A
MIDDLE-ENGLISH DICTIONARY

H. BRADLEY

Oxford University Press

OXFORD LONDON NEW YORK

GLASGOW TORONTO MELBOURNE WELLINGTON

CAPE TOWN IBADAN NAIROBI DAR ES SALAAM LUSAKA ADDIS ABABA

DELHI BOMBAY CALCUTTA MADRAS KARACHI LAHORE DACCA

KUALA LUMPUR SINGAPORE HONG KONG TOKYO

ISBN 0 19 863106 5

A

MIDDLE-ENGLISH
DICTIONARY

CONTAINING

WORDS USED BY ENGLISH WRITERS
FROM THE TWELFTH TO THE FIFTEENTH CENTURY

BY

FRANCIS HENRY STRATMANN

A NEW EDITION, RE-ARRANGED, REVISED, AND ENLARGED

BY

HENRY BRADLEY

OXFORD UNIVERSITY PRESS

FIRST EDITION 1891

Reprinted lithographically in Great Britain by
LOWE AND BRYDONE (PRINTERS) LTD, THETFORD, NORFOLK
from sheets of the first edition
1940, 1951, 1954, 1958, 1963, 1967, 1971, 1974

PREFACE.

———•———

THE Dictionary of Middle-English[1] compiled by Dr. F. H. Stratmann, of
Krefeld, is the only comprehensive dictionary of that stage of the English
language hitherto published in a completed state; the only other existing
work of the same kind being that of Mätzner, of which the first instalment
appeared in 1878, and which has as yet been carried only to the letter J. The
third and latest edition of Dr. Stratmann's dictionary was published in 1878,
and an extensive supplement was issued in 1881. At the time of the author's
death, which took place in 1884, he had apparently almost completed the
preparation of a new edition. The copyright of the work was subsequently
purchased by the Delegates of the Clarendon Press, and the author's materials
for the new edition (consisting of an annotated copy of the printed book) were
placed in my hands to be prepared for the press. The unanimous opinion of
the scholars in this country who were consulted was that simply to reprint the
work in the form contemplated by the author would be wholly unadvisable,
and that an extensive revision was needed in order to adapt it to the needs
of English students. Although the book was marked by great learning and
ability, as well as by extraordinary industry, it had certain serious practical
defects, chiefly, it may be remarked, due to those limitations of scope without
which it would probably have been impossible for the author to bring his work
to completion in any reasonable number of years. The principal respects in
which the present edition aims at being an improvement on the original work
are as follows :—

1. If the definition of a dictionary be that it is a book which explains
the meaning of words, Dr. Stratmann's work can scarcely be regarded as
a dictionary at all. With the *meaning* of Middle-English words, the author
concerned himself but little; his principal care was to identify them etymo-
logically; that is to say, to connect them with their descendants in Modern-

---

[1] In the published editions the work is entitled 'A Dictionary of the Old English Language;'
but in revising it for a new edition the author substituted 'Middle-English' for 'Old English' in
the title-page. He also indicated that in the body of the dictionary the abbreviation 'A.-S.'
(Anglo-Saxon) should be changed into 'O.E.'

English, their antecedents in Old-English, and their cognates in other languages. Such explanations of meanings as were given were merely subsidiary to this object. Not infrequently, indeed, words (and those not merely compounds or obvious derivatives) were left without interpretation altogether; in other cases the sense was indicated only by Latin words, which were often themselves ambiguous. Where a Middle-English word was glossed by a Modern-English one, the latter was commonly the etymological equivalent, even though the word might have undergone a complete change of meaning. In a very large number of instances the only vernacular gloss is an obsolete or dialectal word taken from Halliwell's 'Dictionary of Archaic and Provincial Words,' or some similar source, and quite unintelligible to ordinary English readers. The following examples will illustrate the author's method :—che-owen . . . *chew (chow, chig), mandere, objurgare*; cubbel, *kibble, fustis*; galstron . . . *galster (gauster), crepare*; galwen, *gallow (gally), terrere?*; ӡeolpen, *yelp (yilp), gloriari, jactari*; kitelen . . . *kittle, titillare*; plaigen . . . *plaw, ludere*; schûte . . . *shout, lembus* [the word means a flat-bottomed boat]; scorklin, *scorkle, ustulare*; snobben, *snob, singultire*; slumen, *sloum, dormitare*. In the present edition, I have endeavoured to provide every word with an intelligible explanation in English. When a word has greatly changed in meaning, I have cited the modern form for the sake of completing the etymological history of the word, but in such a manner as not to mislead the reader as to the earlier sense. As the distinctive feature of the book is that the examples are arranged according to the grammatical forms, and not according to the senses, it is obvious that the shades of meaning could not be explained with the same precision and completeness as in larger works like that of Mätzner, or the New English Dictionary, but it is hoped that the explanations given will be found sufficiently full and accurate for ordinary practical needs. The indication of the parts of speech, which was wholly omitted in the original work, has been supplied.

2. Another inconvenient feature of the original dictionary was its perplexing etymological arrangement. The compound words, instead of being inserted in alphabetical order, were placed under their initial element, even though this happened to be an inseparable prefix. Hence, in order to find one of these words, it was necessary to know, or to conjecture its etymology— or rather, Dr. Stratmann's opinion as to its etymology. Suppose, for example, the word amaien is to be sought for in the Dictionary by a student who has no notion of its derivation. The first step of course is to look for it in its alphabetical place. It is not there, nor is there a cross-reference. We next turn to the words under the prefix a-, but amaien is not among them. The cross-reference a-, *see* ab-, ad-, an-, and-, at-, en-, es-, ӡe-, and of-, gives us the choice of ten other places, in any of which the word we are seeking may possibly be found. As a matter of fact, it is placed under es-, so that the reader who takes

the natural and proper course of turning to the several places in their alphabetical order will have looked in ten different pages of the Dictionary before he comes to the word he wants. Even this example, however, does not fully represent the complexity of Dr. Stratmann's method of arrangement. The sub-headings under each of the prefixes are not arranged in the alphabetical order of the words themselves, but in that of the original form of their second element; though this rule is sometimes departed from, apparently by mere oversight. To any one unacquainted with Romanic etymology, the arrangement of the words grouped under com-, for instance, must appear absolutely lawless. The following is an example of the mode of sequence adopted for these compounds: congeien, cogitaciûn, coint, cointise, conjure, collecte, culvert, cullin, couche, couchin, comanden. Romanic philologists will, of course, see at once why this order has been followed, though the position of the word cogitaciûn seems to be due to some mistake. But the majority of those for whom a Middle-English dictionary is intended cannot be expected to find their way readily through such a labyrinth. It has, indeed, often happened that even scholars have erroneously stated a word to be wanting in Stratmann, when in fact it had not been overlooked, but had been recorded in some unlikely place. In this edition I have in the main adopted alphabetical order. Compounds of substantives and adjectives, however, and of words like biforen and anȝēn, are grouped under their first and most important element; and agent-nouns in -ere and nouns of action in -ing follow the verbs from which they are derived. These compounds and derivatives are printed in smaller type and indented, so as not to disturb the general alphabetical arrangement.

3. In cases where a foreign word has come into English by two different channels, and with consequent diversity of form, Dr. Stratmann's usual practice is to treat the word as one. In some instances, where the two forms existed merely as dialectal varieties, without any sense-distinction, this procedure has its advantages; but in others it obscures the facts. The words desk and dish, for example, both descend ultimately from the Latin discus, and therefore in the former editions of this dictionary they are put together as one word. This would be permissible if (as any one would suppose from reading the article) the two forms were employed promiscuously in Middle-English. But the fact is that in that period of the language the words were quite as clearly distinguished as they are now. I have in this and other similar cases separated the two words. It now appears to me that this ought to have been done still more frequently: the verbs chácen and cacchen, for example, would be better separated.

4. The words which in primitive Teutonic began with hl-, hn-, hr- are in the former editions of the dictionary given under the letter H. As the initial h is very generally omitted in Middle-English, this arrangement imposes on

the reader (unless he knows the etymology of the word) the necessity of searching in two different places. I have therefore removed these words to the letters L, N, and R respectively; prefixing, however, an italic *h* to distinguish them from those which have an unaspirated initial.

5. One of the most important changes which have been introduced in this edition relates to the notation of vowel quantities. In the original work the quantities marked were not those which the vowels had in Middle-English, but those of the earliest Old-English, although it is well known that extensive changes in vowel quantity had taken place before the end of the twelfth century, and probably much earlier. The author's procedure in this respect is not, indeed, so wholly irrational as it may at first sight appear. The Old-English quantity admits of being ascertained with tolerable precision, while Middle-English quantity is in many cases uncertain, and in other cases varies according to locality and period. Moreover, the primitive quantity is that which is of greatest etymological importance. I have, however, ventured on the attempt to give a representation of the actual quantity in Middle-English in such a manner as to exhibit its relation to the earlier quantity. Dr. Stratmann employed the circumflex (^) to denote a vowel which was originally long and stressed, original length without stress being indicated by the macron (ˉ). In my notation the macron is placed over an original long vowel which remained long in Middle-English; an original long vowel which has become short by position is marked (˘); and a short vowel which has been lengthened by position is marked with the acute accent. (It might perhaps have been better to use the circumflex for this purpose instead of the acute.) Original short vowels which remained unchanged in quantity are left ·unmarked. It is hoped that this notation will serve to render more intelligible the manner in which changes of quantity influenced the qualitative character of Middle-English vowels. In many points the correctness of my quantitative indications will be somewhat disputable. The vowels a, e, i, o before ld I have marked as 'new long,' except under special conditions which modify the general rules. Similarly I have recognised a lengthening of i and u before nd. The lengthening of e before nd and before n followed by a palatal g, certainly prevailed to a considerable extent, but I doubt whether it was at all so general as the other lengthening, and I have therefore left it unmarked. For similar reasons the lengthening effect of r, followed by a voiced consonant, has not been regarded. There is room for much doubt whether the later positional lengthenings persisted in spite of loss of stress in the second element of a compound. In such cases I have left the vowels unmarked, except where special evidence seems to show that the lengthened quantity was retained. With reference to words of Romanic or Latin origin the question of quantity is very perplexing. Dr. Stratmann correctly recognised that Latin quantity as distinguished from Latin accent had in

general no influence on Middle-English pronunciation, and has therefore treated the unaccented vowels in words derived from Romanic and Latin as (proximately) short, with the exception of those that have compensatory length. I have followed his example in marking as long the vowels in final syllables like -āge, -ōn, -tē, -ūre, &c., which originally were stressed, though owing to a shifting of accent which began, at latest, in the fourteenth century, this notation cannot be regarded as correct for the whole Middle-English period. I should now prefer to mark with the circumflex all Romanic and Latin vowels which were positionally long in early Middle-English, whatever their original quantity, and to leave the rest unmarked.

6. In one instance I have ventured to introduce a diacritic expressive of vowel quality. I have employed the character ü to distinguish the cases in which a Middle-English u expresses either the sound which descended from Old-English y[1], or some sound but slightly different from this. It is clear that in some early loan-words from French, the u was at first sounded in much the same way as the English ü, ū, and I have therefore in these instances marked it with the two points. I now regret not having made a more extensive use of diacritics. The tailed ǫ, ǭ ought to have been used for Middle-English o when representing an Old-English a, ā; the palatal sound of g also might with advantage be distinguished with a special sign.

7. In the etymologies I have attempted to distinguish between direct derivation and collateral relationship of words[2], and have also rectified many errors of detail. Dr. Stratmann's work in this department was eminently scholarly, but the fresh light thrown on many points of comparative Teutonic phonology, by the researches of the past few years, often necessitates an alteration in the form, and sometimes in the substance, of his statements. One error which frequently occurs in the previous editions is that the Old-English verbs in -ian are identified with the continental Teutonic verbs in -jan instead of with those in -ōn, while the umlaut verbs, really corresponding to the continental verbs in -jan, are treated as if peculiar to English.

8. Although the present edition has no pretensions to be regarded as an exhaustive dictionary of Middle-English, it contains a very large number of words not found in the former editions. The author's interest was chiefly in the Teutonic portion of the language, and in his work as originally issued the words of Romanic and Latin origin were almost wholly absent. In deference to the opinion of English scholars, many of these were inserted in later editions, but even in the author's last revision the vocabulary was still conspicuously

[1] In the case of the combination **wur** I have left the u unmarked, since even in Old-English **wyr** often became **wur**.

[2] It should be clearly understood that the Middle-English forms are by no means always the direct descendants of the particular dialectal form in Old-English that happens to be quotable.

defective with regard to the Romanic element. Besides this, many other words, sometimes of considerable literary importance, appear to have been intentionally omitted, because owing to their doubtful derivation they could not be conveniently brought under the etymological arrangement adopted in the work. In the selection of new words for insertion I do not profess to have followed any systematic method. It is probable that many really important words have been overlooked, while some have been inserted that might well have been spared. While I have not expunged the proper names inserted by the author himself, I have not added any others. By further abbreviating the references, which though occupying much space were not full enough to be understood without the help of the table of contractions, it has been found possible to compress the much augmented work into about the same number of pages as were contained in the third edition, together with the supplement.

There are several features of the dictionary which I do not myself regard with approval, though for various reasons I have refrained from altering them. One of these is the normalized orthography of the quotations. My own view is that the quotations, though *not* the headings of the articles, ought to be given in the spelling of the MSS., or at least in that of some standard critical edition; but the gain of textual correctness would not be commensurate with the expenditure of time and labour which would be needed to conform every quotation to the original. Some special points in Dr. Stratmann's system of normalized spelling now appear to me objectionable; for example, he uses the letter **v** not only when the **u** or **v** of the MSS. had the sound of **v**, but also where it had a different consonantal power (=**w**), as in **qvailen, gverdōn**, and **zvo** (Ayenbite). In words taken from early texts which use **g** for ȝ, the former letter is retained, even in the lemmas, although the process of 'normalizing' would seem to have more justification here than in most other cases. The complaint often made that Dr. Stratmann chooses as the 'dictionary forms' of Middle-English words not those of the period of Chaucer and Langland, but those nearest to the Old-English, does not seem to me to be of great moment; the inconveniences of this course being counterbalanced by the advantage of presenting the various forms in something like genealogical sequence. The author, however, did not sufficiently recognise the fact that the orthographical coincidence of Middle-English with Old-English is often merely accidental, especially when the digraphs **æ, ea, eo** are concerned; and most of the articles on words containing these digraphs would need to be rewritten in order to make their arrangement correctly expressive of the facts. A peculiarity of Dr. Stratmann's method, which may cause some embarrassment to the student who is not aware of it, is that he treats 'improper compounds' not as single words but as phrases. Thus the preposition **bisīde** does not occur in its alphabetical place, but examples of it (written **bī sīde** in two words) are found in the article **side**. I have in many cases inserted cross-references to obviate the difficulty arising from this cause.

I am painfully conscious of many serious shortcomings both in the plan and the execution of this edition. The numerous errata will probably seem to some critics to imply extreme carelessness on my part. To this fault I cannot plead guilty; though for many of the oversights I am able to account only by the state of ill-health from which I was suffering when the earlier sheets were passed for press; and if I had then possessed the experience I have subsequently acquired in lexicographical work, I should have known how to take more effectual precautions to ensure accuracy[1]. An assiduous study of Middle-English during the progress of the work has shown me many points in which my knowledge was at the outset materially defective. Perhaps I may reasonably hope for some degree of indulgence on account of the special difficulties of my task. Dr. Stratmann's references are often to books or editions which were not easily accessible to me—some of them, indeed, not to be found even at the British Museum; hence it was frequently not until the article had gone to press that I discovered that he had mistaken the meaning of a word, or had confused together words of different origin and meaning; or, in other cases, that I had given a wrong interpretation to his Latin gloss. The difficulty of accurate proof-reading will be obvious to any person, at all experienced in such work, merely from an inspection of the pages; but it was greatly augmented by the fact that the author's large mass of MS. corrections and additions were made in such a manner that it constantly required the closest attention to discover where the additional words were intended to be inserted. If my attempt to indicate Middle-English vowel-quantity be regarded as in itself justifiable, I may fairly claim that its occasional errors and inconsistencies may be judged with the leniency due to pioneer work. However, although this edition is very far from being such as I should have wished to make it, I venture to hope that in spite of its defects the book in its new form will be found far more useful than if it had been re-issued with merely such slight revision as was contemplated by the original author.

HENRY BRADLEY.

---

[1] One thing which I especially regret is that a few quotations and references have been inserted at second hand, matter which I had caused to be transcribed for me having been sent to the printers under the mistaken belief that it had been properly verified. I hope, however, that all the important errors arising from this cause have been duly noted in the supplementary pages printed at the end of the volume.

EXPLANATION OF REFERENCES.

A. D. = Altenglische Dichtungen des MS. Harl. 2253, herausgegeben von K. Böddeker, Berlin, 1878. [? Written in Herefordshire about 1310.] Quoted by page.

A. P. = Early English Alliterative Poems, edited by R. Morris : i. The Pearl ; ii. Cleanness ; iii. Patience. E.E.T.S., No. 1, 1864. [? Lancashire about 1360 ; by the same author as GAW.] Quoted by number of piece and line.

A. R. = The Ancren Riwle, edited by J. Morton, Cam. Soc. No. lvii. 1853. [? Dorsetshire about 1225.] Quoted by page.

A. S. = Ancient Songs from the time of King Henry II to the Revolution, edited by J. Ritson, London, 1790. Quoted by number of piece.

A. S. (H.) = The same re-edited by W. C. Hazlitt, 1877. Quoted by number of piece.

ÆLF. HOM. = Ælfric's Metrical Lives of Saints, edited by W. W. Skeat, E.E.T.S., Nos. 76, 82, 1881-5. Quoted by page.

ÆLFR. = Ælfric's Grammatik und Glossar, herausgegeben von J. Zupitza, in Sammlung Englischer Denkmäler, Berlin, 1880. Quoted by page.

AL. = The Legend or Life of St. Alexius, edited by F. J. Furnivall. [About 1380.] In six versions, viz. L¹. (MS. Laud 622), L². (MS. Laud 108) L³. (MS. Laud 463), C. (MS. Cotton Titus A xxvi), V. (MS. Vernon), T. (MS. Trinity Coll. Oxford, 57). E.E.T.S., No. 69, 1878. Quoted by line.

ALEX. = The Alliterative Romance of Alexander, edited by J. Stevenson, from the Ashmole MS. Roxburghe Club, 1849. [About 1400-50.] Quoted by line.

ALEX. (Sk.) = The same [The Wars of Alexander], edited by W. W. Skeat, from the Ashmole and Dublin MSS. E.E.T.S., No. xlvii., 1866. Quoted by line.

ALEX.² = 'Appendix from the Bodl. MS. [264]' in ALEX. Quoted by line.

ALEX.² (Sk.) = The same [Alexander and Dindimus], edited by W. W. Skeat, E.E.T.S., No. xxxi., 1878. [? About 1375.] Quoted by line.

ALIS. = King Alisaunder, in Weber's Metrical Romances (Edinburgh, 1810), vol. i. pp. 3-327. [About 1300.] Quoted by line.

ALIS.² = þe Gestes of þe worþie King and Emperour Alisaunder of Macedoine, edited by W. W. Skeat, in vol. with WILL. [About 1375.] Quoted by line.

ALT. BL. = Altdeutsche Blätter von M. Haupt und H. Hoffmann. Quoted by volume and page.

AM. & AMIL. = Amis and Amiloun, in Weber's Metrical Romances (1810), vol. ii. pp. 369-473. [About 1330.] Quoted by line.
= The same, herausgegeben von E. Kölbing, in Kölbing's Altenglische Bibliothek, Bd. ii, 1884. Quoted by line.

AMAD. (R.) = Sir Amadace, in Robson's Early English Metrical Romances (Camd. Soc. No. xviii., 1842), pp. 27-56. [About 1420.] Quoted by stanza.

AMAD. (W.) = Sir Amadas [a later text of the above], in Weber's Metrical Romances (1810), vol. iii. pp. 243-275. Quoted by line.

ANGL. = Anglia, herausgegeben von R. P. Wülcker. Quoted by volume and page.

AN. LIT. = Anecdota Literaria, edited by Thos. Wright, London, 1844. Quoted by page.

ANC. COOKERY = Ancient Cookery from a MS. in the Arundel Collection, printed in Ordinances for the Royal Household. Soc. of Antiquaries, 1790. [About 1440.] Quoted from N.E.D.

ANT. ARTH. (L.) = The Awntyrs of Arthure, in Laing's Select Remains of the Ancient Poetry of Scotland (1885), pp. 86-113. [About 1420.] Quoted by line.

ANT. ARTH. (M.) = Awntyrs of Arthure, edited by F. Madden, in Sir Gawayne volume, Bannatyne Club, 1839. Quoted by line.

ANT. ARTH. (R.) = The Anturs of Arther in Robson's Early English Metrical Romances (Cam. Soc. No. xviii. 1842), pp. 1-26. [? Lancashire about 1420.] Quoted by stanza.

APOL. = The Apology of Lollard Doctrines, formerly attributed to Wiclif, edited by J. H. Todd, Cam. Soc. No. xx. 1842. [About 1400.] Quoted by page.

AR. = Arthur, edited by F. J. Furnivall, E.E.T.S., No. 2, 1864. [? About 1440.] Quoted by line.

AR. & MER. = Arthour and Merlin, a Metrical Romance edited by W. D. Turnbull, from the Auchinleck MS., Abbotsford Club, 1838. [About 1330.] Quoted by line.

ARCH. = Archiv für das Studium der neueren Sprachen, herausgegeben von L. Herrig. Quoted by volume and page.

ASS. = Assumpcioun de notre Dame, edited by R. Lumby, E.E.T.S., No. 14, 1866. [From the Cambridge MS., before 1300.] Quoted by line.

ASS. *B* = Assumpcio Beate Marie in the same volume. [From the British Museum MS. about 1330.] Quoted by line.

AUD. = The Poems of J. Audelay, edited by J. O. Halliwell, Percy Soc. vol. xiv. 1844. [Shropshire 1426.] Quoted by page.

AV. ARTH. = The Avowing of King Arther, in Robson's Early English Metrical Romances, Cam. Soc. No. xviii. 1842, pp. 57-93. [About 1420.] Quoted by stanza.

AYENB. = Dan Michel's Ayenbite of Inwyt, edited by R. Morris, E.E.T.S., No. 23, 1866. [Kent 1340.] Quoted by page.

B. B. = The Babees Book, etc., edited by F. J. Furnivall, E.E.T.S., No. 32, 1868. Quoted by page.

B. OF C. = The Boke of Curtasie, from the Sloane MS. 1986 in the British Museum, printed in B.B. [About 1440.] Quoted by line.

B. D. D. = Be Domes Dæge, edited by R. Lumby, E.E.T.S., No. 65, 1876. Quoted by page.

B. DISC. = Li biaus Disconus, in Ritson's Metrical Romances (1802), vol. ii. pp. 1-90. [About 1460.] Quoted by line.

BARB. = Barbour's Bruce, edited by W. W. Skeat, E.E.T.S., Nos. xi. xxi. xxix, lv. 1870-89. [Scottish about 1375.] Quoted by book and line.

BARB. LEG. = Barbour's Legendensammlungen, herausgegeben von C. Horstmann, Heilbronn, 1881-2. [Scottish about 1375.] Quoted by vol. and page.

BARTH. = Trevisa's Englishing of Bartholomæus de Proprietatibus Rerum. [Written in 1398.] Printed by W. de Worde in 1495. Quoted from N.E.D. (which often gives the readings of the Helmingham MS.).

BEK. = The Life and Martyrdom of Thos. Beket, edited by W. H. Black, from the Harl. MS. 2277. Percy Soc., vol. xix, 1845. [? Gloucestershire about 1300.] Quoted by line.

BENS. VOC. = Vocabularium Anglo-Saxonicum, opera Th. Benson, Oxoniae, 1701.

BEN. WB. = Mittelhochdeutsches Wörterbuch, von Benecke, Müller und Zarncke.

BER. = The Tale of Beryn, edited by F. J. Furnivall, Chaucer Soc. 1876. [About 1400.] Quoted by line.

BEV. = Sir Beves of Hamtune, edited by W. D. Turnbull, Maitland Club, 1838. [About 1320.] Quoted by line.

The same re-edited by E. Kölbing, with versions of six other MSS., E.E.T.S., Nos. xlvi. xlviii. 1885-6. Quoted by line.

BL. HOM. = The Blickling Homilies, edited by R. Morris, E.E.T.S., Nos. 58, 63, 73, 1874-6-80. Quoted by page.

BOKEN. = Livis of Seintis, translated by O. Bokenam, A. D. 1447, Roxburghe Club, 1835. (Also forms Band i. of Kölbing's Altenglische Bibliothek.) [About 1447.] Quoted by page.

BOUT. EV. = Die vier Evangelien in altnordhumbrischer Sprache, herausgegeben von K. W. Bouterwek, Gütersloh, 1857.

BRD. = St. Brandan, edited by T. Wright, Percy Soc. vol. xiv. 1844. [? Gloucestershire about 1300]. Quoted by line.

BREM. WB. = Versuch eines bremisch-niedersächsischen Wörterbuches, Bremen, 1767-1771.

C. B. = Two fifteenth-century Cookery Books from Harl. MS. 279 [about 1430] and Harl. MS. 4016 [about 1450], edited by T. Austin, E.E.T.S., No. 91, 1888. Quoted by page.

C. L. = Castel off Love, edited by R. F. Weymouth, Philol. Soc. 1864. [About 1320.] Quoted by line.

C. M. = Cursor Mundi, edited by R. Morris, E.E.T.S., Nos. 57, 59, 62, 66, 68, 1874-78. [About 1250-1340.] Quoted by line.

CATH. = Catholicon Anglicum, edited by S. J. Herrtage, E.E.T.S., No. 75, 1881. [About 1483.] Quoted by page.

CAXT. VIT. PATR. = Caxton's Vitas Patrum, 1495. Quoted from N.E.D.

CH. = The works of Jeffrey Chaucer, edited by Th. Speght, London, 1687.

CH. (T.) = Canterbury Tales, edited by Tyrwhitt, with other Poems from the Black Letter editions, London, 1843.

CH. A. B. C. = Chaucer's A. B. C. ⎫
CH. ADAM = Chaucer's words to ⎬ in CH. M. P.
his Scrivener Adam. ⎪
CH. AN. = Anelida and Arcite. ⎭

CH. ASTR. = A treatise on the Astrolabe, by G. Chaucer, edited by W. W. Skeat, E.E.T.S., No. xvi., 1872. [Middlesex about 1400.] Quoted by book, chapter, and page.

CH. BOET. = Chaucer's translation of Boethius's ' De Consolatione Philosophiae,' edited by R. Morris, E.E.T.S., No. v. 1868. [About 1374.] Quoted by book, prose (metre), and page.

CH. BUK. = Lenvoy to Bukton : in CH. M. P.

CH. C. T. = Chaucer's Canterbury Tales, edited by F. J. Furnivall, Chaucer Soc. [About 1386.] Quoted by group-letter and line; see comparative Table, p. xxii.

CH. C. T. (W.) = The Canterbury Tales of G. Chaucer, edited by T. Wright, published for the Percy Soc. (vols. xxiv-xxvi.) 1847-51, and for general circulation, London, 1853. Quoted by line.

CH. COMP. M. = Compleint of Mars.
CH. COMP. V. = Compleint of Venus. ⎫
CH. D. BL. = Deth of Blanche. ⎪
CH. F. AGE = The Former Age. ⎪
CH. FORTUNE. ⎬ in CH. M. P.
CH. GENT. = Gentilesse. ⎪
CH. H. F. = House of Fame. ⎪
CH. L. G. W. = Legend of Good Women. ⎭

CH. L. G. W. (Sk.) = Chaucer's Legend of Good Women, edited by W. W. Skeat, Oxford, 1889. [About 1385.]

CH. L. S. = Lack of Stedfastnesse : in CH. M. P.

CH. M. P. = Chaucer's Minor Poems, edited by F. J. Furnivall, Chaucer Soc. Quoted by title of piece and line.

CH. M. P. (Sk). = Chaucer's Minor Poems, edited by W. W. Skeat, Oxford, 1888. Quoted by title of piece and line.

CH. P. F. = Parlement of Foules.

CH. PITIE = Compleint to Pitie.
CH. PROV. = Two Proverbs.
CH. PURSE = Compleint to his Purse.
CH. SCOG. = Lenvoy to Scogan.
} in CH. M. P.

CH. TRO. = Parallel Text Print of Chaucer's Troilus and Creseyde, from three MSS. put forth by F. J. Furnivall, Chaucer Soc. 1881-2. [About 1374.]

CH. TRO. (T.) = Chaucer's Troilus and Creseyde in edition 1843; see CH. (T.).

CH. TRUTH = Truth : in CH. M. P.

CHEST. = The Chester Whitsun plays, edited by T. Wright, Shakspeare Soc. Nos. 17 and 35, 1843-7. [About 1430.] Quoted by vol. and page.

CHR. E. = Chronicle of England in Ritson's Ancient English Metrical Romances (1802), vol. ii. pp. 270-313. [About 1325.] Quoted by line.

CL. = Sir Cleges, in Weber's Metrical Romances (1810), vol i. pp. 313-353. [About 1410.] Quoted by line.

CL. M. = Clene Maydenhod, edited by F. J. Furnivall, E.E.T.S., No. 25, 1867. [About 1376.] Quoted by line.

COMP. BL. KN. = Complaint of the Black Knight, by Lidgate, formerly attributed to Chaucer, printed in CH. (T.). [About 1430.] Quoted by line.

CR. K. = The Crowned King, edited by W. W. Skeat, (in vol. containing the C-Text of Langland), E.E.T.S., No. 54, 1873. [About 1415.] Quoted by line.

CT. LOVE. = The Court of Love, wrongly attributed to Chaucer, printed in CH. (T.). [About 1450.] Quoted by line.

D. ARTH. = Morte Arthure ; or the Death of Arthur, edited by Edmd. Brock, E.E.T.S., No. 8, 1871, previously edited by G. G. Perry, 1865. [Yorkshire beginning of 15th century.] Quoted by line.

DAV. DR. = Adam Davy's Five Dreams about Edward II, edited by F. J. Furnivall, E.E.T.S., No. 69, 1878. [14th cent.] Quoted by line.

DEGR. = The Romance of Sir Degrevant, in the Thornton Romances, edited by J. O. Halliwell, pp. 177-256. Cam. Soc. No. xxx. 1844. [Before 1440.] Quoted by line.

DEP. R. = Richard the Redeles, an Alliterative Poem on [the Deposition of] Richard II, edited by W. W. Skeat (in the C-Text volume of Langland) E.E.T.S., No. 54,

1873. [1399.] Quoted by passus and line.

DEP. R. (W.). = The same, edited by Thos. Wright, Cam. Soc. No. iii. 1838, reprinted in P. P. 1859.

DIEFENB. WB. = Vergleichendes Wörterbuch der gothischen Sprache von Lor. Diefenbach, Frankfurt, 1851.

DIEZ WB. = Etymologisches Wörterbuch der romanischen Sprachen von F. Diez. Bonn, 1853.

DIGB. = Ancient Mysteries from the Digby MS. 133, Abbotsford Club, 1835. [About 1485.] Quoted by page.

DIGB. (F.) = The same [Digby Mysterie], edited and added to, by F. J. Furnivall, New Shakspere Soc., 1882. [About 1485.] Quoted by piece and line.

DREAM = The Dreme, formerly attributed to Chaucer, printed in CH. (T.). [15th cent.] Quoted by line.

D. TROY = The Gest Historiale of the Destruction of Troy, edited by G. A. Panton and D. Donaldson, E.E.T.S., Nos. 39, 56, 1869-74. [About 1400.] Quoted by line.

DU CANGE = Caroli du Fresne, domini du Cange, glossarium ad scriptores mediae et infimae latinitatis, Francofurti, 1710.

E. G. = English Gilds, edited by Miss L. Toulmin Smith, F.E.T.S., No. 40, 1870. Quoted by page.

E. T. = The Erl of Tolous, in Ritson's Metrical Romances (1802), vol. iii, pp. 93-144. [About 1450.] Quoted by line.

E. T. (L.) = The Erl of Tolous and the Emperes of Almayn, herausgegeben von G. Lüdtke, in Sammlung Englischer Denkmäler, Berlin, 1881. Quoted by line.

E. W. = The Fifty Earliest English Wills, 1387-1439, edited by F. J. Furnivall, E.E.T.S., No. 78, 1882. Quoted by page.

ED. = St. Editha, sive Chronicon Vilodunense, herausgegeben von C. Horstmann, Heilbronn, 1883. [About 1420.] Quoted by line.

EGILSS. = Lexicon poëticum antiquae linguae septentrionalis, conscripsit Sveinbjörn Egilsson, Hafniae, 1860.

EGL. = The Romance of Sir Eglamour of Artois, edited by J. O. Halliwell, in the Thornton Romances, pp. 121-175, Cam. Soc. No. xxx. 1844. [? Early 15th cent.] Quoted by line.

ELL. ROM. = Specimens of Early English Metrical Romances, edited by G. Ellis, 2nd edition, London, 1811. Quoted by page.

EM. = Emare, in Ritson's Metrical Romances (1802), vol. ii. pp. 204-247. [About 1460.] Quoted by line.

ENG. ST. = Englische Studien, herausgegeben von E. Kölbing. Quoted by volume and page.

ERC. = The Romance and Prophecies of Thomas of Erceldoune, edited by J. A. H. Murray, E.E.T.S., No. 61, 1874. [About 1425.] Quoted by line.

ERC.(B.) = The same, edited by A. Brandl, in Sammlung Englischer Denkmäler. Quoted by line.

ETTM. LEX. = Lexicon Anglosaxonicum, edited by L. Ettmüllerus, Quedlinburgii, 1851.

F. C. = The Forme of Cury, edited by S. Pegge, London, 1780. [? About 1390.] Quoted by page.

FER. = Sir Ferumbras, edited by S. J. Herrtage, E.E.T.S., No. xxxiv, 1879. [About 1380.] Quoted by line.

FL. & BL. = Floriz and Blauncheflur, edited by R. Lumby, E.E.T.S., No. 14, 1866. [About 1275.] Quoted by line.

FL. & BL. (H.) = The same [Floris and Blauncheflur], edited by E. Hausknecht in Sammlung Englischer Denkmäler, Berlin, 1885. Quoted by line.

FL. & LEAF. = The Flower and the Leaf, formerly attributed to Chaucer. Printed in CH. (T.). [15th cent.].

FLOR. = Le bone Florence of Rome, in Ritson's Metrical Romances (1802), vol. iii. pp. 1–92. [About 1440.] Quoted by line.

FRAG. = Fragment of Ælfric's Grammar, Ælfric's Glossary, and poem on the Soul and Body in the orthography of the 12th cent., edited by T. Phillipps, London, 1838. Quoted by page.

FRITZNER = Ordbog over det gamle Norske Sprog of Joh. Fritzner, Kristiania, 1867.

GAM. = The Tale of ʒong Gamelin, edited by W. W. Skeat, Oxford 1884. [About 1400.] Quoted by line.

GAW. = Sir Gawayne and the Green Knight, edited by R. Morris, E.E.T.S., No. 4, 1864. (Previously edited by F. Madden, Bannatyn Club, 1839.) [? Lancashire about 1360.] Quoted by line.

GEN. & EX. = The Story of Genesis and Exodus, edited by R. Morris, E.E.T.S., No. 7, 1865. [? Norfolk or Suffolk about 1250.] Quoted by line.

GENER. = A Royal Historie of the excellent Knight Generides, edited by F. J. Furnivall, from the Helmingham MS. Roxburghe Club, 1865. [About 1430.] Quoted by line.

GENER. (W.) = Generydes, edited by W. A. Wright, from MS. in Trinity College, Cambridge, E.E.T.S., Nos. 55 & 70, 1873–9. [About 1440.] Quoted by line.

GEST. R. = The Old English Versions of the Gesta Romanorum, edited by F. Madden, Roxburghe Club, 1838. [About 1440.] Quoted by page.

GEST. R. (H.) = The Gesta Romanorum, edited by S. J. H. Herrtage from three MS. (H.) Harl. 7333, (H.²) Additional 9066; (H.³) Cambridge (k. k. 1. 6), E.E.T.S., No. xxxiii. 1879. [About 1440.] Quoted by page.

GOLAG. = The Knightly Tale of Golagros and Gawane, printed in Sir Gawayne volume, edited by F. Madden, Bannatyne Club. 1839; re-edited by M. Trautmann in

ANGL. II. 410 ff. [Scottish about 1440.] Quoted by line.

GOW. = Confessio Amantis of John Gower, edited by R. Pauli, London, 1857. [Middlesex about 1400 ; but edition not from coeval MSS.] Quoted by volume and page.

GR. = The Grave, in Thorpe's Analecta Anglo-Saxonica, p. 142, re-edited in ANGL. V. 289. [12th cent.] Quoted by line.

GRAFF = Althochdeutscher Sprachschatz von E. G. Graff, Berlin, 1834–42.

GREG = Die englische Gregorlegende, herausgegeben von F. Schulz, Königsberg, 1876. [Auchinleck MS. ; 13th cent.] Quoted by line.

GREIN = Sprachschatz der angelsächsischen Dichter, bearbeitet von C. W. M. Grein, Cassel and Göttingen, 1861–4.

GR. GRAM. = Deutsche Grammatik von Jac. Grimm. 2. Ausg.

GRL. = Lovelich's History of the Holy Grail, edited by F. J. Furnivall, E.E.T.S., Nos. xx, xxiv, xxviii, xxx, 1874–78. [About 1450.] Quoted by chapter and line.

GR. WB. = Deutsches Wörterbuch von Jac. und Wilh. Grimm.

GUIL. = A prose translation of the first version of Guileville's Le Pelerinage de l'Homme, edited by W. A. Wright from the Camb. Univ. MS., Ff. 5, 30, Roxburghe Club, 1869. [About 1430.] Quoted by page.

GUIL ² = The Booke of the Pylgremage of the Sowle, translated from the French of de Guileville (? by Lidgate, 1413), edited from Caxton's impression by Miss Cust, London, 1859. Quoted by book, chapter, and page.

GUY = The Romance of Sir Guy of Warwick and Rembrun his son, edited by W. D. Turnbull, Abbotsford Club, 1840. [About 1314.] Quoted by line.

GUY (Z.) = Guy of Warwick, edited by J. Zupitza, from the Auchinleck and Caius MSS. E.E.T.S., Nos. xlii, xlxi, 1883–7. [About 1314.] Quoted by line.

GUY²(Z.) = Guy of Warwick (2nd version), edited by J. Zupitza, E.E.T.S., Nos. xxv, xxvi, 1875–6. [15th cent.] Quoted by line.

H. H. = The Harrowing of Hell, herausgegeben von Ed. Mall, from three MSS., Harl. 2253, Digby 86, and the Auchinleck. Breslau, 1871. [About 1310.] Quoted by line.

= The same, edited by J. O. Halliwell from Harl. MS. 2253, London, 1840. [? Herefordshire about 1310.] Quoted by line.

H. M. = Hali Meidenhad, edited by O. Cockayne, E.E.T.S., No. 18, 1866. [About 1230.] Quoted by page.

H. S. = Robert of Brunne's Handling Sinne; edited by F. J. Furnivall from Harl. MS. 1701, Roxburghe Club, 1862. [Lincolnshire about 1303.] Quoted by line.

H. V. = Hymns to the Virgin, etc., edited by F. J. Furnivall, E.E.T.S., No. 24, 1867. [About 1430.] Quoted by page.

HALDORS. = Lexicon Islandico - Latino - Danicum Biornonis Haldorsonii, cura R. K. Raskii editum, Havniae, 1814.

HALLIW. = A Dictionary of Archaic and Provincial Words, by J.O. Halliwell, London, 1850. Quoted by page.

HAMP.PS. = Hampole's Psalms and Canticles with a Commentary, edited by H. R. Bramley. Oxford, 1884. [Before 1340.] Quoted by psalm and verse ; or by page.

HAMP. TR. = Hampole's Prose Treatises, edited by G. G. Perry, E.E.T.S., No. 20, 1866. [About 1340.] Quoted by page.

HAV. = The Lay of Havelok the Dane, edited by W. W. Skeat, E.E.T.S., No. iv. 1868. [? Lincolnshire about 1280.] Quoted by line.

HEYNE = Kleinere altniederdeutsche Denkmäler, mit ausführlichem Glossar, herausgegeben von Moritz Heyne, Paderborn, 1867.

HICKES = Linguarŭm veterum septentrionalium thesaurus, auctore G. Hickesio, Oxoniae, 1705. Quoted by volume and page.

HIPP. = Dictionnaire de la langue française au XIIe et au XIIIe siècle, par C. Hippeau, Paris, 1873.

HIST. EDWD. = Historie of the Arrivall of Edward IV in England, A.D. 1471, edited by John Bruce, Camd. Soc. No. i. 1838. Quoted by page.

HOCCL. = Poems by Thos. Hoccleve, London, 1796. [About 1400.] Quoted by number of piece and line.

HOCCL. REG. PRIN. = De Regimine Principum, by T. Occleve, edited by T. Wright, Roxburghe Club, 1860. [14th cent.] Quoted from N.E.D.

HOLTZM. = Altdeutsche Grammatik von Ad. Holtzmann.

HOM. = Old English Homilies and Homiletic Treatises of the 12th and 13th centuries, edited by Richd. Morris, E.E.T.S., first series, Nos. 29 and 34; and second series, No. 53. 1867-8, 1873. Quoted by series and page.

HORN (H.) = King Horn, herausgegeben von C. Horstmann, in ARCH. vol. i. pp. 41-58. [Before 1300.] Quoted by line.

HORN (L.) = King Horn, edited by R. Lumby, E.E.T.S., No. 14, 1866. Quoted by line.

HORN (R.) = The geste of King Horn, in Ritson's Metrical Romances (1802), vol. ii. pp. 91-155. Quoted by line.

HORN (W.) = Das Lied von King Horn, herausgegeben von T. Wissmann, Strassburg, 1881. Quoted by line.

HORN CH. = Horn child and maiden Rimnild in Ritson's Metrical Romances (1802), vol. iii. pp. 282-320. [Auchinleck MS. about 1320.] Quoted by line.

HUNT. = The Hunting of the Hare, in Weber's Metrical Romances (1810), vol. iii. pp. 279-290. [? About 1375.] Quoted by line.

IHRE = Glossarium Suio-Gothicum, auctore J. Ihre, Upsaliae, 1769.

IPOM. = The Life of Ipomidon, in Weber's Metrical Romances (1810), vol. ii. 281-365. [About 1440.] Quoted by line.

IPOM. (K.) = Ipomedon, in drei englischen Bearbeitungen, herausgegeben von E. Kölbing, Breslau, 1889. [Version A, ? 14th cent. ; B (same as the above) ; C (prose), 15th cent.] A and B quoted by line, C by page.

ISUM. = The Romance of Sir Isumbras, edited by J. O. Halliwell, in the Thornton Romances, pp. 88-120, Cam. Soc. No. xxx, 1844. [Before 1400.] Quoted by line.

IW. = Iwaine and Gawin, in Ritson's Metrical Romances (1802), vol. i. pp. 1-169. [About 1400.] Quoted by line.

J. T. = St. Jeremie's 15 Tokens before Doomesday, edited by F. J. Furnivall, in E.E.T.S., No. 69, 1878. [14th cent.] Quoted by line.

JARB. = Jahrbuch für romanische und englische Sprache, &c., von Ludw. Lemcke. Quoted by volume and page.

JOHN = The Gospel according to St. John, edited by W. W. Skeat, Cambridge, 1878 [Hatton MS. ? Kent about 1150]. Quoted by chapter and verse.

JOS. = Joseph of Arimathie, edited by W. W. Skeat, E.E.T.S., No. 44, 1811. [Vernon MS. about 1350.] Quoted by line.

JUL. = The Liflade of St. Juliana, edited by O. Cockayne and T. Brock, from two MSS. (Royal 17 A xxvii. and Bodl. 34). E.E.T.S., No. 51, 1872. [About 1200.] Quoted by page.

JUL². = The Metrical Life of St. Juliana (Ashm. 43), in the same vol. [About 1300.] Quoted by line.

K. COL. = Three Kings of Cologne. edited from Cam. Univ. Ee. 4 and Royal 18 A. x. by C. Horstmann, E.E.T.S., No. 85, 1886. [15th cent.] Quoted by page.

K. T. = The King of Tars and the Soudan of Dammas, in Ritson's Metrical Romances (1802), vol. ii. pp. 156-203. [About 1330.] Quoted by line.

KATH. = The Legend of St. Katherine of Alexandria, edited by J. Morton, from MS. Cott. Titus D xviii. with v. r. from Royal 17 A xxvii. Abbotsford Club, 1841. [About 1225.] Quoted by line.

KATH.(E.) = The same, edited by E. Einenkel, E.E.T.S., No. 80, 1884. [About 1200.] Also printed with KATH.² Quoted by line.

KATH ². = The Life and Martyrdom of St. Katherine of Alexandria, edited from a MS. of 15th cent. by H. H. Gibbs, Roxb. Club, 1884. [Prose, about 1430.] Quoted by page.

KIL. = Etymologicum Teutonicae linguae, sive

dictionarium Teutonico-Latinum, studio et opera Corn. Kiliani Dufflaei, editio tertia, Antverpiae, 1599.

KN. L. = Knight of La Tour Landry, edited by Thos. Wright, E.E.T.S., No. 33, 1868. [About 1440.] Quoted by page.

L. C. C. = Liber Cure Cocorum, edited by R. Morris, Philol. Soc. 1862. [About 1420.] Quoted by page.

L. F. M. B. = The Layfolk's Massbook, edited by T. F. Simmons, E.E.T.S., No. 71, 1879. [About 1375-1400.] Quoted by page.

L. H. R. = Legends of the Holy Rood, edited by R. Morris, E.E.T.S., No. 46, 1871. Quoted by page.

L. N. F. = Altenglische Legenden, neue Folge, herausgegeben von C. Horstmann, Heilbronn, 1881. Quoted by page.

LA3. = Laȝamon's Brut, edited by F. Madden, Soc. of Antiquaries, London 1847. [? Worcestershire, first text about 1205, second about 1275.] Quoted by line.

LAI LE FR. = Lai le Freine, in Weber's Metrical Romances (1810), vol. i. pp. 357-371. [About 1325.] Quoted by line.

LANGL. = The Vision of William (Langland, or Langley) concerning Piers the Plowman, three texts, edited by W. W. Skeat, E.E.T.S., Nos. 28, 38, 54, 67, 81, 1867-85; afterwards edited for the Clarendon Press, Oxford, 1886. [*A*-text about 1362, *B* 1377, and *C* 1393.] Quoted by passus and line.

LANGL. (Wr.) = The Vision of Piers Ploughman, edited by T. Wright, London, 1856. [A *B*-text of the above; about 1377.] Quoted by line.

LAUNF. = Launfal, in Ritson's Metrical Romances (1802), vol. i. pp. 170-215. [About 1460.] Quoted by line.

LEB. JES. = Leben Jesu, herausgegeben von C. Horstmann, Münster, 1873. [About 1300.] Quoted by line.

LEECHD. = Leechdoms, Wortcunning, and Starcraf of Early England, by O. Cockayne, Rolls Series, London, 1864-1866, 3 vols. Quoted by volume and page.

LE GON. = Dictionnaire breton-français de L. Gonidec, enrichi par Hersart de le Villemarqué, Saint Brieux, 1850.

LEG. = Altenglische Legenden, herausgegeben von C. Horstmann, Paderborn, 1875. [Late 13th cent.] Quoted by page.

LEG. CATH. = Legendae Catholicae, a little Boke of seintlie Gestes, edited by W. D. Turnbull, Edinburgh, 1840. Quoted by page.

LEXER = Mittelhochdeutsches Handwörterbuch von M. Lexer.

LIDG. BOCH. = The Tragedies gathered by I. Bochas of all such princes as fell from their estates, etc., transl. by John Lydgate, edition of 1558. [? Written about 1450.] Quoted chiefly from the N.E.D.

LIDG. M. P. = A Selection from the Minor Poems of

dan John Lydgate, edited by J. O. Halliwell, Percy Soc., vol. ii., 1840. [About 1430.] Quoted by page.

LIDG. TH. = The Story of Thebes by Lydgate, printed in CH. Quoted by line.

LIDG. TROY = The Historie, Sege, and Distruccion of Troye, translated by John Lidgate, London, 1513. [About 1430.] Quoted by book and chapter, chiefly from N. E. D.

LIND. = Lindisfarne Gospels printed in JOHN, LK., MK., MT. [About 950.]

LITTRÉ = Dictionnaire de la Langue française par E. Littré, Paris, 1873.

LK. = The Gospel according to Saint Luke, edited by W. W. Skeat, Cambridge, 1874. [Hatton MS. ? Kent about 1150.] Quoted by chapter and verse.

LUD. COV. = Ludus Coventriae, edited by J. O. Halliwell, Shakespeare Soc., 1841. [Warwickshire 15th century.] Quoted by page.

M. ARTH. = Le Morte Arthur, edited by F. J. Furnivall, London, 1864. [Harl. MS. 2252, about 1460; the date of composition is much older.] Quoted by line.

M. GOD = 'Mother of God,' by Hoccleve, printed in CH. M. P. [About 1400.] Quoted by line.

M. H. = English Metrical Homilies from MSS. of the 14th century edited by John Small, Edinb., 1862. [About 1325.] Quoted by page.

M. T. = Ancient Metrical Tales, edited by C. H. Hartshorne, London, 1829. Quoted by page.

MAL. = Mallory's King Arthur, edited by R. Southey from Caxton's copy, London, 1817. Since re-edited by H. O. Sommer, London, 1889-90. Quoted by book and chapter.

MAN. (F.) = Robert Manning's History of England, edited by F. J. Furnivall, Rolls Series, London, 1887. [About 1330.] Quoted by line.

MAN. (H.) = Peter Langtoft's Chronicle, as illustrated and improv'd by Robert of Brunne, edited by Th. Hearne, Oxford, 1725. [Lincolnshire 1338.] Quoted by page.

MAND. = The Voiage and Travaile of Sir John Maundeville, reprinted from the edition of 1725, with an introduction, by J. O. Halliwell, London, 1839, reissued 1866. [About 1400.] Quoted by page.

MAP = The Latin Poems commonly attributed to Walter Map, edited by T. Wright, Camden Soc., No. xvi. 1841. [Contains some Middle Eng. Poems: Body and Soul, from MS. Laud 108, 13th cent.; the same from Vernon MS., about 1310; Body and Soul (another version), from MS. Harl. 2253, about 1325; Sir Penny, from MS. Cotton Galba E. ix., 14th cent.] Quoted by page.

b

MARG. = Seinte Margarete (in MARH. vol.), edited by O. Cockayne. [Harl. MS. 2277, about 1300.] Quoted by line.

MARH. = Seinte Marherete, þe meiden ant martyr, from MSS. Reg. 17, A xxvii, and Bodl. 34 [about 1200]; printed together with MARG. (pp. 24–33) and Meidan Maregrete (Trin. Coll. Cantab. MS. about 1258) pp. 34–43, reprinted from HICKES i. 224 ff.; London, 1862, afterwards re-issued E.E.T.S., No. 13, 1866. Quoted by page.

MAS. = A Poem on the Constitutions of Masonry, in Halliwell's Early History of Freemasonry in England, 1844, pp. 12–40. [14th cent.] Quoted by line.

MAT. = The Gospel according to St. Matthew, edited by W. W. Skeat, Cambridge, 1887. [Hatton MS. ? Kent about 1150.] Quoted by chapter and verse.

MED. = Meditations on the Supper of our Lord, etc., by Robert of Brunne, edited by J. M. Cowper, from Harl. MS. 1701, collated with Bodl. 415, E.E.T.S., No. 60, 1875. [About 1320.] Quoted by line.

MER. = Merlin, a Prose Romance, edited by H. B. Wheatley, E.E.T.S., Nos. 10, 21, 36, 1865-6-9. [About 1440.] Quoted by chapter and page.

MIN. = Poems of Laurence Minot, edited by J. Hall, Oxford, 1887. [MS. about 1415, composed about 1352.] Quoted by number of piece and line (in the earlier pages, sometimes by page of Ritson's edition, London, 1825.)

MIR. PL. = A Collection of English Miracle-Plays, from the Chester, Coventry, Townley, and Digby Mysteries, together with God's Promises, edited by W. Marriott, Basel, 1838. Quoted by page.

MIRC = Instructions for Parish Priests by John Myrc, edited by Edw. Peacock, E.E.T.S., No. 31, 1868. [Shropshire about 1400.] Quoted by line.

MISC. = An Old English Miscellany, edited by R. Morris, E.E.T.S., No. 49, 1872. Quoted by page.

MK. = The Gospel according to St. Mark, edited by W. W. Skeat, Cambridge, 1871. [Hatton MS., ? Kent about 1150.] Quoted by chapter and verse.

MONE = Quellen und Forschungen zur Geschichte der teutschen Literatur und Sprache von F. J. Mone, Aachen, 1830.

N. E. D. = A New English Dictionary on Historical Principles, edited by J. A. H. Murray (A–Clivy), Oxford, 1884–1890.

N. P. = Nugae Poeticae. Select pieces of Old English popular poetry, edited by J. O. Halliwell, London, 1844. Quoted by page.

O. & N. = An Old English poem of the Owl and the Nightingale, edited by F. H. Stratmann, Krefeld, 1868, previously edited by T. Wright, Percy Soc. vol. xi. 1843. [? Dorset, Wilts, or Hants ; about 1225.] Quoted by line.

O'BRIEN = An Irish-English Dictionary by J. O'Brien, second edition, Dublin, 1832.

OCTAV. (H.) = The Romance of the Emperour Octavian, edited by J. O. Halliwell, Percy Soc. vol. xiv. 1844. [Before 1400.] Quoted by line.

OCTAV.(S.) = The same, edited by G. Sarrazin, in Kölbing's Altenglische Bibliothek, vol. iii. pp. 64 ff. [About 1400.] Quoted by line.

OCTOV. (S.) = Octovian Imperator edited by G. Sarrazin, in Kölbing's Altenglische Bibliothek, vol. iii. p. 1–63. [Before 1400.] Quoted by line.

OCTOV. (W.) = The same in Weber's Metrical Romances (1810), vol. iii. p. 157–239. Quoted by line.

O'REILLY — An Irish-English Dictionary by T. O'Reilly, Dublin, 1817.

ORF. = Sir Orfeo, herausgegeben von O. Zielke, Breslau, 1880. [About 1300.] Quoted by line.

ORIG. = Originals and Analogues of some of Chaucer's Canterbury Tales, Chaucer Soc., 1875. Quoted by page.

ORM. = The Ormulum by Ormin, edited by R. M. White, Oxford, 1852, re-edited by R. Holt, 1878. [? Lincolnshire about 1200.] Quoted by line, or title of piece and line.

OUDEM. = Bijdrage tot een middel- en oudnederlandsch woordenboek door A.C. Oudemans, Arnhem, 1869–1878.

P. = A Poem on the Times of Edward II, edited by C. Hardwick, Percy Soc., vol. xxviii. 1849. [About 1320.] Quoted by stanza.

P. L. S. = Early English Poems and Lives of Saints, edited by F. J. Furnivall, Philol. Soc., 1865. Quoted by number of piece and line (or stanza).

P. P. = Political Poems and Songs, edited by Thos. Wright, Rolls Series, London, vol. i, 1859. Quoted by page.

P. P. II. = The same, vol. ii., 1861. Quoted by page.

P. R. L. P. = Political, Religious, and Love Poems, edited by F. J. Furnivall, E.E.T.S., No. 15, 1866. Quoted by page.

P. S. = The Political Songs of England, from the reign of John to that of Edward II, edited by T. Wright, Camden Soc., No. vi. 1839. Quoted by page.

PALL. = Palladius on Husbondrie, edited by B. Lodge and S. T. Herrtage, E.E.T.S., Nos. 52 & 72, 1872. [About 1420.] Quoted by book and line.

PARTEN. = The Romans of Partenay or Lusignen, edited by W. W. Skeat, E.E.T.S., No. 22, 1866. [About 1500.] Quoted by line.

PARTON. = The Old English version of Partonope of Blois, edited by W. E. Buckley,

Roxburghe Club, 1862. [About 1440.] Quoted by line.

PEN. PS. = A Paraphrase of the Seven Penitential Psalms, in English verse, supposed to have been written by T. Brampton, in 1414, edited by W. H. Black, Percy Soc., vol. vii. 1842. Quoted by page.

PERC. = The Romance of Sir Perceval of Galles, edited by J. O. Halliwell, in the Thornton Romances, pp. 1–87, Camd. Soc., No. xxx. 1844. [Yorkshire the beginning of the 15th cent.] Quoted by line.

PILGR. S. V. = The Pilgrims' Sea Voyage, edited by F. J. Furnivall, in E.E.T.S., No. 25, 1867. Quoted by line.

PL. CR. = Pierce the Ploughman's Crede, edited by W. W. Skeat, E.E.T.S., No. 30, 1867. [About 1394.] Quoted by line.

PL. T. = The Plowman's Tale, printed in CH. pp. 157–168; reprinted in P. P. ['Complaint of the Ploughman'] pp. 304–346. [? About 1400.] Quoted by stanza.

PR. C. = The Pricke of Conscience by R. Rolle de Hampole, edited by R. Morris, Philol. Soc., 1863. [Yorkshire about 1340.] Quoted by line.

PR. P. = Promptorium Parvulorum, sive Clericorum, dictionarius Anglo-Latinus princeps, auctore fratre Galfrido, grammatico dicto, ex ordine fratrum predicatorum, Northfolciensi, circa A.D. 1440, ad fidem codicum recensuit Albertus Way, Camden Soc., Nos. xxv., liv., lxxxix., 1843–65. Quoted by page.

PROC. = The only 'English Proclamation of Henry III, 18 Oct. 1258, edited by Alex. J. Ellis, Philol. Soc., 1868. Quoted by line of original MS.

PS. = Anglo-Saxon and Early English Psalter, edited by T. Stevenson, Surtees Soc., Nos. xvi, xix, 1843–7. [Before 1300.] Quoted by psalm and verse.

QUAIR = The King's Quair, edited by W. W. Skeat, Scottish Text Soc., 1884. [About 1423.] Quoted by stanza.

R. P. = Religious Pieces in Prose and Verse, edited from R. Thornton's MS. [about 1440] by G. G. Perry, E.E.T.S., No. 26, 1867. Quoted by page.

R. R = The Romaunt of the Rose, formerly attributed to Chaucer, printed in CH. (T.). [? About 1400.] Quoted by line.

R. S. = Religious Songs, edited by Thos. Wright, Percy Soc., vol. xi., 1843. [About 1225.] Quoted by number of piece.

R. W. = [Royal Wills.] A Collection of Wills now known to be extant of the Kings and Queens of England from William I to Henry VII. Soc. of Antiquaries, 1780. Quoted by page.

REB. LINC. = Chronicle of the Rebellion in Lincolnshire, 1470, edited by J. G. Nichols,

in Camden Miscellany, vol. i., Camd. Soc. No. xxxix. Quoted by page.

REL. = Reliquiae Antiquae, edited by T. Wright and J. O. Halliwell, London, 1841–3. Quoted by volume and page.

RICH. = Richard Coer de Lion, in Weber's Metrical Romances (1810), vol. ii. 3–278. [About 1325.] Quoted by line.

RICHEY = Idioticon Hamburgense von M. Richey, Hamburg, 1755.

RICHTH. = Altfriesisches Wörterbuch von Karl von Richthofen, Göttingen, 1840.

RIETZ = Svenskt Dialekt-lexikon of J. E. Rietz, Lund, 1867.

ROB. = Robert of Gloucester's Chronicle, edited from Harl. MS. by T. Hearne, London, 1810. [Gloucestershire about 1300.] Quoted by page.

ROB. (W.) = The same, edited by W. A. Wright, from Cotton MS. Caligula A xi., Rolls Series, London, 1887. Quoted by line.

RQF. = Glossaire de la langue romane par J. B. Roquefort, Paris, 1808.

S. A. L. = Sammlung altenglischer Legenden, von C. Horstmann, Heilbronn, 1878. Quoted by page.

S. & C. I. = Songs and Carols, printed from a MS. in the Sloane Collection in the British Museum, edited by T. Wright, Warton Club, No. 3, 1856. [? Warwickshire beginning of the 15th cent.] Quoted by piece.

S. & C. II. = Songs and Carols, from a MS. of the 15th cent., edited by T. Wright, Percy Soc. vol. xxiii. 1847. Quoted by piece.

S. B. W. = King Solomon's Book of Wisdom, edited by F. J. Furnivall, in E.E.T.S., No. 69, 1878. [? About 1400.] Quoted by line.

S. C. D. J. = Solomon's Coronation, Deeds, and Judgment, edited by F. J. Furnivall, in E.E.T.S., No. 69, 1878. [? About 1400.] Quoted by line.

S. J. = A Song of Joy for Christ's Coming, edited by F. J. Furnivall, in E.E.T.S., No. 69, 1878. [? About 1400.] Quoted by line.

S. S. (WEB.) = The Proces of the Sevin Sages, in Weber's Metrical Romances (1810), vol. iii, pp. 3–153. [About 1320.] Quoted by line.

S. S. (WR.) = The Seven Sages, edited by T. Wright, Percy Soc., vol. xvi. 1846. [About 1425.] Quoted by line.

SACR. = The Play of the Sacrament, edited by W. Stokes, Philol. Soc., 1862. [About 1460.] Quoted by line.

SAINTS (Ld.) = Early South-English Legendary from Laud MS. 108, edited by C. Horstmann, E.E.T.S., No. 87, 1887. [About 1290.] Quoted by number of piece and line, or by page.

SAL. & SAT. = The Dialogue of Salomon and Saturnus, edited with introduction by J. M. Kemble, London, 1848. [? About 1300.] Quoted by line.

SAX. CHR. = Two of the Saxon Chronicles, edited by J. Earle, Oxford, 1865. [MS. Laud from 1122–1164, Peterborough]. Quoted by page.

SCH. & LÜB. = Mittelniederdeutsches Wörterbuch von K. Schiller und A. Lübben, Bremen, 1875.

SCHAMB. = Wörterbuch der niederdeutschen Mundart der Fürstenthümer Göttingen und Grubenhagen von G. Schambach, Hanover, 1858.

SCHMELL. = Glossarium Saxonicum e poemate Heliand inscripto et minoribus quibusdam priscae linguae monumentis collectum a J. A. Schmeller, Monachii, Stuttgartiae et Tubingae, 1840.

SCHMID = Die Gesetze der Angelsachsen, herausgegeben von Reinh. Schmid, 2. Auflage, Leipzig, 1858.

SCHUL. = Gothisches Glossar von Ernst Schulze.

SCHÜTZE = Holsteinisches Idiotikon von J. F. Schütze, Hamburg, 1800–6.

SHOR. = The Religious Poems of William de Shoreham, edited by T. Wright, Percy Soc., vol. xxviii, 1849. [Kent about 1315.] Quoted by page.

SPEC. = Specimens of Lyric Poetry composed in England in the reign of Edward I, edited by T. Wright, Percy Soc., vol. iv. 1842 [from Harl. MS. 2253 (see A. D.).] Quoted by page.

SQ. L. DEG. = The Squyr of Lowe Degre, in Ritson's Metrical Romances (1802), vol. iii, pp. 145–192. [? About 1475.] Quoted by line.

ST. COD. = Codicem manuscriptum Digby 86, in bibliotheca Bodleiana asservatum descripsit, excerpsit, illustravit E. Stengel, Halis, 1871. Quoted by page.

ST. R. = The Stacions of Rome, edited by F. J. Furnivall, E.E.T.S., No. 25, 1867. [Vern. MS. about 1375.] Quoted by line.

STOR. = A Selection of Latin Stories, edited by T. Wright, Percy Soc., vol. viii. 1842. Quoted by page.

TEST. CRES. = Testament of Creseide [by Henryson], formerly attributed to Chaucer, printed in Stowe's edition, 1561. [About 1430.] Quoted by stanza.

TEST. LOVE = The Testament of Love [? by Hoccleve, about 1400], formerly ascribed to Chaucer, printed in CH. Quoted from N. E. D.

TODD = Illustrations of the Lives and Writings of Gower and Chaucer, by H. J. Todd, London, 1810. Quoted by page.

TOR. = Torrent of Portugal, edited by J. O. Halliwell, London, 1842. [About 1435.] Quoted by line.

TOR. (A.) = The same [' Torrent of Portyngale '] re-edited by E. Adam, E.E.T.S., No. li. 1887. Quoted by line.

TOWNL. = The Towneley Mysteries, Surtees Soc., No. 3, 1836. [Yorkshire about 1450.] Quoted by page.

TRANS. = Archaeologia : Society of Antiquaries of London. Quoted by volume and page.

TREAT. = Popular Treatises on Science, edited by T. Wright [containing "þe riȝt put of helle" from Harl. MS. 2277, about 1300 ; another copy of this text occurs in SAINTS (L.) xlvi.], Historical Soc. of Science, London, 1841. Quoted by page.

TREAT. F. = Treatyse of Fysshinge with an Angle, attributed to dame Juliana Bernes, edited by T. Satchell, London, 1883. [15th cent.] Quoted by page.

TREV. = Polychronicon Ranulphi Higden, together with the English translation of John Trevisa, edited by C. Babington and R. Lumby, London, Rolls Series. 1865–86. [? Warwickshire, about 1390.] Quoted by volume and page.

TREV. BARTH. See BARTH.

TRIAM. = The Romance of Sir Triamoure, edited by J. O. Halliwell, Percy Soc., vol. xvi. 1846. [About 1430.] Quoted by line.

TRIST. = Die Nordische und die Englische Version der Tristan-Saga, herausgegeben von E. Kölbing, Heilbronn, 1882–3. Also edited by G. P. McNeill, Scottish Text Soc., 1885–6. [About 1320.] Quoted by line.

TUND. = Visions of Tundale, edited by W. D. Turnbull, Edinburgh, 1843. [? Beginning of 15th century.] Quoted by line.

V. & V. = Vices and Virtues, a Middle English Dialogue, edited by F. Holthausen, E.E.T.S., No. 89, 1888. [About 1200.] Quoted by page.

VAN DER SCH. = Teuthonista, of duytschlender, van Gherard van der Schueren, uitgegeven door C. Boonzajer, Leyden, 1804.

VOC. = A volume of vocabularies, edited by Thos. Wright, 1857. Quoted by page.

VOC. (W. W.) = The same, edited by R. P. Wülcker, London, 1884. Quoted by column.

W. & I. = Wills and Inventories from the Register of the Commissary of Bury St. Edmunds, etc., edited by S. Tymms, Camden Soc., No. xlix. 1850. Quoted by page.

WART. = The History of English Poetry, by Thos. Warton, London, 1840. Quoted by volume and page.

WICL. = The Holy Bible in the earliest English versions made from the Latin Vulgate by John Wycliffe and his followers, edited by For-hall and Madden, Oxford, 1850. [About 1380.] Quoted by title of book, chapter, and verse.

WICL. E. W. = The English Works of John Wyclif, hitherto unprinted, edited by F. D. Matthew, E.E.T.S., No. 74, 1880. [About 1380.] Quoted by page.

WICL. S. W. = Select English Works of John Wyclif, edited by T. Arnold, Oxford, 1869-71. [About 1380.] Quoted by volume and page.

WILL. = The Romance of William of Palerne, edited by W. W. Skeat, E.E.T.S., No. i., 1867. [About 1350.] Quoted by line.

WINT. = Đe Orygynale Cronykil of Scotland be Androw of Wyntown, London, 1795.

[About 1425.] Quoted by book, chapter, and line.

WR. DICT. = Dictionary of Obsolete and Provincial English, compiled by T. Wright, London, 1857. Quoted by page.

YORK = The Plays Performed by the Crafts or Mysteries of York on the day of Corpus Christi in the 14–16th cent., edited by Miss L. T. Smith, Oxford, 1885. [About 1430.] Quoted by piece and line.

ZEIT. = Zeitschrift für deutsches Alterthum, herausgegeben von Mor. Haupt und E. Steinmeier. Quoted by volume and page.

ZUP. ÜBUNGSB. = Alt- und mittelenglisches Übungsbuch von J. Zupitza, Oppeln, 1868. Quoted by page.

COMPARATIVE TABLE OF REFERENCES TO CHAUCER'S CANTERBURY TALES.

SIX-TEXT. GROUP.	LINE.	HEADING.	TYRWHITT. LINE.	WRIGHT. LINE.	MORRIS. VOL.	PAGE.	BELL & SK. VOL.	PAGE.
A § 1	1	General Prologue	1	1	ii.	1	i.	73
§ 2	859	Knight's Tale	861	861		27		115
§ 3	3109	Miller's Prologue	3111	3111		96		188
§ 4	3187	,, Tale	3187	3187		98		192
§ 5	3855	Reeve's Prologue	3853	3853		120		216
§ 6	3921	,, Tale	3919	3919		122		220
§ 7	4325	Cook's Prologue	4323	4323		135		235
§ 8	4365	,, Tale	4363	4363		136		236
B § 1	1	Man of Law's Head Link or Prolog.	4421	4421		170		267
§ 2	99	,, Prologue and Tale	4519	4519		173		272
§ 3	1163	,, ,, ,, Link or Shipman's Prologue [or Squire's Prologue]						
§ 4	1191	Shipman's Tale	12903	14384	iii.	106	ii.	90
§ 5	1625	,, End Link or Prioress' Prol.	12931	14412		107		91
§ 6	1643	Prioress' [Prologue and] Tale	13365	14846		121		106
§ 7	1881	,, End Link or Sir Thopas' Prol.	13383	14864		122		107
§ 8	1901	Sir Thopas	13621	15102		130		115
§ 9	2109	,, End Link or Prol. to Melibeus.	13642	15123		131		117
§ 10	2157	Tale of Melibeus	pp. 106–120	pp. 150–166		138		127
§ 11	3079	Melibeus' End Link	13895	15375		139		129
§ 12	3181	Monk's Tale (1)	13997	15477		198		181
	3565	,, ,, (2)* ('Modern' cases)	14685*	15861		201		187
	3653	,, ,, (3).*	14338*	15949		213		199
§ 13	3957	Monk's End Link	14773	16253		216		203
§ 14	4011	Nun's Priest's Tale	14827	16307		227		213
§ 15	4637	,, ,, End Link	15453	[none†]		[none†]		235
C	[none]	Doctor's Prologue	11929	13410		75		56
§ 1		,, Tale	11935	13416	iii.	75	ii.	56
§ 2	287	,, Link or Pardoner Prol.	12221	13702		85		66
§ 3	329	Pardoner's Preamble or Prol. or Tale.	12263	13744		86		68
§ 4	463	,, Tale	12397	13878		90		73
D § 1	1	Wife's Preamble or Prologue	5583	5583	ii.	206	i.	306
§ 2	857	,, Tale	6440	6440		232		334
§ 3	1265	Wife-Friar Link	6847	6847		245		349
§ 4	1301	Friar's Tale	6883	6883		246		351
§ 5	1665	Friar-Summoner Link	7247	7247		258		364
§ 6	1709	Summoner's Tale	7291	7291		259		365
E § 1	1	Clerk's Head Link or Prologue	7877	7877		278		386
§ 2	57	,, Tale	7933	7933		280		388
	1177	Chaucer's Lenvoy	9053	9053		315		421
§ 3	1213	Clerk-Merchant Link or Mcht's. Prol.	9089	9089		317		422
§ 4	1245	Merchant's Tale	9121	9121		318		424
§ 5	2419	,, End Link or Squire's Prol.	10293	10293		354		461
F § 1	1	Squire's Head Link	10315	10315		354		461
§ 2	9	,, Tale	10323	10323		355		463
§ 3	673	Squire-Franklin Link or Frkl.'s Prol.	10985	10985	iii.	1		486
§ 4	709	Franklin's Proem and Tale	11021	11021		2		489
G § 1	1	Second Nun's Tale	15469	11929		29	ii.	6
§ 2	554	Link or Prol. Canon's Yeoman's	16022	12482		46		24
§ 3	720	Canon's Yeoman's Pream. or Tale §	16188	12648		51		29
§ 4	972	,, ,, Tale §	16440	12900		58		38
H § 1	1	Manciple's Head Link or Prologue	16950	16933		249		236
§ 2	105	,, Tale	17054	17037		252		240
I § 1	1	Blank-Parson Link	17312	17295		261		249
§ 2	75	Parson's Tale	pp. 148–172	pp. 185–211		263		254

TYRWHITT's order following the Ellesmere MS., etc.: A; B § 1, 2; D; E; F; C; B § 3–14; G; H; I.
WRIGHT's, MORRIS', and BELL's order: A; B § 1, 2; D; E; F; G; C; B § 3–14; H; I.
In Wright's 3 vol. edition B § 10 is in vol. ii, 323–386, and I § 2 in vol. iii. 82–189.

---

* In Tyrwhitt's edition, Divisions 2 and 3, are transposed (*following the Ellesmere and other MSS.*).
† But inserted as a footnote.   § There is great disagreement in the MSS. and printed editions here.

LIST OF GENERAL ABBREVIATIONS AND SIGNS.

A. Fr. = Anglo-French.
acc. = accusative.
adj. = adjective.
adv., *adv.* = adverb.
card. num. = cardinal numeral.
cf. = *conferatur*, compare.
comp. = comparative.
comp. = compounds.
conj., *conj.* = conjunction.
Dan. = Danish.
Du. = Dutch.
dat. = dative.
dem. = demonstrative.
deriv. = derivatives.
fem. = feminine.
Fr. = French.
fut. = future.
gen. = genitive.
Ger. = German.
Goth. = Gothic.
Gr. = Greek.
Heb. = Hebrew.
imper. = imperative.
impers. = impersonal.
indic. = indicative.
interj. = interjection.
interrog. = interrogative.
Lat. = Latin.
M. Du. = Middle Dutch.
M. H. G. = Middle High German.
M. L. G. = Middle Low German.
m., *masc.* = masculine.

ms. = manuscript, *mss.* manuscripts.
med. Lat. = Medieval Latin.
mod. Eng. = modern English.
n., *neut.* = neuter.
nom. = nominative.
num. = numeral.
O. E. = Old English (Anglo-Saxon).
O. Fr. = Old French.
O. Fris. = Old Frisic.
O. H. G. = Old High German.
O. L. G. = Old Low German (Old Saxon).
O. N. = Old Norse.
ord. num. = ordinal numeral.
pers. = person.
pl. = plural.
pple., *pple.* = participle.
pr. n. = proper noun.
prep., *prep.* = preposition.
pres. = present.
pret. = preterite.
pron., *pron.* = pronoun.
prov. = provincial.
r. = read.
r. w. = riming with.
rel. = relative.
sb., *sb.* = substantive.
sc. = scilicet.
subj. = subjunctive.
superl. = superlative.
Sw. = Swedish.
v. = verb.

The asterisk (*) after a reference indicates that the passage cited is contained not in the text but in a note. In the case of parallel texts it indicates a quotation from the second of the two versions.

For the meaning of the marks over the vowels see Preface, pp. viii, ix.

a, see ꝥ **al, an,** ꝥ **and, at, habben, he, of,** ꝥ **oð.**
[**a-,** prefix, *O.E.* a-, = *O.L.G., O.Fris.* ā-, *O.H.G.*
ar-, er-, ir-, ur-, *Goth.* us-.]

[**a-²,** prefix, = **an-.**]

[**a-³,** prefix, = **and-.**]

[**a-⁴,** prefix, = **ȝe-.**]

[**a-⁵,** prefix, = **of-.**]

ā, adv., *O.E.* ā, ō, = *O.N.* ā, æ, ei, ey, *O.L.G.,*
O.H.G. eo, io, *Goth.* aiw; *ever, always,* LAȝ.
4013; (*miswritten* æ) 1269; a wurþe hire wa
FRAG. 8; ær ic hit a wuste P. L. S. viii. 9;
world a buten ende A. R. 396; JUL. 22; ever
and a C. M. 18010; a [aa] on ecnesse KATH.
664; þat schal ai [aa] stonden 1490; a, aȝȝ
ORM. 265, 409, 1688; o and o HOM. II. 5; o
buten ende MISC. 73; oo, ai GEN. & EX. 111,
451; ay and oo MAP 337; ai REL. I. 210;
HAV. 159; WILL. 2239; CH. C. T. *A* 63;
LIDG. M. P. 46; þe blisse þat ai shal be
SPEC. 75; longe is ai ant longe is o 106;
ai wiþouten ende B. DISC. 531; ai forþ JOS.
126; for ai IW. 1510; in ai PS. ix. 6; ai HORN
(R.) 159; ei and o REL. I. 275; **ā māre**
[= *O.H.G.* eomer, *mod.G.* immer], *evermore,*
HOM. I. 11; a, ai mare MARH. 3, 5; ai
mare KATH. 1719, 2188.

ā², interj. = *M.H.G.* ā; *ah, oh,* LAȝ. 5015; A.
R. 52; JUL. 70; ORM. 12808; BEK. 1177;
C. L. 1077; S. S. (Wr.) 1250; LANGL. *B* xvii.
124; CH. C. T. *A* 1078; a hwi wepest þu
HOM. I. 45; **ā hā,** *aha,* 'evax' PR. P. 8;
FER. 1148; CH. C. T. *D* 586; **ā ȝē** (? *ms.*
aȝe) P. L. S. xvii. 481; **ā wei,** *ah woe,* LAȝ. 7223;
a wai MISC. 192; a wi HOM. II. 193; a wei le
MAP 336.

aa, see **ā.   aas,** see **ās.   a-bac,** see **bac.**

a-bād, sb., *from* **abīden**; *abode, delay,* DEGR.
129; abod, abood CH. C. T. *A* 965; buten
abade *without delay* JUL. 73.

a-bǣlien, v., *O.E.* abæligan; *provoke to anger*;
ah Bruttes weoren bisie and ofte hine abǣi-
leden (*pret. pl.*) LAȝ. 10275.

abai, sb., *O.Fr.* abai; *condition of being at bay*:
þe dogge ... held it at abaie WILL. 46; at
abei DEGR. 238; to abai H. V. 70.

abaie, adv., *at bay*; so hound abaie *as a hound
at bay* ALIS. 3882.

abailen, v., *O.Fr.* abaillier; *reach, arrive at,*
LIDG. TROY v. 36.

abaiss, abaist, see **abassen.**

a-baiten, v., ? = *M.H.G.* erbeizen; *excite to
desire*; his flesshe on here was so abeitede
(*pple.*) H. S. 181.

abandonen, v., *O.Fr.* abandoner, *mod.Eng.*
abandon; *subdue; subject to another's control;
surrender, give up entirely*; abandoune BARB.
xv. 393; abandone (*pres.* 3 *pl.*) CH. C. T.
I 775; abandonit (*pret.*) BARB. iv. 655;
abandouned (*pple.*) GOW. I. 213; abandoned
III. 253.

abandūn, adv., *O.Fr.* a bandun (*see* **bandōn**);
*under one's control; at discretion; unchecked,
freely, recklessly*; his ribbes and scholder fel
adoun, men might see the liver abandoun
AR. & MER. 223; habben him so abandun
(abaundune) HOM. I. 189, 203; folwed him
abandoun GUY 181; in abandoun, *recklessly,*
R. R. 2342; at abandoun BARB. xv. 59.

abandon-lī, adv., *recklessly,* BARB. viii.
461; xi. 629.

a-bánen, v., *from* **báne,** sb.; *afflict with
disease*; abáne (*imper.*) LANGL. *C* ix. 226.

abasch, sb., *from* **abassen**; *fright, shame*;
abassche GOW. II. 46.

abaschance, sb., *O.Fr.* abaïssance *for* esbaïs-
sance; *dismay, confusion*; abasshaunce
GENER. 5515.

abaschement, sb., *dread,* 'terror' PR. P. 5.

abassen, v., *O.Fr.* esbahiss-; *abash,* CH.
BOET. iv. 7 (146); abàiss BARB. viii. 247;
abaist (*pple.*) MAND. 295; CH. BOET. iii.
12 (107); abasched ALIS. 224; abaischt
MAN. (F.) 5062; WICL. MK. v. 42; abaiste
PR. C. 1431.

abashinge, sb., *abashing, confusion*, CH. BOET. iv. I (109); abasing BARB. xvii. 573.

abatailen, v., *O.Fr.* enbatailler; *embattle*; abatailed (*pple.*) FER. 4310.

abatailment, sb., *battlement*, GAW. 790.

abáte, sb., *discomfiture, surprise*, QUAIR 40.

abatement, sb., *O.Fr.* abatement *for* enbatement; *wrongful seizure of property*, MAN. (H.) 278.

abatement², sb., *O.Fr.* abatement; *abatement, diminution*, PR. P. 5.

abáten, v., *O.Fr.* abatre; *abate; cast down; humble; soften; abolish*; abatin '*subtraho*' PR. P. 5; abatie AYENB. 28; abati ROB. 54; abate WILL. 1141; **abáteþ** (*pres.*) TREV. II. 185; abates PR. C. 1672; **abáted** (*pret.*) P. S. 216; **abáted** (*pple.*) C. L. 1334.

abáve, v., *O.Fr.* esbaubir; *astonish; be astonished*; LIDG. M. P. 144; **abáved** (*pple.*) MAN. (H.) 210; abawed CH. D. BL. 614; abawede (*pl.*) H. S. 9536.

abbeie, sb., *O.Fr.* abbeie, *from Lat.* abbātia; *abbey*, CHR. E. 914; abbei BRD. 12; abbei LAȝ. 29717*; abbai BARB. xx. 599.

abbesse, *see* **abbodesse**.

abbod, sb., *O.E.* abbod, *from Lat.* abbātem; *abbot*, A. R. 314; abbed [abbod] LAȝ. 13131; abbot BRD. 12; abbede (*dat.*) LAȝ. 13117; **abbates** (*pl.*) SAX. CHR. 250.

  abbod-rīche, sb., *O.E.* abbodrīce; *office of abbot*, MISC. 145.

abbodesse, abbesse, sb., *O.E.* abbodisse, *O.Fr.* abbesse; *abbess*: abbodesse LANGL. C vii. 128; abbesse CHR. E. 550; abbes M. H. 164; abbese ROB. 370.

abbodie, sb., *abbacy*, (*ms.* abboddie) MISC. 145.

a-be-ce, sb., *alphabet*, ROB. 266; AYENB. I.

a-bēȝen, abūȝen, v., *O.E.* abēgan, abȳgan; *bend, bow* (*trans. and intrans.*): to non herre i-schal abuie ROB. 102; **abūide** (*pret.*) ROB. 106; ni þei abeiȝedon hem no thing to þe kingus hest ED. 3458.

abeie, *see* **abüggen. abel**, *see* **able**.

a-bélden, v., *grow bold*; abelde ALIS. 2442.

a-belȝen, v., *O.E.* abelgan, = *O.H.G.* arbelgan; *make angry, grow angry*; **abalh** (*pret.*) LAȝ. 26359*; abelh HOM. I. 111; **abolȝen**, abolwen (*pple.*) LAȝ. 1565, 6396; þat hi ne be abolke (? *for* abolȝe) SHOR. 22.

abeliche, *see* **able-līche**.

a-bēoden, v., *O.E.* abēodan, = *O.H.G.* ar-, irbiotan; *offer, announce*; abeode, abede, abude FER. 1540, 1833, 3853; **abēd** (*pret.*) FER. 1829; and his ærnde abed LAȝ. 4423; þai abóde (*plur.*) FER. 1985.

a³-bēre, sb., *cf. O.H.G.* antpāra, *M.H.G.* ambære; *gesture*, SHOR. 60.

a-béren, v., *O.E.* aberan; *bear, carry, endure*, '*portare*,' LK. xi. 46; abeoren HOM. I. 35; abeore bličeliche þe derf H. M. 17; **aberečð** (*pres.*) A. R. 158; aber (*pret.*) KATH. 1555; ure drihten aber ure sunnan HOM. I. 121; aber up *bore up* 225.

abernen, *see* **abrennen**.

abesche, v., *O.Fr.* abechier (*feed with the beak*); *feed*; **abeshed** (*pple.*) GOW. III. 25.

abesse, v., *O.Fr.* abessier = abaissier; *abase*, GOW. I. 111.

abet, sb., *O.Fr.* abet; *fraud, cunning; instigation*; CH. TRO. (T.) ii. 357; mid gile and his abette SHOR. 58.

abette, v., *O.Fr.* abeter; *instigate, incite*, FER. 5816.

  abetting, sb., *abetting* [*v.r.* abet] CH. TRO. ii. 357.

abettement, sb., *A.Fr.* abettement; *abetment*, (*spelt* anbettyment) FER. 2364.

a-bidden, v., *O.E.* abiddan, = *O.L.G.* arpittan; (*refl.*) *offer one's prayer*: þe hwile ich go to abidde (? *for* ȝebidde) me MISC. 41.

a-bīden, v., *O.E.* abīdan, = *O.H.G.* irbītan, *Goth.* usbeidan; *abide, remain, await*, A. R. 208; ORM. 1801; ne durste him nan abiden [abide] LAȝ. 1583; abide LANGL. *B* viii. 64; *A* ii. 210; *C* xxi. 479; HAV. 1797; þu ne darst domes abide O. & N. 1695; if he most abide þe dai BEK. 736; **abideþ** (*pres.*) H. M. 17; abit A. R. 338; MARH. 21; REL. I. 115; R. R. 4989; **abīd** (*imper.*) LAȝ. 21623; ORM. 14020; no abide he nævere þære dægen LAȝ. 21555; **abād** (*pret.*) KATH. 719; ORM. 217; IW. 1180; abad [abod] LAȝ. 5681; sevene night he čer **abiden** (*pret. pl.*); abide AYENB. 13; **abiding** (*pple. pres.*) LANGL. *B* xix. 289; *C* xix. 136; xxiii. 143; abiden (*pple.*) CH. C. T. *E* 1888; TOWNL. 315; abide C. L. 1339; so longe ich (h)abbe abide þer after L. H. R. 20.

  abīding, sb., *O.E.* abīdung; *abiding, waiting, delay, expectation*: noght schende þou me fra abiding mine PS. cxviii. 116; abidinge AYENB. 173; abiding A. P. iii. 419; abidinge WICL. GEN. xlix. 10.

abilitē, sb., *O.Fr.* habilité; *ability*, CATH. 2; CH. ASTR. PROL.; ablete WICL. E. W. 331.

abill, *see* **able**.

abīme, sb., *O.Fr.* abisme, abīme; *abyss*, C. M. 22678.

abit, *see* **habit**.

a-bīten, v., *O.E.* abītan, = *O.H.G.* irpīzan; *bite; bite in pieces, to death*; LAȝ. 26995; þu starest so þu wille abiten al þat þu miht mid clivre smiten O. & N. 77; abute ALIS. 5611; **abīteþ** (*pres.*) LANGL. *C* xix. 32; mete ne drinke he nabit (*for* ne abit) FL. & BL. (L.) 40; abite (*pres. pl.*) LANGL. *B* xvi. 26; **abiten** (*pret. pple.*) REL. II. 276.

ablaqveaciōn, sb., *Lat.* ablaqveātiō ; *ablaqueation, laying bare of roots*, PALL. ii. 1.

ablaqveate, v., *Lat.* ablaqveāre ; *lay bare the roots of a tree* ; **ablaqviate** (*imper.*) PALL. iv. 91.

a²-blăsten, v., *O.E.* onblǣstan ; *blow upon* ; venim and fire togider he cast, that he Jason so sore **ablăste** (*pret.*) GOW. II. 251.

a-blāwen, v., *O.E.* ablāwan ; *blow up* : þos he gan hire herte ablowe SHOR. 160 ; **ablēow**, **ablēu** (*pret.*) HOM. I. 99 ; þet was **ablōwe** (*pple.*) þorȝ þe fenim of þe fende SHOR. 166.

able, adj., *O.Fr.* hable ; *able, suitable, apt*, '*habilis, idoneus*,' PR. P. 5 ; CH. C. T. *A* 167 ; abill BER. 3237 ; *see* **hable**.

    able-līche, adv., *ably* ; abeliche MAS. 243.

    ablenes, sb., *ability*, TEST. LOVE (N. E. D.) ; ablinesse BARTH. (N. E. D.).

ablen, v., *from* **able** ; *enable, qualify, empower, dispose* : **ablith** (*pres.*) REL. II. 44 ; **abiled** (*pple.*) HAMP. TR. 20 ; abelled E. G. 316.

a-blenden, v., *O.E.* ablendan, = *O.H.G.* irblendan ; *blind, deprive of sight*, LAȜ. 14657 ; this leme shal Jupiter ablende LANGL. *B*. xviii. 137 ; **ablendeð**, **ablent** (*pres.*) A. R. 86, 214 ; ablent AYENB. 16 ; **ablende** (*pret.*) HOM. I. 121 ; ROB. 208 ; ablente LANGL. *B* xviii. 323 ; **ablend** (*pple.*) MARH. 15 ; ablent P. S. 380.

a-blīken, v., *O.E.* ablīcan, = *M.H.G.* erblīchen ; *grow pale* ; **ablikeþ** (*pres.*) ST. COD. 53.

a-blinden, v., *O.E.* ablindan, = *O.H.G.* erblinden ; *grow blind* ; **ablíndeð** (*pres.*) A. R. 92 ; ablindað HOM. I. 109 ; ha ablindað MARH. 15.

abloi, interj., *?O.Fr.* ablo ('*courage, allons*') ; GAW. 1173.

ablūcioun, sb., Lat. ablūtiō ; *ablution*, CH. C. T. *G* 856.

a-blūȝen, v., = *M.H.G.* erbliugen ; *frighten, dismay* : þa iwarð þat folc swiðe **ablūied** (*pple.*) HOM. I. 89.

abobben, v., *O.Fr.* abober ; *astonish* ; **abobbed** (*pple.*) AR. & MER. 1969.

abōd, *see* **abād**.

abof, **aboȝe**, *see* **abufen**.

a-boȝien, v., *bow* ; abouwie LEB. JES. 849 ; **aboȝede** (*pret.*) FER. 2070.

abominable, adj., *O.Fr.* abominable ; *abominable*, AYENB. 49 ; CH. C. T. *D* 2006.

abominacioun, sb., *O.Fr.* abomination ; *abomination*, CH. C. T. *D* 2179.

abortif, adj. & sb., *Lat.* abortīvus ; *abortive*, WICL. JOB iii. 16 ; abortiif PL. CR. 485 ; abortives (*pl.*) C. M. 22849.

a-bosten, v., *address boastfully* ; **abostede**, -ed (*pret.*) LANGL. *A* vii. 142 ; *B* vi. 156.

---

abound, adj., *O.Fr.* abonde, *Lat.* abundus, *cf.* abúndin ; *overflowing*, GENER. 311 ; habound TUND. 92.

about, **abouten**, **aboutestonding**, *see* **abūten**.

aboutien, v., *? O.Fr.* abouter, *? mod.Eng.* abut ; *lean forward, stick out* ; **aboutie** (*imper.* 3 *pers.*) A. R. 62.

above, **aboon**, *see* **abufen**.

a-brǣden, v., = *M.H.G.* erbreiten ; *dilate, open wide* : þe ule hire eȝen **abrǎd** (*?pret.*, *for* abradde) O. & N. 1044.

abrege, v., *O.Fr.* abregier ; *abridge*, PR. C. 4571 ; abregge CH. C. T. *A* 2999 ; **abregge** (*imper.*) MIRC 1629 ; **abreggid** (*pple.*) WICL. E. W. 1.

a-breiden, v., *O.E.* abregdan ; *draw out, spring up, awake ; blame, upbraid* ; A. R. 214 ; abreide '*exprobrare*' REL. I. 7 ; **abreid** (*pret.*) A. R. 238 ; abred MAT. xxvi. 51 ; his sweord he ut abræid LAȜ. 26553 ; abraid GEN. & EX. 231 ; Ulixes out of slepe abraid GOW. III. 54 ; **abroiden**, abrōden (*pple.*) HOM. I. 239 ; þe abroidene bureh HOM. II. 175, 209.

a-brēken, v., *O.E.* abrecan ; *break in pieces ; break forth, away* : ȝif we mai owhar abreke AR. & MER. 7903 ; **abræc** (*pret.*) LAȜ. 3532 ; **abrēken** (*pret. pl.*) MAP 338 ; **abróken** (*pple.*) HALLIW. 11 ; ich am ... abroken ut of þon benden LAȜ. 721 ; nu hafeð he mine ban alle abrokene 25928.

a-brennen, v., = *M.H.G.* erbrennen ; *burn up* ; **abernð** (*pres.*) HOM. I. 239.

a-brēoðen, v., *O.E.* abrēoðan ; *degenerate, decay* : þat teonðe werod **abrēað** (*pret.*) HOM. I. 219 ; si al swa swið(e) abreað and adile-ȝede 235.

abróche, adv., *see* **bróche**.

abróche, v., *O.Fr.* abrochier ; *broach (a cask)*, CH. C. T. *D* 177 ; A. P. i. 1122.

absence, sb., *O.Fr.* absence ; *absence*, PR. P. 5 ; CH. C. T. *G* 1214.

absent, adj., *O.Fr.* absent ; *absent*, PR. P. 5 ; CH. C. T. *H* 203.

absoile, *see* **assoilen**.

absolūciūn, sb., *O.Fr.* absolution ; *absolution*, HOM. II. 99 ; A. R. 146 ; absolucioun GAW. 1882.

absolūt, adj., *Lat.* absolūtus ; *absolute, unlimited*, CH. BOET. v. 6 (175).

absténe, v., *O.Fr.* abstenir ; *abstain*, FER. 3761 ; absteine WICL. NUM. vi. 3.

abstinence, sb., *O.Fr.* abstinence ; *abstinence*, AYENB. 236 ; CH. C. T. *I* 682.

abstinent, adj., *Fr.* abstinent ; *abstinent*, CH. C. T. *I* 798.

abstract, pple., *Lat.* abstractus ; *derived*, TREV. I. 21.

a²-bufen, adv. & prep., *O.E.* abufan (*for* on-

bufan); *above*; (*ms.*abufenn) ORM. 588; abuven
A. R. 304; HOM. II. 81; GEN. & EX. 10; þat
is us alle above MISC. 38; aboven HAV.
1700; MAND. 10; CH. C. T. *B* 1344; above
LAȝ. 10302*; AYENB. 35; upe þe doune
above Baþe ROB. 174; ase ich sede above
SHOR. 142; oboven PR. C. 849; abowen
(?=abouen) M. H. 105; aboȝe FER. 2972, 4319.

a-būȝen, v., *O.E.* abūgan; *bow, bend, submit*
(*intrans. and reflex.*): al him scal abuȝe
[abouwe] LAȝ. 18854; þat ich to eni mon
schule abouwe L. H. R. 53; abowe ALIS. 188;
EM. 981; abowe to þe kinges wille BEK. 867;
abūȝþ (*pres.*) O. & N. 782; **abēah** (*pret.*)
HOM. I. 227; abeh '*inclinavit*' II. 111; into
castle he abeh (*went*) LAȝ. 5210.

abūȝen, *see* **abēȝen.**

a-büggen, v., *O.E.* abycgan; *buy, pay for,*
atone for, HOM. I. 41; A. R. 188; SPEC. 112;
þu me smite bi þon rugge ah sare þu hit salt
abuggen LAȝ. 8158; abugge HORN (R.)
1081; C. L. 394; abugge, abeie HORN
(L.) 110, 1075; abiggen ORM. 6907;
abigge ALIS. 901; MIRC 2010; abugge,
abigge LANGL. *A* iii. 236; abegge AYENB. 73;
abouhte (*pret.*) MISC. 64; aboughte CH.
C. T. *A* 2303; abouȝte LANGL. *B* ix. 142;
abouȝte (*pret. plur.*) LANGL. *B* x. 281;
aboht (*pple.*) H. H. 59; þat we habbeþ deore
aboȝt BEK. 1312; about AN. LIT. 91; abouȝt
LANGL. *C* xxi. 433; abouȝte (*pl.*) *B* xviii. 386.

abundance, sb., *OFr.* abundance; *abundance,*
AYENB. 261.

abundaunt, adj., *O.Fr.* abundant; *abundant,*
D. TROY 1695; abundand 5205; habundant
MAND. 179.

abúndin, v., *O.Fr.* abonder; *abound,* PR. P. 5;
abounde REL. I. 265.

abuschid, abussede, *see* **enbusche.**

abūsen, v., *Fr.* abuser; *abuse, misuse, misre-*
present, LIDG. BOCH. (*in* N. E. D.); abūsed
(*pple.*) GUIL.² i. 15 (*in* N. E. D.).

abūsioun, sb., *O.Fr.* abusion; *abuse,* WICL. PS.
xxxi. 18; abusion CH. TRO. iv. 991.

a²-būten, adv. & prep., *O.E.* a-,
on-būtan, = *O.Fris.* abūta; *about, around,*
MK. iii. 34; LK. xvii. 12; abuten A. R. 234;
þe deofel þe geð abutan al swa þe gredie
leo HOM. I. 127; his riht erm schal bicluppen
me abuten 213; abuten [aboute] Eowerwic
LAȝ. 16734; abuten midnihte 7983; abuten
hise lendes ORM. 3211; whan se þat preste
floc ... an siþe þeowted hafden al abuten
550; ever þu were abuten to echen þin ahte
MISC. 176; he foren abuten bi Adad
GEN. & EX. 2482; o and o abuten (*for* buten)
ende HOM. II. 7; abute midniht KATH. 1749;
hwanne oþre slepeþ hire abute O. & N. 1593;
abouten C. L. 703; abouten nine HAV. 1010;
aboute AYENB. 30; LANGL. *C* i. 193; *A* viii.
30; CH. C. T. *A* 2189; Iw. 2055; aboute al

þe urþe heo goþ TREAT. 137; (hi) beoþ aboute
hire to schende BEK. 763; obout PR. C. 1905.

aboute-stondinges, sb., pl., *circumstances,*
AYENB. 174, 175.

aboute-warde, adv., *on the way, trying* (*to*),
TRIAM. 65; EGL. 658.

abuven, *see* **abufen.**

ac, conj., *O.E.* ac, ah, oc, = *O.L.G.* ac, *Goth.*
ak, *O.H.G.* oh; *but,* ROB. 2; AYENB. 6;
HORN (H.) 860; ALIS. 236; S. S. (Web.) 1313;
FER. 3761; TRIST. 220; LANGL. *A* i. 119;
ORM. 63; ac þær biþ sor idol FRAG. 5; ac
aȝen riȝte hi beoþ BEK. 1624; nis nauȝt malice
ac hit is laȝe SHOR. 153; ac, ak C. L. 444;
WILL. 106, 110; ak P. S. 211; ac, ach, ah
HOM. I. 123, 145; ac, ah O. & N. 177, 1176;
LAȝ. 252, 13132; (*misprinted* at) 6901; ak
KATH. 5; SPEC. 46; HORN (R.) 841; CHR. E.
675; auh, auch A. R. 46; auȝ HOM. II. 258;
oc SAX. CHR. 253; GEN. & EX. 213; oc he
fleð fro him REL. I. 212; ok HICKES I. 231;
P. L. S. iii. 7.

a-cálden, v., *O.E.* acealdian, = *O.H.G.* ercal-
ten; *grow cold, make cold*; acoaldeð (*pres.*)
A. R. 404; accólded (*pple.*) GEST. R. 83;
acold P. 23; S. S. (Wr.) 2518.

a-cále, pple., *from O.E.* acalan; *stiff with*
cold: he was so sore acale GOW. III. 296;
acale [akale] LANGL. *B* xviii. 392; *C* xxi. 439;
akale S. S. (Web.) 1512.

a-cangen, v., *? become foolish*; **acanges** tu
KATH. 2112; acanget (*pple.*) 2045.

a-casten, v., *cast down*; akeasten MARH. 14;
akaste (*pret.*) MARH. 2; deað ne acaste nawt
Crist ah Crist overcom deað KATH. 1127;
acast (*pple.*) P. S. 149; TREV. IV. 95; nou is
mi counfort acast PL. CR. 99; akast H. M. 5;
akest A. R. 318.

accept, pple., Lat. acceptus; *accepted*: in time
accept WICL. 2 COR. vi. 2.

acceptable, adj., *O.Fr.* acceptable; *acceptable,*
GOW. III. 250; CH. C. T. *D* 1913.

accepten, v., *O.Fr.* accepter; *accept*; **accepte**
(*pres. subj.*) CT. LOVE 28.

acces, sb., *O.Fr.* acces; *access, attack of fever,*
HALLIW. 13; axsesse, accesse CH. TRO.
ii. 1578.

accidence, sb., *O.Fr.* accidence; *chance,* GOW.
II. 153.

accident, sb., *Lat.* accidens; *occurrence, attri-*
bute, CH. TRO. iii. 918; CH. C. T. *E* 607;
WICL. E. w. 466.

accidie, sb., *med.Lat.* accidia, *from Gr.*
ἀκηδία; *indolence, negligence,* A. R. 208;
sleauþe þet me clepeþ ine clergie accidie
AYENB. 16.

acciōn, sb., *O.Fr.* action; *action,* LANGL. *C* ii.
94; accioun MAN. (H.) 196.

accompt, sb., *O.Fr.* accompt, *cf.* **acount**;
account, GOW. II. 11, 286; acompte E. G. 8.

accomptable, adj., *accountable*, E. G. 379.

accompten, v., *account, reckon, esteem*; accompteth Gow. III. 298.

accustomaunce, sb., *O.Fr.* accoustumance; *custom* : againe our old accustomaunce DREAM 256.

acēmen, v., *O.Fr.* acesmer ; *adorn*, A. R. 360.

achápe, *see* escápen.

achat, sb., *O.Fr.* achat, acat ; *purchase, thing bought*, CH. BOET. i. 4 (15).

acháte, sb., *O.Fr.* acate ; *from Gr.* ἀχάτης ; *agate*, A. R. 134.

achatour, sb., *O.Fr.* achateur, acateur ; *buyer*; catour GAM. 321 ; **achatours**, acatouris (*pl.*) CH. C. T. *A* 568.

achaufen, *see* eschaufen.

áche, sb., *O.E.* æce, ece, *from* áken ; *ache*, MISC. 65 ; TREV. IV. 119 ; WICL. MAT. iv. 23 ; LUD. COV. 223 ; ache, ake *'dolor'* PR. P. 8 ; eche P. L. S. viii. 100 ; A. R. 282 ; HOM. II. 165 ; H. M. 35 ; hache WILL. 905 ; *comp.* hēaved-éche.

áche², sb., *O.Fr.* ache, *cf. mod.Eng.* (small-) age ; *parsley*, VOC. 139 ; PR. P. 6 ; SPEC. 26.

a-chēoken, v., *choke, suffocate* : achoken CH. BOET. ii. 5 (47) ; **achēked** (*pple.*) HOM. II. 181 ; CH. L. G. W. 2008 ; achoked HALLIW. 15.

a-chēosen, v., = *OH.G.* irchiesen, *O.L.G.* akiosan ; *choose* ; **æcēas** (*pret.*) HOM. I. 229.

achēsoun, sb., *O.Fr.* acheison (*for* ocheison) ; *occasion, motive*, AR. & MER. 132 ; *see* encheisūn.

achéven, v., *O.Fr.* achever ; *achieve ; bring to an end ; come to an end* ; GAW. 1838 ; achéved (*pple.*) R. R. 2049.

a-clēoven, v.,=*M.H.G.* erklieben ; *be cleft* ; þin herte aclēf (*pret.*) P. R. L. P. 252.

acloien, v., *O.Fr.* encloer, *cf. mod.Eng.* cloy ; *lame (a horse) with a nail ; lame, hinder, obstruct ; fill to excess* ; acloiȝen, acloiin '*acclaudico, acclavo*' PR. P. 6 ; **acloieth** (*pres.*) CH. P. F. 514; acloied (*pple.*) P. S. 335; acloied '*acclaudicatus*' PR. P. 6.

a-clomen, v., *cf. Du.* verkleumen ; *become torpid* ; aclommid (*pple.*) '*eviratus, enervatus*' PR. P. 6.

aclōsen, v., *O.Fr.* aclōre *for* enclōre, *see* clōsen ; *enclose* ; aclōsed (*pple.*) SHOR. 145.

a-clumsen, v., *grow clumsy* : oure hondis ben aclumsid (*pple.*) '*dissolutae sunt manus nostrae*' WICL. JER. vi. 24.

acoie, v., *O.Fr.* aqvoier ; *quiet, appease* ; acoie R. R. 3564 ; acoied (*pret.*) WILL. 56.

[**acoint**], **aqvointe**, adj. & sb., *O.Fr.* acoint ; *acquainted, acquaintance*, ROB. 465 ; aqvente R. R. 5023 ; acqveintis (*pl.*) CH. C. T. (W.) 7573 (*v. r.* acqveintances *D* 1991).

[**acointance**], **aqveintance**, sb., *O.Fr.* acointance ; *acquaintance* ; aqveintance CH. C. T.

A 245 ; aqveintaunce ALIS. 7258 ; aqvaintonce AYENB. 143.

acointen, v., *O.Fr.* acointer, *from med.Lat.* adcognitāre ; *acquaint* ; aqveintin '*notifico*' PR. P. 13 ; aqveinte CH. D. BL. 532 ; a-cointede (*pret.*) ROB. 15 ; **akointed** (*pple.*) A. R. 218 ; aqveinted GUY 57.

acólden, *see* acálden.

acólen, v., *O.Fr.* acoler ; *embrace* ; **accóles** (*pres.*) GAW. 1936 ; acólen (*pl.*) 2472.

a-cōlen, v., *O.E.* acōlian ; *cool (intrans.)* A. R. 118 ; acōlede (*pret.*) HOM. I. 237 ; acōled (*pple.*) O. & N. 205.

acolit, sb., *med.Lat.* acolytus ; *acolyte*, P. S. 329 ; acolit SHOR. 45 ; Onesimus, the acolit WICL. COL. PROL.

acombren, v., = encombren ; *encumber* ; acombre LANGL. *B* ii. 50 ; acumbri MISC. 33 ; acumbre (*subj.*) GUY 138 ; **acombrede** (*pret. pl.*) DEP. R. ii. 28 ; acombred (*pple.*) ALIS. 8025.

acombring, sb., *encumbering, perplexity* ; accombringe AYENB. 182 ; acomeringe or **acomerment** (*cf. O. Fr.* encumbrement) '*vexatio*' PR. P. 6.

acomplissen, v., *O.Fr.* acomplir ; *accomplish*, CH. BOET. iv. 12 (118) ; **acomplise** (*pres. subj.*) iii. 10 (92).

acord, sb., *O.Fr.* acord ; *accord, agreement*, BEK. 718 ; WILL. 2964 ; MAN. (F.) 5276 ; þeȝ we ne beon at one acorde O. & N. 181.

acordable, adj., *proportionate, harmonious*, CH. BOET. ii. 8 (62).

acordance, sb., *O.Fr.* acordance ; *agreement*, LANGL. *C* iv. 339 ; acordaunce CH. BOET. ii. 8 (62) ; accordaunce iv. 6 (143).

acordant, accordant, adj., *agreeable*, CH. C. T. *A* 37 ; acordant SHOR. 89.

acordin, v., *O.Fr.* acorder ; *accord, reconcile, agree*, PR. P. 6 ; acordi BEK. 731 ; AYENB. 151 ; acordeþ (*pres.*) SHOR. 139 ; acordiþ WICL. LK. v. 36 ; acordede (*pret.*) SAX. CHR. 261 ; acorded (*pple.*) P. L. S. xxiv.

a-córien, v., *taste, feel the pain of*, MISC. 75 ; þu schalt acorien þe rode A. R. 60 ; þu hit schalt acore sore P. L. S. xv. 120 ; þou ne auȝtest nouȝt mi deþ acore M. T. 112 ; acórede (*pret.*) HOM. II. 45 ; þat acorede al þis lond ROB. 75.

acorn, *see* akern. **acost**, *see* cost.

acount, sb., *cf. O.Fr.* conte (*sb.*), aconter (*v.*) ; *account*, E. G. 8 ; acounte SHOR. 47 ; ȝelde trewe acounte E. G. 357 ; **accountes** (*pl.*) BEK. 164.

acounten, v., *O.Fr.* aconter ; *account* : accountie P. L. S. xxiv. 86 ; acounteþ (*pres.*) AYENB. 137 ; acounted (*pple.*) P. S. 337.

acounting, sb., *accounting, calculation*, CH. BOET. i. 2 (8).

acountre, *see* encontre.

acoupen, v., *O.Fr.* encoulper ; *inculpate, accuse* ; **acoupede** (*pret.*) ROB. 544 ; acouped LANGL. *B* xiii.459; acopede BEK.773; acouped (*pple.*) PR. C. 2947.

acoupen[2], v., *cf. O.Fr.* acoper ; *strike, deliver blows* ; **acoupede** (*pret. pl.*) FER. 1594.

acouping, sb., *collision*, WILL. 3438.

a-coveren, v., *O.E.* acofrian, = *O.H.G.* irkoborōn ; *recover*, H. M. 33 ; akoveren A. R. 364, 412 ; **acoverd** (*pret.*) AR. & MER. 315 ; acovered (*pple.*) H. M. 11.

   acoveringe, sb., *recovery*, H. M. 27 ; acovring C. L. 572 ; acoverunge HOM. I. 251.

acróche, v., *O.Fr.* acrochier ; *encroach*, A. P. i. 1068 ; accroche GOW. I. 314.

actif, adj., *Fr.* actif ; *active*, CH. BOET. i. 1 (6).

actuel, adj., *Fr.* actuel ; *actual* ; (*printed* accuel) SHOR. 107, 108.

acumbri, v., *see* **acombren.**

a-cumen, v., *O.E.* acuman ; *attain, come to* : ase ȝef hi hiȝt miȝt wel acome to letten other wile SHOR. 73 ; **acōm** (*pret.*) ROB. 126.

a-cumlen, v., *become cramped ?* acumblid (*pple.*) WICL. JER. vi. 24 ; acomeled '*estomye*' VOC. 161 ; acomelid '*eviratus, enervatus*' PR. P. 6.

acuntren, *see* **encontren.**

acupement, sb., *O.Fr.* acoupement ; *accusation*, FL. & BL. (H.) 1108.

a-cursien, v., *curse* ; acursi O. & N. 1704 ; **acurseþ** (*pres.*) LANGL. *B* xviii. 107 ; **acursede** (*pret.*) HOM. I. 31 ; acorside AYENB. 57 ; acursed (*pple.*) BRD.24 ; WILL. 901 ; acorsed SHOR. 161 ; þe akurside gost A. R. 234.

acusacioun, sb., *O.Fr.* acusation; *accusation* ; accusaciouns TREV. V. 157 ; CH. BOET. i. 4 (15).

acūsen, v., *O.Fr.* acūser ; *accuse* ; acusi SHOR. 34 ; ' acūsede (*pret.*) BEK. 369.

acūsour, sb., *O.Fr.* accuseur ; *accuser*, CH. BOET. i. 4 (16) ; CH. L. G. W. 353.

a-cwákien, v., *quake* ; aqvake HALLIW. 76.

a-cwecchen, v., *O.E.* acweccan (*set in motion*); *move, shake* (*intrans.*) : þe wode **aqveiȝte** (*pret.*) so hi sunge ALIS. 5257.

a-cwellen, v., *O.E.* acwellen, = *O.H.G.* arqvellen ; *kill*, A. R. 334 ; KATH. 1826 ; aqvellen [acwelle] LAȝ. 8773 ; aqvellen S. A. L. 153 ; aqvelle ALIS. 80 ; S. S. (Web.) 2758 ; **aqvelde** [acwelde] (*pret.*) LAȝ. 21 ; aqvelde HORN (R.) 881 ; þu acwaldest MARH. 12 ; **aqveald** [acweld] (*pple.*) LAȝ. 974 ; aqvold MISC. 41 ; þar mid beoþ men acwalde O. & N. 1370.

a-cwenchen, v., *O.E.* acwencan ; *quench*, FRAG. 6 ; A. R. 224 ; aqvenche TREAT. 136 ; REL. II. 274 ; **acwencheð** (*pres.*) HOM. I. 39 ; aqvencheþ SHOR. 37 ; aqvencþ AYENB. 207 ; **acweinte** (*pret.*) A. R. 124 ; aqveinte TREV. IV. 115 ; þe fur aqveinte sone MARG. 239 ; aqvenched (*pple.*) HOM. I. 81 ;

acwenct JUL. 68 ; aqveint S. S. (Web.) 1991 ; GOW. III. 10.

a-cwéðen, v., *O.E.* acweðan ; *resound* : hornes þer **aqveðen** [acweþen] (*pret.*) LAȝ. 27444.

a-cwikien, v., *O.E.* acwician ; *quicken*, (*intrans.*), *become alive*, A. R. 118 ; aqvikien HOM. I. 81 ; aqvikie LANGL. *C* xxi. 394 ; **acwikeð** (*pres.*) HOM. I. 189 ; **aqviked** (*pple.*) AYENB. 203.

ād, sb., *O.E.* ād, = *O.H.G.* eit (*ignis, rogus*) ; *fire, funeral pile*, KATH. 1365* (*v. r.* fur) ; od '*rogus*' FRAG. 4.

ād, *see* **ēad.**

a-dáȝen, v., = *M.H.G.* ertagen ; *awake up* ; adawe ALIS. 2265* ; **adáwed** (*pple.*) CH. C. T. *E* 2400.

adamant, sb., *O.Fr.* adamant ; *adamant, diamond, loadstone*, ' *adamas*,' PR. P. 6 ; CH. C. T. *A* 1990.

adamantine, adj., *L.* adamantinus ; *made of adamant* : writen .. in an adamantine nail WICL. JER. xvii. 1.

adanten, v., *O.Fr.* adanter ; *subdue* ; adaunte L. H. R. 205 ; **adauntede** (*pret.*) ROB. 61.

a-dásen, v., *daze, dazzle, stupefy* : as a witles man greteli **adásed** (*pple.*) TODD 297.

adden, v., *Lat.* addere ; *add*, CH. BOET. iii. 9 (83) ; **addide** (*pret.*) WICL. LK. xix. 11.

addiciōn, sb., *O. Fr.* addition ; *addition*, PR. P. 6.

addre, sb., = *M.L.G.*, *M.Du.* adre, adere ; *see* naddre ; *adder*, S. S. (Web.) 749 ; ALIS.[2] (Sk.) 1009 ; LANGL. *B* xviii. 333 ; eddre AYENB. 26 ; SHOR. 161 ; WICL. GEN. iii. 4 ; addren (*pl.*) ALIS. 5262 ; eddren AYENB. 61.

   edder-wort, sb., *snakeweed*, LEECHD. III. 323.

a-dēaden, v., *cf. O.E.* adȳdan, = *O.H.G.* irtōdin ; *deaden, put to death, die, decay* ; no þing never nes þer inne þet hit muhte adeaden A. R. 112 ; **adēadie** (*pres. subj.*) HOM. I. 202 ; **adēadet** (*pple.*) KATH. 2048.

a-dēaven, v., *O.E.* adēafian, = *M.H.G.* ertauben ; *deafen* ; **adēved** (*pple.*) SHOR. 103.

[**adele**, sb., *O.E.* adela, = *M.L.G.* adele, *O.Sw.* adel ; *addle, filth*.]

   adel-ēi, sb., *addle-egg*, O. & N. 133.

adese, sb., *O.E.* adesa ; *adze*, WICL. IS. xliv. 13 ; adis N. P. 14.

a-deu, *see* **deus.**   **adīte**, *see* **endīte.**

a-dihten, v., = *M.H.G.* ertihten ; *ordain, dispose* ; and so ich mine song **adihte** [adiȝte] (*pres.*) O. & N. 326 ; adihteþ P. S. 329 ; adiht (*pple.*) KATH. 1382 ; adiȝt L. H. R. 151.

a-dileȝien, v., *O.E.* adilgian, adilegian; *abolish* ; **adilegade** (*pret.*) MISC. 145 ; adiliȝede HOM. I. 235.

ādle, sb., *O.E.* ādl *f.*, = *M.L.G.* ādel *m.* (*ulcus*) ; *disease*, ORM. 4803 ; adle '*langvorem*' MAT. iv. 23 ; ādle (*pl.*) '*aegrotationes*' viii. 17.

adlen, v., _O.N._ öðlask, _prov. Engl._ addle;
earn: adlen (_ms._ addlenn) heofnes blisse
ORM. 4185; thank adile TOWNL. 218 (MIR.
PL. 140); adled (_pple._) ORM. 6235.

adling, sb., _earning, merit_, ORM. 17705.

ādliȝ, adj., _O.E._ ādlig; _diseased_, FRAG. 3.

administracioun, sb., _O.Fr._ administration;
administration, SHOR. 57.

admirail, _see_ amirail.

admitin, v., _Lat._ admittere; _admit_, PR. P. 6.

adopcioun, sb., _Lat._ adoptiō; _adoption_,
AYENB. 101.

adornen, aournen, v., _O.Fr._ adorner, aorner,
Lat. adornāre; _adorn_; adorneþ (_pres._) CH.
TRO. iii. 1; aourned (_pple._) L. H. R. 163.

a-dótien, v., _dote, become silly_; & for dol
adóteþ (_pres._) WILL. 2054; adóted (_pple._)
GOW. III. 4; dusie men & adotede A. R.
222.

adrǽden, _see_ andrǽden.

a-drǽfen, v., _O.E._ adrǽfan; _expel_; adrēfde
(_pret._) HOM. I. 219.

a-drāȝen, v., _draw out_; adrawe ROB. 207;
adrōh (_pret._) LAȝ. 7486; adrouȝ JUL.² 205;
hure swerdes þan þai adrowe HALLIW. 22;
adráȝe (_pple._) FL. & BL. 631; AYENB. 218.

adrēfen, _see_ adrǽfen.

a-drenchen, v., _O.E._ adrencan, = _O.H.G._
ertrenchan; _give to drink, drown, plunge in
water_, A. R. 230; adrenche LAȝ. 1507*;
HORN (R.) 109; SHOR. 139; AYENB.
50; P. S. 217; i schal me adrenche S. S.
(Web.) 1470; adrencte (_pret._) ROB. 489;
adrengte LAȝ. 2568; þat folc he al adrente
25698; & adreinten ham sulven i ðer see A.
R. 230; adreint (_pple._) HOM. I. 141; ROB.
51; S. S. (Web.) 1486; TREV. V. 99; adrenct
HOM. I. 225.

a-drēoȝen, v., _O.E._ adrēogan; _suffer_; adriȝe
HORN (L.) 1035; þat his strok miȝt adrie
HALLIW. 22.

a-drillen, v., _debase, corrupt_? wanne man leteþ
adrille þat he god ȝelde schel SHOR. 114;
þat he nadrille 90.

a-drinken, v., _O.E._ adrincan, = _O.H.G._ er-
trinchan; _be drowned_; adrinke MISC. 77;
HORN (L.) 971; adronc (_pret._) A. R.
58; þer inne he adronc LAȝ. 2205; aðeles
adrunken 7852; adrunken [adronke] (_pple._)
LAȝ. 2224; adronke ROB. 430.

a-drīven, v., _O.E._ adrīfan, = _O.H.G._ artrīban;
drive away; addrivan HOM. I. 115; adriven
(_pple._) JUL. 78.

a-drūȝen, a-drūȝien, v., _O.E._ adrȳgan,
adrūwian; _dry up_; adruwien A. R. 150; a-
drouȝe SHOR. 34; adrūeð, adrūweð (_pres._)
A. R. 150; adrūȝede (_pret._) HOM. I. 133.

a-dumbien, v., _O.E._ adumbian; _become
dumb_; adumbed (_pple._) FRAG. 8.

adūne, _see under_ dūne.

a-dūnen, v., _din_: þu adūnest þas monnes
earen O. & N. 337; adenid (_pple._) AUD. 78.

a-dusten [? adüsten], v., _cast down_; aduste
(_pret._) H. M. 41.

advent, sb., _Lat._ adventus; _advent_, HOM. II. 3;
BEK. 1847.

adversarie, sb., _O.Fr._ adversarie; _adversary_,
AYENB. 238; CH. C. T. _B_ 3868.

adverse, adj., _O.Fr._ advers; _adverse_, GOW. II.
116.

adversitē, sb., _O.Fr._ adversité; _adversity_,
A. R. 194.

advertence, sb., _O.F._ advertence; _attention_,
CH. TRO. v. 1257.

a-dwēlen, v., = _O.H.G._ artwelan (_torpere_);
? _grow stupid_; swo heore wit hi demeþ adwóle
(_pple._) O. & N. 1777.

a-dwēschen, v., _O.E._ adwǽscan; _quench, ex-
tinguish_, KATH. 949; adwēschde & a dun
weorp þe wiðerwine of helle 1196; fir ne beoð
adwēsced (_pple._) ['_ignis non extingvitur_']
MK. ix. 46.

ǽ, _see_ ā, ēa, _and_ ǽew.

[ǽ-, prefix, _O.E._ ǣ-,= _O.Fris._, _O.N._, _O.H.G._ ā.]

ǽ-bere, adj., _O.E._ ǽbere, = _O.Fris._ āber,
āuber; _manifest, evident_; æbǽre ORM. 7189;
ebare LAȝ. 2271; eber C. M. 813.

ǽd, ǽdiȝ, _see_ ēad, ēadi.

ǽfen, sb., _O.E._ ǽfen,= _O.L.G._ āvond, _O.H.G._
āband; _evening_: aven, eaven, even LAȝ. 1116,
5763, 19570, 26696, 26910; efen (_ms._ efenn)
ORM. 1105; even GEN. & EX. 1675; AYENB.
113; GOW. III. 13; CH. C. T. _D_ 750; MAN.
(F.) 2440; eve O. & N. 41; HORN (L.) 364;
an eve A. D. 289; aven REL. I. 194; ave
AR. & MER. 5391; ēfne [ēve] (_dat._) O. & N.
323; on þam efne MAT. xxvi. 20.

ēven-līȝht, sb., _O.E._ ǽfenlēoht; _evening
light_, DEGR. 1601.

ēven-song, sb., _O.E._ ǽfensang; _even-song_,
LANGL. _A_ v. 190; evesong A. R. 22.

ēven-sterre, sb., _O.E._ ǽfensteorra; _evening
star_, MISC. 24; WICL. JOB xxxviii. 32; eve-
sterre '(_h_)_esperus_' PR. P. 144; CH. BOET. i. 5
(22).

ēven-tīd, sb., _O.E._ ǽfentīd; _even-tide_, A. R.
404; WICL. GEN. i. 5.

ǽven-tīme, sb., _O.E._ ǽfentīma; _evening-
time_: æventime [evetime] LAȝ. 12858.

ēven-whīle, sb., _evening-time_, WILL. 1747.

ǽfening, sb., _O.E._ ǽfnung; _evening_; evening
LAȝ. 30419; evening, eveninge WICL. MAT.
xx. 8; eveninge '_vesper_' PR. P. 144; in þe
eveninge ROB. 312.

ǽfne, _see_ efen.

ǽfre, adv., _O.E._ ǽfre; _ever_, ORM. 206; æfre,
efre FRAG. 6, 7; æfre, ævere, æfer [efre,
evere], LAȝ. 547, 1276, 3321, 6552; efre, efere,
HOM. I. 9, 21; efre, evere (_ms._ euere), A. R.
4, 62; evre BRD. 3; AYENB. 71; þe faireste

man þat evre (*ms.* eure) ȝut on þi lond cam
HORN (L.) 788; evre, eavere, ever O. & N.
238, 333, 1282; evre, evere HAV. 207,
424; evere buten ende R. S. vii; ēvere mō
R. S. ii; TREAT. 132; HOCCL. i. 199; evere
more MAND. 241; evere, ever CH. C. T. *A* 832;
ever ant a MARH. 11; ever and o MAP
342; falshede ever ȝite heo souhten C. L.
342; ever þe furþe peni P. S. 149; ever te
ever yet MISC. 31; AYENB. 96; avre (*for*
eavre, *see* ENG. STUD. IV. 100) HOM. II.
37; swottre þen eaver eni haliwei KATH.
1707; ever [ere] more MAN. (F.) 287;
P. L. S. viii. 33; **ēver ǣlc**, ever ulc *every*
(*qvisqve*) LAȝ. 2378, 4599; evere ilc HAV.
1330; ever ilc on *every one* (*unus qvisqve*)
GEN. & EX. 185; ever ilk Iw. 2739; CH.
C. T. *A* 1851; TOR. 2574; ever ich on A. R.
20; K. T. 612; ævric (=ǣvre ic) HOM. I.
137; ever uch SPEC. 26; eaver euch KATH.
735; efrec HOM. I. 7; **ēver ihwǎr** *everywhere* A. R. 200; eaver ihwer KATH. 681.

æftemest, adj., *O.E.* æftemest, = *Goth.* af-
tumists; *aftmost, latest*: bi eftemeste erdede
HOM. II. 23.

æften, prep., *O.E.* æftan, = *O.N.* aptan, *Goth.*
aftana; *after*; ?afte S. S. (Wr.) 701; *comp.*
bi-æften.

æfter, prep. & adv., *O.E.* æfter, = *O.L.G.*
aftar, *O.H.G.* after, *O.Fris.* efter, *O.N.* eptir,
Goth. aftra (πάλιν); *after*: æfter [after] þan
flode LAȝ. 19; (he) ferde æfter (*along*) ane
bache 757; heo æxede æfter Brenne 5001;
æfter þine ræde 14178; his moder Elene him
hafde isend æfter 11221; þe heom after com
1587; heo liðen after sæ 20945; after heore
mihte P. L. S. viii. 89; & biheold after help
KATH. 744; after (*ms.* aftterr) godes lare ORM.
120; after þare longe tale O. & N. 140; word
after word 468; after ðis dede GEN. & EX.
355; after (*along*) ulche strete MISC. 66; after
his wille C. L. 1329; make faste þe dore after
þe P. L. S. xvii. 416; he let after him sende
BEK.·1384; after þan þat seint Brendan furst
ðis ile iseȝ BRD. 11; haveþ ȝirned after þe H.H.
164; do efter þes preostes rede HOM. I. 31;
efter þan *after that* 11; efter þan þet þe mon
bið dead 51; efter þet he haveð idon ANGL.
I. 17; efter mine heorte '*secundum cor meum*'
A. R. 56; efter his deaðe 314; efter þan
AYENB. 24; efter wrake ... it criis C. M.
1130; heo biheld after KATH. 1877; and
after þohte hu heo mihte answere finde O. &
N. 469; hit tidde after WILL. 198; Paul com
efter HOM. I. 41.

after-clap, sb., = *Du.* achterklap; *afterclap*,
AR. & MER. 499.

aftir-comer, sb., WICL. GEN. xxi. 23.

after-dēl, sb., = *Du.* achterdeel; *disadvantage*, WR. DICT. 34.

after-ende, sb., *latter end*, HOM. II.
199.

efter-lið, sb., *cf.* O.E. æftera liða; *July*,
MARH. 23.

after-tale, sb., = *M.Du.* achtertale; *defamation*, BEK. 627.

efter-tellere, sb., AYENB. 58.

after-ward, adv., *O.E.* æfterweard; *afterward, afterwards*, H. M. 37; ORM. 14793;
afterward BEK. 119; þe weren efterward
mine milce *that sought after my mercy*
HOM. I. 45; P. S. 333; efterward AYENB. 24;
afterwardes BRD. 10.

æftere, adj., *O.E.* æftere, æftra, = *O.H.G.*
aftero; *after, following*: on þam æfteren
restedaige LK. vi. 1; on his efter tocome
HOM. I. 95.

æfterling, sb., ?*descendant*; **æfterlinges** (*pl.*)
LAȝ. 19116.

æfte-ward, adv., *O.E.* æfteweard; *afterward*;
on æfte-, aftewarde FRAG. 1.

ægede, sb. (*cf.* ēgede, adj.); ?*folly*; in ægede
(ægæde) & in leȝkes ORM. 2166, 8046.

æhte, *see* eahte.

æhte, sb., *O.E.* æht, = *Goth.* aihts, *O.H.G.* ēht,
from āh; *possession, property, cattle*; æhte,
ahte, aihte, LAȝ. 845, 373, 1078; æihte FRAG.
6; ehte ALIS. 1507; sio ehte LK. xx. 14; be-
tacte heom his ehte (*earlier version* æhta) MAT.
xxv. 14; ehte, echte HOM. I. 147; for his
ehte (*ms.* hehte) lure 103; moni mon nafð
ehta (?*pl.*) 113; ehte, eicte (*printed* eitte)
P. L. S. viii. 21, 131; eihte A. R. 214; MISC.
116; eȝte PROCL. 6; eiȝte P. L. S. xvii.
472; eiȝte [aihte] lure O. & N. 1153; eahte
SAX. CHR. 253; on þere weor(l)dliche eahte
HOM. I. 191; ahte KATH. 1725; SPEC. 46; (*ms.*
ahhte) ORM. 1609; hise agte and erve GEN.
& EX. 742; aght TOWNL. 11; aihte MISC.
69; binimeð hem hwile oref hwile oðer
aihte REL. I. 128 (HOM. II. 161); auhte,
auchte, aucte HAV. 531, 1223, 2215; aught
S. S. (Web.) 738; P. S. 256; *comp.* qvic-ēiȝte.

ēiht-grādi, adj., *greedy of property*, HOM. II.
29.

āihte-lēs, adj., *destitute of property*, HOM.
II. 29.

æi, *see* æni. **æie**, **æiȝe**, *see* eȝe.

æiðer, pron. & adv. (*for* *ā-ȝehweþer*), *O.E.*
æghwæðer, ægðer, = *O.H.G.* eogihwedar, *mod.
Eng.* either; *either* (*each of two, any one of two*);
or; æiðer [aiþer] wende to his hole LAȝ. 15982;
æiðer deade men and qviken (*r.* qvike) 7990;
eiþer 1892; eȝȝþerr heore ORM. 413; eiȝðer
HOM. I. 223; eiȝþer, eiþer WILL. 1010,
1240; eiðer KATH. 1983; eiðer GEN. &
EX. 2855; HOM. I. 21; eour eiþer I. 15; to
eiþer limpeð his dole A. R. 10; on eiþer side
ROB. 17; eiþer ... or CH. C. T. *A* 1645;
in felde eiþer in toune H. S. 996; eiþer, aiþer
O. & N. 7, 796; LANGL. *B* ix. 85; aiþer
TREAT. 139; ayþer MAT. xiii. 30; HAV.
2665; S. S. (Wr.) 787; eider, aider AYENB.

53, 66 ; eþer of ȝou DEGR. 1177 ; er E. G. 31 ; fare he norð er fare he suð REL. I. 211 (MISC. 4) ; ēiþers (*gen.*) WILL. 1014.

ælc, ælch, *see* **aighwilc.**

ælde, *see* **elde. ældre,** *see* **eldre.**

ǣ-lenge, ēlinge, adj., *O.E.* ælenge ALFR. P. C. 41, *prov. Eng.* (*Kent*) ellinge ; *distant, lonely, wretched,* LAȝ. 15190 ; elenge P. R. L. P. 85 ; elenge, elinge TREV. VII. 341 ; elenge [eling] LANGL. *B* PROL. 190 ; elinge LEG. 183 ; S. A. L. 160 ; elenge [alenge] CH. C. T. *D* 1199 ; ye libbeþ an alenge lif S. S. (Web.) 227.

ēlenge-līche, adv., *wretchedly,* LANGL. *B* xii. 45.

æm, *see* **am. ænde,** *see* **ende.**

ǣne, adv., *O.E.* ǣne ; *once* ; ene HOM. I. 15 ; O. & N. 1107 ; MISC. 83 ; P. L. S. xxi. 29 ; REL. I. 113 ; LEG. 81 ; for ene and for evere LAȝ. 20462 ; at ene *at once* ROB. 47 ; LEG. 37 ; MISC. 227 ; MIRC 82.

ǣnes, adv., = *O.H.G.* eines, *O.L.G., O.Fris.* ēnes, ēnis ; *once,* SAX. CHR. 248 ; LAȝ. 29325 ; ORM. 1078 ; enes HOM. I. 37 ; MISC. 78 ; ROB. 197 ; JOS. 25 ; enes oðer twies A. R. 70 ; er enes 420 ; enis P. S. 203 ; enus AUD. 43 ; enes (*aliqvando*) þu sunge O. & N. 1049 ; eanes H. M. 11 ; enes anes HOM. II. 67, 109 ; anes KATH. 126 ; Iw. 292 ; ones GEN. & EX. 3288 ; TREAT. 132 ; MIRC 638 ; WILL. 195 ; MAND. 221 ; at ones HAV. 1295 ; SHOR. 145 ; CH. C. T. *A* 765.

ǣni [? **ǎni**], adj., *O.E.* ǣnig (*ullus, aliqvis*), = *O.Fris.* ēnich, ēnig (*ullus*), *O.L.G.* ēnag (*unicus*), *O.H.G.* einīg, einago (*unicus, ullus, aliqvis*), *Goth.* ainaha (*unicus*) ; *any* ; æni SAX. CHR. 253 ; æny, ani þing 'aliqvid' MAT. v. 23, xxiv. 17 ; anig man MK. ix. 30 ; enig, eni ANGL. VII. 220 ; æni, æi, ei LAȝ. 2392, 4270, 6663 ; eni KATH. 1267 ; TREAT. 133 ; SHOR. 30 ; AYENB. 16 ; WILL. 1077 ; eni þing O. & N. 720 ; eni god man 1015 ; eni, ei A. R. 8, 64 ; FL. & BL. 301, 813 ; eni, eani, ei HOM. I. 13, 33, 185 ; eni, ani, oni CH. C. T. *A* 198 ; aniȝ ORM. 1761 ; ani LEECHD. III. 96 ; GEN. & EX. 2181 ; HAV. 10 ; TRIST. 374 ; ani, oni HOM. II. 19, 43 ; oni PROCL. 6 ; MAND. 2 ; ei R. S. v ; ēnies (*gen. m. n.*) HOM. I. 121 ; SHOR. 95 ; eis [eanies] weis MARH. 13 ; ænigen (*dat. m. n.*) LEECHD. III. 84 ; anigen MAT. xvi. 26 ; eie A. R. 202 ; æine (*acc. m.*) LAȝ. 3692 ; æie (*acc. f.*) LAȝ. 6616 ; ēnie (*nom. pl.*) AYENB. 68 ; anie oðre H. M. 7 ; onie PROCL. 6.

[**ær-**, prefix, = **or-** ; *see* **ær-wēne, -witte.**]

ǣr, adv. & prep., *O.E.* ǣr, = *O.L.G., O.Fris., O. H.G.* ēr, *Goth.* air (πρωί), *O.N.* ār, *mod. Eng.* ere ; *early, before ; rather* ; FRAG. 5 ; P. L. S. viii. 7 ; ær, er, ar (? *for* ǣr) LAȝ. 372, 655, 1581 ; ær þan þat ifel com him to ORM. 8151 ; ær þan he boren wære 814 ; ar & late 6242 ; er TREAT. 136 ; H. H. 222 ; swa er swa (*as soon as*) hi hit

habbeð HOM. I. 105 ; a lutel er 93 ; er crist wes iboren 15 ; er ure drihten come to þisse live 9 ; er dei *before day* 39 ; er timan *before the time* 105 ; er ðon *previously* 27 ; er þisse 11 ; er þonne 37 ; er þon þet heo toferden 93 ; ear 17 ; alse ich er seide A. R. 10 ; er to sone þen to leate 20 ; to er oðer to leate 338 ; ear we weren vorgulte 388 ; er and late SPEC. 99 ; er he were ded CHR. E. 462 ; er dai BEK. 1227 ; er, ere, ear, ar O. & N. 859, 1216, 1309, 1560, 1637 ; er, ar JOS. 122, 243 ; er, or HORN (L.) 535, 553 ; SPEC. 72, 94 ; WILL. 147, 1612 ; CH. C. T. *A* 36 ; MAN. (F.) 2769, 5094 ; er [ar, or] dai LANGL. *A* v. 232, *B* v. 459, *C* viii. 66 ; ear KATH. 368 ; ear, or GEN. & EX. 47, 48 ; ar LEECHD. III. 94 ; HOM. II. 43 ; DEGR. 253 ; AV. ARTH. iii ; aar ALIS. 5033 ; ar, or TREV. I. 131 ; Iw. 66, 1030 ; or HAV. 1043 ; EGL. 256 ; or E. W. 1 ; ȝar, ȝer GUY (Z.) 4570, 10434.

ār-dawes, sb., pl., *O.E.* ǣrdagas = *O.L.G.* ērdagōs, *O.N.* ārdagar ; *former days* : are dawes HAV. 27.

ǣr-dēde, sb., *O.E.* ǣrdǣd ; *former deed* ; erdede HOM. II. 153 ; for þine ǣrdǣden [erdede] (*pl.*) LAȝ. 8745.

ēar-līch, adj., = *O.N.* ārligr ; *early,* A. R. 258 ; **ēarlīche,** [*O.E.* ǣrlīce] (*adv.*) KATH. 116 ; erliche BEK. 905 ; WILL. 1296 ; erli CH. C. T. *A* 33 ; arliche S. S. (Web.) 204 ; orli MAN. (H.) 32.

ār-morwe, sb., *O.E.* ǣrmorgen ; *early morning* : an armorwe HALLIW. 84 ; on ernemarȝen (*for* ermarȝen) HOM. I. 115 ; an ernemorewe MISC. 45.

ærd, *see* **eard. ēre,** *see* **ēare, āre.**

ǣrende, ērende, sb., *O.E.* ǣrende, = *O.L.G.* ārundi, *O.N.* erendi, ǫrindi, eyrindi, *O.H.G.* ārunti ; *cf. Goth.* airus (*messenger*) ; *errand* ; ærnde, ernde, erende, arunde [earende] LAȝ. 1421, 10057, 24838, 25266 ; erende O. &. N. 463 ; HORN (L.) 462 ; erinde A. R. 246 ; erande BEK. 1234 ; CH. D. BL. 134 ; arende REL. I. 130 (HOM. II. 167) ; arunde EM. 8 ; ernde R. S. iv ; ROB. 147 ; P. S. 329 ; AMAD. (R.) xx ; (*ms.* errnde) ORM. DEDIC. 159 ; ernde, erande, arende LANGL. *A* iii. 42 (*B* iii. 41 *has* message) ; erand TRIST. 847 ; erond DEGR. 904 ; erdne GEN. & EX. 787 ; (*ms.* herdne) HORN (H.) 480 ; ērendes (*gen.*) mon LAȝ. 24862.

ěrinde-bēre, sb., *messenger,* A. R. 60 ; erandbere (*ms.* herandbere) REL. I. 146.

ěrinde-beorere, sb., *messenger,* A. R. 60*.

ǣrend-rāke, sb., *O.E.* ǣrendraca = *O.N.* erindreki ; *messenger,* LAȝ. 660 ; erendrake HOM. II. 35 ; ěrendrāken (*pl.*) LK. xiv. 32.

ǣrendien, v., *O.E.* ǣrendian, = *O.L.G.* ārundian, *O.H.G.* ārintōn ; *intercede for, commend* : and bad heom arndien him to hæhȝen þan kingen LAȝ. 23315 ; **ěrnde** (*imper.*) me to þi leve laverd KATH. 2158 ; ernde me þe blisse of heovene A. R. 38; ernde us hevene liht SPEC. 62.

ĕr(e)ndinge, sb., *intercession, mediation*, MISC. 190, 191; erndunge MARH. 23; erndinge HOM. I. 207; HORN (L.) 581; ernding SPEC. 58.

ærer, adv., *O.E.* æror,=*O.H.G.* ēror, *Goth.* airis; *earlier, sooner*, LA3. 25351; al swo hit was erur ƀispeke O. & N. 1738; erer P. R. L. P. 221; þe erur HOM. II. 61.

ārre, adj., *O.E.* ærra,=*O.Fris.* ērra, *O.H.G.* ēriro, *Goth.* airiza; *former*, A. R. 10*; earre H. M. 7; hire erure [erore] freond R. S. v; MISC. 173; þe eror wif ROB. 324.

ærest, adv., *O.E.* ærest, ærist,=*O.L.G.*, *O.H.G.* ērist; *first*, LA3. 18; erest O. & N. 683; MISC. 38; JOS. 56; alra erest þu me scalt don riht HOM. I. 33; þet ich erest seide A. R. 220; þeonne on erest biginneð þe deoflan to weden 264; erst, arst CH. C. T. *A* 776; erst [arst] wil i wite more LANGL. *B* iv. 105; at erst [arst] CH. C. T. *E* 985; on earst MARH. 14; arest LK. ix. 61; arst WILL. 3046; MISC. 198; (a)tarst HAV. 2688; orest GEN. & EX. 2061.

** æreste [ēreste]**, adj., *O.E.* æreste,=*O.H.G.* ēristo; *first, earliest*, LA3. 2646; þet ereste HOM. I. 75; þe ereste dole A. R. 8; ure earste ealdren KATH. 885.

ærh, see **ar3**.   **ærm**, see **earm**.

ær-wēne, adj.,=*O.E.* orwēne; *hopeless, desperate*, LA3. 27537.

ær-witte, adj., *witless*, LA3. 22069.

ēst, see **ēast**.

æt, prep., *O.E.*, æt,=*O.L.G.*, *O.N.*, *Goth.* at, *O.Fris.* et, *O.H.G.* az, ez, *Lat.* ad; *at* : æt þare sæ MK. iv. 1; æt Lundene PROCL. 7; æt þire dure FRAG. 6; et þen fontstone 8; æt Cristes mæsse SAX. CHR. 255; at Wudestoke 251; æt his sadele LA3. 6473; æt þe oðer hælve 6474; he nom ræd æt (*sec. text* of) his monnen 1648; ic wolde iwiten æt (*sec. text* of) þe 9132; þat hit at þe toðen atstod 27631; at hire heonmore læve 1271; hwil he bið at hame H. M. 31; at his borde O. & N. 479; at þan ende 826; at sume siþe herde ich telle 293; at undren MISC. 34; at þon ende 78; sat at borde HORN (L.) 827; þe3 he at diþe laie 1252; leve at (*of*) hire he nam 585; at te laste TREAT. 139; beginne at his heved P. L. S. vi. 13; at te cherche WILL. 1961; at cherche AYENB. 56; ate (=at te) daie of done 13; it was ten at the [te] clokke CH. C. T. *B* 14; at o worde (*with one word*) *G* 1360; she dorst(e) at no wight asken it CH. TRO. iv. 672; al þat þei han a do R. R. 5080; þe word was at (*with*) '*apud*' god WICL. JOHN I. 1; at (*ms.* att) heore rihte time ORM. 216; Crist wolde fulhtned beon at sant Johanes hande 10655; i shal hàfen . . . god læn at god ORM. DEDIC. 143; þer Moises fatte þe lahe at (*from*) ure laverd KATH. 2500; mai he no leve at hire taken GEN. & EX. 2697; es noght at (*to*) laine I w. 703; he rad in et þan est3ete HOM. I. 5; et sume

time 151; up et (*to*) mine chinne 35; þu most bi3eten milce et þine drihtene 33; iwiteð et ower meiden A. R. 64; et te dome MARH. 15; he schal have a do . . . wiþ hem MAND. 132; with wimmen have thou nat a do LIDG. M. P. 66; þer ben ate seve siþe twenti 3ate FL. & BL. (H.) 613; & lokede out þer ate P. L. S. xiii. 340; at after soper CH. C. T. *F* 302.

ēt, sb., *O.E.* ǣt,=*O.L.G.* āt, *M.Du.* aet, *O.H.G.* āz (*n.*), *from* ēten; *food*; ǣte (*dat.*) ORM. 7852; on drunke & on ete P. L. S. viii. 130; on ete oðer on wete HOM. I. 103; ate II. 228; *comp.* ofer-, or-ǣt (-ēt).

æt-béren, v., *O.E.* ætberan,=*Goth.* atbairan (προσφέρειν); *bear away* : a wolf his oþer child atbar (*pret.*) L. N. F. 214.

æt-blenche, v., *escape*, MISC. 79; LEG. 166; from þe dreorie deað ne mai no mon atblenche R. S. v.

æt-breiden, v., *O.E.* ætbregdan; *snatch away*; ætbruden (*pret. pl.*) LK. xi. 52; ætbroiden (*pple.*) MAT. xxi. 43; atbroide O. & N. 1380.

æt-bréken, v., *break away*; atbræc (*pret.*) LA3. 1346; etbrec A. R. 48; etbrēken (*pl.*) 172; etbróken (*pple.*) MARH. 16; er he were him atbroke MISC. 44.

æt-bresten, v., *O.E.* ætberstan; *burst away*; atbresteð (*pres.*) HOM. II. 197; (*ms.* atbrested) REL. I. 224; atbrast (*pret.*) ORM. 14734; atbrosten (*pple.*) REL. I. 222.

æt-créopen, v., *creep away* : þe . . . weoren atcrópene LA3. 5671.

æt-cumen, v., *escape*; atkam (*pret.*) REL. I. 234.

æt-dárien, v., *lurk, escape notice* : nis þer non þat heom atdáreþ MISC. 153.

æt-ealden ? v. : þe weren boðe teames ateald (*? past the age*) HOM. II. 113.

æt-ēowen, v., *O.E.* ætēowan, -ȳwan; *show, appear*; atēowede (*pret.*) MAT. ii. 7; ateawede xvi. 1; atewede ii. 13; atewde MK. ix. 4; þa ateoden (*r.* atsode) him ure drihte LA3. 29620; at(e)awed (*pple.*) HOM. I. 225.

æt-fallen, v., *O.E.* ætfeallan; *fall away*; æt-, atfallen (*pple.*) LA3. 4237, 8995; me is . . . moni crume etfallen A. R. 342; atfalle R. S. vii.

æt-fáren, v., *escape*; atfaren LA3. 27071.

æt-flēon, v., *O.E.* ætflēon; *flee away*; etfleon A. R. 172; etflīhð (*pres.*) A. R. 48; atfliþ, atfliþ O. & N. 37; atflēh (*pret.*) LA3. 5209; atflæh ORM. 19639; etfluwen A. R. 172; atflowen LA3. 2480; etflowen (*pple.*) A. R. 48; atflowen LA3. 27071*.

æt-fōn, v., *O.E.* ætfōn; *receive*; atfēngen (*pret.*) LA3. 15359.

æt-fóren, prep., *O.E.* ætforan; *before*, LK. x. 8; FRAG. 1; ætforen ure isworene redesmen PROC. 8; atforen LA3. 2252; etforen HOM. I. 41; A. R. 226; atvore alle oþer ROB. 351.

æt-gān, v., *O.E.* ætgangan,=*Goth.* atgaggan; *go from* : þar fore ich þenche hire atgo MISC.

161; atgēþ (*pres.*) MISC. 101; (heo) atgaꝺ HOM. I. 35; atgōth (*ms.* atgoht) SPEC. 48 (A. D. 188); þoch his weleþe him atgo [þeih his eihte him ago] REL. I. 176; when mi lif is me atgō (*pple.*) SPEC. 74.

æt-glīden, v., *glide away*; atglide ANGL. IV. 198.

æt-hálden, v., *retain*, LAȝ. 9176; at-, ethalden, -holdan HOM. I. 47, 91; etholden cwide oꝺer fundles oꝺer lone A. R. 208; atholde O. & N. 695; M. T. 96; athálst (*pres.*) ROB. 193; ethalt A. R. 246; athēold (*pret.*) LAȝ. 768; þe nihtegale in hire þohte atheold al þis O. & N. 392; etheold, atheld KATH. 99; athuld ROB. 62; athálden (*pple.*) H. M. 39.

æt-hínden, adv., *behind*, (*ms.* athenden) REL. I. 179 (MISC. 123).

æt-lǽten, v., *? remit*; edie ben alle þo þe here giltes ben atlēten (*pple.*) HOM. II. 69.

æt-lāꝺien, v., *become odious*; atlāꝺed (*pple.*) LAȝ. 2258.

æt-lēden, v., O.E. ætlǽdan; *lead away*; atleden LAȝ. 3200.

æt-líchen, v., *dissemble*: þa lette he hine atlichen LAȝ. 6621.

æt-liggen, v., O.E. ætlicgan, = *Goth.* atligan (παρακεῖσθαι); *lie without*: þat lond þat is longe tilꝺe atleien (*pple.*) REL. I. 128 (HOM. II. 161).

æt-lūtien, v., O.E. ætlūtian; *hide oneself*; etlutian A. R. 400.

æt-rǽchen, v., *reach at; snatch away*; atrauȝt (*pple.*) LEG. 42; who so ever he atraught AR. & MER. 4827; when al mi ro were me atraht SPEC. 37.

æt-rǽden, v., *? surpass in counsel*; atrede CH. C. T. *A* 2449.

æt-rīden, v., *ride away*; atrād, atrǽd (*pret.*) LAȝ. 31439.

æt-rīne, v., O.E. æthrīnan; *touch*; attrine MISC. 53; etrīneꝺ (*pres.*) A. R. 50; ætrān (*pret.*) MAT. viii. 15; ȝif he hine mid sweorde atran LAȝ. 1554; ætrinen (*pple.*) MAT. xiv. 36.

æt-rinnen, v., O.E. ætirnan, = *Goth.* atrinnan; *run away*; atrenne CH. C. T. *A* 2449; æterne HORN (H.) 907; atran (*pret.*) ORM. 1424; atarn SHOR. 149; atarnde ROB. 539; after þan Irisce þæ Uꝺer æturnen (*pl.*) LAȝ. 18267; atærnden 19119.

æt-rūten, v., *? escape*; þu ne miht nowar atrute O. & N. 1768; atroute S. A. L. 160; þer nas prince . . . þat him miȝte atroute ROB. 78.

æt-sáken, v., O.E. ætsacan; *deny, renounce*: ich atsake hine here LAȝ. 28210; atsōk (*pret.*) MISC. 45; ætsōken (*pl.*) LK. viii. 45; atsoken [asoken] LAȝ. 6101; bute hi here laȝe asoke HORN (L.) 65.

æt-scheaken, v., *escape*; atsceken LAȝ. 26516.

æt-schēoten, v., *shoot away* (*intrans.*); atschēt (*pret.*) O. & N. 44; þah mi lif me beo atschóte 1623.

æt-sēchen, v., *seek*; atsechen LAȝ. 13322.

æt-sitte, v., O.E. ætsittan; *resist*, HAV. 2200; ROB. 137; asitte OCTOV. (W.) 1665; þo he godes heste atseet (*pret. subj.*) C. L. 235.

æt-slīken, v., *slip away*; atslīkez (*pres.*) A. P. i. 574.

æt-slūpen, v., *slip away*: al min hope were etslópen (*pple.*) A. R. 148.

æt-springen, v., O.E. ætspringan; *spring up*; atsprong(e) (*pple.*) C. L. 152.

æt-spurnen, v., O.E. ætspurnan; *dash the foot against*; ætsperne (*pres. subj.*) MAT. iv. 6; LK. iv. 11.

æt-standen, v., O.E. ætstandan, = *Goth.* atstandan (ἐπι-, παραστῆναι), O.H.G. azstantan; *stand by, withstand*; etstonden '*stare*' HOM. I. 129; ȝef ha etstonden wulleꝺ mine unwreste wrenches MARH. 14; atstonde O. & N. 750; MISC. 144; þer he wolde atstonde [astonde] LAȝ. 2661; no þing miȝte hem atstonde [astonde] ROB. 20; he nolde nowhar atstonde (*stay*) BEK. 2358; þe king made (h)is stede astonde FER. 4235; etstondest (*pres.*) A. R. 236; atstōd (*pret.*) HORN (R.) 1455; ROB. 15; Arꝺur atstod and biheold LAȝ. 26073; when hem most ned atstod AM. & AMIL. 1728; astod REL. I. 101; CHR. E. 62; atstonden (*pple.*) LAȝ. 19853.

æt-sterten, v., *start away*; etsterten A. R. 48; atsterten, atstirten KATH. 699, 2126; æt-, atsterte, -stürte [ast(e)orte] (*pret.*) LAȝ. 2316, 4264, 12965; etstert (*pple.*) A. R. 48.

æt-stünten, v., *stop, hinder; remain*; ne etstunten ne etstonden þe strencꝺe of mine swenges MARH. 15; þah an etsterte us tene schulen etstunten JUL. 50; he atstünte [astinte] ROB. 168; hu þat ufel wes atstünt LAȝ. 31903.

æt-stutten, v., *tarry*; atstutte [etstutte] (*pret.*) KATH. 23; atstutten H. M. 21.

æt-wappen, v., *? escape*; atwappe (*pres.*) A. P. ii. 1205; atwaped (*pret.*) GAW. 1167.

æt-wenden, v., *escape*; atwende O. & N. 1427; Merlin him ætwende (*pret.*) LAȝ. 18176.

æt-winden, v., O.E. ætwindan; *escape; depart, cease*; atwinden LAȝ. 10003; atwinden ORM. 8004; atwand, atwond (*pret.*) LAȝ. 87, 4243; ꝺis unweder ꝺor atwond GEN. & EX. 3058; atwúnden [atwónde] (*pple.*) LAȝ. 12869.

æt-wīten, v., O.E. ætwītan, *mod. Eng.* twit; *reproach with*; atwite SHOR. 98; AYENB. 198; hwi schal he me his sor atwite O. & N. 1234; awite REL. I. 157; atwīt (*pres.*) LAȝ. 25023; AYENB. 66; etwīteꝺ (*imper.*) A. R. 70; etwāt (*pret.*) KATH. 2364; atwot S. S. (Web.) 1876; (heo) atwiste FL. & BL. 590; M. T. 105; atweste [atwitede] ROB. 33; atwiten LAȝ. 26584; atwiten (*pple.*) LAȝ. 19594; atwiten O. & N. 935; atwite AR. & MER. 9250; atwist HALLIW. 109.

atwītinge, sb., *reproach*, AYENB. 65.

æt-wīten, v., *depart*: and god atwōt (*pret.*) into hise liȝt GEN. & EX. 1049.

æt-wrenchen, v., *escape*; etwrenchen MARH.
15; atwrenche O. & N. 248.

æþ, *see* **ēað.**

æðel, sb., *O.E.* æðel (*pl.* æðelu),=*O.Fris.* ethel,
O.L.G., O.N. aðal, *O.H.G.* edil, adal (*n.*);
natural disposition, nobility, honour: aþel
SPEC. 33; mid æðele help his broðer LA3.
9263; aþele (*pl.*) '*natura*' HALLIW. 105;
aðele, æðelen, aðelen LA3. 3751, 10629, 12915.

ēðel, *see* **ēþel.**

æðele, æðel, adj., *O.E.* æðele,=*O.L.G.* aðal,
eðili, *O.Fris.* ethel, *O.H.G.* adal, edel; *noble,
generous,* LA3. 661, 6528; aþel GAW. 241;
A. P. ii. 761; sum aþel man ORM. 611;
se aðele refe MK. xv. 43; aðel (*sb.*) *prince*
LA3. 10091; aðele (*nobly*) iborene LA3.
26746; æðeles (*gen. n.*) LA3. 16544; aðelere
(*dat. f.*) LA3. 30216; æðele (*pl.*) LA3. 4864;
aþele O. & N. 632; hire eðele vif wittes A. R.
172; aðeles (*sb.*) *noble men* LA3. 7852; aðelere
(*gen. pl.*) LA3. 26695; aðelen (*dat. pl.*) LA3.
27189; mid eðele worde HOM. II. 125; æðelest
(*superl.*) LA3. 2611; atheliste D. ARTH. 1593.

 æþel-boren, adj., *O.E.* æðelboren,=*O.L.G.*
aðalboran; *nobly born,* FRAG. 4; æthelboren
LK. xix. 12.

 æðel-mōd, adj., *noble-minded,* LA3. 23255.

æðelien, v., *O.E.* (ge-)æðelian; *dignify,
honour*; æðelede, aðelede (*pret.*) LA3. 2815,
6650; *comp.* 3e-æðelien.

æþeling, sb., *O.E.* æðeling,=*O.Fris.* etheling,
O.H.G. ediling; *nobleman,* FRAG. 2; aþeling
R. S. vi; REL. I. 172; wo so were next king bi
kunde me clupeþ him aþeling ROB. 354; eþeling MISC. 106; æðelinge (*dat. pl.*) LA3. 874.

æven, *see* **æfen.**   **ævere,** *see* **æfre.**

[æw] æ, sb., *O.E.* æw, æ,=*O.Fris.* ēwa, *O.H.G.*
ēwa, ēha, ēa; *law, right,* HOM. I. 227; ORM.
145; þeo alde e HOM. I. 89; eæ JOHN vii. 19.

æw-bræche, sb., *O.E.* æwbreca, -brica;
'*adulter,*' FRAG. 5; eawbrekere [=*M.H.G.*
ēbrechære] *adulterer* HOM. I. 29.

 ēu-bruche, sb., *O.E.* æwbryce; *adultery,*
HOM. I. 49; eawbruche H. M. 43*; eaubruche
A. R. 204; ewebruche HOM. II. 213.

æx, *see* **eax.**   **af,** *see* **of.**

a-fēeren, v., *O.E.* afæran; *terrify*; afere ROB.
22; FER. 387; LANGL. B xx. 165; RICH.
4104; þu miht mid þine songe afere alle þat
hereþ þine ibere O. & N. 221; afērde (*pret.*)
WART. HIST. I. 16; averde [aferde] LA3.
25554; afæred, aværed, afeared (*pple.*) LA3.
3076, 12730, 17057; afered HOM. I. 53; A. R.
326; SPEC. 85; GOW. II. 21; aferd CH. C. T.
A 628; MAND. 294; TOWNL. 28.

a-fallen, v., *O.E.* afeallan,=*O.H.G.* erfallan;
fall down: ower prude schal avalle O. & N.
1685; avalleð (*pres.*) A. R. 246; afēol (*pret.*)
HOM. I. 223; LA3. 31967; afallen (*pple.*)
HOM. I. 205; þi wal is afallen LA3. 15949;
afalle REL. II. 272; TREV. V. 199.

a-fandien, v., *O. E.* afandian; *try, temp*
afonde P. L. S. xxii. 12; afondeþ (*pres.*) SHOR.
73; afanded (*pple.*) P. L. S. viii. 75*.

a-fangen, v., *receive*; afonge O. & N. 1196;
ALIS. 606; þat seint Michel ous mote afonge
and tofore him lede TREAT. 140; he avangeþ
(*pres.*) SHOR. 50.

a-fáren, v., *O.E.* afaran,=*O.H.G.* arfaran;
depart; afáre (*pple.*) LA3. 13533*; as þ(e)
emperour fram home was afare P. L. S. xix.
177.

a-fēden, v., *O.E.* afēdan; *feed*; afédde (*pret.*)
HOM. I. 227.

afel, sb., *O.E.* afol, *O.N.* afl; *power*: þoh it litel
be it hafeþ mikel afel ORM. 3717.

a-fellen, v., *O.E.* afellan, afyllan,=*O.H.G.*
irfellen; *fell, throw down,* KATH. 689; avel-
len [afallen] A. R. 122; afelle ALIS. 5240;
afelde [afulde] (*pret.*) LA3. 22814; afelled
[avalled] (*pple.*) LA3. 2069; afallet MARH. 11.

a-fermien, v., *O.E.* afeormian; *sweep clean*:
mid besme afermed (*pple.*) LK. xi. 25.

a-ferren, v., *O.E.* afyrran; *remove*; aferrede
(*pret.*) FER. 5565; aferred (*pple.*) MAT.
ix. 15.

affaiten, v., *O.Fr.* afaitier; *affect; fashion,
prepare; fit up, adorn; tame, subdue*; LANGL.
B v. 49; afaiti ROB. 179; SHOR. 111; afeited
(*pple.*) A. R. 284; afaited AYENB. 212.

affecciūn, sb., (? *printed* affectiun), *O.Fr.*
affection; *affection,* A. R. 288; affeccioun
AYENB. 151; CH. C. *B* 586.

affére, v., *O.Fr.* afeir; *pertain*: afféris (*pres.*)
BARB. i. 162; affiered (*pret.*) MER. 225.

afferme, v., *O.Fr.* afermer; *affirm,* MAN.
(H.) 316; LIDG. M. P. 132; affermed (*pple.*)
CH. C. T. *A* 2349.

affiaunce, sb., *affiance,* LANGL. *B* xvi. 238;
GAW. 642.

affiche, v., *Fr.* afficher; *fix,* GOW. II. 211.

affīe, v., *O.Fr.* affier; *trust,* R. R. 5480; afīed
(*pret.*) ALIS. 7351.

affīe, sb., *trust,* FER. 2167.

affīle, v., *Fr.* affiler; *whet,* CH. C. T. *A* 712.

affinitē, sb., *O.Fr.* affinité; *affinity,* SHOR. 70.

afflicciōn, affliccioun, sb., *O.Fr.* affliction;
affliction, WICL. EX. iii. 7.

affluence, sb., *Fr.* affluence; *affluence,* S. A. L.
81.

affluent, adj., *copious, flowing in,* GUIL.[2] (N.
.E. D.).

affodille, affadille, sb., *med.Lat.* affodillus,
Gk. ἀσφοδελός; *daffodil,* PR. P. 7.

afforce, v., *O.Fr.* aforcer, esforcer, *from med.
Lat.* exfortiāre; *force, compel, try,* PR. C. 4253;
devells þat afforces (*pres. pl.*) tham (*endeavour*) to reve fra us the honi of poure life
HAMP. TR. 8; aforcede (*pret.*) ROB. 121;
afforsit (*strengthened*) D. TROY 6593.

afforce, adv., *? Fr.* à force ; *forcibly, by force,* DEP. R. iv. 21.

affrai, sb., *O.Fr.* esfrei ; *attack ; alarm, fright,* MAN. (F.) 9054 ; affrai, afrai CH. C. T. *D* 2156 ; afrai LIDG. M. P. 116.

affraien, v., *O.Fr.* esfraer, -freer, *med.Lat.* exfridāre (*Lat.* ex *out of, Teutonic* friðu-s *peace*) ; *startle, frighten,* P. S. 333 ; afraie, affraie CH. C. T. *E* 455 ; afraied (*pple.*) WILL. 2158 ; M. H. 21 ; affraid SHOR. 158.

afin, adv., *Fr.* à fin ; *at the end,* K. T. 780 ; afine AR. & MER. 50 ; R. R. 3690.

a-finden, v., *O.E.* afindan, = *O.H.G.* irfindan ; *find out,* P. L. S. viii. 29 ; LAȝ. 25775 ; afinde, avinde O. & N. 527 ; avinde REL. I. 183 ; avínde (*pres.*) FER. 757 ; afounde (*pret.*) OCTOV. (W.) 1659 ; afounde (*pple.*) SHOR. 49.

a-fleȝen, v., *O.E.* aflīgan, aflȳgan, = *O.H.G.* arflaugen ; *put to flight* ; avlēieð (*pres.*) A. R. 136 ; aflei fram ham al uvel KATH. 2431 ; alle þe seoven deaðliche sunnen muwen beon avlēied (*pple.*) þuruh treowe bileave A. R. 248.

a-flēmen, v., *O.E.* aflȳman ; *put to flight* ; avlēm (*pres. imper.*) HOM. I. 195 ; aflēmde (*pret.*) LAȝ. 8465 ; aflemden SAX. CHR. 252.

aflen, v., *O.N.* afla ; *get, earn* ; avelen þat men hem blescen HOM. II. 159 ; his mede shal ben þanne garked al se hit beað here aveled (*pple.*) IBID. ; cnapechild is afled wel ORM. 7903.

a-flēon, v., *O.E.* aflēon, = *O.H.G.* arfleohan ; *flee* ; afloȝen (*pple.*) FER. 3132 ; þe king was afloȝe LAȝ. 19076* ; aflowe TREV. V. 429.

aflight, adj., *O.Fr.* afflict ; *afflicted,* OCTOV. (W.) 191 ; GOW. I. 210.

a-flūht, sb., *from* aflēon ; *flight* : aflühte (*dat.*) KATH. 2020 : in afflight TORR. 2043.

afólen, v., *O.Fr.* afoler (*affoler*) ; *befool* ; afóled (*pple.*) O. & N. 206.

a-fōn, avōn, v., *O.E.* afōn ; *receive,* HOM. I. 131, 135 ; LAȝ. 8728, 14939 ; alle þo þat hi 'avoþ O. & N. 843 ; avōuh (*imper.*) HOM. I. 197 ; afēng (*pret.*) B. DISC. 1401 ; aveng ROB. 180 ; heo afengen hine mid sibbe LAȝ. 6580 ; afonge (*pple.*) BEK. 355 ; avonge SHOR. 132.

a-fóren, prep. & adv., *O.E.* onforan, aforn ; *before* ; Fulgenes him wes aforen [aforn] on LAȝ. 10413 ; aforn PR. P. 7 ; bihinde & aforn E. G. 23 ; L. H. R. 150 ; aforn (*ms.* afforn) hem FER. 1080 ; afore þe king LANGL. (Wr.) 2495 ; GOW. II. 88 ; GAM. 656 ; OCTAV. (H.) 1341 ; avore þet he come to ssrifte AYENB. 172 ; **aforne ȝēn** *opposite to* LAȝ. 18528* ; ontrewe avore ie his lhord 18 ; afore iens us CH. TRO. ii. 1139 ; aforn aȝens WICL. MK. xv. 39.

afóre-ward, adv. & prep., *in front* (*of*) : Judas com avoreward MISC. 42 ; afforeward alle FER. 3923.

a⁴-forþen, v., *for* ȝeforþen ; *afford* : and ȝaf hem mete as he miȝte aforþe LANGL. *B* vi. 201 ; aforthe HALLIW. 27.

afoundren, v., *O.Fr.* affondrer, esfondrer ; *founder* (*a horse*) ; afoundred (*pple.*) MAND. 69.

a-frainen, v., *question* : afreine PR. C. (HALLIW. 26) ; affraine FER. 2146 ; **affrained** (*pret.*) LANGL. *B* xvi. 274.

a-freten, v., = *Ger.* erfreszen ; *devour:* þe devel huem afretie P. S. 237.

a-friȝt, *see* afürhten.

a-frounti, v., *O.Fr.* affronter ; *affront,* AYENB. 229 ; afrounted, afronted (*pret.*), LANGL. *B* xx. 5 ; an if a pore man speke a word he shal be foule afrounted P. S. 337.

after, *see* æfter.

a-fūlen, v., *O.E.* afȳlan ; *foul, pollute* : þu heo afūlest FRAG. 7 ; afīled (*pple.*) ALIS. 1064.

a-füllen, v., *O.E.* afyllan, = *O.H.G.* arfullan ; *fill* ; afülled, afeolled (*pple.*) LAȝ. 9811, 19077 ; afelled LK. i. 57.

a-fürhten, v., *O.E.* afyrhtan, = *O.H.G.* erfurhtan ; *affright* ; afriȝt (*pple.*) WILL. 2784 ; AMAD. (R.) xxxvii ; afrigct S. S. (Wr.) 2704 ; þai weren afriȝte FER. 138 ; afrought M. ARTH. 73.

a-fürsien, v., *O.E.* afeorsian ; *remove* : nu þu bist afürsed from alle þine freonden FRAG. 7.

aȝ, *see* ā.

a-gǣten, v., *? enrich* : ageet (*pple.*) SHOR. 119.

a-gān, v., *O.E.* agān, agangan ; *depart, escape, pass* ; agon O. & N. 1280 ; AN. LIT. 90 ; ago R. S. i (MISC. 156) ; ageð (*pres.*) HOM. I. 33 ; A. R. 184 ; ageþ FRAG. 7 ; O. & N. 1453 ; þis worldes wele al agoþ REL. I. 160 ; agān (*pple.*) LAȝ. 337 ; agon is al mi streinþe REL. I. 125 ; out agon BEK. 2193 ; hit is not longe agon K. T. 268 ; agon, agoon CH. C. T. *A* 1782 ; ago P. S. 197 ; þa æstre wes aȝonge [agon] LAȝ. 24195 ; aȝeonge 24241.

a-gāsen, v., *? = Goth.* usgaisjan ; *terrify, bewilder* : þe were so sore agāsed CHEST. II. 85.

a-gǎsten, v. (cf. *mod.Eng.* aghast) ; *terrify* ; agaste ROB. 17 ; CH. TRO. ii. 901 ; hwu þe ate(l)liche deovel schal ȝet agesten ham A. R. 212 ; agastið (*pres.*) H. M. 31 ; agasten MAND. 282 ; agaste (*subj.*) WICL. 2 PARAL. xxxii. 18 ; agaste (*pret.*) LAȝ. 6451 ; KATH. 1256 ; Cristofre him sore agaste P. L. S. xv. 97 ; agast (*pple.*) AN. LIT. 91 ; ALIS. 250 ; WILL. 1778 ; CH. C. T. *A* 2341 ; agest A. R. 372 ; hi were agaste BRD. 8.

āge, sb., *O.Fr.* aage, edāge ; *age,* MISC. 35 ; HORN (L.) 1324 ; ROB. 9 ; SHOR. 63 ; ALIS. 7653.

aȝe, *see* eȝe.

a-ȝefen, aȝeven, v., *O.E.* agifan, = *O.H.G.* ar-, er-, irgeban, *Goth.* usgiban ; *give up, give back, render, deliver,* HOM. I. 29, 31 ; aȝeven, aȝiven LAȝ. 4698, 29052 ; aȝeoven, aȝeven JUL. 58, 59 ; aȝæf (*pret.*) LAȝ. 26356 ; þeos word aȝaf þe nihtegale O. & N.

139; and andswere a3ēven (*pl.*) LA3. 29740; til hi aiaven up here castles SAX. CHR. 265.

a3ein, *see* an3ēn.

a3einen, v. (*from* a3ein),=*M.H.G.* engegenen; *encounter*; a3eineden (*pret. pl.*) þere verde LA3. 17854.

a-3élden, v., *O.E.* agieldan, agildan; *yield up*; a3elde FER. 5026; a3ilde TREV. III. 431; a3ïld (*imper.*; *ms.* ayild) þe to þis knight HALLIW. 126.

ā3en, v., *O.E.* āgan,=*O.L.G.* ēgan, *O.H.G.* eigan, *O.N.* eiga, *Goth.* aigan (ἔχειν),*mod. Eng.* owe; *from* āh; *have, own, possess; owe; be under obligation*; LA3. 4061; a3hen ORM. 6339; owen C. L. 132; awe (*r. w.* lowe) HAV. 1292; such a wif to owe HORN (H.) 440; āwe (*pres.*) C. M. 7149); owe HAV. 1666; ahð LK. xi. 21 (*earlier text* ah); o3eþ LA3. 3058* (*first text* ah); o3þ AYENB. 9; oweþ AN. LIT. 90; CH. C. T. *C* 361; we ageð HOM. II. 41; oweþ LA3. 25320 (*first text* a3en); āwand (*pple.*) AMAD. (R.) iii; āhte (*pret.*) [*mod. Eng.* ought], *owned, owed, ought*, HOM. I. 31; KATH. 248; ahte, aute, ohte LA3. 2228, 2525, 14402; a3te GAW. 767; AMAD. (R.) xiv; agte GEN. & EX. 525; auhte, aucte, aute HAV. 743, 2717, 2800; þis gold aughte sir Isambras ISUM. 659; ohte SPEC. 62; to makien hire cwene of al þet he ouhte A. R. 390; ou3te JOS. 36; LANGL. *B* iii. 68; ou3te, oughte CH. C. T. *A* 1249; þu au3test BEK. 352; ohtest FRAG. 6; þu 3ulde þet tu ouhtest A. R. 406; þet achten we to leven HOM. I. 167; þa Englene lond ærest ahten LA3. 18; men þat ahten to wite KATH. 263; we ahte P. S. 246; ou3ten WILL. 1080.

ā3ere, sb.=*Dan.* ēier, (MOLB. I. 209); *possessor*; ā3eres (*pl.*) SHOR. 99.

ā3en, adj., *O.E.* āgen, ǣgen (ALFR. P. C. 4),=*O.L.G.* ēgan, *O.Fris.* ēgen, ēin, *O.H.G.* eigen, *O.N.* eiginn, *Goth.* aigin (*n.*); *from* āh; *own*; þin a3en heaved HOM. I. 29; ich æm þin a3en [own] mon LA3. 28144; hit wes heora a3en 1945; hire a3e [o3en] lif 3481; a3hen ORM. 363; þin ahen HOM. I. 277; ogen HOM. II. 27; o3en AYENB. 172; o3e, owe, owe O. & N. 100, 259; oghe MISC. 30; awen MAN. (H.) 182; ISUM. 635; AV. ARTH. lvii; awin IW. 583; awen, auen GAW. 293, 836; aun C. M. 6003; owen A. R. 158; GEN. & EX. 120; ouen S. S. (Wr.) 2144; þe kinges o3ene do3ter HORN (L.) 249; his āh3enes þonkes HOM. I. 121; āgenen (*dat. m. n.*) MAT. vii. 3; mid his a3ene muðe HOM. I. 133; a3ene, a3ne [owene] LA3. 335, 14200; ahne KATH. 409; ahene, owe O. & N. 1286; oune LEG. 98; CH. C. T. *A* 804; ā3ere, ā3re (*dat. f.*) LA3. 308, 3941; owere REL. I. 172 (MISC. 106); ā3ene (*acc. m.*) LA3. 314; o3ene ○. & N. 1341; ā3ene (*acc. f.*) HOM. I. 43; a3ene, ahene, o3ene O. & N. 1089, 1542; þurh his ahne nihte KATH. 1049; ā3ene [owene] (*pl.*) LA3. 6877; o3ene AYENB. 33; ā3ene (*dat. pl.*) LA3. 1718; fram his agene

manne SAX. CHR. 255; owene A. R. 158; oune HAV. 2428.

ā3en-sclá3a, sb., *O.E.* āgenslaga; *suicide, self-slayer*, HOM. ⟨ 103, 296.

a-3ēoten, v., *O.E.* agēotan,=*O.H.G.* argiuzan; *pour out*; agoten (*pple.*) HOM. I. 127.

a-gesse, v. (*? for* gesse), *guess, expect*, HORN (L.) 1181.

a-3eten, v.,*O.E.* agetan; *seize, attain, perceive*; to a3ytenne LEECHD. III. 94; a3ite (*ms.* a3itte) (*pres.*) MISC. 193; *comp.* un-a3eten.

a3ien, v., *from* a3e (*see* e3e); *awe, terrify*; agh C. M. 12096; awe P. S. 156; awes (*pres.*) H. S. 1592.

a-gīlen, v., *deceive*; agīled (*pple.*) P. L. S. xiv. 67.

a-ginnen, v., *O.E.* aginnan (?=anginnan); *commence*; aginne O. & N. 1289; HORN (R.) 1285; AYENB. 32; S. S.(Web.) 1410; agunnen [agynne] (*riming with* awinnen) LA3. 18760; (we) aginnen [aginne] LA3. 26572; aginneþ (*pres.*) P. S. 189; aginne (*subj.*) A. R. 74; aginð beaten MAT. xxiv. 49; agan (*pret.*) SHOR. 53; agon [agan] LA3. 6753; agunnen (*pl.*) LK. v. 21; agunnen [agonne] (*pple.*) LA3. 24086; agonne AYENB. 166.

aginninge, sb., *commencement*, AYENB. 16.

aglen, v., *?=Dan.* āgle; *vacillate* [*or perhaps for* a3len, *see* eilen], GEN. & EX. 3809.

aglet, aglot, sb., *O.Fr.* aglet; *tag of a lace,* '*acus*,' PR. P. 8; aglot LUD. COV. 241.

a³-glīden, *O.E.* aglīdan, ?=*andglīdan, cf. M. Du.* ontglīden, *M.H.G.* entglīten; *glide away*; agloōd (*printed* agleed, *riming with* bestrood) (*pret.*) LIDG. M. P. 116; þe dent aglod FER. 3384; agliden (*pple.*) MISC. 93.

a-gliften, v., *? terrify*: as he stod so sore aglifte (*pple.*) .H. S. 3590; þenne were þe Romains al aglifte [oglift] MAN. (F.) 3402.

a-glopnen, v., *? astonish*; aglopned (*pple.*) ALEX. 874.

ā3nien, v.,*O.E.* āgnian,=*O.H.G.* eigenen, *O.N.* eigna; *from* ā3en; *own, appropriate, claim*; ahnien [ohni] þan kaisere Arðures riche LA3. 25359; ahnen ORM. 5649; āhnede [ohnede] (*pret.*) LA3. 11864; *comp.* 3e-ā3nien.

ā3nere, sb.,=*Ger.* eigner; *owner, possessor*; owener W. & I. 22; ō3eneres (*pl.*) AYENB. 37.

ā3ninge, sb., *O.E.* āgnung, āhnung; *possession*; o3ninge AYENB. 37.

agonie, sb., *Fr.* agonie, *from Gr.* ἀγωνία; *agony*, CH. C. T. *A* 3452.

a-grámien, v., *cf. O.H.G.* ergremen; *irritate*; agrámed (*pple.*) ALIS. 3310; B. DISC. 1916; PL. T. 291; agromed O. & N. 933; CHR. E. 863; agremed HALLIW. 33.

a-grāpien, v.,=*O.H.G.* irgreifōn; *apprehend*,

comprehend : after þat þei couþe agrope GOW. I. 254.

agreable, adj., *O.Fr.* aggreable ; *agreeable*, CH. C. T. *B* 767.

agrēen, v., *O.Fr.* agreer ; *agree*, CH. TRO. iii. 131.

a-greiþen, v., *equip, prepare* ; agraiþi AYENB. 148 ; **agreiþed** (*pple.*) WILL. 1598.

agreiþing, sb., *dress* ; **agraiþinges** (*pl.*) AYENB. 216.

agréven, aggreggin, v., *O.Fr.* agrevier, agregier ; *aggrieve*, '*aggravo*,' PR. P. 8 ; **agrévith** (*pres.*) LUD. COV. 41 ; aggregith APOL. 4 ; **agréved** (*pple.*) WILL. 641.

a-grillen, v., *grieve* : ful sore him schal agrille FER. 2195 ; nu ich mai singe hwar ich wulle ne dar me never eft mon agrulle O. & N. 1110.

a-grīsen, v., *O.E.* agrīsan ; *be horrified*, A. R. 306 ; H. M. 31 ; ham schal agrisen KATH. 2317 ; agrise R. S. v ; CH. C. T. *D* 1649 ; i **agrise** (*pres.*) SACR. 902 ; me agriseð LAȝ. 13328 ; **agrās** [agros] (*pret.*) LAȝ. 11976 ; agros ROB. 549 ; HORN (R.) 1326 ; of which she nouȝt agros CH. TRO. ii. 930 ; **agrisen** (*pl.*) JOS. 236 ; ꝧo wurðen he frigti and agrisen GEN. & EX. 667 ; **agrisen** (*pple.*) MARH. 9 ; ALIS. 357 ; MAN. (F.) 9611 ; agrise ROB. 539 ; ich am agrise WILL. 1743.

agroten, agrotnien, v., *surfeit* ; agrotonin, -tone '*ingurgito*' PR. P. 8 ; **agrotied** (*pple.*) CH. L. G. W. 2454 ; agrotonid PR. P. 8.

a-grūwen, v., *?=Ger.* ergrauen (*impers.*) : þet ou **agrūwie** (*? printed* agrupie) *that you may be in dread* A. R. 92.

āȝt, *see* **āht.**

āgt, āgte, *see* **hāht.**

ague, sb., *O.Fr.* (fievre) aigue ; *ague*, PR. P. 8 ; LANGL. *B* xiii. 336.

a-gültan, v., *O.E.* agyltan ; *be in fault*, HOM. I. 17 ; agulte C. L. 335 ; LANGL. *B* xv. 304 ; agelte AYENB. 15 ; hwon me **agülteð** to ou A. R. 186 ; **agülte** (*pret.*) SPEC. 112 ; agilte R. R. 5833 ; **agült** (*pple.*) P. L. S. viii. 6 ; R. S. ii ; MISC. 58 ; ROB. 252 ; agelt WILL. 4391 ; gif man haveð wið us agilt HOM. II. 65.

ah, *see* **ac.**

āh, v., *O.E.* āh, = *O.Fris.* ach, *Goth.* aih (ἔχω), *O.N.* ā (*pret. pres.*) ; *have, owe, ought* : þe mon þe lutel ah LAȝ. 3058 ; god ah þe litel mede ORM. 16529 ; ah, ach HOM. I. 139 ; wif ah [auh] leie sottes lore O. & N. 1471 ; ag MAT. xxiv. 47 ; agh C. M. 7149 ; agh, ogh HOM. II. 17 ; oh HOM. II. 21 ; þe more oh ich to lovie þe SPEC. 70 ; ase mon ouh to donne A. R. 58 ; og GEN. & EX. 1 ; ꝧat us og alle to ben minde REL. I. 216 ; i au (*ms.* aw, *? for* awe) þe honor IW. 720 ; aw he þe ani mare AMAD. (R.) xxiv ; þu **āȝest** HOM. I. 15 ; LAȝ. 16706 ; ahest H. M. 11 ; owest A. R. 126 ; we **āȝen** (*pl.*) [*O.E.* āgon] HOM. I. 21 ;

we agen, ogen II. 57 ; we, heo aȝen LAȝ. 6150, 25320 ; þe treowþe þæt heo us oȝen PROCL. 4 ; ogen GEN. & EX. 15 ; ȝe ohen R. S. v (MISC. 168) ; owen A. R. 82 ; we owen fourti pound CH. C. T. *D* 2106 ; what ȝe owen to do (*v. r.* shulen do) WICL. EX. iv. 15 ; ȝe awe AMAD. (R.) i ; we owe TRIST. 1005 ; **āȝe** (*subjunct.*) LAȝ. 28423 ; þe devel him awe (*ms.* hawe) HAV. 1188 ; *deriv.* āȝen, **æhte.**

a-hebben, v., *O.E* ahebban, = *O.H.G.* arheffen ; *heave up,* LK. xviii. 13 ; ahef(ð) (*pres.*) HOM. I. 113 ; **ahévinde** (*pple.*) A. R. 16 ; ahōf (*pret.*) LAȝ. 21417 ; ahæf 7527 ; ahef MARH. 5 ; **ahéven** (*pple.*) KATH. 2405.

a-herden, v., *O.E.* ahyrdan ; *grow hard* ; aherd (*pple.*) MAT. xiii. 15.

āhnien, *see* **āȝnien.**

a-holken, v., *thrust out* : aholeke (*imper.*) hit ut MAT. v. 29.

a-hōn, v., *O.E.* ahōn, = *O.H.G.* ar-, er-, irhāhan, *Goth.* ushāhan ; *hang up,* LAȝ. 20878 ; ahon him seolven JUL. 49 ; **ahēngen** (*pret.pl.*) SAX. CHR. 253 ; MK. xv. 24 ; (*printed* ahongen) MARH. 3 ; **ahangen** (*pple.*) MK. xv. 15 ; ahonge HOM. I. 41 ; ahon MARH. 3.

āht, adj., *O.E.* āht (ÆLFR. GRAM. 296) ; *brave, valiant, worthy,* FRAG. 2 ; ȝef he is wurþful and aht man O. & N. 1481 ; aht, oht, ocht LAȝ. 4317, 4863, 18355 (*mis-written* æht 7063) ; aȝt [ought] ROB. 101 ; oht KATH. 1727* ; oȝt GAW. 2215 ; **ōhtne** (*acc. m.*) LAȝ. 23387 ; **ōhte** (*nom. pl.*) LAȝ. 4743 ; gode men & aȝte ROB. 459 ; **ōhtere** (*gen. pl.*) LAȝ. 18013 ; **āhtere** (*compar.*) LAȝ. 4348.

āht-, ōht-līche, adv., *O.E.* āhtlīche ; *bravely,* LAȝ. 797, 31141.

ōht-scipe, sb., *valour,* LAȝ. 24671.

āht, *see* **āwiht.**

ahte, *see* **eahte. āhte,** *see* **æhte.**

[**ahtien,** v., *O.E.* eahtian, = *O.Fris.* achtia, *O.L.G., O.H.G.* ahtōn, *O.N.* akta ; *consider, estimate* ; *comp.* ȝe-ahtien.]

ahtlien, v., *O.N.* ætla, *prov.Eng.* ettle ; *think, esteem, purpose ; arrange, set out* ; atlien LAȝ. 29063 ; atteli WILL. 404 ; **ahtil** (*pres.*) M. H. xvii ; attle to schawe GAW. 27 ; & þer to ettleð (*aims*) HOM. II. 79 ; atlien (we) to þan kinge *let us turn to the king* LAȝ. 25996 ; **eghtild** (*pret.*) PR. C. 5784 ; Brien him atlede (*tended*) to LAȝ. 30846 ; þat Alixaundre wiþ his ost atlede þidire WART. II. 104 ; **aghteld** (*pple.*) S. S. (Web.) 3053 ; hire teþ aren white ase bon of whal evene set and atled al SPEC. 35.

ahtlinge, sb., *estimation,* LAȝ. 25761 ; etlunge H. M. 39 ; wiðuten ei etlunge [eatlunge] HOM. I. 263.

a-hungren, v., = *M.H.G.* erhungern ; *famish* ; ahungred (*pple.*) LANGL. *B* x. 59*.

ā-hwǣr, adv., *O.E.* āhwær, = *O.H.G.* eohwār ; *anywhere* ; awher HALLIW. 122 ; awer, owar L. H. R. 30 ; owwhær, owwhar ORM. 833,

8472; owher CH. C. T. *A* 653; owhere Gow. II. 349; ower ALIS. 5629; owhar AR. & MER. 7903; ouhwar A. R. 60; ouwhar (*printed* onwhar) WILL. 1820.

a-hwēnen, v., *O.E.* ahwǽnan; *trouble*: her of þe lavedies to me meneþ and wel sore me **ahwēreþ** (*pres.*) O. & N. 1564; þah ich mid soþe ,heo awene 1258.

a-hweorfen, v., *O.E.* ahweorfan; *turn away*; **awharf** (*pret.*) GAW. 2220.

ā-hwider, adv., *O.E.* āhwider; *anywhither*; ohwider HOM. I. 247; ouhwuder A. R. 172; owhidre '*qvoqvam*' WICL. 4 KINGS v. 25.

ai, *see* **ā** *and* **ēi**.

aiel, sb., *O.Fr.* aiel; *ancestor*, CH. C. T. *A* 2477; **aieles** (*pl.*) LANGL. *B* xv. 317.

aighwanen, adv. (*for* *ā-ȝehwanen), *O.E.* ǽghwanan; *from all sides,* '*undiqve,*' MK. i. 45.

aighwilc (*for* *ā-ȝehwilc), **ǽlch** (*for* *ā-ȝelīc), pron., *O.E.* ǽghwylc, ǽlc, = *O.H.G.* eogihwelih, eogilīh, *O.Fris.* elk, ek; *each,* MAT. v. 22, vi. 34; ǽlc hefde his iwillen LAȝ. 1996; ǽlc an *each one* 13145; alch [ech] mon 2512; elc, elch P. L. S. viii. 54, 56; ǽlch, elch HOM. II. 31, 149; ewc, euch KATH. 20, 1231; euch A. R. 14; euch, ech O. & N. 434, 975; ech TREAT. 132; CH. C. T. *A* 791; TOR. 2758; hi lovieþ ech oþren AYENB. 268; **ǽlches**, ēlches (*gen. m.*) LAȝ. 7644, 9921; elches monnes HOM. I. 99; eches BRD. 3; **ēwilcum**, ēwilche (*dat. m. n.*) HOM. I. 37, 93; ǽlchen MAT. xviii. 19; elchen, ǽlche [eche] LAȝ. 2511, 13826; echen AYENB. 13; eche O. & N. 800; **ǽlchere**, elchere (*dat. f.*) LAȝ. 621, 1723; elchere HOM. I. 103; echere MISC. 116; **ǽlene** [echne] (*acc. m.*) LAȝ. 4386; **ǽlche**, ēlche (*acc. f.*) MAT. iv. 23, ix. 35; *see* ȝe-hwilc, ilc.

aihte, *see* **ǽhte**.

[**aihwa,** pron. (*for* *ā-ȝehwa), *O.E.* ǽghwa, = *O.H.G.* eogahwer]; **eiwat**, *O.E.* ǽghwæt; *? everyone, everything, anything,* O. & N. 1056*.

aihwǽr, adv. (*for* *ā-ȝehwǽr), *O.E.* ǽghwǽr, = *O.H.G.* eogihwār; *everywhere*; ǽwher, ǽiwær LAȝ. 13372, 17827; eȝȝwhǽr ORM. 645; eihwer KATH. 1728; aihwer HOM. I. 271; aiwhere A. P. ii. 228; aihwar HICKES I. 223; aihware HOM. II. 222; aiwhare PR. C. 8199; aiware O. & N. 216; aiwhare, -whore MAN. (F.) 721, 6285.

aimen, *see* **ámen**.

aimont, sb., *Fr.* aimant; *adamant,* AYENB. 187.

air, sb., *O.Fr.* air, aer; *air,* MAND. 312; air, aer TREV. IV. 139; air, eir A. R. 104; eir TREAT. 138.

aire, sb., *Fr.* aire; *aerie, eyry*: haukes of nobule eire DEGR. 46.

airen, *see* **eiren**.

airiss, adj., *aerial,* CH. H. F. 964.

aise, sb., *O.Fr.* aaise, eise; *ease*; eaise, eise A. R. 108, 114; eise AYENB. 48; HORN (R.) 1265; ese BEK. 2233; ese, eese, ease CH. C. T. *A* 969.

aise[2], adj., *Fr.* aise; *at ease*: ȝif ȝe beoð eise [aise] A. R. 20.

aisel, *see* **eisil**.

aisiē, adj., *O.Fr.* aisié, aaisié; *easy,* HOM. II. 47; eise MAP 336; aise, ese MAN. (F.) 6438. **ēsē-līche,** adv., *easily,* BRD. 18.

aisiement, sb., *O.Fr.* aisement; *easement*: esement CH. C. T. *A* 4179; **aisiamentis** (*pl.*) WINT. vii. viii. 772; eisementes MAND. 214.

aisien, v., *O.Fr.* aaisier, *from med.Lat.* adagiāre; *ease*; eisi AYENB. 82; esie FER. 1946; to esen hem and don hem alle honour CH. C. T. *A* 2194; **ēsed** (*pple.*) WILL. 1632.

aiþer, *see* **ǽiðer**.

ajoinen, v., *O.Fr.* ajoindre; *adjoin*; **ajoined** (*pple.*) WILL. 1753.

ajournen, v., *O.Fr.* ajurner; *adjourn*; **a-journed** (*pple.*) D. ARTH. 340.

ajugen, v., *O.Fr.* ajugier; *adjudge*; **ajuged** (*pple.*) CH. BOET. i. 4 (15).

ak, *see* **ac**.

āk, sb., *O.E.* āc, = *O.Fris.* ēk, *O.N.* eik, *O.H.G.* eih, eich *f.*; *oak,* '*qvercus,*' VOC. 228; ake 191; oc FRAG. 3; ok ROB. 22; ook, ok FER. 4561; CH. C. T. *A* 1702; **āke** (*dat.*) REL. I. 52; PERC. 773; **ōkes** (*pl.*) LANGL. *A* v. 18.

ōc-werne, sb., *O.E.* ācwern; *squirrel*; (*mss.* ocquerne, aquerne) P. L. S. viii. 182.

a-kēlen, v., *O.E.* acēlan, = *M.H.G.* erkuelen; *cool*; akele Gow. II. 91; (h)is blod scholde sone akele FER. 4492; **akēlþ** (*pres.*) MISC. 30; **akēlde** (*pret.*) ROB. 442.

áken, v., *O.E.* acan [*? allied to O.N.* aka, *drive, Lat.* agere]; *ache*: aken & smerten HOM. II. 207; akin '*doleo*' PR. P. 8; ake HOM. I. 149; MISC. 95; FER. 1831; LUD. COV. 232; **ákeþ** (*pres.*) REL. I. 111; LANGL. *A* vii. 243; þet heaved me akþ AYENB. 51; his bones akeþ SHOR. 2; þine banes akeð þe H. M. 31; aken CH. C. T. *B* 2113; **æke,** eke (*pres. subj.*) A. R. 360, 368; **ékinde** (*pple.*) A. R. 360; **ōc** (*pret.*) LAȝ. 6707; HOM. II. 21; ok ROB. 68; ook S. A. L. 42; oken, oke LANGL. *B* xvii. 194; **ōke** (*pret. subj.*) P. L. S. xx. 66; *deriv.* **áche**.

āken, adj., *O.E.* ācen, = *O.H.G.* eichīn; *oaken*: an oken bord ALIS. 6415; oken wode (? okenwode) D. ARTH. 272.

a-kennednesse, sb., *O.E.* acennedness, acenniss; *generation,* HOM. I. 209; aken(n)esse H. M. 45.

a-kennen, v., *O.E.* acennan; *beget*; akende (*pret.*) LAȝ. 21240; akenned (*pple.*) LK. i. 35; HOM. I. 89; ORM. 7141; he was akennet of Marie KATH. 332.

a-kennen[2], v., = *O.H.G.* archennan; *recognize, reconnoitre*; **akende** (*pret.*) LAȝ. 7243*

(*first text* ikende); **akenninge** (*pple.*) ALIS. 3468.

a-kēpen, v., = **kēpen** : þar hii wolde akepe (? a kepe) LA3. 26937* (*earlier text* kepen).

aker, sb., *O.E.* æcer, = *O.N.* akr, *Goth.* akrs, *O.H.G.* achar, *Lat.* ager, *Gr.* ἀγρός; *acre, field,* '*ager*,' FRAG. 3 ; MAT. xiii. 38; akir WICL. 1 KINGS xiv. 14; **ækeres**, akeres (*gen.*) MAT. vi. 28, 30; **akere** (*dat.*) MAT. xiii. 24; **akeres** (*pl.*) FRAG. 2 ; acres REL. I. 173.

> **aker-land**, sb., = *G.* ackerland; *ploughed land,* VOC. 270 ; akerlond CHR. E. 16.

> **aker-man**, sb., = *G.* ackermann; *husband-man,* HALLIW. 36 ; **acremen** (*pl.*) L. LE. F. 176.

aker, akir, sb., *tidal wave,* '*impetus maris,*' PR. P. 8.

akern, sb., *O.E.* æcern, = *O.N.* akarn, *Goth.* akran; *acorn*; **akernes** (*ms.* hakernes) (*pl.*) WILL. 1811; acornes CH. BOET. ii. 5 (50); acharns TREV. I. 195.

aketoun, sb., *O.Fr.* aqueton, auqueton; *a sort of quilted jacket,* CH. C. T. B 2050; acketoun B. DISC. 1175.

a-kīmen, v.,? = *M.H.G.* erkūmen; *? grow faint,* **akīmed** (*pple.*) LA3. 26354.

al, eal, adj., *O.E.* eall, eal, = *O.L.G.* al, *O.N.* allr, *O.H.G.* all, *Goth.* alls; *all* ; al world HOM. I. 35 ; al þas wrake 15 ; al folc REL. I. 49; al blisse LANGL. B xvi. 190; al þi leave KATH. 787 ; al þe lare 939 ; all ure blisse ORM. 708 ; al þat 3er O. & N. 1259; al a 3er BRD. 3 ; al þis lond ALIS. 886 ; & swa hit al iwearð LA3. 290; mid al his hirede LA3. 3378 ; al (*for* alne) dai O. & N. 373; al þe winter ROB. 59; al (*for* alle) þe twelf dahes KATH. 1844 ; a (*? for* al) Denmark and England HAV. 610 ; **alles** (*gen. m. n.*) HOM. I. 121 ; alle volkes A. R. 431 ; alle kunnes FL. & BL. 793 ; **allen** (*dat. m. n.*) LK. xxiv. 19 ; allen, alle LA3. 135, 1686; **alre** (*dat. f.*) HOM. I. 123; in alre blisse LA3. 6065 ; mid alle (*for* alre) mare mihten (*r.* mihte) 699 ; mid alle mine mihte HOM. I. 191 ; **ealne**, alne (*acc.m.*) MAT. xviii. 32, LK. ix. 215; alne HOM. I. 225 ; alle dæi heo sungen LA3. 20981; **alle** (*acc. f.*) HOM. I. 99 ; O. & N. 433; heo ferden alle nihte LA3. 26922 ; **ealle**, alle (*nom. acc. pl. m. f.*) HOM. I. 23, 97 ; ealle, alle P. L. S. viii. 88 ; alle O. & N. 222 ; alle heo weren ease LA3. 751 ; alle [al] his men 3858 ; alle þine þreates KATH. 2133 ; us alle 548; alle five HAV. 2128 ; alle þreo C. L. 563 ; we alle habbeþ enne vader AYENB. 145 ; alle men MAND. 10 ; alle and some LEG. 45 ; SPEC. 42 ; C. L. 489; FER. 1513; RICH. 5846; CH. C. T. A 2187; alle (*for* alre) monne A. R. 384 ; **al** (*nom. acc. pl. n.*) LA3. 10556; þurh al (*in all things*) swa wel idihte 25280 ; þurh alle þing 2722 ; **alre**, [*O.E.* ealra] (*gen. pl.*), MAT. xiii. 32 ; KATH. 254; C. L. 232; FER. 2884 ; alre þinge A. R. 398 ; allre shafte ORM. 346; here alre fet BRD. 17 ; þe alre worst(e) SPEC. 104; alre mæst ORM. 2595 ; alre mest O. & N. 684 ; on alre erest A. R. 136; alre best REL. I. 116; alre, aller LANGL. B xix. 468, C xxii. 473 ; alre erest *first of all* HOM. I. 33 ; alra 37 ; ælra 221 ; aller CH. C. T. A 586; S. S. (Wr.) 365 ; aldre GEN. & EX. 322 ; alder WILL. 3345; alþer HAV. 1978; AYENB. 27 ; S. S. (Web.) 3560; LIDG. M. P. 20 ; **eallen** (*dat. pl.*) MAT. ii. 16 ; allen HOM. I. 125, 225 ; AYENB. 145 ; allen, alle LA3. 344, 6756; ealle SAX. CHR. 251 ; alle KATH. 125 ; mid alle his wrenche O. & N. 827; best of alle CH. C. T. A 796.

> **at al** (*? for* alle), *at all,* CH. C. T. E 1045.

> **in alle**, *in all*: he lived in alle þre and sixti 3ere TREV. III. 363.

> **mid alle** (*O.E.* mid ealle, = *M.L.G.* mit alle), *wholly,* KATH. 656; O. & N. 1458; P. L. S. xx. 27 ; mid alle fordon HOM. I. 17 ; iwepned mid alle LA3. 26339.

> **wiþ alle**, *withal,* ORM. 2174.

al, adv., *all, wholly* : water geð al to gedere (*altogether*) ut A. R. 320; and was mid ivi al bigrowe O. & N. 27 ; al mid wisdome LA3. 443; al an oþir rule HOCCL. iii. 7 ; al stilliche A. R. 82; al hali3 ORM. 8871 ; al mi3tful (migtful) ASS. 219; GEN. & EX. 2694; all to longe BEK. 774; CH. L. G. W. 824; **al** to fewe SHOR. 93 ; al beo þu meiden *though thou be a virgin* H. M. 43; al be it (*albeit*) þat HOCCL. I. 209; al þus LA3. 3669 ; A. R. 238; FL. & BL. (H.) 817.

> **al āne** *alone* : noht ne ma3 þe man bi bræd al ane libben ORM. 11344; al one WILL. 659; hue wonede al one HORN (R.) 80.

> **al mōst, al mēst**, *almost,* A. R. 222 ; BRD. 7; al mest dead LA3. 19328*; H. H. 47 ; CH. C. T. A 155.

> **al swa, al se, al so, as** (*mod.Eng.* also, as), LA3. 70, 468, 4308, 24579 ; al swa S. S. (Web.) 3945 ; al so sone se he mighte 569; þat ne an oþir es al swa PERC. 1452; als 684 ; al swa al se HOM. I. 153; al swa se 159; al swa KATH. 1986; al swa as 289; **ase** wod wulf 31; as ha set in a bur 139 ; þer as tis blisse is 1723; al swa se ORM. 3536; al se 14801; **als** if it wære 589; al swo al se hit is biforen iseid PROCL. 5; al so ase A. R. 62 ; al se ofte al se 8 ; al se ofte ase 322 ; þere ase (*where*) . . . is 80; al so fer so a boge mai ten GEN. & EX. 1238; als swilc als he is dede ðor in 3836 ; als he cam 1785 ; al so ros þe mone HAV. 1955; al so fresch as þe hauk JOS. 595 ; al se AYENB. 15 ; and mi wif al se [als] LANGL. A v. 144; als AV. ARTH. lxv.; & þe ladi als GAW. 933 ; als mani as men list to have MAND. 209; als sir Iwain made his mane IW. 2303 ; ase MISC. 48; drau3tes as me draweþ in poudre P. L. S. xvii. 225 ; þer as he 3ut is xx. 106; as it þoughte me CH. C. T. A 385 ; as freendli as he were his owen broþer 1652 ; as feondes remeþ MISC. 76; þer as burnes were busi bestes to hulde WILL. 1708; no more of þis as nou

C

CH. C. T. *B* 1242 ; as þan GOW. III. 285 ; as swiþe LANGL. *A* iii. 96 ; i schal ou elle as blive K. T. 1040 ; as for þe time MAND. 122 ; os E. T. 475 ; EGL. 247.

al þāh, *although,* SPEC. 23 ; al þaȝ he bi AYENB. 19 ; al þoȝ FER. 321.

al-, *see* el. āl, *see* awel.

alabastre, sb., *O.Fr.* alabastre ; *alabaster,* CH. C. T. *A* 1910.

a-lacchen, v., *catch* : alle þat þai þan alacche miȝt FER. 3098 ; alehte (*pret.*) hine betwux his earmes SAX. CHR. 249.

a-lēnen, v., *O.E.* alǣnan ; *lend* ; aleane LAȝ. 31603* ; alēneþ (*pres.*) 24000*.

a-lǣten, v., *O.E.* alǣtan ; *let off* : let alǣten þis wæter LAȝ. 15932.

alai, sb., *Fr.* aloi ; *alloy,* LANGL. *B* xv. 342.

alaien, v., *Fr.* aloyer ; *allay, alloy* ; alaied (*pple.*) LANGL. *B* xv. 346 ; alaid ' *temperatus, permixtus* ' PR. P. 9.

Alamanie, pr. n., *Alamannia,* SAX. CHR. 264 ; Alemaine LAȝ. 1977.

alambic, sb., *Fr.* alambic ; *alembic,* CH. TRO. iv. 520.

a-lámen, v., = *M.H.G.* erlamen ; *grow lame* ; alamed, alemed (*pple.*) O. & N. 1604.

a-langien, v. (*cf.* ǣlenge), *?make dreary* ; alangeþ (*pres.*) AR. & MER. 4212.

alant, sb., *Fr.* alan, = *Sp.* alano ; *an Alan dog* ; alantz (*pl.*) CH. C. T. *A* 2148.

alarge, v., *O.Fr.* alargir ; *enlarge,* WICL. GEN. xxxii. 12.

alaski, v., *O.Fr.* alascher, *Lat.* elaxare ; *let loose, release,* LAȝ. 8838*.

a-lāðien, v., = *M.H.G.* erleiden ; *become odious* ; alōþeþ (*pres.*) O. & N. 1277 ; mi lif me is alāðed LAȝ. 25930.

albe, sb., *Fr.* aube, *Lat.* alba ; *alb,* REL. I. 129 ; aube AYENB. 236.

alcali, alkali, sb., *Fr.* alcali ; *alkali,* CH. C. T. *G* 810.

ald, *see* eald.   alder, *see* ealder, aler.

aldien, *see* ealdien.

ále, sb., *O.E.* ealu, ealo, = *O.L.G.* alo(fat), *O.N.* öl *n.; ale,* A. R. 114 ; HAV. 14 ; CH. C. T. *A* 832 ; MAND. 251 ; aille TOWNL. 90 ; of þan ale LAȝ. 24440 ; to then ale R. S. 83 ; at þen ale, at þe nale, atte nale, atte ale LANGL. *A* 42 ; CH. C. T. *D* 1349 ; *comp.* brūd-ale ; *see also* aleð.

    ále-brē, sb., *bread sopped in ale,* L. C. C. 53 ; alberei, alebrei, albri ' *alebrodium, fictum est* ' PR. P. 9.

    ále-hūs, sb., *alehouse* ; álehūse (*dat.*) HOM. II. 11 ; aillehowse TOWNL. 310.

    ále-stake, sb., *a sign before an alehouse,* CH. C. T. *A* 667.

a-lēapen, v., *O.E.* ahlēapan ; *leap up* ; alēop (*pret.*) LAȝ. 30683.

a-leggen, v., *O.E.* alecgan, = *O.H.G.* arleccan ; *lay aside, put down, refute,* · LK. v. 19 ; LAȝ. 7714 ; þu miht lihtliche . . . al mi sor aleggen HOM. I. 197 ; heo ne mihte noht alegge þat þe ule hadde hire ised O. & N. 394 ; alegge (*pres. subj.*) HOM. I. 91 ; uvele lawes . . . bote if he hem alegge BEK. 1638 ; aleide (*pret.*) LAȝ. 7125* ; WILL. 5240 ; aleigd (*pple.*) MAT. xxviii. 6 ; aleid FRAG. 6 ; LAȝ. 31966 ; SPEC. 105 ; TREV. IV. 449 ; alaid is Darie þin honour ALIS. 2386.

aleggen, v., *O.Fr.* alleguier ; *allege,* LANGL. *B* xi. 88 ; alegge CH. C. T. *E* 1658 ; allege PR. C. 5584.

aleggin, v., *O.Fr.* alegier ; *relieve, lighten,* ' *allevio,*' PR. P. 9 ; allege PR. C. 3894 ; alegget (*pple.*) WILL. 1034.

alei, sb., *O.Fr.* alee ; *alley* ; aleis [aleies, alaies] (*pl.*) CH. C. T. *E* 2324.

a-lēinen, v., = *O.H.G.* arlougnan ; *conceal* ; alained (*pret.*) HALLIW. 38.

a-lēomen, v., *illumine,* HOM. II. 7 ; a-lūm(e)ð -limeð, -lemeð (*pres.*) 107, 109, 141.

aler, sb., *O.E.* aler, alr *m. cf. O.H.G.* elira, erila, *M.L.G.* elre *f., O.N.* elri *n.; alder* ; aller PALL. ix. 90 ; alder CH. C. T. *A* 2921 ; aldir ' *alnus* ' PR. P. 9 ; olrr FRAG. 3.

    alder-ker, sb., *alder-carr, grove of alders,* ' *alnetum,*' PR. P. 9.

a-lēsen, v., *O.E.* alēsan, alȳsan, = *O.H.G.* erlōsan ; *release, deliver, redeem,* LAȝ. 1084 ; A. R. 124 ; MISC. 140 ; alise fram helle wite HOM. I. 229 ; alēs (*imperat.*) MARH. 20 ; alēse (*pres. subj.*) L. H. R. 180 ; alēsde (*pret.*) HOM. I. 19 ; JUL. 40 ; he hine alesede mid his blode R. S. vii (MISC. 186) ; alisden (*for* alisde) LAȝ. 11167 ; alēsed (*pple.*) KATH. 1150 ; alused P. L. S. viii. 68 ; ȝe ne beoð ne alesde of deofles anwalde HOM. I. 127.

    alēsednesse, sb., *O.E.* alȳsedness ; *release, redemption,* HOM. I. 129 ; alysednesse LK. xxi. 28.

    alēsend, sb., *releaser, redeemer,* HOM. I. 125 ; alesent JUL. 66.

    alēsendnesse, sb., *?for* alēsednesse ; *release,* HOM. I. 87 ; H. M. 11* ; alisendnesse HOM. I. 227 ; alisendnisse MK. x. 45.

    alēsnesse, sb., *O.E.* alēsness, alȳsness, = *O.H.G.* arlōsnessi, ANGL. VII. 220 ; *release, deliverance,* HOM. I. 15 ; (*printed* alefnesse) H. M. 11.

    alēsunge, sb., *O.E.* alȳsing, = *O.H.G.* irlōsunga ; *release, deliverance,* KATH. 1153 ; alesinge HOM. I. 143.

aleð, sb., *O.E.* ealoð (*gen. and dat. of* ealu ; *see* ale) ; REL. I. 132 (HOM. II. 13) ; ealaþ LEECHD. III. 118.

alfe, sb., *O.E.* ælf, elf, ylf, *pl.* ylfe, = *M.L.G.* alf, *O.N.* alfr, *O.H.G.* alp (*pl.*, elbe) *m. ; elf,* AUD. 77 ; alve [alfe] LAȝ. 19268 ; elf CH. C. T. *D* 373 ; elfe ' *lamia* ' PR. P. 138 ; alven

[alvene] (*pl.*) LA3. 19255; elvene (*for* elven)
ROB. 130; JARB. XIII. 162; **ælvene** (*gen.
pl.*) LA3. 21747.

  elf-qvēne, sb., *elf-queen*, CH. C. T. *B* 1978.
  Alf-rēd, pr. n., O. & N. 235; REL. I. 170.
  Alf-rīch, pr. n., REL. I. 170.

alfin, sb., *O.Fr.* aufin, *L.Lat.* alphinus (*from
Arab.* al-fīl); *bishop at chess, foolish person*:
awfin of the chekar '*alfinus*' PR. P. 18; D.
ARTH. 1343; **aufins** (*pl.*) GEST. R. 61.

algorisme, sb., *Fr.* algorithme; *algorism*,
AYENB. 1; algrim HALLIW. 42; augrim A. R.
214; DEP. R. iv. 53; awgrim stoones CH. T.
A 3210.

aliance, alliance, sb., *O.Fr.* aliance; *alliance*,
ROB. 89; CH. C. T. *A* 2973.

a-libban, v., *O.E.* alibban; *live*, HOM. I. 109.

alie, allie, sb., *O.Fr.* alié; *ally*, WICL. EX.
xviii. 5; TREV. V. 221; CH. C. T. *G* 297.

alie, sb., *O.Fr.* alie; *service-berry*; **aleis** (*pl.*)
R. R. 1377.

alien, allien, v., *O.Fr.* alier; *ally*, CH. C. T.
E 1414; **alied** (*pple.*) ROB. 65.

alien, adj., *O.Fr.* alien; *alien*, PR. P. 10; ALIS.
3918; TREV. III. 447.

aliénen, v., *Fr.* aliéner; *alienate*, WICL.
ECCLES. xi. 36; aliene CH. BOET. i. 6 (27).

a-liggen, v., *O.E.* alicgan, = *O.H.G.* irliccan;
subside, MARH. 5; aligge LA3. 26298; a
muchel wind alið mid a lutel rein A. R. 246;
þis lutle pine þat alið i lute hwile KATH.
2183; **alei** (*pret.*) MARH. 12; þe loudinge
alai (*ms.* alay) LA3. 24873*.

a-līhten, v., *O.E.* alihtan; *alight, descend;
make light; alleviate*, LA3. 26618; no wonder
þe3 hit smite harde þer hit doþ ali3te TREAT.
136; **alīhte** (*pret.*) HOM. I. 79; LA3. 21121;
MISC. 37; god almihti . . . alihte a dun to helle
A. R. 248; ali3te BEK. 1895; SHOR. 129;
ali3ted (*pple.*) '*alleviata*' WICL. IS. ix. 1;
ali3t FL. & BL. 21; so sone so he was ali3t
ALIS. 4490.

a-līhten, v., *O.E.* alȳhtan, = *O.H.G.* irliuhten;
enlighten; ali3te AYENB. 109; FER. 1261;
ali3te (*pret.*) SHOR. 84; alight Gow. II.
183; þu . . . havest aliht mi þester heorte
HOM. I. 185.

  ali3tinge, sb., *O.E.* alȳhting; *illumination*,
AYENB. 221.

a-limpen, v., *O.E.* alimpan; *happen*; **alomp**
(*pret.*) LA3. 18053.

a-līsen, v., *? O.E.* ahlīsian; *mention by report*;
alīsed (*pple.*) P. L. S. 67.

a-liðen, v., *pass away*, LA3. 12041; **aliðen**
(*pple.*); þa seove 3er weoren aliðene 3970.

aliþien, v., *O.E.* aleoðian; *dismember*: nou
haveþ he . . . mine leomes aliþede LA3.
25929*.

alkamie, alkenamie, sb., *O.Fr.* alkemie, al-
quemie; *alchemy*, LANGL. *A* x. 212.

alkatran, sb., *O.Fr.*, *Span.* alquitran, *Port.*
alcatrão (*from Arab.* alqatrān); *bitumen*,
MAND. 99; alka[t]ran A. P. ii. 1034.

aller, *see* aler.

alles, adv. (*gen. of* al), *wholly, altogether*,
A. R. 64; KATH. 796; ROB. 17; alles to swiðe
HOM. I. 103; þa he alles spac LA3. 488; þo
heo was alles þider icome BEK. 73; whon he
wolde alles bicome mon C. L. 659.

allunge, adv., *O.E.* eallunga, eallinga; *wholly,
altogether*, LA3. 8797; A. R. 164; H. M. 47;
turn me allunge to þe HOM. I. 185; eallunge
MAT. iii. 10, allinge P. L. S. xiii. 218; JOS.
440; P. S. 214; M. T. 108; allinges MAND. 189.

almande, sb., *Fr.* amande; *almond*, '*amyg-
dalum*'; almaunde PR. P. 10; **almandes** (*pl.*)
C. M. 6895; almondes L. C. C. 9.

almander, sb., *O.Fr.* almandier; *almond-tree*;
almaunder WICL. ECCLES. xii. 5; **almanders**
(*plur.*) WICL. GEN. xxx. 37; almandres R. R.
1363.

almarie, sb., *O.Fr.* almarie, almaire; *ambry,
cupboard, clothes-press, bookcase*; almari
'*armarium*' PR. P. 10; **almaries** (*pl.*)
LANGL. *B* xiv. 246; the almeries of Neemie
WICL. 2 MACC. ii. 13.

almesse, *see* elmesse.

al-mi3t, adj., *O.E.* ælmiht; *omnipotent*, MIRC
461; A. P. i. 497; almight Gow. I. 190; SACR.
996; to god almi3te FER. 256.

al-mihtin, adj., *? for* almiht; *omnipotent*,
HOM. I. 15, 97; almigtin GEN. & EX. 30;
almihten LA3. 16783; almicten REL. I. 234;
almihtinde 282.

al-mihti3, adj., *O.E.* ealmihtig, ælmeahtig, =
O.H.G. almahtig, *O.N.* almättigr; *almighty*,
HOM. I. 217; ælmihti 221; almihti A. R. 430;
SPEC. 73; almi3ti AYENB. 1; þe almihti god
FRAG. 8; god almihti HOM. I. 5; almahti3
ORM. 95.

  almighti-hēde, sb., '*omnipotentia*,' PR. P. 10.

almoner, sb., *O.Fr.* almosnier, aumosnier;
almoner; (*? printed* amoner), AYENB. 190;
aumoner C. M. 15219.

along, *see* 3elang, -long.

alōsen, v., *O.Fr.* aloser; *praise*; alosi AYENB.
183; **alōsed** (*pple.*) R. R. 2354; ALIS. (Sk.)
139.

alouaunce, sb., *allowance*, LANGL. *B* xiv.
109.

alouin, v., *O.Fr.* allouer; *allow*, '*alloco*,'
PR. P. 10; alowe SHOR. 137; þis stat is
moche to alowe vor his dignete AYENB. 227;
aloue (*pres.*) CH. C. T. (Wr.) 10988; aloue
(*subj.*) RICH. 4662.

alp, sb., *cf.* Norfolk *dial.* olp; *a sort of finch,
'ficedula*,' PR. P. 10; **alpes** (*pl.*) R. R. 658.

alpi, *see* anlepi3. **als**, *see under* al.

alsene, sb., *cf. M.Du.* alsene, *M.H.G.* alansa;
awl, VOC. 150; elsin '*sibula*' PR. P. 138.

alter, sb., *O.Fr.* alter, auter, *Lat.* altāre; *altar*, ORM. 1061; alter, auter GEN. & EX. 758, 1297; auter HAV. 369; JOS. 295; MAN. (F.) 1375.

alter-clŏᵹ, sb., *altar cloth*, REL. I. 129.

alum, sb., *O.Fr.* alum; *alum*, A. P. ii. 1035; CH. C. T. *G* 813.

alūre, sb.; *O.Fr.* alure, aleure; *place to walk in, passage, gallery*, ALIS. 7210; GUY p. 85; PR. P. 10; alūr(e)s (*pl.*) ROB. 192; throu the aleris of his soler '*per cancellos coenaculi sui*' WICL. 4 KINGS i. 2.

a-lūten, v., *O.E.* alūtan; *incline*; alute MISC. 48; aloute BEK. 2140; PL. CR. 750; B. DISC. 1254; alouted (*pret.*) WILL. 3721.

alve, *see* alfe.

alvisch, adj., = *M.H.G.* elbisch; *elfish, like an elf*, GAW. 681; on alvisc smiᵹ LAᵹ. 21131; elvisch CH. C. T. *B* 1893; an elvish knight ELL. ROM. II. 90.

al-wealdende, adj., *O.E.* ealwealdend; *all-ruling*, KATH. 615; alwaldinde LAᵹ. 19548; alwældend ORM. 153; alweldand IW. 2199.

am, v. (1 *pers. sing. pres.*), *O.E.* eom, eam, am, = *O.N.* em, *Goth.* im; *am*, KATH. 463; A. R. 430; ORM. 205; O. & N. 170; ROB. 115; WICL. GEN. xv. 1; i am come her FER. 354; eam FRAG. 6; eam, æm, em, am LAᵹ. 720, 8033, 8899, 10524; æm, eom P. L. S. viii. 1, 2; eom MISC. 192; ich hit eom *it is I* MAT. xiv. 27; em MK. vi. 50; HOM. I. 115.

art (2 *pers. sing. pres.*), *O.E.* eart, = *O.N.* ert; *art*, KATH. 450; ORM. 2206; O. & N. 38; ROB. 31; WICL. JOHN viii. 12; eart HOM. I. 29; eart, ært, ert FRAG. 6, 7; LAᵹ. 1442, 1499, 2237; ært, ert MAT. xiv. 28, LK. i. 4; ert A. R. 26; HOM. II. 29.

is (3 *pers. sing. pres.*), *O.E.* is, = *O.L.G.* is, *Goth.*, *O.H.G.* ist; *is*, ORM. 667; wa is me HOM. I. 35; hwet is us to donne 91; lutel me is of ower luve JUL. 27; me is þe wurs O. & N. 34; what is þe *what is the matter with thee* ASS. 227; what is ᵹou BRD. 21; he . . . is riden to þe feldes CH. C. T. *A* 1503; is, us LAᵹ. 745, 5354; es C. M. 5880; Iw. 34; E. W. 39.

aren (*pl. pres.*), *O.E.* earun, = *O.N.* eru; *are*, HOM. II. 73; MARH. 8; HAV. 619; SPEC. 85; LANGL. *B* iii. 80; arn HOM. I. 269; ORM. 4555; GEN. & EX. 16.

sind, **sinden** (*pl. pres.*), *O.E.* sind, sindon, = *O.L.G.* sind, sindun, *O.H.G.* sint, sintun, *Goth.* sind; *are*, MAT. x. 31, xxii. 8; sinden ORM. 389; sindon, sindan MISC. 146; senden HICKES I. 223; REL. I. 210; senden, sunden P. L. S. viii. 144; sende HOM. II. 173; sunden, sunde LAᵹ. 4359, 16796.

sī (*subj. pres.*), *O.E.* sī, sie, seo, = *O.L.G.*, *O.H.G.* sī, *Goth.* sijau; HOM. I. 245, II. 91; ORM. 3378; si, seo LAᵹ. 8545, 14893; sēon (*pl.*) (*O.E.* sīen, sēon) LAᵹ. 13837; sion (*ms.* syon), sien MAT. vi. 25, JOHN xvii. 22.

a-mǣden [*for* ᵹemǣden, *O.E.* gemǣdan], *? distract, derange*; amǎd (*pple.*) HORN (L.) 574; P. S. 156; S. A. L. 154; amad, amead H. M. 37; amed A. R. 324; MISC. 48; heo weoren amadde LAᵹ. 4438.

amaien, v., *O.Fr.* esmaier, amaier, = *Ital.* smagare; *dismay*: þou miᵹt not me amaie FER. 485; ne amai þe nouᵹt S. S. (Wr.) 1536; amaied (*pple.*) ALIS. 1749; CH. TRO. iv. 641.

a-maistren, v., = *Ger.* ermeistern; *master*, LANGL. *B* vi. 214 (*A* vii. 200); ameistren A. R. 140; amaistreþ (*pres.*) AYENB. 129; amaistrede (*pret.*) P. L. S. xxiv. 60.

a-mānsien, v., *O.E.* amānsumian, *O.H.G.* armeinsamōn; *excommunicate*; amansi BEK. 1744; amonsi P. S. 196; amānsede (*pret.*) P. L. S. xvii. 512; amānsed (*pple.*) HOM. I. 45; KATH. 2101; O. & N. 1307; ROB. 335.

amānzing, sb., *O.E.* amānsumung; *excommunication*, AYENB. 189.

a-māsen, v., *O.E.* amasian; *amaze*; amásed (*pple.*) A. R. 270; WILL. 686; GOW. II. 25.

ambesas, sb., *Fr.* ambesas; *both aces*, ROB. 51; H. H. 108 (110).

ambiciōn, sb., *Fr.* ambition; *ambition*, AYENB. 22.

amblen, v., *Fr.* ambler; *amble*; amblinde (*pple.*) HALLIW. 53; amblinge CH. C. T. *E* 388.

ambler, sb., *Fr.* ambleur; *ambler*, CATH. 9.

amboht, sb., *O.E.* ombeht; *O.H.G.* ampaht; ambaht, *Goth.* andbahts; *servant*, ORM. 2329.

áme, sb., *from* ámen; *aim, guess*, D. TROY 7088; GENER. 5959; A. P. iii. 128.

amel, sb., *O.Fr.* esmal; *enamel*, HALLIW. 54.

amelen, v., *O.Fr.* esmailler; *enamel*; amelid (*pple.*) HALLIW. 54.

ámen, v., *O.Fr.* aesmer, esmer, = *mod.Eng.* aim; *estimate*: amin '*aestimo*' PR. P. 190; hou manie þer deide i mai nought ame MAN. (F.) 4449; aime WILL. 1596; eimeþ (*pres.*) WICL. LEV. xxvii. 8.

amende, sb., *O.Fr.* amende, *from Lat.* ēmendāre; *reparation, fine, compensation*; amendes (*pl.*) AYENB. 37; PR. C. 1589.

amenden, v., *O.Fr.* amender, *from Lat.* ēmendare; *correct, reform; give satisfaction for an offence, amend, mend*, C. L. 1117; amende BEK. 1094; schon amende JOS. 423; amendi R. S. ii; amendeᵹ (*imp.*) A. R. 420.

amendment, sb., *O.Fr.* amendement; *correction; improvement; reparation*, ROB. 32; SHOR. 51; AYENB. 32.

amenusen, v., *O.Fr.* amenuisier; *diminish*; amenusi E. W. 28; amenused (*pple.*) CH. BOET. ii. 4 (40).

a-mēoken, v., *render meek*; **amēkid** (*pple.*) GEST. R. 177.

amerciment, sb., *O.Fr.* amerciment; *forfeit, fine*; **amerciments** (*pl.*) CH. C. T. *I* 752.

amercin, v., *O.Fr.* amercier; *amerce, fine*, PR. P. 11; amerci LANGL. *B* vi. 40.

a-merien, v., *O.E.* amerian; *purify*; **amered** (*pple.*) S. S. (Web.) 2266.

a-merran, v., *O.E.* amerran, -myrran; *mar, destroy, corrupt*, HOM. I. 23; amerre MISC. 74; amærre, -marren [amorre] LA3. 5356, 19469; **amerreþ** (*pres.*) AYENB. 203; a-merde, -mærde [amorde] (*pret.*) LA3. 3825, 11725; þu amerdest FRAG. 8; **amerred** (*pple.*) SHOR. 105.

amerveillen, v., *O.Fr.* esmerveiller; **amerveilled** (*pple.*) *struck with wonder, surprised*, L. R. H. 160; amervailed WILL. 3857.

amēsen, v., *O.Fr.* amesir (*from Lat.* mitire); *appease*; amese you D. TROY. 12842; TOWNL. 194.

amesūren, v., *O.Fr.* amesurer; *moderate*; **amesūreþ** (*pres.*) AYENB. 252.

ămete, sb., *O.E.* æmete, = *M.Du.*emte, eempte, *O.H.G.* āmeiza; *ant*; amote AYENB. 141; amte, emte WICL. PROV. vi. 6; ampte LIDG. M. P. 88; ematte 'formica' VOC. 177; **ǎmeten** (*pl.*) ROB. 296; emeten BEK. 2241; emoten ALIS. 6566.

ămete-hül, sb., *O.E.* æmethyll; *ant-hill*, ROB. 296.

a-méten, v., *O.E.* ametan; *measure*; **améten** (*pple.*) FRAG. 2.

ametist, sb., *Fr.* amethyste; *amethyst*, P. L. S. xxxv. 93; amatiste MISC. 98.

ami, sb., *Fr.* ami; *friend*, ALIS. 1966.

amiable, sb., *Fr.* amiable; *friendly*, CH. C. T. *A* 138.

amice, sb., *O.Fr.* aumuce; *amice*, PR. P. 11; amise WICL. IS. xxii. 17.

amidon, sb., *Fr.* amidon; *starch*, L. C. C. 8.

aministren, v., *O.Fr.* aministrer; *administer*; **aministreþ** (*pres.*) CH. BOET. iv. 6 (135).

amirail, sb., *O.Fr.* amiral, *mod.Eng.* admiral; *emir, Saracen ruler*, ROB. 409; admiral FL. & BL. 201; admiral, admirail LA3. 27668, 27680; admiraud HORN (H.) 96.

amit, sb., *O.Fr.* amit; *garment, amice, hood*, 'amictus,' VOC. 231; WICL. EXOD. xxxix. 21.

amonesten, v., *O.Fr.* amonester; *admonish*; **amonesteþ** (*pres.*) AYENB. 8; CH. BOET. v. 5 (135).

amonicioun, sb., *O.Fr.* amonition; *admonition*, CH. BOET. i. 4 (135).

amorous, adj., *O.Fr.* amorous; *amorous*, GOW. I. 89; R. R. 83.

amortisen, v., *O.Fr.* amortir; *deaden; alienate in mortmain*; amorteise LIDG. M. P. 207; **amortiseden** (*pret.*) LANGL. *B* xv. 315; a-mortised (*pple.*) CH. C. T. *I* 247.

amounten, **amounti**, v., *O.Fr.* amonter; *amount, ascend, rise*, L. H. R. 38; **amuntet** (*pres.*) MISC. 28; **amounted** (*pret.*) A. P. ii. 395.

amóven, v., *O.Fr.* esmovoir; *set in motion; excite*; **amóved** (*pple.*) GOW. I. 296; ameved [amoved] CH. BOET. i. 1 (6).

ampairi, v., *see* enpeiren.

ampre, sb., *O.E.* ampre, *prov.Eng.* amper; *tumour*; **ampres** (*pl.*) HOM. I. 237.

ampulle, sb., *Lat.* ampulla; *phial, ampulla*, LA3. 14986; **ampulles**, ampolles (*pl.*) LANGL. *A* v. 527.

amte, *see* amete.

a-mürðrin, v., *O.E.* amyrðrian; *murder*, HOM. I. 247; **amürðred** (*pple.*) LA3. 16147.

an, v., *O.E.* ann, an, = *O.N.* ann, *O.H.G.* an (*pret.-pres.*); *favour, grant*, O. & N. 1739; as i þe loue and an TRIST. 839; 3if god hit an LA3. 14851; þat he mire dohter wel on 11928; on MISC. 116; ich on wel þat 3e witen KATH. 1761; we **unnen** PROCL. 2; 3if 3e me blisse unnen [unneð] KATH. 2376; þe þe ufel unnen LA3. 28117; þet me god unnen MARH. 21; unne (*subj.*) P. L. S. viii. 158; LA3. 25114; god almihti unne me . . . þet ich mote þe iseo HOM. I. 199; (*ms.* hunne) GEN. & EX. 2249; *comp.* 3e-an; *deriv.* unnen, ēstə.

an, *see* and.   **an-**, *see* en-.

[an-, on-, prefix, *from* an prep.]

[an-², on-², prefix, = and-.]

an, **on**, **o**, **a**, prep. & adv., *O.E.* an, on, = *O.Fris.* an, on, a, *O.L.G.*, *O.H.G.* an, ana, *Goth.* ana, *O.N.* ā, *Gr.* ἀνά; *on*: an are halfe LA3. 20717; an horse and an [a] fote 502; riden an (*sec. text* uppe) horse 12979; he bar an [on] his hande 25809; an þan ilke time 3890; an, on [a] liue 3537, 10785; þa spæche him eode an (*sec. text* at) wille 13076; Brennes cuðe an (*sec. text* of) hauekes 4896; an (? = and) long 138; he havede ane honde 3793; þe king wes ane (*in*) France 23690; hevede . . . kinebearn on wombe 199; on [a] slæpe 1159; on [a] godes nomen 10136; he com on lond 116; þe king nom þat writ on hond 484; þu hit halst on (*sec. text* mid) unriht 7374; inoh he havet on (*of*) þirti 3313; muchele men on mihte 23919; ofte he hire lokede on 18538; hardliche heo on slo3en [an hewen] 1529; he heom on leide þat weoren lawen gode 2077; þah he hefde brunie on 1553; on heo duden heore iweden 9450; one leoden (*r.* leode) 3718; a londe & a watere 550; a (*sec. text* bi) dæi(e) oðer a nihte 19651; to dai a seoven nihten 11929; and swor a (*sec. text* bi) seint BRD. 17314; a (? = and) þat her com liðen ma of heore leoden 6040; ileid an (*in*) horde P. L. S. viii. 6; on helle 116; wel lange ic habbe child ibeon a worde & ec a dede 2; a watere & a londe 41; an

[on] ende A. R. 146; an Englisch 130; on ever iche halve 50; enes a wike 344; bringeð a vlihte 248; o ðe rode 262; an elle REL. I. 132; þare manie rotes one (*ms.* onne) wacseð 128; on ende 1487; þat tu havest on heorte 2142; he heold on (*continued*) to herien 434; o [on] worlde 527; a (*v. r.* til) ðet 719*; on (*ms.* onn) eorþe ORM. 422; on Abrahames time 4089; he wolde ... us ... bringen on to folʒhen þeʒre bisne 7717; þat naht þat Crist was boren one (*ms.* anne) 3753; þat standeþ o þe godspelboc 315; ænes o þe ʒer 1078; an efne O. & N. 323; nis on þe non holinesse 900; and þar one sulieþ 1240; þah no preost a londe nere 1314; a worlde 1363; a sumere 416; a dai 219; heo hadde gode þrote and schille and fale manne song a wille 1722; an horn hue ber an honde HORN (R.) 1111; he sloh a felde him þat (h)is fader aqvelde 997; an eve BRD. 31; ich wondri on mi þoʒt 15; none oþer cloþes nadde he on 29; a godes name 7; hi wende a lond 6; an, on ende C. L. 822, 973; on English 74; a live 1422; a last *at last* 457; o (? = oð) þat 152; an haste AYENB. 31; him on to loki 244; ane his left half 156; a þe left half 156; he valþ a grund 91; ones a dai 73; ones a ʒer E. G. 359; an erþe SHOR. 102; a ʒere 34; an eve, a morwe K. T. 468; he fel a slepe 397; on þinen namen '*in nomine tuo*' MAT. vii. 22; seofan siðan ... on daig '*septies in die*' LK. xvii. 4; on (*in*) þine huse FRAG. 6; on to lokienne *to look on* 8; on (*v. r.* in) ðer sæ HICKES I. 223; on londe HOM. I. 13; on heofene & on eorðan 11; beon on worlde 35; on þisse deie 99; to dei on fowertene niht 123; men cwelað on (*of*) hungre 111; & him on bleow (*breathed into*) gast 221; a londe 17; a domes deie 239; a (? = and) þet drihten com 15; alle hire luveden þat hire on lokeden MARH. 2; don us mare wa on JUL. 43; þe reve het on live ant o leomen 58; on eorðe H. M. 25; þat mahten bringe þe on (*induce thee*) mis for to donne 17; on þe fot of the dune HOM. II. 89; lustliche on to siene 163; þe þrop þe preste(s) one wunien 89; enes o dai, 67; a dai 109; be þu wis on þi word & war on þine speche REL. I. 186; ilc fis on water GEN. & EX. 162; on morgen 1093; on live 3595; so faiger he was on to sen 2659; king on Englene loande PROCL. 1; on godes nome MISC. 39; he is feir & briht on heowe 96; Aþulf sede on hire ire HORN (L.) 309; prut on herte 1389; wiþ swerde ihc þe an hitte (? anhitte) 712; on live HAV. 363; on horse riden 370; þis cloþes þat ich one (*ms.* onne) have. 1145; on eiþer side ROB. 17; o live 40; þre siþe .. a ʒer 376; þat Peteres pans .. to þe pope nere not on isend BEK. 618; þeʒ al þe toun were a fure 1049; on wicchecraft nout i ne con AN. LIT. 7; on fote gon SPEC. 90; a naht *by night* 34; when elde him comeþ on A. D. 252; on felde WILL. 173; a daie 610; a live 129; on haste ALIS. 1855; on þe dai MAND. 288; on live JOS. 707; GOW. II. 372; on [a] grounde and on [a, o] lofte LANGL. *A* i. 88 (90); þai felle on slepe S. S. (Wr.) 2745; he rod on hunting E. T. 931; ronne bournes al on blode FLOR. 609; so semeli on to see 2071; i ran on [a] blode H. H. 55; o þat come domes dai 148; & Danmark on (*for* of) him wan MAN. (H.) 57; o fote 163; sixteen hundred þei brought on (*for* of) live RICH. 2059; what shal worthe on me MIR. PL. 151; a bedde S. S. (Web.) 1513; on fōn *begin* LAʒ. 21194; fon on 5630, 17393, 17423; & feng on þus to speken KATH. 315; we voð on to spekene A. R. 74; fo we on O. & N. 179; **taken on** *begin* LAʒ. 25965; on take ROB. 170.

an ān (*printed* anan), **anon**, KATH. 31; þeʒ wisten sone an an (*ms.* anan) ORM. 225; fowertiʒ daʒhes aʒ on an (*ms.* onnan) 11331.

ān, ăn, a, ōn, ō, card. num., *O.E.* ān, = *O.L.G.* ēn, *O.N.* einn, *O.H.G.* ein, *Goth.* ains; *an (a),* one: þæt þi sien an '*ut sint unum*' JOHN xvii. 22; ða nam man an and an (*one by one*) SAX. CHR. 253; on an to eastren 256; a dæis fare 262; an child HOM. I. 77; ure an 21; a scep 121; an [a] preost LAʒ. 1; an hore 15579; þe an 22757; an efter ane *one after another* 6969; an and twenti 9541; þa weoren heo al an 29080; a wif 2528; þe an KATH. 576; an swa swote smeal 1600; ʒette me an hwat 768; a meiden 66; a crune 1585; in a londe 21; an hwet wite þu MARH. 5; al me is an þin olhnung ant tin eie 5; an an forð riht 15; meiden an eadiest '*virgo beatissima*' 13; an man ORM. 11213; an duhtiʒ wif 113; an an 225; an (*for* ane) cribbe 3321; an (*for* anes) mannes 5813; a mikel here 3370; a cribbe 3366; an hors O. & N. 773; an an, on 488, 1554, 1658; on old stoc 25; on hare 383; ʒif me hit halt evre forþ in on 356; a word 502; a tonge 156; o song 333; an wis man GEN. & EX. 2649; and kiste (h)is breðere on and on 2266; an hare ROB. 457; an (? *for* ane) fourti ʒer 230; an vewe wilde hinen 540; heo were al at on 113; a doʒter 12; at o time 306; an hound WILL. 10; an five mile 5110; a stounde 159; on a time 8; an hors MAND. 249; on geð ... in one sliddrie weie A. R. 252; heo is ever on 6: an on 20; in on 6; ʒif a mon is god 86; þet o mon beo vor one þinge twien idemed 308; a sihðe 48; o luve 12; no qvodh on HAV. 1800; but on 962; ilk an 1770; an hundred 2126; a king 27; of a tale 3; habbe velaʒrede þe on wiþ þe oþre AYENB. 43; bieþ al on 259; a kniʒt 45; o þing 46; whenne on hath don a sinne MIRC 71; and boþe on god 465; þre daies slep he al on on REL. I. 226; an on to þe þrote '*usqve ad guttur*' TREV. III. 101; þei wepen evere in on [oon] CH. C. T. *A* 1771; a so gret best BRD. 31; a fourten night S. S. (Web.) 2363; an (*for* ane) oðer speche A. R. 100; an [on] oþer tale O. & N. 544; **āne**

[*O.E.* āna (*masc.*), āne (*fem.*)] *alone* : þa he
ane wæs MK. iv. 10 ; he ane HOM. I. 165 ;
KATH. 224 ; ane he gon riden LA3. 6466 ;
we hine læteð ane 25702 ; nu þu ært al ane
[one] (*alone*) 17934 ; as ha . . . wes . . . hire
ane (*sola*) JUL. 31 ; he cuþe him ane ben ORM.
3194 ; he was king one ROB. 315 ; heo wende
alone BEK. 59 ; betere is þat ich one deie SPEC.
81 ; left was he one WILL. 211 ; him ane ich
luvie MARH. 4 ; let þe gome one GAW. 2118 ;
nawt ane . . . ah MARH. 14 ; nout one A. R.
46 ; no3t one ROB. 193 ; ānes (*gen. m. n.*)
LA3. 15574 ; anes eorðliches monnes HOM. I.
33 ; anes mannes ORM. 22 ; wið his anes wit
KATH. 589 ; ane kinges 73 ; ever ich ones
every one's A. R. 134 ; mid his ones mihte 160 ;
one monnes 202 ; to þan anes (*for* ane) LA3.
17304 ; for þen anes *for the nonce* JUL. 71 ;
(*ms.* þe naness) ORM. 7160 ; for þen ones
SHOR. 125 ; ALIS. 1624 ; WILL. 2015 ; ānre
(*gen. f.*) LK. xviii. 25 ; ānun (*dat. m. n.*) HOM.
I. 245 ; anan JOHN xx. 1 ; anen MAT. vi. 24 ;
ana LEECHD. III. 88 ; biforen þam preoste ane
HOM. I. 77 ; buton ane treowe 221 ; ane, one
A. R. 78, 124, 150 ; liveð . . . bi him one 12 ; on
ane da3e LA3. 82 ; an one dæie 6443 ; i had
no broþir bot him ane PERC. 2043 ; alle heo
weoren bi ane (*together, cf. M.H.G.* bi ein)
LA3. 22947 ; na3t o del to onen and þet oþer
del to an oþren AYENB. 175 ; alle we bieþ of
one kende 186 ; mid one worde FRAG. 8 ; he
sat up one vaire bo3e O. & N. 15 ; mid mine
one songe 789 ; bute þe one HOM. I. 201 ; as
hit were for þan one BRD. 18 ; þei weoren at
one K. T. 277 ; āre (*dat. f.*) JOHN xx. 7 ; HOM.
I. 173 ; LA3. 555 ; ore O. & N. 17 ; in ore ni3t
FL. & BL. 83 ; ænne (*acc. m.*) HOM. I. 221 ;
anne 93 ; ænne, enne, anne [one, on] LA3. 88,
394, 556 ; ænne, an ORM. 3374, 8118 ; enne
A. R. 56 ; enne gost AYENB. 145 ; enne, anne
O. & N. 799, 831 ; SHOR. 93, 95 ; enne, en
MISC. 50, 51 ; en P. S. 150 ; anne stroc he 3ef
him ROB. 223 ; ane man AYENB. 58 ; āne
(*acc. f.*) HOM. I. 91 ; ane [one] dohter LA3.
2400 ; (*miswritten* ane) 2247 ; ane hwile
KATH. 183 ; ane, ana LEECHD. III. 96 ; one
O. & N. 199 ; one boc A. R. 54 ; one herte
AYENB. 145 ; of ane (*for* are) sunne HOM. I.
23 ; bi one halve O. & N. 109 ; āne (*pl.*) ORM.
19764 ; for hio ane JOHN xvii. 20 ; ane feue
LA3. 11752* ; ane fewe P. S. 194 ; þa kinges
. . . ane þer wuneden LA3. 23880 ; þer heo
weren one bi ham sulven A. R. 154 ; bi ham
ane MARH. 14.

an-and, prep., ?=anefen ; *nearto* ; anant (? *ms.*
an-ante) TREAT. 137 ; onond, anont, onont,
anonde, ononde A. R. 4, 6, 124, 164, 426 ;
anont H. M. 9 ; C. L. 1076 ; MIRC 1961 ; onont
te under KATH. 2531 ; anende A. P. i. 1135.

a-napped, pple., *sleepy* ; anapped he was P.
L. S. xvii. 278.

an-ar3ien, v., *make cowardly* ; anarwiþ (*pres.*)
ALIS. 3346.

†

an-bel3en, v., *be angry* ; anbælh (*pret.*) LA3.
26359 ; onbol3en (*pple.*) LA3. 1696.

an-bīden, v., *O.E.* onbīdan ; *await* : an-
bīdende (? *ms.* & bi diemde) [abidinge] heore
wille LA3. 8622.

an²-bürsten, v., *burst out* : and he anbursten
agon swulc (he) weore a wilde bar LA3.
25831 ; cnihtes anburste [aborst] (*pple.*)
weoren 25241.

ancessour, sb., *O.Fr.* ancessour ; *ancestor* ;
auncessour MAN. (F.) 945 ; ancestre BEK. 429.

ancestrie, sb., *O.Fr.* ancesserie ; *ancestry* ;
auncestrie MAN. (H.) 14.

ancheisoun, *see* encheisōn.

ancien, adj., *O.Fr.* ancien ; *ancient* ; auncien
MAND. 93.

anclēou, sb., *O.E.* anclēow, = *O.Fris.* onclewe,
cf. O.H.G. anchal, *O.N.* ökla ; *ankle* ; oncleou
'*talus*' FRAG. 2 ; ancle VOC. 148 ; ancle,
anclee CH. C. T. *A* 1660 ; anclowe (*dat.*)
AR. & MER. 5206.

an²-cnāwen, v., *O.E.* oncnāwan, = *O.H.G.* int-,
incnāhan ; *recognize, acknowledge* : ge . . .
oncnāweð '*cognoscitis*' MK. viii. 17 ; on-,
oknāun (*pple.*) *aware* C. M. 8627 ; þe aknawe
AM. & AMIL. 2099 ; þou art soþes aknowe
LEB. JES. 332 ; ich am aknowe WILL. 4391.

ancre, sb., *O.E.* ancra ; *anchorite,* A. R. 6 ;
ROB. 380 ; LANGL. *C* iv. 141 ; ancren (*pl.*)
A. R. 4 ; *see* anker².

ancre-hūs, sb., *house of anchorites,* A. R. 88.

ancren, v., *anchor ;* iancred (*pple.*) A. R. 142.

an-cume, sb., *accident* ; oncume HOM. I.
147 ; oncome C. M. 5927.

an-cumen, v., *come upon* : onkumen (*pple.*)
GEN. & EX. 841.

and, prep., *O.E.* and, ond, = *O.N.* and-, *O.Fris.*
and, anda, end, enda, ond-, *Goth.* and, anda-,
O.L.G. and, ant, *O.H.G.* ant-, ent-, int-, *Gr.*
ἀντί ; *against, towards, near, with regard
to* ; þe holie pistle þe me ret to dai and
(*in*) ech holie chirche HOM. II. 117 ; his
bærd and (? *for* an) his chinne LA3. 20305 ;
doð up and (? an) waritreo 5704 ; þe brude þat
briht is and bleo MISC. 91 ; þe mon þe spareþ
ieorde and ionge childe REL. I. 184 (MISC.
130) ; and ende C. L. 1177* ; and last 127 ;
from heih and (*ms.* &) to herre HOM. I. 207 ;
and þat hit wes dæi liht LA3. 5667 ; he is end
longe feouwer & sixti munden 21993 ; biwinneð sone sucurs & help and (*from*) ure
loverd A. R. 244 ; ent domis dai MISC. 216 ;
ande long (*lengthwise*) nouht over þwert HAV.
2822 ; ende long MARH. 10 ; *comp.* an-and.

and², conj., *O.E.* and, ond, = *O.Fris.* and, ande,
end, ende, *O.L.G.* endi, *O.H.G.* anti, enti, inti,
unti ; *and,* ANGL. VII. 220 ; HOM. I. 29 ; KATH.
289 ; (*ms.* annd) ORM. 1849 ; æfre & æfre
206 ; æver tweie and tweie tuhte to somne LA3.
24749 ; ærneð ævere vorð & vorð 16441 ;

and ȝef þu miht æine finden 3692; ant 104; & (*for* & ȝif) þu hit nult ileven 8313; an 2297, 5780; and if AN. LIT. 12; and (*for* and if) HAV. 2862; JOS. 389; PERC. 1156; ANT. ARTH. xvi; and i cacche miȝte LANGL. *B* ii. 192 [ȝif i mihte cacche *A* ii. 167]; and, ant A. R. 174, 230; and, an O. & N. 4, 7; C. L. 715, 789; AYENB. 15; and, ant, an HAV. 29, 36; ant JUL. 2, 3; KATH. 1977; SPEC. 23; an S. S. (Web.) 730; ande E. G. 20; ænd, end MK. i. 40, iii. 26; ent HOM. I. 241; ? a P. L. S. iii. 13; wen he is holden a mai APOL. 56.

ande, sb., *O.E.* anda, onda (*indignation, envy*), = *O.L.G.* ando, *O.H.G.* ando, anado, *O.N.* andi (*breath, spirit*); *breath, spirit,* IW. 3555; PR. C. 3054; and TOWNL. 154; onde '(*h*)*aleine*' VOC. 146; PR. P. 364; ALIS. 3501; FER. 2242; GOW. II. 260; ne drageð ge non onde REL. I. 217; onde (*envy*) LAȝ. 22756; A. R. 274; O. & N. 419; ROB. 40; C. L. 211; P. S. 196; P. L. S. v.² 51; R. R. 148; þah men to me han onde SPEC. 29; for anden MAT. xxvii. 18; þurh ande MK. xv. 10; þurh onden [onde] KATH. 893.

  ond-ful, adj., *full of envy,* A. R. 68*.

anden, v., *O.N.* anda; *breathe*; ande CATH. 9; ondin PR. P. 364; andes (*pres.*) C. M. 21075.

[ander-, ande-, prefix, = ender, = *O.E.* end; *formerly.*]

  ander-sīth, andesīth, adv., *formerly,* C. M. 2110, 24268.

andetted, *see* endetted.

andetten, v., *O.E.* an-, ondettan; *confess*: hio andetten (*pret.*) hiora sinnan MAT. iii. 6.

and-ȝǣten, v., *confess*: andȝæten ure sinnes ORM. 15163.

  andȝǣtinge, sb., *confession,* ORM. 18027.

  andȝǣtnesse, sb., *confession,* ORM. 2762.

andi, adj., *O.E.* andig; *envious*; ondi MAP 336.

andīren, sb., *? a corruption of O.Fr.* andier; *andiron*; aundirin PR. P. 19; aundire VOC. 176.

and-lǣt, sb., = *M.L.G.* antlāt, *O.Fris.* andlēte; *face*; onlæt ORM. 12939; onlete HOM. I. 59; ne turne þine anleth me fra PS. xxvi. 9.

an²-drǣden, v., *O.E.* an-, ondrædan, = *O.L.G.* and-, ant-, andrādan, *O.H.G.* intrātan; *dread,* MK. vi. 50; ondræden MAT. i. 20; adreden HOM. I. 111; LAȝ. 8744; he mai him sore adreden HICKES I. 223; he mai him adrede grame O. & N. 1484; ac sore þu miȝt þe adrede R. S. i; adrēde (*pres.*) P. L. S. viii. 3; ondrædde, ondredde (*pret.*) MAT. ii. 22, xiv. 30; adredde HORN 1170; adrǽd [adred] (*pple.*) LAȝ. 10952; adred HAV. 1258; SPEC. 110; TOWNL. 25; adrad HAV. 278; WILL. 1980; MAND. 282; TOR. 289; we weren adredde KATH. 1345; aren adradde [adrad] LANGL. *B* xix. 21.

and-sǣte, adj. & sb., *O.E.* andsǣte, = *Goth.*

andasēts; *odious; enemy*; ORM. 16071; and-sete HOM. II. 115, 183; ansete I. 107.

and-seche, sb., *O.E.* andsæc; *denial,* HOM. II. 147.

and-springen, v., *O.E.* onspringan, = *O.L.G.* ant-, anspringan, *O.H.G.* intspringan; *spring up, rise*; andsprong(e) (*pple.*) C. L. 152*.

and-sware, sb., *O.E.* andswere, answare [answere], andswaru; *answer*; LAȝ. 4412, 11925, 26355; andsware, -swere ORM. 2404, 12016; andswere MAT. ii. 12; GEN. & EX. 3081; ondswere JUL. 11; andsvare, ondsware, answare, onsware, answere, ansvere, onswere O. & N. 55, 149, 470, 1176; answare ISUM. 674; answere SHOR. 123; ansvere ROB. 197: onswere A. R. 8.

and-swerien, v., *O.E.* and-, ondswerian; *answer,* LAȝ. 22421; andsweren ORM. 2036; onswerien A. R. 94; onswerie HOM. I. 73; ansverie AYENB. 67; answere CH. C. T. *D* 1077; onswere JOS. 377; ich ansverie BEK. 531; and-, answerede (*pret.*) MAT. viii. 8, xii. 39; answerede GEN. & EX. 2728; ansverede BRD. 9; answereden MAND. 229.

[and-wane], awane, adj., *O.N.* andvana, andvani; *wanting, lacking*: to dei he mei to marȝan hit him is awane HOM. I. 21; hit is onwane of his hele 29.

and-weorc, sb., *O.E.* andweorc; *material*: of þissen andweorke alle þing he iwrouhte FRAG. 8.

and-wurden, v., *O.E.* andwyrdan, = *O.H.G.* antwurtan, *Goth.* andwaurdjan; *reply*; and-wurde (*pret.*) HOM. I. 91.

ane, *see* an.

an-efen, prep., *O.E.* on efen, = *O.L.G.* aneban, *mod.Eng.* anent; *near to, against*; anen MAND. 80; onefent (t *paragogic, see* GR. GRAM. III. 217) A. R. 164*; anent ALEX. 735; anentis, anemptis WICL. JOHN v. 45; REL. II. 47; anemptes HALLIW. 61; onence PR. C. 1355.

ān-ēȝed, adj., *one-eyed*; oneyid PR. P. 365; oniȝed A. P. ii. 102.

a²-nēh, adv., *near*; al þat heo aneh comen LAȝ. 3827.

anélen, v., *O.Fr.* aneler, *from Lat.* anhelāre; *pant after, puff at*: berez and borez and etaines ... þat him anélede (*pret.*) of þe heȝe felle GAW. 722 [*Morris renders '* attacked, worried*']; of linage a gret gentilman anelid to þis bischaprik WINT. VIII. xxxviii. 230.

an-ēlen, v., *O.E.* an-, onǣlan; *set on fire, burn*; anēlend (*ms.* anhelend) HOM. I. 219; onealde (*pret.*) HOM. I. 97; anēled (*pple.*) GOW. III. 96; anelid WICL. IS. xvi. 7.

an-élie, v., *anoint with oil,* SHOR. 44; anéleþ (*pres.*) 43; anélid (*pple.*) '*inunctus*' PR. P. 11; anelet MIRC 1812.

anélinge, sb., *anointing with oil,* '*inunctio*,' PR. P. 11.

a-nemnen, v., *O. E.* anemnan, = *M. H. G.*

ernennen ; *name* ; **anemnede** (*pret.*) P. L. S.
xv. 20.

ānerlī, adv., *? from* **ān** ; *alone*, BARB. vii. 59 ;
all anerli v. 281 ; HAMP. TR. 4.

ānes, *see* **ǣnes.**

a nēuste, *see* **nēhweste** under **nēh.**

aneus, sb., pl., *O.Fr.* aniaus, *pl. of* anel ; *rings,
links, fetters,* MAN. (H.) 167, 278.

a-nēwen, v., = *O.H.G.* irniuwōn ; *renew* ;
anewe DEP. R. iii. 24 ; ? ennēwiþ WICL.
ECCLUS. xxxviii. 30 ; ? enewes TOWNL. 314.

ān-fald, adj., *O.E.* ānfeald ; *simple,* HOM. I.
151 ; MARH. 11 ; ORM. 18668 ; C. M. 6342 ;
ofeald HOM. II. 187.

  ānfalde-līche, adv., *simply,* HOM. I. 5.

an-felt, sb., *O.E.* an-, onfilt, = *O.H.G.* anevalz,
M.Du. aenbelt, -bilt ; *anvil* ; anvelt '*incus*'
REL. I. 6 ; TREV. III. 207 ; CH. D. BL.
1165 ; anefelt WICL. ECCLUS. xxxviii. 29 ; anvilt
FER. 1308.

an²-fōn, v., *O.E.* an-, onfōn, = *O.H.G.* antfā-
han ; *receive* ; onfon LAȜ. 1069 ; ORM. 17948 ;
onfōþ (*pres.*) ORM. 4224 ; heo onfoþ FRAG.
7 ; onfēng (*pret.*) LAȜ. 134 ; onfengen MAT.
ii. 12 ; ORM. 14396 ; anfōn (*pple.*) LAȜ. 8827.

an-gard, sb., *arrogance,* D. TROY 9745 ;
ongart M. H. 49 ; angardez (*gen.*) GAW. 681.

ange, sb., *O.E.* ange, = *O.H.G.* ango ; *anguish,
distress* : him was waȝ & ange ORM. 11904 ;
unæþe & al wiþ ange 16289 ; & dide him
mikel ange 19804.

angel, sb., *O.E.* angel, = *O.H.G.* angul, *O.N.*
öngull ; *angle, fish-hook,* '(h)amus,' FRAG. 2 ;
MAT. xvii. 27 ; angil PR. P. 12.

  angil-hōc, sb., *fish-hook,* WICL. IS. xix. 8.

  ongel-twæcche, sb., *O.E.* angeltwicce ;
angle-twitch (*worm used as a bait*) ; ' *lu(m)-
bricus*' FRAG. 3.

angel, *see* **engel.**

an²-ȝēn, prep. & adv., *O.E.* an-, ongegn, -gēn,
-gean, = *O.L.G.* angegin, *O.H.G.* ingagan,
O.N. īgegn ; *again, against,* LEB. JES. 316 ;
HOM. I. 219 ; heo urnen onȝein him HOM. I. 3 ;
dunt aȝein dunt 15 ; aȝein ȝeven *give back,
return* 31 ; hwan ic aȝen cherre (*return*) 79 ;
oȝein 7 ; onȝean FRAG. 6 ; ongen, -geanes
MAT. xii. 30, 32 ; onȝein, -ȝean, aȝein, -ȝein,
-ȝen, -ȝæn, -ȝeines, -ȝenes LAȜ. 1590, 1667, 1900,
5917, 6178, 8837, 16223, 23157 ; for þi was
mikel wræche set onȝæn þat woh ORM. 18 ;
onȝænes kinde 249 ; aȝein, aȝean, aȝeines A. R.
14, 50, 216 ; and eiþer aȝen [ayein] oþer
swal O. & N. 7 ; aȝeines riht 1371 ; aȝeines
þare hete nere hit al noht R. S. v (MISC. 182) ;
þat aȝen us is so hende ROB. 109 ; aȝein even
159 ; he was aȝen idrive BEK. 678 ; aȝein
kuinde JOS. 106 ; cominge him aȝeines 562 ;
aȝein [ageines] his wille LANGL. *A* vii. 70 ;
i shal bringe þe aȝein ['*reducam te*'] into
þis lond WICL. GEN. xxviii. 15 ; aȝens his
broþer ['*adversus fratrem suum*'] iv. 8 ; aȝen,
aȝenst MAND. 12, 220 ; aȝen, agein HAV. 489,

493 ; agein TOR. 2709 ; aȝe cunde BRD. 9 ;
ayen MISC. 27 ; zelleþ ham ayen AYENB. 36 ;
aye, ayens þane holi gost 28, 29, ; oȝain
TRIST. 712 ; a-, egain C. M. 1537, 1538 ;
agen (*r. w.* ben) GEN. & EX. 405 ; ageon 3912 ;
agon (*r. w.* fon) 438 ; on-, aȝenes PROCL. 5, 6 ;
aȝenest LAȜ. 22476*.

aȝein-bíe, v., *redeem,* WICL. LK. xxiv. 21 ;
aȝenboghte (*pret.*) MAND. 2 ; **aȝeenboȝt**
(*pple.*) WICL. ISA. lxii. 12.

  aȝein-bíere, sb., *redeemer,* WICL. JOB xix.
25.

  aȝein-bite, sb., *remorse,* AYENB. 1.

agein-char, sb., *for* ȝeinchar ; *return,*
ANGL. II. 233.

aȝein-cüme, sb., *coming again,* S. A. L. 155.

ayēn-yefþe, sb. (*for* ȝēnȝefþe) ; *a gift made
in return,* AYENB. 120.

aȝēn-rīsing, sb., *resurrection,* WICL. JOHN
xi. 25.

aȝein-sawe, sb., *for* ȝeinsawe ; C. M. 8382.

aȝēn-turn, sb., *for* ȝēnturn ; WILL. 4182.

aȝein-ward, adv., *O.E.* ongeanward ; *backwards* ; onȝeinwærd, aȝeward LAȜ. 1673 ;
aȝeinward JUL. 72 ; aȝeanward A. R. 274 ;
agenward GEN. & EX. 1782 ; aȝein-, aȝenward
WICL. MK. iv. 35 ; ayeward AYENB. 48.

ayēn-wiȝte, sb. (*for* ȝēnwiȝte), = *Ger.* ge-
gengewicht ; *counterpoise,* AYENB. 247.

anger, sb., *O.N.* angr *n. ; anger, grief,* WILL.
552 ; GOW. I. 282 ; IW. 1529 ; E. T. 914 ;
anger and tene GEN. & EX. 2992 ; anger and
wa PR. C. 3517 ; in anger and in care EGL.
150 ; in anger, pine, and mekile wo MIR. PL.
161 ; angres ful A. R. 370 ; angre (*dat.*) CH.
BOET. ii. 4 ; angres (*pl.*) LANGL. *B* xii. 11.

  [**anger-lich**, adj., *O.N.* angrligr] **angerlíche**
(*adv.*) *painfully, angrily,* TREV. III. 81 ;
GOW. I. 292 ; angerli MAN. (F.) 3759.

an²-gin, sb., *O.E.* an-, ongin, = *O.L.G.*, *O.H.G.*
anagin ; *commencement,* HOM. I. 217 ; angun
II. 107 ; **anginne** (*pl.*) MAT. xxiv. 8.

an²-ginnen, v., *O.E.* an-, onginnan, = *M.Du.*
ontginnen, *O.H.G.* inginnan ; *commence* ; on-
gan (*pret.*) ORM. 2801 ; ongon JUL. 13 ;
ongunnen HOM. I. 89.

an²-ȝit, sb., *O.E.* and-, ondgit ; *intellect* ;
anȝite (*dat.*) HOM. I. 99.

an²-ȝiten, v., *O.E.* ongitan ; *perceive,* LAȜ.
26623 ; aȝitte (*pres.*) MISC. 193 ; onȝeat
(*pret.*) HOM. I. 223 ; anȝæt LAȜ. 15726.

angle, sb., *O.Fr.* angle, *Lat.* angulus ; *angle,
corner,* CH. C. T. *B* 304.

Angle, pr. n., *Anglia,* MAN. (F.) 4097.

angren, v., *O.N.* angra ; *anger, vex, distress,*
ORM. 428 ; angre TREV. V. 355 ; þat angreth
me sore LANGL. *B* v. 117 ; **angred** (*pple.*)
MED. 744 ; PR. C. 302 ; þe king was angred
FER. 262.

angri, adj., *troublesome* ; *troubled, angry,*

'*molestus*,' REL. I. 8 ; GOW. I. 283 ; CH. C.
T. *A* 3157.

angrom, sb. (*cf.* agrámien), *affliction*, '*an-
gustia*,' PS. cxviii. 143.

ang-sum, adj., *O.E.* angsum (*from* *ang,
adj.=O.H.G. ang, *O.N.* öngr, *Goth.* aggws,
narrow) ; *strait, narrow*, '*angustus*,' MAT.
vii. 14.

    anxumnesse,sb.,*O.E.*angsumness; *anxiety*,
ORM. 10457.

anguisse, sb., *O.Fr.* anguisse ; *anguish;* ROB.
177 ; CH. BOET. iii. 7 (79) ; MAN. (F.) 4835 ;
anguise A. R. 234 ; angusse TREAT. 140.

anguissen, v., *O.Fr.* anguisser ; *anguish* ;
anguisseþ (*pres.*) CH. BOET. iii. 8 (80) ; an-
guisi (*? printed* angrisi) (*subj.*) AYENB. 146.

anguissous, adj., *O.Fr.* angoissous ; *full of
anguish*, ROB. 157 ; CH. BOET. ii. 4 (41) ; þe
anguisuse deaðe A. R. 112.

ān-hād, sb., *unity*, KATH. 932 ; onhod C. L.
10 ; anhede PR. C. 16 ; onhede AYENB. 79 ;
TRANS. XIX. 333; WICL. I TIM.iii.2 ; APOL. 35.

an-hangen, v., *O.E.* onhangian ; *hang up* ;
anhongen LA3. 13166 ; anhonge BEK. 724 ;
anhongeþ (*pres.*) AYENB. 51 ; **anhanged,**
anhonged (*pple.*) CH. C. T. *B* 3945 ; anhonged A. R. 126 ; AYENB. 241 ; an-, on-
honged WILL. 1564, 4773.

an²-healden, v.,=*O.Fris.* onthalda, *M.H.G.*
enthalten ; *retain* ; **anhialde** (*pple.*) AYENB.
152.

an²-hebben,v.,*O.E.*anhebban,=*O.H.G.*ant-,
intheffan ; *heave up, lift, exalt*, LA3. 12627 ;
onhefð (*pres.*) HOM. I. 113 ; **anhōf** (*sec. text*
ahof) (*pret.*) LA3. 21625 ; anhefden, onheve-
den HOM. II. 177 ; **onheven** (*ms.* onhevene)
(*pple.*) HOM. I. 117.

an-hē3en, v., *? exalt* ; anhe3i (*? for* ahe3i),
AYENB. 23.

ān-hende, adj., *one-handed* ; onhende '*mancus*' FRAG. 3.

an-hēten, v., *O.E.* onhǣtan ; *heat* ; **anhēt**
(*pres.*) MISC. 30 ; AYENB. 108.

an-hitten, v., *hit* : wiþ swerde ihc þe **anhitte**
(*?* an hitte) HORN (L.) 712 ; anhitte (*pret.*)
ROB. 185.

an-hōn, v., *O.E.* onhōn ; *hang up*, LA3. 729 ;
(hi) **anhōð** (*pres.*) O. & N. 1646 ; **anhēng**
(*pret.*) LA3. 29358 ; anhongen (*pple.*) LA3.
1023 ; anhonge O. & N. 1195 ; P. L. S. x. 7 ;
SHOR. 86 ; P. S. 213 ; anhōn (*pple.*) LA3.
30839 ; A. D. 261.

an-hungren, v., *be hungry* ; **anhungred**
(*pple.*) ALIS. 1230 ; OCTOV. (W.) 467 ; P. 23 ;
TREV. III. 471 ; LANGL. *B* x. 59*, *C* xii. 43* ;
anhungrid PR. P. 172.

ān-hūrned, adj., *O.E.* ānhyrned ; *one-horned,
'unicornis* ;' þe anhurnde MARH. 7.

[**ān-hwerfed,** adj., *constant*.]

anhwerfed-lēk, sb., *constancy* ; anwher-
fedle3c ORM. 11124.

ǎni, *see* ǣni.

aniente, anientice, v., *O.Fr.* anientir, anien-
tiss- ; *annihilate*, LANGL. *B* xvii. 285 ; aniin-
tischin PR. P. 12 ; **anientised** (*pple.*) E. G. 6 ;
cf. enentischt.

a-nimen, v., *O.E.* animan ; *receive, take up* ;
anymeð (*imper.*) MAT. xxv. 28 ; anam (*pret.*)
HOM. I. 229.

ānin, v.,=*O.H.G.* einōn ; *unite* ; onin PR. P.
365 ; ōneþ (*pres.*) AYENB. 88 ; ōned (*pple.*)
TREV. V. 341 ; CH. C. T. *D* 1968.

āninge, sb., = *O.H.G.* einunga ; *union* ;
oninge PR. P. 365 ; AYENB. 65 ; TREV. V. 9.

an-innen, prep., *O.E.* oninnan ; *within* : ic inc
habbe . . . aninne mine benden LA3. 5617 ; he
lette al þat wel weorpen þer aninnen 6428.

anious, *see* anoious.

anīs, sb., *Fr.* anīs ; *anise*, SPEC. 26.

a-niðerien, v., *O.E.* aniðerian,=*Ger.* ernie-
dern ; *abase* ; aniðeri [aneoþeri] LA3. 14861 ;
aneþered (*pple.*) ROB. 217.

anjoini, *see* enjoinen.

ān-kenned, adj., *O.E.* āncenned ; *only-be-
gotten*, ORM. 17003.

anker, sb., *O.E.* ancor,=*O.H.G.* ancher *m.,
from Lat.* ancora *f.; anchor*, HAV. 521 ; WILL.
568 ; TRIST. 366 ; LIDG. M. P. 50 ; ankir
OCTAV. (H.) 433 ; **ankeres** (*pl.*) LA3. 25539 ;
ancres ALIS. 1416 ; *deriv.* ancren.

    anker-mon, sb., *anchor-man*, '*proreta*,'
FRAG. 2.

anker², sb., *O.E.* ancor ; *anchorite*,BEK. 1155 ;
R. R. 6348 ; **ancres** (*pl.*) LANGL. PROL. *A* 28 ;
cf. ancre.

anlace, sb., *a sort of dagger*, D. ARTH. 1148 ;
anlas CH. C. T. *A* 357 ; aunlaz HAV. 2554.

ān-lěpi3, sb., *O.E.* ānlēpig, ǣnlȳpig ; *single,
sole*, ORM. 11 ; anlepi KATH. 74 ; anlæpi LA3.
13359 ; onlepi HAV. 1094 ; AYENB. 13 ; APOL.
38 ; onlepi, elpi [anlepi] A. R. 116, 366 ; enlepi,
enlipi, alpi HOM. I. 23, 75 ; olupi E. G. 350 ;
alpi REL. II. 275 ; ænne ǣlpine (*acc.*) broðer
LA3. 31451.

    ōnlěpi-hēde, sb., *singleness*, AYENB. 21.

    ōnlěpi-līche, adv., *singly, alone*, MISC. 28 ;
AYENB. 55.

ān-līch, adj., *O.E.* ān-, ǣnlīc,=*M.Du.* eenlīc ;
only : min anliche sune LK. ix. 38 ; in on-
liche stude A. R. 152 ; ōnlīche (*adv.*) AYENB.
265 ; WILL. 3155 ; MIRC 656 ; PL. CR. 534 ;
anli IW. 3744 ; ōnlūkest (*superlat.*) A. R. 90.

an-līch, adj.,*O.E.* an-, onlīc,=*M.H.G.*anelīch,
Goth. analeiks ; *alike, like*, MAT. xviii. 23 ; þet
is him anlich AYENB. 186 ; þet stat makeþ þaire
þet hit wel lokeþ anlike to þe angles of hevene
227 ; alike PR. P. 10 ; **alīche** (*adv.*), *alike* GOW.
I. 268 ; A. P. ii. 1477 ; alle are þei aliche longe
LANGL. *B* xvi. 57 ; olike PR. C. 7560 ; al it was

him olike loð GEN. & EX. 2024; Saxons . . . hight(e) alle oliche MAN. (F.) 41.

    anlīcnesse, sb., *O.E.* anlīcness; *likeness, image*, AYENB. 88; an-, onlicnesse, onlichnesse HOM. I. 59, 95, 223; onlicnesse A. R. 18; ORM. 5056; onlicnesse [anlichnisse] LA3. 1141, 21154; anlicnisse MAT. xxii. 20.

an-līcnen, v., = *M.H.G.* anelīchen; *resemble*, anlikni AYENB. 101; **anlīkneþ** (*pres.*) 16; anlīcned (*pple.*) *compared*, 61.

an²-līhten, v., *O.E.* onlȳhtan, = *O.H.G.* intliuhten; *enlighten*; onlīhteð (*pres.*) HOM. I. 97; **onlīht** (*imper.*) PS. cxviii. 135; he onlihte Daviðes heorte HOM. I. 97.

ān-mōd, adj., *O.E.* ānmōd, = *O.H.G.* einmuot; *unanimous*, HOM. I. 101; onmod II. 183.

ānnesse, sb., *O.E.* ānness, = *O.H.G.* einussa, einnissa; *unity*, HOM. I. 93; onnesse A. R. 12.

anniversarie, sb., *Fr.* anniversaire.; *anniversary* anniversaries (*pl.*) A. R. 22.

[**anoilen**, v., *O.Fr.* enuiler; *anoint with oil*].   anoiling sb., *anointing*, H. S. 844; anoilinge AYENB. 14.

anoint, anointin, *see* enoint, enointin.

anoious, adj., *O.Fr.* anoious, anuieus (=ennuieus); *noxious*, CH. BOET. i. 2 (7); anious GAW. 534; noious WICL. I TIM. vi. 9.

anour, -ouren, *see* honour, -ouren.

anourne, v., *for* aournen; *see* adornen; *adorn*, WICL. GEN. xxiv. 47.

anournement, sb., *ornament*, A. P. ii. 1290.

ān-rēd, adj., *O.E.* ānrǣd, *O.H.G* einrāt; *constant, persevering*, HOM. I. 115.

    ānrǎd-liche, adv., *persistently*, HOM. II. 61.

    ānrēdnesse, sb., *O.E.* ānrǣdness; *constancy*, HOM. I. 107; an-, onrednesse A. R. 12.

an-sēchen, v., *O.E.* onsēcan; *attack*; onsēken (*pres.*) GEN. & EX. 851; onsōhte (*pret.*) LA3. 5657.

ansel, anser, sb., *a sort of balance*; aunsel, auncel, aunser, auncer LANGL. *A* v. 132, *B* v. 218, *C* vii. 224.

an-sēne, sb., *O.E.* an-, onsīon, -sīen, -sȳn, = *O.L.G.* asiun; *countenance*; onsene '*facies*' FRAG. 2; HOM. I. 191; O. & N. 1706; ansine LEECHD. III. 132; ansiene '*aspectus*' MAT. xxxviii. 3.

an-sīne, sb., *O.E.* an-, onsȳn; *?swoon*; as povre wif þat falleþ in ansinę AN. LIT. 10.

an²-standen, v., *O.E.* andstandan; *resist*: a3en þe Deneis to anstond(e) ROB. 267.

ān-standende, pple., *standing alone*, HOM. I. 221.

ant, *see* and.

ante, sb., *O.Fr.* ante, *from Lat.* amita; *aunt*; aunte ROB. 37; LANGL. *B* v. 153.

antefne, sb., *O.E.* antefn, *from med.Lat.* antiphona; *anthem*, A. R. 34; antein P. L. S. ix. 185.

anticrist, sb., *antichrist*, PR. C. 4065.

antiphoner, sb., *O.Fr.* antiphonier; *antiphoner*, VOC. 230; CH. C. T. *B* 1709.

antre, *see* aventūre.

an-ufen, adv., *O.E.* onufan; *above*; anufene LA3. 16432; onufen LK. xi. 44; onuven A. R. 236; JUL. 53; anoven HORN (L.) 624; P. S. 188.

anuven-an, prep. & adv., *on from above*, (? *for* uvenan), LA3. 15493; anovenon FL. & BL. (H.) 643.

    anufen-ward, adv. & prep., *towards the top (of)*, P. L. S. xiii. 341; FER. 622; þa com se fir onufenweard þone stepel SAX. CHR. 249.

anui, sb., *O.Fr.* anui, anoi (*see* ennui); *annoyance, trouble*, A. R. 94; anoi AYENB. 267; nui C. L. 442.

anuien, v., *O.Fr.* anoier, =ennoier, ennuier; *annoy*; anuie HAV. 1735; anoie SHOR. 36; anoie, noie WICL. EX. xxii. 22; anuied (*pple.*) LA3. 2259*; BEK. 1001.

anunciacioun, sb., *annunciation*, TREV. VIII. 351.

an-under, prep., *under*, HOM. I. 193; MISC. 44; REL. I. 48; A. P. i. 1080; þat alle anunder him were BRD. 1; anunder bis SPEC. 35; anonder HORN (L.) 567; anondir OCTOV. (W.) 550.

an-uppen, prep., *O.E.* onuppan; *upon*, HOM. II. 107; þe he onuppen falð MAT. xxi. 44; anuppon, onuppen, anuppe HOM. I. 3, 43, 133; anuppe þisse dune MISC. 85; anoppe LA3. 1916*.

anūri, *see* honouren.

anvenime, *see* envenime.

an-wald, sb., *O.E.* an-, onweald, = *O.H.G.* anawalt; *power, dominion*; **anwalde**, onwalde (*dat.*) LA3. 8456, 13950; whæt heo hæfden on anwolde 13182; under mire onwalde HOM. I. 13; of deofles onwalde 15.

au-weald, adj., *O.E.* onweald; *powerful*; anwealda gastes HOM. I. 219.

an-wende, v., *O.E.* onwendan; *avert*, LA3. 31438.

ān-, ōn-wil, adj., *obstinate*, A. R. 56, 400.

ān-wille, adj., *O.E.* ānwille, = *O.H.G.* einwilli; onwille '*obstinatus*' FRAG. 5.

an-wiðre, prep., *against*: þe wind him com onwiðere [onwiþere] LA3. 2884; haveð þeos wind & þeos weder awiðer [awiþer] him istonde 12060.

anxum, *see* angsum.

aouren, v., *O.Fr.* aourer; *adore*; **aourede** (*pret.*) P. L. S. xix. 32.

aournen, *see* adornen.

apaie, v., *O.Fr.* apaier; *satisfy, appease*, HICKES I. 228; apaied (*pple.*) ROB. 117; WILL. 1883; LANGL. *A* vi. 110.

apainen, v., *O.Fr.* apainer; *pain*: him **apaineþ** (*pres.*) mani a screwe SHOR. 146.

apaisen, v., *O.Fr.* apaisier; *appease*; apai-sede (*pret.*) CH. BOET. iv. 7 (148).

ápe, sb., *O.E.* apa,= *O.N.* api, *O.H.G.* affo; *ape,* VOC. 188; A. R. 248; O. & N. 1325.

    ápe-warde, sb., *ape keeper,* LANGL. *B* v. 640.

    ápe-ware, sb., *counterfeit wares,* A. R. 248.

apēche, v., *O.Fr.* empecher, *Lat.* impedicāre; *hinder; impeach,* SACR. 302; apēched (*pple.*) SHOR. 38.

apēchour, sb., *impeacher,* PR. P. 12.

apeiren, *see* empeiren.

apel, sb., *O.Fr.* apel; *appeal, peal,* ' *appellatio, classicum,*' PR. P. 12, 13; ROB. 473; REL. II. 227; LANGL. *C* xx. 284.

apélin, v., *O.Fr.* apeler; *appeal*; appelin ' *appello*' PR. P. 13; appele P. R. L. P. 157; apéliden (*pret.*) WICL. 1 MACC. x. 64.

aperceive, v., *O.Fr.* apercevoir; *perceive,* CH. ASTR. ii. 35; aparceiveþ (*pres.*) AYENB. 57.

apéren, v., *O.Fr.* aparoir; *appear*; apeeren LANGL. *A* iii. 109; apere REL. II. 228; appere CH. C. T. *A* 2346; PR. C. 5243; apérede, apierede (*pret.*) MISC. 26, 27.

apert, adj., *O.Fr.* apert; *apert, open, manifest,* ROB. 501; AYENB. 203; ALIS. 2450; M. T. 70; PR. C. 4490.

    apert-líche, adv., *apertly,* ROB. 375; JOS. 276; apertli WILL. 4706.

apointen, v., *O.Fr.* apointer; *appoint*; apoint GOW. II. 151; apointed (*pple.*) GOW. II. 265.

apoisonen, *see* enpoisonen.

apostáte, sb., *O.Fr.* apostate; *apostate,* AYENB. 19.

apostel, sb., *O.E.* apostol, *O.Fr.* apostle, *Lat.* apostolus; *apostle,* HOM. I. 41; AYENB. 41; apostles (*pl.*) HOM. I. 75; apostle (*gen. pl.*) MAT. x. 2; apostlene SHOR. 51.

apostem, sb., *O.Fr.* aposteme, apostume, *Gr.* ἀπόστημα; *imposthume,* CATH. 11; apostim, apostume PR. P. 13; apostims (*pl.*) PR. C. 2995.

apostolie, sb., *O.Fr.* apostolie, apostoile; *pope*; þan holi appostolie (*sec. text* pope) LAȜ. 29614; apostoill MAN. (H.) 130.

apotecarie, sb., *O.Fr.* apotecaire; *apothecary*; **apotecaries** (*pl.*) CH. C. T. *A* 425; MAND. 51.

appallen, v., *? O.Fr.* apallir, *mod.Eng.* appal; *grow or make pale or feeble*; **appalleþ** (*pres.*) SHOR. 91; **apalled** (*pple.*) CH. C. T. *A* 3053.

appareil, sb., *O.Fr.* appareil; *preparation, apparatus, apparel,* WICL. 1 MACC. ix. 35; apparail MAN. (F.) 657.

appareilen, v., *O.Fr.* apareiller; *prepare, fit out, apparel*; apparaile, apparaille LANGL. *A* ii. 148; **aparailde** (*pret.*) WILL. 1146; aparailed (*pple.*) MISC. 26.

apparence, sb., *O.Fr.* apparence; *appearance,* CH. C. T. *F* 218.

appel, sb., *O.E.* æppel, æpl, *m. n.,* = *O.Fris.* appel, *O.H.G.* aphol *m., O.N.* epli *n., Ir.* abhal, ubhal, *Gael.* ubhal, *Welsh* afal; *apple,* ORM. 8118; O. & N. 135; TREAT. 132; þen appel R. S. vii; H. H. 10; appel of the eie VOC. 145; eppel HOM. I. 25; AYENB. 84; þes eppel A. R. 52; **epple** (*dat.*) HOM. I. 123; of þen epple A. R. 66; **apples** (*pl.*) GEN. & EX. 1129; SPEC. 35; æpple LEECHD. III. 118; applen ROB. 283; **applen** (*dat. pl.*) P. L. S. xxiii. 78.

    appul-ȝerd, sb., *apple-yard,* PR. P. 13.

    appul-hord, sb., *stock of apples,* PR. P. 13.

    appul-mōs, sb., = *M. H. G.* apfelmuos; *stewed apples,* ' *pomacium,*' PR. P. 13.

    appul-seller, sb., PR. P. 13.

    appel-trē, sb., *apple-tree,* H. H. 93 (95).

    æpel-tūn, sb., ' *pomarium,*' FRAG. 4.

appenden, v., *O.Fr.* apendre; *appertain, belong*; **appendeþ** (*pres.*) LANGL. *A* i. 43; appendith CH. C. T. *I* 1024; apendez GAW. 623.

appertenaunce, sb., *O.Fr.* apertenance; *appurtenance,* ALIS. 1829.

appertenent, adj., *appertinent,* CH. C. T. *E* 1010.

appertienen, v., *O.Fr.* apertenir; *appertain,* CH. BOET. iii. 4 (73).

appetit, sb., *Ó.Fr.* appetit, apetit; *appetite,* CH. C. T. *D* 1218.

applíen, v., *O.Fr.* aplier; *apply,* CH. BOET. v. 4 (161); applie LUD. COV. 34.

aprentis, sb., *O. Fr.* aprentis; *apprentice,* LANGL. *C* iv. 281; prentis *A* iii. 218; **aprentis** (*pl.*) WICL. E. W. I. 382.

apróchen, v., *O.Fr.* aprocher; *approach,* CH. BOET. i. 1 (6); aprochi P. L. S. xxi. 118.

aproprien, v., *O.Fr.* aproprier; *appropriate*; **apropred** (*pple.*) AYENB. 40; appropried PR. C. 9346.

apróve, v., *O.Fr.* aprover; *approve,* AM. & AMIL. 803; **apróved,** appreeved (*pple.*) TREV. III. 265; appreved CH. L. G. W. 21.

apte, adj., *Fr.* apte; *apt,* PALL. iii. 858; apt S. A. L. 68.

aqveintance, *see* acointance.

aqvére, v., *O.Fr.* aqverre; *acquire*; acqwere S. S. (Wr.) 1081.

aqvitance, acqvitance, sb., *O.Fr.* acqvit-ance; *acquittance,* LANGL. *B* xiv. 189; CH. C. T. *A* 4411.

aqvite, v., *O.Fr.* aqviter; *acquit,* R. R. 6742; aqvitti AYENB. 137; acwiten A. R. 124.

ar, sb., *O.N.* ör *n.; scar,* ' *cicatrix,*' HALLIW. 77; arre WICL. LEV. xxii. 22; an érre (*ms.* a nerre) VOC. 209; **erres** (*pl.*) PR. C. 5327.

ār, *see* ǣr *and* ōr.

Arabisz, sb., *O.Fr.* Arabis, Arabit, Arabi ; *Arab*, HOM. I. 5.

aráce, v., *O.Fr.* esracier, *Lat.* exradicare ; *pull up by the roots, tear away*, CH. C. T. *E* 1103 ; **aráched** (*pple.*) SHOR. 156.

a-rēchen, v., *O.E.* arǣcan, = *O.H.G.* erreichen ; *reach* ; arechin '*attingo*' PR. P. 14 ; arechen, areachen A. R. 128, 166 ; areachen to þe heovene MARH. 12 ; areche HORN (L.) 1220 ; P. L. S. xx. 94 ; SHOR. 154 ; GOW. I. 150 ; WICL. JOHN xiii. 26 ; al þat his ax areche might RICH. 7039 ; arǣhte [arahte] (*pret.*) LA3. 10539 ; to Daris . . . twenti pund he ara3te FL. & BL. 812 ; sturne strokes that ara3te HALLIW.·77 ; ara3t (*pple.*) FL. & BL. 687 ; o best him was arau3t (*? for* atrau3t) S. S. (Web.) 895.

a-rēden, v., *O.E.* arǣdan, = *O.H.G.* irrāten ; *conjecture, interpret, explain* ; aredan HOM. I. 121 ; of whiche no man ne couþe areden þe nombre ALIS. 5115 ; arede MISC. 45 ; GOW. I. 24 ; hi þat arēdeþ (*pres.*) þise redeles SHOR. 24 ; arēde (*imper.*) WICL. MAT. xxvi. 68 ; arǎd (*pret.*), þe sweven . . . þat Daniel an one arad GOW. I. 25.

a-rǣren[arere], v., *O.E.* arǣran ; *rear up, raise up*, LA3. 2023 ; areran ane buruh HOM. I. 93 ; þet heo muhten þe deade arearen vrom deaðe to live A. R. 390 ; arere BEK. 901 ; TRIST. 2834 ; OCTOV. (W.) 21 ; LUD. COV. 132 ; arere þet heved AYENB. 31 ; arǣreþ (*pres.*) FRAG. 7 ; arēr (*imper.*) HOM. I. 211 ; arērde (*pret.*) LA3. 14229 ; arearde KATH. 1043 ; þeo þat . . . uvele lawe arerde MISC. 64 ; arēred (*pple.*) ROB. 7 ; he hedde arered and imad manie werren AYENB. 239 ; þe rode is up arered SPEC. 85.

arainen, v., *O.Fr.*, arainier, araigner ; *arraign* ; areine P. L. S. xxv. 85 ; arene L. H. R. 147 ; areined (*pret.*) TREV. IV. 303.

aranie, eranie, sb., *O.Fr.* araigne ; *spider*, '*aranea*,' PR. P. 14, 140.

a-ráten, v., *reprove*, LANGL. *B* xi. 98 ; rebuked and arated xiv. 163.

arber, sb., = herber ; *arbour* : in þe garden . . . was an arber fair and grene and in þe arber was a tre SQ. L. DEG. 28.

arbitres, sb., *arbitress*, AYENB. 154.

arbitrour, sb., *arbitrator* ; arbitrouris (*pl.*) WICL. 3 ESDR. viii. 26.

arblaste, sb., *O.Fr.* arbaleste, *Lat.* arcubalista ; *arbalast*, ROB. 377 ; arblaste, arblast AYENB. 47, 71 ; arblast RICH. 1867.

arblaster, sb., *O.Fr.* arbalestier ; *arbalaster* ; arblasters (*pl.*) R. R. 4196.

arc, sb., = *O.Fr.* arc, arche ; *arc, arch* ; ark CH. C. T. *E* 1795 ; arche TREV. I. 215 ; arches (*pl.*) LEG. 157.

archangel, sb., *Lat.* archangelus ; *archangel*, HOM. I. 41.

arche, *see* **arke.**

archebischop, sb., *O.E.* arcebiscop, *Lat.* archiepiscopus ; *archbishop*, BEK. 493 ; ærchebiscop LA3. 12636 ; erchebischop MAN. (F.) 6873.

  archebischop-rīche, sb., *archbishopric*, P. L. S. xiii. 71.

  ærchebiscop-stōl, sb., *see of an archbishop*, LA3. 12657.

archedēkne, erchēdekne, sb., *O.E.* arcediacon, *Lat.* archidiaconus ; *archdeacon*, CH. C. T. *D* 1302 ; arcedekne BEK. 170.

archer, sb., *O.Fr.* archier ; *archer*, LEG. 220 ; AYENB. 45 ; archer, archier CH. C. T. *B* 1929 ; archērs (*pl.*) ROB. 199.

archerie, sb., *archery*, LUD. COV. 44 ; PR. P. 14.

architriclin, sb., *O.Fr.* architriclin, *Lat.* architriclīnus ; *ruler of the feast (often mistaken for a proper name)*, MISC. 29 ; WICL. JOHN ii. 8.

ardaunt, adj., *Fr.* ardent ; *ardent*, CH. BOET. iii. 12 (106).

āre, sb., *O.E.* ār f., = *O.N.* ār ; *oar*, FLOR. 1878 ; TRIST. 366 ; ore PR. P. 368 ; HAV. 1871 ; WILL. 2754 ; S. S. (Wr.) 3153.

  āre-hole, sb., *oar-hole*, VOC. 239.

āre, sb., = *O.E.* ār, *O.N.* ǣra, *O.L.G.*, *O.H.G.* ēra, *O.Fris.* ēre ; *honour, grace, mercy*, M.T. 38 ; are & milce ORM. 1476 ; þin are (*ms.* þi nare) TRIST. 2135 ; in menske and are C. M. 4245 ; swa bide ich godes are [ore] LA3. 2972 ; he dude mare to Peteres are 31956 (*miswritten* ære 5336) ; crie him 3eorne þer of merci & ore [are] A. R. 136 ; ore FRAG. 8 ; O. & N. 886 ; HAV. 211 ; ROB. 340 ; TOR. 916 ; M. ARTH. 1344 ; bi godes ore FL. & BL. 173 ; he 3af alle þe kni3tes ore HORN (L.) 1509 ; for to habbe þin(e) ore BEK. 533 ; in goddes ore MIRC 585 ; ore, oore CH. C. T. *A* 3726 ; aore HOM. I. 187.

  ōr-fest, adj., *O.E.* ārfæst ; *dutiful, 'pius,'* FRAG. 3.

  āre-ful, adj., *O.E.* ārfull ; *merciful*, ORM. 1460.

  āre-lēs, adj., *O.E.* ārleas ; *merciless*, HOM. I. 173 ; arelies, oreles P. L. S. viii. 100 ; arelæs ORM. 9881 ; orleas FRAG. 6 ; oreleas HOM. II. 226.

  ār-würðe, adj., *O.E.* arwyrðe ; *honourable* ; ārwurðore (*comp.*) MAT. xx. 28.

  ār-würðen, v., *O.E.* ārwurðian, -wyrðian, -weorðian ; *honour*, SAX. CHR. 30.

  ār-w(ü)rðlice, adv., *honourably*, HOM. I. 229.

a-recchen, v., *O.E.* areccan, = *O.H.G.* irrecchan ; *explain, expound, enounce*, LA3. 28097 ; areche us þis bispel MAT. xv. 15 ; þis sweven (heo) aræhten 25629 ; araht (*pple.*) MARH. 1 ; arou3t HALLIW. 86.

a-redden, v., *O.E.* ahreddan, = *O.H.G.* arrettan ; *deliver, save*, LA3. 26909 ; A. R. 390 ; aredde FL. & BL. 689 ; M. T. 97 ; arudden HOM. I. 253 ; KATH. 1142 ; arüde (*imper.*)

MARH. 6 ; **aredde** (*subj.*) O. & N. 1569 ; **aredde** (*pret.*) HOM. I. 87 ; II. 33 ; LA3. 831 ; **ared** (*pple.*) A. R. 392 ; MISC. 86.

a-réde, v., *? O.E.* aredian ; *reach, hit* : wham he mai with þe hache arede FER. 1706.

a-rēde, v., *O.E.* arǣdan ; *read*, LA3. 22719 ; ALIS.² (Sk.) 573.

a-reimen, v., *rescue, redeem* : þet is ure raunsun þet we schulen areimen us mide A. R. 126.

a-reisen, v., *raise* ; areise WILL. 4342 ; **areisid** (*pple.*) WICL. JOHN iii. 14.

āren, v., *O.E.* ārian, = *O.H.G.* ēren, *O.N.* æra ; *show mercy*, ORM. 1462.

ārende, *see* **ǣrende**.

a-rēosen, v., *O.E.* ahrēosan ; *rush upon* ; **aruren** (*pret. pl.*) MAT. vii. 27.

a-rēowen, v., *? for* 3e-hrēowen ; *commiserate, repent* ; **arewen** P. S. 240 ; hit shal him never þenne arewe MAS. 15 ; me arēoweð þi sar JUL. 35 ; **arēowe** (*subj.*) A. R. 66 ; þe deore drihtin arēaw us KATH. 1379 ; areu H. H. 29.

a-repen, v., *touch* : þet heo muwen arepen & arechen A. R. 128 ; areppen LA3. 26034.

arerage, sb., *O. Fr.* arrerage, arrierage ; *arrearage*, LANGL. C x. 274 ; arrerage, arerage CH. C. T. A 602.

arēre, adv., *O.Fr.* arere ; *backwards*, GOW. I. 315 ; LANGL. A v. 198 ; arrere B v. 354.

a-resien, v., *O.E.* ahrysian ; *shake, totter* : þe tre aresede as hit wold falle S. S. (Web.) 915.

aresonen, v., *O.Fr.* araisoner ; *call to account* ; **aresoneþ** (*pres.*) R. R.6220 ; **aresunede** (*pret.*) MISC. 35 ; **aresoned** (*pple.*) PR. C. 2460.

ārest, *see* **ǣrest**.

arest, sb., *O.Fr.* arest ; *arrest*, CH. C. T. E 1282.

areste, v., *O.Fr.* arester ; *arrest*, TREV. VIII. 283 ; CH. C. T. A 827 ; D. ARTH. 633.

aretten, v., *esteem, account*, CH. BOET. ii. 4 (40) ; arette WICL. LEV. xi. 4.

āre-thēde, sb., *for* ǣr-þēode ; *people of long ago*, ISUM. vi ; arthede DEGR. 7.

arewe, *see* **arwe**. **arf**, *s.e* **erfe**.

arfeþ, *see* **earfeþ**.

ar3, adj., *O.E.* earg, earh, = *O.Fris.* arg, erg, *O.N.* argr, *O.H.G.* arg ; *cowardly, idle, bad*, ar3e GAW. 241 ; arh HOM. I. 277 ; eær3h LA3. 4336 ; æruu 3455 ; are3, areh O. & N. 407 ; areu 1498 ; aru [arwe] as an hare ROB. 457 ; arwe, arwhe '*timidus, pavidus*' PR. P. 14 ; arwe ALIS. 3340 ; eruh [arch] A. R. 274 ; þe arege, arewe, erewe REL. I. 176 (MISC. 116, 117) ; ar3e, ærwe (*pl.*) P. L. S. viii. 10 ; er3e HOM. I. 161 ; erewe MISC. 58 ; arwe RICH. 3821 ; arwe men and kene HAV. 2115 ; ar3(e) (*adv.*) SHOR. 22 ; ærhest (*superl.*) LA3. 21733.

arh-hēde, sb., *cowardice*, MISC. 74.

ærh-scipe, sb., *cowardice*, LA3. 12411.

---

ar3nesse, sb., *cowardice, sloth*, AYENB. 31 ; arghnesse HOCCL. i. 435.

ar3ien, v., *cf. M.Du., M.H.G.* argen ; *make or grow cowardly* or *bad* : þet eower heorte er3ian (*pres.*) swiðe HOM. I. 13 ; me ar3es of mi selfe ALEX. 537 ; **ar3e** (*pres. subj.*) GAW. 2301 ; **ar3ed** (*pret.*) A. P. ii. 572 ; þat arghede alle þat þer ware PERC. 69 ; þou . . . hast **arwed** meni hardi men LANGL. C iv. 237.

argoile, sb., *dregs of wine*, CH. C. T. G 813.

ar3ðe, sb., *O.E.* yrgð, = *O.H.G.* argida ; *cowardice, sloth* ; ærhðe [arhþe] LA3. 23546 ; are3þe, arehþe O. & N. 404.

arguen, argue, v., *Fr.* arguer ; *argue*, LANGL. A xi. 130 ; argu *convict* APOL. 31.

arguere, sb., *arguer* ; **argueres** (*pl.*) LANGL. B x. 116.

argument, sb., *Fr.* argument ; *argument*, SHOR. 142.

argumenten, v., *Fr.* argumenter ; *argue, reason*, GOW. I. 16.

arh, *see* **ar3**.

a-rihten, v., = *O.H.G.* arrichtan ; *raise up* : such gestening he aright (*pret.*) TOR. 1366 ; ariht (*pple.*) P. S. 325.

a-rīmen, v., *O.E.* arīman ; *number* : þa lette þe kaisere arimen (*sec. text* telle) al þæne here LA3. 25392 ; **arīmed** (*pple.*) 28937.

a³-rīnen, v., *O.E.* onhrīnan, ahrīnan, = *O.L.G.* anthrīnan ; *touch* : he scal . . . metes ne arinan HOM. I. 115 ; **arīneð** (*pres.*) A. R. 408 ; **arān** MARH. 20.

a-rīsen, v., *O.E.* arīsan, = *O.L.G.* arīsan, *O.H.G.* arrīsan, *Goth.* urreisan ; *arise*, HOM. I. 141 ; LA3. 1248 ; A. R. 324 ; up arise BEK. 1395 ; til þe sonne gan arise CH. C. T. B 791 ; **arīst** (*pres.*) O. & N. 1397 ; ALIS. 5458 ; **aris** (*imper.*) HOM. I. 45 ; ariseþ MISC. 42 ; **arās** (*pret.*) HOM. I. 91 ; KATH. 337 ; aras, aros LA3. 404, 11476 ; aros BEK. 451 ; AYENB. 173 ; he aros of deaðe HOM. II. 23 ; (þu) arise JUL. 63 ; (heo) **arisen** (*pret. pl.*) LA3. 1983 ; hi arise BRD. 10 ; **arise** (*pret. subj.*) LA3. 25988 ; **arisen** (*pple.*) C. L. 1430 ; K. T. 442 ; arise WILL. 1297.

arīzinge, sb., *arising*, AYENB. 13.

arisnesse, sb., *state of being risen*, REL. I. 282.

ā-rīst, sb., *O.E.* ǣrīst, *cf.Goth.* urrists *f.* ; *rising, arising* ; ariste MAT. xxii. 23 ; HOM. I. 87 ; A. R. 258 ; JUL. 62 ; ærist ORM. 16236 ; arist GOW. I. 320 ; **ǣristes** (*gen.*) HOM. I. 229 ; at the sonne āriste CH. ASTR. ii. 12 ; on þam æriste MK. xii. 23.

arke, sb., *O.E.* earc, = *Goth.* arka, *O.Fris.* erka, *O.N.* örk, *O.H.G.* archa, *from Lat.* arca ; *ark*, '*arca*,' VOC. 199 ; ORM. 1690 ; MAN. (H.) 136 ; in arke or in kiste HAV. 2018 ; arche LA3. 8965 ; A. R. 334 ; GEN. & EX. 561.

arm, *see* **earm**.

armee, sb., *O.Fr.* armee ; *army*, CH. C. T. A 60.

armen, v., *O.Fr.* armer ; *arm,* MAN. (F.) 1174; armi (*ms.* ærmi) LAȝ. 15313; arme ROB. 63 ; armeð (*pres.*) A. R. 262 ; armeþ AYENB. 111 ; armede (*pret.*) LAȝ. 8655*; iarmed (*pple.*) HORN (L.) 1215.

armes, sb. pl., *O.Fr.* armes, *from Lat.* arma ; *arms,* A. R. 60; HAV. 2925 ; ROB. 63 ; AYENB. 170.

armþe, *see* earmðe.

armūre, sb., *Fr.* armure, *from Lat.* armatūra ; *armour,* AYENB. 170; WILL. 3769; armure, armoure PR. P. 14 ; armure, armour ROB. 139, 402 ; armour MAN. (F.) 5117.

armurer, sb., *Fr.* armurier ; *armorer* ; **armurers** (*pl.*) CH. C. T. *A* 2507.

armurie, sb., *armour,* MAN. (H.) 194 ; **armories** (*pl.*) *armed forces* TRIAM. 49.

arn, *see* earn.

arnement, sb., *O.Fr.* arrement ; *ink,* ALIS. 6418 ; S. S. (Web.) 2776.

aromat, sb., *Fr.* aromate ; *aroma,* L. H. R. 224 ; **aromaz** (*pl.*) A. R. 372.

aromatik, adj., *aromatic,* MAND. 174.

arouten, v., *? O.Fr.* arouter; *? send away* ; þou shelt her to aroute HALLIW. 86 ; **aroutid** (*pple.*) DEP. R. iii. 221.

arrai, sb., *O. Fr.* arrei, arroi ; *array* ; arai CH. C. T. *D* 927 ; MAND. 252 ; WILL. 1597.

arraie, v., *O.Fr.* arreier, arreer ; *array,* CH. C. T. *B* 1202; araie WILL. 3561; araieþ (*pres.*) SHOR. 108 ; **arraie** (*imper.*) MAND. 214.

arre, *see under* ærer.

arren, *see* irren.

arrivaile, sb., *arrival,* GOW. II. 4.

arrīve, v., *O.Fr.* arriver ; *arrive,* MAND. 31 ; arive BRD. 11 ; **arīved** (*pple.*) LAȝ. 16063.

arrive, sb. [*? mistake for* armee]: at manie a nobil arive [*v. r.* armee] hadde he be CH. C. T. *A* 60.

arrogance, sb., *arrogance,* AYENB. 21 ; CH. C. T. *D* 1112.

arrogant, adj., *O.Fr.* arrogant ; *arrogant,* CH. C. T. *I* 396.

ars, sb., *O.E.* ears,= *O.Fris.* ers, *O.H.G.* ars, *O.N.* ars, rass ; *arse, hinder parts,* '*anus*,' VOC. 179; PR. P. 14 ; TOWNL. 9 ; ers, ars LANGL. *B* v. 175; ers VOC. 208 ; P. S. 153 ; TREV. V. 171 ; CH. C. T. *A* 3734.

   ars-rop, sb., WICL. I KINGS v. 9 ; **ars-þarm** (*ms.* harsþarme), VOC. 186 ; *the lower bowel.*

   ars-wisp, sb., '*manpirium*,' VOC. 179 ; '*anitergium*,' PR. P. 14.

arsmetike, sb., *O.Fr.* arismetique; *arithmetic,* GEN. & EX. 792 ; arsmetrike P. L. S. xvii. 224.

arsoun, sb., *Fr.* arçon ; *saddle-bow, saddle,* ALIS. 4251 ; B. DISC. 1172; his **arsouns** weren off iren RICH. 5539.

arst, *see* ærest.

arstable, sb., *? astrolaðe,* ALIS. 287.

art, sb., *Fr.* art ; *art,* ROB. 145 ; AYENB. 65 ; **ars** (*pl.*) ALIS. 665.

arten, v., *Lat.* artāre ; *confine, restrain, press,* CH. TRO. i. 388 ; artin '*arto*' PR. P. 14; arte HOCCL. iv. 8.

article, sb., *Fr.* article ; *article,* AYENB. 12 ; **articles** (*pl.*) A. R. 262.

artificial, adj., *O.Fr.* artificial ; *artificial,* ÇH. C. T. *B* 2.

aru, *see* arȝ. **arveð,** *see* earfeð.

arwe, sb., *O.E.* earh, (*dat. pl.* arewan), = *O.N.* ör (*pl.*) örvar *f.* ; *arrow,* GEN. & EX. 478; arwe AYENB. 66 ; CH. C. T. *E* 1673; S. & C. I. xxxv ; on arwe [arewe] LAȝ. 2476; an arewe JUL. 72 ; arewe BRD. 15 ; WILL. 885; earewe A. R. 60 ; **arewen** (*pl.*) ALIS. 5283; arewen, arwen [arewe] LAȝ. 5510, 12576; earewen A. R. 60 ; H. M. 15 ; arwes MAND. 190.

as, *see* al swa *under* al. **as,** *see* es.

ās, aas, sb., *Fr., Lat.* ās ; *ace,* CH. C. T. *B* 3851. [**ās,** sb., *O.N.* āss, = *Goth.* ans (δοκός) ; *plank* ; *comp.* bēt-, wind-ās.]

a-sadien, v., = *M.H.G.* ersatten ; *satiate* ; asad (*pple.*) REL. I. 122 ; P. S. 212.

asailen, assailen, v., *O.Fr.* assaillir ; *assail,* A. R. 246, 362 ; asailȝe H. M. 47 ; assaille HORN (L.) 637 ; asaileþ (*pres.*) AYENB. 17 ; assilled (*pple.*) SHOR. 141.

a³-sáken, v., *O.E.* an-, onsacan, = *O.L.G.* antsacan, *O.N.* andsaka ; *deny, renounce* : bute hi here laȝe asōke HORN (L.) 65.

asaumple, sb., *O.Fr.* esample ; *example,* A. R. 112.

asaut, sb., *O.Fr.* assaut, asalt ; *assault,* ROB. 175 ; assauz (*pl.*) A. R. 196.

ascance, adv., *as it were, as though,* CH. C. T. *D* 1745 ; askauns LIDG. M. P. 35 ; ascaunces CH. TRO. i. 292.

ascapie, *see* escapen.

ascenden, v., *O.Fr.* ascendre ; *ascend* ; **ascendid** (*pple.*) WICL. JOHN xx. 17.

ascensioun, sb., *O.Fr.* ascension ; *ascension,* SHOR. 126.

asch, sb., *O.E.* æsc, = *O.H.G.* asc, asch, *O.N.* askr *m.*, *M.Du.* esch, *M.H.G.* esche *f.* ; *ash* (*-tree*), VOC. 162 ; P. L. S. xiii. 171; CH. C. T. *A* 2922 ; asche VOC. 181 ; LAI LE FR. 168 ; MAN. (F.) 12438; asche, esche PR. P. 15, 143. **esch-kēi(e),** sb., *ash-key,* PR. P. 143. **esch-trē,** sb., *ash-tree,* VOC. 228.

a-scháken, v., *O. E.* asceacan ; *shake off* : ascakeð þæt dust of eowren foten MAT. x. 14 ; ær þe dai weore al asceken [asake] LAȝ. 19154.

a-schámien, v., *O. E.* ascamian ; *make ashamed, put to shame* ; aschámed (*pple.*) LEG. 70 ; WILL. 1035 ; S. S. (Wr.) 2314; ALIS. 3309; aschomed CHR. E. 864.

a-scheden, v., = *O.H.G.* erscuttan; *shed*; **asced** (*pple.*) HOM. I. 127.

asche, sb., *O.E.*, asce, æsce, axe, = *O.H.G.* asca, ascha, asche, asga, *Goth.* azgo, *O.N.* aska; *ashes*; ashe GOW. III. 345; aske, asche '*cinis*' PR. P. 15; aische, aske WICL. GEN. xviii. '27; esche SHOR. 107; axe MISC. 78; asse HALLIW. 95; esse AYENB. 137; pale as ashe [aische, aschen] colde CH. C. T. *A* 1364; **escan** (*dat.*) LK. x. 13; **asken** (*pl.*) A. R. 214; HAV. 2841; asshen HOM. II.65; askes LANGL. *B* iii. 97; ORM. 1001; askes, axes TREV. V. 167, VII. 5; asches MAND. 107; axen LA3. 25989; P. S. 203; (*ms.* acxen) HOM. II. 95.

  aske-baðie, sb., *one who sits in the ashes*, A. P. 214.

  aske-fīse, sb., = *Swed.* askefīs (RIETZ 139); *one who blows the ashes*, '*ciniflo*,' PR. P. 15.

ascheler, sb., *ashlar*, AUD. 78.

a-schenden, v., *? for* 3eschenden; *disgrace*: þat lond he wole aschende BEK. 1354; asende LA3. 18067*; **assend** (*pple.*) ROB. 263.

a-schēwelen, v., *scare away*: þar ich aschēwele [ascheule] pie and crowe O. & N. 1613.

aschi, adj., = *Ger.* aschig; *ashy*, CH. C. T. *A* 2883; aski '*cinereus*' CATH. 14.

aschien, *see* askien.

a-schortien, v., *O.E.* asceortian; *shorten*: (þ)i tunge is ascorted FRAG. 8.

a-schrenchen, v., *O.E.* ascrencan; *deceive*: ech oþren aschrencheþ SHOR. 17; **ashreint** (*pple.*) ALIS. 4819; ich was aschreint S. S. (Web.) 1485.

a-schunchen, v., *? terrify*; ne mei hit me ashunche SPEC. 38.

a²-schunian, v., *O.E.* onscunian; *shun*; aschoune (*? printed* aschonne) DEP. R. ii. 185.

ascrī, sb., *O.Fr.* escri; *cry, shout*; A. P. ii. 1296; askri GOW. II. 386.

ascrīen, v., *O.Fr.* escrier; *cry out*, WILL. 3814; **ascrīed** (*pret.*) A. P. iii. 195.

a-scüren, v., = *askerren; ? scare, drive away*; ascür (*imper.*) him so scheomeliche A. R. 296.

ase, *see* alswa *under* al.

a-sēarien, v., *O.E.* asēarian, = *O.H.G.* arsōren; *grow dry*; **asēred** (*pret.*) S. S. (Web.) 606.

a-sēchen, v., *O. E.* asēcan; *seek out*, LA3. 27866.

asēgen, v., *O.Fr.* aseger; *besiege*; **asēged** (*pret.*) ROB. 184; **asēged** (*pple.*) WILL. 4224.

asēle, v., *O.Fr.* enseeler, *med.Lat.* insigillare; *seal*, LANGL. *A* ii. 37; åsele ALIS.² (Sk.) 829; asēled (*pple.*) MISC. 228; LANGL. *C* iii. 114.

a-senchen, v., *make to sink*; **asencte** (*pret.*) ROB. 416; azenkte AYENB. 49; ure scipen he aseingde LA3. 25697; asenchtest JUL. 33; al here atil and tresour was al so aseint ROB. 51.

a-senden, v., *O.E.* asendan, = *M.H.G.* ersen-

---

den; *send out*, HOM. I. 91; **asende** (*pret.*) HOM. I. 225.

aserchen, *see* enserchen.

aserve, v., *? for* deserve: L. H. R. 147.

a-setten, v., *O.E.* asettan; *appoint*; azet (*pres.*) AYENB. 140; on **asette** *pple.* tidan HOM. I. 115.

Asie, pr. n., *Asia*, MARH. 2.

a-sī3en, v., *O.E.* asīgan; *sink, fall*; asien MISC. 90; **asēh** (*pret.*) SAX. CHR. 249; HOM. II. 109.

a-sinken, v., = *M.H.G.* ersinken; *sink down*: **asinkeþ** (*pres.*), þat hi nasinkeþ SHOR. 136; **asonken** (*pret.*) MAP 345.

a-sitten, v., *O.E.* asittan, = *M.H.G.* ersitzen; *sit*: aset (*sec. text* sat) þe kaisere swulc he akimed weore LA3. 26353.

aske, sb., *? O.E.* åðexe, = *O.H.G.* egedehsa; *lizard*; **askes** (*pl.*) ARCH. LVII. 254; askes (*? ms.* arskes) and oþer wormes M. H. 141.

aske, *see* asche.

åske, sb., *O.E.* æsce, = *O.H.G.* eisca; *demand*; axe LA3. 1053.

aske-baðie, -fīse, *see under* asche.

åskien, åxen [*?* åxen], v., *O.E.* åscian, åcsian, åxian, åhsian, = *O.Fris.* åska, åschia, *O.N.* æskja, *O.L.G.* ēscōn, *O.H.G.* eiscōn; *ask*; asken A. R. 242; C. L. 1023; (*ms.* asskenn) ORM. 10278; askin '*peto, posco*' PR. P. 15; aske AUD. 79; aske leve HAV. 2952; asche EGL. 1090; ashe ED. 2775; asche, esse ROB. 16, 374; eskien HOM. I. 25; asken, axen CH. C. T. *B* 101; axe CH. D. BL. 1276; aske, axe LANGL. *A* iv. 90; esche MISC. 139; axien MAT. xxii. 46; axien [axi] LA3. 7193; acsi, oxi AYENB. 110, 114; axe GOW. I. 47; WICL. JOHN i. 19; asse TOWNL. 58; **aske,** axe (*pres.*) CH. C. T. *A* 1347; ich acsi þe a qvestioun SHOR. 136; eski MARH. 13; aischest, aishest, aissest, axest O. & N. 473, 707, 995; oxist P. S. 200; acseð A. R. 8; nou esche we TREAT. 136; **aske** (*imper.*) H. M. 9; esca hine hwet he habbe bi3eten HOM. I. 35; **askede** (*pret.*) A. R. 66; MISC. 26; GEN. & EX. 3191; askede LA3. 29229; he axede gavel of þan lond 6122; heo æxede [axede] æfter Brenne 5001; askede, aschede ROB. 16, 33; acsede REL. I. 30; HORN (H.) 43; easkede JUL. 15; escade HOM. I. 43; þu askedest O. & N. 1310; *comp.* of-askien.

åsker, sb., *asker*, WICL. IS. ix. 4; þe axere, (*plaintiff*) and þe defendaunt E. G. 361.

åskunge, sb., *O.E.* åscung; *asking*, A. R. 70; askinge '*petitio*' PR. P. 15; ascinge REL. I. 131 (HOM. II. 11); escunge MARH. 16; acsinge AYENB. 209.

a-slǣped, pple., *overcome with sleep*; heo is asleped sviþe FL. & BL. 582; aslepid ('*sopitus*') with ful miche drunkenesse WICL. JUDITH xiii. 4.

a-slǎkien, v., *O.E.* asleacian; *slacken*; aslake

LEG. 137 ; LIDG. M. P. 231 ; PALL. iv. 769 ;
asláki (*pres. subj.*) AYENB. 253 ; **asláked**
(*pple.*) TREV. III. 143 ; til atte laste aslaked
was his mood CH. C. T. *A* 1760.

a-slēan, v., *O.E.* aslēan,=*O.H.G.* arslahan ;
slay ; aslæn,-slan LA3. 22271, 22576 ; aslen FL.
& BL. 105 ; S. A. L. 155 ; **aslō3** (*pret.*) LA3.
22319 ; aslou3 P. L. S. xii. 88 ; (þu) aslo3e LA3.
24812 ; heo aslo3en [aslowe] 20956 ; aslowe
ARCH. LII. 37 ; **asla3e** (*pple.*) HORN (L.)
88 ; asla3e, aslawe FER. 84, 321 ; aslawe
ROB. 170 ; SHOR. 120 ; LEG. 13 ; AL. (T.)
165 ; aslæ3en LA3. 21892.

a-slīden, v., *O.E.* aslīdan ; *slide away* ; aslide
HALLIW. 93.

a³-slüppen, v.,=*ansluppen, *cf. M.H.G.* ent-
slupfen ; *slip away* : ne miht i him asluppe
SPEC. 38 ; **aslipped** (*pple.*) A. P. iii. 218.

a-smellen, v., *smell out* : þe bor hem gan
ful sone asmelle S. S. (Web.) 891.

a-snǣsen, v., *O.E.* asnǣsan ; *strike, thrust* :
þene horn þet he asnēseð mide alle þeo þet
ha areacheð A. R. 200.

a-softien, v., *soften* ; asofte HALLIW. 94.

asottie, v., *O.Fr.* asoter ; *become foolish*, HOM.
I. 17 ; assote GOW. I. 68 ; **asotid** (*pple.*) FER.
2007.

a-spárien, v., = *M.H.G.* ersparn ; *spare* ;
aspare LANGL. *B* xv. 136.

aspe, sb., *O.E.* æspe, æpse,= *O.H.G.* aspa,
O.N. ösp ; *aspen*, VOC. 181, 285 ; CH. C. T.
A 2921 ; aspe, espe PR. P. 15, 143 ; espe VOC.
228.

aspect, sb., *aspect*, CH. C. T. *A* 1087.

aspen, adj.,= *M.H.G.* espin ; *aspen* : aspen lef
CH. TRO. iii. 1200.

a-spenden, v., *O.E.* aspendan ; *spend* ; a-
spenen (*pres. subj.*) HOM. I. 123.

a-spēowen, v., *spue out* ; aspēweð (*pres.*)
HOM. II. 199.

aspīe, aspīen, *see* spīe, spīen.

a-spillen, v., *spill, waste, destroy* ; aspille O.
& N. 348 ; mi soule to aspille BEK. 1024 ;
aspillað (*pres.*) HOM. I. 13 ; þeo þet forleoseð
& aspilleð al hore god A. R. 148.

aspre, adj., *O.Fr.* aspre ; *harsh, cruel*, CH. AN.
23 ; aspere CH. BOET. ii. 1 (32).

　asper-liche, adv., *harshly*, GUY 84 ; asperli
A. P. iii. 373.

　asprenesse, sb., *harshness*, CH. BOET. iv. 4
(127).

aspretē, sb., *O.Fr.* aspreté ; *asperity*, A. R. 354.

a-springen, v., *O.E.* aspringan, = *O.H.G.*
arspringæn ; *spring up* ; aspringe SHOR. 120 ;
asprang (*pret.*) HOM. I. 227 ; **asprungen**
(*pple.*) ANGL. I. 17 ; asprongun DIGB. 118.

assai, sb., *O.Fr.* essai, *Lat.* exagium ; *trial,
test*, P. L. S. xxx. 166 ; FLOR. 1500.

assaien, v., *O.Fr.* es-, assaier ; *try, test*, LANGL.

A iii. 5 ; asaie WILL. 3754 ; **asaid** (*pple.*)
AYENB. 117.

asse, sb., *O.E.* assa, *cf. O.N.* asni, *Lat.* asinus ;
ass, A. R. 32 ; SHOR. 122 ; MAND. 70 ; an asse
TREAT. 138 ; þane asse AYENB. 156 ; **asse**
(*gen.*) LANGL. *B* xviii. 11 ; asse earen A. R.
172 ; was le3d in asse cribbe ORM. 3711 ; an
asse ban C. M. 7171 ; **assen** (*acc.*) LK. xiii. 15 ;
assen (*pl.*) ALIS. 1866 ; assen, asses CH. C. T.
D 285.

asse, sb., *O.E.* asse ; *she-ass*, LEG. 7 ; ane asse
... mid hire colt HOM. I. 3 ; **assen** (*gen.*) MK.
xi. 2.

assemble, v., *O.Fr.* asembler ; *assemble*, CH.
BOET. iii. 7 (80) ; **assemblede** (*pret.*) P. S.
190 ; asembleden WILL. 3815.

assemblē, sb., *O.Fr.* asemblee ; *assembly*,
LANGL. *B* PROL. 217 ; A. P. i. 759 ; IW. 19.

assene, sb., *O.N.* asna ; *she-ass*, MAT. xxi. 2.

assent, sb., *assent*, CH. C. T. *A* 817 ; asent
SHOR. 57 ; WILL. 1300.

assenten, v., *O.Fr.* assentir ; *assent*, LANGL.
C xx. 264 ; assente *B* iii. 113 ; **assentede**
(*pret.*) P. L. S. xxiv. 150.

assetz, sb., *O.Fr.* assez ; *satisfaction*, LANGL.
B xvii. 237 ; assethe PR. C. 3610 ; aseþ WICL.
S. W. II. 237.

assigne, asigne, v., *O.Fr.* assegner ; *assign*,
LANGL. *A* iv. 109 ; **assignede** (*pret.*) P. L.
S. xii. 131 ; asegned (*pple.*) WILL. 581.

assīse, sb., *O.Fr.* assise ; *assize, size*, S. S.
(Web.) 2490 ; OCTOV. (W.) 81 ; R. R. 1237 ;
asise ROB. 53 ; LEG. 46 ; WILL. 4451.

assoigne, *see* essoigne.

assoilen, v., *O.Fr.* assoldre, *from Lat.* absol-
vere ; *absolve, pardon ; solve, explain* ; LANGL.
A PROL. 67 ; asoile MIRC 1968 ; asoili AYENB.
172 ; absoile MER. 11 ; **assoili** (*pres. subj.*)
BEK. 2027 ; **asoilede** (*pret.*) ROB. 476 ; a-
soiled (*pple.*) MISC. 32.

assoinen, v., *O.Fr.* essoigner ; *excuse* ; asunien,
asonien, A. R. 64 ; assoini AYENB. 242 ; a-
soinede ROB. 539.

assumpcioun, sb., *assumption*, SHOR. 126.

assūrance, sb., *O.Fr.* asseurance ; *assurance*,
CH. C. T. *B* 341.

assūre, v., *O.Fr.* asseurer ; *assure*, CH. C.
T. *B* 1231 ; asseūred (*pple.*) BOET. i. 4 (16).

a-standen, v., *O.E.* astandan, = *O.L.G.*
astandan, *O.H.G.* arstantan ; *stand up* : þat
deor up astōd LA3. 6495.

astat, *see* stat.

a-stellen, v., *O.E.* astellan ; *constitute, establish* ; **astalde** (*pret.*) HOM. I. 19 ; astalden
LA3. 8960 (*misspelt* astalleden 8950) ; **astald**
(*pple.*) LA3. 8116 (*misspelt* astalled 8391).

a-stenchen, v.,= *Ger.* erstenken ; *annoy with
stench* : stute nu . . . to astenchen me wið þe
stench þet of þi muð stiheð MARH. 12 ; **asteinte**
(? *stinking*) horesone ALIS. 880.

a-steorven, v., *O.E.* asteorfan, = *O.H.G.* arsterban; *die*, MARH. 12; A. R. 326; asterve FER. 3058; **astorven** (*pple.*) A. R. 310; astorve O. & N. 1200.

a-stēren, v., *? O.E.* astēran; *excite*; **astērde** (*pret.*) HOM. I. 95; þat he . . . þe storvene astearde (*v. r.* arearde) KATH. 1043; mani a Sarzin•þar was **astēred** FER. 3899.

a-sterte, v., *start up, start away*, LANGL. *B* xi. 392; MED. 570; CH. C. T. *A* 1595; **asterte** (*pret.*) FER. 1468; asturte MISC. 42; þe eotend up asturte LAȜ. 26045; astirte til him HAV. 893; *see* **ætsterten**.

a-sterve, v., *O.E.* asterfan, astyrfan, = *M.H.G.* ersterben, *mod.Eng.* starve; *kill*, AYENB. 240; **asterved** (*pple.*) 240.

a-stīȝen, v., *O.E.* astīgan, = *O.H.G.* erstīgan; *ascend, descend*; astien MISC. 55; astie ROB. 317; **astīghð** (*pres.*) HOM. II. 107; **astāh**, asteh (*pret.*) HOM. I. 17, 91; asteh II. 111; asteih, astei MISC. 53, 55; **astiȝen** (*pret. pl.*) LAȜ. 8691; **astiȝe** (*pret. subj.*) HOM. l. 227; **astoȝe** (*pple.*) FER. 2971.

    astīunge, sb., *ascension*, HOM. I. 209.

astoffe, v., *O.Fr.* estouffer; *suffocate*, JARB. XII. 161.

astonie, v., *O.Fr.* estoner; *astound, stun*, AYENB. 126; astone GOW. III. 54; **astoned** (*pple.*) CH. TRO. i. 274; astouned SHOR. 88; hors and man astuned lai AR. & MER. 6297.

astōren, *see* **stōren**.

astranglen, *see* **stranglen**.

a-strecchen, v., *O.E.* astreccan, = *M.H.G.* erstrecken; *stretch out, extend*; astretchin '*attingo*' PR. P. 16.

a-strengþi, v., *strengthen*, ROB. 180; **a-strengþed** (*pple.*) MISC. 32.

astrolabie, sb., *Fr.* astrolabe, *Lat.* astrolabium; *astrolabe*, CH. ASTR. PROL.

astrologie, sb., *O.Fr.* astrologie; *astrology*, CH. ASTR. PROL.

astrologien, sb., *O.Fr.* astrologien; *astrologer*, CH. ASTR. PROL.

astronomer, sb., *astronomer*, TREV. IV. 209.

astronomie, sb., *O.Fr.* astronomie; *astronomy*, PR. P. 16; LAȜ. 24298; LANGL. *A* x. 207.

astronomien, sb., *O.Fr.* astronomien; *astronomer*, GOW. II. 230.

astruien, v., *for* **destruien**; **astrúþ** (*pres.*) AYENB. 17; **astruid** (*pple.*) HOM. II. 147.

astudien, *see* **studien**.

a-stünten, v., *O.E.* astyntan (*obtundere*); *cease*: astunten hore cleppe A. R. 72; he nel nevere astinte' FER. 1842; astente WILL. 1527; **astünte** (*pret.*) LAȜ. 31891*; astunte, astinte ROB. 128, 546; **astint** (*pple.*) MAP 341.

a-stürien, v., *O.E.* astyrian; *stir up*; **astireden** (*pret.*) MK. xv. 11; **astired** (*pple.*) LEECHD. III. 130; astured weoren Romweren

alle mid sterclichere wræðe LAȜ. 25329; astered DEGR. 757; *cf.* **astēren**.

asūr, adj., *Fr.* azur; *azure*, JOS. 195.

aswágen, v., *O.Fr.* assouagier; *assuage*, LANGL. *A* v. 100; aswagi BEK. 1452; aswage CH. C. T. *B* 3834.

a-swebben, v., *O.E.* aswebban; *stupefy*: astonied and asweved CH. H. F. 549.

a-swelȝen, v., *devour*; **aswolwe** (*pple.*) BEVES 786.

aswelten, v., *O.E.* asweltan; *faint, die*: þet he **aswelte** (*pres. subj.*) wiðinnen A. R. 216; þe fuir . . . **aswelt** (*pres.*) ALIS. 6639; cnihtes þar **aswalten** (*pret.*) LAȜ. 27474*.

a-swīken, v., *O.E.* aswīcan (*deceive*); *cease, fail*; **aswīkeð** (*pres.*) KATH. 2187; **aswīkeð** (*imper.*) LAȜ. 15327; **aswīkie** (*subj.*) MAT. v. 29; **aswāc** (*pret.*) LAȜ. 16112.

a-swinden, v., *O.E.* aswindan, = *O.H.G.* arswintan; *vanish, decrease, waste, decay*, MARH. 13; aswinde ANGL. I. 10; O. & N. 1574; **aswint** (*pres.*) LAȜ. 17940; as imet sweven aswindeð hire murhðen JUL. 74; **aswond** (*pret.*) HOM. I. 133; ærst aswond þat corn LAȜ. 31793; **aswunden** [aswonde] (*pple.*) LAȜ. 19599; aswunde O. & N. 1480; aswounde L. H. R. 52.

a-swinken, v., *procure by labour*; aswinke BEK. 1663.

a-swoȝen, pple., *O.E.* aswogen; *in a swoon*; aswoȝe AL. (T.) 141; he fell aswowe to grounde LAUNF. 755.

a-swolkenesse, sb., *O.E.* aswolcenness; *sloth, indolence*, HOM. I. 83.

at, conj. & rel. pron., *O.N.* at; *that*, S. S. (Web.) 3824; M. H. 73; ANT. ARTH. xiv; TOWNL. 2; the beste bodi at thar war PERC. 150; þat at him fel to knaw(e) PR. C. 171.

at, *see* **æt**. **at-** (*compounds*), *see* **æt-**.

atache, attache, v., *O.Fr.* atacher; *attach*, LANGL. *A* ii. 174; **attached** (*pple.*) MAN. (H.) 158.

atachement, sb., *attachment*, E. G. 360.

a⁵-taken, v., *? =* **oftaken**; *take*; atake BEK. 1961; CH. C. T. *G* 556; AM. & AMIL. 2070; al þat þai atake miȝt(e) FER. 3128; shal atake '*apprehendet*' WICL. LEV. xxvi. 5; **atōk** (*pret.*) AR. & MER. 468.

a-tastin, v., *taste*, PR. P. 16; **ataste** (*pres. subj.*) CH. BOET. ii, 1 (30).

āte, sb., *O.E.* āta; *oat*; ote '*avena*' PR. P. 372; MIRC 1483; MAP 350; **ōten** (*acc.*) A. R. 312; **āten** [ote] (*pl.*) LAȜ. 24438; ooten RICH. 6004; otin VOC. 177; oten, otes LANGL. *A* iv. 45 (58); *comp.* **sǣd-āte**.

atel, adj., *O.E.* atol, eatol, = *O.N.* atall; *dreadful, cruel*; atell ORM. 4803; atele hurtes HOM. I. 275.

    atel-līch, adj., *O.E.* ate(l)līc; *cruel*; atelich A. R. 6; MISC. 183; MAP 343; in ateliche (*ms.* atteliche) pole LAȜ. 21748; mid þine ateliche sweore O. & N. 1125; ateliche [eateliche] wihtes JUL. 46, 47; þe alle weren eateliche to bihaldene HOM. I. 41; eatelukest (*superl.*) H. M. 41.

a-tellen, v., *O.E.* atellan, = *M.L.G.* ertellen, *O.H.G.* arzellan; *narrate, tell,* HOM. II. 113.

a-temien, v., *O.E.* atemian; *tame:* we ne muȝen atemien (*afford*) to wurðen godes bord HOM. II. 199; atemie LANGL. *C* xx. 240; atamed (*pple.*) AYENB. 153; hors wel atamed DEP. R. iii. 27.

āten, adj., *O.E.* āten (LEECHD. III.), 292; *oaten:* otene grotes L. C. C. 47.

a-tende, v., *O.E.* atendan; *kindle, inflame,* BER. 2728; a candle he atendeþ (*ms.* attendeþ) FER. 2413; atende (*pres. subj.*) MISC. 52; ðere heðene monnan heortan . . . muhten beon atende HOM. I. 95; so sone so hi weren of þe holi goste atende (*ms.* attende) MISC. 56.

a-tēon, v., *O.E.* atēon; *treat;* atoȝen (*pple.*) HOM. II. 205; weoren . . . reouliche atoȝene LAȜ. 24846.

a ⁴-tēonen, ᷐v., *? for* ȝetēonen; *irritate;* atēoned (*pple.*) CHR. E. 61; atened FER. 114; BEVES 2601.

a-teorien, v., *O.E.* ateorian; *fail;* atiereð (*pres.*) HOM. II. 29.

atiffen, v., *O.Fr.* atiffer; *decorate,* A. R. 360.

atil, sb., *O.Fr.* atil; *equipment, gear,* ROB. 51.

atilen, v., *O.Fr.* atillier; *adorn;* atiled (*pple.*) ROB. 184.

atīr, sb., *attire,* LAȜ. 3275*; WILL. 4465; RICH. 519.

atiren, v., *O.Fr.* atirier; *attire,* S. S. (Wr.) 3218; atíred (*pret.*) ROB. 547; atíred (*pple.*) WILL. 1228.

atlien, *see* ahtlien.

a-treȝen, v., *afflict, grieve;* atraid (*pret.*) S. S. (Web.) 1867; atraiyed (*pple.*) K. T. 605.

a-trukien, v., *? for* ȝetrukien; *fail;* atroketz (*pres.*) P. R. L. P. 221; atrokien (*ms.* attrokien) (*subj.*) LEB. JES. 11; nu beoþ þine teþ atru(kied) FRAG. 8.

attámin, v., *O.Fr.* atamer; *pierce, broach (a vessel), begin; 'attamino,'* PR. P. 16; atame L. H. R. 210; attamede (*pret.*) LANGL. *C* xx. 68; D. ARTH. 2175; attamed, atamed (*pple.*) CH. C. T. *B* 4008.

atteigne, v., *O.Fr.* ateindre; *attain,* GOW. I. 131; atteine CH. BOET. iv. 1 (118); ateine P. L. S. xxv. 87.

attemperaunce, sb., *temperance,* CH. BOET. iv. 6 (138); LIDG. M. P. 209.

attemprē, adj., *attempered,* AYENB. 153; attempree MAND. 276.

attempren, v., *O.Fr.* atemprer; *attemper,* GOW. I. 87; attempered (*pple.*) LIDG. M. P. 3.

attencioun, sb., *Fr.* attention; *attention,* CH. BOET. ii. 1 (29).

attendance, attendaunce, sb., *O.Fr.* attendance; *attendance,* CH. C. T. *D* 933.

attenden, v., *O.Fr.* atendre; *attend, listen,* GRL. xxii; atend C. M. 21803; atende LUD. COV. 259; attende (*pres.*) PALL. i. 511.

attente, sb., = entente, A. R. 252.

ǎtter, āter, sb., *O.E.* ātor, āttor, ǣtor, = *M.L.G.* etter, *O.H.G.* eitar, eittar, *O.N.* eitr; *poison, pus,* HOM. I. 23, 51; atter ORM. 15383; GEN. & EX. 372; C. L. 1150; LANGL. *B* xii. 256; þat atter LAȜ. 14995; þet atter A. R. 208; attir 'sanies' PR. P. 16; ǎttere (*dat.*) LAȜ. 16089.

ǎtter-coppe, sb., *O.E.* ātorcoppa, *cf. Scotch dial.* ettercap; *spider,* O. & N. 600; K. L. 63; attircoppe 'aranea' PR. P. 16.

ǎtter-lāðe, sb., *O.E.* āttorlāðe LEECHD. III. 34; *antidote to poison, betony:* drinc þeonne atterloðe (atterlaðe) & drif þene swel aȝeanward A. R. 274; atterloþe 'more(l)le' REL. I. 37 (VOC. 141).

ǎtter-lich, adj., *O.E.* ātorlīc; *venomous;* atterluch A. R. 212*.

ǎttern, adj., *O.E.* ǣtren, ǣttren, *M.H.G.* eiterin; *venomous* (*ms.* atterne), FRAG. 8; þurh atterne drench LAȜ. 16084.

atternesse, sb., *venomousness,* A. R. 196.

attíse, v., *O.Fr.* atiser; *entice,* HALLIW. 109.

ǎttrien, v., *O.E.* ǣtterian, = *M.H.G.* eitern, *O.N.* eitra; *venom, poison;* attreð (*pres.*) HOM. I. 153; A. R. 84; attred (*pple.*) HOM. II. 19; ORM. 15376; iattred A. R. 208.

ǎttriȝ, adj., = *O.H.G.* eitrig; *venomous,* ORM. 9785; attri A. R. 82; MARH. 14; H. M. 15; ðat attrie ðing REL. I. 215.

aturn, sb., *O.Fr.* aturn (*atour*); *dress, attire,* A. R. 426; H. M. 23; attour R. R. 3718.

aturnē, sb., *O.Fr.* atorné; *attorney,* E. G. 58.

aturnen, v., *O.Fr.* aturner, atorner (*atourner*); *turn; dress, attire:* he atornde as vaste as he miȝte ROB. 419; er he . . . habbe hire swuch aturned and imaked A. R. 284; leofliche aturnet HOM. I. 257.

a²-tween, prep., *between,* LIDG. M. P. 263.

a²-twix, prep., *between,* W. & I. 32; atvix AM. & AMIL. 865; atwixen CH. TRO. 417; atwixe TREV. VII. 167; atwixin, atwixt, *betwixt, 'inter'* PR. P. 17.

a³-twinnen, v., *separate, divide;* atwinne HALLIW. 109.

āþ, sb., *O.E.* āð, = *Goth.* aiþs, *O.L.G.*, *O.Fris.* ēth, ēd, *O.N.* eiðr, *O.H.G.* eid; *oath,* IW. 2264; ænne að [oþ] LAȜ. 19981 (*miswritten* æð 704); oað (*r. w.* wrað) 653; oþ AYENB. 64; þen oþ ROB. 446; ooþ, ooth CH. C. T. *A* 120; oth HAV. 1118; ōþe (*dat.*) PROCL. 5; wiþuten oþe HORN (L.) 347; āþes (*pl.*) ORM. 4479; oðes A. R. 198; āðen (*dat. pl.*) LAȜ. 5165; *comp.* **mān-āþ.**

aþel, aþeling, see æðel, æðeling.

a-þeostren, v., *O.E.* aþeostrian; *be darkened*: se mone aþēstreð HOM. I. 239; aþēostrede (*pret.*) LAȝ. 2860; aþēostred (*pple.*) MK. xiii. 24.

aþinken, see ofþünken.

a-þrāwen, v., *throw off*: of his horse he aþrēu (*pret.*) LAȝ. 807*.

a-þrüsemen, v., *O.E.* aþrysman; *suffocate*, A. R. 40; aþrüsmeð (*pres.*) HOM. I. 251.

āþum, sb., *O.E.* āðum, = *O.Fris.* āthum, āthom, *O.H.G.* eidum, eidam; *son-in-law* (*ms.* aþumm) ORM. 19832; aðum, oðem [oþom] LAȝ. 3619, 23106; oþam *'gener'* FRAG. 2; oðem REL. I. 130 (HOM. II. 165); oþom ROB. 182; odam ALIS. 2081.

a-þürsten, v., ?=*M.H.G.* erdursten; *cf.* ofþürsten; *suffer thirst*; aþürst (*pple.*) P. L. S. xiii. 329; LEG. 7; TREV. III. 471; he werþ aþurst wel sore REL. II. 273; aþurst, aþrust, aþrist LANGL. *B* x. 59*, *C* xii. 43*; athrest LUD. COV. 190.

aube, see albe.

aubel, ebel, sb., *Fr.* aubel; *abele, white poplar*, PR. P. 17.

auburn, adj., *L.Lat.* alburnus; *auburn*, PR. P. 17.

audience, sb., *O.Fr.* audience; *audience*, LANGL. *B* xiii. 434; CH. C. T. *E* 104.

auditour, sb., *O.Fr.* auditeur; *auditor*, SHOR. 96; GOW. II. 191.

aufin, see alfin. augrim, see algrim.

august, sb., *August*, CH. ASTROL. i. 10 (6); augst GOW. III. 370.

auh, see ac. āuhte, see æhte.

auk, adj. (= *avek, as hauk = *havek), *O.N.* öfugr, *O.H.G.* abuh; *awkward, left-handed, perverse,* (*ms.* awke) *'sinister, perversus,'* PR. P. 18; auke dedis D. ARTH. 13.

auk-li, adv., *awkwardly, 'sinistre, perverse,'* PR. P. 18.

auk-ward, adv., *awkward, awry*, D. ARTH. 2247; þe world þai al aukward sett(e) PR. C. 1541.

aul, see awel.

aumperour, see emperour.

aun-, see an-.

austēre, adj., *O.Fr.* austere; *austere*, MAN. (H.) 54; austerne WICL. S. W. I. 1.

austeritē, sb., *O.Fr.* austerité; *austerity*, PR. C. 5376.

auter, see alter.

autoritē, sb., *O.Fr.* autorité, auctorité; *authority*, AYENB. 147; MAN. (F.) 14596; autorite, auctorite CH. C. T. `*B* 4165.

autour, sb., *O.Fr.* auteur; *author*, LANGL. *B* x. 243; auctour CH. C. T. *E* 1141.

autumpne, sb., *Fr.* automne; *autumn*, CH. BOET. i. 2 (8).

āuþer, pron. & conj. (*for* *ā-hweþer), *O.E.* āhwæðer, āwðer, = *O.H.G.* ioweder, *mod.Eng.* or; *either* (*one of two*); *or*; GAW. 88; auþir. ANT. ARTH. xvi; heore owwþerr ORM. 124; ouðer, ouþer SAX. CHR. 32, 38; ouðer of þeos two A. R. 286; ouþer of þam IW. 3589; ouþer . . . or MAND. 184; ouþer bi daie or bi nighte PERC. 87; ouþer [oþer] elles LANGL. *A* i. 151; ouþer [oþer] . . . or CH. C. T. *A* 1593; oiþer ALIS. 4899; AL. (L¹.) 408; oiþer . . . or MAN. (H.) 2; oðer on him sulf oðer on his freond A. R. 202; oþer AYENB. 9; oþer dairim oþer daisteorre O. & N. 328; oþer catel oþer cloþ PL. CR. 116; oþer . . . or ROB. 17; clusung oþer iending FRAG. 1; fader oðer moder LAȝ. 26092; twelf winter oþer more HAV. 787; liþer oþer god TREAT. 133; oþþr, or ORM. 480, 6255, 10882; other [or] . . . other [or] LANGL. *B* xiii. 369; a maister or a pope CH. C. T. *A* 261.

availen, v., *avail*: ne mai mi strengþe not availle [availe] CH. C. T. *A* 2401; H. V. 81; availle MAN. (F.) 5084; avail C. M. 90; awailȝe BARB. i. 336; awail C. M. 7992; aveile CH. TRO. i. 20; availeþ (*pres.*) CH. H. F. 363.

avālen, v., *O.Fr.* avaler; *lower, cause to descend*, CH. C. T. *A* 3122.

avancen, v., *O.Fr.* avancer; *advance*; avance MIN. 39; avaunce CH. C. T. *A* 246; MAN. (F.) 7246; avonci AYENB. 82; avaunceð (*pres.*) A. R. 156; avanced (*pple.*) P. L. S. xvii. 383.

avancement, v., *O.Fr.* avancement; *advancement*, SHOR. 148; avauncement ROB. 312.

avant, sb., *from* avanten; *vaunt, boast, promise*, KATH.² 56; avaunt A. P. ii. 664; FER. 355; ALEX.² (Sk.) 570.

avantáge, sb., *O.Fr.* avantage; *advantage*, CH. C. T. *A* 1293; avauntage GOW. I. 194; avontage AYENB. 209.

avanten, avaunte, v., *O.Fr.* avanter; *vaunt, boast*, CH. C. T. *D* 1014; avanti SHOR. 118.

avantwarde, sb., *O.Fr.* avantgarde; *vanguard*, D. ARTH. 324; avauntwarde LANGL. *C* xxii. 95.

avarice, sb., *O.Fr.* avarice; *avarice*, AYENB. 16; CH. BOET. ii. 5 (45).

aveien, v., *O.Fr.* avier; *teach, inform*; aveide (*pret.*) SHOR. 19; avaid (*pple.*) 158; awaied A. P. i. 709.

aveiment, sb., *O.Fr.* aveiement; *instruction*, SHOR. 77; awaimentis (*pl.*) WINT. (N. E. D.).

ave-long, adj., = *Dan.* aflang, *Swed.* aflång, *from Lat.* oblongus; *oblong, 'oblongus,'* PR. P. 17.

avelen, see aflen. āven, see æfen.

avenaunt, adj. & sb., *Fr.* avenant; *suitable, graceful; convenience*; S. S. (Web.) 1419; MAN. (F.) 7532; ilka man after his avenaunt 4729.

avence, sb., *O.Fr.* avence; *the plant avens geum,* 'avencia,' VOC. 139; PR. P. 17.

Avene, pr. n., *O.E.* Afen; *Avon,* LAȝ. 21267; Avene striem P. L. S. viii. 126.

avener, sb., *O.Fr.* avenier, *Lat.* avenārius (*an officer of the stable, charged with the supply of provender*); VOC. 176; PR. P. 18.

avenge, v., *? for* revenge; *avenge,* LANGL. B v. 128*.

aventaille, sb., *O.F.* aventaille; *movable front of a helmet,* FER. 238; aventaile WILL. 3608.

avente, v., *O.Fr.* esventer; *refresh with air,* FLOR. 1941; awent BARB. vi. 305; **aventid** (*pret.*) D. TROY 7090; **aventid** (*pple.*) *escaped,* WICL. S. W. I. 219.

aventūre, sb., *O.Fr.* aventure; *adventure,* A. R. 340; MISC. 29; WILL. 254; ALIS. 7837; be aventure MAND. 185; to do his lif an auntre [in aventure] ROB. 65.

aventūren, v., *O.Fr.* aventurer; *adventure, venture*; aventure MAN. (H.) 70; auntre CH. C. T. *A* 4209.

aventurous, adj., *adventurous,* GOW. I. 93; RICH. 271.

aver, sb., *horse*; **averes** (*pl.*) HALLIW. 117.

avēr, sb., *O.Fr.* aveir; *property*; avere MAN. (H.) 124; aveer MAND. 292; havoir R. R. 4720.

averil, sb., *O.Fr.* avril; *April,* LAȝ. 24196; TREAT. 136; SPEC. 27.

averous, adj., *cf. Prov.* averōs (*avaricieux*); *greedy,* LANGL. B xv. 132, C xvii. 279; avarous GOW. II. 147.

avetrol, sb., *? corrupted from O.Fr.* avoltre; *? bastard,* ALIS. 2693; S. S. (Web.) 1107.

avīli, v., *O.Fr.* avilir; *make vile,* P. L. S. xi. 10; **aviled** (*pret.*) A. P. ii. 1713.

avīs, sb., *O.Fr.* avis; *advice,* ROB. 89; CH. C. T. *A* 1868; MAND. 180; MAN. (F.) 5268.

avīse, v., *O.Fr.* avīser; *advise, devise, consider,* GOW. I. 132; with alle ... þe mirþe þat men couþe avise GAW. 45; of þine answere avise þe right wel CH. L. G. W. 335.

avisement, sb., *O.Fr.* avisement; *advisement,* GOW. I. 147; CH. C. T. *E* 1531; E. W. 27.

avisiōn, sb., *O.Fr.* avision; *vision,* ROB. 255; avisioun MAND. 114.

avocat, sb., *O.Fr.* avocat; *advocate,* AYENB. 127; avocat, avoket TREV. VIII. 253.

avoi, int., *O.Fr.* avoi (*expressing surprise*); A. P. ii. 863; BEK. 2066.

avoidance, sb., *emptying,* 'evacuatio,' PR. P. 19.

avoiden, v., *O.Fr.* esvuidier, évuider; *empty; make void; avoid;* 'evacuo,' PR. P. 19; avoide WICL. 1 COR. xv. 24.

avou, sb., *vow,* TREV. IV. 49; WILL. 532; (h)is avou (*ms.* auow) he haveþ imad FER. 1818.

avouē, sb., *O.Fr.* avoé, = avocat; *advocate, patron saint*; avowe ALIS. 3160; **avowēs** (*pl.*) BEK. 2118.

avoueisoun, sb., *advowson,* BEK. 575.

avouen, v., *O.Fr.* avoer, *from Lat.* advocāre; *avow, acknowledge, declare*; avowe MAN. (H.) 320; avowin 'advoco' PR. P. 19; **avouh** (*imper.*) HOM. I. 197; **avoud** (*pple.*) AYENB. 101.

avouen, v., *O.Fr.* avoer, *from Lat.* *advōtāre; *vow, bind by a vow,* PL. CR. 847; avoue WICL. GEN. xxxi. 13.

avouerie, sb., *O.Fr.* avouerie, avoerie; *patronage, protection,* P. S. 189; avouerie, avoerie AYENB. 101, 146.

avouter, sb., *O.Fr.* avoutre, avoltre; *adulterer,* PR. P. 19; CH. C. T. *D* 1372.

avoutrie, sb., *O.Fr.* avoutrie; *adultery,* LANGL. B ii. 175; CH. C. T. *D* 1306; MAND. 54.

await, sb., *O.Fr.* await, agvait; *ambush,* CH. C. T. *D* 1657; **awaites** 'insidiae' TREV. IV. 165.

awaite, v., *O.Fr.* agvaitier; *await,* LANGL. B x. 333 (*A* xi. 222 *has* aspie); out he ȝeode for te awaite what þat wonder were BEK. 86; **awaiteð** (*pres.*) A. R. 174; awaiteþ C. L. 767; **aweited** (*pple.*) WICL. EX. xxi. 13.

a-wāken, v., *O.E.* *awacan (*pret.* awōc); *awake*: awake (? =awakie) P. L. S. xiii. 323; CH. L. G. W. 1332; **awōc** (*pret.*) LAȝ. 1253; awok ROB. 15; awok, awook LANGL. *A* viii. 128, *C* x. 293; CH. C. T. *A* 3364; Olimpias of slepe awok ALIS. 355; þe clerkes awoke P. L. S. x. 61; **awáke** (*pple.*) C. L. 473; REL. I. 144.

a-wākien [awákie], v., *O.E.* awacian, = *O.H.G.* erwachen; *awake,* LAȝ. 17914; he mai awakien LEB. JES. 692; ich wolde awakien þe A. R. 238; awakie MISC. 192; **awáke** (*imper.*) LANGL. B v. 399; **awákede** (*pret.*) A. R. 236; awakede AYENB. 128; awakede, awaked LANGL. B xiv. 332; (hi) awakede hine MISC. 32; **awáked** (*pple.*) CH. BOET. iii. 11 (106); after he was awaked, WILL. 679.

a-wākenen, v., *O.E.* awacnian, awæcnian; *awaken (intrans.),* JUL. 11; þet god muwe þer of awakenen A. R. 44; **awákeneð** (*pres.*) H. M. 27; **awákned** (*pret.*) LANGL. B xix. 478; **awákened** (*pple.*) A. R. 58.

a-wálden, v., = *M.H.G.* erwalten; *govern, subdue,* LAȝ. 4345; awealden (? = awelden) JUL. 69; awolde P. L. S. xvii. 335; **aweald** (*imper.*) KATH. 652.

awane, *see* andwane.

a-wappen, v., *? astonish*; **awaped** (*pple.*) ALIS. 899; AR. & MER. 3240; awhaped CH. L. G. W. 814.

awarden, v., *O.Fr.* awarder; *award*; **awarde** (*pres.*) CH. C. T. *C* 202.

a-warien, v., *O.E.* awergian, awyrgian; *curse,*

ban, A. R. 284; **awariede** (*pret.*) LAȝ. 13946; **awaried** (*pple.*) R. S. IV; awaried MISC. 45; awirigd HOM. I. 223; awariede wihtes KATH. 1066; his beoden beoð aweriede HOM. I. 49; ge aweregede gostes II. 69; ine awarȝede glednesse AYENB. 27; *see* **awürien.**

a-warnie, v., *? for* ȝewarnie; *defend*, FER. 4027.

a-waschen, v., *O.E.* awascan, = *O.H.G.* arwaskan; *wash away*; **awescen** (*pple.*) HOM. I. 37.

awe, *see* eȝe *and* eowe.

a-wecchen, v., *O.E.* aweccan, = *O.H.G.* arwekkan. uswakjan; *arouse, awaken* (*trans.*), HOM. I. 267; awecche REL. II. 278; awechche LEB. JES. 691; **awecð** (*pres.*) HOM. II. 179; **aweccheð** (*imper.*) MAT. x. 8; **awahte** (*pret.*) KATH. 1044; aweiȝte ALIS. 5858; hio awehten hine MK. iv. 38; awehten, aweihten LAȝ. 811, 17805; **awaht** [aweht] (*pple.*) LAȝ. 4520.

a-wēden, v., *O.E.* awēdan; *rave, be mad,* LAȝ. 22020; awede SHOR. 163; WILL. 45; TRIST. 3181; B. DISC. 957; **awēdeþ** (*pres.*) O. & N. 509; **awědde** (*pret.*) ROB. 162; **awěd** (*pple.*) A. R. 96; he was neȝ awed BEK. 1486; awedde hundes LAȝ. 19538.

a-wéȝen, v., *O.E.* awegan; *bear up*; **aweigeð** (*pres.*) HOM. II. 181; þu awiðhst HOM. I. 233.

awei, *see under* wei.

awel, sb., *O.E.* awul, awel, āl, æl, eal, *cf. O.H.G.* āla *f.*, *O.N.* alr *m.*, *awl*; owul '*fuscinula*' FRAG. 4 (VOC. 93); owel, ewel O. & N. 80; al '*subulam*' WICL. DEUT. xv. 17; el A. R. 324; **eawles** (*pl.*) KATH. 2209; eaules MISC. 153; aules A. R. 212; oweles LEG. 159; oules BRD. 22; MAP 339; CH. C. T. *D* 1730; eles AYENB. 66; eaule (*dat. pl.*) MISC. 83.

a-wélden, v., *govern, subdue*, HOM. I. 81; awelden, awilden LAȝ. 4083, 27327; awelde REL. I. 184 (MISC. 128); TREV. I. 91; **awéldeð** (*pres.*) A. R. 144; awalde, awælde, **awelde** (*pret.*) LAȝ. 1643, 4663, 6167; **awald** (*pple.*) LAȝ. 4146; awild 12084*; awealt KATH. 555.

a-weleȝen, v., *? make* or *becomerich, luxurious*; heo wolde elles awilegen A. R. 176; hwit awilegeð (?) þe eien 282; nes þer nan swa wræcche Brut þat he nes **awælȝed** LAȝ. 22718.

a-welten, v., *O.E.* awyltan; *roll away*; awelte (*pret.*) MAT. xxviii. 2.

a-wemmen, v., *injure*; awæmmen LAȝ. 21290; **awemmed** (*pple.*) HOM. I. 83; REL. I. 128; LAȝ. 2212.

āwen, *see* āȝen.

a-wendan, v., *O.E.* awendan, = *O.H.G.* irwendan, *Goth.* uswandjan; *turn away*, HOM. I. 109; awende (*pret.*) HOM. I. 219; **awend** (*pple.*) LAȝ. 1973*; Thomas . . . out of mi lond is awend BEK. 1240; awent TREV. V.

229; hi alle wurðon awende of þan feȝre hiwe þe hi an ȝescapen were to loðlice deoflen HOM. I. 219.

awene, awne, sb., *O.N.* ögn, *O.H.G.* agana, *Goth.* ahana; *awn, ear of corn*, '*arista*,' PR. P. 18; agunes (*pl.*) VOC. 155.

a-weorpen, v., *O.E.* aweorpan; *cast out, cast down*; aworpen A. R. 122; awarpen KATH. 590; **awerpeð** (*pres.*) HOM. I. 25; aworpeð 113; awarpeþ REL. II. 210 (P. L. S. xxxii. 2); **awurpen** (*pret.*) MAT. xxi. 39; aworpen (*pple.*) A. R. 278.

awer-mōd, sb., *? pride*, (ms. awwermod) ORM. 4720.

a-wēsten, v., *O.E.* awēstan; *lay waste*; **awēsteð** (*pres.*) HOM. I. 13; **awēste** (*pret.*) LAȝ. 6452; **awēst** (*pple.*) LAȝ. 2149.

ā-wiht, pron., *O.E.* āwiht, āuht, āht, = *O.H.G.* eowiht; *aught, anything*, HOM. I. 103; þat he . . . him awiht hafde kiþed ORM. 16979; þat miht oht (ms. ohht) angren oþre 432; ȝif heo wes awiht hende LAȝ. 7027; þat ævere aht cuðe þer on 11986; þe heom oht lufeden 25346; eawiht HOM. I. 3; owiht, out [eawicht] A. R. 88, 124; auȝt LANGL. (Wr.) 3331; aught RICH. 2460; aȝt AYENB. 194; SHOR. 95; eawt KATH. 1193; ouht, oȝt O. & N. 662; ouȝt, out WILL. 952, 1823; TREV. III. 39; oht SPEC. 71; oȝt HORN (L.) 976; BEK. 1422; ogt GEN. & EX. 1793; am i oght that shreu TOWNL. 180.

a-wilden, v., *O.E.* awildian; *grow wild*, A. R. 176*.

[**a-wildgen**], **a-wileȝen**, v., = **awilden**; *grow wild; bewilder, dazzle*, A. R. 176; hwit awilegeð þe eien 282.

a-winnen, v., *O.E.* awinnan, = *O.H.G.* arwinnan; *win, acquire*, LAȝ. 18761; HORN (L.) 1071; awinne FL. & BL. 132; P. L. S. xvii. 99; AYENB. 85; REL. II. 243; S. S. (Web.) 1822; **awonne** (*pple.*) LAȝ. 605*.

a-wīten, v., *know; watch over*; **awiste** (*pret.*) LAȝ. 27264; awuste HOM. I. 288.

a-wláten, v., *disgust*; **awláted** (*pple.*) ROB. 485.

a-wlenchen, v., *make splendid*; **awlencð** (*printed* awleneð) hire mid cloðes REL. I. 129 (HOM. II. 163).

awnen, v., = *M.H.G.* ougenen; *show*; (ms. awwnenn) ORM. 979; & sone an an se þis was þær þurh godes engles awned 3385.

a-wrāðien, v., *become wroth* : ne noht so glad þat hit ne **awrōþeþ** O. & N. 1278; **awrāððed** (*pple.*) LAȝ. 24834.

a-wrecchen, v., *pull out* : & habbeð al mi kinelond **awræht** (*pres.*) ut of mire hond LAȝ. 15435; whan þe limes beeþ **awreiȝt** ('*dislocantur*') out of here oune places TREV. II. 181.

a-wréken, v., *O.E.* awrecan, = *O.H.G.* errechan; *reject*; *avenge* : þat ilke þat Howel

haf(ð) ispeken ne scal hit na man awreken
LAȝ. 25194 ; awreken sunne A. R. 334 ; awreke
AYENB. 9 ; RICH. 1771 ; he mihte . . . ful sone
hine awreke MISC. 44 ; **awrec** (*imper.*) þe nu
on me R. S. ii. (MISC. 162) ; **awrēc** (*pret.*)
A. R. 334 ; **awréken** (*pple.*) TRIST. 2446 ;
awreke O. & N. 262 ; awreke ROB. 142 ;
ALIS. 2519 ; he wolde ben awreke FER. 1821 ;
awreken, awroken CH. C. T. *A* 3752, *F* 784 ;
awroke LANGL. *B* vi. 204.

a-wrīten, v., *O.E.* awrītan ; *write out* ; **awrīt**
(*pres.*) FRAG. 2 ; **awrāt** (*pret.*) HOM. I. 13 ;
awriten (*pple.*) FRAG. 7 ; hit is awriten on
boken HOM. I. 113.

a-wundrien, v., *O.E.* awundrian ; *amaze, surprise* : seint Gregorie awundreð him A. R.
146 ; **awundrede** (*pret.*) KATH. 312 ; awon-
drede MISC. 32 ; **awundred** (*pple.*) LAȝ.
15972 ; awundred KATH. 1249 ; awondred
FER. 3197 ; WILL. 872 ; of mi tale ne beoþ
noght awondred HALLIW. 123.

a-würien, v., *O.E.* awyrgan, = *O.H.G.* er-
wurgen ; *worry, choke, devour* ; awirien PL.
CR. 662 ; þe vox **awürieð** al enne floc A. R.
202 ; awurieþ (*? printed* awarieþ) MISC. 149 ;
awüried (*pple.*) A. R. 252.

a-würðen, v., *O.E.* aweorðan, = *O.H.G.*
erwerdan ; *turn (into)* ; *come to nothing* ;
awürðeð (*pres.*) A. R. 200 ; to blisse hit
awürðe (*subj.*) LAȝ. 25580 (*sec. text has*
teorne).

a-würðien, v., *O.E.* aweorðian ; *honour* ; a-
würðede (*pret.*) LAȝ. 9529.

ax, axe, *see* eax. **axe,** *see* asche.

axe, axen, -ere, *see* aske, etc.

[**axel,** sb., *O.N.* öxull ; *axle.*]

  axel-trē, sb., *O.N.* öxultrē ; *axle-tree,* C. M.
21268 ; axille tre MAND. 181 ; axiltre PR. P.
20 ; exiltre VOC. 278.

axes, *see* acces.

axien, *see* askien.

b.

bā, *see* bēȝen.

ba, v., *kiss,* CH. C. T. *D* 433.

baban, sb., *infant,* A. R. 234.

bábe, sb., *cf. Swed.* babe (IHRE) ; *babe,* GOW.
I. 344 ; H. V. 1 ; LUD. COV. 84 ; bab TOWNL.
149.

babel, sb., *bauble,* P. S. 335 ; GOW. III. 224 ;
babul, babil ' *librilla* ' PR. P. 20.

babelavante, sb., *? jester,* CHEST. II. 34.

babeuri, sb., *?* = babwinerie ; *grotesque
absurdity* ; baberi D. TROY 1563 ; **babeuwries**
[*v. r.* babeuries, rabewiures] CH. H. F. 1189.

babi, sb., *baby,* LANGL. *B* xvii. 95 ; **babies**
(*pl.*) GOW. I. 268.

babir-lippid, adj., *large lipped,* ' *labrosus,* '
PR. P. 20 ; baberlipped LANGL. *B* v. 190.

babischen, v., *O.Fr.* baubiss-, baubir ; *scoff
at* ; **babished** (*pret.*) TOWNL. 78.

bablen, v., = *Ger.* babeln, *Dan.* bable ; *babble* ; babelinde (*pple.*) A. R. 100 ; **bablede**
(*pret.*) LANGL. *A* v. 8 (Wr. 2487).

babuin, babewin, sb., *O.Fr.* babuin ; *baboon,*
PR. P. 20 ; **babewines** (*pl.*) MAND. 210 ;
baboines A. P. ii. 1409.

babwinrie, sb., *cf. Fr.* babouinerie ; *from
babouin* (*baboon*) ; *grotesque ornamentation,*
WICL. E. W. 8.

bac, adv., *back,* MISC. 228.

bac, sb., *O.E.* bæc, = *O.L.G.* bac, *O.N.* bak
n., *O.Fris.* bek *m.; back,* A. R. 290 ; HAV.
556 ; þe deor feol a bac (*aback*) LAȝ. 6493 ; teh
hine a bac ward 20086 ; a bac TREAT. 134 ;
bac & side ORM. 4776 ; sant Johan droh
him o bach 10656 ; bak MAN. (H.) 133 ;
TOR. 2590 ; have he turned þe bak P. S. 339 ;
on bak ANT. ARTH. xlv ; a bak O. &
N. 877 ; JOS. 496 ; RICH. 4716 ; LUD. COV.
211 ; bæch LK. ix. 62 ; a bak (bakcloþ ?)
to walken in CH. C. T. *G* 881 ; a bec HOM.
I. 239 ; **bake** (*dat.*) H. H. 54 ; **bakkes** (*pl.*)
clothes, LANGL. *B* x. 362 ; and alle his bakkes
rente WILL. 2096.

  bac-bīten, v., *backbite* ; bacbite MIRC 1267 ;
bakbite LANGL. *B* ii. 80 ; **bacbite** (*pple.*)
P. S. 157.

  bac-bītere, sb., *Swed.* bakbītare (RIETZ) ;
backbiter, R. S. vii ; WICL. LEV. xix. 16 ;
bacbitare A. R. 82.

  bac-bītunge, sb., *backbiting,* HOM. I. 205 ;
A. R. 82 ; bakbitinge LANGL. *B* v. 130.

  bac-bōn, sb., = *Swed.* bakbēn ; *backbone,*
VOC. 146 ; bakbōn L. H. R. 190.

  bak-clōþes, sb. pl., *clothes,* LANGL. *B* x.
362*.

  bac-dünt, sb., *blow on the back,* A. R.
290.

  bac-half, sb., *farther side,* WICL. GEN. xix.
6 ; bakhalf CH. ASTR. i. 4 (4).

  bak-sīde, sb., = *Swed.* baksīda ; *back,* CH.
ASTR. i. 15 (8).

  bac-ward, adv., *backward,* MAN. (F.) 5052 ;
bakward B. DISC. 1637 ; CH. C. T. *A* 4281.

  bac-warde, sb., *rear-guard,* LAȝ. 23814.

  bak-water, sb., *backwater,* TREV. I. 57.

bacche, sb., = *Swed.* bak *n.; batch, baking* ;
bahche, batche (? bacche) ' *pistura* ' PR. P. 21.

bach, *see* bæch.

bacheler, sb., *O.Fr.* bacheler ; *bachelor,* ROB.
30 ; bacheler, bacheleer CH. C. T. *A* 80 ;
wheþer he be kniȝt or bachiler WILL. 840.

bachelerie, sb., *O.Fr.* bachelerie ; *the body of
bachelors, state of a bachelor,* ROB. 76, 191 ;
CH. C. T. *E* 270.

bacin, sb., *O.Fr.* bacin ; *basin,* FL. & BL. 563 ;
SHOR. 51 ; S. S. (Web.) 2242 ; RICH. 2557 ;
basin ' *pelvis* ' VOC. 178 ; ALIS. 2333 ; JOS.

697; **basins** (*pl.*) of bras P. S. 189; bascins MARH. 9.

bacinet, basinet, sb., *O.Fr.* bacinet (*bassinet*); *helmet,* RICH. 403, 5266.

backe, sb., = *Dan.* (aften)bakke; *bat,* WICL. LEV. xi. 19; bakke '*vespertilio*' PR. P. 21; LIDG. M. P. 152.

backen, v., *from* bac; *clothe*; bakken [bak] LANGL. *A* xi.• 185.

bacōn, bacoun, sb., *O.Fr.* bacon [*M.Du.*, *M.L.G.* bake, *O.H.G.* bacho *acc.* bachun]; *bacon,* LANGL. *B* v. 194; CH. C. T. *D* 1753; bacun '*petaso*' PR. P. 20; **bacouns** (*pl.*) FER. 2696.

badde, sb., *? cat*; **baddis** [*v. r.* baddeʒ] ALEX. 1763.

badde, adj., *bad,* WILL. 5024; ALIS. 2118; S. S. (Web.) 643; TREV. VII. 343; CH. C. T. *E* 1608; our biʒete was badde REL. II. 242; þis badde king ROB. 108; two badde prestes LANGL. *B* x. 281; **baddere** (*compar.*) CH. C. T. *F* 224.

> **badde-līche,** adv., *badly,* ROB. 566.

, **baddenesse,** sb., *badness,* LANGL. *B* xii. 49.

bāde, sb., = *M.H.G.* beit *f.*; *from* bīden; *dwelling, delay,* TRIST. 345; bode HOM. II. 185; **bāde** (*dat.*) C. M. 7399; PERC. 41; bode FLOR. 374; CH. AN. 119; boute bod(e) WILL. 149; *comp.* a-bād.

bādien, v., = *M.L.G.* beiden, *O.H.G.* beiton; *wait*: badien þe king wolde þat his folc come LAʒ. 25649.

bæch, sb., *valley,* LAʒ. 2596; **bæche,** bache (*dat.*) 757, 5644; over baches and hulles LANGL. *C* viii. 159; **bæcchen** (*dat. pl.*) LAʒ. 21776; *comp.* cou-bache.

bæften, *see* biæften.

bæl, bāl, sb., *O.E.* bæl, *O.N.* bāl, *mod.Eng.* bale (-fire); *funeral pile, bonfire*; bæl LAʒ. 17130; bal WR. DICT. 153; **bāle** (*dat.*) M. H. 169; in a bale of fiir PL. CR. 667.

bǣlen, v., = *O.N.* bǣla; *burn*: bolne & bele A. P. i. 18.

[**bǣlien,** v., *O.E.* (a)bǣligan, bylgan, *from* belʒen; *comp.* a-bælien.]

bēre, sb., *O.E.* bǣr, = *O.L.G.,* *O.H.G.* bāra, *O.Fris.* bēre; *from* beren; *bier,* '*feretrum,*' FRAG. 4; ORM. 8167; bære [bere] LAʒ. 19481; bere VOC. 193; MISC. 101; GEN. & EX. 2481; FL. & BL. 14; BEK. 899; P. L. S. xvii. 196; GOW. III. 263; LIDG. M. P. 34; ANT. ARTH. xiv; TOWNL. 232; bere, beere CH. C. T. *A* 2900; beere PR. P. 32; **bēren** [beres] (*pl.*) LAʒ. 27876; *comp.* hors-bēre.

> **bǣr-disc,** sb., = *O.E.* bǣrdisc; '*ferculum,*' FRAG. 4.

bēr-lēp, sb., *basket,* WICL. MK. viii. *in* SEL. W. I. 17.

bēr-man, sb., *O.E.* bǣrman; *porter,* LAʒ. 3317; HAV. 868.

bærm, *see* bearm.

bærn, *see* bearn.   **bærnen,** *see* brennen.

bǣwen, v.? = *M.H.G.* bæn (*fovere*); *foment* : to clensen & to bæwen ORM. 15153; **bǣweþ** (*pres.*) ORM. 19719.

baffen, v., *beat, strike*: apon þair brestes fast þai beft M. H. xviii; **beft** (*pple.*) GOLAG. & GAW. 870.

baffin, v., = *Du.* baffen, *Ger.* bäffen; *bark,* '*latro,*' PR. P. 20.

> **baffing,** sb., *barking,* P. P. II. 53; baffinge PR. P. 20.

baften, *see* bi-æften.

bagage, sb., *O.Fr.* bagage; *luggage,* P. R. L. P. 18; baggage DREAM 101.

bagge, sb., *cf. O.N.* baggi *m.* (*pakke, bylt,* FRITZN.); *bag,* '*sacculus,*' PR. P. 21; TRANS. XVIII. 28; TREV. I. 257; PL. CR. 223; H. S. 501; beren bagge on bac A. R. 168; a bagge ful of eiren P. L. S. xii. 57; a beggeres bagge LANGL. *B* xiv. 248; **baggen** (*pl.*) & packes A. R. 168; bagges P. S. 150.

> **bagge-ful,** sb., *bagful,* P. L. S. xii. 57.

> **bagge-pīpe,** sb., *bagpipe,* PR. P. 21; CH. C. T. *A* 565.

bagge, bage, sb., *badge,* PR. P. 20; **bagges** (*pl.*) HALLIW. 132; bages ALEX. 4180.

baggen, v., *squint, look away*: þat baggeth (?) foule CH. D. BL. 623; þat þei baggen (?) not þer fro WICL. S. W. I. 191.

> **bagging-lī,** adv., *with a side glance,* R. R. 292.

baggin, v., *become baggy,* '*tumeo,*' PR. P. 21.

> **bagged,** pple., *pregnant*: þe mere was bagged wiþ fole PERC. 717.

baghel, sb., *O.N.* bagall, *from Lat.* baculus; *bishop's crosier,* P. S. 307.

bai, adj., *O.Fr.* bai, *Ital.* baio, *Lat.* badius; *bay* (*colour*), CH. C. T. *A* 2157; B. DISC. 1044.

bai, sb., ? = abai; *bay, bark,* GAW. 1582.

bai, sb.? = *O.N.* bagr; '*obstaculum,*' PR. P. 21.

baiard, sb., *O.Fr.* baiart; *bayard, bay horse,* CH. C. T. *A* 4115.

baie, sb., = abai, *barking; holding at bay*: bai of bor ALIS. 200; bestis baie LUD. COV. 180; baie WILL. 35.

baie, sb., *Fr.* baie, *Ital.* baia; *bay* (*of the sea*), TREV. I. 57; bai (*opening in a wall*) A. P. ii. 1392; PALL. ii. 198.

baie, sb., *Fr.* baie; *berry*; bai '*bacca*' PR. P. 21; **baies** (*pl.*) PALL. ii. 414.

> **bai-trē,** sb., *bay-tree,* VOC. 181.

baiin, v., *O.Fr.* baier; *bay, bark at,* PR. P. 21; baie EGL. 286; **baien** (*pres.*) GAW. 1909.

baile, bailí, sb., *O.Fr.* baille (*barrière, porte avancée*); *bailey,* C. M. 10023; bali A. P. i. 1082; wern put . . . in baile (*prison*) FER. 1211; **bailes** (*pl.*) C. L. 687.

baili, sb., *O.Fr.* baili; *bailiff,* WILL. 5387; C.

M. 11195; bailif ROB. 499; baillif CH. C. T. A 603; **bailifs** (*pl.*) AYENB. 122.

bali-schepe, sb., *office of bailiff,* '*balliatus,*' PR. P. 22; bailliship WICL. S. W. I. 22.

baillie, sb., *O.Fr.* baillie; *jurisdiction, administration,* BEK. 202; P. L. S. xxiv. 78; ALIS. 7532; R. R. 4302; MAN. (F.) 5243; bailie JARB. XIII. 166; **bailies** (*pl.*) AYENB. 26.

bain, adj., *O.N.* beinn, *prov. Eng.* bain; *direct, prompt,* TRIST. 708; IW. 766; GAW. 1092; AMAD. (R.) xlvii; AV. ARTH. lviii; TOWNL. 28; LUD. COV. 173; bein AMAD. (Web.) 514; beane CHEST. I. 50; beaine II. 181; *comp.* un-bain.

bained, adj., *burst*; bainid, as benis or pesin, '*fresus,*' PR. P. 21.

baisk, adj., *O.N.* beiskr; *harsh,* WR. DICT.152; beȝsc ORM. 6698; bask WICL. S. W. III. 42.

baisment, sb., = abashment; *fear,* A. P.i.174.

baissen, v., = abassen; *abash, be abashed*; **basshede** (*pret.*) WICL. JOS. ii. 11; **baist** (*pret.*) GAW. 376; **baiste** (*pple.*) D. ARTH. 2856; basschede 2121.

baite, sb., *O.N.* beita; *from* bīten; *bait, food,* C. M. 16931; GOW. I. 310; beite HALLIW. 172; beit LIDG. M. P. 218.

baiten, v., *O.N.* beita, = *O.H.G.* beizen; *bait, make to bite*: if knight let his eien baiten (*feast*) on ani woman CH. TRO. i. 192; baite WILL. 1723; TOR. 1566; beiton '*commor- deo*' PR. P. 29; þe bere beite HAV. 1840; wilde bueres bete A. D. 237; þeȝ dursten beȝten (*punish*) men for æþelike gilte ORM. 10171; baite (*pres. subj.*) MAND. 243; **baited** (*pple.*) A. P. ii. 55; *comp.* a-baiten.

baiting, sb., *O.N.* beiting; *baiting,* ALIS. 199; beitinge PR. P. 29.

baiþen, v., *O.N.* beiða; *? make to obtain*; i shal baiþen þi bone GAW. 327; I baiþe hit ȝow neuer to grante 1840; þat bai(þ)eþ me mi bone SPEC. 27.

bak, *see* bac.

báken, v., *O.E.* bacan, = *O.N.* baka, *O.H.G.* pachan, bachan; *bake*; bakin, bake PR. P. 21; bake CH. C. T. A 384; **bákest** (*pres.*) ORM. 1566; **book** (*pret.*) WICL. IS. xliv. 19; P. R. L. P. 191; boke REL. I. 83; **báken** (*pple.*) RICH. 3613; baken ORM. 993; bakin, bake PR. P. 21; bake mete CH. C. T. A 343; ibake LANGL. A vi. 285; LIDG. M. P. 85; tiles wel ibake CH. L. G. W. 709.

bákere, sb., *O.E.* bæcere, = *O.N.* bakari; *baker,* VOC. 176; E. G. 354; **bákares** (*pl.*) MISC. 189; bakers LANGL. A iii. 70.

bakestre, sb., *O.E.* bæcestre (*fem.*); *baker,* '*pistor,*' FRAG. 4; baxtere HALLIW. 152; **bakesteres** (*pl.*) LANGL. B iii. 79.

bak-ern, sb., *O.E.* bæcern; *bakehouse,* '*pis- trinum,*' FRAG. 4 (VOC. 93).

bak-hous, sb., *O.E.* bæchūs, = *M.H.G.*

bach-, bachūs; *bakehouse,* '*pistrina,*' PR. P. 21; bachous '*pistrinum*' VOC. 204; E. W. 73.

bákinge, sb., *baking,* '*pistura,*' PR. P. 21.

bal, sb., *O.N.* böllr, = *M.H.G.* bal, *gen.* balles; *ball,* TREAT. 137; TREV. III. 425; S. S. (Wr.) 2066; RICH. 4506; (balle?) CH. C. T. A 2614; þane little bal AYENB. 179; **balle** (*dat.*) TREAT. 134; **balles** (*pl.*) LAȝ. 17443; ALIS. 6481.

bal-pleiere, sb., *ball player,* PR. P. 22.

bal-pleowe, sb., *ball-play,* A. R. 218.

bāl, *see* bǽl.

baláde, sb., *O.Fr.* balade; *ballad,* GOW. I. 133.

balan, sb., *? O.Fr.* balin; *a strong stuff made of tow,* ALEX. 4851.

balance, sb., *O.Fr.* balance; *balance,* ROB. 200; AYENB. 30; belaunce WILL. 948.

bald, balden, *see* beald, bealden.

baldemoin, sb., *gentian,* GOW. I. 99; PR. P. 22.

bále, sb., *O.Fr.* bale; *bale,* '*bulga,*' PR. P. 22; A. P. iii. 157; **báles** (*pl.*) FER. 4202.

bále, adj., *O.E.* bealu, = *Goth.* *balws; *baleful, evil, pernicious,* MIRC 1383; bale drinch HOM. I. 283; wið se bale bere KATH. 2370; bale deeþ LANGL. A xviii. 35; bale stour A. P.˙iii. 426; bale biernez D. ARTH. 1483.

bále, balu, bælu, sb., *O.E.* bealu, bealo, = *O.L.G.* balu, *O.H.G.* balo, *O.N.* böl *n.; bale, destruction, calamity,* LAȝ. 1455, 2597, 19519; þat baleu MISC. 42; & tet beali [bali] blencte JUL. 72, 73; bale H. M. 3; GEN. & EX. 68, 1984; HAV. 327; SPEC. 26; S. S. (Web.) 702; WILL. 75; LANGL. *B* xviii. 200; GAW. 2419; IW. 3062; PR. C. 6465; AMAD. (R.) iv; hwone þe bale is alre hecst þonne is þe bote alre necst O. & N. 687, 699; **balewes** (*gen.*) MISC. 97; **balewe,** balwe, baluwe (*dat.*) LAȝ. 310, 1618, 27065; balewe [bale] KATH. 551; al þi blisse to balewe [bale] schal iwurþen REL. I. 183, 185 (MISC. 126, 127); after bale comeþ bote FL. & BL. 821; þat is falle bale leche AL. (T.) 179; **baluwen** (*gen. pl.*) LAȝ. 9685; **balewen** (*dat. pl.*) FRAG. 5.

bale-band? sb., *cruel bond*: **balebondes** (*pl.*) (? bale bondes) MARH. 13.

bale-dünt? sb., *mortal blow*; **balidüntes** HOM. I. 281.

balu-feht, sb., *mortal fight,* LAȝ. 5943.

balu-ful, adj., *O.E.* bealofull; *baleful,* LAȝ. 24938; HOM. I. 215; baluh-, baleful A. R. 114.

bale-lēas, adj., *O.E.* bealulēas; *innocent*; balelez A. P. iii. 227.

balu-ræs [baloureas], sb., LAȝ. 25936.

baluh-sið, sb., *O.E.* bealusið; *evil fate,* HOM. I. 200; balesið MARH. 23; to his bale- siðe LAȝ. 567; **balesíþes** (*pl.*) REL. I. 179.

bale-stour, sb., *death struggle*: bed me bilive mi balestour A. P. iii. 425.

bælu-wīs, adj., *O.N.* bölvīss (*ms.* bæliwis); *wicked, baleful,* LAȝ. 17130.

balei, sb., *O.Fr.* balai; **baleis** (*pl.*), *rods,* SHOR. 47; LANGL. *A* x. 176.

baleisen, v., *flog with rods*; **baleised** (*pret.*) LANGL. *B* v. 175.

balēne, sb., *O.Fr.* baleine, *Lat.* bàlaena; *whale,* VOC. 222; **baleines** (*pl.*) TREV. II. 13.

bales, sb., *O.Fr.* balais; *balas, ruby,* CT. LOVE 78; balas QUAIR 46.

balȝ, adj., *flat-topped* : balwe '*planus*' PR. P. 22; a balȝ berȝ GAW. 2172; his balȝe haunchez 2032.

bali, *see* **beli.**

balinger, sb., *A.Fr.* balenger, = *O.Fr.* baleinier (*whale-ship*); *a sea-going vessel*; **balingers** (*pl.*) N. E. D.

baliste, sb., *Lat.* ballista; **balistis** (*pl.*) WICL. 1 MACC. vi. 20.

balke, sb., *O.E.* balca, = *O.Fris.* balka, *O.L.G.* balco, *O.H.G.* balco, balcho, *O.N.* bjalki; *balk, beam; ridge in a field; 'trabs, porca,'* PR. P. 21, 22; CH. C. T. *A* 3920; TOWNL. 99; LUD. COV. 343; **balkes** (*pl.*) '*trefs*' VOC. 170; LANGL. *A* vi. 109.

balkin, v., = *Swed.* balka; *balk, plough awry,* '*porco,*' PR. P. 22; but so wel halt no man þe plogh þat he ne **balkeþ** oþer while GOW. I. 296.

ballard, sb., *baldheaded man,* TREV. I. 241; WICL. 4 KINGS ii. 23.

balle, sb., = *M.H.G.* balle; *ball,* '*pila,*' VOC. 240; OCTAV. (H.) 1272; C. M. 16788; *see* **bal.**

balled, adj., *bald,* TREV. I. 283; LANGL. *B* xx. 183; CH. C. T. *A* 198; ballid '*calvus*' PR. P. 22; WICL. LEV. xiii. 41; ballid and bar was þe reson DEP. R. iv. 70.

balled cōte, sb., *bald coot,* '*blarie,*' VOC. 165.

balledness, sb., *baldness,* ROB. 482*; REL. II. 56; ballidnesse '*calvities*' PR. P. 22.

ballen, v., *O.Fr.* baller, baler; *dance*; bale C. M. 13139.

ballok, sb., *O.E.* bealluc, *prov.Eng.* ballock; '*testiculus,*' VOC. 208; **ballokes** (*pl.*) MAND. 162; WICL. LEV. xxii. 24.

ballok-cod, sb., *scrotum,* VOC. 208.

ballok-knīf, sb., *knife with two knobs on the handle,* LANGL. *B* xv. 151.

ballok-stōn, sb., '*genitale,*' VOC. 208.

bealloc-wirt, sb., *orchis,* LEECHD. III. 313.

balmin, *see* **basmen.** **balsme,** *see* **basme:**

baltren, v., = *Dan.* baltre; *stumble about* ; **balteres** (*pres.*) A. P. iii. 459; **baltirde** (*pret.*) D. ARTH. 782.

balu, *see* **bale.** **bame,** *see* **basme, -en.**

ban, sb., *O.E.* bann, = *O.N.* bann, *O.L.G., O.H.G.* ban, *O.Fris.* ban, bon; *ban, edict* : (h)is ban aboute he sende ROB. 187; er þe þe ban es igred SHOR. 71; lat crie þe ban þorgh þe toun E. G. 359; **banne** (*dat.*) A. P. ii. 95; ich folȝi þan ahte manne and fleo bi nihte in heore banne O. & N. 390; *comp.* ȝe-**ban.**

bān, sb., *O.E.* bān, = *O.L.G., O.Fris.* bēn, *O.H.G., O.N.* bein; *bone,* LAȝ. 7559; HOM. I. 253; KATH. 232; TRIST. 274; bon ROB. 126; boon WICL. GEN. xxix. 14; **ban,** banes [bones] (*pl.*) LAȝ. 1603, 1875; bon FRAG. 6; O. & N. 1120; MISC. 54; CHR. E. 417; bone REL. I. 120; banes KATH. 2517; IW. 2052; bones A. R. 350; buones AYENB. 148; **bānen** (*dat. pl.*) LAȝ. 29562; *comp.* bac-, chēak-, hǣfd-, lend-, schin-bān(-bōn).

bān-fīr, sb. (*ms.* banefyre), *bonfire,* '*ignis ossium,*' CATH. 20.

bān-lēs, adj., *O.E.* bānlēas; *boneless,* KATH. 251.

bōn-schawe, sb., *sciatica,* PR. P. 44.

bān-wort, sb., *O.E.* bānwyrt, = *M.H.G.* bein-wurz (*senecio*); *bonewort (name of many distinct plants)*; '*osmunda*' VOC. (W. W.) 556; bonwort or daisie *MS. in* PR. P. 52.

band, sb., = *O.L.G., O.Fris., O.N.* band, *O.H. G.* pant, band *n.; band (bond),* IW. 2394; DEGR. 869; TOWNL. 219; irene band ORM. 19821; bond '*vinculum, ligamen*' PR. P. 43; CL. M. 72; '*instita*' VOC. 182; '*laqueus*' BEK. 1099; TOR. 318; '*foedus*' WICL. GEN. xiv. 13; MAN. (F.) 2851; hwo so brekis this bond AUD. 11; and gat of hem þe band [bond] (*obligation*) CH. C. T. *B* 1558; þe bond þat hiȝte matrimoine *A* 3094; **bande** (*pl.*) SAX. CHR. 254; bonde '*vincula*' VOC. 180; SPEC. 58; ᵹor (h)ise fon he leide in bonde GEN. & EX. 2693; bandes PR. C. 3209; slæpes bandes ORM. 2971; bondes '*bandeaus*' REL. II. 83; MARH. 13; HAV. 538; iren bondes MAND. 163; les me out of bonde SPEC. 29.

band-dogge, sb., = *banddog,* VOC. 187: bonddogge '*molosus*' PR. P. 43.

bande, *see* **bounde.**

bandōn, sb., *O.Fr.* bandon; *disposition, discretion,* R. R. 1163; bandun, baundun A. R. 338; ich am in hire baundoun SPEC. 27; *see* **abandūn.**

báne, sb., = *O.E.* bana, bona, *O.Fris.* bona, *O.L.G.* bano, *O.H.G.* pano, *M.Du.* bane, *O.N.* bani *m.* (*interfector*); *bane, destroyer, destruction,* '*exitium,*' PR. P. 22; HOM. I. 243; REL. I. 219 (MISC. 15); CH. C. T. *A* 1097; MAN. (F.) 8330; TOR. 1691; he slouȝ his fader ban(e) TRIST. 901; bane, bone KATH. 2397; MARH. 13; he wes moni ennes monnes bone [bane] LAȝ. 7554; bone A. R. 222; PERC. 1338; *comp.* hen-**bane.**

bānen, adj., = *M.H.G.* beinīn; *made of bone*; wiþ bonene wal REL. I. 112.

banere, sb., *O.Fr.* banere, baniere, *banner,* A. R. 300; baner ROB. 167; CH. C. T. *A* 966; MAN. (F.) 8445.

banerere, sb., *O.Fr.* banerier; *standard*

bearer, C. M. 12703; banerrere OCTOV. (W.) 1604.

baneret, sb., *O.Fr.* baneret; *banneret*, DEGR. 1017; **banerets** (*pl.*) ROB. 551.

baneur, sb., *O.Fr.* baneor; *standard-bearer*, ROB. 361; baneur, baniour C. M. 12723; banier, baneour LANGL. *B* xv. 428; baniour TREV. II. 215.

banischin, v., *O.Fr.* baniss-; *banish*, PR. P. 23; **banished** (*pple.*) CH. C. T. *A* 1725.

bank, sb., = *O.Fris.* bank, bonk *m.*, *O.H.G.* banch, panch *m. f.*, *O.N.* bakki *m.*; *bank* (*mound, shore*); *bench*; '*ripa*,' VOC. 239; PR. P. 23; bank *scamnum* LUD. COV. 170; bonk GAW. 700; bonk (*mound*) A. P. i. 102; **banke** (*dat.*) LANGL. *A* PROL. 8; upon þe see bonke GAW. I. 165; bonke WILL. 2718; stod uppen ane boncke (*sec. text* benche) LAȝ. 25185; **bankes** (*pl.*) TREV. II. 27; ISUM. 169; þurh bankes & þurh græfes ORM. 9210.

banker, sb., ? *O.Fr.* banquier; *cloth*, or *carpet*, *for a bench*, PR. P. 23; **bankers** (*pl.*) ANT. ARTH. xxxv.

bannen, v., *O.E.* bannan, bonnan (*pret.* bēon, *pple.* gebonnen), = *O.H.G.* bannan, *O.Fris.* banna, bonna (*pret.* bēn, bante, *pple.* bannen), *O.N.* banna (*pret.* bannăĉa); *ban, proclaim, summon, curse*; bannen, bonnien [banni] LAȝ. 7952, 8054, 19907; bannin, '*maledico, exsecro*,' PR. P. 23; banne REL. I. 177; WILL. 476; FER. 5424; LANGL. *A* i. 60; i banne þe birde þat me bar ANT. ARTH. vii; **bannede** (*pret.*) GEN. & EX. 3213; banned PERC. 2123; bonnede [bannede] his ferde LAȝ. 1763; bonneden helmes 22288; *comp.* **for-, ȝe-bannen**.

banne-note, sb., = *prov.Eng.* bannut; *walnut*, VOC. 181.

bantel, sb., ? *pillar*; **bantelez** (*pl.*) A. P. i. 991; bantels i. 1016; bantelles ii. 1459.

baptesme, baptisme, sb., *O.Fr.* baptesme; *baptism*, LANGL. *B* xi. 119, *C* xiii. 58; baptim WICL. EPH. iv. 5; baptem A. P. i. 626.

baptisen, v., *Fr.* baptiser; *baptize*; baptize ROB. 86; **baptísed** (*pple.*) JOS. 686.

    baptízing, sb., *action of baptism*, ROB. 86; C. M. 171.

baptiste, sb., *Fr.* baptiste; *baptist*, A. R. 160.

baptiste, sb., *baptism*, B. DISC. 212.

baptize, baptis, sb., *baptism*, C. M. 12754; baptise B. DISC. 1360.

bar, sb., *O.Fr.* bar; = **baron**, ROB. 544.

bar, adj., adv. & sb., *O.E.* bær, = *M.Du.* baer, *O.H.G.* bar, par, *O.N.* berr; *bare, naked, unarmed, mere*, BRD. 24; IW. 2470; þat heo miȝt(e) of þe hexte men þat lond make bar ROB. 125; bare [bar] LAȝ. 3420; bare SPEC. 50; þe bare O. & N. 56; bare eorðe & chos to bedde HOM. II. 139; beaten hire beare bodi JUL. 27; for one bare sunne P. L. S. viii. 105; mid one bare worde O. & N. 547; beð ... bare of euch blisse KATH. 847; outtaken bare

(*adv.*) two A. P. ii. 1573; on hire bare foten LAȝ. 4997; *comp.* **þrēd-bare.**

bar-fōt, adj., *O.E.* bærfōt, = *O.Fris.* berfōt, *M.H.G.* barvuoz; *barefoot*, LAȝ. 8843; HAV. 862; ORF. 226; M. H. 90; barfoot LUD. COV. 59; barvot A. R. 420.

bar-hēved, adj., = *M.H.G.* barhoubet; *bareheaded*, MAN. (F.) 3252; barhed S. & C. II. xvii; bareheed WICL. LEV. xiii. 45.

bar-legged, adj., *bare-legged*, WILL. 2767.

bare-lī, adv., *barely, plainly*, GAW. 548; MIN. iii. 38.

barenesse, sb., *bareness*, '*nuditas*,' PR. P. 24.

bare-vīs, adv., *with uncovered face*; MAN. (H.) 122.

bār, sb., *O.E.* bār, = *M.L.G.* bēr, *O.H.G.* bēr, pēr; *boar*; bar, bor LAȝ. 7503 (*miswritten* bær 1697); bare IW. 241; DEGR. 43; AV. ARTH. iii; bor A. R. 280; HAV. 1867; ROB. 132; AYENB. 69; WILL. 203; boor MAND. 238; boor, bore LANGL. *B* xi. 333; bore P. S. 151; **bāres** (*gen.*) LAȝ. 30391; **bōres** (*pl.*) HAV. 2331; bores, boores CH. C. T. *A* 1658.

bōr-sper, sb., *O.E.* bārspere; *boar-spear*, '*vena(bul)um*,' FRAG. 2.

barain, adj., *O.Fr.* baraigne; *barren*, A. R. 158; baraine GAW. 1320; barein CH. C. T. *A* 1977.

baratour, sb., *O.Fr.* barateor (*swindler*); *quarrelsome person, rioter*, PR. P. 115; *champion*, PERC. 263.

barbar, sb., *Gr.* βάρβαρος; *barbarian*, WICL. I COR. xiv. 11; barbre CH. C. T. *B* 281; **barbaris** (*pl.*) WICL. DEEDS xxviii. 1.

barbe, sb., *O.Fr.* barbe; *barb* (*of an arrow*); *sort of veil used by women*; P. L. S. xxxi. 350; GAW. 1457, 2310.

barberi, sb., *O.Fr.* berberis; *barberry*, PR. P. 23; barbere AV. ARTH. vi.

barbet, sb., *Fr.* barbette; *sort of veil*, P. S. 154.

barbican, sb., *O.Fr.* barbacane; *barbican, outwork to a fortress, '* antemurale,*'* PR. P. 23; ALIS. 1591; GAW. 793; barbecan FL. & BL. 244.

barbour, sb., *O.Fr.* barbeor; *barber*, S. S. (Web.) 1874; TREV. IV. 285; CH. C. T. *A* 2025.

barbre, see **barbar**.

bare, see **bar** *and* **barewe**.

bāre, sb., *O.N.* bāra, = *O.E.* bēre, *M.Du.* baere, *cf. mod.Eng.* bore (*tidal wave*); *wave*, M. H. 135; TRIST. 356; *cf.* **bēre**.

bareȝ, see **barȝ**.

bareinesse, sb., *barrenness*, WICL. GEN. xxvi. 1.

bareintē, sb., *barrenness, '*sterilitas,*'* PR. P. 24; WICL. GEN. xxvi. 1.

barel, sb., *O.Fr.* bareil; *barrel, '*cadus,*'* VOC. 176; CH. C. T. *B* 3083; MAN. (F.) 3616; barail P. L. S. xxiii. 23.

barelle-ferrers, sb. pl., *cf.* *O.Fr.* ferriere (*travelling-bottle*); *vessel for carrying wine on journeys* : barelle ferrers thei brochede D. ARTH. 2715 ; barell-ferreris BARB. xv. 39.

baret, sb., *O.Fr.* barat ; *debate, trouble, fraud,* A. R. 154 ; HAV. 1932 ; GAW. 353 ; baret & strif P. L. S. xxiii. 137 ; in gret baret and bale WILL. 5517 ; barat AYENB. 39.

barewe, sb., *O.E.* (meox-)berewe, ?=*M.Du.* berve (*cf.* *M.Du.* gherve = gherwe) ; *from* **béren** ; *barrow, handcart, bier* : þeȝ ich scholde beo þider ibore in barewe oþer in bere BEK. 899 ; barowe PR. P. 25 ; FLOR. 2031 ; bare GREG. 196 ; me leiden hem in bare & burden hem ful ȝare HORN (L.) 891 ; *comp.* **hand-, whēl-barowe.**

barȝ, sb., *O.E.* bearg, bearug, bearh, = *M.Du.* bargh, *O.N.* börgr, *O.H.G.* parch, parh, parc, paruc, parug ; *barrow (castrated pig)* : bareȝ [bareh] O. & N. 408 ; bæruh FRAG. 3 ; baru ROB. 207 ; **barowes** (*pl.*) D. ARTH. 191.

bargaine, sb., *O.Fr.* bargaine ; *bargain,* PR. P. 24 ; bargain AYENB. 9 ; bargein GOW. II. 223 ; **bargaines** (*pl.*) CH. C. T. *A* 282.

barganer, sb., *bargainer* : barganers and okerars and lufars of simonee TOWNL. 313.

barganiin, v., *O.Fr.* bargaignier ; *bargain,* PR. P. 24 ; bargaine WICL. S. W. II. 213 ; bargeine E. W. 472 ; bargane (*contend*) BARB. ix. 224.

barge, sb., *O.Fr.* barge, *L.Lat.* barca ; *barge,* WILL. 2767 ; CH. C. T. *A* 410.

barin, v., *O.E.* barian, = *M.Du.* baren, *O.H.G.* (gi)paron, *O.N.* bera ; *bare, strip, uncover,* PR. P. 24 ; his fader he it gan unhillen & baren GEN. & EX. 1912 ; **bared** (*pple.*) A. P. ii. 1149.

bark, sb., *O.N.* börkr, = *M.Du.* barke *m.* ; *bark (of a tree),* ‘*cortex,*’ VOC. 229 ; MAND. 189 ; IW. 741 ; barke PR. P. 24 ; LANGL._*B* xi. 251 ; **barke** (*dat.*) R. R. 7267.

barkin, v., = *Dan.* barke, *Swed.* barka ; *bark (leather), tan ; become covered with a bark ;* ‘*funio, tanno,*’ PR. P. 25 ; he barked over as a turfe C. M. 11824 ; **barkid** (*pple.*) LIDG. M. P. 53.

barkere, sb., = *Swed.* barkare ; *tanner,* PR. P. 24 ; barkare VOC. 194.

barlic, sb., *cf.* bere (*O.N.* barlak *is probably from O.E.*) ; *barley,* REL. I. 215 ; bærlic SAX. CHR. 252 : barlich, barli WICL. EX. ix. 31 ; berlei MAND. 272 ; barliche (*dat.*) TREV. I. 405.

barli-brēd, sb., *barley-bread,* VOC. 198 ; LANGL. *A* vi. 137 ; barlibreed CH. C. T. *D* 144.

barli-cake, sb., *barley-cake,* GOW. III. 216.

barli-corn, sb., *barley-corn,* WICL. 2 KINGS xiv. 30.

barliȝ-lāf, sb., *barley-loaf,* ORM. 15511.

barli-méle, sb., *barley-meal,* PR. P. 25.

barli-sēle, sb., *barley-season,* ‘*tempus ordeacium,*’ PR. P. 25 ; *cf.* **méte-sēle.**

barli-water, sb., *barley-water,* S. S. (Web.) 1574.

barm, *see* **bearm.**

barmeken, barnekinch, sb., = **barbican** : Balaan in þe barmeken sa bitterli fiȝtis ALEX. 1301 ; at the barnekinch he abad DEGR. 375.

barn, *see* **bearn.**

baronáge, sb., *O.Fr.* baronage, barnage ; *baronage,* HORN (L.) 1282 ; AYENB. 58 ; JOS. 62 ; barnage FL. & BL. 639 ; HAV. 2947 ; WILL. 4797.

baronesse, sb., *O.Fr.* baronesse ; *baroness,* PR. P. 25.

baronet, sb., *inferior baron,* ‘*barunculus,*’ VOC. 262.

baronie, sb., *O.Fr.* baronie ; *barony,* ROB. 349, 479 ; **baronies** (*pl.*) AYENB. 38.

barre, sb., *O.Fr.* barre ; *bar,* ‘*vectis,*’ PR. P. 24 ; VOC. 261 ; HAV. 1794 ; a barre of iron CH. C. T. *A* 1075 ; **barren** (*pl.*) HOM. I. 131 ; barres LANGL. *B* xviii. 319 ; þe barres of his belt GAW. 162 ; þe barren of þe burhe KATH. 2349.

barrēre, sb., *O.Fr.* barriere ; *barrier,* PR. P. 24 ; **barēres** (*pl.*) A. P. ii. 1239 ; barers FER. 4668.

barrin, v., *O.Fr.* barrer ; *bar, close with a bar,* PR. P. 24 ; **barred** (*pret.*) WILL. 2046 ; **barrid** (*pple.*) TRIAM. 1188.

Barrocschire, pr. n., *O.E.* Bearrucscire ; *Berkshire,* P. L. S. xiii. 48.

bars, sb., *O.E.* bears, bærs, = *M.Du.*, *O.Fr.* bars ; *bass* (*fish*) ; base B. B. 167 ; bace PR. P. 20.

bartrin, v., *O.Fr.* bareter ; *barter,* PR. P. 25.

baru, *see* **barȝ.**

barūn, sb., *O.Fr.* bar *acc.* baron ; *baron,* HOM. II. 35 ; HAV. 31 ; P. L. S. ii. 108 ; **barūnes** (*gen.*) LAȝ. 5319 ; **barouns** (*pl.*) AYENB. 38.

bas, adj., *O.Fr.* bas, *from late Lat.* bassus ; *base, low,* GOW. I. 98 ; bas (*low*) voice MER. 572.

baschen, *see* **baissen.**

base, sb., *Fr.* base, *Lat.* basis ; *base ;* bas CH. ASTR. ii. 41 (52) ; basse A. P. i. 999 ; **bases** (*pl.*) ii. 1278.

baselard, sb., *a sort of dagger,* ‘*pugio,*’ PR. P. 25 ; ‘*sica,*’ VOC. 263 ; baslarde LANGL. *B* iii. 303 ; MIRC 48.

baser, baser, sb., *executioner,* N. E. D.

basilisk, sb., *Gr.* βασιλίσκος ; *basilisk,* PS. XC. 13 ; **basilicok** CH. C. T. *I* 853.

basin, *see* **bacin.** **bāsk,** *see* **baisk.**

basken, v., *? O.N.* baˇcask ; *bask* : he baskeþ him about þer ine GOW. I. 290.

basket, sb., *basket,* ‘*corbis,*’ PR. P. 26 ; C. M. 4489 ; **bascates** (*pl.*) LEB. JES. 21 ; basketes CH. C. T. *C* 445.

basme, sb., *O.Fr.* basme, *Lat.* balsamum ; *balm,* H. M. 13 ; balsme GOW. III. 315 ;

bame A. R. 164; baume TREV. I. 107; MAND. 50.

basmen, v., *embalm*; balmin, baumin '*balsamo*' PR. P. 27; bame P. R. L. P. 216; balmbe CHEST. I. 165; baumed (*pple.*) ALIS. 4670; bawmede (*smeared*) WICL. JOHN ix. 6.

bass, sb., *Lat.* basium; *kiss*, CT. LOVE 797.

bast, sb., *O.E.* bæst, = *O.N.*, *M.L.G.*, *O.H.G.* bast *m. n.*; *bast*: take a stalworþe baste and binde mi handes bihind me faste HALLIW. 148; bare as a bast D. TROY 4773.

baste-trē, sb., *lime-tree*, '*tilia*,' VOC. 192.

bast, sb., *O.Fr.* bast (bāt); *bastardy*: an sone ... þat was a bast ibore ROB. 425; he was bigeten o bast AR. & MER. 7644.

bastard, sb., *O.Fr.* bastard; *bastard*, CHR. E. 890; L. H. R. 50*; TREV. I. 251.

bastile, sb., *Fr.* bastille; *tower of a castle*, PR. P. 26.

bastin, v., *O.N.* basta, = *O.H.G.* besten, *from* bast; *baste*, '*subsuo*,' PR. P. 26; *basting* (*pple.*) R. R. 104.

bastōn, sb., *O.Fr.* baston, bastun; *stave*, *stanza*, REL. II. 175; bastoun, bastun C. M. 14923.

bat, sb., ? = *Ger.* batze *m.* (*massa*); *clod*; a bat of erþe LANGL. *C* xix. 92.

bat, sb., *for* debat; *bate*, LUD. COV. 12; bate C. M. 9684.

bāt, sb., *O.E.* bāt, *cf. O.N.* bāt̔r; *boat*, JUL. 60; þene bat [bot] LAȜ. 23865; bot TRIST. 354; S. S. (Wr.) 3146; boot WICL. JOS. vi. 22; bāte (*dat.*) DEGR. 919; bote ALIS. 3497; bōtes (*pl.*) MAN. (H.) 241; bāten (*dat. pl.*) LAȜ. 14746.

bōte-swain, sb., *boatswain*, ST. R. 21.

[bāt, sb., *comp.* grist-bāt.]

bataile, sb., *O.Fr.* bataille; *battle*, O. & N. 1197; AYENB. 83; S. S. (Web.) 2011; bataile, bataille CH. C. T. *A* 879; bataille HORN (L.) 574; SHOR. 109; batel C. M. 3463.

batailed, adj., *O.Fr.* bataillie; *battlemented*, C. M. 9902; baitailed A. P. ii. 1183.

batailen, v., *O.Fr.* batailler; *engage in battle*, *fight*, CH. BOET. i. 4 (18).

batailous, adj., *battaillous*, GOW. III. 118.

[batailour], batelur, sb., *O.Fr.* batailleor; *warrior*, ALIS. 1433.

batails, sb. pl., *? battels*, APOL. 76.

batelment, sb., *battlement*, A. P. ii. 1459.

[bātien, v., *comp.* grist-bātien.]

batildōre, sb., *? O.Fr.* batadoir, *mod. Eng.* battledore; *mallet*, VOC. 269; batildoure PR. P. 27.

bátin, v., *for* abáten; *bate*, PR. P. 26; batede (*pret.*) A. P. ii. 440.

bátin, v., *for* debáten; PR. P. 26.

batren, v., *batter*; batride, batred (*pret.*)

LANGL. *A* iii. 192*; ibatrid (*pple.*) FER. 896.

batte, sb., *Fr.* batte; *bat, cudgel*, '*fustis*,' VOC. 263; PR. P. 26; botte (*dat.*) A. R. 366; battes (*pl.*) TREV. I. 381; ALIS. 78; battis WICL. MAT. xxvi. 47; botten [battes] LAȜ. 21593; bottes MISC. 43.

battin, v., *beat with clubs*, '*fustigo*,' PR. P. 26; battede (*pret.*) LANGL. *A* iii. 192.

bature, sb., *batter*, '*batura*,' CATH. 24; batere L. C. C. 38.

baðð, sb., *O.E.* bæð, = *O.N.* bað, *O.L.G.* bath, *O.Fris.* beth, *O.H.G.* bad, pad; *bath*, HOM. II. 226; R. S. v; baþþ ORM. 18044; beð A. R. 394; þet softeste beð HOM. I. 35; beþ AYENB. 74; baðe (*dat.*) LAȜ. 2852; beðe HOM. I. 23; beðes (*pl.*) A. R. 396.

bāðe, adj. & prn., *O.N.* bāðir, = *O.L.G.* bēthia, *O.Fris.* bēde, *O.H.G.* beide; *both*, SAX. CHR. 253; KATH. 1622; hio ... baðe (*earlier text* bu tu) forð eoden LK. i. 7; þis meiden was baðe faderles & moderles KATH. 77; baþe ISUM. 400; for to deme baþe þe gode & þe uvele HOM. I. 143; baþe wæren alde ORM. 250; Crist is baþe god & man 1360; baðe, boðe [boþe] LAȜ. 17014, 27305; baþe, boþe, beþe HAV. 1680, 2543, 2816; boþe C. L. 497; boþe we habbeþ stefne brihte O. & N. 1681; boþe heo were noble men ROB. 48; þat ... is boþe meke & milde 57; þe oþere boþe were him agein MAN. (F.) 610; boþe two MAP 336; CH. C. T. *A* 1716; boþe two his hondes JOS. 697; boþe tvo FL. & BL. 525; hi deieþ boþe tro TREAT. 139; boðe, beoðe A. R. 10, 162; boþe, beþe LANGL. *C* xiii. 73; beþe ALIS. 4847; bāðre (*gen.*) KATH. 1790; þurh þeȝre baþre bisne ORM. 2794; boþer LANGL. *B* ii. 66; Iw. 3556; bāðen (*dat.*) MARH. 21; boþen HAV. 471; PL. CR. 224; baðe KATH. 1791; baþe ORM. 7636; boþe SHOR. 57; mid boðe honden A. R. 338.

Baðen [Baþe], pr. n., *O.E.* Baðan; *Bath*, LAȜ. 21026; Baþe MISC. 145; CH. C. T. *A* 445.

bāðie, sb., *? one that bathes*, A. R. 214.

bāðien, v., *O.E.* baðian, = *O.N.* baða, *O.H.G.* badon; *bathe*, LAȜ. 6657; baþien P. L. S. viii. 124; R. S. v (MISC. 180); baþie LEG. 164; baþi ROB. 146; baþen MAND. 88; báþeþ (*pres.*) AYENB. 167; þe (heo) baðieð LAȜ. 17189; he ... ofte hine þer inne báðede (*pret.*) LAȜ. 17028; baþede WILL. 98; and baþede (h)is swerd in hure blod(e) FER. 3100; báthed (*pple.*) CH. C. T. *A* 3.

báthinge, sb., *bathing*, PR. P. 26.

bau, int., *cf. Ger.* ba; *bah*, LANGL. *B* xi. 135.

baubelet, sb., *O.Fr.* baubelet; *bauble*; beaubelez (*pl.*) A. R. 388*.

baud, adj., *O.Fr.* baud (*hardi, gai*); *joyous, boastful*, R. R. 5674; of wordes he was swiþe baud HORN (H.) 96.

baude, sb., *bawd,* '*leno*,' PR. P. 27; LANGL.
A iii. 124; CH. C. T. *D* 1354.

baudekin, sb., *O.Fr.* baudekin; *a precious
silk stuff,* PR. P. 27; ALIS. 759.

bauderie, sb., *gaiety, mirth,* CH. C. T. *D* 1305.

bauderik, sb., = *M.H.G.* balderich; *baldrick,*
GAW. 2486; baudrik CH. C. T. *A* 116;
baudri ALIS. 4698; **bauderikis** (*pl.*) LIDG.
M. P. 8.

baudi, adj., *dirty,* LANGL. *B* v. 197; baudi
and totore CH. C. T. *G* 635.

baudstrot, sb., *O.Fr.* baudetrot; *bawd,* VOC.
217; LANGL. *A* iii. 42*.

baume, baumen, *see* **basme, -en.**

bauson, sb., *badger,* PR. P. 27; bausin '*castor*'
VOC. 177; **bausons** (*pl.*) TREV. I. 327;
baucines WILL. 2299.

baxtere, *see* **bakestre.**

be, *see* **bī.**

bēacne, *see* **bēkne.**

bēah, sb., *O.E.* bēag, bēah, = *O.H.G.* pouc,
boug, *O.N.* baugr; *from* **būȝen**; *ring, collar,
bracelet,* '(ar)milla,' FRAG. 2; beeȝ '*torquem*'
WICL. GEN. xli. 42; enne beh LAȝ. 24520;
beiȝe TRIST. 265; begh REL. I. 160; bei
HALLIW. 171; ? beey (*printed* beeth) SHOR.
109; **bēie,** biȝe (*dat.*) L. H. R. 28, 29; **bēȝes,**
bēhȝes [bēȝes] (*pl.*) LAȝ. 7425, 21640; beges
and ringes GEN. & EX. 1390; beies HOM. I.
193; beren beȝes [beiȝes, biȝes] ful briȝte
abouten here nekkes LANGL. *B* PROL. 161;
comp. **sweor-bēah.**

beaiell, sb., *O.Fr.* besaiel; *great-grandfather,*
D. TROY 13474.

beald, báld, adj., *O.E.* beald, bald, = *O.L.G.*
bald, *O.H.G.* pald, *Goth.* *balþs, *O.N.* ballr;
bold; bald HOM. I. 257; ORM. 2185; TRIST.
997; SHOR. 123; IW. 1402; bald [bold] LAȝ.
6342; bald, bold AYENB. 100, 105, 158;
bold O. & N. 405; AN. LIT. 3; TOR. 1; be
thou bold it shal be bought M. ARTH. 3483;
beld S. S. (Web.) 2042; B. DISC. 2123; **bálde**
(*dat. m.*) LAȝ. 9613; **báldere** (*gen. dat. f.*)
LAȝ. 10387, 16045; **báldne** (*acc. m.*) 6594;
bealde (*pl.*) SHOR. 100; alle his bigginges
bolde ISUM. 78; belde HORN (L.) 602; ALIS.
5004; þurh belde worde O. & N. 1715;
báldere (*gen. pl.*) LAȝ. 17256; **bálden**
(*dat. pl.*) 25163; **báldore** (*compar.*) ROB.
509; SPEC. 27; baldore, boldare LANGL.
A iv. 94; **báldest** (*superl.*) LAȝ. 26178; *comp.*
un-báld.

bálde, adv., *M.H.G.* balde; *boldly, quickly,*
M. H. 23, 121.

bóld-hēde, sb., *boldness,* O. & N. 514; MAN.
(F.) 13465.

bálde-líche, adv., *O.E.* bealdlíce; *boldly,*
LAȝ. 24673; KATH. 719; A. R. 62; balde-,
boldeliche O. & N. 401, 1707; boldliche
AYENB. 63; **báldelíke** (*adv.*) HAV. 53; lok

hou it is and tel me boldeli (*? quickly*) CH. C.
T. *A* 3433; **báldelíker** (*compar.*) P. L. S. xi.
69.

bóldness, sb., *boldness, impudence, confidence,* D. TROY 226; LANGL. *B* xviii. 386;
MAN. (H.) 40.

báld-sipe, sb., *boldness,* LAȝ. 24943*.

bealden, bálden, v., *O.E.* bealdian, *cf. O.H.G.*
balden, *Goth.* balþjan; *embolden*: to balden
[boldi] (*encourage*) þine leode LAȝ. 4385; þe
brandes to balde PERC. 792; þe wenche bigan
to bolde S. S. (Wr.) 1679; his harte began
to bolde OCTAV. (H.) 975; **báldede** (*pret.*)
HOM. I. 273; LAȝ. 16327*; boldede hire hertes
LANGL. (Wr.) 1756; *comp.* **un-bálden.**

bealtē, sb., *O.Fr.* bealté, beauté; *beauty,* SPEC.
53; beaute WILL. 4534; beaute, beute CH.
C. T. *A* 2385.

bēam, sb., *O.E.* bēam, = *O.Fris.* bām, *O.L.G.*
bōm, *O.H.G.* baum, paum, boum, poum,
Goth. bagms, *O.N.* baðmr; *beam,* '*trabs*,'
FRAG. 4; al swa great swa a beam LAȝ.
2848; beem '*trabs, liciatorium, radius*' PR.
P. 30; LANGL. *B* x. 264; MAN. (H.) 103;
W. & I. 39; bem JUL². 46; A. P. ii. 603;
bēmes (*pl.*) ALIS. 7664; MAND. 131; CH.
D. BL. 337; grete bemes of liȝt TREV. I.
235; þe leome þa strehte west riht a seoven
bǣmen wes idiht LAȝ. 17887; *comp.* **glēo-,
mōr-, sunne-, web-bēam.**

[bēami], bēmi, adj., *radiant,* TREV. BARTH.
(N. E. D.).

bēamien, v., *O.E.* bēamian; *beam, shine*;
beme '*radio*' PR. P. 30; beeming KATH. 46.

bēane, sb., *O.E.* bēan, = *O.N.* baun, *O.H.G.*
bōna; *bean*; beene VOC. 177; bene PR. P. 30;
LANGL. *B* iii. 141*; WICL. EZ. iv. 9; al nas
wurþ a bene ROB. 497; nouȝt a bene worþ
WILL. 4754; þis Absolon ne roghte nat a
bene of al his pleie CH. C. T. *A* 3772; **bēne**
(*dat.*) MAND. 158; þe worþ of a bene P. 47;
bēana, bēane, bēanen (*pl.*) LEECHD. III.
86, 108; bene RICH. 6004; benes HAV. 769;
bene strau CH. C. T. *E* 1422; **bēnen** (*dat.
pl.*) PL. CR. 762.

bēn-codde, sb., *bean-pod,* VOC. 233; **bean-
coddan** (*dat. pl.*) LK. xv. 16.

bēanen, adj., *O.E.* bēanen; *made of beans*:
bene bred VOC. 198; WICL. E. W. 61.

béard, sb., *O.E.* beard, = *O.Fris.* berd, *M.Du.*
baerd, *O.H.G.* bart, part; *beard,* HOM. I. 279;
bærd [beord] LAȝ. 10753; berd MARTH. 9;
ALIS. 1597; GOW. II. 367; CH. C. T. *A* 332;
DEGR. 819; berd, beerd WICL. 2 KINGS
xix. 24; beerd '*barba*' PR. P. 31; bi niȝinges
beard(e) [beorde] LAȝ. 1672; uppe niȝinges
bærde 10703; **bérdes** (*pl.*) GOW. I. 111.

bérd-heer, sb., *beard-hair,* TREV. III.
325.

bérd-lēs, adj., *O.E.* beardlēas; *beardless,*
GOW. II. 369; DEP. R. iii. 235.

bếarded, adj., *bearded*; berdid '*barbatus*' PR. P. 32; beerdid WICL. E. W. 308.

bearm, sb., *O.E.* bearm, = *O.L.G.*, *O.H.G.* barm, *O.N.* barmr, *Goth.* barms; *from* beren; *bosom, lap* : in his bærm (*sec. text* lappe) he hit læide REL. I. 141; LA3. 30261; barm '*gremium*' PR. P. 25; ALIS. 4203; **barme** (*dat.*) HORN (L.) 706; B. DISC. 577; CH. C. T. *B* 3256; A. P. iii. 510; slepeð i ðe deofles berme A. R. 212; **barmes** (*pl.*) MARH. 22.

   barm-clōþ, sb., *apron*, CH. C. T. *A* 3236.

   barm-fel, sb., *apron*, REL. I. 240.

   barm-hatres, sb. pl., *aprons*, REL. II. 176.

   barm-skin, sb., *leather apron*, '*melotes*,' PR. P. 25.

bearn, sb., *O.E.* bearn, = *O.Fris.* bern, *O.L.G.*, *O.H.G.*, *O.N.*, *Goth.* barn, *Scotch* bairn; *from* beren; *child*, A. R. 38; H. M. 35; þet bearn FRAG. 5; þat bearn, bærn, bern LA3. 91, 298, 5024; barn, bærn MAT. xxii. 24; barn WILL. 9; LANGL. *A* ii. 3; MIRC 1688; MAN. (H.) 310; ANT. ARTH. xviii; bern HÓM. II. 21; MARH. 10; MISC. 128; HAV. 571; SPEC. 81; P. L. S. xxiii. 22; AM. & AMIL. 164; **bearnes** (*gen.*) FRAG. 5; barnes ORM. 8044; **bearne** (*dat.*) MAT. xxii. 25; **bearn** (*pl.*) '*sobole(s vel) liberi*' FRAG. 2; LA3. 5104; bearn, bern HOM. I. 99, 225; bærn, barnes ORM. 6808, 8040; bearnes A. R. 272; **bearne** (*gen. pl.*) MAT. xx. 20; FRAG. 8; berne SPEC. 58; **bearnan,** bearnen, bernen, bearne (*dat. pl.*) MAT. vii. 11, 17, 25; MK. iii. 28; LK. i. 17; bearnen HOM. I. 241; bernen LA3. 8061; *comp.* **helle-, kine-, mōder-, stēop-bearn** (-bern).

   barn-āge, sb., *childhood*, A. P. ii. 517.

   barn-hēde, sb., *infancy*, C. M. 166; PR. C. 8252.

   bærn-lēs, adj., *childless*, LA3. 8990; barnles C. M. 7086.

   barn-tēam, sb., *O.E.* bearntēam; *brood of children*, H. M. 31; bernteám GEN. & EX. 3748; barntem C. M. 17792.

beast, *see* bēst.   **be-,** *see* bi-.

bēaten, v., *O.E.* bēatan, = *O.N.* bauta (*pple.* bautinn), *M.H.G.* bōzen; *beat, strike, correct*, A. R. 184; KATH. 1183; mid festen hine beaten MK. xiv. 65; beate and wesse AYENB. 236; beten MISC. 45; beten and bouken LANGL. *B* xiv. 19; bete BEK. 758; MAN. (H.) 328; hire sire and hire dame þreteþ hire to bete R. S. vii; **bēateð** (*pres.*) H. M. 31; beateð, bet A. R. 184, 186; ha beat and smit AYENB. 30; biat 100; hit bet BRD. 26; (heo) beteþ MISC. 83; þei . . . beten þe stretis *they lounge about* WICL. E. W. 152; **bēt** (*imper.*) A. D. 289; **bētende** (*pple.*) GEN. & EX. 2713; **bēot** (*pret.*) A. R. 364; BRD. 24; beot, bet, bette LANGL. *B* x. 176 (*C* xii. 124); biet HOM. II. 169; beet, but TREV. VIII. 229; bet GEN. & EX. 3958; TOR.

1365; bet a doun burwes WILL. 1073; beet MAND. 83; LAUNF. 751; beet, bette WICL. NUM. xxii. 23; beoten HOM. I. 121; JUL. 26; ALIS. 7565; beoten, bieton MK. xiv. 65; LK. xx. 11; beten HAV. 1876; MAND. 46; þe wawes beote him BRD. 24; biete AYENB. 156; beete CH. C. T. *A* 4308; bute FER. 2907; þai bette M. H. 72; **bēote** (*subj.*) A. R. 364; BEK. 759; **bǣten** (*pple.*) ORM. 8168; beten GOW. III. 247; beten wiþ rede golde BEV. 1159; beten wiþ besantus ANT. ARTH. xxix; a grip of golde richeli beton on þe molde EGL. 1031; bete SPEC. 97; beete H. V. 76; bet HAV. 1916; bett TOWNL. 227; *comp.* **a-, 3e-, to-bēaten;** *deriv.* **bētel.**

   [**bēatere,** sb., *O.E.* bēatere; *beater;* *comp.* **market-bētere.**]

   bēatunge, sb., *beating*, A. R. 326; beatinge KATH. 1616.

beautē, *see* bealtē.

bec, sb., *O.Fr.* bec; *beak*, REL. I. 210; bek MAND. 48; **beke** (*dat.*) A. P. ii. 487; **bekes** (*pl.*) CH. L. G. W. 148.

bēche, sb., *O.E.* bēce, boecae (EPIN.), *M.L.G.* boke, *O.H.G.* buocha, *Lat.* fāgus, *Gr.* φηγός (*oak*) *f.*; *beech*, '*fagus*,' VOC. 181; PR. P. 27; **bēches** (*pl.*) AYENB. 23; ALIS. 5242; LANGL. *A* v. 18; *see* bōs.

bēchen, adj., *O.E.* bēcen; *made of beech*, CH. C. T. *G* 1196; in 3one bēchene wode D. ARTH. 1713.

bēcnien, *see* bēknien.

bed, sb., *O.E.* bed *n.*; *prayer*; **bede** (*pl.*) MK. ix. 29 (*earlier text* gebedu); *comp.* 3e-bed.

   bed-hūs, sb., *O.E.* bedhūs; *house of prayer*, MAT. xxi. 13.

   bed-rēp, sb., *O.E.* bedrīp; *reaping on request;* **bedrēpes** (*ms.* bederpes) (*pl.*) E. W. 27.

bed, sb., *O.E.* bedd, = *O.L.G.*, *O.Fris.* bed, *Goth.* badi, *O.H.G.* beti, peti; *bed*, LA3. 19037; MAND. 88; þet bed AYENB. 31; bedd ORM. 4418; at mi **beddis** (*gen.*) side CH. P. F. 98; **bedde** (*dat.*) O. & N. 967; CH. C. T. *A* 3269; þonne men gað to bedde LA3. 711; he lið i bedde 31591; **beddes** (*pl.*) BRD. 6; *comp.* **child-, feðer-, gras-bed.**

   bed-bere, sb., *? L.G.* bedbüre; *toral*, E.W. 41.

   bed-chambre, sb., *bedchamber*, TREV. II. 201; LANGL. *B* v. 222.

   bed-clōth, sb., *bedcloth*, PR. P. 27.

   bed-fēre, bed-ifēre, sb., *bedfellow*, GOW. II. 229; bediver (*printed* bedyner) SPEC. 491.

   bed-lawir, adj., *bedrid*, PR. P. 28.

   bed-lawir, sb., '*clinicus*,' PR. P. 28.

   bed-rēaf, sb., *O.E.* bedrēaf; *bedclothes*, FRAG. 4.

   bed-rede, adj. & sb., *O.E.* bedrida, -reda, -ryda, *cf. M.L.G.* bedderede; *bedrid*, '*clinicus*,'

VOC. 267 ; P. L. S. xxx. 57 ; CH. C. T. *D* 1769 ; MAN. (F.) 8964 ; **bedreden** (*pl.*) LANGL. *A* vii. 131.

bed-stede, sb., = *M.Du.* bedstede ; *bedstead,* PR. P. 28.

bed-strāu, sb., *bedstraw,* FRAG. 7 ; bedstre WICL. PS. vi. 7.

bed-süster, sb., *husband's concubine,* ROB. 27.

bed-tīme, sb., *bedtime,* O. & N. 324.

bedde, sb., = *O.Fris.* bedda, *M.H.G.* bette ; ? = ȝebedde ; O. & N. 1500*.

beddien, v., *O.E.* beddian, = *O.H.G.* bettōn ; *put to bed,* LAȝ. 6658 ; bedden H. M. 43 ; ORM. 2712 ; HAV. 1235 ; LANGL. *B* ii. 97 ; ne muhen ha . . . bedden in a breoste H. M. 43 ; **bedded** (*pple.*) HAV. 2771 ; **bedde,** i-bedde (*f.*) O. & N. 968.

   bedding, sb., *O.E.* bedding, = *M.H.G.* bettunge ; *bedding,* 'stramentum,' FRAG. 4 ; beddinge CH. C. T. *A* 1616.

béde, sb., = *O.L.G.* beda, *O.Fris.* bede, *O.H.G.* beta, bita, *Goth.* bida *f., mod.Eng.* bead ; *from* **bidden** ; *prayer,* 'oratio,' PR. P. 28 ; HOM. II. 135 ; LAȝ. 25514 ; ORM. 1156 ; GEN. & EX. 1375 ; HAV. 1385 ; GOW. II. 372 ; AM. & AMIL. 2351 ; þ(e)os b(e)ode HOM. I. 65 ; **béden** (*pl.*) P. L. S. xvii. 56 ; nu we & heore beden [bedes] singeð LAȝ. 19688 ; beoden biddeð LAȝ. 19722 ; beoden HOM. I. 49 ; A. R. 140 ; MARH. 8 ; bede MISC. 56 ; bedes ORM. 1748 ; AYENB. 141 ; WILL. 3024 ; if i bidde ani bedes LANGL. *B* v. 407 ; bedis *beads* PL. CR. 323 ; a peire of bedes LANGL. *B* xv. 119 ; CH. C. T. *A* 159 ; **béden** (*dat. pl.*) CHR. E. 494.

béde-hūs, sb., = *O.H.G.* betehūs ; *house of prayer,* MISC. 39.

béde-man, sb., = *M.H.G.* beteman ; *beadsman,* AL. (L.) 658 ; bedeman, beodemon *A* iii. LANGL. 46 ; beodemon A. R. 356.

béde-sang, sb., ORM. 6746.

béde-woman, sb., 'oratrix,' PR. P. 28.

Bedeford, pr. n., *O.E.* Bedanford ; *Bedford,* ROB. 4.

bedel, *see* **büdel.**

béden, v., *O.L.G.* bedōn, *O.H.G.* betōn ; *pray*: na man nalde sel bede beoden for heore saule LAȝ. 25514 ; bede OCTAV. (H.) 910 ; ic . . . bidde ȝuw & bede ORM. 18337 ; bedes HAV. 2392 ; he . . . beod & bid O & N. 1437 ; biet & bit HOM. I. 167 ; **béde** (*subj.*) LANGL. *A* viii. 102 ; his bedes **bédand** (*pple.*) night and dei HALLIW. 156.

bēden, *see* **bēoden.**

bee, *see* **bēo. beel,** *see* **būle.**

been, *see* **bēon. bēf,** *see* **boef. beffen,** *see* **baffen. bēȝ,** *see* **bēah.**

beft, *see* **baffin.**

[**bēȝe,** ? = *M.H.G.* (ge)bouge (*pliant*) *comp.* leoðebēie.]

bēȝen, adj., *O.E.* bēgen, *cf. Goth.* bai, *Lat.* ambō, *Gr.* ἄμφω ; *both* ; beiȝen, beien, beine [beje] LAȝ. 2531, 2543, 32137 ; þa kinges beie tweien (*Ital.* ambedue) 30038 ; beien SAX. CHR. 265 ; bege MAT. xv. 14 ; beie BEV. 1212 ; S. B. W. 174 ; þat . . . ȝonge were beie (*ms.* beye) ROB. 47 ; beie þe kinges P. L. S. xi. 44 ; baie AR. & MER. 1528 ; of hem beȝen (*for* bam) ORM. 15091 ; **bā** [*O.E.* ba, *cf. Lat.* ambae *f., O.E.* ba, bu, *Goth.* ba, *Lat.* ambo *n.*] ORM. 373 ; ba tunge & tale KATH. 636 ; his blod & his brain ba weoren todascte LAȝ. 1469 ; mid childe heo weren ba twa (*cf. Ital.* ambedue) 2399 ; boa 281 ; ba twa MK. ii. 22 (*earlier text* bu tu) ; ba twa his honden JUL. 48 ; ba bi dei & bi niht HOM. I. 247 ; bo FL. & BL. 547 ; P. S. 157 ; ALIS. 6763 ; S. S. (Web.) 304 ; bo þin(e) eȝe O. & N. 990 ; þat godes hii were bo ROB. 229 ; his fader & his moder bo AL. (T.) 235 ; bo to JOS. 300 ; bo, boa A. R. 60, 212 ; boo MIRC 3 ; **bēire** (*gen.*) HOM. I. 99 ; KATH. 1790* ; O. & N. 1584 ; FL. & BL. 534 ; ROB. 197 ; BEK. 128 ; LANGL. *B* ii. 66 ; for heore beire nome LAȝ. 5283 ; **bā** (*dat.*) LAȝ. 9804 ; bo MIRC 1360 ; of hem bo C. L. 1454 ; in bo two his honden A. R. 396.

beȝen, būȝen, v., *O.E.* bēgan, bȳgan, = *O.Fris.* bēja, *O.N.* beygja, *M.H.G.* böugen ; *from* **būȝen** ; *incline, bend* ; beien JUL. 27 ; he mot nede beien LAȝ. 1051 ; whan i ne mai his hurte so buye BEK. 1529 ; uneþe he miȝte bye his rug P. L. S. xvii. 167 ; mine kneon ich beie HOM. I. 191 ; beieð A. R. 266 ; he beieð a dun toward þe his deorewurðe heaved HOM. I. 203 ; ant buȝeþ me to grounde REL. I. 122 ; **bēi** (*imper.*) LAȝ. 5068 ; MARH. 20 ; beih þe to me HOM. I. 211 ; beie (*subj.*) SPEC. 81 ; þat hit ne breke ne beie H. M. 15 ; **bēide** (*pret.*) JUL. 77 ; and beide (*drew*) heom to somne LAȝ. 29089 ; þe hwil(e) þat ha buhde hire KATH. 2400 ; byde P. L. S. xvii. 168 ; & hire cneow beigden MK. xv. 19 ; beighed ALIS. 4373 ; Scottes & Bruttes beiden (*sec. text* droȝen) to gaderes LAȝ. 5178 ; **būhed** (*pple.*) HOM. I. 277 ; *comp.* a-, ȝe-beȝen ; *deriv.* **bēi-sum.**

   bēȝend-lich, adj., *O.E.* (ge)bēgendlīc ; *humble* : mid beienliche worden LAȝ. 4930.

beggen, *see* **büggen.**

beggin, v., *beg,* 'mendico,' PR. P. 28 ; beggen A. R. 356 ; begge LANGL. *B* vi. 195 ; WICL. E. W. 352 ; CH. C. T. *A* 4525 ; ISUM. 150 ; **beggide** (*pret.*) WICL. JOHN ix. 8.

   beggere, sb., *beggar,* HORN (R.) 1133 ; CH. C. T. *A* 242 ; hit is beggares rihte vor te beren bagge on bac A. R. 168 ; **beggeres** (*pl.*) AYENB. 36 ; LANGL. *A* PROL. 40.

   beggerie, sb., *beggary,* LANGL. *B* vii. 88 ; WICL. PROV. xxiv. 34.

   beggestere, sb., *beggar,* CH. C. T. *A* 242.

   beggeþ [begged], sb., *begging, mendicancy,*

LANGL. *C* ix. 138; goon a begged (*v.r.* bigged, beggere, begger) CH. C. T. *F* 1580.

beggild, begenild, sb., *one given to begging*; **beggilde** (*gen.*) A. R. 168*; **begeneldes** LANGL. *C* xi. 263.

begginge, sb., *begging,* '*mendicatio,*' PR. P. 28; goþ on begginge MAND. 207; begging WICL. E. W. 353; bigging S. W. 128.

beȝsc, *see* baisk.   **bēh,** *see* bēah.

behave, -havinge, *see* bihabben.

bēi, *see* bēah.   **bēien,** *see* bēȝen.

beiken, *see* bēken.

beil, sb., *cf. Swed.* bögel, bygel, *Ger.* bügel; *? handle,* W. & I. 23.

bein, *see* bain.   **beiten,** *see* baiten.

bēi-sum, adj.(from bēȝen,v.),? = *Ger.* beugsam; *obedient*; buhsume & beisume KATH. 1805.

bek, sb., *? O.E.* bec (*dat.* bæce), *O.N.* bekkr *m.,*=*M.L.G.* beke, *O.H.G.* pah, pach, bach *m.f.*; *beck, brook,* '*rivulus,*' VOC. 239; PR. P. 29; L. H. R. 82.

bek, sb., *?*=bekne; *beck, sign, nod,* PR. P. 29; WICL. JOB xxvi. 11; **bekkes** (*pl.*) TREV. VIII. 221.

bek, *see* bec.

beken, v., *Fr.* becquer,=*Ital.* beccare; *peck*; **bekeð** (*pres.*) A. R. 84; **bekede** (*pret.*) 102; ? biked ALIS. 2337.

bēken, v., *? from* bǣwen; *steep, soak, warm*: to beike his boones FLOR. 99; a softe bē-kinde (*pres. pple.*) bað HOM. I. 269; and ligges bekeand in his bed IW. 1459; bekand BARB. xix. 552; i **bēked** (*pret.*) me TEST. CRES. 36.

bekken, v., *?*=bēknien; *beck, nod*; **bekke** (*pres.*) CH. C. T. *C* 396; beckes PS. xxxiv. 19; **bekkis** (*pl.*) TOWNL. 319 (MIR. PL. 193); **bekked** (*pret.*) CH. TRO. ii. 1260.

bēkne, beekne, sb., *O.E.* bēacen, bēcn,=*O. Fris.* bāken, bēken, *O.L.G.* bōcan, *O.H.G.* pauchan *n.; beacon,* '*pharus,*' PR. P. 29; bekne, bekene LANGL. *B* xvii. 262, *C* xx. 228; *comp.* **fore-bēacne.**

bēknien, v., *O.E.* bēacnian, bēcnan, *O.L.G.* bōcnian, *O.H.G.* pauhnen; *beckon*; þe unc becnien scu(len) . . . to drihtenes dome FRAG. 7; bǣcnien LAȝ. 21938; becnen ORM. 223; beknin, bekin '*annuo, nuto*' PR. P. 29; **bēkneð** (*ms.* bekned) (*pres.*) REL. I. 215; **bēacniende** (*pple.*) '*innuens*' LK. i. 22; bē-kenide, bikenede (*pret.*) WICL. DEEDS xxi. 40; hio becneden ('*annuerunt*') heore ge-feren LK. v. 7.

bēkninge, sb., *O.E.* bēacnung, bēcnung; *beckoning,* '*nutus,*' PR. P. 29; bikening WICL. GEN. xlii. 6; becnunge FRAG. 7.

bel, beu, adj., *O.Fr.* bel; *fair, beautiful*; bel ost GUY 68; a bele babees B. B. 3; beau A. P. i. 197; beaus amis GUY 4.

bel-ami, sb., *Fr.* bel ami; *fair friend,* A. R. 306.

bel-dame, sb., *grandmother,* PR. P. 29.

bel-father, sb., *grandfather,* PR. P. 30.

bel-līche, adv., *beautifully,* PL. CR. 344.

beu-pere, sb., *good father*; beau pere, hi seide to the pope BEK. 1299; boke hiȝte þat beupere LANGL. *B* xviii. 229.

beau-sire, sb., *fair sir,* BEK. 768; beusher D. TROY 1863; bewsher TOWNL. 66.

beld, *see* beald.

bélde, sb., *O.E.* bældo, byldo,= *O.H.G.* paldi, baldi, *Goth.* balþei; *from* beald; *fortitude, courage, comfort,* P. L. S. xxx. 104; FLOR. 762; GAW.650; M.H. 37; PERC. 1412; bi a childe of litil belde overcomen i am in min elde C. M. 12237.

bélden, v., *O.E.* bældan, byldan, = *O.L.G.* beldian; *comfort, encourage*; to belden & to forfren ORM. 662; to beolden (*invigorate*) it & strengen 2614; bealden KATH. 1622; belde LAI LE FR. 231; as he bigon to belde (*grow bold*) AL. (V.) 49; IW. 1220; **béldest** (*pres.*) C. L. 348; (he) beldeð [bealdeð] A. R. 162; bilde (*imper.*) EGL. 3; **bálde** (*pret.*) JUL. 8; belden LAȝ. 8636; **bélded** (*pple.*) ORM. 2746; *comp.* **a-, un-bélden.**

belden, *see* búlden.

bēle, *see* būle.

beleevable, *see under* bilēven.

belef, in phr. *O.Fr.* à belif, beslif; *from med. Lat.* bis-līquus; *obliquely*; a belef GAW. 2486, 2518; *see* embelif.

bēlen, *see* bǣlen.

belȝen, v., *O.E.* belgan,= *O.L.G., O.H.G.* bel-gan, *O.N.* belga; *? swell with anger*; **.bælh** (*pret.*) LAȝ. 15839; þa balh he hine ('*indignatus est*') LK. xv. 28; **bolȝhen** (*pple.*) ORM. 7145; *comp.* **a-, an-, ȝe-belȝen.**

beli, sb., *O.E.* belg, bælg, bylg, bylig,= *O.N.* belgr, *O.H.G.* palc (*follis*), *Goth.* balgs, (ἀσκός), *mod.Eng.* belly, bellows; þe brest wiþ þe beli (*belly*) PR. C. 679; beli '*fou*' VOC. 171; belou '*follis*' 180; beli, belu WICL. JER. vi. 29; beli '*venter*' belou '*follis*' PR. P. 30; beli [bali] CH. C. T. *C* 534 (*belly*), *I* 351 (*bellows*); bali R. S. v; M. T. 200; his bali for to fillen PL. CR. 763; **belies** (*pl.*) *bellows* A. R. 284; belies (*bellows*) ISUM. 412; belies, balies (*bellies*) LANGL. *A* PROL. 41; balez (*bowels*) GAW. 1330; a peir belwis (*bellows*) W. & I. 23; bulies BRD. 22.

beli-naked, adj., *stark naked,* CH. C. T. *E* 1326.

belken, v., *? O.E.* bealcian, = *? M.L.G.* belken (*mugire*); *belch*; belke CATH. 27; TOWNL. 314; haþ belkid (*pple.*) out '*eructavit*' WICL. PS. xliv. 2.

belle, sb., *O.E.* belle,=*M.Du., M.L.G.* belle; *bell,* '*campana,*' PR. P. 30; LAȝ. 29441; LANGL. *B* PROL. 165; bere the belle CH. TRO. iii.

149; **bellen** (*pl.*) FRAG. 7; LAȝ. 16929; HORN (H.) 1294; ROB. 509; AL. (L.³) 586; belle SAX. CHR. 259; belles ORM. 901; MAND. 243.

belle-drǣm, sb., *sound of a bell*, ORM. 922.

belle-ȝēter, sb., *bell-founder*, PR. P. 30.

belle-house, sb., '*campanile*,' VOC. 273.

belle-weder, sb., *bell-wether*; belwedir PR. P. 30; belleweder (*noisy person*) TOWNL. 86; belweather (*leader*) LIDG. BOCH. in N. E. D.

belle, sb., *tunic*, TREV. I. 403; AN. LIT. 12; ANT. ARTH. xxix (*printed* belte; *see errata*).

bellen, v., *O.E.* bellan, = *O.H.G.* pellen, bellen (*latrare*); *roar*, *bellow*; bellin '*mugio*' PR. P. 30; þe bole bigan to belle REL. II. 19; as loude as belleþ wind CH. H. F. 1803; **belling** (*pple.*) as a bole WILL. 1891; **bollen** (*pple.*) *inflated* MAP 334; WICL. I COR. v. 2; bollen hertes COMP. BL. KN. 101; a bleddre ibollen ful of winde A. R. 282; *comp.* **to-bellen**; *deriv.* **belle, bolle**?

bellinge, sb., = *O.H.G.* pellunga (*latratus*); *roaring, lowing*, '*mugitus*,' PR. P. 30.

belt, sb., *O.E.* belt, = *O.N.* belti, *O.H.G.* balz, *Lat.* balteus; *belt*, '*zona*,' VOC. 231; IW. 253; **belte** (*dat.*) CH. C. T. *A* 3929.

belte, sb., *axe*, '*securis*,' PR. P. 30; '*coing*,' VOC. 163; N. P. 13.

beltane, sb., *May-day*; (N. E. D.).

belten, v., *gird with a belt*; **belted** (*pret.*) C. M. 15285.

belwen, v., *O.E.* bylgean; *bellow*; belwe GOW. II. 72; L. H. R. 145; cou . . . þat wolde belwe after boles LANGL. *B* xi. 333.

belwinge, sb., *bellowing*; bellewing GOW. III. 203.

bēm, *see* **bēam**.

bēme, sb., *O.E.* bēme, bȳme; *trumpet*, MAP 348; PR. C. 4677; beeme ? *noise* AR. 108; **bēman** (*pl.*) HOM. I. 87; bemen FRAG. 7; A. R. 210; bemen [bemes, bumes] LAȝ. 4462, 5107, 5874; beme HOM. II. 115; MISC. 163; bemes WILL. 1154; K. T. 499; ISUM. 429; **bēmene** (*gen. pl.*) LAȝ. 19926; HOM. II. 113; **bēman** (*dat. pl.*) MAT. xxiv. 31.

bēmen, v., *O.E.* bȳmian; *sound the trumpet*, A. R. 210.

bēmare, sb., *O.E.* bȳmere; *trumpeter*, '*tubicen*,' FRAG. 2; A. R. 210.

bēmen, *see* **bēamien**. **bēn**, *see* **bēon**.

benche, sb., *O.E.* benc *f.*, ? = *O.L.G.*, *O.Fris.* benk, *O.N.* bekkr *m.*; *bench*, '*scamnum*,' PR. P. 30; P. L. S. xxxv. 11, xxxi. 114 (*bank*); LANGL. *A* iv. 32; AR. & MER. 2325; **benche** (*dat.*) LAȝ. 14963; ORM. 14087; REL. I. 187; HORN (L.) 369; on þine benche FRAG. 6; benke MAN. (H.) 58; iwstices atte benche ROB. 570; **benche** [benches] (*pl.*) LAȝ. 24872; benkes ORM. 15231; *comp.* **kine-, side-benche**.

benched, pple., *furnished with*, *seated on benches*; **benched** (*pple.*) CH. TRO. ii. 822; ibenched PL. CR. 205; benked ORM. 15231.

benchinge, sb., *benching*, TREV. BARTH. (*in* N. E. D.); benkinge ORM. 15232.

bend, sb., *O.E.* bend *m. f.*, = *O.L.G.* *bend (*acc. pl.* bendi), *O.Fris.*, *M.Du.* bende, *O.N.* benda, *Goth.* bandi *f.; from* **binden**; *tie, ribbon, band, bond*, '*vitta*,' VOC. 197; GOW. III. 11; CH. C. L. 810; W. & I. 35; his tunge bend MK. vii. 35; nenne bend AYENB. 48; mid æne (*r.*ane) bende of golde LAȝ. 24747; band (*stripe*) IW. 2394; **bendes** (*pl.*) P. L. S. viii. 95; A. R. 382; HOM. II. 63; MISC. 53; ROB. 58; bendes, bende O. & N. 1428, 1472; þe king heom dude i(n) bende LAȝ. 11862; **benden** (*dat. pl.*) MAT. xi. 2; LAȝ. 587; bende BEK. 15; AYENB. 220; B. DISC. 252; alused . . . of bende P. L. S. viii. 68; fram deaþes harde bende SHOR. 124; *comp.* **blōd-bend**.

bendel, sb., *O.Fr.* bendel, = *O.H.G.* bendel, *O.N.* bendill; *bendlet* (*in heraldry*), *little band*, RICH. 2964; **bendeles** MAN. (F.) 10043.

benden, v., *O.E.* bendan, = *O.N.* benda; *bend*; bende OCTAV. (H.) 1020; bende hire browes TREV. I. 9; bende '*tendo*' PR. P. 30; **bende** (*pret.*) FLOR. 859; GOW. I. 234; EGL. 294; bende (h)is bowe ROB. 16; þe amiral bende (*knitted*) (h)is browes FER. 1954; brente him brimli FER. 545; **bend** (*pple.*) SPEC. 34; ibent C. L. 743; AYENB. 174; *comp.* **un-benden**.

bēne, sb., *O.E.* bēn, = *O.N.* bœn; *prayer, request*, LK. i. 13; HOM. I. 67; ORM. 1459; GEN. & EX. 2511; SHOR. 47; AYENB. 99; SPEC. 58; REL. I. 113; þreting ne bene R. S. i (MISC. 156); grante me a bene HORN (L.) 508; **bēnen** (*dat. pl.*) MARH. 14; *see* **bōne**.

bēne, adj., ? *gracious*; A. P. iii. 418; þene wei bene P. L. S. viii. 170; *comp.* **un-bēne**.

bēne, adv., *pleasantly*, ANT. ARTH. vi.

bēne, *see* **bēane**.

benedight, pple., *from Lat.* benedictus; *blessed*, C. M. 18705; TOWNL. 91.

Benediht, sb., *Lat.* Benedictus; *Benedict*, LAȝ. 13159; A. R. 162.

benefet, benfeet, bienfait, sb., *O.Fr.* bienfait; *benefit*, LANGL. *B* v. 436, *C* viii. 42; benefet WICL. ESTH. xvi. 16; bienfait GOW. I. 304; (*noble deed*) III. 187.

benefice, sb., *O.Fr.* benefice; *favour, free gift, kindness; benefice*; CH. C. T. *A* 291; **benefices** (*pl.*) AYENB. 42; MIRC 317; WICL. S. W. III. 200.

beneisūn, sb., *O.Fr.* beneison; *benison, blessing*, HAV. 1723; benison CH. C. T. *E* 1365; benisoun MAN. (F.) 6961; benissoun PR. C. 3405; benzown S. S. (Web.) 3485.

Benet, sb., *O.Fr.* Benoit; *Benet, Benedict*, PR. P. 31; **benetis** (*pl.*) WICL. S. W. III. 285.

bengere, sb., *corn-bin*, PR. P. 31.

benigne, adj., *Fr.* benin; *benign,* CH. C. T. *A* 518; MED. 1103.

benignitē, sb., *Fr.* benignité; *benignity,* CH. C. T. *B* 1668.

benke, *see* **benche.**

bent, sb., *field,* GAW. 253; over the brode bent DEGR. 200; **bente** (*dat.*) JOS. 450; CH. C. T. *A* 1981 (*hillside*); opon þe bente bare ANT. ARTH. xliv; i walk on his bent TOWNL. 101; **bentis** (*pl.*) FLOR. 1040.

bent, sb., *O.E.* beonet, = *O.H.G.* binuz; *bent* (*sort of grass*), VOC. 191; A. P. iii. 392.

bēo, sb., *O.E.* bēo, bīo, bī, = *O.H.G.* bia, *M.Du.* bie *f.,* *O.N.* by *n.*; *bee,* '*apis*,' FRAG. 3; bee VOC. 177; PR. P. 27; bi LIDG. M. P. 88; **been** (*pl.*) MISC. 148; PL. CR. 727; been, [bees] WICL. DEUT. i. 44; a swarm of been CH. C. T. *F* 204.

bēod, sb., *O.E.* bēod, = *O.L.G.* biod, *Goth.* biuds, *O.H.G.* biet *m.,* *O.N.* biŏðr *m.,* bjŏð *n.; from* **bēoden**; *table*; **beode** (*dat.*) P. L. S. viii. 132; biede HOM. II. 228; ANGL. I. 23.

bēode, *see* **bede.**

bēoden, bēodan, v., *O.E.* bēodan, bīodan, = *O.L.G.* biodan, *O.Fris.* biada, bieda, *Goth.* (ana-, faur)biudan, *O.N.* bjŏða, *O.H.G.* biutan, piotan, bieten; *bid, announce, proclaim, invite, command,* HOM. I. 13; beoden A. R. 114; and lette beoden alle þa bocares LA3. 32124; þu scalt . . . beode (*preach*) þer godes godspel 29507; beoden & bodien KATH. 1480; beode SPEC. 42; min erande for to beode BEK. 1408; beode, bode O. & N. 530; beode, bude FER. 1793, 1825; bedin '*offero*' PR. P. 28; bede HORN (L.) 462; HAV. 1665; MAN. (F.) 11774; GAW. 374; MIN. 19; PR. C. 5958; and wolde him batail bede TRIAM. 972; **bēde** (*pres.*) AV. ARTH. xix; beodeð, beot HOM. I. 47, 55; A. R. 194, 208; al þat te godspel beodeþ ORM. 4917; bedeþ 15745; to thilke honour þat 3e me bede CH. C. T. *E* 360; **bēad** (*pret.*) A. R. 230; bead, bed GEN. & EX. 909, 1069; biad AYENB. 41; bæd, bed [bead, bad] LA3. 18821, 23407; bæd ORM. 11799; bed LK. viii. 29; SAX. CHR. 250; KATH. 441; ROB. 106; CHR. E. 122; TRIST. 2601; ANT. ARTH. i; he him bed risen HOM. II. 147; and bed alle godne dai P. L. S. ix. 200; beed AN. LIT. 11; bad '*mandavit*' PR. C. 6275; buden MK. x. 48; & buden here ferde LA3. 5146; boden HICKES I. 225; AM. & AMIL. 896; þai boden hem landes brade TRIST. 2636; he boden him bringen ut . . . ðo men GEN. & EX. 1067; **bude** (*subjct.*) HOM. I. 13; **boden** (*pple.*) CH. L. G. W. 366; C. M. 10683; haveð hem boden godun dai GEN. & EX. 1430; we ben alle boden (? *for* beden) þider HOM. II. 159; were boden to þe bridale LANGL. *B* ii. 54; bodin LIDG. M. P. 258; bode GOW. I. 12; *comp.* a-, bi-, for-, 3e-, on-bēoden; *deriv.* bod, bode, büdel.

bīdinge, sb., *M.H.G.* bietunge; *oblation,* WICL. E. W. 167.

beoden, *see* **beden.** **beof,** *see* **boef.**

bēon, v., *O.E.* bēon, bīon; *be,* FRAG. 7; KATH. 502; þe scal beon þe bet LA3. 701; þat ich king beon mæi 13082; beon iseien *videri* A. R. 94; beon, bion (*ms.* byon) MAT. xx. 26, 27; beon, bon (= beon, *see* ENG. STUD. IV. 99); HOM. I. 5, 77; O. & N. 262, 1195; þat 3e ower fihtlac leteþ beo (*cease*) 1699; beon, ben ORM. 21, 10255; beon, beo BEK. 105, 107; been PR. P. 30; ben MAN. (F.) 24; bien ANGL. I. 9; HOM. I. 219; MISC. 26; buen SPEC. 25; H. H. 67; buen, ben P. L. S. viii. 20, 87; ben, bi GEN. & EX. 15, 1195; beo, be FER. 517; bee EGL. 366; lat þi sorwe be (*cease*) HAV. 1265; lat þe somnour be CH. C. T. *D* 1289; by SHOR. 128; AYENB. 5; S. S. (Wr.) 536; bue REL. I. 110; boe II. 244; to beonne HOM. I. 193; to bienne AYENB. 131; **bēon,** beo, [*O.E.* bēom, = *O.L.G.* bium, *O.H.G.* pim, bim, bin] (*pres.*) LA3. 3945, 9843; bute ich habbe þine help ne beo ich never bliðe HOM. I. 197; swa muche . . . ich beo (*v. r.* iwurðe) him þe leovere JUL. 17; byo MAT. ix. 21; be HOM. II. 17; SPEC. 30; þu bist FRAG. 7; MISC. 96; SPEC. 72; bist, beost LA3. 3053, 9837; best HOM. II. 29; ORM. 2455; GEN. & EX. 2884; S. S. (Web.) 1663; bes REL. I. 181; þou bes (*shalt be*) honged DEGR. 839; við HOM. I. 13; R. S. v; swa us w(u)rse við (*will be*) LA3. 972; to mær3e wæne hit dæi buð 19484; ase softe as he is her ase herd he við þer A. R. 304; beoþ, beþ ORM. 167, 2683; beð GEN. & EX. 182; beth, bes HAV. 1261, 2007; (we, 3e, 3it, heo) beoð A. R. 8; beoþ PROCL. 2; ALIS. 20; beoð, beoþ, boþ O. & N. 75, 911, 1227; beoð, boð, beað HOM. I. 73, 89; beoð, buð [beoþ] LA3. 5093, 5099; þa hwile 3e buð a life P. L. S. viii. 12; bieþ HICKES I. 222 (ANGL. I. 8); AYENB. I; bieth, biedh MISC. 28, 31; beþ ROB. 7; TRANS. XIX. 320; PL. CR. 254; buþ E. G. 354; beoð, beon KATH. 493, 505; ben LANGL. *A* i. 6; beth, *B* iii. 27; beth, been CH. C. T. *A* 178; buþ, ben WILL. 946, 4447; ben GOW. II. 370; MAND. 3; WICL. JOHN viii. 10; al þat hie bi ben HOM. II. 179; b(ē)o, be (*imper.*) O. & N. 546; hal beo (*earlier text* wes) þu MK. xv. 18; beoð LA3. 19172; beth WILL. 3797; bes HAV. 2246; bēo (*subj.*) O. & N. 1225; þu beo FRAG. 6; liðe him beo drihten LA3. 4; wit beon HOM. I. 33; heo beon A. R. 8; O. & N. 1221; beo 3e swiðe bisie LA3. 7915; **bēon,** bēn (*pple.*) ORM. 2311, 8399; ben LANGL. *B* xi. 236; bien MAN. (H.) 34; bin TOR. 359; ibeon (beon) LA3. 8325; wel lange ic habbe child ibeon [iben] P. L. S. viii. 2; ibien HICKES I. 222; ibeo BRD. 4; ibee RICH. 5138; ibie MISC. 34.

bīinge, sb., *being, essence,* AYENB. 103.

bēor, sb., *O.E.* bēor, = *O.Fris.* biar, bier, *O.H.G.* pier, bier *n.; O.N.* biorr *m.; beer,* FRAG. 4; HORN (R.) 1108; beor, bor O. & N. 1011; ber GAW. 129; bor O. & N. 1009; **bēore** (*dat.*) LAӡ. 13542; bere HORN (H.) 1148.

beórd, *see* beard.

beore, *see* bere. **beoren,** *see* beren.

beorӡ, *see* berӡ.

berk, sb., *bark, bark (of a dog),* ANGL. IV. 197.

beorken, v., *O.E.* beorcan; *bark (as a dog),* JARB. XIII. 163; berkin '*latro*' PR. P. 32; berke AYENB. 179; WILL. 35; ALIS. 1935; **berkest** (*pres.*) A. R. 122; beorkeð [borkeþ] LAӡ. 21340; berketh LIDG. M. P. 29; **berkinde** (*pple.*) REL. I. 266; **bark** (*pret.*) GOW. I. 221; **burken, borken** (*pret. pl.*) ED. 912, 928; *comp.* be-berken.

　berkere, sb., *barker,* S. A. L. 70; berkar '*latrator*' PR. P. 32.

　berkinge, sb., *barking (of a dog),* TREV. I. 83; berking WILL. 55.

beorme, *see* berme.

beorn, bearn, bern, sb., *O.E.* beorn, biorn; *hero, warrior, man,* LAӡ. 1552, 7645, 8083; þu liþere bern P. L. S. xx. 57; beurn, bern ALIS². (Sk.) 9, 219; biurn AV. ARTH. xlv; buirn, burn LANGL. *B* xi. 353; burn WILL. 332; birn ANT. ARTH. iii; **beornes** (*pl.*) MISC. 92; biernes, bernes, burnes LANGL. *A* iii. 256; bernes M. H. 39; bernus DEGR. 500; bernis (*printed* berins) ISUM. 454; burnes GAW. 259; buirnes JOS. 29; **berne** (*gen. pl.*) SPEC. 58; **beornen** (*dat. pl.*) LAӡ. 23408; berne MISC. 93.

beornen, *see* brinnen.

bēot, sb., *O.E.* bēot; *menace,* LAӡ. 16338; *comp.* ӡebēot.

bēotien, v., *O.E.* bēotian; *menace;* **bēoteden** (*pret.*) LAӡ. 20521.

beouste, *see* biwiste.

beovien, *see* bivien.

beqvarre, sb., *O.Fr.* bequarre; *the note B natural,* REL. I. 292.

bercel, sb., *archer's butt,* '*meta,*' PR. P. 32.

berd, *see* beard *and* brerd.

bére, sb., *O.E.* bera, = *O.H.G.* pero, bero; *bear (animal),* '*ursus,*' PR. P. 32; HOM. II. 211; HAV. 573; AYENB. 15; LANGL. *B* xv. 294; CH. C. T. *A* 1640; MAND. 291; bere, beore ROB. 202; beore LAӡ. 25590; ALIS. 1821; beore, bore A. R. 198, 202; O. & N. 1021; beore, beare JUL. 72, 73; **beore** (*gen.*) L. H. R. 140; þes laste bore hweoːp A. R. 202; berse LAӡ. 22282; **béres** (*pl.*) GEN. & EX. 191; bueres A. D. 237.

　bére-fel, sb., *bearskin,* WILL. 2430; **bere-skin,** WILL. 1735; CH. C. T. *A* 2142.

bére-ward, sb., *bear-keeper,* PR. P. 32; P. P. 364.

bére, sb., *O.E.* bere, = *O.N.* barr *m.,* Goth. *baris, whence* barizeins (κρίθινος), *Lat.* far; *bear(-corn), barley,* VOC. 264; AYENB. 141; C. M. 13506; *comp.* pol-bére.

　bére-méle, sb., *barley meal,* LEECHD. III. 98.

bere-, *stem of* beren, v.

　bere-bag, sb., *bag-bearer,* MIN. ii. 20.

　bere-blisse, sb., *bringer of bliss,* AYENB. 72.

bere, sb., *pillow-case,* CH. D. BL. 254; *comp.* bed-, pilwe-bere.

bēre, sb., *O.E.* (ge)bǣre, = *O.Fris.* bēre, *O.L.G.* (gi)bāri, *O.H.G.* (ge)bāre; *gesture, noise, uproar,* FL. & BL. 468; ROB. 208; ALIS. 550; MIRC 276; MAS. 623; his untohe bere H. M. 31; þe ӡonge man . . . spak wit wel faire bere AL. (T.) 129; bere and noise TREV. IV. 367; who makis sich a bere TOWNL. 109; beere MISC. 231; beare LAӡ. 28163*; **bēre** (*dat.*) KATH. 2370; WILL. 43; ANT. ARTH. x; and warni men mid mine bere O. & N. 925; *comp.* a-, ӡe-bēre.

bēre, adj., *O.E.* (leoht-, wæstm-)bǣre; *bearing, fruitful;* childre bere GEN. & EX. 1465; *comp.* ǣrende-, liht-bēre.

bere, sb., *bearing;* for bere of þine childe SHOR. 133.

bēre, sb., *O.E.* bǣre, = *O.N.* bāra, *M.Du.* bǣre; *wave;* **bēres,** beares, bieres (*pl.*) LAӡ. 1341*, 28077*, 28625; cf. bāre.

beren, v., *O.E.* beran, = *O.L.G.* beran, *O.H.G.* peran, beran, *O.Fris., O.N.* bera, *Goth.* bairan, *Lat.* ferre, *Gr.* φέρειν; *bear,* A. R. 4, 230; M. T. 88; þis scal beren eower saule to hevene riche HOM. I. 39; beoren nið 125; beren [bere] wapen LAӡ. 499; witesse beren 13231; we sculleð bere þin ær(n)de 24803; beren child ORM. 2031; beoren KATH. 2491; (*ms.* boren) HOM. I. 69; bueren REL. I. 267; bere HAV. 623; PERC. 190; bere fals witnesse LANGL. *B* ii. 623; beore ALIS. 771; beare LAӡ. 24566*; buere SPEC. 104; to berene HOM. I. 147; **bereð** (*pres.*) HOM. I. 109; hue bereþ þe pris SPEC. 30; þe bitte þat beoreð forð (*projects*) as a water-bulge H. M. 35; (hi)bereð, bereþ O. & N. 1372; bereð me genge (*bear me company*) MARH. 21; **bere** (*subj.*) GEN. & EX. 3513; þat ti lust ne beore þe to þat te lef were H. M. 15; þat heo ure erende beore [bere] to þin hevonkinge R. S. iv (MISC. 168); beoren A. R. 382; **berinde** (*pple.*) AYENB. 96; berande ALIS. 5109; **bær,** ber, bear, bar (*pret.*) LAӡ. 1180, 1825, 12279, 14771; ber HOM. I. 131; MARH. 22; A. R. 368; P. S. 248; SPEC. 60; bar ORM. 12615; GEN. & EX. 338; HAV. 557; BRD. 20; LEG. 6; CH. C. T. *A* 105; þe erthe ne bar nan corn SAX. CHR. 262; þat bar þat blisful barn LANGL. *A* ii. 3; bare TRIST. 4; EGL. 340; þu bere MISC. 38; bere, beere LANGL. *A* iii. 189; (þeӡ) **bæren** ORM. 7576; **bæren, beren** LK. vii. 14, 24; beren HOM. I. 59; KATH. 2496; alle þa him

beren onde LAȜ. 31473; hue beren huem so
swiþe heȝe SPEC. 106; beeren MAND. 172;
bere ROB. 177; til þei bere [beere] leves
LANGL. *B* v. 139; baren 365; so boldeli þei
þem bere TRIAM. 1506; boren GEN. & EX.
684; **bǣre** (*subjct.*) ORM. 2029; bere HOM.
I. 39; BEK. 193; bere [biere] LAȜ. 2513;
boren (*pple.*) ORM. 161; H. H. 196; HAV.
1878; S. S. (Wr.) 73; born MAND. 274; he
hadde … born him wel CH. C. T. *A* 87; bore
AYENB. 221; *comp.* a-, æt-, for-, ȝe-beren;
deriv. **bǣre** (**bēre**), **bearm, berme, bürðe,
bürðene.**

berere, sb., *bearer,* MAND. 83; *comp.* **kāi-
berere.**

beringe, sb., *bearing, bringing forth; support,* MISC. 26; AYENB. 259; TREV. II. 171;
bering GEN. & EX. 2178; ALIS. 484.

beren, adj., *O.E.* beren, *Goth.* barizeins; *made
of barley* : fif berene hlafes JOHN vi. 9.

bēren, v., *O.E.* (ge)bǣran; = *O.L.G.* (gi)bārian,
M.Du. baeren; *gesticulate, shout;* bere HAL-
LIW. 166; M. T. 99; mi selve so to bere (*r. w.*
lere) SPEC. 93; **bēreþ** (*pres.*) L. H. R. 215;
beeringe (*pple.*) L. H. R. 140; **bērde** (*pret.*)
JUL. 53; *comp.* ȝe-bēren.

berfrai, sb., *O.Fr.* berfroi; *belfry,* A. P. ii.
1187; belfrai PR. P. 30; **berfreis** (*pl.*) MAN.
(F.) 1031.

berȝ, sb., *O.E.* beorg, beorh, = *O.L.G., O.Fris.*
berg, *O.H.G.* berg, perg, *O.N.* biarg, *mod.Eng.*
barrow; *from* berȝen; *hill, mountain,* GAW.
2172; beoruh FRAG. 3; **berhȝe** (*dat.*) LAȜ.
12311; berghe, berwe LANGL. *B* v. 589;
berges (*pl.*) REL. I. 222; **beorȝen** [b(e)orewe]
(*dat. pl.*) LAȜ. 20854; bergen LK. xxiii. 30.

berg, sb., *O.E.* (hēafod) beorg, = *O.N.* biörg
f.; *protection,* GEN. & EX. 926; *comp.* **hals-
berȝ.**

beru-ham, sb., *horse-collar,* PR. P. 33; ber-
hom VOC. 278; **beruhames** (*pl.*) 168.

berg-lēs, adj., *without protection,* GEN. & EX.
3048.

[**berȝe,** sb., *? shelter; comp.* her-berȝe.]

berȝen, v., *O.E.* beorgan, = *O.L.G.* bergan,
O.H.G. pergan, *O.N.* biarga, *Goth.* bairgan;
protect, preserve, HOM. II. 209; bergen REL.
I. 208; berȝhen ORM. 1559; berȝe AYENB.
197; berwen HAV. 697, 1426; **bergeð** (*ms.*
berged) (*pres.*) HOM. II. 197; birgeð 195;
barh (*pret.*) ORM. 14572; and barg ðe child
fro ðe dead GEN. & EX. 1330; bargh REL. I.
22; barw HAV. 2022; **borgen** (*pple.*) HOM. II. 187;
borgen, borwen GEN. & EX. 886, 1102;
berȝhed HICKESI.233; berȝhed (*ms.* beryhed)
PS.xxxii. 17; *comp.* bi-, ȝe-berȝen; *deriv.* berȝ,
borȝ, burȝ.

beriher, sb., *protector,* PS. lxi. 7.

berhles, sb., *from* berȝen; *salvation,* ORM.
5115.

berie, sb., *O.E.* berie, berige, = *M.Du.* bere,
besie *f., O.H.G.* beri, *O.N.* ber, *Goth.* basi *n.*;
berry : as broun as is a berie CH. C. T. *A* 207;
berien (*pl.*) A. R. 276; beries GEN. & EX.
2062; *comp.* **blac-, ivi-, mūl-, mūr-, strau-,
win-berie.**

berien, v., *O.N.* berja, = *O.H.G.* perien, *Lat.*
ferire; *beat,* A. R. 188*; (*v. r.* bunsen) bere
TOR. 1045; as hit on ehe bereð KATH. 1056;
wiþ a swere þat Arcita me þurgh þe herte bere
CH. C. T. *A* 2256; bi þe berid weie ['*per
tritam viam*'] WICL. NUM. xx. 19.

berien, *see* **bürȝen.**

beril, sb., *Gr.* βήρυλλος; *beryl,* MISC. 98; P. L.
S. xxxv. 92; beryl SPEC. 25; A. P. i. 1010.

Berkeleie, pr. n., *Berkeley,* ROB. 439.

berken, *see* **beorken.**

berling, sb., (*from* bere) *young bear;* **ber-
lingis** (*pl.*) DEP. R. iii. 40.

berm, *see* **bearm.**

bērman, *see under* **bǣr.**

berme. sb., *O.E.* beorma, bearma, = *M.Du.,
M.L.G.* berme, barme *m.*; *barm, yeast, 'spuma,'*
VOC. 258; PR. P. 32; ORM. 997; R. S. vii;
CH. C. T. *G* 813; L. C. C. 39; beorme MISC.
189.

bermin, v., *work out, 'spumo,'* PR. P. 32;
bermi SHOR. 15.

bern, *see* **bearn** *and* **beorn.**

bern, berne, sb., *O.E.* bere ern (LK. iii. 17);
berern, bern; *barn, 'horreum,'* WICL. DEUT.
xxviii. 17; LANGL. *B* xix. 340; berne ORM.
10487; TREV. I. 173; **berne** (*dat.*) O. & N.
607; GOW. I. 162; CH. C. T. *A* 3258;
berne (*pl.*) LK. xii. 18; bernes ROB. 348;
AYENB. 30.

bernake, bernakile, sb., *Fr.* barnaque, bar-
nache, barnacle, bernacle; *barnacle* (*bit for a
horse*), '*camus,*' PR. P. 33; bernacle '*camus*'
WICL. 4 KINGS xix. 28; **barnacles** (*pl.*) 353.

bernake, sb., *barnacle-goose;* barnakylle, ber-
nak '*barnacus*' PR. P. 32.

bernen, *see* **brennen.**

berslet, sb., *from O.Fr.* berseret; *hound,* D.
TROY 2196; berselette ANT. AR. iii; barslett
ALEX. (*Dubl. MS.*) 786; **bercelettus** (*pl.*)
AV. ARTH. vii.

bersten, *see* **bresten.**

berten, v., *? perspire* : i bert(e) in bedde
REL. II. 211.

berðene, *see* **bürðene. beru,** *see* **berȝ.**

berwe, sb., *? O.E.* bearu (*grove*); '*umbraculum,*' PR. P. 33.

berwen, *see* **berȝen.**

besagew, sb., *O.Fr.* besaiguë; *double-edged
axe,* PARTON. 1936.

besant, sb., *O.Fr.* besant; *Byzantine coin,*
ALIS. 3115; besaunt R. R. 1106; besaunt
'*minam*' WICL. EZ. xlv. 12; **beȝsanz** (*pl.*)
ORM. 8102; besans ALIS. 1572.

besi, *see* **büsi.**

besme, sb., *O.E.* besma, besema, = *O.Fris.* besma, *O.H.G.* pesamo, besamo; *besom,* '*scopa,*' PR. P. 33; AYENB. 172; WICL. IS. xiv. 23; **besmen** (*dat. pl.*) MARH. 5; JUL. 17; mid besme '*scopis*' LK. xi. 25; beseme HOM. II. 87.

besquite, sb., *biscuit,* MAN. (H.) 171; bisqwite PR. P. 48.

best, sb., *O.Fr.* beste, *Lat.* bestia *f.*; *beast,* A.R. 48; BRD. 31; þat ilke best (*r. w.* nest) O. & N. 99; þat best WILL. 3113; beest C. L. 121; beast KATH. 2067; best, beste HAV. 279, 944; beste GEN. & EX. 194; **bestes** (*gen.*) A. R. 58.

best-lich, adj., *beastly*: þe bestliche mon A. R. 58; beastliche men H. M. 25; beestli WICL. I COR. xv. 44.

bestial, adj., *pertaining to, like a beast,* BOKENH. 95; bestiall GOW. I. 140; **bestaille** (*pl.*) folk MAND. 224.

bestial-liche, adj., *belonging to the lower animals,* **bestiallī** (*adv.*) GEST. R. (H.) xlvi. 366.

bestaile, sb., *O.Fr.* bestaille (= *mod.* bétail); *cattle,* GOW. II. 138; bestaille MAND. 284; beastial (*single beast*) LIDG. *in* N. E. D.; **bestailes** (*pl.*) KN. T. 103.

bestialitē, sb., *O.Fr.* bestialité; *nature of a beast,* CH. TRO. i. 735.

bēstinge, sb., *O.E.* bȳsting, *from* bēost, = O. H.G. biost; *biesting* (*first milk after calving*), '*colostrum,*' CATH. 30.

bet, ? adv.; go bet (? *go quickly*) CH. C. T. C 667; go bete S. S. (Wr.) 1005.

bet, adv., *O.E.* bet, = *O.Fris.* bet, *O.N.* betr, *O.L.G.* bet, bat, *O.H.G.* paz, baz (*compar.*); *better,* A. R. 416; GEN. & EX. 2366; AYENB. 195; OCTOV. (W.) 1191; CH. C. T. A 242; HOCCL. i. 178; bett ORM. 4660; me luste bet O. & N. 39; as wel or bet WILL. 172; to love þat lond þe bet ROB. 107; þe bet LANGL. C ix. 42; bet, bæt LA3. 13448, 28560.

best, adv., *O.E.* betst, = *O.Fris.* best, *O.L.G.* bezt, best, *O.N.* bezt, *O.H.G.* bezest (*superl.*); *best* (*adv.*), HOM. I. 7; A. R. 6; god loved he best CH. C. T. A 533; godes give is best HOM. II. 107; bezst [best] LA3. 26606.

betere, adj., *O.E.* betera, betra *m.*, betere, betre *f. n.*, = *O.Fris.* betera, *O.L.G.* betara, *O.N.* betri, *Goth.* batiza, *O.H.G.* peziro, bezero'; *better,* A. R. 310; SPEC. 28; ælc wende to beon betere þene oðer LA3. 24536; betere is (*is preferable*) 5343; betere lif leden P. L. S. viii. 137; betere [beter] is o song O. & N. 713; betere bezet AYENB. 102; betre CHR. E. 382; betre hit is HOM. I. 49; þeh him s(e)olf þe betre nere 153; bettere, bettre, better LANGL. A iv. 80; bettre, better CH. C. T. A 256; beter K. T. 320; ge sende beteren LK. xii. 7; þa weren hire beteren (*her betters*) LA3. 3749.

betste, adj., *O.E.* betsta *m.*, betste *f. n.*, = *O.Fris.* besta, *Goth.* batista, *O.H.G.* pezisto, bezesto; *best* (*adj.*) ORM. 2948; beste BEK. 1447; þat beste þat we habbeð HOM. II. 221; here beste for to do ROB. 111; þe beste mon A. R. 332; his bezte [beste] cnihtes LA3. 8844.

bētās, sb., *O.N.* beitiāss; *sail-yard,* MAN. (F.) 12088.

bēte, sb., *O.E.* bēte, *Lat.* bēta; *beet,* VOC. 190; beetes (*pl.*) LUD. COV. 22.

bētel, sb., *O.E.* bētel, bȳtel, = *M.L.G.* bōtel, *M.H.G.* bōzel; *from* **bēaten**; *beetle, mallet,* STOR. 29; betil PR. P. 34; bitil VOC. 180; 3e schulen iseon bunsen ham mit tes deofles **bētles** (*v. r.* bettles) (*pl.*) A. R. 188.

bēten, v., *O.E.* bētan, = *Northumb.* bœta, *O.L.G.* bētien, *O.Fris.* bēta, *O.N.* bœta; *from* bōte; *mend, remedy,* A. R. 92; ORM. 4440; & gunnen here gultes beten P. L. S. viii. 137; beten alle owre bruchen KATH. 1406; betan HOM. I. 113; beten, bæten LA3. 5941, 18397; bete O. & N. 865; MISC. 41; ROB. 291; S. S. (Web.) 2123; AMAD. (R.) iv; bete his nede ALIS. 5065; REL. II. 278; oure serewes to bete SPEC. 89; to bete þe sinne H. 225; 3our bales to bete WILL. 3167; bete [beete] his hunger LANGL. B vi. 239; fires bete CH. C. T. A 2253; nettes bete 3927; to betene, to betende HOM. I. 7, 29; **bēte** (*pres.*) HOM. I. 21; **bēt** (*imper.*) HOM. I. 17; **bĕtte** (*pret.*) HOM. I. 83; KATH. 1210; ah ever me þat fur bette LA3. 2862; þat her bales bette B. DISC. 2088; greet fir þei under betten CH. C. T. G 518; **bĕt** (*pple.*) SPEC. 30; WILL. 3960; bett ISUM. 764; a brighte fir wel bett PERC. 439; *comp.* 3e-bēten.

bēten, *see* **bēaten. betere,** *see* **bet.**

bētil, sb., *O.E.* bētel; *beetle* (*insect*), VOC. 255; bitil PR. P. 37.

betren, v., *O.E.* betrian, = *O.Fris.* beteria, *O.N.* betra, *O.H.G.* peziron, bezeron; *better, improve*; **betred** (*pret.*) PS. xii. 5; beterid (*pple.*) WICL. S. W. III. 349.

beteringe, sb., *act of making better,* WICL. S. W. I. 55.

betternes, sb., '*benignitatem,*' PS. li. 5.

beð, *see* **bað. beþe,** *see* **bāðe.**

bēðen, v., *? O.N.* beiča; *ask,* GEN. & EX. 2498.

bēþien, v., *steep, foment,* GEN. & EX. 2447; **bethed** TR. F. 8.

beupere, bewsher, *see under* **bel.**

beutē, *see* **bealtē.**

béver, sb., *O.E.* beofer, = *M.L.G.* bever, *O.H.G.* biber, piber, *O.N.* bjorr, *Lat.* fiber; *beaver,* '*castor,*' VOC. 220; P. L. S. viii. 182; PL. CR. 295.

béver-hat, sb., *beaver-hat,* CH. C. T. A 272.

béver-h(e)wed, pr. n., *beaver-coloured,* GAW. 845.

Beofer-lai, pr. n., *Beverley*, FRAG. 5; Beverlēi MISC. 146.

beveráge, sb., *O.Fr.* bevrage; *beverage*, ROB. 26; RICH. 4365; GAW. 1112; A. P. ii. 1433; LANGL. *A* v. 189.

béveren, adj., = *M.H.G.* biberin, *?Lat.* fibrinus; *beaver coloured*; his beveren berd ANT. ARTH. xxviii; beverine lokkes D. ARTH. 3630.

beveren, v., = *L.G.* bevern; *tremble* : mani knightes shoke and **bevered** MAL. i. 15.

[**bi-,** prefix, *O.E.* bi-, be-, *mod.Eng.* be-, *for* bī, *prep.*]

bī, prep., *O.E.* bī, big, be, = *O.L.G.* bī, be, *O.H.G.* pī, bī, pe, be, *Goth.* bi; *by* : whan a man is an urþe ded and his soule bi (*with*) god TREAT. 134; þulke faire womman þat stent bi him P. L. S. x. 10; þe qvene þat by þe stod SPEC. 70; þat þe bi stod R. S. v; SPEC. 58; stod hem bi ORM. 3340; to stonden ðe bi GEN. & EX. 3666; bi þare see O. & N. 1754; þe wuneð bi westen KATH. 591; þat woneþ hire bi SPEC. 96; bi mire side LAȜ. 7879; bi eche side BRD. 22; bi þi side JOS. 42; bi are halve heo riden LAȜ. 9460; an oþer abbei is þer bi P. L. S. xxxv. 147; alle þat ter bi gað KATH. 1482; miȝt(e) we . . . com(e) bi tvo skinnes WILL. 1688; bi þone toppe he hine nom LAȜ. 684; he tok his child bi þe hande ISUM. 325; liveð . . . bi him one A. R. 12; ben . . . bi him selfen (*by himself, alone*) ORM. 822; þo heo were al bi hem selve ROB. 104; bi (*in*) þan dagen HOM. II. 47; bi olde dawe MISC. 47; bi þon time JUL. 4; bi time (*betimes*) GEN. & EX. 1088; HORN (L.) 965; bi daie LAȜ. 2082; O. & N. 241; bi niȝte BEK. 59; þat ich ne mai iseo bi [be] lihte O. & N. 366; ȝif he selleþ meche bi ȝere E. G. 355; al swa hit is bi mine songe O. & N. 1373; þer com a pr(e)ost bi þe weie HOM. I. 79; þe fuheles þe fleon bi ðe lufte MARH. 9; þat flihþ bi grunde O. & N. 506; as hi wende bi þe wode P. L. S. xiii. 153; he þat comeþ not in bi þe dore WICL. JOHN x. 1; bi watere & bi londe LAȜ. 10511; þe children ȝede to tune bi dales (*r.* dale) & bi dune HORN (L.) 154; bi (*on*) heore life hehte heom LAȜ. 24317; bi (*per*) god HORN (L.) 165; þet me ne zverie . . . by þe hevene AYENB. 6; hit is awriten bi (*of*) him HOM. I. 129; al hit wes lesinge þat heo· seiden bi þan kinge LAȜ. 19079; seið uvel bi an oðer A. R. 86; as þe Giwes dude bi oure loverd BEK. 2161; he deþ wel bi me A. D. 294; þe king havede bi his wive twene sonen LAȜ. 4288; heo schal libben bi (*on*) elmesse A. R. 414; noht ne maȝ þe man bi bræd al ane libben ORM. 11344; bi ði leve (*by thy leave*) GEN. & EX. 2865; bi ȝoure leve CH. C. T. *A* 3916; ne talde þeȝ noht teȝre kin . . . bi wimmen ac . . . bi wepmen ORM. 2061; sunne likeð him bi (*according to*) þine tale A. R. 334; ðe mone is more bi mannes tale ðan al ðis erðe GEN. & EX. 141; and dude al bi his rede BEK.

169; it semed bi here speche LANGL. *B* PROL. 52; bi hu muchel þe an passeð þe oðre H. M. 23; þo was it bi a fot to schort L. H. R. 30; a cite bi nome Nazareth icalled JOS. 78; i clepede þe bi þi name WICL. IS. xliii. 1; he wente hous bi hous CH. C. T. *D* 1765; ȝer bi [be] ȝere *E* 402; his doghter hadde a bed al bi hir selve right in þe same chambre bi and bi *A* 4142; up ros oure swete ladi and kist þe apostles bi & bi ASS. *B* 360; sende hit bi his sonde LAȜ. 3151; alle þingis ben maad bi him WICL. JOHN i. 3; as sche sat be þe welle OCTAV. (H.) 327; þe castel be þe see TOR. 871; þet me offrede wilem be þo ialde laghe MISC. 27; be dai FER. 2129; be middai AUD. 75; be þis time WILL. 1031; (hi solde) be an oþer weje wende into hire londes MISC. 27; zvereþ . . . be god AYENB. 6; be the rode GENER. 2445; sio manigeo wundrede be (*at*) his lære MK. xi. 18; be us foure þis i telle C. L. 495; be þet he heþ ofguo AYENB. 13; hi ssolden bi iborȝe be him 12.

bī, sb., *O.N.* bȳr; *mod.Eng.* -by *in local names; from* būen; *town*, HALLIW. 172; C. M. 13290; GUY· 267.

bi-æften, -aften, bæften, adv. & prep., *O.E.* beæftan, bæftan; *behind, after*, LAȜ. 7570, 8680, 26057; bieften HOM. I. 39; he let bi-aften ðe more del GEN. & EX. 3377; bæften FRAG. 1; baften HOM. II. 199; ORM. 14688; beften LK. xxii. 6; befte me MAT. xvi. 23; baften, baft PS. lxxvii. 66; on bafte [baft] *abaft* C. M. 22150.

bi-barren, v., *bar*; **bibarred** (*pple.*) A. R. 170.

bibben, v., *drink*; **bibbes** (*pres.*) A. P. ii. 1499; this miller haþ so wisli **bibbed** (*pple.*) ale CH. C. T. *A* 4162.

bi-bēoden, v., *O.E.* bi-, bebēodan, = *O.H.G.* pipiutan; *enjoin, command*; **bibēad** (*pret.*) HOM. I. 225.

bi-berȝen, v., *O.E.* bebeorgan, = *O.H.G.* pi-pergan; *protect* : þene scute biberh LAȜ. 1461.

bi-berken, v., *bark at* : þe felle dogge þet bit and **beberkþ** (*pres.*) alle þo þet he mai AYENB. 66.

bible, sb., *Fr.* bible; *bible*, LANGL. *B* ix. 41; GOW. I. 15.

bi-blēden, v., *cf. Ger.* beblüten; *cover with blood*; **biblëd** (*pple.*) A. R. 118*; BEK. 2210; S. S. (Wr.) 810; TREV. VIII. 155; þe ground . . . bibled was mid his blode SHOR. 87; **bebled** MIRC 1934; L. H. R. 190.

bi-blōdegien, v., *?for* *ȝeblōdegien; *make bloody*; **biblōdege** (*imper.*) A. R. 292; **biblōdeget** (*pple.*) MARH. 3.

bi-blotten, v., *blot*; **beblotte** (*imper.*) CH. TRO. ii. 1027.

bi-bóde, sb., *O.E.* bibod; *command, commandment*, HOM. I. 221, 125; be-, **bibode** (*dat.*) P. L. S.

viii. 131*; **bebode** (*pl.*) MAT. xix. 17; biboda, -bode, -bodan, -boden HOM. I. 119, 131.

bi-būfen, adv., *above*, ORM. 17970.

bi-būȝen, v., *O.E.* bi-, bebūgan; *escape, turn aside*; **bibǣh** (*pret.*) LAȝ. 8193; heo hine bibuȝen 12252; **biboȝen** (*pple.*) 10569.

bi-būrien, v., *O.E.* bi-, bebyrgan; *bury*, LAȝ. 10436; A. R. 216; beburie ROB. 166; to beberienne MAT. viii. 21; **bibüriede** (*pret.*) KATH. 2227; biburiȝede LAȝ. 1728; heo . . . biburȝeden 15071; **bibüried** (*pple.*) LAȝ. 10833; bebered AYENB. 263.

bi-cachen, v., *catch, ensnare*, HOM. II. 35; bikache REL. I. 183; **bicahte** (*pret.*) HORN (R.) 663; **bikaht** (*pple.*) ORM. 11621; bicauȝt S. S. (Web.) 638; becaȝt AYENB. 54; bicauhte, -kehte (*pl.*) P. L. S. viii. 160.

bi-callen, v., *call, address, accuse*; becalle GUY (Z.) 7596; **bicalle** (*pres.*) A. P. i. 912; becalle ANT. ARTH. xxxii; bicalleð GEN. & EX. 2314.

bi-cáren, v., *? be careful about*; to ofte man bicareð þis lif ANGL. I. 27.

bi-casten, v., *cast around*; *clothe, invest*: hi . . . hit al bicaste wiþ bolehuden BRD. 5; wiþ deope diches beoþ bicaste C. L. 694.

bicche, sb., *O.E.* bicce, = *O.N.* bikkja; *bitch*, REL. I. 187 (MISC. 137); AN. LIT. 11; TREV. III. 141; LANGL. *A* v. 197; CHES. I. 181; bicche, bicke PR. P. 35; bichche S. S. (Web.) 1802; **bicchen** (*pl.*) ALIS. 5394.

bicched, adj., *? cursed*; that bichid body LUD. COV. 395; so breme and bicchid in himselfe ALEX. 165; bicched bones (*dice*) CH. C. T. *C* 656; biched bones TOWNL. 241.

bi-charmen, v., *charm*; becharmeþ (*pres.*) AYENB. 257.

bi-chaunten, v., *enchant*, AR. & MER. 721.

bi-cherren, v., *O.E.* becerran, = *O.H.G.* becheren; *seduce, deceive*, A. R. 368; bicharren LAȝ. 5355; bicheorre 969*; bicherre MISC. 46; þu bicherrest [bich(e)orrest] LAȝ. 3837; he bicherreð monie mus to þe sto(c)ke HOM. I. 53; ȝe bicherreð (bichearreð) JUL. 42, 43; **bicherde** (*pret.*) HOM. II. 59; KATH. 1188; **bichærred** (*ms.* bischærred) (*pple.*) LAȝ. 7980; bicherd P. L. S. viii. 160; A. R. 224; S. A. L. 149.

bi-chirmen, v., *scream around*: hi me bichirmeþ [bichermet] and bigredeþ O. & N. 279.

bi-clappe, v., *clap about*, CH. C. T. *G* 9.

bi-clarten, v., *soil*: þat spatel þat swa biclarted ti leor HOM. I. 279.

bi-cleopien, v., *O.E.* becleopian; *appeal, accuse*, A. R. 344; MISC. 61; biclepiean, biclupien P. L. S. viii. 54; þat he scholde fram þulke curt biclipie to þe kinge BEK. 606; (*hie*) bicleepieð (*pres.*) HOM. II. 173; we beclepeþ þe dom C. L. 498; **bicl(e)oped**, bicleped

(*pple.*) O. & N. 550; a preost . . . þat of manslaȝt was bicliped BEK. 365.

bi-clüppen, v., *O.E.* bi-, beclyppan; *embrace*, A. R. 90; beclippe GOW. II. 95; MAND. 52; biclippe '*complectimini*' WICL. PS. xlvii. 13; **biclüpte** (*pret.*) LAȝ. 26776; ROB. 309; biclipte P. L. S. xii. 59; beclepte AYENB. 240; **biclüpped** (*pple.*) HOM. I. 201; biclipped TREV. I. 179.

bi-clūsen, v., *O.E.* beclȳsan; *inclose*, LAȝ. 15023; biclosi ROB. 558; **biclūsde** (*pret.*) LAȝ. 18614; **biclūsed** (*pple.*) FRAG. 7; A. R. 378; biclūset KATH. 598.

bi-clūtien, v., *dress up*; biclūte (*imper.*) A. R. 316.

bi-cnāwen, v., *O. E.* becnāwan, = *O.H.G.* pechnāhan; *acknowledge, know*; biknowen RICH. 4476; LANGL. *B* PROL. 204; beknowen GOW. I. 62; beknawe AYENB. 132; mon hwi nultu þe bicnowe R. S. i; biknowe CH. C. T. *A* 1558; him one to biknowe SHOR. 95; **biknōwe** (*pres.*) GAW. 2385; ȝif i beknowe mi name AM. & AMIL. 1279; **beknēu** (*pret.*) AYENB. 215; biknewe LANGL. *B* 145; **biknāwe** (*pple.*) AR. & MER. 425; he was biknowe þat Rimenild wes (h)is owe HORN (R.) 993; he was beknowe WILL. 2172.

   bi-cnāwinge, sb., *knowledge*; beknawinge, AYENB. 126.

[**bi-cnāwlēchen**, v., *acknowledge*.]

   bicnāwlēchinge, sb., *confession*, AYENB. 32.

bi-colmen, v., *cover with soot*; bicolmede (*pret.*) HORN. (L.) 1064.

bi-comsen, v., *commence*; **bikomsed** (*pret.*) WILL. 2523.

bi-cráven, v., *crave*, GEN. & EX. 1388.

bicume-līc, adj., *M.H.G.* bekomenlīch; *becoming*, HOM. I. 129; bicumelich II. 127; **becumeliche** daȝes ['*tempus acceptabile*'] I. 11; **bicumelīche** (*adv.*) REL. I. 131.

bi-cumen, v., *O.E.* bi-, becuman, *O.L.G.* becuman, *O.H.G.* pichoman, -qveman, *Goth.* biqiman; *become, come into being, happen, befit*, A. R. 64; ne mei nan man bicumen to mine heovenliche federe HOM. I. 119; þat ich wulle bicumen [bicome] þin mon LAȝ. 8482; wær sullen we bicumen 21913; þat him walde bicumen a freoboren burde JUL. 6; alle worldaihte schulle bicumen to nouhte REL. I. 183 (MISC. 126); bicome BEK. 40; bikimeð, bicumeð (*pres.*) JUL. 54, 55; so hit bicumeþ to havekes cunne O. & N. 271; it bicumeð him swiðe wel MISC. 23; **bicam** (*pret.*) LANGL. *B* v. 493; it so bicam GEN. & EX. 2007; becam MAND. 231; þer wiste non wher he becam GOW. II. 120; bicom KATH. 1563; H. H. 48; þe him bicom swiðe wel LAȝ. 18169; hit bicom me wurse A. R. 316; þo hit bicom þat he haȝte O. & N. 105; bicomen HAV. 2257; heo to londe bicomen

LAȝ. 105; **bicumen** (*pple.*) A. R. 340; michel sorge is me bicumen GEN. & EX. 2227; bicomen P. L. S. xxx. 22; bicome WILL. 222.

bi-cweste, sb., *bequest*; bi-qveste LANGL. *B* vi. 87; MAN. (H.) 86.

bi-cwesten, v., *bequeath*; biqvest PL. CR. 69.

bi-cwéðen, v., *O.E.* bi-, becweðan; *bequeath*; beqwethin '*lego*' PR. P. 31; liches . . . smeren and winden and biqveðen (*bewail?*) GEN. & EX. 2448; ic ou wile seggen . . . hwat þat word biqveþ (*means*) HOM. I. 75; **biqveð** [bicwaþ] (*pret.*) LAȝ. 9190; biqveþ ROB. 381; biqvaþ GAM. 99; **beqveþen** (*pple.*) PL. CR. 409; biqveþe TREV. I. 5.

bi-cwide, sb., *legacy*; biqvide ROB. 381; beqvide AYENB. 112.

bid, v.,=**bud,** *must*; **bidde** HAV. 1733.

bi-dǣlen, v., *O.E.* bi-, bedǣlan, = *O.L.G.* bedēlian, *O.H.G.* piteilan; *deprive*; **bidǣled** (*pple.*) FRAG. 6; ORM. 4677; witte bidǣled LAȝ. 10283; bideled of þe hore HOM. II. 49; bidelid REL. I. 185 (MISC. 134).

bi-daffen, v., *befool*; beþ nought **bidaffed** CH. C. T. *E* 1191.

bi-daggen, v., ?=bidēawen;=*M.Du.* bedauwen; *splashed with clay*; **bidagged** (*pple.*) ALIS. 5485.

bi-daȝien, v., *dawn upon*; **bedaweþ** (*pres.*) GOW. II. 193; **bedaghe** (*subj.*) D. TROY 758.

bidden, v., *O.E.* biddan, = *O.Fris.* bidda, *O.L.G.* biddian, *Goth.* bidjan, *O.N.* biðja, *O.H.G.* bittan, pittan; *pray, beg, bid, command,* A. R. 228; GEN. & EX. 1802; HAV. 529; & bidden (*pray*) for heom HOM. I. 7; & h(e)om to gode bidden 139; bidden hine mildce LAȝ. 12271; bidden his milce R. S. iv; bidden þes bone KATH. 611; bidden for þe ORM. 6130; to bidden us to Criste 8990; biddin '*oro, mando*' PR. P. 35; bidde AYENB. 5; if heore sunnen bidde bote O. & N. 858; mi mete for to bidde HORN (H.) 1218 (R. 1183); to biddene HOM. I. 39; bidde (*pres.*) WILL. 4754; ich . . . bidde þin ore HOM. I. 205; ihc bidde þe þi blessing ASS. 158; ȝif i **bidde** (*subj.*) ani bedes LANGL. *B* v. 407; biddest MARH. 11; bist MISC. 84; biddeð LAȝ. 16834; biddeþ CH. C. T. *A* 3641; he þat biddeþ borweþ LANGL. *B* vii. 81; þe wikked he biddeþ (*for* beodeþ) to gon heore wai P. L. S. xxvii. 11; biddes ISUM. 270; biddez S. A. L. 153; beddez GAW. 1374; bit A. R. 156; ORM. 5454; bit, bid O. & N. 441, 445, 1352; & helpes me biddað HOM. I. 13; hii biddiþ vor ur(e) fon ROB. 235; heo biddeþ heom to gode MISC. 85; bide (*imperat.*) REL. I. 186; bide hine HOM. I. 17; bide [bid] LAȝ. 4378; bide [bidde] his grace H. M. 11; **bidden** (*subj.*) H. M. 11; þat ȝe bidden for me JUL. 75; **biddinde** (*pple.*) AYENB. 219; biddinde [biddinge] LAȝ. 16754; **bæd, bed** (*pret.*) LK. vii. 3, viii. 37; bæd, bad, bead,

bed LAȝ. 2454, 3971, 4767, 4879; bed A. R. 156; AYENB. 191; P. S. 248; & bed hine witen þene forwundede mon HOM. I. 85; bad ORM. 2385; HORN (L.) 79; BEK. 2256; FER. 2127; AN. LIT. 12; MAND. 228; TOR. 1245; he bad hise bede GEN. & EX. 1375; heo bad for Horn HORN (L.) 79; so god it bad (*? for* bead) 57; þu bede LAȝ. 31604; O. & N. 550; ROB. 133; bone þat þou bede P. S. 153; þou me bede singe SHOR. 118; bæden LK. iv. 38; ORM. 697; beden FRAG. 6; heo beden [bede] hine come to helpe LAȝ. 666; bede WILL. 2102; AL. (T.) 370; GOW. I. 111; non oþer hevene hi ne bede FL. & BL. 553; **bǣde** (*subj.*) ORM. 9595; bede LAȝ. 29884; A. R. 222; O. & N. 1678; and bede ane bone KATH. 2401; þat þe watur bede LEB. JES. 317; bede, beede LANGL. *B* viii. 102; þat hi for him bede BEK. 2256; **beden** (*pple.*) HOM. II. 63; WICL. Lk. xiv. 7*; bede WILL. 2410; GOW. I. 70; were bede to þe bridale LANGL. *A* ii. 36; & dide al als him beden (*for* boden) was ORM. 3138; *comp.* ȝe-**bidden**; *deriv.* **bed, bede, beden.**

biddere, sb., *O.E.* biddere; *one who orders or begs*; bidder PR. C. 3679; **bidderes** (*pl.*) and beggers LANGL. *B* vi. 206.

biddunge, sb.,=*M.H.G.* bittunge; *bidding, praying,* HOM. I. 69; A. R. 108; biddinge AYENB. 194.

bīde, sb., *O.E.* bīd,=*M.H.G.* bīt; *delay,* TOR. 1463; C. M. 1761; wiþouten bide L. H. R. 113.

bi-dēawen, v., *M.H.G.* betouwen; *bedew*; **bedēaweþ** (*pres.*) AYENB. 95; **bidēwed** (*pple.*) PALL. vi. 110.

bidel, *see* **büdel.**

bi-delven, v., *O.E.* bedelfan,=*O.Fris.* bidelva, *O. L. G.* bedelƀan, *O. H. G.* bitelban; *dig round, dig*; bidelve REL. I. 116; S. S. (Web.) 1374; AR. & MER. 1026; **bedalf** (*pret.*) MAT. xxv. 18; þai bidolven S. S. (Web.) 2044; bidolven (*pple.*) CH. BOET. v. 1 (151); bedolve WILL. 5252.

bi-dēmen, v., *condemn,* HOM. I. 167.

bīden, v., *O.E.* bīdan,=*O.L.G.* bīdan, *Goth.* beidan (προσδέχεσθαι), *O.N.* bīða, *O.H.G.* bītan, *mod.Eng.* bide; *await, expect, endure,* MISC. 152; AN. LIT. 5; bide P. S. 204; S. S. (Web.) 2864; OCTOV. (W.) 874; GAW. 374; LANGL. *B* xviii. 307; MIN. v. 52; TOR. 1023; bide an dwelle P. L. S. iii. 40; i mai no lenger bide CH. C. T. *A* 4237; bjde an answare FLOR. 264; muche chele he bid SPEC. 110 (A. D. 176); bide (*imper.*) H. M. 11; bide (*subj.*) HOM. II. 149; **bād** (*pret.*) PERC. 569; Herode king bad after þeȝre come ORM. 6507; bod GOW. III. 292; þe navre wowe ne bod HOM. II. 33; bod [bood] ['*mansit*'] WICL. 1 KINGS xxiii. 18; bood CH. C. T. *A* 4399; **biden** (*pret. pl.*) LAȝ. 12690; REB.

LINC. 17; hi nefre ne bide nane niede HOM.
II. 221; **biden** (*pple.*) ALIS. 897; MAN.
(F.) 3403; *comp.* **a-, an-, ȝe-, over-bīden**;
deriv. **bīde, bāde.**

bīding, sb., *O.E.* bīding; *delay,* FER. 4026;
PR. C. 4708.

bidēne, adv., *at once, at the same time,* ORM.
4793; HAV. 730; L. H. R. 63; MAN. (H.)
45; MIN. iv. 53; (*ms.* by dene) MIRC 1870;
342; bifore, hure knele ȝe alle bi dene (*ms.*
bi dene) ASS. 347; þe orf deiede aɬ bidene
P. S. 342; hit faleweþ al bidene (*ms.* by dene)
SPEC. 61; bedene (*ms.* bedene, be dene) GUY
(Z.) 2408; EGL. 1128; ANT. ARTH. xxiv;
MIR. PL. 117.

bi-derken, v., *darken*; **biderked** (*pple.*)
GOW. I. 81.

bi-dernan, v., *O.E.* bi-, bedyrnan; *conceal,*
FRAG. 8.

bi-didren, v., *O.E.* bedidrian; *delude,* ORM.
15391; þe defel hafde hem al bididred &
forblended 19138.

bi-dihten, v., *bedight*; **bidight** (*pple.*) DEGR.
144.

bi-dōn, v., = *M.L.G.* bedōn, *M.H.G.* betuon;
defile with ordure, A. R. 130.

bi-dóte, v., *make to dote, make foolish*; bedote
CH. L. G. W. 1547.

bidowe, sb., *? kind of weapon,* LANGL. *A* xi.
211.

bi-dravelen, v., = *L.G.* bedrabeln; *drabble*;
bedrabilid (*pple.*) PR. P. 28; his berd was bi-
draveled LANGL. *B* v. 194.

bi-drenchen, v., *soak, steep*; **bedreint**
(*pple.*) CT. LOVE 577.

bi-driven, v., *O.E.* bedrīfan; *drive about*:
we beoð . . . mid wedere **bidrivene** LAȝ.
6206.

bi-droppe, v., = *Ger.* betropfen; *bedrop*; be-
droppe GOW. III. 254; **bidropped** (*pple.*)
LANGL. *B* xiii. 321.

bi-dütten, v., *shut up*: ðær þu bist feste
bidytt GR. 27.

bi-dwelien, v., *seduce, confound*: heo bid-
weolieð (*pres.*) simple men A. R. 128; as he
bidweoled [*v. r.* bidweolet] (*pple.*) were
KATH. 1258.

bied, *see* **bēod.**

bien, *see* **bēon.**   **bien,** *see* **büggen.**

bi-ēode, v., *O.E.* bi-, beēode; *went round,*
LAȝ. 1188; biȝede SHOR. 145; bieoden
KATH. 1614.

bieren, *see* **beren.**   **biern,** *see* **beorn.**

bi-fallen, v., *O.E.* befeallan, = *O.L.G.* (*incidere*)
bifallan, *O.H.G.* pifallan; *befall, fall,* C. L.
293; bifalle LAȝ. 4508*; HAV. 2981; WILL.
547; in seknesse bifalle HOM. I. 157; bi-
valleð (*pres.*) A. R. 246; bifēol (*pret.*) LEG.
4; bifel GEN. & EX. 963; LANGL. *A* PROL. 6;
befel HOM. I. 231; hit biful LAȝ. 1818*; BEK.

219; þe oreisouns . . . þat bifulle (*pertained*)
þer to P. L. S. xxi. 170; **bifallen** (*pple.*) HOM.
I. 281; bevalle AYENB. 49; þo was hit bivalle
(*first text* ilumpen) LAȝ. 7195*.

bi-fallinge, sb., *happening, occurrence,* CH.
TRO. iv. 990; **bifallinges** (*pl.*) WICL. WISD.
viii. 8.

bi-fealden, v., *O.E.* bi-, befealdan, = *O.H.G.*
pifaldan; *envelop*; **befēold** (*pret.*) MK. xv.
46; bi-, **bevealde** (*pple.*) AYENB. 8, 188; bi-
folde CL. M. 27; LAI LE FR. 172.

bi-felen, v., *O.E.* bi-, befeolan, *O.L.G.* bi-
felhan, *O.H.G.* pifelhan, -felahan; *commit*:
helle þe we weren in **bifolen** (*pple.*) HOM. I.
123; oft ic habbe bevolen an oðer senne V. &
V. 9.

bi-fellen, v., *O.E.* befyllan; *fell, cut down*;
biveollen (*? ms.* biveolen) LAȝ. 27069.

bi-finden, v., *O. Fris.* bifinda, = *O.H.G.* pi-
findan; *find*; **bivond** (*pret.*) ROB. 267; **bi-
funden** (*pple.*) HOM. II. 47; H. M. 31; ORM.
748.

bi-flēan, v., *O.E.* beflēan; *flay*; bevlaȝe
AYENB. 73; þo þet bevlēaþ þe povre volk 218;
beflain (*pple.*) GOW. III. 183.

bi-flēden, v., = *Ger.* beflüten; *flow round*:
þæ sæ hine **biflédde** LAȝ. 25738.

bi-flēon, v., = *O.E.* beflēon, = *O.H.G.* pifleo-
han; *flee, escape*; biflien HOM. I. 169; bifluen
P. L. S. viii. 77; bivli SHOR. 127; be-, bivli
AYENB. 9, 134; **bevliȝþ** (*pres.*) AYENB. 73;
biflē (*subj.*) SHOR. 36; **bevloȝe** (*pret.*)
AYENB. 77.

bi-flīing, sb., *shunning*; bevliinge AYENB.
121.

bi-flēten, v., *cause to be surrounded*: mid see
him **biflétte** (*pret.*) HORN (L.) 1396.

bi-flōwen, v., *flow around*; **bi-flōwe** (*pple.*)
HORN (H.) 612.

bi-fóle, v., *befool,* GOW. I. 10.

bi-fōn, v., *O.E.* bi-, befōn, *O.L.G., O.H.G.* bi-
fāhan; *contain, encompass*: þe loverd þet al þe
world ne muhte nōut bivon A. R. 76; Brutus
bifeng al þat him biforen wes LAȝ. 829; mid
æne (*r.* ane) bende of gold ælc hafde his hæfd
bivonge 24748; mid deofle bivon HOM. I. 9.

bi-forcen, v., *force, ravish*; **beforst** (*pret.*)
BARB. 536.

bi-fóren, adv. & prep., *O.E.* bi-, beforan, =
O.L.G. biforan; *before,* ORM. 117; PROCL.
.5; SPEC. 24; JOS. 28; Mihhal eode biforen
HOM. I. 41; eour eiþer sunegað biforan
drihten 15; biforen [bifore] and bihinden
LAȝ. 439; biforen þon kinge 8820; biforen
alle oðre H. M. 19; ðor biforen GEN. & EX.
665; biforen, -voren FRAG. 7; bivoren A. R.
190; swo we biforan qveþen MISC. 146;
biforen, -fore ['*ante*'] MAT. v. 12; LK. i. 8;
biforn HAV. 1022; CH. C. T. *A* 100; biforn,
bifore MAND. 1, 225; seen beforn, [bifore
se] ['*praevidere*'] WICL. ECCLES. iv. 13; bi-

fore P. L. S. xviii. 136 ; þe sterre þet iede bifore hem MISC. 27 ; bifore þat oure king wes ded P. S. 246.

bifore-casting, sb., *pre-calculation, malice aforethought*: of biforecasting ['*de industria*'] WICL. JER. xxxviii. 4.

bifore-cnāwen, v., *foreknow, know beforehand*: God biforeknēw (*pret.*) alşo the thingis to cominge of hem WICL. WISD. xix. 1* ; **bifore-knōwinge** know thow '*scito praenoscens*' WICL. GEN. xv. 13.

bifore-cumen, v., *come before* : as he knew₃ him strongli biforecummen ['*se a viro praeventum*'] of the man WICL. 2 MACC. xiv. 31.

bifore-cütten, v., *cut beforehand*: biforekitte ₃e ['*praecedite*'] the braunchis therof WICL. DAN. iv. 11.

bifore-gān, v., *go before* : merci and treuthe shul biforgo ['*praecedite*'] thi face WICL. PS. lxxxviii. 15.

bifore-gōere, sb., *forerunner* : I schal sende thi biforegoere ['*praecursorem*'] an aungel WICL. EX. xxxiii. 2 ; nise ₃e and sue the **biforgoeris** ['*praecedentes*'] JOSH. iii. 3 ; sentist waspis beforgoeres of thin ost WISD. xii. 8.

bifore-greiðen, v., *prepare beforehand* : I shal beforgreithe ['*praeparabo*'] thi seed WICL. PS. lxxxviii. 5 ; the vois of the Lord beforgreithende herttis ['*praeparantis cervos*'] xxviii. 9 ; he . . upon the flod is beforgreithide it xxiii. 2.

bifore-greiðinge, sb., *preparation*: ri₃twisnesse and dom beforgreithing ['*praeparatio*'] of thi sete WICL. PS. lxxxviii. 15 ; so is the beforgreithing of it lxiv. 10.

bifor-gürden, v.: *gird before;* God that **beforgirte** [(*pret.*) me ('*praecinxit me*')] with vertue WICL. PS. xvii. 33 ; thou beforgirtist me with vertue xvii. 40 ; the Lord beforgirte himself lxxxii. 1.

bifore-had, pple., *previously held*; Pharao shal restore thee to the biforehad ['*pristinum*'] gree WICL. GEN. xl. 13 ; after their visage befornhadde WICL. 4 KINGS xvii. 40.

bifore-hond, adv., *see* hand.

before-ocupien, v., *occupy beforehand* : it befornocupieth that couegteth it WICL. WISD. vi. 14 ; beforocupie wee his face in knouleching PS. lxxxxiv. 2 ; if he were befornocupied ['*praeoccupatus*'] bi deth WICL. WISD. iv. 7.

biforn-passend, pple., *excelling* : in alle thi werkes befornpassende be thou ['*praecellens esto*'] WICL. ECCLUS. xxxiii. 23.

bifor-rēdiinge, sb., *preparation* ; the beforrediing ['*praeparationem*'] of the herte of hem herde thin ere WICL. PS. ix. 17.

bifore-renner, sb., *fore-runner* : i shal

sende an aungel thi beforerenner ['*praecursorem*'] WICL. EX. xxxiii. 2.

bifore-rīpe, adj., sb., *early* : the beforeriip grapes ['*praecoquae uvae*'] WICL. NUM. xiii. 21.

bifore-scēawen, v., *foreshow* ; **beforeshēwinge** (*pple.*) WICL. GEN. xxxxi. 11.

bifore-sēn, v., *foresee* : that cannot beforese ('*praevidere*') in to tyme to comynge WICL. ECCLUS. iv. 13 ; alle my weies thou befornsē₃e (*pres.*) WICL. PS. cxxxviii. 4.

bifore-senden, v., *send before* ; **biforesente** (*pret.*) ['*praemisit*'] WICL. 2 MACC. xii. 21 ; **biforsent** (*pple.*) WICL. WISD. xix. 2.

biforn-setten, v., *set before* ; into eche folc of kinde he befornsette ['*praeposuit*'] a gouernour WICL. ECCLUS. xvii. 14 ; **beforsettende** ['*praeponens*'] WICL. ESTH. i. 8.

bifore-singen, v., *lead the chant*; Chinonye . . . was souereyn to biforsynge melodie ['*ad praecinendam melodiam*'] WICL. 1 PARALIP. xv. 22 ; with the whiche she beforesonge WICL. EX. xv. 21.

bifor-spékere, sb., *spokesman* ; profete, that is, interpretour ether bifor-spekere [forspekere] WICL. EX. vii. 1 *note*.

bifore-spékinge, sb., *preface* : prefacioun or biforespeking WICL. 2 MACC. ii. 33.

bifore-strecchen, v., *hold out, extend* : beforstrecche ['*praetende*'] thi merci to men knowende thee WICL. PS. xxxv. 11.

bifore-táken, v., *anticipate* : soone shul befortaken ['*anticipent*'] us thi mercies WICL. PS. lxxviii. 8.

bifore-tellen, v., *declare* ; heuenes shulen his ri₃twisnes beforetelle ['*annuntiabant*'] WICL. PS. xxxxix. 6 ; **befortolde** (*pret.*) thi ri₃twisnesse xxxix. 10.

bifor-wallinge, -wal, sb., *outwork* ; ['*antemurale*'] WICL. IS. xxvi. 1.

bifore-warnen, v., *warn before* ; **bifornwarneden** (*pret. pl.*) ['*praemonebant*'] WICL. WISD. xviii. 19.

bifore-wēven, v., *spread over* ; the cloudis . . . that beforeweuen (*pres.*) ['*praetexunt*'] alle thingus theraboue WICL. JOB xxxvi. 28.

bifore-wrīten, v., *prescribe* : as in the law it is beforn writen '*praescriptum*' WICL. 2. PARALIP. xxx. 5.

bi-frapen, v., *beat about* ; **bifraped** (*pple.*) FER. 2987.

bi-frēosen, v.,=*M.H.G.* bevriesen; *freeze* ; **bifroren** (*pple.*) ORM. 13856 ; befrose GOW. I. 220.

bi-fūlen, v., *O.E.* befȳlan, = *Du.* bevuilen ; *befoul, defile*, A. R. 128 ; befilin VOC. 143 ; bevēlþ (*pres.*) AYENB. 178 ; **bifoulet** (*pple.*) C. L. 1147.

big, adj., *big*, HAV. 1774 ; WILL. 173 ; CH. C. T. *A* 546 ; PR. C. 1460 ; bigge (*pl.*)

LANGL. *B* vi. 216; **biggore** (*compar.*) Jos. 452; bigger GAW. 2101.

big-lī, adv., *strongly, firmly, loudly* : ful bigli me haldes A. P. iii. 321 ; bound hom full bigli GAW. 1141, 1162, 1584 ; ALEX. (Sk.) 1371, 1138, 423 ; D. TROY 6035.

big, sb., *O.N.* bygg, = *O.E.* bēow, *O.L.G.* beu (*corn*), *prov.Eng.* big *n.* ; *barley*, VOC. 233.

bīȝ, see **bēah.**

bi-gabben, v., = *Swed.* begabba ; *mock, deceive* : i nam noȝt **bigabbed** ROB. 458.

bi-ȝǣte, sb., *acquisition, gain, generation, product*, ORM. 16835 ; for þære muchele biȝæte [biȝeate] LAȝ. 609 ; biȝete ROB. 65 ; P. L. S. xiii. 358 ; REL. I. 265 ; WILL. 2303 ; P. S. 200 ; biyete MISC. 76 ; bigete GEN. & EX. 896 ; AR. & MER. 1437 ; biȝete, biȝeate A. R. 154, 166 ; KATH. 472 ; biȝeate H. M. 27 ; beiete OCTOV. (W.) 848.

bi-gálen, v., *O.E.* begalan, = *O.H.G.* begalōn ; *enchant* ; **bigáleð** (*pres.*) HOM. II. 197 ; heo **bigolen** (*pret.*) þat child mid galdere swiðe stronge LAȝ. 19256.

bi-galwen, v., *? terrify* ; þat hors was swift as ani swalwe no man might þat hors begalowe ELL. ROM. II. 171.

bigamie, sb., *Fr.* bigamie ; *bigamy*, GEN. & EX. 448.

bi-gān, v., *O.E.* bi-, begangan, -gān, = *O.L.G.* bigangan, *O.H.G.* pigangan, -gān ; *go round* : mid golde bigon LAȝ. 7623 ; al þat þe sæ **bigǣð** (*pres.*) 11200 ; wo þe **bigō** REL. II. 273 ; þe mid watere is **biȝeonge** [biȝonge] (*pple.*) LAȝ. 23702 ; þa wes þa welle an an al mid attre bigon *tainted* 19773 ; bigan HOM. I. 149 ; þe feld with oten was al bigon *over-run* LEG. 34 ; wel begon R. R. 580 ; wo was þis wrecched womman þo bigon CH. C. T. *B* 918 ; þe erþe it is whiche ever mo wiþ mannes labour is bego *cultivated* Gow. I. 152.

bi-gápen, v., = *M.Du.* begapen ; *gape at* : þes kelser **bigápede** (*pret.*) ham KATH. 1262.

bi-ȝelpen, v., *boast of* : hou shulde i thanne me beyelpe . . . of thi largesse Gow. III. 155 ; biȝelp GUY 1455.

bīȝels, sb., *O.E.* bȳgels ; *from* **būȝen** ; *arch, '(arcus) vel fornix,'* FRAG. 3.

bi-gēmen, v., *O.E.* begȳman, *O.H.G.* pigoumen ; *observe, take care of*, SAX. CHR. 258; **bigēmeð** (*pres.*) HOM. II. 183.

bī-genge, sb., *O.E.* bīgeng ; *service, worship* : ȝif we his bigenge haldeð HOM. I. 119.

bi-ȝeonden, -ȝeonde, -ȝende, adv. & prep., *O.E.* begeondan ; *beyond*, LAȝ. 1231, 2639, 4884 ; begenden JOHN iii. 26 ; biȝonden ORM. 10603; beiondin S. & C. I. xxix; biȝonde MAND. 1 ; biȝonde þe broke A. P. i. 141 ; biionde (*ms.* biyonde) MISC. 145 ; biihonde PR. C. 4458 ; beionde SAX. CHR. 266 ; beȝonde M. H. 78 ; biȝonde, -ȝende ROB. 2, 181 ; biȝonde, -ȝunde TREV. VIII. 81 ; biȝende AN. LIT. 5 ; GEST.

R. 1 ; beiende þe ze AYENB. 165 ; biȝunde BRD. 3 ; biȝendis WICL. MK. iii. 8 ; biȝondes WICL. 2 COR. x. 16.

bi-ȝēoten, v., *O.E.* begēotan, *O.H.G.* pigiuzan; *sprinkle* ; **bigoten** (*pple.*) JUL. 27 ; biȝoten HOM. I. 261.

bi-gernen, v., *O.E.* bigrinian ; *ensnare* ; **bigernin** (*pres. pl.*) APOL. 64.

bi-getel, adj., *? advantageous* : he maden swiðe bigetel forward GEN. & EX. 1992.

bi-ȝeten, v., *O.E.* bi-, begitan, -gietan, -gytan, = *O.L.G.* bigetan, *Goth.* bigitan, *O.H.G.* pigezan ; *beget, acquire*, HOM. I. 27 ; biȝeten, biȝiten, biȝeoten LAȝ. 434, 1206, 13496 ; biȝeten, biȝeoten KATH. 1633 ; biȝiten A. R. 142 ; biȝete O. & N. 1629 ; bigeten GEN. & EX. 1532 ; **biȝeat** (*pret.*) LAȝ. 3179 ; bejæt, bejet SAX. CHR. 256 ; biȝet A. R. 160 ; ich biȝet hit iwriten MARH. 2 ; þe fader þat me biȝat BRD. 24 ; bigat ORM. 13986 ; WILL. 177 ; (þu) bigete HOM. II. 29 ; heo biȝæten LAȝ. 19366 ; heo biȝeten Rome 7368 ; **biȝeten** (*pple.*) ORM. 1645 ; S. S. (Web.) 1066 ; beieten SAX. CHR. 256 ; biȝeten, biȝiten LAȝ. 4245, 8865 ; beyete AYENB. 130 ; biȝite P. L. S. xxiii. 18 ; biȝute BEK. 119 ; þat ha beoð biȝetene KATH. 264.

begetare, sb., *begetter, 'genitor,'* PR. P. 28.

begetinge, sb., *begetting, 'genitura, generatio,'* PR. P. 28.

bi-giften, v., *entrust* ; **begifte** (*pret.*) OCTOV. (W.) 675.

biggen, see **büggen.**

biggen, v., *O.N.* byggja, byggva ; *from* **būen** ; *build, cultivate, inhabit* ; bitwenen men to biggen (*dwell*) ORM. 1611 ; biggin '*aedifico*' PR. P. 35 ; bigge WICL. S. W. II. 50 ; PL. T. 475 ; **bigge** (*pres.*) APOL. 71 ; **biggede** (*pret.*) GEN. & EX. 1137 ; bigged MAN. (F.) 8725 ; planted and bigged PR. C. 4850 ; **bigged** (*pple.*) GAW. 20 ; LIDG. M. P. 264 ; biggit ANT. ARTH. vi.

bigger, sb., *builder* ; biger FLOR. 8.

bigginge, sb., *O.N.* bygging, ; *habitation, building, 'aedificium,' 'aedificatio,'* PR. P. 35 ; bigging GEN. & EX. 718 ; ISUM. 78 ; bugging P. S. 151.

bi-gīlen, v., *beguile*, HORN (R.) 328 ; bigile PR. C. 1264 ; **bigīleð** (*pres.*) A. R. 330 ; begileþ AYENB. 16 ; biȝuled KATH. 1054 ; **bigīled** (*pple.*) P. L. S. xiii. 323 ; LANGL. *B* xviii. 290 ; bigiled, biȝuled A. R. 268, 270.

bi-gīlere, sb., *deceiver* ; begilere WICL. JOB xii. 16.

bi-ginnen, v., *O.E.* beginnan, = *O.L.G.* biginnan, *O.H.G.* piginnan ; *begin*, KATH. 281 ; ORM. 7822 ; biginne CH. C. T. *A* 42 ; we willen biginne ure larspel HOM. I. 75 ; biginne, bigunnen LAȝ. 20739, 28130 ; bigunnen H. M. 11 ; **bigünþ** (*pres.*) SHOR. 71 ; **bigan** (*pret.*) ROB. 10 ; CH. C. T. *A* 44 ;

began HOM. I. 219; bigon [bigan] him to speken LA3. 25765; bigon HOM. I. 43; O. & N. 13; þu bigunne HOM. II. 85; begonne AYENB. 71; CH. C. T. *G* 442; bigunnen KATH. 294; GEN. & EX. 536; HAV. 1011; and bigunnen . . . ræsen to somne LA3. 27493; begonnen MAND. 41; bigonne BEK. 527; **bigunnen** (*pple.*) JUL. 28; ORM. 7823; PR. C. 6476; begonnen MAND. 40; bigonne SHOR. 58.

    biginnare, sb., = *M.L.G.* beginner; *beginner*; biginner A. P. i. 436; beginnare PR. P. 28.

    biginninge, sb., *M.H.G.* beginnunge; *beginning,* KATH. 290; biginning ORM. 706; biginninge, biginnunge A. R. 54, 206; biginnunge JUL. 27.

bi-ginnen, v., *deceive*; biginne MISC. 79; C. M. 3880.

bigli, adj. & adv., *O.N.* byggiligr; *inhabitable; pleasant; great, big*; bigli blis FLOR. 1681; bigli landis 220; bigli bild A. P. i. 962; bigli and swiþe JOS. 571; *see also under* **big.**

bi-glīde, v., *O.E.* beglīdan; *creep past,* SPEC. 87.

bi-glūed, pple., *caught with birdlime,* LIDG. M. P. 115.

bi-grǣden, v., *cry out against*; bigrede O. & N. 1413; bigraden HOM. II. 173; **bigrēdeð** (*pres.*) HOM. II. 69; **begrědde** (*pret.*) M. ARTH. 1812; bigradde O. & N. 1144; bigradden *wept* ALIS. 5175; **bigrǎd** (*pple.*) S. S. (Web.) 1518.

bi-gráven, v., *O.E.* bi-, begrafan, = *O.H.G.* bigraban; *bury*; **bigráven** (*pple.*) AR. & MER. 98; begrave GOW. I. 189; (*engraved*) I. 127.

bi-grīpen, v., *O.E.* begrīpan, = *O.Fris.* bigrīpa, *O.H.G.* begrīfen; *apprehend, comprehend, reprehend,* REL. I. 220; begripe PALL. ii. 278; **begrīpeþ** (*pres.*) GOW. III. 102; & ec Farisewishe men **bigrǎp** (*pret.*) he þus wiþ worde ORM. 9754; sant Johan hafde þe king **bigripen** (*pple.*) of his sinne 19857; mid senne begripe HOM. I. 237.

bi-grippen, v., = *M.H.G.* begripfen; *grip, seize*; **bigripte** (*pret.*) GAW. 214.

bi-grōwen, v., *grow around*; **begrōwe** (*pple.*) GOW. II. 358; was mid ivi al bigrowe O. & N. 27.

bi-grucchen, v., *grumble at*; **bi-gruccheþ** (*pres.*) LANGL. *A* vii. 62.

bi3t, sb., *O.E.* byht; *from* **bū3en**; *bight, bending,* REL. I. 190; þe bi3t of þe þi3es GAW. 1349; *see* **bo3t.**

bī-gürdel, sb., *O.E.* bi-, biggyrdel, = *O.Fris.* bigerdel, *O.H.G.* pīgürtel; *girdle,* A. R. 124; **bīgirdles** (*pl.*) LANGL. *A* viii. 87; **bīgerdlen** (*dat. pl.*) MAT. x. 9.

bi-gürden, v., *O.E.* begyrdan, = *O.H.G.* bigurtan; *begird*; **bigürt** (*pres.*) A. R. 378; bi-

gerte (*pret.*) SHOR. 51; **bigirt** (*pple.*) HAL-LIW. 174.

bi-habben, v., *behave*; behave FLOR. 1567.
    behaving, sb., *behaviour,* MER. 49.

bi-hǣs, bihǣst, sb., *O.E.* behǣs *f.*; *behest, promise,* HOM. II. 61; bihas ALIS. 7840*; bihæste, biheste LA3. 1263, 18751; biheste HOM. I. 33; A. R. 208; BEK. 1406; CH. C. T. *B* 41; beheste AYENB. 225.

bi-hǣsten, v., *from* **bihǣst**; *promise*; **bihastest** (*pres.*) HOM. I. 185; **behested** (*pple.*) LIDG. BOCH. in N. E. D.

bi-ha3e, sb., = *O.Fris.* bihach, *M.Du.* behaegh, *M.H.G.* behage *f.*; *?pleasure*; bi his biha3e HOM. I. 137.

bi-ha3ien, v., = *O.L.G.* bihagon, *O.Fris.* bihagia; *?please*; **biha3eð** (*pres.*) HOM. I. 135.

bi-half, sb., *for* **half**; *behalf*: on mi **behalfe** (*dat.*) CH. TRO. ii. 1458; on goddes behalve MAND. 225; in þe kinges bihalf (*v. r.* half) TREV. VIII. 347.

bi-halven, v., = *O.H.G.* behalben; *surround*; bihalve HAV. 1834; **bihalven** (*pres.*) GEN. & EX. 3355.

bi-hangien, v., *hang round*; bihonge ALIS. 758; **bihonged** (*pple.*) LA3. 20125; behonged WILL. 5015.

bi-hāt, sb., *O.E.* behāt, = *O.L.G.* bihēt, *O.H.G.* piheiz; *promise*; **behāte** (*dat.*) HOM. I. 225; **bihātes** (*pl.*) HOM. I. 89; behotes PS. cxv. 14.

bi-hāten, v., *O.E.* behātan, = *O.H.G.* piheizan; *promise,* HOM. I. 25; LA3. 18396; bihoten A. R. 6; behoten WICL. I. PARAL. xxix. 9; behotin PR. P. 29; **behōte,** bihōte (*pres.*) LANGL. *A* v. 235; behote, bihoote, biheete CH. C. T. *A* 1854; behet(e) MIR. PL. 15; bihateð LA3. 20626; bihat HOM. I. 39; A. R. 6; behat, behot AYENB. 64, 97; (3e) bihateð LA3. 12487; (heo) bihoteð A. R. 430; **bihēt** (*pret.*) HOM. I. 99; KATH. 415; A. R. 176; ORM. 5574; GEN. & EX. 1884; HORN (L.) 470; BEK. 1010; biheihte [behehte] LA3. 1263; beheighte MIR. PL. 20; bihi3te LANGL. *A* v. 47; WICL. GEN. xxxviii. 23; (þu) bihete LA3. 18925; H. H. 197 (199); behete FER. 1402; bihet ORM. 7621; ie bihi3ten CH. C. T. *F* 1327; (heo) bihæten LA3. 5165; bihete R. S. v; P. L. S. xvi. 164; **bihāten** (*pple.*) KATH. 756; ORM. 13823; bihoten A. R. 182; HAV. 564; bihote O. & N. 1745; GOW. II. 28; WICL. ROM. i. 2; bihiht SPEC. 37.

    bi-hētere, sb., *promiser,* WICL. 2 MAC. x. 28; biheter HEB. vii. 22.

    behōtinge, sb., = *O.H.G.* piheizunga; *promise,* AYENB. 207.

bi-hátien, v., = *M.L.G.* behaten; *hate*; **bi-háted** (*pple.*) CH. BOET. iii. 4 (75); bihatid H. V. 89.

bi-hāwen, v., *O.E.* behāwian; *look at, behold,*

þe folk to bihowe MAN. (F.) 11165 ; **behāwe**
(*imper.*) MAT. vii. 5.

bi-hēafdin, v., *O.E.* behēafdian, = *M.H.G.*
behoubeten; *behead*, JUL. 41°; bihefden KATH.
2273 ; bihafdi LAȝ. 26296 ; **behǣfdede** (*pret.*)
MAT. xiv. 10; **bihēaveded** (*pple.*) MISC.
154 ; biheveded P. L. S. xv. 171 ; biheveded,
bihedid WICL. LK. ix. 9.

   bihēfdunge, sb., *O.E.* bihēafdung ; *beheading*, A. R. 184.

bi-healden, bihálden, v., *O.E.* bi-, behealdan, *O.L.G.* bihaldan, *O.H.G.* pihaltan ; *behold,
hold*, HOM. II. 35 ; bihalden ORM. 15663 ;
bihalde IW. 1426 ; biholden A. R. 54 ; bihalde,
biholde O. & N. 71, 1325 ; bihelde S. S. (Web.)
744 ; þe alle weren eate(l)liche to bihaldene
HOM. I. 41; **biháldeð** (*pres.*) LAȝ. 21322 ; bihalt
A. R. 214 ; þe pater noster bihalt (*avails*) me
noht HOM. I. 65; hwet bihalt (*signifies*) . . .
þet tu ne buhest to me MARH. 7; we muȝhen
sen what it bihalt ORM. 13408 ; behalt WICL.
JOB xxxix. 29 ; **biháld** (*attend*) unto mi
bede stevene PS. v. 3 ; **bihēold** (*pret.*) KATH.
744 ; hit naht ne beheld (*availed*) SAX. CHR.
250; biheeld LANGL. (Wr.) 25 ; bihuld BEK.
710 ; biheolden, bihulden LAȝ. 5737, 26321 ;
bihielden HOM. II. 23 ; bihelden KATH. 741 ;
behólden (*pple.*) MAND. 13 ; E. T. 358; i
am beholde (*obliged*) GOW. III. 354.

   behóldere, sb., *beholder*, '*inspector*,' PR. P.
28.

   biháldunge, sb., *beholding*, MARH. 14;
biholdunge A. R. 52; beholdinge '*inspectio*'
PR. P. 28.

bi-hēawen, v., *O.E.* bi-, behēawan, = *O.H.G.*
pihauwan ; *hew around* ; **bihēwe** (*pple.*)
LEG. 46.

bi-hēden, v., = *O.H.G.* behuotan ; *guard, observe, prevent* : hwat can þat ȝongling hit
bihede ȝif hit misdeþ hit mot nede O. & N.
635; **bihĕdde** (*pret.*) REL. II. 225; his nest
noȝt wel he ne bihedde O. & N. 102 ; þe king
hit wel bihedde LAȝ. 19046; þou art mis
bihĕd (*pple.*) LEG. 17.

bi-hélden, v., *wash by pouring* ; wið halewende wattres biheolden ham alle KATH.
1400.

bi-helien, v., = *O.H.G.* behelian, = *O.L.G.* bihelian, *O.H.G.* pihellen ; *cover* ; al þes world
is **biheled** (*pple.*) mid heþene hode MISC. 91 ;
behelid RICH. 5586.

bi-hemmen, v., *hem round* ; bihemmen and
bilegge O. & N. 672.

bi-hengen, v., = *Ger.* behengen ; *hang round* :
bihenged (*pret.*) al wiþ belles ORM. 951.

bihēten, *see* **bihāten.**

bi-hēve, adj., *O.E.* behēfe ; *profitable*, HOM. I.
213, II. 7 ; **bihēvest** (*superl.*) A. R. 298 ;
comp. un-bihēve.

bi-hēve, sb., *? O.E.* behēfe ; *profit* ; moni uvel
ich iseo þer inne & none biheve A. R. 96 ;

biheve, biheeve C. L. 1425 ; to hare biheve
(*printed* byhove) S. S. (Wr.) 2301.

bi-hiȝen, v., *make haste* ; **bihiȝe** (*imp.*) C. M.
5087 ; **bihīde** (*pret.*) S. S. (Wr.) 952.

bi-hinden, adv. & prep., *O.E.* behindan, =
O.L.G. bihindan ; *behind*, FRAG. 8 ; HOM. I.
143 ; JUL. 49 ; ORM. 401 ; BEK. 1374 ; AR.
& MER. 331 ; bihinden [bihinde] LAȝ. 439 ;
jif he loke bihinden [bihinde] him CH. BOET.
iii. 12 (108) ; bihinde REL. I. 183 ; hwo liþ
bihinde O. & N. 528 ; bihinde þe bure 937 ;
behinde AYENB. 130; bihindis WICL. PHIL.
iii. 13.

bi-hōf, sb., *O.E.* behōf, = *O.Fris.*, *M.L.G.* behōf, *M.H.G.* behuof ; *behoof, profit, use* ; **bihōve** [bihofe] (*dat.*) LAȝ. 1050 ; bihove HAV.
1764 ; to his owene bihove A. R. 70; to ure alre
bihove REL. I. 173 (MISC. 108) ; bi-, behove C. M.
7100; serveþ to our behove TRANS. XIX. 330.

   behōf-fol, adj., *profitable*, HIST. EDW. 13 ;
WICL. PS. cxliv. 15.

   behōf-lich, adj., *O.E.* behōflīc ; *profitable* ;
behoveli '*opportunus*' PR. P. 29 ; CH. TRO.
ii. 261 ; KATH.[2] 90.

   behōf-sam, adj., *profitable*, AYENB. 99 ; bihovesum AR. & MER. 2804.

bi-hōflen, v., *O.E.* bi-, behōfian, = *O.Fris.* bihōvia ; *behove* ; bihoven [' *portere*'] WICL.
WISD. xv. 12; **bihōfeþ** (*pres.*) ORM. 16706; bihoveð REL. I. 83; LAȝ. 945; (*ms.* bioueð) GEN.
& EX. 1159 ; bihoveþ LANGL. *B* xvii. 313); hem
bihoveþ muche mete ROB. 177 ; þe bihoveþ
godes helpe SHOR. 95; behoveþ AYENB. 58 ;
behoviþ TOR. 162; bi-, behoves WILL. 729,
2349 ; behoves IW. 3022; S. S. (Wr.) 1545 ;
þe ne be(h)ofiað nanre dedbote HOM. I. 245;
bihōfede, bihovede (*pret.*) LAȝ. 657, 1081 ;
bihovede A. R. 394 ; JUL. 50 ; hir bihoved þer
to bide PERC. 2228.

bi-hōfiȝ, adj., = *M.L.G.* behōvig, *M.H.G.* behuofec (*egenus*) ; *? profitable* ; hwet is elde bihovig HOM. II. 109.

bi-hōfðe, sb., = *O.Fris.*, *M.L.G.* behōfte ; *use*,
HOM. I. 19; bihofþe ROB. 26 ; (*ms.* biofþe)
LAȝ. 4378* ; bihofþe, bihefthe, bihofte LANGL.
C xiii. 187 ; bihofte (*ms.* biofte) GEN. & EX.
1408.

bi-hoȝien, v., *O.E.* behogian ; *care for* ; þe
laverd scal bihoȝian þet he habbe godes fultum HOM. I. 113 ; heore gode wepnen wurðliche **bihoȝeden** (*pret. pl.*) LAȝ. 17369.

bi-hōn, v., *O.E.* behōn, = *O.H.G.* pihāhan ;
clothe ; mid gode ræve bihon LAȝ. 5620 ; bi
hēngen (·*pret.*) HOM. II. 89 ; **bihangen,**
-hongen (*pple.*) LAȝ. 3637, 24465.

bi-horȝen, v., *defile* ; **behorewed** (*pple.*)
AYENB. 237.

bi-hrīnen, v., *touch, fall upon* ; birine HORN
(L.) 11.

bi-hūden, v., *O.E.* bi-, behȳdan, *O.H.G.* behuotan; *conceal* ; **bihüt** (*pres.*) HOM. I. 109;

bihū̆d (*imper.*) REL. I. 177 ; **bihū̆de** (*subj.*) HOM. I. 109 ; **behĕdde** (*pret.*) MAT. xxv. 25 ; bihū̆ded (*pple.*) FRAG. 7 ; bihud A. R. 100.

bi-hürnen, v., *hide in a corner*; **biherneþ** (*pres. pl.*) PL. CR. 642.

bi-jápe, v., *make sport of*, GOW. III. 167 ; bijapeþ (*pres.*) PL. CR. 46 ; **bijáped,** bejaped (*pple.*) CH. C. T. *A* 1585 ; god wil nouȝt be bigiled . . . ne bijaped LANGL. *B* xviii. 290.

bīke, sb., *bee's nest*, C. M. 76 ; GOL. & GAW. 406 ; a bike of waspes HALLIW. 175 ; as bees dos in the bike TOWNL. 325.

bi-kēmen, v., *? grow faint*; **bikēmet** (*pple.*) KATH. 1298*.

bi-kennen, v., = *O.H.G.* bechennan; *indicate, consign*; bikenne MAN. (H.) 274 ; **bikenne** (*pres.*) LANGL. *A* ii. 32 ; i bikenne ȝou to krist WILL. 5424 ; þat bikenneth þat croiz HAV. 1268 ; **bikende** (*pret.*) GAW. 596 ; bekende D. ARTH. 2340 ; his modir in keping to þe he bekende HALLIW. 160.

bi-kennen, v., *for* ȝe-kennen ; *? beget*; **bikenned** (*pple.*) REL. I. 234.

bi-keorven, v., *O.E.* beceorfan, = *O.Fris.* bikerva; *amputate*, JUL. 67 ; **bikorven** (*pple.*) A. R. 362 ; alle weren . . . hefdes bicorven MARH. 19.

biker, sb., *Lat.* bicarium, *O.N.* bikarr, = *O.L.G.* biker, *O.H.G.* becher; *beaker*, '*cymbium*,' PR. P. 35.

biker, sb., *fight*, ROB. 538 ; CH. L. G. W. 2659 ; [*v. r.* beker, bikre, bikkir] MAN. (F.) 9234 ; bikir '*pugna*' PR. P. 35 ; ALIS. 1661.

bikerin, v., *mod.Eng.* bicker; *fight*, '*pugno*,' PR. P. 36 ; bikere MAN. (H.) 256 ; beker ANT. ARTH. iv ; **bikere** (*pres.*) LANGL. *B* xx. 78 ; **bekerde** (*pret.*) D. ARTH. 2096.

bikering, sb., *bickering*, ROB. 540 ; ALIS.² (Sk.) 390.

bil, sb., *O.E.* bil, bill, = *O.L.G.*, *M.H.G.* bil, *mod.Eng.* bill (*axe*) ; *sword, axe, pickaxe*, P. S. 151 ; W. & I. 37 ; bill Iw. 3225 ; bille '*ligo, marra*' PR. P. 36 ; **billes** (*pl.*) ALIS. 1624 ; *comp.* **twī-bil.**

bil-ibēat, sb., *conflict of swords*, LAȝ. 1740.

bi-lacchen, v., *catch* ; **bilaucte** (*pret.*) HORN (H.) 681 ; **bilagt** (*pple.*) GEN. & EX. 773.

bi-lǣden, v., *O.E.* belǣdan, = *O.H.G.* pileitan; *? conduct, treat*; and wel narewe þe **bilēdeþ** O. & N. 68 ; wes uvele bilĕd (*pple.*) MISC. 45 ; lodliche was bilad C. L. 1136 ; bilad (*lead astray*) ALEX.² (Sk.) 906.

bi-lǣfen, -lēven, v., *O.E.* belǣfan, *Goth.* bilaibjan ; *relinquish, remain*, LAȝ. 1055, 2254, 19777; bileafen FRAG. 7 ; bileaven þe sunne A. R. 340 ; bilefen ORM. 8380 ; þei . . . shule wiþ Satanas bileven H. H. 232 (236) ; bleve ROB. 14 ; S. S. (Wr.) 48 ; WILL. 2577 ; MIN. 10 ; bi-, beleve CH. C. T. *F* 583 ; bileve, -leave O. & N. 464, 1688 ; blevin PR. P. 39 ; þer he ssel bleve AYENB. 172 ; **bilēf** (*imper.*) BEK.

141 ; **bilēfde,** bilæfde (*pret.*) LAȝ. 389, 1613 ; bilefde A. R. 372 ; bilefte DEGR. 1885 ; bileafden HOM. I. 93 ; bileuid MIN. iii. 66 ; blefede OCTOV. (W.) 507 ; blefte AYENB. 12 ; **belǣfed** (*pple.*) FRAG. 6 ; bilefed ORM. 8914 ; bileved HOM. I. 81 ; ALIS. 5310 ; bileaved A. R.' 168.

bi-lǣfing, sb., *remaining; patience, perseverance*; bileveing AR. & MER. 8611 ; blevinge AYENB. 72.

bi-lǣve, sb., *? remnant*; **bilēven** (*pl.*) GEN. & EX. 3154.

bi-lǣwen, v., *O.E.* belǣwan; *betray*; belewen HOM. I. 229 ; beleawan MAT. xxvi. 16.

bi-laggen, v., *clog with mud*; **bilagged** (*pple.*) '*esclaboté*' VOC. 173 ; belaggid '*madidatus*' PR. P. 29.

bi-langen, v., = *M.Du.* belanghen; *belong*; belonge GOW. II. 351 ; **belongeþ** (*pres.*) AYENB. 12 ; wat belongeth hit to me MISC. 29 ; belongeþ LANGL. *C* ii. 43 ; **bilongende** (*pple.*) GOW. I. 121.

bi-lappen, v., *wrap about* ; **bilapped** (*pple.*) ORM. 14267 ; S. S. (Web.) 2210 ; AM. & AMIL. 1014 ; hit is bilepped & bihud A. R. 100.

bi-lasten, v., *? O.E.* behlæstan ; *? charge*; **belast** (*pple.*) HALLIW. 161.

bi-láven, v., *wash all over*: **biláved** (*pple.*) mid blode MISC. 140.

bild, *see* **büld.**　　**bīle,** *see* **būle.**

bile, sb., *O.E.* bile ; *bill, beak*, A. R. 84 ; HOM. II. 49 ; O. & N. 79 ; REL. I. 210 ; P. S. 333 ; TREV. V. 291 ; bile, bille LANGL. *B* xi. 349 ; CH. C. T. *B* 4051 ; bille GOW. II. 106 ; S. S. (Wr.) 2196.

bil-fōder, sb., *? food for young birds, provender*, WILL. 81 ; bilfodur 1849.

bi-lēafe, sb., *for* ȝelēafe; *belief*, HOM. I. 101 ; bileave A. R. 2 ; SHOR. 142 ; beleave, beliave MISC. 27 ; bileave, beleave AYENB. 12, 14 ; bileve BEK. 30 ; bileve, bileeve CH. C. T. *A* 3456 ; beleve MAND. 131 ; *comp.* **un-bilēave.**

bilēf-ful, adj., HOM. II. 19.

bi-lēan, v., *O.E.* belēan ; *prohibit, hinder by blame* : for to **bilēande** þat no man werpe þe gilt . . . anuppen god HOM. II. 107.

bi-leggen, v., *O.E.* bilecgan, *O.H.G.* pileggan; *lay round, surround, besiege*; bilegge (*gloze*) O. & N. 672 ; & þus þa wordes we **bileggeð** (*explain*) HOM. I. 57 ; **bilæ(i)de** (*pret.*) LAȝ. 14223 ; þe bære was bileȝd (*pple.*) wiþ bæten gold ORM. 8167.

bi-lenge, adj., *? for* ȝelenge ; *belonging, related to*, ORM. 2230.

bi-lēoȝen, v., *O.E.* belēogan, = *O.H.G.* beliugan; *belie, caluminiate*; bilie LANGL. *B* v. 414; biliveth x. 22 ; witnesses false and fele **bilowen** (*pret. pl.*) hine SHOR. 84 ; **bilowen** (*pple.*) LANGL. *B* ii. 22 ; & te sakelease ofte bilowen A. R. 68.

bi-lēovien, v.,= *Ger.* belieben ; *love, delight in* : & þis wel **bilēovede** (*? for* bilovede) LA3. 5204.

bi-lēsnien, v. *cf.* **lēas** (*falsehood*)*; delude*; þus þe deofel wule bilesnien þe wreche HOM. I. 23.

bi-lēven, v. *for* 3elēven; *believe* ; beleven MISC. 56; beleve AYENB. 151 ; MAND. 53 ; belive FER. 398 ; ich **bilēve** (*pres.*) on god HOM. I. 217; (we) **bilīveð** LA3. 13966; beleve we MISC. 28 ; bileving THREE KINGS 2 ; **bi-lēfde** (*pret.*) LA3. 2856*; bilufde 13415 ; bilefden HOM. I. 19; LEB. JES. 397 ; belevit (*expected*) D. TROY 10919.

    beleevable, adj., *credible*, WICL. PS. xcii. 5.

bī-levinge, sb., *cf.* **bī-live** ; *means of living, food*; he het heom . . . heore in and heore bilevinge greiþi þat heo schulde MISC. 84.

bile-wit, adj., *O.E.* bilewit; *simple, innocent* ; bilehwit HOM. I. 95 ; MARH. 22 ; bilewhit ORM. 6654 ; **bilewitte** (*pl.*) HOM. I. 223.

    bile-hwitnesse, sb., *simplicity*, HOM. I. 95.

bilibre, sb., *Lat.* bilíbra ; *weight of two pounds*, WICL. APOC. vi. 6.

bi-liggen, v.,*O.E.*belicgan,= *O.H.G.*piliggan; *lie by, lie about, lie with*, LA3. 506 ; þe six werkes . . . þe bilige to nihte REL. I. 132 ; **bilai** (*pret.*) ROB. 19 ; bilæi [bilai] þa burh LA3. 9433; þat bilai (*appertained*) to his kinedom C. L. 286; **bilein** (*pple.*) SPEC. 46 ; beℓein GOW. I. 338 ; his doughtir þat was bilain RICH. 1119.

bi-li3e, v.,*O.E.*bihlehhan,-hlyhhan; *laugh at*, P. L. S. xix. 235 ; **bilō3** (*pret.*) P. L. S. xiii. 358 ; SHOR. 102 ; bilowe ROB. 299 ; bilaute HORN (L.) 681.

bi-līhten, v., *illumine* : godes brihtnesse bilihte hem HOM. II. 31.

bi-līken, v., *simulate, feign* ; **bilīked** (*pple.*) O. & N. 842.

bi-limien, v., *dismember*, LA3. 29364; bilime ROB. 301 ; **bilimeð** (*pres.*) KATH. 2160 ; A. R. 360; bilimeden (*pret.*) AR. & MER. 5775 ; **bilimed** (*pple.*) BEK. 560.

bi-limpen, v.,*O.E.*belimpan; *happen, pertain to*, ORM. 11085; **bilimpeð** (*pres.*) HOM. I. 51; to Mercene lawe bilimpeþ viii schiren MISC. 146 ; bilimpes REL. I. 216; **belamp** (*pret.*) HOM. I. 219; **bilumpen** (*pple.*) ORM. 2905.

bī-live, sb., *O.E.* bīgleofa, = *O.H.G.* pīlibi ; *livelihood, means of living*, ROB. 496 ; bilive, -leve HOM. II. 27, 99; hiF LANGL. *B* xix. 230 (*C* xxii. 235); bileove A. R. 168.

bi-līven, v., *O.E.* bēlīfan, = *O.H.G.* pi-, beliban, *mod.G.* bleiben ; *remain* ; bilive REL. I. 274 ; **belāf**, belæf, beleaf (*pret.*) SAX. CHR. 257, 259 ; 3ho bilæf wiþ hire frend ORM. 2391; bilef GEN. & EX. 671.

bille, sb., *O.Fr.* bulle, *from Lat.* bulla ; *bill, note*, LANGL. *A* iv. 34 ; CH. C. T. *C* 170 ; **billes** [' *libellos* '] TREV. V. 144.

bille, v., *enter* (*in a bill, etc.*) ; *libel*, P. P. II.

228 ; pardoun in book is **billed** (*pple.*) L. H. R. 138.

billen, v., *bill, peck with the beak*, ' *rostro*,' PR. P. 36 ; **billeð** (*pres.*) REL. I. 210 ; **billid** (*pple.*) DEP. R. iii. 37.

billing, sb., *pecking*, REL. I. 218 (MISC. 13).

billet, sb., *small note* ; bilet [*v. r.* bille] '*matricula*' PR. P. 36.

billet, sb., *piece of wood* ; bilet '*tedula* ' PR. P. 36.

billin, v.,= *M.H.G.* billen ; *hack with a bill*, '*marro*,' PR. P. 36.

bi-lōken, v., *look at, look about*, ORM. 2917 ; beloken HOM. II. 77 ; **bilōkeð** (*pres.*) A. R. 132 ; biloken REL. I. 220 ; **belōked** (*pple.*) OCTOV. (W.) 1046.

bi-long, adj.,= *O.L.G.* bilang ; *pertaining* ; ðe reching wurð on God bilong GEN. & EX. 2058.

bi-looghe, *see under* **lāh**.

bi-lūken, v., *O.E.* bi-, belūcan, = *O.L.G.* bi- lūcan, *O.H.G.* pilūchan ; *shut in, close around*, REL. I. 183 (MISC. 129) ; bilouken SHOR. 121 ; biloke FER. 2127 ; **bilūkeð** (*pres.*) H. M. 37 ; bilucþ FRAG. 1 ; biluken ORM. 12126 ; **belēac** (*pret.*) HOM. I. 225 ; bilek O. & N. 1081 ; **bi-lōken** (*pple.*) A. R. 160 ; R. S. vii ; REL. I. 184 ; bilokenn ORM. 11186; biloke C. L. 992 ; beloke AYENB. 97.

bi-lürten, v., *deceive* ; bilirten REL. I. 217 (MISC. 13) ; GEN. & EX. 316 ; **bilürt** (*pple.*) A. R. 280* (*v. r.* bichard).

bi-luvien, v. impers., *delight, please* ; 3ef ham **biluveð** (*pres.*) to heren him HOM. I. 257; 3if hit eow biloveð LA3. 989 ; 3if . . 3e alle biluvien gode mine lare 921 ; 3ef me swa **biluvede** (*pret.*) JUL. 24; alle hit biluveden LA3. 1013 ; **beloved** (*pple.*) GOW. II. 273 ; LANGL. *B* iii. 211 ; he was bet biloved P. L. S. xxiv. 33 ; so wel bilovede [beloved] CH. C. T. *A* 1429.

bi-mænen, v., *O.E.* bimǣnan, = *O.H.G.* pi-meinen ; *bemoan* ; *mean, signify* : mid þine muþe bimænen þine neode FRAG. 8 ; bimenen REL. I. 174 (MISC. 115) ; sunne bimenen (*moan*) HOM. I. 13 ; bimene ROB. 490 ; M. ARTH. 856; wat mai this bimene (*mean*) HAV. 1259; **bi-mēneð** (*pres.*) GEN. & EX. 2226 ; wat þis mon-taigne bimeneth [bemeneþ] LANGL. *A* i. 1 ; þer fore we ne bimeneþ þe no3t BEK. 983 ; men . . . bimeneþ goode mete3iveres and in mane haveþ LANGL. *B* xv. 143 ; **bimēnte** (*pret.*) LANGL. *B* xviii. 18 ; hi bimende P. L. S. xvii. 426 ; gie is moche **bemoonid** (*pple.*) of all(e) HALLIW. 163.

bi-mālen, v., *from O.E.* māl (*spot, mole*) ; *soil, sully*, bimolen LANGL. *B* xiv. 4.

bi-mang, *see under* **mang**.

bi-mannen, v., = *M.H.G.* bemannen ; *man, furnish with men* : þah an castel beo wel bemon(n)ed (*pple.*) HOM. I. 23.

bi-másen, v., *confuse* ; **bimásed** (*pple.*) A. R.

270 ; and lefte us liinge . . . al bemased in a
soune CHEST. II. 93.

bi-melden, v., = Ger. bemelden ; *denounce* ;
bimelde (*pres. subj.*) AN. LIT. 3.

bi-mēn, sb., *from* **bimǣnen** ; *complaint,* GEN.
& EX. 2894.

bi-moderen, v., = M.Du. bemoderen ; *cover
with dirt* ; **bimodered** (*pple.*) P. S. 158.

bi-mowe, v., *make a scornful grimace,* APOL.
81 ; the lord schal bemowe them WICL. PS.
ii. 4 ; **bimowiden** (*pret. pl.*) WICL. 2 PARAL.
xxx. 10.

bi-murnen, v., O.E. bi-, bemurnan ; *bemourn* ;
bimurneð (*pres.*) HOM. I. 149 ; II. 23 ; **bi-
murneden** (*pret.*) HOM. I. 155 ; bimorneden
WICL. LK. xxiii. 27 ; bemournet D. TROY
3279.

binacle, sb., *pinnacle,* RICH. 4150.

bī-náme, sb., = M.H.G. bīname ; *byname,
cognomen* ; **bīnámes** (*pl.*) CH. BOET. iii. 9
(84).

binde, sb., O.E. binde, = O.H.G. binta ; *anything that binds ; bine, stem of climbing plant ;
ribbon* ; PR. P. 36 ; A. P. iii. 444 ; *comp.*
wude-binde.

bind-balk(e), sb., = Ger. bindebalken ; *tie-
beam,* '*trapecula,*' VOC. 203.

binden, v., O.E. bindan, = O.L.G., Goth. bindan, O.Fris., O.N. binda, O.H.G. bintan,
pintan ; *bind,* LAȝ. 18458 ; HAV. 1961 ; to
bindene HOM. I. 37 ; **bindeþ** (*pres.*) SPEC.
104 ; bint A. R. 6 ; AYENB. 33 ; CH. COMPL.
M. 47 ; bind (*imper.*) JUL. 37 ; **band** (*pret.*)
Iw. 1776 ; band, bond CH. C. T. A 4082 ;
bond LAȝ. 767 ; HAV. 537 ; TREAT. 136 ;
bond his wunden HOM. I. 79 ; bond, bounde
WICL. JER. xxxix. 7 ; MAT. xiv. 3 ; bounde
N. P. 20 ; luve bonde hire GEN. & EX. 2692 ;
bunden (*pret. pl.*) LAȝ. 10006 ; ORM. 15820 ;
summe þer weren þet his eȝan bundan
HOM. I. 121 ; bunden, bounden HAV. 2442,
2506 ; bounden WILL. 1219 ; bonde ROB.
126 ; **bunden** (*pple.*) HOM. II. 11 ; HAV.
1428 ; IW. 289 ; MAN. (H.) 138 ; bunden,
bonden PR. C. 3208, 3210 ; bounden H. H.
53 ; CH. C. T. E 704 ; WICL. JOHN xi. 44 ;
bounde WILL. 2091 ; bonden WILL. 2238 ;
PERC. 1830 ; *comp.* **for-,** ȝe-, **un-binden** ;
deriv. **binde, band, bend.**

bindere, sb., O.E. bindere ; *binder* ; **binderes**
(*pl.*) HAV. 2050 ; *comp.* **bōk-binder.**

bindunge, sb., *binding,* HOM. I. 207 ; bind-
inge '*ligatio*' PR. P. 36 ; WICL. S. W. III.
481.

bi-nēden, v. impers., = M.H.G. benœten ;
need ; to mine gode ne **beniedeð** þe '*bonorum
meorum non indiges*' HOM. I. 217.

bi-neoðen, adv. & prep., O.E. beneoðan,
-nyðan, = O.Fris. binetha, binitha ; *beneath,*
A. R. 304 ; JUL. 56 ; bineoðen, -neoðe [bi-
neoþe, beniþe] LAȝ. 1325, 14985 ; bineoþen,

-neþen C. L. 1304 ; bineþen P. L. S. viii. 44 ;
ORM. 10729 ; her bineðen GEN. & EX. 10 ;
beneþen MAND. 158 ; bineþen, -neþe WICL.
GEN. vi. 16 ; bineþe TREAT. 132 ; beneþe
AYENB. 108 ; biniþe FER. 3260.

bing, sb., O.N. bingr ; *heap,* '*cumera,*' PR. P.
36 ; M. H. 97.

bi-nimen, v., O.E., bi-, beniman, = O.L.G.
biniman, Goth. biniman, O.H.G. pi-, beneman ;
take away, A. R. 194 ; CHR. E. 292 ; binimen þe
þine rihte LAȝ. 3694 ; binimen, -nime P. L. S.
viii. 24 ; binime BEK. 1008 ; WICL. ECCLUS.
xxviii. 19 ; bi-, benime C. L. 1054 ; AYENB.
39, 59 ; binime, -neme TREV. I. 73 ; **bineme**
(*pres.*) MISC. 32 ; **binim** (*imper.*) S. S. (Wr.)
705 ; **bineome** (*subj.*) HOM. I. 245 ; **binam**
(*pret.*) TREV. IV. 143 ; binam, binom LAȝ.
15989, 26112 ; binomen LAȝ. 215 ; MISC. 56 ;
binumen (*pple.*) FRAG. 7 ; ORM. 7299 ; GEN.
& EX. 198 ; binume O. & N. 1226 ; binomen,
-nome LANGL. B iii. 312 ; benomen R. R.
1509 ; binomen (*benumbed*) C. M. 22829.

binne, sb., O.E. binn (*manger*), O.Fr. benne
(*basket*) ; *bin, manger, chest,* '*capsa,*' VOC. 180 ;
L. H. R. 145 ; CH. C. T. A 593 ; fram þare
binne ['*a praesepio*'] LK. xiii. 15 ; in a bestis
binne LUD. COV. 159.

binnen, adv. & prep., O.E. binnan, beinnan,
= M.L.G., M.Du. binnen ; *for* **bi-innen** ;
within, SAX. CHR. 249 ; ORM. 6292 ; GEN. &
EX. 1032 ; REL. I. 182 ; binnen heo iwenden
LAȝ. 5920 ; binnen lut ȝearen 221 ; binne HAV.
584 ; SHOR. 40 ; MAN. (H.) 7457.

bin-ward, adv., *within,* LAȝ. 30773.

bi-nōten, v., *use* ; **benōteþ** (*pres.*) AYENB. 90.

bī-paþ, sb., *bypath,* MED. 486 ; bipath LIDG.
M. P. 114.

bi-pēchen, v., O.E. bepǣcan ; *deceive,* LAȝ.
5301 ; REL. I. 180 ; bipeche MISC. 72 ; P. L. S.
v (a). 10 ; **bipēhte** (*pret.*) HOM. I. 91, II.
191 ; **bipēiȝt** (*pple.*) P. L. S. v (b). 22 ; be-
paht SAX. CHR. 260 ; we beoð . . . bipahte
R. S. v (MISC. 176).

bipēching, sb., *deception,* HOM. II. 213.

bi-pennen, v., *pen up, shut in* ; bipenned
(*pple.*) A. R. 94.

bi-pilien, v., *peel* ; bipiled (*pple.*) A. R. 148.

bipilunge, sb., *peeling,* A. R. 148.

bi-prēonen, v., O.E. beprēnan ; *pin up* ; bi-
prēoneþ (*pres.*) MISC. 101.

bi-qvaschin, v., *be shivered in pieces* ; bi-
qvashed (*pret.*) LANGL. B xviii. 246.

biqveste, *see* **bicweste.** **biqvide,** *see*
bicwide.

bir, *see* **bür.**

bi-rǣden [bireaden], v., O.E. berǣden, =
M.H.G. berāten ; *advise, deliberate,* LAȝ.
31072 ; birede M. T. 98 ; hvo þet wille wisliche
him berede AYENB. 172 ; **birēde** (*pres. subj.*)
MISC. 78 ; **birĕdde** (*pret.*) SHOR. 124 ; bi-

radde, bireadde KATH. 1237 ; biradden CHR. E. 40 ; **birăd** (*pple.*) SPEC. 41 (A. D. 185).

birche, sb., *O.E.* beorc, byrc,≐*M.Du.* berk, *O.H.G.* bircha, pircha, *O.N.* biörk *f.* ; *birch,* PR. P. 36 ; birche, berche CH. C. T. *A* 2921 ; burche VOC. 181 ; birke FLOR. 1518 ; PERC. 773 ; birches (*pl.*) ALIS. 5242.

birde, *see* **bürðe** *and* **bürde.**

bi-rēavien, v., *O.E.* berēafian, *Goth.* biraubōn, = *O. H. G.* pi-, beroubōn, *mod.Eng.* bereave ; *deprive, rob* ; birævien [bireave] LAȝ. 30311 ; bireaven H. M. 29 ; bireven JOS. 356 ; bireve LANGL. *B* vi. 248 ; **birēaveð** (*pres.*) A. R. 120 ; **birēvede** (*pret.*) C. L. 1349 ; oþre beræfedest [birefedest] rihtes istreones FRAG. 7, 8 ; bireveden HOM. I. 79 ; **berēaved,** beræfed (*pple.*) FRAG. 5, 6 ; biræfed ORM. 2832 ; þus wes þas kineriche of heora kinge biræved LAȝ. 2897 ; bireved O. & N. 120 ; HORN (L.) 622 ; SPEC. 101.

bi-rēdien, v., ? *make ready, prepare,* LAȝ. 4198.

bi-reinen, v., = *M.H.G.* beregenen ; *rain upon, wet* ; birine HORN (R.) 11 ; **bireined** (*pple.*) A. R. 344 ; WICL. EZ. xxii. 24 ; bereinid '*complutus*' PR. P. 32.

bi-rēmen, v., *cry out at,* HOM. II. 29.

bi-rennen, v., = *M.H.G.* berennen ; ? *run around* ; biarnde (*pret.*) LAȝ. 26775.

bi-rēowen, -rēwen, v., = *M.H.G.* beriuwen ; *repent, commiserate,* ORM. 4506, 7784 ; bir(e)ue TREAT. 139 ; **birēwe** (*pres. subj.*) LANGL. *B* xii. 250.

birēounesse, sb., = *M.H.G.* beriuwenisse ; *commiseration* (? *printed* bireaunesse), A. R. 66.

bi-rēowsen, v., *O.E.* behrēowsian, = *M.H.G.* beriuwesen ; *repent,* JUL. 44 ; ORM. 8800 ; birewsien, bireusien HOM. I. 21, 23 ; bireusie MISC. 78 ; bireusi (*printed* birensi) SHOR. 43.

birēousunge, sb., *O.E.* behrēowsung ; *repentance,* FRAG. 8 ; A. R. 372 ; bireowsinge H. M. 21.

bi-rīden, v., *O.E.* berīdan ; *ride around,* LAȝ. 10739.

birie, sb., ? = *M.Du.* berie ; *bier,* AYENB. 258.

biriȝen, *see* **büriȝen.**

bi-rinnen, v., *O.E.* birinnan, = *Goth.* birinnan, *O.H.G.* pirinnan ; *flow over ; run round ; surround* ; **bieorn** (*pret.*) LAȝ. 26064 ; biurnen 26066 ; wiþ tieres al **birunne** (*pple.*) HORN (L.) 654 ; bironne L. H. R. 137 ; þat lond is biurnan [biurne] mid þære sæ LAȝ. 1233 ; wit eild . . . berunnun C. M. 8351.

bi-rīpan, v., *pluck the fruit of* ; **birīpe** (*imp.*) PS. lxxix. 13.

bi-rīsen, v., ? *for* ȝerīsen ; *befit, become* : wisdom **biriseð** (*pres.*) weran & clenesse birisað wifan HOM. I. 111 ; þe biriseð (*sec. text* bicomeþ) to ælche kinge LAȝ. 9821.

birke, *see* **birche.**

birkin, adj., *O.N.* birkinn, = *M.Du.* berken, *O.H.G.* pirchīn ; *made of birch,* GOL. & GAW. 31.

birle, sb., *O.E.* byrle, byrele, *cf. O.N.* byrli ; *cup-bearer,* '*pincerna,*' FRAG. 2 ; birle [borle] LAȝ. 24164 ; **birles** (*pl.*) ORM. 14023 ; **birlen** (*dat. pl.*) LAȝ. 19942.

birlen, v., *O.E.* byrlian, = *O.N.* byrla ; *act as cup-bearer, pour out,* A. R. 114 ; ORM. 15418 ; brillin PR. P. 51 ; **birlide** (*pret.*) WICL. JER. xxv. 17 ; birled M. H. 120 ; sche birlid whit win TOR. 292 ; **birled** (*pple.*) ORM. 15225.

birler, sb., = *O.N.* byrlari ; *one who pours out drink,* '*exelerarius,*' VOC. 212 ; brillare PR. P. 51.

bi-robben, v., *rob* ; **berobbeþ** (*pres.*) AYENB. 39.

bi-rollen, v., *roll over* : al **birolled** (*pple.*) wiþ þe rain A. P. ii. 959.

bi-rōwen, v., *O.E.* berōwan ; *row about* ; birouwen [birowe] LAȝ. 21028.

birst, *see* **brust.** **birþe,** *see* **bürðe.**

birðel, *see* **bürðel.**

bis, adj., *dark grey, neutral blue* : he lis toumbed in a marble bis MAN. (H.) 230 ; bis and asure LIDG. BOCH. (*in* N. E. D.).

bi-sæȝen, v., *let down, lower* ; biseið (*pres.*) HOM. II. 215 ; **biseid** (*pple.*) in þe grune HOM. II. 211 ; bisaid 213.

bi-sálen, v., *assail* ; **besále** (*pres. pl.*) P. R. L. P. 103.

bi-samplen, v., ? *for* **examplen** (*sample*) ; *moralize about* ; **bi-saumpleð** (*pres.*) A. R. 88 ; **bisaumpled** (*pret.*) 316.

bī-sawe, sb., *proverb,* A. R. 88 ; bisawe TREV. III. 309.

bi-sāwen, v., = *M.H.G.* besæjen ; *sow* ; **bisaweð** (*pres.*) HOM. I. 107.

bi-sceadin, v., *O.E.* besceadian ; *overshadow* ; **beshádeth** (*pres.*) GOW. III. 111 ; **besháded** (*pple.*) GOW. II. 109.

bi-schadwe, v., = *M.Du.* beschaduwen, *M.H. G.* beschatewen ; *shadow around,* LEG. 83 ; **bi-schadeweþ** (*pres.*) S. S. (Web.) 586.

bi-scheden, v., = ? *M.H.G.* beschuten ; *besprinkle* : he **bischedde** (*pple.*) it wiþ watir WICL. 4 KINGS viii. 15.

bi-schenden, v., *ruin* ; **bischent** (*pple.*) C. M. 14838.

bi-schéren, v., *O.E.* bisceran ; *shear* ; **bi-schorn,** biscorn (*pple.*) C. M. 12231.

bi-schilden, v., ? *for* ȝeschilden ; ? *protect* : þine feond (? *r.* freond) þu **biscildest** (*pres.*) HOM. I. 7.

bi-schīne, v., *O.E.* bescīnan, = *O.H.G.* bi-scīnan, *Goth.* biskeinan ; *shine on,* HORN (H.) 12 ; beshine BER. 1113 ; **bishīneþ** (*pres.*) ORM. 18851 ; half þe eorþe þe sonne bi-schīneþ JARB. XIII. 167 ; beshineþ GOW. III.

242; **bischōn** (*pret.*) TREV. I. 393 ; **beschine** (*pple.*) TREV. I. 325.

bi-schīten, v., *O.E.* bescītan,=*M.H.G.* beschīzen ; *defile with excrement* ; **bishiten** (*pple.*) ALIS. 5485 ; besshit TOWNL. 235.

bischop, sb., *O.E.* biscop, bisceop, biscep, *from Lat.* episcopus ; *bishop*, R. S. vi ; biscop MISC. 43 ; **bispes** (*pl.*) MISC. 39 ; **bischopen** (*dat. pl.*) O. & N. 1761 ; *comp.* **arche-bischop**.

    bischop-hād, sb., *O.E.* bisceophād ; *office of bishop* ; bischophood WICL. PROL. I TIM. ; bischophade C. M. 21248.

    bischop-rīche, sb., *O.E.* biscoprīce ; *bishopric*, BEK. 1008 ; biscoprīche FRAG. 2.

    biscop-stōl, sb., *O.E.* bisceopstōl ; *episcopal see*, MISC. 145 ; **biscopstōle** (*dat.*) LAȝ. 18218.

    biscop-würt, sb., *O.E.* bisceopwyrt ; *bishop's wort, wood betony*, FRAG. 3.

bischopen, v., *O.E.* biscopian ; *confirm* (*as a bishop*) ; **bisschopeþ** (*pres.*) SHOR. 5 ; **busshoppede** (*pret.*) LANGL. C xviii. 268 ; bisbede, ibisbede MIRC 158.

    bisceopunge, sb., *bishoping, confirmation*, HOM. I. 101.

bi-schrēwen, v.,*beshrew, corrupt* ; **bischrēwe** (*pres.*] CH. C. T. *D* 845 ; beshrewiþ ['*depravat*'] WICL. PROV. x. 9 ; **bischrēwed** (*pret.*] LANGL. *B* iv. 168 ; **bischrēwed** (*pple.*) TREV. I. 113 ; beshrewed GOW. I. 76 ; þis world is al beshrewed P. 45.

bi-schrīchen, v., *screech about* : and þe **bischrīcheþ** [biscricheþ] (*pres.*) and bigredeþ O. & N. 67.

bi-schütten, v., *shut up* ; **bishette** (*pret.*) ALIS. 5709 ; bischitten, bishetten LANGL. *A* ii. 189 ; **bischet** (*pple.*) WILL. 2014 ; bishet OCTOV. (W.) 1280 ; beshet R. R. 4488 ; besset AYENB. 94.

bi-scorn, sb., *scorn*, TREV. I. 179.

bi-scornen, v., *cover with scorn* ; **biscornd** (*pret.*) C. M. 16611 ; bescorned [*v. r.* biscorned] CH. C. T. *I* 278.

bi-scunien, v.,*shun*, P. L. S. viii. 77 ; bischune (*imper.*) REL. I. 183.

bise, *see* **bisse**.

bīse, sb., *O.Fr.* bīse (*? from O.H.G.* bīsa) ; *north wind* : bigan a wind to rise out of þe north men calleth bise HAV. 724.

bi-sēchen, v., *O.E.* besēcan,= *O.Fris.* bisēka, *O.H.G.* besuochen, *mod. Eng.* beseech ; *seek, visit ; beseech* : nu ich mot bisechen [biseche] þat þing þat ich ær forhowede LAȝ. 3494 ; þæt heo walden bisechen þene king & bidden hine mildce 12272 ; þu scholdest bisechen me cosses A. R. 102 ; biseken GEN. & EX. 2492 ; HAV. 2994 ; **bisēche** (*pres.*) KATH. 2375 ; LANGL. *A* i. 58 ; ALIS. 4631 ; beseche MAND. 316 ; beseke LUD. COV. 129 ; bisekþ, bisehþ O. & N. 1439 ; bezekþ AYENB. 117 ; **bisēc** (*imper.*) HOM. I. 17 ; biseke we nu godes

migt GEN. & EX. 4155 ; **bisecheð** (*procure*) eow wepnen LAȝ. 12513 ; **bisōhte** (*pret.*) KATH. 2395 ; þa bisohte he nutescalen LAȝ. 29265 ; bisoghte PERC. 2144 ; bisouhten A. R. 230 ; **bisōht** (*pple.*) R. S. v ; SPEC. 72 ; bisouht A. R. 234 ; bisoȝt FL. & BL. 127.

    bisēchere, sb., *one who beseeches* ; **bisēcheris** (*pl.*) WICL. ZEPH. iii. 10.

    bisēchinge, sb., *beseeching*, WICL. 2 PARAL. vi. 19 ; bezechinge AYENB. 98.

    besēkand-lik, adj., *propitious*, PS. cxxxiv. 14.

bi-sēgen, v., *for* **asēgen** ; *besiege* ; bisegi ROB. 399 ; bisege TREV. I. 91 ; **bisēged** (*pret.*) WILL. 2843 ; **bisēged** (*pple.*) MAN. (F.) 5108.

bi-seggen, v., *O.E.* besecgan, = *O.L.G.* biseggean, *O.H.G.* pisagen ; *signify* ; biseið (*pres.*) HOM. II. 173.

bi-sēme, v., *for* **ȝesēmen** ; *beseem, seem, befit*, GOW. I. 110 ; A. P. i. 310 ; **bisēmeð** (*pres.*) JUL. 55 ; þat ivele hem bisemeþ PL. CR. 58 ; bisemez GAW. 1612 ; as þei besemen WILL. 2529 ; bisemen '*conveniunt*' WICL. ROM. i. 28 ; **bisēmede** (*pret.*) A. R. 148 ; þine wordes beoþ . . . so **bisēmed** (*pple.*) O. & N. 842 ; besemid TRIAM. 720.

bi-senchen, v., *O.E.* bisencan, = *O.L.G.* bisenkian ; *submerge*, A. R. 400 ; bisenken ORM. 19690 ; hi **bisencheð** (*pres.*) us on helle HOM. I. 107 ; **bisencte** (*pret.*) HOM. I. 87 ; biseinte [bisencte] A. R. 334 ; ure ifan . . . beoð **bisencte** (*pple.*) in to helle HOM. I. 87.

bi-senden, v.,=*M.L.G.*, *M.H.G.* besenden ; *send to* : erl Jon . . . **bisende** (*pret.*) him al so & bisouȝte (h)is grace ROB. 491.

bī-sēne, adj., *O.E.* bīsēne, *early mod.Eng.* bisson ; *blind* ; bisen C. M. 16595 ; bisne GEN. & EX. 472 ; þu art blind oþer bisne O. & N. 243.

bi-sengen, v., *O.E.* besengan ; *singe* ; bezenge AYENB. 230.

biseninge, pple. *cf. O.N.* býsna ; *ill-boding, monstrous*, BARB. LEG. II. 129.

bi-sēon, v., *O.E.* besēon,= *O.L.G.* bisehan, *O.H.G.* pisehan, *Goth.* bisaihvan ; *see, provide*, A. R. 344 ; biseon to me HOM. I. 197 ; bisen GEN. & EX. 1313, 2141 ; S. S. (Web.) 507 ; biseo BEK. 105 ; to **bisēonne** MARH. 15 ; þonne **besihþ** (*pres.*) þeo soule sorliche to þen lich(ame) FRAG. 5 ; **bisih** (*imper.*) H. M. 33 ; bisih, bisiȝ [biseh] þe LAȝ. 16048, 24157 ; þe bet he hine biseo [beseo] O. & N. 1272 ; **bisæh** (*pret.*) LAȝ. 4907 ; biseh JUL. 21 ; bisai ROB. 192 ; þa eorles . . . biseȝen heom bihinde LAȝ. 26506 ; **bisēhe** (*subj.*) MISC. 193 ; **biseȝe** (*pple.*) P. L. S. xii. 103 ; besein GOW. I. 341 ; biseie HOCCL. i. 142 ; elde . . . haþ me biseie LANGL. *B* xx. 201 ; so richeli biseie CH. C. T. *E* 984.

    bisēing, sb., *circumspection* ; beziinge AYENB. 184.

bisēken, see bisēchen.

bi-seowen, v., *O.E.* besiwian; *sew*; bisǫwe WILL. 1688; **besewed** (*pple.*) WILL. 3117; GOW. III. 312.

bi-serve, v., *serve diligently*; beserve C. M. 23053.

bi-setten, v., *O.E.* besettan, = *O.H.G.* pisezzan, *Goth.* bisatjan, *mod.Eng.* beset; *place, arrange, set; surround*, REL. I. 213; bisette LANGL. *B* v. 264; CH. C. T. *A* 279; besette þe tour FER. 3228; (h)is catel to bisette LEB. JES. 116; wel bezette þane time AYENB. 214; þe biset his iþonc on his ehte HOM. I. 101; bisette (*pret.*) GEN. & EX. 2687; WILL. 1214; and bisette þis schip al aboute BRD. 21; biset (*pple.*) A. R. 58; KATH. 1578; it was eȝwhær biset wiþ deorewurþe stanes ORM. 8169; þou hast wel beset mi god OCTAV. (H.) 956.

bisi, see büsi.

bī-sibbe, sb., *? relative*, SHOR. 70; '*affinis*' VOC. (W. W.) 562.

bi-sīchen, v., *sigh, sigh for*; bisīcheð (*pres.*) HOM. II. 201.

bi-sīde, -sīden, -sīdes, see sīde.

bi-siht, sb., = *M.H.G.* besiht, *f.; provision*; besiht C. L. 311; besiȝte PROCL. 6.

bi-sinken, v., *O.E.* besincan, = *O.L.G.* bisincan, *M.H.G.* besinken; *sink*: bisinkeð (*submerges*) HOM. II. 177; hwa se lið i leiven deope bisunken H. M. 33.

bi-sitten, v., *O.E.* besittan, = *O.L.G.* bisittian, *O.H.G.* pisizzan; *sit upon, beset, oppress*; hit schal bisitten [bisitte] oure soules sore LANGL. *A* ii. 110.

bi-slaberen, v., = *L.G.* beslabern; *beslaver*; bislabered (*pple.*) LANGL. *B* v. 392.

bi-sloberen, v., *bedabble*; beslobered (*pple.*) LANGL. *B* xiii. 401.

bi-slomered, pple., *covered with mud*: his hosen al beslomered in fen PL. CR. 427.

bismer (? bīsmer), sb., *O.E.* bi-, bysmer, bismor, = *O.L.G.* bismer, -smar, *O.H.G.* pi-, bismer, *n.; mockery, contumely, insult*, HOM. I. 107; þe bismar ROB. 12; þe þe bismar haveð idon LAȝ. 4403; **bismere** (*dat.*) KATH. 551; LANGL. (Wr.) 2651 [bismer *B* v. 89]; H. S. 7400; bi-, busmere TREV. I. 179; lauhwen him to bismare A. R. 270; bismare, bisemare CH. C. T. *A* 3965; bisemare ALIS. 648.

   bismere-spǣche, sb., *cf. O.E.* bysmorsprǣc; *blasphemy*, MAT. xxvi. 65.

bismerien (? bīsmerien), v., *O.E.* bismerian, bismrian, = *O.H.G.* pismerōn, bismarōn; *mock, insult*; **bisemereþ** (*pres.*) and scorneþ þe guode men AYENB. 22; **bismereden** (*pret.*) LK. xxiii. 36.

bismerwen, v., *O.E.* besmyrian, = *M.H.G.* besmiren; *besmear*; bismeoruwed (*pple.*) A. R. 214.

bi-smīten, v., *O.E.* besmītan, = *O.H.G.* pi-

smīzan; *contaminate, soil*; besmiten MK. vii. 15; ne bismīt (*printed* bi sunt) (*imper.*) þu þe mid drunkenesse HOM. I. 13.

bi-smitten, v., = *M.Du.* besmetten, *O.H.G.* bismizzen; *contaminate, pollute*; **bismitted** (*pret.*) '*contaminaverunt*' PS. liv. 21; bismitted (*pple.*) A. R. 214; besmetted, -smet AYENB. 32, 229.

bi-smōken, v., *cover with smoke*; bismōked (*pple.*) CH. BOET. i. 1 (5).

bi-smoteren, v., *stain, defile*; bismotered (*pple.*) CH. C. T. *A* 76.

bi-smudden, v., *besmut, soil*; bismuddet (*pple.*) A. R. 214*.

bi-smudelen, v., *make muddy, soil, defile*; bismuðeled (*pple.*) A. R. 214*.

bīsne, sb., *O.E.* bȳsen, bīsen, = *O.N.* bȳsn, *O.L.G.* (an)būsn, *Goth.* (ana)būsns *f.; example*, ORM. 8987; HALLIW. 179; sette us bisne HOM. I. 5; ure laverd ... teacheð us þurh a bisne 245; **bīsne** (*dat.*) H. M. 45; þa he to bisne nom LAȝ. 30; *comp.* for-bīsne.

bīsne, see bīsene.

bīsnien, v., *O.E.* bȳsnian, = *O.N.* bȳsna; *give an example*; **bīsenes** (*pres.*) M. H. 124.

   bīsnunge, sb., *O.E.* bȳsnung; *example*, HOM. I. 93; bisening L. H. R. 118.

bi-snīwen, v., *O.E.* besnīwan, = *M.H.G.* besnīen; *cover with snow*; **bisniwe** (*pple.*) LANGL. *C* vii. 266; bisnewed LANGL. *B* xv. 110; besnewed AYENB. 81; GOW. I. 111.

bi-sōcne, sb., *petition, request*, A. R. 376; (*spelt* bisockne) LAȝ. 30219*; bisokne S. A. L. 155; þoru bisokne of þe king delaied it was ȝute ROB. 495; bisōcnen (*dat. pl.*) HOM. I. 261.

bi-soilen, v., *besoil*; bisuiled (*pple.*) SHOR. 108; bisoiled MISC. 225.

[**bi-sorȝe-līch,** adj., = *Ger.* besorglīch (*solicitous*); *comp.* un-bisorȝelīch.]

bisp, see bischop.

bi-sparren, v., *O.E.* besparrian, = *O.H.G.* pisparran, -sperran; *shut up, enclose*; **bi-sperred,** bisperede (*pret.*) LANGL. *B* xv. 139; **bisparrede** (? *ms.* bisparred, *printed* bisparreð) (*pple. pl.*) A. R. 94*.

bi-spéken, v., *cf. O.E.* besprecan, = *O.H.G.* pi-sprechan, *O.L.G.* besprekan; *talk of, talk with*: þi soster bispekeþ þi deþ P. L. S. xiii. 144; **bispac** [bispek] (*pret.*) L. H. R. 32; sge ne bispac him nevere GEN. & EX. 1444; bispek ROB. 524; heo bispeken heom bitweonen þet heo walden ibuȝen to þere apostlan fer(r)eden HOM. I. 91; bispoken AR. & MER. 1240; **bispéke** (*pple.*) O. & N. 1738; ROB. 55; ȝe habbeþ a mong ȝu ... bispeke þat ilome BEK. 919.

bī-spel, sb., *O.E.* bigspell, = *M.H.G.* bīspel; *parable, example*, HOM. I. 233; O. & N. 127; **bīspellen** (*dat. pl.*) MAT. xiii. 3.

bi-spēten, v., *spit upon* ; bispete WICL. MK. x. 34 ; bispēteð (*pres.*) A. R. 288 ; **bispätten** (*pret.*) WICL. MK. xv. 19 ; **bispět,** bispat (*pple.*) WICL. LK. xviii. 32 ; bispat MISC. 140.

bi-spitten, v., *spit upon* ; bispitte WICL. MK. xiv. 65 ; **bispit** (*pple.*) P. R. L. P. 240.

bi-spotten, v., *bespot,* CH. BOET. iii. 4 (73).

bi-sprǣden, v., = *M.H.G.* bespreiten ; *bespread* ; **bisprǣdde** (*pret.*) LA3. 16521 ; **besprěd** (*pple.*) K. T. 762.

bi-sprengan, v., *O.E.* besprengan, = *M.H.G.* besprengen ; *besprinkle,* HOM. I. 127 ; **bispreng** (*imper.*) HOM. I. 73 ; **bisprenge** (*subj.*) ROB. 128 ; **bisprengde** (*pret.*) BEV. 350 ; bispreinde WICL. HEBR. ix. 19 ; **bispreint** (*pple.*) ST. R. 502 ; bespreint wiþ bloode LIDG. M. P. 91.

bisprengil, v., *sprinkle over* ; **besprengild** GEST. ROM. 26.

bi-spūsen, v., *for* spūsen ; *espouse* ; **bispūsed** (*pple.*) HOM. II. 159 ; bispusede II. 13.

bisse, sb., *O.Fr.* bysse, *from Gr.* βύσσος ; *a sort of fine stuff,* GOW. III. 34 ; bis SPEC. 26 ; biis GUY (Z.) 2835 ; WICL. LK. xvi. 19.

bi-standen, v., *O.E.* bestandan, = *Goth.* bistandan, *O.H.G.* pistantan ; *stand about, stand around* ; bistonden HOM. II. 173 ; **bistant** (*pres.*) O. & N. 1438 ; wenne hundes hine bistondeð LA3. 1698 ; **bistōd** (*pret.*) R. S. v (MISC. 180) ; GEN. & EX. 3857 ; whan hem bistode nede WILL. 175 ; stormes hem bistode TRIST. 367 ; **bistonden** (*pple.*) LA3. 30323 ; A. R. 264 ; JUL. 31 ; & is mid storme faste bistonden HOM. II. 43.

bi-stáren, v., *stare at* ; þe keiser **bistárede** (*pret.*) hire KATH. 309.

bi-stéden, v., = *M.Du.* besteden, *O.H.G.* pistaten ; *appoint, place* ; **bested** (*pple.*) TRIAM. 1461 ; bistad CH. C. T. *B* 649 ; whar was þe child ... bistad GREG. 455 ; bestad HOCCL. i. 129 ; LUD. COV. 329 ; wofulli bestad GOW. I. 198 ; bistaðed A. R. 264 ; MARH. 3.

bi-stéken, = *M.Du.* besteken ; *shut,* A. R. 62 ; **bisteken** (*pple.*) HOM. I. 247.

bi-stélen, v., *O.E.* bi-, bestelan ; *steal, steal away, steal on* : he wenden wel to bistelene in to þare burh3e LA3. 9756 ; and **bistál** (*pret.*) from þan fihte 28422 ; elde me is **bistólen** (*pple.*) on P. L. S. viii. 9 ; HOM. I. 161 ; & is bistole a wai FER. 3876.

bī-sti3, sb., *by-path* : **bisti3es** (*pl.*) SHOR. PS. xxii. 3 (*in* WICL. PREF. 4).

bi-stōwen, v., *bestow, place* ; bistowe SHOR. 95 ; LANGL. *B* vii. 75 ; CH. C. T. *A* 3981.

bi-steppen, v., *? for* 3esteppen ; *? step* ; beo heo **bistepped** (*pple.*) þer ute A. R. 174.

bi-strēwen, v., *O.E.* bestrēowian, = *M.H.G.* bestreuwen, -striuwen ; *bestrew* ; **bistrē-weden** (*pret.*) HICKES I. 167 (HOM. I. 5).

bi-strīden, v., *O.E.* bestrīdan, *? = M.L.G.*

bestrīden ; *bestride,* LA3. 28020 ; bistride HAV. 2060 ; ALIS. 706 ; **bistrīdeþ** (*pres.*) LANGL. *B* xvii. 78 ; **bistrōd** (*pret.*) LEB. JES. 803 ; þe stede he bistrood GAM. 189 ; bestrod TOR. 504.

bi-strüpen, v., *O.E.* bestrȳpan, = *M.H.G.* bestroufen ; **bestrēpþ** (*pres.*) þe zeve zennes vram þe herte AYENB. 123.

bi-stünten, v., *check, rein in* ; **bistint** (*pret.*) ALIS. 1183.

bi-stürien, v., *O.E.* bestyrian ; *bestir* ; bistere ALIS. 5837 ; bestere GOW. III. 295 ; **bestireþ** (*imper.*) 3ou AR. & MER. 6025.

bi-stüring, sb., *emotion* ; besteriinge AYENB. 263.

bi-sülien, v., *O.E.* besylian, = *M.H.G.* besüln ; *soil, defile* ; **bisülieð** (*pres.*) HOM. II. 37.

bi-sulpen, v., *? pollute* ; **bisulpez** (*pres.*) A. P. ii. 575.

bi-swǣten, v., = *M.H.G.* besweizen ; *cover with sweat* ; **biswǣt** (*pple.*) LA3. 9315.

bi-swāpen, v., *O.E.* beswāpan ; *envelope, entangle* : bi s(c)andlice senne **beswāpen** (*pple.*) HOM. I. 239.

bi-swenken, v., *? = M.H.G.* beswenken (*dolo capere*) ; *toil, labour* ; **biswenkez** (*pres.*) D. ARTH. 1128.

bi-swīkeŋ, v., *O.E.* bi-, beswīcan, = *O.L.G.* beswīcan, *O.H.G.* piswīchan ; *deceive,* LA3. 754 ; ALIS. 4609 ; biswiken A. R. 224 ; ORM. 11678 ; besviken SAX. CHR. 264 ; biswike O. & N. 158 ; HORN (L.) 667 (R. 669) ; beswike RICH. 5918 ; **biswīkeð** (*pres.*) HOM. I. 25 ; P. L. S. viii. 7 ; **biswāc** (*pret.*) LA3. 2126 ; biswak ORM. 12478 ; besvak C. M. 3734 ; biswoc HOM. I. 213 ; biswok MAN. (F.) 11398 ; biswiken (*pple.*) FRAG. 8 ; A. R. 54 ; ORM. 11640 ; GEN. & EX. 3561 ; biswiken SPEC. 45 ; bisviken C. M. 8632 ; biswike HAV. 1249 ; biswiķe ROB. 272 ; **biswīked** '*deceptus*' P. S. lxxvi. 3.

bezvīkere, sb., *deceiver,* AYENB. 171.

bezvīkinge, sb., *deceit,* AYENB. 28.

bi-swingen, v., *O.E.* beswingan ; *flog* ; **bi-sw(u)ngen** (*pple.*) SPEC. 81.

bi-swinken, v., *O.E.* beswincan ; *procure by labour* : þat mowen her bred biswinke LANGL. *B* vi. 216 ; **biswinkeþ** (*pres.*) PL. CR. 722 ; **biswonk** (*pret.*) L. H. R. 27 ; biswunken LEB. JES. 361 ; biswonke LANGL. *B* xx. 290 ; **beswonke** (*pple.*) HALLIW. 169.

biswinc-ful, adj., *toilsome* ; biswincfule A. R. 188.

bit, sb., *M.L.G.* bit, bet, = *M.H.G.* biz *n.* ; (*bi idle-*)*bit,* bitt '(*frenum*) *lupatum*' PR. P. 37 ; **bitte** (*dat.*) CH. L. G. W. 1208.

bi-tācnen, v., = *O.Fris.* bitēkna, *O.H.G.* bezeichenen ; *betoken, signify,* ORM. 919 ; bitokene TREV. V. 9 ; **bitācneð** (*pres.*) HOM. I. 79 ; bitocneð A. R. 170 ; betocneþ AYENB. 15 ; what bitacnieð [bitocneþ] þa draken LA3.

16008; **bitācnede** (*pret.*) HOM. I. 89; bitoknede ROB. 288; þat he has lord and king of kinges wel bitoknede þe sterre SHOR. 123; **bitācned** (*pple.*) H. M. 5; bitocned A. R. 374.

bitācnunge, bitācninge, sb., = *O.H.G.* pizeihinunga; *signification,* HOM. I. 47, 51; bitocnunge A. R. 270; bitokninge BEK. 102.

bi-tæchen, v., *O.E.* betæcan; *assign, commit, deliver,* ORM. 6126; bitæchen, bitechen LA3. 5314, 11502; bitechan ANGL. VII. 220; bitechen A. R. 300; HAV. 203; biteachen MARH. 5; biteche WILL. 5184; biteche (*pres.*) AL. (T.) 178; Crist i þe biteche HORN (R.) 577; **bitæc, -tēc** (*imper.*) LA3. 16847, 18922; **bităhte**(*pret.*) HOM. I. 11; KATH. 608; (*ms.* bitahhte) ORM. 14774; bitahte, bitæhte LA3. 247, 27246; bitagte GEN. & EX. 212; bitauhte HAV. 2212; betau3t(e) S. S. (Wr.) 324; betaughte R. R. 4438; þe ring þat i him bitaughte PERC. 2156; betohte HOM. I. 221; bitei3te BEK. 1827; **bităht** (*pple.*) SPEC. 42; bitau3t WILL. 5289; biteiht A. R. 166.

bi-táken, v., *commit, deliver,* LA3. 6251; HAV. 1226; bitake S. S. (Wr.) 30; þei schulen bitake ['*tradent*'] WICL. MAT. xxiv. 9; **bitáke** (*pres.*) ROB. 475; mi soule bitake [betake] i to Sathanas CH. C. T. *A* 3750; hi betakeþ AYENB. 36; bitak (*imper.*) HORN (R.) 785; **bitōk** (*pret.*) LA3. 22464*; WILL. 66; S. S. (Wr.) 412; CH. C. T. *G* 541; he bitoc it in warde ROB. 183; betoke TOR. 2250; **betōke** (*subj.*) AYENB. 89.

bī-tále, sb., *parable,* LEB. JES. 243.

bi-talten, v., *shake*; **bitalt** (*pple.*) A. P. i. 1160.

bi-tavelen, v., *overcome in gaming*; an anlepi meiden, wið hire anes muð, haveð swa bitevelet (*v. r.* bitaulet ow) . . . alle KATH. 1288.

bite, sb., *O.E.* bita, = *O.N.* biti, *O.H.G.* bizzo; *from* **biten**; *bit, morsel,* ROB. 207; C. L. 1343; þane bite (*earlier text* bitan) ['*buccellam*'] JOHN xiii. 30; vor þe bite of one epple A. R. 334; an bite brædes ORM. 8640; þe bite þat þei eten LANGL. *B* xviii. 200.

bite, sb., *O.E.* bite, = *O.L.G.* biti, *O.Fris.* biti, bite, *O.H.G.* biz; *bite, act of biting,* A. R. 288; S. S. (Web.) 760; MAND. 49; þin bite ['*morsus tuus*'] HOM. I. 123; þane bite AYENB. 61; þe bite of swordes egge KATH. 2437; mid steles bite LA3. 11327; bite, bitte C. M. 8500; bite, bitt(e) '*morsus*' PR. P. 37; bitte '*morsus*' VOC. 198; bitt ALIS. 5436; bit IW. 94; bite *cutting edge* AR. & MER. 4808; bit, bitte GAW. 212, 2224; biten (*dat. pl.*) LA3. 21364.

bitel, adj., *biting, keen,* ORM. 10074; bitele duntes LA3. 26967.

bītel, see **bētil.**

bītel-brouwed, adj. [*v. r.* biterbrowed], *beetle-browed,* LANGL. *A* v. 109.

bi-tellen, v., *O.E.* betellan (*defend*) = *O.Fris.*

bitella, *M.H.G.* bezellen (*acquire*); *defend, conquer* : bitellen wel & weren ORM. 2045; ðet tu beo mi motild a3eines mine soule fon þet heo hire ne muwen bitellen HOM. I. 205; to bitellen eowere rihtes LA3. 7894; bitelle LEG. 169; TOWNL. 217; and lust hu ich con me bitelle O. & N. 263; **bitelleð** (*pres.*) A. R. 226*; **bitóld** (*pret.*) GEN. & EX. 920; nu þu havest Brutlond al **bitáld** (*pple.*) (*sec. text* awonne) to þire hond LA3. 18099.

bitellunge, sb., *? defence, excuse;* A. R. 392.

bīten, v., *O.E.* bītan, = *O.L.G.* bītan, *O.Fris., O.N.* bīta, *Goth.* beitan, *O.H.G.* pīzan, bīzan; *bite,* A. R. 364; biten mete LA3. 15340; on his lippe . . . bite CH. C. T. *A* 3745; þer was no knif þat wolde him bite EGL. 491; **bīteþ** (*pres.*) ORM. 9954; bit HOM. I. 123; A. R. 186; MISC. 70; AYENB. 66; heo biteð LA3. 20173; biteþ AYENB. 70; **bītinde** (*pple.*) A. R. 428; AYENB. 143; **bāt** (*pret.*) ORM. 12422; IW. 2029; bot HOM. II. 181; GEN. & EX. 2926; C. L. 1343; L. H. R. 135; S. S. (Web.) 773; OCTOV. (W.) 329; LUD. COV. 29; Charlis bot his lippen FER. 141; he bot [boot] his lippes LANGL. *A* v. 67; biten (*pret. pl.*) ALIS. 5435; þa scipen biten on þat sond LA3. 1788; bite LEG. 161; **biten, beten** (*pple.*) CH. L. G. W. 2318; beten L. H. R. 74; ibite ALIS. 5438; ibete TREV. VII. 299; *comp.* a-, and-, for-**bīten**; *deriv.* bite, bitel, biter, bait.

[**bītere,** sb., *biter*; *comp.* bac-**bītere.**]

bīting, sb., *biting,* LIDG. M. P. 101; **bītinga** (*? plur.*) HOM. I. 33.

bi-tēon, v., *O.E.* betēon, = *Goth.* bitiuhan, *O.H.G.* piziuhan; *bestow, draw over, betray,* LA3. 9117; þi luve . . . hwar meiht tu biteon hire betere þen upon me A. R. 398; here swinc he wel biten GEN. & EX. 3626; **bitēg** (*pret.*) GEN. & EX. 2878; **bitó3en** (*pple.*) HOM. I. 31; LA3. 19902; ðor (h)aveth a skie hem . . . bitogen (*covered*) GEN. & EX. 3796; bitogen [bitowe] REL. I. 174; þe nihtegale al hire ho3e mid rede hadde wel bito3e O. & N. 702; or elles ich hevede uvele bitowen (*employed*) muchel of mine hwule A. R. 430.

biter, adj., *O.E.* biter, = *O.N.* bitr, *O.H.G.* biter, pitter, *cf. Goth.* baitrs; *from* **bīten**; *bitter,* HOM. II. 33; GEN. & EX. 3300; MISC. 27; S. S. (Web.) 849; biter wind HOM. I. 167; biter ine smac AYENB. 82; bitter A. R. 302; þet bittere FRAG. 6; nawt bittres KATH. 1704; to þan **bittre** (*dat.*) deðe HOM. I. 27; of bitere speche 95; **bittre** (*pl.*) HOM. I. 43; bittere [bitere] swipen LA3. 21247; wið bittre besmen MARH. 5; **bitere** (*adv.*) [*O.E.* bitre], bitere LA3. 16381*; bitere, bittere, bittre LANGL. *B* iii. 249; **bitterer** (*compar.*) PR. C. 7272; biterure HOM. II. 173; bittrore SPEC. 99; **bitterest** [biterest] (*superl.*) LA3. 9685; bittrest HOM. I. 283.

biter-browed, bitter-browid, adj. [*v. r.* bitelbrouwed], LANGL. *A* v. 109.

biter-hēde, sb., *bitterness*, AYENB. 28.

biter-līche, adv., *O.E.* biterlīce ; *bitterly*, LAȝ.13626* ; LANGL. *Civ.*144 ; bitterliche A.R. 364 ; bitterli WILL. 2083 ; biterlüker (*comp.*) MISC. 92 ; bitterlükest (*superl.*) H. M. 35.

biternesse, sb., *O.E.* biterness ; *bitterness*, MISC. 28 ; SHOR. 91 ; bitternesse A. R. 372.

bitere, sb. = *O.H.G.* bitteri ; *bitterness* : for bitter of mi galle LANGL. *A* v. 99.

bi-tīden, v., *for* ȝetīden ; *betide, happen*, A. R. 278 ; bitide HORN (L.) 543 ; WILL. 730 ; (*ms.* bityde) LAȝ.2236* ; shame þe mai bitide SPEC. 90 ; bitīdiþ, bitit (*pres.*) LANGL. *B* xi. 393 ; ȝif us eni ufel bitit HOM. I. 71 ; bitīdde (*pret.*) O. & N. 1107 ; GEN. & EX. 3861 ; BEK. 1511 ; GAW. 2522 ; tel me hou þat bitidde S. S. (Wr.) 589 ; betidde ANT. ARTH. i ; bitīd (*pple.*) GEN. & EX. 1978.

bitīding, sb., *happening*, CH. BOET. v. 3 (155).

bi-tilden, v., *O.E.* bi-, beteldan ; *cover (as with a tent), envelop* : and lette hine bitillen (?=bitilden) mid goldfaȝe pallen LAȝ. 27852 ; bitild (*pple.*) JUL. 8.

bi-tīmen, v., *for* ȝetīmen ; *happen* : ȝif sunne bitīmed bi nihte A. R. 324 ; bitīmde (*pret.*) MARH. 2.

bitor, sb., *O.Fr.* butor ; *bittern*, CH. C. T. *D* 972 ; bittor CHEST. I. 51 ; botors (*pl.*) AR.& MER. 3130.

bi-tōveren, v., *perhaps from M.L.G.*, *M.Du.* betōveren, = *M.H.G.* bezoubern ; *fascinate*, GEN. & EX. 2962.

bi-traie, bi-traisse, v., *for* traie ; *betray*, LAȝ. 8923* ; BEK. 316 ; SHOR. 112 ; LANGL. *A* i. 37 ; bitraiede (*pret.*) MISC. 38 ; betraied, betraised CH. D. BL. 1120 ; bitraied (*pple.*) CH. L. G. W. 137 ; bitrasshed R. R. 1520.

bitraing, sb., *betrayal* ; betraȝing WICL. WISD. xvli. 11.

bitraising, sb., *betraying* ; betraisinge [betraiing] CH. L. G. W. 2460.

bi-trappe, v., *O.E.* betræppan, -treppan, = *M.Du.* betrappen ; *entrap, take in a trap*, GOW. III. 257 ; bitrappet, bitreppet (*pple.*) A. R. 174* ; betrapped D. ARTH. 1630.

bi-traut, v., *betray* ; betraut D. TROY 731 ; betrautid (*pret.*) 11767.

bi-travailen, v., *work at*, or *for ; compose (a book)* ; *torment* : for no bred þat ich bitravaile (*pres.*) LANGL. *C* xvi. 210 ; thei bitraveliden (*pret.*) mi wiif WICL. JUDG. xx. 5 ; þis storie is bitravailled (*pple.*) bi cause of Britaine TREV. I. 27.

bi-tréden, v., *tread over* ; bitredeth (*pres.*) TREV. BARTH. (*in* N. E. D.).

bi-trenden, v., *surround* : sorwe him gan betrende FER. 4006 ; betrende (*pret.*) LEG. 127 ; bitrent (*pple.*) CH. TRO. ii. 1231.

bi-trēouþen, v., *betroth* ; bitrēuþede (*pret.*) SHOR. 66 ; betreuþede FER. 2105.

bi-trēuþinge, sb., *betrothing*, SHOR. 59.

bi-tristen, v., *trust* ; betrost (*pple.*) GENER. 3615.

bi-trüflen, v., *delude* ; bitrüfleð (*pres.*) A. R. 106.

bi-trümen, v., *O.E.* betrymian ; *encompass*, MARH. 20 ; bitrümede (*pret.*) HOM. II. 87 ; bitrümet (*pple.*) MARH. 6 ; betremed LK. xxi. 20.

bitte, *see* bite *and* bütte.

bitted, pple., *having a bit* ; bited PALL. i. 1162.

bitter, *see* biter.

bittren, v., *O.E.* biterian, = *O.H.G.* bitarōn ; *make bitter*, A. R. 308 ; bitteret (*pres.*) HOM. I. 23.

bi-tühten, v., *clothe* ; bitüht [bitiȝt] (*pple.*) mid ruȝe felle O. & N. 1013 ; wiþ fair pal it was betight AL. (L³.) 596 ; bitought GUY 232.

bi-tūnen, v., *O.E.* bi-, betýnan, = *M.Du.* betuinen, *M.H.G.* beziunen ; *hedge in*, A. R. 164 ; bitūnde (*pret.*) HOM. I. 83 ; bitunden LAȝ. 4287 ; bitūned (*pple.*) A. R. 164 ; bi-, betuned FRAG. 6 ; bituined KATH. 1659.

bi-turnen, v., *turn about* : biturn þe and cum aȝean A. R. 394 ; biturnde (*pret.*) MARH. 12 ; ROB. 210.

bi-twechen, v., *? exorcize, rid* ; betweche (*pres.*) P. R. L. P. 23.

bi-twēonen, adv. & prep., *O.E.* betwēonum, -twēonan, bitwēon ; *between*, MAT. xxv. 32 ; bitweonen A. R. 72 ; bitweonen, -tweone HOM. I. 91, 141 ; bitweonen [bitwine] LAȝ. 2080 ; bitwene þissen broðeren 10461 ; bitweone twom monnen 22968 ; bitwene 24274 ; betwune 4307 ; bitwenen ORM. 384 ; bitwinen HOM. II. 51 ; bitweone, -twene O. & N. 1379 ; bitwene LANGL. *A* iii. 68 ; so mikel love was hem bitwene HAV. 2967 ; bitvene TREAT. 134 ; betvene AYENB. 66 ; bitwen GEN. & EX. 760.

bi-twih, adv. & prep., *O.E.* betwih, -tweoh, -twyh, -twuh ; *between* : bitwiȝe Ængle londe & Normandie LAȝ. 20947 ; bitveiȝen, bitwihan HOM. I. 37 ; bitwihen O. & N. 1747 ; bituhen KATH. 1526.

bi-twix, adv. & prep., *O.E.* betwix, -tweox, -twyx, -twux, = *O.Fris.* bitwischa, -twiskum, *mod.Eng.* betwixt ; *between, betwixt*, AM. & AMIL. 142 ; betwix D. ARTH. 801 ; bitwix, -twixen CH. C. T. *D* 880, *B* 1075 ; betwex, -tweox, -tweoxe, -twexe, -twuxe MAT. xx. 26, MK. vi. 51, x. 26, xiv. 19 ; bitwux, -twixe, -twixen, -tuxe, -tuxen [bitwixte] LAȝ. 2329, 2335, 2553, 5010, 10886 ; bitwex HORN (L.) 346 ; bitwuxan HOM. I. 91 ; betwuxe 225 ; bitwuxen FRAG. 1 ; bituxen O. & N. 1747 ; bitvixte SHOR. 77.

biþ, *see* wið.

bi-þarf, v. (*pret.-pres.*), *O.E.* beþearf, = *O.L.G.* bitharf, *O.H.G.* pidarf ; *neéd, want* : þæt ge

... beþurſen MAT. vi. 32 ; **beþurfe** (subj.)
LEECHD. III. 114.

bi-þecchen, v., O.E. bi-, beþeccan, = O.H.G.
bedecchen ; cover : mid pælle **biþæht** (pple.)
LA3. 19215 ; **biþehte** (pl.) 22338.

bi-þenchen, v., O.E. biþencan, = O.H.G. piden-
chan, Goth. biþagkjan ; think upon, remind,
bethink, MARH. 5 ; biþenche BEK. 43 ; SHOR.
33 ; he mot hine ful wel biþenche O. & N.
471 ; þis word ... þe deþ deþenche AYENB.
101 ; biþenken ORM. 2917 ; biþenke WILL.
3057 ; biþinken HOM. I. 61 ; bethinkin ' cogi-
to, meditor' PR. P. 34 ; **biþenche** (pres.)
HOM. I. 161 ; he li𐌳 & bi𐌳enche𐌳 him hwonne
he wule arisen A. R. 324 ; biþenkiþ WICL.
LK. xiv. 31 ; biþencþ O. & N. 1509 ; beþengþ
AYENB. 18 ; **biþenc** (imper.) LA3. 9816 ; bi-
þenk þe AN. LIT. 91 ; biþench R. S. i (MISC.
156) ; biþench þe bet P. L. S. xix. 27 ; **biþōhte**
(pret.) LA3. 670 ; O. & N. 939 ; þene mon he
lufede & wel biþohte HOM. I. 59 ; bi𐌳ouhte
A. R. 200 ; biþo3te HORN (L.) 411 ; **bi𐌳ōuht**
(pple.) A. R. 342 ; biþouht C. L. 698 ; heo
was ... wel biþouht LA3. 2510 ; ich habben
(r. habbe) me biþoht wh(a)t ich don wulle
16745 ; biþo3t FL. & BL. 469 ; þat þu beo bet
biþo3t P. L. S. xx. 54 ; in his wisdom was al
bi𐌳ogt ear 𐌸anne it was on werlde bro3t GEN.
& EX. 37 ; to bidden his milce to late we beo𐌳
biþohte R. S. iv.

biþenchinge,, sb., action of thinking ; be-
þenchinge AYENB. 233.

bi-þrāwen, v., torture : þat i wiþ love am so
beþrōwe (pple.) GOW. III. 5.

bi-thrēten, v., threaten ; **bithrĕtt** (pple.)
C. M. 10102.

bi-þringen, v., O.E. bi-, beþringan, = O.H.G.
piðringan ; oppress : þer ich wes o þon fihte
.. hærde **bi𐌳rungen** [biþrongen] (pple.) LA3.
8814 ; wiþ wandraþ biþrungen ORM. 14825.

bi-þünchen, v., = O.H.G. piðunchen ; seem :
3if him so bi𐌳unched (pres.) A. R. 346.

bi-þurfen, v., ? O.E. beþurfan (pret. bēþorfte),
= O.H.G. bedurfin, (pret. piðorſti) ; be needful,
need ; biþorten (earlier text beþorfton) (pret.)
LK. ix. 11.

bi-uven, bufan, adv., O.E. beufan, bufan, =
M.Du., M.L.G. boven ; above, HOM. I. 95, 159 ;
bufen, bufenn eorþe ORM. 4773 ; buven KATH.
2396 ; HOM. II. 111 ; he is buven a P. L. S.
viii. 44 ; ærest wes þe white buven & seo𐌳𐌳en
he wes bineo𐌳en LA3. 15978 ; buven, buve
A. R. 46, 106 ; buve O. & N. 208 ; bove AN.
LIT. 5 ; al þat is bove and under molde
SHOR. 117 ; comp. a-bufen.

bivien, beovien, büvien, v., O.E. bifian, beo-
fian, = O.L.G. bivon, O.Fris. beva, O.N. bifa,
O.H.G. piben, biben ; tremble, LA3. 23530,
25242, 27718 ; **bive𐌳** (pres.) GEN. & EX. 2280 ;
beoveden (pret.) LA3. 28357.

bi-wæven, v., „O.E. bewæfan, = Goth. bi-
vaibjan ; clothe ; **biwēfde** (pret.) LA3. 28474 ;

biwēved (pple.) MISC. 55, 140 ; ALIS. 1085 ;
biweved hii were bothe mid Welsse mantles
ROB. 539 ; biwæived LA3. 22132 ; mid wintre
he wes biweaved 130.

bi-wailen, biweilen, v., bewail, CH. C. T.
B 26 ; biweile WICL. GEN. xxiii. 1 ; **biweileþ**
(pres.) ALIS. 4395 ; **biwailinge** (pple.) CH.
TRO. iv. 1223 ; **biweilddest** (pret.) CH. BOET.
i. 6 (26).

bi-waine, be-waine, be-wan3e, sb., profit,
gain, BARB. LEG. II. 130, 136.

bi-wáken, v., M.H.G. bewachen ; watch about,
GEN. & EX. 2444 ; biwake S. S. (Web.) 2764 ;
AL. (T.) 565 ; **biwákeden** (pret.) HOM. II. 35 ;
hine biwakeden in þere nihte þritti hundred
cnihten (r. cnihte) LA3. 7576 ; biwoken ORM.
3339 ; **bewáked** (pple.) GOW. II. 350.

bi-wallen, v., = M.L.G. bewallen ; surround
with a wall ; **biwalled** (pple.) LA3. 18607 ;
LEG. 177.

bi-walwien, v., O.E. bewealwian ; wallow
about : **biwalede** (pret.) hine a blode LA3.
27744 ; **biwaled** [biwalewed] (pple.) on axen
25989.

bi-wappen, v., ? astonish ; **bewaped** (pple.)
BEVES 1689 ; bewhapped and assoted GOW.
III. 4 ; bevapid FER. 3037.

bi-wárien, v., invest (money), lay out, expend :
if þe clerk beware his faiþ in chapmanhode at
such a faire GOW. I. 262 ; þi wit is wel bi-
wáred (pple.) CH. TRO. i. 636.

bi-wedden, v., O.E. beweddian, = O.Fris.
biweddia ; wed ; **biwedded** (pple.) FRAG. 8 ;
HOM. I. 149 ; a king of Britaine havede heo
biwedded LA3. 4500 ; his biweddede wif
31960.

bī-wei, sb., = M.H.G. bīwec ; by-way, MAN.
(F.) 10145.

biweilen, see **biwailen.**

bi-wélde, v., ? for 3ewélden ; wield ; bewelde
GOW. I. 312 ; biwalt (pres.) HOM. II. 25 ;
biwálden (pret.) HOM. II. 87.

bi-wellen, v., well up ; bewel TREV. I. 111.

bi-wenden, v., O.E. bewendan, = O.H.G.
bewenden, Goth. biwandjan ; turn away, turn
round : to hwuche of þeos foure mei he him
biwenden A. R. 306 ; (hit) biwent (pres.) him
ofte 132 ; biwende (pret.) KATH. 2362 ; Jesus
hine biwende MISC. 45 ; biwente HORN (L.)
321 (R. 329) ; MAP 334 ; bewent K. T. 1026 ;
heo biwenden heom sone LA3. 26576.

bi-weorpen, v., O.E. beweorpan, = O.L.G.
bi-, bewerpan, O.H.G. piwerfan ; overwhelm :
mid holiwatere beworpen (pple.) FRAG. 7.

bi-wēpen, v., O.E. bewēpan, = O.Fris. bi-
wēpa ; weep for, HOM. I. 39 ; A. R. 108 ; ORM.
15136 ; biwēpen MISC. 78 ; biwepe O. & N.
980 ; biwepe SHOR. 102 ; R. R. 5121 ; biwepe
(moisten) PALL. iv. 61 ; **biwēop** (pret.) A. R.
278 ; biweop ALIS. 3655 ; biwiep HOM. II.
145 ; biweopen MARH. 21 ; biwepte GOW.

III. 182; WICL. GEN. 1; bewōpen (*pple.*)
CH. TRO. iv. 916; here visage al biwope
(*covered with tears*) S. S. (Web.) 1186;
biweped WILL. 661.

biwēpere, sb., *one who beweeps*; biwē-
peris (*pl.*) WICL. WISD. xviii. 10.

bi-werian, v., *O.E.* bewerian, *O.L.G.* biwerien, *O.H.G.* piwerian; *defend*, HOM. I. 15;
biwerien P. L. S. viii. 168; þu . . . biwerest
manne corn from deore O. & N. 1126.

bi-wēten, v., *wet, moisten*; biwĕt (*pple.*)
PALL. vi. 110; wiþ teres al bewet HALLIW.
171.

bi-wēven, v., *whirl away*; biwĕft (*pret.*)
C. M. 24109; biwafðe LAȝ. 30856.

bi-wicchen, v., *bewitch*; bewicche MAND.
159; biwicched (*pret.*) LANGL. *B* xix. 151;
biwicched (*pple.*) TREV. II. 423; biwucched
(*sec. text* biwicched) LAȝ. 24276.

bi-wīȝelien, v., *delude, deceive*, LAȝ. 969; bi-
wīhelin [biwihelin] JUL. 56, 57.

bi-wīlen, v., ? = biwīȝelien; REL. I. 182 (MISC.
123); biwīled (*pple.*) SHOR. 67; GAW. 2425.

bi-wimplen, v., = *M.L.G.*, *M.Du.* bewim-
pelen; *cover with a wimple*; bewimpled
(*pple.*) GOW. II. 359.

bi-winden, v., *O.E.* bi-, bewindan, = *Goth.*
biwindan, *O.H.G.* piwintan; *wind around*:
biwinden þe rapes HOM. I. 47; biwindeð
(*pres.*) HOM. II. 95; biwond (*pret.*) LEG.
88; biwunden (*pple.*) JUL. 76; mid seorwe
biwunden FRAG. 5; wiþ þornis al bewonde
P. L. S. vi. 3; bewounde GOW. II. 295.

bi-winnen, v., = *M.H.G.* bewinnen, *for* ȝe-
winnen; *win*, LAȝ. 472; biwinne FL. & BL.
374; CHR. E. 465; biwinneð (*pres.*) H. M.
29; biwan (*pret.*) ROB. 420; biwan, -won
LAȝ. 29, 4674; biwon HOM. I. 41; þu biwunne
LAȝ. 7878; (heo) biwunnen [biwonne] 3782;
biwunnen (*pple.*) A. R. 228; biwunnen [bi-
wonne] LAȝ. 1311; biwonne AN. LIT. 11.

bī-wist, -west, sb., *O.E.* bī-, bīgwist *f.;
being, living*, HOM. II. 161, 169; bewiste
M. H. 69; C. M. 13832; bewiste, beowust,
beouste A. R. 156, 160; hu beoð þine beouste
LAȝ. 17809; whar beo heore beouste (*printed*
beonste) 26090.

bi-wīten, v., *go away*, ALIS. 5203.

bi-witen, v., *O.E.* bewitan; *pret.* bewiste; *take
care of*, HOM. I. 245; biwiten, -witten LAȝ.
6659, 13782; biwiste, biwuste (*pret.*) LAȝ.
207, 12920; biwistest JUL. 33; biwusten
MISC. 52; biwust (*pple.*) A. R. 104.

bi-wīten, v., *O.E.* bewītan (*pret.* bewāt); *take
care of*; biwit (*pres.*) MARH. 3; (ha) biwiteð
H. M. 29; biwāt (*pret.*) LAȝ. 13028; bi-
witen (*pple.*) 19535.

bi-witie, v., *O.E.* bewitian, bewitigan; *take
care of*, LAȝ. 6659*; biwiteȝen 25010; to bi-
witiende (= biwitienne) HOM. II. 195; bi-
witede (*pret.*) LAȝ. 31013*.

bi-wlappe, v., *bewrap*, WICL. JOB xviii. 11.

bī-word, sb., = *M.H.G.* bīwort; *byword*, SAX.
CHR. 259; biword CH. TRO. iv. 769; TOWNL.
96.

bi-wrappen, v., *wrap up*; schulen bewrappe
'*involvent*' WICL. JOB xviii. 11; bewrapped
(*pple.*) WILL. 1735; biwrabbet A. R. 260*.

bi-wrēȝen, v., = *O.Fris.* biwrōgia, *O.H.G.*
beruogen; *bewray, accuse*, CH. C. T. *A* 2229;
bewrie WILL. 2435; bewrēie (*pres. subj.*)
M. T. 102; biwrēide (*pret.*) ALIS. 4562;
bewrēied (*pple.*) GOW. II. 356.

biwrēire, sb., *one who reveals*; bewraier
'*recelator*' PR. P. 34.

biwrēiing, sb., *action of revealing*, CH. C.
T. *G* 147.

bi-wréke, *avenge*; bewreke RICH. 6283;
LIDG. TROY ii. 16.

bi-wrenchen, v., *O.E.* bewrencan; *deceive*;
biwrencheð (*pres.*) A. R. 92.

bi-wrēon, v., *O.E.* bewrēon; *cover*; biwreo,
-wro O. & N. 673; biwrīeth (*pres.*) ALIS.
6453; biwriȝen [biwreȝe] (*pple.*) LAȝ. 5366;
biwrien A. R. 262.

bi-wrixlen, v., *for* ȝewrixlen; *change*; bi-
wrixled (*pple.*) A. R. 310.

bi-würchen, v., *O.E.* bewyrcan; *work*; be-
wroght (*pple.*) EGL. 1152.

blā, adj., *O.N.* blār, = *O.E.* blāw, blǣw, *O.Fris.*
blāu, *O.H.G.* blāw, ? *Lat.* flāvus; *blue-black*,
livid, PR. C. 5261; blaa ISUM. 311; blo HICKES
I. 230; SPEC. 86; RICH. 2644; TRIST. 2976;
MAN. (H.) 173; reed and blo GEN. & EX. 637;
blo and blac MAP 335; þe water so blo EM.
318; as blo as led TOWNL. 224; blo [bloo]
askes LANGL. *B* iii. 97; bloo '*lividus*' PR. P.
40; bloo '*fulvus*' REL. I. 8; blodi and bloo
LUD. COV. 335.

blō-erþe, sb., '*argilla*,' PR. P. 40.

blā-mon, sb., *O.N.* blāmaðr; *Moor*, MARH.
10; blac as a bloamon [blamon] A. R. 236;
bloman OCTOV.(W.)1405; blā-, blōmen (*pl.*)
C. M. 2118; blomen TREV. I. 45; of Ethiope
he brohte þa bleomen LAȝ. 25380.

blānesse, sb., *blueness, lividity*; blonesse
'*livor*' PR. P. 40; blones WICL. EX. xxi. 25.

blabbe, sb., *blabber, tell-tale*, CH. TROIL. iii.
251; PR. P. 37; blab (*chatter*) BER. 3022.

blaberen, v., cf. *Dan.* blabbre, *Ger.* blappern;
blabber, WICL. PROL. 1 ESDR.; blaberin '*blatero*' PR. P. 37; þou blaberest HALLIW. 180;
þei blaberin LUD.COV. 164; i blaberde LANGL.
A v. 8.

blabir-lippid, adj., *with swollen lips*, CATH.
33; DIGB. iii. 927; cf. babir-lippid.

blac, adj., *O.E.* blæc, = *M.Du.* black, *O.N.*
blakkr; *black*, LAȝ. 3070; A. R. 234; HAV.555;
TREAT. 133; FER. 2437; blak '*niger, ater*'
PR. P. 38; CH. C. T. *A* 2130; MAND. 51; K.
T. 773; blak and blo SPEC. 68; TOWNL. 224;

þe blake deovil HOM. I. 251 ; **blake** (*pl.*) HOM.
I. 51 ; blake monekes ROB. 433 ; þa blake
claƚes LAȝ. 13102 ; **blaccre** (*compar.*) MARH.
10 ; blakkure MISC. 149.

    blak-berie, sb., *O.E.* blæcberige ; *blackberry*,
HALLIW. 180 ; (*ms.* blake berie) VOC. 140 ;
WILL. 1809.

blac, adj.,ʼ*O.E.* blæc, *M.Du.* black, *O.H.G.*
plach, blach, *O.N.* blek *n. ; ink* ; blek ʻ*atra-
mentum*ʼ PR. P. 39 ; blacche, blacchepot
VOC. 181.

blāc, adj., *O.E.* blāc, blǣc,=*O.L.G.* blēc, *O.N.*
bleikr, *O.H.G.* pleich, *mod.Eng.* bleak ; *pale*,
LAȝ. 19888 ; HOM. I. 249 ; blac ant won SPEC.
74 ; blak L. H. R. 150 ; ANT. ARTH. li ; bloc
A. R. 332 ; bleche AYENB. 53 ; GOW. II. 210 ;
bleik ʻ*pallidus*ʼ PR. P. 39 ; LIDG. TH. 2332 ;
blāke (*pl.*) LAȝ. 1888 ; weren for hunger grene
and bleike HAV. 470.

blad, sb., *O.E.* blæd, = *O.L.G.* blad, *O.N.* blaȝ,
O.H.G. blat ; *blade* (*of grass, of a sword*),
FER. 4433 ; þat blad AUD. 12 ; blad, blade
PR. P. 37 ; blade CH. C. T. *A* 618 ; **bláde** (*dat.*)
MAN. (F.) 10039 ; *comp.* windel-blad.

    blad-smith, sb., PR. P. 37 ; D. TROY 1592.

bládin, v.,=ʻ*M.L.G.* bladen, *M.H.G.* blaten ;
blade (*in two senses*) : bladyn herbys ʻ*detirso*,ʼ
bladyn haftys ʻ*scindulo*ʼ PR. P. 37.

blǎdre, sb., *O.E.* blǣdre, blǣddre,=*M.L.G.*
bladere, bledere, *M.Du.* blaere, *O.N.* blaȝra,
O.H.G. plātra, blātra ; *from* **blāwen** ; *bladder* ;
blader REL. I. 190 ; bladdre CH. C. T. *G* 439 ;
bleddre A. R. 282 ; SHOR. 2 ; bledder, bledder
PL. CR. 222 ; bleddir ʻ*vesica*ʼ PR. P. 39 ; bloure
(*ms.* blowre) TOWNL. 62 ; blore YORK xxvi.
188 ; **blǎdran**, blǎdre (*dat.*) LEECHD. III.
82, 84.

bladren, v., *? swell like a bladder* ; **bladdirþ**
(*pres.*) N. P. 66.

blǣd, sb., *O.E.* blǣd, blǣȝ, TREAT. 17, *cf. O.
H.G.* plāt ; *from* **blāwen** ; *blast, breath* ;
blēade (*dat.*) HOM. I. 97.

    blǣȝ-fæst, adj., *O.E.* blǣdfæst ; *? pros-
perous, famous* LAȝ. 10100, 6986.

blǣs, sb.,=*M.Du.* blaes, *M.H.G.* blās ; *blast*,
bles A. R. 82 ; windes bles MISC. 93 ; CL. M.
30 ; blas FER. 2648 ; **blase** (*dat.*) LAȝ. 27818.

blǣsen, v.,=*O.E.* blǣsan, *O.N.* blāsa, *O.H.G.*
blāsen ; *blow* : he gan to blasen out a soun
CH. H. F. 1802.

blǣst, sb., *O.E.* blǣst,=*O.H.G.*, *M.Du.* blāst,
O.N. blāstr ; *from* **blǣsen** ; *blast* : bemene blæst
LAȝ. 19926 ; blest AYENB. 203 ; blast ʻ*flatus*ʼ
PR. P. 38 ; TREAT. 136 ; SPEC. 106 ; CH. TRO.
ii. 1387 ; hornes blast GEN. & EX. 3464 ; REL. I.
223 ; windes blast ALIS. 235* ; S. & C. I. lii ;
blaast TREV. IV. 371 ; **blǎste** (*dat.*) BRD. 19 ;
wiþ strengþe of his blast(e) þe white brent þan
rede AR. & MER. 1538 ; *comp.* þoner-blǎst.

blǣsten, v.,=*O.E.* blǣstan ; *blast, breathe heavily* ; blaste BRD. 27 ; LEG. 167 ; and gan to

puffen and to blaste CH. H. F. 1866 ; he grunte
and blaste P. L. S. xv. 99 ; blasten ALIS. 5348
comp. **a-blǣsten.**

blǣten, v., *O.E.* blǣtan,=*O.H.G.* plāzan ;
bleat ; bletin ʻ*balo*ʼ PR. P. 39 ; blete LEG.
152 ; LANGL. *C* xviii. 38 ; BER. 3245 ; TOWNL.
106 ; **blǣteþ** (*pres.*) ORM. 1315 ; **blētinge**
(*pple.*) TREV. IV. 153.

    blētinge, sb.,=*O.H.G.* plāzunga ; *bleating*,
ʻ*balatus*,ʼ PR. P. 39.

blǣȝ, *see* **blǣd, blēȝ,** *and* **blēaȝ.**

blaffen, v.,=*M.Du.* blaffen ; *stammer* ; wlaffes
ʻ*balbeie*ʼ VOC. 173 ; **wlaffing** (*pple.*) LEG.
119.

    blaffere, sb., *stammerer*, ʻ*traulus*,ʼ PR. P.
37 ; wlaffere AYENB. 262.

blafford, sb., *? M.Du.* blaffaert ; *stammerer*,
ʻ*traulus*,ʼ PR. P. 37.

blak, *see* **blac.**

blake, sb., *blackness* : biholden ever his blake
& nout his hwite A. R. 282.

blakeberied [*v. r.* **blakberied, blakburied**],
sb., *cf.* beggeþ ; hir soules goon a blakeberied (*? a-blackberrying*) CH. C. T. *C* 406.

blākien, v., *O.E.* blācian,=*O.N.* bleikja, *O.
H.G.* bleichen ; *turn pale*, MARH. 9 ; his neb
bigon to blakien [blokie] LAȝ. 19799 ; blaken
MAN. (H.) 183 ; blake AV. ARTH. xv ; þanne
gan bleiken here ble P. S. 341 ; hwenne þin
heou **blōkeþ** (*pres.*) MISC. 101 ; **blākede**
(*pret.*) LAȝ. 7524.

blakin, v., *O.E.* blacian ; *blacken*, PR. P. 38 ;
blaken, blake C. M. 14747 ; **blakeþ** (*pres.*)
SHOR. 155 ; **blakede** (*pret.*) JUL. 48 ; **blaked**
(*pple.*) CH. C. T. *B* 3321.

blaknen, v. : he bigan . . . wit hem ful sore to
blakne (*look black, grow angry*) HAV. 2165.

blāknen, v., ?=**blākien** ; of deithe nammore
to blokne SHOR. 4.

blāme, sb., *O.Fr.* blasme ; *blame*, H. M. 33 ;
HORN (L.) 1265 ; P. L. S. xxii. 4 ; AYENB. 23.

    blāmable, adj., *blameable*, TREV. VI. 25.

    blāme-ful, adj., *imputing blame, blame-
worthy*, WICL. ESTH. xvi. 6 ; blamefull KATH.[2]
106.

        blāmeful-(1)i, adv., APOL. 112.

    blāme-lēs, adj., *blameless*₁ TREV. I. 251 ;
GOW. III. 220.

blāmen, v., *O.Fr.* blasmer ; *blame*, A. R. 64 (*v.
r.* lastin) ; blame MARG. 200 ; **blāmeȝ** (*pres.*)
HOM. II. 73 ; blameþ AYENB. 17 ; **blāmede**
(*pret.*) ROB. 272.

    blāmer, sb., *one who blames* ; **blāmeres**
(*pl.*) WICƚ. IS. l. 6.

blanc, adj., *O.Fr.* blanc, *from O.H.G.* blanch ;
mod.Eng. blank ; *shining, white* ; blonc REL.
I. 37 ; his blonk sadel GAW. 2012.

blanchard, adj., *O.Fr.* blanchart ; *whitish*,
GENER. 2458 : stedis . . . baith blanchart and
bai GAW. & GOL. II. 19.

blanchet, sb., *Fr.* blanchet; *a sort of paint*, HOM. I. 53.

blanchin, v., *Fr.* blanchir; *blanch*, PR. P. 38; **blawnchede** (*pple.*) M.ARTH. 3040; blanched P. P. II. 50.

blanc-mangere, sb., *O.Fr.* blanc-manger; *blanc-mange*, LANGL. *B* xiii. 91; blankmanger CH. C. T. *A* 387.

bland, sb., *O.E.*, *O.N.* bland; *mixture*; in bland *together* ALEX. 2786; GAW. 1204; A. P. ii. 885.

blanden, v., *O.E.* blandan, *O.N.* blanda, = *Goth.* blandan, *M.Du.* blanden, *O.H.G.* blantan; *mix*: **blonde** hit with milke L. C. C. 24; **blande** (*pple.*) GAW. 1931.

blanden, v., *O.Fr.* blandir; *blandish, flatter*; **blondeþ** (*pres.*) SHOR. 73; **blaundissinge** (*pple.*) CH. BOET. ii. 1 (30).

  blandere, sb., *flatterer*; blondere AYENB. 61.

  blandinge, sb., *blandishing*; blandinge AYENB. 75; blandis(s)inge P. L. S. xix. 165.

blandissen, *see* **blanden**.

blānesse, *see under* **blā.**

blanke, sb., *O.E.* blanca, blonca; *cf. O.H.G.* blanc ros, *also* planchaz (*white horse*); (*white, grey*) *horse*; blonke WILL. 3326; ANT. ARTH. xliii; **blanken**, blonken (*pl.*) LA3. 5862, 13512.

blanket, sb., *O.Fr.* blanquete; *coarse woollen cloth, blanket*, BEK. 1167; hir belle was of blonket (? *printed* blenket) ANT. ARTH. (R.) xxix (M. plonket); *see* **blanchet.**

blāren, v., = *M.Du.* blaeren (*vagire, balare*); *weep noisily*; blorin '*fleo*' PR. P. 40.

blās, *see* **blǣs.**

blāse, sb., *O.E.* blǣse; *blaze, flame, torch*, HOM. I. 185; A. R. 254; HAV. 1254; LANGL. *B* xvii. 212; CH. TRO. iv. 184; BER. 2352; enne blase of fure LA3. 2859; þei setten al on blase GOW. II. 244; blase, blese PR. P. 39; C. M. 8877; **blāsen** (*pl.*) A. R. 254; blesen ['*facibus*'] JOHN xviii. 3.

blāsen, v., *blaze*, '*flammo*,' blasin PR. P. 38; brennen & blasen LANGL. *B* xvii. 232; **blāseþ** (*pres.*) TREV. I. 5; GOW. I. 258; vort al þet hus blasie vorð A. R. 296; **blāsed** (*pret.*) ALIS.[2] (Sk.) 729.

  blāsinge, sb. & adj., *flaming*, '*flammacio*,' PR. P. 38; blasing starre (*cynosure*) P. R. L. P. 54.

blāsen, *see* **blǣsen.**

blasfēme, sb. & adj., *blasphemer, blasphemous*: shulden siche blasfemes be stoned to deeþ WICL. S. W. III. 347; heithen and blasfeme men WICL. 2 MACC. x. 4.

blasfēmen, v., *Fr.* blasphemier, *Gr.* βλασφημεῖν; *blaspheme*; **blasfēmeþ** (*pres.*) AYENB. 30; **blasfēmed**, blasphemed (*pple.*) CH. C. T. *D* 2183.

blasfemie, sb., *Fr.* blaspheme, *Gr.* βλασφημία; *blasphemy*, A. R. 198; AYENB. 57.

blasfemour, blasphemour, sb., *blasphemer*, CH. C. T. *D* 2213.

blasoun, sb., *O.Fr.* blason; *blazon, shield*, GAW. 828; RICH. 5727.

blǣst, *see* **blǣst.**

blaunner, sb., *kind of fur*, GAW. 155; blaunner B. DISC. 117 [blaundemere *F*].

blāwen, v., *O.E.* blāwan, = *O.H.G.* plāen, plāhen, *Lat.* flāre; *blow* (*as the wind*), ALIS. (Web.) 5630; blawen, blauwen, blowen LA3. 790, 794, 4462; blauwen HOM. I. 81; blawe HAV. 587; ISUM. 393; blaue ANT. ARTH. xxvi; bloawen A. R. 210; blowe HAV. 913; ROB. 7; CH. C. T. *A* 565; blowe & blaste MARG. 213; bragge or blowe LUD. COV. 183; **blāuþ** (*pres.*) AYENB. 25; bostus and blawus AV. ARTH. xxiii; (hi) blaweþ AYENB. 24; þe engles ... bloweþ heore beme MISC. 163; **blāwe** [blou] (*imper.*) LA3. 25789; blou HAV. 585; SPEC. 51; ær þeo bemen blowen FRAG. 7; **blāwende** (*pple.*) HOM. I. 87; blowand WILL. 3358; **blēow** (*pret.*) HOM. I. 221; bleow ALIS. 491; bleou, bleu LA3. 808, 1762; atter þet þe alde deovel bl(e)ou on Adam HOM. I. 75; blew GEN. & EX. 3085; bleu HOM. II. 19; MISC. 55; BRD. 24; LEG. 165; C. M. 12540; (*ms.* blew) ROB. 13; LANGL. *B* v. 515; blu AV. ARTH. xv; bleowen MAT. vii. 27; bleowen, blewen [bleuwen] LA3. 5107, 22061; blewe CH. C. T. *B* 4589; **blāwen** (*pple.*) PR. C. 685; TOWNL. 14; blowen LANGL. *B* v. 18; S. S. (Wr.) 2181; *comp.* a-blāwen; *deriv.* **blǣd**, **blǣdre.**

  blāwere, sb., *O.E.* blāwere; *blower*; blower TRIST. 535; *comp.* horn-blāwere.

  blāwing, sb., *O.E.* blāwung; *blowing, breathing*, HOM. I. 75; blawing GAW. 1601; blowinge A. R. 426; blowinge PR. P. 41.

blē, *see* **blēo.**   **blēad**, *see* **blǣd.**

blēat, adj., *O.E.* blēat; *wretched*, = *O.N.* blautr (*soft*), *O.Fris.* blāt, *M.Du.* bloot, *M.H.G.* blōz (*naked*); *naked, wretched*: we habbeð ... moni ænne gode wifmon iworht to bletere widewe LA3. 23620; treon ... mid þicke boȝe no þing blete O. & N. 616.

blēað, sb., *O.E.* blēað (*ignavus, timidus*), = *O.N.* blauðr, *O.L.G.* blōth, *O.H.G.* blōd; *timid; soft*: dre(d)ful and bleð GEN. & EX. 2590; blæð (? *for* blæt) LA3. 18737.

blecchen, v., *from* blac; *blacken*; blekkin PR. P. 39; iblæcched (*pple.*) LA3. 17700; bleckid WICL. JOB xxx. 30.

blecen, v., *O.Fr.* blecier, blecher (= *mod.Fr.* blesser); *hurt, wound*; **blecheþ** (*pres.*) AYENB. 40, 238; **blissid** (*pple.*) RICH. 546.

blēche, *see* **blāc.**

blēchen, v., *O.E.* blǣcan (*pple.* blǣced); *from* blāc; *bleach*, HOM. II. 57; **blēched** (*pple.*) CH. BOET. APP. i. 45; wule a weob beon ... mid one watere wel ibleched A. R. 324.

blecken, *see* **blecchen.**

bleddre, *see* **blǎdre.**

blēde, sb., *O.E.* blēd, = *O.H.G.* bluot *f.; from* **blōwen;** *flower, bloom,* HOM. II.256; LEG.85; þar never gras ne springþ ne bled O. & N. 1042; withoute eni blede (*offspring*) AL. (T.) 27; **blede, blæde, blæden** (*earlier text* bleda) (*pl.*) JOHN xv. 2, 4, 16; bleden [bledes] LAȝ. 28832.

blēden, v., *O.E.* blēdan, = *O.Fris.* blēda, *O.H.G.* bluoten; *from* blōd; *bleed*; blede HAV. 103; GOW. I. 188; TRIAM. 379; **blĕdde** (*pret.*) LAȝ. 7523; A. R. 114; LANGL. *B* xix. 103; his woundes bledden ALIS. 5845; *comp.* **bi-blēden.**

blee, *see* **blēo.**

bleik, bleiken, *see* **blāc, blākien.**

blēine, sb., *O.E.* blēgen, = *Dan.* blēgn, *M.Du.* blēine *f.; blain,* '*papula,*' PR. P. 39; blein GEN. & EX. 3027; **bleine** (*dat.*) R. R. 553; **blēines** (*pl.*) WICL. EXOD. ix. 9; pokkis and bleines FLOR. 2024; **blēinen** (*dat. pl.*) MARH. 18.

blēininge, sb., *blistering,* PL. CR. 299.

blek, *see* **blac.**

blēlī, adv., *for* **blēthelī** = **blīþlīche;** *willingly,* WICL. E. W. 417; PARTON. 771.

blemissen, v., *O.Fr.* blesmir; *blemish;* blemishen, blemschin '*obfusco*' PR. P. 39; **blemissed** (*pple.*) CH. BOET. i. 4 (20); al blemischid is thi ble TOWNL. 223.

blench, sb., *trick, deceit,* R. S. i (MISC. 156); AYENB. 130; A. P. ii. 1202; blenk MAN. (H.) 201; blink H. S. 4185; **blenches** (*pl.*) O. & N. 378; blenkes HAV. 307.

blenchen, blenken, v., *O.E.* blencan, = *O.N.* blekkja; *blench, flinch, steal away, deceive,* HOM. I. 55: a vleih mei eilen þe & makien þe to blenchen A. R. 276; ich am war and blinke can wel blenche O. & N. 170; ȝif þou . . . fram him nelt blenche LEG. 157; þanne shaltou blenche at a berghe LANGL. *B* v. 589; his ie gan blenche H. S. 3576; blinche a strok BEK. 2172; blenke MAN. (H.) 115; D. TROY 2483; D. ARTH. 2857; **blenchiþ** (*pres.*) LIDG. M.P.215; **blencte, blenchte** (*pret.*) JUL. 72,73; bleinte LAȝ. 1460; CH. C. T. *A* 1078; (*ms.* blenyte) ROB. 338; bleint MAN. (F.) 1522; Tristrem bleint bi side TRIST. 2779; þei lokede aboute & bleinte bihinde þe busch WILL. 3111; bot dauid sagh and blenked (*v. r.* blenkid, blenched) lau C. M. 7668; *comp.* **at-blenchen.**

blenden, v., *O.E.* blendan, = *O.Fris.* blenda, *O.H.G.* blenden; *make blind;* blende MISC. 88; LANGL. *B* x. 129; CH. TRO. iii. 207; **blendeþ** (*pres.*) AYENB. 33; blendeþ mannes heorte ORM. 4525; blent HOM. I. 251; CH. P. F. 600; **blende** (*pret.*) LAȝ. 29361; R. S. iv; A. P. ii. 1788; *comp.* **a-, for-, ȝe-blenden.**

blenden, v., *O.E.* blendan; *from* blanden; *blend, mingle;* blende TOWNL. 225; **blende** (*pret.*) GAW. 1361; **blent** (*pple.*) D. TROY 3493.

blenk, *see* **blench.**

blenken, *see* **blenchen** *and* **blinken.**

blēo, sb., *O.E.* blēo, blēoh, blēo, *M.Du.* blie, *O.Fris.* blie, blī, *O.L.G.* blī *n.; colour, complexion,* MARH. 9; SPEC. 33; feire on bleo MISC. 95; hit . . . changeþ his bleo TREAT. 139; bleo, blo O. & N. 152, 1547; blee OCTAV. (H.) 50; bright of blee EGL. 33; ble GEN. & EX. 457; WILL. 3083; A. P. i. 212; TOWNL. 223; blie SHOR. 103; **blēos** (*gen.*) FRAG. 3; ones bles GEN. & EX. 1725.

[bler, sb., = *Ger.* blerr; *soreness of eyes.*]

bler-ēiȝed, adj., *cf. L.G.* bleer-oged; *bleareyed;* blereyed '*lippus*' PR. P. 39; VOC. 225; blereiȝed WICL. LEV. xxi. 20; blereighed LANGL. *B* xvii. 324.

blēren, v., *cf. Swed.* blira, plira; *blear (the eyes); blink, wink;* blere RICH. 3708; P. 41; ȝit schal i blere here ie CH. C. T. *A* 4049; i bleri REL. II. 211; þou blerest noȝt so min eȝe FER. 507; **blēred** (*pret.*) LANGL. *A* PROL. 71; blerid S. S. (Wr.) 2952; **blēred** (*pple.*) R. R. 3912; GAW. 963.

blēren [? **blēren**], v., *protrude the tongue:* grin on him and blere PR. C. 2226; with þi tunge on folk þou bleere H. V. 60; **blēride** (*pret.*) D. ARTH. 782.

blēri, adj., *bleared:* bleri [*v. r.* blered] eien LANGL. *C* vii. 198.

blēs, *see* **blǣs.**

bleschin, v., *cf. M.Du.* bleschen (? = beleschen), *O.L.G.* leskian, *O.H.G.* lescan; *quench,* '*extinguo,*' PR. P. 39; blissen GEN. & EX. 553; and čis fier blessede 3803.

bleschinge, sb., *quenching,* '*extinctio,*' PR. P. 39.

blese, *see* **blase.**

blesmin, v., *from O.N.* blǣsma (*maris appetens*); *desire the ram,* CATH. 34*; **blismed** (*pple.*) PS. lxxvii. 70.

blessen, *see* **bleschin. blessien,** *see* **bletsien.**

blest, *see* **blǣst. blēt,** *see* **blēat.**

blēten, *see* **blǣten.**

blĕtsien, v., *O.E.* blētsian, blēdsian, blǣdsian; *from* blōd, *orig. meaning* '*sprinkle with blood*', *see* ANGL. III. 156; *bless;* (*printed* blecsien) FRAG. 7; bletseiȝen LAȝ. 32157; blescien HOM. I. 215; blessi MISC. 55; blecen HOM. I. 57; blessen, blissen CH. C. T. *A* 3448; blisse LUD. COV. 50; **blesceð** (*pres.*) HOM. I. 137; **blesse** (*imper.*) AN. LIT. 7; blesce þe . . . mid te eadie rode tocne A. R. 290; **bletsede** (*pret.*) MK. viii. 7; blessede ROB. 338; heo blescede hire MARH. 17; bliscede GEN. & EX. 1546; & blessed so wiþ his briȝt bront aboute WILL. 1192; blesseden WILL. 196; **blettsedd** (*pple.*) ORM. 4826; bledsed HOM. II. 25; blessed SPEC. 83; blissed HAV. 2873; and wolde have blissid (*stricken*) him bet RICH. 546; his blissede zone AYENB. 87; *comp.* **ȝe-blessien.**

blessed-ful, adj., *full of blessing;* bl(i)ssidful C. M. 11234; blestful N. E. D. 917.

blessed-hēd, sb., *blessedness*, S. A. L. 82 ; blissedhede AYENB. 97.

blessid-lī, adv., *blessedly*, WICL. GEN. xxx. 10 ; **blessedlocurre** (*comp.*) ED. 2711.

blĕtsing, sb., *O.E.* blētsung ; *blessing*, ORM. 10661 ; blessunge HOM. I. 71 ; A. R. 154 ; blessinge LAȜ. 13261 ; blissinge AYENB. 97 ; bliscing GEN. & EX. 1532.

blēð, *see* blēað.

bleu, adj., *O.Fr.* bleu ; *blue*, C. L. 712 ; C. M. 9920 (*sub.*) ; blew PR. P. 41 ; TRIST. 2404 ; ALIS. 5272 ; GOW. III. 158 ; CH. C. T. *E* 2219 ; bleu as inde MAND. 48 ; blewe askes LANGL. *C* iv. 125.

blēven, *see* bi-lǣfen.

bliand, sb., *O.Fr.* bliaut, *L.Lat.* blialdus ; *a sort of rich stuff, or garment made of it* ; blihand TRIST. 410 ; bleaunt GAW. 1928.

blichening, sb., *mildew*, PALL. į. 827.

blie, *see* blēo.

blīken, v., *O.E.* blīcan, = *O.N.* blīkja, *pret.* bleik, *M.H.G.* blīchen ; *grow pale* : his lippes shulle bliken REL. I. 65 ; *comp.* a-blīken ; *deriv.* blāc.

blikien, v., *O.N.* blika, = *M.H.G.* blicken ; *shine, gleam*, LAȜ. 27360 : schal blikien schenre þen þe sunne A. R. 362 ; hire bleo blikieþ (*for* blikeþ) so briht SPEC. 52 ; **bli-kinde** (*pple.*) MARH. 13 ; **blikede** (*pret.*) KATH. 2396 ; blikked GAW. 429.

bliknen, v., *O.N.* blikna ; *glitter ; grow pale* : bliknande perles A. P. ii. 1467 ; þenne blikned þe ble ii. 1759.

blin, sb., *O.E.* blinn ; *for* *bi-linne ; *cessation* : widuten blin C. M. 942 ; wit-outen blin 881, 1897.

blind, adj., *O.E.* blind, = *O.L.G.* blind, *O.N.* blindr, *Goth.* blinds, *O.H.G.* plint, blint ; *blind*, ORM. 1859 ; BEK. 1695 ; sum blind mon O. & N. 1237 ; þe blinde man GOW. II. 210 ; **blindne** (*acc. m.*) HOM. I. 111 ; ænne blindne MK. viii. 22 ; þes blinden hand 23 ; **blinde** (*pl.*) KATH. 1062 ; **blindere** (*gen. pl.*) MAT. xv. 14 ; *comp.* stāre-blind.

blind-fellen, v., *strike blind, put out the eyes ; blindfold* ; PR. P. 40 : þauh þu þin eien **blindfellie** (*subj.*) A. R. 106 ; **blindfellede** (*pret.*) A. R. 106 ; MISC. 45 ; **blindfeld** (*pple.*) *struck blind*, C. M. 19615 ; blinfeld, *blinded*, MAN (H.) 54.

blind-fellunge, sb., *blindfolding*, A. R. 188.

blin-hēde, sb., *blindness*, HAMP. PS. xcvi. 2.

blind-hwarven, v., *blind by spinning round* : his eȝen weore **blintwharved** (*pple.*) C. L. 1146.

blindlunge, adv., = *O.H.G.* blintilingo ; *blindly*, MARH. 15.

blindnesse, sb., *O.E.* blindness ; *blindness*, PR. P. 40 ; WICL. 4 KINGS vi. 18 ; blindnisse MK. iii. 5.

blind-worm, sb., = *Swed.* blindorm ; *blind-worm*, VOC. 223.

blindin, v., *cf. O.N.*, *O.Fris.* blinda, *Goth.* gablindjan, *O.H.G.* plinden, blinden ; *make blind*, PR. P. 40 ; S. S. (Wr.) 2994 ; **blindiþ** (*pres.*) WICL. S. W. I. 379 ; H. V. 92 ; *comp.* a-blinden.

blink, sb., = *Dan.*, *Swed.* blink ; *glimpse* : þe leste þoghte þat of godenesse hadde ani blinke H. S. 4449 ; **blinkes** (*pl.*) TEST. CRES. 226.

blink, *see* blench.

blinken, v., = *Du.* blinken (*pret.* blonk), *Swed.* blinka (*pret.* blinkade), *Dan.* blinke (*pret.* blinkede), *mod.Eng.* blink ; *shine, look at* ; before þi iȝen hit **blenkis** (? *for* blinkes) (*pres.*) HALLIW. 185 ; **blenked** (*pret.*). GAW. 799 ; qvenne þe balefulle birde blenked on his blod ANT. ARTH. xlii ; *deriv.* blanc.

blinnen, v., *O.E.* blinnan *for* *bi-linnan, = *O.H.G.* pilinnan ; *cease*, ORM. 4505 ; MISC. 195 ; blinnen, blinne GEN. & EX. 289, 1963 ; blinne ROB. 302 ; WILL. 55 ; P. S. 212 ; MIR. PL. 3 ; belinne GAM. 557 ; **blinneth** (*pres.*) HAV. ·329 ; **blan** (*pret.*) FER. 1626 ; IW. 178 ; DEGR. 1117 ; AV. ARTH. lviii ; blann ORM. 14565 ; blunne (*ms.* blinne, *r. w.* sunne) HAV. 2670.

blinnunge, sb., *cessation*, KATH. 1694.

blischen, *see* blüschen.

blismen, *see* blasmen.

blisse, sb., *O.E.* bliss, blīðs ; *from* blīðe ; *bliss, joy* ['*gaudium*'], LK. xv. 10 ; HOM. I. 7 ; KATH. 1723 ; A. R. 192 ; ORM. 719 ; GEN. & EX. 11 ; BEK. 689 ; SPEC. 61 ; GOW. II. 145 ; blisse, blis O. & N. 420, 1280 ; blis SPEC. 60 ; ALIS. (Web.) 2638 ; TRIST. 2166 ; CH. C. T. *B* 33 ; blisse wes on daie LAȜ. 858 ; mid muchelere blisse 2100 ; to þere blisse MISC. 37 ; **blisse** (*pl.*) R. S. v (MISC. 170) ; blissen [blisses] LAȜ. 9027 ; blissen A. R. 192 ; **blissene** (*gen. pl.*) AN. LIT. 96.

blis-ful, adj., *blissful*, HOM. I. 191 ; KATH. 1879 ; WILL. 1055 ; CH. C. T. *A* 17 ; blisvol AYENB. 148 ; þe **blisfulle** (*pl.*) tidinge HOM. I. 77 ; blisfulle songes LAȜ. 9539.

blisful-hēde, sb., *blessedness*, PR. C. 7836 ; C. M. 6852 ; blisfulhed HAMP. PS. i. 1.

blisfu(l)-līche, adv., *blissfully*, A. R. 40 ; blisvolliche AYENB. 94.

blisfulnesse, sb., *blissfulness*, CH. BOET. iv. 2 (113).

blissien, v., *O.E.* blissian, blīþsian ; *make happy* ; ne mei nan mon ... blissien him mid þisse wordle ['*gaudere cum saeculo*'] HOM. I. 33 ; blissien mire duȝeðe LAȜ. 19041 ; blissen HOM. II. 93 ; ORM. 444 ; bliscen A. R. 360 ; ich wiht ... blisseþ hit hwanne ich cume O. & N. 435 ; **blissie** (*subj.*) HOM. I. 105 ; blisse O. & N. 478 ; þah ȝe blissen ow þrof KATH. 848 ; þu havest **blissed** (*pple.*) min(e) soule HOM. II. 71 ; *comp.* ȝe-blissien.

blissien, *see* **bletsien.**

blister, sb.,=*M.Du.* bluister; *blister, boil,*
C. M. 6011 ; **blistres** (*pl.*) FL. & LEÁF 408.

blīðe, adj., *O.E.* blīðe, = *O.L.G.* blīði, blīthi,
O.N. blīðr, *Goth.* bleiþs, *O.H.G.* blīdi ; *blithe,*
glad, cheerful, KATH. 1879 ; A. R. 348 ; blīðe
and fagen GEN. & EX. 1343 ; blīðe [bliþe] LAȝ.
1379 ; bliþe ORM. 668 ; O. & N. 418 ; ROB.
114 ; HAV.651 ; AYENB. 87 ; ALIS. 443 ; bliþe
bode ihc þe bringe ASS. 104 ; bliþe stones
GAW. 162 ; bliþe, blithe CH. C. T. *A* 846 ; mi
blithe fader D. TROY 2342 ; blive REL. I. 48 ;
blīðre (*compar.*) FRAG. 6 ; LAȝ. 2527 ; ich
am þe bliþure O. & N. 1108 ; **blīðest** (*superl.*)
LAȝ. 31147.

  blīðe-līche, adv., *O.E.* blīðelīce ; *blithely,*
A. R. 68 ; bliþeliche BRD. 5 ; bluðeliche HOM.
I. 31 ; LAȝ. 3304 ; bliþeliche, bleþeli WILL.
819, 1144 ; bleþeliche S. S. (Web.) 503 ; bliþe-
like ORM. 935 ; bliþeli MAN. (H.) 163 ;
blīðelüker (*compar.*) H. M. 3.

  blīðe-mōd, adj., *O.E.* blīðemōd ; *of cheerful*
mood, LAȝ. 29701.

blīþenesse, sb., *O.E.* blīðness, *O.H.G.* blīd-
nissa ; *blitheness, joy,* BOET. ii. 3 (37).

blīþe, sb.,=*O.N.* blīða ; *joy,* A. P. i. 354 ;
L. C. C. 36.

  blīþe-ful, adj., *joyful,* PS. ciii. 34* ; C. M.
17887.

blīthin, v., *O.N.* blīða, = *O.L.G.* blīthon,
O.H.G. blīden ; *make blithe, 'exhilaro,'* PR.
P. 40 ; blith(e) C. M. 17870 ; **blithid** (*pret.*)
DEP. R. iii. 94.

blīve, *see under* **līf.**

blō, *see* **blā.   blobure,** *see* **blubber.**

blōc, *see* **blāc.**

blōd, sb., *O.E.* blōd, = *O.L.G.*, *O.Fris.* blōd,
O.N. blōð, *Goth.* blōþ, *O.H.G.* bluot, pluot ;
blood, O. & N. 1127 ; þat blod LAȝ. 23973 ;
ORM. 1770 ; ROB. 172 ; Sarra ðat faire blod
GEN. & EX. 1192 ; þet blod AYENB. 218 ;
blod, blood CH. C. T. *A* 635 ; blud WICL.
JOHN vi. 54 ; blud IW. 2071 ; **blōdes** swot
sweat of blood A. R. 360 ; blodes flod ORM.
15516 ; **blōde** (*dat.*) HOM. I. 87 ; AYENB.
107 ; heo were of on(e) blode ROB. 15 ; *comp.*
heorte-blōd.

  blōd-bend, sb., *ligature for stopping bleed-*
ing, A. R. 420 ; TRIST. 2208.

  blōd-güte, sb., *O.E.* blōdgyte ; *bloodshed-*
ding, LAȝ. 630.

  blōd-hound, sb., *bloodhound,* WILL. 2183 ;
blodhond, D. ARTH. 3640.

  blood-īrin, sb., *O.H.G.* pluotīsarn ; *lancet,*
fleam, PR. P. 40.

  blōd-lēs, ad'., *O. E.* blōdlēas ; *bloodless,*
KATH. 251.

  blōd-lēs, sb., *letting of blood,* TREV. VI. 115.

  blōd-lĕten, adj., *cf. O.N.* blōðlātinn ; *having*
been bled ; blodletene (men) A. R. 260.

  blōd-lĕtere, sb., *bloodletter,* VOC. 212 ;
blodelater 195.

  blōd-lĕtunge, sb., *blood-letting,* A. R. 114 ;
blodleting HOM. I. 283.

  blōd-rēad, adj., *O.E.* blōdrēad ; *blood-red* :
of blodrede scarlet ROB. 313.

  blōd-rüne, sb., *O.E.* blōdryne, = *O.Fris.*
blōdrene ; *issue of blood,* HOM. I. 207.

  blōd-ssedinge, sb., *bloodshedding,* ROB.
548.

  blōd-strēam, sb.,=*Ger.* blūtstrōm ; *stream*
of blood ; **blōdstrēmes** (*pl.*) LAȝ. 28359.

  blood-soukere, sb., *bloodsucker,* TREV. IV.
243.

  blōd-swētunge, sb., *blood-sweating,* HOM.
I. 207.

  blōd-wīte, sb., *O.E.* blōdwīte ; *fine for*
bloodshedding, 'quite de sanc espandu,' REL.
I. 33.

  blōd-wrékere, sb., *avenger of blood* ; blood-
wreker JOS. xx. 5.

  blōd-würt, sb.,=*M.H.G.* bluotwurz ; *blood-*
wort, REL. I. 36 ; blodwerte LEECHD. III.
314.

blōdede, pple., *stained with blood,* LAȝ.
26811.

blōdeȝien, blōdechen, v., *O.E.* (ge)blōde-
gian,=*O.N.* blōðga ; *render bloody* ; **blōdeke**
(*subj.*) A. R. 418 ; *comp.* bi-blōdeȝien.

blōdi, adj., *O.E.* blōdig, = *O.L.G.* blōdag,
O.N. blōðugr, *O.H.G.* pluotag ; *bloody,* A. R.
112 ; SPEC. 62 ; AYENB. 46 ; PR. C. 5261 ; bi his
blodie swote HOM. I. 207 ; blodie veldes LAȝ.
26811.

blōding, sb., *black pudding* ; **bloodinges**
(*pl.*) TOWNL. 89.

blok, sb., *O.Fr.* bloc,=*M.Du.* bloc, *O.H.G.*
bloch *n.* (*cippus*), piloch *n.* (*clausura*) ; *? block,*
'*truncus, codex,*' PR. P. 49 ; L. H. R. 141 ; a
blok (*inclosed space*) as brod as a halle A. P.
iii. 272 ; **blockes** (*pl.*) GOW. I. 314 ; blokkes
S. & C. II. lxvi.

blōkien, *see* **blākien.**

blōme, sb., *O.N.* blōmi,=*O.L.G.* blōmo, *Goth.*
blōma, *O.H.G.* bluomo ; *from* **blōwen** ; *bloom,*
flower, 'flos,' PR. P. 40 ; ORM. 10773 ; HAV.
63 ; H. V. 50 ; MAN. (F.) 4166 ; PS. cii. 15 ;
FLOR. 686 ; ISUM. 176 ; TOWNL. 89.

blōmin, v.,=*M.H.G.* blōmen, *M.Du.* bloe-
men ; *bloom, flower, 'floreo,'* PR. P. 40 ; blomen
ORM. 10769 ; blome LUD. COV. 56 ; **blōmede**
(*pret.*) GEN. & EX. 2061 ; blomed L. H. R.
135.

blonc, *see* **blanc.   blonden,** *see* **blanden.**

blonder, *see* **blunder.   blonke,** *see* **blanke.**

blōre, *see* **bladre.   blōren,** *see* **blāren.**

blose, sb., *?=Icel.* blossi ; *flame, torch,* A. P.
i. 910.

blōsme, sb., *O.E.* blōsma, blōstma,=*M.L.G.,*
M.Du. blosem ; *blossom,* H. M. 11 ; SPEC. 26 ;

CH. C. T. *A* 3324; PR. P. 41; uppe þare blosme
HOM. II. 217; blosme, blostme MARH. 10;
blostme A. R. 192; O. & N. 437; REL. II. 8;
(*ms.* blosstme) ORM. 1928; **blosmen** (*pl.*)
A. S. 31; blostmen FRAG. 2; A. R. 276; blosman, blostme HOM. I. 5, 109; blosme FL.
& BL. 294; blosme, blostme O. & N. 16;
blostme, blostmes HOM. II. 89, 161; blosmes
SPEC. 45; LANGL. *B* xvi. 7; mid blostmen
HOM. II. 89; wiþ blosmen SPEC. 43.

blŏsmi, adj., *blossomy*, CH. P. F. 183.

blŏsmin, v., *O.E.* blōstmian; *blossom, 'floreo,'*
PR. P. 40; blosme WICL. PS. lxxi. 16; **blŏs**
meth (*pres.*) CH. C. T. *E* 1462; **blŏsmede**
(*pret.*) LANGL. *B* v. 140 (Wr. 2751); **blŏsmed**
(*pple.*) R. R. 108.

blŏstme, *see* **blŏsme.**

blot, sb., *blot*, CH. C. T. *I* 1010; A. P. i. 781;
blott TOWNL. 106; *see* plot.

blote, adj., *? soft with moisture*: your blote
hides P. L. S. xiii. 154.

blottin, v., *blot*, ' *oblitero,*' PR. P. 41.

blout, adj., *O.N.* blautr, ? = *O.E.* blēat; *soft*:
he maden here backes al so bloute as here
wombes HAV. 1910.

blowe, sb., *cf. O.H.G.* bliuwan, *Goth.* bliggvan (*beat*); *blow, act of striking*; blaw
TOWNL. 195; **blowes** (*pl.*) S. & C. II. lx.

blōwen, v., *O.E.* blōwan, *pret.* blēow, = *O.L.G.*
blōjan (*pple.* geblōid), *O.H.G.* pluōn (*pret.*
plūhita); *blow, blossom*; blowe O. & N. 1133;
blowe REL. I. 262; ROB. 352; **blōweð** (*pres.*)
HOM. I. 193; A. R. 30; **blēou** [bloude] (*pret.*)
LA3. 2013; blewe ANGL. III. 280; **blōwe**
(*pple.*) O. & N. 1636; *comp.* 3e-**blōwen**; *deriv.*
blēde, blōme.

blōwen, *see* **blāwen.**

blubber, sb., *mod.Eng.* blubber; *bubble*,
TEST. CRES. 192; blobure PR. P. 40.

blubren, v., = *L.Ger.* blubbern; *bubble up*;
blubrande (*pple.*) A. P. ii. 1017; bloberond
D. TROY 9642; þe borne blubred (*pret.*)
GAW. 2174.

[**blü3en**, v., *O.E.* blȳgan, = *M.H.G.* blūgen,
bliugen; *terrify; comp.* a-**blü3en.**]

blunder, sb., *blunder*, GAW. 18; blundur
FLOR. 1330; blunder, blonder TOWNL. 30,
98.

blundren, v., *blunder*: we blundren [blondren] CH. C. T. *G* 670; thou blondirs þi selfen
D. ARTH. 3975; blondringe (*pple.*) TREV.
II. 169.

    blunderer, sb., *blunderer*, PR. P. 40.

    blunderinge, sb., *blundering*, PR. P. 40.

blunt, adj., *blunt,* ' *hebes,*' PR. P. 41; A. P. i.
176; blunt & blind ORM. 16954.

blunten, v., *? stagger*; he blunt(e) in a bloc
A. P. iii. 272.

blūre, sb., *? blister, swelling*; grete loppis
ouere all þis lande þei flye, that with byting

makis mekill blure YORK xi. 294 (TOWNL.
blowre); so mani thus broght i on blure
(*r. w.* cure) TOWNL. 310.

blusch, sb., *blush*, JOS. 657; GAW. 520.

blüschen, v., *O.E.* blyscan (*rutilare*) (MONE
355), = *M.L.G.* bloschen (*candere, rubere*)
(SCHILLER I. 363); *blush, shine forth, cast a*
glance; blischis (*pres.*) ALEX. 872; bluschande (*pple.*) GAW. 1819; **blusched** (*pret.*)
A. P. i. 979; blushit the sun D. TROY 4665;
the king blischit on the berin D. ARTH. 116;
blushed red CT. LOVE 1199.

blüsing, sb., *from O.E.* blysian (*to blaze*),=
Swed. blysa (*to shine,* RIETZ 42); *blazing*:
the blising of the cole BER. 561.

blüsnen, v., = *shine*; **blüsnande**, blisnande
(*pple.*) A. P. i. 163, 1404; **blisned** (*pret.*) i.
1047.

blustren, v., *rush wildly about*; **blustered**
(*pret.*) A. P. ii. 886; blustreden LANGL. *B* v.
521.

bō, *see* **bē3en.**

boban, -ant, sb., *O.Fr.* boban; *pomp*; so
prout he is and of so great boban GUY 2816;
the bobant of the worlde K. L. 38.

bobance, sb., *O.Fr.* bobance; *pomp, ostenta*
tion, P. L. S. xii. 45; bobance, bobaunce
CH. C. T. *D* 569; bobaunce P. S. 189; WILL.
1071.

bobbe, sb., *bob, bunch*, GAW. 206; a bob of
cheris TOWNL. 118; **bobbis** (*pl.*) of grapes
HALLIW. 190.

bobbe, sb., *cf. Dan.* bobbe, *Swed.* bobba
(*buprestis*); *some kind of insect*: spiders,
bobbs, and lice HALLIW. 190.

bobbin, v., *cf. Swed.* bobba (*knock against*,
RIETZ 44); *bob*, L. H. R. 179; **bobbende**
(*pple.*) '*illudentes*' WICL. 3 ESDR. i. 51;
bobbed (*pret.*) S. S. (Web.) 2246; bobbiden
REL. II. 45; **bobbed** (*pple.*) H. V. 126;
bobbid LIDG. M. P. 261.

bobet, sb., *slap, box on the ear,* '*colaphus,*'
PR. P. 41; *see* **buffet.**

bobetin, v., *slap,* '*colaphizo,*' PR. P. 41.

bōc, sb., *O.E.* bōc, *pl.* bēc (*cognate with* bōc,
a beech-tree), *cf. O.N.* bōk, *pl.* bœkr, *Goth.*
bōka (γράμμα), *pl.* bōkos f. (βίβλος), *O.Fris.*
bōk, *pl.* bōka, bōk f. n., *O.L.G.* bōk, *pl.* bōk
n. f., *M.L.G.* bōk n., *O.H.G.* puoh, buoh,
buoch, *pl.* puoh n.; *book*, ORM. 599; SPEC. 41;
þeos boc LA3. 10110; A. R. 12; þa boc '*li*
brum' LK. iv. 17; one boc A. R. 54; þissere
boc (*earlier text* bec) FRAG. 1; boc, bok
O. & N. 1325; þis boc, bok AYENB. 1, 42; bok,
buk PR. C. 51, 336; **bēch**, [*O.E.* bēc] (*dat.*),
LK. iii. 4; on lives bec MISC. 71; bæch MK.
i. 2; on þere boc HOM. I. 89; leorni bute boc
learn without book (*learn by heart*) MISC. 99;
boke CH. C. T. *B* 190; þat læred was o boke
ORM. 8932; buke IW. 1947; **bēc** (*nom. acc.*
pl.) FRAG. 8; HOM. I. 101; bæc, boc [bokes]

LA3. 53, 7263; **bōkes** ORM. 5811; in owre bokes (*for* boken) KATH. 839; **bōken** (*dat. pl.*) LA3. 6298; on boken FRAG. 6; HOM. I. 113; lered on . . . boken GEN. & EX. 4; in boke O. & N. 350.

bōk-binder, sb., *bookbinder*, D. TROY 1589.

bōc-fel,sb., *O.E.* bōcfell, = *O.H.G.* puohfel; *parchment*, LA3. 50; MARH. 23.

bōc-hūs, sb., *O.E.* bōchūs; *library*; **bōc-house** (*dat.*) AYENB. I.

bōc-ilǣred, adj., *book-learned*, LA3. 16915; bokilered REL. I. 170; boklered A. P. ii. 1551.

bōc-leden, sb., *O.E.* bōclæden; *book Latin*, HOM. I. 107.

bōc-leorned, adj., *book-learned* : boke-lornut birnus ANT. ARTH. lv.

bōc-lōre, sb., *O.E* bōc-lār; *book-learning*, HOM. II. 155.

bōc-rūne, sb., *letter*, LA3. 4496.

bōc-spel, sb., *tale in a book*, LA3. 17487.

bōc-staf, sb., *O.E.* bōcstæf, = *O.N.* bōkstafr, *O.H.G.* buohstab; *letter*, ORM. 4308; **bōc-staven** (*pl.*) LA3. 7637.

[**bōc**, sb., *O.E.* bōc, = *O.N.* bōk, *O.H.G.* buohha, *mod.G.* buche, *beech*.]

bōc-mast, sb., = *M.L.G.* bōkmast; *beech mast*, HALLIW. 67.

bōc-trēow, sb., *O.E.* bōctrēow; *beech-tree*, 'fagus,' FRAG. 3.

bōcare, *see* **bōkere**.

bocchen, v., *cf. M.Du.* butsen, *O.Fr.* boucher, boucier; *botch, patch, mend*; bocchin ['sarcirent'] WICL. 2 PARAL. xxxiv. 10.

bochchare, sb., *botcher, cobbler*, PR. P. 42; bochoure BARB. LEG. I. 117.

boce, boche, sb., *O.Fr.* boce, boche; *boss*; *hump, swelling*; *botch*; boce or boos of a booke 'turgiolum' PR. P. 41; C. M. 8087; bos CH. C. T. *A* 3266; boche *botch*, 'gibbus' VOC. 224; bohche, botche (? bocche) 'ulcus' PR. P. 42; bocche WICL. LEV. xiii. 18; LIDG. M. P. 31; **boces** (*pl.*) WICL. IS. iii. 18; bocches LANGL. *B* xx. 83.

bocher, sb., *O.Fr.* bochier; *butcher*, P. S. 192; ALIS. 1885; CH. C. T. *A* 2025; boucher LANGL. *C* vii. 379; **bochiers** (*pl.*) LANGL. *B* ii. 187; bouchers LANGL. *C* i. 221.

bocherie, sb., *O. Fr.* boucherie; *butchery*, AYENB. 64; OCTOV. (W.) 733.

bochi, adj., *botchy*, 'gibbosus,' VOC. 225.

bocin, v., *swell out, project* ; *make to project*; bocyn owte 'turgeo' PR. P. 41; men þat **boosen** hor brestis WICL. S. W. III. 124; **bost** (*pret.*) D. TROY 3022; **bost** (*pple.*) 1564.

bocle, sb., *O.Fr.* bocle; *buckle*, 'pluscula,' PR. P. 41; SPEC. 35; AYENB. 236.

boclen, v., *Fr.* boucler; *buckle*, bokelin, PR. P. 42; **bokelinge** [bocling] (*pple.*) CH. C. T.

A 2583; **bucled** (*pple.*) PL. CR. 299; bocult ANT. AR. xxix. 20.

bocler, sb., *O.Fr.* bocler, bucler; *buckler, small shield*, CH. C. T. *A* 112; boklēr P. 20; bookeler WICL. S. W. III. 265.

bod, sb., *O.E.* bod, = *O.Fris.* bod, *O.N.* boð, *O.H.G.* pot *n.; from* bēoden; *message, command, offering, invitation*, LA3. 8510, 23407; C. M. 4096; þat ilche bod MISC. 142; bode 'nuncium' PR. P. 41; HOM. II. 213; ORM. 5255; GEN. & EX. 395; ASS. 104; WILL. 2154; RICH. 3298; TRIST. 2082; AM. & AMIL. 1254; MAN. (H.) 22; CH. P. F. 343; GAW. 1824; þis bode SAX. CHR. 38; sende heast(e) & bode (*message*) KATH. 48; **bode** (*dat.*) P. L. S. viii. 131; A. R. 400, 416; PL. CR. 716; for no bode (*offer*) of pans GREG. 438; **bode** (*pl.*) HOM. I. 57; II. 209; he wende þat he brohte boden swiðe gode LA3. 27999; comp. bi-, for-, 3e-, mis-bod.

bod-word, sb., = *O.N.* boðorð; *word of command, message*, M. H. 44; A. P. ii. 473; TOWNL. 75; bodeword ORM. 7; GEN. & EX. 361; bodeword 'praeceptum' PS. ii. 6; MAN. (F.) 9810.

bōd, *see* **bād**.

bóde, sb., *O.E.* boda, = *O.Fris.* boda, *O.L.G.* bodo, *O.H.G.* poto, boto *m.; from* bēoden; *messenger*, RICH. 1359; **bóden** (*pl.*) 'nuncios' LK. ix. 52; **bóden** (*pl.*) ['angeli'] HOM. I. 219.

bode, *see* **bod**.   **bodekin**, *see* **boidekin**.

bodi, *see* **bodi3**.

bódien, v., *O.E.* bodian, = *O.Fris.* bodia, *O.N.* boða; *bode, announce, proclaim*, LK. i. 19; KATH. 1481; he lette . . . his cume bodien LA3. 23290; bodian þa(n) soðen ileafen HOM. I. 97; **bódest** (*pres.*) O. & N. 1152; bodeþ FRAG. 5; GOW. I. 153; bodieð A. R. 212; **bódiende** (*pple.*) HOM. I. 95; **bódede** (*pret.*) LA3. 27123; bodeden, bodedan HOM. I. 95, 97; comp. 3e-bódien.

bódunge, sb., *O.E.* bodung; *boding, announcement, prediction*, HOM. I. 89; bodinge ROB. 428.

bodi3, sb., *O.E.* bodig *n.,* = *O.H.G.* boteh, potah *m.; body*, ORM. 1555; bodi LA3. 4908; O. & N. 73; GEN. & EX. 4134; TREAT. 138; PERC. 1166; þet bodi A. R. 102; AYENB. 235; þe soule loveþ þe bodi so C. L. 1169; an duc . . . þat hardi bodi was ROB. 183; **bodies** (*gen.*) H. M. 35; **bodie** (*dat.*) HOM. I. 209; AYENB. 72; þat me hire heavet . . . totwemde from þe bodie MARH. 19; **bodies** (*pl.*) KATH. 1444.

bodi3-lích, sb., *body*, ORM. 16294.

bodi-lích, adj., *bodily*, AYENB. 72; bodilíche (*pl.*) workes 212; bodili C. M. 428; bodilíche (*adv.*) PL. CR. 619.

boef, bēf, sb., *O.Fr.* boef, buef, beuf, *mod.Eng.* beef; *ox, beef*, CH. C. T. *D* 1753; beof S. A. L.

156; P. S. 334; bef P. 43; beef ALIS. 5248; bouf WILL. 1849; biffe PR. P. 28.

bōȝ, sb., *O.E.* bōg, bōh (*arm, bough*),=*O.N.* bōgr (*arm, bow of a ship*), *M.Du.* boech, *O.H.G.* buog (*arm*), *Gr.* πῆχυς, *mod.Eng.* bough, bow (*of a ship*); *not allied to* būȝen, *but to Sanskr.* bahu-s (*arm*); *arm, bough, bow of a ship*; O. & N. 242; AYENB. 17; bog GEN. & EX. 608; bōgh IW. 392; TRIAM. 410; bouȝ BRD. 9; bouȝ or braunche ['*ramus*'] WICL. MAT. xxiv. 32; bouh P. S. 341; þe bouh hwon he adeadeð A. R. 150; bough CH. C. T. *A* 1980; bou S. S. (Wr.) 610; bōȝe, bōwe (*dat.*) O. & N. 15, 125; he is under wude boȝe HORN (L.) 1227; bōȝes (*pl.*) AYENB. 17; boȝhes ORM. 10002; bowes [boȝes] A. R. 30; bowes WILL. 23; bowes of wilde bores D. ARTH. 188; bowis WICL. MK. xi. 8; boes ANT. ARTH. iii; bughes PR. C. 680; bōgen (*dat. pl.*) MAT. xiii. 32; boȝe, bowe O. & N. 616.

bōu-līne, sb.,=*O.N.* bōglīna, *M.Du.* boechlijne; *bowline*, MAN. (F.) 12061.

bōu-sprēt, sb.,=*M.Du.* boechspriet; *bowsprit*, MAN. (F.) 12061.

bóȝe, sb., *O.E.* boga,=*O.Fris.* boga, *O.L.G.*, *O.H.G.* bogo, *O.N.* bogi; *from* būȝen; *bow* (*for shooting*); *arch*; AYENB. 45; boge GEN. & EX. 483; *enne* boȝe, bowe LAȝ. 1453, 6471; bowe FRAG. 6; A. R. 250; bowe ROB. 16; CH. C. T. *A* 108; bowe [bouwe] '*fornicem*' WICL. PROV. xx. 26; seint Marie cherche at þe bowe ['*de arcubus*'] TREV. VIII. 153; *comp.* el-, rein-, sadel-bóȝe; *deriv.* bowȝere.

   bówe-draught, sb., *the distance a bow will carry*; a bowedraughte MAND. 240; bowdraucht BARB. vi. 58.

   bów(e)-man, sb., *bowman*, ROB. 378.

   bówe-schote, sb., *bow-shot*, ALIS. 3491.

boge, *see* bulge.

bōȝen, v., *? O.E.* bōgan (*jactare*); *utter, tell of*: the minde of the abundaunce of thi swetnesse þei shul bowen out ['*eructabunt*'] WICL. PS. cxliv. 7.

boggisch, adj., *boastful*, '*tumidus*,' PR. P. 42.

   boggische-lī, adv., *boastfully*, '*tumide*,' PR. P. 42; bogeisliche WILL. 1707.

bóȝien, v.,=*M.L.G.*, *M.Du.* bogen; *bow, bend, turn away*; boȝe (? =bouȝe) A. P. i. 196; boȝe fro þis benche GAW. 344; bowin '*flecto, curvo, inclino*' PR. P. 46; bowe S. S. (Wr.) 2536; bóȝed (*pret.*) GAW. 481; David bowide fro þe face of Saul ['*declinavit David a facie Saul*'] WICL. I KINGS xix. 10; boweden here hedes MAND. 235; summe beouweden speren LAȝ. 22297; iboued (*pple.*) TREAT. 139; *comp.* a-bōȝien.

bóȝien, v., *O.E.* bogian,=*O.Fris.* bogia; *dwell*; bugian heou (*r.* hou) longe wult þu boȝie (*ms.* beoȝie) abuten þissere burȝe LAȝ. 29233;

Brennes þer boȝede (*sec. text* abod) 5974; heo þer boȝeden þe wile þe heo luveden 7903.

boglen, v., *boggle*; bogelid (*pple.*) HALLIW. 959.

boȝ-sam, adj., *cf.* būhsum *and M.Du.* booghsaem (*flexible*); *obedient*, AYENB. 59; bousom Iw. 1155; PR. C. 85; boxom M. H. 54.

   boȝsam-līche, *obediently*, AYENB. 70; buxumlier (*comp.*) WILL. 723.

   boȝsamnesse, sb., *obedience*, AYENB. 33; bōcsomnesse ROB. 234.

boȝt, sb.,=*Du.* bogt, *Dan.*, *Swed.* bugt; *from* būȝen; *bending*; bought TOR. 558; of þeose bouȝt was heore croune ALIS. 4712; *see* biȝt.

boȝted, adj., *vaulted*: hit watz . . . boȝted on lofte A. P. iii. 449.

boȝtling, sb., *from* būggen; *one ransomed*; boghtlinges (*pl.*) C. M. 17262.

boidekin, **bodekin**, sb., *bodkin, dagger*; '*subucula, perforatorium*,' PR. P. 42; boidekin CH. C. T. *A* 3960.

bōie, sb.,=*M.Du.* boeve, *? M.H.G.* buobe (*see* GR. WB. II. 457); *boy* (*ms.* boie), HORN (L.) 1075; S. S. (Web.) 960; WILL. 464; LANGL. *A* PROL. 77; CH. C. T. *D* 1322; bōiȝes (*gen.*) WILL. 1705; bōies (*pl.*) HAV. 1899; BEK. 79.

boile, v., *O.Fr.* boillir; *boil*, GOW. II. 265; C. M. 11886; boile, buile '*scatere*' WICL. EXOD. xvi. 20; boilinde (*pple.*) LEG. 164; boiland S. S. (Web.) 2460; boilinge WICL. IS. lvii. 21; boillede (*pret.*) MARG. 247; boilid (*pple.*) L. C. C. 43.

   boiling, sb., *boiling*, WICL. S. W. II. 202; builing JONAH i. 15.

boiloun, sb., *O.Fr.* boillon; *bubble*; boilouns (*pl.*) S. S. (Web.) 2488.

boinard, sb., *O.Fr.* buinard; *simpleton*, AN. LIT. 9; DEP. R. ii. 164.

boiste, sb., *O.Fr.* boiste (*boîte*); *box*, '*pyxis*,' PR. P. 42; CH. C. T. *C* 307; boiste [bost] C. M. 14003; boist Iw. 1835; MAND. 85; buist GOW. II. 247; bouste BARB. LEG. I. 124; bustes, boistes (*pl.*) A. R. 226.

boistous, adj., *rough, clumsy, boisterous*, JARB. XIII. 176; TREV. II. 311; ALIS. 5660; CH. C. T. *H* 211; LIDG. M. P. 91; boistois AYENB. 103; buistous '*rudis*' WICL. MAT. ix. 16.

   boistous-līche, adv., *boisterously*, TREV. II. 147; boistousli CH. C. T. *E* 791; BER. 134, 163.

bōk, *see* bōc.

bokeram, sb., *O.Fr.* bouqueran; *buckram*, P. P. II. 171; PR. P. 42; bougeren AYENB. 258.

bōkere, sb., *O.E.* bōcere,=*Goth.* bōkareis, *O.H.G.* buochari; *scribe* ['*scriba*'], MAT. viii. 19; bōcares (*pl.*) LAȝ. 32125.

boket, sb., *bucket*, '*situla*,' PR. P. 42; REL. I.

9; CH. C. T. *A* 1533; bucket C. M. 3306; buket QUAIR 70.

bōkien, v., *O.E.* bōcian; *book*; boke GOW. I. 3; ibōked (*ms.* ibocked) A. R. 158; GOW. III. 319.

ból, sb., *Gr.* βῶλος; *bole*; boole armoniak (*astringent earth*), CH. C. T. *G* 790.

ból, sb., *O.N.* bolr,=*M.Du.* bol; *tree trunk*; bóle (*dat.*) GAW. 766; A. P. ii. 622; bóles (*pl.*) HALLIW. 193.

bóle-ax, sb.; *O.N.* bolöx; *axe*, REL. II. 176; bulaxe ORM. 9281; boleaxis (*pl.*) OCTOV. (W.) 1039.

bul-rische, sb., *bulrush*, PR. P. 244.

bol-werk, sb.,=*M.L.G.*, *M.Du.* bolwerk; *bulwark*, LIDG. M. P. 237.

bolace, sb., *O.Fr.* beloce; *bullace*; bolaces (*pl.*) VOC. 162; WILL. 1809.

bóld, sb., *O.E.* bold,=*O.Fris.* bold *n.* [=*O.E.* botl]; *house*, HOM. I. 253; LA3. 7094; KATH. 1664; MISC. 96; ROB. 116; REL. II. 217; bóldes (*pl.*) JUL. 72; LEB. JES. 830; MAP 343; bólde (*gen. pl.*) LA3. 25882; *comp.* büri-, bürŏe-bóld.

bold, bolden, *see* beald, bealden.

bole, *see* bule. **bolge,** *see* bulge.

bol3ien, v., *? elate*; bole3eŏ, boluweŏ, bolhes (*pres.*) A. R. 214; 3ebol3en (*pple.*) MAT. xxvi. 8.

bolgit, *see* bulgit.

bolk, sb., *prov.Eng.* boak; *belching*; bolke (*dat.*) LANGL. *B* v. 397.

bolkin, v., ?=*M.L.G.* bolken (*mugire*); *belch*, '*ructo*,' PR. P. 43; bolked, balked (*pret.*) TREV. II. 195; bolkede out ['*eructavit*'] WICL. WISD. xix. 10.

bolkinge, sb., *belching*, '*eructatio*,' PR. P. 43.

bolknen, v.,=bolkin; bolkenand (*pple.*) PS. cxliii. 13.

bolle, sb., *O.E.* bolla,=*O.Fris.* bolla, *O.N.* bolli *m.*, *M.L.G.*, *M.Du.* bolle, *O.H.G.* polla *f.*, *mod.Eng.* bowl, boll; *globe, boll, bowl, cup,* VOC. 199; PR. P. 43; HORN (L.) 1123; LANGL. *A* v. 213; CH. C. T. *G* 1210; A. P. ii. 1145; LIDG. M. P. 52; þene bolle LA3. 14994; a bolle with meede TREV. VII. 528; þe bolle of þe popi HALLIW. 193; bollen [bolles] (*pl.*) LA3. 19781; TREAT. 138; *comp.* chēse-, flax-, þrote-bolle.

bolle, sb., *? O.N.* bolli; *measure for grain*, BARB. iii. 211.

bollen, v.,=*Du.* bollen; *?strike*: whiche man ... bolled (*pret.*) þe HALLIW. 193.

bollen,v., *? for* bolnin; bolleþ (*pres.*)LANGL. *A* v. 99; bollinge (*v. r.* bolnande) C. M. 6011; iswolle and ibolled TREV. I. 299; bollid [*v. r.* bolnid] ['*inflati*'] WICL. I COR. v. 2.

bollen, *see under* bellen.

bollere, sb., *drunkard* ['*ganeo*'], TREV. III. 359; bollars (*pl.*) TOWNL. 242.

bolling, sb., *? for* bolning; LANGL. *A* vii. 204.

bolnin, v.,=*Dan.* bolne, *O.N.* bolgna; *swell*, '*tumeo*,' PR. P. 43; bolne (*pres.*) GAW. 512; TOWNL. 197; al mi bodi bolneth LANGL. *B* v. 119; bolnes HAMP. PS. i. 5*; bolnand (*pple.*) C. M. 6011; bolnede (*pret.*) MAN. (F.) 8204; bolned A. P. ii. 363; bolned (*pple.*) C. M. 4726; YORK xlvi. 45; bolnid WICL. I COR. v. 2.

bolning, sb., *swelling*, LANGL. *B* vi. 218; C. M. 12083; M. H. 130; bolninge '*tumor*' PR. P. 43; VOC. 224.

bolster, sb., *O.E.* bolster, = *O.N.* bolstr, *O.H.G.* bolstar; *bolster*, MK. iv. 38; bolstir '*culcitra*' PR. P. 43; bolstar '*cervical*' VOC. 178; bolstre (*dat.*) FRAG. 6; HOM. II. 139; R. S V; L. H. R. 210.

bol-strau3t, adj., *? for* bolt-strau3t; *prostrate*, WILL. 1852.

bolstren, v.,=*Ger.* bolstern, polstern; *bolster*; bolsterid (*pple.*) LIDG. M. P. 200.

bolt, sb., *O.E.* bolt (*cross-bow bolt*),=*O.H.G.* bolz *m.*, polz (*catapult*),=*M.Du.* bolt (*headed arrow*); *bolt* (*arrow*), bolt (*of a door*), REL. I. 183 (MISC. 128); P. L. S. xix. 54; A.D. 291; LUD. COV. 136; here bolt is sone ischote S. S. (Web.) 989; upright as a bolt CH. C. T. *A* 3264.

bolt, adv., *bolt upright*, PALL. i. 967.

bolten, v., *bolt*: bolted (*pple.*) with iren LANGL. *B* vi. 138.

bomblen, *see* bumblen.

bommen, *see* bumbin.

bon, adj., *O.Fr.* bon; *good*, TOR. 2143; boon RICH. 1540.

bon, *see* bune. **bōn,** *see* bān.

bonchef, sb., *O.Fr.* bon (*good*), chef (*head*); *cf.* meschief; *good luck, happiness*, GAW. 1764; boonchief TREV. I. 87.

bonchen, *see* bunchen. **bond,** *see* band.

bondáge, sb., *med.Lat.* bondagium; *bondage*, TREV. I. 389; MAN. (H.) 71.

bonde, sb., *O.E.* bonda, bunda, = *M.L.G.* bunde, *O.N.* bōndi=būandi; *husbandman, servant,* '*servus*,' PR. P. 43; MISC. 77; P. S. 150; MIRC 1520; CH. C. T. *D* 1660; H. V. 77; an oht bonde LA3. 15291; bounde S. S. (Web.) 582; AR. & MER. 691; bonde (*pl.*) MISC. 89; barouns, burgeis & bonde WILL. 2128; *comp.* hūs-bonde.

bonde-hēde, sb., *servitude*, C. M. 5404.

bonde-man, sb., *bondman*, '(*servus*) *nativus*,' VOC. 182; O. & N. 1577; HAV. 32; BEK. 552; LANGL. *B* v. 194 (W. 2859); bondemen '*servi*' TREV. I. 246.

bond-schepe, sb., *condition of a bondman*, PR. P. 43.

bond-womman, sb., *bondwoman*, TREV. II. 97.

bone, *see* bane. **bōne,** *see* bān.

bōne, sb., *O.N.* bōn; *boon, request, prayer,* HOM. I. 37; A. R. 234; KATH. 611; ORM. 7606; HAV. 1659; AN. LIT. 11; BRD. 30; WILL. 1095; S. S. (Wr.) 546; GAW. 327; TOR. 1578; AV. ARTH. xiii; bidden þeos bone JUL. 30; bone, boone CH. C. T. *A* 2269; boone EGL. 101; **bōne** (*dat.*) LAȝ. 14912; **bōne** (*nom. pl.*) AUD. 72; **bōnen** (*dat. pl.*) A. R. 142; MISC. 63.

bōnen, v., *pray*; **bōne** (*pres.*) ORM. 5223; **bōned** (*pple.*) 694.

bonēr, adj., *O.Fr.* bonnaire; *gentle, courteous*: so boner and þewed A. P. ii. 733; with wordes bonere ALIS. 6732; sche is meke and boneire B. DISC. 1727; to bonour with hem þat þou ne be B. B. 41; be þou bonaire TRIST. 2731.

   bonaire-līche, adv., *courteously*, AYENB. ·bonerli 265; [debonerli] C. M. 23872.

   bonērnesse, sb., *gentleness*, WICL. 1 COR. iv. 21; S. W. II. 357.

bonertē, sb., *gentleness*: he calde me to his bonerte A. P. i. 761; of so moche bonerite H. S. 1927.

bonet, sb., *O.Fr.* bonet; *bonnet (of a sail)*, 'artemo,' PR. P. 43; VOC. 268; DEP. R. iv. 72.

boni, sb., *O.Fr.* bugne, *cf. mod.E.* bunion; *swelling;* 'tumor,' PR. P. 43.

boni, adj., *bonny*: bonie londes ALIS. 3903.

bonk, *see* **bank. bonnien,** *see* **bannen.**

bor, sb., *O.E.* bor,=*M.L.G., O.N.* bor; *gimlet* (*ms.* bore), TOWNL. 219.

bōr, *see* **būr. bŏr,** *see* **bār.**

borace, sb., *Ital.* borace; *borax*; boras CH. C. T. *A* 630.

borage, sb., *O.Fr.* borrage; *borage,* PR. P. 44; REL. I. 51; burage 'borago' REL. I. 37.

bord, sb., *O.E.* bord *n.* (*board, border, shield*), *cf. O.L.G., O.Fris.* bord *m.* (*board, border, shield*), *Goth.* baurd *n.* (*board*), *O.N.* borð *n.* (*board, border, shield*), *O.H.G.* bort *m.* (*border*); *board, table, shield, border, ship's side*: HAV. 1722; BEK. 691; ALIS. 1270; WICL. EXOD. xxv. 27; heo letten to somne sæiles gliden, bord wið borden (*r.* borde) LAȝ. 20935; þis wes þat ilke bord þat Bruttes of ȝelpeð 22953; þat halȝhe bord ORM. 1096; þet bord AYENB. 167; as stif as eni bord P. L. S. xvii. 334; the theef fel over bord CH. C. T. *B* 922; for bord ne cloþing *G* 1017; burd AMAD. (R.) lxvi; **bordes** (*gen.*) BEK. 1186; **borde** (*dat.*) HOM. I. 105; A. R. 142, 324; O. & N. 479; GAW. 481; heo seten to borde LAȝ. 14950; at godes borde AYENB. 235; breken þe schipes bord [**bordes**] (*pl.*) JUL. 78, 79; bordes (*sec. text* sceldes) þer scænden LAȝ. 5186; **borden** (*dat. pl.*) LAȝ. 22778; *comp.* mete-, sid-, stēr-bord.

   bord-clōð, sb., *tablecloth ornamented*, REL. I. 129; bordcloth E. W. 101.

   bord-knīf, sb., *table-knife,* VOC. 253; boorde-knife PR. P. 44.

bord-fēlāwe, sb., *companion*; bordfelauis (*pl.*) WICL. JUDG. xiv. 11.

bord-lēas, adj., *without a table*: sitte as a beggar bordlas on ground LANGL. *C* xv. 141.

burd-wōgh, sb., *wall made of boards,* 'tabellatum,' VOC. 237.

borde, sb.=*M.L.G.* borde, *O.N.* borði (*carpet*), *O.H.G.* borto; *? border*; on brode silkin borde GAW. 610.

bordel, sb., *Fr.* bordel; *brothel, prostitution,* P. L. S. xxi. 92; WICL. LEV. xix. 29; **bordels** (*pl.*) CH. C. T. *I* 885.

   bordel-hūs, sb., *brothel*; bordelhous WICL. EZ. xvi. 24, 39.

   bordeler, sb., *O.Fr.* bordelier; *keeper of a brothel,* GOW. III. 322.

bordure, sb., *Fr.* bordure; *border,* PR. P. 44; CH. BOET. i. 1 (6); bordoure D. ARTH. 4211; **bowerdurs** (*pl.*) D. TROY 12861.

borduren, v., *border*; **bordured** (*pple.*) MAND. 217.

bóre, sb., *O.N.* bora; *bore, perforation,* 'foramen,' PR. P. 44; M. H. 57; TRIST. 2539.

borel, *see* **burel.**

bórin, v., *O.E.* borian,=*M.Du.* boren, *O.N.* bora, *O.H.G.* borōn; *bore, perforate,* PR. P. 44; **bóre** (*pres.*) FRAG. 4; (heo) borieþ 6; **bórien** (*subj.*) KATH. 1949; **bóriinde** (*pple.*) AYENB. 66; **bóred** (*pret. pple.*) L. H. R. 203; ibored TREV. IV. 397; *comp.* þurh-bóreſ.

borȝ, sb., *O.E.* borg, borh, = *M.H.G.* borc; *from* berȝen; *pledge, caution-money, money borrowed*: þorȝ borȝ and ȝemer ȝeld SHOR. 113; buten þe eorl Adionard ... walde heore borh beon LAȝ. 11913; ȝeldeð hire ȝarew(e) borh JUL. 72; borgh, boruȝ, boru (*ms.* borw) LANGL. *A* iv. 76; borgh (*ms.* borghe) TOWNL. 231; burugh RICH. 3259; borwe, borowe 'vas' PR. P. 45; borwe (*dat.*) LANGL. *B* xiv. 190; i dar take god to borwe GOW. II. 34; ech of hem had leid his faiþ to borwe CH. C. T. *A* 1622; **boreges** (*pl.*) HOM. II. 17; borwes LANGL. *A* i. 75; AM. & AMIL. 899; borwes BEK. 585; borewis 'vades' WICL. PROV. xxii. 26; *comp.* in-borȝ.

   borghe-gage, sb., *pledge,* H. S. 9576.

   borghe-gang, sb., *suretyship,* H. S. 9582.

   boru-hēd, sb., *suretyship*: þe boruheed of Crist WICL. S. W. III. 10.

   borwe-schepe, sb., *suretyship,* PR. P. 44*.

borȝ, *see* **burȝ.**

borgeis, borgesie, *see* **burgeis, burgesie.**

borȝien, v., *O.E.* borgian,= *O.Fris.* borgia, *O.H.G.* borgēn, *O.N.* borga, *mod.Eng.* borrow; *receive on pledge, borrow; be surety for, ransom, save, preserve*: to borȝe (*save*) him fram (h)arme LAȝ. 21268*; borwe ALIS. 1243; CH. C. T. *B* 105; wolde borwe monei of him TREV. VII. 413; to borewen us alle he wes ibore SPEC. 25; us alle from bale to borowe MIR. PL. 149;

borȝeþ (*pres.*) AYENB. 36; **borwe** (*subj.*) LANGL. *B* iv. 109; þet ever ich ... boruwe et tisse vrakele worlde so lutel so heo ever mei A. R. 204; **borwede** (*pret.*) ROB. 393; borwede, borwed LANGL. *B* vi. 101; borwed WILL. 1705; **borwid** (*pple.*) HOCCL. i. 369; iborwed C. L. 822.

borware, sb., *borrower, surety, 'mutuator, sponsor,'* PR. P. 44.

borȝing, sb., *borrowing*: if ye ȝyuen borwynge to hem WICL. LK. vi. 34; he that taketh borewing, servant is of the usurer WICL. PROV. xxii. 7; borwynge '*mutuatio*' PR. P. 44; borewinge ANGL. IV. 198.

borlīch, *see* **burlīch.** **borne**, *see* **burne.**

boru, borw, *see* **borȝ** *and* **burȝ.**

borwage, sb., *from* **borȝ**; *suretyship*, PR. P. 44.

borwien, *see* **borȝien.**

bōs, sb.,=*L.G.* bōs (SCHÜTZE), *O.N.* bāss, *Goth.* bansts; *boose, cow-stall*; bose '*bostar*' VOC. 235; boos '*bostar, bucetum*' PR. P. 41.

bosard, *see* **busard.** **bosch**, *see* **busch.**

bōsem, sb., *O.E.* bōsm, bōsum,=*O.L.G.* bōsom, *O.H.G.* buosum, buosem; *bosom*, ORM. 19391; M. H. 125; bosum A. R. 146; GEN. & EX. 2809; A. P. iii. 107; bosom CH. C. T. *D* 1993; **bōsme** (*dat.*) HOM. I. 53; AYENB. 163; bosme [bosome] LAȝ. 14984; he ne was boren of wifes bosme HOM. II. 133; boseme A. R. 146; in Abrahames boseme sitte LEB. JES. 163; in his boseme he bar a thing LANGL. *B* xvi. 254; **bōsmes** (*pl.*) LAȝ. 7849.

bosine, sb., *O.Fr.* bosine, busine, *Lat.* bucina; *trumpet*, AYENB. 137.

boskage, sb., *O.Fr.* boscage; *boscage*, IW. 1671.

bosken, *see* **busken.** **bosse**, *see* **boce.**

bost, sb., *boast, glory, noise*, ROB. 209; AYENB. 71; WILL. 1141; ALIS. 4069; SPEC. 94; bost, boost LANGL. *B* xiv. 247 (Wr. 9397).

bost-ful, adj., *boastful*, WICL. S. W. I. 2; FLOR. 270.

bosten, v., *boast*, LANGL. *B* ii. 80; boste IW. 70; boste & blowe TRANS. XVIII. 25; **bosteþ** (*pres.*) CH. C. T. *D* 1672.

bostere, sb., *boaster*, TREV. IV. 295; boster DEP. R. ii. 80.

bostinge, sb., *boasting*, SHOR. 112.

bosti, adj., *boastful*, TREAT. 138.

bōsum, *see* **bōsem.**

bot, *see* **būten.** **bōt**, *see* **bāt.**

bote, *see* **būten.**

bōte, sb., *O.E.* bōt,= *O.N.* bōt, *O.Fris.* bōte, *O.L.G., Goth.* bōta (ὄφελος), *O.H.G.* puoza, *mod.E.* boot, *in phrase* 'to boot;' *repair, remedy, redress*; buoza FRAG. 6; LAȝ. 21926; O. & N. 688; HAV. 1200; ROB. 448; RICH. 3086; ANT. ARTH. xvi; aftir bale comeþ bote FL. & BL. 821; þat is bote of alle bale AMAD. (R.) xvii; vorȝiveð ham hore gultes hwon heo

ham iknoweð and bihoteð bote A. R. 430; do bote WILL. 1378; bote þou do me bote S. A. L. 155; ȝeoven bale i bote stude H. M. 33; gan to bote HOM. I. 15; til ure sawle bote ORM. 2692; to ȝoure bote SHOR. 159; bote, boote CH. C. T. *A* 424; to bote [boote], *to boot, in addition*, LANGL. *B* xiv. 268; bute MIN. i. 4; **boten** (*pl.*) A. R. 120; *comp.* **brigge-, dǣd-, sün-bōte.**

bōt-dai, sb.=*Ger.* büsztag; *day of atonement*, HOM. II. 69.

bōt-lēs, adj., *O.E.* bōtleas; *bootless*, WILL. 134; a botles bale LANGL. *C* xxi. 208.

bōte, sb., *O.Fr.* bote; *boot*, '*ocrea*,' VOC. 197; PR. P. 45; **bōtes** (*pl.*) HORN (H.) 522; P. S. 330; MAND. 250; bootes TREV. I. 353; CH. C. T. *A* 273.

bōte-lees, adj., *without boots*, LANGL. *B* xviii. 11.

botel, sb., *O.E.* botl *n.* ?=*O.L.G.* bodal *m.*; *house*; botle (*dat.*) MAT. xxvi. 3; bottle ORM. 2788; *cf.* **bōld.**

botel, sb., *? O.Fr.* bottel; *bottle (of hay), bundle*: a botel hei CH. C. T. *H* 14.

botel, sb., *O.Fr.* bouteille; *bottle*, PR. P. 45; botelle LUD. COV. 138.

boteler, sb., *O.Fr.* bouteillier; *butler*, ROB. 187; WICL. GEN. xl. 1; CH. C. T. *B* 4324; buteler GEN. & EX. 2092.

botelerie, sb., *O.Fr.* bouteillerie; *wine-cellar*, ROB. 191.

botem, *see* **boþem.**

botemai, sb., *bitumen*, ALIS. 6190; **botemeis** (*pl.*) 4763.

bōten, *see* **būten.**

bōten, v., *mod.Eng.* boot *in phrase* 'it boots not;' *cf.* **bēten**, *from* **bōte**; *make better*; bote AM. & AMIL. 2340; it bōted (*pret.*) not K. L. 66; he was bōtid (*pple.*) of mekile care EGL. 188.

boteras, sb., *buttress*, '*fultura*,' PR. P. 45; **boteraces** (*pl.*) WICL. EZ. xli. 15.

boterase, v., *furnish with a buttress*; **boterased** (*pple.*) LANGL. *B* v. 598.

botere, *see* **butere.**

boterel, sb., *O.Fr.* boterel; *toad*, AYENB. 187.

boterie, sb., *ale-cellar, pantry, buttery,* '*cellarium*,' PR. P. 45; botrie '*promptuarium*' VOC. 178; boteri E. G. 98.

botew, sb., *sort of boot*, '*coturnus, botula*,' PR. P. 45; **botwez** (*pl.*) E. G. 332.

bot-forke, sb., *? for* batte-forke; *? forked stick*, SPEC. 110.

bōtnen, v., *make* or *become better, amend, repair*; **bōtneð** (*pres.*) KATH. 2523; bōtnede (*pret.*) CHR. E. 768; botnede ek of uvel(e) P. L. S. xii. 151; bōtned (*pple.*) WILL. 1055; botened, bootned LANGL. *B* vi. 194.

bōtnere, sb., *healer*: heil botenere of everie bodi blinde WART. II. 109.

bŏtninge, sb., *amendment*, SHOR. 96.

Botolf, pr. n., *Botolf*: Botolfs to(u)n, *Boston*, AN. LIT. 4.

botōn, sb., *O.Fr.* boton ; *button*, VOC. 238 ; botoun AYENB. 86 ; botouns (*pl.*) P. S. 239 ; botones LANGL. *B* xv. 121 ; *see* bothom.

botonin, v., *Fr.* boutonner ; *button*, PR. P. 46.

botor, *see* bitor.

botraille, sb. ? paterfamulias . . . shulde sette botraille atweine derk and lighte LIDG. M. P. 170.

botte, *see* batte. botum, *see* boþem.

bōþe, sb., *O.N.* bōð, būð = *M.Du.* boede, *M.L.G.* bōde, *M.H.G.* buode ; *booth*, PR. P. 46 ; ORM. 15817 ; AYENB. 215 ; P. 45 ; GOW. III. 281 ; A. P. iii. 441 ; bōþes (*pl.*) ALIS. 3457.

bōðe, *see* bāðe.

boþel, sb., *?chrysanthemum* ; boþul, bothil '*vaccinia*' PR. P. 46.

boþem, sb., *O.E.* botm, = *O.L.G.* bodom, *O.H.G.* podam, bodem, *O.N.* botn ; *bottom*, VOC. 159 ; JOS. 15 ; GAW. 2145 ; A. P. iii. 144 ; boþum S. S. (Wr.) 809 ; botum, botim '*fundus, glomus*' PR. P. 45 ; botme (*dat.*) AYENB. 140 ; GOW. I. 108 ; CH. C. T. *C* 1321 ; MAND. 52 ; PARTEN. 4480.

   boþem-lēz, adj., = *M.H.G.* bodemlōs ; *bottomless*, A. P. ii. 1022.

bothom, sb. (*? form of* botōn) ; *bud* : the bothom bright of hewe R. R. 2959 ; the bothom faire and swote 3008 ; botheum 1719.

bothon, sb., *O.E.* boðen ; *chrysanthemum*, LEECHD. III. 315.

bōu, *see* bōȝ.

bouche, sb., *Fr.* bouche ; *allowance of victual granted by king or noble to his household, etc.*, DEGR. 998.

boude, *see* bude.

bouel, sb., *O.Fr.* bouel, buele ; *bowel*, A. P. iii. 293 ; bouelle '*viscus*' VOC. 247 ; bouele GOW. II. 265 ; boueles (*pl.*) P. S. 221.

bouelen, bowailen, v., *disembowel*, '*eviscero*,' PR. P. 46 ; boweld (*pret.*) MAN. (H.) 329.

bouȝ, *see* bōȝ.

bouge, *see* bulge. bouȝen, *see* būȝen.

bougerie, sb., *heresy* ; bugerie MAN. (H.) 320.

bougette, sb., *budget* ; bowȝettes (*pl.*) TREV. VII. 385.

bougouns, sb. pl., *an instrument of music* : simbales and sonetes sware þe noise, and bougouns busch batered so þikke A. P. ii. 1416.

Bougre, pr. n., *O.Fr.* Bougre (*Bulgarian*) ; *heretic*, AYENB. 19.

bouȝt, *see* boȝt. bōuh, *see* bōȝ.

bouk, *see* būc.

bouked, pple., *having a protuberance*, ALIS. 6265.

bouken, v., = *M.Du.* buiken, *M.H.G.* būchen ; *buck* (*i.e. steep in lye*), LANGL. *B* xiv. 19 ; bouke (*imper.*) REL. I. 108.

boule, sb., *Fr.* boule ; *bowl* : round as a boule GUIL.[2] (N. E. D.) ; plei withe bowlis PR. P. 46.

bounde, *see* bonde.

bounde, sb., *O.Fr.* bodne, bonde, bonne, *mod.Eng.* bound ; *boundary*, GOW. III. 92 ; boundes (*pl.*) AYENB. 206 ; CH. C. T. *E* 46 ; bandus AV. ARTH. iii ; bunnen (*dat. pl.*) LAȝ. 1313.

bounden, v., *O.Fr.* bonder, bonner ; *bound, set bounds to* ; bounded (*pple.*) GOW. I. 218.

bounen, v., *from* būn ; *prepare*, JOS. 414 ; boune D. TROY 827 ; boune (*pres. subj.*) M. ARTH. 3257.

bounsen, *see* bunsen.

bountē, sb., *O.Fr.* bonté ; *bounty, goodness*, C. L. 1214 ; SPEC. 28 ; CH. C. T. *B* 1664 ; C. M. 9531 ; bunte MISC. 36.

bountevous, adj., *from O.Fr.* bontif ; *bounteous*, CH. TRO. i. 883.

bour, *see* būr.

bourde, sb., *O.Fr.* bourde ; *jest*, CH. C. T. *H* 81 ; GAW. 1409 ; þe bourdes (*pl.*) and þe trufles AYENB. 58.

   bourde-ful, adj., *full of jesting*, WICL. WISD. i. 11 (*margin*) ; bourdfull D. TROY 3952.

   bourdful-li, adv., REL. II. 45.

bourden, v., *O.Fr.* bourder ; *jest* ; bourde (*pres.*) CH. C. T. *C* 778 ; bourdinge (*pple.*) LANGL. *B* xv. 40 ; bourded (*pret.*) GAW. 1217 ; bourdedest AYENB. 20.

bourding, sb., *jesting*, GAW. 1404.

bourdēs, sb., *O.Fr.* behordeis ; *tournament*, H. S. 4662 ; bourdys or tournamentys WILL. 1477.

bourdíse, v., *joust* ; burdised (*pret.*) S. S. (Web.) 740.

bourdour, sb., *O.Fr.* bourdeor ; *joker*, MAN. (H.) 204.

bous, sb., *? drink, ? drinking-cup*, SPEC. 111.

bouten, *see* būten. bouwen, *see* būȝen.

bove, *see* biuven. bowe, *see* boȝe.

bōwen, *see* bōȝen. bowien, *see* boȝien.

bowȝere, sb., *from* boȝe ; *bow maker ; archer*, '*arcuarius, architenens*,' PR. P. 46 ; bowere VOC. 195 ; bowiares (*pl.*) ROB. 541.

box, sb., *O.E.* box, = *O.H.G.* buhsa, *from Gr.* πύξις ; *box*, HOM. II. 145 ; ROB. 456 ; CH. C. T. *A* 4390.

box, sb., *O.E.* box, from *Lat.* buxus, *Gr.* πύξος ; *box(-tree)*, VOC. 163.

   box-trē, sb., *O.E.* boxtrēow ; *box tree*, VOC. 228 ; CH. C. T. *A* 1302.

box, sb., *box* (*on the ear*) ; '*alapa*,' PR. P. 46 ; CH. L. G. W. 1388 ; D. ARTH. 1111.

brā, sb., *O.N.* brā, = *O.E.* brēaw ; *see* brēu ; brae, *river-bank* ; under ane bra thair galai

drench BARB. iv. 372 ; bra *'ripa'* CATH. 39 ;
WINT. VIII. xxvi. 74 ; bro MAN. (H.) 310.

brac, sb., *O.E.* (ge)bræc,=*M.L.G., O.N.* brak,
O.H.G. (ge)breh ; *loud noise* ; *(ms.* bracc)
ORM. 1178.

bráce, sb., *Fr.* bras, *from Lat.* brāc(c)hium ;
arm (of the sea), MAND. 15, 21, 126.

bráce, sb., *O.Fr.* brace, brache, brase, *from
Lat.* brāc(c)hia (*pl.*) ; *armour for the arms ;
clasp, clamp, hook ; pair, couple* ; wel bornist
brace upon his boþe armes GAW. 582 ; paire
brase LIDG. TROY iii. 22 ; *'uncus'* PR. P. 46 ;
brase (*pair*) LIDG. TROY i. 6.

brácen, v., *O.Fr.* bracier ; *embrace, brace* : in
armes for to brace HALLIW. 203 ; brace
(*imper.*) it in irine D. ARTH. 1182 ; ibraced
(*pple.*) JOS. 265.

brácer, sb., *O.Fr.* brasseure ; *bracer,* CH. C.
T. *A* 111 ; brasere VOC. 263 ; brasers (*pl.*)
D. ARTH. 1859.

brache, sb., *cf. O.Fr.* brachet, *M.H.G.* bracke ;
brach (hound) ; braches (*pl.*) GAW. 1142 ;
REL. I. 151.

brad, sb., ? =*O.H.G.* prat, prart (*ora*) ; *brad,
nail, 'aculeus,'* VOC. 234 ; *see* **brerd** *and* **brod.**

brād, adj., *O.E.* brād,=*O.L.G., O.Fris.* brēd,
Goth. braids, *O.N.* breiðr, *O.H.G.* breit ; *broad,*
ORM. 3431 ; S. S. (Web.) 2784 ; IW. 163 ;
PERC. 269 ; AV. ARTH. iii ; brad [brod] LAʒ.
7635 ; brād 1320 (*miswritten* bræd 14219) ;
brod A. R. 102 ; HAV. 1647 ; TOR. 1252 ;
brood CH. C. T. *A* 155 ; ā brād=*abroad :* þat
blod stod al a brod ROB. 261 ; her winges
boþe a brod she spradde GOW. II. 105 ; and
held hir lappe a brod [brood] CH. C. T.
F 441 ; a brood LANGL. (Wr.) 9031 ; MAND.
45 ; þe brode [brade] stret(e) P. L. S. viii.
172 ; **brāde** (*dat. n.*) LAʒ. 4541 ; brādne
(*acc. m.*) LAʒ. 5087 ; brāde (*pl.*) LAʒ. 4212 ;
brāden (*dat. pl.*) LAʒ. 30272 ; brādder
(*compar.*) ROB. 43 ; bradder, broddere *'latior'*
WICL. AMOS vi. 2 ; brōde (*adv.*) WILL.
754 ; CH. C. T. *A* 739.

brōd-arwe, sb., *broad arrow, 'catapulta,'*
PR. P. 53.

brood-axe, sb., *O.E.* brādæx,=*O.N.* breiðox ;
broad-axe, 'dolabra,' PR. P. 53 ; brodax VOC.
234.

brād-ēged, adj., *broad-eyed,* SAX. CHR. 256.

brāde, *see* **brēde.**

brādien, v., *O.E.* brādian ; *from* **brād ;** *cf.*
brædan ; *widen, enlarge* ; brōdes (*pres.*)
ALIS.² (Sk.) 122 ; brōdid (*pret.*) DEP. R. ii.
141 ; HAMP. PS. cxviii. 32 ; ibrōded (*pple.*)
O. & N. 1312.

brǣde, sb., *O.E.* brǣdu,=*O.Fris.* brēde, *Goth.*
braidei, *O.H.G.* breiti ; *from* **brād ;** *breadth* :
a bræde [in brede] LAʒ. 21995 ; brede PR. P.
49 ; O. & N. 174 ; ROB. 353 ; AYENB. 105 ;
GOW. II. 388 ; MAND. 117 ; CH. C. T. *A* 1970 ;
WICL. GEN. vi. 15 ; ANT. ARTH. xlv ; al

aboute on brede þei spredde RICH. 4434 ; þe
knight ligges þer on brede PERC. 797 ; in
breede and lengþe LIDG. M. P. 98 ; breade
A. R. 102 ; MISC. 145.

brǣden, v., *O.E.* brǣdan, = *Goth.* braidjan,
O.L.G. brēdean, *O.N.* breiða, *O.H.G.* breiten ;
from **brād ;** *broaden, expand ;* breden LAʒ.
14283 ; bredin PR. P. 49 ; brede A. P. i. 813 ;
PALL. xi. 101 ; blosmes **brēdeþ** (*expand*)
(*pres.*) SPEC. 45 ; **brǣdde** (*pret.*) JOS. 642 ;
GAW. 1928 ; bordes heo **brǣdden** (*pl.*)
LAʒ. 18523 ; *comp.* **a-brǣden.**

brǣdling, adv., *broadwise* : he wile smite
mid bredlinge swuerde HOM. II. 61.

brag, sb., *brag,* P. L. S. xxix. 92 ; DEGR. 231 ;
brag and boost TREV. V. 109.

brag, adj., *bragging, boastful, proud, valiant* :
þat makeþ us so brag and bolde SPEC. 24 ;
wordes bragge SHOR. 110 ; hi schulde nouʒt
beren hem so bragg(e) PL. CR. 706 ; braggest
(*superl.*) WILL. 3048.

bragance, sb., *boasting* : with boste and
bragance TOWNL. 99.

braggen, v., *mod.Eng.* brag ; *boast, sound
loudly* : he bosteþ and brageþ (*pres.*) LANGL.
B xiii. 281 ; whanne þe voice of þe trompe
. . . in ʒoure eeris braggiþ WICL. JOSH. vi. 5 ;
braggede (*pret.*) D. ARTH. 3657.

braggere, sb., *bragger,* LANGL. *B* vi. 156.

bragot, braget, braket, sb., *Welsh* bragawd,
=*Ir.* bracat (*malt liquor*) ; *bragget (a liquor
made of ale and mead fermented together),*
PR. P. 46 ; bragot, braket CH. C. T. *A* 3261
(Wr. bragat) ; braget TREV. I. 399.

brai, sb., *O.Fr.* brait ; *bray,* ALIS. 2175.

braid, sb., *O.E.* brægd, bregd,=*O.N.* bragð
n. ; from **breiden ;** *throw, twist ; stratagem ;*
GOW. II. 21 ; CH. L. G. W. 1164 ; TOR. 1611 ;
he abod moni a bitter braid TRANS. XVIII.
24 ; scho braid hit a doun at on braid S. S.
(Wr.) 483 ; þat was a þeves braid MAN. (H.)
164 ; breid ROB. 22 ; REL. I. 224 ; he went
to þe uscher in a breide CL. 448 ; habben
bares heorte and remes brede LAʒ. 30392 ;
braides IPOM. 1834 ; ivele breides REL. I.
218 (MISC. 14) ; *comp.* **up-breid.**

braiin, v., *O.Fr.* braire ; *bray, neigh, resound
harshly, 'barrio,'* PR. P. 47 ; braie MAN. (F.)
3488 ; braieþ (*pres.*) GOW. I. 144 ; braien
GAW. 1163 ; braiinde (*pple.*) AYENB. 73.

braiin, v., *O.Fr.* breier, broier ; *bray (in a
mortar), 'pinso, tero,'* PR. P. 47 ; braiid
(*pple.*) WICL. 1 KINGS xxv. 18.

brail, sb., *O.Fr.* braiel ; *brail* ; brailes (*pl.*)
GUIL.² (*in* N. E. D.).

brain, adj., *furious,* GAW. 286.

brain [braʒen], sb., *O.E.* brægen, brægn,=
M.L.G. bregen, bragen, *O.Fris.* breinn *n. ; brain,*
LAʒ. 1468 ; brain TREAT. 138 ; brein FER.
1901 ; BEVES 1558 ; braines (*pl.*) JOS.
501.

brain-lēs, adj., *brainless,* PR. P. 47 ; N. P. 9.

brain-panne, sb., = *M.L.G.* bregenpanne ; *brainpan, skull,* MAND. 234.

brain-wōd, adj., '*brain-mad,*' *insane,* WILL. 2096 ; LANGL. *A* x. 61 ; GAW. 1461.

brainid, sb., *having a brain,* PR. P. 47.

brainin, v., *brain,* '*excerebro,*' PR. P. 47 ; braineþ (*pres.*) WICL. IS. lxvi. 3.

brake, see **braken.**

bráke, sb., *cf. M.L.G.* brake, *M.Du.* braeke, *Swed.* bråka : *flax-brake, baker's brake,* '*rupa, vibra,*' VOC. 217, 276.

bráke, sb., *? O.Fr.* brac (*accus. of* bras) ; *lever, winch* : gunnes & boȝes of brake FER. 3263.

braken, sb., *cf. Swed.* bräken ; *brake, bracken,* A. P. ii. 1675 ; brakin '*felix*' VOC. 191 ; brake '*feugere*' VOC. 156 ; '*filix*' [*sic*] PR. P. 47 ; LEECHD. III. 315 ; LUD. COV. 22.

brákin, v., = *M.Du.* braeken, *M.L.G.* braken ; *vomit,* '*vomo,*' PR. P. 47 ; as an hounde that et gras so gan ich to brake LANGL. *C* vii. 431 ; brákez (*pres.*) A. P. iii. 340.

brallen, v., = *Ger., Du.* brallen, *Dan.* bralle, *?mod.Eng.* brawl ; *cry out* ; bralle TOWNL. 91.

bráme, sb., *cf. M.Du.* braeme, *O.H.G.* brāma, prāma ; *bramble,* '*tribulus, vepres,*' VOC. 192.

bran, see **bren.**

branche, sb., *O.Fr.* branche ; *branch,* '*ramus,*' PR. P. 48 ; braunche LEB. JES. 596 ; branches (*pl.*) ROB. 152 ; bronches AYENB. 9.

branched, adj., *branched* ; braunched WILL. 754.

branchen, v., *Fr.* brancher ; *put forth branches ; ramify ; branch out* ; brauncheth (*pres.*) WICL. ECCLUS. xxxix. 19 ; braunchis D. TROY 8750 ; **braunching** (*pple.*) trees WICL. JER. xvii. 2.

brancher, sb., *O.Fr.* *brancher, *Fr.* branchier ; *young hawk,* M. ARTH. 190.

branchi, adj., *bearing branches* ; braunchi WICL. 4 KINGS xvii. 10.

brand, sb., *O.E.* brand, brond, = *O.Fris.* brand, *O.N.* brandr, *O.H.G.* brant, *mod.Eng.* brand ; *from* **brinnen** ; (1) *sword* ; (2) *burning wood, firebrand* ; C. M. 2873, 3170 ; IW. 1933 ; AV. ARTH. xiv ; TOWNL. 71 ; þene brand [brond] LAȝ. 7544 ; his brond he up ahæf 7527 ; brond HOM. I. 81 ; KATH. 2395 ; MISC. 154 ; WILL. 1244 ; nou tak þou þe brond FER. 2238 ; þe Sarsin þanne adrou (h)is brond 580 ; þei . . . sette . . . wilde bround an on in king Daries lond ALIS. 1855 ; **brandes** (*pl.*) PERC. 774 ; brondes A. R. 368 ; AYENB. 205 ; CH. C. T. *A* 2338 ; *comp.* **fūr-brand.**

brond-īr(e), sb., *O.E.* brandīsen ; *gridiron,* '*tripus,*' VOC. 178.

brand-rīþe, sb., *O.N.* brandreiᵹ, *cf. O.E.* brandrēda, *M.L.G.* brandrēde, *O.H.G.* brand-reita ; *?brandreth ; gridiron,* '*tripus,*' VOC. 232.

branden, v., = *M.Du.* branden ; *brand* ; brondin '*cauterizo,*' PR. P. 53 ; **brondit** (*pple.*) APOL. 103.

brandishen, v., *O.Fr.* brandiss- ; *brandish* ; braundishen WICL. JOB xxxix. 23 ; **brandissende** (*pple.*) WILL. 2322.

branken, v., ? = *M.H.G.* brangen, 'prangen ; *prance* : **brankand** (*pple.*) stedez D. ARTH. 1861.

brant, adj., *O.E.* brant, = *O.N.* brattr ; *steep,* ALEX. 3649.

bras, sb., *O.E.* bræs ; *brass,* '*aes,*' PR. P. 47 ; GEN. & EX. 467 ; TREAT. 136 ; brass ORM. 17417 ; breas, bres JUL. 30, 31 ; bres FRAG. 4 ; A. R. 216 ; AYENB. 203.

bras-pot, sb., *brass-pot,* PR. P. 47 ; VOC. 256.

brase, sb., *cf. O.Swed.* brasa, *M.Du.* brase ; *burning coal* : in brasse and in brinstone ANT. ARTH. xv.

brásen, adj., *O.E.* bræsen ; *brazen,* ALIS. 5585 ; C. M. 12193 ; o þat brasene neddre ORM. 17424.

brasiere, sb., *brazier,* '(*faber*) *aerarius,*' PR. P. 47 ; **brasiers** (*pl.*) D. TROY 1589.

brasil, sb., *Fr.* bresil ; *brasil,* PR. P. 47.

brásni, adj., *like brass* ; braasni WICL. DEUT. xxviii. 23.

brassik, sb., *from Lat.* brassica ; *cabbage,* PALL. ix. 53 ; brasik 137.

brastlien, v., *O.E.* brastlian, = *M.H.G.* brasteln ; *rattle, clash,* LAȝ. 26970 ; sceldes brastleden (*pret. pl.*) 27463.

brat, sb., *O.E.* bratt (*pallium*), MAT. v. 40* ; *Irish* bratt (*cloth*) ; *cloak,* CH. C. T. *G* 881.

brāþ, adj., *O.N.* brāᵹr ; *violent,* ORM. 7164 ; C. M. 4003 ; broþ GAW. 2233 ; **brōþe** (*pl.*) wordez A. P. ii. 1409.

brāþ-lī, adj., *O.N.* brāᵹligr ; *violent* : a bra(þ)li braid ANGL. II. 240 ; broþelich A. P. ii. 847 ; **brōþlíche** (*adv.*) MAP 343 ; brathli C. M. 63 ; broþeli MAN. (H.) 166 ; PERC. 2123 ; A. P. ii. 1256.

brāþþe, sb., *from* **brāþ** (*cf.* wræþþe, wráþþe, *from* wráþ) ; *violence,* ORM. 4561 ; brathe D. TROY 5075.

brāþ-ful, adj., *violent* : brōþefulle (*pl.*) wordes MAN. (H.) 55.

braulin, v., *cf. Swed.* bravla (RIETZ 51) ; *brawl,* '*jurgo,*' PR. P. 48 ; he **braules** (*pres.*) him (*boasts*) for his brighte wedes D. ARTH. 1349.

braulere, sb., *brawler,* PR. P. 48 ; **braweleres** [braleres] (*pl.*) LANGL. *B* xvi. 43 ; brawlers TREV. IV. 209.

braulinge, sb., *brawling,* '*jurgium,*' PR. P. 48 ; brawelinge, braulinge LANGL. *B* xv. 233, *C* xvii. 360 ; YORK xxx. 142.

braun, sb., *O.Fr.* braon ; *brawn,* '*lacertus, pulpa,*' PR. P. 48 ; þe braun of a bore VOC. 267 ; neiþer bacon ne braun LANGL. *B* xiii. 91, *C* xvi. 100 ; ful big he was of braun and ek of bones

CH. C. T. *A* 546; braun (*meaning* '*boar*') D. ARTH. 1095; **braunes** '*lacertos*' WICL. JOB xxii. 9; braunes harde and stronge CH. C. T. *A* 2135.

braunche,　braunched,　*see*　**branche, branched.**

brē, *see* **brēie** *and* **brēu.**

brēad, sb., *O.E.* brēad,= *O.Fris.* brād, *O.L.G.* brōd, *O.N.* brauð, *O.H.G.* brōt; *bread,* A. R. 192; SHOR. 30; bread, brad HOM. I. 241; þet bread, briad AYENB. 110; bread, bred GEN. & EX. 364, 2079; bræd, bred LAȝ. 19717, 31800; bræd ORM. 996; HOM. II. 27; R. S. vii; HAV. 633; BRD. 6; þat bred MISC. 40; for evere mi bred had(de) ben bake FER. 577; breed SPEC. 104; CH. C. T. *A* 341; **brēades** (*gen.*) A. R. 262; bredes SHOR. 20; **brēade** (*dat.*) SHOR. 30.

　brēd-corn, sb., *bread-corn,* LANGL. *B* vi. 64.

　brēd-lees, adj., *breadless,* LANGL. *B* xiv. 160.

　brēad-lēp, sb., *bread-basket,* GEN. & EX. 2078.

brēade, *see* **brǣde** *and* **brēde.**

brēch, sb., *O.E. pl.* brēc,= *O.N.* brōk (*pl.* brœkr), *O.H.G.* bruoch, pruoh; *breech* (*breeches*), A. R. 420; BEK. 260; LANGL. *B* v. 175; breech AL. (L.[1]) 731; MAND. 250; brech, brek '*braccae*' PR. P. 48; brek VOC. 238; IW. 1770; MAN. (H.) 161; **brēchen** (*dat. pl.*) LAȝ. 16749; breche FL. & BL. 258; breke LIDG. M. P. 114.

　brēk-belt, sb., *waist-belt,* '*lumbo,*' VOC. 238.

　brēch-gerdel, sb., *girdle,* AYENB. 205; breigurdel MISC. 193; bregirdil WICL. JER. xiii. 1.

　brēch-lēs, adj., *breechless*; breklesse D. ARTH. 1048.

[**breche, breke,** sb., *O.E.* (æw-, wiðer)breca, = *O.H.G.* (hūs)prehho; *one who breaks*; *comp.* æw-, spūs-breche, wed-breke.]

brēche, sb., *O.Fris.* breke,= *M.H.G.* breche; *from* **brēken**; *breach, act of breaking,* ALIS. 2168; GOW. II. 138; C. M. 8220; (?=bruche) SHOR. 133; *comp.* **schip-breche.**

brēche, sb., *cf. M.L.G.* brāke, *M.H.G.* brāche; *? fallow field* O. & N. 14.

bred, sb., *O.E.* bred,= *M.Du.* bred, berd, *O.H.G.* bret; *board,* PR. P. 48; C. M. 16578; nailid on a bred of tre REL. I. 63; brede (*dat.*) O. & N. 965; A. P. iii. 184; *comp.* tǣvel-, wax-bred.

　bred-chēse, sb., *some kind of cheese,* '*junctata*' [i. e. *juncata*], PR. P. 48.

bred, *see* **brerd. brēd,** *see* **brēad** *and* **braid. brēde,** *see* **brǣde** *and* **brūde.**

brēde, sb., *O.E.* brǣde *f.* ?=*M.L.G.* brāde *f.* (SCHILLER I. 412), *O.H.G.* prāto, brāto *m.* (*assatura*); *from* **brēden**; *roast meat,* P. L. S. viii. 73; O. & N. 1630; HAV. 98; SPEC. 27; AV. ARTH. xxxi; swines brede . . . al to diere he hi beið ANGL. I. 15; þeo brǣde LEECHD. III. 98; brade HOM. II. 224; of þere

brede he æt LAȝ. 30596; þa . . . **brāden** (*acc.*) LEECHD. III. 98; brede [breade] LAȝ. 30583; **bredes** (*pl.*) ALIS. 5249; **breden** (*gen. pl.*) LAȝ. 30589.

brēden [**brēade**], v., *O.E.* brǣdan (*pple.* gebrǣdd),= *O.Fris.* brēda, *O.H.G.* prātan, brātan, (*pret.* priat); *roast,* LAȝ. 25986; brede R. S. v (MISC. 180); JUL.[2] 170; RICH. 1493; **bret** (*pres.*) HOM. I. 53; **brēde** (*imper.*) L. C. C. 25; **brĕdde** (*pret.*) LAȝ. 30584; bradden 20978; bredden AR. & MER. 7305; **brēde** (*pple.*) RICH. 3613; brad GAW. 891; bred GEN. & EX. 3147; ibred MISC. 54.

brēden, v., *O.E.* brēdan, = *M.Du.* broeden, *O.H.G.* pruotan, bruoten; *from* **brōd**; *breed,* ANGL. III. 61; brede SPEC. 101; **brēdeth** (*pres.*) LANGL. *B* xi. 339; bredeð, bret A. R. 200, 222; (hie) **brēdeð** HOM. II. 55; bredeþ ROB. 177; **brĕdde** (*pret.*) O. & N. 101; bredden LANGL. *B* xi. 347; **brĕd** (*pple.*) REL. I. 211; ibred O. & N. 1724; heo weoren ibredde LAȝ. 30071.

　brēding, sb., *breeding,* C. M. 3479.

brēden, *see* **brǣden** *and* **breiden.**

bref, sb., *O.Fr.* bref, brief; *brief, official document, letter,* S. S. (Wr.) 3213; MAN. (H.) 237; breve C. M. 19606.

bref, adj., *O.Fr.* bref, *Lat.* brevis; *brief, short,* A. P. i. 268; in breff time BER. 871; wiþ wordis **breve** (*pl.*) H. V. 55.

　brefli, adv., *shortly*; bre[fl]i C. M. 130; breveli C. M. 18199; brefeli SQ. L. D. 873.

　briefnes, sb., *shortness,* LUD. COV. 79.

bregand, sb., *O.Fr.* brigand; *irregular footsoldier*; **bregaundeȝ** (*pl.*) D. ARTH. 2096.

brēȝe, *see* **brēu. bregge,** *see* **brügge.**

bregge, v., *for* **abregen**; *abridge,* WICL. S. W. II. 407; briggen HALLIW. 211; **briggid** (*pple.*) MAN. (H.) 247; breiggid WICL. MK. xiii. 20.

　breggere, sb., *abridger*; perlipominon, that is . . . word bregger WICL. PREF. EP. I. 72.

breid, *see* **braid.**

breiden, v., *O.E.* bregdan, brēdan,= *O.L.G.* bregdan, *O.Fris.* breida, brīda, *O.N.* bregða, brigða, *? O.H.G.* brettan, *? M.H.G.* brīten; *twist, wrench, hurl, draw* (*a sword, etc.*), *plait,* HOM. I. 201, II. 217; breiden . . . ane crune A. R. 124; breden verliche a dun 222; a gret ok he wolde breide a doun as it a smal ȝerde were ROB. 22; he schal breide a wei ['*evellet*'] WICL. PS. xxiv. 15; ȝif þou **brēdest** (*pres.*) wod S. S. (Web.) 1907; þe ancre . . . þene gult ne up breide hire A. R. 426; bredeþ wode P. 63; **breid** (*pret.*) A. R. 280; breid he . . . a sweord LAȝ. 1548; þe king hine bræid [breid] sæc 6667; he braid . . . efter his alderen 6895; bræid hine of his stede 26777; braid [breid] him aȝeinward JUL. 72, 73; braid IW. 3248; AV. ARTH. xliii; of his slep an on he braid HAV. 1282; þe þef braid out (h)is knif CHR.

E. 670; sche braid hure wel neʒ wod FER. 2099; þan braide [*for* braid] he up of his bed WILL. 686; braided ALIS. 5856; and brudden up Baldolf LAʒ. 20335; faste hi drowe & breide P. L. S. xxi. 108; broʒden (*pple.*) MAP 339; broiden O. & N. 645; broidin '*laqueatus*' PR. P. 53; breided, broided CH. C. T. *A* 1049; *comp.* a-, æt-, for-, ʒe-, to-, wið-breiden *deriv.* braid, (up-)brüd.

brēie, sb., =*M.Du.* broeye, *M.H.G.* brüeje; *broth, gravy*; bre L. C. C. 17; *comp.* ale-brēi.

brein, *see* brain.

breiþen, v., *? O.N.* bregða, =breiden; *rush, fly*; breiþed (*pret.*): wine breiþed uppe into his braine A. P. ii. 1421.

brek, sb., *O.E.* (ge)brec, brece; *act of breaking*, ALIS. 2168*; MAN. (F.) 14662; C. M. 6344; breke PR. P. 49.

breke, *see* breche.

brēken, v., *O.E.* brecan, = *O.L.G.* brecan, *O.Fris.* breka, *Goth.* brikan, *O.H.G.* prechan, brechen, *Lat.* frangere; *break*, A. R. 242; KATH. 2028; GEN. & EX. 3147; breoken JUL. 58; breke O. & N. 1080; AYENB. 52; MAND. 228; him þouhte þat his herte wolde breke CH. C. T. *A* 954; to brekene HOM. I. 43; brékest (*pres.*) ORM. 1548; ʒif hit itit þet þu brekest godes heste HOM. I. 21; brekeþ TREAT. 136; brec (*imperat.*) MARH. 18; brekeð on (*begin*) KATH. 1301; bræc, bræc, breac (*pret.*) LAʒ. 1558, 2623, 10978; brac P. L. S. viii. 93; GEN. & EX. 3100; he brac here bendes HOM. II. 113; brak Iw. 3778; CH. C. T. *A* 1468; WICL. NUMB. xi. 8; a water brak out TREV. I. 191; brak, brek ROB. 22, 126; brec HOM. I. 171; MARH. 5; brec on (*v. r.* bigon) KATH. 2295; brek AYENB. 16; (þu) breke LAʒ. 5037; LANGL. *C* xxi. 383; brǣken, brēken (*pret. pl.*) JOHN xix. 32, 33; breken LAʒ. 27506; JUL. 59; AYENB. 64; ALIS. 5891; S. S. (Wr.) 1218; breeken, braken WICL. 2 PARAL. xxxi. 1; brake ISUM. 652; bróken (*pple.*) HAV. 1238; CH. TRO. v. 1204; brokin OCTAV. (H.) 944; wiðuten brokene breade A. R. 342; brokun WICL. IS. xxxvi. 6; brokene mete MK. viii. 20; *comp.* a-, æt-, for-, ʒe-, to-bréken; *deriv.* brac, brake, brek, breche, briche, brüche, brüchel.

brékere, sb., =*M.H.G.* brechære; *breaker*, HOM. I. 83; *comp.* ǣw-brékere.

brékinge, sb., *breaking*, AYENB. 48.

brekil, adj., *brittle*, '*fragilis*,' PR. P. 177; TOWNL. 101.

brembel, sb., *O.E.* brembel (? brēmbel) brēmel, brǣmel *m.*, = *O.H.G.* brāmal(busc) *bramble*, C. M. 924; brembil, brimbil WICL. ECCLUS. xliii. 21; JOB xxxi. 40; brimbil, bremmil '*tribulus, vepres*' PR. P. 49; brembles (*pl.*) HOM. I. 223; brimbles II. 129.

brembel-flour, sb., *bramble-flower*, CH. C. T. *B* 1936. [*Harl. ms.* brember-flour.]

brēme, adj., *O.E.* brēme, brȳme; *valiant, spirited, famous*, ORM. 7197; GREG. 260; WILL. 18; GAW. 1142; A. P. iii. 430; M. ARTH. 229; TOWNL. 197; þu art wel modi and wel breme O. & N. 500; breme and bold K. T. 835; þis strif þat is so breme C. L. 538; þilke feste was wel breme FL. & BL. 792; breme as a bare DEGR. 1240; a breme bare AMAD. (Web.) 171; a bor so brime S. & C. II. xx; brim '*ferus, ferox*' PR. P. 51; MAN. (H.) 28; þe se þat was ful brim HAV. 2233; breme wittes LANGL. *B* xii. 224; her breme blastis DEP. R. iii. 365; bremme heed DEP. R. ii. 130; when briddes singeþ breme SPEC. 44; foughten breme [breeme] CH. C. T. *A* 1699.

brēme-li, adj., *fierce, furious*: mani bremli blast C. M. 24847; brēmeli (*adv.*), WILL. 23; GAW. 1598.

brēmnes, sb., *fury*, D. TROY 4665.

brēme, sb., *O.Fr.* bresme; *bream (the fish)*, VOC. 222; PR. P. 49; CH. C. T. *A* 350.

bren, sb., *O.Fr.* bren; *bran, 'furfur,'* PR. P. 49; CH. C. T. *A* 4053; AYENB. 210; bren, bran LANGL. *B* vi. 184; bran VOC. 201.

bran-brēd, sb., *bran-bread*, VOC. 198.

brende, adj., *from* brennen; *of tawny colour*, LIDG. M. P. 202.

brend-fier, *see under* brüne.

brene, *see* brüne.

brengen, v., *O.E.* brengan, = *O.Fris.* brenga, branga, *O.L.G.* brengean, *M.H.G.* brengen; *from* bringen; *bring*; brenge AYENB. 87; S.S. (Web.) 2169; ED. 3; (*? ms.* bringe, *r. w.* genge) LAʒ. 15094; brengeþ (*pres.*) SHOR. 6; brōhte (? brŏhte), [*O.E.* brōhte, = *O.Fris.* brōchte, *O.L.G.* brāhta, *O.H.G.* prāhta] (*pret.*), *brought*, H. M. 45; P. S. 221; (*ms.* brohhte) ORM. 14361; brochte HOM. I. 171; me heom brohte drinken LAʒ. 13585; brouhte 63; broute [brofte] 36; broʒte, brouhte O. & N. 107; broʒte AYENB. 118; þat moni mon broʒte to deþe ROB. 17; brouhte me a slep LANGL. *A* v. 8; brouʒte *B* v. 8; broghte, brouʒte, broughte CH. C. T. *B* 522; broute REL. II. 277; brohtest JUL. 61; brohton HOM. I. 101; brohten MAT. ii, 11; HORN (R.) 44; brouhten A. R. 114; brŏht (*pple.*) *brought*, HOM. I. 283; SPEC. 29; broʒt LANGL. *B* xii. 136; brouht, brouct, brouth, browt HAV. 336, 513, 1979, 2412; *comp.* ʒe-brengen.

brenie, *see* brünie.

brenne, sb., *O.N.* brenna, *cf. O.H.G.* brinna; *from* brinnen; *burning, conflagration*: hwan he ... haveden ... þe fir brouth on brenne HAV. 1239.

bren-stōn, sb., *O.N.* brenna-, brenni-, brennusteinn; *sulphur*, HICKES I. 228; *see* brimston, brünstōn, *under* brüne.

brennen, bernen, v., *O.N.* brenna, *O.E.* bernan, bærnan, = *O.L.G.*, *O.H.G.* brennan,

M.Du. bernen (*urere, cremare*), *Goth.* (ga)-brannjan (καίειν); *from* **brinnen** ; *burn*, HAV. 916 ; his tail schal brennen in þe glede CH. C. T. *B* 111 ; brenne MAND. 48 ; brenne his londes WILL. 1133 ; shal brenne '*ardebit*' WICL. LEV. vi. 12 ; bernen A. R. 306 ; bærnen ORM. 1529 ; lim heo gunnen bærnen LAȝ. 15466 ; berne ROB. 81 ; TREAT. 136 ; AYENB. 173 ; **brenneþ** (*pres.*) C. L. 1258 ; **brenne** (*subj.*) FL. & BL. 5 ; **brenning** (*pple.*) KATH.[2] 41 ; LANGL. *C* xx. 83 ; **brende** (*pret.*) ORM. 1086 ; GEN. & EX. 1108 ; P. L. S. xiii. 121 ; FER. 2243 ; LANGL. *B* xvii. 326 ; GOW. II. 164 ; WICL. 4 KINGS xiv. 4 ; he brande, barnde LAȝ. 3824, 30459 ; bernde A. R. 242 ; as þah ha bernde MARH. 19 ; barnde ROB. 300 ; brenden SAX. CHR. 262 ; RICH. 6057 ; þer brenden cerges inne HAV. 594 ; þe fires brende upon þe auter CH. C. T. *A* 2425 ; **brend** (*pple.*) ORM. 1000 ; HAV. 2832 ; brend gold CH. C. T. *A* 2162 ; RICH. 3349 ; brend, brent WILL. 2646, 4367 ; brent HOCCL. i. 390 ; *comp.* **a-, for-brennen.**

brennar, sb., *burner*, PR. P. 49 ; **brenneris** (*pl.*) WICL. S. W. III. 329.

brenninge, sb., *O.E.* berning ; *burning*, GEN. & EX. 3654 ; CH. C. T. *A* 996 ; berninge AYENB. 205.

brēost, sb., *O.E.* brēost, = *O.L.G.*, *O.N.* briost, *O.Fris.* briast *n.; breast*, P. L. S. xvii. 66 ; ALIS. 621 ; þat breost LAȝ. 27556* ; þi breost P. L. S. xvii. 66 ; briest AYENB. 175 ; brest ORM. 4774 ; GEN. & EX. 370 ; breest TREV. III. 341 ; brist LUD. COV. 14 ; breoste A. R. 18 ; moni breoste LAȝ. 4540 ; þa breste [þe breost] 6497 ; **brēostes** (*gen.*) HOM. I. 183 ; a þan breoste LAȝ. 27553 ; ut of þire breoste 15831 ; þire br(e)oste GR. 20 ; upon þe breeste of our lord MAND. 92 ; **brēost** (*pl.*) LEECHD. III. 120 ; þa breost þe þu suke LK. xi. 27 ; þine breoste FRAG. 6 ; breosten MISC. 151 ; brustes FER. 1072 ; **brēostam,** breostan, breosten (*dat. pl.*) LEECHD. III. 120, 122, 128 ; breosten KATH. 2130 ; briesten AYENB. 247.

brēst-bōn, sb., *O.E.* breostbān ; *breast-bone*, CH. C. T. *A* 2710 ; brustbon FER. 1623.

brēst-clūt, sb., VOC. 173.

brēst-līn, sb., *O.E.* brēostlīn ; *breast-linen*, ORM. 955.

brēst-plate, sb., *breast-plate*, CH. C. T. *A* 2120.

brēost-þonc, sb., *O.E.* brēostgeþanc ; *mind*, LAȝ. 1936.

brēoste-wunde, sb., *breast-wound*, A. R. 194.

brēosted, pple., *having a breast*, GUY 261.

brēoðen, v., *O.E.* (a)brēoðan ; *perish* : þer fore ȝe sculleð breoðen LAȝ. 5807 ; heou **brēoðeð** [breþiþ] þa Frensce 5196 ; a broþin (? *degenerate*) (h)eir P. L. S. i. 44 ; *comp.* **a-brēoðen.**

brēowen, v., *O.E.* brēowan, = *M.H.G.* briuwen, *M.L.G.* brūwen, *M.Du.* brūwen, brouwen, *O.N.* brugga ; *brew* ; breowe HOM. II. 257 ; brewen GEN. & EX. 4054 ; brewe MAN. (F.) 1245 ; S. S. (Wr.) 1490 ; E. G. 355 ; bruwin, browin PR. P. 54 ; **brēu** (*ms.* brew) (*pret.*) P. S. 69 ; breu, breuh LANGL. *A* v. 134 ; breuȝ HOM. II. 256 ; brewed S. S. (Wr.) 1285 ; þei browe ROB. 26 ; **browen** (*pple.*) MAP 337 ; RICH. 4365 ; browe LANGL. *B* xviii. 361 ; ibrowen P. 29 ; ibrowe S. S. (Web.) 1494 ; brewid FLOR. 687 ; *deriv.* broþ, brüþen.

brēwere, sb., = *M.H.G.* briuwer, *M.L.G.* brūwere ; *brewer*, LANGL. *B* xix. 394 ; **brēowares** (*pl.*) MISC. 189 ; brueres R. S. vii (MISC. 188).

brēwestere, sb., *female brewer*, LANGL. *B* v. 306 ; E. G. 355 ; browstere LUD. COV. 132 ; browstar or brewere PR. P. 54.

brēr, sb., *? sprout, young shoot* ; brer [breer] on ris C. L. 123.

brerd, sb., *O.E.* brerd, breard, = *O.H.G.* prart, prat, *Swed.* brädd ; *margin, brink*, ORM. 14040 ; WICL. EXOD. xxxvii. 11 ; berd '*margo*' PR. P. 32 ; **brerde** (*dat.*) TREV. II. 173 ; upon helle brerde A. R. 324 ; breorde LAȝ. 23322.

brerd-ful, adj., *cf. Swed.* bräddfull ; *full to the brim*, ORM. 14529 ; bredful, bratful LANGL. *B* PROL. 41 ; bretful HOM. II. 167 ; P. S. 333 (P. 41) ; PL. CR. 223 ; CH. C. T. *A* 687.

brēre, sb., *O.E.* brēr *m.; brier*, '*tributus, vepres*,' VOC. 181, 229 ; PR. P. 49 ; '*eglanter*' VOC. 163 ; LANGL. *C* iii. 28 ; WICL. JOB xxxi. 40 ; TOWNL. 100 ; S. & C. I. xxix ; scharp as brere CH. C. T. *E* 1825 ; breere LIDG. M. P. 218 ; **brēre** (*dat.*) MIN. vii. 128 ; briȝt so blosme on brere GREG. 18 ; rede brere L. C. C. 42 ; **brēres** (*pl.*) A. R. 276 ; ORM. 9212 ; WILL. 1809 ; MAND. 115 ; CH. C. T. *A* 1532 ; a burþen of brere SPEC. 110.

bres, *see* **bras.**

brése, sb., *O.E.* briosa, breosa (*asilus, tabanus*) ; *gadfly*, '*oestrum*,' VOC. 255 ; '*asilus*' PR. P. 49 ; '*bruchus*' CATH. 43 ; PS. civ. 34.

bresed, adj., *? bristly*, A. P. ii. 1694 ; GAW. 305.

brēsen, *see* **brüsen.**

brest, berst, sb., *O.N.* brestr, *O.E.* berst, = *M.H.G.* brest, breste, *M.Du.* berst, berste ; *damage, defect*, PR. P. 49 ; A. P. ii. 229 ; C. M. 16561 ; breste HALLIW. 209 ; berst BEV. 1930 ; *see* **brüst.**

brēst, *see* **brēost.**

bresten, berstan, v., *O.N.* bresta, *O.E.* berstan, = *O.L.G.* brestan, *O.H.G.* prestan, brestan, *O.Fris.* bersta ; *burst* ; brestin PR. P. 50 ; bresten, berste CH. C. T. *A* 1980 ; breste, berste, WICL. MK. ii. 22 ; bersten A. R. 80 ; berste LAȝ. 27683 ; HORN (R.) 662 ; **berste** (*pres. subj.*) O. & N. 990 ;

brast (*pret.*) CH. C. T. *B* 697; TRIST. 191; TOR. 687; DEGR. 1622; brast, barst LANGL. *B* vi. 180; barst LAȝ. 1467; ALIS. 625; þet meari bearst ut JUL. 59; brosten OCTOV. (W.) 1088; MAN. (F.) 1833; broston ED. 3449; bursten JUL. 59; RICH. 2892; stanes þer bursten LAȝ. 27719; þe bees bursten out RICH. 2892; hire limes burste P. L. S. xv. 193; borsten K. T. 1049; **brosten** (*pple.*) CH. C. T. *A* 3829; WICL. GEN. vii. 11; iborsten HICKES I. 228; *comp.* æt-, to-bresten; *deriv.* brest, brüst.

bretasce, sb., *O.Fr.* bretesche; *brattice, bartizan,* '*propugnaculum,*' PR. P. 50; brutaske ROB. 536.

bretasce, v., *fortify with a brattice*; bretexed (*pple.*) LIDG. TROY ii. 11.

   briteising, sb., *bratticing*, WICL. S. W. I. 191.
brēð, sb., *O.E.* brǣð *m., allied to O.H.G.* prādam; *breath*, A. R. 80; unl(e)ofne brēð HOM. I. 153; breþ TREAT. 136; SHOR. 102; P. L. S. v.² 41; TREV. II. 185; þe hote breþ O. & N. 1454; breþ, breeþ CH. C. T. *C* 552; bretz R. L. P. 221; **brēðe** (*dat.*) HOM. II. 145; for þan ufele breðe I. 43; breþe CH. C. T. *A* 5; LIDG. M. P. 62; brethe PR. C. 613.

   brēþ-man, sb., *? blower of a wind instrument*; **brēthemen** (*pl.*) D. ARTH. 4108.
brēthe, sb., *? O.N.* brǣði; *from* brāþ; *? fury* D. ARTH. 2213; breþ '*furorem*' PS. xxxvi. 8; **brēthe** (*dat.*) TOWNL. 197.
breþel, sb., *? from* brēoðen; *wretch, worthless person*; (*printed* breyel) '*miserculus*' PR. P. 50; briþel (*? printed* briȝel) REL. II. 119; **brethelis** (*pl.*) LUD. COV. 308.
breþeling, sb., *? =* breþel; AR. & MER. 64; briþeling R. S. vi (MISC. 184, 185).
brēthin, v., *breathe, smell, exhale perfume,* '*spiro,*' PR. P. 50; breþi TREAT. 136; **brēþiþ** (*pres.*) WICL. JOHN iii. 8; **brēþede** (*pret.*) TREV. VIII. 195; D. TROY 8777.
breu, sb., *O.Fr.* breu; *broth*; brewe '*brodium*' VOC. 241.

[**brēu-**, *stem of* brēowen, *brew.*]

   brēú-arn, sb., *O.E.* brēawern; *brewhouse,* VOC. 178.
   brēu-hous, sb., *? =* M.H.G. briuhūs; *brewhouse,* VOC. 204; CH. C. T. *A* 3334.
brēu (*?* **brēwe**), sb., *O.E.* brēaw, brǣw *m.,* = *O.N.* brā *f., O.H.G.* brā, prā *n. f., O.Fris.* brē *n.; mod.Eng.* brow; *eyelid, eyebrow; bank, river-side* (mod. brae); breȝe SPEC. 34; brie '*cilium*' VOC. 76; bank ne bre MAN. (F.) 10333; **brēie** (*ms.* breye) (*dat.*) R. S. v (MISC. 182); **brēwas** (*pl.*) LEECHD. III. 98; **brēwis** (*pl.*) WICL. LEV. xiv. 9; breowen LAȝ. 18374; brees AV. ARTH. xv; briȝes MISC. 226; *comp.* ēȝe-brēu; *see* brā.

breue, *see* bref.
brēven, v., = *O.N.* brefa, *M.H.G.* brieven; *from* bref; *commit to writing*, P. S. 156;

brēved (*pple.*) GAW. 2521; A. P. ii. 197; ibrevet HOM. I. 249; MARH. 16.
brevet, sb., *Fr.* brevet; *brevet*, LANGL. *A* PROL. 71.
brevetour, sb., *carrier of brevets,* '*brevigerulus,*' PR. P. 50.
brēwe, *see* brēu. **brēweu**, *see* brēowen.
brī, sb., *O.E.* brīw, = *M.Du., O.H.G.* brī, *M.L.G.* brig; *broth, sops,* '*puls,*' FRAG. 4.
brībe, sb., *O.Fr.* bribe (*morsel of bread*), mod.Eng. bribe; *? plunder*; feyning a cause for he wolde han a bribe [*v. r.* wolde bribe] CH. C. T. *D* 1378 (Wr. 6960).
briberie, sb., = *O.Fr.* briberie (*mendicancy*), mod.Eng. bribery; *pilfering*, TREV. VIII. 81; **briberies** (*pl.*) CH. C. T. *D* 1367.
brībeþ, sb.: gōn a bribeþ [bribed] (*go a-begging*) LANGL. *C* ix. 246.
brībin, v., *O.Fr.* briber (*to beg*), mod.Eng. bribe; *pilfer,* '*manticulor,*' PR. P. 50; bribe CH. C. T. *A* 4417.
bríbour, sb., *O.Fr.* bribeur; *pickpocket, thief, rascal,* LUD. COV. 183; **bríboures** (*pl.*) TREV. II. 147.
briche, adj., *O.E.* brice, bryce; *from* brēken; *frail, poor,* H. S. 5821.
bríche, *see* brüche.
brid, sb., *O.E.* bridd *m., mod.Eng.* bird; *young bird; bird in general;* '*pullus*' FRAG. 3; '*avis*' PR. P. 50; A. R. 102; O. & N. 124; S. S. (Wr.) 3035; MAND. 48; WICL. LEV. xvii. 13; culfre bridd ORM. 7887; Judas was a liþer brid (*youth*) P. L. S. xxii. 1; mi brid i fed L. H. R. 133; **briddes** (*gen.*) P. L. S. xxv. 13; **briddes** (*pl.*) A. R. xii. 118, 128; KATH. 64; LANGL. *B* xii. 131; WILL. 179; CH. C. T. *A* 2929; ANT. ARTH. xxvii; birdes D. ARTH. 2510; birdus TOR. 2044; **bridde** (*dat. pl.*) O. & N. 644.
   brid-līm, sb., *bird-lime*, PR. P. 50; birdlīm VOC. 221.
brīde, sb., *O.Fr.* brīde; *bridle*, ALIS. 2626; (*ms.* bridel, *r. w.* ride) HORN (L.) 772 (R. 778, H. 801).
brīde, *see* brüde.
brīdel, sb., *O.E.* brīdel, = *O.H.G.* brītel, brīdel, *M.L.G., M.Du.*breidel; *from* breiden; *bridle,* A. R. 74; O. & N. 1028; ROB. 396; SHOR. 6; CH. C. T. *A* 2506; þane bridel AYENB. 253; bridil RICH. 5817; **brīdle** (*dat.*) AYENB. 204; **bridles** (*pl.*) WICL. JAMES iii. 3.
brīdelin, v., *O.E.* (ge)brīdlian, = *M.H.G.* brītelen; *bridle,* '*freno,*' PR. P. 50; **brīdleð** (*pres.*) A. R. 74; bridlen GOW. I. 110; **brīdled** (*pple.*) ORM. 11664; brideled MAND. 253; ibridled CH. L. G. W. 111.
brīe, *see* brēu. **brīest**, *see* brēost.
brige, sb., *O.Fr.* brigue; *intrigue, quarrel,* '*briga, dissensio,*' PR. P. 50; CH. C. T. *B* 2872.
   brige-less, adv., *without cavil*, ANGL. V. 28.

brige, v., *? Fr.* briguer; *entrap* : þo fende hafs cast þis snare for to brige men WICL. S. W. III. 416.

Brigge, pr. n., *O.E.* Brycge, *M.Du.* Brugghe; *Bruges,* SAX. CHR. 254.

brigge, *see* **brügge.**

brigous, adj., *O.Fr.* brigeus, *from Lat.* brīgosus ; *disputable,* TREV. III. 203.

briȝt, *see* **briht.**

briht, adj., *O.E.* bryht, breoht, beorht, berht, byrht, = *O.L.G.* berht, bereht, *O.H.G.* beraht, *O.N.* biàrtr, *Goth.* bairhts ; *bright,* HOM. I. 61; LAȝ. 7239; MISC. 88; SPEC. 25; K. T. 240; *(ms.* brihht) ORM. 3431 ; briht, briȝt O. & N. 623; bricht HOM. II. 255; MISC. 27; briȝt AYENB. 74; DEGR. 1483; briȝt and schene SHOR. 149; briȝt win GAW. 129; brigt GEN. & EX. 132 ; briȝt, bright CH. C. T. *A* 1062; brith HAV. 589; brict HORN (H). 14 ; *printed* britt HICKES I. 228 ; brist REL. I. 48 ; briht, breoht LK. xi. 34, 36 ; breost MAT. vi. 22 ; his brithe bride HAV. 2131 ; þet brihte [br(e)ohte] i-cunde HOM. I. 147, 149; þat . . . brehte . . . kinde HOM. II. 205 ; **brihte** *(dat. m.)* MARH. 19 ; **brihte** *(pl.)* H. M. 19; HAV. 2610 ; his eȝen b(e)oð swa brichte HOM. I. 165; **brihtre** *(compar.)* KATH. 1680; **brihtre,** brictere HOM. I. 39, 139 ; brihtere, brihture A. R. 38, 182; brihtore SPEC. 57 ; **brihtest** *(superl.)* MARH. 21 ; brictest GEN. & EX. 1910.

  briȝt, sb., *beautiful woman* : breue me briȝt quatkin of priis beres þe perle so maskellez, A. P. i. 754 ; brigt ; *brightness, splendour,* GEN. & EX. 143 ; bright TOWNL. I.

  brihte, adv., *brightly,* CHR. E. 770 ; *(ms.* brihhte) ORM. 2138 ; briȝte BRD. 2.

  briht-līce, adv., *O.E.* beorhtlīce ; *brightly,* MK. viii. 25 ; brihtliche A. R. 154; briȝtliche AYENB. 150 ; **brihtlüker** *(compar.)* A. R. 96.

  brihtnesse, sb., *O.E.* bryhtness, beorhtness; *brightness,* HOM. I. 217, II. 31 ; brictnesse I. 239 ; briȝtnesse AYENB. 27 ; brihtnisse, brehtnisse JOHN v. 41, xvii. 22.

brihten, v., *O.E.* (ge)brihtan, beorhtan, byrhtan, = *O.N.* birta; *brighten,* A. R. 148; **brihte** *(imper.)* HOM. I. 200 ; til on morwen þe dai bright S. S. (Wr.) 1997.

brike, sb., *Fr.* brique, *cf. M.L.G., M.Du.* bricke ; *brick,* E. G. 372 ; brik W. & I. 37.

bríke, sb., *O.North.Fr.* brique ; *snare, trap* : Genilon Oliver brought this worthy king in such a brike CH. C. T. *B* 3580.

brillin, *see* **birlen.**

brim, sb., *O.E.,* = *O.N.* brim; *stormy sea, waves,* L. H. R. 125 ; MIN. vi. 57 ; **brimme** *(dat.)* GAW. 2172 ; and lepith dune into the brimme P. L. S. xxxv. 157.

  brün-swin, sb., *? O.N.* brimsvīn; *?porpoise;* ' *delphinus,*' PR. P. 54.

brim, sb., *O.E.* brymm *(pl.* brymmas) ; *brim, brink, margin* ; **brimme** *(dat.)* Gow. II. 293; TREV. I. 423 ; CH. L. G. W. 2451 ; EM. 694 ; A. P. i. 232 ; bi þan brimme LAȝ. 4472 ; to þis londes brimme HORN (L.) 190; **brimmes** *(pl.)* TREV. viii. 139.

brim, brime, *see* **brēme.**

brime, sb., *O.N.* brimi ; *? burning heat, fire* : ant þurh þe brime [*? read* brune] ablindeð MARH. 15.

brim-fīr, *see* **brinfīr,** *under* **brüne.**

brimmen, v., *O.E.* bremman *(rage, roar),* = *O.H.G.* breman, *M.H.G.* bremen, *M.Du.* bremen ; *(of swine)* be in heat, copulate ; *(of trees) bear fruit* : the sonner wol thei brimme aȝein and bringe forth pigges moo PAL. iii. 1070 ; brimen and beren GEN. & EX. 118 ; brime ' *subare* ' CATH. 44.

brimse, sb., *O.E.* brimse, = *M.Du.* bremse ; *gadfly,* WR. DICT. 257.

brim-stōn, sb. (= **brinstōn,** *from* **brinnan**) ; *brimstone, sulphur,* CHR. E. L181 ; brimstoon, bremston CH. C. T. *A* 629 ; brim-, brumston WICL. DEUT. xxix. 23, JOB xviii. 15 ; bremston BEVES 3277 ; brumston MAP 339 ; MISC. 227 ; *see* **brenstōn,** *under* **brenne,** *and* **brün-stān,** *under* **brüne.**

brīn, sb., *O.N.* brūn *(pl.* brȳnn), *Dan., Swed.* brȳn ; *brow,* ' *supercilium,*' PR. P. 51 ; MAN. (F.) 12344 ; bryne AV. ARTH. xv.

brīne, sb., *O.E.* brȳne *(for* brīne), = *M.Du.* brijne ; *brine,* ' *salsugo,*' PR. P. 51 ; C. M. 6348 ; PALL. iii. 39.

brinfīr, *see under* **brüne.**

bringe, sb., *bringing* ; wel bið him þere bringe LAȝ. 743.

bringen, v., *O.E.* bringan *(pple.* brungen), = *O.L.G.* bringan, *O.H.G.* pringan, bringan *(pret.* prang, *pple.* prungan), *Goth.* briggan, *O.Fris.* bringa ; *bring,* A. R. 268; GEN.& EX. 312; H. H. 6; bringen on *(induce)* ORM. 7717 ; bringen heom to gadere MISC. 37 ; bringen [bringe] LAȝ. 741 ; ne mihten heo . . . nenne a dun bringe 26588 ; bringe O. & N. 1029 ; HAV. 72 ; to deþe bringe ROB. 335 ; to ende bringe C. L. 288 ; fruit . . . bringe LEB. JES. 566 ; **bringeð** *(pres.)* HOM. I. 103 ; H. M. 33 ; & bringeð hire on to gederen A. R. 222 ; þeonne bringeð hy up sum luðer word 426 ; bringeþ O. & N. 524 ; **bring** *(imper.)* MARH. 19 ; **bringe** *(subj.)* P. L. S. xix. 284; *for the pret. see under* **brengan** ; *comp.* **ȝe-bringen** ; *deriv.* **brengen.**

  bringere, sb., *bringer,* MAND. 243 ; bringare PR. P. 51 ; CH. C. T. *D* 1197.

brinie, *see* **brünie.**

brink, sb., *cf. M.L.G., Dan., Swed.* brink, *Icel.* brekka ; *brink,* ' *margo, ripa,*' PR. P. 52 ; **brinke** *(dat.)* A. R. 242*; LEG. 130 ; ALIS. 3491 ; MAN. (F.) 4623 ; bi þe se brinke HORN (L.) 141 ; on mi pittes brinke CH. C. T. *E* 1401 ; in þe

brinke ['*in litore*'] WICL. JOHN xxi. 4;
brinkes (*pl.*) TREV. VIII. 349.

brinnen, v., *O.E.* brinnan, beornan, byrnan,
= *O.L.G.*,*O.H.G.*,*Goth.* brinnan, *O.N.* brinna;
burn, (*intrans.*); brinne C. M. 5749; ISUM.
695; brinne, brenne CH. C. T. *D* 52; beornen
FRAG. 7; beornen, bernen LA3. 2858, 16218;
beornen (*r.w.* 3eorne) HOM. I. 197; birne 239;
TOR. 555; berne AR. & MER. 2320; **brinneð**
(*pres.*) REL. I. 215; bearneð H. M. 37; bir-
nende, beorninde (*pple.*) HOM. I. 41, 95;
berninde R. S. iv (MISC. 166); **born** (*pret.*)
LA3. 16217; burnen 4579; **burne** (*subj.*) LA3.
28085; *comp.* for-brinnen; *deriv.* **brenne,
brennen, brüne, brand.**

brīsel, *see* **brūsil. brīsen,** *see* **brūsen.**

brīsewort, *see under* **brūse.**

brist, *see* **brüst.**

Bristōwe, pr. n., *O.E.* Brycgstōwe; *Bristol,*
ROB. 277; TREV. II. 35; Bristouwe [Brus-
touwe] LA3. 21027.

briteising, *see* **bretasce.**

britel, *see* **brütel. britnen,** *see* **brütnen.**

britoner, *see* **brütiner.**

briþel, briþeling, *see* **breþel, breþeling.**

brixle, sb., *O.N.* brigsli (*for* *brig̃sli); *from*
breiden; *reproach, upbraiding*; brixel, bricsl,
brixil C. M. 10319, 24044, 28196; *comp.* up-
brixle.

brixlen, v., *O.N.* brigsla; *reprove*; **brüxlez**
(*pres.*) A. P. iii. 345.

brō, *see* **brā** *and* **brēu.**

broc, sb., *? menace*; þis was hire broc (*first
text* ibeot) LA3. 21029*.

brōc, sb., *O.E.* brōc *m.* (*brook*), = *M.Du.* broek,
O.H.G. bruoch *n.* (*marsh*); *brook,* HOM. I. 187;
ænne broc LA3. 10828; brok, brook CH. C. T.
A 3922; **brōke** (*dat.*) LA3. 10831; LANGL.
B vi. 137; A. P. i. 141; **brōkes** (*pl.*) LA3.
31228; A. R. 258.
    brōc-minte, sb., *O.E.* brōcminte; *brook-
mint,* VOC. 140.

brocáge, sb., *brokage, trade or action of a
broker; jobbery,* LANGL. *B* xiv. 267, *C* xvii.
109; CH. C. T. *A* 3375.

brōche, sb., *O.Fr.* broche; *broach* (*spit*);
brooch, '*veⁱu,*' VOC. 178; '*monile*' 199; A. R.
420; WICL. JER. iv. 30; (*spear*) L. L. R. 133;
haþ set a broche (*a broach*) þe tonne LIDG.
M. P. 164; **broches** (*pl.*) AYENB. 229; LANGL.
B xvii. 245; bruchez D. ARTH. 3257.

brōchin, v., *Fr.* brocher; *broach* (*a cask*), *spur*
(*a horse*), PR. P. 52; broching FER. 3657;
brōchede (*pret.*) D. ARTH. 2714.

brocke, sb., = *M.Du.* brocke, *O.H.G.* brocco
m.; from **brēken**; *fragment*; broc (*broken
skin*) TRANS. XXX. 381; **broccan** (*dat. pl.*)
MAT. xv. 37.

brocour, brokour, sb., *cf. med.Lat.* bro-
carius; *broker,* LANGL. *B* ii. 65, *C* iii. 66.

brod, sb., *O.E.* brord, *O.N.* broddr (*spiculum,
ora*), *cf. O.H.G.* prot, prort *m.* (*ora*); *spike;
brad; ear of corn,* '*aculeus,*' VOC. 202; '*clavus
acephalus,*' PR. P. 53; brodd & blome ORM.
10773; **brurdes** (*pl.*) A. P. ii. 1474.
    brurd-ful, adj., *cf.* **brerdful**; *full to the
brim,* A. P. ii. 383.

brōd, sb., *O.E.* brōd, = *M.Du.* broed, *M.H.G.*
bruot; *brood,* O. & N. 1633; netes brod GEN. &
EX. 3712; þe luþer brod ROB. 70; **brōde** (*dat.*)
O. & N. 518; C. M. 1507; he sit a brode P. L.
S. v.² 35.

brōd, *see* **brād.**

brodden, v., *from* **brod**; *shoot, sprout,* ORM.
10769.

brōdien, *see* **brādien.**

brōdin, v., *brood,* '*foveo,*' PR. P. 53; *see*
brēden.

broiderer, *see* **broudiour.**

broiderie, *see* **brouderi. broidin,** *see*
breiden.

broilen, v., *O.Fr.* bruiller; *broil,* '*ustulo,
torreo,*' PR. P. 53; broile CH. C. T. *A* 383.

brok, sb., *O.E.* broc, *cf. Dan.* brok, *Ir., Gael.*
broc, *Welsh* broch; *badger,* '*taxus,*' PR. P.
53; '*teissoun*' VOC. 166; Iw. 98; **brockes**
(*pl.*) LA3. 12817; brokkes TREV. I. 327.

brok, sb., *O.E.* broc, = *O.N.* brokkr; *? nag*;
hait brok CH. C. T. *D* 1543.

brōke, *for* **brōken,** *pple. of* **brēken**; leep-
fullis of broke meat WICL. S. W. II. 14.
    brōke-bakkid, adj., *hunch back,* '*gibbosus,*'
PR. P. 53; GAM. 720.
    brōke-footid, adj., *broken footed* ['*fracto
pede*'], WICL. LEV. xxi. 19.
    brōke-legged, adj., *bandy-legged,* LANGL.
B vi. 138.
    brōke-rügget, adj., *hump-backed,* H. M. 25.
    brōke-shankid, -schonket, adj., *broken
legged,* LANGL. *A* vii. 131.

brokel, *see* **brüchel. brōken,** *see* **brüken.**

brokken, v., *? murmur*; a3e the crokkere to
brokke SHOR. 106; **brokkinge** (*pple.*), as a
nyghtyngale CH. C. T. *A* 3377.

brol, sb., *offspring, child*; brolle LANGL. *B* iii.
204; brol PL. CR. 745; REL. II. 177; **brollis**
(*pl.*) WICL. S. W. 195.

brōm, sb., *O.E.* brōm *m.*, = *M.Du.* broem;
broom, '*genista,*' FRAG. 3; VOC. 140; A. P. iii.
392; broom MAND. 130; **brōme** (*dat.*) ALIS.
2492; **brōmes** (*pl.*) ['*myricae*'] WICL. JER.
xvii. 6; CH. H. F. 1226.
    Brōm-holm, pr. n., CH. C. T. *A* 4286.

brond, *see* **brand. brōsen,** *see* **brüsen.**

brosnien, v., *O.E.* brosnian; *crumble, break*
(*intrans.*): þu scalt rotien & brostnian (*? for*
brosnian) FRAG. 7.

brotel, *see* **brütel.**

broþ, sb., *O.E.* broð, = *O.H.G.* prod *n.; ? from*

brēowen; *broth*, ROB. 528; P. S. 334 (P. 43); MAND. 250; AL. (V.) 310.

brōþ, *see* **brāþ.**

broþel, adj., *?from* **brēoðen**; *wretched*, LANGL. *A* xi. 61; **broþel** (*subst.*) GOW. III. 173; LUD. COV. 217; **broþels** (*pl.*) PL. CR. 772; *see* **breþel.**

brōðer [**brōþer**], sb., *O.E.* brōðor, brōður,= *Goth.* brōþar, *O.L.G.* brōthar, bruother, *O.Fris.* brōther, *O.N.* brōðir, *O.H.G.* pruoder, bruoder, *Lat.* frāter, *Gr.* φρατήρ; *brother*, LAȝ. 2105; broþer O. & N. 118; (*ms.* broþerr) ORM. 296; broþer, brother CH. C. T. *A* 529; broder FLOR. 718; **brōðer**, [*O.E.* brōðor,= *O.N.* brōður, *O.H.G.* bruoder, *Goth.* ʽbrōþrs, *Lat.* frātris] (*gen.*) LAȝ. 6741; broþer ROB. 291; þis was na broþer dede (? broþerdede) C. M. 3750; broðres HOM. II. 147; **brōðer**, [*O.E.* brēðer, *O.N.* brōður, brœˇr] (*dat.*) REL. I. 223; broðer MAT. xxii. 25; broþer LAȝ. 4294; **brōðre**,[*O.E.* brōðor, brōðru, brōðra, = *O.Fris.* brōthera, brōthere, *O.N.* brœˇðr, *Goth.* brōþrjus, *O.H.G.* bruoder, bruodra, *Lat.* frātres] (*pl.*) MK. x. 30; broþre MISC. 55; broðre, breðre, breðren HOM. I. 7, 9, 11; breðre SAX. CHR. 32; breþre ORM. 8269; breþer WILL. 2641; MAN. (H.) 51; breþir AMAD. (R.) lx; breder AUD. 35; breˇðeren GEN. & EX. 823, 2271; briþer S. S. (Wr.) 1899; broþren AYENB. 102; broþeren SHOR. 45; broðeren, breðren, breðeren [broþers] LAȝ. 2101, 2137, 4292; breðren A. R. 54; JUL. 40; breþeren BEK. 463; MAND. 222; breþeren, briþren WICL. GEN. xxix. 4; **brēðere**, [*O.E.* brōðra,= *O.N.* brœˇðra] (*gen.pl.*) GEN. & EX. 2213; breþerne BRD. 26; **brōðren** (*dat. pl.*) MAT. xviii. 35; broþren MISC. 53.

broþer-hēde, sb., *brotherhood*, AYENB. 110; CH. C. T. *B* 1232; AM. & AMIL. 362; GAW. 2516; C. M. 1159.

brōþer-rēdene (*ms.* -reddene), sb., *O.E.* brōþorrǣden; *fraternity*, HOM. I. 41; broþerrede AYENB. 110.

broþ-fal, sb., *?cf. O.N.* brottfall; *?epilepsy*; ORM. 15504.

brouderi, broiderie, sb., *Fr.* broderie; *embroidery*, WICL. EXOD. xxviii. 39; browdrie xxxv. 33.

broudin, v., *Fr.* broder; *embroider*, PR. P. 53; **brouded** (*pple.*) CH. C. T. *A* 3238.

broudinge, browding, sb., *embroidery*, CH. C. T. *A* 2498.

broudiour, brouderere, sb., *Fr.* brodeur; broiderer, PR. P. 53; broiderer WICL. 2 KINGS xxi. 19.

broue, *see* **brūwe.**

brouet, sb., *Fr.* brouet,= *Ital.* brodetto; *broth*, PR. P. 54; DEP. R. ii. 51; L. C. C. 22; **broues,** [*O.Fr.* broues, broez, *prov.Eng.* browis]; 'adipatum' PR. P. 53; brouis VOC. 199; brues 266; and soupid of þe brouwis a sope RICH. 3077; brois HAV. 924.

brouken, *see* **brūken.** **broun,** *see* **brūn.**

brousch-, *see* **brusch-.** **brouwe,** *see* **brūwe.**

[**browe,** sb. ?= *M.Du.* brūwe (*broth*); *comp.* fisc-browe].

brüche, sb.,*O.E.* bryce,= *M.Du.* breuk, *O.H.G.* pruh, bruch; *from* **brēken**; *breach; violation; crime*; H. M. 21; ROB. 26; REL. I. 241; at þe furmeste bruche þat he fond he lep in II. 272; for monnes bruche KATH. 1210; breche LK. xxiv. 35; **brüches** (*pl.*) SPEC. 30; þe bruches of hire bodi KATH. 1615; **brüchen** (*dat. pl.*) MARH. 21; i bote of þeos bruchen A. R. 28; *comp.* ǣ-, burh-, griþ-, mund-, schip-, spūs-brüche (-briche).

brüche, adj., *O.E.* bryce,= *Goth.* brūks, *O.H.G.* prūch; *from* **brüken**; *useful*; briche REL. I. 225; MAN. (F.) 11406.

brüchel, adj.,= *M.Du.,M.L.G.* brokel; *from* **brēken**; *brittle*; KATH. 2029; A. R. 164; brukel, brokel M. H. 120, 154; brokil 'fragilis' PR. P. 53; LANGL. *C* xi. 47*; H. V. 86; of brokele kende SHOR. 3; **brüchelure** (*compar.*) A. R. 164.

brüchelnesse, sb., *fragility*, P. L. R. P. 251; brukilnesse QUAIR cxciv.

[**brüd,** sb., *O.E.* bryd, brygd; *from* **breiden**; *comp.* up-brüd].

brüde, sb., *O.E.* brȳd,= *O.L.G.* brūd, *O.H.G.* prūt, brūt, *O.N.* brūðr, *Goth.* brūþs; *bride*, A. R. 164; H. M. 5; MISC. 91; HORN (L.) 1058; P. S. 239; brede MAT. xxv. 1; AL. (T.) 129; brid ORM. 14194; bride PR. P. 50; HAV. 2131; CH. C. T. *E* 1818; brude (?for burde) LAȝ. 294.

brüd-ale, sb., *O.E.* brȳdealo; *bridal*, HORN (L.) 1032; REL. I. 265; bruid-, bridale LANGL. *B* ii. 43; bridale ORM. 14003; GEN. & EX. 1674; OCTAV. (H.) 1702; bredale MISC. 29; AYENB. 118.

brüd-gifte, sb., *giving in marriage*, 'nuptias,' MAT. xxii. 2.

brüd-gume, sb., *O.E.* brȳdguma,= *O.L.G.* brudigumo, *O.H.G.* brutigomo; *bridegroom*, MARH. 19; bridgume ORM. 10422; bredgume MAT. ix. 15; MISC. 29; bredgome AYENB. 233; **bredgumen** (*gen.*) MK. ii. 19.

brüd-lāc, sb., *O.E.* brȳdlāc; *nuptials*, JUL. 7; H. M. 3.

brüd-þing, sb. *O.E.* brȳdþing; *nuptials*, MISC. 99.

brügge, sb.,*O.E.* brycg,= *O.N.* bryggja, *O.Fris.* bregge, *O.H.G.* brucca; *bridge*, LAȝ. 19242; WILL. 2140; LANGL. *B* v. 601; brigge CH. C. T. *A* 3922; MAN. (H.) 241; MAND. 7; þe brigge . . . al of marbre imad is sche FER. 1680; þe brigge of þe nose VOC. 183; bregge B. DISC. 1252; **brügge** (*gen.*) ROB. 539; **brügge** (*dat.*) A. R. 242; brigge HOCCL. I. 190; **brüggen** (*pl.*) ROB. 555; brigges SAX. CHR. 254; *comp.* draht-, tu-brügge.

brigge-bōte, sb., *O.E.* brycgbōt ; *repairing* or *restoration of a bridge*, REL. I. 33.

brigge-ward, sb., *bridge keeper*, FER. 1700.

brüggen, v., *O.E.* brycgian, *M.L.G.* bruggen ; *bridge* ; **briggeden** (*pret.*) HOM. II. 91 ; **ibrügged** (*pple.*) LAȜ. 21276.

brük, sb., *Lat.* brūchus ; *a kind of locust*, WICL. LEV. xi. 2 ; bruik HAMP. PS. civ. 32.

brüken, v., *O.E.* brūcan (*pret.* brēac),= *O.L.G.* brūcan, brūkan, *O.Fris.*, *O.N.* brūka, =*Goth.* brūkjan, *O.H.G.* prūhhan, prūchen ; *enjoy, use*, A. R. 202 ; H. M. 33 ; ORM. 2154 ; þat bred bruken HOM. II. 95 ; bruken [brouke] he heo þohte LAȜ. 4800 ; bruken nanes drenches 19755 ; broken 15334 ; ælra þara þinge þe on paradis beoð þu most bruce HOM. I. 121 ; bruke MISC. 76 ; brouke LANGL. *B* xi. 117 ; CH. C. T. *E* 2308 ; so mote ich brouke finger or to HAV. 1743 ; broke P. L. S. xiv. 51 ; FER. 3484 ; to brukene HOM. I. 105 ; **brouke** (*pres.*) P. S. 332 ; brukest HOM. I. 111 ; **brūc** (*imper.*) GEN. & EX. 1831 ; bruc his [brouk hit] on wunne LAȜ. 24180 ; broke AMAD. (R.) lxi ; *deriv.* **brüche**.

brūn, adj., *O.E.* brūn,=*O.N.* brūnn, *O.H.G.* prūn ; *brown*, C. M. 18833 ; broun CH. C. T. *A* 207 ; mi brune her MISC. 193 ; of broune here ROB. 429 ; hire browe broune SPEC. 28 ; þe brune (*sb.*) HAV. 2181 ; this feire broune is sone to the kinge MER. xxi. (373).

brüne, sb., *O.E.* bryne, byrne, *cf. O.N.* bruni, *m. ; from* **brinnen** ; *burn, burning*, HOM. I. 203 ; LAȜ. 8255 ; H. M. 9 ; þe brune of golnesse A. R. 254 ; huses brune O. & N. 1155 ; bringen on brune *set on fire* KATH. 1364 ; brene AYENB. 264 ; *? comp.* **herte-bren**.

brin-fīr, sb., *burning fire* ; brimfir GEN. & EX. 754 ; **brinfīres** (*gen.*) 1164.

   brendfier-rein, sb., GEN. & EX. 1110.

   brün-stān, sb., '*sulphur*,' PS. x. 7 ; PR. C. 4853 ; brunston MISC. 150 ; brun-, brin-, brenston, WICL. GEN. xix. 24, PS. x. 7 ; brinston P. L. S. xxi. 143 ; A. P. ii. 967 ; ANT. ARTH. xv ; bren-, bernston AYENB. 49, 73 ; bronston MAN. (F.) 14684* ; LUD. COV. 308 ; bornston VOC. 211 ; *see* **bren-stōn**.

brünen, v.,=*O.H.G.* brūnen ; *become brown* ; brouniþ (*pres.*) ALIS. 3293.

brünie, sb., *O.E.* byrne,=*O.N.* brynja, *O.L.G.*, *Goth.* brunjo, *O.H.G.* prunja ; *coat of mail*, A. R. 382 ; R. S. iv ; HORN (L.) 591 ; brunie, burne LAȜ. 1553, 6718 ; bruni ALIS. 1869 ; GAW. 2012 ; brinie HAV. 1775 ; brenie HOM. I. 243 ; breni ANT. ARTH. xli ; burne HOM. I. 155 ; **bürnan** (*pl.*) LAȜ. 1701.

brünied, adj., *provided with a coat of mail* : **breniede** (*pl.*) knightes D. ARTH. 316 ; *see* **ȝe-brünied**.

Brunne, pr. n., *Bourn*, MAN. (F.) 135.

brunswin, *see under* **brim**.

brunt, sb., *brunt*, '*impetus*,' PR. P. 54 ; A. P. i. 176 ; bront LIDG. M. P. 261 ; bronte S. B. W. 3166.

brurd, *see* **brod**.

brusch, sb., *Ital.* brusco ; *brush, brushwood*, '*bruscus*,' PR. P. 54 ; MAN. (F.) 8338.

bruschalle, sb., *Fr.* broussaile ; *brushwood* ; PR. P. 54.

brusche, sb., *Ital.* brusca ; *brush* : wiped it with a brushe LANGL. *B* xiii. 460.

brusche, broush, sb., *forcible rush*, ALEX. 783 ; at a brush ALEX. 2133.

bruschen, v., *brush* (*clothes*) ; brusche B. B. 180.

bruschen, v., *rush with force* ; **brusches** (*pres.*) ALEX. 963 ; bruschese D. ARTH. 1681 ; brusshet D. TROY 1192 ; **bruschid** (*pret.*) WINT. viii, xiii. 93.

bruschet, sb., *thicket*, HALLIW. 215 ; FER. 800.

[**brüse**, sb., *? bruise*]

   brīse-wort, sb., *O.E.* brȳsewyrt (LEECHD. I. 374) ; *bruise-wort*, '*anagallis*,' LEECHD. II. 373 ; '*consolida major*,' III. 316.

brüsen, v., *O.E.* brȳsan (B. D. D. 49), *O.Fr.* bruisier, brisier ; *bruise, break* ; bruse WICL. DEUT. ix. 3 ; brisen HOM. II. 61 ; HAV. 1835 ; brisin '*quasso*' PR. P. 52 ; brese PALL. i. 913 ; **brüsede** (*pret.*) JOS. 501 ; brusden L. H. R. 40 ; **brīsid** (*pple.*) WICL. LK. xx. 18 ; bresid FLOR. 103 ; TRIAM. 237 ; *comp.* **for-, to-brüsen**.

brüsil, adj., *fragile* ; brisill, '*fragilis*,' CATH. 44 ; H. S. 8571 ; bresil HAMP. PS. ii. 11.

brüst, sb., *O.E.* byrst (*loss*),=*M.L.G.* burst, borst (*want, defect*), *O.H.G.* brust (*breaking*) ; *from* **bresten** ; *damage, defect, injury* ; burst LAȜ. 1610 ; SPEC. 24 ; brist E. T. 833 ; PR. C. 6205 ; brist, birst C. M. 6344 ; **bürstes** (*pl.*) LAȜ. 2461.

brüste, sb., *O.E.* byrst, = *O.N.* burst, bust, *O.H.G.* purst ; *bristle* : þe brust(e) of a swin ELL. ROM. II. 322.

brüsten, v., *cf. Ger.* bursten, *Dan.* börste ; *beat* : ofte hi him bete and burste (*r. w.* nuste) AL. (T.) 331 ; **bürsted**, ibirsted (*pple.*) LAȜ. 18950 ; þou were betin and brist MAP 336 ; *see* **büsten**.

brüsten, v., *O.N.* byrsta (*pple.* byrstr) ; *? clothe with bristles* ; **brüst** ase a bore P. S. 151 ; ibrusted (*sec. text* ibrustled) mid golde LAȜ. 3639.

brüstle, sb., =*M.Du.* borstel ; *bristle* ; brustel K. T. 777 ; brustile, bristile, burstil '*seta*' PR. P. 52 ; bristile TOWNL. 100 ; **brüstelis** (*pl.*) ALIS. 6621 ; brustles, bristles, berstles CH. C. T. *A* 556 ; bristles TREV. II. 217 ; brostles BEV. 748.

brüstlien, v., *bristle* ; **bristled** (*pple.*) ALIS. 5722 ; þe bristlede boor CH. BOET. iv. 7

(148); **ibürstled** (ibrustled) mid stele LAӠ. 16095.

brustlien, v., *clatter, clash* : and **brustleþ** (*pres.*) as a monkes froise GOW. II. 93; **brustleden** sceldes LAӠ. 20143.

brüsūre, sb., *O.Fr.* brisure ; *bruise, breach*, WILL. 2461; WICL. LEV. xxiv. 20; brissure '*qvassatio*' PR. P. 52; brosure WICL. 2 ES. vi. I.

Brüt, pr. n., *O.E.* Bryt; *Briton*, LAӠ. 6401; **Brütten**(*dat. pl.*) 7360.

Brüt-lond, pr. n., *Britain*, LAӠ. 2194.

Brüt-lēoden, pr. n. (*pl.*) *British nation*, LAӠ. 14626.

Brütaine, Britaine, pr. n., *Fr.* Bretagne; *Britain*, LAӠ. 1395.

brütel, adj., *from O.E.* brēotan = *O.N.* briota; *brittle*, TREV. IV. 91; brutel, britel GOW. I. 33 (HALLIW. 215); brutel, britel, brotel LANGL. *B* viii. 42; brutel, brotel CH. C. T. *E* 1279; britil, brotil WICL. 2 COR. iv. 7; brotel SHOR. 5; AYENB. 129.

brotel-hēde, sb., *frailty*, AYENB. 130.

brütelnesse, sb., *brittleness*, TREV. II. 219; CH. C. T. *E* 1279.

Brütene, pr. n., *O.E.* Bryten, Breoten; *Britain*; Brutenne, Bruttene ,LAӠ. 2509, 7130; on Breotene FRAG. 5.

Brütiner, sb., *inhabitant of Brittany*, LANGL. *A* vii. 142; **britoner** *B* vi. 178; britonere *B* vi. 156.

Brütisc, adj., *O.E.* Brytisc; *British*, LAӠ. 6318.

brütlen, v., = **brütnen**, *cf. Sc.* brittle; *hew in pieces* : seint Thomas wes biscop, and barunes him qvolde; heo **brütlede** (*pret.*) him MISC. 92.

brütnen, v., *cf. Swed.* brytning (*fracture*); *? for* **brütten**; *break up, divide, hew in pieces*; bruttene WILL. 1133; brittene HAV. 2700; D. ARTH. 963; **brütned** (*pret.*) WILL. 1073; bretinid HALLIW. 210; **britned** (*pple.*) ORM. 14178; britned, brittened GAW. 2, 680 ; *comp.* **for-, to-brütnen.**

brütten, v., *O.E.* (for-, to)bryttan, brittan, *O.N.* brytja; *break in pieces* : þe dede bodi þei britten MAN. (H.) 244 ; *comp.* **for-, to-brütten.**

Brütūn, pr.n., *Fr.* Breton; *a Briton*; **brütūns** (*pl.*) LAӠ. 1958.

[**brüþen**, sb., *O.E.* bryÞen, *from* **brēowen**; *act of brewing*.]

brüþen-lēd, sb., *brewing-lead*, R. S. v (MISC. 182).

brüwe, sb., *? O.E.* brū, brūw (*pl.* brūa, brūwa), VOC. 42, 64, 282; *brow*; bro(u)e '*supercilium*' VOC. 179, 183; PR. P. 53; B. DISC. 883; **brüwen** (*pl.*) LAӠ. 22283; bruwes HOM. II. 213; bro(u)wen SPEC. 39; brouwes K. T. 417; MISC. 225; (*ms.* browes) SPEC. 34;

TOR. 1454; brues, broues C. M. 8079; broues REL. I. 54; **bro(u)wen** (*dat. pl.*) BRD. 19.

būc, sb., *O.E.* būc, = *O.L.G.* būc, *O.N.* būkr, *O.H.G.* būh, pūch; *belly, paunch, body*, FRAG. 6; JUL. 71; O. & N. 1132; a bouk of a motoun MAN. (H.) 174; **būkes** (*gen.*) REL. I. 218; **būke** (*dat.*) HOM. I. 25 ; A. R. 134; bouke CH. C. T. *A* 2746; **boukes** (*pl.*) ALIS. 3946; **būken** (*dat. pl.*) LAӠ. 17319.

bucched (? **būcched**), adj., ? = **bicched**; arh ich was . . . and neh dun fallen, and mine fan derve swa bucched and swa kene HOM. I. 277.

bucke, sb., *O.E.* bucc, bucca, *cf. O.N.* bukkr, *O.H.G.* poch, boch; *buck, he-goat*, A. R. 100; bucke, bukke CH. C. T. *B* 1946; bukke TREV. I. 339; MAN. (F.) 15749; bucc ORM. 1140; **buckes** (*gen.*) buckes [bukkes] horn CH. C. T. *A* 3387; **buckes** (*pl.*) HOM. II. 37; bukkes ORM. 1330; *comp.* **goot-, rā-bukke.**

Bu(c)kingham, pr. n., *O.E.* Buccinga hām; *Buckingham*, MISC. 146.

budde, sb., *bud*, '*gemma*,' PR. P. 54; PALL. iii. 1144; **buddis** (*pl.*) LIDG. M. P. 217.

buddun, v., *bud*, '*gemmo*,' PR. P. 54.

bude, sb., *weevil*, '*polumita*,' VOC. 255; boude '*gurgulio*' PR. P. 46; *comp.* **scharn-bude.**

büdel, sb., *O.E.* bydel, = *O.H.G.* butil; *from* **bēoden**; *beadle*, HOM. I. 117; KATH. 1928; O. & N. 1169; SPEC. 22; bidel ORM. 97; budul LANGL. *A* ii. 77; bedel LANGL. *B* ii. 109; LK. xii. 58; WICL. GEN. xii. 43; **büdeles** (*gen.*) HOM. I. 99; **büdeles** (*pl.*) JUL. 17; P. S. 151; bedeles HOM. I. 237; AYENB. 37.

büen, *see* **bēon.**

būen, v., *O.N.* būa (*pple.* būinn), *O.L.G.* būan, *O.H.G.* pūan, būan, būwan, *Goth.* bauan; **būn** (*pple.*), *cf. prov.Eng.* boun; *ready, prepared*, ORM. 523; C. M. 14992; boun SPEC. 100; WILL. 1138; JOS. 461; MAN. (F.) 4766; TOR. 2044; EGL. 583; sche was boun to goon CH. C. T. *F* 1503; *see* **busken**; *comp.* **ӡe-būen**; *deriv.* **būr, bī, biggen.**

bueren, *see* **bēren.** **bufen**, *see* **biuven.**

buffard, sb., *? fool*, LIDG. M. P. 32.

buffe, sb., *blow, buffet*, AV. ARTH. iv.

buffen, v., *cf. L.Ger.* buffen, *Du.* boffen; *stutter*; **boffing** (*pple.*) ROB. 414; *see* **puffen.**

buffere, sb., *stutterer* : of **bufferes** (*pl.*) '*balborum*' WICL. IS. xxxii. 4.

buffet, sb., *O.Fr.* buffet, *mod.Eng.* buffet; *blow with the hand; a sort of stool*, A. R. 182; WILL. 4700; PR. C. 5203; buffat, boffat WICL. JOHN xviii. 22; buffet, boffet '*alapa, tripes*' PR. P. 41, 54, 55; **buffetes** boffetes (*pl.*) C. L. 1148.

buffetin, v., *O.Fr.* buffeter; *buffet, 'alapo,'* PR. P. 54; **buffetede**, buffeted (*pret.*) LANGL. *B* vi. 178; buffeteden A. R. 106.

buffetunge, sb., *buffeting*, HOM. I. 207.

būӡen, v., *O.E.* būgan, = *M.Du.* buighen, *Goth.*

biugan, *O.H.G.* piugan, *O.N.* biuga, *Lat.* fugere, *Gr.* φεύγειν, *mod.Eng.* bow ; *bend, turn, turn aside, bow* ; to godes bord bugen (*go*) HOM. II. 92 ; buȝen [bouȝen] ut of telde LAȝ. 5386 ; þe him buwen [bouwe] wolden 3709 ; buȝhen ORM. 6627 ; ȝef þu nult to ure wil buhen (*submit*) & beien JUL. 27 ; buwen A. R. 22 ; bouȝe AYENB. 8 ; he sal boughe him ['*inclinabit se*'] PS. ix. 31 ; to bowe into þe bent DEGR. 55 ; **būwe** (*pres.*) HOM. I. 191 ; **būhð** [*O.E.* byhð] A. R. 266 ; buhþ MISC. 96 ; bouweþ C. L. 1305 ; bouwes JOS. 571 ; WILL. 948 ; heo buheð to him KATH. 365 ; **būh** (*imper.*) MARH. 20 ; buh from uvele ['*declina a malo*'] HOM. I. 117 ; bouȝ AYENB. 194 ; boweth LANGL. *B* v. 575 ; **būhe** (*subj.*) KATH. 1484 ; **būinde** (*pple.*) A. R. 18 ; **bēaȝ** (*pret.*) AYENB. 239 ; beah JUL. 77 ; beh, bæh LAȝ. 4745, 15740 ; þa beh (*bent*) ha þe swire MARH. 22 ; and beh him to me over bord SPEC. 54 ; ich beih to þe deofle A. R. 304 ; beiȝ, beih C. L. 358 ; beiȝ AL. (T.) 433 ; bei REL. II. 276 ; bæh ORM. 8961 ; **buȝen** (*pret. pl.*) LAȝ. 24683 ; buȝen to fulehte HOM. I. 91 ; boȝen AYENB. 84 ; GAW. 2077 ; buhe (*subj.*) KATH. 2400* ; *comp.* **a-, an-, bi-, for-, ȝe-būȝen** ; *deriv.* **bēaȝ, bēȝen, bīȝels, biȝt, boȝe, boȝt.**

bouȝinge, sb., *bending*, AYENB. 153.

būȝen, *see* **bēȝen.**

bugge, sb., *?Welsh* bwg ; *bugbear, spectre*, PR. P. 55.

būggen, v., *O.E.* bycgan, bicgan,=*O.L.G.* buggean, *Goth.* bugjan ; *buy*, HOM. I. 163 ; KATH. 1633 ; A. R. 208 ; O. & N. 1368 ; AN. LIT. 9 ; bugge C. L. 1112 ; beggen LK. ix. 13 ; ANGL. I. 10 ; biggen ORM. 15793 ; buggen [bigge, beggin] LANGL. *A* iv. 76 ; bigge ALIS. 5494 ; P. 41 ; E. G. 353 ; bigge, begge TREV. II. 19 ; bigge, bie WICL. GEN. xli. 41, 57 ; begge AYENB. 23 ; bigen GEN. & EX. 2166 ; buie P. L. S. xxv. 49 ; bien, bie PS. cxxix. 8 ; bien, beie CH. C. T. *C* 1246, *C* 845 ; bie HOCCL. iii. 31 ; beie HAV. 53 ; OCTOV. (W.) 388 ; **bihð** (*pres.*) HOM. II. 157 ; buð A. R. 148 ; buþ HOM. I. 185 ; MISC. 63 ; buiþ H. H. 85 ; beið ANGL. I. 15 ; baiþ AYENB. 23 ; (þe) buggeþ HOM. I. 185 ; beȝe JOHN xiii. 29 ; (heo) buggeð A. R. 190 ; biggeþ PL. CR. 360 ; **būȝe** (*imper.*) LAȝ. 30810 ; **būgge** [bigge] (*subj.*) LAȝ. 3556 ; **bohte** (*pret.*) LAȝ. 4799 ; (*ms.* bohhte) ORM. 711 ; boȝte AYENB. 133 ; bouȝte WICL. GEN. xxv. 10 ; bohtest SPEC. 58 ; bouhten A. R. 362 ; P. S. 339 ; boht (*pple.*) H. H. 97 ; boght OCTAV. (H.) 589 ; *comp.* **a-, for-, ȝe-büggen.**

büggere, sb., *buyer* ; biggere, beggere HOM. II. 213 ; biggere E. G. 355 ; **büggares** (*pl.*) LEB. JES. 856 ; biggeres MAND. 86 ; *comp.* **land-büggere.**

büggunge, sb., *O.E.* bycgung ; *buying*, A. R. 362 ; bugginge [*v. r.* bigginge] LANGL. *B*

xix. 230 ; begginge AYENB. 38 ; biing (*redemption*) PS. cxxix. 7.

būgil, sb., *O.Fr.* būgle ; *the plant called bugle*, LIDG. M. P. 199 ; bugle VOC. (W. W.) 554.

būgle, sb., *O.Fr.* būgle, *Lat.* būculus, *prov. Eng.* bugle ; *young bull ; buffalo ; 'bubalus'* PR. P. 55 ; WICL. DEUT. xiv. 5 ; bugelle (*bugle-horn*) TOR. 142 ; **būgles** (*pl.*) ALIS. 5112 ; MAND. 269.

būgle-horn, sb., ALIS. 5282 ; CH. C. T. *F* 1253.

būh-sum, adj., *? O.E.* būhsum,=*Du.* buigzaam, *Ger.* biegsam, *mod.Eng.* buxom ; *from stem of* **būȝen** ; *flexible, obedient*, HOM. I. 57 ; KATH. 1805 ; A. R. 356 ; ORM. 6176 ; MISC. 139 ; buxum GEN. & EX. 980 ; buxom WILL. 3085 ; buxome LANGL. *B* i. 110 ; FLOR. 1725 ; M. H. 62 ; *comp.* **un-būhsum.**

būhsum-līche, adv., *obediently*, HOM. I. 215 ; buxomelich LANGL. *B* xii. 114 ; buxomli MAND. 82 ; buxumli HOCCL. vi. 52.

būhsumnesse, sb., HOM. I. 73 ; H. M. 41 ; buxumnèsse '*obedientia*' PR. P. 57.

builden, *see* **bülden.** **buile,** *see* **būle.**

buine, sb., *? purchase, buying* ; þauh clennesse ne beo nout buine [bune] ['*non ematur*'] A. R. 368 ; wiðuten bune (*v. r.* buggunge) 362* ; wiþuten bune HOM. I. 185.

buist, *see* **boiste.** **buistous,** *see* **boistous.**

buket, *see* **boket.** **bul,** *see* **bol.**

bul, bule, sb., *O.Fr.* boul ; *? falsehood*, C. M. 21270.

bulch, *see* **bulke.**

bulchin, sb., *diminut. of* **bule** ; *young bull*, MAN. (H.) 174.

büld, sb., *house* ; bild HALLIW. 175 ; PL. CR. 157 ; bilde (*dat.*) A. P. i. 962 ; **büldes** (*pl.*) '*aedificia*' TREV. II. 71.

bülden, v., *O.E.* byldan ; *build*, LAȝ. 2656 ; KATH. 1657 ; builden LANGL. *B* xii. 228 ; bilden [bulden, builden] CH. C. T. *D* 1977 ; bulde FER. 2377 ; bilde LUD. COV. 20 ; belden PL. CR. 706 ; belde TRIST. 2810 ; MAN. (H.) 135 ; to belde (*live*) and to bide in blisse D. ARTH. 8 ; **bülde** (*pret.*) LAȝ. 29218 ; ROB. 21 ; TREV. II. 75 ; bilded (*pple.*) MAND. 58 ; ibüld FL. & BL. 643 ; þi bur is sone ibuld R. S. V.

buildinge, sb., *building*, TREV. I. 19 ; a buildinge heo was LEB. JES. 871 ; bildinge, beldinge '*aedificatio*' PR. P. 30 ; belding PL. CR. 501 ; belding GENER. 244.

bulder-stōn, sb.,=*Swed.* bullerstēn ; *boulderstone*, HAV. 1790.

bule, sb., =*M.L.G.* bulle, *M.Du.* bulle, bolle, *O.N.* boli ; *bull*, ORM. 990 ; bulle '*taurus*' VOC. 187 ; bole '*taurus*' VOC. 177 ; PR. P. 43 ; HAV. 2438 ; TREV. VII. 445 ; MAN.

(F.) 478; **bule** (*gen.*): anes bule hude LA3. 14187; **bule** (*pl.*) KATH. 61; bules, boles PS. xlix. 13; boles LANGL. *B* xi. 333; CH. C. T. *A* 2139; bulles WICL. GEN. xxxii. 15; boolis HEB. ix. 13.

bole-hŭde, sb., *bull-hide*, LA3. 14187*; BRD. 5; bolehide ROB. 116.

bŭle, sb., *O.E.* bȳl, = *O.Fris.* bēl, *M.L.G.* bŭle, *M.Du.* buile, *M.H.G.* biule, *prov.Eng.* bile; *boil*; bile '*gibbus*' VOC. 207; '*ulcus*' 267; '*pustula*' PR. P. 36; beel WICL. LEV. xiii. 18*; **bŭla** (*pl.*) LEECHD. III. 84; bules [*v.r.* biles, belis, boilus] LANGL. *C* xxiii. 84; biles '*ulcera*' WICL. EXOD. ix. 9; beles AYENB. 224.

bulge, sb., *O.Fr.* boulge, bouge, *cf. mod.Eng.* bulge, bilge; *bag; swelling; bubble*, IW. 263; bouge '*bulga*' PR. P. 46; bouge '*uter*' WICL. PS. lxxvii. 13; '*gibbus*' LEV. xxi. 20*; boge MAN. (F.) 11197; **bulges** (*pl.*) TREV. VII. 385; *comp.* **water-bulge.**

bulgit, adj., *? bulged*; they com . . . with bolgit schipis REL. II. 24.

bŭli, see **beli**.

bulke, sb., *O.N.* bulki, *cf. mod.Eng.* bulk; *heap, hump*; bolke '*cumulus*' PR. P. 43; bowk GRAIL xxviii. 189; **bulches** (*pl.*) *humps* MAP 338.

bulken, v., *strike, beat*; on her brestes gon thei bulk (*pres.*) C. M. 18511.

bulle, sb., *Lat.* bulla; *bull* (*episcopal or papal*), ROB. 473; AYENB. 62; LANGL. *A* PROL. 66.

bulloke, sb., *O.E.* bulluc; *bullock*, '*buculus*,' VOC. 177; bulluc A. S. 4; bullok PR. P. 55.

bulten, v., = *Swed.* bulta (? = pulten); *knock*, A. R. 366*; **boltid** (*pret.*) WINT. IX. viii. 162; bult D. TROY 7476.

bulten, v., *O.Fr.* buleter, *Ital.* burattare; *bolt* (*sift*) *meal*; bulte PR. P. 55; CH. C. T. *B* 4430; **bulted** (*pple.*) bræd ORM. 992; bulted flour LIDG. M. P. 98.

bultinge, sb., *bolting, sifting*, PR. P. 55.

bumbin, bummin, v., = *Ger.* bummen, *Du.* bommen; *hum, make a booming sound,* '*bombizo, bombilo*,' PR. P. 55; **bumbith** (*pres.*): as a bitor bumbith (*v.r.* bumbliþ) CH. C. T. *D* 972; bommeþ (*drinks*) LANGL. *A* vii. 139; **bummede** (*pret.*) LANGL. *A* v. 137; bummed *B* v. 223.

bumblen, v., = *Swed.* bumbla, bumla (RIETZ 65); *bumble*: as a bitor bumbliþ [bumleþ, bombleth] (*pres.*) in þe mire CH. C. T. *D* 972 (Wr. 6554).

bŭme, see **bēme**.

bunche, sb., *hump*; **bunches** (*pl.*) MAP 344.

bunchon, v., = *Du.* bonken; *beat, strike,* '*tundo*,' PR. P. 55; þei **bonchen** (*pres.*) þeire brestis HALLIW. 194; **bonchede** (*pret.*)

LANGL. *A* PROL. 71; bonched [*v.r.* bunchede] *B.* PROL. 74.

bundel, sb., *cf. M.Du.* bundel; *from* **binden**; *bundle*, '*fasciculus*,' PR. P. 55; WICL. S. SOL. i. 12; JER. ii. 32; **bundels** (*pl.*) MISC. 212.

bune, sb., *O.E.* bune; *hollow stem*; red(e) bunne '*calamus*' PR. P. 55; he nimþ verst his pricke and his bonne (*? printed* boune) AYENB. 150; voiding of bunis (*? barrels*) (*pl.*) HALLIW. 219; a deed sparcle of bonis (*ether of herdis of flex*) ['*favilla stuppae*'] WICL. IS. i. 31.

bŭne, see **buine**.

[**bŭnen**, v., *comp.* **to-bŭnen.**]

bunge, sb., *bung*, '*lura*,' PR. P. 55.

bunne, sb., *bun*, '*placenta*,' PR. P. 55; B. B. 130; **bonn(e)s** (*pl.*) N. P. 10.

bunne, see **bune** *and* **bounde**.

bunsen, v., *cf. L.Ger.* bunsen, *Du.* bonzen, *mod.Eng.* bounce; *knock, beat*, A. R. 188; tundere þat is bete and bounse TREV. I. 281.

bunt, sb., *bunt*; the bunt of the saile HALLIW. 219.

bunten, v., ? = **bulten**; *? sift, bolt*; **bonteþ** (*pres.*) þet mele AYENB. 93.

bunting, sb., *bunting* (*a bird*), '*pratellus*,' PR. P. 56; bounting SPEC. 40.

bŭr, sb., *O.N.* byrr (*gen.* byrjar), *m.; impetus, strong wind, speed*, GAW. 290; A. P. iii. 148; bir M. H. *page* xvii: to him he stirt wiþ bir ful grim IW. 1661; **bire** (*dat.*) ALEX. 711; bere FLOR. 659; wiþ bure '*cum impetu*' WICL. JUDITH xiv. 2.

bŭr, sb., *O.E., O.N.* bŭr *n.* = *O.H.G.* pūr, būr, *m.; from* **bŭen**; *bower* (*lady's room*), A. R. 34; MARH. 21; R. S. v (MISC. 178); þe bur O. & N. 958; ænne bur LA3. 29218; bour HAV. 2072; H. H. 32; bor A. P. i. 963; **bŭres** (*gen.*) O. & N. 652; bures [boures] LA3. *A* 1062; boures CH. C. T. *A* 3367; **bŭre** (*dat.*) ORM. 3323; HORN (L.) 286; boure AYENB. 226; WILL. 1760; GAW. 853; in boure ant in halle P. S. 193; **boure** (*pl.*) FL. & BL. (H.) 660*; bures LA3. 5982; **bouren** (*dat. pl.*) LA3. 2025.

bŭr-cniht, sb., *bower-servant*, LA3. 18960.

bŭr-maiden, sb., *bower-maiden*, PR. P. 56; bourmaiden REL. II. 175.

bŭr-ðein, sb., *O.E.* bŭrþegn; *bower-servant, chamberlain*; burðeine LA3. 15357; **burh-þeines** (*pl.*) 13716.

bŭr-ward, sb., *guardian of the chamber*, LA3. 19176.

bour-woman, sb., *maid*, WICL. S. W. II. 9.

bŭr, sb., *O.E.* (ge)bŭr, = *O.H.G.* būr, *mod. Eng.* boor; *husbandman*; *?* beuir REL. I. 187; *?* bouer LIDG. M. P. 192; *comp.* **3e-bŭr.**

burble, burbulle, sb., *bubble*, '*bulla*,' PR. P. 56; **burbels** (*pl.*) S. A. L. 168.

burblon, burbelin, v., *cf. Span.* borbollar; *bubble,* PR. P. 56 ; **burbelit** (*pret.*) D. TROY 3697.

burch, *see* burȝ. **burd,** *see* bord.

bürde, sb. (? *O.E.* byrdu, *fem. of* byrde, '*of high rank*') ; *lady,maiden,* LAȝ. 19271 ; MARH. 17 ; WILL. 683 ; GAW. 752 ; buirde C. L. 863 ; P. L. S. XXX. 41 ; K. T. 374 ; buirde LANGL. *A* iii. 14 ; berde *C* iv. 15 ; birde PERC. 1289 ; berde DEGR. 759 ; beerde H. V. 13 ; bürde (*gen. pl.*) MARH. 21.

[**burde,** adj. ; *comp.* þóle-bürde.]

bürde, *see* bürðe.

burdōn, sb., *Lat.* burdō ; *mule* ; **burdōnes** (*pl.*) (*earlier ver.* burdowns) WICL. 4 KINGS v. 17.

burdǫn[2], sb., *O.Fr.* bourdon ; *book stud,* PR. P. 56.

burdōn[3], sb., *pilgrim's staff; staff* ; burdon HORN (L.) 1061 ; bordon BEV. 2063 ; bordon [burdoun, bordon] LANGL. *A* vi. 8 ; P. S. 150 ; (*ms.* burdowne) WINT. VIII. xxviii. 56.

burdōn[4], sb., *O.Fr.* bourdon, bordon, *mod. Eng.* burden (*of a song*) ; *bourdon,undersong* ; burdoun [bordoun] CH. C. T. *A* 673.

bureh, *see* burȝ.

burel, sb., *O.Fr.* burel ; *a coarse woollen stuff,* P. S. 221 ; borel clerke (*lay-clerk*) GOW. I. 5 ; borel man (*plain man*) CH. C. T. (W.) 11028 ; borel men (*laymen*) 13691 ; þanne shal borel [burel] clerkes ben abasched LANGL. *B* x. 286.

burȝ, bureh, sb., *O.E.* burg, burh, buruh, =*O.L.G.* burg, *O.H.G.* burg, purg, *O.N.* borg, *Goth.* baurgs *f.* (πόλις), *mod.Eng.* borough ; *fort, castle, borough ; burrow* ; O. & N. 766 ; burȝ, boruȝ L. H. R. 54 ; burȝ, borȝ GAW. 2, 259 ; burg GEN. & EX. 1837 ; burh KATH. 46 ; ORM. 7262 ; ane heȝe burh [borȝ] LAȝ. 218 ; þas burh [borh] 6050 ; þer he (*the boar*) burh hafveð ichosen 12312 ; burch SAX. CHR. 249 ; HOM. I. 225 ; burgh ALIS. 2056 ; ISUM. 547 ; buruh '*urbs*' FRAG. 4 ; HOM. I. 93 ; buruh wiðuten wal A. R. 74 ; of þere buruh 54 ; borȝ, boru ROB. 47, 72 ; bourgh GOW. I. 30 ; Lincolne þe gode boru HAV. 773 (*spelt* borw 847) ; burwȝ PL. CR. 118 ; burwhe PR. P. 56 ; buri HOM. I. 253 ; **büregh,** [*O.E.* byrig] (*dat.*) HOM. II. 31, 33 ; bureȝ FL. & BL. 213 ; buri FRAG. 4 ; beri (*ms.* buriase) MAT. xxiii. 34 ; biri GEN. & EX. 2257 ; birie HOM. I. 225 ; burȝe, burhe, burie [borwe, borewe] LAȝ. 293, 2168, 3553, 4264, 9888 ; burie REL. I. 182 ; bureh, burewe MISC. 39, 55 ; borghe LANGL. *B* vi. 308 ; borwe WILL. 1889 ; **bürh,** [*O.E.* byrig] (*pl.*) HOM. I. 13 ; burh, buriȝe, burhȝes [borewes] LAȝ. 2657, 9928, 15364 ; burȝhes ORM. 6982 ; burghes CH. C. T. *D* 870 ; bourghes LIDG. M. P. 210 ; burgan, burgen (*dat. pl.*) MAT. xi. 1, xiv. 13 ; burȝen [borewe] LAȝ. 6165 ; *comp.* hæfed-, kine-, mōder-burh.

buri-bold, sb., *palace,* KATH. 440.

burch-briche, sb., *O.E.* burg-, burh-, buruh- bryce, -brice ; '*quite de forfesture,*' REL. I. 33 ; burghbreche TREV. II. 95.

burh-cnave, sb., *town lad* ; **burhcnaven** [borhcnaves] (*pl.*) LAȝ. 15555.

burh-folc, sb., *borough folk,* LAȝ. 9758 ; HOM. II. 89.

burh-ȝat, sb., *O.E.* burhgeat ; *castle gate,* LAȝ. 17670 ; KATH. 1679.

burh-mon, sb., *O.E.* burhman ; *townsman,* LAȝ. 12441 ; buruhmon A. R. 350.

borgh-mōt, sb., *meeting of burgesses,* E. G. 350.

burh-rēve, sb., *O.E.* burhgerēfa ; *town reeve,* KATH. 1927.

burȝ-toun, sb., *O.E.* burhtūn ; *fortified town,* WICL. JOSH. vii. 2 ; borwton [borwtoun, burghtoun, borȝtown] LANGL. *C* iv. 112 ; burȝhes tun ORM. 6538.

burh-wal, sb., *O.E.* burhweall ; *town wall,* LAȝ. 22091.

burh-ware, sb. (collective), *O.E.* burgwaru *f.* ; *town dwellers,* '*civitas,*' MAT. xxi. 10 ; burh- weren [borhmen] LAȝ. 28368 ; burȝeweren 28392.

burgāge, sb., *O.Fr.* burgage, bourgage (*héri- tage roturier dans un bourg*) ; *burgage* ; **borgāges** (*pl.*) LANGL. *A* iii. 77.

burgeis, sb., *O.Fr.* burgeis ; *burgess,* FL. & BL. 115 ; WILL. 1889 ; bourgeis ROB. 479 ; borgeis AYENB. 162 ; burgeis (*pl.*) LANGL. *A* iii. 150 ; burjas (*ms.* buriase) MIN. v. 15 ; burges [burgeises] (*magistrates*) C. M. 16060.

bürȝels, sb., *O.E.* byrgels, =*O.L.G.* burgisli ; *tomb, sepulchre* ; buriels C. L. 1284 ; TREV. V. 153 ; an buriels al niwe imad ROB. 204 ; buriles '*sepulcrum*' FRAG. 4 ; buryles MISC. 53 ; burieles LANGL. *B* xix. 142 [buriels, *C* xxii. 146] ; biriels MAN. (F.) 9026 ; beriels '*sepulcrum*' VOC. 178 ; berieles AYENB. 228 ; birigeles GEN. & EX. 2474 ; biriel, beriel '*sepulcrum, tumulus*' PR. P. 37 ; beriel S. S. (Wr.) 2561.

bürȝen, sb., *O.E.* byrgen, byrigen ; *tomb* ; berien HOM. I. 241 ; beregen JOHN xix. 41 ; **büriene** (*dat.*) HOM. I. 111 ; **beriene,** bere- gene (*dat. pl.*) MAT. viii. 28, xxiii. 27 ; beri- enne, berigenne JOHN xi. 38, xx. 1.

bürȝen[2], v., *O.E.* byrgan, byrigan ; *bury* ; burien (? burȝen) LAȝ. 27872 ; berien, berigen LK. ix. 59, 60 ; birȝen (*ms.* birrȝenn) ORM. 15254 ; birien GEN. & EX. 2424 ; biriin, beriin '*sepelio*' PR. P. 36 ; burie [buriȝe, berie] ROB. 252 ; birie P.S. 197 ; WICL. JOHN xix. 40 ; berie CH. C. T. *C* 884 ; EGL. 488 ; to **büriene** (*ger- und*) HOM. I. 37 ; **bürie** (*imper.*) ALIS. 4628 ; birie S. S. (Wr.) 1257 ; **büriede** (*pret.*) HOM. I. 93 ; HORN (R.) 906 ; ROB. 50 ; burede BEK. 2219 ; bureden LAȝ. 19822 ; **büried**

(*pple.*) LANGL. *B* xi. 66 ; beried MAND. 15 ; *comp.* bi-, ȝe-bürȝen.

   bürier, sb., *one who buries* ; biriers (*pl.*) WICL. EZ. xxxix. 15.

   büryinge, büriȝing, sb., *burying*, P. L. S. xxv. 80.

bürȝen³, v., ? = *O.Fris.* burgia, *M.H.G.* bürgen ; *protect* : to burȝen him seolven LAȜ. 21268 ; burege HOM. II. 191 ; burhen JUL. 26 ; to burewen ham wiþ þe FRAG. 7 ; buruwen from þes deofles botte A. R. 366 ; burwe HAV. 2870 ; büreȝe (*pres. subj.*) HOM. I. 25 ; *comp.* ȝe-bürȝen.

burgesie, sb., *O.Fr.* borgesie ; *citizenship* ; borgesie AYENB. 161.

burgoize, sb., *O.Fr.* borgeise ; *wife of a burgess*, KN. T. 12.

burh, *see* burȝ.   buri, *see* burȝ.

bürien, *see* bürȝen.

bürien, v., *O.E.* (ge-)byrian, *O.N.* byrja, = *O.L.G.* (gi-)burian, *O.H.G.* (gi-)burren ; *be due, beseem* ; bürþ (*pres.*) AN. LIT. 4 ; ȝuw birþ understanden ORM. 89 ; bers M. H. 10 ; (hi) buriaþ FRAG. 1 ; bürde (*pret.*) GAW. 2278 ; A. P. I. 316 ; burd MAN. (H.) 76 ; als it birde ORM. 2472 ; þat birde wel to him ben grim HAV. 2761 ; him bird M. H. 17 ; *comp.* ȝe-bürien.

[bürizen, v., *O.E.* byrigan, byrgean, byrgan, = *O.N.* bergja, *taste* ; *comp.* a-bürizen.]

bürinesse, sb., *O.E.* (be-)byrigniss ; *act of burying*, HOM. I. 35 ; burinæsse [burinisse] LAȜ. 25852 ; berines D. TROY 4336.

burjounen, v., *O.Fr.* burjoner (*bourgeonner*) ; *bourgeon, bud*, WICL. LEV. xix. 23 ; burjoneþ, [burgeouneth] (*pres.*) LANGL. *B* xv. 73 ; burgounende (*pple.*) WICL. WISD. xix. 7.

   burjoningis, sb., pl., *buds*, HAMP. PS. 513.

burjoun, sb., *O.Fr.* burjon (*bourgeon*) ; *bud*, (*ms.* burioun) C. M. 10735 ; WICL. S. W. III. 30.

burle, sb., *O.Fr.* bourel ; *flock of wool*, '*tomentum*,' PR. P. 56.

burlen, v., *burl* : burle, ' *extuberare*,' CATH. 48.

burlen², v., = *L.Ger.* burreln ; *welter* ; burland (*pple.*) E. T. 99 ; Betres lai burling in hur blode FLOR. 1639 ; *see* burblen.

burlīch, adj., *mod.Eng.* burly, *cf. O.H.G.* burlīch ; *tall, stately* ; burli PERC. 269 ; burli, borli C. M. 8541 ; borlic REL. I. 222 ; borlich ANGL. III. 280 ; DEGR. 759 ; þi burliche bodi ANT. ARTH. xvi ; burlīche (*pl.*) knightes D. ARTH. 586.

burn, *see* beorn.

burne, sb., *O.E.* burna *m.* burne *f.* (*rivus*), *cf. O.Fris.* burna, *Goth.* brunna (πηγή), *O.L.G.* brunno, *O.H.G.* prunno, brunno, *O.N.* brunnr, *m.* (*fons*), *prov.Eng.* burn ; *brook, streamlet*, O. & N. 918 ; REL. I. 1 ; burne, borne MAN. (F.) 8164 ; C. M. 8964 ; borne GAW. 731 ; bourne (*gen.*) LANGL. *A* PROL. 8 ; bornes *B* PROL. 8 ; bournes (*pl.*) FLOR. 609.

bürne, sb., *for* bürðene.; *load, bundle*, BARB. LEG. II. 82 ; CHEST. I. 65.

bürne, *see* brünie.

burnen, v., *O.Fr.* burnir ; *burnish* ; bornith (*pres.*) CH. TRO. i. 327 ; burned (*pple.*) GOW. II. 231 ; CH. C. T. *A* 1983.

burnesh, v., *prov. Eng.* barnish ; *grow plump*, GENER. 780 ; barnish BARTH. (*in* N. E. D.).

burnet, adj. & sb., *Fr.* brunet ; *brown, brown cloth*, PR. P. 56 ; HOM. II. 163 ; REL. II. 108, 19 ; burnettes (*pl.*) R. R. 226.

burnissen, v., *O.Fr.* burnir, burniss- ; *burnish* ; burnist (*pple.*) BARB. viii. 225 ; A. P. ii. 1085 ; GAW. 212 ; bornist A. P. ii. 554 ; burneisshed P. R. L. P. 102 ; bright bornished gold GUIL.² v. 5 ; borniste A. P. i. 220.

   bornishour, sb., *burnisher*, VOC. (W. W.) 604.

burre, sb., = *Swed.* burre, borre *m.*, *Dan.* borre ; *bur*, '*lappa*,' PR. P. 56 ; AR. & MER. 8290.

bürst, bürsten, *see* brüst, brüsten.

[bursten ? v. ; *comp.* an-, ȝe-bursten.]

burton, v., ? *O.Fr.* borter ; *butt (with the horns), strike*, ' *arieto*,' PR. P. 56 ; þe mastes faste to gidere burte (*pret.*) MAN. (F.) 4626.

bur-trē (? bür-trē), sb., *bur-tree, elder*, '*sambucus*,' VOC. 228 ; CATH. 49 ; HALLIW. 221.

bürðe, sb., *O.E.* (ge-)byrd, = *O.Fris.* berde, berthe, *O.L.G.* (ge-)burd, *O.H.G.* burt, *Goth.* (ga-)baurþs *f.; from* béren ; *birth, race, nation, nature*, HOM. I. 273 ; burðe KATH. 84 ; burde, burðe A. R. 158 ; athalt hire burðe (*nature*) H. M. 13 ; burde, burþe HOM. II. 47 ; þe burþe (*fruit*) þat þou beere L. H. R. 146 ; firme birðe GEN. & EX. 1484 ; Rachel non birðe ne nam 1697 ; birþe GOW. II. 76 ; CH. C. T. *B* 192 ; WICL. JOHN ix. 1 ; þat was of hire kin & al of hire birde ORM. 2052 ; birthes [' *nationes* '] PS. CV. 27.

   bürðe-, bürde-bold, sb., *ancestral mansion*, KATH. 140.

bürð-tīd, sb., *O.E.* gebyrdtīd ; *nativity*, HOM. I. 277 ; (bü)rdtīd FRAG. 5.

bürþ-tīme, sb., *time of birth*, LEG. 75 ; birdetime REL. I. 211.

bürðel, adj., ? *fertile* ; ilc birðhel tre GEN. & EX. 119.

bürðen, v., ? *be born* ; birðen GEN. & EX. 1471.

bürðene, sb., *O.E.* byrðen, = *O.L.G.* burthinnia, *cf. O.H.G.* purdīn, burdīn, *mod.Eng.* burden ; *from* béren ; *burden, load, birth*, LAȜ. 25970 ; one burðene A. R. 232 ; burþene C. L. 958 ; burþen SPEC. 110 ; TREV. I. 73 ; burþene, berþene L. H. R. 56, 57 ; at on burdene (*birth*) GEN. & EX. 1467 ; birþene HAV. 807 ; birþen PS. xxxvii. 5 ; at oon birthen (*birth*) TREV. I. 205 ; birdene MISC. 12 ; berðene MAT. xi. 30 ; berdene JOHN xvi. 33 ; berþene WICL. ECCLUS. xxxiii. 25 ; berdene MISC. 34.

bürðer, sb., *O.E.* byrðor; *bearing (of children)*: i þe burðerne (*? for* burðere *or* burðene) of bearn H. M. 37.

buruȝ, buruh, *see* burȝ.

burwen, *see* burȝen.

bus, v., 3 sing., *see* behōfian ; *it behoves,* C. M. 10639; bos 9870; boes [*v. r.* bihowes, bihoveþ] CH. C. T. *A* 4027 ; bud (*pret.*) MIN. ix. 28 ; **bove** (*subj.*) D. TROY 5115.

busard, sb., *O.Fr.* busart; *buzzard,* ALIS. 3049; bosard PR. P. 45 ; bosarde R. R. 4033 ; busherd VOC. 220.

busch, busk, sb., = *O.H.G.* busc, *M.H.G.* busch, bosch, bosche, *M.Du.* bosch, *O.Fr.* bosc, bosche, *med.Lat.* boscus ; *bush,* WILL. 819, 3069 ; busk '*dumus*' PR. P. 56 ; GEN. & EX. 2779; GAW. 182 ; bosche SHOR. 131 ; **busche, bushe** (*dat.*) CH. C. T. *A* 1517 ; busse AYENB. 28 ; **boskes** (*pl.*) ROB. 547 ; buskes [busches] LANGL. *B* xi. 336 ; bushes MAND. 115 ; buskes PERC. 758.

buschaile, sb., *O.Fr.* boschaille ; *ambush;* busshaile GENER. 9189; buscaile D. ARTH. 895 ; buskaile 1634 ; boschaile OCTOV. (W.) 1607.

buschel, sb., *O.Fr.* buissel, boissel ; *bushel,* P. 30 ; buschel, bushel CH. C. T. *A* 4093 ; buischel, boischel WICL. LK. xi. 33.

buschen, v., *place in ambush; lie in wait*; buske YORK xiii. 8 ; **bussed** (*pple.*) MAN. (H.) 187 ; **busket** (*pret.*) D. TROY 1168.

buschen[2], v., *? make bushy*; ibusched (*pple.*) wiþ þornes MAN. (F.) 9194.

busching, sb., *training on bushes*; bosshing PALL. xi. 33.

busehen[3], v., *?= M.H.G.* büschen, *M.Du.* buischen ; *strike, beat; but with the head; rush; gush out*; buschen on felde WILL. 173 ; busche TREV. II. 191 ; **busched** (*pret.*) A. P. iii. 143 ; bosshet D. TROY 11120.

busshinge, sb., *striking,* DEP. R. i. 99.

buschi, adj., *bushy*; busshi WICL. IS. vii. 19.

büsi, adj., *O.E.* bysig, = *M.L.G., M.Du.* besich ; *busy, anxious, careful,* WILL. 588 ; busi, bisi, besi CH. C. T. *A* 1491 ; bisi [busi] LAȝ. 2837 ; bisi A. R. 142 ; MARH. 16 ; C. L. 787 ; M. H. 108 ; bisi, besi '*assiduus, solicitus*' PR. P. 37 ; besi GOW. II. 42 ; MAND. 3 ; **bisine** (*acc. m.*) LAȝ. 10596 ; **bisie** (*pl.*) LAȝ. 19557; AYENB. 58 ; **bisegure** (*compar.*) A. R. 182 ; **bisegæste** (*superl.*) LAȝ. 10476.

bisi-hēde, sb., = *M.Du.* besicheit ; *industry,* AYENB. 55 ; bisihed ALIS. 3.

bisi-līche [büsilīche], adv., *busily,* LAȝ. 4473; bisiliche A. R. 146 ; SHOR. 55 ; besiliche GOW. II. 43 ; busili WILL. 650 ; **bisiloker** (*compar.*) WICL. 1 PET. i. 22 ; **bisilükest** (*superl.*) JUL. 44.

bisi-schipe, sb., *business,* A. R. 384; besiship GOW. II. 39.

büsie, sb., *O.E.* bysigu, bisigu, bysgu; *labour, occupation* ; besie D. ARTH. 3630.

bisi-vol, adj., *full of business,* AYENB. 226.

büsien, v., *O.E.* bysigan, bysgian, bisgian, = *M.Du.* beseghen ; *busy, occupy; be busy*; bisien CH. BOET. i. 2 (8) ; besien GOW. II. 43 ; **bisiede** (*pret.*) '*satagebat*' WICL. LK. x. 40 ; bisied GAW. 89 ; **bisied** (*pple.*) PS. xxxix. 18.

bisinesse, sb., *business,* '*assiduitas, diligentia, solicitudo,*' PR. P. 37 ; SHOR. 92 ; AYENB. 56 ; PL. CR. 727 ; besinesse '*diligentia*' TREV. I. 5 ; GOW. II. 60.

busk, *see* busch.

busken, v., *O.N.* būask (*get oneself ready*), *mod.Eng.* busk; *prepare; adorn*; buske WILL. 2210 ; **buskeþ** (*pres.*) P. L. S. xxx. 20 ; þe king boskes lettres JOS. 414 ; hue boskeþ huem wiþ botouns P. S. 239 ; **buske** (*imper.*) þe forþ to fare EGL. 348 ; buskeþ [buske] ȝou to þat bote LANGL. *B* ix. 133 ; **busked** (*pret.*) GAW. 1411 ; þai busked and maked hem boun TRIST. 144 ; hit was **buskid** (*pple.*) above wiþ besauntus DEGR. 1427.

busking, sb., *fitting out, setting out*; C. M. 3245 ; bosking TRIST. 92.

busse, sb., *O.Fr.* busse, *cf. O.E.* butsa-carlas (*boatmen*) SAX. CHR. 201 ; *a kind of boat,* MAN. (F.) 187 ; MAN. (H.) 169.

buste, *see* boiste.

büsten, v., *? beat, bruise*; beateð þe & **büsteð** (*pres.*) þe H. M. 31 ; þah þu me **büste** (*subj.*) ant beate JUL. 24 ; ofte þei him bete & buste (*r. w.* niste) AL. (L.[3]) 331 ; *comp.* to-büsten.

bustlen, v., *wander blindly*: busteling (*sb.*) forþ as bestes over valeies & hulles LANGL. *A* vi. 4.

but, sb., *Fr.* but, *mod.Eng.* butt; *target, boundary*; '*meta,*' PR. P. 56 ; a but of lond '*amseges*' VOC. 270 ; that might the ston to his but bring(e) OCTOV. (W.) 899.

but[2], sb., *act of '*putting*' a stone,* HAV. 1040.

büten, būt, bǔt, prep., adv., conj., *O.E.* beūtan, būtan, būton, būte, = *O.L.G.* biūtan, būtan, *mod.Eng.* but ; *outside, without, except, unless, but* : ȝif þe wisa mon bið butan gode wercan HOM. I. 109 ; buton ane treowe 221 ; buten ende 147 ; bute (*unless*) he hine drive a wei 21 ; nawiht for ure ernunge bute (*but*) for his muchele mildheortnesse 19 ; buten wæstme ['*sine fructu*'] MAT. xiii. 22 ; buto ['*nisi*'] xi. 27 ; bute JOHN xix. 11 ; bute þu hefdest unifouh FRAG. 7 ; buten (*v. r.* wiðute) live KATH. 252 ; nis buten an god 367 ; nefde ha bute iseid swa þat al þe eorðe ne bigon to cwakien MARH. 19 ; et te laste vers buten an A. R. 20 ; þis world nis buten a wei to heovene 150 ; buten leave 238 ; he ne mei no þing don us bute bi godes leave 230 ; beon buten [boute] LAȝ. 3749 ; buten [bote] Noe & Sem 23 ; bute mochelere ferde 3679 ; bute (*sed*) nele he . . . þe Evelin . . . bitæchen 8263 ; nefede he boten (*? bōten for* bouten) anne sune 88 ; beute ȝif (*unless*) 26433 ; buten

†

childre ORM. 204; but (*ms.* butt) if 45; bute neste *outside the nest* O. & N. 1386; þat ich bute anne craft ne can 794; bute (*unless*) þu wille bet aginne ne schaltu bute schame iwinne 1289; bute here 1790; bute lese wordes þu me lenst 756; nowiȝt bote sorwe 884; non oðer wile ge ... buten (*sed*) one goð & one sit MISC. 22; bute it were bi hire wille HAV. 85; fro londe woren he bote a mile 721; but if WILL. 472; as schip boute mast 567; but LANGL. *B* PROL. 194; but what ben þes þingis a mong so mani(e) men WICL. JOHN vi. 9; but (*yet*) CH. C. T. *A* 74; þei conne meche of holi writ but þei undirstonde it not MAND. 136; beouten PL. CR. 651; bouten ende TREAT. 132; boute, bote SHOR. 18, 40; bote þat þou me Wilekin bringe AN. LIT. 12; boute gile SPEC. 38; boute, bout, bot GAW. 361, 1285, 1782; boten P. S. 156; þou ne sselt habbe god bote me AYENB. 5; bot ISUM. 150; þer nas bot o wei BEK. 637; and was bot seven winter old S. S. (Wr.) 14; *comp.* **a-būten.**

butere, sb., *O.E.* butere, *Lat.* butirum, *from Gr.* βούτυρον; *butter,* SAX. CHR. 259; HAV. 643; buttere GEN. & EX. 1014; buttire VOC. 198; butter, botere WICL. GEN. xviii. 8; botere E. G. 356.

buter-flȝe, sb., *O.E.* buttorflēoge, *cf. M.Du.* botervlieghe; *butterfly,* FL. & BL. 473; butterflie PALL. iv. 946; boterflie CH. C.T. *B* 3980.

butle, sb., *cf. O.L.G.* butli (FÖRSTEM. NAMENB. II. 319); *house;* buttle HOM. II. 185; *see* **botel.**

butte, sb., *cf. L.Ger., Swed.* butte, *Du.* bot; *butt* (*kind of fish*), PR. P. 56; HAV. 759.

bütte, sb., *O.E.* bytt, *O.N.* bytta; *butt, wineskin, cask,* '*uter*,' FRAG. 4; butte, bitte *?uterus* H. M. 35; bollen as a bitte (*ms.* bite, *r. w.* pite=pitte) MAP 334; bit ['*uter*'] PS. cxviii. 83*; **bütta,** betta, bitte (*ms.* bytte), bitton, bitten (*pl.*) MAT. ix. 17, MK. ii. 22, LK. v. 37; bittes E. G. 382.

butten, v., *cf. Swed.* butta, botta (RIETZ 48), *M.Du.* botten, *? M.H.G.* butzen, *mod.Eng.* butt; *drive, thrust,* AR. & MER. 5175; he bigan ... to stiren & to butten ORM. 2810; button '*pello*' PR. P. 56; but (*pple.*) HAV. 1916; *see* **putten.**

buttinge, sb., *thrusting, tilting,* HAV. 2322.

buttok, sb., *buttock,* '*nates*,' VOC. 183; PR. P. 56; P. L. S. xvii. 163; CH. C. T. *A* 3803; **bottokes** (*pl.*) SAINTS (Ld.) xlvi. 725.

buven, *see* **biuven. būvien,** *see* **bivien.**

būwen, *see* **būȝen. büxum,** *see* **būhsum.**

bȳ, *see* **bī.**

c.

cā, sb., *cf. Dan.* kaa, *M.Du.* kā, kae, *O.H.G.* cāha; *chough;* kaa VOC. 188; coo, keo '*monedula*' PR. P. 84; *see* **chōȝe.**

cā-dāwe, sb., *jackdaw,* '*monedula*,' PR. P. 57.

cabache, sb., *cabbage,* ANC. COOKERY (1440), CAXT. VIT. PATR. (*in* N. E. D.).

caban, sb., *O.Fr.* caban, cabane; *cabin,* '*capana*,' PR. P. 57; LANGL. *A* iii. 184; cabane D. ARTH. 757.

cabil, sb., *? Lat.* caballus; *cf.* **capel;** *horse, nag,* (*ms.* cabylle) VOC. (W.W.) 697.

cáble, sb., *Fr.* cable, *late Lat.* capulum; *cable,* PR. P. 57; GOW. II. 142; cabille VOC. (W.W.) 731; **káblen** [cables] (*pl.*) LAȝ. 1338.

caboche, v., *cut off the head of* (*a deer*) *close behind the horns:* ms. (1425) *in* N. E. D.

cacchen, chácen, v., *O.Fr.* cachier, chacier; *catch, chase:* ȝif he me mihte cacchen [cache] LAȝ. 31501; cacche WILL. 2266; cacche, cache CH. C. T. *A* 4105; chacche LANGL. *A* ii. 167; kecchen A. R. 324; kecche HORN (R.) 1377; cachin [kacchin] a wei '*abigo*' PR. P. 57, 269; vor to cachie and verri þane di(e)vel vram him AYENB. 178; chacen out alle misbeleevinge men MAND. 3; chace PR. C. 4316; chace a wai LIDG. M. P. 139; þe oþre ... **chácieþ** (*pres.*) forþ Olivere FER. 955; alle ... kache me a wai M. H. 151; **cahte** (*pret.*) LAȝ. 4547; caȝte ROB. 14; kagte is (a) wei GEN. & EX. 949; cauȝte, caughte CH. C. T. *A* 498; keihte A. R. 154; keiȝt(e) L. H. R. 134; keȝte ANT. ARTH. xlix; chaced LANGL. *B* xvii. 51; ha cahten KATH. 1990; **caht** (*pple.*) P. S. 152; cauȝt LANGL. *B* xvii. 219; hu he havede þene nome icaht LAȝ. 10843; ikeiht A. R. 88; haþ cachid ['*comprehendit*'] WICL. MIC. iv. 9; þus am i cachet from kiþe ANT. ARTH. (R.) xii; *comp.* **bi-cacchen.**

cachere, chácere, sb., *catcher, hunter;* þise **cacheres** (*pl.*) .. coupled hor houndez GAW. 1139; chaseris BARB. vii. 91.

chácinge, sb., *chasing, hunting,* PR. P. 68; TREV. II. 359.

cache, sb., *O.Fr.* cache, cace, chace; *catch, chase;* chace MISC. 199.

cachepol, sb., *O.Fr.* chácepol; *catchpoll,* HOM. I. 97; kachepol LANGL. *C* xxi. 46; **catchepollis** (*pl.*) '*apparitores*' WICL. I KINGS xix. 20.

cacherel, sb., *catchpole;* **cachereles** (*pl.*) P. S. 151; kachereles AYENB. 263.

cad, sb., *pet lamb,* '*cenaria*,' VOC. 219.

kod-lomb, sb., *? cade-lamb,* '*ricus*,' VOC. 245.

cadas, sb., *O.Fr.* cadas, *caddis, floss silk,* '*bombycinum*,' PR. P. 57.

cáde, sb., *Fr.* cade, *from Lat.* cadus; *barrel,* PR. P. 57; **kádes** (*pl.*) PALL. xi. 331.

cáde[2], sb., *?membrum virile,* AR. & MER. 934.

cadence, sb., *cadence,* GOW. II. 82; (*ms.* cadens) LUD. COV. 189.

cader, sb., *? Welsh* cadair; *cradle,* A. R. 82*; kader 378.

caf, *see* **chaf.**

cǎf, adj., *O.E.* cǎf; *swift, eager;* cof REL. I. 212; cof & qvik A. P. ii. 624; kene he was &

kof MAN. (H.) 66 ; þe cove A. R. 66 ; biforen kafe & kene ORM. 19962.

cōve, adv., *swiftly*, O. & N. 379; cōfer (*compar.*) HOM. I. 231.

cōf-līche, adv., *swiftly*, LA3. 1705; cofli GAW. 2011 ; ALIS.² (Sk.) 1009.

caf, sb., *? cf. O.E.* cofa, *mod.Eng.* cove ; *?box* : of wod dri as teindire þa mad a caf BARB. LEG. II. 194.

caft, keft, sb., *? harlotry*, A. R. 206.

cáge, sb., *O.Fr.* cage, *from Lat.* cavea ; *cage*, A. R. 102; ALIS. 3555; P. 19; CH. C. T. *A* 1294.

caggen, *draw, bind, fasten* ; caggis [caches] ALEX. (Sk.) 1521 ; þai . . . keruen & caggen (*pres. pl.*) & man it clos A. P. i. 512 ; þai wer cagged (*pple.*) & ka3t on capeles al bare ii. 1254; caget D. TROY 3703.

cai, adj., *cf. Fris., Dan.* kei, *? M.Du.* kei, kai (*foolish*) ; *left* : þe kai fote GAW. 422.

cail, *see* keil.

cailewei, sb., *O.Fr.* (poire de) cailloel ; *a sort of pear*, LANGL. *B* xvi. 69; caleweis (*pl.*) R. R. 7093.

cainard, sb., *O.Fr.* caignard, cagnard ; *coward*, SPEC. 110 ; A. S. 36 ; CH. C. T. *D* 235.

[cair, adj., *O.N.* kǣrr ; *dear.*]

kager-le3c, sb., *O.N.* kǣrleikr ; *love* ; (*ms.* kaggerrle33c) ORM. 2187.

cairen, v., *? O.N.* keyra (*drive*) ; *go* : in contre to cairen [kairen] aboute LANGL. *A* PROL. 29 ; kaire GAW. 1048 ; caired (*pret.*) A. P. ii. 85 ; þe kouherde kaired to his house WILL. 373 ; þe candelstik . . . watz caired (*pple.*) þider A. P. ii. 1478 ; icaired SPEC. 37.

caiser, sb., *O.H.G.* kaisar, keiser, = *O.E.* cāsere, *Goth.* kaisar, *O.N.* keisari ; *from Lat.* Caesar ; *emperor*, ALIS. 1409; MAN. (F.) 5120; kaiser A. R. 138 ; kaisere, kæisere, keisere, keiser [caiser, kaisere] LA3. 5220, 5965, 7331, 27366 ; kaisere LANGL. *B* xix. 134 ; kaser ORM. 8329; keiseres (*gen.*) KATH. 3 ; (*ms.* ke33seress) ORM. 3519; caisere (*dat.*) MK. xii. 17.

caitif, sb. & adj., *O.Fr.* caitif, chaitif, *mod. Eng.* caitiff ; *captive ; wretch, wretched, despicable*, '*captivus*,' WICL. I PARAL. v. 6 ; WILL. 710; caitif, [chaitif] LANGL. *C* xxiii. 236.

caitif-dōm, sb., *captivity*, WICL. EZ. xx . 3 ; TOWNL. 156.

caitive-hēde, sb., *wretchedness ; baseness*, C. M. 22382 ; caitef hede 7353.

caitif-lī, adv., *wretchedly*, R. P. 38.

caitifnes, sb., *captivity, wretchedness*, TOWNL. 315.

caitifen, v., *make captive ; render wretched*, WICL. PROL. JER. ; R. P. 36.

caitivitē, sb., *O.Fr.* caitiveté ; *captivity, wretchedness*, C. M. 7353, 18191, 22382 ; PR. C. 455 ; çaitifte WICL. EPH. iv. 8.

cáke, sb., *cf. Swed.* kaka ; *cake*, '*libum, placenta*,' PR. P. 58 ; VOC. 198; H. M. 37 ; L. H. R. 137 ; CH. C. T. *A* 668.

cáke-brēad, sb., *cake (of bread)* ; cakebrede LANGL. *B* xvi. 229.

cakel, adj., *cackling*, A. R. 66.

cakelin, v., = *Du.* kakelen, *Swed.* kakla ; *cackle*, '*gracillo*,' PR. P. 58 ; kakelen A. R. 66 ; cacleþ (*pres.*) GOW. II. 264.

cakelinge, sb., *cackling*, '*gracillacio*,' PR. P. 58.

cakkin, v., '*caco*,' PR. P. 58.

cal, sb., *O.N.* kall ; *call, act of calling*, A. P. ii. 61 ; calle (*dat.*) C. M. 3022.

cāl, *see* caul.

caladrie, sb., *kind of bustard* ; ['*charadrius*'] WICL. DEUT. xiv. 18.

calament, sb., *Fr.* calament, *med.Lat.* calamentum ; *sort of herb*, TREV. BARTH. (N. E. D.) ; '*calamenta balsamita*,' PR. P. 58.

calamus, sb., *Lat.* calamus ; *reed, an aromatic plant*, TREV. BARTH. (N. E. D.) ; WICL. EZ. xxvii. 19.

calami, sb., *sort of aromatic herb (earlier text* chaalami) WICL. EX. xxx. 24.

calcatori, sb., *Lat.* calcatorium ; *wine press*, PALL. i. 461.

calch, sb., *O.E.* calic, *from Lat.* calix (*acc.* calicem) ; *cf. O.Fr.* calice, chalice ; *chalice*, MISC. 43 ; calch, calice HOM. II. 91, 215 ; caliz A. R. 284; HAV. 187; BRD. 14; chalice SHOR. 20; chalis AYENB. 167 ; calices (*pl.*) AYENB. 41.

calcidoine, sb., *O.Fr.* calcedoine ; *chalcedony*, GOW. III. 133.

calcininge, sb., *calcination* ; calceniinge CH. C. T. *G* 771.

calcinatiōn, sb., *Fr.* calcination ; *calcination*, CH. C. T. *G* 804.

calculaciōn, sb., *Lat.* calculatiōnem ; *calculation*, GOW. II. 345.

calculatour, sb., *Lat.* calculator ; *one who calculates*, WICL. S. W. II. 408.

calculen, v., *Fr.* calculer ; *calculate* ; kalcule CH. ASTR. i. 22 (14) ; calculed (*pple.*) LANGL. *B* xv. 364.

kalkuler, sb., *calculator*, CH. ASTR. i. 23 (14).

calculinge, sb., *calculation*, TREV. I. 39 ; CH. TRO. i. 71.

calculōse, adj., *Lat.* calculosus ; *stony*, PALL. ii. 274.

cáld, adj., *O.E.* ceald, cald, = *O.L.G.* cald, *Goth.* kalds, *O.N.* kaldr, *O.H.G.* chalt ; *from* cálen ; *cold*, HOM. I. 97 ; IW. 360; PR. C. 767 ; cald, cold LA3. 4519, 19756; cold A. R. 400 ; O. & N. 622 ; cold red REL. I. 182 ; child [cold] LANGL. *C* xviii. 49 ; cheald JOHN xviii. 18 ; chald MISC. 30 ; AYENB. 153 ; cheld SHOR. 9 ; L. N. F. 214 ; GREG. 798.

cáld, sb., *cold*, HAV. 856 ; P. S. 330 ; WILL.

908; MAND. 256; **chealdes** (*gen. n.*) MAT.
x. 42; **cálde** (*dat.*) JUL. 21; IW. 2974; colde
GOW. II. 38; **cálde** (*pl.*) HOM. I. 95; cares ful
colde P. S. 152; **cáldore** (*compar.*) L. H. R. 143.

káld-hēd, sb., '*refrigerium,*' PS. lxv. 12.

cálde-līche, adv., *coldly*, HOM. I. 277.

báld-mǣwe, sb., = **mǣw**; *? a sort of gull*,
LIDG. M. P. 202; **calmēwe** M. T. 133.

cálde, sb., ?=**kelde**; *cold*: kalde, PS. cxlvii. 17.

cálden, v., *O.E.* cealdian,= *O.H.G.* chalten;
grow cold; colde REL. I. 120; CH. TRO. iv.
362; M. ARTH. 3647; his blod bigan to colde
S. S. (Wr.) 1678; hire heorte bigan to chelde
HORN (L.) 1148 (kolde H. 1185); keld(e) REL.
II. 210; coldeþ (*pres.*) FRAG. 5; MISC. 101;
keldeþ ANGL. III. 279; *comp.* **a-, for-, of-cálden**.

cále, sb., *? cold* : in hete & in cale LEG. 148.

[**cálen**, v., *O.E.* calan, *O.N.* kala (*pret.* kōl,
pple. kalinn); *be cold, grow cold*; *comp.* **a-cálen**; *deriv.* **cáld, cōl**.]

calende, kalendes, sb., *O.E.* calend, *from
Lat.* calendæ; *first of the month; month*;
GOW. III. 123; PALL. x. 23; kalendes of
chaunge CH. TRO. v. 1647.

calender, sb., *Lat.* calendarium; *calendar*;
kalender LA3.7219; CH.ASTR. i. 11 (7); (*guide*)
CH. L. G. W. 542; (*register*) D. ARTH. 2641.

calenge, see **chalenge**.

cales, sb., *name of a fabulous creature*; a
feolle worm, cales ALIS. 7093.

caleweis, see **cailewei**.

calf, sb., *O.E.* cealf, cælf, celf,= *O.L.G.* calf,
O.H.G. chalb *n.*, *O.N.* kālfr *m.*; *calf*, H. M. 37;
ORM. 5858; P. S. 332; calf [kalf] LANGL. *A* vii.
274; kælf FRAG. 3; kelf A. R. 136; an fet
chalf LK. xv. 27; **kalves** (*gen.*) fleis GEN. &
EX. 1013; **calve** (*dat.*) A. S. 4; **calveren** (*pl.*)
MAND. 105; calveren, calves WICL. NUM.
xxix. 32.

calf-flēsh, sb., = *Ger.* kalbfleisch; *veal*, VOC.
200.

calf-lees, adj., *without a calf*, WICL. JOB
xxi. 10.

calf, sb., *O.N.* kālfi = *M.Du.* kalf (*pulpa*),
calf (of the leg), '*sura,*' PR. P. 58; CH. C. T.
A 592.

calice, see **calch**.

caliōn, sb., *pebble*, '*rudus,*' PR. P.58; **caliouns**
(*pl.*) MER. xx. 329.

caliphe, sb., *caliph*, GOW. I. 245; califfe
MAND. 36; caliphee 230.

caliz, see **calch**.

calke, sb., *shoe*; **calken** (*dat. pl.*) MK. vi. 9.

calke-trappe, sb., *O.E.* calcatrippe (*hera-clea*), VOC. 68; *O.Fr.* chaucetrappe; *caltrop*,
'*tribulus,*' REL. I. 37 (VOC. 140); caltrap
saliunca PR. P. 58; caltrap of irin, fote
hurtinge '*hamus*' PR. P. 59; **calketrappen**
(*pl.*) ALIS. 6070; vol of . . . calketreppen

and of grines AYENB. 131; kalketrappes
LANGL. *C* xxi. 296.

calkin, v., *? for* **calculen**; *calculate*, P. P. II.
61; '*calculo*' PR. P. 58.

calle, sb., *O.Fr.* calle, cale; *caul; net*, P. S.
158; CH. C. T. *D* 1018; K. T. 365; þe calle of
þe liver ['*reticulum jecoris*'] WICL. LEV. iii. 4;
kelle '*reticulum*' VOC. 196; EM. 303; **kelles**
(*pl.*) TOWNL. 313.

callin, v., *O.E.* ceallian,= *M.Du.* callen, *O.N.*
kalla, *O.H.G.* challōn; *call*, '*voco,*' PR. P. 58;
callen MARH. 3; so shulen men callen it ai
HAV. 747; calle HOM. II. 257; GOW. I. 148;
PALL. i. 1042; **calle** (*pres.*) SPEC. 59; (þei)
calles WILL. 239; callen MAND. 1; **callede**
(*pret.*) PERC. 610; called [kallid] LANGL. *B*
xviii. 93; **cald** (*pple.*) HOM. I. 271; GEN. &
EX. 1700; icalled JOS. 78; PL. CR. 574; *comp.*
bi-callen.

callinge, sb., *calling*, '*vocatio,*' PR. P. 58.

calme, adj., *Fr.* calme; *calm*, PR. P. 58; GOW.
III. 230; D. TROY 2011.

calmen, v., *Fr.* calmer; *calm*; calme DEP. R.
iii. 366; D. TROY 4587.

calstok, see under **caul**.

caltrappin, v., *catch with a caltrop*, PR. P. 59.

calu, adj., *O.E.* calu (*gen.* calwes), = *M.Du.*
calu, *O.H.G.* chalaw; *mod. Eng.* callow; *bald*
['*calvus*'] WICL. LEV. xiii. 40; calu3 ALIS.
5950; þe calewe P. L. S. ix. 89; þe calouwe
mous (*the bat*) [*Fr.* la chauve-souris] AYENB.
27.

calven, v., *O.E.* cealfian, = *M.Du.* kalven,
M.H.G. kalben; *calve*: þe cou **calvide** (*pret.*)
WICL. JOB xxi. 10.

calving, sb., *calving*, PALL. viii. 66.

calver, adj., *? fresh*; calvur PR. P. 59; cal-war F. C. 19.

cámb, sb., *O.E.* camb, comb,= *O.L.G.* camb,
O.N. kamb, *O.H.G.* champ; *comb*, ORM.6340;
comb P. 16; a cokkes comb MAND. 207; č̣e
dikes comb GEN. & EX. 2564; komb CH. C. T.
B 4049; kokis coom PR. P. 281; **kómbe** (*dat.*)
AYENB. 258; **cómbes** (*pl.*) P. L. S. xix. 249; *comp.* **horg-, huni3-comb**.

cambok, sb., *M.Lat.* cambuca, *cf. Ir., Gael.*
camog; *fetter*, '*pedum,*' VOC. 232.

camēl, sb., *O.Fr.* camel, chamel; *camel*, P. L. S.
i. 22; C. M. 6001; kamel GEN. & EX. 1398;
camēls, chamēls (*pl.*) WICL. JUDG. viii. 21;
camels [chameiles, camailes] C. M. 3304;
chamoilis (*later vers.* camelis) 1 WICL. PARAL.
xii. 40.

camēl-hār, sb., *camel-hair*, M. H. 41.

camēl-hīde, sb., *camel skin*, C. M. 2249.

cameline, sb., *O.Fr.* cameline; *camlet*, R. R.
7366.

cameline, sb., *kind of sauce*, L. C. C. 30.

cameliōn, camle, sb., *Gr.* χαμαιλέων; *chame-leon*, GOW. I. 133; **camles** (*pl.*) MAND. 289.

camelĭōn, sb., *used for camelopard*, WICL. DEUT. xiv. 5; camelion is a flekked best in colour liche to a lupard TREV. I. 159.

cammid, adj., *snub-nosed*, '*simus*,' PR. P. 59; **cammede** (*pl.*) REL. I. 240.

     cammednesse, chammidnesse, sb., '*simitas*,' PR. P. 59.

cammoc, sb., = *O.E.* cammoc; *rest-harrow* (*plant*); **cammokes, kammokes** (*pl.*) LANGL. *B* xix. 309.

camoca, cammoca, sb., = *O.Fr.* camoca; *a rich silken fabric* [cammoka, camaca, camaka, cammaka], LANGL. *C* xvii. 299.

cammock, = **cambok**.

camomille, sb., *Fr.* camomille; *camomile*, P. L. S. xxxi. 114.

camp, sb., *O.E.* camp, comp, = *O.Fris.* kamp, komp, *O.H.G.* champh *m.*, *cf. O.N.* kapp *n.*; *battle*; þat comp LA͡Ʒ. 23889; comp & iʃiht 4347; **kampe** (*dat.*) D. ARTH. 3670; *comp.* ͡Ʒe-camp.

     komp-ĭfēre, sb. (*? printed* kemp ifere), *comrade in battle*, A. R. 274.

campar, sb., *football player*, PR. P. 60.

campesōn, sb., *for* gambisōn; *doublet worn beneath the armour*, RICH. 376.

campin, v., *O.E.* campian, = *O.Fris.* kampa, kempa, *O.H.G.* chamfan, chemfan, *O.N.* keppa; *fight, contend; play at football*, '*pedipilo*,' PR. P. 60; **kempe** D. ARTH. 2633.

     campinge, sb., *football match*, '*pedipiludium*,' PR. P. 60; LIDG. M. P. 200.

campiōn, *see* champiōn.

camus, camois, adj., *Fr.* camus; *low, flat* (*nose*), FER. 4437; CH. C. T. *A* 3934.

camused, adj., *flatnosed*, GOW. II. 210.

can, v., *O.E.* cann (can), conn (con), = *O.L.G.* can, *Goth.* *O.N.* kann, *O.H.G.* chan, (*pret. pres.*); *mod.Eng.* can; *know; know how to; can*; GEN. & EX. 309; ROB. 10; AYENB. 21; can ['*scit*'] WICL. JOHN vii. 15; nou can i no red TRIAM. 595; can, kan ORM. 1314, 5282; i can [kan] a noble tale CH. C. T. *A* 3126; ʃhe . . . can [kan] hem þer fore as muche thank as me *A* 1808; kan ['*novi*'] MK. xiv. 68; can, con LA͡Ʒ. 3291, 13332; can, kan, con, kon O. & N. 197, 263, 680, 757; con A. R. 18; KATH. 817; C. L. 30, 555; H. H. 26; AUD. 16; ne con crist him nenne þonc HOM. I. 31; nis nan sunne þet he ne con 35; þu canst O. & N. 1182; thou canst . . . þe ricthe gate HAV. 846; const [canst, kanst] LANGL. *A* iii. 166; hwet so þu conʃt . . . biþenchen JUL. 66; cunnen ORM. 5514; MIRC 237; þe wel cunnen a [conne of] speche LA͡Ʒ. 7301; cunne O. & N. 911; BEK. 1325; connen, conne MAND. 5, 213; conne SPEC. 103; ED. 3453; **cunne** (*subj.*) A. R. 64; HORN (L.) 568; cunne, kunne O. & N. 47, 188; cunne [conne] LA͡Ʒ. 28644; conne AYENB. 118; þou konne LANGL. *B* xix. 26; ͡Ʒef þou conne MIRC 1356; ͡Ʒe conne [konne, cunne,

kunne] CH. C. T. *A* 3118; *deriv.* **cunnen, kennen.**

can, v., = **gan**; *began, set to* : can [gan, con, gon] on him to stare C. M. 13557; can him to confort QUAIR iv; Moises on þe roche kan stand C. M. 6300; the croune that Ihesus couth (*did*) ber BRUCE iii. 460.

canacle, conacle, sb., *? mistake for* covercle; *? lid of a cup*, A. P. ii. 1461, 1515.

cancre, sb., *O.Fr.* cancre, chancre; *canker, cancer*, A. R. 330; canker [kankir] WICL. 2 TIM. ii. 17; er the rose, as it come into canser D. TROY 2344; cankir, *canker-worm* '*teredo*' PR. P. 60.

canker-dort, sb., *state of suffering*, CH. TRO. ii. 1752.

cancre-frete, adj., *eaten away with canker*, ROB. 299.

cancren, v., *from med.Lat.* cancerāre; *canker*, kankir HOCCL. REG. PRIN. (*in* N. E. D.); **cancringe**, (*pple.*) WICL. PREF. EP. 69; **cancred (concred)**, *ulcerated*, TREV. BARTH. (*in* N. E. D.); here is a cankerd (*depraved*) company YORK vii. 97.

candele, sb., *O.E.* candel, *from Lat.* candēla; *candle*, HOM. II. 47; AYENB. 102; candle BEK. 1969; condle CHR. E. 505.

     candel-bēm, sb., *candle-stick*, '*lucernarium*,' PR. P. 60.

     condel-lĭht, sb., *candle-light*, LA͡Ʒ. 23752.

     kandel-messe, sb., *candlemas*, ORM. 7706; condelmesse A. R. 412.

     candel-staf, sb., *O.E.* candelstæf; *candlestick*, WICL. EXOD. xxv. 33; candelstef MK. iv. 21.

     candel-quencher, sb., *candle-extinguisher*, WICL. EX. xxv. 28.

     candel-stikke, sb., *candlestick*, PR. P. 60; TREV. III. 187; kandelstike H. S. 9374.

candeler, chaundeler, sb., *Fr.* chandelier, *mod.Eng.* chandler; *candle-maker*, '*candelarius*,' PR. P. 60, 71; chaundeler WICL. JUDG. iv. 4*.

candrede, sb., *Welsh* cantred, *from* cant (*hundred*); *cantred* : a candrede is a countrai þat conteineþ an hondred townes TREV. I. 343.

cane, sb., *O.Fr.* cane; *cane*, '*canna*,' VOC. 191; can (*slender glass tube*) LIDG. TROY (*in* N. E. D.).

cāne, chāne, sb., *khan*, MAND. 188, 221.

canel, sb., chanel, *O.Fr.* canel, chanel; *canal, channel, pipe*; '*canalis*,' PR. P. 60, 69; kanel (*throat*) GAW. 2298; chanel TREV. I. 133; **canels** (*pl.*) C. M. 1866; canel [chanel] 22577; canel WICL. S. W. II. 335; canels or pipes PALL. i. 464; canel of a bell '*canellus*' PR. P. 60.

     canel-bōn, sb., *collar-bone*, CH. D. BL. 943; ANT. ARTH. xl; kanelbon REL. II. 78.

canele, sb., *O.Fr.*canele (*cannelle*); *cinnamon,* LA3. 17744; canelle MAND. 187; canel P. L. S. xxxv. 76; WICL. EXOD. xxx. 23.

cānen, v., *Ger.* kānen; *become mouldy*; cāned (*pple.*) '*acidus*' CATH. 53.

canevas, sb., *Fr.* canevas; *canvas,* CH. C. T. *G* 939; canvas PR. P. 60.

cang, adj., *cf. Swed.* kång; *foolish,* KATH. 260 A. R. 62; (? *wanton*) 56.

    cang-līche, adv., ? *foolishly,* A. R. 56.

    kang-schipe, sb., ? *folly,* A. R. 338.

cang, sb., ? *fool,* A. R. 270; **kanges** (*pl.*) 362.

cangen, v., ? *become foolish*: we arn cangede (*pple. pl.*) A. R. 362*; *comp.* a-cangen.

cangun, ? sb.,=**cang**: beo he cangun o$\v{c}$er crupel H. M. 33; (*v. r.* cang) A. R. 62*.

canne, sb., *O.E.* canne, = *O.H.G.* channa; *can, drinking vessel*; kan LUD. COV. 259; cannes (*pl.*) WICL. JOHN ii. 6.

canōn, canoun, sb., *Gr.* κανών; *canon, rule,* CH. ASTR. iii. 32 (42); C. M. 26290; APOL. 73; canon WICL. APOC. PROL.; C. M. 21190.

canon, adj. *from* sb.; *canonical*; lawe canoun TREV. II. 117.

canonie, sb., *O.Fr.* canonie, chanoine, *from Lat.* canonicus, *cf. O.N.* kanunkr; *canon,* SAX. CHR. 250; canoun BEK. 386; chanon VOC. 194; chanoun 209; **kanunkes** (*gen.*) ORM. DEDIC. 9; **canones, canunes** (*pl.*) LA3. 21861, 29852; canunes O. & N. 729; chanounes HAV. 360.

canoniel, adj., *Fr.* canonial; *canonical,* A. R. 8.

canonistre, sb., *canonist,* WICL. S. W. I. 32; **canonistres** (*pl.*) LANGL. *A* viii. 135.

canonisid, pple., *catholic*; the seuene epistolis that ben clepid canonisid WICL. JAMES PROL.

canonizacioun, sb., *the act of canonizing,* PR. P. 60; MAND. 176.

canonizen, canonisen, v., *med.Lat.* canonizare; *canonize,* WICL. S. W. II. 387; GOW. I. 254; III. 280.

    canonizing, sb., *canonisation,* WICL. S. W. III. 456.

canopē, sb., *canopy,* PR. P. 60; WICL. JUD. xiii. 10.

cant, adj., *cf. M.Du.* kant; ? *eager, brave*: cant and kene MIN. vii. 107; C. M. 8943; kant & cof MAN. (F.) 1886; mani kant knight HALLIW. 488.

    cant-lī, adv., *eagerly,* MIN. v. 64.

cantaride, sb., *cantharides,* PALL. i. 128.

Cantebrügge, pr. n., *Cambridge,* P. L. S. xiii. 66; Cantebrîgge CH. C. T. *A* 3921.

cantel, sb., *O.Fr.* cantel; *bit, piece cut off,* CH. C. T. *A* 3008; B. DISC. 346; ANT. ARTH. xli; a cantel of brede VOC. 258; kantel SHOR. 33; GENER. 5934.

    cantel-cāpe, sb., *O.E.* cantelcāpa; *cloak,* LA3. 29749.

canticle, sb., *Lat.* canticulum; *canticle,* GEN. & EX. 4124.

cantin, v., *divide, share,* '*partior,*' PR. P. 60.

[Cantware, pr. n. pl., *O.E.* Cantware (*pl.*); *inhabitants of Kent.*]

    Cantware-büri, pr. n., *O.E.* Cantwara burg; *Canterbury,* LA3. 2821; Cantoreburi FRAG. 5; Canterburi BEK. 240.

canunk, *see* **canonie.**

capacitē, sb., *Fr.* capacité, *from Lat.* capacitatem; *capacity,* W. & I. 66.

capcioun, sb., *capture,* WICL. 2 PET. ii. 12.

capcious, adj., *captious,* WICL. S. W. II. 13; capciows BOKEN. 7.

cape [? **cāpe**], sb., *O.Fr.* cape, chape, *cf. O.N.* kāpa; *cape, cope*: he nom ane cape of his ane cnihte LA3. 13097; and nomen tailes of reh3en and hangede on his cape [cope] 29559; cape, cope '*capa*' PR. P. 91; chape of a schethe 69; cope SHOR. 110; 3if he have$\v{d}$ enne widne hod & one ilokene cope A. R. 56; kope HAV. 429; coope WICL. EXOD. xxv. 7; **cōpes** (*pl.*) LANGL. *A* PROL. 53; *comp.* **cantel-cāpe.**

capel [**capil, capul**], sb., *O.N.* kapall, ? *from Ir.* capall, capull, *Gael.* capull, *Lat.* caballus; *horse, nag,* CH. C. T. *A* 4088; capul LANGL. *A* iv. 22 [capel *C* v. 24]; capul PR. P. 61; capel (*hen*) TOWNL. 99.

capen [? **cāpen**], v., *provide with a cape* or *cope*; cōpeþ (*pres.*) LANGL. *A* iii. 138; **chaped** (*pple.*) CH. C. T. *A* 366.

capitain, sb., *O.Fr.* capitain; *captain,* GOW. I. 360; CH. C. T. *H* 230; captein WICL. S. W. III. 360; capitane BARB. viii. 52; **capteins** (*pl.*) WICL. S. W. I. 323; *see* **chevetain.**

capital, adj., *O.Fr.* capital; *capital*: we writeþ capital lettres TREV. I. 129; capitale WINT. VI. xix. 37; capitale ennime (*deadly foe*), BARB. iii. 2; **capitalen** (*pl.*) A. R. 258.

capital, sb., *O.Fr.* chapitel; *capital*: pilers . . . with . . . capitale P. S. L. xxxv. 69; **capitals** (*pl.*) N. E. D.

capitle, *see* **chapitle.**

capitolie [**capitoile, capthole**], sb., *O.Fr.* capitolie; *capitol, citadel,* CH. C. T. *B* 3893; capitole BARB. i. 543.

capoun, sb., *O.E.* capūn, *O.Fr.* capon, chapon; *capon*; capun REL. I. 217; capoun [chapoun] P. S. 334 (P. 43); **capōns** (*pl.*) AYENB. 38; capones LANGL. *B* xv. 466.

capparis, sb., *Lat.* capparis; *caper* (*the plant*); [*ear. vers.* caperis] WICL. ECCLES. xii. 5.

cappe, sb., *O.E.* cæppe, *O.Fris.* cappa, *O.N.* kappa, *O.H.G.* chappa; *cap,* '*pileus,*' PR. P. 60; '*caleptra*' (καλύπτρα) VOC. 182; P. S. 330; MAND. 247; þis manciple sette here aller cappe CH. C. T. *A* 586; keppen (*dat. pl.*) A. R. 420; *comp.* **fleil-, niht-cappe.**

capped, pple., *wearing a cap*: ask thi cappid maistres P. P. II. 107.

cappere, sb., *capmaker,* E. G. 12.

capret, sb., *cf. Ital.* capretto ; *gazelle,* WICL. DEUT. xii. 15.

capricorn, sb., *O.Fr.* capricorn ; *Capricorn,* CH. ASTR. i. 17 (9).

caprifíe, v., *Lat.* caprifīcāre ; *ripen by caprification,* PALL. iv. 592.

caprifig, sb., *wild fig,* PALL. iv. 589.

capstan, sb., *capstan,* A. P. ii. 418.

captive, sb., *captive,* CH. TRO. iii. 333.

captive, v., *make captive,* LIDG. M. P. 38.

captivitē, sb., *O.Fr.* captivité ; *captivity (ms.* captyuide), A. P. ii. 1612.

capul, *see* capel.

car, adj., *Gael.* cearr ; *left (hand)* : þe car honde (*printed* carhonde) ANT. ARTH. (R.) xlviii.

car, *see* ker.

caracte, sb., *O.Fr.* caracte ; *character, letter* ; carecte LANGL. *B* xii. 90 ; caracter, carecter WICL. APOC. xiii. 16, 17 ; carecte GOW. I. 57 ; **carectus** [*v. r.* carrectis] (*pl.*) þat Crist wroot LANGL. *B* xii. 80.

caraine, *see* caroine.

carake, sb., *Fr.* caraque ; *a sort of ship* ; carike CH. C. T. *D* 1688 ; **carackes** (*pl.*) SQ. L. DEG. 819.

caraldes, sb., pl., ?=**caroles** ; *? chains,* A. P. iii. 157.

carawai, sb., = *Ger.* karvei, *from Arab.* karawiȳā ; *caraway,* PR. P. 62.

carbuncle, sb., *O.Fr.* carbuncle, charboucle ; *carbuncle* (*gem*), TREV. II. 235 ; charbucle H. M. 43 ; charbocle CH. C. T. *B* 2061 ; charbugle FL. & BL. 234 ; charbokill D. TROY 3170 ; charboncle MAND. 239 ; charboncle of armes LIDG. BOCH. (*in* N. E. D.).

    charbuncle-stōn, sb., *carbuncle,* FER. 1741.

carcais, *see* carkeis.

carde, sb., *Fr.* carde ; *card* (*for wool*), PR. P. 62.

carde², sb., *O.Fr.* carte ; *playing-card* : **cardes** (*pl.*) CHEST. II. 83.

    cardinge, sb., *playing at cards,* REL. II. 224.

carden, v., *Fr.* carder ; *card* (*wool*) ; cardin PR. P. 62 ; karde LANGL. *C* x. 80 ; **carded** (*pple.*) LANGL. *B* x. 18.

cardiake, sb., *Fr.* (passion) cardiaque ; *heart disease,* HALLIW. 232 ; cardiacle LANGL. *B* xiii. 335 ; CH. C. T. *C* 313.

cardinal, adj., *O.Fr.* cardinal ; *chief, cardinal* : cardenale PR. P. 62 ; **cardinals** (*cardinal virtues*) (*pl.*) C. M. 10008 ; AYENB. 123.

cardinal², sb., *cardinal (ecclesiastic),* LAȝ. 29497.

cardoun, sb., *O.Fr.* cardon ; *thistle, ' cardo,'* VOC. 191.

cardue, sb., *Lat.* carduus ; *thistle,* WICL. 4 KINGS xiv. 9.

†

cáre, sb., *O.E.* cearu, caru, = *O.L.G.* cara, *Goth.* kara, *O.H.G.* chara ; *care,* LK. x. 40 ; LAȝ. 23051 ; H. M. 27 ; ORM. 4852 ; HAV. 2062 ; C. L. 217 ; CH. C. T. *A* 1569 ; PR. C. 7263 ; god sente on him seknesse & care GEN. & EX. 775 ; care and sorwe ROB. 301 ; kare O. & N. 1590 ; WILL. 726 ; kare and howe S. S. (Web.) 1493 ; on kare & on pine HOM. I. 129 ; **cárun,** caren (*dat. pl.*) LK. viii. 14 ; *comp.* mōd-care.

    car-ful, adj., *O.E.* cearful ; *careful,* BEK. 639 ; WILL. 2201 ; chærful LAȝ. 21572 ; **karefulle** (*pl.*) cnihtes 16760.

      carfu(l)-līche, adv., *carefully,* C. L. 203 ; carfulli WILL. 4347.

      carfulnesse, sb., *carefulness,* HOM. I. 115.

    káre-lēas, adj., *O.E.* cearlēas ; *careless,* A. R. 246 ; careles LAȝ. 12478.

carf, sb., ?=**kirf** ; *cut, wound* ; **carffes** (*pl.*) D. ARTH. 2713.

cari, sb., *some rough cloth,* PL. CR. 442.

cariáge, sb., *O.Fr.* cariage ; *carriage, ' vectura,'* PR. P. 62 ; TREV. III. 391 ; WICL. GEN. xlv. 19 ; CH. C. T. *I* 677 ; BARB. viii. 275 ; K. COL. 40.

carie, sb., *? thing of price, treasure* ; *? cartload* : i wol geve þe gimmes and bighes, ten þousand **caries** (*pl.*) ALIS. 6695.

cárien, v., *O.E.* cearian, = *Goth.* karōn, *O.H.G.* charōn ; *care, be anxious,* A. R. 48 ; he wile carien for hire H. M. 5 ; karien HOM. I. 193 ; carie ROB. 312 ; care MIN. viii. 1 ; **cáre** (*pres.*) SPEC. 54 ; carest WILL. 3182 ; careþ P. 28 ; chareð MAT. vi. 34 ; xxi. 34 ; (we) carieþ P. S. 149 ; **cárande** (*pple.*) GAW. 674 ; **cárede** (*pret.*) BEK. 125 ; S. A. L. 158 ; heo careden LEB. JES. 25.

carien, v., *O.Fr.* carier ; *carry,* LANGL. *A* ii. 132 ; E. G. 20 ; carie GOW. II. 293 ; CH. C. T. *A* 130 ; MAND. 49.

    cariare, sb., *carrier, ' vector,'* PR. P. 62 ; carier BARTH. (*in* N. E. D.).

caritēd, sb., *O.Fr.* carité, *Lat.* cāritātem ; *charity,* SAX. CHR. 263 ; kariteþ ORM. 2998 ; charite AYENB. 79 ; CH. C. T. *A* 532 ; cherite HOM. I. 39.

cark, sb., *A.Fr.* carc ; *grief, trouble,* AR. & MER. 3952 ; cark [carke] GAM. 760 ; karke (*dat.*) MAN. (H.) 135.

carkeis, sb., *? O.Fr.* carcois ; *carcase,* PR. P. 62 ; HOCCL. (*in* N. E. D.) ; karkeis WICL. EXOD. xxi. 35 ; carcais PR. C. 874.

carken, v., *O.Fr.* carkier, *mod.Eng.* cark ; *be anxious* ; carke PL. T. 198 ; carke and care SQ. L. DEG. 924 ; **carke** (*pres.*) SPEC. 54 ; JOS. 30 ; **carked** (*pple.*) AR. & MER. 4464.

carken, *see* chargin.

carl, sb., *O.N.* karl, = *O.H.G.* charl, karl (*allied to* cheorl, *q. v.*) ; *husbandman ; man, male ; ' rusticus,'* PR. P. 62 ; HAV. 1789 ;

CH. C. T. *A* 545; C. M. 13808; TOWNL. 213; karl IW. 559; *deriv.* kerling.

carl-man, sb., *O.N.* karl-, karmačr; *male person*; carman C. M. 2937; **carlmen** (*pl.*) and wimmen SAX. CHR. 261.

carlisch, adj., *manly, masculine*; karlische þinges HOM. I. 273.

carme, sb., *Carmelite*; **carmes** (*pl.*) WICL. S. W. III. 353.

carnaciōn, sb., *incarnation*, MISC. 216.

carnal, adj., *O.Fr.* carnel, charnel; *carnal*, LUD. COV. 84; LIDG. M. P. 44.

carnalitē, sb., *carnality*, LUD. COV. 114; GEST. R. (H.¹) 195.

carnel, *see* kernel.

caroine, sb., *O.Fr.* caroigne, charoigne; *carrion, carcase*, ROB. 216; AYENB. 86; LANGL. *B* PROL. 193; carein APOL. 105; careine CH. P. F. 177; carion PR. C. 847; caraine A. P. ii. 459; karin D. TROY 13025; caren 1971; caraing P. S. 203; **charoines** (*pl.*) A. R. 84; careins WICL. GEN. xv. 11.

carole, sb., *O.Fr.* carole; *carol, dance, song; chain*, ROB. 53; S. S. (Wr.) 2885; **caroles** (*pl.*) CH. C. T. *A* 1931.

carolin, v., *O.Fr.* caroler; *carol*, PR. P. 62; carole and singe CH. D. BL. 849; karold (*pret.*) C. M. 7600; karoled (*pple.*) MAN. (F.) 1777.

carp, sb., *cf. O.N.* karp, (*boast*), Swed. karp ('*prat*' RIETZ); *? talk*, A. P. ii. 23, 1327; S. A. L. 83; carpe A. P. i. 882.

carpe, sb., *O.Fr.* carpe, *O.H.G.* charpho; *carp* (*a fish*), '*carpus*,' PR. P. 62.

carpenter, sb., *O.Fr.* carpentier; *carpenter*, ROB. 537; CH. C. T. *A* 3189; **carpenters** (*pl.*) L. H. R. 30; carpentours D. TROY 1597.

carpentrie, sb., *O.Fr.* carpenterie; *carpentry*, WICL. EX. xxxv. 33.

carpin, v., *O.N.* karpa, *mod.Eng.* carp (at); *speak, talk, say,* '*fabulor*,' PR. P. 62; carpe WILL. 4581; JOS. 440; LANGL. *B* xix. 65; CH. C. T. *A* 474; DEGR. 9; LIDG. M. P. 191; to carpe wiþ 3our qwene ANT. ARTH. xi; the king . . . carpis (*pres.*) þes wordes D. ARTH. 639; carpe (*imper.*) HOM. I. 287; carpende (*pple.*) GOW. III. 195; karpede (*pret.*) ISUM. 234; carped GAW. 1088.

carpare, sb., *talker*, PR. P. 62.

carping, sb., *talk*, WILL. 4660; karping IW. 127.

carr, *see* ker.

carre, sb., *O.Fr.* car, char; *car*, '*carrus*,' PR. P. 62; chare '*currus*' 69; carre MAND. 130; chare 175; char, chare CH. C. T. *A* 2138; chaar, chare ['*currum*'] WICL. GEN. xli. 43; carris ['*plaustra*'] NUM. vii. 9.

cart, sb., *? O.E.* cræt; *cf. O.N.* kartr; *cart*, '*rheda*,' PR. P. 62; LEG. 169; CH. C. T. *D* 1539; E. G. 353; kart VOC. 167; cart,

carte LANGL. *A* ii. 154; carte VOC. 181; WICL. 3 KINGS x. 29; karte ORM. PREF. 48; kert '*currus*' FRAG. 4; carte (*dat.*) LA3. 11396; carten (*pl.*) AYENB. 35; cartes MAND. 250; cartes and waines GEN. & EX. 2362; *comp.* dung-cart.

cart-clout, sb., *iron plate to protect the axle-tree*, '*epuscium*,' VOC. 278.

carte-bodi, sb., VOC. 167.

carte-bondes, sb., pl., '*bendes de les roes*' VOC. 167.

cart-full, sb., *cart-ful*, DÉP. R. ii. 158.

cart-hors, sb., *cart-horse*, VOC. 187.

carte-lōde, sb., *cart-load*, HAV. 895.

cart-mare, sb., *cart-mare*, LANGL. *B* vi. 289.

carte-nave, sb., '*timpana*,' VOC. 180.

cart-sadel, sb., '*dorsilollum*,' VOC. 202; REL. I. 81.

carte-staf, sb., *cart-shaft*, ROB. 99; cartstaf GAM. 590.

cart-wei, sb., *cart-way*, LANGL. *A* iii. 127; TREV. I. 63.

cart-wheel, sb., *cart-wheel*, CH. C. T. *D* 2255.

kart-wright, sb., *cart-builder*, '*carpentarius*,' VOC. 194.

carte, sb., *O.Fr.* carte; *? treatise,* GOW. III. 130.

cartin, v., *cart, carry in a cart*, PR. P. 62; cart LANGL. *C* vi. 62.

cartere, sb., *carter*, AYENB. 160; CH. C. T. *D* 1542; S. & C. I. lxi; kartere [cartare] O. & N. 1186; cartare PR. P. 62.

cas, sb., *Fr.* cas; *case, chance*, ROB. 9; CHR. E. 662; AYENB. 36; WILL. 326; C. M. 1407; GEST. R. (H.¹) lii. 230.

casbalde, sb., *a term of reproach*, YORK xxxiv. 194; TOWNL. 123.

cask, adj., *O.N.* kaskr, karskr; *swift, lively*; kaske (*pl.*) HAV. 1841; *see* crask.

casse, sb., *O.Fr.* casse, *cf. mod.Eng.* case, *box*, PR. P. 269; cas CH. C. T. *A* 2080.

cassen, v., *Fr.* casser; *quash*; cassed (*pple.*) E. G. 311.

cassidoin, sb., *O.Fr.* cassidoine; *chalcedony*; **cassidoines** (*pl.*) A. P. ii. 1471.

cast, sb., *O.N.* kast; *cast, throw*, WILL. 4652; MAND. 92; CH. C. T. *A* 3605; FLOR. 1406; WICL. NUM. xxxv. 17; ANT. ARTH. xlviii.

caste, *see* chastien.

castel, sb., *O.E.*, *O.Fr.* castel; *castle*, BEK. 2089; þæt castel LK. xix. 30; þene castel LA3. 191; castel and bur3 me mai iwinne O. & N. 766; kastel ORM. 19941; **castelle** (*dat.*) MAT. xxi. 2; castele AYENB. 121; castele, castle LA3. 14138, 30771; **castles** (*pl.*) SAX. CHR. 251.

castel-büri, sb., *castle precincts*, LA3. 6714.

castel-3at, sb., *castle-gate*, LA3. 18652.

castel-toun, sb., ALIS. 5131.

castel-tour, sb., *castle tower*, TRIST. 158.

castel-wal, sb., *castle wall*, HOM. I. 141; MAN. (F.) 5078.

castel-werk, sb., *castellated work*, WILL. 2220.

castelet, sb., *O.Fr.* castelet, chastelet; *small castle*, S. S. (Web.) 2754; chastelet LANGL. *B* ii. 84.

castellain, sb., *O.Fr.* castellain; *master of a castle*, GOW. I. 184.

castellion, sb., *castellan*, GENER. 3128.

casten, v., *O.N.* kasta; *cast, throw*, SPEC. 37; CH. TRO. iii. 711; to casten him in irens LANGL. *A* iv. 72; kasten KATH. 946; caste GOW. I. 334; WICL. JOHN viii. 59; PERC. 682; anker . . . kaste HORN (H.) 1053; keasten JUL. 67; kesten A. R. 56; kesten in feteres HAV. 1784; keste AYENB. 99; MAN. (H.) 54; casteð (*pres.*) HOM. II. 177; a . . . rote þet kest vele kveade boȝes AYENB. 31; caste (*pret.*) HOM. I. 47; LAȝ. 1919*; ROB. 151; BRD. 8; MAND. 93; kaste KATH. 1360; it kaste a cri MAP 339; keste ISUM. 608; hi casten (*pl.*) heore lot MISC. 50; kesten GAW. 1649; cast (*pple.*) FLOR. 2181; icast HOM. I. 51; BEK. 450; C. L. 807; TRANS. XVIII. 25; ikest A. R. 228; *comp.* a-, bi-, for-, over-, ümbe-casten.

    castere, sb., *caster*, WICL. PROV. xxviii. 7.

    castinge, sb., *casting*, WICL. S. W. I. 351; casting (*vomit*) WICL. 2 PET. ii. 22; kasting WILL. 942.

castigatiōn, sb., *chastisement*, CH. L. S. 26.

castracioun, sb., *castration; cutting out a hive*; PALL. vii. 113; xi. 267.

casuel, adj., *Fr.* casuel, *mod.E.* casual; *accidental*, CH. TRO. iv. 391.

casuel-lī, adv., *casually*, CH. H. F. 679.

casueltē, sb., *Fr.* casualité; *chance*, N. P. 23.

cat, sb., *O.E.* catt, cf. *O.N.* köttr *m.*, *O.Fris.* katte, *O.H.G.* chaza *f.*, *Lat.* catus, cattus, *Ir.*, *Gael.* cat *m.*, *Welsh* cath *f.*; *cat*, A. R. 102; H. M. 37; AYENB. 179; LANGL. *B* PROL. 149; CH. C. T. *D* 350; AR. & MER. 8726; cat, kat O. & N. 810, 831; ofte museþ þe kat after hire moder REL. I. 180; cattes (*gen.*) LANGL. *B* PROL. 179; cattes drit P. S. 240; kattes minte *catmint*, '*nep(e)ta*' REL. I. 37; cattes (*pl.*) MAND. 129.

catapuce, sb., *O.Fr.* catapuce, *from Lat.* catapotium; *a plant*, CH. C. T. *B* 4155.

catecumeling, sb., *catechumen*, LANGL. 6728 (W.); catekumelinges (*pl.*) *B* xi. 77.

catel, sb., *O.Fr.* catel, chatel, *Lat.* capitāle, *mod.Eng.* cattle *and* chattel; *property*; *chattel, cattle*; LAȝ. 30673*; HAV. 225; C. L. 990; AYENB. 36; RICH. 1546; LANGL. *B* vii. 22; S. S. (Wr.) 1210; wiþ heore catel and heore gode TRANS. XVIII. 24; catel hadde þei inough and rente CH. C. T. *A* 373; ćatell

C. M. 6002; chatel HOM. I. 271; chetel A. R. 224 (*v. r.* feh); chateus (*pl.*) ROB. 18.

cateracte, sb., *Fr.* cataracte; *cataract*, TOWNL. 29; cateractes (*pl.*) TOWNL. 32; cateractes LIDG. (*in* N. E. D.).

caterwrawed, v., *caterwauling*: gon a caterwrawed [wawed] CH. C. T. *D* 354.

cathedral, adj., *O.Fr.* cathedral; *cathedral*, ROB. 282.

catirpel, sb., *caterpillar*, PR. P. 63.

catour, *see* achatour.

caucē, sb., *O.Fr.* caucié, chaucié; *causeway, paved road*, PR. P. 64; kauce MAN. (H.) 183.

    caucĕ-wei, sb., *causeway*, PR. P. 64.

cauciōn, sb., *Fr.* caution; *caution (bail, security, pledge)*, ROB. 506; caucioun WICL. LK. xvi. 6; kaucion ALIS. 2811.

caudel, sb., *O.Fr.* caudel, chaudel; *caudle*, ROB. 561; cawdel LANGL. *A* v. 205.

caudrōn, sb., *O.Fr.* cauderon, chauderon; *caldron*, BRD. 17; LEG. 41; TREV. V. 323.

cauken, v., *O.Fr.* cauquer (*trample*); *tread, copulate as a cock*, LANGL. *B* xii. 229; caukede (*pret.*) xi. 350.

caul, sb., *O.E.* caul, cawl, = *O.N.* kāl, *O.H.G.* chōl, cōl; *from Lat.* caulis; *cabbage*, FRAG. 3; kaul VOC. 141; coul MIR. PL. 8; kal REL. I. 52; cal [col] C. M. 12526; col L. C. C. 48; cool PALL. i. 879.

    caul-, cōl-plante, sb., *cabbage*; kōle-plantes (*pl.*) LANGL. *B* vi. 288.

    cāl-stok, sb., *O.N.* kālstokkr; *cabbage-stem*, VOC. 190; PR. P. 58.

cauri-mauri, sb., *a sort of coarse stuff*, LANGL. *A* i. 62; *see* cari.

cause, sb., *Fr.* cause; *cause*, A. R. 316; AYENB. 224; CH. C. T. *A* 419; bī cause, *because* 174; be cause MAND. 15, 46.

    cause-lēs, adj., *causeless*, CH. TRO. iv. 1448; GENER. 724; causelesse CH. AN. 229.

causen, v., *Fr.* causer; *cause*; causeþ (*pres.*) GOW. I. 262; R. R. 4235.

causer, sb., *one who causes*, M. GOD 12.

cautēle, sb., *Fr.* cautele; *cunning*, PR. P. 64; GOW. III. 140; cautēlis (*pl.*) WICL. S. W. I. 6; cautelis GEST. R. (H.[1]) 123.

cautelous, adj., *cautious, crafty*; cautelouse WICL. JOB v. 13; cautellous WICL. S. W. I. 223.

cáve, sb., *Fr.* cave; *cave*, GEN. & EX. 1137; WILL. 25; CH. C. T. *B* 500.

cavel, *see* kevel.

cavern, sb., *O.Fr.* caverne; *cavern*, CH. BOET. iii. 9 (82).

cavillacioun, sb., *O.Fr.* cavillacion; *act of cavilling*, CH. C. T. *D* 2136.

cedre, sb., *Fr.* cedre; *cedar*, AYENB. 131; C. M. 1377.

ceint, sb., *O.Fr.* ceint, *Lat.* cinctus; *girdle*, CH. C. T. *A* 329; saint IW. 1772.

celebrable, adj., *O.Fr.* celebrable; *celebrated, famous*, CH. BOET. iii. 9 (84).

celestial, adj., *O.Fr.* celestial; *celestial, heavenly*, GOW. III. 301.

celidoine, sb., *O.Fr.* celidoine; *celandine*, SPEC. 26.

cēlin, v., *ceil (a room)*; *cover (a wall)*, '*celo*,' PR. P. 65.

  cēlinge, sb., *ceiling*, FER. 1330.

celle, sb., *O.Fr.* celle; *cell*, P. L. S. ix. 60 ; PL. CR. 739; celle [selle] CH. C. T. *A* 172; **celles** (*pl.*) A. R. 152; cellen AYENB. 267.

celler, sb., *O.Fr.* celier, *Lat.* cellārium; *cellar*, GOW. III. 12; celer PR. P. 65; celere A. R. 214; **selers** (*pl.*) WICL. PS. cxliii. 13.

cellerer, sb., *Fr.* cellerier; *cellarer*, REL. I. 273.

celsitūde, sb., *O.Fr.* celsitude; *loftiness*, CT. LOVE 610.

celūre, sb., *canopy, screen*; selure, PL. CR. 201; GAW. 76; a celour or a coverlit, '*cela-torium*' VOC. (W.W.) 571; selour E. W. 76; silour 36; silure '*celatura*' PR. P. 456.

cementing, sb., *binding together*, CH. C. T. *G* 817.

cempe, *see* kempe.

cendal, sb., *O.Fr.* cendal, sendal; *a rich stuff*, MISC. 43; sendal P. L. S. i. 11; CH. C. T. *A* 440.

cēne, sb., *O.Fr.* çaine; *the Lord's Supper*, WICL. APOC. PROL.

cenefectorie, adj., *Lat.* scēnifactōria (ars); cenefectorie craft (*tent-making trade*), WICL. DEEDS xviii. 3.

cenith, senith, sb., *O.Fr.* cenith; *zenith*, CH. ASTR. i. 19 (11).

cense, sb., *for* encens; *incense*, WICL. LEV. ii. 1.

cense [2], sb., *O.Fr.* cense, *Lat.* census; *tribute*; ? cence (*printed* tence) ALIS. 3025.

censen, sensen, v., *offer incense*, PR. P. 66; **sensinge** (*pple.*) CH. C. T. *A* 3341.

  censinge, sb., *the act of offering incense*, PR. P. 66.

censer, sb., *for* encenser; *O.Fr.* encensier; *censer*; censer [sencer, senser, sensure] CH. C. T. *A* 3340.

cent, sb., *O.Fr.* cent, *Lat.* centum; *hundred*, OCTOV. (W.) 1463.

centre, sb., *Fr.* centre; *centre*, GOW. III. 92; CH. ASTR. i. 21 (12).

centure, sb., *O.Fr.* centoire; *centaury (a herb)*, [*v. r.* centaure, sentaurie] CH. C. T. *B* 4153.

cepter, ceptre, sb., *O.Fr.* sceptre; *sceptre*, PR. P. 66; ALIS. 6716.

cerchen, v., *O.Fr.* cerchier; *search*; cerche LIDG. M. P. 159; serche D. TROY 1537; seer-gin PR. P. 453; serchit (*pret.*) D. TROY 1533; cerched (*pple.*) MAND. 315.

  ceerchinge, sb., *searching*, PR. P. 67.

cercle, sb., *Fr.* cercle; *circle*, L. H. R. 28; GAW. 615; TREV. V. 171.

cerclen, serclen, v., *O.Fr.* cercler, *from Lat.* circulāre; *encircle*, CH. TRO. iii. 1717.

cerge, sb., *O.Fr.* cerge; *wax candle*, ED. 1274; **cerges** (*pl.*) HAV. 594; *see* serge.

cerial, adj., *Ital.* cereale, *from* cerro *evergreen oak* : ook cerial *evergreen oak*, CH. C. T. *A* 2290.

cerimoin, ceremoin, cerimonie, sb., *O.Fr.* cerimonie; *ceremony*, WICL. DEUT. xi. 32.

cert, adv., *O.Fr.* cert; *certainly*, ALIS. 5802.

certein, adj. & adv., *Fr.* certain; *certain*, ROB. 52; MAN. (F.) 4875.

  certain-līche, adv., *certainly*, LEG. 47.

certeintē, *O.Fr.* certaineté; *certainty*, MAN. (H.) 278.

certes, adv., *Fr.* certes; *certainly*, O. & N. 1769; FL. & BL. 523; P. L. S. xvii. 441; WILL. 732.

certifie, v., *Fr.* certifier; *certify*, PR. C. 6546; we ... certifie (*pres.*) GOW. I. 192.

cerūce [serūce], sb., *Fr.* ceruse; *ceruse (white lead)*, CH. C. T. *A* 630.

cessaciōn, sb., *O.Fr.* cessation; *cessation*, LUD. COV. 107.

cessen, cēsen, v., *O.Fr.* cesser; *cease*, LANGL. *A* ii. 122; cesse CH. C. T. *B* 1066; cese TRANS. XIX. 329; sese WILL. 1516; **sēsed** (*pple.*) GAW. 1.

  cessinge [cēsinge], sb., *discontinuing*, WICL. LEV. xxiii. 2.

cestred, cestered, adj., *darkened*, PS. cxxxviii. 12; lxxiii. 20.

cete, cethe-grande, sb., *Lat.* cētus, *pl.* cētē (grandia); *whale*, MISC. 16.

cetewale, cetuale, *see* zedewal.

ceðen, *see* cüðð̄e.

chaalami, *see* calami.   **cháce**, *see* cache.

chaceable, adj., *fit to be hunted*, GOW. II. 169.

chácen, chácinge, *see* cacchen.

chaceour, chasur, sb., *hunter (horse)*, P. L. S. ii. 109.

chaere, sb., *O.Fr.* chaere, *Lat.* cathedra, *Gr.* καθέδρα; *chair*, HORN (L.) 1261; ROB. 321; chaire P. L. S. xvii. 256; chaiere HORN (R.) 1271; C. M. 9954.

chaf, sb., *O.E.* ceaf, cef, = *M.L.G.*, *M.Du.* kaf; *chaff*, LAȜ. 29256; ORM. 1528; GEN. & EX. 2889; chaf [caf] C. M. 4751; chef HOM. I. 85; A. R. 270; þet chef AYENB. 62; caf PR. C. 3148; APOL. 54; **cheve** (*dat.*) HOM. I. 85; AYENB. 210; þa chefu ['*paleas*'] MAT. iii. 12.

  chaf-finch, sb., *chaffinch*, PR. P. 68.

cháfen, *see* chaufen.

chaffare, *see* chēapfare *under* chēap.

chaft, sb., *O.N.* kiaptr; *jaw*, '*maxilla*,' CATH. 57; **chaftis** (*pl.*) ANT. ARTH. (M.) xi.

  chaft-bān, sb., C. M. 1073.

chaine, sb., *O.Fr.* chaene; *chain*; cheine MAN. (F.) 8661; **chaines** (*pl.*) AYENB. 264.

chainin, v., *chain*; ?cheenin PR. P. 72; cheine (*pres.*) LANGL. *C* xxi. 287.

chaire, *see* **chaere**.

chaisel, sb., *O.Fr.* chaisel, chainsil; *linen, linen garment,* ALIS. 279; S. S. (Web.) 1814; cheisil LA3. 23761; MISC. 51.

chalami, *see* **calami**.

chalandre, sb., *Fr.* calandre; *sort of lark,* P. L. S. xxxv. 97.

cháld, *see* **cáld**.

chalenge, sb., *O.Fr.* chalonge, calenge; *challenge, claim, injustice,* AYENB. 34; calenge SHOR. 131.

chalengeable, adj., *that may be challenged,* LANGL. *C* xiv. 117.

chalengen, v., *O.Fr.* chalengier, calengier; *challenge; accuse; claim*: challenge ['*calumniari*'] WICL. I PARAL. xvi. 21; CH. C. T. *D* 1200; chalange M. H. 3; PR. C. 2253; calangi ROB. 451; **kalengest** (*pres.*) A.R. 54; calengeþ AYENB. 43; **chalanged** (*pple.*) LANGL. *B* v. 174.

chalanger, sb., *challenger, accuser,* REL. I. 8; WICL. PS. lxxi. 4; **chalengeris** (*pl.*) JOB xxxv. 9.

chalenginge, sb., *challenging, contradiction,* WICL. GEN. xliii. 18; LANGL. *B* xv. 338; chalanginge (*accusing*) *B* v. 88.

chalf, *see* **calf**. **chalice**, *see* **calch**.

chalk, sb., *O.E.* cealc, = *O.H.G.* calc, chalc, *from Lat.* calx; *chalk,* CH. C. T. *G* 806; calk, chalk, *calx, ' creta'* PR. P. 58.

> **chalk-whit**, adj., *white like chalk,* GAW. 798.

> **chalc-stōn**, sb., *O.E.* cealcstān; *chalkstone,* FRAG. 4; chalkston CH. C. T. *G* 1207.

chalōn, sb., *O.Fr.* chalon, *cf. mod.Eng.* shalloon; *a sort of woollen stuff, ' tapetum,'* VOC. 178; chalun '*toral*' PR. P. 68; chaloun E. G. 351; **chalōns** (*pl.*) CH. C. T. *A* 4140.

champaine, sb., *O.Fr.* champaigne; *plain, open country,* DREAM 2064.

champartie, sb., *O.Fr.* champart; *partnership,* CH. C. T. *A* 1949; LIDG. M. P. 131; champertie E. G. 400.

chamberling, sb., *cf. O.Fr.* chamberlan, *O.H.G.* chamarlinc; *chamberlain*; chamberling, A. R. 410; chamberlein ROB. 390.

chambre, sb., *O.Fr.* chambre; *chamber,* BEK. 452; MAND. 24; chaumbre A. R. 92; chombre AYENB. 215.

> **chamber-dore**, sb., *chamber door,* IW. 63; chambirdore ISUM. 652.

> **chaumber-wouh**, sb., *chamber-wall,* JOS. 204.

chambrēre, sb., *Fr.* chambriere; *chamberwoman,* MAND. 102.

chamēl, *see* **camēl**.

chamlit, sb., *some eastern fabric,* N. E. D.; chamelot QUAIR clvii.

champiōn, **campiōn**, sb., *O.Fr.* champion; *champion,* PR. P. 69; champioun CHR. E. 49; campion C. L. 906; **champiūns** (*pl.*) A. R. 236.

chance, *see* **cheance**.

chancel, sb., *Fr.* chancel; *chancel*; chauncel PR. P. 71.

chanceler, sb., *Fr.* chancelier; *chancellor,* P. L. S. xvii. 240; canceler SAX. CHR. 251.

chancelerie, sb., *Fr.* chancellerie; *chancellorship,* BEK. 359; chauncelrie LANGL. *B* PROL. 93.

chancerie, **chauncerie**, sb., *cf.* **chancelerie**; *chancery,* LIDG. M. P. 104.

chandeler, *see* **candeler**.

chanel, *see* **canel**.

change, sb., *O.Fr.* change; *change*; chaunge REL. I. 268; chonge AYENB. 104.

changeable, adj., *changeable,* TREV. II. 201; PR. C. 1473.

changen, v., *O.Fr.* changier; *change*; chaungen A. R. 6; H. M. 7; chaungi HORN (L.) 1052; changeþ (*pres.*) TREAT. 139; changede (*pret.*) MARH. 3; LA3. 3791*.

> **changinge**, sb., *change*; PEN. PS. 38; chaunginge A. R. 6; TREV. II. 201.

changeour, sb., *Fr.* changeur; *changer,* REL. I. 7; **chaungeris** (*pl.*) WICL. MAT. xxi. 12*.

chanoun, *see* **canonie**.

chantable, adj., *worthy to be sung,* WICL. PS. cxviii. 54.

chanten, v., *Fr.* chanter; *chant*; **chaunteþ** (*pres.*) CH. C. T. *A* 3367.

chantement, sb., *enchantment,* ROB. 243; B. DISC. 1900; WILL. 653.

chanterie, sb., *O.Fr.* chanterie; *chantry*; chaunterie CH. C. T. *A* 510.

chantūr, sb., *O.Fr.* chanteur; *chanter*; chauntur S. A. L. 60; chauntour TREV. II. 349.

chape, *see* **cape**.

chapelein, sb., *O.Fr.* chapelain; *chaplain,* BEK. 961; LANGL. *C* xiv. 127.

chapelet, sb., *O.Fr.* chapelet; *chaplet,* PR. P. 69.

chapelle, sb., *O.Fr.* chapelle, chapele; *chapel,* GAW. 2186; chapele MARH. 20; ROB. 473; AYENB. 56.

chapitle, sb., *O.Fr.* capitle, chapitre; *chapter; summary,* P. S. 254; AYENB. 220; capitle, chapitle PR. P. 61; capitle WICL. HEB. viii. 1; chapitre P. L. S. xvii. 435; CH. ASTR. ii. 9 (22); **cheapitres** (*pl.*) A. R. 14; chapitles P. S. 332; capiteles AYENB. 43.

> **chapitel-, chapitre-hous**, sb., *chapterhouse,* LANGL. *B* v. 174.

chappen, v., = *M.Du.* kappen; *split, chap (the hands)*; **chappid** (*pple.*) RICH. 4550; TOWNL. 98.

char, chare, *see* carre.

charbucle, *see* carbuncle.

char-cole, sb., *charcoal, 'carbo,'* PR. P. 69; GAW. 875.

charen, *see* cherren.

charge, sb., *O.Fr.* charge, carche; *charge,* A. R. 140; ROB. 416; charge, charche TREV. VII. 405; charche E. G. 358; charges (*pl.*) MAND. 302.

chargeaunt, adj., *O.Fr.* chargeant; *burdensome,* CH. C. T. *I* 692.

chargeous, adj., *burdensome,* CH. C. T. *ed.* Morris, p. 160 (*B* 2433 *has* chargeant); WICL. 2 COR. xi. 9; charjous WICL. PROV. xxvii. 3.

chargeour, sb., *charger (large dish),* H. R. 136; L. C. C. 21; B. B. 142.

chargin, v., *O.Fr.* chargier, carchier, carquier; *charge,* PR. P. 69; chargeþ (*pres.*) AYENB. 97; chargede (*pret.*) BRD. 11; charged (*pple.*) PR. C. 2949; charged, karked C. M. 8253; icharged A. R. 204; icarked AYENB. 138.

charien, *see* carien.

charieter, sb., *charioteer,* WICL. 3 KINGS xxii. 34; 4 KINGS ii. 12.

cháriʒ, adj., *O.E.* cearig, = *O.H.G.* charag, (*? mod.Eng.* chary); *from* cáre; *sad, sorrowful*: turtle ledeþ chariʒ lif ORM. 1274.

chariot, sb., *Fr.* chariot; *chariot,* TREV. III. 391.

charitable, adj., *Fr.* charitable; *charitable,* CH. C. T. *A* 143.

charitē, *see* caritēd.

charitous, adj., *full of charity,* GOW. I. 172.

charkin, cherkin, v., *O.E.* cearcian (?= cracian); *creak, 'strido,'* PR. P. 70, 76; charke GOW. II. 102; WICL. AMOS ii. 13.

charme, sb., *O.Fr.* charme; *charm,* LANGL. *B* xiii. 342; charmes (*pl.*) AYENB. 43.

charmin, v., *O.Fr.* charmer; *charm, 'incanto, fascino,'* PR. P. 70.

charmere, sb., *charmer,* AYENB. 257.

charmeresse, sb., *enchantress, witch,* AYENB. 19; charmeresses (*pl.*) CH. H. F. 1261.

charming, sb., *enchantment,* ALIS. 404.

charnel, sb., *O.Fr.* carnel; *charnel-house,* LANGL. *B* vi. 60.

charre, *see* cherre.

charrei, sb., *O.Fr.* carrei, charei; *vehicles, carriages,* ALIS. 5096.

chartre, sb., *O.Fr.* chartre; *charter,* HAV. 676; LANGL. *B* xi. 122; cartre ROB. 77.

chartre, sb., *O.Fr.* chartre; *prison,* GEN. & EX. 2043.

chaste, adj., *O.Fr.* chaste; *chaste,* A. R. 368; HAV. 288.

chast-hēd, sb., *chastity,* GEN. & EX. 2022; chasthede AYENB. 230.

chaste-līche, adv., *chastely,* AYENB. 239.

chásteine, chestein(e), sb., *O.Fr.* chastaigne; *chesnut,* CH. C. T. *A* 2922; chesteine PR. P. 73; chesteines (*pl.*) MAND. 307.

kestein-tree, sb., *chestnut tree,* WICL. IS. xliv. 14.

chastetē, sb., *O.Fr.* chasteté; *chastity,* A. R. 6; AYENB. 159; chastite MISC. 227.

chastiement, sb., *O.Fr.* chastiement; *chastisement,* A. R. 72; chastisement AYENB. 17.

chastien, v., *O.Fr.* chastier, *from Lat.* castigare; *chasten, chastise,* REL. I. 131 (HOM. II. 11); chasten A. R. 218; chastie LEG. 16; chasti ROB. 428; AYENB. 8; PR. C. 3549; chasten, chastiʒen (? chastizen), chastisen LANGL. *B* v. 34; chastizin PR. P. 70; chasteþ (*pres.*) AYENB. 17; castede (*pret.*) HOM. II. 137; chastised WILL. 54.

chastinge, sb., *chastisement,* AYENB. 68; LANGL. *B* iv. 117.

chastísinge, sb., *chastising,* PR. P. 70; GOW. II. 44; WICL. LEV. xxvi. 18.

chastísen, *see* chastien.

chasur, *see* chaceour.

chat, sb., *Fr.* chat; *catkin*: þe chattes (*pl.*) of hasele MAND. 168.

chatel, *see* catel.

chatere, sb., *chatter (dat.),* O. & N. 284.

chaterin, v., *chatter, 'garrio,'* PR. P. 70; cheateren A. R. 152; chatre LIDG. M. P. 150; chaterest (*pres.*) O. & N. 322; chetereþ TREV. I. 239;) chatere (*subj.*) LANGL. *C* xvii. 69; chatre LANGL. *B* xiv. 226; chetering (*pple.*) H. S. 359.

chaterer, sb., *chatterer,* CATH. 60.

chaterestre, sb., (*female*) *chatterer,* O. & N. 655.

chateringe [chatering], sb., *chattering,* O. & N. 576, 744.

chaud, adj., *O.Fr.* chaud; *hot,* LANGL. *B* vi. 313.

chauncelrie, *see* chancelerie.

chauncerie, *see* chancerie.

chaufen, v., *O.Fr.* chaufer; *chafe, make warm, become warm,* ANT. ARTH. xxxv; chaufe PARTEN. 224; chaffin 'calefacio' PR. P. 68; chaufed (*pple.*) WICL. IS. xliv. 15.

chaufur, sb., *chafer,* E. W. 101; chafur 46; chafour 'calefactorium' PR. P. 68; W. & I. 23.

chaul, chaulen, *see* chavel, chavelen.

chavel, sb., *O.E.* ceafl, ?= *O.L.G.* kafal (*dat. pl.* kaflun), *M.L.G.* kavel; *mod.Eng.* jowl, *for* chowl; *jaw,* TRIST. 1468; chavil IW. 1991; chaul ALIS.[2] 1119; WICL. 1 KINGS xvii. 35; chewil 'mandibulum' VOC. 187; choule (*dat.*) AUD. 77; chole PL. CR. 224; þen chin him of swipte mid alle þan chevele LAʒ. 26056; chæfles [choules] (*pl.*) LAʒ. 6507; chaveles REL. I. 220; chavelis [chaulis] C. M. 7510; chaules MAP 338.

chavil-bōn, chaulbōn, sb., *jawbone*; '*mandibula*,' PR. P. 70; chavilbon LUD. COV. 37.

chavelen, v., *cf.* L.Ger. kavelen, *Du.* kevelen; *chatter*; chaule P. S. 240; chefleð, cheofleð (*pres.*) A. R. 70, 128.

chavelinge, sb., *chattering, gabble*: mid chavling(e) and mid chatere O. & N. 284; chevelunge A. R. 100.

cheafer, sb., *O.E.* ceafor,= *M.Du.* kever, *O.H.G.* chevar, kever; *cockchafer*; **cheaffers** (*pl.*) ['*scarabaei*'] TREV. II. 211.

cheafle, sb., *from* chavelen; *? chatter, gabble*, A. R. 72.

chēake, sb., *O.E.* cēace, cēoce, = *O.Fris.* tziake, ziake, *M.L.G.* kēke, kāke, *M.Du.* cāke *f.*, *Swed.* kēk *m.* (*cheek, jaw*); *cheek*, AYENB. 248; his overe cheoke MARG. 159; cheke '*mala, gena*' PR. P. 72; '*gena*' VOC. 185; '*joue*' 145; ['*maxillam*'] WICL. JUDG. XV. 15; SPEC. 34; GOW. II. 14; choke L. H. R. 218; **chēoken** (*pl.*) A. R. 70; chekes LANGL. *A* iv. 37; chekes, cheekes CH. C. T. *A* 633; chekes, chokes MAN. (F.) 1820; **chēken** (*dat. pl.*) HOM. II. 73; AN. LIT. 11.

chēk-bōn, sb., *O.E.* cēacbān; *cheek-bone*, FER. 5650; cheekboon ['*maxillam*'] WICL. JUDG. XV. 15 : chekebon PR. P. 72.

cheance, sb., *O.Fr.* cheance; *chance*, ROB. 14; cheaunce SHOR. 60; chaunce PR. C. 3768; god ȝeve þe good chaunce CH. C. T. *G* 593.

chēap, sb., *O.E.* cēap, cēp, *cf.* O.L.G. cōp, *O.Fris.* kāp *m.*, *O.N.* kaup *n.*, *O.H.G.* chouf *m.*; *mod.Eng.* cheap; *purchase, bargain*: hire cheap wes þe w(u)rse LAȝ. 385 ; no mihtest þu þurh nen(n)e chep finde neouwer na bred 31799 ; deore cheap hefdes tu on me HOM. I. 281 ; so liht cheap A. R. 398 ; begge as guod cheap ase me mai AYENB. 44 ; þet is þe wunderlukeste chep (*bargain*) þet eni mon efre funde HOM. I. 163 ; cheep '*pretium*' PR. P. 72; god chep of corn P. S. 341 ; good chep (*bon marché*) or dere GOW. II. 169 ; light chep TOWNL. 102 ; gret chep (*grand marché*) CH. C. T. *D* 523 ; MAND. 208 ; bien gret chep (*for god chep*) WICL. E. W. 185 ; in chepe '*in foro*' TREV. VIII. 319.

chĕap-fare, sb., *mod.Eng.* chaffer; *merchandise, trade*, AYENB. 36; cheffare A. R. 310; MISC. 40; HAV. 1657; P. L. S. xxi. 67; chaffare C. L. 1112; LANGL. *A* PROL. 31; ['*negotiatio*'] WICL. IS. xxiii. 3.

chĕp-fari, v., *chaffer, trade*, AYENB. 162; chaffare CH. C. T. *B* 139.

chĕp-mon, sb., *O.E.* cēap-, cēp-, cȳpman, = *O.H.G.* choufman ; *chapman, merchant*, A. R. 208 ; chep-, chapmon LAȝ. 30681, 30690; chapmon O. & N. 1575; chapman '*mercator*' PR. P. 69 ; CH. C. T. *A* 397 ; **chĕpmen** (*pl.*) LAȝ. 11356; chepmen R. S. vii; chep-, chapmen MISC. 39, 76 ; chapmen HAV. 51 ; (*ms.* chappmenn) ORM. 15783.

chăpman-hōd, sb., '*mercatus*,' PR. P. 69 ; chapmanhode CH. C. T. *B* 143.

chĕap-scamel, sb., *O.E.* cēapsceamul; *counter*, '*teloneum*,' LK. v. 27.

chēapien, v., *O.E.* cēapian, = *O.L.G.* cōpōn, *O.Fris.* kāpia, *O.N.* kaupa, *Goth.* kaupōn, *O. H.G.* chaufan, choufōn ; *purchase, buy*: wið chatel mon mai luve cheape HOM. I. 271 ; ch(e)api [chepin] ... & sullen JUL. 63 ; chepen GAW. 1271 ; chepin '*licitor*' PR. P. 72 ; chepe CH. C. T. *D* 268; **chēpe** (*pres.*) P. S. 159; hwon he vor so liht wurð . . . cheapeð þine soule A. R. 290 ; heo cheapeð (*sells*) hire soule þe chepmon of helle 418; **chēpede** (*pret.*) OCTOV. (W.) 389 ; cheped LANGL. *B* xiii. 380; cheapeden MAT. xxi. 12 ; *cf.* **chēpen** *and* **coupen**.

chēapild, adj., *fond of bargaining*, A. R. 418.

chēpinge, sb., *O.E.* cēping, cēaping, cēapung; *act of buying; market;* '*forum*,' HOM. II.211; LAȝ. 23413; WILL. 1822; LANGL. *A* iv. 43 ; cheping JUL. 52; cheepinge ['*foro*'] WICL. MAT. xi. 16.

chēping-bōþe, sb., *market-booth*, ORM. 15573.

chēpeing-toun, sb., *market town*, AM. & AMIL. 1700, 1816.

chēaste, sb., *O.E.* cēast, *cf.* O.Fris. kāse, *O.H.G.* kōsa ; *dispute, contention, strife*: þe vormest(e) is cheaste oðer strif A. R. 200 ; cheast HOM. II. 163 ; cheaste, chiaste AYENB. 65, 67 ; cheste HOM. I. 111 ; SHOR. 113 ; OCTOV. (W.) 754 ; þeos cheste O. & N. 177 ; cheste and sake 1160 ; wiþouten cheste RICH. 5143; S. S. (Wr.) 1638 ; cheeste LANGL. *B* xiii. 109; strif and cheste [cheeste] CH. C. T. *I* 556 ; chost P. S. 151 ; MIRC 338.

chec, *see* **chek**.

chéf, sb. & adj., *O.Fr.* chef, chief ; *chief*, ROB. 74, 212; MAN. (F.) 940; chief BEK. 1003.

chief-citē, sb., *chief city*, P. L. S. xiii. 42.

cheef-mete, sb., *? chief meat*, LANGL. *A* vii. 281.

chéf-stīward, sb., *chief steward*, WILL. 3841.

chef, *see* **chaf**.

chĕffare, *see* **chĕapfare**, *under* **chēap**.

cheine, *see* **chaine**.   **cheisil**, *see* **chaisel**.

chek, sb., *O.Fr.* eschec ; *check*, PR. P. 71 ; **chek mat** *check mate* S. A. L. 213 ; CH. TRO. ii. 754 ; **chekkes** (*pl.*) *defeats* TREV. III. 231 ; **ches** [*O.Fr.* eschecs] *chess*, CH. D. BL. 619 ; plaie ate ches AYENB. 52.

chēke, *see* **chēake**.

chékeful, -lew, *see under* **chéoke**.

cheker, v., *O.Fr.* eschekier ; *chess-board; exchequer* ; AYENB. 45; escheker FL. & BL. 345; esscheker [*v. r.* cheker] LANGL. *A* iv. 26.

chekered, adj., *checkered*, PR. P. 72.

chekkin, v., *check*, PR. P. 72 ; cheke (*pres.*) LANGL. *C* xxi. 287.

chélde, see **kélde.**   **chele,** see **cheole.**

chele, sb., *O.E.* cele, ciele, cyle, cile *m.; cold, coolness,* HOM. I. 33; LA3. 30811; ORM. 1615; MISC. 73; ROB. 7; SPEC. 110; AYENB. 75; P. S. 256·; GOW. II. 369; LANGL. *A* i. 23; TREV. I. 135; now͜er heate ne chele KATH. 1701; in þe norþhalf . . . no man ne woneþ for chele TREAT. 137; chele, chule, P. L. S. viii. 100, 118; chile BRD. 3.

chelle, sb., *O.E.* cille, cylle, = *O.H.G.* chella; *censer,* HOM. I. 193.

chemenē, chimenee, sb., *O.Fr.* cheminee; *chimney, furnace,* CH. C. T. *A* 3776; chimenee LANGL. (Wr.) 5803; chimnei WICL. MAT. xiii. 42.

chemise, sb., *O.Fr.* chemise; *chemise,* REL. I. 129 (HOM. II. 162).

chéne, see **chíne.**

[**chéoke-, chóke-,** *stem of* chéoken.]

    chéke-, chókeful, adj., *choke-full,* D. ARTH. 1552, 3604.

    chéke-, -chóke-lew, adj., *apt to choke*: stelthe is medid with a cheklew [*v. r.* chokelew, chekelew] bane HALLIW. 243.

chéoken, v., *choke*; chekin '*suffoco*' PR. P. 72; cheke H. S. 3192; **chóke** (*pres. subj.*) TOWNL. 91; **chókede** (*pret.*) TREV. IV. 139; *comp.* **a-chéoken.**

cheole, sb., *O.E.* ceole, = *M.Du.* kele, *O.H.G.* chela; *throat*: martres cheole P. L. S. viii. 182; herte him . . . on þe chel(e) FER. 3194.

cheorl, sb., *O.E.* ceorl, = *Ger.* kerl, *mod. Eng.* churl; *husbandman, rustic, churl, fellow,* FRAG. 2; LA3. 4260; MISC. 108; þene cheorl A. R. 86; clipe þinne cheorl (*husband*) JOHN iv. 16; cherl ['*rusticus*'] WICL. WISD. xvii. 16; ORM. 14788; GEN. & EX. 2715; AYENB. 76; ALIS. 5598; WILL. 4; IW. 268; CH. C. T. *D* 2182; þe cherl is def SPEC. 111; a gret cherl LANGL. *A* v. 204; cherl, chorl PL. CR. 221; chorl LIDG. M. P. 182; **cheorles** [chorles], cherles (*gen.*) O. & N. 512, 1494; **cheorles** (*pl.*) JOHN iv. 18; þa eorles & þa sweines & þa cheorles LA3. 11904; chorles A. P. ii. 1258; cherles ['*servi*'] TREV. III. 415; þine cherles þine hine HAV. 620; **cheorlen** (*dat. pl.*) LA3. 21536; *cf.* **carl.**

    cherl-līch, adj., *belonging to a husbandman or rustic*; cherl(l)īche travel WICL. ECCLUS. xxvii. 7.

    cheorlisch, adj., *O.E.* ceorlisc; *churlish, rustic*; cherlish WICL. ECCLUS. vii. 16.

chēosen, v., *O.E.* cēosan, = *O.L.G.* kiosan, kiesan, keosan, *O.N.* kiōsa, *O.Fris.* kiasa, tziesa, *Goth.* kiusan, *O.H.G.* chiusan, chiosan; *choose, distinguish,* LA3. 15147; KATH. 1894; A. R. 242; SPEC. 62; cheose O. & N. 1343; HORN (R.) 666; ALIS. 1659; chiesen HOM. I. 219; chiese AYENB. 86; cesen SAX. CHR. 250; chesen ORM. 9218; GEN. & EX. 3429; M. T. 92; þat men mouthe . . . a peni chesen

(*discern*) HAV. 2147; che̦sin PR. P. 73; chese LANGL. *B* xv. 38; PR. C. 79; Charles het Richard chuse his kni3tes FER. 4367; **chēose͡ð** (*pres.*) H. M. 39; chiest AYENB. 126; chest SHOR. 109; (hi) chieseþ AYENB. 45; **chēos** (*imperat.*) A. R. 102; ches [chees] CH. C. T. *A* 1595; chus MARG. 103; **chēas** (*pret.*) JOHN vi. 70; H. M. 15; AYENB. 77; chæs LA3. 12175; ORM. 2541; ches MAND. 2; PR. C. 2132; PERC. 1207; AMAD. (R.) xlvi; and his weig ͜eðen ches (*took*) GEN. & EX. 2736; such strengþe he him þo ches C. L. 1317; chalkwhit(e) chimnees þer ches (*discerned*) he ino3e GAW. 798; chees WICL. JOHN xv. 16; CH. C. T. *E* 2148; þu chure MARH. 19; (heo) curen [chosen] LA3. 6889; cusen SAX.CHR.250; chosen GEN. & EX. 543; HAV. 372; LANGL. *A* PROL. 31; MAND. 225; chose BEK. 1986; S. A. L. 5; **córen** (*pple.*) LA3. 16358; koren BEV. 770; corn AM. & AMIL. 1431; core ED. 4317; corn, chosen PS. xvii. 27; cosen SAX. CHR. 250; chosen ORM. 9623; MAND. 225; *comp.* 3e-chēosen; *deriv.* **cost, cūre, cüste, chūse.**

    chēosere, sb., *chooser*; **chēosers,** chēsers (*pl.*) TREV. V. 309.

    chēseresse, sb., (*female*) *chooser,* WICL. WISD. viii. 4.

    chēsunge, sb., *choosing,* HOM. II. 19; chiezinge AYENB. 42.

chēowen, v., *O.E.* cēowan (*pret.* cēaw), = *O.H. G.* chiuwan (*pret.* chou), *O.N.* tyggva, tyggja (*pret.* tögg); *chew, eat, devour*; chiewe AYENB. 111; chewen (*ms.* chewwenn) ORM. 1241; chewe HOM. II. 183; upon þe bridel chewe GOW. I. 334; late hem chewe as þei chose LANGL. *B* xviii. 199; al þat tu **cheowest** (*pres.*) H. M. 35; cheowe͡ð, cheouwe͡ð A. R. 80, 84; chit te & cheowe͡ð (*printed* cheope͡ð) þe H. M. 31; cheweþ CH. C. T. *A* 3690; chewiþ code WICL. DEUT. xiv. 6; bestes þat chewen not her code MAND. 20; **chēwe** (*subj.*) MIRC 255; *comp.* **to-chēowen;** *deriv.* **chēu.**

    chēwinge, sb., *chewing,* PR. P. 74; AYENB. 111.

chēp, see **chēap.**

chēpen, v., *O.E.* cēpan, cȳpan; *sell;* **chēptan** (*pret.*) MK. xi. 15.

cher, sb., *O.E.* cerr, cierr, cyrr *m., cf. O.H.G.* cher *m.,* chera *f., M.L.G.* kere *f., mod.Eng.* chare (*turn of work*); *turn, act of turning; occasion, time; piece of work, job*; one cherre A. R. 314; sum chearre JUL. 41; don a char P. S. 341; **cherre** (*dat.*) BEV. 3461; at þa(n) latere cherre LA3. 8356; makeden hine þridde charre (*for the third time*) king 6844; et one cherre A. R. 324; at ane chere H. M. 23; þe him de͡ . . . wiken & **cherres** (*pl.*) HOM. I. 137; charres GAW. 1674; charis TOWNL. 106; *comp.* 3ein-char.

chēr, adj., *O.Fr.* cher, chier; *dear*: a most cheere hinde WICL. PROV. v. 19.

chĕre-līche, adv., *dearly*, PL. CR. 582; cherli WILL. 62.

cherche, *see* chireche.

chĕre, sb., *O.Fr.* chĕre, chiere, *mod.Eng.* cheer; *countenance*, MARH. 3; MISC. 40; BRD. 6; PR. C. 1636; cheere '*vultus*' PR. P. 72; makien him glede chere A. R. 190; wiþ riȝt a merie chere [cheere] CH. C. T. *A* 857; chiere AYENB. 155.

cheri, sb., *O.Fr.* cerise, *Lat.* cerasum; *cherry*, PR. P. 72; chiries (*pl.*) LANGL. *A* vi. 296.

  cheri-feire, sb., *cherry-fair*, GOW. I. 19; LIDG. M. P. 231.

  cheri-stōn, sb., *cherry-stone*, PR. P. 72.

  cheri-tīme, sb., *cherry-time*, LANGL. *B* v. 161.

  cheri-trē, sb., *cherry-tree*, VOC. 181; chiritree PALL. iii. 1147.

chĕrin, v., *O.Fr.* chĕrer; *cheer*, '*hilaro*,' PR. P. 72; chĕrid (*pple.*) WICL. RUTH iii. 7.

cherissen, v., *Fr.* cherir; *cherish*; chersin PR. P. 73; cherische [cherisse] (*imper.*) CH. C. T. *E* 1388; cherisched (*pple.*) A. P. ii. 543.

  cherissing, sb., *cherishing*, LANGL. *C* v. 112; cherisshinge PALL. x. 24.

cherl, *see* cheorl.

cherre, *see* cher.

cherren, v., *O.E.* cerran, cirran, = *O.L.G.* keran, *O.Fris.* kera, *O.H.G.* cherran (*pret.* cherta); *turn*; chearre JUL. 33; charre MAP 348; wiþ wives and childre ȝeȝen charen GEN. & EX. 1712; chare H. S. 2062; charin a wai *turn away* PR. P. 70; LUD. COV. 325; charre (*pres.*) GAW. 1678; hwan ic aȝen cherre HOM. I. 79; churreþ REL. I. 172; cher (*imper.*) me from sunne HOM. I. 215; ne cherre he ongean MK. xiii. 16; chearren KATH. 2261; cherde (*pret.*) O. & N. 1658; cherde, chærde, charde LAȝ. 7234, 28744, 31354; cherden [charden] . . . aȝein MARH. 3; chirden (*ms.* chyrden) LK. xxiii. 56; charde A. P. i. 607; *comp.* bi-, ȝe-cherren.

chervelle, sb., *O.E.* cerfille, *cf. M.L.G.* kervele, *O.H.G.* chervola; *from Lat.* caerefolium; *chervil*, PR. P. 73; cherveles [chervelles] (*pl.*) LANGL. *B* vi. 296.

cherwin, v., '*torqueo*,' PR. P. 73.

  chervinge, sb., '*torcio*,' PR. P. 73.

ches, *see* chek.

chēse, sb., *O.E.* cēse, cȳse, *from Lat.* cāseus; *cheese*, HAV. 643; MAND. 272; E. G. 356; he bindeȝ uppon þa swike chese & bret hine HOM. I. 53; chese [chise] TREV. I. 405.

  chēse-bolle, sb., *poppy*, '*papaver*,' PR. P. 73; chesbolle VOC. 190.

  chēs-fat, sb., *O.E.* cȳsefæt; *cheese-vat*, VOC. 202.

  chēse-kake, sb., *cheese-cake*, PR. P. 73.

  chēs-lēpe, sb., *cheese-basket*, VOC. 222.

chēs-lippe, sb., *O.E.* cȳslybb, ?=*M.H.G.* kæseluppe; *rennet*, VOC. 202.

chesel, *see* chisel. chēsen, *see* chēosen.

chesible, sb., *O. Fr.* chesuble, chasuble; *chasuble*, PR. P. 73; chesibles (*pl.*) LANGL. *B* vi. 12.

chesōn, sb., *for* achesōn; *cause, account*, S. S. (Wr.) 680; chesoun ALIS. 3930.

cheste, *see* chiste. chĕste, *see* chĕaste.

chesteine, *see* chasteine.

chestre, sb., *O.E.* ceaster, *Lat.* castrum; *city, fortified town*, '*civitas*,' FRAG. 4; MAT. v. 35; ORM. 8479; Chestre *Chester* P. L. S. xiii. 27.

chete, sb., *O.E.* cete, cyte; ? *cell*; hende as hake in chete SPEC. 31.

chĕte, sb., *for* eschēte; *confiscation*, PR. P. 73.

chetel, *see* catel.

chetel, sb., *O.E.* cetel, cytel, citel, = *O.Fris.* ketel, szetel, tsetel, *O.N.* ketill, *O.H.G.* chezil, *Goth.* katils; *kettle*, '*cacabus*,' FRAG. 4; JUL.² 54; chetil W. & I. 23; ketil, chetil PR. P. 273; ketel '*lebes*' PS. cvii. 10; cheteles [ketels] (*pl.*) WICL. LEV. xi. 35.

  ketel-hat, sb., *helmet*, '*pelliris*,' PR. P. 273; ketille-hatte M. ARTH. 3516; ketelle-hates (*pl.*) 2993.

chēten, v., ? *O.N.* kǣta; *console, cheer*: hi hit chēteð (*pres.*) & blissið HOM. I. 233.

chētinge, sb., *for* eschĕtinge; *escheating*, '*confiscacio*,' PR. P. 73.

chĕtour, sb., *for* eschētour, *O.Fr.* eschetour; *escheator*, '*confiscator*,' PR. P. 73.

chĕðen, *see* cüððe.

chēu, sb., *from* chēowen; *act of chewing*: wurmene cheu HOM. II. 121; chest & chew (*printed* chep) HOM. II. 13; *comp.* ȝe-chēu.

chevachie, sb., *O.Fr.* chevauchie; *expedition on horseback*; cheuache CH. COMPL. M. 144; chivachie CH. C. T. *A* 85.

chevaler, sb., *O.Fr.* chevaler; *chevalier*; chivaler LANGL. *B* xviii. 99.

chevalerie, *O.Fr.* chevalerie; *chivalry*, CHR. E. 225; chevalrie [chivalrie] CH. C. T. *A* 45.

chevalrous, adj., *O.Fr.* chevalereus; *chivalrous*, D. ARTH. 3604.

chevaunce, sb., *O.Fr.* chevance; *gain, profit*, GOW. II. 275.

chevel, *see* chavel.

chéven, v., *O.Fr.* chevir; *attain an end, succeed; happen; be subject*; cheeven LANGL. *A* PROL. 31; cheve BEK. 856; chéve (*pres. subj.*) DEGR. 465; *see* chevise.

chevesaile, sb., *O.Fr.* cheveçaille; *necklace*, R. R. 1081.

chevese, sb., *O.E.* cefes, cyfes, cyfese, = *O.H.G.* chebisa, chebis; *concubine*, MARH. 3; chevese, chivese LAȝ. 384, 6356.

cheves-boren, adj., *born of a concubine*, LA3. 4334.

chevetain, sb., *O.Fr.* chevetain; *chieftain*, SHOR. 112; MAN. (F.) 9779; chevetein BEK. 251; WILL. 3379; *see* **capitain.**

chevise, v., *O. Fr.* cheviss-, chevir, *see* **cheven**; *procure, obtain; succeed; help, profit*; CH. COMPL. M. 289; chevisin '*provideo*' PR. P. 74; **chewise** (*imper.*) D. ARTH. 1750; **chevesed** (*pple.*) ALEX.² (Sk.) 966.

chevissance, sb., *O.Fr.* chevissance; *agreement, bargain*, CH. C. T. *B* 1519.

chēw, chēwen, *see* **chēu, chēowen.**

chibolle, sb., *O.Fr.* ciboulle; *small onion*, PR. P. 74; **chibolles** (*pl.*) LANGL. *B* vi. 296.

chiche, sb., *O.Fr.* chiche, *Lat.* cicer (? *mod. Eng.* chick-pease); *vetch*, PALL. iv. 57; **chichis** (*pl.*) WICL. 2 KINGS xvii. 28.

chiche, adj., *O.Fr.* chiche; *parsimonious*, R. R. 5588; A. P. i. 604.

Chichestre, pr. n., *O.E.* Ciceaster; *Chichester*, ROB. 2; Cicestre MISC. 145.

chīde, sb., *O.E.* (ge)cīd; *contention*, MAP 342.

chīden, v., *O.E.* cīdan; *chide, contend*, LA3. 8149; GEN. & EX. 2722; to chiden agen oni scold REL. I. 183; chidan HOM. I. 113; chidin '*contendo*' PR. P. 74; chide O. & N. 287; AYENB. 67; LANGL. *A* iv. 39; P. 51; **chist** (*pres.*) þu . . . me chist O. & N. 1331; chideð A. R. 198; chideð, chit H. M. 31, 39; chit . . . wið gode HOM. I. 103; chid O. & N. 1533; **chidde** (*pret.*) O. & N. 112; ROB. 390; CH. C. T. *A* 3999; þis holi man him chidde P. L. S. xvii. 414; chidden '*litigabant*' WICL. JOHN vi. 53; GEN. & EX. 1927; hi chidde BEK. 1910; **chidde** (*pple.*) AR. & MER. 7237.

   chīdere, sb., *chider*; chidar PR. P. 74; **chīderes** (*pl.*) LANGL. *B* xvi. 43.

   chīderesse, sb., *chider* (*fem.*), R. R. 150, 4265.

   chīdestere, sb., *chider* (*fem.*), CH. C.T. *E* 1535.

   chīdinge, sb., *O.E.* cīdung; *chiding*, AYENB. 30; TREV. II. 97.

chiere, *see* **chēre.   chiesen**, *see* **chēosen.**

chiken, sb., *O.E.* cicen, ciccen MAT. xxiii. 37*, cycen VOC. 63, *cf. M.L.G.* küken, kuken; *chicken*, '*pullus*,' FRAG. 3; chekin PR. P. 74; **chikene** (*pl.*) MAT. xxiii. 37; LEB. JES. 439; chikenes [chiknes] CH. C. T. *A* 380.

   chicne-mete, sb., *O.E.* cicena mete; ? *chickweed*, FRAG. 3; chikne mete REL. I. 37.

   chekin-wēd, sb., *chickweed*, PR. P. 74.

chikkin, v., ? *produce chickens*; '*pullulo*,' PR. P. 74.

chilce, *see* **childse.**

child, sb., *O.E.* cild, = *O.L.G.*, *O.Fris.* kind, *O.H.G.* chind, chint; *child*, MAT. ii. 9; HOM. I. 37; LA3. 295; þat child O. & N. 1463; ROB. 10; þet child AYENB. 84; þat worþi child WILL. 541; **childes** (*gen.*) ORM. 8056;

it is no childes plei CH. C. T. *E* 1530; **childe** (*dat.*) GEN. & EX. 966; þer hvile þet hi is mid childe (*pregnant*) AYENB. 224; wiþ childe gon SPEC. 82; **child** (*pl.*) MAT. xxi. 15; childre HOM. I. 73; LA3. 5317; ORM. 8005; GEN. & EX. 715; ALIS. 4638; PL. CR. 756; TOR. 2643; childere LUD. COV. 30; childre, children HOM. II. 17, 19; O. & N. 631; children HOM. I. 7; A. R. 334; BEK. 24; MAND. 249; childre (*gen. pl.*) HOM. I. 7; childrene LA3. 15563; A. R. 422; childrene, children HAV. 449; (*ms.* chyldren) MAT. xxi. 16; LANGL. *A* iv. 103; childon, childen (*dat. pl.*) MAT. xiv. 21, xv. 38; children LA3. 12992; *comp.* cnáve-, mōder-, stēop-child.

   child-bed, sb., *childbed*, HOM. II. 47; liþ . . . a childbedde ROB. 379.

   child-bering, sb., *child-bearing*, WICL. GEN. xxv. 24.

   child-gered, adj., *childish-mannered*, GAW. 86.

   child-hēde, sb., *O.E.* cildhād; *childhood*, AYENB. 82; **childhāde** (*dat.*) MK. ix. 21; KATH. 79; childhade (*ms.* -haden) [childhode] LA3. 20312; childhode A. R. 314; childhede GEN. & EX. 2652; CH. C. T. *B* 1691.

   childish, adj., *O.E.* cildisc, *cf. O.H.G.* chindisc; *childish*, LANGL. *B* xv. 145.

   child-lēs, adj., *childless*, GEN. & EX. 930; childlæs ORM. 2312.

   children-lēs, adj., *without children*, TREV. I. 183.

   child-lī, adj., *childlike*, '*puerilis*,' WICL. TOB. i. 4; GOW. II. 228; HOCCL. i. 64.

childin, v., = *O.H.G.* chindōn; *bring forth (children)*, '*pario*,' PR. P. 74; þe shal . . . þin wif an sune childen ORM. 156; childi AYENB. 224; childe ALIS. 604; **childide** (*pret.*) WICL. GEN. iv. 1; childed PARTEN. 1157; **childed** (*pple.*) MAND. 133.

   childinge, sb., *child-bearing*, WICL. GEN. xxv. 24; childing ALIS. 623; GOW. I. 69; C. M. 11059; þurh hire childþinge [chilðinge] MISC. 158, 159.

childse, sb., ? *puerility*; idelnesse and chilce P. L. S. viii. 7 (HOM. II. 220; I. 288; MISC. 58).

chile, *see* **chele.**

chilindre, sb., *Fr.* cylindre; *cylinder*, CH. C. T. *B* 1396.

chillin, v., ? = *Du.* killen, *M.Du.* kilden; *chill, be cold*, PR. P. 75; **chillande** (*pple.*) D. ARTH. 2965; **chilled** (*pret.*) A. P. iii. 368.

   chillinge, sb., *cf. Du.* killing; *chilling*, PR. P. 75; LANGL. *B* vi. 313 (Wr. 4424).

chimbe, sb., *cf. M.L.G.*, *M.Du.* kimme; *edge, brink*: the streem of lif nou droppeþ on þe chimbe CH. C. T. *A* 3895.

chimbe, chime, sb., *cymbal*, C. M. 12193; chime PR. P. 75; **chimbes** (*pl.*) PS cl. 5; MAN. (F.) 11387.

chimben, v., *cf. Swed.* kimba, *Dan.* kime; *chime*; chimbe CH. C. T. *A* 3896; chimin PR. P. 75; chime GOW. II. 13.

chimblen, v., *O.N.* kimbla; *bind up, bundle*: **chimbled** (*pple.*) over hir blake chin with milkqvite vailes GAW. 958.

chime, *see* **chimbe. chimenee,** *see* **chemenē.**

chin, sb., *O.E.* cin, VOC. 43, cinn *?m.* (*mentum*), *cf.* M.L.G. kin, kinne *m. n.,* M.Du. kinne, *O.N.* kinn (*gena*), *Goth.* kinnus *f.* (σιαγών), *O.H.G.* chinni *n., Lat.* gena, *Gr.* γένυς *f.*; chin, '*mentum,*' VOC. 146, 183, 207; PR. P. 75; MAND. 107; þane chin LAȝ. 8148; **chinne** (*dat.*) HOM. I. 35; O. & N. 96; TREAT. 139; TREV. I. 257; swor bi his chinne LAȝ. 18764; þet nimþ þane viss bi þe þrote and bi þe chinne AYENB. 50.

chinche, adj., *? O.Fr.* chinche; *stingy,* '*perparcus,*' PR. P. 75; S. S. (Web.) 1244; HAV. 1763 (*the rime requires* chiche).

chincherie, sb., '*parcimonia,*' PR. P. 75.

chinchin, v., *be stingy,* '*perparceo,*' PR. P. 75.

chīne, sb., *O.Fr.* eschīne; *chine, backbone,* '*spina,*' PR. P. 75; ALIS. 3245; D. ARTH. 3390.

chine, sb., *O.E.* cinu,=*M.Du.* kene, *mod.Eng.* chine (*in local names*); *from* **chīnen**; *fissure, crack, cleft,* PARTEN. 4343; chine, chene ['*hiatus*'] TREV. I. 233,415; LANGL. *C* xxi. 287; chene PALL. i. 450; **chínes** (*pl.*) '*scissuras*' WICL. IS. ii. 21; chinis BER. 2353; chines, kines MAN. (F.) 1720, 13976; *comp.* erð-**chíne.**

chīnen, v., *O.E.* (to)cīnan, = *O.L.G.* kīnan, *Goth.* keinan,*O.H.G.* chīnan; *split, burst open* (*intrans.*): þat makeþ heorte chinen MISC. 73; chine FER. 212; PALL. vi. 199; **chīneð** (*pres.*) HOM. II. 199; þet gles ne brekeð ne chineð I. 83; þe corn... chineth LEB. JES. 564; **chān** (*pret.*) ALIS. 2228; deþ her hert(e) chon AR. & MER. 7764; **chīned** (*pple.*) L. H. R. 142; *comp.* **to-chīnen**; *deriv.* **chíne.**

chippe, sb., *? O.E.* cipp (*dentale*); *chip,* '*assula,*' PR. P. 75; **chippes** (*pl.*) GOW. I. 106; CH. C. T. *A* 3748.

chippen, v.,=*M.Du.* kippen; *? chip (with a hatchet, &c.*); chippe '*dolare*' CATH. 64; HALLIW. 250.

chippinge, sb., *chipping, piece,* PR. P. 75.

chireche, sb., *O.E.* cirice, cyrice, = *O.L.G.* kirika, kerika, *O.N.* kirkja, *O.H.G.* chiricha, *? from Gr.* κυριακή (ἐκκλησία); *church*; circe SAX. CHR. (E.) 262; LAȝ. 26140; REL. I. 129; chirche, churche A. R. 22, 28; O. & N. 608; chereche MISC. 31; cherche AYENB. 7; GOW. I. 63; kirke, chirch'e LANGL. *B* vi. 93, *A* vii. 34; kirke ORM. 3531; HAV. 36; **chirechen** (*pl.*) LAȝ. 20971; chirchen CHR. E. 923; churchen BEK. 1414; *comp.* **hēved-chirche.**

chirche-clōðes, sb., *church vestments,* A. R. 420.

chireche-dure, sb., *church-door,* HOM. I. 73; MISC. 151; **kirkedure,** ORM. 1327.

kirke-flōr, sb., *church floor,* ORM. 9015.

chirch-gang, sb., *church-going; churching* (*of women*), HOM. II. 47; chirchegong GEN. & EX. 2465; ROB. 380.

chirche-ȝeard, sb., *churchyard,* A. R. 318*; chircheȝard LEG. 153; churchȝerd BEK. 2217; kirkegærd ORM. 15254.

chireche-griþ, sb., *O.E.* ciricgriþ; *sanctuary of the church,* LAȝ. 22322; chirchegrið A. R. 174.

chirche-hai, sb., *churchyard,* '*coemeterium,*' VOC. 178; AR. & MER. 6738; chirchhai MIRC 330*; chirchhei E. W. 26; **chircheie** (*dat.*) A. R. 318.

chirche-hawe, sb., *churchyard,* S. S. (Web.) 2625; TREV. V. 65; E. W. 81; CH. C. T. *I* 964; chirchau E. W. 80.

chirche-kei, sb., *church key,* GOW. I. 10, 12.

chiric-lond, sb., *church-land,* LAȝ. 14855.

chirche-rēaf, sb., *robbery of churches*; chirchereve CH. C. T. *D* 1307.

chirche-song, sb., *O.E.* cyricsang; *church song,* O. & N. 984.

chirch-sōcne, sb., *O.E.* ciricsōcn; *parish,* HOM. II. 89.

cherch-toun, sb., *O.E.* cirictūn; *churchyard,* AYENB. 41.

chirche-wardein, sb., *churchwarden,* E. G. 145.

chirche-weorc, sb., *church work,* HOM. I. 31.

chirche-wīce, sb., *office of the church*: þe landes þat lien to þe **circe wīcan** (*dat. pl.*) SAX. CHR. 263.

chirchin, v., *church* (*women*), '*purifico,*' PR. P. 75.

chiri, *see* **cheri.**

chirkin, v., *creak, twitter,* '*strideo, sibilo,*' PR. P. 76; **chirkeþ** (*pres.*) as a sparwe CH. C. T. *D* 1804; *see* **charkin.**

chirking, sb., *screaming,* CH. C. T. *A* 2004.

chirm, sb., *O.E.* cirm, cyrm; *noise, shouting*; chirme (*dat.*) O. & N. 305.

chirmen, v., *O.E.* cyrman, cirman,=*M.Du.* kermen, *M.L.G.* kermen, karmen; *cry out, shout,* A. R. 152; **chirmed** (*pret.*) PARTEN. 878; *comp.* **bi-chermen.**

chirne, sb.,=*M.Du.* kerne; *churn,* PR. P. 76; kirne VOC. 202.

chirnen, v.,=*M.Du.* kernen; *churn*; chirne PR. P. 76.

chirpen, v., *chirp.*

chirpinge, sb., *chirping,* '*garritus,*' PR. P. 76.

chīse, *see* **chūse.**

chisel, sb., *O.E.* cisel, ceosel,=*M.Du.* kesel, *O.H.G.* chisili; *pebbles, shingle,* '*sabulum,*' PR. P. 76; **cheselis** (*pl.*) LUD. COV. 56; *comp.* **sand-chisel.**

chisel, sb., *O.Fr.* cisel; *chisel,* PR. P. 76; SHOR. 137.

chiste, sb., *O.E.* cist, cyst, cest,=*O.N.,O.H.G.* kista, *? from Lat.* cista; *chest, box,* ' *arca sive sarcophagus,*' TREV. II. 307; BEK. 2458; chiste [cheste] CH. C. T. *D* 317; cheste '*loculus*' FRAG. 3; þa cheste LK. vii. 14; ane cheste LA3. 27858; **chiste,** kiste (*dat.*) HAV. 222, 2018; kiste PERC. 2109.

chiteren, [chitren], v., *chatter,* CH. C. T. *G* 1397; chitre GOW. II. 318.

     chiteringe, sb., *chittering,* TREV. II. 159; chitering WICL. DEUT. xviii. 10.

chitirling, sb., *chitterling,* PR. P. 76.

chitte, sb., *cf. L.G.* kitte, *H.G.* kitze; *whelp,* '*catulus,*' VOC. 177; **chittes** (*pl.*) ['*catulos*'] WICL. IS. xxxiv. 15.

chivachē, *see* **chevachie.**

chivaler,chivalrie, *see* **chevaler,chevalerie.**

chivelen, v., *? shiver, tremble*; his chekes ... þei chiveled (*pret. pl.*) for elde LANGL. *B* v. 193; ichiveled (*pple.*) for elde *C* vii. 200.

chiverin, v., *shiver, quake,* PR. P. 76; chiveren [chiverin] in ise R. S. v (MISC. 177); chivere & schake L. H. R. 144; **chivered** (*pret.*) LEG. 165; **cheverid** (*pple.*) for chele D. ARTH. 3391.

chō3e, sb., *O.E.* ceō; *chough*; choghe CH. P. F. 345; chou3e '*monedula*' VOC. 177; chou3he TREV. IV. 307; choughe PR. P. 57; **chō3en** (*pl.*) P. L. S. xvii. 185; þe crowes and the choughes MAND. 59.

chois, sb., *O.Fr.* chois; *choice,* ROB. 111; TREV. IV. 385.

     chois-lī, adv., *choicely,* WILL. 1753.

chóken, *see* **chéoken.**

chol, sb., *? double chin*; his chin wiþ a chol lollede so greet as a gos ei PL. CR. 444.

chold, *see* **cáld.**

chop, sb., *blow, knock, dispute*; **choppes** (*pl.*) LANGL. *C* xi. 275 (*A* x. 187).

choppen, v., *?=M.Du.* koppen; *chop, hew, strike*: þei choppen (*pret.*) alle þe bodi in smale peces MAND. 201; chop (*imper.*) LANGL. *A* iii. 253; **choppid** (*pple.*) D. ARTH. 1026.

     chopping, sb., *striking, fighting,* LANGL. *B* ix. 167.

chōst, *see* **chéaste. chou3e,** *see* **chō3e.**

choul, *see* **chavel.**

chucken, v., *chuckle, boast*: he chukketh (*pres.*) whan he hath a corn ifounde CH. C. T. *B* 4372.

chuffe, choffe, sb., *clown,* '*rusticus,*' PR. P. 77.

chuffer, sb., *? boaster,* TOWNL. 216.

chuffing, sb., *? boasting,* ORM. 12177.

chullen, v., *? drive*; and chulles (*pres.*) him

as men don a balle HALLIW. 249; chased to daie and chulled (*pple.*) as hares D. ARTH.1444.

chürche, *see* **chireche. chürren,** *see* **cherren.**

chūse, adj., *? O.E.* cȳs, cīs, *from* **chēosen**; *elegant; fond, loving* : þat was chuse [chise] of þe childe ALIS.² 49; gent ich wes ant chis REL. I. 123; þe ladi is of lemon chis ALIS. (Web.) 3294; knightis ... chosin ful **chīse** (*pl.*) LUD. COV. 180.

     chīs-lī, adv., A. P. ii. 543.

chūsen, *see* **chēosen.**

cicatrice, sb., *Fr.* cicatrice; *cicatrix, scar,* PALL. iii. 51.

ciclatoun, sb.,*cf.Span.* ciclaton; *a sort of garment or stuff,* HICKES I. 225; CH. C. T. *B* 1924; ischrud mid hwite ciclatune HOM. I. 193.

cicle, sb., *Fr.* cycle; *cycle,* TREV. V. 377.

ciconie, sb., *stork,* WICL. JER. viii. 7; **sikonies** (*pl.*) MAND. 45.

cider, *see* **sider.**

cimbale, sb., *Lat.* cymbalum; *cymbal*; simbale PR. P. 456; **cimbalis** (*pl.*) WICL. 2 KINGS vi. 5.

ciment, sb., *O.Fr.* ciment; *cement,* PALL. i. 449; siment WICL. GEN. xi. 3.

cinamome, sb., *Fr.* cinnamome, *Lat.* cinnamōmum; *cinnamon,* CH. C. T. *A* 3699; cinamum PR. P. 78.

cindre, sb., *? Fr.* cendre; *cinder*; **cindres** (*pl.*) MAND. 101; *see* **sinder.**

cink, sb., *O.Fr.* cinc (*five*); *cinque* (*ports*); sink pors ROB. 515.

cinoper, *see* **sinopir.**

cipres, sb., *O.Fr.* cipres; *cypress,* AYENB. 131; C. M. 1377.

     cipre-trē, sb., *cypress-tree,* WICL. S. SOL. i. 13.

cipres, sb., *fine gauze,* LANGL. *B* xv. 224.

circe, *see* **chireche.**

circuit, sb., *circuit,* CH. C. T. *A* 1887.

circumcīden, v., *Lat.* circumcīdere; *circumcise,* WICL. GEN. xvii. 11; WICL. S. W. I. 335.

circumcīs, pple., *Lat.* circumcīsus; *circumcised,* GEN. & EX. 999.

circumcīse, sb., *circumcision,* GEN. & EX. 2847.

circumcīse, v., *circumcise,* C. M. 2668; circumcīsed (*pple.*) GEN. & EX. 1200.

     circumcīsing, sb., *act of circumcision,* C. M. 2681.

circumcisiōn, sb., *Lat.* circumcisio, *cf. Fr.* circumcision; *circumcision,* HOM. I. 83; GEN. & EX. 991; WICL. ROM. iii. 1; S. W. I. 335; PR. P. 78, 456.

circumference, sb., *Lat.* circumferentia, *cf. Fr.* circonférence; *circumference,* GOW. III. 90.

circumscrīven, v., *Lat.* circumscrībere; *circumscribe,* CH. TRO. v. 1877.

circumstance, sb., *circumstance*; **circumstaunces** (*pl.*) A. R. 316.

Cirencestre, pr. n., *O.E.* Cirin-, Cyrenceaster; *Cirencester*, ROB. 2.

cirurgian, sb., *O.Fr.* cirurgien; *surgeon*, ROB. 566; surgien LANGL. *B* xx. 308; surgienes, surgiens (*pl.*) LANGL. *B* xiv. 88, xx. 177*.

cisar, *see* sider.

cisoure, sb., *O.Fr.* cisoires; *scissors*, 'forfex,' PR. P. 78.

cisterne, sb., *O.Fr.* cisterne; *cistern*, PR. P. 67; S. A. L. 136; MAND. 106.

cisternesse, sb., *pit*, GEN. & EX. 1942.

citaciōn, sb., *Fr.* citation; *citation*, ROB. 473.

citē, sb., *Fr.* cité; *city*, A. R. 228; MISC. 26; GEN. & EX. 2669; MAND. 244; CH. C. T. *A* 2701; PS. xlv. 5.

citeien, sb., *O.Fr.* citeien; *citizen*; citezein (? citeȝein) GOW. I. 75; citeȝens (*pl.*) WILL. 3850; citeȝenis E. G. 23; citeseins TREV. IV. 129.

citenere, sb., *citizen*; cittenere VOC. 211.

citir-trē, sb., *Lat.* citrus; *citron*, PR. P. 78; citur tre PALL. viii. 8, xi. 66.

citole, sb., *O.Fr.* citole; *a musical instrument*, CH. C. T. *A* 1959.

citrinacioun, sb., *med.Lat.* citrinātio; *a process in alchemy*, CH. C. T. *G* 816.

ciūn, sb., *O.Fr.* cion; *scion*, PR. P. 79; siouns (*pl.*) P. L. S. xxxv. 74.

cive, sb., *Fr.* cive; cíves (*pl.*) *garlic*, PR. P. 78.

civilitē, sb., *O.Fr.* civilité, *from Lat.* cīvīlitas, *mod.Eng.* civility; *freedom (of a city)*, *citizenship*, WICL. DEEDS xxii. 28.

clacke, sb., = *M.Du.* klacke; *clack (of a mill)*; klakke VOC. 180; PR. P. 79.

clacken, v., = *M.Du.* klacken, *O.N.* klaka; *clack, gabble*; clake L. C. C. 54; þu clakes(t) (*pres.*) O. & N. 81.

clǣmen, v., *O.E.* clǣman, = *M.Du.* kleemen, *O.H.G.* chleimen; *smear*; cleime LIDG. M. P. 53; Þolōme (*pres.*) SACR. 708; clēme (*imper.*) PALL. iii. 871; cleme hit with clai A. P. ii. 312; clēmede (*pret.*) ['*linivit*'] WICL. EX. ii. 3; *comp.* ȝe-clǣmen.

clǣne, adj., *O.E.* clǣne (*clean*), = *O.L.G.* clēne, (*small*), *M.Du.* clēne (*small, thin, clean*), *O.H.G.* chleini (*fine, thin*); *clean, pure, chaste*: clǣne [cleane] mon LAȝ. 6290; clene FRAG. 7; O. & N. 584; GEN. & EX. 605; CH. C. T. *A* 504; clene lif HOM. II. 85; clene brǣd ORM. 1590; clene water ROB. 435; of ure sunne make us clene HOM. I. 63; cleane inwit A. R. 2; clene, klene, cliene AYENB. 5, 159, 224; clane M. T. 206; ne mai no man clene telle BEK. 128; Vortiger ane hit dude for alle clane [cleane] LAȝ. 13264; bursten hire bondes & breken alle clane JUL. 58; clĕnnere (*compar.*) P. L. S. xxi. 98; LANGL. *B* xiii. 296; clĕnneste (*superlat.*) H. M. 43; clennest LANGL. *B* xiv. 43.

clǣn-leȝc, sb., *purity*, ORM. 4622.

clǣn-lĭch, adj., *O.E.* clǣnlīc; *cleanly*: clenliche cloþinge AYENB. 216; clēnlīche (*adv.*) A. R. 344; BEK. 130; PL. C. 229; cleniike (*ms.* clennlike) ORM. 1644; clanliche P. L. S. xv. 65.

clǣnnesse, sb., *O.E.* clǣnness; *cleanness*, ORM. 4598; clennesse A. R. 164; O. & N. 491; AYENB. 201; castitas þet is clenesse on Englisc HOM. I. 105; cleannesse H. M. 11; clannesse ROB. 332.

clēan-schipe, sb., *purity*, H. M. 21.

clǣnsien, v., *O.E.* clǣnsian; *cleanse, purify*; clensien FRAG. 7; clensen A. R. 314; (*ms.* clennsenn) ORM. 1126; clensi SHOR. 50; AYENB. 137; clense LANGL. *B* vi. 106; clansi O. & N. 610; clanse MIRC 259; clĕnseð (*pres.*) HOM. I. 83; clĕnsede (*pple.*) HOM. I. 237; clĕnsed (*pple.*) HOM. II. 87; PR. C. 3705; *comp.* ȝe-clǣnsen.

clĕnsunge, sb., *O.E.* clǣnsung; *cleansing*, HOM. II. 141; (*ms.* clennsinng) ORM. 15006; clansinge LEG. 99.

clai, sb., *O.E.* clǣg, = *O.Fris.* klai, *M.L.G.* klei *m.*; *clay*, REL. II. 210; clay SPEC. 85; ALIS. 915; PR. C. 411; clei FRAG. 5; '*argilla*' PR. P. 80; TREV. V. 129; CH. C. T. *G* 807; cleiȝ LEG. 14; S. A. L. 178.

clei-clot, sb., *clod of clay*, FRAG. 5; R. S. V. (MISC. 172).

clai-daubed, pple., *daubed with clay*, A. P. ii. 492.

clei-pit, sb., *clay-pit*, PR. P. 80.

cleȝi, adj., *clayey*, WICL. ECCLUS. xxii. 1.

claien, adj., *made of clay*, WICL. JOB iv. 9.

claim, sb., *O.Fr.* claim; *claim*, GOW. I. 250.

claimen, v., *O.Fr.* clamer, claimer; *claim*: claimen [cleimen] and asken LANGL. *B* xiv. 259; cleimeþ (*pres.*) he after cloþes WILL. 4481; cleimed (*pret.*) þeim quit of her servise MAN. (F.) 4548; clame TOWNL. 200.

clak, sb., *O.E.* clæc (-lēas), = *M.H.G.* klac, *O.N.* klakkr; *spot, stain*: ȝif þat ȝe wel ȝuw loken fra clake & sake ORM. 9317.

clakke, *see* clacke.

clam, adj., *cf.* Ger., Du., Dan. klam; *sticky*, '*glutinosus, viscosus*,' PR. P. 79.

clām, sb., *O.E.* clām, = *M.Du.* kleem; *clay*; clom LANGL. *A* xii. 100.

clamance, sb., *med.Lat.* clamantia; *claim*, MAN. (H.) 186.

clamerin, v., *clamber, climb with hands and feet; cluster, crowd*; '*repto*,' PR. P. 79; clamberande (*pple.*) cliffes GAW. 1722; whanne ani clambrede (*pret.*) up LEG. 194; clombred (*pple.*) so þik GAW. 801.

clammen, v., *smear*, WICL. S. W. II. 93.

clamour, sb., *O.Fr.* clamor; *clamour*, GOW. I. 21; CH. C. T. *A* 995.

cland, sb., *O.N.* kland *n.*; *difficulty, calamity* : he makede him selven muchel clond LA3. 11704.

clāne, *see* **clǣne.**

clap, sb., = *M.Du.* klap, *O.H.G.* chlaph ; *clap, stroke,* A. R. 102* ; **clappes** (*pl.*) MAN. (H.) 175 ; *comp.* 3ain², þonder-clap.

claper, sb., *Fr.* clapier ; *rabbit-burrow* ; **clapers** (*pl.*) R. R. 1405.

clappe, sb., *? Ger.* klapfe, klaffe ; *clap, noise,* '*strepitus,*' PR. P. 79.

clappe, sb., = *M.Du.* klappe, kleppe, *M.H.G.* klaffe ; *clapper, clack* (*of a mill*), CH. C. T. *A* 3144 ; clappe of a mille PR. P. 79 ; hold þou þi clappe (*tongue*) ED. 3357 ; cleppe A. R. 70.

clappin, v., *O.E.* clæppan (LEECHD. III.88), = *O.Fris., O.N.* klappa, *M.L.G.* klappen, *O.H.G.* chlaphōn ; *clap, beat, clatter ; babble ; 'pulso,'* PR. P. 79 ; clappen to gidre ['*complodere*'] WICL. JUDG. vii. 19 ; clappe OCTOV. (W.) 569 ; **clappeþ** (*pres.*) SHOR. 135 ; clappeþ at þe ... gate CH. C. T. *D* 1581 ; and clappeþ as a mille *I* 406 ; what so men clappe (*babble*) or crie *G* 965 ; **clappide** (*pret.*) D. ARTH. 956 ; **clappíd** (*pple.*) HOCCL. i. 394 ; *comp.* **bi-clappen.**

  clapper, sb., *cf. L.G.* klapper, klepper ; *clapper, rattle,* GOW. II. 13 ; coppe & claper he bare ... as he a mesel ware TRIST. 3173 ; þe cleper of þe melle AYENB. 58.

  clappinge, sb., *sounding* (*of a bell*), '*tintilla-cio,*' PR. P. 79.

clap-wīpe, sb., *carrot,* '*daucus,*' VOC. 190.

clarē, sb., *O.Fr.* claré, claret ; *claret,* HAV. 1728 ; LAUNF. 344 ; claret PR. P. 79 ; clarre CH. C. T *A* 1471.

claretee, sb., *O.Fr.* clarté ; *brightness, light,* MAND. 86 ; clarité WICL. S. W. I. 405 ; *see* clertē.

clarifíen, v., *Fr.* clarifier ; *clarify,* WICL. 3 ESDR. viii. 82 ; **clarifíeth** (*pres.*) LUD. COV. 103.

clarine, sb., *Fr.* clarine ; *trumpet,* PR. P. 80.

clárionere, sb., *trumpeter,* PR. P. 80.

clarioninge, sb., *trumpeting* ; **clarioninges** (*pl.*) CH. H. F. 1242.

clarioun, sb., *O.Fr.* clarion ; *clarion,* MAN. (F.) 11384 ; CH. C. T. *A* 2600.

[**clarten,** v., *soil, dirty ; comp.* **bi-clarten.**]

claspe, = *L.G.* klaspe ; *clasp, brooch, 'offendix,'* CATH. 65 ; RICH. 4084 ; clospe PR. P. 83 ; clespe VOC. 238 ; **claspes** (*pl.*) P. S. 222 ; D. ARTH. 1108.

claspen, v., *clasp* ; claspe CATH. 65 ; **clasped** [clapsed] (*pple.*) CH. C. T. *A* 273.

clastren, v., *?clatter* ; ant **clastreþ** (*pres.*) wiþ heore colle P. S. 157.

clater, sb., *M.Du.* klater ; *clatter* : that thou had holden stille thi clater TOWNL. 190.

clateren, v., = *M.Du.* klateren ; *clatter, rattle,*

KATH. 2026 ; þe arwes ... clatren (*pres.*) fast and ringe CH. C. T. *A* 2359 ; **clatered** (*pret.*) GAW. 731 ; clattered LIDG. M. P. 106 ; *comp.* **tō-clateren.**

claterer, sb., *clatterer,* D. TROY 11375 ; **claterers** (*pl.*) P. P. 271.

clatering, sb., *rattling,* A. P. ii. 1515.

clatte, sb., *? rattle,* PR. P. 79.

clāð, sb., *O.E.* clāð *m.*, *cf. O.Fris.* klāth, *O.N.* klǣði, *M.H.G.* kleit *n.; cloth, garment,* KATH. 1428 ; þu hefdest clað to werien HOM. I. 33 ; inete & clað LA3. 16853 ; claþ ORM. 3208 ; no linene cloð A. R. 418 ; cloþ O. & N. 1174 ; MISC. 48 ; BEK. 691 ; AYENB. 45 ; CH. C. T. *D* 1633 ; cloth '*pannus*' PR. P. 83 ; HAV. 185 ; **clāðes** (*pl.*) HOM. I. 37 ; he rende his claðes JUL. 70 ; claþes IW. 1803 ; claðes [cloþes] LA3. 2367 ; cloðes A. R. 418 ; cloþes FRAG. 6 ; ROB. 463 ; **clāðen** (*dat. pl.*) LA3. 3187 ; claþe MISC. 75 ; *comp.* **barm-, bord-, hand-, hēved-, seil-clāð.**

  clōth-lēs, adj., *O.N.* klǣðlauss ; *without clothes,* CH. C. T. *I* 343.

  clāð-makinge, sb., *cloth-making,* CH. C. T. *A* 449.

clāþen, clēþen, v., *O.E.* clāðian, *O.N.* klǣða, = *M.H.G.* kleiden ; *clothe* ; claþen ORM. 2710 ; clathe PR. C. 3553 ; cloðen GEN. & EX. 2630 ; cloþen GOW. II. 227 ; GAM. 72 ; cloþe HAV. 1138 ; cleþe SPEC. 37 ; A. P. ii. 1741 ; IW. 1787 ; heo ... clāþeð (*pres.*) heom mid 3eoluwe claþe HOM. I. 53 ; clōþeþ (*imper.*) ROB. 36 ; **clōðinde** (*pple.*) A. R. 16 ; clōþede (*pret.*) AYENB. 133 ; cloþeden LANGL. *A* PROL. 53 ; cladde HAV. 1354 ; RICH. 3097 ; cledde M. H. 87 ; **clōðed** (*pple.*) REL. I. 212 ; cloþed, icloþed CH. C. T. *A* 911, 1048 ; icloþed JOS. 295 ; *comp.* **un-clāþen.**

  clāþere, sb., *clothier* ; **clōtheres** (*pl.*) LANGL. *B* x. 18.

  clāþing, sb., = *M.H.G.* kleidunge ; *clothing,* ORM. 19074 ; cloþing LA3. 3187* ; cloþinge P. L. S. xvii. 506 ; kleþinge '*vestimentum*' PS. ci. 27 ; cleþing ANT. ARTH. ix.

clau, sb., *O.E.* clāwu, clēa, clēo, = *O.L.G.* clāwa, *M.Du.* klauwe, *O.Fris.* klewe, *O.H.G.* chloa, cloa (*pl.* chlawen), *O.N.* klō (*pl.* klœr); *claw,* '*ungula,*' FRAG. 2 ; claw, cle PR. P. 80 ; clawe, cle WICL. EX. x. 26 ; *?* cleo MISC. 95 ; cleu HICKES I. 228 ; clee VOC. 221 ; **clawe** (*pl.*) O. & N. 153 ; clawes (*ms.* clawwess) ORM. 1225 ; clawes ALIS. 2186 ; clauen AYENB. 61 ; clees GOW. II. 39 ; MAND. 198 ; cleen PALL. i. 937.

clause, sb., *Fr.* clause ; *clause,* TRANS. XVIII. 22 ; GOW. III. 87 ; DEP. R. PROL. 92, i. 83.

clausūre, sb., *Lat.* clausūra ; *closure,* PR. P. 83.

claustre, sb., = *O.N.* klaustr, *L.G.* klauster ; *from Lat.* claustrum ; *cloister* ; **claustres** (*pl.*) AYENB. 267 ; *see* **cloistre** *and* **clūster.**

  clawstre-man, sb., *monk,* ORM. 6352.

claver, sb.,=*M.H.G.* klaber; *claw, finger-nail*; clawres (*?for* clavres) (*pl.*) A.P. ii. 1697.

clāver, sb., *O.E.* clâfre, clæfre,=*M.Du.* klāver, *L.G.* klåver, klēver; *clover*, D. ARTH. 3241; wite clovere VOC. 140; *comp.* heort-clēvre.

clavren, v., *cf. Du.* klaveren, *Dan.* klavre; *climb*: claverande (*pple.*) on heghe D. ARTH. 3324; hweʒer þe cat of helle clavrede (*pret.*) ever toward hire A. R. 102.

claw, *see* clau.

clāwin, v., *O.E.* clāwian, = *O.H.G.* klawen, *M.L.G.* klawen, clauwen, klowen, klouwen, *M.Du.* klauwen; *claw, scratch, 'scalpo,'* PR. P. 80; CH. C. T. *D* 940; clowe [clawen, clawe] LANGL. B PROL. 154; clowe FER. 463; clāweþ (*pres.*) PL. CR. 365; GOW. II. 93; clāwe (*subj.*) O. & N. 154; clāwed (*pret.*) CH. C. T. *A* 4326; cleu (*ms.* clew) FER. 5339; S. S. (Web.) 925.

clāwinge, sb., *clawing, scratching, 'scalpi-lacio,'* PR. P. 80; þe bore likide þe clavinge wel S. S. (Wr.) 978.

clō, *see* clau.

clēafer, sb.,=? *M.H.G.* klouber; *claw, finger-nail*; clēafres (*pl.*) A. R. 102; *see* claver.

clēche, sb., *claw, finger-nail*; clēches (*pl.*) A. R. 174*.

clēchen, v., *seize* (*with claws, hands*); cleche HORN (R.) 963; C. L. 734; cleke IW. 2478; þenne sir Gauan bi þe coler clēches (*pres.*) þe kniʒt ANT. ARTH. xlviii; clekes D. ARTH. 1865; clekis ALEX. 842; clāhte (*pret.*) A. R. 102*; clāht (*pple.*) SPEC. 37.

clef, *see* clif. clei, *see* clai.

cleken, v., *O.N.* klekja; *?hatch, bring forth*; ? clekit (*pple.*) TOWNL. 311.

clēken, *see* clēchen. clemben, *see* climben.

clēmen, *see* clǣmen.

clemeren, *see* climbren.

clemmen, v., *O.E.* (be)clemman, = *O.L.G.* (ant-, bi)klemmian, *M.Du.* clemmen, *O.H.G.* (pi)chlemmen; *from* clam; *compress*; clem-(m)id (*pple.*) HALLIW. 254; *comp.* for-clemmen.

clenchen, v.,=*M.H.G.* klenken; *make to clink*: (h)is harpe he gan clenche HORN (R.) 1498; clenchin aʒen *'obgarrio'* PR. P. 80; clenken (*? printed* blenken, *v. r.* clinken) . . . a belle CH. C. T. *B* 1186; clinken (*? for* clenken) P. S. 189; þat aʒen clenkeþ (*pres.*) SHOR. 113.

clenchin, v.,=*O.H.G.* (gi-, in)clenken; *clench, 'retundo,'* PR. P. 80; ich wot ʒef smiþes schule uvele clenche O. & N. 1206; cleint (*pple.*) L. H. R. 138; iclenched CH. C. T. *A* 1991; *comp.* un-clenchen.

clēne, *see* clǣne.

clengen, v. (*properly causative from* clingen); *cling*: þer clengez (*pres.*) þe colde GAW. 2078; þe forst clenged (*pret.*) 1694.

clenken, *see* clenchen. clēnsien, *see* clǣnsien.

clēo, *see* clau. cleof, *see* clif.

cleopien, *see* clépien.

clēoven, v., *O.E.* clēofan, = *O.N.* kliufa, *O.H.G.* chliuban; *cleave, split*; cleove ALIS. 7702; cleven KATH. 2027; HAV. 917; cleve MAN. (H.) 188; clefe PR. C. 6736; clēf (*pret.*) HAV. 2643; P. L. S. xxiv. 254; ALIS. 3708; TRIST. 2384; LANGL. B xviii. 61; cleef MAND. 86; clæf LAʒ. 21390; (þai) clove FER. 2724; clofe TOWNL. 255; clóve (*subj.*) LEG. 66; clófen (*pple.*) ORM. 1224; cloven FER. 1623; CH. C. T. *A* 2934; MAND. 95; cloven [cleft] WICL. MAT. xxvii. 51; clovin LUD. COV. 45; iclove ROB. 49; TREV. VIII. 223; *comp.* a-, for-, to-clēoven.

clēvinge, sb., *act of splitting, cleft*, TREV. II. 383; MAND. 86.

cléovien, *see* clévien.

clēowe, sb., *O.E.* cliwe, = *O.H.G.* cliwa, chliwa; *clew*; cl(e)owe, clewe O. & N. 578; clewe VOC. 180; a clewe of þrede TREV. II. 385; a klewe of yarn VOC. 157; clewe [cliwe] CH. L. G. W. 2016.

clēowen, v., *O.E.* clēowen, cliewen, cliwen, Ger. kleuen; *ball, clew*, FRAG. 4.

clépien, v., *O.E.* cleopian, cliopian, clypian, cli-pian; *call*, MK. x. 47; KATH. 2199; (*ms.* clep-yen) MISC. 51; clepien, clepian, cleopien, cluplien [clepie, cleopie] LAʒ. 852, 2498, 5945, 20357; ʒif ei wolde cleopien him to mete A. R. 260; cleopien, clipien HOM. I. 185, 189; clepen ORM. 15325; clepin *'voco'* PR. P. 81; S. S. (Wr.) 1452; clepen, clepe CH. C. T. *A* 643; clepe WILL. 1299; cleopie LEG. 33; bischopes he let clipie BEK. 472; me to clupie king ROB. 322; to þe i clépie (*pres.*); ant calle SPEC. 59; cleopeþ O. & N. 1315; cleopaʒ HOM. I. 119; we cleopiaʒ to gode 113; ʒif we clepieʒ þine feder 55; ʒe me þenne clepiaʒ (*invoke*) 13; (hie) clepieʒ II. 21; þerne diaþ hi clepieþ lif AYENB. 72; clipieþ SHOR. 107; cleopiende (*pple.*) HOM. 117; clépede (*pret.*) LAʒ. 191; GEN. & EX. 1274; OCTOV. (W.) 759; cleopede toward heovene MARH. 18; clupede HORN. (L.) 225; clipede FER. 142; clepeden ORM. 12978; WICL. JOSH. xxii. 34; clépéd (*pple.*) ORM. 14008; MAND. 305; TOR. 341; clepid LANGL. B x. 21; clept HOCCL. i. 225; *comp.* bi-, ʒe-, in-, wiʒ-clépien.

clépinge, sb., *O.E.* cleopung, clypung; *calling*, WICL. EPHES. iv. 1; ED. 1123; cleopinge LAʒ. 10287.

cleppe, *see* clappe. cleppen, *see* clüppen.

cleppin, v.,=*M.L.G.*, *M.Du.* kleppen; *strike, sound, 'tinnio,'* PR. P. 81.

clēr, adj., *O.Fr.* clēr, clier; *clear*, FL. & BL. 224; cleer as water TREV. I. 123; TREAT. 133; clier AYENB. 24; þe cliere light LIDG. M. P.

215; þe sunne þat schines clēre (*adv.*) C. M. 291; clērore (*compar.*) LEG. 98; clierer AYENB. 267.

clēr-līche, adv., *clearly*, BRD. 21; clierlīche AYENB. 155.

clērnesse, sb.,*clearness*,LEG. 170; AYENB. 95.

clēre-wort, sb., *name of a red herb*, D. ARTH. 3241.

clerc, sb., *O.E.* cleric, *O.Fr.* clerc; *clerk,clergyman, scholar*, SAX. CHR. 250; LAȝ. 39; P. L. S. xvii. 50; clerk A. R. 318; CH. C. T. *A* 480; clerkes (*pl.*) O. & N. 729.

clerke-lī, adv., *like a clerk, 'clericaliter,'* PR. P. 81.

clerematin, sb., *kind of fine bread*, LANGL. *A* vii. 292 (*B* vi. 306; *C* ix. 328).

cleretē, *see* clerte.

clērin, v., *O.Fr.* clairer; *make clear, clear*, PR. P. 81; clēreþ (*pres.*) GOW. III. 313; cleerid (*pple.*) WICL. EX. xix. 16.

clergeal [clergial], adj., *learned*, CH. C. T. *G* 754.

clergial-lī, adv., *learnedly*, D. ARTH. 200.

clergeōn, sb., *O.Fr.* clergeon (*petit clerc*); *clerk in minor orders*, GOW. I. 255; CH. C. T. *B* 1693.

clergesse, sb.,*O.Fr.*clergesse; *learnedwoman*, A. R. 6; KATH. 75.

clergie, sb., *O.Fr.* clergie; *clergy*, P. L. S. xxv. 155; P. S. 324; JOS. 171; clergies (*pl.*) KATH. 538.

clertē, sb., *splendour*, AUD. 45; WICL. TOBIT xiii. 20; clerete WICL. DEEDS xxii. 11.

clōt, clōt, sb.,=*M.L.G.* klōt *m.*, *M.H.G.* klōz *m. n.; wedge, 'cuneus,'* PR. P. 81.

clēte, sb., *? O.H.G.* chletta, chledda; *burdock, 'lappa,'* VOC. 191; of al France ȝaf nouȝt a clete TRANS. XVIII. 24.

clēþen, *see* clāþen.

clēve, sb., *O.E.* cleofa, clyfa,=*O.N.* klefi *m.; chamber*, HAV. 557; ANGL. IV. 189; kleve *'cubile'* PS. xxxv. 5.

clēven, *see* clēoven.

clēvi, adj., *O.E.* clifig, *cf. Du.* klevig, *Ger.* klebig; *adhesive* : clivi as clide A. P. ii. 1692.

clēvien, v., *O.E.* cleofian, cliofian, clifian,= *O.L.G.*clivōn,clevōn;*cf. O.H.G.*chlebēn; *from* clīven; *cleave, adhere*, (ms. cluyen, *printed* clenyen), SPEC. 37; clivin, cleve PR. P. 82; cleve [klive] PS. xciii. 20; cleovien JARB. XIII. 174; cliven GEN. & EX. 372; clive FER. 1901; clēveþ (*pres.*) TREAT. 137; cliveth CH. BOET. iii. 12 (101); cleovieð LAȝ. 9389; clēve, clive (*subj.*) LANGL. *B* xi. 219; clēviinde (*pple.*) AYENB. 107; clēfede (*pret.*) LK. x. 11; clevede BEK. 2242; WICL. JOB xxix. 10; cleved WILL. 734; HOCCL. i. 31; cliveden HAV. 1300; clēved (*pple.*) HOM. II. 73.

clēviinde-līche, adv., *adheringly*, AYENB. 103.

clēvre, *see* clāver.   clēwe, *see* clēowe.

clib, adj., *? from* clīven; *?obstinate*, JARB. XIII. 163; clibbe (*pl.*) MISC. 144; ȝee mote boe wel clibbe (*? printed* chybbe) to floe ham REL. II. 243; clibbest (*superl.*) JARB. XIII. 163.

clicchen, v., *?* = clucchen; *clutch* : to folden and to clicchen (*v. r.* clucche) LANGL. *C* xx. 120; clihte (*pret.*) L. H. R. 145; cliȝte BER. 2515; hit cliȝt (*adhered*) to geder A. P. ii. 1692; clighte CHEST. I. 115; in ȝoure armour so fast icliȝt (*pple.*) ASS. *B* 719; he fond the . . . finger with the ring icliȝt into the paume of the hond TREV. VII. 537; cleȝt A. P. ii. 1655.

clide, sb., *O.E.* cliþe; *bur*, A. P. ii. 1692.

client, sb., *Fr.* client; *client*, GOW. I. 284.

clif, sb., *O.E.* clif, cleof, = *M.Du.*, *O.N.* klif *n.; cliff* : þat clif [clef] LAȝ. 1926; klif GAW. 713; clif, clef TREV. III. 455; clive (*dat.*) P. L. S. viii. 175; B. DISC. 1163; clive, cleve TREV. VII. 461; clife, clefe EGL. 379, 415; to þan cleove LAȝ. 20861*; clives (*pl.*) LAȝ. 21807; kliffes PS. cxiii. 8; cleves CH. L. G. W. 1466; cliven (*dat. pl.*) ALIS. 5429; *comp.* sǣ-clif.

clift, *see* clüft.

cliket, sb., *Fr.* cliquet; *clicket, sort of latch, latch-key, 'clavicula,'* PR. P. 82; LANGL. *A* v. 613; CH. C. T. *E* 2046; clekett *'clavis'* CATH. 66.

cliketed, pple., *fastened with a latch* [*v.r.* clikated, cliketted, i-clicated], LANGL. *B* v. 623.

climat, sb., *Fr.* climat; *climate*, LANGL. *C* xviii. 106; CH. ASTR. ii. 39 (48).

climben, v., *O.E.* climban,=*O.H.G.* chlimban; *climb*, A. R. 140; ORM. 11860; CH. C. T. *E* 106; climben [clemben] LAȝ. 851; climbe O. & N. 833; WILL. 707; climme SHOR. 3; clembe P. L. S. xxiii. 72; clamb (*pret.*) PERC. 1223; clam ROB. 333; S. S. (Wr.) 851; clamb, clomb LEG. 74, 75; clomb A. R. 354; þu clumbe LAȝ. 21439; heo clumben [clemde] 9420; A. R. 244; clomben, clumben CH. C. T. *A* 3636; clomben GAW. 2078; clomme ROB. 410; him þaȝte he clemde upon þis treo P. L. S. xiii. 123; clombe (*pple.*) CH. C. T. *B* 12; iclumben LAȝ. 21432; HOM. I. 211; A. R. 178.

climare, sb., *climber*, PR. P. 82.

climbren, v.,=*O.H.G.* klimmern, klemmern, *climb*: clemeri LEG. 166.

clinen, v., *O.Fr.* cliner;=declīnin; *bow down, 'declino,'* PR. P. 82; i clīne (*pres.*) to þis acorde LUD. COV. 114.

clingen, v., *O.E.* clingan; *cling, adhere; wither away*; clinge O. & N. 743; ALIS. 915; clinge & drie FER. 2524; clinge & cleve LUD. COV. 54; i clinge (*pres.*) as cleiȝ S. A. L. 178; whan þou . . . clingest for drie LANGL. *B* xiv. 50; ant clingeþ so þe clai SPEC. 85; his beli clinges

PR. C. 823; **clang** (*pret.*) C. M. 4699; his limes clonge a wei P. L. S. xxiv. 215; **clongen** [clungen] (*pple.*) C. M. 4581; clungen was his chek M. H. 88; clunge L. H. R. 142; clongin M. ARTH. 751; clonge PALL. ii. 46; *comp.* **for-clingen**; *deriv.* **clengen.**

clinken, = *M.Du.* clinken, *? O.N.* klykkja; *clink*; clinkin '*tinnio*' PR. P. 82; þei herde a belle clinke CH. C. T. *C* 664; *deriv.* **clenchen.**

clint, sb., *O.N.* *klintr,=*Icel.* klēttr, *L.Ger.,* *Dan., Swed.* klint; *rock*; **clintes** (*pl.*) C. M. 17590; klintes HALLIW. 497.

clipien, *see* **clépien.**   **clippen,** *see* **clüppen.**

clippen, v., *O.N.* klippa; *clip* (with shears), ORM. 4106; clippin '*tondeo*' PR. P. 82; clippe WICL. GEN. xxxi. 19; clippe and schave CH. C. T. *A* 3326; **clippid** (*pple.*) WICL. GEN. xxix. 16.

   clippare, sb., *O.N.* klippari; *clipper, shearer,* '*tonsor,*' PR. P. 82; clippere WICL. IS. liii. 7; **clippers** (*pl.*) MAN. (H.) 238.

   clippinge, sb., *clipping,* '*tonsura,*' PR. P. 82.

     clipping-tīme, sb., *shearing-time,* GEN. & EX. 1739.

clips, sb., *for* **eclipse**; LANGL. *B* xviii. 135; clippice PR. P. 82.

clipsi, adj., *eclipsed,* R. R. 5351.

clisterie, sb., *O.Fr.* clistēre; *clyster,* TREV. IV. 393.

[**clīve,** sb., *O.E.* clīfe,=*M.L.G.* klīve, *M.Du.* klijve, *O.H.G.* chlība; *bur; comp.* **gār-clīve.**]

clīven, v., *O.E.* clifan,= *O.H.G.* klīban (*adhere*), *M.Du.* clīven, *O.N.* klīfa (*climb*); *climb*: þet wilneþ heʒe to clive AYENB. 23; hi cliven (*pret. pl.*) into þe helle AYENB. 126; hi biþ . . . heʒe iclive (*pple.*) 26.

clīven, v., = **clēoven**; *split, cleave*; clivin '*findo*'. PR. P. 82; þe helm he clāf MAN. (F.) 10895.

cliver, sb., *O.E.* clifer (*gen. pl.* clifra); *claw*; **clivres,** clevres (*pl.*) O. & N. 155, 270; clivre (*dat. pl.*) 78; *cf.* **claver.**

cliver, adj., *? O.E.* clibbor; *? from* clīven *? mod.* *Eng.* clever; *tenacious*: & te devel (is) cliver on sinnes REL. I. 213.

clivi, *see* **clévi.**   **clivien,** *see* **clévien.**

clobbed, *see* **clubbed.**

cloche, *see* **clucche.**

clocher, sb., *Fr.* clocher; *belfry,* PR. P. 82; LIDG. M. P. 201.

clodde, *see* **clotte.**   **cloddres,** *see* **cloter.**

cloftunge, sb., *O.E.* clufþung; *ranunculus,* LEECHD. II. 377.

clogge, sb., *clog, lump, obstacle,* '*truncus,*' PR. P. 83; **clogges** (*pl.*) P. S. 154; TOWNL. 313.

cloistre, sb., *O.Fr.* cloistre; *cloister,* BEK. 2077; AYENB. 151; CH. C. T. *G* 43.

cloistrer, sb., *O.Fr.* cloistrier; *monk*; **cloistrers** (*pl.*) AYENB. 67.

clóke, sb., *claw,* TOWNL. 324; **clókes** (*pl.*) A. R. 102; PR. C. 6936; klokes D. ARTH. 792.

clóke, sb., *cloak,* LAʒ. 13097*; CH. C. T. *A* 157; clooke PR. P. 83.

clōked-lī, adv., *from* **clōke**; *disguisedly,* S. & C. II. lvii.

cloken, v., *Fr.* cloker, clocher; *limp, walk lame*: þere conninge clerkes shul clocke [clokke] bihinde LANGL. *B* iii. 34.

clokke, sb.,=*M.Du.* klocke, *O.N.* klokka, klukka; *clock,* PR. P. 83; a clokke CH. C. T. *B* 4044; ten of þe clokke it was *I* 5; at vii. of the clokke W. & I. 17.

clokkin, v., *O.E.* cloccian = *Swed.* klokka, klukka, *M.H.G.* klucken, glucken; *cluck,* PR. P. 83.

clom, sb., *? silence,* AYENB. 264, 266; clum [clom] CH. C. T. *A* 3639.

clōm, *see* **clām.**

[**clomen,** v., *Du.* kleumen, *? be stiff; comp.* **a-cleomen.**]

clomsen, v., *? be numb, stiff*; whan þou clom-sest (*pres.*) for colde LANGL. *B* xiv. 50; he is outher clomsed (*pple.*) or wod PR. C. 1651; clumsid hondis '*manus dissolutas*' WICL. IS. xxxv. 3; *comp.* **a-clumsen.**

clond, *see* **cland.**

clōs, adj., *Fr.* clōs; *close, shut up,* GOW. II. 94; A. P. i. 183; cloos TREV. II. 75.

clōs, sb., *O.Fr.* clōs; *close, field, enclosed place,* ROB. 7; cloos or ʒerde '*clausura*' PR. P. 83.

clōsen, v., *from* **clōs**; *close, enclose*; close ROB. 28; clōsede (*pret.*) HAV. 1310; clōsed (*pple.*) CHR. E. 629.

   clōsingis, sb. pl., *gates,* WICL. PROV. viii. 3.

closet, sb., *? O.Fr.* closet; *closet,* GAW. 934; TREV. V. 313.

closūre, sb., *O.Fr.* closure; *closure,* '*clausura,*' PR. P. 83.

clot, sb.,=*M.Du.* klot (*lump, clod*), *O.H.G.* cloz *m.* (*lump*); *clod,* P. L. S. xxxiii. 25; H. V. 13; ane clot of hevi eorče A. R. 140; **clotte** (*dat.*) A. R. 254; **clottes** (*pl.*) TREV. II. 23; clottis '*glebae*' WICL. JOB xxviii. 6; *comp.* **clei-clot.**

clōt, *see* **clēt.**

clōte, sb., *O.E.* clāte, *f.*; *burdock,* '*lappa,*' FRAG. 3; PR. P. 83; REL. I. 37; clote [cloote] WICL. HOS. ix. 6.

   clōte-leef, sb., *leaf of the burdock,* CH. C. T. *G* 577.

cloter, sb., *cf. Swed.* klotr, *n.* (*globus*), *M.Du.* kloter (melk); *clot*: in **cloddres** (*pl.*) of blod his her was clunge L. H. R. 142.

cloterin, v., *cf. M.Du.* cloteren; *clot,* '*coagulo,*' PR. P. 83; **clotred** [clotered] (*pple.*) blood CH. C. T. *A* 2745.

clotte, sb., *cf. M.Du.* klotte, *? Ger.* klotze; *clod,* CATH. 68; REL. II. 192; clodde '*gleba*' PR. P. 83; PALL. i. 73.

clōð, *see* **clāð.**

clou, sb., *Fr.* clou (de girofle); *clove,* L. C. C. 17; clowe (? = clou) CATH. 68.

  clowe-gilofre, sb., *clove-gilly-flower,* MAND. 263.

clouchin, sb., *? M.Du.**klouwken; *from* klouwe, cloue; *little ball,* '*glomicellus,*' PR. P. 83.

cloude, *see* **clūde.**

cloue, sb., *M.Du.* klouwe; *ball*; (*ms.* clowe) '*glomus*' PR. P. 83.

clouȝ, sb., *clough,* TRIST. 1761; **cloughis** (*gen.*) M. ARTH. 893; clous C. M. 17590.

clouse, *see* **clūse.**   **clout,** *see* **clūt.**

cloute, sb., *blow on the head,* TRIAM. 781; ISUM. 619; LUD. COV. 98.

clóve, sb., *O.E.* cluf (*acc. pl.* clufe LEECHD. II. 36); *clove*: clove of garlek PR. P. 84.

clōvere, *see* **clāver.**   **clōwen,** *see* **clāwin.**

club, sb., *cf. Swed.* klubb, *m.* (*lump*); (*ms.* glubbe) '*globus*' WICL. NUM. xvi. 11.

clubbe, *O.N.* klubba *f.*; *club,* '*fustis,*' PR. P. 84; LAȝ. 26062; HAV. 1927; '*pedum*' VOC. 240; klubbe A. P. i. 1348; clobbe TRIST. 2338; **clubbes** (*pl.*) Iw. 3200; **clubben** (*dat. pl.*) LAȝ. 21504; *comp.* **mæin-clubbe.**

clubbed, adj., *Swed.* klubbed; *clubbed,* CH. C. T. *B* 3088; clubbid '*rudis*' PR. P. 84; clobbed CH. C. T. *B* 3088.

clucche, sb., *clutch*; **cloches** (*pl.*) LANGL. *B* PROL. 154; cloches MAP 338.

clucchen, v. *clutch*: to cluche [clucche] or to clawe LANGL. *B* xvii. 188; **cluche** (*pres.*) REL. II. 211; cluchches A. P. ii. 1541; upe here ton heo seten icluȝt LEG. 192; *comp.* **for-clucchen.**

clūde, sb., *O.E.* clūd *m.*, *cf. M.L.G.* klūde *f.*, (*stone, as a weight*); *rock*: þa clude (*sec. text* cleve) LAȝ. 1915; cloude SPEC. 44; wormes woweþ under cloude SPEC. 44; **clūdes** (*pl.*) ORM. 2656; knarres & cludes O. & N. 1001; clude LAȝ. 21849; **clūden** (*dat. pl.*) LAȝ. 31880.

clūde, sb. (*prob. same word as the preceding*); *cloud,* C. M. 18402; cloude '*nubes*' PR. P. 84; BEK. 1415; ALIS. 777; CH. C. T. *B* 3956; **clouden** (*pl.*) TREAT. 136; cloudes AYENB. 108; PS. xvii. 12; cloude (*dat. pl.*) A. D. 243.

clūdi, adj., *cloudy*; cloudi BRD. 23; LIDG. M. P. 139.

clūdiȝ, adj., *O.E.* clūdig; *rocky,* ORM. 2734.

clüft, sb., = *Swed.* klyfta, *Dan.* klöft, *M.H.G.* kluft *f.* (*fissure*); *from* clēoven; *cleft*; clift '*scissura, rima*' PR. P. 81; '*furchure*' VOC. 148; clift [clifte] C. M. 19834; **clifte** (*dat.*) CH. C. T. *D* 2145; to helle clifte L. H. R. 205;

clüftes, cliftes '*fissurae*' TREV. IV. 347; cliftus ['*scissuras*'] WICL. IS. xxii. 9.

clum, *see* **clom.**   **clumsen,** *see* **clomsen.**

clüpien, *see* **clépien.**

clüppen, v., *O.E.* clyppan, *cf. O.Fris.* kleppa; *embrace,* HOM. I. 185; A. R. 424; **cluppe** SPEC. 38; clippe HORN (R.) 1362; LANGL. *B* xvii. 188; PARTON. 621; cleppen FL. & BL. 594; cleppe OCTOV. (W.) 585; **clippeþ** (*pres.*) CH. C. T. *E* 2413; **clüpte** (*pret.*) LAȝ. 5011; BEK. 288; clipte WILL. 1265; MAN. (F.) 12118; **clipped** (*pple.*) WILL. 859; *comp.* **bi-clüppen.**

  clüppunge, sb., *act of embracing,* A. R. 324; cluppunge, cluppinge H. M. 3; cluppinge HOM. I. 201.

clūse, sb., *O.E.* clūs, = *O.H.G.* chlūsa, *from med.Lat.* clūsa; *mill dam*: et ter mulne cluse A. R. 72; clouse PR. P. 84.

clūsen, v., *O.E.* (be)clȳsan; *close*; **clūsden** (*pret.*) LAȝ. 9760; *comp.* **bi-, for-clūsen.**

  clūsung, sb., *O.E.* clȳsung, clȳsing; *closing,* FRAG. 1.

cluster, sb., *O.E.* cluster, clyster, = *L.G.* kluster; *cluster,* TREV. IV. 151; WICL. DEUT. xxxii. 32; clustir, closter '*botrus*' (βότρυς), *racemus*' PR. P. 84; closter '*botrus*' VOC. 277; **clustres** (*pl.*) MAND. 168.

[**clüster,** sb., *O.E.* clūstor; *cloister.*]

  clüster-loke, sb., *O.E.* clūstorloca; *cloister*; **clüsterlokan** (*pl.*) HOM. I. 43.

clusteren, v., *L.Ger.* klustern; *cluster*: moni **clustered** (*pple.*) clowde clef alle in clowtez A. P. ii. 367.

clūt, sb., *O.E.* clūt, = *O.N.* klūtr; *clout, rag,* A. R. 256; R. S. v (MISC. 172); clout '*panniculus, pittacium*' PR. P. 84; REL. I. 65; MAND. 293; þane clout SHOR. 15; a wollen clout ALIS. 4459; ich nabbe clout ne lappe P. R. L. P. 227; **clūtes** (*pl.*) HOM. I. 277; ORM. 3327; HAV. 547; cloutes PL. CR. 244; sche rente it al to cloutes CH. C. T. *E* 1953; *comp.* **pilche-, sweþel-, winde-clūt.**

clūte, sb., = *M.L.G.* klūte; *clod,* O. & N. 1167.

clūtien, v., *O.E.* (ge)clūtian; *patch, mend*; cloutin '*sarcio*' PR. P. 84; she wolde cloute mi cote LUD. COV. 98; **clūtie** (*pres. subj.*) A. R. 256; **clouted** (*pple.*) PL. CR. 424; i-clouted LANGL. *B* vi. 61; *comp.* **bi-clūtien.**

clouter, sb., *patcher,* '*sartor,*' PR. P. 84.

cnag, sb., *cf. Dan.* knag, *Swed.* knagg *m.*; *knot, peg*; knagg FLOR. 1795.

cnaggen, v., *fasten with a knot*; **knagged** (*pple.*) D. TROY 4972; knaged GAW. 576.

cnak, sb., *O.N.* knakkr; *crack, noise*; knak FER. 4599; **knackis** (*pl.*) WICL. E. W. 156; knakkes (*v. r.* crekes) CH. C. T. (Wr.) 4049.

cnakken, v., *cause to sound*; **knackus** (*pres.*) WICL. S. W. III. 482.

cnal, sb., *? =* **cnül**; *knock,* FER. 463*.

cnap, sb., *O.E.* cnæpp (*vertex*),= *O.N.* knappr (*knob*) ; *knob* ; **knappes** (*pl.*) of golde LANGL. *B* vi. 272.

cnap, sb.,= *Du.* knap ; *blow, knock* : þe knight, under **knappis** (*pl.*), uppon knes fell D. TROY 6437.

cnápe, sb., *O.E.* cnapa, = *O.Fris.* knapa, *O.L.G.* knapo, *O.N.* knapi ; = **cnáve** ; *boy*, ['*puer*'], MAT. viii. 6 ; knape GEN. & EX. 477 ; S. S. (Web.) 930 ; GOW. III. 321 ; L. H. R. 136 ; GAW. 2136 ; S. & C. I. lxi ; **cnápes** (*gen.*) ORM. 4106 ; **cnápen** (*acc.*) LK. ix. 47 ; *comp.* herde-cnape.

    cnápe-child, sb., *male child*, ORM. 7895 ; knapechild GEN. & EX. 2585.

cnarre, sb., = *L.Ger.* knarre ; *knot* ; knarre GAW. 1434 ; CH. C. T. *A* 549 ; **knarres** (*pl.*) & cludes O. & N. 1001 ; ful of knarres '*verticibus plenum*' WICL. WISD. xiii. 13.

cnarri, adj., *knotty, gnarled* ; knarri CH. C. T. *A* 1977.

cnáve, sb., *O.E.* cnafa, = *O.H.G.* chnabe, *mod.Eng.* knave ; *boy, servant* (*ms.* cnaue), LA3. 292 ; PL. CR. 288 ; knave GEN. & EX. 1151 ; HAV. 450 ; TREAT. 139 ; WILL. 2394 ; FLOR. 2095 ; CH. C. T. *A* 3434 ; TOR. 2008 ; þe kokes knave A. R. 380 ; knight, swain and knave ALIS. 7968 ; is youre child a knave TOWNL. 113 ; knafe M. H. 131 ; **cnáven** (*pl.*) LA3. 15559.

    knáve-barn, sb., *male child*, C. M. 2668.

    cnáve-child, sb., *male child*, HOM. II. 47 ; LA3. 15526* (*first text* cnihtbærn) ; knave-child P. L. S. xxiii. 45 ; WICL. EXOD. i. 16 ; CH. C. T. *B* 715.

cnávisch, adj., *knavish* ; a knavisch speche CH. C. T. *H* 205.

cnáwe, sb., *O.E.* (ge)cnáwe ; *conscious* : he nolde be knowe for no þing þat hit wes a maide A. D. 258.

    cnáwenesse, sb., *knowledge* ; cnow(e)nesse HOM. II. 25.

cnáw-lēche, sb., *from stem of* **cnáwen** ; *knowledge* ; knauleche C. M. 12162 ; knoweleche TREV. III. 216 ; knouleche MAND. 313 ; CH. C. T. *B* 1220 ; LUD. COV. 93 ; souзten him a mong hise cosins and hise knouleche (*acquaintances*) WICL. LK. ii. 44 ; knoulache AUD. 19 ; knaulage PERC. 1052.

cnáwelēchi, v., *from* **cnáwlēche** ; *acknowledge* ; cnowlechi LEB. JES. 198 ; knoulechin '*fateri*' PR. P. 280 ; knouleche MAP 335 ; MAND. 120 ; knoulache MIRC 1670 ; **cnáwlēcheð** (*pres.*) KATH. 1352 ; зif ha . . . cnawlecheð soð H. M. 9 ; **knōulēchide** (*pret.*) WICL. ESTH. viii. 1 ; knowleched LANGL. *B* v. 481 ; knoulecheden WILL. 4782.

    cnáwlēchinge, sb., *acknowledging*, KATH. 1388 ; knaulechinge AVENB. 132 ; cnoulechunge A. R. 92 ; knowleching GOW. II. 23.

cnāwen, v., *O.E.* (ge)cnáwan (*pret.* cnēow), *O.H.G.* (int-, ir-)chnāhan, knāhan (*pret.* chnāta), *O.N.* *pres.* knā (*pret.* knātta) ; *know*, ORM. 1314 ; knawe HAV. 2785 ; AYENB. 26 ; PR. C. 429 ; knaue ANT. ARTH. xix ; cnowen HOM. II. 81 ; knowen WILL. 577 ; knowe LANGL. *B* xix. 198 ; cnouen REL. I. 172 ; knoue AUD. 43 ; **cnāweð** (*pres.*) HOM. I. 249 ; **knōu** (*imper.*) H. H. 165 (167) ; **cnēow** (*pret.*) ORM. 12584 ; cneou LA3. 17069 ; kneow ALIS. 113 ; cnew HOM. II. 127 ; kneu HORN (R.) 1151 ; ROB. 315 ; C. M. 11032 ; kneew ALIS.[2] (Sk.) 541 ; þou cnewe REL. I. 48 ; knewen GEN. & EX. 2904 ; knewen AYENB. 246 ; knewe LANGL. *B* xii. 154 ; зef þu cneowe JUL. 22 ; **knāwen** (*pple.*) lw. 2666 ; PR. C. 444 ; knowen GEN. & EX. 3037 ; knowen MAND. 2 ; knowe WILL. 726 ; *comp.* **an-, bi-, зe-cnāwen.**

    cnāwing, sb., *knowing* ; knawing PR. C. 145 ; cnowunge A. R. 280.

cnāwes, adj., *? for* **cnāwe** ; beo nu ken & cnawes KATH. 2070 ; *comp.* **sōð-cnāwes.**

cnē, *see* **cnēo.**

cnéden, v., *O.E.* cnedan, = *M.L.G.* kneden, *O.H.G.* chnetan ; *knead* ; knedin PR. P. 279 ; knede CH. C. T. *A* 4094 ; **cnédest** (*pres.*) ORM. 1486 ; **knéden** (*pple.*) R. R. 217 ; knodon PR. P. 280 ; *comp.* **зe-cnéden.**

    cnédare, sb., *kneader* ; knedare PR. P. 279.

    [**cnéding**, sb., *kneading.*]

      knéding-trogh, sb., *kneading-trough,* CH. C. T. *A* 3548.

      knéding-tubbe, sb., *kneading-tub,* CH. C. T. *A* 3563, 3594.

cnēlien, *see* **cnēolien.**

cnēo, sb., *O.E.* cnēo, cnēow,= *O.L.G.* kneo, knio, *Goth.* kniu, *O.Fris.* knē, knī, *O.N.* knē, *O.H.G.* chneo, chniu, *Lat.* genu, *Gr.* γόνυ ; *knee* ; cneow FRAG. 6 ; knee '*genu*' VOC. 186 ; to þe niþe cne (*degree*) HICKES I. 226 ; kne GEN. & EX. 444 ; in (þe) teþe kne ROB. 228 ; **cn(ē)owe** (*pl.*) LA3. 5388* ; cneon HOM. I. 191 ; MARH. 9 ; falleð a cneon A. R. 16 ; kneon TREAT. 139 ; cnewes (*ms.* cnewwess) ORM. 6467 ; knewes AR. & MER. 6551 ; knes HORN (L.) 383 ; knees [knowes] LANGL. *A* v. 203 ; **cnēowen** (*dat. pl.*) LK. v. 8 ; cneowen LA3. 12685 ; þe king læi on cneouwen 32046 ; cneowe 29573 ; cneon 4996 ; kneon MISC. 149.

    knēo-bōn, sb., *knee-bone*, CHR. E. 758.

    cnē-hole, sb., *O.E.* cnēowholen (VOC. 68) ; *knee-holly* (*printed* cue-hole), REL. I. 37.

    knee-panne, sb., *knee-pan*, VOC. 183.

    cnēo-res, sb., *O.E.* cneōress, -ryss ; *generation*, MAT. xvii. 17.

cnēolien, v.,= *M.L.G.* knēlen, *M.Du.* knielen, *kneel* : cneolin MARH. 20 ; cneolen, cnelen ORM. 6138, 11392 ; knele HAV. 1320 ; MAND. 14 ; EGL. 1279 ; **cnēwlest** (*pres.*) HOM. II.

25 ; cneuleð 83 ; **cnēole** (*imper.*) LAȝ. 24163 ;
cneoleð A. R. 18 ; **knēoli** (*subj.*) R. S. v (MISC.
168) ; **knēolinde** (*pple.*) A. R. 18 ; **cnēoulede**
(*pret.*) LAȝ. 29654 ; kneulede BEK. 540 ;
knelede SPEC. 106 ; TREV. V. 455.

     knēlare, sb., *kneeler*, PR. P. 279.

     knēuling, sb., *kneeling*, HORN (H.) 810 ;
cneling̃ ORM. 1451.

cnēowien, v., *O.E.* cnēowian, *cf. O.H.G.* chni-
wen, chniuwen ; *kneel* ; **knēwede**(*pret.*) HOM.
I. 121 ; knowede MISC. 48.

     knēouwunge, sb., *O.E.* cnēowung ; *kneeling*,
HOM. I. 199.

cnicche, *see* **cnücche**.

cnīf, sb., *O.E.* cnīf, = *M.Du.* knijf *n.*, *O.N.*
knīfr *m.* ; *knife*, KATH. 1953 ; þat cnif ORM.
4128 ; enne longne cnif LAȝ. 3775 ; knif A. R.
282 ; HAV. 479 ; SPEC. 102 ; ALIS. 1061 ;
MAND. 51 ; **cnīfes** (*gen.*) ORM. 4257 ; **knīve**
(*dat.*) HOM. I. 69 ; cnifes 22805 ; cnives
12282 ; **cnīve** (? *ms.* cniues, *r. w.* live) (*pl.*)
LAȝ. 18027 ; knifes [knives] CH. C. T. *A* 233 ;
knives A. R. 212 ; HAV. 1769 ; **cnīven** (*dat.*
pl.) LAȝ. 4009 ; *comp.* **bord-knīf.**

     knīf-w(e)**orpare**, sb., *knife-thrower*, A. R.
212.

cniht, sb., *O.E.* cniht, cnyht, cnieht, cnioht,
cneoht, = *O.Fris.* knecht, *O.H.G.* chneht, *mod.*
Eng. knight ; *boy ; servant ; knight*,
HOM. II. 181 ; cniht, knit [cniþt] LAȝ. 346,
2800 ; cniht, kniȝt O. & N. 1087, 1575 ; kniht
HORN (R.) 508 ; knicht WART. II. 2 ;
kniȝt DEGR. 1506 ; ich am þi kniȝt ROB. 115 ;
Lazar þe kniȝt LEB. JES. 678 ; knigt GEN. &
EX. 283 ; knight CH. C. T. *A* 43 ; knith, knict,
knicth, knit HAV. 32, 77, 1650, 2427 ; knict,
knit HORN (H.) 524, 1038 ; **cnihtes** (*gen.*)
LAȝ. 19950 ; **cnihte** (*dat.*) LAȝ. 7703 ; knihte
A. R. 86 ; **cnihtes** (*pl.*) SAX. CHR. 252 ; (*ms.*
cnihhtess) ORM. 8185 ; knihtes A. R. 358 ;
cnihtes, cnihte LAȝ. 3617, 26780 ; kniȝte
P. L. S. xix. 263 ; **cnihte** (*gen. pl.*) LAȝ.
8172, 18007 ; cnihtene (*printed* cinhtene)
3346 ; cnihtene KATH. 1571 ; **cnihten** (*dat*
pl.) MK. iii. 9 ; LAȝ. 714 ; mid cnihten HOM.
I. 231 ; *comp.* **būr-, in-, leorning-cniht.**

     cniht-bærn, sb., *male child* (*sec. text* cnave-
child), LAȝ. 15526.

     kniȝt-child, sb., *male child*, L. N. F. 214.

     kniht-hōd, sb., *O.E.* cnihthād ; *knighthood*,
HORN (R.) 543 ; kniȝthōd AYENB. 83.

     kniȝt-lī, adj., *O.E.* cnihtlīc ; *knightly*, WICL.
2 MACC. viii. 9.

     cniht-scipe, sb., *knighthood*, '*militia*,' HOM.
I. 243 ; LAȝ. 3801 ; knihtshipe P. S. 335.

     knight-weede, sb., *attire of a knight*, ALIS.[2]
(Sk.) 544.

     cniht-weored, sb., *troop of knights*, LAȝ.
26766.

cnihten, v., *knight, create a knight* ; knihten
HORN (R.) 640 ; kniȝted (*pple.*) WILL. 1354.

cnobbe, sb., *M.L.G.* knobbe ; *knob* : knobbe
PR. P. 280 ; þe **knobbes** (*pl.*) sittinge ·on his
chekes CH. C. T. *A* 633.

cnobbel, sb., = *M.Du.* knobbel ; *knob* ;. knoble
PR. P. 280.

cnobbid, adj., *knobbed* ; cnobbid PR. P. 280.

cnok, sb., *knock, blow* ; knok LANGL. (Wr.)
6262 ; knok *C* vi. 178 ; knokke *B* x. 327 ; RICH.
491.

cnokien, v., *O.E.* cnucian, *O.N.* knoka ; *knock* ;
knokeþ (*pres.*) AYENB. 116 ; cnoken A. P. i.
726 ; **knoke** (*imper.*) CH. C. T. *A* 3432 ;
cnokieð MAT. vii. 7 ; **knokede** (*pret.*) S. S.
(Wr.) 1420 ; **knoked** (*pple.*) MIN. iii. 68.

cnokil, sb., = *M.Du.*, *M.L.G.* knokel, *M.H.G.*
knuchel ; *knuckle*, '*nodus*,' VOC. 186 ; ' *con-*
dylus,' PR. P. 280 ; **knokelis** (*pl.*) REL. I. 190.

cnol, sb., *O.E.* cnoll ; *knoll* ; knol GEN. & EX.
4129 ; **cnolles** (*pl.*) '*colles*' HICKES I. 168
(HOM. II. 111) ; knolles PS. lxiv. 13.

cnollin, v., = **cnüllen** ; *knock* ; knollin, '*pulso*,'
PR. P. 280 ; **knolled** (*pple.*) E. G. 401.

cnop, sb., = *O.Fris.* knop, *O.H.G.* chnoph,
cnopf ; *knop, knob, button ; bud ;* knop LUD.
COV. 245 ; as hit were a gilden knop L. C. C.
39 ; þe knop of þe kne VOC. 208 ; **knoppe**
(*dat.*) W. & I. 35 ; **knoppes** (*pl.*) R. R. 1080,
1683 ; knoppis DEGR. 1494 ; knoppis of bars
WICL. EXOD. xxvi. 11.

     knop-weed, sb., *centaurea nigra*, LEECHD.
III. 319.

cnoppe, sb., = *M.Du.*, *M.L.G.* knoppe *m.* ; *knob,*
bud ; knoppe '*nodus, gemma*' PR. P. 280.

cnoppe, v., *stud, provide with knobs*, '*bullare*,'
CATH. 205 ; **knopped** (*pple.*) R. R. 7260 ;
knopped schon PL. CR. 424.

cnorre, sb., = *M.Du.*, *M.L.G.*, *M.H.G.* knorre ;
knot, excrescence ; knor BER. 2514.

cnotte, sb., *O.E.* cnotta, = *M.Du.* knutte, *M.H.*
G. knotze ; *knot*, H. M. 33 ; knotte '*nodus*' PR.
P. 280 ; C. M. 8411 ; CH. C. T. *F* 401 ; þane
knotte AYENB. 253 ; wiðute knotte A. R. 2 ;
knotten (*pl.*) BEK. 1479 ; cnotten, cnottes
KATH. 1157 ; knottes MAND. 190 ; PL. CR. 161.

     knot-lēs, adj., *without a knot or obstacle*,
CH. TRO. v. 768.

cnottel, sb., = *M.Du.* knuttel ; *little knot* ;
knottiles (*pl.*) HALLIW. 498.

cnotti, adj., *knotty*, HOM. I. 281 ; knotti '*no-*
dosus' PR. P. 280 ; CH. C. T. *A* 1977.

cnottien, v., *knot* ; **cnotted** (*pple.*) SAX. CHR.
262 ; knotted MAND. 197 ; iknotted A. R. 420 ;
cnottede schurgen KATH. 1551.

cnōwen, *see* **cnāwen**.

cnücche, sb., = *M.L.G.* knucke, knocke ; *bun-*
dle ; knitche ['*fasciculus*'] WICL. AMOS ix. 6 ;
knohches (*pl.*) of hai RICH. 2985 ; bindeþ hem
in knucchen (*printed* knucchenus) MISC. 225.

cnül, sb., *O.E.* cnyll ; *knell*, REL. II. 277 ; (þe)
laste knel 79.

cnüllen, v., *O.E.* cnyllan, = *M.H.G.* knüllen; *knock, beat; ring bells*; knille PERC. 1348; MAS. 689; lete al so þe belles knille MIRC 779; **knülled** (*pple.*) P. S. 193; *see* **cnollin.**

cnurned, adj., *? gnarled, knotty*; to his cnurnede cneon MARH. 10; knorned stonez GAW. 2166.

cnütten, v., *O.E.* cnyttan, = *M.L.G.* knutten; *knit*; knittin '*nodo, necto*' PR. P. 279; knitten LANGL. *B* PROL. 169; knitte his browes CH. C. T. *A* 1128; **knüt** (*pres.*) A. R. 396; as me knit a net P. L. S. xvii. 157; **knütte** (*pret.*) ALIS. 2251; knette GOW. II. 30; **knitted** (*pple.*) PR. C. 7215; icnut KATH. 1525; to gedir hit is iknit AUD. 10; *comp.* **un-cnütten.**

cō, *see* **cā.**

cobarde, sb., *? baker's tongs*; '*vertebra*,' VOC. 201.

[cobbel, sb., *cf. Swed.* kobbel (*boundary-stone*, RIETZ).]

cobil-stōn, sb., *cobble-stone*, '*rudus*,' VOC. 256; '*petrilla*' PR. P. 84.

cobben, v., *cf. Swed.* kubba; *strike*; **cobbit** (*pret.*) D. TROY 8285.

cobelere, sb., *cobbler*, '*sartor*,' PR. P. 84; LANGL. *A* v. 170.

coc, sb., *O.E.* coc, cocc, *O.Fr.* coc, *O.N.* kokr; *cock*, '*gallus*,' MK. xiv. 68 (*the earlier text has* hana); A. R. 140; O. & N. 1679; REL. I. 217; AYENB. 258; cok MAND. 91; **cockes** (*gen.*) BEK. 1090; a **cokkes comb** MAND. 207; **cockes fōt** *? columbine*, LEECHD. III. 319; **cokkes** (*pl.*) TREV. I. 339; *comp.* **pā-, weder-, wude-coc.**

cōc, sb., *O.E.* cōc, *cf. O.H.G.* coch; *Lat.* coqvus, cocus; *cook*; LAȜ. 19948; cok, kok HAV. 967, 2898; kok SPEC. 101; cook LANGL. *B* v. 155; CH. C. T. *A* 2020; LIDG. M. P. 217; kuk VOC. 211; **kōkes** (*gen.*) A. R. 380; **cōkes** (*pl.*) LAȜ. 3315; TREV. V. 173.

cocard, sb., *O.F.* cocart, cocard; *fool*, GOW. II. 221.

cocatrice, sb., *O.Fr.* cocatrice; *cockatrice*, WICL. PS. XC. 13; kocatrice '*basiliscus*' PR. P. 281.

coccou, *see* **cuccu.**

cock, sb., *fight, strife*, PS. cxliii. 1.

[cocke, sb., = *Ger.* kocke *m.*; *cock, heap*; *comp.* **hai-cocke.]**

cockere, sb., *one who makes hay-cocks*; **coker(e)s** [cokares] (*pl.*) LANGL. *C* vi. 13.

cocken, v., *? fight*; he wole grennen cocken & chiden REL. I. 188; cocke wiþ knif P. S. 153; **cocken** (*pres. pl.*) ALEX. 2042; mon þat siþ briddes **cokkinde** (*pple.*) REL. I. 262.

cocker, sb., *? fighter*; cocker þef & horeling REL. I. 188; **cokkers** (*pl.*) TOWNL. 242; fiȝters and cokkers TREV. IV. 173.

cockunge, sb., *? fighting*; wið strong cockunge overcume hire flesch H. M. 47; cokkinge TREV. II. 83.

cocle, sb., *cockle*, '*cochlea*,' PR. P. 86.

cocodril, sb., *Lat.* crocodīlus; *crocodile*; cokedril WICL. LEV. xi. 29; cokadrill ALIS. 6620; **cocodrilles** (*pl.*) MAND. 198.

cocumber, sb., *O.Fr.* cocombre; *cucumber*, PALL. i. 981.

cod, sb., *O.E.* codd (*bag*), *cf. Du.* kodde (*coleus*) (KIL. DICT.), *O.N.* koddi *m.*; *bag; cod, husk*; MAT. x. 10; CH. C. T. *C* 534; (*pillow*) TOWNL. 84; codde *husk, scrotum* PR. P. 85; mid þine . . . codde (*skin*) O. & N. 1124; winberian **coddes** (*pl.*) LEECHD. III. 112; coddis ['*siliqvis*'] WICL. LK. xv. 16; coddis of silke HALLIW. 262; *comp.* **ballok-, bēn-, pés-cod.**

cod, sb., *cod (the fish)*, B. B. 174.

code, *see* **cude.**

codling, sb., *codling (fish)*, '*morus*,' PR. P. 85; '*mullus*' VOC. 189.

codule, sb., *O.E.* cudele; *cuttle*, '*sepia*,' PR. P. 85.

co-empciōn, sb., *Lat.* co-emptiō; *co-emption*; coempcioun, þat is to sein, comune achat, or biing togidere CH. BOET. i. 4 (15).

coeterne, adj., *Lat.* co-aeternus; *co-eternal*, CH. BOET. v. 6 (172).

cōf, cōfe, cōflīche, *see* **cāf.**

cofin, sb., *O.Fr.* cofin, *mod.Eng.* coffin; *basket*, '*cophinus*,' PR. P. 85; WICL. PS. lxxx. 7; cofine (*pie-crust*) L. C. C. 34, 39, 41; **coffins** (*pl.*) PALL. iv. 672.

cofre, sb., *O.Fr.* cofre; *coffer*, BEK. 1923; C. L. 992; SPEC. 27; CH. C. T. *E* 585.

cofren, v., *Fr.* coffrer; *store in a coffer*, PL. CR. 68.

cog-boote, *see* **cok-boot** *under* **coke.**

cogge, sb., = *M.L.G.* kogge, *M.Du.* kogghe, *O.N.* kuggr; *small boat*, CH. L. G. W. 1477; D. ARTH. 476; **cogges** (*pl.*) RICH. 4784; MAN. (F.) 12068.

cogge [2], sb., *cf. Swed.* kugge *m.*; *cog, tooth of a wheel*, PR. P. 85; frogge þat sit at mulne under cogge O. & N. 86.

coghen, *see* **coughen.**

cogitaciūn, sb., *O.Fr.* cogitaciun; *cogitation*, A. R. 288.

cognacioun, sb., *O.Fr.* cognation, *Lat.* cognātiō; *kindred*, WICL. GEN. x. 31, xxiv. 4.

cognisaunce, sb., *O.Fr.* conissance; *knowledge*, DREAM 3091.

coi, adj., *O.Fr.* coi, qvoi, *mod.Eng.* coy; *quiet*, PR. P. 86; CH. C. T. *A* 119.

coien, v., *make quiet*, CH. TRO. ii. 801; **coies** (*pres.*) ALIS.[2] (Sk.) 1175.

coife, sb., *O.Fr.* coife; *coif*, PR. P. 86; MAN. (F.) 10896; helm & coife FER. 1605.

coilōn, sb., *O.Fr.* coilon; *testicle*; **coilōns,** coillons (*pl.*) CH. C. T. *C* 952.

coin, sb., *O.Fr.* coin, coing; *coin*, CH. F. AGE 20; coigne LANGL. *C* ii. 46; coni, cuni PR. P. 282.

coin², **qvoin**, sb., *Fr.* coin; *quince*, VOC. 163, 192.

coinen, v., *O.Fr.* coigner; *coin* : coigne þe monei GOW. II. 83; **coined** (*pple.*) MAN. (H.) 239.

coint, adj., *O.Fr.* coint, *from Lat.* cognitus; *mod. Eng.* quaint; *famous, well-known; knowing, clever, skilful; neat, elegant*; WILL. 653; GAW. 1526; cointe, cwointe A. R. 140; of se cointe (*v. r.* icudd) keiser KATH. 580; qvoint MAP 337; A. P. i. 888; þe qvointe swike ROB. 105; qveinte CH. C. T. *B* 1426.

cointe-liche, adv., *skilfully, gracefully*, ROB. 25; qveintliche WILL. 3233; qveinteli R. R. 783.

cointen, v., *acquaint*; **cointed** (*pret.*) him WILL. 4644; qvainted him MAN. (H.) 225.

cointise, sb., *O.Fr.* cointise; *skill; wisdom; ornament*; qvointise A. P. iii. 39; ROB. 51; cointice, qveintise WILL. 1665, 4220; qveintise R. R. 838; qvaintise A. P. ii. 1632.

coise, sb., =?*cweise*; ?*ulcer (term of abuse for a woman)* : þis foule great coise GOW. I. 100.

coisi, adj.,?=*qvasi*, *or*=*cweisi from* **cweise**; ?*queasy*; codling, cungur, and suche coisi fishe M. T. 118.

coite, sb., *quoit*, PR. P. 86.

coiten, vb., *play at quoits, 'petriludo,'* PR. P. 86.

   coiter, sb., *quoit-player, 'petriludus,'* PR. P. 86.

cok, *see* **coc**.   **cōk**, *see* **cōc**.

[cok, coke ?]

   cok-belle, sb., *small bell, 'nola,'* PR. P. 86; cokebelle TREV. I. 219.

[coke, sb., *O.Fr.* coque; *cf.* **cogge.]**

   cok-boot, sb., *cockboat*, LIDG. M. P. 152; cogboote *'scafa'* PR. P. 86.

cokel [cokkel], sb., *O.E.* cocel, coccel; *cf. Ir.* cogal; *cockle (weed)*, CH. C. T. *B* 1183; kokil [*'zizania'*] WICL. MAT. xiii. 25; REL. II. 80; cockel I. 36; cokkil LIDG. M. P. 149.

cōken, v., *from* **cōc**; *cf. M.Du.* koken, *O.H.G.* cochōn, *from Lat.* coqvere; *cook*; **cōked** (*pple.*) LANGL. *C* xvi. 60.

cokenei, sb., *mod. Eng.* cockney; ?*hen's egg* (*see* MURRAY *in* ACADEMY 10 *May* 1890); *spoiled child, milksop, simpleton*, PR. P. 86; cokenai CH. C. T. *A* 4208; ne no kokenei . . . coloppes for to maken LANGL. *B* vi. 287.

coker, sb., *O.E.* cocor, = *O.Fris.* koker, *O.H.G.* chochar; *quiver for arrows; kind of half-boot or gaiter*; enne koker LAȝ. 6470; cocur, cokir *'cothurnus'* PR. P. 84; **cokeres** (*pl.*) LANGL. *B* vi. 62.

cokerel, sb., *cockerel*; kokerel PR. P. 281.

cōkerie, sb, = *M.Du.* kokerie; *cookery*, GOW. II. 83.

cokerin, v., *keep warm*, PR. P. 85.

   cokeringe, sb., *warming, 'fotio,'* PR. P. 85; O. & N. 504.

coket, sb., *sort of fine bread*, LANGL. *C* ix. 326.

cokewold, *see* **cukeweald**.

cól, sb., *O.E.* col, *cf. O.N.* kol *n.*, *O.H.G.* chol *n.*, cholo *m.; coal*; col AYENB. 205; WILL. 2520; dude þer inne muchel col LAȝ. 2366; col groweþ under lond TREV. I. 399; col [cole] CH. C. T. *A* 2692; cole HORN (L.) 590; S. S. (Wr.) 2170; PR. C. 6762; a qvic cole BEV. 1548; as blak as cole P. L. S. ii. 42; **cóle** (*dat.*) HOM. I. 27; LAȝ. 17701; **cóles** (*pl.*) MAND. 101; **cólen** (*dat. pl.*) HOM. I. 251; brennen to colen GEN. & EX. 2653; *comp.* **char-cole**.

cól-blak, adj., *coal-black*, O. & N. 75; CH. C. T. *A* 2142.

cól-fox, sb., = *Ger.* kohlfuchs, *Du.* koolvos; *brant-fox*, CH. C. T. *B* 4405.

cól-mōse, sb., *O.E.* colmāse, = *Ger.* kolmeise; *titlark*, PR. P. 88.

cól-püt, sb., *O.E.* colpytt; *coal-pit*, MISC. 183.

cól-rake, sb., *coal-rake*, VOC. 276; PR. P. 88.

col, **colle**, ?adj. *or* pr. n.; ?*cunning*; ?*nickname given to rogues*: colle tregetour CH. H. F. 1277.

cōl, sb., *O.E.* cōl, = *M.Du.* koel, *O.H.G.* chuol; *from* **cálen**; *cool*, MAND. 305; cool MISC. 149; his red was to coul SHOR. 105; þat cole red ROB. 131.

cōl, *see* **caul**.

colcase, sb., *Lat.* colocāɔia; *Egyptian bean*, PALL. iv. 25.

cóld, **cólden**, *see* **cáld**, **cálden**.

cólder, sb., *prov.Eng.* colder (*chaff*); ?*rubbish* : coolder, coldir, *'petrosa, petro,'* PR. P. 86.

cōle, sb., *coolness* : in hete or cole S. & C. II. xxx.

coler, sb., *O.Fr.* coler; *collar*, ROB. 223; LANGL. *B* PROL. 169; coller PR. P. 87.

colere, sb., *O.Fr.* colere; *choler*, WICL. ECCLUS. xxxi. 23; coler GOW. III. 99.

colered, pple., *having a collar*, CH. C. T. *A* 2152.

colerik, adj., *choleric*, CH. C. T. *A* 587; colrik AYENB. 157.

colfre, *see* **culfre**.

coliaundre, *see* **coriandre**.

cōlien, v., *O.E.* cōlian, = *O.L.G.* cōlōn; *cf.* **kélen**; *cool, become or make cool*; colin *'frigefacio'* PR. P. 87; colen GAM. 540; colen her carez GAW. 1254; **cōlede** (*pret.*) JUL. 70; **coolid** (*pple.*) PR. P. 87; *comp.* **a-**, **for-cōlien**.

coliere, sb., *cf. M.H.G.* kolære; *from* **cól**; *collier, coal-dealer*; **colieres** (*pl.*) OCTOV. (W.) 495; kolieres WILL. 2523.

colike, sb., *Fr.* colique; *colic*, PR. P. 87.

colirie, sb., *O.Fr.* colire, *Lat.* collȳrium; *eye-salve*, WICL. APOC. iii. 18.

colis, sb., *O.Fr.* coleis; *sort of broth*; kolis L. C. C. 20.

colit, sb.,=acolit; *acolyte*, WICL. 2 THESS. PROL.; PR. P. 88.

colke, sb., *core of fruit*, '*interior pars pomi*,' CATH. 71; appel . . . þat even in middes has a colke PR. C. 6445; colke (*dat.*) TOWNL. 281.

cōlknīf, sb., *see* caul.

collacioun, sb., *Lat.* collātio; *discourse; consideration; comparison*, WICL. 2 MACC. xii. 43; CH. C. T. (W.) 8199; collacion GOW. I. 237; colasioun CH. BOET. iv. 4 (125); collaciōns (*pl.*) AYENB. 155.

colle, sb., *? O.E.* (morgen-) colla; *? rage*, P. S. 157.

collecte, sb., *med.Lat.* collecta; *collection; collect* (*short prayer*); þe collecte of euerich tide A. R. 20; collect WICL. 2 PARAL. vii. 9; colect 2 ESD. viii. 18; collectis (*pl.*) or gaderingis of moneie I COR. xvi. 1.

collen, v., *? for* acollen; *embrace*; colle WICL. ECCLES. iii. 5; kolled (*pret.*) WILL. 69; colled LANGL. *B* xi. 16.

collinge, sb., *embracing*; collingis (*pl.*) WICL. GEN. xlvi. 29.

collen², v., *cf.* Swed. kulla (*behead*); *? kill*; cole C. M. 3135; *see* cüllen.

colmi, *see* culmi.

colok, sb., *can, drinking vessel*, '*cantharus*,' VOC. 257.

colop, sb., *cf. O.Swed.* kollops; *collop, mincemeat*; colloppe '*frixatura, carbonacium*' PR. P. 88; colopes [coloppes] (*pl.*) LANGL. *A* vi. 287.

colōr, sb., *O.Fr.* colour; *colour*; colour, SHOR. 26; WILL. 764; som colour of riȝte ROB. 313; colur HORN (L.) 16.

colōrin, v. *O.Fr.* colorer; *colour*, PR. P. 88; coloureth (*pres.*) GOW. III. 139; colourid (*pple.*) WICL. GEN. xxx. 32; colored A. P. ii. 456.

colpōn, sb., *Fr.* coupon; *shred*; culpoun PR. P. 108; colpōns [culpōns] (*pl.*) CH. C. T. *A* 679; culpons IW. 642.

colt, sb., *O.E.* colt; *colt*, '*pullus*,' PR. P. 88; HOM. I. 3; JUL. 55; CH. C. T. *A* 3263; LIDG. M. P. 255; þat colt LEB. JES. 798; þet colt AYENB. 185; i hadde al wei a coltes (*gen.*) toþ CH. C. T. *D* 602; coltis fōt *colt's foot* tussilago LEECHD. III. 319.

coltisch, adj., *coltish*, CH. C. T. *E* 1847.

columbīne, sb., *Fr.* colombīne; *columbine* (*plant*), SPEC. 26.

columne, sb., *O.Fr.* columne; *column*, PR. P. 88.

colvere, colvre, *see* culfre.

colward, adj., *O.Fr.* culvert; *? perverse*, A. P. ii. 181; *cf.* culvert.

colwen, v., *blacken with coal*; colwid (*pple.*) PR. P. 88; (h)is kollede snoute HORN (R.) 1088.

colwie, adj., *grimy*, HORN (L.) 1094.

comandement, sb., *O.Fr.* co-, commandement; *commandment*, CH. C. T. *D* 67; commandement MISC. 33.

comanden, v., *O.Fr.* co-, commander; *command*; commaunde GOW. I. 2; comaunde (*pres.*) C. L. 956.

commaunding, sb., *commandment*, GOW. I. 3.

comandour, sb., *O.Fr.* comandeor, comandour; *commander*, PR. P. 88.

comb, *see* camb.

combrance, sb., *trouble, annoyance*, D. TROY 9169; combraunce A. P. ii. 3; DEP. R. iii. 113; cumbraunce D. TROY 12076.

combren, v., *O.Fr.* combrer, *from Lat.* cumulus; *encumber, annoy* : to combren þe chirche PL. CR. 461; hit combreȝ (*pres.*) uchone A. P. ii. 1023; cumbred (*pple.*) *entangled* TRANS. XVIII. 28; cumbered EM. 483; þat cumbered was wiþ parlesi M. H. 129; combred CH. BOET. iii. 10 (94); A. P. ii. 901.

comburment, sb., *cf. O.Fr.* encombrement; *encumbrance, perplexity*, ALIS. 471, 7764.

combust, adj., *Lat.* combustus; *burnt up* (*said of planets near the sun*), CH. TRO. iii. 686; ASTR. ii. 4 (31).

come, *see* cüme. comel, *see* cumel.

comen, *see* cumen.

comende, v., *Lat.* commendāre; *commend*, LANGL. *C* xv. 35.

comēte, sb., *Fr.* comete; *Lat.* comēta; *comet*, LAȝ. 17871; ROB. 416.

comin, *see* cumin.

cominalt *see* comonaltē.

comlen, *see* cumlen.

commedi, sb., *O.Fr.* comedie, *Gr.* κωμῳδία; *comedy*, TREV. I. 315.

commencement, sb., *O.Fr.* commencement; *commencement*, MISC. 30.

commendaciōn, sb., *commendation*, GOW. III. 145.

commissarie, sb., *commissary*, LANGL. *A* ii. 154.

commixen, v., *mix together*, PALL. ii. 21, iii. 3.

commixtioun, sb., *Lat.* commixtio, *O.Fr.* commistion; *mixing together*, TREV. II. 159.

commodious, adj., *convenient*, PALL. ii. 22.

commoeven, v., *O.Fr.* commovoir; *excite, move*; comméveþ (*pres.*) CH. TRO. v. 1797; commoevede (*pret.*) CH. BOET. iii. 12 (107); comméved (*pple.*) GOW. III. 205.

commoevinge, sb., *excitement*, CH. BOET. iv (12).

commūn, adj., *O.Fr.* comun; *common*, AYENB. 37; comun P. L. S. xvii. 404; commun [comoun] C. M. 7176.

comūn-līche, adv., *commonly*, AYENB. 145.

K

communi, v., *O.Fr.* communier; *commune*, AYENB. 102; comune [comine] WICL. PS. cxl. 4.

  comuner, sb., *partaker; citizen*, WICL. 1 PET. v. 1; TOWNL. 210; **cominers,** *members of town council*, E. G. 380.

  comuninge, sb., *participation; public meeting*; comeninge WICL. 1 COR. x. 16; E. G. 380.

communiōn, sb., *O.Fr.* communion; *(holy) communion*, PR. P. 89.

comonaltē, sb., *O.Fr.* communalté; *commonalty*, MAND. (H.) 54; cominalte WICL. PROV. xxix. 2*.

comp, *see* camp.

compacient, adj., *compassionate*, WICL. 1 PET. iii. 8.

compaignable, adj., *O.Fr.* compaignable; *sociable*; [*v. r.* companable, com-, cumpaniable] CH. C. T. *B* 1194; GENER. 2260; PR. P. 109.

compaignie, sb., *O.Fr.* com-, cumpaignie; *company*, MAN. (F.) 3131; cumpaignie P. L. S. xiii. 185; companie MISC. 138; cumpanie WILL. 1124.

compainoun, sb., *O.Fr.* compaignon; *companion*, ROB. 252.

companāge, sb., *O.Fr.* companage; *anything eaten with bread*, P. S. 240; WICL. S. W. I. 19.

comparer, sb., *? checking-clerk*, AYENB. 243.

comparisōn, sb., *O.Fr.* compareson; *comparison*, CH. C. T. *B* 846; comparisoun BOET. ii. 7 (58); AYENB. 81.

comparisounen, v., *compare*, WICL. MK. iv. 30; **comparisunez** (*pres.*) A. P. ii. 161; **comparisoun** (*imper.*) we ['*comparemus*'] WICL. 1 MACC. x. 71; **comparisound** (*pret.*) CH. BOET. ii. 7 (58).

compas, sb., *O.Fr.* compas; *compass*; cumpas C. L. 739; compas [cumpas] CH. C. T. *A* 1889; C. M. 8797.

compassement, sb., *contrivance, stratagem*, MAN (H.) 255; compacement WILL. 1981.

compassen, v., *O.Fr.* compasser; *compass*, LANGL. *C* xxii. 241; **cumpassede** (*pret.*) LEG. 12.

  compassinge, sb., *boundary*, R. R. 1349; *contrivance* CH. H. F. 1188; CH. C. T. *A* 1996.

compassiōn, sb., *O.Fr.* compassion; *compassion*, AYENB. 148; CH. C. T. *A* 1110.

compellen, v., *O.Fr.* compeller; *compel*; **compellede** (*pret.*) TREV. IV. 95.

compensen, v., *Lat.* compensāre; *compensate, requite*; **compensed** (*pple.*) GOW. I. 365.

compēr, sb., *O.Fr.* compair; *compeer*, '*compar*,' PR. P. 89; CH. C. T. *A* 670; **compērs** (*pl.*) WILL. 370; cumpers MISC. 212.

compilatour, sb., *Fr.* compilateur; *compiler*, CH. ASTR. PROL. (2).

compilen, v., *O.Fr.* compiler; *compile*; compiled (*pple.*) GOW. III. 48.

compilour, sb., *compiler*, MAN. (H.) 6.

complainte, sb., *O.Fr.* complainte; *complaint*, PR. P. 89; compleinte GOW. I. 111.

compleine, v., *O.Fr.* complaindre; *complain*, CH. C. T. *A* 908; **compleignende** (*pple.*) GOW. I. 88.

  compleininge, sb., *act of making a complaint*, GOW. I. 327.

complēt, adj., *Lat.* complētus; *complete*, WICL. S. W. I. 323.

complexioun, sb., *O.Fr.* complexion; *constitution of body, temperament*, AYENB. 157; **complexiōns** (*pl.*) GOW. III. 97.

complī, sb., *O.Fr.* complie; **complīn** BEK. 2078; cumplie A. R. 24, 428; cumpelie 22.

complie, v., *O.Fr.* complir; *accomplish*, SHOR. 100; *see* complissen.

complīn, sb., *last service of the day in monastic establishments*, CH. C. T. *I* 386; cumpline '*completorium*' PR. P. 109.

complissen, v., *accomplish*, CH. BOET. iv. 4 (124); MER. I. ii. 61; *see* complie.

compositiōn, sb., *O.Fr.* composicioun; GOW. I. 31.

compoune, v., *Lat.* compōnere; *compose, put together*, CH. BOET. iii. 9 (87).

comprehende, v., *Lat.* comprehendere; *comprehend*, CH. C. T. *E* 223; PR. C. 7463.

compte, sb., *O.Fr.* compte; *counting*, R. R. 5028.

compten, v., *O.Fr.* compter; *count*, GOW. III. 31, 44.

compuncciōn, sb., *O.Fr.* compunction; *compunction*, WICL. PS. iv. 5.

compunct, adj., *Lat.* compunctus; *having compunction*, WICL. DEEDS ii. 37.

comsen, v., *O.Fr.* co-, cumencer; *commence*; comse WILL. 2244; **comseth** [cumseþ] (*pres.*) LANGL. *A* i. 128; cumseð JUL. 2.

  comsinge, sb., *beginning*, WILL. 4868.

comūne, sb., *Fr.* commune; *community, people*, LANGL. *B* PROL. 115; '*commons*,' *food*, ROB. 528.

comunetē, sb., *O.Fr.* communité; *community*, DEP. R. iv. 41.

comūnlīche, *see under* commūn.

con, *see* can. **conabill,** *see* covenable.

conand, *see* convenant.

conceit, sb., *notion, conception*, '*conceptus*,' PR. P. 89; þe word was liche to þe conceipte GOW. I. 7; **conseites** (*pl.*) CH. TRO. iii. 755.

conceitate, sb., *conception*, TOWNL. 75.

conceive, v., *O.Fr.* conçoivre; *conceive*, LEG. 83; **conceiveþ** (*pres.*) AYENB. 136.

  conceivinge, consēwinge, sb., *conception*, PR. P. 89; WICL. GEN. iii. 16.

concēlement, sb., *concealment*, MAN. (H.) 297.

concēlen, v., *O.Fr.* conceler; *conceal*; concēled (*pple.*) GOW. II. 282.

concentrik, adj., *O.Fr.* concentrique; *concentric*; consentrik CH. ASTR. i 17 (9).

concepcioun, sb., *conception*, SHOR. 119.

concience, *see* conscience.

concilie, sb., *O.Fr.* cuncilie, concile; *council*, SAX. CHR. 254.

conclāve, sb., *Lat.*, *O.Fr.* conclāve; *conclave*, GOW. I. 254.

conclūden, v., *conclude*, CH. C. T. *A* 1358.

conclusiōn, sb., *O.Fr.* conclusion; *conclusion*, GOW. I. 23.

concordable, adj., *Lat.* concordābilis; *agreeing*, GOW. I. 253, 361, III. 204.

concordaunce, sb., *O.Fr.* concordance; *agreement*, LIDG. M. P. 48.

concorden, v., *make to agree*, CH. TRO. iii. 1702.

concubīne, sb., *O.Fr.* concubīne; *concubine*, ROB. 27; CH. C. T. *A* 650.

concupiscence, sb., *O.Fr.* concupiscence; *carnal desire*, GOW. III. 267, 285.

concurbite, sb., *Lat.* cucurbita; *flask used in distilling*; concurbites (*pl.*), CH. C. T. *G* 794.

condescenden, v., *O.Fr.* condescendre; *condescend*, CH. C. T. *E* 407; condecendre AYENB. 157.

condiciōn, sb., *O.Fr.* condicion; *condition*, AYENB. 173; condicioun CH. C. T. *A* 38.

condite, v., *from* conduit; *conduct*, MERL. I. ii. 50, iii. 577; condīted (*pret.*) I. ii. 144.

conditour, sb., *conductor*, MERL. III. 549.

condūen, v., *O.Fr.* conduire, *from Lat.* condūcere; *conduct*; coundue GAW. 1991; condūeþ (*pres.*) AYENB. 122; condīe (*subj.*) MAN. (H.) 182.

conduit, sb., *O.Fr.* conduit; *conduit, conduct*, TREV. IV. 365; condut ROB. 212; AYENB. 202; cundut O. & N. 483; saf conduit *safe-conduct* GOW. II. 160; condwis (*pl.*) *channels*, AYENB. 91; condites MAND. 217; coundutes (*carols*) of krist masse GAW. 1654.

conestable, sb., *O.Fr.* conestable; *constable*, HAV. 2286; cunestable HOM. I. 247; constable ROB. 538.

conestablesse, sb., *constable's wife*, CH. C. T. *B* 539.

confeccioun, sb., *O.Fr.* confeccion; *preparation*, TREV. I. 221.

confederacie, sb., *Lat.* confoederātio; *confederacy*, CH. BOET. ii. 6 (53).

confermen, v., *O.Fr.* confermer; *confirm*, L. H. R. 27; confermi BEK. 481; AYENB. 121.

   conferminge, sb., *confirmation*, ROB. 277; SHOR. 13.

confessen, v., *O.Fr.* confesser; *confess*; confesse (*pres.*) GOW. I. 107.

confessioun, sb., *O.Fr.* confession; *confession*, MAND. 119.

confessour, sb., *confessor*, P. L. S. xii. 1; AYENB. 172; confessours (*pl.*) AYENB. 267; confessōren (*dat. pl.*) HOM. I. 239.

confirmaciōn, sb., *O.Fr.* confirmacion; *confirmation*, GQW. I. 258; MAN. (H.) 209.

confirmement, sb., *O.Fr.* confirmement; *confirmation*, SHOR. 15.

confit, sb., *O.Fr.* confit; *preserve*: dates in confite B. B. 167.

confonde, v., *O.Fr.* confundre, -fondre; *confound*, MARG. 173; confundi SHOR. 112.

conforme, v., *O.Fr.* conformer; *conform*, GOW. II. 106.

confort, sb., *O.Fr.* con-, cunfort; *comfort*, H. M. 27; WILL. 1408; cunfort HOM. I. 185; kunfort [cumfort] A. R. 236 (*v. r.* elne).

   confort-lēs, adj., *comfortless*; comforteles, coumfortles GENER. 987, 3075.

confortable, adj., *helpful, comforting*, GENER. 2212; comfortabill HAMP. TR. 2.

conforten, v., *O.Fr.* conforter; *strengthen, comfort*, LANGL. *B* i. 201; conforti P. L. S. xv. 116; conforted (*pple.*) WILL. 380.

confūs, adj., *O.Fr.* confūs; *confused*, LANGL. *B* x. 136; confous GREG. 644.

confusiōn, sb., *O.Fr.* confusion; *confusion*, AYENB. 229; confusion [confusioun] CH. C. T. *G* 23.

congeien, v., *O.Fr.* congier (*congédier*); *bid farewell to, dismiss*, CH. TRO. v. 478; congeie LANGL. *A* iii. 167; congeide (*pret.*) GOW. II. 238.

congelatiōn, sb., *O.Fr.* congelation; *congelation* (*in alchemy*), GOW. II. 86.

congēlen, v., *O.Fr.* congeler; *congeal*; congēled (*pple.*) GOW. III. 86.

congen, v., = congeien, MAN. (H.) 323.

conger, sb., *O.Fr.* congre; *conger*; cunger VOC. 97; kunger P. L. S. xxxiv. 2.

congie, sb., *O.Fr.* congié (*congé*); *farewell*; congeie LANGL. *B* xiii. 202.

congraffen, v., *cf. O.Fr.* graffe; *write down*; congraffet (*pple.*) C. L. 1055.

congruitē, sb., *O.Fr.* congruité; *concord*, GOW. II. 90, III. 136.

coning, sb., *O.Fr.* connin; *cony, rabbit*, 'cuniculus,' VOC. 177; WILL. 182; ne koning ne hermine MISC. 70; coning [coni] WICL. LEV. xi. 5; cunig P. L. S. viii. 182; conig MIN. viii. 75; cunin HOM. II. 231; coninges (*pl.*) LANGL. *B* PROL. 193; conies CH. P. F. 193.

coningere, sb., *rabbit burrow, warren*; coningere 'cunicularium' PR. P. 90; coningeris (*pl.*) LIDG. M. P. 174.

conjecte, v., *Lat.* conjectāre; *conjecture, guess*, WICL. EZ. xxi. 19; conjecte (*pres.*) CH. BOET. i. 6 (27).

conjectere, sb., *diviner*, WICL. xxiii. 7.

conjectinge, sb., *divination* : he shal take conjectinge or suspicioun WICL. Ez. xxi. 19.

conjoignen, v., *O.Fr.* conjoindre, *mod.Eng.* conjoin ; *join together* ; **conjoignen** (*pres. subj.*) CH. BOET. iii. 10 (92).

conjoint, pple., *joined together*, GOW. III. 101, 127.

    conjoininge [**conjoigninge**], sb., *conjoining*, CH. C. T. *G* 95.

conjōn, congeōn, sb., *dwarf*, PR. P. 90; conjoun AYENB. 76 ; conjoun (*printed* coinoun) ALIS. 1718; congioun DEP. R. iii. 45.

conjuncciōn, sb., *conjunction (in astronomy)*, CH. ASTR. ii. 32 (41) ; GOW. III. 67.

conjūre, sb., *conjuration, magic*, GOW. II. 247.

conjūre, v., *O.Fr.* conjurer ; *conjure*, CH. C. T. *B* 1834 ; **conjūrede** (*pret.*) BEK. 2313.

conjuresōn, sb., *O.Fr.* conjureison ; *conspiracy; enchantment*, FL. & BL. 312 ; conjuracioun CH. BOET. i. 4 (18).

connen, *see* cunnen.

conquére, v., *O.Fr.* conqverre; *conquer*, FER. 537 ; MAN. (F.) 3127; **cuncweari** (*pres.*) H. M. 33.

conquerour, sb., *O.Fr.* conqvereur ; *conqueror*, P. S. 250; CH. C. T. *A* 862.

conqueste, sb., *O.Fr.* conqveste ; *conquest*, SHOR. 148 ; MAN. (F.) 4599.

conquesten, v., *O.Fr.* conquester ; *conquer* ; **conquest** (*pple.*) A. P. ii. 1304.

conrai, sb., *O.Fr.* conroi ; *provision*, C. M. 11513 ; conrei (*troop*) MAN. (H.) 304.

conscience, sb., *O.Fr.* conscience ; *conscience*, A. R. 306; concience (*consciousness*) GAW. 1194.

conseil, sb., *O.Fr.* con-, cunseil ; *counsel*, LAȝ. 2324*; MISC. 26 ; ROB. 48; HOCCL. i. 76 ; consail BEK. 881 ; cunsail L. N. F. 217 ; counsail A. R. 70 ; PR. C. 5943.

conseilere, sb., *O.Fr.* conseller ; *counsellor*, MAN. (H.) 54 ; L. N. F. 216 ; kunsiler A. R. 410 ; **counseilors** (*pl.*) GOW. II. 223 ; counsellers MER. iii. 37.

conseille, v., *O.Fr.* conseillier ; *counsel*, LANGL. *B* PROL. 115 ; CH. C. T. *D* 66 ; conseili ROB. 214.

consecracioun, sb., *O.Fr.* consecration ; *consecration*, APOL. 48.

consence, sb., *O.Fr.* consense, kunsence ; *consent*, A. R. 288.

consentemént, sb., *O.Fr.* consentement ; *consent*, AYENB. 11, 19.

consenten, v., *O.Fr.* con-, cunsentir ; *consent*, kunsenten A. R. 272 ; consent GOW. II. 95 ; **consentede** (*pret.*) BEK. 1871.

    consentinge, sb., *consenting*, AYENB. 117, 176.

consequent, sb., *consequence, inference*, CH. BOET. iii. 9 (84).

conserve, sb., *preservation*, GOW. III. 22, 86.

conserven, v., *O.Fr.* conserver ; *keep*, CH. TRO. v. 309.

consideratiōn, sb.. *O.Fr.* consideration ; *consideration*, GOW. III. 178.

consistorie, sb., *O.Fr.* consistorie ; *consistory*, LANGL. *B* ii. 177 ; constorie *A* iii. 32.

consonans, sb., *consonant*, P. L. S. xxxiv. 3.

conspiracie, sb., *conspiracy*, TREV. III. 33 ; CH. C. T. *C* 149.

conspiracioun, sb., *O.Fr.* conspiracion ; *conspiracy*, WICL. 2 PARAL. xxxiii. 24 ; **conspiraciōns** (*pl.*) AYENB. 23.

conspīre, v., *O.Fr.* conspīrer ; *conspire*, GOW. I. 232.

conspīrement, sb., *conspiracy*, GOW. I. 216.

constance, sb., *O.Fr.* constance ; *constancy*, AYENB. 167.

constellāciōn, sb., *O.Fr.* constellacion ; *constellation*, CH. C. T. *A* 1088.

constitūciōn, sb., *O.Fr.* constitution ; *constitution*, GOW. II. 75.

constorie, *see* consistorie.

constreinin, v., *O.Fr.* constraindre ; *constrain*, PR. P. 91 ; constreinen MAND. 188 ; **constreined** (*pple.*) CH. BOET. iii. 2 (97).

    constreiner, sb., *overseer of workmen* ; **constreiners** (*pl.*) WICL. EX. v. 6.

    constreigning-lī, adv., WICL. 1 PET. v. 2.

constreinte, sb., *constraint*, CH. TRO. iv. 741.

construcciōn, sb., *O.Fr.* construction ; *construction*, PR. P. 91.

construin, v., *O.Fr.* construire ; *construe*, PR. P. 91 ; construe LANGL. *A* iv. 133.

consuetūde, sb., *Lat.* consuetūdo ; *custom*, WICL. 1 KINGS xx. 25.

consul, sb., *O.Fr.* consul ; *consul*, TREV. I. 239.

consūmén, v., *Fr.* consumer ; *consume* ; **consūmed** (*pple.*) WICL. GEN. xli. 36.

consumpt, adj., *Lat.* consumptus ; *consumed*, CH. BOET. (60).

contac, sb., *contest, strife*, AYENB. 15 ; contak S. S. (Wr.) 1718 ; contek ROB. 90.

contacki, v., *debate, strive*, AYENB. 57 ; **conteckede** (*pret.*) P. L. S. xiii. 310.

contē, sb., *O.Fr.* conté, comtat ; *county* ; counte LANGL. *B* ii. 85 ; **countēs** (*pl.*) MAN. (H.) 133.

conteini, v., *Fr.* contenir ; *contain*, ROB. 547 ; **conteined** (*pple.*) MAND. 1.

contekour, sb., *debater*, E. G. 4 ; **conteckours** (*pl.*) BEK. 196.

contemplaciōn, sb., *O.Fr.* contemplacion ; *contemplation*, AYENB. 204 ; contemplaciun A. R. 142.

contemplatif, adj., *Fr.* contemplatif; *contemplative*, AYENB. 247.

contempt, sb., *O.Fr.* contempt; *contempt*: in contempte of regalie GOW. I. 217.

contenance, sb., *O.Fr.* contenance; *countenance*, ROB. 333; CH. C. T. *B* 320; cuntenaunce WILL. 1397.

content, adj.,*Fr.*content; *content*, GENER. 368.

contesse, sb., *O.Fr.* cuntesse, contesse, cuntesse; *countess*, SAX. CHR. 264; contas ROB. 159.

continuele, adj., *O.Fr.* continuel; *continual*, PR. C. 8947.

   continue(l)-līche, adv., *continually*, P. L. S. xvii. 220.

continence, sb., *O.Fr.* continence; *continence*, GOW. I. 19.

continuance, sb., *continuance,patience*, GOW. II. 7, 14.

continuin, v., *O.Fr.* continuer; *continue*, PR. P. 91; continue LANGL. *B* xii. 39; contune APOL. 12; **continued** (*pret.*) TREV. I. 73.

contourben, v., *Lat.* conturbāre; *agitate*; contourbed (*pple.*) GOW. I. 49.

contrarie, adj., *O.Fr.* contraire; *contrary*, MISC. 30; AYENB. 136; contrare (*sb.*) A. P. ii. 3; contraire GOW. I. 22.

   contrari-lī, adv., APOL. 101.

contrarien, v., *O.Fr.* contrarier; *contradict*, CH. C. T. *F* 705.

contrarious, adj., *cf. Ital.* contrarïōso; *contrary*, PR. P. 91; AYENB. 28; contrarius MISC. 228; PR. C. 1414.

contrait, sb., *contract*, SHOR. 62.

contre-, *see* countre-.

contrē, sb., *O.Fr.* contree, cuntree; *country*, LA3. 1282*; MAND. 244; contre [cuntre] CH. C. T. *A* 216; cuntre WILL. 6.

   contrē-man, sb., *countryman*; contraimen (*pl.*) P. L. S. xii. 56.

contrevore, sb., *cf., O.Fr.* troveure, *and* contróven; *contrivance*, MAN (H.) 334.

contriciōn, sb., *O.Fr.* contriciun; *contrition*, LANGL. *B* xi. 130.

contróven, v., *O.Fr.* controver; *contrive*; contreove, contreve MAN. (F.) 7146, 8986; contreve LANGL. *B* x. 19; **contróves** (*pres.*) PR. C. 1561.

contubernial, adj., *Lat.* contubernālis; *having the same dwelling*: humble folk ben Cristes frendes; thei ben contubernial [*ed. Morris* conturnialli] with the Lord CH. C. T. *I* 760.

contumelie, sb., *O.Fr.* contumelie; *contumely*, CH. C. T. *I* 556.

conveie, v., *O.Fr.* conveier; *convey; accompany*, WILL. 5111; MAN. (F.) 13123.

convenant, **covenant**, sb., *O.Fr.* co-, convenant; *covenant*, ROB. 179, 185; covenant,

covenaunt SHOR. 64; covenaunt ALIS. 2036; conand BARB. i. 561; cunnand iii. 759.

convenient, adj., *Lat.* conveniens; *convenient, suitable*, CH. BOET. iii. 11 (97).

conventicul, sb., *O.Fr.* conventicule; *assembly*,WICL. PS. xv. 4.

convers, sb., *O.Fr.* convers; *proselyte*, convert, WICL. I PARAL. xxii. 2; DEEDS ii. 11.

conversaciōn, sb., *O.Fr.* conversation; *conversation, manner of life*, AYENB. 96, 241.

conversen, v., *O.Fr.* converser; *spend one's time*; conversand (*pple.*) PR. C. 4197.

converten, v., *O.Fr.* convertir; *convert*, GOW. II. 58; converte TREV. V. 445.

convicten, v., *convict*; convict (*pple.*) CH. BOET. i. 4 (19); WICL. I COR. xiv. 24; convicte [convicted] WICL. DAN. xiii. 61.

cop, sb., *O.E.* copp,=*O.Fris.*kop, *O.N.* koppr, *O.H.G.* choph; *summit, head*, P. S. 70; REL. I. 144; þa turres cop LA3. 7781; þe cop of þe hille MAND. 17; **coppe** (*dat.*) CH. H. F. 1166; PARTEN. 5911; bi þe coppe he him nam LA3. 684*; from þe tures coppe A. R. 228; **coppes** (*pl.*) TREV. III. 451; þe coppis of þe hillis WICL. GEN. viii. 5.

cōpe, sb.,=*O.L.G.* cōpa, *O.H.G.* chuopha; *coop*; coop '*ciphus*' (κύφος) VOC. 178.

cōpe, *see* **cāpe**. **cōpen**, *see* **coupen**.

cōpen, v., *put on the cope*, LANGL. *C* vii. 288; cōpeþ (*pres.*) *B* iii. 142; cōpide (*pret. pl.*) *C* iii. 240; cōped (*adj.*) *C* iv. 38.

copenere, *see under* **copnien**.

coper, sb., *O.E.* copor (LEECHD. III. 16), *from Lat.* cuprum; *copper*, TREV. I. 261; CH. C. T. *G* 829.

coperōn, **coperūn**, sb., *O.Fr.* couperon; *ornament on the lid of a vessel, 'capitellum,'* PR. P. 91; **coperounes** (*pl.*) A. P. ii. 1461; coprounes GAW. 796.

coperōse, sb., *Fr.* couperose, *mod.Eng.* copperas; *sulphate of copper*, PR. P. 91.

copful, *see under* **cuppe**.

copi(e), sb., *Fr.* copie; *copy*, MAN. (H.) 293; PR. P. 92; (*abundance*) TREV. I. 301.

copiin, v., *Fr.* copier; *copy*, PR. P. 92.

copious, adj., *O.Fr.* (*Prov.*) copiōs; *copious*, PR. P. 92; TREV. I. 399.

cople, *see* **couple**.

copmaker, *see under* **cuppe**.

copnien (? cōpnien), v., *? await longingly*; ich copni þi cume MARH. 21; **copniδ** (*pres.*) KATH. 802; copneδ 2378.

   copenere (? cōpenere), sb., *O.E.* copenere; *lover*, TREV. II. 199; copenere [copinere] O. & N. 1342; copinere MISC. 150; ROB. 335; copiner S. S. (Web.) 2225.

coppe, sb. = *M.Du.* koppe, *L.G.* kobbe; *spider*, MIRC 1937; (*? printed* loppe) CH. ASTR. i. 3 (4); *comp.* **atter-coppe**.

   cop-web, sb., *cobweb*, TREV. VII. 343.

coppe, *see* ꞓuppe.

copped, adj., *from* ꞓop; *having a crest*: ꞓoppid as a lark HALLIW. 269.

cops, *see* ꞓosp.

corāge, sb., *O.Fr.* corage; *courage,* AYENB. 164; MAN. (F.) 5384.

corageus, adj., *O.Fr.* corageus; *courageous,* ROB. 359.

coral, sb., *O.Fr.* coral; *coral,* SPEC. 25; CH. C. T. *B* 4049.

ꞓorbel, sb., *O.F.* corbel, corbeil; *raven*; corbial A. P. ii. 456; **ꞓorbeles** (*gen.*) GAW. 1355.

corbet, sb., *O.Fr.* corbet; *some architectural ornament, ? corbel*; **corbetz** [corbettes] (*pl.*) CH. H. F. 1304.

corbin, sb., *O.Fr.* corbin (*corbeau*); *raven,* A. R. 84; C. M. 3332; corbun [corboun] C. M. 1892.

corde, *O.Fr.* corde; *cord,* P. L. S. xviii. 165; AYENB. 58; CH. C. T. *A* 1746; binde him ... wiþ strong **corden** (*pl.*) P. L. S. xx. 68.

corden, v., *accord, agree,* CH. C. T. (W.) 17142; [accorde, acorde *H* 208].

cordioustē, sb., *courageousness,* WICL. PS. liv. 9.

corduaner, sb., *O.Fr.* cordouanier; *cordwainer, shoemaker,* PR. P. 92.

cordwane, sb., *O.Fr.* cordouan; *cordovan leather,* PR. P. 92.

córe, sb., *O.Fr.* cœr; *core* (*of fruit*); 'arula,' PR. P. 93; LIDG. M. P. 43; SACR. 757; kore REL. II. 79.

corfeu, sb., *O.Fr.* covrefeu, cuevrefu; *curfew,* CH. C. T. *A* 3645; corfu S. S. (Web.) 1429.

coriandre, coliaundre, sb., *O.Fr.* coriandre; *coriander,* WICL. EXOD. xvi. 31.

[**corien,** v.,=*O.H.G.* corōn, chorōn, *M.Du.* coren; *taste*; *comp.* **a-corien.**]

coriour, [**curiour**], sb., *O.Fr.* conreeur; *currier,* WICL. DEEDS ix. 43; coriowre PR. P. 93.

coriūn, sb., *a musical instrument,* LAȝ. 7002.

cork, sb.,=*M.Du.* kork; *cork,* 'suberies, cortex,' PR. P. 93.

    cork-bark, sb., 'cortex,' PR. P. 93.

    cork-trē, sb., 'suberies,' PR. P. 93.

corkes, sb. pl., [*? misreading for* toskes]; *? tusks*; grassed as a mereswine wiþ corkes fulle huge D. ARTH. 1090.

ꞓorleu, sb., *Fr.* corlieu; *curlew,* LANGL. *B* xiv. 43; ꞓurlew WICL. PS. civ. 40.

cormoraunt, sb., *Fr.* cormoran; *cormorant,* WICL. LEV. xi. 18.

corn, sb., *O.E.* corn,=*O.L.G.* corn, *O.Fris.,* *O.N.* korn, *O.H.G.* chorn, *Goth.* kaurn; *corn, grain,* ORM. 1500; O. & N. 1126; SHOR. 32; HAV. 699; P. S. 342; a corn JUL. 79; þet corn me deˣ in to gerner HOM. I. 85; heo freten þet corn and þat græs LAȝ. 3905; þet corn AYENB. 140; korn, koren HAV. 1167, 1879; coren

GEN. & EX. 2134, 2237; corne (*pl.*) CH. H. F. 698; cornes PR. C. 3420; *comp.* barli-, hwēte-corn.

    corn-ēr, sb., *ear of corn*; corn eres (*pl.*) TREV. II. 305.

    corn-lond, sb., *corn-land,* TREV. II. 43.

    corn-stak, sb., *stack of corn*; cornstakkes (*pl.*) MAN. (F.) 14690.

corn, sb., *O.E.* corn (LEECHD. III. 62), *? O.Fr.* corn; *corn* (*on the foot*), PR. P. 93.

cornardie, sb., *O.Fr.* cornardie; *deception,* AYENB. 130.

cornemūse, sb., *Fr.* cornemūse; *bagpipe,* CH. H. F. 1218; GOW. III. 358; R. R. 4350; cormuse PR. P. 93.

corner, sb., *O.Fr.* cornier; *corner,* MAND. 217; cornieres (*pl.*) AYENB. 124.

cornered, adj., *cornered, having corners,* TREV. I. 179, 305.

cornet, sb., *O.Fr.* cornet; *from* corn; *cornet* (*wind instrument*), OCTOV. (W.) 1069, 1189.

corni, sb., *corny,* CH. C. T. *C* 315.

corniculēre, sb., *Lat.* corniculārius (*a Roman official*); CH. C. T. *G* 369.

Cornwaile, pr. n., *Cornwall,* LAȝ. 2246; HAV. 178.

corolarie, sb., *Fr.* corollaire, *Ital.* corollārio; *corollary,* CH. BOET. iii. 10 (91).

coronación, sb., *coronation,* PR. P. 93.

coronal, sb., *from Lat.* coronālis (*adj.*); *diadem,* WICL. JUD. xvi. 10; coronall GOW. II. 46; curonalle ANT. AR. xlix; corounal *point of a lance* RICH. 6218; cornall B. DISC. 929, 1603; **coronales** (*pl.*) MAND. 209.

corōne, *see* crūne.

coroner, sb., *coroner*; coroners (*pl.*) E. G. 350.

corosif, adj. & sb., *corrosive,* CH. C. T. *G* 853; corsies (*pl.*) A. P. ii. 1034.

corour, sb., *O.Fr.* coreor; *courier,* WICL. JOB ix. 25.

corporal, sb., *Fr.* corporal; *corporal* (*altar-cloth*); corporeals (*pl.*) HOM. II. 163.

corps, *see* cors.

correcciōn, sb., *correction,* PR. C. 9594.

correcte, v., *from Lat.* correctus (*pple.*); *correct,* PR. C. 9595; corette DEP. R. PROL. 59; corretted (*pple.*) CH. C. *D* 66.

correpcioun, sb., *Lat.* correptio; *rebuke,* WICL. 2 PET. ii. 16.

corrin, sb., *Lat.* corōna; *? tonsure*; ȝe holi monkes wiþ ȝur corrin REL. II. 175.

corrumpe, v., *O.Fr.* corrumpre (*corrompre*); *destroy, defile, become corrupt,* WICL. HOS. ii. 12; CH. BOET. iii. 11 (96); corompen (*pres.*) CH. BOET. iii. 11 (98); corrumped (*pple.*) PR. C. 2558.

corrupciōn, sb., *O.Fr.* corruption; *corruption,* AYENB. 227; corrupcioun TREV. V. 155.

corrupt, adj., *Lat.* corruptus; *corrupt*, GOW.
I. 34; corupt AYENB. 82.

co(r)rupten, v., *corrupt*, WICL. 2 COR. iv. 16.

cors, sb., *O.Fr.* cors; *body; corpse*; ALIS. 2195;
GAW. 1237; WICL. GEN. xxiii. 3; þet cors
MISC. 28; ded cors MAN. (F.) 4492; corps
[cors] LANGL. *B* xv. 23; CH. C. T. *F* 519;
corps SHOR. 88; cors (*pl.*) ROB. 154.

cors, *see* curs *and* cours.

corsaint, sb., *Fr.* corps saint; *a holy body,
relic of a saint*, LANGL. *C* viii. 177; *B* v. 539;
DREAM 941; E. G. 97; D. ARTH. 1164.

corser, sb., *horse dealer*, OCTOV. (W.) 818;
coresur E. T. 977; cosir '*mango*' VOC.
(W.'W.) 684.

corserie, sb., (*horsedealing*), *cheating trade*,
WICL. S. W. III. 283; coseri D. ARTH. 1582.

corsete, sb., *O.Fr.* corset; *under-garment*,
'*ventrale*,' VOC. 259; corsette TREV. II. 361.

corsies, *see* corosif.

corsing, sb., *dishonest trading*, M. H. 139.

cort, corteis, *see* curt, curteis.

cortin, sb., *O.Fr.* cortine; *curtain*, GAW. 1185;
curtin CH. TRO. iii. 10; curtine CH. C. T. *D*
1249; cortines (*pl.*) WILL. 2055.

cortined, adj., *curtained*, GAW. 1181; ALIS.
1027.

corumpable, adj., *corruptible*, CH. C. T. *A*
3010.

corûne, *see* crûne.

corvee, sb., *Fr.* corvée; *forced labour*;
corvees (*printed* tornees) (*pl.*) AYENB. 38.

corvesers, corvisers, sb. pl., *O.Fr.* corvoi-
sier; *shoemakers*, E. G. 371, 384.

cos, sb., *O.E.* coss, = *O.N.* koss, *O.Fris.* kos,
O.L.G. cos, cus, *O.H.G.* chus, cus; *kiss*, WICL.
GEN. xxvii. 26; AUD. 60; þes cos A. R. 102;
cos of pes JUL. 74; cos of pees TREV. V.
215; cus PR. P. 111; H. V. 12; kus HOCCL.
i. 155; cosse (*dat.*) MISC. 42; REL. I. 29;
cosses (*pl.*) A. R. 102; kosses WILL. 1011;
cossen (*dat. pl.*) LA3. 30452.

cosche, sb., *hut*, '*tugurium*,' PR. P. 94.

cosīn, *O.Fr.* cosīn; *cousin*, HORN (L.) 1444;
ROB. 38; MAN. (F.) 4753.

cosināge, sb., *O.Fr.* cosinage; *cousinship*,
CH. C. T. *B* 1329; MIRC 168.

cosīne, sb., *O.Fr.* cosīne; *female cousin*, ROB.
330; WILL. 602.

[cosp, sb., *O.E.* cosp, cops; *fetter; comp.*
fōt-, hand-, sweor-cops.]

cost, sb., *O.E.* cost (*manner*) = *O.N.* kostr,
O.H.G. chost, *Goth.* kustus *m.* (δοκιμή); *from*
chēosen; *choice, quality, manner*, GAW. 546;
LANGL. (Wr.) 1491; þer næs cost nan oðer
LA3. 18166; knewen he nogt ðis dewes cost
GEN. & EX. 3327; nēdes costes *necessarily*
ANGL. III. 61; ARCH. LII. 33; þe king þat
kinde was of coste C. M. 8179; costes (*pl.*)
HOM. I. 29; SHOR. 10; M. H. 147; son sum

ic was waxen man þa flæh i childes costes
ORM. 8056; ðe culver haveð costes gode
REL. I. 226; costus DEGR. 364; we ne ma3en
alre coste halden crist(es) bibode HOM. I. 21;
nawiht heo nusten of heore vare costen [coste]
LA3. 31914; *comp.* far-, or-, un-cost.

cost, sb., *Lat.* costus; *a herb*, PR. P. 94.

cost, sb., *O.Fr.* coust, *cf. M.Du., M.H.G.*
kost; *cost, expense*, ROB. 33; P. 40; LIDG. M. P.
208; hver lite profit liþ and moche cost
AYENB. 83; upe his coust MARG. 281; bi
lives coste MARH. 7; at min oune coste CH.
C. T. *A* 804; costes (*pl.*) HOCCL. i. 362;
costis ['*impensas*'] WICL. 2 PARAL. viii. 16.

cost-ful, adj., *expensive*, GEN. & EX. 3878;
cost-volle (*pl.*) AYENB. 229.

cost-lewe, adj., *costly*, '*sumptuosus*,' PR. P.
95; TREV. IV. 213; CH. C. T. *I* 418.

costāge, sb., *O.Fr.* costage; *expense*, ROB.
391; CH. C. T. *B* 1562.

costard, sb., *costard* (*apple*), '*quirianum*,'
PR. P. 94.

coste, sb., *O.Fr.* coste, *Lat.* costa; *coast*, GOW.
III. 296; TOR. 1419; þei knew(e) ful wel þe
cost(e) (*region*) FER. 1552; costes (*pl.*)
LANGL. *B* ii. 85.

costeien, v., *O.Fr.* costeer; *coast, go by the
side of*; costez (*pres.*) GAW. 1695; costei-
ing (*pple.*) R. R. 132; costinge MAND. 127.

costen, v., *O.Fr.* couster, *from Lat.* constāre;
cf. M.Du., M.H.G. kosten, *O.N.* kosta; *cost,
expend*: he schal costen þe ful deore C. L.
1092; costiþ (*pres.*) LIDG. M. P. 152; coste
['*impende*'] WICL. DEEDS xxi. 24; coste
(*pret.*) CH. C. T. *A* 1908; and 3et wolde i
swere þat it coste me moche more LANGL.
B xiii. 383; if þou have o3t on hur cost (*pple.*)
(*expended*) AV. ARTH. xxviii.

coster, sb., *med.Lat.* coster; *curtain, tapestry*;
coostre '*subauleum*' PR. P. 94; costerdes
(*pl.*) SQ. L. DEG. 833.

costevous, adj., *O.Fr.* costeous; *expensive*,
WICL. 2 PARAL. xxxv. 24; costius D. TROY
3777; costuous '*sumptuosus*' PR. P. 95;
AYENB. 228.

costnien, v., *O.E.* costnian; *try, tempt*: bute
he icostned (*pple.*) weoren (*r.* weore) þrie
inne compe LA3. 24669.

costnunge, sb., *O.E.* costnung; *temptation*,
MK. iv. 17; ne led us noht in to costnunga
(*pl.*) HOM. I. 67.

costnien, v., = costen; *cost, expend*; costneþ
(*pres.*) AYENB. 75; costnede (*pret.*) A. R.
290; AYENB. 145; hii costnede ROB. 390;
þou3 it hadde costned (*pple.*) me catel
LANGL. (Wr.) 406.

costninge, sb., *expense*: mid his a3ere
costninge LA3. 22547; costnigge AYENB.
151.

costrelle, sb., *large bottle*, VOC. 176; PR. P.
95; CH. L. G. W. 2664.

costume, sb., *O.Fr.* custume; *custom*; custome REL. I. 131 (HOM. II. 11) ; P. L. S. xvii. 204 ; kustume MISC. 47 ; custome 29.

cot, cóte, sb., *O.E.* cot *n.*, cote, *cf. M.L.G.* kote *f. m.,* hut, *O.N.* kot *n.,* *M.Du.* kot *n.,* kote *f.; cot, cottage, hut*; cot '*cella*' FRAG. 4 ; mene mei nout wiŧuten swink a lutel kot areren A. R. 362 ; seþþe i . . . cot hade to kepe P. S. 152 (A. D. 104) ; cote '*casa*' VOC. 273 ; HAV. 737 ; cóte (*dat.*) CH. C. T. *A* 2457 ; coten MAT. xxi. 13 ; cótes (*pl.*) LANGL. *B* viii. 16 ; WICL. WISD. xi. 2 ; *comp.* salt-, schēp-, swīn-cóte.

cot-līf, sb., *O.E.* cotlif (SAX. CHR. 136), *O.N.* kotlīfi ; *cottage, dwelling* : wo is him þat uvel wif bringeþ to his cotlif REL. I. 178 (MISC. 118).

cotāge, sb., *O.Fr.* cotage ; *cottage,* VOC. 232 ; CH. C. T. *B* 4012.

cóte, sb., *O.Fr.* cote, cotte, *cf. O.H.G.* cozo, cozzo ; *coat,* '*tunica,*' VOC. 238 ; GOW. II. 47 ; GAW. 152 ; cote [coote] CH. C. T. *A* 103 ; WICL. GEN. xxxvii. 3.

cōte, coote, sb., *Du.* koet ; *coot,* '*mergus,*' VOC. 188, 253 ; '*mergus, fulica*' PR. P. 95 ; ['*larum*'] WICL. LEV. xi. 16 ; a balled cote VOC. 165.

cóte-armure, sb., *cf. O.Fr.* cote a armure ; *coat of arms*; a cote armure CH. H. F. 1326 ; GAW. 586 ; PR. P. 95 ; CH. C. T. *A* 1016 ; cote armur TRO. v. 1665.

coteler, sb., *O.Fr.* coutelier ; *cutler,* PR. P. 96 ; BER. 3296.

cóten, v., *coat, provide with coats*; cóteþ (*pres.*) LANGL. *A* iii. 138.

cotidian, adj., *O.Fr.* cotidian, *from Lat.* cotīdiānus ; *daily,* GOW. II. 142 ; a fever cotidiene PR. C. 2987 ; cotidien (*sb.*) *quotidian fever* R. R. 2401.

cotier, sb., *O.Fr.* cotier ; *cottier*; cotiers (*pl.*) LANGL. *C* x. 97.

cotoun, sb., *O.Fr.* coton ; *cotton,* MAND. 212 ; coton, cotun PR. P. 96.

cōðe, sb., *O.E.* cŏðu ; *disease* : coðe oðer qvalm HOM. II. 177 ; coþe ALEX. 2815 ; cothe '*syncope*' PR. P. 96 ; TOWNL. 31.

cou, see **cū.**

couard, sb. *and* adj., *O.Fr.* cuard, couard ; *coward,* H. H. 138 (140) ; coward (*v. r.* kouard) ROB. 285 ; ALIS. 2108 ; cueard A. R. 288*.

  couard-lī, adv., *cowardly,* WICL. S. W. I. 192 ; WILL. 3336 ; A. P. ii. 1629.

couarden, v., *O.Fr.* coarder ; *cow, terrify*; cowardiþ (*pres.*) ALIS. 3344.

couardie, sb., *O.Fr.* couardie ; *cowardice,* CH. C. T. *A* 2730.

couardíse, sb., *O.Fr.* couardise ; *cowardice,* CH. TRO. v. 412.

couche, sb., *O.Fr.* couche ; *couch, bedchamber,* AYENB. 171 ; GOW. II. 132 ; entre into þi couche ['*cubiculum*'] WICL. MAT. vi. 6 ; cowche (*den of a boar*) AV. ARTH. xii.

coucheor, sb., *jeweller,* D. TROY 1597.

couchin, v., *O.Fr.* colcier, colchier, couchièr ; *lay, lay down; place together; lie down;* '*colloco,*' PR. P. 96 ; couchid (*pple.*) MISC. 217 ; kouchid WILL. 2240.

coufel, see **cuvel.**

cough, sb., = *M.Du.* kuch ; *cough,* '*tussis,*' PR. P. 97 ; PALL. viii. 123 ; couȝ [kouȝhe] TREV. IV. 287 ; coughe [coghe] CH. C. T. *E* 1957.

coughen [coghen], v., = *M.Du.* kuchen, *M.H.G.* kūchen ; *cough,* CH. C. T. *E* 2208 ; coughen, couhin '*tussio*' PR. P. 97 ; couwe (*pres.*) REL. II. 211 ; coghiþ FLOR. 248 ; koghe (*imper.*) MIRC 891 ; couhede [coughed] (*pret.*) LANGL. *A* v. 205 ; coughed, kouhed TREV. II. 195.

coul, sb., *? M.H.G.* kobel ; *? cattle-pen,* '*saginarium,*' PR. P. 97 ; see **cuvel.**

coulte, see **qvilte.** **coume,** see **cumbe.**

counte, sb., *O.Fr.* conte, cunte ; *account,* MAN. (H.) 136.

countesse, see **contesse.**

countin, v., *O.Fr.* cunter ; *count,* '*computo,*' PR. P. 98 ; counteþ (*pres.*) LANGL. *A* iii. 137 ; countes A. P. ii. 1685 ; counted (*pple.*) A. P. ii. 1730.

countour [counter], sb., *Fr.* comptoir ; *counting-house,* CH. C. T. *B* 1403.

countour[2], sb., *O.Fr.* conteor ; *calculator*; cowntere, counter '*computator*' PR. P. 98, 99 ; countour (*treasurer*) ROB. 538.

countour[3], sb., *counter*; gold and silver and countours (*pl.*) RICH. 1939.

countreféte, sb., *imitation, counterfeit,* MAND. 218 ; countirfete '*conformale*' PR. P. 99.

countreféten, v., *counterfeit*; contrefeten CH. BOET. v. 6 (173).

countrepeis, sb., *counterpoise*; countirpeis, counterpois PR. P. 99.

countrepeise, v., *O.Fr.* contrepeser ; *counterpoise,* CH. TRO. iii. 1357 ; counterpese GOW. III. 135 ; counterpeised (*pple.*) GOW. III. 190.

countrepléten, v., *plead against*; countrepléted (*pple.*) CH. L. G. W. 475.

countrevailen, v., *O.Fr.* contrevaloir ; *prevail against*; contrevaile GOW. I. 28 ; countrevaileth (*pres.*) 270.

countre-, counterwaiten, v., *be on one's guard against,* CH. C. T. *B* 2509.

countrin, v., = **encontren** : *harmonise; encounter* : cowntrin in songe '*occento*' PR. P. 99 ; i wiłłe countur (*encounter, combat*) wiþ þe kniȝte ANT. ARTH. 36 ; countred (*pret.*), wiþ þe erle of Kent þei countred at Medeweie MAN. (H.) 38.

coupable, adj., *O.Fr.* coupable ; *culpable,* LANGL. *B* xvii. 300.

coupare, sb., *cooper,* '*cuparius,*' PR. P. 99.

coupe, sb., *O.Fr.* cupe, coupe ; *cup, drinking vessel,* ' *crater,*' PR. P. 99 ; ROB. 118 ; mid gildene coupe (*first text* bolle) LA3. 24612* ; a coupe wiþ win sche hadde in hande MAN. (F.) 7563 ; coupes (*pl.*) AYENB. 35 ; coupes of clene golde LANGL. *A* iii. 22.

coupe, sb., *O.Fr.* colpe, *from Lat.* culpa ; *fault, sin, guilt,* LANGL. *B* v. 305.

coupe, sb., ?=cūpe ; *coop for poultry,* REL. I. 4.

coupen, v., *?Fr.* couper ; *slash, slit* ; coupes (*pres.*) D. ARTH. 798 ; couped (*pple.*) shon TOR. 193 ; galoches icouped LANGL. *B* xviii. 14.

coupen, v., *O.N.* kaupa ; *see* chēapien ; *buy* ; coupe HAV. 1800 ; copen LIDG. M. P. 105.

couple, sb., *O.Fr.* couple, cople ; *couple,* LEG. 70 ; LANGL. *B* ix. 140 ; coples (*pl.*) FER. 1328.

couplin, v., *O.Fr.* cupler, copler ; *couple,* ' *copulo,*' PR. P. 99 ; cople LANGL. *C* iii. 190 ; kupleð (*pres.*) boðe togederes A. R. 78 ; coupleþ [' *copulatis* '] WICL. IS. v. 8 ; icuplet (*pple.*) KATH. 1059.

courbe, adj., *O.Fr.* courbe ; *curved, bent,* GOW. I. 99.

courbe, sb., *curve, bend,* GOW. II. 159.

courben, v., *Fr.* courber ; *bend* ; courbed (*pret.*) LANGL. *B* i. 79.

couren, v., *cf. Dan.* kūre, *Swed.* kūra, *Ger.* kauern ; *cower* ; coure P. S. 329 ; FLOR. 785 ; PL. T. 155 ; kouriþ (*pres.*) ALIS. 2053 ; koured (*pret.*) WILL. 47.

cours, sb., *O.Fr.* curs, cours ; *course, running,* TREAT. 132 ; P. L. S. xxv. 125 ; ALIS. 712 ; GOW. I. 130 ; CH. C. T. *A* 8, *F* 76 ; cors FER. 1108 ; GAW. 116.

courser, sb., *O.Fr.* corsier, coursier ; *courser,* CH. C. T. *A* 1704 ; curser PR. P. 110.

court, *see* curt.

courtepī, sb., *short jacket* ; CH. C. T. *A* 290 ; R. R. 219 ; VOC. 196.

couschote, *see under* cū.

cousloppe, cousokulle(s), *see under* cū.

coutere, sb., *from O.Fr.* cute (*elbow*) ; *elbow-piece* (*of armour*), D. ARTH. 2566 ; GAW. 582.

couþen, *see* cūðen. cōve, *see* cāf.

cóve, sb., *O.E.* cofa, *cf. O.N.* kofi, *M.L.G.* kove, *M.H.G.* kobe ; *cove, small room, cell,* C. M. 11617.

covē, *Fr.* couvée ; *covey,* PR. P. 96.

coveiten, *see* cuveiten. covele, *see* cuvele.

covenable, adj., *O. Fr.* covenable (*convenable*) ; *suitable,* WICL. GEN. xl. 5 ; cuvenable MISC. 27 ; conabill BARB. v. 266.

covenab-lī, adv., *suitably,* WICL. MK. xiv. 11.

covenant, *see* convenant.

covent, sb., *Fr.* couvent ; *convent,* P. L. S. xvi. 160 ; AYENB. 219 ; kuvent A. R. 12.

coverchief [kerchef], sb., *O.Fr.* couvre-, cuevrechief ; *kerchief,* CH. C. T. *B* 837 ; kovercheef LIDG. M. P. 48 ; keverchief P. L. S. xxxiv. 126.

covercle, sb., *O.Fr.* covercle ; *lid of a cup,* CH. H. F. 792 ; PR. P. 97 ; VOC. 198.

coveren, v., ?=recoveren, *gain, attain; heal, recover* ; covere HAV. 2040 ; PR. C. 811 ; keoveri P. L. S. xxiii. 136 ; kevere ROB. 256 ; kuvere, kevere WILL. 128, 1521 ; kevered (*pret.*) GAW. 1755 ; *see* acoveren.

coveren, v., *O.Fr.* covrir, cueuvrir, *Lat.* co-operīre ; *cover* ; covere MAN. (F.) 14101 ; covere, kevere LANGL. *B* iii. 60 ; kevere WICL. HABAK. ii. 17 ; kevered (*pple.*) JOS. 176.

coverlit, sb., *Fr.* couvre-lit ; *coverlet,* C. M. 11239 ; keverlet E. W. 4.

covert, *O.Fr.* covert ; *covert,* WILL. 2217 ; PR. C. 4489.

covertūre, sb., *O.Fr.* coverture ; *covering, bedclothes; horse-cloth,* HORN (L.) 696 ; kuvertur(e) A. R. 214.

covine, sb., *O.Fr.* co-, convine ; *intrigue* WILL. 3147 ; CH. C. T. *A* 604.

Covintrē, pr. n., *O.E.* Cofantrēo ; *Coventry,* ROB. 6.

crabbe, sb., *O.E.* crabba, =*M.Du.* krabbe, *O.N.* krabbi ; *crab,* ' *cancer,*' VOC. 189 ; PR. P. 99 ; HOM. I. 51 ; LIDG. M. P. 58 ; crabben (*pl.*) HOM. I. 51 ; ALIS. 4943.

crabbe, sb., *crab (apple),* PALL. i. 569.

crabbe-appule, sb., *cf. Swed.* krabbäpple ; *crab-apple,* ' *macium,*' PR. P. 99.

crabbe-trē, sb., *crab-tree,* ' *macianus, arbutus,*' PR. P. 99 ; crabtre ' *arbutus* ' VOC. 192.

crabbed, adj., *cf. Swed.* krabbed ; *crabbed,* CH. C. T. *E* 1203 ; LIDG. M. P. 132 ; crabbede wordes LANGL. *B* x. 104.

cracche, sb., ?*itch* ; crache C. M. 11823.

cracche, sb., *O.Fr.* crache, creche ; *manger, crib,* ' *praesepe,*' PR. P. 99 ; WILL. 3233 ; WICL. JOB vi. 5 ; cratche WICL. LK. ii. 7 ; crecche A. R. 260.

cracchin, v., *cf. Swed.* kratsa, *Du.* kratsen, krassen, *L.G.* kratsen, krassen, kraschen ; *scratch,* PR. P. 99 ; cracchen [cracche] us or clawen LANGL. *B* PROL. 154 ; cracche TREV. VIII. 37 ; crechen JUL. 35 ; cracching (*pple.*) PARTEN. 5892 ; crached (*pret.*) ORF. 78 ; with taseles cracched (*pple.*) LANGL. *B* v. 446.

cracchinge, sb., *scratching,* CH. C. T. *A* 2834.

cradel, sb., *O.E.* cradol, *cf. Ir.* craidhal, *Gael.* creathall ; *cradle,* A. R. 82 ; ALIS. 3655 ; CH. C. T. *A* 2019 ; credel, cradel PR. P. 101 ; credil S. S. (Wr.) 789 ; cradele (*dat.*) O. & N. 631 ; credile MAN. (H.) 243.

cradel-barn, sb., *child in cradle* ; kradel-barnes (*pl.*) HAV. 1912.

cradel-bond, sb., *swaddling-cloth*, L. H. R. 134.

cræft, sb., *O.E.* cræft *m.*, *cf. O.L.G.* craft, *O.Fris.* kreft, *O.H.G.* chraft *f.*, *O.N.* kraptr *m.*, *mod.Eng.* craft ; *strength* ; *art*, LAӠ. 8287 ; þe craft is ihate astronomie 24297 ; he cuðe þene uvele craft 2840 ; craft O. & N. 757 ; REL. Į. 211 ; WILL. 635 ; CH. C. T. *A* 401 ; LIDG. M. P. 162 ; (*ms.* crafft) ORM. 18809 ; creft KATH. 872 ; AYENB. 35 ; þene creft FRAG. I ; þough he criede wiþ alle þe craft þat he coude MAND. 305 ; **craftes** (*gen.*) men *crafts-men* (*cf.* **craftmon**) TREV. I. 113 ; **crafte** (*dat.*) O. & N. 787 ; **cræftes,** craftes, cræfte (*pl.*) LAӠ. 2395, 10105, 15789 ; craftes O. & N. 568 ; PERC. 560 ; merci schal hir craftez kiþe A. P. i. 356 ; **cræften** (*dat. pl.*) LAӠ. 11797 ; *comp.* driӠ-, dweomer-, lēche-, lēoð-, wicche-, wunder-craft.

[**craft-ful,** adj., *crafty*].

craftful-līche, adv., *craftily*, REL. II. 176.

creft-lēas, adj., *O.E.* cræftlēas ; '*iners*,' FRAG. 2.

craft-mon, sb., *craftsman*, CH. C. T. *A* 1899 ; **craft-monnen** (*pl.*) LAӠ. 28944.

cræften, v., *O.E.* cræftan, = *M.L.G.* kreften, *O.Fris.* (ur)krefta ; *? accomplish* ; crefte SHOR. 157 ; crafte PALL. i. 428 ; **crefteþ** (*pres.*) SHOR. 2.

crafti, adj., *O.E.* cræftig, = *O.H.G.* chreftig ; *crafty, skilful*, M. H. 2 ; a crafti weorcman LAӠ. 22892 ; crafti and strang PR. C. 9088 ; crefti KATH. 125 ; MISC. 91 ; **craftier** (*compar.*) WILL. 1680 ; C. M. 8753.

crafti-līche, adv., *craftily*, LANGL. *A* x. 5 ; craftili Iw. 3104 ; **creftiluker** (*compar.*) KATH. 260.

crag, sb., *Ir.*, *Welsh* craig, *Gael.* creag ; *crag*, WILL. 2850 ; **cragge** (*dat.*) R. R. 4156 ; REL. I. 126 ; **cragges** (*pl.*) PR. C. 6393.

crahien, v., *?* = crakien ; *crack, crash* ; to crahien ant to crenchen mit swire MARH. 9.

craier, sb., *O.Fr.* craier ; *a small ship* ; **craiers** (*pl.*) D. ARTH. 738.

crak, sb., = *M.Du.* crac, *M.H.G.* krach ; *crack, crash*, ALIS. 641 ; TREV. I. 409 ; C. M. 18953 ; krak B. DISC. 962 ; **krakkes** (*pl.*) A. P. ii. 1402.

cráke, sb., *cf. O.N.* krāka (*crow*) ; *crake* (*bird*) ; **crákes** (*pl.*) S. S. (Web.) 3532 ; *comp.* **night-crake.**

crakenelle, sb., *Fr.* craquelin ; *cracknel*, PR. P. 100.

crakien, v., *O.E.* cracian, = *M.L.G.* kraken, *M. Du.* craken, *M.H.G.* krachen ; *crack, break, make a cracking noise* ; crake RICH. 5423 ; FLOR. 92 ; TRIST. 887 ; L. H. R. 144 ; REL. I. 2 ; krake HORN (H.) 1118 ; **crakeþ** (*pres.*) SHOR. 99 ; so chaunteþ he and crakeþ CH. C. T. *E* 1850 ; (þei) craken Gow. III. 94 ; and **craked** (*pret.*) boþe hire legges LANGL. *B* xviii.

73 ; banes þer **crakeden** (*pl.*) LAӠ. 1875 ; **kraked** (*pple.*) HAV. 1238 ; icraked GRĘG. 602.

crakinge, crakkinge, sb., *cracking, boasting*, '*crepor, jactancia*,' PR. P. 100.

crallen, v., *Ger.* krallen ; *? twist, curl* ; **cralle** (*imper.*) L. C. C. 35 ; **cralled** (*pple.*) IBID. ; curious harneis qvaintli crallit PL. T. 134.

cram-cake, sb., *? pancake*, WICL. EX. xxix. 2.

crammel, sb., *crumb* ; **crammeles** (*pl.*) AYENB. 253.

crammien, v., *O.E.* crammian ; *cram* ; cremmin PR. P. 101 ; **crommeþ** (*pres.*) P. S. 238 ; crammid (*pple.*) WICL. HOS. xiii. 6 : beo ur mouþ crommed with clai P. L. S. xxv. 112 ; icrammed [icrommed] CH. C. T. *C* 348 ; i-crammed [icrommet] LANGL. *A* PROL. 41.

cramminge, cremminge, sb., *stuffing*, '*farcinacio*,' PR. P. 101.

cramp, sb., = *O.H.G.* chramph, *mod.Eng.* cramp ; *hook, claw* ; cromp L. H. R. 139.

crampe, sb., = *M.L.G.*, *M.Du.* krampe, *O.H. G.* chrampho ; *cramp* (*spasm*), '*spasmus*,' PR. P. 100 ; CH. TRO. iii. 1071 ; crampe [crompe] LANGL. *B* xiii. 335.

crampishen, v., *cramp*, CH. AN. 174.

cramsin, v., *? scratch*, PR. P. 100.

cráne, ? cräne, sb., *O.E.* cran, *?* crān, *cf. O.L.G.* krān, krōn, *M.Du.* craen, krane (*?* kräne) *f.* ; *crane*, '*grus*,' VOC. 188 ; PR. P. 100 ; REL. II. 79 ; LIDG. M. P. 154 ; crane, cran CHˊ. P. F. 344 ; cron [crane] LAӠ. 20163 ; **cráne** (*dat.*) AYENB. 56 ; **kránes** (*pl.*) HAV. 1726 ; cronez A. P. ii. 58.

crani, sb., *cranny*, '*rima*,' PR. P. 100 ; **cranis** (*pl.*) SACR. 710.

cranke, sb., *crank, winch*, '*girgillus, haustrum*,' PR. P. 100.

crapaute, sb., *a precious stone* ; crepawnde, crapawnde, crepaud PR. P. 101 ; grapond VOC. 256 ; **crapawtes** (*pl.*) EM. 142 ; crepawdis SACR. 171.

crappe, sb., *? chaff*; '*acus criballum*,' PR. P. 100 ; **crappis** (*pl.*) '*curalis*' VOC. 201.

craschin, v., *crash*, PR. P. 100 ; **craschede** (*pple.*) brainez D. ARTH. 2114.

crásen, v., *cf. Dan.* krase, *Swed.* krasa, *mod. Eng.* craze ; *crack, break* : þe cablis crasen M. T. 128 ; þe pot was **crásed** (*pple.*) CH. C. T. *G* 934 ; coveitise haþ crasid Ӡoure croune DEP. R. I. 8.

crask, adj., *?* = cask ; *? burly*, '*crassus*,' PR. P. 100.

cratche, see **cracche.**

cräu, sb., *O.E.* crāw ; *from* crāwen ; *crow* (*of a cock*) ; **crōwe** (*dat.*) BEK. 1090 ; CH. C. T. *A* 3675 ; *comp.* hane-cräu.

crauke, see **croke.**

craulen, v., *O.N.* krafla, *cf. M.H.G.* krabelen ; *crawl* ; **craulande** (*pple.*) C. M. 6612.

cravant, see **creant.**

cráven, v., *O.E.* crafian,=*O.N.* krefja (*pret.* krafča); *crave,* GEN. & EX. 1320; crave C. L. 249; AN. LIT. 11; S. S. (Web.) 52; IW. 543; HOCCL. i. 432; **cráve** (*pres.*) SPEC. 72; craveþ SHOR. 61; crave LANGL. *B* iii. 221; we craven HOM. I. 59; **crávede** (*pret.*) GEN. & EX. 1418; HAV. 633; **icráved** (*pple.*) A. R. 2*; *comp.* **bi-cráven.**

crawe, sb., = *Dan.* krave, *O.H.G.* chrago; *craw, throat,* PR. P. 101; WICL. 4 KINGS vi. 25.

cráwe, sb., *O.E.* cráwe,= *O.H.G.* cráwa, chrája, chrãa, *O.L.G.* krája; *crow (thè bird),* '*cornix,*' VOC. 188; crowe O. & N. 1130; crowe ROB. 490; CH. C. T. *A* 2692; **crōwen** (*pl.*) P. L. S. xvii. 185; crowe O. & N. 304; *comp.* **niȝt-crōwe.**

   crōwe-fōt, sb., *crow-foot,* PR. P. 105.

   crōwe-lēc, sb., *O.E.* cráwan léac; *name of a plant,* FRAG. 31.

cráwen,v.,*O.E.*cráwan(*pret.*crēow),*cf.O.H.G.* chráhan (*pret.*chráta); *crow (as a cock);* crowe MISC. 45; er þe cok crawe P. S. 238; **crēow** (*pret.*) MAT. xxvi.74; creu S. A. L. 162; C. M. 15992; crew WICL. MAT. xxvi. 74; CH. C. T. *B* 4387; crewe S. S. (Web.) 2536; **crōwen** (*pple.*) TOWNL. 182; crowe LUD. COV. 278; icrowe CH. C. T. *A* 3357; *deriv.* **cráu, crēd.**

   crōwing, sb., *crowing,* TREV. VII. 535.

creaciōn, sb., *O.Fr.* creation; *creation,* GOW. II. 158.

creance,sb.,*O.Fr.*creance,*med.Lat.*crēdentia; *credence,* CH. C. T. *B* 915.

creant, adj., *O.Fr.* creant; *craven*; creaunt A. R. 288; cravant KATH. 133.

creatour,sb.,*O.Fr.* creatour; *creator,* LANGL. *B* ix. 26; CH. C. T. *C* 901.

creatūre, sb., *O.Fr.* creature; *creature,* BEK. 2238; CH. C. T. *A* 2769.

crecche, *see* **cracche.**

Crekkelāde, pr. n., *O.E.* Crecca gelád; *Cricklade* (*printed* Creskelade), ROB. [*v. r.* creck-, creoke-, kreke-, kyrke-] 269.

crēd, sb., *? O.E.* crēad; *from* **crūden**; *?crowd*; for . . . mikel crede MAN. (F.) 11244.

[**crēd,** sb.,*from* **cráwen**; *comp.* **han-crēd.**]

crēde, sb., *from Lat.* crēdo; *creed,* HOM. II. 17; SHOR. 138; REL. I. 211.

credel, *see* **cradel.**

credence, sb., *O.Fr.* credence; *belief, trust, faith,* GOW. I. 59, 65, 152; S. A. L. 60.

credible, adj., *Lat.* crēdibilis; *credible,* GOW. I. 23.

creft, *see* **cræft.**

crei, sb., *crowing,* O. & N. 335.

créke, sb., *basket, crib,* PR. P. 101; M. H. 64.

créke, sb., *?creak*; **crékes** (*pl.*) CH. C. T. *A* 4051 [knakkes (Wr.) 4049].

crékin, v., *? = M.Du.* kreken, *Ger.* krechen; *creak, croak,* '*gracillo,*' PR. P. 101; **crékes** (*pres.*) REL. II. 79; a crowe . . . **créked** (*pret.*) TREV. VII. 535.

crelle, sb., *?Gael.* criol; *creel,* '*sporta,*' PR. P. 101.

crēme, sb., *cf. Fr.* crēme, *med.Lat., Ital.* crema; *cream,* LANGL. (Wr.) 4365.

crēme, *see* **crisme.**

cremelen, v., *cf. L.G.* krömeln; *crumble,* L. C. C. 36.

cremmin, *see* **crammien.**

crempen, v., = *M.Du.* krempen, *M.H.G.* krempfen; *contract, restrain*; crempe O. & N. 1788; *comp.* **for-crempen.**

crenchen, v., *?cringe*; crenchen [crenge] mit swire MARH. 9.

crenclen, *see* **crinklen.**

crēopen, v., *O.E.* crēopan,= *O.N.* kriupa, *O. Fris.* kriapa, *M.L.G.* krúpen; *creep*; kreopen P. L. S. xvii. 107; creope ALIS. 576; crepen LAȝ. 29313; cr(e)ope [crepe] O. & N. 819; crepe SHOR. 146; **crēopeð** (*pres.*) HOM. I. 23; creopþ, cropþ O. & N. 826; criepeð HOM. II. 199; **crēop** (*imper.*) A. R. 292; **crēo-pinde** (*pple.*) MARH. 11; **crēap** (*pret.*) KATH. 908; crep GEN. & EX. 2924; ALIS. 571; S. S. (Wr.) 479; crep [creep] CH. C. T. *A* 4226; krep S. & C. I. i; crap (*for* cræp) LAȝ. 29282; crop (*ms.* crope) AV. ARTH. lxv; crepte GOW. III. 258; CH. C. T. *A* 4193; (þou) creptest LANGL. *A* iii. 184; heo crupen [crope] LAȝ. 18472; cropen GEN. & EX. 2974; crope BEK. 2241; crepten WILL. 2235; **crópen** (*subj.*) LANGL. *B* PROL. 186; **crópen** (*pple.*) WICL. E. W. 296; H. V. 84; crope GOW. I. 198; CH. C. T. *A* 4259; LIDG. M. P. 240; icrope TREV. III. 397; *comp.* **æt-crēopen.**

   crēpinge, sb., *creeping,* '*repcio,*' PR. P. 101.

   crēpere, sb., *O.E.* crēopere; *creeper,* PR. P. 101.

crepel, *see* **crüpel.**

cresce, v., *Lat.* crēscere; *increase*: bad hem cresce and multipli GOW. III. 276; cresen WICL. 4 KINGS xx. 10; **cressing** (*pple.*) WICL. S. W. I. 338.

crēsen, v., *?crush*; her seed if that me **crese** (*subj.*) PALL. v. 77.

cresse,sb.,*O.E.*cresse,cerse,= *O.H.G.*chresso, cresso, cressa, *O.L.G.* kerse; *cress,* '*nasturtium,*' PR. P. 102; A. P. i. 343; kerse GOW. I. 299; is noȝt worþ a kerse LANGL. *B* x. 17; *comp.* **water-, wel-cresse.**

cresset, sb., *?Fr.* creuset, *mod.Eng.* cressit; *crucible; lamp*; GOW. III. 217; cressit '*crucibulum*' PR. P. 102.

creste, sb., *O.Fr.* creste; *crest,* '*crista,*' A. P. i. 855; PR. P. 102; L. H. R. 212; creeste of the broche WICL. EX. xxviii. 23.

crete, sb., *cf. M.Du.* kratte (*basket*), *O.H.G.* crezzo (*calathus*); *? cradle:* þe kinges zone . . . þet wepþ ine his crete AYENB. 137.

creulen, v., *?=Du.* krevelen ; *crawl* ; **creuland** (*pple.*) C. M. 6612.

crevace, sb., *O.Fr.* crevace ; *crevice,* CH. C. T. *I* 363 ; crevas S. S. (Wr.) 768 ; crevice PR. C. 9186 ; crevisse GAW. 2183.

crevil, sb.,=*M.Du.* krevel ; *? itching,* LIDG. M. P. 31.

crevis, sb., *O.Fr.* crevice ; *crayfish* ; krevis LIDG. M. P. 154.

crī, sb., *O.Fr.* crī ; *cry,* LAȝ. 11991* ; MAP 339 ; BEK. 81 ; kri WILL. 2174.

cribbe, sb., *O.E.* cribb, crybb,=*O.L.G.* cribbia, *O.H.G.* crippa, chripia ; *crib,* '*praesepe,*' PR. P. 103 ; HOM. I. 277 ; ORM. 3321 ; MAND. 70 ; PR. C. 5200 ; ['*fiscella*'] TREV. IV. 353.

crīen, v., *O.Fr.* crīer ; *cry,* HAV. 2443 ; crie BRD. 22 ; **crīe** (*imper.*) A. R. 136 ; **crīande** (*pple.*) WILL. 4347 ; **crīede** (*pret.*) LANGL. *B* v. 451 ; **crīeden** (*pl.*) HOM. I. 279.

crīing, sb., *crying,* FER. 2433.

crike, sb., *O.Fr.* crique, *cf. M.Du.* kreke ; *creek,* '*scate(b)ra,*' PR. P. 103 ; CH. C. T. *A* 409 ; *?*(*printed* trike) GEN. & EX. 2947 ; krike HAV. 708.

crike, sb., *O.N.* kriki ; *? thigh :* his nese went unto þe crice (*r. w.* swike) HAV. 2450.

criket, sb., *Fr.* criquet ; *cricket* (*animal*), VOC. 164 ; LANGL. *B* xiv. 42.

crikke, sb., *crick, cramp,* '*spasmus,*' PR. P. 103 ; thou might stombie and take the crik REL. II. 29.

crīme, sb., *Fr.* crīme ; *crime,* CH. C. T. *G* 455 ; WICL. DEEDS xxiii. 29.

crimpil, sb., *wrinkle,* '*ruga,*' PR. P. 103.

crimplin, v., *crumple,* '*rugo,*' PR. P. 103.

crinklen, v., *?=Du.* krinkelen ; *crinkle:* the hous is crinkled [crencled] (*pple.*) CH. L. G. W. 2012.

criour, sb., *O.Fr.* crieur ; *crier,* LEG. 220 ; TREV. I. 247.

crious, adj., *clamorous,* WICL. PROV. ix. 13.

crip, sb., *pouch, scrip,* P. R. L. P. 156.

cripel, *see* crüpel.

crippen, v., *break, cut* ; **crippid** (*pple.*), WICL. LEV. xxii. 24.

crips, *see* crisp.

crisme, sb., *O.Fr.* cresme, *Gr.* χρῖσμα ; *chrism,* PR. P. 103 ; GEN. & EX. 2458 ; creme MIRC 634 ; creime SHOR. 13 ; AYENB. 41.

    crisme-child, sb., *child that dies within the month of birth,* MISC. 90.

    crisme-clōð, sb., *chrism or chrisom cloth,* HOM. II. 95.

crisolíte, sb., *O.Fr.* crisolite ; *chrysolite,* MAND. 276.

crisopace, sb., *O.Fr.* crisopace ; *chrysoprase,* MISC. 98.

crisp, adj.,*O.E.* crisp, *from Lat.* crispus ; *crisp, curly,* MAND. 168 ; his crispe her CH. C. T. *A* 2165 ; crips TREAT. 138.

    crisp-heed, sb., '*crispitudo,*' PR. P. 103.

    crispenesse, sb., '*crispitudo,*' PR. P. 103.

crispen, v., *curl* ; **crispedde** (*pple.*) TREV. I. 53 ; cresped GAW. 187.

crist, sb., *O.Fr.* crist ; *Christ,* KATH. 612 ; **cristes messe,** *Christmas,* AYENB. 213.

cristal, sb., *Fr.* cristal ; *crystal,* FL. & BL. 274 ; CHR. E. 638 ; SPEC. 52.

cristen, adj. & sb., *O.E.* cristen ; *christian :* a cristen man LANGL. *B* xi. 120 ; cristene men FRAG. 7 ; **cristene** (*sb.*) SHOR. 56 ; þis gode king þat þus cristene bicom ROB. 75 ; **cristine** [cristene] (*pl.*) LAȝ. 29773.

    cristen-dōm, sb., *christendom,* A. R. 30 ; BEK. 108 ; þane cristendom SHOR. 5.

    cristen-lī, adv., *O.E.* cristenlīce ; *as a christian,* CH. C. T. *B* 1122.

cristianitē, sb., *Lat.* christiānitātem ; *christianity,* CH. C. T. *B* 544.

cristnien, v., *O.E.* cristnian ; *christen:* cristnie SHOR. 9 ; **cristnede** (*pret.*) P. L. S. xv. 177 ; **cristned** (*pple.*) ORM. 1782.

cristninge, sb., *christening,* AYENB. 14.

critouns, sb., pl., *O.Fr.* cretons ; *greaves* (*refuse of tallow*), WICL. PS. ci. 4.

criðe, sb., *crib, manger,* M. H. 63.

crive, sb., *? grave* ; krive MAN. (H.) 91.

croce, croche, sb., *O.Fr.* croce, croche ; *crutch, crozier :* '*pedum,*' PR. P. 103, 104 ; a bischopes crosse [croce] LANGL. *B* viii. 94 ; croche '*sustentaculum*' VOC. 277.

crochet, sb., *Fr.* crochet ; *crotchet* (*in music*), '*semiminima,*' PR. P. 104 ; TOWNL. 116.

crocke, sb., *O.E.* crocca ; *crock, pot* ; '*olla,*' FRAG. 4 ; A. R. 214 ; crokke SHOR. 106 ; TREV. III. 85 ; LANGL. *B* xix. 275.

crockere, sb., *potter* ; '*figulus,*' WICL. WISD. xv. 7 ; crokkere SHOR. 106.

crocodile, sb., *O.Fr.* crocodile ; *crocodile* ; a crocodilles (*gen.*) hide PALL. i. 960 ; *see* cocodrill.

croft, sb., *O.E.* croft,=*M.Du.* kroft, crocht ; *croft, field,* LANGL. *B* v. 582 ; TOWNL. 199 ; LUD. COV. 36 ; **crofte** (*dat.*) S. A. L. 160.

crōh, sb., *O.E.* crōg, = *M.Du.* kroeg, *O.H.G.* cruag *m.; pitcher, jug,* H. M. 39 ; **croos** (*pl.*) MISC. 29.

crois, sb., *O.Fr.* crois, croiz, cruiz ; *cross,* C. L. 1183 ; P. L. S. iii. 88 ; JOS. 446 ; CH. C. T. *E* 556 ; croiz HORN (R.) 1314 ; HAV. 1263 ; creoiz A. R. 18 ; croiz, croice BEK. 959, 971 ; croice WILL. 350 ; *see* cros.

croisen, v., *O.Fr.* croiser, cruisier ; *sign with the cross :* to croici þrie his foreheved P. L. S.

xvii. 73 ; creosieð (*imper.*) A. R. 64 ; croised (*pple.*) L. H. R. 133.

croiserie, sb., *O.Fr.* croiserie ; *crusade*, ROB. 486.

crōk, sb., = *Du.* croec, crōc (*curl*), *O.N.* krōkr m. (*hook*) ; *crook, hook,* '*uncus,*' VOC. 237 ; '*pedum*' PR. P. 104 ; on his bake he bar a crok TOR. 1018 ; crok *ringlet* P. S. 327 ; crook ALIS. 4819 ; LIDG. M. P. 217 ; croc *fraud* ORM. 11635 ; crōkes (*pl.*) *hooks* LEG. 169 ; LANGL. *C* xxi. 296 ; þoȝ ȝur crune be ischave fair beþ ȝur crokes (*locks*) REL. II. 175 ; crokez *reaping hooks* A. P. i. 40 ; crokes *frauds* A. R. 268 ; SPEC. 105 ; crefti crokes KATH. 125.

crōke, sb., *core* (*of fruit*), *coke,* '*arula,*' VOC. 267 ; crauke '*cremium* (*quod restat in frix-orio*)' PR. P. 101 ; mi banes als kraukan (*pl.*) dried þa PS. ci. 4.

crōke, sb., = *Du.* krōke (*curl*) ; *? for* crōk ; kroke D. ARTH. 3352.

crōken, v., *croak* ; crouken, '*coaxo,*' PR. P. 105.

crōken, v., = *Du.* kroeken, krōken ; *crook* ; crokin '*curvo*' PR. P. 104 ; S. S. (Wr.) 609 ; croki S. A. L. 161 ; AYENB. 177 ; croke HICKES I. 229 ; GOW. II. 144 ; i crōke *I am bent* REL. II. 211 ; crokeð HOM. II. 61 ; þe stature boweþ and crokeþ TREV. II. 185 ; crookeden (*pret.*) WICL. PS. lvi. 7 ; crōked (*pple.*) GAW. 653 ; CH. C. T. *B* 560 ; croked, icroked O. & N. 80, 1676 ; þe crokede hond P. L. S. xvii. 345 ; þe crōkede (*pl.*) AYENB. 224 ; *comp.* for-crōken.

crōkidnesse, sb., *crookedness*, WICL. S. W. I. 273.

croket, sb., *O.Fr.* croquet, = crochet *lock of hair*, P. S. 329 ; GOW. II. 370 ; HALLIW. 281.

crokke, *see* crocke.

crol, crollen, *see* crul, crullen.

crombe, *see* crumbe.

crome, *see* crume. crommen, *see* crammien.

cromp, *see* cramp.

crompid, adj., *crisp* : a crompid cake of the leepe of therf looves WICL. EX. xxix. 23.

crōn, *see* crāne.

crone, sb., *? O.Fr.* caroigne, *cf. M.Du.* kronie ; *crone*, CH. C. T. *B* 432.

crōnen (?cronen), v., *? = M.Du.* kronen, kreunen (*groan, complain*), *O.H.G.* chrōnen (*chatter*) ; *croon, moan* ; ȝe croine (*r. w.* mone) TOWNL. 116.

cronesank, sb., *name of a plant* ; '*persicaria, saucheneie*' VOC. 140 ; REL. I. 36.

croniclen, v., *chronicle* ; cronculd (*pple.*) EGL. 1339 ; cornicled P. R. L. P. 10.

croniclere, sb., *historian* ; '*historicus*' PR. P. 104.

cronike, sb., *O.Fr.* cronique ; *chronicle*, TREV. III. 47 ; cronicle PR. P. 104 ; cronique GOW. I. 31, 89 ; TREV. I. 5, II. 77.

crop, sb., *O.E.* cropp, = *M.Du.* krop, *O.N.* kroppr, *O.H.G.* chroph ; *crop* (*of a bird*), *throat* ; *top of a tree* ; *crop of corn* ; '*cyma,*' VOC. 229 ; P. S. 238 ; LANGL. *C* xix. 75 ; CH. BOET. iii. 2 (69) ; P. L. S. ii. 98 ; L. H. R. 69 ; hit hadde crop so an heort ALIS. 688 ; crop and more LEG. 105 ; a tre ... of whilk þe crop es turned do(u)nward PR. C. 663 ; þe crop of þe rede brere L. C. C. 42 ; croppe '*annona*' PR. P. 104 ; croppe (*dat.*) C. M. 8458 ; in crop ant rote SPEC. 100 ; croppes (*pl.*) R. R. 1396.

cropere, sb., *O.Fr.* cropiere ; *crupper*, PR. P. 105 ; croperes [cropiers] (*pl.*) LANGL. *B* xv. 453.

cropon, sb., *O.F.* cropion, crepon ; *haunch, buttock* ; '*clunis,*' PR. P. 105 ; cropins (*pl.*) B. B. 140.

croppen, v., *O.N.* kroppa ; *crop, pluck* ; croppe M. T. 167 ; þe wiði þet sprutteð ut þe betere þet me hine ofte croppeð (*pres.*) A. R. 86 ; suche (foules) cometh to mi croft and croppeth mi whete LANGL. *B* vi. 33 ; hast þou ... crop-ped (*pple.*) ȝerus (*r.* ierus) of corne MIRC 1502.

cros, sb., *O.N.* kross ; *cross*, LAȜ. 31386 ; KATH. 727 ; LANGL. *A* vi. 13 (*B* v. 529 *has* cruche) ; cros or crouche HALLIW. 282 ; one crosse, *across*, D. ARTH. 3667.

    cros-līne, sb., *cross-line*, CH. ASTR. i. 12 (7).

    cros-sail, sb., *cross-sail*, A. P. iii. 102.

crosse, sb., *madman*, LAȜ. 20325 (2*nd text* fol).

crosselet, croslet, sb., *O.F.* croisel ; *crucible*, CH. C. T. *G* 1147, 1191, 1198.

crossen, v., *O.N.* krossa ; *cross* ; crossid (*pple.*) '*cruce signatus*' PR. P. 105.

crōte, sb., *? O.Fr.* crote, *cf. Ger.* krotze ; *clod, lump,* '*glebula,*' PR. P. 105 ; croote ['*cremium*'] WICL. PS. ci. 4 ; ilk a crote MAN. (F.) 2102 ; crot (*v. r.* grot) C. M. 2528.

crouche, *see* crúche. crouchen, *see* crúchen.

crouden, croudinge, *see* crúden, crúdinge.

crouke, sb., *O.E.* crúce, = *O.L.G.* crúca ; *? pitcher* ; CH. C. T. *A* 4158.

crouken, *see* crúchen. croune, *see* crúne.

croupe, sb., *O.Fr.* croupe ; *hinder parts*, ALIS. 2447 ; CH. C. T. *D* 1559 ; MAND. 289.

crouþ, sb., *Welsh* crwth, *prov.Eng.* crowd ; *fiddle* : fiele ne crouth SPEC. 53 ; harpe fidele and crouthe B. DISC. 137 ; he herde a sym-phonie and a crowde ['*chorum*'] WICL. LK. xv. 25 ; croudis (*pl.*) ['*chori*'] JUDG. xi. 34.

crōwe, *see* crāwe, *and* crāu.

crúcche, sb., *O.E.* crycc, = *O.H.G.* krucka, *M.Du.* krucke ; *crutch*, H. V. 81 ; mid his crucche LAȜ. 19482.

crúche, sb., = *O.L.G.* cruce ; *cross*, KATH. 1171 ; crepe to cruche HOM. II. 95 ; cruche [crouche] LANGL. *B* v. 529 ; crouche SHOR. 15 ; AYENB. 111 ; E. G. 54.

crúchen, v., *sign with the cross* ; crouche

(*pres.*) CH. C. T. *A* 3479; crouche**þ** SHOR. 15; heo ... crúchede (*pret.*) hire KATH. 628.

crūchen, v., *cf. Ger.* krauchen; *crouch* : i can nauther crouke ne knele TOWNL. 163; kniȝtes crouke**þ** hem to & crūche**þ** (*pres.*) full lowe PL. CR. 751; **þ**ei so lowe crouchen 302.

crucifie, v., *Fr.* crucifier; *crucify*, C. M. 18273.

crucifix, sb., *crucifix*, A. R. 16.

crucken, v., *cf. M.L.G.* krucken; *bend, twist* : ich foile awaie thai crucke (*pres.*) PALL. ii. 210.

crūd, sb., *crowd*, AR. & MER. 5510.

crudde, curde, sb., *curd*, '*coagulum*,' PR. P. 105; **cruddes** (*pl.*) LANGL. *B* vi. 284; curdes VOC. 178.

cruddin, v., *curdle,* '*coagulo*,' PR. P. 105; **crudded** (*pple.*) WICL. JOB x. 10.

crūde, sb., *barrow, handcart* ; *crowde,* '*cenivectorium*,' PR. P. 105.

 croude-wain, sb., *cf. M.Du.* kruiwaghen; *handcart*, AM. & AMIL. 1858.

crūden, v., *O.E.* crūdan (*pret.* crēad), *M.Du.* kruiden, (*pret.* krood); *crowd, thrust, push*; crude HORN (L.) 1293; croudin PR. P. 105; croude CH. C. T. *B* 801; out croude (*? printed* treud**ě**) SHOR. 33; croude (*imper.*) ALIS. 609; crodin (*pret. pl.*) P. R. L. P. 245.

 crūdinge sb., *thrusting, pushing*; croudinge CH. C. T. *B* 299; crowdinge '*cenivectura*' PR. P. 105.

cruel, adj., *O.Fr.* cruel; *cruel,* A. R. 100; ROB. 57.

 cruel-liche, adv., *cruelly,* CH. TRO. iv. 1274; crewelli MERL. i. 2 (127).

 cruelness, sb., *cruelty,* TREV. I. 177; SACR. 769.

crueltē, sb., *cruelty,* A. R. 268; MAN. (F.) 5186.

cruet, sb., *cruet,* '*phiala*,' REL. I. 7; crouet SHOR. 49; **cruetz** (*pl.*) BRD. 14.

crul, adj., = *M.H.G.* krul, *M.Du.* krol; *curly*; (*ms.* crolle) ALIS. 1999; **crulle** (*pl.*) CH. C. T. *A* 81.

crullen, v., = *M.L.G.* krullen, *Du.* krollen; *curl*; **crollid** (*pple.*) FER. 1354; curlid PR. P. 111.

crumb, adj., *O E.,* crumb, = *O.L.G.* crumb, *O.H.G.* chrumb, *cf. Ir.* crum, crom, *Gael.* crom, *Bret.* kroumm, kromm; *curved,* ORM. 9207; **croume** (*dat. n.*) S. S. (Wr.) 2477.

crumbe, sb., *hook, bend*; crombe, crome, croumbe, '*uncus, bucus*,' PR. P. 104.

crumben, v., *cf. O.H.G.* (ge)chrumben, *M.Du.* krommen; *make curved, bend*; cromin PR. P. 104; **crumpt** (*pple.*) HALLIW. 284.

crume, sb., *O.E.* cruma (*?m.*), *cf. Ger.* krume, *Swed.* kruma, krumme *m.*; *crumb,* '*mica*,' VOC. 258; A. R. 342; crumme PR. P. 106; GOW. III. 35; crome TREV. IV. 399; MIRC 2013; ane crou'ne LEB. JES. 155; **cruman** (*pl.*) MAT. xv. 27; crumen A. R. 344; crumes (*ms.* crummess) ORM. 1475; cromes CH. C. T. (T.) 15528.

crom-bolle, sb., *crumb bowl,* PL. CR. 347.

crümelen, v., *Ger.* krümeln; *crumble*; **kremelid** (*pple.*) L. C. C. 36.

crummin, *cf. L.G.* krömen; *crumble, break into crumbs,* PR. P. 106.

crumpe, sb., *cramp, spasm,* TOWNL. 308.

crumplen, v., *crumple*; **crumpled** (*pple.*), C. M. 8087; was crompild and crokid FLOR. 1979.

crundel, sb., *O.E.* crundel (*? mound*); *cave,* HOM. II. 139.

crūne, sb., *cf. O.N.* krúna, *O.Fr.* corōne, corūne; *crown,* ORM. 8158; MISC. 48; þe þornene krune A. R. 258; crune [croune] LAȝ. 4251, 13110; crune, croune, corune HAV. 568, 1319, 2734; crune, corune GEN. & EX. 2638, 2642; his croune was ... of ismite, blodi was his heved BEK. 2227; corone [croune] LANGL. *B* ii. 10; coroune AYENB. 168; **crūnen** (*pl.*) A. R. 238; crounen ROB. 242; *see* corrin.

crūnen, v., *cf. O.N.* krúna, *O.Fr.* corōner, corūner; *crown*; crouni BEK. 1728; to croune þe to king(e) ROB. 105; krūne**ð** (*pres.*) A. R. 392; coroune**þ** AYENB. 72; **crūneden** (*pret.*) LAȝ. 31935; crūned (*pple.*) ORM. 5462; icruned HOM. I. 121.

crūninge, sb., *coronation,* PROCL. 8; crouning ROB. 430.

crüpel, sb., *O.E.* (eorð-) crüpel, cryppel, = *O.H.G.* crupel, *O.N.* krypill, kryppill, *M.Du.* krepel, kreupel, *O.Fris.* kreppel; *from* crēopen (*cripple,* H. M. 33; REL. I. 243; cripel CHR. E. 771; cripel [crepel] C. M. 12260; cripil, crepil '*claudus*' PR. P. 103; crepel GOW. III. 147.

crüpelen, v., *? go like a cripple*; cripelande (*pple.*) REL. I. 211 (MISC. 5).

crūs, sb., = *M.H.G.* krūs, *M.Du.* kruis; *curly, angry*; HAV. 1966; crus [crous] C. M. 14740. [**cruschel,** sb., *cf. M.H.G.* kroschel, krossel, krustel, kruspel.]

 cruschil-bōn, sb., '*cartilago*,' PR. P. 106.

cruschin, v., *O.Fr.* cruisir, croisir; *crush,* '*quasso*,' PR. P. 106; cruschen D. ARTH. 1134.

crūse, crouse, sb., *M.Du.* kruise, = *O.N.* krūs, *M.H.G.* krūse; *cruse, drinking-vessel, bowl*; '*cantharus*,' PR. P. 105; croos MISC. 29; cruce PALL. xi. 51; **cruses** (*pl.*) '*cyathi*' PALL. i. 584.

crūskin, sb., *a little* crūse, PR. P. 106.

cruste, sb., *cf. Ger.* kruste *f.,* krust *m., from Lat.* crusta, crustum; *crust,* PR. P. 106; crust P. S. 204.

crusted, adj., *having a crust* : a crustid cake WICL. EXOD. xxix. 23.

cū, sb., *O.E.* cū, = *O.Fris.* kū, *O.N.* kȳr (*acc.* kū), *O.L.G., M.Du.* kō, *O.H.G.* chuo; *cow,* A. S. 4; ku '*vacca*' FRAG. 3; A. R. 418; ALIS. 5956; cou P. L. S. xiii. 221; AYENB. 56; LANGL. *B* vi. 289; **kī** [*O.E.* cȳ] (*pl.*) MAN. (F.) 4732; C. M. 4564; kie PS. lxvii. 31; kuin (*r. w.* slen)

ALIS. 760 ; kyn P. L. S. xiii. 235 ; CH. C. T. *B* 4021 ; kiin [kien] LANGL. *B* vi. 142 ; kien WICL. GEN. xxxii. 15 ; kin, ken WILL. 6, 244 ; ken AYENB. 191 ; OCTOV. (W.) 672.

Cou-bache, pr. n., P. L. S. xiii. 244.

cou-calf, sb., *cow calf*, LANGL. *B* xv. 462.

cou-herde, sb., *O.E.* cūhyrde; *cowherd*, WILL. 4.

cou-schote, sb., *O.E.* cūsceote, cuscote, *prov. Eng.* cushat ; *wood-pigeon*, '*palumbus*,' CATH. 79 ; couscot '*palumbus*' VOC. 221 ; qvisht PALL. i. 758.

cou-sloppe, sb., *O.E.* cūsloppe, cūslyppe ; *cowslip*, VOC. 162.

cou-sokulle, sb., *some plant*, *? cowberry*, *? cowslip ;* '*vaccinium*' VOC. 176.

cubbel, sb., *staff*, A. R. 140.

cubicularie, sb., *Lat.* cubiculārius ; *chamberlain*, WICL. JUD. xii. 6.

cubit, sb., *Lat.* cubitus ; *cubit*, WICL. JER. xxxviii. 12 ; A. P. ii. 319 ; TREV. II. 235 ; TOWNL. 28 ; WICL. MAT. vi. 27 ; cubits (*pl.*) WICL. GEN. vi. 15 ; cupidez A. P. ii. 315.

cuccu, sb., *O.Fr.* cucu, coucou ; *cuckoo*, A. S. iii. ; cukkou TREV. I. 229 ; coccou AYENB. 59 ; CH. P. F. 358.

cüchene, [kichene], sb., *O.E.* cycene, *from Lat.* coquīna, *cf. O.H.G.* cuchina, chuchina ; *kitchen*, LA3. 3316 ; kuchene A. R. 214; kichene, kechene WILL. 1681, 1707 ; kechene AYENB. 171 ; kechine PERC. 455.

[**cucking**, sb., *kicking ?*]

cucking-stōl, sb., *cuckingstool*, P. S. 345 ; co(c)kingstōl REL. II. 176.

cuc-stool, sb., *cucking-stool*, '*cadurca*,' PR. P. 107 ; cokstol P. 72.

cude, sb., *? = O.H.G.* cuti (*gluten*) ; *? pitch* : hwit cude *? mastic* LEECHD. III. 136 ; code, sowters wex PR. P. 85 ; DIGB. M. 35 ; coode ['*pice*'] WICL. EX. ii. 3.

cude, sb., *O.E.* cudu, cwudu, cweodu, cwidu ; *cud*, ORM. 1237 ; cude [code] C. M. 1958 ; chewiþ kude [code, qvide, qvede] ['*ruminat*'] WICL. LEV. xi. 3, DEUT. xiv. 6.

cūf, sb., *O.E.* cȳf ; *tub*, FRAG. 4 ; kīve (*dat.*) REL. II. 191 ; keve (*? printed* kove) BEV. 2415.

cuffe, sb., *cuff*, *sleeve*, '*c(h)irot(h)eca*,' PR. P. 106 ; coffes (*pl.*) LANGL. *B* vi. 62.

cüggel, sb., *O.E.* kycgel (*dart*), kuggel, *cf. Welsh* cogeil (*distaff*, *truncheon*) ; *? cudgel*, A. R. 292.

cukeweald, sb., *cuckold* ; cukeweld O. & N. 1544 ; kukewald D. ARTH. 1312 ; cokewold LANGL. *A* iv. 140 ; CH. C. T. *A* 3152.

cul, sb., *stroke* : þe cul of þer eax A. R. 128.

cūl, *see* kēl.   **cūle**, *see* cuvele.

culfre, sb., *O.E.* culfre, culufre, *mod. dialects* culver ; *pigeon*, *dove*, ORM. 1258 ; kulvre A. R. 98; culver WICL. GEN. viii. 8 ; colvre AYENB.

142 ; colvre, colvere MARG. 293, 299 ; colvere [colver] LANGL. *B* xv. 396 ; culfren, culfre (*gen.*) HOM. I. 95, 141 ; culfre brid ORM. 7887 ; þet tu . . . habbe kulvre kunde A. R. 292 ; culveren [colvere] (*pl.*) LA3. 24521 ; colfren ROB. 190 ; colveren ALIS. 5405 ; colveren, colveres MAND. 118 ; culvere briddis ['*pullos columbarum*'] WICL. LEV. x. 7.

colver-hous, sb., *pigeon-house*, AYENB. 142.

[**culien**, v., *? burn ; comp.* for-culien.]

cüllen, v., *cf. M.Du.* killen, kellen (*kill*), *M.H.G.* kellen, *mod.Eng.* kill; *? =* cwellen; *strike*, *beat ; kill* : oþur to culle oþur to bete P. L. S. xxx. 146 ; counseilede Caim to cullen [killen] his broþer LANGL. *A* i. 64 ; killin '*occido*' PR. P. 274 ; culles (*pres.*) JOS. 545 ; kulle al ut þet is i þe krocke A. R. 346; cülle (*subj.*) A. R. 126 ; kille CH. C. T. *D* 1041 ; cülde (*pret.*) S. A. L. 150 ; K. T. 179 ; ofte me hine culde LA3. 20319 ; culde [kilde] LANGL. *A* iii. 180; (h)a kulden hem doun FER. 2660 ; keld (*pple.*) OCTOV. (W.) 1063 ; þei were ikilde [ikeld] TREV. VIII. 5.

cullin, v., *O.Fr.* coillir, *from Lat.* colligere ; *cull*, *gather*, *select*, PR. P. 107 ; coile DEP. R. iii. 200; sex hundred of hise he colede (*pret.*) out MAN. (F.) 2731.

culm, sb., *soot*, '*fuligo*,' PR. P. 108.

[**culmen**, v., *? cover with soot ; comp.* bi-colmen.]

culmi [colmi], adj., *? sooty*, LANGL. *B* xiii. 356 ; his colmie snute HORN (L.) 1082.

cülne, sb., *O.E.* cyln, *cf. O.N.* kylna ; *kiln*, *furnace* ; kulne REL. II. 81 ; kilne PR. P. 274.

cülpe, sb., *mod.Eng.* kelp ; *sea-weed* ; cülpes (*pl.*) TREV. II. 81.

culrāge, sb., *Fr.* curage (*? O.Fr.* *culrage) ; *arse-smart*, *water-pepper*, M. T. 133 ; '*persicoria*' PR. P. 108.

culter, sb., *O.E.* culter, *from Lat.* culter ; *coulter*, LANGL. *B* vi. 106 ; WICL. PROV. xxiii. 2 ; CH. C. T. *A* 3763 ; colter MAP 338.

cultūre, sb., *Fr.* culture ; *culture*, PALL. i. 21.

culvere, *see* culfre.

culvert, adj., *O.Fr.* culvert, *Lat.* collibertus ; *villanous*, A. R. 96 ; FL. & BL. 329 ; *cf.* colward.

cumb, sb., *O.E.* cumb *m.* ; *coomb*, *a measure of capacity* ; coume '*cumba*' PR. P. 97.

Cumberland, pr. n., *O.E.* Cumbra land ; *Cumberland*, MISC. 146.

cumblen, *see* comblen.

cume, sb., *O.E.* cuma, = *O. H. G.* qvemo, *stranger*, MAT. xxv. 35 ; þu bist him cume deore LA3. 4373 ; cumene (*gen. pl.*) LK. ii. 7; *comp.* wil-cume.

cüme, sb., *O.E.* cyme, *m.*, *cf. O.L.G.* cumi, *O.H.G.* chumi, qvemi *f.*, *Goth.* qums *m.* ; *advent*, *arrival*, A. R. 188 ; O. & N. 436 ; cume, kume, kime [come] LA3. 3962, 6646,

14827; cume, come KATH. 413, 671; come ORM. 718; GEN. & EX. 2267; HORN (L.) 530; C. L. 1031; WILL. 4192; D. ARTH. 1203; *comp.* ȝein-, hām-, hider-, on-, tō-, ūt-cüme.

[**cüme**, adj., *O.E.* cyme; *becoming.*]
  cüme-līch, adj., *O.E.* cymlīc; *comely*, H. M. 25; kumli IW. 2886; comli ALIS. 6055; LANGL. *B* xv. 444; cümelīche (*adv.*) MARH. 19; comeliche SPEC. 37; **comloker** (*compar.*) GAW. 869; **comlokest** (*superl.*) GAW. 53.
  comli-hēde, sb., *comeliness*, GOW. II. 354.
  comlinesse, sb., *comeliness*, PR. P. 89.

cumel, sb.; *? ambush, shelter under tents*; on cumelan, comelan, comlen, comele LAȝ. 6630, 11009, 20273, 30400.

cumeling, sb., = *O.H.G.* chomeling; *from* cumen; *stranger, new comer*, GEN. & EX. 834; cumling IW. 1627; cumling [comeling] WICL. LEV. xxv. 47; comeling PR. P. 89; **komlinges** (*pl.*) ROB. 18.

cumen, v., *O.E.* cuman, = *O.L.G.* cuman, O. *Fris.* kuma, koma, *O.N.* koma, *O.H.G.* choman, qveman, *Goth.* qiman; *come*, LK. xviii. 16; SAX. CHR. 250; HOM. I. 19; JUL. 53; ORM. 2179; GEN. & EX. 305; REL. I. 223; cumen [come] LAȝ. 3025; kumen A. R. 394; cumæn, cumæn FRAG. 6, 7; cume KATH. 694; cume, come O. & N. 611; comen HAV. 413; come CH. C. T. *D* 1684; þa sunde to cumene LAȝ. 16029; makeþ to comene AYENB. 106; to cumende HOM. II. 117; þat is to cominge TREV. III. 471; cümest, kimest [*O.E.* cymest, cymst], (*pres.*) JUL. 62, 63; kymst LK. xxiii. 42; comest O. & N. 585; cumeð FRAG. 5; cumeð, kimeð A. R. 350; kumeð, kimeð, kimð HOM. I. 33, 81, 153; kemð LK. ix. 26; cumeþ, comeþ O. & N. 302, 1531; comeð KATH. 2460; comeþ LANGL. *A* i. 59; sche comeþ ride LAUNF. 867; (we) cumeð HOM. I. 51; ȝe comeþ P. L. S. xii. 111; hi comeð viii. 118; **kume** (*subj.*) A. R. 424; cume HORN. (L.) 143; cumen we *let us come*, LAȝ. 13006; **cum** (*imper.*) LAȝ. 12944; cum, kum A. R. 98, 292; cumeþ WICL. MAT. iv. 19; **cominde** (*pple.*) AYENB. 264; **cam** (*pret.*) GEN. & EX. 494; LEB. JES. 285; CH. C. T. *A* 547; S. S. (Wr.) 460; MAND. 12; kam HAV. 863; cam LANGL. *B* ix. 151; com (? cōm), come *A* x. 166, *B* ix. 137 (Wr. 5449, coom); com (? cōm) FRAG. 8; MARH. 2; KATH. 154; O. & N. 133; BEK. 4; WILL. 39; TOR. 662; (*ms.* comm) ORM. 101; (*r. w.* hom) L. N. F. 216; a vuhel com fl(e)on from h(e)ovene HOM. I. 81; þet com wel forð (*sprang up*) 133; s(e)oð-ðen he com tó monne (*he became man*) 167; þa com þer a mon irnen LAȝ. 5748; seinte Katerine of noble cunne com P. L. S. xix. 1; þo hit com to even AYENB. 191; so sone so she to him come AL. (L.³) 511; (þu) come ORM. 2812; O. & N. 1058; HORN (R.) 1178; JOS. 434; comen SAX. CHR. 249; MAT. ii. 2; ORM. 7492; GEN. & EX. 1979; S. S. (Wr.)

318; comen [coomen] LANGL. *B* xx. 219; coomen K. T. 189; come O. & N. 1671; ROB. 15; komen, keme HAV. 1012, 1208; þai kemen FER. 3130; kemin S. & C. I. xlvi; cōme (*subj.*) O. & N. 1015; ROB. 220; **cumen** (*pple.*) ORM. 134; GEN. & EX. 365; IW. 2956; comen MISC. 197; CH. C. T. *A* 671; *comp.* a-, an-, bi-, ȝe-, over-cumen; *deriv.* cume, cüme, cwēme.

  cominge, sb., *coming*, TREV. I. 29.

cumin, sb., *O.Fr.*, cumin, coumin (*cf. O.E.* cymen), *Lat.* cumīnum, cymīnum; *cummin*, MAT. xxiii. 23; comin PR. P. 89; SPEC. 27; ALIS. 6797; WICL. IS. xxviii. 25.

cumlen, v., *? O.N.* kumla; *? become cramped*; i **comble** (*pres.*) REL. II. 211; **cumblid** (*pple.*) [comelid] hondis ['*manus dissolutas*'] WICL. IS. xxxv. 3; comelid '*eviratus*' PR. P. 88; *comp.* a-cumlen.

cumsen, *see* comsen.

cün, sb., *O.E.* cynn, = *O.N.* kyn, kynni, *O.Fris.* ken, *O.L.G.* cunni, *O.H.G.* chunni, *Goth.* kuni *n.; kin, kind, species*, KATH. 445; monna cun HOM. I. 97; cun, kun LAȝ. 319, 328; kun A. R. 308; O. & N. 714; fader moder ant al mi kun SPEC. 91; i have no kinne [kin, kun] þere LANGL. *B* v. 639; kin ORM. 2053; HAV. 393; his kin and his frendes MAND. 309; ken AYENB. 42; WILL. 722; GOW. III. 332; **cünnes** (*gen.*) JUL. 11; alles cunnes wilde d(e)or HOM. I. 79; on ælches cunnes [kinnes] wise LAȝ. 8072; heora nexta **cünnes** mon (*kinsman*) 2727; Daviþes kinnes man ORM. 12528; his cunnes men BEK. 1657; kunnes C. L. 855; ones kunnes treou A. R. 150; for nones kunnes mede P. S. 193; kinnes GEN. & EX. 756; eches kinnes chapman HOM. II. 193; **cünne** (*dat.*) FRAG. 7; O. & N. 271; he is of aðele cunne LAȝ. 11465; of þises wineȝeardes kynne ['*genimine*'] MK. xiv. 25; **cünna** (*pl.*) LEECHD. III. 94; cunne A. R. 128; þreo cunne fan MARH. 1; on feole kunne wisen (*r.* wise) *in many ways*, LAȝ. 1717; a þre cunne wise MISC. 38; kinne HOM. II. 63; alle kinne sinnes LANGL. *B* xx. 370; tweire kunne salve O. & N. 888; tweire kinne HOM. II. 95; of alle kinne cwike der ORM. 14558; *comp.* dēor-, fuȝel-, man-, wīf-cün.

  kün-rēd, sb., *kindred*, DEGR. 835; kenred SHOR. 45; **künrēde** (*dat.*) O. & N. 1677; kinrede MAND. 67; **cünrēdes** (*pl.*) HOM. I. 261.

  cün-rēden, sb., *O.E.* cynrǣden; *race, kindred*, JUL. 60; cunraden (*? printed* cum raden) HICKES I. 225; kunrede ROB. 284; kinrede '*prosapia*' PR. P. 275; CH. C. T. *A* 2790; kinrede, -rade TREV. VII. 471; kinrade WILL. 522; þa twelf kunreden HOM. I. 141.

  cün-rün, sb., *O.E.* cynryn; *generation*, '*generatio*,' FRAG. 2.

[**cünd**, adj., *comp.* god-, gram-cünd.]

[cund, adj., ?=cūð; *comp.* náme-cund.]

cünde, sb., *O.E.* cynd, = *O.N.* kyndi *n.*, *cf. Lat.* gens *f.*; *kind, nature,* HOM. I. 99; MARH. 8; þene wæi þe toward his cunde (*sec. text* cuþþe) læi LAȝ. 21492; þat beoþ of þine cunde O. & N. 251; he haþ angles cunde TREAT. 134; þeos (beastes) doð hare cunde . . . in a time of þe ȝer H. M. 25; þeo kunde A. R. 14; none nerre kunde þer nas ROB. 169; kuinde JOS. 131; kinde ORM. 663; WILL. 2506; CH. C. T. *A* 2451; GAW. 5; þe kinde [kuinde] þat of him com C. L. 1 9; as kinde askeþ LANGL. *B* ii. 27; ðe hertes haven an oðer kinde REL. I. 216; to forsake his wedded wif and do his kinde oþer wai MIRC 230; of kinde ['*naturae*'] WICL. DEUT. xxiii. 12; þe lawe of kinde GOW. III. 204; aȝenst kinde MAND. 223; kende, kinde, '*genus, progenies, prosapia*' PR. P. 271; kende SHOR. 9; AYENB. 28; þat it were comen of riche kende LAI LE FR. 138; after kindes wune GEN. & EX. 1652; *comp.* god-, ȝe-, man-cünde.

cünde, adj., *O.E.* cynde, *mod.Eng.* kind; *natural, native; kind,* TREAT. 136; he þat is so cunde ROB. II. 227; þe cunde folk of þe lond ROB. 40; kuinde c. L. 1044; kinde CH. C. T. *E* 602; (h)is kinde lond GEN. & EX. 1279; his kinde fader WILL. 309; kuinde wit LANGL. *A* i. 53; kende, kinde PR. P. 271; kende SHOR. 90; *comp.* ȝe-, un-cünde.

künd-[h]ēde [kindhēde], sb., *kindness,* ROB. 452; kuindhede LEB. JES. 122.

cünde-lich, adj., *O.E.* cyndelīc, *mod.Eng.* kindly; *natural,* KATH. 964; hit is a kindelich þing AVOW. ARTH. xi.; kindeli WICL. WISD. xii. 10; kendelich AYENB. 47; kündelīche (*adv.*) A. R. 120; kindelike GEN. & EX. 2500; kuindeli LANGL. *A* iii. 15; kindeli PR. C. 1686.

kündenesse, sb., *kindness,* LEG. 185; kindenes WILL. 321.

kinde-ship, sb., *benignity,* GOW. II. 292.

cündel, sb., *offspring*: þesne kundel A. R. 200; kündles (*pl.*) A. R. 200; kindles REL. I. 222; kindlis ['*genimina*'] WICL. LK. iii. 7.

cündlen, v., *prov.Eng.* kindle; *bring forth*; kindlin '*feto*' PR. P. 275; kündleð (*pres.*) A. R. 194; wanne he is ikindled (*pple.*) REL. I. 209.

cüne, *see* kine. cunger, *see* conger.

cunin, *see* coning.

cunnen, v., *O.E.* cunnan, = *Goth.* kunnan, *O. Fris. O.N.* kunna, *O.H.G.* chunnan; *see* can; *know; be able;* HOM. I. 73; ORM. 2958; alle þe creftes þat clerke ah to cunnen KATH. 524; cunnen . . . þonc [*cf. Germ.* dank wissen] A. R. 124; cunne LAȝ. 25113; kunne WICL. ECCLES. vii. 26; knowe and kunne [konne] LANGL. *B* xv. 45; konnen C. L. 1071; conne AYENB. 21; GOW. I. 200; S. S. (Wr.) 33; PARTEN. PROL. 104; Creseide shal not conne

knowe me CH. TRO. v. 1404; konne ROB. 45; cune (*imper.*) HOM. II. 29; cunnand (*pple.*) M. H. 93; þat heeld him self so kunninge [konning] LANGL. *B* xi. 70; cūðe (*pret.*) [*mod.Eng.* could] LAȝ. 904; cuðe, kuðe GEN. & EX. 289, 1659; kuðe KATH. 1544; A. R. 66; kuþe O. & N. 714; couþe HAV. 112; S. S. (Wr.) 2319; couthe [kouþe] LANGL. *A* PROL. 182; coude SHOR. 79; koude CH. C. T. *A* 713; cuðest A. R. 280; cuðen KATH. 1330; heo cuþen alle spechen HOM. I. 93; þa . . . þa weiȝes cuðen LAȝ. 26915; couþen P. S. 248; couthe MIR. PL. 95; cuþe (*subj.*) O. & N. 663; couþe AYENB. 105; ? coud (*pple.*) CH. D. BL. 787.

cunnen, v., *O.E.* cunnian, = *O.L.G.* (gi-)kunnōn, *cf. O.H.G.* chunnēn; *try, test,* ORM. 834; kunnen HOM. I. 151; cunneð (*pres.*) HOM. II. 87; heom i folhi . . . þet cunnið to beon cleane MARH. 13; smeihte ant cunnede (*pret.*) þer of A. R. 114; *comp.* ȝe-cunnen.

cunninge, sb., *O.E.* cunning, *mod.Eng.* cunning; *experience, skill, knowledge, wisdom,* '*scientia*,' PR. P. 109; kunninge LANGL. *B* x. 446; WICL. GEN. xliv. 15; kunning LIDG. M. P. 31; conninge AYENB. 115.

cunte, sb., *O.N.* kunta, = *O.Fris.* kunte; '*vulva*,' VOC. 186; ANGL. IV. 190; counte REL. II. 282.

cūpe, sb., *O.E.* cȳpe, *cf. M.Du.* cūpe; ? *from Lat.* cūpa; *basket,* FL. & BL. 447; cupe, kipe '*sporta*' TREV. IV. 359; cūpen (*pl.*) FL. & BL. 435; twelf cūpe ful C. L. 1278.

cupidez, *see* cubit.

cuppe, sb., *O.E.* cuppa; *cup,* HAV. 14; LANGL. *A* v. 184; MAND. 52; cuppe ['*calicem*'] WICL. JOHN xviii. 11; þa cuppe LAȝ. 14996; cuppe, kuppe GEN. & EX. 2047, 210; cuppe [coppe] CH. C. T. *A* 134; coppe ROB. 117; SHOR. 20; MAN. (F.) 1364; cuppes (*pl.*) ORM. 14043; coppis S. S. (Wr.) 1795.

cup-bord, sb., *cupboard,* A. P. ii. 1440.

cop-ful, sb., *cupful,* REL. I. 52.

cop-maker, sb., *cup maker,* '*cipharius*,' VOC. 213.

cuppe-mēle, -māle, adv., *by cups,* LANGL. *A* v. 139

cuppe-schoten, adj., *drunk* ['*enivré*'], MAN. (F.) 7560.

curat, sb., *med.Lat.* cūrātus; *curate,* CH. C. T. *A* 219.

curatour, sb., *Fr.* curateur; *curator,* MIRC 860.

cūre, sb., *O.E.* cyre, = *M.H.G.* küre *m.; from* chēosen; *choice, custom*: þet wes þe bezste cure LAȝ. 8077; hefde he þene cure 9467; wind god of cure HORN (R.) 1446; cire HOM. I. 221; rigt and kire GEN. & EX. 451; after londes kire 1693.

cūre, sb., *O.Fr.* cūre; *cure,* '*cura*,' PR. P. 110; FER. 1062; LANGL. *B* PROL. 88; CH. C. T. *B* 188.

L

cüren, v.,=*Ger.* küren; *from* **cüre**; *select*, KATH. (E.) 1870; **icüred** (*pple.*) A. R. 56*.

cürin, v., *O.Fr.* cürer, *from Lat.* cūrāre, *mod. Eng.* cure; *take care of*, '*curo*,' PR. P. 110; **cürede** (*pret.*) WICL. 3 KINGS xviii. 30.

curiosité, sb., *Fr.* curiosité; *curiosity*, GOW. III. 383.

curiour, *see* **coriour.**

curious, adj., *O.Fr.* curious; *curious*, TREV. II. 169.

cürnel, sb., *O.E.* cyrnel; *kernel*: þene curnel HOM. I. 79; kurnel '*granum*' FRAG. 3; kirnel '*granum*,' PR. P. 276; '*nucleus*' VOC. 267; kirnel [kernel] LANGL. *B* xi. 253; **cürneles** (*pl.*) LAȝ. 29266; L. H. R. 26; L. C. C. 25; curnles A. R. 260*; kerneles MAN. (F.) 14682.

cürnen, v.,=*Ger.* körnen; *form grain*: curne ROB. 490; kurne [kerne] LANGL. *C* xiii. 180.

curraiin, v., *O.Fr.* conreer; *curry*, '*strigillo*, *cociodio*,' PR. P. 110; curri [currei] PL. CR. 365.

curre, sb., *cf. Swed.* kurre, *M.Du.* corre; *cur*, A. R. 140*; CH. L. G. W. 396; LIDG. M. P. 29; **kurres** (*pl.*) PL. CR. 644.

cürre, adj., *O.N.* kyrr, kvirr,=*M.H.G.* kürre, *Goth.* qairrus; *?quiet, gentle*; kurre A. R. 288*.

curre, sb., *O.Fr.* curre; *chariot*; cure JUL. 9.

curs, sb., *O.E.* curs; *curse*, REL. I. 129; C. L. 1385; CH. C. T. *A* 661; LIDG. M. P. 53; curs [cors] P. S. 333 (P. 38); cors SHOR. 106; FER. 303.

 curs-ful, adj., *accused*, WICL. PROV. xxviii. 9.

cursien, v., *O.E.* cursian; *curse*; cursi BRD. 26; cursen GEN. & EX. 4005; cursen [curse] CH. C. T. *D* 1624; curse GOW. II. 364; cursest, kursest (*pres.*) O. & N. 1178; kursed A. R. 198; curseþ [corseþ] LANGL. *B* xv. 166; **cursede** (*pret.*) SAX. CHR. 264; REL. II. 277; **cursed** (*pple.*) HOM. II. 11; ORM. 16059; icorsed MIRC 1261; *comp.* **a-cursien.**

 cursed-hēde, sb., *cursedness*, WICL. BAR. ii. 33.

 cursednesse, sb., *cursedness*, CH. C. T. *G* 1101.

cursunge, sb., *O.E.* cursung; *cursing*, HOM. I. 205; cursinge REL. I. 131 (HOM. II. 11); GEN. & EX. 3926; corsinge AYENB. 97.

curt, sb., *O.Fr.* curt, cort; *court*, SAX. CHR. 266; HAV. 1685; kurt A. R. 210; cort AYENB. 137; curt, court BEK. 165, 747.

curteis, adj., *O.Fr.* cortois, courtois; *courteous*, MISC. 155; HAV. 2916; BEK. 1190; curteis, corteis WILL. 194, 231.

 kurteis-līche, adv., *courteously*, WILL. 873.

curteisie, sb., *O.Fr.* courtoisie; *courtesy*, HAV. 194; P. L. S. xxiv. 229; CH. C. T. *A* 46; kurteisie A. R. 70.

cürtel, sb., *O.E.* cyrtel,=*O.N.* kyrtill; *kirtle*, *tunic*, MISC. 49; BEK. 2361; SPEC. 40; anne curtel LAȝ. 4993; kurtel A. R. 10; kurtel [kirtel] TREV. VII. 355; kirtel ORM. 10137; PL. CR. 229; CH. C. T. *A* 3321; kertel AYENB. 191; S. S. (Web.) 1883; **cürtle** (*dat.*) HOM. II. 139; **cürtles** (*pl.*) TREV. VII. 401; kirtles ORM. 9292; kertles AYENB. 267; **kertlen** (*dat. pl.*) MAT. vii. 15.

curteour, sb., *courtier*; kourteour WILL. 342.

curtiler, sb., *O.Fr.* courtilier (*gardener*); *monastery gardener*, REL. II. 277.

curtine, *see* **cortin.**

cus, *see* **cos.**

cüssen, v., *O.E.* cyssan,=*O.N.* kyssa, *O.Fris.* kessa, *O.H.G.* cussan; *kiss*, A. R. 424; C. L. 552; cusse SPEC. 38; kissin '*osculor*' PR. P. 277; kisse M. H. 83; CH. C. T. *A* 3680; kesse WILL. 5045; OCTOV. (W.) 585; **cüs** (*imper.*) A. R. 102; **cüste** (*pret.*) LAȝ. 1194; MISC. 42; HORN (L.) 405; kiste GEN. & EX. 1652; LANGL. *B* xviii. 420; keste AYENB. 240; custen BEK. 288.

 cüssinge, sb., *cf. M.H.G.* küssung; *kissing*, SPEC. 70; kessinge FL. & BL. 513.

cussin, sb.,=*O.H.G.* cussin, *O.Fr.* cuissin, coissin; *cushion*; cushin PR. P. 111; **kussinis** (*pl.*) HALLIW. 500.

cüste, sb., *O.E.* cyst,=*O.Fris.* kest, *O.L.G.* cust, *O.H.G.* cust, chust, *Goth.* (ga-)kusts *f.*; *from* **chēosen**; *virtue, quality*; nane custe LAȝ. 21498; **cüste** (*pl.*) REL. I. 178; LAȝ. 31019; MISC. 118; **cüste** (*dat. pl.*) O. & N. 9; na non nuste of Baldulfes custe LAȝ. 20324.

cüsti, adj., *O.E.* cystig,=*O.H.G.* chustīg; *liberal, munificent*, FRAG. 3; LAȝ. 6366: kistiȝ ORM. 4698: *comp.* **mete-cüsti.**

 cüstinesse, sb., *O.E.* cystignesse; *liberality*, HOM. I. 105.

custume, *see* **costume.**

cut, sb., *lot*, '*sors*' PR. P. 111; CH. C. T. *A* 835; kut REL. I. 7.

cutte, sb., *M.Du.* kutte; *pudendum muliebre*; kutte LUD. COV. 218.

cuttin, v., *cut*, '*scindo, seco*,' PR. P. 111; kutte ['*scindere*'] WICL. LEV. x. 6; cutte [kitte] CH. C. T. *C* 954; cotte MIRC 1929; **kutte** (*subj.*) MAND. 49; **cutte** (*pret.*) LAȝ. 8182*; kutte ALIS. 2336; kutten [kitten, ketten] LANGL. *B* vi. 191; citte HAV. 942; **cutted** (*pple.*) PL. CR. 434; PR. C. 3715; *comp.* **for-cutten.**

 cuttere, sb., *cutter*, VOC. 195.

 cuttinge, sb., *cutting*, PR. P. 111.

 cutte-pors, sb., *cut-purse*, *pick-pocket*, LANGL. *A* vi. 118.

cū̆ð, adj., *O.E.* cū̆ð, = *O.L.G.* cūth, *Goth.* kunþs, *O.H.G.* chund, *O.N.* kunnr; *known*, LAȝ. 747; ȝef þu cneowe ant were cuð (*acquainted*) wið þe king JUL. 22; cuþ ORM. 9240; MISC. 96; in ever euch londe ich am cuþ [cuuþ] (*known*) O. & N. 922; couþ BEK. 1198; GOW. II. 310; couþ, kouth CH. C. T. *E* 942; kouth LIDG. M. P. 25; heo beoð . . . al to kuðe A. R. 204; þe **cuddeste** [cuðest,

cuddest] (*superl.*) KATH. (E.) 821; *comp.*
for-, náme-, un-cūð.

cūð-lǣchen, v., *O.E.* cūðlǣcan; *make acquainted, familiar*; cūðlēchen HOM. II. 45;
he hine cūðlæhte LA3. 17103.

kūðlēchunge, sb., *acquaintance*, A. R. 68.

cūð-līch, adj., *familiar*: mid cūðliche [couþ-
liche] worden LA3. 19679; and cuðliche speken 18809; cuþli3 ORM. 2204; cuthli C. M.
17696; couþli GAW. 937; kouþli WICL. PROL.
I KINGS.

cūð-rēdne, sb., *familiarity*, A. R. 168*.

cūðen, v., *O.E.* cȳðan, = *O.Fris.* kētha,
O.L.G. cūthian (*pret.* cūthda), *Goth.* (gasvi-)
kunþjan, *O.H.G.* chunden (*pret.* kundta,
chundita); *make known, show*, LA3. 1157;
HOM. II. 223; kuðen his strencðe A. R. 222;
cuþe P. L. S. viii. 50; REL. I. 183 (MISC. 124);
cuiþe LEG. 38; JOS. 484; þat so much love
him kuiþe wolde C. L. 590; kiðen HOM. II.
139; REL. I. 209 (MISC. 3); kiþen ORM. 210;
kiþe WILL. 1184; MAN. (F.) 6690; PERC.
1234; þe soþ i wille þe kiþe OCTAV. (H.) 609;
keþen SHOR. 20; couþe PL. CR. 17; kēðe
(*pres.*) ANGL. VII. 220; he cuðeð MARH.
19; kiþe [couþe] LANGL. B v. 181; þu cuþest
O. & N. 90; cuþeþ REL. I. 178 (MISC. 118);
kiþeþ ORM. 1131; his craft he ðus kiðeð REL.
I. 211 (MISC. 5); he cuþ O. & N. 132; kiþ
HOM. II. 79; (we) cuðeð KATH. 1348; cūþ
(*imper.*) MISC. 90; cuið HOM. I. 215; kīþe
(*subj.*) CH. C. T. B 636; cūðde [cudde] (*pret.*)
LA3. 15194; cudde 2462; cuþede 4494; cudde
HOM. I. 123; KATH. 1354; ROB. 56; C. L. 756;
muchele luve he us cudde P. L. S. viii. 97;
kidde GEN. & EX. 1651; P. S. 342; LIDG.
M. P. 125; kidde þat he was hende TRIST.
2415; · kedde SHOR. 48; cuddest JUL. 62;
kidden ORM. 17822; kidden [kidde] LANGL.
B v. 440; kīþed (*pple.*) ORM. 16979; cud
KATH. 814; kud FER. 1401; kid GOW. II.
299; MAN. (F.) 1621; it was ful loude kid
HAV. 1060; of alle kudde & kuðe sunnen
A. R. 342; *comp.* ȝe-cūðen.

cūði, adj., *O.E.* cȳðig, = *O.H.G.* kundīg, *O.N.*
kunnigr; *knowing, familiar*: cuðie meyes
LA3. 5098.

cūððe [cüþþe], sb., *O.E.* cȳðð, cȳððu, *O.H.G.*
chundida, *M.H.G.* künde; *kith, kindred,
native country*, LA3. 2435; his cuþþe and
(h)is kun LEG. 107; þe king þat þis kuþþe
auȝte JOS. 434; cuþþe [kiþþe] LANGL. A iii.
197; keþþe SHOR. 19; kiþþe TREV. IV. 267;
kiþe ANT. ARTH. xii; fro kiþe and kinne
GOW. II. 267; cēðen, cheðen (*pl.*) HOM. I.
230, 235; *comp.* un-cūðde.

kiþþe-lī3, adv., *familiarly*, ORM. 16532.

cuveiten, v., *O.Fr.* cuveiter; *covet*, A. R. 60
[*v. r.* ȝirni]; coveite GOW. II. 136; coveiteþ
(*pres.*) ROB. 306.

cuveitise, sb., *O.Fr.* coveitise; *covetousness*;
coveitise SPEC. 49; AYENB. 137.

cuveitūs, adj., *O.Fr.* coveitūs; *covetous*;
coveitus FL. & BL. 367; coveitous S. S.
(WEB.) 1591; covaitous AYENB. 80; covetous
MAN. (F.) 4529.

cuvel, sb., *? cf. M.H.G.* kübel; *basket*; coul
'*cupa, tina*' PR. P. 97; E. G. 382; in lepes
& in coufles (*pl.*) so muche viss hii ssolde
him bringe ROB. 265.

cuvel-staf, sb., *pole for carrying two-handled
vessels*, GEN. & EX. 3710.

cuvele, sb., *O.E.* cufle, *cf. O.N.* kufl *m.*,
M.Du. covele, covle *f.*, kovel *m.; cowl*;
kuvele A. R. 10; ane cule [covele] LA3. 17698;
covele BEK. 2228; under covele and cope
SHOR. 110; cuvel, covel HAV. 768, 2904;
coule (*ms.* cowle) PR. P. 97.

cwákien, v., *O.E.* cwacian; *quake, tremble*,
MARH. 19; A. R. 116; qvakien [cwakie] LA3.
17915; qvakie MISC. 193; A. D. 241; ? hwa-
kien HOM. I. 143; whakin, qvakin PR. P.
523; qwake, whake PR. C. 5411, 7343;
cwákede (*pret.*) JUL. 21; qvakede HICKES
I. 228; ROB. 24; OCTOV. (W.) 1713; qvakide
WICL. I KINGS xxviii. 5; qvaked LANGL.
B xviii. 246; qvōk CH. C. T. A 1576; qvoc
M. H. 9; MAN. (H.) 292.

qvákinge, sb., *O.E.* cwacung; *quaking*,
ROB. 336; WICL. MK. xvi. 8.

cwále, sb., *O.E.* cwalu, = *O.L.G., O.H.G.* qvala,
O.N. kvöl; *from* cwélen; *death, slaughter*;
ȝef hine to cwale for us alle HOM. I. 121;
qvale com on orve LA3. 31809.

qvále-hūs, sb., *torture-chamber*, LA3. 727.

qvále-sīð, sb., *mortality*, LA3. 31900.

cwalm, sb., *O.N.* cwealm, cwelm, = *O.L.G.*
qvalm, *O.H.G.* qvalm, chvalm; *from* cwélen;
death, slaughter; torture; (*printed* cwalm)
H. M. 29; cwalm, cvalm, qvalm O. & N. 1157,
1199; qvalm ROB. 252; REL. II. 244; P. S.
342; C. M. 4721; þe qvalm muchele þe wes
on moncunne LA3. 31877; qvelm P. L. S. iv.
9; cwalme (*dat.*) HOM. I. 115; qvalme CH.
C. T. A 2014; cwalmes (*pl.*) LK. xxi. 11;
comp. man-, orf-cwalm.

cwalm-hūs, sb., *torture-chamber*, KATH.
600; A. R. 140.

cwalm-steou, sb., *O.E.* cwealmstōw; *place
of execution*, A. R. 106; cwalm-stōwe (*dat.*),
HOM. I. 283.

cwarterne, sb., *O.E.* cweartern; *prison*,
MAT. v. 25; KATH. 670; ORM. 6168; qvarterne LA3. 19293; HOM. II. 213; qvarterne
(*pl.*) SAX. CHR. 262.

[cwátien, v., *cf. M.H.G.* qvāzen; *satiate*;
comp. over-qvatien.]

cwávien, v., *tremble*, MARH. 48*; qvavin
'*tremo*' PR. P. 419; qvávide (*pret.*) WICL. I
KINGS xxviii. 5; qvaved LANGL. B xviii. 61.

cwecchen, v., *O.E.* cweccan, = *M.Du.* qvicken;
twitch, shake; qvecchen mid hafde LA3.
25844; qvecchen [cwecche] to cuchene

3316; qveche ALIS. 4747; qvicchin '*moveo*'
PR. P. 421; **qvaȝte** (*pret.*) FER. 607; þat þa
eorðe aȝæn qvehte [cwehte] LAȝ. 20141; he
qveite toward þe qvene WILL. 4344; heo
qvehten [cwehten] heore scaftes LAȝ. 23907;
comp. **a-cwecchen.**

cwēd, adj., *cf.* O.E. cwēd (*dung*), BLICKL.
GLOSSES, = O.Fris. qvād, M.Du. qvaed, M.H.
G. qvāt, kāt, kōt (*bad ; dung*); *bad*; cved
LAȝ. 29600; qved O. & N. 1137; ROB. 414;
ALIS. 1243; P. S. 256; he dude more qved
CHR. E. 443; þai þoghte to do him qved
ISUM. 611; kved, kvead AYENB. 11, 14; qvead
SHOR. 147; qvad GEN. & EX. 536; Gow. II.
246; þe qvede ROB. 314; MIRC 1658; þene
qvede MISC. 44; he deð al to cweade A. R.
72; **qvēde** (*pl.*) P. L. S. iv. 31; qvade ȝere
CH. C. T. B 1628.

    kvēad-hēde, sb., = O.Fris. qvaedheed; *bad-
ness*, AYENB. 101.

    qvēd-līc, adj., *badly*, PS. xvii. 22*; kvead-
liche AYENB. 8; cwedli (*sparingly*) HOM. I. 269.

    kvēadnesse, sb., *iniquity*, AYENB. 10.

    qvēd-schipe [cwēdsipe], sb., *evil*, LAȝ.
5066; cweadschipe A. R. 310.

cweise, sb., O.N. kveisa, *cf.* M.L.G. qvese;
ulcer, A. R. 328.

cwélen, v., O.E. cwelan, = O.L.G. qvelan,
O.H.G. chelan; *die ; be tormented*; **cwélað**
(*pres.*) HOM. I. 111; **qvéle** (*subj.*) REL. I. 174;
(MISC. 112); **qvēlen** (*pret.*) LAȝ. 31826;
deriv. **cwále, cwalm, cwüld.**

[cwelle, sb., *killer ; comp.* **man-cwelle.]**

cwellen, v., O.E. cwellan, = O.L.G. qvellian,
O.H.G. qvellen, chellen, O.N. kvelja; *from*
cwale; *mod. Eng.* quell; *kill, torture* ; (*ms.*
cwellenn, *printed* cwellen) ORM. 1843; qvellen [cwelle] LAȝ. 644; qvellin '*suffoco*'
PR. P. 419; cwelle R. S. v. (MISC. 168); ne
mai him no feond cwelle S. A. L. 153; qvelle
ROB. 10; BEK. 767; WILL. 1246; LIDG.
M. P. 109; M. H. 96; EGL. 521; the dokes
criden as men wolde hem qvelle CH. C. T. B
4580; and with here axes out the braines
qvelle (*knock*) TRO. iv. 46; **qvelde** (*pret.*)
HORN (L.) 988; cwalden ORM. 15526;
qvalden [cwelden] LAȝ. 1752; qvolde MISC.
92; **cwelled** (*pple.*) HOM. I. 279; qvelled
GAW. 1324; iqvald [icwelled] LAȝ. 10928;
comp. **a-cwellen.**

    cwellere, sb., O.E. cwellere; *killer*, MK. vi.
27; MARH. 22; qvellere WICL. TOB. iii. 9;
qveller M. H. 40; *comp.* **man-qvellere.**

    qvelling, sb., *killing*, ROB. 296.

cwellen, v., = M.Du., O.H.G. qvellen; *spring
up as a well* : þe welle . . . þet alne wai kvelþ
(? *ms.* kvelȝ) AYENB. 248.

cwelm, *see* **cwalm.**

cwelmen, v., O.E. cwelman, cwylman, = O.
L.G. qvelmian; *from* **cwalm**; *kill* : qvelm(e)
['*trucident*'] PS. xxxvi. 14

cwēme, adj., O.E. (ge)cwēme, *cf.* O.H.G. (bi-)
qvām; *from* cumen ; *convenient, agreeable*,
ORM. 466; qveme HAV. 393; GAW. 2109;
TOWNL. 2; hit wes him swiðe qveme LAȝ.
2427; qweme C. L. 500; D. TROY 633; wheme
FLOR. 145; **qvēmere** (*compar.*) HOM. II. 63;
qvēmest (*superlat.*) H. M. 11; GEN. & EX.
3764; *comp.* **ȝe-cwēme.**

cwēme, sb., O.E., ? *pleasure* ; to cweme HOM. I. 23;
almahtiȝ god to cweme ORM. 1661; þan
folke to qveme [cweme] LAȝ. 347; to qveme
C. L. 1018; WILL. 3404; MAN. (F.) 7390;
to qveme [qweme] P. S. 325 (P. 9).

    qvēme-ful, adj. ['*placabilis*'], WICL. EXOD.
xxxii. 12; qvemfull(e) qveene ALIS.² (Sk.) 582.

cwēmen, v., O.E. cwēman; *please*, HOM. I.
247; H. M. 5; R. S. V. (MISC. 184); hu þu
miht drihtin cwemen ORM. 1217; cweme
LAȝ. 13486*; qvemen MISC. 185; RICH.
3432; qveme O. & N. 209; GREG. 258;
Gow. I. 166; PS. cxiv. 9; MAN. (H.) 286;
qvemin PR. P. 420; kveme AYENB. 26:
cwēmeð (*pres.*) : servises inedde ne cwemeð
nout ure loverde '*coacta servitia deo non
placent*' A. R. 338; **cwěmde** (*pret.*) FRAG.
8; cwemmde ORM. 2595; *comp.* **ȝe-cwēmen.**

    cwēmnesse, sb., O.E. cwēmness ; *pleasure*,
HOM. I. 213; qvemnesse II. 55.

cwēn, sb., O.E. cwēn, cwene, *cf.* O.N. kvān,
kona, O.L.G. qvān, qvena, O.H.G. qvena,
Goth. qēns, qinō (γυνή), *mod. Eng.* quean *and*
queen ; *woman ; queen*, MARH. 19; ORM.
2159; qven TRIST. 1193; qven, qvene [cwene]
LAȝ. 182, 3729, 25843; qven, qvene MISC.
195, 196; kven, kvene AYENB. 80, 130; qwen
[qwean], qwene C. L. 881, 1192; qwen, qwene
LUD. COV. 80, 215; cwene A. R. 170, 296;
kwene HOM. I. 193; qvene R. S. i. (MISC. 156);
FL. & BL. 818; SHOR. 131; ROB. 26; CH.
C. T. B 161; qwene PR. C. 4461; qviene S.
A. L. 154; **qvēne** (*gen.*) ROB. 26; qvene red
REL. 182 (MISC. 122); **qvēne** (*dat.*) LAȝ. 140;
TOR. 766; **qvēne** (*pl.*) LAȝ. 24713.

cwenchen, v., O.E. (a-)cwencan; *quench*,
qvenchen HOM. I. 175; REL. I. 132; qvenche
SHOR. 19; AYENB. 204; TREV. I. 187; LIDG.
M. P. 151; cwenken ORM. 1191; **qvencheþ**
(*pres.*) TREAT. 136; **cwenchte** (*pret.*) JUL.
68; þat fur qveinte an on P. L. S. xv. 204;
qvenched (*pple.*) MAND. 70; qvenchid WICL.
MK. ix. 44; cwenked ORM. 4417; qveint
Gow. II. 201; LANGL. B xviii. 344; CH.
C. T. A 2321; ikvenct AYENB. 186; iqveint
MIRC 1194; *comp.* **a-cwenchen.**

    qvenchinge, sb., *quenching, extinction*,
TREAT. 135; qvenching MISC. 7.

cwerkin, v., = O.Fris. qverka, O.N. kyrkja;
choke, strangle, '*suffoco*'; qverkin PR. P. 420.

cwerne, sb., O.E. cweorn, cwyrn, = O.N. kvern,
O.L.G. qvern, O.H.G. qvirn, churne, *Goth.*
qairnus; *quern, hand-mill*; qverne '*mola*'
PR. P. 420; P. L. S. xix. 233; AYENB. 181;

CH. H. F. 1798; WICL. EX. xi. 15; **qvernes**
(*pl.*) TREV. III. 391; *comp.* **pepir-qwerne**.

cweorn-stŏn, sb., *O.F.* cweornstān; *mill-
stone*, FRAG. 4; qwernston VOC. 233.

cwessen, see **qvaschen**.

cweðen, v., *O.E.* cweðan, = *O.L.G.* qvethan,
qvedan, *O.N.* kveða, *O.Fris.* qvetha, qveda,
O.H.G. qvedan, chvedan, chedan, *Goth.* qiþan;
speak, say, MAT. iv. 17; cweþen HOM. I. 37;
qveðen II. 29; GEN. & EX. 3525; **cwěðe**
(*pres.*) HOM. I. 99; KATH. 869; i qveþe him
qvite R. R. 6999; & cweðeð þe al cwite H. M.
41; cweþeð, cweð HOM. I. 31, 113; qveð
HOM. II. 5; (we) cweðað HOM. I. 113;
qvéðende (*pple.*) REL. I. 129; cweðinde
HOM. I. 99; qwething LUD. COV. 362;
cwaþ (*pret.*) quoth ORM. 5214; qvaþ WILL.
251; TRIST. 607; PL. CR. 98; qvað, qvad
GEN. & EX. 755, 2881; qvaþ LANGL. *C* iv. 490;
cwað, cwaþ, quaþ, qvad, qveþ O. & N. 117,
187, 1186; qvað, qveð [cwaþ] LAȝ. 696, 6203;
qvaþ, qveþ MISC. 40, 86; cweð KATH. 379;
A. R. 122; cweð, qvoð MARH. 4, 22; qvath,
qvoth, qvodh, qvod, hwat HAV. 606, 909, 1650,
1800, 1888; qvoþ [qvod] CH. C. T. *A* 788;
qvod GOW. I. 47; S. S. (Wr.) 668; coþ ANT.
ARTH. xxxvii; hio **cwǣðen** [*O.E.* cwǣ-
don] (*pl.*) JOHN i. 22; cweþen HOM. I. 5;
cweðen KATH. 134; cweðen, ? cweden JUL.
26, 27; qveðen LAȝ. 893; qveþen MISC. 146;
qvoðen GEN. & EX. 3267; **qvéðen** [*O.E.*
cweden] (*pple.*) GEN. & EX. 1496; *comp.*
a-, bi-, for-, ȝe-, mis-cweðen; *deriv.* **cwide**,
cwiste.

cwic, adj., *O.E.* cwic, cwyc, cuc, = *O.L.G.* qvic,
O.N. kvikr, kykr, *O.H.G.* qvech, chech,
Goth. qius (*accus.* qivana), *mod.Eng.* quick;
living, KATH. 1890; ORM. 1370; qvic [cwic]
LAȝ. 22; slou to fiȝte & qvic to fle ROB.
455; qvic, qvik HAV. 1405, 2210; WILL.
1212, 1564; qvic CHRON. E. 762; CH. C. T.
A 1015; qvik oþer ded BEK. 1822; qvik
water WICL. GEN. xxvi. 19; kvic, qvic
AYENB. 67, 205; qwik PR. C. 6390; qvek
M. H. 26; qvik, whik PR. P. 421, 524; whik
M. T. 156; TOWNL. 163; (þat) cwuce bread
HOM. I. 241; nowiht **cwikes** (*gen.*) *nothing
alive* A. R. 334; nawiht qvikes LAȝ. 25758;
elc þing cuces HOM. I. 225; bi mine qvike
[cwike] live LAȝ. 14595; þa andswarede þe
king **qvickere** (*dat. fem.*) stefne 16874; mid
qvickere [cwickere] stevene 15873; **cwike**
(*pl.*) JUL. 62; cwike briddes KATH. 64;
cwike & grene boȝhes ORM. 10002; demen
. . . cwike & deade HOM. I. 209; of qvike
and of dede LANGL. *B* xix. 53; qvike [cwike]
(*dat. pl.*) LAȝ. 27317; qvic eiȝte ROB. 537;
cwickure (*compar.*) A. R. 112; **cwikest**
(*supl.*) A. R. 112; *comp.* **sām-cweoc**.

cwic-liche, adv., *quickly*, A. R. 246; qvicliche
[cwicliche] LAȝ. 4697; qvic-, qvikliche WILL.
908, 2127; qvicliche ALIS. 2607; **cwicluker**
(*compar.*) A. R. 270.

qvik-mīre, sb., *quickmire*, PL. CR. 226.

cwicnesse, sb., *quickness*, A. R. 150; qvik-
nesse PR. P. 421.

qvik-silver, sb., *O.E.* cwicseolfor; *quick-
silver*, CH. C. T. *A* 629; MAND. 52.

cwidden, v., *O.E.* cwiddiar, *cf. O.N.* kviðja,
O.H.G. qvettan, chvettan; *say*, ORM. 19358;
cwiddest (*pres.*) KATH. 2172; ase hire
nome cwiddeð A. R. 174; qviddieð LAȝ.
25325; qviddeþ MISC. 85; **cwiddeden** (*pret.*)
ORM. 8613; qviddeden (*sec. text* cwiddede)
LAȝ. 13775; **cwidded** (*pple.*) ORM. 282;
icwiddet HOM. I. 261.

cwide, sb., *O.E.* cwide, cwyde, = *O.L.G.*,
O.H.G. qvidi, *O.N.* kviðr *m.; from* **cweðen**;
saying; promise; legacy, A. R. 208; þes
witeȝan cwide . . . is ifulled HOM. I. 91; ich
forȝive ælchere widewe hire laverdes qvide
[cwide] LAȝ. 14857; qvide O. & N. 685; qvede
GEN. & EX. 1463; qvethe GAW. 1150; **qvides**
[cwides] (*pl.*) LAȝ. 9141; on ealden cwiden
MAT. v. 27; *comp.* **bi-cwide**.

qveþe-word, sb., *promise*, ‘*legatum*,’ PR. P.
420.

cwīe, sb., *O.N.* kvīga : *heifer*; qwie ‘*juvenca*’
VOC. 218.

qwīe-calf, sb., *O.N.* kvīgukālfr, *prov.Eng.*
whycalf; *female calf*, ‘*vitula*,’ VOC. 219.

cwikien, v., *O.E.* cwician, = *O.H.G.* qvichan,
chichan; *enliven, make or be lively*; qwiken
PR. C. 1723; qvikin PR. P. 421; qvikin [qvike]
C. M. 8622; qvikie S. A. L. 159; **qvĭkie** [qvike]
(*pres.*) LANGL. *B* xviii. 344; þe brond þe is
al aqvenched . . . ne qvikeð he nevre HOM. I.
81; qviken HOM. II. 177; **cwikede** (*pret.*)
H. M. 43; **qviked** (*pple.*) CH. C. T. *F* 1050;
comp. **a-cwikien**.

cwiknen, v., *O.N.* kvikna; *quicken, give life,
receive life*; qviknin ‘*vivifico*’ PR. P. 421;
qvikene WICL. LK. xvii. 33; **qvikne** (*pres.*)
LANGL. *B* xviii. 344*; aȝen qvikenid (*v. r.*
qvikide) [‘*revixit*’] WICL. GEN. xlv. 27*.

cwiste, sb., *O.E.* (ge)cwiss, = *Goth.* (missa-)
qiss; *from* **cweðen**; *bequest*; qviste HAV.
219; *comp.* **bi-qveste**.

cwiver [**cover**], adj., *O.E.* cwifer; *brisk,
lively*, A. R. 140.

cwüld, sb., *O.E.* cwyld, cwild *m. f. n., cf.
O.H.G.* qvelida *f.; from* **cwelen**; *death,
slaughter*, ‘*pestis*,’ FRAG. 3.

d.

dā, sb., *O.E.* dā; *doe*, IW. 2027; do ‘*dama*’
FRAG. 3; MAN. (F.) 15749; doo PR. P. 124;
WICL. PROV. vi. 5.

dabbe, sb., *stroke, slap*, ALIS. 2306.

dabben, v., *cf. M.Du.* dabben; *slap, strike*;
dabbeþ (*pres.*) P. S. 192.

dæd, see **dēad**.

dæde, sb., *O.E.* dēd, = *O.L.G.* dād, *O.N.* dāð,

Goth. dēds, *O.H.G.* tāt *f.*; *from* dōn; *deed*;
dede LAȝ. 26556; A. R. 62; ORM.. 2267;
ROB. 73; PR. C. 2485; or he dide ani oþer
dede HAV. 1356; bitvex him & þe was mad
a prive dede MAN. (H.) 259; in dede CH. C.
T. *A* '659; dade HOM. II. 187; dēde (*pl.*)
FRAG. 2; BRD. 27; dede, deden P. L. S.
viii. 5, 45; dæde, dede, deden [dedes] LAȝ.
4864, 6564, 14177; dede, dade, deden, dedes
HOM. II. 9, 15, 131, 222; dæden MAT. xviii.
31; deden C. L. 938; his uvele deden A. R.
86; mi gode deden SPEC. 99; in . . . mine
dearne deades KATH. 575; dēdan (*gen. pl.*)
HOM. I. 117; dēdan, dede (*dat. pl.*) HOM. I.
99, 159; deden FRAG. 7; LAȝ. 13976; dede
P. L. S. viii. 2; O. & N. 1376; *comp.* elmes-,
gōd-, harm-, mis-, oþer-, üvel-, wel-dæde.

dæd-bōte, sb., *O.E.* dædbōt; *atonement,
reparation for a crime,* ORM. 6025; þa dedbote HOM. I. 23; penitence þet is dedbote
A. R. 348; satisfac(c)ioun dedbote AYENB.
32; deedbote REL. II. 243.

dæi, sb., *O.E.* dæg, = *O.L.G.* dag, *O.Fris.* dei,
O.N. dagr, *Goth.* dags, *O.H.G.* tag; *day,* SAX.
CHR. 260; dæi, dai LAȝ. 833, 10246; deȝ HOM.
I. 223; dei & niht 65; dei KATH. 747; tō dai
to-day, 788; dei A. R. 20; to dai 278; daȝ HOM.
II. 257; þe latste daȝ (*ms.* daȝȝ) ORM. 4168;
godes gast of hefne com . . . an daȝ at undern
time 19458; to daȝ 5415; daig MAT. vi. 34;
dai FRAG. 7; GEN. & EX. 83; (*ms.* day) PROCL.
7; **a dai** *by day* O. & N. 219; hit was neȝ
dai liȝt P. L. S. xvii. 277; as hi fischede a dai
bi þe se P. L. S. xx. 3; al dai [*O.E.* ealne dæg]
BEK. 1458; þet habbeþ al dai hare eȝe to hevene
AYENB. 75; nou a dai GOW. I. 64; **dæiges,**
dæies [*O.E.* dæges] (*gen.*) SAX. CHR. 249, 251;
dæiȝes LAȝ. 30032; dæies [daiȝes] and nihtes
3255; þes daȝes [daies] lihte 26464; deiȝes
HOM. I. 97; deies & nihtes 7; deies A. R. 256;
daies KATH. 1078; (*ms.* dayes) HAV. 2353; bi
daies lihte O. & N. 1431; haveþ daies kare
1590; be dæies (*for* dæie) SAX. CHR. 261;
nou a dais CH. C. T. G 1396; **daȝe,** dæie, daie
[*O.E.* dæge] (*dat.*) LAȝ. 82, 2775, 4923; on his
dæie [daiȝe] 9090; æfter his deie [daiȝe] 2942;
en Edwardes kinges deȝe ANGL. VII. 220;
deȝe, deiȝe, deie, daie HOM. I. 99, 191, 229;
daiȝe MK. x. 34; deie JUL. 10; bi deie A. R.
32; daie FL. & BL. 259; bi daie O. & N. 241;
dages [*O.E.* dagas] (*pl.*) MAT. xxviii. 20;
daȝes HOM. I. 11; daȝen 119; daȝhes ORM.
1902; daȝes, dæȝes, dæies, dawes [daȝes,
daiȝes, daies, dawes] LAȝ. 1113, 3895, 8724,
19216; daȝen 3615; daȝes, daies AYENB. 198,
214; dages, daies REL. I. 225, 226; daies,
dahes KATH. 1562, 1844; daies MISC. 28;
WILL. 3843; WICL. JOHN xi. 6; daies, dawes
BRD. 5; 6; daiges GEN. & EX. 2455; dawes A. R.
70; dawes MISC. 38; HAV. 2344; ALIS. 1436;
GOW. II. 113; LAUNF. 1; dæges LEECHD. III.
84; deaȝes FRAG. 5; daiȝe [*O.E.* daga] (*gen.
pl.*) MAT. xxiv. 29; fifti daȝa HOM. I. 87;

daȝene, dawene LAȝ. 3615, 4605; twenti dahene
gong KATH. 2502; daȝen SHOR. 126; **dagon**
(*dat. pl.*) LK. iv. 25; dagen MAT. ii. 1; daȝen
HOM. I. 11; an his daȝen LAȝ. 10932; after his
daȝen 19848; o þon dawen 1284; dagen HOM.
II. 47; daghen MISC. 35; on þan dæge
LEECHD. III. 84; dawen ALIS. 5631; of dage
brogten GEN. & EX. 3545; bi olde dawe MISC.
47; P. L. S. xiii. 19; þat he had broȝt of dawe
FER. 2143; mani a bold burn was sone brouȝt
of dawe WILL. 3817; *comp.* dēaþ-, ēaster-,
ende-, ȝōl-, hāli-, lif-, love-, messe-, reste-,
somer-, wā-dæi (dei, dai).

daies ēiȝe, sb., *daisy,* ALIS. 7511; daies ie
CH. C. T. *A* 332; daies eȝes SPEC. 43.

dæi-hwam-līc, adj., *O.E.* dæghwamlīc;
daily; ure daihwamliche bred HOM. II. 27;
heore daȝwhamlike swinc ORM. 6238; **dei-
hwamlīche** (*adv.*) HOM. I. 17.

dai-lī, adj., *O.E.* dæglīc; *daily,* A. P. i.
313.

dai-līht, sb., *O.E.* dæglēoht; *daylight,* LAȝ.
III. 89; TOWNL. 2.

daig-rēd, sb., *O.E.* dæg-, dægerēd, -ræd, =
M.Du. dagheraed; *dawn,* JOHN viii. 2; R. S.
iv. (MISC. 162).

daiȝ-rēwe, sb., *dawn,* HOM. II. 255; dairewe
O. & N. 328*; daierewe MISC. 163; dairawe
A. P. ii. 893; ALEX. 392.

dai-rieme, sb., *O.E.* dægrima; *daybreak,*
'*aurora,*' REL. I. 130 (HOM. II. 167); dairim
O. & N. 328; dæiríme (*dat.*) SAX. CHR. 249.

dag-sang, sb., *day-singing*; wid daȝsang &
wid uhtensang ORM. 6358.

dai-spring, sb., *day-spring,* ALIS. 4290.

dai-steorre, sb., *O.E.* dægsteorra; *day-star,*
O. & N. 328; daisterre WICL. JOB xxxviii. 32.

dæl, dāl, sb., *O.E.* dǽl *m.*, dāl *n.*, *cf.* *O.L.G.*,
O.Fris. dēl *m. n.*, *O.N.* deili *n.*, *Goth.* dails *f.*,
O.H.G. teil *m. n.*, *mod.Eng.* deal; *part, portion*:
þe moare dæl of heom PROCL. 2; iwhilc dæl
ORM. 1722; del 2715; muchel dæl of his cunne
LAȝ. 9436; þat beste del [deal] 2951; mesten
dæl (*? ms.* dal) [del] alle *nearly all* 2153;
muche deal goldes 6079; nenne dāl (*? for* dæl)
JOHN xiii. 8; del SHOR. 145; LANGL. *B* xv.
480; AMAD. (R.) xiv; þane del HOM. I. 229;
þe meste del A. R. 330; þet feirest del 276;
þat beste del HOM. II. 143; sum del O. & N. 870;
ri song were ispild ech del 1027; sum del
GEN. & EX. 353; BEK. 688; ha was sum del
offruht KATH. 668; mesten del MARH. 17;
il(c) del HAV. 818; þat oþer del BRD. 3; þe
furþe del WILL. 1284; deel TREV. I. 153;
þe heþ grat del ine his kende AYENB. 17; eche
daies dol 112; hi nolden þer of makie nones
cunnes dol MISC. 50; he hæfde to dæle þat
suðland LAȝ. 2111; him þat lond to dale
com 2117; from þe oþer dele of Engelond
TREV. II. 85; **dæles** (*pl.*) FRAG. 1; dæles,
dales MAT. ii. 22, xvi. 13; deles, doles AYENB.

17 ; he a fif dæle [deale] dælde his ferde LAȝ.
21125 ; **dæle** (*gen. pl.*) FRAG. 1 ; *comp.* **after-,**
for-, ȝe-, or-, tō-dæl (-dāl, -dōl) ; *see also*
dāle.

   dāl-neominde, sb., *O.E.* dælneomende ;
sharer, HOM. I. 47.

dælen, v., *O.E.* dǣlan, = *O.L.G.* dēlian, *O.Fris.*
dēla, *O.N.* deila, *Goth.* dailjan, *O.H.G.* teilan,
mod.Eng. deal ; *divide, share,* ORM. 6175 ;
heo wolden al þis lond dælen [deale] heom bitwenen LAȝ 4053 ; to dælen and to dihten þis
kinelond 23627 ; þer heo gunnen dælen (*sepa-*
rate) 18897 ; nu wit scullen delen (*fight about*)
þene dæð of mire maȝen 26041 ; heo gunnen
delen (*fight*) 30417 ; þet ich mote delen
(*take part*) ine ham A. R. 28 ; dealen lif &
soule 224 ; to dealen his feder chetel to
neodful(l)e 224 ; delen MISC. 49 ; to delen þat
uvel from þe gode C. L. 139 ; to delen with
swich poraille CH. C. T. *A* 247 ; dele AYENB.
76 ; S. S. (Wr.) 3442 ; PR. C. 3460 ; dele, deale
O. & N. 954 ; hi dǣleð (*pres.*) FRAG. 6 ; þei
deleþ LANGL. *B* x. 28 ; þe womman þat ȝe with
delen (*cohabit*) *B* vii. 90 ; on four doles delen
he ðe ger GEN. & EX. 151 ; dēl (*imper.*) GEN.
& EX. 3239 ; deleð elmesse ' *date ele(e)-*
mosynam' HOM. I. 111 ; þet ðu ne dēle (*subj.*)
(*share*) noht þer inne 67 ; dælde (*pret.*) LAȝ.
21126 ; delde BEK. 1191 ; delden HOM. I. 91 ;
dæled (*pple.*) ORM. 598 ; deled ROB. 23 ;
CHR. E. 370 ; þis þing was deled and diȝt so
hem þouȝt best AR. & MER. 5349 ; *comp.* **bi-,**
ȝe-, tō-dælen.

   dēlare, sb., *O.E.* dǣlere ; *divider,* '*partitor,*
distributor,' PR. P. 117 ; FRAG. 2.

   dēlende, sb., *O.E.* dǣlend (Lk. xii. 14) ;
partner : non to him ne come ne delende
nere of his eadinesse HOM. I. 217.

   dēlinge, sb., = *O.H.G.* teilunga ; *dealing,*
partition, division, TREV. I. 247 ; ['*distributio*']
WICL. NUM. xxxvi. 4 ; *comp.* **tō-dēlinge.**

dæþ, *see* **dēað.**

daffe, sb., *fool, idiot,* PR. P. 111 ; deffe 116 ;
P. S. 328 ; LANGL. *A* i. 129 ; CH. C. T. *A* 4208.

[**daffen,** v., *render foolish ; comp.* **bi-daffen.**]

daft, adj., *O.E.* (ge-)dæft, *mod.Eng.* deft, daft ;
mild, gentle ; stupid : doumb and daft MAP
343 ; dafte ORM. 2175 ; CHEST. I. 134 ; deft
'(*h*)*ebes*' PR. P. 116 ; ðat defte meiden REL. I.
209 ; *comp.* **ȝe-daft.**

   dafte-leȝc, sb., *gentleness,* ORM. 2188.

   dafte-līke, adv., *meekly, becomingly,* ORM.
1215 ; deftli LAI LE FR. 360.

[**daftiȝ,** adj., *cf. L.Ger., Du.* deftig.]

   daftiȝ-līke, sb., *decently,* ORM. 15921.

daȝ, *see* **dæi.**

dāȝ, sb., *O.E.* dāg, dāh, = *M.Du.* deegh, *O.N.*
deigr, *Goth.* daigs, *O.H.G.* teig ; *from root of*
Goth. deigan (πλάσσειν) ; *dough* ; dah ' *massa* '
FRAG. 4 ; dagh VOC. 201 ; doȝ AYENB. 205 ;
dogh MIRC 1882 ; doh LEECHD. III. 88 ; dou

REL. II. 277 ; **dōghe** (*dat.*) L. C. C. 41 ; *comp.*
sūr-dāȝ.

   dōu-ribbe, sb., *baker's scraper,* ' *sarpa, costa*
pasthalis,' PR. P. 129 ; VOC. (W. W.) 575.

   dōu-trough, -trou, sb., *kneading trough,*
PR. P. 129.

dagen, *see* **daȝien.**

dagge, sb., *O.Fr.* dague ; *strip of cloth,* ' *frac-*
tillus,' PR. P. 111 ; DEP. R. iii. 193 ; highe
shoos knopped with **dagges** (*pl.*) R. R.
7260.

daggen, v., *O.Fr.* daguer ; *tear into rags :*
leet dagge his cloþes LANGL. *B* xx. 142 ; **dagges**
(*pres.*) D. ARTH. 2102 ; **dagged** (*pple.*) CH. C.
T. *I* 421.

   dagging, sb., *act of slitting* ; dagging of
sheres CH. C. T. *I* 418.

[**daggen,** v., ?=**dēawen** ; *comp.* **bi-daggen.**]

dagger, sb., *dagger,* CH. C. T. *A* 392.

daggered, adj., *wearing a dagger :* now
swerded, now daggered MAND. 137.

daȝien, v., *O.E.* dagian, = *O.N.* daga, *O.H.G.*
tagen ; *from* **dæi** ; *dawn, become day* ; (*ms.* dai-
ȝen) [daȝeȝe] LAȝ. 26940 ; dagen GEN. & EX.
16 ; daiin, dawin PR. P. 112 ; dawen SPEC. 96 ;
S. S. (Web.) 2249 ; dawe WILL. 3261 ; LANGL.
B xviii. 179 ; **dageð** (*pres.*) HOM. II. 103 ;
daweð A. R. 352 ; þer daweþ him no dai CH.
C. T. *A* 1676 ; dawe (*subj.*) P. S. 238 ; **daȝede,**
dawede (*pret.*) LAȝ. 1694, 8523 ; dawed
DEGR. 597 ; *comp.* **a-, of-daȝen.**

   dawunge, sb., *O.E.* dagung, = *M.H.G.* tag-
unge ; *dawning,* A. R. 20 ; dawinge ROB. 208 ;
dawing TREV. VIII. 343* ; dagheinge HAMP.
PS. xlv. 5 ; lxxiii. 17.

daȝnien, v., *for* **daȝien** ; *dawn* ; dawnin
PR. P. 114.

   daigening, daiening, sb., = *Dan., Swed.*
dagning ; *dawning,* GEN. & EX. 77, 3264 ;
dawening ALIS. 403 ; daweninge CH. C. T.
B 4372 ; TREV. II. 9 ; dauninge PR. P. 115.

dagoun, sb., *piece* (*of cloth, etc.*), CH. C. T. *D*
1751.

dagswaine, sb., *blanket* ; daggisweine PR. P.
112 ; dubbide wið **dagswainnes** (*pl.*) dow-
blede they seme D. ARTH. 3610.

dāh, *see* **dāȝ.**

dahet, sb., *O.Fr.* dahait, deshet ; *misfortune* ;
dahet [*v. r.* dehaet] habbe þat ilke best . . .
þat fuleþ his owe nest O. & N. 99 ; daþet
(? *printed* dayet) MAN. (H.) 143 ; daþeit S. S.
2395 ; daþet (*interj.*), *woe to, evil befall,* AM.
& AMIL. 1569 ; TRIST. 1875 ; daþeit HAV. 300,
926 ; dait JUL.² 202 ; dai þat P. L. S. xix. 357.

daheð, sb., *from* **daȝien** ; ? *dawn* ; euch
daheðes (*gen.*) dei *every day at dawn* JUL. 7.

dai, daig, *see* **dæi.**

daierie, *see* **dēierie.**

daigening, *see under* **daȝien. dainen,** *see*
deinen.

dalc, sb., *O.E.* dalc, dolc,=*O.N.* dalkr; *pin, brooch,* '(*spi*)*nter*,' FRAG. 2; dalke *firmaculum, monile* CATH. 89; dolc sor and blein GEN. & EX. 3027.

dále, sb., *O.E.*dæl,*cf.O.L.G.,Goth.*dal, *O.H.G.* tal *n., O.N.* dalr *m.; dale, valley,* ORM. 9203; ROB. 362; LANGL. *A* PROL. 15; PR. C. 1044; þat dale LAȝ. 27166; **dále** (*dat.*) O. & N. 1; GEN. & EX. 19; ALIS. 2550; **dáles** (*pl.*) LAȝ. 20860; MAP 348; deales A. R. 282; **dálen** (*dat. pl.*) LAȝ. 21775.

dále, sb.,=*M.L.G.* dēle, *Goth.* daila (μετοχή), *O.H.G.* teila, *mod.Eng.* dole; *division, part;* ane dale HOM. I. 123; KATH. 99; ne schaltu habben ... dale of heovene riche MARH. 22; þe feorþe dale ORM. 8273; ne scalt þu næver halden dale of mine lande LAȝ. 3084; þa feorðe dale 3311; ane dæle (? *for* dale) 17754; dole GEN. & EX. 152; C. L. 291; A. P. i. 136; þeos laste dole A. R. 412; heo eoden to þære dale LAȝ. 19666; **dōlen** (*pl.*) A. R. 10; *see* **dǽl.**

dali, sb., ? *O.Fr.* daille; *die,* '*tessera, alea*,' PR. P. 112.

daliance [daliaunce], sb., *dalliance,* CH. C. T. *A* 211; daliaunce GAW. 1012.

dalien, v., *dally, play;* dali GAW. 1253; daile MAN. (H.) 116; disours dalie ALIS. 6991; **dailieden** (*pret.*) GAW. 1114.

dalke, sb., ? *dim. of* **dále;** *little dale,* '*vallis,*' PR. P. 112; REL. II. 78; dalk PR. C. 6447.

dalke, *see* **dalc.**

dalle, sb., *mod.dial.* doll; *hand;* put furthe þi dalle TOWNL. 118.

dalmatik, adj., *O.Fr.* dalmatique, *Lat.* dalmatica; *dalmatic,* PR. P. 112; VOC. 249.

dam, sb., *O.N.* damr,= *O.Fris.* dam, dom, *M.H.G.* tam; *dam,* '*agger,*' PR. P. 113; '*stagnum*' VOC. 239; PS. cvi. 35*.

dam², sb.,*O.Fr.*dam, dan, dans,*Lat.*dominus; *sir* (*prefixed to names*), HAV. 2468; þis boc is dan Michelis AYENB. 1; þe ersebisshop of Anxus, danz Guard MAN. (H.) 147.

damáge, sb., *O.Fr.* damage; *damage,* ALIS. 959; TREV. II. 391; MAN. (F.) 4552.

damaging, sb., *damage, injury:* withoute damagingue LEG. 45.

damasin, sb., *damson* (*plum*), PR. P. 113; VOC. 192; **damisins** (*pl.*) wiche withe þer taste delite LIDG. M. P. 15.

dáme, sb., *O.Fr.* dame; *dame, lady,* KATH. 2111; A. R. 230; R. S. vii; MAND. 302; PERC. 336.

damesele, sb., *O.Fr.* damisele, dameisele; *damsel,* HORN. (L.) 1169; damisele WILL. 401.

damnable, adj., *Fr.* damnable; *condemnable;* dampnable SHOR. 152.

damnacioun, sb., *Fr.* damnation; *damnation,* CH. A. B. C. 23.

damnen, v., *Fr.* damner; *condemn;* damni AYENB. 137.

dampen, v., *drown, suffocate;* damped (*pple.*) A. P. ii. 989.

dan, *see* **dam².** **dáne,** *see* **déne.**

danger, sb., *O.Fr.* dangier, daunger, *mod.Eng.* danger; *control, power; resistance,* A. R. 356; SHOR. 162; CH. C. T. *A* 663; now wolde God, i had her all wiþoute daunger at my wille GOW. II. 40.

dangerūs, adj., *O.Fr.* dangereus; *dangerous, hard to please,* A. R. 108; CH. C. T. *B* 2129.

dank, sb., *dampness:* on þe danke (*dat.*) of þe dewe many dede liggis D. ARTH. 3750.

dank, adj., *damp;* dannke D. ARTH. 313.

danken, v., *cf. Swed.* dänka; *bedew, damp;* dankit (*pret.*) D. TROY 7997.

danz, *see* **dam.**

daper, adj., ? *M.Du.* dapper (*brave*),=*O.H.G.* tapher (*heavy, brave*); *dapper;* dapir '*elegans*' PR. P. 113.

[dappel, ? = *O.N.* depill (*macula*).]

dappel-grai, adj., *dapple-gray,* CH. C. T. *B* 2074.

dar, sb., *dace* (*the fish*), '*capito*,' VOC. 253; darce B. B. 156, 174.

dar², sb., ? *O.N.* darr (*spear*); *dart,* '*jaculum,*' VOC. 253; (VOC. W.W. 558 *reads* dart, *but the index has* dar).

dar, *see* **dear** *and* **þarf.**

darial, sb., *O.Fr.* dariole; *kind of pastry,* darials (*pl.*) L. C. C. 38; with darielles endordide D. ARTH. 199.

dárin, v., ?=*L.Ger., Du., Fris.* (be-)daren; *lurk, be concealed,* '*lateo,*' PR. P. 113; LIDG. M. P. 153; dare SPEC. 50; TRANS. XVIII. 26; WICL. MK. vii. 24; MAN. (F.) 5183; GAW. 2258; A. P. i. 838; **dárie** (*pres.*) MARH. 16; dare E. T. 553; i droupe and dare MIN. i. 9; dareð so ge ded were REL. I. 217; and dariþ þer for drede TRIAM. 321; we dearieð JUL. 43; deovelen þat in ham dearieð [daried] KATH. 553; daren ANT. ARTH. iv; dare CH. C. T. *B* 1293; **dáre** (*subj.*) O. & N. 384; **dearede** (*pret.*) KATH. 1135; dared WILL. 4055; *comp.* **at-dárien.**

dark, *see* **dearc.**

darnel, sb., *Fr.* darnel; *darnel,* TREV. VII. 525; M. H. 145; dernel WICL. MAT. xiii. 25, 29; '*zizania, lolium*' PR. P. 119; **dernels** (*pl.*) [*v. r.* derneiles, darnels] WICL. MAT. xiii. 30.

dart, sb., *O.Fr.* dart [*cf. O.E.* daroð, *O.N.* darraðr, *O.H.G.* tart]; *dart,* '*telum,*' PR. P. 114; MAND. 117; CH. C. T. *A* 1564.

darten, v., *cf. O.Fr.* darder; *pierce with a dart,* H. R. 143.

daschen, v., *cf. Dan.* daske, *Swed.* daska; ?= **dwēschen;** *dash, rush, strike;* dasche ALIS. 2887; dasche a doun P. L. S. xvii. 366;

dasse (*pres. pl.*) AR. & MER. 9135; da-schande (*pple.*) A. P. iii. 312; daschte (*pret.*) ROB. 51; dassed AR.& MER. 6293; dast 9271.

dasewen, *see* dásin.

[**dasi**, adj., *cf. Swed.* dasig.]

dasi-berd, sb., *cf. L.G.* dösbart; *simpleton*, VOC. 217; dosebeirde, dossiberde CHEST. II. 34, I. 201.

dásin, v., *O.N.* dasa (*be sluggish*),=*M.Du.* dasen, dwasen (*rave, be foolish*), *O.E.* dwæsigean (*be foolish*), *mod.Eng.* dase; *grow dim, grow dizzy; stupefy, bewilder, 'caligo,'* PR. P. 114; dase M. T. 311; þe iȝen schulen not dasewe ['*non caligabunt*'] WICL. IS. xxxii. 3; i dáse (*pres.*) and i dedir TOWNL. 28; þe eiȝen dasewetz P. R. L. P. 221; dasen [dasewen, daswen] CH. C. T. *H* 31; dásed (*pret.*) A. P. iii. 383; dásed (*pple.*) CH. H. F. 658; L. H. R. 216; PR. C. 6647; *comp.* a-dásen.

    dásed-lī, adv., *coldly, indifferently*: when a man God dasedli loves PR. C. GLOSS. 289.

    dásednes, sb., *coldness, indifference*, PR. C. 4904.

dastard, sb., *?from O.N.* dæstr (*weary*), *mod.Eng.* dastard; *stupid person*; dastard or dullard PR. P. 114.

daswen, *see* dásin.

dáte, sb., *Fr.* date; *date (point of time)*, CH. C. T. *G* 1411; MAN. (F.) 1747.

dáte, sb., *O.Fr.* date; *date (fruit), 'dactylus,'* PR. P. 114; dátes (*pl.*) MAND. 57.

daþeit, daþet, *see* dahet.

dau, daw, adj., *? melancholy*: if god helpe amang i mai sit downe daw TOWNL. 26.

    dau-lī, dawlī, adv., *? sadly, miserably*, D. TROY 728, 870.

      dawli-lī, adv., *for* dawlī, D. TROY 5359.

daubin, v., *daub, plaster, 'limo* [*? linio*], *muro,'* PR. P. 114; daauben (*pres.*) ['*liniunt*'] WICL. EZ. xiii. 11; daube (*imper.*) A. P. ii. 313.

    dauber, sb., *dauber, 'argillarius'* PR. P. 114; VOC. 181.

    daubinge, sb., *plaster*, LANGL. *C* ix. 198.

daunce, sb., *O.Fr.* dance, danse; *dance*; þere was no daunce GOW. III. 291; daunse [daunce] CH. C. T. *A* 476; daunces (*pl.*) ALIS. 6990.

dauncen [daunsen], v., *O.Fr.* dancer, danser; *dance*, CH. C. T. *A* 2202; daunce GOW. II. 95; daunecen (*pres.*) ALIS. 5213; þoȝ þou daunce A. P. i. 345; daunsid (*pret.*) RICH. 185; daunsed GOW. 1026; þise damseles daunsede LANGL. *C* xxi. 471.

    dauncere, sb., *dancer, 'tripudiator, tripudiatrix,'* PR. P. 114.

    daunceresse, sb., *(female) dancer*; daunseresse WICL. ECCLUS. ix. 4.

    dauncing, sb., *dancing*, S. & C. II. xxii; dawncinge, idem est quod dawnce PR. P. 114; **dawncinge pipe** '*carola*' PR. P. 114.

daunselen, v., *fondle, caress, cajole*; daunseled (*pple.*) LANGL. *A* xi. 30.

daunsen, v., *fondle, caress (v. r.* daunte), WICL. IS. lxvi. 12.

daunt, sb., *check, stoppage*: þe crosses dunt ȝaf him a daunt L. H. R. 145.

daunten, v., *O.Fr.* danter, donter, *from Lat.* domitāre; *subdue, tame*, LANGL. *A* iii. 268; daunte WICL. MK. v. 4; daunted (*pret.*) LANGL. *B* xv. 393.

daunten, v., *? fondle*; upon þe knes men shal daunte ȝou ['*blandientur vobis*'] WICL. IS. lxvi. 12; daunt LIDG. M. P. 35.

dāwe, sb.,=*O.H.G.* tāha; *daw, 'monedula,'* CATH. 91; *comp.* cā-dāwe.

dawien, *see* daȝien. **dawunge**, *see under* daȝien. **day**, *see* dæi.

de-, prefix, *imitating words from O.Fr. with* de- *for* des-, dis-; *see* debreiden, debréken.

dē, sb., *O.Fr.* dé; *die (for playing)*; dee GOW. II. 209; dēs (*pl.*) AYENB. 45; dees GOW. II. 39; dees [dis] CH. C. T. *C* 467; dise TOWNL. 240; dise (*used as sing.*) '*alea*' VOC. 202; dice PR. P. 120; dicis (*pl.*) PR. P. 121.

dēad, adj., *O.E.* dēad,=*O.Fris.* dād, dāth, *O.L.G.* dōd, *O.N.* dauðr, *Goth.* dauþs, *O.H.G.* tōt; *dead*, HOM. I. 51; A. R. 134; s. v.; LIDG. M. P. 35; dead, dæd LAȝ. 196, 2540; dead [ded] O. & N. 1619; GEN. & EX. 217, 2493; diad, diead AYENB. 12, 79; after þat he was dæd for us ORM. 11551; ded HAV. 634; ded [deed] CH. C. T. *A* 148; ðe dede se GEN. & EX. 750; þe dede slep *the sleep of the dead* CH. D. BL. 127; þe dide slep FL. & BL. 66; þa brude deade iweard LAȝ. 294; þas dædan FRAG. 5; dēadne (*acc. m.*) LAȝ. 7993; dēade (*pl.*) JUL. 62; hes beoð deade A. R. 310; þa deden HOM. I. 37; deden, dede [deade] LAȝ. 9279, 22873; þe deade A. R. 390; wiþ þe dede PERC. 155; dēadre (*gen.*) manne ['*mortuorum*'] MAT. xxii. 32; *comp.* sām-dēd.

    dēad-līc, adj., *O.E.* dēad-, dēaðlīc,=*O.H.G.* tōd-, tōtlīh; *deadly, deathly, mortal*, HOM. I. 223; deadlich KATH. 1104; A. R. 206; diadlich MISC. 27; a dedlich mon MISC. 46; deaðlic HOM. I. 221; deadliche senne SHOR. 99; deadliche ifoan PROCL. 6; þine dædliche ivan LAȝ. 8550; deadli [deedli] WICL. ESTH. xiii. 2; dedli GOW. I. 355; LANGL. *B* xiv. 78; dōdlīche (*adv.*) MARH. 22; he zeneȝeþ diadliche AYENB. 86.

    deadlīcnesse, sb., *O.E.* deadlīcness; *deadliness*, A. R. 382.

    dēde-stoure, sb., *death-throe*, PR. C. 1820.

dēaden, *see* dēden.

dēaf, adj. & sb., *O.E.* dēaf, = *O.Fris.* dāf, *O.L.G.* dōf, *O.N.* daufr, *Goth.* daubs, *O.H.G.* toub; *deaf*, MARH. 20; diaf AYENB. 1; dæf ORM. 9887; def MAP 335; def [deef] CH. C. T. *A* 446; deef PR. C. 782; LIDG. M. P.

189; LUD. COV. 254; **dēave** (*pl.*) MARH. 6; deve HOM. II. 129; defe [deve] LANGL. *B* xix. 126; (þe) deave KATH. 1061.

dēf-li, adv., *deafly*, TOWNL. 100.

dēfnes, sb., *deafness*, '*surditas*,' VOC. 224.

dēaȝ, sb., *O.E.* dēag, dēah; *dye, colour*; **dēhe** (*dat.*) MISC. 193.

dēaȝen, v., *O.E.* dēagan (*pret.* dēog), dēagian (MONE 356); *dye*; deie [die] CH. C. T. *F* 725 (Wr. 11037); diin '*tingo*' PR. P. 124; **dīed** [idīȝed] (*pple.*) TREV. II. 331.

    dēier, [dīere], sb., *dyer*, CH. C. T. *A* 362 (364); **dīhȝeres** (*pl.*) E. G. 359.

    dīestere, sb., *dyer*, LEG. 40.

dēah, v., *O.E.* dēah, dēag, = *O.L.G.* dōg, *Goth.* daug (συμφέρει), *O.H.G.* toug (*pret.-pres.*); *is powerful; is befitting* (*oportet*); deah, deh KATH. 1446, 1853; deh MARH. 1; deih HOM. I. 302; þet ou me deih A. R. 420; Thebald nouht ne deih MAN. (H.) 133; þat þing þat noht ne dæh ORM. 4872; hwile þine dages dugen REL. I. 184; *deriv.* **duȝen**, **duȝeðe**, **duhtiȝ**.

deakne, *see* **diacne.**   **dēal,** *see* **dǣl.**

deale, interj. [*Stratm. regards this as imper.* of **dǣlen**; *others compare O.Fr.* deu le set (*God knows*)], *? expressing wonder*: lo, deale, hwat he seið A. R. 362; 286; o, dele, þis is a fole Briton MAN. (H.) 167.

dear, v., *O.E.* dearr, dear, = *O.L.G.* (gi-)dar, *Goth.* dars, *O.H.G.* (ge-)tar, (*pret.-pres.*); *dare*, HOM. I. 187; dar MARH. 16; ORM. 10659; O. & N. 1110; H. H. 142; P. S. 324; AYENB. 32; WILL. 564; LANGL. *B* xv. 108; MAND. 249; SACR. 316; der HOM. I. 27; LAȝ. 6639; A. R. 206; SHOR. 5; **derst**, dærst, darst LAȝ. 20375, 24779, 24785; derst HOM. I. 27; JUL. 47; darst ORM. 5615; O. & N. 853; C. L. 1081; hou þerstou . . . nemne his name BRD. 27; **durren** (*pl.*) JUL. 42; ne durren [dorren] heo me abiden LAȝ. 15063; durre ALIS. 1314; men durre (*? for* þurve) selde here orf in hous a wintre bringe ROB. 43; dorren, dorre AYENB. 22, 78; doren P. S. 325; WICL. GEN. xliv. 26; dur MAND. 271; dur [dore] CH. C. T. *G* 661; we ne þore abide noȝt BEK. 1375; **durre** (*subj.*) LAȝ. 24783; O. & N. 1706; dorre ROB. 112; durren A. R. 128; *deriv.* **durren, dursti ȝ**.

dearc, adj., *O.E.* deorc; *dark*, GR. 26; darc, dorc JUL. 30, 31; derk '*tenebrosus, obscurus*' PR. P. 119; ROB. 560; C. L. 71; GOW. II. 120; MAND. 237; LUD. COV. 21; deorc LK. xi. 34; FRAG. 5; deork BEK. 1411; dorc HOM. I. 253; durk TREAT. 134; P. L. S. ii. 43; dirk ANT. ARTH. vi; LIDG. M. P. 144; darke niȝt P. 57; dorcke nipt (*r.* niht) LAȝ. 7563*; **derker** (*compar.*) LANGL. *B* x. 182; **darkest** (*superl.*) MARH. 8.

    deork-hēde, sb., *obscurity*, HALLIW. 298; derkhede LEG. 86; durchede BRD. 2.

derk̄-līche, adv., *O.E.* deorclīce; *darkly*, LANGL. (Wr.) 6363.

derknesse, sb., *darkness*, LANGL. *B* xvi. 85; CH. C. T. *A* 1451.

deark, sb., *darkness*; derk, WILL. 1285; ALIS.[2] (Sk.) 714; bi þe derke HORN (L.) 1431.

    derk-ful, adj., *O.E.* deorcfull; *dark* ['*tenebrosus*'], WICL. LK. xi. 34.

dearkien, v., *? O.E.* (a-)deorcian; *make dark; lurk in the dark*: derkin '*obscuro*' PR. P. 119; derken CH. BOET. i. 4 (20); i durk(e) and dare C. M. 25432; **derkede** (*pret.*) B. DISC. 1379; þe child . . . darked in his den WILL. 17; **derked** (*pple.*) GOW. III. 307; derkid WICL. 3 KINGS xviii. 45; dirkid LIDG. M. P. 138; idurked BEK. 1414.

dearknien, v., = *O.H.G.* tarchnēn (*conceal*); *darken, grow dark*; durcnin HOM. I. 259*; **darkneþ** (*pres.*) REL. I. 241; dirkins ANT. ARTH. v; **dürcninde** (*pple.*) KATH. 2049*.

dearne, *see* **derne.**

dēað, sb., *O.E.* dēað, = *O.Fris.* dāth, dād, *O.L.G.* dōþ, dōd, *Goth.* dauþus, *O.H.G.* tōd, *cf. O.N.* dau˘ði; *death*, KATH. 1127; deaþ, diaþ SHOR. 25, 28; diaþ AYENB. 12; diath MISC. 27; dieð P. L. S. viii. 93; deað, deð, dæð [deaþ, deþ] LAȝ. 284, 2587, 20677; deþ [deeth] CH. C. T. *A* 964; dæþ ORM. 17450; deth, ded PR. P. 115; dead GEN. & EX. 508; ded HAV. 1687; MIRC 247; IW. 1262; PR. C. 815; **dēaðes** (*gen.*) KATH. 1143; deaðes swot A. R. 110; ȝiveð deaðes dunt 274; deaþes (*in death*) O. & N. 1632; deaþes drench *deadly potion* AYENB. 130; deðes dom *sentence of death* HOM. I. 279; deþes wonde ROB. 49; deþes wounde FER. 851; diþes wunde HORN (L.) 640; hure deþes stounde (= *Ger. ire todesstunde*), 953; dæþes pine ORM. 1381; **dēaðe** (*dat.*) HOM. II. 97; Crist aras of deaðe HOM. I. 91; þa þe fader wes on deaðe LAȝ. 318; efter his deaðe A. R. 314; deaþe [deþe] O. & N. 1617; diaþe AYENB. 7; diþe HORN (L.) 58.

dēþ-dai, sb., *O.E.* dēaðdæg; *death-day*, LANGL. iii. 104.

dǣþ-shildiȝ, adj., *O.E.* dēaðscyldig; *worthy of death*, ORM. 10436.

    dæðshildiȝnesse, sb., *worthiness of death*, ORM. 10430.

dēaþ-sīþ, sb., *death*, LAȝ. 6348*.

dēþ-þrōwe, sb., *death-throe*, P. L. S. xv. 192; dedthrawe D. ARTH. 1150.

dēað-üvel, sb., *fatal sickness*, A. R. 314; deþüvel BEK. 2211; deethevel TREV. VIII. 9.

dēau, sb., *O.E.* dēaw, *? = O.Fris.* dau (*dat.* dawe), *O.H.G.* tau, tou, *O.N.* dögg; *dew*: deau, diau AYENB. 136, 144; daw (*? for* deaw) HOM. I. 233; dæw ORM. 13865; dew GEN. & EX. 3325; deu TREAT. 137; MAP 347; SPEC. 72; M. H. 26; þene deu HOM. I. 159; deuȝ II. 256; **dēwes** (*gen.*) GEN. & EX. 3327; dewes dropes

SPEC. 114 (A. D. 179); dēwe (dat.) TREAT. 137; dēowes (pl.) SPEC. 44; comp. mil-dēu.

dēu-lappe, sb., cf. Dan. dōglæp; dew-lap, 'palear' PR. P. 120; VOC. 221; dēulappes (pl.) PALL. iv. 711.

dēu-water, sb., cf. M.H.G. touwazzer; dew water: rein water oÞer deu water HOM. II. 151.

dēaven, v., O.E. (a-)dēafian, = O.Fris. dāva, cf. O.N. deyfa; from dēaf; deafen, grow deaf, make deaf; deve GAW. 1286; him deaveþ þa æren FRAG. 5; dēvid (pple.) LUD. COV. 348; comp. a-dēaven.

dēawen, v., O.E. dēawian, cf. M.Du. dauwen, O.H.G. touwen (touuen), O.N. döggva; dew, bedew: dewin 'roro' PR. P. 120; dæwen ORM. 13848; deweth (imp.) ['rorate'] WICL. IS. xlv. 8.

debat, sb., O.Fr. debat; debate, strife, WILL. 4380; LANGL. A v. 181.

debáte, v., O.Fr. debatre; debate, do battle, GOW. I. 40; debateþ (pres.) I. 124; debetande (pple.) wiþ himselfe GAW. 2178; debated (pret.) 67.

debonaire, adj., O.Fr. de bon aire; meek, gentle: þet debonere child . . . cussed þe 3erd A. R. 186; ROB. 167; fortune · was hem debonaire GOW. I. 22; deboneire CH. BOET. i. 5 (22); a maiden . . . ful debonere A. P. i. 162.

debonair-lī, adv.: up his look debonairli he cast CH. TRO. ii. 1259; BOET. iv. 3 (122); bere me debonureli WILL. 730.

debonernesse, sb., mildness: for treuthe and debonernesse WICL. PS. xliv. 5.

debonairetē, sb., meekness; debonairiete CH. D. BL. 984; C. T. I 540; Þuruh his debonerte A. R. 390; debonerte A. P. iii. 418.

de-breiden, v., tear apart; debreidinge [v. r. to breidinge] (pple.) WICL. MK. i. 26; ix. 25.

de-brēken, v., cf. bréken, debríse; break asunder; debrékinge (pple.) WICL. MK. i. 26; debróken (pple.) WICL. EZ. xxxi. 15.

debríse, v., O.Fr. debrisier, debruisier; break in pieces, WICL. EZ. xxxiv. 27; debrüsede (pret.) L. H. R. 40.

decas, sb., cf. cas; decay, ruin, GOW. I. 32.

deceit, sb., deceit, TREV. II. 391; deceite CH. C. T. G 1367.

deceivable, adj., O.Fr. decevable; deceptive, GOW. I. 233; the juge was made favourable, thus was the lawe deceivable I. 216; R. R. 4836; desaivabel and trecherous PR. C. 4231.

deceivance, sb., O.Fr. decevance; deception: here of a desceivaunce þei conseild him to do MAN. (H.) 133; desceivaunce 195.

deceivaunt, adj., deceitful: ne be nought deceivaunt GOW. I. 82, 222.

deceive, v., O.Fr. decevoir; deceive, CH. C. T. C 918; deceivi AYENB. 82.

deceivar, sb., deceiver, 'fraudator,' PR. P. 115.

decembre, sb., O.Fr. decembre; December, ROB. 441; december GOW. III. 236.

decepcioun, sb., O.Fr. deception; deception, LIDG. M. P. 76, 260.

deces, sb., O.Fr. deces, from Lat. decessus; decease, MAN. (F.) 5352.

decimacioun, sb., Lat. decimātio; paying tithes, LIDG. M. P. 135.

deciple, see disciple.

declaracioun, sb., Lat. dēclārātio; declaration, TREV. I. 243.

declāren, v., O.Fr. declarer; declare, CH. C. T. A 3002; declāringe (pple.) CH. BOET. i. 1 (5).

declīn, sb., O.Fr. declīn; decline, decay, P. S. 154.

declīne, v., O.Fr. decliner; decline, A. P. i. 333; 3if þou canst decline þilke tweie names and speke Latin TREV. I. 327; wiþ heed declīnid (pple.) lowe LIDG. in HALLIW. 295.

decoccioun, sb., Fr. décoction; decoction: the coke bi mesour sesonith his potages by decoccioune to take their avauntages LIDG. M. P. 82.

decollacioun, sb., med.Lat. decollātio; beheading: of the decollacioun of Saint John TREV. V. 49.

decópen, v., O.Fr. decoper; cut, slash; with shoon decoped (pple.) R. R. 842.

decourren, v., O.Fr. decoire, decourre; ? depart from: of pride the parchemin decourreth (pres.) LANGL. (W.) 9302 [decorreth B xiv. 193].

decrē, sb., O.Fr. decré; decree, CH. C. T. A 640.

decrees, sb., O.Fr. decreis; decrease, GOW. III. 154.

dēd, see dēad and dēaÞ.

dēde, see dæde.

dēden, v., O.E. dēdan, dȳdan (kill), = O.H.G. tōden, Goth. dauþjan (kill); cf. O.H.G. tōdēn (die); become dead: þe olde tre bigan to dede S. S. (Wr.) 623; dēadid [dedid] (pple.) WICL. 1 KINGS xxv. 37; comp. a-dēden.

dederen, see dideren.

deduit, O.Fr. deduit; amusement, LEG. 6; CH. C. T. A 2177; dedut BRD. 2; WILL. 4998; see dūt.

dēef, dēf, see dēaf.

defáce, v., deface, CH. C. T. E 510; defáced (pple.) AYENB. 191.

defáden, v., fade away; diffade P. L. S. xxx. 8; es my face defádide (pple.) D. ARTH. 3305.

defailed, pple., O.Fr. defaillir; grown feeble, weak: recreid and defailed AYENB. 33.

defaute, sb., O.Fr. defaute; default, fault, want, ROB. 2; AYENB. 73; defaut MAN. (F.) 5113; i hungered and had defaute of mete

PR. C. 6190; defaulte GOW. II. 145; **defautes** (*pl.*) A. R. 136; defauȝtes TREV. I. 9.

defaut-lēs, adj., *without defect, faultless*: ȝat any man might ordaine defautles PR. C. 8699.

defauten, v., *cf. It.* difaltare; *want*: greetlich thei **defauten** (*pres.*) WICL. JUDG. viii. 5; **defautide** (*pret.*) WICL. NUM. xi. 33.

defauting, sb., *failure*, WICL. WISD. xi. 5.

defautif, adj., *faulty, defective*, WICL. EX. vi. 12*.

defende, v., *O.Fr.* defendre; *defend*, MAN. (F.) 14636; defendi AYENB. 157; we **defendeþ** (*pres.*) ROB. 198; **defendide** (*pret.*) WICL. JUDG. xv. 1.

defendour, sb., *O.Fr.* defendeour; *defender*, PR. P. 115; defendor ROB. 198.

defensable, adj., & sb., *O.Fr.* defensable; *defensible*: in hillis and moost defensable placis WICL. JUDG. vi. 2; hii hulde hem there **defensables** (*pl.*) to libbe oþer to deie ROB. 549.

defense, sb., *O.Fr.* defense; *defence*, ROB. 253.

defensin, v., *Lat.* dēfensāre; *defend*, PR. P. 115.

defensioun, sb., *O.Fr.* defension, *Lat.* dēfensio; *prohibition*: domes of defensioun WICL. ECCLUS. xlviii. 7.

defensour, sb., *protector*: yow be my defensour SAC. 854.

deffe, see **daffe**.

defféted, pple., *enfeebled*: þou languissed and art deffeted for talent and desiir of þi raþer fortune CH. BOET. ii. 1 (30); defet TRO. v. 627.

defformen, v., *fashion, engrave*: ministracioun of deeth **defformid** (*pple.*) by lettris in stones WICL. 2 COR. iii. 7.

deffūse, sb., *? prohibition*: fore gret dule of deffuse of dedes of armes D. ARTH. 255.

defiaunce, sb., *O.Fr.* deffiance; *defiance*, LIDG. M. P. 92.

defīe, v., *O.Fr.* de-, desfier; *defy*, MAN. (F.) 5975; **defied** (*pret.*) LANGL. *B* xx. 65.

defīen, v., *digest; be digested, decay*, LANGL. *A* v. 219; þe bodi sal defien ANGL. III. 61; defiin '*digero*' PR. P. 115; to defie þe mete TREV. IV. 441.

deflōren, v., *O.Fr.* deflourer; *deflower*: þe which book Robert bisshop of Hereforde **deflōrede** (*pret.*) ['*defloravit*'] TREV. I. 39; she hadde be **defloured** (*pple.*) GOW. II. 322.

defoil, sb., *trampling*: under hors knightes defoile AR. & MER. 7999; defoil 9191.

de-foul, sb., *dirt, filth*, TREV. I. 109; defoule A. P. iii. 289.

defoule, v., *O.Fr.* defouler; *tread under foot*;

insult, defoul, ROB. 57; WICL. PS. cxxxviii. 11; **defoiled** (*pret.*) WILL. 4614; **defouled** (*pple.*) AYENB. 167.

defoulinge, sb., *treading upon*: power of defoulinge on serpents WICL. LK. x. 19.

de-foulen², v., *make foul, 'inquino, deturpo,'* PR. P. 116; A. P. ii. 1146; þei schulleþ ... defoule clerkes and holi places TREV. I. 129; defoule GOW. III. 20; þere mighte no envie ... defoule þe saule HAMP. TR. 38; **defouled** (*pple.*) CH. C. T. *F* 1398.

defoulinge, sb., *pollution, 'deturpacio, maculacio,'* PR. P. 116; WICL. JUD. iv. 10; men forsakinge the **defoulinges** (*pl.*) or unclennesses of the world 2 PET. ii. 20.

defourme, adj., *ill formed*: other seven oxen in as miche defourme and leene WICL. GEN. xli. 19.

defrai, sb., *O.Fr.* defroi; *disorderly living*: thurch mi sinne and mi defrai ich am comen to mi last dai AR. & MER. 9695.

defrauden, v., *O.Fr.* defrauder; *defraud*; **defraudeþ** (*pres.*) LANGL. *A* vii. 71; *C* x. 64.

deft, see **daft**.   **deft**, see **dæi**.

dēȝen, v., *O.N.* deyja, = *O.Fris.* dēja, *Goth.* (af-)daujan, *O.L.G.* dōjan, *O.H.G.* touwan (touuan); *die*, ORM. 8090; deȝen [deiȝen] LAȝ. 283, 31796; deigen GEN. & EX. 3127; deien A. R. 134; KATH. 958; deien WILL. 3353; deiin PR. P. 117; deȝe H. H. 54; GAW. 996; deiȝe SHOR. 124; deie HORN (L.) 109; L. H. R. 22; (*ms.* deye) HAV. 168; ROB. 494; S. & C. I. lii; deie [die] CH. C. T. *A* 1797, *E* 364; deghe PR. C. 1939; diȝe [die] LANGL. *A* ii. 187; dien C. L. 218; die MAND. 201; M. H. 79; **dæide**, deȝede [deide] (*pret.*) LAȝ. 4280, 7150; deide KATH. 1101; LANGL. *B* xi. 200; diȝede JOS. 134; deieden A. R. 301; dieden MAND. 286; **idæied** (*pple.*) LAȝ. 3737.

degīse, **degīsen**, *see* **disguīse** *and* **disguīsen**.

degráden, v., *O.Fr.* degrader, *Lat.* dēgradāre; *degrade*, TOWNL. 20; **degráded** (*pple.*) TREV. V. 35; Agaminon degrated of his degre D. TROY 12574.

degrē, sb., *O. Fr.* degre, degret; *degree*, H. M. 15; CH. C. T. *A* 1891; **degrēz** (*pl.*) A. R. 288.

dēh, see **dēaȝ**.   **dei**, see **dæi**.

dōie, sb., *O.N.* deigja; *dairy-maid*; (*ms.* deye) PR. P. 116; A. D. 259; CH. C. T. *B* 4036; (*ms.* deie) P. S. 327 (P. 15); daie REL. I. 129; (HOM. II. 163).

dēien, see **dēaȝen** *and* **dēȝen**.

dōierie, sb., *dairy*, CH. C. T. *A* 597; deirie PR. P. 117.

deificatiōn, sb., *O.Fr.* deification; *deifying, apotheosis*: the boke of his deification GOW. II. 166; deificacion 158.

deignous, adj. [*for* desdeinous]; *disdainful*, CH. TRO. i. 289; ye have to longe be deignous

R. R. 353 ; his name was hoote deinous Sime-kin CH. C. T. *A* 3941.

deinen, v., *O.Fr.* deigner, *from Lat.* dignāri ; *deign* ; daineþ (*pres.*) AYENB. 18 ; **deined** (*pret.*) LANGL. *B* vi. 310.

deintē, sb., *O.Fr.* deinté, deintet (*dignity*), *mod.Eng.* dainty ; *dignity, importance, value,* A. R. 412 ; WICL. E. W. 220 ; cloþes of gret deinte ALIS. 7070 ; deintee LANGL. *B* xi. 47 ; dainte GAW. 1250 ; daintethe TOWNL. 245 ; **deintēs** (*pl.*) *articles of value, dainties,* BEK. 1202.

    deintē-full, adj., *pleasant,* GOW. III. 28.

deis, sb., *O.Fr.* deis (*table*) ; *dais,* ROB. 349 ; SHOR. 102 ; LAUNF. 899 ; CH. C. T. *A* 370 ; des WILL. 4312 ; dees MAN. (F.) 9282.

deitē, sb., *O.Fr.* deité ; *deity,* CH. C. T. *F* 1047.

dēkene, *see* diacne.   **dēl,** *see* dǣl.

delai, sb., *O.Fr.* delai ; *delay,* ROB. 421 ; wiþoute delaie LAȝ. 17480*.

delaien, v., *O.Fr.* delaier ; *delay,* ROB. 156.

delaiement, sb., *O.Fr.* delaiement ; *delay,* GOW. II. 9, 297.

delāten, v., *Lat.* dīlātāre ; *stretch out, increase* : it nedeth nought that i delāte (*subj.*) the pris GOW. III. 190.

déle, *see* dále.

delectaciūn, sb., *O.Fr.* delectation ; *delectation,* S. A. L. 74.

dēlen, *see* dǣlen.

delf, sb., *O.E.* delf, dælf, = *M.L.G.* delf *m.* ; *ditch, hole dug in the ground, quarry* ; **delfe** (*dat.*) TOWNL. 230 ; **delves** (*pl.*) ['*lapici-dinis*'] WICL. 2 PARAL. xxxiv. 11.

delfen, v., *O.E.* delfan, = *O.Fris.* delva, *O.L.G.* (be-)delban, *O.H.G.* (pi-)telpan ; *delve, dig, bury,* LK. xvi. 3 ; delven (*ms.* deluen) GEN. & EX. 2452 ; LANGL. *B* xix. 359 ; delven [delve] diches LAȝ. 9238 ; delvin PR. P. 118 ; LIDG. M. P. 145 ; delve P. L. S. xii. 144 ; AYENB. 108 ; CH. C. T. *A* 536 ; delve pittis WICL. GEN. xxvi. 25 ; delfeþ (*pres.*) ORM. 6485 ; delved HOM. I. 49 ; **dalf** (*pret.*) GEN. & EX. 2718 ; ROB. 131 ; S. S. (Web.) 2419 ; SHOR. 165 ; dulfen [dolve] LAȝ. 21998 ; dulven A. R. 292 ; dulve MARG. 219 ; dolven GEN. & EX. 3189 ; LANGL. *A* vi. 178 ; dolve [delveden] WICL. PS. xxi. 17 ; **dulve** (*subj.*) A. R. 384 ; dolve AYENB. 263 ; dolven (*pple.*) WILL. 2630 ; MAND. 62 ; R. R. 4070 ; *comp.* bi-, ȝe-, ümbe-delven.

    delvere, sb., *O.E.* delfere ; *delver, digger,* LANGL. *A* PROL. 102.

    delvinge, sb., *digging,* '*fossura,*' PR. P. 118 ; LANGL. *B* vi. 250.

delfin, *see* dolfin.   **delful,** *see under* doel.

deliberen, v., *Lat.* dēlīberāre ; *deliberate, re-solve* ; for which he gan deliberen for the beste CH. TRO. iv. 141 ; **delibered** (*pple.*) iv. 183.

deliberacioun, sb., *O.Fr.* deliberation ; *de-liberation* : whan i this supplicacion with good deliberacioun . . . had after min entente write GOW. III. 352 ; deliberacioun LIDG. M. P. 72.

delicacie, sb., *med.Lat.* dēlicācia ; *delicacy, luxury,* GOW. I. 14 ; III. 21.

delicat, adj., *Fr.* delicat ; *delicate,* CH. C. T. *E* 927.

    delicat-lī, adv., *luxuriously* : ne doth him nouȝte dine delicatli LANGL. *B* xiv. 250.

delice, sb., *O.Fr.* délice *from Lat.* dēlicia ; *pleasure,* WICL. GEN. ii. 8 ; ROB. 195 ; P. R. L. P. 248 ; **delices** (*pl.*) A. R. 368 ; AYENB. 24.

delicious, adj., *O.Fr.* delicieus ; *delicious* ; GOW. III. 24 ; PR. C. 9291 ; LEG. 204.

    delicious-līche, adv., *luxuriously* : of a riche man . . . þat ladde is liif wel delicious-liche LEB. JES. 150.

delīt, sb., *O.Fr.* delīt ; *delight,* A. R. 102 ; H. M. 3 ; ALIS. 2096 ; CH. C. T. *A* 335.

delitable, adj., *O.Fr.* delitable ; *delectable,* BRD. 2 ; WICL. GEN. iii. 6 ; CH. BOET. ii. 1 ; PR. C. 5239 ; places **delitables** (*pl.*) CH. C. T. *F* 899.

delīten, v., *O.Fr.* deliter ; *delight,* A. R. 52 ; **delited** (*pret.*) PR. C. 8336.

    delītinge, sb., *pleasure* ; **delītingus** [*later ver.* delitingis] (*pl.*) in thi riȝtt hond WICL. PS. xv. 10.

deliverance, sb., *O.Fr.* delivrance ; *deliver-ance,* PR. C. 3585 ; deliveraunce TREV. II. 291.

delivre, adj., *O.Fr.* delivre ; *lively, active,* P. L. S. xvii. 250 ; delivere CH. C. T. *A* 84 ; deliver WILL. 3556 ; deliver *delivered of a child* A. P. ii. 1084 ; M. H. 168.

    deliver-līche, adv., *promptly,* WILL. 1245 ; CH. TRO. ii. 1086 ; delivérli he dressed up GAW. 2009 ; deliverli R. R. 2283.

    delivernes, sb., *suppleness* : delivernes and bewte of bodi PR. C. 5900.

delivren, v., *O.Fr.* delivrer ; *deliver,* P. S. 216 ; delivere ROB. 93 ; MAND. 2 ; delivere, dilivere C. L. 1124 ; **delivrede** (*pret.*) A. R. 234.

delle, sb., = *M.Du.* delle ; *dell, valley,* REL. II. 7 ; dellun (*dat. pl.*) ANT. ARTH. iv.

delūge, sb., *O.Fr.* deluge, *from Lat.* dīluvium ; *deluge, flood,* LIDG. M. P. 251.

delven, *see* delfen.

demaien, *see* desmaien.

demaine, sb., *O.Fr.* demain ; *domain, de-mesne, possession,* GOW. III. 349 ; demeigne [demeine] CH. C. T. *B* 3855.

demande, sb., *O.Fr.* demande ; *demand,* ROB. 500 ; BEK. 817 ; CH. C. T. *G* 430.

dēme, sb., *O.E.* dēma, = *O.H.G.* tuomo ; *judge,* MAT. v. 25 ; LAȝ. 363 ; ORM. 650 ; O. & N. 1783 ; **dēman,** demen (*dat.*) MAT. v. 25, xxvii. 2 ; demen P. L. S. vii. 48 ; **dēmen** (*pl.*) MAT. xii. 27 ; **dēmen** (*dat. pl.*) LK. xxi. 12.

demean, sb., *demeanour*: somewhat strange and sad of her demeane she is CT. L. 734.

demeine, sb., *O.Fr.* demeine, *from Lat.* dominium; *dominion*, CH. C. T. *B* 3855.

dēmen, v., = *O.E.* dēman, = *O.Fris.* dēma, *O.N.* dœma, *Goth.* dōmjan, *O.H.G.* tuoman; *from* dōm; *mod.Eng.* deem; *judge, condemn; think, suppose*, ORM. 652; MAND. 41; þet tu schalt demen þi sulven wod A. R. 120; and demen (*condemn*) me to deaðe JUL. 24; MAT. vii. 1; deman HOM. I. 95; deme O. & N. 188; WICL. GEN. xv. 14; LIDG. M. P. 12; mi fader of heven has me doun sent to deme ioure dedes MIR. PL. 188; to demene HOM. I. 45; dēmeð (*pres.*) HOM. I. 105; he demað stiðne dom 95; dēm (*imper.*) A. R. 12; dēmde (*pret.*) FRAG. 8; LA3. 4936; A. R. 306; SHOR. 85; demden KATH. 330; HAV. 2820; dēmed (*pple.*) M. H. 7; PR. C. 1995; *comp.* for-, 3e-dēmen.

  dēmend, sb., *O.E.* dēmend; *judge*; dēmendes(*gen.*) HOM. II. 171.

  dēmere, sb., *O.E.* dēmere; *judge*, HOM. I. 279; AYENB. 12; demare A. R. 286.

  dēmester [dempster], sb., *cf. mod.Eng.* deemster (*in the Isle of Man*); *judge*, C. M. 7005.

  dēminge, sb., *judgment*, '*judicium*,' PR. P. 118; in valse dēminges (*pl.*) AYENB. 27; MAN. (H.) 86.

demēnen, v. *O.Fr.* demener; *behave, manage*: to demene well his shaft RICH. 455; demene GOW. I. 196; William whi3es . . . demēned (*pret.*) hem dou3tili WILL. 1222.

demeoren, v., *O.Fr.* demorer, -murer, -meurer, *mod.Eng.* demur; *tarry, linger*: demeore (*imp.*) 3e þe lengre A. R. 242; *see* demúre.

demer, *see* dwimer. **demére**, *see* demúre.

demmen, v., *? O.E.* (for-)demman, = *O.Fris.* demma, *Goth.* (faur-)damjan (*dam, stop up*); *? meet with a check, come to its boundary*: a mannez dom mi3t dri3li demme, er minde mo3t malte in hit mesure A. P. i. 223; uche a dale so depe þat demmed (*? pret.*) at þe brinkez ii. 384.

demoniak, adj. & sb., *Lat.* dæmoniacus; *possessed by a devil*: i hold him . . . demoniak CH. C. T. *D* 2240; he nas no fool ne no demoniak 7874.

demonstraciōn, sb., *Lat.* dēmonstrācio; *proof*, GOW. II. 368; demonstratiōns (*pl.*) III. 46.

demonstratif, adj., *Fr.* démonstratif, *from Lat.* dēmonstrātīvus; *irrefragable, incontestable*, CH. C. T. *D* 2272.

demonstraunce, sb., *O.Fr.* demonstraunce; *show*, LYDG. M. P. 60.

demplen, v., *wrangle, dispute*: no more of þis to demple, tak þat þat 3e first ches MAN. (H.) 196.

demúre, sb., *O.Fr.* demoere, -meure; *demur, delay*, FL. & BL. 591; demere (H.) 1011.

den, sb., *O.E.* denn (*cubile*), = *M.Du.* den, denne (*area, antrum*), *M.L.G.* denne, danne, *O.H.G.* tenni *n.* (*area*); *den, cave,* '*specus*,' PR. P. 118; MISC. 152; þet den GR. 24; denne '*spelunca*' VOC. 241; WICL. JOHN xi. 38; denne (*dat.*) FRAG. 8; LA3. 22028; ALIS. 5400; WILL. 20; AV. ARTH. vi.

dēn, sb., *O.Fr.* dean, deien, *Lat.* decānus; *dean*, P. S. 332; deen E. G. 64; dēnes (*pl.*) LANGL. *A* ii. 150.

denaien, *see* deniin.

·**déne**, sb., *O.E.* dene *m.*, denu *f.*; *cf. M.Du.* dan *m.*; *valley,* '*vallis*,' FRAG. 3; '*convallem*' PS. lix. 8*; A. P. i. 295; dane ['*vallis*'] LK. iii. 5; þe dene of tieares AYENB. 160; dánes (*pl.*) 59.

Dene, sb. pl., *O.E.* Dene; *Danes*: Dene (*gen. pl.*) lawe MISC. 146.

  dane-geld, sb., *Dane-tribute*, MAN. (H.) 110.

  Dene-marke [Denemarche], pr. n., *O.E.* Dena mearc; *Denmark*, LA3. 4563; Denemark, Denemarch ROB. 3.

dene, denien, *see* düne, dünien.

dēnerie, sb., *from* dēn; *deanery*, PR. P. 118.

deniin, v., *O.Fr.* denier; *deny*, PR. P. 118; denie WICL. MAT. xxvi 34; deni TOWNL. 182; denaie GAW. 1497; deníede (*pret.*) WICL. GEN. xviii. 15; D. TROY 8009; denoied (*pple.*) CH. BOET. iii. 10 (88).

denne, *see* dene.

dennien, v., *shelter as in a cave or den, lodge*: wu he dennede (*pret.*) him in ðat defte maiden REL. I. 209 (MISC. 2); þu was . . . dennet (*pple.*) in a beastes cribbe HOM. I. 277.

denoumbren, v., *Fr.* denombrer; *count out; tell* ['*dinumerare*'], WICL. PS. lxxxix. 11.

denounce, v., *O.Fr.* denoncer, *from Lat.* dēnunciāre; *command*, WICL. 2 THESS. iii. 6.

Densc, adj., *O.E.* Denisc; *Danish*; Densce (*pl.*) LA3. 12854; Denshe HAV. 2575.

dent, *see* dünt.

denten, v., *Lat.* dentāre; *indent; cf. O.Fr.* endenter; '*indento*' PR. P. 118; dent (*for* dented) (*pple.*) GOL. & GAW. vi. (65).

dentinge, sb., *morticing*: twei dentingis (*pl.*) schulen be in the sides of a table WICL. EX. xxvi. 17*.

dēofel, sb., *O.E.* dēofol, *m. n.*, *Lat.* diabolus; *devil*, HOM. I. 21; ORM. 671; deovel A. R. 208; deovel [deavel] LA3. 17669; dievel AYENB. 15; devel HAV. 446; WILL. 1976; divel REL. I. 209 (MISC. 2); dēofles (*gen.*) P. L. S. viii. 98; deovles, deoveles H. M. 17, 41; dēofle (*dat.*) A. R. 84; dievle MISC. 28; dēofles (*pl.*) P. L. S. viii. 98; dēofle MAT. ix. 34; HOM. I. 27; MISC. 152; deovlen JUL. 22; dievlen AYENB. 86; develen ROB. 506; dēoflene (*gen. pl.*) JUL. 38; deovlene O. & N. 932; **dēoflan,** d(e)ovelen (*dat. pl.*) HOM. I. 73, 103.

dēofel-ʒild, sb. (*ms.* -ʒyld), *O.E.* dēofolgild, = *O.L.G.* diobolgeld; *tribute to the devil,* HOM. I. 227.

dēofel-līc, adj., *O.E.* dēofollīc; *devilish :* þa deoflīche ʒītsunge HOM. I. 105.

dēofel-schīn, sb., *devilish magic*; **dēofel-shīne** (*dat.*) ORM. 8110.

dēvel-hēd, sb., *nature of the devil :* no devel-hede i ne habbe in me LEB. JES. 499.

dēvelnesse, sb. *demon :* alle goddes of genge **dēvelnesses** (*pl.*) ['*dæmonia*'] ere þa PS. XCV. 5.

deol, *see* doel.

dēop, adj., *O.E.* dēop, dīop, = *O.Fris.* diap, *O.L.G.* diop, diap, *O.N.* diupr, *Goth.* diups, *O.H.G.* tiuf; *deep,* A. R. 224; MARH. 15; deop [deap] LAʒ. 647; deop, deep LANGL. *A* PROL. 15; diep AYENB. 264; dep ORM. 16697; MAP 335; so dep oþ ROB. 233; dup FER. 1687; hu deope is þe eorðe A. R. 232; þe deope diʒhelnesse ORM. 5501; a dupe dich BRD. 27; **dēoppere** (*dat. f.*) FRAG. 8; **dēope** (*pl.*) SPEC. 62; **dēopen** (*dat. pl.*) LAʒ. 21776; **dēoppre** (*compar.*) HOM. I. 49; seoven voten deopere [deoppere] LAʒ. 15895; deoppere [depper] LANGL. *A* x. 182; depper CH. C. T. *G* 250; **dēppest** (*superl.*) TREV. I. 45.

dēope, adv., *deeply,* H. M. 33; deope leared KATH. 388; deope in his flesch hi wode BEK. 2232; depe lǣred ORM. 7207; þat ʒe ne falle to depe SHOR. 104.

dēop-līche, adv., *O.E.* dēoplīce; *deeply,* A. R. 154; swor swiðe deopliche JUL. 13; deplike HAV. 1417.

dēopnesse, *O.E.* dēopness; *deepness,* HOM. I. 49; diepnesse AYENB. 211; depnisse KATH. 980; depnesse D. ARTH. 746; **dūpnesse** (*dat.*) LEECHD. III. 84.

dēope, sb., *O.E.* dȳp, dȳpe, *cf. Goth.* diupei, *O.H.G.* tiufī; *deep,* P. L. S. xv. 84; depe CH. C. T. *A* 455; on deopan ['*in altum*'] LK. v. 4; ine þe depe SHOR. 146.

dēopien, v., *cf. Goth.* (ga-)diupjan; *deepen, penetrate deeply:* dēopeð (*pres.*) into þe soule A. R. 288; þa þe dic wes ... allunge idēoped (*pl.*) LAʒ. 15473.

dēor, sb., *O.E.* dēor, dior, = *O.Fris.* diar, *O.L.G.* dier, *O.N.* diur, dȳr, *Goth.* dius, *O.H.G.* tior, tier, *mod.Eng.* deer; *beast; deer :* þat deor LAʒ. 6495; deor, der ORM. 1177, 1312; der REL. I. 215; dur MISC. 199; **dēores** (*gen.*) A. R. 196; **dēore** (*dat.*) HOM. II. 37; **dēor** (*pl.*) HOM. I. 43; II. 35; LAʒ. 1125; MARH. 10; libbeþ al so wilde deor [dor] O. & N. 1012; duer CHR. E. 30; der GEN. & EX. 178; **dēoren** [deore] (*gen. pl.*) LAʒ. 20852; **dēoran** (*dat. pl.*) HOM. I. 115; deoren MK. i. 13; deoren LAʒ. 1128; dere GAW. 1322.

dēor-cen, sb., *O.E.* dēorcynn; *race of beasts,* HOM. I. 221.

dēr-fald, sb., *enclosure for deer,* SAX. CHR. 249.

dēor-frið, sb., *O.E.* dēorfrið; *deer forest,* LAʒ. 1436.

dēor, adj. & adv., *O.E.* dēor (*heavy, severe,* GREIN); *grievous, grievously :* nou fele we it dere PR. C. 1469; dēre (*pl.*) GAW. 564; sor-wes dere GEN. & EX. 3742; **dērure** (*compar.*) KATH. 948.

deorc, *see* dearc.

dēore, adj., *O.E.* dēore, dīore, dȳre, = *O.Fris.* diore, diure, *O.L.G.* diuri, *O.N.* dȳrr, *O.H.G.* tiuri; *dear, precious, noble, beloved,* A. R. 408; deore, dure LAʒ. 143, 2963; deore, dere ORM. 2206, 2356; (*ms.* dyere) LK. vii. 2; þane diere time AYENB. 36; deere, dere CH. C. T. *A* 2453; dere MAND. 252; P. 78; of dere pris GEN. & EX. 2247; what him dere þouʒt WILL. 1268; þe cuntre was dere JOS. 37; a dere dai GAW. 92; Arthur þe dere PERC. 508; of duʒti men and of dere AV. ARTH. i; dure frend FER. 282; duere REL. I. 110; **dēorre** (*compar.*) A. R. 392; derre ORM. 18221; his **dēoreste** (*superl.*) mon 8975; of derrest pris LANGL. *B* ii. 13.

dēore, adv., *dearly,* A. R. 392; deore aboʒt BEK. 112; diere ANGL. I. 15; dere HAV. 1637; RICH. 5312; deere LIDG. M. P. 217; dere, duere H. H. 59; duere P. S. 214; **dēorre** (*compar.*) A. R. 190; derrer AYENB. 36; heo him hadde **dēorest** iboʒt (*superl.*) H. M. 13; P. L. S. xiii. 136; durest LAʒ. 3081.

dēor-līche, adv., *O.E.* dēorlīc; *dearly,* LAʒ. 31752; JUL. 76; drinke to him deorli SPEC. 111; derli WILL. 1421; **dēorluker** (*compar.*) LAʒ. 30060.

dērenesse, sb., *dearness,* S. S. (Web.) 3144.

dēor-würðe, adj., *O.E.* dēorwyrðe; *precious,* FRAG. 6; deorewurðe JUL. 10; deorewurþe ORM. 6689; durewurðe LAʒ. 16686; derwurþe AV. ARTH. xxii; derewerþe FL. & BL. 289; derworþe C. L. 27; WILL. 1538.

dēorewurð-līche, adv., *preciously,* H. M. 11; durewurðliche LAʒ. 15151.

dēore, sb., *cf. O.H.G.* tiurī; *dearness* ; dere HAV. 824 : ROB. 416; MAN. (F.) 16419.

Dēorebi, pr. n., *O.E.* Dēora bȳ; *Derby,* MISC. 146; Derbi ROB. 3.

deork, *see* dearc.

dēorling, sb., *O.E.* dēorling, dȳrling, = *O.N.* dȳrlingr; *from* dēore; *darling,* LAʒ. 6316; A. R. 56; derling ORM. 9219; WILL. 1538; PR. C. 8791; LUD. COV. 53; durling REL. I. 170 (MISC. 102); BRD. 3.

deorne, *see* derne.

dēovel, *see* dēofel.   **dēp,** *see* dēop.

depart, v., *O.Fr.* despartir, *from Lat.* dis-partīre; *mod.Eng.* depart; *divide, distribute ;*

separate (trans. and intrans.) ; depart, WILL. 2334 ; departi TREAT. 140 ; þe Rede see . . . **departeþ** (*pres.*) þe southside of Inde from Ethiopia TREV. I. 63 ; **departede** (*pret.*) BEK. 483 ; ROB. 424 ; departed (*pple.*) PR. C. 3710.

departable, adj., *separable, distinct* : the Trinite, thre persones in parcelles departable fro other LANGL. *B* xvii. 26.

depeinten, v., *depict, paint* ; **depeinted** (*pple.*) C. L. 704 ; WILL. 3217.

dēpen, v., *O.E.* dēpan, = *O.Fris.* dēpa, *O.L.G.* dōpean, *Goth.* daupjan, *O.H.G.* toufen ; *baptize* : depe SHOR. 11 ; AYENB. 107.

depict, pple., *Lat.* dēpictus ; *depicted, painted* : a liknesse depict upon a wal LIDG. M. P. 177.

depōse, v., *O.Fr.* depōser ; *depose, put down*, LANGL. *B* xv. 514.

depost, sb., *O.Fr.* depost ; *deposit*, WICL. I TIM. vi. 20.

Deppeford, pr. n., *Deptford*, CH. C. T. *A* 3906.

deprāven, v., *O.Fr.* depraver ; *depreciate, defame* : deprave LANGL. *B* iii. 178 ; **deprāveden** (*pret. pl.*) WICL. PROV. i. 30.

depresen, v., *cf. Lat.* dēpressus *from* dēprimere ; *depress, vanquish* : þou con . . . fro þat mariag al oþer depres A. P. i. 777 ; þat prince of pris **depresed** (*pret.*) him so þikke GAW. 1770 ; depreced prouinces 5.

depresen, v., *? O.Fr.* despresser ; *release* : wolde ȝe . . . despresce ȝour prisoun GAW. 1218.

depressioun, sb., *O.Fr.* dépression, *from Lat.* dēpressio ; *depression*, CH. ASTR. ii. 28 (34).

deprīve, v., *O.Fr.* deprīver ; *deprive*, A. P. ii. 185.

dēpþe, sb., *from* dēop ; *cf. O.N.* dȳpt, *M.Du.* diepte ; *depth* : derknessis weren on the face of **dēpthe** (*dat.*) WICL. GEN. i. 2 ; thei descenden into the depthe as a stoon EX. xv. 5 ; **dēpthis** (*pl.*) of watris PS. cxlviii. 7.

deputātiōn, sb., *Fr.* députation ; *deputation* : he shall . . . ordeigne his deputation of suche juges as ben lerned GOW. III. 178.

depúten, v., *Lat.* dēputāre ; *ascribe* : al what evere to be **depúte** (*pple.*) to the grace of God WICL. ROM. PROL.

der, see **dear**.   **dēr**, see **dēor**.

derai, sb., *O.Fr.* desroi ; *disorder, tumult*, ALIS. 1177 ; derai B. DISC. 386 ; ANT. ARTH. xl ; disrai ALIS. 4353 ; drai MIN. viii. 34.

deraien, v., *O.Fr.* deraier ; *throw into disorder* : he **derāied** (*pret.*) him as a devel WILL. 2061 ; draied 1210.

deraine, ᵇsb., *O.Fr.* deraine, deresné : *claim* : this dereine by the barouns is imad ALIS. 7353.

deraine, v., *O.Fr.* derainier, -raisnier ; *defend one's cause, vindicate a claim*, ALIS.² (SK.) 124 ; dereini ROB. 285 ; dreinen C. L. 974.

dére, sb., *O.E.* daru, = *M.Du.* dare, ᵈere,

M.L.G. dere, *O.H.G.* tara ; *injury, harm,* GEN. & EX. 3214 ; FER. 4613 ; IW. 2577 ; RICH. 1696 ; E. T. 642 ; EGL. 513 ; TOWNL. 28 ; þis hete gret dere me doþ LEG. 5.

dēre, see **dēore**.

derf, adj., *O.E.* dearf, *O.N.* diarfr, *O.L.G.* dervi ; *from* **derven** ; *firm, hard, heavy ; valiant ;* MARH. 5 ; ORM. 16780 ; GAW. 1492 ; A. P. ii. 862 ; for þi derfe dede AV. ARTH. liv ; **derfe** (*pl.*) TOWNL. 317 ; derfe dintes ALEX. 2091 ; destines derf(e) & dere GAW. 564 ; derve (*ms.* derue) MISC. 9 (*printed* derne REL. I. 215) ; **derfre**, dervre (*compar.*) JUL. 16, 17 ; H. M. 21 ; **derveste** (*superl.*) KATH. 565 ; derf(e) (*adv.*) M. H. 23 ; *comp.* **un-derf.**

derf-līc, adj., *cf. O.N.* diarfligr ; *hard, brave* : þi derfli dede C. M. 1143 ; **derflīche** (*adv.*) ipined A. R. 114 ; deorflike ORM. 9752 ; derflike REL. I. 218 ; derfli ANT. ARTH. xxiv ; A. P. iii. 109 ; derffli D. TROY 1339.

derfnes, sb., *cf. O.E.* gedeorfnyss ; *severity, gravity*, C. M. 3996 ; dervenesse HOM. I. 21.

derf-schipe, sb., *obstinacy*, KATH. 978.

derf, sb., *cf. O.E.* gedeorf ; *tribulation, affliction* : nas na man . . . þat durste him derf makien LAȝ. 10943 ; vor te drien derf ine godes servise A. R. 80 ; þe derf of deaðૂ KATH. 2426.

der(f)-ful, adj., *strict, severe*, A. R. 348.

derfful-līche, adv., *painfully*, KATH. 1090.

dérien, v., *O.E.* derian, = *O.L.G.* derian, *O. Fris.* dera, *M.Du.* deren, daren, *O.H.G.* terian, terran ; *from* **dére** ; *injure, hurt*, P. L. S. viii. 168 ; derian, derien HOM. I. 13 ; derie HORN (L.) 786 ; AYENB. 126 ; derie MISC. 76 ; SHOR. 28 ; ALIS. 3657 ; deren HOM. II. 73 ; REL. I. 212 ; LANGL. *B* vii. 50 ; dere HAV. 488 ; MAND. 13 ; CH. C. T. *A* 1822 ; PS. lxxxviii. 23 ; PR. C. 2290 ; MIN. i. 52 ; PERC. 1171 ; AV. ARTH. iii ; **déreð** (*pres.*) MISC. 14 ; þas twa þing deriað oft þan alden HOM. I. 109 ; derieþ MARG. 226 ; **dérede** (*pret.*) LAȝ. 9657 ; GEN. & EX. 242 ; deredest FRAG. 7.

déringe, sb., *O.E.* derung ; *injury, 'nocumentum,'* PR. P. 119.

derk, see **dearc**.

derne, adj., *O.E.* derne, dyrne, dierne, = *O. L.G.* derni, *O.H.G.* tarni ; *dark, secret, hidden, retired*, MARH. 8 ; ROB. 289 ; C. L. 1030 ; WILL. 860 ; LANGL. *B* ii. 175 ; GAW. 1012 ; M. H. 166 ; ᵗᵉᵒȝ he be derne hunte MISC. 2 ; derne love CH. C. T. *A* 3278 ; bi a derne sti ISUM. 40 ; godes derne runes A. R. 96 ; derne stedes AYENB. 143 ; derne, deorne LAȝ. 304, 13604, 13624 ; dearne KATH. 574 ; dærne ORM. 2004 ; durne BEK. 23 ; **derne** (*adv.*) SPEC. 23 ; A. P. ii. 697 ; luvie derne O. & N. 1357 ; dierne HICKES I. 222 (ANGL. I. 11) ; **derne** (*sb.*) O. & N. 608 ; in derne ['*in abscondito*'] PS. ix. 29 ; AV. ARTH. lii ; let us halde us in derne DEGR. 607 ; *comp.* **un-derne.**

derne-līche, adv., *secretly,* A. R. 146; O. & N. 1423; M. T. 103; derneliche [deorneliche] LA3. 4392; dearneliche KATH. 407; durne-liche BEK. 27; dernelike MISC. 14; dernelike and stille AN. LIT. 5; dærnelike, -li3 ORM. 6914, 7370; dernli TOWNL. 141; **dernelüker** (*compar.*) A. R. 128.

dearne-schipe, sb., *privity,* A. R. 152*; darnscipe (? *for* dærn-) LA3. 258.

dernel, *see* **darnel.**

dernen, v., *O.E.* dernan, dyrnan, = *O.L.G.* der-nian, dernean, *O.H.G.* tarnan; *conceal*: dernen, dærnen, deorne, LA3. 6660, 18549; derni SHOR. 79; **dærnden** (*pret.*) LA3. 7694; *comp.* **bi-dernan.**

derner, sb., *door-head*; derner [dernere] C. M. 6075; dirner 6103.

dersten, *see* **dreste.**

Dertemūðe, pr. n., .*Dartmouth,* LA3. 1786; Dertemouþe CH. C. T. *A* 389.

dērþe, *see* **dierþe.**

derven, v., *O.E.* deorfan (*labour*), = *O.L.G.*(far-) dervan, *O.Fris.* (for-)derva, *O.N.* (for-)diarfa; *grieve, afflict, injure,* A. R. 382; C. L. 676; to derven mine soule HICKES I. 227; derve JOS. 47; derveð (*pres.*) KATH. 1684; derve (*subj.*) MARH. 12; derfde (*pret.*) LA3. 8731; *comp.* 3e-derven; *deriv.* **derf.**

dervenesse, *see* **derfnes** *under* **derf.**

des, *see* **deis.**

desaivabel, *see* **deceivable.**

desalī, *see under* **düsi.**

desaraien, v., *O.Fr.* desarroyer; *fall into disorder*; **desaraied** (*pret.*) TREV. VII. 243.

desblāmen, v., *O.Fr.* desblâmer; *free from blame, exonerate*; **deblāmeth** (*imper.*) CH. TRO. ii. 17.

desceivaunce, *see* **deceivance.**

descenden, v., *O.Fr.* descendre; *descend*: descend MAN. (H.) 134; **descendeþ** (*pres.*) AYENB. 123; GOW. II. 280; **descendid** (*pret.*) LIDG. M. P. 79.

descensioun, sb., *O.Fr.* descension; *descent,* CH. ASTR. ii. 4 (32); discencioun ii. 4 (33).

descensorie, sb., *apparatus for extracting oil*; **descensories** (*pl.*) CH. C. T. *G* 792.

descent, sb., *O.Fr.* descente; *descent, lineage,* GOW. III. 207; dissent LIDG. M. P. 2.

deschargen, dischargen, v., *O.Fr.* deschar-ger; *discharge*; dischargen LANGL. *B* xv. 528; deschargen *C* xviii. 231; dischargen MAND. 302; discharge WICL. JAS. v. 15.

de-sclandre, sb., *O.Fr.* esclandre, *from Lat.* scandalum; *disgrace,* BEK. 2061; disclaundre CH. TRO. iv. 537.

de-sclandren, v., *for* **sclandren**; *disgrace, slander*: thu desclandrest (*pres.*) thin owe loverd BEK. 2050; disclaundrid (*pret.*) CH.

C. T. *I* 623; for disobedience disclaundrid (*pple.*) is perpetualli mi name LIDG. M. P. 143.

descolouren, v., *O.Fr.* descoulourer; *dis-colour*; **descoloured** (*pple.*) GOW. III. 339.

descrīen, v., *O.Fr.* descrier; *describe,* '*describo*,' PR. P. 120; descrie EGL. 1177; discrighe ALIS. 137; discri TOWNL. 203; discrieth MAS. 323; **discrīed** [descried] (*pple.*) TRIAM. 781.

descripcioun, sb., *O.Fr.* description; *descrip-tion,* TREV. I. 29; WICL. S. W. I. 316; de-scription GOW. III. 144.

descrīven, v., *O.Fr.* descrivre; *describe, draw,* CH. TRO. v. 1314; descrive WILL. 5005; GOW. I. 264; descrife PR. C. 2304; discrive OCTOV. (W.) 629; diskrive CH. D. BL. 915; **descrīveþ** (*pres.*) AYENB. 168; **discrīven** (*pl.*) LANGL. *B* xvi. 53; **descrīved** (*pret.*) WILL. 3041; discrivede WICL. NUM. xxxiii. 2; **descrēved** [descrived] (*pple.*) TREV. I. 171; discrived WICL. LK. ii. 1; discrived [descrived] WICL. NUM. xi. 26.

descrīving, sb., *describing,* ROB. 60; dis-criving WICL. LK. ii. 2.

descuren, descuveren, *see* **discoveren.**

deseriten, *see* **disheriten.**

desert, sb., *O.Fr.* desert; *desert, wilderness,* GEN. & EX. 1227; AYENB. 240; deseert WICL. NUM. xiii. 1; desart MAND. 42; dezert AYENB. 131; disert MAND. 42; diserd GEN. & EX. 973; desertes (*pl.*) MAND. 63.

desert [2], sb., *O.Fr.* desert; *desert, merit,* PR. P. 120; WICL. JOB xxi. 4; desserte A. P. i. 594; dissert A. P. iii. 12; deserte (*dat.*) ROB. 253.

deserven, v., *O.Fr.* deservir; *deserve*: deservin '*mereor*' PR. P. 120; diserve GOW. I. 66; des-serve CH. TRO. v. 273; disserve TREV. V. 213; deserve (*pres.*) LANGL. *B* xiv. 86; **disserved** (*pret.*) A. P. ii. 613; **deserveden** (*pl.*) MISC. 225; **deserved** (*pple.*) GOW. I. 29; disserved GAW. 452; deservet MISC. 225; deservid DEGR. 1131; is disservid (*well pleased*) WICL. HEB. xiii. 16.

deserving, sb., *deserving,* IPOM. 452.

desēse, *see* **disēse. desi,** *see* **düsi.**

desirable, adj., *O.Fr.* desirable, *from Lat.* dēsīderābilis; *desirable,* WICL. PS. xviii. 10.

desīre, sb., *O.Fr.* desier; *desire,* GOW. I. 65; desir CH. C. T. *G* 671; **desīres** (*pl.*) LIDG. M. P. 187, 221.

desīre-ful, adj., *desirous,* WICL. DAN. x. 3.

desīren, v., *O.Fr.* desirer; *desire,* PR. C. 8031; desiri AYENB. 244; **desīri** (*pres.*) ROB. 309; desires A. P. ii. 545; desīreþ (*pl.*) TREV. I. 137; desiren LANGL. *B* ix. 104; desire (*imp.*) LIDG. M. P. 187; desīrede (*pret.*) BEK. 225; desirid LIDG. M. P. 109; desīred (*pple.*) ROB. 253.

desīringe, sb., *desiring,* IPOM. 204.

desīrous, adj., GOW. I. 244; desirus CH. C. T. *A* 1674.

M

de-skateren, v.; *scatter about*; **deskatered** (*pple.*) P. S. 337.

deske, sb., *? Du.* disch (*table*); *see* **disch**; desk, *pulpit*, '*pluteum*,' PR. P. 120; '*lectrinum*' 299; W. & I. 39.

deslaien, v.; *cf.* delaien; *hinder*; **deslaied** (*pple.*) GOW. II. 60, 115.

desolacioun, sb., *O.Fr.* desolation; *solitude*, WICL. EZ. xii. 19*; LIDG. M. P. 114.

desolat, adj., *desolate, lonely*, CH. TRO. v. 540; desolate GOW. I. 248; dissolate TREV. III. 181.

desolaten, v., *make desolate*; **desolatid** (*pple.*) WICL. EZ. xii. 19.

des-ordeinen, v., *deprive of clerical orders*: if a clerk hadde misdo ... that me scholde him anon desordeinen BEK. 721; desordeini ROB. 473.

despeir, sb., = **desespeir**; *despair*, MAND. 93; in despeire GOW. III. 217; dispair LANGL. *B* xx. 163; in dispaire PR. C. 6293.

despeirable, adj., *hopeless*; ['*desperabilis*'] WICL. JER. xv. 18.

despeiren, v., *O.Fr.* desperer; *despair*; despeire GOW. I. 272; despeired (*pret.*) WICL. S. W. III. 135; GOW. I. 318; CH. TRO. v. 713; despaired AYENB. 34; **dispeired** (*pple.*) GOW. I. 281; LIDG. M. P. 76.

desperacion, sb., *Lat.* dēsperātio; *desperation*, LANGL. *C* xx. 289; disperacion MISC. 215.

desperischen, *see* **dispereschen**.

desperance, sb., *O.Fr.* desperance; *despair*; desperaunce CH. C. T. *I* 421 (*Pet. ms.*); GOW. II. 119; A. R. 8.

desperate, *see* **disesperat**.

despers, *see* **dispers**.

despisabil, adj., *despicable*, HAMP. PS. xlviii. 19.

despisen, v., *cf. O.Fr.* despire, *from Lat.* dēspicere; *despise*; despisin PR. P. 120; despise MAS. 326; despice PR. C. 9426; despise PR. C. 4252; LIDG. M. P. 118; **despisest** (*pres.*) ROB. 31; despiseth LANGL. *B* xv. 54; **despise** (*pl.*) GOW. I. 87; dispisen MAND. 295; dispise LANGL. 10682 (W.); **despised** (*pret.*) TREV. I. 11; despised (*pple.*) RICH. 1837; TREV. I. 13.

despeisere, sb., *despiser*, D. ARTH. 538.

despising, sb., *despising*, TREV. III. 287.

despit, sb., *O.Fr.* despit, *from Lat.* despectus, *mod.Eng.* despite; *contumely, scorn, spite*, ROB. 547, 464; WILL. 3335; despite MAN. (H.) 54; RICH. 2177; dispit A. P. iii. 50; JOS. 580; dispite M. H. 70; **despites** (*pl.*) S. S. (Wr.) 1897; dispitis WICL. ROM. i. 24.

despitous, adj., *spiteful*, CH. C. T. *A* 516, *D* 762; dispitouse TRO. v. 199; dispitous TREV. V. 87.

despitous-liche, adv., *scornfully*, TREV. III. 389; despitousli WILL. 1136; despitusli 1210;

dispitousli D. TROY 7650; dispetusli MED. 615.

desplaien, *see* **displeien**.

despoilen, v., *O.Fr.* despoiller; *despoil, take away*; despoilen [despoillen] CH. C. T. *E* 374; dispoili AYENB. 45; MED. 615; dispuile WICL. I KINGS xxxi. 8; **dispuilide** (*pret.*) WICL. I KINGS xviii. 4; **despuiled** (*pple.*) A. R. 260; despuled AR. & MER. 1403; dispoiled GAW. 860; dispoilid WICL. 2 COR. v. 4; despuiled GOW. I. 116.

despoilinge, sb., *spoil*, CH. BOET. iv. 7 (147).

destance, *see* **distance**.

desten ?=**dusten**, v.; *throw*: ne non harm hine do(u)n deste SHOR. 52; over þe brigge he deste TRIST. 2393.

desténen, v., *O.Fr.* destiner; *destine*; **desteined** (*pple.*) D. TROY 2673.

desténing, sb., *destiny*, ALIS. 6866.

dester, sb., *O.Fr.* destre, *from Lat.* dextera (manus); *right hand*: þi stedes þat þouȝ haddest in dester leddes MAP 334.

destinal, adj., *pertaining to destiny*, CH. BOET. iv. 6 (135).

destinee, sb.. *O.Fr.* destinee; *destiny*; destinee LANGL. *B* vi. 276; destine A. P. i. 757; iii. 49; GOW. II. 94; destene WILL. 315; destegne CH. TRO. iv. 931; destinie CH. BOET. iv. 6 (135); destenie TREV. V. 237; destanee CH. L. G. W. 952; destanie TREV. III. 401.

destitūte, adj., *Lat.* dēstitūtus; *destitute*, P. L. S. xxxi. 97; LIDG. M. P. 34.

destrer, sb., *O.Fr.* destrier; *war-horse*, B. DISC. 569; CH. C. T. *B* 2103.

destructiōn, sb., *destruction*, MAN. (H.) 202; GOW. II. 37; destructioun S. S. (Web.) 2761; destruccion PR. C. 4049; destruccioun PR. P. 120; destruccioun S. S. (Web.) 393.

destruien, v., *O.Fr.* destruire; *destroy*; destruiie WILL. 2929; destrui C. M. 22348; destrue ROB. 236; destroie PR. C. 4453; destrie WICL. MAT. ii. 13; destrei P. L. S. iv. 37; distruie WICL. I MAC. ix. 73; distroie PR. C. 4472; distrie [destroie] LANGL.*B* ii. 14; disstrie A. P. ii. 907; **destruie** (*sb.*) ROB. 46; **destruiþ** (*pres. pl.*) AYENB. 35; **destruide** (*pret.*) P. L. S. xxiv. 196; destruied LANGL. (W.) 1139; destrude ROB. 242; destroied TREV. I. 97; destriede ROB. 55; distruide [distruiede] WICL. MAT. xxii. 7; distriede I MAC. ix. 73; **destruied** (*pple.*) JUDG. vi. 30; destruied PR. C. 4073; destrued WILL. 2646; destruit 2847; destrud AYENB. 30; destroied MAND. 95; destroiid RICH. 1358; destried WICL. MAT. xxiv. 2; distruiȝed L. H. R. 33; distruied GOW. III. 15; distroied LIDG. M. P. 85; distried ALIS. 130; disstried A. P. ii. 1159.

destroiere, sb., *destroyer*, PR. P. 120; distriere WICL. JUD. viii. 25; **destrieris** (*pl.*) JUDG. iii. 17.

destroiing, sb., *destroying*; destroienge TREV. III. 449.

determinable, adj., *determinable*, A. P. i. 594.

determinat, pple., *Lat.* dēterminātus; *determinate*, CH. ASTR. i. 21 (11), ii. 18 (29).

determine, v., *O.Fr.* determiner; *determine*, LIDG. M. P. 66.

detraccīōn, sb., *O.Fr.* detraccion; *detraction*, AYENB. 10.

detractour, sb., *O.Fr.* detracteor; *detractor*, PR. P. 120.

dette, sb., *O.Fr.* dette, dete, debte; *debt*, PR. P. 120; A. R. 126; AYENB. 35.

dette-lēs, adj., *free from debt*: in honour dettles CH. C. T. *A* 583.

detted, adj., *from* dette; *owed*: to whom oni thing is detted, ethir owid WICL. DEUT. xv. 2.

dettūr, sb., *O.Fr.* detor, deteur, detteur; *debtor*, A. R. 312; dettour HOCCL. vi. 7; dettours (*pl.*) AYENB. 113.

dēð, *see* dēáð. **dēu,** *see* dēáu *and* due.

deus, sb., *O.Fr.* deus, deu (*dieu*); *God*, HAV. 1312; a deu (*ms.* dew) TOWNL. 94; par de CH. C. T. *A* 1312.

deutē, *see* duetē. **dēvel,** *see* dēofel.

dēven, *see* dēaven *and* dūven.

Devene [**Devenes**], pr. n., pl., *people of Devon*, LAȝ. 30880.

Devene scīre, pr. n., *O.E.* Defena scire; *Devonshire*, LAȝ. 21015; Devene (*printed* Denene) schire ROB. 5.

devēr, sb., *O.Fr.* devoir; *duty*, SHOR. 54; WILL. 474; LANGL. *A* xii. 2; D. ARTH. 1940; E. G. 5.

devīn, devīnen, devīnour, *see* div-.

devīs, sb., *O.Fr.* devīs, vīse, *cf. Ital.* devisa; *device*, WILL. 3222; to maken aftur his devis LEG. 46; at his aun devise C. M. 11576.

devīsement, sb., *O.Fr.* devisement; *description*: i knew hit bi his devisement in þe apocalippez A. P. i. 1018.

devīsen, v., *from Lat.* dīvīsus (*pple.*); *divide*: þis buk . . . in seven partes devised (*pple.*) es PR. C. 348, 986.

devīsen², v., *O.Fr.* deviser, devisier; *tell; devise, contrive; plan, determine*: devisi AYENB. 144, 103; nis no nede heore armes to devise ALIS. 7377; devise GOW. I. 31; device WILL. 1602; CH. TRO. iii. 407; ne devise hi miȝte best out of chambre wende BEK. 875; TREV. I. 293; ordain and divise a gene S. S. (Wr.) 2033; devīsede (*pret.*) LANGL. *C* xxii. 331; LEG. 46; devīsed (*pple.*) MAN. (H.) 24.

devīsinge, sb., *describing*, MAND. 5.

divisīōn, *see* divisīōn.

devociūn, sb., *O.Fr.* devocion; *devotion*, A. R. 286; devocion ROB. 330; AYENB. 107; devocioun P. L. S. xii. 139; PR. C. 5906.

devoide, v., *O.Fr.* desvuidier (*dévider*); *put away*, WILL. 2044; A. P. ii. 908; devoided (*pple.*) R. R. 2929.

devoidinge, sb., *putting away*: in devoidinge þe vilanie þat venkqvist his þewez A. P. ii. 543.

devors, *see* divors.

devōt, adj., *O.Fr.* devōt; *devout*, A. R. 376; devout ROB. 369.

devōte-liche, adv., *devoutly*, WILL. 1245; devouteliche 2974; devoutliche AYENB. 211; devotli P. L. S. xxx. 6; A. P. ii. 814; CH. C. T. *A* 482; devoutli LYDG. M. P. 72.

devouren, v., *O.Fr.* devorer; *devour*, CH. C. T. *G* 21.

devowrar, sb., *devourer*, '*devorator*,' PR. P. 120.

devouresse, sb., *a female devourer*: a devouresse of men WICL. EX. xxxvi. 13.

dēw, dēwen, *see* dēáu, dēawen.

dia, sb., *medicine*: dragge nor dia was none in buri towne LIDG. M. P. 49; with **dias** (*pl.*) and drogges LANGL. *C* xxiii. 174.

diacne, sb., *O.Fr.* diacne, *O.E.* diacon; *from Lat.* diaconus, *Gr.* διάκονος; *deacon*, AYENB. 190; HOM. I. 71, 81; deakne SHOR. 51; dēkenes (*pl.*) WICL. I TIM. iii. 18.

diad, *see* dēad.

diadēme, sb., *Fr.* diademe; *diadem*, BEK. 2153; CH. C. T. *F* 43.

diadēmed, adj., *crowned*: David shall be diademed LANGL. *A* iii. 268.

dial, sb., *dial*, PR. P. 120.

dialatik, sb., *Fr.* dialectique; *dialectics, logic*, WICL. PREF. EP. vii.

dialoge, sb., *Lat.* dialogus, *Gr.* διάλογος; *prophecy*: Daniel in his **dialokez** (*pl.*) devised A. P. ii. 1157.

diamaund, sb., *Fr.* diamant; *diamond*, SPEC. 25; **diamantz** (*pl.*) CH. C. T. *A* 2147.

diapenidion, sb., *cf. Fr.* pénide, *mod.Eng.* pennet: *barley sugar, medicated sweetmeat*, LANGL. *B* v. 123.

diaper, sb., *O.Fr.* diaspre, diapre, *med.Lat.* diasprus; *diaper* (*kind of cloth*): towelle of diaperi B. B. 129; diaper 268.

diapred, adj., *variegated like diaper*: cloth of gold diapred wel CH. C. T. *A* 2159; dighte in **diaperde** (*pl.*) wedis D. ARTH. 3251.

diaþ, *see* dēáð.

dīc, sb., *O.E.* dīc, *m. f., cf. O.L.G.* dīc, *O.Fris.* dīk, *M.H.G.* tīch *m., O.N.* dīk, dīki *n.; dike, ditch*: enne dic, ane dich [one dich] LAȝ. 646, 6425; þa dich [þe dich] 12570; dich A. R. 246; ROB. 86; BRD. 27; AYENB. 57; ALIS. 6632; þeovene dich (*den of thieves*) MISC. 39; dik GEN. & EX. 281; in a dīk [dike, diche] falle LANGL. *B* xi. 417; to delven a dich [diche] xix. 359; diche MAND. 29; dīche (*dat.*) CH. C. T. *A* 4106; ORF. 347; to þare diche

O. & N. 1239; dike RICH. 6021; **dīches** (*pl.*)
LA3. 9238; ROB. 409; dikes HAV. 1923;
diche P. L. S. viii. 21; HOM. I. 163; MISC. 59.

dīcen, v., *from* dēs, *see* dē; *play with dice;
cut into cubes*: dicin, or pley with dicis
'*aleo*' PR. P. 121; dicin, as men do brede
'*quadro*' PR. P. 121.

dīcer, sb., *dicer, gambler*: she makis dīsers
(*pl.*) to selle thare corne and thare catelle
TOWNL. 243; disars 242.

dīcing, sb., *dice-playing*: at the dising he
dos us no wrang TOWNL. 240.

dīch, *see* dīc.

dictour, sb., *? Lat.* dictātor, *or*=dihtere;
steward, viceroy, D. ARTH. 709.

dīderin, v., *cf. prov.Eng.* didder, dither;
shiver, tremble, '*frigutio,*' PR. P. 121; dintus
gerut him to dedur AV. ARTH. xxv; dedir
(*pres.*) TOWNL. 28.

[**dīdren,** v., *O.E.* dyderian (*cheat*); *cf. Dorset
dial.* dather (*bewilder*); *comp.* bi-didren.]

dīen, *see* dēa3en *and* dē3en.

dīep, *see* dēop. **dīere,** *see* dēore.

dīerþe, sb., = *O.N.* dȳrð,=*O.L.G.* diurtha,
diuritha, *O.H.G.* tiurida; *from* dēore; *dearth*;
dierþe AYENB. 256; derþe P. S. 323; TREV.
II. 55; LANGL. *B* vi. 330; MAND. 38; (*beauty*)
A. P. i. 99.

dīestere, *see under* dēa3en.

diēte, sb., *Fr.* diète; *diet,* A. R. 112.

diēten, v., *from* diēte, sb., *diet*: if thow
diēte (*subj.*) thee thus LANGL. *B* vi. 270.

dieð, *see* dēað.

diffamacioun, sb., *Fr.* diffamacion. *from Lat.*
diffāmātio; *cf. mod.Eng.* defamation; *dishonour*: it were a greet diffamacioun for a
man to use more ringes þan oon TREV. II.
313.

diffāme, defāme, v., *O.Fr.* defamer; *defame,*
TREV. IV. 97, 99.

differen, v., *O.Fr.* differer; *defer*: **differre**
(*pres.*) ['*differat*'] WICL. DEUT. vii. 10.

difference, sb., *O.Fr.* differance; *difference,*
AYENB. 10; TREV. V. 401.

difficultē, sb., *O.Fr.* difficulté; *difficulty,* CH.
C. T. *B* 218.

diffūse-lī, adv., *cf. Lat.* diffūsus (*adj.*); *at
length, in detail*: Luk ... telliþ more diffuseli
how man stieþ up to God WICL. S. W. I. 391.

dī3el, adj., *O.E.* dȳgol, dīgol, dēgol, dīegol,=
O.H.G. tougal; *secret, hidden*: in one swiþe
di3ele hale O. & N. 2; on dīglen (*O.E.*
dīglum) ['*in abscondito*'] MAT. vi. 4; di3elen
[di3ele] bi hælves LA3. 26935.

dī3el-līc, adj., *O.E.* dīgollīc; *hidden*: (his)
digeliche, dieliche tocume HOM. II. 5, 7;
dī3ellīche (*adv.*) FRAG. 8; di3elliche & stille
LA3. 13539.

dī3elnesse, sb., *O.E.* dȳgelness; *secrecy*
LA3. 2391; di3helnesse ORM. 5501.

[**dī3en,** adj.,=*O.H.G.* tougan; *secret.*]

dī3en-līc, adj., *secret*: mid dī3enlīche rūnen
LA3. 415; di3enliche ... biwitie 6659*.

dī3en, *see* dē3en *and* dēa3en.

digest, sb., *med. Lat.* dīgestum; *the Pandects*:
þe lawes of digest TREV. III. 255.

digest, adj., *Lat.* dīgestus; *digested*: when
Phebus entrith in the Ariete, digest humoures
upward doon hem dresse LIDG. M. P. 195.

digestible, adj., *O.Fr.* digestible; *easy of
digestion,* CH. C. T. *A* 439.

digestioun, sb., *O.Fr.* digestion; *digestion*:
cleer eir and walking makiþ good digestioun
B. B. 54; to digestioun repastis be nat goode
LIDG. M. P. 155.

digestive, sb., *O.Fr.* digestif; *a help to digestion*: digestives (*pl.*) CH. C. T. *B* 4151.

digge, sb., *duck*; digges (*pl.*) CHEST. I. 52
(MIR. PL. 9); L. C. C. 9.

diggin, v., *?*=dīkin; *dig,* '*fodio,*' PR. P. 121;
diggen him up MAND. 107; digge TREV. IV.
329; LIDG. M. P. 145; ston ... to digge WILL.
2243; digge (*v.r.* delve) pittis WICL. GEN.
xxvi. 25; digg(e) it up S. S. (Wr.) 631; **digged**
(*pret.*) up LANGL. *B* vi. 109 [*A* vii. 100
dikeden]; diggide WICL. IS. xxxiv. 15; deggid
LIDG. M. P. 113; **diggiden** (*pl.*) WICL. NUM.
xxi. 18*.

diggar, sb., *digger,* '*fossor,*' PR. P. 118.

digne, adj., *O.Fr.* digne, *from Lat.* dignus;
worthy, dignified, ROB. 132; JOS. 252; WILL.
583; CH. BOET. ii. 4 (43); PR. C. 74; E. G.
29.

digne-līche, adv., *with dignity,* AYENB. 20;
WILL. 520.

dignitē, sb., *O.Fr.* dignité; *dignity,* A. R. 140;
BEK. 244; PR. C. 3872.

dihten, v., *O.E.* dihtan, = *O.H.G.* dihtōn,
dictōn, *O.N.* dikta, *mod.Eng.* dight; [*from
Lat.* dictāre]; *prepare, set in order*: þa setten
heo biscopes þan folken (*r.*folke) to dihten LA3.
10201; KATH. 1471; dihtin, ditin '*dicto,
paro*' PR. P. 123; hi leten hem di3te a gret
schip BRD. 5; him to dethe di3te AV. ARTH.
iv; di3t(e) a baþ EGL. 526; dighte PERC. 652;
al þat he dihteð (*pres.*) & demeð HOM. I. 267;
dihte (*pret.*) MARH. 2; he dihte feole domes
LA3. 7221; he hit (his kineriche) dihte and
delde 9192; þus he us diste 25907; þe king
dihte him for to wende K. T. 954; di3te
SHOR. 10; and di3te me dereli LANGL. *B*
xix. 2; di3t (*pple.*) WILL. 151; dight RR. C.
448; a lettre has he dight DEGR. 153; idiht
KATH. 1607; O. & N. 641; idi3t P. L. S. xxxv.
106; *comp.* a-dihten.

dihtende, sb., *disposer, guider*: dihtende of
alle shafte HOM. II. 123.

di3tere, sb., *O.E.* dihtere; *disposer, governor,* AYENB. 100.

di3tinge, sb., *disposition; decoration,* AYENB.
1724.

dihtnare, sb., *O.E.* dihtnere; *steward,* '*dispensator,*' FRAG. 2.

dīk, *see* dīc.

dīkin, v., *O.E.* dīcian,=*M.Dut.* dīken, *O.Fris.* dīka, dītsa; *dig, make a ditch or dike,* '*fosso,*' PR. P. 121;ᵗ diken, diche LANGL. *B* vi. 143; xix. 232; dike GOW. I. 15; CH. C. T. *A* 536; **dīked** (*pple.*) WILL. 2233; idiched C. L. 674.

   dīkere, sb., *O.E.* dīcere; *ditcher*; **dīkers** (*pl.*) LANGL. *A* PROL. 102.

dil, adj., ?=dul; *dull, foolish*: dill & slaw ORM. 9885; dille GAW. 1529; A. P. i. 679; TOWNL. 136; domb & dille HALLIW. 303.

dilataciōn, sb., *Lat.* dīlātātio; *dilatation; enlargement, diffuseness,* CH. C. T. *B* 232.

dile, sb., *O.E.* dile,=*O.H.G.* tilli; *dill,* '*anet(h)um,*' VOC. 140; PALL. iv. 167; dille PR. P. 121; diles (*gen.*) LEECHD. III. 92.

dilȝien, v., *O.E.* (a-)dilgian, dilegian,=*O.Fris.* (ur-)diligia, *O.L.G.* (far-)diligōn, *O.H.G.* tiligōn; *efface, extirpate, extinguish*; dilȝhen (*ms.* dillȝhenn) ORM. 4083; *comp.* a-, for-dilȝien.

diligence, sb., *O.Fr.* diligence, *Lat.* dīligentia; *diligence*: AYENB. 238; GOW. II. 61, 88; CH. C. T. *B* 1729; oon hath slewthe anothir diligence LIDG. M. P. 161.

diligent, adj., *O.Fr.* diligent, *from Lat.* dīligens; *diligent*, AYENB. 32, 220; GOW. II. 39; diligent travaile LIDG. M. P. 89.

   diligent-līche, adv., *diligently*, AYENB. 70; **diligentlier** (*comp.*) WICL. 3 KINGS iii. 21 (P.); 2 MAC. xi. 36.

dill, *see* dil.   **dille**, *see* dile.

dillen, v., ?=dullen; *render dull, become dull*: dille C. M. 202; TOWNL. 136; dillen (*pres. subj.*) REL. I. 217.

diluvie, sb., *Lat.* dīluvium, *cf. O.Fr.* diluve; *cf.* delūge; *deluge, flood*: the diluvie or greet flood WICL. 2 PET. ii. 5; diluvie WICL. GEN. vi. 17 *.

dim, adj., *O.E.* dimm,=*O.Fris.* dim, *O.N.* dimmr; *dim, obscure*, O. & N. 577; GEN. & EX. 286; AYENB. 159; MAND. 60; PR. C. 1166; PERC. 1994; dim and derk C. L. 71; þe dim bicom SHOR. 86; his dimme eiȝen FRAG. 5; dimme (*adv.*) GOW. II. 292; þauȝ i loke dimme LANGL. *B* x. 179; dimme (*pl.*) REL. I. 210 (MISC. 3).

   dim-līche, adv., *cf. O.E.* dimlīc (*adj.*); *dimly*: dimli A. P. iii. 374; A. R. 210.

   dimnesse, sb., *O.E.* dimness; *dimness,* '*obscuritas,*' FRAG. 7.

dīme, sb., *Fr.* dīme, *O.Fr.* disme, *from Lat.* decima; *tenth, tithe,* LIDG. M. P. 137; **dīmes** [dismes] (*pl.*) LANGL. *B* xv. 526.

diminuen, v., *Fr.* diminuer; *Lat.* dēminuere; *diminish, disparage*; diminued (*pple.*) ['*derogastis*'] or spoken ivel WICL. EZ. xxxv. 13.

diminutiōn, sb., *Fr.* diminution, *from Lat.* dēminūtio; *diminution*, GOW. III. 89.

dimmin, v., *O.E.* dimmian,=*O.N.* dimma; *dim,* '*obscuro,*' PR. P. 121; dimmen REL. I. 65; **dimmiþ** (*pres.*) P. L. S. vi. 7; him dimmeþ (þa) eiȝen FRAG. 5; **dimmede** (*pret.*) TREV. VII. 411; dimmede [dimmed] LANGL. *A* v. 200.

   dimminge, sb., *dimming*, RICH. 6977; bi (þe) dimminge of þe dai TOR. 514*.

dindelin, v., *cf. M.Du.* tintelen; *tinkle,* '*tinnio,*' PR. P. 121.

dine, dinen, *see* dūne, dūnien.

dīnen, v., *O.Fr.* disner; *dine*; dinen [dine] LANGL. *A* v. 58; **dīnede** (*pret.*) ROB. 558.

dīner, sb., *O.Fr.* disner; *dinner*, RICH. 655; LANGL. *B* vi. 293.

dinge, *see* dünge.

dingen, v., *cf. O.N.* dengja,=*M.H.G.* tengen, *prov.Eng.* ding; *beat, strike*, FER. 717; LANGL. *B* vi. 143; dinge HAV. 215; ALIS. 1732; PERC. 1967; **dang** (*pret.*) EGL. 550; wiþ his tail þe erþ(e) he dang IW. 3167; dong HAV. 1147; RICH. 5270; **dongen** (*pl.*) MAP 338; **dungen** (*pple.*) HAV. 227; PR. C. 3256; dongen MIN. vii. 74.

   dinging, sb., *blow, stroke*: dongun wiþ mani dingings (*pl.*) APOL. 37.

dingle, sb., ?=*O.H.G.* tunculla (*whirlpool*); *mod.Eng.* dingle; *? sea-bottom*: deopre þen eni sea dingle HOM. I. 263.

dint, *see* dünt.

diocesan, sb., *O.Fr.* diocesian, *med.Lat.* dioecēsānus; *diocesan, bishop*: yow be mi defensour in our diocesans (*gen.*) sight SACR. 854.

diocīse, sb., *O.Fr.* diocise, diocese; *diocese,* CH. C. T. *A* 664.

dippen, *see* düppen.   **dirk,** *see* dearc.

directe, v., *from Lat.* dīrectus; *direct*, LIDG. M. P. 247; **directe** (*pres.*) 149; directiþ 244.

dirige, sb., *Lat.* dīrige (*the first word of a funeral hymn*); *dirge*: a placebo and dirige for all the sowles of the bredern and susters E. G. 190, 191; dirige, office for ded men '*exequie*' PR. P. 121; deregi E. G. 145; say his dorge and masse LIDG. M. P. 111.

disalouen, v., *O.Fr.* desalouer; *disallow*: disalowed (*pple.*) LANGL. *B* xiv. 130.

   disalowing, sb., *disapproval*, LANGL. *B* xiv. 139.

disarmen, v., *O.Fr.* desarmer; *disarm*, CH. BOET. i. 4 (13).

dis-asenten, v., *disagree, refuse consent*; disasent (*pple.*) D. TROY 8016; dissaisent 9368.

disavauncen, v., *O.Fr.* desavancer; *hinder damage*: disavaunce CH. TRO. ii. 511.

disavauntāge, sb., *disadvantage, misfortune*: TREV. II. 161; **disadvauntāges** (*pl.*) III. 287.

disaventūre, sb., *misadventure, mischance,* CH. TRO. iv. 269, 725.

disc, *see* **disch.**

discerne, v., *O.Fr.* discerner; *discern,* CH.
C. T. *A* 3003.

disch, sb., *O.E.* disc (*board, table*), = *O.N.* diskr
(*dish*), *O.H.G.* tisc (*table*), *from Lat.* discus;
see **deske**; *dish,* JOS. 297; (*quoit*) WICL. 2
MAC. iv. 14; ibroken nep oðer disch A. R.
344; disc LAȝ. 19692; dish '*discus, scutela*'
PR. P. 122; **dische** (*dat.*) MAND. 52; disse
HORN (L.) 1144; **disches** (*pl.*) MISC. 175:
comp. **bær-disc.**

    disch-ful, sb., *dishful*: thre dischful of blod
S. S. (Web). 1906; disschfol 1918; disseful
1900; dissvol AYENB. 119; **dischfolles** (*pl.*)
S. S. (Web.) 1892.

    disc-þein, sb., *table servant,* FRAG. 4.

dischargen, *see* **deschargen.**

dischere, sb., *from* disch; *dishmaker, dish-
seller,* LANGL. *A* v. 166; *C* vii. 372.

    dissheres, sb., (*female*) *dishseller,* LANGL.
B v. 323.

dischevelen, v., *O.Fr.* descheveler; *dishevel*:
disheveled (*pple.*) CH. P. F. 235.

disciple, sb., *O.Fr.* disciple; *disciple,* P. S. L.
xvii. 390; **disciples** (*pl.*) HOM. I. 141; MISC.
40.

disciplesse, sb., *female disciple*: a disciplesse
(*v. r.* sum disciplisse) WICL. DEEDS ix. 36.

disciplīne, sb., *O.Fr.* disciplīne; *discipline,*
P. S. L. xvii. 114; PR. C. 5556.

disciplīne, v., *O.Fr.* disciplīner; *discipline,*
BEK. 2367.

disclandre, *see* **desclandren.**

disclōse, adj., *disclose,* GOW. I. 294; desclos
GOW. II. 354.

disclōsen, v., *disclose, reveal,* GOW. I. 294;
diclose LIDG. M. P. 26; **desclōsed** (*pple.*)
GOW. I. 157.

discomfiting, sb., *defeat,* CH. C. T. *A* 2721.

discomfitūre, sb., *discomfiture,* CH. C. T. *A*
1008; AN. 326.

discomfort, sb., *O.Fr.* descomfort; *discomfort;* TREV. I. 363; disconforte CH. TRO. iv.
283; discoumfort WICL. EZ. vii. 19.

discomforten, v., *O.Fr.* desconforter; *discomfort, defeat,* CH. C. T. *A* 2704; desconfite
MAN. (F.) 13944; descowmforten PR. P. 122;
disconforted (*pret.*) WICL. NUM. PROL.; **dis-
coumfortid** (*pple.*) JOSH. v. 1; disconfite
GOW. I. 362; discomfit III. 200.

disconsolat, adj., *disconsolate,* CH. TRO. v.
542; LIDG. M. P. 205.

disconvenience, sb., *discord, unsuitableness*:
LIDG. M. P. 82.

discordable, adj., *O.Fr.* descordable, *from
Lat.* discordābilis; *not in accord with, in-
compatible,* GOW. II. 225; elements that ben
so discordable CH. TRO. iii. 1704.

discordance, sb., *O.Fr.* descordance; *discordance,* AYENB. 259.

discordant, adj., *O.Fr.* discordant; *discordant,* GOW. III. 163.

discorde, sb., *O.Fr.* des-, discorde; *discord,*
PR. P. 122; discord AYENB. 43; descord ROB.
196; SHOR. 141.

discordin, v., *O.Fr.* descorder; *discord,* PR. P.
122; **discordeþ** (*pres.*) TREV. I. 123; discor-
dede (*pret.*) TREV. III. 97.

    discording, sb., *disagreement*: bituene hem
nas non discording ROB. 255.

discoveren, v., *O.Fr.* descouvrer; *discover*:
discurin PR. P. 122; H. V. 63; diskoverez
(*pres.*) GAW.418; discuretz E. G. 55; **dīscover**
(*imp.*) I. 305; **discoveringe** (*pple.*) WICL.
2 KINGS vi. 20; diskeveret (*pple.*) JOS. 350.

    discovering, sb.; discuringe of cownselle PR.
P. 122; descuvering WILL. 1024, 1044.

discovert, adj., *O.Fr.* descovert; *uncovered,*
DREAM 6.

discoverte, sb., *uncovered part*: Alisaunder
smot him in the discoverte ALIS. 7417.

discreciōn, sb., *O.Fr.* discretion; *discretion,*
AYENB. 155; discrecioun CH. C. T. *A* 2537.

discreet, adj., *O.Fr.* discret; *discreet,* CH. C.
T. *A* 312; LIDG. M. P. 155; discrete TREV.
V. 459; GOW. III. 167.

    discrēt-lī, adv., *discreetly,* WICL. S. W. III.
170.

discrēsen, v., *med.Lat.* discrescere; *decrease,*
LIDG. M. P. 244; GOW. II. 189.

discrien, *see* **descrien.**

discussen, v., *from Lat.* discussus (*pple.*);
mod.Eng. discuss; *examine, search into*:
Crist sal discusse alle thing PR. C. 6247.

disdainen, v., *O.F.* desdeigner, desdaigner,
from Lat. dēdignārī; *disdain*; disdeine LIDG.
M. P. 37; **desdaineþ** (*pres.*) GOW. III. 227;
disdaineth [disdeintith] R. R. 5688; disdeig-
neth GOW. I. 84; **dedeineden** (*pple.*) WICL.
MAT. xxi. 15.

disdainous, adj., *O.Fr.* desdeinous; *disdainful*; desdainous CH. TRO. ii. 1216.

[**dīse,** sb., *M.L.G.* dise (*flax prepared for
spinning*).]

    dīse-stæf, sb., *O.E.* distæf; *distaff,* '*colus*'
FRAG. 4; distaf VOC. 157; disestafe 269;
distaf CH. C. T. *A* 3774; **distaves** (*pl.*) TREV.
VIII. 211.

disdein, sb., *O.Fr.* desdaing, = *Ital.* disdegno;
disdain; disdein C. M. 11309; desdein [dis-
dain] CH. C. T. *G* 41; dedein ROB. 172; JOS.
244; MIRC 1159.

diséren, v., *cf. O.Fr.* deseriter; *disinherit*:
deséredin (*pret. pl.*) treu airs unriȝtfulli MISC.
211.

diseritē, sb., *O.Fr.* deserité; *cf.* **disheriten**;
a disinherited person: so þat þe **deseritēs**
(*pl.*) into þis lond come ROB. 85; þe kniȝtes
were diserites in þe lond aboute wide 563.

disēse, sb., *O.Fr.* desaise; *disease, trouble,*

GOW. I. 139; CH. C. T. *B* 3961; desese TREV. IV. 203; disease LIDG. M. P. 30; diseese WICL. 2 COR. i. 4*; **disēses** (*pl.*) TREV. II. 185; disesis WICL. S. W. III. 10.

disēse-ful, adj., *troublesome*, WICL. GEN. xxxix. 10; EX. xxiii. 9; this widowe is hevi, or diseseful, to me WICL. LK. xviii. 5.

disēse, v., *O.Fr.* desaisier; *trouble, annoy*, CH. TRO. iv. 1272; **disēseþ** (*pres.*) GOW. III. 3; **disēsed** (*pple.*) GOW. I. 89; disesid WICL. I KINGS xxviii. 15*.

disespeir, sb., *O.Fr.* desespeir; *despair*; in desespeire GOW. II. 125; dessespeir CH. TRO. i. 605.

disespeire, v.,=despeiren; *Fr.* désespérer; *despair, be without hope*, LIDG. M. P. 236; **disespeired** (*pple.*) of our owne offence LIDG. M. P. 179.

disesperat, adj., *hopeless*: disesperat [desperate] of alle blis CH. H. F. 2015.

disesperaunce, sb., *hopelessness*; desesperaunce CH. TRO. ii. 1307.

disfigūre, sb., *deformity*: schulde tellen of his disfigure CH. C. T. *D* 960.

disfiguren, v., *O.Fr.* desfigurer; *deform*; defigurd (*pple.*) PR. C. 2338.

disguīse, adj., *O.Fr.* desguisé; *fashionable*: thaire degise atire HAMP. PS. cxlvi. 11; degise CH. C. T. *I* 417; **degīse** (*sb.*) PR. C. 1518; **degīses** (*sb. pl.*) 1524.

disguīsen, v., *O.Fr.* desguīser; *disguise*; desguise GOW. II. 227; **desguīsede** (*pret.*) MAN. (F.) 4744; **disguīsed** (*pple.*) WILL. 1677.

disgīsing, sb., *covering*: the wrecchid swollen membres that thai schewe thurgh desgising [disguising] CH. C. T. *I* 425.

[**disguīsi**, adj.]

disgīsi-lī, adv., *strangely*: desparaged were i disgisili ʒif i dede in þis wis WILL. 485.

disgīsines, sb., *strange appearance*: precious clothing is coupable for his straungenes, and disgisines CH. C. T. *I* 414.

disheriten, v., *O.Fr.* desheriter; *disinherit*; desherite HAV. 2547; A. P. ii. 185; **ideserited** (*pret.*) ROB. 375; see **diseritē**.

dishonōr, *O.Fr.* deshonor; *dishonour*; deshonor FL. & BL. 655; MAN. (F.) 6242.

dishonouren, v., *dishonour*; **deshonoured** (*pple.*) of that she hadde be defloured GOW. II. 322; deshonoured 377.

dishoneste, adj., *O.Fr.* deshoneste; *dishonest*, CH. C. T. *H* 214.

dishonesten, v., *O.Fr.* deshonester; *dishonour*: whan thou hast **dishonestid** (*pple.*) thi frend WICL. PROV. xxv. 10.

disi, *see* **düsi**.

disjoint, sb., *O.Fr.* desjoinct; *dilemma, awkward position*: of what wight that stant in swich **disjointe** (*dat.*) CH. TRO. iii. 447; i stonde in this disjoint G. T. *B* 1601.

diskumfit, *see* **discomfite**.

dismaien, v., *cf. Span.* desmayar; *? for esmaien*; *dismay*; demaien JOS. 84; **desmaie** (*imper.*) ʒou no longer WILL. 3040; dismaie LIDG. M. P. 263; demai GAW. 470; **demaide** (*pret.*) ROB. 156; dismaid (*pple.*) GAW. 336.

disme, *see* **dime**.

dismembrin, v., *O.Fr.* des-, de-membrer; *dismember*, PR. P. 122; **demembred** (*pple.*) ROB. 559.

dismitte, v., *for Lat.* dīmittere; *dismiss*, WICL. JER. xxxii. 11; **desmittiden** (*pret. pl.*) DEEDS xvii. 10; **dismitted** (*pple.*) or relesid I MAC. x. 43.

disobedience, sb., *disobedience*, LIDG. M. P. 143.

disobeie, v., *O.Fr.* desobeir; *disobey*, GOW. I. 86; **disobeide** (*pret.*) I. 338.

disobeisaunce, sb., *O.Fr.* desobeisance; *disobedience*, GOW. I. 86.

disobeisaunt, adj., *O.Fr.* desobeissant; *disobedient*, CH. P. F. 428; GOW. I. 248; LIDG. M. P. 143.

disordéne, adj., *from O.Fr.* ordener; *inordinate, excessive*, AYENB. 34; desordene CH. BOET. ii. 2 (36); AYENB. 46; disordein CH. C. T. *I* 818.

desordéne-līche, adv., *inordinately*, AYENB. 55.

disordinat, adj., *immoderate, excessive*, CH. C. T. *I* 422; R. R. 4818; disordinat of language LIDG. M. P. 258.

disour, sb., *O.Fr.* disour; *speaker, teller of a story*, B. DISC. 139; **disours** (*pl.*) ALIS. 6991.

disparāge, sb., *want of equality*, CH. C. T. 908.

disparagen, v., *O.Fr.* desparagier; *disparage*; **disparagide** (*pret.*) TREV. VII. 385; **desparaged** (*pple.*) WILL. 485.

disparplin, v., *for* esparplin; *scatter*; disparplen PR. P. 122; disparple WICL. EZ. xii. 14; **disparplith** (*pres.*) JOHN x. 12; desparpleth MAND. 4; **disparpoilide** (*pret.*) GEN. xi. 9; **disparplid** (*pple.*) JOHN xvi. 32.

dispencer, sb., *O.Fr.* despensier; *steward*: Hue þe Dispencer ROB. 559.

dispenden, v., *O.Fr.* despendre; *spend, dissipate*; dispendin PR. P. 122; dispendi AYENB. 53; despende PR. C. 125; **despendeth** (*pres.*) GOW. I. 106; **dispended** (*pret.*) PR. C. 2435; despended LANGL. *B* xii. 49; despent GOW. II. 162; dispendede AYENB. 128; **dispended** (*pple.*) CH. C. T. *E* 1403; dispended, dispent GOW. I. 197, 309.

dispendour, sb., *steward*: desspendoure AYENB. 190; dispender [*later ver.* dispendere] WICL. LK. xii. 42; **dispenderis** (*pl.*) WICL. I COR. iv. 1; despendours CH. C. T. *B* 2843.

dispensacioun, sb., *O. Fr.* dispensacion;

dispensation, TREV. III. 469; dispensacioun wiþ þis lawe winnes miche money WICL. S. W. III. 162.

dispense, sb., *O.Fr.* despense; *expense, outlay*; dispence ALIS. 2616; despense AYENB. 21; **despenses** (*pl.*) CH. C. T. *B* 2842.

dispensin, v., *O.Fr.* despenser; *dispense,* PR. P. 122.

disperaciōn, *see* **desperaciōn.**

disperishen, v., *Lat.* disperīre; *disperse,* WICL. JUD. xi. 3; WISD. xvi. 29; **disperisht** (*pret.*) WICL. LAM. v. 18.

dispers, adj., *Lat.* dispersus; *dispersed,* GOW. II: 185; III. 175; **dispers** (*pple.*) II. 177.

displeien, v., *O.Fr.* despleier, -ploier; *display*; **displeied** (*pret.*) CH. C. T. *A* 966.

displesaunce, sb., *O.Fr.* desplesance; *displeasure*: CH. TRO. iii. 430; iii. 1245; LIDG. M. P. 71; displesaunce DREAM 1461.

displésen, v., *O.Fr.* desplaisir, *from Lat.* displicēre; *displease*, '*displiceo*,' PR. P. 123; WICL. EZ. xx. 43; displease LIDG. M. P. 169; **displése** (*pres.*) LANGL. *B* xiii. 135; **displésed** (*pret.*) LIDG. M. P. 91; **desplésed** (*pple.*) GOW. III. 173.

displésing, sb., *displeasure*, TREV. II. 411.

dispōnen, v., *Lat.* despondre, *Lat.* dispōnere; *dispose*; **dispōneþ** (*pres.*) CH. BOET. iv. 6 (134); TRO. iv. 935; **dispōne** (*imper.*) v. 300.

disport, sb., *O.Fr.* desport; *disport, sport*; disporte GOW. II. 175; desport CH. C. T. *A* 775; sport LUD. COV. 185; **disportes** (*pl.*) DREAM. 104.

disporte, v., *O.Fr.* desporter; *sport, divert,* GOW. I. 75, 119; CH. TRO. ii. 1673.

dispōsen, v. *O.Fr.* dispōser; *dispose*: LIDG. M. P. 149; **dispōsith** (*pres.*) LIDG. M. P. 197; **dispōsed** (*pret.*) LIDG. M. P. 20, 176; TREV. I. 109; **dispōsid** (*pple.*) WICL. S. W. III. 68, 161; disposed LIDG. M. P. 159.

dispositiōn, sb., *O.Fr.* disposition; *disposition*, CH. TRO. ii. 526; disposicion of Venus GOW. I. 50; after the disposition of natural complexion GOW. I. 92; dispocicioun LIDG. M. P. 174; **disposiciouns** (*pl.*) TREV. I. 51.

dispreisin, v., *O.Fr.* desprisier; *blame*, '*culpo, vitupero*,' PR. P. 123; dispreise LIDG. M. P. 84.

dispreisinge, sb., *blaming*, CH. C. T. *B* 2876.

dispréven, v., *disprove*: schul . . . nouȝt hem despreve by despitusoun TREV. I. 17; þis **dispréveð** (*pres.*) nouȝt Gaufrede his storie TREV. V. 339.

dispúten, v., *O.Fr.* des-, disputer; *dispute,* LANGL *B* viii. 20; dispute [dispuite] C. L. 1082; dispuite MIRC 673; desputi P. L. S. xix. 74; sputs KATH. 1315; spute JOS. 148; **des-pútede** (*pret.*) AYENB. 79.

dispúting, sb., *disputation, contention*: me þuncheð betere þa ha beo ear overcomen wið

desputinge KATH. 558; whether mi disputing is aȝens man WICL. JOB xxi. 4.

disputesoun, sb., *O.Fr.* desputoison; *disputation*, CH. C. T. *B* 4428; disputacioun LEG. 139; desputeson GOW. I. 90.

[**dis-rewli**, adj., *from* **rewle**; *unruly*.]

disrewli-li: lede his lif disrewlilie R. R. 4902.

disseisen, v., *O.Fr.* dessaisir; *disseize*; **disseised** (*pple.*) C. L. 1088.

disseisine, sb., *disseizin*, E. G. 361.

dissenciōn, sb., *O.Fr.* dissension, *Lat.* dissensio; *dissension*, PR. C. 4061; dessencioun WICL. S. W. III. 133; discencioun III. 141; dissenciun C. M. 22221, 22238.

dissenten, v., *from Lat.* dissentīre; *disagree*: if i dissente (*subj.*) LIDG. M. P. 44.

disseveraunce, sb., *O.Fr.* desseverance; *separation*, CH. TRO. iii. 1375; BOET. iii. 11 (96).

disseveren, v., *O.Fr.* dessevrer; *divide, sever*: i wolde him fro the court desever GOW. I. 332; **dissever** (*refl.*) GOW. II. 97; **deseverd** (*pple.*) MISC. 31; **dissevert** (*pret.*) D. TROY 1602; A. P. iii. 314.

dissimulatiōn, sb., *Fr.* dissimulation; *dissimulation, pretence*, GOW. I. 74.

dissimulen, v., *Lat.* dissimulāre; *dissimulate, hide, pretend*, CH. BOET. v. 6 (178): som can dissimele and blow the bukkis horn by apparence of feined kindenesse LIDG. M. P. 160; **dissimelide** (*pret.*) WICL. I KINGS x. 27; dissimilide I KINGS xxiii. 13.

dissimulinge, sb., *dissembling*, CH. TRO. v. 1625; **dissimulinges** (*pl.*) CH. C. T. *F* 285.

dissimulour, sb., *Lat.* dissimulātor; *dissembler*: CH. C. T. *B* 4418.

dissolucioun, sb., *Fr.* dissolution; *dissipation*, R. R. 4900; LIDG. M. P. 247.

dissolūt, adj., *Lat.* dissolūtus; *dissolute*: now passing besi, now dissolut, now idil LIDG. M. P. 245; dissolute laughters B. B. 26.

dissolven, v., *dissolve*: **dissolved** (*pple.*) S. A. L. 80; dissolvid WICL. S. W. III. 163; dissolved 68.

distaf, *see under* **dīse.**

distance, sb., *O.Fr.* distance; *distance*: '*distantia*' PR. P. 123; a distance (*disagreement*) there is ispronge BEK. 1285; þer was at londone a lute distance (*tumult*) ROB. 570; without destaunce RICH. 1669.

disteine, v., *O.Fr.* desteindre; *stain*, CH. L. G. W. 269; steine GOW. I. 225; **desteined** HOCCL. i. 340.

distempraunce, sb., *O.Fr.* destemprance; *unfavourableness*: the destempraunce of the hevéne CH. BOET. iii. 11 (97).

distemperen, v., *O.Fr.* destemprer; *disorder*: **distemperes** (*pres.*) WICL. S. W. III. 157; distemperin '*distempero*' PR. P. 123.

distempring, sb., *want of due proportions,*

disorder : þe destempringe of þise vour qualites AYENB. 153.

distemperūre, sb., *cf. O.Fr.* tempereure; *intemperance, want of moderation*, WICL. S. W. III. 156.

distemprē, adj., *intemperate* : CH. BOET. iv. 3 (121).

distillatiōn, sb., *O.Fr.* distillation, *Lat.* distillātio ; *act of distilling*, GOW. II. 86.

distillen, v., *O.Fr.* distiller, *from Lat.* distillāre ; *drop, flow* ; **distilleth** (*pres.*) GOW. I. 3 ; **distilling** (*pple.*) LIDG. M. P. 262.

[**distinct**, adj., *Lat.* distinctus ; *distinct.*]

distinct-lī, adv., *distinctly*, LIDG. M. P. 63.

distincte, v., *O.Fr.* distincter ; *distinguish*, R. R. 6199.

distincten, v., *from Lat.* distinctus ; *distinguish* : to distincti betvene þe guode þinges and þe kveade AYENB. 152.

distinctiūn, sb., *O.Fr.* distinction ; *distinction*, A. R. 12 ; distinccioun TREV. I. 111 ; CH. BOET. ii. 5 (46) ; **distinctiuns** (*pl.*) A. R. 12.

distinguen, v., *Fr.* distinguer, *from Lat.* distinguere ; *distinguish* : wilde men **destingeð** (*pres.*) nouȝt noþer to sette der feeldes by boundes noþer by meres TREV. I. 135 ; art þou **distingwed** (*pple.*) and embelised bi þe springing floures CH. BOET. ii. 5 (47).

distract, adj., sb., *Lat.* distractus ; *distracted* : distracte were thai stithli D. TROY 3219; distrauhte in thouhte LIDG. M. P. 206.

distracten, v., *med.Lat.* distractāre ; *distract* : bodeli þingis **distractiþ** (*pres.*) men WICL. S. W. III. 84.

distractioun, sb., *detraction, slander* : suffre no distractioun LIDG. M. P. 67.

distreinen, v., *O.Fr.* destraindre ; *constrain, compel* ; destrigni BEK. 711 ; **destreine** (*imp.*) CH. TRO. v. 596 ; **destrainede** (*pret.*) i. 354 ; distreigned BEK. 751.

distresse, sb., *O.Fr.* destrece ; *distress*, GOW. II. 268 ; RICH. 2763; distrès A. P. ii. 1159; LIDG. M. P. 76 ; distresce BEK. 755 ; destresse ROB. 442 ; GOW. I. 333 ; LIDG. M. P. 227.

distresen, v., *distress* : (þay) **distressed** (*pret.*) him wonder strait A. P. ii. 879.

disturbance, sb., *O.Fr.* destourbance ; *commotion* ; desturbance REL. I. 265 ; destourbance ROB. 436.

disturben, v., *O.Fr.* destourber ; *disturb* ; desturben A. R. 162 ; desturbi P. L. S. xvii. 360 ; **destorbeþ** (*pres.*) AYENB. 179.

disturbing, sb., *disturbing* ; destourbinge LEB. JES. 428 ; destorbinge AYENB. 225.

disturblen, v., *trouble* ; disturbelin ' *turbo, conturbo* ' PR. P. 123 ; disturble WICL. PS. ii. 5 ; **destorblist** (*pres.*) HALLIW. 307 ; **distourblen** (*pl.*) WICL. S. W. III. 122 ; **disturblinge** (*pple.*) WICL. DEEDS xvii. 13 ; **distourblede** (*pret.*) R. R. 1713 ; disturblid

(*pple.*) JOHN xxii. 27 ; distourbled S. W. III. 135 ; distroubled CH. D. BL. 522 ; distrobled WICL. MAT. xxiv. 6.

disturbling, sb., *troubling, disturbance* : fro disturbling of men WICL. PS. xxx. 21 ; distrobelinge of pece ' *disturbrum, turbacio* ' PR. P. 123 ; distourbling of witte WICL. S. W. III. 135 ; disturbling LK. xxiii..19.

disturbour, sb., *O.Fr.* destourbeur; *disturber* ; desturbour BEK. 1110.

disturnen, v., *O.Fr.* destourner, *Fr.* détourner ; *turn away* : thi fader prei al thilke harme desturne CH. TRO. iii. 669.

disūsen, v., *become unaccustomed, misuse,* ' *obsoleo, dissuesco, abutor,*' PR. P. 123 ; virtues of God may no mon disuse WICL. S. W. III. 157 ; a riche mon þat **disūside** (*pret.*) his richesse II. 1.

dis-were, sb., *cf.* **were** ; *doubt,* ' *dubium,*' PR. P. 123 ; withoute diswaire L. C. C. 51 ; withouten diswaire 25.

ditane, sb., *Fr.* dictame ; *dittany* (*a plant*), PR. P. 123.

ditē, sb., *O.Fr.* dité, dicté, *mod.Eng.* ditty ; *poem, verse; singing*, PR. P. 123 ; WICL. DEUT. xxxi. 19; **ditees** (*pl.*) TREV. IV. 309.

dīten, *see* **dihten**. **diþ**, *see* **dēað**.

dīten, v., *O.Fr.* dicter, *Lat.* dictāre ; *indite, indict* : ditin or inditen letters and speche ' *dicto* ' PR. P. 123 ; ditin or inditen for trespace ' *indicto* ' PR. P. 123.

dīting, sb., *composition,* ' *dictamen, indictatio,*' PR. P. 123 ; (he) in his diting . . . tellus D. TROY 3850 ; his diting of his dedis 7392.

dītour, sb., *historian, orator*, TREV. III. 163 ; advoketes and **dītoures** (*pl.*) II. 373.

dīven, *see* **dūven**.

diveren, v., = *M.Du.* daveren; *tremble*, KATH. 619 ; divere HOM. I. 283 ; speoken i ne dar nawt, ah diveri ant darie MARH. 16.

divers, adj., *O.Fr.* divers ; *divers*, SHOR. 142 ; AYENB. 15 ; C. M. 11054.

divers-līche, adv., *diversely*, TREV. V. 339 ; diversli pined PR. C. 7473 ; LIDG. M. P. 119.

diversitē, sb., *O.Fr.* diversité ; *diversity*, TREV. III. 467 ; CH. C. T. B 220.

diversorie, sb., *Lat.* dīversōrium ; *inn*, LEG. 90*.

divíden, v., *Lat.* dīvidere ; *divide* ; **divíden** (*pl.*) CH. ASTR. i. 7 (5) ; **divíde** (*imper.*) WICL. EX. xiv. 16 ; **divídeden** (*pret.*) WICL. PS. xxi. 19 ; **divídid** (*pple.*) WICL. 4 KINGS ii. 8 ; divided S. W. III. 130 ; CH. ASTR. i. 15 (8) ; GOW. I. 7.

dīvīn, sb., *O.Fr.* di-, devīn, *Lat.* dīvīnus ; *divine, augur* ; di-, **devīnes** (*pl.*) TREV. III. 219 ; þe devines and þe wichen AYENB. 19.

dīvīn², sb., *cf.* **divīnen**, **divīne** ; *prophecy;*

divination ; theology : Merlin, in his devin, of him has said MAN. (H.) 282 ; a maister of **divīne** (*dat.*) R. R. 6490.

divīne, adj., *Lat.* dīvīnus, cf. *O.Fr.* divīn, devin ; *divine,* LIDG. M. P. 95 ; devine 233.

[**divīnen**], **devīnen,** v., *O.Fr.* devīner, *Lat.* dīvīnāre ; *divine, prophecy ; interpret* : it is nought goode a sleping hounde to wake, ne yeve a wight a cause to devine CH. TRO. iii. 715 ; he wole devine soone LANGL. *B* xiii. 98 ; TREV. I. 159 ; IV. 385 ; Daniel of hire un-doinge **devīned** (*pret.*) and seide LANGL. *B* xv. 589.

> **divīninge,** sb., *divination,* TREV. III. 57 ; CH. BOET. v. 3 (157).

> [**divīneresse**], **devīneresse,** sb., *a female diviner* : thou wenest ben a grete devineresse CH. TRO. v. 1535.

divinitē, sb., *O.Fr.* divinité ; *divinity,* P. L. S. xvii. 237.

[**divīnour**], **devīnour,** sb., *O.Fr.* devineor ; *diviner,* WICL. NUM. xxii. 5.

divīsen, *see* devīsen.

divisiōn, sb., *O.Fr.* division ; *division,* CH. C. T. *A* 2476 ; GOW. I. 30 ; **divisiouns** (*pl.*) LIDG. M. P. 89 ; devisiouns CH. ASTR. i. 8 (6).

divors, sb., *O.Fr.* divorce, *Lat.* dīvortium ; *divorce,* TREV. I. 251 ; be ech man war þat he procure no fals devours WICL. S. W. III. 192 ; **divorces** (*pl.*) LANGL. *B* ii. 75.

dō, *see* dā.

dobben, *see* dubben. **doble,** *see* duble.

doc, dok, sb., *O.N.* dockr ; *tail,* GAW. 191.

docke, sb., *O.E.* docce ; *dock* (*plant*), CH. TRO. iv. 461 ; sour(e) dokke *sour dock,* ' *surel*(*l*)*e*,' rede dokke ' *parel*(*l*)*e* '. VOC. 162.

docken, v., *dock, cut short* ; dokkin ' *decurto* ' PR. P. 126 ; docke WICL. E. W. 430 ; his top was docked [dokked] CH. C. T. *A* 590.

doctour, sb., *O.Fr.* doctur ; *doctor,* CH. C. T. *A* 411.

doctrīne, sb., *O.Fr.* doctrine ; *doctrine,* MAND. 167 ; CH. ASTR. PROL. (2).

dodden, v., *shave, clip* : ȝe shulen dodde [' *attondebitis* '] WICL. LEV. xix. 27 ; **doddeþ** (*pres.*) P. S. 192 ; **doddid** (*pple.*) PR. P. 125 ; idodded A. R. 422.

> **doddunge,** sb., *shaving,* A.R. 14.

doder, sb.,= *M.H.G.* toter ; *dodder* (*a plant*), REL. I. 37 ; ' *cuscuta* ' VOC. (W.W.) 577.

doel, sb., *O.Fr.* doel, duel, deul, *from Lat.* (cor-)dolium, cf. *prov.Eng. and Sc.* dool, dule ; *grief, mourning,* ALIS. 5121 ; MAND. 202 ; deol LANGL. *A* v. 216 (doel *B* v. 386) ; doel, dol, duel, dul, del WILL. 564, 781, 1510, 1909, 2757 ; doel [duil, deol, deil] WICL. 2 KINGS xiv. 2 ; duel SPEC. 34 ; deol HORN (L.) 1048 ; ROB. 166 ; dool DEGR. 1736.

> **deol-ful,** adj., *doleful,* LAȝ. 6901* ; dul-, delful GAW. 560, 1517.

dulful-līche, adv., *dolefully,* P. L. S. xxiv. 219.

dōȝ, *see* dāȝ.

dogerel, adj., *doggerel* : rim dogerel CH. C. T. *B* 2115.

doȝeþe, *see* duȝeðe.

dogge, sb., *O.E.* docga, = *M.Du.* dogghe ; *dog,* ' *canis,*' VOC. 187 ; PR. P. 125 ; A. R. 290 ; ROB. 69 ; WILL. 46 ; MAND. 66 ; CH. C. T. *E* 2014 ; **doggen** (*pl.*) P. S. 239 ; *comp.* **band-, helle-dogge.**

> **doggid,** adj., *dogged,* P. S. 199 ; so dogget (*base*) a dede D. TROY 10379.

> **dogged-līche,** adv., *doggedly,* FER. 1289 ; doggetli (*meanly*) had done D. TROY 13071.

doggisch, adj., *doggish* ; **doggische** (*dat. m.*) WICL. ECCLUS. xii. 22*.

dohter, sb., *O.E.* dohtor,= *O.L.G.* dohtar dohtor, dohter, *O.Fris.* dochter, *O.N.* dottir, *Goth.* dauhtar, *O.H.G.* tohter, *Gr.* θυγάτηρ ; *daughter,* KATH. 74 ; (*ms.* dohhterr) ORM. 128 ; dohter, doȝter, douter LAȝ. 142, 211, 3018 ; doȝter AYENB. 26 ; dogter GEN. & EX. 1646 ; doghter PR. C. 2130 ; doghtir EGL. 58 ; douhter FRAG. 8 ; A. R. 54 ; douhter, douȝter S. L. 301, 347 ; douȝter BEK. 22 ; dowter GEN. & EX. 1847 ; doughter, doutir, doster PR. P. 129 ; doȝter, doster HORN (L.) 249, 697 ; douter AN. LIT. 10 ; doftir (? dostir) B. DISC. 689 ; douȝter (*gen.*) WILL. 3152 ; doster [dohter] LAȝ. 2982 ; his dohtres ræd SAX. CHR. 254 ; **dohter,** dochter, docter (*dat.*) LAȝ. 3368, 3373, 3433 ; douter REL. I. 184 ; **dohter** (*pl.*) LK. xxiii. 28 ; dohtere, dohtren [dohtres] LAȝ. 2924, 24509 ; douhtren, douȝtren C. L. 289 ; dohtren (? dehtren) H. M. 19 ; dehtren HOM. I. 247 ; CHR. E. 545 ; deȝter A. P. ii. 270 ; dohtres ORM. 6383 ; douhtres HAV. 350 ; **dohter,** dohtrenne [dohterne] (*gen. pl.*) LAȝ. 2689, 2701 ; **dohtren** (*dat. pl.*) LK. i. 5 ; LAȝ. 3085 ; *comp.* **god-, steop-dohter.**

> **doughter-lēs,** adj., GOW. III. 305.

dohti, *see* duhtiȝ.

dok, *see* doc. **doke,** *see* dūke.

doket, sb., *? from* doc ; *? rag, clout,* TOWNL. 313.

doket [2], sb., *O.Fr.* ducat ; *ducat* (*a coin*) ; do-kettis (*pl.*) SACR. 315.

dol, *see* dul *and* doel. **dōl,** *see* dǣl.

dolc, sb., *ulcer, ? wound* ; dolc sor and blein GEN. & EX. 3027 ; wiðute knotte and dolke A. R. 2.

dōle, *see* dāle.

dolfin, sb., *O.Fr.* ? doulphin, cf. *Prov.* dalfin, *from Lat.* delphīnus ; *dolphin,* DEGR. 1038.

[**dolken** ? *comp.* **for-dolken.**]

dollen, *see* dullen.

dolour, sb., *O.Fr.* dolour, dolur ; *grief,* MISC. 212 ; S. S. (Web.) 1270 ; FER. 2175 ; dolur LEG. 72.

dōm, sb., _O.E._ dōm,=_O.L.G._, _O.Fris._ dōm, _O.N._ dōmr, _Goth._ dōms, _O.H.G._ tuom, _mod. Eng._ doom; _judgment, decision, decree; authority_, ORM. 1472; C. L. 491; HAV. 2487; WILL. 1220; WICL. GEN. xviii. 19; þe dom ... ne lest he nawiht longe HOM. I. 169; þat rihtne dom us ʒive wolde O. & N 1692; þen dom ROB. 337; þane laste dom AYENB. 74; dōmes (_gen._) O. & N. 1695; **dōmes dæi** LAʒ. 24274; domes dai R. S. iv; **dōmes man** _doomsman_ (_judge_) A. R. 156; LANGL. _B_ xix. 302; TREV. IV. 49; **dōme** (_dat._) O. & N. 179; mid unimete dome heo ferden ut of Rome LAʒ. 9228; imong þissen dome dæd iwarð þe gode king 15064; þe dai of dome PR. C. 1859; **dōmes** (_pl._) HOM. I. 103; LAʒ. 7221; þah þine domes derne beon MARH. 8; _comp._ **alder-, cristen-, eorl-, frēo-, hæðen-, hāli-, kine-, king-, martir-, pāpe-, rīche-, þēow-, wis-dōm**; _deriv._ **dēmen.**

dōm-hūs, sb., _judgment-hall_: dome-howse '_pretorium_' PR. P. 126.

dōm-kēte, adj., _swift of judgment_, HOM. II. 83.

dōm-place, sb., _place of judgment, 'forum'_: he disputide in the sinagoge ... and ... in the chepinge or domplace WICL. DEEDS xvii. 17; xvi. 19.

dōm-seotel, sb., _O.E._ dōmsetl; _judgment seat_, JUL. 55.

dōm-stōl, sb., _judgment-seat_, A. R. 306.

domb, _see_ **dumb.**

dominacioun, sb., _Fr._ domination; _domination_, CH. C. T. _B_ 3409.

domlen, v., _be dull or cloudy_; domland (_pple._) PR. C. 1443.

dompen, _see_ **dumpen. domping**, _see_ **dopping. don**, _see_ **dun.**

dōn, v., _O.E._ dōn,=_O.L.G._ dōn, doan, duan, duon, _O.Fris._ dua, _O.H.G._ tuon; _do_, KATH. 784; A. R. 16; MAND. 3; þe sculde þas ernde dōn LAʒ. 1421; don on _don, put on_ 1701; harm don HOM. I. 213; don hire schome H. M..17; don t(o) understandin ORM. 3067; don (_commit_) sunne MISC. 72; to deþe don (_put_) CHR. E. 908; ALIS. 730; he scholde ... in strong warde him do (_put_) BEK. 379; do (_commit_) þeof þe 396; schame do 1488; he ssel do a wai þe brondes AYENB. 205; do bote WILL. 1378; shal do (_make_) þe noʒt drede LANGL. _B_xx. 153; þei schule of do her cloþes TREV. IV. 1; to donne KATH. 2201; A. R. 6; ORM. 2949; what him weore to donne [donde] LAʒ. 4769; to done BEK. 248; i have to done JOS. 161; what is best to done CH. C. T. _A_ 3544; to donde MISC. 193; dō (_pres._) A. R. 200; ich do þe wel to witene LAʒ. 3163; dest LAʒ. 13154; KATH. 754; A. R. 124; SPEC. 104; SHOR. 96; AYENB. 129; þat fruit þat þou dest bere LEG. 6; dest, dost O. & N. 49, 237, 321; dost ORM. 5258; deð A. R. 174; deþ HOM. II. 27; FRAG.

7; P. L. S. viii. 11; ROB. 321; AYENB. 5; deð [doþ] LAʒ. 674; deþ, doþ O. & N. 156, 564; he deþ [doþ] me gret wo C. L. 899; doþ ORM. 1042; bet heo heolden heore wurðing dei þene we doð HOM. I. 9; we doð sunne 55; ʒe doð LAʒ. 1434; heo doð A. R. 328; gret harm heo me doþ C. L. 896; ne do þu þeofðe HOM. I. 13; do sei me JUL. 41; do nim þe wreche R. S. ii. (MISC. 162); do of ðin son GEN. & EX. 2781; don (=do on) þin helm FER. 460; dof(=do of) WILL. 2343; do (_make_) me love þe so SPEC. 71; do out all þoʒtes AYENB. 210; do a wei þi maumetes JOS. 102; do me sikernesse þer to 623; qvoth Ubbe doth him swiþe fete HAV. 2037; þat ich do ham wurðschipe KATH. 506; **dōende** (_pple._) MAT. xxiv. 46; doinde AYENB. 194; **dude** (_pret._) A. R. 102; O. & N. 1637; BEK. 1050; he dude him on þe wei P. L. S. xii. 119; he dude to dethe WILL. 3427; dude heom to understonde MISC. 86; dude [dede] LAʒ. 431; dude [dide] KATH. 1552; dide [dede] CH. C. T. _A_ 2262; dede AYENB. 78; ISUM. 67; dede mankinde bote GEN. & EX. 24; god dede (_caused_) ðat he on swevene cam 224; alle þe Englis dede he swere HAV. 254; he ... dede him on gate WILL. 1119; dudest A. R. 306; deodest [dudest] LAʒ. 2294; duden A. R. 330; and duden hine i benden (_r._ bende) LAʒ. 31011; duden of claðes 16759; heore brugge heo duden a dun 19242; diden ORM. 429; and hi swa diden SAX. CHR. 250; þei diden dedli sinne LANGL. _B_xiv. 78; he deden hem binden GEN. & EX. 2193; we þin heste dude forleten H. H. 169 (171); **dōn** (_pple._) ORM. 237; K. T. 813; he ben don ut of paradis GEN. & EX. 381; don [doon] CH. C. T. _A_ 2092; _comp._ **bi-, for-, ʒe-, tō-, un-dōn**; _deriv._ **dǣde.**

[**dōere**, sb., _doer; comp._ **gōd-, mis-, üvel-dōere.**]

dōinge, sb., _doing_, TREV. I. 27.

donat, sb., _Lat._ Dōnātus; _a grammar; elementary book of instruction_: Donat and Dindimus GOW. II. 90; PR. P. 126; donet S. S. (Web.) 179; LANGL. _B_ v. 209.

donge, _see_ **dunge.**

dongeōn, donjoun, sb., _O.Fr._ dongon, donjon; _dungeon_, CH. C. T. _A_ 1057; dungun, dongeon LANGL. _A_ PROL. 15, _B_ PROL. 15; donguon FER. 2301; A. P. ii. 1224.

donken, v., _moisten_; donkeþ (_pres._) SPEC. 44; donkande (_pple._) GAW 519; **donkit** (_pret._) D. TROY 9639.

donkinge, sb., _moisture_, D. ARTH. 3249.

doppar, sb.; _cf._ **doupar**; _didapper_, '_mergulus_,' PR. P. 127.

doppe, sb., ? _O.E._ (dūfe-)doppa; ? _sea gull_, ALIS. 5776; _comp._ **dēve-doppe.**

dopping, sb., _the dabchick_; doppinges (_v. r._ dompinges) (_pl._) LANGL. _C_ xiv. 169.

dorc, _see_ **dearc.**

dord, sb., *?whey* : **dordus** (*pl.*) '*serum*' VOC. 178.

dore, sb., *O.E.* dora, *?=Du.* tor ; *dor* (*beetle*) ; **dorren** (*pl.*) AR. & MER. 6428.

dore, *see* **dure.**

Dorkcestre, sb., *O.E.* Dorca ceaster ; *Dorchester*, P. L. S. xiii. 36 ; Dorchestre ROB. 2.

Dorchestre-sēten, sb., pl., *inhabitants of Dorchester*, LAȝ. 29615.

dorlot, sb., *O.Fr.* dorelot, dorlot ; *ornament for a woman's dress* ; dorlott '*trica, caliendrum*' PR. P. 127 ; vair dorilot AYENB. 177.

dormant, adj., *fixed*: his table dormant in his halle alway stood redi covered al the long dai CH. C. T. *A* 353 ; table dormounte HALLIW. 311.

dormous, sb., *dormouse*, '*glis*,' VOC. 220 ; PR. P. 127.

dornen, v., *deceive* : þer husbondes þei wil so dorne S. & C. II. lvii.

Dorsēte, pr. n., *O.E.* Dorsǣte ; **Dorsǣtan**, Dorsǣtum (*dat.*); *Dorset*, LAȝ. 21014; O. & N. 1753.

dortour, sb., *O.Fr.* dortoir ; *dormitory*, PL. CR. 211 ; CH. C. T. *D* 1855.

dosc, *see* **dusc.**

doseper, sb., *O.Fr.* douze pairs (*pl.*) ; *one of the twelve peers ; champion* : ferst they sent out a doseper in blake armes OCTOV. (W.) 949 ; RICHD. 9 ; inne Franse weren italde twelfe iferan, þe Freinsce heo cleopeden **dusze pers** (*pl.*) (*sec. text* dosseperes) LAȝ. 1623 ; dosse pers of France ROB. 188 ; dozepers HALLIW. 315 ; duze pers MAN. (H.) 81 ; duseperis ANT. AR. 22 ; D. ARTH. 66 ; dusseperez 2029.

dosil, sb., *O.Fr.* dosil, dousil, dusil ; *spigot, faucet* ; dosiile VOC. 198 ; dosele [doseil] S. S. (Web.) 1150 ; **dosils** (*pl.*) ROB. 542.

dosser, sb., *Fr.* dossier ; *pannier, basket*; docer, dorcer PR. P. 125, 127 ; doser GAW. 478 ; **dossers** (*pl.*) CH. H. F. 1940.

dotāge, sb., *dotage*, CH. C. T. *A* 3898 ; A. P. ii. 1425.

dotard, sb., *dotard*, CH. C. T. *D* 331 ; do-**tardes** (*pl.*) PL. CR. 825.

dóte, adj., *doting, foolish*, REL. I. 184 ; TRIST. 1912 ; BEV. 217 ; doote YORK xxxv. 5.

dotel, sb.,=**dotard** : þenne þe dotel on dece drank þat he miȝt A. P. ii. 1517.

doteren, v.,*?* = **totiren** ; *totter* ; dotur AV. ARTH. xvi ; þe duc **dotered** (*pret.*) to þe ground DEGR. 1109.

dótien, v.,=*M.Du.* doten *O.Fr.* (re-)doter ; *dote* ; dotie LAȝ. 3294 ; doton '*deliro, desipio*' PR. P. 128; doten [dote] CH. C. T. *G* 983 ; dote LUD. COV. 98 ; AV. ARTH. xvi ; **dótest** (*pres.*) LANGL. *A* i. 129 ; dotes tu KATH. 2111 ; he dotes PR. C. 785 ; þei doteþ TREV. IV. 403 ; **dótie** (*subj*) A. R. 224 ; he

dared as **dóted** (*pple.*) man WILL. 4055 ; doted and dased L. H. R. 216 ; *comp.* a-**dótien**.

dottipol, sb., *cf.* **dodden** *and* **pol** ; *baldhead* : fi, dottipols, with youre bookes TOWNL. 145.

dōu, *see* **dāȝ.**

douaire, sb., *O.Fr.* doaire ; *dower*, CH. C. T. *E* 848 ; dowarie E. G. 361 ; dowere TREV. IV. 73 ; dower WICL. EX. xxii. 16.

douce, adj., *O.Fr.* dols, doux, *Lat.* dulcis ; *sweet* : he drawes into dowce Fraunce D. ARTH. 1251.

doucen, v., *from* **douce** ; *sweeten* ; dowce L. C. C. 7 ; PR. P. 129.

doucet, adj., *O.Fr.* doucet ; *sweet* : doucet drinkes HALLIW. 313 ; dowcet mete '*dulceum*' PR. P. 128.

doucette, sb., *a sweet pie or custard* ; dou-**cettes** (*pl.*) B. B. 60 ; dowcetes 148.

doucette[2], sb., *a musical instrument* : to pipe both in doucet and in riede CH. H. F. 1220.

doude, sb., *slut*, MAN. (F.) 11255 ; if she be never so foule a dowde TOWNL. 312.

douen, v., *O.Fr.* douer, doer ; *endow* ; dowe WICL. EX. xxii. 16 ; dowe [douwe] LANGL. *C* iv. 322 ; **dūed** (*pret.*) PL. CR. 1528.

dowinge, sb., *endowment*, WICL. S. W. III. 171 ; TREV. III. 37.

douȝter, douhter, *see* **dohter.**

douȝti, douhti, *see* **duhtiȝ.**

douken, *see* **dūken.**

doul, *see* **duvel.**

doun, *see* **dūn.**

dounen, v., *? O.N.* deyna ; *smell* : swet to downen (*ms.* dowwnenn) ORM. 6745.

doupar, sb. ; *cf.* **doppar** *and* **dopping** ; *the dabchick* : dowpar, brid [dooper, H.] '*mergus*' PR. P. 129.

douren, v.,=*L.Ger.* dūren, *M.H.G.* tūren ; *?grieve* ; **doured** (*pret.*) A. P. iii. 372.

douse, sb.,=*? M.Du.* duise (*concubina*) ; *harlot*, SPEC. 111 ; dowse TOWNL. 104.

dousour, sb., *O.Fr.* dulçor, douçor ; *sweetness* : now for singlerti o hir dousour we call her fenix of Arrabi A. P. i. 429.

doust, *see* **dust.   doute,** *see* **dūte.**

douve, *see* **dūve.**

douvre, sb., *O.Fr.* douvre (*fossé, canal*) ; *rabbit-warren* ; douwere '*cuniculus*' PR. P. 128.

Dóvere, pr. n., *? O.E.* Dofere (*dat.* Doferan) ; *Dover*, HAV. 139 ; MAN. (F.) 4335 ; Dovere, Dovre LAȝ. 7415, 8583.

dowen, *see* **douen** *and* **duȝen.**

dowrie, sb.,=**douaire** ; *dowry* : disherite & deprive dowrie of widoes A. P. ii. 185 ; MAN. (H.) 151 ; doure PR. P. 128.

dozeine, sb., *O.Fr.* dozaine ; *dozen*, LANGL. *B* iv. 37 ; CH. C. T. *A* 578.

drabelen, *see* **dravelen.**

drǣd, sb., *dread, terror*; dred, drede A. R.
6, 364; i þon castle wes muchel dred [drede]
LA3. 1682; heo hefden muchele drede
2088; he hadde of water dred GEN. & EX.
660; drede HOM. II. 71; MISC. 193; HAV.
828; ROB. 337; AYENB. 32; LANGL. *B* xvi.
85; CH. C. T. *B* 680; RICH. 5515; PR. C.
5263; hwone hit is alre mest on drede O. & N.
684; drade ALIS. 5740.

   drēd-ful, adj., *dreadful*, A. R. 302; C. L.
454; H. H. 200 (202); dredvol AYENB. 14.

   drēdful-lī, adv., *dreadfully*, LANGL. *B*
xvii. 62.

   drēd-lēs, adj., *dreadless*, GAW. 2334; PL. CR.
524.

   drēd-lich, adj., *terrible*, A. R. 58; **drēdliche**
(*adv.*) HOM. I. 14.

   drēdnesse, sb., *terror*, HOM. I. 235; JUL.
69.

drǣden, v., *O.E.* drēden, (on-)drǣdan (*pret.*
drēd),=*O.L.G.* (and-)drādan (*pret.* drēd), *O.
H.G.* (in-)trātan (*pret.* triat, triet); *dread,* ORM.
1218, 5907; dreden HOM. I. 21; LA3. 31164;
A. R. 136; drede GOW. I. 344; LANGL. *B* xx.
153; K. T. 429; **drēdest** (*pres.*) JUL. 24; dredeþ
ORM. 6179; P. 47; dret MISC. 95; AYENB.
26; (hi) dredeþ AYENB. 74; ne dred þe nou3t
LEG. 81; ne drede 3e ow nawiht KATH. 1403;
drĕdde (*pret.*) ORM. 19965; GEN. & EX. 767;
SHOR. 49; dradde BEK. 127; GOW. I. 341;
comp. **an-, for-, of-drǣden.**

   drǣdung, sb., *dreading*, ORM. 5602.

drǣdi, adj., *timid*; dredi GEN. & EX. 872;
WICL. JUDG. vii. 3; L. H. R. 140.

drǣfen, v., *O.E.* drǣfan,=*O.N.* dreifa, *Goth.*
draibjan, *O.H.G.* treiben, *Dorset dial.* dreve;
from **drīfen**; *drive*; **drēved** (*pret.*) A. P. i.
979; *comp.* **a-, tō-drǣfen.**

drǣm, *see* **drēam.**

draf, sb., *cf. M.L.G., M.Du., Swed.* draf,
m. n. O.H.G. pl. treber; *draff, dregs,* 'se-
gisterium,' VOC. 233; LA3. 29256; draffe
LANGL. *B* x. 11; draf CH. C. T. *I* 35; þet draf
AYENB. 93; drof GEN. & EX. 3582; *comp.*
wīn-draf.

   draf-sak, sb., *sack of dregs*, CH. C. T. *A*
4206.

drāf, sb., *O.E.* drāf,=*M.H.G.* treip; *from*
drīfen; *drove, driving*: it hiled al ðis werldes
drof GEN. & EX. 102; wiþ duntes drof him al
todraf L. H. R. 141; **drōve** (*dat.*) WICL.
GEN. xviii. 7; drove of bestis WILL. 181.

draft, sb.,=**draf**; *dregs*, WICL. PS. xxxix. 3.

dra3-, *stem of* **dra3en,** v.

   draw(e)-brügge, sb., *drawbridge,* ALIS.
1205; drawebrigge D. ARTH. 2474.

dragē, draggē, sb., *O.Fr.* dragee; *meslin,
mixed corn,* 'mixtilio,' PR. P. 130; dragge
'dragetum' (*? comfit, sweetmeat*) VOC. 178;

? **draggēs** (*pl.*) *? medicated sweetmeats*, CH.
C. T. *A* 426 [6-*text has* drogges; *see* **drogge**].

dra3en, v., *O.E.* dragan, = *O.L.G., Goth.*
dragan, *O.N.* draga, *O.Fris.* draga, drega,
M.L.G. dragen, dregen, *O.H.G.* tragan;
draw, carry, LA3. 10530; for to dra3en hine
ut of þisse wutte HOM. I. 47; dragen HOM.
II. 29; GEN. & EX. 2378; dra3hen ANGL. I.
9; nile i noht dra3hen upon me . . . þat
maht ORM. 18406; drahen, draien, drehen, dreien KATH. 33, 1193, 2237; dreaien,
dreien, drahen HOM. I. 257; dra3e AYENB.
12; to schupe we mote dra3e HORN (L.)
1420; draie TOWNL. 106; drahe, drawe
O. & N. 1375; drawen A. R. 160; drawe P. S.
153; CH. C. T. *A* 1416; WICL. JOHN iv. 7;
drawe his her P. L. S. xix. 266; qveþer i wit
riht and lawe mai him wit me til helle drawe
M. H. 56; **dra3e** (*pres.*) O. & N. 970; (hi)
drageþ, dregeþ MISC. 150, 151; swa drieðð
his erme saule in eche pine HOM. I. 27; **drah**
(*imper.*) HOM. II. 149; **drōh** (*pret.*) MARH.
2; ORM. 769; þes duc . . . to þare sæ him
droh LA3. 93; droh (*was obedient*) to his
ræde 9527; purpre and pal he droh REL. I.
119; dro3 FL. & BL. 683; GAW. 1188; oþer
folk þider dro3 ROB. 41; drog, drug GEN. &
EX. 1746, 2717; drogh PR. C. 2249; ISUM.
364; AMAD. (R.) xxvi; drogh [drou3, drou]
LANGL. *B* xiv. 106; dro3, drou3, drou, dreu3
WICL. EX. iv. 7, JUDG. v. 21, IS. x. 13; drouh
A. R. 102; he drouh hine a bak MISC. 43;
drou3 . . . his swerd MARG. 309; to falshede
evere he drou3 LEB. JES. 772; drou3, drou
WILL. 1068, 2208; drou HAV. 719; EGL.
316; hu þu deað drohe KATH. 2467; þu . . .
drowe me to þe FRAG. 8; dro3en LA3. 818;
GAW. 1463; dro3hen ORM. 8704; drohen
KATH. 2155; drowen A. R. 110; drowen
LANGL. *B* xi. 330; þe ministris . . . þat drowen
þe water WICL. JOHN ii. 9; **dra3hen** (*pple.*)
ORM. 7413; ut of latin ðis song is dragen
GEN. & EX. 13; drain TRIST. 706; drauhen
MAN. (H.) 183; drawen HAV. 1925; MAND.
40; þar for þis buk es on Ingles drawen PR. C.
336; *comp.* **a-, 3e-, tō-, wið-dra3en.**

   drawere, sb., *drawer,* WICL. S. W. I. 193.

   drawinge, sb., *drawing,* P. L. S. xxi. 133.

draggē, *see* **dragē.**

draggin, v., *cf. Swed.* dragga; *drag,* PR. P. 130.

dragme, sb., *O.Fr.* drame, dragme, *from Lat.*
drachma, *Gr.* δραχμή; *dram (weight, measure,
coin),* WICL. LK. xv. 8; drame PR. P.
130; three hundrid **dragmes** (*pl.*) of silver
WICL. 2 MAC. iv. 19.

dragonce, sb., *Lat.* dracontium; *serpentine
(herb),* REL. I. 301; dragaunce PR. P. 130;
dragauns VOC. 256; dragans HALLIW. 315.

dragonet, sb., *from* **dragoun**; *a young dragon*: that signifieth the dragonet ALIS. 602.

dragoun, sb., *O.Fr.* dragon; *dragon,* MARG.
158; dragun GEN. & EX. 2924.

draȝþe, sb., ?=**draht,** AYENB. 251.

drahen, see **draȝen.**

draht, sb.,=*M.H.G.* dracht *f., O.N.* drāttr *m.; from* **draȝen**; *draught, pull; load; current of air;* drinkeð a draht HOM. II. 199; to þe is al mi draucht 256; agen he maden here dragt GEN. & EX. 3745; draght [draȝt] C. M. 21266; i woldę have drawe the same draught (*move in chess*) CH. D. BL. 682; a draught [drauht, drauȝt] of win C. T. *A* 396; draut (*ms.* drawte) PR. P. 131; a þan vorme drahte swið(e) monie he ilahte LAȝ. 29259; **drahtes** (*pl.*) P. S. 153; drauȝtes as me draweþ in poudre P. L. S. xvii. 225; drawe draghtes and (bere) berþenes HALLIW. 315.

     drauht-brigge, sb., *draw-bridge,* MAN. (H.) 183; drautbrigge PR. P. 131; draȝt-brigge FER. 1113.

     draut-welle, sb., *draw-well,* PR. P. 131.

drai, see **derai.**

dráke, sb., *O.E.* draca, *from Lat.* draco; *dragon,* ORM. 1843; R. S. v; GEN. & EX. 283; K. T. 408; M. ARTH. 2607; **dráken** (*gen.*) LAȝ. 17876; þe drake heaved A. R. 246; **dráken** (*pl.*) LAȝ. 15935; see **dragoun.**

dráke², sb.,=*L.G.* drake, *M.L.G.* antdrake, *O.H.G.* antrache; *drake,* HAV. 1241; CH. C. T. *A* 3576; **drákes** (*pl.*) SPEC. 44; *comp.* **schel-dráke.**

dráne, sb., *O.E.* dran, drǣn, *cf. O.L.G.* *dran (*pl.* drani), *N.L.G.* drone *f., O.H.G.* treno *m.; drone, 'fucus,'* PR. P. 130; VOC. 177; LIDG. M. P. 154; þa drane SAX. CHR. 256; dro(ne) FRAG. 3; **dránes** (*pl.*) PL. CR. 726.

dráper, sb., *O.Fr.* drapier; *draper,* LANGL. *C* vii. 250.

draperie, sb., *O.Fr.* draperie; *drapery,* REL. II. 175; (P. L. S. xxxiv. 11).

drappen, v.,=**droppen**; *drop, sink:* neigh to dede we gan drappe OCTOV. (W.) 567.

draste, drasti, see **dreste, dresti.**

drauȝt, drauht, see **draht.**

drauk, sb.,=*M.Du.* dravick; *weed,'zizanium,'* VOC. 265; PR. P. 130; REL. II. 80; drauc (*printed* dranc) M. H. 152.

dravelen, v.,=*L.G.* drabeln; *drivel, slaver, talk foolishly:* þei don but dravele þer on LANGL. *A* xi. 11; [drevele, drivele *B* x. 11]; drinken and drivelen [dravelen] *B* x. 41 (W. 5683); drabelin PR. P. 129; **draveled** (*pret.*) GAW. 1750; **draveled** (*pple.*) P. 55; *comp.* **bi-dravelen.**

drawebrügge, see under **draȝ-.**

drawen, see **draȝen.**

drēam, sb., *O.E.* drēam (*song, delight*),=*O.L.G.* drōm (*dream, feasting*), *O.Fris.* drām, *O.N.* draumr, *O.H.G.* traum, troum (*dream*), *mod.Eng.* dream; *sound, music; dream,* KATH. 1498; A. R. 210; þu iherdest þene dream FRAG. 7; englene dream H. M. 19; dream, drǣm, drem LAȝ. 1010, 14286, 24554;

---

drǣm ORM. 923; drem BEV. 1339; ALÍS.² (Sk.) 781; AL. (T.) 487; þe bemenę drem HOM. II. 115; þe drem . . . of harpe and pipe O. & N. 21; belles (? belle) drem REL. I. 223; dreem SPEC. 7; drem *dream* GEN. & EX. 2095; HAV. 1284; GOW. I. 24; PR. C. 8076; mid te dredful dreamę of þe englene bemen A. R. 214; mid fulle dreme O. & N. 314; **drǣmes** (*pl.*) FRAG. 7; **drēmen** (*dat. pl.*) LAȝ. 22876; *comp.* **belle-, glēo-, man-drēam.**

drēm-rēdere, sb., *interpreter of dreams,* TREV. III. 143; dreemreder WICL. GEN. xl. 23.

drecche, sb., ? *delay;* dreche HALLIW. 317; drechch(e) GAW. 1972.

drecchen, v.,*O.E.* dreccan,=*M.H.G.* trecken; (? *draw*), *harass, delay;* drechen GEN. & EX. 1420; drecche FER. 1997; GOW. II. 5; CH. TRO. ii. 1264; P. R. L. P. 85; ac Sathanas þe frecche þe saule wule drecche (*afflict*) MISC. 75; **dreccheþ** (*pres.*) SPEC. 113; PL. CR. 504; wiles he drecched (*tarries*) ðore REL. I. 211 (MISC. 4); (ha) dreccheð HOM. I. 251; drecche (*subj.*) R. S. ii (MISC. 162); if it so betide þis night þat þe in slepe dreche ani wight Iw. 480; **drecchand** (*pple.*) PS. cviii. 10; **draihte** (*pret.*) ALIS.² (Sk.) 752; *comp.* **ȝe-drecchen.**

drecchunge, sb., *harassing, delay,* H. M. 7; drecchinge FER. 493; GOW. 218.

drēd, drēden, drēdi, see **drǣd, drǣden, drǣdi.**

drēf, sb. (? *or* adj.),=*O.H.G.* truobī; ? *trouble*; ? *or*=*O.H.G.* truobī; *painful*; idolatrie utwrogte hem sorges dref GEN. & EX. 4144.

drēf-ful, adj., *sorrowful*: dreful and bleð GEN. & EX. 2590.

drēfful-lī, adv., *sorrowfully,* HALLIW. 317.

drēfen, v., *O.E.* drēfan,=*O.H.G.* truoben, *Goth.* drōbjan; *from* **drōf**; *trouble, disturb*; dreven HOM. II. 195; ic and Eve sulen alle (h)is blisse dreve GEN. & EX. 318; **drēveð** (*pres.*) REL. I. 220; **drēfed** (*pple.*) ORM. 147; *comp.* **for-, ȝe-drēfen.**

dreg, sb., *O.N.* dregg *f.,*=*Swed.* drägg *m.*; *dregs,* PS. xxxix. 3; **dregges** (*pl.*) VOC. 198; LANGL. *B* xix. 397; dreggis PR. P. 131.

dreȝ, adj., *O.N.* driugr, *prov.Eng.* dree; *from* **drēoȝen**; *long; tedious; patient,* GAW. 1750; driȝe 724; dreȝ, dreiȝ, drih HOM. II. 49, 256, 258; his moder leof and driȝe LEG. 38; bi a river brode and dreghe HALLIW. 317; **on drēghe,** *at a distance,* D. ARTH. 786; drogh him o dreghe ANT. ARTH. xliv; a dreȝ A. P. ii. 71; a dreigh [drei] MAN. (F.) 1042, 6048; on driȝe GAW. 1031; whi draghes þou on dregh *why dost thou delay* D. TROY 11647; a drigh GOW. II. 46; derfe dintes and **drēȝe** (*pl.*) ALEX. 2091; þe sinnez driȝe A. P. i. 822; þou lovest Tristrem **dreiȝe** (*adv.*) TRIST. 3035; *comp.* **un-drēȝ.**

drēȝ-lī, adv., =*O.N.* driugliga; *at great*

length, tediously, GAW. 1026 ; dre3li, dri3li A. P. ii. 74, 476 ; dreli TOWNL. 90 ; dreghli D. TROY 2379.

drē3e, *see* drü3e.

drē3e, sb., *length* : alle the dreghe of the daie D. ARTH. 2915 ; dregh of the derke night D. TROY 678.

drē3en, *see* drēo3en.

dreggi, adj.,=*Swed.* dräggig ; *from* dreg ; *dreggy,* '*faeculentus,*' PR. P. 131.

drehen, *see* dra3en.

dreie, sb., ?=*Swed.* drög (*traha*); *dray* ; dreie ['*rheda*'] TREV. III. 145.

dreien, *see* dra3en. drēm, *see* drēam.

dreight, sb., *from* drē3, adj., *length* : the dai of þe dreight (*the longest day*) drivin uppo long D. TROY 10633.

drēmels, sb., *dream,* LANGL. *A* viii. 138.

drēmen, v., *O.E.* drēman, drȳman (*be joyful*), = *O.N.* dreyma, *O.H.G.* troumen (*dream*), *mod.Eng.* dream ; *from* drēam ; *sound, resound, sing aloud; dream*; heogunnen dremen LA3. 13586 ; harpen gunnen dremen 22885 ; dremen (*dream*) GEN. & EX. 2067 ; drēme (*pres.*) GOW. I. 281 ; murie dreâmeð engles HOM. I. 191 ; þet ower beoden . . . dreamen wel ine drihtenes earen A. R. 430 ; a selkuth drem drēmde (*pret.*) me nou HAV. 1284 ; him drempte GEN. & EX. 1941 ; þeines þer dremden LA3. 11575.

drēmere, sb., *O.E.* drēmere (*musician*),= *O.H.G.* troumāri (*dreamer*) ; *dreamer,* WICL. GEN. xxxvii. 19 ; dremare '*somniator*' PR. P. 131 ; dremer C. M. 4111.

drēminge, sb., *dreaming,* '*sompniacio,*' PR. P. 131 ; to hir he tolde of his dremeing S. S. (Web.) 3089.

drench, sb., *O.E.* drenc, *cf. O.L.G.* dranc, *O. H.G.* tranch *m., Goth.* dragk *n., mod.Eng.* drench ; *from* drinken ; *drink, potion,* R. S. I ; drench, drænc LA3. 10234, 16084 ; drench, drenche ROB. 68, 69 ; drenche (*dat.*) MISC. 82 ; HORN (H.) 1199 ; drenches (*gen.*) LA3. 19755 ; drenche (*gen. pl.*) LA3. 19759.

drenchen, v., *O.E.* drencan (*pret.* drencte), *O.Fris.* drenka, *O.N.* drekkja, *O.H.G.* tren- chen, *Goth.* dragkjan, *mod.Eng.* drench ; *drown, plunge in water,* LA3. 1507 ; HAV. 583 ; ich wille me drenchen in þe welle S. S. (Web.) 1464 ; i schal drenchen in þe depe CH. C. T. *B* 455 ; drenche O. & N. 1205 ; LANGL. *B* viii. 50 ; GOW. III. 295 ; heo let . . . drenche boþe two ROB. 27 ; i schal drenche min arewis in blood ['*inebriabo sagittas meas sanguine*'] WICL. DEUT. xxxii. 42 ; to'drench-ende HOM. II. 39 ; drengte (*pret.*) LA3. 12111 ; dreinte P. S. 217 ; MAN. (F.) 2008 ; dreint (*pple.*) ROB. 52 ; GOW. III. 10 ; idreint CH. C. T. *A* 3520 ; *comp.* a-, for-drenchen.

drenching, sb., *drowning,* TREV. III. 395 ; WICL. S. W. III. 158.

dreng, sb., *O.E.* dreng, *? from O.N.* drengr ; *warrior, man* (*male person*), HAV. 31 ; dring LA3. 12713 ; C. M. 15414 ; dringen (*dat. pl.*) LA3. 12933 ; *comp.* here-dring.

drēnien, v., *O.E.* drehnian,=*M.H.G.* trehenen (*from* trahen *drop, tear*) ; *strain out* : ge drēnieð (*pres.*) þanne (*r.* þane) gnet a weig MAT. xxiii. 24.

drenklen, *see* drinklen.

drēo3en, v., *O.E.* drēogan (*accomplish, carry through, suffer*) *cf. O.L.G.* (bi-)driogan (*deceive*), *Goth.* driugan (στρατεύεσθαι), *O.H.G.* triugan (*deceive*) ; *endure ; carry through to the end, accomplish,* P. L. S. viii. 145 ; drie3en ANGL. I. 24 ; dregen GEN. & EX. 512 ; dre3hen eche þine ORM. 1505 ; drei3en, driæn FRAG. 6, 8 ; drehen KATH. 626 ; deað drehen JUL. 7 ; pine to þolie ant dre3e SPEC. 62 ; þat þou ne mist þis foreward holde ne dre3e 105 ; dreye HORN (R.) 1047 (H. 1078) ; drei M. H. 59 ; dree ISUM. 379 ; dri3en, drien [dre3en] LA3. 370, 6228 ; drighe, dreghe PR. C. 2044, 2235 ; dri3e GAW. 560 ; A. P. ii. 372 ; drien A. R. 80 ; drie MARG. 34 ; WILL. 459 ; R. R. 7484 ; K. T. 235 ; he smot as faste as he might drie HALLIW. 318 ; penance for to drie P. 27 ; drie (*pres.*) HICKES I. 228 ; DEGR. 560 ; drihð A. R. 356 ; þa sunfulle monne þe dre3eð (*perform*) a heore uvele werkes HOM. I. 23 ; þat man his licames lust drige (*satisfy*) HOM. II. 31 ; drēg (*pret.*) GEN. & EX. 429 ; dre3 AL. (T.) 337 ; þe sorewe þat he dre3 (*? printed* dregh) S. S. (Web.) 2660 ; dreih A. R. 136 ; drei3h, drei WILL. 2796, 2864 ; al þat ha druhen [drehden] KATH. 628 ; drógen (*pple.*) GEN. & EX. 2402 ; *comp.* a-, 3e-drēo3en ; *deriv.* drihte.

drēopen, v., *O.E.* drēopan,=*O.L.G.* driopan, *O.Fris.* driapa, *O.N.* driupa, *O.H.G.* triufan ; *drip*: drepe LUD. COV. 170; dreepith (*pres.*) LIDG. M. P. 161 ; *deriv.* drope, drippe.

drēori3, drēri3, adj., *O.E.* drēorig,= *O.H.G.* trūreg ; *from* drēosen ; *dreary, sad,* ORM. 4752, 4838 ; dreori '*moestus*' FRAG. 4 ; REL. I. 178 ; ich am dreori MARH. 15 ; dreori þo3t BEK. 1273 ; dreori fustes A. R. 106 ; of dreori mod(e) SPEC. 70 ; dreri KATH. 2049 ; CH. C. T. *E* 514 ; PR. C. 791 ; ISUM. 125 ; AV. ARTH. lxvi ; dreri, druri WICL. 2 ESDR. ii. 2 ; heo weoren . . . an heorte druri [dreri] LA3. 14547 ; druri EM. 808 ; A. P. i. 323 ; drērier (*comp.*) WICL. GEN. xl. 7.

drīri-hēd, sb.,=*O.H.G.* trūregheit ; *dreariness, sadness,* GEN. & EX. 1122.

drēori-, drēri-līche, adv., *drearily,* KATH. 1898, 2316 ; dreriliche LEG. 173.

drēri3-mōd, adj., *O.E.* drēorigmōd ; *sad-minded,* ORM. 6541 ; drerimod AN. LIT. 6 ; ANGL. IV. 199 ; drurimod FER. 1103.

drēorinesse, sb., *O.E.* drēorigness ; *dreari-ness* ; drerinesse ['*tristitia*'] WICL. ECCLUS. iv. 8.

[**drēoɛen**, v., *O.E.* drēosan,= *O.L.G.* driosan, *Goth.* driusan, *fall*; *comp.* ӡe-, tō-drēosen; *deriv.* drēoriӡ.]

drepen, v., *O.E.* drepan,= *O.N.* drepa, *O.H. G.* trefan; *strike, beat*, HOM. I. 283; HAV. 1865; drepe D. TROY 9854; drepeð (*pres.*) REL. I. 221; drepez A. P. ii. 246; **drap** (*pret.*) MAP 343; drop HAV. 2229; drapen SAX. CHR. 262; drape PS. xciii. 6; **drēpe** (*subj.*) HAV. 506; **dropen** (*pple.*) GEN. & EX. 2648.

   dreping, sb., *smiting, killing*: þer was swilk dreping of þe folk HAV. 2684.

drēpen, v., *O.E.* drȳpan, = *O.N.* dreypa, *M.H.G.* troufen, tröufen; *besprinkle*; heom me **drēpeþ** (*pres.*) mid þe piche MISC. 151; and elles where hem **dripe** (*imper.*) PALL. ii. 277.

drēpen, *see* drēopen. **drēri**, *see* drēoriӡ.

drēried, adj., *from* drēoriӡ; *saddened, wretched*: alle dai dreried i in went PS. xxxvii. 7.

dressen, v., *O.Fr.* dresser, *from med.Lat.* dīrectiāre, *from* dīrectus (*straight*); *mod.E.* dress; *make straight, direct, set in order, prepare, equip; dress; direct one's course, go*: CH. C. T. *A* 3468; **dressinge** (*pple.*) WICL. NUM. xxiv. 1; **dressede** (*pret.*) D. ARTH. 786; **dressed** (*pple.*) A. P. ii. 92.

   dressinge, sb., *mod.Eng.* dressing; *making straight, directing, preparing*, '*directio*,' PR. P. 131; **dressingis** (*pl.*) WICL. PS. xcviii. 4.

dressour, sb., *O.Fr.* drecoir, drechoir; *dresser, table for preparing food*; dressure '*dressorium*' PR. P. 131; dressore L. C. C. 20.

dreste, sb., *O.E.* drestan, derstan, dærstan,= *O.H.G.* trestir; *dregs, dross*, ['*faex*'] WICL. PS. lxxiv. 9; dersten '*amurca*' FRAG. 4; drestis PR. P. 131; drastis '*faeces*' VOC. 178.

dresti, adj., *full of dregs*, '*faeculentus*,' PR. P. 131; þi drasti [darsti] speche CH. C. T. *B* 2113.

drevèlen, *see* dravelen.

drēven, *see* drǣfen *and* drēfen.

drī, sb., *O.E.* drȳ; *wizard*, '*magus*,' FRAG. 2.

   drīӡ-craft, sb., *O.E.* drȳ-, drēocræft; *magical art*, ORM. 16053.

   drīӡ-man, sb., ·*O.E.* drȳman; *magician*, ORM. 16051.

drīe, *see* drūӡe.

drīen, *see* drēoӡen *and* drūӡen.

drīfen, v., *O.E.* drīfan,= *O.N.* drīfa, *O.Fris.* drīva, *Goth.* dreiban, *O.H.G.* trīban; *drive*, ORM. 16982; drive CH. TRO. v. 1179; gomen . . . driuen LAӡ. 24706; ich isæh þæ uþen i þere sæ driuen 28073; drife M. H. 31; drive BEK. 197; toward þe halle he gan drive S. S. (Web.) 967; **drīvest** (*pres.*) A. R. 230; swa hund þene heort driveð LAӡ. 26762; þat te deovel . . . driveð ow to donne JUL. 26; drifþ AYENB. 75; hi driveþ IBID.; heo þe driveþ

heonne O. & N. 66; þus þei drive forþ (*pass*) þe dai GOW.·I. 16; **drīve** (*subj.*) HOM. I. 21; **drif** (*imper.*) LAӡ. 17613; A. R. 274; none cheffare ne drive ӡe 418; **drīvende** (*pple.*) A. R. 244; **drāf** (*pret.*) MARH. 8; ORM. 8260; draf, drof JUL. 76, 77; LAӡ. 309, 7843 (*miswritten* dræf 9367); drof BEK. 675; MAND. 86; A. P. i. 1152; a wind . . . drof hem intil Enge(l)lond HAV. 725; (H)avelok it sau and þider drof 1793; drof [droof] CH. C. T. *D* 1540; dref MAN. (F.) 1590; þou drive HORN (R.) 1279; heo driven balles LAӡ. 24703; heo driven [drive] in to þan castle 1675; driven ALIS. 5731; drive ROB. 21; driven [dreven] WICL. 2 KINGS vi. 3; dreve RICH. 5092; **drifen** (*pple.*) ORM. 8239; driven MAND. 67; to longe we haven driven ure dusischipes KATH. 1816; drive WILL. 979; dreven WICL. EX. viii. 9; L. ʽH. R. 68; *comp.* bi-, for-, ӡe-, tō-, þurh-drifen; *deriv.* drāf, drifte.

   drīfer, sb.,= *M.L.G.* drīver, *O.H.G.* trīpāri; *driver*, VOC. 213.

drifte, sb.,= *M.Du.*, *M.L.G.* drift, *O.N.* dript, *M.H.G.* trift; *from* drifen; *drift, act of driving; herd*, '*armentum*,' VOC. 279; PR. P. 132; drift ALIS². (Sk.) 897; þe drift of þe snouh S. A. L. 221.

drīӡ, *see* drī *and* drēӡ. **drīӡe**, *see* drūӡe.

drīӡen, *see* drēoӡen *and* drūӡen.

drih, adj., *tame, ·gentle*: lomb is drih and milde HOM. 49.

drihte, sb., *O.E.* dryht, driht,= *O.Fris.* drecht, *O.L.G.* druht, *O.N.* drott, *Goth.* drauhts *f.*; *from* drēoӡen; *retinue, host*, LAӡ. 92.

   driht-fare, sb., *retinue*, KATH. 1852.

   driht-folc, sb., *O.E.* dryhtfolc,= *O.L.G.* druhtfolc; *dependants*, LAӡ. 111.

   driht-ful, adj., *glorious*; þe drihtfule [*sec. text* kinewurðe] godd Apollo mi lauerd JUL. 13.

   driht-līc, adj., *O.E.* dryhtlīc; *noble*: þis drihtliche lond LAӡ. 3784; drihtliche men 1623.

   driht-mon, sb., *vassal*, LAӡ. 14715.

   drihtnesse, sb., *majesty*, HOM. I. 101; KATH. 1346.

drihten, sb., *O.E.* dryhten, drihten,= *O.L.G.* drohtin, *O.Fris.* drochten, *O.N.* drōttin, *O.H.G.* truhtīn; *lord*, HOM. I. 91; drihten [driste] LAӡ. 4; drihten, dristen REL. I. 171 (SAL. & SAT. 227); drihtin KATH. 1095; ORM. 2; driӡten REL. I. 186; driӡtin GAW. 996; drihte, dricte (*printed* dritte), drichte (*printed* drithte) P. L. S. viii. 40, 55, 60; drihte C. L. 27; M. H. 1; dryhte MISC. 83; driӡte HORN (L.) 1310; SHOR. 61; LANGL. *B* xiii. 269; (? *printed* drighte) ALIS. 6402; MAP 348; drihtenes (*gen.*) FRAG. 6; A. R. 430; drihtenes, drihtnes HOM. I. 11, 105; **drihtene** (*dat.*) LAӡ. 20; drihtene, drihtne HOM. I. 33, 105.

drillen, v., = *M. Du.* drillen, *Swed.* drilla, *Dan.* drille; *? act deceitfully; elude*: selcuð vs thinc o þe, pilate wit drightin for to drill C. M. 16390; þe ded ai wen we for to drille 23715; *comp.* **a-drillen.**

drinc, sb., *O.E.* drinc (? =drync); *drink,* M. H. 24; þene drinc LAȜ. 1303; drinc, drinch ORM. 165, 15388; drinch HOM. I. 283; drink MAND. 248; P. 42; DEGR. 1739; **drinke** (*dat.*) BRD. 11; drinches [dringes] (*pl.*) LAȜ. 3558; drinkes HOM. II. 179; HAV. 1738; *comp.* **love-, over-drinc.**

   drinke-lēs, adj., *without drink,* HOM. I. 141.

dring, *see* **dreng.**

drink, *see* **drinc.**

drinkelen, *see* **drinklin.**

drinken, v., *O.E.* drincan, = *O.L.G.* drincan, *Goth.* drigkan, *O.Fris.* drinka, *O.N.* drekka, *O.H.G.* trinchan, trinkan; *drink,* MAT. xx. 22; LAȜ. 5804; A. R. 44; HAV. 800; (*ms.* drinnkenn) ORM. 165; drinke SHOR. 26; drinca LEECHD. III. 128; **drinkeð** (*pres.*) HOM. II. 199; **drinc** (*imper.*) LAȜ. 14348; **dranc** (*pret.*) GEN. & EX. 1660; (*ms.* drannc) ORM. 1374; drank HOM. II. 111; RICH. 3090; CH. C. T. *B* 743; LIDG. M. P. 91; dronc A. R. 210; dronc, dronk [drong] LAȜ. 6928, 19796; drank LEB. JES. 323; heo drunken LAȜ. 14285; drunken GOW. III. 21; (*ms.* drunnkenn) ORM. 4797; dronken WILL. 1906; CH. C. T. *A* 822; dronken [dronke] LANGL. *B* xiv. 64; þu drunke (*subj.*) ORM. 14482; **drunken** (*pple.*) ORM. 14057; RICH. 661; dronken CH. C. T. *A* 135: *comp.* **a-, ȝe-drinken;** *deriv.* **drinc, drench, drünc, drunken.**

   drinkere, sb., *O.E.* drincere, = *O.H.G.* trinchāre; *drinker,* TREAT. 138; AYENB. 52; MAN. (F.) 3986; **drinkeres** (*pl.*) HOM. II. 55; drinkares A. R. 216.

   drinkinge, sb., *drinking,* HOM. II. 37; drinking ORM. 19063.

drinklin, v., *plunge, drown,* '*mergo,*' PR. P. 132; ðe ðe childre so drinkelen bead GEN. & EX. 2768; **drenkled** (*pret.*) C. M. 1236; drenkled (*pple.*) MAN. (F.) 997.

drīpen, *see* **drēpen.**

drippe, sb., = *Dan.* dryp, *M.Du.* druppe, *O.H.G.* trupha; *from* drēopen; *drip, drop,* '*gutta, stilla,*' PR. P. 132.

drippin, v., = *Dan.* dryppe, *M.L.G., M.Du.* druppen, *M.H.G.* trüpfen; *drip,* '*stillo,*' PR. P. 132.

drīri, *see* **drēoriȝ.**

drit, sb., *O.N.* dritr, *cf. M.Du.* drijt; *dirt, excrement,* '*stercus,*' PR. P. 132; TREV. IV. 423; P. L. S. i. 7; cattes drit P. S. 240; drit and donge ALIS. 4718; LANGL. *A* vii. 178; muk or drit WICL. E. W. 22.

   drit-cherl, sb., *cf. L.G.* dritkerl; (*term of contempt*), HAV. 682.

drīte, v., *O.E.* (ge-)drītan, = *M.Du.* drijten, *O.N.* drīta; '*cacare,*' CATH. 109.

drivel, sb., = *M.L.G., M.Du.* drevel; *bondservant,* H. M. 29; drivil PR. P. 132; **driveles** (*pl.*) KATH. 2154.

drivelen, *see* **dravelen.** **drīven,** *see* **drīfen.**

drobli, droblin, *see* **drubli, drublin.**

drōf, adj., *O.E.* drōf, = *M.L.G.* drōve, *M.Du.* droeve, *O.H.G.* truobi; *troubled, sad*: drof he wes on mode LAȜ. 1040.

   drōf-līc, adj., *O.E.* drōflīc, = *M.L.G.* drōflīk; LAȜ. 1026.

drōf, *see* **drāf.**

drogge, sb., *O.Fr.* drogue; *drug*; **drogges** [drugges] CH. C. T. *A* 426.

drogman, sb., *med.Lat.* dragumānus, *from Arab.* tarjumān; *interpreter,* ALIS. 3401.

dromedarie, sb., *O.Fr.* dromadaire; *dromedary,* MAND. 243; **dromedaries** (*pl.*) LEG. 97.

dromelus, sb., *dromedary,* ORM. 6966.

dromound, sb., = *O.N.* dromundr, *from O.Fr.* dromont, *Lat.* dromō, *Gr.* δρόμων; *a kind of war-vessel*: dromoun RICH. 1407; droumund 2525; droumound 2458; **dromouns** (*pl.*) ALIS. 90; AR. & MER. 113

drōne, *see* **dráne.**

drōnen, v., = *M.L.G., M.Du.* dronen; *roar, bellow*: he **drouned** (*pret.*) as a dragon ALIS². (Sk.) 985.

dronken, *see* **drunken.**

dronklen, v., *? =* **druncnen;** *submerge, be submerged, drown*: þe schip þat was so grete, it **dronkled** (*pret.*) in the flode MAN. (H.) 170; dronkeld 289; four & twenti þousand in Temse . . . wer **dronkled** (*pple.*) of Danes MAN. (H.) 43.

drope, sb., *O.E.* dropa, = *O.N.* dropi, *O.L.G.* dropo, *O.H.G.* tropho, tropfo; *from* drēopen; *drop,* '*gutta, stilla,*' FRAG. 3; HOM. I. 211; MAND. 51; CH. C. T. *A* 131; HOCCL. i. 415; PR. C. 3065; ænne drope blod [one blodes drope] LAȜ. 7650; a lutel deawes drope A. R. 184; a watres drope GEN. & EX. 1018; þer ne ful noȝt a reines drope P. L. S. xvii. 369; o drope of blode xix. 294; no drope of water S. S. (Web.) 1153; þane drope AYENB. 92; **dropen** (*pl.*) A. R. 220; TREAT. 136; AYENB. 84; droppe LK. xxii. 44; *comp.* **rein-drope.**

   drope-falling, sb., *rain*: as dropefalling droppende upon erthe WICL. PS. lxxi. 6; **dropefallingus** (*pl.*) lxiv. 11.

   drope-mēle, adv., *O.E.* dropmǣlum; *by drops,* A. R. 282.

dropesie, *see* **idropesie.**

droppen, v., *O.E.* droppan, = *M.L.G., M.Du.* dropen, *M.H.G.* tropfen; *drop*; **droppeþ** (*pres.*) MAND. 50; droppe (*subj.*) MIRC 1938; droppinde (*pple.*) SAINTS (Ld.) 317; **dropped** (*pret.*) CH. C. T. *G* 580; þe cloþes droppede TREV. IV. 429; as þei dropped a doun LANGL. *B* xvi. 79; *comp.* **bi-droppen.**

droppinge, sb., *dropping,* '*stillacio*,' PR. P. 132; dropping of rein CH. C. T. *B* 2276; droppinge of flesshe, or fishe in the rostinge '*cadula*' PR. P. 133.

dros, sb., *O.E.* dros,=*M.L.G.* dros, *O.H.G.* gitros; *? from* drēosen; *dross,* A. R. 284; PR. C. 3339; L. H. R. 147.

drosne, sb., *? O.E.* drosn (*pl.* drosna), *O.H.G.* trosena, trusana, truosana (*pl.*); *dregs,* '*fex*,' FRAG. 4; **drosenes** (*pl.*) and dregges LANGL. *C* ix. 193.

drotin, v., *? O.N.* dratta; *drawl, lisp,* '*traulo*,' PR. P. 133.

droukinge, sb.,=**drouping**; *slumber,* PR. P. 113: als i lay in a winteris nit in a droukening (?=droupening) bifore the dai MAP 334; fordolled in a droukning dred L. R. H. 141.

drounen, *see* druncnen.

droupin, *see* drūpen.

droupnen, *see* drūpnen.

drōven, v., *from* drōf; *trouble, disturb*; drove C. M. 11974; **drōves** (*pres.*) PR. C. 1319; **drōved** (*pple.*) ['*turbata*'] PS. vi. 4; *see* drēven.

drōving, sb., *trouble, affliction* ['*tribulatio*'], PS. xxi. 12; C. M. 22384; **drōvinges** (*pl.*) xxxiii. 20.

drōvi, adj.,=*M.L.G.* drōvich; *turbid,* REL. I. 220; CH. C. T. *I* 816; A. P. ii. 1016.

drubli, drobli, adj., *muddy,* '*turbidus*,' PR. P. 132.

drublinesse, sb., *muddiness,* '*turbulencia*,' PR. P. 133.

drublin, v., ? =**trublen**; '*turbo*,' PR. P. 133.

drǖerie, sb., *O.Fr.* druerie; *love, friendship,* FL. & BL. 382; drurie HOM. I. 271; LANGL. *A* i. 85; driwerie A. R. 250; **drǖries** (*pl.*) PR. C. 7825; GAW. 1506.

drǖʒe, adj., *O.E.* drȳge, drīge, drēge=*L.G.* drüge, *M.L.G.* drüge, drōge, *M.Du.* droogh; *dry*; mid druʒe fotan HOM. I. 87; druiʒe L. H. R. 142; druie A. R. 276; druie ROB. 531; CHR. E. 346; LEG. 12; þoleð ... hwile druie (*sb.*) & hwile wete HOM. II. 123; drue P. S. 193; druʒe, driʒe A. P. ii. 385, 412; (*ms.* driʒʒe) ORM. 9883; þe drige stedes HOM. II. 85; driʒe wude FRAG. 3; driʒe TRANS. XVIII. 23; drige, drie GEN. & EX. 616, 3910; drie REL. I. 226; TREAT. 135; GOW. II. 266; drie [dreie] CH. C. T. *A* 420; CH. P. F. 380; for drie C. T. *E* 409; drege MAT. xii. 43; dreʒe SHOR. 145; dreie HOM. I. 227; draie AYENB. 240.

drīnesse, sb., *dryness,* PR. P. 132; LIDG. M. P. 194.

drǖʒen, drǖʒeṅ, v., *O.E.* drȳgan, drūgian, drūwian, = *L.G.* drügen, *M.L.G.* drügen, drōgen; *dry*; drigen HOM. II. 87; driin '*sicco*' PR. P. 132; dreie LUD. COV. 230; whan þou druiʒst (*pres.*) LANGL. *A* i. 25; drieþ CH. C. T. *A* 1495; **drūide** (*pret.*) TREV.

V. 113; dride HOM. II. 155; driede MAND. 68; dreide LK. vii. 38; driʒʒed (*pple.*) ORM. 8625; dried GEN. & EX. 3681; PS. xxi. 16; *comp.* a-, for-drǖʒen, -drūʒien (drūwien).

druggen, v., *drag*: drugge and drawe CH. C. T. *A* 1416.

druggunge, sb., *dragging*: his owune rodé on his softe schuldres so herde druggunge HOM. I. 207.

drūhþe, sb., *O.E.* drūgað,=*L.G.* drügde, *M.Du.* drooghte; *drought*; druhþe ORM. 8626; druʒþe AYENB. 68; druiþe [drouʒte] TREV. VII. 473; drouʒþe SAINTS (Ld.) 315; drouhþe, drouthe, drouʒte, droʒþe A. P. ii. 524; drouthe LANGL. *B* x. 296; drugte GEN. & EX. 2348; droughte [droghte] CH. C. T. *A* 2.

drūnc, drǖnch, sb., *O.E.* drync,=*O.H.G.* trunch, *O.N.* drykkr; *from* drinken; *drink, drinking,* A. R. 14, 114; drunch MARH. 8; **drǖnche** (*dat.*) HOM. II. 41; mid unmete drunche HOM. I. 103.

drūnc-wil, adj., *greedy of drink,* A. R. 216*.

druncnen, v., *O.E.* druncnian, = *O.N.* drukkna, *Swed.* drunkna, drukkna, *mod.Eng.* drown; *be drowned*; *be, make drunk*: i shal drunkne ['*inebriabo*'] thee with mi teres WICL. IS. xvi. 9; **drunkenes,** dronkenes (*pres.*) M. H. 138; ha druncneð þer in MARH. 15; druncnen (*ms.* drunncnenn) ORM. 15398; **druncnie** (*subj.*) A. R. 58; **drunknede** (*pret.*) WICL. IS. lxiii. 6; þe swin urnen ... into þe sæ & druncnede hem selven HOM. II. 39; dronkened PS. lxiv. 10; drouned A. P. ii. 372; **druncned** (*pple.*) ORM. 6795; drouned MAND. 57; LIDG. M. P. 149; *comp.* of-druncnen.

druncning, sb., *drowning,* ORM. 14547.

drunken, adj., *O.E.* drunken,=*O.N.* drukkin, *O.H.G.* trunchan; *from* drinken; *drunken,* ORM. 14065; GEN. & EX. 871; þa heo weore swa drunken Laʒ. 13465; dronken MAND. 250; dronke AYENB. 75; CH. C. T. *A* 3128; þa drunkene cnihtes Laʒ. 13526; *comp.* for-, win-drunken.

drunken-hēd, sb., *drunkenness,* SAX. CHR. 209; WICL. IS. v. 11; dronkenhede GOW. III. 20; dronkehede AYENB. 260.

dronke-lēc, sb., *drunkenness,* MIRC 31.

drunke-shepe, sb., *drunkenness,* '*ebrietas*,' PR. P. 134; dronkeshipe GOW. III. 20.

drunken, v., *drown, plunge in water*; **drunkede** (*pret.*) WICL. ECCLUS. xxxix. 28; dronked (*pple.*) WILL. 3516; *comp.* of-drunken.

drunken, sb., *O.E.* druncen,=*Goth.* drugkanei, *O.N.* drykkni, *O.H.G.* trunchenī; *drunkenness,* HOM. I. 175 (HOM. II. 228; MISC. 67); Laʒ. 6070; the millere þat for drunken [dronken] was al pale CH. C. T. *A* 3120; drunke P. L. S. viii. 128.

drunken-lēwe, sb., *drunkard*, WICL. MAT. xxiv. 49; drunkelewe, dronkęlewe GOW. III. 5; dronkelewe LANGL. *B* viii. 83; drunke-, dronkenlewe CH. C. T. *D* 2043; drunkeleu '*ebriosus*' PR. P. 133; drunkleu LIDG. M. P. 68.

drunkenesse, sb., *O.E.* drunceness, druncenness; *drunkenness*, HOM. I. 13; O. & N. 1399; drunkennesse ORM. 166; dronkennesse [dronkenesse] CH. C. T. *C* 549.

drunken-som, adj., *given to drink*, A. R. 216*.

drūp, adj., *humble, drooping*; **drūpest** (*superl.*) MARH. 16; drupest alre monne KATH. 2050.

drūpen, v., *O.N.* drūpa, *droop; decline, sink down*: droupin '*latito*' PR. P. 113; droup(e) & dare TRANS. XVIII. 26; M. ARTH. 2575; **droupe** (*pres.*) MIN. i. 9; TOWNL. 223; droupus ANT. ARTH. v; **drupand** (*pple.*) [drupand, droupand] . . . he sagh þeir chere C. M. 4457; **drouped** (*pret.*) his arwes droupĕd nought wiþ feþeres lowe CH. C. T. *A* 107; dropede FER. 1103.

droupinge, sb., *drooping*, '*latitatio*,' PR. P. 113; drouping GAW. 1750; D. TROY 3291.

drūpi, adj., *downcast*, HOM. I. 205; HALLIW. 321; makeð drupie chere A. R. 88.

drūpnin, v., = *O.N.* drūpna; *droop, be humble*, HOM. I. 259; i droupne ant dare SPEC. 54; **droupninde** (*pple.*) HORN (H.) 1126.

droupninge, **droupening**, sb., *dejection*, MAP 340.

drūri, *see* drēoriʒ.

drūð, sb., *O.Fr.* drūd, *from O.H.G.* drūt; *darling*, HOM. I. 269.

drūwien, *see* drūʒien.

dubben, v., *O.E.* dubbian, SAX. CHR. 219, = *O.N.* dubba; *dub, adorn*: to cnihte hine dubben [dobben] LAʒ. 22497; dubbe him knight ISUM. 484; þeo kniʒtis heore bodi **dubbeþ** (*pres.*) ALIS. 4311; **dubbede** (*pret.*) HAV. 2314; **dubbed** (*pple.*) GAW. 75; L. H. R. 127; dubbed and dight PR. C. 8790; dubbed wiþ precious stones MAND. 241; idubbed [idobbed] LAʒ. 19578; idobbed AYENB. 83.

dubbing, sb., *dubbing of a knight*, HORN (L.) 438; dobbinge SHOR. 15.

dubbement, sb., *O.Fr.* adobement; *embellishment*: the dubbement dere of doun & dalez A. P. i. 121.

duble, adj., *O.Fr.* duble, doble; *double*, A. R. 70; P. L. S. ii. 176; doble BEK. 417; double GOW. I. 7; i will have three sithe **double** (*sb.*) of his RICH. 2096; doublechinned TREV. I. 299.

doublenesse, sb., *duplicity, falseness*, CH. C. T. *G* 1300.

dublen, v., *O.Fr.* doubler, dobler; *double*: dublin, dobelin PR. P. 125; dubli BRD. 28;

doubleþ (*pres.*) SHOR. 34; dobleþ AYENB. 22; **dubled** [doubled] (*pret.*) C. M. 6954.

dobler, sb., *O.Fr.* doublier; *platter*, A. P. ii. 1146; **dobeleres** (*pl.*) LANGL. *B* xiii. 81, *C* xvi. 91.

dublet, sb., *O.Fr.* doublet; *doublet*; doblette, dobbelet PR. P. 125; GAW. 571; doublett LIDG. M. P. 54; dobelat VOC. 182; doplit 238.

dūc, sb., *O.Fr.* duc; *duke*, LAʒ. 86; A. R. 300; FL. & BL. 697; duc [duk] CH. C. T. *A* 893; duk [duik] WICL. MAT. ii. 6; duik L. H. R. 149.

duchē, sb., *Fr.*, duché; *duchy*, LANGL. *C* iv. 245; MAN. (F.) 3150.

ducherie, sb., *duchy, dukedom*: TREV. IV. 145; ducherie of Kent 187; that daie **ducheries** (*pl.*) he delte D. ARTH. 3615.

duchesse, sb., *O.Fr.* duchesse; *duchess*, CH. C. T. *A* 923.

dudde, sb., *? sort of hood or cloak*, '*birrus*,' PR. P. 134.

due, adj., *O.Fr.* deut, *from Lat.* dēbitus; *due* ʒif me be diʒt a destine due to have A. P. iii. 49; TREV. IV. 189; GOW. I. 19; in dewe time TREV. V. 407; LIDG. M. P. 207; þe most holi werke, and most duwe to prelatis WICL. S. W. III. 179; **due** (*sb.*): to singe againe as was hir due LIDG. M. P. 182.

due-līche, adv., *duly, properly*, GOW. III. 245; WICL. NUM. xxix. 24; GOW. III. 354; dewli LIDG. M. P. 253.

duel, *see* doel.

duer, *see* dēor.

duetē, sb., *duty*, GOW. III. 176; deute AUD. 24; dewtee LIDG. M. P. 141; dute PR. P. 135.

duʒe, adj., *from* dēaʒ; *? useful*; duwe P. L. S. xxv. 123; *comp.* ʒe-duʒe.

duʒen, v., *? O.E.* dugan = *O.N.* duga, *? O.H.G.* tugan; *from* dēaʒ; *avail, be valiant, be worth*; **dowes** (*pres.*) A. P. iii. 50; **duʒende** (*pple.*) LAʒ. 4123; duʒende þewas HOM. I. 109; **doh(t)e** [*O.E.* dohte, *cf. O.N.* dugða, *O.H.G.* tohta] (*pret.*) M. H. 149; al he solde þat outh douthe HAV. 703; dought TRIST. 1125; MAN. (F.) 7550; *comp.* ʒe-duʒen.

dūʒen, v., *O.E.* (ge-)dēgan, dȳgan, ?= *O.L.G.* dōgen; *? lack*; wel mei duhen ancre of oðer wim(p)lunge A. R. 420*.

duʒeðe, sb., *O.E.* duguð, dugoð, = *O.Fris.* duged, *O.N.* dygð, *O.H.G.* tuged, tugend; *virtue, virility, nobility, power, riches*: unʒearu to elchere duʒeðe HOM. I. 103; Brutus & his duʒeðe [doʒeþe] LAʒ. 1819; duweðe 4945; þat hi forleteþ in heore duʒeþe [duhþe] O. & N. 634; þi dugeþe gin þu delen REL. I. 184 (MISC. 133); duheðe JUL. 4; duþe MISC. 91; douthe GAW. 1365; doweþes (*pl.*) MISC. 112.

duʒeðe-king, sb., *prince*: a dun hem ferdę Bedver to his duʒeðekinge LAʒ. 25956.

duʒeþe-cniht, sb., *retainer,* LAȝ. 10166.

duʒeð-līche, adv., *virtuously, nobly,* LAȝ. 16844.

duʒeðe-mon, sb., *retainer,* LAȝ. 14066.

duggen, v., *? cut;* mai he dug him a doket TOWNL. 313.

duȝti, *see* **duhtiȝ.**

dūhen, duheðe, *see* **dūȝen, duʒeðe.**

duhtiȝ, adj., *cf. O.E.* dyhtig, = *M.H.G.* tühtic; *doughty, powerful, good,* ORM. 113; duhti KATH. 782; þine domes . . . beoð duhti MARH. 8; duhti [dohti] mon LAȝ. 14791; dohti 7652; duȝti GAW. 724; AV. ARTH. xiv; douhti TRIST. 1467; TRANS. XVIII. 25; AM. & AMIL. 207; doȝti A. P. ii. 1182; doghti ISUM. 273; douȝti WILL. 1101; WICL. EX. xvii. 25; duhtie men LAȝ. 7257; doȝtiere (*compar.*) FER. 531; douȝtiore LANGL. *A* v. 84; douȝtier *B* v. 102; doȝtiest (*superl.*) A. P. ii. 1306; þe doȝtieste FER. 1448; *comp.* **un-duhti.**

   dughti-hēde, sb., = *M.H.G.* tüchtikeit; *bravery, goodness,* C. M. 10166.

   duhtiȝ-leȝc, sb., *goodness;* (*ms.* duhhtiȝleȝ3c) ORM. 4904.

   douhti-līche, adv., *doughtily,* JOS. 495.

   duhtiȝnesse, sb., *doughtiness,* ORM. 17582; doȝtinisse FER. 3197; doughtinesse MAN. (F.) 2868.

dūke, sb., *O.E.* dūce, *from* dūken; *duck,* '*anas,*' VOC. 220; duke [doke] LANGL. *B* xvii. 62; CH. C. T. *A* 3576; doke PR. P. 125; **dokes** (*pl.*) MAND. 216.

   dokelinge, sb., *duckling,* PR. P. 125.

   doke-wēd, sb., *duckweed,* PR. P. 125.

dūken, v., = *M.L.G.* dūken, duiken, *M.H.G.* tüchen; *duck* (*plunge under water*); **doukand** (*pres. pple.*) ALEX. 4091; **dūked** (*pret. pple.*) C. M. 23203.

   dūkere, sb., = *O.H.G.* tüchāri; *diving bird;* doukere '*plounjoun*' REL. II. 83; dokare VOC. 253.

dul, adj., *O.E.* dol, = *O.L.G.* dol, *M.L.G., M.Du.* dul, dol, *O.H.G.* tol; *?* = **dwal;** *dull,* '*hebes, obtusus,*' PR. P. 135; CH. C. T. *F* 279; dul wit KATH. 1268; þi brain is dul LIDG. M. P. 191; **dulle** (*pl.*) LANGL. *A* i. 129; dulle (*v. r.* dulte) neiles A. R. 292*; **duller** [doller] (*compar.*) TREV. III. 409; *see* **dil.**

   dulnesse, sb., *dullness,* PR. P. 135; CH. D. BL. 879.

   dul-, dol-witted, adj., *dull-witted,* TREV. III. 467.

dullard, sb., *cf. M.Du.* dullard; *dullard,* '*duribuccius, agrestis,*' PR. P. 135.

dullen, v., = *M.Du.* dullen, dollen (*be insane*); *dull, make dull, grow dull;* dullin, dollin '*hebeto, tepefacio*' PR. P. 126, 135; it dulleþ me to rime CH. C. T. *G* 1093; min heed dullith LUD. COV. 343; **dullid** (*pple.*) LIDG.

M. P. 217; i am ner hande dold so long have i nappid TOWNL. 98; *comp.* **for-dullen.**

dult, adj., *cf. mod.Eng.* dolt; *dull, blunt,* HOM. I. 203; KATH. 1268*; dulte neiles A. R. 292.

dumb, adj., *O.E.* dumb, = *O.L.G., O.Fris.* dumb, *O.N.* dumbr, *Goth.* dumbs, *O.H.G.* tumb; *dumb,* HOM. II. 125; ORM. 211; O. & N. 416; MAP 335; domb, doumb AVENB. I, 51; a dombe best P. L. S. xiii. 220; domb [doumb] CH. C. T. *A* 774; doumbe LANGL. *B* x. 137; dum PR. P. 135; dom PR. C. 49; L. H. R. 148; doum LUD. COV. 126; **dumbe** (*pl.*) MISC. 39; oþer godes þet dumbe beoð ant deave MARH. 6; doumbe ROB. 131; **dumben** (*dat. pl.*) HOM. I. 229.

   dumnesse, sb., = *O.H.G.* tumpnessi; *dumbness,* PR. P. 135.

dumbien, v., *O.E.* (a-)dumbian, *cf. O.H.G.* tumben; *become dumb;* i doumbed ['*obmutui*'] PS. xxxviii. 3; *comp.* **a-dumbien.**

dumpen, v., = *O.N.* dumpa; *beat, ? drive;* dumpe A. P. iii. 362; C. M. 22643.

dumping, sb., *diving bird;* **dumpinges** [dompinges] (*pl.*) LANGL. *C* xiv. 169.

dun, adj., *O.E.* dunn, *? Ir.* dunn, *Gael.* donn; *dun, dark-coloured,* R. R. 1213; **dunne** (*pl.*) A. P. i. 30; LIDG. M. P. 152; donne LAUNF. 988; L. H. R. 144; Dun (*pr. n.*) is in the mire CH. C. T. *H* 5.

dūn, sb., = *M.L.G.* dūn, *O.N.* dūnn *m.; down* (*of birds*); doun '*pluma*' PR. P. 128; GOW. II. 103.

dūn, *see* **dūne.**

dunch, adj., = *O.Fris.* diunk; *dark,* HALLIW. 24.

dunch, sb., = *Dan.* dunk; *stroke, blow,* '*percussio,*' PR. P. 135.

dunchin, v., = *Dan.* dunke, *Swed.* dunka; *beat,* '*tundo,*' PR. P. 135; **dunchen** (*pres.*) HOM. I. 283.

dūne, sb., *O.E.* dyne, dyn, = *O.N.* dynr, *M.L.G., M.Du.* done; *din:* þene dune LAȝ. 8642; as a þunres dune KATH. 2024; dine PR. P. 121; HOM. II. 117; HAV. 1860; dene TREV. I. 415; dene [din] MAN. (F.) 9917; din A. P. ii. 862; PR. C. 7332; **dine** (*dat.*) GEN. & EX. 3467; dine [dene, deone] LANGL. *B* xviii. 62, *C* xxi. 65; *comp.* **erðe-dine.**

dūne, sb., *O.E.* dūn, *cf. O.Ir.* dūn, *mod.Eng.* down; *hill,* HOM. I. 143; O. & N. 919; GEN. & EX. 587; dun ORM. 14568; uppen þere dune LAȝ. 27256; doune ROB. 362; SPEC. 85; bi doune and dale ALIS. 1767; a **dūne** [*O.E.* a, of dune]; *adown, down,* LAȝ. 30297; a dun 551; a dun ward 9297; a dun A. R. 60; O. & N. 1454; duste him a dun riht (*downward*) to þer eorðe MARH. 12; a dune, a doun HAV. 567, 2735; a doun ROB. 208; CH. C. T. *A* 393; þe piler fel a doun S. S. (Wr.) 2045; knele a doun AUD. 79; **dūne** (*adv.*) *for* a dune *down* LAȝ. 26546; he com dun ORM. 631; dune, dun, doun HAV. 888, 925, 1815; he fl(e)ah **dūn rihte** 25613; hewe him to the grounde dounright S. S. (Web.) 621; OCTOV. (W.) 1095; PS. CV. 18; ALIS. 2299; WILL. 1165; downrightes slew there our English knightes RICH. 1761; **dūn ward,** *downward,* LAȝ. 13106; & draf þir to dun riht KATH. 2023; dun ward ORM. 2056; dunewardes KATH. 1191; **dūnes** (*pl.*) A. R. 380; **dūnen** (*dat. pl.*) LAȝ. 21775; dune HORN (L.) 154.

Dūn-holm, pr.n., *O.E.* Dūnholm; *Durham:* on Dunholme MISC. 146.

dunge, sb., *cf. M.Du.* dunge, *Swed.* dynga (*dung*), *O.N.* dyngja (*heap*), *M.H.G.* tunge *f.*; *dung,* A. R. 142; dunge, donge '*fimus*' PR. P. 127; donge ALIS. 4718; donge [dounge] LANGL. *A* iv. 130; ? dung MARH. 15; dong AYENB. 61; MAND. 49; CH. C. T. *A* 807; dunge (*dat.*) LK. xiii. 8 (*the earlier text has* meoxe); ROB. 310; ding(e) P. L. S. i. 7.

dung-, dong-cart, sb., *dung-cart;* CH. C. T. *B* 4226.

dung-heep, sb., *dung-heap,* SPEC. 103.

dong-hel, sb., *dung-hill,* AYENB. 81; S. S. (Web.) 2411; donghil CH. P. F. 597.

dungen, v., *cf. O.E.* dyngan, = *Swed.* dynga, *M.Du.* dungen, *M.H.G.* tungen; *dung;* dunge [donge] CH. C. T. *B* 4226; dongen PL. CR. 785.

dūnien, v., *O.E.* dynian, *O.N.* dynja, duna, = *O.L.G.* dunian, *M.Du.* donen; *din, sound, resound,* LAȝ. 30410; denie HORN (L.) 592; **dūnede** (*pret.*) MARH. 20; þe erþe dunede under hem ROB. 459; dinede an migtful hornes blast GEN. & EX. 3464; þondir dined(e) shille AL. (L.¹) 561; denede WILL. 5014; REL. II. 7; *comp.* **a-dūnien.**

dūninge, sb., *cf. O.E.* dunung, = *M.Du.* doninge; *sound, din,* '*bombus,*' PR. P. 135.

dunnien, v., *O.E.* dunnian; *become dark:* þin heu **dunnet** (*pres.*) P. R. L. P. 221; **dunnid** (*pple.*) '*subniger*' PR. P. 135.

dunnir, sb., *? for* þuner; *thunder:* þer nis dunnir, slete, no hawle ALT. BL. I. 319 (P. L. S. xxxv).

dünt, sb., *O.E.* dynt, *cf. O.N.* dyntr, *Swed.* dunt, *mod.Eng.* dint; *blow, stroke,* HOM. I. 15; LAȝ. 8420; KATH. 22; ALIS. 1505; GAW. 452; sweordes dunt A. R. 60; þene dunt SAINTS (Ld.) 315; dint ORM. 4290; LANGL. (Wr.) 12124; dint dent WILL. 1234, 3750; dent HORN (L.) 152; LAUNF. 332; **dinte, dente** (*dat.*) FER. 617, 1843; **düntes** (*pl.*) LAȝ. 22780; A. R. 284; O. & N. 1227; C. L. 1161; dintes HOM. II. 95; PR. C. 7017; M. H. 135; **dünte** (*gen. pl.*) A. R. 424; **dünten** (*dat. pl.*) LAȝ. 27051; *comp.* þonder-dent.

dünten, v., = *O.N.* dynta, *Swed.* dunta; *strike, give a blow;* (the kinges men) **dünten** him (*pret.*) HAV. 2448; (ha) dintede HOM. I. 281; **dünted** (*pple.*) L. H. R. 138; dunet HOM. I. 281.

---

dūp, *see* **dēop.**

duplicitē, sb., *O.Fr.* duplicité, *from Lat.* duplicitas; *duplicity,* LIDG. M. P. 165, 170.

düppen, v., *O.E.* dyppan, = *Dan.* dyppe; *dip, immerse;* dippin '*intingo*' PR. P. 121; dippest (*pres.*) ORM. 1551; duppeþ TREV. I. 117; duppes JOS. 534; düpe (*imper.*) LEECHD. III. 118; dippe ['*intingat*'] WICL. LK. xvi. 24; **dipped** (*pple.*) PR. C. 8044; idept AYENB. 106.

dippere, sb., *diver* (*bird*) ['*mergulus*'], WICL. LEV. xi. 17.

dūr, *see* **dēor.**

duracioun, sb., *Lat.* dūrātio; *duration:* iaf hem eke duracioun, some to wexe & wane some CH. H. F. 1022.

durcnien, *see* **dearknien.**

dúre, sb., *O.E.* duru, = *O.Fris.* dure, dore, = *M.L.G.* dore, *Goth.* daurō, *O.N. pl.* dyrr (*gen.* dura), *O.H.G.* turi, tura, *Gr.* θύρα, *Lat.* foris; *door,* LK. xi. 7; GEN. & EX. 1082; pore [dore] LAȝ. 25889; dore HAV. 1788; AYENB. 210; CH. C. T. *A* 550; þe dore . . . sche fleȝ out of þe hokes FER. 2182; **dúre** [*O.E.* duru, dura, duran] (*dat.*) MK. i. 33; HOM. I. 167; O. & N. 778; þere dure MISC. 44; dure [dore] LAȝ. 10181; wiþuten dore 2382; **dúren** [*O.E.* dura, duru, duro, duran] SAX. CHR. 217 (*pl.*) JOHN XX. 19; HOM. I. 87; MISC. 153; duren [dores] LAȝ. 2363; doren ROB. 495; SPEC. 110; SHOR. 55; **dúren** (*dat. pl.*) MAT. xxiv. 33; *comp.* **chamber-, chireche-dure.**

dúr-herre, sb., *door-hinge,* VOC. 261.

dóre-nail, sb., *M.H.G.* türnagel; *door-nail:* ded as a dorenail WILL. 628; LANGL. *A* i. 161.

dúre-pin, sb., GEN. & EX. 1078; HORN (L.) 973.

dor-stode, sb., *O.E.* durustod, *O.N.* durastoð; *door-post,* VOC. 170.

dóre-trē, sb., *cf. Dan.* dörtræ; *door bar,* HAV 1806; LANGL. *B* i. 185.

dúre-ward, sb., *O.E.* duruweard; *door-keeper,* LAȝ. 17672; MISC. 43; doreward SHOR. 46.

düre, *see* **dēore.**

düren, v., *O.Fr.* dürer; *last, continue;* dure ROB. 70; CH. C. T. *A* 2770; deore P. L. S. ix. 63; **dürede** (*pret.*) LAȝ. 26708*; whi þat tempeste so longe time durede [dured] LANGL. *B* xviii. 63.

düringe, pple. & prep., *during,* TREV. I. 261.

duresse, sb., *O.Fr.* duresce, *cf. mod.Eng.* duress; *hardship, harm,* WILL. 1074; S. S. (Web.) 2189; HOCCL. i. 12; LIDG. M. P. 118.

durk, *see* **dearc. durne,** *see* **derne.**

durren, v., *? O.E.* durran = *O. H. G.* (ge-)turren, *Goth.* (ga-)daursan; *from* dear; *dare;* durn, dorn, darn PR. P. 114; dur PR. C. 4135; **durste** [*O.E.* dorste, *cf. O.H.G.* (ge-)torsta, *Goth.* (ga-)daursta] (*pret.*) durst, KATH. 1315;

A. R. 158; GEN. & EX. 2593; SPEC. 104; ISUM. 618; durste [dorste] LA3. 357; dorste LANGL. *A* viii. 118; CH. C. T. *A* 227; dorste AYENB. 143; JOS. 664; PERC. 1966; dorste, dorte HORN (L.) 388, 928; i ne þerste . . . do such a fol dede MARG. 304; dorstest AYENB. 73; dorsten LK. ix. 45; dursten HOM. I. 93; HAV. 1866; MAND. 282; durst (*pple.*) MAN. (F.) 6362.

dürsti, adj., *O.E.* dyrstig, = *O.H.G.* (gi-)tur-stīg; *daring, bold.*

dürsti-līche, adv., '*audacter*,' FRAG. 1; dir-sti3like ORM. 16152.

dusc, adj., *O.E.* dosc (ACADEMY No. 849); *dusk*; dosc HOM. I. 259; dosk LEG.203; deosc A. R. 94.

dusknesse, sb., *darkness* ['*caligo*'], WICL. JOB xxiii. 17.

duschen,v.,? = *M.L.G.*duschen; = **dwēschen**; *strike, beat*; dusched (*pret.*) AR. & MER. 5624; A. P. ii. 1538; dusched a doun . . . hure fon FER. 3068; *see* **daschen.**

dusching, sb., *tumbling*; . . . dinning and dusching of sinfulle PR. C. 7350.

düsi, adj., *O.E.* dysig, = *M.Du.* deusig, *O. H.G.* tusīg, *mod.Eng.* dizzy; *stupid, foolish*, KATH. 782; A. R. 208; dusi luve ne last noht longe O. & N. 1466; than waxes . . . his heved feble and disi PR. C. 771; desi REL. I. 179; þe dusie mon HOM. I. 117; **desien** (*dat. m.*) MAT. vii. 26; **düsie** (*pl.*) A. R. 222; þa dusian HOM. I. 111; **düsien** (*dat. pl.*) LA3. 2811; **düsigest** (*superl.*) A. R. 182.

düsi-lēc, sb., *folly*, KATH. 425.

düse-līche, adv., *foolishly*, HOM. I. 119; desali BARB. vii. 210.

desinisse, sb., *O.E.* dysigness; *folly*; desy-nysse MK. vii. 22.

düsi-schipe, sb., *folly*, A. R. 182.

düsi, sb., *O.E.* dysig; *folly*, HOM. I. 111.

düsien, v., *O.E.* dysegian, = *O.Fris.* dusia; *behave foolishly*; **desige**ð (*pres.*) MK. ii. 7; desiet REL. I. 177.

dusken, v., *from* dusc; *darken, make or grow dark*: dusken his herte PL. CR. 563; þine ehnen schulen doskin H. M. 35; **dusked** (*pret.*) CH. C. T. *A* 2806.

dust, sb., *O.E.* dust, = *M.L.G.*, *M.Du.*, *O.N.* dust; *dust*, '*pulvis*,' PR. P. 135; HOM. I. 223; A. R. 122; ORM. 1636; LUD. COV. 225; þat dust [doust] LA3. 27646; doust REL. I. 28; þet doust AYENB. 108; **duste** (*dat.*) JUL. 41; douste P. L. S. xviii. 26.

dusten, v., ? = *O.N.* dusta (*dust, beat*); *strike, beat, throw down*, A. R. 212; 3if hit dusteð 314; duste KATH. 1094; JUL. 48; he dust(e) him doun FER. 854; ich habbe a dun . . . idust MARH. 11; *comp.* **a-dusten.**

dusti, sb., *cf. O.E.* dystig; *dusty*, '*pulverulen-tus*,' PR. P. 135; HICKES I. 224.

dūt, sb., = **dedūt**; *pleasure, joy*, SHOR. 129; P. L. S. xxxv. 9; GAW. 1020.

dūtable, adj., *doubtful* : doutable R. R. 5416.

dūtance, sb.,*O.Fr.* dutance; *doubt, fear* : have you no doutance of all these English cowards RICH. 1862; he saide he hadde thereof do-taunce ALIS. 582; withouten **doutance** (*dat.*) RICH. 5872; out of doutaunce CH. TRO. iv. 933.

dūte, sb., *O.Fr.* doute; *doubt, fear*, A. R. 220; MISC. 44; doute MARG. 107; C. L. 1425.

dūte-lēs, adj., *doubtless* : nade he ben du3ti . . . douteles he hade ben ded GAW. 724.

dūten, v., *O.Fr.* douter; *doubt, fear*, MISC. 148; **dūte**ð (*pres.*) A. R. 244; **doutede** (*pret.*) HAV. 708; **douted** (*pple.*) C. L. 382.

dūtous, adj., *O.Fr.* doutous; *doubtful*; dout-ous CH. BOET. i. 1 (5); dotous RICH. 4839.

dütten, v., *O.E.* dyttan, ? = *M.H.G.* tützen; *shut*, SPEC. 110; me schulde dutten [ditten] his muð A. R. 82; dettan LK. xi. 53; ditten ORM. 18590; dutte [ditte] out LANGL. *A* vii. 178; ditte C. M. 19452; **ditte**ʒ (*pres.*) A. P. ii. 588; we ditteð HOM. II. 199; **düte** (*imper.*) A. R. 106; **dütte** (*pret.*) LA3. 19812*; MARG. 208; dutten A. R. 106; dit (*pple.*) GAW. 1233; idut REL. I. 90; C. L. 31; **iditte** (*pl.*) S. B. W. 242; *comp.* **for-, un-dütten.**

dūþe, *see* **du3eðe.**

dūve, sb., *O.N.* dūfa, = *O.L.G.* dūva, *Goth.* dūbo, *O.H.G.* tūba; *dove*, HOM. II. 49; GEN. & EX. 695; douve MAND. 87; RICH. 5671; LANGL. *B* xv. 393; H. S. 304; doufe TOWNL. 33; doufe, douve, dofe (? dofe), dove C. M. 1895, 1901; (*ms.* dove, dowe) PR.P. 128; dove LUD. COV. 48; **dovene** A. P. ii. 481.

dūve-brid, sb., *young dove*, HOM. II. 47.

[**dūve**, *stem of* **düven.**]

dīve-dap, sb., *didapper*, '*mergulus*,' PR. P. 127; divedap, devedoppe, devedep WICL. DEUT. xiv. 7, LEV. xi. 1, 17.

dúvel, sb., *cf. M.L.G.* dovel, *M.H.G.* tübel; *felly of a wheel*; **dowlis** ['*modioli*'] WICL. 3 KINGS vii. 33.

düvel-rihtes, adv., *headlong* : feollen ba (a) duvelrihtes KATH. 1599.

düvelunge, adv., *headlong, with a plunge*, ant te meiden duvelunge feol dun to þe eorðe MARH. 20; develing AR. & MER. 7762; BEV. 649.

düven, v., *O.E.* dūfan (*pret.* dēaf); *dive, plunge*; þet þet sweord in dēæf (*pret.*) LA3. 6505; & def . . . dun to þer eorðe JUL. 76.

düven, v., *O.E.* dȳfan (*pret.* dȳfde), = *O.N.* deyfa; *dive*; diven LANGL. *B* xii. 163; **dūve**ð (*pres.*) A. R. 282; he diveð dun to grunde REL. I. 221; **dēf** (*imperat.*) MARH. 17; duve ['*demergat*'] HOM. II. 43; **dēfde** (*pret.*) MARH. 22; euch dunt defde in to hire liche JUL. 28.

duwe, duweðe, *see* duʒe, duʒeðe.

dwal, adj. & sb., *O.E.* dwol,=*M.L.G.* dwal (*fatuous*), *Dan.*dval, *Goth.*dwals (μωρός); *from* **dwélen**; *bad, foolish*: þat devel dwale GEN. & EX. 20; dwale *heretic* ORM. 7454; ne chid þu wiþ nenne **dwáles** (*? fools*) REL. I. 183; *see* dul.

    dwal-kenned, adj., *misleading*: þurh dwal-kende lare ORM. 7441.

dwále, sb.,=*O.N.* dvöl (*delay*), *Swed.* dvala, *Dan.* dvale (*stupor*), *O.H.G.* twala (*delay*); *delay; stupefaction,* GEN. & EX. 1220; dwale [dvale] C. M. 14197; stod as he were in dwale MAN. (F.) 9059; dwale *soporiferous drink* GOW. III. 14; CH. C. T. *A* 4161; REL. I. 324.

dwéle, *see* dweole.

dwélen, v., *O.E.* *dwelan (*pple.* gedwolen), *cf. O.H.G.* twelan; *be stupefied*: þe child lai **dwéling** (*pple.*) S. S. (Web.) 770; *comp.* a-dwélen.

dwélien, v., *O.E.* dwelian,=*M.L.G.* dwelen; *err, be deceived*: ʒif þe larðeu **dwélað** (*pres.*) HOM. I. 109; ge dwelieð MAT. xxii. 29; dwele ['*errant*'] PS. xciv. 10; **dwéled** (*pret.*) PS. cxviii. 176; *comp.* bi-dwélien.

dwelle, sb., *delay*: withouten dwelle LEG. 57; dvelle FER. 648; dvell(e) C. M. 2831.

dwellen, v., *O.E.* dwellan, = *O.N.* dvelja (*pret.* dvalda), *O.H.G.* twellan (*pret.* twalta); *dwell, delay,* ORM. 9938; dwellen HAV. 54; i nille nouʒt longe dwellen S. S. (Wr.) 1625; dwelle ALIS. 225; SHOR. 19; WILL. 701; CH. C.T. *A* 2354; dwelle SPEC. 82; him reoweþ þat he **dwelleþ** (*pres.*) her swa swiþe lange on eorþe ORM. 5576; **dwelde** (*pret.*) LANGL. *B* xx. 342; mani winter he dwelde þere S. S. (Wr.) 3204; dwalde ORM. 13218; dwelden LEG. 4; **dwelled** (*pple.*) ORM. 226; *comp.* ʒe-dwellen.

    dwelling, sb., *dwelling,* HAV. 1352; dvelling, dwelling LEG. 43; GREG. 395.

dweole, sb., *O.E.* gedweola; *error*: þat is dweole R. S. ii. (MISC. 160); dweole [dwele] O. & N. 1239; wend a wei mine eien vrom þe worldes dweole A. R. 62; dwele LEB. JES. 149; (*ms.* dwelle, *r. w.* wele) P. L. S. i. 13.

    dweole-, dwéle-song, sb., O. & N. 926.

dweoleð, sb., *? error*; heo was igon a dweoleð (*v. r.* o dweole) A. R. 224.

dweolðe, sb., *? = Goth.* dvaliþa; *error, ? delirium*; þet ... dweolðe me ne derie HOM. I. 199; dweoluhðe 195; he spek o dwelthe ED. 3149.

dweomer, *see* dwimer.

dwer, sb., *cf.* diswére; *? doubt*; MAP 361; S. & C. I. liii; dwere LUD. COV. 17, 117; TOWNL. 302.

dwerʒ, sb., *O.E.* dweorg, dweorh, = *O.N.* dvergr, *O.H.G.* tverg; *dwarf*; dwergh Iw. 2390; dweri (*? ms.* dwery, *printed* dwerþ) WILL. 362; dwerou, dweruh '*nanus*' PR. P.

134; dverw TRIST. 2062; dwerf TREV. I. 231; IPOM. 1746; **dwerghes** (*pl.*) MAND. 205; dwerewes [dv(e)rwes] ALIS. 6266.

[**dwēschen,** v., *O.E.* dwǣscan; *? shake*; *comp.* a-dwēschen].

dwild, sb., *O.E.* (ge-)dwild, dwyld; *error*; dwyld SAX. CHR. 258; dwilde ORM. 9736; *comp.* ʒe-dweld.

[**dwimer,** sb., *O.E.* dwimor (*phantom*); *comp.* ʒe-dwimor.]

    dweomer-cræft, sb., *magic art,* LAʒ. 30634.

    dweomer-lác, sb., *magic play,* LAʒ. 270; demerlaik ALEX. 414; A. P. ii. 1561.

dwīnen, v., *O.E.* dwīnan (*pret.* dwān),= *M.Du.* dwīnen, *O.N.* dvīna (*pret.* dvīnaða), *prov.Eng.* dwine; ['*tabesco*'] WICL. PS. cxviii. 139; dwinin '*evanesco*' PR. P. 134; dwine PR. C. 703; as gres in medowe i drie and dwine HALLIW. 326; dwineþ GOW. II. 117; **dwīned** (*pret.*) WILL. 578; dwine (*pple.*) ORF. 259; dwined R. R. 360; *comp.* for-dwīnen.

dwole, sb., *O.E.* dwola,= *O.H.G.* (ga-)twola; *error,* O. & N. 825.

dwólunge, sb., *O.E.* dwolung; *delusion*: þer þe dwolunge rixað HOM. I. 117.

e.

e, *see* en. **e-,** *see* ʒe-. **ē,** *see* ǣ. **ē-,** *see* ǣ-.

ēa, int., *O.E.* ēa, ēaw; *ah*; e ale MAT. xxiii. 17.

ēa, sb., *O.E.*ʹēa, *O.Fris.* ā, ē,= *O.N.* ā, *O.L.G.*, *O.H.G.* aha, *Goth.* ahwa, *Lat.* aqua; *water, river*; eaa FRAG. 3; þas æ LAʒ. 2506; an æ Saba ʒehaten ORM. 7091; in are swiðe feire æ LAʒ. 1400.

    ēa-frosk, sb., *waterfrog*; **ēafroskes** [*v. r.* eafraskes, eafreskes] (*pl.*) HOM. I. 251, 326.

ēac, conj., *O.E.* ēac, ēc,= *O.Fris.* āk, *O.L.G.* ōc, *O.N.*, *Goth.* auk, *O.H.G.* ouh, *later Eng.* eke; *also*; æac HOM. I. 221; ec, æc, eke [eak, eke] LAʒ. 3745, 8001, 25682; ec HOM. I. 3; MARH. 14; ORM. 159; þe is nemed ec arm-herfnesse HOM. II. 5; ec, eke A. R. 50, 168; ek HAV. 1025; ROB. 200; ALIS. 2164; and ek ich frovri fele wihte O. & N. 535; ek [eek] CH. C. T. *A* 41; ek ōc.

ēacnen, v., *O.E.* ēacnian,=*Goth.* auknan; *increase, grow*: ekni hor robberie ROB. (W.) 2092*; þe beggere **ēcneð** (*pres.*) his bode HOM. II. 213.

[**ēad,** sb., *O.E.* ēad, = *O.N.* auðr; *wealth, happiness.*]

Ēd-gār, pr. n., *O.E.* Ēadgār; *Edgar,* P. L. S. ix. 144.

Ēd-mund, pr. n., *O.E.* Ēadmund; *Edmund,* P. L. S. xvii. 1.

ēdnesse, sb., *O.E.* ēadness; *beatitude,* HOM. II. 75; (*ms.* hednesse) SAL. & SAT. 230.

Ēad-ward, pr.n., *O.E.* Eadweard; *Edward,* MISC. 145.

ēadi, adj., *O.E.* ēadig, *cf. O.L.G.* ōdag, ōdog, *Goth.* audags (μακάριος), *O.N.* auðigr, *O.H.G.* ōtag; *rich, happy,* '*beatus,*' FRAG. 1; MARH. 2; A. R. 28; SHOR. 129; eadi, eædi, edi LAȝ. 2361, 6638, 16559; edi HOM. I. 173; MISC. 160; REL. II. 228; ȝedi P. L. S. viii. 115*; ædiȝ ORM. 2333; þeo ædie burde LAȝ. 16271; **ēadine** (*acc. m.*) HOM. I. 115; **ēadiȝe,** ædie (*pl.*) HOM. I. 47, 113; **ēadiure** (*comp.*) H. M. 13; **ēadieste** (*superlat.*) H. M. 45.

ǣdiȝ-leȝc, sb., *happiness,* ORM. 5706.

ēadi-līche, adv., *happily,* A. R. 328.

ēadinesse, sb., *O.E.* ēadigness; *happiness,* HOM. I. 217; A. R. 28.

ēadmēde, -mōd, etc., *see under* ēað.

ēaȝe, sb., *O.E.* ēage, = *O.Fris.* āge, *O.L.G.* ōga, *O.H.G.* ouga, *O.N.* auga, *Goth.* augō; *eye:* anre nedle eage LK. xviii. 25; eȝe O. & N. 426; þet eȝe HOM. I. 15; AYENB. 147; eie LANGL. *B* x. 123; eȝhe ORM. 9393; eghe PR. C. 2234; PERC. 691; ehe KATH. 1056; eie A. R. 60; HAV. 2545; þe neldis ei(e) P. L. S. i. 22; eiȝe [iȝe] WICL. MAT. v. 29; iȝe TREV. VIII. 159; MAN. (H.) 330; ie MIRC 62; CH. C. T. *A* 2680; **ēagen** (*dat.*) MAT. vii. 4; LEECHD. III. 96; eȝan, eȝen [= *O.E.* ēagan, *Goth.* augōna] (*pl.*) HOM. I. 23, 43; eȝen P. L. S. viii. 38; BRD. 26; AYENB. 81; egen REL. I. 210; eȝen, eien O. & N. 75; eiȝen FRAG. 5; eiȝe ROB. (W.) 6716*; WILL. 463; eien A. R. 64; eien [ȝeȝen (=ieȝen)] C. M. 17837; ighen [ihen, eien] CH. C. T. *A* 152; ehne, ehnen KATH. 497, 2002; eghne ISUM. 620; **ēagene,** eagen (*gen. pl.*) LEECHD. III. 94, 96; **eȝan,** eȝen, æȝen, eȝenen [*O.E.* ēagum, *Goth.* augam] (*dat. pl.*) LAȝ. 1885, 5736, 7247, 17076; egen, eien HOM. II. 23, 25; eien A. R. 16; AN. LIT. 91; eȝe BRD. 3; eȝhne ORM. 370; ene (*ms.* een) (*r. w.* kene) ANT. ARTH. xlvii.

ēȝe-brēu, sb., = *O.H.G.* ougebrā; *eyebrow;* **ēȝebrēwe** (*dat.*) P. S. 239.

ēhe-lid, sb., = *M.H.G.* ougelit; *eyelid,* HOM. I. 265; **eiȝeliddes** (*pl.*) TREV. V. 189.

eȝhe-salfe, sb., *O.E.* ēagsealf; *eye-salve,* ORM. 1852.

ǣh-sēone, eæhsēne, sb., = *O.H.G.* ougsiuni; *eyesight,* LAȝ. 3092, 9703; exsene REL. I. 103; out of his eighesene TRIST. 2450.

ēiȝe-siȝt, sb., *eyesight,* S. A. L. 164; iesiȝte MIRC 325.

ēh-sihðe. sb., *evesight,* MARH. 17; eihsihðe A. R. 58; eȝhesihþe ORM. 1868.

ēie-sōr, sb., *eye-sore,* REL. I. 110.

ēh-þürl, sb., *O.E.* ēag-, ēahþyrl; *window,* HOM. I. 83; eiþurl A. R. 62; (*ms.* eyþurl) MARH. 8.

eahte, ahte, card. num., *O.E.* eahta, = *O.Fris.* achta, *O.L.G., O.H.G.* ahto, *Goth.* ahtau, *O.N.* ātta, *Lat.* octo, *Gr.* ὀκτώ; *eight,* JUL. 60, 61; æhte [eahte, ehte] LAȝ. 3919, 26502; ehhte ORM. 580; eihte A. R. 12; eihte MISC. 145; eȝte AYENB. 45; egte GEN. & EX. 1349; eiȝte WICL. JOHN xx. 26; eiȝte and nienti ROB. 62; fifti & eiȝte 533; eighte [auht] MAN. (F.) 3714; eighte and twentithe CH. C. T. *A* 5; eite LUD. COV. 129; ahte CHR. E. 324; aughte D. ARTH. 278; aȝt A. P. iii. 11; aght IW. 1438.

ehte-tēne, card. num., *O.E.* eahtatȳne; *eighteen,* LAȝ. 18014*; æhtene 14252; eiȝtetene ROB. 407; eiȝte-, eiȝtene TREV. III. 17; eiȝtene PR. P. 137.

eiȝte-tēoþe, ord. num., *O.E.* eahtatēoða; *eighteenth,* P. L. S. xii. 5; eȝtetenþe PROCL. 7; eiteteþe, eiȝteteþe, eiȝttene ROB. (W.) 1860*, 6490, xx. 409*.

eiȝte-ti, card. num., *O.E.* eahtatig, = *O.H.G.* ahtozug; *eighty,* ROB. 292; WICL. I KINGS xxii. 18; eiȝti '*octoginta*' PR. P. 137.

eahtuðe, ehtuþe, ord. num., *O.E.* eahtoða, = *O.H.G.* ahtodo, *Goth.* ahtuda, *O.Fris.* achtunda, *O.N.* āttundi; *eighth,* HOM. I. 81, 107; eihtuðe, eihteoðe A. R. 14, 236; ehteðe, eihteðe HOM. II. 47, 137; ehtende 87; eihteoþe MISC. 145; eiȝteþe ROB. 213; eighteþe LANGL. (Wr.) 9538; eghþe MIRC 498; (*ms.* ehhtennde) ORM. 4196; eȝtende AYENB. 10; SHOR. 99; egtende GEN. & EX. 1199; aghtend PR. C. 6895.

ēake, sb., *O.E.* ēaca, = *O.N.* auki; *addition:* tō ēken, *in addition,* HOM. II. 51; teken [to eke] MARH. 4; teken ORM. 747; teke [teken, to eke] A. R. 78; teke H. M. 25; ? on ēke SPEC. 34; *comp.* over-ēake.

ēke-name, sb., = *O.N.* aukanafn; *nickname:* a vile ekename H. S. 1531; ekename, nekename '*agnomen*' PR. P. 352.

eal, *see* al.

eald, adj., *O.E.* eald, ald, = *O.L.G.* ald, old, *O.H.G.* alt, *cf. Goth.* alþeis; *old,* P. L. S. viii. 2; ald HOM. I. 159; ORM. 747; AYENB. 104; Iw. 1556; MIN. iii. 19; he was fiftene ȝer ald LAȝ. 301; an ald mon 18707; ald englisch MARH. 23; twelf winter ald SHOR. 44; old A. R. 6; O. & N. 25; eld, old PR. P. 137; CH. C. T. *D* 1213; þa ealde laȝe HOM. I. 9; eld OCTOV. (W.) 656; ED. 1656; S. & C. I. li.; an eld cloth HAV. 546; þe ealde lawe SHOR. 44; þe ialde (*ms.* yalde) laȝe AYENB. 7; iealde (*ms.* yealde) 97; þu alde monslahe MARH. 12; þe alde king LAȝ. 2959; olde 3001; þe olde man BRD. 2; **ealden** (*dat. m. n.*) MK. ii. 21; alde [olde] O. & N. 1183; be þo ialde laghe MISC. 27; alde (*pl.*) ORM. 126; þa alde LAȝ. 28444; elde SPEC. 95; LUD. COV. 73; **ealden** (*dat. pl.*) MAT. v. 21; alden LAȝ. 6774; to olde [elde] men WICL. MAT. v. 21; **eldre** (*compar.*) *elder* ORM. 13215; MISC. 58; þe eldre [uldre] P. L. S. viii. 162; ældre, eldere [eldre] LAȝ. 3926, 4290; uldere ROB. (W.) 750*; ealdre HOM. I. 23; **eldren** (*pl.*) *elders, parents* ROB. 65;

SHOR. 97 ; ælderen, ældere [eldre] LAȝ. 7354, 12516; ulderne ROB. (W.) 1495* ; elder ROB. (W.) 907* ; þe ealdren H. M. 27 ; eldren LANGL. *A* iii. 248; eldre MISC. 46 ; ure eldre HOM. II. 43 ; ildre REL. I. 175 ; eldres CH. C. T. *E* 65 ; eldrene (*gen. pl.*) KATH. 81 ; ælderne LAȝ. 13922 ; elderne (*ancestors*) DEP. R. i. 65 ; eldest (*superl.*)' MK. x. 43 ; eldeste CHR. E. 423 ; AYENB. 104 ; ældeste [eldeste] LAȝ. 2930 ; ealdeste H. M. 41 ; uldest ROB. (W.) 732*.

eld-fader, sb., *O.E.* eald-, ealdefæder, = *O. Fris.* ald-, aldafeder, *O.L.G.* aldfader, *O.H.G.* altfater; *ancestor, grandfather; father-in-law, stepfather;* '*avus*' VOC. 205 ; el(d)fadir '*socer*' PR. P. 137; aldfader M. H. 122 ; ealdefader ANGL. VII. 220 ; aldevader LAȝ. 31009.

eald-līc, adj., *O.E.* ealdlíc; *oldish*: an oldli man WICL. JOB xli. 23 ; aldelīke (*adv.*) ORM. 1229 ; aldeliȝ 2553.

eld-mōder, sb., *O.E.* ealdmōder; *ancestress, grandmother; mother-in-law;* '*avia*' VOC. 205 ; C. M. 1189; eldmodir '*socrus*' PR. P. 137.

eldnesse, sb., *O.E.* ealdnyss; *oldness*; eldnesse WICL. DEUT. viii. 4.

ealde, sb., = *M.L.G.* olde, *O.N.* öld (*gen.* aldar), *Goth.* alþs (*gen.* aldais) *f.*; *old age*: þis weorldes ald ORM. 8831 ; þat him schal on ealde (? = elde) sore reowe REL. I. 184 ; for þas kinges alde (? *for* ælde) LAȝ. 19411 ; in olde S. S. (Wr.) 641 ; in to olde SHOR. 2.

ealder, sb., *O.E.* ealdor, aldor ; *prince, chief*; ælder, alder LAȝ. 1365, 16562 ; alder HOM. I. 219; ealdres (*pl.*) ['*principes*'] MAT. ii. 4; **aldren** (*dat. pl.*) HOM. I. 231.

alder-dōm, sb., *O.E.* ealdordōm ; *principality*, ORM. 18278.

alder-man, sb., *O.E.* ealdorman, = *O.Fris.* aldirmon, *mod.Eng.* alderman ; *ruler, viceroy*, ORM. 14061 ; aldermon '*princeps*' FRAG. 2 ; LAȝ. 1420.

alder-scipe, sb., *rank of a ruler* ['*principatus*'], HOM. I. 219.

ealdien, v., *O.E.* ealdian, *cf. O.H.G.* alten, altēn, *Goth.* (us-)alþan ; *grow old*; iealdi AYENB. 97; elden R. R. 396; aldeð (*pres.*) HOM. I. 35 ; eldeþ ORM. 18826; elde (*subj.*) HOM. I. 21 ; ældede (*pret.*) LAȝ. 2937 ; (mi bones) eldeden WICL. PS. xxxi. 3 ; *comp.* forealdien.

ealdinge, sb., *O.E.* ealdung ; *growing old*, AYENB. 95.

ēam, ǣm, sb., *O.E.* ēam, = *O.Fris.* ēm, *M.Du.* oom, *O.H.G.* oheim ; *uncle*, LAȝ. 8831, 8897 ; eam, em GEN. & EX. 1758, 3747 ; eom SAX. CHR. 261 ; em HICKES I. 227 ; HAV. 1326; OCTOV. (W.) 1379; GOW. II. 267 ; WICL. I PARAL. xxvii. 32 ; MAN. (F.) 4123 ; GAW. 356 ; M. ARTH. 1681 ; min em (*ms.* mi nem) TRIST. 921 ; eem PR. P. 139 ; B. DISC. 397 ; iem (*ms.* yem) LEECHD. III. 82 ; CHEST. II. 55 ; ēmes

(*gen.*) WILL. 3426 ; emis S. S. (Wr.) 1054 ; for the manere of thee her ēme (*dat.*) CH. TRO. (BELL) i. 1022 (3-*Text* 1015 *has* em) ; ēmes (*pl.*) PERC. 1050.

eam, *see* **am**.

ēan, adj., *O.E.* (ge-)ēan ; *filled* ; ēne [eene, ȝene] (*pl.*) WICL. PS. cxliii. 13.

ēanen, v., *O.E.* ēanian, = *Du.* oonen ; *yean*; enin '*feto*' PR. P. 140; ēnede (*pret.*) TREV. IV. 451.

ēar, sb., *O.E.* ēar, *Northumb.*, eher, = *M.L.G.* ār, *O.H.G.* ehir, ahir, *Goth.* ahs, *O.N.* ax, *Lat.* acus (*gen.* aceris) *n.*; *ear (of corn); er* '*spica*' VOC. 233 ; PR. P. 141 ; CH. L. G. W. 76 ; ēare (*dat.*) MK. iv. 28 ; ere ALIS. 797 ; iere (*ms.* yere) AYENB. 28 ; ēar (*pl.*) MAT. xii. 23 ; eares A. R. 260; GEN. & EX. 2104 ; eres ROB. 490 : *comp.* corn-ēr.

ēar, ēarlīche, *see* **ǣr**.

eard, erd, sb., *O.E.* eard, = *O.L.G.* ard, *M.H.G.* art; *land, country, dwelling* ['*regio*'], HOM. I. 13, 59; ærd LAȝ. 29174 (*miswritten* ard 13473) ; ORM. 1416 ; erd GEN. & EX. 210 ; ure erd is on hevene HOM. II. 149 ; earde (*dat.*) A. R. 358 ; þe king of þissen earde LAȝ. 7417 ; þa com he to þan ærde þer læi his ferde 19202 ; erde HOM. I. 115 ; LANGL. *B* vi. 203 ; A. P. i. 248 ; ich fare hom to min(e) erde O. & N. 460 ; *comp.* kine-, middel-, ūten-eard.

erd-folc, sb., *O.E.*..., *people*, GEN. & EX. 1880.

eardien, v., *O.E.* eardian, = *O.L.G.* ardōn, *O. H.G.* artōn ; *dwell, inhabit*: ha ne muhen nawt somen earden in hevene H. M. 43 ; erden WILL. 5260 ; erdest (*pres.*) PS. xxi. 4 ; eardið MARH. 9 ; erdeden (*pret.*) HOM. I. 89.

earding, sb., *O.E.* eardung ; *habitation*; erding HOM. II. 159.

earding-stōwe, sb., *dwelling-place*, MAT. xvii. 4 ; eardingstowe, erdingstowe O. & N. 28 ; erdingstouwe HOM. II. 173.

ēare, sb., *O.E.* ēare, = *O.Fris.* āre, *O.L.G.* ōra, ōre, *O.H.G.* ōra, *O.N.* eyra, *Goth.* ausō *n., cf. Lat.* auris *f.*; *ear*, KATH. 1734 ; þet eare, ieare AYENB. 177, 189 ; iare MISC. 36 ; ære ORM. 2800 ; ere P. L. S. xv. 118 ; ere '*auris, ansa*' PR. P. 141 ; ēaren (*dat.*) LEECHD. III. 90 ; ēaran, earen (*pl.*) MAT. xiii. 16 ; MK. iv. 9 ; earan, earen, eren HOM. I. 23, 49, 127 ; earen MARH. 20 ; A. R. 90 ; O. & N. 338 ; HORN (R.) 969 ; earen, æren FRAG. 5, 7 ; earen, ieren AYENB. 204, 257 ; eren ALIS. 6448 ; P. S. 154 ; eres MAND. 205 ; PR. C. 782 ; ires HORN (L.) 959 ; ȝeres (= ieres) HALLIW. 946 ; ēaran, earen (*dat. pl.*) MAT. xiii. 15 ; LK. iv. 21 ; it com þe kinge to ere ROB. 492.

ēre-lappe, sb., *O.E.* ēarlæppa ; *earlap*, REL. I. 54.

ēar-prēon, sb., *O.E.* ēarprēon ; *earring, 'inauris,'* FRAG. 2.

ēre-ring, sb., *earring*, WICL. JOB xlii. 11.

ȝēr-wigge, sb., *O.E.* ēarwicga; *earwig*, PR. P. 143*.

ēared, adj., *from* ēare; *eared* (*having a handle*); erid WICL. NUM. iv. 7.

ēaren, v., *from* ēar; *come into ear*: atte crestemasse barlich beginneþ eere K. COL. 30, 31.

earewe, *see* arwe.

[earfeð, sb., *O.E.* earfoð, earfeðe, = *O.N.* erfiði, erfaði, *O.L.G.* arbed, arbid, arbedi, arbidi, arvithi, *O.Fris.* arbeid *n.*, *cf. Goth.* arbaiþs, *O.H.G.* arbeit *f.*; *labour, tribulation.*]

earfeþ-, earveþ-sīþ, sb., *O.E.* earfoðsīð; *calamity*, FRAG. 5.

earfeð, adj., *O.E.* earfoðe, earfeðe, = *O.N.* erfiðr; *difficult*; earf(e)ð [erveð] KATH. 999; erveð [arveð] for te paien A. R. 108; arfeþ ORM. 17334; erveðer (*compar.*) HOM. II. 63.

arveð-finde, adj., *difficult to find*, HOM. II. 27.

arveð-forþe, adj., *difficult to effect*, HOM. II. 131.

erveð-helde, adj., *difficult to hold*, P. L. S. viii. 157; arefeðheald(e) HOM. II. 229.

erveð-līche, adv., *O.E.* earfoðlīce; *with difficulty*, A. R. 328; arveðliche HOM. II. 123.

earveð-, ærfeð-, erfeðnesse, sb., *O.E.* earfoðness; *difficulty*, HOM. I. 21, 105, 223.

ærfeð-telle, adj., *difficult to count, innumerable*, HOM. I. 231.

arveð-winne, adj., *difficult to win*, HOM. II. 49.

earm, sb., *O.E.* earm, = *O.Fris.* erm, *O.L.G.*, *O.H.G.* arm, *O.N.* armr, *Goth.* arms, *Lat.* armus; *arm*, HOM. I. 189; erm 213; ærm [arm] LAȜ. 28040; arm ROB. 17; MAND. 172; an arm of þe se TRIST. 2246; arum HAV. 1982; earme (*dat.*) A. R. 112; earmes (*pl.*) LAȜ. 1872; SHOR. 123; earmes, ermes A. R. 394, 402; armes TREAT. 139; (*ms.* arrmess) ORM. 7616; ærmen, armen (*dat. pl.*) LAȜ. 2233, 21869; arme MISC. 150.

erm-ēddre, sb., *brachial vein*, A. R. 258.

arm-hole, sb., *arm-hole, armpit*; (*ms.* harm-) VOC. 245.

earm², arm, sb., *O.E.* earm, = *O.L.G.* arm, *O.Fris.* erm, *O.H.G.* arm, *O.N.* armr, *Goth.* arms; *poor, miserable*, H. M. 9; ærm on his mode LAȜ. 6608; earme steorve MARH. 12; þet ærme folc SAX. CHR. 253; þe arme gume REL. I. 186; his erme saule HOM. I. 27; earme (*dat. n.*) HOM. I. 35; ermne (*acc. m.*) 115; earme (*pl.*) P. L. S. viii. 115; arme O. & N. 537; ærmest (*superl.*) LAȜ. 14893; earmest JUL. 53; *deriv.* ermen·

arm-heorted, adj., *tender-hearted, merciful*, HOM. II. 95.

armhertnesse, sb., *mercy*, HOM. II. 95.

erm-līc, adj., *O.E.* earmlīc; *poor, miserable*,

HOM. I. 115; ærmliche biþrungen LAȜ. 9435; ærmlīche, armliche (*dat. pl.*) LAȜ. 593, 1046.

earming,·sb., *O.E.* earming, erming, yrming, = *M.Du.* arminc, erminc; *poor, miserable creature*; þa erming sauten HOM. I. 41; erming FRAG. 7; makede him . . . erming (*a pauper*) HOM. II. 61; O. & N. 1111; ærming LAȜ. 16690; earminges (*pl.*) MISC. 69 (erminges HOM. II. 230).

earmðe, sb., *O.E.* earmðu, ermðu, yrmðu, = *O.H.G.* armida; *misery*, H. M. 27; ermðe MK. xii. 44; HOM. I. 113; ærmðe LAȜ. 16143; armþe REL. I. 182.

earn, sb., *O.E.* earn, = *M.L.G.* arn, *O.N.* örn, (*gen.* arnar), *cf. O.H.G.* aro (*gen.* aren), *Goth.* ara (*gen.* arins); *eagle*, A. R. 134; ærn LAȜ. 2826; ORM. 5880; ern HAV. 572; WILL. 3105; OCTOV. (W.) 196; LAUNF. 268; arn AYENB. 61; ernes (*gen.*) REL. I. 209; ROB. 177; earnes (*pl.*) A. R. 196; ærnes, arnes [earnes (*ms.* hearnes)] LAȜ. 21753, 21755; ernes MAN. (F.) 10202.

earnien, v., *O.E.* earnian, *cf. M.Du.* arnen, ernen, *O.H.G.* arnēn; *earn*; ernien HOM. I. 7; arne N. P. 14; earneð (*pres.*) H. M. 19; earnie ȝe HOM. I. 41; ernede (*pret.*) HOM. II. 5; earned (*pple.*) KATH. 2255; *comp.* ȝe-, of-earnien.

earninge, vbl. n., *O.E.* earning, earnung; *earning, merit*, MISC. 60; erninge P. L. S. viii. 32; ernunge HOM. I. 19; earninges (*pl.*) P. L. S. viii. 161.

ease, *see* aise.

ēast, sb., *O.E.* ēast, = *O.Fris.* āst, *M.Du.* oost, *O.N.* aust(-maðr); *east*, H. M. 11; east (*in the east*) and west feor and neor i do wel faire mi mester O. & N. 923; þet ieast AYENB. 124; þat æst & west & suþ & norþ þis middelærd biluken ORM. 12125; verden heo æst [est] (*eastward*) LAȜ. 10590; est 'oriens' PR. P. 143; GEN. & EX. 829; toward þan est BRD. 2; est and west P. S. 331; i shall walk eest MIR. PL. 172.

Ǣst-ængle, pr. n., *O.E.* Eastengle; *East-Angles*, LAȜ. 12253; Estangle ROB. 4.

æst-dale, sb., *O.E.* ēastdæl; *eastern part*, ORM. 16400.

æst-ende [ēasteande], sb., *O.E.* eastende; *east-end*, LAȜ. 28305.

ēst-ȝet, sb., *east gate*, HOM. I. 5; estȝate TREV. IV. 451.

æst-halve, sb., *O.E.* ēasthalf; *east part*, LAȜ. 29287.

ēast-lond, sb., *O.E.* ēastland; *the east*: alle þe meistres weren in ēstlonde (*dat.*) KATH. (E.) 535.

ēast-partie, sb., *east, east division*, MAND. 156.

ēst-rīche, sb., *the east*, HOM. II. 45.

ēst-sēe, pr. n., *O.E.* ēastsǣ; *east sea*, P. L. S. xiii. 18.

Ēast-sexe, pr. n., _O.E._ Eastseaxe; _Essex,_ MISC. 146; Æstsæx [Estsex] LAȝ. 15390; Estsex ROB. 3.

ǣst-sīde, sb., _east-side,_ LAȝ. 21798; estside ROB. 2.

ēast-ward, adv., _O.E._ ēastweard; _eastward,_ FRAG. 5; estward BRD. 2.

ēst-wind, sb., _east-wind,_ PR. P. 143.

ēasten, adv., _O.E._ ēastan,=_O.N._ austan, _O.L.G._ ōstan, ōstane, ōstana, _O.H.G._ ōstana; _eastward_; esten fro ȝa(n) GEN. & EX. 1264; A. R. 232; MAN. (H.) 45; bi este HORN (L.) 1325; P. L. S. xviii. 2; fram þe easte SHOR. 122.

ēaster, sb., _O.E._ ēaster (_gen._ ēastres, _pl._ ēastro) _n.,_ ēastre (_pl._ ēastran, ēastron), _cf._ _O.H.G._ ōstra (_pl._ ōstrun) _f._; _Easter_; ester BRD. 7; þa æstre [easter] wes aȝonge LAȝ. 24195; aster MIRC 143; iēstre (_dat._) AYENB. 213; an ane æstre LAȝ. 24143; astere AUD. 41; ED. 3140; ēastre (_pl._) MAT. xxvi. 2; eastren MK. xiv. 1; eestern '_pascha_' PR. P. 143; ēastren (_dat. pl._) SAX. CHR. 256; befor estren E. W. 40; in estern (_ms._ esterne) GEN. & EX. 3290.

ēaster-dei, sb., _O.E._ ēasterdæg, = _O.H.G._ ōstertag; _Easter-day,_ HOM. I. 123; esterdei A. R. 412.

ēster-ēven, sb., _O.E._ ēasterǣfen; _Easter-eve,_ A. R. 70; HOM. II. 95; estereve TREV. V. 189.

ēster-feste, sb., _Easter-feast,_ ROB. 441.

ēaster-līc, adj., _O.E._ ēasterlīc, = _O.H.G._ ōstarlīch; _Easter_: fram þan esterliche (_dat._) deie HOM. I. 89.

ǣster-lomb, sb., =_Ger._ ōsterlamm; _paschal lamb,_ ED. 88.

ēster-morewe, sb., = _Ger._ ōstermorgen; _Easter morning,_ SPEC. 96.

ēster-tīd, sb., _O.E._ ēastertīd (TREAT. 6); _Easter-tide,_ HOM. I. 87; estertide WILL. 1417; AR. & MER. 2849.

ēasterne, adj., _O.E._ ēasterne; _eastern_: of æsterne (? _ms._ æstene) weorlde LAȝ. 27393.

eatel, _see_ atel. eaten, _see_ eten.

ēað, adj., _O.E._ ēaðe, ēðe, ȳðe (_easy, desert_), = _O.L.G._ ōthi (_easy_), _O.H.G._ ōdi (_easy, desert_), _O.N._ auðr (_empty_), _Goth._ auþs (ἔρημος); _easy,_ KATH. 626; HOM. I. 279; æþ ORM. 13012; eð him wes on heorten LAȝ. 2234; þa wes his hurte æðe 8178; an his mode him wes þa æðe 22359; is . . . eð to overkesten A. R. 274; eþ ROB. 327; GOW. I. 60; MIN. v. 47; MAN. (H.) 194; eth WILL. 3571; eþe GAW. 676; A. P. i. 1201; CH. TRO. v. 850; bliðe hi bieþ and eaðe HICKES I. 224; ēaðe (_adv._) HOM. II. 173; eaðe meiht tu beon prut A. R. 276; eaðe, ieþe P. L. S. viii. 143; æþe ORM. 17676; eþe HORN (R.) 61; C. L. 1281; iþe HORN (L.) 57; eþere (_compar._) LK. v. 23; hwæðer is eðre to seggene MK. ii. 9; mid eþran

ledene FRAG. 1; þe æð (_easier_) þe scal iwurðen LAȝ. 22100; _comp._ un-ēaðe.

ēð-bēte, adj., _easy to amend,_ HOM. II. 63.

ēāð-falle, adj., _easily overcome_: ho is eaðfalle (_v. r. for_ eð-fallen), A. R. 62.

ēð-fēle, adj., _easily felt,_ A. R. 194.

ēþ-gete, adj., _O.E._ ēaðgete; _easily got,_ MISC. 74.

ēð-healde, adj., _O.E._ ēðhylde; _easily held_ ['_contenti_'], LK. iii. 14.

ēð-lēte, adj., _lightly esteemed,_ P. L. S. viii. 37; þe oþre mannes wif was lef, his awene eðlete 130.

ēðe-līc, adj., _O.E._ ēaðe-, ēðe-, ȳðelīc,=_O.N._ auðligr; _easy, light, trifling,_ HOM. I. 21; eðelich JUL. 13; on eðelich stiche A. R. 282; þurh eðeliche dede LAȝ. 24921; for æþelike gilte ORM. 10172; eðeliche & wake HOM. I. 255; eðeliche men II. 35; æþeliȝ (_adv._) ORM. 12534; þe eþelikeste (_superl._) ston is cristal FL. & BL. 274.

ēð-lūke, adj., _easily pulled,_ JUL. 70.

ēad-mēde, sb., _O.E._ ēad-, eaðmēdu,=_O.L.G._ ōdmōdi, _O.H.G._ ōt-, ōdmuotī; _grace_: þurh his ædmeden LAȝ. 10013; and ȝeorneden Arðures grið & his aðmeden 21866.

ēad-mēde, adj., _O.E._ ēadmēde,=_O.L.G._ ōdmōdi, _O.H.G._ ōtmuoti; _humble_: þe edmeda riche HOM. I. 115.

ēd-mōd, adj., _O.E._ ēad-, ēaðmōd; _humble, gracious,_ '_humilis,_' FRAG. 3; A. R. 158; MISC. 141; ed-, admod HOM. II. 89, 187; he was ædmod & milde ORM. 2887; ādmōde (_pl._) HOM. I. 115; mid ædmode worden LAȝ. 22422; we ahte to beon þe ēdmŏddre (_comp._) HOM. I. 5.

ǣdmōd-leȝc, sb., _humility,_ ORM. 19297.

ēadmōd-līche, adv., _O.E._ ēadmōdlīce; _humbly, graciously,_ HOM. II. 141; edmodliche A. R. 94; æðmodliche LAȝ. 26288; ædmodliȝ ORM. 1108.

ēadmōdnesse, sb., _O.E._ ēadmōdness; _humility, graciousness,_ HOM. I. 17; edmodnesse A. R. 8; MISC. 193; ædmodnesse ORM. 19218.

ēd-mōdi, adj., _O.E._ ēadmōdig; _humble_: þa edmōdies (_gen._) monnes bonen A. R. 246.

ēð-sēne, adj., _easily seen, plain,_ KATH. 381; eðsene, eðcene A. R. 116.

ēð-winne, adj., _easy to win,_ HOM. II. 49.

ēaðien, v., _ease_; ēðede (_ms._ eððede) (_pret._) his sorge GEN. & EX. 1439.

ēað-mōdien, v., _O.E._ ēadmōdian; _humble_: ēadmōdieð (_imper._) A. R. 278*; (he) ēadmōde(de) (_pret._) hine seolfne HOM. I. 17.

ēau, _see_ ǣ. eaul, _see_ awel.

eaver, _see_ ever. ēaw, _see_ ǣ.

[ēawen, v., _O.E._ ēawan, ēowan, īowan, īewan, ȳwan, = _O.Fris._ auwa, _O.H.G._ ougen, _Goth._ augjan; _show_; _comp._ æt-ēowen.]

eax, sb., _O.E._ eax, æx, acas (MAT. iii. 10),=

O.N. öx, *O.L.G.* acus, *O.H.G.* achus, *Goth.*
aqizi; *axe*, A. R. 384; eax, æx LAȝ. 2312,
6473; æx '*securis*' FRAG. 4; axe VOC. 196;
ORM. 10063; ROB. 17; ax [axe] WICL. MAT.
iii. 10; CH. L. G. W. 2000; ax VOC. 234;
O. & N. 658; HAV. 1894; LEG. 47; S. S.
(Web.) 384; MAND. 251; P. S. 222; ex PR. P.
144; B. DISC. 1180; **eaxe** (*dat.*) LAȝ. 2310;
axen (*dat. pl.*) LAȝ. 7478; *comp.* **bol-, brād-,
hand-, wī-eax(-ax).**

[**eax,** sb., *O.E.* eax, = *O.H.G.* ahsa, *Lat.* axis;
axle.]

   ax-trēo, sb., *axle-tree,* JUL. 56; axtre MIRC
334; ax-, extre WICL. ECCLUS. xxxiii. 5; extre
VOC. 180; PR. P. 145; CH. ASTR. i. 14.

eaxle, sb., *O.E.* eaxl, = *O.Fris.* axle, *O.N.*
öxl, *O.L.G.* ahsla, *O.H.G.* ahsla, ahsela,
ahsala, *Lat.* axilla; *shoulder*; æxle, exle LAȝ.
2263, 18032; **eaxle** (*pl.*) LEECHD. III. 120;
LK. xv. 5; exlan MAT. xxiii. 4; axeles
['*scapulis*'] PS. xc. 4; he hit berð an his eaxlun
HOM. I. 245.

eban, sb., *Lat.* ebenus, *Gk.* ἔβενος; *ebony,*
PR. P. 135.

ebbe, sb., *O.E.* ebba, *O.Fris.* ebba, ebbe; *ebb,*
ALIS. 6184; CH. C. T. *F* 259; MAN. (F.)
12317.

ebbin, v., *O.E.* ebbian; *ebb,* PR. P. 135; ebbe
LUD. COV. 56; ebbe or flowe M. T. 131;
ebbeð (*pres.*) REL. I. 128; ebbeþ TREAT. 137.

   ebbinge, vbl. n., *ebbing,* HOM. II. 177.

ēbure, *see* **ǣbere.**

ēc, *see* **ēac.**

ecco, sb., *Lat.* ēchō; *echo,* PR. P. 135.

ēch, *see* **aighwilc.** **éche,** *see* **áche.**

ēche, adj., *O.E.* ēce; *eternal,* MAT. xix. 16;
KATH. 302; A. R. 108; REL. I. 130; ORM.
971; O. & N. 1279; eche reste HICKES I.
224; þet eche fur FRAG. 7; in eche (*for ever*)
JUL. 35; **ēches** (*gen. m.*) MK. iii. 29; **ēchere**
(*dat. f.*) ANGL. VII. 220; *comp.* **efen-ēche.**

   ēche-līc, adj., *O.E.* ēcelīc; *eternal*: ðeo
echeliche riche HOM. I. 139; II. 23; KATH.
2387.

   ēche-liche, adv., *eternally,* FRAG. 7; HOM.
I. 139.

   ēcnesse, sb., *O.E.* ēcness, ēcnyss; *eternity,*
MARH. 7; KATH. 664; echnesse LK. i. 33;
HOM. I. 109, 251; echenesse A. R. 144.

ēche, sb., ?=**ēake,** ?*increase*: wiþoute wane
and eche SHOR. 10.

eched, sb., *O.E.* eced, æced, = *O.L.G.* ecid,
from Lat. acētum; *vinegar*; æched LK. xxiii.
36; echede (*dat.*) MAT. xxvii. 48.

ēchen, v., *O.E.* ēcan, ȳcan, = *O.L.G.* ōkian, *cf.*
O.H.G. auchōn; *from O.E.* *ēacan (*pple.*
ēacen), *Goth.* aukan (*pret.* aiauk), *Lat.* augēre,
mod.Eng. eke; *increase, augment,* A. R. 124;
to echen þin ahte R. S. v; echan LK. xii. 25;
æchen [eche] LAȝ. 13065; eche ROB. 199;

SPEC. 92; SHOR. 38; ALIS. 6026; WICL.
ECCLUS. xviii. 5; echen, eken HOM. II. 31,
57; eken ORM. DED. 57; eke MAN. (H.)
176; PR. C. 3256; ēche (*pres.*) CH. TRO. i.
705; echeð HOM. I. 103; ekes ['*adjiciet*'] PS.
xl. 9; ēked (*pple.*) PL. CR. 244.

   ēchinge, sb., *addition*; eking WINT. VIII.
viii. 53; GOW. II. 22; i shal putte upon
Dibon **ecchingus** (*pl.*) WICL. IS. xv. 9.

echin, sb., *Lat.* echīnus, *Gr.* ἐχῖνος; *echinus
sea-urchin*; **echinnis** (*pl.*) CH. BOET. iii. 9
(82).

ēchnesse, *see* **ēcnesse** *under* **ēche.**

ēchte, *see* **ǣhte.**

eclipse, sb., *Fr.* eclipse; *eclipse,* LANGL. *C* xxi.
140; esclepis [clipes, clippis] C. M. 16814.

eclipse, v., *eclipse*; **eclipsen** (*pres.*) GOW. II.
153.

ecliptik, adj. *and* sb., *ecliptic*: the ecliptik
line CH. ASTR. i. 21 (12); a latitude fro the
ecliptik ii. 4 (19).

ēcnen, *see* **ēacnen.**

[**ed-,** *prefix, O.E.* ed-, *Goth.* id-, *O.N.* ið-, *O.
H.G.* it-; *again, back.*]

ēd, *see* **ēad.**

edder-wort, *see under* **addre.**

eddest, edist, adj. superl., ? *for* oddist, *see*
od; ? *distinguished in war,* D. TROY. 5324,
5950.

eddre, *see* **addre.**

ēddre, sb., *O.E.* ǣdre, ǣddre, = *O.Fris.* ēddre,
O.H.G. ādra; *vein*; (*ms.* heddre) REL. II 273;
ēddren (*pl.*) A. R. 258; *comp.* **erm-ēddre.**

ēde, *see* **ēode.**

edera, sb., *Lat.* hedera; *ivy*: crownid with
edera WICL. 2 MACC. vi. 7; eder JON. iv. 6.

ed-gān, v.; ich schal ... undervongen deaðes
wunde and ich hit wulle heorteliche vorto
ofgan [edgan C] þine heorte A. R. 390.

ed-grō(w)e, sb., *prov.Eng.* edgrow, ed-
growth; *aftermath,* PR. P. 135.

ed-healden, v., ? *retain*; edhalde (? *for* et-
halde) HOM. I. 149.

ēdi, *see* **ēadi.**

edificatiōn, sb., *a building*; edificaciouns
(*pl.*) TREV. I. 405.

edifíen, v., *Fr.* edifier, *from Lat.* ædificāre,
mod.Eng. edify; *build, make, form* ['*edifi-
care*'], WICL. GEN. xi. 8*; edifie LANGL. *B*
xviii. 43; WICL. DEEDS xx. 32; edifíede
(*pret.*) the rib into a woman GEN. ii. 22;
edefiede an auter viii. 12.

[**edisch,** sb., *O.E.* edisc, edesc; *park.*]

   edis(h)-henne, sb. ['*coturnix*'], PS. civ. 40.

edit, sb., *O.Fr.* edit, *Lat.* ēdictum; *edict*:
ROB. 568.

[**ed-lǣchen,** v., *O.E.* edlǣcan; *repeat; comp.*
ȝe-edlǣchen.]

ed-lēn, sb., *O.E.* edlēan,=*O.H.G.* itlōn; *recompense*, HOM. I. 103.

ēdmōd, *see* ēadmōd *under* ēað.

ēdnesse, *see under* ēad.

ed-wīt, sb., *O.E.* edwīt,=*O.L.G.* edwīt, *Goth.* idveit, *O.H.G.* it-, itawīz; *reproach, blame*: þet edwit A. R. 108; edwit ROB. 379; ædwit LAȜ. 5827; falle in edwite O. & N. 1233.

ed-wīte, v., *O.E.* edwītian, *cf. Goth.* idveitjan, *O.H.G.* itwīzon; *reproach, blame,* LANGL. B v. 370; P. L. S. iv. 8; **edwiteð** (*pres.*) HOM. I. 253; edwiteð, edwiten H. M. 37; **edwīte** (*subj.*) A. R. 426; edwite WICL. ECCLUS. viii. 6; **edwīted** (*pret.*) LANGL. C vii. 421.

ed-wīting, sb., *reproach, blaming*; edwiting (*var. r.* upbreiding) WICL. WISD. v. 3*; ECCLUS. vi. 1*.

ee, *see* ēaȝe. **eek,** *see* ēac. **eest,** *see* ēast.

eem, *see* ēam. **eete,** *see* ēten. **ef,** *see* if.

efen, adj. *and* adv., *O.E.* efen, emn, em,= *O. Fris.*even, iven, *O.N.* iafn, *O.H.G.* eban, epan, *Goth.* ibns; *even, equal,* ORM. 1840; efne & smeðe A. R. 2; þet is efne wið þe HOM. I. 209; þole wið efne heorte þe dom 265; þer scal þe hehȝe beon æfne [efne] þan loȝe LAȜ. 22928; his æfne wiht of golde 30835; ete ... enes o dai & evene (*just*) fille HOM. II. 67; evene dom C. L. 490; evene juge CH. C. T. *A* 1864; þet makeþ þane wal emne AYENB. 151; an emne *equally* SHOR. 75; al an evene (*quietly*) ROB. (W.) 9567; efne þissen worden ... Beduer heo gon hirten LAȜ. 25939; he heom wes leof æfne al swa heore lif 13924; æfne forð rihten (*r.* rihte) 22773; ich singe efne O. & N. 313; evene GEN. & EX. 331; WILL. 747; þat is evene above þin heved TREAT. 132; to bere him evene CH. C. T. *A* 1523; **em forth** (*according to*) mi might 2235; em forth mi wit CH. TRO. ii. 997; evene forþ wiþ þi selve LANGL. (Wr.) 11637; evene weren þe endes two LEG. 48; whan þe dai and þe niȝt beþ evene TREV. I. 325; þei ben evene wiþ aungels ['*aequales angelis sunt*'] WICL. LK. xx. 36; *comp.* an-efen.

efen-ald, adj., *O.E.* efeneald,=*O.H.G.* epanalt; *of the same age,* ORM. 18605; evenald (*ms.* -hald) H. M. 41; evenold, -eld '*coaetaneus*' PR. P. 143.

even-cristene, sb., *fellow-christian,* HOM. I. 53; LANGL. (Wr.) 3353; evencristen C. L. 976; emcristene HICKES I. 223; emcristen AYENB. 10.

efen-ēche, adj., *O.E.* efenēce; *coeternal,* ORM. 18582.

even-hēde, sb., *equity* ['*aequitas*'], WICL. PS. x. 8.

efen-hēh, adj., *O.E.* efenhēah, = *O.H.G.* ebenhōh; *equally high,* ORM. 15720.

even-kniȝt, sb., *fellow-soldier* ['*commilito*'], WICL. PHILIP. ii. 25.

even-lēchen, v., *O.E.* efenlǣcan; *resemble*;

evenlēcheð (*pres.*) HOM. I. 113; *comp.* ȝeefenlǣchen.

geve-lengðe, sb., *O.N.* iafnlengð; *? equinox,* GEN. & EX. 147.

even-līc, adj., *O.E.* efenlīc, *O.N.* iafnligr,= *Goth.* ibnaleiks; *equal,* ORM. 1837; eveliche MISC. 90; gevelic REL. I. 213; GEN. & EX. 282; **evenlīche** (*adv.*) REL. I. 172; E. W. 5; evenli WICL. JOB ix. 32.

efen-mahtiȝ, adj., *equally mighty,* ORM. 18571.

efen-mēte, adj.,=*O.H.G.* ebenmāzi; *of the same size,* ORM. 3099; evenmete PS. xlviii. 13.

efennesse, sb., *O.E.* efennyss, *emness*; *equity*; evennesse ['*aequitas*'] WICL. DEUT. ix. 5; evennes PS. x. 8.

efen-nēxta, sb., *neighbour,* HOM. I. 17.

efen-rīke, adj., *equally powerful,* ORM. 11868.

efen-servaunt, sb., *fellow-servant,* WICL. APOC. xix. 10; **evenservanntis** (*pl.*) MAT. xviii. 28.

efen-soukere, sb., *brother,* WICL. 2 MACC. ix. 29.

even-worþ, adj., *equivalent,* WICL. JOB xxviii. 19.

ēfen, *see* ǣfen.

efenling, evenling, sb., *O.E.* efenling; *equal,* HOM. I. 57, 67.

ēfeste, *see* ōveste.

effect, sb., *Lat.* effectus; *effect, fruit, purpose*; the swete effecte of Aprelle floures P. L. S. xxxi. 67; mi purchas is th' effect of al mi rent CH. C. T. *D* 1451; to tellen hem theffect of his entent *E* 1398; took effect CH. BOET. i. 4 (15).

effectual, adj., *effectual*; effectueel CH. C. T. *D* 1870.

effectual-lī, adv., *effectually*: effectuali WICL. E. W. 388.

effectuōs, adj., *Lat.* effectuōsus; *effectual,* APOL. 55.

effectuous-lī, adv., *effectually,* LUD. COV. 380.

efficāce, sb., *efficacy,* A. R. 246.

efflouren, v., *go out of bloom*: effloureth (*pres.*) ['*deflorescit*'] PALL. iii. 82.

effrai, sb.,=affrai; *terror,* BARB. xi. 250.

effraien, v.,=affray; effraiit (*pple.*) BARB. vii. 610.

effusciōn, sb., *dispersion,* CHEST. 92.

efne, sb., *O.N.* efni; *ability, nature,* A. R. 6, 372; on mannes efene HOM. II. 137; after his evene *according to his ability* HOM. I. 187; bi his evene KATH. 57; bi here evene P. S. 157.

efnen, v., *O.E.* efnan, *cf. O.N.* iafna, *O.H.G.* ebanōn, *Goth.* (ga-)ibnjan; *render even, equalise,* ORM. 1396; evenen H. M. 19; WICL. IS. xxviii. 25; evenin '*aequo*' PR. P. 143; emni AYENB.

16; efneð, eveneð (*pres.*) A. R. 82, 132; oure loverd him silf him eveneþ to soþnisse þere BEK. 1629; **efnede** (*pret.*) A. R. 126; **efned**, evened (*pple.*) HOM. II. 103, 211; evened PS. lxxxviii. 7; *comp.* 3e-efnen.

efning, sb., *O.N.* iafningi; *equal*, HOM. I. 191; A. R. 182; efning, efening ORM. 10702, 10991; evening KATH. 119; O. & N. 772; MIRC 1229; þet neveðe on eorþe non evening MISC. 95.

ēfre, see **æfre**.

eft, adv., *O.E.* eft, æft,=*O.L.G.*, *O.Fris.* eft; *afterwards; again*, KATH. 1449; ORM. 181; REL. I. 225; GEN. & EX. 1169; C. L. 751; WILL. 882; LANGL. *B* xvii. 111; GOW. II. 264; CH. C. T. *A* 1669; MAND. 14; WICL. GEN. xvii. 9; HOCCL. i. 408; ævric mon scal eft mowen bi þon þe he nu saweð HOM. I. 137; dude hit eft & eft A. R. 266; wanne he come eft a3e ROB. 220; and ever eft he was lame B. DISC. 477; eft, æft LA3. 3264, 6445; efte ROB. (W.) 3628*; and **eft sōne** (*eftsoons*) cume þar to O. & N. 821; ef sone ROB. (W.) 1896.

eft, sb., ?=*O.E.* æfest, æfst; *? malice*; eft and niþe and felounie M. H. 130; eft and nithe and felonni 125; iowes hauis eft and nithe at me for the ferlikes that i kithe 35.

efter, see **æfter**.

eft-sōne, adv., see under **eft**.

eft-white, v., *? repay*, TOWNL. 106.

ēg, see **ēi**.

egal, adj., *O.Fr.* esgal, egual, igal, *from Lat.* æquālis; *equal*; under ioure ierde egal to min offence CH. TRO. iii. 88; **egalle** (*pl.*) CT. LOVE 1041; E. G. 401.

egal-lī, adv., *with equanimity, impartially*, CT. LOVE 365; agreableli or egali CH. BOET. ii. 4 (43).

egalitē, sb., *O.Fr.* egauté, egalité, *Lat.* æquālitas; *evenness of mind, equanimity*; she is as these martires in egalite CH. C. T. *I* 949; egalite CH. BOET. ii. 4 (42).

e3e, **ei3e**, sb., *O.E.* ege, *O.N.* agi=*O.H.G.* egi *m.*, for *Goth.* agis *n.*; *awe, terror*, HOM. I. 99, 113; lufe ne for e3e (*ms.* e33e) ORM. 4481; a3he 7185; eige MAT. xiv. 26; eige, age GEN. & EX. 432, 2550; ei3e, æi3e, eie, eaie LA3. 4733, 9126, 9129, 22881; þene muchelne æie [eie (*ms.* eye)] 16142; eie P. L. S. viii. 140; MARH. 5; A. R. 418; FL. & BL. 302; P. S. 196; eie BEK. 282; REL. I. 116; RICH. 3609; H. S. 38; þou haddest of him non eie MAN. (F.) 2894; awe 1850; eie, æie SAX. CHR. 261, 266; aghe TOWNL. 305; ah3e LEG. CATH. 88; aighe AR. & MER. 465; of wham men stondeþ aie FER. 408; aie, awe ALIS.² (Sk.) 1243; awe '*pavor, terror*' PR. P. 17; HAV. 277; CH. C. T. *A* 654; HOCCL. i. 383; for love ne for awe WILL. 5430; àue C. M. 18050.

ei-fold, adj., *haughty*, ROB. (W.) 7729*.

ei-ful, adj., *O.E.* egefull; *awful*, KATH. 40; a3heful ORM. 7172; **eifulle** (*pl.*) word HOM. II. 81; ahefulle develes I. 271.

---

e3e-lǣs, adj., *O.E.* egelēas,=*O.N.* agalauss; *aweless, fearless*, ORM. 6191; a3lez GAW. 2335; **æielēse** (*pl.*) LA3. 19410.

æie-lēste, sb., *fearlessness*, LA3. 19291.

ei-līch, adj.,=*O.H.G.* egilīh; *terrible*, HOM. II. 5; aghlich GAW. 136; a3elīche (*adv.*) A. R. 56*; a3li A. P. ii. 937.

ē3e, see **ēa3e**.

ēgede, adj. (*etymology unknown, cf.* **ǣgede**); *foolish, silly*; for nawt ne þunche hit hire egede H. M. 39; his egede orhel MARH. 11; hwu egede þing is orhel A. R. 282; (*printed* a gade) S. S. (Web.) 2638.

ē3ern, adj., ?=*O.N.* āgiarn; *lascivious*; ðis oref is swiðe egerne HOM. II. 37.

[**e3ese**, **eise**, sb., *O.E.* egesa, egsa,=*O.L.G.*, *O.H.G.* egiso *m.; terror.*]

æi3es-ful, adj., *O.E.* egesfull; *fearful, terrible*, LA3. 17972; eisful HOM. I. 111; MARH. 9; (*printed* eifful) A. R. 190.

eis-līc, adj., *O.E.* egeslīc,=*O.L.G.* egislīc; *terrible*, FRAG. 3; HOM. I. 87; eisliche wihte P. L. S. viii. 142; eiseliche HOM. II. 67.

egg, see **ēi**.

egge, sb., *O.E.* ecg,=*O.Fris.* eg, *O.N.* egg, *O.L.G.* eggia, *O.H.G.* ekka, *Lat.* aciēs; *edge*, '*acies*,' PR. P. 136; LA3. 5202; ORM. 6639; ALIS. 1271; B. DISC. 1923; WICL. HEBR. xi. 34; L. H. R. 136; mid sweordes egge A. R. 60; under a wode egge JOS. 475; þe hi3est of þe eggez A. P. ii. 451.

eghelinge, adv., *edgewise*, M. ARTH. 3676.

egge-tōl, sb., *edge-tool*, WILL. 3755; GOW. II. 251.

egged, adj., *O.E.* ecged,=*O.N.* eggiaðr; *edged*, MAN. (F.) 7862; a . . . knif iegged in eiþer side ROB. 310.

eggement, sb., *from* **eggen**; *incitement*: thurgh wommanes eggement [egment] CH. C. T. *B* 842.

eggen, v., *O.N.* eggia, *cf. mod.Eng.* egg on; *provoke, incite*, A. R. 146; ORM. 11819; eggin '*incito*' PR. P. 136; eggeð (*pres.*) REL. I. 131; **eggede** (*pret.*) LEB. JES. 897; TREV. IV. 93; egged WILL. 1130; PR. C. 5483; Adam and Eve he eggede (heggen) to (don) ille LANGL. *A* i. 63; eggiden ['*provocaverunt*'] WICL. DEUT. xxxii. 16.

eggunge, vbl. n., *egging on, instigation*, A. R. 82; eggunge, egginge H. M. 5; egging ORM. 11675; A. P. ii. 241; PR. C. 5487.

eggen, v.,=*M.Du.* egghen, *O.H.G.* eckan; *cf.* eiþe; *harrow*: eggen oþer harwen LANGL. *C* vi. 19.

eghelinge, see under **egge**.

Egipcián, sb., *Egyptian*, WICL. DEEDS vii. 24.

égipstone, sb., *gypsum*, VOC. 94.

Egipte, pr. n., *Egypt*: of Egipte londe HOM. I. 11.

egiptisch, adj., *O.E.* egyptisc; *Egyptian*: þam egiptissen folche (*v.* folce) HOM. I. 87.

egle, sb., *O.Fr.* egle, aigle; *eagle*, MAND. 48; CH. C. T. *A* 2178; egle is ern MAN. (F.) 13757.

egleche, adj., *O.E.* æglæca, aglæca; *? bold, valiant*; knihtes egleche REL. I. 170 (MISC. 102); sleiȝe men and egleche S. A. L. 148; þe lefdi was egleche ROB. (W.) xx. 125.

eȝlen, *see* eilen.

eglentier, sb., *O.Fr.* eglenter, eoglentier; *eglantine*, MAND. 14.

egre, adj., *O.Fr.* egre, aigre; *eager*, ROB. 80; REL. II. 277; WILL. 3636; CH. C. T. *E* 1199; WICL. PS. cvi. 11.

  egre-lich, adv., *eagerly*, LANGL. *B* xvi. 64.

egren, v., *O.Fr.* aigrier; *urge, excite*: egren him to done felonies CH. BOET. iv. 6 (141); he ... egerd (*pret.*) him with D. TROY 7329.

eȝte, *see* eahte. **ēȝte**, *see* æhte. **ēhe**, *see* ēaȝe. **ehte**, *see* eahte. **ēhte**, *see* æhte.

ēhten, v., *O.E.* ēhtan; *persecute*; ? egte GEN. & EX. 470.

  ēhtere, sb., *O.E.* ēhtere; '*persecutor*,' FRAG. 2.

ehtende, *see* eahtuðe.

ei, int.,=*M.H.G.* ei; *alas*; ei CH. C. T. *A* 3768; ei god AYENB. 105; ei what is me HOCCL. i. 393.

ēi, egg, sb., *O.E.* æg, *O.N.* egg,=*O.H.G.* ei; *egg*; ey, O. & N. 104; ROB. 404; GOW. III. 76; CH. C. T. *B* 4035; WICL. LK. xi. 12; LIDG. M. P. 204; S. & C. I. xxix; ei, eiȝ LEG. 124, 125; non eiȝ (*ms.* no neiȝ) WILL. 83; ei, ai GAM. 610; aig LK. x. 1, 12; ai ALIS. 568; TRIST. 3167; MAN. (H.) 175; ei, eg PR. P. 136; ēie (*ms.* eye) (*dat.*) TREAT. 132; æȝere [*O.E.* ægru,=*O.H.G.* eigir] (*pl.*) LEECHD. III. 134; eiren A. R. 66; P. L. S. xii. 60; eyren AYENB. 178; MAND. 49; LIDG. M. P. 29; L. C. C. 7; egges LANGL. *B* x. i. 343; S. & C. II. lvi; ēir (*gen. pl.*) monger, *egg-seller*, P. L. S. xii. 69; eire (*dat. pl.*) O. & N. 106; *comp.* adel-, gōs-ēi.

  ai-schelle, sb., *egg-shell*, ALIS. 577.

ēi, *see* æni.

[**ēie**, sb., *O.E.* ēg, īg,=*M.Du.* ei, *O.N.* ey *f.*; *island*].

  ēi-lond, sb., *O.E.* ēg-, īgland,=*M.Du.* eiland, *O.N.* eyland; *island*, REL. I. 220; eland MAN. (H.) 77; ilond LAȝ. 1133*; OCTOV. (W.) 539; iland TRIST. 1024.

eie, *see* eȝe. **ēie**, *see* ēaȝe. **eier**, *see* heir. **ēiȝ**, *see* ēi.

eiȝe, *see* eȝe. **ēiȝe**, *see* ēaȝe.

eiȝte, eihte, *see* eahte. **eiȝtēne**, *see* ehtetēne *under* eahte. **ēihte**, *see* æhte.

ēihwēr, *see* aihwēr. **ēiland**, *see under* ēie.

eile, sb., *O.E.* egl *f.* (*point, bit of straw*) egla (*ear of corn*) VOC. 38, *cf.L.G.* aile *f.*(*arista*); *ear of corn, bit of straw*, '*acus*,' HALLIW. 343; þa eigle ['*festucam*'] LK. vi. 41; eilen (*pl.*) A. R. 270.

eile, adj., *O.E.* egle,=*Goth.* aglus; *grievous, painful*: eile and hard C. L. 223; eile BEV. 513; þat water wes sturne and eile (*ms.* eille, *r. w.* seile) heore þuhte swiþe eille [eil] of æðelene hire fædere LAȝ. 3281; eil H. M. 25.

eile, sb.,=*Goth.* aglō (θλίψις); *pain*: deð lesse eile to þen eien A. R. 50.

eilen, v., *O.E.* eglan,=*Goth.* (us-)agljan; *ail, annoy, be painful*, A. R. 276; RICH. 3629; eȝlen ORM. 4767; eili H. M. 47; eileð (*pres.*) KATH. 1699; what eileþ þe FER. 1560; CH. C. T. *B* 1081; i wot wel what siknesse ȝou eileþ LANGL. *B* vi. 259; eilie (*subj.*) A. R. 418; þis eisful wiht þet hit ne eile [eili] me nawiht MARH. 9; eilede (*pret.*) WILL. 951; ailed (*pple.*) GAW. 438; MIN. ix. 28.

eimeri, eimbre, sb., *O.E.* æmerge, *O.N.* eimyrja; *ember*, '*pruna*,' PR. P. 136.

eir, *see* air, eire *and* heir.

eire, *see* īren.

eire, sb., *O.Fr.* erre, *cf. mod.Eng.* '*justiçes in eyre*;' *swiftness, journey*: alle the flote com wiþ gret eire MAN. (F.) 1486; as þe ssipes wiþ gret eir [heir, ire] come toward londe ROB. (W.) 1183; þe eire of iustize ROB. (W.) 10647.

eirēde, *see* ērēde.

eiren, v., *O.Fr.* errer, oirrer; *journey*; aires (*pres.*) D. ARTH. 1329.

eischste, *see* askien.

eise, *see* aise *and* eȝese. **eiselich**, *see* aisie.

eisien, v., *O.E.* egesian, egsian, = *O.H.G.* egisōn; *terrify*: þet he his men eisian ne der HOM. I. 111.

eisil, sb., *O.Fr.* aissel; *vinegar*, A. R. 402; C. L. 1150; eisel R. R. 217; MIRC 1884; AUD. 64; eisel, aisel WICL. NUM. vi. 3, MK. xv. 36 (*v. r.* vinegre); C. M. 18019; aisiles (*gen.*) JOHN xix. 29; mid eisile (*earlier version* ecede) MK. xv. 36.

ēit, sb., *? O.E.* īgoð; *ait* (*eyot*), *small island*: þan æit LAȝ. 23873.

  ēit-lond, sb., *small island*, LAȝ. 1117.

eiteteþe, eiteti, eiteþe, *see* eiȝte-tēoþe *under* eahte.

eiþe, sb., *O.E.* egeð, ?=*M.Du.* egede, *M.L.G.* egede, eide, *O.H.G.* egida; *harrow*; eiþes (*pl.*) LANGL. *C* xxii. 273.

ēiþer, *see* æiðer.

ēk, *see* ēac. **ēke**, *see* ēake. **ēken**, *see* āken.

ēken, *see* ēchen. **ēking**, *see* ēching.

eker, sb., *some sea-monster*: thei woneth in water iwis with eker [*v. r.* iker] and fisch ALIS. 6172; ekeris (*pl.*) 6202.

ēkni, *see* ēacnen.

[**el-**, *O.E.* el-, æl-, ele-, *cf. O.L.G.* eli-, *O.H.G.* eli-, el-, *Goth.* alja-, *Lat.* ali-, *Gr.* ἀλλο-.]

  el-hēowed, adj., *changed in colour*: his leor deaðlich ant blac ant elheowit HOM. I. 249.

ele-lende, sb., *O.E.* ele-, ellende, *O.H.G.* eli-, ellenti; (*exiled*) *foreigner*; aliande '*extraneus*' PR. P. 9.

ele-lendis, elendis, adj., *O.E.* elelendisc; *foreign*; (*ms.* helendis) HOM. I. 79, 81.

el-reordi, adj., *O.E.* elreordig; *speaking another language, barbarous* : elreordi (*? ms.* welreordi) feond LA3. 25658.

el-þēod, sb., *O.E.* elþēod; *foreign nation*: for alþeodene (*gen. pl.*) gold LA3. 2327.

el-ðēodi, adj., *O.E.* elþēodig; *foreign*, A. R. 348*; alþeodi FRAG. 2.

al-þēodisc, adj., *O.E.* elþēodisc; *foreign*, LA3. 2301 ; ealðēodisce (*pl.*) MAT. xxvii. 7.

ēl, *O.E.* æl, = *O.H.G.* āl, *O.N.* āll; *eel*, '*anguilla*,' PR. P. 137 ; ēles (*pl.*) ALIS. 5792; MAND. 161 ; elis L. C. C. 50.

ēl-ger, sb., *M.Du.* aelgheer; *eel-spear*, '*fuscina*,' PR. P. 138.

ēl, *see* awel.

elbowe, *see under* elne.

ēlch, *see* aighwilc.   **eld**, *see* eald.

ēld, sb., *O.E.* æled, = *O.L.G.* ēld, *O.N.* eldr; *fire*, HOM. II. 258.

elde, sb., *O.E.* eldo, ældo, yldo, *cf. O.L.G.* eldī, *O.H.G.* eltī, altī, *O.N.* elli *f.*; *from* eald ; *age*, HOM. II. 35 ; ORM. 201 ; GEN. & EX. 705; ROB. 11 ; SPEC. 48 ; SHOR. 168 ; AYENB. 69 ; WILL. 5227 ; LANGL. *B* xi. 43 ; CH. C. T. *A* 2447 ; MAND. 293 ; WICL. LK. i. 36 ; A. P. ii. 657 ; PR. C. 1513 ; elde me haþ overcome L. H. R. 22 ; fiftene 3er þan was his elde MAN. (F.) 9744 ; on hire elde LK. i. 36 ; ich was of swuche elde A. R. 318 ; wimman of elde (*of age*) HAV. 2713 ; elde, ulde P. L. S. viii. 9, 188 ; ælde [ealde] LA3. 25913 ; ulþe (*for* ulde) P. L. S. xix. 6.

elden, *see* ealdien.

ēlden, v., *cf. O.N.* elda; *from* ēld; *light a fire*; ? eilden GEN. & EX. 2892.

eldern, adj., *? from, compar. of* eald ; *elderly, ancient*: elderne mannes late ORM. 1213 ; miņ eldrin folk of juen lede C. M. 18016.

eldern-liche, adv., *of old time*, E.G. 352.

ēlding, sb., *cf. O.N.* elding, *Scotch* eldin ; *fuel*, C. M. 3164 ; eilding '*focale*' PR. P. 136.

eldnesse, *see* ealdnesse.

eldringes, sb. pl., *parents*, AYENB. 35.

éle, sb., *O.E.* ele *m.*, ele, œle, = *O.H.G.* oli, olei *n.*, *O.L.G.* olig, *Goth.* ale, *O.Fr.* oile, uile ; *from Lat.* oleum ; *oil*, FRAG. 8 ; ORM. 13252; SHOR. 41 ; eli HICKES I. 229; eole KATH. 2512 ; eoile H. M. 45 ; eoli MARH. 11 ; A. R. 320; olie GEN. & EX. 1548 ; oli PS. iv. 8; oile, uile AYENB. 93.

éle-sǣw, sb., *oil*, ORM. 8667.

elebre, sb., *O.Fr.* ellebore; *hellebore*, PR. P. 138 ; PALL. i. 1044.

eleccioun, sb., *O.Fr.* election, *Lat.* ēlectio ;

election, *choice of fit times*, H. S. 10990; MAND. 155; CH. C. T. *B* 312 ; eleccióuns (*pl.*) of times ASTR. ii. 4 (18) ; election GOW. I. 18.

elect, sb., *election* ; elect WINT. IX. xxvii. 141 ; electes N. E. D.

elect, adj., *elect* ; eleckte CHEST. I. 212 ; elect (*sb.*) *bishop elect*, WINT. IX. xxvii. 121.

electre, sb., *O.Fr.* eleutre, *Lat.* ēlectrum ; *a metal*, WICL. EZ. i. 4, 27.

elegeance, *see* enlegeance.

element, sb., *O.Fr.* element; *element* ; elementz (*pl.*) TREAT. 138.

[**ēlen**, v., *O.E.* ǣlan ; *burn* ; *comp.* an-ēlen.]

ēlenge, ēlenglīch, ēlengenes, *see* ǣlenge.

ēlet, sb., *O.E.* ǣled ; *fuel*, HOM. II. 119 ; BEV. 3088.

elevacioun, sb., *O.Fr.* elevacion, eslevation, *Lat.* ēlevātio, *mod.Eng.* elevation ; *altitude above the horizon*, CH. ASTR. ii. 23 (32).

elevat, adj., *Lat.* ēlevātus ; *elevated*, CH. ASTR. ii. 23 (32).

elfe, *see* alfe.

Eli, pr. n., *O.E.* Elīg ; *Ely*, ROB. 6.

élien, v., *M.H.G.* ölen, olen ; *anoint* ; elie SHOR. 41 ; élede (*pret.*) ROB. (W.) 5329; iéled (*ms.* iheled) (*pple.*) ROB. 457 ; he wes icruned and ieled LA3. 31941.

élinge, sb., = *M.H.G.* ölunge, olunge ; *oiling* : þe laste elinge *extreme unction* MIRC 533 ; eliinge SHOR. 41.

elifaunt, *see* olifänt.

elite, sb., *election* : þe pape wild not consent, he quassed þer elite MAN. (H.) 209 ; elítes (*pl.*) IBID.

eliten, v., *from* elíte ; *elect* : Creusa . . . þat Eneas afterward elit (*pret.*) to wed D. TROY 1490.

ellarne, sb., *O.E.* ellarn, = *M.L.G.* elhorn, alhorn ; *elder (tree)*, '*sambucus*,' VOC. 140, 181 ; helren (*for* elren) 163 ; ellerne [helderne, illern, eller] LANGL. *A* i. 66 ; eldir PR. P. 137 ; *see* hiller.

elle, *see* elne.   **ellen**, *see* elne.

elles, adv., *O.E.* elles, = *O.Fris.* elles, *O.H.G.* elles, alles, *Goth.* aljis ; *else*, ORM. 7983; WILL. 1132 ; MAND. 5 ; PR. C. 1281 ; HOCCL. ii: 23 ; in lenten and elles [ellis] LANGL. *B* PROL. 91 ; summe qveðen ælles [elles] LA3. 893 ; ellis SACR. 211 ; elles hwar A. R. 78 ; 3if heo ouht elles understode O. & N. 662 ; elles hwider lokin HOM. I. 261 ; elles whoder P. L. S. xvii. 549 ; helles-hwat HOM. 215.

elleven, *see* endleven.

elm, sb., *O.E.* elm, = *O.H.G.* elm, *cf. O.N.* almr; *elm*, '*ulmus*,' PR. P. 138 ; TRANS. XVIII. 24 ; CH. C. T. *A* 2922.

elmesse, sb., *O.E.* ælmesse, *from Gk.* ἐλεη-μοσύνη ; *alms*, A. R. 222 ; AYENB. 17 ; don elmesse HOM. I. 37 ; elmesse to wurchen II.

59; almesse LANGL. *B* iii. 75; CH. C. T. *B* 168.

elmes-dēde, sb., *almsdeed*, HOM. I. 135; almesdede R. S. iv; P. L. S. xvii. 385; C. M. 11350.

almes-disch, sb., B. B. 139; almesdisches (*pl.*) TREV. VIII. 183.

elmes-ful, adj., *charitable*, HOM. I. 143; almesful HOM. II. 85; ORM. 9931; BEK. 1674; LANGL. *C* vii. 48.

elmes-ȝeorn, adj., *charitable*, HOM. I. 43.

ælmes-mon, sb., *almsman*, LAȜ. 19662.

almes-werc, sb., *almsdeed*, ORM. 10118.

elne, sb., *O.E.* ellen, *cf. O.H.G.* ellan, ellen, ellin, *O.L.G.* ellen, ellean, ellien, *Goth.* aljan *n.*, *O.N.* eljan, eljun *f.*; *strength, comfort*, HOM. I. 215; MARH. 10; A. R. 106; M. H. 27.

ellen-lǣs, adj., *comfortless*, ORM. 10908.

elne, sb., *O.E.* eln, = *O.H.G.* elna, elina, *O.N.* öln, alin, *Goth.* aleina, *Lat.* ulna, *Gk.* ὠλένη; *ell*, ALIS. 2750; an elne long SPEC. 35; an elne (*printed* a nelne) brod D. TROY 170; elne, elle PR. P. 138; elle OCTOV. (W.) 934; elne (*pl.*) GEN. & EX. 563; elnen ROB. 429.

el-bowe, sb., *O.E.* elnboga, = *O.H.G.* elinbogo, *O.N.* öln-, ölbogi; *elbow*, FRAG. 2; elbowe TREAT. 139.

eln-ȝerde, sb., *ell yard*, GAW. 210; elneȝerdes (*pl.*) ROB. (W.) 8834.

elle-wand, sb., *ell-wand*, PR. P. 139.

elnen, v., *O.E.* elnian, = *O.N.* elna, *O.H.G.* ellinōn, *Goth.* aljanōn; *comfort*, HOM. I. 215; A. R. 10; elnede (*pret.*) KATH. 672.

elning, elnung, sb., *O.E.* elnung; *comfort*, HOM. I. 185, 201.

elongacioun, sb., *Lat.* ēlongātiō; *elongation, angular distance*, CH. ASTR. ii. 25.

elp, sb., *O.E.* elp, elpend, *cf. M.H.G.* elfant; *elephant*, REL. I. 223; alpes (*gen.*) ban (*ivory*) L. N. F. 248; *see* olifant.

ēlpi, *see* ānlĕpiȝ.

elring, sb., *? = M.H.G.* erling, *cf. Ger.* elritze; *a sort of fish, ? minnow*: als the quale fars wit the elring M. H. 136.

elsin, *see* alsene.

elten, v., *O.N.* elta; *knead*, GEN. &EX. 2892.

elvene, *see* alfe. em, *see* am *and* efen.

ēm, *see* ēam. · emang, *see* en.

ēmatte, *see* ǎmete.

embassade, sb., *O.Fr.* embassade; *embassy*; *business or message of an ambassador*; DREAM 1993; now mi imbasset i · have seid to you LUD. COV. 77.

embassadrie, sb., *embassy*, CH. C. T. *B* 233.

embe, *see* ümbe.

embelif, adv. & adj., *O.Fr.* en belif; *obliquely*; *oblique; not at right angles*: under the cercle of cancer, ben ther 12 devisiouns embelif CH. ASTR. i. 20 (11); embelif cercle

ii. 26; embelif orisonte . . . embelif angles ii. 26.

embeþonk, *see* ümbe-þanc.

embir-dai, *see* ümbri-.

emb-snīðen, *see* ümbesnīðen.

emel, emell, *see* mel.

emeraude, sb., *Fr.* emeraude; *emerald*, SPEC. 26.

emeraudes, sb. pl., *O.Fr.* esmeroides, *from Lat.* haemorrhoides, *Gk.* αἱμοῤῥοῖδες; *emerods*, REL. I. 190; PR. P. 139.

ěmete, *see* ǎmete.

emforþ, *for* efen forþ: *see* efen.

emne, emnen, *see* efen, efnen.

[emnetten, v., *O.E.* emnettan; *from* efnen; *make equal; comp.* ȝe-emnetten.]

emparāle, sb., *med.Lat.* imperiālis; *imperial* (*a coin*): for emperāles (*pl.*) that were not smale i bought him thare OCTOV. (W.) 1911.

empericě, sb., *empress*, SAX. CHR. 264; WILL. 5343; TREV. IV. 189; emperisse SHOR. 117; emperesse (emperasse) ROB. (W.) 9064.

emperie, sb., *O.Fr.* emperie; *empire*, ROB. 85; TREV. III. 73; enpir MAN. (F.) 3068.

emperour, sb., *O.Fr.* empereor; *emperor*, ROB. 46; WILL. 5251; S. S. (Wr.) 1; aumperour LEB. JES. 393; SAINTS (Ld.) 344; aumperoures SAINTS (Ld.) 248.

empiri, sb., *Fr.* empyree, *? from Gr.* ἔμπυρος; *empyrean*: þis heven is cald empiri PR. C. 7761.

emplastre, sb., *O.Fr.* emplastre, *from Gr.* ἔμπλαστρον; *plaster*: make an enplaster [*later ver.* plaster] upon the wounde WICL. Is. xxxviii. 21.

emplastren, v., *Lat.* emplastrāre; *plaster over, gloss*: als fair as ye his name emplastre (*v. r.* implastre) (*pres.*) he was a lecchour CH. C. T. *E* 2297.

emprīse, sb., *O.Fr.* emprīse, emprinse; *enterprise*, CH. C. T. *A* 2540; C. M. 9802; OCTOV. (W.) 1060, 769; GOW. I. 111.

ěmti, adj., *O.E.* ēmtig, ǣmtig, ǣmetig; *empty*, '*vacuus*,' PR. P. 139; HOM. II. 87; KATH. 393; AYENB. 143; emti [empti] CH. C. T. *A* 3894; emti, amti ROB. 17, 50; empti A. R. 156; M. H. 151.

ěmtien, v., *O.E.* ǣmtian, ǣmtigian; *empty*; empte CH. C. T. *G* 741; þo þet ěmteþ (*pres.*) þe herte of hire guode AYENB. 58; ěmptede (*pret.*) TREV. VIII. 89; ǣmteden LAȜ. 30408.

ěmtinge, sb., *emptying*, '*evacuatio*,' PR. P. 139.

ēn, *see* ēan.

en, prep., *cf. O.Fris.* en, = in; *in, on*: en Edwardes kinges deȝe ANGL. VII. 220; en mang *among* SAX. CHR. 261; en ende HOM. I. 227; en Enge(l)londe CHR. E. 336, 480; en middes PARTEN. 870; ene þe þridde dai MISC. 199; e mang TOWNL. 22; C. M. 7364; e mange C. M. 14342.

O

[en-, pref., *O.Fr.* en-,=in-.]

en-amelen, v., *for* amelen; *enamel*; enameled (*pple.*) MAND. 219; enamild PR. C. 8902.

enamūren, v., *O.Fr.* enamourer; *enamour*; enamūred (*pple.*) MAN. (F.) 9367.

enarmen, v., *O.Fr.* enarmer; *arm*; enarmed (*pplé.*) C. L. 1351.

enbaissing,=abashinge, CH. BOET. iv. 1 (109).

enbanen, v., *? fortify*; enbaned (*pple.*) under þe abatailment GAW. 790; enbaned under batelment A. P. ii. 1458.

en-batailen, v., *from O.Fr.* batailler; *make battlements; entrench*: walled welle with highe walles enbatailed (*pple.*) R. R. 136; whan that he was embatailed GOW. I. 221.

enbaumen, v., *O.Fr.* en-, embaumer; *embalm, anoint*; enbaumed (*pret.*) LANGL. *B* xvii. 70.

enbelise, v., *O.Fr.* embelir; *embellish*, GAW. 1033.

enbénen, v., *? baste* (*in cookery*): with ȝolkes of eiren enbéne (*imper.*) it L. C. C. 27; enbene hit wele withe þi riȝt honde 26.

enbībing, pple., *imbibing*, CH. C. T. *G* 814.

enblaunchen, v., *O.Fr.* enblanchir; *render white*: enblaunched (*pple.*) with bele paroles LANGL. *B* xv. 113.

enblemischen, v., *A.Fr.* emblemiss-, emblemir; *make ill, infect*: oo lepre left unheelid mai enblemissche mani folc WICL. S. W. I. 281.

en-blōwen, v., *inspire*; enblōwid (*pple.*) with the spirit of retorik WICL. PREF. EP. I. 77.

enbrácen, v., *O.Fr.* embraċier; *embrace*, P.L.S. xxv. 113.

enbréven, v., *O.Fr.* embrever; *record*; enbréved (*pple.*) A. R. 344.

enbróche, v., *O.Fr.* embrocher; *put on the spit*, L. C. C. 35; enbróchid (*pret.*) 34.

enbrouden, v., *embroider*; embrouded (*pple.*) FER. 553; CH. C. T. *A* 89; enbrouden QUAIR 152.

enbusche, v., *O.Fr.* enbuscher; *lay in ambush*, MAN. (F.) 986; abuschid (*pple.*) WILL. 3633.

enbuschement, sb., *ambushment*, HALLIW. 333.

encens, sb., *O.Fr.* encens; *incense*, CH. C. T. *A* 2277.

encensen, v., *O.Fr.* encenser; *incense*; ensensede (*pret.*) LEG. 81.

encenser, sb., *O.Fr.* encensier; *censor*; encenseres (*pl.*) WICL. NUM. iv. 7; encensers EX. xxxvii. 16.

encenti, *see* ensenten.

enchácen, v., *O.Fr.* enchacier; *drive away*: enchase P. R. L. P. 45; enchácede (*pret.*) FER. 2906.

enchantement, sb., *O.Fr.* enchantement; *enchantment*, ROB. 10; an enchauntement SAINTS (Ld.) 248.

enchanten, v., *O.Fr.* enchanter; *enchant*; enchaunted (*pple.*) LANGL. *B* ii. 41.

enchaunteresse, sb., *female enchanter*, CH. BOET. iv. 3 (123).

enchanterie, sb., *enchantment*: þo þe clerc adde iseid his enchaunterie ROB. 10; enchanterie 148.

enchantour, sb., *O.Fr.* enchanteor; *enchanter*; enchanteor ROB. (W.) 668; enchantours (*pl.*) P. L. S. xxi. 122.

enchargen, v., *O.Fr.* encharger; *charge, enjoin*: vertue is encharged (*pple.*) . . . to you CH. BOET. v. 6 (178).

enchaufen, v., *cf. O.Fr.* eschaufer; *become warm*: enchaufe (*pres.*) WICL. JOB vi. 17.

encheisōn, sb., *O.Fr.* enchaisoun; *occasion, cause*, WILL. 3697; encheson, enchesun ROB. 111, 452; en-, ancheisoun AYENB. 47, 205; ancheisun A. R. 158; enchesoun M. GOD 61.

enclinaunt, adj., *inclining, prone*, WICL. EX. xxxiii. 22; MAN. (F.) 5320.

enclīne, sb., *O.Fr.* enclin; *salutation*: made his enclīnes (*pl.*) D. ARTH. 82.

enclīne, v., *O.Fr.* enclīner; *incline*, CH. C. T. *B* 3092; A. P. i. 629; enclīnede (*pret.*) P. L. S. xvi. 159.

encloien, v., *O.Fr.* enclouer; = acloien; *lame, obstruct*; encloie (*pres.*) LANGL. *C* xxi. 296*; encloied (*ms.* encloyde) (*pple.*) HALLIW. 333; *see* acloien.

enclōse, v., *enclose*, MAND. 84; enclōsed (*pple.*) WILL. 2220; aclosed SHOR. 145.

enclōser, sb., *a setter in metal*: alle craftese men and enclōsere (*pl.*) WICL. 4 KINGS xxiv. 14, 16.

enclūden, v., *Lat.* inclūdere; *include*; enclūdeth (*pres.*) PALL. iv. 435, 337.

encomberment, sb., *O.Fr.* encombrement; *trouble*: saved þam alle fro þer encumberment MAN. (H.) 148.

encombrance, sb., *encumbrance*, MAND. 284.

encombre, sb., *O.Fr.* encombre; *trouble*: withouten encombre MAN. (H.) 189.

encombren, v., *encumber, tire*, PL. CR. 483; encombriþ (*pres.*) HOCC. i. 355; encombred (*pple.*) LANGL. *C* ii. 31; CH. C. T. *A* 508; *see also* acombren.

encumbering, sb., *trouble*: allas this is a grete encumbering MAN. (H.) 117.

encombrous, adj., *O.Fr.* encombros; *cumbrous*: hard matere is encombrouse for to here CH. H. F. 861; ful encomberouse is the usinge COMP. M. 339.

encontre, sb., *O.Fr.* encontre; *encounter*, ROB. 391; acountre HALLIW. 17.

encontre, v., *O.Fr.* encontrer; *encounter*, D. ARTH. 1320; encountrede (*pret.*) BRD. 19; acuntred WILL. 3602.

encorporen, v., *Lat.* incorporāre; *incorporate*; encorpored (*pret.*) TREV. VIII. 285; encorporing (*pple.*) CH. C. T. *G* 815.

encorsife, adj., *from O.Fr.* encorser (*grow fat*); *fattened* ['*incrassatus*'], HAMP. PS. 516.

encrees, sb., *increase,* CH. TRO. ii. 1332; encresse iv. 1229; encrese P. R. L. P. 27; encres GOW. I. 215; encres or incres '*incrementum*,' PR. P. 139; **encrécis** (*pl.*) WICL. PROV. iv. 9.

encrésen, v., *O.Fr.* encreistre, *Lat.* increscere; *increase*; encrecin PR. P. 129; to encresse in Goddis service WICL. S. W. I. 358; encresse CH. C. T. *B* 1068; wel fele i alwai mi love encrease CH. TRO. ii. 549; A. P. i. 958; ever enduire and ever encres MISC. 218; for to encrecen her richesse R. R. 5739; encrese WILL. 683; mai God encrese our feith WICL. S. W. I. 94; encresce LANGL. *B* xi. 389; **encrésede** (*pret.*) TREV. IV. 191; encreest P. R. L. P. 211; incresed CH. BOET. iii. 6 (77).

encróchen, v., *O.Fr.* acrocher; *approach*: þai schal comfort encroche in kiþes ful moni A. P. iii. 18; D. ARTH. 3213; delit þat his come **encróched** (*pret.*) A. P. i. 1116.

end, *see* **and.**

endamagen, v., *O.Fr.* damagier; CH. BOET. i. 4 (15).

endauntūre, sb., *taming*: huo þet tekþ colte endaunture AYENB. 220.

ende, sb., *O.E.* ende, *cf. O.L.G.* endi *m.,* *O.N.* endi, endir *m.,* *O.H.G.* endi, enti *m. n.,* *Goth.* andeis *m.* (ἄκρον, πέρας, τέλος); *end, district, territory,* ORM. 19326; LANGL. *B* ii. 100; se ende MISC. 35; hit ne me(i) . . . neome nan ende HOM. I. 265; þen ende A. R. 208; þene ende REL. I. 175; þane ende FL. & BL. 819; þe ende þet tu ant heo habbeð in ierdet MARH. 16; ech man . . . wende hom in his ende P. L. S. x. 41; have an ende CH. P. F. 495; hwonne come ich to ende A. R. 204; at þen ende LEG. 181; in eche ende ROB. 34; þo hit was ibrouȝt to ende P. L. S. x. 64; ende, ænde [eande] LAȝ. 243, 27505; to ende bringe C. L. 288; ænde LK. i. 33; PROCL. 3; ȝende (*for* iende) H. S. 2347; ED. 1846; ÉGL. 961; **endes** (*pl.*) LAȝ. 7835; **enden** (*dat. pl.*) MAT. xxiv. 31.

ende-daȝ, sb., *O.E.* endedæg; *day of one's death,* ORM. 7665; endedæi LAȝ. 3720; endedai BEK. 1963 (1973).

ende-land, sb., *O.N.* endiland; *? adjacent land,* ORM. 17916.

ende-lēas, adj., *O.E.* endelēas; *endless,* A. R. 146; endeles HOM. II. 169; GOW. II. 231.

endelēsnes, sb., *O.E.* endelēasnyss; *endlessness,* PR. C. 8122.

end-lī, adj., *M.H.G.* endelīch; *final,* PARTEN. 4011.

ende-sīð, sb., *death-journey,* GEN. & EX. 3777.

ende, *see* **and** *and* **enede.**

endeine, v., *Lat.* indignāri; *be angry, indignant,* WICL. IS. lvii. 6; ne endein

(*imper.*) mi lord EX. xxxii. 22; **endeineden** (*pret. pl.*) WISD. xii. 27; endeined (*pple.*) JOB xxii. 2.

enden, v., *O.E.* endian, = *O.N.* enda, *O.L.G.* endōn, *O.H.G.* endōn, entōn; *end,* A. R. 24; ORM. 3254; endie MISC. 192; endi AYENB. 110; ende H. H. 243 (247); GREG. 243; A. P. ii. 402; **endeð,** endeþ (*pres.*) O. & N. 943; endede (*pret.*) LAȝ. 8187; ROB. 32; *comp.* ful-, ȝe-enden.

endunge, sb., *O.E.* endung, = *O.H.G.* endunga; *ending,* MAT. xiii. 39; HOM. I. 71; endinge A. R. 428; KATH. 292; AYENB. 31; ending SPEC. 59; PR. C. 3772.

ending-dai, sb., *? last day,* S. B. W. 14.

endenten, v., *med.Lat.* indentāre; *fix in*: towrres endentid (*pple.*) with presios stonis TOR. (A.) 227; þen arne þai into the vine anon the dai with derk endente A. P. i. 627; endent 1012.

endentinge, sb., *joint, rabbeting*; **endentingis** (*pl.*) WICL. I PARAL. xxii. 3.

endentūr, sb., *med.Lat.* indentūra; *crevice,* A. P. ii. 313.

ender, adj., *latter, last past*: þis ender dai SPEC. 60; AN. LIT. 11; P. L. S. xxvi. 2; S. S. (Web.) 1080; WILL. 3042; GOW. I. 45; PL. CR. 239; þis ender niht C. M. 4561; ender niȝt AR. & MER. 917.

endetten, v., *O.Fr.* endeter; *indebt*; **endetted** (*pple.*) CH. C. T. *G* 734; andetted A. R. 126.

en-dirken, v., *make obscure*: no wickednesse shal endirken it CH. BOET. iv. 3 (120).

endīte, v., *O.Fr.* endicter, endīter; *indict; indite,* TREV. IV. 213; CH. C. T *A* 2741; adite RICH. 1174; indite '*indicto*' PR. P. 261; and doggez to dethe endite GAW. 1600.

endīter, sb., *inditer, writer, secretary*; **endīters** (*pl.*) and writeris WICL. ESTH. viii. 9*.

endītinge, sb., *style of composition,* '*dictamen, indictacio*,' PR. P. 139; ALEX.[2] (Sk.) 243; enditing CH. ASTR. PROL. (2).

endive, sb., *Fr.* endive; *endive,* PR. P. 140.

endleven, card. num., *O.E.* endleofan, ellefan = *O.Fris.* andlova, elleva, *O.L.G.* ellevan, *O.N.* ellifu, *cf. Goth.* ainlif, *O.H.G.* einlif; *eleven,* ROB. 298; endluve GEN. & EX. 1921; enlevene [ellevēne] WICL. LK. xxiv. 33; onleven RICH. 2725; enleven (eolleve) LAȝ. 23364; bi æellevene *by elevens* 14531; enleve DEGR. 342; elleven C. M. 18870; elleovēne FRAG. I; MISC. 55; elleve BEK. 489; aleven P. R. L. P. 216.

endleofta, ord. num., *O.E.* endleofta, ellefta, = *O.N.* ellifti; *eleventh,* HOM. I. 117; endleft (*misprinted* -lest) ALIS. 57; endlefte ROB. 270; eollefta MISC. 146; elleft M. H. 26; ellevend PR. C. 4798.

endōren, v., *O.Fr.* endorer; *varnish*; **endōre** (*imper.*) hit with ȝolkes of egges L. C. C. 37; endoured (*pple.*) pigions B. B. 278.

O 2

endōren, v.,=adōren; *adore*: i forloine mi
dere endōrde (*pple.*) A. P. i. 368.

endōsen, v., *O.Fr.* endosser; *protect, guard*:
ye and i agans the feinde are welle endoost
(*pple.*) TOWNL. 166.

en-douten, v., *see* dūten; *doubt, fear*: if i
ne hadde endouted (*pple.*) me R. R. 1664.

endowen, v., *endow*; the gode vertues that
the bodi is endowed (*pple.*) with MAND. 252.

en-drīen, v., *from O.E.* drēoȝen; *suffer*: shuld
. . . shame in all mi life endri CT. LOVE 725.

endung, *see under* enden.

endūre, v., *O.Fr.* endurer; *endure*, P. L. S.
xxix. 74; MAND. 256; enduire MISC. 218.

ēne, *see* ǣne.

enede, sb., *O.E.* ened,=*O.N.* önd, *O.H.G.*
anut, anit, *Lat.* anas (*gen.* anātis); *duck*, REL.
I. 302; ende '*anas*' PR. P. 139; (*ms.* hende)
HAV. 1241; ende mete *ducks' meat* '*lenticula*' PR. P. 140.

enemi, sb., *O.Fr.* enemis, *Lat.* inimīcus;
enemy, CH. C. T. *A* 1643.

  enemi-ful, adj., *hostile*: it shal be desolat
  as in enemiful wastete WICL. IS. i. 70.

  enemi-liche, adv., *hostilely*, WICL. NUM. xxv.
  18; enmili (*later v.* enemili) 2 MACC. xiv. 11.

enemiable, adj., *O.Fr.* anemiable; *hostile*:
a bure he made aȝen the enmiable (*v. r.* enemiable) fole WICL. ECCLUS. xlvi. 7.

enemien, v., *from* enemi; *be hostile*: overjoȝe
not to me that enemien (*pres.*) to me wickeli
WICL. PS. xxxiv. 19.

enemitē, sb., *O.Fr.* enemistié; *enmity*, WICL.
GEN. iii. 15.

ēnen, *see* ēanen.

enentischt, enentist, pple., *from O.Fr.* ane-
antiss-, aneantir; *exhausted* ['*exinanita*']
HAMP PS. lxxiv. 8.

en-erden, v., *dwell in*; enerdend (*pres. pple.*)
D. TROY 12856; enerdond (*pple.*) 4117.

en-faminen, v., *O.Fr.* famine; *famish, hunger*;
enfamined (*pret.*) CH. L. G. W. 2429; faste
failed hem þe fode, enfaminied monie A. P. ii.
1194.

enfarsen, v., *O.Fr.* enfarcir; *stuff*: rost hit
on broche of irne bigge enfarsed (*pple.*) L.
C. C. 36.

en-fatten, v., *make fat*: the herte of this
peple is enfattid (*pple.*) WICL. MAT. xiii.
15; infattid DEEDS xxviii. 27.

enfēblen, v., *O.Fr.* enfeblir; *render feeble*:
we bi sinne enfēblen (*pres.*) oure feiþ WICL.
S. W. I. 94; þe strengþe is enfēblid (*pple.*)
I. 117.

enfēblishen, v., *O.Fr.* enfebliss-; *make
feeble*; enfēblished (*later ver.* febled) (*pple.*)
WICL. EX. xxii. 14.

enfecte, adj., *O.Fr.* infect, *from Lat.* infectus;
tainted, injured, PALL. i. 294.

enfecten, v., *O.Fr.* infecter; *infect; taint*:

a chanoun of religioun . . . wold infecte [enfect
T.] al a toun CH. C. T. *G* 973; enfecte
(*pret.*) þi faire love to þe fairist of Grise
D. TROY 2979; enfecte [*v. r.* infect] CH.
C. T. *A* 320; he enfecte the firmament with
his felle noise 936.

enfeffement, sb., *grant, fief*: ful riche was
þin enfeffement P. R. L. P. 163.

enfeffen, v., *enfeoff; endow*: a wille þat
stant enfeffid (*pple.*) in fraunchise P.R.L. P.64.

enfermen, v., *strengthen, fortify*: to enfermi
þen toun ROB. 552.

enfermer, sb., *O.Fr.* enfermier; *super-
intendent of the infirmary in a monastery*,
M. H. 29; *see* fermerere.

enflāmen, v., *O.Fr.* enflamer; *inflame*: pride
covetise and envie han so enflaumed (*pret.*)
the hertes MAND. 3.

enforce, v., *O.Fr.* enforcier; *enforce*, MIRC
1362.

  enforsinge, sb., *fury*, GEST. R. 116.

enformaciōn, sb., *information*, GOW. I.
253; informacion I. 272.

enforme, v., *O.Fr.* enformer; *inform*, D. TROY
654; enformede (*pret.*) TREV. V. 451.

enfoublen, v., *O.Fr.* afublir; *wrap up*: hir
frount folden in silk, enfoubled (*pple.*) ai
qvere GAW. 959.

enfrai, sb., *cf. O.Fr.* esfrei, esfroi, effrei, effroi;
affray: for ferdnes of a foulle enfrai TOWNL.
179.

eng, sb., *O.N.* eng; *cf. prov. Eng.* ing;
meadow, CATH. 115.

engeilen, v., *O.Fr.* engieler; *congeal, freeze*:
stones engeiled (*pple.*) falleth doun arow
HALLIW. 334.

engel, sb., *O.E.* engel, ængel, *from Lat.*
angelus; *angel*, HOM. I. 27; KATH. 666;
A. R. 146; ORM. 143; ængel MAT. i. 29;
HOM. I. 227; angel GEN. & EX. 67; SHOR.
120; engle (*dat.*) A. R. 66; engles (*pl.*)
LAȝ. 29689; angles BRD. 9; engle (*gen. pl.*)
ORM. 67; engle kinde 666; englene A. R.
92; englen (*dat. pl.*) MISC. 83; *comp.* hēh-
engel.

  engel-wird, sb., *angel-host*, GEN. & EX. 4140.

engendren, v., *O.Fr.* engendrer; *engender*,
CH. C. T. *E* 1272: engendrede (*pret.*) SHOR.
168.

  engendring, sb., *begetting, production*: thei
  made a fulle good engendring . . . thei engendrede the devel of helle R. R. 6117.

engendrūre, sb., *O.Fr.* engendrūre; *generation*, SHOR. 167.

engin, sb., *O.Fr.* engin, *from Lat.* ingenium;
engine; *understanding, skill*, FL. & BL. 759;
SPEC. 58; ALIS. 1213.

  engine-ful, adj., *skilful*: hi ben . . . gode
  and engineful to fighth ALIS. 4868.

enginen, v., *O.Fr.* engignier; *contrive; tor-*

ture, MAP 337; **engínid** (*pret.*) LANGL. *B* xviii. 250; thei hem have **engíned** (*pple.*) GOW. I. 79; engined CH. C. T. *B* 4250.

enginour, sb., *O.Fr.* engigneur; *engineer, contriver*, MAN. (F.) 4571.

enginous, adj., *O.Fr.* engignos; *skilful*, GOW. III. 99.

Engle, pr. n., *O.E.* Engle; *Angles, Englishmen*; **Engle** (*gen.*) lond MISC. 145; Engle land SAX. CHR. 250; Engle lond (*printed* Engelond) A. R. 82; Enguelonde SAINTS (Ld.) 26, 47, 73; mang **Engle þēode** ORM. 4231; on **Engle spēche** GEN. & EX. 814; Ængle ANGL. VII. 220; Engle, Englene LAȜ. 17, 1980; Englene REL. I. 170 (MISC. 102); PROCL. 1; he wolde of Engle þa æˀelæn tellen LAȜ. 13.

Englisch, adj. & śb., *O.E.* Englisc; *English*: an (on) Englisch HOM. I. 77; A. R. 130; BEK. 73; an Englisc LAȜ. 18231; Englich, Englischs, Engliss, Englisse ROB. (W.) 4802, 7543, 1450, 61; **Englisce** (*pl.*) men LAȜ. 1973; Englische men SAINTS [Ld.] xx. 81.

en-gleimen, v., *make clammy; besmear; 'visco, invisco,'* PR. P. 198; the man that muche honi eteth his mawe it **engleimeth** (*pres.*) LANGL. *B* xv. 56; alle englaimez þe gresse D. ARTH. 1130; **engleimed** (*pple.*) WICL. S. W. III. 150; englaimed in glotenie ALEX.² (Sk.) 676.

en-glūting, sb., *? smearing over*: of þe pot and glassis engluting (*v. r.* enluting) that of the aier mighte passe no thing CH. C. T. *G* 766; *see* enlūting.

en-gólden, v., *cover or deck with gold*; **engóldid** (*pple.*) with gold WICL. APOC. xvii. 4; engoldid and ensilvered WICL. BAR. vi. 7.

engreggen, v., *O.Fr.* engregier, *from Lat.* ingravāre; *make heavy*: everich of hem encreseth and **engreggith** (*pres.*) other CH. C. T. *I* 979; his herte engredgide (*pret.*) WICL. EX. viii. 15; engregid (*pple.*) vii. 14.

engreinen, v., *O.Fr.* engrainer; *dye in grain, of a fast colour*: of red scarlet **engreined** (*pple.*) LANGL. *B* ii. 15.

engréven, v., *O.Fr.* engrever; *trouble, burden*; no thing **engréveth** (*pres.*) me R. R. 3443.

en-grinen, v., *from O.E.* grinian; *ensnare*: þe herten **engrined** (*pple.*) ine þe dievles nette AYENB. 154.

enhabiten, v., *O.Fr.* enhabiter; *inhabit*; **enhabited** (*pple.*) MAND. 150; inhabited 187.

enhansen, v., *A.Fr.* enhauncer, *O.Fr.* enhaucer; *enhance, exalt*; enhaunse CH. C. T. *A* 1434; anhansi ROB. 198; enhauncen CH. C. T. *I* 614; enhaunce (*later ver.* enhaunse) WICL. EX. xv. 2; she deceivable **enhaunseth** (*pres.*) up the humble chere CH. BOET. ii. 1 (33); **enhaunside** (*pret.*) WICL. LK. i. 52; **enhaunsid** (*pple.*) D. TROY 13378; enhaunced

MAND. 95; enhaused CH. ASTR. ii. 26 (37); enhansed LANGL. *C* xii. 58.

enhaunsere, sb., *exalter*, WICL. EX. xvii. 15.

enhaunsing, sb., *elevation, exaltation*, CH. ASTR. ii. 39 (16); WICL. EX. xvii. 15.

en-haunten, v., *exercise*: **enhaunten** (*pres. pl.*) WICL. 2 PAR. xix. 6; bisili **enhaunte** (*imper.*) thi feeld WICL. PROV. xxiv. 27; en-**hauntide** (*pret.*) ix. 26; enhaunteden WICL. TOB. ix. 12.

en-hēȝen, v., *raise, exalt*; **enhīed** (*pple.*) GEST. R. (H.) 123.

enherite, v., *O.Fr.* enheriter; *inherit*, A. P. ii. 240.

enhorten, v., *O.Fr.* enorter; *exhort*: ful wisli to enhorte the peple CH. C. T. *A* 2851; **enhurte** (*imper.*) hem WICL. 2 KINGS xi. 25; **enorteth** (*pres.*) the studie of lessoun WICL. PREF. EPIS. iii (63).

ēni, *see* ǣni.

enjoien, v., *O.Fr.* enjoir; *enjoy*, PR. P. 140; enjoie WICL. LK. i. 14.

enjoinen, v., *Fr.* enjoindre; *enjoin*; anjoini AYENB. 172; **enjunȝe**, angeonni (*pres.*) A. R. 346*; i ajoine þee D. TROY 2197; en-**joinede** (*pret.*) ROB. 234; **anjoint** (*pple.*) (*printed* amoynt) A. P. i. 893.

enke, sb., *O.Fr.* enque, *Gr.* ἔγκαυστον; *i k*, HICKES I. 229; REL. II. 227; WICL. JER. xxxvi. 18; inke PR. P. 261.

enk-horn, sb., *ink-horn*, WICL. EZ. ix. 11.

enker, adj., *O.Fr.* (vert) encré; enker grene, *dark green*, GAW. 150.

[**ēnker-**, *? O.N.* einkar-; *? very*.]

ēnker-lī, adv., *? O.N.* * einkarliga (*specially, particularly*); *earnestly*, D. ARTH. 507, 2222; BARB. i. 92; encreli i. 301, 425.

enlácen, v., *O.Fr.* enlacer; *bind, entangle, perplex*: þat man . . . **enláceþ** (*pres.*) him in þe cheine CH. BOET. i. 3 (13); **enláced** (*pret.*) iii. 8 (80); v. 1 (149).

enlangoured, adj., *O.Fr.* enlangouré, *from* langor, *Lat.* languor; *sickly*, R. R. 7399.

enlargen, v., *O.Fr.* enlarger; *enlarge*, MAND. 45;

enlegeance [elegeance], sb., *O.Fr.* alegeance; *alleviation*, ROB. (W.) 1884.

enleven, *see* endlefen.

enliance, sb., *O.Fr.* aliance, alliance; *alliance in marriage*, ROB. 12.

enluminen, v., *O.Fr.* enluminer; *illumine*; **enlumined** (*pret.*) CH. C. T. *E* 33.

en-lūting, sb., *? med. Lat.* lūtāre; *daubing with clay*, CH. C. T. *G* 766; *see* englūting.

en-nēwen, v., *renew*; **ennēwith** (*pres.*) WICL. ECCLUS. xxxviii. 30; **ennēwe** (*imper.*) thou signes xxxvi. 6.

enoint, pple., *O.Fr.* enoint, enuint; *anointed*; CH. C. T. *A* 199; anoint R. R. 1889; TREV. III. 25; enoignt P. P II. 12; inoint L. H. R. 224.

enointin, anointin, v., *anoint*, PR. P. 12, 140;

†

anointe WICL. GEN. l. 2; **anointed** (*pple.*)
WILL. 139; enointed A. P. ii. 1446.

enointing, sb., PR. P. 140; anointing WICL.
EX. xxx. 31.

enoumbre, v., *O.Fr.* enombrer, *med.Lat.*
inumbrāre; *conceal in shadow*, MAND. 1;
enumbred (*pret.*) 136.

en-ourlen, v., *? from O.Fr.* ourler; *?clothe*;
enourled (*pret.*) in alle þat is clene A. P. ii. 17.

enournen, v., *cf. O.Fr.* aorner; *adorn*: enournen and araien WICL. ESTH. viii. 9; with
vertuez **enn(o)urned** (*pple.*) GAW. 634, 2027;
anurned A. P. i. 1027.

enourning, sb., *adornment*: wimmenis enourning WICL. ESTH. ii. 12, 9.

enpeirement, sb., *O.Fr.* empeirement; *im-
pairment*,AL.(L[1])255;emparementAYENB.148.

enpeiren, v., *O.Fr.* enpeirer, *cf. Sp.* empeorar;
impair, CH. BOET. iv. 3 (120); empeire
MAND. 316; ampairi AYENB. 10; apeire
LANGL. *B* v. 573; **apeirede** (*pret.*) ROB. 279.

enpir, *see* **emperie.**

enpoisōne, sb., *poison*, D. ARTH. 213.

enpoisonen, v., *O.Fr.* empoisoner; *poison*;
apoisonede (*pret.*) CHR. E. 781; enpoisoned
[apoisounde, apoisoned] (*pple.*) LANGL. *B* iii.
127, *C* iv. 164.

enpoisoner, sb., *poisoner*, CH. C. T. *C* 884.
enpoisoning, *poisoning*, CH. C. T. *C* 891.

enpressen, v., *O.Fr.* enpresser; *? oppress*:
þere as povert **enpresses** (*pres.*) A. P. iii. 43;
enpreces 528.

enprisonen, v.,*O.Fr.*enprisonen; *imprison*;
emprisonede (*pret.*) ROB. 94.

enprisonment, sb., *imprisonment*, A. P. ii. 46.

en-prūden, v̇., *become proud*; enprīdid (*ms.*
enpriuide) (*pret.*) GEST. R. (H.) 174.

enqveintance, sb.,=**acointance**; *acquaintance*, ROB. 330.

enqvére, v., *O.Fr.* enqverre; *inquire*, BEK.
333; P. S. 325; TREV. V. 11.

enqvíringe, sb., *inquiry*, CH. C. T. *B* 888.

enqveste, sb.,*O.Fr.* enqveste; *inquest*, BEK.
333; SHOR. 94.

enresonen, v., *converse with*; enresonede
(*pret.*) his men ROB. 321.

enrolli, v., *O.Fr.* enroller; *enroll*, E. G. 359.

ensample,sb.,*O.Fr.*ensample; *example*,MISC.
27; ensaumple WICL. 1 THESS. i. 7; ansaum-
ple SAINTS [Ld.] 65; *see* **asaumple, sample.**

ensampler, sb., *O.Fr.* essamplaire, exam-
plaire; *pattern, copy*: the whiche oure en-
saumpleris (*pl.*) [*earlier ver.*saumpleers] deliten WICL. JOSH. PROL. *page* 555.

enségen, v.,*besiege*; he...castelles enseggez
(*pres.*) D. ARTH. 623; þei **enségen** (*pl.*) þe
soulis of men WICL. S. W. II. 155; **enségid**
(*pret.*) WICL. DEUT. xxviii. 52.

ensent, sb., *consent*; ensent [*v. r.* assent]
ROB. 317.

ensenten, v., *assent*; encenti ROB. (W.) 2118;
ensenteþ (*pres.*) ROB. (W.) 6204; **ensentede**
(*pret.*) þerto ROB. 446; encented 171.

enserchen, aserchen, v., *O.Fr.* encerchier;
search into, WICL. PROV. xxviii. 11, 3 KINGS
xx. 6; **enserches** (*pres.*) D. ARTH. 4311.

enserchour, sb., *searcher*, GEST. R. 55.

en-silveren, v., *silver over*: engoldid and ensilvered (*pple.*) ben fals WICL. BAR. vi. 7.

ensoinen, v., *O.Fr.* ensoignier; = **assoinen**;
excuse: mi sone from þe schulde beon en-
soinet (*pple.*) L. H. R. 132.

en-speren, v., *from* spürien; *inquire*; en-
spered (*pret.*) GEST. R. 317.

enspīre, v., *O.Fr.* inspirer; *inspire*, P. L. S.
xxix. 81; **enspīrud** (*pple.*) CH. C. T. (W.) 5.

enstōre, v., *cf. Lat.* instaurāre; *restore; store
up*; enstore, instore WICL. EPH. i. 10; en-
stooride (*pret.*) WICL. 4 KINGS xii. 14; in-
stōrid (*pple.*) in this word WICL. ROM. xiii. 9.

enstōringe, sb., *reparation*, WICL. 4 KINGS
xii. 12.

enstranglen, v., *O.Fr.* estrangler, *from Lat.*
strangulāre; *strangle*; **enstrangled** (*pple.*)
MAND. 194.

ensue, v., *O.Fr.* ensuire; *ensue*, P. R. L. P. 43.

ensūre, v., *for* assūre; *ensure*, CH. C. T. *B*
1231; D. ARTH. 2324.

entached, *see* **entechen.**

entaile, v., *O.Fr.* entailler; *engrave, carve*,
R. R. 608; **entailled** (*pple.*) with mani riche
portraitures PL. CR. 139, 398.

entaille, sb., *O.Fr.* entaille; *carving; quality,
fashion, sort*; FER. 730; MAN. (F.) 15846; of
entaille *persons of quality* 14886.

entalenten, v., *O.Fr.* entalenter; *excite*, CH.
BOET. v. 5 (168).

entāme, v., *O.Fr.* entamer, *for* *attamer, *Lat.*
attāmināre; *begin*, CH. A. B. C. 79; **entámede**
(*pret.*) D. ARTH. 2203.

enteche, sb., *symptom* (*of illness*): þe en-
tecches (*pl.*) of min evele WILL. 557.

entechen, v., *O.Fr.* entechier; *infect; imbue
with good or bad qualities*; **enteched** (*pple.*)
and defouled CH. BOET. iv. 3 (120); entached
wel ROB. (W.) *G* 165; on of the best enteched
creature CH. TRO. v. 832.

entemperate, adj., *temperate*, ROB. (W.)
8848*.

entemprē, adj., *temperate*, ROB. 429.

entempri, v., *moderate*, TREAT. 289.

entencioun, sb., *O.Fr.* entencion; *intention*,
R. R. 5361; intencion LUD. COV. 240, 247.

entendable, adj., *O.Fr.* entendable; *mindful*,
Gow. III. 157.

entendement, sb., *O.Fr.* entendement; *dis-
cernment, instruction*, Gow. III. 142, 143;
a tale of great entendement I. 179.

entenden, v., *O.Fr.* entendre, *Lat.* intendere
intend, be intent, look steadfastly, P. L. S. xvii.

227; ech to his owen nedes gan entende CH.
TRO. iii. 375; to love min enemies i wolde not
entende P. R. L. P. 199; this love to vertu alle
intende (*pres.*) LUD. COV. 361; entendeth R. R.
5312; CH. BOET. i. 2 (8); entende (*imper.*)
P. L. S. xvii. 234.

entente, sb., *O.Fr.* entente; *intent*, A. R. 252*;
MARG. 277; in ful good intent CH. C. T. *A* 958.

ententen, v., *O.Fr.* ententer; *pay attention
to; intend, attend*, CH. BOET. v. 1 (150);
ententid (*pret.*) ALIS. 2833.

ententif, adj., *O.Fr.* ententif; *attentive*, WICL.
2 PARAL. vi. 3.

  ententif-liche, adv., *attentively*, BEK. 460;
AYENB. 210; ententiifli WICL. 2 PARAL.
xxxiii. 13; ententifli P. CR. 624, 2550.

entēr, adj., *Fr.* entier; *entire*, L. H. R. 196.

  entēr-lī, adj., *entire, whole*: besechinge you
ever with min enterli hert P. R. L. P. 41.

enterchaunge, sb., *interchange*, CH. BOET.
iii. 2 (65).

enteren, v., *Fr.* enterrer; *inter*; entered
(*pple.*) MAND. 94.

enterlace, sb., *cf.* entrelácen; *alternate
rhyme*, MAN. (F.) 86.

enterlūde, sb., *O.Fr.* entrelude, *med.Lat.* in-
terlūdium; *interlude*; enterlūdez (*pl.*) GAW.
471; plaiers of enterludes and plaies WART.
III. 173; interlodies REL. II. 47.

entermedlen, v., *O.Fr.* entremesler, -meller;
intermeddle; entermedled (*pple.*) CH. BOET.
ii. 6 (54).

enterment, sb., *O.Fr.* enterrement; *interment,
burial*: at the entierment of þe tru prinse D.
TROY 9105; enterment MAN. (H.) 327.

enterpéle, v., *Fr.* interpeller; *question*, WICL.
DEUT. xxv. 7.

enteune, sb., *from O.Fr.* entoner, *see* en-
tūnen; *melody, music*; enteunis (*pl.*) CH.
D. BL. 307.

enticement, sb., *enticement*, ALIS. (L.²) 264;
entisment S. S. (Wr.) 3418; entisement CH.
C. T. *I* 967.

entirférin, v., *O.Fr.* entreferir; *interfere*, PR.
P. 140.

entisen, v., *O.Fr.* enticier; *catch an infection;
entice*: entise teches of filþe GAW. 2436; (he)
entísid (*pret.*) man to glotoni TOWNL. 21;
entisede ROB. 32.

entitelen, v., *entitle*; entitelid (*pple.*) CH.
P. F. 30.

entraile, sb., *Fr.* entraille; *entrails*, CH. C. T.
E 1188; intraile PR. P. 262.

entrecommunen, v., *from O.Fr.* communier;
mix together: men most entrecommunen ifere
CH. TRO. iv. 1325; entercomuninge (*communication*) of merchaundise CH. BOET. ii. 7 (57).

entredit, sb., *interdict*, ROB. 495.

entredīten, v., *interdict*; enterdīted (*pret.*)
MAN. (H.) 209; (hi) entreditede (*pl.*) ROB. 495.

entree, sb., *O.Fr.* entree; *entry*, LANGL. (Wr.)
6823; entre ROB. 158; MAN. (F.) 14091; en-
treie ROB. (W.) 3311.

entrelácen, v., *O.Fr.* entrelacier; *entangle*:
so entreláced (*pple.*) þat it is unable to ben
unlaced CH. BOET. iii. 12 (105).

entremes, sb., *O.Fr.* entremes; *entremets*:
trufles vor entremes AYENB. 56; this entre-
messe is dressed for iow all CH. P. F. 665.

entremeten, v., *O.Fr.* entremetre; *meddle
with; busy one's self about; interpose*, A. R.
414; to entremete hir of sich vice R. R. 5948;
entremetti AYENB. 152; entermet MAN. (F.)
12615; i entremete (*pres.*) me of brokages
6973; entremetteth CH. C. T. *B* 2732; entre-
metiþ CH. BOET. iii. 12 (104).

entreparten, v., *from* parten, *O.Fr.* partir;
share, CH. TRO. i. 591.

entréten, v., *O.Fr.* entreater; *treat, handle;
entreat*, P. R. L. P. 168; smitten and vilentli en-
tréted (*pple.*) MAND. 91; vileinsli entreted 95.

entriken, v., *O.Fr.* entriquer, *Lat.* intricāre;
entangle; deceive; engage (in an affair):
wherof that he the worde entríketh (*pres.*)
GOW. I. 359; entríked (*pret.*) R. R. 1642.

entrin, v., *O.Fr.* entrer; *enter*, PR. P. 140
entren MAND. 273; entre PR. C. 5340; entri
SHOR. 46; intered (*pret.*) LIDG. M. P. 10.

  entring, sb., *entering*: þis entriinge to þe
sepulcre is cominge to þe service of Crist
WICL. S. W. I. 132; entering P. L. S. xxxi. 197.

entunen, v., *O.Fr.* entoner; *intone, sing*;
entuned (*pple.*) CH. C. T. *A* 123.

enui, sb., *O.Fr.* enoi, ennui, ani; *annoy*; anui
[ennui] A. R. 94; anoi AYENB. 267; nui C. L.
442.

envelupen, v., *O.Fr.* envoluper, enveloper;
envelope: he is most envoluped (*pple.*) in
sinne CH. C. T. *C* 942.

envenime, v., *O.Fr.* envenimer; *envenom*, CH.
C. T. *D* 474; en-, avvenimeþ (*pres.*) AYENB.
26, 27; envenemed (*pple.*) WILL. 4428.

  enveniminge, sb., *poisoning*, CH. C. T. *E*
2060.

envie, sb., *O.Fr.* envie; *envy*, HORN (L.) 687;
P. L. S. ix. 69; C. L. 833; anvie MISC. 33.

envie, v., *Fr.* envier; *envy, vie*, CH. C. T. *D*
142; D. BL. 406.

envīnen, v., *O.Fr.* enviner; *store with wine*:
a bettre envīned (*pple.*) man was nowher
noon CH. C. T. *A* 342.

envious, adj., *O.Fr.* envieus; *envious*, AYENB.
27; envius FL. & BL. 356.

enviroun, adv., *O.Fr.* environ; *round about*:
the erthe is . . . aboute enviroun . . . 20425 miles
MAND. 185; aboute the king stonden environ
Attendaunce, Diligence and . . . mani one CT.
LOVE 1031; environ GOW. III. 76.

enviroun, sb., *O.Fr.* environ; *environs, circuit, compass*, WICL. GEN. xxiii. 17; alle the

Egipciens delviden bi enviroun of the floode EX. vii. 24.

enviroune, v., *Fr.*environner; *environ,*WICL. PS. lviii. 7.

envirouninge, sb.,*circumference,*CH. BOET. v. 4 (164); longe **envirouningis** (*pl.*) of weie WICL. JOSH. v. 4.

ēode, v., *O.E.* ēode, = *Goth.* iddja (*pret.*); *went,* HOM. I. 41; A. R. 52; SPEC. 42; C. L. 320; he eode a lond LA3. 122; eode, 3eode ROB. 53, 79; L. H. R. 26; ede HORN (R.) 115; MAP 335; geode (? = *O.E.* geeode) MAT. ix. 25; 3eode, 3ede HORN (L.) 381, 1025; 3ede HOM. I. 229; WILL. 1767; S. S. (Wr.) 1964; MAN. (F.) 7552; CH. C. T. *G* 1281; 3ede into þe temple ORM. 136; gede GEN. & EX. 618; iede SACR. 299; to þe kirke iede HAV. 1355; for hem ne iede gold ne fe 44; giede HOM. II. 175; 3ode AMAD. (R.) lxvi; iode EGL. 531; M. H. 72; vde ROB. (W.) 8953*; þu eodest MISC. 139; eoden HOM. I. 3; beggeres faste aboute eoden [3eden, 3ede] LANGL. PROL. 40; iedon SAX. CHR. 250, 262; ieden (*ms.* yeden) AYENB. 74; *comp.* **bi-, for-, 3e-, of-, ofer-, under-ēode.**

eole, *see* **éle. eom,** *see* **am** *and* **ēam.**

eont, sb., *O.E.* ent; *giant;* **eontas** (*pl.*) HOM. I. 93.

ēored, sb., *O.E.* ēored, ēorod, = *O.L.G.* eorid; *legion, troop:* twelf **ēorde** (*pl.*) ængle MAT. xxvi. 53.

eorl, sb., *O.E.* eorl, = *O.L.G.* erl, *O.N.* iarl; *earl,* LA3. 7551; REL. I. 170; BEK. 507; erl AYENB. 71; CH. C. T. *D* 1157; 3orl DEGR. 1881; **eorle** (*dat.*) LA3. 8927; HOM. II. 35; **eorles** (*pl.*) P. L. S. viii. 161; ORM. 3989; 3ierles HOM. II. 230; **eorlene** (*gen. pl.*) LA3. 4766; **eorlen** (*dat. pl.*) LA3. 24966; ærlen HOM. I. 231.

eorl-dōm, sb., *O.E.* eorldōm; *earldom,* LA3. 11560; erldom HAV. 2909.

erl-marschal, sb., *earl marshal,* ROB. 152.

eorne, adv., (*? for* 3eorne); *earnestly:* crieð him eorne merci A. R. 44.

eornen, *see* **rinnen.**

eornest, sb., *O.E.* eornost, = *O.H.G.* ernust, *mod.Eng. earnest; battle; earnest, seriousness:* þer wes feht swiðe stor eornest ful sturne LA3. 16480; on eornest ic segge eow MAT. v. 18; on eornest S. A. L. 161; ernest '*seriositas*' PR. P. 142; GOW. I. 297; CH. C. T. *A* 3186; mid gode ernest ROB. 175; an ernest TREV. VII. 335; more for erneste ðan for gamen GEN. & EX. 411.

ernest-ful, adj., *earnest,* HOCCL. i. 293.

earnestful-lī,adv.,*earnestly,*BARB.viii.144.

eornest, adj., *earnest;* earnest C. M. 26351; D. TROY 2713.

ernest-lī, adv., *O.E.* eornostlīce; *earnestly,* A. P. ii. 277.

eorre, *see* **irre.**

eorðe, sb., *O.E.* eorðe, iorðe, earðe, = *O.L.G.* ertha, erðe, *O.Fris.* erthe, erde, *O.H.G.* erda, *Goth.* airþa, *O.N.* iörð; *earth,* A. R. 122; þeo eorðe HOM. I. 53; þa eorðe [þe earþe] LA3. 27817; eorþe ORM. 2596; C. L. 95; TREV. V. 163; eorþe [erþe, erthe] LANGL. *A* vii. 2; erþe GOW. III. 92; PS. xi. 7; erþe (*printed* erye), erthe, erde PR. P. 141; eerþe REL. I. 282; oerþe II. 226; erþe, ierþe AYENB. 8, 12; 3orþe N. P. 50; 3orthe GUY (Z.) 2825; yerthe D. TROY 8345; urþe TREAT. 132; (*ms.* vrþe) MIRC 483; eorðan (*gen.*) MAT. xi. 25; eorðan (*dat.*) HOM. I. 13; on eorðen LA3. 1150; to þere eorðe 25616; to þære eorþe FRAG. 6; þa cheorles... þa tileden þa eorðen LA3. 20958; & leide hit on urþen (*ms.* vrþen) 3863.

erð-chine, sb., *cleft in the earth,* REL. I. 217.

eorþe-qvake, sb., *earthquake,* TRANS. XVIII. 28; ertheqwake PR. P. 141.

erthe-qvave, sb.,*earthquake;* **ertheqvaves** (*pl.*) WICL. ESTH. xi. 5.

eorð-dine, sb., *O.E.* eorðdyne; *earthquake,* SAX. CHR. 258; erðedine GEN. & EX. 1108; erthedin PR. C. 4790; erþedene PR. P. 141.

erð-grine, sb., *earthquake:* erðgrine strong inou ROB. 530.

erþ-griþe, sb., *earthquake,* ROB. 414.

eorð-hole,sb.,*cave in the earth,*HOM. II. 139.

erðe-horn, sb., *a contrivance said to have been used at the battle of the Standard:* in ilke strete and wai þei ordeind an erþehorn MAN. (H.) 118; erþehornes (*pl.*) 118.

eorð-hūs, sb., *O.E.* eorðhūs; *house in the earth,* LA3. 2360.

eorð-īvi, sb., *O.E.* eorðīfig; '*hedera terrestris,*' LEECHD. III. 88; oerþivi REL. I. 37.

eorð-līc, adj., *O.E.* eorðlīc; *earthly,* LA3. 25875; eorþlic, -li3 ORM. 7036, 10933; eorþlich ALIS. 429; urþlich P. L. S. xvii. 269.

erthe-moving, sb., *earthquake,*WICL. MAT. xxviii. 2; erthemovinge AMOS i. 1; erthemovingis (*pl.*) ESTH. xi. 5; MAT. xxiv. 7.

eorþe-rīche, sb., *O. E.* eorðrīce, = *O.H.G.* erdrīche; *earthly kingdom,* ORM. 12132.

erthe-schaking, sb., *earthquake,* WICL. MAT. xxviii. 2.

eorðe-tilie, sb., *O.E.* eorðtilia; *tiller of the ground,* LA3. 22117; A. R. 414; erþetilie CHR. E. 93; eorðetilien (*later ver.* erþe-tilies) (*pl.*) LA3. 22118.

erthe-tiliere, sb., *tiller of the ground,*WICL. GEN. iv. 2; erthtiliers [*earlier ver.* erthetilieris] (*pl.*) 4 KINGS xxv. 12.

erþe-tilþe, sb., *O.E.* eorðtilð; *agriculture* ['*agricultura*'] WICL. 2 PARAL. xxvi. 10.

eorðe-ware, sb.pl.,*O.E.* eorðwaru; *inhabitants of the earth,* A. R. 322; HOM. II. 69.

erthe-werching, sb., *agriculture:* as the erthewerching ['*rusticatio*'] of a tree shewith the frute of him WICL. ECCLUS. xxvii. 7.

eorðen, adj.,=*O.H.G.* erdīn, *Goth.* airþeins; *earthen*; erþen CH. C. T. *G* 761; wiðinnen one eorðene castle A. R. 388.

[**eorþien**, v., *bury*] ; **erdit** (*pple.*) BARB. xiii. 666 ; erdide WINT. IX. xii. 7.

   erþing, sb., *burial*, C. M. 1190, 18041.

eotand, eotend [**eatant**], sb., *? for* eten ; *? giant*, LAȝ. 1375, 1866.

eoten, *see* eten.   **ēou** *see* **ēow**.

Eoverwīc, ȝeorc [**ȝorc**], pr. n., *O.E.* Eoforwīc ; *York*, LAȝ. 2669, 2673 ; Everwik ROB. 2 ; Everwich MISC. 146 ; Yerk HAV. 1178.

eovese, *see* **evese**.

ēow, pron., *O.E.* ēow, iow (*dat.*), ēowic, iowih (*acc.*), *cf. O.Fris.* io, iu, *O.L.G.* iu, eu, *O.H.G.* iu (*dat.*), iuwih, iwih, iuh (*acc.*), *Goth.* izvis (*dat. acc. pl.*) ; *you* (*dat. & accus. plur.*), P. L. S. viii. 79 ; ȝe hit ieseð eow selven HOM. I. 35 ; eou 105 ; eow, geau ANGL. VII. 220 ; eou, eow, ȝeow [ȝou, ou] LAȝ. 737, 5453, 5455 ; eou R. S. vii ; S. A. L. 153 ; ow KATH. 277 ; ow, ou, eu O. & N. 114, 1683, 1793 ; eu, giu HOM. II. 17, 35 ; ou H. H. 2 ; JOS. 73 ; ʋu [ow] C. L. 567 ; non of ou A. R. 256 ; demeð ou sulven 406 ; ȝou MARG. 136 ; WILL. 262 (*spelt* ȝow 238) ; AUD. 6 ; iou (*ms.* you) HAV. 169 ; ȝou [ou] LANGL. *A* i. 14 ; ȝou [iou] CH. C. T. *A* 728 ; ȝuw ORM. 89 ; ȝeu ANGL. I. 16 ; (*ms.* ȝew) PROCL. 7 ; ȝeu, ȝiu HOM. I. 223, 237 ; ȝu SPEC. 136 ; gu MAT. xxv. 45 ; GEN. & EX. 325 ; iu (*ms.* yu) MISC. 33 ; HAVEL. 2595 ; ȝo ANT. ARTH. xxv.

ēowde, sb., *O.E.* ēowde,=*O.H.G.* ēwit, *Goth.* aweþi ; *flock of sheep* : to þara eowde oðer falde HOM. I. 245.

ēowe, sb., *O.E.* ēowu=*O.H.G.* awi, ouwi, *M. Du.* ouwe, oie, *O.Fris.* ei, *O.N.* ær ; *ewe* ; ewe REL. I. 161 ; an ouwe (*ms.* a nouwe) MAP 335 ; awe A. S. 3.

   ewe-lamb, sb.,=*M.Du.* oilam ; *ewe-lamb*, WICL. GEN. xxi. 28.

ēowen, *see* **ēawen**.

ēower, pron., *O.E.* ēower,=*O.L.G.* iuwar, *O.H.G.* iuwer, *Goth.* izvara (*gen. pl.*) ; *your*, *of you* : an eower MAT. xxvi. 21 ; eower ealre ix. 7 ; hwilc eower HOM. I. 245 ; eower alre sunne 21 ; eour 15 ; ower KATH. 1283 ; ower FL. & BL. 534 ; ȝour HORN (L.) 815 ; iour AM. & AMIL. 852 ; ȝure nan ORM. 9271 ; ilc gure GEN. & EX. 3471 ; ȝoure LANGL. *C* xxii. 473 ; ȝoure on BRD. 5.

ēower, ēuwer, pron., *O.E.* ēower,=*O.H.G.* iuwer, *O.L.G.* iuwar, iuwa, *O.Fris.* iuwe, *Goth.* izvar ; *your*, HOM. I. 13 ; eouwer LAȝ. 3652 ; S. A. L. 153 ; ower, eur O. & N. 1699 ; ower ROB. 500 ; ower, ouwer, our A. R. 8, 46, 70 ; ower (*for* owres) swinkes lan KATH. 806 ; giwer HOM. II. 65 ; gur silver GEN. & EX. 2260 ; he is ȝour CH. TRO. ii. 587 ; i wolde it were ȝoures C. T. *B* 1464 ; ȝor P. L. S. xxx. 58 ; or SPEC. 37 ; eowre þeode LAȝ. 12526 ;

eoure sunne R. S. vii. ; oure soule A. R. 50 ; eure saule REL. I. 171 ; ȝiure hierte HOM. I. 217 ; yure levedi HAV. 171 ; kristes ore and ioures 2798 ; ȝure (*for* ȝur) preost ORM. 934 ; oure [ȝoure] LANGL. *A* iii. 64 ; **ēowres** (*gen. m. n.*) MAT. x. 30 ; **ēowren**, eowre (*dat. m. n.*) MAT. x. 29, xxv. 8 ; in eowre londe LAȝ. 12524 ; **ēowre** (*pl.*) HOM. I. 91 ; eoure godes LAȝ. 13947 ; ȝiure HOM. I. 223 ; giure HOM. II. 115 ; iure sennen MISC. 32 ; oure douȝtris and ȝoure WICL. GEN. xxxiv. 16 ; in owre bokes KATH. 839 ; **ēowren** (*dat. pl.*) MAT. ix. 4.

ēpen, v., *O.N.* œpa (æpa),=**wēpen** ; *call, call upon* ; ēpeþ (*pres.*) ORM. 9198.

epetīte, sb., *Lat.* hēpatītis, *Gr.* ἡπατίτης ; *a precioys stone*, P. S. L. xxxv. 94.

epicicle, sb., *Fr.* epicycle, *Gr.* ἐπίκυκλος ; *epicycle*, CH. ASTR. ii. 35 (44).

epiphani, sb., *Gr.* ἐπιφάνεια ; *epiphany*, SPEC. 96.

epistle, sb., *O.Fr.* epistle ; *epistle*, A. R. 8.

epitaf, sb., *Lat.* epitaphium ; *epitaph*, MAN. (F.) 16682.

eppel, *see* **appel**.

eqvaciōn, sb., *Lat.* aeqvātio ; *equation, equal partition; calculation*, CH. ASTR. ii. 37 (46) ; **eqvaciouns** (*pl.*) i. 22 (14) ; ii. 36 (44).

eqval, adj., *Lat.* aeqvālis ; *equal* : the 24 houres eqvals (*pl.*) of the clokke CH. ASTR. i. 16 (8) ; eqvales ii. 8 (21).

eqvator, sb., *Lat.* aeqvātor ; *equator* : same cercle is cleped also the weiere, ' eqvator,' of the dai CH. ASTR. i. 17 (9).

eqvinoxial, adj., *equinoctial*, CH. ASTR. i. 17 (9).

eqvinoxium, sb., *equinox* : upon the hour of mid dai whan it is eqvinoxium MAND. 183 ; are thise two signes called the eqvinoxiis (*pl.*) CH. ASTR. i. 17 (9).

eqvipolence, sb., *O.Fr.* eqvipollence ; *equivalent*; eqvipolences (*pl.*) R. R. 7077.

eqvitē, sb., *O.Fr.* equité ; *equity*, LANGL. *B* xvii. 304 ; CH. C. T. *E* 439.

er, *see* **æiðer**.   **ēr**, *see* **ær, ēar**.

eranie, *see* **aranie**.

erbe, erber, *see* **herbe, herber**.

erber, sb., *O.Fr.* (h)erbiere ; *first stomach of ruminants* ; þe erber diȝt he ȝare TRIST. 486 ; siþen þai slit þe slot, sesed þe erber GAW. 1330 ; and begin in firste to make the erbere BK. ST. ALB. F. iii.

erchebischop, *see* **archebischop**.

erchedēkne, *see* **archedēkne**.

erd, erdien, *see* **eard, eardien**.

ērdēd, *see under* **ær**.   **erdit**, *see* **eorþien**.

ērdne, *see* **ærende**.   **ēre**, *see* **ēare**.

ē-rēde, adj., *destitute of counsel, at a loss*, O. & N. 1295 ; eirede MISC. 192.

erege, sb., *O.Fr.* herege, *Lat.* hæreticus ;

heretic: huanne ne dra3þ voulliche þet bodi of oure lhorde, ase doþ þe **ereges** (*pl.*) AYENB. 40.

eremin, *see* ermine.

eremite, sb., *O.Fr.* ermite, hermite, *from Lat.* erēmīta; *hermit*, HOM. II. 85; A. R. 12; ermite, hermite P. L. S. xv. 64, 78; þou schost have ben ermite or frere GREG. 907; eremite (heremites) (*pl.*) LA3. 29851; *deriv.* ermitāge.

ērende, *see* ǣrende.

eresie, *see* heresie.

erewe, *see* ar3.

erfe, erve, sb., *O.E.* erfe, ierfe, yrfe,= *O.Fris.* erve, *O.H.G.* erbe, erbi, arbe, *Goth.* arbi *n.* (κληρονομία); *heritage; cattle*, GEN. & EX. 169, 183; of þat erfe þat tær was drihtin to lake 3arked ORM. 1068; cwealm on men and on erve SAX. CHR. 254; erfes ['*pecora*'] PS. cxlviii. 10.

erfe-blōd, sb., *blood of animals*, ORM. 1788.

erf-kin, sb., *cattle*, GEN. & EX. 3177.

arf-name, sb., *cf. M.Du.* arfname, *O.Fris.* erfnama, -noma; *heir*, ORM. 17744.

er(f-)ward, sb., *O.E.* erfe-, yrfeweard; *heir*, GEN. & EX. 934.

erfeð, *see* earfeð. **er3**, *see* ar3.

éri, adj., *mod.Eng.* eerie; *timid*, C. M. 17685.

érines, sb., *eeriness*, BARB. ii. 295.

ērid, *see* ēared.

érien, v., *O.E.* erian, = *M.Du.* eren, *O.N.* erja, *O.Fris.* era, *O.H.G.* erren, *Goth.* arjan, *Lat.* arāre; *plough*: erien and sowe O. & N. 1039; ærien LA3. 10030; eriin '*aro*' PR. P. 141; erie AYENB. 214; LANGL. *B* vi. 4; eeren, ere WICL. EXOD. xxxiv. 21; ere MAND. 44; CH. C. T. *A* 886; érede (*pret.*) A. R. 384; ROB. 21.

érere, sb., *ploughman* ['*arator*'], WICL. AMOS ix. 6.

éringe, sb., *cf. O.H.G.* erunga; *ploughing*, LANGL. *A* PROL. 21.

erigant, *see* herigant.

ērinde, *see* ǣrende.

eritāge, sb., *O.Fr.* heritāge; *heritage*, ROB. 381; **eritāges** (*pl.*) 454.

eritāgen, v., *inherit, endow*, WICL. PS. xxxvi. 11; thi sed jentiles shal eritagen (*later ver.* enherite) IS. liv. 3; the lawe of lif he **eritāgede** (*pret.*) them ECCLUS. xvii. 9.

erite, sb., *from Lat.* hæreticus; *heretic*; þe erites (*pl.*) HOM. I. 143.

erl, *see* eorl.

erles, sb., ?=erres, *O.Fr.* erre (*pl.* erres); *earnest-money*: þis ure laverd 3iveð ham her as on erles H. M. 7; giffen on erles D. ARTH. 2687.

ērlī, ērlīk, *see under* ǣr.

erm, *see* earm.

er-man, sb., *from* érien; *cultivator*; hear-man H. M. 47.

ermen, v., *O.E.* yrman, ierman; *from* earm; *make miserable*; erme CH. C. T. *C* 312.

erming, sb., *O.E.* erming; *misery*: theo bisscheop weop for erming ALIS. 1525.

ermight, sb. ? if thow be a gentill man let thi beiting and thi ermight be, and come prove thi strenghe on me TOR. (A.) 1008.

ermine, sb., *O.Fr.* ermine; *ermine*, P. L. S. viii. 182; S. S. (Web.) 473; hermine MISC. 70; eremin MAN. (F.) 11315.

erming, *see* earming.

ermitāge, sb., *O.Fr.* ermitage, hermitage; *hermitage*, IW. 1672; ermitage, hermitage C. M. 8161; **heremitāges** (*pl.*) MAND. 93.

ermite, *see* eremite. **ermðe**, *see* earmðe.

[**ern**, sb., *O.E.* ern, ærn, ?=ran; *house; comp.* bak-, ber-, brēu-, cwart-, hēd-, shǣw-, slǣp-ern (-erne).]

ern, *see* earn.

ērnde, *see* ǣrende.

erne, *see* hürne.

ernen, *see* rennen *and* rinnen.

ernes, sb., ?=erles; *earnest-money*, P. R. L. P. 231; ernes, ernest ['*pignus*'] WICL. 2 COR. i. 22, v. 5; ernest '*arra*' PR. P. 142.

ernest, *see* eornest *and* ernes.

ernien, *see* earnien.

Ernlē3e, pr. n., *Arley* (*dat.*) LA3. 5.

erratik, adj., *O.Fr.* erratique, *Lat.* errāticus; *erratic*: the erratik sterres CH. TRO. v. 1825.

erraunt, adj., *O.Fr.* errant, *from* errer; *arrant*: an outlawe, or a thef erraunt CH. C. T. *H* 224; that erraunte jewe D. ARTH. 2895.

erre, sb., *O.Fr.* erre; *earnest-money*; **erres** (*pl.*) L. H. R. 217; *see* erles *and* ernes.

erre, *see* ar *and* irre.

erren, v., *O.Fr.* errer; *err*, PL. CR. 846; **erren** (*pres.*) CH. C. T. *G* 449; **errede** (*pret.*) TREV. IV. 473; erride WICL. GEN. xxi. 14.

erring, sb., *error*, PR. C. 5728, 5974.

errour, sb., *O.Fr.* errour; *error*, SHOR. 30; AYENB. 69; LEG. 213; PR. C. 4266.

ers, *see* ars.

ertin, v., *O.N.* erta; *? provoke*, '*irrito*,' PR. P. 142.

erþe, sb., *O.E.* yrð; *ploughing*; **erthes** (*pl.*) PALL. iv. 68.

erþe, *see* eorðe *and* eardien.

erþing, *see under* eorþien.

eruh, *see* ar3.

erūke, sb., *caterpillar*, WICL. JOEL i. 4.

erve, *see* erfe. **ervest**, *see* hervest.

erveð, *see* earfeð.

es, ?=se, pron., (*acc. f.*); *her*, HOM. II. 163; hes 159; is, his SHOR. 77, 136; his ROB. 349;

AYENB. 58; **es**, is (*acc.pl.*) GEN. & EX. 135, 154, 1768 (*printed* it 2974) ; HAV. 970, 1174 ; is ROB. 338 ; hes, his HOM. I. 55, 237 ; his MISC. 34 ; þe þe ehte wile healden wel þe wile he mai his welden ȝive his for godes luve P. L. S. viii. 28 ; his, hise AYENB. 5, 13.

ēs, sb., *O.E.* ǣs, = *M.L.G.*, *M.H.G.* ās, *M.Du.* aes ; *corpse, carrion ; bait, decoy* ; ees '*esca*' PR. P. 143 ; þet es HOM. I. 123 ; ierne to þe mete ase deþ þe hond to þe es (*ms.* hes) AYENB. 55 ; **ēses** (*gen.*) HOM. I. 123 ; **ēse** (*dat.*) IBID.

escápen, v., *O.Fr.* escaper ; *escape* ; eschape PR. C. 2678 ; aschape WILL. 1671 ; ascapie AYENB. 131 ; askape LANGL. *A* ii. 180 ; scapen MISC. III ; ARCH. LII. 37 ; **askápeþ** (*pres.*) MISC. 41 ; **ascápede** (*pret.*) LAȜ. 1611*.

eschaufen, v., *O.Fr.* eschaufer (*échauffer*) ; *warm, heat* ; eschaufeþ (*pres.*) CH. BOET. i. 5 (22) ; achaufe [chaufe] (*subj.*) LANGL. *C* xv. 68* ; achaufed (*pple.*) A. P. ii. 1143.

    eschaufing, sb., *heating*, CH. C. T. *I* 537.

eschaunge, sb., *O.Fr.* eschange ; *exchange*, LANGL. *C* vii. 280 ; eschaunge [eschaung, eschange], CH. C. T. *A* 298.

esche, *see* asche *and* áskien.

escheker, *see* cheker.

eschel(e), sb., *O.Fr.* eschele (*échelle, troupe rangée en échelle*) ; *troop, battalion*, WILL. 3379 ; eschelle TOWNL. 47 ; **escheles** (*pl.*) MAN. (H.) 288.

eschend, *see* ȝeschenden.

eschēte, sb., *O.Fr.* eschoete, escheete ; *escheat* ; **eschētes** (*pl.*) LANGL. *B* iv. 175.

eschewen, v., *O.Fr.* eschiver, *mod.Eng.* eschew ; *avoid, escape*, CH. BOET. v. 6 (177) ; for to eschewe that place MAND. 61 ; for to eschewe folie 293 ; WICL. LEV. xi. 11 ; P. L. S. xxv. 121 ; what to done best were, and what eschuwe CH. TRO. ii. 696 ; escheue GOW. I. 19 ; the puple eschewide (*pret.*) to entre in to the citee WICL. 2 KINGS xix. 3.

    eschuing, sb., *avoiding*, CH. BOET. iii. 11 (99).

eschien, *see* áskien.

eschieu, adj., *O.Fr.* eschiu ; *unwilling, disinclined* ; eschiewe CH. C. T. *I* 971.

escrīen, v., *O.Fr.* escrier ; *cry out* : thei being aferd escrīed (*pret.*) L. H. R. 169 ; *see also* ascrīen.

escūsen, v., *O.Fr.* escuser ; *excuse*, A. R. 304* ; excusi AYENB. 7 ; escūsede (*pret.*) SHOR. 40 ; excūsed (*pple.*) PR. C. 6077.

ēse, *see* aise. **ēsen**, *see* aisien.

eskien, *see* áskien.

esking, sb., *cf. prov. Engl.* esking ; *penthouse, shed,* '*grunda*,' VOC. 236.

esmaien, v., *O.Fr.* esmaien ; *dismay* : esmaie GOW. II. 239 ; *see* amaien.

espe, *see* aspe.

espeir, sb., *O.Fr.* espeir ; *hope* : and for that suche is min espeir GOW. I. 370.

esperaunce, sb., *O.Fr.* esperance ; *hope*, CT. LOVE 1032.

espiaille, sb., *from O.Fr.* espier ; *spying* : ful subtilli he hadde his espialle CH. C. T. *D* 1323 ; espiailles (*pl.*) C. T. *B* 2509.

espie, *see* spíe. **espien**, *see* spíen.

espirituel, adj., *O.Fr.* espirituel ; *spiritual, heavenly* : a place espirituel R. R. 649, 671.

espleit, sb., *O.Fr.* espleit ; *exploit, advantage*, LEG. 25 ; esploit GOW. II. 258.

essai, *see* assai.

esschien, essien, *see* áskien.

essel, sb., ? *O.Fr.* aissel ; ? *bar of a gate*, LAȜ. 18992.

essoigne, sb., *O.Fr.* essoigne, essoine ; *cf.* assoinen, ensoinen ; *essoin, excuse*, HALLIW. 340 ; assoigne P. S. 191 ; essoine, assoine E. G. 361 ; asoine ROB. 539.

ēst, *see* ēast.

establissen, v., *O.Fr.* establiss-, establir ; *establish*, CH. BOET. i. 4 (15) ; *see* stáblen.[2]

estat, *see* stat.

ĕste, sb., *O.E.* ēst, = *O.Fris.* ēst, enst, *O.N.* āst, *O.L.G.*, *O.H.G.* anst, *Goth.* ansts ; *from* an ; *favour, grace, pleasure*, P. L. S. viii. 181 ; HOM. II. 99 ; O. & N. 1504 ; flesches este A. R. 364 ; **ĕste** (*dat.*) H. M. 29 ; TRIAM. 1416 ; for goddis este EGL. 904 ; **ĕsten** (*pl.*) ['*deliciae*'] HOM. I. 241 ; estes (*ms.* esstess) *dainties* ORM. 7542.

ĕst-dēde, sb., *deed of kindness*, GEN. & EX. 2758.

ĕst-ful, adj., *O.E.* ēstful ; *full of kindness*, A. R. 108.

ĕst-mete, sb., *O.E.* ēstmete ; *dainty food*, HOM. II. 37.

ĕste, adj., *O.E.* ēste ; *favourable, delicious*, A. D. 292 ; þat lond nis god ne hit nis este O. & N. 999 ; nas nan este (*ms.* esste) mete þær ORM. 829.

ĕst-līch, adj., *O.E.* ēst-, ēstelīc ; *delicious* : estliche ınetes HOM. II. 179 ; esteliche eten II. 31 ; **ĕstlīche** (*adv.*) A. R. 204.

estellaciōn, sb., *astrology*, ALIS. 587.

ēster, *see* ēaster.

estimacioun, sb., *O.Fr.* estimation ; *estimation* : withouten estimacioun MAND. 149, 239 ; bi estimatioun 41.

estrange, adj., *O.Fr.* estrange ; *strange* : his manere estraunge CH. TRO. i. 1077 (1084) ; *see* strange.

ĕstre, sb., *O.Fr.* estre ; *state, quality ; locality, inner part*; ALIS. 5468 ; MAN. (F.) 5398 ; **ĕstres** (*pl.*) and the herberes WILL. 1768 ; the estres [eestres] of the grisli place CH. C. T. *A* 1971 ; talked of othir estres (*matters*) PERC. 1559.

et, *see* æt *and* it. **ēt**, *see* ǣt.

etel, adj., *O.E.* etol, *cf. O.H.G.* ezal; *gluttonous*; ætul MAT. xi. 19; *comp.* ofer-etel.

éten, v., *O.E.* etan,=*O.L.G.* etan, *O.N.* eta, *Goth.* itan, *O.Fris.* ita, *O.H.G.* ezan, *Lat.* edere; *eat*, ORM. 7806; CH. C. T. *E* 1890; eten, eoten, eaten HOM. I. 31, 45, 53; eten, æten [eaten] LAȝ. 13456, 23037; eoten MARH. 14; ete O. & N. 108; MISC. 38; WICL. GEN. iii. 14; AYENB. 51; (*spelt* ethe) 205; to etene A. R. 54; AYENB. 51; éte (*pres.*) M. H. 70; ette PR. C. 4675; etest O. & N. 599; REL. I. 264; est ROB. 238; AYENB. 54; eteþ SHOR. 23; et HOM. I. 103; MARH. 17; AYENB. 135; hi eteþ O. & N. 1007; et (*imper.*) C. L. 174; æt (*pret.*) LAȝ. 8725; et HAV. 656; P. L. S. xvii. 307; et ORM. 11549; eet C. L. 235; M. H. 39; at GEN. & EX. 342; BEK. 274; eet [ete] LANGL. *B* vi. 298; þu ete HOM. II. 181; æten MAT. xxvi. 21; LAȝ. 13444; eten ORM. 4797; H. H. 172; eten, eeten LANGL. *B* v. 612; eeten RICH. 113; ēte (*subj.*) C. L. 177; eete TREV. VIII. 67; éten (*pple.*) GEN. & EX. 329; HAV. 657; LANGL. *B* xvii. 97; CH. C. T. *A* 4351; IW. 763; WICL. LEV. vii. 16; etten S. S. (Wr.) 837; *comp.* fr(a)-, ȝe-éten; *deriv.* æt.

étere, sb., *O.E.* etere; *eater*, TREV. IV. 297; **éteres** (*pl.*) AYENB. 47.

étinge, sb., *eating*, HOM. II. 37; AYENB. 56; eting ORM. 19063.

eten, sb., *O.E.* eoton,=*O.N.* iötunn; *giant*, TRIST. 950; C. M. 7443; etin L. H. R. 118; yhoten (=ioten), eten PS. xviii. 6; getenes (=ietenes) (*pl.*) GEN. & EX. 545; *see* eotand.

etenisch, adj., *O.E.* eotonisc; *gigantic*: getenisse men GEN. & EX. 3715.

eterne, adj., *O.Fr.* eterne; *eternal*, CH. C. T. *G* 44.

eternitē, sb., *O.Fr.* eternité; *eternity*, CH. BOET. v. 6 (171).

etfóren, ethálden, *see* ætfóren, æthálden.

etlen, etlunge, *see* ahtlien.   ēð, *see* ēað.

ēþel, sb., *O.E.* ēðel, œðel, *m. n.*, *cf. O.N.* ōðal *n.*, *O.H.G.* uodal; *property, country, 'patria,'* FRAG. 3; HOM. I. 113, 115; æðel (*earlier version* eðel) MK. vi. 1; þat æðel heo biwunnen LAȝ. 4744; þat æðel (*ms.* aðel) wes his aȝene 20201.

ēþel-þēowe, sb., *homeborn slave*, H. M. 31.

eðele, eþeling, *see* æðele, æðeling.

ēþem, sb., *O.E.* æðm, eðm, *cf. O.L.G.* āthom, *O.H.G.* ātum, ādum; *breath*, HOM. I. 43.

ēþen, *see* hæþen.

ēþen, v., *cf. M.H.G.* eiden; *? from* āð; *? adjure*: i ēþe (*pres.*) þe GAW. 379.

ēðien, v., *O.E.* ēðian; *breathe*; eði MARH. 13.

ethique, sb., *O.Fr.* ethique; *ethics*: wherof the firste ethique is named GOW. III. 140.

ēu, *see* ēow *and* æw.   **ēuch,** *see* aighwilc.

ēur, ēuwer, *see* ēower.

evangelie, sb., *Lat.* evangelium; *gospel*, WICL.

ROM. i. 1; R. R. 5453; **evaungeles** (*pl.*) CH. C. T. *B* 666.

evangelise, v., *O.Fr.* evangelizer; *evangelise, preach*, WICL. LK. i. 19; **evangelísist** (*pres.*) WICL. IS. xl. 19; **evangelísinge** (*pple.*) viii. 1; **evangelíside** (*pret.*) WICL. LK. iii. 18.

evangelíser, sb., *preacher, evangelist*; **evangelíseris** (*pl.*) WICL. PS. lxvii. 12.

evangeliste, sb., *Lat.* evangelista; *evangelist*; ewangeliste HOM. I. 81; A. R. 94; ewangelíst SAINTS [Ld.] 77.

evel, evelien, *see* üvel, üvelien.

ēven, *see* æfen.

evene, evenen, *see* efen, efnen.

ever, sb., *O.E.* eofor, *cf. O.N.* iöfurr, *O.H.G.* eber, *Lat.* aper; *boar*; **eaveres** (*pl.*) H. M. 13.

ever-vern, sb., *O.E.* eoforfearn; *polypody*, REL. I. 36; **everferne** (*dat.*) A. P. iii. 438.

ēver, *see* æfre.   **everne,** *see* ȝe-fürn.

evese, sb., *O.E.* efese, yfese, *cf. M.L.G.* ovese, *O.Fris.* ọse, *mod.Eng.* eaves; *edge, border, eaves; 'stillicidium, imbrex,'* PR. P. 144; þe evese of þe hil WICL. TOB. xi. 5; the wode evese (*ms.* hevese) VOC. 159; evis (*ms.* euis *printed* enis) D. ARTH. 3376; evez GAW. 1178; ovese (*ms.* ouese) REL. I. 219 (MISC. 15); ovise PARTEN. 5504; **eovesen** (*pl.*) LAȝ. 29279; eveses LANGL. *B* xvii. 227; MAN. (F.) 14689; *comp.* hūs-evese.

evesien, v., *O.E.* efesian, efsian; *clip the edges*; **evesede** (*pret.*) A. R. 398; A. D. 258; **evesed** (*pple.*) PL. CR. 166; (*printed* enesed) GAW. 184.

evesunge, sb., *? O.E.* efesung; *eavesing*, A. R. 142; evese or evesinge of a house PR. P. 144; **evesinges** (*pl.*) LANGL. *B* xvii. 227*.

evest, sb., *O.E.* æfest; *envy, malice*: in niþe and evest [eust] (*printed* enest [enst]) C. M. 23138; evist [eust] (*printed* enist [enst]) and hete C. M. 23279; *cf.* eft².

evete, sb., *O.E.* efete; *eft, n-ewt, 'lacerta,'* FRAG. 3; *'lesard'* VOC. 157; ['*stellio*'] WICL. PROV. xxx. 28; LEG. 151; eute, neute PR. P. 355; **eveten** (*pl.*) P. L. S. viii. 138; evetes BEV. 1539; (*printed* enettes) TREV. I. 335; ewtes MAND. 61.

evidence, sb., *O.Fr.* evidence; *evidence*, LANGL. *B* xvii. 195; GOW. I. 14.

evident-lī, adv., *by observation*, ['*evidenter*'] CH. ASTR. ii. 23 (32).

ēvre, *see* æfre.   **ēw,** *see* īw.

ewāge, sb., *cf. Lat.* aqvāticus; *beryl*; **ewāges** (*pl.*) LANGL. *B* ii. 14.

ewangelist, *see* evangeliste.

ēwe, *see* ēowe *and* ǣ.

ewer, sb., *O.Fr.* aiguier, ewer; *ewer*, MAN. (F.) 11425; eware PR. P. 143.

ēwilch, *see* aighwilc.   **ex,** *see* eax.

exactour, sb., *torturer*; **exactours** (*pl.*) ben thei that enqveren the truthe bi mesurable betingis WICL. *marg. gloss to* DEUT. xvi. 18.

exalacioun, *O.Fr.* exhalation ; *exhalation*: of exalation i finde fire kinlid of the same kinde Gow. III. 96 ; **exalatiōns** (*pl.*) 95.

exaltacioun, sb., *exaltation* (*in astrology*), CH. C. T. *D* 702, *E* 2224.

exaltate, adj., *Lat.* exaltātus ; *highest*: Pisces, wher Venus is exaltate CH. C. T. *D* 704.

examinen, v., *O.Fr.* examiner ; *examine*; examini AYENB. 153 ; examine CH. C. T. *B* 2391 ; **examined** (*pret.*) CH. C. T. *B* 2400.

examining, sb., *examination*: i mi self shalle make examining TOWNL. 193 ; examining CH. C. T. *B* 2392.

excellence, sb., *O.Fr.* excellence ; *excellence*, CH. C. T. *E* 408.

excellent, adj., *O.Fr.* excellent ; *excellent*, MAND. 133.

excepte, prep., *except*, LANGL. *B* xv. 53.

exces, sb., *O.Fr.* exces ; *ecstasy of mind*, WICL. PS. XXX. 23 ; excess WICL. DEEDS xi. 5 ; axcess x. 10.

Exchestre [Excestre], pr. n., *O.E.* Exanceaster ; *Exeter*, LA. 30942 ; Excestre MISC. 145 ; ROB. 4.

exciten, v., *O.Fr.* esciter ; *excite*; **exciteþ** (*pres.*) CH. C. T. *G* 744 ; **excitede** (*pret.*) TREV. III. 33.

exclūde, v., *Lat.* exclūdere ; *exclude*, WICL. PROL. ROM. ; **exclūdid** (*pple.*) PR. P. 144.

excusable, adj., *O.Fr.* excusable ; *excusable*, Gow. I. 76.

excusaciōn, sb., *O.Fr.* escusation, excusation, *from Lat.* excūsātiō ; *excusation*, PR. P. 145.

excūsement, sb., *excuse*, Gow. I. 76.

execuciōn, sb., *O.Fr.* execution ; *execution*, P. L. S. xxv. 37.

executen, v., *O.Fr.* executer ; *carry out*: to execute her lordis bidding Þ. L. S. xxxi. 8.

executour, sb., *Fr.* executeur ; *executor*, N. P. 11,

executrice, sb., *female executor*: fortune, executrice of wierdes CH. TRO. iii. 568.

exempt, adj., *Fr.* exempt ; *exempt*, PR. P. 145.

exempten, v., *exempt*; exemptide (*pple.*) '*exemptus*' PR. P. 145.

exercen, v., *O.Fr.* exercer ; *exercise*, *practise*: mai exercen or haunten CH. BOET. ii. 6 (52).

exercisen, v., *O.Fr.* exercise ; *observe*: the new fest of whiche iij. in the ȝere we **exercise** (*pres. pl.*) LUD. COV. 71.

exercitacioun, sb., *O.Fr.* exercitation ; *exercise*, CH. BOET. iv. 6 (140).

exhortacioun, sb., *O.Fr.* exhortation ; *exhortation*: taketh gode hede to this exhortacioun P. L. S. xxxi. 373.

exil, sb., *Fr.* exil ; *exile*, AYENB. 131.

exiltrē, *see* axel-trē.

exilin, v., *Fr.* exiler ; *exile*, PR. P. 145 ; **exiled** (*pple.*) TREV. III. 235.

exilinge, sb., *banishment*: þe exilinge of Anaxagore CH. BOET. i. 3 (11).

exinatores, sb. pl., *senators*, YORK xxx. 21.

existence, sb., *reality*, CH. H. F. 266 ; to se him that is freend in existence from him that is by apparence R. R. 5550.

exle, *see* **eaxle.**

exorcisacioun, sb., *from Lat.* exorcīzāre, *Gr.* ἐξορκίζειν ; *exorcism* ; **exorsisaciouns** (*pl.*) CH. H. F. 1263.

expans, adj., *Lat.* expansus ; *separate* (*in astrology*): his expans yeeres CH. C. T. *F* 1275 ; mine **expanse** (*pl.*) yeris ASTR. ii. 45 (56).

expectant, adj., *O.Fr.* expectant ; *expectant*: expectant ai telle i mai mete R. R. 4571.

expelle, v., *Lat.* expellere ; *expel*, CH. C. T. *A* 2751.

experience, sb., *O.Fr.* experience ; *experience*, CH. C. T. *B* 4168.

experiment, *O.Fr.* experiment, *from Lat.* experīmentum ; *experiment*, PR. P. 145 ; D. TROY 13215.

expert, adj., *O.Fr.* expert ; *experienced in*: (he) shal not ben expert ani thing of evel WICL. ECCLES. viii. 5 ; '*expertus*' PR. P. 145.

exposicioun, sb., *O.Fr.* exposition, *from Lat.* expositiō ; *exposition*, PR. C. 3856, 4715 ; exposicioun of dremes MAND. 44.

expounen, v., *O.Fr.* expondre, *Lat.* expōnere ; *expound, interpret*, WICL. JOSH. xx. 9 ; writ PR. C. 4271 ; expoune (*imper.*) þe þis speche A. P. ii. 1729 ; expoundeth (*pres.*) Gow. I. 26 ; expounden (*pl.*) CH. TRO. v. 1276 ; expouned (*pret.*) MED. 735 ; exponid N. P. 2 ; expounede (*pple.*) WICL. GEN. xli. 8.

expounere, sb., *interpreter*, WICL. GEN. xl. 22 ; expouneris (*pl.*) GEN. xli. 8.

expouning, sb., *explanation*: expouning of speche A. P. ii. 1564 ; interpretinge or expouning of wordis WICL. 1 COR. xii. 10.

expresse, adj. & adv., *O.Fr.* expres ; *express*, *expressly*: a þinge expresse A. P. i. 909 ; proved expresse A. P. ii. 1158.

expresse-lī, adv., *expressly*, Gow. I. 357.

expresse, v., *express, tell*, CH. C. T. *B* 1666.

exstasie, sb., *ecstasy*, WICL. DEEDS iii. 10.

extainten, v., *cf. O.Fr.* estendre, *from Lat.* extinguere ; *extinguish, subdue*, D. TROY 4925.

extenden, v., *Lat.* extendere ; *survey, seize*, *extend*: extendours he sette for to extend þe land MAN. (H.) 83 ; wille King Richard alle his lond extende 202.

extendours, sb. pl., *surveyors*, MAN. (H.) 83.

extorciōn, sb., *extortion*, PR. P. 145.

extorcionere, sb., *extortioner*, '*extortor, exactor, predator, angarius*,' PR. P. 145.

ēy, *see* **ēi.** **eye,** *see* **eȝe.**

f.

fā, adj. & sb., *O.E.* fāh, fāg,=*O.H.G.* (gi-)fēh;
hostile ; *foe* : hold oðer fa HOM. I. 231 ; a
foo cragge GAW. 1430 ; **fā** (*subst.*) Iw. 874 ;
faa PR. C. 1453 ; fo HAV. 1363 ; BEK. 870 ;
he wole be fo to þine fon SHOR. 90 ; fo [foo]
CH. C. T. *A* 63 ; vo, foa [fa] A. R. 62, 274 ;
fāwe (*pl.*) D. ARTH. 747 ; fawe [fau] fellis
ANT. ARTH. vii ; **fōn** (*subst. pl.*) BEK. 711 ;
C. L. 683 ; von AYENB. 255 ; voan [fan] A. R.
388 ; faas, fais M. H. 58, 137 ; *comp.* ʒe-fā.

   fā-mon, sb., *O.E.* fāhman ; *foe,* JUL. 24 ;
H. M. 41 ; foman PR. P. 169 ; SHOR. 41 ;
A. P. ii. 1175 ; TOWNL. 134 ; voamen [fa-
men] (*pl.*) A. R. 186.

   fō-shipe, sb., *enmity,* HOM. II. 45 ; fooschip,
foschip A. P. ii. 918, 919.

fā, *see* fēa.

fable, sb., *O.Fr.* fable ; *fable,* PR. P. 145 ;
A. P. i. 591 ; fabull GUY (Z.) 3254.

fablen, v., *O.Fr.* fabler ; *fable, talk* ; fableden
(*pret. pl.*) ['*fabularentur*'] WICL. LK. xxiv.
15 (*v. r.* talkiden.)

   fabler, sb., *O.Fr.* fableor ; *fabler* ; fablers
(*pl.*) WICL. BAR. iii. 23 ; fabulers LANGL.
A ii. 157.

   fabling, sb., *saying, discourse* : wicked
fablinges (*pl.*) ['*fabulationes*'] talde to me
bot noght als þe lagh of þe PS. cxviii. 85.

fāc, adj., *O.N.* feikr (*monstrous*) ; *treacherous* :
Vortiger wes fac (*ms.* fæc) for her he his
laverd biswac (*ms.* biswæc) LAʒ. 13507.

facchen, *see* fecchen.

fáce, sb., *O.Fr.* face ; *face,* P. L. S. xvii. 171 ;
ARCH. LII. 38 ; AYENB. 88 ; ALIS. 1314 ;
everi signe [of the zodiac] is departid in
3 evene parties bi [10] degrees, and thilke
porcioun thei clepe a face CH. ASTR. ii. 4 (19) ;
faas P. L. S. xxviii. 22 ; faz M. H. 85 ; **fáces**
(*pl.*) WICL. IS. xiii. 8.

fácen, v., *from* fáce ; *soil, deface, disfigure* ;
be unabashed : facin or shewin boolde face
'*effronto*' PR. P. 145 ; all **fácid** (*pret.*) hir
face with hir fell teris D. TROY 9129.

   fácinge, sb., *disfigurement* : he ... wesshed
him anone, refreshing his face for facing of
teres D. TROY 9215.

facioun, *see* fasoun.

facound, sb., *O.Fr.* faconde, *from Lat.* fa-
cundia ; *eloquence* : of eloquence was never
founde so swete a souning facounde CH. D.
BL. 924 ; D. TROY 3791 ; CH. P. F. 557 ; facunde
HAMP. PS. xi. 4* ; '*facundia, eloquencia*' PR.
P. 145.

facound, adj., *O.Fr.* facond, *from Lat.* fa-
cundus ; *eloquent* : nature with facound vois
P. F. 521.

facrēre, sb., ? *Fr.* faire croire ; ? *make-believe,*
art of dissembling : a craft which cleped
is facrere GOW. I. 230.

facultē, sb., *O.Fr.* faculté ; *faculty,* PR. P. 145.

fade, adj., ? *great, powerful,* GAW. 149 ;
TRIST. 153 ; fede 2474 ; fadde PERC. 615.

fáde, adj., *O.Fr.* fade, *from Lat.* fatuus ;
withered, SAINTS (Ld.) 318 ; GOW. II. 117 ;
H. S. 3220 ; ['*marcidus*'] WICL. ECCLUS. xi.
12 ; þi faire hewe is al fade WILL. 891.

fáden, v., *fade,* CH. BOET. iv. 3 (119) ; fade
P. L. S. xxx. 42 ; WILL. 579 ; D. TROY 784 ;
fáden (*pres.*) GOW. II. 117 ; min herte
fádide (*pret.*) ['*emarcuit*'] WICL. IS. xxi. 4.

fader, *see* fæder.

fadien, v., *O.E.* fadian, ?=*O.H.G.* (keun-)va-
tōn ; ? *dispose, suit* ; PARTEN. PROL. 164.

fadme, *see* faðme.

fæc, sb., *O.E.* fæc,=*O.Fris.* fek, *O.H.G.* fah ;
space : bin(n)e fece HOM. I. 235.

fæder, fader, sb., *O.E.* fæder (*gen. dat.* fæder),
=*O.L.G.* fader (*gen. dat.* fader), *Goth.* fadar,
O.Fris. feder (*gen.* feder, federes), *O.N.* faðir
(*gen.* föður), *O.H.G.* fater (*gen.* fater, fateres) ;
father, LK. i. 32, xxii. 42 ; LAʒ. 212, 3199 ;
fader HAV. 1224 ; LANGL. *A* i. 14 ; TOR.
2058 ; vader (*ms.* uader) AYENB. ·8 ; fadir
ISUM. 313 ; feder REL. I. 1 ; feder, feader
HOM. I. 11, 137 ; feder, veder, veader A. R.
10, 26, 54 ; fader (*gen.*) HOM. I. 171 ; LAʒ.
398 ; ORM. 186 ; ROB. 302 ; TOWNL. 182 ;
feder, federes MARH. 2, 6 ; þi fader love FER.
1351 ; faderes LK. xi. 47 ; P. L. S. viii. 99 ;
GEN. & EX. 1536 ; federes A. R. 406 ; fader
(*dat.*) LAʒ. 2279 ; fader [fadir] CH. C. T.
A 100 ; **faderes** (*pl.*) ROB. 127 ; **fæderen**
(*dat. pl.*) LK. i. 55 ; faderen LAʒ. 5724 ;
comp. eld-, fore-, fōster-, god-, hēh-fæder.

   fader-cwellare, sb., *slayer of a father* ;
fader-qwellare '*patricida*' PR. P. 145.

   fadir-hōd, sb., *fatherhood* ; besoughte his
holi fadirhode MAND. 315.

   fadir-in-lawe, sb., *father-in-law,* '*socer*,'
PR. P. 145.

   fader-lēs, adj., *O.E.* fæderlēas ; *fatherless,*
KATH. 78 ; federleas A. R. 10.

fæʒe, adj., *O.E.* fǣge,=*O.L.G.* fēgi, *O.N.*
feigr, *O.H.G.* feigi, *Scotch* fey ; *doomed to
die* ; *cowardly* : fæiʒe LAʒ. 9705 ; folc þer
wes fæie 4694 ; his fæie blod 16647 ; þat
bearn nas ne wi(h)t feie 298 ; feiʒe, feie JOS.
558, 569 ; LANGL. *C* xvi. 2 ; feie SPEC. 28 ;
SHOR. 124 ; GAW. 1067 ; AV. ARTH. iv ; (h)is
feie blod him gan to pinge FER. 2430 ; ðis
feie folk ALIS. (Sk.) 397 ; veie BEV. 3032 ;
mani fei schalle be first appone the felde
levide D. ARTH. 517 ; MISC. 112.

   fæie-scipe, sb., *slaughter,* LAʒ. 21258.

   fæie-sīð, sb., *death,* LAʒ. 304 ; **fæiesiþe**
[feaisiþe] (*dat.*) 26040.

fæʒen, adj. & adv., *O.E.* fægen, fægn,=*O.L.G.*
fagan, fagin, *O.N.* feginn ; *allied to O.E.*
gefēon (*pret.* fēah), *O.H.G.* (gi-)fehan (*pret.*
fah) (*rejoice*), *mod.Eng.* fain ; *joyful* ; *joyfully* ;
fæin, fain LAʒ. 4891, 11619 ; ic walde fein

pinian HOM. I. 35 ; vein A. R. 192 ; fagen REL. I. 220 ; GEN. & EX. 15 ; faȝe FER. 308 ; fain WILL. 1783 ; fain P. L. S. xvii. 412 ; PR. C. 4552 ; ISUM. 556 ; fain, fawe CH. C. T. *A* 2437, *D* 220 ; fain, vawe ROB. 120, 183 ; fain, faun PR. P. 146, 152 ; fawen HAV. 2160 ; fawe HOM. I. 199 ; fawe ALIS. 1956 ; K. T. 1058 ; with herte ful faine LUD. COV. 25 ; **feinre** (*compar.*) JUL. 47 ; **fainest** (*superl.*) WILL. 3933 ; A. P. ii. 1219 ; *comp.* ȝe-, un-faȝen.

fain-hēd, sb., *gladness,* D. TROY 2446.

fainnes, sb., *gladness* ['*laetitiam*'] PS. l. 10.

fæȝer, adj., *O.E.* fæger,=*O.N.* fagr, *O.L G.* fagar, *O.H.G.* fagar, *Goth.* fagrs ; *fair, beautiful* : fæiȝer, feiȝer, fæier, fair, feier, fæir LAȝ. 152, 3886, 4825, 7594, 7764, 31905 ; feier, feir HOM. I. 131, 137 ; (*ms.* faȝȝerr) ORM. 6392 ; faiger, faier, fair GEN. & EX. 126, 1058, 2636 ; faier REL. I. 226 ; (*ms.* fayer) S. S. (Wr.) 2533 ; fair and god HAV. 500 ; fair, vair (*ms.* feyr) O. & N. 579, 584 ; vair (*ms.* uayr) AYENB. 16 ; fair [feir] LANGL. *A* i. 10 ; CH. C. T. *A* 3977 ; feir A. R. 206 ; MISC. 172 ; R. S. v. ; CHR. E. 603 ; K. T. 445 ; **fæire** [faire] (*adv.*) LAȝ. 4842 ; feire CHR. E. 612 ; (*ms.* faȝȝre) ORM. 1215 ; faire and wel HAV.224 ; he grette him faire BEK. 34 ; feȝre (*dat. n.*) HOM. I. 219 ; veirne (*acc. m.*) A. R. 236 ; **faire** (*pl.*) PR. C. 9249 ; **væiren** (*dat. pl.*) LAȝ. 15163 ; mid faire worde O. & N. 158 ; **fehere** [*O.E.* fægerra] (*compar.*) KATH. 2323 ; fairre WILL. 4437 ; feirure [fairere] LAȝ. 2405 ; fairere BRD. 33 ; feirore C. L. 737 ; fairer HORN (L.) 13 ; veȝereste, veierste, feireste, feiruste (*superl.*) LAȝ. 2217, 4080, 25305, 29663 ; fæirest 13894 ; *comp.* un-fair.

fair-hēde, sb., *beauty,* BRD. 9 ; ALIS. 6392 ; fairehede · MAN. (F.) 696 ; feirhede FLOR. 1670 ; vairhede ROB. (W.) 2515.

feier-lēc, sb., *O.N.* fagrleikr ; *beauty,* HOM. I. 261 ; feirlec MARH. 19 ; feirlek C. L. 145.

vair-līche, adv., *fairly,* AYENB. 59.

feir-schipe, sb., *beauty,* C. L. 690 ; fairsipe LAȝ. 3139*.

fæȝer, sb., *O.E.* fæger, *? cf. O.H.G.* fagarī ; *beauty* : feir, feire JUL. 6, 7 ; heo never ne beoð sead þi veir to iseonne HOM. I. 193 ; fair ALIS. 164.

fæȝernesse, sb., *O.E.* fægernesse,=*O.H.G.* fagarnessi ; *fairness, beauty* : væȝernesse, fæiernesse, feirnusse LAȝ. 3272, 22724, 24266 ; fæirnesse FRAG. 1 ; veirness(e) A. R. 52 ; faȝernesse ORM. 12253 ; faiernesse GEN. & EX. 1233 ; fairnesse CH. C. T. *A* 1098.

fæȝnien, v., *O.E.* fægnian, fagenian,=*O.N.* fagna, *O.L.G.* faganōn, faginōn, *O.H.G.* faginōn, feginōn, *Goth.* faginōn ; *mod.Eng.* fawn ; *from* fæȝen ; *rejoice ; flatter* ; fele shule fagenien on his burde HOM. II. 135 ; fainen LAȝ. 3588 ; faine sal ['*laetentur*'] alle þat

hope in þe PS. v. 12 ; faunin ' *blandio*' PR. P. 152 ; þeonne **veineð** (*pres.*) he mid ou A. R. 194 ; wulves þat fainen of hare preie HOM. I. 277 ; he **fagnede** (*pret.*) hire GEN. & EX. 1441 ; fageneden MK. xiv. 11 ; heo væineden þas LAȝ. 18807 ; fainede JOS. 243 ; fauhnede [faunede] wiþ þe tailes LANGL. *B* xv. 295 (*C* xviii. 31).

fainare, sb., *fawner,* ' *adulator*,' PR. P. 146.

fæȝnunge, sb., *O.E.* fægnung ; *fawning* : vawenunge A. R. 290 ; faghning [fauning] C. M. 12350 ; fauning SPEC. 23.

fæȝrien, v., *O.E.* fægrian,=*O.N.* fegra ; *from* fæȝer ; *make fair, become fair* : þus heo doð for to feiren heom seolven HOM. I. 53 ; **fairiþ** (*pres.*) ALIS. 2903 ; þis trau greneþ and vaireþ be his virtue AYENB. 95 ; **feirede** (*pret.*) C. L. 876 ; fairede (*ms.* fairhede) A. R. 302*.

fæie, see **fæȝe.**   **fæien,** see **feȝen.**

fæiȝe, see **fæȝe.**   **fæin,** see **fæȝen.**

fæine, sb., *?from* fæȝen ; *?joy* ; faine FLOR. 844.

fæl, see **fal.**

fēle, adj., *O.E.* fǣle, *?*=*O.H.G.* feili (*venalis*); *proper, estimable* : fele LAȝ. 3290* ; O. & N. 1378 ; MAP 350, 352 ; PERC. 729 ; fele [feele] TREV. I. 399 ; *comp.* un-fēle.

fēllen, see **fellen.**

fēmen, see **fāmien.**

fǣr, sb., *O.E.* fǣr, fēr, *cf. O.L.G.* fār *m.,* *O.N.* fār *n.,* *M.H.G.* vār *m.,* vāre *f.* ; *fear* : fer MAP 338 ; feer ' *timor, terror*' PR. P. 156 ; fere HORN (H.) 1285 ; BRD. 8 ; TREV. VII. 369 ; S. B. W. 42 ; WICL. 2 PARAL. xiv. 14 ; CH. C. T. *D* 1022 ; fere [feere] LANGL. *B* xiii. 162 ; wiðutan f(e)ore HOM. I. 97.

fēr-ful, adj., *fearful,* GOW. II. 13 ; P. L. S. xxx. 93.

fēr-lāc, sb., *terror,* A. R. 306 ; ferlak C. L. 672 ; fearlac KATH. 39 ; (*ms.* farlac) MARH. 10.

fēr-līc, adj., *O.E.* fǣrlīc ; *terrible, sudden,* PS. cxxxviii. 6* ; ferlich A. R. 362 ; ALIS. 5948 ; feorlic [feorlich] LAȝ. 25553 ; ferli WILL. 3934 ; A. P. i. 1083 ; ISUM. 235 ; an swiðe ferli sigt GEN. & EX. 2774 ; þe færliche dæð LAȝ. 6835 ; þurh ferliche deð HOM. II. 61 ; ferliche frechen KATH. 732 ; **fērlich** (*sb.*), *miracle, wonder,* LAȝ. 5381 ; FL. & BL. 456 ; feorlich KATH. 2086 ; ferlik HAV. 1849 ; ferli LANGL. *A* PROL. 6 ; PERC. 530 ; PR. C. 2955 ; ANT. ARTH. xli ; **fǣrlīche** (*adv.*) LAȝ. 10376 ; ferliche HOM. I. 89 ; H. M. 21 ; verliche A. R. 178 ; ROB. 299 ; AYENB. 55 ; færlike ORM. 3074 ; ferlike GEN. & EX. 2799 ; ferli A. P. ii. 269.

fērli-ful, adj., *wonderful, horrible* : a ferliful frute D. TROY 179 ; a ferliful folke 13306.

vērlīch-hēde, sb., *haste,* AYENB. 55.

fērli-līc, adv., *wondrously,* C. M. 11424 ; ferlili A. P. ii. 962.

færde, *see* **ferde.**

fērde, sb., *?* = *M.H.G.* (ge-)vǣrde, fērde ; *terror,* WICL. GEN. ix. 2 ; A. P. iii. 215 ; KN. L. 36 ; for ferde GAW. 2130 ; MIN. iv. 27 ; PR. C. 6864 ; PERC. 911 ; MIR. PL. 175.

　　fērd-ful, adj., *terrible,* '*terribilis,*' PR. P. 156 ; WICL. EX. xix. 18 ; feerdful H. V. 14.

　　fērd-laik, sb., *terror,* PR. C. 6427.

　　fērd-lī, adj., *terrible,* N. P. 65.

　　fērdnes, sb., *terror* ; ferdnes PR. C. 2231.

fērdi, adj., *fearful* : be not ferdi (*v. rr.* sori, eri) C. M. 17685.

fēren, v., *O.E.* fǣran (*terrify*), = *M.Du.* vēren, vaeren (*fear*), *O.L.G.* fāran (*observe*), *O.H.G.* fāren (*observe, lie in wait*); *fear, terrify* : he wile him fǣren . . & skerren ORM. 675 ; fearen A. R. 230* ; ferin '*terreo*' PR. P. 156 ; fere LANGL. *B* vii. 34 ; WICL. 2 COR. x. 9 ; PR. C. 2291 ; D. TROY 1929 ; fērde (*pret.*) GAW. 1588 ; fēred (*pple.*) SPEC. 24 ; M. H. 134 ; ferd MAN. (H.) 191 ; PERC. 925 ; þeʒh ʒe be ferd of ʒour fon WILL. 3366 ; *comp.* a-, for-, of-fǣren.

fēren, *see* **fēren.**

fēringe, adv., *O.E.* fǣringa, fǣrunga, = *O.H.G.* fāringa ; *suddenly* ['*subito*'], LK. ii. 13 ; fēringes MAN. (H.) 185.

　　fēring-lī, adv., *suddenly* : ferinkli, bathe mare and læsse waned þai PS. lxxii. 19.

fǣrlīc, *see under* **fǣr.**

fæst, adj., *O.E.* fæst, = *O.L.G.* fast, *O.N.* fastr, *cf. O.Fris.* fest, *O.H.G.* festi ; *fast, firm* : fast '*fixus*' PR. P. 151 ; fast (*ms.* fasst) & findiʒ laf ORM. 1602 ; fest HOM. I. 111 ; vest AYENB. 189 ; make faste þe dore P. L. S. xvii. 416 ; he makede his castles treowe & swiðe væste LAʒ. 11898 ; **faste** (*adv.*) O. & N. 796 ; BRD. 5 ; LANGL. *A* i. 42 ; RICH. 3331 ; PERC. 1020 ; stike(de) feste [faste] LAʒ. 7533 ; heore grið heo setten fæste 9562 ; heo riden vaste [faste] 9754 ; holdeð hine veste A. R. 34 ; so faste shal i binde þe H. H. 127 ; he slepen faste HAV. 2128 ; þanne bihelden he him faste 2148 ; vaste ROB. 568 ; faste bi (*close by*) Radistone LAʒ. 9* ; faste bi þe Belle CH. C. T. *A.* 719 ; on fest Radestone LAʒ. 9 ; & fusden an veste (*quickly*) 7986 ; þær on fast ORM. 3334 ; **fæstre** (*compar.*) LAʒ. 9775 ; vastre O. & N. 656 ; faster [fastor, fastore] ROB. (W.) 3392 ; *comp.* ār-, blǣð-, héte-, rōt-, scháme-, sōþ-, stéde-, trēow-, word-fæst.

　　fæst-līce, adv., *O.E.* fæstlīce ; *firmly, quickly* : ʒierneliche listede and fastliche hield HOM. II. 163 ; fastliche LAʒ. 13001, 23609 ; hine hiealde vestliche ine his wille AYENB. 166, 27775 ; his horn he vastliche (*violently*) bleu LAʒ. 808 ; festliche heom feht wið 11897 ; festliche HOM. II. 61, 77 ; I. 115 ; wunnen aʒean þe **vestlūcher** (*comp.*) A. R. 238 ; te vestluker holde 234.

　　fastnesse, sb., *O.E.* fæstness ; *support, firmness, stability* ['*firmamentum*'], PS. xxiv.

14 ; of fastnesse ['*firmitatis*'] WICL. GEN. xli. 32 ; vestnesse AYENB. 104.

　　fast-rēde, adj., *O.E.* fæstrǣd, = *O.N.* fastrāðr ; *wise in counsel,* O. & N. 211.

　　fest-schipe, sb., *firmness* ; vest-chipe A. R. 202.

fæste, sb., *? O.E.* fæst, *cf. O.N.* festr, *O.L.G.* festi, *O.H.G.* fasti ; *bond, fetter* ; fest '*ligamen*' PR. P. 158.

fæsten, v., *O.E.* fæstan, *cf. O.H.G.* festan, *O.N.* festa, *O.L.G.* festian ; *fasten, make fast* : in feteres ful faste festen HAV. 82 ; festin '*figo*' PR. P. 158 ; fasten GREG. 240 ; faste M. ARTH. 3324 ; and þat ich him wolde . . . faste on honde ROB. 150 ; wo so festeð (*pres.*) hope on him REL. I. 221 ; fasten (*pret.*) P. S. 214 ; fested (*pple.*) PR. C. 5295 ; fest [fast] LANGL. *B* ii. 123 ; WILL. 1650 ; fest M. H. 79 ; *comp.* ʒe-, hand-, līf-, staðel-, un-fæsten.

　　festinge, sb., *O.E.* fæsting ; *fastening,* '*fixura*' PR. P. 158.

fæsten, *see* **fasten.**

fæstnien, v., *O.E.* fæstnian, = *O.Fris.* festna, *O.L.G.* fastnōn, *O.H.G.* festinōn ; *fasten, make fast* : væstnien LAʒ. 29061 ; festnen, festhi KATH. 1180, 2011 ; (*ms.* fesstmenn) ORM. 1718 ; vestni AYENB. 117 ; festne CH. C. T. *A* 195 ; fastnede (*pret.*) LAʒ. 29118 ; fastned (*pple.*) ANT. ARTH. xxxix ; festened WILL. 447 ; ifestned HOM. I. 121 ; *comp.* ʒe-, un-fæstnien.

　　fastnare, sb., *fastener,* PR. P. 151.

　　fastning, sb., *O.E.* fæstnung ; *fortress, confirmation* ; ['*firmamentum*'] WICL. PS. xvii. 2 ; festning HOM. I. 67.

fāʒ, *see* **fāh.**

fagen, fagenien, *see* **fæʒen, fæʒnien.**

faʒer, *see* **fæʒer.**

fāʒien, v., *O.E.* fāgian, = *O.N.* fā, *cf. O.H.G.* (gi-)fēhen ; *colour, stain* : fāʒeden (*pret.*) LAʒ. 16413 ; **ifāwed** [ivowet] mid bloðe 4165 ; *see* **fōwin.**

fágin, v., *?* = **fæʒnien** ; *flatter, coax* : fagin or flatterin '*adulor*' PR. P. 146 ; to fage and to speke plesantli to men WICL. S. W. I. 44 ; **faage** (*imp.*) to thi man WICL. JUDG. xiv. 15 ; by plesaunce him **fáged** (*pret.*) LIDG. M. P. 27 ; **fáginge** (*pple.*) wordis WICL. S. W. I. 56.

　　fáginge, sb., *flattery, adulation,* '*adulacio,*' PR. P. 146 ; passe not the boondis of soþ for faging of men WICL. S. W. II. 6.

fagot, sb., *O.Fr.* fagot ; *fagot,* TREV. III. 259 ; C. M. 3164.

fāh, adj., *O.E.* fāh, fāg, = *O.N.* (blā-)fār, *O.H.G.* fēh, *Goth.* faihus ; *coloured, variegated,* LAʒ. 24653 ; fah [fou] P. L. S. viii. 182 ; foh and grei R. S. iv ; foh (*? printed* feh) ant gris REL. I. 109 ; fouh FRAG. 3 ; fouh, vouh, fou MISC. 70, 96, 165 ; fowe and griis TRIST. 12209 ; fou and grene REL. I. 243 ; þe foaʒe neddre HOM. I. 53 ; faʒe neddren I. 51 ; vaʒe LAʒ. 30999 ; *comp.* góld-fāh.

fai, sb., *O.Fr.* fee, faie, *cf. Ital.* fata ; *fay*, *fairy*, Gow. I. 193.

fai, *see* fei. **faier**, *see* fæ3er.

faierie, sb., *O.Fr.* faerie (*féerie*), *mod.Eng.* fairy; *fairyland; enchantment; fairy*, CH. C. T. *E* 1743 ; fairie ALIS. 6924 ; feirie WILL. 230; faieries (*pl.*) CH. C. T. *D* 872.

faile, sb., *O.Fr.* faille ; *fail, failure* ; wituten faile HAV. 179; wiþouten faile C. L. 1162; faille SHOR. 109 ; al was elles turnd to faille MAN. (F.) 16428.

failen, v., *O.Fr.* faillir ; *fail*, Gow. I. 337; faile [failie] LANGL. *B* vi. 326 ; faili GAW. 1067; faileþ (*pres.*) AYENB. 186; him ne failleþ no3t BRD. 13 ; þei sein we failen MAND. 134; þe king ... failede (*pret.*) his mihte LA3. 2938*; and failede ... of his hure A. R. 404; win failede MISC. 29.

failing, sb., *failure, lack*: þurgh failing of fode þat faintes þe pepull D. TROY 11162.

fain, fainen, *see* fæ3en, fæ3nien.

fainen, *see* feinen.

faint, faintise, *see* feint, feintise.

fair, *see* fæ3er. **faire**, *see* fáre *and* feire.

fait [feet], sb., *O.Fr.* feit, feet, *from Lat.* factum ; *feat, action done, deed*, LANGL. *A* i. 160.

faiten, v., =affaiten ; *mortify, tame* : forte faiten hire flesch þat frele was to sinne LANGL. *A* v. 49 [*B* 67 affaiten].

faiten[2], v., *dissemble*, LANGL. *B* vii. 94 ; faite H. V. 76 ; fait(e) TRIST. 3054 ; (hio) faiteden (*pret.*) LANGL. *B* PROL. 42.

faiterie, sb., *deceit*, '*fictio, simulatio, ficticium*,' PR. P. 147 ; Gow. I. 47.

faitour, sb., *O.Fr.* faiteor ; *deceiver, begging impostor*, '*fictor, simulator*,' PR. P. 146 ; faitoures (*pl.*) LANGL. *B* x. 71.

fāken, sb., *O.E.* fācen, =O.L.G. fēcan, *O.N.* feikn, *O.H.G.* feichan ; *fraud, stratagem, crime*, ORM. 12797 ; facne ['*nequitia*'] MAT. xxii. 18.

fāken, adj., *cf. O.E.* fǣcne, *O.L.G.* fēcni, *O.N.* feikn ; *fraudulent* : heo was faken biforen & atterne (*r.* attern) bihinden FRAG. 8 ; faken trowþe ORM. 12655 ; foken MAN. (H.) 194.

fāken-līche, adv., *O.E.* fācenlīce ; *craftily*, FRAG. 8.

fāknien, v., = *O.H.G.* feihnōn; *? deceive*; fōken (*pres.* 3 *plur.*) REL. I. 179.

fal, sb. (*from* fallen), =O.L.G., O.H.G. fal, *O.Fris.* fal, fel, *O.N.* fall ; *fall*, '*casus, lapsus, ruina*,' PR. P. 147 ; KATH. 2322 ; H. M. 15 ; MÁP 337 ; MAN. (F.) 10875 ; LIDG. M. P. 29 ; fall ORM. 11862 ; ALIS. 4412 ; val A. R. 326 ; þat heo wes al forfaren þurh þere leodene væl LA3. 2921 ; þæt ne mihte þes kinges folc of heom fæl (*slaughter*) makien 635 ; for his leodene valle (*ms.* uallen) 29994; *comp.* on-fal.

fáld, sb., *O.E.* falod, fald, *cf. Dan.* fold, *Swed.* fálle ; *fold* (*for cattle*) ; '*ovile*' VOC. (W.W.)

670 ; alle folkes to fald sal falle PR. C. 4637 ; fold LEB. JES. 544; WICL. JOHN x. 16; folde REL. I. 209; CH. C. T. *A* 512; L. H. R. 141 ; **fálde** (*dat.*) HOM. I. 245; goð al se shep to volde HOM. II. 37; **fáldes** (*pl.*) ORM. 3339 ; *comp.* dēr-fald, pund-fold.

fáld[2], sb., *cf. M.H.G.* valt, valde, *Swed.* fáll ; *from* fealden ; *fold* (*of a garment, etc.*) ; ase hit his (*r.* is) in holie boke iwriten ine mani a fe(a)lde (*r. w.* tealde) SHOR. 91 ; a mong þe volde of harde rinde O. & N. 602 ; ne vint he red in one volde 696 ; robes in folde PERC. 32 ; vestement of riche fold M. H. 113 ; under felde SPEC. 24.

fálden, *see* fealden.

falding, sb., *a sort of cloth*, PR. P. 147 ; CH. C. T. *A* 391.

fale, *see* fele, falu.

fale ? adj. *or* sb. : þa3 þe fader þat him formed were fale of his hele A. P. iii. 92.

fal3e, sb., = *Ger.* falge; *fallow* : falghe, falowe '*novale*' CATH. 121 ; falowe WICL. JER. iv. 3 ; falwes (*pl.*) HAV. 2509 ; CH. C. T. *B* 656.

fal3en, v., = *Ger.* falgen, felgen ; *fallow* ; falwe '*novo*' PR. P. 148 ; hi falleweden (*pret.*) erþe CHR. E. 94.

fallace, sb., *O.Fr.* fallace, *cf. mod.Eng.* fallacy ; *deceitfulness*, WICL. MAT. xiii. 22 ; fallas Gow. I. 63.

falle, sb., *? O.E.* feall, =O.H.G. falla ; *trap*, '*decipula*,' PR. P. 147 ; felle CATH. 126 ; *comp.* mūs-falle.

fallen, v., *O.E.* feallan, =O.L.G., O.H.G. fallan, *O.Fris.*, *O.N.* falla ; *fall*, LA3. 801 ; ORM. 7640 ; vallen (*ms.* uallen) A. R. 214 ; þu schalt falle O. & N. 956 ; al þat heom wule þar falle (*befall*) 630 ; falleð (*pres.*) LA3. 1401 ; valleð (*ms.* ualleð) HOM. I. 49 ; al so sone ase eni mon valleð into luðer speche A. R. 96 ; (heo) falleð HOM. I. 53 ; fallið, fallid ROB. (W.) 8601*, 132*; falle (*subj.*) A. R. 280: þet fallinde (*pple.*) üvel (*epilepsy*) 176; fallinge eville MAND. 69 ; fallinge eville or londe eville PR. P. 148 ; fēol (*pret.*) MISC. 187 ; HORN (L.) 428; ALIS. 563 ; SPEC. 62; feol, veol A. R. 226, 280; feol, vel HOM. I. 79, 93; feol, fel HORN (L.) 334, 428; feol, ful LA3. 177, 10416; fel GEN. & EX. 72 : WICL. GEN. xvii. 17 ; K. T. 1086; LAUNF. 755 ; hem fell to þolen wa ORM. 897 ; as vel (*it was due*) to an king ROB. 446; fel, ful HOM. II. 155; fel [fil] LANGL. *B* xiv. 79 ; CH. C. T. *A* 845 ; vil AYENB. 16 ; ful BRD. 1 ; FER. 591 ; feollen MAT. xiii. 4 ; LA3. 578 ; feollen, felle KATH. 1598 ; fellen LANGL. *B* xviii. 309; fullen C. L. 100; villen AYENB. 67 ; fēolle (*subj.*) A. R. 58 ; felle HAV. 1673 ; GREG. 968 ; fallen (*pple.*) MISC. 22 ; H. H. 110 (108) ; MAND. 52 ; *comp.* a-, æt-, bi-, 3e-, ofer-, tō-fallen ; *deriv.* fal, falle, fellen.

fallinge, sb., *falling*, A. R. 280 ; falling PR. C. 1558 ; vallinge ROB. (W.) 4241.

falowe, *see* fal3e. falowen, *see* falwen.

fals, adj. *O.Fr.* fals, faus, *from Lat.* falsus; *false*, A. R. 74; O. & N. 210; BEK. 1004; fals and fikel C. M. 786; false qverne P. L. S. xix. 232; false godes HOM. I.237; false domes MISC. 149; þane valsne (*acc.*) peni AYENB. 24; falsest, valsest (*superl.*) LA3. 30182, 31422.

fals, sb., *O.E.* fals,= *O.N.* fals, *from Lat.* falsum ; *falsehood*, SAX. CHR. 253 ; fals and flærd ORM. 7334, 10027 ; faus JOS. 208.

fals-dōm, sb., *falsehood* : ne con ich saien non falsdom AN. LIT. 4 ; falsedom ROB. (W.) 852*.

fals-hēde, sb., *falsehood*, ROB. 225 ; PL. CR. 419 ; GOW. II. 226 ; PR. C. 1176.

fals-lēk, sb., *falsehood* : to get i falle hem feole, for falsleke fifti folde SPEC. 32.

fals-lī, adj., *false, deceptive* : mi fikel fleische, mi falsli blod SPEC. 32.

fals-līche, adv., *falsely*, LANGL. B xviii. 349 ; valsliche AYENB. 28.

fals-(s)chipe, sb., *falsity*, H. M. 27 ; falsshipe P. S. 150.

falsetē, sb., *O.Fr.* fauseté ; *falsity*, AN. LIT. 5.

falsien, v., *O.Fr.* falser ; *cf. O.N.* falsa ; *falsify; deceive; break, damage; fail* : his hereburne gon to falsie [fausie] LA3. 23967 ; falsen CH. TRO. iv. 964 ; falsin '*falsifico*' PR. P. 148; valseþ (*pres.*) AYENB. 40; þet ower bileave falsie A. R. 270.

falsere, sb., *falsifier* : alse me valseþ þe kinges sel, ha ssel bi demd valsere ate daie of dome AYENB. 62; falsere . . . of holi scriptures WICL. PROL. CATH. EPP.

falsing, sb., *falsehood* : falsing and flateri D. TROY 11330.

falsnesse, sb., *falseness, fraud*, P. S. 220; CH. C. T. G 1304 ; falsnessis (*pl.*) WICL. JOB xiii. 9.

faltrin, v., *cf. Ital.* faltare ; *falter, stumble, tremble; 'cespito'* PR. P. 148; þi limes faltren CH. C. T. B 772 ; faltered (*pret.*) GAW. 430 ; falterde D. ARTH. 1092.

falu, adj., *O.E.* fealo (*pl.* fealwe),= *O.L.G.* falu, *O.H.G.* falo, *O.N.* fölr (*pl.* fölvar), *mod.Eng.* fallow (deer) ; *brown, yellow*, A. D. 252 ; falwe CH. C. T. A 1364; falewe REL. I. 266 ; þe falewe feld RICH. 461 ; to þe fale erþe GAW. 728 ; falewe lockes LA3. 18449 ; veldes falewe wurðen 27468.

falwe, see fal3e.

falwen, v., *O.E.* fealwian, fealuwian, *cf. O.H.G.* falewen ; *turn brown, fade* : valuwen HOM. I. 193; his heu shal falewen REL. I. 65 : falewen shule þi floures SPEC. 89; falewi [falwi] R. S. i. (MISC. 156, 157); valeweð (*pres.*) A. R. 132 ; faleweþ MISC. 93 ; falowiþ P. L. S. vi. 7 ; valouweþ AYENB. 81; falewede (*pret.*) LA3. 4163 ;

falewed A. P. ii. 1539; falwed (*pple.*) L. H. R. 132.

falwen, *see* fal3en.

falwi, sb., *? brown colour* ; falewi O. & N. 456.

fām, sb., *O.E.* fām,= *O.H.G.* feim; foam, DEGR. 1386 ; fom BRD. 19; SPEC. 102 ; FER. 699 ; fom [foom] CH. C. T. G 564.

famaciōn, sb., *cf. Lat.* (dif-)fāmātio ; *defamation* : fals famaciōns (*pl.*) and fained lawez A. P. ii. 188.

fāme, sb., *O.Fr.* fāme ; *fame*, A. R. 222 ; ROB. 29 ; S. S. (Web.) 1232.

fāmen, v., *spread fame of, make famous* : he watz fāmed (*pple.*) for fre þat fe3t loved best A. P. ii. 275.

fámen, v., *O.Fr.* afamer ; *starve* : Steven wille us traveille and famen us to dede MAN. (H.) 122; thin fámit (*pple.*) folke LUD. COV. 105.

fāmi, sb., *O.E.* fāmig ; *foamy* ; fomi CH. C. T. A 2506.

fāmien, fǣmen, v., *O.E.* fǣman,= *O.H.G.* feimen ; *'spumo'* ; fomin '*spumo*' PR. P. 169; fome TREV. VII. 377 ; femin JUL. 68 ; feme LEB. JES. 223 ; fōmeþ [vomeþ] (*pres.*) WICL. MK. ix. 17 ; fāmed (*pret.*) S. S. (Wr.) 959 ; vemde ROB. 208.

familaritē, sb., *O.Fr.* familiarité ; *familiarity* : she useþ ful flatring familarite CH. BOET. ii. 1 (30).

familier, adj., *O.Fr.* familier ; *familiar*, CH. C. T. A 215.

famine, sb., *O.Fr.* famine ; *famine*, LANGL. C ix. 347 ; CH. C. T. C 451.

famlen, v.,= *Dan.* famle, *Swed.* famla, *? O.N.* falma ; *grope, blunder* : stameren oþer famelen REL. I. 65 ; famplen H. M. 37.

famous, adj.,= *O.Fr.* fameus ; *famous*, TREV. V. 215.

fan, sb., *O.E.* fann, *from Lat.* vannus ; *fan*, '*vannus*,' PR. P. 148; ['*ventilabrum*'] WICL. MAT. iii. 12 ; fanne VOC. 233 ; CH. C. T. A 3315 : i shal scatere them with a fan in the 3atus of the lond WICL. JER. xv. 7.

fān, v.,= *M.H.G.* vēhen ; *set at enmity* : he fōde (*pret.*) man wið god HOM. II. 209.

fanc, sb., *O.Fr.* fanc (*fange*) ; *mud* (*printed* fauc), AYENB. 251.

fand, sb., *? from* finden ; *experiment, trial* : of ðis fruit wile ic haven fond GEN. & EX. 336 ; after his owen fond 3150.

fandien, v., *O.E.* fandian,= *O.Fris.* fandia, *O. L.G.* fandōn, *O.H.G.* fantōn ; *from* finden ; *try, tempt, prove* ; fandie HOM. I. 239 ; fanden ORM. 11336 ; fande Iw. 316 ; MIN. vi. 45 ; PR. C. 2228 ; fondian godes HOM. I. 93 ; fondien [fondi] LA3. 2949 ; on sæ and on londe his win he lette fonden 30679 ; fonden H. H. 68 (70) ; S. S. (Wr.) 1489 ; let Adam fonden ðe tre ðe Noe barg GEN. & EX. 3476 ; vonden A. R. 194 ; hit þohte fondi þas O. & N. 1442 ; vondi

AYENB. 15 ; fonde WILL. 1019 ; CH. C. T. *B* 347 ; PERC. 1466 ; he schal fonde þe dent of mine honde HORN (L.) 151 ; i wol fonde to helpe þe A. D. 258 ; **fande** (*pres.*) D. ARTH. 984 ; fondeþ LANGL. *B* xii. 180 ; GOW. II. 106 ; fonde we þis toun to were RICH. 4402 ; **fandede** (*pret.*) MAT. xxii. 35 ; fondede BEK. 717 ; JOS. 505 ; his fandeden MK. viii. 11 ; fondeden KATH. 121 ; **fanded** (*pple.*) PS. xi. 7 ; *comp.* ȝe-**fandien**.

fandere, sb., *tempter* ; vondere AYENB. 116.

fanding, sb., *O.E.* fandung ; *temptation*, ORM. 12262 ; PR. C. 314 ; fondunge A. R. 178 ; fondinge MIRC 240.

fáne, sb., *O.E.* fana, = *Goth.* fana, *O.L.G.*, *O.H. G.* fano ; *banner ; vane*, PR. P. 148 ; chaung- ing as a fane [vane] CH. C. T. *E* 996 ; of red gold was his fane EGL. 1192.

fang, sb., *O.E.* fang, *cf. O.Fris.* fang *m.*, *O.N.* fang *n.*, *M.H.G.* fang, vanc *m.; from* **fōn** ; *catching*, C. M. 3728.

[**fangel**? *comp.* **newe-fangel**].

fangelen? v., hold ȝou stil and **fangel** (*imp.*) noȝth REL. I. 257.

fangen, v., *O.N.* fanga, = *O.H.G.* (furi-)fan- gōn ; *seize, catch*, ORM. 10799 ; fangin '*appre- hendo*,' PR. P. 149 ; fange PS. xxiii. 5 ; GAW. 391 ; M.H. 80 ; i nolde fange [fonge] a ferþing LANGL. *B* v. 566 ; fongen R. S. ii (MISC. 162) ; PL. CR. 786 ; fonge FL. & BL. 300 ; ROB. 73 ; SHOR. 42 ; CH. C. T. *B* 377 ; M. ARTH. 3503 ; ne dar no fuȝel þar to fonge [vonge] O. & N. 1135 ; no cold þat þu ne fonge HAV. 856 ; *comp.* a-, an-, under-**fangen** ; *see* **fōn**.

fangere, sb., = *O.H.G.* (in-)fangare ; ['*sus- ceptor*'] PS. iii. 4.

fannien, v., *from* **fan** ; *winnow, fan; fly* : vanni his corn AYENB. 139 ; fanne '*vanno*' PR. P. 149 ; fannez (*pres.*) on the windez A. P. ii. 457 ; fannand (*pple.*) GAW. 181.

fanōne, sb., *O.Fr.* fanon ; *priest's maniple*, '*fanula*,' VOC. 193 ; fanun or fanen '*fanula*, *manipulus*' PR. P. 149 ; fanoun MIRC 1917.

fant, *see* **font**.

fantasie, sb. *O.Fr.* fantasie ; *fancy*, ALIS.[2] (Sk.) 384 ; CH. C. T. *A* 3191 ; fantesie PARTEN. 358.

fantise, *see* **feintise**.

fantosme, sb., *O.Fr.* fantosme, *Gr.* φάντασμα ; *phantom*, GOW. III. 172 ; (*? printed* fantesme) A. R. 62 ; fantome PR. C. 1197 ; **fantums** (*pl.*) WICL. S. W. III. 20.

far, sb., *O.E.* fær, = *O.N.* far *n.; navigation* : on þam fare SAX. CHR. 248 ; *comp.* **in-fer**.

far-cost, sb., *O.N.* farkostr ; *means of con- veyance, boat, ship*, M. H. xix ; *see* **fercost**.

farand, adj., *pleasing, handsome, joyous* : at uch farand fest among his fre meni GAW. 100 ; þat farand fest A. P. ii. 1757 ; lest les þou leve mi talle farande i. 864 ; if thai wer farande and fre and faire ii. 607 ; fair and farand BARB. ii. 514.

farande-lī, adv., *pleasantly* : farandeli on a felde he fettelez him to bide A. P. iii. 435.

farcen, v., *O.Fr.* farcir ; *stuff* ; farce '*farcio*' PR. P. 150 ; farsed (*pple.*) B. B. 23 ; farsud CH. C. T. *A* 233.

fardel, sb., *O.Fr.* fardel ; *bundle* ; **fardels** (*pl.*) WICL. JUDG. xix. 17 ; C. M. 5004.

fardung, *see* **ferding** *under* **ferdien**.

fáre, sb., *O.E.* faru, = *O.N.* för *f.*, *mod.Eng.* fare ; *journey ; conveyance; equipment ; be- haviour ; success, fortune*, H. M. 19 ; HAV. 1337 ; ALIS. 7072 ; LANGL. *B* xviii. 18 ; CH. C. T. *B* 569 ; S. S. (Wr.) 842 ; MIN. x. 5 ; A. P. ii. 861, iii. 98 ; M. ARTH. 945 ; PERC. 1037 ; ANT. ARTH. iv. ; LUD. COV. 183 ; fare [faire] ROB. 52 ; se firste fare SAX. CHR. 257 ; swulc fare of fleoȝen her was LAȝ. 3904 ; her wes unimete fare 10236 ; þat durste þene king fræine of his fare coste 25562 ; þat na mon under criste of heore vare [fare] muste 25718 ; al most redi was here fare GEN. & EX. 3179 ; whar fore was as þis fare formest bigunne WILL. 4580 ; song and cri and suche fare MIRC 332 ; þei buskid þem wiþ riale fare EGL. 1054 ; *comp.* **chēap-, driht-, forð-, hām-, schip-, wel-fare**.

fáren, v., *O.E.* faran = *O.L.G.*, *O.H.G.*, *Goth.* faran, *O.Fris.*, *O.N.* fara, *mod.Eng.* fare ; *go, travel*, LAȝ. 2512 ; KATH. 1564 ; ORM. 3456 ; HAV. 264 ; JOS. 506 ; wolde hom faren GEN. & EX. 1418 ; faran HOM. I. 115 ; farin '*meo*' PR. P. 150 ; fare WILL. 5142 ; MAN. (F.) 6081 ; PR. C. 1863 ; PERC. 1113 ; fare wið H. M. 27 ; to chirche fare MIRC 265 ; after him for to fare S. S. (Wr.) 290 ; to **fárene** in to heovene FRAG. 6 ; **fáre** (*pres.*) O. & N. 460 ; þou farest ek bi me Pandarus as he CH. TRO. iv. 463 ; fareð, vareð A. R. 120, 214 ; swa hit fareð (*goes*) H. M. 7 ; & fareþ ouer þe flod TRIST. 1304 ; he fariþ wel ['*valet*'] WICL. GEN. xxix. 6 ; he ferð HOM. I. 103 ; þer alle zaulen vareþ wel AYENB. 1 ; feareð HOM. I. 247 ; far (*imper.*) LAȝ. 17994 ; **fáre** (*subj.*) LANGL. *B* x. 405 ; ear we faren henne KATH. 1393 ; **fárinde** (*pple.*) LAȝ. 4262 ; wel farande ISUM. 333 ; **fōr** (*pret.*) ORM. 1807 ; O. & N. 1474 ; he for neh ut of his iwitte MARH. 6 ; he for to Lundene LAȝ. 11396 ; foor MAN. (F.) 7923 ; foren, voren LAȝ. 100, 5555 ; foren GEN. & EX. 2482 ; fore ANT. ARTH. lv ; þei fore forþward PERC. 1425 ; **fáren** (*pple.*) REL. I. 225 ; LANGL. *A* v. 5 ; sinden faren ut of life ORM. 8361 ; hu habbe ȝe fare HORN (L.) 1355 ; *comp.* **for-, ȝe-, ofer-faren** ; *deriv.* **far, fáre, ferde, fēre, férien, fōre**.

farȝen, v., *farrow* : zoȝe huanne hi heþ **ivarȝed** (*pple.*) wel bleþeliche bit men icloþed mid huit AYENB. 61 ; ifarwed TREV. III 213 ; þus beoð þeos pigges iveruwed A. R. 204.

farh, far, sb., *O.E.* fearh, = *O.H.G.* farch, *Lat.* porcus ; *farrow (young pig)* ; **faren** (*pl.*) ALIS. 2441.

farme, *see* feorme.   **farsen,** *see* farcen.

farsūre, sb., *Lat.* farsūra ; *stuffing,* PR. P. 150 ; of alle þo þinges þou make farsure L. C. C. 26.

fart, sb., = *Ger.* farz ; *crepitus ventris,* PR. P. 150 ; CH. C. T. *A* 3806 ; LUD. COV. 21 ; *see* fert.

farten, v., = *Ger.* farzen ; *break wind,* LANGL. *B* iii. 231 ; fartin ' *pedo* ' PR. P. 150 ; *see* ferten.

fartare, sb., *one who breaks wind,* PR. P. 150.

fáse, sb., *Lat.* phase ; *passover* : offre ӡe fase WICL. EX. xii. 21 ; religioun of fase 43.

faselin, v., = *Ger.* faseln ; *unravel,* PR. P. 150.

fasil, sb., *Ger.* fasel ; *fringe,* PR. P. 150.

fasoun, sb., *O.Fr.* faceon, fazon ; *fashion,* P. L. S. xxv. 41 ; fasoun, facioun WILL. 500, 934 ; fasceon MAND. 205 ; fazoun S. S. (Web.) 1384.

fast, *see* fæst.

fastare, *see under* fæsten.

faste, sb., = *ON., O.L.G. O.H.G.,* fasta, *O.Fris.* festa *f.* ; *fast ; abstinence,* ' *jejunium,* ' PR. P. 151 ; (*ms.* fasste) ORM. 11333 ; **fasten** (*dat. pl.*) P. L. S. viii. 74 ; festen A. R. 138.

fasten, v., *O.E.* fæstan, = *O.H.G.* fasten, *Goth.* fastan, *O.N.* fasta, *O.Fris.* festia ; *fast, abstain from food* ; (*ms.* fastenn) ORM. 11326 ; fæsten MK. ii. 19 ; festen HOM. I. 29 ; faste P. L. S. xv. 69 ; veste AYENB. 50 ; **fasteð** (*pres.*) REL. I. 211 ; **fastinde** (*pple.*) HAV. 865 ; **faste** (*pret.*) BRD. 5 ; veste A. R. 126 ; **ivast** (*pple.*) LAӡ. 22310 ; MISC. 38.

    fasting, sb., *fasting,* P. L. S. xiii. 136 ; vastinge MISC. 28 ; festunge HOM. I. 207.

fasten, sb., *O.E.* fæsten *n.* ; *fast* ; festene (*dat.*) HOM. I. 7 ; *comp.* mid-festen.

    festen-dei, sb., *O.E.* fæstendæg ; *fast day* ; vestendei A. R. 318.

fasten, *see* fæsten.   **fastnien,** *see* fæstnien.

fat, sb., *O.E.* fæt, fat, = *O.L.G., O.N.* fat, *O.H.G.* faz, *mod.Eng.* vat ; *vessel, cask, vat,* ' *cupa,* ' VOC. 233 ; PR. P. 151 ; ' *cuve,* ' REL. I. 81 ; fet MK. xi. 16 ; þet fet HICKES I. 229 ; feat, veat JUL. 68, 69 ; þet feat MARH. 17 ; a vet 18 ; **vete** (*dat.*) AYENB. 203 ; **fate** (*pl.*) MAT. xii. 29 ; faten MISC. 29 ; fattis RICH. 1491 ; *comp.* lēoht-, water-fat.

fāt, adj., *O.E.* fætt, = *M.Du.* vet, *O.N.* feitr, *O.H. G.* feizt, feizit ; *from* *fǣten, = *O.N.* feita, *M.H.G.* veizen ; *fat,* TREAT. 138 ; RICH. 6003 ; P. S. 332 ; CH. C. T. *A* 200 ; fatt ORM. 995 ; fat, fet ' *pinguis* ' PR. P. 151 ; fet LK. xv. 23 ; GEN. & EX. 2098 ; vet A. R. 136 ; AYENB. 53 ; þe vatte LAӡ. 19445 ; þet fette swin HOM. I. 81 ; **fāttere** (*comp.*) P. S. 330.

    fett-hēd, sb., *fatness,* GEN. & EX. 1547.

    fattisch, adj., *fat, plump* : armes eueri lith fattissh, flesshi, nat grete therwith CH. D. BL. 951.

    fatnes, sb., *fatness,* ' *pinguedo,* ' VOC. 247 ; PR. P. 151 : with fatnes of whete he fed am ai PS.

lxxx. 17 ; fattnes cxlvii. 14 ; fattenes lxxii. 7 ; the fatnes of (his flok) WICL. GEN. iv. 4.

fätten, v., *O.E.* fǣttan, = *M.Du.* vetten, *O.H.G.* feizten ; *fatten, make fat, grow fat,* WICL. PS. cxl. 5 ; fatte CH. C. T. *D* 1880 ; ase swin ipund ine sti vor te vetten A. R. 128 ; **vetteþ** (*pres.*) AYENB. 246 ; **fatted** (*pple.*) PS. xxii. 5 ; ivetted A. R. 136.

faðme, sb., *O.E.* fæðm *m. f., cf. O.L.G.* fathom, faðom, *O.N.* faðmr, *O.H.G.* fadem *m. ; fathom* ; fadum, fadme ' *ulna* ' PR. P. 145 ; ane veðme LAӡ. 27686 ; fedme OCTOV. (W.) 1656 ; twenti feþme FER. 2312 ; fadme CH. C. T. *A* 2916 ; þreo hund fedme lang HOM. I. 225 ; five fedme long ALIS. 546.

faðmien, v., *O.E.* fæðmian, = *O.N.* faðma ; *fathom ; embrace* ; fadmin ' *ulno* ' PR. P. 145 ; mine armes weren so longe that i **fadmede** (*pret.*) . . . Denemark HAV. 1295 ; frendez fellen in fere faþmed to geder A. P. ii. 399.

[**fauc,** *see* fanc.]

fauchōn, sb., *O.Fr.* fauchon ; *falchion,* LANGL. (Wr.) 9622 ; fauchoun OCTOV. (W.) 105.

faucūn, sb., *O.Fr.* faucon ; *falcon,* Ө. & N. 111 ; faucoun SPEC. 26.

faun, sb., *O.Fr.* faon ; *fawn* ; (*ms.* fawne) VOC. 188.

faunin, *see* fæӡnien.

faunt, sb. *O.Fr.* fant ; *child* : godes sone and maidens faunt L. H. R. 145 ; faunt A. P. i. 101 ; WICL. EX. ii. 3 ; LEV. xii. 3 ; faunte LIDG. M. P. 576 ; **fauntes** (*pl.*) LANGL. (W.) 4367 ; fauntis WICL. JOB xxi. 11.

    faunteken, sb., *little child* ; conformen **fauntekines** (*pl.*) LANGL. *B* xiii. 213 ; fele fauntekins of freeborne childire D. ARTH. 844.

fauntee, sb., *childishness,* LANGL. *B* xv. 146*.

fauntelet, sb., *little child ; infancy,* LANGL. *C* xii. 310.

faunteltē, sb., *childishness,* LANGL *B* xi. 41.

faur, *see* fēower.   **faus,** *see* fals.

fauset, sb., *Fr.* fausset ; *faucet,* L. H. R. 211 ; faucet WICL. JOB xxxii. 19.

faute, sb., *Fr.* faute ; *fault,* JOS. 208 ; ALEX.² (Sk.) 303 ; faut GAW. 2433 ; **fautez** (*pl.*) A. P. ii. 177.

    faut-lēs, adj., *faultless* : fautles of hir fetures GAW. 1760 ; fautlez A. P. ii. 794 ; fautlez in his five wittez GAW. 640.

fauten, v., *cf. O.Fr.* falte (*faute*) ; *be wanting in ; be lacking in* : fools that **fauten** (*pres.*) inwit, i finde that holi chirche sholde finden hem that hem fauteþ LANGL. *B* ix. 66.

fauti, adj., *faulty,* MAND. 175 ; GAW. 2386.

fautour, sb., *O.Fr.* fauteur *from Lat.* fautor ; *favourer, follower, adherent* : fautour, or meintinore ' *fautor* ' PR. P. 152 : him and his **fautours** (*pl.*) he cursed MAN. (H.) 209 ;

Simon and Iohn with here **fautores** (*dat.*)
TREV. IV. 443.

favel, sb., *O.Fr.* favele, fauvel; *flattery, deceit*:
both fals and favel LANGL. *B* ii. 6; favel with
his fikel speche LANGL. *B* ii. 78.

faverous, adj., *favourable*: the time is þan so
faverous R. R. 82.

favorabel, adj., *O.Fr.* favorable; *favourable*:
til þam þe world es favorabel PR. C. 1344.

favorin, v., *favour*, '*faveo*,' PR. P. 152: whi
favure (*pres.*) ȝe þanne falcej godus ALEX.[2]
(Sk.) 740; **faveret** (*pple.*) full wele D. TROY
3868.

favour, sb., *O.Fr.* favor; *favour*, AYENB. 230.

fawe, *see* **fæȝen** *and* **fēa.**

fawen, *see* **fæȝen.** **fāwien,** *see* **fāȝien.**

fawn, *see* **faun.**

fax, sb., *O.E.* feax, = *O.Fris.*, *O.N.* fax, *O.L.G.*,
O.H.G. fahs; *hair*, WILL. 2097; REL. II. 176
(P. L. S. xxxiv. 15); FLOR. 1545; GAW. 181;
A. P. ii. 790; ANT. ARTH. xxix.; MIR. PL. 183;
vax (*ms.* uax) JUL. 29; væx (*sec. text* her)
LAȝ. 21873; **faxes** (*gen.*) GR. 45; **vaxe** (*dat.*)
LAȝ. 24843.

fé, *see* **feoh.**

fēa, adj., *O.E.* fēa, = *O.Fris.* fē, *O.L.G.* fāh,
O.N. fār, *O.H.G.* fōh, *Goth.* faus, *Lat.* pau-
cus; *few, scanty*: þair clothing was . . . fa C.
M. 8599; fo GEN. & EX. 2403; **fēawe** (*pl.*) KATH.
950; feawe, fewe, feuwe P. L. S. viii. 52, 176;
feawe SHOR. 16; veawe þer bieþ zuiche AYENB.
162; fewe MAND. 255; a fewe men ROB. 18;
fawe MISC 70; vawe 177; mid fewe worde
REL. I. 183; feaue MISC. 34; ine feue ȝere LAȝ.
387*; fæwe ORM. 424; fowe, foue REL. I. 85;
fone PS. cvi. 39; fone, fune MIN. ii. 28,
29; PR. C. 764; vewe contreies ROB. (W.)
7586.

  fēunesse, sb., *O.E.* fēaniss; [*v.r.* fonenesse]
fewness, PS. ci. 24.

feaȝen, *see* **fēȝen.**

[-feald, *O.E.* -feald, *O.L.G.*, *O.Fris.* -fald, *O.N.*
-faldr, *O.H.G.* -falt, *Goth.* -falþs; *-fold; see*
ān-, feole-, fēower-, fīfhund-, mani-, twā-,
þrēo-, þūsen-feald (-fald, -fold).]

fealden, fálden, v., *O.E.* fealdan (*pret.*
fēold), = *O.H.G.* faldan, *M.L.G.* valden, volden,
O.N. falda, *Goth.* falþan (*pret.* faifalþ); *fold,*
clasp; *bend, give way, falter*; falde IW.
1425; ISUM. 307; foldin '*plico*' PR. P. 168;
leves . . . folde O. & N. 1326; þe feendes
strengþe to folde K. T. 1118; sixti sir Roger
doun can (*r.* gan) folde TRIAM. 326; for no
fa(i)ntise Arthur nold folde M. ARTH. 2547;
his herte began to folde EGL. 726; his legges
gunne to volde FER. 4567; ant mei no finger
felde SPEC. 48; aiþer oþer in armez con (*r.*
gon) felde GAW. 841; care and kunde of elde
makeþ mi bodi felde REL. I. 120; þi tunge
vóldeþ (*pres.*) MISC. 101; falt mi tunge
O. & N. 37; ope ane viealdinde (*pple.*)

stole AYENB. 239; (h)is bodi a side he felde
FER. 841; **fólden** (*pple.*) PL. CR. 227; MAN.
(F.) 11889; þe folden fust LANGL. *B* xvii. 176;
folden in silk GAW. 959; volden to grunde
LAȝ. 15885; mi nest falde (? *allied*) cun JUL.
32; *comp.* **bi-, ȝe-, ümb-, un-fealden (-fál-**
den, -fólden).

feale, *see* **fēle.** **fearen,** *see* **fáren.**

fearn, sb., *O.E.* fearn *n., cf. O.H.G.* farn *m.;*
fern, '*filex*' (*sic*), FRAG. 3; fern, ferne CH.
BOET. iii. 1 (64); ferne AR. & MER. 8876;
værne (*dat.*) LAȝ. 12819; *comp.* **ever-, fen-**
fern.

fēawe, *see* **fēa.**

fēble, adj., *O.Fr.* fēble, fieble; *feeble,* HOM. I.
47; A. R. 56; R. S. vii.; GEN. & EX. 1072; P.
L. S. xvii. 547; fieble HOM. II. 191; AYENB.
31; **fēblore** [feblor] (*comp.*) ROB. (W.) 3424;
fēblest (*superl.*) GAW. 354.

  fēble-līche [fīblelīche], adv., *feebly,* ROB.
(W.) 11771; feblelike HAV. 418.

  fēblenis, sb., *feebleness,* ROB. (W.) 9110*;
febilnesse PR. P. 152.

fēblesse, sb., *O.Fr.* feblesce; *feebleness,* ROB.
442; fieblesse CH. BOET. iv. 2 (112).

fēblin, v., *O.Fr.* febloier; *enfeeble,* PR. P. 152;
febli P. L. S. xvii. 546; **fēblie** (*pres. subj.*) A. R.
368; **fēbleden** (*pret.*) WICL. JUD. xv. 4.

fec, *see* **fæo.**

fecche, sb., *Lat.* vicia; *vetch*; fehche '*vicia*'
PR. P. 153; feche VOC. 201; ficche WICL.
IS. xxviii. 25.

fecchen, v., *O.E.* feccan, = *Ger.* facken (GR.
WB. III. 1229); *fetch,* JUL. 69; ORM. 8634;
LANGL. *B* ii. 180; CH. C. T. *G* 411; he wile
hem fecchen into chirche HOM. II. 61; water
fecchan JOHN iv. 15; vechchen A. R. 368;
fecche HORN (R.) 357; AN. LIT. 10; P. L. S.
xiii. 222; WILL. 26; hwuch este mihtestu
þar vecche O. & N. 1504; facche P. S. 191;
vacche FER. 2517; **feccheð** (*pres.*) REL. I.
217 (MISC. 13); **fecheð** (*imper.*) JUL. 67; fæche
(*printed* sæche) [feche] LAȝ. 3571; **fecche**
(*subj.*) R. S. ii; if he fer fecchen fode REL.
I. 216; **feight(e)** (*pret.*) MAN. (F.) 9058;
þer he dæd fæhte LAȝ. 6460; foched MAN.
(F.) 12690.

fecht, *see* **fehten.**

fēden, v., *O.E.* fēdan, = *O.Fris.* fēda, *O.N.*
fǿða, *Goth.* fōdjan, *O.L.G.* fōdean, *O.H.G.*
fuotan; *from* **fōde**; *feed,* LAȝ. 3253; ORM.
2080; feden, fede HAV. 621, 906; veden
A. R. 214; fede MAND. 58; M. H. 43; vede
AYENB. 199; **fēdeð,** fet (*pres.*) REL. I. 215,
225; fedeþ [fet] LANGL. *B* PROL. 194; fet
HOM. I. 63; A. R. 198; **vēd** (*imper.*) A. R.
100; **fédde** (*pret.*) KATH. 1843; GEN. &
EX. 2630; CH. C. T. *A* 146; feddest JUL.
61; **fēd** (*pple.*) C. L. 1276; ifed P. L. S. xiii.
135; ived A. R. 206; O. & N. 1529.

  fēdinge, sb., *O.E.* fēding, = *O.N.* fǿðing:

feeding, PR. P. 152; feding TREV. III. 183; MED. 39.

feder, *see* **fæder. fedme,** *see* **faðme.**

fee, *see* **feo** *and* **feoh. feel,** *see* **fēle.**

feend, *see* **fēond. feer,** *see* **fēre.**

feet, *see* **fait.**

fefering, sb., *from O.E.* feferian; *fever*; **fēveringes** (*gen.*) MAN. (F.) 3925.

feffement, sb., *O.Fr.* fieffement; *enfeoffment*, LANGL. *B* ii. 72.

feffen, v., *O.Fr.* feffer, fieffer; *enfeoff, endow*; feffe WILL. 1061; **feffede** (*pret.*) ROB. 368.

fefre, sb., *O.E.* fefer (MAT. viii. 15), *from Lat.* febris; *fever*, A. R. 112; fevre AYENB. 29; fevre [fevere] LANGL. *B* xiii. 336; fivere PR. P. 163; fiver PR. C. 2989.

[**fēȝe,** adj., *O.E.* fēge; *fit; comp.* ȝe-fēȝe.] **fēȝe,** *see* **fæȝe.**

fēȝen, v., *O.N.* fegja, fægja, = *M.Du., M.H.G.* vegen; *polish, adorn; cleanse*: fæien heore steden LAȝ. 8057; **feaȝeð** (*pres.*) A. R. 58*; feg ðe ðus of ði brest filde (*r.* filðe) REL. I. 213; feahe þi meidenhad wið alle gode þeawes H. M. 45; **feiede** (*pret.*) D. ARTH. 1114; fæȝeden heore wepnen LAȝ. 7957.

    feahunge, sb., *adornment*, H.M. 43.

    fēȝer, sb.,= *M.H.G.* vegære; *cleanser*; feiar PR. P. 203.

fēȝen, v., *O.E.* fēgan,= *O.Fris.* fōga, *O.H.G.* fuogen; *adapt, fit, join*; veien A. R. 396; feȝe P. S. 154; fie S. S. (Wr.) 2981; vie TREAT. 139; **vieþ** (*pres.*) BEK. 658; [**fēȝede** (*pret.*)] feide þe lemes to ure licame HOM. II. 25; fiede LAȝ. 50; mannes bodiȝ fēȝed (*pple.*) is of fowre kinne shafte ORM. 11501; *comp.* ȝe-fēȝen.

    fēiunge, sb., *O.E.* fēgung, = *M.H.G.* vüegunge, veiunge; *joining*, A. R. 78; feinge HOM. II. 25.

fēȝer, *see under* **fēȝen.**

feȝt, *see* **feht. feh,** *see* **feoh.**

feht, sb., *O.E.* feoht, fyht,= *O.H.G.* (gi-)feht *n.; fight*, KATH. 608; þet fecht, þis fecht HOM. I. 151; feȝt A. P. ii. 275; feht, fæht, feiht, feoht, fiht LAȝ. 172, 407, 1744, 4170, 18693; feight, feit PR. P. 153; fiht HAV. 2668; þat fiht C. L. 973; viht A. R. 358; figt GEN. & EX. 870; viȝt AYENB. 180; **fehtes** (*gen.*) LAȝ. 8609; **fehte** (*dat.*) HOM. II. 115; fihte, fuhte LAȝ. 605, 6807; *comp.* ȝe-feht.

    fiht-lāc, sb.. *O.E.* feohtlāc; *battle*, O. & N. 1699.

    ficth-wīte, sb., *O.E.* fiht-, fyht-, feohtwīte; *fine for fighting, 'quite de medlee de la merci,'* REL. I. 33.

fehten, v., *O.E.* feohtan, = *O.Fris.* fiuchta, fiochta, *O.H.G.* fehtan; *fight*, KATH. 603; feohten, fihten LK. xiv. 31, 32; fechten, fihten HOM. I. 153, 155; fehten, feihten, fihten, fuhten LAȝ. 5765, 1491, 1580, 3939; feȝte ANT. ARTH. xxxii; feightin, feitin PR. P. 153;

fihten K. T. 1078; (*ms.* fihhtenn) ORM. 1842; figten GEN. & EX. 3227; fihte, vihte, fiȝte, viȝte O. & N. 172, 1069, 1679; fiȝte WILL. 3825; viȝte AYENB. 15; fighten LANGL. *C* xxii. 65; fiht, fiȝt (*pres.*) O. & N. 176, 1072; (we) fehtað HOM. I. 107; (heo) vihteð A. R. 268; **feaht,** faht fæht, feht (*pret.*) LAȝ. 1591, 4353, 8806, 11278; faht MARH. 2; faht, vaȝt O. & N. 1071; faȝt ROB. 274; FER. 465; faȝt GEN. & EX. 3386; faght OCTAV. (H.) 535; fauh MAN. (H.) 170; fauȝt WILL. 3426; faught RICH. 5677; faut TOR. 668; **fuhten** A. R. 196; fuhten [fohte] LAȝ. 576; fohten (*ms.* feohten) 7482; fuȝten HORN (L.) 1375; fouhten HAV. 2661; fouȝten WILL. 3414; foȝtun ANT. ARTH. xliv; **foghten** (*pple.*) MIN. v. 62; foghtin EGL. 439; *comp.* for-, ȝe-, wið-fehten.

    feightare, sb., *O.E.* feohtere, = *O.H.G.* fehtāri; *fighter, 'pugnator,'* PR. P. 153; fiȝttere SHOR. 109; fiȝteres (*pl.*) TREV. I. 351; fechtaris BARB. xi. 106.

    fihtinge [fiȝtinge], sb., *fighting*, O. & N. 1704; viȝtinge AYENB. 37.

    fechting-sted, sb., *battlefield*, BARB. xv. 378.

fei, sb., *O.Fr.* fei, feid; *faith*, HAV. 255; BEK. 2062; JOS. 245; SACR. 589; fai SHOR. 139; DEGR. 428; feið GEN. & EX 2687; feiþ P. L. S. xxv. 136; MAND. 18; EGL. 440; feiȝþ WILL. 275.

    feiþ-ful, adj., *faithful*, WILL. 5434; MAND. 139.

    feiþful-lī, adv., *faithfully*, MAND. 246.

    feith-hēd, sb., *fidelity*, ROB. (W.) APP. G. 59*.

    feith-liche, adv., *in faith, truly*, WILL. 2732; feiþli 777; feiþeli 201; feiȝþli 4793; feiȝþeli 230; feþli 132, 209; feȝtli 703; faitli ALIS. 804; faithli D. TROY 11447; faitheli GAW. 1635.

feide, sb., *med.Lat.* faida; *feud*, ALIS. 96.

fēie, *see* **fæȝe. fēien,** *see* **fēȝen** *and* **feoȝen.**

feier, *see* **fæȝer. feiȝ, feih,** *see* **feoh.**

feild, *see* **feld. fein,** *see* **fæȝen.**

feinin, v., *O.Fr.* feindre; *feign, 'fingo,'* PR. P. 153; feine CH. C. T. *A* 736; PL. CR. 273; **feignede** (*pret.*) BEK. 42; feinede ROB. 336; **feined** (*pple.*) PR. C. 2556; fained A. P. ii. 188.

    feinere, sb., *feigner*: the feinere in mouth desceiveth his frend WICL. PROV. xi. 19; teinare *'fictor, simulator'* PR.P. 153; **feineres** (*pl.*) WICL. JOB xxxvi. 13.

    feining, sb., *pretending, pretence*, WICL. IS. xxxii. 6; love withoute feininge WICL. ROM. xii. 9; *'fictio, simulacio'* PR. P. 153.

feint, adj., *O.Fr.* faint; *faint, 'segnis,'* PR. P. 153; SPEC. 25; MIRC 1193; faint PL. CR. 47.

    feint-liche, adv. *faintly*, P. S. 325; feinteliche ROB. (W.) 10595.

    feintnesse, sb., *faintness, segnities,'* PR. P. 153; MED. 594.

feintin, v., *faint,* '*fatisco,*' PR. P. 153 ; feinte WILL. 3638 ; DAV. DR. 118.

feintise, sb., *O.Fr.* feintise, faintise ; *dissimulation, cowardice, languor,* MARG. 34 ; WILL. 436 ; MIRC 1207 ; faintise AYENB. 26 ; PL. CR. 251 ; IW. 79 ; PR. C. 3519 ; fantise ROB. (W.) 460*.

feir, *see* fæȝer.

feire, sb., *O.Fr.* feire ; *fair, market,* FL. & BL. 216 ; WILL. 1822 ; feire, faire LANGL. *A* v. 119 ; *B* v. 205 ; faire AYENB. 76 ; GAM. 271.

feiren, *see* fæȝrien. **feirie,** *see* fairie.

feið, *see* fei.

fekel, *see* fikel.

fel, sb., *O.E.* fel, fell, = *O.L.G., O.Fris., O.H.G.* fel (*gen.* felles), *O.N.* (bōk-)fell, (ber-)fiall, *Goth.* (þruts-)fill *n., mod.Eng.* fell ; *hide, skin,* KATH. 1617 ; REL. I. 211 ; WILL. 1720 ; S. S. (Wr.) 2483 ; K. T. 752 ; (*ms.* fell) ORM. 10441 ; fel ['*pellem*'] WICL. JOB ii. 4 ; þet fel HICKES I. 227 ; fel, vel A. R. 120, 364 ; vel O. & N. 830 ; felle (*dat.*) ROB. 287 ; GAW. 1359 ; PR. C. 82 ; velle AYENB. 210 ; a feier þing of flesche and felle EGL. 26 ; **fellan,** fellen (*dat. pl.*) HOM. I. 225 ; *comp.* **barm-, bōc-fel.**

fel, adj.. *O.E.* (wæl-)fel, = *M.Du.* fel, *O.Fr.* fel ; *fell, fierce, cruel,* P. L. S. xxiv. 35 ; CHR. E. 791 ; AYENB. 66 ; RICH. 4419 ; CH. C. T. *D* 2002 ; LIDG. M. P. 145 ; a fel wind LANGL. *B* xvi. 31 ; fel and grim M. H. 123 ; felle (*pl.*) WICL. JOB xxxvi. 13 ; felle frekes LANGL. *B* v. 170 ; mid hire felle wrenches LAȜ. 5302* ; wit wordez felle A. P. i. 367 ; he lokede felle OCTAV.(H.)936 ; nas never feller(*comp.*) fever WILL. 897 ; **felliste** (*superl.*) AYENB. 61.

fel-hēde, sb., *cruelty,* AYENB. 29, 159.

fel-lī, adv., *cruelly,* WILL. 3274 ; A. P. ii. 571 ; PR. C. 4449 ; PERC. 613 ; felle TRIST. 97.

felnes, sb., *atrocity, astuteness* ['*astutia*'], WICL. S. W. I. 55 ; fellnesse WICL. JOB v. 13.

fel, sb., *O.N.* fiall, *mod.Eng.* fell ; *mountain,* M. H. 25 ; DEGR. 1149 ; felle (*dat.*) GAW. 723 ; D. ARTH. 2489 ; felles (*pl.*) ANT. ARTH. i.

felaȝe, sb., *O.N.* fēlagi ; *fellow, partner, companion,* HORN (L.) 996 ; SHOR. 161 ; felage GEN. & EX. 1761 ; felaghe PR. C. 5485 ; velaȝe AYENB. 36 ; feolahe JUL. 49 ; felaȝe, felawe HOM. I. 187, 201 ; felawe P. L. S. x. 42 ; ALIS. 2255 ; LANGL. *B* vii. 12 ; CH. C. T. *A* 395 ; PERC. 460 ; feo-, felawe A. R. 114 ; felowe MAND. 24 ; **felaȝes** (*pl.*) GAW. 1702 ; felahes KATH. 2339 ; velaghes MISC. 34 ; felawes, felaus ROB. (W.) 1376, 2335 ; *comp.* **bord-fēlawe.**

fēolau-līch, adj., *brotherly,* A. R. 282.

fēlaȝ-līche, adv., *brotherly,* A. R. 38.

fēolau-rēden, sb., *company* ; veolauredden A. R. 106 ; velaȝrede AYENB. 9 ; felaurede RICH. 3137 ; K. T. 956 ; felaurade HORN (R.) 174.

fēlaȝ-schip, sb., *fellowship,* A. P. ii. 271 ; felawscipe HOM. I. 185 ; feolauschipe A. R. 160 ; felauschupe JOS. 165.

fēlawschipen, v., *associate with, unite in* : ne **fēlawship** (*imper.*) [*later vers.*felowschipe] with hem mariagis WICL. DEUT. vii. 3 ; **fēlawshipten** [*Purv.* felouschipiden] (*pret.*) hem selven with hem in the batail I KINGS xiv. 22 ; **fēlowshipte** [*Purv.* felouschipid] (*pple.*) thurȝ mariage GEN. xxv. 7 ; felaschipid [felouschipid] PROV. xxvi. 23.

fēlaȝen, v., *make a partner* ; **vēlaȝeþ** (*pres.*) AYENB. 102.

fēld, sb., *O.E.* feld, feald, *cf. O.Fris.* feld, field *m., O.L.G.* feld, *O.H.G.* felt, feld *n.;* *field, plain,* ORM. 14568 ; SPEC. 105 ; LANGL. *B* i. 2 ; MAND. 50 ; DEGR. 269 ; þene feld LAȜ. 15285 ; a veld [on feold] 17156 ; ænne bræne (*r.* bradne) fæld 14202 ; þen feld ROB. 213 ; veld (*ms.* ueld) A. R. 102 ; AYENB. 131 ; feld [feeld] CH. C. T. *A* 886 ; feild ALIS. 3403 ; fild PERC. 1679 ; TOR. 3 ; AMAD. (R.) xvi ; **fēlde** (*dat.*) O. & N. 1714 ; REL. I. 217 ; vælde LAȜ. 15188 ; fielde S. A. L. 157 ; **fēldes,** felde (*pl.*) LAȜ. 4164, 24694 ; **fēlden** (*dat. pl.*) LAȜ. 18404.

fēld-bēst, sb., *cattle* ; oure . . . feeldbeestes (*pl.*) and housbeestis WICL. NUM. xxxii. 26.

fild-man, sb., = *Ger.* feldmann ; *countryman,* '*rusticus,*' VOC. 213.

fēld-mous, sb., *field-mouse,* PALL. i. 918.

field-wode, sb., *name of a plant* ; ? = fieldwort ; ? *gentian,* GOW. II. 262.

fēlde, fēlden, *see* fald², fealden.

fēldi, adj., *champain* : the feeldi places of Moab WICL. NUM. xxii. 1 ; feeldi GEN. xiv. 6 ; LK. vi. 17.

fēldȝefare, fēldefare, sb., *O.E.* feldefare, feolufor, fealafor (*turdus pilaris*) ETTM. 344, ? = *O.H.G.* felefor (*onocrotalus*) ; *fieldfare,* REL. II. 78 ; CH. P. F. 364 ; feldefare, felfare PR. P. 154 ; **fēldfares** (*pl.*) WILL. 183.

féle, adj., *O.E.* fela, feola, feala, feolu, feolo, = *O.Fris.* felo, *O.L.G.* filu, filo, *O.H.G.* filu, filo, file, *Goth.* filu, *Gr.* πολύς ; *much, many,* SHOR. 16 ; WILL. 5 ; A. P. i. 21 ; M. ARTH. 6 ; ANT. ARTH. xxi ; hu fele hlafe MAT. xv. 34 ; feole þinge MK. v. 26 ; fele ȝere HOM. I. 35 ; feole oðre godere werke 9 ; fele wundre 227 ; fele shulen fallen ORM. 7640 ; ne ilef þu nouht to fele uppe þe see þat floweþ REL. I. 174 ; þær wæren fele gode men ORM. 55 ; þe wise mon mid fewe word(e) can fele biluken REL. 183 ; hu fele ger GEN. & EX. 2400 ; fele wordes LANGL. *B* iii. 338 ; fele siþe HAV. 778 ; vele manne MISC. 75 ; fela þefas SAX. CHR. 253 ; feole huntes 256 ; feale tacne 249 ; þer weore feondes to feole [fale] LAȜ. 1286 ; feola wintre 2046 ; fele ȝer 108 ; fele [fale] ȝere 13733 ; feole cnihtes 24537 ; væle þusend 22350 ; feole KATH. 89 ; JOS. 18 ; feole domes FRAG. 8 ; feole [veole] siðen

A. R. 236, 320; fele, feale P. L. S. viii. 5, 35; fele [veole] þeode O. & N. 387; fele [vele] wrenche 813; of so feole [fele] þinge 1772; swa vale wise 1663; fele, vale ROB. 9, 200; feel ROB. (W.) 208; fale HICKES I. 232; to velen AYENB. 102.

feole-fold, adj., *O.E.* felafeald; *manifold*; féolevold A. R. 180; feolefolde mede HOM. I. 135; feolevalde sorȝe LAȝ. 27129.

féle-faldien, v., *multiply*; félefaldes (*pres.*) PS. xi. 9.

féle, sb., *O.E.* féle(-léas), = *O.Fris.* féle; *feeling*, C. M. 17017; reaves me fele of pine HOM. I. 285.

[**féle**, sb., *? O.E.* (unge-)féle; *? sensible; comp.* eð-féle.]

fele, *see* **fǽle**.

félen, v., *O.E.* feolan (*pret.* fealh), = *O.N.* fela, *O.Fris.* (bi-)fella, *O.L.G.* (bi-)felhen, *O.H.G.* felahan, *Goth.* filhan (κρύπτειν, θάπτειν); *hide, conceal*; fele D. ARTH. 3237; þat godes gift fra man wil fele (*printed* sele) M. H. 3; féle (*pres.*) A. P. ii. 914; ie fele TOWNL. 67; *comp.* bi-félen.

félen, v., *O.E.* félan, = *O.Fris.* féla, *O.L.G.* (gi-)fōlian, *O.H.G.* fuolan, fualan; *feel*, Gow. III. 281; fele A. P. ii. 107; APOL. 1: féleð (*pres.*) HOM. II. 93; veleð A. R. 178; fielez S. A. L. 148; ȝif þou fele þe siker P. L. S. xxix. 45: félde (*pret.*) ROB. 16; SHOR. 86; felte CH. C. T. *G* 521 (Wr. 12449); veiled ROB. (W.) 6962*; féled (*pple.*) SPEC. 68; WILL. 638; *comp.* ȝe-félen.

félend-lī, adv., *sensibly*: he spak felendli or wiseli, weel WICL. ECCLUS. xiii. 27.

féling, sb., = *Ger.* fühlung; *feeling*, PR.C. 3030; velunge A. R. 110; felunge H. M. 13; velinge AYENB. 241.

fel-fare, *see* **féldȝefare**.

felȝe, sb., *O.E.* felg, felge, = *M.Du.* velge, *O.H.G.* felga; *felloe (of a wheel)*; veli VOC. 180; felwe '*canthus*' PR. P. 154; felowe VOC. 234; felien (*pl.*) KATH. 1944; felies WICL. 3 KINGS vii. 33.

felȝen, *see* **folȝen**.

felicité, sb., *O.Fr.* felicité; *felicity*, CH. C. T. *A* 1266.

felie, *see* **felȝe**.

feliole, sb., *round tower*; felioles (*pl.*) SQ. L. DEG. 835; filioles (*pl.*) A. P. ii. 1462; faire filiolez þat fiȝed and ferlili long GAW. 795.

fellen, v., *O.E.* fellan, fyllan, = *O.L.G.* fellian, fellean, fellien, *O.H.G.* fellen, *O.N.* fella; *from fallen*; *fell, cut down, knock down*, P. S. 189; fellin '*dejicio*' PR. P. 154; vellen heo to grunde MISC. 89; felle JOS. 368; M. H. 45; walles felle O. & N. 767; felle [fille] an oke CH. C. T. *A* 1702; fellen, fællen, feollen LAȝ. 4204, 5632, 8600; velle, fulle L. H. R. 30; he let fulle þe grete tren ROB. 170; fel (*imper.*)

LAȝ. 3697; felde, fælde, feolde, fulde (*pret.*) LAȝ. 1716, 1917, 3710, 4089; felde ROB. 70; HAV. 1859; fulde FER. 3734; felden ALIS. 5844; feolden CHR. E. 94; *comp.* **a-, bi-, ȝe-fellen**.

fallinge-ax, sb., *felling-axe*, WICL. PS. lxxiii. 6.

fellen, *see* **fúllen**.

feloun, sb., *O.Fr.* felon, felun; *felon*, BEK. 565; AYENB. 29; felun FL. & BL. 329.

felon-líche, adv., *like a felon, cruelly*: falsliche and felonliche LANGL. *B* xviii. 349; felonlike MAN. (F.) 493; **felonloker** (*comp.*) 3028.

felonie, sb., *O.Fr.* felonie; *felony*, FL. & BL. 331; P. L. S. xiii. 97; AYENB. 149; felonie [feloniȝe, velonie] ROB. (W.) 1446.

felonous, adj., *O.Fr.* feloneus; *vicious, cruel, wicked*: felonouse and foule MAND. 65.

felschen, v., *? renovate*: gerte felschene his firez D. ARTH. 1975.

felt, sb., *O.E.* felt, = *M.Du.* vilt, *O.H.G.* filz; *felt, 'filtrum,'* PR. P. 154; A. P. ii. 1689.

felt-wort, sb., *O.E.* feltwyrt; *verbascum*, LEECHD. GLOSS. II. 384.

felten, v., = *M.H.G.* vilzen; *felt*: **felted** (*pple.*) botes P. S. 330 (P. 26).

feltren, v., = *Fr.* feutrer, *Ital.* feltrare; *felt; join together*; filter, filtir (*pres.*) A. P. ii. 696, 224; **felterd** (*pple.*) TOWNL. 85; filtered D. ARTH. 1078.

félðe, *see* **fúlðe**. **felwe**, *see* **felȝe**.

femelle [**femele**], adj. & sb., *O.Fr.* femelle; *female*, CH. C. T. *D* 122; femelles [femeles] (*pl.*) LANGL. *B* xi. 331.

fémen, *see* **fámien**.

feminin, adj., *O.Fr.* feminin, *from Lat.* féminínus; *feminine, womanly*: the feminin gender TOWNL. 309; feminine shamefastness KN. CUR. 97.

feminité, sb., *womanhood*: o serpent under feminitee CH. C. T. *B* 360; a thing contrarie to feminite LIDG. M. P. 47.

fen, sb., *Arabic* fan; *section of Avicenna's Canon*, CH. C. T. *C* 890.

fen, sb., *O.E.* fenn *n. m.*, *cf. O.N.* fen *n.*, *O.Fris.* fenne *m.*, *? O.H.G.* fenni *n.* fenna *f.*, *Goth.* fani *n.*; *fen*, GEN. & EX. 490; SPEC. 54; PR. C. 566; þet fule fen HOM. I. 81; a ful deop fen MISC. 150; ven A. R. 138; **fenne** (*dat.*) LAȝ. 21741; FER. 2331; venne O. & N. 832; fenne (*pl.*) P. S. 70; *comp.* láȝe-, mōr-fen.

fen-fern, sb., *O.E.* fenfearn; fen vern '*salvia,*' VOC. 139.

fen-līc, adj., *marshy*; venliche A. R. 206.

fénd, fōnd, *see* **féon, féond**.

fence, sb., *O.Fr.* defense; *defence, bulwark* '*protectio, defensio,*' PR. P. 155; '*municio, defensaculum*' PR. P. 155; he stood at fence agein them HALLIW. 352 ; the fens & the fin stuff of all the tulkes D. TROY 7363.

fencen, v., *defend, fence* : fast stand we to fence TOWNL. 308 ; fensin PR. P. 155 ; fencid (*pple.*) or defencid ' *defensus, munitus, defensatus*' PR. P. 155.

fenden, v., *for* defenden ; *defend,* LANGL. *B* xix. 61 ; fende MAN. (F.) 6491.

fenel, fenkil, sb., *O.E.* finul, *from Lat.* feniculum ; *fennel,* PR. P. 155 ; fenil SPEC. 44 ; finkil VOC. 191.

 fenel-seed, sb., *fennel seed,* LANGL. *A* v. 156.

fenestral, sb., *window* ; fenestralle VOC. 237 ; PR. P. 155.

fenestre, sb., *O.Fr.* fenestre ; *window,* BRD. 15 ; LANGL. *B* xviii. 15.

feng, sb., *O.E., O.Fris.* feng, = *O.N.* fengr ; *from* fōn ; *capture, seizing,* HOM. I. 39 ; **fenges** (*gen.*) LAȝ. 8610 ; **fenge** (*dat.*) O. & N. 1285 ; *comp.* for-, here-feng.

fēnix, sb., *O.E.* fenix, *from Lat.* (*Gr.*) phoenix ; *phœnix,* MAND. 48 ; A. P. i. 430 ; fenex of Arabie CH. D. BL. 980 ; fenis VOC. 189.

fenken, v., = **vencuse** ; *vanquish* : Philip **fenkes** (*pres.*) in fight ALIS.² (Sk.) 323 ; haddest þou **fenked** (*pple.*) þe fon þat in þi flech dwellen ALEX.² (Sk.) 339 ; hee had fenked þe folke too fare ALIS.² (Sk.) 111 ; ifenked 117.

fenkil, *see* **fenel.**

fenni, adj., *O.E.* fennig, fenneg, = *O.H.G.* fennīg ; *dirty, vile,* A. P. ii. 1113.

fenni-līche, adj., MARH. 15 ; H. M. 11.

fente, sb., *O. Fr.* fente ; *the slit in a robe closed by a brooch or trimmed with fur, &c.,* ' *fibulatorium,*' fimbria, PR. P. 156.

feo, sb., *O.Fr.* feu, fiu, *from med. Lat.* feudum ; *fee,* ALIS. 7963 ; fee PR. P. 152 ; **fees** (*pl.*) A. P. ii. 960.

feo, *see* **feoh. fēoȝen,** *see* **fēon.**

feoh, sb. *O.E.* feoh, = *O.N.* fē, *O.Fris.* fia, *O.L.G.* fehu, *O.H.G.* fihu, *Goth.* faihu, *Lat.* pecu, *mod. Eng.* fee ; *cattle, property, money,* ' *pecunia,*' FRAG. 4 ; feoh (feo) LAȝ. 4429 ; feh P. S. 152 ; (*ms.* fehh) ORM. 3279 ; þet feh HOM. I. 91 ; feih A. R. 326 ; feo MISC. 95 ; FL. & BL. 25 ; feiȝ, fei C. L. 467 ; gold ne fe HAV. 44 ; and gaf him lond and agte and fe GEN. & EX. 783 ; fee C. M. 11679 ; CH. C. T. *D* 630 ; ISUM. 90 ; EGL. 294 ; hirdès that kepe thar fee MIR. PL. 95 ; mi gold and fee SACR. 382.

feoht, *see* **feht. feole,** *see* **féle.**

fēon, v., *O.E.* fēon, fēoȝan, = *O.H.G.* fīen, *O.N.* fiā; *Goth.* fijan ; *hate* ; f(ē)ode (*pret.*) HOM. II. 209 ; wiþ freomen þu art ferli feid MISC. 86 ; said þe laverd to þe fede [fedde, qvede, fende] (*devil*) C. M. 12948 ; *comp.* ȝe-fēoȝen ; *deriv.* fēond.

fēond, sb., *O.E.* fēond, fiond, fȳnd, = *O.L.G.* fiond, *O.Fris.* fiand, fiund, *O.H.G.* fiant, fient, *Goth.* fijands, *mod. Eng.*

fiend ; *from* fēon ; *hater, enemy, fiend ;* ' *osor, hostis,*' FRAG. 5 ; R. S. vii ; SPEC. 93 ; L. H. R. 137 ; þe feond hine ferede LAȝ. 237 ; feond he wes þes kinges 11352 ; veond A. R. 66 ; feond [fend] C. L. 893 ; fend ORM. 12354 ; EGL. 564 ; SACR. 884 ; feend MAND. 87 ; voend REL. I. 243 ; fiend HOM. II. 191 ; viend, vend AYENB. I, 158 ; find TOR. 991 ; AUD. 2 ; **fēonde** (*dat.*) LAȝ. 25791 ; **fēond** (*pl.*) FRAG. 6 ; LAȝ. 10311 ; HOM. I. 13 ; fond 67 ; feondes MARH. 8 ; viendes AYENB. 79 ; **fēonda** (*gen. pl.*) HOM. I. 33 ; **fēonden** (*dat. pl.*) LAȝ. 373 ; feonde HOM. I. 13 ; P. L. S. viii. 111 ; viende ANGL. I. 20 ; *comp.* ȝe-fēond.

fēnd-, feend-līch, adj., *O.E.* fēondlīc ; *fiendish,* CH. C. T. *B* 783 ; fendli WICL. JAS. iii. 15 ; **fēondlīche** (*adv.*) stor LAȝ. 85 ; feondliche wunden 28579.

fēond-rǣs, sb., *hostile course,* LAȝ. 23960.

fēond-scaðe, sb., *O.E.* fēondsceaða, -scaða ; *monster,* LAȝ. 26039.

fēond-scipe, sb., *O.E.* fēondscipe, = *O.L.G.* fiundscipi, *O.H.G.* fiantscaf ; *enmity,* LAȝ. 7714.

fēond-slæht, sb., *slaughter of foes,* LAȝ. 16456.

fēond-ðēwes, sb. pl., *evil conduct* ; feonðewæs LAȝ. 579.

feor, sb., *O.Fr.* feur ; *price,* HOM. I. 287 ; A. R. 398 ; **feore** (*dat.*) C. L. 1091.

feor², sb., *O.E.* feorh (*gen.* feores), = *O.N.* fiör, *O.L.G., O.H.G.* ferah ; *life* : þa ich wes an vore (? = veore) fiftene ȝere LAȝ. 15700.

feor, adj. & adv., *O.E.* feorr, fiorr, (*remote*), = *O.L.G.* ferr, fer, *O.Fris.* fir, *O.H.G.* fer, ferro, ferr, *O.N.* fiarri, fiarr, *Goth.* fairra (πόῤῥω), *Lat.* porro, *Gr.* πόῤῥω ; *far,* C. L. 1304 ; ALIS. 3851 ; þe nes noht feor [for] from heom LAȝ. 543 ; he ferde to feor 1720 ; feor, veor A. R. 76, 104 ; feor ihc am iorne HORN (L.) 1146 ; feor, fer ORM. 6986, 8473 ; feor, veor, for, vor O. & N. 398, 653, 710 ; fer ' *procul*' PR. P. 156 ; HAV. 359 ; IW. 1815 ; A. P. i. 334 ; PR. C. 3895 ; fur P. L. S. i. 50 ; as fur forþ as we mowe BEK. 880 ; of feor *from afar* HOM. I. 247 ; A. R. 250 ; of feor iboren MISC. 98 ; of se feor KATH. 825 ; on fer *afar,* C. M. 12352 ; a fer ['*procul*'] WICL. GEN. xxii. 4 ; GOW. I. 314 ; a ver AYENB. 91 ; hi stode o fur P. L. S. xviii. 43 ; bi fer JOS. 592 ; be vee AYENB. 112 ; vram ver 164 ; in so feorre londe MISC. 85 ; of ferre londe ROB. 331 ; ferre londes WILL. 155 ; verre stedes AYENB. 204 ; **fir** [*O.E.* fyrr, = *O.N.* firr, *O.Fris.* firra, ferra] (*compar.*) MARH. 10 ; þes þe fir HOM. I. 103 ; fir & neor 137 ; fur A. R. 192 ; no fur BEK. 1172 ; fur and ner TREAT. 133 ; feor [fur] and ner O. & N. 923, 1657 ; ner and feorre 386 ; ferre WILL. 2613 ; ferre, firre A. P. ii. 97, 766 ; ac furre fleeþ P. L. S. xxxv. 132 ; verrore ROB. (W.) 2292 ; **ferrer** [*O.H.G.* ferror] LK. xxiv. 28 ; MAN. (H.) 44 ; IW. 1813 ; þanne walkede i ferrer PL. CR. 207; ferrer forth PR. C. 2329; ferror, ferrore ROB. (W.) 8458*, 9011*; **ferrest** [*cf. O.L.G.* ferrist, *O.H.G.* ferrost], (*superl.*) WILL. 2433; *ms.* farreste *corrected to* ferreste ROB. (W.) 3072*; it ŝtreccheþ ferrest ALIS. 4826; þe ferreste CH. C. T. *A* 494; þe ferreste ende KATH. 1565; verroste ROB. (W.) 3072.

fer-caster, sb., *a far-seeing person*: forwise & a fercaster D. TROY 3949.

fer-casting, sb., *foresight, suspicion*: of malice and of **fercastinge** (*dat.*) thei passen alle men undre hevene MAND. 219.

für-wei, sb., = *O.L.G.* ferweg; *long way*: er he a furwei com BEK. 1226.

fēor, *see* **fǣr.** **fēore,** *see* **fēre.**

fēorling, sb., ? = *M.H.G.* vierling; *farthing*; ferling P. S. 69.

feorme, veorme, sb., *O.E.* feorm; *repast, banquet,* LAȝ. 14426, 14432; ferme LK. xiv. 12; REL. I. 131; TREV. VII. 217; farme DREAM 1752.

feorren, v., *O.E.* feorran, fyrran, = *O.H.G.* firren, *O.N.* firra; *remove*; ferre PS. li. 7; verri (*printed* nerri) SHOR. 164; **verreþ** (*pres.*) AYENB. 178; firres (*? printed* firnes) A. R. 76*; iverred (*pple.*) AYENB. 240; *comp.* **a-ferren.**

feorren, ferrene, adv. & adj., *O.E.* feorran, feorrane, = *O.L.G.* ferran, ferrane, *O.H.G.* ferron, ferrana; *afar, from afar* [' *a longe* '], MAT. xxvi. 58, MK. v. 6; þa iseȝen heo nawiht feorren [vorre] a muchel fur smokien LAȝ. 25733; þa isehȝen heo feorre [a vorre] ænne selcuði ste(o)rre 17864; þa com þe ilke Belial ... feorren to JUL. 71; þet stod er veorren to A. R. 288; is feorrene ikumen 70; þe sonne and mone and mani(e) sterren bi easte ariseþ swiþe ferren SHOR. 137; figeren (*for* fieren) GEN. & EX. 3519; feorre, veorre, ferre O. & N. 327, 1322; furre REL. II. 175; of feorren icumen HOM. I. 249; of ferrene ifat KATH. 1296; ferne MAN. (F.) 5040; fro ferne HAV. 1864; ROB. 193; fro ferre GOW. III. 95; ferde on ferren [*earlier text* fyrlen] land LK. xix. 12; on ferrum GAW. 1575; C. M. 5751; o ferrum MIN. vii. 70; o ferrom MAND. 271; of ferrene [forne] londe 5328; of furrene londe P. L. S. xix. 20; of ferne londe HAV. 2031; ferrene kinges LAȝ. 3331; verrene londes MISC. 27; to ferne halwes CH. C. T. *A* 14.

feorsien, v., *O.E.* feorsian, fyrsian; *remove*; fersien HOM. II. 205; fersen ORM. 14198; firsin JUL. 16; **fürseð** (*pres.*) A. R. 76; *comp.* **a-fürsien.**

feorst, *see* **frist.**

fēorðe, adj., *O.E.* fēorða, = *O.L.G.* fiortho, *O.N.* fiorði, *O.H.G.* fiordo, fierdo; *fourth,* LAȝ. 3311; veorðe A. R. 228; feorþe MISC. 145; feorþe, ferþe ORM. 530, 4318; feorðe HOM. I. 105; forðe (*for* feorðe) HOM. I. 397;

fierðe 237; ferþe GOW. III. 113; CH. C. T. *D* 364; PERC. 1118; ferþe [fierþe] LANGL. *B* xix. 293; furþe BEK. 587; CHR. E. 429; ALIS. 52; FER. 1466; feurðe TREV. III. 19*; fourþe TREV. III. 13; MAND. 234.

fērðing, sb., *O.E.* fēorðing, fēorðung, = *O.N.* fiorðungr; *farthing,* LK. xii. 59; ferþing LANGL. *A* iv. 41; verþing ROB. 507; ferthing (*bit*) CH. C. T. *A* 134; ferþing noke H. S. 5812.

feoutē, *see* **feutē.**

fēower, card. num., *O.E.* fēower, feðer-, fyðer-, = *O.Fris.* fiower, fiuwer, *O.L.G.* fiuwar, fiar, fior, *O.H.G.* fior, fier, fiari, fiere *m. f.,* fioru *n., O.N.* fiorir, fiorar, fiögur, *Goth.* fidwōr (*dat.* fidwōrim), fidur-; *four*; feower, feouwer, fower feour, four LAȝ. 25, 194, 1902, 2092, 25395; feower, fouwer, fuwer HOM. I. 49, 159; fuwer, ? fuȝer II. 85, 211; fowwerr, fowre ORM. 507, 11275; vor, vour [fouwer] A. R. 160, 172; four & twenti ROB. 22; vour AYENB. 124; fowre JUL. 59; fowre, foure GEN. & EX. 559, 816; alle foure BEK. 2395.

fowre-cornered, adj., *four-cornered,* 'quadrangulus,' PR. P. 175; fourcornarde S. S. (Wr.) 139.

four-fald, adj., *O.E.* feowerfeald; *four-fold,* four-vald LAȝ. 1356*.

feðer-fōted, adj., *four-footed,* HOM. I. 43; fourfoted WILL. 191.

fēower-nōked, adj., *four-nooked,* LAȝ. 21999.

four-score, sb., *fourscore*: þe ȝer of grace a þousand ... and fourscore & sevene ROB. 382.

foure-square, adj., *square,* PR. P. 175.

fēower-tēne, card. num., *O.E.* feowertēnē, tȳne, = *Goth.* fidwōrtaihun; *fourteen,* LAȝ. 25675; fowertene HOM. I. 123; fourtene WILL. 1337; vourtene, fourtine ROB. 271, 279; fourte(n) niȝt BRD. 19; furten FL. & BL. 210.

four-tēoþe, ord., *O.E.* fēowertēoða; *fourteenth,* BRD. 15; fourteþe ROB. 300; fourtende GEN. & EX. 3143; fourtend PR. C. 4808.

fēower-ti, card. num., *O.E.* fēowertig, = *O. L.G.* fiwar-, fiartig, *Goth.* fidwōrtigjus; *forty,* SAX. CHR. 253; feowerti LAȝ. 31932; fowertiȝ ORM. 15594; feortig MAT. iv. 2; feortiȝ HOM. I. 227; fourti TREAT. 139.

fēower-tiȝþe, ord., *O.E.* fēowertigoða; *fortieth,* PROCL. 8; fower-, fuwertiðe HOM. II. 47, 113; furteohte HOM. I. 229; vourtaȝte AYENB. 13.

fer, *see* **far, feor, for,** *and* **fers.**

fēr, *see* **fǣr** *and* **fūr.**

feraunt, *see* **ferraunt. ferce,** *see* **fers.**

ferche, ferchs, *see* **fresh.**

fercost, sb., = **farcost**; *a kind of ship*: in floines and fercestez (*pl.*) D. ARTH. 742.

ferde, sb., *O.E.* ferd, fierd, fyrd, = *O.Fris.* ferd, *O.N.* ferð, *O.L.G.* fard (*gen.* ferdi), *O.H.G.* fart (*gen.* ferti); *from* **fǣren**; *military expedition, march,* HOM. II. 51; O. & N. 1668; HAV. 2535; WILL. 5326; ferde, verde, verd A. R.

74, 232, 264; ferde, færde LAȝ. 1413, 4429 (*misprinted* færð 1310); ferde, furde FER. 85, 95; ferd ORM. 14850; ROB. 19; WILL. 386; **ferde** (*dat.*) ALIS. 5579; þere ferde LAȝ. 4999; verde [varde] ROB. (W.) 7920; with grete furde hi come to Engelonde P. L. S. xviii. 10; **ferde** (*pl.*) LAȝ. 5146; **ferden** (*dat. pl.*) LAȝ. 5105; *comp.* **leod-, schip-, þrüm-ferde.**

ferd-wīte, sb., *O.E.* ferdwīte; *fine in lieu of military service*, ' *quite de morance de ost*,' REL. I. 33.

fērde, fērdful, fērdlaik, fērdli, fērdnes, *see* **færde.**

fērdi, *see* **færdi.**

ferdien, v., *O.E.* fyrdian; *from* **ferde**; *go on a military expedition*; furdien (*ms.* furðien) HOM. II. 189.

ferding, sb., *O.E.* fyrding; *expedition,* GEN. & EX. 842; furding FRAG. 2; fardung HOM. II. 189.

fēre, sb., *O.E.* (ge-)fēra; *from* **fáren**; *companion*, A. R. 114; ORM. 8608; O. & N. 223; GEN. & EX. 338; HAV. 1214; SHOR. 69; H. H. 67; MIRC 190; AMAD. (R.) xii; þou þeves fere GREG. 897; hir fere·[feere, fiere] fleeþ fro hire LANGL. B xvii. 318; feere S. S. (Wr.) 244; L. H. R. 147; HOCCL. vi. 23; **fēren** (*pl.*) LAȝ. 26135; KATH. 1252; HORN (L.) 19; veren A. R. 86; alle mine fere P. L. S. xv. 58; *comp.* **ȝe-, plei-, scōle-fēre.**

fēr-hēde, sb., *society, company*, ALIS. 3060; ROB. 138; verhede ROB. (W.) 2917.

fēr-rēden, sb., *O.E.* gefērrǣden; *society, company*, KATH. 703; HOM. II. 33; feorreden A. R. 106*; ferrede P. L. S. xvi. 145; RICH. 2278; foredene ROB. (W.) G. 59; **ferrǣdene** (*pl.*) LAȝ. 6020.

fēre, sb., *O.E.* gefēr *n.*; *from* **fáren**; *companionship, company* : al the fere þat him folowes D. TROY 1132; in fere *in company, together*, MISC. 210; CH. C. T. *B* 328; A. P. i. 89; M. ARTH. 3702; MIR. PL. 79; in fere, i fere WILL. 1429, 2180; in (i) fere MIRC 25; i fere TREAT. 136; DEGR. 244; twein kinges þere ævere weoren i fere LAȝ. 27435; in feer ROB. (W.) APPEN. xx. 268; in feere S. S. (Wr.) 263; HOCCL. iii. 40; i veore LAȝ. 3286.

fēre, adj., *O.E.* (ge-)fēre, = *O.Fris.* fēre, *O.N.* fœrr (fǣrr), *O.H.G.* (gi-)fuori; *able to go, in health*, HOM. I. 25; HORN (R.) 155; ALIS. 3510; þat makiþ þe sike þus fere FLOR. 2006; hal and fere ISUM. 601; hol & fere WILL. 1583; MAN. (F.) 9650; fere, feore LAȝ. 17618, 30601; feere FER. 517; GEN. & EX. 2812; *comp.* **un-fēre.**

fēre, sb., = *O.Fris.* fēre, *O.N.* fœri, *O.L.G.* (gi-)fōri, *O.H.G.* (gi-)fuori; *ability, power* : after þine fere ORM. 1251; þus is þes world of false fere MISC. 95; *comp.* **un-fēre.**

[**fēre**, sb., *O.N.* (tor-)fœra; *comp.* **tōr-fēre.**]

fēre, *see* **fēr** *and* **fǣr.**

fēren, v., *O.E.* fēran (*go*), = *O.Fris.* fēra, *O.N.* fœra, *O.L.G.* fōrien, *O.H.G.* fuoran (*carry*); *go, travel*, HOM. I. 11; færen LAȝ. 4401; **fērð**, fearð (*pres.*) HOM. I. 103; **fērde** (*pret.*) KATH. 5; ORM. 2661; MISC. 139; WILL. 30; S. S. (Wr.) 573; AMAD. (R.) xix; þa hu to wode ferde LAȝ. 302; for to lokien hu hit þer ferde HOM. I. 41; hwou Robert with here loverd ferde (*behaved*) HAV. 2411; and glad was he if ani wight wel ferde CH. TRO. iii. 1791; as a wod mon he ferde K. T. 98; ferde, verde FER. 3228, 3598; fierde S. A. L. 148; (heo) ferden FRAG. 7; LANGL. *B* ix. 143; forþ þei ferden ALIS. 181; furde BEK. 20; *comp.* **for-, ȝe-, þurh-fēren.**

fēren, v., *O.Fr.* aferir; *belong, be due* : þe fēres (*pres.*) lofsang PS. lxiv. 2.

fēren, *see* **færen.**

feret, *see* **foret.**

feri, sb., *O.N.* ferja, = *M.H.G.* vere; *from* **ferien**; *ferry*, '*pormeus*,' PR. P. 156.

feri-boot, sb., *ferry-boat*, PR. P. 156.

feriāge, sb., *ferryage*, '*feriagium, naulum*,' PR. P. 156.

ferialle, adj., *festival*, '*feriālis*,' PR. P. 156.

fērie, sb., *O.Fr.* fērie, *from Lat.* fēria; *festival, holiday*, LANGL. *B* xiii. 415; **fēries** (*pl.*) WICL. LEV. xxiii. 2, 4.

ferien, v., *O.E.* ferian, = *O.N.* ferja, *O.H.G.* ferren, *Goth.* farjan, *mod.Eng.* ferry; *carry, transport*; verien LAȝ. 32098; fere FLOR. 2086; fereð (*pres.*) HOM. I. 111; feriiþ P. L. S. v. (b) 42; fere (*imper.*) LAȝ. 26109; **ferede** [verede] (*pret.*) LAȝ. 237; **feried** (*pple.*) A. P. i. 945.

fēringe, *see* **fǣringe.**

ferkien, v., *O.E.* fercian; *carry, proceed*; ferke WILL. 3630; D. ARTH. 984; ferkeþ (*pres.*) GOW. III. 295; ferkis ALEX. 926; ferkand (*pple.*) D. ARTH. 1451; he ferked (*pret.*) over þe flor A. P. ii. 133; þei ferkid hem forþ DEP. R. iii. 90; ferkid D. TROY 1036.

fērlien, v., *from* **fǣrlīc**; *wonder*; fērliede (*pret.*) LANGL. *C* xiv. 173*.

fērling, *see* **fēorling.**

fermacie, sb., *Gr.* φαρμακεία; *medicine*: fermacies (*pl.*) of herbes CH. C. T. *A* 2713.

ferme, adj., *Fr.* ferme; *firm*, '*firmus*,' PR. P. 156; CH. BOET. iii. 6 (78).

ferme, sb., *Fr.* ferme; *farm, rent*, '*firma*,' PR. P. 156; CH. C. T. *A* 252b; he sette his tounes & his londes to ferme ROB. 378; he sette to ferme ['*ad firmam tradidit*'] TREV. VII. 413.

ferme, *see* **feorme.**

fermen, v., *O.Fr.* fermer, *from Lat.* firmāre; *strengthen*: to fermie love bitwene hem two FER. 2113; ferme MAN. (F.) 15507.

fermerere, sb., *the officer in charge of the infirmary*: our sextein and our fermerere that han ben trewe freres CH. C. T. *D* 1859; fermerer *'enfermarius'* CATH. 127; *see* enfermerere.

fermerie, sb., *O.Fr.* enfermerie, *M.Lat.* infirmāria (*infirmary*: if ʒe fare so in ʒowre fermorie LANGL. *B* xiii. 108.; rewfulnes salle make the fermorie HALLIW. 353; *'valetudinarium'* a fermeri VOC. 231; *'refectorium'* a ferñori 236.

fermien, v., *O.E.* feormian, fearmian; *cleanse*; ferme HALLIW. 353; he fermeð (*ms.* fermed) (*pres.*) his bernes flore LK. iii. 17; *comp.* a-fermien.

[**fermere,** sb., *cleanser*; *comp.* **gong-firmar.**]

fermin, v., *from* ferme; *farm, offer for hire,* *'firmo, vel ad firmam accipio,'* PR. P. 157.

fermisoun, sb., *med.Lat.* firmātiō; *close time* (*when animals are not allowed to be killed*); in fermisoun time GAW. 1156; flesch fluriste of fermisone (*venison*) with frumentee noble D. ARTH. 180; faire bi fermesōnes (*pl.*) by frithis and felles ANT. ARTH. i.

fermour, sb., *farmer, bailiff*: a fermour, eithir a baili WICL. LK. xvi. 1; a fermour as keper of a toun S. W. I. 22; fermour CH. L. G. W. PROL. 376.

fern, adj. & adv., = *O.L.G.* fern, *Goth.* fairnis (παλαιός); *past*: þe snou of ferne yere CH. TRO. v. 1176; i dar nat speke a word of ferne yeer HOCCL. i. 423; *see* **fürn.**

fern, *see* **fearn.**

ferne, ferren, *see* **feorren.**

ferrai, sb., *cf. Scot.* foray; *plundering*: som of ferrai TOWNL. 310; *see* **forrai.**

ferraunt, adj., *O.Fr.* ferrant; *? iron-gray*: on a sted ferraunt DEGR. 371; ferant, feraunte, feraunt D. ARTH. 22591, 1811, 2451; ferauns (*sb.*) DEGR. 1245.

ferriour, sb., *? O.Fr.* fourrageur; *plunderer*: ferriours (*pl.*) fers unto þe firthe ridez D. ARTH. 2753; *see* **forrier.**

ferrum, *see under* **feorren.**

fers, adj., *O.Fr.* fiers, fier; *fierce, wild,* BEK. 5; A. P. ii. 101, 217; so fers & so breme WILL. 3641; fers [feers, fiers] CH. C. T. *A* 1598; a feers man ['*ferus homo*'] WICL. GEN. xvi. 12; fers, fier, feer FER. 351, 409, 2073; wiþ ferse folke EGL. 42; ferce ROB. (W.) 5717; fierse bestes LANGL. *B* xv. 300.

fers-līche [fresliche], adv., *fiercely,* P. L. S. xxv. 32; fersli WILL. 1190, 1766.

fersnesse, sb., *fierceness,* WICL. JUD. iii. 1; D. RICH. ii. 7.

fers², sb., *O.Fr.* fierce; *the queen at chess,* CH. D. BL. 654; ferses (*pl.*) 723.

fers, *see* **fresch** *and* **vers.**

fersch, *see* **fresch.** **fersien,** *see* **feorsien.**

ferst, *see* **frist** *and* **fürst.**

fert, sb., = *O.N.* fretr, *O.H.G.* firz; *peditum,* VOC. 209; REL. I. 260.

ferten, v., = *O.N.* freta, *O.H.G.* firzen, *Gr.* πέρδειν; *break wind*; bucke verteþ (*pres.*) A. S. 4.

ferteren, v., *from* fertre, sb.; *deposit*; fertered (*pret.*) thaim at a nunrie M. H. 143.

fertre, sb., *O.Fr.* fiertre, *Lat.* feretrum; *bier, shrine, tomb*: the fertre of alabastre where the bones liʒn MAND. 60; thei setteñ him upon a blak fertre 225.

ferþe, ferþing, *see* **feorðe.**

fervent, adj., *Fr.* fervent; *fervent,* AYENB. 121.

fervent-līche, adv., *fervently*: ferventliche him preie CH. TRO. iv. 1356.

fervour, sb., *O.Fr.* fervour; *fervour,* WICL. DEUT. xxix. 20.

fesant, sb., *O.Fr.* faisan, faisant; *pheasant,* '*phasianus,*' VOC. 189; fesaunt PR. P. 158; CH. P. F. 357; fesauns (*pl.*) WILL. 183.

fēse, sb., *hurry*; vese [*glossed* '*impetus*'] CH. C. T. *A* 1985.

fēsin, v., *O.E.* fēsian, = **fūsen**; *from* **fūs** *drive,* PR. P. 158; fese HALLIW. 354; fese a wei þe cat BER. 351; fēseþ, vēseþ (*pres.*) TREV. I. 339, II. 17; fēsed (*pple.*) MAN. (H.) 192, 274; D. ARTH. 2842; fesid P. R. L. P. 198; ifesid into helle P. L. S. ii. 172; *see* **fūsen.**

fest, *see* **fæst.** **fest,** *see* **fūst.**

feste, sb., *O.Fr.* feste, *from Lat.* fēstum; *feast,* AYENB. 156; þat feste HAV. 2354; þe feste HORN (L.) 521; feste (*pl.*) A. R. 222.

festinge, sb., *feasting,* PR. P. 158.

fest-lich, adj., = *Ger.* festlich; *festive,* CH. C. T. *F* 281.

festeien, v., *O.Fr.* festier; *feast*: to bringe to me to festeie WICL. WISD. viii. 9; his lordes festeing (*pple.*) CH. C. T. *F* 345.

festen, *see* **fæsten** *and* **fasten.**

festin, v., *O.Fr.* fester; *feast, make a feast,* PR. P. 158; he festeþ (*pres.*) hem CH. C. T. *A* 2193.

festival, sb. & adj., *O.Fr.* festival; *festival,* A. P. ii. 136; festifulle (*pl.*) daies MAND. 137, 208.

festival-lī, adv., *gaily*: how a man scornede festivali and mirili swiche vanite CH. BOET. iii. 7 (59).

festnen, *see* **fæstnien.**

festour, sb., *feaster*: never festour fedde bett L. H. R. 220.

festre [fester], sb., *suppuration,* C. M. 11824; fester N. P. 65; *? comp.* **gúte-festre.**

festrin, v., *fester, suppurate*; feestrin PR. P. 158; so festred (*pple.*) ben his woundes LANGL. *B* xvii. 92.

festu, sb., *O.Fr.* festu; *bit of straw,* LANGL. *B* x. 278; WICL. MAT. vii. 3.

fet, *see* **fat** *and* **fait.**

fet, adj., *O.Fr.* fait, *from Lat.* factus; *skilful, handsome*: so she goth in patens faire ¦and féte (*pl.*) CT. LOVE 1087; ye fele ther fete so fete ar thai HALLIW. 354.

fet-lī, adv., *from O.Fr.* fait ; *gracefully* : he þat fetli in face fettled alle eres A. P. ii. 585 ; fetli him kissed GAW. 1757.

'fetel, fetil, sb., *O.E.* fetel, = *O.N.* fetill, *O.H.G.* fezel ; *girdle, fetter,* M. H. 140 ; **fetteles** (*? printed* festeles) (*pl.*) PS. cxlix. 8 ; *comp.* **fōt-fetel.**

feten, v., *O.E.* fetian, fettan, = *O.N.* feta, *M.Du.* vaten, vatten, *M.H.G.* vazzen ; *fetch,* GEN. & EX. 2744 ; to fetten watur LEB. JES. 310 ; i shal þe fete bred an chese HAV. 642 ; fette Fer. 1260 ; JOS. 167 ; GOW. II. 198 ; K. T. 363 ; OCTOV. (W.) 549 ; A. P. ii. 802 ; to fette me to helle MAP 338 ; fette, vette ROB. 147, 325 ; his fo fetteþ him in uche ende C. L. 430 ; **fette** (*pret.*) HOM. II. 61 ; FL. & BL. 790 ; GEN. & EX. 1535 ; LANGL. *B* ii. 64 ; fette wif *took a wife* MAT. xxii. 25 ; and mete inouȝ hem fette BEK. 888 ; he vette feþerhome HOM. I. 81 ; fætte, vætte [vette] LAȝ. 9177, 29673 ; fette [fatte] TREV. III. 259 ; fatte KATH. 720 ; vatte HOM. I. 83 ; fet (*pple.*) REL. I. 128 ; MAP 339 ; LANGL. (Wr.) 14569 ; CH. C. T. *A* 819 ; *comp.* ȝe-fetten.

feter, sb., *O.E.* fetor, feotor *f., cf. O.L.G.* feter, *O.N.* fiōturr *m., O.H.G.* fezera *f. ; fetter* ; fetir PR. P. 159 ; **fetteres** (*pl.*) HAV. 82 ; BEK. 15 ; GAM. 384 ; veteres AYENB. 128 ; vetres FER. 1313.

fetir-lok, sb., *chain-lock,* 'sera compeditalis,' PR. P. 159 ; fetirlok of gold W. & I. 37.

feterien, v., *O.E.* (ge-)feterian, fetrian, = *O.Fris.* fitera, *O.N.* fiōtra, *O.H.G.* (ka-)fezarōn ; *fetter* ; feterie LANGL. *C* iii. 212 ; feteren GAM. 384 ; binde and fetere HAV. 2758 ; **fetered** (*pple.*) MAN. (F.) 8568 ; fettred LANGL. *B* xvii. 110 ; *comp.* ȝe-, un-feterien.

fētfolk, *see under* fōt.

fēþ, *see* fei.

fetis, adj., *O.Fr.* fetis, faitis, *from Lat.* factīcius ; *skilfully made, neat; skilful* ; WILL. 126 ; FER. 5883 ; CH. C. T. *A* 157.

fetis-līche, adv., *skilfully, neatly,* WILL. 96 ; fetiseli A. P. ii. 1462 ; fetisli R. R. 577 ; fetousli CH. C. T. *A* 273.

fetlak, sb., *? = Ger.* fiszlach, -loch ; *fetlock* ; **feetlakes** (*pl.*) RICH. 5816 ; fitlokes AR. & MER. 5902.

fetlen, v., *fettle, make ready* ; **fetled** (*pple.*) GAW. 656 ; feteled D. ARTH. 2149 ; fetild TOWNL. 309.

fētles, sb., *O.E.* fætels (LEECHD. III. 16) ; *vessel,* ORM. 14450 ; þis bruchele vetles [fetles] A. R. 164 ; a strong vetles MARG. 207 ; fetles, feteles GEN. & EX. 561, 1225 ; veteles LEB. JES. 314.

fetten, *see* fatten *and* feten.

fetūre, sb., *O.Fr.* feture, faiture ; *feature,* P. L. S. xxv. 41 ; **fetūres** (*pl.*) WILL. 857 ; GAW. 145.

feðer, *see* fēower.

feðere, sb., *O.E.* feðer, fiðer *f. n., cf. O.N.* fiōðr, *O.L.G.* fethera, *O.H.G.* federa *f., Gr.* πτέρον ; *feather,* LK. xvi. 6 ; feþere LAȝ. 49 ; feþer O. & N. 1688 ; feþer [feþere] CH. C. T. *A* 2144 ; fedir PR. P. 152 ; VOC. 221 ; fiþer A. P. ii. 1026 ; fiþera (*pl.*) MAT. xxiii. 37 ; veðeren A. R. 140 ; veþeren AYENB. 270.

feþer-bed, sb., *O.E.* feðerbed ; *feather-bed,* A. P. iii. 158.

feðer-home [feþerhame], sb., *O.E.* feðerhama, -homa ; *covering of feathers, winged body,* LAȝ. 2874 ; feþerham [feþerhome] ROB. (W.) E 2 ; feþerhome HOM. I. 81.

feðeren, adj., = *M.H.G.* vederen ; *made of feathers* : veðerene (*pl.*) balles LAȝ. 17443.

feðerien, v., *O.E.* (ge-)fiðerian ; *feather* : he feþered (*pret.*) his feet and his hondes TREV. VII. 223 ; feþered (*pple.*) WILL. 191 ; *comp.* ȝe-fiðerien.

feðme, *see* faðme.

fēðren, v., *from* fōðer ; *charge, load* ; veðren A. R. 140 ; iveðred [ifeðered] (*pple.*) þet is icharged 204.

feuer, *see* fefre.   feunesse, *see under* fēa.

feurþe, *see* feorþe.

feutē, sb., *O.Fr.* feeltē ; *fealty* : alle deden him feute ALIS. 2910 ; '*omagium, fidelitas*' PR. P. 159 ; feoute ROB. (W.) APP. xx. 530*.

feute, *see* fuite.

feuter, sb., *Fr.* feutre ; *? a rest for the spear,* WILL. 3437 ; feutire D. ARTH. 1366.

feverer, sb., *February* : þe monþe of feuerer ROB. 399 ; feverȝere D. ARTH. 436 ; feverel ALIS. 57 ; februar GOW. III. 126 ; februarie LIDG. M P. 2 ; i þe sixtenðe dei of **feoverreres** (*gen.*) moneð JUL. 79.

fevre, *see* fefre.   fēwe, *see* fēa.

fī, interj., = *M.H.G., O.Fr.* fi ; *fie,* FER. 1578 ; WILL. 481 ; S. S. (Web.) 1599 ; K. T. 612 ; CH. C. T. *A* 1773.

fiaunce, sb., *O.Fr.* fiance ; *affiance, trust* ; fiaunce AYENB. 164 ; nor in her ieftis have fiaunce R. R. 5482.

fīblelīche, *see* fēblelīche *under* fēble.

fibreches, sb. pl., *? contrivances* : the fibreches in forcerers LANGL. *B* x. 211.

ficche, *see* fecche.

ficchen, v., *O.Fr.* ficher ; *fix,* CH. BOET. iii. 9 (88) ; fichede (*pret.*) D. ARTH. 4239.

ficching, sb., *print, impress* : the ficching of nailis WICL. JOHN xx. 25.

ficheu, sb., *? O.Fr.* fissel ; *fitchew, polecat* ; fitchewes (*pl.*) PL. CR. 295.

fichs, *see* fisch.   fieble, *see* fēble.

fīen, *see* fēȝen.   field, *see* fēld.

fíen, v., *see* defíen ; *digest* : fiin or defíin mete and drink '*digero*' PR. P. 159 ; þi flesche foode þe wormes wol fie P. L. S. xxix. 65.

fīend, *see* **fēond.** **fier,** *see* **fūr.** **fierðe,** *see* **fēorðe.**

fiers, *see* **fers.**

fīf, card. num., *O.E.* fīf, fīfe, = *O.L.G.* fīf, fīvi, *O. Fris.*fīf, *O.H.G.* fimf, finf, fimfi, finfi, *Goth.* fimf, *O.N.* fimm ; *five,* K. T. 1017 ; A. P. i. 848 ; fif ʒere LAʒ. 11675 ; mid fif hundred cnihten 1425 ; fif moneþ ORM. 233 ; þe fif & þrittude KATH. 43 ; vif A. R. 26 ; þurh fife grimme wundes ORM. 1443 ; betere is on almesse before þanne ben after vive (*ms.* vyue) P. L. S. viii. 14 ; vif pors (*cinque ports*) ROB. (W.) 7888 ; cnihtes sunen vive LAʒ. 13993 ; þe vormeste vive A. R. 22 ; vive þe verste ROB. (W.) 10671.

 fīf-feald, adj., = *O.H.G.* finffalt ; *fivefold,* HOM. II. 35 ; fiffald ORM. 7836.

 fīf-lēf, adj., *O.E.* fīflēaf ; *fiveleaf, 'quinquefolium,'* REL. I. 36.

 fiff-sum, adj., *five in all,* BARB. vi. 149.

 fīf-tēne, card. num., *O.E.* fīftēne, -tȳne ; *fifteen,* LAʒ. 301 ; ROB. 70.

 vīf-tēoþe, ord. num., *O.E.* fiftēoða ; *fifteenth,* P. L. S. xii. 142 ; fifteþe DEGR. 1869 ; fiftened ORM. 9170 ; fiftend PR. C. 4812.

 fīf-teoʒaðe, ord. num., *O.E.* fiftigoða ; *fiftieth,* HOM. I. 89 ; fiftugeðe, -tuðe II. 117.

 fīf-tiʒ, card. num., *O.E.* fiftig ; *fifty* ; (*ms.* fiftiʒ) ORM. 8102 ; fifti HOM. I. 225 ; LAʒ. 1285.

fifte, ord. num., *O.E.* fifta, = *O.Fris.* fifta, *O.H. G.* finfto, fimfto, *O.N.* fimti ; *fifth,* HOM. I. 103 ; LAʒ. 6532 ; ORM. 4320 ; GEN. & EX. 159 ; WILL. 1322 ; LANGL. *B* xiv. 298 ; PR. C. 3088 ; vifte A. R. 110 ; fifþe LAʒ. 6532 ; MIRC 157.

fïgarde, sb., *Lat.* pygargus ; *? bison, ? roebuck,* WICL. DEUT. xiv. 5.

fīge, sb., *O.Fr.* fige ; *fig,* MAND. 50 ; **fīges** (*pl.*) A. R. 150.

 fīge-trē, sb., *fig-tree,* PR. P. 159 ; figetree WICL. JOHN i. 48.

fïger, sb., *O.Fr.* figier ; *fig-tree,* A. R. 148 ; LEB. JES. 842.

 figer-trē, sb., *fig-tree,* TRIST. 3082.

fïger, *see* **fūr.** **fighten,** *see* **fehten.**

fīʒti, adj., *valiant in fight* : migti men and figti GEN. & EX. 546.

figūren, v., *from* **figūre,** sb. ; *make* ; gemini which is figured (*pple.*) redeli lich to two twinnes Gow. III. 119 ; as felon figured is a scorpion III. 122 ; the faireste figured folde that figurede was ever D. ARTH. 2151.

figūre, sb., *O.Fr.* figūre ; *figure,* AYENB. 234 ; WILL. 447 ; CH. C. T. *B* 3412 ; **figūres** (*pl.*) A. R. 214 ; GEN. & EX. 1006.

fïht, fïhten, *see* **feht, fehten.**

fïhsscheþ, *see* **fischoð.**

fïkel, adj., *O.E.* ficol, *mod.Eng.* fickle ; *deceitful,* HAV. 1210 ; ALIS. 2661 ; LANGL. *A* iii. 117 ; TRANS. XVIII. 22 ; A. P. iii. 283 ; PR. C. 1088 ; fikel and fals HOM. I. 185 ; vikel MISC. 93 ; fikil S. S. (Wr.) 985 ; þis fïkele

world H. M. 7 ; **fïkele** (*pl.*) LEG. 15 ; fikele & swikele reades A. R. 268.

 fïkel-līche, adv., *deceitfully* : fike(l)li þai dide [*'dolose agebant'*] PS. V. 11.

fïkelien, v., *deceive, flatter* ; fikele [*v. r.* vikeli] ROB. 31 ; fekel ROB. (W.) 713* ; **vikeleð** (*pres.*) A. R. 84, 198 ; **fïkelinde** (*pple.*) HOM. I. 253 ; fïkeled (*pret.*) TREV. V. 169.

fïkelare, vikelare, sb., *deceiver, flatterer,* A. R. 84, 86.

fïkelunge, sb., *flattery,* A. R. 82 ; vikelinge [*v. r.* fekeling, fikeling] ROB. (W.) 711.

fīken, v., = *O.Swed.* fikja (RIETZ 136), *fidget ; hurry about ; trifle ; flatter* ; fikin aboute *'vagor'* PR. P. 162 ; a wai fike RICH. 4749 ; nulle we nout mit te fike (*? trifle*) HICKES I. 225 ; fike SPEC. 46 ; þu **fīkest** (*pres.*) MARH. 13 ; fikeð mid te heaved & stingeð mid te teile A. R. 206 ; fikeð & fondeð al his migt REL. I. 223 ; **vīkiinde** (*pple.*) A. R. 256 ; fīked (*pret.*) GAW. 2274.

fïkenung, sb., *deceit* ; mid fikenunge (*dat.*) fearð HOM. I. 103 ; *cf.* **fïkelunge.**

fīl, *see* **vīl.**

fīlateri, sb., *Gr.* φυλακτήριον ; *phylactery* : **fīlateries** (*pl.*) WICL. MAT. xxiii. 5.

filberd, sb., *filbert* ; filberde, *'fillum,'* PR. P. 160 ; philliberd Gow. II. 30 ; **filbirdes** (*pl.*) hanging to the ground SQ. L. DEG. 37.

 filberd-trē, sb., *filbert-tree,* PR. P. 160.

fïld, *see* **fēld.**

fïldor, sb., *O.Fr.* fil d'or ; *gold thread* : rich riverez as fildor fin her bonkes brent A. P. i. 106 ; folden in with **fïldore** aboute þe faire grene GAW. 187.

fīle, sb., *O.E.* fēol, = *O.H.G.* fīla, fihala, *? O.N.* þēl ; *file, 'lima,'* PR. P. 160 ; CH. C. T. *A* 2508 ; vile, file A. R. 184.

 fīle-feste, adv., *? fast, securely* : ibunden vile veste [*v. r.* hetefaste, heleveste] mid te holie monnes beoden A. R. 244.

fīle², sb., *worthless person,* MAN. (H.) 95 ; MIN. vii. 139 ; REL. I. 194 ; fule file HAV. 2499 ; fals file HALLIW. 356 ; file (*concubine*) LANGL. *B* v. 60 ; **files** (*pl.*) MAN. (F.) 15215.

fīlen, *see* **fūlen.**

filet, sb., *O.Fr.* filet ; *fillet,* P. S. 154 ; WICL. S. SOL. iv. 3 ; CH. C. T. *A* 3243.

fïlʒen, *see* **folʒen.**

fīlin, v., = *O.H.G.* fīlōn ; *file, 'limo,'* PR. P. 160 ; file CH. P. F. 212 ; **vīleð** (*pres.*) A. R. 184 ; **fïlede** (*pret.*) ED. 1421 ; fīled (*pple.*) GAW. 2225 ; iviled A. R. 284.

filiole, sb., *see* **feliole.**

fīling, *see under* **fūlen.**

fille, sb., *O.E.* fille (VOC. 79 ; LEECHD. III. 34) ; *chervil* ; the fenil ant the fille SPEC. 44 ; worþ a fille ROB. 297 ; BEK. 946.

fille, *see* **fülle.** **fillen,** *see* **fellen** *and* **füllen.**

filme, sb., *O.E.* fylmen; *film,* '*folliculus,*' PR. P. 160.

filor, sb., ? *grindstone*: a denez axe nue dizt . . . filed in a filor GAW. 2225; filoure, of barbours crafte '*acutecula, filarium*' PR. P. 860.

filosofe, sb., *Fr.* philosophe; *philosopher,* AYENB. 120; philosophre CH. C. T. *A* 297.

filosofie, sb., *Fr.* philosophie; *philosophy*; filozofie AYENB. 164; philosophie ROB. 130; SHOR. 137; CH. BOET. i. 3 (10).

filsnen, v., *lurk*: þar filsnez (*pres.*) þat fende D. ARTH. 881.

filsten, *see* fülsten. **filtren,** *see* feltren.

fīlþe *see* fülðe.

fīn, sb., *O.E.* fīn; *end,* MISC. 144; HAV. 22; FL. & BL. 441; ROB. 91; mad his fin (*made his peace*) MAN. (F.) 15966.

fīn, adj., *O.Fr.* fīn, *from* ·*M.H.G.* fīn, vīn, *cf. O.N.* finn; *fine,* CHR. E. 628; SPEC. 52; ALIS. 2657; OCTOV. (W.) 1754; fīneste (*superl.*) LANGL. *B* ii. 9.

fīn-līche, adv., *finely,* C. L. 1132; WILL. 768.

final, adj., *Fr.* final; *final,* PR. C. 3990.

final-lī, adv., *finally*: MAND. 38; CH. C. T. *B* 1781.

finch, sb., *O.E.* finc, *cf. O.H.G.* finco; *finch,* CH. C. T. *A* 652; *comp.* **chaf-, góld-finch.**

[**finde,** adj., *O.E.* (ēað-)fynde,=*O.N.* (tor-)fyndr; *comp.* **arveð-finde.**]

finden, v., *O.E.* findan, = *O.L.G.* findan, fīthan, *O.H.G.* findan, *O.Fris.* finda, *Goth.* finþan, *O.N.* finna; *find,* FRAG. 6; HOM. I. 107; LAʒ. 1232; ORM. 383; AN. LIT. 3; finde O. & N. 470; hi heten him sikernisse finde BEK. 800; to vinde hem mete ROB. 378; þu finst (*pres.*) MAT. xvii. 27; vindeð LAʒ. 21780; fint HOM. II. 87; vint O. & N. 696; AYENB. 74; hi vindeþ AYENB. 120; **fand** (*pret.*) LK. xiii. 6; ORM. 13; TRIST. 787; Iw. 1705; ISUM. 374; vand AYENB. 57; fond LAʒ. 11362; GEN. & EX. 440; HORN (L.) 597; FER. 4849; WILL. 293; LANGL. (Wr.) 9926; CH. C. T. *A* 653; MAND. 220; ich him fond ligge BRD. 27; vond A. R. 258; fuṅde MAT. xx. 6; HOM. I. 199; LAʒ. 22; MISC. 42; founde REL. II. 274; BEK. 168; where he it founde WILL. 396 (*see* ENG. STUD. III. 472); sunne ne fond(e) [foundest] þou H. H. 77; fundest JUL. 28; funden LAʒ. 277; ORM. 3398; GEN. & EX. 2948; HAV. 602; funde HORN (L.) 882; fonden MISC. 33; fonde BRD. 6; founden MAND. 17; founde P. L. S. xviii. 12; WILL. 2094; wen hi þen deþ founde LEG. 66; **funde** (*subj.*) P. L. S. viii. 34; ORM. 833; founde WILL. 2176; **funden** (*pple.*) HOM. II. 161; IW. 1734; PR. C. 4608; AV. ARTH. lvi; funden, funde HAV. 1427, 2376; founden CH. TRO.

iv. 944; founde WILL. 394; fonden PERC. 1902; *comp.* **a-, ʒe-finden**; *deriv.* **fand, fandien, fundien.**

findere, sb., *finder*; findare '*inventor*' PR. P. 161; beholdeth me therof no finder, her bokes ben mi shewer ALIS. 4794.

findinge, sb., = *O.H.G.* (ar-)findunga; *finding,* '*inventio,*' PR. P. 161; finding WICL. WISD. xiv. 12; new **findinges** (*pl.*) of vanite PR. C. 1556; findingis WICL. DEUT. xxviii. 20.

findiʒ, *see* **fŭndi.**

findles, fŭndles, sb.,=*Dan.* findelse; *invention,* A. R. 6, 8, 208.

findling, *see* **fŭndeling.**

fīnen, finissen, v., *O.Fr.* finer, finir; *finish*; fine HORN (L.) 262; fīnede (*pret.*) ROB. 140; fined A. P. ii. 450; finischid (*pple.*) WILL. 5398.

fīning, sb., *death, end of life*: God geve alle good fining ALIS. 8015.

fīnen², v., *from* fīn; *pay as a fine*: ROB. (W.) 10520; hii fīnede (*pret.*) 9521.

fīnen², v.,=*M.H.G.* fīnen, *O.Fr.* (a-)finer; *refine*; fine PR. C. 4913; clenzeþ and fīneþ (*pres.*) þet gold AYENB. 106.

finger, sb., *O.E.* finger, = *O.H.G.* finger, fingar, *O.N.* fingr, *Goth.* figgrs; *finger,* C. L. 1104; vingre (*dat.*) AYENB. 5; **fingres** (*pl.*) BEK. 1194; vingres A. R. 290; fingre MK. vii. 33; **fingren** (*dat. pl.*) LAʒ. 49.

finger-neil, sb., *finger-nail,* HOM. I. 281.

fingirling, sb.,=*Ger.* fingerling; *thimble,* '*digitabulum,*' PR. P. 161.

finisment, sb., *end, finish*: þe forme to be finisment foldes ful selden GAW. 499.

finissen, *see* **fīnen.** **finkel,** *see* **fenel.**

finne, sb., *O.E.* finn,=*M.Du.* vinne, *Lat.* pinna; *fin,* VOC. 221; PR. P. 161; A. P. ii. 531; **finnes** (*pl.*) MARH. 9; **vinnes** (*r. w.* inne) ALIS. 6591.

finned, adj., *having fins*: of þe finnede (*pl.*) fihcs our fode to lacche ALEX.² (Sk.) 298.

fiole, sb., *O.Fr.* fiole, phiole, *Lat.* phiala; *phial, viol,* WICL. NUM. vii. 13; viole JOS. 290.

fīr, fīren, fīri, *see* **fŭr, fŭren, fŭri.**

firmament, sb., *Fr.* firmament; *firmament,* GEN. & EX. 95; ROB. 112.

firme, firmest, *see* **forme, formest.**

firmþe, sb., *O.E.* fyrmð, frymð; *entertainment*: frimtha ANGL. VII. 220.

firnen, v., ?=*M.L.G.* vernen; *remove*; **firnes** (*v. r.* furseð) (*pres.*) A. R. 76.

firre, sb., *O.E.* furh,=*O.H.G.* forha, foraha, *O.N.* fura, fyra; *fir,* '*abies,*' VOC. 228; CH. C. T. *A* 2921; **firris** (*pl.*) WICL. 4 KINGS xix. 23.

firre-trē, sb., *fir-tree,* '*abies,*' VOC. 192.

firren, adj., *made of fir*: a firrene wowe HAV. 2078.

firren, *see* **feorren.**

firs, sb., *O.E.* fyrs (*rhamnus*) VOC. 80; fyrsas (*rubos*) GREIN I. 364; *furze*; firse ['*saliuncula*'] WICL. IS. lv. 13; with a wispe of firsen [firses] LANGL. *A* v. 195.

firsien, *see* **feorsien.**

first, sb., *O.E.* first, fyrst, = *O.H.G.* first; *ceiling*, '*laquear*,' FRAG. 4; firste, virste MISC. 178, 179 (ST. COD. 100).

first, *see* **frist, fürst. firþ,** *see* **friþ** [2].

firþer, firþren, *see* **furðer, furðrien.**

fis, *see* **fisch.**

fīs, sb., = *Swed.* fīs, *Dan.* fiis; *flatus ventris*, VOC. 209; *see* **fīst.**

fisch, sb., *O.E.* fisc, = *O.L.G.*, *O.H.G.* fisc, *O.N.* fiskr, *Goth.* fisks, *Lat.* piscis; *fish*, BRD. 8; ase fisch hal JUL. 59; fisch ... whon þe worm he swoleweþ C. L. 1129; þar is fuȝel þar is fisc [fis] LAȝ. 1235; inne þisse venne is fisc unimete 21964; fis HOM. I. 51; GEN. & EX. 162; fis, fihs O. & N. 1007; viss ROB. 265; þane viss AYENB. 50; **fisches** (*gen.*) BRD. 11; **fisces,** fisches (*pl.*) LAȝ. 2009, 21324; fisches MARH. 9; SHOR. 146; fiṣhes LANGL. *B* xv. 581; fiskes ORM. 13297; fisses HICKES I. 223; GEN. & EX. 2945; fisches, fische TREV. III. 67, VIH. 81; **fiscen** (*dat. pl.*) LAȝ. 21743; fische MAND. 103; *comp.* **sæ-, schel-, stok-fisch.**

    fisc-browe, sb., *a fish sauce,* '*garum*,' VOC. 241.

    fisshe-dai, sb., *fast-day*: hasteletes on fisshedai L. C. C. 37; fihshedai P. S. 151.

    fisch-hook, sb., *fish-hook,* TREV. II. 295.

    fisch-leep, sb., *fish-basket,* '*nassa*,' PR. P. 163.

    fi(s)ch-manger, sb., *fishmonger,* VOC. 212.

    fish-net, sb., *O.E.* fiscnett; *fishing-net,* HOM. II. 175.

    fisch-pōl, sb., *O.E.* fiscpōl; *fish-pond*; **fischpōle** (*dat.*) S. S. (Wr.) 883.

    fisch-ponde, sb., *fish-pond,* PR. P. 163.

    fisch-wer, sb., *O.E.* fiscwer (LK. v. 4); *fish-weir*; **fischweris** (*pl.*) TREV. I. 423.

fischin, v., *O.E.* fiscian, = *O.N.* fiskja, *O.H.G.* fiscōn, *Goth.* fiskōn; *fish, 'piscor,*' PR. P. 163; fisken ORM. 13297; fissen HORN (L.) 1136; **fischede** (*pret.*) P. L. S. xx. 3.

    fishere, sb., *O.E.* fiscere, = *O.H.G.* fiscāri; *fisher,* HORN (R.) 1134; HAV. 2230; fiscære LAȝ. 9083; fischare PR. P. 162; fixere HOM. I. 97; vissere AYENB. 50; **fischeris** (*pl.*) WICL. IS. xix. 8; vissares ROB. (W.) 5359.

      fisher-bōt, sb., *fishing-boat,* S. S. (Web.) 3568.

    fischinge, sb., *fishing,* PR. P. 163; HORN (H.) 676; þat beþ ago ... a vissinge ROB. 265; fuschinge ROB. (W.) 2198*.

fischoð, sb., *O.E.* fiscað, fiscoð; *act of fishing*: gan on fissoð JOHN xxi. 3; (h)is men were iwend a visseþ ROB. 264; a fihsscheþ [visseþ] ROB. (W.) 5341*.

fisicien, sb., *O.Fr.* fisicien, phisicien; *physician,* LANGL. *A* vii. 170; fizicien AYENB. 174; phisicien CH. D. BL. 39; **fisiciens** (*pl.*) A. R. 370; P. L. S. xxi. 7.

fisike, sb., *O.Fr.* fisique, phisique; *physic,* AYENB. 53; S. S. (Web.) 186; phisik CH. C. T. *B* 4028.

fisiologet, sb., *Physiologus*: ðus it is on boke set, ðat man clepeð fisiologet MISC. 10.

fisk, *see* **fisch.**

fisken, v., *? cf. Swed.* fjeska; *scamper about*: fiskin aboute '*vagor*' PR. P. 162; **fiskeþ** (*pres.*) ... aboute LANGL. *C* x. 153; he fiskez hem bifore GAW. 1704.

fisonomie, sb., *O.Fr.* phisonomie; *physiognomy*: the childe couthe of fisenamie S. S. (Wr.) 1072; thou schuldest bi phisonomi be shapen to that maladi of love drunk GOW. III. 5; the faireste of fissnanni D. ARTH. 3332; fisnamie 1114.

fist, sb., = *M.Du.* vijst; fiist PR. P. 163.

fist, *see* **fūst.**

fīstin, v., = *Ger.* fisten, *M.Du.* vijsten, *mod. Eng.* foist; *break wind silently,* PR. P. 163.

    fīsting, sb., '*liridacio*,' PR. P. 163.

fit, adj., *fit, 'congruus,*' PR. P. 163; þat oþer þing nis non his fitte (*? equal*) O. & N. 784.

fit [2] [? **fitte**], sb., *O.E.* fitt (GREIN I. 300); *contest,* CH. C. T. *A* 4184, 4230; fitt EGL. 254.

fitchew, *see* **ficheu.**

fiten, v., *quarrel with*: feng to fiten his maumez JUL. 70.

fiteren, v., *? adorn*; **fitered** (*pple.*) cloþes MIRC 1146.

fitin, fittin, sb., *falsehood,* '*mendacium*,' PR. P. 163.

fitte, sb., *OE.* fitt; *song, stanza,* LANGL. *A* i. 139; fitt(e) EGL. 344.

fitten, v., = *M.Du.* vitten; *fit*; **fittes** (*pres.*) D. ARTH. 1989; **fittid** (*pple.*) 2455.

fiðele, = *O.N.* fiðla, *from med.Lat.* fidula, *from Lat.* fidis; *fiddle,* LAȝ. 7002; fiþele BRD. 9; BEV. 3735; LANGL. (Wr.) 5374; fithele CH. C. T. *A* 296; fiþele [feþele] ROB. H. 244; viþele AYENB. 105; **fiþeles** (*pl.*) MAN. (F.) 11386; **fiþelen** (*dat. pl.*) LAȝ. 3642.

fiþelen, v., *fiddle*: ich can nat tabre ... ne fiþelen at festes LANGL. *C* xvi. 206; fiþele *C* viii. 107; fithel *B* xiii. 447.

    fiþelere, sb., *fiddler,* LANGL. (Wr.) 5791; **fiþeleres** (*pl.*) HORN (H.) 1522.

fiðelinge, sb., *fiddling,* LAȝ. 22701.

fiþer, *see* **feþere. fivere,** *see* **fefre.**

fix, adj., *Lat.* fixus; *fixed*: with eye fix looke

on his visage LIDG. M. P. 235; latitudes of sterres fixe CH. ASTR. PROL. (3); sterres fixes i. 21 (11).

fixation, sb., *fixing*: that there be fixation with tempred hetes of the fire GOW. II. 86.

fixen, v., *Lat.* fixāre; *fix, make stationary; establish*: hir eien she **fixethe** (*pres.*) on him LYDG. M. P. 35; stelle eratice, nat **fixed** (*pple.*) 153; the wehle of fortunat fixid fast S. & C. II. vi.

fixene, sb., = *O.H.G.* fuchsin; *from* fox; *vixen*: the fixene fox HALLIW. 359.

fiz, sb., *O.Fr.* fiz; '*Fitz*,' *son of*, ROB. 431; P. L. S. ii. 179.

flā, sb., *O.E.* flā, = **flān**; *arrow, dart*: ane fla [flo] LA3. 1456; flaa A. R. 60*; flo '*telum*' FRAG. 4; CH. C. T. *H* 264; S. &. C. I. xxxv; **flān** (*pl.*) LA3. 1844; flon ROB. 394; floon GAM. 648.

flacken, v., = *M.Du.* vlacken, *Swed.* flacka; *palpitate*: her colde brest began to hete her herte al so to flacke and bete GOW. III. 315.

flǣrd, sb., *? O.E.* fleard, *? O.N.* flǣrd; *fraud, deceit*: fals & flærd ORM. 12177; flerd (*r. w.* erd) REL. I. 219 (MISC. 14).

flǣsc, sb., *O.E.* flǣsc, = *O.L.G.* flēsc, *O.Fris.* flēsk, flāsk, *O.H.G.* fleisc; *flesh*, FRAG. 6; flǣsh ORM. 3532; flesc HOM. I. 81; MISC. 193; flesch BEK. 263; ALIS. 694; ban & flesch KATH. 1103; þet flesch, vleschs A. R. 80, 112; flesch, flehs, fleis O. & N. 1007, 1399; fleisch [flehs] WICL. GEN. ii. 23; bred and fles GEN. & EX. 3316; fless ROB. 243; þet vless AYENB. 47; **flēsches** (*gen.*) O. & N. 1390; flesches lust A. R. 274; þes fleisces HOM. I. 85; **flēsce** (*dat.*) MISC. 28; þa wes he . . . . oflust after deores flǣsce LA3. 30555; flesche, fleshe, vleisse O. & N. 83, 1387; *comp.* **calf-flēsch**.

   flēsch-flīe, sb., *flesh-fly*, PR. P. 166; flesh-fle3e [flessflie] C. M. 5956.

   flēsch-hēwere, sb., = *Ger.* fleischhauer; *butcher*, OCTOV. (W.) 750.

   flēsch-hook, sb., *flesh-hook*, PR. P. 166.

   flēschi, adj., = *Ger.* fleischig; *fleshy*, PR. P. 166.

   flēsch-līch, adj., *O.E.* flǣsclīc; *fleshly, carnal*, A. R. 194; A. P. ii. 265; flǣshlike man ORM. 17276; **flēschlīche** (*adv.*) KATH. 1101; C. L. 1388.

      flēschlinesse, sb., *O.E.* flǣsclīcness; *fleshliness*, PR. P. 166; fleischlinesse WICL. DEUT. xvii. 17*.

   flēsc-mete, sb., *O.E.* flǣscmete; *flesh-meat*, SAX. CHR. 259.

   flēsch-schamil, sb., *meat-shambles, market*, CATH. 135.

fla3e, sb., = *M.L.G.* vlage, *M.Du.* vlaeghe (*? nimbus*); *flake*; a flai(e) (*v. r.* flake) of snow(e) VOC. 160; flawe '*floccus*,' PR. P. 164; a flawe of fire CATH. 133; a flaw(e) of snawe ALEX. 1756; **flawes** (*pl.*) of fire D. ARTH. 2556.

flagelle, sb., *Lat.* flagellum; *whip, scourge*: tak this flagel devoutli in thi hand LIDG. M. P. 146; **flagelle** (*dat.*) of eqvite and resoun IB.

flaget, *see* **flaket**.

flagge, sb., *sod; flag; 'cespes,'* PR. P. 163; **flagges** (*pl.*) and spraies TREV. IV. 157.

flaggi, adj.: flaggi place *place where flags grow*, WICL. EX. ii. 3.

fla3t, sb., *sod*, A. P. i. 57; flaghte CATH. 132.

fla3t², sb., *flake (of snow, fire)*: flaghte of snawe '*floccus*' CATH. 133; flaght (*printed* slaght) of fire C. M. 17372.

flaien, *see* **fle3en**.

flailen, v., *O.Fr.* flaieler, *from Lat.* flagellāre; *scourge*: thei him nailid and il **flailid** (*pret.*) S. & C. II. lx.

flair, sb., *O.Fr.* flair; *odour*, PR. C. 9017.

flāke, sb., *cf. O.N.* fleckr, *Ger.* fleck; *fleck, blemish*: his flok is withouten flake A. P. i. 946.

flāke², sb., *O.N.* flaki; *flake, 'floccus,'* PR. P. 163; LUD. COV. 140; a flake of snowe REL. II. 81; **flākes** (*pl.*) CH. H. F. 1192; flakes of soufre A. P. ii. 954.

flāken, v., *O.Fr.* flaquer; ? = **flasken**; *splash*: **flāke** (*printed* slake) (*imper.*) PALL. ii. 311.

flakeren, v., = *O.N.* flökra, *M.Du.* flackeren; *flutter*: flakereð (*pres.*) A. R. 222; **flakerande** (*pple.*) A. P. ii. 1410.

flaket, sb., *flagon*, TREV. III. 171; flaget VOC. 257; **flaketes** (*pl.*) WILL. 1888; flagetes 1893.

flamme, sb., *O.Fr.* flamme; *flame*; flamme [flaumbe] CH. BOET. ii. 6 (51); flaume PR. C. 6737; flaumbe LANGL. *B* xvii. 207.

flammen, v., *flame*; **flaumeþ** [flaumbeth] (*pres.*) LANGL. *B* xvii. 225.

   flamming, sb., *gleaming, flaming*: the flamming of þe flese was ferli to see D. TROY 970.

flān, sb., *O.E.* flān, = *O.N.* fleinn; *arrow*, A. R. 60*; H. M. 15; C. M. 4314; flon LA3. 311*; ALIS. 785; GAW. 1161; *see* **flā**.

flān, *see* **flēan**.

Flandre, pr.n., *M.Du.* Vlandere; *Flanders*, LA3. 7588.

flanke, sb., *cf. Swed.* (snö-)flanka; = **flāke²**; **flaunkes** (*pl.*) of fir A. P. ii. 149.

flanke, sb., *O.Fr.* flanc; *flank, 'ilium,'* PR. P. 163; alle þe flesche of þe flanke he flappes in sondire D. ARTH. 2782; flaunke 1156.

flap, sb., = *Du.* flap; *flap*, TOWNL. 206; flappe '*alapa, flabellum,'* PR. P. 163; **flappes** (*pl.*) AR. & MER. 8084; flappes of scourges LANGL. *B* xiii. 67.

flappin, v., = *Du.* flappen; *flap, clap, applaud*, PR. P. 164; **flappes** (*pres.*) D. ARTH. 2781; **flapped** (*pret.*) S. S. (Web.) 766; L. H. R. 176; flappiden [flappeden] wiþ hondes WICL. IS. iii. 16.

flashe, sb., *? O.Fr.* flache; *shallow pool,* '*lacuna,*' PR. P. 164; flosche GAW. 1430; PS. lxxxvli. 5.

flasken, v., *O.Fr.* flasquer; *splash* : heo **vlaskeð** (*pres.*) water þer on A. R. 314; flascheþ TREV. I. 63.

flat, adj., = *O.N.* flatr, *O.H.G.* flaz; *flat,* '*planus,*' PR. P. 164; WILL. 4414; BEV. 1040; IW. 259; flat (*subst.*) GAW. 507.

> **flatlinge,** adv., *prostrate*: and leide him flatling on þe grounde HALLIW. 360.

> **flat-mouthed,** adj., *flatmouthed*; flatmouthede as a fluke D. ARTH. 1088.

> **flatnesse,** sb., *flatness,* PR. P. 164.

flat, sb., *cf. Ger.* flatz; *slap,* AR. & MER. 4910; BEV. 3256; RICH. 5265.

flateren, v., *cf. M.Du.* flattern, *? O.N.* flaðra; *flatter,* LANGL. *B* xx. 109; flaterin '*adulor*' PR. P. 164; vlateri AYENB. 61; **flattereð** (*pres.*) A. R. 222*; **flateringe** (*v. r.* flikeringe) (*pple.*) CH. C. T. *A* 1962.

> **flaterere,** sb., *flatterer,* LANGL. *B* ii. 165; **flattereres** (*pl.*) WILL. 5480.

> **flatterunge,** sb., *flattering,* A. R. 320; flateringe PR. P. 164; CH. D. BL. 639.

flaterie, sb., *O.Fr.* flaterie; *flattery,* S. S. (Web.) 2155; CH. C. T. *B* 4514; vlaterie AYENB. 197.

flatour, sb., *O.Fr.* flateor; *flatterer,* AYENB. 257; CH. C. T. *B* 4515.

flatrour, sb., *flatterer,* GOW. III. 170; **flatrours** (*pl.*) III. 159.

flatten, v., *pour, smite,* AR. & MER. 9562; fette water . . . and flatte it on his face LANGL. *A* v. 224; flatten (*pret. pl.*) (*v. r.* flapten) on with flailes *C* ix. 180*.

fláþe, sb., *cf. O.H.G.* flado (*acc.* fladōn); *flat cake, flat fish*: flathe PR. P. 164; **fláthen** (*pl.*) VOC. 127.

flaumbe, *see* **flamme.**

flaun, sb., *O.Fr.* flaon ; = **fláþe** ; *pancake,* VOC. 241; PR. P. 164; **flaunes** (*pl.*) HAV. 644.

flaunke, *see* **flanke.**

flavor, sb., *O.Fr.* flaveur; *flavour*; so frech **flavorez** (*pl.*)of fritez were A. P. i. 87.

flawe, *see* **flaȝe.**

flax, sb., *O.E.* fleax, = *O.Fris.* flax, *O.H.G.* flahs; *flax,* '*linum,*' PR. P. 164; VOC. 156; flax [flex] CH. C. T. *A* 676; flex MAT. xii. 20; LANGL. *B* vi. 13; **vlexe** (*dat.*) AYENB. 236.

> **flax-bolle,** sb., = *Ger.* flachsbolle; *head of flax*; (*? printed* filaxlolle) VOC. 156.

> **flax-top,** sb., *head of flax*: flaxtop WICL. ECCLUS. xxi. 10; flax top WICL. IS. ii. 31.

flēa, sb., *O.E.* flēa, flēah, *cf. O.N.* flō, *M.L.G.* vlō *f.,* *O.H.G.* flōh, flōch *m.*; *flea* ; flee '*pulex*' PR. P. 165; fle HICKES I. 231; flee (*pl.*) P. L. S. i. 5.

> **flē-wort,** sb., *O.E.* flēawyrt; *flea-wort,* HALLIW. 362.

flēam, sb., *O.E.* flēam, = *O.H.G.* floum, *O.N.* flaumr; *from* **flēon**; *flight, escape* : fleam makian LAȝ. 1577; flæm 577; he nam flem [fleom] 8857; vlem [fleom] 24070; þei hadde take flem (*ms.* flem) ALIS. 4341; we ben . . . on fleme HOM. II. 149; he turnde to fleme LAȝ. 6407; torne . . . in fleme PS. lxxxviii. 24.

flēan, v., *O.E.* flēan, = *M.Du.* vlaen, vlaeghen, *O.N.* flā; *flay, skin* : he heom lette qvic flan [flean] LAȝ. 6418; flon HORN (H.) 92; flo HAV. 612; HORN (R.) 92; P. S. 191; flen '*excorio*' PR. P. 166; flen HORN (L.) 86; ALIS. 1734; WILL. 1682; **flēþ** (*pres.*) C. L. 1308; **flēa** (*subj.*) JUL. 24; riȝt his (*r.* is) þat fendes flea þe (*printed* fleathe) SHOR. 97; **flōgh** (*pret.*) Iw. 1699; flow HAV. 2502; MAN. (F.) 12452; vloȝen LAȝ. 20957; **flawen** (*pple.*) HAV. 2476; flawin, flain PR. P. 163; flain REL. I. 152; PR. C. 9520; MIR. PL. 119; vlaȝe AYENB. 73; iflawe ALIS. 893; **ivlaȝene** (*pl.*) LAȝ. 27377; *comp.* bi-flēan; *deriv.* ? flaȝe ? flaȝt.

> **flēar,** sb., = *M.Du.* vlāre; *flayer,* '*excoriator,*' PR. P. 165.

flecchen, v., *Fr.* flechir; *bend; flatter,* P. S. 344; **fleccheþ** (*pres.*) L. H. R. 137; **fleeching** (*pple.*) BARB. v. 622*; **flechchi** (*subj.*) AYENB. 253; **flecchede** (*pret.*) P. L. S. xv. 116.

flecchere, sb., *O.Fr.* flechier, *mod.Eng.* fletcher; *arrowmaker,* '*petularius, flectarius,*' PR. P. 165; **flecchours** (*pl.*) D. TROY 1593.

flecked, adj., = *O.H.G.* flecchōt; *flecked, spotted*: a flecked [flekked] pie CH. C. T. *E* 1848; flekked TREV. I. 159; **fleckede** (*pl.*) feþeres LANGL. (Wr.) 7234.

flēden, v., = *O.N.* flœda, *M.Du.* vloeden, *M. H.G.* vluoten; *from* flōd; *flood, flow,* MARH. 9; whænne þa sæ **vlēdeð** (*pres.*) LAȝ. 22019; þe eadie flod þet of ham **flēdde** (*pret.*) HOM. I. 209; *comp.* bi-flēden.

flēden, v., = *M.Du.* vlieden, *cf. O.N.* flȳja (*pret.* flȳða), *Swed.* flȳ (*pret.* flȳdde); *take flight, flee* ; flede MIRC 1374; **flēde** (*pret.*) M. H. 73; fleede D. ARTH. 1432; **flēdde** GEN. & EX. 3384; HAV. 1431; vledde AYENB. 206; fledden LANGL. *A* ii. 209; CH. C. T. *A* 2930; MAND. 61; **flēd** (*pple.*) RICH. 2301; TRIST. 2223; LIDG. M. P. 186; DEGR. 319.

flee, *see* **flēa.**

fleen, *see* **flēan,** *and* **flēon.**

flees, *see* **flēos.**

flēȝe, sb., *? O.N.* fley *n.; ship*; **flēin** (*ms.* fleyne) (*pl.*) OCTOV. (W.) 1671.

flēȝe, *see* **flēoȝe.**

flēȝen, v., *O.E.* flēgan, flȳgan, = *O.N.* fleyja, *O.H.G.* (ar-)flaugen; *put to flight*: **fleide** (*pret.*) a wei þat fearlac KATH. 1602; flaied A. P. ii. 960; M. H. 69; **flaied** (*pple.*) MIR. PL. 122; *comp.* a-flēȝen.

> **flaiing,** sb., *flight*: þe dai of flaiing and of afrai PR. C. 6112.

flēჳen, see flēoჳen.

flegge, sb., *flag (plant)*, '*acorus*,' PR. P. 165; VOC. 226.

flēie, flēien, see flēoჳe, flēoჳen, flēჳen.

fleil, sb., = *O.H.G.* flegil, *M.Du.* vleghel, *O.Fr.* flaiel; *flail*, '*flagellum, tribulum*,' PR. P. 165; fleჳl ORM. 1500; fleiles [flailes] (*pl.*) LANGL. *A* vii. 174.

　fleil-cappe, sb., *cf. Ger.* flegelkappe; *clown's cap*, PR. P. 165.

fleiðing, sb., ? = *M.Du.* vleidinghe (*flattery*); *instigation*, GEN. & EX. 692.

fleke, sb., *O.N.* fleki; *flake, hurdle*, '*crates*,' VOC. 201; PR. P. 165; flekes (*pl.*) MAN. (H.) 321.

flēken, v., ? *O.N.* flœkja; ?*entangle*; þei flēked (*pret.*) þam over thvert MAN. (H.) 241.

flekeren, see flikeren.　flēm, see flēam.

flem, sb., = flum; *river*: þe grate flem of þi flod A. P. iii. 309; i þe flem Jurdan HICKES I. 229; to þe flim Jordan S. A. L. 158.

flēme, sb., *O.E.* flēma, flȳma; *from* flēon; *fugitive*, HOM. I. 157; HORN (L.) 1271; six ჳer and a monþ he was fleme BEK. 1850; flēmen (*pl.*) LA3. 5952; flēamene (*gen. pl.*) ANGL. VII. 220.

flēmen, v., *O.E.* flēman, flȳman, *O.N.* flæma; *put to flight*; flemen, fleman LA3. 1579, 6574; i shal flēmen þe HAV. 1160; fleme ROB. 562; MAN. (H.) 82; flēmeþ (*pres.*) ALIS. 3348; flēmde (*pret.*) ORM. 8243; heo hine flemden out of þane londe LA3. 323; flēmed (*pple.*) MISC. 196; CH. C. T. *G* 58; PR. C. 2977; flemid M. ARTH. 3560; TOWNL. 320; DEGR. 899; *comp.* a-, ჳeflēmen.

　flēmere, sb., *one who puts to flight*, CH. C. T. *B* 460.

flēmen, v., *flow*: blode flēmit (*pret.*) o fer in flattes aboute D. TROY 10004.

flēn, see flēan *and* flēon.

flēoჳe, sb., *O.E.* flēoge, flȳge, = *O.H.G.* fliuga, *M.Du.* vlieghe; *fly (insect)*; vleჳe (*ms.* uleჳe) AYENB. 136; (*ms.* fleye) P. S. 238; vliჳe A. R. 8; flie MAND. 61; CH. C. T. *A* 4352; noჳt worþ a flie ROB. 428; flēoჳen, fleჳen (*pl.*) LA3. 3900, 3904; vleჳen AYENB. 58; vlien [fleჳen] A. R. 290; flien RICH. 2918; vliჳe, vlie O. & N. 600; *comp.* buter-, flǣsc-, hors-, húnd-flēoჳe.

flēoჳen, v., *O.E.* flēogan, flīogan, = *O.H.G.* fliugan, fliogen, *O.N.* fliuga, *M.Du.* vlieghen; *fly (as a bird)*; fleჳhen ORM. 5991; flegen GEN. & EX. 610; fleghe PR. C. 7947; fleighe [flee] LANGL. *B* xii. 241; fleie HAV.1813; MAN. (F.) 2257; fleon LA3. 2870; ALIS. 544; (*ms.* flon) HOM. I. 81; brid hwon hit wule vleon A. R. 130; fleen MAND. 48; fleo BRD. 9; flee EGL. 194; fliჳe GAW. 524; fliin PR. P. 167; flie MAP 335; ISUM. 375; vli AYENB. 254; flēo, flo (*pres.*) O. & N. 372; þu flihst [fliჳst] 227;

fliჳeð [flieþ] LA3. 21356; flegeð REL. I. 210 (MISC. 3); fleჳheþ ORM. 5889; flihð, vlið A. R. 132, 142; flihþ O. & N. 506; vliჳþ AYENB. 206; flið (*ms.* flid) HOM. I. 85; we fleoþ BRD. 9; (heo) fleoð LA3. 21759; fleoþ [floþ] O. & N. 278; vlēon (*subj.*) A. R. 130; vlēoinde (*pple.*) A. R. 130; vliinde AYENB. 71; fl(ē)ah fleh [fleah] (*pret.*) LA3. 25613, 27788; fleh HOM. I. 81; he flæh up intil heofne ORM. 5885; fleჳ BRD. 9; FER. 2182; flegh LAUNF. 473; fleiჳ P. L. S. xxv. 18; flegh [fleigh, flei] CH. C. T. *B* 4607; fleigh MAND. 27; flei HAV. 1305; EGL. 311; M. H. 101; fluჳen HOM. I. 129; fluჳen, fluwen [floჳe, flowen] LA3. 813, 3901; floჳe FER. 2249; flowen CH. C. T. *B* 4581; RICH. 4366; A. P. i. 89; his eiჳen flowen out of his hed JOS. 362; fluwe (*subj.*) A. R. 132; flówen (*pple.*) CH. H. F. 905; flowe GOW. II. 252; iflowe ROB. 29; *deriv.* flēoჳe, flie, flügge, flüht.

flēon, v., *O.E.* flēon, flīon, = *O.H.G.* fleohan, fliohan, *O.Fris.* flia, *Goth.* þliuhan; *flee*, LA3. 1578; A. R. 160; HORN (R.) 887; fleon, flen ORM. 4144, 9803; C. L. 519; flen GEN. & EX. 1086; fleen PR. P. 166; fléo, flo O. & N. 406 1231; to fleo alle his fon BEK. 1072; flee PERC. 876; to flēonne (*ger.*) JUL. 44; to fleonne, flenne, flæinde [fleonde, flende] LA3. 1570, 4662, 5561; flihst (*pres.*) O. & N. 405; fliihð, flicð LA3. 21339, 21343; A. R. 246; fliht [fliჳt] O. & N. 176; whan he fliჳþ out of londe BEK. 1489; fleoþ, fleþ ORM. 815, 9812; fleð HOM. II. 73; (heo) fleoð MARH. 13; fleð [fleoþ] LA3. 5894; vleþ AYENB. 41; flīh (*imper.*) A. R. 162; flih, fliჳ LA3. 3092, 16078; þu flēo (*subj.*) LA3. 16080; flēonde (*pple.*) HOM. II. 175; fleoinde A· R. 288; flēah (*pret.*) KATH. 16; vleaჳ AYENB. 129; fleh SAX. CHR. 264; flǣh, fleh, fleih [fleჳ] LA3. 552, 1612, 9360; flǣh ORM. 823; fleg GEN. & EX. 430; fleih JOS. 98; he fleih his holi A. R. 160; hit fleih vrom him 374; fleih [fleiჳ] to þe freeres LANGL. *A* ii. 186; fleiჳ TRIST. 2357; fleigh CH. C. T. *B* 3879; fleu LEB. JES. 559; (*printed* flen) ROB. 221; flew EGL. 786; flu-gen (*pl.*) MK.v.14; heo hine fluჳen LA3. 12251; fluwen [floჳen] 5564; flowen 817; baþe fluჳhen fra þe folc ORM. 893; fluhen JUL. 52; flogen GEN. & EX. 861; floჳen FER. 1620; fluwen A. R. 154; flowen LANGL. *A* ii. 209; JOS. 18; MAND. 226; S. S. (Wr.) 822; flógen (*pple.*) GEN. & EX. 1750; flowen RICH. 2203; *comp.* a-, æt-, bi-, ჳe-flēon; *deriv.* flēam, flēme, flēmen, flüht.

flēon, see flēoჳen.

flēos, sb., *O.E.* flēos, flȳs, = *M.Du.* vlies; *fleece*; flees '*vellus*' PR. P. 166; APOL. 104; flees [flus] LANGL. *C* x. 270; flēose (*dat.*) A. R. 66; flese WICL. GEN. xxx. 35.

flēot, sb., *O.E.* flēot *m.*, = *O.N.* fliōt, *M.Du.* vliet, *O.Fris.* fleet, *M.L.G.* vlet, *M.H.G.* vliez *n., mod.Eng.* -fleet (*in place-names*); *creek, rivulet*; fleet '*aestuarium*' PR. P. 166.

flēote, sb. *O.E.* flēot; *fleet* : al his fleote LAȝ. 2155 ; flete '*classis*' PR. P. 166 ; RICH. 1653 ; MAN. (F.) 1462 ; **flēte** (*dat.*) GOW. I. 197 ; D. ARTH. 1189.

flēoten, v., *O.E.* flēotan,=*O.L.G.* fliotan, *O.N.* fliōta, *O.H.G.* fliozan; *flow, swim,* HORN (R.) 159 ; L. H. R. 33 ; vleoten [fleote] LAȝ. 22010 ; flete LANGL. *B* xx. 44 ; MAND. 100 ; TRIST. 350 ; fleete LIDG. M. P. 98 ; swa se water fræm aȝ **flēteþ** (*pres.*) forþ ORM. 18093 ; fleteþ [flet] LANGL. *B* xii. 168 ; CH. C. T. *B* 463 ; flet REL. I. 220 ; FER. 4311 ; þe fisches þet i þe flodes fleoteð MARH. 9 ; **flēte** (*subj.*) HAV. 522 ; **flēotinde** [vleotinde] (*pple.*) A. R. 46 ; fletande L. C. C. 54 ; fletende water HOM. II. 177 ; fletinge CH. C. T. *A* 1956 ; **flēt** (*pret.*) LAȝ. 28960 ; GEN. & EX. 3187 ; P. L. S. xxiv. 251 ; it flæt up i þe lift ORM. 3466 ; fluten LAȝ. 32033 ; ꝥe fisses . . . floten abuven GEN. & EX. 2946 ; flote BRD. 21 ; fletide WICL. 4 KINGS vi. 6 ; **flōten** (*subj.*) LAȝ. 1032 ; *comp.* over-, tō-flēoten ; *deriv.* flēot, flēote, flot, flóte.

flērd, *see* flērd. **fleren,** *see* fliren.

[**flēringe,** sb., *O.E.* flēring ; *flooring ; comp.* up-flēringe.]

flēs, *see* flēos.

flēsc, flēsch, flēss, *see* flǣsc.

flet, sb., *O.E.* flett,=*O.L.G.*, *O.Fris.*, *O.N.* flet, *M.H.G.* vletze, *O.H.G.* flazzi ; *from flat ; floor ; apartment,* WILL. 5368 ; AN. LIT. 9 ; P. S. 337 ; LAUNF. 979 ; me wule swopen þine flor & þet flet clensien FRAG. 7 ; **flette** (*dat.*) ALIS. 1807, 2884 ; FER. 853 ; MIRC 273 ; GAW. 1374 ; he com to þan vlette þer þe feond lai LAȝ. 26023.

flēt, flēte, *see* flēot, flēote.

[**flēte,** sb., *O.E.* flēt ; *cream ; deriv.* flētin.]

[**flēten,** v.,?=*M.H.G.* flœzen ; *comp.* bi-flēten.]

flēten, *see* flēoten.

fleteren, v., *? fit for flight* : with flonez **flete-rede** (*pret.*) þai flitt fulle frescli þer frekes D. ARTH. 2097.

flētin, v.,=*Dan.* flöde ; *skim the cream,* PR. P. 167.

flēu, adj.,=*M.Du.* flauw ; *thin,* '*bassus,*' PR. P. 167.

fleumatike, adj., *O.Fr.* fleumatike ; *phlegmatic,* AYENB. 157.

fleume, sb., *O.Fr.* fleume ; *phlegm,* GOW. III. 99 ; (*ms.* flewme) LIDG. M. P. 195.

flēwort, *see under* flēa. **flex,** *see* flax.

flexpeng, sb., *? gudgeon,* '*fundulus,*' VOC. 253.

flēye, *see* flēȝe *and* flēoȝe.

flicche, sb., *O.E.* flicce,=*M.L.G.* flicke, *O.N.* flikki *n.* ; *flitch* : þei don hem to Donmowe. . . to folwen after þe flicche LANGL. *B* ix. 169 ; [flucchen, flicche *A* x. 189] ; fliche H. M. 37 ; flikke PR. P. 167.

flicht, *see* flüht.

flie, sb., *O.E.* flyge, *O.H.G.* flug ; *act of flying* : in flie L. H. R. 221.

flīe, flīen, *see* flēoȝe, flēoȝen.

flīȝe, *see* flēoȝe. **fligge,** *see* flügge.

fliht, *see* flüht.

fliker, adj., *unsteadfast, wavering* : man of fliker þoht M. H. 36.

flikeren, v., *O.E.* flicerian,=*M.Du.* fliggeren, vleggeren ; *flicker, waver ; flatter* ; flekerin '*volito*' PR. P. 165 ; **flikereð** (*pres.*) so mit þe A. R. 290 ; flekirs ALEX. 505 ; her gost þat **flikered** (*pret.*) ai a loft CH. TRO. iv. 1221 ; þis bischop flekerid in his þoht M. H. 92.

flinder, sb.,=*Ger.* flinder ; *splinter* ; **flendris** (*pl.*) GOL. & GAW. 915.

flindre, sb.,=*M.Du.* vlinder ; *moth* : zvo long vliȝþ þe vlindre aboute þe candle þet hi bernþ AYENB. 206.

flinder-mous, sb., *bat* : winges like a backe or flindermouse HALLIW. 363.

flingen, v., ? = *O.N.* flengja (*flog*) ; *fling* ; flinge FER. 583 ; B. DISC. 338 ; TOR. 2027 ; Alisaundres folk forþ gon fling(e) ALIS. 1111 ; **flingande** (*pple.*) D. ARTH. 2757 ; þoruȝ þe heorte þe launce **flang** (*pret.*) ALIS. 2749 ; flunge ALIS. 5892.

flint, sb., *O.E.* flint, = *M.Du.* vlint ; *flint,* '*silex,*' FRAG. 4 ; PR. P. 167 ; HAV. 2667 ; TRIST. 1452 ; þene vlint A. R. 220 ; **flinte** (*dat.*) HOM. I. 129 ; LANGL. *B* xiv. 64 ; **flintes** (*pl.*) PR. C. 7018 ; vlintes AYENB. 136.

vlint-sex, sb., *flint knife,* HOM. I. 81.

flint-stōn, sb.,=*M.L.G.* vlintstēn ; *flint-stone,* MAND. 150.

flipen, v., *pull off* : **flipit** (*pret.*) of the flese D. TROY 954.

fliren, v.,=*Dan.* flire, *Norw.* flira ; *fleer, sneer* : lagh(e) and flire (*printed* flerye, *r. w.* atire) FLOR. 1769 ; **fleriande** (*pple.*) D. ARTH. 1088.

flischen, v., *slash, cut* ; **flisches** (*pres.*) D. ARTH. 2768 ; **flischande** (*pple.*) D. ARTH. 2141.

flit, sb., *O.E.* flīt *n.,* *cf. O.L.G.* flīt, *O.H.G.* flīz *m.* ; *contest, strife,* KATH. 688 ; IW. 93 ; PS. cvi. 40* ; A. P. ii. 421 ; þat flit LAȝ. 24966 ; flit & win REL. I. 130 (HOM. II. 165) ; flit ne strif AL. (L.³) 20 ; **flīte** (*dat.*) HOM. II. 55.

fliten, v., *O.E.* flītan,=*O.H.G.* flīzan ; *quarrel* ; KATH. 721 ; fliten ne chiden HOM. I. 113 ; fiitin '*contendo*' PR. P. 167 ; fliten HORN (R.) 855 ; flite WILL. 2545 ; IW. 1027 ; TRIAM. 1227 ; A. P. i. 353 ; **flīteð** (*pres.*) & winneð HOM. II. 55 ; flit HOM. I. 113 ; **flōt** (*pret.*) MAP 344 ; AMAD. (R.) xxxvi ; hio fliten betweoxe heom hwilc heore wære ildest LK. xxii. 24 ; **fliten** (*pple.*) GEN. & EX. 3689 ; C. M. 13867.

flītere, sb., *O.E.* flītere ; *one who quarrels* ; **flīteris** (*pl.*) PR. P. 106 ; flitars MIR. PL. 179.

fliting, sb., *contention,* HOM. II. 13.

fliteren, v.,=*Ger.* flittern; *flutter*: it flitterid (*pple.*) al a brode MAL. i. 137.

flitten, *see* flütten.

flix-rop, sb., *? spleen, milt,* '*lien,*' VOC. 179.

flō, *see* **flā.**

floc, sb., *O.E.* flocc,=*O.N.* flokkr; *flock, troop,* HOM. I. 3; LAȜ. 4192; enne floc A. R. 202; (*ms.* flocc) ORM. 510; flok AR. & MER. 3866; a flok of briddis ALIS. 566; þane flok SHOR. 109; floc or drove of many hoggis WICL. MAT. viii. 30; **flocke,** flokke (*dat.*) LAȜ. 18005, 23801; **flockes** (*pl.*) O. & N. 280; GEN. & EX. 1637; C. M. 3992; and delde a þreo vlockes [flockes] his duhtie cnihtes LAȜ. 30964; flokkes TREV. V. 219.

　floc-mēle, adv., *O.E.* floccmælum; *by flocks, in companies,* WICL. 2 MACC. xiv. 14; flokmele CH. C. T. *E* 86.

flocke, sb.,=*M.Du.* vlocke, *O.H.G.* floccho, *? Lat.* floccus; *flock* (*of wool*); **flockes** (*pl.*) O. & N. 427; flokkes LUD. COV. 241; flokkis PR. P. 167.

flocken, v., *flock together, congregate*; flokkin PR. P. 167; þei **flocken** (*pres.*) ANT. ARTH. xxvi; **flockede** (*pret.*) LAȜ. 4729*; flokked A. P. ii. 386.

flōd, sb., *O.E.* flōd m. n., *cf. O.L.G.* flōd m. f. n., *Goth.* flōdus f., *O.N.* flōd n., *O.H.G.* fluot f.; *from* **flōwen**; *flood, river,* '*flumen,*' VOC. 239; ORM. 6793; O. & N. 946; GEN. & EX. 644; SPEC. 70; ALIS. 6185; þat flod HOM. I. 225; LAȜ. 20171; of a drope waxeð a muche flod A. R. 74; þe flod MISC. 150; al þe stret(e) a watere orn as hit were a gret flod P. L. S. xvii. 371; vlod AYENB. 247; **flōde** (*dat.*) HOM. I. 93; æfter who so roweþ aȝein þe flod P. S. 254; þan flode LAȜ. 19; **flōdes** (*pl.*) LANGL. *B* vi. 326; PR. C. 4706; *comp.* **sǣ-, water-flōd.**

　flōd-ȝat, sb., *floodgate,* VOC. 233; flodȝet A. R. 72.

　flōd-drī, sb., *mire*: hie secheð to þe fule **flōddri** (*dat.*) HOM. II. 37.

　flōde-wombe, sb., *river bed*: nakened shal be the flodwombe ['*alveus rivi*'] WICL. IS. xix. 7.

floiȝene, sb., *O.Fr.* flouin (*vaisseau léger*); *a kind of small ship*: ther were **floigenes** (*pl.*) on flote HALLIW. 365; in floines and fercostez and flemesche schippes D. ARTH. 743.

flōke, (**? flōk**), sb., *O.E.* flōc (*plaice* VOC. 65, 281), *O.N.* flōki m. (*sole*); *flook* (*sort of fish*), '*phoca,*' VOC. 254; fluke D. ARTH. 1088.

　flōke-mouthed, adj., *mouthed like a flatfish*: thou wenes for to fiai us **flōke**-mouthede schrewe D. ARTH. 2780.

flōn, *see* **flān** *and* **flēan.**

flōr, sb., *O.E.* flōr m. f., *cf. O.N.* flōrr m., *M.Du.* vloer m., *M.H.G.* vluor m. f.; *floor,* ORM.

15566; ROB. 288; MAP 349; A. P. ii. 133; i þene flor LAȜ. 22809; **flōre** (*dat.*) AN. LIT. 5; ISUM. 653; to þen flore FRAG. 5; **flōres** (*pl.*) D. ARTH. 3249.

flōr, *see* **flour.**

florin, sb., *Fr.* florin; *florin*; floring RICH. 5865; florein GOW. II. 138; florence ISUM. 551; D.TROY 1365; **florins,** florens(*pl.*) OCTOV. (W.) 392, 794; florencis OCT. (W.) 1907.

florischen, *see* **flourin.**

flosen, v., *? = Dan.* flosse; *? be shaggy*; felt **flosed** (*pret.*) him umbe A. P. ii. 1689.

flosche, *see* **flasche.**

flot, sb., *O.E.* flot n., *M.L.G.* vlot m.,=*O.N.* flot (*act of floating*); *from* **flēoten**; *float* (*a brewer's utensil*), E. W. 22; *grease* A. P. ii. 1011; and tagte . . . ilc fis . . . his **flōtes** (*gen.*) migt GEN. & EX. 162; when alle were . . . o flōte (*afloat*) MAN. (F.) 12079.

　flot-gres, sb., = *M.Du.* vlotgras; *float-grass,* '*ulva,*' PR. P. 168.

flōte, sb., *O.E.* flota=*O.N.* floti, *M.L.G., M.Du.* vlote; *fleet,* LAȜ. 2155*, 4530*; MAN. (F.) 1737; BARB. iii. 601; *comp.* **sǣ-flote.**

flōte [2], sb., *O.F.* flote; *troop, flock,* HAV. 738; JOS. 28; þe lombe þer . . . hatz feried þider his faire flote A. P. i. 945.

floteren, v., *O.E.* flotorian, *mod. Eng.* flutter; *be tossed in the waves; flutter* (*as a bird*): to floteri in þe grete se SAINTS (Ld.) 303; **floterin** (*pres.*) þo on heȝe C. M. 1781; þe foules flotereð CH. BOET. iii. 11 (96); **floterep** ['*fluctuate*'] WICL. IS. xxix. 9; **floterande** (*pple.*) HALLIW. 364; **flotterede** (*pret.*) HORN (R.) 129.

　flotering, sb., *rolling as the waves*: he shal not ȝive into withoute ende flotering to the riȝtwise WICL. PS. liv. 23.

floteri, adj., *flying about*: wiþ floteri [flotri] berde CH. C. T. *A* 2883.

flōti, adj., *waving*: þe firre i folȝed þose floti valez A. P. i. 127.

flōtien, v., *O.E.* flotian,=*M.Du.* vloten, *O.H.G.* flozen; *? float*: amanges ous to flotie SHOR. 21.

flotise, sb., *froth, scum*: flotise or flotice of a pott PR. P. 168.

flotten, v., *cf. M.Du.* vlotten; *swim*: a . . . whal . . . bi þat bot **flotte** (*pret.*) A. P. iii. 248.

floðer, sb., *cf. O.E.* flæðer; *flake* (*of snow*); **floþre** (*? gen. pl.*) MISC. 149.

flounder, sb., *cf. O.N.* flyðra; *flounder* (*the fish*); flondire REL. I. 85; **floundurs** (*pl.*) B. B. 171; flounders 282.

flour, sb., *O.Fr.* flour, flūr, flōr; *flower*; *flour,* HAV. 1719; C. L. 919; AYENB. 93, 95; CH. C. T. *A* 4; flour [flur] MISC. 195; **flūres** (*pl.*) A. R. 340; flures [flores] O. & N. 1046.

　floure-lēs, adj., *flowerless*: an herbe he broughte flourelesse, all grene DREAM 1862.

flourette, sb., *O.Fr.* florete ; *small flower* : alle in floures and **flourettes** (*pl.*) R. R. 890.

flouri, adj., *O.Fr.* flōri ; *flowery,* A. P. i. 57.

flourin, florischin, v., *O.Fr.* florir ; *flower, flourish; brandish,* PR. P. 167, 168 ; floure [florishe] WICL. HABAK. iii. 17 ; florishe L. C. C. 9 ; flourep, florissep (*pres.*) AYENB. 28, 95 ; florischep TREV. IV. 173.

flouroun, sb., *O.Fr.* floron ; *flower-ornament;* a white coroune shee beer with **flourouns** (*pl.*) smale CH. L. G. W. PROL. 215.

floute, sb., *O.Fr.* flaute ; *flute,* PR. P. 168.

floutour, sb., *O.Fr.* flauteur, fleusteor ; *player on the flute;* **floutours** (*pl.*) minstrales R. R. 763.

floutin, v., *play on the flute,* PR. P. 168 ; **floutinge** [floitinge] (*pple.*) CH. C. T. *A* 91.

flōwen, v., *O.E.* flōwan (*pret.* flēow), *cf. O.N.* flōa (*pret.* flōda), *Du.* vloeijen (*pret.* vloeide) ; *flow,* ORM. 4783 ; flowen CH. TRO. III. 1758 ; al þe leor schal vlowen o teares A. R. 64 ; flowe LUD. COV. 56 ; þe se bigan to flowe HORN (L.) 117 ; **flōweð** (*pres.*) KATH. 2519 ; flowep O. & N. 946 ; þe see þat flowep REL. I. 174 ; **flēaw** (*pret.*) HOM. I. 211 ; flew GEN. & EX. 3875 ; vleau A. R. 112 ; fleoweden 110 ; *comp.* **over-flōwen.**

flōwinge, sb., *flood, stream;* 'fluxus, venilia,' PR. P. 168 ; flodiʒ rereden up ther **flōwingis** (*pl.*) WICL. PS. xcii. 3 ; the flowingis of Jordan JOSH. v. 1*.

floxen, v., *Swed.* flaksa (*shake the wings*); *leap;* (þat child) **floxede** (*pret.*) and pleide togenes hire ['*exultavit infans in utero ejus*'] HOM. II. 127.

flügge, adj.,=*M.Du.* vlugge, *O.H.G.* flukke ; *able to fly;* fligge PR. P. 167 ; flegge M. T. 124.

flüht, sb., *O.E.* flyht *m.,*=*M.Du.* vlucht *f.;* *from* flēoʒen ; *flight, act of flying;* vluht A. R. 132 ; fligt GEN. & EX. 161 ; REL. I. 210 ; fliʒt (*?printed* flight) ALIS. 348 ; flicht (flith) LANGL. *C* xv. 172 ; **flühte** (*dat.*) LAʒ. 2880 ; mid þisse fluhte he fleh in to h(e)ovene HOM. I. 81 ; **flühtes** (*dat. pl.*) *wings* LAʒ. 2885.

flüht², sb., *O.E.* flyht,=*O.H.G.* fluht, *M.Du.* vlucht ; *from* flēon ; *flight, act of fleeing,* A. R. 48 ; fluiʒt JOS. 506 ; fliht ORM. 19683 ; fliht makie *take flight* LAʒ. 21405* ; fliʒt A. P. ii. 377 ; flight CH. C. T. *A* 988 ; we . . . turneð to fluhte JUL. 45 ; he deð him o fluhte A. R. 248 ; bringeð him o fluhte 294.

fluke, *see* **flōke.**

flum, sb., *O.Fr.* flum ; *river,* LAʒ. 542 ; MISC. 38 ; GEN. & EX. 806 ; H. H. 206 (208) ; (*ms.* flumm) ORM. 10342 ; flum [flom] AYENB. 202 ; *see* **flem.**

flumbarding, sb., *O.Fr.* flambard, *med.Lat.* flambardus ; *fiery, ardent person* : hit is an hardi flumbarding ALIS. 1788 ; **flumbardinges** (*pl.*) 6700.

flunder, sb., *cf.* flinder ; *spark;* **flunderis** (*pl.*) REL. I. 240.

flūr, *see* **flour.** **flūs,** *see* **flēos.**

flüschen, v.,?=*L.Ger.* flüschen ; *flush* : flush(e) for anger DEP. R. ii. 166.

flüsen, v., ? =*L.Ger.* flüsen ; *fly* : þat hit open **flüste** (*pret.*) HORN (R.) 1080 ; ?fliste FL. & BL. 473.

flütte, sb., *flitting, migration;* flitte REL. I. 235 ; flette 160.

flütten, v., *O.N.* flytja,=*Swed.* flytta, *Dan.* flytte ; *flit, carry, migrate* : þet moni þusunt muhten bi flutten (*? subsist*) A. R. 202 ; flittin '*amoveo, transfero*' PR. P. 167 ; flitten GEN. & EX. 1522 ; to flitten men til hefnes ærd ORM. 15648 ; leden hem fra land to land ʒif þat teʒ flitten (*migrate*) sholden 2082 ; flitte PR. C. 3762 ; fele times have ich fonded to flitte it fro þouʒt WILL. 623 ; þannes nolde he flitte FER. 495 ; he xal nevir flitte out of þis grave LUD. COV. 341 ; ha **flütteð** (*pres.*) from þe heate to þe chele HOM. I. 251 ; **flüte** (*imper.*) HOM. I. 265 ; **flitte** (*subj.*) HORN (L.) 711 ; **flittinge** (*pple.*) LANGL. *B* xi. 62 ; **flütte** (*pret.*) LAʒ. 30503 ; flit (*pple.*) TRANS. XVIII. 23 ; iflut FRAG. 5 ; iflut hider KATH. 826 ; ivlut LAʒ. 27934 ; *comp.* **for-flütten.**

flüttunge, sb., *flitting, removal, ? subsistence;* MARH. 22 ; flittinge ORM. 18023 ; flitting C. M. 5227.

fl(u)we, flowe, sb.,=*M.Du.* vlouwe ; *flue, fluff,* PR. P. 168.

flüx, flix, sb., *Fr.* flux ; *flux,* PR. P. 167 ; WICL. MAT. ix. 20 ; **flüxes** (*pl.*) LANGL. *B* xx. 80.

fnæd, sb., *O.E.* fnæd ; *fringe,* '*fimbria,*' FRAG. 4 ; fned MAT. ix. 20 ; **fnæde** (*pl.*) MAT. xxiii. 5.

fnǣst, sb., *O.E.* fnǣst ; *breath,* O. & N. 44.

fnästen, sb., = *O.H.G.* fnāstōn ; *breathe;* fnaste HAV. 548 ; **fnästed** (*pret.*) GAW. 1702.

fnästing, sb., *snorting* : þe fnesting ['*fremitus*'] of his hors WICL. JEREM. viii. 16.

fnēosen, v., *O.E.* fnēosan, = *Du.* fniezen, *Dan.* fnȳse, *Swed.* fnȳsa (*snort*) ; *sneeze;* fnese BER. 42 ; **fnēsep** (*pres.*) CH. C. T. *H* 62.

fnēsinge, sb., *O.E.* fnēosung ; *sneezing,* ['*sternutatio*'], WICL. JOB xli. 9.

fō, *see* **fā.**

fobbe, fobbere, *see* **foppe, foppere.**

focchen, v., *cf. O.E.* feccan ; *fetch* : am i sent to foche the childre of Israelle TOWNL. 60 ; fotche 199 ; **fochchez** (*pres.*) GAW. 1961 ; foch (*imper.*) 395.

fōdder, *see* **fōder.**

fōde, sb., *O.E.* fōda ; *food; offspring;* 'ali-

mentum,' PR. P. 168; HOM. I. 63; ORM. 828;
O. & N. 94; MISC. 159; GEN. & EX. 176;
ISUM. 134; fode, vode A. R. 142; mi child
mi oune fode HORN (R.) 1350; þis freli fode
EGL. 1254; to wedde þat freli fode AMAD.
(R.) liv; vode P. L. S. i. 57; a liþer vode P. L. S.
xxiii. 12; fude PERC. 1326; þo unlede fode
AL. (T.) 333.

fōd-rēde, sb., *sustenance, means of living*:
ðe spinnere . . . festeð atte hus rof hire fōd-
rēdes (*pl.*) MISC. 15.

fōder, sb., *O.E.* fōdor, fōddor, fōddur, =
M.Du. voeder, *O.N.* fōdr, *O.H.G.* fuotar, *mod.Eng.* fodder; *food, fodder; child, offspring*; fodir '*pabulum*' VOC. 264; þou hast
born a sori foder (*child*) ALIS. 645; foder
[fodder] CH. C. T. *A* 3868; foddur, foodir
'*pabulum*' PR. P. 168; fodder LAʒ. 27031;
fōddre (*dat.*) A. R. 416.

fōdien, v., *from* fōde; *feed*: fōdes (*pres.*)
hire wiþ faire wordes WILL. 2050; fōded
(*pret.*) it wiþ floures 57.

fōdinge, sb., *feeding*, PR. P. 168.

fōdme, sb., *? product*, GEN. & EX. 124.

fōdneð, sb., *? O.E.* fōdnoð: *sustenance*; to
fōdneðe (*dat.*) & to scrude HOM. I. 137.

fōdrien, v., *M.Du.* voederen, = *O.N.* fōðra,
O.H.G. fuoteren; *fodder, feed*; fōddred
(*pple.*) WICL. I KINGS xxviii. 24; foþer PS.
xxx. 4.

fōʒ, sb., = *M.H.G.* vuoc *m.*, *mod.Ger.* fug;
fitness, decency: mid foʒe and mid rihte O. &
N. 184.

foʒel, *see* fuʒel.

fogge, sb., *rank grass*, A. P. ii. 1683.

fōh, *see* fāh.

foil, sb., *O.Fr.* foil, fuil; *leaf*, D. ARTH. 2704;
PALL. ii. 194.

foilen, v., *? O.Fr.* fuler, fouler, foler; *? trample
under foot, subdue*; in vein þu **foilist** (*pres.*)
þi flesch wiþ abstinens APOL. 44.

foilen, v., *defile*: a sinner that . . . foilithe
(*pres.*) the commandments of God GEST. R.
151; hast thowe **foiled** (*pret.*) mi dowter 143.

foine, sb., *O.Fr.* foine; *beech-martin*, PR. P.
168; furris of foine DEP. R. iii. 150.

foinen, v., *fence, thrust*: thei foinen (*pres.*)
ech at oþer CH. C. T. *A* 1654; þei foine wiþ
daggers & wiþ swerdes LIDG. TH. 4325;
foinede (*pret.*) DEGR. 274; D. ARTH. 1898;
foined GAW. 428.

foining, sb., *thrusting*: fell was the fight,
foining of speires D. TROY 10287.

foisōn, foisoun, sb., *O.Fr.* foison, fuison;
abundance, power, CH. C. T. *A* 3165; foisoun
CHR. E. 711; fuisoun FER. 4457.

fōkel, adj. *or* sb., *?* = fāken; REL. I. 179.

fōken, *see* fāken.

fol, adj., *O.Fr.* fol; *foolish*, LAʒ. 1442*; A. R.
164; AYENB. 26; fol (*subst.*) *fool*, P. L. S.

xvii. 415; A. P. ii. 750; fool CH. C. T. *B* 3271;
fule REL. 110; fōles (*pl.*) HAV. 2100.

fol-herdi, adj., *foolhardy*, A. R. 62.

folhardi-lī, adv., *foolhardily*: if i hadde
doon . . . foolhardili WICL. 2 KINGS xviii.
13.

foolhardinesse, sb., *foolhardiness*, CH.
BOET. i. 3 (10).

folherdi-schipe, sb., *foolhardiness*, A. R.
182.

fol-līch, adj., *foolish*: þerof wexeþ vele
zennes . . . simulacion, follich yeve AYENB. 22;
ani foli þing WICL. JOB i. 22; foli thing
I COR. i. 25; foli dedis R. R. 5009.

foli-līche, adv., *foolishly*: thou dest
foliliche AR. & MER. 5064; folili MAND. 134;
WICL. NUM. xii. 11; A. P. ii. 696.

fol, *see* ful.

fólc, sb., *O.E.* folc, = *O.L.G.* folc, *O.Fris.*,
O.N. folk, *O.H.G.* folc, folch, *mod.Eng.* folk;
people, troop, ORM. 141; þat folc com to
gadere LAʒ. 859; heo letten forð bi siden an
oþer folc [folk] riden 5499; þat folc HORN
(L.) 618; þat folc of Kent ROB. 269; folc, volc
(*ms. uolc*), volk A. R. 156, 362; volc AYENB.
43; folc, folk HAV. 89, 438; WICL. JOHN xi.
50; folk CH. C. T. *A* 744; LIDG. M. P. 12;
swa mikel folk com never to gider PR. C.
6013; **folkes** (*gen.*) MAT. xiii. 15; GEN. &
EX. 2785; **folke** (*dat.*) HOM. I. 91; þa alre
feireste wifman þe þa wunede on folke LAʒ.
2218; **fólc** (*pl.*) HOM. I. 9; volke LAʒ.
23058; **folke** (*gen. pl.*) ORM. 12149; volke
[folke] LAʒ. 19995; *comp.* **burh-, driht-, fōt-,
land-, Suð-folc.**

fólc-king, sb., *O.E.* folccyning; *king of a
people*, LAʒ. 9501.

fólk-rēde, sb., *O.E.* folcræden; *people*;
folkerede AYENB. 196.

fóld, *see* fáld *and* feald.

fólde, sb., *O.E.* folde, = *O.L.G.* folda, *O.N.*
fold; *ground, land*, LANGL. *B* i. 153; FLOR.
342; GAW. 1694; M. ARTH. 3549; AV. ARTH.
i; þe volde dunede aʒen LAʒ. 23946; uppen
þissere **vólden** (*dat.*) 26120; heo fullen to
folde 5390; on folde SPEC. 24; A. P. ii. 403;
fólden (*acc.*) LAʒ. 7938.

fólden, *see* fealden.

fóldin, v., *cf. Dan.* folde; *from* fáld[1]; *fold
(cattle)*, put in a fold, PR. P. 168.

fóle, *see* fol.

fóle, sb., *O.E.* fola, = *O.H.G.* folo, *O.N.* foli,
Goth. fula, *cf. Lat.* pullus, *Gr.* πῶλος; *foal,
'pullus,'* FRAG. 3; PR. P. 168; HORN (R.)
587; LANGL. *B* xi. 335; MAND. 250; PERC.
717; ANT. ARTH. xlii; foole (*horse*) D. TROY
1245; foale 8341; þan **fólen** (*accus.*) MK. xi. 5.

folebairie, sb., *O.Fr.* fole-baerie; *ambition*:
folebairie, þet we clepieð ambicion AYENB. 17.

foleien, v., *O.Fr.* foloier, folier; *play the fool*,
CH. BOET. iii. 2 (67).

fólen, v.,=*Ger.* fohlen, *Dan.* fole ; *foal, bring forth a foal* ; **fóled** (*pple.*) CH. C. T. *D* 1545.

folet, folt, sb., *O.Fr.* folet ; *fool, 'stolidus,'* PR. P. 168 ; MAN. (F.) 4527, 7229.

folt-hēd, sb., *folly, madness* : in fersnesse ne in foltheed DEP. R. ii. 7.

folȝen, v., *O.E.* folgian, fylgan, fylgean,= *O.Fris.* folgia, fulgia, *O.L.G.* folgōn, *O.H.G.* folgen, *O.N.* fylgja ; *follow,* HOM. I. 33, 147 ; (*ms.* folien), folȝhen ORM. 944 ; folȝen, (*ms.* fulien) folien, fulien [folȝen] LAȜ. 1002, 4140 ; folgen, folwen GEN. & EX. 401, 3272 ; folwin PR. P. 169 ; LUD. COV. 268 ; folwe WILL. 189 ; GOW. I. 50 ; CH. C. T. *D* 1124 ; volewen, voluwen A. R. 28, 102 ; filghe PS. xvii. 38' ; filowe REB. LINC. 13 ; felowe MAN. (H.) 291 ; **felgie** (*pres.*) MAT. viii. 19 ; folȝeð P. L. S. viii. 7 ; volȝeþ SHOR. 40 ; AYENB. 12 ; folȝez A. P. ii. 6 ; folewes DEGR. 1561 ; folȝeþ, fulieð (*ms.* fulied), voleweþ O. & N. 307, 1239 ; (hes) folȝiað HOM. I. 119 ; hornes him fulieð LAȜ. 21338 ; folhið, folhen KATH. 2340 ; buten we fulien þissen ræde LAȜ. 16756 ; **folȝede,** fulede (*pret.*) LAȜ. 95, 24328 ; folgede MISC. 31 ; folwede [folwed] CH. C. T. *A* 528 ; vulede MISC. 38 ; foleȝeden, fulieden HOM. I. 3, 151 ; folweden LANGL. *A* iv. 25 ; felgden MAT. viii. 1.

folwere, sb., *O.E.* folgere ; *follower* ; folwar LANGL. *B* v. 549 ; folware PR. P. 169 ; feoleware A. R. 364 ; **folwers** (*pl.*) LANGL. *C* xviii. 103*.

folh-sum, adj.,=*Ger.* folgsam ; *from* folȝen ; *obsequious,* ORM. 7750.

folie, sb., *O.Fr.* folie ; *folly,* LAȜ. 3024* ; A. R. 52 ; P. L. S. xiv. 3 ; SHOR. 139.

foliot, sb., *?foolish matter* ; ne singe ich heom no foliot for all mi song is of longinge O. & N. 868.

folk, *see* folc.

folmarde *see* fūlmart *under* fūl.

folt, *see* folet.

folt-hēd, *see under* folet.

foltin, v., *from* folet ; *act like a fool* : foltin or doon as a foole '*stultiso*' PR. P. 169 ; lad folted (*pple.*) men to sinne C. M. 2303 ; this is a foltid man MAN. (H.) 164.

foltisch, adj., *foolish* : wher God hath not maad the wisdom of this world foltisch WICL. I COR. i. 20 ; foltisch chaffering WICL. S. W. I. 309 ; a foltissh face LIDG. M. P. 166 ; suche **foltisshe** (*pl.*) foolis 168.

foltrie, sb., *folly, 'fatuitus, stoliditas, follicia,'* PR. P. 169.·

folwen, *see* folȝen *and* fulwen.

fōm, fōmen, *see* fām, fāmien.

fon, adj. & sb., *? cf. Swed.* fåne ; *foolish ; fool* : þis folted fon MAN. (F.) 4051 ; MIRC 358 ; like a fon TOWNL. 80 ; fonne shepe GEST. R. 218 ; **fonnis** (*pl.*) LUD. COV. 367.

fōn, v., *O.E.* fōn,=*O.Fris.,* *O.N.* fā, *O.L.G.,* *O.H.G.,* *Goth.* fāhan (*for* *fanhan) ; *seize, take,* HOM. I. 61 ; ORM. 3733 ; SPEC. 41 ; mede fon HOM. II. 93 ; fon ich hit wulle LAȜ. 5437 ; it wullen (*r.* wulle) on fon (*begin*) 21194 ; fo MISC. 80 ; AL. (L.³) 539 ; fēð [foð] (*pres.*) A. R. 88 ; foþ REL. I. 183 ; on me heo foþ C. L. 895 ; **fōð** (*imper.*) HOM. II. 187 ; **fēng** [*O.E.* fēng, = *O.L.G.* fēng, *O.H.G.* fiang, fieng, *O.N.* fēkk, fēngum] (*pret.*) LAȜ. 22878 ; s. s. (Web.) 924 ; ich hem feng of þe BEK. 783 ; feng on to tellen JUL. 10 ; feng to þe fliȝt A. P. ii. 377 ; veng to deliten A. R. 52 ; þu fenge FRAG. 6 ; þer fengen feole to LAȜ. 659 ; **fonge** [*O.E.* (ge-)fangen,=*O.L.G.,O.H.G.* (ge-)fangan] (*pple.*) FLOR. 1831 ; fo we on (*let us begin*) O. & N. 179 ; fon MARH. 3 ; *comp.* **a-, bi-, ȝe-, on-, under-fōn** ; *see* **fangen.**

fōn, *see* fān. **fond,** *see* fand.

fond, pple., *see* fonnen.

fondien, *see* fandien *and* fundien.

fondling, *see* fundeling.

fondour, *see* foundeōr.

fonel, sb., *funnel, 'fusorium,'* PR. P. 170 ; fonel, funel C. M. 3306.

fongen, *see* fangen. **fonke,** *see* funke.

fonnen, v., *be foolish, infatuated* : fonne and dote CT. LOVE 458 ; þou **fonnist** (*pres.*) LUD. COV. 36 ; **fonned** (*pple.*) *fond, infatuated,* R. R. 5367 ; WICL. DEUT. xxxii. 21 ; fond GEST. R. 20 ; *comp.* **for-fonnen.**

fonnednesse, sb., *foolishness* : WICL. JER. xxiii. 13.

font, sb., *O.E.* font, fant, *from Lat.* font-em ; *cf. O.Fr.* font ; *font,* HOM. I. 59 ; MARH. 1 ; funt ORM. 17208 ; GEN. & EX. 3290 ; fant FER. 548 ; founȝt SHOR. 11.

funt-fat, sb., *font,* MISC. 4.

font-stōn, sb., *font,* FRAG. 8 ; SHOR. 97 ; JOS. 7 ; CH. C. T. *B* 723 ; fantstane PR. C. 3309 ; fanston HOM. II. 95 ; funtstane PR. C. 3124 ; vanstone ROB. (W.) 5898*.

foppe, sb., *fop, fool* : foppe idem quod folet PR. P. 170 ; **fobbes** (*pl.*) LANGL. *C* iii. 193.

foppere, sb., *?=Ger.* fopper ; *cheat* ; **fobberus** [*v. r.* fobberes, fobbes, freres] and faitours (*pl.*) LANGL. *C* iii. 193, *B* ii. 182.

[for-, prefix, *O.E.* for-, = *O.H.G.* far-, fir-, *Goth.* fair-, fra-, faur-.]

[for-², prefix, = for, *prep.*]

[for-³, prefix, = fōre, *prep.*]

for, prep. & conj., *O.E.* for, fær, *cf. O.L.G.* for, far, fur, *O.Fris.* for, fore, fori, *O.H.G.* fur, fure, furi, *O.N.* fur, fyr, *Goth.* faur ; *for* : þrouwede deð for al moncun HOM. I. 17 ; her to onswerede an for ham alle KATH. 1134 ; þe vor king halden LAȜ. 14613 ; al þat þu me seist for schame O. & N. 1283 ; he shulde ben iknowe ... for treitour P. S.

218; stingeþ for wod MISC. 148; hit leide on for wod P. L. S. xvii. 370; morned neiȝh for mad (*as if she were mad*) WILL. 1761; zelþ ... vor þaꝺe valsne peni AYENB. 24; for (*by*) mi lif JUL. 16; for gode H. M. 25; P. L. S. xv. 153; GAW. 1822; heo ne dursten for gode don þer þa misbode LAȝ. 13249; for þire lufe 18890; god nam swa muchele wreche for ane misdede P. L. S. viii. 104; for twam þingen HOM. I. 135; for are misdede 173; þet heo ne gruchie ... vor none mete ne vor none drunche A. R. 108; he ne mihte for his live iseo þat man wiþ hire speke O. & N. 1078; his care was al for hire BEK. 131; he weren for (*of*) Ubbe swiþe adrad HAV. 2304; he died for hungre MAND. 230; for chele qvake LANGL. B x. 59; for þon [for þan] *therefore* LAȝ. 989; he scholde never die for þon C. L. 1072; for þon þe *because* 123; for þan þe he sunne atend HOM. II. 107; for þen þe FRAG. 1; for þon þet he scolde swote smelle HOM. I. 53; for þan þat hit no wit not O. & N. 780; vor þan þet he let him overcome AYENB. 181; for þan ['*qvia*'] LK. iv. 32; for þan þe ['*enim*'] MAT. xiv. 24; for hwon *wherefore* HOM. I. 85; vor hwon (*v. r.* for hwi) A. R. 314; for þi *because* LAȝ. 22691; for þi þe HOM. I. 41; for þi þat he is so with HAV. 2043; for þi þat PR. C. 375; for þi þat he was wroht of eorþe ORM. 77; for þat he was godes preost 131; for þi þat ich nule þe forsaken (*v. r.* for ich nule forsaken þe) JUL. 32, 33; for þi þat he hadde (*v. r.* for he hadde) slain Asahel WICL. 2 KINGS iii. 10; for þat he deleð þe sowle & þe lichame HOM. II. 7; for þat ich fleo bi nihte O. & N. 365; for þat he hade þe Sarazenes slain ISUM. 489; for ich hit wot O. & N. 1248; al folc him luvede for he dide god justise CHR. E. 268; for he nefde nenne sune LAȝ. 148; for eȝþer here ȝede swa riht after godes lare ORM. 119; for þu art lutel and unstrong HOM. I. 113; for alle ure goden LAȝ. 13919; to holden and to knowe for evr(e) ech(e) oþere wiȝte HORN (L.) 671; vor þe feste A. R. 22; for alle oþer BEK. 721; for to (*until*) þet heo comen to his (? *read* heore) lives ende HOM. I. 155; ALIS. 5363; PL. CR. 311; for to, for te HOM. II. 23, 33; LAȝ. 7563*, 25803*; for te þe hete a morwe come TREAT. 137;

vort midniht A. R. 236; fort þet hit was non MISC. 50; fort hit was eve O. & N. 41; þat ece fer þe ham ȝearcod was fer (*for*) hare prede HOM. I. 221.

for, *see* feor³.

forāge, sb., *Fr.* fourrage; *forage,* SHOR. 122; CH. C. T. *E* 1422.

forager, sb., *O.Fr.* fourrageur; *messenger*: frenesies and foule iveles foragers (*pl.*) of kinde LANGL. *B* xx. 84.

for-arnen, v., *ride hard*; vorarnd (*printed* uor arnd) ROB. (W.) 7490 (H. 362).

for-bannen, v., = *M.Du., M.H.G.* verbannen; *banish*: let forbonne O. & N. 1093.

for-barren, v., *bar*; forbarre WILL. 3333; P. R. L. P. 60; C. M. 8213.

for-bēaten, v., *beat down*; forbete LANGL. *B* xviii. 35; forbētun (*pple.*) ANT. ARTH. li.

for-bēoden, v., *O.E.* forbēodan, = *O.Fris.* for-, furbiada, *Goth.* faurbiudan, *O.N.* furbióða, *O.H.G.* for-, far-, fer-, firpiotan; *forbid,* LAȝ. 27408; verbeoden MISC. 46; forbedin '*prohibeo, interdico*' PR. P. 170; forbēdeþ (*pres.*) MAND. 71; CH. C. T. *C* 643; forbedes A. P. ii. 1147; forbiet MISC. 28; vorbiet AYENB. 6; forbēode (*subj.*) A. R. 8; god forbeode A. D. 259; forbeode [forbede] LANGL. *A* iii. 107; forbēad (*pret.*) GEN. & EX. 311; SHOR. 157; vorbead A. R. 70; forbæd ORM. 1955; furbed MAN. (F.) 9158; vorbed RQB. 496; forbæd [*sec. text* forbed] LAȝ. 20579; furbed MAN. (F.) 9158; forbude (*subj.*) HOM. I. 245; forbóden (*pple.*) ORM. 12021; GEN. & EX. 325; MAND. 20; R. R. 6616; PL CR. 147.

for-bēren, v., *O.E.* forberan, = *O.H.G.* furi-, for-, far-, fer-, firberan; *forbear*: unðewes to forberen HOM. II. 39; he wule vorberen ou lesse A. R. 218; forbere ASS. 60; GOW. I. 164; CH. C. T. *A* 885; hi ne miȝte forbere no more BEK. 1426; vorbere AYENB. 148; verbere ROB. (W.) 6552; forbére (*pres. subj.*) P. L. S. xxxi. 362; vorber (*pret.*) A. R. 218; forbar HAV. 764; for baren SAX. CH. 262; þauh þu me vorbere HOM. I. 197; forbóren (*pple.*) HOM. I. 197; forbore R. S. vii.

vorbéringe, sb., *forbearing,* AYENB. 148.

forbernen, *see* forbrennan.

forbersten, *see* forbresten.

for³-bī, prep., *cf. Ger.* vorbei; *beside,* ALIS. 5487; MAND. 11; CH. C. T. *C* 125; AV. ARTH. xxv.

for-binden, v., *O.E.* forbindan, = *O.H.G.* furi-, ferpintan; *bind up*; forbíndeþ (*pres.*) FRAG. 5; ORM. 4524; forbúnden (*pple.*) ORM. 13775.

forbischin, v., *O.Fr.* forbir; *furbish,* PR. P. 170.

for³-bīsne, sb., *O.E.* forebȳsen; *example, parable,* KATH. 698; O. & N. 98; REL. I. 218; þere ... forbisne HOM. I. 133; vorbisne A. R. 76; AYENB. 46; forbisne, -bisene C. L. 555, 980; forbisene, forebisene, forbusne

LANGL. *B* viii. 29, *A* ix. 24, *C* xviii. 277 ; for-būsene (*dat.*) LANGL. *B* xv. 555.

for³-bīsnen, v., *exemplify* : ðis forbīsnęde (*pple.*) ðing REL. I. 222.

forbīseninge, sb., *parable* : i sal helde mi erĕ in forbiseninge PS. xlviii. 5 ; i sal open mi mouth in forbīseninges (*pl.*) lxxvii. 2.

for-bīteň, v., = *Ger.* verbeiszen ; *bite in pieces*; forbīteþ (*pres.*) LANGL. *B* xvi. 35 ; forbāt [forbot] (*pret.*) LAȝ. 6497.

for-blak, adj., *very black*, CH. C. T. *A* 2144.

for-blāwen, v., *O.E.* forblāwan ; *blow to pieces* ; forblōwe (*pple.*) GOW. I. 160.

for-blēden, v., *cf. Ger.* verblüten ; *bleed to exhaustion* ; forblĕd (*pple.*) L. H. R. 191 ; TRIAM. 1243 ; M. ARTH. 3434.

for-blenden, v., = *Ger.* verblenden ; *blind*, ORM. 15391.

for-bod, sb., *O.E.* forbod, = *M.H.G.* verbot ; *prohibition, interdiction*, REL. I. 177 ; forbode KATH. 2232 ; GEN. & EX. 324 ; HORN (L.) 76 ; forbóde (*dat.*) LAȝ. 1446.

for-breiden, forbrēden, v., *O.E.* forbregdan, forbrēdan ; *pervert, corrupt* ; forbreideþ, for-brēdeþ (*pres.*) O. & N. 510, 1383 ; forbroiden (*pple.*) REL. I. 211 ; forbroide men ROB. 21 ; forbrode O. & N. 1381.

for-brĕken, v., *O.E.* forbrecan, = *O.H.G.* fer-, firprechan ; *break to pieces*; forbrac (*pret.*) GEN. & EX. 3049 ; Richard … vorbrec þere his necke ROB. 375 ; forbróken (*pple.*) REL. I. 211 ; forbroken man SAX. CHR. 256.

for-brennen, v., *O.E.* forbernan, = *O.H.G.* fur-, for-, fer-, firbrennan ; *burn up* ; forbrennen, -bærnen, furbernen [forberne] LAȝ. 645, 7389, 16658 ; forbrenne LANGL. *A* iii. 88*, *C* iv. 125 ; forbernen HOM. I. 143 ; forberne HORN (R.) 692 ; furberne LEG. 18 ; forbrende (*pret.*) GEN. & EX. 3784 ; vorbernde AYENB. 186 ; forbarnde LAȝ. 23202 ; verbarnde ROB. 296 ; forbrend (*pple.*) MAP 339 ; for-, vorbrend L. H. R. 26 ; furbrend MISC. 152 ; forbrent ALIS. 4669 ; M. ARTH. 1925 ; vorbernd A. R. 54.

for-bresten, v., *O.E.* forberstan ; *burst* ; forberst (*pret.*) LAȝ. 1912.

for-brinnen, v., *O.E.* forbeornan, -byrnan, = *O.H.G.* far-, fer-, firprinnan ; *be burnt up* ; forbearnen JUL. 67 ; þu forbernest (*pres.*) wel neȝ for onde O. & N. 419 ; forbearn (*pret.*) SAX. CHR. 249 ; forburnen LAȝ. 29297.

for-brūsen, v., *break in pieces* ; forbrūsed (*pple.*) CH. C. T. *B* 3804.

for-brūtnen, v., *break in pieces*; forbrittened (*pple.*) D. ARTH. 2273.

for-brütten, v., *O.E.* forbryttan ; *break in pieces* ; forbrett (*pres.*) MAT. xii. 20.

for-būgen, v., *O.E.* forbūgan ; *decline, avoid*, MK. vi. 48 ; forbuȝhen ORM. 7514 ; forbuwen A. R. 206 ; forbūh (*imper.*) H. M. 17 ; vorbuwen (*pret.*) A. R. 306.

for-büggen, v., *redeem* ; forbugge C. L. 1090 ; forbigge WICL. EX. vi. 6 ; hi couþen .hire zennen vorbegge AYENB. 78 ; forbouȝt (*pple.*) SHOR. 164 ; forbought ELL. ROM. II. 355.

forbügger, sb., *redeemer* : i wote that mi forbier (*v. r.* forbigger) liveth WICL. PREF. EP. vii.

for³-bürþe, sb., = *Ger.* vorgeburt ; *previous birth*, C. M. 3545.

[for-cálden ?], v., forcóld (*pple.*) ; *chilled*, S. S. (Web.) 2623.

for-casten, v., = *Dan.* forkaste, *Swed.* förkasta ; *cast away, reject* ; vorkest (*pple.*) AYENB. 186.

forkesting, sb., *rejection*, A. R. 278.

force, sb., *O.Fr.* force ; *force*, WILL. 1217 ; CH. C. T. *C* 133 ; C. M. 13044 ; forss BARB. x. 784 ; fors iv. 132 ; ma na fors *make no account* v. 85 ; make no fors CH. C. T. *H* 68 ; mast fors *most especially* BARB. viii. 11 ; no fors *it is no matter* CH. C. T. *B* 285 ; *G* 1019.

forcelet, sb., *O.Fr.* forcelet ; *strong place*, PR. P. 170.

forcement, sb., *fortifications* ; forsemens (*pl.*) or strengthes WICL. IS. xxv. 12.

forcen, v., *O.Fr.* forcer ; *compel* : his fader us forset (*pret.*) with his foule wille D. TROY 1924 ; forcede D. ARTH. 1070 ; i forced (*reflex.*) me for to luve Jesu HAMP. TREAT. 6 ; he has forsede (*pple.*) her and filede D. ARTH. 978.

forcinge, sb., *compulsion, violence* : whanne forsinge he made, he assentide not WICL. 4 KINGS v. 16.

forcer, sb., *O.Fr.* forcier ; *chest, coffin*, WILL. 4432 ; S. S. (Wr.) 2949.

forceress, *see* fortress.

for-chĕosen (? for³-), v., *? choose* : þat sche was forchósen (*pple.*) from þe beginning of þe world MAND. 132.

forchūre, sb., *O.Fr.* forcheure ; *fork or dividing of the legs* : Firumbras … was a … long man in forchure FER. 549.

for-clemmen, v., = *Ger.* verklemmen ; *? compass* ; forclemmed (*pple.*) A. P. iii. 395.

for-clēve, v., *cleave in pieces*, FER. 543 ; forclēaf (*pret.*) LAȝ. 26455* ; forclef BRD. 19 ; FER. 1619 ; verclef ROB. (W.) 8269*.

for-clingen, v., *O.E.* forclingan ; *wither* ; forclonge (*pret.*) P. L. S. xxiv. 216 ; forclungen (*pple.*) ORM. 13851.

for-clōsen, v., *cf. O.E.* forclȳsan ; forclōsede (*pret.*) SAINTS (Ld.) 303.

for-clucchen, v., *? cramp* : up hor ton hi set al vorcluȝt (*pple.*) & qvaked al vor fere LEG. 165.

for-cóld, *see* forcáiden.

for-coveren, v., *cover wholly* : she forcoveride (*pret.*) the nakid of the nak WICL. GEN. xxvii. 16.

for-crásen, v., *break to pieces ; be in ruins* :

old werk, **forcrásed** (*pple.*) alle S. S. (Web.) 721.

for-crempen, v.,=*M.H.G.*verkrempfen; *contract*; for-, **vorcrempeþ** (*pres.*) O. & N. 510.

for-cróken, v., *become crooked*: þe hond was ek **forcróked** (*pple.*) P. L. S. xvii. 336.

for-culien, v., *?set on fire, burn*; **vorkuliinde** (*pple.*) hire sulven mid þe fure of sunne A. R. 306; þet te soðe sunne . . . haveð wiðuten **vorkuled** (*pple.*) ou 50.

for-cumen, v., *O.E.* for-, forecuman,=*O.H.G.* furiqveman; *come before*; **forcom** (*imper.*) PS. xvi. 13.

for-cursien, v., *curse*; **forcursed** (*pple.*) SAX. CHR. 262.

for-cutten, v., *cut to pieces*; **forkutteþ** (*pres.*) CH. C. T. *H* 340; **forkutte** (*pret.*) TREV. III. 405.

for-cūð, adj., *O.E.* forcūð; *wicked*: sum forcuð [forcouþ] kempe LA₃. 28240; **forcūðere** (*compar.*) HOM. II. 83; **forcūðest** (*superl.*) KATH. 2242.

for-cweðen, v., *O.E.* forcweðan, = *O.H.G.* far-, ferqvedan, *Goth.* faurqiþan; *forsake, renounce*: hwen þu **forcwiðest** (*pres.*) . . . ure . . . godes KATH. 389.

for³-cwiddare, sb., *foreteller*, A. R. 212.

for-cwikien, v., *revive*; **forqvichieð** (*pres.*) þan here time cumeð HOM. II. 77.

ford, sb., *O.E.* ford,=*M.Du.* vord, *O.H.G.* furt; *ford, 'vadum,'* FRAG. 3; ALIS. 4343; TRIST. 2818; þene vord [ford] LA₃. 20159; ford [forþ] WICL. GEN. xxxii. 22; forde, forth LANGL. *A* vi. 57, *B* v. 576; forþ A. P. i. 150.

for³-dēde, sb., *previous deed*, A. R. 394; WILL. 5182.

for³-dēl, sb., = *M.H.G.* vorteil; *advantage*, HALLIW. 370.

·for-dēmen, v., *O.E.* fordēman,=*O.N.* fordœma, *O.H.G.* fur-, fortuomen; *condemn*, HOM. I. 95; LA₃. 3457; KATH. 428; fordeme O. & N. 1098; **fordēmde** (*pret.*) MARH. 2; **fordēmed** (*pple.*) C. L. 447.

for-dil₃ien, v., *O.E.* fordilegian,=*O.L.G.* fardiligōn, *O.H.G.* fertiligōn; *exterminate, destroy*; **fordil₃hed** (*pple.*) ORM. 14541.

for-dolken, v., *? wound severely*; **fordolked** (*pple.*) A. P. i. 11.

for-dōn, v., *O.E.* fordōn,=*O.H.G.* fertuon; *ruin, destroy*, LA₃. 2567; LANGL. *B* v. 20; MAND. 56; fordo PR. C. 3395; furdo MAN. (F.) 11663; to vordonne A. R. 210; **fordeð** (*pres.*) KATH. 214; vordeþ AYENB. 121; **fordüde** (*pret.*) MARH. 2; fordede GEN. & EX. 426; furduden MISC. 151; fordeden PL. CR. 495; **fordōn** (*pple.*) ORM. 18923; CH. C. T. (T.) 17239.

for-drǣden, v., *terrify; be afraid of*; **vordrǎdde** (*pret.*) ROB. (W.) 2088; **fordrěd** (*pple.*) GEN. & EX. 1557; SPEC. 88; (*ms.* forrdredd) ORM. 147.

for-dra₃en, v., *draw along*: þe devel þe **fordrawe** (*pres.*) FER. 1796; with flateringe of lippis she **fordro₃** (*pret.*) him WICL. PROV. vii. 20.

for-drēfen, v., *perturb*; **fordrēfed** (*pple.*) ORM. 2194.

for-drenchen, v., *O.E.* fordrencan,=*M.L.G.* vordrenken; *drown; intoxicate*, HOM. II. 213; fordrenche P. L. S. viii. 166; **fordrenct** (*pple.*) KATH. 2343; þas men beoð mid miste (*r.* muste) fordrencte HOM. I. 91.

for-drīfen, v., *O.E.* fordrīfan,=*O.H.G.* far-, fertrīban; *drive asunder*; **fordriven** (*pple.*) REL. I. 220; fordrive R. R. 3782; mid wedere fordreven LA₃. 6206*.

for-dru₃en, v., *O.E.* fordrūgian; *dry up*; fordruie, fordrue O. & N. 919; **fordried** (*pret.*) TREV. VIII. 139; **furdrüit** (*pple.*) LEB. JES. 596; **vordrüwede** (*pl.*) A. R. 148.

for-drunken, adj., *O.E.* fordruncen,=*M.L.G.* vordrunken; *drunken*; vordrunken A. R. 214; þa **fordrunkene** (*pl.*) cnihtes LA₃. 13518.

for-dullen, v.,=*M.L.G.* vordullen; *become dull*; **fordolled** (*pple.*) L. H. R. 141.

for-dütten, v., *O.E.* fordyttan; *shut up*; forditten RICH. 4170; **forditteð** (*pres.*) HOM. II. 197; fordut LA₃. 17139; **fordütted** (*pple.*) FRAG. 7; fordit MAP 339; **fordütte** (*pl.*) FRAG. 7.

for-dwine, v.,=*M.Du.* verdwijnen; *fade away, decay*, P. L. S. xxiv. 215; **fordwinnen** (*pple.*) REL. II. 211; **fordwined** R. R. 366.

fōre, sb., *O.E.* fōr, = *O.H.G.* fuora; *from fáren*; *course, march, proceeding, expedition*, GEN. &. EX. 2984; ROB. 386; ALIS. 2355; REL. I. 160; CH. C. T. *D* 110; Brennes . . . heom beh æfter mid starkere fore LA₃. 5568; þe cnihtes weoren on fore 5858; heo nomen heore vore into þas kinges bure 13667; and so forleost þe hund his fore O. & N. 817; he folwed the fore of an oxe TREV. I. 197; *comp.* forð-fóre.

fóre, prep. & adv., *O.E.* fore,=*O.L.G.*, *O.H.G.* fora, *O.Fris.* fara, *Goth.* faura; *before*: þet tu fore ibeden havest MARH. 21; þat ich fore habbe imurned LA₃. 13472; non no spak him on word fore ALIS. 4038; þis . . · perle . . . þe joueler ₃ef fore alle his god A. P. i. 733; (h)is sone þat al þe sorwe is fo re WILL. 2941; þu ne derst cumen biforen him fore þine gulte HOM. I. 27; þer fore LA₃. 316; fore þe miracle þet hi seghe was here beliave þe more istrengþed MISC. 30; bote ief hit bi vore zome niede AYENB. 7; þis Gilbert hire tolde fore BEK. 31; he tolde ous fore hou hit scholde beo P. L. S. xx. 37; he tolde him fore TREV. III. 143; nis nout . . . god so grim ase ₃e him vore makieð A. R. 334; so stille and derne he was þe fend fore C. L. 1030; þa þe fore stopen ['*praeibant*'] LK. xviii. 39; se time . . . þe god fore scewede

HOM. I. 227; ich com ... for te warnin ow fore 253; ich warni þe vore [v. r. biforen] JUL. 47; for þe welþe ... i have him wrouȝt fore WILL. 2076; as þei fore seiden JOS. 208; as he fore tolde LANGL. *A* xi. 165; þe vore izede Jon AYENB. 190; god fore [for] wot it CH. C. T. *B* 4438.

for-eald, adj., *very old*; forold CH. C. T. *A* 2142.

for-ealdien, v., *O.E.* forealdian, *cf. O.H.G.* faralten; *grow old*; **vorólded** (*pple.*) L. H. R. 24.

fore-arsoun, sb., *saddle-bow* : his ax on his forearsoun hing RICH. 5053; on his stede ful þe dent, biside þe forarsoun FER. 3385.

fore-bēcne, forbēacne, sb., *O.E.* forebēacn; *omen* ; (*pl.*) MK. xiii. 22; JOHN iv. 48.

foredene, *see* **fērrēden** *under* **fere.**

fore-fader, sb., *O.N.* forfaðir ; *forefather*, TREV. VIII. 9; forefadres (*pl.*) LANGL. *B* v. 501.

[fore-feng], forfeng, sb., *O.E.* fore-, forfeng; *'quite de avant prise,'* REL. I. 33.

fore-finger, sb., *forefinger*, VOC. 179.

fore-fōt, sb., *forefoot*; forefēt (*pl.*) WILL. 3284.

fore-ganger, -gōere, sb., = *Ger.* vorganger, -genger, -geer; *a goer before* : i go before bodword to bere, and as forgangere am i TOWNL. 165; foregoere LANGL *A* ii. 162; WICL. HEB. vi. 20; als anticrist lims and his **forgangers** (*pl.*) PR. C. 4151.

fore-ȝeard, sb., *forecourt, hall*; foreȝerd WICL. EZ. x. 3; in her stretis ether **forȝerdis** (*pl.*) 2 ESDR. viii. 16.

fore-gengle, sb., *predecessor*; **vorgenglen** (*pl.*) LAȝ. 25082.

fore-gessen, v., *anticipate, expect* : a conscience disturblid presumeth | *gl. marg.* that is bi **forgessing** (*pple.*) grete ivels to cominge on it self] evere wickid thingis WICL. WISD. xvii. 10.

fore-gōere, *see* **fore-ganger.**

fore-hēved, sb., *O.E.* forhēafod, *cf. M.L.G.* forehouvet ; *forehead*, BEK. 2200; forheafod HOM. I. 127; vorheaved A. R. 18; forheved PERC. 495; forheed CH. C. T. *A* 154.

forein, adj. & sb., *O.Fr.* forain; *foreign, foreigner*, MAND. 183; E. G. 361; some hadde **foreines** (*pl.*) to frendes LANGL. *C* x. 199.

forel, sb., *O.Fr.* forrel, fourrel; *sheath*, PR. P. 171; C. M. 15791; furel (*v. r.* scheþe) WICL. JOB xx. 25.

fore-land, sb., *foreland*; **forlondes** (*pl.*) TREV. I. 305; forlondez GAW. 699.

fóren, prep. & adv., *O.E.* foran, forne, = *O.L.G.* foran, forana. *O.H.G.* forna; *before*; foren ongen eow MAT. xxi. 2; forenan his hafde [*Ger.* vorn an seinem haupte] LAȝ. 23968;

foren [forn] to his breoste 20121; foren wenden 26899; forn at þere burȝe 24032; forn aȝan (*v. r.* aȝæn), (*opposite him*) 18528; min seoruwe is forn on FRAG. 6; þeȝ token eft forn on to serfen ORM. 553; forn aȝeins WICL. GEN. xvi. 12; swim þu foren me HOM. I. 51; foren us to gon GEN. & EX. 3541; *comp.* a-, æt-, bi-, tō-fóren.

for-ēode, v., (*pret.*) *forwent, let go*, SPEC. 23; forȝode HALLIW. 375; forȝede CH. TRO. ii. 1330; voreoden A. R. 406; furȝeden MAN. (F.) 15711.

fore-partie, sb., *forepart, front* : foreparti of the hede '*sinciput*' VOC. 183.

fore-said, pple., *aforesaid* : this forsaid stret M. H. 52; this ilk forsaid nonnrie 58; þe foresaide folk ALEX.² (Sk.) 111; tho forseide lake MAND. 199; þe vorzede manne AYENB. 190.

fore-scēwing, sb., *O.E.* forescēawung; *foreshowing*; forsceuing C. M. 5745.

fore-schip, sb., *O.E.* forscip; *foreship*, RICH. 2618.

fore-siht, sb., = *O.H.G.* forasiht; *foresight*; forsiȝt, forsighte PR. P. 171.

fore-slēve, sb., *front of the sleeve* : of a freres frokke were the **forslēves** [foresleves] (*pl.*) LANGL. *B* v. 81.

fore-spēche, sb., *O.E.* fore-spræc, -spræc; *preface* ; vorespeche AYENB. 5.

forest, sb., *O.Fr.* forest ; *forest*, ROB. 375; CH. C. T. *A* 1975; LAI LE FR. 145.

fore-stain, sb., *prow of a ship* : frekes one þe forestaine fakene þeire coblez D. ARTH. 742; '*prora*' forstanig VOC. 274.

forester, forster, sb., *O.Fr.* forestier; *forester*, ROB. 499; LANGL. *B* xvii. 112; CH. C. T. *A* 117.

foret, feret, sb., *O.Fr.* furet; *ferret*, PR. P. 171.

fore-tāken, sb., *O.E.* foretācn, = *O.H.G.* forazeichan ; *foretoken*; fortaken ORM. 16157; fortacne HOM. I. 87; foretoken GEN. & EX. 2994; GOW. I. 137; fortocne HOM. II. 81.

fore-tēþ, sb. pl., *O.E.* foretēþ; *front teeth*, LANGL. *C* xxi. 386; FLOR. 1609.

fore-top, sb., *foretop*, VOC. 183; fore-, fortop SAINTS (Ld.) 317 (TREAT. 137).

fore-þouht, sb., *forethought*, L. H. R. 145.

fore-wal, sb., *O.E.* foreweall; *front wall, bulwark* : the wal and the fore wal [biforeval either a strengthe bifore the wall] WICL. IS. xxvi. 1.

fore-ward, sb. & adv., *O.E.* fore-, forweard!; *front part, first part; forward, forth* : on forward *in the beginning* HOM. I. 73; furst and a foreward BEK. 492; fro þat dai forward AM. & AMIL. 155; **forewardes** (*adv.*) forward MAND. 61.

fore-ward², sb., *O.E.* foreweard (SAX. CHR.

67), *cf. O.N.* forvörꝺr; *agreement, covenant,*
A. R. 62; foreward C. L. 1055; voreward
O. & N. 1689; foreward, forward LAȝ. 1091,
23657; LANGL. *A* vi. 36; forward CH. C. T.
A 829; M. ARTH. 3695; **forewardes** (*pl.*)
LAȝ. 31133; SPEC. 37.

fore-warde, sb., *vanguard,* CATH. 139;
P. P. ii. 280.

for-færen, v., = *M.L.G.* vorvēren, *M.Du.*
vervēren, -vaeren; *terrify;* **forfæred** (*pple.*)
ORM. 665; forfered CH. C. T. *F* 527; IW.
1678.

for-fal, sb., *O.N.* forfall (*impediment*); *sickness;* forval (*? ms.* forwal) LAȝ. 31590.

for-fáren, v., *O.E.* forfaran, = *O.H.G.* fur-,
for-, ferfaran, *O.Fris.* forfara; *perish, destroy,* GEN. & EX. 1087; let forfaren al
mankin ORM. 14582; þeȝ wolden him forfaren
al & cwenken 19632; forfare WILL. 2762;
A. P. ii. 1168; GAM. 74; EGL. 119; LUD. COV.
47; al þis lond he· wule forfare LAȝ. 25683;
forvare REL. I. 176; furfare LEG. 30; for-
várꝺ (*pres.*) A. R. 138; **forfōren** (*pret. pl.*)
LAȝ. 12380; **forfáren** (*pple.*) HAV. 1380;
þeos utlaȝen þa þisne wei us habbeoꝺ forvaren
(*intercepted*) LAȝ. 27373.

[**? for-fehten,** v.], **forfohte** (*pple.*), *exhausted
with fighting,* LAȝ. 26189*; forfouȝten JOS.
579; forfouten WILL. 3686.

[**for-fehten,** v., *fight for, defend.*]

forfiȝtere, sb., *defender,* WICL. IS. xix. 20.

for³-fenden, v., *keep, protect, prohibit* : wile
thou not forfenden him that mai wel do WICL.
PROV. iii. 27; thee also i shal not forfende
JOB xxxiv. 31; forfendin PR. P. 171; **forfende**
(*imp.*) thi foot from the sties of hem WICL.
PROV. i. 15; forfende (*pl.*) WICL. JOB xv.
11; from alle evel weie i **forfendede** (*pret.*)
mi feet WICL. PS. cxviii. 101; **forfendid** (*pple.*)
WICL. 2 KINGS xxii. 24; WICL. S. W. I. 241.

for-fēren, v., *O.E.* forfēran; *perish, destroy;*
forfērde (*pret.*) LAȝ. 7280; A. D. 298; GAW.
1617; A. P. ii. 571; vorverden A. R. 334.

forfēren, *see* **forfæren.**

forfet, sb., *O.Fr.* forfait, *med.Lat.* foris factum,
mod.Eng. forfeit; *offence, forfeit* : he schalle
finde no forfete amonge us MAND. 294; that
we falle into forfet with our fre wille D. TROY
2295; of oulde **forfetis** (*pl.*) upbraide not þi
felawe B. B. 29.

forfeten, v., *forfeit; transgress, commit sin* ;
forfetin '*forefacio, delinquo*' PR. P. 172; forfet
D. TROY 4446; forfette his landez D. ARTH.
554; ȝef he forfete (*subj.*) in eni of hem MAS.
331; þaȝ faurti forfete A. P. ii. 743; Jesus Crist
that never **forfeted** (*pret.*) CH. C. T. *I* 273;
ne forfeted . . . þe mede A. P. i. 616; oure
forme fader hit con forfete A. P. i. 636; we
mone be **forfetede** (*pple.*) and flemide for
ever 1155.

forfetour, sb., *forfeiter,* '*forefactor,*' PR. P.
172.

forfetūre, sb., *O.Fr.* forfaiture; *forfeiture,*
LANGL. *B* iv. 131; forfeiture RICH. 131; '*forefacio, forefactura,*' PR. P. 172; this shall be
thi forfeiture GOW. II. 268.

for-flütten, v., = *Dan.* forflytte, *Swed.* för-
flytta; *remove;* **forflitte** (*pple.*) AMAD. (Web.)
381.

for-fonnen, v., *become silly, senseless* : he
might faithli **forfonnet** (*pple.*) be a fole
holden D. TROY 632.

for-freten, v., = *Du.* vervreten; *devour, corrode;* **forfreteþ,** -fret (*pres.*) LANGL. *B* xvi.
29; vorvret A. R. 138; **forfrēten** (*pret. pl.*)
MISC. 226; forfrete (*pple.*) WILL. 2376.

forfriȝten, *see* **forfürhten.**

for-fruschen, v., *dash to pieces;* **furfrusched**
(*ppl.*) MAN. (F.) 1180.

for-fürhten, v., *frighten* : ꝺis **forfrigted**
(*pple.*) folc GEN. & EX. 3519.

for-gabben, v., *mock* : who so **forgabbed**
(*pple.*) a frere PL. CR. 1257.

for-gān, v., *O.E.* forgangan, -gān, = *O.L.G.*
for-, fargangan, *Goth.* faurgaggan, *O.H.G.* fer-
gangan, -gān; *forgo, pass by; escape; dispense
with; abstain from;* forga PR. C. 1842;
forgon A. R. 8; SPEC. 88; his rihtes forgon
BEK. 726; in whiche manere we mowe best
þe develes poer forgon P. L. S. xvii. 301;
forgon, -go CH. C. T. *E* 171; forgo R. S. vii;
WILL. 5187; vergon ROB. (W.) 5903*; **furgō**
(*pres.*) REL. I. 103; vorgeꝺ al þet he luveꝺ
A. R. 364; forgang (*imper.*) þu ones treowes
westm HOM. I. 221; þe his lond hafde **forgān**
(*pple.*) LAȝ. 22130.

for-garen, v., *O.N.* fyrirgöra; *ruin, destroy;
oppose* : þat home þat aungelez **forgart** A. P. ii.
240; hit watz **forgarte** (*pple.*) A.·P. i. 321;
full of sinne and all forgart ȝæn God ORM.
14582.

forge, sb., *O.Fr.* forge; *forge,* '*fabrica,*' PR. P.
172; CH. C. T. *A* 3762.

forgeable, adj., *that can be forged or made* :
forgeable thingis WICL. PREF. EPIS. vi.

for-ȝearwien, v., = *Dan.* forgiöre, *Swed.* för-
göra; *destroy, corrupt;* **forgart** (*pple.*) ORM.
14584, 17531; A. P. i. 321.

for-ȝélden, v., *O.E.* forgildan, = *O.L.G.* far-
geldan, *O.H.G.* for-, far-, fergeltan; *requite,*
HOM. I. 65; and þu mi muchele swinc mid
sare **forȝéldest** (*pres.*) LAȝ. 2298; forȝhelde
PS. xvii. 21; **forȝélde** (*subj.*) A. R. 428; LANGL.
B vi. 279; GAW. 839; god forȝelde þe DEGR.
860; (*ms.* foryelde) M. ARTH. 1548; forȝilde
AV. ARTH. xlii.

for-yhéldinge, sb., = *M.H.G.* vergeltunge;
retribution, PS. xxvii. 4; **foryhéldinges** (*pl.*)
PS. cii. 2.

for-ȝelwen, v., *? = M.H.G.* vergilwen; *turn
yellow, fade* : ꝺu . . . **forgelwes** (*pres.*) (*ms.*
-gelues) REL. I. 212 (MISC. 6).

for-ȝēmen, v., *O.E.* forgȳman ; *neglect* ; vorȝemen A. R. 372 ; **foriēmeþ** (*ms.* -yemeþ) (*pres.*) MISC. 114 ; (ȝe) forȝemeð HOM. I. 13 ; **forȝĕmden** (*pret.*) (*ms.* forrȝemmdenn ORM. 7502 ; forȝēmed (*pple.*) C. L. 947.

forȝerd, *see* fore-ȝeard.

for-ȝeten, v., *O.E.* forgitan, -gytan,= *O.L.G.* for-, fargetan, *O.H.G.* fergezan ; *forget,* HOM. I. 161 ; forȝetin, -ietin (*ms.* -yetin) '*obliviscor*' PR. P. 174 ; forȝiete HOM. II. 221 ; forȝete, forȝuten LANGL. *B* xvii. 242, *C* xx. 208 ; forȝete WICL. GEN. xli. 51 ; A. P. i. 86 ; forȝiten P. L. S. viii. 17 ; vorȝiten A. R. 272 ; furgete MAN. (F.) 14119 ; forȝeteþ (*pres.*) ORM. 2965 ; voriet (*ms.* -yet) AYENB. 18 ; forȝut P. L. S. viii. 13 ; (ha) forȝeoteð MARH. 15 ; forȝete (*subj.*) L. C. C. 24 ; **forȝat** (*pret.*) P. L. S. xvii. 216 ; LANGL. *B* xi. 59 ; WICL. GEN. xl. 23 ; foriat (*ms.* -yat), -gat HAV. 249, 2636 ; forgat S. S. (Wr.) 1273 ; forȝæt [forȝet] LAȝ. 25833 ; forȝet KATH. 1835 ; forȝeten LAȝ. 5514 ; A. R. 330 ; P. S. 342 ; WILL. 1909 ; forȝēte (*subj.*) CH. C. T. *A* 1882 ; **forȝeten** (*pple.*) C. L. 998 ; forȝeten is god ['*oblitus est deus*'] WICL. PS. x. 11 ; forȝetin [-ieten, -geten] CH. C. T. *A* 1914 ; forgeten IW. 1584 ; PR. C. 3909 ; forȝete WILL. 4934 ; vorȝiten A. R. 100 ; verȝite ROB. (W.) 6428* (*errata*) ; forȝute BRD. 26.

forȝetinge, sb., '*oblivio*,' PR. P. 174 ; MAND. 5 ; voryetinge AYENB. 18.

forȝetingnes, sb., *forgetfulness,* WICL. 2 PET. i. 9.

forȝet-ful, adj., *forgetful,* CH. C. T. *E* 472 ; WICL. JAS. i. 25.

for-ȝetil, adj., *O.E.* forgytol, = *M.L.G.* vorgetel ; *forgetful,* '*obliviosus,*' PR. P. 174 ; forȝetel (? *printed* foryetel) GOW. III. 98.

forgetelnesse, sb., *O.E.* forgytelness ; *forgetfulness,* HOM. II. 71 ; forȝetelnesse GOW. II. 19 ; forȝetelnesse CH. C. T. *I* 827 ; forgetelnes PS. ix. 19.

forȝetil-schip, sb., *forgetfulness* : for a forgetilschip Richard and he boþe les MAN. (H.) 176.

forghe, *see* furh.

[**for-ȝife-,** *stem of* forȝifen.]

vorȝive-lich, adj., = *M.H.G.* vergebelich, ? *O.E.* forgifendlīc ; *pardonable,* A. R. 346.

for-ȝifen, v., *O.E.* forgifan, = *O.H.G.* fur-, for-, far-, fergeban ; *forgive,* ORM. 1464 ; forȝive WICL. NUM. xxx. 9 ; forȝefen, -ȝeven HOM. I. 23, 37 ; forȝeve P. L. S. xvii. 214 ; forȝive, -give LANGL. *B* iii. 8 ; **forgieve** (*subj.*) MK. xi. 25 ; þou forȝeve A. D. 261 ; **forȝaf** (*pret.*) ORM. 19294 ; P. L. S. xxv. 47 ; voriaf (*ms.* -yaf) AYENB. 114 ; forȝef LAȝ. 4273 ; **forȝǣfe** (*subj.*) ORM. 1143 ; forȝiven (*pple.*) A. R. 124 ; vorieve (*ms.* uoryeue) AYENB. 29 ; þine sunnen þe beð forgivene HOM. II. 145.

forȝifenesse, sb., *O.E.* forgifeness, -gifenness ; *forgiveness,* ORM. 1477 ; vorȝivenesse,

forȝevenesse HOM. I. 29, 217 ; forȝivenesse LEG. 12 ; forȝeovenesse H. M. 43 ; voryevenesse AYENB. 32.

foryevinge, sb.,= *M.H.G.* vergebunge ; '*remissio,*' PR. P. 174.

forgin, v., *O.Fr.* forgier ; *forge,* PR. P. 172 ; forge WICL. EX. xxvii. 3 ; **forged** (*pple.*) CH. C. T. *A* 3256.

forgere, sb., *forger, fabricator, smith* : God that is forgere of alle thinges WICL. ECCLES. xi. 5 ; **forgers** (*pl.*) WICL. PREF. EP. vi.

forging, sb., *fabrication* ; forging of gold WICL. ECCLUS. xxxii. 8.

for-glopnen, v., *terrify* ; **forglopned** (? *ms.* forrgloppnedd) (*pple.*) ORM. 670.

for-gnaȝe, v., *gnaw to pieces,* FER. 1148 ; forgnaȝeþ (*pres.*) SHOR. 98 ; forgnaweð A. R. 338 ; **vorgnōwe** (*pret.*) LEG. 161.

for-gnīden, v., *O.E.* forgnīdan, = *O.H.G.* fargnītan ; *bruise* ; **forgnōd** (*pret.*) ['*contrivit*'] PS. cvi. 16.

for-greiðen, v., *prepare, make ready* : in þair levinges forgraiþe lickam of þa PS. xx. 13.

forgraiðing, sb., *preparation* : forgraiþinge ['*præparationem*'] of þair herte herd ere þine PS. ix. 38.

for-gremen, v., *deprive of (by displeasing God)* : þe mihtes þe Adam forgremede (*pret.*) us alle HOM. II. 35.

for-grōwen, v., *grow to excess* : **forgrouwen** (*pple.*) mid brimbles HOM. II. 129 ; forgrowe P. P. I. 363 ; GENER. (Wr.) 3667.

for-gülten, v., *O.E.* forgyltan ; *render guilty ; forfeit by guilt* : to forgilten us ȝæn god ORM. 2619 ; **forgilt** (*pres.*) HOM. II. 211 ; **forgülte** (*pret.*) HOM. I. 23 ; ear we weren **vorgülte** (*pple.*) A. R. 388.

for-hacchen?, v., *despise* ; **forheccheð** (*pres.*) H. M. 41* ; **forhaht** (*pple.*) SPEC. 37.

for-hardien, v., *O.E.* forheardian, *cf. O.H.G.* far-, fer-, furihertan ; *harden* ; **forha(r)dede** (*pret.*) GEN. & EX. 3338.

for-hāren, v., *grow hoary* : thou olde and **forhōrid** (*pple.*) man GUY 11089.

for-hāten, v., *O.E.* forhātan ; *promise ; renounce* : ȝif þu wult . . . vorhoten þine sunne A. R. 340 ; **forhōteð** (*pres.*) HOM. II. 199 ; **forhēt** (*pret.*) P. L. S. xvii. 86 ; þat he lang forhight C. M. 11334 ; vorheten A. R. 192 ; **forhōte** (*pple.*) SHOR. 162.

for-healden, v., *O.E.* forhealdan, = *M.L.G.* vorholden ; *detain, retain* ; **forhēld** (*pret.*) GEN. & EX. 2026 ; **forhōlde** (*pple.*) MISC. 149.

for-hēawen, v., *O.E.* forhēawan, *O.H.G.* furhouwan ; *hew in pieces* ; **forhēow,** -heou [forhew] (*pret.*) LAȝ. 4593, 28032 ; **forhēwe** (*pple.*) FER. 899.

for-hēden, v., *neglect* : wimmen he **forhĕdde** (*pret.*) LAȝ. 2579*.

forheed, *see* forehēved.

for-hefednesse, sb., *O.E.* for-, færhæfedness; *continence,* HOM. I. 101.

for-hélen, v., *O.E.* forhelan,=*O.H.G.* far-, ferhelen; *conceal, hide, cover, protect,* GEN. & EX. 2593; forhele O. & N. 798; forhile him i sal for mi name knewe he PS. xc. 14; forhéle (*imper.*) REL. I. 177; forhéle (*subj.*) LAȝ. 4360; forheole HOM. I. 37; forhal (*pret.*) ORM. 233; forhólen (*pple.*) GEN. & EX. 2317; AN. LIT. 8; MAP 337; it sholde wurþen wel forholen wiþ þe defel ORM. 2468; forhole ALIS. 6967; S. S. (Web.) 250; nis him nout forhole P. L. S. viii. 39.

forhiler, sb., *protector*: PS. xvii. 31, xxxvi. 39, lxxxiii. 10, cxiii. 9.

forhiling, sb., *cover, shelter, protection*: PS. xvii. 36, xc. 1, civ. 39.

for-hengen, v.,=*M.H.G.* verhengen; *hang*; forhenge HAV. 2724.

for-hoȝien, v., *O.E.* forhogian; *contemn, despise*; forhohien KATH. 993; forhowien A. R. 166; forhoȝhen ORM. 3959; forhoȝeð (*pres.*) LAȝ. 31565; forhoweþ FRAG. 5; forhoȝie (*subj.*) HOM. I. 49; forhoȝeden [forhowede] (*pret.*) LAȝ. 19421.

for-hoht, -hoght, sb., *contempt,* PS. cxviii. 22.

for-hōnen, v., *? M.Du.* verhoonen; *despise*; forhone PS. xliii. 6.

for-hōren, v.,=*O.H.G.* for-, farhuorōn; *debauch*; vorhoren A. R. 394; forhōred (*pple.*) HOM. II. 81; ORM. 2043.

forhōren, *see* forhāren.

for-hūden, v., *O.E.* forhȳdan,=*L.G.* ferhüden; *hide*; forhid (*pple.*) GEN. & EX. 1875.

for-huȝien, v., *contemn, despise*; forhugien (*pres. subj.*) MAT. xviii. 10; forhugede (*pret.*) GEN. & EX. 3814.

for-hungren, v.,=*Ger.* verhungern; *famish*; forhungred (*pple.*) ORM. 5679; WILL. 2515; forhungrid PR. P. 172; forhongered LAȝ. 23562*.

for-huschen, v., *? deride*; forhuste (*printed* forhusce) (*pret.*) LAȝ. 3171; forhusten 29021.

for-īdlen, v.,=*M.H.G.* verītelen; *become idle*; forīdled (*pple.*) A. R. 116.

for-irken, v., *disgust*: of manna he ben forirked (*ms.* forhirked) (*pple.*) to eten GEN. & EX. 3658.

for-justen, v., *overthrow, defeat*: jolili this gentille forjustede (*pret.*) anoþer D. ARTH. 2088; so mani groundes he forjustede D. TROY 296; forjusted A. P. ii. 1215; forjustid (*pple.*) DEGR. 1897; thus es þe geante forjuste 2896.

forke, sb., *O.E.* forc [ÆLF. HOM. I. (Sk.) 430], *from Lat.* furca, *O.Fris.* forke, *O.Fr.* furke; *fork,* 'furca,' VOC. 180; SPEC. 110; forken (*pl.*) LAȝ. 21102; forkes (*gallows*) DEP. R. i. 108; *comp.* pic-forke.

forked, adj., *forked,* CH. C. T. *A* 270; C. M. 18843.

for-keorven, v., *O.E.* forceorfan; *cut to pieces*; vorkeorven A. R. 360; forkerveþ (*pres.*) CH. C. T. *H* 340; forcarf (*pret.*) LAȝ. 8186*; forkarf B. DISC. 1325; forcorven (*pple.*) H. M. 11; forcorvin FER. 617.

forkin, sb., *baker's shovel,* 'pala,' VOC. 276.

for-lǣden, v., *O.E.* forlǣdan, = *O.L.G.* farlēdean, *O.H.G.* for-, far-, ferleiten; *seduce*; forlēdeþ (*pres.*) LAȝ. 1333*; forlēdde (*pret.*) SHOR. 164.

for-lǣren, v., *O.E.* forlǣran; *seduce*; forlēreð (*pres.*) HOM. II. 29; þat þi dweolesong heo ne forlēre (*subj.*) O. & N. 926.

for-lǣten, v., *O.E.* forlǣtan, = *O.L.G.* for-, farlātan, *O.H.G.* fur-, for-, far-, ferlāzan; *forsake, leave,* ['*dimittere*'] MAT. xv. 32; ORM. 3769; forletan, -leten HOM. I. 19, 107; forleten LAȝ. 30599; A. R. 406; forlete HORN (L.) 218; CH. C. T. *C* 864; he scholde þe lif forlete C. L. 178; for-, furlete O. & N. 966; furlete HICKES I. 225; ferleten HOM. I. 219; forlatin '*desolo*' PR. P. 172; forletēð (*pres.*) HOM. II. 27; forlet HOM. I. 119; alle þo þe here sinnen forleteð HOM. II. 41; forlēt (*pret.*) HOM. I. 19; Adam god forlet ORM. 355; al folc wel neh forlet to þenken oht of hefne 18875; forleten GEN. & EX. 4068; alle þe breme bestes ... forlete us & folwed him WILL. 2311; forlēten (*pple.*) ORM. 3119; forlete ALIS. (Sk.) 679; vorlete L. H. R. 34.

forlētenesse, sb., *O.E.* forlǣtennyss; *contempt, scorn,* PS. cxxii. 3.

forlēting, sb., *neglect, scorn, contempt,* PS. cxxii. 4.

for-lǣven, v.,=*O.H.G.* firleiben; *leave*; forleve WICL. LEV. xxv. 46; forlēaf (*imper.*) JUL. 33; forlēft (*pple.*) WILL. 2497; forlaft P. S. 340.

for-laȝe, sb., *? O.N.* forlög (*fate*); *? opportunity,* HOM. I. 19.

for-lange, adj., *very long,* ORM. 6996.

for-langen, v.,=*M.H.G.* verlangen; *be possessed with longing*: ȝif þat tu forlanged (*pple.*) art to cumen up til Criste ORM. 1280.

vorlonginge, sb., *longing,* A. R. 274.

for-laped, pple., *? satiated with drinking,* P. S. 238.

for-lēden, v., *lead away, bring forth*: þethen sal i forlede David horn PS. cxxxi. 17.

for-leȝernesse, sb., = *O.L.G.* farlegarnessi; *adultery,* ORM. 2032; forleirnisse MAT. v. 32; *see* forliȝer.

for-lēinen, v., = *O.L.G.* farlōgnian, *O.H.G.* fur-, for-, ferlougnen; *deny*: þat mai he (*misprinted* be) nouȝt forlain TRIST. 1586.

for-lengþen, v., *lengthen*; forlenghþed (*pret.*) ['*prolongaverunt*'] PS. cxxviii. 3.

for-lēosen, v., *O.E.* forlēosan,=*O.L.G.* for-,

farleosan, -liosan, *O.H.G.* fur-, for-, far-, fer-liosan, -leosan ; *lose*, LAȝ. 9486 ; KATH. 347 ; A. R. 108 ; ne let tu neaver mi sawle for-leosen MARH. 3 ; forlesen ORM. 2278 ; for-leose HORN (R.) 665 ; þu forlēost [-lost] (*pres.*) O. & N. 1649 ; vorleoseð A. R. 120 ; forleost MK. ix. 41 ; for-, vor-, furleost O. & N. 619, 693, 949 ; forliest MISC. 31 ; we for-leosað HOM. I. 105 ; (hi) vorlieseþ AYENB. 57 ; forlēose (*subj.*) SPEC. 94 ; forlēas [forles] (*pret.*) LAȝ. 213 ; vorleas A. R. 232 ; AYENB. 181 ; forles SAX. CHR. 264 ; GEN. & EX. 189 ; (þu) forlure FRAG. 7 ; vorlure A. R. 310 ; verlore ROB. (W.) 4658* ; (heo) forluren HOM. I. 93 ; ORM. 1412 ; forloren GEN. & EX. 241 ; for-lóren (*pple.*) [*mod.Eng.* forlorn] A. R. 146 ; GEN. & EX. 1886 ; HAV. 580 ; H. H. 243 ; forlorn CH. C. T. *F* 1557 ; forlore O. & N. 1391 ; furlore MISC. 153.

forlīezinge, sb., *loss*: vor þet me bevliȝt hire folie and hire vorliezinge AYENB. 156.

forlórennesse, sb., *O.E.* forlorenness ; *lost condition* ; vorlorenesse A. R. 66, 110.

for-lēve, *see* forlǣven.

for-liȝer, sb., *O.E.* forliger *n.*= *O.H.G.* for-ligiri ; *fornication*, HOM. I. 103 ; forleigre (*pl.*) MK. vii. 21.

for-liggen, v., *O.E.* forlicgan, = *O.H.G.* fur-ligan ; *commit adultery* ; forlīð (*pres.*) H. M. 41 ; forlai (*pret.*) GOW. I. 347 ; vorlai AYENB. 206 ; forlæiȝen LAȝ. 15375 ; forleȝen (*pple.*) ORM. 1988 ; forleien SHOR. 65 ; forleie FL. & BL. 301 ; furlain MAN. (F.) 12271.

forliȝer, sb., *O.E.* forligere ; *adulterer*, '*adulter*,' FRAG. 5 ; forlier MAT. xii. 39 ; forliȝeres (*pl.*) HOM. I. 117.

for-lītenen, v., *? lessen*: we hafe as losels liffide … and forelītenede (*pple.*) the loos þat we ar laitande (*printed* layttede) D. ARTH. 254.

for-livien, v., *?*= *M.H.G.* verleben ; eni for-lived (*pple.*) (*decrepit*) wrecche FL. & BL. *B* 98.

for³-lōkien, v., *forelook, look forward* ; for-luke PR. C. 1946.

forlōkere, sb., *provider, looker-out* : God is the forlookere of him that ȝeldeth grace WICL. ECCLUS. iii. 34.

forloinen, v., *O.Fr.* forlonger ; *forsake, depart, go astray, err* : þaȝ i forloine (*subj. pres.*) A. P. i. 368 ; forloine ii. 750 ; uch freke forloined (*pret.*) fro þe riȝt waies A. P. ii. 282 ; þai forloine her faith and folȝed oþer goddes A. P. ii. 1165.

forlond, *see* foreland.

forlórenesse, *see under* forlēosen.

for-lost, pple., *utterly lost*: she held hire self a forlost creature CH. TRO. iv. 728 ; she forloste iii. 228.

for-lüst, adj., *very anxious* : forlist after þe blisse of heofene ORM. 11475.

formaille, *see* formel.

formal, adj., *Lat.* formālis ; *formal*: the cause formal is the maner of here werking CH. *B* 2590.

for-manging, sb., *change* : noght es to þam formanging, and þarwith drede þai na thing PS. liv. 20.

forme, sb., *O.Fr.* forme ; *form*, '*forma*,' PR. P. 172 ; CH. C. T. *B* 1294 ; hi makede a forme (*agreement*) P. L. S. xiii. 314 ; an hare whan he in forme liþ TREAT. 139 ; forme *bench* ['*scamnum*'] of þe chirche TREV. VII. 369 ; an fourme hii made þat eiþer helde is owe lond in his hond ROB. (W.) 8766 ; in þis fourme (*on this condition*) ROB. (W.) 8907.

forme-liche, adv., *according to proper rule* : that kanst so wel and formeliche arguwe CH. TRO. iv. 468.

forme, sb., *O.E.* forma, feorma, *O.L.G.*, *O.Fris.* forma ; ?= frume (*see* GR. GRAM. III.627) ; *first*, JUL. 60 ; ORM. 14431 ; O. & N. 820 ; ALIS. 1292 ; S. S. (Web.) 1424 ; JOS. 685 ; ure forme fader gult HOM. I. 171 ; ȝiefe him his forme mete 231 ; þe forme mon C. L. 1477 ; oure forme fader A. P. i. 638 ; PR. C. 483 ; forme, vorme LAȝ. 4942, 25151 ; þe vorme dole A. R. 10 ; furme P. S. 155 ; firme GEN. & EX. 59 ; formen (*dat.*) HOM. I. 243.

formel, formaille, sb., *cf. O.Fr.* (faucon) formel ; *female* (*of birds of prey*) : his formel, or his make CH. P. F. 369 ; a formel egle 372 ; faukone ne formaille appone fiste handille D. ARTH. 4003.

for-melden, v., = *O.H.G.* for-, fermelden ; *announce* ; formelt (*pple.*) C. M. 13496.

for-melten, v., *O.E.* formeltan ; *melt away*, FRAG. 6 ; formelteð (*pres.*) HOM. II. 151 ; formealten (*subj.*) H. M. 13 ; a ðet ha beon formealte (*pple.*) HOM. I. 251.

formen, v., *Fr.* former, *from Lat.* formāre ; *form* ; formin PR. P. 172 ; formede (*pret.*) HAV. 1168 ; fourmede (*pple.*) D. ARTH. 1061 ; fourmit D. TROY 1540.

former, sb., *former, creator*, MED. 795 ; former, formour, formeour, fourmour LANGL. *C* xx. 133, ii. 14, *B* xvii. 67, ix. 27 ; formiour of alle the world MAND. 2.

formere, comp. adj., *former*, MAT. xxi. 36 ; WICL. GEN. xxxii. 17 ; formere fader (*ancestor*) MAND. 2.

formest, adj. & adv., *O.E.* formest, fyrmest, = *O. Fris.* formest, *? Goth.* frumist ; *foremost, first*, LEG. 49 ; WILL. 939 ; LANGL. *B* x. 215 ; seþþen Adam formest sunne bigon C. L. 1140 ; a vormest LAȝ. 24611 ; formest, fermest MK. ix. 35, x. 44 ; formast A. P. ii. 494 ; furmest P. S. 154 ; firmest REL. 213 ; GEN. & EX. 1472 ; formeste MISC. 31 ; þe formeste LAȝ. 6855 ; þet formeste liht HOM. I. 139 ; vormest(e) A. R. 200 ; formeste MK. xii. 30.

formour, *see under* formin.

forn, *see* fóren.

fornaise, sb., *O.E.* fornaise; *furnace*; fornais AYENB. 131; forneis MAND. 35; WICL. DEUT. iv. 20; fornis A. P. ii. 1010; VOC. 233; fornis 276; furneis WICL. GEN. xv. 17; PR. P. 183; furnasse VOC. 200; fourneise WICL. MAT. vi. 30; fournes MISC. 211; **fornaise** (*dat.*) AYENB. 205; furneise JUL. 32; **forneises** (*pl.*) WICL. LEV. xi. 35.

forn-cast, ppl., *forecast*, CH. C. T. *B* 4407.

for-nean, adv., *O.E.* fornean; *almost*, '*pene*,' FRAG. 1; fornean en ende þissere wurold HOM. I. 227.

for-nēh, adv., *O.E.* forneah; *nearly, almost,* H. M. 41; forneh wod MARH. 7.

forneis, *see* fornaise.

fornicari, adj., *Lat.* fornīcārius; *given to fornication*, WICL. PREF. EPIS. vii; fornicarie womman WICL. APOC. xvii. 16.

fornicatioun, sb., *O.Fr.* fornication, *from Lat.* fornīcātiō; *fornication*: thei seie that fornication is no sinne dedli MAND. 19; fornicacioun MAND. 69; fle ʒe fornicacioun WICL. I COR. vi. 18; fornicacion AYENB. 47, 206.

fornicatour, sb., *fornicator*: fornicatouris, fornicatours (*pl.*) WICL. HEB. xiii. 4, APOC. xxi. 8.

for-nimen, v., *O.E.* forniman; *take away*; fornumen (*pple.*) GEN. & EX. 2228; fornomen PS. cviii. 23.

foróld, *see* foreald.   **foroure**, *see* furre.

for-oute, prep., *without*: foroute alle greves of sauʒtes to the cite WILL. 2681.

for-painen, v., *torture severely*; forpained (*pple.*) A. P. i. 246.

for-pampren, v., *pamper*; forpampred (*pple.*) CH. APPEND. BOET. 5 (180).

for-pinchen, v., *pinch to pieces*: hit shal be so forpinched (*pple.*) totoilled, and totwiht P. S. 337.

for-pīnen, v., *torture, pine, famish*; forpīnede (*pret.*) LAʒ. 29130; vorpīned [forpeined] (*pple.*) ROB. (W.) 1159; forpined CH. C. T. *A* 205.

for-possen, v., *toss up and down*: as in ballaunce forpossed (*pret.*) up and doun HALLIW. 373; forpossed with woe LIDG. BOCH. i. 5 (f. iii.)

for-pricken, v., *spur violently*; verpriked (*pret.*) ROB. (W.) 7490*; vorpriked (*pple.*) ROB. 362.

for-rǣden, v., *O.E.* forrǣdan, = *O.H.G.* for-, ferrāten; *deceive, seduce*; forrēadeð (*pres.*) JUL. 18; vorrādde [forradde] (*pret.*) LAʒ. 14867; forrěd (*pple.*) GEN. & EX. 2192; forrad MAP 337.

forrai, sb., *foray*: sum war set to the forrai BARB. xvi. 612; *see* ferrai.

forraien, v., *O.Fr.* forrer; *forage, plunder*: to forrai no goudes A. P. ii. 1200; **forraise** (*pres.*) thi landez D. ARTH. 1247; **forraiede** (*pret.*) 3017.

for-raken, v., *overdo with walking*: i am weri forraked (*pple.*) and run in the mire TOWNL. 105.

for-recchen, v., = *O.H.G.* ferrecchan; *? pervert, corrupt*; forraht (*pple.*) ORM. 14540.

forreier, sb., *O.Fr.* forrier; *forager, harbinger*; forrier AYENB. 195; forreiers (*pl.*) MAN (F.) 13228; *see* ferriour.

forred, *see* furrin.

for-rennen, v., *cf. Ger.* verrennen; *bar by running*: foraernen þa wateres LAʒ. 12861.

for-rīdel, sb., *O.E.* forrīdel; *outrider, precursor*; vorrīdeles (*pl.*) A. R. 206.

for-rīden, v., *ride before, anticipate, intercept*: þat þa Romleoden heom forriden (*pple.*) haveden LAʒ. 26931.

forrier, *see* forreier.

for-rotien, v., *O.E.* forrotian, = *M.L.G.* vorroten; *rot away*, R. S. v.; vorrotien A. R. 344; forroti P. L. S. v. I, 4; **vorrotede** (*pret.*) AYENB. 205; forroteden H. M. 13.

forrour, *see* furrour.

for-sāken, v., *O.E.* forsacan, = *O.H.G.* fur-, for-, farsahhan; *forsake*, LAʒ. 23687; MAND. 226; forsake HAV. 2778; WILL. 1358; S. S. (Wr.) 65; PR. C. 4406; þu me schalt þries fursake MISC. 41; **forsáke** (*pres.*) *B* v. 431; fursake HICKES I. 226; forsakeð A. R. 226; forsakeþ ORM. 17243; **forsōc** (*pret.*) GEN. & EX. 1833; BEK. 1982; forsok ALIS. 7715; þu forsoke A. R. 108; ha forsoken H. M. 21; vorsok ROB. (W.) 289; **forsáken** (*pple.*) GEN. & EX. 3811; fursaken REL. I. 103; forsake BEK. 661; vorsake [versake] ROB. (W.) 6374*.

for-sákien, v., *cf. O.E.* sacian; *forsake; deny*: of thi frendes forsákid (*pple.*) thou be S. & C. I. xlv.

forsákinge, sb., *denial*, '*refutacio, derelictio, desercio*,' PR. P. 172.

for-schāden, v., *O.E.* forscēadan, *cf. O.H.G.* forsceiden; *depart*: al bið ðes faxes feirnes forscēden (*pple.*) GR. 47.

for-scháken, v., *drive away*: forscháken (*pple.*) als gressop with gram PS. cviii. 23; swa sones of forschaken in land cxxvi. 4.

for-schalden, v., *scald*; forscaldede, -schaldede (*pret.*) JUL. 70, 71; **forschalded** (*pple.*) A. R. 246.

for-schámen, v., *O.E.* forsceamian; *put to shame*; forshámed (*pple.*) ORM. 2183.

for-s(ch)ápe, v., *transform*, AN. LIT. 11; forscháped (*pple.*) WILL. 2639.

for-schenden, v., *disgrace*; forshend (*pple.*) PARTEN. 3306.

for-scheppen, v., *O.E.* forsceppan; *deform, transform*; forschüppeð (*pres.*) A. R. 120; forschōp (*pret.*) WILL. 4394; forshepte, -shupte (*pret.*) LANGL. *B* xvii. 288, *C* xx. 270; forshápin (*pple.*) TOWNL. 115; forschipt C. L. 634.

forschipping, sb., *deformity*, C. L. 640.

forschüppild, sb., *transformer*; vorschuppild A. R. 120.

for-schrinken, v., *O.E.* forscrincan; *shrink*; forscrunken (*pret.*) MAT. xiii. 6.

for-schütten, v., *O.E.* forscyttan; *shut up*; forscütted (*pple.*) FRAG. 7.

for-senchen, v., *O.H.G.* farsenkan; *submerge*; forsenctest, -senchtest (*pret.*) JUL. 60, 61.

for-sēon, v., *O.E.* forsēon, = *O.L.G.* forsehan, *O.H.G.* far-, fersehan; *despise*, ORM. 9619; vorsien ANGL. I. 31; forsihð (*pres.*) HOM. I. 111; (ha) forseoð MARH. 15; forsēh (*pret.*) HOM. I. 111; forsēwen, forseȝen (*pple.*) HOM. I. 227, 113.

for-sēoðen, v., *boil away*; forseþe SHOR. 165; leste ure owune teares vorsēoðen (*subj.*) us ine helle A. R. 312.

for-set, sb., *from* forsetten; *obstacle, stop*, MAN. (F.) 2912.

for-setten [? for-³], v., *O.E.* forsettan, = *M.H.G.* versetzen; *set before; bar, impede*; forsettez (*pres.*) A. P. ii. 78; forsette (*pret.*) M. ARTH. 3046; forset (*pple.*) JOS. 487.

　forsetting, sb., *purpose*: i sal open in sauter mi forsettinge PS. xlviii. 5; speke sal i fra biginninge forsettinges (*pl.*) PS. lxxvii. 2.

for-singen, v., *tire with singing*: forsongen (*pple.*) R. R. 663.

for-sinken, v., = *O.H.G.* forsinchan; *sink away*; forsanc (*pret.*) GEN. & EX. 1114.

for-sitten, v., *O.E.* forsittan; *supersede, neglect*; forsēte (*pret. subj.*) LAȝ. 28518.

for-slǣpen ? v., *overcome with sleep*; forslēpt (*pple.*) ['*sopitus*'] WICL. PROV. xxiii. 34.

for-slāwen, v., *O.E.* forslāwian; *neglect*; forslēwiþ (*pres.*) SHOR. 114; forsleweth CH. C. T. *I* 685; wanne hii vorsoke is & vorslēwede [*v. r.* forsleuþed] (*pret. pl.*) ROB. (W.) 4055.

for-slēuþen, -slōuþen, v., *lose through sloth*, CH. C. T. *B* 4286; forslēuþed(e) (*pret.*) ROB. 197; forslēuþed (*pple.*) LANGL. *B* x. 445, *C* viii. 52.

for-sluggen, v., *neglect*; forsluggeth [forsloggeth] (*pres.*) CH. C. T. *I* 685.

for-smīten, v., *smite to pieces*; forsmāt (*pret.*) LAȝ. 1598; forsmiten (*pple.*) L. N. F. 213; forsmitten PALL. iii. 272.

for-spánen, v., *O.E.* forspanan, = *O.L.G.* forspanan, *O.H.G.* far-, ferspanan; *entice, seduce*: warþ þa þat wif forspannen (*pple.*) þurh þe deofles lare HOM. I. 223.

for-spékin, v., = *M.L.G.* vorspreken; *charm, enchant, 'fascino,'* PR. P. 173; forspókin (*pple.*) TOWNL. 115.

for²-spéker, sb., = *O.H.G.* fersprechāre; *advocate*; vorspeker HOM. I. 83; vorspekere REL. II. 229.

for-spenden, v., *O.E.* forspendan, *cf. O.H.G.*

forspentōn; *spend*; forspende (*pret. subj.*) HOM. I. 31.

for-spillen, v., *O.E.* forspillan, -spildan, = *O.H.G.* farspildan; *spill*; forspille PS. xciii. 23; wile ... forspille ['*perdiderit*'] MAT. xvi. 25.

for-sprǣden, v., *extend, spread*; forsprēde (*imper.*) þi merci thorgh the land PS. xxxv. 11.

forst, *see* **frost**.

for-stal, sb., *O.E.* forsteall; *forestalling*, REL. I. 33; forstalles (*pl.*) ANGL. VII. 220.

for-stallen, v., *forestall, intercept*: forstalleþ (*pres.*) mi feire LANGL. *A* iv. 43; forstalle (*pres. subj.*) E. G. 396.

　forstallinge, sb., *forestalling*, TREV. II. 95.

for-standen, v., *O.E.* forstandan; *hinder*: and forstōd (*pret.*) heom þene vord LAȝ. 20159.

for-stélen, v., *O.E.* forstelan, = *M.L.G.* vorstelen, *O.H.G.* for-, fur-, ferstelan; *steal away*, HOM. II. 35; forsteolan HOM. I. 109; forstóle (*pple.*) MISC. 192; furstole 53.

for-stoppen, v., = *M.L.G.* vorstoppen, *M.Du.* verstoppen; *stop up*; forstoppeð (*pres.*) A. R. 72; forstoppid (*pret.*) RICH. 4843.

for-stormen, v., *toss about with storms*: the ship which on the wawes renneth and is forstormed (*pple.*) and forblowe GOW. I. 160.

for-süneȝen, v., *O.E.* forsyngian, *cf. M.H.G.* versündigen; *sin*; forsineged (*pple.*) HOM. II. 143; þam forsünegede (*who has sinned*) HOM. I. 95; þo forsinegede II. 93.

for-swǣlen, v., *O.E.* forswǣlan; *burn up*; forswǣlde, -welde (*pret.*) LAȝ. 16228, 29127; sie sunne hit forswǣlde MK. iv. 6; forswǣled (*pple.*) LAȝ. 16271.

for-swǣten, v., *spoil with sweating*: swart ant forswǣt (*pple.*) P. S. 158.

for-swarten, v., *blacken*; forswarted (*pple.*) P. L. S. xxiv. 227.

for-sweiȝen, v., *go astray*: þei mowe nought forswei GOW. III. 275; if so be that a king forswei (*subj.*) GOW. III. 224.

for-swelȝen, v., *O.E.* forswelgan, -sweolgan, *M.L.G.* vorswelgen, *cf. O.H.G.* farswelgan; *swallow up*; vorzuelȝe AYENB. 56; forswalȝe [forsw(e)olȝe] LAȝ. 28453; forswolhen MARH. 9; forswolehen R. S. v; vorsuolwe ROB. 205; vorswoluwen A. R. 164; forswoleȝeð (*pres.*) HOM. I. 123; vorswoluweð [forsweolhes] A. R. 66; forswealh (*pret.*) HOM. I. 123; forsvalȝ MARG. 161; vorzualȝ AYENB. 67; (hi) vorzuolȝe 206; forswolgen (*pple.*) HOM. II. 43; forswolhen MARH. 13.

for-swelten, v., *O.E.* forsweltan; *faint, die*, JUL. 19, 79.

for-swelten², v., *kill*, MARH. 5; JUL. 18; forswelt (*pple.*) ALIS. 7559.

for-swérin, v., O.E. forswerian, = M.L.G. vorsweren, O.H.G. fersweren; forswear, commit perjury, 'perjuro,' PR. P. 173; he him vorzuerþ AYENB. 6; ne forswerie þu þe HOM. I. 13; vorsuerie ROB. (W.) 7945; **forswóren** (pple.) HAV. 1423; LANGL. B xix. 367; heo beoð forsworene LAȝ. 21185; versuore ROB. (W.) 7442.

    forswérere, sb., perjurer, PR. P. 173.

    forswéringe, sb., perjury; 'perjurium' PR. P. 173; slaghter and forswering PR. C. 3367.

    forswórenesse, sb., O.E. forsworennyss; perjury, HOM. I. 103.

for-swǽten, see forswǽten.

for-swingen, v., harass, bruise: when þou were so forswong (pple.) among the jues þei did þe hong L. H. R. 194.

for-swinken? v., exhaust with labour;" forsw(u)nken, forswunke (pple.) REL. I. 180.

for-swiðen, v., burn up; forswiðande (pple.) A. R. 306*.

[**for-swunden,** pple. of *for-swinden, v.,= O.H.G. farswindan (swoon, fail).]

    forswunden-lecg, sb., indolence: unlust & forswundenlecg is drihtin swiþe uncweme ORM. 2623.

    forswundennesse, sb., indolence, remissness : him is idelnesse laþ & all forswundennesse ORM. 4736.

fort, see under for.

for-tateren, v., tear in tatters: fortatird (pple.) and torne TOWNL. 239.

for-taxen, v., tax heavily : fortaxed (pple.) and ramid TOWNL. 98.

forteletes, sb. pl., O.Fr. fortelesse; forts, MAN. (F.) 4822.

for-tēon, v., O.E. fortēon, = O.H.G. farziuhan; seduce, draw aside; fortēð (pres.) HOM. II. 87; fortēoþ (? are seduced) REL. I. 182.

for-travaillen, v., tire out; fortravailled (pple.) hi were sore P. S. L. xiii. 313.

for-treden, v., O.E. fortredan, = O.H.G. furtretan; tread down; fortrede HALLIW. 374; PS. xc. 13; fortreden (pple.) HOM. II. 155; fortroden CH. C. T. I 190.

fortress, sb., O.Fr. forteresse; fortress; forceresses (pl.) MAN. (F.) 7143.

for-trūsten, v., cf. Dan. fortröste, Swed. förtrösta; trust, entrust : lest sum fortrūste (subj.) him HOM. I. 249.

for-tühten, v., O.E. fortyhtan; seduce; fortehten (pret.) HOM. II. 87.

    fortühting, v., seduction, HOM. II. 107.

fortunáte, adj., Lat. fortūnātus; happy; auspicious : that house mai not be fortunate P. L. S. xxxi. 198; borne, bi influence hevenli so fortunate LIDG. M. P. 37.

fortúne, sb., O.Fr. fortune; chance, fortune, CH. C. T. A 1252; the fortune of everi chaunce

GOW. III. 12; fortune did your flesch to diȝe A. P. i. 306; þe guodes of fortune AYENB. 76; be fortune MAND. 266; þe levedi fortune AYENB. 26; dame fortune 181; dam fortone PR. C. 1273; governed bi foolisshe happes and **fortúnes** (pl.) CH. BOET. i. 6 (26).

fortunel, adj., O.Fr. fortunel; fortuitous: fortunel happes CH. BOET. v. 1 (152).

fortúnen, v., O.Fr. fortuner; make fortunate; give fortune; occur by chance : wel coude he fortune the ascendent CH. C. T. A 417; obedience mai well fortune a man to love GOW. I. 104; o stronge god that ... hem **fortúnest** (pres.) CH. C. T. A 2377; suche merveiles **fortúnede** (pret.) than CH. D. BL. 288.

fortunous, adj., fortuitous : be moeved bi fortunouse fortune CH. BOET. i. 6 (26); fortunous þinges ii. 3 (38).

forþ, see ford.

forð, adv., O.E. forð, = O.L.G. forth, forð, ford, O.Fris. forth, ford, M.H.G. vort; forth, forward, away : evre forð to domes dei HOM. I. 45; forth to þe wedding will i wende ALEX. (Sk.) 827; forð þe king wende LAȝ. 25751; he droh þat witness(e) forþ ORM. 3078; fareþ forþ 6406; he droȝ forþ a riche ring FL. & BL. 683; þat him hadde ifed & ... forþ ibroȝt P. L. S. xiii. 135; (hi) bere forþ boþe xviii. 84; þe pomp of pride ai forth shawis BARB. iv. 121; furth in wiþevin far into the evening ALEX. (Sk.) 3055; and so forþ TREV. II. 351; ase þet child wext and comþ vorþ AYENB. 119; hennes forþ WILL. 1050; ȝif þu wult þeonne vorð more vorhoten þine sunne A. R. 340; wulle forð rihte (straightway) makien king of ane Peohte LAȝ. 13673; forð riht 6372; forð riht an an 8515; forþ riht an an he flæh a weȝ ORM. 3196; forswoleȝeð þene hoc forð mid (together with) þan ese HOM. I. 123; & swipte þat hæfved of forð mid þan helme LAȝ. 21426; forþ wiþ [for wit] (forthwith) C. M. 11001; & let it eornen forþ wiþ ORM. 1336; so fer forþ fram his men WILL. 209; so forð ant so feor MARH. 15; se forð KATH. 243; ase forð as (as far as) i mei JUL. 47; he wule to vorð A. R. 294; hire of þuhte þat heo hadde þe speche so feor forþ iladde O. & N. 398; vorþ to ROB. 7; forð þet ic alegge þine fortened HOM. I. 91; vorþ niȝte (late in the night) hit was LEG. 91; whanne it was forþ daies (late in the day) WICL. MK. vi. 35; bilevede vorþ (continued to remain) ROB. (W.) 7100; smite vorþ (continued smiting) 6206; dude vorþ (went on doing) 6576.

forð-béren, v., O.E. forðberan; bear forth : þo hi þat heved forþbēre (pret. pl.) P. L. S. xviii. 84.

forð-bringen, v., O.E. forðbringan; bring forth : i seiȝ a clerk a boke forthe bringe

P. L. S. xxvi. 5; geres the erthe froit forth-bring M. H. I; frute furthbring TOWNL. 2; doghtirs þat þai forthebroght PR. C. 5866; an englisch þuse wordes forþbroȝte (*pret.*) P. S. L. xvii. 570; the childe was **forthe-broght** (*pple.*) S. S. (Wr.) 3116; after mi wille this is furthbroght TOWNL. I.

forð-callen, v., *provoke*: in þar graves at nithe forþkalled (*pret.*) him PS. lxxvii. 58.

forð-casten, v., *cast away*: forthkeste mi saghes hindward þare PS. xlix. 17.

forð-clepien, v., *O.E.* forðclipian; *call forth*: HOM. I. 231; as an egle **forthclepinge** (*pple.*) his briddis to flee WICL. DEUT. xxxii. 11.

forð-come, sb., *coming forth; departure*: fained es Egipt in forthcome of am PS. civ. 38.

forð-cutten, v., *cut up*: forthkutten and purge his erthe WICL. IS. xxviii. 24.

forð-dōn, v., *O.E.* forðdōn; *put forth, utter*: sal ic non wurd mugen forðdon uten ðat gode leið on GEN. & EX. 3993.

forð-draȝen, v., *draw forth or out*: þe fischer þan þe child forðdrouȝ (*pret.*) wiþ salt and wiþ þe crismecloþ GREG. 347.

forþen, *see* forthien, furþen.

for-þenchen, v., *O.E.* forþencan,=*O.H.G.* fordenchen; *repent*: (he) him ssel þer of vorþenche AYENB. 5.

forð-ēode, v. pret., *passed away*: ileiden þa untrummen men bi þere stret þere Petrus forðeode HOM. 91; dæi þer forðeode LAȝ. 13455; forthȝeden a litil WICL. GEN. xliv. 4; vii niȝt forðgeden GEN. & EX. 1755.

forðer, forðerlike, *see* furðer.

forð-fare, sb., *departure, decease*, LAȝ. 6009.

forð-fáren, v., *depart, die*: i sal forthfare PS. x. 2; care that dai ther letten forthfare ALIS. 6936; he moste deien and forðfare C. L. 218; alle þe forðfárinde (*pple.*) vondeþ to vordonne A. R. 210; ȝif Sikelin is forðfáren (*pple.*) LAȝ. 23164; þemperour was forðfare faire to Crist WILL. 5266.

forð-fadres, sb. pl., *O.E.* forð-fæderas; *forefaɥers*, KATH. 94; forðfederes MARH. 4.

forð-fetten, v., *fetch forth*: anon his sone was **forthefete** (*pple.*) S. S. (Wr.) 2440.

forð-fōr(e), sb., *O.E.* forðfōr; *departure*, GEN. & EX. 3158.

forð-gān, v., *O.E.* forðgān; *come forth; pass away*: xiij. ger gan so forðgon GEN. & EX. 845; forthgoo S. S. (Wr.) 761; she was in feste daȝes **forthgōende** (*pple.*) with gret glorie WICL. JUDITH xvi. 27; **forthegōne** (*pple.*) '*profectus*' PR. P. 173.

forð-gōing, sb., *progress*: forðgoing of pro-fitende men WICL. PS. PROL.

forð-gang, sb., *O.E.* forðgang,=*O.Fris.* forth-gong; *going forth, progress*: forthgang PS. cxliii. 14; ich was þe beginnunge hwi such þing hefde vorðgong A. R. 318; i vorðgong of gode live 374.

forð-glīden, v., *glide away, pass*; **forð-glōd** (*pret.*) ðis oðer dais nigt GEN. & EX. 113; forðglod 129, 157.

forþien, v., *O.E.* forðian; *promote, effect, execute*; forðen HOM. II. 49; GEN. & EX. 341; REL. I. 212; forðest (*pres.*) JUL. 54; min erdne þu forðe (*imper.*) GEN. & EX. 1372; forþed (*pple.*) ORM. 212; *comp.* a-, ful-, ȝe-forðien.

forð-læden, v., *O.E.* forðlædan; *lead away*: there was mani a weping heie as the childe was **forthlådde** (*pple.*) S. S. (Wr.) 2442.

forð-lēapen, v., *leap forth*: moni ladde þer forth-lēp (*pret.*) to lave & to kest A. P. iii. 154.

forð-līch, adj., *healthy, likely to live*: ha haveð of þe forschuppet bearn sar care and shome, & fearlac of þat forðlich H. M. 35.

forð-lōkien, v., *O.E.* forðlōcian; *look forth*; **forthlōked** (*pret.*) over mennes sonnes PS. xiii. 2; forthloked lxxxiv. 12.

forð-nimen, v., *seize*: þes fures icunde is þet hit forðnimeð (*pres.*) swa hwet him neh bið HOM. I. 95; forðnam ðis folc GEN. & EX. 3351.

forð-passen, v., *pass over*: go and forth-passe (*imper.*) into Mesopotani WICL. GEN. xxviii. 2; he **forthpassid** (*pret.*) com into Mesopotanie 5.

for-þræsten, v., *O.E.* forþræstan; *distort*; forþræst (*pple.*) A. P. ii. 249; MISC. 224.

for-þrāwen, v., *overthrow*; vorþrawe AYENB. 86.

forðriht, *see under* forð.

for-þringen, v., *O.E.* forþringan; *oppress*; forþrungen (*pple.*) ORM. 6169.

for-þrüsmen, v., *O.E.* forþrysmian; *suffocate*; forþresmede (*pret.*) MAT. xviii. 28.

forð-schēwen, v., *show forth*: sal þi might forthschewe PS. cxliv. 4.

forð-sið, sb., *O.E.* forðsið; *departure, decease*; (ms. feorþsiþ), FRAG. 5; vorðsiðe (*dat.*) HOM. I. 197.

forð-tēon, v., *lead away; betray*: man mid is gele eggeð us and fondeth, and **forðtēð** (*pres.*) to idele þonke HOM. II. 199; on ech of hise deden [he] is iefned to þe deore wuas geres he forð teoð HOM. II. 37; ðes feawe word þe ich nu here forðtēah (*pret.*) (*pro-duced, alleged*) HOM. II.185; forðtegh II.145.

for-þunchen, v.,=*O.N.* fyrirþykkja, = *M.H.G.* verdunken, *M.L.G.* vordunken; *be dis-pleasing; repent*; vorðenche HICKES I. 224; hit ne ssel þe vorþenche AYENB. 159; for-þuncheð (*pres.*) JUL. 16; forþinkeþ LANGL. *B* ix. 129*; þat forþinkeþ me MAND. 303; me forþinkeþ þat i so did(e) M. ARTH. 3849; vorþingþ AYENB. 27; forþinkes WILL. 5422; hit him forþouȝte (*pret.*) sore inou ROB. 11; his dedus him sore forþoȝte AMAD. (Rob.)xviii.

forðinchinge, sb., *repentance*, A. R. 110*; vorþenchinge AYENB. 20.

for-þürsten, v., *cf. Ger.* verdursten; *be thirsty*; **forthirst** (*pple.*) PR. P. 173; forþrist (*ms.* forrþrisst) ORM. 5679.

forð-ward, adv., *O.E.* forðweard; *forward*, MISC. 11; swimmen forðward HOM. I. 51; þannen forðward HOM. II. 189; vorðward A. R. 242; forþward ORM. 5226; forþward BEK. 674; WILL. 3630; vorþward ROB. (W.) 3534; forðward [forþwardes] heo wenden LAȝ. 5370; heonne forðwardes KATH. 2099.

forð-werpen, v., *throw away*: **forthwerpand** (*pple.*) PS. xvi. 11.

foru, see furh.

[for-wáken?] v., **forwáke** (*pple.*) *exhausted with watching*, SPEC. 28.

[for-wákien?] v., **forwáked** (*pple.*), *exhausted with watching*, WILL. 790; CH. C. T. B 596; forwakid TOWNL. 104.

for-wal, see forfal.

[for-walken?] v., **forwalked** (*pple.*), *exhausted with walking*, WILL. 2236; LANGL. B xiii. 204.

for-wallen, v., *boil away*; **forwalleð** (*pres.*) HOM. I. 251.

for-wandren, v., *exhaust oneself with wandering*; **forwandreþ** (*pres.*) WILL. 737; **forwandred** (*pple.*) LANGL. B PROL. vii; R. R. 3336.

forwanien, see forwenian.

forward, see foreward.

for-warȝien, v., *cf. O.H.G.* for-, furwergēn; *curse*; **forwarȝed** (*pple.*) ORM. 8048.

forwe, see furh.

for-weaxen, v., *O.E.* forweaxan, *cf. M.H.G.* verwahsen; *grow to excess*; **forwoxen** (*pple.*) PARTEN. 2990.

[for-welhen, ? = forwelken] v., ? furweolewede (*pret.*) LEB. JES. 846; **vorwelwed** (*pple.*) L. H. R. 22.

for-welken, v., = *Ger.* verwelken; *become withered*; **forwelked** (*pple.*) R. R. 361.

for-wenden, v., *cf. M.L.G.* vorwenden, *M.H.G.* verwenden; *convert*; **forwent** (*pple.*) GEN. & EX. 1121.

for-wénien, v., = *M.H.G.* verwenen; *spoil by indulgence*; forweni, -wene, forwanie LANGL. B v. 35, C vi. 138, A v. 33*; **forwéned** (*pple.*) HOM. II. 41; forweined DEP. R. i. 27.

[for-wēpen?] v., **forwēped** (*pple.*), *exhausted with weeping*, CH. D. BL. 126; forwept WILL. 2868.

for-wēri, adj., *excessively weary*, WILL. 2443; R. R. 3336; M. ARTH. 2901.

for-wérien [forwérie], v., *wear out*; vorwerien LAȝ. 14479; **forwéred** (*pple.*) R. R. 235; forwerd PL. CR. 736; ha weren alle forwerede HOM. I. 251.

for-wērien, v., *weary out*; **forwēried** (*pple.*) GEN. & EX. 3894.

for-wernen, v., *O.E.* forwyrnan, -wiernan; *prohibit, refuse*: hwi wolde god swa litles þinges him forwerne HOM. I. 221; forwurnen LAȝ. 3497; **forwerneþ** (*pres.*) FER. 2809.

for-werpen, v., *O.E.* forweorpan, = *O.L.G.* for-, farwerpan, *O.H.G.* for-, far-, ferwerfan; *reject*, ORM. 1320; REL. I. 216; vorw(e)orpen A. R. 120; **forwarp** (*pret.*) PS. lxxvii. 60*; forworpen (*pple.*) A. R. 218; ORM. 1393.

for³-wīse, adj., *O.E.* forewīs; *prescient, far-seeing*: a foole shulde be forwise soche ferlies to know D. TROY 2539; in fele thinges forwise 3949.

for-wit, sb., *foreknowledge*, LANGL. B v. 166.

[for-wīten, v., *O.E.* forewītan,] **forwāt** (*pret.-pres.*); *foreknows*: what that God forwot most needes be CH. C. T. B 4424; forwoot TRO. 1042.

for-witer, sb., *foreknower*: biholder and forwiter of alle thinges CH. BOET. v. 6 (178).

for-witing [for-weting] sb., *foreknowledge*, CH. C. T. B 4434.

for-wiþ, adv., *before*; forwit C. M. 14941.

for-wlenchen, v., *O.E.* forwlencean; *render proud*; **forwleint** (*pple.*) SPEC. 24.

for-woundien, see forwundien.

for-wrappen, v., *cover up*; **forwrapped** (*pple.*) CH. C. T. C 718.

for-wrēien, v., *O.E.* forwrēgan; *accuse*, P. L. S. viii. 49; JUL. 46; **forwrēied** (*pple.*) A. R. 172*.

for-wúndien, v., *O.E.* forwundian, = *M.H.G.* verwunden; *wound*; **forwúndede** (*pret.*) LAȝ. 14713; forwundeden HOM. II. 33; **forwúnded** (*pple.*) HOM. I. 83; forwounded ROB. 56; forwounded WILL. 3686; R. R. 1830; vorwounded 1287.

for-wundrien, v., = *M.H.G.* verwundern; *amaze, astonish*; **forwundred** (*pple.*) ORM. 3417; forwondrid M. T. 19; forwondret ANT. ARTH. xxvi.

for-würchen, v., *O.E.* forwyrcan, = *O.L.G.* for-, farwirkean, *O.H.G.* far-, ferwirchen, -wurchen; *ruin, destroy*; **forworhte** (*pret.*) HOM. I. 221; **forworht** (*pple.*) LAȝ. 16694; forwroht ORM. 17547; forwrogt GEN. & EX. 266.

for-wurðen, v., *O.E.* forweorðan, -wurðan, *O.H.G.* fur-, for-, farwerdan; *perish, go wrong*, LAȝ. 28425; forwurðan HOM. I. 13; vorwurðen A. R. 210; forworþe P. S. 328; PS. i. 6; **forwurþeþ** (*pres.*) ORM. 18826; forworthes PR. C. 780; **forwurðe** (*subj.*) KATH. 2191; **forwarð** (*pret.*) HOM. II. 155; forworðen LAȝ. 1777; **forwurþen** [*O.E.* forworden] (*pple.*) ORM. 19216; forwurden REL. I. 211; forworþe, -wurde, -wurþe (*r. w.* borde) O. & N. 573, 1492.

forwurþen-līke, adj., *perverse*, ORM. 6245.

fosse, sb., *O.Fr.* fosse; *fosse,* MAND. 32.

fŏster, sb., *O.E.* fōster, fōstor, fōstur,=*O.N.* fōstr; *connected with* **fōde**; *nursing, protection; foster-child;* MARH. 8; þet foster FRAG. 8; freo monne foster (*child*) KATH. 451; heo was a seli foster S. A. L. 148; Eleine min aȝen voster LAȝ. 25921; in oure fostre MIR. PL. 194.

    fŏster-child, sb., *O.E.* fōstercild; *foster-child,* FRAG. 2.

    fŏster-fæder, sb., *O.E.* fōsterfæder; *foster-father,* FRAG. 2; fosterfader ORM. 8855.

    fŏsterling, sb., *O.E.* fōstorling; *fosterling, nursling*; **fŏsterlinges** (*pl.*) LAȝ. 28574.

    vŏster-mōder [**fŏstermōder**], sb., *O.E.* fōster-, fōstormōder; *foster-mother,* LAȝ. 25899; vostermoder MARH. 8.

fŏstre, sb., *O.E.* fōstre,=*O.N.* fōstra; *nurse,* GEN. & EX. 2624.

fŏstrien, fŏstren, v., *O.E.* fōstrian, *O.N.* fōstra, = *M.Du.* voedsteren, *mod.Eng.* foster; *nurse,* FRAG. 6, 8; fostrin MARH. 2; (*ms.* fosstrenn) ORM. 1558; fostre CH. C. T. *E* 593; PS. liv. 23; **fŏstreð** (*pres.*) A. R. 296; **fŏstrede** (*pret.*) LAȝ. 25858*; **fŏstred** (*pple.*) GEN. & EX. 2618; HAV. 1434; HOCCL. i. 229.

    fŏstrunge, sb., *fostering,* H. M. 37; fostring CH. C. T. *D* 1845.

fŏstrild, sb., *nurse,* A. R. 72.

fōt, sb., *O.E.* fōt,=*O.L.G.,* *O.Fris.* fōt, *O.N.* fōtr, *Goth.* fōtus, *O.H.G.* fuoz; *foot,* ORM. 11369; treden under fot ORM. DEDIC. 73; broht to grund(e) & under fot 11773; ϸo dunes fot GEN. & EX. 1303; (*ms.* uot) A. R. 390; þan vot ROB. 490; and him deþ þe wordle onder vot AYENB. 85; & i þe leitende fur het warpen euch (*i.e.* '*each man*') fot KATH. 1371; and ever ilk fot of hem slowe HAV. 2432; com ofer drīȝe **fōt** (*dry foot*) ORM. 10338; drie foot ['*pede sicco*'] TREV. V. 239; drui fot JUL. 32; **hōt fōt** (*immediately*) MAP 339; **fōt hōt** S. S. (Web.) 843; K. T. 104; CH. C. T. *B* 438; we beoð under vote [fote] LAȝ. 21925; heo . . . is under vote MISC. 89; on fote A. P. ii. 79; gon on fote HAV. 113; on fote gon SPEC. 90; ten þusend men a vote [fote] LAȝ. 25211; **fēt** (*pl.*) LAȝ. 31973; ORM. 4775; CHR. E. 630; vet A. R. 122; fiet S. A. L. 150; fit S. & C. I. liv; **fōte** (*gen. pl.*) LK. xx. 43; 8 fote long MAND. 75; thre fote brad C. M. 7447; fif foten [fote] he is deop LAȝ. 21997; foote JOS. 14; **fōtan** (*dat. pl.*) FRAG. 6; mid druȝe fotan HOM. I. 87; fotan, foten MAT. vii. 6; JOHN xi. 32; foten LEECHD. III. 132; HOM. II. 207; foten, voten LAȝ. 4089, 14820; fote FL. & BL. 393; mid dreie fote HOM. I. 227; fallen him to fote (*Ger.* im zu füszen) SPEC. 51; fel to his fote MISC. 53; falleþ him to fete CH. C. T. *B* 1104; *comp.* **crāwe-, hāre-fōt.**

    fōte-knave, sb., *footman*: of mi lioun no

help i crave, i ne have none other foteknave Iw. 2575.

    fōt-cops, sb., *O.E.* fōtcops; *fetter*; **fōt-copses** (*pl.*) MK. v. 4.

    fōt-fetel, sb.,=*Ger.* füszfeszel; *fetter*; **fōt-fetteles** (*? printed* -festeles) (*pl.*) PS. civ. 18*.

    fōte-fest, sb., *prisoner*: laverd fotefest unleses he PS. cxlv. 7; he herd sighinge of fōtefeste sone ci. 21; sighinge of **fōte-festes** (*pl.*) lxxviii. 11.

    fōt-folk, sb.,=*M.H.G.* vuozvolc; *infantry,* MAN. (F.) 12019; fotvolc ROB. 398; fetfolk ROB. (W.) 8210.

    fōt-folower, sb., *follower on foot*: footfolowers (*pl.*) of the princis WICL. 3 KINGS xx. 14.

    foot-grene, sb., *snare for the foot,* ['*pedica*'] WICL. JOB xviii. 10.

    fōt-lame, adj.,=*Ger.* füszlam; *lame in the foot,* P. S. 335; fotlome 194.

    fōt-man, sb., *footman,* ALIS. 1611; votman ROB. 199.

    fōt-scamel, sb., *O.E.* fōtsceamol,= *O.L.G.* fōtscamel, fuozscamal; *footstool,* ['*scabellum*'] MAT. v. 35; **fōtsceomele** (*dat.*) HOM. I. 91.

    fōt-schēte, sb., *footcloth, carpet*: make redi youre foteschete aboue þe coschin B. B. 177.

    fōt-stake, sb., *support, base*: thre pilers, and so feele footstakes (*pl.*) WICL. EX. xxvii. 14; footstakis of the seintuarie xxxviii. 26; tablis with fourti silveren footstakis xxxvi. 24.

    fōt-steppe sb., *footstep,* PR. P. 174.

    foot-trappe, sb., *snare*: the foottrappe of him is hid in the erthe WICL. JOB xviii. 10.

    fōt-welm, sb., *O.E.* fōtwelm, -wylm; *sole oɟ the foot,* '*planta*' (*pedis*), FRAG. 2.

    fōt-wunde, sb., *Ger.* füszwunde; *wound in the foot*; votwunde A. R. 194.

fōtinge, sb., *footing,* PR. P. 174.

fōþer, sb., *O.E.* fōðer,=*O.L.G.* vōther, *M.H.G.* vuoder, *mod.Eng.* fother; *burden, load*: everi strok . . . falleþ doun as a foþer FER. 641*; on his head falleþ þe foþer RICH. 1732; þat hadde ilad of dong ful mani a foþer [fother, fothir] CH. C. T. *A* 530; foþir ALIS. 1809; voðer [foðer] to hevi vor te veðren mide þe soule A. R. 140; þritti voðere [foþer] LAȝ. 25762.

fōðere, v.,=? feðren; A. R. 140*.

fou, *see* **fāh.**

fouaille, sb., *O.Fr.* fouaille; *fuel,* '*focale*,' PR. P. 174.

foudre, sb., *O.Fr.* fuldre, fuildre, foudre, *from Lat.* fulgur; *lightning*: dint of thonder, ne . . . foudre CH. H. F. 535.

fouh, *see* **fāh.**   **foul,** *see* **fuȝel, fūl.**

foun, sb., *O.Fr.* faon; *fawn,* '*hinnulus*,' PR. P. 175; WICL. PROV. vi. 5; fowen MAN. (F.) 15750; **founes** (*pl.*) MAND. 290; CH. D. BL. 429.

foundacioun, sb., *O.Fr.* fondation ; *foundation,* MAND. 81.

foundement, sb., *O.Fr.* fondement, fundement ; *foundation,* L. H. R. 119 ; PL. CR. 250 ; WICL. 1 TIM. vi. 19 ; foundement, fundement ROB. 131, 310.

founden, *see* fúndien.

founden, v., *O.Fr.* funder, fonder ; *found* ; founded (*pple.*) P. S. 330 ; MAND. 30.

foundeōr, sb., *O.Fr.* fondeor ; *founder,* JOS. 68 ; fondooure MAN. (H.) 84 ; foundours (*pl.*) LANGL. *A* xi. 213.

foundrin, v., *O.Fr.* fondrer ; *founder, stumble* ; *ruin,* PR. P. 175 ; foundred (*pret.*) CH. C. T. *A* 2687 ; foundret ANT. ARTH. xlii.

four, *see* fēower.

fourche, sb., *O.Fr.* forche ; *fork (of the body),* MAN. (F.) 1824.

fourched, adj., *forked* : sat on the fourched tre TRIST. 503.

fourchen, v., *O.Fr.* fourchier ; *fork (of harts)* : first whan an hert hath fourched (*pple.*) REL. I. 151.

fourches, sb. pl., *forks or haunches of deer* : bi hoȝes of þe fourches (*pl.*) GAW. 1356.

fourme, *see* forme.

fous, *see* fūs.   **foute,** *see* fuite.

fōw, *see* fāh.   **fowel,** *see* fuȝel.

fowen, *see* foun.

fowesōn, sb., *advowson,* ROB. (W.) 9678*.

fōwin, v., *O.N.* fāga ; *cf.* fāȝien ; *cleanse, polish,* '*purgo,*' PR. P. 175 ; fowen BEV. 1120 ; fōwid (*pple.*) PR. P. 175.

    fōware, sb., *O.N.* fāgari ; '*purgator,*' PR. P. 174.

    fōwinge, sb., '*emundacio, purgacio,*' PR. P. 175.

fox, sb., *O.E.* fox, = *O.H.G.* fuhs, fuchs, *M.L.G., M.Du.* vos *m., connected with O.H.G.* foha, *Goth.* fauhō *f.* (ἀλώπηξ) ; *fox,* REL. I. 217 ; fox, vox O. & N. 812, 819 ; vox, wox REL. I. 272 ; foxes (*gen.*) glōve, *foxglove,* 36 ; voxe [foxe] (*dat.*) LAȝ. 20840 ; voxes (*pl.*) A. R. 128.

    fox-liche, adv., *cunningly,* HOM. I. 31.

foxerie, sb., *foxish manners* : wrie me in mi foxerie under a cope of papelardie R. R. 6795.

foxing, sb., *foxlike trick* : in ure skemting he doð raðe a foxing MISC. 14.

foxish, adj., *foxish,* '*vulpinus,*' PR. P. 175.

frā, prep. & adv., *O.E., O.N.* frā, = *O.H.G.* frā ; *from* : swa fer fra godes riche ORM. 1265 ; drihtin takeþ it fra me 4820 ; fra þat Adam god forlet 355 ; þe limes þat er cutted fra þe bodi PR. C. 3715 ; fro hevene to helle HOM. II. 111 ; god ledde hem fro helle nigt GEN. & EX. 89 ; fro ðan ðat he singen bigan 188 ; he ros fro dede REL. I. 209 ; dede hire fete ... fro Winchestre

HAV. 318 ; wende hire fro HORN (L.) 367 ; nevere heo nule him wende fro C. L. 1170 ; fro lond(e) to londe ROB. 9 ; to schilde hem fro schame 80 ; ne do min huerte þe turne fro SPEC. 71 ; comen ... fro þemperour of Grece WILL. 1424 ; fro whennes he come LANGL. *B* v. 532 (*A* vi. 16 *has* from) ; no þing þou mai take fro us MAND. 294 ; he þat cam doun fro hevene WICL. JOHN iii. 13 ; god schilde us fro schame EGL. 508 ; i am sent fro (*by*) god LUD. COV. 85 ; passen to and froo LIDG. M. P. 122.

fracchin, v., *cf. prov.Eng.* fratch (*speak peevishly*), '*strido,*' PR. P. 175.

frēēt, *see* frēt.

fragilitē, sb., *O.Fr.* fragilité ; *fragility,* LIDG. M. P. 44.

fraȝnen, *see* freinen.

fraht, sb., *M.Du.* vracht ; *freight* : frauȝt H. V. 76 ; fraght D. TROY 1118 ; fraght of a shipe CATH. 141 ; fraught LUD. COV. 137 ; freight, freit '*vectura,*' PR. P. 177.

frahten, v., *M.Du.* vrachten ; *freight* ; frauȝt(e) CH. C. T. *B* 171 ; frauȝt (*pple.*) WILL. 2732.

frai, sb., *for* afrai ; *terror,* L. H. R. 192 ; BARB. xv. 255.

fraiel, sb., *O.Fr.* fraiel, frael ; *basket* ['*cala-thus*'] WICL. JER. xxiv. 2 ; RICH. 1549 ; fraiel [freiel] LANGL. *B* xiii. 94.

fraik, *see* freke.   **frainen,** *see* freinen.

fraisen, v., *? O.E.* frāsian, = *O.L.G.* frēson, *O.H.G.* freisōn, *Goth.* fraisan ; *? tempt* ; fraisez (he) (*pres.*) the pople D. ARTH. 1248 [*but perhaps an error for* fraiez].

fraisten, v., *O.N.* freista ; *inquire, examine, try* ; fraist(e) MAN. (F.) 8391 ; fraiste (*pres.*) A. P. i. 169 ; fraistes PR. C. 1090 ; ISUM. 669 ; fraisted (*pret.*) IW. 3253 ; fraisted (*pple.*) GAW. 1679 ; silver fraisted wiþ þe fire ['*argentum igne examinatum*'] PS. xi. 7.

fraked, adj., *O.E.* fraced, fracoð ; *wicked* : þis frakede folc HOM. II. 83.

frakel, adj., *base, evil, ugly,* MISC. 93 ; AL. (L.3) 279 ; he bið wiðuten feire & frakel wiðinnen HOM. I. 25 ; hu vrakel is þe worldes blisse A. R. 182 ; þis fikele world & frakele H. M. 7 ; seiþ man vaire bivore and frakele (*adv.*) bihinde REL. I. 183 (MISC. 122).

frakin, sb., = *Swed.* fräkne, *cf. Dan.* fregne ; *freckle, spot,* '*lenticula,*' VOC. 209 ; frakine, fraken, frakne, freken PR. P. 176 ; a fewe freknes (*pl.*) in his face ispreind CH. C. T. *B* 2169.

fraklī, *see under* frec.

frakni, frekeni, adj., = *Swed.* fräknig ; *freckly,* '*lentiginosus,*' PR. P. 176.

fraknid, pple., *freckled,* '*lentiginosus,*' PR. P. 176 ; frakned D. ARTH. 1081.

fram, prep., *O.E.* fram, from, = *O.L.G., O.H.G., Goth.* fram ; *from* : ne mahtu ... wende min herte fram him KATH. 1506 ; icomen ... fram þe se side HORN (L.) 203 ; þo þrie kinges ... þet comen fram verrene londes MISC. 27 ; hit is so fur fram þe sonne TREAT. 137 ;. fram ȝere to ȝere BEK. 331 ; oure lhord aros vram diaþe to live AYENB. 7 ; vram þe ginninge of þe worlde 12 ; a litil fram his halle S. S. (Wr.) 750 ; þat freo beoð fram ham H. M. 37 ; to wite þe fram þe fende SHOR. 90 ; ofslagen ... fram (by) his agene manne SAX. CHR. 255 ; fram god ich am isend ROB. 264 ; from souþ to norþ I ; þei konne noȝt don it from [fram] hem LANGL. B i. 194 (Wr. 853) ; steorren sculen from heovene falle HOM. I. 143 ; þe from [fram] drihtene com LAȝ. 20 ; fare we from þisse londe 991 ; keccheð ei god from þe A. R. 324 ; vrom (ms. urom) deie to deie 218 ; ne schal me firsen him from JUL. 17 ; wenden heom from MISC. 152 ; he can schede from [vrom] þe rihte þat woȝe O. & N. 197 ; þat from hevene dude alihte C. L. 656 ; from morwe til eve WILL. 763 ; butan þe beon clene from all sake HOM. I. 113 ; from loðen alesen LAȝ. 1084 ; þet he wite & wardie ou vrom (v. r. wið) alle þeo þet ou awaiteð A. R. 174 ; schild þi suikeldom from [vram] þe lihte O. & N. 163 ; shild us from helle SPEC. 61.

fram-ward, prep. & adv., O.E. fromweard ; *away from, away* : þat evene framward þan est toward þane west drouȝ BRD. 3 ; fromward MISC. 94 ; fromward, frommard [framward] LAȝ. 1899, 26168 ; fleo þe swuðere vrommard ham A. R. 92 ; heo makieð frommard hore nest 134.

fráme, sb., O.N. frami m.,=M.Du. vrame f. ; *profit, advantage,* ORM. 961 ; REL. I. 169, 209 ; GREG. 998 ; MAN. (F.) 2483 ; he dede ðe Ebri(u)s frame GEN. & EX. 2540 ; note ant frame SPEC. 71 ; ȝif hit be for mi frame HALLIW. 378 ; frame *frame 'fabrica'* PR. P. 176.

frámien, v., O.E. framian,=O.Fris. framia, O.N. frama ; *be profitable ; frame, construct* : al ... þat ham wule framien H. M. 29 ; he sulden ... here orf framen GEN. & EX. 1642 ; framin *'dolo'* PR. P. 176 ; **frámes** (*pres.*) A. R. 126* ; litel hit frames MAN. (F.) 11112 ; **frámed** (*pple.*) RICH. 1859.

fráminge, sb., O.E. fremung ; *advantage, profit ; hewing* : framinge or afframinge, or winninge, *'lucrum, emolumentum'* PR. P. 176 ; framinge of timbri *'dolatura'* PR. P. 176.

France, pr. n., O.Fr. France ; *France,* LAȝ. 1231.

franchise, sb., O.Fr. franchise ; *franchise,* ROB. 47 ; PARTEN. 3745 ; fraunchise WICL. I MACC. x. 34.

franchisen, v., *make free* : to fraunchise our mortalite LIDG. M. P. 249 ; thus standen alle men **fraunchísed** (*pple.*) GOW. I. 269 ; kinges ... were not fraunchised to make resistence LIDG. M. P. 128.

frangebill, sb., O.Fr. frangible, *from Lat.* frangere ; *breakable, brittle,* S. & C. II. lvii.

frank, sb., O.Fr. franc ; *pen for fattening cattle or birds,* 'saginarium,' PR. P. 177.

Franke, sb., O.E. Franca, cf. O.H.G. Franko ; *Frank, Frenchman* ; **Francene** (*gen. pl.*) LAȝ. 3715.

Fronc-lond, sb., O.E. Francland ; *Frankland, France,* KATH. 7.

frankelein, sb., *franklin, freeholder,* '*libertinus*,' PR. P. 177 ; CH. C. T. A 331 ; **frankeleines** (*pl.*) LANGL. B xix. 39.

franken, v., *fatten* ; **frankid** (*pple.*) '*saginatus*,' PR. P. 177.

frankinge, sb., *fattening,* 'saginacio,' PR. P. 177.

frankincens, sb., *frankincense,* 'francum incensum,' PR. P. 177 ; frank ensens L. H. R. 218.

frapaille, sb. pl., *attendants,* MAN. (F.) 13319 ; *see* **frape.**

frape, sb., ? O.Fr. frape ; *troop,* AL. (L.²) 390 ; CH. TRO. iii. 410 ; MAN. (F.) 9825.

frarie, sb., *fraternity, brotherhood* : nombre of this frari is 63 LIDG. M. P. 164.

fraternité, sb., O.Fr. fraternité ; *fraternity,* CH. C. T. A 364.

fraude, sb., O.Fr. fraude ; *fraud,* TREV. IV. 9.

fraud-ful, adj., *fraudulent* : he is a þef and fraudful revar APOL. 112.

frauden, v., O.Fr. frauder, *Lat.* fraudāre ; *defraud* : the hiire of ȝoure werkmen ... that is fraudid (*pple.*) of ȝou WICL. JAS. v. 4.

fraudulent, adj., *fraudulent, deceptive* : undir floures of fraudulent fresshenesse the serpent darethe LIDG. M. P. 160 ; dispoosid of kinde for to be fraudulent 197.

frauȝt, *see* **fraht.**

fraunchise, *see* **franchise.**

frā-ward, adv., prep. & adj., *mod.Eng.* froward ; *turned away ; averse from ; unfavourable* ; SAX. CHR. 256 ; þa turnest tu þe fraward god ORM. 4672 ; fraward WICL. DEUT. xxi. 18 ; if he fraward be PR. C. 87 ; froward HOM. II. 121 ; GEN. & EX. 3322 ; þe weder was cold and froward S. S. (Web.) 2622.

frōward-līche, adv., *frowardly,* TREV. I. 379.

frāwardnes, sb., *frowardness,* PR. C. 1173 ; frowardnesse PR. P. 180 ; LIDG. M. P. 145.

frē, *see* **frēo.**

frec, adj., O.E. frec, fric,=O.N. frekr, *Goth.* friks, O.H.G. freh, frech ; *insolent, daring* ; frek MIN. iv. 54 ; DEGR. 1365 ; vrech [frec] A. R. 128 ; frik *'crassus'* PR. P. 179 ; LIDG. M. P. 230 ; þe frecche MISC. 75 ; **fréche** (*pl.*) LAȝ. 9419 ; freke D. ARTH. 2821 ; **frikest** (*superl.*) L. H. R. 221 ; *see* **fréke.**

frec-līche, adv., *daringly,* FRAG. 6 ; vrecliche LAȝ. 31772 ; vrechliche A. R. 204 ; frekli D. ARTH. 1360 ; frakli BARB. vii. 166.

frēden, v., _O.E._ (ge-)frēdan, = _M.Du._ vroeden, _O.H.G._ fruotan; _be wise, perceive_; frede Gow. II. 119; **frēdeþ** (_pres._) SHOR. 7; **frĕdde** (_pret._) MK. v. 29; SHOR. 120; S. S. (Web.) 1514; _comp._ ȝe-frēden.

frōfren, see **frōfren.**

freinen, v., _O.E._ fregnan, frignan (_pret._ frægn, gefrægnde MK. xv. 2 LIND.), ? = _O.L.G._ frignan (_pret._ fragn), _O.N._ fregna (_pret._ frā), _Goth._ fraihnan (_pret._ frah); _inquire, question, ask,_ KATH. 1645; JUL. 36; H. M. 33; GEN. & EX. 1047; freine Gow. II. 384; OCTOV. (W.) 1381; Brien hine gon fræine of his fare coste LAȝ. 30734; ? fræinien 17116; fraȝnen (_ms._ fraȝȝnenn) ORM. 5664; fraine FER. 1216; PERC. 1293; **freineð** (_pres._) A. R. 264; frainez A. P. i. 129; freine (_imper._) MIRC 912; fraine (_subj._) AMAD. (R.) lvi; **frain** (_pret._) LANGL. _B_ xvi. 174*; þaih þu fraini after freond MISC. 92 (_sec. text_ axede); freinede LAȝ. 7244; he fræinede þis leodfolc æfter heore kineleoverde (_r._ -loverde) 6627; freinede GEN. & EX. 2759; freined WILL. 1303; M. ARTH. 678; frainede LANGL. _A_ i. 56; fraȝned (_pple._) ORM. 12973; _comp._ ȝe-, of-freinen.

Freinsse, see **Frenkisch.**

freitour, sb., _hall for meals,_ '_refectorium,_' PR. P. 177; BRD. 13; PL. CR. 220.

frek, see **frec.**

frēke, sb., _O.E._ freca; _cf._ **frec**; _bold, insolent man; warrior, man_; ALIS. 2161; FER. 113; WILL. 402; A. P. ii. 6; ANT. ARTH. xxxii; freke, fraik, freek, freik, freike LANGL. _B_ iv. 12, _C_ viii. 21*, _C_ xvi. 80, _B_ xiv. 105*, _A_ iv. 13; fréchen, freoken (_pl._) KATH. 732.

freken, see **frakin.**

frekled, pple., _freckled:_ ifracled wiþ whit & rede FER. 3659.

frēle, adj., _O.Fr._ frēle, fraile, _from Lat._ fragilis; _frail,_ P. L. S. xxv. 136; LANGL. _A_ iii. 117; WICL. WISD. xiv. 1.

frēlnesse, sb., _frailty:_ the frelnes of mankende LUD. COV. 108; freilneesse '_fragilitas,_' PR. P. 177.

frēlēs, see _under_ **frīe.**

frēletē, sb., _O.Fr._ frailetē; _frailty,_ LANGL. _B_ iii. 55; freeltee MAND. 5; frealte GEST. R. 25*; freilte TOWNL. 165.

frēlsen, see **frēolsen. fremde,** see **fremede.**

freme, adj., _O.E._ freme; _mod.Eng._ frim; _good, tasty; beautiful, fresh, vigorous_; L. C. C. 5; A. P. i. 1078.

[**frem-som,** adj., _O.E._ fremsum; _useful._]

fremsomnes, sb., _O.E._ fremsumness; '_benignitas,_' PS. lxiv. 12*.

fréme, sb., _O.E._ fremu; _profit, advantage,_ HOM. I. 217; II. 195; for þe freme of þe loande PROCL. 3; vreme AYENB. 69; freome FRAG. 5; A. R. 124*; he deð him selva freoma LAȝ. 674; _comp._ un-fréme.

frem-ful, adj., _O.E._ fremfull; _advantageous,_ HOM. II. 149; vremvol AYENB. 80; frem-, freomful HOM. I. 109, 135.

fremede, adj., _O.E._ fremede, fremde, fremðe, = _O.L.G._ frĕmidi, fremithi, _O.H.G._ fremidi, _Goth._ framaþis; _strange, foreign, not one's own,_ HOM. I. 185; HORN (R.) 68; a vreomede [fremde] child A. R. 184; fremde HAV. 2277; MIRC 1352; of fremde londe CH. C. T. _F_ 429; **fremde** (_pl._) ANGL. I. 9; **fremden** (_dat. pl._) MAT. xvii. 25; fremde (_ms._ fremmde) men ORM. 1250; fremde & sibbe ROB. 346; wiþ fremid(e) and sibbe P. S. 202.

fremed-lī, adv., _like a stranger, strangely:_ fer floteþ fro his frendez fremedli he ridez GAW. 714; fremedli the Frenche tunge fei es belefede D. ARTH. 1250; fremidli 3405.

fremeðe, sb., = _M.H.G._ vremede, vremde; _strangeness_; flesses fremeðe GEN. & EX. 349.

frémien, v., _O.E._ fremian, fremman, = _O.L.G._ fremmean, _O.N._ fremja, _O.Fris._ fremma, _O.H.G._ (gi-)fremen; _promote, perform, profit,_ KATH. 288; þet havelese monnam meie fremian HOM. I. 111; vreomien A. R. 284; fremen GEN. & EX. 173; he sholden his wille freme HAV. 441; freme ['_proficere_'] PS. lxxxviii. 23; heo scullen me monradene . . . fremmen (_do_) LAȝ. 24010; **frémede** (_pret._) MAT. xxvii. 24; KATH. 2399; fremede ['_profuit_'] HOM. I. 127; **frémed** (_pple._) HOM. II. 29; _comp._ ful-fremien.

fremmen, see **fremien.**

fremðe, see **frümþe.**

frēn, see **frēoȝen. frēnd,** see **frēond.**

frenesie, sb., _Fr._ frenesie; _frenzy,_ CH. C. T. _D_ 2209; Gow. I. 287.

frenetike, adj., _Fr._ frenetique; _frantic, silly,_ PR. P. 178; frantik [frenetike, frentik] LANGL. _B_ x. 6; frentik _C_ xii. 6, _A_ xi. 6, _C_ xii. 82*.

frenge, sb., _Fr._ frange; _fringe,_ PR. P. 178; **frenges** (_pl._) GAW. 598; CH. C. T. _D_ 1383.

frengen, v., _Fr._ franger; _fringe_; ifrenget (_pple._) L. N. F. 507.

Frenkisch, adj., _O.E._ Frencisc; _French,_ GAW. 1116; Frensc LAȝ. 7663; Frensch C. L. 25; Freinsse [_v. r._ Frensse, Frensche] ROB. (W.) 7890; Frenchs, Franche 7539*; Frenisch 7542*.

frēo, adj., _O.E._ frēo, frio, frī, frȳ, _O.Fris._ frī, _O.H.G._ frī, _Goth._ freis _m._ frija _f._; _free_; LAȝ. 23574; KATH. 1179; O. & N. 1507; BEK. 554; ALIS. 192; moni freo ȝeve HOM. I. 19; freo iboren JUL. 7; freo of heorte MISC. 96; of peines freo SPEC. 62; freo, vreo A. R. 220, 370; freo, fre ORM. 2974, 12201; fre GREG. 42; fre man (? freman) HAV. 628; þi merci is wel fre SPEC. 72; fre [free] CH. C. T. _A_ 852; frie HOM. II. 53; fri SHOR. 102; vri AYENB. 86; þa barn senden frie MAT. xvii. 26; freo men (? freomen) MISC. 86; freo [fre] men

TREV. III. 417; þan freo cnihten LAȝ. 20474; **frērre** (*compar.*) HOM. I. 271; freore C. L. 243; **vrēoest** (*superl.*) A. R. 398.

frēo-brōðer, sb., *O.E.* frēobrōðor; *younger brother* (*? ms.* frobroder), ALIS.[2] (Sk.) 23.

frēo-dōm, sb., *O.E.* freodōm; *freedom*, H. M. 5; freo-, fredom TREV. III. 61; fredom HAV.*631; MAND. 193; MAN. (H.) 71.

frē-hertid, adj., *freehearted*, PR. P. 177.

frēo-lāc, **-laic**, **vrēoleic**, **frēolēc**, sb., *O.E.* frēolāc (*oblation*); *liberality*, A. R. 192, 222, 386; freolaic KATH. 2398.

frēo-lich, adj., *O.E.* frēolīc; *free*, MARH. 13; KATH. 68; freli M. ARTH. 3121; **frēolīche** (*adv.*) *freely* LAȝ. 185; freliche TREV. VIII. 103; freli CH. C. T. *A* 1207; alle his **frēlīche** (*pl.*) felawes WILL. 428; **frēlokest** (*superl.*) WILL. 2634.

frēo-man, sb., *O.E.* frēoman; *freeman*: **frēmannes** (*gen.*) REL. I. 183; MISC. 128; **frēomen** (*pl.*) (? freo men) LAȝ. 19390.

frēness, sb., *liberality*, PR. P. 178; frenesse LANGL. *B* xvi. 88.

frēo-scipe [**frēsipe**], sb., *liberty*, LAȝ. 459; vreoschipe *liberality* A. R. 386; freshipe HOM. II. 121.

free-stōn, sb., *freestone*, E. W. 59.

frēo, sb., *O.E.* frēo, = *O.L.G.* frī; *noble woman*; fre GAW. 1549; DEGR. 745; fri FER. 3441.

frēoȝen, v., *O.E.* frēogan; *free*, *liberate*; freoiȝen, freoien LAȝ. 882, 5619; freoin MARH. 3; fren GEN. & EX. 2787; **vrī** (*imper.*) AYENB. 190; **frēode** (*pret.*) LAȝ. 6175; **frīed** (*pple.*) HOM. II. 103; *comp.* ȝe-frēoȝen.

frēolsen, v., *O.E.* frēolsian, = *O.N.* frelsa, frialsa; *? for* *frēohalsen, *cf. O.Fris.* frī-, friahals, *Goth.* freihals (*liberty*)*; keep free from labour*, *celebrate* : messedaȝ to freolsen (*ms.* freollsenn) ORM. 2720; **frēls** (*imper.*) REL. I. 22; **frēlsed** (*pple.*) C. M. 16942.

freome, **freomien**, *see* **frēme**, **frēmien**.

frēond, sb., *O.E.* frēond, frīond, frȳnd, = *O.Fris.* friond, friund, *O.L.G.* friund, *O.H.G.* friont, friunt, *Goth.* frijōnds, *O.N.* frændi; *from O.E.* frēogan, frēon, *Goth.* frijōn (*to love*); *friend*, LAȝ. 708; H. M. 11; freond, frend ORM. 1609, 17960; froend REL. I. 243; vriend (*ms.* uriend) AYENB. 158; frend WILL. 727; freend PR. P. 178; frende HAV. 375; frind AUD. 22; **fr(ē)ondes**, vrendes (*gen.*) O. & N. 1154; & dude us frendes dede SPEC. 112; **frēonde** (*dat.*) LAȝ. 14329; **frēond** [*O.E.* frȳnd, freondas] (*pl.*) LAȝ. 703; A. R. 250; MARH. 8; BEK. 367; freond, frend LK. vii. 6, xiv. 12; fr(e)ond, frend O. & N. 477; frend ORM. 443; GEN. & EX. 1775; MAP 334; freondes LAȝ. 1987*; frendes GREG. 953; **frēondene** (*gen. pl.*) LAȝ. 14901; **frēondan** (*dat. pl.*) SAX. CHR. 248; freondan, freonden JOHN xv. 13, 15; freonden FRAG. 6; LAȝ. 197; freonde P. L. S. viii. 111; frienden HOM. I. 231; frende SHOR. 90.

frēndesse, sb., *a female friend* : clepe thou prudence thi frendesse WICL. PROV.* vii. 4; mi frendesse WICL. S. SOL. i. 8.

frēnd-lēs, adj., *O.E.* frēondlēas; *friendless*, A. D. 299.

frēnd-līch, **frēndlī**, adj., *O.E.* frēondlīc; *friendly*, CH. C. T. *A* 2680; **frēondlīche** (*adv.*) LAȝ. 14845; **frēndeloker** (*compar.*) LANGL. *B* x. 225; frendloker *A* xi. 171.

frēndli-hēde, sb., *friendliness*, GOW. II. 286.

frēond-mon, sb., *friend*; vreondmon HOM. I. 199; **frēndmen** (*pl.*) HOM. II. 183.

frēond-rēde, sb., *O.E.* freondrǣden; *friendship*, ALIS. 1488; vrendrede AYENB. 149.

frēond-scipe, sb., *O.E.* freondscipe, = *O.L.G.* friundscipi, -skepi, *O.H.G.* friuntskeph, -scaf; *friendship*, LAȝ. 5106; freondschipe A. R. 98; frendshepe PR. C. 1884.

frēndsom, adj., *friendly* : frendsome es thi merci PS. lxviii. 17.

frēonden, v., = *M.H.G.* friunden, vreonden; *make friends*, A. R. 420.

frēosen, v., *O.E.* frēosan, = *O.N.* friōsa, *O.H.G.* friusan; *freeze*; fresin PR. P. 178; frese ORF. 245; **frēoseþ** (*pres.*) TREAT. 137; fr(e)ost O. & N. 620; hi freoseþ TREAT. 136; **freese** (*subj.*) LANGL. *C* xiii. 192; **frēs** (*pret.*) MAN. (H.) 121; GAW. 728; frees WICL. ECCLUS. xliii. 22; **fróren** (*pple.*) MISC. 151; frore SPEC. 25; *comp.* bi-, ȝe-frēosen.

frēre, sb., *O.Fr.* frēre; *friar*, BEK. 1124; LANGL. *B* iii. 35; CH. C. T. *A* 208; **frēres** (*pl.*) *C* xi. 12; freeres *A* ii. 186.

fresch, adj., *O.E.* fersc, = *O.N.* ferskr, *O.H.G.* frisc; *fresh*, P. L. S. xxx. 49; als fresch for to fiȝt(e) as þei were on morwe WILL. 3640; fresch flesch LANGL. *B* vi. 312; fresh ORM. 6348; CH. C. T. *A* 92; PR. C. 1254; versc HOM. I. 175; fersch TREAT. 136; fersch, verss ROB. 7, 217; ferche [versche], ferchs, fers ROB. (W.) 4424*, 1291*, 4424*; on fresche steden ALIS. 2405; of fresche watir MAND. 115.

fresch-lī, adv., *freshly*, *rashly*, '*recenter*, *noviter*,' PR. P. 178; GAW. 1294.

freshnesse, sb., *freshness*, LIDG. M. P. 245.

fresche, sb., *mod.Eng.* fresh; *flood of a river*; *onrush of men* : the fresshe was so felle of the furse grekes D. TROY 4730.

freschen, v., = *M.H.G.* vrischen; *refresh*: the water him **freschid** (*pret.*) HALLIW. 380.

frēse, sb., = *O.L.G.* frēsa, *M.Du.* vreese, *O.H.G.* freisa; *peril*, *fear*; no frese TOWNL. 30.

frēsen, *see* **frēosen**.

freslīche, *see* **ferslīche**, *under* **fers**.

frest, **fresten**, *see* **frist**, **fristen**.

frete, sb., *O.E.* frǣtu (*pl.* frǣtwe), = *O.L.G.* fratoh, *mod.Eng.* fret(-work); *ornament* : her fret of gold CH. L. G. W. 225; wiþ frette of perle GOW. II. 228.

[frēt, frǣt, sb., *cf. M.H.G.* frāz (*food*).]

frēte-wil, adj., *voracious*, A. R. 128*.

freten, v., *O.E.* fretan, = *O.H.G.* frezan, *Goth.* fraītan, *mod.Eng.* fret; *eat, devour*, LAȝ. 31677; fretin '*corrodo*' PR. P. 179; frete SHOR. 161; PR. C. 6570; freteþ (*pres.*) ORM. 16133; vret A. R. 184; frat (*pret.*) PS. lxxix. 14; fret ROB. 203; ALIS. 703; WILL. 87; freet LANGL. (Wr.) 12468; fræten MK. iv. 4; freten ALIS. 5171; freeten MISC. 229; freten (*pple.*) GEN. & EX. 2101; ALIS. 5179; CH. C. T. *A* 2068; ifreten FRAG. 7; MISC. 147.

fretinge, sb., *devouring, wasting; rust; ? pain in the bowels*: she died, as bi freting of the addres GEST. R. 278; fretinge WICL. 3 KINGS viii. 37; '*corrosio, torcio*' PR. P. 179.

fretien, v., *O.E.* frætwian, *O.L.G.* fratohōn; *fret, adorn*; fretted (*pple.*) LANGL. *B* ii. 11; A. P. ii. 1476; *comp.* ȝe-fratewen.

frette, sb., *release, redemption*: with him mak þer frette or with his bodi pas tille Gascoin MAN. (H.) 290.

freþ, Freþeric, *see* frið. freþien, *see* friðien.

frēvren, *see* frōfren.

frī, sb., *O.N.* frio, frǣ, = *? Goth.* fraiw, *mod.Eng.* fry; *seed*: to the and to thi fri mi blissing graunt i TOWNL. 24.

frī, *see* frēo.

Frīe, pr. n., *O.E.* Frig, = *O.N.* Frigg; *Venus*, ROB. 112; Frēon (*dat.*) LAȝ. 13931.

Frī-dæiȝ, *O.E.* Frigdæg, Frigedæg, = *O.N.* Frīadagr, *O.H.G.* Frīatag; *Friday*, SAX. CHR. 258; Fridæi LAȝ. 13932; Fridai KATH. 2531; ROB. 112; Vridei A. R. 422.

frī-niht, sb., *O.N.* frjānātt; *? night before Friday*: he þolede oðe longe friniht A. R. 122.

[frīe, sb., *O.N.* frȳja; *reproach, taunt*.]

frē-lēs, adj., *cf. O.N.* frȳja laust (*blamelessly*); *blameless*, A. P. i. 272.

frīen, v., *O.N.* frȳja; *reprove, blame*: frie HAV. 1998; to friȝȝenn & to tælen ORM. 16513.

frīen², v., *O.Fr.* frīre, *from Lat.* frīgere; *fry*; friin '*frigo*' PR. P. 179; frie CH. C. T. *A* 383; ifrīd (*pple.*) AYENB. 111; ifriȝet LANGL. *A* vii. 298.

frīinge, sb., *frying*, '*frixatura*,' PR. P. 179; þe dieveles panne . . . huerinne he makeþ his frīinges (*pl.*) AYENB. 23.

frīing-pan, sb., *frying-pan*, '*frixorium*,' VOC. 199; '*sartago, fricsorium*' PR. P. 179, 256.

frien, *see* frēoȝen. friend, *see* frēond.

friȝt, -en, friȝti, *see* fürht, -en, fürhti.

frik, *see* frec.

frikien, v., *O.E.* frician; *leap, dance*; frikieð (*pres.*) HOM. II. 211.

---

frim, *see* freme. frīnd, *see* frēond.

[Frise, pr.n., *O.E.* Frisa, Frysa, Fresa, = *O.Fris.* Frisa; *Frisian*.]

Fris-lond, pr. n., *O.E.* Frys-, Fresland; Frisia, LAȝ. 23377.

frise, sb., *Fr.* frise; *frieze*, PR. P. 179.

frīsen, v., *Fr.* friser; *put a nap on* (*cloth*); '*villo*,' frise (*pres.*) PR. P. 179; frīsed (*pple.*) IBID.

frīsare, sb., '*villator*,' PR. P. 179.

frist, sb., *O.E.* first, fyrst *m.*, *cf. O.Fris.* first, ferst *n.*, *O.N.* frest *n.*, *O.H.G.* frist *f.*; *space of time, term, respite*; frest PR. P. 178; ORM. 261; C. M. 15954; do þou nouth on frest þis fare HAV. 1337; first HOM. I. 161; first, feorst [forst] LAȝ. 4506, 12242; furst HOM. I. 19; BEK. 630; FL. & BL. *A* 638 (*B* 401); ferst P.L.S. viii. 19; & ferst ['*spatium*'] næfden þæt hio æten MK. vi. 31; he þet . . . deþ hit a verst AYENB. 32; firste (*dat.*) HOM. II. 145; a þisses dæies ferste [furste] LAȝ. 15853.

frist, *see* fürst.

fristel, sb., *O.Fr.* frestel; *flute*: with trompes, pipes, and with fristele Iw. 1395.

fristen, v., *? O.E.* firstan, = *O.Fris.* fersta, *O.N.* fresta, *O.H.G.* fristan; *respite, delay*; friste CATH. 143; frestin, or lende to freste PR. P. 178; no langer wold he freste HALLIW. 380; frist(e) (*pres.*) A. P. ii. 743; verste (*imper.*) AYENB. 173; firsten (*subj.*) MARH. 15; firstede (*pret.*) KATH. 2399.

frītūre, sb., *O.Fr.* fritūre; *fritter*, B. B. 167.

friȝð, sb., *O.E.* friȝð *n.*, friȝðo, freoðo *f.*, *cf. O.N.* friðr (*gen.* friðar), *O.L.G.* fritho, fridu, *O.Fris.* fretho, frede, ferd, *O.H.G.* fridu, frido (*gen.* frido, frides) *m.*; *peace, safety, protection*; HOM. II. 101; GEN. & EX. 681; bihahte to halden friȝð LAȝ. 2549; friþ ORM. 3380; on griðe & on friðe HOM. I. 13; *comp.* un-friȝð.

friþ-sōcne, sb., *O.E.* friðsōcn; *asylum*, REL. I. 33.

Frēþeric, pr. n., *O.E.* Freoðerīc, = *O.H.G.* Fridurīh; *Frederick*, ROB. 480.

friȝð², firð, sb., *O.E.* friȝð; *but cf. O.E.* fyrhð (*? fir-wooa*); *game-preserve, forest*, WILL. 822; IW. 157; GAW. 1912; A. P. ii. 534; friþ or forest toun or fild AMAD. (R.) lxxi; freþ DEGR. 486; ȝe huntieð i þes kinges friðe LAȝ. 1432; þe floures in þe frith(e) LANGL. *B* xii. 219; in freþe and in feld TOR. (A.) 586; freth P. S. 334; firthe D. ARTH. 1409; TOWNL. 131; *comp.* dēor-friȝð.

friðien, v., *O.E.* friðian, freoðian, = *O.N.* friða, *O.Fris.* frethia, ferdia, *O.L.G.* frithōn, *Goth.* (ga-)friþōn, *O.H.G.* (ge-)fridōn; *keep in peace, preserve, secure from disturbance*: eower lond ic wulle friþian HOM. I. 15; ne þat . . . deade bodi nulen ha nawt friðie 283; friðen GEN. & EX. 786; friþe MAN. (F.) 8733; fruðien LAȝ.

16804; vreþie AYENB. 7; friðe me ane hwile JUL. 48; friðede (*pret.*) KATH. 2399*; friþed (*pple.*) ANT. ARTH. i; friþed [freþed] in LANGL. *C* viii. 228; *comp.* ȝe-friðien.

frō, *see* frā.

frobrōðer, *see* frēobrōðer *under* frēo.

frōde, sb., *? O.N.* frauði; *? frog, toad*; frōden (*pl.*) HOM. I. 326; frude P. L. S. viii. 138.

frōfre, sb., *O.E.* frōfor, = *O.L.G.* frōfra, *O.H.G.* fluobara; *consolation, comfort*, ORM. 8786; frovre (*ms.* froure) JUL. 11; MISC. 90; GEN. & EX. 54; frofre, frovre [frovere] LAȝ. 9075, 21898; vrovre A. R. 92; HOM. II. 258; frover(e) REL. I. 171; frēfre (*dat.*) HOM. I. 11.

frōfer-gōst, sb., *O.E.* frōfergāst; *Paraclete, Comforter, ‘paracletus,’* FRAG. 3; frofregast HOM. I. 97; ORM. 10554.

frōfren, frēfren, v., *O.E.* frōfrian, frēfrian, *cf. O.L.G.* frōfrean, *O.H.G.* fluobiren; *console, comfort*; frofren ORM. 150; frovren KATH. 287; vrovren A. R. 72; froveren REL. I. 186; frefrien HOM. II. 215; vroæfrien LAȝ. 14844; freveri S. A. L. 150; frevre BEV. 183; frōvri, froveri (*pres.*) O. & N. 535; frevreð HOM. I. 145; freverep SHOR. 7; froreþ SPEC. 73 (A. D. 203); frōvere (*? subj.*) C. L. 889; frōvrede (*pret.*) MARH. 8; frōfred (*pple.*) ORM. 2198; *comp.* ȝe-frōfren.

frēfring, frōfring, sb., *O.E.* frēfrung, frōfrung; *consolation*; frefringe HOM. II. 117; froring H. H. 164 (166).

frogge, sb., *O.E.* frocga, frogga; *frog, toad, ‘reine,’* VOC. 154; O. & N. 146; ROB. 69; TREV. IV. 397; [‘*ranam’*] WICL. PS. lxxvii. 45; frogge, frugge ‘*bufo’* PR. P. 180; froggen, frogge (*pl.*) HOM. I. 51, 53; wroggen REL. II. 277.

frōh, adj., *O.N.* frār (*swift*)? *light, ? swift, capricious*; frogh PALL. iii. 671; frouh MISC. 94; fals and frouȝ MAP 337; CL. M. 36; fals or frow H. S. 2305; frow(e) (*adv.*) ‘*festine,’* 216.

froise, sb., *pancake*, PR. P. 180; VOC. 242; GOW. II. 93; L. C. C. 50.

froit, *see* früt.

frok, sb., *O.Fr.* froc; *frock*, PR. P. 179; A. P. ii. 136; frokke (*dat.*) LANGL. *A* v. 64.

from, *see* fram.   frome, *see* frume.

front, *see* frount.

frosch, sb., *O.E.* frox, forsc, = *O.H.G.* frosc, *O.N.* froskr; *frog, ‘rana,’* VOC. 223; frosh WICL. PS. lxxvii. 45; frosch, frosk PR. P. 180; froske (*dat.*) D. ARTH. 1081; froskes (*pl.*) GEN. & EX. 2977; PS. civ. 30; TOWNL. 62; froskis C. M. 5928; *comp.* ēa-frosk.

frosch, *see* frusch.

frost, sb., *O.E.* frost, forst, = *O.Fris.* forst, *O.N., O.L.G., O.H.G.* frost; *from* frēosen; *frost, ‘gelu,’* PR. P. 180; ORM. 12655; MAND. 130; frost [forst] WICL. PS. lxxvii. 47; forst TREAT. 137; GAW. 1694; vorst (*ms.* uorst)

HOM. I. 193; ROB. 265; forste (*dat.*) HOM. I. 35; vrostes (*pl.*) AYENB. 108; forstes O. & N. 524; *comp.* hōr-frost.

forst-tīme, sb., *winter*, TREV. IV. 467.

frosti, adj., = *O.H.G.* frostag; *frosty*, CH. C. T. *A* 268.

froten, v., *O.Fr.* froter; *rub*; frote WICL. DEUT. xxiii. 25; froteþ (*pres.*) AYENB. 155; CH. C. T. *A* 3747; frotede (*pret.*) TREV. IV. 25.

frotunge, sb., *friction*: þuruh so monie duntes & frotunges (*pl.*) A. R. 284.

froþe, sb., *O.N.* froða; *froth*, GAW. 1572; frothe ‘*spuma’* PR. P. 180.

froþen, v., *froth*: froþiþ (*pres.*) or fomeþ [‘*spumat’*] WICL. MK. ix. 17; froþen [frothen] CH. C. T. *A* 1659; froþande (*pple.*) A. P. ii. 1721.

frouh, *see* frōh.

frounce, sb., *O.Fr.* fronce; *wrinkle, flounce*: þe lappe of hir garment iplitid in a frounce CH. BOET. i. 2 (9); frounces (*pl.*) LANGL. *B* xiii. 318.

frounce-lēs, adj., *without a wrinkle*: hir forheed frounceless al pleine R. R. 860.

frouncen, v., *O.Fr.* froncier; *wrinkle, gather up*: he frounces (*pres.*) boþe lippe & browe GAW. 2306; þe front ... fronces C. M. 3571; his nose frounced (*pple.*) full kiked stoode R. R. 3137; with pais fronst ALIS. 1630.

frounin, v., *O.Fr.* frognier; *frown*, PR. P. 180; froune LIDG. TH. 254.

frount, front, frunt, sb., *O.Fr.* front; *front*, PR. P. 180, 181; frount BEK. 1195; MAN. (F.) 3360; frount (*forehead*) ROB. (W.) 4426.

frountel, sb., *O.Fr.* frontel, fronteau; *frontlet, headband*: the bout and the barbet with frountel shule feȝe P. S. 154; fruntel PR. P. 181; frontels (*pl.*) E. W. 5.

frounter, sb., *O.Fr.* frontier; *front, façade*: to make a frounter for a loveris herte P. R. L. P. 57; frikis þe frountere welle a five hundreth D. ARTH. 2899; att frounter of thees welles clere LIDG. M. P. 16; was writen in the frontures (*pl.*) 18.

frōvre, *see* frōfre.

frōw, *see* frōh.

fructifien, v., *O.Fr.* fructifier, *from Lat.* fructificāre; *fructify*; fructifie MAND. 50; no þing frutefiing (*pple.*) nor profitable destroien þe cornes plentevouse of frutes of reson CH. BOET. i. 1 (6); þet zed þet vil into þe guode londe fructefide (*pret.*) AYENB. 234.

fructuous, adj., *Lat.* fructuōsus; *fruitful*, MAND. 3.

fructuous-lī, adv., *fruitfully*: prechithe fructuouslie the word of God GEST. R. 233.

frūde, *see* frōde.

[frume, adj., *O.E.* fruma, = *Goth.* fruma, *cf. Gr.* πρόμος.]

frum-berdling, sb., *O.E.* frumbyrdling ; *young man* ; **frumberdlinges** (*pl.*) HOM. II. 41.

frum-ken(n)ed, adj., *O.E.* frumcenned ; *first-begotten*, HOM. I. 87 ; hire frumkennede sune MAT. i. 25.

frum-ræs ? sb., *first onset* ; ᷆ æt þon **frum-ræsen** (*dat. pl.*) LAȜ. 8665.

frum-schaft, -scheft, sb., *O.E.* frumsceaft ; *first creation*, JUL. 2, 3.

frum-schapen, adj., *O.E.* frumsceapen ; *first created* : Adam þene frumscepene mon HOM. I. 123.

frume, sb., *O.E.* fruma ; *beginning*, LAȜ. 13265 ; FL. & BL. 135 ; at þe frume MISC. 87; from þe worlde frume [frome] O. & N. 476 ; at te frume 1513 ; frome C. L. 595 ; ALIS. 5356 ; FER. 1104 ; *comp.* **ord-frume**.

frumentee, sb., *O.Fr.* fromentee ; *frumenty* (*wheat boiled in milk*), D. ARTH. 180 ; fur-mente L. C. C. 6.

frumpil, sb., *cf. Du.* frommel ; *wrinkle*, ' *ruga*,' PR. P. 181.

frumplen, v., *cf. Du.* frommelen ; *frumple*, *wrinkle* ; **frumplid** (*pple.*) PR. P. 181.

frümþe, sb., *O.E.* frumð, frymð ; *beginning*, ORM. 18558 ; MISC. 142 ; on frumþe FRAG. 8; i þe frumðe A. R. 8 ; on fremðe MAT. xix. 8.

frunt, fruntel, *see* **front, frontel.**

frunten, v., *O.Fr.* afronter ; *strike, kick, rush, drive, tumble* : þe freke him **frunt** (*pret.*) with his fot A. P. iii. 187 ; with a sword . . . frunt him in the fase D. TROY 6923 ; frunt him evin in the fase with a fin arow 6984 ; frunt unto Ector 6887 ; he frunt of his fol flat to þe ground 6890.

frusch, sb., *rush* : so fuerse was the frusshe when þai first met D. TROY 1196 ; þe freike, with the **frusshe** (*dat.*) fell of his horse 5830 ; ferkes one a frusche D. ARTH. 2900 ; frosch MAN. (F.) 12882.

fruschen, v., *O.Fr.* fruisser (=*mod.Fr.* froisser) ; *rub* : þei fruschen to gidere MAND. 238 ; frouschen DEGR. 1087 ; **fruschede** (*pret.*) JOS. 505 ; **frusht** [froust] (*pple.*) ALIS. 1630.

frŭt, sb., *O.Fr.* fruit ; *fruit*, HOM. I. 135 ; A. R. 150 ; TREAT. 133 ; S. S. (Web.) 559; fruit GEN. & EX. 327 ; frit A. P. ii. 1043 ; froit HAMP. PS. cxxv. 6 ; **frŭtis** [froitis] (*pl.*) BARB. x. 191.

  fruite-ful, adj., *fruitful* : hilles . . . tries fruitefull and cedres alle PS. cxlviii. 9.

  fruitestere, sb. (*female*) *fruiterer* : tombe-steris fetis and smale and yonge **fruitesteres** (*pl.*) CH. C. T. *C* 478.

  fruit-lēs, adj., *fruitless*, PR. C. 5666.

frŭten, v., *bear fruit*, ' *fructifico*,' PR. P. 182 ; i as vine **frŭted** (*pret.*) swotnesse of smel WICL. ECCLUS. xxiv. 23.

fuel, *see* **fuȝel.**

fuȝel, sb., *O.E.* fugel, fugol, = *O.N.* fugl, *Goth.* fugls, *O.H.G.* fogel ; ? *for* *fluȝel ; *fowl, bird*, O. & N. 1097 ; fuȝel, foȝel [fowel] LAȜ. 2832, 20174 ; fugel, fuel GEN. & EX. 160, 221 ; fuhel MARH. 3 ; fuel A. R. 126 ; voȝel AYENB. 71 ; foghel PR. C. 7075 ; fowel HORN (R.) 1414 ; TREAT. 139 ; fowel [foul] CH. C. T. *A* 190 ; foul ALIS. 1982 ; MAND. 48 ; TOR. 2050 ; **fuȝeles** (*pl.*) HOM. I. 133 ; fugeles MAT. xiii. 4 ; fuȝeles, voȝeles [foȝeles] LAȜ. 21754, 21759 ; fuȝeles, fuheles, foweles O. & N. 1144, 1660 ; fuheles KATH. 2124 ; foȝeles SHOR. 146 ; foules LANGL. *A* v. 199 ; wilde foules ROB. 52 ; **fuwelene** (*gen. pl.*) A. R. 298 ; **foulen** (*dat. pl.*) WILL. 805 ; fuȝele, foȝle, vowele O. & N. 64, 277 ; fouweles LANGL. *C* xviii. 11 ; *comp.* **henne-, sēd-, water-fuȝel.**

  fuȝel-cün, sb., *O.E.* fugelcynn ; *bird-tribe*, HOM. I. 95 ; foȝelcun LAȜ. 8109 ; fuelkun [fowelcun] O. & N. 65.

  fugel-fliht, sb., *flight of birds*, GEN. & EX. 3321.

fugitif, adj. & sb., *fugitive* : he preieth for the seruaunt fugitife WICL. PREF. EPP. vii ; he is like a fugitif LIDG. M. P. 167 ; alle manere of **fugitifes** (*pl.*) MAND. 66.

foulin, v., *O.E.* fugelian ; *fowl, catch birds*, ' *aucupor*,' PR. P. 175.

  foȝelere, sb., *O.E.* fugelere ; *fowler* ; voȝe-lere AYENB. 254 ; foulere WICL. PROV. vi. 5 ; **fuheleres** (*gen.*) MARH. 3.

  foulinge, sb., *fowling*, ' *aucupium*,' PR. P. 175.

fuerse, *see* **fers.** **fuhel**, *see* **fuȝel.**

fuir, *see* **für. fuisoun**, *see* **foisōn.**

fuite, sb., *O.Fr.* fuite (*voie du cerf qui fuit*) ; *track, trace*, GAW. 1425 ; feute WILL. 90 ; foute 33 ; feute ' *vestigium* ' PR. P. 159.

ful, adj. & adv., *O.E.* full, = *O.N.* fullr, *Goth.* fulls, *O.L.G.* ful, *O.Fris.* ful, fol, *O.H.G.* foll ; *full*, GEN. & EX. 110 ; LANGL. PROL. *A* 17 ; CH. C. T. *A* 90 ; hundred is ful tel A. R. 372 ; euch godes ful JUL. 3 ; a scip ful [fol] of golde LAȜ. 23694 ; ful neh HOM. I. 29 ; LAȜ. 124 ; ful iwis ORM. 687 ; O. & N. 1189 ; ful [fol] iwis LAȜ. 1324 ; ful wel HOM. I. 17 ; MARH. 4 ; A. R. 6 ; ORM. PREF. 17 ; ful briht HICKES I. 222 ; ful faste HAV. 82 ; ful hard TRANS. xix. 321 ; a ful bolde bacheler LANGL. *B* xvi. 179 ; ful wis A. P. i. 747 ; ful clene ANT. ARTH. xxx ; ful late PR. C. 789 ; vol AYENB. 15 ; fol wel SHOR. 159 ; RICH. 1550 ; þe fulle give HOM. II. 109 ; þe fulle tale ORM. 11270 ; þat is mi fulle ivo MISC. 42 ; at þe fulle *at full* [' *plene* '] TREV. III. 443 ; mid fulle dreme O. & N. 314 ; at þe fulle flod WILL. 2745 ; mid fullere strenðen (*r.* strenðe) LAȜ. 29047 ; ane hand fulle LEECHD. III. 92 ; an(e) hand folle S. S. Web.) 1588 ; enne koker fulne LAȜ. 6470 ; **fulle** (*pl.*) GEN. & EX. 2952 ; hi nefre ne beoð fulle HOM. I. 103 ; hare fulle fan

H. M. 31 ; twelf wilien fulle MK. vi. 43 ; fifti scipen fulle LAȝ. 1285 ; fulle seoven nihte 1632 ; weres fulle fowertiȝ ORM. 11210 ; fulle six ȝer HORN (R.) 926 ; ahte ant tventi folle ȝer CHR. E. 416 ; *comp.* aȝe-, bále-, cáre-, drēd-, eȝes-, ĕst-, fēr-, ȝeorn-, grið-, hih-, ȝelēaf-, luf-, măn-, miht-, milds-, nēod-, niþ-, rēd-, sīde-, sorh-, sün-, tīr-, þŏht-, wā-, wil-, wurþ-ful.

ful-hēde, sb., *fulness,* ['*plenitudo*'] PS. xxiii. 1 ; volhede AYENB. 119.

ful-līche, adv., *O.E.* fullīce ; *fully,* HOM. I. 73 ; LAȝ. 14124 ; A. R. 124 ; O. &. N. 1687 ; fullike ORM. 16302 ; fulli PR. C. 4570.

fulnesse, sb., = *O.H.G.* folnassi, folnessi, folnissi, folnussi ; *fulness,* PR. P. 182 ; volnesse AYENB. 266.

ful-sum, adj., *plentiful,* GEN. & EX. 2153 ; fulsom D. TROY 3068.

fulsum-hēd, sb., *copiousness,* GEN. & EX. 1548.

fulsum-lī, adv., *copiously,* WILL. 4325.

fulsomnesse, *copiousness,* CH. C. T. *F* 405.

ful, sb., *O.E.* ful ; *bumper,* LAȝ. 14335.

fūl, adj., *O.E.,* fūl,= *O.N.* fūll, *Goth.* fūls, *O. H.G.* fūl ; *foul,* A. R. 118 ; ORM. 2032 ; a poke ful and blac HAV. 555 ; ful, vul (*ms.* wl) O. & N. 94, 236 ; foul ROB. 69 ; SPEC. 103 ; IW. 1332 ; voul AYENB. 32 ; his fule saule LAȝ. 27634 ; his fule stench MARH. 11 ; mid þine fule holde FRAG. 7 ; fūlne (*acc. m.*) O. & N. 1196 ; ȝif þine vo enne fulne nome A. R. 316 ; fule lustes HOM. I. 243 ; foule airen ALIS. 4719 ; mid fule worde O. & N. 285 ; fūle (*adv.*) HOM. II. 37 ; ORM. 1201 ; stinkeð . . . fule A. R. 138 ; foule ALIS. 1118 ; LANGL. *A* v. 66 ; fūlre (*compar.*) HOM. I. 25 ; A. R. 84 ; ORM. 15831 ; fūluste (*superl.*) A. R. 216 ; fouleste P. L. S. xiii. 365.

fūl-ȝetóȝen, pple., *ill-educated, trained ill* ; fulitowen A. R. 108, 186 ; never so angresfule, ne so fulitowune 244.

fūlȝetóȝen-schip, sb., *vulgarity of manners* : alle hise fūlitóhe-schipes (*pl.*) & hise unhende gomenes H. M. 33.

voul-hēde, sb., *foulness,* AYENB. 17.

fūl-līc, adj., *O.E.* fūllīc ; *foul-seeming,* FRAG. 1 ; fūllíche (*adv.*) HOM. II. 173 ; A. R. 124.

fūl-mart, sb., *foulmart, polecat,* MIR. PL. 8 ; fulmard PR. P. 182 ; folmarde A. P. ii. 534.

fūlnesse, sb., *O.E.* fūlness, *O.Fris.* fūlnisse, *O.H.G.* fūlnussi ; *foulness* ; foulnes WICL. PREF. EPIST. 7.

fūl-som, adj., *foul* : sotte there thou ligges, for the fūlsomeste (*superl.*) freke that fourmede was ever D. ARTH. 1059.

ful-brengen, v., *cf. O.Fris.* fulbranga ; *accomplish* ; fulbröht (*pple.*) ORM. 16335.

fulcnen, v., *from* *fulc, *O.E.* fulloc (*baptism*) ; *baptize,* HOM. II. 137.

fulcnere, sb., *baptist,* HOM. 131.

fulcninge, sb., *baptism,* HOM. II. 9.

ful-dēde, sb., *perfection,* A. R. 372.

ful-dōn, v.,= *O.H.G.* folletuon ; *accomplish* A. R. 372 ; voldō (*pple.*) AYENB. 28.

ful-doun, adv., *flat down, completely down* : the wallis of the cite shulen fuldoun falle WICL. JOSH. vi. 5.

fūlen, fūlen, v., *O.E.* fūlian, fȳlan, = *O.H.G.* fūlen ; *M.Du.*vuilen ; *foul,defile* ['*inquinare*'], A.R.124 ; MARH. 13 ; filen HOM.II.127 ; ORM. 4493 ; file PR. C. 1210 ; APOL. 22 ; foulin '*turpo, maculo*' PR. P. 175 ; fūleð (*pres.*) HOM. I. 53 ; fuleþ O. & N. 100 ; fouleth LANGL. *B* xix. 310 ; (þu) vilest ROB. (W.) 802 ; fouled (*pret.*) A. P. ii. 269 ; foulid (*pple.*) ['*polluta*'] WICL. LEV. vii. 20 ; ifuled FRAG. 5 ; iviled ROB. 435 ; *comp.* a-, bi-fūlen.

fīling, sb., *defouling,* PR. C. 2345.

ful-enden, v., *O.E.* fullendian, = *Ger.* vollenden ; *finish, achieve, complete,* P. L. S. viii. 123 ; fulendi MISC. 66 ; fulende BEK. 2305 ; fulended (*pple.*) HOM. II. 61.

ful-fair, adj., *handsome* : us mai falle here a fulfaire gifte D. TROY 3155.

ful-fattid, pple., *quite fattened* : ful fat maad is the loved, and aȝen winsed ; fulfattid, fulgresid WICL. DEUT. xxxii. 15.

ful-fēding, sb., *overfeeding* : no merke of fulfeding ȝovun WICL. GEN. xli. 20.

ful-festning, sb., *confirmation,* HOM. I. 67.

ful-forðien, v., *accomplish, achieve* ; fulforðie HOM. I. 237 ; fulforþed (*pple.*) ORM. 15597.

ful-fremien, v., *O.E.* fulfremian ; *accomplish, perfect* ; fulfremed (*pple.*) ORM. 2530.

fulfremed-līke, adj., *O.E.* fullfremedlīce ; *perfectly,* ORM. 10751.

fulfremednesse, sb., *O.E.* fullfremedness ; *perfection,* ORM. 6083.

ful-füllen, v., *O.E.* fulfyllan ; *fulfil,* A. R. 288 ; fulfillen GEN. & EX. 1222 ; fulfille A. P. ii. 264 ; fulfille, folfulle LANGL. *A* vi. 36 ; volvelle AYENB. 64 ; vulvelden (*pret.*) MISC. 29 ; fulfild [folfuld] (*pple.*) C. L. 561 ; folfeld SHOR. 147.

fol-vellinge, sb., *fulfilling,* AYENB. 260.

ful-grésid, pple., *quite fattened,* WICL. DEUT. xxxii. 15.

fulhten, v., *baptize,* HOM. I. 73 ; fullehteð (*pres.*) LAȝ. 29769 ; fullehtede (*pret.*) LAȝ. 29693.

fulhtnen,= **fulhten,** v., ORM. 94 ; fuluhtnie HOM. II. 17 ; ifulhtnet (*pple.*) KATH. 1391.

fulohtninge, sb., *baptism,* HOM. II. 15 ; fulhtning ORM. 10798.

fulien, *see* folȝen.

ful iwis, *see under* ful.

fülle, sb., *O.E.* fyllo, = *O.H.G.* fullī, follī, *Goth.* fullei ; *fill, fulness,* FRAG. 7 ; HORN (L.) 402 ; EGL. 560 ; to eten hire fulle HOM. I. 53 ; þe

vulle (*ms.* uulle) of endelese blisse HOM. II.
35; fille HAV. 954; P. L. S. xvii. 391; WILL.
768; S. S. (Wr.) 967; volle AYENB. 247; *comp.*
ofer-fülle.

füllen, fullen, v., *O.E.* fyllan, fullian, *cf. O.N.*
fylla, *O.Fris.* fella, fullia, *O.H.G.* fullen, *O.L.G.*
fullian, fullōn, *Goth.* fulljan; *fill*, MARH. 18;
AN. LIT. 8; þu woldest fullen al þet was his
wille FRAG. 7; fulle LAȝ. 23352; ROB. 296;
godes hese to fulle P. L. S. viii. 175; fille RICH.
6039; fellen LK. xv. 16; velle AYENB. 92;
fülleð (*pres.*) HOM. I. 39; luve fulleð þe lawe
A. R. 386; fulleþ R. S. vii; C. L. 731; me
calleþ me fulle flat, *they call me Fill-the-hall,*
SPEC. 47; fülde (*pret.*) LAȝ. 14965; A. R. 40;
HORN (R.) 1155; þe swotnesse. . . fulde al
þat lond L. H. R. 29; filde GOW. II. 204;
feled ROB. (W.) 6220*; felden JOHN vi. 13;
velden AYENB. 233; fülled (*pple.*) LAȝ. 12076;
filled A. P. ii. 104; mine wordes þat filled
shulen ben ORM. 215; *comp.* a-, ful-, ȝe-, of-,
ofer-füllen.

fillinge, sb., *filling,* PR. P. 160.

fullen, v., *O.Fr.* fuler, fouler, *from med.Lat.*
fullāre; *full (cloth)*; fulle clothe '*fullo*' PR.
182; fulled [foulled] (*pple.*) LANGL. *B* xv. 444.

fullere, sb., *O.E.* fullere; *fuller* ['*fullo*']
MK. ix. 3; fullare VOC. 194; PR. P. 182; vol-
leres (*pl.*) AYENB. 167; fullaris P. S. 188.

fulling, sb., *fulling,* '*fullatura,*' PR. P. 182.

fulling-stokkes, sb. pl., *fulling frame,*
LANGL. *B* xv. 444.

füllen, *see* fellen. fullien, *see* fulwen.

fulling, *see under* fulwen.

fulluht, sb., *O.E.* full-, fulwiht, -wuht, fulluht;
baptism, ORM. 4054; MISC. 90; fulluht, ful-
leht [folloht] LAȝ. 36, 9607; fulht MARH. 19;
fullouht, fullouȝt JOS. 682, 693; fulloȝt REL.
II. 243; follouht LANGL. *C* xviii. 76; folht P.
S. 157; fulhtes (*gen.*) HOM. I. 23.

ful-máken, v., *accomplish, perfect*; volmad
(*pple.*) AYENB. 260.

fülmart, -mard, *see under* fūl.

fulnað, sb., *O.N.* fullnaðr; *fulness*: of his
fulnaþe (*ms.* full naþe) of haliȝ gast ORM.
18362.

fülsenen, v., *? for* fülstnen; *aid, increase;*
support, maintain: as fortune wolde fulsun
hom the frairer to. have GAW. 99; i shall
filsin þis forward D. TROY 9243; filsin her
strenght 9381; þat he filsened (*pret.*) þe
faithful in the falce lawe A. P. ii. 1167.

ful-serven, v., *serve fully*: me can zigge,
hus þet serveþ, and naȝt volserveþ (*pres.*)
his ssepe he liest AYENB. 33.

ful-sounen, v., *sound clearly*: whanne the
trompe fulsouneþ (*pres.*) WICL. JUDG. vii.
18; thei fulsouneden (*pret. pl.*) 20.

ful-spēde, adv., *prepared, quick*: fulspede
goo ȝe before the Lord into fiȝt WICL. NUM.

xxxii. 20; fulspeed goo ȝe before ȝoure
britheren DEUT. iii. 18.

fülst, sb., *O.E.* fylst, fullēst, -æst, = *O.H.G.*
folleist; *help, assistance,* HOM. I. 213; H. M.
17; fülste (*dat.*) LAȝ. 1747.

fülsten, v., *O.E.* fylstan, fullēstan, fullæstan,
= *O.L.G.* fullēstian, *O.H.G.* folleisten; *help,*
assist, LAȝ. 5581; vulsten (*ms.* wlsten) 11302;
filsten REL. I. 186; fülste (*pres.*) O. & N.
889; vulsteð (*ms.* uulsteð) HOM. II. 29; fülste
(*pret.*) LAȝ. 1148; felsten LK. v. 7.

fülstnen, v., = fülsten; filstnen (*ms.* fill-
stnenn) ORM. 6170; filstnede (*pret.*) REL. I.
209; *see* fülsenen.

ful-timbren, v., *build completely*: þat was i
sexe ȝeres all & fowertiȝ fulltimbred (*pple.*)
ORM. 16321.

fultum, sb., *O.E.* fultum; *help*; fulteam
ANGL. III. 151; HOM. I. 105; GEN. & EX.
2824; þurȝ godes fultume PROCL. I; fultume
[foltome] (*dat.*) LAȝ. 14618; beon heom a
fultume 16424.

fülþe, sb., = *O.L.G.* fullitha, *O.H.G.* fullida;
fulness, M. H. 7; fulthe D. TROY 5414.

fülðe², sb., *O.E.* fylð, = *O.H.G.* fūlida; *filth,*
A. R. 128; fulþe '*ordure*' VOC. 159; MISC. 159;
C. L. 1138; MIRC 1942; filðe REL. I. 213;
(MISC. 7); filthe '*sordes*' PR. P. 160; felðe
MAT. xxxiii. 27; felþe SHOR. 4; velþe AYENB.
81; fülðen (*pl.*) A. R. 128.

fülþ-hēde, sb., *filth,* ROB. (W.) 5900.

[ful-þēon], v., fulþógen (? ful þógen)
(*pple.*); *adult, grown up,* HOM. II. 41.

[ful-þrīfen], v., fulþrifen (*pple.*), *full-*
thriven, ORM. 5130.

ful-waxen, adj., = *O.H.G.* follewahsan; *full-*
grown, ORM. 10884; fulwaxen P. L. S. xxx.
38; volwoxe ROB. 430; volwexe AYENB. 26.

fulwen, v., *O.E.* fulwian, fulligan; *? for* *ful-
wīhan* (*wīhan = Goth.* weihan ἁγιάζειν); *bap-*
tize, LAȝ. 2402; folewen HICKES I. 229; folowe
MIRC 85; fulli MISC. 56; fulwede, folwede
(*pret.*) JOS. 683, 694; folewede H. H. 22; fullede,
[follede] TREV. V. 401; follede [fulled] LANGL.
B xv. 440; fulȝed (*pple.*) A. P. ii. 164; fulled,
ifulled FER. 1084, 5835; folut ANT. ARTH.
xviii; ifulwed C. L. 1457; ivulȝed LAȝ. 16881;
ifulhet MARH. 1; ivolled MISC. 38; ROB. 239.

fulling, sb., *baptism*: there is fullinge of
fonte and fulling in blood shedinge, and þorugh
fuir is fulling LANGL. *B* xii. 282; follinge *C* xv.
208.

ful-würchen, v., *O.E.* fullwyrcan, = *O.H.G.*
fol(l)awurchan; *accomplish*; fulwroht (*pple.*)
ORM. 15597.

fūme, sb., *O.Fr.* fum; *fume, smoke*: from the
fire depertith fume HALLIW. 385.

fumetere, sb., *O.Fr.* fumeterre, *Lat.* fūmus
terræ; *fumetory*: fumus terre VOC. 140;
'*fumus terre,*' PR. P. 182; centuri and fumetere
CH. C. T. *B* 4153.

fumōse, adj., *O.Fr.* fumos; *causing flatulence*; metes þat ar fumose in þeire degre B. B. 139.

fumositē, sb., *O.Fr.* fumosité; *indigestibility; heady quality (of wines)*: the fumosite of alle maner skinnes B. B. 140; win of Spaine . . . of which ther riseth such fumosite CH. C. T. *C* 567; þe **fumositees** (*pl.*) of fisch, flesche and fowles B. B. 139.

fundeling, fündeling, sb., = *M.H.G.* vundelinc, *M.Du.* vondeling, *Swed.* fyndling; *foundling,* PR. P. 182; WILL. 481; fundling HORN (R.) 226; fondling GREG. 472; findling P. L. S. xxiii. 56; **foundelinges** (*pl.*) ALIS. 4604; LANGL. *B* ix. 193; foundlinges PS. lxvii. 6.

fundement, see **foundement.**

fúnden, v., *O.E.* fundian, = *O.L.G.* fundōn, cf. *O.H.G.* funden; *from* finden; *seek, endeavour,* GEN. & EX. 2953; funde P. L. S. vii. 10; HORN (L.) 103; founden SPEC. 82; to fluiȝt founden JOS. 506; founde HORN (R.) 107; GREG. 462; TRIST. 1880; M. ARTH. 1965; AMAD. (Web.) 52; TOWNL 216; i mot heþen fonden (*go*) PL. CR. 408; **fúndie** (*pres.*) MISC. 192; when þou an hunting foundes EGL. 259; whænne swa æi ferde fundeð to þan ærde LAȝ. 21758; he fundeð (*seeks*) swiðe me to forswolhen MARH. 10; þider fundeþ evrich man O. & N. 719; **founding** (*pple.*) WILL. 1749; **fúndede** (*pret.*) HOM. II. 117; & fundede to varene LAȝ. 17858; foundeden JOS. 596.

fúndi, adj., *O.E.* (ge-)fyndig (*capable*), findig (*heavy*), cf. *Swed.* fyndig, *M.H.G.* vündic (*inventive*); *firm, solid*; findiȝ ORM. 4149; findige speche HOM. II. 119; he hem makede fundie on worde 117.

fundiing, sb., *later Eng.* foundering; *benumbment with cold*: his maill eiss of ane fundiing begouth BARB. xx. 75.

funding, sb., = **fundeling**; *foundling,* HORN (L.) 220; M. H. 167.

fündles, see **findles.**

fundling, see **fundeling.**

funel, sb., *funnel, 'fusorium, infusorium,'* PR. P. 170; ne mele, ne bucket, ne funell (*v. r.* fonel) C. M. 3305.

funerāl, adj., *med.Lat.* funerālis; *funeral,* CH. C. T. *A* 2942.

funke, sb., = *O.H.G.* funcho, *M.Du.* vonke; *spark, 'igniculus,'* PR. P. 182; þat ilke firi funke GOW. III. 18; a fonk(e) of fuir LANGL. *C* vii. 335; þat was not worth a fonk(e) MAN. (H.) 172.

funt, see **font.**

fur, see **feor** *and* **furh.**

fur-, see **for-.**

fūr, sb., *O.E.* fȳr, = *O.L.G.*, *O.H.G.* fiur, *O.Fris.* fiur, fior; *fire,* FRAG. 7; HOM. I. 39; P. L. S. viii. 125; LAȝ. 1187; A. R. 124; BRD. 8; þat fur R. S. iv; fuir C. L. 1253; ALIS. 547;

JOS. 260; MAND. 35; fuir LANGL. *A* iii. 88; fir SAX. CHR. 249; ORM. 10452; HAV. 585; CH. C. T. *A* 1246; fir, fier GEN. & EX. 99, 103; fier S. S. (Wr.) 2505; WICL. LK. xxii. 25; LIDG. M. P. 35; figer GEN. & EX. 3522; fer HOM. I. 239; SHOR. 99; TRIST. 1475; RICH. 781; SACR. 494; þet for MISC. 30; ver ANGL. I. 22; hit his zet alle a ver AYENB. 205; **fēres** (*gen.*) MAT. v. 22; **fūre** (*dat.*) HOM. I. 89; mid fure & mid here LAȝ. 2159; *comp.* helle-fír.

fïir-belowis, sb., '*reposilium,*' VOC. 261.

fūr-berninde, *pple.,* = *Ger.* feuerbrennend; *burning as fire*: þe furberninde drake R. S. V. (MISC. 180).

fūr-brond, sb., = *Ger.* feuerbrand; *firebrand,* LAȝ. 25608; fuirbrond ALIS. 6848; firbrand PR. C. 7421.

fïir-drake, sb., *O.E.* fȳrdraca; *fire-drake,* GOW. III. 95.

fïir-forke, sb., *fire-fork,* PR. P. 161.

fïir-herth, sb., = *Ger.* feuerherd; *fire-hearth,* PR. P. 161.

fūr-ïre, sb., = *M.H.G.* viurīsern; *fire-iron,* BRD. 30; furirin PR. P. 161.

fïir-, fïer-panne, sb., *O.E.* fȳrpanne; *fire-pan,* WICL. EX. xxxviii. 4.

fïir-reed, adj., *M.H.G.* viurrōt; *red as fire,* CH. C. T. *A* 624; fierreed WICL. LEV. xiv. 19.

fïir-sticke, sb., *fire-stick,* HAV. 966.

fïir-stōn, sb., *O.E.* fȳrstān = *Ger.* feuerstein; *firestone, flint,* PR. P. 161.

fïer-vessel, sb., *fire-vessel*; **fïervessels** (*pl.*) WICL. EX. xxvii. 3.

fürde, fürdien, fürding, see **ferde, ferdien, ferding.**

furel, see **forel.**

fūren, adj., *O.E.* fȳren, = *O.H.G.* fiurīn; *fiery*; firen C. M. 995; fiiren WICL. 4 KINGS ii. 11; of swilc firene kinde ORM. 17412; **fürene** (*pl.*) LAȝ. 18863; furene tungen HOM. I. 89; furene hweoles A. R. 356.

fūren, v., *O.E.* fȳrian, = *M.L.G.* vüren, *O.H.G.* fiuren; *set on fire*; firin '*ignio,*' PR. P. 162; **fūre** (*imper.*) MARH. 18; **fïrid** (*pple.*) WICL. PROV. xxx. 5.

furgōn, sb., *O Fr.* fourgon, cf. *Ital.* forcōne; *poker,* PR. P. 182; frugon CATH. 144.

furh, sb., *O.E.* furh (*dat.* fyrh), = *O.H.G.* furh, furuh, *O.Fris.* furch, *M.L.G.* vore *f.*; *furrow*; furg REL. I. 217; furgh GOW. II. 63; furghe H. S. 2448; forghe CH. BOET. v. 5 (170); forȝ, forwe FER. 3720, 5593; vorouȝ, forowe TREV. VII. 535; forwe AR. & MER. 3470; LANGL. *B* xiii. 372; forw HAV. 1094; forowe, fore '*sulcus*' PR. P. 171; foure (*ms.* fowre) FLOR. 1300; forȝes (*pl.*) A. P. ii. 1547; forwes LANGL. *B* vi. 106; forwis WICL. JOB xxxi. 38; fores MAN. (F.) 13024; vores FER. 1565.

fur-land, sb., *ploughland*, ROB. (W.) 2198.*

furȝ-long, sb., *O.E.* furlang; *furlong*; furlong, furlonge LANGL. *A* v. 5, *B* v. 5; furlong WILL. 13; MAND. 190; forlang B. DISC. 306.

[fürht, adj., *O.E.* fyrht, forht, = *O.L.G.* forht, *O.H.G.* forht, *Goth.* faurhts; *fearing* : comp. god-fürht, un-frigt.]

fürht, sb., *O.E.* fyrhtu, fyrhto, *cf. O.L.G.*, *O.H.G.* forhta, *Goth.* faurhtei, *O.Fris.* fruchta; *fright, fear*; friȝt MAP 338; S. S. (Wr.) 948; frigt GEN. & EX. 1234.

frig(t)-ful, adj., *O.E.* forhtfull; *fearful, timid*, GEN. & EX. 3459.

fürhten, furhtien, v., *O.E.* fyrhtan, forhtian, *cf. O.H.G.* furhten, forhten, *Goth.* faurhtjan, *O.Fris.* fruchta; *be afraid, make afraid*; forhtigen ['*pavere*'] MK. xiv. 33; frigtede (*pret.*) GEN. & EX. 1861; *comp.* a-, for-, of-fürhten.

fürhti, adj., = *O.H.G.* forhtīg; *timid*; frigti GEN. & EX. 667, 984.

frigti-hēd, sb., *timidly*, GEN. & EX. 2222.

frigti-līke, adv., *timidly*, GEN. & EX. 1617.

füri, adj., = *M.L.G.* fürich, *M.Du.* vierigh, *M.H.G.* viurec; *fiery*, BRD. 15; firi CH. COMPL. M. 96; PR. C. 7762; fürie (*pl.*) LAȝ. 18863*; fuirie LEG. 189.

furial, adj., *Lat.* furiālis; *raging, furious* : in that furialle pein of helle CH. C. T. *F* 448.

furie, sb., *O.Fr.* furie; *fury*, LIDG. M. P. 206.

furious, adj., *Lat.* furiōsus; *furious*, LIDG. M. P. 157; N. P. 2.

furme, see forme.

furmente, see frumente.

[fürn, adv., *O.E.* fyrn (*of old*), = *O.L.G.* furn; *comp.* ȝe-fürn.]

fürn-daȝes, sb., *former days, days of old* : i furndaȝen [*O.E.* in fyrndagum, *cf. O.L.G.* an furndagon] LAȝ. 27118.

fürn-ȝēr, sb. & adv., *O.E.* fyrngēar; *years past; in former years* : þe kindenesse þat mine evencristene kidde me fernȝere LANGL. *B* v. 440; þe lost of vernȝere AYENB. 92.

furneien, v., *O.Fr.* furnir; *procure*; furneie (*imp.*) a tree RICH. 5517.

furre, sb., *O.Fr.* fuerre, forure; *fur*, PR. P. 183; furres (*pl.*) of armine SQ. L. DEG. 839; shepe ... berithe furres P. R. L. P. 16; forours MAN. (F.) 11193..

fürren, see feorren.

furrin, v., *O.Fr.* forrer; *fur*, PR. P. 183; furred (*pple.*) GAW. 1737; furred, forred LANGL. *B* xiii. 227, *B* xx. 175.

furrour, forrour, sb., *O.Fr.* fourreur; *furrier*, MAN. (F.) 12453.

fürsien, see feorsien.

fürst, adv., *O.E.* fyrst, = *O.N.* fyrst, *O.H.G.* furist; *first*, P. L. S. xii. 6; furst þer sit an old cherl P. S. 156; first SAX. CHR. 253; for þi comen þeȝ him first to seken i þat ende

ORM. 6876; first, verst ROB. 9, 196; first [ferst] CH. C. T. *A* 725; ferst WILL. 648; frist DEP. R. i. 107; a furst JOS. 553; a verst AYENB. 46.

fürste, adj., *O.E.* fyrsta, = *O.N.* fyrsti, *O.L.G.*, *O.H.G.* furisto; *first*, C. L. 300; firste LANGL. *B* xvi. 184; þe firste preost ORM. 293; ferste SHOR. 45; DEGR. 1522; verste AYENB. 5; friste M. ARTH. 149; E. G. 80; i þan virste ænde LAȝ. 20863.

fürst, see frist.   fürþe, see feorðe.

furþen, adv., *O.E.* furðon, furðum; *in the first place* : þer ne mot beon furþan on stef ofer itel FRAG. 2; forðon, forþon, forðe HOM. I. 5, 7, 23; forþen ORM. 825; forðen, forðe HOM. II. 13, 137.

furðer, adv., *O.E.* furður, furðor, = *O.L.G.* furthor, *O.Fris.* further, forther, *O.H.G.* furdor, furder, furdir, *M.H.G.* vurder, vurder; *further*, LAȝ. 4880; ne schalt tu gon no furðer A. R. 228; furþer REL. I. 173; EGL. 373; furþer mo MISC. 149; vorþeremore (*furthermore*) ROB. (W.) 2464; forþer, ferther LANGL. *A* ii. 176, *B* v. 6; furþer, firþer H. S. 337; flæh it ta na forþer mar ORM. 7338; forþir PR. C. 440; forþer, ferþer CH. C. T. *A* 36, 4117.

furþer(e), adj., *O.E.* furðra, = *O.L.G.* furthero, forthero, *O.H.G.* fordero; *earlier, former ; fore*; forther(e) WICL. GEN. xxxii. 17; mines furþur ealde fader ANGL. VII. 220; his forþere fet BRD. 30; IW. 2004.

furðer-līc, adv., *further*; als hi com swa forþerliȝ (*far forth*) ORM. 14812; þet he furðerlüker (*comp.*) echeð his pine A. R. 236.

furþest, adj., *furthest, farthest*; forþest CH. BOET. iv. 6 (136); þe fertheste ende of Northfolke LANGL. *B* v. 239 (Wr. 4950).

furðrien, v., *O.E.* fyrðrian, *cf. M.L.G.* vorderen, *O.H.G.* fordarōn; *further, promote, help forward* : þo þe leveð þat swilch þing hem muge furðrie oðer letten REL. I. 131; (HOM. II. 11); furþren GOW. I. 228; firþren ORM. 5084; forþren SPEC. 99; fürðreð (*pres.*) A. R. 156; forþered (*pple.*) CH. BOET. ii. 4 (41).

fortheringe, sb., = *M.L.G.* vorderunge, *O.H.G.* forderunga; '*promotio*,' PR. P. 174; furþering GOW. I. 228.

füs, adj., *O.E.* füs, = *O.L.G.* füs, *O.N.* füss, *O.H.G.* funs; *eager, ready*, ORM. 9065; FL. & BL. 368; C. M. 14089; fous S. A. L. 148; SPEC. 50; B. DISC. 288; vous REL. II. 243; füse (*pl.*) LAȝ. 4399; heo beoþ fuse to bringen þe ut of huse FRAG. 6; vouse MISC. 46.

füsen, v., *O.E.* fȳsan, = *O.L.G.* füsian, *O.N.* fȳsa; *from füs; hasten, set out, prepare* : þa com þer fusen unimete ferde LAȝ. 5778; alle we mote fusen 13534; hem to scipe füsde (*pret.*) 1511; Brutus ... fusde to fihte 1735; fusde hine sulfne 1865; *comp.* ȝe-füsen.

S

fusible, adj., *O.Fr.* fusible; *fusible*: oiles ablucioun and metal fusible CH. C. T. *G* 856.

fūst, sb., *O.E.* fȳst,=*M.Du.* vuist, *O.Fris.* fest, *O.H.G.* fūst; *fist*, o. & N. 1538; fust [fist] LANGL. *B* xvii. 138 (Wr. 11644); fist PR. P. 163; **fūste** (*dat.*) FER. 1901; **fūstes** (*pl.*) A. R. 82; TREAT. 139; vustes [fustes] LA3. 22785; vestes ROB. (W.) 7081; **fēsten** (*dat. pl.*) MK. xiv. 65.

fustain, sb., *O.Fr.* fustaine; *fustian*, REL. I. 129; blankettes shall be of fustiane SQ. L. DEG. 841; of fustian he werede a gepoun CH. C. T. *A* 75; ȝe meshakele [is] of medeme fustane HOM. II. 163; fustian (*v. r.* fustein) '*furesticus*' PR. P. 183.

fuwel, *see* fuȝel.   **fūwer,** *see* fēower.

g and ȝ.

ȝa, *see* ȝe.   **gā,** *see* gangen.   **ȝā,** *see* ȝēa.

gabbe, sb., *O.N.* gabb, *mod.Eng.* gab; *jest, imposture*, FL. & BL. 489; wiþouten gabbe C. L. 507; gabbe ant gile SPEC. 49; **gabbes** (*pl.*) KATH. 2269.

gabben, v., *O.N.* gabba, = *O.Fris.* gabbia (*accuse*), *M.L.G.* gabben, *mod.Eng.* gab; *ridicule; mock, deceive; babble, chatter*: lauhwen oȝer gabben A. R. 200; we agen to gabben us selven for þat we synegeden HOM. II. 65; gabbin '*mentior*' PR. P. 183; gabbe LEG. 52; LANGL. *B* iii. 179; CH. C. T. *A* 3510; ne noþing nolleþ do er þan me **gabbeth** (*pres.*) of ham AYENB. 69; þo burnes þat so bleþeli gabbe WILL. 1994; þat i gabbe [*v. r.* lie] not WICL. 2 COR. xi. 31; **gabbed** (*pret.*) M. ARTH. 1105; gabbit BARB. iv. 290; **gabbid** (*pple.*) M. H. 57; igabbet KATH. 2305; *comp.* bi-gabben.

  gabbere, sb., *mocker, deceiver*, CH. C. T. *I* 784; gabbar(e) '*mendax*' PR. P. 183; **gabberes** (*pl.*) MAND. 160.

  gabbinge, sb., *mocking, lying, 'mendacium,'* PR. P. 183; O. & N. 626; LANGL. *B* xix. 451; **gabbingis** (*pl.*) BARB. iv. 768.

gáble, sb., *O.Fr.* gable (SCHMIDL. CATHOL.), *cf. O.H.G.* gibil, *O.N.* gafl, *Goth.* gibla; *gable*, LANGL. *B* iii. 49; CH. C. T. *A* 3571; gabil LIDG. M. P. 204.

gáble[2], sb.; *cable* : gabulle EGL. 1193; **gábulle-rope** N. P. 18; *cf.* cáble.

gad, sb., *O.N.* gaddr (*nail, sting*), = *O.H.G.* gart (*goad*), *Goth.* gazds *m.* (κέντρον); *rod, staff, bar, spike, 'pertica,'* PR. P. 184; VOC. 234; GEN. & EX. 3185; HAV. 279; a gad of stele L. C. C. 6; gadde TREV. VII. 199; **gaddes** (*pl.*) HAV. 1016; grete gaddes of iren TREV. VI. 199; gaddes of stele D. ARTH. 3621.

[**gáde,** sb., *O.E.* (ge-)gada,=*M.L.G., M.Du.* gade, *M.H.G.* gate; *companion; comp.* ȝe-gáde.]

gāde, sb., *O.E.* gād; *goad*; gode '*stimulus*' FRAG. 2; PL. CR. 433; WICL. ECCLUS. xxxviii. 26.

  gād-wand, sb., *carter's whip*, BARB. x. 232.

gadeling, sb., *O.E.* gædeling,=*O.L.G.* gaduling, *Goth.* gadiliggs, *O.H.G.* gateling, *M.H.G.* getelinc; *companion, fellow*, MISC. 44; HAV. 1121; LANGL. *B* xx. 156; a luþer gadeling ROB. 310; gadeling [gadling] GAM. 102; þat covetous gadling S. S. (Wr.) 1589; þou berdles gadling TOR. 1015; gadeling, gedeling REL. I. 180; **gadelinges** (*pl.*) LA3. 12335; stalworþe gadelinges ALIS. 1192.

gaderen, *see* gæderien.

[**gæder,** sb., *? O.E.* gador(-wist) ;] **tō gædere,** gadere [gedere], [*O.E.* tō gædere, gadore,= *O.Fris.* tō gadera, *M.L.G.* tō gadere]; *together*, LA3. 52, 10963; al to gædere ic am forlor(en) FRAG. 8; te gædere SAX. CHR. 262; to gaderē O. & N. 807; MISC. 37; TREAT. 138; SHOR. 92; to gedere ROB. 32; al to gedere A. R. 320; to gedre ORM. 991; MAND. 70; to gidere AYENB. 43; to gidere [gidir] LANGL. *B* PROL. 46; **to gideres** xvi. 80; þe an is qvenched al to gederes HOM. I. 81; to gederes KATH. 114; A. R. 354; **æt gadere** [*O.E.* æt gædere] *together* [' *simul*'] JOHN xx. 4.

gæderien, v., *O.E.* gædrian, gaderian,= *O. Fris.* gadria, *M.L.G.* gaderen, *M.H.G.* gatern; *gather*; gædere(n), gaderen LA3. 25463, 30183; gederen KATH. 990; A. R. 146; (*ms.* gaddrenn) ORM. 13407; gaderen GEN. & EX. 2134; gaderin PR. P. 184; gadrie FER. 3338; gaderi AYENB. 120; gadere WILL. 30; gadre [gedere] WICL. EX. xxiii. 10; **gadereð** (*pres.*) MAT. vii. 16; gadereþ BEK. 617; **gederinde** (*pple.*) A. R. 128; **gæderede,** gaderede (*pret.*) LA3. 1480, 9255; gederede P. S. 70; he gederede michel ferde HOM. II. 51; gadred [gadrede] LANGL. *B* xvi. 80 (W. 10960); þu gæderedest FRAG. 6; **gadred** (*pple.*) HAV. 2577; P. 14; gedret, gedrit D. TROY 1174, 1275; *comp.* ȝe-gæderien.

  gaderere, sb., *gatherer*, TREV. I. 13; **gaderares** (*pl.*) MISC. 67.

  gederunge, gederinge, sb., *O.E.* gaderung, = *M.L.G.* gaderinge; *gathering*, A. R. 206, 286; gedering C. L. 643; gaderinge AYENB. 192; (*? ms.* garderynge) MAN. (F.) 4354; gadering PR. C. 8831.

gadie, sb., *? spike*; gadien (*pl.*) KATH. 1945; mid irnene gadien JUL. 56.

ȝēn, *see* ȝein.   **ȝēp,** *see* ȝēap.

ȝēr, *see* gār *and* ȝēar.

gærsume, sb., *O.E.* gersume (*acc.* gersuman SAX. CHR. 177),=*O.N.* gersemi *f.*; *treasure*, FRAG. 8; gærsume, gersume, garsume LA3. 2642, 4428, 4561; gersume HOM. I. 91; FL. & BL. 405; gersum MARH. 3; KATH. 799; gersom REL. II. 217; garsum A. R. 350.

ȝæru, ȝærwen, *see* ȝearu, ȝearwien.

ȝæt, *see* ȝeat.

gǣte, sb., *care* : nimen gete FRAG. 6.

    gǣte-lǣs, adj., *without care*, ORM. 6190.

gǣten, v., *O.N.* gǣta ; *guard, keep* : he sholde wel gæten hire & hire child ORM. 2079 ; gete HAV. 2762 ; AV. ARTH. lix ; an angel has þe ȝate to gete C. M. 997 ; gētes (*pres.*) M. H. 23 ; þou geetis D. TROY 11746 ; gēateð (*imper.*) A. R. 50* ; gĕtte (*pret.*) C. M. 7503 ; gate D. TROY 566.

    gǣter, sb., *keeper, guardian* ; Mars . . . geter of his good D. TROY 972 ; geeter of the god 11739.

[gǣten, adj., *? O.N.* gǣtinn ; *diligent.*]

    gǣten-līche, adv., *? diligently* ; gētenlüker (*compar.*) HOM. II. 121.

ȝǣten, *see* ȝáten.

gafel, *see* gavel.

gaffe, sb., *Fr.* gaffe ; *gaff*, REL. II. 174.

gagates, sb., *Lat.* gagātes, *Gr.* γαγάτης ; *jet*, TREV. II. 17 ; gagates [gogathes] I. 337 ; a gagate REL. I. 53.

gáge, sb., *for* cáge ; *enclosure* ; (*v. r.* cage) MAN. (F.) 1343.

gáge, *see* wáge.

gaȝel, sb., *O.E.* gagel, gagol,=*M.H.G.* gagel, *M.Du.* gaghel ; *the plant gale,* ' *myrtus*,' VOC. 141 ; gawil, gaul PR. P. 189.

gagelin, v.,=*M.L.G.* gagelen, *M.Du.* gag- helen ; *cackle* (*as a goose*), PR. P. 184 ; gagult (*pret.*) REL. I. 86 ; gaglide DEP. R. iii. 101.

    gagelinge, sb., *cackling*, PR. P. 184.

gágen, *see* gaugin. ȝaȝhen, *see* gain.

gaggin, v., *gag*, ' *suffoco*,' PR. P. 184.

gai, adj., *O.Fr.* gai ; *gay*, PR. P. 184 ; CH. C. T. *A* 3339 ; A. P. ii. 830 ; þe gaie gerles WILL. 816.

    gai-līche, gai-lī, adv., *gaily*, WILL. 1625, 2591.

    gaienesse, sb., *pleasure*, LANGL. *C* xii. 66 ; gainesse *B* x. 81.

gaignáge, *see* gaináge.

gaile, sb., *O.Fr.* gaole, geole ; *gaol, jail* ; jaiole (*ms.* iaiole), jaile (*ms.* Iaile) C. M. 13174.

gailer, sb., *O.Fr.* gaiolier ; *gaoler*, P. L. S. xxiv. 218 ; CH. C. T. *A* 1064 ; gailer, jailer LANGL. *A* iii. 133, *C* iv. 175.

gaillard, adj., *Fr.* gaillard ; *gay, merry*, CH. C. T. *A* 4367.

gain, sb., *O.N.* gagn ; *gain, advantage* ; TRIST. 614 ; gaȝhen ORM. 13923 ; gein MARH. 18.

    gaȝhen-lǣs, adj., *O.N.* gagnlauss ; *gainless*, ORM. 2019.

gain, gainen, *see* gein.

gaináge, sb., *O.Fr.* gaignage ; *profit from cultivation of the soil* : the gainage of the ground PL. CR. 391 ; gaignage HALLIW. 389.

gaine, *see* ȝeinen.

gāinge, *see* ganging *under* gangen.

gainlīche, *see under* ȝein.

gaire, *see* geare.

gaisein, sb., *marsh* [' *lacuna* '], PALL. i. 36.

gaite, *see* gāt.

gaitre, sb., *dog-wood tree* : laxatives . . . of catapus or of gaitre beriis CH. C. T. *B* 4155.

gal, sb.,=*M.H.G.* gal ; *from* gálen ; *singing*, HOM. II. 197 ; gleo and gal MISC. 97.

    gal-ful, adj., *tuneful, melodious* : ȝe holden him [Mercurie] galful & god of the tounge ALEX.[2] (Sk.) 667.

gāl, sb., *O.E.* gāl,=*M.H.G.* geil ; *luxury* : þat gal MISC. 151.

gāl[2], adj., *O.E.* gāl, = *O.L.G.* gēl, *O.H.G.* geil ; *luxurious, lascivious*, ORM. 1201 ; MISC. 148 ; þe gole men HOM. II. 31.

    gōl-hēd, sb., *luxury*, GEN. & EX. 534.

    gāl-lich, *O.E.* gāllīc ; *luxurious* : ga(l)liche dede HOM. I. 149 ; golliche deden REL. I. 132.

    gālnesse, sb., *O.E.* gālness ; *luxury, lascivi- ousness*, HOM. I. 103 ; ORM. 8015 ; golnesse A. R. 198 ; O. & N. 492.

    gālschipe, sb., *O.E.* gālscipe ; *luxury*, REL. I. 132 ; golsipe (*ms.* kolsipe) and ȝiscing MISC. 11 ; so kolde of kinde ðat no golsipe is hem minde 610.

galaunt, sb., *O.Fr.* galant, gallant ; *gallant* ; galauntes (*pl.*) B. B. 136.

galamelle, sb., *med.Lat.* calamella ; *sugar- cane* : thei drinken gode beverage and swete and norisshinge, that is mad of galamelle MAND. 142.

galantine, *see* galentine.

galaxie, sb., *Fr.* galaxie ; *galaxy*, CH. P. F. 56.

galban, sb., *Lat.* galbanum ; *the gum gal- banum* : take galban (*ear. text* galbantum) of good odour WICL. EX. xxx. 34.

galder, sb., *O.E.* gealdor, galdor, *O.N.* galdr ; *from* gálen ; *enchantment* ; galdere (*dat.*) LAȝ. 19257.

[gále, sb., *from* gálen ; *? singer* ; *comp.* nihte-gale.]

gále[2], sb., *O.Fr.* gale (*gaiete*), = *Ital.* gala ; *merriment*, SPEC. 26 ; þorȝ his oȝene gale SHOR. 107 ; it made michel gale S. A. L. 168 ; þe niȝtingale in wode makiþ miri gale ALIS. 2548.

galegale, sb., *from* gálen ; *noisy fellow* : thu hattest niȝtingale, thu miȝtest bet hoten galegale O. & N. 256.

galeie, sb., *O.Fr.* galee ; *galley,* ' *galea*,' FR. P. 185 ; HORN (L.) 185 ; gale TOR. 1315 ; galai men MIN. iii. 57 ; galaies (*pl.*) MIN. iii. 51.

gálen, v., *O.E.* galan (*pret.* gōl),=*O.N.* gala ;

sing, cry out; gale CT. LOVE 1356; **gáles** (*pres.*) YORK xxxiii. 23 ; *comp.* **bi-gálen.**

gálere, sb., *O.E* galere ; *enchanter, 'incantator,'* FRAG. 2.

galentine, sb., *O.Fr.* galentine ; *a kind of sauce*: sause of galentine CH. BOET. APP. 16 (180) ; B. B. 281 ; galantine 174.

galewes, *see* **galȝe.**

galȝe, sb., *O.E.* gealga, galga,=*O.Fris., Goth.* galga, *O.L.G., O.H.G.* galgo, *O.N.* galgi ; *gallows*; **galwe** (*dat.*) BEV. 1217; **galwes** (*pl.*) MAN. (H.) 172; L. H. R. 132; galewes P. S. 221; galous, galowes, galos D. TROY 12890, 12885, 13116.

galhe-forke, sb., *gallows,* A. R. 174*.

galwe-trē, sb., *O.E.* galgtrēow, galgatrē ; *gallows,* HAV. 43.

galian, sb., *a beverage named after Galen* ; **galianes** (*pl.*) CH. C. T. C 306.

gálin, v., *O.E.* (be-)galian ; *sing, make a loud noise, 'crocito,'* PR. P. 185; **gálieð** (*pres.*) A. R. 128*; hundes þer galieð LAȝ. 20858 ; þar **gálede** (*pret.*) þe gouk D. ARTH. 927.

galingale, sb., *O.Fr.* galingal ; *a kind of spice,* P. L. S. xxxv. 73 ; CH. C. T. *A* 381 ; L. C. C. 8.

galiote, sb., *O.Fr.* galiot *m.,* galiote *f.*; *small galley*: mani of **galiotes** (*pl.*) MIN. iii. 81 ; galietes TREV. VIII. 552.

galle, sb., *O.E.* gealla *m., cf. O.N.* gall *n., O.L.G., O.H.G.* galla ; *gall, bile ; anything bitter ; bitterness, anger,* A. R. 106 ; ORM. 15383 ; REL. I. 226 ; HAV. 40; A. D. 294; ALIS. 5073 ; LANGL. *B* xvi. 155 ; A. P. ii. 1022; MIR. PL. 112 ; se swerta gealle LEECHD. III. 82 ; **gallen** (*dat. pl.*) MAT. xxvii. 34.

galle², sb., *O.E.* gealla, *cf. O.N.* galli *m., M.L.G., M.H.G.* galle *f. ; gall, sore place,* PR. P. 185 ; A. P. i. 1059 ; ȝif eni wiȝt wil clawe us on þe galle CH. C. T. *D* 940.

gallin, v., *O.E.* geallian ; *gall, 'strumo,'* PR. P. 185 ; þe hors was . . . **galled** (*pple.*) upon þe bak(e) GOW. II. 46 ; gallid LIDG. M. P. 168.

gallinge, sb., '*strumositas,*' PR. P. 185.

galloc, sb., *O.E.* galluc ; *comfrey,* REL. I. 36 (VOC. 139).

galoche, sb., *O.Fr.* galoche ; *a sort of patten, 'crepida,'* PR. P. 184; CH. C. T. *F* 555; **galoches** (*pl.*) LANGL. *B* xviii. 14.

galōn, galoun, sb., *O.Fr.* galon ; *gallon,* LANGL. *B* v. 346, *A* v. 191 ; galun R. S. vii ; HORN (L.) 1123.

galopen, v., *O.Fr.* galoper ; *gallop*; **galopiþ** (*pres.*) ALIS. 460.

galoxie, *see* **galaxie.**

galpen, v., ?=*O.L.G.* galpōn (*cry out*), *M.Du.* galpen ; *gape, yawn,* LANGL. *B* xiii. 88; **galpeth** (*pres.*) ['*oscitat*'] TREV. V. 389;

galping (*pple.*) '(*h*)*iante,*' VOC. 126; CH. C. T. *F* 354.

galper, sb., *one who gapes,* REL. I. 291.

galpinge, sb., *yawning,* TREV. V. 389.

galstren, v., = *O.H.G.* galstrōn (*bewitch*), *L.G.* galstren (*demand*) ; *boast* ; ȝelstreð [*v. r.* galstres] (*pres.*) . . . & ȝelpeð of hore god(e) A. R. 128.

galte, sb., *O.N.* galti (*boar*), *cf. O.H.G.* galza (*sow*) ; *boar,* PR. P. 185 ; VOC. 204 ; D. ARTH. 1101.

ȝalu, *see* **ȝeolu.** **galwe,** *see* **galȝe.**

galwen, v., *O.E.* (a-)gælwian ; *scare, terrify* ; **galwed** (*pple.*) L. H. R. 132; *comp.* **bi-galwen.**

gambesōn, sb., *O.Fr.* gambaison, gambeson ; *quilted doublet*: gaumbisoun ALIS. 5146 ; gomes with **gambassoune** (*dat.*) HALLIW. 390; his gloves and his **gamesuns** (*v. r.* gambesouns) (*pl.*) gloet as the gledes ANT. ARTH. xxxi.

gamelos, sb., *Gr.* χαμαιλέων; *chameleon*: þe gamelos þet leveþ bi þe eir AYENB. 62.

gamen, gáme, sb., *O.E.* gamen, gomen, = *O.L.G., O.H.G., O.N.* gaman, *O.Fris.* game, gome, *mod.Eng.* game, gammon ; *sport, jest, game,* GEN. & EX. 2015; HAV. 2250; TRIST. 2406; AM. & AMIL. 710; C. M. 3522; DEGR. 3; gamen & gleo P. L. S. viii. 145 ; gomen, gome, game LAȝ. 7014, 15856, 18594 ; gomen A. R. 214 ; R. S. ii ; MAP 347 ; game '*ludus, jocus*' PR. P. 185 ; HORN (L.) 198 ; ROB. 16, 26 ; CH. C. T. *A* 853 ; in ernest or in game GOW. I. 297 ; gammin *joy* BARB. xix. 804; gaume TREV. VII. 111 ; the gammin ga *affairs may turn out* BARB. xi. 319 ; game, gome O. & N. 521, 1443, 1649; gome CHR. E. 456; þat gome R. S. vii; þet geme AYENB. 34; **gamene** (*dat.*) PERC. 1689 ; gomene LAȝ. 3045; **gomenes** (*pl.*) A. R. 318; SPEC. 24; gaumes D. TROY 1620.

gome-ful, adj., *jesting, sportive,* MARH. 10; mid gomenfulle worden LAȝ. 21430.

gamful-li, adv., *jocosely,* TREV. VII. 111.

game-gobelin, sb., *demon who plays with men,* VOC. (W. W.) 597.

gamen-līch, adj., *sportive*: mi gode gameliche game gurte to grounde REL. II. 8.

game-līche, adv., *O.E.* gamenlīce ; *jokingly* ; gameli WILL. 427 ; gomenli GAW. 1079.

geme-man, sb., *gambler* ; **gememen** (*pl.*) AYENB. 63.

gamensum, adj., *O.N.* gamansamr ; *sportive*: gam(e)sum & glad WILL. 4193.

gamenien, gámen, v., *O.E.* gamenian, gamnian, = *O.N.* gamna ; *joke, play ; amuse, delight*; gomenin MARH. 14; **gamenen** (*pres.*) ALIS. 5461 ; **gamenede** (*pret.*) FL. & BL. 31 ; ne gamnede hire þat gle riȝt nouȝt BEV. 3016 ; gamede LEG. 68 ; gomede LAȝ. 4588 ; A. R. 368.

ȝamer, *see* ȝeomer.

gamme, sb., *Fr.* gamme; *gamut,* PR. P. 185.

gan, *see* ginnen.   **gān,** *see* gangen.

gānde, sb., = *Swed.* gående ; *journey, going* : abute furten niȝt gonde FL. & BL. 210.

gandre, sb., *O.E.* gandra; *gander,* REL. I. 217; **gandres** (*pl.*) [' *anseres* '] TREV. III. 297; MAND. 216.

gang, sb., *O.E.* gang, gong, geong, = *O.L.G., O.H.G.* gang, *O.N.* gangr, *Goth.* gaggs ; *walk, act of going, gang, troop; passage; privy, sewer,* HOM. I. 229; ORM. 8910; gong H. M. 19; gong, ȝong KATH. 569, 2502; ȝong A. D. 252; feole dawen· ȝong LAȝ. 1298; ȝeong 4605; þis. proute ȝong ARCH. LII. 35; gong ' *latrina* ' PR. P. 203; S. S. (Web.) 1315; *comp.* chirche-, forð-, hand-, here-, in-, mis-, üm-, üt-gang.

   ȝong-dawes, sb. pl., *O.E.* gangdagas ; *rogation days* (? *ms.* ȝoing-), A. R. 412.

   gong-firmar, sb., *scavenger,* PR. P. 203.

   gong-hūs, sb., *privy,* A. R. 84.

   gong-men, sb. pl., *scavengers,* A. R. 84.

   gong-þurl, sb., *hole of a privy* : þes fikelares mester is to wrien & te helien þet gongþurl A. R. 84.

gangen, gān, v., *O.E.* gangan, gongan, geongan, gān (*pret.* gēong, gieng, gang), = *O.L.G.* gangan, gān (*pret.* gēng), *O.H.G.* gangan, gān (*pret.* giang, gieng), *O.N.* ganga, gā (*pret.* gēkk), *Goth.* gaggan (*pret.* gaggida) ; *go* ; gangen REL. I. 211 (MISC. 5); ORM. 1076; gan 7767; gange AN. LIT. 9; MIR. PL. 122; gange LANGL. *B* ii. 167; go slepe *B* vi. 303; gonge HAV. 1185; gon 1045; ȝeonge LAȝ. 27764; ȝonge 9061*; gan [gon] 597; ne scal him neoðer gon fore gold ne na gærsume 22853; ȝonge, ionge (*ms.* yonge) H. H. 132 (134); gan Iw. 800; nouþer gold ne seolver ne moste gan for þe HOM. I. 9; gan, gon, go O. & N. 214, 952, 1431; gon TREAT. 137; gon ... slepen A. R. 270; goṅ seken GEN. & EX. 3598; so ne mai hit nouȝt gon C. L. 1084; he mai gon al gelde SPEC. 24; hou longe schalt þou maidin gon (*continue a maiden*) GUY (Z.) 7020; gon [goon] CH. C. T. *A* 12; ga PERC. 1462; PR. C. 4100; go MAND. 4; i wil go sittin LUD. COV. 20; guon AMAD. (Web.) 670; guo AYENB. 32; to ganne LAȝ. 22279; MARH. 4; **gæst** (*pres.*) LAȝ. 26437; gest A. R. 86; AYENB. 129; þu gest al to mid swikelhede O. & N. 838; gast ORM. 4660; gangeð REL. I. 213 (MISC. 7); gangeþ, gaþ ORM. 1228, 1236; ȝeongeþ, geð [goð] LAȝ. 23499, 23665; ȝongeþ P. S. 216; hit geþ to naht 220; goð A. R. 364; geþ O. & N. 528; ROB. 205; C. L. 409; AYENB. 56; þus hit geþ bitvene hem tvo SPEC. 103; JOS. 394; geth TOR. 2142; HOM. I. 27; gað [goþ] LAȝ. 711; goð A. R. 10; guoþ AYENB. 34; ga we nu til þat ilke tun ORM. 3390; go we (*let us*) kiþe oure kniȝthod WILL. 1184; **gā** (*imper.*)

HOM. I. 35; LAȝ. 26107; gað HOM. I. 33; KATH. 1466; **gonge** (*subj.*) HAV. 690; **gangende** (*pple.*) MAT. xiv. 25; ganninde [goinde] LAȝ. 1582.; goinde S. A. L. 148; guoinde AYENB. 120; **gān** (*pple.*) ORM. 4352; PERC. 1062; PR. C. 1995; gon EGL. 375; ga HOM. I. 21; ȝegan 33; *comp.* a-, æt-, bi-, for-, ȝe-, ofer-, þurh-, under-, ümbegangen (-gān); *deriv.* gang, genge; *see* ēode.

   [**ganger,** sb., = *M.L.G.* ganger, genger; *goer*; *comp.* **fore-ganger** ;] gōere (*ms.* goare) ' *ambulator* ' PR. P. 200; gōeres (*pl.*) MAND. 277.

   ganging, gāing, sb., *going* : þat þe poveral get sum bote, and ganging þat ar lame o fote C. M. 12259; goinge WICL. JOB xxxi. 7; til þe time of þe son dounganging PR. C. 4778; þei lepith als ligh(t)li a þe longe goinge (*death on the gallows*) out of the domes carte D. RICH. iii. 136; **gāinges** (*pl.*) PS. xvii. 37; goingis WICL. JOB xiv. 16; guoinges AYENB. 231.

ganglen, *see* janglen.

gānien, ȝānien, v., *O.E.* gānian, = *O.H.G.* geinōn, *Gr.* χαίνειν; *yawn*; gone GOW. II. 263; ȝanin ' *hio, oscito,* ' PR. P. 536; **gāneth,** [goneth] (*pres.*) CH. C. T. *H.* 35; ganes, gonis ' *baaile* ' VOC. 152; gones AV. ARTH. xii ; **gōninde** (*pple.*) HICKES I. 228; ȝaninge WICL. 2 MACC. vi. 18; gapinge nor ganinge B. B. 135; ȝāned (*pret.*) RICH. 276.

   gāninge, ȝāninge, sb., *O.E.* gānung; *yawning,* PR. P. 185.

gannok, sb., *banner,* MAN. (H.) 113.

ganokir, ganneker, sb., *alehouse-keeper,* PR. P. 185.

gant, *see* gaunt.

gante, sb., *O.E.* ganot, ganet (*swan*), *cf. M.L.G.* gante, *O.H.G.* ganazo ; *gannet,* ' *bistarda,* ' PR. P. 186.

gap, sb., *O.E.* geap, = *O.N.* gap; *gap,* ' *intervallum,* ' PR. P. 186; gappe FER. 4989; **gape** (*dat.*) LIDG. M. P. 114; gappe CH. C. T. *A* 1639.

ȝāpe, *see* ȝeap.

gápin, v., *O.E.* geapian, = *O.N.* gapa, *M.L.G., M.Du.* gapen ; *gape, desire,* ' *hio,* ' PR. P. 186; gape LANGL. *B* x. 41; gapen and desiren CH. BOET. ii. 2 (36); **gápeð** (*pres.*) REL. I. 220; gapes AV. ARTH. xii; **gápande** (*pple.*) WILL. 2875; ȝeápede (*pret.*) MARH. 46; i gaped aboute PL. CR. 156; gapeden CH. BOET. i. 4 (15); *comp.* **bi-gápen.**

   gápinge, sb., *gaping, desire,* ' *hiatus, hiacio,* ' PR. P. 186; her cruel ravine devouringe al þet þei han geten, shewiþ oþer **gápinges** (*pl.*) CH. BOET. ii. 2 (36).

gar, *see* ȝearwien.

gār, sb., *O.E.* gār, = *O.L.G., O.H.G.* gēr, *O.N.* geirr *m., cf. Goth.* gairu *n.*; *spear*; gare, gære[gar]LAȝ.5079,15225; gāre(*dat.*)OCTAV. (H.) 1527; ISUM. 453; **gāres,** gæres (*pl.*) LAȝ. 1751, 1847; geres HOM. II. 35; gōren (*dat. pl.*) GEN.& EX. 3458; *comp.* ēl-gēr, nave-gār.

găr-olīve, sb., *O.E.* gārclīfe ; *agrimony* (*plant*), 'agrimonia,' VOC. 139.

găr-fangil, sb., *spear*, 'anguillaria, anguillare,' PR. P. 186.

găr-fisch, sb., *garfish*, PR. P. 186.

găr-lēk, sb., *O.E.* gārlēac ; *garlick*, HICKES I. 232 (P. L. S. xxxv. 105) ; L. C. C. 53 ; garlek, -leek CH. C. T. *A* 634 ; garlec REL. I. 37 ; (*sauce*) B. B. 152 ; grene garlike B. B. 278.

gărlēk-monger, sb., *garlick dealer*, LANGL. *C* vii. 373.

garant, sb., *some dark coloured gem* : the blake [clustres of grapes] ben of onichez and **garantez** (*pl.*) MAND. 219.

garbāge, sb., *garbage*, 'exta,' PR. P. 186.

garce, garcen, *see* **garse, garsen.**

ȝard, *see* **ȝeard.**

gard-brace, sb., *O.Fr.* garde-bras ; *armour for the arm*, CH. D. BL. 1554.

[**garderinge**, *see* **gederunge** *under* **gæderien.**]

gardevian, sb., *O.F.* gardeviant ; *meat safe*, B. B. 196.

gardin, sb., *O.Fr.* gardin ; *garden*, AYENB. 94 ; S. S. (Web.) 419 ; CH. C. T. *A* 1051 ; A. P. i. 260.

gardin-ȝate, sb., *garden gate*, MAND. 210.

gardiner, sb., *O.Fr.* gardenier ; *gardener*, AYENB. 94 ; gardeneer (*later ver.* gardinere) WICL. JOHN xx. 15 ; garthinere TOWNL. 267.

gardwin, sb., *cf. O.Fr.* gueredun ; *reward*, *guerdon* : gersoms, and golde and **gardwines** (*pl.*) D. ARTH. 1729 ; *see* **gverdon.**

gare, *see* **geare.** **ȝare**, *see* **ȝearu.**

gāre, sb., *O.E.* gāra, = *M.Du.* gheere, *O.H.G.* gōro, *O.N.* geiri ; *gore, strip*, TRIST. 2868 ; gore 'lacinia' PR. P. 203 ; VOC. 238 ; CH. C. T. *A* 3237 ; a gore (*printed* agore) AR. & MER. 6405 ; under gore (*gown, per synecdochen*) O. & N. 515 ; SPEC. 26 ; AN. LIT. 2 ; REL. II. 210 ; under þe wode gore SPEC. 91.

ȝāre, *see* **ȝeāre.**

gáren, ȝáren, *see* **ȝearwien.**

gargate, sb., *O.Fr.* gargate (*gorge*) ; *throat*, CH. C. T. *B* 4525 ; *see* **gorget.**

gargulie, sb., *Fr.* gargouille ; *gargoyle*, PR. P. 186.

gargulūn, sb., *part of the inwards of a deer* : þai griped to the gargulun GAW. 1335 ; he tiȝt þe mawe on tinde, and eke tho gargiloun TRIST. 508.

garisōn, waresōn, sb., *O.Fr.* garisōn, gvarisōn, warisōn ; *protection ; remedy, cure ; treasure, store*, ROB. 114, 409 ; garisoun C. L. 870 ; GAW. 1837 ; garison, garisoun ROB. (W.) 8461, 8558 ; garisun, warison WILL. 2259, 5073 ; warison FER. 1099 ; warisoun CHR. E. 836 ; A. D. 288 ; MAN. (F.) 1284.

garison, *see* **gærsume.**

garissen, warisse, v., *O.Fr.* garir, guarir, warir ; *protect*, CH. C. T. *C* 906 ; warishen [warschen] LANGL. *B* xvi. 105 ; warische WILL. 4283 ; **guarisshed** (*pple.*) L. H. R. 155.

garite, sb., *O.Fr.* garite, *mod.Eng.* garret ; *watch-tower, upper story*, 'specula,' PR. P. 187 ; **garites** (*pl.*) PL. CR. 214.

ȝarkien, *see* **ȝearkien.**

garland, *see* **gerland.** **gărlēk**, *see under* **gār.**

ȝarm, sb., *outcry*, A. P. ii. 971.

ȝarmen, *see* **ȝermen.**

garment, *see* **garnement.**

ȝarn, sb., *O.E.* gearn, = *O.N.*, *O.H.G.* garn ; *yarn*, PR. P. 536 ; (*ms.* yarn) VOC. 157 ; ȝern N. P. 18.

ȝarn-windel, sb., *yarn-winder*, 'girgillus,' PR. P. 536 ; (*ms.* yar[n]windel) VOC. 157.

garnement, garment, sb., *O.Fr.* garnement ; *garment*, LANGL. *B* xiv. 24 ; S. S. (Web.) 2775 ; **garnimentz** (*pl.*) FER. 1395.

ȝarnen, *see* **ȝeornen.**

garnet, sb., *O.Fr.* grenat ; *garnet* ; gernet SPEC. 25 ; granate ALEX. (Sk.) 3344 ; **grenaz** (*pl.*) MAND. 219 ; garnettes EM. 156.

garnet-appil, sb., *pomegranate* : the garnet appille of coloure golden hewid HALLIW. 392.

garnischin, v., *O.Fr.* garnir, guarnir, warnir ; *garnish*, PR. P. 188 ; warnische MAN. (F.) 15907 ; **warnished** (*pple.*) WILL. 1083.

garnisōn, sb., *O.Fr.* garnison ; *garrison* : a garnison she was of alle goodnesse P. R. L. P. 57 ; TREV. VIII. 522 ; CH. C. T. *B* 2217 ; **garnisōns** (*pl.*) MAN. (F.) 7127.

garrai, sb., *? armed force* ; *commotion* : after us send his garrai TOWNL. 64 ; he made all the garrai 113.

garren, v., *cf. M.Du., M.L.G., M.H.G.* garren ; *roar, chatter* ; **garrende** [garringe] (*pple.*) WICL. 3 JOHN 10 ; ȝarrande GAW. 1595.

garring, sb., *chirping, chattering, roaring* : the garring and fliing of briddus APOL. 95.

garse, sb., *gash*, CATH. 150 ; garce 'caesura, incisio' PR. P. 186 ; **garses** [garcen] (*pl.*) A. R. 258.

garsen, v., *O.Fr.* garser ; *gash* ; garse 'scarificare' CATH. 150 ; garcin PR. P. 186.

garsing, sb., *pricking the skin with a lancet*, HALLIW. 393 ; garcinge 'scarificacio, incisio' PR. P. 186.

garsoun, sb., *Fr.* garçon ; *boy*, S. S. (Web.) 1428.

garsume, *see* **gærsume.**

[**gart ?** *comp.* **an-, ofer-gart.**]

gartere, sb., *O.Fr.* jartiere ; *garter*, VOC. 196 ; PR. P. 188.

garteren, v., *garter*, 'subligo,' PR. P. 188.

garth, *see* **ȝeard.** **garthiner**, *see* **gardiner.**

ȝaru, *see* **ȝearu.**

ȝarwe, ȝarowe, sb., *O.E.* gearwe, gearewe,

gearuwe, = *O.H.G.* garwa, garawa; *yarrow,* '*millefolium,*' PR. P. 536; ȝarou VOC. 226; yarou REL. I. 53.

ȝarwien, *see* ȝearwien.

gāsen (?gásen), v., *cf. Swed.* gasa (*gape*) .(RIETZ 188); *gaze*: loke and gase LIDG. M. P. 203; a gase B. B. 39; gāsed (*pret.*) CH. C. T. E 1003; *comp.* a-gāsen.

gāspin, v., *O.N.* geispa; *gasp,* PR. P. 188; gaspe GOW. II. 263; gaspe S. & C. I. lxi; gaispande (*pple.*) D. ARTH. 1462.

gāst, sb., *O.E.* gāst, gæst, = *O.L.G.* gēst, *O.H. G.* ȝeist; *ghost, spirit,* ORM. 259; GEN. & EX. 202; FLOR. 778; MAN. (H.) 185; mi gast ant mi bodi MARH. 20; gast, gæst, gost LAȝ. 9071, 17130, 17136; gost A. R. 368; O. & N. 1401; JOS. 49; god is a gost AYENB. 211; þane gost 246; moni sori gost (*man, by synecdoche*) P. S. 70; a grimli gost FER. 539; so grisli a gost WILL. 1730; þe holi gost MAND. 18; gost [goost] CH. C. T. A 205; gāstes (*gen.*) GEN. & EX. 1486; gostes O. & N. 1398; gāste (*dat.*) HOM. I. 113; gāstes (*pl.*) ORM. 9046; gāsten (*dat. pl.*) MK. i. 27.

gāst-līc, adj., *O.E.* gāstlīc, = *O.H.G.* geistlīch; *spiritual,* M. H. 2; gastlic, -liȝ ORM. 14244, '14398; gostlich A. R. 194; þi gostlich fader BEK. 757.

gāstlines, sb., *spirituality*; gastlines [goostlines] C. M. 6448; in gastlines [goostlines, gasteli þingus] 14508.

gāst, sb., *terror,* ED. 3110.

gāst-ful, adj., *terrified,* WICL. DEUT. xx. 8.

gāst-lich, adj., *ghastly,* P. L. S. xv. 147; AR. & MERL. 1494; gastli CH. C. T. A 1984.

gāstnes, sb., *terror,* WICL. JOSH. ii. 9; gastnesse CH. BOET. iii. 5 (75).

gastan, *see* wastan.

gāsted, adj., *endowed with a spirit; spiritual*; gostid ALEX. (Sk.) 1874.

gastarios, sb., *a fish,* B. B. 234.

gāsten, v., *O.E.* gæstan; *terrify, scare*: gaste crowen from his corn LANGL. A vii. 129; gāste (*pres. subj.*) WICL. 2 PARAL. xxxii. 18; gāst (*pple.*) GAW. 325; WICL. DEUT. xx. 8; igast A. R. 372; *comp.* a-, of-gāsten.

ȝat, *see* ȝeat.

gāt, sb., *O.E.* gāt, *cf. Goth.* gaits, *O.N.* geit, *O.H.G.* geiz *f., Lat.* haedus *m.*; *goat,* '*capra,*' VOC. 219; ORM. 988; gat [got] LAȝ. 21315; got FRAG. 3; A. R. 100; GEN. & EX. 940; got [goot] CH. C. T. A 688; goot '(*h*)*edus, capria*' PR. P. 205; MAND. 47; geit WICL. LEV. xxiii. 19; gōte (*dat.*) GOW. III. 96; gǣt [*O.E.* gǣt] (*pl.*) ORM. 1206; gæt [geat] LAȝ. 25682; get HOM. II. 37; CHR. E. 34; REL. II. 275; E. G. 354; geet P. S. 198; geat A. R. 100; gaite ['*hircos*'] HAMP. PS. xlix. 10; gāten (*gen. pl.*) LAȝ. 21310; efter gate horden A. R. 100.

goot-bukke, sb., *O.E.* gātbucca; *goat-buck,* TREV. III. 129.

goot-herde, sb., *O.E.* gāthyrde; *goat-herd,* PR. P. 206.

gáte, sb., *O.N.* gata (*gen.* götu), = *Goth.* gatwō, *O.H.G.* gaza, *mod.Eng.* gait, *prov.Eng.* gate; *road, street; gait, way of going;* '*via, iter,*' PR. P. 188; MAP 338; LANGL. B iv. 42; PERC. 1961; A. P. i. 395; SACR. 629; thou canst ful wel þe ricthe gate to Lincolne HAV. 846; for lord knew the wai of rightwis and the gate of wicked sall perisch HAMP. PS. i. 7; whilc gate *in what way* ORM. 2281; schal no gom(e) . . . oþer gate it make WILL. 3761; alle gate [*O.N.* alla götu] *by all means, always* BEK. 1136; A. D. 260; MIR. PL. 154; þet heo hit ȝelde alle gate A. R. 58; þær he bi gate ȝede ORM. 12749; in his gate GOW. III. 196; nou er þei alle on gate MAN. (H.) 209; other gátis (*gen.*) *otherwise* HAMP. PS. cxviii. 39*; gátes (*pl.*) H. H. 43; gatis HAMP. PS. lxvii. 26*, xvi. 6.

gáte-lǣs, adj., *without a road,* ORM. 9211.

gáte-schādil, -shōdell, sb., *parting of roads,* '*compitum,*' PR. P. 188; gateschedel VOC. 271.

gáte, ȝáte *see* ȝeat.

ȝāten, v., *O.E.* geātan (*pret.* gētte, gātte), ? *O.N.* jātta, jāta, *cf. O.Fris.* gētta; *grant, concede,* ORM. 9819; ȝate (*ms.* yate) C. M. 26950; ȝetten LAȝ. 10052; ȝete A. P. i. 557; gete HICKES I. 230; ic gǣte (geate, iete) SAX. CHR. 39; ȝetteð A. R. 170; ȝētte (*subj.*) A. R. 26; god ȝate PR. P. 201; iǣtte (*pret.*) SAX. CHR. 256; ȝatte ORM. 2372; gatte GEN. & EX. 1574; þe king him ȝette al þat he ȝirnde LAȝ. 4426; ȝetede JUL. 7; ȝettede KATH. 1590; ȝetteðest HOM. I. 209; getton [ietten] SAX. CHR. 32; ȝāted (*pple.*) ORM. 154; iȝetted MARH. 16; hwanne he havede his wille yat (*printed* wat) HAV. 1674; *comp.* and-ȝǣten.

ȝēttunge, sb., *concession,* A. R. 204.

gat-tōðed [? gat-], adj., ? *from* gat = ȝeat; *or from* gāt; ? *having interstices between the teeth;* ? *lascivious,* CH. C. T. A 468.

gaude, sb., *gaud, toy, finery,* '*nuga,*' PR. P. 188; CH. C. T. C 389; gaudes (*pl.*) S. S. (Web.) 3957; L. H. R. 134; LIDG. M. P. 92.

[gaude², sb., *O.Fr.* gaude (*weld*); ? *a green dye.*]

gaudi, adj., *light green*: gaudi grene '*subviridis*' PR. P. 189; HALLIW. 394; gaudi (*sb.*) of grene GAW. 169.

gauden, v., ? *dye with gaude*; ? *adorn*: gauded (*pple.*) al with grene CH. C. T. A 159.

gaugin, v., *Fr.* jauger; *gauge,* PR. P. 189; gáge (*imp.*) PALL. iv. 427.

gauginge, sb., *act of measuring,* '*dimencionatus,*' PR. P. 189.

gaul, *see* gaȝel.

ȝaulen, gauling, *see* goulen.

gaunt, adj., *gaunt*: gaunt or lene '*macio-lentus*' PR. P. 189; gaunte or swonge [or slender] '*gracilis*' PR. P. 189.

gauren, v., *? look, stare*, CH. C. T. *A* 3827.

gavel, sb., *O.E.* gafol, gaful, gafel; *from* ȝifen; *tribute, tax, rent, usury*, LAȝ. 6105; A. R. 202; MISC. 46; þet gavel AYENB. 35; gafel MAT. xvii. 24; govel GEN. & EX. 844; goul '*fenus*' PR. P. 206; **gavele** (*dat.*) MISC. 150; gafele ['*usura*'] MAT. xxv. 27; *comp.* lond-gavel.

> **gaveler**, sb., *usurer*, PS. cviii. 11; **gavelers** (*pl.*) MISC. 150; gaveleres AYENB. 35.

> **gavelinge**, sb., *usury*, MISC. 31; AYENB. 34.

gavelen, v., *make into sheaves*; gavelin corne '*manipulo*' PR. P. 189.

gavelle, sb., *O.Fr.* gavelle; *sheaf of corn*; gavel PR. P. 189.

gavelok, sb., *O.E.* gafoluc, = *O.N.* gaflok, gaflak, *cf. Welsh* gaflach, *Gael.* gobhlag; *gave-lock, spear*, ALIS. 1620; MAN. (H.) 297; **gavelokes** (*pl.*) PS. liv. 22; gavelockes AYENB. 207.

gawel, *see* gaȝel.

gāwen, v., *look*: ne make þou na mo men gawen on me JUL.[2] 125; *see* gōwen.

gazafilace, sb., *Gr.* γαζοφυλάκιον; *treasury*: þe golde of þe gazafilace A. P. ii. 1283.

ȝe, pron., *O.E.* ge, gie, = *O.L.G.* gi, *Goth.* jus, *O.N.* jer, er, *O.H.G.* ir; *ye*, LAȝ. 741; A. R. 4; ORM. 366; BEK. 26; WILL. 269; LANGL. *A* i. 16; PR. P. 537; ȝe (*ms.* ye) O. & N. 1730; CH. C. T. *A* 726; (*ms.* ye) HAV. 161; EGL. 4; ȝe, ge, gie, ȝie HOM. II. 5, 21, 143; ge GEN. & EX. 330; ȝie HOM. I. 217; yhe PR. C. 68; ȝe (yhe) ALEX. (Sk.) 1034; *see* ēow *and* ēower.

ȝe², conj., *O.E.* ge, = *O.L.G.* ge, gie, ja, *O.H.G.* joh, *Goth.* jah; *and*: þeos sunne fordeð eiðer ȝe saule ȝe lichoma HOM. I. 103; ȝa læwed folc ȝa læred ORM. 846.

ȝe, *see* he *and* ho.   **ȝē**, *see* ȝēa.

[ȝe-, ȝi-, i-, e-, a-, prefix, *O.E.* ge, = *O.L.G.* ge-, gi-, *O.Fris.* ge-, gi-, je-, e-, i-, *O.H.G.* ga-, ge-, gi-, *Goth.* ga-; *see* GR. GRAM. II. 733; III. 50; WB. IV. 1, 1594.]

ȝēa, adv., *O.E.* ȝēa, jā,=*O.L.G.*, *O.H.G.*, *O.N.* jā, *Goth.* jā, jai, *O.Fris.* jē, gē; *yea*, KATH. 1234; H. M. 27; ȝea, ȝe FER. 1575, 2318; ȝea, ȝa, ȝe LANGL. *A* iii. 107; *B* iii. 111; ȝea ȝhe WICL. MAT. v. 37; ȝa (*ms.* yaþ), '*ita, etiam, imo*,' PR. P. 536; ȝa ORM. 2411; WILL. 258; GREG. 909; (*ms.* ya) HAV. 1888; ALIS. 3571; Iw. 1004; LIDG. M. P. 40; iaa FLOR. 1736; ioo EM. 888; ȝe HOM. I. 47; MARH. 4; KATH. 991; FL. & BL. 585; BEK. 36; MAND. 292; ȝe, ie CH. C. T. *E* 1345; ie TOR. 829; ȝe, ȝui A. R. 136, 334; ȝē ich [*cf. M.H.G.* jā ich (*see* GR. GRAM. III. 765)], A. R. 408; ȝui [ȝeoi, ȝei] he 334; ȝe [ȝea] hit 52.

ȝü-se, adv., *O.E.* gese, gise, gyse (*see* GR.

GRAM. III. 764); *yes*, LAȝ. 17208; A. R. 392*; ȝise A. P. iii. 117; ȝus, ȝis FER. 1573, 3799; ȝis PR. P. 539; WILL. 697; EGL. 55; ȝis [ȝus] LANGL. *A* v. 103; ȝes SHOR. 153.

ȝēac, sb., *O.E.* gēac, = *O.N.* gaukr, *O.H.G.* gauh, gouh, *prov. Eng.* gowk; *cuckoo; fool*; ȝek HALLIW. 951; gouk D. ARTH. 927; gok REL. I. 291.

ȝēk-pintel, **gaukpintil**, sb., *cuckoo-pint* (*arum maculatum*), LEECHD. III. 319.

[ȝe-æðelien] i-æðelien, v., *O.E.* geæðelian; *ennoble*, LAȝ. 22496; iaðeled LAȝ. 23333.

[ȝe-āhnien] i-āhnien, v., *O.E.* geāgnian; *own*, LAȝ. 3743; iāhnede (*pret.*) LAȝ. 2483; iahned 1932.

[ȝe-ahtien] i-ahtien, v., *O.E.* geeahtian; *reckon*: moni mon for aihte uvele iauhteþ (*pres.*) MISC. 118.

[ȝe-an] i-an, v., *O.E.* gean (*pret. pres.*); *favour, grant*; iunne (*subj.*) HOM. I. 125.

ȝeant, sb., *O.Fr.* geant; *giant*, ROB. (W.) 349; geaunt ROB. (W.) 4158; HORN (L.) 802; MAN. (F.) 1471; jeant 287 (*spelt* yeant 301); giaunt WICL. JOB xvi. 15; jeaunt LANGL. *A* vii. 219; giaund D. TROY 5303; **geauntes** (*pl.*) LANGL. *C* xxiii. 215; jeauntez A. P. ii. 272; geans (*v. r.* geandes) ROB. (W.) 333*.

ȝēap, adj., *O.E.* gēap, *cf. O.N.* gaupa (*lynx*); *cunning, prompt, vigorous, bold*: ȝep and … iwær LAȝ. 7581; ȝæp and war 13095; ȝæp & wis ORM. 8937; ȝiep HOM. II. 195; ȝep S. A. L. 158; GAW. 105; ȝape ALEX. (Sk.) 3304; ȝope 2201; C. M. 5370; ȝeep SPEC. 39; noþer ȝep ne wis O. & N. 465; þou art ȝong and ȝep LANGL. *C* xi. 287; ȝop MIRC 1763; a ȝop knight D. TROY 6642; þe ȝepe knight 902; þat iēpe (*sb.*) was with child D. TROY 13231; ȝēape (*pl.*) A. R. 362; ȝæpe wordes ORM. 13503; ȝepe MAN. (H.) 320; *comp.* hinder-ȝǣp.

ȝēp-hēde, sb., *cunning*, O. & N. 683.

ȝǣp-leȝc, sb., *cunning*, ORM. 2523.

ȝēp-līche, adv., *O.E.* gēaplīce; *readily, eagerly, cunningly*, LANGL. *B* xv. 183; ȝepli WILL. 3346; A. P. ii. 1708; ȝapeli ALEX. (Sk.) 80, 2406, 4866; ȝopeli 1393.

ȝēap-scipe, sb., *O.E.* gēapscipe; *cleverness, cunning*, LAȝ. 2760; giap-, ȝiep-, ȝepshipe HOM. II. 195.

ȝeápen, *see* gapen.

ȝēar, sb., *O.E.* gēar, gēr, = *O.L.G.* gēr, jār, *O.Fris.* jēr (gēr), jār, *O.H.G.* jār, *Goth.* jēr, *O.N.* ār, *year*; gear, gær SAX. CHR. 252, 253; ȝer A. R. 412; ORM. 9503; O. & N. 101; ȝer [ȝier] LAȝ. 3672; ȝeer WICL. JOHN xi. 49; ger GEN. & EX. 152; jer (*ms.* yer) AYENB. 110; ȝar (*printed* yare) S. S. (Web.) 568; ȝere BARB. vi. 188; ȝēres (*gen.*) LAȝ. 29055; ȝeres ȝive LANGL. *B* viii. 52; ēare (*dat.*) PROCL. 8; to ȝere *in this year* LAȝ. 8039; BRD. 11; CH. C. T. *D* 168; to iere TOWNL. 231 ; ʒēr (*pl.*) ORM. 8753; WILL. 5369 ; two and þritti ʒer A. R. 404; BARB.i. 39 ; ʒeer MAND. 12 ; ʒer [ieer] CH. C. T. *A* 601 ; ieer HOCCL. i. 110 ; ʒer, ʒier SHOR. 63, 71 ; ʒier ANGL. I. 15 ; gier REL. I. 130 ; ʒer, ʒere LAʒ. 301, 3416; HORN (L.) 524, 732; ʒere L. C. C. 34 ; ʒeres MARH. 2 ; A. R. 190 ; ʒēare, ʒiere (*gen. pl.*) HOM. I. 225 ; ʒera 93 ; ʒere 35 ; fif & twenti ʒere LAʒ. 7058 ; yhere PR. C. 741 ; ʒēren (*dat. pl.*) LAʒ. 377 ; *comp.* furn-, lēp-ʒēr.

ʒēr-dai, sb.,=*M.H.G.* jārtac ; *anniversary*, PR. P. 537 ; E. G. 281.

ʒeer-lī, adv., *O.E.* gēarlīc ; *yearly*, HOCCL. i. 421.

ʒeard, garþ, sb., *O.E.* geard, *O.N.* garꝺr,= *O.L.G.* gard, *Goth.* gards, *O.H.G.* gart, *Lat.* hortus, *Gr.* χόρτος, *mod.Eng.* yard, *prov.Eng.* garth; *piece of land, yard, garden* ; ʒeard, ʒerd, ʒord, *hortus'* PR. P. 537 ; ʒard P. L. S. xxiii. 73 ; ʒerd VOC. 270 ; CH. TRO. ii. 820 ; a ʒerd or a gardin WICL. JOHN xviii. 1 ; (*ms.* yerd) HAV. 702 ; garth PALL. i. 778 ; ʒarde (*dat.*) HALLIW. 950; appils garthis (*pl.*) HAMP. PS. lxxviii. 1*; throne garthis ['*sepes*'] HAMP. PS. lxxxviii. 39 ; *comp.* chirche-, fóre-, middel-, wīn-, wort-ʒeard.

geare, sb., *O.E.* gearwe,=*O.L.G.* garewi, *O.H.G.* garawi, garewi, garwi, *O.N.* gervi *f.*, *mod.Eng.* gear; *apparatus, clothing* ; gere R. S. iv. (MISC. 164) ; SPEC. 36 ; WILL. 1716 ; GOW. II. 227 ; A. P. ii. 1811 ; PERC. 189 ; TOR. 652 ; gare TRIST. 2868 ; EM. 198 ; geir (*equipment*) BARB. ix. 709 ; ger (*property*) xviii. 160 ; (*provisions*) viii. 458 ; (*armour*) v. 110 ; gaire D. TROY 905 ; *comp.* messe-gere.

ʒeare, *see* ʒearu.

ʒēare, ʒāre, adv., *O.E.* gēara, *mod.Eng.* yore ; *long ago* ; LAʒ. 2671, 3415 ; ʒare hit is þet ich wuste her of A. R. 88 ; ase was ʒeare iseid 298 ; ʒare JUL. 17 ; BRD. 31 ; hit is ful ʒare HORN (L.) 1356 ; in helle ich habbe yare ibeo MISC. 147 ; ʒore FL. & BL. 653 ; SPEC. 28 ; WILL. 1503 ; MIRC 1304 ; GAW. 2114 ; S. S. (Wr.) 2153 ; so ʒore MAP 339 ; ʒore [yoore] CH. C. T. *B* 174 ; þat ʒore hedden him abide C. L. 1339.

ʒearkien, ʒarkien, v., *O.E.* gearcian, *from O.E.* gearc (*ready*) ; *prepare, proceed*, LAʒ. 2631, 19513; ʒarken KATH. 1752 ; (*ms.* ʒarrkenn) ORM. 96 ; ʒarki SAINTS (Ld.) 301 ; iong . men iepeli iarke (*pres.*) into eld D. TROY 414 ; ʒerkeꝺ A. R. 410 ; ʒarkeꝺ H. M. 47 ; ʒarkeþ [ʒarketh] LANGL. *B* vii. 80 ; ʒarkie 23275 ; ʒarke (*imper.*) LAʒ. 32070 ; ʒarkede (*pret.*) ROB. 12 ; garkede GEN. & EX. 3261 ; þai ʒarkit to þe iates D. TROY 10738 ; ʒarkeden ROB. (W.) 260* ; giarked (*pple.*) HOM. II. 5 ; ʒarked A. P. ii. 1708 ; ʒarked, yarked C. M. 8982 ; iarked EM. 329 ; iarket to end D. TROY 5595: with ʒep men at þe ʒatis ʒarkit full þik D. TROY 11265 ; *comp.* ʒe-ʒearkien.

ʒarking, sb., *O.E.* gearcung ; *preparation*, ORM. 10800.

ʒearnien, ʒearnung, *see* ʒeearnien, -ung.

ʒearu, ʒāre, adj., gearu, gearo,=*O.L.G.* garu, *O.H.G.* garo, garewo, *O.N.* görr ; *ready* ; ʒaru HOM. I. 153 ; ʒaru, ʒæru LAʒ. 7783, 22506; ʒeruh A. R. 394 ; geare MK. xiv. 38 ; ʒeare, ʒare FER. 4193, 5186 ; ʒare O. & N. 488; ROB. 52; WILL. 3265 ; MAN. (F.) 8831 ; AV. ARTH. lxiv ; sorewe eou schal beon ful ʒare S. A. L. 153 ; ʒare (*ms.* yare) GAM. 90; iare GOW. II. 237 ; RICH. 6727 ; S. S. (Wr.) 305 ; M. ARTH. 218 ; TOWNL. 37 ; ʒærewe (*pl.*) LAʒ. 9457 ; ʒarewe KATH. 1750 ; O. & N. 378 ; iarwe RICH. 6751 ; he woren iare into Denemark for to fare HAV. 2954 ; ʒarewere (*compar.*) HOM. I. 213 ; ʒárest (*superl.*) WILL. 2729 ; *comp.* un-ʒearu.

ʒeáre, adv., *O.E.* geare, gearo, gearwe; *readily*, HOM. I. 223 ; þe he ful ʒeare (*sec. text* wel) wuste LAʒ. 525 ; tolde him ful ʒare hu he hadde ifare HORN (L.) 467 ; writen ful ʒare (*plainly*) A. P. i. 833 ; gare GEN. & EX. 390.

ʒǽre-witele, adj., *prudent* : næs þe king noht . . . swa ʒǽre witele LAʒ. 18547.

ʒearwien, garen, v., *O.E.* gearwian, gearuwian, gerwan, gyrwan, *O.N.* görva, gerva, göra, gera, =*O.L.G.* garuwian, gerewian, *O.H.G.* garewen, garen, *prov.Eng.* gar; *prepare; make* ; ʒærwen LAʒ. 12450 ; yaren HAV. 1350 ; garen AN. LIT. 13 ; garen, geren GEN. & EX. 1417, 2441 ; ʒare MAN. (H.) 294 ; gare PERC. 1411 ; HAMP. PS. xii. 4* ; gere ANT. ARTH. xxi ; ger BARB. vii. 19 ; ʒarewieꝺ eow to fihte LAʒ. 7473 ; gers (*pres.*) him sink HAMP. PS. i. 1* ; þou gers ix. 3* ; gare (*imper.*) iii. 6* ; gerand (*pple.*) HAMP. PS. xxii. 7* ; ʒæᵣwede [gurede] (*pret.*) LAʒ. 16197 ; ʒared MAN. (H.) 58 ; garte RICH. 6037 ; ISUM. 402 ; garte, gerte LANGL. *B* x. 175, vi. 303 ; ʒarweden, ʒareweden, gereden LAʒ. 1873, 9761, 9782 ; gered (*pple.*) GAW. 179 ; A. P. ii. 1568 ; gert MIN. vii. 53 ; gert LANGL. *B* v. 130 ; HAMP. PS. xvii. 5* ; *comp.* for-, ʒe-ʒearwien.

[ʒe-áskien] i-áskien, v., *O.E.* geáscian,= *O.H.G.* geeiscōn ; *ask* ; iéscad (*pple.*) HOM. I. 35.

ʒeat, sb., *O.E.* geat, *cf. O.L.G.*, *M.Du.* gat, *O.Fris.* jet *n.*, *O.N.* gatt *f.*; *gate* : þæt narewe geat MAT. vii. 13 ; ʒæt [ʒet] LAʒ. 6059 ; ʒet A. R. 74 ; BEK. 2250 ; þet ʒet HOM. I. 5 ; ʒat ROB. 540 ; C. L. 699 ; þe giate HOM. II. 105 ; ʒate ORM. 4122 ; þe ʒate KATH. 2454 ; LANGL. *B* xi. 108 ; ʒate, yate, gate PR. P. 188, 536 ; þe gate AYENB. 189 ; si gate HOM. I. 237 ; þa nærewe gate Lk. xiii. 24 ; ʒeáte (*dat.*) A. R. 424 ; FER. 1720 ; ʒete [ʒeate] LAʒ. 20442 ; þan ʒate 19137 ; ʒate A. D. 260 ; WILL. 3757 ; ʒate HOCCL. i. 178 ; EGL. 281 ; æt þam gate JOHN x. 1 ; þare gate HOM. I. 239 ; ʒeáte (*pl.*) HOM. I. 127 ; gate MAT. xvi. 18 ; ʒate, ʒaten [ȝeate] LAȝ. 1604, 11241; ȝeten HOM. I. 141; A. R. 222; giaten HOM. II. 113; ȝates BRD. 4; yhates PR. C. 2127; *comp.* burȝ-, ēst-, flōd-, lid-ȝat.

ȝáte-herd, sb., *gate-keeper*; gáteherden (*pl.*) A. R. 100.

ȝǽte-ward [ȝeateward], sb., *O.E.* geat-weard; *gate-keeper*, LAȝ. 18998; ȝeteward A. R. 270; ȝateward H. H. 137 (139); LANGL. *A* vi. 85; C. M. 1245; gateward LANGL. *C* viii. 243; gatwarde *C* xiv. 92.

[ȝe-ban] i-ban, sb., *O.E.* gebann; *ban, 'edictum,'* FRAG. 2.

[ȝe-bāned] i-bōned, adj., *boned*: wel iboned & strong ROB. 414.

[ȝe-bannen] i-bannan, v., *O.E.* gebannan; *summon*; ibonned [ibanned] (? *for* ibonnen, ibanne *r. w.* monnen, manne) (*pple.*) LAȝ. 20206; havestu ... ibanned ferde O. & N. 1668.

[ȝe-bēat, sb., *O.E.* gebēat, = *M.H.G.* gebōz; *beating, stroke;* comp. bil-ibēat.]

[ȝe-bēaten] i-bēaten, v., *O.E.* gebēatan, = *M.H.G.* gebōzen; *beat*; ibēaten (*pple.*) A. R. 176; ibeten SPEC. 70; ALIS. 3848; ibeate, ibiate AYENB. 236, 239; ibete CH. C. T. *A* 3759.

[ȝe-bed] i-bed, sb., *O.E.* gebed, = *O.L.G.* gibed, *O.H.G.* ge-, gibet; *prayer*; ibéde (*dat.*) HOM. I. 7; ibéde (? *pl.*) P. L. S. viii. 150; ibeoden (*dat. pl.*) HOM. I. 89.

[ȝe-bedde] i-bedde, sb., *O.E.* gebedda, = *M.H.G.* gebette; *bed-fellow*, O. & N. 1570.

ȝe-bēȝen, v., *O.E.* gebēgan, -bȳgan; *inflect, incline*; ibēiȝed (*pple.*) FRAG. 1; ibuid TREAT. 139.

ȝe-belȝen, v., *O.E.* gebelgan; *swell, be indignant*; gebulge [gebolgen] (*pret.*) MAT. xx. 24; MK. x. 41; ibolwe (*pple.*) O. & N. 145.

ȝe-bēoden, v., *O.E.* gebēodan, = *O.L.G.* gibiodan, -beodan, *O.H.G.* gi-, gepiotan; *command*; ibóden (*pple.*) LAȝ. 22122.

[ȝe-bēot] i-bēot, sb., *O.E.* gebēot; *menace*, LAȝ. 21029.

[ȝe-bérded] i-bérded, adj., = *M.H.G.* ge-bartet; *bearded*, TREV. II. 197.

[ȝe-bēre] i-bēre, sb., *O.E.* gebǽre (*pl.* ge-bǽru), = *O.L.G.* gibári, *O.H.G.* gebâre n.; *gesture; shouting*: of faire ibere MISC. 100; ibēre (*pl.*) O. & N. 222, 1348; reuliche iberen LAȝ. 15067.

[ȝe-bēren] i-bēren, v., *O.E.* geberan, = *O.L.G.* giberan, *O.H.G.* ge-, giberan, -peran; *bear*, LAȝ. 27850; iber (*pret.*) A.R. 194; MISC. 37; (þou) ibere REL. I. 49; ibóren (*pple.*) R. S. v; C. L. 21; CH. C. T. *E* 626; bezst iboren LAȝ. 16357; wel ibore 678; ibore BEK. 32, 1172; LIDG. M. P. 116; abore HALLIW. 9.

[ȝe-bēren] i-bēren, v., *O.E.* gebǽran, = *M.*

H.G. gebāren, -bæren; *behave*: þe mon þe swa ibēreð (*pres.*) LAȝ. 21010; ibērde (*pret.*) JUL. 52.

[ȝe-berȝen] i-berȝen, v., = *O.H.G.* gi-, ge-, gabergan, *Goth.* gabairgan; *protect, preserve*; iborȝen (*pple.*) HOM. I. 71; LAȝ. 4265; iborȝe AYENB. 5; iborȝe [iborewe] O. & N. 883; ibor-hen JUL. 56; iboruwen A. R. 8; iborewe S. S. (Web.) 826.

ȝe-bēten, v., *O.E.* gebētan; *mend, repair*; ibeten HOM. I. 11; ibete MISC. 66; ȝebětte (*pret.*) HOM. I. 35; ibět (*pple.*) KATH. 1219; A. R. 272; WILL. 4613.

[ȝe-bidden] i-bidden, v., *O.E.* gebiddan; *bid*; ibidde MISC. 144; ibidde (*pres.*) LAȝ. 20655; ibiddeð (*imper.*) A. R. 144; ibiddeþ eu MISC. 42; ibed (*pret.*) MISC. 42; ibéden (*pple.*) KATH. 2464; ibede LAȝ. 18433; P. L. S. xvii. 142; AYENB. 117; to colde gistninge he was ibede REL. II. 277.

[ȝe-bide] i-büde, sb., = *M.H.G.* gebite; *abode*: þu nahtes(t) i nane stude habben freo monnes ibude LAȝ. 15577.

[ȝe-bīden] i-bīden, v., *O.E.* gebīdan, = *Goth.* gabeidan, *O.H.G.* gebītan; *bide, remain; endure*, LAȝ. 4721; ibide FL. & BL. 175; þat þu bern ibīdest (*pres.*) MISC. 128; ibād (*pret.*) HOM. I. 35; he ibad þes wederes LAȝ. 9734; þer fore he sorȝen (*r.* sorȝe) ibad 13268; þe bedeles ... ibide þe kinges heste BEK. 988; ibiden (*pple.*) LAȝ. 3419.

ge-bínden, v., *O.E.* gebindan, = *Goth.* gabin-dan, *O.H.G.* gibintan; *bind*, MK. v. 3; ibinden LAȝ. 2487; ȝebúnden (*pret.*) HOM. I. 229; ȝebúnden (*pple.*) HOM. I. 3; ibunden A. R. 254; mid sorȝen ibunden LAȝ. 12635; ibunde O. & N. 1354; þou art ibounden ... to helpen me CH. C. T. *A* 1149; ibounde AYENB. 145.

[ȝe-bīten] i-bīten, v., *O.E.* gebītan, = *M.H.G.* gebīzen; *bite, eat*; ibite HOM. I. 233.

[ȝe-blenden] i-blenden, v., *O.E.* geblen-dan; *blind*; iblend (*pple.*) MARH. 13; TREV. V. 249; iblent SHOR. 111; ALIS. 3956; S. S. (Web.) 2523.

[ȝe-blěssien] i-blěssien, v., *O.E.* geblētsian; *bless*; iblěssi (*pres. subj.*) AN. LIT. 6; iblěssed (*pple.*) C. L. 1441; SHOR. 125; PL. CR. 520; CH. C. T. *D* 323; iblescsed A. R. 376; REL. I. 159; iblesset MARH. 5; ibletsede HOM. II. 5.

ȝe-blissien, v., *O.E.* geblissian; *be glad, rejoice; make happy*, JOHN v. 35; þa engles ... ham iblissieð (*pres.*) HOM. I. 41; þat folc wes al iblissed (*pple.*) LAȝ. 27841.

[ȝe-blōwen] i-blōwen, v., *O.E.* geblōwan; *blossom*; iblōwen (*pple.*) MARH. 10; þat evre stont iliche iblowe O. & N. 618.

[ȝe-bod] i-bod, sb., *O.E.* gebod; *command, mandate, message*, REL. I. 184; ibóde (*pl.*) HOM. I. 11.

[ȝe-bódien] i-bódien, v., O.E. gebodian; announce; ibóded (pple.) HOM. I. 121; LAȝ. 23027.

[ȝe-bōn] i-bōn, see under ȝe-būen.

ȝe-bōned, see ȝe-bāned.

[ȝe-bórennesse] i-bórenesse, sb., birth, HOM. I. 205; A. R. 158; H. M. 37.

[ȝe-breiden] i-breiden, v., O.E. geþregdan; braid; ibræid (pret.) LAȝ. 15274; ibroiden (pple.) LAȝ. 23764, 29252; ibroide P. L. S. xvii. 156; (ms. ibroyde) O. & N. 645.

[ȝe-bréken] i-bréken, v., O.E. gebrecan, = Goth. gabrikan, O.H.G. gebrechan; break; ibróken (pple.) KATH. 1218; LANGL. A PROL. 68; TRIST. 2901; ibroke O. & N. 1558; BEK. 1005; SHOR. 161.

[ȝe-brengen] i-brengen, v., O.E. gebrengan; ibrōht (pple.) LAȝ. 1852; MARH. 2; O. & N. 1559; SPEC. 39; ibrouht A. R. 144; C. L. 580; ibroȝt ROB. 473; ibrout ANGL. III. 63.

[ȝe-brennen] i-brennen, v., M.H.G. gebrennen; burn; ibrend (pple.) P. S. 213; ANGL. IV. 199; þe ibernde ver dret AYENB. 116.

[ȝe-brid] i-brid, sb., young brood; ? ibridde (dat.) O. & N. 123*.

[ȝe-bringen] i-bringen, v., O.E. gebringan; bring; LAȝ. 16993; ibringe O. & N. 1539.

ge-brōðre, sb., pl., O.E. gebrōðor, gebrōðru, = M.H.G. gebruoder; brothers, MAT. xii. 47; ibrodere, ibroðeren LAȝ. 3880, 10446; ibroðran HOM. I. 125.

[ȝe-brúcen] i-brúcan, v., O.E. gebrúcan; enjoy; ibrúce (pres. subj.) HOM. I. 233.

[ȝe-brúnied] (i-)brúnied, ibúrned, adj., O.E. gebyrnod; provided with a coat of mail, LAȝ. 26277.

[ȝe-būen] i-būen, v., O.E. gebūan (pple. gebūn); make ready, prepare; ibōn (? = ibūn) (pple.) LAȝ. 12805, 25788; iboen AN. LIT. 13.

[ȝe-brústed] i-brústed, pple., made bristly, LAȝ. 14296.

ȝe-būȝen, v., O.E. gebūgan, = Goth. gabiugan, O.H.G. ge-, gapiugan; bow, bend; ibuȝen HOM. I. 91; (ȝe-)ibūȝað (pres.) 13; ibóȝen (pple.) LAȝ. 11751.

ȝe-büggen, v., O.E. gebycgan: buy; ibie TOR. 1223; gebeið (pres.) MAT. xiii. 44; iboht (pple.) SPEC. 62; ibouht A. R. 398; iboȝt FL. & BL. 118; AYENB. 145; ibouȝt PL. CR. 569.

[ȝe-búhsum] i-búhsum, adj., obedient, HOM. I. 75, 113.

ibúhsumnesse, sb., obedience, HOM. I. 109.

[ȝe-búr, sb., O.E. gebūr, = O.H.G. gipūr; settler; comp. nēhe-búr.]

[ȝe-búrde, sb., O.E. gebyrd, = O.L.G. giburd, O.H.G. ga-, ge-, gipurt, -burt, Goth. gabaurþs; birth.]

geberde-tīde, sb., birth time, MK. vi. 21.

[ȝe-bürȝen] i-bürȝen, v., O.E. gebyrgan; bury; ibüried (pple.) HICKES I. 166; LAȝ 6014; KATH. 335; ROB. 23; ibiried GEN. & EX. 2520; ibured BEK. 321; ibered SHOR. 89.

[ȝe-bürȝen] i-bürȝen, v. (? for iberȝen) protect, LAȝ. 9988; þæt me ham ibureȝe from þam uvele pinan HOM. I. 43.

ȝe-būrien, v., O.E. gebyrian, = O.L.G. giburian, O.H.G. giburren; happen, fall to, behove; ibüreþ (pres.) REL. I. 172; iburd A. R. 420*; hit iburð (ms. iburd) breke þas word HOM. I. 79; (heo) iburiaþ FRAG. 2; geberede (pret.) MAT. xviii. 33.

[ȝe-bursten] i-bursten, v., ? enrage; iburst (pple.) HOM. I. 255; JUL. 69; heo weoren ... laðliche iburste LAȝ. 1889.

[ȝe-camp] i-camp, sb., O.E. gecamp; combat; icompe (dat.) HOM. I. 107.

ȝe-cende, adj., O.E. gecynde; natural: si ȝecende lage HOM. I. 235.

ȝe-cēosen, v., O.E. gecēosan, = Goth. gakiusan, O.H.G. ga-, ge-, gichiosan; choose, HOM. I. 227; icheose REL. I. 182; gechēas (pret.) MAT. xii. 18; iches HOM. I. 97; ichæs LAȝ. 6356; icure 11555; icóren (pple.) MARH. 16; C. L. 203; wel icoren HOM. I. 195; icorn TRIST. 2280; icore HOM. II. 143; MISC. 38; BRD. 33; SPEC. 25; P. S. 247; FER. 766; ED. 23; icoren, ichosen LAȝ. 1539, 12312; KATH. 836, 1624; A. R. 56, 160; ichosen PROCL. 2; ichose BEK. 227; AYENB. 42; E. G. 349.

ȝe-cherren, v., O.E. gecerran, = O.H.G. gikerran; turn; icherren HOM. I. 117; gecherreð (pres.) MAT. x. 13.

[ȝe-chēu] i-chēu, sb., chewing, HOM. II. 35.

[ȝe-clǣmen] i-clǣmen, v., O.E. geclǣman; smear; iclēm (imper.) HOM. I. 225.

ge-clǣnsien, v., O.E. geclǣnsian, -clānsian; cleanse, MAT. viii. 2; iclense PL. CR. 760; iclǣnsed (pple.) LAȝ. 10835; ROB. 43; iclensed HOM. I. 59; iclenzed AYENB. 74.

ȝe-clepien, v., O.E. geclypian; call, HOM. I. 231; icleped (pple.) C. L. 1357; SHOR. 45; TRIST. 1674; CH. C. T. A 867; WICL. DEEDS i. 23; iclepet KATH. 2480; icliped BRD. 7; icluped FL. & BL. 140.

ge-cnāwe, adj., O.E. gecnāwe; conscious, LK. iv. 22.

[ȝe-cnōulēchen] i-cnōulēchen, v., for cnōulēchen; acknowledge; ich icnoulēche (ms. -ie) (pres.) HOM. I. 205.

ȝe-cnāwen, v., O.E. gecnāwan; know, recognise, HOM. I. 121; icnawen [icnowe] LAȝ. 24805; ȝecnowen HOM. I. 223; to icnowen him sulven A. R. 182; iknowen C. L. 36; iknawe AYENB. 82; iknowe HORN (R.) 1213; ALIS.² (Sk.) 607; CH. C. T. (Wr.) 6952; fleschliche iknowe SHOR. 63;

icnōwe (*pres.*) REL. I. 144; ich icnowe me gulti HOM. I. 205; icnōwe (*subj.*) O. & N. 477; icnēou, icneow (*pret.*) LAȝ. 6625, 9727; ȝecnew HOM. II. 143; icnew HOM. I. 93; ikneu FL. & BL. 509; BEK. 87; (þu) icneowe FRAG. 6; þou iknēwe (*subj.*) FER. 358; icnāwen (*pple.*) MARH. 22; KATH. 424; ich nes neaver ... wið him icnawen (*acquainted*) JUL. 15; beute ȝif þu wulle icnawen beo þat Arður is king over þe LAȝ. 26433; beon icnowen of his pliht 18567; icnowen A. R. 64; beo iknawe ALIS. 724; wel ich am þer of iknowe BEK. 783.

[ȝe-cnēden] i-cnēden, v., O.E. gecnedan,= O.H.G. gichnetan; *knead*; iknéde (*pple.*) REL. II. 277.

[ȝe-cnütten] i-cnütten, v., O.E. gecnyttan; *knit*, LAȝ. 29272; icnüt (*pple.*) KATH. 1525.

[ȝe-crimpled] i-crimiled, pple., *? with hair crimpled*, LANGL. C xvii. 351; *see* crimpil.
ȝecðe, *see* ȝicðe.

[ȝe-cumen] i-cumen, v., O.E. gecuman,= Goth. gaqiman, O.H.G. kaqveman; *come*; icōm(e) (*pret. subj.*) GREG. 271; icumen (*pple.*) LAȝ. 320; MARH. 4; A. R. 64; icume O. & N. 138; icomen, icome GREG. 53, 541.

[ȝe-cünde] i-cünde, sb., O.E. gecynd *n. f.*, *cf.* O.H.G. kikunt; *kind, nature, genus, property*: þet faire icunde HOM. I. 147; þereto nes him nout incunde L. N. F. 215; al swa tacheð his icunde LAȝ. 22004; to munien his ikunde 2033; þat ich mote ... biȝite mine ikunde 16279; icunde, ikunde O. & N. 114, 1383.

[ȝe-cünde] i-cünde, adj., O.E. gecynde; *natural, native*: ure icunde lond LAȝ. 22155; his icunde speche MISC. 56; hit is him ikunde FRAG. 5; icündere (*v. r.* icundur) (*comp.*) O. & N. 85; *comp.* un-icünde.

icünde-līche, adv., O.E. gecyndelīce; *naturally*, HOM. I. 99; O. & N. 1424.

[ȝe-cunnen] i-cunnen, v., O.E. gecunnian, = O.H.G. cachunnen; *try*; icunned (*pple.*) LAȝ. 26237.

[ȝe-cüðen] i-cüðen, v., O.E. gecyðan; *make known*; icüd (*pple.*) LAȝ. 8196; KATH. 540; A. R. 64; BEK. 2347; SPEC. 25; iked SHOR. 24.

icüþnesse, sb., O.E. gecyðness; *testimony*, '*testimonium*,' FRAG. 3.

ȝe-cwēme, adj., O.E. gecwēme; *convenient, pleasing*, HOM. I. 225; icweme FRAG. 8; A. R. 120; R. S. iv; iqveme LAȝ. 117; HOM. II. 9.

ȝe-cwēmen, v., O.E. gecwēman; *please*, HOM. I. 25; icweme 267; iqvemen [icweme] þan kinge LAȝ. 13288; icweme [qveme] O. & N. 1784; iqveme REL. I. 174; AYENB. 228; iqvēmeþ (*pres.*) HORN (L.) 485; icwěmde (*pret.*) HOM. I. 117; iqwēmed (*pple.*) C. L. 1394.

ȝe-cwéðen, v., O.E. gecweðan,=O.L.G. ge-,

giqvethan, Goth. gaqiþan, O.H.G. giqvedan; *speak, say*; iqvéþe (*pres.*) O. & N. 502; iqveð (*pret.*) LAȝ. 2267; icwede (*pple.*) O.& N. 1653; icweþen FRAG. 1; iqveðen [icweþe] LAȝ. 9140.

ȝed, sb., O.E. gedd, gydd, gidd; *saying*; ȝeddes (*pl.*) LAȝ. 25853.

[ȝe-dǣlen] i-dǣlen, v., O.E. gedǣlan,=O.L. G. gedēlian, Goth. gadailjan, O.H.G. geteilen; *part, divide*, FRAG. 5; idǣled [idealed] (*pple.*) LAȝ. 12196; ideled A. R. 204; TREV. VIII. 169; ideld BEK. 332.

ȝe-dafte, adj., O.E. gedæfte; *gentle, mild*, ['*mansuetus*'], MAT. xxi. 5.

[ȝe-dāl] i-dāl, sb., O.E. gedāl; *separation, division, distribution*, HOM. I. 135; idol FRAG. 5.

ȝe-daven-līc, adj., O.E. gedafenlīc; *proper, fit*; (*? printed* ȝedanfenlic) HOM. I. 221.

gedd, sb., O.N. gedda; *pike (fish)*; geddis (*pl.*) BARB. ii. 576.

ged-dēde, sb., *? mistake for* god-dēde [*Stratm. compares Dan.* gied, *M.Du.* gade *something pleasing*]: love is geddede AN. LIT. 96.

ȝeddien, v., O.E. geddian, gyddian, giddian; *speak, sing*, LAȝ. 21429; ȝedde LANGL. A i. 138; (*ms.* ȝede) S. S. (Wr.) 215; ȝeddede (*pret.*) LAȝ. 7873; ȝeoddede (*printed* ȝeoðde) FRAG. 8.

ȝeddinge, sb., O.E. geddung, giddung; *proverb, saying*, PR. P. 537; geddinges (*pl.*) CH. C. T. A 237; geddingus DEGR. 1421.
ȝēde, *see* ēode.

gedeling, *see* gadeling.

[ȝe-delven] i-delven, v., O.E. gedelfan; *delve, dig*; idolven (*pple.*) HOM. I. 49; A. R. 292.

[ȝe-dēmen] i-dēmen, v., O.E. gedēman,= Goth. gadōmjan; *deem, judge*; iděmden (*pret.*) LAȝ. 4054; idēmed (*pple.*) A. R. 170; C. L. 114; idemid SHOR. 2.

ȝeder, adj., *? cf.* O.E. ædre, edre (*quickly*), O.L.G. adro (*quick*); *great; vehement*: with ȝedire ȝoskinges ALEX. (Sk.) 5042.

ȝeder-lī, (*adv.*), *? promptly*, GAW. 453; A. P. ii. 463.

geder, *see* gæder.

[ȝe-derven] i-derven, v., O.E. gedeorfan; *grieve, injure*; idorven (*pple.*) MARH. 16; idorve O. & N. 1158; idorwen [iderved] A. R. 106; idervet of deovlen MARH. 20.

gedi, *see* gidi.

[ȝedihten] i-dihten, v., O.E. gedihtan; *ordain, dispose*: to dæðe idihte LAȝ. 19671; idihte (*pret.*) FRAG. 5; HOM. I. 13; idiht (*pple.*) KATH. 1607; O. & N. 641; þu us havest ful wel idiht R. S. ii; idight M. ARTH. 610.

ȝedire, *see* ȝeder.

ȝe-dōn, v., O.E. gedōn,=O.H.G. ga-, ge-, gituon; *do, make*, HOM. I. 29; idon LAȝ. 24378;

idüde (*pret.*) LAȝ. 18432; idōn (*pple.*) A. R.
316; MAND. 10; þa burh wes swiðe wel idon
LAȝ. 2029; idon [idoon] CH. C. T. *A* 1025;
idoon HOCCL. vi. 58; edon MISC. 218; AUD.
16; ido O. & N. 463; ROB. 470; AYENB. 30;
(h)is lifdawes wern ido FER. 994; idoo LIDG.
M. P. 223; TRIAM. 411.

[ȝe-draȝen] i-draȝen, v.,=*Goth.* gadragan,
O.H.G. ketragan; *draw*; idrahen (*pple.*)
KATH. 1200; idraȝe [idrawe] O. & N. 586;
idrawe BEK. 1625; CH. C. T. *G* 1440.

gedre, *see* gædrien.

[ȝe-drecchen] i-drecchen, v., *O.E.* gedrec-
can; *trouble*; idreiȝt (*pple.*) P. L. S. xix. 45;
idraht LAȝ. 4521; idrecched 5054; þat mi
saule ne beo idriȝt ASS. 190; idrecchet H. M. 29.

[ȝe-drēfen] i-drēfen, v., *O.E.* gedrēfan,=
O.H.G. ge-, ketruoban; *trouble, disturb*; id-
rēfed (*ms.* idrefeð) (*pple.*) LAȝ. 171; id-
reaved HOM. I. 193.

[ȝe-drēȝen] i-drēȝan, v., *O.E.* gedrēogan;
perform, endure, suffer, HOM. I. 29; idrehen
JUL. 27; hire willen idriȝen LAȝ. 1270; þat he
ne mihte idriȝen to ihæren þene muche drem
6708; idróhen (*pple.*) MARH. 21; þu havest
for mi luve muchel idrohen & idrahen JUL.
35; idrowe REL. I. 174 (MISC. 112).

[ȝe-drēosen] i-drēosen, v., *O.E.* gedrēosan,
·=*Goth.* gadriusan; *beat* : he was to deþ id-
róre (*pple.*) GREG. 155.

[ȝe-drinken] i-drinken, v., *O.E.* gedrincan,
=*Goth.* gadrigkan, *O.H.G.* getrincan, -trin-
chan; *drink*; idrunken (*pple.*) LAȝ. 6692;
idrunke MISC. 29; idronken LEG. 58; idronke
AYENB. 51; idronken (*printed* ydroked) PALL.
xi. 334.

[ȝe-drīven] i-drīven, v., *O.E.* gedrīfan,=
O.H.G. getrīban; *drive, endure*; idriven
(*pple.*) KATH. 1816*; idriven [idreve] LAȝ.
6213; we habbeoð idriven þat swa longe
25701; idrive BEK. 678; C. L. 199; idreve
TREV. V. 147.

[ȝe-duȝe] i-duȝe, iduwe, v., *? useful*, O. & N.
1582.

[ȝe-duȝen] i-duȝen, v., *O.E.* gedugan, ?=*M.
H.G.* getugen; *thrive* : honoure oure godes
. . . while ȝou is wel idoȝt (*pple.*) P. L. S. xv.
182; þo þe king was hool and wel idouȝth
ALIS. 5906.

ȝe-dweld, sb., *O.E.* gedwyld, -dwild; *error*,
HOM. I. 227; gedwel(d) MAT. xxvii. 64.

[ȝe-dwellen] i-dwellen, v., *O.E.* gedwellan,
=*O.H.G.* getwellan; *dwell upon, delay* :
hevene is al idveld (*pple.*) SHOR. 147.

ȝe-dwilð, sb., *error* : þa asprang þis ȝedweld
ofer all middenard HOM. I. 227; on monie
gedwilþan (*dat. pl.*) 119.

[ȝe-dwimor] i-dwimor, sb., *O.E.* gedwimor;
'*phantasma*,' FRAG. 3.

ȝe-dwole, sb., *O.E.* gedwola,=*O.H.G.* catvola;
error; gedwolen (*acc.*) MAT. xxiv. 24.

ȝe-earnien, v., *O.E.* geearnian, = *O.H.G.*
gearnēn; *earn, merit*; iernien HOM. I. 19;
ȝearnien, ȝearnie 221, 223; ȝearnest (*pres.*)
HOM. I. 221; ȝearnede (*pret.*) HOM. I. 233;
iærned (*pple.*) LAȝ. 24154.

ȝe-earnung, sb., *O.E.* gearnung; *earning,
merit* : efter his ȝearnunge HOM. I. 231;
iearnunge 99.

ȝe-edlǣchen, v., *O.E.* geedlǣcan; *repeat*;
ȝeedlēhte (*pret.*) HOM. I. 229.

ge-efenlǣchen, v., *O.E.* geefenlǣcan; *make
like, be like*, MAT. vi. 8.

ge-efen-līc, adj., *equal*; o geuelike *on
equal terms* MISC. 10.

[ȝe-efnen] i-efnen, v., *O.E.* geefnan, *cf. Goth.*
gaibnjan, *O.H.G.* geebenōn, caepanōn; *make
equal*; iefned (*pple.*) HOM. II. 37; A. R. 128.

[ȝe-emnetten] i-emnetten, v., *O.E.* geemn-
ettan; *make equal*; iemnette (*pple. pl.*)
FRAG. 2.

ȝe-enden, v., *O.E.* geendian,=*O.H.G.* gi-
entōn; *end, finish*; iendað (*pres.*) HOM. I.
129; ȝeendode (*pret.*) HOM. I. 223; iended
(*pple.*) FRAG. 2; BEK. 1768; SHOR. 150.

ȝiendunge, sb., *O.E.* geendung; *ending*,
HOM. I. 217; iendung FRAG. 1.

[ȝe-ēode] i-ēode, v., *O.E.* geēode (*pret.*) ;
conquered, LAȝ. 4253.

ȝeer, *see* ȝēar.

geet, *see* gēt *and* gāt.

geeten, *see* gǣten.

ȝe-eten, v., *O.E.* geetan, = *O.H.G.* geezan,
gezan; *eat*; ȝete C. M. 7116; ȝeetst (*pres.*)
HOM. I. 221; ieten (*pple.*) LAȝ. 31773; iete
ROB. 418; BRD. 14; C. L. 338; BER. 2782;
iȝeten (*for* ieten, *as Germ.* gegeszen *for* ge-
eszen) LAȝ. 6691; iȝete SHOR. 23; (*ms.* iyete)
AYENB. 13; OCTOV. (W.) 757; iȝette P. L.
S. iii. 33.

ȝef, *see* ȝif.

ȝe-fā, sb., *O.E.* gefāh; *foe*; ifa LAȝ. 15855;
ȝefo HOM. I. 231; ivo O. & N. 1716; AYENB.
171; ifān (*pl.*) MARH. 5; ifon REL. I. 175;
ifoan PROCL. 6.

[ȝe-fæȝrien] i-fæȝrien, v., *? make fair*; i-
feired (*? ms.* ileired) (*pple.*) mid golde LAȝ.
23954.

[ȝe-fēren] i-fēren, v., *terrify*; ifēred (*pple.*)
LAȝ. 27140.

[ȝe-fæsten] i-fæsten, v., *O.E.* gefǣstan,=
O.H.G. gifestan; *make fast*; ivaste (*pret.*)
LAȝ. 22551; ifast (*pple.*) P. L. S. iii. 95; ifest
GREG. 114.

ȝe-fæstnien, v., *O.E.* gefastnian; *fasten*; ȝe-
festnede (*pret.*) HOM. I. 221; ivæstned
[ifastned] (*pple.*) LAȝ. 31132; ifestned KATH.
1523; ivestned A. R. 218; ifastened ALIS.
179.

[ȝe-faȝen] i-faȝen, adj., *O.E.* gefægen; *fain,
gladly*; ifaȝe SHOR. 67; ifaie REL. II. 276.

[3e-fallen] i-fallen, v., O.E. gefeallan,=O.
H.G. gefallan; fall; ivēl (pret.) MISC. 29;
ifallen (pple.) O. & N. 514; ivallen A. R. 58;
ifalle TRIST. 1937; CH. C. T. G 61.

[3e-fandien] i-fandien, v., O.E. gefandian;
try, tempt; ifonded (pple.) LA3. 16314;
ivonded A. R. 58; MISC. 38.

[3e-fáren] i-fáren, v., O.E. gefaran; fare, go;
ifare O. & N. 400; ivōr (pret.) LA3. 6090;
iváren (pple.) A. R. 366; hu heo ivaren weren
LA3. 561; ifare HORN (L.) 468; she was fro
þis world ifare CH. TRO. iv. 1169.

[3e-fealden] i-fealden, v., O.E. gefealdan,
=O.H.G. kifaldan; fold, bend; ivólden (pple.)
A. R. 122; ifolde AL. (T.) 446; LANGL. B xvii.
166; ivolde FER. 5795; a doun he fel ifold(e)
TRIST. 2790.

[3e-fēden] i-fēden, v., O.E. gefēdan; feed;
ifēd (pple.) SPEC. 110; WILL. 768; R. R. 471;
P. 26; ived LA3. 13573.

[3e-fē3e, adj., O.E. (un-)gefēge,=O.H.G. ka-
fōgi, M.H.G. gevüege; suitable; comp. un-
ifēie.]

[3e-fē3e, sb., O.E. gefēg,=O.H.G. kifōgi;
junction; comp. stef-ifēi3e.]

[3e-fē3en] i-fē3en, v., O.E. gefēgan, cf. O.H.
G. ga-, kafōgan; join; ivēied [ife3et] (pple.)
to gederes A. R. 26; er he be swo iveid (disposed) HOM. II. 11.

3e-feht, sb., O.E. gefeoht,=O.H.G. kafeht;
fight, battle; ifiht LA3. 4347; gefehte (pl.)
MK. xiii. 7.

[3e-fehten] i-fehten, v., O.E. gefeohtan,=
O.H.G. gefehtan; fight; ivohten [ifohte]
(pple.) LA3. 25693; ivo3te AYENB. 176.

[3e-feiþed] i-feiþed, pple., treated with
enmity, LA3. 14133*.

[3e-fēlen] i-fēlen, v., O.E. gefēlan,=O.L.G.
gifōlien, O.H.G. gafuolan; feel; ivelen A. R.
232; ifele P. L. S. xxiii. 18; ivēlþ (pres.)
AYENB. 31; ifēlde (pret.) HOM. I. 123; HORN
(L.) 54.

[3e-fellen] i-fellen, v., O.E. gefellan, -fyllan,
M.H.G. gevellan; fell; ifelde (pret.) HORN
(R.) 58; ifelled (pple.) LA3. 988; iveld AYENB.
50.

3e-fēo3en, v., O.E. gifīoge, gefiage (JOHN vii.
7*); set at variance: mine sunnen habbeð
... ivēied [iveed] me toward þe HOM. I.
187, 202; wið þine sune þu beost ivæid LA3.
9837; heo wusten heom ifæied 21214; ? ivet
C. L. 310.

[3e-fēond] i-fēond, sb., pl., ? for fēond;
enemies, LA3. 16077.

[3e-fēre] i-fēre, sb., O.E. gefēra, -fǣra; com-
panion, JUL. 4?; R. S. v; HORN (R.) 227;
REL. II. 275; ivere HOM. I. 173; LA3. 26020;
MISC. 66; ifēran, -feren (pl.) LA3. 588, 1621;
3eferen HOM. I. 231; iveren A. R. 392; ifere
P. L. S. viii. 116.

ifēr-rēd, sb., company, HOM. I. 243; 3efered
HOM. I. 231.

3efēr-rēden, sb., .O.E. gefērrǣden; company; gefērrēdene (dat.) MAT. xviii. 17.

[3e-fēren] i-fēren, v., O.E. gefēran; go: þeo
þat child wes vorð ivǣred (pple.) (advanced)
& wes of twealf 3ere LA3. 11064; 3et ne beoð
fif dæi3es allunge iverede (pret.) 8797.

[3e-feterien] i-feterien, v., O.E. gefeterian,
=O.H.G. kafezarōn; fetter; ifetered (pple.)
P. S. 217; CH. C. T. A 1229; iveotered
A. R. 32.

[3e-fetten] i-fetten, v., O.E. gefetian, -fettan,
=M.H.G. gevazzen; fetch; ifet (pple.) HOM.
I. 147; P. L. S. xxi. 71; JOS. 428; CH. C.
T. F 174; MAN. (F.) 1624; ifat KATH. 1296.

[3e-finden] i-finden, v., O.E. gefindan,=
M.H.G. gevinden; find, LA3. 6219; KATH.
516; ivinden A. R. 156; ifinde S. S. (Web.)
2371; ifindeð (pres.) MARH. 20; ifond
(pret.) KATH. 161; ivond A. R. 66; ivunden
LA3. 12303; ifúnden (pple.) A. R. 48; ifunde
O. & N. 705; ifonden MISC. 27; ivonde,
-vounde AYENB. 92, 186; ifounden, -founde
LANGL. B xv. 225; ifounden PL. CR. 631.

[3e-fiðerien] i-fiðerien, v., O.E. gefiðerian;
feather: earewe þet is iviðered (pple.) A. R. 60.

[3e-flēmen] i-flēmen, v., O.E. geflēman,
-flȳman; put to flight; iflēmed (pple.) LA3.
1363; C. L. 113.

[3e-flēon] i-flēon, v., O.E. geflēon,=O.H.G.
gefliehen; flee; iflō3en (pple.) LA3. 4764;
ivlowen A. R. 168; iflowen C. L. 470; iflowe
ROB. 311; ALIS. 1907.

[3e-flīt, sb., O.E. geflīt; contention.]
iflīt-ful, sb., O.E. geflītfull; contentious,
'conten(tios)us,' FRAG. 2.

[3e-fōg, adj.,O.E.(un-)gefōg,=M.H.G.gevuoc;
suitable; comp. un-ifōh.]

3e-fol3en, v., O.E. gefylgan, cf. O.H.G. ge-
folgen; follow; 3efolged (pple.) HOM. I. 237.

[3e-fōn] i-fōn, v., O.E. gefōn,=O,H.G. gifāhan,
Goth. gafāhan; seize, LA3. 4555; ifo FL. &
BL. 694; ivo O. & N. 612; (he) ifēhð (pres.)
HOM. I. 131; (hi) ifoð, -voþ O. & N. 1645;
ifēng (pret.) HOM. I. 77; LA3. 820; ivongen,
-fon (pple.) LA3. 26848, 28137; ifonge ALIS.
4573; ivonge MISC. 50.

3e-forðian, i-forðien, v., O.E. geforðian; perform, execute, HOM. I. 31, 37; þine heste
ivorðen LA3. 31561; ivorðed (pple.) A. R.
408; þat worc wes iforðed LA3. 8709; his
heast was iforðet KATH. 2278.

3e-fratewen, v., O.E. gefrætwian; adorn;
gefratewed (pple.) MAT. xii. 44; ifretid BER.
3926.

[3e-frēden] i-frēden, v., O.E. gefrēdan;
perceive, experience; ifrede, ivrede MISC. 78,
82; ivrede (printed inrede) SHOR. 7; he
heom ivrědde (pret.) LA3. 27138.

[ȝe-freinen] i-freinen, v., *O.E.* gefrignan, gifregna, -frægna,=*Goth.* gafraihnan ; *inquire,interrogate* ; ivreined (*pple.*) A. R. 338 ; ifreined CH. C. T. (Wr.) 12361.

[ȝe-frēoȝen] i-frēoíȝen, v., *O.E.* gefrēogan ; *set free, liberate,* LAȝ. 29474 ; ifrēoȝed (*ms.* i-freoied) (*pple.*) LAȝ. 20895 ; ifreoed FRAG. 6 ; ivrid AYENB. 86.

[ȝe-frēonde] i-frēonde, sb., pl., *O.E.* gefrȳnd, =*M.H.G.* gevriunde ; *friends,* LAȝ. 11591.

[ȝe-frēosen] i-frēosen, v., = *O.H.G.* gefriusan ; *freeze* ; ifróre (*pple.*) MISC. 152 ; TREAT. 137.

[ȝe-friðien] i-friðien, v., *O.E.* gefriðian, -freoðian,=*Goth.* gafriþon, *O.H.G.* gefridōn ; *preserve* ; ifreþed (*pple.*) LANGL. *C* viii. 228*.

[ȝe-frōfren] i-frōfren, v., *cf. O.E.* gefrēfrian ; *console, comfort* ; ifrēfrað (*pres.*) HOM. I. 97 ; ifrōvred [ifrēvered] (*pple.*) LAȝ. 19545.

ȝeft, *see* gift.

ȝe-füllen, ifülle, v., *O.E.* gefyllan,=*O.H.G.* gifullan, *Goth.* gafulljan ; *fill,* HOM. I. 57, 119 ; ivullen LAȝ. 32184 ; ifülled (*pple.*) A. R. 156 ; MISC. 149 ; C. L. 1280 ; ifuld FL. & BL. 644 ; gefelled, gefeld MAT. ii. 14, xiii. 14.

ȝefüldnesse, sb., *O.E.* gefylledness ; *plenitude* ; ȝefildnesse HOM. I. 219.

ȝe-fulwen, ȝe-fullen, v., *O.E.* gefulwian ; *baptize* : buton her richlice ȝelife on god, and þat he beo ȝefulled (*pple.*) HOM. I. 229 ; ifulled 241 ; MISC. 151 ; ifulled FER. 5835 ; LANGL. *B* xix. 40 ; ifulwed C. L. 1457 ; ifulȝed HOM. I. 37 ; ifulehȝad 113 ; ifulȝede men HOM. I. 101 ; ifulhet MARH. 1 ; ifolwed JOS. 7 ; ifolled FER. 5829 ; ivolled ROB. 239.

[ȝe-fülsta] i-fülsta, sb., *O.E.* gefylsta ; *helper,* HOM. I. 113.

ȝe-fürn, adv., *O.E.* gefyrn ; *of old, formerly* ; ifurn MISC. 193 ; O. & N. 1306 ; REL. I. 182 ; ivurn, ifeorn [iv(e)orn] LAȝ. 24017, 25024 ; gefern ['*olim*'] MAT. xi. 21 ; ifern FER. 3207.

ifürn-daȝes, sb., pl., *days of old* : ine ivurn-daȝen LAȝ. 32206.

ifürn-ȝēr, sb., pl., *years of old* : ine þan ivurn-ȝere(n) LAȝ. 25139.

[ȝe-fürne] i-fürne, sb., *? antiquity* : a vor-bisne is of olde ivurne (*ms.* iwrne, iwurne) O. & N. 637.

[ȝe-füsen] ifüsen, v., *O.E.* gefȳsan ; *set out, hasten* ; LAȝ. 22123 ; forð he ivüsde (*pret.*) 10492.

ȝe-gada, sb., *O.E.* gegada,=*O.L.G.* gigado, *M.H.G.* gegate ; '*complex,*' FRAG. 5.

ȝe-gæderien, v., *O.E.* gegæderian ; *gather* ; igederian HOM. I. 95 ; igædered (*pple.*) FRAG. 6 ; igedered A. R. 76 ; C. L. 1418 ; ȝegadered HOM. I. 229 ; igadered LAȝ. 1501 ; PL. CR. 189 ; P. 52.

ȝegæderung, sb., *O.E.* gegaderung, igæde-rung ; *gathering,* '*congregatio, concio,*' FRAG. 2 ; igederunge (*dat.*) HOM. I. 89.

[ȝe-gān] i-gān, v., *O.E.* gegān, -gangan, = *O.H.G.* gigān, -gangan, *O.L.G.* gigangan, *Goth.* gagaggan ; *go* ; igon A. R. 208 ; igān, -gon (*pple.*) LAȝ. 2064, 6046 ; igon LEG. 35 ; CH. C. T. *B* 1402.

ȝegōinge, sb., *return,* ['*regressus*'] TREV. III. 111.

[ȝe-gǽst ?].

igǽst-līche, adj., *terrible,* LAȝ. 17869.

[ȝe-ȝearkien] i-ȝearkien, v., *O.E.* gegear-cian ; *make ready* ; iȝarked (*pple.*) LAȝ. 14259.

[ȝe-ȝearwien] i-ȝearwien, v., *O.E.* gegear-wian, -gerwan, -gyrwan,=*O.H.G.* gigarawan ; *make ready, equip* ; iȝearwed, -ȝarwed (*pple.*) LAȝ. 220, 2076 ; iȝarwed HOM. I. 139 ; uvela heo weoren igarede LAȝ. 6272.

[ȝe-ȝel] i-ȝel, sb., *yell,* LAȝ. 17799.

[ȝe-ȝelden] i-ȝélden, v., *O.E.* gegyldan ; *yield, give back* ; iȝólden, iȝolde C. L. 343, 1480 ; iȝolde FL. & BL. 809 ; iyolde AYENB. 120.

[ȝe-ȝēmen] i-ȝēmen, v., *O.E.* gegēman ; *take care of,* A. R. 44* ; iȝēmed (*pple.*) C. L. 448.

ȝēȝen, v., *O.N.* geyja (*bark*) ; *? shout* : ȝeien KATH. (E.) 205 ; A. R. 288 ; to ȝeȝen ne to reme ST. COD. 97 ; ȝēȝe (*pres.*) SPEC. 111 ; GAW. 1215 ; ȝeiȝeð HOM. I. 43 ; ȝēiede (*pret.*) A. R. 152 ; ȝeide MARH. 6 ; ȝeiden LAȝ. 27750.

ȝe-genge, i-geng(e), sb., *company,* HOM. I. 237, 243.

[ȝe-ȝernen] i-ȝernen, v., *O.E.* gegyrnan, gegernan (ZEIT. XXI. 36) ; *yearn, wish for* ; iȝerned (*pple.*) A. R. 192 ; iȝirned SPEC. 28.

ȝegge, *see* gigge.

[ȝe-ȝifen] i-ȝifen, v.,=*O.H.G.* gegeben ; *give* ; iȝiven (*pple.*) A. R. 82 ; C. L. 60 ; iȝefen HOM. I. 19 ; iȝeven LAȝ. 3167 ; C. L. 608 ; CH. BOET. iv. 6 (141) ; iȝive O. &. N. 551 ; (*ms.* yyeue) AYENB. 79.

[ȝe-gladien] i-gladien, v., *O.E.* gegladian ; *gladden* ; igladed (*pple.*) LAȝ. 19587 ; SHOR. 132 ; WILL. 850 ; P. L. S. xxiv. 130.

ȝe-glengen, v., *O.E.* geglengan ; *adorn* ; ȝeglenge (*pres. subj.*) HOM. I. 229.

[ȝe-glīden] i-glīden, v., *O.E.* geglīdan ; *glide* ; igliden (*pple.*) LAȝ. 9159* ; MISC. 95.

gegnen, *see* ȝeinen.

[ȝe-gōdien] i-gōdien, v., *O.E.* gegōdian ; *do good* ; igōded (*pple.*) A. R. 386.

ȝe-grámien, v., *O.E.* gegremian, *cf. O.H.G.* kagreman ; *irritate* ; ȝegrámeden (*pret.*) HOM. I. 225 ; igrámed, -gremed (*pple.*) O. & N. 1603 ; ure godes þat tu igremed havest KATH. 1467.

[ȝe-grēde] i-grēde, sb., *clamour,* O. & N. 1643.

[ȝe-grēden] i-grēden, v., *cry out*; igrǎd (*pple.*) O. & N. 1149.

ȝegrémien, *see* ȝegrámien.

ȝe-grēten, v., *O.E.* gegrētan, = *O.H.G.* gigruozzan, -gruozan; *greet, salute*, KATH. 2303*; igræten LAȝ. 20523; ȝegrĕtte (*pret.*) HOM. I. 227; igrette LAȝ. 1260; igretten R. S. v; igrĕt (*pple.*) SHOR. 119.

igrētinge, sb., *greeting*, PROCL. 1.

[ȝe-grínden] i-grínden, v., *O.E.* gegrindan; *grind*; igrounde (*pple.*) LANGL. *B* xviii. 78; CH. C. T. *A* 3991.

[ȝe-grīpen] i-grīpen, v., *O.E.* gegrīpan,= *O.H.G.* kagrīfan; *gripe, seize*, LAȝ. 21513; igrīpe (*pres. subj.*) JUL. 72; igrāp (*pret.*) MARH. 12; igrap [igrop] LAȝ. 1566; igripen 4559; igripen (igripe) (*pple.*) LANGL. *A* iii. 175.

[ȝe-griðien] i-griðien, v., *O.E.* gegriðian; *grant peace*; igriðed (*pple.*) LAȝ. 18263.

[ȝe-grōwen] i-grōwen, v., *O.E.* gegrōwan; *grow*; igrōwen (*pple.*) CH. C. T. *A* 3973.

[ȝe-grūre] i-grūre, sb., *? terror*, LAȝ. 812.

[ȝe-gülten] i-gülten, v., *O.E.* gegyltan; *sin*; igült (*pple.*) HOM. II. 200; igelt ANGL. I. 7.

[ȝe-gürden] i-gürden, v., *O.E.* gegyrdan; *gird*; igürd (*pple.*) A. R. 418; ROB. 174; igert AYENB. 236.

[ȝe-habben] i-habben, v., *O.H.G.* gehabēn, *Goth.* gahaban; *have*; ihaved (*pple.*) KATH. 467; BEK. 292; ihaved [ihafd] LAȝ. 4501; iheved A. R. 108; ihad P. L. S. xii. 108.

[ȝe-hādien] i-hādien, v., *O.E.* gehādian; *ordain*; ihāded [ihoded] (*pple.*) LAȝ. 15643; ihoded O. & N. 1177, 1311; AYENB. 49.

ge-hǣlen, v., *O.E.* gehǣlan, = *O.H.G.* geheilan, *Goth.* gahailjan; *heal*, MAT. ix. 28; ȝehēlde (*pret.*) HOM. I. 229; ihēled (*pple.*) LAȝ. 29991; A. R. 274.

ȝe-hǣse, sb., *command*; ȝehese HOM. I. 221.

[ȝe-hāl] i-hāl, adj., *O.E.* gehāl, = *M.Du.* gheheel, *M.H.G.* geheil; *whole*; ihal [ihol] & isund LAȝ. 821; ihol A. R. 256; SHOR. 27; ihōle (*pl.*) LANGL. *A* vii. 55.

ihōl-līche, adv., *wholly*, AYENB. 109.

ȝə-hǎlȝien, v., *O.E.* gehālgian,=*O.H.G.* geheilegōn; *hallow, sanctify*; gehāleged (*pple.*) LK. xi. 2; ihaleȝed LAȝ. 29443; ihalȝed MISC. 193; ihalewed [ihaleȝed] A. R. 18; ihalewed P. L. S. xx. 75; ihalȝede stedes AYENB. 40.

ȝe-hāten, v., *O.E.* gehātan,=*O.L.G.* gihētan, *Goth.* gahaitan, *O.H.G.* ca-, geheizan; *call, bid*; ȝehāten (*pple.*) ORM. 112; ȝehaten, -hoten HOM. I. 219, 225; ihaten MARH. 2, 19; KATH. 2202; ihaten, -hoten, -hoaten [ihote] LAȝ. 15, 219, 558; ihoten FRAG. 8; A. R. 58; JOS. 291; i(h)oten GEN. & EX. 2416; ihote BEK. 701; DAV. DR. 5; LANGL. *A* i. 61.

[ȝe-healden] i-healden, v., *O.E.* gehealdan, = *O.H.G.* gehaltan; *hold, keep*, P. L. S. viii. 28*; his scrift ihalden HOM. I. 9; þet he mihte ... us ihalden (*preserve*) 95; feht ihalden LAȝ. 5505; ihēold (*pret.*) LAȝ. 31544; þa wes Vortigernes hired for hehne ihalden LAȝ. 13988; ihólden (*pple.*) A. R. 250; iholde ROB. 470; CH. C. T. *A* 2958; HOCCL. i. 184; þe wrenne was wel wis iholde O. & N. 1723; ihealde SHOR. 34; ihialde MISC 30; þet me is ihialde to done AYENB. 8.

[ȝe-hēawen] i-hēawen, v., *O.E.* gehēawan; *hew*; ihēawen (*pple.*) FRAG. 3; ihewe TREV. V. 263.

[ȝe-hēden] i-hēden, v., *O.E.* gehēdan; *heed*: heo wēren wel ihĕdde (*pple.*) LAȝ. 2725*.

[ȝe-hēȝen] i-hēȝen, v., *O.E.* gehēan,=*O.H.G.* gihōhan; *exalt, elevate*; ihǣȝed [hihēȝed] (*pple.*) LAȝ. 1251; iheied A. R. 154; iheiet MARH. 11.

[ȝe-hélen] i-hélen, v., *O.E.* gehelan; *conceal*; ihel (*pret.*) KATH. 2055; ihólen (*pple.*) A. R. 146; ihole AYENB. 26.

[ȝe-helmed] i-helmed, adj., *O.E.* gehelmod, = *O.H.G.* gehelmōt; *helmeted*, LAȝ. 26277.

[ȝe-helpen] i-helpen, v., *O.E.* gehelpan,= *Goth.* gahilpan, *O.H.G.* gehelfan; *help*; i-holpen (*pple.*) HOM. I. 81; A. R. 242; iholpe ROB. 405; AYENB. 184; TREV. V. 447.

[ȝe-hende] i-hende, adv., *O.E.* gehende; *at hand*; O. & N. 1131; MISC. 28; AYENB. 212.

[ȝe-hēnen] i-hēnen, v., *O.E.* gehēnan, = *O.H.G.* kihōnan, *Goth.* gahaunjan; *humble, oppress*; ihēned (*pple.*) HOM. I. 115.

[ȝe-henten] i-henten, v., *O.E.* gehentan; *take, receive, catch*; to ihente þe unstrong H. M. 21; ihent (*pple.*) SPEC. 28.

[ȝe-heorted] i-heorted, adj., *hearted*: swete iheorted A. R. 282: ha is hardre iheorted H. M. 37.

[ȝe-hēowien] i-hēowien, v., *O.E.* gehēowian, -hīwian; *colour*; iheouwed (*pple.*) A. R. 356; ihewed BEK. 2206.

ȝe-herbereȝen, v., *dwell, lodge*: ine zwo poure house iherberȝed (*pple.*) AYENB. 130; seven awergede gostes ware on hire ȝeherbereȝede HOM. II. 143.

[ȝe-hēre] i-hēre, adj., = *M.H.G.* gehǣre; *? obedient*: þeȝ he ne be nauȝt ihere ac wikke SHOR. 27.

ihĕrsum, adj., *O.E.* gehȳrsum, = *O.H.G.* gahōrsam; *obedient*; HOM. I. 109; ihersam (*ms.* hihersam) 223.

ihĕrsumnesse, sb., *O.E.* gehȳrsumnyss; *obedience*, HOM. I. 109.

ihĕrsumien, v., *O.E.* gehȳrsumian,=*O.H.G.* gihōrsamōn; *obey*; (*ms.* hihersamian) HOM. I. 221; ihĕrsumed (*pple.*) HOM. I. 141.

ȝe-hēren, v., *O.E.* gehēran, -hȳran,=*O.H.G.*

gihōran, *Goth.* gahausjan ; *hear, obey*, MARH.
6 ; A. R. 78 ; C. L. 67 ; iheren [ihiren] LAȝ.
3039 ; ihere O. & N. 224 ; SPEC. 46 ; S. S.
(Web.) 1871 ; DEGR. 27 ; ihure REL. I. 170
(MISC. 102) ; TREAT. 135 ; ȝehiran HOM. I. 243 ;
ihiere ANGL. I. 23 ; ahere OCTOV. (W.) 23 ;
gehīre (*pres.*) LK. ix. 9 ; ihered A. R. 90 ;
we ihereð siggen HOM. I. 19 ; iheer (*imper.*)
LANGL. (Wr.) 11643 ; ihereð R. S. v (MISC.
168) ; ihiereth MISC. 27 ; ihĕrde (*pret.*) MARH.
2 ; KATH. 149 ; O. & N. 1667 ; ihurde BEK.
41 ; FER. 112 ; iherden HICKES I. 167 ; P. S.
189 ; ihērd (*pple.*) A. R. 92 ; AYENB. 97.

[ȝe-herien] i-herien, v., *O.E.* geherian;
glorify, celebrate ; ihered (*pple.*) BEK. 1213 ;
AYENB. 23 ; P. 2 ; ihered & iheied beo þu
KATH. 2413 ; iherid OCTOV. (W.) 865.

ȝehēse, *see* ȝehǣse.

[ȝe-heviȝen] i-heviȝen, v., *O.E.* gehefigian,
-hefegian ; *make heavy* ; iheveged (*pple.*)
A. R. 332 ; ihevied CH. BOET. v. 5 (171).

ge-hlūd, sb., *O.E.* gehlȳd, *cf. M.H.G.* gelūt ;
clamour, [' *tumultus* '] MAT. xxvii. 24 ; gehled
MK. v. 38.

[ȝe-hōked] i-hōked, adj., *hooked*, O. & N.
1675.

[ȝe-hōld] i-hōld, sb., *O.E.* geheald ; *hold,
castle,* O. & N. 621.

[ȝe-hōn] i-hōn, v., *O.E.* gehōn, = *O.H.G.*
gihāhan ; *hang* : ȝif þu art þar over ihonge
(*pple.*) O. & N. 1136 ; ihon LAȝ. 2086.

[ȝe-horsed] i-horsed, pple., *O.E.* gehorsod ;
horsed, WILL. 1950.

[ȝe-hósed] i-hósed, pple., *hosed,* MISC. 91 ;
TREV. I. 29.

[ȝe-hudeked] i-hudeket, pple., *provided
with a hood,* A. R. 424*.

[ȝe-hūden] i-hūden, v., *O.E.* gehȳdan ; *hide* ;
ihede S. S. (Web.) 1314 ; ihūd (*pple.*) KATH.
1187 ; A. R. 146 ; ROB. 87 ; C. L. 32 ; ALIS.
2403 ; ihed AYENB. 109 ; ihid PL. CR. 643.

[ȝe-hūren] i-hūren, v., *O.E.* gehȳran ; *hire,*
ihērede (*pret.*) MISC. 34 ; ihūred (*pple.*) H.
M. 29.

[ȝe-hūrned] i-hūrned, pple., *O.E.* gehyrned,
= *O.H.G.* gehurnt ; *horned,* MARH. 9 ; i-
horned TREV. III. 397.

[ȝe-hwa, pron., *O.E.* gehwa,= *O.L.G.* gihve ;
each one ;] ihwat [*O.E.* gehwæt,*O.L.G.* gih-
vat ;] *each thing; ?something,* O. & N. 1056 ;
comp. aihwa.

ȝe-hwǣr, adv., *O.E.* gehwǣr ; *everywhere* ;
ihwær, -whær, -were [iwar, -ware] LAȝ. 260,
2968, 13765 ; ihwer KATH. 681 ; ant te worldes
weldent is ihwer mi warant MARH. 12 ; ihwar
A. R. 200 ; ȝewer, ihwer, uwer HOM. I. 165, 189,
231 ; ihware O. & N. 216* ; *comp.* aihwǣr.

[ȝe-hwēnen] i-hwēnen, v., *vex, grieve?*
god so þou schelt iwene (*?printed* i-wenne)
SHOR. 33.

[ȝe-hwilc] i-hwilc, pron., *O.E.* gehwilc,= *O.
L.G.* gihwilīc, *O.H.G.* gahwelīh ; *each* ; iwilc,
-wilch, uwilc, wilch HOM. I. 13, 59, 125, 129 ;
iwhilc, ilc an *each one* ORM. 503, 513 ; iwil
LAȝ. 18956 ; uich O. & N. 1378 ; i-, üwilche
(*dat. m. n.*) HOM. I. 45, 133 ; uiche dai MISC.
43 ; ihwülche (*pl.*) oþre *some others* FRAG. 1 ;
comp. aihwilc.

[ȝe-hwīlen] i-hwīlen, v.,= *Goth.* gahweilan
(πάνεσθαι) ; *have leisure* : ne mei heo nout
ihwulen vor to hercnen mine lore A. R.
422.

ȝei, *see* ȝēa.

ȝeie, sb., *?for* *ȝe-eȝe ; *awe, fear* : ȝeie and
drednesse HOM. I. 233, 239.

gein, adj., *O.N.* gegn, *prov.Eng.* gain ; *convenient, ready,* ' *directus,*' PR. P. 189 ; a gein
paþ LIDG. TH. 2148 ; gain A. P. ii. 749 ; M.
ARTH. 1904 ; þat gain(e) weede JOS. 299 ;
geiner (*compar.*) MAN. (F.) 3376 ; geinest
(*superl.*) SPEC. 29 ; gainest Iw. 1979 ; þe
geinest(e) gatis WILL. 4189 ; *comp.* un-gein.

geine, adv., *readily, quickly; near* ; full
gaine FLOR. 1986 ; hoo gainest (*superl.*) hem
miȝt ride WILL. 3465.

gain-līch, adj., *mod.Eng.* gainly ; *gracious* :
mi gainlich god A. P. iii. 83 ; gainli ii. 728 ;
gein-, gainlīch (*adv.*) WILL. 369, 391 ; geȝn-
like ORM. 18084.

ȝein, prep. & adv., *O.E.* gegn, gēn, gean,=
O.N. gegn, gögn, *O.L.G.* (an-)gegen, *O.H.G.*
gagan, gagen ; *against* ; gein LIDG. M. P.
162 ; gain A. P. i. 138 ; gain þe dore PARTEN.
2791 ; þer yen ... strive HAV. 2271 ; if he ...
bi ðor gen (*r. w.* ben) GEN. & EX. 2797 ; gon
(*r. w.* gon) 1148 ; ȝæn kinde ORM. 2322 ;
ȝǣnes 2320 ; yeines þi wille MISC. 90 ; gains
C. M. 3958, 8108 ; *comp.* an-, in-, tō-ȝein.

ȝein, adv., *O.E.* gegn ; *for* anȝēn ; *again,
back* ; to habbe ȝein his owe LAȝ. 22137* ; yee
sal gain to yur maisturs wend C. M. 12809.

ȝein-call, sb., *counter demand* : oþer lettes
him with gaincall C. M. 28778.

ȝein-char, sb., *O.E.* geancyrr ; *turning
again,* SPEC. 46.

ȝein-cherringe, sb., *return* ; þar nis no iein-
cherringe ne ... non endinge MISC. 74.

ȝein-clap, sb., *blow in return* ; ȝainclappes
(*pl.*) KATH. 129.

ȝein-cüme, sb., *O.E.* geancyme ; *again-
coming, return,* A. R. 234 ; ȝancume 394 ; ȝein-
come C. M. 2719.

ȝein-cumen, v., *return* : hit þohte him sove
ȝere are he ȝeincüme (*pret.*) LAȝ. 4437*.

gaine-cumming, sb., *attack,* BARB. ii. 450.

ȝein-giving, sb., *giving back* ; ganegeving
BARB. i. 115.

ȝein-rās, sb., *onrush* ; gainras ['*occursus*']
HAMP. PS. xviii. 7.

ȝein-sawe, sb., *gainsaying* ; gainsawe M. H.
75 ; C. M. 8382.

ʒein-seggen, v., *gainsay, contradict, deny*: al þis mai she not ʒeinsei [gainsai] C. M. 883; þus we gainsaie (*pres.*) ʒoure gilt ALEX.² (Sk.) 396; þei gainsain hure saviour 420.

ʒein-segging, sb., *contradiction*: in tokenalso of ʒeinseiing [ienseiing, gainsaiing] C.M. 11361; ʒenseiing MED. 635; ganesaiing BARB. i. 580.

ʒein-standing, sb., *opposition*: gainstanding and restrening MASSBK. 69.

ʒein-turn, sb., *counterturn*, KATH. 2118; ʒaintorn WILL. 3552.

gein, *see* gain. ʒeinde, *see* joinen.

geinen, ʒeinen, v., *O.N.* gegna, gagna,=*O. H.G.* gaʒanen, gagenen; *go to meet; be profitable*, KATH. 179; geʒnen ORM. 12929; þer mai no miht ne ʒiftes ʒeine P. L. S. xxv. 89; me schal þe ʒene O. & N. 845; gaine WILL. 598; gaune TOWNL. 319; geineð (*pres.*) A. R. 362; H. M. 45; geineþ MISC. 92; P. S. 324; (h)is bone him geineþ noht A. D. 260; gaineþ SHOR. 41; us geineþ [gaineth] no raunsoun CH. C. T. *A* 1176; gaines M. ARTH. 1071; ʒained (*pple.*) GAW. 1724; *comp.* a(n)-ʒeinen.

ʒe-innian, v., *O.E.* geinnian; *retrieve, repair*: þa wolde god ʒefyllan & ʒeinnian þone lere HOM. I. 221.

ʒeir, *see* ʒear *and* geare.

geit, *see* gāt. ʒēk, *see* ʒēac.

[ʒe-kēlen] i-kēlen, *O.E.* gecēlan; *cool*; ikēled (*pple.*) HOM. I. 149; ikelet KATH. 2291.

[ʒe-ken] i-ken, adj., *known*: iken & icnawen KATH. 424.

[ʒe-kennen] i-kennen, v., *O.E.* gecennan; *learn, know*, LAʒ. 28559; ikende (*pret.*) LAʒ. 25430; ikenned (*pple.*) LAʒ. 4844.

ʒe-kennen, v., *O.E.* gecennan,=*O.H.G.* kichennan; *be born*; gekenned (*pple.*) MAT. i. 20; ikenned LAʒ. 26128; TREAT. 138; ikend AYENB. 12.

[ʒe-kēpen] i-kēpen, v., *keep; watch for;* ['*praestolari*'] A.R.156; KATH.399*; ikēpeþ (*pres.*) O. & N. 1228; ikēpte (*pret.*) H. M. 19.

ʒel, sb., *yell*, MAP 339; ʒelles (*pl.*) ANT. ARTH. vii; *comp.* ʒe-ʒel.

[ʒe-lacchen] i-lacchen, v., *O.E.* gelæccan; *seize, take*; ilecche (*pres. subj.*) JUL. 73; ilahte (*pret.*) LAʒ. 29260; JUL. 39; ilaht (*pple.*) P. S. 214; ilaut AN. LIT. 90.

[ʒe-lāden] i-lāden, v., *O.E.* gehladan,=*O. H.G.* cahladan; *load, burden*; ilāde (*pple.*) TREV. III. 231; E. G. 358; (*ms.* ilade) LAʒ. 15952.

ʒe-lǣden, v., *O.E.* gelǣdan,=*O.H.G.* gileitan; *lead*; ʒelĕdde (*pret.*) HOM. I. 221; ʒelǣd[ilǣd] (*pple.*) P. L. S. viii. 3; ilad LAʒ. 24989; KATH. 2233; BEK. 374; MAND. 13; CH. C. T. *A* 530; huere lif (was) al wiþ joie ilad SPEC. 106 (alad ARCH. LII. 34); iled A. R. 54.

[ʒe-lǣnen] i-lǣnen, v.,*lend*; ilēned (*? printed* ileued), ilēnet, ileaned (*pple.*) KATH. 1648, 2177; ileaned A. R. 208; ilend AYENB. 19; ilent CH. C. T. *G* 1406.

ʒe-lǣr, adj., *empty*; ilær LAʒ. 15961; iler ROB. 541; ʒelǣre (*acc. f.*) LEECHD. III. 92.

[ʒe-lǣren] i-lǣren, v., *O.E.* gelǣran,=*O.H. G.* gilēran, *Goth.* galaisjan; *teach*; ilǣred (*pple.*) LAʒ. 6898; ilered A. R. 64; BEK. 158; SHOR. 136; ilearet KATH. 1314; ilærede men FRAG. 6.

[ʒe-lǣsten] i-lǣsten, v., *O.E.* gelǣstan,= *O.L.G.* gilēstean, *O.H.G.* geleistan; *last, endure; fulfil*, LAʒ. 17001; ilaste P. L. S. viii. 122; HORN (L.) 660; ROB. 246; ileste O. & N. 341; P. L. S. xii. 108; AYENB. 68; ilēsteð (*pres.*) A. R. 320; ilasteþ TREAT. 132; ilast J. T. 32; ʒelĕste (*pret.*) HOM. I. 235; ilaste BRD. 8; þa while his daʒes ilasten LAʒ. 27863; ·lǣsted, ilasted, -last, alast (*pple.*) LANGL. *A* iii. 185*, *B* iii. 191.

ʒe-lǣte, sb., *O.E.* gelǣte, *cf. M.L.G.* gelāt, *M.H.G.* gelāz, gelǣze *n.*; *visage*; ilete O. & N. 403, 1446.

·i-lǣted, adj., *visaged*; ilatet H. M. 33; ilatet se lučere JUL. 33.

[ʒe-lǣten] i-lǣten, v., *O.E.* gelǣtan,=*O.H. G.* gelāzan; *let*; ilēten (*pple.*) A. R. 112; ilete LAʒ. 18980; AYENB. 193; CH. TRO. iii. 781.

[ʒe-lǣven] i-lǣven, v., *O.E.* gelǣfan; *leave, relinquish*; ilēved (*pple.*) LAʒ. 28583; þer nas noʒt ileved BEK. 2228.

[ʒe-lang] i-lang, adv., *O.E.* gelang,=*O.L.G.* gelang, *O.H.G.* gilang, *mod.Eng.* along (of); *related, dependent*; ilong HOM. I. 197; BEK. 1642; C. L. 229; SPEC. 61; whær on hit weore ilong LAʒ. 15502; on hire is al mi lif ilong R. S. ii; along CH. C. T. *G* 930; al is on mi self along GOW. II. 22.

[ʒe-lāð] i-lāð, adj., *O.E.* gelāð; *loath*; (*ms.* ylað) O. & N. 1607.

ʒe-lāðie, v., *O.E.* gelaðian,=*Goth.* galaþōn, *O.H.G.* galadōn; *invite*, HOM. I. 231; ilāðet (*pple.*) KATH. 1918*; ilaðede (*pl.*) LAʒ. 19966.

ʒelāðere, sb., *inviter*; ʒelāðieres (*pl.*) HOM. I. 235.

ʒelāðunge, sb., *O.E.* gelaðung; *invitation*, HOM. I. 101.

ʒĕld, sb., *O.E.* geld, gield, gi·d, gyld,=*O.L.G.* geld, *O.Fris.* jeld, *Goth.* gild, *O.N.* giald, *O.H.G.* gelt; *tribute, retribution*, KATH. 212; A. R. 58; JUL. 75; (*? printed* yeld) ALIS. 2959; ʒeld [ʒield] LAʒ. 7194; ʒĕlde (*dat.*) HOM. I. 163; ʒielde ANGL. I. 9; ʒilde (*? printed* gilde) LAʒ. 4784; ʒĕldes (*pl.*) ORM. 10170; *comp.* dēofel-ʒild.

ʒild-wite, sb., *fine*: he shall be in the gild-wit of half a bushel of barlei E. G. 185.

gelde, adj., *O.E.* gelde, *O.N.* geldr,=*M.H.G.* galt; *sterile, barren*, SPEC. 24; elde makiþ me geld(e) REL. II. 210; þat geld (*sb.*) in houses

makes wenand PS. cxii. 9; ȝif ha ne mai nawt
teamen ha is iclepet gealde H. M. 33.

ȝelde-halle, *see under* **gild.**

gelden, v.,= *O.N.* gelda, *Ger.* gelten; *geld,
castrate*; geldȝen PR. P. 190; gelde [gilde]
CH. C. T. *B* 3342; **geldid** (*pple.*) VOC. 250;
gelded, gilded TREV. V. 173.

geldere, sb., *gelder,* ' *castrator,*' PR. P. 190.

geldinge, sb., *O.N.* gelding; ' *castratio,*' PR.
P. 190.

ȝélden, v., *O.E.* geldan, gieldan, gildan, gyldan, *O.L.G.* geldan, *O.Fris.* jelda, *Goth.* (fra-,
us-)gildan, *O.N.* gialda, *O.H.G.* geltan, keltan,
mod.Eng. yield; *pay tribute; recompense,*
KATH. 568; A. R. 58; ne scalt þu ȝelden uvel
onȝein uvel HOM. I. 15; his mede ȝelden
ORM. 19903; ȝelden, ȝulden LAȝ. 7372,
13481; gelden HOM. II. 5; GEN. & EX. 6;
ȝelde SPEC. 69; WILL.-321; WICL. LK. x.
35; (ich) wole heore while ȝelde ROB. 115;
here wed . . . ȝelde L. N. F. 216; he wolde . . .
ȝelde him to godes servise A. D. 257; þe
soule . . . op ȝelde SHOR. 86; acountes ȝelde
(*give*) E. G. 357; (*ms.* yelde) HAV. 2712;
AYENB. 18, 38; ȝielde S. A. L. 157; ȝelde,
ȝilde FER. 1638, 4015; ȝilde TREV. IV. 9;
ȝulde FL. & BL. 176; P. L. S. xiv. 44; ȝulde
acountes BEK. 778; þou **ȝéldest** (*pres.*) noȝt
but borwest LANGL. *B* v. 296; yeldeþ þonkes
ᴧAYENB. 18; ȝelt A. R. 302; he me ȝilt [ȝelt]
mede LAȝ. 21071; ha ȝeldeð him his gretunge
HOM. I. 257; **ȝéld** (*imper.*) ORM. 5214; he
ȝélde (*subj.*) ORM. 7378; god ȝilde you EGL.
271; **ȝeald** (*pret.*) KATH. 128; geald, ȝiald
HOM. II. 45, 169; geald GEN. & EX. 2581;
ȝald LANGL. *B* xii. 193; (*ms.* yald) M. H.
32; þu ȝulde A. R. 406; ȝolde BEK. 775; heo
ȝulden A. R. 404; gulden GEN. & EX. 1062;
ȝolden MAND. 105; yolde MISC. 39; ȝulde
(*subj.*) KATH. 217; **ȝólden** (*pple.*) ORM.
6239; MAND. 251; GAW. 453; yholden PR.
C. 5672; ȝolde FER. 3684; (*ms.* yolde) AYENB.
73; *comp.* a-, for-, ȝe-ȝélden.

ȝéldere, sb., = *M.H.G.* geltære; *yielder,
debtor*; (*ms.* yeldere) AYENB. 163.

ȝéldinge, sb.,= *M.L.G.* geldinge, *M.H.G.* gel-
tunge; *payment of debt*; ' *redditio*' PR. P. 537;
yéldinges (*pl.*) AYENB. 262.

gelding, sb., *O.N.* geldingr; *gelding,* ' *equus
castratus,*' PR. P. 190; E. W. 53; [' *eunuchus*']
WICL. GEN. xxxvii. 36.

[ȝe-lēaded] i-lēaded, adj., *from* lēad; *loaded
with lead,* A. R. 418.

[ȝe-lēafe] i-lēave, sb., *O.E.* gelēaf; *leave,
permission,* SHOR. 46; AYENB. 50.

ȝe-lēafe[2], sb., *O.E.* gelēafa,= *O.L.G.* gilōbo,
-lōvo, *O.H.G.* giloubo; *faith*; ileafe MAT.
xv. 28; HOM. I. 5; ileave AYENB. 50; ilæfe
LAȝ. 2974; gielefe HOM. II. 143; ȝelēafen (*acc.*)
HOM. I. 229; ileafen 97. ·

ȝelēaf-ful, adj., *O.E.* gelēaffull; *faithful,*

HOM. I. 5; ileafful FRAG. 3; ileaffulle men
HOM. I. 101.

[ilēaf-lēas, adj., *O.E.* gelēaflēas; *incre-
dulous.*]

ilēflēaste, sb., *O.E.* gelēaflēast; *incredulity,* HOM. I. 95.

[ȝe-lēapen] i-lēapen, v., *O.E.* gehlēapan,=
O.H.G. giloufan; *run away, leap*: þe wicke
giu (was) a wei ilópe (*pple.*) LEG. 33; (he)
was ilope TREV V. 5.

[ȝe-lēfed] i-lēfed, pple., *O.E.* gelȳfed; *believing*: ileafen ilefede men HOM. I. 91.

ȝe-lēfen, v., *O.E.* gelēfan, -lȳfan, = *O.L.G.*
gilōvian, ge-, gilōbian, -lōbean, *O.H.G.* ge-
louban, *Goth.* galaubjan; *believe,* HOM. I. 229;
he scal ileafan on þa halȝa þreomnesse 99;
ileven LAȝ. 5762; A. R. 224; ileeven C. L.
1424; ileve KATH. 342; HORN (R.) 559;
AYENB. 11; S. S. (Web.) 1198; TREV. VIII.
177; ilive FER. 1804; ilēveð (*pres.*) A. R.
68; iluvet P. L. S. viii. 66; ilēf (*imper.*) A. R.
266; SHOR. 31; ileveð A. R. 56; ilēove
(*subj.*) TREAT. 134; ȝalife HOM. I. 229; ilēfde
(*pret.*) LAȝ. 29012; MARH. 18; O. & N. 123;
AYENB. 191; ilefden A. R. 110; ilēved (*pple.*)
BEK. 2328; SPEC. 47.

ȝe-leggen, v., *O.E.* gelecgan,= *O.H.G.* gileg-
gan, *Goth.* galagjan; *lay, place*; ileggen LAȝ.
26890; ileide (*pret.*) FRAG. 5; geleigd (*pple.*)
MK. xv. 47; ileid SHOR. 126; PL. CR. 263;
CH. C. T. *A* 3568; þe hen hwon heo haveð
ileid A. R. 66.

gēlen, v., *O.E.* gǣlan (*delay*), *cf. O.N.* gǣla
(*delight*), *Goth.* gailjan (εὐφραίνειν), *M.H.G.*
geilen (*revel*); þer i con (*r.* gon) gele (*? delay*)
A. P. i. 930; he gēlð (*? dallies*) wið his herte
HOM. II. 31; *? deriv.* gēlunge.

[ȝe-lengen] i-lengen, v., *O.E.* gelengan,=
O.H.G. gilengan; *prolong; reach, attain*:
lēteð me ilenge riht to Stanhenge LAȝ. 17834;
ilenged (*pple.*) FRAG. 2; þenne beoð þine
daȝes ilenged HOM. I. 13; ilengd AYENB. 198.

[ȝe-lēoden] i-lēoden, v., *O.E.* gelēodan;
grow, increase: al ure **iledene** (*? for* ilodene
or ilidene) (*pple.*) folc LAȝ. 13857.

[ȝe-lēof] i-lēof, adj.,= *M.H.G.* gelieb; *dear
to one another*: þar two ileove [ilove] in one
bedde liggeþ O. & N. 1047.

[ȝe-lēoȝen] i-lēoȝen, v., *O.E.* gelēogan; *tell
lies, deceive*; ilóȝen (*pple.*) HOM. I. 91; iloȝe
[ilowe] O. & N. 847; treuþe iloȝe AYENB. 65;
ilowe S. S. (Web.) 2272.

[ȝe-leornen] i-leorne, v., *O.E.* geleornian,
= *O.H.G.* gelernen, -lirnen; *learn,* MISC. 93;
ileorned (*pple.*) LAȝ. 13279; KATH. 386;
ileorned [ilorned] O. & N. 216; ilierned AYENB.
70; ilerned LANGL. *B* vii. 49.

[ȝe-lēoten] i-lēoten, v., *O.E.* gehlēotan;
cast lots: þer vore him scal ileoten (*fall
to his lot*) bitterest alre baluwen LAȝ. 31306;
nas hit noht swa ilóten (*pple.*) (*destined*) 7819;
þu . . . þat art iloten [ilote] to him H. M. 11.

T 2

[ȝe-lēsen] i-lēsen, v., = M.H.G. gelœsen; *release, deliver*; ilēsed (*pple.*) HOM. I. 71; MISC. 62; ilus(e)d P. L. S. viii. 68*.

ȝe-lēsten, *see* ȝelǣsten. ȝe-lēte, *see* ȝelǣte.

[ȝe-leðered], i-leðered, adj., *from* leðer; *leathered, loaded with leather*: ne beate [ȝe] ou mid schurge ileðered A. R. 418.

[ȝe-letten] i-letten, v., O.E. gelettan, = *Goth.* gelatjan, O.H.G. gilezzan; *hinder*; ilet (*pple.*) A. R. 164; PROCL. 6; SHOR. 71; ilat ALIS. 1776; heo beoð ilette A. R. 350.

[ȝe-lēven] i-lēven, v., O.E. gelȳfan; *permit, concede*; ilēvet (*pple.*) (*? printed* ilenet) KATH. 771*.

ȝe-līc, adj., O E. gelīc, = O.L.G. ge-, gilīc, O. *Fris.* gelīk, O.N. glīkr, *Goth.* galeiks, O.H.G. ga-, ge-, gilīch; *like*, HOM. I. 219; gelic, -lich LK. xiii. 18; ilik REL. I. 218; ALIS. 6995; ilik, -like LANGL. *B* i. 91; ilich H. M. 11; FL. & BL. 49; AYENB. 15; ilich ham [heom iliche] KATH. 502; þu wenest þat ech song beo grislich þat þine pipinge nis ilich O. & N. 316; godes riche þat evre is svete and evre iliche 358; iliche A. R. 4; S. S. (Wr.) 3451; þe him iliche (*earlier text* gelic) beo iliche [O.E. gelice] (*adv.*) A. R. 112; BRD. 33; CH. C. T. *A* 2526; LIDG. M. P. 15; iliche A. P. ii. 228; eliche (?= aliche) DEP. R. i. 66; elike TOWNL. 57; iliche (*pl.*) LA3. 19705; ilicchest (*superl.*) LA3. 29482; *comp.* un-ilic.

ȝelīcnesse, sb., O.E. gelícness; *likeness*, HOM. I. 127; A. R. 230; KATH. 992.

[ȝe-līche] i-līche, sb., O.E. gelíca, = O.H.G. gilícho; *equal, peer*; ilike HOM. I. 151; his ilike, iliche LA3. 12789, 25378; þine ilike [iliche] O. & N. 157; P. L. S. xxiii. 66; nas non his iliche HORN (L.) 18; godes iliche SHOR. 167.

[ȝe-līche²] i-līche, sb., = O.H.G. gelíchī; *likeness*: in Hornes ilike HORN (L.) 289; iliche KATH. 1843; his iliche A. R. 136.

[ȝe-lided] i-lided, pple., *provided with a lid*, A. R. 58.

ȝe-līf-fæsten, v., O.E. gelíf-fæsten; *vivify*; ȝelíffeste (*pret.*) HOM. I. 219.

[ȝe-liggen] i-liggen, v., O.E. gelicgan; *lie*; ileien (*pple.*) A. R. 326; ileien [ileie] LA3. 24923; ileie HORN (L.) 1139; (*ms.* ileye) ROB. 282; TREV. VII. 353; ileiȝen [ileien, ileie] LANGL. *A* v. 65, *B* v. 82.

[ȝe-līhten] i-līhten, v., O.E. gelíhtan; *alleviate*; ilíhted (*pple.*) A. R. 356; iliȝt P. L. S. ix. 20.

[ȝe-līhten²] i-līhten, v., O.E. gelýhtan; *illumine*; ilíht (*pple.*) C. L. 794.

[ȝe-līke] i-līke, adj., *? pleasing, agreeable*: ræd þat heom weore ilike LA3. 15117.

ȝelīke, *see* ȝelíche.

[ȝe-līkien] i-līkien, v., O.E. gelícian, *cf.* O.H. G. gilíchan; *like, have pleasure*; ilíki (*pres. subj.*) AYENB. 109.

[ȝe-limed] i-limed, adj., *limbed*, C. L. 624.

[ȝe-limpe, i-limpe, sb., O.E. gelimp; *accident, chance*; *comp.* un-ilimpe.]

ȝe-limpen, v., O.E. gelimpan, = O.H.G. gilimphan; *befall*; ilimpen LA3. 4508; ȝelamp (*pret.*) HOM. I. 231; ilomp A. R. 54; hit ilomp LA3. 386; on his daȝen ilompen reoðen on leoden LA3. 31791; ilumpe (*subj.*) HOM. I. 91; ilumpen (*pple.*) HOM. I. 93; þa wes hit ilumpen (*sec. text* bivalle) LA3. 7195.

[ȝe-līðe] i-līðe, adj., = M.H.G. gelinde; *? agreeable*, LA3. 4917.

[ȝe-līðen] i-līðen, v., O.E. gelíðan, = *Goth.* galeiþan, O.H.G. gilídan; *go*; ilíðen (*pple.*) LA3. 26094.

[ȝe-līðeȝen] i-līðeȝen, v., O.E. gelíðegian; *mitigate*; ilíðegað (*pres.*) HOM. I. 97.

[ȝe-livien] i-livien, v., O.E. gelifian, -leofian, *cf.* O.H.G. gilebēn; *live*; ileve AYENB. 93; ilived (*pple.*) ROB. 195; A. D. 237.

ȝelke, *see* ȝeolke.

gellē, sb., O.Fr. gelee; *jelly*, PR. P. 190.

ȝellen, v., O.E. gellan, giellan, gyllan, = O.H. G. gellan, O.N. gella, gialla, M.L.G. gellen, gillen; *yell, shout, scream, sound*, KATH. 2040; þe se ... bigan to ȝellen and to bellen HALLIW. 951; shal ȝellen [*ululabit*] WICL. Is. xv. 2; ȝellin *'vociferor'* PR. P. 537; ȝelle MAN. (F.) 10207; S. S. (Wr.) 805; GAW. 1453; EGL. 411; hellehoundes here i ȝelle MAP 338; foulis herde sche ȝelle OCTAV. (H.) 330; (*ms.* yhelle) PR. C. 7341; helmes gunnen gullen [ȝelle] LA3. 16407; ȝulle and rore P. L. S. xix. 241; ȝelle (*pres.*) CH. C. T. *A* 2672; gelð HOM. II. 31; (hi) yelleþ AYENB. 71; ȝeollinde (*pple.*) LEG. 183; ȝal (*pret.*) O. & N. 112; BRD. 23; S. S. (Wr.) 801; gullen, gollen [ȝollen] LA3. 9797; ȝulle BRD. 27; ȝollen LEG. 43; ȝelled ROB. (W.) 4239*.

ȝellinge, sb., [*'ululatus'*] WICL. EXOD. xxxii. 17; ȝullinge BRD. 23.

ȝe-lōȝien, v., O.E. gelógian; *place*; ȝelōgode (*pret.*) HOM. I. 221; ilōgod (*pple.*) FRAG. I.

[ȝe-lōken] i-lōken, v., O.E. gelócian; *look, see, observe*, HOM. I. 45; ilōked (*pple.*) LA3. 5277; A. R. 372; AYENB. 7; þe lahen þet tu ham havest iloket MARH. 10.

[ȝe-lōme] i-lōme, adv., O.E. gelóme, = O.H. G. kilómo; *frequently*, P. L. S. viii. 24, 46; ilome LA3. 16500; A. R. 136; O. & N. 49; ROB. 3; C. L. 435; A. D. 290; ilome and ofte BRD. 24; ilōmest (*superl.*) O. & N. 595.

[ȝe-lósien] i-lósien, v., O.E. gelósian; *lose*; ilósed (*pple.*) LA3. 12492; KATH. 2046.

ȝelp, sb., O.E. gelp, gielp, gilp, gylp, = O.L.G. gelp, O.H.G gelph, gelf; *boasting*, HOM. I. 103; ORM. 12041; C. L. 1364; MAN. (H.) 264; ȝelp [ȝeolp] LA3. 21007; (*? printed* yelp)

LAUNF. 718; IW. 2765; ȝelpe (*dat.*) A. R. 210.

ȝelpen, v., *O.E.* gelpan, gielpan, gilpan, gylpan,=*M.H.G.* gelfen; *boast*, A. R. 146; ȝelpen, ȝeolpen LAȜ. 12072, 22949; ȝelpen, ȝilpen ORM. 2042, 4925; ȝelpe MAN. (F.) 7463; BER. 3268; S. & C. I. xxxv; (*printed* yelpe) ALIS. 1065; GOW. II. 116; ȝilpe FER. 694; ȝelpest (*printed* ȝeolpest) [ȝulpest] (*pres.*) O.&N. 1299, 1650; ȝelpeð KATH. 1287; ȝealp [ȝalp] (*pret.*) LAȜ. 2870; ȝelp KATH. 476; (*ms.* yelp) MISC. 46; yalp AYENB. 208; þu ȝulpe LAȜ. 26835; iȝolpe (*pple.*) ALIS. 3368.

ȝelpere, sb., *boaster*; yelpere AYENB. 22.

ȝelpunge, sb., *boasting*, A. R. 330; (? *ms.* ȝelwunge) HOM. I. 11; ȝelpinge (ȝulpinge) & bost ROB. (W.) 4266; ȝelping GAW. 492.

ȝelstren, *see* ȝàlstren.

gēlsunge, sb.,=*O.H.G.* geilsunge; *luxury*, A. R. 100*.

gelt, *see* gült.

ȝelte, sb., *cf. M.L.G.* gelte (*castrated sow*); *O.H.G.* gelza; '*scrofa*,' *sow*, VOC. 177; *see* gilte.

ȝelu, *see* ȝeolu.

[ȝe-lūken] i-lūken, v., *O.E.* gelūcan; *shut*, *lock*; iloken (*pple.*) LAȜ. 32202; iloke BEK. 824; ALIS. 2769; one ilokene cope A. R. 56; þe ȝeten weren ilokene HOM. I. 141.

gēlunge, sb., ?*from* ȝēlen; ?*luxury*; A. R. 100.

gelūs, adj., *O.Fr.* gelōs; *jealous*, A. R. 90; O. & N. 1077; GEN. & EX. 3495; jelous WICL. ECCLUS. xxvi. 8; CH. C. T. *A* 1329.

gelousnes, sb., *jealousy*: the spirit of gelousnes WICL. NUM. v. 14; jelousnesse WICL. PROV. vi. 34.

gelusie, sb., *O.Fr.* gelosie; *jealousy*, A. R. 90; gelousie WICL. NUM. v. 14*; jelousie, jelosie CH. P. F. 252; jelesie, jalousie LANGL. *A* x. 184, *B* ix. 166.

gelustē, sb., *jealousy*, WICL. NUM. v. 15; to jelouste WICL. PS. lxxvii. 58.

[ȝe-luvien] i-luvien, v., *O.E.* gelufian; *love*; ilufed (*pple.*) FRAG. 7; iluved LAȜ. 13644; R. S. ii.

ȝe-mæcche, sb., *O.E.* gemæcca,=*O.H.G.* gimahho; *wife*; Marian þine gemæcchen MAT. i. 20; *cf.* ȝemáke.

ȝe-mæht, sb.,=*O.H.G.* kamaht; *might*; mid alle heore imæhte LAȜ. 30979; gemihte (*dat.*) LK. i. 17.

[ȝe-mælen] i-mælen, v., *O.E.* gemǣlan; *speak*; imelen HOM. I. 193.

ȝemǣne, adj. & adv., *O.E.* gemǣne,=*O.Fris.* gemēne, *O.H.G.* ge-, gimeini, *Goth.* gamains; *in common, universal, common, commonly*; imæne, imene, imone LAȜ. 978, 11037, 13614; þeȝ sungen alle imæn ORM. 3376; ȝemene HOM. I. 101; imene REL. I. 216; B. DISC. 608; þet heo alle habbeð imene A. R. 12; speken

of alle imene 90; hit is fale oþer wiȝtē imene O. & N. 628; more his (*r.* is) blisse god and clene a monge frendes to habbe imene SHOR. 125; þe imeane blisse HOM. I. 261; þet is þin wið me imeane KATH. 1867; mi kinedom is imone ROB. 311; hi gonne to singe imone BRD. 17.

imēnnesse, sb., *O.E.* gemænness; *community*, HOM. I. 217.

[ȝe-mēne] i-mēne, sb.,=*M.H.G.* gemeine; *companion*; imene HOM. II. 31; Aþulf wes his imone HORN (R.) 530; þeh þe wulf beon (*r.* beo) ane buten ælc imane [imone] LAȜ. 21308.

[ȝe-mēne] i-mēne, sb., *cf. O.E.* gemāna, *O.H.G.* kimeinī, *Goth.* gamainei; *community, company*: þat wiþ þe fule haveþ imene O. & N. 301; imone HOM. I. 245; MISC. 85; wiþoute mannes imone SHOR. 118.

ȝe-mǣre, adj., *O.E.* gemǣre; *boundary, district*; gemæren ['*finibus*'] MAT. ii. 16.

[ȝe-mǣrsien] i-mǣrsien, v., *O.E.* gemǣrsian,=*O.L.G.* gimārsōn; *celebrate*; imērsed (*pple.*) HOM. I. 243.

[ȝe-mǣte] i-mǣte, adj., *O.E.* gemǣte,= *O.H.G.* gemāzi, *M. H. G.* gemǣze; *meet, moderate, suited for*; imete HOM. I. 105; LAȜ. 6584; L. H. R. 30; S. A. L. 158; *comp.* un-imǣte.

imētnesse, sb., *moderation*, HOM. I. 101.

[ȝe-mǣten] i-mǣten, v., *O.E.* gemǣtan; *appear in a dream*: me imǣtte (*pret.*) a sweven LAȜ. 28016; imēt (*pple.*) JUL. 74.

[ȝe-māh] i-mōh, adj., *O.E.* gemāh; *bold*; imouh FRAG. 5.

imōuhnesse, sb., *O.E.* gemāhness; *boldness*, '*procacitas*,' FRAG. 5.

[ȝe-mained], adj., *having strength*; *comp.* stīð-imained.]

[ȝe-mak] i-mak, adj., *O.E.* gemæc,=*O.H.G.* gemach; *adapted for, suitable*: wiþ middel smal ant wel imake SPEC. 28.

[ȝe-máke] i-máke, sb., *O.E.* gemaca,= *O.L.G.* gimaco, *O.H.G.* kamahho; *wife, partner, equal*, FRAG. 8; in worle nes nere non þin imake FL. & BL. *B* 77; imáken (*pl.*) LAȜ. 18206.

ȝe-mákien, v., *O.E.* gemacian,=*O.H.G.* ka-, gimahhōn, -machōn; *make*; ȝemáked (*pple.*) HOM. II. 145; imaked A. R. 54; HAV. 5; TREAT. 138; SPEC. 111; SHOR. 62; LANGL. *A* PROL. 14; imad ROB. 470; imaad CH. C. T. *B* 693; HOCCL. i. 203.

ȝeman, *see* ȝoman.

ȝe-mancünn, sb., *O.E.* manncynn; *mankind*: þa wes þa sume hwille godes ȝei on ȝe mancinne efter the flode HOM. I. 225.

[ȝe-mang] i-mang, sb., *O.E.* gemang, -mong, =*O.L.G.* gimang, *M.H.G.* gemane; *mixture*,

crowd; **imonge** (_dat._) LAȝ. 10868; **imong**
(_adv. and prep._) _for_ on imong [_O.E._ on gemang,
_O.L.G._ an gimang] _among_, HOM. I. 27; imong
þon scipmonnen LAȝ. 2229; her wes harpinge
imong 22702; heom imong (_printed_ imang)
FRAG. 6; imong moncun MARH. 12; ꝥer
imong MISC. 19; imong (_ms._ ymong) GEN. &
EX. 1349.

ȝēme, sb., _O.E._ giēme, gȳme; _care, heed_,
LAȝ. 6981; A. R. 32; FL. & BL. 38; ROB.
194; C. L. 502; SHOR. 11; LANGL. _B_ x. 195;
P. 9; MAN. (F.) 7125; mid his ȝeme (_earlier
text_ gymene) HOM. I. 117; nim ȝeme of þi
ȝuheꝥe KATH. 1462; ȝeme, ieme (_ms._ yeme)
O. & N. 649; yeme AYENB. 24; _see_ **gōme.**

     ȝēme-lēas, -lēs, adj., _O.E._ gēme-, gȳmelēas,
= _O.H.G._ goumilōs; _negligent_, HOM. I. 109,
205.

     ȝēmelēas-līche, adv., _heedlessly_, A. R. 92;
gemeslesliche REL. I. 129.

     ȝēmelēas-(s)chipe, sb., _negligence_, A. R.
202.

     ȝēmelēaste, sb., _O.E._ gȳmelēast; _negli-
gence_ A. R. 46; ȝemeleste HOM. I. 117; ȝeme-
læste ORM. 2913.

[ȝe-mearkien] i-merkien, v., _O.E._ gemear-
cian; _mark, note_; imerked (_pple._) FRAG. 8;
A. R. 42.

[ȝe-medemien] i-medemien, v., _O.E._ ge-
medemian, _cf._ _O.H.G._ gemetemēn; _humble_;
þe almihti god hine seolf imedemede (_pple._)
HOM. I. 97.

gemels, sb., pl., _Lat._ gemellī; _twins_, WICL.
GEN. xxxviii. 27.

[ȝe-melten] i-melten, v., _O.E._ gemeltan;
melt; imolt(e) (_pple._) TREV. VIII. 129*.

[ȝe-melten²] i-melten, v., _O.E._ gemyltan;
melt; imulten REL. I. 183; imelt (_pple._)
A. R. 284; imealt H. M. 45.

[ȝe-membred] i-membred, adj., _cf._ _O.Fr._
membre; _parti-coloured_, A. R. 420.

gemen, _see_ gamen.

ȝemēne, _see_ ȝemǣne.

ȝēmen, v., _O.E._ gēman, giēman, gȳman, = _O.N._
geyma, _O.L.G._ gōmean, _O.H.G._ goumen,
Goth. gaumjan (βλέπειν); _take care of, care,
observe, regard_, A. R. 98; ORM. 2911; ȝemen
þes fehtes LAȝ. 8609; ȝeme P. S. 215; H. H.
24; WILL. 2734; LANGL. _B_ viii. 52; S. S.
(Wr.) 99; M. ARTH. 2512; (_ms._ yeme) PS. xi.
8; D. TROY 136; (_ms._ yeme) and gete HAV.
2960; ȝime LAȝ. 12581*; ȝime in store
P. L. S. i. 18; ȝēmes (_pres._) A. P. ii. 1493;
þet ȝe ȝemen þenne (_r._ þene) halie sunne dei
HOM. I. 11; ȝēmande (_pple._) PERC. 1136;
ȝēmde (_pret._) LAȝ. 30845; ȝemede CHR. E.
152; gemden ['_observabant_'] LK. vi. 7;
Bruttes her of ȝemden LAȝ. 9168; ȝemde (_ms._
yemede, _r. w._ fremde) HAV. 2276; ȝemit
BARB. xi. 112; ȝemit (_pple._) BARB. viii.
494; _comp._ **bi-, for-, ȝe-ȝēmen.**

ȝēmer, sb., = _O.N._ geymari; _guardian_;
yemer M. H. 114; emer TUND. 224.

gēming, sb., _care_, GEN. & EX. ·2783.

[ȝe-mengen] i-mengen, v., _O.E._ gemengan;
mix, trouble; imenged (_pple._) LAȝ. 28561;
ROB. 119; imenget KATH. 605; imengd HOM.
I. 185; BEK. 2147; SHOR. 9; AYENB. 196;
imeind A. R. 332; O. & N. 18; CH. C. T. _A_
2170.

ȝēmer, _see_ ȝēomer.

[ȝe-merren] i-merren, v., _O.E._ gemerran,
-myrran; _mar_; imerred (_pple._) H. M. 11.

Gemes-tīd, sb., _St. James' day_ (_July_ 25),
ROB. (W.) 10278.

[ȝe-met] i-met, sb., _O.E._ gemet, = _O.L.G._
gimet, _O.H.G._ ca-, kimez; _measure; moder-
ation_; þet rihte imet FRAG. 5; bi ꝥon ilke
imet ꝥe ȝe meteꝥ HOM. I. 137; best is ever
imete A. R. 286; _comp._ **un-imet.**

     [ȝemet-fæst], adj., _O.E._ gemetfæst; _moder-
ate_].

     imetfest-līche, adv., _moderately_, FRAG. 1.

     imet-līc, adj., _O.E._ gemetlīc, = _O.H.G._ ki-
mezlih; _moderate_: an imetliche broc LAȝ.
21783.

[ȝe-met²], adj., _O.E._ gemet (_fit_); _comp._ **un-
imet.**]

[ȝe-mēten] i-mēten, v., _O.E._ gemetan, = _Goth._
gamitan, _O.H.G._ gemezan; _measure_; imēten
(_pple._) LAȝ. 21995; REL. II. 217; imete FER.
2092.

ȝe-mēten, v., _O.E._ gemētan; _meet_; imeten
FRAG. 7; P. L. S. viii. 67; LAȝ. 19187; ȝemete
HOM. I. 231; imete R. S. v; HORN (L.) 940;
REL. I. 144; FER. 72; imǣtte (_pret._) A. R. 154;
ROB. 179; CHR. E. 611; imēt (_pple._) K. T.
960; SACR. 237; such pine hi habbeþ imet
MISC. 152.

     imēting, sb., _O.E._ gemēting; _meeting_, '_con-
ventus_,' FRAG. 2.

[ȝe-mildsen] i-mildsen, v., _O.E._ gemiltsian;
show clemency: imilze þu Octa & his iveren
al swa LAȝ. 16837; ȝe imilcien (_subj._) HOM.
I. 39.

[ȝe-missen] i-missen, v., = _M.H.G._ gemissen;
miss: þat he me ne imisse REL. I. 102 (MISC.
195); imist (_pple._) A. R. 78; O. & N. 581.

gemmaire, sb., _Lat._ gemmārius; _jeweller_:
the graving of the gemmaire WICL. EX. xxviii.
11; gemmaire (_gen. pl._) craft EX. xxxix. 6.

gemme, sb., _O.Fr._ gemma; _gem_, MISC. 98;
gemis (_pl._) ne gewellis D. TROY 1368; _see_
gim.

[ȝe-mōded], adj., _having a mood_; _comp._
stiꝥ-imōded.]

ȝe-mōt, sb., _O.E._ gemōt; _convention, assem-
bly_, '_synodus_;' imot FRAG. 2; witene imot
LAȝ. 11545; gemōten (_dat. pl._) MAT. x.
17; _comp._ **hāli-mōt.**

gēmsle, sb., _O.N._ geymsla; _care_, ORM. 5095;

ʒemsall BARB. xx. 231 ; to ʒemsall *under his care* BARB. xvii. 231 ; ʒeimseill BARB. xi. 329 ; ʒhemsell ii. 136 ; ʒeemsell xx. 231.

ʒe-münde, sb., *O.E.* gemynd,=*Goth.* gamunds, *O.H.G.* gimunt ; *mind* ; imunde MISC. 37 ; of lihte nabbeþ hi none imunde O. & N. 252 ; on his geminde ['*in memoriam ejus*'] MK. xiv. 9.

　imünd-lēas, adj., *O.E.* gemyndlēas ; *without a mind, 'amens,'* FRAG. 3.

[ʒe-münde²] i-münde, adj., *O.E.* gemynde ; *? mindful* ; þine beoden þe beoð þe so imunde MARH. 12.

[ʒe-münden] i-münden, v., *O.E.* gemyndan ; *mind, be mindful of* ; imende (*imper.*) AYENB. 262.

[ʒe-mündi] i-mündi, adj., *O.E.* gemyndig ; *mindful* : ʒif we imundie beoð godes bibode HOM. I. 89.

ʒe-müneʒen, v., *O.E.* gemynegian ; *remember, remind* ; imüneget (*pple.*) MARH. 12 ; þanne beo ic ʒemeneʒed mines weddes HOM. I. 225 ; imeneʒed SHOR. 69.

[ʒe-munan] imunen, v., *O.E.* gemunan ; *remember,* LAʒ. 16309.

[ʒe-münnen] imünnen, v., *O.E.* gemynnan ; *remember,* LAʒ. 8037 ; iminne S. A. L. 169.

[ʒe-münten] i-münten, v., *O.E.* gemyntan ; *intend* ; imünt (*pple.*) A. R. 408 ; REL. II. 277 ; imint HOM. II. 101 ; FER. 576 ; iment HORN (L.) 795 ; ALIS. 5942 ; OCTOV. (W.) 1953.

ʒēn, *see* ʒein.

[ʒe-námen] i-námen, v., *O.E.* genamian ; *name* ; inámid (*pple.*) WICL. I MAC. x. 1.

genciane, sb., *O.Fr.* gentiane ; *gentian* (*the plant*), PR. P. 190 ; VOC. 226.

ʒend, *see* ʒeond.

ʒende, *see* ende.

gender, sb., *O.Fr.* gendre, genre ; *gender, kind, 'genus,'* PR. P. 190.

gendrin, v., *O.Fr.* gendrer ; *gender, beget,* PR. P. 190 ; gendre WICL. GEN. xvii. 20 ; gendrith (*pres.*) ALIS. 6589 ; gendride (*pret.*) WICL. MAT. i. 2.

　gendrer, sb., *parent* ; gendrers [*later ver.* gendreris] (*pl.*) WICL. ZECH. xiii. 3.

　gendringe, sb., *engendering,* CH. H. F. ii. 457.

gendrūre, sb., *O.Fr.* (en-)gendrūre, *med.Lat.* generātūra ; *begetting,* WICL. JOB xl. 12.

ʒene, *see* ʒeʒen.

genealogi, sb., *O.Fr.* genelogie, *Gr.* γενεαλογία ; *genealogy* ; (*v. r.* genologi) C. M. 7846 ; genelogie MAN. (H.) 111 ; withoute genelogi (*later ver.* genologie) WICL. HEB. vii. 3 ; genolgie, genolagie YORK 271, 208 ; nethir ʒive tent to fables and genologies (*pl.*) WICL. I TIM. i. 4.

[ʒe-nēden] i-nēden, v.,*O.E.* genēdan,-nȳdan ; *need ; compel* ; inēd (*pple.*) A. R. 72.

ʒe-nemnen, v., *O.E.* genemnan, = *O.H.G.* ginemnan, -nennan ; *nominate, name* ; inemnen LAʒ. 24366 ; ʒenamned (*pple.*) HOM. I. 229 ; inemned A. R. 252 ; P. S. 215 ; AYENB. 66 ; E. W. 27 ; inempned C. L. 61 ; inemd HOM. II. 39.

ʒeneow, sb., *yawn, gape* : (he) ʒeonede mid his wide geneow MARH. 9.

generacioun, sb., *O.Fr.* generacion ; *generation,* TREV. II. 231 ; MAND. 206, 223.

general, sb., *O.Fr.* general ; *general,* AYENB. 14 ; in generalle MAND. 296.

　general-līch, adj., *universal, catholic* : holi cherche generalliche AYENB. 263.

　generalliche, adv., *generally* : generalliche ech manere of volk, ac specialliche þe greate lhordes AYENB. 16 ; generali C. M. 29118 ; generalli CH. C. T. *I.* 371.

generūs, adj., *O.Fr.* generous, *from Lat.* generōsus ; *free, generous* : ʒineris, ʒinerus, jenerus D. TROY 1242, 357, 3917.

[ʒe-nēwen], v., *O.E.* genīwian ; *renew, renovate* ; inēwed (*pple.*) AYENB. 107.

genge, sb., *O.E.* genge (SAX. CHR. 207), *O.N.* gengi, *cf. O.H.G.* (ana-)genge ; *company, retinue, people,* HOM. I. 87 ; LAʒ. 15095 ; ORM. 3918 ; MAP 348 ; WILL. 1600 ; A. D. 241 ; PS. ii. 8 ; þe . . . bereð me genge MARH. 21 ; D. TROY 1225 ; him and his genge wel he fedde HAV. 786 ; ginge ALIS. 1509 ; RICH. 4978 ; genge ['*gentes*'] HAMP. PS. ii. 1, 8 ; ging ALEX. (Sk.) 3484 ; gingis (*pl.*) ALEV (Sk.) 1648, 2435 ; *comp.* ʒe-genge.

genge², adj., *O.E.* genge, = *O.H.G.* gengi, *M.L.G.* genge, *M.Du.* ghenge ; *from* gangen ; *current, usual,* O. & N. 804, 1002, 1065.

gengen, v., *O.E.* gangan,=*M.H.G.* gengen ; *pass, move* : swa þat it muʒhe gengen us to berʒhen ure sawle ORM. 4160 ; make þi clerk before þe ʒinge MIRC 1963 ; gengþ (*pres.*) O. & N. 376 ; geinde (*pret.*) LAʒ. 4568 ; forð gengden (geinde) þa qvenen 12865.

[gengle, sb., = *O.H.G.* gengil ; *comp.* foregengle.]

ʒenien, *see* ʒeonien.

genillere, sb., *O.Fr.* genouillere ; *armour for the knees* : þe strok . . . ful opon is genillere bar awai his chaunceore FER. 5631.

ʒe-nimen, v., *O.E.* geniman,=*Goth.* ganiman, *O.H.G.* gineman ; *seize* ; ʒenam (*pret.*) HOM. I. 223 ; inom LAʒ. 71 ; inumen (*pple.*) HOM. I. 29 ; MARH. 2 ; inumen, -nomen [inome] LAʒ. 186, 5383 ; inume O. & N. 541 ; inomen ALIS. 4668 ; inome BEK. 655 ; SHOR. 126 ; ASS. 5 ; MIRC 495 ; inumene (*pl.*) A. R. 42.

genital, adj., *O.Fr.* genital, *Lat.* genitālis ; *genital,* WICL. NUM. xxv. 8.

genitrise, *see* genterise.

ʒe-niþerien, v., *O.E.* geniðerian, -neðerian, *cf. O.H.G.* ginidiran ; *debase, humble* ; geneþered (*pple.*) MAT. xxiii. 12; iniðered, -niþered LAȝ. 10218, 25235; iniþered HOM. I. 117.

ȝe-nōh, adj. & adv., *O.E.* genōh, -nōg, = *O.L.G.* ginōg, *O.N.* gnōgr, *O.H.G.* genuog, *Goth.* ganōhs; *enough*, LAȝ. 2558; KATH. 347; ORM. 731; SPEC. 28; H. H. 51; inoh, inouh O. & N. 1182; inoȝ ROB. 18; AYENB. 22; inogh PR. C. 1466; inouh FRAG. 7; A. R. 86; inouh [-nouȝ] C. L. 408; inouȝ LIDG. M. P. 226; inouȝ, inou WILL. 100, 714; (*ms.* inowȝ, inow) WICL. LEV. xiii. 28; inou TREAT. 140; HAV. 563 (*spelt* inow 706); MAND. 59; PL. CR. 230; inogh [-nouȝ, -nough, -nou] CH. C. T. *B* 1477, BOET. iii. 3 (71); inog, anog GEN. & EX. 600, 3670; i-, anouȝ TRIST. 1535, 1598; i-, enogh C. M. 18226; enogh PERC. 952; inōhe (*pl.*) KATH. 514; inohȝe, inowe LAȝ. 2367, 3388; inoȝe HOM. I. 13; O. & N. 16; treos inoȝhe ORM. 14; inowe PL. CR. 215; inowe men RICH. 5905; anowe S. S. (Web.) 921; AM. & AMIL. 872.

gent, adj., *O.Fr.* gent; *gentle, well-born,* HAV. 2139; ROB. 24; SPEC. 89; þis is þe ladi so gent and fre C. L. 859; gente (*pl.*) O. & N. 204.

ȝent, *see* ȝeond.

gentelerie, sb., *gentility, gentry* : þuru þis lore and gentelerie he amendit huge companie MISC. 138.

genterie, sb., *gentry,* CH. C. T. *D* 1146; gentrie (*v. r.* genterise) LANGL. *C* xxi. 21*.

genterise, sb., *O.Fr.* genterise; *genteel carriage, gentility,* A. R. 168; gentrise JUL.[2] 52; gentrise C. M. 27568; gentris D. TROY 131; genterise, gentrise LANGL. *B* xviii. 22, *C* xxi. 21; genitrise [gentrise] ROB. (W.) 1313.

gentil, sb., *noble,* D. TROY 128; (*lady*) 437; gentiles (*pl.*) MAND. 251; gentils EGL. 1000; gentiles TREV. VII. 99; gentilles TREV. VII. 87; gentillis B. B. 22.

gentil[2], adj., *O.Fr.* gentīl; *gentle, noble, generous,* A. R. 358; AYENB. 75; CH. C. T. *A* 72; a gentil tree PALL. iii. 709; gentil (*thoroughbred*) maris PALL. iv. 779; gentiller (*comp.*) HOM. I. 273; gentileste (*superl.*) A. P. i. 1014.

gentil-līche, adj., *gently,* LANGL. (Wr.) 1381.

gentil-man, sb., *mod.Eng.*gentleman; *nobleman, 'generosus, proceres,'* VOC.182; jentilman TOR. 91; gentilmen (*pl.*) P. S. 213; gentilmen children beeþ itauȝt to speke Frensche TREV. II. 159.

gentilnesse, sb., *nobility (of rank and of manners),* CH. TRO. iii. 1364.

gentil-woman, sb., *noble-born woman, 'generosa,'* PR. P. 190; gentilwomman CH. C. T. *I* 838.

gentilesse, sb., *O.Fr.* gentillise, gentilece (*gentillesse*), *gentility,* AYENB. 89; gentilesse [gentillesse] CH. C. T. *B* 3441; PALL. xi. 28.

gentiletē, sb., *gentility,* AYENB. 89.

[ȝe-nūht, sb., *O.E.* genyht, *cf. O.H.G.* ca-, genuht; *abundance.*]

[ȝenūht-sum, adj., *O.E.* genyhtsum, *cf. O.H.G.* canuhtsam; *abundant.*]

inihtsumnesse, sb., *O.E.* genyhtsumniss, '*abundantia,*' FRAG. 4.

ȝēo, *see* hēo.   ȝeōȝeðe, *see* ȝūȝeðe.

ȝēol, *see* ȝōl *and* ȝōle.

ȝeolke, sb., *O.E.* geolca; *yolk,* SAINTS (Ld.) 317; ȝolke TREAT. 137; D. ARTH. 3283; yholke PR. C. 6448; ȝelke '*vitellus*' PR. P. 537; VOC. 267; yelke VOC. 150.

ȝeolp, *see* ȝelp.

ȝeolu, adj., *O.E.* geolo, = *M.Du.* ghelu, *O.H.G.* gelu; *yellow,* P. S. 158; ȝeoluh A. R. 88; geleu REL. I. 129; ȝeleu (*? printed* yeleu) RICH. 2644; ȝelou, ȝalou MAND. 48; ȝelu [yelou] as wex CH. C. T. *A* 675; ȝolou ALIS. 1999; yalu HALLIW. 943; yalow ALEX. (Sk.) 607; þe ȝeolewe clað HOM. I. 53; ȝelwe (*? printed* yelwe) GOW. III. 255; hir ȝelwe [ȝelowe, yolwe] heer CH. C. T. *A* 1049; ȝeoluwe (*pl.*) S. A. L. 150; ȝeluwe froggen HOM. I. 51; ȝealwe FER. 5881.

ȝalou-souȝt, sb., *jaundice,* HALLIW. 950.

ȝeolu-mon, sb., *man dressed in yellow, bumbailiff* : ther stont up a ȝeolumon P. S. 158.

ȝelhwnesse, sb., *yellowness,* PR. P. 537; ȝalownes CATH. 425.

ȝeoman, *see* ȝoman.

geomancie, sb., *O.Fr.* geomancie *from Lat.* geōmantia; *mensuration* : geomaunce GOW. III. 45; gemensie, geomesie LANGL. *A* xi. 153, *B* x. 208; MAND. 234; geomancie CH. C. T. *I* 605.

ȝēomer, adj., *O.E.* gēomor, gīomor, *O.L.G.* jāmar, *O.H.G.* jāmar; *lamentable, miserable,* HOM. I. 193; ȝemer SHOR. 113; yemer MISC. 30; hare ȝeomere bileave KATH. 1831; ȝeomere spelles LAȝ. 24942; þa ȝemere scipen 12039; iemere (*ms.* yemere) þoȝtes AYENB. 215; ȝōmere (*adv.*) O. & N. 415; ȝēomerest (*superl.*) LAȝ. 16566.

ȝēomer-līc, adj., *O.E.* gēomorlīc; '*lamentable,* LAȝ. 29564; ȝomerli A. P. ii. 971.

ȝēomer-līche, adv., *lamentably,* MARH. 6; gemerliche HOM. II. 35; ȝomerli GAW. 1453; ȝamerli ANT. ARTH. vii.

ȝēmernesse, sb., *misery,* MISC. 28.

[? ȝēomer-rēd] iumpred, sb., *mourning,* A. P. ii. 491.

ȝēomerien, v., *O.E.* gēomerian, = *O.H.G.* jāmarōn; *lament*; ȝōmers (*pres.*) D. TROY 543; ȝēomerinde (*pple.*) HOM. l. 97; gemerinde MK. v. 38; ȝēomerede (*pret.*) LAȝ. 23492; ȝamurt ANT. ARTH. vii.

ȝēomerunge, sb., *O.E.* gēomerung; *lamentation,* H. M. 35; ȝomering D. TROY 1722.

geometrie, sb., *O.Fr.* geometrie; *geometry,* LANGL. *B* x. 208; gemetrie '*geometria*' PR.

P. 190; gemettrie and gramer PR. C. 7801;
gemetri MAS. 53; gematri D. TROY 8394.

geometrien, sb., *geometrican*; **gemetriens**
(*pl.*) LIDG. TREAT. I, 2.

ӡeon, pron., *O.E.* geon (ALFR. P. C. 443), *cf.*
O.Fris. jene, *O.H.G.* jener, *O.N.* enn, *Goth.*
jains; *yon*; ӡeon [ӡon, ӡone] liӡt LANGL. *B* xviii.
145, *C* xxi. 149; ӡon A. P. i. 672; C. M. 654;
in ӡone stede Iw. 2549; on ӡone hil MAN.
(F.) 12300; ӡone kniӡtes WILL. 4572; gune
men GEN. & EX. 3135.

ӡeond, adv. (adj.), *O.E.* geond, = *L.G.* gind,
giend, gond, *Goth.* jaind (ἐκεῖ); *yonder*; ӡond
O. & N. 119; WILL. 846; ӡond [yond] in the gardin CH. C. T. *A* 1099; ӡond hoves a rial kniӡt
AMAD. (R.) xlvi; þe holi man was ӡund of
Irlande BRD. 1; her and gund GEN. & EX.
3851; ӡent rid Maximion A. D. 250; (*ms.* yend
for ӡen, ӡeon) AYENB. 256; ӡonde (*for* ӡone)
liӡt LANGL. *B* xviii. 187; o ӡond half ORM.
10588; a ӡund half BRD. 33; ӡendis WICL.
I ESDR. iv. 16.

ӡeond-ward, adv., *to that place,* LAӡ. 30781.

ӡeond², prep., *O. E.* geond, giond, 'gind;
through, across: ӡeond þa eorðe HOM. I. 91;
ӡeond, ӡuond, ӡond, ӡend LAӡ. 28, 113, 494,
23200; ӡeond, ӡeont A. R. 72, 164; ӡend
HORN (R.) 1012; ӡent al þis world P. S. 246;
ӡond al þe world C. L. 1448; ӡont MARH. 23.

ӡeonde[n], adv., *O.E.* geondan, = *M.L.G.* gen-
ten; *yonder*: þat ӡeonde liggeþ LEG. 21; yeonde
O. & N. 119*; ӡeonde, ӡeonde, ӡonde LANGL.
C xxi. 263; ӡonde (*? for* ӡond) CH. C. T. *A*
1099*; *comp.* **bi-ӡeonden.**

ӡeonder, adv., = *M.L.G.* gender, ginder, *M.Du.*
ghender, ghinder, *Goth.* jaindre (ἐκεῖ); *yonder*;
ӡonder FER. 1557; MAN. (F.) 12299; C. M.
3148; þat faire child þat stondeþ ӡonder
[yonder] CH. C. T. *B* 1018; yonder TOR.
2070; jondur EGL. 290; ӡender A. P. ii. 1617;
ӡendir HALLIW. 952; go þu yunder HAV. 922.

ӡeong, *see* gang *and* ӡung.

ӡeonien, v., *O.E.* geonian, ginian, = *O.H.G.*
ginen; *yawn,* A. R. 242; **ӡeoneþ** (*pres.*)
R. S. v; (MISC. 182); ӡeneþ ST. COD. 101;
ӡeniþ (*? printed* yenith) ALIS. 485; yoneþ
MISC. 183; ӡyonie [vonie] (*subj.*) O. & N. 292;
ӡeoniinde (*pple.*) LEG. 185; ӡeonede (*pret.*)
MARH. 9; ӡenede MAP 338; LEG. 126;
yenede BEV. 787; ӡeneden WICL. PS. xxxiv. 21.

ӡe-openen, v., *O.E.* ӡeopenian, = *O.H.G.* gi-
ofanōn; *open*; ӡeopenede (*pret.*) HOM. I.
225; iopened (*pple.*) A. R. 242; P. S. 221;
iopenet MARH. 12.

ӡeor, ӡur, sb., *cry, wail,* A. R. 306; ӡur JUL.
51; *see* ӡeorre.

ӡeorde, *see* ӡerde.

ӡeoren, v., *? = O.L.G., O.H.G.* gerōn, *O.Fris.*
jeria; *yearn for, desire*; ӡeore (*? for* ӡeorne)
(*pres. subj.*) LANGL. *A* i. 33; ӡirde (*? for*
ӡirnde) (*pret.*) LAӡ. 11514; ӡirden [ӡeorden]
10297.

ӡeorn, adj., *O.E.* georn, gyrn, = *O.L.G.* gern, *O.
N.* giarn, *O.H.G.* gern, *Goth.* gairns; *willing,
desirous, eager*; ӡerne CH. C. T. *A* 3257;
ӡeorne (*adv.*) HOM. I. 11; MARH. 2; A. R.
108; ORM. 2718; LEG. 17; he þonkede hire
ӡeorne [ӡerne] LAӡ. 1261; ӡeorne, ӡorne O. & N.
538, 1352; who se . . . ӡerne [ӡeorne] biholdeþ
þis ilke writ C. L. 76; ӡerne MIRC 70; CH. C.
T. *D* 993; MAN. (F.) 7064; DEGR. 605;
gierne HOM. II. 9; drepen him he wolden
yerne HAV. 1865; ӡorne AV. ARTH. xxiii;
ӡurne BRD. 5; CHR. E. 52; ӡirne ED. 727;
ӡarne BARB. iii. 547; þe **gernere** *the more
willingly* HOM. II. 63; *comp.* **elmes-, lof-,
mēd-, sŏft-ӡeorn.**

ӡeorn-liche, adv., *O.E.* geornlīce; *eagerly,*
HOM. I. 259; ӡeorneliche A. R. 82; gierneliche
REL. I. 129; ӡeornelüker (*compar.*) A. R.
234; ӡernloker JOS. 593.

[**ӡeorne,** = *O.H.G.* gernī, *Goth.* gairnei; *desire*;
comp. **sŏft-ӡeorne.**]

ӡeorn-ful, adj., *O.E.* geornfull; *desirous,
diligent,* ORM. 1631; ӡornful L. H. R. 224; we
schulde . . . beon yeornfulle MISC. 144.

ӡeornfulnesse, sb., *diligence,* ORM. 2690;
ӡernvolnesse ANGL. I. 31.

ӡeornen, v., *O.E.* gyrnan, = *O.L.G.* gernean,
girnean, *O.H.G.* gernen, *O.N.* girna, *Goth.*
gairnjan; *yearn, desire,* ORM. 3579; ӡeornen,
ӡernen, ӡirnen LAӡ. 3434, 15409, 23128; irne
(*ms.* yrne) MISC. 144; ӡerne WILL. 58; yherne
PR. C. 4663; ӡarne BARB. ii. 507; ӡirne
(*pres.*) A. R. 400; AN. LIT. 3; ӡerne HORN
(L.) 915; ӡeorneð LAӡ. 8250; iwhilc man
þat ӡeorneþ godes are ORM. 1363; ӡernes
(*? printed* yernes) Iw. 1242; yhernes, yhornes
PS. li. 2; heo ӡirneð HOM. I. 207; ӡarnis
BARB. ix. 20; ӡerne (*subj.*) LANGL. *A* i. 33;
(*ms.* yerne) HAV. 299; ӡurne we **his** dohter
LAӡ. 934; ӡarnand (*pple.*) BARB. vii. 11;
ӡernede (*pret.*) KATH. 1591; ӡernde, ӡirnde
[ӡeornde] LAӡ. 4427, 4720; ӡirnde MARH.
2; yeornden SAX. CHR. 250; gerneden GEN.
& EX. 3657; ӡirnden SPEC. 43; ӡarnit BARB.
xx. 209; ӡe ӡarnit xii. 282; ӡerned (*pple.*)
HOM. I. 271; A. P. i. 1189; ӡerned, ӡirned
H. H. 162 (164); *comp.* **ӡe-ӡernen.**

ӡirnunge, sb., *O.E.* geornung; *yearning,* A.
R. 114; ӡirning SPEC. 72; ӡerning C. M. 7310;
yherning PR. C. 1127; ӡarning BARB. xviii. 220.

ӡeorre, sb., *O.E.* ӡerre; *outcry,* ALEX. (Sk.)
5042; *see* ӡeor.

ӡeorren, v., *O.E.* georran, girran, gyrran, =
M.H.G. girren; *roar, rattle*; ӡurren (*ms.*
ӡuren) MARH. 16; ӡellen & ӡur(r)en KATH.
2040; he(o) **yeorreþ** (*pres.*) & heo gredeþ
MISC. 83; ӡurren þa stanes mid þan blod-
stremes LAӡ. 28358; ӡerand (*pple.*) ALEX.
(Sk.) 4745; ӡürde (*pret.*) JUL. 50.

ӡüring, sb., *crying, shrieking*: þe un-
wiht ӡurde þat monie weren awundret hwet
te ӡurung [*sec. text* ӡur] mahte beon JUL. 50.

ȝēoten, v., *O.E.* gēotan, = *O.L.G.* geotan, giotan, *Goth.* giutan, *O.H.G.* giozan ; *pour out, shed*: blod ȝeoten (*shed*) LAȝ. 17318 ; ȝeoten . . . teres HOM. I. 39 ; he sholde melten bras & ȝeten (*found, cast*) him a neddre ORM. 17418; geten GEN. & EX. 3548 ; ȝetin PR. P. 538 ; ȝete [ȝeete] WICL. EXOD. xxvi. 19; ȝēoteð (*pres.*) A. R. 282 ; gēt (*imper.*) GEN. & EX. 2815 ; ȝēat (*pret.*) FRAG. 6 ; þat te blod ȝeat [ȝet] (*poured*) a dun of þe ȝerden JUL. 16, 17 ; gette, get GEN. & EX. 582, 585 ; ȝote MAP 338 ; ȝote JUL.[2] 55 ; yhotten PS. lxxviii. 3 ; ȝóten (*pple.*) (*ms.* ȝotenn) ORM. 1773 ; (*? printed* yoten) RICH. 371 ; AM. & AMIL. 2024; yhotin PS. xliv. 3 ; igoten JUL. 38 ; iȝotten H. M. 45 ; iȝote P. S. 250 ; TREV. I. 233; *comp.* a-, bi-ȝēoten; *deriv.* góte, güte.

ȝēoter, sb., *O.E.* gēotere ; *pourer, founder, caster;* (*printed* geoter) ALIS. 6735 ; ȝeetere WICL. JER. vi. 29 ; *comp.* belle-ȝēter.

ȝētinge, sb., *pouring, shedding,* ' *fusio,*' PR. P. 538 ; ȝeting WICL. JER. li. 17.

ȝeove, ȝeoven, *see* ȝife, ȝifen.

ȝeoxe, sb., *O.E.* geocsa, gicsa, gisca ; *hiccup* ; ȝoxe HALLIW. 954.

ȝeoxen, v., *O.E.* giscian, = *O.H.G.* geskōn, *M. L.G.* gischen ; *hiccup;* sob ; ȝixin, yexen ' *singultio*' PR. P. 539 ; ȝiske CATH. 426 ; ȝexeþ [ȝeskeþ, ȝoxeþ] (*pres.*) CH. C. T. *A* 4151 ; ȝoxide (*pret.*) WICL. 4 KINGS iv. 36 ; ȝoxede and siȝte ofte BEK. 1568.

ȝixinge, sb., *O.E.* giscung, gicsung ; *hiccup, sobbing,* ' *singultus,*' PR. P. 539 ; ȝoxing ROB. 34 ; WICL. LAM. iii. 56 ; ȝosking WR. DICT. 1038.

ȝēp, *see* ȝēap.   gepoun, *see* gipōn.

ȝēpse, sb., = *M.L.G.* gǣpse, gēspe ; ȝēspe ' *vola*' PR. P. 537 ; (? yepsen, *printed* thepsen) (*pl.*) VOC. 147 ; yespon PALL. xii. (p. 220).

[ȝe-pünden] i-pünden, v., *O.E.* gepyndan ; *impound*; ipünd (*pple.*) A. R. 128.

ger, *see* geare.   gēr, *see* gār.

ȝēr, *see* ȝēar.

[ȝe-rād] i-rēd, adj., *O.E.* gerād, -rǣd, = *Goth.* garaids, *M.H.G.* gereit; *ready, prepared*: hi beoþ alle irede MISC. 40.

irād-, irǣd-līche, adv., *readily,* LAȝ. 11532, 29631 ; iredliche AYENB. 1.

ȝe-rēchen, v., *O.E.* gerǣcan, -rǣcean, = *O.H.G.* gereichan ; *reach*: sige gerechen SAX. CHR. 201 ; irāut (*pple.*) M. T. 88.

[ȝe-rǣden] i-rǣden, v., *O.E.* gerǣdan, = *Goth.* garaidjan, *M.H.G.* gereiten; *ordain* ; irǣd (*pple.*) LAȝ. 18100 ; as biforen irad wes C. L. 654.

[ȝe-rǣden[2]] i-rǣden, v., *O.E.* gerǣdan, = *Goth.* garēdan, *O.H.G.* girātan ; *counsel*; ireden HOM. I. 15 ; irǣd [irad] (*pple.*) LAȝ. 11622 ; irad BRD. 20 ; FER. 4083.

ȝerǣdnesse, sb., *O.E.* gerǣdness ; *counsel*:

of one þinge sei me irednesse (*miswritten* iredynesse) MISC. 85.

ȝe-rǣdi, adj., *ready* ; iredi O. & N. 488 ; BRD. 30 ; AYENB. 173 ; FER. 354; ireadi JUL. 8; aredi LAȝ. 17947* ; SHOR. 81 ; AYENB. 121 ; aredili WILL. 5006 ; ȝerēdie (*pl.*) HOM. I. 239.

[ȝe-rǣsen] i-rǣsen, v., *O.E.* gerǣsan ; *? rush, make an onset* ; irǣsed (*pple.*) LAȝ. 27296.

gerard, sb., *? villain*: þat gredi gerard [Herod] als a gripe C. M. 11811 ; gerard 11905 ; gerard [*other texts* devil, fend] 22307.

ȝerd, *see* ȝeard.

ȝerde, sb., *O.E.* gerd, gierd, gird, gyrd, = *O.L.G.* gerda, *O.H.G.* gerta, *mod.Eng.* yard (measure) ; *rod, staff,* ' *virga,*' PR. P. 537 ; MARH. 11 ; ORM. 16181 ; ROB. 22 ; MAN. (H.) 292 ; WICL. EX. xiv. 6, DEUT. xxiii. 1 ; gerde GEN. & EX. 2851 ; a ȝerde [yerde] long CH. C. T. *A* 1050 ; ȝerde, ȝerd A. R. 184 ; ȝerd FRAG. 3 ; HOM. I. 243 ; O. & N. 777 ; SHOR. 132 ; yerd AYENB. 95 ; anne ȝerd [ȝeord] LAȝ. 14771 ; yeorde REL. I. 184 ; ȝurd BEK. 2368 ; ȝerde (*dat.*) MAND. 85 ; mid one ȝerde FRAG. 5 ; mid ter ȝerde A. R. 184 ; if men smot it with a ȝerde CH. C. T. *A* 149 ; ȝerden (*pl.*) A. R. 186 ; ȝerden [ȝurden] L. H. R. 26 ; ROB. (W.) 6985 ; ȝerdes LANGL. *A* v. 128 ; ȝerden (*dat. pl.*) LAȝ. 20318 ; JUL. 17 ; *comp.* eln-, kine-, līm-, seil-ȝerde(-ȝerd).

gerdel, *see* gürdel.   gerden, *see* gürden.

ȝerdien, v., *? beat with rods*: hwil me ȝerdede (*pret.*) hire MARH. 6.

gerdōn, *see* gverdōn.   gere, *see* geare.

gēre (? gēre), sb., *? = *geare ; *or ? Lat.* gȳrus (*turn*); *manner behaviour; change, vicissitude*: in his geare CH. C. T. *A* 1372 ; as don þes lovers in hir qveinte géres [geeres] (*pl.*) CH. C. T. *A* 1531 ; HOM. II. 35, 165 ; wilde geris PERC. 1351 ; *deriv.* géri, gérish.

gér-ful, adj., *changeable, unlucky*: gerful violence CH. TRO. iv. 256 ; right as hir dai (*Friday*) is gerful [*v. r.* grisful, gerful, gereful] CH. C. T. *A* 1538.

ȝēre, *see* ēare, ȝēar.

ȝerēdie, *see* ȝerǣdi.

ȝe-reard, sb., *O.E.* gereord ; *language,* HOM. I. 225 ; irord 97.

[ȝe-rēaven] i-rēaven, v., *O.E.* gerēafian ; *seize*; irēaved (*pple.*) AYENB. 143 ; irǣved LAȝ. 10508.

[ȝe-recchen] i-recchen, v., *O.E.* gereccan, -reccean, *O.H.G.* kirechan ; *expound, narrate* ; irǣht (*pres.*) FRAG. 2 ; ireht (*pple.*) HOM. I. 89 ; iraht LAȝ. 10842.

ȝerecchednisse, sb., *O.E.* gereccednyss ; *relation, narration*; (*ms.* gerechednysse) MAT. i. 1 ; irecednesse *interpretation* HOM. I. 97 ; irǣcchednesse ' *historia*' FRAG. 2.

ȝe-rēchen, v., *cf. O.L.G.* rōkian, *M.H.G.* geruochen ; *? reck, care*: wham so hit ireche

HORN (R.) 358 ; (reche H. 364, recche L. 352).

Ʒe-rēfe, sb., *O.E.* gerēfa ; *steward, sheriff* ['*villicus*'] LK. xvi. 3 ; ireve P. L. S. viii. 25* ; irēfen (*dat. pl.*) HOM. I. 115.

Ʒe-rēfen, v., *O.E.* gehrēfan ; *roof, cover* ; Ʒerēf (*imper.*) HOM. I. 225.

[Ʒe-rekenen] i-rekenen, v., *O.E.* gerecenian, = *O.H.G.* gerechenōn ; *reckon* ; irekened (*pple.*) H. M. 33 ; irikened A. R. 82.

Ʒeren, *see* Ʒearwien.

Ʒe-rennen, v., *run* ; Ʒerne HORN (L.) 705.

Ʒe-reordung, sb., *O.E.* gereordung ; *dinner.*
 ireordung-hūs, sb., *refectory,* '(*refecto*)-*rium,*' FRAG. 4.

[Ʒe-reste] i-reste, sb., *rest, repose,* HOM. I. 47.

[Ʒe-resten] i-resten, v., *O.E.* gerestan, = *O.H.G.* girestan ; *rest* : on heore iræste (*pple.*) steðen LAƷ. 26646.

gērfaucun, sb., *Lat.* gȳrofalcōnem ; *gerfalcon* (*a kind of hawk*), PR. P. 190 ; gerfaukon WICL. DEUT. xiv. 16.

gerfaunt, sb., ? there also ben mani bestes that ben clept orafles ; in Arabie thei ben clept gerfauntz (*pl.*) MAND. 289.

géri, adj., *cf.* gére ; *changeable, changeful,* CH. C. T. *A* 1536 ; DEP. R. iii. 130 ; LIDG. M. P. 24.

[Ʒe-rīden] i-rīden, v., *O.E.* gerīdan ; *ride* ; iriden (*pple.*) LAƷ. 24855 ; iride AR. & MERL. 3103.

[Ʒe-riht, adj., *O.E.* geriht, *cf. O.H.G.* gerehti, *Goth.* garaihts ; *right* ;] irihte (*adv.*) HOM. I. 15.
 iriht-lēchan, v., *O.E.* gerihtlǣcan ; *make right, correct,* HOM. I. 115.

Ʒe-rihte, sb., *O.E.* geriht *n.* ; *right,* LAƷ. 7906 ; furleosen his irihte MISC. 97 ; þas gerichtan (*pl.*) ANGL. VII. 220.

[Ʒe-rihten] i-rihten, v., *O.E.* gerihtan, = *O.H.G.* ka-, gerihtan, *Goth.* garaihtjan ; *make right, correct* ; iriht (*pple.*) LAƷ. 14889 ; þou hast iriƷt þat was a mis SHOR. 131 ; iriht up *erected* A. R. 364.

[Ʒe-rīmé, adj., *able to be counted* ; *comp.* un-Ʒerīm.]

[Ʒe-rīmen] i-rīmen, v., *O.E.* gerīman, = *O.H.G.* girīman (*number*) ; *count* ; irīmed (*pple.*) AYENB. 99.

[Ʒe-rīnen] i-rīnen, v., *O.E.* gehrīnan ; *touch* ; irineð (*pres.*) HOM. I. 77 ; alle heore vingeres irinen (*pple.*) mid goldringes LAƷ. 24600.

[Ʒe-rīp] i-rīp, sb., *O.E.* gerīp ; '*messis,*' FRAG. 2.

Ʒe-rīsen, v., *O.E.* gerīsan, = *O.L.G.* girīsan, *O.H.G.* kirīsan ; *become, befit* ; gerīseð (*pres.*) HOM. II. 141.

gérish, adj., ? *from* gére ; ? *impetuous* ; gerisshe LIDG. M. P. 243.

gerken, *see* Ʒearkien.

gerl, gurl, sb., *cf. L.G.* gör, *mod.Eng.* girl ; *boy or girl,* CH. C. T. *A* 3769 ; girl A. P. i. 205 ; girles (*pl.*) ALIS. 2802 ; P. S. 337 ; gerles WILL. 816 ; LUD. COV. 181 ; gerles, gurles LANGL. *B* i. 33, *C* ii. 29.

gerland, sb., *O.Fr.* gvirlande, garlande, gerlaunde ; *garland* : þe garland Roberd tok MAN. (H.) 331 ; a garland (*v. r.* garlond, gerland) had he set upon his heed CH. C. T. *A* 666 ; in garlande gai A. P. i. 1185 ; garland, garlant, garlond VOC. 196, 262, 268 ; garlonde PR. P. 187 ; a gerland of leves P. S. 218 ; gerlande LANGL. *B* xviii. 48 ; gerlondes (*pl.*) AYENB. 71.

germain, adj., *Fr.* germain, *from Lat.* germānus ; *german* : broþer germain AYENB. 146 ; sister germain WICL. 3 KINGS xi. 19.

germaunder, sb., (*herb*), '*germandra,*' PR. P. 190.

Ʒermen, v., *O.E.* gyrman ; *from* Ʒarm ; *cry out, bellow* : þe fend bigan to crie and Ʒarme HALLIW. 951 ; Ʒermis (*pres.*) D. ARTH. 3911 ; Ʒarmand (*pple.*) ALEX. (Sk.) 4745.

germine, v., *Lat.* germināre ; *bud,* PALL. xii. 48.

Ʒern, *see* Ʒeorn. Ʒerne, *see* Ʒerennen.

Ʒerne-mouþe, Ʒernemūðe, pr. n., *Yarmouth,* LAƷ. 30543 ; ROB. 227.

Ʒernen, *see* Ʒeornen.

gerner, sb., *O.Fr.* gerner, gernier, grenier, *med.Lat.* grānārium ; *garner,* HOM. I. 85 ; MAND. 52 ; þet gernier AYENB. 191 ; gernere LANGL. *B* vii. 129.

gernet, *see* garnet.

Ʒerneseie, pr. n., *Guernsey,* ROB. 243.

[Ʒe-róbed] i-róbed, pple., *robed* : (þ)us irobed in russet LANGL. *A* ix. 1.

Ʒerre, *see* Ʒeorre.

gers, *see* græs. Ʒersten-, *see* Ʒestren-.

gersume, *see* gærsume. gerþ, *see* gurd.

[Ʒe-rūmen] i-rūmen, v., *O.E.* gerȳman, = *O.H.G.* garūmman ; *clear, evacuate* ; irūm (*imper.*) LAƷ. 3698 ; Ʒerīmed (*pple.*) HOM. I. 231.

[Ʒe-rüsten] i-rüsten, v., *O.E.* gehyrstan, *cf. O.H.G.* kihrusten ; *adorn* : irūst (*pple.*) al mid golde LAƷ. 25812.

ges, sb., *O.Fr.* get, giet ; *jess* : ase me of halt þane voƷel be þe ges þet he ne vli to his wille AYENB. 254.

Ʒes, *see* Ʒü-se *under* Ʒēa.

[Ʒe-sǣle] i-sǣle, adj., *happy,* LAƷ. 7666 ; *comp.* un-isǣle.

[Ʒe-sǣli] i-sǣli, adj., *O.E.* gesǣlig ; *happy* ; iseli HOM. I. 117 ; LAƷ. 28861 ; A. R. 60.

ȝe-sēlhðe, sb., *felicity*, HOM. I. 109; iseluhðe A. R. 382.

ȝe-sǣlþe, sb., *O.E.* gesǣlð; *felicity*; ȝeselþe, -selþi P. L. S. viii. 8; HOM. I. 161; HOM. II. 220; ANGL. I. 8; (*ms.* hiselþe) REL. I. 181 (*printed* his elþe MISC. 124).

[ȝe-saht] i-sauȝt, pple., *reconciled*, HOM. II. 256; isahte R. S. v (MISC. 176).

[ȝe-sahtnien] i-sahtnien, v., *reconcile*; isehtnede (*pret.*) HOM. I. 83.

[ȝe-samme] i-samme, adv., *alike*, LANGL. *A* x. 193.

[ȝe-samnien] i-samnien, v., *O.E.* gesamnian, -somnian, = *O.H.G.* gisamanōn; *assemble*; isomnie LAȝ. 24248; isomned (*pple.*) HOM. I. 135.

isomnunge, sb., *O.E.* gesamnung, -somnung; *assembly*, HOM. I. 93.

ȝe-sāwen, sb., *O.E.* gesāwan; *sow*; ȝesāwen (*pple.*) HOM. I. 241; isawe SHOR. 27; isowen LANGL. *B* v. 550; isowe O. & N. 1614.

ȝe-scēad, sb., *O.E.* gescēad, = *O.H.G.* gascait, -sceit; *distinction, discretion*, HOM. I. 97; þet iscead FRAG. 1; ȝesceod HOM. I. 231.

iscēad-wīs, adj., *O.E.* gescēadwīs; *reasonable*: þa isceadwise mon HOM. I. 105.

iscēadwīsnesse, sb., *O.E.* gescēadwīsness; *discretion*, HOM. I. 105.

ȝe-sceafte, sb., *O.E.* gesceaft, = *O.H.G.* gascaft, *Goth.* gaskafts *f.*; *creature*, FRAG. 8; ȝescefte HOM. I. 219; iscefte 75.

ȝe-schēaden, v., *O.E.* gescēadan, = *O.H.G.* gesceidan, *Goth.* gaskaidan; *shed, separate*; ȝescēodeð (*pres.*) HOM. I. 237; isched, -schod (*pple.*) TREV. V. 369.

ȝe-scheapien, v., = *O.H.G.* gescafōn; *shape, form*; ischeaped (*pple.*) A. R. 200.

ȝe-schēawen, v., *O.E.* gesceawian, = *O.H.G.* gascauwōn; *show*; ischēawed (*pple.*) KATH. 1573; A. R. 154; ischeawed SHOR. 161.

[ȝe-schenden] i-scenden, v., *O.E.* gescendan, -scyndan, = *O.H.G.* gescentan; *disgrace*, LAȝ. 25508; iscend (*pple.*) FRAG. 7; ischend MARH. 15; A. R. 248; O. & N. 1336; MIRC 1113; eschend MISC. 221.

iscendnesse, sb., *O.E.* gescendnyss; *disgrace*, HOM. I. 93.

[ȝe-schēoten] i-schēoten, v., *O.E.* gescēotan, = *O.H.G.* gisciozan; *shoot*; iscoten (*pple.*) LAȝ. 6487; REL. I. 183 (MISC. 129); ischoten JUL. 72; ischote, -shote, -scote O. & N. 23, 1121; ischote CHR. E. 928; ishote ALIS. 5953.

ȝe-scheppen, v., *O.E.* gesceppan, -scyppan, = *Goth.* gaskapjan, *O.H.G.* kisceffan; *shape, make, form*; ȝesceōp, -scop (*pret.*) HOM. I. 219; iscop HICKES I. 223; HOM. I. 77; LAȝ. 21090; ischop MISC. 57; ischeápen (*pple.*) HOM. I. 203; JUL. 5; iscæpen [isape] LAȝ.

15857; ischapen KATH. 221; R. S. iv; REL. I. 174; ALIS. 6465; WILL. 2634; CH. C. T. *G* 1080.

[ȝe-schéren] i-schéren, v., *O.E.* gesceran, = *O.H.G.* gisceran; *shear*; ischorn, -shorn (*pple.*) CH. C. T. *A* 589; ischore LEG. 182; S. S. (Web.) 597.

[ȝe-schilden] i-schilden, v., *O.E.* gescildan, -sceldan; *shield*; ischilde, -shilde O. & N. 781; iscilt (*pres.*) HOM. I. 111; iscilde (*subj.*) HOM. I. 53; LAȝ. 23735; ischilde A. R. 84; god almihti þat ischilde S. S. (Web.) 1461; ishilde REL. I. 245; ischeld (*imper.*) SHOR. 86.

[ȝe-schīren] i-schīren, v., *pronounce*: ne dar heo noht a word ischire O. & N. 1532; loke þat þu na mare swulc þing ne iscire (*pres. subj.*) LAȝ. 17129.

[ȝe-schrīven] i-schrīven, v., *O.E.* gescrīfan; *shrive*; iscrifen (*pple.*) HOM. I. 23; ischriven A. R. 332; MISC. 152; ischrive BEK. 2188; ishrive LANGL. *B* v. 91.

ȝe-schrŭden, v., *O.E.* gescrȳdan; *clothe*; ischrŭd (*pple.*) KATH. 1187; A. R. 66; R. S. v; ischrid REL. II. 191; P. L. S. v. (b.) 11; ȝescridde (*pl.*) HOM. I. 225.

[ȝe-scōle] i-scōle, sb., *shoal, school, army*, HOM. I. 243; *comp.* heri-scōle.

ȝescung, *see* ȝiscung.

ȝe-sēchen, v., *O.E.* gesēcan, -sēcean, = *Goth.* gasōkjan, *O.H.G.* gisuochan; *seek*; isechen HOM. I. 31; LAȝ. 28827; ȝeseche HOM. II. 145; iseche O. & N. 741; isōhte (*pret.*) LAȝ. 24556; isoute ST. COD. 97; isōht (*pple.*) P. S. 216; isoȝt R. S. v; BEK. 68; isouht MISC. 46.

ȝesecðe, *see* ȝesihðe.

[ȝe-seȝd] i-sēd, sb., = *M.H.G.* gesegede; *?dictum*, P. L. S. xvii. 439.

[ȝe-seȝen] pple., *? seen*; *see* ȝesēon.]

iseȝen-līc, adj., *O.E.* gesewenlīc; *visible*: alle iseienliche ðing HICKES I. 166; *comp.* un-iseȝenlīc.

[ȝe-seggen] i-seggen, v., *O.E.* gesecgan, = *O.H.G.* gisagen; *say*; isegge (*pres. subj.*) HOM. I. 15; isegd (*pple.*) HOM. I. 233; iseid MARH. 19; MISC. 26; C. L. 1123; iseid, ised, isæd LAȝ. 4150, 5428, 31706; iseid, ised O. & N. 395, 1037.

[ȝe-sellen] i-sellen, v., *O.E.* gesellan, -syllan, = *O.L.G.* gisellien; *sell*; iseald (*pple.*) HOM. I. 13; isold FRAG. 7; FL. & BL. 192; P. L. S. xxi. 77; C. L. 344.

ȝesēlðe, *see* ȝesǣlðe.

[ȝe-sēme ?]

isēme-līche, adj., *seemly*, LAȝ. 21785.

[ȝe-sēmen] i-sēmen, v., *O.E.* gesēman; *seem, accommodate, conciliate, befit*: on name þe hire mihte isemen LAȝ. 9587; þa he hafde al iset and al hit isēmed (*pple.*) 21128; isemet MARH. 23.

gēsen, adj., *O.E.* gǽsne, gēsne (*barren*), *cf. O.H.G.* keisenī (*barrenness*); *scarce, rare; barren*: mi wafres þere were gesene LANGL. B xiii. 271; trew love is full geson S. & C. II. viii; when eggis and craime be gesoun B. B. 170; for faute of witt her liif is gesoun H. v. 64.

[ʒe-senden] i-senden, v., *O.E.* gesendan,= *O.H.G.* gisentan, *Goth.* gasandjan; *send*; isende (*pres. subj.*) MISC. 196; isend (*pple.*) FRAG. 7; PROCL. 11; BEK. 1328; isent CH. C. T. *E* 2220.

[ʒe-sēne] i-sēne, adj., *O.E.* gesēne, -sȳne; *visible*, LAʒ. 9548; O. & N. 166; MARG. 122; CH. C. T. *A* 592; asene CHR. E. 44; ALIS. 847.

ʒe-sēon, v., *O.E.* gesēon, = *O.L.G.* gisehan, *O.H.G.* gasehan, *Goth.* gasaihwan; *see*; iseon MARH. 7; A. R. 92; R. S. iv; iseo BEK. 125; C. L. 556; iseo, iso O. & N. 371; iseon, iseen ALIS. 144, 527; gesien (*ms.* gesyen) MAT. xxviii. 1; isien ANGL. I. 8; isuen SPEC. 101; ise TRIST. 1337; CH. C. T. *E* 2402; isee P. L. S. i. 14; isi SHOR. 107; isihst (*pres.*) A. R. 178; isihst, isixt (*ms.* isixst) LAʒ. 5195, 21959; isihst, -sist O. & N. 1225; isixt LEG. 193; isihð A. R. 6; isihþ C. L. 708; isihð, isið HOM. I. 125; isikth LEB. JES. 539; iziʒþ AYENB. 81; (heo) iseoð A. R. 196; iseoþ TREAT. 132; SHOR. 56; izig (*imper.*) AYENB. 161; iseiʒe (*subj.*) FRAG. 6; isæh, iseh, iseih (*pret.*) LAʒ. 553, 1408, 4671; iseh HOM. I. 43; iseih A. R. 166; iseʒ, -seih (*ms.* iseyh) O. & N. 29; iseʒ BEK. 93; iseiʒ C. L. 319; isegh S. S. (Web.) 966; þou iseʒe P. L. S. x. 55; iseʒe AYENB. 20; iseʒen [isehe] LAʒ. 6420; (*ms.* iseien) A. R. 190; isēhe, iseʒe (*subj.*) HOM. I. 31, 253; iseʒe O. & N. 425; iseie L. H. R. 28; isehen (*pple.*) KATH. 1746; iseʒen HOM. I. 97; LAʒ. 6626; SHOR. 160; iseien A. R. 92; iseiʒen [iseʒe, iseien, iseie] LANGL. *A* v. 4; iseʒe BRD. 1; iseie WILL. 1874.

[ʒe-seowien] i-seowien, v., *sew*; iseouwed (*pple.*) A. R. 200; isowed BEK. 1838; isewed PL. CR. 229.

ʒe-setnesse, sb., *O.E.* gesetnyss; *constitution, statute*; isetnesse HOM. I. 87; þas isetnesse (*pl.*) 115; isetnesses PROCL. 4.

ʒe-setten, v., *O.E.* gesettan,= *Goth.* gasatjan, *O.H.G.* kesezzan; *set, constitute*; ʒesette (*pret.*) HOM. I. 227; isette LAʒ. 22054; iset (*pple.*) KATH. 1997; A. R. 254; BEK. 1366; isette hure A. R. 428; ænne isetne dæi LAʒ. 15184.

[ʒe-sib] i-sib, adj. & sb., *O.E.* gesibb; *related, relation*, ROB. 315; þe weren his moder isib LAʒ. 411; isibbe (*pl.*) MISC. 144; heo weren isibbe LAʒ. 30533; menie of hem him were isibbe P. L. S. xvi. 85; *comp.* un-isibbe.

isibsum, adj., *peaceable*, HOM. I. 95.

[ʒe-sīʒen] i-sīʒen, v., *O.E.* gesīgan,= *O.H.G.* gesīgan; *glide*; isiʒen [iseʒe] (*pple.*) LAʒ. 12537; isihen hider KATH. 2084.

ʒe-sihðe, sb., *O.E.* gesihð; *sight*; isihðe LAʒ. 13990; HICKES I. 224; ʒesichðe (*ms.* ʒesychðe), ʒesecðe HOM. I. 229, 239.

gesīne, sb., *O.Fr.* gesine; *childbed*, C. M. 3906; LUD. COV. 150.

ʒe-sinhēowen, sb., pl., *O.E.* gesinhīwan; *husband and wife,* '*conjuges,*' FRAG. 2.

[ʒe-sitten] i-sitten, v., *O.E.* gesittan,=*Goth.* gasitjan, *O.H.G.* gasizzan; *sit*: he þa iset (*pret.*) (*possessed, occupied*) þet seld xxx ʒera HOM. I. 93; iseten (*pple.*) LAʒ. 18532; MISC. 154; whan we beoþ isete BRD. 13.

[ʒe-slǽpen] i-slǽpen, v., *O.E.* geslǽpan,= *Goth.* gaslēpan; *sleep*; islēpe (*pple.*) BRD. 6; P. S. 343; AYENB. 128; islape AR. & MER. 2377.

[ʒe-slēan] i-slēan, v., *O.E.* geslēan,=*O.H.G.* geslahan; *slay*; islaʒen, -slæʒen, -slawen (*pple.*) LAʒ. 322, 965, 7678; islaʒe O. & N. 1142; AYENB. 58; islawe BEK. 2189; ALIS. 725; JOS. 96; (*ms.* isleien) A. R. 156; islein KATH. 201; islain TRIST. 1081; CH. C. T. *B* 605.

gēsling, *see under* gōs.

[ʒe-smakien] i-smakien, v., *taste*; i-smacked (*pple.*) A. R. 316*; þet hedde wel ... ismacked þe ilke zvetnesse AYENB. 93.

[ʒe-smecchen] i-smecchen, v., *O.E.* gesmeccan, = *O.H.G.* kesmecchan; *taste*; i-smaht (*pple.*) HOM. I. 189; ismeiht A. R. 92.

[ʒe-smérien] i-smérien, v., *O.E.* gesmerian, -smirian, *cf. O.H.G.* gesmiran; *smear*; i-sméred (*pple.*) MISC. 28; AYENB. 93; ismiret H. M. 13*.

[ʒe-smīten] i-smīten, v.,=*Goth.* gasmeitan, *O.H.G.* gasmīzan; *smite*; ismāt (*pret.*) LAʒ. 25615; ismite (*pple.*) BEK. 2143; LEG. 28; REL. I. 267; CH. BOET. iii. 7 (80); ismete P. 73; TREV. III. 283.

[ʒe-sōm] i-sōm, adj., *O.E.* gesōm; *agreed, concordant*, R. S. v; isōme (*pl.*) O. & N. 1735; ROB. 40; to maken us isome S. J. 20; *comp.* un-isōm.

ʒe-sōðien, v., *O.E.* gesōðian; *verify*, LAʒ. 29011; isōðet (*pple.*) HOM. I. 261.

[ʒe-spēde] i-spēde, sb., *? for* spēde; *success*, LAʒ. 23417.

[ʒe-spēden] i-spēden, v., *O.E.* gespēdan, *cf. O.H.G.* gespuotōn; *speed*; ispēde (*pres.*) MISC. 40; ispědden (*pret.*) LAʒ. 3399; ispěd (*pple.*) BEK. 1485; SPEC. 110; CH. C. T. *A* 4220.

[ʒe-spéken] i-spéken, v., *O.E.* gesprecan,= *O.H.G.* gisprechan; *speak*; ispac (*pret.*) LAʒ. 19504; ispeken 3263; ispéken (*pple.*) HOM. I. 51; LAʒ. 3136; A. R. 108; ispeke O. & N. 1293; BRD. 15; C. L. 1062; AYENB. 44.

[ȝe-spillen] i-spillen, v., *O.E.* gespillan, = *O. H.G.* gispillan ; *spill, destroy, ruin* ; ispilled *(pple.)* HOM. I. 17 ; ispild MISC. 144 ; ASS. 18 ; AYENB. 75.

[ȝe-spinnen] i-spinnen, v., = *O.H.G.* kispinnan ; *spin* ; isponne *(pple.)* P. L. S. xvii. 156 ; SHOR. 95 ; ALIS. 7251.

ȝespon, *see* ȝepse.

[ȝe-sprǣden] i-sprǣden, v., *O.E.* gesprædan, = *O.H.G.* gispreitan ; *spread* ; isprĕd *(pple.)* A. R. 390 ; isprad BRD. 6 ; CH. BOET. iii. 6 (78) ; espred AUD. 78.

[ȝe-springen] i-springen, v., *O.E.*, = *O.H.G.* gispringan ; *spring* ; isprunge *(pple.)* O. & N. 300 ; HORN (L.) 548 ; ispronge BEK. 1285.

gesse, sb., = *M.Du.* ghisse ; *guess, supposition* : up gesse LANGL. *B* v. 421 ; bi gesse LIDG. M. P. 58.

gesseraunte, *see* jesseraunt.

gessin, v., = *M.L.G.* gissen, *M.Du.* ghissen, *Dan.* gisse, *Swed.* gissa ; *guess, suppose, have an opinion, expect,* '*aestimo, arbitror, opinor,*' PR. P. 190 ; gesse RICH. 4482 ; A. P. i. 498 ; he seide he wolde gesse to arive at Westnesse HORN (R.) 1187 ; gess BARB. xiv. 270 ; gesse *(pres.)* TRANS. XIII. 26 ; CH. C. T. *A* 82 ; gessist WICL. MK. iv. 41 ; gesside *(pret.)* WICL. I KINGS i. 13 ; gessiden WICL. MK. vi. 49.

gessare, sb., = *M.L.G.* gisser ; *guesser,* '*aestimator,*' PR. P. 190.

gessinge, sb., = *M.L.G.* gissinge, *M.Du.* ghissinge ; *supposition, expectation,* '*aestimatio,*' PR. P. 190 ; TREV. II. 59 ; CH. BOET. i. 4 (21).

gest, sb., *O.E.* gest, gæst, giest, gist, gyst, = *O.N.* gestr, *O.L.G.*, *O.H.G.* gast, *Goth.* gasts, *Lat.* hostis ; *guest,* '*hospes,*' PR. P. 191 ; SPEC. 30 ; AYENB. 249 ; gist A. R. 68 ; REL. I. 130 (HOM. II. 165) ; gesse B. B. 342 ; gheste TREV. IV. 81 ; geste *(dat.)* GEN. & EX. 1054 ; gestes *(pl.)* WILL. 4904 ; PR. C. 1374 ; gistas HOM. I. 109 ; gistes A. R. 414 ; TREV. III. 461 ; geste GEN. & EX. 1070 ; HORN (L.) 1217 ; gustes ROB. (W.) 5788, 8853*.

gest-halle, sb., *guest-hall,* LAI LE FR. 258.

gest-hūs, sb., *O.E.* gæst-, gesthūs, = *O.H.G.* gasthūs ; *guest-house, inn,* ORM. 7040.

gest-, (gist-)wise, adv., *in manner of a guest,* TREV. III. 161.

ȝest, sb., *O.E.* gist, = *M.L.G.* gest, *M.H.G.* jest, gest ; *yeast* ; ȝeest '*spuma*' PR. P. 537.

[ȝe-standen] i-standen, v., *O.E.* gestandan, = *Goth.* gastandan, *O.H.G.* gistantan ; *stand* ; istonden LAȝ. 15505 ; istonde *(pple.)* HOM. I. 47 ; istond(e) TRIST. 973.

ȝe-staðel-fæsten, v., *O.E.* gestaðelfæstan ; *make stable* ; ȝesteþelfaste *(pret.)* HOM. I. 221.

ȝe-staðelien, v., *O.E.* gestaðolian ; *make stable* ; istaþeleð *(pres.)* HOM. I. 115 ; ȝe-staþeled *(pple.)* HOM. I. 225 ; þis lond wes istaðeled LAȝ. 6777.

geste, sb., *O.Fr.* geste, *mod.Eng.* jest ; *achievement, story,* HORN (R.) SUPERSC. ; A. P. i. 277 ; gestes *(pl.)* LANGL. *C* xii. 23 ; jeestes *A* xi. 23 ; gestis HAMP. PS. xlvii. 12* ; of gret gestis *(? of large dimensions)* ane sow thai maid BARB. xvii. 597.

[ȝe-stefnen] i-stefnan, v., *O.E.* gestefnan ; *constitute* ; istefned *(pple.)* A. R. 310.

ȝesten, v., *? overflow* : þai ȝeȝd & ȝolped of ȝestande *(pple.)* sorȝe A. P. ii. 846.

gesten, v., *lodge, feast* : to gesten in hir leif licam C. M. 10079 ; his men wer wel gested *(pple.)* with brede MAN. (H.) 160.

gestinge, sb., *entertainment, lodging* : his niȝtes gestinge FL. & BL. 125 ; gesting C. M. 11443.

gesten[2], v., *from* gest, sb., *story ; tell a story* : gestin in romaunce '*gestio*' PR. P. 191 ; i kan nat geeste, rum ram ruf, bi lettre ne, god woot, rim holde i but litel bettre CH. C. T. *I* 43.

gesting, sb., *telling a tale* : gestinge or romauncinge '*gesticulatus, rythmicatus*' PR. P. 191.

[ȝe-stēoren] i-stēoren, v., *O.E.* gestēoran, -stȳran, = *O.H.G.* gistiuran ; *steer, govern,* HOM. I. 95.

[ȝe-stīȝen] i-stīȝen, v., *O.E.*, = *O.H.G.* gestīgan, *Goth.* gasteigan ; *ascend* ; istien *(pple.)* A. R. 400 ; istihe H. M. 47 ; *?* istoȝen HOM. I. 107.

ȝe-stillen, v., *O.E.* gestillan, = *O.H.G.* gastillan ; *still, allay* : he ȝestilde *(pret.)* windes HOM. I. 229 ; istild *(pple.)* SHOR. 133.

[ȝe-stinken] i-stinken, v., *O.E.* gestincan, = *O.H.G.* gestinchan ; *smell* ; istinckeð *(pres.)* A. R. 84.

gestnen, v., = *M.L.G.* gesten, *O.N.* gista ; *to entertain ; lodge, be entertained* : he wule gistnen mid ou A. R. 402 ; gestened *(pret.)* C. M. 2712 ; DEGR. 935 ; igistned *(pple.)* SHOR. 13.

gestning, sb., = *Swed.* gästning ; *entertainment, banquet,* GEN. & EX. 1507 ; ALIS. 1779 ; gestninge FL. & BL. 201 ; gistninge LAȝ. 14262 ; C. L. 1265 ; gistninges *(pl.)* A. R. 414.

gestonie, sb., = gestninge ; *entertainment* : a gestonie with alle maner of minstralsie TOR. (A.) 2374.

gestour, sb., *mod.Eng.* jester ; *a story-teller* ; gestoure '*gesticulator*' PR. P. 191 ; poetes and gestoures *(pl.)* TREV. IV. 101 ; gestiours CH. H. F. iii. 108.

[ȝe-strangien] i-strongien, v., *O.E.* gestrangian ; *comfort* ; istronged *(pple.)* LAȝ. 10301.

[ȝe-strecchen] i-strecchen, v., *O.E.* gestreccan, = *O.H.G.* gestrecchan ; *stretch* ; i-strǣhte *(pret.)* LAȝ. 26778 ; istreiht *(pple.)* FRAG. 5 ; istreiȝt TREAT. 140 ; and fond þe dede ligge istreiȝht LEG. 52 ; istraht CHR. E. 756 ; istrauȝt JOS. 269 ; TREV. III. 211.

[ʒestren-, *O.E.* geostran-, gystran-, gyrstan-,
= *Goth.* gistra-, *O.H.G.* gestren, kestre, *Lat.*
hesternus ; *yester-*.]

ʒersten-, ʒursten-dæi [ʒorstendai], sb. &
adv., *yesterday*, LAʒ. 17063, 21299 ; ʒurstendai
AN. LIT. 4 ; ʒisterdai TREV. III. 145 ; PERC.
509 ; gisterdai GEN. & EX. 2732 ; yhiʒtredai
PS. lxxxix. 4 ; ʒusterdai JOS. 330.

ʒistern-ēve, sb. & adv., *yestereven*, WILL.
2160 ; yistereven MIR. PL. 122 ; ʒusturevin AV.
ARTH. xxxvii.

ʒester-, ʒister-, ʒüster-niʒt, sb. & adv.,
yesternight, FER. 5765 ; C. M. 15988 ; ʒersten-
niʒt (*?printed* yerstenenight) BEV. 2912.

ʒe-strengen, v., *confirm* ; ʒe-strenð (*pres.*)
HOM. I. 239 ; istrenget (*pple.*) KATH. 2197.

[ʒe-strēon] i-strēon, sb., *O.E.* gestrēon, *cf. O.
L.G.* gestriuni, *O.H.G.* kastriuni ; *progeny,
procreation, gain,* HOM. I. 19 ; LAʒ. 22597 ;
REL. I. 175 ; MISC. 151 ; ·istrēones (*gen.*)
FRAG. 7.

ʒe-strēonen, v., *O.E.* gestrēonan, -strȳnan,
O.H.G. gastriunan ; *generate, procreate* ; i-
strēonieð (*pres.*) HOM. I. 133 ; ʒestrīende,
-strienede (*pret.*) HOM. I. 225, 227 ; gestre-
onede, -strenede MAT. i. 3 ; istrēoned (*pple.*)
A. R. 66 ; C. L. 1380.

[ʒe-súnd] i-súnd, adj., *O.E.* gesund, = *O.L.G.*
gisund, *O.H.G.* gesunt ; *sound, healthy* ; i-
sund FRAG. 3 ; isund [isond] LAʒ. 821 ; i-
sunde (*pl.*) O. & N. 1102 ; izounde AYENB.
205.

[ʒe-sünde] i-sünde, sb., = *M.L.G.* gesunde,
M.Du. gesonde, *O.H.G.* gisuntī ; *health :* þine
isunde LAʒ. 9112 ; mid isunde MISC. 140 ;
comp. un-isünde.

isünd-ful, adj., *O.E.* gesundfull ; *healthy,*
HOM. I. 115.

[ʒe-sundren] i-sundrin, v., *O.E.* gesun-
drian ; *sunder, separate* ; isundred (*pple.*)
A. R. 252.

[ʒe-süneʒen] i-süneʒen, v., *O.E.* gesyngian ;
sin ; isüneged (*pple.*) A. R. 306 ; A. D. 260 ;
isuneʒet MISC. 193 ; iseneged (*printed* i-
senoged) SHOR. 6 ; izeneʒed AYENB. 173 ;
isinewed P. L. S. xiv. 19.

[ʒe-sutelien] i-sutelien, v., *O.E.* geswute-
lian ; *manifest* ;· isuteled (*pple.*) A. R. 8.

ʒe-swebben, v., *O.E.* geswebban ; *sleep,
swoon* ; ʒeswefede (*pret.*) HOM. I. 221 ; i-
sweved (*pple.*) LAʒ. 3075.

ʒe-swelʒen, v., *O.E.* geswelgan, = *O.H.G.*
· geswelgan, -swelhan ; *devour* ; isuolʒe, -swol-
we (*pple.*) O. & N. 146.

ʒe-swenchen, v., *O.E.* geswencan ; *fatigue,
afflict* ; iswenchet (*pres.*) HOM. I. 13 ; ge-
swenched (*pple.*) LK. iv. 38.

[ʒe-swérien] i-swérien, v., *O.E.* geswerian,
= *O.H.G.* gesverran ; *swear* ; iswóren (*pple.*)
A. R. 96 ; R. S. v ; isworn CH. C. T. *A* 3301 ;

iswore P. 48 ; S. S. (Wr.) 2935 ; þa aðes
weoren iswórene LAʒ. 12171.

[ʒe-swīken] i-swīken, v., *O.E.* geswican,
= *O.H.G.* gisvīchan ; *cease* ; iswiken, iswican
HOM. I. 15, 17 ; iswīkeð (*pres.*) unrihtwisra
dedan HOM. I. 117 ; iswīke (*subj.*) O. & N. 929 ;
ʒesweac (*pret.*) HOM. I. 223 ; iswiken JUL. 43.

ʒe-swinch, i-swinch, sb., *O.E.* geswinc ;
labour, P. L. S. viii. 18 ; iswinc HICKES I.
223 ; iswinc, -swinch HOM. II. 181, 221 ;
iswink MISC. 59 ; iswinche (*dat.*) HOM. I.
129.

[ʒe-swingen] i-swingen, v., *O.E.* geswingan,
= *O.H.G.* gesvingan ; *swing, flog* ; iswungen
(*pple.*) WART. I. 24 ; iswongen SPEC. 84 ;
iswonge REL. II. 278.

[ʒe-swingle] i-swingle, sb., *flogging* ; i-
swica(n) þenne weorð þa iswingla HOM. I. 13.

ge-swüstre, sb., pl., *O.E.* gesweostor, = *M.H.
G.* geswester ; *sisters,* MAT. xix. 29 ; isustren
HOM. II. 219.

get, sb., *?gain* ; GAW. 1638 ; *comp.* an-ʒit,
for-ʒet.

get[2], sb., *O.Fr.* get ; *throw ; quality ;* 'machina,
modus,' PR. P. 191 ; a singere of þe beste get
MAN. (F.) 4024 ; jet P. S. 329 ; jette DEP.
R iii. 159 ; ges (*pl.*) AYENB. 254.

get[3], sb., *O.Fr.* jaiet ; *jet (mineral),* S. & C. I.
xxvi ; geet CH. C. T. *B* 4052 ; jete LIDG.
M. P. 201 ; geete (*dat.*) DEGR. 1461.

ʒet, *see* ʒit *and* ʒeat.

ʒet, ʒete, adv., *O.E.* get, giet, git, gyt, geta,
gita, gyta, *cf. O.Fris.* jeta, eta, ita ; *yet,* LAʒ.
109, 28636 ; þa ʒet leovede Corineus 2245 ;
ʒet sæiþ þeo soule FRAG. 6 ; et þan est ʒete
. . . þet ʒet me hat speciosa porta HOM. I. 5 ;
þe formeste dei þet eaver ʒiete was iseʒen
buven eorðe 139 ; ʒet, ʒete, ʒut A. R. 66, 92,
356 ; ah ʒet KATH. 70 ; þa ʒet 561 ; þe ʒet
MARH. 1 ; godes engel seʒde þær of sant
Johan ʒet mare ORM. 780 ; þat tu nart noht
ʒet in heofne 7951 ; þeʒ ne wæren ʒet noht
tahte 9802 ; ʒet ʒif we wole repente P. L. S.
xxv. 141 ; get GEN. & EX. 313 ; giet HOM. II.
21 ; ʒet þu me seist of oþer þinge O. & N.
309 ; artu ʒut [jet] (*ms.* yet) inume 541 ; abid
ʒete 747 ; ʒut þu aischest 995 ; þah mi lif me
beo atschote þe ʒet ich mai do gode note
1624 ; (*ms.* yete) HAV. 495 ; ʒet [ʒit] LANGL. *B*
PROL. 210 ; ʒet [yet], ʒit [yit] CH. C. T. *A* 70,
255 ; þo ʒit ne mihten heo . . . ileeven hit C. L.
1423 ; þat never ʒute iholde nas 266 ; ʒite
DEGR. 986 ; ʒit GOW. III. 242 ; WICL. JOHN
ii. 4 ; MAN. (H.) 7 ; EGL. 76 ; he is sekere
þan evere ʒit he was P. S. 333 ; ʒette ROB. (W.)
6639* ; (*ms.* yhit) PR. C. 105 ; ʒit, ʒut WILL.
186, 515 ; ʒut er þe worldes ende BRD. 3 ; þat
he þe ʒut of londe wende BEK. 680 ; ʒute 2144 ;
ʒute more HORN (L.) 70 ; ʒiet ANGL. I. 7 ; giet
(*ms.* gyet), giot, geot MAT. xv. 16 ; MK. viii.
17 ; LK. xxi. 9 ; ʒeot, ʒuit SAINTS (Ld.) 302.

[ʒe-tācnien] i-tācnien, v., *O.E.* getācnian,

cf. O.H.G. gizeichnan ; *give a token, signify* ; itācneden (*pret.*) HOM. I. 101 ; itācned (*pple.*) LAȜ. 32115.

itācnunge, sb., *O.E.* getācnung ; *significa-tion,* HOM. I. 97.

ȝetōcnisse, sb., *speech* : dumben he forgeaf getocnisse HOM. I. 229.

[ȝe-tǣchen] i-tǣchen, v., *O.E.* getǣcan, -tǣcean ; *teach ; show,* LAȜ. 10395 ; iteiht (*pple.*) A. R. 170; itaȝt FL. & BL. 404 ; AYENB. 54; itaiht LAȜ. 758; itauȝt [-taught] CH. C. T. *A* 127.

[ȝe-tǣse] i-tǣse, adj., *O.E.* getǣse; *con-venient, at hand*; itase LAȜ. 6502.

[ȝe-támien] i-támien, v., *O.E.* getemian,= *Goth.* gatamjan, *O.H.G.* gizeman ;`*tame* ; i-támed (*pple.*) SHOR. 133 ; itemed LAȜ. 8834 ; KATH. 1291.

[ȝe-tawen] i-tawen, v., *O.E.* getawian, = *Goth.* gataujan (ποίειν), *O.H.G.* gizawan ; *taw, tan* : velles wel itauwed (*pple.*) A. R. 418.

[gete (? gēte), adj., *O.E.* (ēaδ-)gete, ? *O.N.* (tor-)gætr ; *comp* ēþ-gete.]

ȝete, *see* ȝet. gēte, *see* gǣte.

getēe, sb., *Fr.* jetēe, *cf. mod.Eng.* jetty ; *projecting upper story,* PR. P. 192.

ȝe-tēȝen, v., *O.E.* getīgan ; *tie* ; getēȝed (*ms.* getegeδ) (*pple.*) LK. xix. 30 ; (*ms.* iteied) KATH. 1292 ; A. R. 14 ; iteied [itiȝed] LAȜ. 25972 ; (*ms.* iteyed), iteid O. & N. 778 ; CH. C. T. *A* 457 ; itiȝed C. L. 1130.

[ȝe-tǣiled] i-teiled, adj., *having a tail,* A. R. 206 ; iteilede draken HOM. I. 251.

ge-tel, sb., *O.E.* getæl,= *O.L.G.* gital ; *number,* MAT. xiv. 21 ; itel FRAG. 2 ; þat itel of þan scipen LAȜ. 7805 ; itele (*dat.*) FRAG. 2..

[ȝetel ? *comp.* bi-, for-ȝetel.]

[ȝe-tellen] i-tellen, v., *O.E.* getellan,= *O.L.G.* gitellean, *O.H.G.* gizellan; *tell ; count ; esteem,* HOM. I. 133; itelle LAȜ. 24627 ; itáld (*pple.*) LAȜ. 8392; KATH. 1293 ; AYENB. 24 ; itold A. R. 198; SPEC. 26; WILL. 1493.

geten, v., *O.Ē.* (bi-, for-, on-)gitan, gietan, geotan, gytan,= *O.N.* geta, *O.L.G.*. (bi-, for-) getan, *Goth.* (bi-)gitan, *O.H.G.* (ar-, bi-, fer-) gezan ; *get, obtain, beget,* HAV. 792 ; GOW. II. 242 ; to geten him gilte spores LANGL. *B* xviii. 14 (12098) ; ȝete TRANS. XVIII. 23; gete ROB. 152; JOS. 23; ISUM. 151; gete (*pres.*) GEN. & EX. 1497; þou getist EGL. 347 ; geteþ P. S. 334 ; get ORM. 10219; gat (*pret.*) WILL. 2895; gat ['*genuit*'] WICL. GEN. xi. 10; and gat upon 'hire Constantin MAND. 12; (þou) gete LANGL. *B* xviii. 332; (þei) geten xx. 156; geeten, goten WICL. GEN. vi. 4; JER. xvi. 3; goten ˈMAND. 67; gaten WILL. 1592; geten (*pple.*) JOS. 523; MAND. 19; CH. C. T. *A* 291; þe good þat þou hast geten LANGL. *B* v. 295; befor þat he was geten and forth broght PR. C. 443; geten, gete WILL. 799, 1030; ȝeten, igete

GREG. 132, 339; *comp.* a-, and-, bi-, for-, of-, under-ȝeten (-ȝeoten, -ȝiten).

ȝeting, sb., *getting ; begetting ; gains,* C. M. 22035; in rightwise getinge es he ai PS. xiii. 6; thou shalt ete of the getingus (*pl.*) of it WICL. ECCLUS. vi. 20.

geten, *see* eten. ȝeten, *see* ȝeeten.

gēten, *see* gǣten *and* gāten.

ȝēten, *see* ȝēoten.

[ȝe-tēon] i-tēon, v., *O.E.* getēon,=*Goth.* ga-tiuhan, *O.H.G.* geziohan ; *lead, draw,* LAȜ. 2418; itóȝen, -towen (*pple.*) O. & N. 1725; itohen JUL. 9; on itowen sweord A. R. 324; wel itohe (*educated*) H. M. 25 ; þa wimmen... alre bezst itoȝene LAȜ. 24646.

[ȝe-tēonen] i-tēonen, v., *vex*; itēned (*pple.*) P. S. 149; MIRC 1258.

[ȝe-tīd] i-tīd, sb.,= *O.L.G.* getīd, *O.H.G.* gizīt ; *time, hour* : on oþer two itide ANGL. I. 15..

ȝe-tīden, v., *O.E.* getīdan; *betide, happen* ; i-tiden H. M. 31; itide O. & N. 1733; itīt (*pres.*) P. L. S. viii. 63; A. R. 186; itĭdde (*pret.*) LAȜ. 27898; A. R. 152.

[ȝe-tilien] i-tilien, v., *O.E.* getilian; *till, cultivate* ; itiled (*pple.*) LAȜ. 10026; A. R. 78.

[ȝetilδe] i-tilδe, sb., *O.E.* getilδ; *tilth,* ANGL. I. 10.

[ȝe-timbren] i-timbren, v., *O.E.* getimbrian, = *O.H.G.* gizimbrōn; *construct, build* ; i-timbred (*pple.*) LAȜ. 26226; A. R. 124; i-timbret KATH. 1972.

[ȝe-tīmien] i-tīmien, v., *O.E.* getīmian ; *be-fall, happen ; ? use opportunities,* HOM. I. 53; itīmode (*pret.*) 93; itimede 123; i-tīmed (*pple.*) LAȜ. 27978.

ȝe-trēowe, -trēouwe, adj., *O.E.* getrēowe; *true, faithful,* LAȜ. 4451, 7395.

ge-trukien, v., *O.E.* getrucian ; *run short, fail* : þa þat win getrukede (*earlier text* geteorode) (*pret.*) JOHN ii. 3.

ȝetrümen, v., *O. E.* getrymian ; *confirm, strengthen* ; getremede (*pret.*) MK. xvi. 14.

getten, *see* ȝāten.

gettin, jettin, v., *O.Fr.* geter, jeter, *from Lat.* jactāre ; *swagger ; make a great show ; 'lascivo, gesticulo,'* PR. P. 192, 258 ; jett N. P. 46; jettand (*pple.*) perles ALEX. (Sk.) 4444.

gettere [jettere], sb., *boaster,* WICL. E. W. 243; getteris (*pl.*) 23.

[ȝe-tühten] i-tühten, v., *O.E.* getyhtan; *train, discipline* ; itüht (*pple.*) P. S. 220; ituht efter wittes wissunge HOM. I. 267; hwon he ... haveδ ituht hit wel A. R. 184; þei buþ to þe king itiȝt (*drawn*) ALIS. 7164.

[ȝe-turnen] i-turnen, v., *? for* turnen, LAȜ. 25574; iturnde (*pret.*) HOM. I. 97.

ȝe-tüδen, v., *O.E.* getīδian,=*M.H.G.* gezwī-den; *grant* ; itüþed (*pple.*) HOM. I. 157; itudet MARH. 9.

ȝe-twinnes, sb. pl., *O.E.* getwinnas ; *twins,* LAȜ. 12256.

ȝe-þanc, sb., *O.E.* geþanc, -þonc,=*O.H.G.* gidanc, -danch; *cogitation, thought*, HOM. I. 243; iþanc P. L. S. viii. 54; iþonc HOM. I. 101; geþankes (*pl.*) MAT. xii. 25.

[ȝe-þanked], i-þoncked, adj., *minded*, A. R. 210.

ge-þávian, v., *O.E.* geþafian; *permit*, ANGL. VII. 220; iðavien HOM. I. 199; iþeven LAȝ. 15279; iþévað (*pres.*) HOM. I. 113; ȝeðáfode (*pret.*) HOM. I. 229; iþeafede 121.

geþe, sb., ? *hurry* : before hur bedd lai a stone, the ladi toke it up anon and toke it in a gethe, on the mouth sche him hit FLOR. 1605.

[ȝe-þeawed] i-þeawed, adj., *mannered* : wel iþeawed SHOR. 3; iðæwed [iþeuwed] LAȝ. 6536; iþewed CH. BOET. iv. 6 (139).

[ȝe-þeinen] i-þeinen, v., *O.E.* geþegnian; *serve*; iþeinet (*pple.*) MARH. 23.

[ȝe-þenchen] i-þenche, v., *O.E.* geþencan, -þencean,=*O.L.G.* githenkēan, *O.H.G.* giden-chan; *think on, consider, remember*, HICKES I. 223; MISC. 36; iþench (*imper.*) LAȝ. 8257; iþench þu þes HOM. I. 197; iþenche (*subj.*) O. & N. 723; iþóht, -þouht (*pple.*) O. & N. 1560; iþouht A. R. 164; iþoȝt BEK. 110; iþouȝt SHOR 115.

ȝe-þeon, v., *O.E.* geþēon,=*O.L.G.* gethīhan, *Goth.* gaþeihan, *O.H.G.* gedīhan; *thrive, flourish, grow*, HOM. I. 221; iþeon LAȝ. 9116; iþeo MISC. 57; BRD. 7; iþi SHOR. 102; ȝeðihð (*pres.*) HOM. I. 223; iðeȝ (*pret.*) BEK. 151; iþæh, -þai [iþeh] LAȝ. 299, 9622; iþeu (*printed* yþen) ROB. 346; iðoȝen LAȝ. 30074; iþóȝen (*pple.*) HOM. I. 107; iðoȝe [iþoȝe] LAȝ. 19903.

[ȝe-þingen] i-þingen, v., *O.E.* geþingan; *thrive, grow*; iþungen (*pple.*) HOM. I. 107.

[ȝe-þóht] i-þóht, sb., *O.E.* geþōht,=*O.H.G.* gedāht; *thought*; iðōhte (*dat.*) HOM. I. 99; iþōhtas (*pl.*) HOM. I. 109.

[ȝe-þólien] i-þólien, v., *O.E.* geþolian,=*O.L.G.* getholōn, *cf. Goth.* gaþulan; *suffer*, LAȝ. 482; A. R. 122; iðolien HOM. I. 43; iþóled (*pple.*) MISC. 34; ROB. 24; AYENB. 182.

[ȝe-þræsten] i-þræsten, v., *O.E.* geþræstan; *twist, afflict, torment*; iþraste LAȝ. 28581; iþrăst (*pple.*) M. T. 110.

ȝe-þrean, v., *O.E.* geþrēan,=*O.L.G.* githrōon, *O.H.G.* gedrōwan; *afflict*; geþrēd (*pple.*) MAT. viii. 6.

[ȝe-þrēaten], v., *O.E.* geþrēatian; *compel*; iðrăt (*pple.*) A. R. 304*.

[ȝe-þringen] i-þringen, v., *O.E.* geþringan, = *O.L.G.* gethringan, *O.H.G.* gedringan; *press*; iþrunge (*pple.*) O. & N. 38; R. S. v; iþrongen CH. BOET. ii. 7 (57).

[ȝe-þüden] i-þüden, v., *push* : þe iþüd (*pple.*) beoð to hellen LAȝ. 9159.

[ȝe-þüld] i-þüld, sb., *O.E.* geþyld, *cf. O.L.G.* githuld, *O.H.G.* gedult; *patience*, HOM. I. 105

[ȝe-þüldiȝ] i-þüldi, adj., *O.E.* geþyldig,= *O.H.G.* gedultig; *patient*, HOM. I. 105.

[ȝe-þünchen] i-þünchen, v., *O.E.* geþyncan, =*O.H.G.* gedunchan; *seem, appear*; iþüht (*pple.*) A. R. 412; elche men wes iþuht HOM. I. 93.

[ȝe-þwære] i-þwære, adj., *O.E.* geþwære; *united, accordant, 'concors,'* FRAG. 2.

[ȝe-unnen] i-unnen, v., *O.E.* geunnan,=*O.H.G.* giunnan, gunnan; *favour, grant*, LAȝ. 21083; iuðe (*pret.*) LAȝ. 16549; ȝif hit me Crist ȝiuðe (*ms.* iȝuðe) MISC. 193; iunned (*pple.*) A. R. 30.

ȝeve, *see* ȝife.

ȝeven, *see* ȝifen *and* efen. ȝever, *see* ȝifer.

[ȝe-wær] i-war, adj., *O.E.* gewær,=*O.L.G.* gewar, *O.H.G.* giwar; *wary, cautious*, LAȝ. 7261, 7581; iwar A. R. 104; heo wel wiste and was iwar þat heo song hire a bisemar O. & N. 147; iwar LANGL. *B* i. 42; þe king was þer of iwar ROB. 168; iwer AYENB. 100; awar (*ms.* a ware) P. L. S. iv. 9; *comp.* un-iwar.

iwarnesse, sb., *wariness*, O. & N. 1228.

[ȝe-wákien] i-wákien, v., = *M.H.G.* ge-wachen; *wake*, LAȝ. 28082; iwáked (*pple.*) KATH. 1760.

[ȝe-wáld] i-wáld, sb., *O.E.* geweald, -wald, = *O.L.G.* giwald, *O.H.G.* gewalt; *power, sway*, LAȝ. 5064; HOM. I. 103; MARH. 5; iwold FRAG. 6; iweld O. & N. 1543; *comp.* un-iweald.

ȝe-wálden i-wálden, v., *O.E.* gewealdan, = *O.L.G.* giwaldan, *Goth.* gawaldan, *O.H.G.* giwaltan; *rule, govern*, LAȝ. 17213; iwolde ST. COD. 95; alre sceafte ȝewalt (*pres.*) HOM. I. 231; þe deað him wes iwealde (*pple.*) (*decreed*) HOM. II. 45.

[ȝe-walken] i-walken, v., *O.E.* gewealcan; *walk, travel*; iwalken (*pple.*) LAȝ. 112; muchel ic habbe iwalken bi water ant bi londe HICKES I. 228.

[ȝe-wān] i-wān, sb., *resource, device; fortune; property*, LAȝ.7706; his freondes striveð to gripen his iwon R. S. v; þeȝ ich have nou liþer iwon BEK. 1022.

[ȝe-wánien] i-wánien, v., *O.E.* gewanian, -wonian; *diminish, wane*; iwáned (*pple.*) HOM. II. 33.

[ȝe-warnien] i-warnien, v., *O.E.* gewearnian, = *O.H.G.* kiwarnōn; *warn*; iwarned (*pple.*) A. R. 318.

ȝe-wassen, v., *wash*; ȝewasse (*pple.*) HOM. I. 239; mi bed bie iwasshen mid mine teares HOM. II. 65; i-waschen A. R. 288; iwesscen, iweschen HOM. I. 37, 159; iwesse be zoþe sscrifte AYENB. 112.

ȝe-waxen, v., *O.E.* geweaxan,=*O.H.G.* ka-wahsan; *wax, grow*, HOM. I. 13; iwaxen (*pple.*) A. R. 380; iwaxen LAȝ. 12902; iwoxe P. L. S. xvii. 143.

U

[ʒe-wēde] i-wēde, sb., *O.E.* gewǣde,=*O.L. G.* giwādi, *O.H.G.* kawāti; *garment, clothes,* LAȝ. 26754; MISC. 193; alle his iwede LAȝ. 9333; on iwedan HOM. I. 109.

ʒe-wélde, adj., *O.E.* (un-)gewylde; *? potent; comp.* un-iwélde].

ʒe-wélden, v., *O.E.* geweldan, -wyldan; *rule, restrain, tame,* ['*domare*'] MK. v. 4; iwelt (*pres.*) HOM. I. 111; ʒewilde (*pret.*) HOM. I. 229; þis lond he iwalde LAȝ. 9029; iwáld (*pple.*) KATH. 190*.

[ʒe-weleʒen] i-weleʒen, v., *O.E.* gewelegian, -welgian; *? make rich, luxurious;* hie beð iwilegede (*? printed* iwilegeð) (*pple.*) HOM. II. 209.

ʒe-wemmen, v., *O.E.* gewemman, = *O.H.G.* giwemman; *defile, pollute;* iwemmed (*pple.*) KATH. 1427; iwemmed ROB. 339; CL. M. 100; TREV. V. 213.

iwemmednesse, sb., *O.E.* gewemmednyss; *defilement, corruption,* FRAG. 1.

ʒe-wende, sb., = *M.H.G.* gewende; *? contrivances* : men habbeþ a mong oþre iwende a rumhus at heore bures ende O. & N. 651.

ʒe-wenden, v., *O.E.* gewendan, = *Goth.* gawandjan, *O.H.G.* giwentan; *turn, return, go,* MARH. 2; iwenden LAȝ. 527; iwende FL. & BL. 61; ʒe-, ʒiwende (*pret.*) HOM. I. 225, 229; iwende A. R. 260; iwende LAȝ. 325; MISC. 41; LEB. JES. 125; iwend (*pple*). FRAG. 8; O. & N. 1519; iwend-, -went ·BEK. 974, 1198; iwent KATH. 1300.

[ʒe-wénen] i-wénen, v., *O.E.* gewenian, = *O.H.G.* giwennan; *wean* : þeo þet hefde iwist and iwénet (*pple.*) hire MARH. 2.

[ʒe-wēnen] i-wēnen, v., *O.E.* gewēnan, = *Gòth.* gawēnjan, *O.H.G.* gewānan; *ween, hope, think;* iwēnes (*pres.*) MAP 336; iwēnde (*pret.*) LAȝ. 20237.

[ʒe-wēpen] i-wēpen, v., *O.E.* gewēpan; *weep;* iwōpen (*pple.*) L. N. F. 215; iwope TREV. IV. 1.

ʒewer, *see* ʒehwǣr.

[ʒe-werc, v., *O.E.* geweorc,=*O.H.G.* giwerch, *Goth.* gawaurki; *work; comp.* hand-ʒewerc.]

ʒe-werʒen, v., *O.E.* gewergian,=*Goth.* gawargjan; *curse, condemn;* ʒewerged (*pple.*) HOM. I. 243.

[ʒe-werpen] i-werpen, v., *O.E.* geweorpan, = *Goth.* gawairpan, *O.H.G.* gewerfan; *throw;* iworpen (*pple.*) A. R. 368; iworpe O. & N. 1121; iworpe JOS. 221; iwarpen JUL. 49.

ʒe-wēsten, v., *to make empty, vacant;* iwēsten (*pret.*) LAȝ. 4116.

[ʒe-wéven] i-wéven, v., *O.E.* gewefan,= *O.H.G.* geweban; *weave;* iwéve (*pple.*) P. L. S. xvii. 156; TREV. V. 369; iwoven '*contextus*' REL. I. 7; CH. BOET. i. 1 (6).

[ʒe-wiht] i-wiht, sb., *O.E.* gewiht,=*M.H.G.* gewiht; *weight;* iwichte (*dat.*) HOM. I. 173.

[ʒe-wiht[2]] i-wiht, adj., *valiant, brave;* iwihte (*dat. pl.*) LAȝ. 12175.

[ʒe-wil] i-will, sb., *O.E.* gewill; *will,* HOM. I. 61; P. L. S. viii. 7; iwil LAȝ. 5935; MISC. 34; AYENB. 94.

[ʒe-wil[2]] i-wil, adj., *? agreeable* : þat him wes ful iwil LAȝ. 29515.

ge-wille, sb., *will,* MAT. vi. 10; iwille P. L. S. viii. 56; HOM. I. 101; iwille LAȝ. 6229; iwillan (*acc.*) HOM. I. 93; iwillen LAȝ. 1996.

[ʒe-willed] i-willed, pple., = *M.H.G.* gewillet; *willed,* LANGL. C ii. 189*.

[ʒe-wilnien] i-wilnien, v., *O.E.* gewilnian; *desire;* iwilned (*pple.*) A. R. 60; iwilned LEG. 8.

iwilnunge, sb., *O.E.* gewilnung; *desire,* HOM. I. 95.

[ʒe-win] i-win, sb., *O.E.* gewinn,=*O.L.G.* giwin, *O.H.G.* gawin; *contest, strife; ? gain,* LAȝ. 9044; (*ms.* iwyn) MISC. 144.

[ʒe-winden] i-winden, v., *O.E.* gewindan, = *O.H.G.* gawintan; *wind, wrap up;* iwonden (*pple.*) WART..II. 2; iwounden LANGL. B v. 525; iwounde LEB. JES. 696.

[ʒe-winne, sb., *O.E.* gewinna; *adversary, rival; comp.* wiðer-iwinne.]

[ʒe-winnen] i-winnen, v., *O.E.* gewinnan, =*O.L.G.* giwinnan, *O.H.G.* gawinnan; *win,* LAȝ. 2194; iwinne O. & N. 766; iwinne ROB. 519; SPEC. 113; iwon (*pret.*) LAȝ. 2560; (þou) iwonne FER. 478; iwunnen (*pple.*) LAȝ. 7233; iwonne SHOR. 94; LANGL. B v. 93; CH. C. T. A 2659; ewonne AN. LIT. 87.

[ʒe-wis] i-wis, adj., *O.E.* gewis, -wiss,=*O.H. G.* gewiss; *certain,* MARH. 6; A. R. 270; O. & N. 35; GEN. & EX. 159; (*ms.* iwiss) ORM. 687; iwis LAȝ. 29481; TREAT. 132; PL. CR. 555; CH. C. T. A 3277; TRIAM. 216; TOR. 1213; M. H. 17; MIR. PL. 130; L. C. C. 5; SACR. 483; ʒit was þe sonne þo seve siþe iwis brihtore þen heo nou is C. L. 101; iwisse [?=*O.H.G.* gewisso], adv., *certainly,* GEN. & EX. 91; iwisse LAȝ. 19315*; WILL. 697; M. H. 41; A. P. ii. 84; TOR. 392; M. ARTH. 3633; mid iwisse P. L. S. viii. 71; mid iwisse LAȝ. 7607; SPEC. 61; SHOR. 23; to iwisse LAȝ. 3545.

ʒewis-līche, adv., *O.E.* gewislíce; *certainly,* HOM. I. 111, 115; iwisliche LAȝ. 26184; iwislūcur (*compar.*) FRAG. 1.

[ʒe-wīsien] i-wīsien, v., *O.E.* gewīsian, -wissian,=*O.H.G.* kawīsan; *direct, teach, show;* iwisse (*pres. subj.*) REL. I. 102; iwīsed (*pple.*) LAȝ. 1525; SHOR. 122.

ʒewissunge, sb., *cf. O.E.* gewīsung; *instruction, direction,* HOM. I 93.

[ʒe-wist] i-wist, sb.,=*O.H.G.* giwist; *existence, living;* iwust A. R. 156*.

[ʒe-wit] i-wit, sb., *O.E.* gewitt,=*O.L.G.* giwit, *O.H.G.* gewizzi; *wit, senses, intellect,* FRAG.

5 ; HOM. I. 105; O. & N. 774 ; iwit LAʒ.
17138 ; (he) for neh ut of his iwitte MARH. 6.

[ʒe-wita] i-wita, sb., O.E. gewita; witness,
'testis,' FRAG. 3 ; gewiten (pl.) LK. xxiv. 48 ;
gewitene (gen. pl.) SAX. CHR. 253.

[ʒe-wīteʒen] i-wīteʒen, O.E. gewītgian,=
O.H.G. gewīzegōn ; prophesy ; iwītegede
(pret.) HOM. I. 5.

[ʒe-witen] i-witen, v., O.E. gewitan ; under-
stand, know, LAʒ. 15800; C. L. 67 ; ALIS. 763 ;
iwiten, -wite LANGL. B viii. 124 ; iwite FL. &
BL. 206 ; ROB. 487 ; SHOR. 128 ; AYENB. 29 ;
iwiteð (imper.) A. R. 64 ; iwiste (pret.) HOM.
II. 220 ; iwist (pple.) MARH. 2 ; iwist P. L.
S. x. 38 ; CH. BOET. v. 3 (156) ; iwist, -wust
C. L. 261, 1473 ; iwust A. R. 48.

[ʒe-wīten] i-wīten, v., O.E. gewītan ; depart ;
see, see to, watch over, LAʒ. 17235 ; hit cumeþ
weopinde & woniende iwīteþ (pres.) FRAG.
5 ; þa feol he a dun & iwāt (pret.) HOM. I.
93 ; þe mete forð iwąt LAʒ. 658 ; summe to
þere sæ iwiten 18116 ; iwiten (pple.) FRAG. 6 ;
wes i þere ilke wike þe ærchebiscop forð iwiten
LAʒ. 13244; þe haveð iwiten al þis ærd
13572 ; he sculde beon iwite (punished) 2084 ;
iwite LAI LE FR. 321.

[ʒe-witnesse] i-witnesse, sb., O.E. gewit-
ness ; witness, ' testimonium,' FRAG. 3 ; HOM.
I. 91 ; KATH. 2491*.

[ʒe-wīven] i-wīven, v., O.E. gewīfian ; take
a wife, marry : so him is a live þat uvele
iwiveþ (r. iwive) REL. I. 178 (MISC. 118).

[ʒe-wrāðien] i-wrāðien, v., O.E. gewrāðian ;
make wroth ; iwrāðede (pret.) LAʒ. 27698 ;
i-wreþed (pple.) HOM. I. 149 ; iwreþed
AYENB. 161.

[ʒe-wreʒen] i-wreʒen, v., O.E. gewrēgan ;
accuse ; iwrēied (pple.) A. R. 172 ; iwreied
TREV. V. 117.

[ʒe-wréken] i-wréken, v., O.E. gewrecan,
= Goth. gawrikan, O.H.G. gerechan ; avenge ;
iwréken (pple.) LAʒ. 3661 ; AN. LIT. 7 ; GEN.
& EX. 1856.

[ʒe-wrench,=? M.H.G. geranc ; comp. un-
iwrench.]

[ʒe-writ] i-writ, sb., O.E. gewrit ; writing,
letter, treatise ; iwrite (dat.) P. L. S. viii. 51 ;
iwriten (pl.) REL. I. 173 (MISC. 108).

ʒe-wrīten, v., O.E. gewrītan ; write ; ʒe-,
iwriten (pple.) HOM. I. 11 ; iwriten MARH.
2 ; A. R. 410 ; R. S. v ; iwriten SHOR. 91 ;
JOS. 317 ; iwrite C. L. 24.

[ʒe-wrīðen] i-wrīðen, v., O.E. gewrīðan,=
O.H.G. kirīdan ; bind together, restrain ;
iwrāð (pret.) HOM. I. 123 ; iwriðen (ms.
ywriðen) (pple.) LAʒ. 25974 ; iwriþen R. R.
160 ; iwriþe REL. II. 18.

ʒe-wúnden, v., O.E. gewundian,=O.H.G. gi-
wuntōn ; wound ; gewúnden (pret.) MK. xii.
4 ; iwúnded (pple.) A. R. 240 ; iwonded
AYENB. 148 ; iwounded TRIST. 85.

[ʒe-wune, adj., O.E. gewuna, = O.L.G. ge-
w(u)no, -wono, O.H.G. gewon ; accustomed.
habituated.]

iwune-līche, adv., O.E. gewunelīce ; usually,
HOM. I. 131.

ge-wune, sb., O.E. gewuna,=O.L.G. giwono,
O.H.G. giwona ; custom, habit ['consvetudo'],
JOHN xviii. 39 ; iwune HOM. I. 55 ; iwune LAʒ.
18805 ; iwune, -wune, -wone O. & N. 475, 1320.

[ʒe-wunien] i-wunien, v., O.E. gewunian,=
O.L.G. gi-wonōn, M.H.G. giwonen ; inhabit ;
habituate, accustom ; iwuned (pple.) A.R.230;
iwuned MISC. 44; don al se heo ær weoren
iwuned [iwoned] LAʒ. 12611 ; þa Bruttes hafden
iwuned here wel feole wintre 31655 ; iwoned
R. S. v (MISC. 172) ; iwoned BEK. 1409 ;
AYENB. 106 ; a kniʒt . . . þat was iwoned to
pleie wiþ him TREV. VIII. 187 ; ase hui iwo-
nede were LEG. 35.

ʒe-wurchen, v., O.E. gewyrcan, -wyrcean,=
O.L.G. giwirkean, Goth. gawaurkjan, O.H.G.
giwurchan, -wirchan ; work, make, perform,
iwurchen LAʒ. 28995 ; iw(u)rche, iwerche
(pres. subj.) REL. I. 173 ; ʒeworhte (pret.)
HOM. I. 219 ; iworhte, -wrohte LAʒ. 2066, 4208 ;
iworht [iwroht] (pple.) LAʒ. 21139 ; iwraht
MARH. 9 ; iwroht CH. E. 755 ; SPEC. 36 ;
iwrouʒt PL.CR. 162 ; iwroght, [-wrouht, iwrouʒt]
CH. C. T. E 1324.

ʒe-wurht, sb., O.E. gewyrht,=O.L.G. ge-
w(u)rht ; work, deed, merit : after heore i-
wurhte LAʒ. 24189 ; buton gewerhtan JOHN
xv. 25.

ʒe-wurðen, v., O.E. geweorðan,-wyrðan,-wur-
ðan,=O.L.G. giwerthan, O.H.G. giwerdan ; be
become, be made, happen ; iwurðen MARH. 7 ;
al þe wo þet nu is . . . & ever schal iwurðen A. R.
52; leteð hine iwurðen (let him alone) 96 ;
freond iwurðen [iworþe] LAʒ. 5441 ; to reouþe
iwurðen 24944 ; ær heom mihte iwurðen
what heo don wolde 25333 ; ʒif þu þis
nult iwurðen (allow) 8910 ; ʒif ʒe wolden
iwurðen & don mine iwille 19318 ; iwurþen
REL. I. 185 ; iwurþe (r. w. eorþe) 178 ; i-
worþen PL. CR. 665 ; iworþe BEK. 1252 ; P.
L. S. v. A 4 ; iworþe [aworthe] LANGL. B vi.
228 ; (heo) iwurðeð (pres.) HOM. I. 135 ;
iwurþe (let be done) godes wille MISC. 134;
iworþe 41 ; iworþe þi wil AYENB. 262 ; ʒewarð
(pret.) HOM. I. 227 ; iwearð A. R. 236 ; þa
iwearð þer muchel eie on godes folke HOM.
I. 93 ; iwearþ FRAG. 8 ; iwearð, iwarð, iwerð
LAʒ. 259, 290, 6769 ; iwurðen KATH. 2487 ;
alle þing iworþen FRAG. 8 ; iworðen [iworþe]
(pple.) LAʒ. 3733 ; iworþe (r. w. worde) O. &
N. 660.

[ʒe-wurðien] i-wurðien, v., O.E. gewurðian,
-wyrðian, -weorðian, cf. O.H.G. gawerdōn ;
honour, LAʒ. 29687 ; iwurðed (pple.) LAʒ.
13422.

[ʒe-wurðiʒen] i-wurðiʒen, v.,=M.L.G. ge-

werdigen; *honour*; **iwurdget**, iwurget (*pple.*)
JUL. 62, 63; **iwurðegede** (*pl.*) HOM. I. 137.

ȝexen, *see* ȝeoxen.

ȝh-, *see* ȝ-.

Giane, pr. n., *Guienne*, ALEX. (Sk.) 5667.

giaund, *see* geant.

gibbe, sb., *O.Fr.* gibbe, *Lat.* gibba; *hump*:
knobbe in a beestis back or breste, þat is
clepid a gibbe '*gibber, gibbus*' PR. P. 28c.

gibelet, sb., *O.Fr.* gibelet; *giblet*, '*idem quod
garbage*,' PR. P. 193; gibelet of fowlis '*pro-
fectum*,' '*exta*,' VOC. 179; **giblettes**, giblotes
(*pl.*) *additions* ['*appendicia*'] TREV. VII.
403.

gibet, sb., *O.Fr.* gibet; *gibbet*, A. R. 116;
ROB. 519; AYENB. 128; gibite [gebat] WICL.
GEN. xl. 22; **gibetis** (*pl.*) WICL. JOSH. x. 27.

ȝicche, sb., *O.E.* gyccæ (ZEIT. V. 196),=
M.H.G. jucke; *itch*, C. M. 11823; WICL. SEL.
W. IV. 91; icche, ȝiche ['*pruritus*'] PR. P.
259.

ȝicchen, v., *O.E.* giccan (LEECHD. III. 50),=
O.H.G. jucchan, juckan; *itch*; ȝichin, ȝikin,
ȝekin, ichin, ikin, ekin '*prurire*' PR. P. 258,
538; icche H. V. 80; **ȝicchinde** (*pres. pple.*)
A. R. 80; ȝitchinge WICL. 2 TIM. iv. 3; **ȝechid**,
icched (*pret. pple.*) CH. C. T. *A* 3682.

   ȝicchinge, sb.,=*M.H.G.* juckunge; *itching*,
A. R. 238; ȝokkin SPEC. 50.

ȝicðe, sb., *O.E.* gicða, giocða, giecða,=*O.H.G.*
juchido; *itch*; ȝikthe '*prurigo*' PR. P. 538;
ȝecðe H. M. 9.

gīde, sb., *Fr.* guide; *guide*, LANGL. *B* vi. 1;
CH. C. T. *A* 804; gie WILL. 2727.

gīde², sb., ?=gīte; ? *dress*: dame Gainour
he ledus ine a gliderand gide ANT. ARTH.
ii; gide xxix.

gīden, gīen, v., *Fr.* guider, *O.Fr.* gvier; *guide*;
gie CH. C. T. *A* 1950; S. S. (Wr.) 5; **gieþ**
(*pres.*) FER. 4921; gides D. ARTH. 3005; gidis
ALEX. (Sk.) 5387; **gīed** (*pret.*) LANGL. *A* ii.
162; **gīdid** (*pret.*) ALEX. (Sk.) 4425; giit
BARB. xix. 708.

   gīder, sb., *guider, master*, S. & C. II. lvii.

   gīderesse, sb., *female guide*, CH. BOET.
I (108.)

   gidinge, sb., *guidance*: o sterre . . . the
giding of thi bemes bright CH. TRO. v. 642.

gidi, adj., *giddy, foolish*, (*v. r.* gudi) ROB.
(W.) 1542; L. H. R. 58; þe gidie O. & N. 291.

   gidi-hēde, sb., *madness*, MISC. 143; P. L. S.
ix x. 13

   gidi-līche, adv., *giddily*, O. & N. 1282.

   gidinesse, sb., *giddiness*, ST. COD. 54.

gie, *see* ȝé. **gīe, gīen**, *see* gīde, gīden.

ȝield, *see* ȝeld. **ȝiēr**, *see* ȝēar.

ȝierne, *see* ȝerne. **giet**, *see* ȝet.

gieste, *see* giste.

ȝif, conj., *O.E.* gif, gyf,=*O.Fris.* jef, *cf. Gotn.* ja-

bai (*see* GR. GRAM. III.283); *if*, FRAG. 8; HOM.
I. 7; A. R. 22; ORM. 1438; PROCL. 6; C. L.
181; WILL. 99; LANGL. *A* PROL. 37; MAND.
3; PL. CR. 62; H. S. 665; A. P. ii. 758; L.
C. C. 32; ȝif (*ms.* yif) CH. C. T. *A* 145;
(*ms.* yif) HAV. 126; gif EGL. 495; ANT.
ARTH. xx; ȝif, ȝef LAȝ. 356, 5318; O. & N.
51, 1180; ȝef HOM. I. 221; MARH. 2; HORN
(L.) 1094; ROB. 8; SPEC. 36; SHOR. 27;
MIRC 85; gef GEN. & EX. 311; (*ms.* yef)
MISC. 27; AYENB. 5; DEGR. 185; M. H. 30;
ȝief HOM. I. 219; (*ms.* gyef) JOHN xx. 15; gife
ALEX. (Sk.) 565; *see* if.

[**ȝife**, sb., *O.E.* gifa, geofa, = *O.H.G.* gebo;
giver; *comp.* **rǣd-ȝive**.]

ȝife², sb., *O.E.* gifu, geofu, giofu, giefu, gyfu,=
O.Fris. jeve, *O.H.G.* geba, *Goth.* giba; *gift*,
ORM. 5482; ȝife, ȝefe, ȝeve (*ms.* ȝeue) HOM.
I. 17, 19; ȝive (*ms.* ȝyue) P. L. S. viii. 32;
give HOM. II. 105; HAV. 357; MAN. (H.)
155; geve (*ms.* geue) LK. ii. 40; ȝefe, ȝeve,
ȝeove LAȝ. 401, 1790, 7704; ȝeove A. R. 202;
ȝeove, ȝeoven, ȝeven, ȝiven (*pl.*) LAȝ. 930,
5445, 5464, 20494; ȝeoven A. R. 368; KATH.
37; gifes ORM. 5361; *comp.* **morȝen-ȝive**.

   ȝeve-cüsti, adj., *O.E.* cystig; *liberal in
giving*; (*ms.* ȝeue-, *printed* geue-) LAȝ. 4862.

ȝifen, v., *O.E.* gifan, giefan, giofan, geofan,
gyfan,=*O.N.* gefa, *O.Fris.* jeva, *Goth.* giban,
O.H.G. geban, keban; *give*, ORM. 1864;
ȝiven (*ms.* ȝiuen) FRAG. 6; (*ms.* ȝyuen) PL. CL.
54; ȝiven . . . vorbisne A. R. 68; given GEN.
& EX. 11; (*ms.* iuen) SAX. CHR. 264; ȝive
(? *printed* yive) ALIS. 5526; gife AMAD.
(R.) liii; ȝifen, ȝifven, ȝiven, ȝeven, ȝeoven
LAȝ. 4921, 17013, 20879, 28273, 29235; ȝiven,
ȝeoven KATH. 636; H. M. 19; ȝiven, ȝeven
LANGL. *B* ix. 161; þat rihtne dom us ȝive
wolde O. & N. 1692; ȝefe [yeve] answere
1710; ȝive, ȝeve CH. C. T. *A* 225, 505;
ȝeve, ȝive C. L. 1111; jeve (*ms.* yeue),
ive (*ms.* yue) AYENB. 7, 46; ȝefen, ȝevan
HOM. I. 13, 19; ȝieven ANGL. I. 10;
ȝieven HOM. II. 9; ȝieve MISC. 193; ȝeven
GEN. & EX. 1508; ȝevin PR. P. 537; ȝeve
ROB. 13; SPEC. 39; MIRC 138; ȝeve souke
give suck, '*lacto*' PR. P. 537; TREV. III.
141; yeve HOCCL. iii. 30; (*r. w.* live) HAV.
485; ȝeve, geve TOR. 933, 2088; geve EGL.
265; to ȝivene HOM. II. 179; to yevene
AYENB. 193; mi hert . . . mai nocht gif (*incline*)
me till dwell BARB. xix. 107; ȝive, yeve
(*pres.*) O. & N. 1686; al þis ic ȝife (*will give*)
þe ORM. 11383; ȝiveð, ȝifð A. R. 60; ȝieveð
HOM. II. 161; ȝeveþ, ȝifþ TREAT. 133, 134;
iveþ (*ms.* yueþ) MISC. 60; AYENB. 45; giffs
BARB. i. 227; ȝif (*imper.*) A. R. 294; ȝif me
. . . iseon A. R. 38; ȝif LAȝ. 8902; ȝef us
drinken 13580; ȝef SPEC. 104; (*ms.* yef)
MISC. 33; yeveþ LIDG. M. P. 48; ic ȝife (*subj.*)
ORM. 5220; (þu) ȝeve LAȝ. 8253; god ȝuve
þat ure end(e) beo god P. L. S. viii. 61;
gifand BARB. xiii. 160; **geaf**, gæf, iaf, iæf,

(*pret*.) SAX. CHR. 255, 260; ȝeaf, ȝæf, ȝef LAȝ.
1089, 7170, 9179; ȝiaf HOM. I. 223; yeaf, yaf
AYENB. 5, 81; ȝaf SHOR. 21; MAND. 97;
(*ms*. ȝaff) ORM. 761; ȝaf, yaf CH. C. T. *A*
177, 496; yaf M. ARTH. 269; yaf, gaf WILL.
395, 5381; HAV. 218, 315; Goldeboru gret
and yaf (*printed* was) hire ille 1129; gaf
GEN. & EX. 232; EGL. 407; gaf a mikel cri
C. M. 8630; þe ladi gret and gaf hire ille ISUM.
315; ȝaf [yaf], ȝef [yef] O. & N. 55, 1176; ȝaf,
ȝef SPEC. 43, 106; ȝef A. R. 92; ROB. 58;
H. H. 89; ȝef þan folke drinken *gave to
drink* HOM. I. 129; ȝef him . . . onswere KATH.
357; þet te dunt ȝef MARH. 22; gef GAW.
370; (þou) ȝeve JUL. 61; L. N. F. 222; ȝeve,
ȝave BEK. 782, 784; H. H. 167; gæfen ORM.
6476; ȝeven LAȝ. 352; ȝeven, ȝeeven C. L. 616;
ȝeve ROB. 269; ȝeven, ȝaven, ȝoven WICL.
GEN. xix. 33, 2 ESDR. ix. 17, MAT. xxvi. 67;
goven GEN. & EX. 844; SACR. 39; he . . .
goven hem ille HAV. 164; gove WILL. 4781;
ȝæfe (*subj*.) ORM. 3281; ȝefe HOM. I. 25; ȝeve
BEK. 308; ȝifen (*pple*.) ORM. 2111; gifen
IW. 3540; ȝiven HOM. I. 69; given HAV.
365; ȝive LANGL. *B* v. 390; ȝeven WILL.
2857; yeve M. ARTH. 88; yeoven REL. I.
168; ȝoven GOW. I. 79; MAND. 13; yoven
HAV. 224; HOCCL. i. 99; ȝove, ȝovin PR. P.
538; *comp*. a-, for-, ȝe-ȝifen(-ȝefen).

givere, sb., = *O.H.G.* gebāre; *giver*, LANGL.
B vii. 70; yevere AYENB. 120.

ȝevinge, sb., *cf. M.L.G.* gevinge; *giving*, PR.
P. 538; yevinge AYENB. 120.

ȝīfer, *see* ȝifre.

ȝifveðe, ȝeveðe, adj., *O.E.* gifeðe, gyfeðe;
given, LAȝ. 8160, 8366.

ȝifveðe², sb., *O.E.* gifeðe; *gift*, LAȝ. 8118;
yefþe (? *for* yefte) AYENB. 13.

ȝiveness, sb., *O.E.* gifness; *indulgence,
forgiveness*; gevenesse HOM. I. 33; II. 107;
ANGL. I. 25.

ȝifre, adj., *O.E.* gīfre, *O.N.* gīfr; *greedy, 'gu-
losus,'* FRAG. 5; ȝifer ORM. 10218*; ȝiver
A. R. 84; ȝever (*ms.* ȝeuer, *printed* ȝener)
D. TROY 3955; þe ȝivre glutun A. R. 214;
ȝēfere [ȝifre] (*pl.*) LAȝ. 7337.

ȝīver-līche, adv., *greedily*, A. R. 240; ȝiver-,
yeverli D. TROY 902, 13231.

ȝifernesse, sb., *O.E.* gīfernyss; *greediness*,
HOM. I. 83; ORM. 9318; ȝivernesse LAȝ.
9491; A. R. 198; R. S. vii; givernesse REL.
I. 216; ivernesse (*ms.* yuernesse) MISC. 38;
ȝivernes, ȝevernes (*printed* ȝinernes, ȝenernes)
D. TROY 869, 1275.

ȝift, sb., *O.E.* gift, *O.Fris.* jeft, = *O.H.G.* gift,
O.N. gipt *f.*; *from* ȝifen; *gift*, BEK. 570; þet
ilke ȝifte HOM. I. 71; ȝift [ȝifte] CH. C. T. *D*
2146; ȝifte PR. P. 538; P. S. 331; ȝeft LAȝ.
1790*; yeft HAV. 2336; ȝifte (*dat*.) WICL.
NUM. iii. 9; mid þare ȝift(e) LAȝ. 7705*;
ȝiftes (*pl*.) A. R. 28; WILL. 5357; LANGL.
B iii. 99; giftes GEN. & EX. 1416; ȝeftes

SHOR. 45; giftis, giftes, giftez ALEX. (Sk).
3777, 3128, 2027.

gigante, gigaunte, sb., *giant*, TREV. II. 231,
367; *see* geant.

gīge, sb., *O.N.* gīgja, = *M.H.G.* gīge, *O.Fr.*
gīge, *mod.Eng.* jig; *fiddle*, LEG. 140; gigges
(*pl*.) CH. H. F. 1942.

gigge, sb., *? frivolous woman*; A. R. 204;
gigges PL. T. 707; gigges [gegges] TREV. I.
403; gegges FL. & BL. 439; *? comp.* hwirl-
gigge.

gigging, sb., *clattering*; gigginge of shealdes
CH. C. T. *B* 2504.

gigelot, sb., *a romping girl*, P. S. 154; giglot
PR. P. 193; giglote TREV. VII. 409.

gigelotri, sb., *romping*; geglotries (*pl*.)
BARB. *page* 530.

gigour, sb., *O.Fr.* gigeour; *fiddler*, HORN
(L.) 1472.

gil, sb., *? cf. O.N., M.L.G., O.H.G.* gil; *gill (of
a fish)*; *throat*; (*ms.* gille) FLOR. 1419; gille
'*branchia*' PR. P. 194; VOC. 222; gile WICL.
TOB. vi. 4; giles (*pl*.) VOC. 159; A. P. iii.
269.

gīl, sb., *Du.* gijl; *new ale, 'celia,'* PR. P.
193.

ȝild, *see* ȝeld.

gilde, sb., *O.E.* gild, = *M.L.G.* gilde, *M.Du.*
ghilde, *O.Fris.* jelde; *gild (guild), 'frater-
nitas,'* PR. P. 193; CATH. 155; LIDG. M. P.
207; E. G. 29; yelde 370; ylde MISC. 96;
ȝilden (*pl*.) LAȝ. 32001; gildes MAN. (F.)
14746.

ȝilde-, ȝelde-halle, sb., *guild-hall*, CH. C.
T. *A* 370; gildehalle '*praetorium*' PR. P.
193.

ȝilden, *see* ȝelden.

gilden, *see* gülden.

gildinge, sb., *see after* gülden.

gildre, sb., *O.N.* gildra; *snare*; gildire
HAMP. PS. xxxvi. 33*; gildert PS. ix. 31;
gilders (*pl*.) M. P. 69; gildirs HAMP. PS.
cxxxix. 4*.

gildren, v., *O.N.* gildra; *catch in a snare*;
gildirs (*pres*.) HAMP. PS. xxx. 10*; gilderd
(*pple*.) C. M. 23307; gildird HAMP. PS.
lxviii. 27*.

gīle, sb., *O.Fr.* gvīle; *guile, treachery*, LAȝ.
16382*; A. R. 128; ROB. 310; SPEC. 49;
SHOR. 164; s. s. (Wr.) 1346; PERC. 1035;
to do þe gile REL. I. 182; gile (gille) ROB.
(W.) xx 61*; *see* wīle.

gīle-ful, adj., *guileful*, TREV. VII. 59; WICL.
PS. v. 7.

gīleful-līche, adv., *guilefully*, TREV.
VIII. 5; gilefulli WICL. PROV. xx. 19.

gīlen, v., *O.Fr.* gīler, gvīler; *beguile, deceive*,
A. R. 128; gile S. S. (Wr.) 2637; gili AYENB.
15; gīleþ (*pres*.) REL. I. 116; gīled (*pret*.)

ROB. (W.) XX 254; **gīled** (*pple.*) HORN (L.)
1452; WILL. 689; *comp.* **a-, bi-gīlen.**

gīling, sb., *cunning*; mid grete gilinge ROB.
(W.) XX 59.

gīling-liche, adv., *with guile*: thi brothir
com gilingliche WICL. GEN. xxvii. 35; who
goth gilendeli, shewith prive thingus WICL.
PROV. xi. 13.

gilerie, sb., *O.Fr.* gillerie; *deceit,* S. A. L.
225; gileri, gilri HAMP. PS. ix. 31*; **gilres,**
gilris (*pl.*) HAMP. PS. cxiii. 17*, ix. 25*.

[**gill,** sb., *O.N.* gil;] **gille-strēmes,** sb., pl.;
cascades, ALEX. 3231 (*miswritten* gille-
stormes *in Dubl. MS.*).

gille, sb., *O.Fr.* gelle; *gill* (*measure*), PR. P.
194; gille, jille LANGL. *A* v. 191.

Gille, sb., *Fr.* Gille; *clown*: þ(o)u art a
grameful Gille LEG. 212.

gillen, v., *disembowel* (*fish*): gillin or gille
fische '*exentero*' PR. P. 194.

gillinge, sb., '*exenteracio,*' PR. P. 194.

gilmins, sb., pl., *an abusive name for monks*:
hail be ȝe gilmins wiþ blake gunes P. L. S.
xxiv. 7.

gilofre, sb., *Fr.* girofle, *from Gr.* καρυόφυλλον';
gilliflower, SPEC. 27; R. R. 1368; A. P. i.
43; giloffre ALEX. (Sk.) 5426.

gilour, sb., *O.Fr.* gileor; *deceiver,* LEB. JES.
873; LANGL. *A* ii. 89; WICL. PROV. iii. 32;
CH. C. T. *A* 4321; gilours (*pl.*) WICL. PS.
liv. 24.

gīlous, adj., *deceitful, treacherous*: gilous
falshede WICL. COL. ii. 8; trecherous, or
gīlous (*pl.*) workmen WICL. 2 COR. xi. 13.

ȝilpen, *see* ȝelpen. **gilt,** *see* gült.

gilte, sb., *O.E.* gilte, *O.N.* gilta; *sow,* PR. P.
194; VOC. 204.

gilten, *see* gülten.

ȝim, sb., *O.E.* gimm; *gem*; ȝimmes (*pl.*)
HOM. I. 259; A. R. 342*; ROB. 489; gimmes
ALIS. 3152: ȝimmen (*dat. pl.*) FRAG. 2;
ȝimme LAȜ. 6081; *see* gemme.

ȝim-stān, sb., *O.E.* gimstān; *gem-stone,*
KATH. 1662; ȝimston A. R. 134; ȝimston
(ȝemston) LAȜ. 21143; (*ms.* ymston) MISC. 98;
ȝimstōnes (*pl.*) HOM. I. 193.

gimbire, gimbüre, sb., *O.N.* gymbr; *young
sheep,* VOC. 187, 219.

gimelot, sb., *O.Fr.* gimbelet; *gimlet,* PR. P.
194; gimlet B. B. 121.

ȝīmen, *see* ȝēmen.

ȝimmen, v., *O.E.* gimmian; *set with gems;
bud*; wintreowe ȝimmeþ *pres.*) FRAG. 2; þi
ȝimmede (*pple.*) bur HOM. I. 273.

gimowe, gimew, sb., *Fr.* jumeau, jumelle;
mod.Eng. gimbals; *anything consisting of a
pair of jointed pieces; pair of tongs; hinge,
etc.;* '*vertinella, gemella,*' PR. P. 194.

gin, sb., *? shortened from, of* engin; *but cf.*
O.N. ginna (*deceive*), *mod.Eng.*gin; *ingenuity,
contrivance; snare,* KATH. 1980; þurh snoter
gin (*ms.* gyn) ORM. 7087; feondes gin MISC.
144; ginne FL. & BL. 195; **ginne** (*dat.*) C. L.
93; ALIS. 1219; LANGL. *B* xviii. 250; ALIS.[2]
(Sk.) 1135; M. ARTH. 3037; mid ginne O. & N.
669; P. L. S. xix. 222; windowes corven wiþ
ginne MAN.(F.) 4580; **ginne** (*pl.*) LAȜ.18839;
vele ginnes heþ þe dievel AYENB. 54; **ginnen**
(*dat. pl.*) LAȜ. 1336.

gin-ful, adj., *cunning, treacherous,* LANGL.
B x. 208.

[**gin**[2], sb., *O.E.* (an-)gin, ginn, = *O.L.G., O.H.G.*
(ana-)gin, ginni; *from* **ginnen**; *comp.*an-gin.]

ȝing, *see* ȝung.   **ging(e),** *see* genge.

gingebreed, sb., *cf. med.Lat.* gingibrētum;
gingerbread, CH. C. T. *B* 2044; (*ms.* gynge-
bred) VOC. (W. W.) 587.

gingelin, v., *jingle,* PR. P. 195; ȝingle CH.
C. T. *A* 170.

gingelinge, sb., *jingling, resounding; 'reso-
nancia'* PR. P. 195.

gingen, *see* ȝüngen.   **gingere,** *see* gingivere.

gingivere, sb., *O.E.* gingifre, *O.Fr.* gengibre,
Gr. ζιγγίβερις; *ginger,* LAȜ. 17745; A. R.
370; gingivre SPEC. 27; gingere ALEX. (Sk.)
5426; B. B. 121; ginger sauce B. B. 152.

giniper, sb., *Lat.* jūniperus, *O.Fr.* geneivre;
juniper, MAND. 28.

ginnen, v., *O.E.* (an-, be-)ginnan, = *M.L.G.*
ginnen, *O.L.G., O.H.G.* (bi-)ginnan, *Goth.*
(du-)ginnan; *begin*; ginne MAN. (F.) 201;
ginne (*pres.*) SACR. 502; ginnest BEK. 760;
þe blostme ginneþ springe O. & N. 437; **gan**
(*pret.*) ORM. 2805; HAV. 2443; CH. C. T. *A*
301; he gan þenche P. L. S. ix. 173; he gan
to sike WILL. 691; gon LANGL. *A* PROL.
11; Laȝamon gon [gan] liðen LAȜ. 27; þe
mone gon to scine 17861; gon C. L. 209;
heo gunnen LAȜ. 24695; GEN. & EX. 534;
gunne HORN (L.) 51; K. T. 1037; ISUM.
448; gunne, gonne WILL. 1164, 1961; gonnen
LANGL. *B* xiii. 267; gonne CH. C. T. *A* 1658;
gonnen (*pple.*) MAN. (H.) 167; *comp.* **a-,
an-, bi-ginnen.**

ginninge, sb., *beginning,* AYENB. 16.

ginnen[2], v., *? for* enginen; *but cf. O.N.*
ginna (*deceive*); *devise, contrive*; iginned
[*older text* idihte] (*pple.*) LAȜ. 28627*; þis
pinful(l)e gin was o swuch(e) wise iginet
KATH. 1981; *comp.* **bi-ginnen**[2].

ginour, sb., *for* enginour; *workman*; ginnur
FL. & BL. 323; **ginours** (*pl.*) RICH. 2913.

gīour, sb., *O.Fr.* gvieour; *conductor,* TREV.
I. 349; giour [gwiour] LANGL. *B* xx. 71;
gioure DEP. R. PROL. 29; **gīours** (*pl.*) MAN.
(F.) 3379.

gipe, sb., *O.Fr.* gipe, jupe; *cassock*: beggars
with ... botis reveling as a gipe R. R. 7262.

Gipeswich, pr. n., *O.E.* Gipes wic ; *Ipswich*, ROB. 297.

gipōn, sb., *O.Fr.* gipon ; *a short cassock, petticoat*, OCTOV. (W.) 1029 ; gipon, gepoun CH. C. T. *A* 75 ; gipoun TREV. I. 403.

ʒippin, v., *peep*, *chirp*, PR. P. 538.

gipsiere, gipsere, sb., *Fr.* gibeciere (*gamebag*); *pouch*, PR. P. 195 ; gipsere CH. C. T. *A* 357.

gird, girdel, girden, *see* **gürd**, *etc.*

girdiller, *see* **gürdlere** *under* **gürdel**.

ʒiren, *see* **ʒeoren**.   **girl**, *see* **gerl**.

ʒirn, *see* **grin**.   **ʒirnen**, *see* **ʒeornen**.

ʒirnen, v., *O.E.* geirnan, gerinnan, = *O.H.G.* girinnan, *Goth.* garennan ; *run together; congeal*; þus ʒirneʒ (*pres.*) þe ʒere in ʒisterdaies moni GAW. 529 ; a ʒere ʒernes ful ʒerne 498 ; þe ʒonge men . . . ʒornen þer oute A. P. ii. 881.

ʒirnunge, *see under* **ʒeornen**.

girþ, *see* **griþ**.

ʒis, *see* **ʒü-se** *under* **ʒēa**.

gisarme, sb., *O.Fr.* gisarme ; *halberd*, LAʒ. 1567* ; GEN & EX. 4084 ; ALIS. 2303 ; gisarmes (*pl.*) MAN. (F.) 12000.

ʒiscen, *see* **ʒitsen**.

gīse, *see* **gvīse**.

gīsel, sb., *O.E.* gīsel, = *O.N.* gīsl, *O.H.G.* gīsal ; *hostage, 'obses,'* FRAG. 2 ; ʒīsles (*pl.*) LAʒ. 5324 ; ʒīslen (*dat. pl.*) LAʒ. 16853.

gīsen, v., *O.Fr.* (de-)gviser ; *dress, prepare* : þou gīsed (*pret.*) the gerne R. P. 91 ; thei gised them fulle gai TRIAM. 660 ; Triamoure was gīsed (*pple.*) fulle gai TRIAM. 710.

gīser, sb., *Fr.* gesier ; *gizzard*, CH. BOET. iii. 12 (107).

ʒissen, *see* **ʒitsen**.

gist, gistnir, *see* **gest, gestnen**.

giste, sb., *O.Fr.* giste, *from* gesir (*to lie*); *joist; lodging; 'trabes,'* PR. P. 196 ; LEG. 91 ; SHOR. 4 ; gieste TREV. VII. 35 ; hieð toward his giste (*couch*) A. R. 350.

giste[2], sb., *? refreshment* : more him likede þat ilke giste þane ani flechs isode oþar irost LEG. 8.

gisten, v., *O.Fr.* gister ; *agist; lodge* (*cattle*); joist (*pple.*) A. P. ii. 434.

ʒistren–, *see* **ʒestren–**.

ʒit, pron., *O.E.* git, gyt, = *O.L.G.* git, *O.N.* it, *M.H.G.* ez ; *you two*, HOM. I. 93 ; ORM. 4498 ; mine sunen ʒit beoð beien LAʒ. 5020 ; ʒet 5618 ; *see* **inc**.

ʒit, *see* **ʒet**.

gite, sb., *a sort of garment*, CH. C. T. *A* 3954.

giterne, sb., *O.Fr.* gviterne ; *guitar*, LANGL. *B* xiii. 233 ; CH. C. T. *A* 3333 ; WICL. IS. v. 12 ; giternes (*pl.*) CH. C. T. *C* 466

giternen, v., *O.Fr.* gvisterner ; *play on the lyre or guitar* : to harpe and gitterne R. R. 2321.

giternere, sb., *guitar-player*, A. P. i. 91.

giterning, sb., *playing on the lyre* : giterninge & daunsinge WICL. E. w. 8 ; giterning CH. C. T. *A* 3363.

gitōn, sb., *O.Fr.* gvidon ; *pennon, streamer, 'conscisorium,'* PR. P. 197.

ʒītsen, v., *O.E.* gītsian, = *M.H.G.* gītsen, gītesen ; *covet* ; ʒiscen A. R. 196 ; ʒisceð (*pres.*) þah after muchele deale mare H. M. 29 ; ne gisce þu nog(t) ðin nestes ðing GEN. & EX. 3515.

ʒītsere, sb., *O.E.* gītsere, = *M.H.G.* gītesære ; *one who covets, a miser*, HOM. I. 101 ; ʒissare [ʒiscere] A. R. 202 ; ʒisceres (*pl.*) P. L. S. viii. 135.

ʒītsunge, sb., *O.E.* gītsung ; *coveting*, HOM. I. 103 ; ʒitsunge, ʒitsinge LAʒ. 6560, 9490 ; ʒittsunng ORM. 4560 ; ʒissunge A. R. 258 ; ʒissinge P. R. L. P. 222 ; giscing GEN. & EX. 1874 ; (*misprinted* gisting) REL. I. 216 ; issing MISC. 38 ; icinge AYENB. 16.

gith, sb., *Lat.* gith, git ; *gith*, PALL. x. 155.

Giu, sb., *O.Fr.* Giu, Jui ; *Jew*, MISC. 47 ; LEG. 13 ; Jeu AYENB. 43 ; **Gius** (*pl.*) A. R. 114 ; Giwes MARH. 3.

Giugne, sb., *June*, ALEX. (Sk.) 3537.

giugou, sb., *? Fr.* joujou ; *gewgaw, toy* ; gugau PR. P. 218 ; giuegouen (*pl.*) A. R. 196.

ʒive, ʒiven, giveness, *see* **ʒife, ʒifen, ʒifeness**.

gīven, v., *fetter* ; **gīvede** (*pret.*) TREV. VI. 203 ; gived LANGL. *B* xx. 191 ; **gīved** (*pple.*) WICL. GEN. xxxix. 22 ; here the weilingus of the givede PS. ci. 21.

ʒiver, *see* **ʒifre**.

gīves, sb., pl., *gives, fetters*, LAʒ. 15338 ; OCTOV. (W.) 222 ; wiþ feteres ant wiþ gives P. S. 221.

ʒixen, *see* **ʒeoxen**.

giwerie, pr. n., *Jewry*, A. R. 394 ; jewerie CH. C. T. *B* 1679.

glace, glas, sb., *ice*, ALEX. (Sk.) 3032.

glacen, v., *O.Fr.* glacer, glacier ; *glance aside; flash* : gladande glori con to me glace A. P. i. 171 ; wheþer it wole glase or glente H. V. 109 ; his swerde **glasedde** (*pple.*) lowe GUY (Z.) 5067.

glacing, sb., *glancing* : glacinge or wrong glidinge of bolts or arowis PR. P. 197.

glácin, *see* **glásen**.

glad, *see* **glæd**.

gladene, sb., *O.E.* glædene ; *iris plant*, REL. I. 36 ; gladen ALEX. (Sk.) 4094 ; *see* **gloden**.

gladien, v., *O.E.* gladian, = *O.N.* glaða, gleðja ; *make glad, be glad*, LAʒ. 16947 ; to gladien & to blissen us HOM. II. 93 ; gladie BEK. 1204 ; gladie (gladiʒe) ROB. (W.) 3350 ; gladen ORM. 1128 ; CH. TRO. ii. 979 ; glade WILL. 827 ; LANGL. *B* x. 43 ; IW. 1440 ; MIN. vi. 53 ; gleadien HOM. I. 259 ; gledien A. R. 178 ; MARH. 21 ; gledie AYENB. 266 ; glathe LUD. COV. 171 ; gledie (*pres.*) MARH. 6 ; gladieþ REL. I. 123 ; he gledie ou and frovre ou A. R.

430; **gladede** (*pret.*) ROB. 265; þa gladede (h)is mod LAȝ. 4410; þat gladede his herte LANGL. (Wr.) 14267; Abraham ȝour fader gladide þat he schulde se mi dai WICL. JOHN viii. 56; **gladed** (*pple.*) WILL. 600; *comp.* ȝe-gladien.

glader, sb., *gladdener*: thou glader of the mount of Citheroun CH. C. T. *A* 2223.

gladinge, sb., *gladdening, gladness,* C. L. 841; WICL. I KINGS iv. 8; gleadunge HOM. I. 261; gledunge A. R. 94.

gledunde, sb., *? for* **gledunge**: mi gleo ant mi gledunde MARH. 3.

glæd, glad, gled, adj., *O.E.* glæd, = *O.L.G.* glad, *O.Fris.* gled, *O.N.* glaðr, *O.H.G.* glat; *brilliant, glad,* LAȝ. 3962, 7013, 9374; glað O. & N. 434; GEN. & EX. 4051; BRD. 6; P. L. S. iii. 34; (*ms.* gladd) ORM. 2811; glad under gore SPEC. 26; glead JUL. 71; gled MARH. 10; A. R. 282; REL. I. 180; AYENB. 265; glad GEN. & EX. 3671; wiþ glade mode SPEC. 80; **glade** (*pl.*) LAȝ. 12349; KATH. 1683; ORM. 160; O. & N. 424; LEG. 32; beoð glede A. R. 380; **gladder** (*compar.*) GOW. II. 53; þe gladdere ȝe was on heorte LAȝ. 2407*; gladere WICL. RUTH iii. 7; gladdere ROB. (W.) 7377; **gladdest** (*superl.*) ALIS. 5261; SPEC. 26.

glad-chēred, adj., *good-humoured,* MAN. (F.) 9752.

glad-līche, adv., *O.E.* glædlīce; *gladly,* LAȝ. 22305; P. L. S. xix. 145; gladlike ORM. 10463; HAV. 805; gledliche A. R. 126; **gladloker** (*compar.*) GAW. 1064; gledluker A. R. 188; gladliere HAMP. PS. lxxvi. 6*.

gladnesse, sb., *O.E.* glædness; *gladness,* PR. P. 197; gladnisse LAȝ. 12329*; glednesse A. R. 94; AYENB. 27.

glad-shipe, sb., *O.E.* glædscipe; *gladness,* ORM. 784; SPEC. 38; gladscipe LAȝ. 8397; gledscipe HOM. I. 45; gledschipe A. R. 180; glaidschip BARB. v. 298*.

gladsum, adj., *gladsome,* WICL. PS. ciii. 15; gladsom CH. C. T. *B* 3968; LUD. COV. 16.

gladsum-lī, adv., *joyfully,* WICL. WISD. vi. 17; BARB. xvi. 20.

glæd, sb., *O.E.* glæd; *gladness*; glad GENER. (Wr.) 1255; a gladen he waitis (*watches for a lucky moment*) ALEX. (Sk.) 131; glath MAN. (F.) 3260; **glade** (*dat.*) TOR. 1143.

gled-ful, adj., *joyful,* A. R. 242.

glæs, sb., *O.E.* glæs, = *O.L.G.* gles, *O.H.G.* glas, *cf. O.N.* gler; *glass*; glas, gleas KATH. 1676, 2029; glas HORN (L.) 14; SPEC. 31; CH. C. T. *A* 198; A. P. i. 989; gles A. R. 164; AYENB. 82; þet gles HOM. I. 83.

glæs-fat, sb., *glass vessel* [*sec. text* urinal], LAȝ. 17725.

glaire, sb., *Fr.* glaire, *O.Fr.* clēre; *albumen; glossiness,* CATH. 157; glaire [gleire] of an ei CH. C. T. *G* 806; þat glent(e) as glaire

A. P. i. 1025; with a glaire hoge D. TROY 5926.

glaive, sb., *O.Fr.* glaive; *sword,* A. P. i. 653; gleive HAV. 1770; ROB. 203; JOS. 497.

glam, sb., *O.N.* glam; *word, talk, noise,* GAW. 1426, 1652; A. P. ii. 499, 830, 849; glaam ALEX. (Sk.) 5504.

glárin, v., = *M.L.G.* glaren; *glare, shine,* PR. P. 198; **gláreth** (*pres.*) MISC. 27; **gláringe** (*pple.*) eien CH. C. T. *A* 684.

glas, see **glace** *and* **glæs.**

glásen, v., *? O.N.* glæsa, *mod.Eng.* glaze; *polish, cause to shine*; glacin PR. P. 197.

glásing, sb., *polishing*: glacinge, or scouringe of harneis '*pernitidacio, perlucidacio*' PR. P. 197.

glasi, adj., = *Ger.* glasig; *glassy,* LIDG. M. P. 53.

glásien, v., *glaze*; glasin '*vitro*' PR. P. 198; glase LANGL. *A* iii. 49; iglásed (*pple.*) CH. D. BL. 323.

glásinge, sb., *glasswork*: the stori of Troye was in the glasinge iwrought CH. D. BL. 326.

glasin, adj., *O.E.* glæsen, = *M.Du.* glasen, *O.H.G.* glesīn; *made of glass,* '*vitreus,*' PR. P. 198; a glasen houve LANGL. *B* xx. 171; þe glesne-ehþurl HOM. I. 83.

Glastingbüri, pr. n., *O.E.* Glæstinga byrig (SAX. CHR. 159); *Glastonbury,* ROB. 478.

glath, see **glæd.**

glaumen, v., *? O.N.* gleyma; *make a noise, yelp*; **glaumande** (*pple.*) GAW. 46.

glavir, sb., *chattering,* ALEX. (Sk.) 5504.

glaveren, v., *? Welsh* glafru; *chatter, flatter, deceive*; **glaverez** (*pres.*) A. P. i. 688; **glaver** (*subj.*) WICL. E. W. 330; **glaverande** (*pple.*) GAW. 1426; glaverande gomes D. ARTH. 2538; wiþ glaveringe wordes PL. CR. 51.

glē, see **glēo.** **glead,** see **glæd.**

glēam, see **glēm.**

glēaw, gleg, adj., *O.E.* glēaw, *O.N.* glöggr, = *O.L.G., O.H.G.* glau, *Goth.* glaggwus; *sagacious, skilful,* FRAG. 3; gleaw MISC. 104; gleu REL. I. 181; he is of worde swiþe gleu O. & N. 193; gleuȝ S. A. L. 158; gleg of ei C. M. 13448; **glēawun,** gleawen (*dat. pl.*) LK. x. 21, xi. 25; **glǣwest** (*ms.* glæuest) (*superl.*) LAȝ. 16237.

glēbe, sb., *O.Fr.* glēbe; *glebe,* TREV. VIII. 335.

gled, see **glæd.**

glēde, sb., *O.E.* glida, *O.N.* gleða; *glede, kite,* '*milvus,*' VOC. 188, 220; A. P. ii. 1696; OCTAV. (H.) 680; **glēdes** (*pl.*) MAND. 309; P. I. 344.

glēde, sb., *O.E.* glēd, = *O.Fris.* glēd, *M.Du.* gloed, *O.N.* gloð, *O.H.G.* gluot; *from* **glōwen**; *burning coal,* MISC. 43; HAV. 91; FER. 2241; GOW. I. 280; PERC. 756; M. ARTH. 780; ane berninde (*pple.*) glede HOM. I. 27; glede [gleede] LANGL. *B* ii. 12; CH. C. T. *A* 1997; gleede HOCCL. i. 159; **glēden** [*O.E.* glēda] (*pl.*) JOHN xxi. 9; HOM. I. 43; A. R. 406; gleden [gledes] LAȝ. 18863; gledes ORM. 1067; M. H. 133; glodez A. P. i. 79; **glēdan** (*dat. pl.*) JOHN xviii. 18.

glēd-rēad, adj., *O.N.* glōðrauðr; *red like hot coals*, HOM. I. 249.

glēdien, *see* **gladien.**   **gleg**, *see* **glēaw.**

glei, *see* **glíen.**

gleim, sb., ? =**cleim** (**clām**); '*limus, gluten*,' PR. P. 198; PL. CR. 479.

gleimen, v., ? =**cleimen**; '*visco*' PR. P. 198.

gleimous, adj., *slimy*, PR. P. 198.

   gleimousnesse, sb., *clamminess, stickiness,* '*limositas*,' PR. P. 198.

gleire, *see* **glaire.**   **gleive**, *see* **glaive.**

glēm, sb., *O.E.* glǣm [*from* *glīmen,=M. H.G.* glīmen v.]; *gleam, splendour, brightness,* HICKES I. 227; HAV. 2122; A. P. ii. 218; gleem SPEC. 57; glem, gleam MARH. 9, 12; gleam A. R. 94.

glēmin, v., *gleam*, '*radio*,' PR. P. 198; **glēaminde** (*pple.*) KATH. 1668; glemand BARB. viii. 226; **glēmede** (*pret.*) SPEC. 36; glemed FLOR. 379; GAW. 598.

   glēminge, sb., *gleaming;* '*conflagracio, flammacio*,' PR. P. 199.

glen, sb., *glen*: in a glen thair galai dreuch BARB. iv. 372*.

glēne, sb., *O.Fr.* glene, glaine; ? *handful of corn ears;* '*spicatum*,' PR. P. 199.

glēnen, v., *O.Fr.* glener, glainer; *glean, gather*; glene A. P. i. 955.

   glēnar, sb., *gleaner*, PR. P. 199.

   glēninge, sb., *gleaning*, PR. P. 199.

[**glengen**, v., *O.E.* glengan; *adorn, trim* (*a lamp*); *comp.* ȝe-glengen.]

glent, sb.,=*Swed.* glänt; *glimpse; shining*; MAN. (F.) 12782; GAW. 1290.

glenten, v.,=*Swed.* glänta, ?*M.H.G.* glenzen; *shine; look; move quickly*; glent(e) H. V. 109; **glente** (*pret.*) GEN. & EX. 1029; MAN. (F.) 12423; als he bi Wawain glent (*passed quickly*) MAN. (F.) 12729; þe wal of jasper þat glent(e) as glaire A. P. i. 1025; her eie glent(e) a side CH. TRO. iv. 1223; when þe swird out glente OCTAV. (H.) 892; whiderward his iȝen glente H. S. 6186; þe strok a doun him glente (*glanced*) FER. 616; gle(i)ves glētering glent(e) opon geldene scheldus DEGR. 279.

glēo, sb., *O.E.* glēow, glīw, glīg,=*O.N.* glȳ, *mod.Eng.* glee; *joy, song*, KATH. 146; MARH. 3; A. R. 210; MISC. 97; FL. & BL. 477; ALIS. 191; þat gleo LAȝ. 7006; glew GEN. & EX. 459; glew MAN. (H.) 295; gleu HAV. 2332; MAP 347; M. H. 23; glu '*musica*' PR. P. 200; gle WILL. 824; DEGR. 34; gle ant gome CHR. E. 456; hii ȝeve him selver for (h)is glē ROB. 272; glie ANGL. I. 24; HOM. II. 229; SHOR. 103; **glēwe** (*dat.*) TRIST. 290.

glēo-bēam, sb., *O.E.* glēobēam; *harp*; **glēobēames** (*pl.*) HOM. I. 193.

   glēo-cræft, sb., *O.E.* glēocræft; *art of music*, LAȝ. 7012.

   glēo-drēm, sb., *O.E.* glēodrēam; *glee, song*, LAȝ. 1823.

   glēo-man, sb., *O.E.* glēomann; *glee-man*, ALIS. 1152; gleomon LAȝ. 7004; glewman HOM. II. 29; glemon SPEC. 49; gleman LANGL. *B* ix. 101.

glēowien, v., *O.E.* glēowian, glīwian; *make music, play, amuse*, LAȝ. 20315; glewen MAP 340; glewe SPEC. 38; TRIAM. 108; **glēowede** (*pret.*) A. R. 368; glewed (? *called loudly*) A. P. iii. 164.

   glēowinge, sb., *music*, HORN (L.) 1468.

gles, *see* **glæs.**

glēsen, v., *O.E.* glēsan; *gloss*; **glēsþ** (*pres.*) FRAG. I.

   glēsing, sb., *O.E.* glēsing; *glossing*, FRAG. I.

glete, sb., *O.Fr.* glete; *gore; mud, clay*; glet(e), glette A. P. i. 1059, ii. 306; PR. C. 459; glett(e) ALEX. 4491; kastis out of thaire hert all glet HAMP. PS. *page* 512.

glēu, *see* **glēo.**   **glēu**, *see* **glēaw.**

glēw, *see* **glēo.**   **glēwen**, *see* **glēowien.**

glewishe, adj., *from* glū; *glutinous*: glewide it with **glewishe** clei WICL. EX. ii. 3.

[**glīde**, ? *stem of* glīden, v.]

   glīde-worm, sb., ? *glow-worm*, '*incedula*,' VOC. 190.

glīden, v., *O.E.* glīdan,=*O.L.G.* glīdan, *O. Fris.* glīda; *glide*, LAȝ. 800; GEN. & EX. 370; glide ALIS. 2245; SHOR. 161; P. L. S. iii. 10; **glīdeð** (*pres.*) JUL. 3; glit LAȝ. 22043; MISC. 94; C. L. 1454; **glidende** (*pple.*) HOM. I. 43; **glād** (*pret.*) LAȝ. 15714; PERC. 2116; glod CH. C. T. *F* 393; AMAD. (Web.) 761; GAW. 661; forð glod ðat firme ligt GEN. & EX. 75; gliden LAȝ. 24753; GEN. & EX. 733; þe tieres glide of hire lere FL. & BL. 501; **gliden** (*pple.*) GEN. & EX. 3460; CH. C. T. *E* 1887; *comp.* a-, ȝe-, tō-glīden,

   glīdare, sb., *glider*, '*serptor, serptrix*,' PR. P. 199.

[**glider**, adj.,=*O.L.G.* glidir (HEINE 120); *from* glīden; *slippery*.]

   glethur-lī, adj., *smoothly, quickly*, HALLIW. 404.

gliderin, v., ?=gliteren; *flash, glitter*, '*rutilo*,' PR. P. 199; wiþ scharp sverdes on helmes gan glidre (*smite*) MAN. (F.) 4372; **gliderande** (*pple.*) ANT. ARTH. ii.

glie, *see* **glēo.**

glien, v., *O.N.* gliā; *squint*; glie SHOR. 100; **glíande** (*pple.*) VOC. 225; glide (*pret.*) LEG. 169; gleied [gliȝed] C. M. 3862; **gliȝt** (*pret.*) *shone* A. P. i. 114; (*looked*) iii. 453; **gleiit**, gliet (*pple.*) D. TROY 3772, 3943.

glíere, sb., *squinter,* '*strabo,*' VOC. 225; gliare PR. P. 199.

glíinge, sb., *squinting,* '*strabositas,*' PR. P. 199.

gliffen, *see* **glüffen.** **gliften,** *see* **glüften.**

glim, sb., *cf. Swed.* glim (RIETZ); *splendour*; **glimme** (*dat.*) A. P. i. 1087.

glimerin, v., *cf. Ger.* glimmern, *Dan.* glimre, *Swed.* glimra; *glimmer,* '*radio,*' PR. P. 199; **glimerand** (*pple.*) WILL. 1427; glemirring TOR. (A.) 426; **glemered** (*pret.*) & glent GAW. 172.

glimering, sb., *glimmering,* WICL. E. W. 339.

glimsen, v., ? = *M.Du.* glinsen; *mod.Eng.* glimpse; *shine*: (snou) þat **glimsede** (*pple.*) on hare eiȝen ROB. (W.) xx. 252.

glimsing, sb., *splendour,* CH. C. T. *E* 2383.

glisien, v., *O.E.* glisian, *cf. O.Fris.* glisa; *shine*; **glisiinde** (*pple.*) MISC. 91; glisiande ALIS. (Sk.) 180; sceldes þer **gliseden** (*ms.* cliseden) LAȝ. 21725; glised HORN CH. 173.

glisnien, v., *O.E.* glisnian; *glisten*; **glissinand** [*v. r.* glasinand] (*pple.*) ALEX. (Sk.) 3015*; glisnande A. P. i. 165; glisninge WICL. HABAK. iii. 11; **glissenede** (*pret.*) LAȝ. 21725*.

glissen, v., *O.E.* glissian (GREIN I. 516); *glance*; he **glisset** (*pret.*) up with his ene ANT. ARTH. xxviii; FLOR. 1659.

glisten, glistnen, v., *glisten*; **glistinde** (*pple.*) MARH. 9; glistinde & gleaminde KATH. 1668; **glistnede** (*pret.*) SPEC. 36; glistnede as gleam de δ MARH. 9.

glistren, v., = *M.L.G.* glistern, *M.Du.* glisteren; *glister, shine*; **glistering** (*pple.*) EM. 350; **glistred** (*pret.*) GOW. II. 252; glistrid FER. 4438.

glitenien, v., *O.E.* glitinian, = *O.H.G.* glizinōn; *shine*: glitene on golde LAȝ. 15715.

glitren, sb., *O.N.* glitra, = *M.H.G.* glitzern; *glitter*; glitren [gliteren] (*pres.*) CH. C. T. *A* 977; glitteren ['*rutilent*'] WICL. JUDG. v. 31; **gliterand** (*pple.*) PS. xliv. 14; glitterand BARB. viii. 233; gliteringe DEGR. 1839; glitered (*pret.*) & glent GAW. 604.

glob, sb., *Lat.* globus; *mass, assembly,* WICL. JOSH. iii. 13; the men ... togidre gedred in a glob JUDG. ix. 47; glub 4 KINGS iv. 17; glubbe NUM. xvi. 11.

globbare, *see under* **gulpen.**

globben, *see* **gulpen.**

glocken, v., *shake violently,* PALL. viii. 131.

glōde, sb., (?) ALEX. (Sk.) 1334.

gloden, sb., '(h)eliotropium,' REL. I. 37 (VOC. 149); *see* **gladene.**

gloffare, sb., '*devorator,*' PR. P. 199; *see* **globbare** *under* **gulpen.**

gloghen, v., *O.N.* glugga; *gaze, stare*: to glogh uppon gomes D. TROY 2922.

gloiere, sb., = **glíere**; *squinter*: gloiere or gogileie '*strabo*' PR. P. 199.

glomben, glommen, v., ? = *Swed.* glomma, *L.G.* glummen; *observe; frown; be gloomy*; glombe R. R. 4356; **gloumbes** (*pres.*) A. P. iii. 94; gloumes ALEX. (Sk.) 4142; and **glommede** (*pret.*) als he were wraþ HALLIW. 404.

glop, sb., *surprise*: mi hert is risand now in a glope TOWNL. 146.

glōpen, v., ? = *M.Du.* gleopen, gluipen, *O.Fris.* glupa; *terrify*; **glōped** (*pret.*) A. P. ii. 849.

glopnen, v., *cf. O.N.* glúpna; *be astonished, be terrified,* A. R. 212*; **glopnede** (*pret.*) D. ARTH. 1074; **glopnid** (*pple.*) C. M. 11611; *comp.* **a-, for-glopnen.**

glōpned-lī, adv., *fearfully,* A. P. ii. 896.

gloppning, sb., *fright, amazement*: gloppning [*v. r.* gloppening] C. M. 19633; gloppininge D. ARTH. 3864.

gloppen, gloppinge, *see* **gulpen.**

glōren, v., = *Swed.* glōra (RIETZ 200), *M.Du.* gloeren; *glare, stare*: til þat his lippes þar of dude glore S. A. L. 47; **glōrand** (*pple.*) ALEX. (Sk.) 4552; than glopnede we gloton and **glōrede** (*pret.*) D. ARTH. 1074.

glōrie, sb., *O.Fr.* glorie; *glory,* A. R. 358; SHOR. 5; glori ALEX. (Sk.) 1730.

glōrien, v., *Lat.* gloriāri; *glory,* WICL. PROV. xx. 14.

glorifíin, v., *Fr.* glorifier; *glorify,* PR. P. 199; **glorifieþ** (*pres.*) AYENB. 25.

gloriūs, adj., *O.Fr.* glorius, glorios; *glorious,* MISC. 27; glorious TREV. IV. 201; **gloriōsest** (*superl.*) ALEX. (Sk.) 1611.

glorious-liche, adj., *gloriously,* LANGL. *C* xx. 15.

gloriousnesse, sb., *glory,* '*gloriositas,*' PR. P. 199.

glōse, sb., *O.Fr.* glōse; *gloss, flattery, falsehood,* ROB. (W.) 2381; AYENB. 187.

glōsen, v., *O.Fr.* glōser; *gloss, explain; pervert; flatter*; **glōseþ** (*pres.*) PL. CR. 345; glose ['*blandire*'] WICL. JUDG. xiv. 15; **glōsinde** (*pple.*) ROB. 497; **glōsed** (*pret.*) WILL. 60; MAN. (H.) 34.

glōsare, sb., *glosser, glozer,* PR. P. 199.

glōsinge, sb., *explanation, comment, flattery, deceit;* '*glosacio, adulacio*' PR. P. 200; CH. C. T. 7374; LANGL. *B* xiii. 74; **glōsinges** (*pl.*) and gabbinges LANGL. *B* xx. 124.

gloten, glotoun, *see* **gluten, glutūn.**

Gloucestre, pr. n., *O.E.* Glēawan ceaster; *Gloucester,* LAȝ. 16236; ROB. 2.

gloum, *see* **glomben.**

glouten, v., *look sullen*; **gloutid** (*pret.*) RICH. 4771.

glouting, sb., ? *looking sulky,* B. B. 134.

gloðeren, v., ? *flatter*: ne noȝt i kepe to gab ne gloþer C. M. 8401.

glōve, sb., *O.E.* glōf, glōfa, *O.N.* glōfi ; *glove,*
'*chirotheca,*' PR. P. 200 ; GOW. II. 370 ; C. M.
8116 ; LIDG. M. P. 110 ; glōven (*pl.*) LA3.
28581 ; A. R. 420 ; HORN (R.) 800 (H. 823) ;
glove HORN (L.) 794.

glōver, sb., *glover,* VOC. 181 ; glovare PR. P.
200.

glōw-berd, sb., *glow-worm,* '*nocticula,*' VOC.
255.

glōwin, v., *O.E.* glōwan (*pret.* glēow), = *O.N.*
glōa, *M.Du.* gloejen, *O.H.G.*.gluojan ; *glow,*
'*cando,*' PR. P. 200 ; glowe LANGL. *B* xvii.
222 ; PR. C. 7360 ; glōwende (*pple.*) ORM.
1067 ; glowinge MARG. 214 ; glōwid (*pret.*)
S. S. (Wr.) 2688 ; gloweden CH. C. T. *A* 2132.

　glōwinge, sb., *glowing,* '*candor, coruscacio,*'
PR. P. 200.

glū, sb., *O.Fr.* glū, glūs ; *glue,* AYENB. 246 ;
TREV. III. 451 ; *deriv.* glewishe, glüwi.

glubbere, *see under* gulpen.

glucchen, *see* gulchen.

glüffen, v., *dazzle, look bright:* no gai gere
to gliffe in 3our e3en ALEX. (Sk.) 4599 ; 3if 3e
þurh 3emeleaste glüffeð [gliffen] (*subj.*) of
wordes A. R. 46 ; gliffed (*pret.*) ANT. ARTH.
(M.) xxviii. (R. glisset) ; gliffnit BARB. vii. 184.

glüften, v., *look, gaze* ; gliftes (*pres.*) D.
ARTH. 2525 ; glifte (*pret.*) GAW. 2265 ; A. P.
ii. 849 ; (*? printed* glyste) FLOR. 1659 ; glifted
þei þer eiene MAN. (F.) 3399 ; 3if þat he
glüfte (*subj.*) in ani half LEG. 199 ; *comp.* a-
gliften.

glug, sb., *clod:* place of safir is stones and
the gluggis (*pl.*) of him gold WICL. JOB
xxviii. 6.

glüin, v., *O.Fr.* gluer ; *glue,* PR. P. 200.

　glüinge, sb., *gluing, sticking,* '*conglutinacio,*
conviscacio,*' PR. P. 200.

glummen, *see* glommen.

glut, sb., *O.Fr.* glout ; *glutton,* PL. CR. 67 ;
gluttis (*pl.*) ALEX. (Sk.) 4532.

gluten, v., *O.Fr.* glotir, gloutir ; *glut* ; glotie
LANGL. *C* x. 76*.

gluterie, sb., *gluttony:* gloterie MISC. 217 ;
glotori, glotteri, glutiri, glutturi HAMP. PS.
lxxxvi. 4, xiii. 5*, lxviii. 19, lxxxii. 6* ; glotries
(*pl.*) ['*comessationibus*'] WICL. DEUT. xxi. 20.

gluternesse, sb., *? from O.N.* glutr ; *? luxury,*
ORM. 11653.

glutūn, sb., *O.Fr.* glutun, gluton ; *glutton ;*
A. R. 214 ; glotoun AYENB. 50.

　glotoun-līche, adv., *like a glutton:* þet þou
sselt ete zviþe and glotounliche AYENB. 110.

glutunerie, sb., *O.Fr.* gloutonerie ; *gluttony* ;
(*? ms.* glutenerie) HOM. I. 49.

glutunie, sb., *O.Fr.* gloutonie ; *gluttony,* A.
R. 194 ; MISC. 31 ; glotonie [gloteni3e] ROB.
(W.) 6782 ; SPEC. 49 ; LANGL. *A* PROL. 22 ;
glotonies CH. C. T. *C* 514.

glutunous, adj., *gluttonous, greedy* ; gloton-
ous CH. BOET. i. 5 (26) ; glotounius ALEX.[2]
(Sk.) 790.

glüwi, adj., *from* glü ; *gluey :* pittis of gluwi
clei WICL. GEN. xiv. 10.

gnacchen, v., *gnash,* REL. I. 240.

gnack, sb., *O.N.* gnak ; *trick :* laweieris
wiþ here gnackis (*pl.*) & japes WICL. E.
W. 184.

gna3en, v., *O.E.* gnagan, = *M.L.G.* gnagen,
O.H.G. gnagan, nagan, *O.N.* gnaga, naga ;
gnaw ; gnaghe HAMP. PS. cxviii. 40* ; gnawen FRAG. 6 ; gnawe ROB. 404 ; CH. TRO.
iv. 621 ; PR. C. 6908 ; gna3(e)st (*pres.*) SHOR.
97 ; gnaweþ MISC. 151 ; gnawe (*subj.*) M. H.
142 ; gnōgh (*pret.*) H. S. 3581 ; gnou3 (*ms.*
gnow3e), gnou (*ms.* gnow), gneu C. M. 6043 ;
gnou (*ms.* gnow) RICH. 5074 ; CH. C. T. *B*
3638 ; gnowen WICL. JOB xxx. 3 ; MISC. 225 ;
gnowe TREV. IV. 137 ; gnōwe (*subj.*) ROB.
539 ; gnawen (*pple.*) PR. C. 864 ; ignawe
HICKES I. 227 ; *comp.* for-, tō-gna3en.

gnawinge, sb., *gnawing :* gnawinge, or foule
biting '*corrosio*' PR. P. 200 ; bi gnawing of
rattis CH. C. T. *I* 605.

gnaggen, v., *gnaw* ; gnaggid (*pple.*) up bi
the gomis LUD. COV. 384.

gnaisten, *see* gnāsten.

gnappen, v., *? for* cnappen ; *? = M.Du.* knap-
pen ; *? bite, snap at* ; sum(e) gnappede (*pret.*)
here fete and handes H. S. 10208.

gnāre, sb., *snare :* be the boord of hem maad
bifore hem into a gnare [*later ver.* gin] WICL.
ROM. ii. 9 ; gnare of the devil WICL. I TIM.
vi. 9 ; he hangede hem with a grane or a
gnare WICL. MAT. xxvii. 5 ; gnare [grane]
WICL. LK. xxi. 35* ; gnáris (*pl.*) WICL. 2
KINGS xxii. 6 ; WICL. S. W. II. 363.

gnāren, v., *choke, strangle* ; *snare, entraþ* ;
to gnare men in his net WICL. E. W. 437 ;
as taren gnaren corn WICL. S. W. I. 96 ;
gnárid (*pple.*) with the woordis of the mouth
WICL. PROV. vi. 2 ; gnared WICL. IS. xxviii. 13.

gnāst, sb., *O.E.* gnāst, *cf. O.N.* gneisti, *O.H.G.*
gneist ; *burning ashes* ['*favilla*'] WICL. IS.
i. 31 ; þene ene gnast HOM. I. 81 ; gnast,
knast '*emunctura*' PR. P. 277 ; gnōstis (*pl.*)
HALLIW. 406.

gnāstin, v., *O.N.* gneista ; *gnash,* '*fremo,*
strido,*' PR. P. 200 ; he shal gnasten ['*frendet*']
WICL. IS. v. 29 ; he with his tethe gnaistes
ALEX. (Sk.) 5321 ; gnaiste PR. C. 7338 ;
gnaisted (*pret.*) '*fremuerunt*' PS. ii. 1 ; þeir
teþ gnaisted MAN. (F.) 1821 ; gnaistid HAMP.
PS. ii. 1.

　gnāstere, sb., *gnasher,* '*fremitor,*' PR. P.
200.

　gnāstinge, sb., *gnashing,* '*fremitus,*' PR. P.
200 ; gnastinge of horsis WICL. JER. viii.
16 ; þer endeles gnaisting is of tooth C. M.
26760 ; þe voices of gnastinge TREV. VII. 81.

gnat, sb., *O.E.* gnætt, ?=*Swed.* gnadd (RIETZ 204); *gnat,* MIRC 1937; CH. C. T. *D* 347; gnet MAT. xxiii. 24; þene gnet A. R. 8; **gnattes** (*pl.*) GEN. & EX. 2988; PS. civ. 31; TREV. V. 159.

gnawen, gnawing, *see* **gnaȝen.**

'gnēde, adj., *O.E.* gnēde, (un-)gnȳde; *stingy,* MISC. 92; HAV. 97; MIRC 319; GREG. 823; C. M. 5392; TRIST. 2838; her beþ blisse gnede REL. I. 120; of gifte was he never gnede DEGR. 1159; a gnedi gloton with to grete chekes LANGL. *C* xvi. 85; (g)nede A. P. ii. 146; gnede, knede (*ms. in gloss.*), A. P. I. 154.

  gnēde-līche, adv., *O.E.* gnēdelīce; *stingily,* A. R. 202.

gnēden, v., ?*grudge*: these, ful securli mete and drinke at thi nede, non that dai shal thee gnede MAS. 667.

gnet, *see* **gnat.**

gnīden, v., *O.E.* gnīdan,=*M.L.G.* gnīden, *O.H.G.* gnīten; *rub, brush,* A. R. 238; gnide (*ms.* gnyde, *printed* guyde) REL. I. 175; **gniden** (*pret.*) A. R. 260; *comp.* **for-gnīden.**

gnitte, sb., *cf. M.L.G.* gnitte; ?*gnat*; **gnittus** (*pl.*) REL. I. 86.

gnof, sb., ?*churl; miser*; a riche gnof CH. C. T. *A* 3188.

gnōst, *see* **gnāst.**

gnudden, v., *rub, brush, crush,* A. R. 238*; **gnodde** (*imper.*) HALLIW. 406; **gnodded** (*pret.*) CH. APPEND. BOET. i. 11; moni kniȝt ... gnodded (*printed* guodded) gras on þe grounde ALIS. 2374; gnuddeden A. R. 260*.

ȝo, *see* **hēo.**

gobbe, sb., *lump, mass,* ['*moles*'] WICL. IS. xl. 12.

gobben, v., *cut in lumps*; goben, gobone (*imp.*) B. B. 155, 281.

gobelin, sb., *Fr.* gobelin; *goblin,* WICL. PS. xc. 6; goblin ALEX. (Sk.) 5491.

gobet, sb., *Fr.* gobet; *small piece, 'massa, frustum,'* PR. P. 200; TREV. IV. 399; WICL. ROM. ix. 21; CH. C. T. *A* 696.

  gobet-līche, adv., *piecemeal* ['*membratim*'], TREV. IV. 103.

  gobet-mēle, adv., *piecemeal,* WICL. 2 MAC. xv. 33.

goblet, sb., *goblet*; **goblets,** goblettis, gobletis, goblettes ALEX. (Sk.) 2935, 3701, 5131.

goboun, sb., *fragment*; **gobouns** (*pl.*) TREV. IV. 155; gobins B. B. 161.

ȝoc, sb., *O.E.* geoc, gioc, joc, juc,=*Goth.* juk, *O.N.* ok, *O.H.G.* joh, *Lat.* jugum; *yoke,* A. R. 156; (*ms.* ȝocc) ORM. 4045; ȝok VOC. 234; C. L. 957; CH. C. T. *E* 113; APOL. 80; ioc FRAG. 5; yok AYENB. 255; yok [ȝhoc] PS. ii. 3; **giókes** (*pl.*) HOM. II. 195; ȝockis WICL. JOB i. 3.

god, sb., *O.E.* god,=*O.L.G., O.Fris.* god, *O. H.G.* got, cot, *O.N.* goð, guð, *Goth.* guþ; *god,*

LAȝ. 5354; BEK. 2; god woof [goddot] *God knows* C. M. 3729; goddot HAV. 606; AN. LIT. 13; **gódes** (*gen.*) O. & N. 357; CHR. E. 630; for godes luve H. M. 35; godes (*ms.* godess) sune (*the son of God*) ORM. 267; þe godes peni AYENB. 91; góde (*dat.*) HOM. I. 87; ORM. 1952; GEN. & EX. 2816; swa ich mote gode iþeon LAȝ. 9116; vor gode (*by God*) ROB. (W.) 7000; gódes (*pl.*) KATH. 53; **goden** (*dat. pl.*) LAȝ. 8378.

god-child, sb., *godchild,* A. R. 210; **godchildere** (*pl.*) MIRC 152.

god-cund, adj., *O.E.* godcund, = *O.L.G.* godcund, *O.H.G.* gotchund; *divine,* ORM. 5873; sumne godcunde mon LAȝ. 10139.

  godcund-hēde, sb., *divinity,* REL. I. 222.

  godcund-leȝc, sb., *divine nature,* ORM. 1389.

  godcundnesse, sb., *divinity,* KATH. 985; godcun(d)nesse P. L. S. viii. 196; HOM. II. 73.

god-cünde, sb.,=*O.H.G.* gotchundī; *divinity*; goddcunde LAȝ. 24960.

god-doȝter, sb., *O.E.* goddohter; *god-daughter,* AYENB. 48.

goddot, *see* **god.**

god-fader, sb., *O.E.* godfæder; *godfather,* HOM. II. 17; SHOR. 69.

god-ful-hēd, sb., *divinity,* GEN. & EX. 53.

god-fürht, adj. & sb., *O.E.* godfyrht; *god-fearing,* HOM. I. 27; godfruht, -friht II. 19, 167, 187.

  godefrigti-hēd, sb., *piety,* GEN. & EX. 493; frigtihed 540.

god-hōd, sb.,=*O.H.G.* gotheit; *godhead,* A. R. 112; godhede C. L. 1335; AYENB. 12; MAND. 144.

god-mōder, sb., = *O.E.* godmōder; *god-mother,* AYENB. 48; ED. 588.

godnesse, sb., = *O.H.G.* gotnissa; *deity,* HOM. I. 101.

god-sib, sb., = *O.E.* godsibb, *m.*; *gossip,* LANGL. *A* v. 152; gossib CHAUC. C. T. *D* 243; of his godzib oþer of his godzibbe AYENB. 48.

  godsib-rēde, sb., SHOR. 69; gosibrede TREV. I. 357; gossiprede LIDG. M. P. 36.

god-sune, sb., *O.E.* godsunu; *godson,* VOC. 214; godsone BRD. 2.

gōd, adj. & sb., *O.E.* gōd,=*O.L.G., O.Fris.* gōd, *Goth.* gōds, *O.N.* gōðr, *O.H.G.* guot; *good; goodness, goods,* LAȝ. 1782, 29203; good MAND. 56; on good man MISC. 33; goed AN. LIT. 96; guod SAINTS (Ld.) 312; AM. & AMIL. 16; gud PR. C. 80; ISUM. 177; gude MIN. iii. 37; þet gōd (*sb.*) A. R. 248; al þat god of þisse londe LAȝ. 999; al hire god WILL. 3523; ȝif ȝe maȝen eni oðer god don HOM. I. 37; þu dest me god A. R. 124; hit doþ me god ASS. 210; guod AYENB. 6; AL. (T.) 190; se gode man HOM. I. 237; þat gode wif O. & N. 1578; þet gode & þet ufele HOM. I. 7; i praise þe gode man FER. 1729; guode man LEG. 32; one gode stunde MISC. 42; þe gode mannes ORM. 1459; gōdes (*gen. m.*) LAȜ. 15790; in so gode kinges londe A. & N. 1095; hwat dostu godes O. & N. 563; gōde (*dat. m.*) LAȜ. 27733; a gode Fridai BRD. 17; hu scolde oðer monnes goddede comen him to gode [= *Ger.* ihm zu gute kommen] HOM. I. 9; al þat þu maht don to gode 23; þe þe mest deð nu to gode P. L. S. viii. 31; what se we don to gode ORM. 426; and singe . . . for hire gode O. & N. 1596; a ðat ha hit leose for gode H. M. 35; wiþ heore caˆtel and heore gode TRANS. XVIII. 24; gōdere (*dat. f.*) LAȜ. 3597; mid godere heorte HOM. I. 3; efter godere lore A. R. 428; to godre heale 194; goder hele (*for the prosperity of*) ROB. (W.) 7570; gōdne (*acc. m.*) O. & N. 812; SHOR. 135; godne dai HORN (R.) 731; P. L. S. ix. 200; AN. LIT. 6; godne, goudne [godne] LAȜ. 3538, 8299; guodne AYENB. 238; gōde (*acc. f.*) LAȜ. 19195; gode (*pl.*) LAȜ. 521; O. & N. 605; gode men ORM. 3954; þeo goden FRAG. 7; þe gode A. R. 178; gode men HOM. I. 5; gōdere (*gen. pl.*) MAT. xv. 49; LAȜ. 5705; HOM. I. 9; gōdan (*dat. pl.*) HOM. I. 107; goden his gode dide P. L. S. xvii. 485; HOM. I. 229; LAȜ. 15206; of ealle gode men SAX. CHR. 257; to þe guoden AYENB. 72.

gōd-dēde, sb., *O.E.* gōddæd, = *M.H.G.* guottāt; *good deed*, HOM. I. 9; LAȜ. 21072; A. R. 150; MISC. 81.

guod-dōere, *good-doer*, AYENB. 135.

gōd-ful, adj., *good*, LAȜ. 17038.

gōdful-līche, adv., *kindly*, MISC. 90.

gōd-hēde, sb., = *M.H.G.* guotheit; *goodness*, LAȜ. 21072*; O. & N. 351; ALIS. 7060; FLOR. 1682; guodhēde AYENB. 233.

gōd-lies, adj., *O.E.* gōdlēas; *devoid of good; having no possessions, poor,* P. L. S. viii. 173; godles KATH. 846; H. M. 31.

gōd-leic, adj., *O.N.* gōðleikr; *goodness,* A. R. 136; godleic, -lec KATH. 290, 899; godlec MARH. 10; AN. LIT. 8; godleȝc ORM. 1768.

gōd-lich, adj., *O.E.* gōdlīc, = *O.L.G.* gōdlīc, *O.H.G.* guotlīch; *goodly,* ANGL. III. 280; A. P. ii. 753; godli IPOM. 75; gudlīche cnihtes LAȜ. 860; mid godlīche scruden 5362.

gōdlīche (*adv.*) LAȜ. 18857; DEGR. 675; goodliche C. L. 1396; godli WILL. 169.

goodlī-hēde, sb., *goodness,* R. R. 4604; CH. TRO. ii. 841.

goodlinesse, sb., *goodness,* 'benignitas, benevolencia,' PR. P. 201.

gōd-man, sb., *master, sir,* ALEX. (Sk.) 2407; gudman 436; goodman CH. C. T. *C* 361; gudmans (*gen.*) (*hero's*) ALEX. (Sk.) 3433; goodmen (*pl.*) (*wealthy men*) DEP. R. i. 66.

gōdnesse, sb., *O.E.* gōdness, = *O.H.G.* guotnassi; *goodness,* LAȜ.6897; ORM.2109; TRANS. XVIII. 28; guodnese AYENB. 18; godnisse P. L. S. xi. 20; goodnesse, gudnes ALEX. (Sk.) 1676, 3103.

good-schipe, -schüpe, sb., *goodness*; C. L. 503.

gōd-spel, sb., *O.E.*, godspell, (*whence O.L. G.* godspell, *O.H.G.* gotspel, *O.N.* guðspiall); *gospel,* O. & N. 1270; BEK. 1072; M. H. 5; þat godspel (*ms.* godd-) LAȜ. 29525; gospel CH. C. T. *A* 481; þet gospel A. R. 388.

gōdspel-bōc, sb., *O.E.* godspellbōc; *gospel,* ORM. 6458.

gōdspellere, sb., *O.E.* godspellere; *evangelist,* HOM. I. 89; A. R. 94*; AYENB. 12; M. H. 47; gospellere R. R. 6887.

gōdspel-wrihte, sb., *evangelist,* ORM. 5635.

godard, sb., *? O.Fr.* godart; *gutter, drain*; the water gosshet through godardis (*pl.*) D. TROY 1607.

goddesse, sb., *goddess,* TREV. III. 73; CH. C. T. *A* 2063; godas TREV. IV. 65.

ȝode, *see* ēode. gōde, *see* gāde.

godelen, v., *slander* : hvanne þe on godeleþ (*pres.*) þane oþrene AYENB. 66.

godelinge, sb., *slandering,* AYENB. 6; godelinges (*slanderers*) AYENB. 66.

gōden, v., *O.E.* gōdian, = *M.Du.* goeden, *M.H. G.* guoten, güeten; *make good, become good; endow with goods,* ORM. 10866, 15909; & gōded (*pret.*) (*improved*) it svithe SAX. CHR. 263; þat þing þat gōded (*pple.*) is & eked ORM. 2124; *comp.* ȝe-gōden.

gōdlēas, *see under* gōd.

gōere, *see under* gangen.

gōfisch, adj., *foolish*; goofish [gofisshe] CH. TRO. iii. 535.

gogathes, *see* gagates.

[gogel? sb. *or* adj.]

gogil-ēiid, adj., *goggle-eyed,* 'limus,' PR. P. 201; gogiliȝed ['luscus'] WICL. MK. ix. 46.

gogelen, v., *look aside*; gogelen (*pres. pl.*) fer fro goddis lawe WICL. E. W. 341.

ȝoȝelinge sb., *? monotonous song,* O. & N. 40; ȝuhelinge JUL. 57.

goie, *see* joie.

gojon, sb., *O.Fr.* goujon, goignon; *pivot*; gojone of a polein 'vertibulum, cardo' PR. P. 201.

gojon, sb., *Fr.* goujon; *gudgeon,* 'gobio,' PR. P. 201, 202.

ȝok, *see* ȝoc. gōk, *see* ȝēac.

ȝokel, sb., *O.N.* iökull; *icicle*; yokle PR. P. 259.

ȝōken, v., *cf. Ger.* jochen, *Lat.* jugāre; *yoke,* GOW. II. 63; yoken 'jugo' PR. P. 539; heo ȝeókeden (*pret.*) heora earmes LAȜ. 1872;

iȝóked (*pple.*) TREV. VII. 445; iȝoket REL. II. 210.

gōki, sb., *mod.Eng.* gawky; *fool*, LANGL. *B* xi. 299; goukou LANGL. *C* xiv. 120*; *see* ȝēac.

ȝokkin[g], *see* ȝicching.

gol, *see* gul.

gōl, sb., *? O.Fr.* gaule; *goal*, SHOR. 145.

[gōl? adj., *cf. O.N.* gōla (*well*), gōlastr (*best*), *O.H.G.* urguol (*famous*).]

gōl-lī, adj., *? O.N.* gōligr, = *O.H.G.* guollīch; *? noble, famous*, REL. I. 85; golike tun ORM. 15662.

gōl, *see* gāl.

ȝōl, sb., *O.E.* geōl, *O.N.* jōl; *yule, Christmas*, ORM. 1915; AN. LIT. 5; MAN. (F.) 10370; GAW. 284; ȝoill, yule, ȝhule BARB. ix. 204; ȝōles (*gen.*) ORM. 1910; ȝōle (*dat.*) PERC. 1803; *cf.* ȝōle.

ȝeōl-dæi, sb., *Yule day*, LAȝ. 22737; ȝoldaȝ ORM. 11063; ȝoldai S. & C. I. liii.

ȝoill-ēvin, sb., *Christmas-eve*, BARB. ix. 204.

iōl-niht, sb., M. H. 101.

góld, sb., *O.E.* gold, = *O.L.G.*, *O.H.G.* gold, *Goth.* gulþ, *O.N.* gull; *gold*, LAȝ. 4779; ORM. 6474; þat gold REL. I. 173; þet gold AYENB. 167; golde (*dat.*) SPEC. 38; gulde Iw. 888.

góld-fæt, sb., *O.E.* goldfæt, = *O.L.G.* goldfat; *golden vessel*, FRAG. 6.

góld-fāh, adj., *O.E.* goldfāh; *gold coloured*, LAȝ. 31406; góldfóhne (*acc. m.*) FRAG. 6; góldfāȝe [goldfawe] (*pl.*) LAȝ. 26706.

góld-finch, sb., *O.E.* goldfinc; *goldfinch*, O. & N. 1130; ALIS. 783; CH. C. T. *A* 4367.

góld-flūr, sb., *? the aurelia*; góldeflūrs (*pl.*) VOC. 162.

góld-hord, sb., *O.E.* goldhord; *gold-hoard*, HOM. I. 109; MARH. 17; A. R. 152; S. S. (Wr.) 2004.

góld-hous, sb., *treasury*, HALLIW. 408.

góldisch, adj., *like gold*: al is not golde that shinethe goldisshe hewe LIDG. M. P. 190.

góld-mestling, sb., *O.E.* goldmæstling; '*aurichalcum*' FRAG. 4.

góld-plante, sb., góldplantes (*pl.*) VOC. 162.

góld-qvarelle, sb., '*aurifodina*,' VOC. 271.

góld-ring, sb., *O.H.G.* goldring, *O.N.* gullhringr; *gold ring*, HOM. I. 193; LAȝ. 24600.

góld-smið, sb., *O.E.* goldsmið, *O.H.G.* goldsmid; *goldsmith*, A. R. 236; goldsmith CH. C. T. *G* 1337.

góld-smiðri, sb., *goldsmith's work*: herneis so . . . riche wrought and wel of goldsmithri of brouding, and of steel CH. C. T. *A* 2498.

góld-þrēd, sb., *M.H.G.* goltdrāt; *gold-thread*, CH. C. T. *B* 3665.

góld-webbe, sb., *gold tissue*, AMAD. (R.) xlv.

góld-wīr, sb., *gold-wire*, LAȝ. 7048; LANGL. (Wr.) 901; ALIS.² (Sk.) 180; PR. C. 9107.

góld², sb., *the turnsol* (*a plant*): [Leuchothoe] sprong up out of the molde into a flour, was named golde GOW. II. 356; góldes (*pl.*) *marigolds* PALL. iv. 174; gouldes [*intubis*] *endive* PALL. i. 702.

gólden, *see* gülden.

góle, sb., *O.Fr.* gole, goule; *throat*, CHEST. I. 229; D. ARTH. 3725; goulez (*pl.*) *gules* GAW. 610; gols ALEX. (Sk.) 4819.

ȝōle, sb., *O.E.* geōla; *December*, MAN. (H.) 47; H. S. 815.

golee, sb., *O.Fr.* goulee; *mouthful* (*of words*), CH. P. F. 556.

golet, sb., *O.Fr.* goulet; *gullet; gorget*, PR. P. 202; CH. C. T. *C* 543; D. ARTH. 1772.

golf, sb., *O.N.* golf; *heap of sheaves*, PR. P. 202.

goliardeis, sb., *O.Fr.* goliardois, *from med. Lat.* goliardus; *buffoon; teller of ribald stories*: he was a jangler, and a goliardeis CH. C. T. *A* 560; a goliardeis a gloton of wordes LANGL. *B* PROL. 139.

golíke, *see under* gōl.

golanand, pple., *? cf. O.N.* gola, gula (*breeze*); *tempestuous*, ALEX. (Sk.) 4796.

goliōn, *see* guliōn.   ȝolke, *see* ȝeolke.

goll, sb., *? mod.Eng.* gull; *young bird*; gollis (*pl.*) ['*pulli*'] WICL. DEUT. xxii. 6.

ȝollen, v., = ȝeollen; ȝellen; *? howl*; ȝolle O. & N. 972; ȝolle (*v. r.* ȝelle) TREV. IV. 395; yolle (*pres.*) CH. C. T. *A* 2672; ȝollinge (*v. r.* ȝullinge, ȝellinde) (*pple.*) L. H. R. 44; ȝollide (*pret.*) ['*ululabat*'] WICL. JUDG. v. 28.

ȝollinge, sb., *yelling*, WICL. NUM. x. 6; yollinge O. & N. 1643.

ȝolpen, = ȝeolpen, ȝelpen; *? boast*; ȝolped (*pret.*) A. P. ii. 846.

golsoght, *see under* gul.

ȝolu, *see* ȝeolu.

golven, v., *stack corn*; golvin or golvon, '*arconiso*,' PR. P. 202; golve VOC. 154.

ȝoman, sb., *? for* ȝung man, *mod.Eng.* yeoman; *young man; manservant, steward*, WILL. 3649; qven he throded was to yoman, he was archer C. M. 3077; '*effebus, valleta*' CATH. 427; Heliceus, ȝoman and despencer of Abraham MAND. 123; kniȝte, squiere, ȝoman and knave AMAD. L.; yoman AM. 635; ȝomane D. ARTH. 2629; yeman [*v. r.* ȝoman] CH. C. T. *A* 102; ȝeman S. & C. II. xxx; ȝheman BARB. v. 235; ȝomans (*pl.*) ISUM. 408; ȝomen (*pl.*) WILL. 3649; MAN. (H.) 297; ȝhomen BARB. v. 257; ȝemen A. P. i. 534; CH. C. T. *A* 2511.

gomie, *see* gume.

gōme, sb., *O.N.* gaum, = *M.L.G.* gōm, *M.Du.* goom, *O.H.G.* gauma, gouma; *care*, R. S. v;

MISC. 45 ; P. L. S. ii. 5 ; (*printed* ʒome) GREG. 987; nime gome ROB. 38 ; nemaþ gome FER. 1745 ; gom ORM. 916 ; *see* ʒēme.

gōme², sb., *O.E.* gōma, = *O.H.G.* guomo, *cf. O.N.* gōmr ; *gum, palate,* 'gingiva,' PR. P. 202 ; VOC. 245 ; L. H. R. 218 ; gōmes (*pl.*) REL. II. 78 ; *see* gūme.

gomen, *see* gamen.　ʒōmer, *see* ʒēomer.

gomme, *see* gumme.　ʒon, *see* ʒeon.

gōn, *see* gān.

ʒond, ʒonder, ʒone, *see* ʒeond, ʒeonder, ʒeon.

gonel, sb., *O.Fr.* gonelle, gonele ; *gown* ; gonels (*pl.*) FER. 4345, 4369.

gonfanoun, *see* gunfanoun.　gong, *see* gang.　ʒong, *see* ʒung.　gongen, *see* gangen.

ʒonien, *see* ʒeonien.　gōnien, *see* gānien.

goot, *see* gāt.　ʒōp, *see* ʒēap.

góre, sb., *O.E.* gor, = *O.N., O.H.G.* gor, *mod. Eng.*'gore ; *mud,* '*limus,*' PR. P. 203 ; A. P. ii. 306 ; (*ms.* gorre) AL. (L.) 1005 ; gore and fen B. DISC. 1471 ; gorred (*pret.*) ALEX. (Sk.) 4645.

gōre, *see* gāre.　ʒōre, *see* ʒēare.

gorge, sb., *O.Fr.* gorge ; *throat, neck,* LANGL. B x. 66 ; D. ARTH. 3760 ; gorg ALEX. (Sk.) 4985.

gorgen, v., *O.Fr.* gorger ; *gorge* : all hei gorgen (*pres. pl.*) as a ravene ALIS. 5625.

gorgere, sb., *O.Fr.* gorgiere ; *armour for the throat* : RICH. 321 ; B. DISC. 1616 ; gorger (*neckerchief*) GAW. 957.

gorst, sb., *O.E.* gorst ; *gorse,* '*juniperus,*' REL. I. 37 ; WICL. IS. lv. 13 ; gorstez (*pl.*) A. P. ii. 534.

ʒorþe, *see* eorþe.

gōs, sb., *O.E.* gōs, = *O.N.* gās, *O.H.G.* gans, *f., Gr.* χήν *m. f.* ; *goose,* HAV. 1240 ; REL. I. 217 ; P. S. 333 ; goos CH. C. T. *A* 4137 ; MAND. 49 ; guos AYENB. 32 : gēs (*pl.*) A. R. 128 ; gees HAV. 702 ; MAND. 49 ; LIDG. M. P. 118 ; gies S. A. L. 156.

　gōs-ei, sb., *goose-egg,* PL. CR. 225.

　gōs-flēsch, sb., '*caro aucina,*' goseflesche VOC. 300 (W. W.) 661.

　gōs-hauk, sb., *O.E.* gōshafoc ; *goshawk,* ALIS. 483 ; WICL. JOB xxxix. 13 ; CH. C. T. *B* 1928.

　gōs-herde, sb., *goose keeper,* PR. P. 204.

　gōsis gres, sb., *a plant* : gosis gres or cameroche or wilde tanzi PR. P. 204.

　gōsling, sb., *gosling,* '*ancerlus,*' VOC. 177 ; PR. P. 204 ; gesling VOC. 187 ; geslinge VOC. 220 ; gusling 252 ; whan . . . gōslinges (*pl.*) hunt the wolfe to overthrow S. & C. II. lviii.

gosibrēde, *see* godsib *under* god.

ʒosking, *see* ʒixing *under* ʒeoxen.

gŏspel, *see under* gŏd.

gosshien, v., *wag, shake* ; þih and shonkes

and fet oppieʒ, wombe gosshieʒ (*pres.*) and shuldres wrenchieʒ HOM. II. 211 ; *see* guschen.

gossib, *see* godsib *under* god.

gossomor, sb., *gossamer* ; gosesomer '*filaundre*' VOC. 147, 164 ; gossomire 239 ; gossummer 273 ; gossomer '*filandria, lanugo*' PR. P. 205 ; CH. C. T. *F* 259 ; LIDG. M. P. 26.

gōst, *see* gāst.　gōstid, *see* gāsted.

gŏt, *see* gāt.

góte, sb., = *M.L.G., M.Du.* gote ; *prov. Eng.* goit ; *from* ʒēoten ; *channel, stream,* PR. P. 205 ; gótez (*pl.*) A. P. i. 607 ; gotis ALEX. (Sk.) 4796.

gotere, sb., *O.Fr.* goutiere ; *furrow, gutter,* PR. P. 206 ; '*aquaductum, guttarium aquaductile,*' VOC. 237 ; gotir 272 ; goter CH. TRO. iii. 737 ; goteres (*pl.*) TREV. I. 181 ; gutters HAMP. PS. xli. 9 ; goteris WICL. PS. xliv. 11.

gotous, adj., *O.Fr.* gutus, goteux ; *gouty* ; gotous mann or woman '*guttosus*' PR. P. 206.

gottes, *see* gut.

goþelen, v., = *Ger.* godeln, gudeln ; *sputter, bubble* : hise guttes bigonne to goþelen LANGL. *B* v. 347 ; a slab of ire þat glowinge a fure were in water hit wolde goþeli loude TREAT. 135.

ʒou, *see* ēow.

gouge, sb., *Fr.* gouge ; *gouge,* (*ms.* gowge) N. P. 18.

ʒouhþe, *see* ʒuʒeðe.　gouk, *see* ʒēac.

gouki, *see* gōki.　goul, *see* gavel.

goules, *see* góle.

goulen, ʒoulen, v., *O.N.* gaula ; *howl* ; goule MISC. 228 ; PR. C. 477 ; i shal ʒoule ['*ululabo*'] WICL. MIC. i. 8 ; ʒaulé (*pres.*) GAW. 1453 ; goul (*imper.*) WICL. EZ. xxi. 12 ; goulende (*pple.*) REL. I. 291 ; gouleden (*pret.*) HAV. 164 ; ʒaulit ANT. ARTH. vii.

gound, *see* gund.　goune, *see* gūn.

goupen, sb., *O.N.* gaupn, = *G.* gaufen ; *double handful* ; goupines (*pl.*) VOC. 147*.

ʒour, *see* ēower.

gourde, sb., *Fr.* gourde ; *gourd* ; CH. C. T. *H* 82 ; PALL. iv. 456 ; gurds (*pl.*) ALEX. (Sk.) 3701.

goute, sb., *O.Fr.* goute ; *gout,* ROB. 564 ; SPEC. 48 ; SHOR. 2 ; the goute lette hire nothing for to daunce CH. C. T. *B* 4030.

gouten, v., *Fr.* goutter ; *drop, drip* ; gouton as candelis '*gutto*' PR. P. 206.

goutous, adj., *O.Fr.* gutus ; *gouty* : a qvene goutus and croket REL. I. 196.

ʒouþe, *see* ʒuʒeðe.　govel, *see* gavel.

govern, sb., *O.Fr.* governe ; *government, management* : his bischopriche hadde ibeo withoute govern and rede BEK. 1789.

governail, sb., *O.Fr.* governail, *from Lat.* gubernāculum ; *rudder; power to govern* ;

governaile CH. C. T. *E* 1192; GEST. R. 263;
schippes . . . born aboute of a litel governaile
WICL. JAS. iii. 4; knawin war of gud governale
BARB. xi. 161; of the marchis than had he
the governale xvi. 358; **governailis** (*later
ver.* governails) (*pl.*) WICL. DEEDS xxvii. 40.

governance, sb., *O.Fr.* gouvernance; *government*, CH. C. T. *E* 23; MAND. 38.

governen, v., *O.Fr.* governer, guverner;
govern; governe ROB. 106; governi SHOR.
109; how i mai governe me TOR. (A.) 779;
governe LANGL. *A* iii. 271; **governeþ** (*pres.*)
AYENB. 124; **governit** (*pret.*) BARB. xx. 603;
governit (*pple.*) BARB. vi. 366.

governour, sb., *O.Fr.* guverneur; *steersman,
ruler, governor,* WICL. DEEDS xxvii. 11; D.
ARTH. 1201; WICL. PROV. xxii. 34; **governers** (*pl.*) ALEX. (Sk.) 3552.

ʒow, see **ēow.**

gōwen, ? = **gāwen,** v., tu (= þu) **gōwes** (*pres.*)
þær onne ORM. 12233.

ʒoxe, ʒoxen, see **ʒeoxe, ʒeoxen.**

grā, see **grai.**

grāce, sb., *O.Fr.* grace; *grace,* HOM. I. 49;
LAʒ. 6616*; A. R. 8; KATH. 298; HORN (L.)
571; BEK. 421; as him grace (*chance*) fell D.
TROY 76; grass BARB. xiv. 361; þe borde
was leid . . . þe **grāces** (*pl.*) seid MAN. (F.)
16086.

    grāce-dōing, sb., *thanksgiving*: WICL.
Is. li. 3.

    grāce-lēs, adj. & sb., *graceless*, '*ingraciosus*',
PR. P. 206; **grācelees** (*sb.*) CH. C. T. *G* 1078.

gracious, *O.Fr.* gracieus; *gracious, beautiful,*
WILL. 312; graceux ALEX. (Sk.) 3667, 798*;
gracieux 1964; **graciouser** (*compar.*) AYENB.
24; **graciousest** (*superl.*) ALEX. (Sk.) 4909.

    gracious-līche, adv., *graciously,* TREV. VII.
35; graciouslie PENIT. PS. 28; graciousli A. P.
ii. 488.

gradeuable, adj., *having a degree*; clerkes
that ben gradeuable B. B. 284.

grádid, pple., for **degrádid**; *degraded,* ALEX.
(Sk.) 2430.

grēd, sb., *? clamour*; gred GEN. & EX. 3230,
3717; grad ALIS. 5204.

grēden, v., *O.E.* grǣdan, grēden; *cry out,* A.R.
236; H. M. 47; greden LANGL. *A* iii. 63; grede
O. & N. 308; HAV. 96; BEK. 129; AYENB.
22; grede (*v. r.* grete) ant grone H. H. 82;
crie and grede P. L. S. i. 32; blowe and grede
M. ARTH. 791; wepe and grede LUD. COV.
361; graden ALIS. 2751; **grēdeþ, grēdeð**
(*pres.*) HOM. II. 129; ALIS. 142; gret A. R.
330; AYENB. 56; gred O. & N. 1533; (heo)
gredeþ O. & N. 1671; AYENB. 71; alle þo þat to
me grede (*cry*) SPEC. 82; **grēd** (*imper.*) A. R.
290; HOCCL. vi. 36; **grēde** (*subj.*) A. R. 284;
grēdde [gradde] (*pret.*) LAʒ. 8634; gredde
A. R. 244; GEN. & EX. 3585; HORN (R.)

1202; gradde O. & N. 1662; ALIS. 1067;
gredden SHOR. 84; gradde AL. (L.¹) 364;
comp. **bi-grēden.**

grēdinge, sb., *clamour,* LAʒ. 23564*; AYENB.
212; greding S. J. 156.

grēdi, adj., *O.E.* grǣdig, = *O.L.G.* grādag,
Goth. grēdags, *O.N.* grādugr, *O.H.G.* grātag;
greedy; grediʒ ORM. 10217; gredi FRAG. 7;
P. L. S. viii. 133; GEN. & EX. 1494; A. D.
292; þenne bið he gredi þes eses HOM. I.
123; þe gredie leo 127; **grǣdie** (*pl.*) FRAG.
6; gredie hundes A. R. 324; **grēdiure**
(*compar.*) A. R. 416.

grēdiʒ-leʒc, sb., *greediness,* ORM. 3994.

    grǣdi-līche, adv., *greedily,* FRAG. 6; gre-
diliche A. R. 240*; gredilike REL. I. 215;
gredeliche HOM. II. 173; gredeli ALEX. (Sk.)
1435.

grēdiʒnesse, sb., *O.E.* grǣdignyss; *greediness,* ORM. 4522; gredinesse HOM. I. 103;
A. R. 416.

grǣf, sb., *O.E.* grǣf; *carved work,* '*sculptura*,'
FRAG. 2; **gráves** (*pl.*) PS. lxxvii. 58.

grǣfe, sb., *O.E.* grǣfe (*cavern,* MAT. xxi.
3 LIND.); *trench, quarry*; **grǣfes** (*ms.* grǣ-
fess) (*pl.*) ORM. 9210; leien in greaves (*older
text* stangraffen) LAʒ. 31881*; greves D. ARTH.
1872; *comp.* **stān-grǣfe.**

grǣnen, v., *? cf. M.Du.* graenen (*touwen*)
? prepare; heo **grǣneden** [greinede] (*pret.*)
heore steden LAʒ. 23909.

grǣs, sb., *O.E.* grǣs, gǣrs, = *O.L.G., O.H.G.,
O.N., Goth.* gras, *M.Du.* gras, gars, *O.Fris.*
gers; *grass,* FRAG. 3; grǣs, gras LAʒ. 3905,
16414; gres H. M. 35; gras O. & N. 1042;
TREAT. 137; WILL. 644; GOW. I. 152;
MIRC 1499; gres PR. P. 210; GEN. & EX.
3049; REL. I. 173; HAV. 2698; MAN. (H.)
336; PERC. 1192; gress BARB. ii. 361; þat
gras HORN (L.) 130; þet gers AYENB. 111;
girss BARB. viii. 445, xi. 572, xli. 582; **grases**
(*pl.*) SPEC. 114; graces ROB. (W.) 1011;
greses ORM. 8193; *comp.* **flot-gres.**

    gras-bed, sb., *grass bed*; **grasbedde** (*dat.*)
LAʒ. 23985.

    gras-grēne, adj., = *M.H.G.* grasgrüene;
grass-green, ALIS. 299.

    gres-hoppe, sb., *O.E.* gǣrshoppa; *grass-
hopper,* '*locusta*,' FRAG. 3; VOC. 165; ORM.
9224.

    grass-hopper, sb., = *M.Du.* grashopper;
grasshopper, VOC. 177.

gráfe, sb., *O.E.* grǣf, *cf. O.N.* gröf, *Goth.* graba
f., *O.L.G.* graf, *O.Fris.* gref, *O.H.G.* grab
n.; grave, L. H. R. 79; '*sepulcrum*' VOC.
249; TOWNL. 225 (MIR. PL. 150); grave VOC.
231; PR. P. 207; GEN. & EX. 3184; HAV.
408; TREV. VII. 21; WICL. JOHN xix. 41;
graffis, gravis (*pl.*) ALEX. (Sk.) 4451, 2101.

    graf-makere, sb., '*bostarius*,' VOC. 231.

gráve-stān, sb., *stone-coffin*, MARH. 22; graveston *gravestone* '*cippus*' PR. P. 208.

gráfe, sb., *O.N.* greifi, = *O.H.G.* grāvo, *O.Fris.* grēva; *steward*, '*villicus*, *praepositus*,' VOC. 211; greive HAV. 1771; his bidel & his greȝfe (*ms.* greȝȝfe) ORM. 18365.

gráfere, *see under* **gráven**.

graffe, **griffe**, sb., *Fr.* greffe; *graft*, '*sur-culus*,' PR. P. 212; graffe LANGL. *C* ii. 201.

graffen, v., *Fr.* greffer; *graft*, LANGL. *B* v. 137; griffin, graffin '*insero*' PR. P. 212; graffid (*pple.*) WICL. ROM. xi. 17.

graffare, sb., *cf. O.Fr.* graffeur; *one who engrafts*, '*insertor*,' PR. P. 212.

gragge, sb., *? throat*, '*vicecolla*,' VOC. 222.

grai, **grā**, adj., *O.E.* grǣg, grēg, *O.N.* grār, *cf. O.Fris.* grē, *O.H.G.* grāu; *gray*; grai REL. II. 210; (*ms.* gray) SPEC. 34; grei H. M. 43; R. S. iv; GEN. & EX. 1723; grei as glas CH. C. T. *A* 152; his greie fel O. & N. 834; i ðe greȝe kuvele A. R. 12; greye monekes ROB. 440; grei, grai (*subst.*) P. L. S. viii. 182 (HICKES I. 224); (*ms.* gray) MISC. 96; B. DISC. 839; grei LANGL. *B* xv. 215; grai [gra] M. H. 42; gro LAUNF. 237; in gro ant in gris SPEC. 26.

grai, sb., *O.N.* grey; *badger*, '*taxus*'; grei PR. P. 209; ant tu grisliche gra þu luþere liun MARH. 6; *see* GRIMM'S MYTHOL. 945.

grai-hond, sb., *O.E.* grǣghund, ? = *O.N.* grey-hundr; *greyhound*, S. S. (Web.) 755; grei-hound TRIAM. 473; greahund A. R. 332; grehound LIDG. M. P. 29; grohund PARTEN. 1389; grihond AYENB. 75; greuhond S. S. (Wr.) 738.

graid, **graiden**, *see* **greiþ**, **greiþen**.

graiel, sb., *O.Fr.* grael, = *Ital.* graduale; *gradual* (*part of the service of the mass, or the book containing this*), VOC. 193; REL. I. 291; graiel [grael] TREV. V. 231.

graien, v., = *O.H.G.* grāwen; *become gray*; graies (*pres.*) GAW. 527.

grain, sb., *Fr.* grain; *grain*, *small portion*: sum graine of godhede ALEX. (Sk.) 5622; grein LANGL. *A* vi. 121; grains (*pl.*) AYENB. 230; grains [granes] ALEX. (Sk.) 2024.

grain, *see* **grein**.

graine, sb., *? skins dyed in grain*; *? tanned leather*; sum graine to be neþire gloves ALEX. (Sk.) 2767; grein B. B. 178.

graiþ, *see* **greiþ**.

gram, adj., *O.E.* gram, grom, = *O.L.G.* gram, *O.N.* gramr, *O.H.G.* gram; *angry*; AL. (T.) 406; gram (*ms.* gramm) & gril ORM. 7145; hete gram GEN. & EX. 1228; god was him gram HAV. 2469; bliþe oþer grom O. & N. 992; gráme (*pl.*) AM. & AMIL. 214; LAȜ. 24774*.

grome-lĭch, adj., *O.E.* gramlīc, = *O.N.* gram-ligr; *dreadful*, *harmful*, MARH. 9.

gramaire, sb., *O.Fr.* gramaire; *grammar*, GOW. III. 136; grammere LANGL. *C* xii. 123;

CH. C. T. *B* 1726; gramire TREV. II. 161; gramere ALEX. (Sk.) 631.

gramer-scole, sb., *grammar school*: he held a gramerscole TREV. V. 51.

gramarian, sb., *O.Fr.* gramarien; *gram-marian*, TREV. V. 161; gramariens (*pl.*) LANGL. (Wr.) 8173.

gráme, sb., *O.E.* grama, *cf. M.L.G.* gram, *O.N.* gremi; *anger*, *harm*, P. S. 199; WILL. 2200; JOS. 539; CH. C. T. *G* 1403; A. P. iii. 53; MIN. v. 18; DEGR. 1372; TRIAM. 1223; TOR. 2029; he qvakede for grame FL. & BL. 712; & heore grame dærnden LAȜ. 7694; þer fore ȝe sculen han grome 1435; þu dest me grame [grome] O. & N. 49; al him turnde hit to grome 1090; grome MARH. 17; A. R. 180; of grome & of teone KATH. 1363; þer of hi hedde grome MISC. 39; þene eche grome 83; greme GAW. 312; A. P. ii. 947; **grámen** (*acc.*) HOM. I. 223; **grómen** (*pl.*) H. M. 7.

gram-cund, adj., *given to anger*, ORM. 1545.

gramcundnesse, sb., *irascibility*, ORM. 3833.

gráme-ful, adj., *irritating*, *vexatious*, LEG. 212; gremful P. L. S. ii. 156.

gramerci, *see* **grant merci** *under* **grand**.

grámien, v., *O.E.* gremian, = *Goth.* gramjan, *O.N.* gremja, *O.H.G.* greman; *anger*, *irritate*, HOM. II. 69; gromien [gramie] LAȜ. 25216; grame MED. 548; LUD. COV. 27; AUD. 77; gremien MARH. 12; gremen A. P. ii. 1347; greme HAV. 442; **grómeð** (*pres.*) MARH. 17; gremeð A. R. 334; summe laverdes ... god gremiað HOM. I. 111; ne grim (we) ALEX. (Sk.) 4653; **grámande** (*pple.*) MAN. (F.) 14423; **grámede** (*pret.*) SPEC. 36; him gro-mede [gramede] wið þene king LAȜ. 18552; him gremed(e) [gromede] wið ham KATH. 2106; gramed AL. (L.¹) 734; gremed ['*irritavit*'] PS. ix. 25; **grámed** (*pple.*) L. H. R. 132; *comp.* a-, ȝe-, of-grámien.

gráming, sb., *grief*: [in helle] is waning and graming [*?read* granung] and toþen grisbating HOM. I. 33.

grān, v., *from* *grīnen*, = *O.H.G.* grīnan; groan, C. M. 3731; grane, grone ALEX. (Sk.) 726*, 3238*; gron MAP 343; **grānes** (*pl.*) ANT. ARTH. (M.) xlviii.

granate, *see* **garnet**.

[grand, adj., *O.Fr.* grand, grant; *great*.]

gran(d)-dame, sb., *grandam*, MARH. 22.

grand-sire, sb., *grandsire*, P. L. S. xvii. 492; grantsire ROB. (W.) 6353, 7101.

grant-merci, interj., *O.Fr.* grant merci; *many thanks*: grant merci GAW. 1037, 1392; graunt merci LEG. 58; GOW. I. 106; CH. C. T. *E* 1088; gramerci TOWNL. 80.

grane, sb., *? =* **grin**; *snare*: grane (*v.r.* snare), WICL. LK. xxi. 35*.

X

grange, sb., *O.Fr.* grange; *grange, granary,*
VOC. 178; grange [graunge] CH. C. T. *A* 3668;
graunge, gronge PR. P. 208; gronge HAV. 764.

grānien [grōnie], v., *O.E.* grānian; *groan,*
LA3. 25557; granen H. M. 47; gronen vor his
eche A. R. 326; gronin '*gemo*' PR. P. 214;
ne gronie ne sike P. L. S. xvii. 580; grone
H. H. 80; TREV. XVII. 535; greōneþ (*pres.*)
FRAG. 5; granis ALEX. (Sk.) 717; grōni
(*subj.*) O. & N. 872; grāninde (*pple.*) HOM.
I. 43; granand PR. C. 798; grōnede (*pret.*)
B. DISC. 478; he groned(e) as a bere GREG.
620.

grānunge, sb., *O.E.* grānung; *groaning,*
HOM. I. 253; ? granung (*printed* graming) 33;
graning LA3. 17797; groninge MISC. 91;
groning ANT. ARTH. xxvi; greoning FRAG. 5.

grānken, v., *groan*; **grānkis** (*pres.*) TOWNL.
155.

grant, sb., *O.Fr.* grant; *grant,* P. L. S. ix.
148; graunt A. R. 238.

Grantebrügge, pr. n., *O.E.* Grantan brycg;
Cambridge, ROB. 6.

granten, v., *O.Fr.* granter; *grant; confess*;
grante ROB. 115; granti P. L. S. xvi. 175;
graunti (*pres.*) O. & N. 745; graunte CH.
C. T. *C* 327; granteþ AYENB. 7; **granti3eþ**
(*imp.*) ROB. (W.) 6896*; **grantede** (*pret.*)
LA3. 4789*; GEN. & EX. 1423; grantit BARB.
xix. 61; **igranted** (*pple.*) A. R. 34.

grantinge, sb., *granting, yielding,* AYENB.
10; grauntinge '*concessio, stipulacio*' PR. P.
208; **grantinges** (*pl.*) AYENB. 134.

grantise, sb., *grant, concession*: MAN. (H.)
130, 134.

grāp, sb., *O.E.* grāp, = *O.N.* greip; *from* **grīpen**; *art*; Galienes **grāpes** (*pl.*) KATH.
(E.) 853.

grāpe, sb., *O.Fr.* grape, crape; *grape,* '*uva,*'
PR. P. 207; VOC. 240; LANGL. *B* xiv. 30;
grāpes (*pl.*) BRD. 19; MAND. 219.

grāpin, v., *O.E.* grāpian, = *O.N.* greipa, *O.H.G.*
greifōn; *grope, handle*; grapin HOM. I. 251;
grape HAMP. PS. cxiii. 15; gropin PR. P. 214;
grope ALIS. 1957; LANGL. *B* xiii. 347; CH.
C. T. *A* 644; FLOR. 1633; **grōpeþ** (*pres.*)
O. & N. 1496; (hi) gropieþ SHOR. 125;
grōpe (*imper.*) MIRC 912; **grōpie** (*subj.*) A. R.
368; **grāpede** [gropede] (*pret.*) LA3. 30269;
graped PR. C. 6566; gropede FER. 1388;
groped here hondes (*gave a bribe*) TREV.
VII. 7; **grōpid** (*pple.*) WICL. EX. x. 21;
comp. a-grāpien.

grōpinge, sb., *groping, handling,* TREV.
III. 467; **grōpunges** (*pl.*) A. R. 206.

grapnell [grapenell], sb., *? from O.Fr.*
grapin; *grapnel,* CH. L. G. W. 640.

gras(e), *see* **grǣs**.

grāsen, v., *O.E.* grasian, = *M.H.G.* grasen;
graze; grase LIDG. M. P. 121; gresin PR. P.
210; **grāseþ** (*pres.*) GOW. I. 142.

graspen, v., *cf. L.Ger.* grapsen; *grasp,* WICL.
JOB v. 14; **graspid** (*pret.*) FLOR. 678;
grasped bi þe walles CH. C. T. *A* 4293.

grāt, sb., *O.N.* grātr, = *Goth.* grēts; *weeping*;
grot GEN. & EX. 1577, 1978, 2288, 3717.

grāte, sb., *cf. Ital.* grāte, *from Lat.* crātes =
grate, PR. P. 207; P. R. L. P. 136.

grāten, v., *O.Fr.* grater, *from med.Lat.* crā-
tāre; *grate*; **grāte** (*pres.*) PR. P. 207.

grāten, *see* **grēten**.

grāve, *see* **grāfe**.

gravelle, sb., *O.Fr.* gravele; *gravel,* VOC.
271; PR. P. 207; gravel HORN (L.) 1465;
WICL. GEN. xxii. 17; MAND. 234.

gravel-lī, adj., *made of gravel, like gravel
or sand,* WICL. ECCLUS. xxv. 27; MAND.
272.

gravel-pitte, sb., *gravel-pit,* '*arenarium,*'
PR. P. 207.

grāven, v., *O.E.* grafan, = *O.N.* grafa, *M.L.G.*
graven, *O.H.G.,* *Goth.* graban, *mod.Eng.*
grave; *engrave; dig; bury*: he lette þer on
graven sǣlcuðe runstaven LA3. 9960; gold
graven ANGL. IV. 194; gravin '*fodo, sculpo*'
PR. P. 208; in erþe grave CH. C. T. *E* 681;
grave (*for* bigrave) HAV. 613; **grāves** (*pres.*)
HAMP. PS. vi. 5*; **grōf** (*pret.*) PS. vii. 16*;
þai grofe L. H. R. 108; **grāven** (*pple.*) A. P.
ii. 1332; a cave ðe was ðor in roche graven
GEN. & EX. 1138; þere he sholde be graven
(*for* bigraven) LANGL. *B* xi. 67; graivin ALEX.
(Sk.) 423; igraven LA3. 7636; igrave HORN
(L.) 566; TREV. III. 391; MED. 987; *comp.*
bi-grāven.

grāfere, sb., *O.E.* grafere; *engraver,* '*scul-
ptor,*' FRAG. 2; grafer VOC. 213; graver LIDG.
M. P. 28; **grāvers** (*pl.*) MAS. 208.

grāvinge, sb., *digging; engraving,* '*fossio,
fossatura; sculptura,*' PR. P. 208; smithes
craft and gravinge TREV. II. 229; HALLIW.
414; grafeinge, gravinge HAMP. PS. lxxix. 17*,
xxi. 17*.

grē, sb., *O.Fr.* grē; *step, degree,* '*gradus,*'
PR. P. 208; ST. R. 23; M. ARTH. 48; þe gree
(*prize*) 3ut haþ he geten LANGL. *C* xxi. 103;
when the Grekes hade the gre & the ground
won(n)en D. TROY 1352; grew ALEX. (Sk.)
3270; gre *victory* ALEX. (Sk.) 3296; *prize*
818; gree *superiority,* 3651; grēce, grecis
(*pl.*) *stairs* ALEX. (Sk.) 5050,332; grece PALL.
i, 463.

grē², sb., *O.Fr.* gret, grē, *from Lat.* grātuṁ;
favour, goodwill: resave in gre . . . this rolle
AN. LIT. 83; receive in gre CH. C. T. *E* 1154;
at þi gre A. P. iii. 347; to gree MAND. 295; at
gre LIDG. M. P. 22; to Josepe he made is
gre with guode wille LEG. 48; to his freont
make þi gre 455.

grēable, adj., *suitable,* B. B. 129.

grēade, sb., *O.E,* grēada; *bosom, lap*: þet
me do þe elmesse into þe greade of þe povre

AYENB. 196; grede S. S. (Web.) 1802; þe coppe he putte undur his grede ALIS. 4187.

grēat, adj., *O.E.* grēat,=*O.Fris.* grāt, *O.L.G.* grōt, *O.H.G.* grōz; *great,* ALIS. 935; swa great swa a beam LA3. 2848; great, griat MISC. 31, 33; græt ORM. 2479; gret GEN. & EX. 2098; BEK. 71; MIRC 1173; 3eo was gret mid childe LA3. 19248*; gret [greet] CH. C. T. *A* 318; hire heorte was so gret (*swollen with anger*) O. & N. 43; hath þi herte be . . . gret MIRC 1173; gret wiþ childe K. T. 538; grat AYENB. 18; mid **grǣten** (*dat. n.*) ane huxe LA3. 27881; mid **grēatere** (*dat. f.*) heorte 569; **grēatne** (*acc. m.*) LA3. 2284; gratne AYENB. 238; **grēate** (*acc. f.*) BEK. 11; þurch griate luve MISC. 34; bur^xene grete LA3.25970; **grēate** (*pl.*) A. R. 194; AYENB. 21; grete oþes HAV. 2852; grete lordes MAND. 13; alle þe grete of Grece WILL. 1595; grettis *great men* ALEX. (Sk.) 3651; **grēaten** (*dat. pl.*) AYENB. 139; of greate (*coarse*) heorden A. R. 418; **grŏtture** (*compar.*)A.R. 194; gretture, grettere O. & N. 74; grettere HAV. 1893; gretter MAND. 2; grettere [grettore, grettor] ROB. (W.) 4794; greter BARB. xx. 463; **grŏttest** (*superl.*) BRD. 2; WILL. 928; hir gretteste oþ CH. C. T. *A* 120; gretteste [grettest] ROB. (W.) 8371; greste A. R. 66; SHOR. 8.

grēt-līche, adv.,*greatly,* HOM. II. 13; WILL. 2444; gretli ALEX. (Sk.) 472; **grētlüker** (*compar.*) A. R. 426.

grētnisse, sb.,*greatness,* BRD. 8; gretnesse TREV. II. 181; MAND. 297; R. R. 552.

grētum-lī, adv., *greatly,* BARB. i. 365, iii. 668, xiii. 210, xviii. 322, xix. 113.

grēaten, v., *O.E.* grēatian, *cf. O.H.G.* grōzen; *grow great,* A. R. 128; grete [greti] ROB. 68; **grēteþ** (*pres.*) ALIS. 464; greteth 'grandescunt' PALL. iii. 1025.

grēatnen, v., *become pregnant;* **grētnede** (*pret.*) JOS. 88.

Grēc, Grīc, sb., *O.E.* Grēc; *Greek;* Gric [Greck] LA3. 382; **Grickes** (*pl.*) ORM. 17560; grekis [grekez] ALEX. (Sk.) 986; **grēken** [grecen] (*gen. pl.*) ALEX. (Sk.) 3216.

Grīc-lond [Grēclond], pr. n., LA3. 327; Gricland (*ms.* Griccland) ORM. 16423.

Grēce, pr. n., *Fr.* Grēce; *Greece,* ALIS. 997; WILL. 1424.

grēce, *see* **gresse.**

[gred? *cf.* **gredel.]**

gred-īren, sb., *gridiron* ['*craticula*'] WICL. EX. xxvii. 4; gredire L. H. R. 58; gridire P. L. S. xv. 204; gredirne A. P. ii. 1277.

grēd, *see* **grǣd.** **grēde,** *see* **grēade.**

gredel, sb., *cf. Welsh* greidell, *Ir.* greideal; *griddle,* L. C. C. 13; gredil A. R. 122; L. H. R. 58*; as gridell full gai gretful of fiche D. TROY 13826.

grēden, *see* **grǣden.**

grēdi, grēdi3, *see* **grǣdi.**

grēf, sb., *O.Fr.* grief; *grief, harm,* A. R. 392; FL. & BL. 187; CHR. E. 738; WILL. 2473; greef [grief] CH. C. T. *G* 712; grefe ALEX. (Sk.) 4157.

grēfe, adj., *O.Fr.* grief, gref, *from Lat.* gravis; *grievous, harmful*: als a felon grefe MAN. (H.) 138; þou ert so grefe 259; greve 288.

grēf-ful-lī, adv., *grieviously;* grefulli ALEX. (Sk.) 973.

grēf-hēd, sb., *? harmfulness*: [*v.r.* grenehede] CH. C. T. *B* 163.

grēf-lī, adv., *sadly*: greefli bigo ALIS.² (Sk.) 490; greefli bigo with a grim peeple 994.

grege, grig, sb., *cricket,* ALEX. (Sk). 1753.

gregeis, sb., *O.Fr.* gregois; *Greek*: eche of his men a Gregeis ALIS. 2431; Gregoise WILL. 5104; GOW. II. 253; Gregious HALLIW. 416; Gregies ALIS. 3700, 3734; *see* **Grēc.**

gregen, v., *O.Fr.* greger; *make heavy, aggravate*: grege PR. C. 2990; gregge WICL. E. W. 319; gregge (*subj.*) ECCLUS. viii. 18.

greiles [greilles], sb., pl., *from O.Fr.* graile, graille; *fifes,* MAN. (F.) 3521.

grēi, *see* **grāi.** **greife,** *see* **grāfe.**

grein, sb., *? socket*: scharpe as a sweord boþe bi the grein and at ord ALIS. 6534; þe grain al of grene stele GAW. 210.

grein, *see* **grain.**

greiþ, adj., *O.N.* grei˘r; ?=3erād; *prepared, equipped;* graiþ M. H. 99; AV. ARTH. xxxvi.; þe graiþe gate LANGL. *A* i. 181; **greithe** (*adv.*) MIRC 346; graid *excellently* ALEX. (Sk.) 3689; **greiðe** (*pl.*) A. R. 16; LEB. JES. 635; þe graiþest (*superl.*) gate MIN. vii. 48; graithest [grethest] *richest* ALEX. (Sk.) 1865; alþire graithist ALEX. (Sk.) 162.

greiþ-līc, adj., *O.N.* grei˘ligr; *suitable, well-planned*; mid grei˘licre speche LA3. 445; mid græi(˘)lichen (*sec. text* faire) worden 10039; **graiþliche** (*adv.*) PL. CR. 529; greiþli WILL. 984; graiþli M. H. 44; PERC. 491.

greiþe, sb., *O.Fr.* grei˘i; *preparation, contrivance,* JOS. 66; graiþe C. M. 3523; graithe ALEX. (Sk.) 5578.

greiþen, v., *O.N.* grei˘a; *prepare, equip, build,* ALIS. 170; græi˘ien, græiþen [greiþi] LA3. 8058, 17288; gre3þen ORM. 98; greiþe HAV. 1762; WILL. 1719; MAP 334; greiþi MISC. 84; **greiðeð** (*pres.*) A. R. 236; graithes M. ARTH. 3530; greiþe me selver P. S. 151 (A. D. 103); **graithand** (*pple.*) HAMP. PS. lxiv. 7; **greiðede** (*pret.*) LA3. 1079; greiþede BEK. 931; greiþide ['*paravit*'] WICL. GEN. xxvii. 14; graiþed PERC. 123; graithid HAMP. PS. xxxii. 14; graithd lxvii. 11; graith lvi. 10; graid xxiii. 2; **greiþed** (*pple.*) WILL. 1945; igreiþed SPEC. 48; PL. CR. 196; igrei˘et KATH. 1993; graithid (*ms.* grauithid) HAMP. PS. xlviii. 9; graid D. TROY 1664; *comp.* **a-greiþen.**

greiðing, sb., *preparation* : in the firste grei-
thing of oure werc WICL. ECCLES. PROL. ;
cloth of oþer colur grene of swiþe guod grei-
þingue LEG. 42 ; there is putte to hem grete
greithinge of melis WICL. 4 KINGS vi. 23 ;
graithinge HAMP. PS. lxiv. 10.

greive, *see* grāfe.

Grēke, pr. n., *the Greek language*, ALEX. (Sk.)
5009.

Grēkin, sb., *Grecian*, ALEX. (Sk.) 5504.

Grēkisc, adj., *O.E.* Grēcisc ; *Greek*, FRAG. 1;
HOM. I. 97 ; þis Grickische fur A. R. 402 ; þe
Grickishe še S. A. L. 156 ; mid Grickisce fure
LAȝ. 628.

grēme, grēmien, *see* grāme, grāmien.

gremþe, sb., = *M.H.G.* gremde ; *from* gram ;
anger, WILL. 2080 ; ALIS.² (Sk.) 279.

grenat, *see* garnet. grene, *see* grin.

grēne, sb., *O.N.* grein *f.* ; *discord, quarrel* ;
in game ne in grene HAV. 996.

grēne, adj., *O.E.*, grēne, = *O.Fris.* grēne, *O.N.*
grœnn, *O.L.G.* grōni, *O.H.G.* gruoni ; *green*,
LAȝ. 24652 ; GEN. & EX. 2776 ; ROB. 352 ;
SPEC. 90 ; ALIS. 638 ; MAND. 54 ; greene CH.
C. T. *E* 2037 ; grēne (*subst.*) MISC. 92 ; JUL.²
206 ; grene of conscience CH. C. T. *G* 90 ;
grene boȝhes ORM. 10002 ; þe woundes grene
C. L. 1433 ; grēnnere (*compar.*) BRD. 15 ;
comp. gras-grēne.

  grēne-fish, sb., *cod*, B. B. 174 ; (*printed*
gronefish) 154.

  grēn-hēde, sb., *greenness*, REL. II. 84;
AYENB. 28 ; youthe withoute grenehede [*v. r.*
grefhede] or folie CH. C. T. *B* 163.

grēnnesse, sb., *O.E.* grēnness ; *greenness*,
PR. P. 210.

Grēne-wĭch, pr. n., *O.E.* Grēne wīc ; *Green-
wich*, CH. C. T. *A* 3907.

grēne, sb., *cf. M.H.G.* grüene *f.* ; *green, grass-
plot*, HORN (L.) 851 ; to pleien o þe grene P. S.
222 ; HAV. 2828 ; FER. 693 ; CH. C. T. *D* 998.

grēnin, v., *O.E.* grēnian, *cf. O.H.G.* gruonen ;
become green, '*viresco*,' PR. P. 210 ; grenen
H. M. 35 ; greni AYENB. 95 ; grēneð (*pres.*)
A. R. 150.

grēning, sb., *becoming green*, C. M. 16867.

grenien, *see* grinien.

grēnling, sb., *cf. Ger.* grünling ; *? codfish*, PR.
P. 210.

grennien, v., *O.E.* grennian, = *M.H.G.* grennen, *O.N.* grenja (*howl*) ; *grin ; gnash the teeth*,
FRAG. 7 ; gren nen A. R. 212 ; REL. I. 188 ;
grennin '*ringi*' PR. P. 210 ; grenni with (h)is
teth LEB. JES. 223 ; grenne PS. xxxvi. 12 ;
grenne and berke ALIS. 1935 ; grenninde
(*pple.*) LEG. 185 ; þe teeþ grennand RICH.
3406 ; grennede (*pret.*) HOM. I. 227 ; D. ARTH.
1075 ; he siȝte & grente sore AL. (T.) 217 ;
heo grenneden hin hon (*read* on) LAȝ. 29550 ;
see grünnien.

grennunge, sb., *O.E.* grænnung (EPIN.
GL.) ; *grinning*, A. R. 212 ; grenning TOR.
(A.) 1126 ; grenninge PR. P. 210.

grēot, sb., *O.E.* grēot, = *O.L.G.* greot, griot,
O. N. griōt (*stone*), *O.H.G.* grioz (*gravel*);
grit, sand, gravel, A. R. 70 ; me graveþ . . .
in greot (*ms.* greote) and in ston MISC. 92 ;
gret '*gravele*' VOC. 159 ; neiðer ston ne gret
GEN. & EX. 3774 ; TRIST. 2501 ; greet WICL.
JOB xxi. 33 ; REL. II. 81 ; as grein þat lith in
þe greot(e) [grete] LANGL. *C* xiv. 23 ; his fless
lai under grete (*v. r.* stone) C. M. 16935.

grēpe, sb., *O.E.* grœp (*scrobis*) (EPIN. GL.) ;
? furrow : hi nabbeþ never a grepe ORIG. 214.

gres, *see* græs.

grēs, grēselīche, *see* grīs, grīslīche.

grēsen, v., *O.Fr.* gresser ; *grease* ; gresin
PR. P. 210.

  grēsinge, sb., *greasing*, '*saginacio*,' PR. P.
210.

gresse, grēse, sb., *O.Fr.* gresse, graisse,
craisse ; *grease*, L. H. R. 58* ; gresse [grese]
CH. C. T. *C* 60 ; grese AYENB. 205 ; grece
ROB. (W.) 8485.

grēt, *see* grēat *and* grēot.

grēte, sb., *? O.N.* grǣti ; *weeping* ; (*dat.*) C. M.
189 ; ANT. ARTH. xxv. ; MIR. PL. 150.

[grēte, sb., = *O.H.G.* grōzī ; *greatness* ; *comp.*
un-grēte].

  grēt-full, ådj., *quite full*, D. TROY 13826 ;
(*printed* grecfull) 331.

grēten, grāten, v., *O.E.* grētan, *O.N.* grāta
(*pret.* grēt), = *Goth.* grētan (*pret.* gaigrōt),
prov. Eng. greet ; *weep, cry* ; gretin '*ploro,
fleo*' PR. P. 210 ; grete HORN (L.) 889 ; TRIST.
730 ; A. P. i. 331 ; PR. C. 7099 ; grete LANGL.
A v. 216* ; grete and grone H. H. 82 ; TOWNL.
227 ; graten HAV. 329 ; groten GEN. & EX.
1984 ; grēte (*pres.*) ANT. ARTH. vii. ; greten
AN. LIT. 11 ; grōtinde (*pple.*) HAV. 1390 ; gre-
tand MAN. (F.) 6941 ; grēt (*pret.*) GEN. & EX.
2287 ; HAV. 615 ; MAN. (F.) 3613 ; M. H. 18 ;
ISUM. 315 ; thou grete TOWNL. 292 ; greten
GEN. & EX. 3207 ; he greten and gouleden
HAV. 164 ; grēten (*pple.*) MAN. (F.) 15613 ;
grete EGL. 926 ; graten, igroten HAV. 241,
285 ; *deriv.* grāt, grēte.

  grētinge, sb., *weeping*, '*ploratus, fletus*,' PR.
P. 211 ; greting HAV. 166 ; PR. C. 496.

  grēting-ful, adj., *sorrowful*, HAMP. PS.
xvii. 10.

grēten², v., *O.E.* grētan, = *O.Fris.* grēta, *O.L.G.*
grōtean, *O.H.G.* gruozzan, gruozan ; *greet*,
['*salutare*'] MARK xv. 18 ; ORM. 2805 ; wult
greten ure godes KATH. 2303 ; gretin PR. P.
211 ; greten, grǣten LAȝ. 8293, 13302 ; greten,
grete LANGL. *A* v. 187, *B* v. 343 ; grete WILL.
1430 ; CH. C. T. *E* 1014 ; MIR. PL. 152 ; grē-
teð (*pres.*) LAȝ. 3157 ; GEN. & EX. 2382 ;
grēt (*imper.*) KATH. 1466 ; greteð þe lefdi
mid one ave Marie. A. R. 430 ; grētte (*pret.*)

LAȝ. 133; HORN (R.) 788; BEK. 452; LANGL.
B x. 218; LAUNF. 252; gretten WILL. 4554;
þat him and hise wiþ swerd(e) gretten ALIS.
5696; grēt (*pple.*) HAV. 2290; *comp.* ȝe-
grēten.

grētunge, sb., *O.E.* grēting, *cf. M.H.G.*
grüezunge; *greeting*, LK. i. 29; A. R. 250;
gretinge LAȝ. 4512; greting BEK. 1238.

grēting-word, sb., *salutation*, ORM. 2799.

grēt-wurt, sb., *elecampane*, '*elna, enula,*' VOC.
(W. W.) 554.

Greu, adj. & sb., *O.Fr.* Greu; *Greek*;
writen in grew MAND. 76; Griu MISC. 50;
Gru C. L. 24; þe Grewes (*pl.*) for gremþ
ginneþ on me werre WILL. 2080.

greu, *see* grēuwen *and* grē.

grevance, sb., *O.Fr.* grevance; *grievance,*
IW. 126; grevaunce ALIS. 965; MAN. (F.)
9495; PR. C. 3019.

grēve [? gréve], *cf.* grōf, grǣfe; *grove,*
WILL. 3634; GAW. 1355, 1898; A. P. i. 320;
grēves (*pl.*) TRIST. 14; GOW. II. 261; CH.
C. T. *A* 1495; B. DISC. 551; PERC. 172;
grevis ANT. ARTH. xxxiii; grevez D. ARTH.
927.

grēve, *see* grǣfe.

gréven, v., *O.Fr.* grever, *from Lat.* gravāre;
cf. gregen; *grieve, burden, injure*; grevin,
grevi (*ms.* greui, *printed* greni) H. M. 31; greve
HAV. 2953; WILL. 689; MAND. 11; grevi
BEK. 1522; AYENB. 39; gréveð (*pres.*) A.
R. 236; GEN. & EX. 3818; greves ALEX. (Sk.)
2912; gréve (*subj.*) LANGL. *B* xvii, 111.

grévinge, sb., *grieving, hardship*: þer doel
ever dwellez, greving & greting & grisping
harde of teþe A. P. ii. 158.

gréves, sb., *O.Fr.* greves; *greaves, armour
for the legs*; ALEX. (Sk.) 3893; grévez (*pl.*)
GAW. 575.

grevous, adj., *O.Fr.* grevous; *grievous,* P. L.
S. xxi. 5; TREV. IV. 103; CH. C. T. *A* 1010;
grevousere (*comp.*) WICL. JOB vi. 3.

grevous-līche, adv., *grievously,* AYENB. 47;
TREV. II. 327; grevousli MAN. (F.) 4274;
grevosliere (*comp.*) HAMP. PS. ii. 6*.

grevousnesse, sb., *grievousness,* '*grava-
men,*' PR. P. 211.

Grīc, *see* Grēc.

grīde, sb., *? O.Fr.* gride; *outcry*; grídis (*pl.*)
ALEX. (Sk.) 544.

gridlen, v.: sa gret plente als a grideld
(*pple.*) [griddeled] frost (*? hoar frost*) C. M.
6518.

grīen, v., ? =grūwen; grīed (*pret.*) GAW.
2370.

griffe, *see* graffe.

griffōn, sb., *O.Fr.* griffon, *from Lat.* grȳphus;
griffin, CH. C. T. *A* 2133; ALIS. 494; EGL.
848; griffoun '*grifo, grifes*' PR. P. 212;
MAND. 269; griffin, griffun WICL. LEV. xi.

13; DEUT. xiv. 12; griffon, greffon, TOR.
(A.) 1971, 1981; grefine VOC. 220; grif-
founes (*pl.*) MAND. 268; griphonnes TREV.
I. 135.

griffon, sb., *O.Fr.* grifon; *a name applied to
Greeks*: his baner upon the wall he pulte,
mani a griffin it bihulte RICH. 1921; pleighed
at the chesse with o griffoun of hethenesse
ALIS. 3133; griffons (*pl.*) RICH. 1767;
griffouns WILL. 1961.

grig, *see* grege. gright, *see* grucchen.

griking [griging], sb.; *morning twilight,*
PS. lxxvii. 34; þe griginge of þe daie D.
ARTH. 2510.

gril, adj.,=*M.Du.* gril, *M.H.G.* grel; *keen,
fierce,* '*horridus,*' PR. P. 212; AM. & AMIL.
1275; gril (*ms.* grill) & gram ORM. 7196; þis
sper þat is so gril P. R. L. P. 226; grille
(*adv.*) AL. (L.[1]) 564; grille (*pl.*) R. R. 73;
wordes grille MAP 334; woundes . . . grille
SPEC. 91; grones full grille ANT. ARTH. xlviii;
griller, [grèlere] (*comp.*) MAN. (F.).4578.

gril-līch, adj., *horrible*; grilich D. ARTH.
1101.

grille, sb.,=*M.L.G.* grille; *harshness,* GUY
(Z.) 11488; E. T. 279; wiþouten grucching or
grille S. A. L. 207.

grillen, v., *O.E.* grillan, griellan, = *M.L.G.*
grellen, *M.H.G.* grellen, grillen, *Du.* grillen;
irritate, be grievous to; grille REL. II. 19;
ANT. ARTH. xlix; grill(e) SACR. 786; i shall
. . . never more the greeve ne grill(e) MIR. PL.
4; so gretli can (*r.* gan) he grille E. T. 165;
þat ich seide ouȝt him for to grulle (*rhyming
with* wille) LEG. 38; me grulleð aȝean mine
pine A. R. 366; ȝef hire herte ther to grille
MIRC 103; grülde (*pret.*) FRAG. 6; riht so
me grulde (*twanged*) schille harpe ʊ. & N.
142; igrüld (*pple.*) HOM. II. 259; *comp.*
a-grillen.

grim, adj., *O.E.* grimm,=*O.L.G.* grim, *O.N.*
grimmr, *O.H.G.* grimm; *grim, cruel,* '*austerus, horridus,*' PR. P. 212; GR. 30; LAȝ.
15566; ALIS. 754; MIRC 1561; IW. 2183;
MAN. (F.) 5208; PR. C. 2250; a grim word
A. R. 100; he is to þe swiþe grim HAV. 2398;
grim and wroþ S. S. (Web.) 818; grim and
sturne TREV. V. 101; þe grimme wrastlare
A. R. 280; grimmere (*dat. f.*) LAȝ. 18315;
grimne (*acc. m.*) LAȝ. 2283; grimme (*pl.*)
S. A. L. 152; grimme wundes ORM. 1443;
grimmen (*dat. pl.*) LAȝ. 683; grimmest
(*superlat.*) A. R. 202.

grim-leic, sb., *O.N.* grimmleikr; *cruelty*;
grimmeleȝc ORM. 4719.

grim-līch, adj., *O.E.* grimlīc; *grim,* MISC.
63; þe grimli gost SPEC. 111; grimlīche
(*adv.*) LAȝ. 1914; A. R. 104; O. & N. 1332;
grimlīche (*dat. pl.*) LAȝ. 8176; grimlokest
(*superl.*) ANT. ARTH. viii.

grimnesse, sb., *O.E.* grimness; *grimness,*
'*austeritas,*' PR. P. 212; CH. C. T. *I* 864.

grim[2], sb.,=*M.H.G.* grim ; *fury, rage,* ALIS. (Sk.) 904.

 grim-cund-le3c, sb., *cruelty* : þwert ut of grimcundle3c ORM. 4704.

 grim-ful, adj., *cruel,* HOM. I. 253.

grīm, sb., *cf. Dan.* grim ; *grime, soot* : grim or gore HAV. 2497.

Grīmes-bi, pr. n., *Grimsby,* LA3. 22692 ; HAV. 745.

grimin, *see* gramien.

grin, sb., *O.E.* grin, gryn *f. n.; snare, trap,* AYENB. 47 ; grine, grune O. & N. 1056 ; grune [' *laqueum* '] HOM. II. 209 ; **grüne** (*dat.*) MARH. 3 ; grene JUL. 73 ; & mid grine hine selfne aheng MAT. xxvii. 5 ; in one grine O. & N. 1059 ; **grines** [grenes] (*pl.*) WICL. JOB xxii. 10 ; JER. xviii. 22 ; grenes (*printed* greues) TREV. II. 385 ; girns BARB. ii. 576* ; *see* grane *and* grone.

grind, sb., *O.E.* gegrind ; *act of grinding,* HOM. II. 183.

 grind-stōn, sb., *grind-stone,* A. R. 332 ; grindstoon [' *mola* '] WICL. NUM. xi. 8.

[**grindel**, ? sb. ; *from* grinden.]

 grindel-stān, sb., *prov.Eng.* grindlestone ; *grind-stone,* A.R. 332* ; grindelston GAW. 2202.

grindel[2], adj., *angry, impetuous,* GAW. 2338 ; A. P. iii. 524.

 grindel-laik, sb., *anger, impetuosity,* GAW. 312.

 grindel-lī, adv., *wrathfully* : Gawain ful grindelli with greme þenne said GAW. 2299.

grinden, v., *O.E.* grindan ; *grind,* A. R. 70 ; grindin ' *molo* ' PR. P. 212 ; grinde A. P. i. 81 ; grinde ate qverne AYENB. 181 ; **grindest** (*pres.*) ORM. 1486 ; grint A. R. 70 ; WILL. 1242 ; grindeð HOM. II. 181 ; grinde CH. C. T. *C* 538 ; **grindand** (*pple.*) ALEX. (Sk.) 4552 ; **grónd** (*pret.*) JUL. 56 ; gronde [' *contrivit* '] WICL. EXOD. xxxii. 20 ; grunden GEN. & EX. 3339 ; **grúnde** (*subj.*) A. R. 70 ; **grúnden** (*pple.*) IW. 676 ; grunden stel HAV. 2503 ; grounden VOC. 155 ; GAW. 2202 ; groundin L. C. C. 30 ; *comp.* 3e-grínden.

 gríndere, sb., *grinder* ; WICL. ECCLES. xii. 4 ; **grínderis** (*pl.*) WICL. ECCLES. xii. 3.

 gríndinge, sb.,*grinding,* PR. P. 212 ; AYENB. 265.

grinien, v., *O.E.* (ge-)grinian, (be-)grynian ; *catch in a snare* ; **grenede** [' *illaqueavi* '] WICL. JER. l. 24.

grīp, sb., *O.N.* grīpr,=*O.H.G.* grīf ; *raven, vulture,* HAV. 572 ; BRD. 20 ; ALIS. 6345 ; OCTOV. (W.) 447 ; gripe TOR. (A.) 1961 ; **grīpes** (*pl.*) LA3. 28062 ; MISC. 151.

grip, sb., = *M.Du.* grippe ; *furrow, ditch* ; grippe PR. P. 212 ; þan birþ men casten hem in poles or in a grip HAV. 2101 ; summe leie in dikes stenget, and summe in **gripes** (*pl.*) bi þe her drawen ware 1924.

gripe, sb., *O.E.* gripe,=*M.L.G.* gripe, grepe, *O.N.* gripr, *M.H.G.* grif ; *grip, act of seizing,* LANGL. *C* xx. 146 ; a growen grape of a gripe (*cluster*) ALEX. (Sk.) 1347 ; **gripen** (*for* gripe) (*dat.*) LA3. 15273 ; **grippis** (*pl.*) 544.

gripel, adj., *O.E.* gripul ; *eager to seize,* LA3. 7336.

grīpen, v., *O.E.* grīpan,=*O.L.G.* grīpan, *O. Fris.,O.N.* grīpa,*Goth.* greipan,*O.H.G.* grīfan ; *grip, seize, grasp,* FRAG. 6 ; REL. I. 174 ; his freondes strived to gripen his iwon R. S. v ; gripe RICH. 4471 ; **grīpeþ** (*pres.*) LANGL. *A* iii. 235 ; **grīpes** (*imp.*) [' *adprehendite* '] PS. ii. 12 ; gripis gripis *vultures seize* ALEX. (Sk.) 5453 ; **grāp** (*pret.*) LA3. 27691 ; ORM. 8125 ; grure grap euch mon KATH. 1994 ; grop HAV. 1776 ; ? grep BEV. 2309 ; gripen LA3. 18027 ; HAV. 1790 ; grepen RICH. 5070 ; **gripen** (*pple.*) HOM. I. 273 ; PS. ix. 16 ; gripen [grepe] LANGL. *A* iii. 175* ; griped LANGL. *B* iii. 181 ; *comp.* bi-, 3e-gripen ; *deriv.* grīpe, grīpel, grippen, grāp.

 grīpinge, sb., *griping, seizing, ' constrictio, compressio,'* PR. P. 213 ; **grīpinges** (*pl.*) PS. cxlix. 6.

grippel, sb.,=*Du.* grippel, greppel ; *trench, ditch,* PR. P. 212.

grippen, v.,=*M.H.G.* gripfen ; *grip, seize, grasp* ; **grippest** (*pres.*) M. T. 118 ; **gripte** (*pret.*) ROB. 22 ; he gript(e) his mantel WILL. 744 ; *comp.* bi-grippen.

grīs, sb., *O.Fr., cf. O.L.G., O.Fris., M.H.G.* grīs ; *gray fur,* SPEC. 26 ; LANGL. *B* xv. 215 ; griis TRIST. 1220 ; gris and grai B. DISC. 839 ; gris and gro LAUNF. 237 ; somme manteles wiþ veir & gris MAN. (F.) 11418.

grīs[2], sb., *O.N.* grīss (*pl.* grīsir) ; *young pig, 'porcellus,'* VOC. 204 ; REL. II. 79 ; LANGL. *A* iv. 38 ; L. C. C. 55 ; wen me bedeþ þe gris opene þe shet ANGL. IV. 195 ; **grises** (*pl.*) A. R. 204*.

[**grīs**[3] (? grīse), sb., *cf. M.L.G.* grēse, *M.H.G.* grūs, grūse ; *horror.*]

 grīse-ful, adj., *horrible,* P. L. S. i. 33 ; **grīs-, grīseful** WICL. DEUT. ii. 10 ; WICL. WISD. xi. 19.

 grīsful-lī, adv., *dreadfully, horribly* : dredende grisfulli WICL. WISD. xvii. 3.

 grīs-lích, adj., *O.E.* grīs-, grȳslic, *cf. O.Fris.* grīslīk, *M.Du.*grijselijk, *Swed.*grȳselig(RIETZ) *M.L.G.* grēselik, ? *M.H.G.* grous-, griuslich ; *grisly, horrible,* MARH. 8 ; A. R. 190 ; O. & N. 224 ; AYENB. 49 ; S. S. (Web.) 756 ; K. T. 168 ; grissli3 ORM. 3842 ; grislich, grisli ALIS. 684, 6633 ; grisli A. P. ii. 1534 ; grisli, griseli PR. C. 1404, 1757 ; gresli BER. 1577 ; greesli MIR. PL. 34 ; greseli N. P. 26 ; **grislíche** (*adv.*) FRAG. 7 ; A. R. 118 ; ROB. 24 ; WILL. 4343 ; PL. CR. 585 ; grisliche [greseliche] ROB. (W.) 8659 ; griseli ALEX. (Sk.) 3879 ; **grislíche** (*pl.*) LA3. 28063 ; **grislüker**

(*compar.*) HOM. II. 171 ; grisloker ROB. 560 ; þe grislikeste weder P. L. S. xvii. 353.

grīslines, sb., *grisliness*, PR. C. 2310; griseli-nesse WICL. DEUT. ii. 11.

grīse, adj. (*? for* grīsi); *horrible*, B. DISC. 597 ; C. M. 23249.

grīsen, v., *O.E.* (a-)grīsan, *cf. M.Du.* grīsen, *M.L.G.* grēsen ; *feel horror*, REL. I. 130 (HOM. II. 165); JUL. 57 ; grise S. A. L. 152 ; TRANS. XVIII. 26 ; S. S. (Wr.) 2674 ; L. H. R. 121 ; M. H. 23 ; LIDG. M. P. 113; MIR. PL. 177 ; me grīses (*pres.*) A. R. 366* ; þet ou grise wið ham 92* ; grōs (*pret.*) HORN (L.) 1314 ; ham gros FRAG. 6 ; hit was no wundir þo3 him gros H. S. 7875 ; his herte .. gros & greu MAN. (F.) 8532 ; *comp.* a-grīsen.

grīsung, sb., *horror*, A. R. 190* ; grising WICL. JOB xviii. 20.

[grīsi, adj., *O.H.G.* griusīg.]

grīsi-lîche, adv., *horribly*, WILL. 4343.

grīsines, sb. ['*horror*'], WICL. GEN. xv. 12.

grīsil, adj., '*horridus, terribilis,*' PR. P. 213.

grīsle, sb., *horror*, MARH. 15 ; ich cwaike of grisle & of grure HOM. I. 253.

grisping, sb.,=gristbiting; *gnashing of the teeth*, A. P. ii. 158.

grist, sb., *O.E.* grist,= *O.L.G.* grist(-grimmo), *M.H.G.* gris(-gram) ; *grist, grinding*, E. G. 336; A. P. i. 465 ; þi mille haþ grounde þi laste grist H. V. 74.

gris-bāt, sb., *gnashing of teeth*, LA3. 5189.

grist-bēatien, v., *O.E.* gristbātian ; *gnash the teeth*, JUL. 69 ; grisbātede (*pret.*) A. R. 326.

grist-bātinge, sb., *hissing noise*, LA3. 1886 ; grisbating HOM. I. 33 ; MISC. 230.

grist-bite, v., *O.E.* gristbitian ; *make a hissing noise* ['*dentibus stridere*'], TREV. VII. 377.

grist-biting, sb., *O.E.* gristbitung ; ['*stri-dor dentium*'] TREV. VII. 501.

gristel, sb., *O.E.* gristel,= *O.Fris.* gristel ; *gris-tle*, VOC. 179 ; gristil '*cartilago*' PR. P. 213 ; gristles (*pl.*) HOM. I. 251.

gristil-bōn, sb., '*cartilago,*' PR. P. 106.

gristen? v., *grind the teeth* : þi tethe be not pik-inge, gris(t)inge (*pple.*) ne gnastinge B. B. 136.

grið, sb., *O.E., O.N.* grið *n.; peace, security*, HOM. I. 45 ; A. R. 284; GEN. & EX. 560 ; hælden grið [griþ] LA3. 1416 ; grið & frið 2816 ; lifves grið (=*O.N.* lifs grið) 8866 ; griþ REL. I. 172 ; WILL. 3899 ;. GREG. 580 ; MAN. (F.) 3659 ; griþ (*ms.* griþþ) & friþ ORM. 3380 ; hi nabbeþ noþer griþ ne sibbe O. & N. 1005 ; þat þider wol fleon ta sechen griþ C. L. 702 ; grith HAV. 511 ; ANT. ARTH. v ; griþ, griht H. H. 126 ; griþes (*gen.*) ORM. 6559 ; girth BARB. ii. 44, iv. 47 ; griðe (*dat.*) LA3. 480 ; on griðe & on friðe HOM. I. 13 ; *comp.* chireche-grið.

griþ-brüche, sb., *O.E.* griðbryce ; *breach*

of the peace, O. & N. 1734 ; grithbriches (*pl.*) ANGL. VII. 220.

grið-ful, adj., LA3. 9171 ; A. R. 406.

griðfulnesse, sb., *peacefulness*, A. R. 416.

grið-lîche, adv., *peaceably*, LA3. 121.

grið-sergeant, sb., *bailiff* : bedels and greives, grithsergeans (*pl.*) HAV. 266.

griðien [griþie], v., *O.E.* griðian ; *make peace, grant peace, keep in peace*, LA3. 5551, 10605, 21908 ; griðede (*pret.*) LA3. 31032 ; *comp.* 3e-griðien.

gro, sb., *horror*, ALEX. (Sk.) 3238 (*Dub. ms. has grone : see grān*).

grō, *see* grai.   grob, *see* grubbe.

grobben, *see* grubben.   grócere, *see* grosser.

grochen, *see* grucchen.

grōf, sb., *? O.E.* grāf *m.* ; *grove*, AR. & MER. 1640 ; grove '*lucus*' PR. P. 215 ; MAND. 271; grōve (*dat.*) O. & N. 380 ; CH. C. T. *A* 1478 ; 3eond þan groven [grove] LA3. 469.

grōfe, sb., *O.N.* grōf,= *M.L.G.* grōve, *O.H.G.* gruoba, *prov.Eng.* (*Derbys.*) grove (*mine*) ; *cf. mod. Eng.* groove ; *cave*, ALEX. (Sk.) 5394.

grogginge, *see* grucchen.

groin, groni, sb., *O.Fr.* groing ; *snout*, '*rostrum,*' PR. P. 214 ; CH. C. T. *I* 156.

groinen, v., *Fr.* grogner ; *grunt* ; groine WICL. IS. xxix. 4 ; groniin '*murmuro*' PR. P. 214 ; groigned [groined] (*pret.*) C. M. 13590 ; groininde (*pple.*) 176.

groinere, sb., *murmurer* : WICL. PROV. xxvi. 20.

groininge, sb., *grunting* (*of swine*), '*grun-nitus,*' PR. P. 214 ; groining CH. C. T. *A* 2460.

groising, sb., *? from O.Fr.* grossir ; *growing thick*, PALL. i. 59.

grom, sb., *O.N.* gromr, *cf. M.Du.* grom; *groom; boy* ; HAV. 790 ; HORN (R.) 971 ; BEK. 148 ; WILL. 1767 ; P. S. 218 ; grom oþer maide A. D. 258 ; moni knight and grom ALIS. 1150 ; grōme (*dat.*) LEG. 32 ; grōmes (*pl.*) A. R. 422 ; O. & N. 1115 ; AYENB. 210.

grōme, *see* grāme.

gromil, sb., *Fr.* gremil ; *gromwell* (*grey-millet*), SPEC. 27 ; gromali PR. P. ; gromilioun A. P. i. 43.

grōn, grōne, *see* grān.   grond, *see* grund.

grone, sb., *?*=grin ; *snare*, A. R. 278 ; gronen [grunen] (*pl.*) A. R. 278.

grōnien, *see* grānien.   gronten, *see* grunten.

grōpe, sb.,= *O.Fris., M.L.G.* grōpe, *M.Du.* groepe ; *channel* ; groupe PR. P. 216 ; (*r. w.* scoupe=scōpe) MAN. (F.) 8165.

grōpien, *see* grāpien.

grōpin, v.,= *M.L.G.* grōpen ; *groove* : groupin wiþ an iren '*runco*' PR. P. 216 ; grave & groupe MAN. (F.) 8166.

grōping-īren, sb., *gouge*, '*strofina*,' VOC. 276 ; the groping-iren than speke to the compas N. P. 14.

grōs, adj., *O.Fr.* gros ; *large* ; groos B. B. 145.

grōsen, v., *heap together*, '*ingrosso*,' PR. P. 214.

grosses, sb., pl., *green figs*, PALL. iv. 633.

grosser, sb., *Fr.* (marchand-) grossier; *grocer*; grocer PR. P. 213 ; **grossers** (*pl.*) 213*.

grot, sb., *O.E.* grot ; *fragment, particle*, A. R. 260 ; LEG. 24 ; WILL. 4257 ; CH. C. T. *D* 1292; (karf) hem al to grótes (*pl.*) HAV. 472.

grōt, sb., *? O.N.* grautr; (*barley*) *groat*; grōtes (*pl.*) L. C. C. 47.

grōt, see **grāt.**

grōte, sb., *M.L.G.* grōte, *cf. O.Fris.* grāta ; *groat* (*coin*), LANGL. *A* v. 31 ; grote, groote LIDG. M. P. 50, 152 ; CH. C. T. *B* 4148.

grōten, see **grēten.** **ground**, see **grúnd.**

groute, sb., *O.E.* grūt,=*M.L.G.* grūt, *M.Du.* gruite ; *wort* (*of beer*), VOC. 178 ; PR. P. 217.

grōve, see **grōf.** **grovelinge**, see **grufelinge.**

grōwen, v., *O.E.* grōwan,=*O.Fris.* grōwa, *O.N.* grōa, *O.H.G.* grōen, grüen; *grow*; growe O. & N. 1134 ; groo D. TROY 1403 ; growe LANGL. *B* xx. 56 ; MAND. 88 ; grōweþ (*pres.*) REL. I. 123 ; grouweþ P. S. 332 ; grōwinde (*pple.*) SHOR. 159 ; groaund D. TROY 11462 ; grēu (*pret.*) LAȝ. 2014 ; HAV. 2333 ; FL. & BL. 483; ROB. 470 ; greowen LAȝ. 8696; gr(e)owe [grewe] O. & N. 136; grewen MISC. 48; grēowe (*subj.*) REL. I. 173 ; grōwen (*pple.*) PL. CR. 221 ; so riche were growen hise sunen GEN. & EX. 1897 ; *comp.* **bi-, ȝe-grōwen.**

grōwinge, sb., *growing*, '*crescentia*,' PR. P. 215.

grubbe, sb., *grub, caterpillar*, TR. FISH. 26 ; grub [grob] ALEX. (Sk.) 1753 ; **grobbes** (*pl.*) PALL. vii. 63.

grubbin, v., *dig up*, '*fodico*,' PR. P. 217 ; grobbe CH. APPEND. BOET. i. 29 ; **grobbe** (*imper.*) PALL. viii. 6 ; **grubbed** (*pret.*) L. H. R. 94.

grubbare, sb., *digger*: grubbare in the erthe '*fossor, confossor, fossatrix*' PR. P. 217.

grucchen, v., *O.Fr.* grouchier, groucier ; *grudge, murmur*, '*murmuro*,' WICL. NUM. xiv. 36; grutche WICL. PS. lviii. 16; grochi AYENB. 67 ; grugge LUD. COV. 228 ; grughe YORK xxx. 473 ; **gruccheð** (*pres.*) A. R. 114; groches PR. C. 297; grogginge B. B. 135; **grucche** (*imper.*) WILL. 1450; **grucched** (*pret.*) MAND. 57 ; grucheden GEN. & EX. 3354 ; gratchiden WICL. PS. cv. 25 ; gruȝt A. P. ii. 810; gright D. TROY 9315, 9367 ; grochgede TREV. VI. 137 ; **groched** (*pple.*) ALEX. (Sk.) 1467 ; gricchit D. TROY 7072.

grucchende-līche, adv., *grudgingly*: mani thingus grucchendeli whistrende WICL. ECCLUS. xli. 19 ; grochindeliche and mid ȝorȝe of herte AYENB. 193.

grucchere, sb., *grumbler*, A. R. 108*.

grucchild, sb., *grumbler*, A. R. 108.

grucchunge, sb., *grudging*, A. R. 114 ; grucching O. & N. 523; GEN. & EX. 3318; WILL. 1461 ; grichching A. P. iii. 53.

grudge, sb., *grudge*, B. B. 93.

gruel, sb., *O.Fr.* gruel ; *gruel*, L. C. C. 47 ; growelle, gruell B. B. 150, 273; gruwel LANGL. *A* vii. 169.

grüen, see **grüwen.**

grüf? o **grüfe**, phrase, *cf. O.N.* ā grūfu ; '*supinus*'; CATH. 259; þei fillen (a)gruf (*on the face*) CH. C. T. *A* 949; and laide her gruf (*face downwards*) upon a tre EM. 656; on groufe fallen D. ARTH. 3944.

grufelinge, adv., *groveling*, '*supinus*,' CATH. 166 ; grovelinge '*supinus*' PR. P. 215 ; ['*pronus*'] WICL. DAN. viii. 18 ; grovelinge to his fete þai felle A. P. i. 1119 ; **gruflinges** C. M. 17709 ; groflinges TOWNL. 40 ; grovelings ALEX. (Sk.) 5276.

grüfnen, v., *from O.N.* grūfa ; *? bow down* ; **grüfnede** (*? printed* grusnede) (*pple.*) and strekede and starf wið ðan GEN. & EX. 481.

grüllen, see **grillen.**

grunching, sb., *cf.* **gränken**; *murmuring*, BARB. xvi. 9.

grúnd, sb., *O.E.* grund,=*O.L.G.* grund, *Goth.* grundus, *O.H.G.* grunt, *O.N.* grunnr ; *ground, soil; bottom; foundation, reason*, LAȝ. 2296, 3191 ; P. S. 196 ; þe grund of helle pit ORM. 12059 ; his hertes grund 13286 ; he valþ a grund AYENB. 91 ; grond PR. C. 7213 ; ground LANGL. (Wr.) 10924; he was leid … in ground CHR. E. 750; the ground and cause of al mi peine CH. COMPL. M. 160; grount PALL. ix. 154; **grúnde** (*dat.*) A. R. 268; O. & N. 506; bringen him to grunde ORM. 12547 ; were … brouht to grunde HAV. 1979; grounde AYENB. 23 ; and caste him into þe derkeste grounde MISC. 230; falle to grounde TREAT. 136; al is folc ȝede anon to gronde (*were beaten*) ROB. (W.) 1766; ground harme 1431 ; a ground hate D. TROY 1403 ; eode to gronde *perished* ROB. 79; on grounde and on lofte LANGL. *A* i. 88; hire coverchiefes ful fine were of grounde CH. C. T. *A* 453; *comp.* **sē-grund.**

ground-ebbe, sb.,=*Ger.* grundebbe ; LIDG. M. P. 50.

grúnd-fulled, pple., *filled throughout*, LAȝ. 1088.

grúnd-hāt, adj., *quite hot*, LAȝ. 5692.

grúnd-laden, pple., *laden throughout*, LAȝ. 1106.

ground-līas, adj., *O.E.* grundlēas ; *groundless*, SHOR. 154.

grúnd-līch, adj.,=*Ger.* gründlich; *profound, radical, solid*: mid grundliche stre(n)gðe LAȝ. 15813 ; **grúndlīche** (*adv.*) JUL. 69;

grundliche onslowen LAȝ. 1739; ete grund-like HAV. 651; ȝroundli BER. 4001.

ground-sope, sb., =*M.Du.*grondsop; '*faex, sedimen,*' PR. P. 216.

grúnd-stalworð, adj., *strong, powerful*: þe ston was mike ... grundstalworthe man he sholde be þat mouthe liften it to his kne HAV. 1025.

grúnde-swülie, sb., *O.E.* grundeswylige (LEECHD. GLOSS.); *groundsel,* FRAG. 3; grun-deswilie REL. I. 37.

grúnd-wal, sb., *O.E.* grundweall; *founda-tion, 'fundamentum,'* FRAG. 3; JUL. 72; ORM. 13372; grundwal MISC. 97; grund-, ground-wal C. M. 8424.

grúnden, gründen, v., *O.E.* gryndan, = *M.H.G.* gründen; *ground*; grounden '*fundo*' PR. P. 216; **grendeþ** (*pres.*) SHOR. 137; ȝe gronde ȝou noȝt on a god ALEX. (Sk.) 4490; **groundid** (*pple.*) WICL. I ESDR. iii. 10; igrounded CH. D. BL. 921; grune (*for* grun-den) ALEX. (Sk.) 338.

grüne, see **grin.**

grünnien, v., *gnash* (*the teeth*); grunni AYENB. 67.; *see* **grennien.**

grunten, v., =*O.H.G.* grunzen; *grunt,* A: R. 326; i grunte ant grone REL. I. 120; þei gronten MAND. 274; **grunte** (*pret.*) P. L. S. xv. 99; MAP 341; nevere grunte [gronte] he CH. C. T. *B* 3899.

gruntare, sb., *grunter,* '*grunnitor,*' PR. P. 217.

gruntinge, sb., *grinding, gnashing,* '*grun-nitus,*' PR. P. 218; gruntinge of teeth CH. C. T. *I* 208; grinting, grenting, grunting WICL. MAT. viii. 2, xxii. 13; LK. xiii. 28; **grintingis** (*pl.*) GEST. R. 6.

grüre, sb., *O.E.* gryre, = *O.L.G.* gruri *m.*; *horror, terror,* LAȝ. 27716; KATH. 1994; MISC. 91; swuc grure he hefde A. R. 112; for muchele grure MARH. 7.

grüre-ful, adj., *horrible,* HOM. I. 271; A. R. 210.

grüreful-lïche, adv., A. R. 320.

grüre-blöd, sb., *blood-stream,* A. R. 294; *see* **güre.**

grüsen, grüselien, v.; *eat, nibble*; **grüse** (*pres. subj.*) [gruselie] A. R. 428*.

grusshen, v., *crush, gash*; grusshet (*pret.*) D. TROY 9482.

grut, sb., *cf. L.Ger., Swed.* grut *n.*; *gravel, silt, glarea, limus,*' PR. P. 218; þe toundikes ... wer(e) ... full(e) of grut RICH. 4339.

grütte, sb., *O.E.* grytt (*pl.* grytta), =*M.L.G.* grutte, *O.H.G.* gruzze; *bran*; grut(t)a '*fur-fures*' FRAG. 4.

grütten, adj., *? made of bran*; þet tu ete grut-tene bread A. R. 186.

grüwen, v., =*M.L.G., M.H.G.* grüwen, *O. Swed.* grüa, *Dan.* grüe; *shudder*; grue SACR.

155; C. M. 23027; **grēu** (*ms.* grew) (*pret.*). MAN. (F.) 8532; growit BARB. xv. 541*, xx. 517; *comp.* **a-grüwen.**

greuing, sb., *horror,* BARB. xix. 555*; grow-ing xix. 555.

[**ȝu,** adv., *O.E.* geo, io, iu, = *O.L.G., O.H.G.* giu, ju; *formerly*; *comp.* nŭ-ȝu.]

ȝu, see **ēow.**

gua-, see **gva-.** **gūd,** see **gōd.**

gubernaciōn, sb., *government,* TREV. IV. 33.

gue-, see **gve-.**

ȝuȝeðe, *O.E.* geoguð, iogoð, iugoð, = *O.L.G.* juguth, jugud, *O.H.G.* jugund; *youth*; iuȝeðe (*ms.* Ivȝeðe), ȝuheðe, ȝeoȝoþe HOM. I. 97, 109, 145; ȝuȝeðe, ȝeoȝeþe LAȝ. 6566, 19837; ȝu-heðe KATH. 1462; ȝuwoþe FRAG. 2; ȝuweþe A. R. 156; guweðe HOM. II. 127; ȝeoȝeðe P. L. S. viii. 188; ȝeuȝeþe HICKES I. 224 (ANGL. I. 30); ieȝeþe (*ms.* yeȝeþe) AYENB. 69; gueþe (*printed* guewe) REL. I. 174; iouhþe (*ms.* youhþe) 173; ȝouþe A. D. 289; ȝoeþe, youhþe O. & N. 633; (*ms.* youþe) HAV. 2988; ȝouþe [youthe] CH. C. T. *A* 461; yhouthe PR. C. 5972; he was guð (*young, per metonymiam*) GEN. & EX. 2665.

ȝuweðe-hōd, sb., *O.E.* geoguðhād, = *O.L.G.* jugudhēd; *youth,* A. R. 342; guðhede REL. I. 209; ȝouthede ISUM. 60; yhouthede PR. C. 5713; ȝoutheid, ȝouthede BARB. v. 277, x. 532.

ȝuhelunge, see **ȝoȝelinge.**

ȝui, see **ȝēa.** **ȝuile,** see **hwīle.**

gul, adj., *O.N.* gulr, golr; *yellow*; **gulle** (*pl.*) HALLIW. 424.

gul-soght, sb., = *Swed.* gulsot, *Dan.* guulsot, *jaundice*; '*aurugo,*' CATH. 168; golsoght HALLIW. 409.

gulchen, v., *cf. Ger.* gulken, kulken, glucken, klucken; *swallow greedily*; gulcheð [gluc-ches] in A. R. 240; gulcheð [culcheð] ut *vomits* 88.

gulche-cuppe, sb., *drunkard*: ȝif þe gulche-cuppe weallinde bres to drinken A. R. 216.

Gúldeford, pr. n., *Guildford*; **Gúldeforde** (*dat.*) O. & N. 191.

gülden, adj., *O.E.* gylden, = *O.H.G.* guldīn, *O.N.* gullinn; *from* góld; *golden*; gilden PR. C. 5360; LUD. COV. 76; a guldene crune KATH. 1582; guldene crune [goldene croune] LAȝ. 4251; þe gildene ȝate MAND. 81; gil-dene crunes ORM. 8180; geldene scheldus DEGR. 280.

gülden, v., *O.E.* gyldan, = *O.N.* gylla; *gild*; gildin '*deauro*' PR. P. 193; gilden WICL. EXOD. xxvi. 29; **gilt** (*pple.*) ALEX. (Sk.) 2384; gild (*gilt-plate*) B. B. 131; iguld FER. 1330; igelt AYENB. 26; *comp.* **over-gülden.**

gildinge, sb., =*O.N.* gylling; *gilding,* PR. P. 193.

ȝülden, *see* **ȝelden.**

gúle, sb., *Lat.* gula ; *gluttony* : this vice . . . is cleped gule Gow. II!. 1 ; the lusti vice is hote of gule the delicaci Gow. III. 22.

gulion, sb., *O.Fr.* goleon ; *collar* : cast on her his gulion which of the skin of a leon was made Gow. II. 358 ; golion '*gunella, gunellus*' PR. P. 202.

ʒ**üllen**, *see* ʒellen.

gulpen, v., *cf. Du.* gulpen, golpen ; *gulp* ; igulped [igolped, iglubbed, iglobbed, igloppid] (*pple*). LANGL. *B* v. 346.

   gloppinge, sb., *gulping, swallowing greedily* : gloppinge of drink PL. CR. 183 ; gloffinge '*devoracio*' PR. P. 199 .

   gulpere, sb., *cf. M.Du.* golper ; *gulper, drunkard, glutton* ; **glubberes** [globbares] (*pl.*) LANGL. *B* ix. 60 ; gloffare PR. P. 199.

ʒ**ülpen**, *see* ʒelpen.

gült, sb., *O.E.* gylt *m.* ; *guilt*, LAʒ. 4272 ; O. & N. 1410 ; MAP 341 ; MIRC 1099 ; & nimeð þene gult uppen hire A. R. 258 ; þisne gult MISC. 49 ; as sone as he hedde þe gult idon C. L. 414 ; i nabbe þer of gult non BEK. 827 ; gilt HOM. II. 65 ; GEN. & EX. 2262 ; MAND. 134 ; (*ms.* gillt) ORM. 13730 ; gelt ['*debitum*'] MAT. xviii. 32 ; ANGL. I. 16 ; SHOR. 166 ; AYENB. 30 ; WILL. 4403 ; **gilte** (*dat.*) PR. C. 5559 ; **gültes** (*pl.*) FRAG. 7 ; HOM. I. 39 ; A. R. 356 ; giltes ORM. 1143.

   gült-lēs, adj., *guiltless*, TREV. III. 389 ; gultelese, gulteles ROB. (W.) 5705, 6867 ; giltles FLOR. 2090 ; giltelæs ORM. 871.

gülten, v., *O.E.* gyltan ; *incur guilt, sin* ; gulte P. L. S. viii. 108 ; gilſten WICL. PS. xxxiii. 22 ; he wolde noht on ane wise gilten (*ms.* gilltenn) ORM. 3111 ; **gülteð** (*pres.*) MARH. 17 ; **gülte** (*subj.*) O. & N. 1523 ; **gülte** (*pret.*) HOM. I. 83 ; **gült** (*pple*) MIRC 419 ; gilt M. ARTH. 1377 ; *comp.* a-, for-, ʒe-**gülten.**

   gültare, sb., *one who incurs guilt*, REL. I. 282 ; **gilteris** (*pl.*) WICL. PS. lxxiv. 5.

gülti, adj., *O.E.* gyltig ; *guilty*, A. R. 58 ; BEK. 2111 ; MIRC 901 ; gilti CH. C. T. *A* 660 ; MAN. (H.) 112 ; PR. C. 2949 ; geltig MAT. xxiii. 18 ; gelti SHOR. 40 ; *comp.* ungülti.

gúme, sb., *O.E.* guma, *O.N.* gumi, = *O.L.G.* gumo, *Goth.* guma, *O.H.G.* gomo, *Lat.* homō ; *man*, M. H. 127 ; REL. I. 186 ; gume [gome] LAʒ. 3590 ; gomé SPEC. 36 ; ALIS. 705 ; WILL. 670 ; JOS. 531 ; S. S. (Web.) 1399 ; GREG. 543 ; LANGL. *B* xiii. 181 ; A. P. i. 231 ; ANT. ARTH. viii ; Havelok was a ful god gome HAV. 7 ; **gúmen** (*pl.*) LAʒ. 4621 ; gomen HORN (R.) 169 ; ALIS. 1384 ; **gumene** (*gen. pl.*) LAʒ. 12178 ; gume JUL. 26 ; *comp.* brüd-, here-, scip-gume.

gūme, sb., *cf. O.H.G.* giumo ; *gum*, '*gingiva*,' VOC. 207 ; *see* gōme.

gumme, sb., *Fr.* gomme ; *gum*, PR. P. 218 ; **gummes** [gommes] (*pl.*) LANGL. *A* ii. 202 ; *see* gome.

gun, *see* ʒeon.

gūn, sb., *gown* ; goun '*toga*' VOC. 238 ; BARB. xix. 352 ; (*ms.* goune) CH. C. T. *A* 93 ; A. P. ii. 145 ; **gūnes** (*pl.*) REL. II. 174 ; gounes DEGR. 1604 ; BARB. viii. 468.

ʒ**ünc,** *see* inc.

gúnd, gound, sb., *O.E.* gund, ?=*Goth.* gund, *O.H.G.* gunt, gund ; *purulent matter*, '*chacie*,' VOC. 145 ; gound PR. P. 206 ; REL. II. 78 ; red gound '*scrophulus*' PR. P. 426 ; r(e)ade **goundes** (*pl.*) LANGL. *B* xx. 82.

ʒ**und,** *see* ʒeond.

gúndi, goundi, adj., = *O.H.G.* gundīg ; *full of sores*, '*chacious*,' VOC. 144.

gunfaneur, sb., *standard bearer*, A. R. 300.

gunfanoun, sb., *O.Fr.* gun-, gonfanon ; *standard*, L. H. R. 118 ; gonfanoun ALIS. 1963 ; gunphanoun [gomfainoun] MAN. (F.) 13758.

ʒ**ung**, adj., *O.E.* giung, iung, geong, giong, geng, = *O.Fris.* jung, jong, *O.L.G.* jung, *O.H.G.* jung, *Goth.* juggs, *O.N.* ungr, *Lat.* juvencus ; *young*, LAʒ. 376 ; KATH. 178 ; A. R. 6 ; ORM. 1212 ; BEK. 124 ; gung MISC. 121 ; ʒung, iung HOM. II. 199, 201 ; ʒiung, iung HICKES I. 222 ; ʒung, ʒing P. L. S. viii. 2 ; (*ms.* yung) HAV. 112 ; yhung, yhong PR. C. 214, 3785 ; ʒong C. L. 151 ; JOS. 437 ; A. P. i. 412 ; yong TOR. 1971 ; ʒing MAN. (F.) 2370 ; AMAD. (R.) xl ; þe yunge eorl SAX. CHR. 265 ; þe ʒinge HORN (W.) 582 ; DEGR. 1807 ; ʒ**ungne** (*acc. m.*) LAʒ. 10566 ; yongne AYENB. 162 ; ʒ**unge** (*pl.*) O. & N. 1134 ; AV. ARTH. vii ; ʒonge HORN (W.) 563 ; ʒinge SPEC. 95 ; ISUM. 144 ; LUD. COV. 73 ; þa ʒeonge LAʒ. 28444 ; þan ʒungen 14158 ; to þe ʒunge A. R. 70 ; ʒ**üngre, ʒungre** (*compar.*) ANGL. I. 27 ; A. R. 424 ; (*ms.* ʒunngre) ORM. 13271 ; gungere GEN. & EX. 1510 ; ʒengere, ʒungere [ʒeongere, ʒeongre] LAʒ. 3927, 9189 ; gingre LK. xv. 13 ; ʒonge [ʒongor, ʒongore, ʒonger] ROB. (W.) 7746 ; ʒ**ungeste**, ʒengeste [ʒeongeste] (*superl.*) LAʒ. 3460, 31257 ; þe ʒungeste JUL. 61 ; gungeste GEN. & EX. 2190 ; ʒongost [ʒongoste] ROB. (W.) 8682.

ʒ**ung-hēde**, sb., *youth*, BEK. 3 ; yonghede MISC. 71 ; ʒonghede ROB. (W.) 2195, 2822.

ʒ**ung-lích**, adj., *O.E.* geonglíc ; *young*, BRD. 33.

ʒ**üngen**, v., = *O.H.G.* jungen ; *make young* ; gingen REL. I. 216.

ʒ**ungling**, sb., *O.E.* geongling, = *O.H.G.* jungeling ; *youth, young man*, MISC. 21 ; gungling, geongling MAT. xviii. 2 ; MK. xiv. 51 ; ʒongling O. & N. 635 ; FL. & BL. 705 ; TRIST. 859 ; M. T. 110 ; ['*adolescens*'] WICL. GEN. xxxiv. 19 ; yongling PARTEN. 3843 ; ʒongeling ALIS. 2366 ; **yunglenges** (*pl.*) HOM. I. 237 ; ʒeonglinges LAʒ. 28681 ; ʒonglinges MAN. (F.) 14281.

ʒ**ungþe**=ʒuʒeðe, sb., *youth*, H. S. 2810 ; ʒungthe PR. P. 539 ; ʒongþe WICL. GEN. xlviii.

15; ȝonkþe [ȝouth] MAN. (F.) 1936; ȝengþe ALIS. 1323.

gunne, sb., *mod.Eng.* gun; *ballista, mangonel*, CH. L. G. W. 637; AV. ARTH. lxv; **gonnes** (*pl.*) ALIS. 3268; FER. 3263; gunnes [gunnez] ALEX. (Sk.) 2227.

gunnare, sb., *gunner*, PR. P. 219.

guod, *see* gōd.

ȝur, *see* ēower, ȝeor.

gürd, sb., *O.E.* gyrd,=*O.N.* giörð *f.*, *cf. M. H.G.* gurt *m.*; *girth*; gird ALIS. 2272; ge þ RICH. 5733; gerthe PR. P. 190.

ȝürd, *see* ȝerde.

gürdel, sb., *O.E.* gyrdel, *cf. M.L.G.* gordel, *O.N.* gyrðill, *O.H.G.* gurtil; *girdle*, FRAG. 4; A. R. 420; SPEC. 35; þan gurdel FER. 2419; girdel ORM. 3210; girdel [gerdel] LANGL. *B* xv. 120; CH. C. T. *A* 358; gerdel MAT. iii. 4; AYENB. 236; gefe thaim up þe girdill (*submit to the inevitable*) ALEX. (Sk.) 131; **gürdle** (*dat.*) LAȝ. 1325; **gerdles** (*pl.*) AYENB. 236; *comp.* brēch-gerdel.

　gürdlere, sb.,=*M.H.G.* gurtelære; *girdle-maker*: girdilhare '*lorimarius*' VOC. (W. W.) 651; **gürdelers** (*pl.*) VOC. 123; girdillers D. TROY 1584.

　gürdel-stede, sb., *waist*, FER. 1707; gurdil-stode B. B. 313.

　gürdel-maker, sb., '*corrigiarius*,' VOC. 212.

gürden, v., *O.E.* gyrdan,=*O.N.* gyrða, *O.H.G.* gurten, *M.L.G.*, *M.Du.* gorden; *gird*; girdin '*cingo*' PR. P. 195; ȝe shulen girde about ['*accingetis*'] WICL. EX. xii. 11; gerde AYENB. 236; gerdeþ (*pres.*) WILL. 1240; **gürdeþ** (*imper.*) LANGL. *A* ii. 176; **gürde** (*pret.*) ROB. 435; wit swerdes he hem girte HORN (H.) 1513; hi gerten wel hare lenden AYENB. 236; **gürd** (*pple.*) GAW. 588; girt GEN. & EX. 3149; girt [gert] CH. C. T. *A* 329; *comp.* bi-, ȝe-gürden.

gürden², v., *strike, cut; rush headlong*: lete gurd of min heved FER. 4065; to girden of his heed CH. C. T. *B* 3736; i schal girde of ['*amputabo*'] WICL. 2 KINGS xvi. 9; he gürdus (*pres.*) to sir Gauane throȝe ventaille and pusane ANT. ARTH. xlv; girdez GAW. 2160; gordez 2062; girdis out (*bursts out*) ALEX. (Sk.) 3159; (þai) girdis ALEX. (Sk.) 796; **gird** (*imper.*) TOWNL. 115; **girdand** (*pple.*) ALEX. (Sk.) 1243; **girdid** (*pret.*) out as gutars ALEX. (Sk.) 3231; gurd FER. 4117; A. R. 106; FER. 3932; he gurde Suard on þat hæfd LAȝ. 1596; girde of his heed WICL. 1 KINGS xvii. 51; þe grounde of Gomorre gorde into helle A. P. i. 911; **gürd** (*pple.*) ALIS. 2299.

[**güre**, sb., *O.E.* gyr (*marsh, mire*).]

　güre-blōd, sb., *?gory blood*; on gureblode (*?printed* gure blode) JUL. 28; (*?printed* grure blode) A. R. 294; *see* grüre.

güren, *see* ȝearwien. **gürl**, *see* gerl.

ȝürn, *see* ȝeorn.

gurnard, sb., *Fr.* gurnard; *gurnard* (*a fish*), PR. P. 219; VOC. 253; whan . . . gurnardes (*pl.*) schot rokes out of a crose bow S. & C. II. lviii.

ȝürren, *see* ȝeorren. **ȝürsten-**, *see* ȝestren-.

guschen, v., *gush, rush*; guschez (*pres.*) D. ARTH. 1130; gosshet (*pret.*) D. TROY 1607.

ȝüse, *see under* ȝēa. **güst**, *see* gest.

gut, sb., *O.E.* gutt; *gut*, A. P. iii. 280; gut, gutte LANGL. *B* i. 36, *C* ii. 34; gutte '*viscus*' PR. P. 220; guttes (*pl.*) MARG. 112; ALIS. 2465; FER. 748; TOR. 192; guttes [gottes] TREV. III. 205; gottes ROB. 289.

ȝut, *see* ȝet.

[**güte**, sb., *O.E.* gyte, = *O.H.G.* guz; *from* ȝēoten; *shedding*; *comp.* blōd-güte.]

　güte-festre, sb., *? running ulcer*, A. R. 328.

gutter, *see* gotere.

gutton, v., *gut*, '*exentero*,' PR. P. 220.

[**gūð**, sb., *O.E.* gūð,=*O.N.* gūðr, gunnr (*war*).]

　gūð-strencðe sb., *?warlike strength*, LAȝ. 1595.

ȝuw, *see* ēow.

gverdonen, v., *O.Fr.* gverredoner; *reward, recompense*: gverdon him with nothinge but with sighte CH. TRO. ii. 1293; ȝe oughte the rathere to gverdoune hem CH. C. T. *B* 2465; gverdoned (*pret.*) *B* 2462.

　gverdoning, sb., *reward*, R. R. 2380.

　gverdōn-lēs, adj., *without reward*: gverdonlesse COMP. BL. KN. 397.

gverdōn, sb., *O.Fr.* gverdon; *guerdon, recompense*, PARTEN. 551; gver-, gerdon CH. C. T. *D* 1878.

gverre, *see* werre.

gvīse, sb., *O.Fr.* gvīse; *guise*, LAȝ. 19641*; P. S. 221; gise '*modus*' PR. P. 195; MISC. 111; CH. C. T. *A* 993; A. P. i. 1098.

gvīsen, *see* gīsen.

guweorn, sb., *spurge*, '*spurgia*,' VOC. 140.

ȝw-, *see* hw-, w-.

h.

ha, *see* hé.

hā, interj.,=*O.Fris.*, *M.H.G.* hā; *ha*, CH. C. T. *B* 4571.

haare, *see* hǽr.

habben, v., *O.E.* habban, hæbban (*pret.* hæfde), = *O.L.G.* habbian, hebbean (*pret.* habde), *O. Fris.* habba, hebba (*pret.* hede), *O.N.* hafa (*pret.* hafði), *O.H.G.* habēn (*pret.* habēta), *Goth.* haban (*pret.* habaida); *have*, FRAG. 1; KATH. 291; A. R. 14; þu scalt . . . habben þine sunne forȝevene HOM. I. 37; habben and owen C. L. 132; habbén, hæbben, han LAȝ. 145, 1435, 23162; habbe O. & N. 258; MISC. 29; ALIS. 312; E. G. 352; man him ssel habbe wel (h)oneste AYENB. 214; habben, hafen, haven SAX. CHR. 250, 255, 256; habben, hafen ORM. 123, 4964; haven C. L. 351;

H. H. 98 (100); haven, have HAV. 78, 614´; have WILL. 72; LIDG. M. P. 29; have ant holde SPEC. 35; havin, han PR. P. 225; have, han LANGL. *B* PROL. 109, vi. 148; CH. C. T. *A* 490, 654; han S. S. (Web.) 294; MIRC 1298; HOCCL. i. 31; hafe PERC. 915; to hæbbene LA3. 10273; þet we a3en to habben deore •HOM. I. 57; **hæbbe**, habbe [*O.E.* hæbbe, hafu] (*pres.*) LA3. 464, 3216; hebbe, habbe ANGL. VII. 220; habbe O. & N. 174; BEK. 353; C. L. 329; ic habbe . . . ibeon P. L. S. viii. 2; ich habbe . . . ibeo MISC. 147; as i befor has borne on hand BARB. xiii. 642; habbe, hafe ORM. 465, 958; have WILL. 519; havest LA3. 2275; A. R. 236; O. & N. 155; H. H. 57; hafest ORM. 12236; hast C. L. 950; JOS. 350; hest MISC. 29; hafeþ ORM. 153; haveð P. L. S. viii. 20; A. R. 10; haveþ O. & N. 113; haveð, hafð, hæfeð, hæfð LA3. 1331, 1926, 16019, 20624; hafð HOM. I. 105; haveþ, haþ BEK. 162, 184; haþ WILL. 477; CH. C. T. *A* 915; HOCCL. i. 437; LIDG. M. P. 28; heþ SHOR. 95; AYENB. 9; we habbeoð, habbeð LA3. 364, 6215; (3e) habbeð A. R. 2; h(e)o habbeð HOM. I. 165; habbeþ PROCL. 2; TREAT. 135; AYENB. 8; we habbeþ be FER. 277; habbet P. L. S. viii. 51; hæbbeð LK. x. 7; hass, has, hes BARB. xi. 273, xii. 179, xvii. 904; **hafe** [have] (*imper.*) LA3. 25787; have KATH. 1585; **habbe** (*subj.*) HOM. I. 21; O. & N. 99; A. D. 262; **hæfde**, hefde, hafde, havede, hevede (*pret.*) LA3. 110, 226, 759, 4887, 25414; hafde (*ms.* haffde) ORM. 6; hafde, hefde KATH. 289, 322; hefde HOM. I. 51; A. R. 124; havede HAV. 649; hevede SPEC. 36; H. H. 7; hadde GEN. & EX. 3392; MAND. 2; CH. C. T. *A* 48; WICL. JOHN v. 6; he hadde icome P. L. S. ix. 84; hadde, hedde O. & N. 216, 395; hedde C. L. 308; MISC. 29; JOS. 503; hæfden [hadden] LA3. 20918; þa ifaren hafden bi live 13994; hi hedden levere liese vour messen AYENB. 31; **haved** (*pple.*) KATH. 446; *comp.* bi-, 3ehabben.

having, hawing(e), sb., *having; behaviour,* BARB. vii. 135, 412; xi. 247.

haberdashere, sb., *haberdasher,* CH. C. T. *A* 361.

habergeon, *see* **hauberjoun.**

habilitē, sb., *O.Fr.* habilitē; *ability,* PARTEN. 2341; *see* **abilitē.**

habit, sb., *O.Fr.* habit; *habit, dress,* A. D. 259; MAN. (F.) 9002; abit A. R. 12; BEK. 265; habet ALEX. (Sk.) 3513.

habitable, adj., *O.Fr.* habitable, *from Lat.* habitābilis; *habitable*: the lond of Sarazin habitable MAND. 132; **habitables** (*pl.*) or trepassables 182.

habitacioun, sb., *O.Fr.* habitacion; •*habitation,* CH. C. T. *B* 3406.

habitacle, sb., *O.Fr.* habitacle, *from Lat.* habitāculum; *dwelling, niche*; habitacle RICH.

4148; ALEX. (Sk.) 4334; WICL. DEEDS xii. 7; in ech of the pinacles weren sondri `habitacles (*pl.*) CH. H. F. 1194.

habiten, v., *O.Fr.* habiter, *from Lat.* habitāre; *inhabit, frequent*: in thilke places as thei habiten (*pres.*) R. R. 657.

hable, habil, adj., *O.Fr.* hable; *fit, suitable,* PARTEN. 2355, 4536; TREV. VIII. 133; *see* **able.**

habi(l)-lī, adv., *fitly,* TREV. III. 237.

hablen, v., *strengthen*; **hable** (*pres. subj.*) ALEX. (Sk.) 1768*; *see* **ablen.**

hacche, sb., *O.E.* hæcc *f., cf. M.L.G., M.Du.* heck *n.*; *hatch, wicket-gate, hatch of a ship; ? hay-rack*; D. TROY 2005; (*? printed* hatche) VOC. 261; hecche, hek '*antica*' PR. P. 231; hek '*ostiolum*' VOC. 203; TOWNL. 106; **hacche** (*printed* hatche) (*dat.*) MARG. 222; to þan hacche O. & N. 1058; **hacches** (*pl.*) WILL. 2770; TREV. II. 233; CH. L. G. W. 648; hachches A. P. iii. 179; haches ANT. ARTH. xxxv.

hacchen, v., *cf. Swed.* häcka, *Dan.* hække, *Ger.* hecken; *hatch (eggs), bring forth,* DEP. R. ii. 143; ha3te, haihte (*pret.*) O. & N. 105; ihaht (*pple.*) P. S. 237.

[**hacchen**², v., ?; *comp.* **for-hacchen.**]

hache, adj., *O.Fr.* hache; *axe,* FER. 1701; JOS. 503.

hache², sb., *Fr.* hachis; *hash,* MAN. (F.) 15759.

hache, *see* **ache.**

hachet, sb., *O.Fr.* hachet; *hatchet, 'securicula,'* PR. P. 220; P. S. 223; LANGL. *B* iii. 304; hachit BARB. x. 114.

hacken, v., = *M.L.G., M.Du., M.H.G.* hacken, *O.Fris.* (to-)hakia; *hack, chop*; (*ms.* acken) HOM. II. 139; hakkin PR. P. 221; hakke and hewe CH. C. T. *A* 2865; hakke (*? strive, struggle*) after holinesse LANGL. *B* xix. 399; hakki ROB. (W.) 2982; hackeð (*pres.*) A. R. 298; hakken JOS. 512; **hacke** (*imper.*) L. C. C. 52; **hackede** (*pret.*) A. R. 298; ihacked (*pple.*) A. R. 298; *comp.* **to-hacken.** ·

hakkinge, sb., *chopping, 'sectio,'* PR. P. 222.

hacken², v., *? be burdensome*; hakand [hakande] (*pres. pple.*) ['*molesti*'] PS. xxxv. 13, liv. 4.

hād, hēd, sb., *O.E.* hād, *O.N.* heiðr *m., cf. Goth.* haidus *m., O.H.G.* heit *m., f., O.L.G.* hēd, hēde *f., mod.Eng.* -hood, -head; *order, condition, quality, rank; person (of the Godhead)*: kanunkes had & lif ORM. DEDIC. 9; had [hod] LA3. 13267; hod AYENB. 235; (*ms.* hode) ['*ordinem*'] PS. cix.4; **hādes** (*gen.*) HOM. I. 101; **hōde** (*dat.*) A. R. 318 (*v. r.* ordre); of hede þus dwineþ he GOW. II. 117; his cote of þe same hede PERC. 1103; **hādes** (*pl.*) MARH. 11; hodes [hades] A. R. 26; on þreom hadan HOM. I. 99; *comp.* **cniht-,**

man-, prēost-, widewe-hād (-hōd), chíld-,
god-, ʒuʒeð-, mǣiden-, þral-hād (-hōd),
-hēde, báld-, biter-, brōðer-, drunken-,
fæʒer-, fals-, ful-, fūl-, fürhti-, gāl-, ʒēap-,
gōd-, ʒung-, hāli-, līht-, lüþer-, mād-,
mēok-, siker-, sōþ-, swike-, wis-, wēri-,
wlanc-, wrecche-hēde.

hādien, v., *O.E.* hādian; *ordain*; **hāded**
(*pple.*) ORM. 10881; hoded LAʒ. 21856*;
þe hodede HOM. II. 31; *comp.* for-, ʒe-,
un-hādien.

hāding, sb., *O.E.* hādung; *ordination*, ORM.
15967.

hadok, sb., *haddock*, REL. I. 81; haddok VOC.
222; PR. P. 220; B. B. 174.

hǣfd, *see* **hēafed.** **hǣh**, *see* **hēah.**

hǣl, sb., *O.E.* hǣl,=*O.H.G.* heil, *O.N.* heill
Goth. haili *n.*; *health, prosperity, salvation*;
(interj.) *hail*: þat sal beon þin heal LAʒ.
17755*; heal us MARH. 22; hail AYENB. 262;
dronk hire hail ROB. 118; hail seint Michel
REL. II. 174; il hail C. M. 7320; hail, heil
MISC. 42, 48; heil LUD. COV. 25; heele TREV.
I. 399, V. 397, VIII. 183.

hēl-ful, adj., *salutary, healthful*, TREV. VII.
502; heelful WICL. ECCLUS. vi. 31.

heilnesse, sb., *O.E.* hǣlness; *health*, GEN.
& EX. 2068.

hāl-wende, adj., *O.E.* hālwende; *healthful*,
HOM. I. 103; KATH. 1412; halwende LAʒ.
2851; healuwinde A. R. 190.

hǣl, *see* **hāl.** **hǣlden**, *see* **hélden.**

hǣle, sb., *O.E.* hǣlu,=*O.L.G.* hēli, *O.H.G.*
heilī *f., O.N.* heill; *from* **hāl**; *health*, ORM.
5378; hǣle [hale] LK. xix. 9; JOHN iv. 22;
heale TREV. VIII. 183; hele S. S. (Web.)
2201; WILL. 597; LANGL. *B* vi. 261; MAND.
11; A. P. i. 16; PR. C. 1326; DEGR. 1766;
þa hele LAʒ. 30551; mi soule hele M. ARTH.
3655; heele *'sanitas'* PR. P. 234; hele, heale
A. R. 180, 312; hale S. S. (Web.) 693; REL.
I. 274; hēle (*dat.*) MIRC 369; CH. C. T.
A 3102; to godere þire hǣle LAʒ. 3597; to
goder (*misprinted* geder) hele come þou hider
AN. LIT. 9; to wroþere hele A. D. 243; heale
KATH. 874; *comp.* un-hǣle.

heale-water, sb., *holy water*, A. R. 106.

hāle-wēi, sb., *cf. M.H.G.* heilwǣge *n.*, heila-
wāc *m., O.N.* heilivāgr *m.*; *balsam*; haliwhei
'balsamum' PR. P. 223*; haliwei *'antidotum,
salutiferum'* PR. P. 223; swottre þen eaver
eni haliwei KATH. 1707; heo sculde mid
hāleweie (*dat.*) helen his wunden LAʒ. 23072;
halewi MARH. 14; healewi HOM. I. 200; A. R.
164; halwei P. S. lxxxv. 83 (ALTD. BL. 398);
ant te deovel beot hire his healewi (*draught*)
drinken A. R. 238.

hǣlen, v., *O.E.* hǣlan,=*O.L.G.* hēlean, *O.
H.G.* heilan; *heal*, MAT. xvii. 16; LEECHD.
III. 94; ORM. 2218; hǣlen, helen [heale]
LAʒ. 23072, 26102; helen, healen A. R. 112,

330; hele HAV. 2058; ISUM. 486; M. H. 127;
heele WICL. JOHN iv. 47; hale S. S. (Web.)
1033; to helenne HOM. I. 95; hēl (*imper.*)
A. R. 26; hēlde (*pret.*) MISC. 39; ROB. 244;
helde [healde] LAʒ. 29541; þu heldest JUL.
62; (hi) helden AYENB. 96; hēled (*pple.*)
WILL. 1329; haled MAN. (H.) 7; PR. C.
8323; *comp.* ʒe-hǣlen.

hǣlend, sb., *O.E.* hǣlend,=*O.L.G.* hēliand,
O.H.G. heilant; *saviour*, LAʒ. 9144; hǣlende
ORM. 2216; helend KATH. 185; helend, he-
lind HOM. I. 3, 83; helinde A. R. 112; hē-
lindes (*gen.*) LAʒ. 10191.

hēlere, sb.,=*O.H.G.* heilāri; *healer*, HOM.
I. 83, II. 257; helare LAʒ. 9145*; MISC. 40.

hǣlunge, sb., *O.E.* hǣling,=*M.H.G.* hei-
lunge; *healing*; healunge MARH. 19; helinge
LANGL. *B* xvii. 115; heelinge *'sanatio'* PR. P.
235.

hǣleð, *see* **heleð.** **hǣlf**, *see* **half.**

hǣlle, *see* **helle.**

hēlðe, sb., *O.E.* hǣlð, = *O.H.G.* heilida;
health; helðe LAʒ. 29992; GEN. & EX. 2344;
helþe AYENB. 172; LIDG. M. P. 69; healeth
TREV. II. 185; helthe PR. P. 235; *comp.* un-
hēlðe.

hǣne, *see* **hēan.**

hǣnen, v., *O.E.* hǣnan; *from* **hāne (hōne)**;
stone; henen JOHN xi. 8.

hǣnen, *see* **hēnen.** **hǣp**, *see* **hap.**

hǣp, *see* **hēap.** **hǣr**, *see* **her.**

hǣr, **hār**, sb., *O.E.* hǣr, *O.N.* hār,=*O.Fris.*
hēr, *O.L.G., O.H.G.* hār; *hair*, ORM. 3208;
hǣr (*miswritten* hǣð) [heer] LAʒ. 7048; her
A. R. 424; HAV. 1924; P. L. S. xix. 266; FER.
1580; mi brune her is hwit bicume MISC.
193; hire her is fair inoh SPEC. 28; he vorleas
his her AYENB. 181; an her KATH. 2288;
hear 1429; her, heer CH. C. T. *A* 589; heer
'capillus, crinis' PR. P. 235; S. S. (Web.)
472; MAND. 81; har Iw. 823; haare ALEX.
(Sk.) 5476; hor HAV. 235; ANT. ARTH. xlv;
hēr (*pl.*) A. R. 398; here ALIS. 4989; hǣre
(*dat. pl.*) MK. i. 6; here, heare O. & N. 428,
1550; *comp.* hors-hēr.

heer-bond, sb., = *M.H.G.* hārbant; *hair-
ribbon*, PR. P. 236.

hēr-seve, sb.,=*Ger.* hārsib; *hair-sieve*, L.
C. C. 7.

hǣrd, *see* **heard.**

hǣre, sb., *O.E.* hǣre,=*O.N.* hǣra, *O.H.G.*
hārre, *O.Fr.* haire; *hair shirt, cilicium*; here
LAʒ. 17697; MISC. 101; BEK. 2231; AYENB.
227; here, heare A. R. 126, 130; haire CH. C.
T. *G* 133; hair TREV. V. 107; here [heire,
haire] ROB. (W.) 8952; hǣren (*dat.*) MAT.
xi. 21; hēren (*dat. pl.*) LK. x. 13; LAʒ. 19707;
heren [hearen] A. R. 10.

hǣren, adj., *O.E.* hǣren,=*M.H.G.* hǣrin;
made of hair; heeren WICL. EX. xxvi. 7.

hǣri, adj., *hairy*; heri TREV. III. 285 ; WICL. GEN. xxvii. 11 ; hari C. M. 8085.

hæring, sb., *O.E.* hæring,=*O.Fris.* hereng, *O.H.G.* harinc ; *herring,* FRAG. 3 ; hering HAV. 758 ; ALIS. 6589 ; REL. I. 83 ; hit nere . . . w(u)rþ on hering MISC. 95.

hærm, *see* **hearm. hærre,** *see* **herre.**

hǣsè, hěste, sb., *O.E.* hæs *f., for* *hǣst, *from* **hāten;** *behest, command* ; hæse ORM. 3537 ; hese SAX. CHR. 251 ; hese, hes HOM. I. 165, 181 ; P. L. S. viii. 46, 174 ; hes MISC. 61 ; has HOM. II. 222 ; ALIS. 444 ; heste BEK. 381 ; C. L. 172 ; SHOR. 98 ; AYENB. 5 ; ALIS. 1330 ; CH. C. T. *A* 2532 ; M. ARTH. 3686 ; ane heste LAӠ. 2494 ; heaste KATH. 235 ; heste, hest HOM. I. 11, 125 ; A. R. 6, 58 ; heest AUD. 17 ; hěste (*pl.*) LAӠ. 31561 ; hesten A. R. 386 ; *comp.* **bi-, ӡe-hēs, ūt-hēs, -hěste.**

hēste-dei, sb., *holy day, festival,* HOM. I. 3.

hǣsne, sb., *command*; hesne HOM. I. 229, 237 ; ANGL. I. 12.

hǣte, sb., *O.E.* hǣto,=*O.Fris.* hēte, *O.H.G.* heizī ; *from* **hāt** ; *heat,* ORM. 13855 ; heit, het BARB. xi. 611, 612 ; hete PR. P. 238 ; P. L. S. viii. 117 ; A. R. 120 ; MISC. 37 ; TREAT. 136 ; AYENB. 75 ; PERC. 862 ; AUD. 46 ; when þe hete is overcome A. D. 289 ; hete [heete] CH. C. T. *G* 1408 ; heate KATH. 1701 ; h(e)ate HOM. II. 227 ; **hǣte,** hǣtan, hǣten (*dat.*) LEECHD. III. 94, 128.

hǣten, v., *O.E.* hǣtan,=*O.N.* heita, *O.H.G.* heizen ; *heat, make or become hot*; heaten A. R. 404 ; his heorte feng to heaten JUL. 21 ; hetin PR. P. 238 ; hete MARG. 247 ; **hětte** (*pret.*) CH. P. F. 145 ; **ihǽt** (*pple.*) MISC. 148 ; *comp.* **an-, of-hǣten.**

hǣðe, sb., *O.E.* hǣð, =*O.N.* heiðr, *Goth.* haiþi, *O.H.G.* heida ; *heath (moor, heather),* LAӠ. 12819 ; heþe A. P. ii. 535 ; heþe, hethe CH. C. T. *A* 6, 3262 ; hethe PR. P. 238 ; heþ P. L. S. xxix. 30 ; heeþ [heth] LANGL. *B* xv. 451 ; heghte D. ARTH. 2295.

hǣþe-līch, adj., *O.N.* hǣðiligr ; *cf.* hāþ ; *contemptuous* ; heþelich TRIST. 2897 ; **hǣþe-liӡ** (*adv.*) ORM. 7408 ; hetheli M. H. 43 ; D. ARTH. 268.

hǣþen, sb., *O.E.* hǣðen,=*O.Fris.* hēthen, *O.L.G.* hēthin, *O.N.* heiðinn, *O.H.G.* heidan ; *from* **hǣðe** ; *heathen,* ORM. 1948 ; heþen LANGL. *B* xv. 451 ; MAND. 3 ; heþin OCTAV. (H.) 1598 ; haiþen M. H. 99 ; PR. C. 5521 ; ISUM. 424 ; þe heðene mon LAӠ. 12103 ; **hǣðene** [heaþene] (*pl.*) LAӠ. 14437 ; heþene ROB. 81 ; **hǣðenen** (*dat. pl.*) LAӠ. 14647.

hǣþen-dōm, sb., *O.E.* hǣðendōm,=*O.H.G.* heidantuom ; *paganism,* ORM. 9673 ; heaðendom KATH. 35.

hǣðene-scipe [heaþensipe], sb., *O.E.* hǣðenscipe, *cf. O.H.G.* heidinscaft; *paganism,* LAӠ. 14849.

hǣðenesse,sb.,*O.E.*hǣðennyss; *heathenism, heathendom,* LAӠ. 29388 ; heþenesse MISC. 26 ;

ST. COD. 57 ; LANGL. *B* xv. 435 ; CH. C. T. *A* 49 ; the custom of his hethenenesse [*later ver.* hethenesse] WICL. JUDITH xiv. 6 ; bothe cristendome and hethinnesse E. T. 196 ; heþenisse FER. 121 ; in al hethenis is no Sarsin FER. 2187 ; hathennes TOWNL. 66.

hǣþen, v., *O.N.* hǣða ; *mock, scorn* ; hēþe SPEC. 37 ; **hǣþen** (*pres.*) ORM. 13682.

hǣþing, sb., *O.N.* hǣðing ; *scorn, contempt,* ORM. 19702 ; heþing S. S. (Web.) 91 ; PS. lxxviii. 4 ; MAN. (H.) 273 ; A. P. ii. 710 ; AMAD. (R.) 11 ; hething TOWNL. 174 ; hetheinge [ethinge] HAMP. PS. ii. 4.

hēthing-full, adj., *contemptuous,* D. TROY 3953.

hǣved, *see* **hēafed. hǣwen,** *see* **hēawen.**

hafen, *see* **habben. hafene,** *see* **havene.**

haft, sb., *O.E.* hæft, *cf. O.H.G.* hefti, *O.N.* hepti ; *haft, handle,* P. S. 339 ; FLOR. 1636 ; WICL. DEUT. xix. 5 ; heft ' *manubrium* ' PR. P. 232 ; VOC. 238 ; under heft and under hand S. S. (Web.) 259 ; othir **haftis** (*pl.*) (*affairs*) in hande have we YORK xx. 76.

hag, sb., *? notch, cleft* : [þis castel] es hei sett apon þe crag, grai, and hard, wituten hag [hagg] C. M. 9885.

hagas, sb., *mod.Sc.* haggis ; *a kind of pudding,* ' *tucetum,*' VOC. 242 ; PR. P. 220 ; L. C. C. 52.

haӡe, hahe, hawe, sb., *O.E.* haga,=*M.L.G.* hage, *M.Du.* haeghe, *O.N.* hagi ; *hedge, enclosure, meadow,* O. & N. 585, 1612 ; hawe CH. C. T. *C* 855 ; hawch BARB. xvi. 336 ; haugh YORK iv. 35.

haӡe[2], sb., *O.E.* haga *for* ***haӡe-berie,**= *M.Du.* haeghbesie ; *haw* (*fruit of hawthorn*); hawe C. M. 505 ; CH. C. T. *D* 659 ; HOCCL. i. 380 ; al nas wurþ an hawe ROB. 524 ; **hawen** (*pl.*) ALIS. 4983 ; hawes WILL. 1811 ; LANGL. *B* x. 10 ; *comp.* **chirche-hawe.**

haӡ(e)-þorn, sb., *O.E.* haga-, hægþorn, = *O.N.* hagþorn, *M.H.G.* hagedorn ; *hawthorn,* P. L. S. xiii. 187 ; haweþorn LANGL. *C* xix. 184.

haӡe[3], sb.,=*M.H.G.* hage *m.*; *pleasure* : mid gode haӡe LAӠ. 17507 ; helde (*r.* elde) wiþhouten hawe ANGL. III. 281 ; *comp.* **bi-haӡe.**

haӡel [hawel], sb., *O.E.* hagol, hægel, hægl, =*O.H.G.* hagal, *O.N.* hagl ; *hail,* LAӠ. 11975 ; snou and haӡel (hawel) O. & N. 1002 ; hawel TREAT. 137 ; hail PR. P. 221 ; GEN. & EX. 3046 ; REL. I. 264 ; LANGL. (Wr.) 8362 ; haul SAINTS (Ld.) 317.

hawel-, hail-stōn, sb.,=*M.L.G.* hagelstēn ; *hailstone,* TREV. IV. 69 ; hailston WICL. WISD. v. 23.

haӡelen, v.,=*M.H.G.* hagelen, haglen ; *hail*; haweli BRD. 32 ; hailin PR. P. 221 ; **hailes** (*pres.*) VOC. 201.

haӡer, adj., *? cf. O.N.* hagr ; *apt, dexterous* : a ful

haȝer [haher, hawur] smið A. R. 52; haȝher
hunte ORM. 13471; haver (*for* hawer) P. S.
155; haȝer GAW. 352; þe haȝer stones trased
aboute hir tressour GAW. 1738.

haȝher-līke, adv., *aptly*, ORM. 1214;
hagherlich(e) A. P. ii. 18.

hagge, sb., *?M.H.G.* hacke; *hag, old woman*,
LANGL. *B* v. 191; heggen (*pl.*) A. R. 216.

haȝhe-līke, -līȝ, adv., = *O.N.* hagliga; *becomingly*, ORM. 1228, 1247.

haȝien, v., *O.E.* (on-)hagian, *cf. O.Fris.* hagia,
O.L.G.(bi-)hagōn, *O.N.* haga, *M.L.G., M.H.G.*
hagen; *? be expedient*: ȝef me haheð (*pres.
subj.*) hit HOM. I. 305; *comp.* bi-haȝien.

hahe, *see* haȝe.

hāht, sb., *? from* hangen; *cf. O.N.* hætta;
danger: hu michel haht hit is godes forbod
te brekene V. & V. 11; hauht 87; his soule
beð mikel hagt GEN. & EX. 486; to liven in
hagt 2044; ic am in sorge and hagt 2082;
Amalechkes folc ffedde for agte of dead
3384.

hai, *see* hei.   hāi, *see* hēi.

haifare, sb., *O.E.* hēahfore, hēafre; *heifer*,
'*juvenca*,' VOC. 177; heffre TREV. IV. 451;
hekfere PR. P. 234; VOC. 250; an haifre
(*gen.*) hude P. S. 239.

hail, haile, *see* haȝel, hæl, hāl.

hailin, v., *O.N.* heilla; *salute, greet*; heilin
'*saluto*' PR. P. 233; heȝlen (*ms.* heȝȝlenn)
ORM. 2814; haile (*pres.*) LANGL. *B* v. 101;
hailede [haillede] (*pret.*) LAȝ. 14968; heilede
LANGL. *A* ix. 10.

hailinge, sb., *salutation*, LAȝ. 14442.

hailsen, *see* hālsien.

hain, sb., *cf. Dan.* hegn, *Swed.* hägn *n., M.H.G.*
hagen, hain *m.; park, enclosure*: grete hertes
in þe haines (*pl.*) HALLIW. 439; faiere
parkes inwiþ hainus DEGR. 70.

hainen, v., *? from Fr.* haine (*hatred*); *? hate*:
þei schullen be hatid & hained (*pple.*) doune
as houndis WICL. E. W. 250.

haire, *see* hǣre *and* hār.

hairif, *see under* hei.

hairōn, sb., = *O.Fr.* hairon; *heron*; '*ardea*,'
VOC. 177; heiroun [heroun] CH. P. F. 346;
heirōne (*dat.*) AYENB. 193.

hairounsew, sb., *O.Fr.* herounceu, heroun-
cel; *young heron*; heironsew B. B. 165;
herunsew REL. I. 88; heronsewe '*ardiola*'
CATH. 184; heronsewes (*pl.*) CH. C. T. *F* 68.

hait, adj., *lively, cheerful*, ROB. (W.) 8650.

hait, heit, interj., *cf. Ger.* hot; *gee up!* CH.
C. T. *D* 1543.

haite, v., *O.N.* heita; *? threaten*: ever þou
art mi fo, febli þou canst haite, þere man
schuld menske do TRIST. 3050.

haiþen, *see* hǣþen.

hak, sb., *cf. Du.* hak; *pickaxe, digging-fork*,
'*bidens, fossorium, ligo*,' CATH. 170.

hāke, sb., *cf. O.E.* hacod; *hake*; '*squilla*,'
PR. P. 222; SPEC. 31; LIDG. M. P. 201.

hakele, sb., *O.E.* hacele, = *Goth.* hakuls, *O.N.*
hökull, *M.L.G.* (mis-)hakel, *O.H.G.* hachul;
cloak; hakel GAW. 2081; *comp.* messe-
hakele.

hakenē, hakenei, sb., *Fr.* haquenee, *cf. Span.*
hacanea; *hackney*; hakenai B. B. 155.

hakenei-mon, sb., *? horse-dealer*, LANGL.
A v. 161; *C* vii. 365, 378; *B* v. 318; hakenei-
mannes (*gen.*) *C* vii. 391.

hakken, *see* hacken.

hal, sb., *O.E.* hal, heal; *? secret place, corner*,
SHOR. 160; hāle (*dat.*) MIRC 1384; in one
swiþe diȝele hale O. & N. 2; from hale to
hurne P. S. 150; hāles (*pl.*) MAN. (F.) 9280.

hāl, adj., *O.E.* hāl, hǣl, = *O.L.G., O.Fris.* hēl,
O.N. heill, *O.H.G.* heil, *Goth.* hails; *well,
healthy, sound, whole, hale*, JUL. 59; ORM.
14818; S. S. (Web.) 1064; hal & fere HOM.
I. 25; hwil þe scheld is hal H. M. 15; hal
[hail] beo þu LAȝ. 1498; hal and hæil 12528;
hæl [hol] & isund 1252; hæil [hol] 6636;
hail wurð þu 3516; hol GEN. & EX. 2776;
HAV. 2075; ROB. 86; AYENB. 51; WILL.
1566; LUD. COV. 374; betere is for te gon
sic touward heovene þen al hol touward helle
A. R. 190; hol & sound MAN. (F.) 9657;
hool '*sanus, integer*' PR. P. 242; WICL.
JOHN v. 6; CH. C. T. *C* 357; (he) was mad
hool MAND. 88; an hool moneþ 134; heil
'*sanus*' PR. P. 233; heil and sund REL. I.
210; heil beo þou LEB. JES. 815; heil, hail
MISC. 42, 48; hail ALIS. 3080; hail be þou
MIRC 422; all haill *entirely* BARB. x. 793;
haill and feir *safe and sound* BARB. xv. 514;
þe hole half A. R. 112; of ane hale yhere
PR. C. 3933; hōle (*pl.*) MISC. 38; LANGL.
B vi. 61; hālen (*dat. pl.*) MAT. ix. 12; *comp.*
ȝe-hāl, un-hāl.

hōl-līche, adv., *wholly*, WILL. 945; JOS.
456; holli M. ARTH. 935; GAW. 1049; holli
[hoolli] CH. C. T. *A* 599; hali BARB. v. 57,
xiv. 79; haleli iii. 45, iv. 772; hoili YORK
viii. 24.

hāli-lī, adv., *wholly*, BARB. i. 316.

hālsum, adj., *O.N.* heilsamr, = *O.H.G.* heil-
sam; *wholesome*, MARH. 21; ORM. 2921;
holsum '*saluber*' PR. P. 244; HOCCL. i. 248;
holsom CH. TRO. i. 947; LIDG. M. P. 66;
helsum APOL. 6.

hōlsum-līche, adv., *wholesomely*, HOM.
II. 107.

hōlsumnesse, sb., *wholesomeness*, '*salu-
britas*,' PR. P. 244; HOM. II. 103.

halata, sb., *?a fabled animal*, B. B. 234.

halchen, v., *fasten, embrace*: ȝet [he] hem
halchez (*pres.*) al hole þe halvez togeder
GAW. 1613; half his armes·þer under were
halched (*pple.*) 185; aither halched oþer
GAW. 939.

háld, hálden, háldere, háldinge, *see* **heald, healden** *and* **hélden, healdere, healding.**

hále, sb., *Lat.* halō ; *halo (of the moon),* ' *halo,*' PR. P. 222.

hāle, hālen, *see* **hǣle, hǣlen.**

hálen, *see* **hálien.**

hālewe, *see* **hāliʒ.**

hālewēi, *see under* **hǣle.**

hālewin, *see* **hālʒien.**

half, adj. (adv.), *O.E.* healf, = *O.L.G., O.Fris.* half, *O.N.* halfr, *O.H.G.* halb, *Goth.* halbs ; *half,* MAND. 99 ; M. H. 91 ; half þat ilke win LAʒ. 15006 ; hit is half mon & half fisc 1330 ; hælf [half] ʒaru 8657 ; half þat gavel 22439 ; þreo ʒer and an half BEK. 147 ; an half ʒer ROB. 492 ; an half ʒer, half a ʒer LANGL. *B* ii. 228, *A* ii. 204 ; half a ʒer P. L. S. xvii. 386 ; an (h)alf peni AYENB. 193 ; half a ʒer 173 ; half here lif BEK. 2424 ; ōðer half [*O.E.* ōðer healf, = *O.L.G.* ōther half, *O.H.G.* ander halp] ; *one and the other (second) half, one and a half,* LAʒ. 7856 ; oþer half BEK. 11 ; CHR. E. 120 ; MAN. (F.) 3667 ; oþer half spanne.FER. 744 ; other half MIR. PL. 141 ; þridde half [= *Ger.* dritte halb] yer *two years and a half* H. H. 45 ; **vifte half** *four and a half* LAʒ. 32195 ; half feorþe ʒer *three years and a half* ORM. 8621 ; half so strong HAV. 2245 ; half so freo SPEC. 40 ; þet halve ʒer LAʒ. 3377 ; þet oðer halve ʒer A. R. 412 ; halve a marke LANGL. *B* v. 31 ; **halfen** (*acc. m.*) MAN. (H.) 48 ; þe halven del MAND. 166 ; **halvendel** [halfendel, halfendele] (*adv.*) C. M. 2099 ; halvindel RICH. 1138 ; ʒif halvendel the child were þin K. T. 783 ; ne halvendel the dred CH. TRO. iii. 657 ; i sawe never wommon halvendell so gai EM. 443 ; **halve** (? *pl.*) þa hundes LAʒ. 22444 ; of halve þan blissen 24231.

half-acre, sb., *small piece of land,* LANGL. *C* vii. 267, ix. 23.

half-brōþer, sb., *half-brother,* LANGL. *C* xxi. 422.

half-dēd, adj., *O.E.* healfdēad ; *half-dead,* ROB. 163 ; hi bieþ halfdeade AYENB. 86.

hælf-ʒaru, adj., *half ready,* LAʒ. 8657.

half-idiʒt, adj., *half dressed* ; helf idiʒt FER. 3569.

halve-mon, sb., *coward,* ROB. (W.) 5793 ; **halvemen** (*pl.*) ROB. (W.) 4093, 4378.

half-peni, sb., *O.E.* healf pening ; *half-penny,* MAN. (H.) 238 ; half-peniworthez E. G. 425 ; halpeni ALIS. 3114 ; WICL. EZ. xlv. 12 ; LK. xii. 6.

half-qvic, adj., *O.E.* healf cwic ; *half alive,* HOM. 79 ; whi hit seið alf qvic and noht alf ded 81 ; half qvik (*later ver.* half alive) WICL. LK. x. 30.

half-wōde, adj., *half mad,* GOW. III. 18.

half-süster, sb., *half-sister,* GAW. 2464.

half, sb., *O.E.* healf, = *O.L.G.* half, *O.N.* halfa, *O.H.G.* halb, halba, *Goth.* halba ; *half, side, part, behalf,* ROB. 168 ; SHOR. 51 ; LANGL. *B* iii. 337 ; þe verste half of þe leave AYENB. 1 ; helve FER. 159 ; **halfe** (*dat.*) HOM. I. 37 ; an are halfe LAʒ. 20717 ; an oðer halve he wes glæd þat his ifon weoren dæd 9374 ; bi are halve 9460 ; æt þe oðer hælve 6474 ; halve JOS. 549 ; on ever iche halve A. R. 252 ; on eiþer halve O. & N. 887 ; on oþer half þes wateres MISC. 146 ; on þis half þe see MAND. 234 ; o godes halfe (*on the part, in the name, of God*) ORM. 624 ; on his halfe CH. TRO. ii. 934 ; in þe kinges half WILL. 4831 ; bi halve (*beside*) (*later ver.* besides) þan castle LAʒ. 27913 ; bi halves France 11893 ; heo stod bi halves P. L. S. xix. 13 ; on boþe halve ALIS. 957 ; on alle halve MAP 338 ; *comp.* **bac-, norð-, sūð-half.**

halfinge, adv., *O.E.* healfunga ; *half, not wholly* ; halving GOW. II. 65.

halflunge, adv., = **halfinge,** A. R. 354 ; **halflinges,** ORM. 16575.

halfter, sb., *O.E.* hælfter (*dat.* hælftre), *cf.* *O. H. G.* halftra, *M. Du.* halfter, halter, *M.L.G.* halter *f. ; halter* ; helfter HOM. I. 53 ; halter BEK. 1174 ; halter ne bridel O. & N. 1028 ; heltir, halter ' *capistrum*' PR. P. 235 ; heltere TOWNL. 313.

hālʒien [hălʒi], v., *O.E.* hālgian, = *O.H.G.* heilagōn, *O.L.G.* hēlagōn, *O.N.* helga ; *from* hāliʒ ; *hallow,* LAʒ. 17496 ; halʒhen (*ms.* hallʒhenn) ORM. 4413 ; haliʒen HOM. I. 45 ; (h)aligen GEN. & EX. 258 ; halʒi AYENB. 237 ; halwin ' *consecro*' PR. P. 224 ; halwe ROB. 154 ; GOW. II. 370 ; LANGL. *B* xv. 557 ; LUD. COV. 61 ; halewe [hallowe] WICL. JOHN xi. 55 ; haloe [' *celebrare,*' ' *consecrare*'] TREV. III. 83, 109.; halowe TREV. V. 231, 269 ; **hăleweð** (*pres.*) A. R. 396 ; halweþ MISC. 232 ; **hălʒe** (*imper.*) SHOR. 93 ; **hăleʒede** [halwede] (*pret.*) LAʒ. 22406 ; halʒed A. P. ii. 506 ; (*ms.* hālyhed) PS. xlv. 5 ; halwide WICL. GEN. ii. 3 ; haloede TREV. V. 87 ; **hălewed** (*pple.*) MAND. 1 ; haloide TREV. VI. 113 ; *comp.* **ʒe-hālʒien.**

hălwinge, sb., *O.E.* hālgung ; *act of making holy,* ' *consecratio,*' PR. P. 224 ; TREV. III. 185 ; haloing V. 401.

halh, sb., *O.E.* healh ; *haugh, meadow* ; halche (*v. r.* hawch, hawgh) BARB. xvi. 336 ; haugh YORK iv. 35.

hāli, *see* **hāliʒ.**

halibut, sb., *a sort of fish,* B. B. 157.

hálien, v., ? *O.E.* (ge-)holian (ALFR. P. C. 209), = *O.Fris.* hālia, *O.L.G.* halōn, *O.H.G.* halōn, holōn ; *hale, haul, drag* ; halin ' *traho*' PR. P. 223 ; to haʒie men fro helle LANGL. *B* viii. 95 ; hale GOW. II. 295 ; CH. P. F. 151 ; A. P. iii. 219 ; **háleð** (*pres.*) REL. I. 214 ; haliþ ALIS. 1416 ; halen ANT. ARTH. xxxv ; **hálede**

(*pret.*) LA3. 16712; **ihauled** (*ppie.*) BEK. 1497.

halier, sb., *carrier, porter* : ne soffir not the **haliers** (*pl.*) to hale it all awei E. G. 425.

háli3, adj. & sb., *O.E.* hālig, hāleg, hǣlig,= *O.L.G.* hēlag, *O.N.* heilagr, *O.H.G.* heilīg, heileg, heilag ; *from* hāl ; *holy*, ORM. 541 ; hali KATH. 2073 ; R. S. v ; PR. C. 3690 ; hali [holi] LA3. 10130 ; hali, heli GEN. & EX. 51, 54 ; holi PR. P. 243 ; A. R. 370 ; C. L. 7 ; SHOR. 13 ; MAND. I ; hooli WICL. JOHN xiv. 26 ; heli REL. I. 23 ; holie lore FRAG. 7 ; se halge gast LK. i. 35 ; þe hal3e gast HOM. I. 99 ; halie 85 ; þe halie [holie] qvene LA3. 11148 ; ðe holie tid HOM. II. 3 ; þæt hilige ['*sanctum*'] MAT. vii. 6 ; **hålghe** (*sb.*) *saint* PR. C. 6087 ; halewe BEK. 1085 ; þes hål3an, hal3en, hal3e (*gen.*) gastes HOM. I. 99 ; one holie monnes A. R. 350 ; of þan **hålgen** (*dat.*) gaste MAT. i. 20 ; in þan halie godspelle HOM. I. 47 ; mid te holie monne A. R. 278 ; on halowe Thurs daie HALLIW. 430 ; **håline** (*acc. m.*) HOM. I. 99 ; þa . . . **hålie** (*pl.*) da3es HOM. I. 47 ; holie men A. R. 350 ; (þa) hål3en (*sb., pl.*) LA3. 10122 ; AYENB. 6 ; hal3hen (*ms.* hall3henn) ORM. 6009 ; crist . . . & his halechen SAX. CHR. 263 ; halewen A. R. 124 ; halighes PS. xlix. 5 ; halewes REL. I. 38 ; halwes [halowes] CH. C. T. *A* 14 ; **håle3ene**, haluwene (*sb., gen. pl.*) A. R. 94, 330 ; halewene P. L. S. ix. 181 ; alle halowene tid E. G. 351 ; **hōlieste** (*superl.*) JOS. 113 ; holiist(e) AYENB. 54 ; *comp.* **lif-háli**.

háli-dei, sb., *O.E.* hāligdæg ; *holiday*, A. R. 20 ; hali3da3 ORM. 4350 ; halidai LANGL. *B* v. 409 ; CH. C. T. *A* 3309 ; A. P. ii. 134.

háli-dōm, sb., *O.E.* hāligdōm,= *O.H.G.* heiligtuom ; *sanctuary, saintly relic*, LA3. 15343 ; BEK. 2273 ; E. G. 36 ; halidam GAW. 2123 ; **háli3dōmes** (*pl.*) ORM. 1031.

háli-hēde, adj.,= *O.H.G.* heilīgheit ; *sanctity*, C. M. 1439 ; holihede AYENB. 247.

hōli-līche, adv., *holily*, AYENB. 74 ; halili3 ORM. 15920.

háli3nesse, sb., *O.E.* hāligness ; *holiness*, HOM. I. 99 ; ORM. 8864 ; halinesse [holinisse] LA3. 1820 ; holinesse O. & N. 900 ; holinesse [hoolinesse] CH. C. T. *B* 713.

hōli-niht, sb., (? holi niht), *the night before a holiday*, A. R. 22.

hōli-schüpe, sb., *sanctity*, C. L. 1320.

háli-water, sb., *O.E.* hāligwæter ; *holy water*, A. R. 324 ; mid holiwatere FRAG. 7.

håliwei, -whei, *see under* **hæle**.

halke, sb., *O.E.* healoc ; ? *from* hal ; *corner*, '*angulus, latibulum*,' PR. P. 223 ; HORN (L.) 1087 ; AYENB. 210 ; CH. C. T. *F* 1121 ; **halkes** (*pl.*) TREV. I. 313 ; A. P. ii. 104 ; *deriv.* **halchen**.

halle, sb., *O.E.* heall,= *O.L.G.*, *O.H.G.* halla,

O.N. höll ; *hall*, '*aula*,' PR. P. 222 ; A. R. 192 ; SPEC. 48 ; S. S. (Web.) 175 ; CH. C. T. *A* 2521 ; MIR. PL. 165 ; þa halle LA3. 28033 ; þare halle dure 20999 ; in halle and in bure KATH. 1470 ; **halle** (*pl.*) MISC. 170, 171 ; hallis (*tents*) ALEX. (Sk.) 4148 ; **hallen** (*dat. pl.*) LA3. 2025 ; *comp.* gest-, 3ilde-, mōt-halle.

halle-dure, sb., *hall-door*, LA3. 30153.

halle-flōr, sb., *hall floor*, B. DISC. 1765.

hal-imōt, sb., *court of the lord of the manor*, LA3. 31997 ; P. S. 154.

halm, sb., *O.E.* healm, halm,= *O.L.G.*, *O.H.G.* halm, *O.N.* halmr, *?Lat.* calamus, *Gr.* κάλαμος ; *straw*, '*stipula*,' PR. P. 223 ; *handle*, GAW. 330.

halou, interj.,= *Ger.* hallo ; *halloo*, PR. P. 223.

halowen, v., *halloo*: halowin or criin as schipmen '*celeumo*' PR. P. 224 ; **halowes** (*pres.*) D. ARTH. 3319 ; thei halowede (*pret.*) an hight DEGR. 233 ; foule halowed him C. M. 15831 ; he was halowid (*pple.*) and ihuntid DEP. R. iii. 228.

halowinge, sb., *hallooing*: halowinge of hundis '*boema*' CATH. 172.

hals, sb., *O.E.* heals, hals,= *O.Fris.*, *O.L.G.*, *O.H.G.*, *O.N.*, *Goth.* hals *m.* ; *neck, throat*, '*collum*,' PR. P. 224 ; ORM. 4777 ; HAV. 521 ; LANGL. *B* PROL. 179 ; CH. C. T. *E* 2379 ; RICH. 2561 ; S. S. (Wr.) 2223 ; GAW. 621 ; MAN. (H.) 279 ; D. ARTH. 764 ; AV. ARTH. lxv ; LUD. COV. 342 ; *comp.* **wik-hals**.

hals-wort, sb., *O.E.* halswyrt ; (*a plant*) '*auricula leporina*,' LEECHD. II. 390.

hålsien, v., *O.E.* hālsian, hǣlsian,= *O.H.G.* heilisōn, *O.N.* heilsa (*salute*) ; *conjure, deprecate, adjure ; salute*, HOM. I. 45 ; þat his qvide durste halsien LA3. 13242 ; hailce GAW. 2493 ; **hålsie** (*pres.*) A. R. 348 ; halsi LA3. 32168 ; ich halsi þe o godes nome MARH. 17 ; halse JOS. 400 ; halse [hailse] CH. C. T. *B* 1835 ; hailse LANGL. *B* v. 101 ; hailses GAW. 972 ; **hailse** (*imper.*) PERC. 404 ; **hålside**, hailside, hailsed (*pret.*) LANGL. *A* viii. 10*, *B* viii. 10 ; hailsed L. H. R. 113 ; **hålsed** (*pple.*) P. R. L. P. 85 ; A. P. ii. 1621.

hålsunge, sb., *O.E.* hālsung ; *supplication*, A. R. 330 ; halsinge ['*adjuratio*'] TREV. III. 11 ; halsinge BARB. vii. 117.

halsin, v.,= *O.H.G.* halsen, *O.N.* halsa ; *seize by the neck, embrace*, '*amplecti*,' PR. P. 224 ; halse CT. LOVE 1289 ; **halseþ** (*pres.*) and kisseþ LIDG. M. P. 32.

halsinge, sb., *embrace* ; ['*complexus*'] TREV. VII. 139.

hålsnien, v.,= **hålsien** ; helsni AYENB. 253 ; ich **hålsni** (*pres.*) þe a godes name S. A. L. 161.

halt, adj., *O.E.* healt, halt,= *O.L.G.* halt, *O.N.* haltr, *Goth.* halts, *O.H.G.* halz ; *lame*, '*claudus*,' PR. P. 224 ; MK. ix. 45 ; CH. D. BL. 622 ; A. P. ii. 102 ; MAS. 154 ; þe **halte** (*pl.*) HAV.

543; LEG. 30; holte MISC. 39; **halten** (*dat.
pl.*) HOM. I. 229; halte ORM. 15499.

halten, v., *O.E.* healtian, = *O.L.G.* haltōn, *cf.
O.N.* helta, *M.H.G.* halzen; *halt, be lame,
limp,* GOW. I. 310; haltin '*claudico*' PR. P. 224;
halte HOCCL. vi. 42; **haltinde** (*pple.*) SPEC.
48; **haltede** (*pret.*) TREV. III. 239; haltide
WICL. GEN. xxxii. 31; halted PS. xvii. 46.

　halter, sb., *one who is lame,* '*claudicarius,*
duplicarius,' CATH. 172; haltare '*claudicator*'
PR. P. 224.

　haltinge, sb., *limping,* '*claudicacio,*' PR. P.
224.

halter, *see* **halfter.**

haltrin, v., *cf. M.Du.* halfteren, halteren;
halter; heltrin '*capistro*' PR. P. 235; **ihal-
tred** (*pple.*) LEB. JES. 808.

halve, *see* **helfe.**

halven, v., *cf. M.H.G.* halben, helben; *halve,*
WICL. PS. liv. 24; seiles þer **helfden** (*pret.*)
LAȝ. 7851; *comp.* **bi-halve.**

hǎlwien, *see* **hǎlȝien.**

hām, sb., *O.E.* hām, = *O.L.G.* hēm, *O.H.G.*
heim, *O.N.* heimr *m., cf.* Goth. haims *f.*; *home,*
C. M. 994; hus & ham ORM. 1608; þa was
Verolam a swiȝe kinewurȝe hom 19455;
nute ȝe ... his hom O. & N. 1751; at
hāme (*dat.*) LAȝ. 2436; H. M. 31; ORM.
12985; PERC. 338; at home CH. C. T. *A* 512;
hāmes (*pl.*) HOM. I. 49; hames [homes] LAȝ.
19537; Rome richest alre **hōme** (*gen. pl.*)
REL. I. 122; **hāmen** (*dat. pl.*) LAȝ. 29416.

　hām, adv., *home*: he brohte ham LAȝ.
31813; ham sho hied fast IW. 1836; hwan
he cumeþ hom [ham] O. & N. 1531; hom to
faren GEN. & EX. 1711; come hom S. S.
(Wr.) 2499; fire hem home PALL. v. 192;
hame BARB. xvi. 667; hom JOS. 609; hem
LUD. COV. 30.

　hām-cūme, sb., *O.E.* hāmcyme; *home-com-
ing,* H. M. 31; homcome AN. LIT. 5; homkome
WILL. 807.

　hām-fare, sb., *O.E.* hāmfaru; *attack on a
house,* ['*insultus factus in domo*'] TREV. II. 95.

　hām-holde, adj., *domestic,* D. ARTH. 1843.

　hām-lī, adj., = *M.L.G.* hēmelik, *O.H.G.*
heimlīch; *homely,* HAMP. PS. iv. 1*; hoomli
'*domesticus*' PR. P. 244; homli A. P. i. 1210;
CH. C. T. *E* 1794; **hāmlī** (*adv.*) BARB. xviii.
345; homli PL. CR. 771.

　hāmli-lī, adv., *in a homely manner,* BARB.
xvii. 4.

　hāmlinesse, sb., *homeliness, familiarity,*
HAMP. PS. PROL. 16; hoomlinesse CH. C. T.
B 2876.

　hām-sōcne, sb., *O.E.* hāmsōcn, = *O.N.* heim-
sōkn, *Sc.* hamesucken; *housebreaking,* ANGL.
VII. 220; hamsokne '*quite de entrer en
autri ostel a force*' REL. I. 33; TREV. II. 95.

　hām-ward, adv., *homeward,* LAȝ. 23170;

BĒK. 254; hamward [homward] LANGL. *A* iii.
187; homward GEN. & EX. 2376; homwardis
BARB. vii. 492.

háme, sb., *O.E.* hama, homa, = *O.L.G.* (feˀar-,
līc-)hamo, *O.H.G.* (līh-)hamo; *skin, covering, garment,* PR. P. 224; ALIS. 391; hámes
(*pl.*) *plumage* ALEX. (Sk.) 4986; *comp.* feðer-,
līc-hame (-home).

háme², sb., = *Ger.* hame, *Du.* haam; *hame,
horse collar:* so are þei bounde in þe fendes
hame H. S. 11496; **hámes** (*pl.*) VOC. 168;
comp. **beru-ham.**

hamelen, v., *O.E.* hamelian, = *O.H.G.* ha-
malōn, *O.N.* hamla; *mutilate*; he heomélede
(*pret.*) þa reven LAȝ. 11206; **hameled** (*pple.*)
CH. TRO. ii. 944; ihamld PL. CR. 300.

hamelet, sb., *O.Fr.* hamelet; *hamlet,* '*villula,*'
CATH. 172; a toun hamelet MAN. (H.) 269;
hamlete 310; tounes and **hamelesse** (*for
hamelez French pl.*) 321.

hámen, v., *? lame, cripple*; **hámid** (*pple.*)
TOWNL. 98.

hamer, sb., *O.E.* hamor, homor, = *O.L.G.*
hamur, *O.H.G.* hamar, *O.N.* hamarr; *hammer,*
HAV. 1877; TREV. III. 205; hamer [hamur]
CH. C. T. *A* 2508; hamur PR. P. 225; **hameres**
(*pl.*) BRD. 22; homeres A. R. 284.

　hamer-bētere, sb., *worker with a hammer,*
blacksmith, WICL. GEN. iv. 22*; JOB xli. 15.

　hamer-smiþ, sb., = *Ger.* hammerschmied;
blacksmith, WICL. GEN. iv. 22.

hameren, v., = *M.H.G.* hameren, hemeren;
hammer; homered (*pret.*) GAW. 2311.

hamlounen, v., *see* **hanelōn;** *double, wina
about*: [þe fox] hamlounez (*pres.*) & her-
kenez bi heggez ful ofte GAW. 1708.

hamme, sb., *O.E.* hamm, homm, = *O.H.G*
hamma, *O.N.* höm; *ham* (*bend of the knee*),
thigh, '*poples,*' PR. P. 225; **hamme** [homme]
(*dat.*) TREV. V. 369; **hommen** [hammes]
(*pl.*) A. R. 122; hammes '*garez*' VOC. 148;
hommes A. P. ii. 1541.

hampren, v., *hamper, clog, hinder*; **hampris**
(*pres.*) WILL. 668; **hamperde** (*pret.*) FLOR.
1175; **hampred,** hampered (*pple.*) WILL.
441, 4694.

han, *see* **habben.**

hanche, sb., *O.Fr.* hanke, hanche; *haunch,*
D. ARTH. 1100; in the haunche riȝt Tristrem
was wounded TRIST. 1088; haunche A. R.
280; honche B. DISC. 268.

hanchen, v., *eat, gnaw*; **hanchid** (*pple.*)
ALEX. (Sk.) 774*.

hand, sb., *O.E.* hand, hond, = *O.Fris.* hand,
hond, *O.L.G.* hand, *O.N.* hönd, Goth. handus,
O.H.G. hant; *hand,* ORM. 14684; IW. 2086;
LIDG. M. P. 49; an hand i ȝou hete DEGR. 1400;
nime an hand AYENB. 143; hond GEN. & EX.
104; BĒK. 299; MAND. 172; þeos hond A. R.
124; hond wið honde fuhten þa heȝe men

LAȝ. 174; þe ufere hond (*superiority, mastery*) habben 1520; þe herre hond habben KATH. 758; he hed(d)e þe heire hond CHR. E. 268; the herre hond have D. TROY 9571; þan deofle þu bist isold on hond FRAG. 7; here tuder ... goð wel on hond HOM. II. 177; he wolde ... luȝere him gon an hond (= *Ger.* an hand gehen) LAȝ. 22268; þeos children weoxen an hond 7165; þe wunde þet ever wursed an hond A. R. 326; forȝif ous þat ȝe ous bereþ an hond (*impute*) BEK. 1002; this false knight ... bereþ hire on hond þat sche haþ don þis þing CH. C. T. *B* 620; an honde sullen LAȝ. 31580; heo eoden an honde (*submitted*) þan kinge Gurmunde 28964; þu me gest an honde O. & N. 1651; an **hande** (*gen.*) brede CH. C. T. *A* 3811; **hande** [*O.E.* handa, = *O.N.* hendi] (*dat.*) LAȝ. 3713; alle þa londes (*r.* londe) stondeða mire honde 9475; wind heom stod an honde 22313; honde CH. C. T. *A* 2376; neih honde MISC. 85; mi net liþ her bi honde HORN (L.) 1137; habbe an honde SPEC. 41; þo he hadde ... under honde ROB. 141; þe hundred thousend marc were ipaid **bivóre hond** (*beforehand*) 489; bivoren hond A. R. 212; neih hond 424; slo with his hend HAV. 1412; **hande** [*O.E.* handa, = *O.L.G.* hendi, *O.N.* hendr] (*pl.*) MK. ix. 43; ORM. 10996; honde MISC. 54; HORN (R.) 200; P. L. S. xvii. 170; SPEC. 80; FER. 2119; CH. C. T. *B* 606; hende IW. 3151; PR. C. 3214; TOWNL. 125; honde, hondan, honden HOM. I. 23, 91, 149; honde, honden HOM. II. 21, 181; handen MAT. xv. 2; honden A. R. 114; **handan** (*dat. pl.*) JOHN xix. 3; LEECHD. III. 114; handen HOM. I. 223; honden FRAG. 5; LAȝ. 31006; AYENB. 235; JOS. 272; þet tu ne meiht ... etfleon hore honden A. R. 390; honden, honde HOM. II. 89, 207.

hand-ax, sb., *hand-axe*, HAV. 2553; hand-ax [hond axe, handex] ROB. (W.) 584.

handax-schaft, sb., *shaft of a hand-axe*, BARB. xii. 57.

hand-bal, sb., *ball to play with*; handballe '*pila manualis*' CATH. 173; handball [-balle] ALEX. (Sk.) 1895.

hand-band, sb., *O.N.* handaband; *marriage-tie; ? covenanted portion*; handband [handbonde] C. M. 13426; God gif the to thi handband the dew of heven TOWNL. 43.

hand-barow(e), **-barwe**, sb., *hand-barrow*, PR. P. 225.

hande-brēde, sb., *O.E.* handbrēd; *palm of the hand, span,* '*palmus,*' PR. P. 225; CH. C. T. *A* 3911; handibrede, handibreede WICL. Ez: xl. 5, 43.

hand-clōð, sb., *O.E.* handclāð; *towel,* REL. I. 129.

hond-cops, sb., *O.E.* handcops; *handcuff,* '*manica,*' FRAG. 5.

handi-dandi, sb., *juggling-trick with the hands; bribe,* LANGL. *C* v. 68.

hand-dēde, sb., *exploit,* HAV. 90.

hand-fæsten, v., *O.N.* handfesta; *betroth*; handfest (*pple.*) ORM. 2389; he heo hæfde ihondfæst LAȝ. 2251.

hand-ful, sb., *handful,* ORM. 8648; GEN. & EX. 1919; an honful ȝerden A. R. 254.

hand-gang, sb., *O.E.* handgang; *laying on of hands,* ORM. 13254.

hand-ȝewerc, sb., *O.E.* handgeweorc; *handi-work*; handiwerc HOM. I. 129; MARH. 10; handiwerc, hondiwerc REL. I. 48; hondiwerk SPEC. 60; handewerc (*ms.* handewerrc) ORM. 5054; hondewerc KATH. 1229.

hand-ȝewrit, sb., *O.E.* handgewrit; *hand-writing*; hande writt ORM. 13566.

hand-habbing, pple., *O.E.* handhæbbende; *having in one's hand,* FL. & BL. 668.

hond-hwīle, sb., *O.E.* handhwīl, = *M.H.G.* hantwīle; *moment,* MARH. 15; KATH. 2183*; handwhile ORM. 12166; LANGL. *B* xix. 267; hondhwule A. R. 94; hondwhile D. TROY 406; handqvile, handwhile ALEX. (Sk.) 5524, 3260*.

hond-legginge, sb., *laying on of hands,* TREV. V. 243.

hand-lēs, adj., *handless,* '*mancus, mancatus,*' CATH. 173; on hanles man served hem all S. & C. II. xxix.

handlinges, adv., *O.E.* handlunga; *hand to hand*: þat sammen handlinges wristeld þai C. M. 3933.

hand-maiden, sb., *hand-maid,* PR. P. 225; PS. cxxii. 2.

hond-sæx, sb., *O.E.* handseax; *dagger,* LAȝ. 6474.

hand-sale, sb., *O.N.* handsal, = *M.Du.* hanseel; *hansel, earnest-money, present; first occurrence*; hansale '*strena*' PR. P. 226; hansel GOW. II. 373; DEP. R. 30; hansel, hansell LANGL. *A* iv. 91, *B* v. 326; hanselle GAW. 491; hondeselle (*pl.*) GAW. 66.

hand-sellen, *O.N.* handselja; *handsel, be-troth*: ha wes him sone **ihondsald** (*pple.*) JUL. 7.

handsom, adj., = *M.Du.* handsaem, *mod. Eng.* handsome; *easy to handle, convenient,* TOR. 1301; handsum '*manualis*' PR. P. 225.

hand-staff, sb., *walking-stick;* '*manuten-tum*' a handestaffe CATH. 133.

hand-tame, adj. & sb., = *O.H.G.* hantzam; *tame; mild, meek,* P. S. 341; right hand-tame ['*mites*'] he sal in dome PS. xxiv. 9.

handtamenesse, sb., *mildness, meekness,* PS. lxxxix. 10; handetamenes cxxxi. 1.

hond-werke, sb., *O.E.* handweorc, = *M.H.G.* hantwerc; *handiwork,* ALEX. (Sk.) 4346; WILL. 929.

hand-wimman, sb., *maid,* C. M. 10905; hand-womman 8381.

hand-worme, sb., *O.E.* handwyrm ; *hand-insect*, '*cirus*,' VOC. (W.W.) 643 ; hondwerm VOC. 177.

hand-wriste, sb., *O.E.* handwyrst,= *O.Fris.* handwirst ; *wrist*, VOC. 147.

handele, sb., *O.E.* handel ; *handle*, VOC. 168 ; handil PR. P. 225 ; **hondlen** (*dat. pl.*) JUL. 59.

handlien, v., *O.E.* handlian,= *M.L.G.* handelen, *O.N.* höndla, *O.H.G.* hantalōn ; *handle*; handlin '*manu tracto, palpo*' PR. P. 225 ; handlen ORM. 18913 ; FL. & BL. 450 ; handlen spere HAV. 347 ; hondlien [handli] LA̤. 1338 ; hondlen A. R. 378 ; þe sternes for to handill ALEX. (Sk.) 2480 ; **handleþ** (*pres.*) AYENB. 235 ; heo hondleþ FRAG. 5 ; handilez D. ARTH. 1156 ; **handil** (*imper.*) H. S. 99 ; hondleþ MISC. 54 ; **handlede** (*pret.*) HOM. II. 47 ; BEK. 1983 ; LEG. 54 ; hondlede A. R. 318 ; **handillit** (*pple.*) BARB. xvii. 416 ; handlit x. 648.

handlinge, sb.,= *M.L.G.* handelinge ; *handling* : fole zi̤þe . . . speche . . . handlinge . . . kesinge . . . dede AYENB. 46 ; hondlunge A. R. 60 ; handling sinne H. S. 80, 86, 94, 114.

handlinges, *see under* **hand**.

[**háne**, sb., *O.E.* hana,= *Goth.* hana,= *M.L.G.* hane, *O.H.G.* hano, *O.N.* hani ; *cock*.]

háne-crau, sb., *cock-crow*, '*galli cantus*,' HOM. II. 39.

han-crēd, sb., *O.E.* hancrēd,= *O.L.G.* hanocrād, *O.H.G.* hanachrāt ; *cock crowing*; on hancrede MK. xiii. 35.

hanelōn, sb., *cf.* **hamlounen** ; *doubling* (*of a fox*) ; (*printed* hauelon) MAN. (H.) 308 ; **hanilouns** [hanelons] (*pl.*) LANGL. B x. 149.

hange-man, sb., *hangman*, LANGL. C vii. 368.

hangien, v., *O.E.* hangian, hongian,= *O.L.G.* hangōn, *cf.* *O.H.G.* hangen (*pret.* hangeta), *O.N.* hanga (*pret.* hang̃i) ; *from* **hōn** ; *hang* : heo sculden hongien [hongie] on he̤e treowen LA̤. 510 ; heo izei̤en heore bearn hangen 5725 ; hongi O. & N. 816 ; AYENB. 31 ; hangen a feloun LANGL. B xviii. 377 ; hangen, honge HAV. 335, 2807 ; **honge̤** (*pres.*) TREAT. 137 ; **hangede** (*pret.*) LK. xxiii. 39 ; hangede, hongede LA̤. 13109, 29559 ; þer ure loverd hongede A. R. 106 ; hongede LANGL. A i. 66 ; hangide ['*suspendi*'] WICL. GEN. xxiv. 47 ; hongeden KATH. 331 ; **hanged** (*pple.*) LANGL. B PROL. 176 ; honged WILL. 2086 ; K. T. 979 ; *comp.* **an-, bi-hangien**.

hanginge, sb., *hanging*, PR. P. 226.

haniper, sb., *O.Fr.* hanapier ; *hamper*, '*canistium*,' PR. P. 226.

hanipeles, sb., pl.,= *ampulles*, LANGL. C viii. 165.

hanken, v., *cf.* *Swed.* hanka (RIETZ) ; *? tie up* ; hank(e) C. M. 16044 ; ihaneked (*?* ihancked) (*pple.*) mid golde LA̤. 25872 ; [þe devel] hadde trecherousli hankid þi chosen WICL. S. W. III. 28.

hanles, adj., *see* **handlēs** *under* **hand**.

hannen, v., *? struggle* : let þu þe hundes hannen togaderes LA̤. 31676.

hans, sb., *? reward* : he gaf theo bischop, to gode hans, riche beighes besans and pans ALIS. 1571 ; to gode hans 2935.

hansale, *see* **handsale**.

hansen, v.= **enhansen** ; haunsen JOS. 232.

hant, sb., *O.Fr.* hant ; *usage ; usual place*, *haunt* : tak of þam homage, as custom is & haunt MAN. (H.) 168 ; in wilde wode is his hont ALIS. 6530 ; þer he held his haunt MAN. (H.) 223.

hanten, v., *O.Fr.* hanter ; *haunt, practise* ; haunten PL. CR. 771 ; exercen or haunten CH. BOET. ii. 6 (52) ; **haunte̔** (*pres.*) H. M. 33 ; þe self hantis *practise the same* ALEX. (Sk.) 4667 ; hants Iw. 1470 ; **hauntede** (*pret.*) ROB. 534 ; **haunted** (*pple.*) PR. C. 6344.

hantinge, sb., *haunting, frequenting ; dwelling* ; hauntinge '*frequentacio, exercitacio*' CATH. 179 ; telle in what place is thin haunting R. R. 6084.

haunting-lī, adv., *frequently*, PR. P. 231.

hap, sb., *O.N.* happ *n. ; hap, fortune*, KATH. 187 ; JUL. 60 ; WILL. 414 ; LANGL. B xii. 108 ; PS. xc. 12 ; his hap wes þe bettere LA̤. 4894 ; wurse hap 11999 ; god hap ROB. 412 ; SHOR. 52 ; al was hap upon he̤e GAW. 48 ; uvel hap LEG. 37 ; up happe (*perchance*) WILL. 2722 ; **happe** (*dat.*) LA̤. 18215 ; wiþ happe and sele C. M. 8884 ; **happes** (*pl.*) SPEC. 24 ; LIDG. M. P. 225 ; harde happes TREV. III. 269 ; *comp.* **un-, wiðer-hap** (-hep).

hap-līche, adv., *haply, perhaps*, LANGL. A vi. 104 ; happeli *luckily* BARB. xvii. 438.

happen, adj., *O.N.* heppinn ; *fortunate*, A. P. iii. 13 ; **hapnest** (*superl.*) GAW. 56.

happen[2], v., *cf. Swed.* happa (RIETZ 243) ; *happen* ; happe LANGL. B vi. 47 ; CH. C. T. A 585 ; HOCCL. vi. 70 ; **happeþ** (*pres.*) MAND. 242 ; LIDG. M. P. 27 ; **happede** (*pret.*) TREV. V. 461 ; MAN. (F.) 5518 ; happed GOW. I. 199.

happen[3], v., *wrap up* : happin or whappine in clothis PR. P. 226 ; happe GAW. 1224 ; FLOR. 112 ; happe (*imper.*) A. P. ii. 626 ; happed (*pret.*) MAN. (F.) 9017 ; PERC. 2244 ; happid (*pple.*) TOWNL. 98.

happinge, sb., *wrapping*, HAMP. PS. 510 ; happing YORK xxix. 82.

happenen, v., *?* = **happen**; *happen* ; **happeneþ** (*pres.*) MAND. 158 ; happene (*subj.*) D. ARTH. 1269.

happening, sb., *chance, luck*, YORK xxix. 39.

happi, adj., *happy*, PR. P. 226 ; PR. C. 1334 ; D. ARTH. 1741.

happi-līche, adv., *happily, luckily*, WILL. 2495.

hār, adj., *O.E.* hār, *O.N.* hārr ; *hoar, hoary* ;

hor BRD. 12; JOS. 648; FER. 1580; ich am
old and hor [hoor] LANGL. *A* vii. 76; hoor
CH. C. T. *A* 3878; hor ilocket LA3. 25845*;
to a forest highe and hore M. ARTH. 314;
hāre *(pl.)* TRIST. 378; PERC. 230; hore GAW.
743; ORF. 212; hore vrostes AYENB. 108;
hi ben hore al so a wolf ALIS. 5031; þe holtis
& þe haire heere ALEX. (Sk.) 776; hāreest
(superl.) ALEX. (Sk.) 1062.

hoor-frost, sb., *hoar-frost,* WICL. EX. xvi.
14; horforst TREAT. 137.

hōre-hūne, sb., *horehound, 'maribium,
maruil,'* VOC. 139; horhoune REL. II. 9.

hārnesse, sb., *O.E.* hārness, hōrnesse;
hoariness, ['canities'] WICL. PROV. xx. 29.

hār, *see* **hēr.**

harageous, adj., *violent;* harageous(e)
knighttez D. ARTH. 1645, 1742.

haras, sb., *O.Fr.* haras, haraz; *stud of horses,
'equitium,'* RR. P. 227; PALL. iv. 820.

haras[2], sb., *? harass, annoyance,* DEP. R. iii.
27.

hard, harden, *see* **heard, hearden.**

hardi, adj., *O.Fr.* hardi; *brave,* KATH. 1745;
MISC. 40; BEK. 70; AYENB. 83; LANGL. *B*
xix. 285; A. P. ii. 143; hardi & bold GEN. &
EX. 2121; herdi A. R. 56; hardiere *(compar.)*
LA3. 4348*; BEK. 964; hardieste *(superl.)*
LA3. 4181*.

   hardi-līche, adv., *bravely,* KATH. 676;
hardili A. P. i. 3.

   hardinesse, sb.; *hardiness,* LANGL. *B* xix.
31; WICL. 2 PARAL. xvii. 6.

   hardi-schipe, sb., *bravery,* HOM. I. 271.

hardien, v., *O.Fr.* hardier; *render hardy,*
WILL. 1156; hardie ALIS. 1264.

hardiesse, sb., *Fr.* hardiesse; *boldness,
hardiness,* LEG. 49; hardiesse strengþe an
stedevestnesse AYENB. 83.

hardiment, sb., *O.Fr.* hardement; *boldness,*
CH. TRO. iv. 505; hardement R. R. 3392;
hardimentis *(pl.) deeds of valour* BARB.
xiii. 179*.

hardissi, v., *O.Fr.* hardiss-, hardir; *encourage,*
ROB. (W.) 4465; **hardissede** [hardesched]
(pret.) ROB. (W.) 8976.

hāre, adj., *O.E.* hara, = *O.N.* héri, *O.H.G.*
haso; *hare,* O. & N. 373; HAV. 1994; ROB.
457; AYENB. 75; CH. C. T. *A* 684; MAN.
(F.) 5184; L. C. C. 21.

   hāre-belle, sb., *harebell, 'bursa pastoris,'*
VOC. 226.

   hāre-fōt, sb., *harefoot (a plant), 'avencia,'*
REL. I. 36.

hāren, v., *O.E.* hārian; *from* **hār;** *become
hoary;* hore REL. I. 121; ALIS. 1597; hōre
(pres.) SPEC. 50; horeþ TREV. I. 81.

   hōrunge, sb., *O.E.* hārung; *hoariness; age,*
FRAG. 2.

harewe, *see* **harwe.**

hariin, v., *O.Fr.* harier; *drag, 'traho,'*
PR. P. 227; **haried** *(pple.)* CH. C. T. *A* 2726.

harlen, v., *O.Fr.* harler, hareler; *drag;*
harli P. L. S. xxi. 131; **harlede** *(pret.)*
BRD. 11; hii harlede him out of churche
ROB. 536; harled *(pple.)* GAW. 744.

harlot, sb., *O.Fr.* harlot, *mod.Eng.* harlot;
disorderly person, 'scurra,' REL. I. 7; A. R.
356; LANGL. *B* xvii. 108; CH. C. T. *A* 647;
A. P. ii. 39; LIDG. M. P. 52; TOWNL. 248;
LUD. COV. 217; daunceinge of tumblers and
herlotes *(pl.)* HAMP. PS. xxxix. 6.

harlotrie, sb., *tale-telling, buffoonery, evil-
doing,* P. L. S. xxv. 132; CH. C. T. *A* 3145;
LANGL. *B* v. 413, *C* viii. 76; harlotri ALEX.
(Sk.) 4484, 4555.

harm, *see* **hearm.**

harmesai, interj., *alas!* BARB. *page* 528.

harneis, sb., *O.Fr.* harneis; *harness, armour,
'arma, phalerae,'* PR. P. 228; AYENB. 24;
WILL. 1582; harnais ALEX. (Sk.) 3791;
herneis *['pudenda']* TREV. III. 453.

harneisin, v., *harness, equip; 'armo,'* PR. P.
228; harnesche ALIS. 4708; **harnas** *(imper.)*
ALEX. (Sk.) 998; harneischeþ FER. 2929;
harnest *(pple.)* ALEX. (Sk.) 3785.

harnette, sb., *O.E.* hyrnet; *hornet,* TREV. II.
211.

haro, interj., *O.Fr.* haro, harou *(a cry for
help);* grede harou AYENB. 31; we, out,
haro, help to blaw TOWNL. 14; out & harrou,
what deville is here in SACR. 671; harrou,
i die CH. C. T. *A* 4307; harrou and help!
LANGL. *B* xx. 87.

harood, *see* **heraud. horowe,** *see* **harwe.**

harpe, sb., *O.E.* hearpe, *O.N.* harpa, = *O.H.G.*
harpha, harfa; *harp,* A. P. i. 880; LIDG.
M. P. 30; þa harpe [hearpe] LA3. 4898;
harpen [harpes] *(pl.)* LA3. 14955; **harpen**
(dat. pl.) LA3. 3642.

   harpe-string, sb., *harpstring, 'lira,'* VOC.
202; *'lira, fidicula'* CATH. 176; *'fides'*
VOC. 240; for harpestringis *(pl.)* his ropis
servithe ichoone P. R. L. P. 17; harpstringis
'fidis' PR. P. 228.

harpien, v., *O.E.* hearpian; *play the harp,*
LA3. 20311; harpe P. L. S. ix. 179; **harpide**
(pret.) WICL. 1 KINGS xviii. 10; harpeden
LANGL. *B* xviii. 405.

   harpare, sb., *O.E.* hearpere, = *O.N.* harpari,
O.H.G. harfere; *harper,* ROB. 272; **har-
peres** *(pl.)* HORN (H.) 1521.

   harpinge, sb., *O.E.* hearpung; *playing the
harp,* LA3. 22702.

harre, *see* **heorre.**

harriinge, sb., *snarling,* TREV. II. 159.

harrold, *see* **heraud. harrou,** *see* **haro.**

harsk, adj., *cf.* Dan. harsk, *M.L.G.* harsch;
harsh, PR. P. 228; D. ARTH. 1084.

harwe, harowe, sb., *cf. O.Swed.* harva *f.* ; *harrow,* PR. P. 228; harwe [haru, harou] C. M. 12388; harewe LEG. 46; harowe VOC. 234; **harwes** [harewes] (*pl.*) LANGL. *B* xix. 268.

harwen, harwin, v., *harrow,* PR. P. 228; harwe, harwen LANGL. *B* xix. 263.

harwen, *see* her3ien.

hās, adj., *O.E.* hās, = *O.N.* hāss, *cf. O.H.G.* heis, *M.Du.* heesch, heersch ; *hoarse, harsh,* IW. 3620 ; a hase man HAMP. PS. lxviii. 4* ; hos K. T. 599 ; TOWNL. 109 ; hoos, hors '*raucus*' PR. P. 248 ; hos [hoos, hors] LANGL. *B* xvii. 324 ; whether [his horne] were clere or horse of soun CH. D. BL. 347 ; mid stefne hose O. & N. 504; hoose, horse ['*raucae*'] WICL. PS. lxviii. 4.

hoos-, hoors-hēde, sb., *hoarseness,* '*rau- :itas,*' PR. P. 248.

hoosnesse, hoorsnesse, sb., *O.E.* hāsnyss ; *hoarseness,* '*raucitas,*' PR. P. 248.

hās, *see* **hǣse.**

hasard, sb., *Fr.* hasard ; *hazard,* HAV. 2326 ; hazard AYENB. 171.

hasarderie, sb., *gambling,* ROB. 195 ; CH. C. T. *C* 590.

hasardour, sb., *O.Fr.* hasardour ; *gambler,* CH. C. T. *C* 596 ; C. M. 26852 ; haserder '*aleator*' VOC. 217 ; **hasardoures** (*pl.*) CH. C. T. *C* 613 ; hacerdouris B. B. 56.

hasel, sb., *O.E.* hæsel, *O.N.* hasl, = *O.H.G.* hasal ; *hazel,* GAW. 744 ; hasil VOC. 181 ; hesil PR. P. 238 ; REL. I. 53 ; **hasles** (*pl.*) LA3. 8696.

hasel-bōu, sb., *hazel-bough,* REL. I. 243.

hasel-note, sb., *O.E.* hæselhnutu; *hazel-nut,* WILL. 1811 ; MAND. 158.

hasel-rīs, sb., *hazel bush,* ALIS. 3293.

hasel-woode, sb., *hazel-wood* : CH. TRO. v. 1170 ; **haselwodes** (*pl.*) iii. 841 ; hasilwode v. 502.

hāsen, v., *be hoarse,* '*raucio,*' CATH. 177.

haspe, sb., *O.E.* hæpse, = *M.Du.,* *M.H.G.* haspe, *M.L.G.* haspe, hespe, *O.N.* hespa ; *hasp,* CH. C. T. *A* 3470 ; GAW. 1233 ; and underneþe is an hasp(e) shet wiþ a stapil and a clasp(e) RICH. 4083 ; hespe '*sera*' FRAG. 3 ; VOC. 237 ; '*pessulum, mataxa*' PR. P. 238 ; **haspes** (*pl.*) MAP 338 ; of grete haspis wer(e) þe reinis OCTOV. (W.) 1442 ; þe haspes of his helme D. TROY 1270.

haspen, v., *O.E.* hæspian ; *fasten with a hasp, clasp* ; **hasppes** (*pres.*) GAW. 1388 ; **hasped** (*pple.*) GAW. 607 ; A. P. ii. 419; a dore . . . haspet ful faste JOS. 205 ; ihaspet LANGL. *A* i. 171.

hasping, sb., *embracing* : hasping in armis D. TROY 367.

hassell, sb., *?retainer* ; **hassellis** (*gen.*) DEP. R. ii. 25.

hassok, sb., *?tuft of grass or rushes,* '*ulphus,*' PR. P. 228.

haste, sb., = *O.Fris., M.L.G.* hast, *M.Du.* haeste, *from O.Fr.* haste ; *haste,* '*festinatio,*' PR. P. 228 ; FER. 403 ; CH. C. T. *A* 3545 ; an haste AYENB. 31 ; on haste HORN (L.) 615 (H. 631) ; in alle haste MAND. 243.

haste-līche, adv., = *O.Fris.* hastelīke; *hastily, quickly,* MISC. 29 ; BEK. 1929; AYENB. 65 ; ALIS. 2825 ; hasteli WILL. 597 ; GOW. II. 262.

hasteler, sb., *one who roasts meat* : hasteler and potagere L. C. C. I.

hastelet, sb., *O.Fr.* hastellet, *mod.Fr.* hâte- lettes (*pl.*) ; *harslet, haslet* ; **hasteletes** (*pl.*) in galentine HALLIW. 436 ; hastletez GAW. 1612 ; hasteletes on fisshe dai L. C. C. 37.

hasten, v., = *M.L.G.* hasten, *M.Du.* haesten, *from O.Fr.* haster ; *make haste, hasten,* GOW. I. 336; CH. C. T. *E* 978 ; **haste** (*imper.*) þe LEG. 58 ; **hastede** (*pret.*) BEK. 272 ; AYENB. 174; hastide WICL. GEN. xxix. 12.

hasteri, sb., *?roasted meat* : potage, hasteri and bakun mete L. C. C. I ; hastere 38.

hasti, adj., = *O.Fris.* haestig, *M.Du.* haestigh ; *hasty,* WILL. 475 ; MAND. 243 ; PR. C. 1548.

hasti-hēd, sb., = *M.Du.* haastigheid ; *hasti- ness* : eche of hem in hastihede shall other slee GOW. II. 245.

hastinesse, sb., *hastiness,* ROB. 475 ; TREV. III. 213.

hastif, adj., *O.Fr.* hastif ; *hasty,* ROB. 119 ; CHR. E. 667 ; AYENB. 183 ; A. P. iii. 520 ; CH. C. T. *A* 545.

hastif-līche, adv., *quickly, rashly* : hastif- liche ant blire P. S. 190 ; hastiliche LANGL. *B* xix. 353 ; hastilich S. S. (Wr.) 317 ; he answerede hem ful hastifli and seid LEG. 54 ; ne so hastifli watz hot A. P. ii. 200.

hastiwes, sb., *?for* *hastivesse* ; *hastiness* : qven, we hald our hert fra wreth and hasti- wes M. H. 159.

haswed, pple., *from O.E.* hasu, = *O.N.* höss ; *marked with brown* : haswed arled or grei GEN. & EX. 1723.

hat, sb., *O.E.* hætt, *O.N.* höttr ; *hat,* BEV. 408 ; RICH. 367 ; CH. C. T. *A* 470, 1388 ; hæt '*capitium*' FRAG. 2 ; **hatte** (*dat.*) LANGL. *B* v. 536.

hāt, adj., sb. & adv., *O.E.* hāt, = *O.L.G.* hēt, *O.N.* heitr, *O.H.G.* heiz ; *hot,* ORM. 1564 ; IW. 2030 ; hat hunger HOM. I. 277 ; þat uvel hate (*hot evil, fever*) LA3. 30550 ; þe hote [hoote] somer CH. C. T. *A* 394 ; hat, hot LA3. 2850, 29281 ; hot A. R. 190 ; O. & N. 1275 ; MISC. 30 ; hot hunger LANGL. *B* xviii. 205, *C* xxi. 213 ; hoot '*calidus, fervidus*' PR. P. 249 ; MAND. 55 ; LIDG. M. P. 193 ; het BARB. iv. 114 ; **hōt** (*sb.*) AYENB. 129 ; hot and cold H. H. 50 ; **hāte** (*adv.*) it schon ALIS. 572 ; loveþ so hote [hoote] CH. C. T. *A* 1737 ; **hāte** (*pl.*) PERC. 1969 ; **hǎttre** (*compar.*) P. L. S. viii. 126 ; A. R. 400 ; hatture MISC. 154 ; hattore ROB. 531.

hāt-heorte, sb., *O.E.* hātheorte, = *O.H.G.*

heizherzı; *anger*: on hatheorte MARH. 19; hatherte KATH. 2147.

hāt², sb., *O.E.* (ge-)hāt,=*M.L.G.* hēt, *O.N.* heit, *O.H.G.* (ge-)heiz; *see* hicht; *command, promise,* ORM. 13822; hot GEN. & EX. 935; MAN. (H.) 69; hēte (*dat.*) AMAD. (Web.) 440; hates, hotes '*vota*' PS. xlix. 14; *comp.* bi-hāt.

hate, *see* hete.   hāte, *see* hǣte.

hatel, adj., *O.E.* hatol, hetol,=*O.L.G.* hatul, *M.Du.* hatel, *O.H.G.* hazal; *full of hate, hostile,* GEN. & EX. 2544; A. P. ii. 227; heatel HOM. I. 253; hetel swerd A. R. 400; hatele seve MARH. 6; (þis) hatele tintreohe KATH. 1971; hatelest (*superl.*) MARH. 22.

hatell, *see* haðel.

hāten, v., *O.E.* hātan,=*O.L.G.* hētan, *O.Fris.* hēta, *Goth.* haitan, *O.N.* heita, *O.H.G.* heizan; *call; command; promise,* HOM. I. 31, 77; ʒet ich wule haten [hote] mare LAʒ. 23384; ich wule . . . haten hine hiʒindliche cumen to mine riche 31607; hotin '*promitto*' PR. P. 249; hote P. L. S. xix. 58; hete MIR. PL. 152; hotcn *be called* O. & N. 256; hote Arcite CH. C. T. (Wr.) 1559 [hiʒte, hiht *A* 1557]; hicht, hecht BARB. v. 209, xii. 384: hōte (*pres.*) BEK. 479; S. S. (Wr.) 511; hote, hete WILL. 572, 1123; hete JOS. 412; PALL. iii. 936; DEGR. 1400; ANT. ARTH. xix; AUD. 28; Sathanas þet tu levest upon & ti feader hatest JUL. 55; hateð LAʒ. 3648; Wælsce men me heom hateð [hoteþ] 2124; he hat [hot] þe faren to Rome 26369; hoteþ ALIS. 1215; hat KATH. 364; A. R. 6; C. L. 1006; AYENB. 78; þenne þe preost hine hot aʒefen þa ehte HOM. I. 31; hit deþ þat mon hit hot O. & N. 779; hot, hoot MISC. 29; hi hoteþ AYENB. 122; hāt (*imp.*) LAʒ. 24001; hǎ'te [*O.E.* hātte, hǣtte,=*Goth.* haitada, *O.H. G.* heizze] (*passive*) am, is called KATH. 22; MISC.38; FL. & BL. 479; ROB. 15; AYENB. 1; S. S. (Wr.) 223; GAW. 381; M. H. 149; ich hatte Godlac LAʒ. 4643; þat nu hatte (*is called*) Normandie LAʒ. 24170; hette A. R. 134; hette AYENB. 41; þe furste douʒter hette Merci C. L. 300; het HAV. 2348; het, higte GEN. & EX. 747, 2589; hiʒte, hiʒt LANGL. *B* xviii. 115; highte CH. C. T. *A* 860; hight TOR. 101; PR. C. 966; þu hattest nihtegale O. & N. 255; what hattestou LANGL. *B* xx. 337; hettest JOS. 155; hēt [*O.E.* hēt, heht,=*O.N.* hēt, *O.L.G.* hēt, hiet, *Goth.* haihait, *O.H.G.* hiaz, hiez] (*pret.*) ORM. 4922; MISC. 27; GEN. & EX. 2365; BEK. 90; P. S. 257; AYENB. 5; het ham hihen KATH. 412; hette 433; heet LANGL. *B* xx. 271; TREV. VII. 357; hat, hæhte, hehte, heihte LAʒ. 424, 2049, 4221, 17496; hiet SAINTS (Ld.) 300; hiet LEG. 45; het, hiʒt WILL. 58₁ 521; Crechanben hecht that montane BARB. x. 27; (he) hecht him full fair rewarding x. 579; to naime he hicht [heght] *was called* x. 153; hiʒte C. L. 176; þu hete KATH. 540; H. H. 224; (heo) hehten LAʒ. 26347; hehten

hine aredan hwa hit were HOM. I. 121; highten GOW. II. 173; hete BEK. 1020; hēte (*subj.*) LAʒ. 28519; hāten (*pple.*) LAʒ. 3156; ORM. 5200; hattin BARB. x. 750; hoten GEN. & EX. 101; HAV. 106; WILL. 405; JOS. 79; as þai wer(e) hoten (*bidden*) FER. 4901; was hoten Marcius S. S. (Wr.) 92; *comp.* bi-, for-, ʒe-hāten; *deriv.* hāt, hǣse.

hōtere, sb., *commander,* AYENB. 109.

hōtestre, sb., *mistress,* AYENB. 53.

hētinge, sb.,=*M.L.G.* hētinge, *O.H.G.* (pi-) heizunga; *promise,* HALLIW. 447; MIR. PL. 135; hetting YORK ix. 22; hētingis (*pl.*) YORK xliii. 187.

hatere, sb., pl., *O.E.* hæteru *pl. of* *hæt, ? cf.* *M.H.G.* hāz; *vestments, garments*; L. N. F. 218; he dude of al hire hatere (*printed* batere) ASS. 149; hater GREG. 433; C. M. 20211; hatere (*as sing.*) LANGL. *B* xiv. 1; alle his hateren weoren totoren LAʒ. 30778; hattren SPEC. 110; hattir ALEX. (Sk.) 4118; hateren (*dat.*) A. R. 104; hatere A. P. ii. 33; naked þei goþ wiþouten hater ALIS. 7054.

haterel, sb., *O.Fr.* haterel; *neck,* ['*cervix*'] WICL. IS. xlviii. 4; PR. C. 1492; betwen the haatreel and the schulders WICL. 2 PARAL. xviii. 33; haterels (*pl.*) MAC. i. 64.

hateringe, sb., *? cf. mod.Sc.* hatter; *state of disorder* (MÄTZN.); *? misery*: into þesse wrecheliche hateringe of þisse worelde ['*in defectus huius mundi*'] HOM. II. 33.

hateringe², sb., *from* hatere; *dress*: foliliche spenden in housinge, in hateringe LANGL. *B* xv. 76.

hátien, v., *O.E.* hatıan, *cf. Goth.* hatjan, *O.N.* hata, *O.L.G.* haten, hatōn, *O.H.G.* hazēn, hazōn; *hate,* HOM. II. 5; A. R. 88; LANGL. *B* x. 93; hatien, hatiʒen LAʒ. 14681, 29781; hatien, heatien JUL. 50, 51; hatie MISC. 28; AYENB. 73; haten (*ms.* hatenn) ORM. 5062; to hetiene HOM. I. 15; háteð (*pres.*) A. R. 224; hateþ O. & N. 230; hateð, heteð, hetað HOM. I. 19, 125; heateð 251; (we) hatieð MARH. 17; (heo) hatieð A. R. 310; hatieþ P. L. S. xi. 1; P. S. 157; AYENB. 72; hátie (*subj.*) A. R. 176; hátede (*pret.*) A. R. 130; HAV. 1188; hateden LAʒ. 6878; háted (*pple.*) SPEC. 37; ihated ALIS. 1544.

hátare, sb., *hater,* '*osor,*' PR. P. 229.

hátunge, sb., *O.E.* hatung; *hating, hatred,* A. R. 200; hatinge LAʒ. 8321; HOM. II. 165; MISC. 144.

hatterliche, *see* haterliche *under* heter.

[hað, sb., *O.N.* hað; *scorn, contempt; deriv.* hǣþe-lich, hǣþen, hǣþing.]

hað-ful, adj., *scornful,* HOM. I. 279.

haðel,=æðel, adj. & sb., *noble, knight, man,* GAW. 221, 2056, 2065; hatill, hatell YORK xviii. 223, xxxiii. 293; hatheles (*pl.*) GAW. 895.

hauberc, sb., *O.Fr.* hauberc; *hauberk,* ROB. (W.) 2200, 6017; AYENB. 171; haubert,

haubrec 3609* ; haubrik 4577* ; haubergh(e)
GAW. 203, 268.

haubergeoun, sb., *Fr.* haubergeon ; *habergeon* ; habergeon CH. C. T. *B* 2051 ; haubergeon, habergeon, habergoun, haburgoun,
haberjon, haburjon(e) TREV. I. 411, III. 243,
IV. 199, 437, VII. 243, 327, VIII. 215 ; hauberjoun WICL. I KINGS xvii. 5 ; **haber-
jounis,** haubrischounis (*pl.*) BARB. xi. 131*.

hauberjounen, v., *array with a habergeon* :
thei sawen . . . men **hauberjounid** (*pple.*)
WICL. I MAC. iv. 7.

haugh, *see* **halh.   hauk,** *see* **havec.**

hauken, v., *hawk* : for to hauke ne hunte
ALEX.² (Sk.) 299 ; **haukid** (*pret.*) DEP. R.
ii. 176.

  hauking, sb., *hawking* : on hauking_e wold
he ride CH. TRO. iii. 1730 ; went hauking bi
the river ORF. 293 ; ' *aucipium* ' VOC. 221 ;
' *aucupatus* ' CATH. 179.

haunsen, *see* **hansen.**

haunt, haunten, -ing, -inglī, *see* **hant,** *etc.*

hautain, adj., *O.Fr.* autain, hautain ; *proud,
haughty* ; PARTEN. 2829 ; hautein WILL. 472 ;
the herti houndes hautein of cries WILL.
2187 ; an hautein speche CH. C. T. *C* 330 ;
hauteine R. R. 6104 ; þei he hautein were
ROB. 66 ; detraccioun makith an hautein man
be the more humble CH. C. T. *I* 614 ; hauten
ALEX. (Sk.) 4255 ; hautand YORK iii. 27.

  hauteine-lī, adv., *haughtily,* R. R. 5823.

hautesse, sb., *O.Fr.* haltece, hautece ; *pride ;
nobility,* ROB. (W.) 687 ; so hiȝe hautesse
GAW. 2454 ; hautes ALEX. (Sk.) 2835 ; hautesse of ȝeris *great age* DEP. R. iii. 13.

have, sb., = *M.L.G., M.Du.* have, *O.H.G.*
haba ; *?possession, property,* HOM. II. 217 ;
see **haven.**

  have-lēs, adj., = *O.H.G.* habelōs ; *poor,*
HOM. I. 111 ; GOW. II. 214 ; P. R. L. P. 74.

  havelĕste, sb., *poverty,* HOM. I. 115.

havec, havek, sb., *O.E.* heafoc, hafoc, heafuc,
= *M.L.G.* havek, *O.Fris.* hauk, *O.N.* haukr,
O.H.G. habuh, hapuh ; *hawk,* O. & N. 303 ;
havek REL. I. 265 ; hauk JOS. 595 ; OCTOV. (W.)
727 ; CH. C. T. *D* 1938 ; **hevekes** (*gen.*) REL.
I. 227 ; **haveke** (*dat.*) REL. I. 125 ; **havekes**
(*pl.*) LAȝ. 3258 ; SPEC. 105 ; haukes LANGL.
B iv. 125 ; *comp.* **gōs-, spar-hauk.**

[**haven,** sb., *O.E.* hæfen, *O.N.* höfn *f. ;
possession, property.*]

  haven-lēas, adj., *O.E.* hafen-, hæfenlēas ;
poor, '*inops,*' FRAG. 2 ; havenlese men HOM.
II. 157.

  havenlĕste, sb., *O.E.* hæfen-, hafenlēast ;
poverty, '*inopia,*' SAX. CHR. 38.

hàven, *see* **habben** *and* **hebben.**

havene, sb., *O.E.* hæfene (*acc.* hæfenan, SAX.
CHR. 162, 226), *cf. M.L.G., M.Du.* havene,
M.H.G. habene, *O.N.* höfn *f. ; haven,* '*portus,*'

PR. P. 230 ; ROB. 64 ; HOCCL. iv. 9 : þa
havene LAȝ. 8566 ; **havene** (*dat.*) BEK. 1160 ;
AYENB. 182 ; WILL. 569 ; to are hævene LAȝ.
9362 ; hafene 21812 ; to þe havene of heale
JUL. 33 ; **havenes** (*pl.*) CH. C. T. *A* 407.

  havin-toun, sb., *haven-town,* D. TROY
1789 ; havintoune CATH. 178.

  havenen, v., *take harbour* ; we **haveneden**
(*pret.*) at Samum WICL. DEEDS xx. 15.

havere, sb., *O.N.* hafri, = *O.L.G.* havoro, ha-
vero, *O.H.G.* habaro, habero ; *oats* ; hafri,
hafir VOC. 233 ; havir CATH. 178.

  haver-brēd, sb., = *O.H.G.* haberbrōt ; *oat-
bread,* '*panis avenacius,*' VOC. 198.

  haver-cake, sb., *oat-cake,* LANGL. *B* vi. 284.

  havir-straa, sb., *oat straw,* HALLIW. 438.

having, *see under* **habben.**

havoir, *see* **avēr.   hawe,** *see* **haȝe.**

hawch, *see* **halch, halh.**

hawe? D. ARTH. 3704.

hāwe, adj., *cf. O.E.* hǣwen ; *dark gray* : hawe as
þe leed TEST. CRES. 257 ; hawe [haa] heu ANT.
ARTH. ii ; wiþ hawe bake CH. C. T. *B* 95.

hawe, *see* **haȝe.   hawel,** *see* **haȝel.**

[**hāwen,** v., *O.E.* hāwian ; *gaze at ; comp.*
bi-hāwen.]

  hāwere, sb., *O.E.* hāwere ; *spectator* ;
hāuweres, hawres, ?haures (*pl.*) LAȝ. 1488,
1492, 30465.

hawer, *see* **haȝer.**

hawing, *see* **having** *under* **habben.**

hé, pron., *O.E.,* = *O.L.G.* he, hi, *O.Fris.* hi,
Goth. is, *O.H.G.* ir, er, *Lat.* is ; *he,* A. R. 6 ;
ORM. 7 ; GEN. & EX. 4 ; HAV. 8 ; he LAȝ. 3 ;
hæ 12677 ; (h)a 1383* ; he, ha MISC. 27 ; SHOR.
90 ; AYENB. 10 ; he, (h)a FER. 45, 127 ;
he [' *aliqvis* '] to delve aboute is diligent
PALL. iv. 32 ; **his** (*gen.*) A. R. 32, 50 ; his
heorte LAȝ. 149 ; hus [his] seolver 3212 ; hes
2956 ; his godes MARH. 4 ; ase ure loverd
lerede alle his A. R. 130 ; aȝen þe king and
his BEK. 1576 ; his, hes JUL. 21 ; **him** (*dat.*)
him O. & N. 194 ; GEN. & EX. 3674 ; seide
him þise wordes JOS. 21 ; sche swor him
(*to him*) CH. C. T. *D* 961 ; liþe him beo
drihten LAȝ. 48 ; mid him he hine lædde
584 ; heom 15079 ; him (*for* hine) SAX. CHR.
250 ; P. L. S. viii. 44 ; HOM. I. 165 ; KATH.
2270 ; ORM. 208 ; HAV. 30 ; ROB. 10 ; **hine**
(*acc.*) FRAG. 5 ; HOM. I. 17 ; HOM. II.
143 ; A. R. 58 ; O. & N. 471 ; HORN (L.)
1028 ; CHR. E. 612 ; SHOR. 4 ; AYENB. 7 ;
hine, hin GEN. & EX. 3004, 3468 ; hine, hin,
hene, hune LAȝ. 323, 728, 4226, 6694 ; hin
HOM. II. 109 ; MISC. 27 ; hine (*for* him) SAX.
CHR. 255.

hēo, pron., *O.E.* hēo, hīo, hie, hȳ, hī, hig, = *O.
Fris.* hio hiu (*accus.*), hia (*nom. acc.*) ; *she, her,*
A. R. 8 ; ROB. 237 ; C. L. 301 ; MIRC 86 ;
LANGL. *B* iii. 29 ; AUD. 59 ; heo, hi FRAG. 6, 7 ;

heo, hu, he, ha [ʒeo] LAʒ. 144, 1147, 1230, 2028, 3186; heo (ho), hi HOM. I. 83, 93, 105; heo, ha KATH. 80, 117; MARH. 2, 6; heo, ho, he, hi O. & N. 19, 32, 199, 401, 939; heo, he JOS. 83, 87; heo LANGL. *A* i. 10; hue LANGL. *C* ii. 10, 12; ʒo *C* ii. 44; hio, hie (*ms.* hyo, hye) hi MAT. i. 19; MK. v. 27; þa bewente hie hi JOHN xx. 14; hie LAI LE FR. 114; hie clensede heo selven HOM. II. 145; hie, hi MISC. 29; hue HORN (R.) 77; ALIS.[2] (Sk.) 34; he P. S. 154; SPEC. 26; hi HOM. I. 227; ANGL. I. 14; SHOR. 136; AYENB. 16, 106; ho PL. CR. 411; MIRC 1353; A. P. i. 177; ANT. ARTH. i; ʒho ORM. 115; ʒho, ʒo P. L. S. iii. 79, 84; ghe, ge GEN. & EX. 237, 1024; ʒhe WILL. 119; **hire** [*O.E.* hire, hyre, hiere, = *O.Fris.* hire] (*gen. dat.*) *her*, MARH. 2; ORM. 235; BEK. 61; hire þuhte god in hire heorte to habbe monie under hire KATH. 85; bi hire side HAV. 127; hire, hir LANGL. *A* iii. 8; CH. C. T. *A* 925; hir PERC. 190; hire, hure, here, heore LAʒ. 192, 3211, 3281, 3999, 29286; al is hire A. R. 46; h(e)ore weaden 314; hire, here WILL. 153, 1716; hire (*for* heo) SAX. CHR. 264; A.R. 48; KATH. 126; O. & N. 1297; GEN. & EX. 971; HAV. 316; BEK. 68; ALIS. 525; i moot been hires (*for* hire) CH. C. T. *B* 227.

hit, pron., *O.E.* hit, = *O.Fris.* hit, *O.L.G.* it, *O.H.G.* iz, *? Lat.* id (*nom. acc.*); *it*, LAʒ. 11; A. R. 4; BEK. 25; SPEC. 34; SHOR. 30; H. H. 160; WILL. 198; ALIS. 166; S. S. (Web.) 277; MIRC 20; EGL. 49; hit, it P. L. S. viii. 7, 9; O. & N. 41, 118; ROB. 1; LANGL. *A* PROL. 32; it SAX. CHR. 262; ORM. 763; HAV. 77; MAND. 13; CH. C. T. *A* 39; PR. C. 40; it, et GEN. & EX. 126, 590; hit ar ladies innoʒe GAW. 1251; **his** (*gen.*) TREAT. 139; it can cnawen swiþe wel his moder ORM. 1315; ʒef þat water his kende lest SHOR. 9; **him** (*dat.*) HOM. II. 119; O. & N. 682.

hēo, pron., *O.E.* hēo, hīo, hīe, hī, hȳ, hig, = *O. Fris.* hia (*nom. acc. pl.*); *they, them*, A. R. 142; C. L. 52; PROCL. 4; heo, hi, he (? hē), ha [hii] LAʒ. 16, 283, 402, 3800, 5365; heo hi dæleþ FRAG. 6; heo, ho, ha HOM. I. 5, 23, 81; heo, ha KATH. 264, 648; heo, ho, hi O. & N. 66, 76, 926, 1257; heo, hii ROB. 11, 166; LANGL. *A* PROL. 43; hio (*ms.* hyo), hie, hi (*ms.* hy) MAT. xxiii. 4, 7; hio, hi HOM. I. 223; hie REL. I. 128 (HOM. II. 161); SHOR. 16, 124; hie, hi MISC. 33; hi ANGL. V. II. 220; P. L. S. viii. 49; AYENB. 9; hue HORN (R.) 38; H. H. 237; hee P. S. 215; hee, hi FER. 303, 1393; he GEN. & EX. 379, 1988; HAV. 162; SPEC. 75; hui SAINTS (Ld.) 300; ARCH. LII. 35; ha MARH. 2; **heore** [*O.E.* heora, hiora, hiera, heara, hyra, = *O.Fris.* hiara, hira] (*gen. pl.*), *of them, their*, P. L. S. viii. 146; heore beire wille HOM. I. 99; feole flowen . . . out of heore cuþþe JOS. 18; heore, here JOHN xvii. 12, 20; heore [hire] four wives LAʒ. 25; heore alre laverd 1805; heoræ 4228;

heora 420; whulc here (*of them*) 2950; heore, hore HOM. I. 157; A. R. 28; heore (hore), hire O. & N. 280, 1566; heore, here ORM. 119, 413; heara ANGL. VII. 220; here SAX. CHR. 251; GEN. & EX. 3673; HAV. 52; MAND. 11; CH. C. T. *A* 588; DEGR. 756; here beire red BEK. 2438; here, hoere L. N. F. 217; huere non HORN (R.) 1260; hura, hure LEECHD. III. 108, 132; hure FER. 305; hire guod AYENB. 6; þe kingdom of hevene is hare 144; hore FRAG. 7; hare HOM. I. 221; P. S. 69; he is hare alre schuppend KATH. 304; **heom** [*O.E.* heom, him, = *O.Fris.* hiam, him] (*dat. pl.*), *to them*, MAT. iii. 7; FRAG. 5; P. L. S. viii. 146; JOS. 367; ALIS. 61; heom, hem, him [ham, ʒam, him] LAʒ. 26, 413, 763, 6191; heom (hom), hem, him O. & N. 62, 539, 1448, 1615; him HOM. I. 89; him, himen FER. 106, 160; hem, him HAV. 1808, 1976; hem GEN. & EX. 305; WILL. 169; MAND. 11; CH. C. T. *A* 18; TOR. 1198; HOCCL. v. 12; (*ms.* hemm) ORM. 145; ech of hem TREAT. 132; ham FRAG. 6; A. R. 30; AYENB. 8; hom GAW. 819; DEGR. 2; ANT. ARTH. x; heom, *for* heo LAʒ. 358; O. & N. 929, 1253; hem HOM. II. 7; ham FRAG. 7; A. R. 8; KATH. 280; **hemen**, himen [*O.E.* heoman] (*dat.*) FER. 1963, 2201, 2559; HALLIW. 444.

hēad, *see* **hēafed**.

hēafden, sb., *O.E.* (be-)hēafdian, = *M.H.G.* houbeten; *behead; grow to a head*; hefedin, hedin '*decollo*' PR. P. 231; hede C. M. 13174; that it mai hede PALL. xi. 156; **hēfdid** (*pret.*) C. M. 171; heveded 13176; **hēded** (*pple.*) TREV. IV. 157; *comp.* **bi-hēafden**.

hēfdare, sb., *executioner* : hedare or hefdare, '*decapitator, lictor*' PR. P. 231; hevedare 238; TREV. IV. 157.

hēafed, sb., *O.E.* hēafod, hēafud, = *O.Fris.* hāved, hāfd, *M.Du.* hoofd, *O.N.* höfuð, *Goth.* haubiþ, *O.H.G.* haubit, houbit; *head*, MK. vi. 24; heaved (*ms.* heaued) KATH. 1360; þet heaved A. R. 130; AYENB. 51; heaved, heved, hefed SHOR. 23, 85, 149; heaved, hæfed, hæved, hæfd [hefved, hefd] LAʒ. 574, 1596, 3856, 15989; hæfed ORM. 1557; heved O. & N. 74; ALIS. 205; SPEC. 62; MAND. 10; PR. C. 771; ANT. ARTH. xlii; hafed HOM. I. 233; heved, hed ROB. 17; heved, heed WICL. MK. vi. 24; hefd M. H. 39; head RICH. 1732; LIDG. M. P. 106; hed [heed] LANGL. *B* xvii. 70; hed TOR. 704; off thine awine heid *as of your own will* BARB. ii. 121; **hēfdes** (*gen.*) MARH. 19; **hēafde** (*dat.*) FRAG. 8; heafde, heafede, heafode, hevede LEECHD. III. 82, 104, 130; **hēafde** (*pl.*) LK. xxi. 28; hæfden [hefdes] LAʒ. 14682; hefden A. R. 188; *comp.* **fore-hēaved**.

hǣfd-bān, sb., *O.E.* hēafodbān; *skull*, LAʒ. 1467.

hēd-borwe, sb., *head of a frankpledge*, PR. P. 231.

hǣfed-burh, sb., *O.E.* hēafodburh, *O.N.* höfuð-borg ; *chief city*, ORM. 8469.

hēved-chirche, sb., *O.E.* hēafodcirice ; *chief church, cathedral*, ROB. 69 ; hevedkirke MAN. (F.) 11060.

hēved-clōð, sb., *O.E.* hēafodclāð ; *head cloth*, A. R. 424.

heed-dēre, sb., *chief deer*, DEP. R. ii. 117.

hēaved-eche, sb., *O.E.* hēafodece ; *headache*, A. R. 370.

hǣfed-folc, sb., *? rulers of the people* : þurh þat Judisken hæfedfolc þat he was boren offe ORM. 14492.

hǣfed-hird, sb., *chief family* : þa hirdes . . . hafden an hæfedhird ORM. 587.

hǣfed-kinedōm, sb., *chief kingdom* : [Romes kinedom] was hæfed kinedom abufen oþre ORM. 9175.

hēved-lārðēow, sb., *chief teacher* : heved lorðeau of alle holie chirechen HOM. II. 7.

hēved-lēs, sb., *O.E.* hēafodlēas ; *headless*, S. S. (Web.) 1321 ; heedles PR. P. 232.

hēved-lich, adj., *principal* : þe zeven **hāvedlīche** (*pl.*) zennes AYENB. 15 ; iche hedli sinne WICL. S. W. III. 162.

hēvedling, sb., *O.E.* hēafodling ; *chieftain*, LAȝ. 9986*.

hēvedlinge, **hēdlinge**, adv. & prep., = *M.H. G.* houbetlingen ; *headlong*, WICL. DEUT. xxii. 8 ; LK. viii. 33 ; hedlinge ALIS. 2261 ; **hēdlinges** D. TROY 10175 ; D. ARTH. 3829.

hēved-lond, sb., *O.E.* hēafudland ; *head-land*, VOC. 154.

hēaved-luve, sb., *chief love* : vour **hēaved** luven me ivint i ðisse worlde A. R. 392.

hǣfed-maht, sb., *chief power* : aȝ is set an hæfed maht onȝæn an hæfedsinne ORM. 4566 ; **hēafodmihtan** (*pl.*) HOM. I. 105.

hēd-maiden, sb., *chief maiden* : [Pantasilia] and hir **hēdmaidons** (*pl.*) D. TROY 10902.

hēd-masse-penni, sb., *payment for masses for the dead* : E. G. 144, 145 ; TOWNL. 104.

hēaved-mon, sb., *O.E.* hēafodmann, = *O.H. G.* haubitman ; *head-man, chief, 'primas,'* FRAG. 2 ; hæfedman ORM. 297 ; hevidman (*ms.* hevisdman) ALEX. (Sk.) 441 ; **hēfdmen** (*pl.*) [*'principes'*] HOM. I. 123 ; hæfdmen LAȝ. 16146.

hǣfed-pening, sb., *poll tax*, ORM. 3292.

hǣfed-pliht, sb., *chief danger* : ideileȝc is hæfedpliht ORM. 4738.

hēaved-ponne, sb., *O. E.* hēafodpanne ; *skull*, FRAG. 7.

hǣfed-prēost, sb., *high priest*, ORM. 299.

hēde-rāp, sb., *head-rope* ; **hēderāpis** (*pl.*) . . . þat helde upe þe mastes D. ARTH. 3668.

hēad-shēte, sb., *kerchief*, SQ. L. DEG. 843 ; hedshete and pillow B. B. 179.

hēde-soime, sb., *rope reaching to the heads of the oxen ; traces*, BARB. x. 180.

hēd-stoupis, adv., *headlong* : hurlit (hurlet) doun hedstoupis to the hard urthe D. TROY 6638, 7249 ; hedstoupis of his horse 7434.

hēaved-sünne, sb., *capital sin*, A. R. 10 ; hevedsunne MISC. 37 ; hæfedsinne ORM. 4567.

hēved-toun, sb., *capital town*, ROB. 23 ; hedtoun MAN. (F.) 14762.

hēaved-þēaw, sb., *chief virtue*, HOM. I. 247 ; hedþeu C. L. 799.

hēde-vale, sb., *? = hede wale* ; *great value*, TOR. (A.) 2135.

heed-warde, adv., *towards the head*, TREV. III. 323.

heed-warke, sb., *headache, 'cephalia,'* PR. P. 232.

hēved-welle, sb., *source, head-spring*, GEN. & EX. 868.

hǣfved-wunde, sb., = *Ger.* hauptwunde ; *head-wound, mortal injury*, LAȝ. 7614.

hēah, adj., *O.E.* hēah, hēh, = *O.Fris.* hāch, *O.N.* hār, *O.L.G.* hōh, *O.H.G.* hōh, *Goth.* hauhs ; *high*, HOM. I. 225 ; heh ORM. 2347 ; inne Griclonde he was heh [heȝ] LAȝ. 559 ; an heh king 2042 ; hæh [hei (*ms.* hey, *printed* heþ)] 1173 ; heih 985 ; hei (*ms.* hey) 128 ; hiȝe *noble* GAW. 120 ; (*loud*) 307 ; (*tall*) 1154 ; (*heights*) 1152 ; hongeð hire on heh (*on high*) MARH. 5 ; he saide an heh (*aloud*) SPEC. 41 ; on heh, heih, hei O. & N. 1405, 1456 ; heȝ BEK. 649 ; an heȝ TREAT. 132 ; SHOR. 82 ; heg GEN. & EX. 2011 ; hegh PR. C. 1872 ; ISUM. 16 ; heih A. R. 130 ; heiȝ JOS. 159 ; on heiȝ WILL. 2020 ; hei (*ms.* hey) PR. P. 232 ; HAV. 1083 ; heigh [high] CH. C. T. *A* 316 ; hiȝ FL. & BL. 151 ; on high S. S. (Wr.) 2830 ; ane heȝe ȝefe HOM. I. 17 ; þe heȝe strete P. L. S. xvii. 367 ; þe heȝe ze AYENB. 182 ; hege HOM. II. 89 ; sing þare þine heiȝe masse SAINTS (Ld.) 302 ; þe heighe wei LANGL. *B* x. 155 ; þæs heiȝe kinges FRAG. 7 ; þis highe qvene GOW. II. 113 ; an hiȝe strete TOR. 1526 ; him þoȝt it was hie time HALLIW. 449 ; **hāhes**, hæȝes (*gen. m. n.*) LAȝ. 15416, 21972 ; hæȝen (*dat. m. n.*) LAȝ. 5849 ; **hēahere** (*dat. f.*) HOM. I. 243 ; mid hæȝere strengðe LAȝ. 2188 ; mid hæhȝere stevene 11994 ; **hēhne** (*acc. m.*) HOM. I. 93 ; hæhne LAȝ. 8082 ; hēȝe (*pl.*) LAȝ. 855 ; eȝe word he spekeð 1503 ; hegedages HOM. II. 113 ; heye men HAV. 958 ; heiȝe and lowe LEG. 33 ; **hēhre**, hæȝere (*gen. pl.*) LAȝ. 22448, 30909 ; **hæȝen**, heȝe (*dat. pl.*) LAȝ. 511, 26278 ; **hēhre** (*ms.* hehhre) (*compar.*) ORM. 6297 ; herre KATH. 758 ; MARH. 6 ; A.R. 178 ; O. & N. 1637 ; LANGL. *A* ii. 21 ; þa herre endes LAȝ. 7835 ; heire CHR. E. 268 ; heȝere TREAT. 134 ; hegher PR. C. 992 ; hiere MAND. 92 ; heer PALL. i. 1058 ; **hēhste**, heihste, hexte, hæhȝeste [heheste] (*superl.*) LAȜ. 1807, 2325, 5733, 24142 ; hexte HAV. 1080 ; P. L. S. xv. 14 ; hexte, heixte A. R. 42, 138´ ; heste JUL. 63 ; heȝeste FL. & BL. 560 ; heȝiste BEK. 291 ; heieste PR. P. 232 ; hecst, hekst O. & N. 687 ; hehest KATH. 4 ; heȝhest ORM. 2146 ; heghest PR. C. 993 ; *comp.* **efen-hēh.**

hēȝe, adv., *high, highly,* O. & N. 989 ; þe cloþ þat heng heȝe BRD. 24 ; he halt hire hed heȝe P. S. 154 ; heȝhe fleȝhen ORM. 5991 ; ich am heie iclumben HOM. I. 211 ; he hit heveð to heie up A. R. 86 ; hæh iboren LAȝ. 4912.

hēh-engel, sb., *O.E.* hēahengel ; *archangel,* HOM. I. 43 ; ORM. 1862.

hēah-feder, sb., *O.E.* hēahfæder ; *patriarch,* HOM I. 227 ; hehfader ORM. 17107.

hēi-fol, adj., *haughty,* ROB. (W.) 4011*.

hiȝ-hēde, sb., *altitude* : bihold of þe tur þe hiȝhede FL. & BL. 327 [(H.) 747].

hēh-līch, adj., *O.E.* hēahlīc ; *powerful, complete* : hiȝlich here GAW. 183 ; hehliche grið LAȝ. 10291 ; hehliche dom AYENB. 264.·

hǣh-līche [**hēhlīche**], adv., *O.E.*· hēalīce ; *highly,* LAȝ. 8088 ; heȝliche P. L. S. xiii. 276 ; heih-, heiliche A. R. 56, 190 ; hehlike ORM. 11875 ; heȝli A. P. ii. 1527 ; (*devoutly*) GAW. 755 ; helich (*loudly*) D. ARTH. 1286.

hēh-messe, sb., *O.E.* hēahmæsse, *cf. M.H.G.* hōchmesse ; *high mass,* SAX. CHR. 254 ; HOM. I. 97.

hēi-mon, sb., *man of rank,* ROB. (W.) 4561.

hēhnesse, [**hēhnisse**], sb., *O.E.* hēahness ; *highness, haughtiness,* LAȝ. 29734 ; hehnisse KATH. 211 ; heihnesse A. R. 412 ; heȝnesse AYENB. 97 ; heiȝnesse C. L. 534·

hēh-rēve, sb., *O.E.* hēahgerēfa ; *high prefect,* JUL. 8.

hēi-schipe, sb., *lofty dignity,* HOM. I. 189 ; H. M. 5 ; heihschipe A. R. 100.

hēh-seotel, sb., *O.E.* hēahsetl ; *throne,* JUL. 50 ; hǣhsetle (*dat.*) LAȝ. 18527 ; hehsetle HOM. I. 113.

hēg-tīde, sb., *O.E.* hēahtīd, = *O.N.* hātīð, *M.H.G.* hōchzīt ; *festival* : at hegtide and at gestning GEN. & EX. 1507.

hēiȝ-wei, sb., *highway,* LEG. 98 ; heiȝwai WILL. 1846.

hēahþe, sb., *O.E.* hēahðo, = *Goth.* hauhiþa, *O.H.G.* hōhida, *O.N.* hǣð ; *height* ; heȝþe AYENB. 24 ; A. P. ii. 317 ; heiȝþe JOS. 192 ; WICL. EPHES. iv. 8 ; heighþe MAND. 40 ; heithe PR. P. 233 ; heghte ANT. ARTH. xli ; heght PR. C. 4760 ; on hiȝte [highte] CH. C. T. *A* 1784 ; highte HOCCL. i. 172 ; into the hicht *openly* BARB. v. 487.

hēal, see **hǣl.**

heald, hāld, sb., *O.E.* (ge-)heald, *O.N.* hald *n.* ; *support, custody, stronghold, castle* ; hald ORM. 5026 ; C. M. 9957 ; hold A. R. 74 ; P. S. 218 ; þei

dide him in hold MAN. (F.) 3913 ; helpe and hold i shal him ȝeve TOR. 2088 ; **hálde** (*dat.*) IW. 170 ; hald(e) TRIST. 991 ; holde WILL. 2836 ; A. P. ii. 1597 ; þat not a peni had(de) in hold(e) R. R. 451 ; þe geant stod in his holde PERC. 2005 ; holde LAȝ. 3861 ; *comp.* **ȝe-hóld.**

healden, hálden, v., *O.E.* healdan, haldan, = *O.L.G., Goth.* haldan, *O.Fris., O.N.* halda, *O.H.G.* haltan, *M.L.G.* holden ; *hold ; keep ; observe ; consider* ; healden P. L. S. viii. 28 ; PROCL. 4 ; halden HOM. I. 105 ; MARH. 5 ; S. S. (Wr.) 1233 ; ðat he nevre ma mid te king ... wolde halden (*side with*) SAX. CHR. 264 ; halden [holde] (*account*) þe for herre LAȝ. 1390 ; halden lond 2612 ; hælden (? *for* halden) grið 1416 ; godes laȝhes halden ORM. 941 ; þat we wolden evre to him holden HOM. II. 61 ; hie ne mai hire muð holden 181 ; holden one riwle A. R. 4 ; ich mei wel holden þene wei toward heovene 78 ; up holden (*uphold*) 140 ; þe devel mai ... holden his halimotes P. S. 154 ; healde, hialde AYENB. 5, 206 ; hialde ANGL. I. 10 ; halde A. P. i. 489 ; ISUM. 304 ; griþ to halde O. & N. 1369 ; he mot mid me holde 1680 ; holde H. H. 124 (126) ; helde HORN (R.) 314 (H. 319) ; his foreward to haldene LAȝ. 9879 ; **háldest,** halst [holdest] (*pres.*) LAȝ. 18733, 26430 ; ȝif þu haldest her on *if thou perseverest herein* JUL. 13 ; halst LEB. JES. 333 ; halde'ð, halt LAȝ. 14333, 29752 ; healdeþ SHOR. 17 ; holdiþ festis P. L. S. i. 23 ; (he) haldes þam wise PR. C. 794 ; halt A. R. 50 ; AYENB. 15 ; GOW. II. 128 ; LANGL. *B* xvii. 105 ; hwet halt (*avails*) þe wreððe HOM. I. 17 ; halt flit wið him HOM. II. 43 ; þeo þat ti stede halt H. M. 37 ; me hi halt ... fule O. & N. 32 ; wat halt it to telle longe ROB. 105 ; what halt hit muche her of to telle FER. 1602 ; what halt it longe to strive TRIST. 918 ; halt on H. M. 47 ; (we) haldeð [holdeþ] LAȝ. 7344 ; (heo) haldað HOM. I. 101 ; holdeð A. R. 130 ; **héldand** (*pple.*) BARB. v. 153 ; **háld** (*imper.*) LAȝ. 3329 ; hold þe to ham HOM. I. 213 ; **hálde** (*subj.*) HOM. I. 109 ; þat weo ... halden [holde] alle ure aðes LAȝ. 2334 ; halde þe godes laȝe HOM. I. 55 ; (ȝe) halden 11 ; **hēold** (*pret.*) A. R. 108 ; JOS. 360 ; ALIS. 323 ; heold, hold O. & N. 144 ; heold, heald, held SAX. CHR. 251, 252, 255 ; Ascanius heold [held] þis ... lond LAȝ. 216 ; wið Grickes he heold moni fiht 407 ; huld [heold] 3914 ; heold, held ORM. 2225, 11330 ; heold [held] on *continued* KATH. 434 ; his wai held doune *continued* BARB. xv. 173 ; hield REL. I. 129 ; (HOM. II. 163) ; LIDG. M. P. 86 ; hueld H. H. 159 ; heuld *maintained* TREV. V. 19 ; (*esteemed*) V. 297 ; þis· Gilbert him huld stille BEK. 95 ; held [heeld] CH. C. T. *A* 182 ; heeld RICH. 1335 ; hild AYENB. 206 ; (þu) heolde A. R. 146 ; helde REL. I. 130 (HOM. II. 165) ; he helde [heelde] TREV. III. 231 ; heolden A. R. 152 ; heolden, hulden LAȝ.

2324, 8082; hielden *were feeding (swine)*
MK. v. 14; wiþ þe false (þei) helden LANGL.
B xix. 368; hēolde (*subj.*) O. & N. 51; C. L.
181; gef he þat hielde sinne HOM. II. 31;
ȝif ha ham wel helden KATH. 890; hálden
(*pple.*) MIR. PL. 153; for forleȝen halden
ORM. 2000; þus es ilk man halden to do
PR. C. 5960; haldin M. H. 49; holden CH.
C. T. *A* 141; þou shalt not be holden to þe
oþ ['*non teneberis juramento*'] WICL. GEN.
xxiv. 8; holde (*obliged*) WILL. 317; GOW.
I. 368; *comp.* æt-, bi-, ed-, for-, ȝe-, of-
healden.

hálder, sb., = *O.Fris.* haldere; *holder*: haldere in fee E. G. 362; holdere up of Troie
CH. TRO. ii. 643; háldaris (*pl.*) BARB. iv. 82.

háldinge, sb., *holding; keeping possession,*
'*detencio, retencio,*' PR. P. 242; halding wit
trecheri C. M. 27844; mid louh holdunge of
hire selven A. R. 176; wrang háldings (*pl.*)
of gudes sere PR. C. 5994.

healden, see **hélden.**

hēale, hēalen, see **hǽle, hǽlen.**

hēan, adj. & sb., *O.E.* hēan, = *O.H.G.* hōn,
Goth. hauns; *base, wretched, poor;* hene, hæne,
? *for* hen, hæn LAȝ. 15401, 30316; þis heane
... tintreohe KATH. 1971; this cursed hein
(*vile person*) CH. C. T. *G* 1319; hēne (*pl.*)
P. S. 150; hēnen (*dat. pl.*) LAȝ. 17829.

hēanling, sb., *? a vile man*; hēanlunges
(*? for* heanlinges) (*pl.*) MARH. 14.

hēanen, see **hēnen.**

hēap, sb., *O.E.* hēap, = *O.Fris.* hāp, *O.L.G.* hōp,
O.H.G. hauf, houf; *heap; troop, number;*
þane greate heap AYENB. 130; þe hiap 159;
hæp [heap] LAȝ. 816; hæp ORM. 4330; heep
'*cumulus, acervus*' PR. P. 235; MAND. 62;
hēape (*dat.*) A. R. 314; LIDG. M. P. 106; þa
weoren ... feouwer hundred þusende cnihtes
a þan hæpe [heape] LAȝ. 25396; hepe [heepe]
CH. C. T. *A* 575; gadereþ ... to hepe [an
hepe] P. S. 325 (P. 9); been ... han brouȝt it
to hepe PL. CR. 727; hēapes (*pl.*) HOM. I.
221; KATH. 1996; hapes HOM. I. 219; hēpen
(*dat. pl.*) LAȝ. 29721; *comp.* dung-heep.

hēapin, v., *O.E.* hēapian; *heap*; hepin
'*cumulo*' PR. P. 235; hēapeð (*pres.*) A. R.
314; hepen CH. BOET. v. 2 (153); hǽped
(*pple.*) ORM. 4331.

hēar, see **hǽr.**

heard, adj., *O.E.* heard, = *O.Fris.* herd, *O.L.G.*
hard, *Goth.* hardus, *O.N.* harðr, *O.H.G.* hart,
hert; *hard; powerful, brave; cruel:* þe ... pine
is se heard HOM. I. 253; þe king hefde enne
þein swiþe heard [hard] LAȝ. 1584; hærd
8164; herd 18958; heard [herd] lif A. R. 126,
222; hard MAT. xxv. 24; O. & N. 1694; GEN.
& EX. 3061; SPEC. 105; AYENB. 189; PR. C.
806; hard & starc ORM. 1596; hard & sharp
9663; herd PARTEN. 2586; herde here A. R.
186; þin harde fader CH. C. T. *B* 857; þis

harde (*sb., hardship*) þou hast al for mi gelt
WILL. 2339; **heardes** (*gen. neut.*) HOM. I.
255; **hardne** (*acc. m.*) P. L. S. viii. 86; pinen
harde ant stronge A. D. 237; mid hearde worde
FRAG. 8; harde (*pl.*) O. & N. 530; hardes (*sb.*)
hard parts ['*duris*'] PALL. viii. 135; harder
(*compar.*) AYENB. 174; hærdere LAȝ. 4349;
herdure A. R. 430; hardest (*superl.*) LAȝ. 4181.

hærde, adv., *hard,* LAȝ. 8814; harde ORM.
14783; he smot Corineus harde inoȝ ROB. 213;
swonken ful harde LANGL. *A* PROL. 21;
chast him herd BARB. xviii. 482.

hard-grem, sb., *hardship,* D. TROY 4897.

hærd-iheorted, adj., *hard-hearted,* LAȝ.
11990; herdiheorted A. R. 400.

hard-laike, sb., *O.N.* harðleiki; *wrong;*
hardship, D. TROY 2212; our vile grem and
hardlaike 2768; hardlaike & harme 3475.

hard-līche [hardelīche], adv., *O.E.* heard-
līce; *hardly,* LAȝ. 7480; hardeliche O. & N.
402; ROB. 126; herdeliche A. R. 290.

hardnesse, sb., *O.E.* heardness, = *O.H.G.*
hartnissa; *hardness,* GEN. & EX. 3067;
AYENB. 29; WILL. 1816; two hardness ['*du-
ramenta*'] (*pl.*) in oon vine PALL. iii. 240;
herdnesse HOM. I. 47.

hard-nolled, adj., *stiffnecked:* if oon hadde
be hardnollid, wondur if he hadde be giltles
WICL. ECCLUS. xvi. 11*.

heard-schipe, sb., *hardship,* A. R. 6, 364;
herdschipes (*pl.*) 354, 384.

hearden, v., *O.E.* heardian, *cf. O.H.G.* hertan;
make hard, A. R. 220; hardin PR. P. 227; to
hardi (h)is men ROB. 218; harde (*pres.
subj.*) LAȝ. 5871; hardide (*pret.*) WICL. EXOD.
xiv. 8; harded, iharded (*pple.*) CH. C. T. *F*
245; iharded ROB. 352; ihert AYENB. 29.

harding, sb., *hardening,* CH. C. T. *F* 243.

hardnen, v., *O.N.* harðna; *harden,* ORM.
18219; hardneþ (*pres.*) ORM. 1574.

hearm, sb., *O.E.* hearm, = *O.L.G., O.H.G.*
harm, *O.N.* harmr; *harm, injury,* HOM. I.
253; MARH. 8; hearm, harm JUL. 14, 15;
heærm, hærm [harm] LAȝ. 2171, 9999; harm
O. & N. 1235; TREAT. 136; C. L. 896; AYENB.
8; GOW. I. 270; harum HAV. 1983; herm
A. R. 116; þene herm HOM. I. 31; hærme
(*dat.*) LAȝ. 6384; harme A. P. iii. 17; harmes
(*pl.*) WILL. 453; WICL. JUDITH iii. 2; hermes
A. R. 244; and hermes he worhte LAȝ. 3823;
hærmen (*dat. pl.*) LAȝ. 13748.

harm-dēde, sb., *injurious action,* REL. I.
217.

harm-ful, adj., *harmful,* WICL. PROV. i. 22;
HOCCL. i. 413.

harm-lēs, adj., *harmless,* BEK. 2113; WILL.
1671.

hearmin, v., *O.E.* hearmian, *cf. M.H.G.* her-
men; *harm, injure,* HOM. I. 263; hermien
A. R. 398; hermie FER. 1295; harmi AYENB.

23; harmi, hermi SHOR. 100, 112; harmen MAND. 190; **harmeð** (*pres.*) KATH. 2435; **hermede** (*pret.*) HOM. I. 97; **hærmed** (*pple.*) LAȝ. 3072; ihermed A. R. 124.

harming, sb.: *harming*; in harminge of miself CH. TRO. ii. 1137.

hearpe, see **harpe.**

hēate, hēaten, see **hǣte, hǣten.**

hēaved, see **hēafed.**

hēawen, v., *O.E.* hēawan,=*O.L.G.* hauwan, *M.L.G., M.Du.* houwen, *O.Fris.* hauwa, hawa, houwa, howa, *O.H.G.* hauwan, howan, *O.N.* höggva; *hew*; hæwen ORM. 10083; hewen CH. C. T. *A* 1422; hewe TRIST. 190; hewe to peses ROB. 142; hewe a doun a grete wode TREV. III. 145; to hewene LAȝ. 28030; **hēweþ** (*pres.*) FER. 1613; hewen MAND. 189; huen GAW. 1346; **hēou** (*pret.*) FRAG. 8; heu (*ms.* hew) HAV. 2729; S. S. (Web.) 592; GOW. II. 263; heowen JOS. 511; heowen, heuwen [hewen] LAȝ. 7480, 9796; heuen AV. ARTH. xlvi; hewe P. L. S. xv. 189; FER. 604; **hēwen** (*pple.*) S. S. (Wr.) 1712; GAW. 477; of golde hewen (*forged*) hewe SPEC. 110; *comp.* bi-, for-, ȝe-, to-hēawen.

[**hēawere**, sb., *hewer;* *comp.* flēsch-hēwere.]

hēwinge, sb., *O.E.* hēawung; *hewing,* 'seccio,' PR. P. 239; 'dolatura' CATH. 185.

hēawien, v., *O.E.* hēawian; *hew, cut:* Samuel **hēwide** (*pret.*) him into gobbetis WICL. I KINGS xv. 33; al to peces thai hewed thair sheldes Iw. 641; PR. C. 3712; **hēwid** (*pple.*) WICL. GEN. xxii. 3.

hebben, v., *O.E.* hebban,=*O.L.G.* hebbien, *O.N.* hefja, *O.H.G.* heffan, *M.H.G.* heben, haben, *O.Fris.* heva, *Goth.* hafjan, *M.Du.* haven; *heave, lift*, A. R. 156; hebben, hæben; LAȝ. 13192, 17396; hebbe ROB. 17; FER. 1248; hefen ORM. 11865; heven A. P. i. 16; heve WILL. 348; CH. C. T. *A* 550; heoven MARH. 6; **have** (*pres.*) ANT. ARTH. xv; hevest MARH. 10; heveð A. R. 86; heveð up JUL. 19; **hef** (*imper.*) A. R. 290; heve WICL. GEN. xiii. 14; **hebbe** (*subj.*) SHOR. 18; **hōf** [*O.E.* hōf,=*O.L.G., O.N., Goth.* hōf, *O.H.G.* huob] (*pret.*) ORM. 16705; HAV. 2750; PS. cxx. 1; D. TROY 5259; hof, hæf, heaf [heof] LAȝ. 1914, 6768, 28018; hef A. R. 122; KATH. 184; ALIS. 2297; CH. BOET. i. 1 (5); haf LEG. 6; TREV. IV. 447; hevede HOM. II. 113; P. L. S. xvi. 168; hefde LAȝ. 16509*; þou hove HORN (R.) 1277; (þeȝ) hofen ORM. 16840; hoven ALIS. 5889; hoven LAȝ. 19928; & heoven [hoven] hine to kinge 9025; Scottes huven up muchelne ræm 11280; heven 6789; heven KATH. 1418; SHOR. 68; **hófen** [*O.E.* hafen, =*O.N.* hafinn, *Goth.* hafans, *O.H.G.* haban, *N.H.G.* (ge-)hoben, *O.Fris.* heven] (*pple.*) ORM. 2649; hoven REL. I. 130 (HOM. II. 167); A. P. ii. 206; PR. C. 3126; ihoven LAȝ. 28771; A. R. 282; ihove TREV. VII. 455;

heven PS. xii. 3; iheven HICKES I. 226; heved HOM. II. 111; *comp.* a-, an-, over-hebben.

Hebrēisch, adj., *O.E.* Hebrēisc, Ebrēisc, =*M.H.G.* Hebrēisch; *Hebrew:* þa Hebreisce men HOM. I. 3; an Ebreische ledene A. R. 302.

Hebreu, sb., *Fr.* Hebreu; *Hebrew:* on Ebreu A. R. 136.

hecche, see **hacche.**

hechele, sb.,=*M.L.G.* hekele, *O.H.G.* ha-chele; *heckle* (*for flax*), VOC. 156; (*printed* hethele) MAP 339; hechil REL. II. 176; hekele PR. P. 234; hekile VOC. 269.

hechelen, v.,=*O.L.G.* hekilōn; *heckle* (*flax*); hekelin PR. P. 234.

hecht, see **hāten.** **hēd**, see **hēafed.**

hēde, sb., *O.Fris.* hōde,=*O.H.G.* huota; *heed, care, attention*, P. L. S. ix. 25; AYENB. 32; WILL. 368; S. S. (Web.) 1169; RICH. 3766; PR. C. 592; taken hede SPEC. 35; nimen hede LANGL. *B* xi. 313; hede [heede] CH. C. T. *A* 303; heede LIDG. M. P. 66; hede, hiede GOW. I. 139.

hĕdd-ern, sb., *O.E.* hĕddern; *storehouse,* 'cellarium,' FRAG. 4; LK. xii. 24.

hēde, see **hād, hēafden.**

hēden, v., *O.E.* hēdan,=*O.Fris.* hōda, *O.H.G.* huoten; *heed, take care, guard*, JUL. 8; hede SPEC. 100; P. L. S. v. (b.) 33; þe hiȝe trone þer moȝt ȝe hede A. P. i. 1050; **hĕdd(e)** (*pret.*) MISC. 193; heo leopen to þan bedde & þene king hedden LAȝ. 17801; *comp.* bi-, ȝe-, for-hēden.

hēden, see **hŭden.** **heder**, see **hider.**

hedgin, see **heggin.**

hēdi, ? adj., ? *heedful* [*perhaps should be read* hendi], SPEC. 22.

hedoine, sb., (? *meaning*): herons in hedoine, hiled fulle faire D. ARTH. 184.

heed, see **hēafed.** **heerdes**, see **heorde.**

hēfd, hēfed, see **hēafed.**

hefen, see **hebben.** **hefene**, see **heofene.**

hefi, hefiȝ, see **hevi.** **hefne**, see **heofene.**

heft, see **haft.**

hēȝ, hēȝe, see **hēah, hēhe.**

heȝen, v., *O.E.* hegian,=*M.H.G.* hegen; *from hei; hedge in;* heghin HALLIW. 443.

hēȝen, v., *O.E.* hēgan, hēan,=*O.H.G.* hōhan, *Goth.* hauhjan; *from* hēah; *raise high; exalt, extol:* his monscipe heȝen LAȝ. 31490; (*ms.* heien) 5408; hæien þa toures 5983; HOM. II. 57; heghen PR. C. 4119; heien KATH. 460; heien ant herien MARH. 3; heȝen [hiȝen, hie] WICL. MAT. xxiii. 12; hiȝen JOS. 226; **hēȝ-heþ** (*pres.*) ORM. 2640; heȝeþ and hereþ AYENB. 136; hiȝeþ GOW. III. 295; **hēȝi** (*subj.*) SHOR. 52; **hēȝede** (*pret.*) LAȝ. 7097; hehede hire stefne JUL. 74; heiden MARH. 1; **hēȝed** (*pple.*) HOM. II. 197; *comp.* ȝe-hēȝen.

hēȝinge, sb., *exaltation*, AYENB. 35; heiunge JUL. 3; A. R. 174.

hegge, sb.,=*M.L.G.* hegge, *M.Du.* hegghe, *M.H.G.* hecke; *hedge,* SPEC. 110; ALIS. 783; LIDG. M. P. 116; hegge, hedge PR. P. 232; heg AYENB. 232; **hegge** (*dat.*) O. & N. 17; **hegges** (*pl.*) TREAT. 137; LANGL. *A* iii. 128, 132, *B* iii.

hegge, *see* **hagge.**

heggin, hedgin, v.,=*M.Du.* hegghen; *hedge,* '*sepio,*' PR. P. 232; hegged (*pple.*) LIDG. M. P. 181.

hegh, *see* **hiȝ.** **hēgte,** *see* **hǣþe.**

hēȝþe, *see* **hēahþe.** **hēh,** *see* **hēah.**

hēhe, sb., = *O.H.G.* hōhī, *Goth.* hauhei; *altitude, height,* H. M. 19; heghe D. ARTH. 1146; in þe heiȝe ['*in excelso*'] WICL. I KINGS ix. 12; harez . . . to þe hiȝe runnen A. P. ii. 391.

hei, sb., *O.E.* hege (*pl.* hegas),=*M.L.G.* hege, *M.H.G.* hac (*gen.* hages); *hedge, enclosure*; hai R. R. 54; **heie** (*dat.*) O. & N. 819; þer is a stoute hare in hir hai(e) HALLIW. 439; **haies** (*pl.*) PS. lxxxviii. 41; *comp.* **chirche-hei.**

hei-hōve, sb., *ground ivy*; (*ms.* heyhowe) '*eire terestre*' VOC. 162; B. B. 184.

hai-net, sb., '*cassis,*' PR. P. 221.

hai-rif, sb., *O.E.* hegerife; *goosegrass, clivers,* '*rubia minor,*' PR. P. 221; heireve LEECHD. III. 391; B. B. 184.

hei-sugge, sb., *O.E.* hegesugge; *? hedge-sparrow,* O. & N. 505; hei-, haisugge CH. P. F. 612.

hei-ward, sb., *farm bailiff,* A. R. 418; heiward '*agellarius*' PR. P. 234; haiward SPEC. 110; ALIS. 5756; LANGL. *C* vi. 16, *B* xix. 329.

hei², interj.,=*O.H.G.* hei; *hey!* KATH. 579; P. L. S. xix. 137; hai GAW. 1445.

hēi, sb., *O.E.* hēg, hīg, =*O.N.* hey, *M.L.G.* hoig, hoi, *M.Du.* hei, hoi, *O.H.G.* hewe, howe, houwe, *M.H.G.* houwe, höuwe, hou, höu, *Goth.* hawi (*dat.* hauja); *hay,* '*foenum,*' FRAG. 3; þat hei .LEG. 92; (*ms.* hey) PR. P. 232; LAȝ. 24441; GOW. III. 357; heiȝ [hei] TREV. V. 177; WICL. PS. ci. 5; haig MK. vi. 39; hei (*ms.* hey) LK. xii. 28; hei [hai] CH. C. T. *A* 3262; hai Iw. 3078.

hāi-cok(k)e, sb., *haycock,* CATH. 170.

hēi-stak, sb., *haystack,* PR. P. 233; haistak VOC. 233.

hēi, hēien, *see* **hēah, hēȝen.**

heid, *see* **hēafed.** **heifre,** *see* **haifare.**

hēiȝ, hēiȝþe, *see* **hēah, hēahþe.**

heil, *see* **hail.**

heine, *see* **hēan.**

hēinin, v., *cf. Swed.* höjna (RIETZ); *exalt,* '*exalto,*' PR. P. 233.

heir, sb., *O.Fr.* heir; *heir,* BEK. 24; TREV. IV. 305; eir HAV. 410; P. L. S. i. 44; WILL. 709.

heiren, v., *inherit*; **heire** (*pres.*) MAN. (F.) 13483.

heirōn, heirounsew, *see* **hairōn, hairounsew.**

heit, *see* **hait.**

hek, *see* **hacche.** **hekfere,** *see* **haifare.**

hekele, *see* **hechele.**

heke, sb., *horse*; **hekes** (*pl.*) D. ARTH. 2284.

hel, *see* **hül.**

hḗld, sb., *? = M.H.G.* helde; *slope, hillside*: þe heiȝe held. TRIST. 3274; **hélde** (*dat.*) ST. COD. 58; þare niðer(e) helde HOM. II. 230; **héldes** (*pl.*) HOM. II. 258; ȝeond hulles & ȝeond heldes LAȝ. 12867; heldes ['*devexa*'] PALL. viii. 22.

hḗld², sb., *O.E.* held, hyld *m.*; *grace,* HOM. I. 69.

hḗlde, sb., *O.E.* hyldo, = *O.Fris.* helde, hulde, *O.H.G.* huldī, *O.N.* hylli *f.*; *grace, faith, allegiance,* ROB. 285; þat he þe bere al þe helde þat man schal to his loverd ȝelde. FL. & BL. 397; ihere þou me . . . in helde SPEC. 37; •mihtes(t) tu do þe in his (h)ilde MISC. 96.

hḗlde², sb., *O.E.* helde (VOC. 79); *tansy,* REL. I. 36.

[**hḗlde³,** adj., *O.E.* (ēað-)helde, hylde; *comp.* erveð-helde.]

hḗlden, v., *O.E.* heldan, hyldan; *hold,* (*? for* healden), MISC. 136; S. S. (Web.) 1567; TRIST. 2797; **hélde** (*pres.*) H. H. 105 (107).

hḗlden², v., *O.E.* heldan, hyldan, *cf. M.Du.* helden, *O.H.G.* heldan (*incline*), halden, *O.N.* hella (*pour*), halla (*incline*); *from O.H.G.* hald, *O.N.* hallr (*inclined*); *incline, pour*: helden hit ut HOM. II. 213; scipen gunnen helden LAȝ. 7848; heo scullen . . . hæiden in to hælle 20539; me schal helden eoli and win beoðe ine wunden A. R. 428; heldin '*inclino*' PR. P. 234; helde M. ARTH. 261; min huerte ginneþ to helde (*sink*) SPEC. 48; of his hors he made him held(e) HORN 674; shal helde ['*fundet*'] WICL. LEV. iv. 17; helde [hilde] TREV. IV. 397; healden KATH. 685; JUL. 30; **héldeð** (*pres.*) A. R. 282; heldis A. P. ii. 1330; TOWNL. 221; helt AYENB. 177; þei heldeþ LANGL. *A* x. 60; **hélde** (*imper.*) L. C. C. 40; helde þine ere to me PS. xvi. 6; if þou . . . helde it in PALL. i. 1132; **héldande** (*pple.*) GAW. 972; **hélde,** hulde, halde (*pret.*) L. H. R. 58, 59; helde WICL. MK. xiv. 3; þat he a dun hælde (*sec. text* ful) LAȝ. 2477; healde 1551; heolde 29642; he halde (*? for* hælde) [held(e)] þa milc in þat fur 1196; a wei ward he halde 8878; held ALIS. 2521; helt Iw. 368; haldit [heldit] ALEX. (Sk.) 2141; hælden [h(e)olle] LAȝ. 28038; hilde TREV. II. 347; AL. (L.³) 257; *comp.* **bi-, tō-hélden.**

héldinge, sb., *bending aside,* HAMP. PS. xlii. 4*.

helder, adv., *O.N.* heldr, = *Goth.* haldis;

rather, preferably, GAW. 376; S. S. (Wr.) 1835; hildire ALEX. (Sk.) 4657.

héle, sb., *cf. O.E.* hel (*pretext*); *hiding-place* : gon to wode and maken hélen (*pl.*) and crepen there inne ALIS. 4958.

hēle, sb., *O.E.* hēla, hǣla, = *O.Fris.* hēla, *O.N.* hǣll *m.*; *heel,* '*calx,*' VOC. 179; PR. P. 234; A. R. 112; BEK. 2234; RICH. 3834; LIDG. M. P. 108; hēlen (*pl.*) MARH. 10.

hēle-spor, sb., *heel,* PS. lv. 7; wiknes of mi helespor sal umgive me ai xlviii. 6.

hēle-wōu, sb., *end wall* : side walles hit hedde to, ac non helewou þer nas LEG. 90; (*? ms.* hilewoþ) LA3. 255887 [*earlier text* halle wah] ; hēle-wāges (*pl.*) GR. 17 ; helewowes (*printed* helewewes) FRAG. 6 ; helewoghes MADDEN'S LA3. III. 506.

hēle, *see* hǣle.

hélen, v., *O.E.* helan, = *O.L.G., O.H.G.* helan ; *cover, conceal,* AN. LIT. 8 ; hele P. L. S. xix. 237 ; AYENB. 175 ; WILL. 960 ; CH. C. T. *D* 950 ; S. S. (Wr.) 1111 ; E. T. 1034 ; M. ARTH. 143 ; hele MIN. vi. 16 ; nule ich . . . lengre heolen hit te JUL. 9 ; heole HOM. I. 57 ; 3if þu wel hiles(t) te under godes wenges H. M. 47 ; heleð A. R. 314 ; MARH. 15 ; hel (*imper.*) AMAD. (R.) iii ; heleþ ASS 188 ; hal (*pret.*) MAP 341 ; (þou) hele AL. (T.) 476 ; heien P. L. S. viii. 81 ; hóle (*pple.*) ALIS. 4203 ; OCTOV. (W.) 1355 ; iholen A. R. 146 ; ihole AYENB. 26 ; *comp.* for-, 3e-hélen.

hēlen, *see* hǣlen.

heleð, sb., *O.E.* heleð, hæleð, = *O.L.G.* helith, *O.H.G.* helid ; *armed man, warrior* ; hæleð LA3. 11989 ; heleðes (*pl.*) 1779.

heleð², sb., *O.E.* heoleð, hæleð, = *O.H.G.* helid ; *shelter* : þa makeden heo hus & hæledes (*pl.*) sikere LA3. 1938.

helfe, sb., *O.E.* helfe (VOC. 35) ; hielf (*dat.* hielfe ALFR. P. C. 166), *cf. M.L.G.* helve *n., M.Du.* helve, *M.H.G.* helb, halb *m., O.H.G.* halb, halbe, helbe ; *handle,* ORM. 9948 ; helve '*manubrium*' PR. P. 235 ; S. S. (Web.) 384 ; ELL. ROM. II. 84 ; hilve FER. 4434 ; hilves (*gen.*) FER. 4655 ; helve (*dat.*) WICL. DEUT. xix. 5 ; halve JOS. 503.

helfln, v., *put a handle on* ; helvin '*manubrio*' PR. P. 235.

helfter, *see* halfter.   hēli, *see* hāli3.

hēlīch, *see* hēahlíc *under* hēah.

hélien, v., *O.E.* helian, *cf. O.L.G.* (bi-)helian, *O.H.G.* hellan ; *conceal, cover, protect* : helien þet gongþurl A. R. 84 ; helien [hilien] hem wiþ leves LANGL. *B* xii. 231 ; helie [heli3e, heli] ROB. (W.) 3170 ; a rof shal hile us boþe HAV. 2082 ; hileð (*pres.*) REL. I. 223 ; héle (*imper.*) A. R. 316 ; hile PS. xvi. 8 ; hélede (*pret.*) HOM. II. 197 ; BRD. 29 ; his sconken he helede mid hosen of stele LA3. 21135 ; & helede hine under capen 30849 ; helid S. S. (Wr.) 1234 ; hilide WICL. GEN. vii. 20 ; héled (*pple.*) GOW.

II. 327 ; helid [hiled] LANGL. *A* v. 599 ; hiled GEN. & EX. 3184 ; A. P. ii. 1397 ; iheled A. R. 70 ; P. L. S. xiii. 333 ; LAUNF. 284 ; ihæled [iheled] LA3. 18405 ; iheoled SAINTS (Ld.) 302 ; *comp.* bi-, un-hélien.

hilere, heiler, sb., *protector,* HAMP. PS. xvii. 3, xxvi. 2.

hélinge, sb., *concealment, covering,* TREV. III. 273 ; MAND. 247 ; hilinge ['*opertorium*'] PS. ci. 27 ; hiling VOC. 170 ; WICL. EX. xxxvi. 19.

hēlinge, *see* hǣlunge *under* hǣlen.

helle, sb., *O.E.* hell, = *O.N.* hel (*gen.* heljar), *O.L.G.* hell, hellia, *O.H.G.* hella, *Goth.* halja *f.*; *hell,* '*infernus, tartarus,*' PR. P. 234 ; A. R. 290 ; ORM. 10223 ; SPEC. 39 ; MIRC 509 ; PR. C. 6437 ; helle (*gen.*) A. R. 324 ; etforen helle 3ete HOM. I. 41 ; helle fur (? hellefur) P. L. S. viii. 76 ; HOM. II. 258 ; helle fir ORM. 1529 ; in helle grunde MISC. 77 ; helle bale GEN. & EX. 2525 ; helle 3ates C. L. 1341 ; H. H. 40 ; helle (*dat.*) HOM. I. 19 ; ferde to helle LA3. 1924 ; hælle 20539.

helle-bearn, sb., (? helle bearn) ; *child of hell,* HOM. I. 281.

helle-dǣð, sb., *O.E. death of hell* : sinnes dra3hen sinful man til helledæþ ORM. 7781.

helle-dogge, sb., *dog of hell,* HOM. I. 273.

helle-3at, sb., *O.E.* hellegeat ; *gate of hell,* HOM. I. 41 ; hellegate TOWNL. 314 ; helle-3ates (*pl.*) C. L. 1341.

helle-grund, sb., *O.E.* hellegrund, (? helle grund), = *O.L.G.* helligrund ; *bottom of hell,* ORM. 10508.

helle-hole, sb., *hell* : hurled into hellebole A. P. ii. 223.

helle-hund, sb., *dog of hell* ; (? helle hund) MARH. 6 ; MISC. 151 ; hellehound MAP 337.

helle-pīne, sb., *hell torment,* ORM. 12060 ; HAV. 405.

helle-pit, sb., *hell* : helle pit is nefre ful ORM. 10215 ; helle pitte PR. C. 6447.

helle-stench, sb., *stench of hell,* HOM. I. 193.

helle-þēod, sb., *people of hell* : all þe gaste floc, all helleþeod ORM. 6546.

helle-wā, sb., *woe of hell,* ORM. 10011.

helle-ware, sb., pl., *O.E.* hellwaru ; *hell dwellers,* HOM. II. 53 ; A. R. 244*.

helle-wīte, sb., *O.E.* hellewīte, = *O.L.G.* helliwīti, *O.N.* helvīti ; *punishment of hell,* FRAG. 6 ; HOM. I. 109.

helle-würm, sb., *worm of hell* : þe laðe hellewürmes (*pl.*) HOM. I. 251.

hellen, adj., *belonging to hell* : out of . . . hellen wombe i calde A. P. iii. 306 ; þe hellene schucke H. M. 41.

hellen², v., *O.N.* hella ; *see* hélden² ; *pour out* ; hel (*pres.*) HAMP. PS. cxli. 2 ; (thai) helles HAMP. PS. PROL. *page* 3 ; summe helle (*subj.*)

water B. B. 5 ; hell (*imp.*) lxviii. 29 ; helles ['*ef-fundite*'] lxi. 8 ; helland (*pple.*) *page* 494 ; helt (*pret.*) xli. 4 ; helt (*pple.*) xxi. 13.

helm, sb., *O.E.* helm,=*O.L.G.*, *O.H.G.* helm, *O.N.* hialmr, *Goth.* hilms ; *?from* hélen ; *helmet*, '*galea, cassis*,' PR. P. 235 ; HAV. 379 ; ROB. 49 ; GOW. II. 233 ; RICH. 2562 ; þane helm AYENB. 265 ; helm [healm] LAȝ. 23965 ; hælm [helm] 21419 ; helme (*dat.*) R. S. iv ; helmes (*pl.*) LAȝ. 4542 ; TRIST. 190 ; helmis PERC. 1225 ; helmen (*dat. pl.*) LAȝ. 7578 ; *comp.* kine-helm.

helme, sb.,*O.E.* helma,*cf. O.H.G.* (joh-)halmo *m.*, *O.N.* hialm *f.*; *helm, rudder*, PR. P. 235 ; VOC. 274 ; TREV. VII. 17 ; A. P. iii. 149.

helmid, pple., *helmeted*, WICL. EZ. xxxviii. 5 ; *see* ȝe-helmed.

help, sb., *O.E.* help *f.*, *m.*, *cf.O.N.* hialp,*O.L.G.* helpa, *O.H.G.* helfa, hilfa *f.* ; *help*, MARH. 6 ; A. R. 290 ; BEK. 162 ; C. L. 1363 ; GREG. 84 ; nenne help MISC. 43 ; help, hælp [healp] LAȝ. 16416, 20730 ; hilp FER. 3208 ; help [helpe] LANGL. *B* xiv. 255 ; helpe SAX. CHR. 265 ; ORM. 966 ; GEN. & EX. 1802 ; MIRC 1679 ; CH. C. T. *E* 1324 ; helpes (*gen.*) HOM. I. 13 ; helpe (*dat.*) MARH. 7 ; þet is ow on helpe HOM. I. 259 ; cumen him to helpe (*to his aid*) LAȝ. 8511.

help-ful, adj., *helpful*, WICL. HEB. viii. 12.

help-lēs, adj., *helpless*, HOM. I. 129 ; BEK. 798 ; helples LANGL. *A* viii. 23.

help-lĭch, adj.,=*M.H.G.* helf-, helfelĭch ; *helpful*, P. L. S. i. 1 ; GOW. iii. 46 ; helpeli WICL. PROV. xii. 10 ; REL. II. 52 ; helplī (*adv.*) D. TROY 3579.

helpe, sb., = *M.L.G.* helpe, *M.H.G.* helſe ; *helper*, D. TROY 10803 ; helpe & hirde ORM. 3193 ; helpes (*pl.*) GEN. & EX. 3409.

helpen, v., *O.E.* helpan,=*O.L.G.* helpan,*O.N.* hialpa, *Goth.* hilpan,*O.H.G.* helfan ; *help*, FRAG. 6 ; HOM. I. 27 ; A. R. 204 ; GOW. II. 78 ; helpen [helpe] CH. C. T. *E* 1453 ; helpin PR. P. 235 ; helpę (*pres.*) O. & N. 484 ; helpeð A. R. 196 ; hit helpeþ noht A. D. 258 ; help (*imper.*) LAȝ. 12768 ; helpe (*subj.*) FL. & BL. 128 ; swa me helpe drihten HOM. I. 33 ; halp (*pret.*) ORM. 1342 ; GEN. & EX. 26 ; WILL. 2206 ; halpe LANGL. *B* vii. 6 ; halp [help] CH. C. T. *A* 1651 ; help [halp] LAȝ. 9263 ; A. R. 88 ; help HOM. I. 81 ; þu me hulpe [holpe] LAȝ. 8931 ; þu . . . hulpe þer to A. R. 320 ; (hi) holpen GEN. & EX. 3382 ; LANGL. *B* ix. 113 ; hulpe (*subj.*) LAȝ. 16181 ; A. R. 220 ; ORM. 12033 ; þat he hulpe him LEG. 169 ; holpe LANGL. *C* xxi. 443 ; holpen (*pple.*) ORM. 6201 ; HAV. 901 ; LANGL. *C* xii. 28 ; LIDG. M. P. 224 ; P. S. 332 ; holpe CH. C. T. *E* 2370 ; hulpen LANGL. *B* xv. 130 ; *comp.* ȝe-helpen.

helpere, sb., = *O.L.G.* helpere, *M.H.G.* helfære ; *helper*, TREV. IV. 411 ; helpare '*ad-jutor*' PR. P. 235.

helping, sb.,=*M.Du.* helpinge, *M.H.G.* helfunge ; *help*, LAȝ. 23748 ; BARB. iii. 148.

hĕlsni, *see* hălsnien.

helter, heltrin, *see* halfter, haltrin,

hēlðe, *see* hǣlðe.

helve, helvin, *see* helfe, helfin.

hem, sb., *O.E.* hemm ; *hem, border* : þane hem LAȝ. 4995 ; hemme '*fimbria, limbus*' PR. P. 235 ; A. P. i. 217 ; hemme (*dat.*) MISC. 98 ; hemmes (*pl.*) WICL. MAT. xxiii. 5 ; D. ARTH. 1359.

hēm, *see* hām.

hēme, sb., *?from* hām ; *?man, head of a family* ; an heme P. S. 156 ; heme and hine O. & N. 1115.

hēme,[2] adj., *?from* hām ; *fitting, suitable,* SPEC. 32.

hēme-lī, adv., *fittingly*, GAW. 1852.

heme-luc, sb., *O.E.* hemlēac, hemlic (VOC. 31), hymlic (LEECHD. III. 50) ; *hemlock*, '*herba benedicta*,' REL. I. 37 ; humlok '*cicuta*' VOC. 226, 265 ; whan brom will appeles bere and humlok honi S. & C. II. viii ; homelok '*cicuta*' VOC. 191.

hemen, *see under* hé.

heming, sb., *O.N.* hemingr (*skin of the shanks*), *cf. O.E.* hemming (*a kind of shoe*) ; *part of the skin of a deer* ; heminges (*pl.*) TRIST. 476.

hemisperie, sb., *Lat.* hemisphaerium, *Gr.* ἡμισφαίριον ; *hemisphere* : so fast ai to oure hemisperie bound CH. TRO. iii. 1388.

hemmin, v., *hem*, '*limbo, fimbrio*,' PR. P. 235 ; hemmid (*pple.*) TOWNL. 311 ; *comp.* bi-hemmen.

hemp, sb., *O.E.* hanep,=*O.N.* hampr, *O.H.G.* hanaf, hanef, *Lat.* cannabis, *Gr.* κάνναβις ; *hemp*, VOC. 217 ; HAV. 782 ; BARB. x. 352.

hemp-seed, sb., *hemp-seed*, VOC. 156.

hempen, adj.,=*O.H.G.* hanafīn ; *hempen*, PR. P. 235 ; hempin BARB. x. 360.

hen, sb., *O.E.* henn, hænn,=*M.L.G.* henne, *O.H.G.* henna,=*hanja ; *cf.* háne ; *hen*, A. R. 66 ; O. & N. 413 ; LEB. JES. 438 ; GOW. II. 264 ; henne '*gallina*' VOC. 177 ; PR. P. 235 ; HICKES I. 229 ; HAV. 1240 ; TREV. V. 163 ; L. C. C. 22 ; henne (*gen.*) A. R. 66 ; henne (*dat.*) MAND. 49 ; henne (*pl.*) P. S. 198 ; hennen REL. II. 272 ; P. S. 151 ; AYENB. 38 ; *comp.* mōr-hen.

hen-ăi, sb., *O.E.* henǣg ; *hen's egg* : RICH. 2839.

hen-bane, sb., *henbane*, PR. P. 235 ; LEECHD. III. 331 ; hennebone REL. I. 37.

hen-belle, sb., *O.E.* henne belle ; *henbane*, HALLIW. 444.

hen-cote, sb., *hencoop*, '*galinarium*,' VOC. 274.

henne-fugel, sb., *O.E.* henfugol ; *hen*, SAX. CHR. 259.

henne-harte, adj., *chicken-hearted*, YORK xxiii. 198.

hēn, *see* hēan.

henche-man, sb., *page of honour*, FL. & LEAF 252.

hende, adj., *O.E.* (ge-)hende,=*M.Du.* hende, *M.H.G.* (be-)hende, *? O.N.*hentr; *from* hand; *near. ready, prompt, gracious,* LA3. 2529; H. M. 7; BRD. 3; SHOR. 21; P. S. 192; GREG. 501; LANGL. *B* xx. 187; S. S. (Wr.) 3002; AM. & AMIL. 1583; A. P. ii. 612; M. ARTH. 332; ISUM. 177; god þat is so hende MISC. 52; fer an hende HAV. 359; fair and hende 1104; þat love min herte mi3te not come hende P. R. L. P. 199; nadde his help hende ben WILL. 2513; his hende wif 184; hinde EGLAM. 1297; AV. ARTH. xli; hendure (*compar.*) A. R. 192; hendore SPEC. 27; hen-der MAN. (F.) 8844; hendest (*superl.*) LA3. 13938; A. R. 398; IW. 74; GAW. 26; *comp.* ān-, 3e-, rūm-, spar-, un-hende.

hende-leic, sb.,?=*O.N.* hentleikr; *gracious-ness,* HOM. I. 269; hendeleik HAV. 2793; hendelaic M. H. 49; hendelaik GAW. 1228; A. P. ii. 860.

hende-lich, adj., *graçious*; mid hendeliche worden LA3. 25942; hendelīche (*adv.*) A. R. 316; LANGL. *B* iii. 29; he servede him so hendeliche BEK. 167; hendeli, hendli GAW. 895, 829.

hendenesse, sb., *gentleness* : of hardinesse of herte & of hendenesse LANGL. *B* xix. 31.

hende-schipe, sb., *graciousness*, MARG. 189.

hende-spēche, sb., *mildness of speech,* LANGL. *C* xxiii. 348.

henden, v., *O.E.* (ge-)hendan,=*O.Fris., O.N.* henda ; *take by the hand, capture* ; hende RICH. 4033; S. S. (Wr.) 2860; hende (*pret.*) LA3. 21365*; *see* henten.

hendi, adj., *O.E.*(list-)hendig,=*Goth.* handugs (σοφός), *O.N.* hentugr; *handy, apt, courteous, gentle,* LA3. 4833; HORN (L.) 1336; SPEC. 28; GREG. 951; curteis and hendi MISC. 155; (h)is hendi bodi P. L. S. vi. 8; his hendi (*v. r.* hende) children A. R. 186.

hendi-līche, adv., *fairly, courteously,* LA3. 1227 [*sec. text* hendeliche].

hénen, *see* heonene.

hēnen, v., *O.E.* hēnan, hȳnan,=*O.Fris.* hēna, *M.Du., O.H.G.* hōnen, *Goth.* haunjan ; *from* hēan; *humiliate, oppress,* JUL. 50; hene HOM. I. 13; hænen LA3. 28870; heane [heanin] KATH. 1020; hēneð (*pres.*) HOM. II. 211; he heanið ant hateð me MARH. 8; hea-neð us & harmeð H. M. 13; hēne (*subj.*) SAINTS (Ld.) 307; hēande (*pret.*) JUL. 5; hænde, hunde LA3. 6874, 14412; hēaned (*pple.*) HOM. I. 279; *comp.* 3e-hēnen.

hēninge, sb., = *M.H.G.* hœnunge; *? con-tempt* : ne kepte heo non hening here SPEC. 36; heo heveden him in hening(e) CHR. E. 1030.

hēnen, *see* hǣnen.

henge, sb., = *M.L.G., M.Du.* henge ; *from*

hangen ; *hinge* ; heeng(e) ['*cardine*'] WICL. PROV. xxvi. 14; henges (*pl.*) FER. 2181.

hengel, sb.,=*M.H.G.* hengel; *hinge*: hen-gel of a dore or windowe '*vertebra, vectis*' PR. P. 235; '*gumser*' a hengille VOC. 261; lokis and henglis (*pl.*) WICL. 2 ESDR. iii. 13.

hengen, v., *O.N.* hengja (*pret.* hengða), = *M.L.G.* hengen, *O.H.G.* hengan, henchan (*pret.* hangta, hancta) ; *hang; suspend; be suspended,* HAV. 43; henge A. P. ii. 1584; henge i wile wið þe HOM. I. 285; hinge EGL. 557; hing C. M. 8905; þou hengest (*pres.*) ... so he3e upon þe rode SPEC. 86; hengde (*pret.*) ORM. 13773; henged (*pple.*) ORM. 1018; henged on a tre HAV. 1429; *comp.* bifor-hengen.

hengest, sb., *O.E.* hengest,=*O.H.G.* hengist, *O.Fris.* hengst; *stallion*: hængest LA3. 3546.

henne, *see* hen.

henne, hennes, hens, *see* heonene.

hent, adv., = *M.L.G.* hent, hente, hento, *M.H.G.* hinz, hinze; *until,* AUD. 15, 74, 76.

henten, v., *O.E.* hentan,?=henden; *seize, catch, take,* HOM. II. 209; SPEC. 32; henten, hente LANGL. *B* xiv. 239; hente ROB. 462; BEK. 2203; CH. C. T. *C* 710; A. P. i. 668; AV. ARTH. xiii. hentin, hintin '*rapio*' PR. P. 240; hentest (*pres.*) WILL. 2787; hentes PR. C. 2722; hente (*pret.*) LA3. 21365; GEN. & EX. 2715; JOS. 382; S. S. (Web.) 758; CH. C. T. *A* 698; hent (*pple.*) LIDG. M. P. 225; *comp.* 3e-henten.

hēnðe, sb., *O.E.* hēnðo, hȳnðo,=*O.H.G.* hōnida; *from* hēan ; *poverty,* HOM. I. 115; hinþ(e) '*damnum*' FRAG. 2.

hēo, *see* hé.

heofene, sb., *O.E.* heofone *f.*, heofon, hefon *m., cf. O.L.G.* heben, *O.N.* hifinn, himinn, *Goth.* himins *m.; heaven,* HOM. I. 59; heovene (*ms.* heouene) SAX. CHR. 259; FRAG. 8; þe heovene HOM. I. 195; A. R. 166; heovene, heofne [heavene] LA3. 27455, 29633; hevene KATH. 1799; GEN. & EX. 40; TREAT. 132; heofne, hefne ORM. 10830, 13842; hevene [heven] CH. C. T. *A* 1090; heofen LK. ix. 16; heven DEGR. 1196; PR. C. 7756; hefen M. H. 98; hefne LUD. COV. 21; heofenen (*gen.*) MAT. vi. 26; þare heofene xxiv. 29; hevene A. R. 374; binimeð him hevene wele HOM. II. 11; heofenes MK. xiii. 27; hevene 3at P. L. S. xix. 292; of hevene blisse heo beoþ iflemed C. L. 113; heofnes ORM. 801; hevones GEN. & EX. 287; heofenan, heofenen, heofena, heofene (*dat.*) MAT. xii. 30, xxiv. 29, 30; heovene A. R. 356; to þere heovene LA3. 29580; under hevene 26127; heovene [hovene], hevene O. & N. 728, 897, 916; hevene AYENB. 6; LANGL. *B* v. 516; toward þare hevene MISC. 55; on hevone GEN. & EX. 270; heofenes (*pl.*) MAT. iii. 16; hefenes HOM. I. 225; heofone (*earlier text* heofona) riche ['*regnum coelorum*'] MAT. iii. 2.

Z

hevene-bēm, sb., *sky* ; from ᚦe erᚦe up til hevenebem GEN. & EX. 1605.

hevene-bowe, sb., *rainbow*, C. L. 743.

heven-cope, sb., *canopy of heaven* : the which under the hevencope begripeth all this erthe round GOW. III. 102.

hevene-cwēne, sb., *queen of heaven* ; hevenqvene SPEC. 93.

hevene-dēw, sb., *dew of heaven,* GEN. & EX. 1547 ; hevendeu and erᚦes smere 1573.

hevene-driht, sb., *lord of heaven* ; iche þing vnder hevene driht C. L. 225.

hevene-engel, sb., *O.E.* heofonengel ; *angel of heaven* : liflade ilich hevene engel H. M. 45.

hevone-hil, sb., *heaven-hill* : twen hevone il and helle dik GEN. & EX. 280.

hevenish, adj. *heavenly* : an hevenissh parfit creature CH. TRO. i. 104 ; hevenisshe melodie v. 1825 ; H. F. 1392.

heoven-king, sb., *O.E.* heofoncyning ; *king of heaven,* HOM. I. 61 ; hevonking R. S. iv. (MISC. 168) ; hevenking REL. I. 224 ; P. L. S. ii. 103 ; heveneking C. L. 244 ; SPEC. 45.

heven-lich, adj., *O.E.* heofonlīc ; *heavenly,* KATH. 1327 ; þe heovenliche hird A. R. 94 ; þan heovenliche deme LAȝ. 32053 ; þa hefenliche qvene 21239.

hevene-līht, hovene-līȝt, sb., *O.E.* heofonleoht ; *heaven light,* O. & N. 732.

heven-lōverd, sb., *heaven's lord,* BEST. 226.

heoven-rīche, sb., *O.E.* heofonrīce ; *kingdom of heaven,* A. R. 150 ; hevenriche REL. I. 209 ; GOW. I. 265 ; S. S. (Wr.) 3450 ; A. P. i. 718 ; heofenrīches (*gen.*) ORM. 2153.

hevene-rōf, sb., *O.E.* heofonhrōf ; *vault of heaven,* GEN. & EX. 101.

hevene-ward, adv., *heavenward,* GEN. & EX. 3025.

heoven-ware, sb. (*collective*), *O.E.* heofonwaru ; *dwellers in heaven,* HOM. I. 143 ; hefneware ORM. 12919.

heolen, *see* **hélen.**

heonene, adv., *O.E.* heonane, heonan, heonone, heonon,= *O.L.G.,O.H.G.* hinana, hinan ; *hence,* HOM. I. 11 ; A. R. 230 ; heonene, heonne, hunne [hinene, hinne, hinnes] LAȝ. 1581, 3365, 19119 ; heonen, henen MAT. ix. 24 ; MK. xiv. 25 ; henen HOM. II. 161 ; heonne REL. I. 175 ; heonne forᚦ MARH. 12 ; hine ALEX. (Sk.) 4456 ; hiene D. ARTH. 2582 ; heonne, honne O. & N. 66, 850, 1673 ; henne P. L. S. viii. 199 ; HAV. 843 ; JOS. 215 ; FER. 2332 ; CH. C. T. *A* 3889 ; whan hi schulle henne wende BEK. 34 ; henne, hennes WILL. 329, 1746 ; hunne, hunnes BRD. 11, 33 ; hennes HICKES I. 224 ; SHOR. 41 ; LANGL. *A* i. 152 ; HOCCL. i. 49 ; hens TOR. 10 ; LIDG. M. P. 220.

hénen-sīᚦ, sb., *O.E.* heonansīᚦ ; *departure, death* : sorehful is ure hidercume, and sorilich urę henensīᚦ HOM. II. 185.

heonone-ward, adv., *O.E.* heononweard ; *hence,* A. R. 98 ; heonneward KATH. 1915.

ʼieorde, sb., *O.E.* heord,= *M.Du.* herde, *Goth.* hairda, *O.N.* hiörd, *O.H.G.* herta ; *herd (of animals), company,* HOM. I. 95 ; ane heorde [hierde] LAȝ. 305 ; heerde PR. P. 236 ; þou forhiled me fra herd of liþerand PS. lxiii. 3 ; he dede him on gate holli wiþ al his herde þat he had asembled WILL. 1119 ; king Pharaon . . . garkede his herd GEN. & EX. 3961 ; **heorden** (*pl.*) A. R. 100.

heorde, herde, sb., *O.E.* heorde, hiorde, hierde, hirde, hyrde,= *M.Du.* herde, *O.L.G.* hirde, *Goth.* hairdeis, *O.N.* hirᚦir (*gen.* hirᚦis), *O.H.G.* hirti (*gen.* hirtes), *mod.Eng.* herd ; *shepherd, pastor,* MAT. ix. 36 ; MK. vi. 34 ; herde WILL. 6 ; S. S. (Web.) 903 ; herde, hirde WICL. JOHN x. 11 ; herde [hierde] CH. C. T. *A* 603 ; hirde MARH. 12 ; ORM. 3193 ; GEN. & EX. 456 ; REL. I. 209 ; PR. C. 4638 ; hurde REL. I. 170 ; **heordes** (*gen.*) MAT. xxvi. 31 ; **heordes,** herdes (*pl.*) HOM. II. 35 ; hirdes M. H. 63 ; hirdis TOWNL. 109 ; heordes, heordan LK. ii. 15, 18 ; **hirde** (*gen. pl.*) ORM. 3372 ; herdene HOM. II. 41 ; **hirden** (*dat. pl.*) JUL. 62 ; *comp.* cū-, gāt-, (*h*)reoᚦer-, nēat-, schēp-heorde, swiᚗ-herde (-hirde, -hürde).

herde-cnape, sb., *herd-boy,* S. S. (Web.) 930.

hirde-floc, sb., *company of shepherds,* ORM. 3372.

heorde-mon, sb., = *M.Du.* herdeman ; *herdsman, shepherd,* A. R. 418 ; herdeman MAND. 255 ; hirde-man ORM. 6852 ; ISUM. 88.

hird(e)nesse, sb., *O.E.* hyrdness ; *flocks and herds,* GEN. & EX. 1664, 1732, 1930, 2771.

hird-sipe, sb., *? guardianship,* MISC. 92.

heorde, sb., *O.E.* heorde (*pl.* heordan) ; *tow* ; herdde ['*stuppam*'] PALL. i. 1122 ; **herdes** (*pl.*) *hards (refuse of flax),* L. H. R. 81 ; R. R. 1233 ; heerdes ROB. (W.) S. 7*, 12* ; herdis '*stuppa*' PR. P. 241 ; WICL. DAN. iii. 46 ; **heorden** (*dat. pl.*) A. R. 418.

heorden, v., *form a flock ; collect together* ; herdeied (*pret.*) LANGL. *C* xiv. 148.

hēored, *see* **hīred.**

heorre, sb., *O.E.* heorr, *m. f.* heorre, hearre (ÆLFR. 317) ; hyrre (*dat.* hyrran) ZEIT. xxi. 40 ; *cf. M.Du.* herre *f.,* *O.N.* hiarri *m.* ; '*cardo,*' FRAG. 4 ; *hinge,* HOM. II. 113 ; REL. I. 292 ; WICL. PROV. xxvi. 14 ; GOW. I. 36 ; herre other heengis WICL. PROV. xxvi. 14 ; 3 KINGS vii. 50 ; '*cardo*' VOC. 261 ; PR. P. 237 ; herris' (*cardinal points*) of the world WICL. PROV. viii. 26* ; harre CATH. 176 ; CH. C. T. *A* 550 ; har VOC. 237 ; **heorren** (*dat. pl.*) LEECHD. III. 84 ; *comp.* **dur-herre.**

heort, sb., *O.E.* heort, heorot = *O.L.G.* hirot, *M.Du.* hert, hirt, *O.N.* hiörtr, *O.H.G.* hirz, hiruz ; *hart,* ALIS. 688 ; þene heort LAȝ. 26762 ; hert '*cervus*' PR. P. 237 ; ROB. 371 ; AYENB. 216 ; WILL. 2569 ; GOW. I. 54 ;

MAND. 238; CH. D. BL. 352; DEGR. 42;
hertis tonge *hart's tongue* (*fern*) PR. P.
238; hertis tounge VOC. 162; hertis lethir
'*nebris*' PR. P. 238; **heortes** (*pl.*) A. R. 398;
hertis PERC. 218; **heorten** (*dat. pl.*) LAȝ. 306.

heort-(c)levre, sb., *O.E.* heortclæfre; *hart-clover,* LEECHD. III. 331; hertclovre IBID.

Hürt-ford, pr. n., *O.E.* Heorotford; *Hert-ford,* PROCL. 9.

heorte, sb., *O.E.* heorte, hearte *f., cf. O.N.*
hiarta, *O.L.G.* herta, *O.Fris.* herte, hirte, *Goth.*
hairto, *O.H.G.* herza, *Lat.* cor (*gen.* cordis),
n.; heart; MARH. 18; ALIS. 20; þeo heorte
A. R. 282; heorte, horte O. & N. 37, 947;
heorte, hurte LAȝ. 149, 8178; heorte, herte
ORM. 1460, 1596; hierte HICKES I. 223
(ANGL. I. 13); HOM. I. 217; huerte HORN (R.)
281; REL. I. 110; SPEC. 58; herte RICH.
2928; GOW. III. 262; MAND. 2; CH. C. T.
A 150; EGL. 153; god help þe dere herte
FER. 324; hurte TREAT. 138; mi swete
hurte P. L. S. xiii. 142; harte FLOR. 2075;
TOR. 689; **heortan** (*gen.*) MAT. xii. 34;
heorte HOM. I. 9; for mine heorte blode
(?heorteblode) LAȝ. 15845; wrappe meinþ þe
heorte blod O. & N. 945; herte SPEC. 81;
FER. 323; his herte wille HAV. 70; his
herte rote CH. L. G. W. 1993; at his hurte
grounde BRD. 2; **heorten,** heorte (*dat.*) MAT.
xxiv. 15, 48; heorten, herten, hurten LAȝ.
663; to þare heorte 12964; mid godere heorte
HOM. I. 3; MK. vii. 19; **heorten** (*acc.*) LAȝ.
3455, 6456; **heorten** (*pl.*) MISC. 37; ALIS.
2464; heortan, heorte HOM. I. 13, 95; heorten
[heortes] LAȝ. 5826; herten AYENB. 46;
heortan, heorten (*dat. pl.*) MK. ii. 6, 8.

heorte-blōd, sb., = *M.H.G.* herzebluot; *heart-blood*: min heorte blod HOM. I. 191; SPEC.
74; herteblod HAV. 1819; herteblod P. L. S.
xxv. 46.

herte-bren, sb., *heartburn,* GEN. & EX. 4054.

hart-ful-lī, adv., *heartily,* BARB. iii. 510.

hĕrte-lēs, adj., *O.E.* heortlēas; *heartless,*
RICH. 4410; WICL. PROV. xii. 8.

heorte-lĭch, adj., *hearty;* herteli GOW. I.
251, II. 267; **herteliche** (*pl.*) D. ARTH. 2551;
heorteliche (*adv.*) *heartily,* A. R. 390; JUL.
74; hertelike HAV. 1347; herte-, hertili CH.
C. T. *A* 762; D. ARTH. 2991, 3642.

hertlinesse, adj., *cordiality,* '*cordialitas,*'
PR. P. 238.

heorte-püt, sb., *heart-pit,* ALIS. 2250.

heorte-rōte, sb., *bottom of the heart*: þe
teares . . . walleþ of þe **heorte-rōtes** (*pl.*)
swo water doð of welle HOM. II. 151.

heorte-sār, sb., = *M.L.G.* hertsēr; *heart-sorrow,* HOM. I. 149; heortesor A. R. 180;
hertesōr HOM. II. 207.

herte-spōn, sb., *? midriff, ? breast-bone,* CH.
C. T. *A* 2606.

herte-tēne, sb., *heart-pain;* mani hertetene
him tideþ ANGL. IV. 193; GREG. 388.

heorte-þeauw, sb., *disposition of the heart*;
heorte-þeauwes (*pl.*), devociun, reoufulnesse
. . . and oðre swiche vertuz A. R. 368.
[**heorte,** ? adj.; ? *comp.* **mild-heorte.**]

heorti, adj., *? hearty;* herti '*cordialis*' PR. P.
238; WICL. DEUT. i. 13.

heortien, *see* **hertin.**

hēou, sb., *OE.* hēow, hīw, = *Goth.* hiwi *n.; hue,
colour, species, form,* A. R. 50; heou, hou O. & N.
619; hew ORM. 19251; hew, hiu HOM. II. 99;
heu FL. & BL. 510; HAV. 2918; TREAT. 140;
heuh, heuȝ C. L. 710; heowe LAȝ. 3071 : hewe
WILL. 891; GREG. 794; heu (*ms.* hew), hieu,
hewe CH. C. T. *A* 394; **hīowes** (*gen.*) LK.
ix. 29; **hēowe** [howe], hewe (*dat.*) O. & N.
152, 577; heo wolden of ane heowen [hewe]
heore claðes habben LAȝ. 24649; heouuwe A.
R. 160; hewe GEN. & EX. 4051; in anes
weres hewe ORM. 2172; hewe CH. C. T. *A*
3255; feir on hewe SPEC. 46; heue ANT.
ARTH. ix; hiwe HOM. I. 219; hiwe P. L. S. ii.
43; **hēowes** (*pl.*) C. L. 705; grene over alle
heowes frovreð mest (þe) eien A. R. 150;
hewes LANGL. *B* xi. 357.

heoven, *see* **hebben. heovene,** *see* **heofene.**

hēow, hēowe *see* **hīw, hīwe.**

hēowe, *see* **hēou.**

hēowier, v., *O.E.* hēowian, hīwian; *colour*;
hēowede (*pret.*) A. R. 392; **hēwed** (*pple.*)
D. ARTH. 3252; *comp.* ȝe-hēowien.

hep, *see* **hap. hēp,** *see* **hēap.**

hepe, sb., *cf. M.L.G.* hepe, *M.Du.* heepe;
scythe: what wiþ hepe and what wiþ croke
GOW. II. 223.

hepe, *see* **hüpe.**

hēpe, sb., *O.E.* hēope, = *O.L.G.* hiopa, *O.H.G.*
hiufo (*tribulus*); *hip* (*fruit of the dog rose*); of
hem ne ȝive i nout an hepe S. S. (Web.) 2532;
þe rede hepe CH. C. T. *B* 1937; **hēpen** (*pl.*)
VOC. 163; ALIS. 4983.

hēpe-trē, sb., *hip-tree,* VOC. 181.

hēr, adj., *O.E.* hēr, = *O.L.G.* hēr, *O.H.G.* hēr;
great, lofty: an hær wude LAȝ. 16372.

hēr-lĭch, adj., *O.E.* hēr-, hǣrlīc, = *O.H.G.*
hērlīch; *excellent*; **hērlīche** (*? for* erliche)
(*adv.*) SAL. & SAT. 240.

hēr, adv., *O.E.* hēr, = *O.N.,* *Goth.* hēr, *O.L.G.*
hēr, hier, hīr, *O.Fris.* hīr, *O.H.G.* hiar, hier;
here, MARH. 6; A. R. 236; GEN. & EX. 184;
HAV. 689; TREAT. 137; H. H. 46; her, heer
CH. C. T. *A* 1610; her, here P. L. S. viii. 50,
51; ORM. 241, 3264; O. & N. 462, 931;
heere LANGL. *A* PROL. 38; her, here, ?hire
LAȝ. 21, 7057, 31765; hier AYENB. 18; hier,
hiere ANGL. I. 12; **hēr after,** *hereafter,*
GEN. & EX. 243; SHOR. 164; her aȝeines
A. R. 94; **hēr on,** *hereon,* LANGL. *B* xiii, 130;
hēr bī, *hereby,* O. & N. 127; BEK. 938; her
biforen HOM. II. 83; **hēr inne,** *herein,* HAV.
458; H. H. 163; **hēr mid,** *here with,* LAȝ.
5355; **hēr of,** *hereof,* REL. I. 217; hier of

Z 2

MISC. 193 ; her offe ORM. 7402 ; HAV. 2585 ; hẽr-till, *hereto*, BARB. xiii. 241* ; hẽr tō, *hereto*, LAȝ. 25321 ; A. R. 388 ; O. & N. 657 ; hẽr þurh, *by this*, ORM. 12710 ; go hẽr ūt (*out of here*), A. R. 290 ; her ute sitteð six men LAȝ. 19704 ; heir and their BARB. vi. 27.

hẽr-anont, adv., *against it* : aȝein missawe oðer misdede, lo, heranont remedie & salve A. R. 124.

hẽr-biforne, adv., *before this, already*, CH. L. G. W. PROL. 73.

hẽr-bīwist, sb., *life here below* : þe erveð-liche herbiwist HOM. II. 125 ; witegede þe childes herbiwist 127.

hẽr, *see* hæ̃r.

herald, heraud, sb., *O.Fr.* heraut ; *herald*, CH. C. T. *A* 2533 ; heraud PL. CR. 179 ; harood TOR. (A.) 1711 ; harroldis (*pl.*) TOR. (A.) 2365 ; herrodis BARB. xii. 371.

heraldie, sb., *heraldry* : the heraldie of hem that usen for to lie GOW. I. 173.

herbe, sb., *Fr.* herbe ; *herb*, ALIS. 5084 ; B. B. 184 ; erbe PR. P. 140.

herber, sb., *O.Fr.* herbier ; *herbary, lawn, flower-bed, orchard, arbour* ['*viridarium*'], TREV. VII. 421 ; ALIS. 331 ; S. S. (Wr.) 329 ; herber [erber] LANGL. *B* xvi. 15 ; erber WILL. 1752 ; FER. 1789 ; TOR. 1968 ; erbere, erber A. P. i. 938 ; eerbir H. V. 6 ; herber CH. L. G. W. 203 ; FL. & LEAF 64 ; herbier P. R. L. P. 56 ; herberes (*pl.*) WILL. 1768 ; *see* arber.

herber(e)we, *see* hereberȝe.

herbergãge, sb., *O.Fr.* herbergage ; *lodging, dwelling* : herbergage of his stedis IPOM. 1349 ; herburgage bi night is perilous 4329 ; herbergage GOW. II. 4 ; so streit of herbergage CH. C. T. *B* 4179 ; Julian that men clepe to for gode herberghage MAND. 97.

herberge, herbergen, *see* hereberȝe, -en.

herbergerie, sb., *O.Fr.* herbergie ; *inn, lodging, chamber* : the hous of herbergrie WICL. GEN. xxiv. 32 ; herbergeri GOW. III. 99 ; where is the herborgerie LK. xxii. 11 ; harburgerie FLOR. 1759 ; herbergeries (*pl.*) WICL. JUDITH xiii. 1.

herbergeour, sb., *harbinger* : in helpe to ben his herbergeour GOW. I. 204 ; herbarjours (*pl.*) D. ARTH. 2448 ; herbreouris BARB. xvi. 465, xviii. 334.

herberȝhlẽs, *see under* hereberȝe.

herberi, herbür̃ȝe, *see* hereberȝe.

herce, sb., *O.Fr.* herce ; *hearse*, PR. P. 236 ; herse M. ARTH. 3532 ; CH. COMPL. PITY 15, 36.

hẽrc-wil, adj., *greedy of hearing* : to hercwile & to spekefule ancren A. R. 100.

hẽrcnien, *see* hẽrknien. herd, *see* heard.

herde, *see* heorde. herdel, *see* hür̃del.

herden, v., *O.E.* (a-)hyrdan, = *O.L.G.* herdian, *O.N.* herða, *O.H.G.* hertan ; *from* heard ;

harden ; ihert (*pple.*) AYENB. 29 ; *comp.* a-herden.

herden, *see* heorden. herdes, *see* heorde.

hére, sb., *O.E.* here, = *O.L.G.* heri, *O.Fris.* here, hiri *m.*, *O.H.G.* heri, hari *n.*, *O.N.* herr, *Goth.* harjis *m.*; *host, army*, H. M. 5 ; ORM. 3370 ; O. & N. 1702 ; GEN. & EX. 1787 ; HAV. 346 ; ALIS. 2101 ; GREG. 614 ; GAW. 59 ; A. P. ii. 902 ; C. M. 13507 ; mid his here LAȝ. 1644 ; hære 7872 ; muchelne here LAȝ. 12174.

hére-būrne, sb., *O.E.* herebyrne ; *armour*, LAȝ. 23966.

hére-dring, sb., *soldier*, LAȝ. 8601.

hére-feng, sb., *spoil, plunder* ; (*ms.* herre-) LAȝ. 11716.

Hére-ford, pr. n., *O.E.* Hereford ; *Hereford*, P. L. S. xiii. 34 ; Herford WILL. 165.

Hére-ford-schīre, pr. n., *Herefordshire*, MISC. 146.

hére-gong, sb., *O.E.* heregang, = *O.Fris.* heregong ; *march of an army, military expedition*, SAL. & SAT. 229 ; GEN. & EX. 848 ; AR. & MER. 4094 ; here-, hergong O. & N. 1191 ; hireȝeong LAȝ. 18194.

hérə-gume, sb., *soldier*, LAȝ. 19164.

hére-kempe, sb., *soldier*, LAȝ. 22573.

hére-mærke [hiremarke], sb., *ensign of an army*, LAȝ. 27469.

hære-scrūd, sb., *war clothing*, LAȝ. 5069.

hére-toȝa, sb., *O.E.* heretoga, = *O.L.G.* heritogo, *O.H.G.* herizogo ; *leader of an army*, HOM. I. 111 ; heretoȝe LAȝ. 10319 ; heretowa FRAG. 2.

hér-iscõle, sb., *army*, HOM. I. 243.

here [2], sb., *O.E.* heoru, = *O.L.G.* heru, *O.N.* hiörr, *Goth.* hairus *m.*; *? sword* : mid here and mid fure, *? with sword and fire*, LAȝ. 8245.

[hére, sb., *O.E.* hére, = *O.H.G.* héri ; (*grandeur, praise, majesty*).]

hére-word, sb., *O.E.* hēreword ; *praise*, HOM. II. 83 ; A. R. 148 ; here-, hæreword, LAȝ. 11917, 24658.

hére-wurðe, sb., *worthy of praise*, JUL. 23.

hére [2], adj., *O.E.* hēore, hȳre, = *O.N.* hȳrr, *O.L.G.* (un-)hiuri, *M.H.G.* gehiure ; *gentle, bland*, LAȝ. 25867 ; AM. & AMIL. 16.

[hẽr-līc, adj., *O.E.* (un-)hierlīc, = *O.N.* hȳrligr, *M.H.G.* gehiurlīch ; *mild* ; *comp.* un-hẽrli.]

[hére [3], adj., *O.E.* (ofer-)hȳre, = *O.Fris.* (ovir-)hére, *M.H.G.* gehœre ; *from* hēren ; *obedient* ; *comp.* ȝe-hére.]

hére, *see* hẽr, hæ̃re *and* hūre.

here-berȝe, herbür̃ȝe, sb., = *O.H.G.* heriberga, *O.N.* herbergi, -byrgi ; *from* hére *and* berȝen ; *harbour, inn, lodging, guest-house*, HOM. I. 37, 69 ; herberge GEN. & EX. 1392 ;

here-, herberȝe, -berwe, beorwe LAȝ. 12054, 22358, 24556, 28878; herberȝhe ORM. 6167; herbergh(e), -berwe, -burhe, -borowe CH. C. T. *A* 765; herberwe LANGL. *B* x. 406; herberewe ALIS. 5748; herberewe, -borwe, -borowe PR. P. 236; herberewe, -borewe, -bore WICL. PHILEM. 22, HEB. xiii. 1; herbŏruwe A. R. 260; herberie M. H. 63; herberi R. P. 28; herbri BARB. xvii. 535; herber IW. 2985; GAW. 812; PR. C. 6153; harbor TOWNL. 247.

herberȝhe-læs, adj., *shelterless, without a lodging*; herberȝhelæs ORM. 6166; herberlesse [*later ver.* herboreles] WICL. MAT. xxv. 38; to herbere þe herberles R. P. 28.

hereberȝen, v., = *O.H.G.* heribergōn, *O.N.* herbergia,-byrgia; *harbour, entertain*; herberȝen HOM. I. 61; herbergen GEN. & EX. 1057; herberȝi AYENB. 199; herberwin PR. P. 236; herberewen WICL. ECCLUS. xxix. 32; herborwen FRAG. 6; herberwe PL. CR. 215; herberwed (*pret.*) LANGL. *B* xvii. 73; herbered GAW. 2481; herberd PR. C. 6154; (he) herberid, (thai) herbreit BARB. i. 599, xix. 390; herberged (*pple.*) GEN. & EX. 1602; herborwed HAV. 742; herbarwed WILL. 1626; herbreit, herberiit BARB. v. 48, ix. 689.

herberȝere, sb., *innkeeper*, AYENB. 39.

herboring, sb., *lodging* : hospitalite, that is herboringe of pore men WICL. ROM. xii. 13; i am glad of mi herbouring IPOM. 1354.

hĕred, *see under* hīrĕd.

heremite, *see* eremite.

hēren, v., *O.E.* hēran, hȳran, hīeran, = *O.Fris.* hēra, *O.L.G.* hōrian, *O.H.G.* hōrran, *O.N.* heyra, *Goth.* hausjan; *hear; obey,* MARH. 15; ORM. 901; heren [heeren] C. L. 521; to luvien god & heren him REL. I. 129 (HOM. II. 163); heren tidinge LAȝ. 24023; þi word ich wulle heren LAȝ. 18888; he nalde for nane dome mare heren to Rome 9199; hæren 4887; ich þe wulle huren 1210; hiren 14151; here HORN (L.) 398; PR. C. 526; TOR. 1307; SACR. 9; here [heere] CH. C. T. *A* 914; huren FER. 304; hure ROB. 165; WILL. 3270; huire LANGL. *C* viii. 22; hure, hire O. & N. 312, 1483; hire P. L. S. ii. 159; P. S. 200; TREV. I. 55; S. S. (Wr.) 457; hiere (*ms.* hyere) AYENB. 70; hĕreð (*pres.*) MARH. 6; þenne hereþ god (h)is bene A. D. 295; alle þo þe hereð one loverd HOM. II. 9; her to hereþ (*belong*) viii store schire MISC. 146; hērande (*pple.*) þise kniȝtes *in the hearing of these knights* GAW. 450; hēr (*imper.*) MARH. 10; hĕrde (*pret.*) LAȝ. 1164; KATH. 2352; O. & N. 293; GEN. & EX. 1285; HAV. 286; LANGL. *B* PROL. 189; CH. C. T. *A* 1123; DEGR. 77; (*ms.* herrde) ORM. 907; herden WILL. 1298; JOS. 2; hĕrd (*pple.*) GREG. 1005; i have herd sain GOW. I. 367; *comp.* ȝe-hēren.

hīerere, sb., = *M.H.G.* hœrære; *hearer,* AYENB. 256.

hērunge, sb., = *O.H.G.* hōrunga; *hearing,* A. R. 48; MARH. 2; heringe PR. P. 237.

heresie, sb., *O.Fr.* heresie; *heresy,* AYENB. 267.

heretike, sb., *Fr.* heretique; *heretic,* AYENB. 19; D. ARTH. 1307.

herfest, *see* hervest.

herȝien, v., *O.E.* hergian, = *O.N.* herja, *O.H.G.* heriōn; *from* hére; *harry, lay waste, plunder*; hærȝien, heriȝen LAȝ. 3741, 5063; heriede (*pret.*) HOM. I. 205; herhede KATH. 336; herȝede þat lond LAȝ. 1640; he heregede helle HOM. II. 23; herȝed, heried A. P. ii. 1179, 1786; heried WILL. 3725; IW. 2874; herid M. H. 14; haried [harwed] CH. C. T. *A* 3512; harewede M. T. 25; harowed E. T. 256; iherȝed (*pple.*) LAȝ. 2210; iheriȝed DEGR. 140.

harwere, sb., *one who harries* : how harwere of helle was born this night LUD. COV. 159.

herȝunge, sb., *O.E.* hergung; *harrying, laying waste,* HOM. I. 115; heriunge REL. I. 172.

herien [? hērien], v., *cf. O.E.* hērian, hærian, herian, *O.L.G.* hērōn, *O.H.G.* hērēn, *? Goth.* hazjan; *glorify, celebrate, praise,* MARH. 2; A. R. 88; REL. I. 242; (*ms.* heryen) SPEC. 51; herien & hersumen KATH. 147; we wulleð ... þine monscipe herien LAȝ. 6234; hærien 16844; to herien god WILL. 1875; herie AYENB. 23; WICL. LK. xix. 37; herie (*r. w.* merie) CH. C. T. *E* 616; (*r. w.* derie) SHOR. 28; to heriane HOM. I. 97; herest (*r. w.* biwerest) (*pres.*) O. & N. 1518; heriȝe (*subj.*) HOM. I. 113; herede (*pret.*) LAȝ. 14062; A. R. 414; hered, heired A. P. ii. 1086, 1527; hereden HICKES I. 167; hi ... herede god BEK. 288; *comp.* ȝe-herien.

herier, sb., *worshipper* : the herieris (*pl.*) of Baal WICL. 4 KINGS x. 89.

heri-ful, adj., *worthy of praise* : heriful, or worthi to be preised WICL. DAN. iii. 26.

herunge, sb., *O.E.* herung; *glorification, celebration,* HOM. I. 107; A. R. 86; heringe HOM. I. 5; CH. C. T. *I* 682.

heriet, sb., *? O.E.* heregeatu; *heriot,* P. L. S. xvii. 464.

herigaut, sb., *O.Fr.* herigaut; *? cloak*; erigaut (*printed* erigant) A. P. ii. 148; P. S. 156; herigaus (*pl.*) ROB. 548.

hering, *see* hæring.

heritāge, sb., *O.Fr.* heritage; *heritage,* H. M. 25; HORN (L.) 1281; P. L. S. xiii. 96; CH. C. T. *D* 1641.

heritāge-līk, adv., *heritable* : heritagelik of þe & of þin heires MAN. (H.) 251.

hĕrkien, v., *? =* O.Fris. hērkia, *M.Du.* herken, horken, *O. H. G.* horechen; *from* hēren; *hearken,* HOM. I. 31; herken WILL. 213;

PL. CR. 155; CH. C. T. *A* 1526; herkin
'*ausculto*' PR. P. 237; herke LAI LE FR.
147; **hĕrke** (*imper.*) LUD. COV. 55.

hĕrknien, v., *O.E.* hērcnian, hyrcnigan; *hearken*; hærcnien (*ms.* harcnien) LA3. 14649;
herknie TREAT. 133; hercnen KATH. 1735;
A. R. 320; (*ms.* herrcnenn) ORM. 7748; herkne HOCCL. i. 263; **hĕrkinand** (*pple.*) BARB.
vi. 107; **hĕrkne** (*imper.*) SPEC. 29; hercnia͞ð
mine lare LA3. 1517; hercnie͞ð nu to me
26731; herkneþ H. H. 1; CH. C. T. *A* 2208;
hĕrcnede, hærcnede (*pret.*) LA3. 10163,
22638; hercneden MARH. 21; **hĕrkned**
(*pple.*) A. P. ii. 193.

    hĕrknere, sb., *listener*; **hĕrkneres** (*pl.*)
AYENB. 58.

    hĕrcnung, sb., *O.E.* hērcnung, hȳrcnung;
giving ear, listening, A. R. 104; hercnunge
HOM. I. 229.

herle, sb., *cf. M.L.G.* herle, harle; *fibre*, GAW.
190; TREAT. FISH. 35.

herlot, *see* **harlot.**

herm, hermien, *see* **hearm, hearmien.**

hermine, *see* **ermine.**

hern, *see* **hairon.**

herne, sb., *O.N.* hiarni *m.*, *cf. M.Du.* herne,
hirne, *M.L.G.* herne, harne, *O.H.G.* hirni
n.; brain; **heɪnis** (*pl.*) '*cerebrum*' PR. P.
237; hernes HAV. 1808; MIN. iii. 68; hernez
A. P. i. 58; hærnes SAX. CHR. 262; harnis
BARB. iv. 625.

    herne-panne, sb., *cf. M.L.G.* hernepanne;
skull, '*cranium*,' PR. P. 237; REL. II. 78;
RICH. 5293; hernpanne HAV. 1991.

herne, *see* **hürne.**

herre, sb., (? **hērre**), sb., *O.E.* herra, heorra,
hierra, hearra, *O.N.* herra, harra, = *O.L.G.,
O.H.G.* herro (? **hērro**), *M.H.G.* herre, hērre;
master, sovereign, A. R. 6; MISC. 46; ROB.
102; herre, hære LA3. 5420, 7178; *comp.*
over-herre.

herre, *see* **heorre.**

herse, *see* **herce.**

hĕrsum, adj., *O.E.* hȳrsum, = *O.H.G.* hōr-
sam; *from* **hēren;** *obedient,* HOM. II. 51;
(*ms.* herrsumm) ORM. 2534; hærsum LA3.
19395; þe hersum (*? devout*) evensong GAW.
932; *comp.* 3e-hĕrsum.

    hĕrsumian, v., *O.E.* hȳrsumian, = *O.H.G.*
hōrsamōn; *obey,* HOM. I. 37; hersumien II.
145; hersumen KATH. 147; **hĕrsumest**
(*pres.*) MARH. 4; hi him hersumie͞ð LK. viii.
25.

    hĕrsum-lecg, sb., *obedience*; (*ms.* herr-
summleccg) ORM. 2519.

    hĕrsumnesse, sb., *O.E.* hȳrsumness;
obedience, LA3. 29731; REL. I. 131; hersam-
nisse HOM. I. 223.

hert, *see* **heort.**   **herte,** *see* **heorte.**

herti, *see* **heorti.**

hertin, v., *O.E.* hyrtan, = *M.Du.* herten,
M.H.G. herzen; *from* **heorte**; *take heart;
encourage;* '*animo,*' PR. P. 238; herte D.
ARTH. 1181; hirten LA3. 25941; **hertedin**
(*pret.*) GEN. & EX. 1980; herted (*pple.*)
WILL. 3417.

    herting, sb., *encouragement*; mai non hert-
ing on me ben wrogt GEN. & EX. 1982; her-
ting L. H. R. 88; YORK xvii. 115.

herþ, sb., *O.E.* heor͞ð, = *O.Fris.* herth, hirth,
O.H.G. herd; *hearth,* PR. P. 237; **heorþe**
(*dat.*) LEECHD. III. 128; *comp.* **fīr-herþ.**

    herþ-stok, sb., *andirons,* PR. P. 237.

hervest, sb., *O.E.* herfest, hærfest, = *M.L.G.*
hervest, *M.Du.* herfst, harfst, *O.H.G.* herbist,
herbest; *harvest,* '*auctumnus,*' PR. P. 238;
ROB.59; AYENB. 86; LANGL. *B*vi. 292; HOCCL.
ii. 27; herfest FRAG. 3; þe monþe of hervest *August* ROB. (W.) xx. 175; **hervestes**
(*gen.*) LA3. 25403.

    hervest-mōneþ, sb., *O.E.* hærfestmōna͞ð;
harvest month (August), ROB. 61.

    herfest-tīd, sb., = *M.H.G.* herbestzīt; *har-
vest time*; ORM. 11254; hervesttid C. M.
4060.

    hervest-tīme, sb., *harvest time,* LANGL. *C*
ix. 121.

    hervesten, v., = *M.Du.* herfsten, *M.H.G.* her-
besten; *harvest, reap,* MAND. 300.

hes, *see* **es.**   **hēs, hōse,** *see* **hǣse.**

hesel, *see* **hasel.**

heslen, adj., *of hazel*; hesline D. ARTH. 2504.

hesmel, sb., *? collar,* A. R. 424.

hespe, *see* **haspe.**

hete, háte, sb., *O.E.* hete, *cf. O.L.G.* heti,
M.Du. hate, *O.H.G.* haz *m., O.N.* hatr, *Goth.*
hatis (*gen.* hatizis) *n.; hate, hatred,* KATH.
2434; O. & N. 167; S. A. L. 153; S. S. (Web.)
1205; B. DISC. 198; GREG. 327; C. M.
18527; þurh hete (*ms.* hĕte) & niþ ORM. 1404;
withoute hete LEG. 5; hunger & hete [hate]
LA3. 20728; hæte 20441; hete, hate SHOR.
84, 161; MAN. (F.) 3956; hate PR. P. 229;
BEK. 1665; AYENB 29; GOW. I. 40; WICL.
JOHN xv. 18; HOCCL. I. 19; (h)ate GEN.
& EX. 373.

    hete-feste, adv., *securely*: bind him hetefeste
[heteveste] JUL. 36, 37; bunden hire þerto
herde & hetefeste [heteveste] 58, 59; hetefeste
[heteveste] ibunden MARH. 10, 12; heteveste
A. R. 306, 35, 378.

    háte-ful, adj., *hateful,* CH. C. T. *B* 99.

    háte-lich, adj., *O.E.* hetelīc, = *M.Du.* hatelīc;
hateful, C. L. 682; hetelīke (adv.) HAV. 2655;
hetelīch(e) ELL. ROM. II. 82; ha þe bunden
swa heteli fasti HOM. I. 281.

    háte-rēden, sb., *hatred,* PS. xxiv. 19; ha-
treden PR. C. 3363; hatredin, hateredin
HAMP. PS. v. 6*, xxx. 11*; hatrede HOM. I.

233 ; MAN. (F.) 8992 ; hattrede LANGL. *A* iii. 136*.

hátesum, adj., *hateful*, WICL. PROV. i. 29 ; ʒe han maad me haatsum GEN. xxxiv. 30.

hēte, *see* hǣte.   hetel, *see* hatel.

hēten, *see* hāten.

heter, adj., = *M.L.G.* hetter ; *quick, rough, cruel* ; hatter ALEX. (Sk.) 490, 702 ; A. P. iii. 373.

   heter-līche, adv., *roughly, quickly, fiercely*, KATH. 777 ; JUL. 47 ; hetterliche MARH. 6 ; A. R. 290 ; heatterlīche JUL. 17 ; hatterli [heterli] ALEX. (Sk.) 803 ; heterli GAW. 1462 ; A. P. i. 402 ; hetterli WILL. 150 ; hitterli ALEX. (Sk.) 5322.

hetian, v., = *O.L.G.* hetian ; *hate* : heore uvel . . . þu aʒest to hetien HOM. I. 15 ; þet þa saule heteð (*pres.*) 19 ; hetieð 65.

   hetunge, sb., *hatred* : þet hetunge habbeð hom bitwone HOM. I. 67.

hēþe, *see* hǣðe.

héþen, adv., *O.N.* heðan ; *hence* ; (*ms.* heþenn) ORM. 15570 ; PL. CR. 408 ; A. P. i. 231 ; PERC. 1904 ; go heþen HAV. 683 ; fra heþen forth PS. cxii. 2 ; heðen GEN. & EX. 1644 ; heðen forð HOM. II. 65 ; hethen PR. C. 509 ; hethun ANT. ARTH. xix.

   héðen-sīð, sb., *departure, death*, HOM. II. 133.

   heoðen-ward, adv., *away*, A. R. 248* ; heþenward ORM. 5490.

hēþen, *see* hǣþen.   hēþing, *see under* hǣþen.

hēu, *see* hēou.   heuke, *see* huke.

hēve, *see* hūve.   hēved, *see* hēafed.

hevek, *see* havec.

[heveld, sb., *? O.E.* hefel, hefeld (*thread for weaving*).]

   heveld-bed, sb., *? curtained bed*, H. M. 21.

heven, *see* hebben.

heven, hevene, *see* heofene.

hevenen, v., *O.N.* hefna ; *avenge* ; hevin M. H. xvi.

   hevening, sb., *O.N.* hefning ; *vengeance*, YORK xxxii. 284 ; HALLIW. 447.

hevi, adj., *O.E.* hefig, = *M.Du.* hevigh, *O.H. G.* hebīg ; *heavy* ; (*ms.* heui) SAX. CHR. 253 ; FRAG. 5 ; LAʒ. 26067 ; A. R. 232 ; HAV. 1050 ; AYENB. 31 ; (*ms.* heuy) WICL. LK. xi. 7 ; A. P. iii. 2 ; PR. C. 4583 ; TOR. 1270 ; APOL. 26 ; hevi & sor GEN. & EX. 2565 ; hefiʒ last ORM. 4522 ; hefi LEECHD. III. 88 ; hevie (*pl.*) P. L. S. xv. 192 ; hevie (*adv.*) A. R. 32 ; heviere (*compar.*) P. L. S. xv. 106.

   hevi-chēred, adj., *of sad countenance*, GOW. III. 360.

   hefiʒ-līke, adv., *O.E.* hefiglīce ; *heavily*, ORM. 4771 ; hevali BARB. vii. 209.

   hevinesse, sb., *O.E.* hefigness ; *heaviness*, A. R. 132 ; AYENB. 31 ; hevinesse TREAT. 134.

hevisum-lī, adv., *grievously*, WICL. ECCLUS. vi. 26.

hēvid, *see* hēafod.

hevie, sb., *heaviness, weight* : tuei gegges þe cupe bere and for hevie wroð hi were FL. & BL. (H.) 853.

heviin, v., *O.E.* hefigian ; *make heavy, 'gravo,'* PR. P. 239 ; heveʒi LAʒ. 18408* ; hevie P. L. S. xv. 96 ; to hevin on þi harme *to think only of your wrong* D. TROY 2083 ; hevegeð (*pres.*) A. R. 424* ; hevieð HOM. II. 29 ; hevieþ CH. BOET. v. 5 (171) ; hevied (*pple.*) WICL. MAT. xxvi. 43 ; *comp.* ʒe-heviʒen.

hēw, hēwe, *see* hēou.

hēwe, hēwen, *see* hīwe, hīwen.

hēwen, hēwien, hēwinge, *see* hēawen, hēawien, hēawinge.

hēy, *see* hēi.   hī, *see* hé.

hiane, sb., *Lat.* hyæna ; *hyena* : hiane, þet ondelfeþ þe bodies of diade men, and hise eteþ AYENB. 61 ; hiene CH. FORTUNE 35.

hicchin, v., = *L.G.* hicken, *mod.Eng.* hitch ; *move* ; hicchin (*printed* hitchin), hichin '*amoveo*' PR. P. 239 ; *see* icchen.

hicht, sb., = hāt ; *promise*, BARB. xiv. 16.

hicht, *see* hēahþe.

hīdāge, sb., *from* hīd ; *payment for land*, TREV. II. 97.

hīde, sb., *O.E.* hīd, hȳd, hīgid, = *hīwid* ; *hide (measure of land)* ; hīde (*dat.*) MAN. (H.) 110 ; eiʒte hīde (*pl.*) lond ROB. 297 ; sixti hiden of londe LAʒ. 18241.

hīde, hīdel(s), *see* hūde, hūdels.

hīden, *see* hūden.

hider, adv., *O.E.* hider, hyder, = *Goth.* hidrē, *O.N.* heðra ; *hither* ; (*ms.* hiderr) ORM. 209 ; O. & N. 462 ; GEN. & EX. 2344 ; HAV. 885 ; SPEC. 42 ; MAND. 44 ; CH. C. T. *A* 672 ; PR. C. 508 ; hider geond ['*illuc*'] MAT. xxvi. 36 ; hider in LAʒ. 36 ; hider on, 25326 ; hider tō, KATH. 447 ; A. R. 80 ; H. H. 116 (118) ; hedir BEK. 2439 ; PERC. 1153 ; LUD. COV. 52.

   hider-cūme, sb., *O.E.* hidercyme ; *hithercoming*, HOM. II. 133.

   hider-ward, adv., *hitherward, -wards*, O. & N. 1690 ; hiderward JOS. 354 ; hiderward, -wardes LAʒ. 10154, 30780.

hidor, sb., *O.Fr.* hidor, hisdor ; *fear* : such a hidor hem hent & a hatel drede A. P. iii. 367 ; thou aʒtest habbe more hidour of þine oʒene unriʒte SHOR. 33.

hidous, adj., *O.Fr.* hideus ; *hideous*, WILL. 3177 ; MAN. (F.) 14268 ; hiduouss BARB. x. 594.

   hidous-līche, adv., *hideously, horribly* ; *greatly*, AYENB. 6 ; hidousli WICL. I KINGS xxxi. 3 ; WICL. 4 KINGS xvii. 18 ; (he) blewe so hidousli and hie CH. H. F. 1599.

hidousnesse, sb., *horror, dread* : orrour, ether hidousnesse WICL. DAN. vii. 15 *; þe hidusnes of paine PR. C. 9485.

hidousen, v., *fear, dread* : a man kindeli **hidousiþ** (*pres.*) derknesse WICL. S. W. I. 269 ; mi spirit **hidouside** [' *horruit* '] (*pret.*) DAN. vii. 15.

hīe, *see* hēȝen.

hīen, *see* hiȝien. **hiene**, *see* hiane.

hieren, *see* hēren. **hierte**, *see* heorte.

hīerþe, sb., *cf. O.L.G.* gihōritha, *O.H.G.* gehōreda ; *hearing*, AYENB. 56.

hīf, sb., *O.E.* hēof, *? cf. Ger.* hief ; *? sound* : minne horn mid grǣte **hīve** (*dat.*) blowen LAȝ. 790.

hīȝ, *see* hēah.

hīȝe, sb., *haste* : wiþ mikel hih ORM. 2686 ; an hiȝe (*printed* highe) S. S. (Web.) 1478 ; in hiȝe LEG. 213 ; in hie CH. C. T. *B* 209 ; TRIAM. 277 ; ANT. ARTH. iv ; hegh [hi] HAMP. PS. i. 1 *.

hih-ful, adj., *hasty*, A. R. 302.

hiȝe, *see* hüȝe.

hīȝe, sb., *O.E.* hīge,=*M.L.G.*, *M.H.G.* hīge, hīe ; *?* =hīwe ; *domestic* ; **hīȝez** (*pl.*) A. P. ii. 67.

hiȝe, **hīȝen**, *see* hēhe, hēȝen.

hiȝende, sb., *? =*hīȝinge ; *haste* : an hiȝende LAȝ. 5496.

hīȝend-līche, adv., *quickly, hastily*, LAȝ. 7312 ; hiendliche KATH. 2141 ; hihe(n)dliche MISC. 52.

hight, *see* hēahþe.

hiȝien, v., *O.E.* higian ; *hie, hasten* ; hiȝie MARG. 267 ; hihȝen HOM. I. 105 ; hiȝhen ORM. 2723 ; hihen KATH. 412 ; hien A. R. 92 ; hiȝe WILL. 1286 ; (*printed* highe) ALIS. 5133 ; hie S. S. (Wr.) 3108 ; SACR. 860 ; híeþ (*pres.*) MISC. 93 ; hiȝez A. P. ii. 538 ; híhe (*imper.*) MARH. 21 ; hiȝe LEG. 33 ; high(e) þe PS. xxx. 3 ; þet god . . . hiȝe ham ut of pine A. R. 30 ; hiȝede (*pret.*) HORN (L.) 968 ; BEK. 2224 ; A. D. 258 ; PL. CR. 155 ; hied [hiȝed] LANGL. *B* xx. 322 ; hiȝeden LAȝ. 2317 ; híȝed (*pple.*) WICL. I THESS. ii. 17.

híȝinge, sb., *haste* : an hiȝinge LAȝ. 2358 ; on hijng MISC. 50.

híȝing-lī, adv., *hastily*, WICL. DEEDS xvii. 15 ; thei wenten hiingli 2 KINGS xvii. 20.

hiȝt, *see* hüht, hihte.

hiȝt, *see* hēahþe. **hiȝten**, *see* hāten.

hiȝtlien, v., *? from* hüht ; *ornament* ; hiȝtild (*pret.*) ALEX. (Sk.) 1541 ; a hatt . . . hiȝtild (*pple.*) o floures 4540 ; a wale wode . . . hiȝtild in þat hill with handis of aungels 4969.

hiht, *see* hüht.

hihte, sb., *?* =hihðe ; *haste* ; (*ms.* hiþte) SPEC. 110 ; hiȝt MIRC 559.

hīhte, *see* hēahþe. **hihten**, *see* hühten.

hihten, v., *? from* hüht ; *adorn*, HOM. II. 7ᵇ ; hiȝte TREV. IV. 37 ; hiuhteð (*pres.*) REL. I. 132 (HOM. II. 13) ; hiȝteþ CH. BOET. i. 2 (8) ; hiȝt SHOR. 155 ; hihten (*pret.*) HOM. II. 89 ; in uch an herd þin aþel is hiht (*pple.*) (*? extolled*) SPEC. 33.

hihðe, sb., *O.E.* hihð ; *? haste*, A. R. 324 ; JUL. 77 ; *see* hihte.

hil, *see* hül. **hilde**, *see* helde.

hilden, *see* hélden *and* hülden.

hilder, *see* hillor.

hildinge, *see under* hülden.

hildire, *see* helder.

hilet (*?* =heleð²), sb., *shady place*, ['*umbraculum*'] (*later ver.* schadewing place), WICL. ECCLUS. xxxiv. 19.

hilien, *see* helian.

hilla, interj., *halloa*, ALEX. (Sk.) 1066.

hiller, **hillern**, sb.,=ellarne, *elder* ; hildir or eldir ' *sambucus* ' PR. P. 239.

hillor-trē, sb., *elder tree* ; hildertre, VOC. 163 ; hillortre 191 ; hillerntre PR. P. 239.

hilt, sb., *O.E.* hilt *m. n.*, hilte *f.*, *cf. O.N.* hialt *n.*, *M.L.G.*, *M.Du.* hilte, *O.H.G.* helza *f.* ; *hilt*, ' *capulus*,' PR. P. 240 ; LAȝ. 6506 ; he smot him on þe scheld igult þoruȝout þe bord þoruȝout þe hilte ALIS. 1270 ; hit him up to þe hult GAW. 1594 ; hilte (*dat.*) HORN (R.) 1434 ; TREV. III. 391 ; þere hilte [þan helte, heolte] LAȝ. 1559, 22509.

hilted, adj., *having a hilt* ; hiltede (*pl.*) swerdez D. ARTH. 2274.

hilve, *see* helfe.

himlande, adj., thorowe hopes and himlande hillis D. ARTH. 2503.

himpne, sb., *Lat.* hymnus ; *hymn*, ' *impnus*,' CATH. 186 ; himpne maker [' *hympnista* '] 186 ; mi lippis shuln tellen out an impne WICL. PS. cxviii. 171 ; impne to his haleghs PS. cxlviii. 14 ; **impnes** (*pl.*) PS. xcix. 4 ; CATH. 186 ; imne A. R. 16 ; PR. P. 259.

himpner, sb., *med.Lat.* hymnārium ; *hymnàl, hymn book*, CATH. 186 ; VOC. 230, 249.

hind, adv.,=*O.H.G.* hint ; *backwards*, ALIS. 5200.

hind-ward, adv., *backwards*, [' *retrorsum* '] PS. cviii. 5 ; WICL. PS. lxix. 4 ; D. TROY 8553.

hinde, sb., *O.E.*, *O.N.* hind,=*O.H.G.* hinta, hinda ; *hind, female deer*, LAȝ. 2589 ; ROB. 376 ; WILL. 2866 ; ALIS. 1889 ; LANGL. *B* xv. 274 ; hinde (*ms.* hindes, *r. w.* finde) (*pl.*) LAȝ. 1448 ; (*ms.* hinden) 8108.

hinde-hēle, sb., *hind-heal*, ' *ambrosia*,' REL. I. 36.

hindene, sb., *? hiding place* : þe scawere, þet is þes deofles hindene HOM. I. 53.

[**hinden**, adv., *O.E.* hindan,=*O.L.G.* hindan, *Goth.* hindana, *O.H.G.* hintana ; *hind ; comp.* at-, bi-hinden.]

hinde-forð, adv., *backward*: hindeforth thei seten ALIS. 4708.

hinde-ward, adv., *O.E.* hindeweard; *backward*, PS. ix. 4, xx. 13; hindward lxix. 4; he had him of horse, hindward anon D. TROY 8553.

hinder, sb.,*cf. O.Fris., M.L.G.* hinder *m., O.N.* hindr *n.*; *hindrance*: of hindre þat is of bipeching HOM. II. 213.

[**hinder,** adj., *O.E.* hinder,=*Goth.* hindar, *O.H.G.* hintar; *after.*]

hinder-arson, sb., *back of the saddle*: bitwix him and his hinder-arsoun IW. 680; aswogh he fell adoun an his hinder-arsoun B. DISC. 1171.

hinder-cræft, sb., *deceitful craft, artifice*: of ane hindercræfte (*? ms.* hindere cræfte) LA3. 10489.

hinder-ful, adj., *deceitful,* HOM. II. 59; SAINTS (Ld.) 319.

hinderful-liche, adv., *deceitfully,* HOM. II. 83.

hinder-3æp,adj.,*O.E.* hinder3eap; *insidious,* ORM. 6646.

hinder-word, sb., *deceitful word*: mid his hinderworde bicherde him HOM. II. 59.

hinderling, adj., *O.E.* hinderling; *laggard, coward,* ORM. 4860; **hinderling** (*sb.*) whiche souneþ icast doun fro honeste, or ani image goinge bakward TREV. VII. 109.

hindre, adj., *O.N.* hindri,=*O.H.G.* hintaro; *hinder,* MAND. 107; WICL. GEN. xvi. 13; BRD. 30; at þe hinder gate R. R. 5850; **hindreste, hindereste** [= *O.H.G.* hintarōsto] (*superl.*) *hindermost* CH. C. T. *A* 622; hinmast BARB. viii. 245.

hendir-mār, adj., compar., BARB. vii. 599; hindermore (*sb.*) *hinder part* WICL. EX. xxxiii. 23; hindirmore WICL. GEN. xvi. 13; **hindir-mōris** (*pl.*) WICL. 3 KINGS xxi. 21.

hindren, v., *O.E.* hindrian, = *O.N.* hindra, *M.L.G.* hinderen, *O.H.G.* hintran; *hinder, impede*: to hindren and to lette LIDG. M. P. 93; hindre CH. C. T. *A* 1135; hindire ALEX. (Sk.) 2497; **hindreð** (*pres.*) HOM. II. 193; hindren MISC. 226.

hindrer, sb., *hinderer,* GOW. II. 97; the sonne . . . the hinderer of the night III. 111.

hindringe, sb., *hindering*; hindringe or harminge '*dampnificacio*' PR. P. 240; hindring(e) GOW. I. 213, II. 64; hinderinge '*detrimentum, derogacio, peioracio*' CATH. 186.

hine,sb.,*O.E.* hīna, *? mod.Eng.* hind; *domestic,* HOM. I. 197; BEK. 263; LANGL. *A* PROL. 39; CH. C. T. *A* 603; PS. xviii. 12; M. H. 145; TOWNL. 181; **hīnen** (*pl.*) FRAG. 6; LA3. 368; H. M. 7; ROB. 540; P. S. 149; ALIS. 1215; hine HOM. II. 51; MISC. 82; HAV. 620; *comp.* in-hīne.

hine-folc, sb., *kinsfolk,* GEN. & EX. 3655.

hīne-hēd, sb., *family; service*: alle hīne-hēdes (*pl.*) of genge PS. xxi. 28; and gresse to hinehed (*service*) of men PS. civ. 14.

hīnes-kin, sb., *kindred*: wið wifes and childre and hineskin GEN. & EX. 3775.

hinene, *see* heonene. **hingen,** *see* hengen.

hingren, *see* hüngren.

hinne, *see* heonene. **hīnþe,** *see* hēnðe.

hipe, *see* hüpe. **hīpel,** *see* hüpel.

hippen, *see* hüppen. **hīrd,** *see* hīrēd.

hirde, hirdefloc, *see* heorde.

hirdel, *see* hürdel.

hirdnesse, hirdsipe, *see* heorde.

hire, pron., *gen. f. and gen. pl. of* hé, *inflected like a possessive pronoun*: of hiren AYENB. 38, 111.

hire, *see* here.) **hīre,** *see* hüre.

hīrēd, hīrd, hïrd, sb., *O.E.* hīrēd,=*O.H.G.* hīrāt *m.* (*marriage*); *from* hīw; *cf.* hīw-rǣden; *family, household, retinue*; hired LA3. 6152; þat halie hired HOM. I. 89; hird A. R. 94; KATH. 81; ORM. 512; GEN. & EX. 3222; SPEC. 32; TRIST. 166; **hīrēdes, herdes** (*gen.*) LA3. 2336, 4342; hirdes (*ms.* hyrdes), heordes MAT. xiii. 52, xx. 1; to hirede (*court*) hes comen LA3. 19443.

hīrēd-cnafe, sb., *attendant, domestic,* LA3. 28824; heoredcnave 20967.

hīrēd-, hīrd-cniht, sb., *O.E.* hīrēdcniht; *courtier,* LA3. 4316, 9856.

hīrd-ifēre, sb., *courtier*: þer he huntede (*ms.* hundede) on comelan wið his hirdi-fēren (*pl.*) LA3. 6631.

hīrēd-gume, sb., *courtier,* LA3. 12889.

hīrdling, sb., *knight,* LA3. 12713*.

hīrēd-, hērēd-mon, sb., *O.E.* hīrēdmann; *retainer,* LA3. 2350, 6877; hirdmon KATH. 2247; heredmon GAW. 302; **hīredmen** (*pl.*) GAW. 302; hirdmen P. S. 157.

hīrēd-plæie, sb., *court-play*: mid haveken & mid hunden hiredplæie luvien LA3. 14481.

herd-swein, sb.,*domestic servant,* LA3. 5662.

hīren, *see* hēren *and* hūren.

hirne, *see* hürne. **hirst,** *see* hürst.

hirten, *see* herten *and* hürten.

his,*gen. of* hé, pron.,*sometimes inflected like a possessive pronoun*: hise write SAX. CHR. 250; vor te leren hise A. R. 114; an of hise men KATH. 406; þurh hise gode dedes ORM. 60; hise limes REL. I. 210; to alle hise holde PROCL. I.

his, *see* es.

hisiaus, sb., pl., *O.Fr.* hiziaus; *? heralds*: þis hisiaus (*printed* hisians) and þise kempen . . . þat vor pans oþer vor timlich profit yeveþ ham to crefte na3t oneste AYENB. 45.

hissin, v., *cf. M.Du.* hissen, hischen; *hiss,* '*sibilo,*' PR. P. 242; **hissit** (*pres.*) '*sibilat*' VOC. 180.

hissing, sb., *hissing*, '[*sibilus*'] WICL.. 2 PARAL. xxix. 8.

histoire, sb., *O.Fr.* histoire, estoire ; *history, story*, GOW. II. 23, 80 ; storie R. R. 154 ; HAV. 1640 ; store WILL. 4805 ; '*argumentum, historia*' CATH. 366 ; **stóris** (*pl.*) C. M. 21 ; the stories of Moises lawe WICL. IS. PROL.

historial, adj., *historical*: historial bookes WICL. IS. PROL. ; no fable but knowen for a storial thing notable CH. C. T. *C* 156.

hit, see **hé.**

hitten, v., *O.N.* hitta ; *hit, fall upon, touch* ; hitte LANGL. *B* xii. 108 ; EGL. 269 ; wham so he hitteþ (*pres.*) wiþ (h)is hond FER. 873 ; þe fevere aguful sore him hatte MAN. (F.) 15729 ; we hitteč A. R. 176* ; **hitte** (*pret.*) LAȝ. 1550 ; WILL. 2822 ; ALIS. 2357 ; A. P. iii. 289 ; hit (*pple.*) '*tactus*' PR. P. 242.

hitterlī, see **heterlī.**

hīþe, sb., *O.E.* hȳč ; *harbour*, '*statio*,' PR. P. 242.

hiu, see **hēou. hīve,** see **hūve.**

[**hīw, hēow,** sb., *O.N.* hiu ; *?family* ; *cf.* Goth. heiwafrauja (οἰκοδεσπότης).]

hēow-rǣden, sb., *O.E.* hīw-, hȳw-, heo-rǣden ; *family*, LUKE xix. 9.

hīw-scipe, sb., *O.E.* hīwscipe ; *family*, HOM. I. 87.

hīwe, sb., *O.E.* hīwa (*nom. pl.* hīwan),= *O.H. G.* hīwo, *? O.N.* hio (*nom. pl.* hion) ; *domestic* ; hewe AYENB. 195 ; AR. & MER. 1165 ; CH. C. T. *E* 1785 ; **hēwen** (*pl.*) LANGL. *B* iv. 55 ; heowes SPEC. 114 ; mid þine hiwun HOM. 1. 225 ; ga . . . to þinen heowen MK. V. 19 ; *comp.* **sin-hēowen.**

hīwen, sb., *O.E.* hīwen *n.* ; *family*, (*ms.* hewenn) ORM. 594.

hl-, hn-, see **hl, hn-** (*in letters* **l, n**).

ho, see **hwá.**

hō, interj., *O.N.* hō ; *ho !* : and cride ho CH. C. T. *A* 1706 ; hoo to him hoo hoo HALLIW. 457 ; how *outcry* ALEX. (Sk.) 4732.

hō², sb., *intermission*: wiþouten ho CH. TRO. ii. 1083 ; withoutten ho BARB. xx. 429.

hō-lī, adv., *O.N.* hōgliga ; *tardily*, HAMP. PS. xxxix. 24.

hūlines, sb., *tardiness, lateness*, HAMP. PS. xxxix. 24*.

hō, see **hēo** (*under* **hé**), **hōh** *and* **houȝ.**

hobelen, v., *cf. M.Du.* hobbelen ; *hobble, limp ; saunter about* ; hobelen (*pres.*) PL. CR. 106 ; hobland (*pple.*) BARB. iv. 447 ; hobled (*pret.*) TRIST. 1161 ; hobleden LANGL. *A* i. 113.

hobeler, sb., *O.Fr.* hobeler (*cavalier qui monte un hobin*) ; *? groom* ; **hobelers** (*pl.*) OCTOV. (W.) 1598.

hobi, sb., *O.Fr.* hobin ; *hobby* (*hawk*), '*alaudarius*,' PR. P. 242 ; **hobbis** (*pl.*) DEP. R. i. 90 ; hobinis BARB. xiv. 68, 400.

hoc, sb., *O.E.* hocc ; *mallow*, '*malva*,' FRAG. 3 ; REL. I. 37 ; hok VOC. 265 ; *comp.* **holi-hoc.**

hōc, sb., .*O.E.* hōc,=*M.Du.* hoek ; *hook*, '*uncinus*,' FRAG. 4 ; þene hoc HOM I. 123 ; hok '*hamus*' VOC. 240 ; C. L. 1128 ; hanged worþe he on an hok HAV. 1102 ; **hōke** (*dat.*) MARH. 3 ; TOR. 2508 ; **hōkes** (*pl.*) P. L. S. xix. 248 ; SPEC. 105 ; AYENB. 264 ; GREG. 744 ; LANGL. *B* v. 603 ; hokis ALEX. (Sk.) 5519 ; *comp.* **angel-, fisch-, flesch-, wēod-hoc.**

hūke-nebbide, adj., *hook-nosed,* D. ARTH. 1082.

hochen, v., *? M.Du.* hutsen, hotsen ; *hack, chop* : (thai) **hotchene** (*pl.*) in holle the heþenne knightes D. ARTH. 3687.

hochepot, sb., *Fr.* hochepot, *cf. L.G.* hutspot ; *hotchpot* : ye han cast all hire wordes in an hochepot CH. C. T. *B* 2247 ; goose in an hoggepot L. C. C. 32 ; hogpoch AUD. 29.

hock, sb., *caterpillar* ; **hockes** (*pl.*) ['*campas*'] PALL. i. 882.

hōcour, see **hōker.**

hōd, sb., *O.E.* hōd, = *O.L.G.* hōd, *O.H.G.* huot ; *hood*, MIRC 883 ; PL. CR. 423 ; GAW. 155 ; þere burne hod LAȝ. 21421 ; enne widne hod A. R. 56 ; als ich evere brouke min hod under min hat P. S. 332 ; hod, hood LANGL. *A* v. 172, *B* v. 329 ; CH. C. T. *A* 195 ; hood LIDG. M. P. 103 ; hud ANT. ARTH. ii ; hude BARB. xviii. 308 ; **hōdes** (*pl.*) TREV. I. 353.

hōd-lēz, adj., *hoodless*, A. P. ii. 643.

hōd, hōdien, see **hād, hādien.**

hōdin, v., *hood, provide with a hood* ; hoodin PR. P. 242 ; ihōdede (*pple. pl.*) LAȝ. 7836.

hodren, v.,=*L.G.* hudren ; *cover up* : hodur and happe FLOR. 112 ; hodred (*pple.*) in þer hottes MAN. (H.) 273.

hōen, v., *O.N.* hōa ; *cry out* : hoen on him LANGL. *B* x. 61.

[**hof,** sb., *O.E.* hof *n.*, *cf. O.L.G.. O.H.G.* hof *m.; court.*]

hóve-daunce, sb., *M.Du.* hofdans, *cf. M.H. G.* hovetanz ; *court-dance* : daunce and singe the hovedaunce & carolinge GOW. III. 6 ; the hovedaunce and the carole III. 365.

hōf, sb., *O.E.* hōf, *O.N.* hōfr,= *O.H.G.* buof ; *hoof*, '*ungula*,' FRAG. 2 ; **hūfe** (*dat.*) PR. C. 4179 ; **hōves** (*pl.*) GAW. 459.

hōf², sb., *O.N.* hōf ; *measure, reason*, ORM. 4742 ; **hōve** (*dat.*) C. M. 11973 ; wituten hove PR. C. PREF. xi ; *? comp.* **bi-hōf.**

hōf-lēas, adj., *unreasonable*, A. R. 108 ; hofles MARH. 17.

hofer, sb., *O.E.* hofer,=*M.L.G.* hover, *O.H.G.* hovar ; *hump*, '*gibb(us)*, *struma*,' FRAG. 5.

hoferede, sb., *O.E.* hoferede,=*M.L.G.* hoverde ; *humpbacked* : nowčer halt ne hoveret MARH. 20 ; healde halte & hoverede (*pl.*) KATH. (E.) 1063.

hōfien, v., *for* bi-hōfien; hōven (*pret.*)
BARB. x. 3͜ *.

hog, sb., *hog, pig*, ALIS. 1885; hogges (*pl.*)
AYENB. 89; LANGL. *B* vi. 183; *comp.* þorn-
hog.

hōȝ, *see* hōh.

hoȝe, sb., *O.E.* hoga (*cf. M.L.G.* hoge (*gen.*
hogen), *O.N.* hugi *m.*; *thought, care*, O. &
N. 601; FER. 4539; howe REL. I. 263; S. S.
(Web.) 1450; howe (*dat.*) ROB. 461; ALIS.
1906; þu scalt faren al to howe FRAG. 6;
comp. over-, ümb-hoȝe.

hoh-ful, adj., *O.E.* hohful; (*v. r.* houh-,
howful) O. & N. 1292, 1295; hoȝheful ORM.
2902; hohfulle (*pl.*) LAȝ. 14096.

hogh, *see* houȝ.

hoȝien, v., *O.E.* hogian, *cf. O.H.G.* hogan
(*pret.* hogeta); *take thought, meditate*;
howien REL. I. 173; hoȝað (*pres.*) HOM. I.
113; hoȝeþ O. & N. 455; howeþ SPEC. 23;
hoȝede (*pret.*) LAȝ. 13416; *comp.* bi-, for-,
over-hoȝien.

hōh, sb., *O.E.* hōh, hō *m.*; *hough, hock*:
bineþe at hire ho MARH. 160; hōȝes (*pl.*)
GAW. 1357; hoghez ALEX. (Sk.) 3151.

hōuȝ-senu, sb., *O.E.* hōhsino, *cf. O.Fris.* hox-
ene, hoxne, *O.N.* hāsin, *M.H.G.* hahse, hehse,
M.L.G. hesse; *hamstring*, WICL. I PARAL.
xviii. 4.

hoxenen (? hōxenen), v., *cf. O.H.G.* hah-
sinōn, *M.H.G.* hehsenen, *M.L.G.* hessen;
hamstring: þou shalt hoxe ['*subnervabis*']
WICL. JOSH. xi. 16; hoxened (*pple.*) TREV.
VII. 139.

hōhin, v., *hough, hamstring*; houhin, houghin
'*subnervo*' PR. P. 251.

hoist, *see* host³.

hōk, *see* hōc.

hōked, adj., *hooked*, O. & N. 79; his . . .
hokede neose MARH. 9.

hōken, v., *? hook; pick one's way*: [the
hare] hōkeð (*pres.*) pathes swithe narewe
O. & N. 376; [cucumber floure] is so ferd of
oiles, that therfroo hit hoketh if me setteth
it nigh ther under PALL. iv. 201; hōkit
(*pret.*) out of havin D. TROY 4621.

hōker, sb., *O.E.* hōcor; *mockery, derision*,
HOM. I. 153; MARH. 5; KATH. 778; ROB.
272; hoker & scarn LAȝ. 17307; hocour
ALEX. (Sk.) 1714*; hōkere (*dat.*) JUL. 52;
CH. C. T. *A* 3965; hōkeres (*pl.*) LAȝ.
29790; A. R. 188.

hōker-ful, adj., *scornful*, REL. I. 188.

hōkerful-līche, adv., *mockingly*: a proude
dame and an envieous, hokerfulliche mes-
segging LAI LE FR. 59.

hōker-lahter, sb., *scornful laughter*, HOM.
I. 283.

hōker-l(ē)oð, sb., *song of mockery*, LAȝ.
28872.

hōker-līche, adv., *derisively*, LAȝ. 19412;
KATH. 742; A. R. 198; ROB. 417; P. S.
204.

hōker-word, sb., *O.E.* hōcorword; *scornful language*, LAȝ. 19595.

hōkerien, v., *mock, scoff*; hōkerest (*pres.*)
KATH. 458; (heo) hokerieð þan folke LAȝ.
15785; hōkereð (*imper.*) A. R. 248; hōke-
rede (*pret.*) MISC. 50; nes hit nan swa
wac mon þat him ne hokerede on LAȝ.
14795.

hōkerere, sb., *mocker, scoffer*, ARCH. LII.
36.

hōkerunge, sb., *mockery, derision*, A. R.
188; hokeringe HOM. I. 281.

hoket, sb., *O.Fr.* hoquet; *? difficulty, obstacle*: moni hoket is in amours ALIS. 7000;
him think no hoket TOWNL. 313; of care and
of curstnes, hething and hoket 311.

hokillen, *? beat*: þe maidens . . . maȝtili hokillen (*pret.*) with þe swaif of þe sworde A. P. ii.
1267:

hokkerie, *see* hukkerie.

hol, adj. & sb., *O.E.* hol, = *O.Fris.*, *M.L.G.*
hol, *O.N.* holr, *O.H.G.* hol; *hollow, hole, cave*,
'*cavus*,' PR. P. 242; hóle (*dat. n.*) O. & N.
965; hol (*subst.*) LAȝ. 30862; BEK. 1152;
þat hol LEG. 25; hole WILL. 95; S. S. (Wr.)
2169; CH. C. T. *A* 3440; hoill BARB. xix.
669; hóle (*dat.*) MARH. 10; O. & N. 826;
HAV. 1813; A. P. iii. 306; hóle (*pl.*) LK. ix.
58; þe haveð fif hole HOM. II. 201; holes A.
R. 128; holis ALEX. (Sk.) 4045; holles BARB.
xi. 153*; howis xi. 153.

hól-lēk, sb., *O.E.* hollēac, = *M.H.G.* hol-
louch; *shalot*, '(*caepa*) ascalonia,' VOC. 225.

hól-rische, sb., '*papyrus*,' PR. P. 244.

hōl, *see* hāl.

holard, sb., = huler; *debauchee*, TOWNL. 149.

hóld, adj., *O.E.* hold = *O.Fris.*, *O.L.G.* hold,
O.H.G. hold, *O.N.* hollr, *Goth.* hulþs;
friendly, faithful, GEN. & EX. 1389; P. S.
214; hold oðer fa HOM. I. 231; hold & trig
ORM. 6177; holde MISC. 38; þin holde mon
LAȝ. 14091; mid holde mode HOM. I. 81;
hólde (*pl.*) P. L. S. viii. 134; ROB. 377;
WILL. 2833; M. H. 102; hulde Iw. 887;
hólde (*adv.*) GAW. 2129; hóldeste (*superl.*)
LAȝ. 16369; *comp.* un-hóld.

hólde-līke, adv., *faithfully*, GEN. & EX.
1546; holdeli GAW. 1875, 2016.

hóld-ōþ, sb., *O.E.* holdāð; *oath of friendship,
allegiance*, AR. & MER. 3588; hold-, holde-oþ
HAV. 2781, 2816; holdeoþ ROB. 383.

hóld², sb., *O.E.* hold, = *O.N.* hold; *flesh, carcase*, ['*cadaver*'] HOM. II. 183; heo wulleþ
freten þin fule hold FRAG. 6.

hóld, *see* heald.

hólde, sb., = *M.H.G.* holde, hulde, *M.Du.*
houde; *fidelity*, ALIS. 2912.

hólden, v., *see* healden.

hóle, *see* hule.

hólen, v., *peel*; hooled (*pple.*) barli WICL. PROV. xxvii. 22.

hólen[2], v., *O.E.* holian, = *O.H.G.* holan, *O.N.* hola, *Goth.* (us-)hulōn; *make a hole*; holin '*cavo*' PR. P. 243; hólieð (*pres.*) A. R. 130; hólede (*pret.*) MAN. (F.) 6836; H. S. 10736; hóled (*pple.*) ALIS. 5929.

hóliinge, sb., *? hollowing, excavation*: holiinge and perforacion GEST. R. 10.

hōlen, v., *O.E.* hōlian (*? calumniate*), *cf. Goth.* hōlōn (συκοφαντεῖν) SCH. GLOSS., *O.H.G.* huolēn; *? defraud*; þat hōleþ (*pres.*) o þe laȝhe leod & rippeþ hem & ræfeþ ORM. 9319.

holer, *see* huler.

holet, sb., *? from* hol; *but cf.* helet; *recess, hole,* WICL. S. W. II. 281; siche placis ... shulden be fled as fendis holetis (*pl.*) WICL. E. W. 322; holettez HALLIW. 455.

holh, adj. & sb., *O.E.* holh, *cf. M.L.G.* holich; *hollow, cave,* LAȝ. 761; holȝ, holeh, holeuh O. & N. 643, 1113; holȝ GAW. 2182; holgh PALL. iii. 257; holu ROB. 251; (*ms.* holw) WILL. 295; holouȝ TREV. III. 395; holou PR. P. 242; þet holh HOM. I. 23; in þe holwe asche LAI LE FR. 209; holȝe (*pl.*) A. P. ii. 1695; holwe ROB. 131; his eien holwe CH. C. T. *A* 1363; holȝes (*subst. pl.*) LAȝ. 20848.

holounesse, holouȝnes, sb., *hollowness,* TREV. III. 395.

holȝen, v., *hollow, excavate*; holwed (*pret.*) ROB. 415*.

holi, sb., = holin; *holly,* FRAG. 3; holie (*dat.*) A. R. 418.

holi-hoc, sb., *? O.E.* holihocc; *holly-hock,* '*althaea,*' REL. I. 36.

hōli, *see* hāliȝ.

holier, *see* huler.

holin, sb., *O.E.* holen, holegn; *holly,* '*hous,*' VOC. 163; A. R. 418*; REL. II. 280; CH. P. F. 178; hollin bobbe *holly bough* GAW. 206; *see* holi; *comp.* cnē-hole.

holk, sb., *O.E.* holc; *hollow part*: et te breoste holke HOM. I. 251; *see* holh.

holke, *see* hulc.

holken, v., = *M.L.G.* holken, *Swed.* holka; *hollow out, thrust out*; holkked (*pret.*) A. P. ii. 1222; holket (*pple.*) ANT. ARTH. ix; *comp.* a-holken.

holm, sb., *O.E.* holm, = *O.L.G.* holm, *O.N.* holmr; *holm, hill, island,* PR. P. 243, 244; into þan hæȝe holme LAȝ. 20712.

holm[2], sb., *holm-oak,* PR. P. 244; CH. C. T. *A* 2921.

holmen, adj., *of the holm tree*: holmen leves ALIS. 4945.

holocaust, sb., *Lat.* holocaustum; *holocaust,* GEN. & EX. 1325.

holou, holouȝ, *see* holh.

holste, sb., (*name of a bird*), '*talendiola,*' VOC. 253.

holt, sb., *O.E.* holt, = *O.L.G., O.Fris., O.N.* holt, *O.H.G.* holz; *holt, wood,* '*saltus,*' FRAG. 3; VOC. 270; PR. P. 244; holte (*dat.*) LAȝ. 826; CH. C. T. *A* 6; AV. ARTH. xix; holtes (*pl.*) TRIST. 378; PERC. 230; holtis M. ARTH. 3029.

holt-wode, sb., *O.E.* holtwudu; *wooded knoll*: into a forest, ful dep ... holtwodez (*pl.*) under GAW. 740; over hilles and hethes into holte woddes D. TROY 1350.

holu, *see* holh. hōm, *see* hām.

homāge, sb., *Fr.* homage; *homage,* ROB. 46.

homager, sb., *one who pays homage*; homagers (*pl.*) D. ARTH. 3147.

hóme, *see* háme. homelen, *see* hamelen.

homicīde, sb., *Fr.* homicīde; *homicide, murder*; that. . . vice of homicide GOW. I. 355.

homicīde, sb., *Lat.* homicīda; *homicide, murderer*: she that was an homicide GOW. I. 346; CH. C. T. *B* 1757, *E* 1994.

homme, *see* hamme.

hōn, v., *O.E.* hōn, = *Goth., O.H.G.* hāhan (*for* *hanhan); *hang,* LAȝ. 10009; hōþ (*pres.*) O. & N. 1123; hēng [*O.E.,* hēng = *O.Fris.* hēng, *O.H.G.* hiang, *O.N.* hēkk] (*pret.*) ORM. 7339; GEN. & EX. 3899; WILL. 734; ALIS. 6750; MAND. 93; he heng an his sweore ænne sceld deore LAȝ. 21149; þa heng heo hire hæfned 15688; hing 18374; heng (heeng) CH. C. T. *A* 160; hieng LEG. 23; hing RICH. 5053; hengen LAȝ. 5722; SHOR. 85; þær hio hine hengen LK. xxiii. 33; þe gostes þat þar on hengen LEG. 191; hengen, heengen LANGL. *A* i. 148; henge HAV. 2510; MAN. (F.) 3624; hengen GOW. II. 47; hangen [*O.E.* hangen, = *O.L.G., O.H.G.* (ar-, bi-, gi-) hangen, *O.N.* hanginn] (*pple.*) GEN. & EX. 4074; *comp.* a-, an-, bi-, ȝe-hōn; *deriv.* hangien, hengen, henge, hengel.

hōn[2], v., *? cease*: þou schuldist hoo of swering P. R. L. P. 195; hō (*pres.*) GOW. II. 103; hoo ALEX. (Sk.) 4437; hoo (*subj.*) E. T. 153; hō (*imper.*) B. DISC. 1938; ho [hoo] CH. C. T. *A* 1706; hoo ALEX. (Sk.) 2835.

hond, *see* hand *and* húnd.

honden, v., *? from* hand; *attend upon*: luk ȝe hondene (*pres. pl.*) theme alle that in mine oste lengez D. ARTH. 3209.

hondertīde, *see* undern-tide *under* undern.

hōne, sb., *O.E.* hān (LEO 584), = *O.N.* hein *f.*; *whetstone*; hoone '*cos*' PR. P. 245.

hōne[2], sb., *delay, ? cessation*: wiþouten hone L. H. R. 109; IW. 3667; C. M. 14316; boute hone GAW. 1285; hone, hoin BARB. vi. 564; hune, hone YORK xxv. 272.

[hōnen, v., *comp.* for-hōnen.]

hōnen², v., *? delay, cease*; hone A. P. i. 920; C. M. 5873; and þar þe her na langer hone TOWNL. 11; M. H. 129.

honest, adj., *O.Fr.* honeste; *seemly, splendid*: arai all þi cite ... stoutli & faire þat it be honest [onest] all over ALEX. (Sk.) 1496; honest & hol A. P. ii. 594.

honeste-līche, adv., *honourably*: þe more qvainteliche and þe more honestliche AYENB. 47; forto be keped honestli L. H. R. 76; as honestli to his degre CH. C. T. *E* 2026.

honestē, sb., *O.Fr.* honesté; *honesty, decorum*, TREV. IV. 25; BARB. i. 548.

honesten, v., *Lat.* honestāre; *honour*: to honesten the pore WICL. ECCLUS. xi. 23.

honestetē, sb., *honour, worthiness*: weddid with fortunat honestete CH. C. T. *E* 422.

honger, *see* hunger. hongien, *see* hangien.

honi, *see* huniȝ.

honissen, v., *O.Fr.* honir, hunir (*honnir*); *degrade*; honesschen [honische, hunsen] LANGL. *A* xi. 48; honisez (*pres.*) A. P. ii. 596; honishid [hunischi(s)t] (*pret.*) ALEX. (Sk.) 3004; honest (*pple.*) *brought to shame* ALEX. (Sk.) 3791.

honour, sb., *O.Fr.* honour; *honour, praise*, P. L. S. xvii. 508; CH. C. T. *A* 46; anour GUY 149; honouris (*pl.*) ALEX. (Sk.) 2121.

honourance, sb., *O.Fr.* honorance; *honour*: in honorance of Iesus Crist sitteþ stille ASS. *B* 1; honuraunce of swete Iesu LEG. 3; in the honouraunce of viftene salmes LEG. 75.

honourable, adj., *O.Fr.* honorable, *from Lat.* honōrābilis; *honourable*, GOW. II. 69; III. 225; honourable, honurable ALEX. (Sk.) 223, 1840.

honourabil-lī, adv., *with honour*, R. P. 15; honorabille BARB. xiii. 664; honurabloker (*comp.*) MAN. (F.) 4477.

honouren, v., *O.Fr.* honōrer; *honour*; honouri MARG. 104; SHOR. 95; honoure LANGL. *A* iii. 204; anuri MISC. 26; aouri AYENB. 135; anouriþ (*pres.*) P. L. S. 6; anourest SHOR. 96; honūrede (*pret.*) BRD. 33; anourede KATH. 32; anourede [*v. r.* honowride] WICL. GEN. xix. 1; anourene R. P. 21; honorid [honourd] (*pple.*) ALEX. (Sk.) 3160; honorit BARB. xvi. 672.

houte, hontien, *see* hunte, huntien.

hop, sb., *O.E.* hop, *cf. Sc.* hope; *valley*: hope ALEX. (Sk.) 5390; thorowe hópes (*pl.*) and himlande hillis D. ARTH. 2503.

hōp, hoop, sb., *O.Fris.* hōp, = *M.Du.* hoep *m.*; *hoop, 'circulus,'* VOC. 276; PR. P. 245.

hōp-ring, sb., = *M.Du.* hoepring; *? hoop ring*, HALLIW. 458.

hópe, sb., *O.E.* hopa *m.*, *cf. M.L.G.*, *M.Du.* hope (*gen.* hopen) *m. f.*; *hope*, LAȝ. 13899;

MARH. 8; A. R. 78; ORM. 3816; AYENB. 5; LANGL. *A* iii. 193; þet to þe habbeð hope HOM. I. 202; hop BARB. iv. 104; *comp.* over-, tō-, un-, wan-hope.

hópe-ful, adj., *hopeful*, A. R. 302.

hópien, v., *O.E.* hopian, = *M.L.G.*, *M.Du.* hopen, *M.H.G.* hoffen; *hope, expect*, A. R. 78; hopen KATH. 1151; hopie AYENB. 89; hope RICH. 3474; hópie (*pres.*) HOM. I. 197; BEK. 137; SHOR. 129; hope CH. C. T. *A* 4029; we hopeþ H. H. 165; hópe (*imper.*) LAȝ. 17936; hópede (*pret.*) HORN (L.) 1394; ROB. 33; hoped Iw. 1675; hopeden WILL. 4308.

hóping, sb., *hope*, ALEX. (Sk.) 4518.

hōpin, v., *hoop*; hoopin PR. P. 245.

[hoppe, sb., *O.E.* (gærs-)hoppa; *comp.* greshoppe.]

hoppe², sb., *cf. M.L.G.* hoppe, *O.H.G.* hopfo; *hop, 'humulus,'* PR. P. 245.

hoppe³, sb., *O.E.* hopp (*bulla*); *head of flax*, *'linodum,'* PR. P. 246; hoppen (*pl.*) *'boccaus'* VOC. 156.

hoppen, v., *O.E.* hoppian, = *M.Du.* hoppen, *O.N.* hoppa, *M.H.G.* hopfen; *hop, jump, dance*, H. M. 21; hoppe LANGL. *A* iii. 193; CH. C. T. *A* 4382; (h)oppieð (*pres.*) HOM. II. 211; hoppen (*subj.*) MISC. 91; hoppede (*pret.*) ROB. 278; hopped (*pple.*) MAN. (F.) 13509.

hoppere, sb., *cf. M.Du.* (gras-)hopper; *jumper, dancer*; (h)opperes (*pl.*) *locusts* GEN. & EX. 3096.

hoppestere, sb., *O.E.* hoppestre; *female dancer*; hoppesteres (*pl.*) CH. C. T. *A* 2017 [TEN BRINK *suggests that Chaucer's ms. of Boccaccio's Teseide had* ballatrici *instead of* bellatrici].

hoppere, sb., *hopper (of a mill)*; *seed-basket*, REL. I. 7; hop(p)er LANGL. *B* vi. 63; hop-(p)er CH. C. T. *A* 4039; hopre PALL. x. 43.

hōr, sb., *O.E.* hōr(-cwene), = *O.Fris.* hōr, *O.H.G.* huor; *fornication, adultery*: & drieð hordom & of þe hore fule stinkeð HOM. II. 37.

hōr-cop, sb., *bastard, 'spurius,'* PR. P. 246; TRIAM. 224.

hōr-dōm, sb., = *O.Fris.* hōrdōm, *O.N.* hōrdōmr; *fornication*, P. L. S. viii. 128; A. R. 204; SHOR. 59; MAN. (H.) 58; PR. C. 8259; horedom ORM. 4632; GEN. & EX. 3509.

hōr-ēie, sb., *lustful eye*, A. R. 204.

hōre-hous, sb., = *O.H.G.* huorhūs; *brothel*, *'lupanar,'* VOC. 274; PR. P. 246; S. S. (Web.) 1504.

hōre-man, sb., *fornicator*, GEN. & EX. 4072.

hōr-pleȝe, sb., *fornication*; horeplage GEN. & EX. 530.

hōr, *see* hār *and* hǣr.

hord, sb., *O.E.* hord, = *O.L.G.* hord, *O.H.G.* hort, *O.N.* hodd, *Goth.* huzd; *hoard, treasury*,

A. R. 224; SPEC. 54; AYENB. 185; þet hord
FRAG. 6; hord of gold ORM. 6732; hord of
apples CH. C. T. *A* 3262; legge on hord O. &
N. 1224; **horde** (*dat.*) LAȝ. 6077; GOW. III.
155; ileid an horde P. L. S. viii. 6; hurde PR.
C. 5567; **horde** (*gen. pl.*) FRAG. 7; ORM.
6733; *comp.* **appel-, góld-hord.**

hord-hous, sb., *treasury*, HALLIW. 459.

horden, v., *O.E.* hordian; *hoard*: & horden
þat tu winnest ORM. 12281; horde WICL. E.
W. 338; **hordeþ** (*pres.*) AYENB. 182; hordes
['*thesaurizat*'] PS. xxviii. 7.

hordere, sb., *O.E.* hordere; *hoarder,* SAX.
CHR. 260; hordare '*cellarius*' FRAG. 4; hor-
dier *treasurer* AYENB. 121.

horder-wice, sb., *treasurership*; **horder-
wican** (*dat.*) SAX. CHR. 263.

hóre, sb., *O.E.* horu (*gen.* horwes),=*O.L.G.*
horu, horo, *O.Fris.* hore, *O.H.G.* horo (*gen.*
horawes); *dirt, filth, mud,* HOM. II. 49; O. &
N. 596; P. L. S. xvii. 8; REL. II. 243;
AYENB. 229; B. DISC. 1477; H. V. 83; wiþoute hore FER. 5725.

hōre, sb., *O.E.* hōre, = *O.N.* hōra, *M.Du.*
hoere, *O.H.G.* huora; *whore,* LAȝ. 15579;
A. R. 54; H. M. 31; GEN. & EX. 4082; SHOR.
62; ALIS. 1000; E. T. 653; LUD. COV. 218;
houre ['*pellex*'] TREV. VII. 61; **hōren** (*dat.
pl.*) LAȝ. 29133.

hōre-sone, sb.,=*M.H.G.* huorensun; *bastard,* ALIS. 880; FER. 2016.

hōre, *see* **ūre.**

[**hōren,** v., *O.N.* hōra,=*O.H.G.* huorōn; *commit fornication; comp.* **for-hōren.**]

horȝ, sb., *O.E.* horg, horh; *filth, mud*; horuȝ
P. L. S. xxxv. 34; **horȝe** (*ms.* horie) (*dat.*)
HOM. I. 302; *see* **hóre.**

horȝen, v.,=*M.H.G.* horgen; *cover with filth,
defile*; horyen MISC. 92; **horegede** (*pret.*)
HOM. II. 201; **horwed** (*pple.*) A. P. ii. 335;
comp. **bi-horȝen.**

hori, adj., *O.E.* horig,=*M.L.G.* horeg, *M.H.G.*
horec, horwêc; *filthy, dirty,* HOM. II. 141;
REL. II. 176; P. L. S. v. (b) 13; hoori WICL.
LEV. xxii. 5; horie (*pl.*) LEB. JES. 418; wiþ
tonges horwe CH. COMI. MARS 206; '*turbo*'
TREV. VIII. 231.

horien, *see* **horȝen.**

horlen, *see* **hurlen.**

horlinge, *see* **hurling.**

hōrling, sb., *cf. O.H.G.* huorlinc (*bastard*); *fornicator,* ['*scortator*'] WICL. DEUT. xxiii. 17;
AYENB. 52; S. S. (Wr.) 2189; P. 6; **hōr-
linges** (*pl.*) P. L. S. viii. 52; R. S. vii.; P. S.
238.

horn, sb., *O.E.* horn *m.*, *cf. O.Fris., O.N., O.
H.G.* horn, *Goth.* haurn, *Lat.* cornu *n.*; *horn,*
LAȝ. 789, 4538; HORN (L.) 1109; P. S. 342;
þene horn A. R. 200; **horne** (*dat.*) O. & N.
318; horne (*hoof*) PALL. iv. 795; **hornes**

(*pl.*) MARH. 7; AYENB. 14; MAND. 47; horn
SAX. CHR. 256; **hornen** (*dat. pl.*) LAȝ.
1424.

horn-blāwere, sb., *O.E.* hornblāwere; *horn
blower,* SAX. CHR. 256.

horn-keke, sb., *a sort of fish,* PR. P. 247.

horn-pīpe, sb., *hornpipe,* PR. P. 247.

horned, pple., *cf. O.E.* hyrned; *horned,*
OCTOV. (W.) 1335; ALEX. (Sk.) 4267.

hornen, adj., *cf. O.H.G.* hurnīn; *made of horn*:
þe hornene trumpe WICL. PS. xcvii. 6.

horologe, sb., *O.Fr.* horlòge, *from Lat.* hōro-
logium; *sun-dial, timepiece,* VOC. 249; diale
or diel or an horlege (horlage, orlage) PR. P.
120; in the orloge (*later ver.* orologie) of
Achaz WICL. 4 KINGS xx. 11; the cok that
orlogge is of thropes lite CH. P. F. 350;
orrelegge DEGR. 1453.

horrible, adj., *O.Fr.* horrible; *horrible*: made
noise horrible FER. 3895; that ilke foul horrible vice of homicide GOW. I. 355; devels
horribel til mans sight PR. C. 5618; orrible
sinne CH. C. T. *I* 886.

horrible-lī, adv., *horribly*: how sculd i ...
sinne so horribli CH. C. T. *I* 881.

hors, sb., *O.E.* hors, = *O.Fris.* hors, *O.N.*
hross, *O.H.G.* hros (*gen.* hrosses); *horse,* A. R.
208; O. & N. 629; TOR. 1171; þet hors
AYENB. 140; **horses** (*gen.*) A. R. 74; **horse**
(*dat.*) REL. I. 180; on horse SPEC. 48; **hors**
(*pl.*) HOM. I. 49; II. 179; ORM. 8704; MAND.
38; CH. C. T. *A* 74; hors, horses LAȝ. 897,
3561; horss BARB. viii. 446; **horss** (*gen. pl.*)
BARB. ii. 359; **horsen** (*dat. pl.*) LAȝ. 1025;
horse HOM. I. 9; *comp.* **rōd-hors.**

hors-bak, sb., *horseback*: to horsbak went
thai TOR. 2565; forth he wente on horsbacke
GOW. III. 256; horsebake I. 260.

hors-bēre, sb., *O.E.* horsbær,=*Dan.* ros-
baare; *horse-litter,* PR. P. 247; MAN. (F.)
9605; hors-, horsebere ROB. 163, 165; horsebere LAȝ. 19431.

hors-brēde, sb., *horse-bread*: no baker shalle
bake horsbrede, kepinge osteri E. G. 376, *cf.*
406.

hors-cam, sb., *O.E.* horscamb; *horse-comb*;
horskam HALLIW. 461; horscomb VOC. 202;
TREV. IV. 25.

horse-charche, sb., *horse toll*: an halpeni
of custome ... and þe horsecharche a ferthinge
E. G. 358.

hors-cnave, sb., *groom*; horsknave P. S.
237; horseknave HAV. 1019.

horse-colt, sb., *colt,* WICL. ECCLUS. xxiii. 30.

hors-fleeȝe, sb., *horse-fly,* WICL. JOSH. xxiv.
12.

hors-hēr, sb., *horsehair,* P. L. S. xvii. 158.

hors-h(e)orde, sb., *O.E.* horshyrde; *horse-
keeper, innkeeper,* HOM. I. 79.

hors-hūs, sb., *inn,* HOM. I. 85.

hors-kēpare, sb., *horse-keeper* or *inn-keeper*, '*eqvarius*,' PR. P. 447.

hors-lī, adj., *like a [living] horse*: [the hors of bras] so horsli and so qvik of ye CH. C. T. *F* 194.

hors-lōde, sb., *horse-load*, E. G. 358.

hors-lāf, sb., *horseloaf*; **horselōfis** (*pl.*) E. G. 337.

hors-man, sb., *horseman*, '*eqvester*,' PR. P. 248; **horsmen** (*pl.*) LA჏. 26617.

hors-minte, sb., =*M.H.G.* rosminze; *horsemint*, '*mentastrum*,' PR. P. 248; REL. I. 36.

hors-monger, sb., *horse dealer*, OCTOV. (W.) 836.

hor(s)-schō, sb., *horse-shoe*, PR. P. 248.

hors-þistel, sb., *endive*, LEECHD. III. 333.

hōrs, *see* **hās.**

horsen, v., *horse, provide with a horse*, MAN. (F.) 11794; how we can hors oure king TOWNL. 218; Pollux ... horsit (*pret.*) him in haste D. TROY 1280; horsid TOR. 2165; M. ARTH. 87; **horsede** (*pple.*) HALLIW. 460; horsed LIDG. M. P. 3; horset D. TROY 6468; horsutte AV. ARTH. xxxviii; ihorsed PARTEN. 886.

horsing, sb., *cavalry*: he hade no horsing AV. ARTH. xxxi; bi noumbre of horsinge arered WICL. DEUT. xvii. 16.

hortling, *see* **hurtling.**

horuþe, horþe, sb., *from* **hori**; *filth*, '*sordes*,' WICL. DEUT. vii. 26.

horwen, *see* **horჳen.**

hōs, *see* **hās.**

hóse, sb., *O.E.* hose, =*M.Du.* hose, *O.H.G.* hosa, hose (*pl.* hosun), *O.N.* hosa *f.*; *hose*, '*caliga*,' VOC. 196, 259; PR. P. 248; LA჏. '15216; ROB. 125; TRIST. 1486; GOW. III. 236; **hósen** (*pl.*) A. R. 420; TREV. V. 369; MAND. 59; CH. C. T. *A* 456; PL. CR. 426; hosen and shon HAV. 969; hosin W. & I. 32; hosen (hoses) ROB. (W.) 8013.

hósen, v., *from O.Fr.* hoser; *put on hose*; hosun '*caligo*' PR. P. 248; hose CATH. 189; **hósed** (*pple.*) pie SPEC. 111; cloþed, osed and shod HAV. 971.

hósiinge, sb., *hose*: ine cloჼinge, and in hósiinge (*dat.*) and in ssoinge AYENB. 154.

hósier, sb., *hosier, hose maker*, '*calicator*,' CATH. 189; hoseare [hoseჳere, hosiare, hoser] PR. P. 248.

hōsel, hōslen, *see* **hūsel, hūslen.**

hospital, sb., *O.Fr.* hospital; *hospital*, BEK. 84; TREV. IV. 465; E. G. 350.

hospitaler, sb., *O.Fr.* hospitalier; *hospitaller*: the ile of Rodes, the whiche ile **hospitaleres** (*pl.*) holden MAND. 26; or prest, or hospitalers CH. C. T. *I* 891; templers and hospitelers R. R. 6694; hospitleres MAND. 81; **hospitelers** (*gen. pl.*) MAN. (H.) 178.

hóst, sb., *O.Fr.* host (*armée*); *host*, **army**, SHOR. 108; MAN. (F.) 3238; ost WILL. 3767.

hóst², sb., *Lat.* hospitium; *hotel, lodging*: an host ... to be eside inne GEST. R. 257; the host or herbore WICL. DEEDS xxviii. 23; an ooste, or hous to dwelle inn WICL. PHILEM. 22.

hóst, hoiste³, sb., *Lat.* hostia; *sacrifice, host (in the mass)*: how God is put in the holi host S. & C. II. xxv; hoste '*hostia*' CATH. 190; þe oost ... is maad goddis bodi WICL. E. W. 357; oste 345; that offreth of ჳou an oost to God WICL. LEV. i. 2; a livinge oost WICL. ROM. xii. 1; hoist TREV. V. 9; to offre spiritual **hoostes** (*pl.*) WICL. I PET. ii. 5.

hostāge, sb., *O.Fr.* hostāge; *hostage*, LA჏. 5317*.

hostager, sb., = **hostāge**; **hostagers** (*pl.*) MAN. (F.) 4983.

hostaien, v., *O.Fr.* hostoier, ostoier; *lead an army; wage war*: to hostaie in Almaine D. ARTH. 555.

hóste, sb., *O.Fr.* hoste; *host, entertainer*, BEK. 1208; hoste [oste] CH. C. T. *A* 747.

hóstes, *see* **hóstesse.**

hōst(e), sb., *O.E.* hwōsta, =*O.N.* hōsti, *M.L.G.* hōste, *O.H.G.* huosto; *cough*, '*tussis*,' VOC. 224; PR. P. 248.

hostel, sb., *O.Fr.* hostel; *hostel, hotel*, GEN. & EX. 1397; GAW. 805.

hostelen, v., *O.Fr.* hosteler; *provide with lodgings*: hope shal ... hostel hem LANGL. *B* xvii. 118; ther **hostild** (*pret.*) thai alle thre TOWNL. 289; boþe þei weoren **hostelled** (*pple.*) þere S. R. 548.

hostelēr, sb., *O.Fr.* hostelier, *mod.Eng.* hostler, ostler; *innkeeper*, CH. C. T. *A* 241; hostiler TREV. VIII. 91.

hostellerie, sb., *O.Fr.* hostellerie; *hostelry*, CH. C. T. *A* 23.

hosterie, sb., *hostelry, hotel*: ordaine for me an honest hosteri GEST. R. 315; he ... herberwed him at an hostrie LANGL. *B* xvii. 73; hostrie E. G. 376; calle to gestening or to osteri all þat went bi the wei GEST. R. 19; ostri 90.

hóstesse, sb., *O.Fr.* hostesse; *hostess*, S. A. L. 151; hostes BARB. iv. 635.

hostiari, sb., *Lat.* ostiārius; *doorkeeper*, ['*ostarius*'] TREV. V. 97.

hōstin, v., *O.N.* hōsta, = *M.L.G.* hōsten, *O.H.G.* huosten; *cough*, '*tussio*,' PR. P. 249.

hōt, *see* **hāt.**

hōten, v., *hoot*: hot him ut A. R. 290; **hōtend** (*pple.*) WILL. 2387; *see* **hūten.**

hōten, *see* **hāten.**

hotte, sb., *O.Fr.* hotte; *mod.Eng.* hod; *basket (to carry on the back)*; hott C. M. 5524; **hottes** (*pl.*) (*mss.* hattes) CH. H. F. 1940.

hotte², sb., (*? meaning*): þi nek, þi hotte, þe

develle it breke MAN. (H.) 282; hodred in
þer hottes (*pl.*) 273.

hou, *see* hwu.

houe, sb., *O.Fr.* houe; *hoe*; houes (*ms.*
howes) (*pl.*) FER. 4993.

hou3, sb., *? O.N.* haugr,=*M.H.G.* houc; *hill*;
hogh C. M. 15826; hoo YORK iv. 36; hōes
(*pl.*) ANT. ARTH. v.

hou3, *see* hōh. houlen, *see* hūlen.

hound, *see* hund.

houpen, v., *Fr.* houper; *whoop, cry out*;
houpide, houped, hoped (*pret.*) LANGL. *A* vii.
159, *B* vi. 174; houped CH. C. T. *B* 4590.

hour(e), *see* ūre.

hourschen, v., *? rush*: alle hoursches (*pres.*)
over hede harmes to wirke D. ARTH. 2110.

hous, *see* hūs.

housel, *see* hūsel. houten, *see* hūten.

houve, sb., *O.E.* hūfe,=*O.N.* hūfa, *M.L.G.*
hūve, *O.H.G.* hūba; *cap, head-dress, 'cidaris ;'*
(*ms.* howe, howue) PR. P. 249; (*ms.* houue)
ANGL. I. 82; (*printed* honne) P. S. 327; a
silk houve LANGL. *A* iii. 276 (*B* iii. 293
howue, *C* iii. 451 houe); though i sette his
houve CH. C. T. *A* 3911.

hove, sb., *? dregs of oil, 'amurca,'* PR. P. 250.

hōve, sb., *O.E.* hōfe; *ground ivy, '(h)edera
terrestris,'* PR. P. 250; *comp.* hei-, tūn-hōve.

hovedaunce, *see under* hof.

hovel, sb. *hovel, hut,* W. & I. 19; hovil PR. P.
250.

hóven, v.,=*M.Du., M.L.G.* hoven; *inhabit,
tarry*; hovin PR. P. 251; hove GOW. II. 370;
TRIAM. 1471; LIDG. M. P. 4; whi nill þou over
us hove CH. TRO. iii. 1427; he hóveð (*pres.*)
in ʒe sunne REL. I. 210; and hoveþ þer a
stunde TREAT. 136; hoves AMAD. (R.) xlvi;
PR. C. 7579; hovez A. P. ii. 458; hoven JOS.
489; hóvande (*pple.*) PERC. 533; howand
(*for* hovand) BARB. xv. 461*; hóvede
(*pret.*) ROB. 172; TREV. VIII. 123; he
hovede and abod S. S. (Wr.) 2825; and
hoveden on heiʒe over þe lake ALIS. 5445;
hoved LANGL. *B* PROL. 210; GAW. 2168;
þai hoved on þe flode MIN. iii. 83; thei hovid
& bihelde M. ARTH. 259.

hover, *see* hofer. how, *see* hō *and* hwu.

howe, *see* hoʒe. hōwen, *see* hāwen.

hox, sb.,=houʒsenu (*q.v. under* hōh): David
kitte the hoxes (*pl.*) of all drawinge beestis
WICL. 2 KINGS viii. 4*.

hoxen, *see* hoxenen *under* hōh.

hr-, *see* hr- (*in letter* r).

hū, sb., *O.Fr.* hu; *hue (and cry); cry, clamour,* MAN. (F.) 11984*; a hue from heven i
herde A. P. i. 873.

hu, *see* hwu.

huche, sb., *O.Fr.* huche, *med.Lat.* hutica;
hutch, box, 'cista, arca,' PR. P. 252; BER.
2510; L. C. C. 33; hucche LANGL. *B* iv. 116;
arke or hucche MAND. 85; *see* hwicche.

hucken, *see* hukken.

hūde, sb., *O.E.* hȳd,=*O.Fris.* hūd, hēd, *O.L.
G.* hūd, *O.N.* hūð, *O.H.G.* hūt; *hide, skin,*
MARH. 18; þe hude þe wæs of þare hinde
LA3. 1213; softe as is ... wummon(n)e hude
A. R. 120; þine hude O. & N. 1114; huide
K. T. 752; of huide ne of hewe L. N. F. 218;
hide TRIST. 1451; PR. C. 5299; MIR. PL.
151; hur sone was ... feir of hide and hewe
TRIAM. 468.

hūdels, sb., *O.E.* hȳdels; *hiding place*; hu-
dels, huidels, hidels ['*latebra*'] TREV. I. 199,
V. 117; hudles A. R. 146; an hudlese LA3.
1817; in hidils [hudlis] ['*in abscondito*']
WICL. DEUT. xxvii. 15; hidlis D. TROY 12304;
hidels, hidils HAMP. PS. ix. 30; hidil HAMP.
PS. (*p.* 511); hiddillis BARB. vi. 382.

hidel-līke, adv., *? secretly*, GEN. & EX.
2882.

hūden, v., *O.E.* hȳdan,=*M.L.G.* hūden; *from
hūd*; *hide, conceal*, A. R. 130; huide K. T.
305; MIRC 1105; hiden ORM. 1019; hide
CH. C. T. *A* 1481; M. H. 54; M. ARTH. 110;
hede AYENB. 44; hūde (*pres.*) O. & N. 265;
hut A. R. 130; ʒef me hut ant heleð hit MARH.
15; þeo þet hudeð ham A. R. 174; hidis ALEX.
(Sk.) 504; hīden (*pl.*) ALEX. (Sk.) 3214;
hūd (*imp.*) A. R. 292; hŭdde (*pret.*) LA3.
8586; KATH. 912; ROB. 226; ALIS. 2489;
hudde, hidde TREV. V. 153; LANGL. *B* xvii.
108; hudden JOS. 13; hidden ORM. 13736;
GEN. & EX. 3028; hŭd (*pple.*) P. L. S. viii.
39; FER. 1400; hid CH. C. T. *E* 1944; hed
WILL. 688; *comp.* bi-, for-, ʒe-hūden.

hūding-clōð, sb., *veil, curtain*: þat huding-
cloþ to delde in the temple a to MISC. 50.

hūdnesse, sb., *O.E.* gehȳdness; *retreat*;
hidnes MAN. (H.) 77.

hūdunge, sb., *hiding*, A. R. 174; hudinge
P. L. S. ix. 53; hedinge AYENB. 196.

hue, *see* hēo.

[hŭen, v., *O.Fr.* huer; *set up a hue and cry;
see* hū.]

hūing, sb., *clamour*, O. & N. 1264.

hūen, *see* hēawen. huerte, *see* heorte.

hūf, sb., *? O.N.* hūfe; *? side of a ship*: i mai
touch with mi hufe the ground evin here
TOWNL. 32.

hufen, v., *?=*hóven; *remain*: long shalle thou
hofe TOWNL. 32; ʒif þou hufe (*pres.*) alle þe
daie D. ARTH. 1688.

húge, adj., *O.Fr.* ahuge; *huge, 'magnus,'*
PR. P. 255; WILL. 2569; FER. 546; TREV.
III. 183; LANGL. *B* xi. 242; MAND. 45; CH.
C. T. *A* 2951; A. P. ii. 1659; houge TOR.
550; hoge [huge] ALEX. (Sk.) 821, 1062; hó-
gere [huger, hugir] (*comp.*) ALEX. (Sk.) 3047,
1368.

húge-liche, adv., *hugely, largely, greatly*;
hugeliche ROB. 482; hugeli WICL. GEN. xvii.

2; WICL. DEUT. ix. 20; hogeli, hugeli ALEX. (Sk.) 269, 3226*.

húgenis, sb., *bulk, greatness*: of strengþe, of schap, of hugenis FER. 51.

hüȝe, sb., *O.E.* hyge, *cf. O.L.G.* hugi, *O.H.G.* hugu, *O.N.* hugr, *Goth.* hugs *m.* (νοῦς); *mind*, LAȝ. 4910; huiȝe, huie 2337, 3033; hige & mihte HOM. II. 119; hiȝ(e) & hope ORM. 2777.

hüȝien, v., *O.E.* hycgan,=*Goth.* hugjan, *O. L.G.* huggien, *O.H.G.* huggan, *O.N.* huga, hyggja; *think, meditate*; **hügiende** (*earlier text* hogiende) (*pple.*) MAT. vi. 34; *comp.* for-, ofer-hüȝien.

hüht, sb., *O.E.* hyht,=*O.H.G.* huht (*feeling*); *hope, joy, delight*, HOM. I. 97; hiht ORM. 3816; hiht & hope HOM. I. 217; agon is al min hiȝt FER. 2782; hihte [hiȝte] O. & N. 272; mid gode huhte HOM. I. 109; hiȝt ALEX. (Sk.) 5313.

hiht-lích, adj., *O.E.* hyhtlíc; *joyous*, HOM. II. 213; hiȝtlí (*adv.*) *fitly* GAW. 1612; *comp.* un-hühtlíc.

hühten, v., *O.E.* hyhtan; *hope, rejoice*; hihteþ [hiȝteþ] (*pres.*) O. & N. 436.

hühten, *see* hihten.

huire, *see* hüre.

hüke-nebbide, *see under* hóc.

húke, heuke, sb., *? M.Du.* huike, *O.Fr.* huque; *cloak*, PR. P. 232.

hukel, sb., *cf. O.Sw.* hukli (IHRE); *cloak*: under vreondes huckel A. R. 88.

hukken, v., *hawk, sell*: to merchaunt and huk '*auccionor*' PR. P. 252*.

hukkerie, sb., *cf. Ger.* höckerei; *huckster's trade*; hokkerie [hukkerie, hukri] LANGL. *B* v. 227.

hukstere, sb., *Du.* heukster; *huckster*, PR. P. 252; hucster '*institorem*' VOC. 123; hokester REL. II. 176; **hucsteres** (*gen.*) ORM. 15817; **hoksters** (*pl.*) TREV. II. 171.

huksterie, sb., *huckster's trade*; hoxterie LANGL. *A* v. 141.

hül, sb., *O.E.* hyll, *?=M.Du.* hil; *hill*, FRAG. 3; A. R. 178; BEK. 1717; ænne hul LAȝ. 1645; hil GEN. & EX. 1293; HAV. 1287; RICH. 6048; (*ms.* hill) ORM. 9205; hel WILL. 2233; **hülle** (*dat.*) P. S. 152; ALIS. 2550; helle AYENB. 5; **hülles** (*pl.*) LAȝ. 5191; helles TREV. I. 399; hulle P. L. S. viii. 175; **hülle** (*dat. pl.*) HORN (L.) 208; *comp.* ămete-, dung-, muk-hül (-hil).

hulc, sb., *O.E.* hulc; *shanty, hut, '*tugurium*,'* FRAG. 4; hulke WICL. IS. i. 8; holke VOC. 178; **hulke** (*pl.*) MAN. (F.) 8288.

hülden, v., *O.E* (be-)hyldan, *O.N.* hylda; *flay, skin*; hulde WILL. 1708; hilde FER. 1639; **hilden** (*pret.*) LAȝ. 20957*; hildiden WICL. MIC. iii. 3; **hüld** (*pple.*) TREV. VIII. 167*; ihuld S. A. L. 155.

hildinge, sb., *skinning*, TREV. II. 359.

hule, sb., *O.E.* hule; *husk, pod*: þese hule [hole LANGL. *B* vii. 194; hole '*siliqua*' PR. P. 242; **hulis** (*pl.*) WICL. S. W. II. 71; holes II. 69; *deriv.* hólen, hullen.

hul-wurt, sb., *fleawort, '*pulegium*,'* REL. I. 36.

hule², sb., *hut, shelter*, REL. I. 214; **hulen** (*dat. pl.*) A. R. 100.

hülen, v.,=*Goth.* huljan, *O.L.G.* (bi-)hullean, *O.H.G.* hullen (*pret.* hulta), *O.N.* hylja; *cover, conceal*: hule & huide HOM. I. 279; **hillin** '*operio*' PR. P. 240; hille DEP. R. iii. 326; **hüles** (*pres.*) A. R. 150*; WILL. 97; hilles IW. 741; **hüle** (*imper.*) MIRC 1872; **hilde** (*pret.*) AM. & AMIL. 2302; **hüled** (*pple.*) REL. I. 39; hulet A. R. 388*; ihulet. LANGL. *A* vi. 80; *comp.* un-hülen.

hillinge, sb., *covering, '*operimentum, tegumentum*,'* PR. P. 240; ANT. ARTH. ix.

hülen, v.,=*M.L.G.* hülen, *M.Du.* huilen; *howl*; houlin '*ululo*' PR. P. 250; houle GOW. II. 265; **houleþ** (*pres.*) CH. C. T. *A* 2817.

huler, sb., *O.Fr.* houlier (*homme débauché*); *lecher*, H. M. 31; holier AYENB. 51; **holers** [holours] (*pl.*) ROB. (W.) 624.

hulke, sb., *great awkward fellow*, D. ARTH. 1058, 1085.

hulke², sb., *O.E.* hulc, *cf. M.Du.* hulke, *M.L.G.* holk *m.*, holke *f.*, *O.H.G.* holcha, *from Gr.* ὁλκάς; *sort of ship, '*hulcus*,'* PR. P. 252.

hulken, v., *from* hulc; *hide*; **hulked** (*pret.*) MAN. (F.) 15888.

hullen, v., *peel, become peeled*: take wete, brai hit a litelle, with water it spring til hit **hulle** (*subj.*) L. C. C. 7; *see* hólen.

hülli, adj., *hilly*; hilli PR. P. 240; TREV. I. 333.

huls, *?*v., *cf. M.Du.* hulse, *O.H.G.* hulsa (*pod*); *? gather pods*: everi puls . . . is hervest nowe to huls PALL. vii. 56.

hülstren, v., *from O.E.* heolstor (*hiding-place*); *hide, conceal*: there i hope best to **hülstred** (*pple.*) be R. R. 6149.

hült, *see* hilt.

hulvir, sb., *O.N.* hulfr; *holly*, PR. P. 253; holvir S. & C. II. xl; **hulfere** (*dat.*) COMP. BL. KN. 129.

humanitē, sb., *Fr.* humanité; *humanity*, MIRC 457.

human-lí, adv., *humanely*: the goudwif ful humanli . . . gave gounis S. & C. II. lvii.

humble, adj., *O.Fr.* humble; *humble*, CH. C. T. *E* 949.

humble-hēd, sb., *humble position*: from humblehede [*v. r.* humble bedd] to roial mageste CH. C. T. 16157 (W.)

humble-líche, adv., *humbly*; humbliche FER. 1041; humelich & faire 2050; **humeli,** humilli BARB. iii. 762, xviii. 404.

humblenesse, sb., *humility* : the apostle for humblenesse . . . sette not his name to-fore WICL. HEB. PROL.

[**humble,** sb.,= *Ger.* hummel, *Du.* hommel.]

hombul-bē, sb., *cf. M.L.G.* hummelbee ; *humblebee,* REL. I. 81.

[**humblen,** v., = *Du.* hommelen ; *make a booming sound.*]

humbling, sb.,= *M.Du.* hommeling ; *rumbling* : like the last humblinge after a clappe of oo thundringe CH. H. F. 1039.

humblesse, sb., *O.Fr.* humblesse ; *humility* : with lowe herte humblesse sue GOW. I. 118 ; humblesse and vertu CH. H. F. 630.

Humbre, pr.n., *O.E.* Humbre ; *Humber* : he ferde over þe Humbre LAȝ. 3822 ; bi þare Humbre 3785.

humectate, pple., *made wet,* TREV. I. 267.

humiliaciōn, sb., *Fr.* humiliation ; *humiliation,* CH. C. T. *I* 480.

humilitē, sb., *O.Fr.* humilité ; *humility,* SHOR. 117 ; CH. C. T. *B* 1665.

hummen, v.,= *Ger.* hummen ; *hum* ; humme CH. TRO. ii. 1199 ; **humme** (*pres.*) PALL. vii. 124.

humour, sb., *O.Fr.* humor ; *humour,* CH. C. T. *A* 421 ; **humours** (*pl.*) AYENB. 129.

hūn, sb., *cf. M.Du.* hūn, *O.N.* hūnn ; *top of a mast* : seil heo droȝen to hune LAȝ. 28978.

húnd, sb., *O.E.* hund,= *O.L.G.* hund, *O.Fris.* hund, hond, *O.N.* hundr, *Goth.* hunds, *O.H.G.* hunt ; *hound, dog, 'canis,'* VOC. 219 ; A. R. 324 ; MARH. 6 ; O. & N. 817 ; HAV. 1994 ; P. S. 197 ; hund [hond] LAȝ. 26762 ; he is an haðene hund 16623 ; an hund (*man, in a contemptuous sense*) him gan bihelde HORN (L.) 601 ; hond AYENB. 55 ; OCTOV. (W.) 1530 ; hound BRD. 6 ; WILL. 10 ; LANGL. *A* xi. 48 ; **húndes** (*gen.*) O. & N. 822 ; **húndes tunge,** *hound's tongue (a plant),* REL. I. 37 ; **húnde** (*dat.*) O. & N. 814 ; **húndes** (*pl.*) A. R. 324 ; ORM. 7405 ; hondes MAN. (H.) 75 ; **húnden** (*dat. pl.*) MAT. vii. 6 ; LAȝ. 1424 ; *comp.* **grāi-, helle-, sē-hund.**

hound-berie, sb., *morel,* LEECHD. III. 333.

hound-fisch, sb., *hound-fish,* CH. C. T. *E* 1825 ; houndfish LIDG. M. P. 201.

hound-flēȝe, sb., *dog-fly,* WICL. PS. lxxvii. 45 ; hundfleghe HAMP. PS. lxxvii. 50 ; hundflee C. M. 5956.

hund, card. num., *O.E.* hund,= *O.L.G., Goth.* hund, *O.H.G.* hunt, *Lat.* centum ; *hundred* : nigon hund ȝeare HOM. I. 225 ; hund þousunt LAȝ. 83.

hund-fald, adj., *hundredfold,* HOM. I. 147.

hundred, card. num., *O.E.* hundred, = *O.Fris.* hundred, hunderd, *O.N.* hundraȝ, *O.H.G.* hundert ; *hundred,* A. R. 42 ; O. & N.

1101 ; P. S. 344 ; hu he sette hundred LAȝ. 31998 ; an hundred (*ms.* hunndredd) ORM. 4333 ; twa hundred LAȝ. 1556 ; fower hundred ORM. 4321 ; hondred ROB. I ; AYENB. 55 ; JOS. 476 ; hundreþ MAN. (H.) 35 ; hundreth PR. C. 4524 ; hondered, hundered ROB. (W.) 450, 10646 ; hunder BARB. xiv. 67*.

hundred-feald, adj., *hundredfold,* HOM. II. 227 ; hundredfald HOM. I. 163 ; ORM. 19903.

[**hund** (? **hūnd**) ; (*? meaning*).]

hund-limen, sb. pl., *? servile limbs* : ilches mannes hundlimen alle swinkeȝ HOM. II. 181 ; mid foten and mid honden & mid alle here hundlimes II. 179.

[**hund-,** prefix, *O.E.* hund-,= *O.L.G.* ant-, *O.Fris., M.Du.* t-, *Goth.* tēhund ; *prefix to the 'tens' in numerals.*]

hund-seventi, card. num., *O.E.* hundseofontig ; *seventy,* HOM. II. 51 ; huntseventi REL. I. 173.

húnding, sb., *cynocephalus* : **houndinges** (*pl.*) men clepeth hem . . . from the brest to the ground men hi ben, aboven houndes ALIS. 4962.

[**hūne,** sb., *O.E.* hūne ; *horehound* ; *comp.* **hōrehūne** (*under* **hār**).]

hūnen, *see* **hēnen.**

hunger, sb., *O.E.* hungor,= *O.L.G., O.H.G.* hungar, *O.Fris.* hunger, honger, *O.N.* hungr, *Goth.* hūhrus ; *hunger,* LAȝ. 4042 ; KATH. 1702 ; ORM. 5682 ; LANGL. *B* vi. 323 ; honger ROB. 378 ; H. H. 50 ; AYENB. 75 ; **hungre** (*dat.*) A. R. 260 ; CH. C. T. *B* 100.

hunger-storven, pple., *perished with hunger* : min eie wolde, as though he faste, ben hungerstorven GOW. III. 28.

hüngren, hungren, v., *O.E.* hyngran, *cf. O.L.G.* (ge-)hungrean, *O.H.G.* hungeren, *O.N.* hungra, *Goth.* huggrjan ; *hunger,* A. R. 214 ; him hüngreð (*pres.*) REL. I. 220 ; us hungreth HAV. 455 ; whan þe hungreþ LANGL. *B* xiv. 49 ; þa þe hingreð MAT. v. 6 ; him hüngrede (*pret.*) A. R. 162 ; ISUM. 557 ; *comp.* **a-, an-, for-, of-hüngren.**

hungerere, sb., *hungry person* : voide he shal make the soule of the hungrere WICL. IS. xxxii. 6.

hungriȝ, adj., *O.E.* hungrig, = *O.H.G.* hungarīg ; *hungry,* ORM. 6162 ; hungri MISC. 154 ; H. H. 87 ; hungri gere GEN. & EX. 2136 ; hungri time CH. BOET. i. 4 (15) ; **hungrie** (*pl.*) FRAG. 6.

huniȝ, sb., *O.E.* hunig,= *O.Fris.* hunig, *O.L.G.* honeg, *O.H.G.* honig, honag, honang, *O.N.* hunang ; *honey,* ORM. 9225 ; huni A. R. 404 ; H. M. 9 ; PL. CR. 726 ; honi C. L. 78 ; SHOR. 90 ; MAND. 251 ; þet honi AYENB. 136.

huni-comb, sb., *O.E.* hunigcamb ; *honey-comb, 'favus,'* MISC. 54 ; honicomb WICL. I KINGS xiv. 27.

honi-drope, sb., *honey-drop,* LEG. 65 ; S. J. 5.

honi-socle, sb., *honeysuckle,* PR. P. 245;

hunisuccles (*pl.*) REL. I. 37 ; ? honisoukis WICL. MAT. iii. 4, MK. i. 6.

huni-tiar, sb., *O.E.* hunigtēar ; *drop of honey,* HOM. I. 217 ; huniter HICKES I. 167.

huniʒen, v.,=*M.H.G.* honigen; *honey* : wiþ honi of hevene **ihonied** (*pple.*) swete S. A. L. 70.

hünne, *see* heonene.

hunte, sb., *O.E.* hunta ; *hunter,* HOM. II. 209; ORM. 13471 ; CH. C. T. *A* 2018; hunte [honte] LAʒ. 21337 ; **hunten** (*pl.*) LAʒ. 2590; REL. I. 120; huntes SAX. CHR. 256 ; huntes [hontes] TREV. VII. 357 ; of þe **huntę** (*gen. pl.*) grune ['*de laqueo venantium*'] HOM. II. 209.

Huntin-dūne, pr. n., *O.E.* Huntan dūn ; *Huntingdon,* MISC. 146 ; Hontindone ṚOB. 4.

Huntendone-schīre, pr. n., PROCL. 1.

hunteð, sb., *O.E.* huntoð ; *hunting,* HOM. II. 209 ; honteþ ṚOB. 283.

huntien [honti], v., *O.E.* huntian; *hunt,* LAʒ. 2586; hunten SAX. CHR. 256 ; ORM. 13460; MISC. I ; CH. C. T. *A* 1674; honti, honte ṚOB. 16; honte DEGR. 50; hount BARB. vii. 399 ; **hunteð** (*pres.*) HOM. I. 203 ; ʒe huntieð [honteþ] LAʒ. 1432; **huntand** (*pple.*) BARB. xx. 21 ; **huntede** (*pret.*) A. R. 344; honted (*pple.*) P. S. 150.

huntere, sb., *hunter,* GEN. & EX. 1481; huntare PR. P. 253.

huntunge, sb., *O.E.* huntung; *hunting,* A. R. 330; huntinge LAʒ. 21342; hunting GREG. 717; hontine BARB. iv. 513.

hüpe, sb., *O.E.* hype, *cf. M.Du.* hupe, heupe, *Goth.* hups *m.*, *O.H.G.* huph, huf *f.*; *hip,* A. R. 280; hipe PR. P. 241 ; hipe [hippe] WICL. GEN. xlvii. 29 ; hippe GOW. II. 159 ; hepe VOC. 179; P. S. 329; **hüpes** (*pl.*) ṚOB. 322 ; ALIS. (Sk.) 190; hipes [hippes] CH. C. T. *A* 472 ; hepis MED. 624; hepes ṚOB. (W.) 6580*.

hüpe-bān, sb., *hip-bone,* ATHENÆUM 2467 (*Roy. MS.* 4, A. xiv. fol. 106 *b*) ; hepeboon HALLIW. 445.

hippe-halt, adj., *lame in the hips,* GOW. II. 159.

hüpel, sb., *O.E.* hypel ; *little heap,* FRAG. 2 ; hipil WICL. GEN. xxxi. 47 ; **hüples** (*pl.*) TREV. IV. 137*; huppels I. 179*.

hīpil-mēlum, adv., *in heaps, summarily,* ['*acervatim*'] WICL. WISD. xviii. 23.

hüppen, v.,=*M.L.G.* huppen, *M.H.G.* hüpfen; =**hoppen**; *bound, jump*; huppe ṚOB. 537; JOS. 14 ; **hüppe** (*pres.*) SPEC. 38 ; hupþ O. & N. 379; (þei) huppe [hippe] LANGL. *B* xv. 557; **hüpe** (*imper.*) SPEC. 111 ; **hipping** (*pple.*) FLOR. 1993; **hüpte** (*pret.*) O. & N. 1636; ṚOB. 208 ; BRD. 8; *comp.* over-hüppen.

hürcheoune, *see* irchoun.

hurde, sb., = *M.Du.* hurde, *O.H.G.* hurd, *O.N.* hurð, *Goth.* haurds (θύρα) ; *hurdle,*

grating ; **hurdes** (*pl.*) D. TROY 13459 ; hurdis MIN. x. 14; GOLAG. & GAW. 470.

hurd-reve, sb., '*centauria,*' REL. I. 36.

hürde, *see* heorde.

hürdel, sb., *O.E.* hyrdel ; *hurdle* ; hirdil '*crates*' PR. P. 241 ; herdil VOC. 279; **hürdles** (*pl.*) ALIS. 6104; herdles ṚOB. 232.

hurdice, sb., *O.Fr.* hurdeis, *med.Lat.* hurdicum ; *palisade* : hurdice or hustilment PR. P. 253 ; one hindire hurdace one highte helmede knightez D. ARTH. 3626; with targes and **hurdices** (*pl.*) thee gregeis heom wried ALIS. 2785.

hure, *see* hwure.

húre, sb., *? O.Fr.* hure (*head*) ; *skull-cap,* '*pileus,*' HALLIW. 470; '*tena*' PR. P. 252 ; BEK. 2099; DAV. DR. 59; an old cherl in a blake hure P. S. 156.

hūre, sb., *O.E.* hȳr, = *O.Fris.* hēr, *M.L.G.* hūre, *M.Du.* huere ; *hire,* LAʒ. 31110 ; SPEC. 42; hure, huire A. R. 404, 428 ; hure [huire] LANGL. *A* ii. 91 ; huire H. M. 7; MAND. 265; hire HAV. 910; MIN. vii. 66; for here swinc hire he nu haven GEN. & EX. 3172; hire or mede WICL. JOHN iv. 36; here ALIS. 5221 ; *comp.* hous-, schip-hire.

hūr-mon, sb., *O.E.* hȳrmann ; *one hired,* '*mercenarius,*' FRAG. 5.

hūre[2], sb., *cf. M.Du.* huere ; *courtesan* : wit hire no wolde (he) leike ne lie HAV. 997.

hūren, v., *O.E.* hȳrian, = *M.Du.* hueren, *M.L. G.* hūren, *O.Fris.* hēra ; *hire,* A. R. 126 ; hirin '*conduco*' PR. P. 241 ; hire A. P. i. 506; LIDG. M. P. 43 ; here MISC. 33 ; **hūrede** (*pret.*) HORN (L.) 752; BEK. 1173; herde MAT. xx. 7; hureden LAʒ. 30696; **huired [hired]** (*pple.*) LANGL. *B* vi. 116 (Wr. 4024) ; *comp.* ʒe-hüren.

hüren, *see* hēren.

hürfte, sb., *O.E.* hwyrft ; *orb* : urthe is a lutel hurfte aʒen hevene TREAT. 4.

hurien, v., *? hurry* ; **horied** (*pret.*) A. P. ii. 883.

hurkelen, v., *cf Du.* hurken, *O.N.* hurka, *prov. Eng.* hurkle (*to squat, nestle*) ; *hang down; overhang; nestle* : þat oþer burne watz abraist and **hurkelez** (*pres.*) doun with his hede A. P. ii. 150; [a litill brid] hurkils and hidis ALEX. (Sk.) 504; over the hiʒest hille þat **hurkled** (*pple.*) on erþe A. P. ii. 405.

hurle, sb. [? *for* hwirl], *wave*: the pure popul- and hurle [*v. r.* perle] passis it umbi ALEX. (Sk.) 1154; the pure poplande hourle plaies on mi heved A. P. iii. 319.

hurlen, v., *? cf. M.Du.* horrelen ; *hurl* ; **hurlest** (*pres.*) CH. C. T. *B* 297 ; hurleð to gederes A. R. 166; horliþ P. S. 211 ; hele over hed **hour-lande** (*pple.*) about A. P. iii. 271 ; þe see him **hurlede** (*pret.*) up and doun P. L. S. xxiii. 25; hurled WILL. 1243; to helle he horlede

L. H. R. 140; **hurled** (*pple.*) A. P. ii. 44; hurlid ['*impulsus*'] WICL. PS. cxvii. 13.

hurlung, sb., *contention*: hurling or strife '*incursio, conflictus*' PR. P. 253; ʒif ʒe weren iче worldes þrunge, mid a lutel **hurlunge** (*dat.*) ʒe muhten al vorleosen A. R. 166.

hurliḍ, adj., *? from O.Fr.* hure; *covered with bristles*: his hede is like a stouke, hurlid as hogges TOWNL. 313.

hūrling, sb., *O.E.* hȳrling; *hireling*; **hīr-lingen** (*ms.* hyrlingen) (*dat. pl.*) MK. i. 20.

hūrne, sb., *O.E.* hyrne,= *O.N.* hyrna, *O.Fris.* herne; *nook, corner,* MARH. 8; O. & N. 14; P. S. 150; REL. I. 264; som hurne of þe londe ROB. 178; huirne JOS. 378; hurne, hirne WILL. 688, 3201; hirne PR. P. 241; ORM. 1677; A. P. iii. 178; herne CH. C. T. *F* 1121; þare hernen [herne] MK. xii. 10; LK. xx. 17; **hürnen** (*pl.*) A. R. 314; huirnes, hirnes, hernis LANGL. *A* ii. 209; hirnes ALEX. 3215.

hirne-stān, sb., *corner stone,* ORM. 6824.

hurren, hurrin, v., = *Ger.* hurren, *Swed.* hurra, *M.Du.* horren; *whirr,* '*bombizo*,' PR. P. 254; *comp.* to-hurren.

hurrok, sb., *cf. prov.Eng.* (*Norfolk*) orruck-holes (*oar holes*); *? oar*: withouten maste . . baweline . . kable . . capstan . . hurrok . . handehelme A. P. ii. 419; (he) on helde bi þe hurrok A. P. iii. 185.

hurst, sb.,= *M.H.G.* hurst, *M.L.G., M.Du.* horst, *mod.Eng.* hurst (*in place-names*); *copse, wooded hill,* S. A. L. 156; **hurste** (*dat.*) TREV. I. 419; upon þe hexte hurste SAINTS (Ld.) xlv. 18; hirste D. ARTH. 3369; **hurstes** (*pl.*) ANT. ARTH. v.

hürt, sb., *O.Fr.*, hurt, *cf. M.H.G.* hurt, *M.Du.* hurt, hort; *hurt,* A. R. 112; hirt MAN. (F.) 12401; hurth YORK xl. 34; **hurtes** (*pl.*) HOM. I. 207; LAʒ. 1837; CH. C. T. *F* 471.

hürt-lēs, adj., *unhurt,* ALEX. (Sk.) 102.

hürte, *see* heorte.

hürten, v., *O.Fr.* hurter, *from M.H.G.* hurten, *cf. M.Du.* hurten, horten; *hurt, offend,* A. R. 8; þat tu noht ne shalt tin fot uppo þe stanes hirten (*ms.* hirʒenn) ORM. 11370; hurten [hirte] ['*laedere*'] WICL. EXOD. xii. 23; herte SHOR. 112; hirtiþ (*pres.*) ['*offendit*'] WICL. JOHN xi. 9; hurtes [hortes] PS. xxxvi. 24; **hürte** (*subj.*) LANGL. *B* x. 366; herte (*pret.*) RICH. 4715; hurten LAʒ. 1878; ihürt (*pple.*) A. R. 98; H. M. 21; ihert S. S. (Web.) 772.

hürtunge, sb., *hurting,* A. R. 344.

hurtlen, v., *run together, collide, stumble*: þei schulen . . . hurtlen to gidere ['*collident*'] WICL. JER. xlviii. 12; hurtelin '*impingo, collido*' PR. P. 253; hurtle (*ms.* hurtel) WILL. 5013; ilk an of þam sal . . . again other hortle (*ms.* hortel) PR. C. 4787; he him hurtleth (*v. r.* hurteþ) with his hors a doun CH. C. T. *A* 2616.

hurtlinge, sb., *collision, clashing together,*

'*collicio, contactus*,' PR. P. 253; TREV. IV. 153; hard was the hurteling tho herti betwene D. TROY 10053; hortling MAN. (F.) 2946.

hūs, sb., *O.E.* hūs, = *O.L.G., O.H.G., O.N.* Goth. hūs, *M.Du.* hūs, huis; *house,* A. R. 134; ORM. 1608; O. & N. 623; GEN. & EX. 1619; HAV. 740 (*spelt* hws 1141); C. M. 14742; huis P. S. 327; hous ROB. 20; MAND. 247; s. S. (Wr.) 2815; PS. lxxxiii. 4; hous and hoom CH. C. T. *H* 229; **hūses** (*gen.*) O. & N. 1155; **hūse** (*dat.*) FRAG. 6; O. & N. 479; hi (*r.* i) schal to house (*home*) þi douter do wel spuse HORN (H.) 1034; **hūs** (*pl.*) LAʒ. 1937; hous AYENB. 43; TREV. III. 417; huses HOM. I. 49; A. R. 296; O. & N. 1203; **hūsen** (*dat. pl.*) MAT. xi. 8; *comp.* bak-, béde-, brēu-, cwalm-, gang-, gest-, hōr-, hord-, rūm-, mōt-, tol-hūs.

hūs-berner, sb., *house burner*: **hūsberners** (*pl.*) bakbiteres MISC. 30.

hŭs-bonde, sb., *O.E.* hūsbonda, -bunda, *O.N.* hūsbōndi; *husband, small farmer,* '*pater-familias*,' PR. P. 254; LAʒ. 31958; HOM. I. 247; MAND. 247; husbond BARB. x. 387.

hŭsbonden, v., *practise thrift, husband* (*one's resources*): husbondin or wiseli dis-pendin wordeli goodes '*dispenso iconomice*' PR. P. 254.

hŭsbond-man, sb., *head of a family*: [maidins] lat liʒt be **hŭsbandmen** (*pl.*) S. & C. II. xxii; '*paterfamilias, hic iconomus*' VOC. 211.

hŭsbonderie, sb., *husbandry,* LANGL. *A* i. 55.

hūs-brenning, sb., *house burning*: mans slaghter and husbren(n)ing C. M. 26235.

hous-evese, sb., *house-eaves,* WICL. PS. ci. 7.

hūs-folc, sb., *cf. Dan.* huusfolk; *family*: everilc husfolc GEN. & EX. 3138.

hous-hīre, sb., *rent of a house,* P. S. 330.

hous-hold, sb.,= *Ger.* haushalt; *household,* LIDG. M. P. 67; **housholdes** (*pl.*) MAND. 209.

hous-holdere, -haldere, sb., *householder,* CH. C. T. *A* 339.

hous-kēpare, sb., *housekeeper,* PR. P. 251.

hūs-lāverd, sb. (*ms.* huse-) HOM. I. 247.

hūse-lĕfdi, sb., *mistress of the house,* A. R. 414.

hous-lēk, sb.,= *M.L.G.* hūslōk; *houseleek,* PR. P. 251.

hous-lēs, adj., *houseless,* J. T. 40.

hūs-lēwe, sb., *O.E.* hūshlēow; *protection of the house,* HOM. I. 277.

hūs-rōf, sb., *house roof,* REL. I. 219.

hūs-shipe, sb., *family,* HOM. II. 197.

hŭs-ting, sb., *O.N.* hūsþing; *husting,* LAʒ. 2324.

hūs-wif, sb., *housewife*, VOC. 215; PR. P.
255; (*ms.* husewif) A. R. 416; houswif LANGL.
B xiv. 3.

hūsewif-schipe, sb., *house management*:
housewifschipe is Marthe dole A. R. 414.

[**hüscen,** v., *O.E.* hyscan, *from* husc (hux);
deride; *comp.* for-hüscen.]

huschen, v., =*L.G.* huschen, hussen; *hush*;
husht [hust] (*pple.*) CH. C. T. *A* 2981 (Wr.
2983).

hūsel, sb., *O.E.* hūsel, =*O.N.* hūsl, *Goth.*
hunsl (θυσία); *eucharist*, A. R. 208; ORM.
6215; þat holie husel HOM. II. 61; housel
LANGL. *B* xix. 390; R. R. 6386; TOR. 1273;
housil MISC. 217; PR. C. 3402; hosel ROB.
419; hosil LIDG. M. P. 248.

hūsen, v., *O.E.* hūsian, =*M.L.G.*, *M.H.G.*
hūsen; *house, dwell*; housin PR. P. 251; house
HALLIW. 464; D. ARTH. 4284: **housede**
(*pret.*) ROB. 21.

hūsing, housing, sb., =*M.L.G.* hūsinge, *M.
H.G.* hūsunge; *housing*, C. M. 8591; housinge
MIRC 1147; DEP. R. iii. 217; housing GOW.
II. 352.

huske, sb., ? *M.Du.* hulsche; *husk*, '*cor-
ticillus*,' PR. P. 254; MAND. 188.

huske [2], sb., *a kind of fish*, '*sqvarus*,' PR. P.
254.

hūslen, v., *O.E.* hūslian, *cf. O.N.* hūsla, *Goth.*
hunsljan; *administer the eucharist*, ORM.
6129; houselin PR. P. 250; housele MAND.
261; hoslen HAV. 212; **housled** [houseled]
(*pple.*) LANGL. *B* xix. 3; ihuseled A. R. 16;
ihouseled SHOR. 22; TREV. V. 417.

huspilin, v., *Fr.* houspiller; *plunder*, '*spolio*,'
PR. P. 255.

hŭsting, *see under* **hūs.**

hūten, v., *cf. Swed.* hūta; *hoot*, ORM. 2034;
houtin '*boo*' PR. P. 251; LUD. COV. 179;
grede & houte FER. 3225; **hūted** (*pple.*)
ORM. 4875; ihouted LANGL. *B* ii. 218; *see*
hōten.

hūting, hūtung, sb., *hooting*, JUL. 52, 53.

hutte, sb., *heap*; ['*gleba*'] PALL. ii. 188.

hŭve, sb., *O.E.* hȳf, =*M.L.G.* hūve; *hive*;
huive REL. II. 84; hive SAX. CHR. 256;
P. L. S. iv. 31; CH. C. T. *A* 4373; A. P. ii.
223; hife VOC. 223; heve BEV. 1408.

hŭven, v., *put into a hive*; hivin PR. P. 242.

hux, sb., *O.E.* hucs (húx), husc, =*O.L.G.*,
O.H.G. hosc; *taunt, mockery*: hux and hoker
LAȝ. 28865.

hux-word, sb., *taunt*, LAȝ. 21682.

hveȝel, *see* **hwēol.** **hvet,** *see* **hwat.**

hwá, pron., *O.E.* hwa, =*O.Fris.* hwa, *O.L.G.*
hwe, *O.N.* hvar, *O.H.G.* hwer, *Goth.* hwas,
Lat. qvis; *who*, MARH. 20; hwa was wurse þen
heo KATH. 170; ȝif hwa (*if any one*; *cf. Lat. si
qvis*) is swa sunful HOM. I. 9; wha LAȝ. 4626;
PERC. 1613; PR. C. 90; wha ORM. 2091; hwao,

hwo MISC. 192; hwo, hwoa A. R. 6, 18; hwa
[hwo], wo, wa O. & N. 113, 1195, 1782; hwo,
wo HAV. 76, 296; hvo (*ms.* huo) AYENB. 5;
who CH. C. T. *A* 831; who, ho C. L. 268,
1159; wo ROB. 215; S. S. (Wr.) 1093; ho
FL. & BL. 634; TREAT. 133; WILL. 1286;
JOS. 466; K. T. 894; DEGR. 1433; ȝwo
MAP 334; qva M. H. 3; qvo GEN. & EX.
359; ANT. ARTH. xxiv; qwo E. G. 30, 95;
quha BARB. i. 391; **hwat** (*neut.*) [*O.E.*
hwæt, = *O.Fris.* hwet, *O.L.G.* hwat, *O.N.*
hvat, *O.H.G.* hwaz, *Lat.* qvid] *what* KATH.
150; H. M. 3; hwat herte is swa hard
HOM. I. 269; hwat mihte . . . was icud A. R.
76; hwat tu dest wel 86; of hwat (*for* hwam)
66; hwat, what, wat O. & N. 60, 393, 1730;
hwat, wat HAV. 117, 1951; hwet seggeð heo
HOM. I. 29; mest hwet 137; hwet wene ȝe
MARH. 6; qvhat kin *of what kind* BARB. iv.
649; of hwet cunde 16; & wite hwet he
us lende ANGL. I. 14; hvet AYENB. 20; whæt
[wat] heo don mihten I.Aȝ. 12262; whet he
þer sohte 2393; what cnihtes we beoð 13845;
al what it seȝþ & meneþ ORM. 5503; what
C. L. 1061; nēȝ what BEK. 1958; what qvod
þe prest LANGL. *A* vii. 130; **what swa**
['*qvodcunqve*'] PS. i. 3; **what lütles** þat
he et P. L. S. xvii. 396; **what for** eie what
for love BEK. 337; wat frend wat fa HOM.
I. 237; wat þe was wo SHOR. 87; wet
MISC. 27; H. H. 194; qvat REL. I. 292; M.
H. 5; **hwás** (*gen.*) *whose* ['*cujus*'] MAT.
xxii. 42; HOM. I. 151; god . . . hwas wreððe
MARH. 9; ancren hwas blisse A. R. 348;
hvas AYENB. 38; whas ORM. 3425; WILL.
1441; hwes MISC. 50; whes sune he weore
LAȝ. 17111; was, hwos HOM. II. 43, 187; whos
WICL. JOHN iv. 46; whos [hos] C. L. 61; wos
REL. I. 226; ȝwas SAINTS (Ld.) lii. 49; **hwam**
[*mod.Eng.* whom] (*dat.*) A. R. 56; þuruh hwam
(*for* hwan) 352; to hwam HOM. II. 181; þurh
hwam (*for* hwan) KATH. 225; hwæt is þes
be hwam (*earlier text* þam) ic þellic gehire
LK. ix. 9; his swete moder . . . of hwam he
vleiss nom MISC. 57; hwam (*for* hwan) þie
seche 53; hvam AYENB. 9; wham (*ms.*
whamm) ORM. 6995; wham WILL. 314; PR.
C. 91; wham [wam] he mihte bitæchen his
dohter & his riche LAȝ. 11550 (*miswritten
whæm* 11404); ȝwam SAINTS (Ld.) lv. 140;
whom C. L. 296; **hwane** (*acc. m.*) ['*qvem*']
JOHN xviii. 4; whan [wan] LAȝ. 27487; þuruh
hwon A. R. 320; make king wan he wole ROB.
502; mid **hwan** (*used as dat. masc., neut.*)
['*qvo*'] MAT. vi. 31; for hwan MK. iv. 30; to
whan hit scal iwurðen LAȝ. 17903; for wan
2679; for hwan [wan] O. & N. 453; wos(t) tu to
hwan [wan] man was ibore 716; bi hwan he lai
1509; to wan [whan] we sculle & of wan we
come P. L. S. viii. 164; þe ston upe whan ich
sitte BRD. 27; to hwon MARH. 16; vor hwon
(*printed* hwou) A. R. 62; mid hwon gremeð he god 334; wite ye for hwon MISC. 38;
comp. ȝe-**hwá.**

hwǽr, *see* **hwár** *and* **hweðer**.

hwǽte, sb., *O.E.* hwǽte, *cf. O.L.G.* hwēti, *Goth.* hwaiteis, *O.H.G.* hweizi *m.*, *O.N.* hveiti *n.*; *wheat*, MK. iv. 28; hwete, hweate A. R. 70, 270; hweate HICKES I. 224; whǽte ORM. 10521; hvete AYENB. 141; whete LANGL. *A* iii. 41; MAND. 189; wete SHOR. 30; qvete M. H. 140; **hwǽtes** (*gen.*) MAT. xiii. 36; *comp.* **sǽd-hwǽte.**

     hwēte-corn, sb., *O.E.* hwǽtecorn, = *O.N.* hveitikorn; *grain of wheat*, HOM. I. 241.

hwǽten, adj., *O.E.* hwǽten, = *M.Du.* weiten, *M.H.G.* weizin; *wheaten*; whete bred LANGL. *B* vii. 120; in ane hvetene lhove AYENB. 82.

hwákien, *see* **cwákien.**

[hwal] whal, sb., *O.E.* hwæl, = *O.N.* hvalr, *O.H.G.* wal; *whale*; whal '*cetus*' VOC. 222; SPEC. 34; WICL. JOB vii. 12; CH. C. T. *D* 1930; whal, qwal PR. P. 523; qwal VOC. 189; qval M. H. 136; of **wháles** [wales] (*gen.*) bone LAȝ. 2363.

hwán, sb., *O.E.* hwōn; *little, few*: lit hwan ['*pusillum*'] MAT. xxvi. 39; hwon ['*pusillum*'] MK. i. 19; a litel wan HOM. II. 69; nusten þa Bruttes na whon whæt Vortiger hæfde idon LAȝ. 13203; a qvon C. M. 19782; thocht thai war qvhein (*few*) thai war worthi BARB. ii. 244; qvhoin [qvhone] ix. 163.

hwanene, adv., *O.E.* hwanone, hwanan, hwonan, = *O.L.G.* hwanan, hwanen, *O.H.G.* (h)wanana, hwanan, (h)wannan; *whence*, HOM. II. 191; whanene, whonene, wonene [wanene] LAȝ. xvi. 1430, 13846; wanene, whonene, hwenene, hwenne O. & N. 138, 1300; wanene ROB. 376; hwanen MK. vi. 2; hweonene, hweonne HOM. I. 249; JUL. 38, 39; hwonne MARH. 16; from hwonne þe engles a dun f(e)ollen HOM. I. 61; whenne HORN (R.) 169; whine ALEX. (Sk.) 834*; **hvannes** AYENB. 115; whannes TREAT. 139; whennes WILL. 478; CH. C. T. *C* 335; whens TRIAM. 431; *comp.* **aighwanen.**

hwanne, adv., *O.E.* hwanne, hwonne, hwænne, = *O.H.G.* hwanne, (h)wenne, *O.L.G.*, *Goth.* hwan; *when*; hvanne AYENB. 6; whanne ORM. 133; whanne WICL. JOHN xii. 12; wanne REL. I. 209; whanne, whan WILL. 80, 303; hwanne, hwenne, hwan HOM. I. 35, 79, 81; hwanne, wanne, wane, wonne, hwenne, hwan, wan, hwon, won, wone, hwen O. & N. 38, 165, 324, 684, 1251, 1264, 1446, 1531, 1566; hwonne, hwon A. R. 6, 144; wonne, wenne, whenne, whænne [wane, wan] LAȝ. 714, 3570, 21757, 27174; whenne, when SPEC. 74; hwen MARH. 6; whan BEK. 53; CH. C. T. *A* 179; whan, whon LANGL. PROL. 1; whon JOS. 25; wen S. S. (Wr.) 994; hwan, qvanne HAV. 162, 312; qwanne (*ms.* quuanne) GEN. & EX. 190; qwan LUD. COV. 256; w. & I. 48; qvan S. & C. I. ii; qven A. P. i. 40; M. H. 2; AV. ARTH. iv; ȝwane

SAINTS (Ld.) lii. 47, lx. 33; ȝwan xxxvi. 295; *comp.* **seld-hwonne.**

hwár, adv., *O.E.* hwǽr, hwár, = *O.L.G.*, *Goth.* hwār, *O.H.G.* wār, wāra, *O.N.* hvār; *where*, FRAG. 6; A. R. 8; hwar, hwar, war, hware, ware O. & N. 64, 892, 938, 1049, 1727; hware, hwere HAV. 549, 1881; whar WILL. 394; PR. C. 357; war ne (*unless*) sin war(e) 2342; hwær LK. ix. 58; whær ORM. 1827; whær, wher [ware] LAȝ. 3320, 4454; hwer MARH. 10; R. S. v; hwer he mei wunian HOM. I. 27; hwer bicomen heo þa 129; wher HORN (L.) 416; wher, where WICL. JOHN vii. 11; wor REL. I. 223 (MISC. 20); **whǽr** on *whereon*, LAȝ. 15502; hver an AYENB. 176; **whár bī** *whereby*, WILL. 2256; **hwár fóre** *wherefore*, A. R. 158*; wher fore MAND. 246; **hvēr inne** *wherein*, AYENB. 23; hwær mid MAT. viii. 25; wer mid REL. II. 274; hver mide AYENB. 23; **hwár of** *whereof*, A. R. 12; whar of TREAT. 136; wor of GEN. & EX. 3530; wher of LANGL. *B* iii. 56; whær ofe ORM. 14052; **hwár tō** *whereto*, A. R. 392; hwar [war] to O. & N. 464; hwer to MARH. 16; **hwēr þurh** *by which*, KATH. 236; wher þurȝ PROCL. 6; hwar þurh A. R. 58; **hvēr onder** *under which*, AYENB. 221; **hvēr oppe** *where upon*, AYENB. 251; **hwēr wið** *wherewith*, H. M. 9; wher wiþ CH. C. T. *D* 131; whar with PR. C. 3835; ȝware with SAINTS (Ld.) lxi. 46; *comp.* **ā-, ȝe-hwár.**

[hwarf], sb., *O.E.* hwearf *m.* (*turn, change, congregation*), *cf. O.L.G.* hwarf *m.* (*congregation, council*), *cf. O.N.* hvarf *n.* (*turning, shelter*), *? O.H.G.* warb *m. n.* (*turn*); *from* **hwerfen**; *change*; *? agreement*: þurh warf of þon folke LAȝ. 2070; he þer wærf makede 17485.

hwarfen, v., *O.E.* hwearfian, = *O.N.* hvarfa (*pret.* hvarfaði); *from* **hwerfen**; *turn, change*; wharfen ORM. 14137; **hwarefeð** (*pres.*) HOM. II. 173; **wharfed** (*pple.*) ORM. 9658; *comp.* **blind-hwarven.**

hwat, adj. *O.E.* hwæt, = *O.L.G.* hwat, *O.N.* hvatr; *swift, active, brave*; spac & hwat HOM. II. 183; wat 127; al se hwat se *as soon as*, HOM. I. 79; al wat (*until*) hi kam MISC. 27; al hvet (*ms.* huet) ich habbe idronke AYENB. 51; hvet hi is ido 87; wat comeþ his ascensioun SHOR. 126; on hwat *quickly*, HAV. 1932; þer weoren eorles swiðe whæte [wate] LAȝ. 7137; **wháte** (*adv.*) ALIS. 2639; B. DISC. 1741; **wáte** (*pl.*) LAȝ. 19136.

     hwat-líche, adv. *O.E.* hwætlíce; *swiftly,* O. & N. 1708; whatlike ORM. 12166; **whatlokere** (*compar.*) BEK. 1249; watloker ROB. 429; **whatlokest** (*superl.*) P. L. S. xiii. 315.

hwat, *see* **hwá.**

hwáte, sb., *O.E.* hwæt *n.* (*omen*); *omen, fortune*: þe man þe leveð upen hwate REL. I. 131 (HOM. II. 11); þe luþur wate ROB. 34; gode wate CHR. E. 163; qvate HALLIW. 656; after

sum geste stod him qvate GEN. & EX. 1054;
comp. un-hwáte.

hwaþer, *see* hweðer.

hwecche, *see* hwicche.

hwei [wei], sb., *O.E.* hwæg,=*M.Du.* wei;
whey, O. & N. 1009; whei '*serum*' PR. P.
523; qwhei REL. I. 9.

hwēl, *see* hwēol.

[hwēle] whēle, sb., *weal*; whele, whelle,
wheel '*pustula*' PR. P. 523; VOC. 267.

[hwēlen] whēlin, v., *O.E.* hwēlian; *blister,
swell;* '*pustulo,*' PR. P. 523.

hwelfen, v., *O.E.* (be-)hwylfan, = *O.N.* hvelfa,
O.L.G. (be-)hwelbean, *M.H.G.* welben; *roll*:
he hwelfde (*pret.*) at þare sepulchre dure
enne grete ston MISC. 51; *comp.* over-
hwelven.

[hwēlke] ˌwhēlke, sb., *diminut. of* whēle;
weal, swelling; whelke, qwelke '*pustula*' PR.
P. 523; whelke WICL. LEV. xiii. 2; whēlkes
(*pl.*) CH. Ç. T. *A* 632.

[hwelmen] whelmen, v., *turn*; [*v. r.* weilen,
whielen] CH. TRO. (T.) i. 139; whelmin
'*supino*' PR. P. 524; whelme (*imper.*)
HALLIW. 926; *comp.* over-hwelmen.

hwelp, sb., *O.E.* hwelp, hwylp,=*O.L.G.* hwelp,
O.N. hvelpr, *O. H. G.* welf; *whelp, puppy* :
hweolp A. R. 198;ˈwhelp ORM. 5838; whelp
'*catulus*' PR. P. 524; P. L. S. xviii. 70; welp
AN. LIT. 11; TRIST. 2399; hwelpes (*pl.*)
MK. vii. 28; whelpes LAȝ. 31679; whelpis
OCTOV. (W.) 314.

hwelpen, v., *bring forth whelps*; whelpede
(*pret.*) OCTOV. (W.) 470; whelped (*pple.*)
ORM. 5839; ihweolped A. R. 200.

[h]wēnen, v., *O.E.* (a-)hwǣnan; *? trouble,
grieve* : þat wēneþ (*pres.*) me umbe while
SPEC. 49; *comp.* a-, ȝe-hwēnen.

hwenene, *see* hwanene.

hwenne, *see* hwanne.

hwēol, sb., *O.E.* hwēol, hwēohl, hwēogul,
hwēowol,=*M.Du.* weel, wiel, *O.N.* hvel, hiŏl
n., Gr. κύκλος; *wheel*, A. R. 322; þat hweol
KATH. 1951; wheol ORM. 3642; hefnes
whel 17531; hwel MISC. 149; hveȝel AYENB.
24; ȝweol, wheol, weol LEG. 163, 190, 191;
wheel CH. C. T. *A* 925; qwel HALLIW. 657;
qvheill BARB. xiii. 637; hwēole, hweoles
(*pl.*) KATH. 1942, 2017; hweoles A. R. 356;
weoles ROB. 408; wheles MAND. 241; ? who-
welen SHOR. 109; *cómp.* cart-wheel.

whēl-barowe, sb., *wheelbarrow*, FLOR.
2031.

whēl-spor, sb., *wheel-track*, '*orbita,*' PR.
P. 524.

whēl-wriȝt(e), sb., *wheelwright*, REL. II.
8; qvelwriȝte ANT. ARTH. xxi.

hwēolen, v., *wheel, rotate*; hwēolinde (*pple.*)
A. R. 356.

hwēr, *see* hwār *and* hweðer.

hwerf, sb., *O.E.* hwerf,=*M.Du.*, *M.L.G.* werf
m.; wharf; wherfe (*dat.*) MAN. (H.) 310.

[hwerfen], v., *O.E.* hweorfan,=*O.N.* hverfa
(*pret.* hvarf), *O.H.G.* werban, *Goth.* hwairban;
turn; comp. a-hwerfen; *deriv.* hwarf,
hwarfen, hwerfen, hwirl, hworvel.]

hwerfen [2], wherven, v., *O. E.* hwerfan,
hwyrfan,=*O.N.* hverfa (*pret.* hverfði), *O.H.
G.* werben (*pret.* warpta); *turn, change*;
whærven LAȝ. 31680; wervende (*pple.*) HOM.
II. 87; whærfde (*pret.*) hire nome LAȝ. 6319;
manege ... agen hwærfden MK. vi. 31; wher-
fed (*pple.*) (*perverse*) folc ORM. 9721.

[hwēsen] whēsen, v., *O.E.* hwēsan (*pret.*
hwēos ÆLFR. HOM. I. 86); *wheeze*: whēse
(*pres.*) TOWNL. 152.

hwēstan, v., *? cough*; (he) hwēst (*pres.*)
LEECHD. III. 122; *see* hōstin.

hwet-stōn, sb. *O.E.* hwetstān,=*M.Du.* wet-
steen, *M.L.G.* wettestēn, *O.H.G.* wezzi-, wezi-
stein; *whetstone*, '*cos,*' FRAG. 5; whetston
PR. P. 524; TREV. I. 13.

hwēte, *see* hwǣte.

[hwetten] whettin, v., *O.E.* hwettan,=*O.N.*
hvetja, *M.Du.*, *M.L.G.* wetten, *O.H.G.* wez-
zan; *from* hwat; *whet, sharpen;* '*acuo,*' PR.
P. 524; whete (*pret.*) TREV. VII. 341; whætte
LAȝ. 14215; wette S. S. (Web.) 911; iwhæt
[iwet] (*pple.*) LAȝ. 30579.

whettinge, sb., *whetting*, '*acucio,*'PR.P. 524.

[hweþen]wheþen,adv.,*O.N.*hvaˇtan; *whence*:
MAN. (H.) 236; whethen [weþen] PS. cxx. 1;
whethen PR. C. 5205; weðen HOM. II. 127;
wheþen, qveþen GAW. 461, 871; C. M. 4819;
qveðen GEN. & EX. 1401; qveþen M. H. 131.

hweþer, pron. & adv. (conj.), *O.E.* hwæðer,=
O.L.G. hwethar, *Goth.* hwaþar, *O.H.G.* hwedar,
O.N. hvārr, *Gr.* κότερος; *whether, which of
two*, HOM. I. 173; A. R. 64; hweðer þe beo leovere KATH. 2312; hweþer unker O. & N. 151;
hweþer deþ wurse 1408; hweþer he schal
forþ þe a bak 824; weþer 1360; hwaþer, hwaþer 1198, 1362; wheðer, whæˇðer, weˇðer, whær
[waþer] LAȝ. 905, 5295, 13504, 20877; hweþer,
weþer, HAV. 292, 294; TRIST. 315; wheþer
SPEC. 59; weþer P. L. S. viii. 19; he nuste in
weþer ende turne ROB. 172; weþer est þe
west 220; wer [whether] ur(e) loverd wole
evere wroþ be 352; wheþr (*ms.* wheþþr) ORM.
526; wheþer, wer CH. C. T. *A* 1856, 2397;
WICL. JOHN vii. 17; ȝweþur, ȝweþer SAINTS
(Ld.) xxx. 13, xlv. 74; qveˇðer GEN. & EX.
1471; qveþer M. H. 18; wher C. L. 1040; WILL.
799; MAND. 219; hweðeres (*gen.*) fere
wult tu beon A. R. 284; to whaþere (*dat.*) heo
faren mihten LAȝ. 25742; *comp.* āuþer.

hweðere, conj., *O.E.* hweðere, hwæˇðere, hwæ-
ðre; *nevertheless, yet*, HOM. I. 225; þeih
hweðere II. 224; þoh wheþre (*ms.* þohh-
wheþþre) ORM. 18510; wheþer PS. xxxviii. 6.

hwī, pron. & adv., *O.E.* hwī, hwȳ (*instrum. of* hwá), =*O.L.G.*, hwī, *O.N.* hvī, *O.H.G.* hwiu, hiu, *Goth.* hwē, *Lat.* qvī; (*with, for*) *what;* *why*: cuð me ... hwi þe worldes weldent wuneð in þe MARH. 16; mid hwi (*ms.* hwy) his hit agulden LK. vii. 42; for hwi HOM. I. 153; to hwi; ANGL. I. 13; hwi, whi, wi O. & N. 150, 905, 909; hwi, hwui FRAG. 6, 7; A. R. 162, 320; for wi (*printed* þi) 2578; whi, wi LAȝ. 1577, 3804; hwi HAV. 454; whi SPEC. 49; WILL. 48; MAND. 295; for whi (*for which reason*) ORM. 219; for whi SHOR. 43; PR. C. 709; ȝe habbeþ iherd ... for whi god þe world maken wolde C. L. 568; ȝwi SAINTS (Ld.) xlv. 175; hve is hit voul dede AYENB 47; qvi GEN. & EX. 1759; GAW. 623; M. H. 31; AV. ARTH. xxxiii.

hwicche, sb., *O.E.* hwæcca; *chest*; whicche [whucche] LANGL. *A* iv. 102; whichche A. P. ii. 362; whiche, hoche (?=huche) PR. P. 242; whucche JOS. 39; wheche E. W. 27; *see* **huche.**

hwider, adv., *O.E.* hwider, hwyder, hwæder, *cf. Goth.* hwadrē; *whither*, KATH. 1299; REL. I. 128; hwider [wider] O. & N. 724; whider SPEC. 61; WILL. 701; PR. C. 2115; hvider (*ms.* huider) AYENB. 115; hweder REL. I. 185; whedir LUD. COV. 38; hwuder MISC. 43; whuder [woder] LAȝ. 1202; wuder FL. & BL. 114; whoder BRD. 32; qvider GEN. & EX. 2600; qvedur ANT. ARTH. xi; ȝwodere SAINTS (Ld.) li. 23; *comp.* **ā-hwider.**

whider-ward, adv., *in which direction*; (*ms.* whiderrwarrd) ORM. 16669; WILL. 105; hwuderward A. R. 168; whoderward BEK. 66.

hwik, *see* **cwic.**

hwilc, hwūlc, pron. (adj.), *O.E.* hwilc, hwylc, hwelc, =*O.L.G.* hwilīc, *O.Fris.* hwelik, hwelk, hwek, *O.H.G.* hwelīch, *O.N.* hvilīkr, *Goth.* hwēleiks; (*for* *hwī-līc); *which*, HOM. I. 15, 99; ORM. 471; whilc, whulc [woch] LAȝ. 2167, 2305; whilk S. S. (Wr.) 23; PR. C. 144; hwic HOM. I. 243; hwich of him FER. 160; which GOW. I. 279; which a liȝt LANGL. *B* xviii. 124; a prest ... which was so plesant CH. C. T. *G* 1014; hwuch MARH. 6; A. R. 8; nat ich hwuch þi þoht beo KATH. 512; hwuch, wuch, hwich O. & N. 1378, 1443; whuch, wȝuch C. L. 110; wuch ROB. 220; hwilche mede MAT. v. 46; hwilche ȝife HOM. I. 19; on wilche wise REL. I. 130; whiche help god hem sente WILL. 2705; whuche JOS. 270; ȝwuch SAINTS (Ld.) lv. 143; **whilkes** (*gen. m. n.*) (*ms.* whillkess) ORM. 5287; whulches LAȝ. 20735; hwuches kunnes FRAG. 1; **whūlche** (*dat. m.*) LAȝ. 2303; a wulche time P. L. S. viii. 66; **wūlchere** (*dat. f.*) LAȝ. 4446; **whūlcne** (*acc. m.*) LAȝ. 10120; **hwūlche**, hwiche (*pl.*) HOM. I. 11, 155; whulche, wulche [woche] LAȝ. 11770, 11772; hvichen (*dat. pl.*) AYENB. 15; *comp.* **ȝe-hwilc.**

hwīle, sb. (adv., conj.), *O.E.* hwīl, *cf. O.L.G.*, *O.H.G.* hwīla, *O.N.* hvīla, *Goth.* hweila (χρόνος, καιρός, ὥρα); *while, time*, MARH. 3; swuþe longe hire is þe hwile O. & N. 1591; þe ule one hwile [wile] hi biþohte 199; ane hwile HAV. 722; while ORM. 142; while ['*momentum*'] PR. P. 524; PR. C. 632; a while in þoȝte he stod BEK. 1243; litel while LANGL. *B* xvii. 46; hwule A. R. 246; qvile GAW. 30; LUD. COV. 73; longue ȝwhile SAINTS (Ld.) lxvi. 118; þā **while** [wile] LAȝ. 11309; þa, þe hwile P. L. S. viii. 11, 12; þeo hwile, hwule FRAG. 6; þe hwile [wile] O. & N. 1141; þeo hwule A. R. 422; hwile *for a while, once on a time*, HOM. I. 17; so hwile [wile] dude sum from Rome O. & N. 1016; while ROB. 9; SPEC. 28; þus menie kinges þer were while in Engelonde P. L. S. xiii. 73; a sonde me cam while fram hevene ASS. 240; er while P. L. S. xx. 91; hwil *whilst* MARH. 12; while WILL. 2537; while [whiles] CH. C. T. *A* 35; whiles PR. C. 3645; qvile GEN. & EX. 2041; hwīle ... hwīle REL. I. 128; while ma while nan LAȝ. 12036; **sume hwīles** (*for* hwile) A. R. 272; **ōðer hwīle** *at another time* HOM. I. 23; oðer while LAȝ. 7062; oþer while REL. I. 110; CH. BOET. iii. 12 (105); oþer wile ROB. 100; oðer hwule [hwiles] A. R. 82; þēr hvīle AYENB. 7; þer while S. S. (Web.) 701; þor wile SHOR. 43; ðer wile he bieð a live HICKES I. 222; þer whiles WILL. 2736; **bī while** ORF. 10; **whīlum** (*dat. pl.*) *whilom, formerly* ORM. 4868; hwilem MISC. 30; whilom CH. C. T. *A* 859; hwilun LEECHD. III. 138; an usurer was ȝwilene SAINTS (Ld.) lxvi. 117; hwilon ['*ad tempus*'] LK. viii. 13; hwilen (*ms.* hwylen) MISC.145; whilen LAȝ. 8279; SPEC. 91; wilen MAP 340; hwulon *sometimes* FRAG. 1; *comp.* **hand-, līf-, mor-ȝen-hwīle.**

hwīl-wende, adj., *O.E.* hwīlwende; *temporary*; hwilende H. H. 25; whilende HOM. I. 7; hwilinde MISC. 94.

whīlwend-līc, adj., *O.E.* hwīlwendlīc, hwīlendlīc; *temporary*, ORM. 18787.

[**hwilen**, v., *cf. O.H.G.* wīlōn, *O.N.* hvīla, *Goth.* hweilan (παύεσθαι); *comp.* **ȝe-hwīlen.**]

hwīlende, *see* **hwīlwende** *under* **hwīl.**

[**hwin**] **whin**, v., *whin;* '*saliunca*,' VOC. 229; whinne '*saliunca, ruscus*' PR. P. 524; qvin Iw. 159.

[**hwīnen**] **whīnin**, v., *O.E.* hwīnan, =*O.N.* hvīna; *whine*; whinin '*ululo, gannio*' PR. P. 524; for chele ... hwine MISC. 82; whine CH. C. T. *D* 386; whīne (*pres.*) REL. II. 245.

whīninge, sb., *whining*, '*ululatus*,' PR. P. 524.

hwinge, *see* **winge.**

[**hwippe**] **whippe**, sb., *M.Du.* wippe; *whip*, '*scutica*'; 'whippe PR. P. 524; MAND. 249; CH C. T. *E* 1671; HOCCL. i. 118.

[**hwippin**] **whippin**, v., *M.Du.* wippen; *whip*, PR. P. 524.

[**hwirl,** sb., *O.E.* hwyrfel, *O.N.* hvirfill, = *M.Du.*, *M.H.G.* wervel, *O.H.G.* werbil ; *from* hwerfen ; *whirl* ; *see* hworvel.]

whirl-bōn, sb., = *M.Du.* wervelbēn ; *vertebra* ; whirlebone, or hole of a joint, '*anca*, *vertebrum*, *condulus*' PR. P. 524 ; whirlebone, wherlbone VOC. 183, 208 ; werelbone 179 ; qwhirlbone CATH. 298 ; qvirlilebone PR. P. 42.

whirl-puff, sb., *whirling blast* : a whirlpuff of wind WICL. WISD. V. 21.

whirl-gig(g)e, sb., *whirligig*, PR. P. 525.

whirl-wind, sb., *O.N.* hvirfilvindr, *cf. M.Du.* wervelwind ; *whirlwind*, '*turbo*,' PR. P. 525 ; TREV. VII. 159 ; qwirlwind REL. I. 6.

[**hwirlen**] **whirlin,** v., *O.N.* hvirfla, = *M.Du.* wervelen ; *whirl*, *rotate* ; whirlin '*roto*' PR. P. 525 ; whirle CH. P. F. 80 ; **whirlide** (*pret.*) WICL. I KINGS xvii. 49 ; wirlede LEG. 167.

[**hwispren**].**whisperin,** v., *O. E.* hwisprian, = *M. Du.* wisperen ; *whisper* ; PR. P. 525 ; whispre LIDG. TH. 695.

whisperinge, sb., *O.E.* hwisprung ; *whispering*; '*musitacio*' PR. P. 525 ; **w(h)ispringes** (*pl.*) CH. H. F. 1958 ; qwisperinge PR. P. 349.

[**hwist**] **whist,** interj., *whist, hist* ; whist WICL. JUDG. xviii. 19.

whiste-lī, adv., *silently* : þe whele of fortun ... þat whisteli chaungez ALEX. (Sk.) 1851*.

hwistle, sb., *O.E.* hwistle ; *whistle*, '*fistula*,' FRAG. 2 ; whistle CH. C. T. *A* 4155 ; whistel LIDG. M. P. 27 ; W. & I. 41.

hwistlen, v., *O.E.* hwistlian ; *whistle* ; whistlen WICL. JER. xix. 8 ; **whistlen** (*pres.*) LANGL. *B* xv. 467 ; whistleden (*pret.*) ALIS. 5348.

whistlere, sb., *O.E.* hwistlere ; *whistler*, LANGL. *B* xv. 475 ; **hwistleres** (*pl.*) MAT. ix. 23.

whistlinge, sb., *O.E.* hwistlung ; *whistling; hissing, contempt* : scorpion with vile whistlinge ALIS. 5262 ; whistlinge LANGL. *B* xv. 466 ; to have hevene þoruȝ her whistlinge 471 ; whistelinge PR. P. 525 ; i shal sette this cite in to stoneing and in to whistling WICL. JER. xix. 8.

[**hwistren**] **whistren,** v., *whisper* ; the **whistrende** (*pple.*) grucchere WICL. ECCLUS. xxviii. 15 ; grucchendli whistrende xii. 19.

hwīt, adj., *O.E.* hwīt, = *O.L.G.* hwīt, *O.N.* hvītr, *Goth.* hweits ,*O.H.G.* hwīz ; *white*, A. R. 396 ; O. & N. 1276 ; hwit HAV. 1729 ; whit TREAT. 139 ; TOR. 458 ; qvit GEN. & EX. 2810 ; GAW. 1205 ; M. H. 43 ; ȝwite IOf SAINTS (Ld.) xxxvi. 283 ; Hwīte bī *Whitby* MISC. 146 ; þe white [wite] LAȜ. 15978 ; þe hvite robe AYENB. 228 ; hwīte sun(n)e dei *Whitsunday*, A. R. 412 ; þan whiten sun(n)en dæi LAȜ. 17492 ; a whiten sunnen dæie 17484 ; on hwite sun(n)e dai HOM. I. 209 ; to þan white sun(n)e tide *Whitsuntide* LAȜ. 31524 ; hwītture (*com-*

par.) A. R. 324 ; whittore SPEC. 28 ; whittere BRD. 7 ; *comp.* milch-wīt.

whīt-brēd, sb., = *M.H.G.* wīzbrōt ; *white bread*, E. G. 354.

whit-flowe, -lowe, SD., *whitlow*, '*panaricium*,' PR. P. 525.

whīt-lēd, sb., *white-lead*, PR. P. 525.

whīt-līmin, v., *white-lime, whitewash*, PR. P. 305 ; **whītlīmed** (*pple.*) LANGL. *B* xv. 111 ; iȝwitlīmede (*ppl. pl.*) LEB. JES. 422.

whīt-mete, sb., *O.E.* hwītmete ; *food prepared with milk*, '*lacticinium*,' PR. P. 525.

whītnesse, sb., *whiteness*, '*albedo*,' PR. P. 525.

wīt-þorn, sb., = *M.H.G.* wīzdorn ; *whitethorn*, REL. I. 38.

hwīte, adj., *O.N.* hwīti, = *O.H.G.* wīzī ; *whiteness* : hu hire hwite [white] like him A. R. 56.

hwītel (?hwitel), sb., *O.E.* hwītel, *cf. O.N.* hvitill ; *blanket*, FRAG. 4 ; A.˙R. 214* ; whitel LANGL. *C* xvii. 76.

hwīten, v., *O.E.* hwītian, *cf. Goth.* hweitjan, *O.H.G.* wīzen ; *whiten, become white* ; whiten LANGL. *B* iii. 61 ; whiton PR. P. 525 ; **hwīteð** (*pres.*) A. R. 150 ; ihvīted (*pple.*) AYENB. 178.

whītinge, sb., *O.E.* hwīting ; *whiting*, '*dealbatio*,' '*albatura, candidacium*,' PR. P. 525.

whītstare, sb., *bleacher*, PR. P. 525.

[**hwīting**] **whīting,** sb., *? Du.* wijting ; *whiting (the fish)*, '*glaucus*,' VOC. 222 ; '*gemmarius, merlingus*' PR. P. 525 ; B. B. 156, 174 ; witing '*clamitus*' VOC. 189.

[**hwītnen**] **whītnen,** v., = *O.N.* hvītna ; *whiten* ; **whītened** (*pple.*) PS. l. 9.

[**hwiðeren,** v., *?shake; comp.* to-hwiðeren.]

hwōn, see hwān. **hwonene,** see hwanene. **hwonne,** see hwanne. **hwore,** see hwure.

[**hworvil**]**whorvil,** sb., *? M.Du.* worvel ; *from* hwerfen ; *whorl* ; whorvil, whorl PR. P. 526.

hwu (?hwū), adv., *O.E.* hwu, hu, = *O.L.G.* hwo, *O.Fris.* hu, ho, *O.H.G.* hweo, wio, *Goth.* hwaiwa ; *how*, HOM. I. 237 ; hwu, hu A. R. 182, 256 ; hwu come þu (h)ider HOM. II. 97 ; hu vele 63 ; hvu (*ms.* hw), hu O. & N. 294, 1493 ; hvu (? huu ; *ms.* hw) wis sho was HAV. 288 ; hwou Robert with here loverd ferde 2411 ; whou PL. CR. 141 ; wu REL. I. 172, 224 ; wou HICKES I. 225 ; hu LAȜ. 561 ; ORM. 938 ; GEN. & EX. 244 ; HORN (L.) 468 ; hu michel ['*qvantum*'] LK. xvi. 5 ; hu woc so hit ever beo (*however weak it be*) A. R. 138 ; hu ich hatte MARH. 13 ; hu nu KATH. 2111 ; hou BEK. 131 ; C. L. 569 ; SPEC. 39 ; AYENB. 17 ; WILL. 97 ; TRIST. 514.

howe-gates, adv., *in what manner*, YORK xxvi. 229.

hwüch, see hwilc. **hwüder,** see hwider.
hwülc, see hwilc. **hwüle,** see hwile.

hwure, adv., *O.E.* huru, *cf. Swed.* huru; *at least* : þet þu heom ȝefe rest ˙la hwure þen sunne dei HOM. I. 45; hwure þinge [*'vel'*] MAT. xiv. 36; hwore MK. vi. 56; hure SAX. CHR. 260; MARH. 16; la hure H. M. 23; (*printed* lanhure) KATH. (E.) 558, 774, 1073; ne hure HOM. I. 131; hure and hure O. & N. 11; hure & hure ȝet he hefde vode A. R. 260.

[hwürf, sb., *?*]

hwürf-bān, sb., *O.E.* hwyrfbän; '*vertebra*,' LEECHD. III. 98.

i.

i-, *see* ȝe- (*for many participles with this prefix see the simple verbs*).

ī, *see* ēie, ic, in.

iald, *see* eald.　**iare,** *see* ēare.

ic, pron., *O.E.* ic, ih,＝*O.L.G.* ic, *Goth.* ik, *O.N.* ek, *O.H.G.* ih; *I,* FRAG. 6; HOM. I. 225; P. S. 199; ic [ich] P. L. S. viii. 1; ic, ich MAT. v. 18, 20; ANGL. VII. 220; ic, ich, ihc LAȝ. 461, 697, 872; i 18886; ic, ich, ihc, ih, i O. & N. 1, 293, 868, 1049, 1698; ic, ich L. H. R. 38; ic, i ORM. 105, 4815; GEN. & EX. 309; M. H. 32; ik 11; ich HOM. II. 11; MARH. 1; KATH. 222; A. R. 12; SHOR. 129; AYENB. 1; ic, ich, ihc, i HAV. 3, 119, 288, 1377; ich, ihc, i FL. & BL. 44, 60, 61; ich, i C. L. 85, 326, 329, 495; WILL. 548; TRIST. 2, 764; LANGL. *A* i. 10, 21; i SAX. CHR. 262; MAND. 6; CH. C. T. *A* 730; PR. C. 1738; hic GEN. & EX. 35; REL. 188; hich MISC. 34; hi MISC. 123; HAV. 487.

icchen, v., *?*＝hicchen; *lift up* : icchen upward ORM. 11833; and evere lai þis maide stille hi ne miȝte hire enes icche P. L. S. xxi. 132; icched (*pret.*) ORM. 8123.

icchen, *see* ȝicchen.　**ich,** *see* ic, ȝehwilc.

iche, *see* issen.

īdel, adj., *O.E.* īdel,＝*O.Fris.* īdel, *O.L.G.* īdal, *O.H.G.* ītal; *idle; empty, unoccupied; vain,* A. R. 44; REL. I. 129, 218; O. & N. 917; SPEC. 41; CH. C. T. *A* 2505; idel ȝelp ORM. 12041; HOM. I. 103; KATH. 471; idel þonc HOM. II. 129; an idel word ORM. 825; idel þoht M. H. 33; he nolde ... no time idel beo P. L. S. ix. 59; idel go SPEC. 104; on idel *in vain* HOM. I. 95; O. & N. 920; P. S. 326; on idel & wiþuten ned ORM. 12514; an idel LANGL. *B* xiv. 195; in idel GEN. & EX. 3497; CH. C. T. *C* 642; idele blisse AYENB. 23; of idele manne LAȝ. 3310; mid idele honden AYENB. 218; o sond ne groweð no god, and bitocneð idel A. R. 404.

īdel-hēd, sb., *vanity,* GEN. & EX. 28.

īdel-leȝc, sb., *idleness, folly,* ORM. 2165, 7847.

īdel-līche, adv., *idly,* AYENB. 80; GOW. II.

42; idelleche SHOR. 96; idillich WICL. DEUT. v. 11; ideli WICL. 2 MAC. vii. 18.

īdel-honded, pple., *empty handed,* AYENB. 218.

īdelnesse, sb., *O.E.* īdelness; *idleness, indolence, vanity,* LAȝ. 24911; ORM. 4736; idelnisse P. L. S. ix. 62; þe wordle is idelnesse AYENB. 164; īdelnesses (*pl.*) speke þai PS. xi. 3.

īdel-schipe, sb., *idleness, uselessness,* REL. I. 180; MISC. 144; idelschepe SHOR. 93; idillchipe R. P. 5.

idiote, sb., *Lat.* idiōta, *mod.Eng.* idiot; *uneducated person,* LANGL. *B* xvi. 170; CATH. 194; idiote, neither foule ne righte wice PR. P. 258; idiotes (*pl. adj.*) LANGL. *B* xi. 308.

īdlen, v., *O.E.* īdlian, *cf. O.H.G.* (ar-)ītalen; *? render idle or vain* ; īdils (*pres.*) TOWNL. 313; *comp.* for-īdlen.

īdolastre [idolaster], sb., *O.Fr.* idolastre; *idolater,* CH. C. T. *I* 860.

idolatrie, sb., *Fr.* idolatrie; *idolatry,* GEN. & EX. 4143; A. P. ii. 1173; LANGL. *C* i. 196.

īdole, sb., *O.Fr.* ydole, ydele; *idol,* CH. D. BL. 626; MAND. 173; idoles (*pl.*) MAND. 164; idolus ALEX.² (Sk.) 754.

idre, sb., *Lat.* hydria; *water pot* ; idres (*pl.*) MISC. 29.

idre², sb., *Lat.* hydra; *hydra,* CH. BOET. iv. 6 (134).

idromancie, sb., *Lat.* hydromantīa; *hydromancy,* MAND. 134; idromaunce GOW. III. 45.

idropesie, dropesie, sb., *O.Fr.* idropisie; *from Lat.* hydropisis; *dropsy,* WICL. LK. xiv. 2; dropsie S. A. L. 5.

idropike, adj., *dropsical,* A. P. ii. 1096.

īe, *see* ēaȝe.　**iēm,** *see* ēam.　**iēr,** *see* ēar.

iernen, *see* eornen.　**iēþe,** *see* ēaðe.

if, conj., *O.N.* ef,＝*O.L.G., O.Fris.* ef, *O.H.G.* ibu, *Goth.* iba, ibai; *?*＝ȝif; *if,* HOM. I. 149; ORM. 589; GEN. & EX. 464; HAV. 513; S. S. (Wr.) 56; PR. C. 85; if (*v. r.* ȝif) O. & N. 51; LANGL. PROL. *A* 37; CH. C. T. *A* 145; if, ef M. H. 33, 52; ef ANGL. I. 14; REL. I. 179; HORN (L.) 537.

ifel, *see* üvel.

igain, *see* inȝein.

īȝe, *see* ēaȝe.

ignobilitē, sb., *O.Fr.* ignobilité; *meanness,* L. H. R. 161.

ignorance, sb., *O.Fr.* ignorance; *ignorance,* A. R. 278; CH. C. T. *B* 482; ignorans LUD. COV. 335.

ignoraunt, adj., *ignorant,* CH. BOET. v. 3 (160); ignorant CATH. 191.

ikil, sb., *O.E.* gicel,＝*O.N.* iökull; *piece of ice,* '*stiria*,' PR. P. 259; *comp.* is-ikel.

iker, eker, sb., *?*＝niker, ALIS. 6175.

īl, sb., *O.E.* īl, igl, igel,＝*O.H.G.* igil, *O.N.*

igull; *hedge-hog*: prikiende so piles (*printed* wiles) on ile FRAG. 8; **illes pīl** *dart of a hedge-hog, for hedge-hog*, P. L. S. xviii. 47; iles (*ms.* yles) pilles felles A. R. 418; il(e)s piles ['*ericii*'] TREV. I. 339.

ilc, ülc, adj., *O.E.* ylc=ǣlc, *mod.Sc.* ilk; *each, every*; ilc ORM. 503, 513; ?ilch REL. I. 131; ilc GEN. & EX. 119; ilk S. S. (Wr.) 638; MAN. (H.) 7; PR. C. 53; ilc on other wirwed lai HAV. 1921; ilk on *each one* 1842; ilk a BARB. ii. 74, viii. 26; il del HAV. 818; ich ever ilc *every* P. L. S. viii. 33; TRIST. 47; PL. CR. 432; LIDG. M. P. 59; M. ARTH. 2036; ich, uch WILL. 332, 1488; uch C. L. 1228; SPEC. 33; **ilkes** (*gen. n.*) ORM. 9199; **ilche** (*dat.*) LAȝ. 7006; iche REL. I. 235; DEGR. 89; **ilchere** (*dat. f.*) A. R. 132; **ilchene** (*acc. m.*) LAȝ. 7091; ulcne HOM. I. 51; ever ichne A. R. 214; *see* aighwilc (ǣlc) *and* ȝe-hwilc.

ilche, *see* ilke.

ilde, sb., *island*, '*insula*,' PR. P. 259; MAN. (F.) 271; idle ALIS. 4856; **ildes** (*pl.*) MAN. (H.) 282; idles ALIS. 5618.

ildre, *see* eldre.

ile, sb., *O.E.* ile *m.*, *cf. ON.* il *f.*; *sole, underside of the foot*: to þe ile (*ms.* yle, *printed* þle) of hire helen MARH. 10.

īle, sb., *O.Fr.* isle, *cf. O.H.G.* isila; *isle*, HORN (R.) 1330; BRD. 8; A. P. i. 692; þe īles (*pl.*) of Anglesai GAW. 698; iles, isles MAND. 15, 23.

īlespīl, *see under* il.

ilke, adj., *O.E.* ilca, ylca *m.*, ilce, ylce *f. n.*, *mod.Eng.* ilk; *same*, A. R. 68; ORM. 755; O. & N. 99; GEN. & EX. 258; SHOR. 12; LANGL. *A* i. 81; CH. C. T. *A* 721; PERC. 340; TRIAM. 514; M. ARTH. 1765; SACR. 495; þis ilke dei KATH. 2315; ilke, ulke HORN (L.) 855, 1199; þat ilke lond LAȝ. 1640; þat ulke ȝer 3668; ilche MISC. 142; **ilcan** (*dat. m.*) HOM. I. 129; on þain ilken huse LK. x. 7; bi þan ilche oþe PROCL. 5; in þer ilke nihte LAȝ. 5754; **ilken** (*dat. pl.*) LAȝ. 29906; ilca HOM. I. 43, 87.

ille, adj. (sb.), *O.N.* illr; *ill, bad*, P. L. S. viii. 37; HOM. II. 258; GEN. & EX. 4038; hit is him ille O. & N. 1536; þe ille 421; god or ille TRIAM. 113; don ille A. D. 294; heo gan him telle hire ille ALIS. 1147; an ille tre PR. C. 660; ille wiles ORM. 6647; **illis** (*sb. pl.*) ALEX. (Sk.) 3267*.

ille, adv., *O.N.* illa; *ill*, R. S. v; ORM. 6245; LANGL. *A* i. 51; þa swiken speken ille LAȝ. 5426; þei hire likede swiþe ille HAV. 1165; Rimenild wep wel ille HORN (R.) 677; wel or ille M. H. 37; ille heowet HOM. I. 325.

ille-willand, sb., *enemy*, PS. xx. 9; lxxxviii. 24; cv. 10; cix. 2.

ille-willed, adj., '*malivolus*,' CATH. 195.

illen, v., ? *O.N.* illa; *make, become evil; do harm to*: þat al þis world shal illen ST. COD. 95; to ill thi foe, doth get to thee hatred B. B.

100; thai **illid** (*pret.*) ['*malignaverunt*'] counsaile HAMP. PS. lxxxii. 3.

illern, *see* ellarne.

illing, sb., *O.N.* īlling; *malice*, REL. I. 218; MISC. 419.

illuminen, v., *illumine*; illuminit (*pret.*) BARB. viii. 228; illuminit (*pl.*) BARB. xx. 229; *see* enluminen.

illūmininge, sb., *enlightening*, WICL. 2 COR. iv. 6.

illusioun, sb., *O.Fr.* illusiun, *from Lat.* illūsiōnem; *illusion*; **illusiouns** (*pl.*) MAND. 159.

imāge, sb., *O.Fr.* imāge; *image*, (*ms.* ymage) KATH. 1476; P. L. S. xxiv. 134; AYENB. 87; CH. C. T. *D* 1642; **imāgis** (*pl.*) ALEX. (Sk.) 4068.

imagerie, sb., *Fr.* imagerie; *imagery*, EM. 168; **imageries** (*pl.*) CH. H. F. 1190.

imageour, sb., *image maker*, HALLIW. 473.

imaginacioun, sb., *Fr.* imagination; *imagination*, CH. D. BL. 14; MAND. 251; AYENB. 158; GOW. I. 74.

imaginatif, adj. & sb., *O.Fr.* imaginatif; *imaginative*; *imagination*; to ben imaginatif CH. C. T. *F* 1094; imaginatif shal answere LANGL. *B* x. 115; considerethe in youre imaginatif LIDG. M. P. 5.

imaginen, v., *O.Fr.* imaginer; *think, reflect*; imagine LANGL. *B* xix. 272; imaginie *C* xxii. 277; imagenen *B* xiii. 289; imaginin (*mss.* imagyn) PR. P. 259.

imagining, sb., *plan*; **imagininges** (*pl.*) LIDG. M. P. 211.

imb, *see* ümbe.

imbrowe, v., *O.Fr.* embrouer, *cf. Ital.* imbrodare, *mod.Eng.* imbrue; *dirty*, B. B. 6.

imelle, *see* mel.

imes, adv., ? *O.N.* ȳmiss; *variously*: what men immess ȝeornen ORM. 11510; ?on imis [*ms.* onuius, *perhaps* on mis *amiss*] HOM. I. 57.

immaculate, adj., *Lat.* immaculātus; *immaculate*, LIDG. M. P. 79; immacalat [inmaculate] LUD. COV. 272.

immortal, adj., *Lat.* immortālis; *immortal*; inmortal CH. C. T. *I* 1078; inmortalle MAND. 227; LIDG. M. P. 8.

immuin, adj., *Lat.* immūnis; *free from*, PALL. vi. 237.

immutable, adj., *Lat.* immūtābilis; *unchangeable*, LIDG. M. P. 25, 69.

imne, *see* himpne.

impacience, sb., *O.Fr.* impatience; *impatience*, A. R. 198; inpacience AYENB. 67.

impacient [inpacient], adj., *O.Fr.* impacient; *impatient*, CH. C. T. *I* 401.

impaiable, adj., *from O.Fr.* paier; *implacable*, HAMP. PS. lxxvii. 7*.

imparfit, adj., *from Lat.* imperfectus; *im-*

perfect, LANGL. *C* iv. 389; inparfit *C* xxi. 136; inperfit CH. BOET. iii. 9 (83).

inparfitnesse, sb., *imperfection*, WICL. ECCLUS. xxxviii. 31.

impassible, adj., *O.Fr.* impassible; *incapable of suffering*: he is in generations **inpassibles** (*pl.*) HAMP. PS. lxxi. 5.

impe, sb., *O.E.* impe,=*Dan.* ympe, *Swed.* ymp (? *cf. M.Du.* inte, ente, *O.Fr.* ente); *scion, graft; young tree; 'surculus,'* PR. P. 259; P. P. 218; S. S. (Web.) 577; **impen** (*pl.*) A. R. 378; AYENB. 94; impes LANGL. *B* v. 137; impis ALEX. (Sk.) 4819.

impe-trē, sb., *young tree*, ORF. 68.

impen, v.,=*Dan.* ympe, *Swed.* ympa, *M.H.G.* impfen, impfeten (*cf. M.Du.* inten, enten, *O. Fr.* enter); *graft*; impin '*insero*' PR. P. 259; **imped** (*pret.*) LANGL. *B* v. 138; **impid** (*pple.*) '*insitus*' REL. I. 9; þe image of ipocricie imped-upon fendes PL. CR. 305; iimped A. R. 360.

impare, sb., *engrafter*, '*insertor, surculator*,' PR. P. 259.

imping, sb., *grafting*, PR. P. 260; CATH. 195.

imperfeccioun, sb., *O.Fr.* imperfection; *imperfection*; inperfeccion CH. C. T. *I* 1007.

imperial, adj., *O.Fr.* emperial, imperial, *from Lat.* imperiālis; *imperial*, CH. BOET. i. 1 (7); imperialle, emperial LIDG. M. P. 11, 237; imperiale GOW. III. 113.

impertinent, adj., *O.Fr.* impertinent; *impertinent, irrelevant*, CH. C. T. *E* 53; impertinente TREV. VI. 335.

impitous, adj., *pitiless*; impitous windes B. B. 248.

implíen, v., *imply*; **implíeþ** (*pres.*) CH. BOET. v. 1 (152).

impne, *see* himpne.

impolut, adj., *O.Fr.* impollut; *from Lat.* impollūtus; *unpolluted*; inpolut WICL. HEBR. vii. 26.

importable, adj., *O.Fr.* importable, *from Lat.* importābilis; *unbearable*, CH. C. T. *B* 3792; TREV. I. 261; VII. 139; inportable CH. C. T. *E* 1144.

importen, v., *Lat.* importāre; *import*; whiche inportithe (*pres.*) LIDG. M. P. 117.

importūne, adj., *O.Fr.* importun; *importunate*, R. R. 5635.

impositiōn, sb., *imposition, tax*, CH. BOET. i. 4 (15).

impossible, adj., *O.Fr.* impossible, *from Lat.* impossibilis; *impossible*, MAND. 264; impossibille TOWNL. 93; inpossible WICL. WISD. xi. 18; ALEX.² (Sk.) 268; LUD. COV. 387; im**possible** (*sb.*) CH. C. T. *D* 688; LIDG. M. P. 134.

impotence, sb., *O.Fr.* impotence, *from Lat.* impotentia; *impotence*, LIDG. M. P. 246.

impotent, adj., *O.Fr.* impotent, *from Lat.* impotens; *impotent*, LIDG. M. P. 39, 154.

impressen, v., *from Lat.* impressus; *impress*; **impresse** (*pres. pl.*) CH. C. T. *G* 1071; impresse (*imper.*) it in the mind MAND. 295.

impressioun, sb., *O.Fr.* impression; *impression*, LIDG. M. P. 51; inpressioun 133.

improberabille, adj., *? irreproachable*, B. B. 170.

[**improper**, adj., *O.Fr.* impropre; *improper.*]

improper-lī, adv., *improperly*, GOW. I. 21.

impudence, sb., *O.Fr.* impudence; *impudence*, CH. C. T. *I* 391.

impudent, **inpudent**, adj., *Fr.* impudent; *impudent*, CH. C. T. *I* 397.

impugne, v., *Fr.* impugner; *fight against*, impugn, WICL. 1 ESDR. vi. 12; inpugnen LANGL. *B* PROL. 109; (he) **impugnide** WICL. PS. lv. 2; inpugneden WICL. 1 MAC. xi. 41; inpugned (*pple.*) TREV. V. 195.

impulsiōn, sb., *Lat.* impulsiō; *dashing of the sea*, TREV. IV. 199.

imstōn, *see* ȝimstān *under* ȝim.

in, **i**, prep., *O.E.* in, *O.N.* ī,=*O.H.G.*, *Goth.*, *Lat.* in, *Gr.* ἐν; *in, into*: þa children pleȝeden in þere strete HOM. I. 7; in (*on*) eorðan 13; god (*r.* goð) in þane castel 3; þe cnave wes iboren in þere burhe LAȜ. 293; in þan londe 397; in ane time 14063; faren in ænne tun 19186; i (*later text* o) þan hulle 21287; þu scalt do þe i þene wæi 29512; in hire breoste A. R. 230; in heovene & in eorðe A. R. 398; i þe worlde 10; i þen ende of al his live 404; four siðen i ðe ȝere 422; in eorðe KATH. 353; in þe marhen 603; in a stevene with one voice 1396; in halle & i bure 1470; i þon time JUL. 5; in his moder wambe ORM. 168; i þis middelærd 3496; in mine neste O. & N. 282; in o dai in sume tide 709; i ne vinde nenne gult in þisse monne MISC. 47; in eorþe 56; in gode weie 97; in godes hond GEN. & EX. 104; in Engeland HAV. 108; he was ... brod in þe sholdres 1647; in þ(e) ēnde HORN (L.) 1378; in eiþer side ROB. 174; in one dai P. L. S. ix. 56; in þe rode xx. 42; slaȝþ þri in one stroke AYENB. 61; in þe nome of þe fader JOS. 10; in þe morwe 26; in godes name FER. 2943; wenten forþ in heore wei LANGL. *A* PROL. 48; in (*v. r.* on) Englisch *B* xiii. 71; in (*v. r.* on) þe cros TREV. VIII. 75; in erþe CH. C. T. *A* 1896; in live MAND. 135; in þe brinke WICL. JOHN xxi. 4; ones in þe ȝeer E. G. 354.

in², adv., *O.E.* in, inn,=*O.N., Goth.* inn, *O.L.G.* inn, in, *O.H.G.* in; *in*: þe hine in laðede LK. vii. 39; þa he in to þam huse eode MK. ix. 28; (he) rad in et þan estȝete HOM. I. 5; hwam ha leote in & tut 247; ne mai na mon cume in to godes riche 73; ne moste þer na mon in cumen LAȜ. 6712; & bad hine ...

gan in to þan kinge 17675; heo comen in to havene 5150; (heo) goð in A. R. 272; ðis cete ðanne . . . ðise fisses alle in sukeð REL. I. 220; fare in to þe se HAV. 1393; he orn in BEK. 88; þoru þe faste ȝat he con (r. gon) in teo C. L. 877; he com in at Newegate P. S. 218; lete in AYENB. 264; in come PS. xxiii. 7.

in[3], sb., *O.E.* inn, = *O.N.* inni; *inn, house,* LAȜ. 14263; A. R. 260; FL. & BL. 20; inne (*dat.*) s. s. (Web.) 1606; WILL. 1485; TRIST. 1239; LANGL. *B* viii. 4; þær he was at inne ORM. 12739; innes (*pl.*) MAND. 34; M. H. 63; innen (*dat. pl.*) LAȜ. 14007.

    in-lēs, adj., *without lodging,* LEG. 88.

inabiliten, v., *med.Lat.* inhabilitāre; *disqualify*; inabilitinge (*pple.*) TREV. VIII. 448.

in-biggen, v., *inhabit*; inbigge PS. lv. 7.

in-blāwan, v., *inflate*; inblōwith (*pres.*) WICL. I COR. viii. I; inblōwin (*pple.*) wiþ pride WICL. I COR. iv. 6.

in-borȝ, sb., *O.E.* inborh; *bail*; inborȝes (*pl.*) HOM. I. 73; inboreges II. 17.

in-bowen, v., *cf. O.E.* onbȳgan; *bend*; inbowe WICL. IS. xxvi. 5; inbowid (*pple.*) WICL. JOHN xx. 5.

in-brēðen, v., *breathe into*; inbrēthede (*pret.*) WICL. ECCLES. iv. 12.

    inbrēðinge, sb., *act of breathing into,* WICL. 2 KINGS xxxii. 16; inbrething PS. xvii. 16; JOB xxxii. 8.

inbuchen, v., *place in ambush*; inbuched (*pple.*) FER. 2879; *see* enbusche.

inbulled, pple., *cf. O.Fr.* embullé; *embodied in a papal bull,* TREV. VIII. 432.

inc, pron., *O.E.* inc, = *O.L.G.* inc, *M.L.G.* enk, *O.N.* ykkr, *Goth.* igqis (σφώ); *you two,* LAȜ. 695; JUL. 19; H. M. 11; of inc baðen MARH. 21; inc, ginc, gunc, MK. i. 17, xi. 2, 3; hinc HOM. I. 93; hunc (*ms.* hunke) O. & N. 1733; ȝunc ORM. 4493; ganc GEN. & EX. 2830; inker (*gen.*) H. M. 31; unker HAV. 1882; unker æiðer (*either of you two*) LAȜ. 32170; ȝunker ORM. 6183.

incantaciōn, sb., *Lat.* incantātiō; *incantation,* GOW. III. 45.

incarnaciōn, sb., *O.Fr.* incarnation; *incarnation,* ROB. 9.

incarnate, adj., *Lat.* incarnātus; *incarnate*: (he) becom man incarnate LIDG. M. P. 79.

incertain, adj., *O.Fr.* incertain; *uncertain,* LIDG. M. P. 121.

[**incessant**, adj., *Lat.* incessans; *incessant.*]   incessant-lī, adv., *unceasingly,* S. & C. II. lx.

incest, sb., *Fr.* inceste; *incest,* A. R. 204; inceste SHOR. 70.

in-changen, v., *change,* WICL. JOB xiv. 20; inchaungid (*pple.*) WICL. I COR. xv. 51.

inche, *see* ünche.

inclen, v., *give an inkling of, hint*: too incle þe truthe ALIS.[2] (Sk.) 616.

in-clepin, inclepe, v., *call upon, invoke,* WICL. ROM. x. 13; inclepen (*pres. pl.*) WICL. ROM. x. 12; inclepinge (*pple.*) WICL. DEEDS vii. 58; inclepiden (*pret.*) WICL. 2 MAC. viii. 2.

    incleping, sb., *invocation,* WICL. 2 MAC. xviii. 15.

inclinatiōn, sb., *Lat.* inclīnātiō; inclinatiōns (*pl.*) LIDG. M. P. 91.

inclīnen, v., *Lat.* inclīnāre; *incline*: to me thou wille incline TOWNL. 324.

inclūden, v., *Lat.* inclūdere; *include*; inclūdithe (*pres.*) LIDG. M. P. 118.

inclōs, adj., *O.Fr.* enclos; *shut in,* MAN. (F.) 5107.

inclōsin, *see* enclōsen.

inclūs, sb., *Lat.* inclūsus; *recluse,* TREV. VII. 81.

in-cniht, sb., *O.E.* incniht, = *O.H.G.* inkneht; *indoor servant*; inkniht '*cliens*' FRAG. 2.

incombrous, adj., *heavy,* CH. H. F. 862; *see* encombrous.

in-come, sb., *coming in, entrance,* D. ARTH. 2009.

in-comen, v., *come in*; income RICH. 3305; incomen (*ppl.*) RICH. 3991.

    incoming, sb., *coming in,* ['*ingressus*'] WICL. ECCLES. i. 7; incominge PERC. 493.

incomperable, adj., *O.Fr.* incomparable; *from Lat.* incomparābilis; *incomparable,* LIDG. M. P. 24.

incomprehensibele, adj., *O.Fr.* incomprehensible; *incomprehensible,* LUD. COV. 288.

inconstance, sb., *O.Fr.* inconstance; *inconstancy,* CH. C. T. *D* 1958.

inconstant, adj., *O.Fr.* inconstant,. *from Lat.* inconstans; *inconstant*; inconstaunte LIDG. M. P. 57.

incontinence, sb., *O.Fr.* incontinence, *from Lat.* incontinentia; *incontinence,* MAND. 161.

inconvenience, sb., *inconvenience,* SACR. 897.

inconvenient, adj., *O.Fr.* inconvenient; *inconvenient,* CH. BOET. v. 3 (158).

incrementacion, sb., *means of increase,* PALL. xii. 294.

in-crēping, sb., *thrusting in, creeping in,* ALIS. 2168.

incrésin, *see* encrésen.

in-crōken, v., *bend*; incrōke (*imper.*) WICL. ROM. xi. 10.

incubus, sb., *Lat.* incubus; *incubus,* CH. C. T. *D* 880.

incūrable, adj., *O.Fr.* incurable; *incurable,* LANGL. *B* x. 327.

incurren, v., *Lat.* incurrere; *incur (a penalty),* LIDG. M. P. 141.

inde, sb., *O.Fr.* inde ; *indigo,* C. L. 712 ; MAND. 48.

in-delven, v., *dig,* WICL. GEN. xxxv. 4.

inderlī, *see* innerlī *under* inner.

indeterminat, adj., *Lat.* indēterminātus ; *indeterminate,* CH. ASTR. ii. 17 (27).

indewen, v., *A.Fr.* endouer ; *endue, endow* ; indewe '*subarrare*' CATH. 195 ; induin '*doto*' PR. P. 261 ; indewed, endewed (*pret.*) LIDG. M. P. 117.

[**indifferent,** adj., *O.Fr.* indifferent ; *indifferent.*]

    indifferent-lī, adv., *indifferently,* CH. BOET. v. 3 (157).

indigence, *O.Fr.* indigence, *from Lat.* indigentia ; *indigence,* LIDG. M. P. 50, 188.

indignacioun, sb., *O.Fr.* indignacioun ; *indignation,* TREV. I. 33 ; WICL. E. W. 4.

indiscrēt, adj., *O.Fr.* indiscrēt ; *indiscreet,* MIRC 825.

[**indistinct,** adj., *Lat.* indistinctus.]

    indistinct-lī, adv., *without distinction,* PALL. iii. 1064.

indīten, *see* endīte.

indōsen, v., *O.Fr.* endosser ; *load the back ; flog* ; indoost (*ppl.*) TOWNL. 201 ; *see* endōsen.

indre, *see* inre.

in-dronkenen, v., *be drunk* ; indronkenand (*pple.*) PS. lxiv. 11.

inducciōn, sb., *O.Fr.* induction ; *induction (of a priest),* WICL. E. W. 248.

inducting, sb., *induction,* WICL. E. W. 450.

indulgence, sb., *O.Fr.* indulgence ; *from Lat.* indulgentia ; *indulgence,* CH. C. T. *B* 2964.

indurat, adj., *Lat.* indūrātus ; *hardened, obstinate,* LIDG. M. P. 140.

indūren, v., = endūren ; *endure* ; induire, indure TOWNL. 24, 191.

in-dwellen, v., *dwell in* ; indwelle WICL. PS. xxxvi. 27, lxvii. 7.

    indwelling, sb., *indwelling,* WICL. WISD. ix. 15.

ine, *see* innen.   **ineward,** *see under* inne.

in-fáren, v., *O.E.* infaran ; *enter* ; infare PERC. 1537.

infatten, *see* enfatten.

infecten, *see* enfecten.

in-fer, sb., *O.E.* infær ; *entrance,* HOM. I. 231.

infernal, adj., *O.Fr.* enfernal, infernal ; *relating to the lower regions, infernal* ; Pluto the god infernal GOW. II. 263 ; [dremes] ben **infernals** (*pl.*) illusions CH. TRO. v. 367.

in-ficchen, v., *fix in, insert* ; inficchid (*pple.*) WICL. PS. xxxvii. 3, lxviii. 3.

in-fihten, v., *fight against* ; infightand (*pple.*) PS. xxxiv. 1 ; infaght (*pret.*) PS. cxix. 7 ; TREV. VII. 403.

infirmitē, sb., *O.Fr.* enfirmeteit, -eté ; *infirmity,* LANGL. *C* x. 230 ; GEST. R. 69 ; LUD. COV. 398 ; LIDG. M. P. 161 ; **infirmits** (*for* infirmites) (*pl.*) ALEX. (Sk.) 4279.

inflacioun, sb., *flatulence,* PALL. xi. 504.

in-flēing, sb., *refuge* ; ['*refugium*'] PS. cxliii. 2.

in-flokken, v., *flock in* ; inflokkes (*pres.*) A. P. ii. 167.

influence, sb., *O.Fr.* influance ; *influence,* LIDG. M. P. 16, 247.

influent, adj., *that is poured on, bestowed,* LIDG. M. P. 241.

in-folewing, sb., *consequence* ; infolewingis (*pl.*) WICL. ECCLUS. xxxii. 23.

informacioun, sb., *Fr.* information ; *information,* GOW. I. 272.

informin, v., *O.Fr.* enformer ; *inform,* PR. P. 261.

infortunat, adj., *unfortunate, unlucky,* CH. ASTR. ii. 4 (19) ; C. T. *B* 302.

infortūne, sb., *O.Fr.* infortune, *from Lat.* infortūnium ; *bad fortune ; misfortune,* GOW. II. 208 ; CH. TRO. iii. 1577 ; R. R. 5496 ; LIDG. M. P. 75.

infortuned, adj., *unlucky* ; infortuned night CH. TRO. iv. 716.

infortuning, sb., *misfortune,* CH. ASTR. ii. 5 (19).

in-gang, sb., *O.E.* ingang, = *O.H.G.* ingang ; *entrance, admission,* PS. cxx. 8 ; ingonge LANGL. *B* v. 638 ; ingang *C* viii. 282 ; ingong HOM. II. 203 ; A. R. 62 ; LEG. 25 ; inȝeong LAȜ. 28370.

in-gangan, v., *O.E.* ingān ; *enter* ; inga PS. xxv. 4, xlviii. 20 ; ingo PS. v. 8 ; **ingaas** (*imper.*) PS. xcix. 2, 4 ; **ingōinge** (*pple.*) WICL. GEN. xxxviii. 16.

    ingōing, sb., *entrance,* WICL. PS. lxvii. 25 ; inguoinge AYENB. 72, 105 ; **ingōingus** (*pl.*) WICL. PS. lxvii. 25.

[**in-ȝein**] i-gein, adv., *cf. O.N.* īgegn (*through*) ; *again, back,* M. H. 47, 89, 91, 114.

ingen, sb., ? = engin : againste jeauntis ongentill have we joined with **ingendis** (*pl.*) YORK xxxi. 13.

in-ȝetten, v., *pour in, infuse* ; inȝettis (*pres.*) HAMP. TR. 2.

    inȝetting, sb., *pouring on, bestowal* ; inȝettinge of grace HAMP. TR. 4.

ingnel, adj., *O.Fr.* ignel ; *quick,* AYENB. 141.

ingot, sb., *Fr.* ingot ; *ingot,* CH. C. T. *G* 1223, 1228, 1233 ; **ingottes** (*pl.*) 818.

ingratitūde, sb., *O.Fr.* ingratitude ; *ingratitude,* AYENB. 18.

ingredientes, sb., pl., *materials,* B. B. 127.

ingroost, pple., *comprehended* :   God in persons thre, alle in oone substance air ingroost TOWNL. 170.

inhabitable, adj., *O.Fr.* inhabitable ; *from Lat.* inhabitābilis ; *uninhabitable*, MAND. 161.

inhabiten, *see* enhabiten.

inhabitting, sb., *dwelling,* ALEX. (Sk.) 3736.

in-hēd, sb., *cf. O.E.* ingehygd, -hȳd ; *? deliberation,* HOM. I. 69.

in-hélden, v., *pour in* ; inhielde (*imper.*) CH. TRO. iii. 43.

inheritauns, sb., *inheritance,* LUD. COV. 242 ; enheritauns 244.

inheriten, v., *inherit* ; inherite, ineritin PR. P. 261 ; *see* enherite.

in-hīne, sb., *? domestic,* JUL. 33.

inhonest, adj., *Lat.* inhonestus; *dishonorable*; inhoneste AYENB. 220.

inhoneste-līche, adj.,*dishonourably*,AYENB. 177.

iniqvitē, sb., *O.Fr.* iniqvité ; *iniquity,* CH. C. T. *B* 358 ; ALIS. 131 ; iniqvitees (*pl.*) CH. C. T. *I* 442.

injoien, v., *see* enjoien.

injūre, sb., *O.Fr.* injure, *Lat.* injūria ; *injury,* CH. TRO. iii. 967 ; injurie CH. C. T. *B* 2922 ; injuries (*pl.*) *B* 3001.

injuste, adj., *O.Fr.* injuste ; *unjust,* LIDG. M. P. 120.

inke, sb., *O.E.* inca ; *? apprehension, misgiving* ; inke (*ms.* hinke) and age GEN. & EX. 432.

inke, *see* enke.

inker, pron., *O.E.* incer, = *O.N.* ykkr, *Goth.* igqar ; *from* inc ; *of you two,* H. M. 3 ; inker (*ms.* incker), unker LAȝ. 5102, 26541 ; gunker GEN. & EX. 398 ; incre (*ms.* yncre) (*dat. m.*) MAT. ix. 29.

inkerlī, *see* enkerlī.

in-lǣden, v., *lead, carry in* ; inlǣd (*pret.*) [' *induxit* '] PS. lxxvii. 54 [inladde WICL.] ; inledde PS. lxxxvii. 8.

in-lǣten, v., *let in* ; inlēte (*pret.*) TRIST. 629 ; he was inlǣte (*pple.*) OCTOV. (W.) 1186.

in-laȝe, sb., ' *sujet a la lei le rei*,' REL. I. 33.

in-lāte, sb., = *Ger.* einlasz ; *inlet, entrance,* M. H. 51 ; C. M. 18078.

in-lawe, v., *inlaw, free from outlawry,* CATH. 196.

in-līche, adv., *cf. O.E.* onlīce ; *alike,* A. P. i. 544, 603.

in²-līche, adv., *O.E.* inlīce, = *O.HG.* inlīho ; *inly, sincerely,* LANGL. *B* xiv. 89 ; inli WICL. DEUT. ii. 30 ; HOCCL. i. 237.

in-liggen, v., *lie upon* ; inlai (*pret.*) [' *incubuit* '] PS. ciii. 38.

in-līhten, v., *O.E.* inlȳhtan, = *O.H.G.* inliuhten, *Goth.* inliuhtjan ; *enlighten, illuminate* ; inlȝted (*pple.*) WICL. 2 COR. iv. 6.

in-līhtnen, v., *enlighten* ; inliȝtne WICL. EPH. iii. 8 ; inlȝtened (*pple.*) WICL. EPH. i. 18.

in-loȝen, v., *inflame* ; inloghed (*pret.*) PS. civ. 19 ; inlowed (*pple.*) PS. lxxii. 21.

inmelle, *see* mel.

in-met, sb., *intestines,* D. ARTH. 1122.

inmoeveable, adj., *O.Fr.* immouvable ; *immovable* ; inmoeveable CH. BOET. v. 6 (173).

inmoeveabletē, sb., *unchangeableness,* CH. BOET. v. 6 (173).

inmong, inmonges, *see* mang.

inmortal, *see* immortal.

inne, adv., *O.E.* inne, = *O.L.G.*, *O.H.G.* inne, *O.N.* inni, *Goth.* inna ; *in,* A. R. 62, 280 ; HAV. 807 ; LANGL. (Wr.) 13597 ; CH. C. T. *F* 578 ; HOCCL. iii. 6 ; þan castle þer Æstrild wes inne [ine] LAȝ. 2485 ; he is inne sone forgeten REL. I. 184 ; þo he wes inne II. 272 ; þulke hous ... as seint Thomas was inne ibore BEK. 83 ; þat hus þat brǣd is inne don ORM. 3530 ; þe put þat he was inne iworpe JOS. 221 ; a cloþ of silk sche wond him inne GREG. 133 ; he went inne PERC. 437.

innemest, adj., *O.E.* innemest ; *inmost* : þe inemaste C. L. 809.

inne-warde, sb., *O.E.* inneweard ; *intestines; womb,* ROB. (W.) 1547 ; ineward MISC. 151.

innen, adv. & prep., *O.E.* innan, = *O.L.G.*, *O.H. G.*,*O.N.* innan, *O.Fris.* inna (ina), *Goth.* innana ; *in* : innen huse MAT. ix. 10 ; innen þe FRAG. 6 ; þe inne 7 ; innan HOM. I. 43 ; inne griðe & inne friðe LAȝ. 184 ; inne [ine] lut ȝere 387 ; inne þe see P. L. S. viii. 42 ; inne, ine A. R. 20, 30 ; ine O. & N. 964 ; ine hevene AYENB. 103 ; ine an oþre stede 194 ; riche ine guodes 162 ; inewið A. R. 38 ; *comp.* **an-, bi-, wið-innen.**

inner, adv., *O.E.* innor, = *O.H.G.* innor, *O.N.* innar; *further in* : þe sparewe inner (*ms.* innere) crap (*read* crǣp) LAȝ. 29282 ; *see* inrest.

inre, adj., *O.E.* inra, innera, = *O.N.* innri, *O. H.G.* innero ; *inner, interior,* A. R. 4 ; innere TREV. I. 53 ; þe innere halle WICL. EZ. xliv. 18 ; hit entird as watir in his inerere [' *interiora* '] (*inward parts*) HAMP. PS. cxviii. 17.

inner-lī, adv., *Ger.* = innerlich, *Dan.* inderlig ; *internally,* WICL. IS. xxxiv. 6 ; in(n)er-, inderli MAN. (F.) 3195 ; inderli GOW. I. 227.

inneste, sb., *? O.E.* innosta ; *intimate friend,* WICL. PROV. xxvi. 22.

inneð, sb., *O.E.* innað, innoð, = *O.H.G.* innōd ; *womb* ; inneþe (*dat.*) HOM. I. 83 ; into ðe maidenes inneðe '*in virginis uterum*' HICKES I. 168.

innien, v., *O.E.* innian, = *O.Fris.* innia, *M.L.G.* innen ; *lodge in an inn; get in* (*a harvest*) ; inni ROB. 336 ; inne mi corn H. v. 69 ; innes (*pres.*) JOS. 174 ; inned (*pple.*) WILL. 1638 ; CH. C. T. *A* 2192 ; were our seed inned HOCCL. ii. 29 ; *comp.* ȝe-innian.

inninge, sb., *O.E.* innung ; *getting in* our seed inninge, HOCCL. ii. 15.

in-nimen, v., *take in* ; innomen (*pple.*) A. P. i. 703.

innocence, sb., *O.Fr.* innocence ; *innocence,* AYENB. 181 ; LANGL. *B* xvii. 286 ; CH. C. T. *I* 325, *B* 2966.

innocent, adj., *O.Fr.* innocent ; *innocent,* PR. P. 262 ; AYENB. 150 ; MAND. 134 ; ennosent .AUD. 60 ; innossente A. P. ii. 672 ; þe innossent (*sb.*) A. P. i. 666 ; þis seli innocent (*babe*) LAI LE FR. 164 ; **innocentz** (*pl.*) CH. C. T *B* 2368.

innumerable, adj., *Lat.* innumerābilis ; *innumerable,* AYENB. 267 ; LUD. COV. 241.

inobedience, sb., *O.Fr.* inobedience ; *disobedience,* A. R. 234 ; GOW. I. 83 ; WICL. ROM. v. 19 ; CH. C. T. *I* 391.

inobedient, adj., *disobedient,* CH. C. T. *I* 292 ; LANGL. *C* vii. 19 ; A. P. ii. 237.

inobeishaunce, sb., *Fr.* inobéissance ; *disobedience,* ['*inobedientia*'] TREV. VI. 123.

inobeishaunt, adj., *disobedient,* WICL. DEUT. viii. 20.

inoint, *see* enoint.

in-over, adv., *also,* PS. viii. 8, xv. 7.

inp-, *see* imp-.

in-parken, v., *impark, enclose* ; **inparkis** ALEX. (Sk.) 5499 ; **inparkid** (*pret.*) 4702.

in-putten, v., *place in, place on* ; **inputtide** (*pret.*) WICL. I MAC. xi. 13 ; **inputtiden** WICL. DEEDS xxviii. 10.

in-räs, sb., *onrush, attack,* ['*irruptio*'] PS. xc. 6.

in-rēad, adj., *very red,* A. R. 402 ; **inred** S. S. (Wr.) 6 L.

inrest, adj.,=*O.H.G.* innerōst ; *from* **inner** ; *innermost,* ORM. 1017 ; CH. BOET. iv. 6 (136) ; **inreste** (*dat.*) PS. lxxxv. 13.

in-rīsen, v., *rise against, rebel* : wicked **inräse** (*pret.*) in me PS. lxxxv. 14 ; **inrisen** WICL. PS. xxvi. 12.

   inrīsere, sb., *rebel, insurgent* ; **inrīseris** (*pl.*) WICL. PS. xliii. 6.

inrollen, v., *O.Fr.* enroller, *med. Lat.* inrotulāre ; *enroll* ; **inrold** (*pple.*) TOWNL. 92 ; *see* **enrollin.**

insame, *see* samne.

in-scheden, v., *pour over ; immerse* : in dewe of heven thou shalt be **inshed** (*pple.*) ['*infunderis*'] WICL. DAN. iv. 22 ; the teris enshed WICL. JUD. vii. 23.

in-schilder, sb., *defender,* PS. cxx. 5.

in-seil, sb., *O.E.* insegele,=*O.H.G.* insigili ; *seal,* HOM. I. 127 ; MARH. 5 ; **inseзles** (*pl.*) ORM. DEDIC. 260.

in-sēken, v., *seek* ; **insēkinge** (*pple.*) WICL. HEB. xi. 6.

in-sēme, adv., *cf. O.E.* gesōme ; *together,* A. P. i. 838.

in-senden, v., *send* ; **insent** (*pple.*) PS. xxxix. 4.

in-setten, v., *O.E.* insettan ; *place over, place*

in ; inset PS. xx. 6 ; **insettinge** (*pple.*) WICL. 2 MAC. vii. 21 ; **insett** (*pple.*) WICL. ROM. xi. 23.

in-siht, sb.,=*Ger.* einsicht ; *insight, intelligence,* LAЗ. 30497 ; ORM. 14398 ; C. L. 558 ; M. H. 2 ; insiзt TREAT. 132 ; WILL. 94 ; A. P. ii. 1659 ; insight PR. C. 253 ; mine insihte O. & N. 1187.

insipiens, sb., *Lat.* insipientia ; *folly* : whan in women be found no incipiens, than put hem in trust S. & C. II. lviii.

in-smīten, v., *smite in* ; dreed is **insmiten** (*pple.*) ['*incussus est*'] WICL. 2 MAC. xii. 22.

insolence, sb., *O.Fr.* insolence ; *insolence,* CH. C. T. *I* 391 ; LIDG. M. P. 90, 94.

insolent, adj., *Fr.* insolent ; *insolent,* CH. C. T. *I* 399.

insolible, adj., *O.Fr.* insoluble ; *that cannot be dissolved or solved ;* ['*insolubilis*'] WICL. HEB. vii. 16 ; this is .an insolible (*sb.*) LIDG. M. P. 43.

inspeccioun, sb., *Lat.* inspectiō ; *inspection,* LIDG. M. P. 137.

inspiracioun, sb., *O.Fr.* inspiration ; *inspiration,* TREV. V. 445 ; MAND. 16.

inspiren, v., *Lat.* inspirāre ; *inspire* ; inspire LIDG. M. P. 140 ; **inspírid** (*pple.*) WICL. 2 TIM. iii. 16.

in-standen, v., *stand near, be near* ; **instonding,** instoondinge, instondende (*pple.*) WICL. GEN. xxxviii. 27, JUDG. xi. 5, 3 ESD. v. 47.

instaunce, sb., *O.Fr.* instance ; *instance,* CH. BOET. v. 6 (174) ; WICL. JUD. iv. 8, 1 MAC. xi. 40* ; LUD. COV. 240.

in-steppen, v., *O.E.* insteppan ; *step in* ; instepped PS. cxxxviii. 3.

institūt, pple., *instituted* : this newe parsoun is institut in his churche P. S. 326.

institutiōn, sb., *O.Fr.* institution ; *institution,* WICL. E. W. 248.

instōren, *see* enstōre.

instrument, sb., *O.Fr.* instrument ; *instrument ; musical instrument,* MAND. 60 ; S. & C. II. 78 ; BEK. 1886 ; enstrument AUD. 55 ; **instrumentes** (*pl.*) MED. 884 ; WICL. E. W. 218 ; PR. C. 9264 ; ALEX. (Sk.) 1564.

instuing, sb., *instituting* (*to office*), WICL. E. W. 450.

insuffisance, sb., *O.Fr.* insuffisance ; *insufficiency,* MAND. 315.

insuffisant, adj., *O.Fr.* insuffisant ; *insufficient,* MAND. 293 ; insufficient LIDG. M. P. 240.

insūre, v., *insure,* TOWNL. 191 ; *see* ensūre.

intelligence, sb., *O.Fr.* intelligence ; *intelligence,* LIDG. M. P. 9 ; intelligens LUD. COV. 248, 273.

intemper, adj., *intemperate,* ROB. (W.) 8848*.

intemperat, adj., *Lat.* intemperātus ; *intemperate,* LIDG. M. P. 258.

intemperance, sb., *O.Fr.* intemperance ; *foulness of weather,* [*'intemperies'*] TREV. II. 291.

intenciōn, *see* entencioun.

intenden, *see* entenden.

intent, *see* entente.

in-tēon, v., *march in* ; inteo (? in teo) C. L. 877.

interen, v.,=entrin ; *enter* : the king is intered (*pple.*) intc this citee LIDG. M. P. 10.

interesse, sb., *Lat.* interesse ; *interest,* LIDG. M. P. 210.

interpretacioun, sb., *O.Fr.* entrepretation ; *interpretation,* LIDG. M. P. 238; interpretacion LUD. COV. 272.

interpretator, sb., *Lat.* interpretātor ; *interpreter,* TREV. II. 221, 241 ; V. 397.

interrupciōn, sb., *O.Fr.* interruption, *Lat.* interruptiō ; *interruption,* GOW. I. 36.

interrupt, adj., *O.Fr.* interrupte ; *deprived of,* N. P. 6.

intervalle, sb., *O.Fr.* entreval ; *interval,* CH. C. T. *B* 2723.

intestate, adj., *O.Fr.* intestat ; *intestate,* LANGL. *B* xv. 134.

in-til, prep., *cf. Swed.* intill ; *into, unto,* HAV. 128 ; TRIST. 1386 ; LANGL. *B* xiii. 210 ; WICL. E. W. 238 ; PR. C. 6644 ; M. H. 14 ; ȝeden þær intil þat hus ORM. 6452.

intīre, adj.,=entēr ; *entire,* LIDG. M. P. 68 ; entier 100.

intīsen, intīsing, *see* entīsen.

in-tō, prep., *into, unto,* MARH. 4 ; KATH. 7 ; A. R. 54 ; ORM. 635 ; O. & N. 996 ; BRD. 32 ; MAND. 7 ; B. DISC. 460 ; heo hatieð þe swiðe into þan bare dæðe LAȝ. 14099 ; from hevene into eorþe alihte C. L. 916 ; more encens·into þe fire he caste CH. C. T. *A* 2429 ; into (*v. r.* unto) þe dai *G* 1045 ; inte MISC. 35.

intréten, *see* entréten.

introductorie, sb., *O.Fr.* introductoire ; *introduction,* CH. ASTR. PROL. (3).

inventen, v., *O.Fr.* inventer ; *invent* ; **inventid** (*pple.*) LIDG. M. P. 64.

inviolate, adj., *Lat.* inviolātus ; *inviolate,* LIDG. M. P. 92.

in-ward, adv. & adj., *O.E.* inweard,=*M.H.G.* inwart ; *inward, internal; earnest,* MARH. 8 ; inward bone JUL. 44 ; wið inwarde heorte HOM. I. 209 ; in inwarde helle KATH. 1815 ; inward (*adv.*) A. R. 272 ; inwardes 92.

   inward-lich, adj., *O.E.* inweardlīc ; *inwardly; earnestly* ; inwardliche bede HOM. II. 45 ; **inwardlīche** (*adv.*) A. R. 282 ; inwardliȝ ORM. 697 ; inwardliche bonen JUL. 45 ; inwardliche (*later ver.* inwardli) bowe þin herte WICL. PROV. ii. 2 ; **inwardlūkest** (*superl.*) lerede alle his icorene A. R. 282.

   inwardnesses, sb., pl., *inward parts* ; [*'in praecordiis'*] WICL. WISD. iv. 14*.

in-wenden, v., *enter* ; inwent (*pret.*) [*'ingrediebar'*] PS. xxxvii. 7.

in-wēten, v., *dip in* ; inwĕt (*pple.*) WICL. PS. lxvii. 24.

in-wit, sb., *conscience,* A. R. 4 ; inwit BRD. 26 ; SHOR. 55 ; AYENB. I ; LANGL. *B* xvii. 278 ; GEST. R. 18 ; PR. C. 5428 ; **inwitte** (*dat.*) MAN. (H.) 155 ; inwittis (*pl.*) WICL. HEB. xii. 3.

in-wið, prep., *within,* KATH. 172 ; inwið his heorte HOM. I. 187 ; inwiþ SPEC. 48 ; inwith CH. C. T. *E* 1944 ; PERC. 611 ; A. P. i. 969 ; JUL. 7 ; inewið A. R. 38 ; iwið H. M. 29.

in-wlappen, v., *envelope* ; inwlappith (*pres.*) himsilf WICL. 2 TIM. ii. 4 ; **inwlappinge** (*pple.*) WICL. EZ. i. 4 ; **inwlappid** (*pret.*) WICL. 2 PET. ii. 20.

in-wonen, v., *O.E.* inwunian ; *inhabit* ; inwone PS. lxviii. 36 ; **inwones** (*pres.*) D. TROY 133 ; **inwonet** (*pret.*) D. TROY 13864.

in-wrappen, v., *wrap up; involve* ; inwrappe WICL. PROV. xxix. 6 ; lest i enwrappe (*subj.*) thee with hem [*'ne involvam te cum eo'*] WICL. I KINGS xv. 6 ; **inwrappide** (*pret.*) WICL. 4 KINGS ii. 8 ; **inwrappid** (*pple.*) WICL. NUM. iv. 15.

in-wrīten, v., *inscribe* ; inwrite (*pple.*) WICL. ECCLUS. xlviii. 10.

ipocras, sb., *O.Fr.* ipocras ; *hipocras,* CH. C. T. *E* 1807 ; B. B. 125, 280 ; of win and spices is maad good ipocras LIDG. M. P. 216.

ipocrisie, sb., *O.Fr.* ipocrisie, *from Lat.* hypocrisis, *Gr.* ὑπόκρισις ; *hypocrisy,* A. R. 342 ; AYENB. 17 ; GOW. I. 63 ; CH. C. T. *I* 391 ; ipocresi AUD. 31.

ipocrite, sb., *O.Fr.* ipocrite ; *hypocrite,* A. R. 128 ; GOW. I. 62 ; epocrite AUD. 15 ; **ipocrite** (*adj.*) manere WICL. E. W. 89 ; **ipocritis** (*pl.*) WICL. S. W. I. 177.

ipotame, sb., *O.Fr.* ypotame ; *hippopotamus,* ALIS. 5179 ; ipotanos 6554 ; ipotamus ALEX. (Sk.) 156 ; **ipotaines** (*pl.*) MAND. 268.

ippen, *see* üppen.

irain, sb., *O.Fr.* iragne ; *spider,* PS. xxxviii. 12 ; ireine, irein WICL. PS. xxxviii. 12 ; lxxxix. 9 ; **ireins** (*pl.*) WICL. JOB viii. 14 ; *see* aranie.

iral, sb., *cf.* iris ; *some precious stone* ; hir paietrelle was of irale fine ERC. 61 ; stones iraille ANT. ARTH. (R.) xlvi ; stones of iral ANT. ARTH. (L.) xlvi.

irchoun, sb., *O.Fr.* ireçon ; *urchin, hedgehog,* WICL. IS. xiv. 23 ; urchon *'ericius'* PR. P. 512 ; hircheoune BARB. xii. 353 ; **irchōnes** (*pl.*) PS. ciii. 18 ; urchounes MAND. 290.

ire, *see* irre.

īre, sb., *O.Fr.* īre ; *ire, anger,* ROB. 185 ; AYENB. 147 ; CH. C. T. *A* 1782 ; RICH. 3633 ; PR. C. 8588.

   ire-ful, adj., *angry, passionate,* RICH. 365 ; S. B. W. 20 ; D. TROY 1330.

īre, *see* ēare *and* eirə.

īren, īsen, sb., *O.E.* īren, īsen, īsern,=*O.Fris.*
īsern, *O.L.G.*,*O.H.G.*,*O.N.* īsarn,*Goth.*eisarn ;
iron ; iren A. R. 74, 160 ; ORM. 4129 ; MAND.
50 ; CH. C. T. *A* 500 ; LIDG. M. P. 54 ; PR. C.
8033 ; þet iren HOM. I. 23 ; irin GEN. & EX. 467 ;
iron RICH. 2529 ; irn JUL. 59 ; ire FL. & BL.
˙6 ; TREAT. 135 ; ise (*r. w.* wise) O. & N. 1030 ;
izen AYENB. 139 ; ise ALIS. 5149 ; irene (*dat.*)
LAȝ. 7831 ; irne KATH. 2209 ; barres of irne
D. TROY 6018 ; bonde fulle fast in irens (*pl.*)
(*fetters*) TOWNL. 307 ; in isnes and ine veteres
AYENB. 128 ; he sleped in his irnes (*armour*)
GAW. 729 ; *comp.* fūr-īre.

iren-smiþ, sb., WICL. I KINGS xiii. 19.

īren², adj., *O.E.* īren, īsern, *cf. O.H.G.* īsarnīn,
īsenīn, *Goth.* eisarneins ; *made of iron* :
irene ponne FRAG. 4 ; irene (*pl.*) MAND. 30 ;
irnene HOM. I. 149 ; KATH. 1947, 2150 ; JUL.
57 ; irenen (*dat. pl.*) HOM. I. 121 ; irene LAȝ.
1019 ; RICH. 3981.

īrened, adj., *iron*, PS. ii. 9, cvi. 16.

īris, sb., *Lat.* īris ; *a sort of precious stone*,
MAND. 219.

Īrisch, adj., *Irish*, O. & N. 322 ; þa Irisce
LAȝ. 18059.

irk, adj., = *M.H.G.* erk(-lich) ; *distasteful*,
MAN. (H.) 297 ; M. H. 23 ; FLOR. 1519 ; ANT.
ARTH. vi ; TOWNL. 155 ; erk R. R. 4870 ;
irke (*pl.*) TRIAM. 463 ; ISUM. 118 ; SACR. 917.

irk-sum, adj., *irksome*, '*fastidiosus*,' PR. P.
266.

irksumnesse, sb., *irksomeness*, PR. P. 266.

irkin, v.,=*M.H.G.* erken ; *be irksome*, '*fasti-
dio*,' PR. P. 266 ; irke MIRC 526 ; MAN. (F.)
11122 ; PR. C. 8918 ; LUD. COV. 178 ; erke
AUD. 74 ; irked (*pret.*) GAW. 1573 ; *comp.*
for-irken.

Īr-lond, pr. n., *O.E.* Īralond ; *Ireland*, LAȝ.
17292 ; Irland BRD. 15.

irnen, *see* rinnen. īrnen, *see* īren².

īroni, adj., *made of iron*, WICL. DEUT. xxviii.
23 ; irunni WICL. DAN. ii. 40.

irour, sb., *O.Fr.* iror ; *anger*, S. S. (Web.) 954.

irous, adj., *O.Fr.* iros ; *wrathful* ; CH. C. T.
D 2014 ; irrous ALIS. 329 ; irus MAN. (H.)
116 ; irose WICL. E. W. 307 ; irous, jrus D.
ARTH. 1329, 1957.

irous-lī, adv.,*wrathfully* ; irouslie D. ARTH.
2530 ; irusli BARB. viii. 144*.

irre, adj., *O.E.* yrre, eorre,=*M.Du.* erre,*O.L.
G.* irre,*O.H.G.* irri,*Goth.* airzeis ; *angry, passionate* : irre (*ms.* ire, yr) on his mode LAȝ.
18597 ; eorre MAT. xviii. 34 ; A. R. 304.

irre², sb., *O.E.* irre, yrre, eorre, ierre,=*M.Du.*
erre *n.* ; *anger, passion*, ORM. 18000 ; eorre
MISC. 67 ; urre REL. I. 175 (MISC. 114) ; erre
MIRC 1225 ; eorre (*dat.*) HOM. I. 83 ; A. R. 116*.

irregularitē, sb., *O.Fr.* irregularité ; *irregu-
larity*, WICL. S. W. I. 87.

irregulēr, adj., *O.Fr.* irregulier ; *irregular*,
WICL. S. W. I. 87 ; irreguleer CH. C. T. *I* 782 ;
ye are irregulere TOWNL. 198.

irreligiositē, sb., *O.Fr.* irreligiosité ; *irre-
ligiousness*, WICL. 3 ESD. i. 52.

[irren] arren, v.,=*M.L.G.* erren, irren, *O.H.
G.* irran, *Goth.* airzjan (πλαυᾶυ) ; *irritate, make
angry* ; arrin LUD. COV. 316 ; arreden (*pret.*)
SAINTS (Ld.) xliv. 296 ; WICL. DEUT. xxxii. 16.

irreverence, sb., *O.Fr.* irreverence ; *irreverence*, CH. C. T. *I* 391, 403 ; HAMP. TR. 10.

irsien, v., *O.E.* yrsian, eorsian ; *become angry* ;
irseden (*ms.* yrseden) (*pret.*) MK. xiv. 5.

is-, *see* es-.

īs, sb., *O.E.* īs, *cf. O.Fris.*, *O.H.G.* īs *n.*, *O.N.*
īss *m.* ; *ice*, HOM. I. 43 ; MISC. 148 ; LEG. 165 ;
(*ms.* ys) GEN. & EX. 99 ; iis WICL. WISD. xvi.
22 ; PL. CR. 436 ; ises (*ms.* yses) (*gen.*) GEN.
& EX. 97 ; īse (*dat.*) R. P. v ; ROB. 463 ;
MAND. 130.

īs-ikel, sb., ? *O.E.* īsgicel ; *icicle* ; īsikles,
-ekeles (*pl.*) LANGL. *B* xvii. 227, *C* xx. 193 ;
isseikkles GAW. 731.

Īs-lond, pr. n., *O.N.* Īsland ; *Iceland*, LAȝ.
22488 ; Island MAN. (F.) 10475.

īsen, *see* īren.	isle, *see* üsle.

isope, sb., *O.Fr.* ysope ; *hyssop*, PR. P. 266 ;
isopp VOC. 265.

issen, v., *O.Fr.* issir ; *go out* ; isshen MAN.
(H.) 3345 ; iche (*pres.*) D. ARTH. 1411 ;
isschewis 4060, 3116 ; isseden (*pret.*) MAN.
(F.) 3466 ; issit D. TROY 6631, 6998 ; issed
LIDG. M. P. 6 ; isshit D. TROY 5784.

issue, sb., *O.Fr.* issue ; *issue*, LANGL. *B* xvi.
196 ; MAN. 5084 ; issu LIDG. M. P. 44, 249 ;
isshue MAN. (H.) 19 ; ischeue D. ARTH. 1943.

it, *see* hit *under* hé.

īþe, *see* ūðe, ēaðe.

iþen [ithand], adj., *O.N.* iðinn ; *diligent*,
BARB. iii. 285.

iþen-lī, adv., *diligently, continually*, M. H.
108 ; C. M. 2871 ; iþenli [iþinlik] 23280 ; ith-
andli BARB. ii. 57, iii. 275, 288, vi. 327, x. 287.

ível, *see* üvel.

iven, sb.,=*M.Du.* ieven ; *ivy*, CATH. 199.

iven-lēf (? iven lēf), sb., *leaf of ivy* ; iven
lefes (*pl.*) S. S. (Wr.) 181.

ívi, sb., *O.E.* ifig,=*O.H.G.* ebah ; *ivy*, '*hedera*,' PR. P. 266 ; O. & N. 27 ; S. S. (Web.)
200 ; WICL. JONAH iv. 6 ; ive P. L. S. xxxi.
358 ; *comp.* eorþ-ivi.

ívi-berie, sb., *ivy-berry*, MAND. 168.

ívi-lēf, sb., *ivy leaf* ; all nis worth an ivi
lefe GOW. II. 21 ; he moot(e) pipen in an ivi
leef CH. C. T. *A* 1838 ; pipe wiþ an ivi lefe
WICL. E. W. 372.

ívi-stalke, sb., *stalk of ivy*, TREV. III. 199.

ivori, sb., *O.Fr.* ivoire ; *ivory*, C. L. 737 ;

ivori [evori] TREV. III. 273; ivori, ivoire CH. C. T. *B* 2066; CH. D. BL. 946; ivore FLOR. 598; evorie ALIS. 7666.

ivoired, adj., *of ivory*, PS. xliv. 9.

īw, sb., *O.E.* īw, ēow, *cf. O.N.* ȳr *m.*, *O.H.G.* īwa *f.*; *yew*, '*taxus*,' FRAG. 3; eu (*ms.* ew) REL. I. 7; CH. C. T. *A* 2923.

j.

jáce, sb., *prov. Eng.* (*Devon*) jace; *kind of fringe*; jáces (*pl.*) DEP. R. iii. 130.

jácen, v., *trot, jog along*; jácede (*pret.*) LANGL. *C* xx. 50.

jacinct, sb., *cf. Prov.* jacint; *from Lat.* hyacinthus; *jacinth*; *stuff dyed blue*, H. M. 43; jacinkte ALIS. 5682; jasinkt WICL. EX. xxv. 3; jacinte WICL. 2 PARAL. ii. 7; jacinctes (*pl.*) FL. & BL. 283; jacinctis WICL. S. SOL. v. 14.

jacinctine, adj., *of the colour of hyacinth*: the coope aa jacinctine WICL. EX. xxxix. 20; skinnes jacinctines (*pl.*) xxv. 5.

jáde, sb., (*ms.* Iade), *wretched horse*, CH. C. T. *B* 4002.

jagge, sb. (*ms.* iagge); *jag* (*of cloth*), PR. P. 255.

jaggen, v., *jag, chop*; jaggede (*pret.*) thame in sonder D. ARTH. 1123; jaggid (*pple.*) '*fractillosus*' PR. P. 225; jaggid hode TOWNL. 319; a jupone . . . jaggede in schredes D. ARTH. 905.

jaggen², v., *thrust*; jogges (*pres.*) D. ARTH. 2892; (thei) jaggede (*pret.*) D. ARTH. 2909.

jagounce, sb., *O.Fr.* jagonce; *a sort of gem*, LIDG. M. P. 188; jagounces (*pl.*) R. R. 1117.

jai, sb., *O.Fr.* jai, gai; *jay*, VOC. 177; SPEC. 52; P. S. 328; ALIS. 142; CH. C. T. *A* 642.

jaile, jailer, *see* gaile, gailer.

jakke, sb., = *O.Fr.* jaque, *or M.Du.* jakke; *jacket*, '*balteus*,' PR. P. 256; (*ms.* iakke) FER. 3689; jackis (*pl.*) TREV. VIII. 550.

jak-tēþ, sb., pl., *? for* *chēak-tēþ; *see* chēake; *double teeth*, LANGL. *C* xxiii. 191*.

jalous, *see* gelūs. **jalousie,** *see* gelusie.

jambē, adj., *O.Fr.* jambé; *active, nimble*; a jambe stede D. ARTH. 2894; a jambi stede 373.

jambeaux, sb., pl., *from O.Fr.* jambe; *armour for the legs*, CH. C. T. *B* 2065; jambeaus FER. 5613.

jamnis, sb., pl., *? jaws*, D. TROY 939.

jangle, sb., *O.Fr.* jangle; *jangling*; *idle talk, slander*, ALEX². (Sk.) 458; LANGL. *C* v. 174*; ful of rouninges and jangles (*pl.*) CH. H. F. 1960; jangeles LANGL. *C* vii. 133.

janglen, v., *cf. Du.* jangelen, *O.Fr.* jangler; *jangle, prate, chatter*; jangelin '*garrio, blatero*' PR. P. 256; jangli (*ms.* iangli) REL. II. 243; AYENB. 214; to jangle and to jape LANGL. *B* ii. 94; jangle [gangle] ALIS. 7413; jangil HAMP. PS. xciii. 4; gangliez (*pres.*) LEB. JES. 862; jangle CH. C. T. *F* 220; ganglinde (*pple.*) AYENB. 226; janglande A. P. iii.

90; jangillande YORK vii. 47; **jangled** (*pret.*) MISC. 226.

jangler, sb., *O.Fr.* jangler, jangleur; *jangler*, CH. P. F. 457; jangeleres (*pl.*) LANGL. *A* PROL. 35.

jangleresse, sb., *female jangler*, CH. C. T. *D* 638; jangleresses (*pl.*) *E* 2307.

janglinge, sb., *jangling*: janglinge is, whan a man speketh to muche CH. C. T. *I* 406; jangling D. TROY 671; janglinge LANGL. *C* xi. 270.

Janiver, sb., *Fr.* janvier; *January*, TREV. VIII. 129; MAND. 77; Jeniver ROB. (W.) 7259; Janevere GOW. III. 125.

jápe, sb., *joke, trick*, '*nuga*,' PR. P. 257; LANGL. *B* xx. 144; CH. C. T. *G* 1312; PARTEN. 5695; jápes (*pl.*) ['*ludicra*'] TREV. III. 459; MIRC 61; your japes ar . . . ille A. P. ii. 864.

jápe-worþi, adj., *ridiculous*, CH. BOET. v. 3 (157).

jáperie, sb., *jesting*, S. A. L. 70; jáperies (*pl.*) CH. C. T. *E* 1656.

jápin, v., *joke, play tricks*, '*illudo*,' PR. P. 257; jápede [japed] (*pret.*) LANGL. *A* i. 65; Jupiter [was] a jettoure þat japid mani ladis ALEX. (Sk.) 4415; jáped (*pple.*) CH. C. T. *A* 1729; *comp.* bi-jápen.

jápere, sb., *jester*, (*ms.* iapere) ['*nugax*'] TREV. VII. 453; CH. C. T. *I* 89: jápers (*pl.*) and jangelers LANGL. *A* PROL. 35.

jápinge-stikke, sb., *plaything*, REL. II. 50.

jappon, sb., *jest, gibe*, YORK xxxi. 304.

jardine, sb., *almond*, B. B. 168.

jargoun, sb., *O.Fr.* jargon; *jargon, uproar*, ALEX.² (Sk.) 462; gargoun S. S. (Wr.) 3147, 3158.

jaspe, sb., *O.Fr.* jaspe; *jasper*; (*ms.* Iaspe) PR. P. 257; (*ms.* iaspe) MISC. 96; SPEC. 25; P. L. S. xxxv. 70; jaspre ALEX. (Sk.) 4444; jasper A. P. i. 999.

jaudewin, adj., *? Fr.* Jean des vignes; *simpleton*: þe jaudewin Jupiter ALEX.² (Sk.) 659; thow jawdewyne (*said to a Lollard*) HALLIW. 483.

jaumbe, sb., *? O.Fr.* jambe (*leg*); *? projection*: þe jaumbe of þe justarm D. TROY 11114.

jaumbers, sb., pl., *O.Fr.* jambiere; *leg pieces*, MAN. (F.) 10026.

jaunis, sb., *O.Fr.* jaunisse; *jaundice*, PR. C. 700; jaundis TREV. II. 113; jandis VOC. 224.

javele, sb., *base fellow* (*term of contempt*), A. P. ii. 1495; javel '*joppus, gerro*' PR. P. 257; javellis (*pl.*) *wranglers* YORK xxx. 235.

javellen, v., *wrangle*; do herke howe þou javell (*pres.*) jangill of lewes YORK xxx. 59.

jeant, *see* geant.

jectour, *see* getter.   **jelous,** *see* gelūs.

jeniver, *see* Janiver.   **jentil,** *see* gentil.

jerine, sb., *? a sort of garment*: a jerine of Acres D. ARTH. 903.

jerodine? sb., *some fabric*: a jupone of jerodine jaggede in schredez D. ARTH. 905.

jeseine, sb.,=gesīne, *childbed*, MIR. PL. 83.

jesseraunt, sb., *O.Fr.* jaserant; *coat of mail*, D. ARTH. 904; gesseraunte 4238; jesserand ALEX. (Sk.) 4961; **jesserantis** (*pl.*) ALEX. (Sk.) 2450.

jeste, see **geste.**

jet, jete, jetten, see **get, getten.**

Jeu, see **Giu.**

jeu, sb., *O.Fr.* geu, jeu; *game*, FER. 2224.

jeuparti, sb., *Fr.* jeu parti; *jeopardy, hazard*; *hazardous plan, feat*, HALLIW. 484; juparti CH. C. T. *G* 743; juperdi BARB. x. 340, 413, 524, 539; jeopardies [*v. r.* punȝeis, punȝe] (*pl.*) *skirmishes* BARB. xii. 373: juperdiss BARB. x. 145.

jobarde, sb., *O.Fr.* jobard; *blockhead*, HALLIW. 485; **jobbardis** (*pl.*) LIDG. M. P. 119.

jobbe, sb., *piece, article*; **jobbes** (*pl.*) of gold D. TROY 11941.

jobbin, v., *job, peck*: jobbin withe the bille PR. P. 263.

jocaunt, adj., *Lat.* jocans; *sportive*, GEST. R. 116.

jocund, adj., *Lat.* jocundus; *merry, lively*, D. TROY 316; jocounde LIDG. M. P. 183.

jocundnes, sb., *mirth, pleasure*, AUD. 26.

jogelin, v., *O.Fr.* jogler, jugler, *from Lat.* joculāri; *juggle*, PR. P. 263; jogele LANGL. *B* xiii. 232.

jogelour, sb., *O.Fr.* jougleor, jugleor; *juggler*, AYENB. 172; jogeloure LANGL. *B* vi. 72; CH. C. T. *D* 1467; jogulour '*mimus, histrio*' PR. P. 263; jugelour P. L. S. xv. 19; jogolour ALIS. 5991; **juglūrs** (*ms.* iuglurs) (*pl.*) A. R. 210; jogulours MAND. 237.

jogelrie, sb., *O.Fr.* joglerie; *jugglery*, CH. C. T. *F* 1265.

joggen, v., *jog, go hastily*; **joggeth** (*pres.*) CH. L. G. W. 2705; **jogged** [jugged] (*pret.*) LANGL. *B* xx. 133.

joggen, see **jaggen.**

joie, sb., *O.Fr.* joie, goie; *joy*, A. R. 218; SPEC. 76; (*ms.* ioie) FER. 2037; (*ms.* ioye) P. L. S. ix. 206; (*ms.* yoye) EGL. 4: joie, goie AYENB. 75, 226; goie CH. BOET. v. 6 (179).

joie-ful, adj., *glad, joyful*, TRIST. 1920; joivol ROB. (W.) 786; joiſol A. P. i. 288; joiſulle N. P. 68.

joiful-lī, adv., *gladly*; joiveli D. TROY 374; joifulli 993.

joi-lēs, adj., *joyless*, A. P. ii. 252; iii. 146.

joiel, see **jouel.**

joien, v., *O.Fr.* joir; *rejoice, be glad*, TRIST. 47; joie WICL. LK. i. 14; **joisseþ** (*pres.*) AYENB. 25; (ho) joien MISC. 223.

joiinge, sb., *gladness*, HAV. 2087.

joifnes, sb., *?youth*, GAW. 86.

joinen, v., *O.Fr.* jungre: *join*; joine A. P. ii. 726; joine (*pres.*) R. R. 2355; joineþ AYENB. 88; joined [joigned] (*pple.*) LANGL. *A* ii. 106.

joining, sb., *joining, encounter*; joinenige D. ARTH. 2133.

joinen², v., *for* enjoinen; *enjoin*; joined, joint (*pple.*) A. P. ii. 62, 355.

joinour, sb., *O.Fr.* joignour; *joiner, 'compaginator, archarius'* PR. P. 264.

joint, sb., *Fr.* joint; *joint*, PR. P. 264; LANGL. *C* x. 215; þe stronge strok of the stonde strained his **jointes** (*pl.*) A. P. ii. 1540.

[**joint,** adj., *O.Fr.* joint (*pple.*); *joint.*]

joint-lī, adv., *together, continuously*, ALEX. (Sk.) 1470; D. TROY 1538.

jointūre, sb., *O.Fr.* jointure; *joining*, CH. BOET. ii. 5 (46); **jointours** (*pl.*) *joints, limbs*, ALEX. (Sk.) 722.

joious, adj., *O.Fr.* joious; *joyous*, P. L. S. xii. 36; SHOR. 120.

jol, sb., ?=chavel, *mod.Eng.* jowl; *head, 'caput,'* PR. P. 264.

jolif, adj., *O.Fr.* jolif; *jolly, beautiful*, P. L. S. xii. 117; CHR. E. 362; SPEC. 52; ALIS. 155; A. P. i. 841; joli WILL. 3479; PR. C. 589; **joliere** (*comp.*) D. ARTH. 4110; **jolieste** (*supl.*) D. ARTH. 1658.

joli-lī, adv., *gaily*, D. ARTH. 245, 2088, 4109.

jolivetē, sb., *O.Fr.* joliveté; *jollity, agreeableness, happiness*, AYENB. 53; jolite P. P. 251; CH. C. T. *A* 680; ALEX. (Sk.) 3537, 4202, 4458.

jolle, v., *knock about*, YORK xxxii. 14.

joncate, sb., *med.Lat.* juncāta; *junket*, B. B. 123, 266.

jopone, see **gipōn.**

jornee, sb., *O.Fr.* jornee, *cf. Ital.* giornata; *mod.Eng.* journey; *day's work*, AYENB. 113; journee CH. C. T. *A* 2738; journe (*task, fight*) MAN. (F.) 13874; BARB. xx. 494; journai A. P. iii. 355; jurneie A. R. 352; **jurnēs** (*pl.*) GEN. & EX. 1291; jornes (*journeys*) MAN. (F.) 15942; **journēs** (*gen. pl.*) (*days' work*) MAN. (F.) 7963.

jorneien, v., *O.Fr.* jornoier; *journey*; **jorneied** (*pret.*) MAN. (F.) 3383.

jottes, sb., pl., *peasants, mean people*, LANGL. *B* x. 460.

jouel, sb., *O.Fr.* jouel; *jewel*, (*ms.* iowel) AYENB. 112; juel A. P. i. 249; juwel S. S. (Wr.) 2811; **juelez** (*pl.*) A. P. i. 278.

jouelēr, sb., *jeweller*; jueler(e) A. P. i. 252, 264.

jouken, v., *O.Fr.* jouquier; *rest, slumber*: jouke in her chambre LANGL. *B* xvi. 92; þai **jouke** (*pres.*) in þa strandis ALEX. (Sk.) 4202; **jouked** (*pret.*) A. P. iii. 182.

joupe, sb., *Fr.* jupe; *tunic*, PR. P. 265; HAV. 1767.

joutes, sb., *O.Fr.* joute (*beet root, vegetables generally*); *pot herbs*, LANGL. *B*v. 158; MAND. 58; L. C. C. 15; joutis, potage '*brassica*' PR. P. 265; joutes C. B. 5.

jowe, jouwe, sb., *? O.Fr.* joue; *jaw*, '*mandibula*,' PR. P. 265; þe over(e) jawe [jowe] TREV. III. 109; jowe ['*branchiam*'] WICL. TOB. vi. 4; jowes (*pl.*) MAND. 107; of þe jowes ['*faucibus*'] CH. BOET. i. 4 (15).

jubulacioun, sb., *Lat.* jūbilātiō; *rejoicing*, WICL. PS. cl. 5.

judēisc, adj., *Jewish*: þa judeiscen (*ms.* iu-) men HOM. I. 89.

juge, sb., *O.Fr.* juge; *judge*, '*judex*,' PR. P. 266; juge [jugge] CH. C. T. *A* 1712.

juge-men, sb., pl., *judges*, ALEX. (Sk.) 3402.

jugement, sb., *O.Fr.* jugement; *judgment*, P. L. S. xiii. 174; PR. C. 2802; juggement ROB. 53.

juggen, v., *O.Fr.* jugier; *judge, regulate*, A. R. 118; jugge C. L. 419; LANGL. *A* ii. 127; S. S. (Wr.) 3099; jugged (*pret.*) A. P. i. 7; gentilnes jugget justli his werkes D. TROY 10360; jugged (*pple.*) MAND. 92.

juise, sb., *O.Fr.* juise; *judgment*, P. R. L. P. 71; A. P. ii. 726; CH. C. T. *A* 1739.

jurdan, sb., *chamber-pot*, LANGL. *B* xiii. 83; jurdon '*urna*' PR. P. 267.

jurdisciōn, sb., *jurisdiction*, WICL. E. W. 57.

jurūr, sb., *juror*; jurouris (*pl.*) WICL. E. W. 63.

jūs, jous, sb., *Fr.* jus; *juice*, PR. P. 265; juis TREV. IV. 121; jous, juse ALEX. (Sk.) 339, 410.

jussel, sb., *O.Fr.* jussel (*jus, potion*); *a dish in cookery*, B. B. 53, 170.

just, juste, adj., *Fr.* juste; *just*, CH. C. T. *B* 4240; þi just werkes D. TROY 214.

just-lī, adv., *properly*, D. TROY 512.

juste, sb., *O.Fr.* jouste; *joust, tournament*; justes (*ms.* iustes) (*pl.*) LANGL. *B* xvii. 74; joustes ROB. 137; PARTEN. 988.

justarme, sb., *tilting weapon*, D. TROY 11114; *see* gisarme.

justen, v., *O.Fr.* jouster; *joust, tilt*, LANGL. *B* xviii. 82; juste WILL. 1237; jouste FER. 105.

justinge, sb., *jousting*, D. ARTH. 1957, 2875.

justifīin, v., *O.Fr.* justifier; *justify*, '*justifico*,' PR. P. 268; justifle (*pres. subj.*) D. ARTH. 663; justifiet (*pple.*) A. P. i. 700.

justilen, v., *? from* justen; *cf. mod.Eng.* jostle; *? sport* (*sensu obscoeno*); justilet (*pple.*) D. TROY 12738.

justise, sb., *O.Fr.* justice; *justice*, SAX. CHR. 261; justice PR. P. 268; HAV. 2202; Juno a justis of joies D. TROY 2385; justis [justice] (*justices*) (*pl.*) ALEX. (Sk.) 1601, 3402.

justisen, v., *Fr.* justicier; *administer justice*, C. L. 298.

justiser, sb., *O.Fr.* justicier; *judge, ruler*, MAN. (F.) 2221.

justour, sb., *O.Fr.* justeur; *jouster*; justours (*pl.*) MAN. (F.) 7657.

jusule, sb., *cf.* justilen; *? lechery*; þe jusule of Jupiter, & of his japis als ALEX. (Sk.) 4411.

k.

ka-, kā-, *see* ca-, cā-.

kāl, *see* caul. kalengen, *see* chalengen.

kastand, kastanid, pple., *made like the chestnut*; ALEX. (Sk.) 1537; *see* chasteine.

kateil, *see* catel.

kebben, v., *cf. Sw.* kebba, käbba; *contend*: ʒef þat kebbede (*pret.*) eni of ous ich woʒt (*r.* wot) wel þat he leʒ(e) SHOR. 111.

kebbinge, sb., *contention*; kebbinges (*pl.*) SHOR. 111.

kecchen, *see* cacchen.

kēchel, sb., *O.E.* cīcel (*for* cȳcel), coecil (*tortum*), = *M.H.G.* chüechel; *small cake*: (*ms.* kechell) ORM. 8662; kechel [kichil] CH. C. T. *D* 1747.

kechene, kechin, *see* cüchene.

kēʒe, sb., *O.E.* .cǣg, cǣge, = *O.Fris.* kei; *key*; keʒe (*later ver.* keie) WICL. IS. xxii. 22; keiʒe LEG. 25; SHOR. 47; E.G. 360; (*ms.* keie) '*clavis*' FRAG. 3; LEG. 153; (*ms.* keye) P. R. L. P. 220; LANGL. *B* v. 613; PL. CR. 30; GOW. I. 10; CH. C. T. *E* 2044; LIDG. M. P. 62; kēie (*dat.*) O. & N. 1557; LUD. COV. 31; kēien (*pl.*) ROB. 186; kaigen MAT. xvi. 19; keies RICH. 4147; A. P. ii. 1438; kaies PERC. 2102.

kāi-berere, sb., *key bearer*, '*claviger*,' VOC. 211.

kēi-herde, sb., *key keeper*, HOM. II. 193.

kēʒen, v., *key, fasten with a key*; kēiʒed (*? ms.* keived, *printed* keiþed) (*pple.*) L. H. R. 205; ikeiʒet LANGL. *A* vi. 103; ikeied *C* viii. 266.

kei, sb., *O.Fr*, cai, caie; *quay*, PR. P. 269; keie E. G. 374.

kēie, kēiʒe, *see* kēʒe.

keil, sb., *M.Du.* kegel, = *O.H.G.* chegil (*cone*): cailis (*pl.*) *skittles* REL. II. 224.

kēkin, v., *?* = kīken; *peep*, '*intueo*,' PR. P. 269; kēked (*pple.*) CH. C. T. *A* 3445.

kēl, sb., *O.E.* cēol, = *M.Du.* kiel, *O.H.G.* chiol (kiol), cheol, *O.N.* kioll, kiöll; *keel* (*of a ship*): from þe kele [cule] to þe hacches TREV. II. 233.

kelde, adj. *O.E.* cyldu, = *O.Fris.* kelde, kalde, *O.H.G.* chaltī; *cold*, SPEC. 37; chelde ALIS. 5501; LANGL. *B* i. 23*.

kelden, *see* cálden.

kēlen, v., *O.E.* cēlan, *from* cōl; *cool*, ORM. 19584; heore þurst kelen HOM. I. 141; kelin PR.

P. 270; kele P. L. S. xv. 102; LIDG. M. P. 31;
ANT. ARTH. iv; TOWNL. 245; L. C. C. 6; to kele
þi lust H. M. 25; kēleþ (*pres.*) GOW. II. 360;
þat he ... kele mi tunge WICL. LK. xvi. 24;
kēled (*pple.*) REL. I. 53; *comp.* a-, ʒe-kēlen.

kēlare, sb., *cooler*, '*frigidarium*,' PR. P.
269.

kēlinge, sb., *cooling, refreshing*: times of
kelinge, or refreischinge ['*tempora refrigerii*']
WICL. DEEDS iii. 20.

kelf, *see* calf.

kēling, sb., *large codfish*, HAV. 757; REL. I.
85; *comp.* lobbe-kēlinge.

kelk, sb., *cf.* Swed. kälk (*marrow*); *ova of
fishes*; kelkes (*pl.*) L. C. C. 19.

kelle, *see* calle.   kelp, *see* kilp.

kemben, v., *O.E.* cemban, = *M.H.G.* kemben,
kemmen, *O.N.* kemba; *from* camb; *comb*, A.
R. 422*; kembe FL. & BL. 562; AYENB. 176;
kembeþ (*pres.*) P. S. 329; kemben LANGL.
B x. 18; kembede (*pret.*) D. ARTH. 3351;
kembed MAND. 24; kembed (*pple.*) CH. C.
T. *A* 2143.

kembstere, sb., = *M.Du.* kemstere; *female
comber*; kempstare '*pectrix*' PR. P. 270;
kemster VOC. 194.

kemeling, sb., = cümeling; *foreigner*, ROB.
(W.) 581*.

kemelin, *see* kimlin.

kemes, sb., *O.E.* cemes; *shirt*, M. H. 124;
withouten kirtele or kemse MAN. (H.) 122.

kempe, sb., *O.E.* cempa, = *O.Fris.*, *O.N.*
kempa, *O.L.G.* kempio, *O.H.G.* chempho,
chemphio; *from* camp; *soldier, champion,
'miles, athleta*,' FRAG. 2; KATH. 803; ORM.
3587; HAV. 1036; AYENB. 50; WILL. 4029;
kempe, kempa LAʒ. 1575, 26022; kampen,
cempen (*pl.*) JOHN xix. 23, 32; cæmpen LK.
vii. 8; kempen MARH. 1; AYENB. 45; cempen HOM. I. 243; kempen [kempes] LAʒ. 463;
kempes PERC. 47; CHEST. I. 259; kempene
(*gen. pl.*) A. R. 196; *comp.* hēre-, lēod-
kempe.

kempe[2], adj., *shaggy*, CH. C. T. *A* 2134;
campe A. P. ii. 1695.

kempen, *see* campin.

ken, adj., *knowing*: beo nu ken & cnawes
KATH. 2070; þu art ... ken & icnawen 2254.

kenbowe, sb.: and set his hond in kenebowe
(*akimbo*) BER. 1838.

ken, *see* cün.

kenchen, v., *laugh loudly*, KATH. 2042;
kenchinde (*pple.*) H. M. 17; *cf.* kinken.

kende, *see* cünde.   kene, *see* kine.

kēne, adj. *O.E.* cēne, = *O.N.* kœnn, *M.Du.*
coene, cōne, *O.H.G.* chuoni; *keen, sharp,
bold, brave*, FRAG. 8; A. R. 140; KATH.
1932; HAV. 1832; P. L. S. xxi. 149; MAND.
173; EGL. 1248; TOR. 2710; ANT. ARTH.
xlvii; S. & C. I. lxi; he wes swa kene LAʒ.

1374; wit so kene O. & N. 681; nis of ou
non so kene þat durre abide mine onsene
1705; a kene ax ROB. 490; a swerd ... þat
wel isteled and kene were C. L. 1248; an
arewe kene ant smert(e) CHR. E. 929; kēnen
(*dat. m.*) LAʒ. 23570; þe þornes beþ kēne
(*pl.*) SPEC. 110; biforen riche & kene ORM.
16139; kēnre (*compar.*) KATH. 1953; kēn-
nest (*superl.*) LAʒ. 21300; kenest KATH. 815;
of þan alre kennuste MISC. 48; kēne (*adv.*)
D. TROY 1467.

kēn-līche, adv., *O.E.* cēnlīce; *keenly, boldly,
bravely*, HOM. I. 107; LAʒ. 1573; REL. I. 172;
ken-, keneliche, -li WILL. 37, 152, 2532;
keneliche KATH. 2240; kēnlükeste [kēnlo-
keste] (*superl.*) LAʒ. 25429.

kēn-schipe, sb., *boldness*, LAʒ. 6364; MARH.
11.

kenel, sb., *Fr.* chenil; *kennel*, '*canicularium*,'
PR. P. 271; GAW. 1140.

kenet, sb., *O.Fr.* chenet; *hound*, '*canicula*,'
VOC. 219; PR. P. 271; GAW. 1701; S. S. (Wr.)
1740.

kennen, v., *O.E.* cennan, = *O.Fris.*, *O.N.*
kenna, *O.L.G.* (ant-)kennian, *O.H.G.* (ar-,bi-)
chennan, *Goth.* kannjan; *from* can; *know, acknowledge; make known; teach*, E. G. 43;
ne der ich noht kennen [kenne] ... þat ich
her king weore LAʒ. 6639; whar swa he
mihte hine kennen 11380; kenne JOS. 158;
GOW. I. 204; A. P. ii. 865; PERC. 407;
TRIAM. 506; SACR. 450; he schal ... kenne
þe & cuðen al þat tu easkest JUL. 37; ne
con him ous no man kenne L. N. F. 216; no
man scholde him kenne AL. (L.[2]) 96; and
þaim þe wai til hevin kenne M. H. 3; kenne
(*pres.*) MIRC 880; we kennið ... & cuðeð
KATH. 1347; we kenneð & cnaweð 2096;
heo kenneþ us care P. S. 159; ken (*imper.*)
MARH. 16; kenne me ... on Crist to bileve
LANGL. *C* ii. 78; kende (*pret.*) BRD. 2;
TOWNL. 245; þas word kende LAʒ. 32066; ase
þe angel hire kende HICKES I. 226; and
kende an oon þat (hit) was he S. S. (Wr.) 2875;
kenned (*pple.*) WILL. 343; kend SPEC. 25;
SACR. 15; *comp.* a-, bi-, ʒe-kennen.

kenninge, sb., '*cognitio*,' PR. P. 271.

kennen[2], v., *O.E.* cennan, = *O.L.G.* kennian,
O.H.G. (ki-)chennan; *bring forth, beget*: child
kennen & beren HOM. II. 179; kende (*pret.*)
['*peperit*'] MAT. i. 25; ST. COD. 97; kenned
(*pple.*) ORM. 19268; þat þorgh þe holi gostes
might kenned was REL. I. 160; kened (*v. r.*
kennede) ROB. (W.) 1547; *comp.* a-, ʒe-
of-kennen.

Kent, pr. n., *O.E.* Cent; *Kent*, LAʒ. 7171.

Kentisc, adj., *O.E.* Centisc; *Kentish*: mid
Kentisce leoden LAʒ. 7441.

kēo, *see* cā.

keorven, v., *O.E.* ceorfan, = *M.L.G.*, *M.Du.*
kerven, *O.Fris.* kerva; *carve, cut, split*, JUL.

56; kerven HOM. II. 87; HORN (R.) 241; kervin '*scindo, sculpo*' PR. P. 273; kerve LANGL. *B* vi. 106; WICL. LEV. i. 17; CH. C. T. *F*158; A. P. ii. 1104; kerveþ (*pres.*) C. L. 1257; kervis ANT. ARTH. xlvii; kerve (*imper.*) L. C. C. 6; kerveð [kerveþ] LA3. 5864; keorvinde (*pple.*) A. R. 250; kervinde AYENB. 66; kervond and kene D. TROY 8640; carf (*pret.*) ROB. 116; S. S. (Wr.) 3011; carf [karf] CH. C. T. *A* 100; karf HAV. 471; ALIS. 3629; heo cærf [carf] him þene swure a twa LA3. 4012; þeo her þe me kerf of A. R, 398; (heo) curven [corve] LA3. 21875; corven CHR. E. 757; corve ROB. 313; kurve (*subj.*) A. R. 384; P. L. S. xviii. 50; corven (*pple.*) CH. C. T. *A* 2696; corvin EGL. 279; icorven H. M. 17; hore her beo ikorven A. R. 424; comp. bi-, for-, to-kerven.

kervere, sb., *carver*, CH. C. T. *A* 1899; E. G. 446; purse kerveris (*pl.*) WICL. S. W. III. 320.

keorfunge, sb., *carving*, A. R. 344; kervinge PR. P. 273.

kerving-kníf, sb., *carving knife*, B. OF C. 673.

kēp, sb., *heed, care*, GEN. & EX. 939; LEG. 116; hire sone to servi was alle hire kep ASS. 72; 3ef þou nimest wel god keep SPEC. 103; take keep CH. C. T. *A* 2688; koep AN. LIT. 90; kepe SHOR. 12; GOW. III. 187; PR. C. 381.

kēpen, v., *O.E.* cēpan, = *M.Du.* kēpen; *keep, preserve*, A. R. 332; GEN. & EX. 3378; LANGL. *B* iii. 287; MAND. 252; þe duntes b(e)oð uvel to kepen (*ward off*) HOM. I. 153; whær he mihte þene kæisere iwisliche kepen (*take*) LA3. 26184; þer heo wolden kepen (*wait*) 26937; kepin '*custodio, servo,*' PR. P. 272; kepe SPEC. 35; WILL. 66; WICL. GEN. xxvi. 5; kepe [keepe] CH. C. T. *A* 130; kinemede kepe (*receive*) KATH. 399; to kepe here fon wen þei to hem drowe ROB. 51; men schuld(e) him kepe and wake GREG. 623; to kepe him when he doun sal come PR. C. 5029; ne kēpe (*pres.*) (*like*) ich noht þat þu me clawe O. & N. 154; of þe ne kepe (*care*) i no3t BEK. 998; ne kepeð (*cares*) he HOM. I. 55; kepeð (*awaits*) & copneð þi come KATH. 2457; gode 3eme kepeþ SHOR. 11; kēp (*imper.*) HOM. I. 71; sei3e heom þat hui kepen (*await*) me S. A. L. 161; kĕpte (*pret.*) LA3. 27714; HAV. 879; & god ne kepte (*ms.* keppte) noht of þat ORM. 3114; ne kepte heo non hening here SPEC. 36; and kepte wel his folde CH. C. T. *A* 512; kepten SAX. CHR. 256; þe ontful(l)e ne kepten nout þat me dealede of hore gode A. R. 248; (he) keppit him swithe D. TROY 1230; mi bede kēpid (*pple.*) has he ['*orationem meam adsumpsit*'] PS. vi. 10; comp. 3e-kēpen.

kēpere, sb., = *M.L.G.* kēpere, *M.Du.* kēper; *keeper*, LANGL. *B* xii. 128; GOW. III. 175; CH. C. T. *A* 172; kepare '*custos, conservator*' PR. P. 272; comp. hous-kēpare.

kēpinge, sb., *keeping*, ASS. 56; D. ARTH. 4205; keping PR. C. 4196.

keppe, *see* cappe.

ker, sb., *O.N.* kiarr, kiörr; *marshy ground*, PR. P. 272; at þe kerre side GAW. 1431; kerss BARB. xii. 392; kerres (*pl.*) MAN. (F.) 14574; comp. alder-ker.

kerling, sb., *O.N.* kerling; *from* carl; *old woman*, C. M. 11056.

kerlok, sb., *O.E.* cer-, cyrlic; *charlock,* '*rapistrum,*' HALLIW. 492; carlok '*eruca*' VOC. 265; PR. P. 62.

kernel, sb., *O.Fr.* crenel (*créneau*); *battlement*, A. R. 62; FL. & BL. 231; carnels (*pl.*) C. L. 695; karnels MAN. (F.) 6046.

kernel, *see* cürnel.

kernelen, v., *O.Fr.* creneler (*munir de créneaux*); *crenellate, embattle*; kerneled (*pple.*) LANGL. *B* v. 597.

kerse, *see* cresse.

kertel, *see* cürtel. kerven, *see* keorven.

kessen, *see* cüssen. kesten, *see* casten.

ket, sb., *O.N.* kiöt; *flesh*, REL. I. 218.

kēte, adj., *brave, strong*, FER. 4596; as a king kete REL. II. 9; kete and beld LEG. 48; a borugh . . . þat michel was & kete AL. (L.[3]) 201; þe kete MISC. 76; of þe sonne . . . þe leomes beoþ so kete TREAT. 138; þe stormes beoth so kete S. A. L. 156; kete lordes WILL. 330; kete men LANGL. *A* xi. 56; comp. dōm-kēte.

kĕt-lī, adv., *quickly*, WILL. 1986.

kēte, *see* kīte. ketel, ketil, *see* chetel.

ketling, sb., *O.N.* ketlingr; *whelp*; ketling ['*catulus*'] of a lion WICL. DEUT. xxviii. 22; kitling '*catellus*' PR. P. 277.

kēþen, *see* cūðen. keþþe, *see* cūððe.

kevel, sb., *O.N.* kefli, kafli (*piece cut off*), cf. *M.Du.* cavel *m.* (*lot*), *O.L.G.* kavele *f.* (*lot*); *bridle-bit; clamp, hook; lot*: a kevel (*gag*) of clutes HAV. 547; kevil, keul '*camus*' PR. P. 274; kevil ['*camo*'] PS. xxxi. 9; þan' kest þai cavel (*lots*) C. M. 18907; keveles PERC. 1426; kiviles (? *clamps*) MAN. (F.) 12062.

kevelen, v., *O.N.* kefla; *put a kevel on*: and keviles (*pres.*) his stede PERC. 424; comp. un-kevelen.

kēven, v., ? *turn*; keve A. P. i. 320; kēved (*pple.*) A. P. i. 980.

keveren, *see* coveren.

keweri, sb., *from O.Fr.* queux (*cook*); *cookery*, TREV. I. 405; curi B. B. 150; cūries (*pl.*) 149.

kex, sb., *kex, reed*, VOC. 157; kex [kix] LANGL. *B* xvii. 219; kix '*calamus*' PR. P. 277.

kēye, *see* kē3e.

kíbe, sb., ? *Welsh* cibi; *chilblain, carbuncle, ulcer*, ['*anthrax*'] TREV. VIII. 227.

kichel, *see* **kĕchel. kichene,** *see* **cü̱chene.**

kid, sb., *cf. mod.Eng.* kit ; *? bundle,* '*fascis,*' PR. P. 274.

kide, sb., *O.N.* kið *n. ; kid (animal)*, '*haedus,*' PR. P. 274 ; ORM. 7804 ; WICL. JUDG. xv. 1 ; CH. C. T. *A* 3260 ; REL. II. 80 ; kides *(gen.)* GEN. & EX. 1697 ; **kides** *(pl.)* GEN. & EX. 1535.

kid-nēre, sb., *kidney,* L. C. C. 10 ; kidenei VOC. 149 ; kednei 179 ; **kidneers** *(pl.)* [*later vers.* kideniris, kideneiren] WICL. EX. xxix. 13 ; kideneris LEV. iii. 4.

kīf, *see* **cü̱f.**

kigge, adj., *cheerful* : kigge or joli '*hilaris*' PR. P. 274.

kiken, v., *cf. Ger.* kiken *(prick)* ; *kick* ; kike ['*calcitrare*'] TREV. V. 355 ; LANGL. *C* v. 22 ; WICL. DEEDS ix. 5 ; CH. C. T. *D* 941 ; **kikide** *(pret.)* WICL. DEUT. xxxii. 15.

kīken, v.,=*L.G.* kīken *(pret.* keik*), Du.* kijken *(pret.* keek*), Swed.* kika : *look, peep* : and to þe roof þei kīken *(pres.)* and þei gape CH. C. T. *A* 3841 ; *see* **kēken.**

kikir, sb., '*tentigo,*' VOC. 186, 246.

kīle, sb., *O.N.* kȳli ; *sore,* '*ulcus,*' VOC. 224 ; **kīles** *(pl.)* REL. I. 53.

killen, *see* **cü̱llen. kilne,** *see* **cü̱lne.**

kilp, kelp, sb., *O.N.* kilpr : *handle of a vessel* : kilpe [kelpe] of a caldron '*perpendiculum*' CATH. 203 ; his swerd he pulte up in his kelp (*? scabbard, ? sword-belt*) L. H. R. 140.

kime, *see* **cü̱me.**

kīme, sb., *? wretch* : þe silli kime PL. T. 643.

kīmen, v., *cf. Du.* kuimen, *O.H.G.* chūman (*? complain*) ; ikīmet *(pple.) ? become faint* KATH. 1298 ; *comp.* **a-kimen, bi-kēmen.**

kimlin, sb., *? from* **cumb** ; *brewing-tub,* PR. P. 274 ; kimelin [kemelin] CH. C. T. *A* 3548.

kin, *see* **cü̱n.**

kinch, sb., *bundle,* '*fasciculus,*' VOC. 229.

kinde, kindel, *see* **cü̱nde, cü̱ndel.**

kindlen, v., *kindle, set fire to,* (ms. kinndlenn) ORM. 13442 ; HAV. 915 ; kindlin '*accendo*' PR. P. 275 ; kindle CH. C. T. *C* 481 ; **kündleð** *(pres.)* A. R. 296 ; **kindled** *(pple.)* MAND. 70 ; kindlid [kindeled] C. M. 6759.

     kindling, sb., *ardour (v.r.* kindilling*)*, ALEX. (Sĸ.) 3292.

kindlen, *see* **cü̱ndlen.**

[**kine,** sb., *O.E.* cyne, cine,=*O.H.G.* chuni(-riche) ; *royal dignity.*]

kine-ærd, sb., *royal land,* LAȝ. 19433.

kine-bearn, adj., *O.E.* cynebearn ; *royal offspring,* HOM. I. 273 ; LAȝ. 199 ; kinebern MARH. 4 ; kine-, cunebern HOM. II. 47, 49.

kine-benche, sb., *royal throne,* LAȝ. 9693.

kine-boren, adj., *O.E.* cyneboren ; *of royal birth,* LAȝ. 10098.

kine-burh, sb., *royal city,* KATH. 1883.

kine-dōm, sb., *O.E.* cynedōm ; *realm,* KATH. 1472 ; A. R. 148 ; ORM. 2227 ; FL. & BL. 802 ; ROB. 173 ; C. L. 286 ; cune-, kenedom [kinedom] LAȝ. 6095, 15079.

kīne-ȝerde, sb., *O.E.* cynegerd, -gyrd ; *sceptre,* ORM. 8182 ; P. S. 215.

kine-helm, sb., *O.E.* cynehelm ; *crown,* FRAG. 2 ; LAȝ. 6766 ; kenehelm JOHN xix. 2.

kine-lāverd, sb., *O.E.* cynehlāford ; *royal lord,* LAȝ. 9831.

kine-līch, adj., *O.E.* cynelīc ; *royal* : ane kineliche burh LAȝ. 14130.

kine-lond, sb., *royal land,* LAȝ. 183.

kine-mēde, sb., *royal reward,* KATH. 399.

kine-merke, sb., *sign of royalty,* FRAG. 8 ; kinemerk HAV. 604.

kine-rīche, sb., *O.E.* cynerīce, = *O.H.G.* chuni-, chunerīche ; *realm, royal power,* KATH. 182 ; ORM. 2236 ; MISC. 47 ; kine-, kuneriche LAȝ. 2896, 28931 ; C. L. 1416 ; P. S. 215 ; kuneriche PROCL. 2 ; kunerike HAV. 2804 ; kinrik BARB. i. 158.

kine-ring, sb., *royal ring,* KATH. 409.

kine-sǣte, sb., *royal seat,* ORM. 2224.

kine-scrūd, sb., *royal robe,* HOM. I. 193.

kine-setle, sb., *O.E.* cynesetl ; *royal seat, throne,* HOM. I. 115 ; KATH. 45.

kine-stōl, sb., *O.E.* cynestōl ; *royal seat,* HOM. I. 191 ; LAȝ. 4517.

kine-þēode, sb., *kingdom,* LAȝ. 2947.

kine-wurðe, adj., *royal,* KATH. 567 ; þis kinewurðe [kineworþe] lond LAȝ. 13384 ; kenewurð(e) 8618 ; kineworþe C. L. 14.

kinewurð-līche, adv., *royally* ; kinewurð-liche iwurðget JUL. 62.

kine, *see* **chine.**

kining, sb., *O.E.* cyning, cining, cyng, cing, = *O.Fris.* kining, kening, kinig, kenig, *O.L.G.* cuning, *O.H.G.* chuning, kuning, chunig, kunig, *O.N.* konungr, kongr ; *king,* MK. xv. 2 *(written* kyning MAT. xxi. 5*)* ; kining *(ms.* kyning*)*, king SAX. CHR. 249, 254 ; king FRAG. 1 ; LAȝ. 5962 ; KATH. 27 ; A. R. 138 ; **kinges** *(gen.)* BEK. 203 ; **kingis** fischare, *king fisher,* '*is(p)ida,*' PR. P. 275 ; **kinge** *(dat.)* LAȝ. 25654 ; **kinges** *(pl.)* KATH. 226 ; kinges, kinge LAȝ. 5331, 29693 ; **kinge** *(gen. pl.)* HOM. I. 33 ; A. R. 398 ; ORM. 3588 ; kinge, kingen, kingene LAȝ. 2611, 5378, 21242 ; HOM. II. 45 ; MARH. 18 ; kingene LANGL. *A* i. 103 ; **kininген** *(ms.* kyningen*) (dat. pl.)* LK. xxi. 12 ; kingen LAȝ. 23915 ; *comp.* **folc-lēod-king.**

king-dōm, sb., *O.E.* cyningdōm, = *O.L.G.* cuningdōm, *O.N.* konungdōmr ; *kingdom,* LANGL. *B* vii. 155 ; MAND. 40 ; PR. C. 8788 ; M. H. 96 ; þane kingdom AYENB. 77.

king-hōd, sb., *kingship,* WILL. 4059 ; kinghed LANGL. *A* xi. 216.

1.

king-lēs, adj., *without a king*, ROB. 105.

king-lī, adj., = *O.Fris.* keninglīk, *O.H.G.* chuninglīch ; *kingly*, LIDG. M. P. 20.

king-lond, sb., *kingdom*, GEN. & EX. 1262.

king-rīche, sb., = *O.Fris.* kiningrīke, *O.H.G.* kuningrīchi ; *realm, royal power*, MISC. 26 ; AYENB. 122 ; ALIS. 5981 ; WILL. 2127 ; LANGL. *B* PROL. 125 ; kingrike D. ARTH. 24 ; PR. C. 5780 ; kingrik BARB. i. 57.

kinken, v., = *M.Du.* kinken ; *pant, gasp* : i laghe that i kinke (*pres.*) TOWNL. 309.

kinnen, v., ? *O.N.* kynda ; *kindle, set on fire* ; kinned (*pple.*) A. P. ii. 915.

kīpe, *see* cūpe.

kippin, v., *O.N.* kippa, *cf. Du.* kippen ; *seize,* '*rapio*,' PR. P. 276 ; kippe HAV. 894 ; S. A. L. 166 ; i kippe (*pres.*) ant cacche P. S. 152 ; he kippis TOWNL. 90 ; (þai) kippe A. P. ii. 1510 ; kippe (*imper.*) MAN. (F.) 642 ; kipte (*pret.*) GEN. & EX. 3164 ; HAV. 1050 ; L. N. F. 218 ; kipte heore longe knives ROB. 125 ; he ne kipte of hem non hure P. L. S. ix. 64.

kip-trē, sb., *from* **kippin** ; *beam of a draw-well,* '*telo*,' PR. P. 276.

kire, *see* cūre.

kirf, sb., *O.E.* cyrf, = *O.Fris.* kerf *n.* ; *cut (with a weapon), blow ; piece cut*, GAW. 372 ; GOW. II. 152 ; twine everi kirf a wei warde from the grape PALL. i. 190.

kirke, *see* chireche. **kirne,** *see* chirne.

kirnel, *see* cürnel. **kirtel,** *see* cürtel.

kissen, *see* cüssen.

kiste, *see* cüste, chiste.

kīte, sb., *O.E.* cȳta (*buteo*, VOC. 29, 63) ; *kite,* '*milvus*,' VOC. 177 ; REL. II. 19 ; DEP. R. ii. 159 ; kite [kete] CH. C. T. *A* 1179 ; kete AYENB. 52 ; ALIS. 3048 ; like to kites (*pl.*) (*printed* tokites) HALLIW. 879.

[**kitelen,** v., *cf. M.Du.* kitelen, ketelen, *O.H.G.* chizilôn ; *tickle*.]

kitelinge, sb., *O.E.* citelung, = *M.Du.* kete-linghe, *O.H.G.* chizelunge ; *tickling*, HALLIW. 496 ; kitlinge HAMP. PS. ii. 4*.

kitling, *see* ketling.

kiton, sb., *cf. O.Fr.* chaton ; *kitten*, LANGL. *C* i. 205.

kitte, sb., *cf. M.Du.* kitte ; *milking-pail* ; kitt '*mul(c)trum*' VOC. 217 ; kit BARB. xviii. 168, 223.

kitten, *see* cüttin. **kīðen,** *see* cūðen.

kiþþe, *see* cūððe. **kix,** *see* kex.

klēþing, *see* clāþing *under* clāþen.

kn-, *see* cn-. **ko-,** *see* co-.

kr-, *see* cr-. **kū,** *see* cū.

kün, *see* cün. **kürtel,** *see* cürtel.

kvēd, *see* cwēd. **kvic,** *see* cwic.

lā, interj., *O.E.* lā ; *lo*, ORM. 741 ; GEN. & EX. 3113 ; la hwure HOM. I. 45 ; ea le MAT. xxiii. 17 ; (wen)dest þu la (*oh*) erming her o to wunienne FRAG. 7 ; wei la, le 6 ; la hwet scal þis beon HOM. I. 89 ; wa la H. M. 29 ; la god hit wot O. & N. 1543 ; la, leo, lou, le [lo] LAȝ. 5031, 15736, 21171, 25859 ; lo A. R. 52 ; ROB. 36 ; WICL. JOHN vii. 26 ; lo sire whar ic am here P. L. S. xix. 255 ; loo LIDG. M. P. 145 ; lo, low KATH. 849, 2458 ; lowr (? = lo hwer) her ich abide KATH. (E.) 2403 ; lour hit her *lo, it is here* A. R. 152.

laak, *see* lac. **laas,** *see* lás.

labbe, sb., *talkative person, blabber*, PR. P. 282 ; CH. C. T. *A* 3509.

labben, v., *cf. M.Du.* labben ; *blab, prate* ; labbe (*imper.*) LANGL. *C* xiii. 39 ; a labbing (*pple.*) shrewe CH. C. T. *E* 2428 ; PARTEN. 3751.

label, sb., *O.Fr.* label ; *label*, CH. ASTR. i. 22 (13) ; WICL. E. W. 331.

labor, sb., *O.Fr.* labour ; *labour* ; labour BEK. 49 ; A. P. i. 633.

laborer, sb., *O.Fr.* laboreor ; *labourer*, LANGL. *A* iii. 282 ; laborere *B* iii. 278.

laborin, v., *O.Fr.* labourer ; *labour,* '*laboro*,' PR. P. 283 ; laboure CH. C. T. *A* 186 ; la-bourde (*pret.*) ALEX. (Sk.) 4814.

lac, sb., = *M.Du.* lac, *M.L.G.* lak ; *lack, fault,* HICKES I. 231 ; HAV. 191 ; AYENB. 62 ; lac ne lest HOM. II. 258 ; lak '*defectus*' PR. P. 285 ; ROB. 412 ; LIDG. M. P. 158 ; lake OCTOV. (W.) 1394 ; S. S. (Wr.) 2146 ; laak PALL. iv. 895 ; wiþoute lake SHOR. 146 ; lackes (*pl.*) AYENB. 32 ; lakkes LANGL. *B* x. 262.

lak-lēs, sb., *faultless*, LANGL. *B* xi. 382.

lac², sb., *O.E.* lac *m.*, *cf. M.Du.* lac *m.*, lake *f.,* *O.H.G.* lacha *f.* ; *lake* : over þen lac [þe lake] of Silvius LAȝ. 1279 ; lak A. P. ii. 438 ; lake '*lacus*' PR. P. 285 ; ich leade ham . . . i þe ladliche lake of þe suti sunne MARH. 14 ; **láke** (*dat.*) P. L. S. xxv. 162 ; MAP 337 ; CH. C. T. *D* 269 ; LIDG. M. P. 153 ; LUD. COV. 309, 350, 387.

lāc, sb., *O.E.* lāc *n.* ; ? *from root of* **līken** ; *offering, gift*: heo nomen þat lac LAȝ. 17748 ; bi þat alter was þe lac o fele wise ȝarked ORM. 1062 ; lutel loc [lac] is gode lef P. L. S. viii. 37 ; loc '*munus*' FRAG. 4 ; þreo kinges . . . lok him broȝte P. L. S. xix. 128 ; loac GEN. & EX. 1798 ; **lāke** (*dat.*) LAȝ. 31953 ; brohten to lake KATH. 62 ; lōc (*pl.*) HOM. II. 45 ; lokes [lakes] A. R. 152 ; lakes ORM. 1157.

[**lāc²**] **leic,** sb., *O.E.* lāc *m.*, *O.N.* leikr, = *Goth.* laiks, *O.H.G.* leich *m.; sport, play, activity* ; leik HAV. 1021 ; JOS. 17 ; laik WILL. 678 ; loveli laik was it nevere bitwene þe longe and þe shorte LANGL. *B* xiv. 243 ; sinful laik M. H. 58 ; leȝkes (*pl.*) (*ms.* leȝȝkess) ORM. 2166 ;

laikes MIN. iii. 64; PERC. 1704; ANT. ARTH.. xlii; *comp.* **brūd-, cair-, clǣn-, dūsi-, dweo-mer-, feir-, fēr-, fiht-, frēo-, gōd-, hende-, īdel-, love-, mēok-, rēaf-, schend-, siker-, trēu-, wed-, wōhlāc** (-leic).

lacche, sb., *latch*, ' *cliket*,' VOC. 170; lahche, latche (? lacche) '*pessulus*' PR. P. 283; lache ED. 2929.

lǣcchen, v., *O.E.* (ge-)lǣccan (*pret.* lǣhte); *catch, seize,* ORM. 12300; lacche O. & N. 1057; JOS. 356; lacche foules LANGL *B* 355; lacche with foules *A* v. 199; lache GREG. 275; (hie) **lǣchēð** (*pres.*) HOM. II. 197; lacche me in þin armes WILL. 666; **leacche** (*subj.*) A. R. 164*; **lǣhte** (*pret.*) A. R. 102*; HORN (R.) 249; lahte ut his tunge MARH. 9; laȝte A. P. i. 1204; lauȝte JOS. 222; LANGL. *B* xviii. 324; DEP. R. ii. 159; laute HAV. 744; **lǣht** (*pple.*) ORM. 11621; P. S. 192; lagt GEN. & EX. 3141; laȝt GAW. 156; REL. I. 226; laght IW. 3622; lauȝt WILL. 671; (*printed* laught) ALIS. 685; lacched MAN. (H.) 120; *comp.* **a-, bi-, ȝe-lacchen.**

lǣcching, sb., *taking,* LANGL. *A* i. 101.

lāce, *see* **lās.**

lacerte, sb., *O.Fr.* lacerte; *fleshy muscle,* CH. C. T. *A* 2753.

lache, *see* **lacche, lasche. lāche,** *see* **lǣche.**

lachesse, sb., *O.Fr.* laschesse; *negligence, laziness,* LANGL. *A* ix. 32; lachesse LANGL. *C* ix. 253, x. 269.

lachet, sb., *Fr.* lacet; *latchet,* PR. P. 284; GAW. 591.

lācin, v., *O.Fr.* lacer; *lace,* PR. P. 283; lace HORN (L.) 717; **lācinge** (*pple.*) CH. C. T. *A* 2504; **lāced** (*pret.*) WILL. 1736; lāced (*pple.*) CH. C. T. *A* 3267; ilaced A. R. 620; MISC. 77.

lacken, *see* **laken. lācnien,** *see* **lǣchnien.**

*h*lad, sb.,? *O.E.* hlæd, = *O.N.* hlað; *from h*lāden; *draught, load*: we loden alle twinne lad of his godnesses welle ORM. 19313; ech on other laid good lade TOR. (A.) 1663.

lād, sb., *O.E.* (ge-)lād (*way*), = *M.H.G.* geleite *n.*; *? act of leading*; **lādes [lōdes] mon** (*sb.*) [= *M.L.G.* leides man, *M.H.G.* leites (geleites) man]; *pilot,* LAȝ. 6245; lodes man AYENB. 140; lodęs mon A. P. iii. 179; loder (*for* lodes) man GEN. & EX. 3723.

[**lād-man,** sb., *pilot.*]

lōdemenāge, sb., *pilotage,* CH. C. T. *A* 403.

ladde, sb., *lad, young man,* HAV. 1786; SPEC. III; P. S. 239; FER. 400; AL. (T.) 260; MAN. (F.) 5895; A. P. ii. 36; DEP. R. iii. 146; **laddes** (*pl.*) LANGL. *B* xix. 32; IW. 2266.

*h*laddre, sb., *O.E.* hlǣder, hlǣdder, = *O.Fris.* hleder (? hlēder), hladder, *O.H.G.* hleitra; *ladder*; laddre ROB. 333; C. L. 915; LANGL.

B xvi. 44; laddre, leddre SHOR. 2, 3; lheddre AYENB. 246; leddre HOM. I. 129; A. R. 354; GEN. & EX. 1607; **ledderis** (*pl.*) BARB. ix. 314: þe two leddre stalen A. R. 354; laddren (*v. r.* laddrene, laddres, ladders) ROB. (W.) 8490.

leddir-staf, sb., *? rung of a ladder,* '*scalare,*' PR. P. 293.

lāde, sb., *O.E.* lād, = *O.N.* leið, *M.Du.* lēde, leide, *O.H.G.* leita *f.*; *from* līðen; *path, road, journey; load*; PR. C.3421; he wile folȝhenaȝ þat ilke steornes lade ORM. 2140; ilc an sholde þrinne lac habben wiþ him o lade 3455; lode A. R. 268; TRIST. 419; TOR. 1676; E. G. 396; þe stede liked wel þe lode WILL. 3292; what þai in lode hadde *what they carried* FER. 2703; of stree . . . mani a lode CH. C. T. *A* 2918; loode '*vectura*' PR. P. 310; *comp.* **carte-, hors-, līf-, plōu-lāde(-lōde).**

lōd-cniht, sb., *guide,* LAȝ. 25729.

lōde-sterre, sb., = *M.Du.* leidsterre, *M.H.G.* leitesterne; *loadstar,* MAND. 180; CH. C. T. *A* 2059; lode-, loodsterre TREV. I. 301; ladesterne D. ARTH. 751.

lād-þēow, sb., *O.E.* lād-, lāttēow, latťēow (ALFR. P. C. 3ò5); *leader*; ladþeow ANGL. III. 152; JUL. 33; latteu HOM. II. 197; **lādtēwes** (*pl.*) MAT. xv. 14.

*h*lādel, sb., *O.E.* hlǣdel; *from h*lāden; *ladle*; ladel LANGL. *B* xix. 274; CH. C. T. *A* 2020.

*h*lāden, v., *O.E.* hladan, = *O.L.G.* hladan, O. *H.G.* hladan, ladan, *O.N.* hlaða, *Goth.* (af-)-hlaþan; *load; draw up*; ladin PR. P. 283; lade HORN (R.) 1409; lhade out þet weter AYENB. 178; **hlādeð** (*imper.*) ['*haurite*'] JOHN ii. 8; lōden (*pret.*) ORM. 19313; **lāden** (*pple.*) ORM. 14054; GEN. & EX. 1800; RICH. 1389; lade A. P. i. 1145; *comp.* **ȝe-, over-lāden**; *deriv. h*lad, *h*lādel, *h*last.

lāden, *see* **lǣden. lādi,** *see* **hlǣfdiȝ.**

lǣche, sb., *O.E.* lǣce, = *Goth.* lēkeis, *O.Fris.* lētza, *O.H.G.* lāchi; *leech, physician,* ORM. 19354; leche A. R. 178; SPEC. 92; SHOR. 36; WILL. 576; LANGL. *B* xx. 302; MAND. 199; CH. C. T. *G* 56; S. S. (Wr.) 446; M. ARTH. 200; PR. P. 291; lache HOM. II. 229; **lǣches** (*gen.*) MAT. ix. 14; **lēches** (*pl.*) P. L. S. xv. 190; lechis BARB. v. 437; **lēchene** (*gen. pl.*) HOM. II. 41.

lǣche-craft, sb., *O.E.* lǣcecræft; *art of healing,* LAȝ. 7616; ORM. 1869; lechecraft A. R. 178; ROB. 141; LANGL. *B* xvi. 104; IW. 2736.

lǣche-dōm, sb., *O.E.* lǣcedōm, = *O.H.G.* lāchitoum; *medicine,* ORM. 1851; lechedom HOM. I. 111.

lǣche², sb., *O.E.* lǣce. = *M.Du.* lāke; *leech*; leche '*hirudo*' PR. P. 291.

lǣche³, sb., *? = O.N.* (mein-)leiki; *? countenance*: ladliche lǣches (*pl.*) LAȝ. 1884; þine leches beoþ grisliche O. & N. 1140; mid his lēchen (*dat.*) he gon liȝen LAȜ. 13703; *comp.* cnāwe-, wōh-lēche.

[lĕche⁴, sb., *O.E.* (æg-)lǣca; *comp.* eglēche.]

[lĕchen, v., *O.E.* lǣcan; *from* lāc²; *play;* *comp.* cnāwe-, cūð-, ed-, even-, nēh-, riht-, þriste-lēchen.]

lĕchen², v., = lĕchnien; *heal, act as physician,* ORM. 1856; leche WICL. IS. lxi. 1; leeche DREAM 854; lēched (*pret.*) LANGL. *B* xvi. 113.

lēchunge, sb., = lĕchnunge; *healing, surgery,* HOM. I. 187; leching D. TROY 10223; BARB. xiii. 46.

lĕchnien, v., *O.E.* lǣcnian, lācnian, *O.N.* lǣkna, = *Goth.* lēkinōn, *O.H.G.* lāchenōn; *act as physician;* lechnien HOM. I. 23; lechnien, lacnien [lechnie] LAȜ. 16589, 19500; lecnen A. R. 330; lēchnede (*pret.*) LANGL. *C* ix. 189*.

lēchnunge, sb., *O.E.* lācnung, *O.N.* lǣkning; *healing, surgery,* HOM. I. 202; JUL. 6.

lĕde, adj., *O.E.* (un-)lǣde, = *Goth.* (un-)lēds; *?rich;* he(o) þat beoþ nu lede (*? ms.* leaþe, *r. w.* grede) MISC. 77; *comp.* un-lēde.

lĕden, v., *O.E.* lǣdan, = *O.L.G.* lēdian, *O.Fris.* lēda, *O.N.* leiða, *O.H.G.* leitan; *from* liðen *lead,* ['*ducere*'] LK. vi. 39; læden, leden, leaden [lede] LAȜ. 358, 828, 1000; laden, leden HOM. II. 7; leden A. R. 78; ORM. 938; S. S. (Wr.) 1636; god lif leden P. L. S. viii. 62; leden song GEN. & EX. 699; lede AYENB. 168; LANGL. *B* xvii. 117; TOR. 1061; ferde lede O. & N. 1684; folc ut lede HAV. 89; liȝte (charge) lede (*carry*) LEB. JES. 417; lǣst (*pres.*) R. S. i (MISC. 158); leadeð KATH. 261; let HOM. I. 111; A. R. 174; AYENB. 185; lat MISC. 101; MAP 336; (heo) ledeð A. R. 162; ledeþ AYENB. 50; lēd (*imper.*) A. R. 40; ne led ous naȝt in to vondinge AYENB. 116; lĕdde, ladde, ledde [ladde] (*pret.*) LAȜ. 584, 5606, 6565; he leadde (*carried*) an his honde enne bowe stronge 1452; ledde HOM. I. 11; A. R. 160; ORM. 3204; GOW. II. 374; leadde KATH. 1592; ladde BEK. 887; ALIS. 1485; WILL. 1609; CH. C. T. *A* 2275; þu leddest [leaddest] JUL. 32, 33; ledden HAV. 2451; ladden REL. I. 129; lĕd (*pple.*) GEN. & EX. 649; (*ms.* ledd) ORM. 9359; lad LANGL. *B* ix. 16; MAND. 14; *comp.* æt-, bi-, for-, ȝe-lǣden.

lēdere, sb., *O.E.* lǣdere, = *O.H.G.* leitāri; *leader,* WILL. 1355; LANGL. *B* i. 157; ledere [ledare] TREV. V. 217; ledare PR. P. 292; lord and ledar of Kintir BARB. iii. 660; lēdaris (*pl.*) BARB. xi. 160.

lēdunge, sb., = *M.L.G.* lēdinge; *leading,* HOM. I. 207; ledinge PR. P. 293; leding PR. C. 4217.

lĕfdi, *see* hlǎfdiȝ.

lǣfe, *see* lēave.

læfel, sb., *O.E.* læfl, = *O.H.G.* label; *cup, bowl,* '*scyphus,*' FRAG. 4; læflen (*dat. pl.*) LAȜ. 22762.

lēn, lān, sb., *O.E.* lēn, *O.N.* lān, lēn, = *O.Fris.* lēn, *O.H.G.* lēhan n.; *from O.E.* lēon, = *L.Ger.,* *O.H.G.* līhan, *Goth.* leihwan (δανείζειν); *loan;* ORM. 1518; lane [leane] HOM. I. 257 (326); lone PR. P. 312; A. R. 208; P. L. S. xvii. 479; REL. I. 113; AYENB. 35; A. D. 247; GOW. II. 284; CH. C. T. *B* 1485; AMAD. (R.) xxxviii; hit is godes lone (*printed* loue) REL. I. 175 (MISC. 114, 115); loone (lene) LANGL. *B* xx. 284.

lēne, adj., *O.E.* lǣne, = *O.L.G.* lēhni; *from* lēn; *transitory, temporary*: ȝif we forleosað þas lēnan (*pl.*) worldþing HOM. I. 105.

hlǣne, adj., *O.E.* hlǣne; *lean, poor, thin;* læne LAȜ. 19445; lene PR. P. 296; REL. I. 211; GEN. & EX. 2099; P. L. S. xiii. 236; P. 34; lene [leene] LANGL. *B* PROL. 123; CH. C. T. *A* 287; leene LIDG. M. P. 132; lene, leane A. R. 118, 368; lhene AYENB. 53; lenie BARB. i. 387; lēnre (*compar.*) HOM. I. 37.

lēnnesse, sb., *O.E.* lǣnnyss; *leanness,* PR. P. 296.

lēnen, v., *O.E.* lǣnan, = *O.Fris.* lēna, *O.N.* lāna, lēna, *O.H.G.* lēhanōn; *from* lēn; *lend;* lenen LAȜ. 31621; A. R. 248; ORM. 15795; GEN. & EX. 3170; PL. CR. 741; hu mihte he leanen [lenen] lif to þe deade KATH. 1086; lene HAV. 2072; AYENB. 35; MIRC 1485; IW. 757; lene hem wimmen ROB. 42; leene ['*commodare*'] WICL. DEUT. xv. 9; TRIAM. 683; leendin '*praesto, concedo*' PR. P. 296; lennen CATH. 213; lese wordes þu me lēnst (*pres.*) O. & N. 756; lord þat lenest us lif P. S. 153; ant leneð (*printed* leueð) his luve in liflese schaften JUL. 24; lenþ AYENB. 6; ge leaneð LK. vi. 34; (hi) leneþ AYENB. 35; lēn [lean] (*imper.*) LAȜ. 11494; len P. L. S. xxiv. 234; (*ms.* lene) DEGR. 1599; len [leen] me ȝoure hond CH. C. T. *A* 3082; lēne (*subj.*) LAȜ. 4383; god lene him grace WILL. 327; lēninde (*pple.*) AYENB. 35; lēnde (*pret.*) P. L. S. viii. 61; BEK. 775; þis lond he hire lende LAȜ. 228; lēned (*pple.*) ORM. 4387; lend REL. I. 113; lent LANGL. *B* xiv. 39; mi love is lent and liht on Alisoun SPEC. 28; *comp.* a-, ȝe-lēnen.

lēnere, sb., *lender,* AYENB. 35; leenere WICL. PROV. xxii. 7.

lenninge, sb., *lending, loan,* HAMP. PS. xxxvi. 27*.

hlǣnin, v., *O.E.* hlǣnian; *grow lean, pale;* lenin '*macero*' PR. P. 296; þi rudi neb schal leanen H. M. 35; lene PALL. iii. 810.

hlǣnsien, v., *?make lean;* lēnsed (*pple.*) his fleis HOM. I. 147.

lēre, adj., = *O.L.G. O.H.G.* lāri; *empty, useless;* lere O. & N. 1528; ROB. 81; BER. 1953;

leer TREV. II. 283; lere B. B. 60; TREV. III. 311; *comp.* ȝe-lǣr.

lǣren, v., *O.E.* lǣran = *O.L.G,* lērian (*pret.* lērde), *O.H.G.* lēran, *O.Fris.* lēra, *O.N.* lǣra, *Goth.* laisjan; *teach; learn,* MK. vi. 2; ORM. 6218; leren A. R. 64; leren GEN. & EX. 354; SPEC. 82; lerin '*doceo, disco*' PR. P. 298; lere O. & N. 1017; HAV. 823, 2592; ROB. 67; MIRC 546; LANGL. *B* xviii. 225; CH. TRO. iv. 441; S. S. (Wr.) 131; PERC. 232; PR. C. 155, 5874; AUD. 13; lere, learen [leore] LAȝ. 15213, 23121; to lerene HOM. I. 131; lēareð (*pres.*) A. R. 64; and lereð prestes lore REL. I. 211; lēr (*imper.*) MISC. 128; lērde (*pret.*) HOM. I. 125; P. L. S. viii. 154; LAȝ. 8613; WILL. 341; and lerdest hi to don schome O. & N. 1053; lerden KATH. 470; lærde LAȝ. 7458; lēred (*pple.*) GEN. & EX. 4; CH. C. T. *C* 283; OCTOV. (W.) 715; M. H. 2; *comp.* ȝe-, for-, mis-lǣren.

   lērare, sb., = *O.H.G.* lērāri, *Goth.* laisareis; *teacher,* '*doctor,*' PR. P. 297; lēreris (*pl.*) WICL. HEB. xii. 9.

lǣs, adv., *O.E.* lǣs, = *O.L.G.* les (*compar.*); *less:* þi læs þe þi wiðerwinne þe selle þam deman MAT. v. 25; les A. R. 30; GEN. & EX. 3595; AYENB. 118; hit nis noht les (*r. w.* wes) HOM. I. 57; þe les *the less* ROB. 26; M. H. 19; no þe les O. & N. 374; iwend þe from uvele þi les þe (*lest*) ðu ... losie HOM. I. 117; les te *lest* LAȝ. 6642; A. R. 58; KATH. 2386; heo is afered les te þeo eorðe hire trukie HOM. I. 53; for drede leste he wolde hire bite K. T. 404; þat treo him was forbode lest he hedde þe miht of gode C. L. 1068; las TRIST. 2508; last WILL. 641; les when [' *ne qvando*'] HAMP. PS. ii. 12.

   lăsse, adj., *O.E.* lǣssa, *m.,* lǣsse *f. n.,*= *O. Fris.* lessa; *less,* LAȝ. 7042; P. L. S. viii. 107; ORM. 7906; a lasse hul ROB. 204; Inde þe lasse and þe more MAND. 4; lesse A. R. 86; GEN. & EX. 994; lesse (*n. absol.*) A. R. 92; þe him ȝiveþ lesse P. L. S. viii. 36; lasse ORM. 4896; LANGL. *B* v. 252; lasse ne mare KATH. 1561; þa læsse [lasse] LAȝ. 12735; lasse O. & N. 482; ALIS. 68; lasse & mare M. ARTH. 687; lesse guodes AYENB. 90.

   lěst, adv., *O.E.* lǣst, lǣsest (*superl.*); *least,* MARH. 13; AYENB. 36; er me lest wene A. R. 178; and lest it costeþ GOW. I. 153; least MARH. 15; lǣste, adj., *least* ORM. 15277; leaste HOM. I. 253; leste MAND. 173; þe leste man LANGL. *A* iii. 24; at the leste HORN (R.) 612; CH. P. F. 452; PR. C. 2322; et te leste A. R. 164; bi þe laste HORN (L.) 616.

lǣs, *see* **lěas.**

lǣssen, lǣssenen, v., *lessen;* lessin '*minuo,*' PR. P. 298; lessi AYENB. 175; lassen GAW. 1800; lasse (*pres.*) REL. II. 175; lasseð KATH. 1718; laꜱꜱeþ GOW. III. 147; lessith S. & C. II. xlvi; (þei) lasseþ [lasseneþ] TREV. VIII. 327;

---

lěssed (*pret.*) PS. lxxxviii. 46; lasned A. P. ii. 438, 441; ilěssed (*pple.*) SHOR. 127.

lessinge, sb., *diminution,* AYENB. 268.

lěst, sb., *O.E.* lǣst, lāst (*footstep*), = *O.H.G.* leist (*forma*), *Goth.* laists (ἴχνος); *bootmaker's last;* lest of a boote VOC. 181; lest PR. P. 298; last VOC. 196; lěstes (*pl.*) P. L. S. xxxiv. 13: sutters with ȝour mani lestes REL. II. 176.

[**lěst**[2], sb., *O.E.* (ful-)lǣst, (ge-)lāst, = *O.L.G.* ful-, (ge-)leist, *O.H.G.* (ful-)leist; *help; comp.* fülst.]

   lăst-ful, adj., *O.E.* (ge-)lǣstfull; *helpful, lasting:* þu ware me lastful HOM. II. 183; lesteful lif GEN. & EX. 304.

lǣsten, v., *O.E.* lǣstan, = *O.L.G.* lēstean, *O. Fris.* lāsta, *O.H.G.* leisten, *Goth.* laistjan (διώκειν); *pursue; continue, last, endure;* lesten PL. CR. 855; hu þine sceal a lesten P. L. S. viii. 74*; heo wenden hit scholde lesten o MISC. 152; ðat ic ðe have hoten wel ic it sal lesten (*perform*) GEN. & EX. 2906; lestin '*duro*' PR. P. 299; leste ASS. 112; in sunne leste C. L. 1058; þat mai to hevene leste SHOR. 3; lasten KATH. 279; SPEC. 49; GOW. III. 239; (*ms.* lasstenn) ORM. 8549; laste HAV. 538; lěsteð (*pres.*) LAȝ. 1959; lesteð A. R. 20; lasteþ MISC. 100; WILL. 5538; leasteð MARH. 5; hit lasteð þre wuke HOM. II. 3; lest, last O. & N. 516, 1450, 1466; þei lasten noght þat þei behoten MAND. 252; lěstende (*pple.*) HOM. I. 159; lestinde PROCL. 7; evre lestinde AYENB. 104; lěste (*pret.*) LAȝ. 6542; leste HOM. II. 117; laste ROB. 9; lǎsted (*pple.*) MAND. 10; *comp.* ȝe-, þurh-lǣsten.

   lǎsting-lī, adv., *constantly,* WICL. DEEDS i. 14.

lǣte, lāt, sb., *cf. O.N.* lāt *n.,* lǣti, *M.L.G.* lāt; *from* lǣten; *appearance, pretence:* feir lete HOM. I. 69; he makeð lete of þoleburdnesse II. 79; lat M. H. 123; þe lot of þe windes A. P. iii. 161; man haldeþ hem for gode men for þeȝre gode late, ORM. 1213; for þine fule lete O. & N. 35; lote ANGL. I. 74; wið reuli lote GEN. & EX. 1162; of lote & h(e)ue A. P. i. 895; weie lōt [*cf. O.E.* wega gelǣte] *meeting of roads:* in þe weie lot(e) [' *in bivio*'] WICL. GEN. xxxviii. 21; lātes (*pl.*) A. R. 90; luveliche lates MARH. 14; þe lates of þe foules ALEX. 149; latis ISUM. 180; his lǣtes weoren alle swulc he lome me weore LAȝ. 30776; ofte he hire loh to & makede hire letes 18543; wei(e) letes MIRC 748; mid swiðe væire læten LAȝ. 15661; summe (wenden) to weien læten 15509; *comp.* an-, ȝe-, in-, ūt-, -lǣte.

[**lǣte**[2], adj., *O.E.* (earfoð-)lǣte; *comp.* ēað-lēte.]

lǣten, v., *O.E.* lǣtan, = *O.Fris.* lēta (*permit*), *Goth.* lētan (ἐᾶν, ἀφιέναι), *O.L.G.* lātan (*permit, send*), *O.N.* lāta (*put, permit, behave*), *O.H.G.* lāzan (*permit*); *let, permit; dismiss, leave; cause; think, esteem,* LAȝ. 8612; læten wel of oþre men ORM. 7527; leten GEN. & EX. 4142;

WILL. 2184; GOW. II. 369; hic willeð here
sinnes leten HOM. II. 27; hwanne (we) ure lif
leten schule REL. I. 175 (MISC. 112); þeos . . .
beoð alle ine freo wille to donne oþer to leten
A. R. 8; þet oðer þu most leten 102; leten to
wel of hire sulven 176; lete [leoten] makie
KATH. 1475; leoten JUL. 12; lete O. & N.
1445; SHOR. 147; AYENB. 56; teres lete
shed tears AL. (L.¹) 716; and lete [late]
him in his prisoun stille dwelle CH. C.
T. *A* 1335; laten HAV. 328; latin, laatin
'*dimitto, relinquo, permitto, puto*' PR. P.
288, 289; late P. S. 338; MAND. 11; late
him inne PERC. 961; lāte (*pres.*) HAV.
1741; swa þu lætest lasse of þe swa læteþ
drihtin mare ORM. 4896; letest ANGL. III.
62; þat tu letest lutel of al þat tu schuldest
luvien JUL. 15; þu lezst [letest] LA3. 18067;
þu lest REL. I. 184; FL. & BL. 365; leteð
REL. I. 211; lateþ modili3 ORM. 1296; lates
PR. C. 1567; he ne let him nout blod A. R.
112; hvanne me let (*omits*) wel to done
AYENB. 26; he let [leet] it soþ LANGL. *B* xv.
168; he . . . lat hi grede O. & N. 308; of him
leteþ wel lihtliche 1774; þo þet leteþ (*relinquish*) al þet hi habbeþ AYENB. 165; thai let as
thai armid to stand with god HAMP. PS. lxxvii.
12; lētande (*pple.*) *appearing* D. ARTH.
3831; lǣt (*imper.*) ORM. 7619; let HOM. I. 57;
MARH. 8; let hine & nim þene oðer LA3. 13156;
lat O. & N. 258; LANGL. *B* iii. 73; leteð HOM.
I. 39; A. R. 42; lǣte (*subj.*) P. L. S. viii. 170;
ORM. 4859; lete BEK. 757; ure loverd me
lete ibide þe dai FL. & BL. 175; lete we a wei
þeos cheste O. & N. 177; lēot (*pret.*) SAX.
CHR. 255; let GEN. & EX. 809; HAV. 314;
BEK. 451; let it eornen ORM. 1336; let
lihtli3 þær offe 16517; and let þat uvele mod
ut O. & N. 8; lette 952; he let þat folk fasten
SAINTS (Ld.) xlv. 78; his swete herteblod he let
for us P. S. 257; hi hedde iwrite in hare testament þet hi him let (*bequeathed*) a þousend
and vif hondred pond WILL. 191; he let
make WILL. 5532; sho let als sho him noght
had sen IW. 1809; let [leet] CH. C. T. *A* 128;
leet MAND. 137; and leet li3t of þe lawe
LANGL. *B* vi. 170; liet LEG. 30; let, lette
LA3. 432, 7102; lette KATH. 792; A. R. 54;
þu lete O. & N. 1308; leten ORM. 9821; BRD.
5; ALIS. 3278; RICH. 3305; þe lutel leten
(*accounted*) of godes bode P. L. S. viii. 131;
here lif hi lete þere HORN (L.) 1246; he letten
after him sende A. D. 261; twe3en hafde he leten . . .
cwellen ORM. 8149; lete [laten] CH. C. T.
A 4346; for a knight i mai be lete PERC.
803; laten HAV. 240; IW. 2423; S. S. (Wr.)
1811; MAN. (H.) 61; *comp.* a-, æt-, for-,
3e-lǣten.

lǣven, v., *O.E.* lǣfan,=*O.Fris.* lēva, *O.N.*
leifa, *O.H.G.* leiben, *Goth.* (bi-)laibjan; *from
līven*; *leave, relinquish; remain, be left*;
(*ms.* læuen) LA3. 994; leaven KATH. 429;
A. R. 162; leven A. D. 243; GOW. I. 86; levin

PR. P. 301; leve WILL. 2358; LANGL. *B* xv.
101; ich wole . . . mi lond leve for love of þe
BEK. 39: þat he ne mi3te leve þer 1246;
leif BARB. iv. 608; lēfeþ (*pres.*) ORM. 8664;
lēf (*imper.*) A. R. 102; BEK. 136; lēve
(*subj.*) DEGR. 933; lǣfde (*ms.* lafde) [lefde]
(*pret.*) LA3. 766; lefde MAT. xxii. 25; leafde
KATH. 480; lafte JOS. 707; laftin ALEX.
(Sk.) 886; LANGL. *B* xx. 250; læfden LA3.
3909; lede B. B. 301; lĕfde (*subj.*) A. R. 70;
lĕved (*pple*) HAV. 225; levid MIN. i. 55;
comp. bi-, for-, 3e-, lǣven.

lēving, sb., *residue*; lēvenges (*pl.*) TREV.
III. 113, 117.

lǣwed, adj., *O.E.* lǣwed, *mod.E.* lewd; *lay,
unlearned*, ORM. 846; læwed [lewed] LA3.
24625; lewed BEK. 378; AYENB. 25; MAND.
164; CH. C. T. *A* 574; OCTOV. (W.) 1715;
lewide TREV. III. 319; leud '*laicus, ignarus*'
PR. P. 301; AUD. 18; læwed, lawed SAX. CHR.
254, 255; lawed M. H. 5; lewed men *laymen* ROB. (W.) 9522; lewede men 9676.

lēudnesse, sb., *meanness, ignorance, lay
manners*, '*ignorantia*,' PR. P. 301; lewednesse
TREV. VI. 239, VII. 263, VIII. 217.

[lǣwen, v., *O.E.* (be-,3e-)lǣwan,=*Goth.*lēwjan,
O.H.G. *(gi-)lāhan (GRAFF II. 295); *betray*;
comp. bi-lǣwen.]

ħlāf, sb., *O.E.* hlāf,=*O.N.* hleifr, *Goth.* hlaifs,
hlaibs, *O.H.G.* leib; *loaf, bread*; hlaf MK.
iii. 20; laf ORM. 1474; lof FRAG. 1; HAV.
653; BRD. 13; TRIST. 382; a lof bred BEV.
1420; loof '*panis*' PR. P. 310; hlāfe (*dat.*)
HOM. I. 227; lhove AYENB. 82; hlāfes (*pl.*)
JOHN vi. 5; laves [loaves] LA3. 22781; loves
WICL. JOHN vi. 5; hlāfe (*gen. pl.*) MAT. xvi.
9; hlāfan (*dat. pl.*) MK. vi 52.

ħlāmmesse, sb., *O.E.* hlām-, hlāfmæsse;
lammas; lammesse PR. P. 286; lammasse
ROB. 201; TREV. VIII. 185.

ħlāfdi3, sb., *O.E.* hlǣfdige; *lady*; (*ms.* laff-
di3) ORM. 1807; lafdi KATH. 88; lǣfdi
[leafdi] LA3. 1256; lef-, leafdi A. R. 4, 38;
lef-, leve-, lavedi O. & N. 959, 1051; lhevedi
AYENB. 24; lavedi REL. I. 102; ISUM. 108;
lade GAW. 1810; our lādeis (*gen.*) evin
Mary *our Lady Mary's eve* BARB. xvii. 335;
ladi SPEC. 113; MAND. 86; CH. C. T. *A*
88; laidis (*pl.*) MAN. (F.) 11233; ladisse D.
ARTH. 3081.

lĕfdi-schipe, sb., *ladyship*, A. R. 108.

ħlāford, sb., *O.E.* hlāford *for* *hlāf-weard;
lord; hlaford HOM. I. 217; laferd SAX. CHR.
249; ORM. 42; PR. C. 416; laverd LA3. 268;
laverd, loverd O. & N. 959; hloverd HOM.
II. 43; lhoaverd PROCL. 1; loverd A. R. 2;
GEN. & EX. 30; HAV. 96; BEK. 490; lhord
AYENB. 1; lord LANGL. *A* i. 89; MAND. 3;
CH. C. T. *A* 837.

lāferd-dōm, sb., *O.E.* hlāforddōm; *lordship*,
ORM. 11851; laverddom H. M. 11,

lāverding, sb., *lording, sir* (*title of address*) ; laverding, lording (E.) C. M. 11769 ; loverding ROB. 431 ; **lāferdinges** (*pl.*) ORM. 918 ; laverdinges LA₃. 27394 ; lhordinges AYENB. 67 ; lordinges PL. CR. 609.

lōverd-lēs, adj., *O.E.* hlāfordlēas ; *lordless, widowed*, BEK. 670 ; moni wif loverdles ROB. 142.

lōverd-lich, adj., *lordly*, HOM. II. 23 ; lordlich LANGL. C iv. 199.

lāferdling, sb., *lordling* ; **lōverdlinges** (*pl.*) LA₃. 12664* ; BEK. 524.

lāver(d)-scipe, sb., *O.E.* hlāfordscipe ; *lordship*, HOM. I. 111 ; lordschipe C. L. 142 ; MAND. 9.

lāverd-swike, sb., *O.E.* hlāfordswica ; *traitor to his lord*, LA₃. 22138 ; loverdsvike CHR. E. 1033 ; lordswike P. S. 220.

la₃e, sb., *O.E.* lagu, *O.N.* lög ; *law*, SHOR. 50 ; AYENB. 97 ; þeos la₃e LA₃. 6307 ; on þere alde la₃e HOM. I. 87 ; la₃e [lawe] O. & N. 969 ; la₃he ORM. 2376 ; laghe PR. C. 2163 ; MIR. PL. 175 ; lahe KATH. 780 ; lawe BEK. 300 ; LANGL. *A* iii. 156 ; K. T. 387 ; broþer in lawe (*O.Fr. frere en loi*) ALIS. 4399 ; fadir in lawe '*socer*' PR. P. 145 ; sone in lawe CH. C. T. *E* 315 ; laue AUD. 39 ; la₃e (*pl.*) HICKES I. 223 ; þa ten la₃e HOM. I. 11 ; þas la₃en 15 ; lawe P. L. S. viii. 156 ; la₃e, lawen [lawes] LA₃. 1167, 18172 ; lawen FRAG. 8 ; lawen C. L. 167 ; lagas SAX. CHR. 254.

la₃he-bōc, sb., *O.N.* lögbōk ; *law book*, ORM. 1953.

lawe-breche, sb., *O.E.* lahbrice ; *breach of law*, WICL. Is. i. 5.

lawe-ful, adj. *O.N.* lögfullr ; *lawful*, TREV. III. 193 ; lawful H. V. 113.

laweful-liche, adv., *lawfully*, LANGL. *C* x. 59 ; lawefolich TREV. V. 297, 313.

lah-, lauhfulnesse, sb., *lawfulness*, O. & N. 1741.

lawe-lēs, adj., *O.N.* löglauss ; *lawless*, TREV. III. 73 ; la₃elēase (*pl.*) HOM. II. 229.

lage-liche, adv., *O.E.* lahlīce, *O.N.* lag-, lögliga ; *lawfully*, REL. I. 131 ; la₃helike ORM. 2374 ; lagelike MISC. 22 ; mid laweliche deden 106.

La₃a-mon [**Laweman**], pr. n., *cf. O.E.* lahmann, *O.N.* laga-, lögmaðr (*man of law*) ; *Layaman*, LA₃. 2.

lage-wīs, adj., *O.N.* lagavīss ; *experienced in law*, LK. xi. 46.

la₃e², adj., *O.E.* lagu = *O.L.G.* lagu, *O.N.* lögr *m.; pool* ; lage MAT. vii. 12 ; laie, leie, ARTH. & MER. 5306, 9486 ; lai LEG. 13 ; GOW. II. 5 ; laies (*pl.*) TREV. III. 367 ; lawes LEG. 12 ; lawen (*ms.* lauen, *r. w.* slawen) (*dat. pl.*) ALIS. 3856.

la₃e-fen, sb., *marsh, swamp* : la₃efen [leiven] LA₃. 22835 ; leifen, leiven (*ms.* leiuen, *printed*

leinen). H. M. 33 ; leiven (*printed* leinen) MARH. 14 ; leievenne (*dat.*) A. R. 328.

lei-mūre, sb., *fen*, A. R. 328*.

[la₃e³, ? sb. ; *comp.* in-, ūt-, wiðer-la₃e.]

lā₃e, *see h*lā̆we.

la₃en, v., *O.E.* lagian, = *O.N.* laga ; *institute, ordain* ; lahede (*pret.*) KATH. 1213 ; for þe was wedlac ilahet (*pple.*) H. M. 21.

lā₃en, v., = *M.Du.* leeghen ; *from* lāh ; *make low* ; la₃hen ORM. 2639 ; lawe S. B. W. 126 ; lo₃i AYENB. 28 ; lowin '*humilio*' PR. P. 315 ; lōweþ (*pres.*) GOW. III. 295 ; þe sonne loweþ ALIS. 5746 ; everi man þat loweþ him self TREV. V. 107 ; lawes PR. C. 8505 ; lo₃e þe SHOR. 107 ; lōwen (*subj.*) WICL. JUDG. xix. 24 ; lōwed (*pret.*) WILL. 695 ; lowudest A. R. 190 ; lō₃ed (*pple.*) A. P. ii. 1650 ; lawed PR. C. 8522.

la₃ere, sb., *lawyer* ; lawere PR. P. 290 ; lawer [lawiere] TREV. III. 275 ; **lawieres** (*pl.*) LANGL. *A* vii. 59.

laggen, v. : theire launces thei lachene . . . laggene with longe speres D. ARTH. 2542.

laggin, v., *? splash, muddy, 'palustro, labefacio'* PR. P. 283 ; **laggid** (*pple.*) IBID. *comp.* **bi-laggen.**

lagt, pple., *see* **lacchen.**

lāh, adj., *O.N.* lāgr, = *M.Du.* laegh, leegh, *O.Fris.* lēg, *M.H.G.* lǣge ; *low, humble*, GR. 15 ; ORM. 15246 ; laih [loh] LA₃. 986 ; lo₃ BEK. 1783 ; SPEC. 73 ; SHOR. 145 ; AYENB. 105 ; a lo₃ [lou] LANGL. *A* ii. 234 ; bi loogh *below* A. P. ii. 116 ; louh A. R. 140 ; lau (*ms. law*) MIN. vi. 50 ; N. P. 17 ; lou (*ms.* low) TREAT. 136 ; þe la₃e [lowe] LA₃. 22135 ; þe la₃he leod ORM. 9319 ; laigh BARB. xiii. 651* ; that held the plain ai & the law BARB. vi. 518 ; **lāge** (*pl.*) GR. 17 ; lowe FRAG. 6 ; lowe HAV. 1324 ; **lāhere** (*compar.*) H. M. 27 ; lahre, lah₃hre ORM. 2664, 13270 ; lowure A. R. 380 ; **lā₃hest** (*superl.*) ORM. 15247 ; laheste MARH. 14 ; lo₃este AYENB. 119 ; loweste BEK. 1187.

lāhe, adv., *low*, MARH. 12 ; lihten swa lahe JUL. 10 ; la₃e, lowe O. & N. 1456 ; lawe EM. 323 ; ich ligge lowe HOM. I. 211 ; heo holdeð . . . þet heaved lowe A. R. 130 ; hie and law *wholly* BARB. iv. 594.

lōh-iboren, adj., *low born*, LA₃. 22041.

lō₃-lī, adj., *O.N.* lāgliga ; *lowly*, A. P. ii. 614 ; louli CH. C. T. *A* 99.

lōuhnesse, sb., *lowness*, A. R. 278 ; lo₃nesse AYENB. 246 ; lounesse PR. P. 314 ; *?* laghnes ALEX. (Sk.) 3293.

lōuh-schīpe, sb., *humility*, A. R. 358.

lahe, *see* **la₃e.**

h₁ahhen, v., *O.E.* hlehhan, hlyhhan, hlihhan (*pret.* hlōh), = *O.H.G.* lahhen, *Goth.* hlahjan (*pret.* hlōh), *O.N.* hlǣja (*pret.* hlō) ; *laugh* ; lahhen REL. II. 4 ; lah₃hen ORM. 8142 ; lahen [lahhe] MARH. 14 ; laghe EGL. 522 ; ANT. ARTH. xxxiv ; lau₃he TREV. V. 73 ; laughen [lauhen, laughwen] CH. C. T. *A* 3849 ; lauhwen

A. R. 200; lauȝwe P. L. S. xxviii. 3; lauhin
PR. P. 290; lauȝe PL. CR. 94; lawe FER. 386;
laughwen [lauȝen, laghen, lauhen (*ms.*lawhen)]
LANGL. *A* iv. 93; laȝhen [leiȝe] WICL.
PROV. xxxi. 25; lehȝen, lihȝen [lahȝe] LAȝ.
22419, 23717; lheȝȝe AYENB. 58; leihe HORN
(H.) 366; leighe RICH. 3451; liȝhe, lihe
ROB. 93, 101; liyhe [liȝhe, liȝe, lawgh, laughe]
ROB. (W.) 3066; lauhweð (*pres.*) A. R. 198;
laghes PR. C. 1092; **lahhinde** (*pple.*) HOM.
I. 257; lahinde KATH. 1556; A. D. 295; lauh-
winde HAV. 946; laȝinge P. L. S. ix. 72; la-
ȝande GAW. 988, 1068; lōh (*pret.*) LAȝ. 18542;
HORN (R.) 361; SPEC. 28; logh TRIAM. 1558;
OCTAV. (H.) 988; EGL. 948; louȝ BEK. 710;
LANGL. (Wr.) 13880; WICL. GEN. xvii. 17;
TRIST. 2870; lough CH. C. T. *C* 961; S. S.
(Wr.) 3154; lou REL. II. 272; (*ms.* low) HAV.
903; loȝen FL. & BL. 477; loȝen [lowe] LAȝ.
12872; louȝe [louhe, loughe] CH. C. T. *A* 3858;
lowen P. S. 341; lowe HICKES I. 228; laȝed
A. P. ii. 653; leiȝede WICL. GEN. xvii. 17;
lauȝeden WILL. 1784; lōwe (*subj.*) P. L. S.
xviii. 46; **laughen** [lauhen, laughwen] [= *O.L.
G.* hlagan] (*pple.*) CH. C. T. *A* 3855; *comp.*
bi-liȝen.

laȝhing, sb., *laughing*; leiȝing WICL. JAS. iv.
9; lheȝ(ȝ)inges (*pl.*) AYENB. 63.

*h*lahter, sb., *O.E.* hleahtor, = *O.H.G.* hlahtar,
O.N. hlātr; *laughter*; lahter HOM. I. 283;
laghter PR. C. 1451; laghtur TRIAM. 1558;
laughter CH. TRO. ii. 1169; leihter '*risus*'
FRAG. 4; leiȝtir WICL. JOB viii. 21; leahtre
lahtre (*dat.*) KATH. 2326; lehtre HOM. II.
175; lehtre [lihtre] LAȝ. 3045; **leihtres** (*pl.*)
A. R. 198.

lahter, *see* lehter.

lai, sb., *O.Fr.* lai; *lay, song,* HOM. I. 199; laies
(*pl.*) LANGL. *A* viii. 66; MAN. (F.) 4027.

lai[2], adj., *O.Fr.* lai; *lay*: lai feo BEK. 556.

lai, *see* lei. laie, *see* laȝe, leiȝe.

laiere, *see* leir. laif, *see* *h*láf.

laik, laiken, *see* lāc[2], lāken[2].

lain, lainen, *see* lēin, lēinen.

lainere, *see* lanere.

laire, sb., *O.N.* leir; *mud, clay,* HAMP. PS.
xvii. 46; lare lxviii. 18; withouten lime or
laire ALEX. (Sk.) 5088; laire (*dat.*) C. M. 519.

lairi, adj., *miry,* HAMP. PS. ii. 9*, xvii. 36.

lait, *see* léte *and* lēit.

laiten, v., *O.N.* leita, = *O.E.* wlātian (*look on*),
Goth. wlaitōn (περιβλέπεσθαι); *seek, look for*;
laite IW. 237; lait TRIST. 3052; ALEX. 152;
PERC. 255; to leȝten & to seken ORM. 3457;
laites (*pres.*) GAW. 355; PR. C. 7535; lai-
tand (*pple.*) PS. xxiii. 6; (*printed* layttede) D.
ARTH. 254; laited (*pret. pple.*) A. P. ii. 1768.

laiten, *see* lēiten. laiþ, *see* lāð.

lak, lake, *see* lac.

lake, sb., = *O.L.G.* lacan, *O.H.G.* lachan *n.*;
linen cloth, HALLIW. 502; CH. C. T. *B* 2048.

laken, v., = *O.Fris.* lakia (*attack*), *M.Du.*
laken (*blame, be wanting*); *from* lac; *lack,
be wanting; blame*: hem gan ðat water laken
GEN. & EX. 1231; whan þai wil ani man lake
S. S. (Web.) 1212; lakkin PR. P. 285; lakke
MAND. 214; S. & C. I. lxvii; to lakke his chaffare LANGL. *B* v. 132; hem sholde lakke no
liflode xi. 280; lackeþ (*pres.*) AYENB. 210;
lakes PR. C. 797; **lackand** (*pple.*) HAMP.
PS. v. 9*; wat lacede (*pret.*) ȝeu HOM. I. 233;
lakkede CH. C. T. *A* 756.

lāken, v., *from* lāc; *make offerings, worship*:
to þeowten god & laken ORM. 973; **lākest**
(*pres.*) 1196; lakeþ 7354; þat tu lāke (*subj.*)
7945; **lākeþ** (*imper.*) 6412; **lākeden** (*pret.*)
7430; **lāked** (*pple.*) 6491.

lāken[2], **leiken**, v., *O.E.* lācan (*pret.* lēolc,
lēc), *O.N.* leika (*pret.* lēk), = *Goth.* laikan
(*pret.* lailaik), *M.H.G.* leichen (*pret.* leichte,
pple. geleichen); *play, sport*: als if he wolde
leȝken ORM. 12044; with him lake MIR. PL.
139 (TOWNL. 218); al so he wolde with hem leike
HAV. 469; laike LANGL. *B* PROL. 172; **laikez**
(*pres.*) A. P. ii. 872; **laiked** (*pret.*) WILL.
1026; GAW. 1554; laikid M. H. 71; leikeden
HAV. 954.

lākin, sb., *cf. O.N.* leikni; *plaything*; laikin,
lakan, leikin PR. P. 284.

lakken, *see* laken.

lallen, v., *cf. Dan.* lalle (*prattle*); *speak*;
lalled, laled (*pret.*) A. P. ii. 153, 913.

lām, sb., *O.E.* lām, = *M.Du.* leem, *O.H.G.*
leim; *loam*, M. H. 1; **lāme** (*dat.*) HOM. I.
221; N. P. 39; C. M. 11985; þu makedest
mon of lame JUL. 60.

lamb, sb., *O.E.* lamb (*pl.* lombru), = *O.N.*, *O.
L.G.* lamb, *Goth.* lamb (*pl.* lamba), *O.H.G.*
lamb (*pl.* lember); *lamb*, ORM. 1312; M. H.
45; þet lamb AYENB. 232; lamb [lomb] CH.
C. T. *B* 459; lomb MARH. 12; A. R. 304; ROB.
57; MAND. 264; MAN. (F.) 1592; **lambes**
(*gen.*) ORM. 7752; ischrud mid lombes flosse
A. R. 66; lombe (*dat.*) AYENB. 236; **lambre**
(*pl.*), lambre ORM. 13329; lambren AYENB.
139; LANGL. *B* xv. 200; WICL. LEV. xxiii.
18; LIDG. M. P. 169; lombe (*r. w.* wombe)
ROB. (W.) 7609; lambron PALL. v. 155; lam-
berne, lambren, lambrin TREV. II. 229, 303.

Lam-hūþe, -hēthe, pr. n., *O.E.* Lambhȳð,
Lambeth, ROB. 326; Lambhiþe TREV. VIII.
69.

lambur, sb., *Fr.* l'ambre; *amber*: ladies with
bedis of coralle and lambur B. B. 315.

láme, adj., *O.E.* lama, = *O.L.G.* lamo, *O.N.*
lami, *O.Fris.* lam(a), loma, *O.H.G.*
lam; *lame, infirm*; ['*paralyticus*'] MAT. viii.
6; '*claudus*' PR. P. 286; HAV. 1938; AN.
LIT. 7; LIDG. M. P. 225; lame, lome O. & N.
364, 1732; lome [lame] LAȝ. 19479, 30777;
he cwæð to þam lamen MK. ii. 5; **lámen** (*pl.*)
MAT. iv. 24; HOM. I. 229; *comp.* **fōt-lame.**

lámin, v., *O.E.* lemian, = *O.H.G.* lemen, *O.N.*

lemja; *make lame, be lame*, PR. P. 286; **lámede** (*pret.*) D. ARTH. 4302; **lámed** (*pple.*) HAV. 2755; MAN. (F.) 1836; *comp.* **a-lámen.**

lamentacioun, sb., *Fr.* lamentation; *lamentation*, CH. C. T. *A* 935.

lámi, adj., = *O.H.G.* leimīg; *loamy*, H. M. 47.

lǎmmesse, *see under* **hlāf.**

lamp, sb., *O.Fr.* lame, *from Lat.* lāmina; *thin plate*, CH. C. T. *G* 764.

lampe, sb., *Fr.* lampe; *lamp*, MARH. 20; laumpe LANGL. *A* i. 163; lompe AYENB. 232; **lampen** (*pl.*) P. L. S. xiii. 121; lampes CH. C. T. *G* 802.

lampreie, sb., *O.Fr.* lamproie; *lamprey*, ROB. 442; lamprei B. B. 166, 174.

lān, *see* **lēan.**

hlanc, adj., *O.E.* hlanc; *lank, slender*: lonc he is & leane HOM. I. 249; þe lonke mon REL. I. 188.

lance, sb., *O.Fr.* lance; *lance*, ROB. 137; launce JOS. 264; D. ARTH. 1379; LANGL. iv. 461; ALEX. (Sk.) 1223; **lances** (*pl.*) 788.

lance², sb., *leap*; launce [lanss] BARB. x. 414.

lancegai, sb., *O.Fr.* launcegai; *a kind of spear*, CH. C. T. *B* 1942, 2011; launcegai PR. P. 290.

lanchin, v., *O.Fr.* lancier; *mod.Eng.* launch; *drop, leap, shoot*; launchin PR. P. 200; (þei) **launceþ** (*pres.*) heiȝe her hemmes PL. CR. 551; launches D. ARTH. 2560; launceþ LANGL. *C* xix. 10; þe leves lancen from þe linde GAW. 526; **launced** (*pret.*) luþerli after him WILL. 2755; lansed (*? quaked*) A. P. ii. 957; lansit BARB. viii. 25; launschide D. ARTH. 194.

land, sb., *O.E.*, *O.Fris.* land, lond, = *O.L.G.*, *O.N.*, *Goth.* land, *O.H.G.* lant, land; *land*, IW. 1023; PR. C. 1404; M. H. 98; þat land ORM. 7092; þet land þet he halt of his lhorde AYENB. 19; land [lond] CH. C. T. *B* 494; lond PR. P. 312; FRAG. 7; A. R. 208; HAV. 64; WILL. 2761; LANGL. *A* iv. 131; MAND. 6; AUD. 1; þat lond LAȝ. 241; he com on lond O. & N. 999; ROB. 5; se wide se þe lond was KATH. 49; water and lond GEN. & EX. 103; **londes** (*gen.*) BEK. 124; lande (*dat.*) PERC. 208; londe C. L. 554; AYENB. 37; A. P. i. 936; sumer com to londe (*cf. Germ.* kam ins land); LAȝ. 24242; a londe & a watre LANGL. *B* xvi. 189; loande PROCL. 5; lond, londe, londes (*pl.*) LAȝ. 1409, 5225, 28022; landes SAX. CHR. 263; lande (*gen. pl.*) ANGL. VII. 220; londen, londe (*dat. pl.*) LAȝ. 13314, 22755; londe BRD. I; *comp.* **ēi-, kine-, ūt-land.**

land-brist, sb., *cf. O.N.* brestr; *surf*, BARB. iv. 444.

lond-bügger, sb., *buyer of land*, LANGL. *B* x. 307.

lond-folc, sb., *O.E.* landfolc; *natives*, LAȝ. 30930; O. & N. 1158; ROB. 173.

lond-gavel, sb., *land tax*, LAȝ. 7789.

lond-spēche, sb., *language*, GEN. & EX. 669.

lond-tilie, sb., *ploughman*; **londtilien** (*dat. pl.*) LAȝ. 14850.

lond-üvel, sb., *epilepsy; epidemic disease*, A. R. 360; londivil '*epilencia*' PR. P. 312.

lond-weig, sb., = *Ger.* landweg; *path*, GEN. & EX. 2681.

lande, sb., *O.Fr.* lande (DIEZ 199), *mod. Eng.* lawn; *shrubbery, heath*; launde '*saltus*' PR. P. 291; AYENB. 216; CH. C. T. *A* 1691.

landen, v., *O.E.* landian, *cf. O.H.G.* lantan; *land*: a god schup . . . þat him scholde londe in . . . HORN (L.) 753; londe '*appello, applico*' PR. P. 312; londede (*pret.*) TREV. VIII. 93; **landed** (*pple.*) ISUM. 231.

londinge, sb., *landing*, PR. P. 312.

landed, pple., *cf. M.H.G.* gelandet; *landed*; londid '*terra dotatus*' PR. P. 312.

landone, sb., *? O.Fr.* landon; *? field*: alle his lele lige mene o landone ascriez D. ARTH. 1768.

láne, sb., *O.E.* lane, = *O.Fris.* lane, lone; *lane*, '*viculus*,' PR. P. 286; '*venella*' VOC. 270; RICH. 4373; PR. C. 8919; lone A. P. i. 1065; PALL. ix. 170; **lánes** (*pl.*) '*venel(l)es*' VOC. 165; CH. C. T. *G* 658; lones, lanes LANGL. *A* ii. 192; *B* ii. 216.

lāne, *see* **lēn.**

laner, sb., *O.Fr.* lanier; *sort of hawk*, VOC. (W. W.) 761.

lanere, sb., *Fr.* laniere; *strap, thong, '*ligula*,'* PR. P. 286; lainer WICL. GEN. xiv. 23; **laineres** [lanieris] (*pl.*) CH. C. T. *A* 2054.

laneret, sb., *O.Fr.* laneret; *a sort of hawk*; **lanerettes** (*pl., printed* lauerettes) TREV. I. 339.

lang, adj., *O.E.* lang, long, = *O.Fris.* lang, lang, *O.L.G.* lang, *O.N.* langr, *O.H.G.* lang, *Goth.* laggs, *Lat.* longus; *long*, ORM. 15210; AYENB. 99; S. S. (Web.) 55; IW. 1479; PR. C. 632; langh PALL. i. 1033; long FRAG. 6; LAȝ. 6366; O. & N. 754; GEN. & EX. 563; HAV. 987; ROB. 22; SPEC. 35; 8 fote long MAND. 75; þat hir hadde holden seven ȝer long (*cf. Ger.* sieben jahre lang) S. A. L. 164; umbe long KATH. 518; drawe . . . a long (*stretch*) LANGL. *A* v. 124; an long (*O.E.* and lang, *along*) þare sea LAȝ. 138; ande long nouht over þwert HAV. 2822; ende long MARH. 10; RICH. 2649; CH. C. T. *A* 1991; end lang PR. C. 8582; end long MAN. (F.) 12385; end long þe bord GOW. I. 182; end longes (*for* long) MAND. 49; **longes** (*gen. n.*) REL. I. 175; on lang(e) Fridæi *Good Friday*, SAX. CHR. 263; on lange Fridai HOM. II. 95; **longne** (*acc. m.*) LAȝ. 287; þenge longe dæi 28316; þene dæi longe 5668; al longe dai BEK. 394; alle longe niht LAȝ. 28000; O. & N. 331; al þe longe yeer LIDG. M. P. 152; al þe ȝer longe O. & N. 790; **longe**

(*pl.*) LAȝ. 28548; **lengre** (*compar.*) CH. C.
T. *A* 330; PR. C. 3932; langir, langar BARB. i.
598; vi. 554; vii. 547; **lengust** (*superl.*)
MISC. 124; lengest TREV. VIII. 65.

lange, adv., ORM. 1264; M. H. 72; lange ic
habbe child iben P. L. S. viii. 2; longe A. R.
268; BEK. 10; AYENB. 205; OCTAV. (H.)
130; **leng** (*comp.*) O. & N. 42; BEK. 161;
ne mihte he na leng libben LAȝ. 11015; þe
leng þe more ALIS. 5864; i dar no leng abide
TRIST. 2598; langar BARB. iv. 43; KATH.
810; A. R. 8; ORM. 13163; LAȝ. 471; WILL.
633; HOCC. i. 288; laingire ALEX. (Sk.)
2195; langir 3311; MISC. 153.

lang-mōd, adj., *O.E.* langmōd, = *O.H.G.*
lancmuet; *patient, longsuffering* ['*longa-
nimis*'] PS. cii. 8.

langnesse, sb., *length*, AYENB. 105.

lon(g)-sum, adj., *O.E.* langsum, = *O.L.G.*
langsam, *O.H.G.* langsam; *long*, HOM. I. 107.

lange-wei, adv., *lengthwise*, TREV. I. 295.

lang[2], adv.,=ȝelang; *belonging to, due to*: al
Cristene folkes hald is lang o Cristes helpe
ORM. 13377; long (*v. r.* along) CH. C. T.
G 930; mi lif is long on þe SPEC. 29.

langāge, sb., *O.Fr.* langage; *language*, BEK.
142; SHOR. 122; ALIS. 6857; CH. C. T.
A 211.

lange, sb., *cf. M.Du.* lange; *length* : he is
end longe (*sec. text* on lengþe) feouwer and
sixti munden LAȝ. 21993; to þere worlde
longe 15533.

langen, v., *O.E.* langian, longian, *cf. O.L.G.*
langōn, *O.H.G.* langēn, *O.N.* langa; *from
lang and* lang[2]; *long, desire earnestly;
lengthen; reach forth, extend; belong* : hem
sholde ... after his come langen ORM. 19364;
him gon longen [longie] LAȝ. 18803; dæȝes
gunnen longen [longi] 30627; (Brien) longien
him to lette his maðmes 30896; longie SHOR.
123; longen CH. C. T. *A* 2278; longin '*per-
tineo, desidero*' PR. P. 312; longe MISC. 197;
longeð (*pres.*) KATH. 1915; to þe me longeð
HOM. I. 197; þe longeð [langeþ] after laðe
spelle LAȝ. 15808; hin longeþ O. & N. 890;
me longeþ him to se ROB. 289; longiþ þe dai
ALIS. 139; þat þer til longes HAV. 396; lunginge (*pple.*) *belonging* TREV. VIII. 446; longede [langede] (*pret.*) LAȝ. 10124; longede
BEK. 45; longed A. P. i. 144, ii. 1747; long-
eden C. L. 1340; longed (*pple.*) WILL. 4570;
comp. bi-, for-, of-langen.

longunge, sb., *O.E.* langung; *longing,
desire*, A. R. 190; longinge O. & N. 869; SPEC.
28; longing GOW. III. 309.

langour, sb., *O.Fr.* langour; *languor, disease;
ill fortune*, AYENB. 93; WILL. 918; ALEX.
(S·.) 3289; langoure D. ARTH. 4268; **lan-
gvores**, langours (*pl.*) WICL. MAT. iv. 24;
LK. iv. 40.

langvish, sb., *languishing*; **langvisches**
(*pl.*) ALEX. (Sk.) 2810.

langvissen, v., *O.Fr.* langvir; *languish*;
langvessande (*pple.*) D. ARTH. 4338; lan-
gvissed (*pret.*) MAN. (F.) 9550; langvishide
WICL. DAN. viii. 27.

hlanke, sb., = *M.Du.* lanke, *O.H.G.* lancha;
flank : stiches i þi lonke H. M. 35; and leiþ
(h)is leg o lonke P. S. 156.

[**lānhure**, see hwure.]

lanterne, sb., *O.Fr.* lanterne; *lantern, lamp*,
FL. & BL. 238; AYENB. 195; WICL. MAT. v.
15; lantrine A. P. i. 1047; **lanterns** (*pl.*)
ALEX. (Sk.) 5398.

lapin, v., *O.E.* lapian,=*M.L.G.*, *M.Du.* lapen,
O.H.G. laffan; *lap; taste, lick up*, VOC. 153;
lape LANGL. *A* v. 207; A. P. ii. 1434; lappin
'*lambo*' PR. P. 287; lappeþ (*pres.*) GOW. III.
215; lapiden (*pret.*) WICL. JUDG. vii. 7;
lapede TREV. VIII. 121.

lappe, sb., *O.E.* lappa, læppa,=*O.Fris.* lappa,
M.H.G. lappe; *lap; border; '*ora*,' FRAG. 1;
PR. P. 287; LEG. 34; P. R. L. P. 227; LANGL.
B ii. 35; CH. C. T. *A* 686; GAW. 936; and
leide (hit) uppe his lappe (*first text* bærm)
LAȝ. 30261; þe lappe of oure loverdes cloþ(e)
P. L. S. xxi. 29; and knit hit in his lappe S. S.
(Wr.) 1262; nou es left me no lappe (*rag*) mi
ligham to hele D. ARTH. 3286; *comp.* dēu-,
ēre-, schürte-lappe.

lappen, v.,=*M.Du.* lappen; *wrap up, embrace*,
WILL. 1712; lappin '*involvo*' PR. P. 287; lappe
FLOR. 113; lappeþ (*pres.*) in his armes CH.
COMPL. M. 76; lapped (*pret.*) GOW. II. 268;
TRIAM. 417; lappid TOR. 557; lappide [wlap-
pide] WICL. MAT. xxvii. 59; lapped (*pple.*)
L. H. R. 69; PR. C. 523; lapped in cloutes PL.
CR. 438; ilappid P. L. S. i. 39; *comp.* bi-
lappen.

lappen, see lapin.

hlæpwinche, see under hlēap.

lar, see laire.

lard, sb., *Fr.* lard; ? *lard*, ? *bacon*, PR. P. 288;
C. B. 19.

lardenere, sb., *larder*, BARB. v. 410.

lardēr, sb., *O.Fr.* lardier; *larder*, P. L. S. xiii.
236.

lardin, v., *Fr.* larder; *lard, garnish with
bacon*, PR. P. 288; larde (*imper.*) C. B. 18, 40,
78; lardid (*pret.*) MAN. (F.) 15756.

lāre, sb., *O.E.* lār,=*O.Fris.* lāre, *O.L.G.*, *O.H.G.*
lēra, *Swed.* lära, *Dan.* lære; *lore, doctrine,
precept*, HOM. I. 109; KATH. 469; ORM. 372;
MIN. x. 28; PR. C. 6469; TOWNL. 59; lare
[lære] MK xi. 18; JOHN vii. 16; lare, lære
(? *for* lare) [lore] LAȝ. 693, 1014; lore, loare
[lare] A. R. 80, 254; lore FRAG. 7; O. & N.
1208; HORN (L.) 412; SHOR. 59; WILL.
328; LANGL. *B* xii. 274; CH. C. T. *A* 527; A.
P. ii. 1556; laire YORK xi. 181; loar(e) GEN.

& EX. 181; lere ELL. ROM. II. 373; M. ARTH. 521; lāre (*pl.*) HOM. I. 115; lære (*? for* lare) LA3. 1014.

lār-fader, sb., *cf. O.N.* lǣrifaðir; *teacher,* ORM. 16625; lārefadirs (*pl.*) HAMP. PS. xlix. 7*.

lār-spel, sb., *O.E.* lārspell; *sermon,* HOM. I. 75; ORM. 12686; larspel [lorspel] LA3. 10162; lorspel REL. I. 129.

lōr-þein, sb., *teacher,* FRAG. 5.

lār-ðēow, sb., *teacher,* JOHN i. 38; larþew, -þeu, lorðeu HOM. I. 81, 135; II. 9; lorþeu MISC. 108; lōrðewes (*pl.*) HOM. II. 163.

lārēow, sb., *? O.E.* lārēow;=lārþēow; *preceptor, instructor,* ANGL. III. 152; lareaw HOM. I. 241; lārēwes (*pl.*) ORM. 7233.

large, adj., *O.Fr.* large, *mod.Eng.* large; *large, wide; liberal, generous,* HOM. I. 143; A. R. 168; HAV. 97; ROB. 183; AYENB. 21; leves ... faire and large GOW. I. 137; a large mile LANGL. *A* xi. 118; hus laies and hus large (*bounteous dealing*) LANGL. *C* xxii. 43; largere (*compar.*) A. R. 168; largest (*superl.*) HOM. I. 271.

large-līche, adv., *liberally,* A. R. 112; H. M. 29; AYENB. 37.

largenesse, sb., *extent; liberality;* PR. P. 288; LANGL. *B* v. 632; TREV. III. 393; ALEX. (Sk.) 68.

largesse, sb., *O.Fr.* largesse; *largess, bounty,* A. R. 166; AYENB. 102; LANGL. *A* vi. 112.

larke, sb., *O.E.* lāwerce, *cf. M.Du.* lawerke, *O.H.G.* lēracha *f., O.N.* lǣvirki *m.; lark, 'alauda,'* PR. P. 288; FER. 1498; LANGL. *B* xii. 262; CH. C. T. *A* 1491; laveroc SPEC. 26; L. C. C. 36; laverok GOW. II. 264.

laroun, sb., *O.Fr.* larron, larun; *robber,* ALIS. 4209.

las, see læs.

lás, sb., *O.Fr.* las, laz; *lace,* ALIS. 7698; C. M. 15880; laas CH. C. T. *A* 392; lace GAW. 217.

lasche, sb., *cf. L.G.* laske, *Ger.* lasche; *lash, 'ligula,'* PR. P. 288; whippes lashe CH. P. F. 178.

lasche², sb., *? = O.H.G.* loskie; *morocco,* HALLIW. 506.

lasche³, adj., *O.Fr.* lasche; *loose, lax; lazy;* PR. P. 288; lache CH. BOET. iv. 3 (122).

laschin, v., *lash; draw swiftly; 'verbero,'* PR. P. 288; laschis (*pres.*) ALEX. (Sk.) 1325; lasched (*pret.*) A. P. ii. 707; lashed EM. 298.

laseir, laser, see leiser. lásen, see lácin.

lasken, v., *O.Fr.* lascher, lasker (*lâcher*); *relax,* AL. (L¹.) 681; lask(e) WILL. 950; laske his peines MIRC 1736.

lasnen, see læssen.

lasse, sb., *lass;* lasce M. H. 39; C. M. 2608. lásse, see læs. lássen, see læssen.

last, sb., *O.N.* löstr; *from lēan; fault, vice,*

LA3. 22974; ORM. 4522; lest HOM. II. 258; it is no lest (*blame*) for hem FER. 459; laste (*dat.*) SPEC. 37; lastes (*pl.*) HOM. I. 145; A. P. iii. 198; lasten (*dat. pl.*) HOM. I. 197.

laste-lēs, adj., *O.N.·* lastalauss; *faultless,* KATH. 105; SPEC. 52; leasteles MARH. 12.

hlast, sb., *O.E.* hlǣst *n., cf. O.Fris.* hlest, *O.H.G.* last *f., O.N.* hlass *n.; from* hláden; *burden; last* (*a weight*); lest PR. P. 299; laste (*dat.*) DEP. R. iv. 74; a þousand last CH. C. T. *B* 1628.

lastāge, sb., *market tax,* TREV. II. 97.

lást, see lǽst.

lasten, v., *O.N.* lasta; *blame,* JUL. 70; þou lastest (*pres.*) hem REL. I. 243; lastes (*becomes faulty*) A. P. ii. 1141; laste (*imper.*) A. R. 352.

lastunge, sb., *blame,* A. R. 66.

hlasten, v., *O.E.* (ge-)hlǣstan; *burden:* þai ... wern laste (*pple.*) & lade A. P. i. 1145; *comp.* bi-lasten.

lásten, see lǽsten.

lat, adj., *O.E.* lǣt,=*O.L.G.* lat, *O.N.·* latr, *Goth.* lats, *O.H.G.* laz; *late, slow,* H. M. 37; JOS. 695; A. P. ii. 1172; slau & let HOM. II. 183; later (*compar.*) *later* P. L. S. viii. 67; O. & N. 963; þe latere (*latter*) A. R. 86; at þa(n) latere cherre 8356; latre ORM. 868; latste (*superl.*) *last* ORM. 4168; lateste HOM. II. 5; laste KATH. 585; A. R. 86; A. D. 237; at þan laste LA3. 3765; at þe laste CH. C. T. *A* 707; S. S. (Wr.) 2672; at þe laste ... seide he HAV. 1677; at te laste TREAT. 134; latst ORM. 11765; þanne i last spak with þe HAV. 678; a last A. R. 18; C. L. 1129; SPEC. 102; AYENB. 69; o last HOM. I. 207; þe latemeste (*superl.*) vers A. R. 20*; þe latemeste dai MISC. 171; þe latemiste read LA3. 11080*.

láte, adv., *late,* LA3. 3075; ORM. 753; O. & N. 1147; LANGL. *A* iii. 56; leate, lete [late] A. R. 20, 240.

láte-ful, adj., *late,* WICL. HOS. vi. 3; WICL. JAS. v. 7.

lat-lī, adv., *late,* HAMP. PS. xli. 2*; xlii. 5*; latlier (*comp.*) HAMP. PS. lxxi. 15*.

lat-rēde, adj., *O.E.* latrǣde; *slow,* CH. C. T. *I* 718.

latsum, adj., *O.E.* lǣtsum; *tardy,* WICL. EX. iv. 10; latsom and slaw PR. C. 793.

latsomnes, sb., *backwardness,* HAMP. PS. xxiii. 3*.

lat², sb., *hindrance,* BARB. xii. 516.

lát, láten, see lǽt, lǽten.

láte, see lǽte.

latin, sb., *O.Fr.* latīn; *Latin,* JUL. 2; GEN. & EX. 13; C. L. 23; AYENB. 145; CH. C. T. *A* 638.

latinier, sb., *O.Fr.* latinier; *interpreter,* MAN. (F.) 7573; (*printed* latimer) LA3. 14319; SPEC. 49; latineres (*pl.*) MAND. 58.

latis, sb., *Fr.* lattis; *lattice*; latiis WICL.
PROV. vii. 6; **latisis** (*pl.*) WICL. S. SOL. ii.
9.

latōn, sb., *O.Fr.* laton; *latten, a mixed metal
similar to brass, 'aurichalcum,'* PR. P. 289;
VOC. 255; GOW. I. 221; laton [latoun] CH.
C. T. *A* 699; latoun PR. C. 4371; latun PL. CR.
196.

latte, sb., *O.E.* lætta (VOC. 26),=*M.Du.* latte,
O.H.G. latta; *lath, 'asser,'* CATH. 209; VOC.
235; latthe, laththe *'tigillum'* PR. P. 288.

lǣttēow, *see under* lāde.

lāð, adj., *O.E.* lāð,=*O.L.G.* lēth, lēð, lēd, *O.N.*
leiðr, *O.H.G.* leid, *mod.Eng.* loath; *hateful,
unwelcome; hideous; reluctant*; HOM. I. 35;
MARH. 14; laðs, loð [loþ] LA3. 244, 399 (*miswritten* lǣð 8803); laþ ORM. 4140; IW. 135;
MAP 360; loð A. R. 168; GEN. & EX. 369;
loþ FRAG. 6; WILL. 5201; FER. 495; GREG.
346; ich was him loþ O. & N. 1088; loth
HAV. 261; loop [loth] LANGL. *B* ix. 57; be
him loþ [looth] or leef CH. C. T. *A* 1837;
laiþ H. S. 563; leith AUD. 31; þat laþe ...
folc ORM. 8521; þe lathe Lazare TOWNL.
248; þe loþe gost P. L. S. viii. 135; lōþe (*pl.*)
HORN (L.) 1060; we beoð heom loaðe LA3.
967; laið(e) 3799; leiðe MISC. 15; laithe
M. H. 51; oure laith(e) sinnes A. P. iii. 401;
lāðen (*dat. pl.*) LA3. 507; lōðre (*compar.*)
A. R. 266; ich am him þe laðere [loþere] LA3.
872; þat he bie ... loðere men HOM. II. 95;
lāthere (*adv.*) HAMP. PS. lxxvi. I*; lāðest
(*superl.*) MARH. 3; þat þe is alre laðest
[loþest] LA3. 2308; loðest A. R. 296; *comp.*
3e-lāð.

lāð-ful, adj., *hateful*; laðfule (*ms. and
printed* haðfule) hokeres HOM. I. 279.

lāð-līc, adj., *O.E.* lāðlīc,=*O.H.G.* leidlīch;
loathsome, GR. 42; ladlich [loþlich] LA3.
4574; loðlic GEN. & EX. 749; lodlich A. R.
118; O. & N. 71; loþli WILL. 50; S. S. (Wr.)
3207; TOR. 963; lothli *'abominabilis'* PR. P.
314; lodli MAP 339; ladlich KATH. (E.) 2288;
lōdlīche (*adv.*) C. L. 1136; lāðlīche (*pl.*)
LA3. 1884; ladliche fend P. L. S. viii. 141;
lāðlüker (*compar.*) H. M. 25; ladluker LA3.
15967; lōþlokest (*superl.*) ALIS. 6312; lodlu-
keste A. R. 66.

lōdlīchen, v., *make loathsome*, A. R. 256.

lōdlīh-hēde, sb., *loathsomeness*, AYENB.
203.

lōthsum, adj., = *O.H.G.* leidsam; *loathsome, 'abominabilis,'* PR. P. 314.

lāð², sb., *O.E.* lāð,=*O.L.G.* lēth, *O.H.G.* leid;
annoyance, injury: þat lað LA3. 16073; laþ
ORM. 11887; lath HAV. 76; he ne wile us
don non loð REL. I. 218; þat þe Sarasins ne
dud(e) him loþ FER. 3510; loth ED. 3276;
PERC. 1935; AV. ARTH. lvii; lōþe (*dat.*)
O. & N. 1146; wiþouten loþe L. H. R. 139;
from loðen (? *for* loðe) alesen LA3. 1084.

lōð-lēas, adj., *O.E.* lāðlēas; *innocent*; A. R.
188; loðles HOM. II. 167; þat ladlese meiden
JUL. 45.

lōðlēsnesse, sb., *innocency*, HOM. II. 33.

lāð-spæl, sb., *O.E.* lāðspel; *evil tidings*,
LA3. 20808.

hlāðe, sb., *O.N.* hlaða, *prov.Eng.* lathe; *barn*;
lathe *'horreum'* VOC. 204; M. H. 146; lāðes
(*pl.*) GEN. & EX. 2134.

[laþer, sb., *O.E.* leaðor, *O.N.* löðr *n.*; *lather.*]
laðer-clūt, sb., *lather-cloth*, H. M. 37*.

lāðien, v., *O.E.* laðian,=*O.Fris.* lathia, ladia,
O.N. laða, *Goth.* laþōn (καλεῖν), *O.H.G.* ladōn;
call, invite, LA3. 6673; lāðeð (*pres.*) HOM.
II. 203; laðieð A. R. 144; lāþez (*imper.*)
A. P. ii. 81; lāðede (*pret.*) LA3. 14427; þe
hine in laðede [*'qui vocaverat eum'*] LK. vii.
39; *comp.* 3e-lāðien.

lāðere, sb., = *M.H.G.* lader; *one who in-
vites*; lāðieres (*pl.*) HOM. I. 237.

lāðunge, sb., *O.E.* laðung,=*O.H.G.* ladunga;
invitation, calling together, HOM. I. 93;
laðunge [laþinge] LA3. 5115.

lāðien, v., *O.E.* laðian, *cf. O.N.* leiða, *O.H.G.*
leidēn; *loathe, detest*; laðin JUL. 16; laði H.
M. 9; loþin, *'abominor, detestor'* PR. P.
316; loþe ROB. 32; thei schulen hit loþe
MIRC 215; þat us loþeþ þe lif LANGL. *B*
PROL. 155; lōðie (*subj.*) A. R. 324; leōðede
(*pret.*) LA3. 6097; loþed ALIS.² (Sk.) 335;
comp. a-lāðien.

lōþinge, sb., = *O.H.G.* leidunga; *loathing*,
PR. P. 316.

lāðnesse, sb., = *O.H.G.* leidnissa; *loathsomeness*, HOM. I. 95; loðnesse A. R. 310.

lāððe, sb., *O.E.* lǣððu; *hatred*, A. R. 310*
(*v. r.* loðnesse); lutel me is of ower luve
leasse of ower laððe JUL. 27; wraþþe & laþþe
ORM. 5451; laððe, lǣðe LA3. 2328, 18680;
muchel leððe to hire sunne HOM. II. 141;
laþe HAV. 2976; loþe SHOR. 57; leþe M. H.
81.

lauchtane, adj., *cf.* lake; *? made of cloth*; a
lauchtane mantill BARB. xix. 672.

laude, sb., *Lat.* laus, laudem; *praise*, CH. C.
T. *B* 1645, 3286.

lau3en, *see* hlahhen.

lauhen, *see* hlahhen. lauhte, *see* laechen.

laumpe, *see* lampe. launde, *see* lande.

laurēr, sb., *O.Fr.* laurier, lorer; *laurel*, CH.
C. T. *A* 1027; lorer C. M. 8235; lorel WILL.
2983; loriel PR. P. 313.

lauriat, adj., *Lat.* laureātus; *laureate*: his
triumphe lauriat [laureate, laureat] CH. C. T.
B 3886.

lausen, v., *loose*, GAW. 1784; *see* lēosen.

lāve, sb., *O.E.* lāf,=*O.Fris.* lāva, *O.L.G.* lēva,
O.N. leif, *Goth.* laiba, *O.H.G.* leiba; *from
līven; remainder*: nas þer na mare ... to

lave LAȝ. 28583; loave A. R. 168; love, loove *widow* ['*relicta*'] TREV. VIII. 75, 173; þa lafe ['*reliquias*'] MAT. xiv. 20; laiff, lafe, laif BARB. v. 370, viii. 507, xvii. 920.

lăvedi, *see* *h*lăfdiȝ.

láven, v., *O.Fr.* laver, *mod.Eng.* lave; *wash; pour out a stream; stream (with blood, etc.)*: to lave & to kest(e) A. P. iii. 154; þe deu of grace upon me lave SPEC. 72; **lávande** (*pple.*) A. P. ii. 366; **lávede** (*pret.*) PERC. 2250; hie his fet lavede mid hire hote teres HOM. II. 145; he lavede a sweote LAȝ. 7489; it lavede a blode REL. I. 144; **láved** (*pple.*) LANGL. B xiv. 5; *comp.* bi-láven.

lavendēre, sb., *Fr.* lavandiere; *washerwoman,* SPEC. 49; lavender CH. L. G. W. 358; landar (*v. r.* lavender, lainder) BARB. xvi. 273; **lavendēres** (*pl.*) TREV. VIII. 191.

lavendre, sb., *lavender,* '*lavendula,*' REL. I. 37; '*lavandria, lavendula*' CATH. 210.

lavendrie, sb., *laundry,* LANGL. B xv. 182.

lāverd, *see* *h*lāford. **laveroc,** *see* larke.

lavor, sb., *O.Fr.* lavoir; *laver, washing vessel,* AYENB. 202; lavour REL. I. 7; B. B. 132; **lavoures** (*pl.*) PL. CR. 196.

lāw, *see* lāh. **lawe,** *see* laȝe.

h*lāwe, sb., *O.E.* hlāw, hlǣw, ?=*O.L.G.* hlēo (*dat.* hlēwe), *O.H.G.* hlēo (*pl.* (h)lēwa), *Goth.* hlaiw (τάφος); *mound, barrow, tomb, hill;* lawe GAW. 2171; A. P. ii. 992; ISUM. 392; ilc an lawe & ilc an hil ORM. 9205; lowe HAV. 1699; MAP 339; ALIS. 4348; REL. I. 120; B. DISC. 1000; TOR. 1967; leouwe [lehe] *den,* A. R. 368; in þe liunes lehe KATH. 1847; **lāȝes,** lages (*pl.*) ['*saltus*'] HOM. II. 211.

lāwed, *see* lǣwed. **lawen,** *see* loȝen.

lawer, lawiere, *see* laȝere.

lawmere, sb., *sluggard (a term of reproach),* YORK xxxi. 180.

lax, sb., *O.E.* leax,=*O.N.* lax, *O.H.G.* lahs; *salmon,* HAV. 754; P. S. 151; lex FRAG. 3.

lay, *see* lēi. **laȝe,** *see* laȝe.

lazar, sb., *leper,* LANGL. C xix. 273; **lazars** (*pl.*) CH. C. T. *A* 245.

lē, *see* lā, lēou.

lēac, sb., *O.E.* lēac,=*M.Du.* look, *O.N.* laukr, *O.H.G.* louch; *leek;* lec FRAG. 3; leek '*porrum*' PR. P. 295; LANGL. *A* v. 65; CH. C. T. *A* 3870; þer dedes ar nought worþ a leke MAN. (F.) 12654; **lēc** (*pl.*) & worten ROB. (W.) 6999; *comp.* ker-, gār-, hem-, hol-, hous-, pel-, schin-lēk (-lōk).

lēad, sb., *O.E.* lēad,=*Du.* lood, *M.H.G.* lōt; *lead; leaden vessel; plummet;* læd [leod] LAȝ. 5692; liad AYENB. 141; led HAV. 924; SPEC. 83; a led vol of water ORIG. 216; leed MAND. 75; (che) doth me rennen under the led S. & C. l. xlix; **lēade** (*dat.*) AYENB. 150; lede LANGL. *B* v. 600; CH. C. T. *A* 202; **lōdes** (*pl.*) MISC. 224; *comp.* brüþen-lēd.

lēaden, adj., *O.E.* lēaden; *leaden;* leden CH. C. T. *G* 728; ledin E. W. 46.

lēadin, v., = *M.H.G.* lœten; *from* lēad; *cover with lead,* '*plumbo;*' PR. P. 292; lede PALL. ix. 175; **ilēaded** (*pple.*) A. R. 418.

leedare, sb., *plumber,* '*plumbarius,*' PR. P. 292.

lēaf, sb., *O.E.* lēaf, *cf.* O.Fris. lāf, *M.Du.* loof, *O.N.* lauf, *O.H.G.* laub *n.*, *Goth.* laubs *m.*; *leaf,* MARH. 1; leaf, liaf, lieaf AYENB. 62, 230, 232; lef TREAT. 137; i falewe so doþ þe lef SPEC. 90; lef [leef] LANGL. *B* iii. 337; lef [leef] CH. C. T. *A* 3177; **lēave,** lieave (*dat.*) AYENB. 1; **lēaf** (*pl.*) MAT. xxi. 19; leaf [leves] LAȝ. 46; lef FL. & BL. 224; treo þe bereð lef & blosman HOM. I. 109; leaves A. R. 322; AYENB. 57; læfes ORM. 13737; leves O. & N. 1046, 1326; boþe leeves of þe ȝate WICL. JUDG. xvi. 3; **lēve** (*dat. pl.*) O. & N. 456; *comp.* ívi-lēf; *deriv.* lēavi, lēaven.

lēf-sel, sb., *cf.* Swed. löfsal *m.*, *Dan.* lövsal; *bower of leaves,* A. P. iii. 448; lef-, leefsel CH. C. T. *A* 4061; lefsal ANT. ARTH. vi; levecel '*umbraculum*' PR. P. 300.

[**lēaf²,** adj., *O.E.* (unge-)lēaf,=*O.H.G.* (ge-)loub; *believing; comp.* un-lēaf.]

lēafe, *see* lēave.

lēaȝe, sb., *O.E.* lēah,=*M.Du.* looghe, *O.H.G.* louga, *lye;* leȝe P. S. 154; AYENB. 145; leihe MISC. 193; lie '*lixivium*' PR. P. 294; lie '*lixivum*' REL. I. 8.

leahter, *see* *h*lahter. **leahtrum,** *see* lehtor.

lēal, adj., *O.Fr.* leal; *loyal;* leel MAN. (F.) 5924; lel WILL. 5119; leill [lele] BARB. v. 293; i. 375; **lēlist** (*superl.*) ALEX. (Sk.) 2877.

lēl-līche, adv., *loyally,* WILL. 999; leli HAMP. PS. *page* 1; leleli, leli BARB. ii. 171, xix. 190.

lealtē, sb., *O.Fr.* lealté; *loyalty,* SPEC. 53; leute MIN. *page* 101; LANGL. *A* iii. 273; (lewte [laute] *B* iii. 289); leaute MAN. (F.) 12892; laute, leaute BARB. v. 162, 530.

[**lēan,** v., *O.E.* lēan (*pret.* lōg), = *Goth.* laian (*pret.* lailō), *O.L.G.*, *O.H.G.* lahan; *reproach, blame; comp.* bi-lēan; *deriv.* lehter, last.]

lēan², sb., *O.E.* lēan,=*O.L.G.*, *O.H.G.* lōn, *O.Fris.* lān, *O.N.*, *Goth.* laun; *reward,* HOM. II. 222; ANGL. I. 10; ure swinkes lean MISC. 60; læn ORM. 1518; þat læn LAȝ. 16691; len GEN. & EX. 2838; lan HOM. I. 163; KATH. 806; len P. L. S. viii. 32; REL. I. 183; *comp.* ed-lēn.

lēane, lēanen, *see* lēne, lǣnen.

lēap, sb., *O.E.* lēap,=*O.N.* laupr; *basket; leep* '*sporta, corbis, nassa*' PR. P. 296; ['*fiscellam*'] WICL. EXOD. ii. 3; **lēpe** (*dat.*) O. & N. 359; S. & C. II. lxix; **lēpes** (*pl.*) ROB. 265; C.

M. 13513; lepus AUD. 81; *comp.* bēr-, brēad-, fisch-, sēd-lēp.

*h*lēap, sb., *cf. O.E.* hlȳp, *O.Fris.* (bec-)hlēp, *M.Du.* loop, *O.H.G.* louf *m.*, *O.N.* hlaup *n.;* *from h*lēapen; *leap*; lepe MAND. 113; leep '*saltus*' PR. P. 297; lupe H. M. 23; lupe [leope] LAȝ. 1928; lipe ANGL. IV. 198; lope GOW. I. 310; loup BARB. xiii. 652; lūpe (*dat.*) P. L. S. xv. 148; lēapes (*pl.*) LIDG. M. P. 32; lupes A. R. 48; *comp.* over-lōp (-lēpe).

lēp-ȝēr, sb., *O.N.* hlaupār; *leap-year*, VOC. 239; lepeȝeer MAND. 77; lepe-, lupeȝer TREV. IV. 199.

lhǎp-winche, sb., *O.E.* hlēapwince (VOC. 62, 280); *lapwing*, AYENB. 61; lapwinke '*upupa*' PR. P. 288; lap-, leepwinke WICL. LEV. xi. 19; lapewink B. B. 143.

*h*lēapen, v., *O.E.* hlēapan, = *O.Fris.* hlāpa, *O.L.G.* hlōpan, *O.N.* hlaupa, *Goth.* hlaupan, *O.H.G.* hlaufan, loufan; *leap; run*; leapen A. R. 106; lepen [leape] LAȝ. 23697; læpen ORM. 11792; lheape AYENB. 89; lepen LANGL. *A* ii. 207; lepe FL. & BL. 465; MAND. 113; CH. C. T. *A* 4378; loupe HAV. 1801; lēapeð (*pres.*) A. R. 224; lepeð REL. I. 215; lēapinde (*pple.*) KATH. 196; lēop (*pret.*) A. R. 52; MISC. 44; ALIS. 1075; K. T. 1065; leop, leoup, leup LAȝ. 1462, 9284, 9331; leop [leep] LANGL. *A* ii. 191; leup JUL. 77; lup FER. 243; lep GEN. & EX. 2726; HAV. 891; WILL. 2756; A. P. iii. 179; S. S. (Wr.) 885; PERC. 479; lep [leep] CH. C. T. *A* 2687; lhip AYENB. 45; lip GREG. 479; the Brus lap on (*took horse*) BARB. ii. 28; leopen, lupen LAȝ. 1836, 2600; lepen MAN. (F.) 4746; lepe OCTAV. (H.) 476; lepe [lope] LANGL. *B* iv. 153; lopen HAV. 1896; lopin S. S. (Wr.) 2417; lepte ROB. 63; lēope (*subj.*) A. R. 140; lupe ORM. 12037; lópen (*pple.*) GEN. & EX. 2647; S. S. (Web.) 739; LANGL. *B* v. 198; A. P. ii. 990; *comp.* a-, ȝe-, over-lēapen.

lēping, sb., *running away, escaping*, YORK xxvi. 292.

lēas, adj., *O.E.* lēas, = *O.Fris.* lās, *O.L.G.* lōs, *O.H.G.* lōs, *Goth.* laus, *O.N.* lauss; *from* lēosen; *void, false, vain*, MARH. 15; leas & lutiȝ FRAG. 6; læs [les] LAȝ. 19235; les ROB. 160; RICH. 3230; P. S. 214; S. S. (Wr.) 2470; REL. I. 123; lees '*falsus*' PR. P. 298; SPEC. 49; E. T. 1086; R. R. 8; lēas (*subst.*) A. R. 82; SHOR. 111; les LIDG. TH. 1777; leaveð þe lease and luvieð þe soðe JUL. 74; buten lese LAȝ. 28150; þi lease wit KATH. 1010; lēase (*pl.*) P. L. S. viii. 129; LAȝ. 751; lease swefnes A. R. 268; lese A. P. ii. 1719; lese wordes O. & N. 756; *comp.* āre-, bērd-, cáre-, eȝe-, ende-, fader-, ȝēme-, grúnd-, help-, herte-, lawe-, līf-, miht-, mōder-, nēad-, rǣd-, rēche-, sak-, scháme-, smēch-, stēor-, þēu-, wem-, wit-lēas.

lēas-līche, adv., *O.E.* lēaslīce; *wrongly*; leasliche icleopode FRAG. 1.

[lēaste, sb., *O.E.* -lēast; *from* lēas; *empti-ness*; *comp.* eȝe-, ȝelēf-, ȝēme-, have-, mōd-, rēche-, scháme-, slǣp-lēaste(-lēste).]

leasten, *see* lǣsten.

lēasunge, sb., *O.E.* lēasung, lēasing; *false-hood*, HOM. I. 229; A. R. 82; læsinge, lesinge LAȝ. 2969, 3068; lesinge MISC. 37; LANGL. *B* iv. 18; leesinge PR. P. 298; lesing O. & N. 848; CH. C. T. *G* 479; OCTAV. (H.) 1660; PR. C. 4274; AUD. 27.

lēsing-mongere, sb., *liar*, WICL. ECCLUS. xx. 27; lēsingmongeris (*pl.*) WICL. 1 TIM. i. 10.

leautē, *see* lealtē.

lēave, sb., *O.E.* lēaf, = *M.H.G.* loube; *leave, permission*; (*ms.* leaue) A. R. 56; aske leve A. P. ii. 316: take leve A. P. ii. 401; leave, læve [leve] LAȝ. 1271, 4879; ich ȝeve þe leave [leve] JUL. 34, 35; lefe ORM. 10229; ISUM. 142; leve '*licentia*' PR. P. 300; GEN. & EX. 784; HAV. 1387; H. H. 175; LANGL. *B* xviii. 264; TOR. 440; ich . . . nime leve O. & N. 457; leve [leeve] CH. C. T. *A* 1217; leve, live FL. & BL. 9, 124; live S. S. (Wr.) 1568; leif BARB. iv. 582; REB. LINC. 6; mid þes kinges leave MISC. 145; lēvis (*pl.*) *leave-takings* BARB. xvi. 689; *comp.* ȝe-lēave.

lēf-ful, adj., *allowed, permissible; lawful*, CH. C. T. *A* 3912; W. & I. 22; leefful WICL. LK. vi. 2; leful, leefful [leffol] TREV. I. 243; IV. 431.

lēfsum, adj., *allowable*, LANGL. *B* xi. 92*.

lēave[2], sb., = *O.Fris.* lāva, *M.L.G.* lōve; *for* ȝelēave; *belief, faith*; (*ms.* leave) HOM. I. 57; KATH. 316; SHOR. 143; læfe ORM. 16809; leve REL. I. 131; MISC. 134; MAN. (H.) 247.

lēaf-ful, adj., *faithful*, KATH. 1038; leaf-, læfful LAȝ. 3033, 10854; læfful ORM. 19242; lefful GEN. & EX. 3447; MISC. 23.

lēaf-līch, adj., *O.E.* (ge-)lēaflīc, = *O.H.G.* (ge-)loublīch; *credible*; levelik PS. xcii. 5; *comp.* un-lēaflīch.

lēaven, v., *cf. O.H.G.* loubēn; *put forth leaves; grow to leaf*; leve PALL. iii. 276; levi P. L. S. xv. 126; lēaved (*pple.*) GEN. & EX. 3839; leved WILL. 22; LANGL. *B* xv. 95; ileved LEB. JES. 842.

lēaven, *see* lǣven.

lēavi, adj., *from* lēaf; *leafy*, PALL. iv. 486.

lebard, *see* leopard. leberalle, *see* liberal. leburd, *see* leopard.

lēc, *see* lēac. lecchen, *see* lacchen.

leccioun, sb., *Lat.* lectiō; *reading*, GREG. 986; *see* lessōn.

lēche, sb., *Fr.* lēche, *O.Fr.* lesche; *slice*; '*lesca*,' PR. P. 292; leche viaundez C. B. 34; leese bi lees (*slice by slice*) B. B. 159; leeches (*pl.*) B. B. 150.

lēche, lēchen, *see* lǣche, lǣchen.

lēchen, v., *from* lēche ; *cut into slices* ; lēche (*imper.*) C. B. 12, 31 ; lesshe 12 ; lēchist (*pres.*) C. B. 35 ; lēchide (*pple.*) D. ARTH. 265 ; ilechid C. B. 35.

lecherie, sb., *O.Fr.* lecherie ; *lechery*, A. R. 60 ; GEN. & EX. 3510 ; P. L. S. ix. 140 ; AYENB. 9.

lecherous, adj., *lecherous*, AYENB. 111 ; ROB. (W.) 7208* ; licherous ALEX. (Sk.) 4328.

lēchnien, *see* lǣchnien.

lechour, sb., *O.Fr.* lechour ; *lecher, unchaste person*, TREAT. 133 ; lechur A. R. 216 ; MISC. 30 ; mid lechōrs (*gen.*) mod ROB. (W.) 7208 ; lechūrs (*pl.*) HOM. I. 53.

lēd, *see* lēad. lēd, *see* lēod.

leddre, *see* *h*lǎddre.   lēde, *see* lǣde.

leden, sb., *O.E.* leden, lǣden, lyden ; ?=latīn ; *speech, language* ; on leden (*Latin*) FRAG. 1 ; an Englisc leoden LA3. 29677 ; ledene, leodene C. L. 32 ; leodene REL. II. 229 ; ledene [ledne] LANGL. *B* xii. 253 ; ludene LANGL. *C* xv. 179* ; ledden A. P. i. 877 ; MIR. PL. 9 (leden CHEST. 52) ; ledene (*dat.*) MARH. 23 ; ledene, leodene A. R. 136, 170 ; ledne [ledene] CH. C. T. *F* 435 ; *comp.* bōc-leden.

lēden, *see* lǣden.

lederōn, sb., *Fr.* laideron ; *ugly person*, ALIS. 3210 ; lidron PR. P. 303 ; lidderon, lidrone YORK xxxi. 167, 187.

lee, sb., *skein* ; lee of threde, ' *ligatura*,' PR. P. 291.

lee², sb., *delight*, ALEX. (Sk.) 5615.

lee, *see* *h*lēou.   lēf, *see* lēaf, lēof.

leege, *see* lige.   leer, *see* lǣre.

leese, *see* lēse.   lěfdi, *see* *h*lǎfdi3.

lēfe, *see* lēave.   lēfen, *see* lēven.

lěffol, *see under* lēave.

lěfsel, *see under* lēaf.

left, leften, *see* lüft, lüften.

leg, sb., *O.N.* leggr (*crus*) ; *leg*, ' *tibia*,' PR. P. 293 ; ALIS. 1808 ; legges (*pl.*) LA3. 1876* (*first text* sconken) ; ROB. 338 ; TREAT. 139 ; CH. C. T. *A* 591 ; þe3es legges fet ant al SPEC. 52.

legāt, sb., *Fr.* legat ; *legate, ambassador*, LA3. 24501 ; MAN. (F.) 11083 ; legātis (*pl.*) WICL. PS. lxvii. 32.

legaunce, *see* ligence.

lege, leghe, *see* leuge.

lege, *see* lige.

lē3e, adj., ?=*M.L.G.* lō ; *lea, fallow* ; leie (*ms.* leye) MAN. (F.) 6983 ; mi lond leie (*ms.* leye) liþ P. S. 152 ; *see* lēi.

le3e, *see* lü3e.   lē3e, *see* lēa3e, lēi.

lē3en, v., *O.E.* lēgan,=*M.H.G.* lougen ; *from* lēi ; *flame* : lie and brenne CH. C. T. *D* 1142 ; ilē3ed (*pple.*) REL. I. 266.

lē3en, *see* lēo3en.

legende, sb., *Fr.* legende ; *legend*, CH. C. T. *A* 3141 ; MED. 51.

legge, sb.,=*M.H.G.* lecke ; *from* liggen ; *ledge*, PR. P. 293.

leggen, v., *for* aleggen² ; *allege* : loke noght ye legge (*pres.*) againe oure lai YORK xvii. 147 ; legge xxvi. 45.

leggen, v., *O.E.* lecgan, = *O.L.G.* leggian, *O.N.* leggja, *O.H.G.* leggan, leccan, *Goth.* lagjan ; *from* liggen ; *lay, place*, LA3. 8192 ; A. R. 346 ; ORM. 5097 ; leggen . . . a dun *lay down* KATH. 773 ; legge BEK. 2278 ; LANGL. *B* ii. 34 ; CH. C. T. *A* 3937 ; lein HAV. 718 ; PR. P. 294 ; Lud king lette legge þane wal LA3. 7085 ; leie MAP 341 ; and sai a knight lei(e) him on wiþ a swerde EGL. 416 ; legge (*pres.*) A. R. 346 ; þu leist KATH. 1895 ; leig8 LK. xiv. 29 ; leie8 (*ms.* leie8) lei8 A. R. 212, 288 ; lei8 HOM. II. 209 ; (we) legge8 P. L. S. viii. 159 ; heo leggeo8 [legge þ] LA3. 15822 ; leggiþ ALIS. 6602 ; lege (*imper.*) LEECHD. III. 86 ; leie LA3. 5069 ; legge (*subj.*) FL. & BL. 376 ; thai legge PALL. i. 583 ; leigde (*pret.*) MK. xv. 46 ; le3de ORM. 1334 ; leide O. & N. 467 ; FLOR. 740 ; he heom leide on mid sweord(e) & mid spere LA3. 547 ; sche leide hur doune be þe childe OCTAV. (H.) 366 ; he heom on leide þat weoren lawen gode LA3. 2077 ; laide þat leidest HAV. 636 ; leiden HOM. I. 101 ; KATH. 2252 ; (*ms.* leyden) JOHN xx. 13 ; heo leiden to gadere LA3. 5904 ; le3d (*pple.*) ORM. 3401 ; leid GEN. & EX. 2426 ; GREG. 156 ; *comp.* a-, bi-, 3e-leggen.

leggere, sb., *layer* ; leggeres (*pl.*) (of stoon) WICL. 1 ESDR. iii. 7.

lē3he, *see* lei3e.

legistre, sb., *O.Fr.* legistre ; *jurist* ; legistres (*pl.*) and lawieres LANGL. *B* vii. 59.

legiūn, sb., *Fr.* legion ; *legion*, LA3. 6024 ; legioun CHR. E. 633.

lēhe, *see* *h*lāwe.   leh3en, *see* *h*lahhen.

lēht, *see* liht.

lehter, sb., *O.E.* leahtor, *cf. M.Du.* lǎchter *m.*, *O.L.G.*, *O.H.G.* lastar, laster *n.* ; *from* lēan ; *vice* ; þesne lehter HOM. I. 137 ; lehtres, lahtres (*pl.*) HOM. II. 55, 149 ; leahtrum (*dat.*) HOM. I. 235.

lehter, *see* *h*lahter.

lehtrien, v., *O.E.* leahtrian, *cf. M.Du.* lachteren, *O.H.G.* lastrōn ; *reprove* ; lehtrie (*pres. subj.*) HOM. II. 215.

lei, sb., *O.Fr.* lei ; *law, religion*, KATH. 321 ; lai (*ms.* lay) MISC. 153 ; GEN. & EX. 1201 ; C. L. 240 ; BEK. 346 ; SHOR. 122 ; K. T. 383 ; C. M. 1428 ; DIGB. 3*.

lēi, sb., *O.E.* lēg, lȳg, līg,=*O.H.G.* loug *m.* ; *flame*, KATH. 1412 ; þene lei (*ms.* ley) JUL. 66 ; lei [leie] TREV. IV. 153 ; lege HOM. II. 49 ; (*ms.* leie) P. L. S. viii. 140 ; R. S. v. leie MAP 338 ; leie LANGL. *B* xvii. 207 ; lie BRD. 23 ; li3e (*printed* lighe) ALIS. 3458 ;

lēie (*dat.*) A. R. 202 ; **lēies** (*pl.*) HOM. I. 41 ;
P. R. L. P. 85.

lēi², sb., *O.E.* lēah,=*M.L.G.* lōh, lō, *O.H.G.*
lō, lōch ; *lea, -ley* (*in local names*)*; untilled
ground, grass land* ; lai '*subsecivum*' PR. P.
285 ; TOR. 165 ; lee GAW. 849 ; **lēie** (*ms.* leye)
(*dat.*) H. V. 95 ; erien his leiȝes [leies] LANGL.
A vii. 5 ; *comp.* **wude-lēi.**

lēi-land, sb., *fallow land,* TOWNL. 101 ;
leilond '*ter*(*r*)*e fre*(*s*)*che*' VOC. 153.

lēih-tūn, sb., *O.E.* lāhtūn (VOC. 285), lehtun
(LK. xiii. 19*), lectun(-ward) (VOC. 27) ;
garden, MISC. 45 ; lahtoun REL. I. 264 ;
lēiȝhtone (*dat.*) LEG. 54.

   lēihtūn-ward, sb., *keeper of a garden,*
MISC. 53.

leie, *see* laȝe, lüȝe.

lōie, *see* lēȝe, lēaȝe, lēi.

leien, *see* leggen. **leif,** *see* lēave.

leiȝe, sb., *O.N.* leiga ; *hire* ; leȝhe ORM. 6234 ;
lai C. M. 11814.

lēȝhe-man, sb., *hireling,* ORM. 6222.

leiȝe, *see* leȝe. **lōiȝe,** *see* lēi. **lēiȝt,** *see* lēit.

leiȝter, leihter, *see* hlahter.

leik, leiken, *see* lāc², lāken².

leimūre, *see under* laȝe.

lōin, sb., *cf. O.Fris.* lēine *f., O.H.G.* lougen *m.,*
lougna *f., O.N.* laun *f.* ; *denial,* AMAD. (Web.)
511 ; **lāine** (*dat.*) M. ARTH. 1964 ; withoute
lain MIRC 810.

lōinen, v., *? O.E.* lēgnian, lȳgnian,=*O.Fris.*
lēina, *O.N.* leyna, *O.L.G.* lōgnean, *O.H.G.*
lougnan, *Goth.* laugnjan ; *deny, hide* ; leine C.
M. 2738 ; laine YORK x. 187 ; laine WILL.
906 ; IW. 703 ; PERC. 1940 ; M. ARTH. 989 :
here liflode **līgneð** (*pres.*) hem selven HOM.
II. 31 ; leines A. R. 314* ; **lāined** (*pple.*)
A. P. i. 244 ; *comp.* **a-, for-lēinen.**

   lāining, sb., *concealment,* YORK xxv. 101.

leinten, sb., *O.E.* lengten, lencten, *cf. M.L.G.*
lenten, *O.H.G.* lenzo, lengiz(o) ; *spring ; season
of Lent ;* '*ver,*' FRAG. 3 ; LAȝ. 30626 ; A. R.
70 ; leinte ROB. 187 ; lenten ORM. 8891 ;
SPEC. 43 ; LANGL. *B* xx. 359 ; M. H. 122 ;
ane lenten AYENB. 91 ; lente PR. P. 296 ; *comp.*
mid-lenten.

   leinte-mete, sb., *Lenten food,* HOM. II. 67.

   lente-rine, sb., *Lent,* BARB. x. 815.

   lenten-tīde, sb., *O.E.* lenctentīd ; *Lent,*
SAX. CHR. 255.

   leinten-tīme, sb., *season of Lent :* in leinten
time uwilc mon gað to scrifte HOM. I. 25.

leinþe, *see* lengðe.

leir, sb., *O.E.* leger, *cf. O.L.G., O.H.G.* legar,
leger, *O.N.* legr *n., Goth.* ligrs *m.* ; *lair ; den ;
camp, sick bed,* L. H. R. 200 ; leire (*dat.*) HOM.
II. 103 ; PALL. i. 52 ; laiere D. ARTH. 2293.

   leir-stōwe, sb., *O.E.* legerstōw ; *sepulchre,*
LAȝ. 22874.

leire-wīte, sb., *O.E.* legerwīte ; *fine for
lying with a bondwoman,* TREV. II. 97.

Leirchestre, pr. n., *O.E.* Lægreceaster ;
Leicester, LAȝ. 2915 ; PROC. 9 ; Leicestre
ROB. 2.

leiren, v.,=*M.H.G.* legern (*stretch out*) ; *lay
prostrate* ; þe rihte bileve & þe soðe luve . . .
ben **leirede** (*pple.*) . . . on his heorte HOM. II.
103.

leiser, sb., *O.Fr.* leisir ; *leisure,* JOS. 164 ;
MAN. (F.) 2753 ; CH. D. BL. 172 ; laseir,
lasare, laser, lasair BARB. xiii. 602 ; v. 390 ;
xiii. 59 ; vi. 660 ; lesure PALL. iii. 733 ; iv.
825.

lēit, sb., *O.E.* lēget, lȳget, līget ; *lightning,*
HOM. I. 43 ; GOW. III. 94 ; MAND. 292 ; WICL.
DEUT. xxxii. 41 ; (*ms.* leyt) MAT. xxviii. 3 ;
MISC. 224 ; leiȝt ROB. 308 ; lait AYENB. 66 ;
GAW. 199 ; laite *flame* TREV. V. 121 ; late
ALEX. (Sk.) 553 ; **lēite** (*dat.*) LAȝ. 25599 ;
A. Ṙ. 306.

leiten, *see* laiten.

leiten, v., *? O.E.* lēgettan,=*Goth.* lauhatjan,
O.H.G. lougezan ; *lighten, flame out,* A. R.
202 ; JUL. 66 ; **lēiteð** (*pres.*) MARH. 13 ;
lēite (*subj.*) GOW. III. 95 ; **lēitinde** (*pple.*)
HOM. I. 185 ; **lēitede** (*pret.*) LAȝ. 18539 ;
KATH. 671 ; laitid S. S. (Wr.) 2228.

leið, *see* lāð. **leithlī,** *see under* līht.

lēk, *see* lēac.

lēken, v.,=*M.Du.* leken, *O.N.* leka, *M.H.G.*
lechen ; *leak, drop* ; leke PALL. vi. 33.

lēl, *see* lēal. **leme,** *see* lim.

lemaille, *see* limaille.

lēme, lēmien, *see* lēome, lēomien.

lēmiren, *see* lēomiren.

lemman, *see* lēofman *under* lēof.

lēn, *see* lēan². **lenāge,** *see* lināge.

hlenchen, v., *O.E.* hlencan, = *? M.H.G.*
lenken ; *? bend, twist* ; **lench**(e) (*pres.*)
REL. II. 211.

lende, sb., *O.E.* lenden (*pl.* lendenu),*cf.O.Fris.*
*lenden (*dat. pl.* lendenum), *O.N.* lend (*pl.*
lendar), *O.L.G.* *lendi (*pl.* lendin), *O.H.G.*
lenti (*pl.* lentin, lendin) ; *loin,* '*lumbus,*' PR. P.
296 ; ORM. 4776 ; leende WICL. GEN. xlvi. 26 ;
lendene (*pl.*) MK. i. 6 ; lenden MARH. 18 ;
SHOR. 44 ; AYENB. 236 ; lendes ORM. 3211 ;
CH. C. T. *A* 3237 ; leendes PALL. iv. 683 ;
lendene (*dat. pl.*) LEECHD. III. 140 ; lenden
A. R. 280 ; lende ROB. 377 ; lindes GAW. 139.

   lend-bōn, sb., *loin-bone,* MISC. 12.

lenden, v., *O.E.* lendan, = *M.H.G.* lenden,
O.N. lenda ; *from land* ; *land, arrive, reside,
remain,* ORM. 2141 ; JOS. 81 ; lende HAV. 733 ;
C. L. 504 ; WILL. 1466 ; IW. 1449 ; TOR. 2600 ;
in Acris gunne þai lende ISUM. 508 ; here i
mai no lenger lende M. ARTH. 565 ; **lende**
(*pret.*) DEGR. 1560 ; neh him he heom lænde
[lende] (*? placed*) LAȝ. 1989 ; lent GAW. 1002.

lēnden, *see* lēnen.

lēne, lēnen, *see* **lǣne, lǣnen.**
lēnere, *see under* **lǣnen.**
lenge, sb.,=*M.Du.* lenge, linge; *ling (fish),*
'*lucius,*' PR. P. 296; HAV. 832.
lenge[2], sb., *O.E.* lengu,=*O.H.G.* lengī, *Goth.*
laggei; *from* **lang**; *length*; lenghe REL. II.
281; A. P. i. 416.
[**lenge,** ? adj., *from* **lang**; *comp.* **ǣ-lenge.**]
lengen, v., *O.E.* lengan, = *O.H.G.* lengan,
O.N. lengja; *from* **lang**; *prolong, linger,*
dwell : þi lif lengen REL. I. 185; lenge HAV.
1734; AYENB. 173; JOS. 162; WILL. 5421;
A. P. i. 261; AMAD. (R.) xxviii; i mai no lenger lenge LANGL. *A* i. 185; lenge and lende
M. ARTH. 3276; lenges (*pres.*) DEGR. 1340;
lengden (*pret.*) PL. CR. 310; *comp.* **ȝe-lengen.**
lengðe, sb., *O.E.*, *O.N.* lengð,=*M.L.G.,M.Du.*
lengde; *length,* HOM. I. 261; on lengðe it
sal him rewen REL. I. 221; lengþe LAȜ. 21993*;
O. & N. 174; in lengþe (lenthe) and in
brede LANGL. *B* ii. 88; lengthe PR. P. 296;
lenkþe S. S. (Wr.) 2696; leinþe CHR. E. 26;
REL. I. 125; lenþe A. P. ii. 425; lenthe PR. C.
5899.
lengþen, v., *lengthen*; lengþe WILL. 957; i
wolde have lengþed (*pple.*) his lif LANGL.
B xviii. 300.
lenien, *see* **hleonien.**
lennen, *see* **lǣnen.** **lent,** *see* **lenden.**
lente, sb., *O.Fr.* lente; *lentil*; (*v. r.* lentes or
fetchis) WICL. 2 KINGS xxiii. 11.
lenten, *see* **leinten.**
lentil, sb., *O.Fr.* lentille, *from Lat.* lenticula;
lentil, GEN. & EX. 1488.
lenþe, *see* **lengðe.**
lēo, sb., *O.E.* lēo, *cf. O.H.G., O.Fr.* leōn, liōn,
from Lat. leō; *lion,* REL. I. 125; leo, le ORM.
5834, 5978; leo, leon, liun LAȜ. 1463, 4085,
28074; leon ROB. 302; leun REL. I. 208;
GEN. & EX. 4025; HAV. 1867; leun, liun A.
R. 120, 164; lioun AYENB. 14; **lēones,** leunes
(*gen.*) ORM. 5927, 6027; liunes KATH. 1847;
lēones (*pl.*) JOS. 222.
lēod, lēode, sb., *O.E.* lēod, liod, = *O.Fris.*
liod, *O.L.G.* liud *f., O.N.* liöðr, lȳðr *m.,*
O.H.G. liut *m. n.;* *from* **lēoden**; *people,*
race, nation; leode, lede, led ORM. 7166,
7446, 11313; over al þas leode LAȜ. 2054;
lede *man* LANGL. *B* 522 (leod *v. r.* man
A vi. 6); leude, lede GAW. 126, 675; uche
leod þat liveþ in londe H. V. 106; lede
FLOR. 716; OCTAV. (H.) 1459; ANT. ARTH.
xxii; lued P. S. 155; lud WILL. 452; **lēodes**
(*gen.*) MARH. 20; þisses ledes king LAȜ.
9656; **lēode** (*dat.*) ORM. 5371; þe king wes
on leode, LAȜ. 3877; i þissere leode 4807;
on leode R. S. vi; in lede OCTAV. (H.) 183;
M. ARTH. 653; in ȝon leede A. P. ii. 772; **lēode**
(*pl.*) *men* ORM. 3237; MISC. 96; CL. M. 81;
leode nere þar nane LAȜ. 1118; þa leoden
[!eode] 1784; wolde ye mi leode lusten eure

loverde REL. I. 171 (MISC. 104); lede K. T.
124; MAN. (F.) 7345; þou losest lond & lede
A. D. 120; his lond and his lede [leede]
GAM. 71; lude GAW. 133; lith(e) HAV.
2515; leoden LK. xix. 14; FRAG. 5; leodes
P. S. 150; JOS. 168; ledes P. 65; ledes,
ludes WILL. 195, 390; lithes D. ARTH. 994;
þere leodene king LAȜ. 892; leof heo wes
þon leoden 3234; inne feole leoden 8320.
lēod-bischop, sb., *O.E.* lēodbisceop,=*O.N.*
lȳðbiskup; *diocesan,* CHR. E. 322.
lēodbiscop-rīche, sb., *jurisdiction of a*
diocesan, MISC. 145.
lēod-ferde, sb., *folk army,* LAȜ. 5685.
lēod-kempe, sb., *warrior,* LAȜ. 6025.
lēod-king, sb., *O.E.* lēodcyning; *king of the*
folk, LAȜ. 867.
lēod-līc, adj., = *O.H.G.* liutlīch; *native,*
national : þat leodliche folc LAȜ. 14698.
lēod-qvide, sb., *national speech,* LAȜ. 2914.
lēod-rūne, sb., *O.E.* lēodrūn; *discourse,*
LAȜ. 14553.
lēod-scome, sb., *disgrace of the people*;
LAȜ. 26297.
lēod-spel, sb., *discourse,* LAȜ. 15757.
lēod-swike, sb., *betrayer of the people,* LAȜ.
12942.
lēod-ðeau, sb., *O.E.* lēodþēow; *servant of*
the people, LAȜ. 2059.
[**lēoden,** v., *O.E.* lēodan,=*O.L.G.* liodan,
Goth. liudan, *O.H.G.* liutan; *grow*; *comp.*
ȝe-lēoden; *deriv.* lēod.]
leoden, *see* **leden.**
lēodisc, adj., *national,* LAȜ. 2144; ledisch A.
P. ii. 1556.
lēof, adj. & sb., *O.E.* lēof, līof,=*O.L.G.* liof,
O.Fris. liaf, *O.N.* liūfr, *O.H.G.* liob, liub,
Goth. liubs, *mod. Eng.* lief; *dear, beloved,*
A. R. 250; BEK. 37; leof & wurð KATH.
2263; leof [lef] LAȜ. 344; leof, lef ORM. 733,
2356; leof, lof O. & N. 203, 1277; lief HOM.
II. 7; ANGL. I. 11; luef REL. I. 110; SPEC.
99; lef GEN. & EX. 340; AN. LIT. 3; lef
[leef, lief] CH. T. *A* 1837; leef [lief]
or looþ LANGL. *B* ix. 57; **lēof** (*subst.*)
A. R. 380; SPEC. 92; leof min LAȜ. 29624;
hire leof ALIS. 2906; lef [leof] KATH.
786; Rimenild þi luef HORN (R.) 564; leef
LANGL. *B* ii. 33; mi swete lef [leef, lief] CH.
C. T. *A* 3792; in his leoves heorte A. R. 400;
ya, leve HAV. 1888; enne leove mon LAȜ.
4486; mi leofve mon 8928; his leve mon AN.
LIT. 12; leve [lieve] broþer CH. C. T. *A*
1136; lieve vader AYENB. 117; **lēove** (*pl.*)
MISC. 178; JOS. 240; leve feren KATH. 1375;
frendes lefe and dere PR. C. 2978; **lēovere**
(*compar.*) A. R. 102; leovere heom his (*r.* is)
to libben LAȜ. 466; levere [leovere] KATH.
2312; levere HAV. 1193; levre P. L. S. viii.
15; lever WILL. 453; levir OCTAV. (H.)
1183; AV. ARTH. xliv; luver TREV. VII. 27;

lēovest (*superl.*) A. R. 26; leovest, leofust LAȝ. 3288, 3766; levest me were PL. CR. 16; his leofeste cnihtes LAȝ. 28419; *comp.* ȝe-lēof.

lēof-līc, adj., *O.E.* lēoflīc,=*O.L.G.* liof-, lioblīc, *O.H.G.* liuplīch, *Goth.* liubaleiks; *lovable, lovely,* LAȝ. 31787; swa leoflic and swa lufsum HICKES I. 167; leoflich A. R. 90; leflich SPEC. 52; lefli GREG. 59; hire leflich(e) [leofliche] lich KATH. 1553; lēoflīche (*adv.*) A. R. 306; MISC. 92; he heom leofliche biheold LAȝ. 47; lefliȝ ȝho him fedde ORM. 3181; lēoflūkest (*superlat.*) MARH. 4.

lēof-mon, sb., *leman, dear one, one beloved,* LAȝ. 18611; A. R. 90; O. & N. 1430; lefmon AN. LIT. 11; lemman HAV. 1283; AYENB. 230; WILL. 663; CH. C. T. *A* 4240; LAUNF. 301; A. P. i. 762; M. ARTH. 586; lemmon HORN (R.) 550; lēfmen (*pl.*) ROB. (W.) 10206.

lēofsum, adj.,=*O.H.G.* liebsam; *lovable,* JUL. 17; lefsum HOM. II. 181.

lēofe, *see* lēove.

lēoȝen, v., *O.E.* lēogan,=*O.L.G.* liogan, *O.H.G.* liogan, liugan, *O.N.* liūga, *Goth.* liugan; *lie, tell lies,* P. L. S. viii. 145; lieȝen ANGL. I. 24; legen REL. I. 175; leȝhen ORM. 4907; leȝe SPEC. 91; lieȝe AYENB. 63; liȝen [leȝe] LAȝ. 3034; liȝen HOM. I. 153; P. S. 324; liȝe, lie (*ms.* lye) O. & N. 853; lihe JUL. 37; lien A. R. 68; liin, liȝin PR. P. 304; lie BEK. 1183; S. S. (Web.) 2409; CH. C. T. *A* 763; ne schal i þe lie HORN (L.) 1451; līest (*pres.*) A. R. 236; O. & N. 367; lext AR. & MER. 912; lixt P. L. S. xx. 28; LANGL. *B* v. 163; CH. C. T. *D* 1618; legeð REL. I. 224; legheþ MISC. 36; ligeð HOM. II. 131; liheð H. M. 39; leȝþ SHOR. 100; lihð A. R. 266; ofte him lieð (*fails*) þe wrench A. R. 338; R. S. i. (MISC. 156); lieþ [likth] LANGL. *B* xviii. 31; (hie) lieð HOM. II. 21; lih (*imper.*) A. R. 336; liȝe (*subj.*) A. R. 142; lihinde (*pple.*) JUL. 2; lēh [læh] (*pret.*) LAȝ. 12942, 17684; læh ORM. 12178; leh A. D. 260; leȝ SHOR. 111; leigh AM. & AMIL. 846; ne luȝe þu na HOM. I. 93; lowe LANGL. *B* xviii. 400; (heo) luȝen LAȝ. 15024; P. L. S. viii. 81; lowen MISC. 64; S. S. (Web.) 799; lógen (*pple.*) HOM. II. 61; lowen LANGL. *B* v. 95; WICL. JUDG. xvi. 15; we han lowen on Marine A. D. 262; *comp.* bi-, ȝe-lēoȝen; *deriv.* lüȝe.

lieȝere, sb., *O.E.* lēogere; *liar,* AYENB. 62; lieȝer R. S. vi; lihȝare HOM. I. 125; ligher PS. cxv. 11; liere MISC. 185; IPOM. 928; AUD. 12.

lieȝinge, sb., *lying,* AYENB. 143.

lēoht, līht, sb., *O.E.* lēoht,=*O.L.G.,* *O.H.G.* lioht, leoht, *O.Fris.* liacht; *light;* leoht MAT. iv. 16; liht P. L. S. viii. 140; KATH. 1594; (*r. w.* riht) O. & N. 230; (*r. w.* niht) R. S. v; (*ms.* lihht) ORM. 1918; liht [licht] A. R. 92; liht [liȝt] C. L. 723; liȝt

AYENB. 270; (*r. w.* riȝt) TREAT. 132; lígt (*r. w.* nigt) GEN. & EX. 44; lith, lict HAV. 576, 588; lict (*printed* litt) HICKES I. 227; lihte (*r. w.* drihte) LAȝ. 18585; lēochtes (*gen.*) HOM. II. 11; lihtes REL. I. 132; lihte (*dat.*) O. & N. 366; go bi lihte hom A. D. 299; bi dæies lihte (*ms.* lihten) LAȝ. 19661.

liȝt-bēre, sb., *O.E.* lēohtbǣre; *Lucifer,* AYENB. 11; lig(t)ber GEN. & EX. 271.

lēoht-berinde, pr. n., *Lucifer,* HOM. I. 219.

lēoht-fet, sb., *O.E.* lēohtfæt; *lamp,* [*'lucerna'*] MK. iv. 21; lihtfat ORM. 13399.

light-lēs, adj., *lightless,* PR. C. 4729.

lēoht², līht, adj., *O.E.* lēoht,=*O.L.G.* lioht, leoht, *O.Fris.* liacht, *O.H.G.* lioht; *light, lucid,* LEECHD. III. 88; er hit were liht SPEC. 42; hit was dæi liht LAȝ. 5667; hit is dai liht [liȝt] (*r. w.* niht) O. & N. 332; liȝt TREAT. 134; S. S. (Wr.) 2037; til . . . þat it were lith HAV. 1755; þe lihte dai HORN (R.) 497; þe sunne lihte (*r. w.* brihte) LAȝ. 17863; bi lihte dæie 6400; bi lihte daie MISC. 44; al þene dæi lihte LAȝ. 19566; līhture (*compar.*) A. R. 94.

lihte, adv., MISC. 149; lighte brenne CH. C. T. (Wr.) 6724 (lie and brenne *D* 1142).

lihtnesse, sb., *lucidity,* SPEC. 96; ligtnesse GEN. & EX. 1559; liȝtnesse CH. BOET. i. 2 (8).

liȝtsum, adj., *light, brightness,* WICL. PS. xviii. 9; lightsum PR. P. 304.

leohten, *see* lǣten.

lēohten, līhten, v., *O.E.* lēohtan, līohtan, lȳhtan, *cf. O.H.G.* liuhten, *Goth.* liuhtjan; *light; shine; kindle;* lihten KATH. 1411; (*ms.* lihht-enn) ORM. 19084; ase þe dæi gon lihte (*r. w.* fihten) LAȝ. 28314; liȝte (*r. w.* riȝte) BEK. 1415; of his feire siȝte al þe bur gan liȝte HORN (L.) 386; lighte [lihte] (*r. w.* brighte) CH. C. T. *A* 2426; lihteð (*pres.*) HOM. II. 111; lihteþ C. L. 718; līht (*imper.*) SPEC. 93; light min(e) eghen PS. xii. 4; lihte (*pret.*) LAȝ. 25595; liȝte P. L. S. xiii. 121; þei lihten two torches JOS. 191; *comp.* a-, an(d)-, bi-, in-lihten.

lihtinge, liȝtinge, sb., *O.E.* lȳhtinge; *illumination, enlightenment; lightning,* L. H. R. 46, 47; liȝtinge TREAT. 135; liȝting WICL. PS. xliii. 4.

lēohtnen, v.,=lēohten; *enlighten, illuminate;* lightenin PR. P. 304; liȝtne (*imp.*) WICL. PS. xviii. 9; liȝtnede (*pret.*) WICL. 2 TIM. i. 10.

liȝtnere, sb., *one who enlightens,* WICL. PROV. xxix. 13.

lightninge, sb.,=lihtinge; *lightning, illumination, 'fulgur,'* PR. P. 304; lightning TOR. 1512; liȝtning WICL. 2 TIM. i. 10.

leom, *see* lim.

lēome, sb., *O.E.* lēoma,=*O.L.G.* liomo, *O.N.* liómi; *light, gleam, brightness,* KATH. 1594; A. R. 94; TREAT. 133; leome [leom] LAȝ.

17873; leome, leom ORM. 658, 7627; mines lives leome HOM. I. 191; leme FL. & BL. 235; LANGL. B xviii. 124; leme [lem] C. M. 8048; lem ALIS. 6848; **lēomen** (*gen.*) LAȜ. 17874; **lēomen** (*pl.*) LAȜ. 17868; lemes CH. C. T. B 4120; **lēomene,** lemene (*gen. pl.*) HOM. II. 107.

lēmer, sb., *beamer*; lemer of light YORK xiv. 112.

ħleomeke, sb., *O.E.* hleomoce; *brooklime*; leomeke REL. I. 36; lemeke LEECHD. II. 393.

lēomen, v., ? *O.E.* lēomian, lȳman, *cf. O.N.* liōma; *give light, shine*; **lēomeþ,** lemeþ (*pres.*) SPEC. 31, 33; swa þat liht leomeȝ ant leiteȝ MARH. 13; lemeþ ELL. ROM. I. 269; lumes SPEC. 52; **lēmand** (*pple.*) BARB. viii. 226*; leming TREV. VIII. 525; **lēomede** (*pret.*) JOS. 687; lemede MAN. (F.) 9036; LAUNF. 942; lemed GAW. 591; A. P. i. 119; lemid M. ARTH. 3308; *comp.* a-lēomen.

lēomiren, v., *glimmer*; lēmired (*pret.*) TOR. (A.) 291.

lēon, see lēo.

lēonesse, sb., *O.Fr.* lionnesse; *lioness*, CH. C. T. D 637; lionesse PR. P. 306.

ħleonien, v., *O.E.* hleonian, hlinian, = *O.L.G.* hlinōn, *cf. O.H.G.* hlinen, *Lat.* (ac-, dē-, in-)clīnāre; *lean, incline*: lenie, line TREV. III. 309; hl(e)oneȝ (*pres.*) HOM. II. 39; leneȝ REL. I. 223; leneþ LIDG. M. P. 29; he leneþ on (h)is forke SPEC. 110; **leonie** (*subj.*) A. R. 142; **lénanȝ** (*pple.*) S. S. (Wr.) 2895; **ħlénede,** lenede (*pret.*) JOHN xiii. 23, 25; leonede LAȜ. 10776; leonede [lenede] LANGL. PROL. A 9; lenede WILL. 2991; a doun hi linede P. L. S. x. 45.

leonīn, adj., *lionlike*, CH. C. T. B 3836.

leopard, sb., *Fr.* leopard; *leopard*; leopart, lebbarde PR. P. 291; lubard MAN. (F.) 13795; LANGL. B xv. 293; lipard AYENB. 14; MIN. xi. 3; lebard WILL. 2935; Iw. 240; **libardes** (*pl.*) PR. C. 1228; lepardes CH. C. T. B 3451.

ħlēor, sb., *O.E.* hlēor, = *O.L.G.* hleor, hlear, hlier, *M.Du.* lier, *O.N.* hlȳr *n., cheek, face*; leor '*maxilla*' FRAG. 2; A. R. 64; KATH. 316; hire lufsum leor MARH. 3; lere OCTAV. (H.) 40; ANT. ARTH. xiii.; lure SPEC. 52; **lēre** (*dat.*) ALIS. 799; WILL. 227; EM. 294; A. P. i. 398; M. ARTH. 3624; leor(e) [lere, lire] LANGL. A i. 3; **lēores** (*pl.*) LAȜ. 5076; leres SPEC. 26; P. S. 330; lires HALLIW. 522; **lēre** (*dat. pl.*) FL. & BL. 501.

leornien [leorni], v., *O.E.* leornian, liornian, = *O.H.G.* lernen, lirnen, *O.Fris.* lerna, lirna; *learn*, LAȜ. 9897; leornen KATH. 110; leornen, lernen ORM. 9309, 13005; lernen HOM. II. 73; leorni [lorni] O. & N. 642; leorne [leurne, lerne] TREV. III. 279; V. 193; lerne ROB. 100; LANGL. B xx. 206; CH. C. T. A 308; TOR. 2202; lierni AYENB. 34; lurnen P. S. 155; **leorne** (*imper.*) A. R. 108; leorniaȝ HOM. I. 117; **leornede** (*pret.*) LAȜ. 6297;

KATH. 833; A. R. 254; lurnede P. L. S. x. 19; lernede PERC. 221; lernde ORM. 7248; leorneden FRAG. 7; *comp.* ȝe-leornen.

leornere, sb., *O.E.* leornere; *learner*; **leorneres** (*pl.*) HOM. I. 7.

lerning, sb., *O.E.* leornung; *learning*, AUD. 10.

leorning-cniht, sb., *O.E.* leornungcniht; *disciple*, ORM. 2234.

lēosen, v., *O.E.* (be-, for-)lēosan, = *O.L.G.*, *O.H.G.* (far-)leosan, liosan, *Goth.* (fra-)liusan, *O.Fris.* (for-)liasa, liesa; *lose*, KATH. 805; A. R. 102; C. L. 1177; he scal lessen [leose] þa hond LAȜ. 15238; l(e)osen, lesen O. & N. 351; leose BEK. 950; SPEC. 111; ihc wene þat ihc schal leose þe fiss þat ihc wolde cheose HORN (L.) 663; liese MISC. 26; AYENB. 214; lesen RICH. 3182; lesin PR. P. 298; lese MIRC 325; CH. C. T. A 1215; PR. C. 2915; ANT. ARTH. xxii; lese [leese] WICL. GEN. xviii. 23; lese [luse] TREV. VII. 49; luse FER. 4469; lise S. S. (Wr.) 899; **lēse** (*pres.*) LIDG. M. P. 189; i leese mi game H. V. 76; þu lust þi bischopriche BEK. 859; (he) leost [lost], lust O. & N. 830, 1159, 1193; lest SHOR. 108; GOW. II. 147; þo sennen þurch wiche me liest þo luve of gode MISC. 31; ha leoseȝ MARH. 14; **lēas** (*pret.*) A. R. 54; leas, les, læs [leos, les] LAȜ. 637, 6931, 18202; leas, lieas AYENB. 85, 203; les ALIS. 951; WILL. 887; les [lees] LANGL. B v. 499; lees JOS. 125; S. S. (Wr.) 3425; las MAP 337; þou lore ROB. 428; (we, ȝe, hi) lure MISC. 151; loren P. S. 190; ALIS. 2842; LANGL. B xii. 122; lore SHOR. 18; **lōre** (*subj.*) P. L. S. xvii. 477; loren AYENB. 235; **lōren** (*pple.*) MAP 334; lorn CH. C. T. A 3536; S. S. (Wr.) 3065; PERC. 159; PR. C. 547; lore WILL. 1360; LANGL. B xviii. 79; GOW. I. 189; HOCCL. i. 349; iloren SPEC. 110; C. L. 204; K. T. 563; ilore LAȜ. 18152; ROB. 160; AYENB. 45; LIDG. M. P. 38; *comp.* for-lēosen; *deriv.* lēas, lēsen, los, lósien, lóre, lüre..

ħlēoten, v., *O.E.* hlēotan, = *O.L.G.* hliotan, *O.N.* hliōta, *O.H.G.* hleozan, liozan; *cast lots,* JOHN xix. 24; *comp.* ȝe-lēoten; *deriv.* lot.

lēoð, sb., *O.E.* lēoð, = *O.N.* liōð, *O.H.G.* liod, leod; *song,* REL. I. 129; spel & leoþ (*printed* leow) HOM. I. 153; leod (? leoȝ) H. M. 21; led GEN. & EX. 27; **lēoðe** (*dat.*) LAȜ. 30054; lede BEV. 2130.

lēoþ-creft, sb., *poetry,* FRAG. I.

lēoð-scōp, sb., *poet,* LAȜ. 30609.

leoð, see lið. **leoðe,** see leðe.

leoþi, leoðien, see leþi, leðien.

ħlēou, sb., *O.E.* hlēow, hlēo, = *O.L.G.* hleo, *O.N.* hlȳ, hlǣ, *O.Fris.* hli, *mod.Eng.* lee, *prov.Eng.* lew; *shelter, protection*; lee GAW. 849; a lee DEP. R. iv. 74; we lurkede undir lee D. ARTH. 1446; le EM. 348; C. M. 23326; *comp.* hūs-lēwe.

leōw-stüde, sb., _O.E._ hlēowstede; _sheltered place_, ' _apricus locus_,' FRAG. 5.

lēove, sb., ? = _M.L.G._, _M.Du._ leve, _O.H.G._ liubī; ? _love_: leove (? = love) heom wes bitwune LA₃. 4307; for cristes leofe 17712.

leovien, _see_ livien.

lēovien, v., _O.E._ lēofian, = _O.H.G._ liuban; _love_; lēovie (? = lovie) (_pres._) LA₃. 4556; lēovede (_pret._) LA₃. 4850; ilēoved (_pple._) LA₃. 4487; _comp._ bi-lēovien.

lēp, _see_ lēap.

lepard, _see_ leopard.

[lēpe, sb., _O.E._ (ān-)lēpe; _comp._ sunder-lēpes.]

lēpe, _see_ lēap, _h_lēap. lēpen, _see h_lēapen.

[lēpi, adj., _O.E._ (ān-)lēpig, lȳpig; _comp._ ān-lĕpi.]

lepre, sb., _O.Fr._ lepre; _leprosy_; MISC. 31; GEN. & EX. 3690; WICL. LEV. xiii. 44; TREV. III. 33.

leprūs, adj. & sb., _O.Fr._ lepros, leprus; _leprous_, A. R. 148; MISC. 31; lepros ALEX. (Sk.) 4593.

lēr, _see h_lēor. lere, _see_ lire _and_ lüre.

lēre, lēren, _see_ lǣre, lǣren.

lerke, sb., _wrinkle_; lerkes (_pl._) D. TROY 3029.

lernien, _see_ leornien. les, _see_ lǣs.

lēs, _see_ lēas.

lesarde, sb., _Fr._ lezard, _from Lat._ lacerta; _lizard_, PR. P. 298; lusarde LANGL. _B_ xviii. 335; lisardes (_pl._) PALL. i. 1056.

lese, _see_ lesse.

lēse, sb., _O.E._ lǣs (_dat._, _acc._ lǣswe) _f._; _pasture land_, ROB. 43; BRD. 7; WILL. 175; TREV. V. 303; GOW. I. 17; ine heovene is large leswe A. R. 94; lēswe [lesewe] (_dat._) WICL. IS. vii. 25; lēswe (_acc._) HOM. II. 37; lēswa [lesewes] (_pl._) LA₃. 2011; lesen ROB. (W.) 15; leseo ROB. (W.) 15*, 1005*, 3887*; lesow 7701*.

lēsen, v., _O.E._ lesan, = _O.L.G._, _O.H.G._ lesan, _O.N._ lesa, _Goth._ lisan; _gather_; lese LANGL. _B_ vi. 68; lēseth (_pres._) PALL. viii. 48; laas (_pret._) TREV. I. 11; lēsid (_pple._) WICL. LEV. xix. 10.

lēzere, sb., _gleaner_, AYENB. 86.

lēsinge, sb., _collection_, P. S. 149.

lēsen, v., _O.E._ lēsan, lȳsan, = _O.Fris._ lēsa, _O.N._ leysa, _O.L.G._ lōsian, _O.H.G._ lōsen, _Goth._ lausjan; _from_ lēosen; _release, deliver_, HOM. I. 171; ORM. PREF. 63; GEN. & EX. 2897; lesen of helle pine H. H. 213 (215); leosen ne leˣien KATH. 1530; lēses (_pres._) M. H. 167; lēse (_subj._) HAV. 333; AV. ARTH. xxiii; lēsande (_pple._) R. S. v; lēsde (_pret._) HOM. II. 69; lesede R. S. v; lesed Iw. 2864; lēsed (_pple._) ORM. 4044; lesed [lesid] C. M. 22049; _comp._ a-, ₃e-lēsen.

lēser, sb., _deliverer_, PS. xvii. 3, 48.

lēsinge, sb., _O.E._ lȳsing; _remedy_, ' _solutio_,' PR. P. 298; B. B. 125, 267.

lēsnesse, sb., _remission_, ROB. 173; AYENB. 14; lisnisse P. L. S. xv. 75.

lēsen, _see_ lēosen. lēsewe, _see_ lēse.

leshe, _see_ lesse.

lēsinge, _see_ lēasunge _and under_ lēsen.

leske, sb., _cf. O.Sw._ liuske, _Dan._ lȳske, _M.Du._ liesche; _loin_, ' _inguen_,' PR. P. 298; lesske ORM. 4776; leskes (_pl._) D. ARTH. 3279.

lesse, sb., _O.Fr._ lesse, lesche (_laisse_), _leash_; lees CH. C. T. _G_ 19; leece PR. P. 291; lese OCTOV. (W.) 767; leeshe HALLIW. 510.

lessen, _see_ lǣssen.

lessōn, sb., _O.Fr._ leçon; _lesson_, PR. C. 3857; lessoun AYENB. 135; lessoun, lessun WILL. 341, 4442; lescun A. R. 282.

lest, = þi lǣs þe; _see_ lǣs.

lest, _see_ last, lüst. lĕst, _see_ lĕst.

lēste, _see_ lēaste. lesten, _see_ lüsten.

lĕsten, _see_ lĕsten. lesure, _see_ leisure.

lēswe, _see_ lēse.

lēswien, v., _O.E._ lǣswian; _pasture, feed_; lēseweð (_pres._) HOM. II. 39; lēswe (_imper._) A. R. 100; lǣsiende (_pple._) MK. v. 11; lēseweden (_pret._) WICL. LK. viii. 34.

let, _see_ lat. lētande, _see_ lǣten.

letanie, sb., _O.Fr._ letanie; _litany_, PR. P. 299; A. R. 20; LEG. 180.

lēte, sb., (_court_) _leet_: amercin in a corte or lete PR. P. 11.

lēte, lēten, _see_ lǣte, lǣten.

letere, _see_ litere.

letron, leteron, sb., _cf. med.Lat._ lectrum; _lectern_, PR. P. 299.

lette, sb., = _M.Du._ lette, ? _M.H.G._ letze; _from_ lat; _let, hindrance, delay_, HOM. I. 239; LA₃. 4572*; SHOR. 71; WILL. 1340; GOW. II. 92; CH. C. T. _E_ 300; S. S. (Wr.) 2439; ANT. ARTH. iii; without(e) lette MIRC 677; let BARB. i. 598; vii. 172.

let-lēas, adj. & adv., _without hindrance_; letless BARB. xvi. 568.

letten, v., _O.E._ lettan, = _O.L.G._ lettian, _O.N._ letja, _Goth._ latjan, _O.H.G._ lezzan; _from_ lat; _obs.Eng._ let; _retard, impede_, A. R. 162; REL. I. 131; GOW. II. 72; lette HAV. 1164; SHOR. 69; WILL. 1253; LANGL. _B_ xvii. 78; MAND. 173; i lette ₃ou no₃t BRD. 27; letteð LA₃. 22009; letteþ ORM. 14117; letteð, let A. R. 14, 156; lettes PR. C. 238; latteth PALL. xi. 135; lette (_pret._) CH. C. T. _E_ 389; lætten [lette] LA₃. 1344; _comp._ ₃e-letten.

lettere, sb., _hinderer_, LANGL. _A_ i. 67; lettir YORK xlvi. 143.

lettinge, sb., = _M.L.G._ lettinge; _impediment_, LA₃. 7820; LANGL. _A_ vi. 7; letting GEN. & EX. 1076; lettunge HOM. I. 187.

lettered, pple., *lettered, learned,* WILL. 4088; lettred TREV. V. 19, VII. 15, 17.

lettre, sb., *O.Fr.* lettre, letre; *letter,* GEN. & EX. 993; P. L. S. xxiv. 156; MAN. (F.) 941; letter BARB. x. 353; lettres *(pl.)* A. R. 422; lettris *(deeds)* BARB. xx. 44; lettres *(as sing.)* BARB. ii. 80.

lettrūre, sb., *O.Fr.* lettreūre, letreūre; *literature,* LANGL. *C* x. 198; lettireure HAMP. PS. lxx. 17; letrure ALIS. (Sk.) 1152.

letuarie, sb., *O.Fr.* lettuaire; *electuary,* PR. P. 300; A. R. 370; CH. C. T. *C* 307.

letuce, sb., *lettuce, 'lactuca,'* PR. P. 300; PALL. ii. 202.

leþ, see **liδ.**

leδe, sb., = *M.L.G., M.Du.* (un-, on-)lede; *alleviation, ease,* HOM. I. 35; leþe A. P. iii. 160; þer heo leoδe [lioþe] hafveden LAȝ. 12022.

leδer, sb., *O.E., O.N.* leẍr, = *M.Du., O.H.G.* leder; *leather:* þet leẍer A. R. 392; leþer A. P. ii. 1581; leþer [lether] CH. C. T. *A* 3250; leþir, ledir *'corium'* PR. P. 293; leder TREV. I. 165; II. 19; VIII. 283, 467; leþeren *(pl.) slings* ROB. (W.) 8124*.

lēþer, see **lǔδer.**

leþeren, adj., *cf. M.Du.* lederen, *O.H.G.* lidrīn; *made of leather,* WICL. 1 KINGS v. 9; a leþerne [leþeren] purs LANGL. *A* v. 110; ietherin MIN. xi. 19.

leδerien, v., *O.E.* leẍrian, lyẍrian; *from* laþer; *lather:* leδeri *(pres. subj.)* JUL. 16; he leþerede *(pret.)* a swote LAȝ. 7489*; liẍerede o blode MARH. 5; KATH. 1554.

leþi, adj., *O.E.* liþig (ÆLFR. L. S. 10, 73), = *O.Fris.* letheg, *O.N.* liẍugr, *M.Du.* ledigh, *M.L.G.* ledich, *M.H.G.* ledec, lidic; *? from* leδe; *loose, flexible, unoccupied:* ȝef eni boȝ *(r.* loȝ) þer leþi *(? printed* lothy) were SHOR. 156; a ful leþi þing LANGL. *B* x. 184; leoþi and lene L. H. R. 147; lethi *'flexibilis'* PR. P. 302; his erelappes waxes lethi REL. I. 54.

leδien, v., *? cf. O.N.* liẍa; *set free, alleviate, assuage,* HOM. II. 149; leosen *(r.* lesen) ne leẍien KATH. 1530; (3)if he him wold(e) leoδien of laẍe his benden LAȝ. 4775; leþe GAW. 2438; A. P. iii. 3; léδeδ *(pres.)* HOM. II. 71; leþes ['*mitigas*'] PS. lxxxviii. 10*; leoδe *(imper.)* MARH. 13; leoẍe ure benden LAȝ. 21922; lethe *(adjust)* PALL. vii. 45; leoẍede *(pret.)* MARH. 13; þat weder leoẍede LAȝ. 12042; leþed M. H. 86.

leδδe, see **laδδe.**

leuge, sb., *league (measure of distance);* leuge, leuke, lege TREV. II. 11; V. 245.

hlēuke, *h*lēuc, adj., = *Du.* leuk, *M.H.G.* lawec, lēwic; *? from* **hlēwe;** *tepid;* lheuc AYENB. 31; leuk WICL. APOC. iii. 16; leuke PR. C. 7481*; *(ms.* lewke) *'tepidus'* PR. P. 302; luke LAȝ. 27557.

lēukenesse, sb., *lukewarmness, 'tepor,'* PR. P. 302.

leūn, see **lēo.** **leurne,** see **leornien.**

levain, sb., *O.Fr.* levain; *leaven,* AYENB. 205; levein *'fermentum'* PR. P. 300.

lēve, see **lāve, lēave, lēaven, lēof.**

lěvedi, see **hlǎfdiȝ.** **lēveful,** see *under* **lēave.**

level, sb., *O.Fr.* livel; *level, 'perpendiculum,'* PR. P. 301; level and line LANGL. *B* x. 179 (livel *A* xi. 135); levele *(dat.)* AYENB. 150.

lēven, v., *O.E.* lēfan, lȳfan, = *O.N.* leyfa, *O.H.G.* (er-)louban, *Goth.* (us-)laubjan; *give leave, permit, concede:* eouer axe ich eou leve *(ms.* leue) LAȝ. 1053; lēf *(imper.)* JUL. 29; leaf me gan MARH. 12; lēve *(subj.)* A. R. 88; HORN (L.) 461; HAV. 334; C. L. 17; ROB. 139; PL. CR. 573; *(printed* lene) SPEC. 57; god leve him in his blisse spilen GEN. & EX. 2532; Crist leve *(printed* lene) þat he so mote P. P. 125; lefe ORM. 8873; lēfde *(pret.)* MK. v. 13; lēved *(pple.)* KATH. 771; *comp.* ȝe-lēven.

lēven², v., *O.E.* lēfan, lȳfan, = *O.L.G.* (gi-)lōvian, *O.H.G.* (ge-)louben, *Goth.* laubjan; *believe,* MARH. 5; REL. I. 131; GEN. & EX. 4141; H. H. 231 (235); SPEC. 37; leven an (in) god JUL. 10, 12; ho scal leven o mine godes HICKES I. 226; leeven C. L. 615; levin PR. P. 301; lefen FRAG. 7; ORM. 1153; leve WILL. 4175; ALIS. 326; WICL. EX. iv. 1; lefe TOWNL. 98; lieven S. A. L. 153; to luvene HOM. I. 75; lēve *(pres.)* AYENB. 89; CH. C. T. *E* 2205; lefe ISUM. 253; leove MARG. 169; þou livest FER. 1050; þe loverd þat man on leves HAV. 1781; (hi) leveþ AYENB. 86; luveþ HORN (L.) 44; leven [liven] LANGL. *B* xv. 403; leeven MAND. 108; lēf *(imper.)* JUL. 13; SPEC. 113; lēve *(subj.)* S. S. (Wr.) 1506; lēfde *(pret.)* HOM. II. 125; KATH. 430; levede ROB. 265; lifde REL. II. 244; lefden LK. xxiv. 41; MARH. 2; leeved *(pple.)* HOCCL. i. 220; *comp.* bi-lēven, ȝe-lēfen.

lēvenesse, sb., *'confidentia,'* PR. P. 301.

lēven, see **lǣven, lēaven.**

lēvene, sb., *cf. Swed.* lȳvna, lȳgna (RIETZ), *Goth.* lauhmuni; *flash of lightning, 'fulgur,'* PR. P. 301; GEN. & EX. 3265; GOW. III. 77; levin HAV. 2690; AV. ARTH. lxv; TOWNL. 116; lēvene *(dat.)* CH. C. T. *D* 276.

lēvenen, v., *flash, lighten;* levenand *(pple.)* HAMP. PS. *page* 510.

lever, sb., *O.E.* læfer, = *M.H.G.* leber; *bulrush, 'scirpus,'* LEECHD. III. 335.

lēver, see **lēof.** **leverē,** see **liveree.**

levien, see **livien.**

levour, sb., *O.Fr.* levier; *lever,* ROB. 126; WICL. IS. xxvii. 1; lever PARTEN. 4177.

hlēwe, adj., *? O.E.* hlēowe, *cf. O.N.* hlǣr, *M. Du.* lauw, *O.H.G.* lāwi; *lukewarm, warm, tepid;* lewe HOM. I. 7; HAV. 498; WICL. APOC. iii. 16; REL. II. 211.

lēwnesse, sb., *moderate warmth,* HOM. I. 21.

lēowþ, sb., *O.E.* hlēowþ; *sheltered warmth,* ' *apricitas,*' FRAG. 5.

[lēwe, ? adj.; *comp.* cost-, chēke-, drunken-, sīk-, þürstlēwe, (-lēu).]

lēwed, *see* lǣwed. lēwke, *see* *h*lēuke.

lēwse, *see* lēse. lewtē, *see* lēaltē.

lex, *see* lax. lēy, *see* lēi.

leye, *see* leȝe *and* lei. lēden, *see* *h*lǎden.

liad, liaf, *see* lēad, lēaf.

liard, adj., *O.Fr.* liard; *gray,* P. S. 71; liarde (*grey horse*) PALL. iv. 806.

libard, *see* leopard.

[libbe, sb., *O.E.* lybb,=*M.Du.* libbe, lebbe, *M.H.G.* lüppe; *rennet*; *comp.* chēs-lippe.]

libben, *see* livien.

libel, sb., *Fr.* libelle, *mod.Eng.* libel; *little book; writing*; [' *libellum*'] WICL. MAT. v. 31; libelles (*pl.*) AYENB. 40.

liberal, adj., *liberal*; leberalle D. ARTH. 2318.

libertē, sb., *Fr.* liberté; *liberty,* CH. C. T. *E* 145; libertēs (*pl.*) ALEX. (Sk.) 4348.

librarie, sb., *Fr.* librairie; *library,* TREV. II. 37.

līc, adj., *O.N.* līkr; ?=ȝelīc; *like, similar,* ORM. 2569; REL. I. 227; lic [liic] WICL. MAT. xx. 1; lik GEN. & EX. 223; MAND. 47; lik [like] LANGL. *A* ix. 6; CH. C. T. *A* 412; leche TREV. II. 187; like [' *clemens*'] PALL. i. 52; like, liche RICH. 5498, 5900; lich HAV. 2155; liche S. S. (Wr.) 2991; WILL. 3678; nas non his liche FL. & BL. 88; it was weill lik (*probable*) BARB. xvi. 324; for þi þat teȝ him sinden swiþe like ORM. 8218; þei beþ no man lich(e) P. 56; *comp.* ǣr-, aȝe-, an-, ān-, atel-, beald-, biter-, blīðe-, brāð-, brēme-, briht-, cāf-, clǣn-, crafti-, cüme-, cünde-, cūð-, dēop-, dēor-, derne-, drēori-, earm-, ēaðe-, ēche-, efen-, eȝes-, fǣr-, fǣst-, fēond-, flǣsch-, frēo-, frēond-, ful-, fūl-, gāl-, gāst-, ȝe-, ȝēap-, ȝeār-, ȝeōmer-, ȝeorn-, glǣd-, gōd-, grǎme-, grēat-, grǣdi-, grim-, gris-, hǣþe-, hāl-, hām-, hēah-, help-, hende-, heorte-, heoven-, hēr-, hōker-, hüht-, hwat-, in-, inner-, kēn-, kine-, laȝe-, lǎð-, lēof-, līf-, līht-, man-, mis-, mōdi-, nǎme-, open-, prūd-, rād-, *h*rēou-, rīche-, sǎr-, schǎme-, schand-, sēm-, smerte-, sōð-, spēd-, starc-, stil-, strang-, stürne-, sumer-, sunder-, swēte-, tīd-, tīme-, trēow-, þrā-, þülde-, ütter-, wǎc-, war-, weorld-, wīs-, wōd-, wrāð-, wün-, wunder-, wurð-līc; *also* ǣlc, hwülc, swülc.

līk-lī, adj., *O.N.* līkligr; *likely,* CH. C. T. *A* 1172; HOCCL. i. 74.

likli-hēde, sb., *likelihood,* CH. C. T. *B* 1786.

līklines, sb., *likeliness, likelihood,* BARB. iii. 88; xi. 244.

liknesse, sb., = *M.L.G.* līknisse; *likeness*; (*ms.* liccness) ORM. 1047; GEN. & EX. 202; AYENB. 92; (*v. r.* liknes) LANGL. *B* i. 113; liknesse WICL. LK. v. 36.

[līc², *stem of* līken.]

līc-vol, adj., *pleasing, agreeable,* AYENB. 217; likful P. L. S. xxxv. 80.

līc-wurðe, adj., *O.E.* līcwyrðe; *agreeable, acceptable,* A. R. 120; H. M. 11; HOM. II. 7.

līc-wurþiȝ, adj., *agreeable,* ORM. 15919.

līc³, līch, sb., *O.E.* līc, *cf. O.L.G.* līc, *O.N.* līk, līki, *Goth.* leik *n.* (σῶμα), *O.H.G.* līh, lích *f.*; *body, corpse,* LAȝ. 6682, 10434; ORM. 8183, 16300; lich JOHN xx. 12; MARH. 5; MISC. 149; hire leflīch lich KATH. 1553; þet rotede lich A. R. 216; þat lich AL. (T.) 565; lich, liche GEN. & EX. 2441, 2488; liche ' *funus*' PR. P. 302; REL. I. 178; ALIS. 3482; whan he liþ a cold liche P. L. S. v.² 48; like ALEX. (Sk.) 2931; liche (*dat.*) BEK. 259; LANGL. *A* x. 2; hu mahtest þu gan to þine aȝene liche (*? be present at thy own funeral*) HOM. I. 29.

līc-hame, sb., *O.E.* līchama, -homa,= *O.L.G.* līchamo, *O.N.* līkami, *O.H.G.* līchamo, līhhamo; *body, corpse,* MAT. v. 29; P. L. S. viii. 153; REL. I. 128; licame LAȝ. 11046; lichome, licome O. & N. 1054; lighame YORK v. 110; D. ARTH. 3281, 3286; līghames (*pl.*) 4269; licame KATH. 215; A. R. 4; R. S. v; þes licome lust HOM. I. 19; licham GEN. & EX. 200; likame MISC. 48; WILL. 227; licam [likam] LANGL. *A* PROL. 30; licam AUD. 17.

līcham-līch, adj., *O.E.* licamlīc; *corporeal*; licomlīche (*pl.*) A. R. 240; licomliche pinen KATH. 42; līchamlīche (*adv.*) REL. I. 130.

lich-hūs, sb., *sepulchre,* HOM. II. 169.

lich-raste, sb. (*? for* -rǣste), *O.E.* līcrest; *sepulchre,* LAȝ. 17225.

lich-þrowere, sb., *O.E.* līcþrowere; *leper*; lichþrowæres (*pl.*) LK. iv. 27.

lich-wake, sb., *funeral-wake,* MIRC 1465; CH. C. T. *A* 2958.

licence, sb., *O.Fr.* licence; *licence,* LANGL. *A* PROL. 82.

licenciāt, sb., *Ital.* licenziāto; *licenciate,* CH. C. T. *A* 220.

lich, *see* līc.

liche, sb., *O.E.* (man-, swīn-)līca, = *Goth.* (man-)leika, *O.H.G.* (man-)līcha; *form, figure,* HOM. I. 141; SHOR. 20; a wifmonnes liche LAȝ. 1142; liche, like A. R. 224, 262; like ORM. 5813; GOW. I. 143; *comp.* man-liche.

[līchen, v., = *M.L.G.* liken, *M.H.G.* līchen; *resemble*; *comp.* at-līchen.]

licherous, *see* lecherous.

Lichet-feld, pr. n., *O.E.* Licetfeld; *Lichfield*; (*? printed* Lycchesfeld) MISC. 146.

licken, v., *O.E.* liccian,= *O.L.G.* liccōn, leccōn, *O.H.G.* lecchōn; *lick*; likkin ' *lingo*' PR. P. 305; licke PS. lxxi. 9; **lickest** (*pres.*) AN. LIT. 90; lickeð H. M. 9; lickede (*pret.*) P. L. S. xviii. 69; likked IW. 2008; (*sipped*) A.

P. ii. 1521; lickeden LEB. JES. 157; **licked**
(*pple.*) REL. I. 114; ilicked HOM. I. 200; *comp.*
lik-pot.

likkare, sb.,=*O.H.G.* lecchāri; *licker*, PR.
P. 305.

lickestre, sb. (fem.), *licker*: þe tonge þe
lickestre AYENB. 56.

likking, ŝb., *licking*, TREV. IV. 435.

licorice, sb., *Lat.* liqviritia, *Gr.* γλυκύρριζα;
licorice, PR. P. 303; licoriz LAȝ. 17745; licoris
SPEC. 26; ALIS. 6794; CH. C. T. *A* 3207.

licorous, adj., *Fr.* liquoreux; *dainty, pleasant*,
KN. L. 22; likerous AYENB. 95; LANGL. *A*
PROL. 30; CH. BOET. iii. 4 (72); CH. C. T. *C*
540; TREV. I. 11; likkerwis GAW. 968.

licorousnesse, likerousnesse, sb., *lecher-
ousness*, CH. C. T. *D* 611.

licour, sb., *O.Fr.* likeur; *liquor*, SHOR. 8;
CH. C. T. *A* 3; PR. C. 6763; licoure (*sauce*)
B. B. 141; licur A. R. 164; licour(e), likoure
C. B. 6, 11, 12.

hlid, sb., *O.E.* hlid,=*O.N.* hlǐ, *M.L.G.* lid,
led, *O.H.G.* lit *n.*; *lid, cover*; lid P. L. S.
xvi. 168; w. & I. 23; þat lid of þe þrouwe
LEB. JES. 723; led CL. 272; **hlide** (*ms.* hlyde)
(*dat.*) MAT. xxvii. 60; **lides** (*pl.*) REL. I. 209;
comp. **ēhe-lid.**

lid-ȝate, sb., *O.E.* hlidgeat; *lidgate, postern*,
MIRC 1497.

lidderon, *see* **lederon. līde**, *see* **lūde.**

hlīden, v., *O.E.* (be-, on-)hlīdan (*pret.* hlād);
cover, close; **līdeð** (*? printed* liðeð), lides
(*pres.*) A. R. 84*; *comp.* **un-līden.**

lie, sb., *Fr.* lie; *lees, dregs,* 'lia,' PR. P. 303.

lie, *see* **lüȝe. līe**, *see* **lēaȝe, lēi.**

lief, *see* **lēaf, lēof. lieȝen**, *see* **lēoȝen.**

līen, *see* **lēan.**

lien, v., *O.Fr.* lier; *thicken soup*: make a
lioure of brede and blode, and lie hit þerwithe
L. C. C. 32; li(e) C. B. 13, 15, 19; liid (*pple.*)
C. B. 13.

līen, *see* **liggen, lēȝen, lēoȝen.**

lier, sb., *O.Fr.* lieure (*liaison*); *thickening for
soup*, B. B. 60; C. B. 33; lioure L. C. C. 32.

lies, *see* **lēas. liesen**, *see* **lēosen.**

lieutenant, sb., *Fr.* lieutenant; *lieutenant*,
LANGL. *B* xvi. 47; leef-, levetenaunt TREV.
VIII. 143; luftenand BARB. xiv. 139, 255.

līf, sb., *O.E.* līf, *cf* *O.L.G.*, *O.Fris.*, *O.N.* līf,
O.N. līfi *n.*, *O.H.G.* līb *n. m.; from* **līven**; *life,
body, person*, ORM. 1625; H. H. 179; MAND.
100; S. S. (Wr.) 2816; lif & saule HOM. I.
71; mi leove lif 195; munechene lif 93; lif
& leomen LAȝ. 702; binom hire seolven þat
lif 3776; lif and time O. & N. 1098; mi leve
lif LEG. 31; his douȝter þat leove lif ALIS.
2502; a pore lif (*man*) P. S. 331; ALEX. (Sk.)
599, 5287; lif (*man*) GOW. I. 362; LANGL. *C*
iv. 450; lif and soule *B* xii. 188; ech lif
hit loþede *C* viii. 50; ani lif MIRC 1300; liif

WILL. 357; WICL. JOHN iii. 15; liff BARB. x.
417; **lifes**, lives, lifves (*gen.*) LAȝ. 229, 2425,
8866; after þis lifes ende ORM. 2708; lives
REL. I. 235; HAV. 509; LANGL. *B* xix. 154;
lives boc *the book of life* A. R. 246; lives bec
MISC. 71; heo beoð lives [*O.N.* līfs] *alive*
HOM. I. 31; ah þu nevre mon to goed lives
ne deaþes stal ne stode O. & N. 1632; ever
uch best þat lives is A. D. 240; a lives mon
C. L. 1422; o lives (*for* live) MAP 338; **līfe**
(*dat.*) P. L. S. viii. 12; live A. R. 152; for his
live O. & N. 1078; on live *alive* FRAG. 6; S.
S. (Wr.) 56; DEGR. 966; sari he was on live
LAȝ. 1037; wa wes him on live JUL. 48; wo
is him a live R. S. v. (MISC. 182); if þu were
a live HORN (L.) 107; to live god him wal-
d(e) bringe(e) GREG. 265; and leten ðe oðre
to live gon (*live*) GEN. & EX. 629; ȝef þu mote
to live go HORN (L.) 97; wheþer our to live
go TRIST. 1022; to bringe þe of live HORN
(R.) 695; bringe fro live R. R. 7260; bi life
quickly ORM. 17943; be life FLOR. 1654;
forð heo gunnen liðen a neouste bi life LAȝ.
25636; fæhten bi live 4191; bi live S. S.
(Wr.) 2693; be live LUD. COV. 230; bi live
[blive] KATH. 2311; bi live, blive WILL. 248,
1705; blive BRD. 20; *comp.* **cot-, munec-
līf.**

līf-daȝes, sb., pl., *O.E.* līfdagas,=*O.N.* līfda-
gar; *life-days, life*, HOM. I. 129; lifdages (*ms.*
liue-) GEN. & EX. 4119; lifdawes FRAG. 5;
lifdawes WILL. 3704; lifdaies LANGL. *A* i. 27;
lifdaȝen (*dat. pl.*) LAȝ. 11295; þe (h)wile þu
art on lifdaȝe O. & N. 1141; lifdawe P. L. S.
xiii. 93; ibroht of lifdawe HORN (R.) 914.

[**līf-fæsten**, v., *O.E.* līffæstan; *vivify;
comp.* **ȝe-līffæsten.**]

līf-ful, adj., *life-giving*, KATH. 867.

līf-hāli, adj., *holy of life*, JUL. 2; lifholi A.
R. 346; HOM. II. 133; liifhooli PR. P. 303.

līf-hōlinesse, sb., *holiness of life*, HOM. I.
207; A. R. 142; LANGL. *C* vi. 80.

līf-(h)wīle, sb., *life time*, AN. LIT. 5.

līf-lāde, sb.,=*O.H.G.* līblëita; *livelihood,
life, means of living*, MARH. 20; liflode A. R.
350; REL. I. 129; ROB. 41; LANGL. PROL.
30; MAND. 293; GAW. 133; A. P. ii. 561;
lif-, liiflode PR. P. 308.

līf-lēas, adj., *O.E.* lifleas,=*O.N.* lifláuss,
O.H.G. lībelōs; *lifeless*, FRAG. 8; KATH.
896; lifles LEG. 151.

līf-lēovie?, sb., *? lifelong lover*, KATH. (E.)
1674.

līf-lich, adj., *O.E.* līflīc,=*O.N.* līfligr, *O.H.
G.* līblīch; *lively, vital, corporeal*; livelich
P. P. 218; A. R. 6; lifli WICL. WISD. xv. 11;
liifli 'vivax' PR. P. 308; **līflī** (*adv.*) CH. C.
T. *A* 2087; liiflīche (*pl.*) TRIST. 2845.

līf-sīð, sb., *lifetime*, H. M. 45.

līf-tīme, sb., *life-time*, REL. I. 224; WILL.
999; BARB. i. 308.

lifisch, adj., *living* : lifishe man ORM. 2463 ; 5140, 6106, 6175, 18985 ; ilc lifishe man mennish 18942.

lifnoþ, *see* livenaðŏ.

lift, liften, *see* lüft, lüften.

lige, adj., *O.Fr.* lige, liege ; *liege,* ROB. 457 ; MAN. (F.) 7455 ; mi lege man WILL. 1174 ; lege, liege, lige, ligge D. ARTH. 1901, 1200, 3080, 2221 ; lege man TREV. VIII. 175 ; ligmane D. ARTH. 420 ; lege pouste *full power* BARB. v. 165 ; ligge mene D. ARTH. 1518.

liȝe, *see* lüȝe, ḥlahhen.   **līȝe,** *see* lēi.

ligen, *see* liggen.   **līȝen,** *see* lēoȝen.

ligence, sb., *O.Fr.* ligence ; *allegiance* ; liegeaunce FER. 1270 ; TREV. VIII. 469, 481 ; legeaunce (*v. r.* legeance) ALEX. (Sk.) 2791.

[**liȝer,** sb., = *O.H.G.* ligiri ; *couch, bed ; comp.* for-liȝer.]

liggen, v., *O.E.* licgan, licgean, = *O.L.G.* liggean, *O.N.* liggja, *O.Fris.* lidza, *O.H.G.* liggan, likkan, *Goth.* ligan ; *lie, lie down, lie* (*with*), LEECHD. III. 88 ; LAȜ. 1437 ; KATH. 2286 ; A. R. 4 ; ligge O. & N. 1200 ; FL. & BL. 754 ; AYENB. 31 ; WILL. 2194 ; LANGL. *B* xi. 418 ; MIRC 219 ; S. S. (Wr.) 144 ; E. G. 350 ; PR. C. 475 ; ligge a doun ROB. 100 ; ligge, lien HAV. 876, 2134 ; ligge [lie] bi hire TREV. III. 257 ; ligge, liȝe MAND. 240, 276 ; liggin, ligin, liin PR. P. 304 ; ligen REL. I. 177 ; liȝe GAW. 1096 ; lien SAX. CHR. 262 ; MAP 335 ; lin ORM. 6020 ; GEN. & EX. 942 ; lie CH. C. T. *A* 3651 ; **líst** (*pres.*) FRAG. 6 ; H. M. 9 ; O. & N. 1502 ; wher þe wrong liggeþ [lihþ] LANGL. *B* iii. 169 ; ligges PERC. 797 ; liȝþ, liþ C. L. 775 ; lið [liþ] LAȜ. 21434 ; A. R. 212 ; R. S. v ; ne lið (hit) nawt to þe to leggen lahe upo me KATH. 779 ; liþ FRAG. 5 ; O. & N. 430 ; P. S. 152 ; al þat to þe bodiȝ liþ (*belongs*) ORM. 4608 ; ȝe liggeð [liggen] 2389 ; (heo) liggeð A. R. 316 ; heo liggeoð, liggeð LAȜ. 12740, 27943 ; þe londes þe þer to liggeð JUL. 13 ; liggeþ O. & N. 1048 ; PL. CR. 83 ; ligges M. H. 29 ; liȝe (*imper.*) LAȜ. 28724 ; lie A. R. 290 ; ligge (*subj.*) A. R. 424 ; SHOR. 30 ; **liggende** (*pple.*) GOW. I. 187 ; ligginde SHOR. 122 ; ligand, liand WILL. 2180, 2246 ; ligging CH. C. T. *A* 2390 ; **læig,** laig [*O.E.* læg, = *O.L.G., O.H.G., Goth.* lag, *O.N.* lā (*pret.*)] lay, MK. ii. 4, LK. v. 25 ; læi SAX. CHR. 251 ; læi, lei leai, lai [lai] LAȜ. 393, 650, 681, 28431 ; lei A. R. 54 ; þat lond þat lei into Rome KATH. 28 ; laȝ ORM. 3692 ; lai, lei (*ms.* ley) O. & N. 1494, 1509 ; lai JOS. 176 ; (*ms.* lay) BEK. 892 ; LANGL. *A* v. 46 ; þu læie [leie] LAȜ. 5030 ; leiȝe FRAG. 7 ; (heo) leien [leiȝen] LAȜ. 9804 ; of alle þon londen þa leȝen [leien] into France 1657 ; leien 16826 ; leiȝen P. S. 190 ; JOS. 418 ; leien A. R. 106 ; (*ms.* leyen) HAV. 475 ; leie WILL. 4307 ; (*ms.* leye) BEK. 2457 ; **læie** [leie] (*subj.*) LAȜ. 22254 ; leie HOM. I. 33 ; O. & N. 134 ; leien [*O.E.* (a-)-

legen, = *O.N.* leginn, *O.H.G.* (gi-)legan] (*pple.*) lain, (*ms.* leien) AN. LIT. 11 ; (*ms.* leyen) CHR. E. 742 ; LANGL. *A* iii. 38 ; RICH. 2477 ; lein MAND. 286 ; *comp.* a-, æt-, bi-, fŏr-, ȝe-, of-, ümbe-, under-liggen ; *deriv.* laȝe, legge, leggen, leir, liȝer.

[**liȝer,** sb., *O.H.G.* (bi-)legāri ; *one that lies* ; *comp.* for-liȝer (*under* forliggen).]

ligginge, leginge, sb., *lying down,* TREV. VI. 93, VII. 11.

ligiament, sb., *act of allegiance,* TREV. VIII. 55.

liȝhen, *see* ḥlahhen.   **lignen,** *see* leīnen.

lignāge, *see* lināge.   **liȝt,** *see* līht.

lihen, lihȝen, *see* ḥlahhen.   **lihen,** *see* leoȝan.

lihnen, v., *? from* lēan ; *? blame, reprehend* : to lihnen (*ms.* lihhnenn) þat lærede folc ORM. 7440 ; to lihnen þeȝre spæche 18618.

līht, adj., *O.E.* līht, leoht, = *O.Fris.* licht, *O.H.G.* līhti, liehti, *Goth.* leihts, *O.N.* lettr ; *light (in weight), easy,* A. R. 220 ; (*r. w.* riht) SPEC. 102 ; (*ms.* lihht) ORM. 4500 ; liht, liȝt C. L. 958 ; liȝt (*r. w.* briȝt) GREG. 428 ; hit is . . . liȝt to zigge AYENB. 99 ; light GOW. II. 95 ; leoht MAT. xi. 30 ; þe lihte heorte A. R. 60 ; licht men and waverand BARB. vii. 112 ; licht and deliver x. 61 ; on licht hors xiii. 56 ; lihte (*adv.*) REL. I. 180 ; liȝte ROB. 25 ; lihte (*pl.*) LAȜ. 5903 ; lihte gultes A. R. 346 ; ne luvede ha nane lihte plahen KATH. 106 ; **lihture** (*compar.*) A. R. 140 ; liȝtere WILL. 2119 ; liȝter LANGL. *B* xvii. 39.

liȝt-hēd, sb., *levity,* (*later ver.* lihtnesse) WICL. JER. iii. 9.

liht-lich, adj., *O.E.* līht-, leohtlīc ; *light, easy* ; O. & N. 1759 ; lihtlīche (*adv.*) A. R. 392 ; liȝtliche ROB. 79 ; LANGL. *B* xviii. 267 ; lehtliche [lihtliche] LAȜ. 26079 ; lehtliche H. M. 17 ; lihtlike ORM. 6487 ; leithli YORK ii. 72 ; **lihtlüker** (*compar.*) A. R. 254 ; liȝtloker ROB. 214 ; LANGL. *B* xii. 158 ; A. P. iii. 47.

lihtnesse, sb., *lightness, levity,* HOM. I. 83 ; liȝtnesse WICL. 2 COR. i. 17 ; lightnesse CH. C. T. *A* 3383.

liht-schipe, sb., *levity, lightness,* HOM. I. ˙261.

lihtsum, adj., = *M.H.G.* lihtsam ; *merry, jovial, 'facilis,'* PR. P. 304 ; lightsom R. R. 936.

līht, *see* lēoht.

lihte, sb., *lights, lung,* FRAG. 6 ; þa lihte LAȜ. 6499 ; liȝtes (*pl.*) TRIST. 498 ; D. TROY 10705.

lihten, v., *O.E.* līhtan, = *M.H.G.* lihten, *O.N.* lētta ; *make light ; alleviate ; descend* ; GR. 41 ; JUL. 8 ; a dun heo gunnen lihten LAȜ. 26337 : hwat god þer of muhte lihten A. R. 96 ; ȝe schulen beon idodded four siðen i ðe ȝere vor to lihten ower heaved 422 ; licten REL. I. 234 ; ligten GEN. & EX.

1983 ; to ligten her on er̃e REL. I. 209 ;
ligten him of his birdene 217 ; lihte JOS. 81 ;
lightin '*allevio*' PR. P. 304 ; lighte GOW. II.
371 ; līhteð (*pres.*) KATH. 1791 ; liȝþeþ hem
of children PL. CR. 79 ; līht (*imper.*) MISC.
50 ; SPEC. 58 ; lihte (*pret.*) H. M. 31 ; liȝte
WICL. GEN. xxiv. 64 ; he liȝte a doun LANGL.
B xvii. 64 ; þou lihtest SPEC. 73 ; lichtit
BARB. xiv. 121 ; lihten KATH. 2494 ; heo
lihten of heore steden LAȝ. 25731 ; līht (*pple.*)
SPEC. 37 ; liȝt A. P. i. 987 ; light TOWNL.
107 ; *comp.* a-, ȝe-līhten.

līhten, *see* lēohten.

līhtnen, v.,=līhten ; *lighten, relieve*: heo
was līhtned (*pple.*) of hire evel JOS. 644.

līhtnen, *see* lēohtnen.

līk, *see* līc. **līke,** *see* līche.

[liken, v.,=? *M.L.G.* līken (lycken), *O.H.G.*
līchōn (*polish*) ; *comp.* bi-liken.]

līken, v., *O.E.* līcian,=*O.L.G.* līcōn, *cf. O.N.*
līka, *Goth.* leikan, *O.H.G.* līchēn ; *like, please,*
A. R. 238 ; liken alle men HOM. II. 29 ; tet
(=þe it) maȝ ille liken ORM. 18279 ; liki O. &
N. 342 ; ROB. (W.) 11819 ; līkeþ (*pres.*) TREAT.
140 ; CH. C. T. *A* 777 ; him likeþ MAND.
209 ; me likeþ wel ȝoure wordes LANGL. *A*
i. 41 ; likes PR. C. 7851 ; we him þa (*r.* þæ)
bet likieð LAȝ. 26738 ; þet þe gode likie HOM.
I. 63 ; līkinde (*pple.*) AYENB. 214 ; līkede
(*pret.*) GEN. & EX. 2299 ; C. L. 214 ; hit þe
likede wel LAȝ. 8746 ; *comp.* ȝe-, mis-līken.

　līcunge, sb., *O.E.* līcung ; *liking, pleasure;*
dainty; A. R. 180 ; likinge HOM. I. 271 ;
AYENB. 23 ; LANGL. *B* xi. 20 ; liking P. L. S.
xxv. 140 ; WILL. 452 ; PR. C. 292 ; likinge
TREV. III. 37, VII. 11, 27, 29.

　līkinge-liche, līkinglī, adv., *attractively,*
TREV. III. 465 ; VI. 251 ; VII. 405.

likken, *see* licken.

līknin, v., = *Swed.* līkna, *M.L.G.* līkenen, *O.
H.G.* (gi-)līchinōn ; *liken, make alike, compare,*
'*similo*,' PR. P. 305 ; likni AYENB. 245 ; likne
MAND. 48 ; licken ALEX. (Sk.) 3095 ; līknez
(*pres.*) A. P. i. 499 ; (he) līknede (*pret.*)
MAN. (F.) 3632 ; līkned (*pple.*) CH. C. T. *A*
180 ; þe water is likned to þe worlde LANGL.
B viii. 39 ; ilikned SHOR. 58 ; liknit BARB. i.
396.

lik-pot, sb., *forefinger of the right hand,*
TREV. VII. 73.

lilie, sb., *O.E.* lilie,=*O.H.G.* lilia, *from Lat.*
līlium ; *lily,* SPEC. 33 ; CH. C. T. *A* 1036 ;
þe lilie mid hire faire wlite O. & N. 439 ; leli
MIN. iv. 91 ; lilien (*pl.*) LK. xii. 27.

　lilie-whīt, adj., *white as a lily,* SPEC. 30.

lillen, v., *shine* ; **lilled** (*pret.*) A. P. iii. 447.

lim, sb., *O.E.* lim *n., cf. O.N.* limr *m.; limb,*
'*membrum*,' PR. P. 305 ; TOR. 2498 ; HOCCL.
i. 31 ; PR. C. 1912 ; þet lim A. R. 360 ; lim
& lið JUL. 13 ; lim and liþ M. T. 80 ; lim, lime
ROB. 17, 208 ; lim, lime [leme] LAȝ. 3011,

4227 ; lime TREAT. 137 ; SHOR. 23 ; LANGL.
B xx. 194 ; leme AYENB. 47 ; WILL. 1736 ;
lime (*dat.*) HAV. 1409 ; wiþ lime and liþ(e)
MAP 335 ; lime (*pl.*) LEECHD. III. 120 ; O.
& N. 1098 ; leome, leomen, leomes [lime] LAȝ.
702, 6275, 19501 ; leme HAV. 2555 ; liman,
limes SAX. CHR. 253, 262 ; leoman HOM. I.
109 ; limes A. R. 298 ; ORM. 4005 ; REL. I.
210 : TREAT. 138 ; CH. C. T. *E* 1465 ; lemes
MAND. 159 ; limen [leomen] KATH. 252 ;
lime (*gen. pl.*) MAT. v. 29 ; limen (*dat. pl.*)
A. R. 110 ; mid alle mine lime MISC. 193.

　lim-mēle, adv., *O.E.* limmǣlum ; *limb by*
limb ; LAȝ. 25618 ; limmele [limemeele] TREV.
v. 281 ; lim-[leme-]mele WICL. 2 MACC. i. 16 ;
todrawe limemele BEK. 1813.

līm, sb., *O.E.* līm,=*O.N.* līm *n., M.L.G.,*
O.H.G. līm *m.; lime ; birdlime ; 'gluten,'*
FRAG. 2 ; (*ms.* lyme) '*viscus, calx*' PR. P.
305 ; LAȝ. 15466 ; GEN. & EX. 2552 ; (*ms.*
lym, liim) O. & N. 1056 ; (*ms.* lym) P. L. S.
xii. 54 ; ALIS. 420 ; CH. C. T. *G* 806 ; þet lim
A. R. 228 ; lim & stan ORM. 16285 ; līme
(*dat.*) P. L. S. ii. 91.

　līm-ȝerde, sb., *limed twig,* PR. P. 305 ;
LANGL. *B* ix. 179 ; PL. CR. 564.

　līme-rōd(e), sb.,=*M.L.G.* līmrōde, *M.H.G.*
līmruote ; *limed-twig,* CH. C. T. *B* 3574.

　līm-twig, sb., *lime-twig,* LIDG. M. P. 189.

limaille, lemaille, sb., *O.Fr.* limaille ; *filings,*
CH. C. T. *G* 853, 1162, 1197.

limbo, sb., *limbo,* YORK xxxvii. 102.

[limien, v., *from* lim ; *comp.* bi-, to-limien.]

līmin, v., *O.E.* līman,=*M.H.G.* līmen ; *smear*
with bird-lime ; whitewash with lime, PR. P.
305 ; lime CH. TRO. i. 353 ; limeð (*pres.*)
KATH. 1792 ; līmed (*pple.*) GEN. & EX. 562 ;
ilimed A. R. 226 ; *comp.* whit-līmen.

　līmunge, sb., = *M.H.G.* līmunge ; *gluing*
together, A. R. 138.

limit, sb., *Lat.* līmes, līmitem ; *limit,* ALEX.
(Sk.) 5069.

limiten, v., *set limits to* ; lemete (*pret.*) D.
ARTH. 457 ; lemett (*pple.*) ALEX. (Sk.) 4283.

limitour, sb., *a friar who begs within certain*
limits, PL. CR. 597 ; CH. C. T. *A* 209.

limous, adj., *Lat.* līmōsus ; *muddy,* PR. P.
305 ; PALL. ix. 139.

limp, limpe, sb., *O.E.* (ge-)limp,=*M.L.G.*
(ge-)limp, limpe ; *accident* ; limpes (*pl.*) HOM.
II. 197 ; *comp.* ȝe-, un-limp (-limpe).

　limp-līch, adj., *O.E.* (ge-)limplīc,=*M.L.G.*
limplīk ; *?convenient, suitable* : limpliche mihte
HOM. II. 25.

limpen, v., *O.E.* limpan,=*O.H.G.* limphan,
limfan ; *happen, be becoming* ; limpe JOS. 213 ;
limpeð (*pres.*) KATH. 471 ; to eiþer limpeð
his dole A. R. 10 ; limpes A. P. ii. 174 ;
limpus ANT. ARTH. xlviii ; lomp (*pret.*)
MARH. 4 ; *comp.* a-, bi-, ȝe-limpen.

[lin, sb., = ? *O.H.G.* lun (*bar, peg*).]

lin-pin(n)e, sb., *linchpin,* VOC. 202.

līn, sb., *O.E.* līn, = *O.L.G., O.H.G., O.N.* līn, *Goth.* lein, *from Lat.* līnum, *Gr.* λίνον; *flax,* '*linum,*' REL. I. 7; **līne** (*dat.*) HAV. 539; in lin(e) SPEC. 46; *comp.* **brēost-līn**.

līn-sēd, -seed, sb., *linseed,* VOC. 156, 217.

lināge, sb., *O.Fr.* lināge; *lineage,* MAN. (F.) 944; PARTEN. 5033; lignage ['*tribus*'] TREV. IV. 33, 247; lenage TOR. (A.) 491.

Lincolne, pr. n., *O.E.* Lindcylne; *Lincoln,* LA3. 20614; ROB. 2.

Lincolne-schire, pr. n., *Lincolnshire,* AN. LIT. 4.

lind, *see* li*ð*e.

linde, sb., *O.E.* lind, = *O.N.* lind, *O.H.G.* linda; *lime-tree,* '*tilia,*' PR. P. 305; VOC. 228; B. DISC. 1039; CH. C. T. *A* 2922; SACR. 389; linde (*dat.*) ALIS. 2489; GOW. I. 119; LANGL. *B* i. 154; TRIAM. 426; in ore linde O. & N. 1750.

linde-wodes, sb., pl., *lime tree woods,* GAW. 1178.

linden, adj., *O.E.* linden; *made of lime-tree wood*: linden spon TRIST. 2050.

Lindes-ēie, pr.n., *O.E.* Lindes īg; *Lindsey,* HAV. 734; ROB. 114.

līne, sb., *O.E.* līne, = *O.N., O.H.G.* līna; *line, rope, string,* '*chorda, funiculus,*' PR. P. 305; HORN (L.) 681; AYENB. 150; lines (*pl.*) LANGL. *A* v. 199; *comp.* **bō3-līne**.

līne², līnie, sb., *O.Fr.* ligne; *line,* '*linea,*' PR. P. 305; line P. L. S. xviii. 98; line bi line *from beginning to end* BARB. xvii. 84.

līnen, līnen, adj. & sb., *O.E.* līnen, = *O.Fris.* linnen, *O.H.G.* linīn; *made of linen; linen*; REL. I. 163; linen [line] WICL. EXOD. xxv. 4; linnen ['*linteum*'] TREV. III. 137; in linnen iclo*þ*ed LANGL. *A* i. 3; here smok o line S. S. (Web.) 474; lufsum under line TRIST. 2816; no linene clo*ð* A. R. 418; *þe* linene kertel AYENB. 236; linnene clo*þ* O. & N. 1174; mid linene cla*ð*e JOHN xix. 40; mid linnene cla*ð*e LA3. 22290; lining BARB. xlii. 422.

līnin, v., *cf. Swed.* līna; *line* (*a garment*), '*duplo, duplico,*' PR. P. 306; pers **līned** (*pple.*) with taffata CH. C. T. *A* 440.

lininge, sb., *lining,* '*deplois,*' PR. P. 306.

ling, sb., *O.N.* lyng; *ling, heather,* PR. P. 305; M. T. 189; DEGR. 336.

ling², sb., = line²; in a ling *straight on* BARB. ii. 417; xix. 285; intill a ling BARB. vi. 560; xii. 49.

lingand, pple., *in a line,* BARB. xix. 356*.

linien, *see* *h*leonien.

liniment, sb., *Lat.* linīmentum; *smearing-stuff for casks,* PALL. xi. 440.

*h*linke, sb., *O.E.* hlence, *cf. O.N.* hlekkr *m.* (*chain*); *? link*; linke '*hilla*' (*i.e.* '*link*' *of sausage*) PR. P. 306.

*h*linken, v., *O.E.* (ge-)hlencian (LEECHD. II. 343); *link, make a chain*; linked (*pple.*) LIDG. TH. 1744.

linkwhite, sb., *O.E.* līnetwige; *linnet*; link-whittes (*pl.*) D. ARTH. 2674.

linnen, v., *O.E.* linnan, = *O.N.* linna, *O.H.G.* (bi-)linnan, *Goth.* (af-)linnan; *cease*; linne SPEC. 103; MAN. (F.) 8347; linne*ð* (*pres.*) KATH. 1717*; linnen HOM. I. 67; **linne** (*subj.*) HORN (L.) 992; lan (*pret.*) TRIST. 38; *comp.* **bi-linnen**.

linnunge, sb., *cessation,* HOM. I. 259; MARH. 11; KATH. (E.) 1679.

linnen, *see* līnen.

lins, sb., *O.E.* lynis (*axle tree*), *cf. Du.* luns, *Ger.* lüns (*linchpin*); *linchpin*; linses (*pl.*) SHOR. 109; *cf.* lin.

linse wolse, sb., *linsey woolsey,* B. B. 248.

lint, sb., = *M.Du.* lint (*linen tape*); *lint,* PR. P. 306; BARB. xvii. 612.

lintel, sb., *O.Fr.* lintel; *lintel,* WICL. EXOD. xii. 22.

liōn, *see* lēo. lipard, *see* leopard.

līpe, *see* lēpe.

lipnen, v., *? for* litnen; *trust*: lipne (*imper.*) *þ*ou nouht to borewinge ANGL. IV. 198; lip-nie (*subj.*) HOM. I. 37; ne lipnie na non to muchel to childe ne to wive 161; lipne HOM. II. 220; MISC. 59.

lipning, lippening, sb., *trust,* BARB. xii. 238.

lippe, sb., *O.E.* lippe, *cf. M.Du.* lippe *f., O.Fris.* lippa *m.; lip,* '*labrum,*' PR. P. 306; ALIS. 6429; biting on mi lippe GOW. I. 283; lippen (*pl.*) FRAG. 5; SHOR. 82; AYENB. 210; FER. 141; lippes LA3. 29359; lippen (*dat. pl.*) A. R. 106.

lippē, sb., *? Fr.* lippée; *morsel*: lene folke *þ*at lese wol a lippe at every noble LANGL. *B* v. 250; Juwes han a lippe of owre bileve xv. 493.

lipsen, *see* lispen.

lire, sb., *O.E.* lira; *calf of the leg, thick flesh,* A. R. 130; JUL. 59; H. M. 21; B. DISC. 1325; ISUM. 262; TOWNL. 55; boon ne lire OCTOV. (W.) 1119; his white lire [lere] CH. C. T. *B* 2047; *comp.* **spar-lire**.

lire, *see* lüre.

līre, sb., *cf. O.H.G.* līra, *from Gr.* λύρα; *lyre,* LA3. 7003.

līre, *see* *h*lēor.

liri? sb.: leide here legges a liri [leri] *across* LANGL. *A* vii. 115.

lirken, v., *O.N.* lerka; *? compel*: i lirke (*pres.*) him up with mi hond HALLIW. 523.

lisard, *see* lesarde.

*h*līse, sb., *O.E.* hlīsa, hlīosa; *sound, fame,* MAT. iv. 24.

lisēre, sb., *O.Fr.* lisiere; *edge of cloth, list,*

LANGL. *B* v. 210, *C* vii. 216 [*A* v. 124 liste] ; lisure PR. P. 307.

lisoun, sb., *? O.Fr.* luision (*shining*) ; *? glimpse* ; A. P. ii. 887.

lispen, *see* wlispen.

lisse, sb., *O.E.* liss, liðs ; *from* liðe ; *lenity, tranquillity*, HOM. I. 15 ; P. L. S. viii. 119 ; LAȜ. 3261 ; MISC. 141 ; C. L. 757 ; SPEC. 57 ; WILL. ·2828 ; LANGL. *A* x. 30 ; CH. C. T. *F* 1238.

lissen, v., *O.E.* lissian, liðsian ; *ease, relieve* : ne lissen heore soruwe S. A. L. 152 ; lisse Gow. III. 82 ; R. R. 4128 ; hire care to lisse WILL. 631.

list, *see* lüst.

liste, sb., *O.E.* list *m.*, *cf. O.H.G.* list *m.f.*, *O.L. G.*, *O.N.* list *f.*, *? Goth.* lists (*acc. pl.* listins) *m.* ; *art, cunning*, LAȜ. 17210 ; KATH. 1527 ; O. & N. 757 ; R. S. i ; S. S. (Web.) 2046 ; betere is liste þen luðer(e) strencðe A. R. 268 ; **liste** (*dat.*) SHOR. 24 ; **listes** (*pl.*) MISC. 136 ; tech him of alle þe liste HORN (L.) 235.

   liste-līche, adv., *cunningly*, REL. I. 188 ; listli WILL. 2742.

liste, *see* hlüst.

līste, *O.E.* līst, = *M.L.G.* līste, *O.N.*, *O.H.G.* līsta ; *border of cloth ; edge of anything* ; LANGL. *B* v. 524 ; he smot me . . . on þe list(e) (*edge of the ear*, = *? Ger.* ohrleiste, *or* = hlüst) CH. C. T. *D* 634 ; liiste PR. P. 307 ; and bond him wiþ a liste LEG. 91 ; þe list of þe lifte A. P. ii. 1761.

listen, *see* lüsten, hlüsten.

listes, sb., pl., *? O.Fr.* lice ; *lists (for a tournament)*, CH. C. T. *A* 1862.

listre, sb., *O.Fr.* listre ; *reader* ; listere '*lector*' PR. P. 307 ; lister TREV. VI. 257 ; listres (*pl.*) LANGL. *B* v. 138 ; listris WICL. E. W. 298.

lit, sb., *O.N.* litr ; *colour*, GEN. & EX. 1068 ; SPEC. 36 ; **littis** (*pl.*) ALEX. (Sk.) 4336.

lit, *see* lüt.

litarge, sb., *Fr.* litharge ; *white lead*, CH. C. T. *A* 629, *G* 775.

līte, sb., *O.N.* lýti (*vice*) ; *flaw, vice* : wiþ-outen lite L. H. R. 112 ; TOWNL. 71 ; þus he haves hir led with lite IW. 1620.

litel, *see* lütel.

litin, v., *O.N.* lita ; *dye*, '*tingo*,' PR. P. 308 ; liteð (*pres.*) A. R. 268 ; **lited** (*pple.*) PS. lxvii. 24 ; ilitet KATH. 1432 ; littid [*intingvatur*'] in blode HAMP. PS. lxvii. 25.

   litster, sb., *dyer*, '*tinctor*,' VOC. 212 ; listare PR. P. 307 ; litestere CH. F. ĀGE 17.

litterate, pple., *Lat.* litterātus ; *well-instructed*, TREV. IV. 81, 83.

litere, sb., *O.Fr.* litiere ; *litter, bed*, PR. P. 307 ; TREV. IV. 461 ; C. M. 13817 ; liter MAN. (F.) 9604 ; letere ALEX. (Sk.) 4910 ; littar BARB. ix. 106 ; **literes** (*pl.*) WICL. IS. lxvi. 20.

litnen, v., *? diminish* : hwan Havelok sau . . . his ferd so swiþe littene HAV. 2701.

litnen[2], v., *cf. Swed.* līta ; *? trust* : þa þat litnen (*ms.* littnenn) to þin fode ORM. 6115 ; ne litnie (*? ms.* litmie) na mon to swiðe to þisse live HOM. I. 7 ; *see* lipnen.

litster, *see under* litin.

lið, sb., *O.E.* lið, *cf. O.Fris.* lith *n.*, *? cf. O.L. G.* lith, lið *m.*, *O.H.G.* lid *n. m.*, *O.N.* liðr, *Goth.* liþus *m. ; limb, joint*, JUL. 13 ; H. M. 21 ; GEN. & EX. 1804 ; REL. I. 223 ; liþ LEG. 204 ; WILL. 1724 ; AR. & MER. 8504 ; GOW. I. 99 ; CH. C. T. (T.) 14881 ; TOR. 2498 ; TRIAM. 467 ; lith MIRC 1365 ; PR. C. 1917 ; brekeþ liþ from liþe FRAG. 7 ; **liðes** (*pl.*) A. R. 262 ; lithes HAV. 2163.

leoðe-bēie, adj., *supple-jointed*, MARH. 16 ; leþebei(e) REL. I. 188.

leoþe-wōk, adj., *O.E.* liðu-, leoðuwāc, = *O. H.G.* lidoweich ; *weak in limbs*, REL. II. 229.

liþe-wurt, sb., *O.E.* liðwyrt ; *some plant, ? dwarf elder*, '*ive*,' REL. I. 37 ; liþewort '*ebulus*' LEECHD. II. 398.

lið, sb., *O.N.* lið (*order, assembly, help*) ; *? help, ? company* : nes þer nan oðer lið LAȜ. 5213 ; we wullet gan a leoðe (*? in a body*) 5307 ; þre leodes in oon liþ [lith] LANGL. *B* xvi. 181.

hlið, sb., *O.E.* hlið *n.*, *cf. O.N.* hlið *f.*, = *O.H.G.* lita ; *slope* : under hevene liðe(?) HOM. II. 117 ; ȝeond wudes & ȝeond liðen (*plur.*) LAȜ. 32219.

līth, *see* lēod.

liðe, adj., *O.E.* liðe, = *O.L.G.* lithi, *O.H.G.* lindi, *cf. O.N.* linr ; *gentle, smooth*, A. R. 428 ; ðet weter . . . wes liðe HOM. I. 129 ; liðe him beo drihten LAȜ. 4 ; þat weder wes swiðe liðe 7242 ; liþe ORM. 1307 ; lind B. B. 21 ; S. S. (Web.) 2517 ; CH. H. F. 118 ; RICH. 4859 ; EM. 348 ; PS. cvi. 29 ; ISUM. 494 ; liþe, lithe '*lenis, mollis, tranquillus*' PR. P. 310 ; (þe) efter(e) lið(e) (*sc.* moneð) *July* MARH. 23 ; līþe (*adv.*) TRIAM. 417 ; *comp.* ȝe-liðe ; *deriv.* lisse.

   liðe-liche, adv., *O.E.* liðelīce ; *gently*, HOM. I. 259 ; A. R. 428.

   liðnesse, sb., *gentleness*, HOM. I. 95.

liþe, līthe, sb., *? = O.H.G.* linde ; *mitigation*, '*malacia*,' PR. P. 310 : he ne mouchte no liþe (*ms.* lyþe) gete HAV. 147.

lið-ful, adj., *gentle* : mid liðfulle (*pl.*) worden LAȜ. 1262.

liðen, liðan, v., *O.E.* līðan, = *O.L.G.* līthan, *O.N.* līða (*go, travel*), *Goth.* (af-, ga-, us-)-leiþan (ἔρχεσθαι), *O.H.G.* līdan (*carry*) ; *go, travel*, LAȜ. 27, 1202 ; he ne durste noht . . . intil þat ende liþen ORM. 8374 ; lað (*? pret.*) LAȜ. 4880 ; over sæ þu liðe 5045 ; þer liðen to somne alle Scotleode 20046 ; liðen (*pple.*) LAȜ. 5783 ; *comp.* a-, ȝe-liðen ; *deriv.* lāde, læden.

hlīðen, v., *O.N.* hlýða; *listen*; liþe ALIS. 5721;
A. D. 238; GREG. 375; MAN. (H.) 93; TOR.
337; OCTAV. (H.) 9; M. ARTH. 676; AMAD.
(R.) xxiii; AUD. 66; liþe LANGL. *B* viii.
66; līþe (*imper.*) HORN. (L.) 336; PS. xxx.
3; liðeð GEN. & EX. 2077; liþes HAV. 1400.

liþer, see **lüðer.**

liþere, sb., *O.E.* liðre, lyðre; *sling*; lü-
þeren (*dat. pl.*) ROB. 394; *comp.* staf-liþere.

liðerien, v., *from* liþere; *hurl with a sling,
throw stones*: þese lourdeines lithereþ [li-
theren] (*pres.*) þer to þat alle þe leves fallen
LANGL. *C* xix. 48; liðere (*imper.*) A. R. 290;
iliþered (*pple.*) ROB. 549.

liðerien, see **leðerien.**

[**liðien,** v., = *M.H.G.* liden; *comp.* to-liðien.]

liðien, v., *O.E.* liðian, *cf. O.H.G.* lindian, *O.N.*
lina; *from* liðe; *mitigate*; liþe ALIS. 433;
CH. TRO. iv. 754; H. S. 2280; A. P. i. 357;
þou schalt . . . mi longe sorwe liþe HORN (H.)
428 (R. 412); liþeþ (*pres.*) our pine SHOR. 19;
liðegede (*ms.* liðegedde) (*pret.*) HOM. I. 95;
comp. ʒe-iliðeʒen.

liūn, see **lēo.**

[**live,** sb., *O.E.* (and-, big-)leofa; *means of
living*; *comp.* bī-live.]

līve, see **lēave.** **livel,** see **level.**

[**līven,** v., *O.E.* līfan, *O.Fris.* (bi-)līva, *O.H.G.*
(bi-)līban; *comp.* bi-līven; *deriv.* līf, livien,
lāve, lǣven.]

līven, see **lēven.**

livenað, sb., *O.N.* lifnaðr; *food, means of
living*, HOM. I. 63: liveneð A. R. 356; H. M.
29; lifnoþ AYENB. 138.

livere, see **livre.**

liveree, liverē, sb., *O.Fr.* livree; *livery*, CH.
C. T. *A* 363; MAN. (F.) 10557; livere (*allowance of food*), MAN. (F.) 3117; liverie BARB.
xix. 36; lufre xiv. 233; levere D. ARTH. 241;
in ech levere (*division of an army*) D. ARTH.
3078.

liveren, v., = *M.L.G.* leveren, *M.H.G.* liberen;
coagulate: livred (*pple.*) blod MISC. 148; þe
liverede see ROB. 39.

liveren², v., *Fr.* livrer; *deliver, liberate*:
livere JOS. 707; C. M. 5943; livers (*pres.*)
ALEX. (Sk.) 3152.

livien, v., *O.E.* lifian, leofian, liofian, lyfian,
libban (*pret.* lifode, lifde), *cf. O.L.G.* libbien,
libbean, *O.Fris.* libba (*pret.* livade, lifde),
O.N. lifa (*pret.* lifða), *Goth.* liban (*pret.* li-
baida), *O.H.G.* lebēn (*pret.* lebēta); *live*;
liyien, leovien, luvien, libben LAʒ. 6902,
6904, 18049, 27779; liven SAX. CHR. 263;
GEN. & EX. 308; HAV. 355; MAND. 24;
binimeð his underlinge þat he sholde bi
liven HOM. II. 179; live CH. C. T. *A* 506;
shalt þou never live (*cf. Ger.* erleben) þat dai
SPEC. 90; life PR. C. 1869; live [libbe] LANGL.
A iii. 220; leven E. G. 75; leve S. S. (Wr.)

534; S. & C. I. iii; leve, libbe SHOR. 2, 70; libben KATH. 1901; A. R. 38; ORM. 372; libben, libban HOM. I. 7, 109; libbe O. & N. 1192;
HORN (L.) 63; ROB. 178; AYENB. 32; H. H.
247; ALIS. 7965; S. S. (Web.) 287; to lib-
benne LAʒ. 11775; **livie** (*pres.*) HOM. I. 211;
life AMAD. (R.) liii; libbe SPEC. 27; lifeð
MISC. 192; lifeþ ORM. 2698; liveð MARH. 4;
A. R. 352; leofeð LK. iv. 4; leveþ AYENB. 54;
leviþ EGL. 120; (heo) luvieð, libbeð, libbað
HOM. I. 53, 115, 119; we livieð (?*printed* luueð)
MARH. 17; (heo) libbeð 10; heo livieð A. R.
310; we libbeð 360; libbeþ O. & N. 1012;
AYENB. 53; livieð, luvieð (*imper.*) JUL. 74,
75; liviende (*pple.*) LAʒ. 27213; KATH.
1227; liveand PS. xxxviii. 6; levand TOR.
315; libbinde A. R. 350; þe libbinde AYENB.
13; libbendre [*'viventium'*] MAT. xxii. 32;
livede (*pret.*) HOM. I. 7; lefede 225; livede,
luvede, leovede, leofede, levede LAʒ. 252, 299,
2680, 3236, 15074; livede, leovede S. A. L.
160; levede AYENB. 71; þu livedest HOM. l.
33; liveden LANGL. *A* PROL. 26; leveden
HOM. I. 129; *comp.* ʒe-livien.

livre, sb., *O.E.* lifer, = *O.N.* lifr, *O.Fris.* livere,
O.H.G. libara, lebara; *liver*, R. S. v; TREAT.
139; ALIS. 2156; FER. 1095; þa livere [livre]
LAʒ. 6499; lifre FRAG. 6; livere CH. C. T. *D*
1839; liver GOW. III. 100; levir LUD. COV.
181.

liver-wort, sb., *liver-wort*, LEECHD. III.
336; livirwort PR. P. 309.

livreisūn, sb., *O.Fr.* livraison; ?*award*: in þe
day of livreisun (*judgment-day*) HOM. I. 85;
lieversōns (*pl.*) *allowances* MAN. (F.) 7432.

lō, see **lā.**

lobbe-kēling, sb., *large codfish*, M. H. 136.

lober, sb., *lubber*; **lobres** (*pl.*) (*v. r.* loburs)
LANGL. *A* PROL. 52.

lobi, sb., *lubber*, DEP. R. ii. 170; **lobies** (*pl.*)
LANGL. *A* PROL. 52 *.

loc, sb., *O.E.* loc, = *O.Fris.*, *O.N.* lok, *O.H.G.*
(bi-)loh; *from* lūken; *lock, 'clausura,'* FRAG.
3; AYENB. 255; lok '*sera*' PR. P. 311; LANGL.
A i. 178; MAND. 179; **loke** (*dat.*) O. & N.
1557; **lokes** (*pl.*) AYENB. 151; *comp.* feter-,
war-lok.

lok-smith, sb., *locksmith*, PR. P. 311.

lok-sonn(e)dai, sb., ? *day of Pentecost*,
SHOR. 127.

loc², sb., *O.E.* locc, loc, = *O.N.* lokkr, *O.H.G.*
locch, lock; *lock (of hair)*: ænne loc MAT.
v. 36; lok '*cincinnus, floccus,'* PR. P. 311;
lockes (*pl.*) MAT. x. 30; LAʒ. 21875; SPEC.
34; lockes [lokkes] CH. C. T. *A* 81; lokkes
ALIS. 4164; **lockan,** locken (*dat. pl.*) LK. vii.
44; JOHN xi. 2

lōc, see **lāc.**

lóche, sb., *O.Fr.* loche; *loach, 'silurus,'* VOC.
189; PR. P. 310; REL. I. 85; **lóches** (*pl.*)
L. C. C. 54.

locuste, sb., *Lat.* locusta ; *locust,* C. M. 6041.

lōd, lōde, *see* **lād, lāde.**

Lodelowe, pr. n., *Ludlow,* ROB. 448.

lof, sb., *O.E.* lof, = *O.L.G., O.N.* lof, *O.H.G.* lob ; *praise,* LAȜ. 8376 ; HOM. II. 91 ; PS. viii. 3 ; drihtin to lofe & wurþe ORM. 1141 ; þe sullere lat sum del of his lofe HOM. II. 213 ; ben at love (*printed* lone) & at bode PL. CR. 716.

   lof-ȝeorn, adj., *O.E.* lofgeorn ; *desirous of praise,* HOM. I. 103.

   lof-līk, adj., *fit to be praised,* [' *laudabilis* '] PS. xcv. 4.

   lof-sang, sb., *O.E.* lofsang, = *O.N.* lofsöngr. *O.H.G.* lob-, lobesang ; *song of praise,* HICKES I. 167 ; (*ms.* loffsang) ORM. 18025 ; lofsong HOM. I. 5, 191 ; lofsong (*v. r.* loft song) C. L. 29 ; loftsong [lovesang] LAȜ. 68 ; loftsong HOM. I. 261 ; (*printed* lostsong) MARH. 19.

   lof-songere, sb., *singer of praise,* HOM. I. 153.

   lof-word, sb., = *M.H.G.* lopwort (*ms.* loveword) ; *word of praise,* C. M. 10614.

lōf, sb., = *Swed.* lōf *m.,* *Du.* loef *f.;* *luff* (*of a ship*) : heo scuven ut heore lof LAȜ. 7859 ; he went þene lof A. R. 104; loof MAN. (F.) 12088 ; her loof and her windas RICH. 71 ; heo wenden heore lofes [loves] & liȝen toward londe LAȜ. 20949 ; (hii) twinde hom to þe lofe ROB. (W.) 9026.

lōf, *see* **lāf.** **lofen,** *see* **loven.**

loft, sb., *O.N.* lopt *n.* (*air, upper region, upper chamber*) ; *cf.* lüft ; *height ; upper room, loft* : ' *solarium* ' PR. P. 311 ; and bereþ hire up into þe lofte S. A. L. 160 ; mi doȝter þat sitteþ on þe lofte HORN (L.) 904 ; in your lofte GAW. 1096 ; o loft *aloft, on high* ORM. 11823 ; on lofte LANGL. *A* i. 88 ; on, a lofte CH. C. T. *B* 277 ; bi loft *on high B* xviii. 45 ; bi lofte K. T. 686.

lòftsong, *see* **lofsang** *under* **lof.**

loȝ, sb., *? O.E.* luh, *? Gael.* loch ; *lake,* A. P ii. 441 ; louh, lough MAN. (F.) 1423, 10211 ; louch BARB. iii. 430 ; **loghe** (*dat.*) ANT. ARTH. vii.

lōȝ, looȝ, sb., *O.E.* lōh, = *O.Fris.* lōch (*dat.* lōge), *M.L.G.* lōg (*place*), *? O.H.G.* luog (*cave, den*) ; *place,* SHOR. 126, 145, 157.

lōȝ, *see* **lāh.**

loȝe, adj., *O.N.* logi, = *O.Fris.* loga, *M.H.G.* lohe, *NorthEng.* lowe ; *fire :* loȝhe ORM. 16185 ; lohe A. R. 356* ; lowe GEN. & EX. 643 ; lowe PR. P. 314 ; JOS. 687 ; Iw. 343 ; MAN. (F.) 8341 ; H. S. 484 ; PR. C. 9431.

[**loȝe, lowe,** sb., *liar ; comp.* **war-, wedlowe.**]

loȝen, v., *O.N.* loga, = *M.H.G.* lohen ; *from* loȝe ; *flame ;* lowin '*flammo*' PR. P. 315 ; **lowande** (*pple.*) GAW. 236 ; lowinge DEGR. 1436 ; **lawid** (*pret.*) ALEX. (Sk.) 226.

lōȝen, *see* **lāȝen.**

logge, sb., *O.Fr.* loge ; *lodge, tent, hut,* SAINTS (Ld.) xlv. 279; S. S. (Web.) 2603 ; B. DISC. 550 ; A. P. iii. 457 ; CH. C. T. *B* 4043 ; loge, luge BARB. xix. 660* ; **luggis,** logis (*pl.*) xix. 392, vii. 550.

loggen, v., *O.Fr.* logier ; *lodge,* MAND. 193 ; **loggede** (*pret.*) A. R. 264; þai logit þaim BARB. ii. 304; **loged** (*pple.*) WILL. 1918 ; logged CH. C. T. *B* 4186.

   logging, sb., *lodging,* MAN. (F.) 9201 ; loging BARB. ii. 282 ; luging BARB. iv. 494.

loȝhe, *see* **loȝ.**

lōȝien, v., *O.E.* lōgian, *O.Fris.* lōgia ; *from* lōȝ ; *place ;* **lōgeden** (*pret.*) MK. i. 19 ; *comp.* ȝe-lōȝien.

loght, v., *? =* lutien ; *? lurk, lie in ambush* : i bide for him ȝou loghte (*subj.*) YORK xix. 181.

lōh, *see* **lāh.** **loht,** *see* **lēoht.**

loine, sb., *O.Fr.* logne ; *loin,* ' *lumbus,*' PR. P. 312 ; P. S. 191.

loitren, v., *cf. M.Du.* leuteren, loteren ; *loiter ;* loitron PR. P. 311 ; **loitrande** (*? ms.* loltrande) (*pple.*) A. P. iii. 458.

lok, *see* **loc.**

lōk, sb., *look,* ' *visus,*' PR. P. 311 ; REL. II. 14 ; lok [look] CH. C. T. *A* 3342.

loke, sb., *O.E.* loca (*enclosed place*), *M.Du.* loke (*fence*) ; *enclosure, locked place,* SHOR. 79 ; loke ' *valva* ' PR. P. 311 ; þer weoren in ane loken fif hundred gaten LAȜ. 21309 ; ich am within thi loke GUY 17 ; **loken** (*pl.*) SHOR. 79 ; locun LAȜ. 5926 ; *comp.* **clüster-loke.**

loken, v., *O.N.* loka, = *M.Du.* loken ; *lock,* GEN. & EX. 3193 ; loke ALIS. 3936 ; CH. C. T. *D* 317 ; LUD. COV. 341 ; **lokede** (*pret.*) SHOR. 46 ; lokid S. S. (Wr.) 1222 ; M. ARTH. 2620 ; **ilokked** (*pple.*) MAND. 265.

loker, sb., = *M.Du.* loker ; *locker,* PR. P. 311.

loket, sb., *O.Fr.* loquet, loquette ; *? curl,* P. S. 154.

lōkien, v., *O.E.* lōcian, = *O.L.G.* lōcōn, *M.Du.* lōken ; *look, keep, observe,* GR. 37 ; LAȜ. 24294 ; þe nefre nalde Cristes laȝen lokien HOM. I. 43 ; lokin MARH. 4 ; loken KATH. 1596 ; C. L. 494 ; to loken what teȝ læren us ORM. 1815 ; him birþ loken him þat he ne gilte noht 7863 ; vor te loken riht bitweonen ou A. R. 286 ; lokie BRD. 9 ; loki SHOR. 55 ; AYENB. 5 ; ich can loki manne wike O. & N. 604 ; loke HAV. 376 ; ROB. 488 ; ALIS. 1823 ; TOR. 1492 ; luke Iw. 849 ; PR. C. 205 ; to lokienne FRAG. 8 ; (heo) **lōkieð** (*pres.*) LAȜ. 23077 ; **lōke** (*imper.*) LAȜ. 11770 ; HORN (L.) 748 ; **lōkiende** (*pple.*) MAT. xiii. 13 ; lokinde JOS. 278 ; **lōkede** (*pret.*) MK. vi. 41 ; LAȜ. 2266 ; KATH. 791 ; lokede, lokéd LANGL. *B* xiv. 47 (Wr. 9004) ; lukit BARB.

iv. 321 ; lōked (*pple.*) GEN. & EX. 193 ; god it hafde loked (*awarded, ordained*) swa ORM. 439 ; *comp.* bi-, ȝe-, over-lōken.

lōkere, sb., *looker*, AYENB. 21.

lōkunge, sb., *looking*, A. R. 102 ; lokinge HOM. I. 145 ; H. M. 31 ; AYENB. 19 ; loking HORN (L.) 342 ; A. P. i. 1048.

lokere, *see* wei-lokker.

lokered, pple., *curled*, D. ARTH. 779.

lollen, v., *M.Du.* lollen ; *loll* : loseliche he lolleþ (*pres.*) þere LANGL. B xii. 213 ; his chekes & his chin . . . lollede (*pret.*) PL. CR. 224 ; lik a leþerne pors lullede [lolled] his chekes LANGL. A v. 110.

lollere, sb., *loller* (*lollard*), CH. C. T. B 1177 ; loller AUD. 37 ; lolleres (*pl.*) LANGL. B xv. 207.

loltrande, *see* loitren. lomb, *see* lamb.

lombard, sb., *a dish in cookery*, B. B. 164, 271.

lóme, *see* láme.

lōme, sb., *O.E.* (ge-)lōma, *mod.Eng.* loom ; *tool, implement, vessel*, PS. ii. 9 ; GAW. 2309 ; A. P. ii. 314 ; PERC. 2032 ; REL. I. 54 ; loome ' *utensile, instrumentum*,' PR. P. 312 ; lōmen (*pl.*) A. R. 384 ; SPEC. 41 ; lomes PALL. xi. 447.

lōme², adv., =ȝelōme ; *frequently*, P. L. S. viii. 6 ; ORM. 2178 ; P. S. 197 ; SHOR. 71 ; OCTOV. (W.) 1944 ; ofte and lome S. S. (Wr.) 1892 ; ED. 98 ; lōmere [lomer] (*compar.*) LANGL. B xx. 237.

lomente, sb., *mash*, [' *lomentum* '] PALL. xi. 366.

lomerande, pple., *mod.Eng.* lumbering ; *cf.* lawmere ; *hesitating, creeping*, A. P. ii. 1094.

lompe, *see* lampe *and* lumpe.

lonc, *see* ħlanc. lond, *see* land.

Londreis, *see* Lundreis.

lóne, *see* láne. lōne, *see* lǣne.

long, longien, *see* lang, langien.

longe, *see* lunge. lonke, *see* ħlanke.

loof, *see* ħlāf. loove, *see* lāve.

loppe, sb., *O.E.* lopp, *cf. Dan.* loppe, *Swed.* loppa ; *flea*, AN. LIT. 84 ; loppis (*pl.*) TOWNL. 62.

loppestre, sb., *O.E.* loppestre ; *lobster*, FRAG. 3 ; loppister VOC. 176 ; lopstere 189 ; lopster REL. I. 81.

lopren, v., *curdle, coagulate* ; lopred (*pple.*) PS. cxviii. 70 ; lopird PR. C. 459 ; HAMP. PS. lxvii. 16.

lopiring, sb., *curdling*, HAMP. PS. cxviii. 70*.

lor, sb., *O.E.* lor, = *O.L.G., O.H.G.* (far-)lor ; *from* lēosen ; *loss* ; lore MAN. (F.) 6731 ; GENER. 5457.

lōrd, *see* ħlāford.

lores, sb., pl., *laurel trees*, ALEX. (Sk.) 4972 ; *see* laurēr.

lordein, lurdein, sb., *O.Fr.* lourdein (*lourdaud*) ; *lazy person*, LANGL. C vi. 163 ; lurdein REL. I. 291 ; lordan, lurdan, lurdaine YORK xi. 226, i. 108, xliv. 77.

lōre, *see* lāre.

lorein, sb., *O.Fr.* lorein (*rêne, bride*) ; *bridlerein* ; loreins (*pl.*) BEK. 190.

lorel, sb., *worthless person*, ' *lurco*,' PR. P. 313 ; LANGL. B vii. 136 ; CH. C. T. D 273 ; *cf.* losel.

lorel, lorēr, *see* laurēr.

lorimer, sb., *O.Fr.* lorimier ; *maker of straps, saddler* ; lorimers (*pl.*) A. R. 184*.

lorken, *see* lurken.

los, sb., *O.E., O.N.* los ; *from* lēosen ; *loss*, ' *perditio*,' PR. P. 313 ; MIRC 1279 ; CH. C. T. B 27 ; A. P. ii. 1589 ; D. TROY 5588 ; ga to lose LAȝ. 22844 ; loos LANGL. C xxii. 292.

lōs, lous, adj., *O.N.* lauss ; *from* lēosen ; *loose* : los in þe haft P. S. 339 ; whan þe hors was los [lous] CH. C. T. A 4064 ; loos ' *solutus* ' PR. P. 310 ; his lose tunge GOW. I. 131 ; þe louse ston A. R. 228* ; loise ALEX. 2505*.

     lōse-liche, adv., *loosely*, LANGL. B xv. 213.

lōs², sb., ? *O.Fr.* lōs ; *praise*, ROB. 181 ; AYENB. 26 ; WILL. 1386 ; loos SHOR. 90 ; D. ARTH. 254 ; lose TREV. VIII. 25.

losel, sb., *worthless man* ; loselle TREV. VIII. 41 ; losels (*pl.*) LANGL. A PROL. 74 ; losellis YORK xi. 78 ; *cf.* lorel.

losen, *see* lēosen.

lōsen, v., *praise* ; lōsed (*pple.*) ALEX. (Sk.) 5316 ; losit 2505.

losengere, sb., *O.Fr.* losengeur ; *flatterer*, ALEX. (Sk.) 1923 ; losangere 3543 ; losengeres (*pl.*) WILL. 5481 ; lozengeours MAN. (F.) 2825.

losengerie, sb., *flattery*, LANGL. B vi. 145 ; losengrie A xi. 36.

lósien, v., *O.E.* losian (*loose, perish*), = *M.Du.* losen, *O.N.* losa (*loose*) ; *also* lōsen, lousen, *from* lōs, lous ; *loose, set free ; lose* ; losien LAȝ. 20538 ; losen (*v. r.* lesen) H. H. 36 ; losin ' *solvo* ' PR. P. 313 ; lose A. P. iii. 198 ; louse WICL. GEN. xxvii. 40 ; PERC. 1871 ; OCTAV. (H.) 1634 ; louse and binde AUD. 45 ; louses (*pres.*) PR. C. 1792 ; losen GOW. II. 150 ; þi les þe ðu steorles losie (*be lost*) HOM. I. 117 ; lósede (*pret.*) KATH. 1120 ; þat an þe him losede HOM. I. 245 ; al his folc he losede þer LAȝ. 2201 ; loused FLOR. 1545 ; loosed (*pple.*) R. R. 4511 ; loused [' *dissolutus* '] TREV. VII. 149 ; lost GOW. I. 153 ; MAND. 273 ; CH. C. T. B 1375 ; *comp.* ȝe-lósien.

lósnen, v., *O.N.* losna ; *become loose* : þe lord lósneþ (*v. r.* loosiþ) (*pres.*) þe givede WICL. PS. cxlv. 7.

lossom, *see* lufsum *under* luve.

lost, sb., = *O.L.G., O.H.G.* (far-)lust, *Goth.* (fra-)lusts ; *loss, perdition*, TREV. IV. 213 ;

LANGL. *C* vii. 275; AUD. 13; goþ to lost Gow. I. 147.

lost, *see* **lust.**

lot, sb., *? O.N.* lot; *? from* lūten; *? bow, inclination of the head*; lote (*dat.*) A. P. i. 238; TOWNL. 109.

*h*lot, sb., *O.E.* hlot, *cf. O.Fris.* hlot, *M.L.G., M.Du.* lot *n., O.L.G.* hlot (? hlōt), *O.N.* hlutr, *O.H.G.* hluz *m.; from h*lēoten; *lot,* MAT. xxvii. 35; lot A. R. 358; MISC. 50; RICH. 4262; (*ms.* lott) ORM. 133; hwa se ever wule habbe lot wiþ þe of þi blisse HOM. I. 187; þat lot AL. (L.¹) 607; hlote (*pl.*) MK. xv. 24; hloten LK. xxiii. 34; loten [lotes] LAȝ. 13858; lotes MAN. (F.) 7343; lottis WICL. PROV. xvi. 33.

lote, sb., *? O.Fris.* (hūs-)lota; *? tribute*: in Englond he arerede a lok (*v. r.* lote HALLIW. 530) of uche hous that come smok CHR. E. 445.

lōte, sb., *cf. Sw.* låt (*sound*); *see* **læte**; *noise,* A. P. 876; (*word*) ii. 668.

lōte, *see* **læte.** **lotien,** *see* **lutien.**

lotebī, sb., *paramour,* S. S. (Web.) 1443; R. R. 6339; AUD. 5; lutbi HAMP. PS. lxxv. 10*; lotebies (*pl.*) LANGL. *A* iii. 146.

lōð, lōðien, *see* **lāð, lāðien.**

lotterel, sb., *a term of opprobrium,* YORK xxxii. 259, 382.

louch, *see* **loȝ.**

louȝ, lough, *see h*lahhen.   **lōuh,** *see* **lāh.**

louken, *see* **lūken.**

lounen, v., *cf.* **lūne**; *shelter*; lounit (*pple.*) BARB. xv. 276.

loupe, sb., *loop, noose*; loupis (*pl.*) D. TROY 2806.

loupe², sb., *mod.Eng.* loop(-hole); *window,* (*dat.*) LANGL. *C* xxi. 288; GAW. 792; loupes (*pl.*) PARTEN. 1175.

loup, *see h*lēap.   **louren,** *see* **lūren.**

lous, *see* **lós, lūs.**

lousen, *see* **lósien** *and* **lūsin.**

louten, *see* **lūten.**   **love,** *see* **luve.**

lōve, sb., *O.N.* lófi, = *Goth.* lōfa; *hand, palm*; (*ms.* loue, *printed* lone) '*vola*' VOC. 207; loove ALEX. 2569; lufe TOWNL. 32; loves (*pl.*) ii. 987.

lōve, *see* **lāve.** **loveache,** *see* **luvestiche.**

lóven, v., *O.E.* lofian, = *O.N.* lofa, *M.Du., M. L.G.* loven, *O.H.G.* lobōn; *from* lof; *praise; point out*: to lofen him & wurþen ORM. 208; love A. P. i. 285; to love god WICL. E. W. 320; lóve (*pres.*) D. ARTH. 369; loveð A. R. 408; þe sullere loveð his þing dere HOM. II. 213; lóved (*pret.*) PR. C. 321; lovid OCTAV. (H.) 727; loovid FLOR. 714: lóved (*pple.*) PS. ix. 24.

lóving, sb., *O.E.* lofung; *laudation,* '*indicatio,*' L. H. R. 75; PR. C. 321; loving &

los D. TROY 4878; bi lovinge (*ms.* louyinge, *printed* lonynge) & bedinge WICL. E. W. 167.

lover, sb., *O.Fr.* lover; *skylight; smoke-hole in a roof*; '*lodium,*' PR. P. 315; LANGL. *C* xxi. 288; lovir VOC. 236; lovers (*pl.*) PARTEN. 1175.

lōverd, *see h*lāford. **lovien,** *see* **luvien.**

lōw, *see* **lāh.**

lowable, adj., *O.Fr.* louable; *praiseworthy,* LANGL. *C* vi. 103, xviii. 130.

lowe, *see* **loȝe.**   lōwe, *see* **lāh, h**lāwe.

lowen, *see* **loȝen.**   lōwen, *see* **lāȝen.**

*h*lōwen, v., *O.E.* hlōwan (*pret.* hlēow), *cf. O.H. G.* hlōhen; *low (as cattle)*; loowen WICL. JOB vii. 5; lowin PR. P. 315; lōweþ [luoweþ] (*pres.*) P. S. 332; lhouþ A. S. 4.

lōwing, sb., *lowing,* KATH. 144.

lowere, sb., *O.Fr.* louier; *reward,* ALEX. (Sk.) 5368.

lubard, *see* **leopard.**

lucchen, v., *? pitch, throw*; luche (*pres.*) A. P. iii. 230.

luce, sb., *O.Fr.* luz; *pike (the fish),* '*lucius,*' PR. P. 316; CH. C. T. *A* 350.

lucre, sb., *Fr.* lucre; *lucre,* CH. C. T. *G* 1402.

lūd, *see* **lēod.**

*h*lūd, sb., *voice*; on hire lud to sing SPEC. 27.

*h*lūd², adj., *O.E.* hlūd, = *O.H.G.* hlūd, *O.H.G.* lūt, *cf. Gr.* κλυτός; *loud*; lud A. R. 210; O. & N. 6; crie lowde CH. TRO. i. 400; lauhinge a loude (*aloud*) LANGL. *C* vii. 23; loude & stille *in all conditions* ROB. (W.) XX. 352; BARB. iii. 745; lhūde (*adv.*) A. S. 3; lahȝen lhude ORM. 8142; lude HOM. I. 43; GEN. & EX. 3585; HORN (L.) 209; heo song so lude O. & N. 141; lude [loude] clepian LAȝ. 852; lhoude AYENB. 212; loude TREAT. 136; ALIS. 283; CH. C. T. *A* 772; **hlūdere** (*dat. f.*) LK. i. 42; KATH. 208; ludere ludere stefne LAȝ. 929; mid ludere stefne 1429; **lŭddre,** luddure (*compar.*) A. R. 266, 290; ludder, luddere TREV. VII. 531, 535.

*h*lūde, sb., *cf. M.L.G.* lūd *m. n.,* lūde *f., M.H.G.* lūde *m.,* lūte *f., O.H.G.* lūta, *f.; sound, noise*: mid muchelen lude [mochelere loude] LAȝ. 2591; lūden (*pl.*) REL. I. 188; þa luden heo iherden LAȝ. 27034.

luddok, sb., *buttock,* '*lumbus,*' PR. P. 316; L. C. C. 43; luddockes (*pl.*) '*nates,*' VOC. 186; luddokkis TOWNL. 313.

*h*lūde, sb., *O.E.* hlȳda; *March*; lude ROB. 571; as wederis don in lide P. 35.

[*h*lūden, v., *O.E.* hlȳdan, = *O.H.G.* hlūtan; *sound, cry out.*]

lōdinge, sb., *clamour,* KATH. 145; luding [loudinge] LAȝ. 24873.

lüdene, *see* **leden.**   **lue-,** *see* **lēo-.**

lufe, *see* **luve.**

[**lüffeten**, v., *O.E.* lyffettan ; *flatter.*]

lüffetung, sb., *O.E.* lyffetung ; *flattery,* FRAG. 4.

lüfrē, *see* **liveree**.

lüft, adj., *cf. M.Du.* luft, lucht ; *left (hand),* HOM. I. 213 ; bi his luft side (luftside ?) LAȝ. 24461 ; lift 15263 ; leoft 28047 ; lufthond [*v. r.* left, lefthalf] LANGL. *A* ii. 5 ; lift HAV. 2635 ; ROB. 22 ; MAND. 31 ; left, lift WILL. 2961, 2965 ; CH. C. T. *A* 2953 ; left ' sinister ' PR. P. 293 ; on his lift half HOM. II. 11 (REL. I. 131) ; over her lifte schulder A. P. ii. 981.

lüft, sb., *worthless person,* LANGL. *B* iv. 62.

lüft², sb., *O.E.* lyft *m. f., cf: O.N.* lopt *n.* (*see* loft), *O.L.G., O.H.G.* luft *f.,* Goth. luftus *m.* (ἀήρ) ; *air, upper region ;* 'aer,' FRAG. 3 ; lift A. P. ii. 212 ; M. H. 26 ; sio lift MK. ix. 7 ; lifte PR. C. 1444 ; lift & land ORM. 17533 ; lüfte (*dat.*) KATH. 2124 ; MARH. 9 ; A. R. 244 ; JOS. 385 ; bi þan lufte HOM. I. 129 ; to þon lufte LAȝ. 2881 ; under lufte 10104 ; lufte, lifte (*printed* liste) ROB. 279, 280 ; lifte BRD. 9 ; LANGL. *B* xv. 351 ; foȝeles on ᵭar lefte HICKES I. 223 (ANGL. I. 12).

lüften, v., *O.N.* lypta, = *M.H.G.* lüften ; *lift ;* leften ORM. 2488 ; liften HAV. 1028 ; MAND. 190 ; liften, lifte LANGL. *A* v. 203, *B* v. 359 ; liftin ' *levo* ' PR. P. 303 ; lifte (*pret.*) FLOR. 767 ; lifte (lefte) TREV. VII. 349.

luftenant, *see* **lieutenant**.

lüȝe, sb., *O.E.* lyge *m., cf. O.H.G.* lug *m.,* lugi *f., O.N.* lygi *f. ; from* lēoȝen ; *lie, falsehood ;* lighe PS. v. 7 ; lie CH. C. T. *A* 3391 ; K. T. 1056 ; leȝe (*printed* leghe) S. S. (Web.) 2887 ; (*ms.* leye) HAV. 2117 ; lee TOWNL. 216 ; liȝes (*pl.*) P. L. S. xxx. 141.

lüȝen, adj., *O.E.* (un- ge-)lygen, = *O.N.* lyginn, *O.H.G.* lugīn ; *deceitful* : he is ... hardi and ofte lie SAINTS (Ld.) xlvi. 688.

luge, *see* **logge**.

luging, *see* **logging** *under* **loggen**.

lugge, sb., *pole* : and me mid stone and lugge þreteþ O. & N. 1609.

luggen, v., *cf. Sw.* lugga ; *lug, drag ;* lugge LIDG. TH. 4597 ; luggeþ (*pres.*) GOW. III. 140 ; luggid (*pret.*) DEP. R. ii. 173 ; *comp.* to-**luggen**.

luis, *see under* **lūs**.   **luitel**, *see* **lütel**.

[**lūke** ? adj. ; *comp.* ēaᵭ-lūke.]

lüke, *see* **lēuke**.

lüken, v., *O.E.* lūcan, = *O.L.G.* (ant-, bi-)lūcan, Goth. (ga-, us-)lūkan, *O.Fris.* lūka, *O.N.* lūka, liuka, *O.H.G.* (ant-, ar-, pi-)lūchan ; *close, lock ; pull ;* MARH. 6 ; REL. I. 209 ; luken & teon KATH. 2128 ; lette luken [louke] þa ȝeten LAȝ. 10736 ; we him sculleᵭ to luken (*draw*) 31685 ; luke MISC. 97 ; louke LANGL. *B* xviii. 243 ; AL. (L.³) 611 ; snakes heore eien lukeþ MISC. 151 ; luken ORM. 16432 ; **loukande**

(*pple.*) A. P. ii. 441 ; lēac, lec (*pret.*) JUL. 52, 53 ; þe ȝates læc [lok] ful feste LAȝ. 15311 ; Arᵭur him læc to 21269 ; hit læc toward hirede folc unimete 28522 ; up he læc (*took*) þene staf 29661 ; þa læc þe water ofer hem ORM. 14814 ; lek AR. & MER. 5686 ; MAP 339 ; he lek his eȝen S. S. (Web.) 929 ; luken ut of scaþe sweordes longe LAȝ. 23211 ; luken heom bi vaxe 24843 ; loken AM. & AMIL. 492 ; & þe doren ... loke vaste ROB. 495 ; in armes ilk oþer þei loken MAN. (F.) 3258 ; loken (*pple.*) GOW. I. 226 ; PL. CR. 31 ; WICL. 3 KINGS vii. 9 ; S. S. (Wr.) 2951 ; DEGR. 552 ; he was loken þær wiþinnen ORM. 1091 ; loken kope HAV. 429 ; lokin M. H. 88 ; a Sarsins legge haþ he lokin OCTAV. (H.) 1274 ; *comp.* bi-, ȝe-, to-, un-**lūken** ; *deriv.* **loc**, **loke**.

louker, sb., ' *runcator,*' VOC. 218.

lukin, *see* **lokien**.

lukke, luk, sb., *cf. O.Fris., M.Du.* luck, *M.L.G.* lucke, *M.H.G.* (ge-)lücke *n., O.N.* lukka *f. ; luck,* PR. P. 316.

lukken, v., *cf. M.Du., M.L.G.* lucken ; *prosper, succeed ;* lukke M. T. 225.

lullin, v., *cf. M.Du.* lullen ; *lull,* ' *lallo,*' PR. P. 317 ; lulled (*pret.*) L. H. R. 133 ; CH. C. T. *E* 553.

lulten, v., *sound* ; loude alarom upon launde lulted (*pple.*) watz þenne A. P. ii. 1207.

lūmen, *see* **lēomien**.

lüminen, v., *for* **enluminen** ; *mod.Eng.* limn ; *illuminate* ; lumine [limne] TREV. VII. 295 ; lüminid [limnid] (*pple.*) ' *elucidatus* ' PR. P. 317.

lüminour, limnore, sb., *mod.Eng.* limner ; *illuminator,* ' *elucidator,*' PR. P. 317 ; limenour E. G. 9.

lumpe, sb., = *M.Du.* lompe ; *lump,* ' *frustum,* VOC. 241 ; PR. P. 317 ; TOWNL. 307 ; lompe REL. II. 79 ; a lompe of chese LANGL. *C* x. 150 ; lump (*heap*) BARB. xv. 229 ; (*company*) 342* ; (*crowd*) xix. 377*.

lumpen, adj., *lumpy, heavy* ; heore hertes were colde as lumping led L. H. R. 141.

lunaire, sb., *moonwort,* CH. C. T. *G* 800.

lunatik, adj., *O.Fr.* lunatic ; *lunatic* : lunatik lollers LANGL. *C* x. 107 ; þanne loked up a lunatik (*sb.*) LANGL. *B* PROL. 123.

lünd, sb., *O.N.* lund ; *nature, disposition,* ORM. 7038.

Lunden, Lundene, Lunde [Londene], pr. n., *O.E.* Lunden ; *London,* LAȝ. 7080, 18477, 18495 ; on Lundene MISC. 145 ; Lunde 94 ; Lounde SPEC. 92 ; Londen, Londe LANGL. *B* xv. 148, *C* xvii. 286.

Lundene-burh, sb., *London,* LAȝ. 4284.

Lundenisc, adj., *of London* ; Lundenisce men LAȝ. 31146.

Lundreis, sb., pl., *Londoners* ; Londreis ROB. (W.) 1126.

lūne, sb., *cf. Dan.* luun, *Swed.* lugn, *O.N.* logn; *quiet, rest*: of mine live ʒif me lune HOM. I. 197.

lunge, sb., *O.E.* lungen (*pl.* lungenu) *n., cf. O. Fris.* lungen, *O.H.G.* lungunna, lunginna, lungun, lunga *f., O.N.* lungu *pl. n.; lung,* '*pulmo*,' PR. P. 317; R. S. v (MISC. 178); FER. 1095; GOW. III. 100; WICL. 3 KINGS xxii. 34; LUD. COV. 181; longe ALIS. 2156; B. DISC. 602; **longene** (*pl.*) LAʒ. 6499*; longes VOC. 183; CH. C. T. *A* 2752.

   longe-woo, sb., *lung disease,* PALL. i. 50.

lung-wurt, sb., *O.E.* lungenwyrt; *lung-wort,* '*elle*(*b*)*ore*,' REL. I. 37.

lunginge, *see* **langen.**   **lüpe,** *see* *h*lēap.

lurdane, *see* **lordein.**

lüre, sb., *O.E.* lyre; *from* lēosen; *loss,* FRAG. 2; LAʒ. 980; KATH. 805; A. R. 58; O. & N. 1153; D. TROY 2244; lur GAW. 355; luere, lere REL. I. 262, 263; lire GEN. & EX. 2920; lire of eorþlike ahte ORM. 5667; lere ROB. 526; AYENB. 36; AR. & MER. 8369; þone lere HOM. I. 221; leore ALIS. 1122; **lüren** (*pl.*) A. R. 298; (*dat. pl.*) H. M. 7.

lure, sb., *O.Fr.* loirre; *lure,* '*lurale*,' PR. P. 317.

lüre, *see* *h*lēor.

luren, v., *O.Fr.* loirrer, *cf. M.Du.* leuren, loren; *allure*; lure CH. C. T. *D* 415; lured (*pple.*) LANGL. *B* v. 439.

lüren, v., *? = M.L.G.* lüren (*examine*); *lower, look sullen*; lure HORN (L.) 270; louren PL. CR. 556; lourin '*oboculo, maero*' PR. P. 316; loure S. S. (Web.) 1539; P. S. 154 (A. D. 106); GOW. I. 47; TRIAM. 1032; loure on mi neʒebore LANGL. *B* v. 132; **lourinde** (*pple.*) AYENB. 256; lourand WILL. 2119; louring CH. C. T. *D* 1266.

   lüring, sb., *louring,* O. & N. 423.

lüri, adj., *cf. L.G.* lürig; *sullen,* LIDG. M. P. 52.

lurken, v., *? from* **lüren,** *as* **hĕrkien** *from* **hēren**; *lurk,* HAV. 68; lürkin '*lateo*' PR. P. 317; i lurk(e) (*pres.*) and dare TOWNL. 137; lurkes A. P. iii. 277; (hi) lurkeþ PL. CR. 83; **lorkinde** (*pple.*) WILL. 2213; lurkand(e) BARB. v. 192; x. 627; lurkinge LANGL. *A* ii. 192* (lorkinge *B* ii. 216); lurkinge CH. C. T. *G* 658; **lurkide** (*pret.*) WICL. 1 PARAL. xii. 8; lurkede LANGL. *A* ii. 192.

   lurkere, sb., *lurker,* REL. I. 133; ALEX. (Sk.) 3543.

   lurking, sb., *hiding-place,* PS. xvii. 12.

lurnien, *see* **leornien.**

[**lürten,** v., *= M.H.G.* lürzen; *deceive; comp.* **bi-lürten.**]

lūs, sb., *O.E.* lūs, *= O.N., O.H.G.* lūs; *louse*; lous P. S. 238; LANGL. *B* v. 198; it nas noʒt worþ a lous FER. 439; **lūse** [*O.E.* lȳs], (*pl.*)

P. L. S. i. 5; luse sed '*psyllium*' REL. I. 37; with luis [luise] TREV. VI. 387; lis LANGL. *B* v. 197; lise PR. C. 651.

lusarde, *see* **lesarde.**

lusch, adj., *cf. L.G.* lusch; *loose,* '*laxus*,' PR. P. 317.

luschen, v., *beat, flog; ? come down with a rush*: schall i . . . lusshe all youre limmis with lasschis YORK xxxi. 10; þeir schipes alle in peril were . . . mast & sail, down hit **lusched** [*v. r.* lussed] (*pret.*) MAN. (F.) 2977; luischede D. ARTH. 2226; lusshed YORK xlvi. 37.

lüschen, lüsken, v., *cf. M.H.G.* lüschen, *M.Du.* luischen, *Dan.* lūske; *lie hid*; lusk(e) MAN. (H.) 9; **lūskand** (*pple.*) TOWNL. 189.

lūsen, *see* **lēosen.**

lūsi, adj., *= M.H.G.* lūsec; *lousy*; lousi LANGL. *B* v. 195; CH. C. T. *D* 1467.

lūsin, v., *cf. M.H.G.* lūsen; *clear off lice*; lousin PR. P. 316; **louseþ** (*pres.*) P. S. 239; **iloused** (*pple.*) TREV. III. 353.

lust, lüst, sb., *O.E.* lust *m.,* lyst *f., cf. O.L.G., O.H.G.* lust *f., Goth.* lustus *m.* (ἐπιθυμία), *O.N.* losti *m.,* lyst *f., mod.Eng.* lust; *desire; pleasure; lust*; MARH. 14; O. & N. 507; HOCCL. iii. 60; M. H. 165; AUD. 4; (*ms.* lusst) ORM. 1628; þene lust A. R. 272; let lust over go MISC. 159; ure lust and ure liking P. L. S. xxv. 140; fleisliche lust SPEC. 72; let lust over gan A. D. 290; we hadden gret lust to see his noblesse MAND. 220; lust '*voluptas, libido*' PR. P. 317; list '*delectatio, libitum*' 306, 307; lust, list, lest CH. C. T. *A* 132, 192, 1318; lost AYENB. 82; list GEN. & EX. 1230; sinful list REL. I. 221; list, liste A. P. i. 907, ii. 843; **lustes** (*gen.*) A. R. 318; **luste** (*dat.*) O. & N. 1397; mid muchelere liste þa childere he biwuste LAʒ. 13731; at here oune list(e) MAND. 50; **lustes** (*pl.*) MARH. 1; A. R. 176; ivele lustes REL. I. 129; listes [lestes] of love WILL. 740, 946; luston ['*voluptatibus*'] LK. viii. 14.

   lost-vol, adj., *O.E.* lustfull; *lustful,* AYENB. 80; **lostvoller** (*compar.*) AYENB. 92.

   lustfulnesse, sb., *pleasure* HOM. I. 21.

   lust-lēs, adj., *without desire,* GOW. II. 111; listles PR. P. 307.

   lust-līche, adv., *O.E.* lustlīce; *pleasantly,* REL. I. 129; lustlike MISC. 115.

   lustsum, adj., *O.E.* lustsum, *= O.L.G., O.H.G.* lustsam, lussam; *pleasant*: lussum lif heo ledes SPEC. 34 (A. D. 156); þe lilie is lossom to seo SPEC. 44 (A. D. 164); a lussum (? lufsum) lond C. M. (*Trin. Coll. MS.*) 604.

lustre, sb., *Lat.* lustrum; *period of five years,* TREV. VIII. 29.

hlüst, sb., *O.E.* hlyst *m. f., cf. O.L.G., O.N.* hlust *f.* (*hearing, ear*); *hearing,* HOM. II. 25; liste 61; lust HOM. I. 75; ALIS. 1916;

Caredoc makede lust & þus spæc LaȜ. 11577 ; (*ear*) A. R. 212 ; luste FER. 1900.

lüsten, v., *O.E.* lystan, *cf. O.N.* lysta, *O.L.G.* lustean, *O.H.G.* lusten, lustōn, (*pret.* lusta, lustōta), *Goth.* lustōn ; *be pleasing* ; (*with refl. dat.*) *delight, be pleased* : al þat ham lüsteð (*pres.*) H. M. 25 ; lusteth E. W. 28 ; þer he him losteþ (*delights*) AYENB. 246 ; lust HOM. I. 103 ; JOS. 41 ; ne lust him nu to none unrede O. & N. 212 ; him lust MIRC 820 ; list PR. C. 795 ; whan hem list MAND. 38 ; if þe list.CH. C. T. *A* 1183 ; ȝef þe luste A. D. 289 ; lüste (*pret.*) KATH. 1588 ; A. R. 238 ; BRD. 6 ; ne luste heom hider varen LaȜ. 28811 ; þo luste him ete MISC. 38 ; luste, [liste lest(e)] LANGL. *B* xvii. 139 ; luste, liste PL. CR. 71, 165 ; liste ORM. 8119 ; GOW. II. 366 ; A. P. i. 146 ; liste, leste CH. C. T. *A* 750, 1052 ; *comp.* for-, of-lüsten.

*h*lüsten, v., *O.E.* hlystan, *cf. Swed.* lysta ; *listen, hear* ; lusten P. L. S. viii. 114 ; ȝif ȝe hit lusten w(u)lle LaȜ. 919 ; þine songes luste O. & N. 896 ; listen (*ms.* listenn) ORM. 8574 ; hlesten ANGL. I. 20 ; lesten WILL. 31 ; lheste AYENB. 70 ; hlisteð (*pres.*) HOM. II. 155 ; listiþ LIDG. M. P. 189 ; lüst (*imper.*) MARH. 3 ; lust mire worden LaȜ. 21985 ; hleste (*subj.*) MK. vii. 16 ; lüste (*pret.*) LaȜ. 29526 ; O. & N. 253 ; lheste AYENB. 199.

lusti, adj., = *M.Du.* lustich, *M.H.G.* lustic, *O.N.* lostigr ; *lusty, pleasant, mirthful*, KATH. 1693 ; GOW. I. 110 ; CH. C. T. *A* 80 ; PR. C. 4231 ; lusti on to loken HOM. I. 269 ; in þi youþ be lusti LIDG. M. P. 68 ; lusti, listi '*delectabilis, voluptuosus*' PR. P. 307, 317 ; lustier (*compar.*) CH. C. T. *G* 1345.

lusti-heed, sb., *M.H.G.* lusticheit ; *hilarity*, CH. C. T. *F* 288.

lusti-liche, adv., *lustily*, JUL. 75 ; lustili CH. C. T. *E* 1802.

lustinesse, sb., *pleasantness*, CH. C. T. *A* 1939.

*h*lüstnen, v., *cf. Swed.* lyssna ; *listen* ; lustnen KATH. 110 ; (*v. r.* hercnen) A. R. 422 ; lüstneþ (*pres.*) SPEC. 24 ; listneþ LANGL. (Wr.) 9534 ; lestne (*imper.*) SHOR. 76 ; lustneþ C. L. 1027 ; listneð REL. I. 217 ; lestneþ ANGL. III. 61 ; lüstnede (*pret.*) LaȜ. 26357 ; listnede GEN. & EX. 2137 ; lisnit BARB. vi. 72 ; ix. 685 ; ilüstned (*pple.*) LaȜ. 25128.

lüt, sb., *O.E.* lyt, *cf. O.L.G.* lut, *O.N.* lītt ; (*a*) *little, few*, LaȜ. 252 ; KATH. 34 ; worldliche men ileveð lut A. R. 66 ; mid lut wordes 70 ; þese lit word HOM. II. 105 ; a lute liste O. & N. 763 ; his sone . . . biȝat fule lute BEK. 2379 ; lute god TREAT. 133 ; lute and lute *little by little* 137 ; luite JOS. 148 ; to luite (*v. r.* lutel) C. L. 632 ; lite AYENB. 31 ; (*r. w.* write) HORN (L.) 932 ; to lite ich habbe an horde MISC. 193 ; a lite CH. C. T. *A* 1450 ; lüte (*adj.*) *little*, LaȜ. 22208 ; þet is lute wunder HOM. I. 187 ; a lute wiht A. R. 72 ; i lute hwile KATH.

2183 ; a lute fowel BRD. 9 ; lite (*r. w.* bite) HAV. 1730 ; þi sorȝe is al to lite SHOR. 32 ; it is lite CH. C. T. *A* 2627.

lüte, sb., *lute*, PR. P. 318 ; lütes (*pl.*) CH. C. T. *C* 466.

lütel, adj., *O.E.* lytel, litel, *cf. O.L.G.* luttil, *O.H.G.* luzil, *O.N.* lītill, *Goth.* leitils ; *little*, O. & N. 561 ; þet is lutel wunder HOM. I. 202 ; a lutel lond KATH. 2178 ; an luttel [lutel] child LaȜ. 9124 ; luitel JOS. 39 ; litel LK. xviii. 3 ; HAV. 481 ; CH. C. T. *A* 490 ; (*ms.* litell) ORM. 3205 ; litil LIDG. M. P. 163 ; lütel (*sb.*) *a little* KATH. 354 ; A. R. 74 ; to lutel HOM. I. 213 ; a lutel MARH. 13 ; lutel, litel P. L. S. viii. 6, 69 ; luitel C. L. 73 ; MIRC 627 ; litel AYENB. 29 ; a litel ORM. 1715 ; REL. I. 211 ; a litel (*v. r.* lite) LANGL. *B* xiii. 268 ; litel woot Arcite CH. C. T. *A* 1525 ; a littel HICKES I. 228 ; ane lutle while LaȜ. 5818 ; lutle hwile O. & N. 1451 ; þis lutle . . . stucchen A. R. 428 ; þis lutle pine KATH. 2182 ; þe litle wereld ORM. 17597 ; litill, littill BARB. i. 173 ; xii. 19 ; sum what litles (*gen.*) *little* ORM. 4681 ; litles what 6952 ; ænne lütelne (*acc.*) sune LaȜ. 6329 ; littlene AYENB. 238 ; þane lutle fuȝel O. & N. 1097 ; lutle luve hi beren MISC. 42 ; lütle (*pl.*) A. R. 220 ; O. A. N. 631 ; litle HAV. 2014 ; lütlen (*dat. pl.*) *little at a time* LaȜ. 3569 ; lutlen and lutlen MARH. 12 ; litlum & litlum LANGL. *B* xv. 599.

lüten, v., *O.E.* lūtan, = *O.N.* lūta ; *bow, incline the head, stoop*, ORM. 6139 ; GEN. & EX. 1926 ; loutin '*conquinisco*' PR. P. 316 ; loute MAP 336 ; L. H. R. 34 ; CH. C. T. (T.) 14168 ; MAN. (H.) 333 ; lüteð (*pres.*) REL. I. 223 ; loutes DEGR. 1225 ; (hi) luteð KATH. 1781 ; lout (*lie*), B. B. 157 ; lüteð (*imper.*) A. R. 13 ; lout B. B. 13 ; lütende (*pple.*) A. R. 426 ; lēat (*pret.*) AYENB. 239 ; heo leat lahe to hire leove laverd MARH. 12 ; ðor ðe grave under let (*lay hid*) GEN. & EX. 3188 ; he ros and lutte 1055 ; lutte ORM. 8961 ; GAW. 2236 ; loutede [louted] LANGL. *A* iii. 111 ; his hed loutede a doun ROB. 115 ; louted WILL. 3485 ; loutid S. S. (Wr.) 406 ; loutit BARB. ii. 154 ; ofte heo luten a doun LaȜ. 1880 ; *comp.* a-lūten.

louting, sb., *inclination*, D. TROY 661.

lutenant, *see* lieutenant.

lutien, v., *O.E.* lutian, *cf. O.H.G.* luzzen ; *from* lüten ; *lurk, lie concealed* : hæhte heom þere lutie [lotie] wel LaȜ. 21509 ; lotie TREV. IV. 396 ; þe hare luteþ (*pres.*) al dai O. & N. 373 ; deþ luteþ in his scho R. S. i ; lotieþ LANGL. *B* xvii. 102 ; lotinge (*pple.*) CH. C. T. *G* 186 ; TREV. II. 83 ; lutede (*pret.*) KATH. 1848 ; loted TREV. II. 229 ; *comp.* æt-lutien.

lütiȝ, lüti, adj., *O.E.* lytig ; *cunning*, FRAG. 6, 7.

lütlien, v., *O.E.* lytlian ; *diminish, decrease*, LaȜ. 8838 ; ne make ȝe nowðer mi luve ne min bileave lutlen JUL. 16 ; hit ne me(i) neaver mare lutlin ne wursin HOM. I. 265; lutli O. & N. 540; i lutle REL. II. 211; lüttulde (*pret.*) JOS. 145; litelid (*pple.*) KN. L. 61.

lütling, sb., *O.E.* lytling; *little one*; litling MAT. xviii. 4.

*h*lütter, adj., *O.E.* hlūtor, hluttor, = *O.L.G.* hlutter, *O.H.G.* hlūter, hlutter, *Goth.* hlūtrs; *pure*; lutter ORM. 5707.

lüðer, adj., *O.E.* lȳðer; *evil, bad*, A. R. 96; luðer (luþer) LAȝ. 6833; luþer and qved O. & N. 1137; luþer ant lees SPEC. 49; luþer [liþer, leþer] LANGL. *A* v. 98; liðer REL. I. 131; GEN. & EX. 369; liþer BEK. 364; WILL. 2169; S. S. (Wr.) 2250; lither PR. C 1059; leþer IW. 599; lider PR. P. 303; þe luðere reve MARH. 6; þat leoðre folc HOM. I. 241; luþre mede MISC. 39; lüðere (*adv.*) JUL. 16; him ilomp wel luðere [luþre] LAȝ. 2785; ȝef æi mon him liðere dude 4270; þine lüþere (*pl.*) deden FRAG. 8; luðre men HOM. I. 205; luðere wordes A. R. 424; luþre [luþere] wrenches MISC. 124, 125; lüðerest (*superl.*) MARH. 18.

lüþer-hēde, sb., *badness*, ROB. 240; liþer-hede P. L. S. xiii. 88.

lüðer-liche, adv., *wickedly*, MARH. 5; A. R. 290; luþerliche FRAG. 7; liþerliche S. S. (Web.) 972; liðerlike GEN. & EX. 1563; luþerli WILL. 1208; liþerli CH. C. T. *A* 3299; letherli MIR. PL. 114.

lüðernesse, sb., *wickedness*, HOM. I. 197; luþernesse P. S. 239; lithernes PR. C. 226.

lüþere, *see* liþere.

luve, sb., *O.E.* lufu (*gen.* lufan), *Goth.* lubō, (*gen.* lubōns) (*see* ANGL. VI. 176); *love*; (*ms.* luue) GEN. & EX. 35; þeo luve A. R. 394; heo ches him to luve ant to leove mon MARH. 2; luve, lufe LAȝ. 413, 31412; luff BARB. ii. 515; lufe ORM. 1572; PR. C. 69; ISUM. 242; lufve AMAD. (R.) lii; luve, love O. & N. 968, 1510; love CH. C. T. *A* 1600; WICL. JOHN xiii. 35: Reginoldes love LANGL. *B* 49 (lemmon LANGL. *A* iv. 36); knichtis for thar luffis (*gen.*) (*sweetheart's*) sak BARB. iii. 349; þi love bonde SPEC. 58; luven (*dat.*) SAX. CHR. 251; for godes luven GEN. & EX. 4081; luven, lufe [lofve] LAȝ. 1257, 9630; for þi luve MARH. 8; luve (*pl.*) P. L. S. viii. 156.

love-dai, sb., *love-day, day for the amicable settlement of differences*, P. L. S. xvii. 414; LANGL. *A* xi. 20; CH. C. T. *A* 258; AUD. 26.

love-knotte, sb., *love-knot*, LANGL. *C* xviii. 127; CH. C. T. *A* 197.

love-drink, sb., *philter*, TRIST. 1710.

luf-ful, adj., *loving*, A. R. 222.

love-laik, sb., *amorous play*, TRIST. 2020.

love-longinge, sb., *love longing*, SPEC. 61.

love-lēs, adj., *loveless*, P. S. 255; LANGL. *A* v. 98.

luve-lich, adj., *O.E.* luf-,· lufelīc; *lovely, affectionate*, A. R. 428; lovelich LANGL. *A* ix. 77; þat lufliche word HOM. II. 5; lufli BARB. i. 389; **luveliche** (*adv.*) A. R. 14; luveliche [lofveliche] LAȝ. 7892; bide hine luveliche þet he þe do riht HOM. I. 17; luvelike REL. I. 217; loveliche LEG. 35; JOS. 305.

lufle-lī, adv., BARB. xvii. 315*.

luf-rēden, sb., *O.E.* lufrǣden; *love*, M. H. 30; loverede AYENB. 146; lovered PS. cviii. 5.

luve-rūne, sb., *love talk*, KATH. 109.

love-song, sb., *love song*, SPEC. 74.

luve-spēche, sb., *love speech*, A. R. 204.

luf-sum, adj., *O.E.* lufsum; *loveable, lovely*, MARH. 3; ORM. 3583; ANT. ARTH. xxvii; þi leor is . . . lufsum KATH. 316; lufsom IW. 214; lossum he is SPEC. 26 (A. D. 145); lofsom LAUNF. 942; luf-, luve-, lovesum C. M. 604; lovesom CH. TRO. v. 465; **lufsumre** (*compar.*) H. M. 11; **lufsumest** (*superl.*) JUL. 13.

lufsum-līke, adv., *lovingly*, ORM. 1663; lusumli BARB. xvii. 315.

love-tēar, sb., *tear of love*; lovetēres (*pl.*) SPEC. 70.

luf-þing, sb., *love*, LAȝ. 169.

luve-wende, adj., *O.E.* lufwende; *loveable*, JUL. 65.

luve-wurðe (? luve wurðe), adj., *worthy of love*, A. R. 112; JUL. 17; **luvewurðest** (*superl.*) HOM. I. 187.

luve-wurði, adj., *worthy of love, loveable*, HOM. I. 269.

lūven, *see* lēven. **lūver**, *see under* lēof.

luvestiche, *O.E.* lufestice, = *M.H.G.* lubestecke, *O.Fr.* livesche, *Ital.* levistico; *from Lat.* ligusticum; *lovage, 'levisticum,'* REL. I. 36; loveache PR. P. 314.

luvien, v., *O.E.* lufian; *love*; (*ms.* luuien) A. R. 14; REL. I. 212; luvien [lovie] LAȝ. 3056; luvian HOM. I. 11; luvien, luvie O. &. N. 1341, 1345; lufen ORM. 940; luven GEN. & EX. 9; lovien SPEC. 57; luvie FL. & BL. 392; MISC. 28; lovie AYENB. 6; (*ms.* lovye) SHOR. 91; loven MAND. 160; love WICL. JOHN viii. 42; **luvie** (*pres.*) MARH. 4; lovie BEK. 890; lovie [love] LANGL. *B* iii. 32; luvest A. R. 282; lufeð LK. vii. 5; lufeð, luveð, luvað HOM. I. 19, 99; luveð KATH. 231; luveþ O. & N. 230; gif ge lufieð þa þe eow lufieð LK. vi. 32; luvieð LAȝ. 998; KATH. 952; A. R. 282; luvieþ O. & N. 791; MISC. 36; luvigeð HOM. II. 37; lovieþ FER. 2003; lufen ORM. 3602; **lufe** (*imper.*) MAT. v. 43; luve HOM. I. 65; (þu) luvie (*subj.*) LAȝ. 14585; ȝe luvian HOM. I. 125; **loviinde** (*pple.*) AYENB. 145; luffand BARB. i. 363; **lufede** (*pret.*) LK. vii. 47; lufode, lufede, luvede [lovede] LAȝ. 156, 256, 17027; lufede ORM. 16712; luvede KATH. 106; lufedest FRAG. 6; lovedest BEK. 781; luveden SAX. CHR. 250;

A. R. 282; lufed (*pple.*) IW. 1164; *comp.*
bi-, ȝe-luvien.

lovere, lover, sb., *lover*, CH. C. T. *A* 1379.

luxurie, sb., *Lat.* luxuria; *luxury*, AYENB.
157; CH. C. T. *C* 484.

luxurious, adj., *luxurious*, CH. BOET. i.4 (21).

lynx, sb., *Lat.* lynx; *lynx*, AYENB. 81.

m.

má, see mákien.

mā, adv., *O.E.* mā, = *O.Fris.* mā, mār, *O.
L.G.*, *O.H.G.* mēr, *O.N.* meir, *Goth.* mais,
Lat. magis (*compar.*); *more*, FRAG. 8;
KATH. 1282; MARH. 8; FLOR. 1104; PR.
C. 3997; ma monna *more men* HOM. I.
27; nefede he bern no ma LAȝ. 91; næfre
ma ne shal he ben o nane wise filed ORM.
4206; mo O. & N. 564; GEN. & EX. 354;
HAV. 1846; LANGL. *B* PROL. 147; MAND. 275;
A. P. i. 869; AUD. 80; evre mo P. L. S. viii. 53;
oðer mo oðer les A. R. 42; monie moa þer
beoð 328; fele oþere mo BEK. 571; no þe mo
TREAT. 138; ne mo ne les AYENB. 118; mo
þan þries ten CH. C. T. *A* 576.

māre, adj., *O.E.* māra *m.*, māre *f. n.*, =
O.Fris. māra, *O.H.G.* mēro *m.*, mēra *f. n.*,
O.N. meiri *m.*, meira *f. n.*, *Goth.* maiza, maizei,
maizō, *Lat.* mājor, mājus (*compar.*); *more,
greater*, PR. C. 1047; PERC. 229; nes nefre
mare mon þenne he HOM. I. 131; mare luve
LAȝ. 30055; ne aras he never mare 1555;
eaver se þu mare wa & mare weane dost me . . .
se þu wurches(t) mi wil & mi weol mare KATH.
2140; þat is mare wurð 70; mare genge ORM.
19566; godes engel seȝde . . . ȝet mare 780;
mare lufest tu þat þing 4662; þe mare 10146;
her ufer mar 1715; þæs þe mare MAT. xx. 31
(*earlier vers.* ma); more P. L. S. viii. 1;
O. & N. 213; HAV. 982; LANGL. *B* v. 289;
MAND. 4; HOCCL. i. 65; more wunder A. R.
54; þe more 194; more oþer lasse TREAT.
138; more grat AYENB. 18; nevre mor 220;
hit was more þan ani stede ALIS. 692; þer
nis no more to seie CH. C. T. *A* 2386; boþe
more and lesse 6516; þe more fishes . . . eten
þe lasse HOM. II. 179; moare PROCL. 2;
mor GEN. & EX. 993.

māst, adj. & adv., *O.E.* mǣst, mǣsta,
mǣste, = *O.L.G.* mēst, mēsta, *O.Fris.* māst,
māste, *O.H.G.* meist, meisto, meista, *O.N.*
mestr (? mēstr), *Goth.* maists; *most, greatest*,
LAȝ. 27601; mest P. L. S. viii. 4; O. & N.
684, 852; ealle mest (*nearly all*) SAX. CHR.
250; mest qvat *for the most part* ALEX.
(Sk.) 5010; ever ich mest haveð on(e) olde
cwene A. R. 88; þesne lehter habbeð mest
hwet alle men HOM. I. 137; on mest 209;
ho so haveþ of urþe mest TREAT. 138; þat
is al mest (*almost*) ibroȝt to grounde BEK.
1966; þet bieþ mest worþ AYENB. 23; mest
manne him gremede HOM. II. 169; mast
mannen 7; ant tu schalt . . . meast wunne

ant weole welden MARH. 6; meast alle
KATH. 29; mast hwat HOM. II. 11; it is
mast & heȝhest ORM. 10734; tær was þeȝre
king aȝ mast 8471; most worþi MAND. 1; þe
mæste LAȝ. 11567; meste ROB. 9; þe meste
del A. R. 330; (h)is meste fo SPEC. 24; þe
meste and leste CH. C. T. *A* 2198; at te
meste BRD. 31; maste IW. 2123; þe maste
lufe ORM. 5328; þe moste GEN. & EX. 189;
þe moste swike HAV. 423; þe moste sinne
MAND. 249.

macche, sb., *O.Fr.* mesche; *match* (*for
striking a light*), LANGL. *B* xvii. 213.

macche, see mæcche.

máce, sb., *O.Fr.* mace (*masse*); *club, mace*:
mid mace mid ax & mid anlas ROB. 48; mace
ALIS. 1901; máces (*pl.*) ALEX. (Sk.) 4100;
maciss BARB. xii. 579.

máce², sb., *Ital.* mace, *Fr.* macis; *mace
(plant, spice)*, ALIS. 6796; macis PR. P. 319;
L. C. C. 13; máces (*pl.*) P. L. S. xxxv. 75.

mácer, sb., *Fr.* massier; *mace bearer*;
máceres (*pl.*) LANGL. *B* iii. 76.

machūn, see masoun.

mād, adj., *O.E.* mād(-mōd), (ge-)mǣd, = *O.L.
G.* gemēd, *Goth.* gamaids (ἀνάπηρος), *O.H.G.*
gameit (*stultus*); *mad*, S. S. (Wr. 2091;
maad B. DISC. 2001; i waxe mod (*ms.* mot,
r. w. blod) SPEC. 31; *cf.* mǎd *pple. of* mǣden.

madame, sb., = *O.Fr.* ma dame, *from Lat.*
mea domina; *lady, madam*, ALIS. 269.

mǣdden, v., *from* mǎd *pret. pple. of* mǣden;
madden, make or become mad; maddin
'*insanio*' PR. P. 319; mǎdde (*pres.*) PL. CR.
280; þou maddes YORK xv. 38; mǎdde (*subj.*)
GAW. 2414; if þou madde CH. C. T. *A* 3156;
þat mǎddid (*pret.*) þi men DEP. R. i. 63.

mǎdding, sb., *madness*, ALEX. (Sk.) 1743,
3546; A. P. i. 1154.

mǣdem, see māðem.

mader, sb., *O.E.* mædere; *madder*, E. G.
358; madir PR. P. 319; VOC. 265.

mæcche, sb., *O.E.* (ge-)mæcca; *from* máke;
match, mate, equal; macche LANGL. *B* xiii.
47; HOCCL. i. 307; Zakariȝes mache ORM.
290; þe schrewe fond his macche þo P. L. S.
xiv. 48; mache M. ARTH. 1607; mecche
ANGL. I. 79; MAN. (F.) 13563; ches . . .
Crist to meche REL. I. 225; mehche '*par*'
PR. P. 331; macche *struggle* D. TROY 1324;
mæcchen (*acc.*) MAT. i. 24; macchis (*pl.*)
ALEX. (Sk.) 831; *comp.* ȝe-mæcche.

mæcchen, v., *match, compare*; mach GAW.
282; mache ALEX. (Sk.) 3607; maccheþ
(*imper.*) LANGL. *A* x. 193 [*B* ix. 193 macche];
mached (*pret.*) YORK xxx. 199; mached (*pple.*)
D. ARTH. 2904; machet ANT. ARTH. xxxiv.

mæche, see mēche.

[**mæden**, v., *O.E.* (ge-)mǣdan, = *Goth.* maidjan
(καπηλεύειν), *O.N.* meiða (*hurt*); *madden*];
mǎd (*pret. pple.*), *driven mad, mad*, SPEC.

29; P.S. 331; GREG. 751; (ms. madde, r. w. glad) M. H. 53; comp. **a-mǣden.**

mǎd-hēde, sb., madness, MIRC 1657.

mǎd-lǐche, adv., madly, KATH. 2114.

mǎd-schipe, madness, KATH. 227; med-schipe A. R. 148; KATH. (E.) 267, 325.

mǣȝe, mā̆ȝe, sb., O.E. mǣge, māge; kins-woman; meȝhe ORM. 3178; mehe [mowe] A. R. 76; mage (earlier text mǣge) LK. i. 36; maȝe, moȝe P. L. S. viii. 15; maȝe, mawe [moȝe, mowe] LAȝ. 257, 25897; moge, mowe HOM. II. 125, 221, 225; mowe R. S. v; mowe MISC. 79; ROB. 316; mow(e) 'glos' PR. P. 345; mēhen (pl.) MARH. 16; mowe P. L. S. v. 3.

mæht, see meaht.

mæi, sb.,=? **mæiden**; maid, virgin; mai P. L. S. xvi. 4; M. H. 39; þat mai FL. & BL. 743; (ms. may) SPEC. 26; WILL. 659; MIRC 1351; GREG. 525; GOW. II. 2; CH. C. T. B 851; AMAD. (R.) liii; TOWNL. 67; þat ilke mai (first text maide) LAȝ. 30486*; maȝ (ms. maȝȝ) ORM. 2489.

mæi², v., O.E. mæg,=O.L.G., O.H.G., Goth. mag, O.Fris. mei, O.N. mā (pret.-pres.); may, am able, FRAG. 6; mei HOM. I. 7; mæi, mai [mai] LAȝ. 754, 1202; mæig, maig, mai MAT. iii. 9; MK. x. 26; JOHN iii. 27; mei, mai KATH. 227, 1527; ne maig he tilen him non fode REL. I. 210 (MISC. 3); mai [maiȝ] HICKES I. 223; mai A. R. 6; O. & N. 228; GEN. & EX. 371; C. L. 206; (spelt may) 1; (ms. maj) LEECHD. III. 104; kepe þe ȝates who so mai H. H. 141; maȝ ORM. 199; meȝ HOM. I. 216; mei A. R. 6; SPEC. 48; (þu) mæht (ms. maht, r. w. fæht) LAȝ. 14568; maht MARH. 4; SPEC. 91; maht, miht HOM. I. 17, 23; ORM. 1488, 1492; maiȝt C. L. 1109; mait, maith HAV. 641, 689; maut AN. LIT. 8; meiht HOM. I. 217; A. R. 102; miht LAȝ. 694; H. H. 80; miht, miȝt, mist O. & N. 64, 79; miȝt BRD. 30; AYENB. 7; LANGL. B xii. 10; **maȝen** [O.E. magon,= Goth. magun, O.H.G. magun, mugun, O.L.G., O.Fris. mugun, O.N. megum] (pl.) HOM. I. 43; maȝen, mawen, muȝen LAȝ. 1333, 5335, 12748; maȝe FL. & BL. 632; mahen KATH. 757; mawen SPEC. 96; mugen MAT. x. 28; muȝen HOM. I. 221; muȝhen ORM. 202; muȝe, mawe O. & N. 182; muwen A. R. 4; muwe MISC. 37; P. S. 328; moȝe, mowe, moun FER. 321, 2002, 4996; mowen CH. BOET. i. 6 (25); mowe ROB. 316; C. L. 23; WILL. 3903; MAND. 5; HOCCL. i. 148; E. W. 27; moun WICL. GEN. vi. 20; S. S. (Wr.) 1340; mǣie, maȝe [mawe], [O.E. mǣge, muge] (subj.) LAȝ. 1520, 3180; (ms. meie), maȝe HOM. I. 37, 111; mahe KATH. 267; mawe P. L. S. viii. 168; muȝe PROCL. 6; muge REL. I. 212 (MISC. 6); muhȝe HOM. I. 17; muȝhe ORM. 200; muhe [mowe] O. & N. 1581; moȝe AYENB. 7; þet he hit muwe iheren A. R. 96; mowe HAV. 183;

BEK. 141; mæȝen LAȝ. 17947; mawen FRAG. 8; mowen CH. BOET. i. 5 (25); deriv. **mæin**, **mæiþ**, **meaht**, **muȝen**.

mæi, **māȝ**, sb., O.E. mǣg, māg,=O.L.G., O.H.G. māg, O.N. māgr, Goth. mēgs; relative, kinsman, 'propinquus,' FRAG. 2; SAX. CHR. 255; mæi [mei] LAȝ. 3838; mæi, mai HOM. II. 221, 225; mei I. 93; þi mei [mæi] ne þi moȝe P. L. S. viii. 15; mei R. S. v (MISC. 178); magh [mau] C. M. 2807; mog GEN. & EX. 1761; mǣie (dat.) LAȝ. 20445; **mǣȝes** mæies, meies [meies] (pl.) LAȝ. 457, 5098, 21812; mæiges, mages MK. xiii. 12; LK. ii. 27; meies ba ant mehen MARH. 16; meiis P. L. S. v (a.) 3; māgen (dat. pl.) LK. xxi. 16; moȝe PROCL. 10; comp. **wine-mæi.**

mæiden, maiden, maide, sb., O.E. mǣgden, =M.H.G. magedin; maiden, virgin, girl, LAȝ. 150, 3059, 3111, 22225; maigden, maiden MAT. xiv. 11, MK. xi. 28; meiden KATH. 66; A. R. 64, 166; maȝden ORM. 2102; maiden TOR. 2008; maide ROB. 206; CH. C. T. D 79; þat maide O. & N. 1459; maid GOW. I. 150; **mæidenes** (gen.) LAȝ. 3202; meidenes A. R. 248; maidene (dat.) MAT. xiv. 11; **mæidene**, maidene (pl.) LAȝ. 2214, 20963; meidene HOM. I. 143; maide [meide] O. & N. 1338; **meidene** (gen. pl.) MARH. 4; maidene REL. I. 128 (HOM. II. 161); maidenen (dat. pl.) LAȝ. 2741; comp. **būr-maiden.**

maȝden-child, sb., O.E. mǣgdencild; female child, ORM. 7897; maidin-, maidechild ['puella'] WICL. GEN. xxiv. 55, 57; mæide-, maidechild LAȝ. 14378, 24529.

maȝden-hād, sb., O.E. mǣgdenhād; maiden-hood, virginity, ORM. 4606; meiden-, meide-hod A. R. 164, 394; maidenhod AYENB. 220 maidenhod, -hede CH. C. T. A 2329, D 69; maidenhed GEN. & EX. 1852; GOW. II. 230; maidenhede TRIST. 2134.

meide-lūre, sb., loss of virginity, A. R. 164, 204.

maȝden-man, sb., O.E. mǣgdenman; virgin, ORM. 2085; maidenmon SPEC. 82.

maide-kin, sb., M.Du. magdeken; little girl; 'puella,' PR. P. 319.

mæin, sb., O.E. mǣgen, mægn, = O.L.G. megin, O.H.G. magin, megin, O.N. magn; megin, megn n., mod.Eng. main; from mæi²; power, force; magen [maigen] MAT. xiii. 54; xxii. 29; main LAȝ. 1541; ALIS. 656; P. L. S. xxxii. 10; LUD. COV. 25; mein MARH. 12; C. L. 1479; SPEC. 47; wiþ al his mein [main] FL. & BL. 17, 621; **mæine** (dat.) LAȝ. 1918; meine MISC. 95; mid alle meine HOM. I. 123; maine O. & N. 760; ROB. 49; GOW. III. 4; TRIAM. 740; **mægene** (pl.) MAT. xxiv. 29.

mæin-clubbe, sb., heavy club, LAȝ. 15292.

main-dint, sb., hard blow; **main dintes** (pl.) GAW. 336.

mein-ful, adj., *powerful*, KATH. 2072; mainful A. P. i. 1092.

main-hors, sb., *powerful horse*, GAW. 187.

main-lēs, adj., *O.E.* magenlēas; *without might*, REL. I. 211.

main-līche, adv., *powerfully*, LAӠ. 808*; mainli A. P. ii. 1427; ALEX. (Sk.) 399, 1033, 1217; maineli 3969.

main-strong, adj., *O.E.* mægenstrang; *powerfully strong*, LAӠ. 27731.

mæӡnan, v., *O.E.* mægnian; *strengthen*; (*? printed* meӡhan) HOM. I. 15.

[mæið, sb., *O.E.* mægeð, mægð, = *Goth.* magaþs, *O.H.G.* magad; *virgin*.]

meið-hād, sb., *O.E.* mægðhād; *virginity*, MARH. 8; maiðhod HOM. II. 21; maӡþhad ORM. 2339.

mæiþ, sb., *O.E.* mægð, *cf. O.H.G.* (ge-)māgeda; *family, race*; '(*proge*)*nies, tribus*,' FRAG. 2; hire fader ... was of Asæres maӡþe ORM. 7678.

mēl, sb., *O.E.* mæl, =*O.N.* mæl, mēl, māl, *O.Fris.* mēl, māl, *O.L.G., O.H.G.* māl, *Goth.* mēl *n.* (καιρός); *meal* (*repast*), LAӠ. 19690; mel KATH. 1839; GEN. & EX. 1020; BEK. 2395; mel [meel] CH. C. T. *D* 1774; mel [meal] A. R. 262; meel '*pastus*' PR. P. 331; P. L. S. xxx. 1; mēle (*dat.*) GAM. 636; et ane mele HOM. I. 31; to þon mele LAӠ. 8105; mēles (*pl.*) ORM. 4959; *comp.* underncuppe-, floc-, hīpel-, lim-, pūnd-, scēaf-, stūnd-, wūke-mēlen, (-mēlum, -mālum, -mēle).

mēl-seotel, sb., *seat at table*, MARH. 11.

mēl-tīd, sb., *O.N.* mæltīð, =*M.H.G.* mālzīt; *meal-time*, REL. I. 132; CH. TRO. ii. 1556.

mēl-tīma, sb., *O.E.* mæltīma; *meal-time*, HOM. I. 115; meeltime LANGL. *B* iii. 223.

mēlen, v., *O.E.* mælan, =*O.N.* mæla, *O.H.G.* mahalōn; *from* māl²; *see* maðelen; *speak*, ORM. 2919; mele MISC. 86; HAV. 2059; SPEC. 34; WILL. 4009; BEV. 1243; TRIST. 168; GAW. 2295; D. ARTH. 990; ANT. ARTH. vi; mēaleð (*pres.*) KATH. 1325; melez A. P. i. 496; melis [mellis] ALEX. (Sk.) 2078, 2953; mēlen (*subj.*) JOS. 46; mēalde (*pret.*) KATH. 1245; melid ALEX. (Sk.) 5113; *comp.* ӡe-mēlen.

mēling, sb., *conversation*, WILL. 760.

mǣne, adj., *cf. O.Fris.* mēne; =ӡemǣne; *mean, common*; þe mene and þe riche LANGL. *B* PROL. 18; þe mene folk MAN. (H.) 168; a mene kniӡtes douӡter TREV. III. 257; of mene lore MIRC 2037; þe mene men P. S. 342; þe mene sinnes PR. C. 3192; þeo beon to alle men ... meane (*v. r.* imeane) H. M. 19; mēner (*compar.*) LANGL. *B* xiv. 166; L. C. C. 7.

mǣne-līch, adj., *common, general*: oure meneliche loverd REL. I. 282; al was mænelike þing ORM. 2503.

mǣne², māne, sb., *for* ӡemǣne; *companion*: Aþulf was his mone HORN (L.) 528.

mǣne³, māne, sb., *for* ӡemǣne; *communion, participation, companionship*, ORM. 1948; mene GEN. & EX. 501; meane HOM. I. 275; wiðuten monnes man(e) MARH. 13; mone SHOR. 61, 70; mone (*sec. text* imone) LAӠ. 25916; wiþoute mone ROB. 315.

mǣnnesse, sb., *O.E.* gemænness; *communion*: þe inennesse of halӡen AYENB. 14.

mǣne⁴, māne, sb., = *O.Fris.* mēne, *O.H.G.* meina (*opinio*); *communication, complaint*: to hire ich make min mene REL. I. 103; in a false mene LANGL. *B* xv. 535; mane REL. I. 60; IW. 535; M. H. 119; MIN. iii. 8; mone O. & N. 1520; ROB. 34; SHOR. 89; ALIS. 1281; A. D. 247; FER. 910; CH. C. T. *F* 920; godes prophete makede swuche mone A. R. 64; and made hir mone to Piers LANGL. *B* vi. 125; bi mine mone HAV. 816; moone TRIAM. 182.

mǣnen, mēnen, v., *O.E.* mǣnan, =*O.Fris.* mēna, *O.L.G.* mēnian, *O.H.G.* meinen, *O.N.* meina; *mean; communicate, indicate, signify; complain, moan*; LAӠ. 8294, 29613; menan HOM. I. 111; menen A. R. 274, 316; ORM. 4817; meanen KATH. 1243; mene C. L. 401; TRIST. 1135; GOW. I. 94; TRIAM. 28, 473; ne scal him no man mene þer of strengþe ne of wrange P. L. S. viii. 85; hwi wulleþ men of me hi mene O. & N. 1257; hwat mai þis mene HAV. 2114; hei(e) men ... wolde hem ... mene þat hii nadde non eir ROB. 330; to mene ant minne SPEC. 37; to mene þe soþe WILL. 2218; to whom miӡt i me mene 493; his biriels for to mene MAN. (F.) 9026; what hit wolde mene 9050; manen REL. I. 274; monen GEN. & EX. 180; mone RICH. 4636; PERC. 128; mēne (*pres.*) CH. C. T. *A* 2063; ӡif þu me dest woh ... ic hit mene to mine laverde HOM. I. 33; an oþer þing of þe ich mene (*indicate*) O. & N. 583; þu ... menis þe for litill ALEX. (Sk.) 2741; hit is a schep-(h)erde þat i of mene M. T. 74; hit ... mænet (*laments*) þeo weowe FRAG. 5; as Marc meneþ in þe gospel LANGL. *B* x. 276; and meniþ him bi name TOR. 759; so so bec mæneþ FRAG. 8; þei menen MAND. 132; mēn (*imper.*) HOM. I. 17; mǣnde, mende (*pret.*) LAӠ. 1904, 2438; mende A. R. 64; REL. II. 243; heo nuste hwat heo (*r.* he) mende MISC. 85; ase ich er mende SHOR. 129; menden MARH. 6; mēned (*pple.*) WILL. 1490; *comp.* bi-mænen.

mēninge, sb., =*O.H.G.* meinunge; *meaning, communication, quest, moaning*, PR. P. 332; HOM. II. 63; MAN. (H.) 25; PERC. 1986; ANT. ARTH. xlvi.

mǣr [mēr], sb., *O.E.* (ge-)mǣre, =*O.N.* mǣri *n.*; *limit, boundary, terminus*, LAӠ. 2133; meer '*limes*' PR. P. 333; mere LUD. COV. 171; mēre (*dat.*) A. P. ii. 778; mǣre (*ms.*

mare) (*pl.*) LAȝ. 31220; meres ['*terminos*'] PS. ii. 8; þe meres and þe merkes TREV. V. 265; mæren ['*finibus*'] MK. v. 17 ; *comp.* ȝe-mǣre.

mēre-stān, sb., *boundary stone*, CATH. 235.

mǣre, *see* mere.

mǣre, adj., *O.E.* mǣre,=*O.N.* mǣrr, *O.L.G.* māri, *O.H.G.* māri, *Goth.* mērs ; *famous, illustrious, glorious*, ORM. 806; mære [mere] LAȝ. 1139 (*miswritten* mare 2028) ; mere P. L. S. viii. 196; SPEC. 26 ; GAW. 878 ; mere, meare JUL. 62, 63 ; mēren (*dat. n.*) LAȝ. 409 ; mēren (*dat. pl.*) HOM. I. 221.

   mǣr-līche, adv., *O.E.* mǣrlīce ; *gloriously*, LAȝ. 2677.

mǣrsien, v., *O.E.* mǣrsian,=*O.L.G.* (gi-)-mārsōn ; *celebrate, magnify* ; mǎrsiað (*pres.*) MAT. xxiii. 5; *comp.* ȝe-mǣrsien.

mǣrðe, sb., *O.E.* mǣrðe,=*O.L.G.* māritha, *O.H.G.* mārida, *Goth.* mēriþa ; *glory* : ne delende nere of his eadinesse nof his merče (? *ms.* merhðe) HOM. I. 217.

mǣschen, v., *mash, beat into a mash* : mid þine clivres woldest me meshe [meisse] O. & N. 84.

mǣsse [**masse**], sb., *O.E.* mæsse, *O.Fr.* messe; *mass (eucharistic service)*, LAȝ. 18520; messe A. R. 32 ; SHOR. 27 ; S. S. (Wr.) 1631; messe, masse PR. P. 334; cristes masse *Christmas* O. & N. 481 ; messen (*pl.*) AYENB. 31.

   messe-bōc, sb., *O.E.* mæssebōc ; *mass-book*, ORM. DEDIC. 31 ; messebok HAV. 186.

   messe-cos, sb., *mass-kiss*, A. R. 34 ; HOM. II. 91.

   messe-daȝ, sb., *O.E.* mæsse dæg ; *mass-day*, ORM. 2720 ; messedaȝes (*pl.*) AYENB. 214.

   messe-gere, sb., *mass-garments*, HAV. 188.

   messe-hakele, sb., = *M.L.G.* mishakel ; *chasuble*, SAX. CHR. 249 ; meshakele HOM. II. 163.

   messe-hwīle, sb., *time of celebrating mass* ; messeqvile GAW. 1097.

   masse-prēost, sb., *O.E.* mæsseprēost ; *mass priest*, LAȝ. 29872.

   messe-rēf, sb., *O.E.* mæsserēaf ; *mass-robe*, HOM. II. 215.

mǣssien, v., *O.E.* mæssian ; *say mass* ; messeð (*pres.*) A. R. 268.

mǣst, sb., *O.E.* mæst, = *O.H.G.* mast ; *mast (of ship)*, LAȝ. 4593 ; mast '*malus*' PR. P. 329 ; CH. C. T. *A* 3264 ; mastes (*pl.*) LAȝ. 1100.

mǣst², sb., *O.E.* mæst, = *O.H.G.* mast ; (*beech*)-*mast* ; thei eten mast hawes and swich pounage CH. F. AGE 7 ; mast or apples 37 ; þe wilde bar þenne he i þan mæste [maste] monie swin imeteð LAȝ. 21263 ; *comp.* bōc-mast.

mast-hog, sb., *hog glutted with berries*, PR. P. 329.

   mast-swīn, sb.,=*M.H.G.* mastswīn; '*maialis*,' PR. P. 329.

mǣst, *see under* mā.

mǣsten, v., *O.E.* mæstan,=*O.H.G.* masten, *M.Du.* mesten, mastin ; *fatten up,* '*sagino*,' PR. P. 329.

mǣsti, adj., *cf. Ger.* mastig ; *over-fed, glutted with berries* : mastie swine CH. H. F. 1777.

mǣstling, sb., *O.E.* mæstling ; *mixture of anything; copper; brass* ; mestling A. R. 284* ; meastling H. M. 9 ; mastling ROB. 87 ; messeline ALEX. (Sk.) 4583 ; mastilyon '*bigermen, mixtilio*' CATH. 230 ; *comp.* gold-mestling.

   mǣstling-smiþ, sb., *coppersmith*, FRAG. 2.

mǣte, sb.,= *M.L.G.* māte, = *M.Du.* maete, *O.H.G.* māza ; *measure* ; mete HOM. I. 247; mesure & mete H. M. 41 ; of his oune meete TREV. II. 27.

mǣte², adj., *O.E.* mǣte, = *O.Fris.* mēte, *M.L.G.* māte, *O.H.G.* (ge-, eben-)māzi ; *meet, moderate, suitable, fit* ; mete SPEC. 36; PL. CR. 428; PERC. 802; LUD. COV. 318; þe tre was al sva mete and qveme C. M. 8809; mete [meete] CH. C. T. *A* 2291 ; meete ALIS. (Sk.) 184 ; þei wepen stille ant mete 85 ; *comp.* efen-, ȝe-, or-, un-mēte.

   mēt-leȝc, sb., *modesty*, ORM. 2659.

   mēte-līke, adv., *meetly*, ORM. 10703 ; meteli GAW. 1414 ; meetli R. R. 822.

   mētnesse, sb., *O.E.* (un-)mǣtnesse ; *moderation*, HOM. I. 105.

mǣtels, sb., *dream* ; metels ['*somnium*'] WICL. GEN. xxxvii. 5; metels [meteles] LANGL. *B* PROL. 208 ; mēteles (*pl.*) AYENB. 165.

mǣten, v., *O.E.* mǣtan ; *paint; dream* : lette . . . a swiče wunderlich hweol meten (*design*) & makien JUL. 57 ; meten [meeten] a . . . swevene LANGL. *A* PROL. 11 ; mete WILL. 658 ; mēteþ (*pres.*) REL. I. 262 ; met AYENB. 128 ; mētte (*pret.*) P. L. S. xiii. 132 ; JOS. 442 ; CH. C. T. *A* 3684 ; mēted (*pple.*) *painted* ORM. 1047 ; met *dreamed* HAV. 1285 ; *comp.* ȝe-mǣten.

   mētere, *O.E.* mētere ; *painter; dreamer;* '*pictor*,' FRAG. 2 ; mēteres (*pl.*) AYENB. 32.

   mētinge, sb., *O.E.* mǣting ; *dream, picture*, AYENB. 143 ; meting ALIS. 327 ; WILL. 706 : LANGL. *B* xiii. 4 ; meting '*pictura*' FRAG. 2.

mǣðe [**mēþe**], sb., *O.E* mǣð *f.*; *measure, moderation, modesty*, LAȝ. 977 ; mæþ ORM. 7515 ; meče GEN. & EX. 3601 ; man ne can his muðes meče REL. I. 131 ; methe ALEX. 4325 ; ED. 1716 ; meþ SPEC. 103 ; wiþouten meþ(e) C. L. 318 ; in mesure & meþe A. P. ii. 247 ; Mari ledd(e) hir lif wiþ meþe M. H. 107 ; *comp.* un-mēþ.

   mēð-ful, adj., *reasonable*, A. R. 430 ; meþful SPEC. 32; methful PS. iii. 6; meaðful HOM. II.257.

mēð-lēas, adj., *immoderate*, A. R. 96; methles GAW. 2106.

mēðlēas-liche, adv., *unceasingly*, A. R. 330.

mē̆þe², adj., *modest, reasonable*; meþe HALLIW. 552; meþ C. M. 12271; *comp.* unmēðe.

mēðe-līche, adv., *moderately*, HOM. II. 7; meðelike GEN. & EX. 1758.

mēðeʒien, v., O.E. (ge-)mǣðehian; *moderate, temper*: we sculleð meðegie heore mod LAʒ. 25231; efter þat þe he mēðegeð (*pret.*) nu his dede shal eft ben mēðeged (*pple.*) his mede HOM. II. 153.

mēðien, v., ? O.E. mǣðian; *moderate, accommodate, temper, spare*: ic sal meðen ðe stede GEN. & EX. 1046; þe man þe hit mēðeð (*pres.*) riht REL. I. 132; mēðede (*pret.*) þo his liflode swo þat he was bicumelich to swiche wike HOM. II. 139; an angel meðede hire þat ned GEN. & EX. 1242; ʒif þei hem self couþe have mēþed (*pple.*) MAN.(F.)13615.

mē̆w, sb., O.E. mǣw,=O.N. mār (*larus*); *mew, gull*; meaw FRAG. 3; mowe PR. P. 346; *comp.* sē-māwe.

maflen, v., *cf. M.Du.* maffelen; *stammer*: he wot nouʒt what he maffleþ (*pres.*) TREV. II. 91; mafflid (*pret.*) DEP. R. iv. 63.

magat, magot, sb., *cf.* maðek; *maggot*, PR. P. 321; maked VOC. 255.

maʒ, *see* mæi. māʒ, *see* mǣi.

maʒe, sb., O.E. maga,=O.H.G. mago, O.N. magi; *maw, stomach*, AYENB. 56; mahe H. M. 35; R. S. v (MISC. 178); maghe ['*vulva*'] *womb* HAMP. lvii. 3; mawe FRAG. 6; A. R. 370; mawe MISC. 179; ROB. 311; ALIS. 1260; LANGL. B xi. 56; P. 43; CH. C. T. B 486; A. P. iii. 255; M. H. 124; mawen (*pl.*) MISC. 151.

[maʒe², adj., O.E. maga, *cf.* O.H.G. (un-)mag; *powerful*; *comp.* un-maʒe.]

māge, *see* mǣʒe.   maʒden, *see* mǣiden.

maggen, v., *cf. Sc.* magil, maggle; *mangle*; magged (*pple.*) ALEX. (Sk.) 1268.

magicien, sb., *Fr.* magicien; *magician*, CH. C. T. B 3397.

magike, sb., *enchantment*, CH. C. T. B 214.

magnanimitē, sb., *Fr.* magnanimité; *magnanimity*, AYENB. 164; magnanimitee CH. C. T. G 110.

magnete, sb., O.Fr. magnete; *magnet*, '*magnes*,' PR. P. 325.

magnificence, sb., *Fr.* magnificence; *magnificence*, AYENB. 164; CH. C. T. B 1664.

magnifien, v., O.Fr.magnifier; *magnify*, PR. P. 325; magnifíe (*pres.*) WICL. MAT. xxiii. 5.

magot, *see* magat.   magrē, *see* maugrē.

māʒþ, *see* mǣiþ.

māh, adj., O.E. māh; *eager*, LAʒ. 11201; þa itimede þan deofle al swa deð mahʒe fisce HOM. I. 123; *comp* ʒe-māh.

mahe, *see* maʒe.

mahimet, maumet, mahum (mahun), sb., *Mahomet, idol*, LAʒ. 230, 14585, 29221; mahounde YORK xi. 401; maumet P. L. S. xv. 226; MAN. (F.) 1346; maumez (*pl.*) JUL. 5; maumetes HAMP. PS. cxiii. 12; *deriv.* maumetri.

maht, *see* meaht.

mai, pr. n., O.Fr. mai; *May*, PR. P. 319; mæi [mai] LAʒ. 24200; ALEX. (Sk.) 3699.

mai, *see* mæi, mǣi.

maiden, maigden, *see* mǣiden.

maien, v., ? *for* amaien; *dismay*; maies (*imper. pl.*) not ʒour(e) hertes ALEX. (Sk.) 3010, 3570; maied (*pple.*) 5399.

mail, sb., O.Fr. mail, *Lat.* malleus; *mall, club, hammer*, FER. 4653; malle WICL. E. W. 351; TOR. 322; maul ROB. (W.) 4229; mealles (*pl.*) HOM. I. 253; melles PR. C. 6572; mellis ALEX. (Sk.) 4100.

maile, sb., O.Fr. maile, maille,=*Ital.* maglia, *Lat.* macula; *coat of mail*, PR. P. 320; maille MAN. (F.) 2725; mailes (*pl.*) ANT. ARTH. xxx; in starand mailis ALEX. (Sk.) 3615.

maillet, sb., O.Fr. maillet; *mallet*, PARTEN. 4698; maliet PR. P. 323.

maim, sb., O.Fr. mehaing; *maimed person*; maimes (*pl.*) AYENB. 135.

maimen, mainen, v., O.Fr. mahaigner, mehaigner; *maim*; maime CH. C. T. D 1132; maimin, mainin '*mutilo*' PR. P. 320; maines (*pres.*) ALEX. (Sk.) 5153; imaimed (*pple.*) ROB. 288; AYENB. 135; LANGL. B xvii. 189; mained ALEX. (Sk.) 1273; (*wretched*) 4544.

maining, sb., *maiming*, ALEX. (Sk.) 4088.

main, adj., *strong, large, chief*, ALEX. (Sk.) 3018, 3777, 3932.

main, *see* mæin.

mainē, sb., O.Fr. maisnee (*maisonnée*); *household*, FL. & BL. 782; AYENB. 30; maine, meine ROB. 108, 167; WILL. 184, 416; meine HAV. 827; MAN. (F.) 8427; meinee MAND. 40; meneyhe [meinʒei] ALEX. (Sk.) 3120; meine *chessmen* MAN. (F.) 11396.

maineal, adj., *menial*; meineal TREV. VII. 413; WICL. ROM. xvi. 5.

mainpernour, sb., O.Fr. mainpreneur; *mainpernor*, LANGL. A iv. 99.

main-, meinprise, sb., ? O.Fr. mainprise, *bail*, LANGL. A iv. 75, B iv. 88; GAM. 744.

mainprisin, v., *mainprise*, '*mancipo*,' PR. P. 320; main-, meinprise LANGL. B iv. 179.

maintenaunce, sb., O.Fr. maintenance; *maintenance*, ALEX. (Sk.) 1179; mentenance SHOR. 100.

main-, meintḗne, v., O.Fr. maintenir; *maintain, defend*, WILL. 2698, 3002; LANGL.

A iii. 178; manteine O. & N. 759; **mainteines** (*pres.*) ALEX. (Sk.) 4522; **mainten** (*pl.*) 1265.

maintenour, sb., *Fr.* mainteneur; *maintainer*; LANGL. *C* iv. 288.

maire, sb., *O.Fr.* maire; *mayor*, LANGL. *B* iii. 87; ALEX. (Sk.) 1557.

maister, sb., *O.Fr.* maistre, *Lat.* magister; *master*, O. & N. 1778; S. S. (Wr.) 36; maister mair ALEX. (Sk.) 1557*; maʒstre ORM. 5236; meister A. R. 236; **masteris** (*pl.*) BARB. iv. 411.

maister, adj., *chief,* ALEX. (Sk.) 2037, 5404.

maisterling, sb., *prince*; **maisterlinges** (*pl.*) HOM. II. 113; maisterlingis ALEX. (Sk.) 481.

maistir-ful, adj., *powerful,* WICL. LK. xii. 58.

maister-líke, adj., *with dignity,* ALEX. (Sk.) 228.

maister-men, sb., pl., *craftsmen,* D. TROY 1599.

maister-shipmon, sb., *captain,* TOR. (A.) 1425.

maistren, v., *master*; meistren KATH. 1280; **maisteréde** (*pret.*) D. ARTH. 2683; masteřit BARB. vii. 211; *comp.* a-, overmaistren.

maistresse, sb., *O.Fr.* maistresse; *mistress,* AYENB. 34; CH. BOET. i. 3 (10); TRIST. 102; maistres ALEX. (Sk.) 3763.

maistrice, sb., *O.Fr.* maistrise; *mastery; feat of skill*; BARB. iv. 524, vi. 566; maistrise *powerful charms* ALEX. (Sk.) 333.

maistrie, sb., *O.Fr.* maistrie; *mastery, lordship, dominion,* PR. P. 320; ROB. 3; TRIST. 72; meistrie A. R. 140; **maistries** (*pl.*) YORK xxxvii. 216.

maiŏ, *see* **mæiŏ.**

maitlēs, *see under* **máte.**

maiþe, sb.; *O.E.* mægˣe, mageˣe; *mayweed,* '*camomilla*,' (*printed* maiwe) VOC. 140, (maithe) 162, 190.

majestē, sb., *O.Fr.* majesté; *majesty,* TREV. IV. 9; majeste [mageste] CH. C. T. *B* 3334; mageste P. L. S. xvii. 264.

māk, *see* **maŏek.**

máke, sb., *make, construction*; make [makke] ALEX. (Sk.) 3218; (*timber*) PALL. ii. 437.

máke², sb., *O.N.* maki, *prov.Eng.* make,=ʒe-máke; *consort, partner, friend, equal,* A. R. 114; ORM. 1276; O. & N. 1159; REL. I. 224; HORN (L.) 1409; ĦAV. 1150; P. L. S. xix. 256; SHOR. 67; WILL. 1898; ALIS. 3314; LANGL. *A* iii. 114; S. S. (Wr.) 239; GOW. I. 45; DEGR. 1116; M. ARTH. 1062; LIDG. M. P. 153; þe turtle þat haþ lost hir make ĊH. C. T. *E* 2080; he wenes his mak(e) mai na man find(e) M. H. 49.

máke-lēs, adj., *without a mate,* MARH. 11; C. L. 819; CH. TRO. ị. 172; ANT. ARTH. xxvii.

máke³, adj., *O.N.* makr; =ʒemáke; *easy,* '*aptus,*' PR. P. 321.

mak-lī, adv.,=*M.L.G.* gemaclīch, *M.H.G.* gemechlīche; *easily,* '*faciliter,*' PR. P. 322.

makerel, sb., *O.Fr.* maquerel; *mackerel,* PR. P. 321; HAV. 758.

mákien, v., *O.E.* macian,=*O.Fris.* makia, *O.L.G.* macōn, *O.H.G.* machōn; *make,* A. R. 6; PROCL. 4; makien us stronge KATH. 1026; makien [maki] inc riche LAʒ. 5621; ane neowe Troie þar makian 1246; make-ʒen sehtnesse and griŏ 31370; heo hine wolden maken [makie] duc 362; maken ORM. 1480; GEN. & EX. 278; C. L. 479; Grim dede maken a ful fair bed HAV. 658; make I.ANGL. *A* ii. 117; maken *B* ii. 147; maki TREAT. 138; AYENB. 27; maa (make) ALEX. (Sk.) 1761; **mákie** (*pres.*) SPEC. 86; ich makie moncun wurchen to wundre JUL. 38; make *write verses* LANGL. *B* xii. 22; makest O. & N. 339; he makeŏ A. R. 224; (þu, he) makis, makes [mase, mas] ALEX. (Sk.) 831, 2908, 2210, 2587; mace GAW. 1885; (ha) makieˣ MARH. 11; we makieþ P. L. S. xviii. 1; makieþ, makeþ O. & N. 650; **mákie** (*subj.*) LAʒ. 5880; **mákede** (*pret.*) A. R. 224; TREAT. 132; & makede mani(e) weorkes SAX. CHR. 263; he makede ane heʒe burh LAʒ. 218; (?) makode 1795; made MAND. 261; and made hem alle to knihte HORN (R.) 522; made [maade] C. L. 34; makeden hine to duke LAʒ. 419; maden GEN. & EX. 1992; made *wrote verses* LANGL. *C* vi. 5; **máked** (*pple.*) ORM. 995; SHOR. 71; WILL. 1951; maked, mad GEN. & EX. 183, 2470; mad [maad] CH. C. T. *A* 394; *comp.* ʒe-mákien.

mákere, sb.,=*O.H.G.* machāre; *maker, author,* AYENB. 262, 269; makere, maker LANGL. *B* x. 240; maker CATH. 226; macare PR. P. 319.

máking, sb.,=*O.H.G.* machunga; *making, verse-making; feature*; LANGL. *A* xi. 32; makinge AYENB. 92; LANGL. *C* xiv. 193.

māl, sb., *O.E.* māl,=*M.Du.* mael, *?O.H.G.* meil, *?Goth.* mail; *mole, spot*; mool REL. I. 108; **mōles** (*pl.*) and spottes LANGL. *B* xiii. 315.

māl², sb., *O.E., O.N.* māl,=*O.L.G., O.H.G.* mahal (*speech*), *mod.Eng.* (black-)mail; *speech, payment, tribute* : o Grickishe mal ORM. 4270; samnen laʒhelike & riht þe kinges rihte male 10188; mol HOM. II. 179; hi token unriht mol MISC. 151; moal GEN. & EX. 81; wiþouten male C. M. 5376.

malàdie, sb., *O.Fr.* maladie; *malady,* MISC. 31; CH. C. T. *A* 419; malidi, maledi ALEX. (Sk.) 4281, 2127*.

malancoli, *see* **melancolie.**

mále, sb., *O.Fr.* male, *from O.H.G.* malha, malaha, *mod.Eng.* mail (*post*); *bag, wallet,* '*mantica*,' PR. P. 323 ; LA3. 3543 ; HAV. 48 ; BEV. 1297 ; MIRC 1342 ; SACR. 179 ; **máles** (*pl.*) LANGL. *B* v. 234.

mále, sb., *O.Fr.* mascle ; *male*, CH. C. T. *D* 122 ; **māles** (*pl.*) LANGL. *B* xi. 330, *C* xiv. 147.

maledight, pple., *O.Fr.* maledict ; *cursed*, C. M. 1134.

malegrēfe, *see* **maugrē.**

mālen, v., *cf. O.E.* (ge-)mǣlan ; *spot* ; **mōled** (*pple.*) LANGL. *B* xiii. 275 ; *comp.* **bi-mōlen.**

malencoli, *see* **melancoli.**

malēse, sb., *O.Fr.* malaise ; *discomfort, sickness*, LANGL. *C* ix. 233 ; **maleese** WICL. MAT. iv. 24 ; **maill-eiss** BARB. xx. 75.

malice, sb., *O.Fr.* malice ; *malice, evil*, AYENB. 26 ; CH. C. T. *B* 363 ; ROB. 570 ; **maleces** (*pl.*) *works of malice* ALEX. (Sk.) *page* 279.

maliciūs, adj., *O.Fr.* malicius ; *malicious*, A. R. 210* ; **malicious** S. S. (Web.) 1503 ; A. P. iii. 508.

malicoli, *see* **melancolie.**

malignitee, sb., *O.Fr.* malignité ; *malignity*, CH. C. T. *I* 573.

malingnen, v., *malign* ; ye **malingne** (*pres.*) YORK xxx. 506.

malisūn, sb., *O.Fr.* malizon, maleiçon ; *malediction*, HAV. 426 ; **malisoun** WICL. GEN. xxvii. 12 ; TRIST. 3057.

malkin, sb., *prov.Eng.* mawkin ; *oven cloth* ; '*dossorium*' PR. P. 323 ; VOC. 276.

mallard, sb., *O.Fr.* malart ; *mallard*, VOC. 177 ; malard '*anas*' PR. P. 323.

malle, *see* **mail.**

mallen, v., *O.Fr.* mailler ; *mall, strike with a mall*, JOS. 508 ; malle D. ARTH. 4037 ; deoflen þat ham **mealliδ** (*pres.*) HOM. I. 251 ; **malling** (*pple.*) þurgh metall D. TROY 9520 ; **malled** (*pple.*) PALL. ii. 17.

malliable, adj., *O.Fr.* malleable ; *malleable* ; CH. *G* 1130.

Malmesbüri, pr. n., *Malmesbury*, FRAG. 5.

malskren, v., *O.E.* *malscran (bewitch ; cf.* malscrung=*O.H.G.* mascrunc *fascination)* ; *bewilder ; wander, be bewildered* ; **malsc-rande** (*pres. pple.*) A. P. ii. 991 ; maskinge ['*aberrantem*'] TREV. II. 67 ; **malskred** (*pret. pple.*) in drede A. P. iii. 255 ; hou he had(d)e . . . màlskrid aboute WILL. 416 ; malscrid (*? ms.* malstrid) and mased ALEX. (Sk.) 1270 ; maskede men BRD. 6.

malt, sb., *O.E.* mealt,=*O.N.* malt, *O.H.G.* malz ; *from* **melten** ; *malt,* '*bracium*,' FRAG. 4 ; PR. P. 323 ; LANGL. *B* PROL. 197 ; CH. C. T. *A* 3991.

maltin, v., =*M.H.G.* malzen ; *malt,* '*brasio*,' PR. P. 323.

maltestere, sb., *maltster,* '*brasiatrix, brasiator*,' PR. P. 324.

maltinge, sb., *malting,* PR. P. 324.

malvesie, sb., *malmsey, wine from Napoli di Malvasia,* D. ARTH. 236 ; CH. C. T. *B* 1260.

malwe, sb., *O.E.* mealwe, *cf. Lat.* malve ; *mallow,* '*malva*,' PR. P. 324 ; **malues** (*pl.*) PALL. v. 206.

mamelen, -ere, *see* **momelen, -ere.**

mamere, v., *? stammer,* '*mutulare*,' VOC. 203.

man, v., *O.E.* man, mon,=*O.N.* man, mon, mum, *Goth.* man, *Lat.* memini, *Gr.* μέμονα (*pret. pres.*) ; *remember, have in mind, think, will (used as sign of future tense)* ; mon PERC. 567 ; PR. C. 96 ; TOR. 1113 ; M. ARTH. 3230 ; mun L. H. R. 95 ; MIN. iii. 119 ; i mun walke on mi wai ANT. ARTH. xxv ; þou mone (*ms.* mowe, *r. w.* wone) REL. I. 113 ; þou mon be ded IW. 703 ; þai mun(e) MIN. i. 48 ; i wene that we deie mone HAV. 840 ; þer mone no dintus him dere AV. ARTH. iii ; we mon N. P. 41 ; ye mon TOWNL. 63 ; **mune** (*subj.*) ORM. 14356 ; þat all his gode dede ne mune him noht beon god inoh 7927 ; *deriv.* **mánien, müne, munen, münien.**

man, sb., *O.E.* mann (man), monn, manna (*gen.* mannes, mannan),=*O.Fris.* man, mon, *O.L.G., O.H.G.* man (*gen.* mannes), *Goth.* manna (*gen.* mans, *dat.* mann, *acc.* mannan, *nom. pl.* mans, *acc.* mannans), *O.N.* mannr, maδr ; *man, human being ; male person ; adult male ; also as quasi-pronoun* [=*Ger.* man, *cf. Fr.* on] *one* ; ROB. 38 ; SHOR. 1 ; AYENB. 21 ; WILL. 377 ; ALIS. 1579 ; MAND. 1 ; IW. 2109 ; PR. C. 42 ; Richard þat was his man (*liegeman, vassal*) BEK. 4 ; man (*ms.* mann) hem halt for gode men ORM. 387 ; gef δer is no man (*no one*) REL. I. 223 ; man, mon O. & N. 455, 477, 573 ; nan man LA3. 6257 ; what mai man vinden 15888 ; a wel ibore mon [man] LA3. 678 ; forδ mon [mè] brohte þat water 14948 ; mon FRAG. 6 ; KATH. 925 ; A. R. 8 ; ROB. 10 ; C. L. 622 ; MIRC 105 ; AUD. 6 ; **mannes** (*gen.*) ORM. 1459 ; CH. C. T. *B* 287 ; **monnes** [mannes] LA3. 1566 ; mannes, monnes HOM. I. 97 ; O. & N. 338, 1476 ; **manne** AYENB. 9 ; **men** (*dat.*) MAT. vii. 26 ; HOM. I. 93 ; II. 5 ; (*ms.* man, *r. w.* Moisen) GEN. & EX. 3110 ; manne LEECHD. III. 84 ; O. & N. 800, 1489 ; monne [manne] LA3. 65 ; monne A. R. 68 ; man P. L. S. viii. 10 ; **men** (*nom. acc. pl.*) LA3. 541 ; O. & N. 385 ; LANGL. *A & B* PROL. 32 ; (*ms.* menn) ORM. 123 ; men (*for* man, *impersonal*) shal of (h)is owen galle shenchen him REL. I. 113 (A. D. 294) ; swa men [me] dide sone KATH. 1552 ; what men [me] telle HORN (L.) 366, (R.) 372, (H.) 378 ; in eorδe me hine sette LA3. 27933 ; ase me seiδ A. R. 54 ; me . . . halt O. & N. 32 ; me him wiþnimþ AYENB. 17 ; mon-nan, monnen, monne HOM. I. 15, 21, 109 ; **manne** (*gen. pl.*) P. L. S. viii. 82 ; HOM. II. 19 ; ORM. 2597 ; ROB. 312 ; feower þusend manne MAT. xv. 38 ; to manne frame REL. I. 209 (MISC. 2) ; oþre manne þing AYENB. 11 ; manne, monne O. & N. 475, 585 ; monna, monne, monnan, HOM. I. 91, 97, 135 ; monne A. R. 384 ; monne,

monnen LA3. 639, 24509; mannen HOM. I.
241; II. 7; menne [men] LANGL. *C* iv. 102;
mennes ORM. 386; WILL. 6; MAND. 104;
monnam, monnan, monnen, mennen, monne
(*dat. pl.*) HOM. I. 43, 111, 117, 129; mannan,
mannen MAT. v. 13; JOHN xvii. 6; HOM. I.
327, 231; monnen FRAG. 7; monnen [menne]
LA3. 344; manne LEECHD. III. 100; P. L. S.
xix. 297; manne, monne O. & N. 234, 389;
men SAX. CHR. 257; LA3. 350*; KATH. 406;
A. R. 2ɛ6; BEK. 310; *comp.* aker-, alder-,
bēr-, bonde-, bur3-, carl-, chĕp-, dri3-,
fā-, fōt-, glēo-, 3o-, hēaved-, heorde-, hors-,
lēof-, mére-, sǣ-, stēor-, þēow-, weorc-,
wēp-, wīf-man.

man-cün, sb., *O.E.* mancynn, *O.N.* mannkyn;
mankind, P. L. S. viii. 153; man-, monkun
O. & N. 973; monkun LA3. 436; H. H. 110;
mankin ORM. 1; GEN. & EX. 240; A. P. i.
636; monkin SPEC. 95.

mon-drēam, sb., *O.E.* mandrēam; *shout
of men*; KATH. 2046; þa aras þe mondrem
þat þe volde dunede a3en LA3. 23945.

mon-ferde [**manverde**], sb., LA3. 10774,
16453.

man-hōd, sb., = *M.H.G.* manheit; *manhood*,
SHOR. 148; CH. C. T. *A* 756; manhede
WILL. 431; monhede ROB. 131; C. L. 649.

man-künde, **monkuinde**, sb., *mankind*,
L. H. R. 18, 19; monkuinde, -kinde C. L.
1066; mankinde SPEC. 81; GOW. II. 83;
mankende AYENB. 1.

man-līch, adj., = *O.N.* mannligr, *M.H.G.*
manlīch; *manly, masculine, human*, P. L. S.
xxiii. 33; LANGL. *B* v. 260; monlich KATH.
1324; A. R. 272; manli Iw. 3207; **monlīche**
(*adv.*) LA3. 26855; **monlüker** (*compar.*) A.
R. 422; **manlokest** (*superl.*) WILL. 3419.

manlīc-hēd, sb., = *M.H.G.* manlīcheit; *humanity*; GEN. & EX. 23; manlihed PARTEN.
4352.

man-līche, sb., *O.E.* manlīca, = *Goth.* man-
leika, *O.H.G.* manlīcha; *human form*, REL.
I. 234.

mon-qvalm [**mancwalm**], sb., *O.E.* man-
cwealm; *mortal disease*, LA3. 3908; **man-
cwalmes** (*pl.*) ['*pestilentiae*'] MAT. xxiv. 7;
manqvalm HAMP. PS. i. 1*.

mon-qvelle, sb., *murderer*, R. S. vii.

manqvellere, sb., *homicide, slayer of man*,
['*homicida*'] WICL. MK. vi. 27; WILL. 993;
monqvellere HOM. I. 279; **manqvellers** (*pl.*)
ROB. (W.) 9333.

man-rēd, sb., ? = **manrēden**; SAX. CHR.
261.

mon-rēdne, **-rādene**, sb., *O.E.* manrǣden;
homage, LA3. 420, 6240; manrede HAV. 484;
P. L. S. xiii. 203; ALIS. 4665; monraden ANT.
ARTH. 1; monrade H. H. 88 (90); manredin
BARB. xvi. 303; manrent v. 296.

man-scipe, sb., *dignity, honour*, HOM. I.

235; monscipe (mansipe) LA3. 345; manshipe
ORM. 19014; man(s)chipe WILL. 3337.

mon-slæht, sb., *O.E.* manslyht, *cf. O.H.G.*
manslaht; *manslaughter; murder*; LA3.
27826; monsleiht A. R. 56; manslagt GEN.
& EX. 485; mansla3t BEK. 365; manslau3t,
monslauht L. H. R. 30, 31; monsla3t MIRC
1535; AUD. 2; Marcure was mansla3t (*? manslayer*) ALEX. (Sk.) 4498.

mon-sla3e, sb., *O.E.* manslaga; *homicide,
manslayer*, HOM. I. 13; monslahe MARH.
11; **monsla3en** (*pl.*) HOM. I. 53.

man-sla3ter, sb., *manslaughter*, C. M. 25749;
manslauter TREV. III. 473; manslatir ALEX.
(Sk.) 4486.

man-sla3þe, sb., *for* **mansla3e**; *homicide*;
AYENB. 8; mansle3þe SHOR. 152.

man-slēer, sb., *manslayer*, TREV. III. 41.

man-þēau, sb., *O.E.* manþēaw; *manly
habits*: ler him **monþēwes** (*pl.*) MISC. 128
(REL. I. 184).

mon-weored, sb., *troop of men*, LA3. 20393.

mān, sb., *O.E.* mān, = *O.L.G.* mēn, *O.N.*,
O.H.G. mein; *wickedness, sin*: þat tu wilt
loke wel fra man in aþes & i witness ORM.
4478.

mān-āð, sb., *O.E.* mānāð, = *O.L.G.* mēneth,
O.N. meineiðr, *O.H.G.* meineid; *perjury*,
HOM. I. 49; manaþ ORM. 4480; monoð HOM.
II. 215.

mān-dēde, sb., *O.E.* māndǣd; *crime*, HOM.
I. 99.

mōn-ful, adj., *O.E.* mānfull; '*profanus*,'
FRAG. 4.

mōn-sware, sb., *O.E.* mānswara, = *O.N.*
meinsvari; *perjurer*, FRAG. 7; monsware
LA3. 4148.

mān-sworn, pple., *perjured*, Iw. 3938;
manswore LA3. 22139; monsworn A. P. ii. 182.

manáce, *see* **menáce**.

manciple, sb., *O.Fr.* mancipe; *manciple*,
A. R. 214; CH. C. T. *A* 544.

mande, sb., *O.E.* mand, = *M.L.G.*, *M.Du.*
mande; *maund, basket*; maund(e) '*sportula*'
PR. P. 330.

mandē, sb., *O.Fr.* mandé, *from Lat.* man-
dātum; *Maunday Thursday observance*, BRD.
17; maunde [maundee] LANGL. *B* xvi. 140.

mandement, sb., *O.Fr.* mandement; *command*, ROB. 194; mandement [maundement]
CH. C. T. *D* 1360; maundement WICL. NUM.
xv. 15; mandment ALEX. (Sk.) 3531, 3541.

manden, v., *O.Fr.* mander, *from Lat.* mandāre;
command, send forth: þe mone **mandeth**
(*pres.*) hire bleo SPEC. 44; þe mone mandeth
hire liht 44.

mandragore, sb., *O.Fr.* mandragore; *man-
drake*; mondrake SPEC. 26; mandragge PR. P.
324; **mandragores** (*pl.*) MISC. 19; man-
dragis WICL. GEN. xxx. 14.

máne, sb., *O.E.* manu,=*O.H.G.* mana, *O.N.* mön; *mane* (*of a horse, etc.*), '*juba*,' PR. P. 324; ALIS. 1957; FER. 244; TREV. III. 403; Iw. 1853; GAW. 187.

máne, *see* **mǣne**.

mánen, v., ? *O.N.* meina (*prohibit*); ? *blame*; mone MAP 336; þer dedes þou not mōnes (*pres.*) MAN. (H.) 219.

mánen, *see* **mǣnen**.

manēr, sb., *O.Fr.* manoir; *manor, farm,* BEK. 480; maner [manoir] LANGL. *A* v. 595.

manēre, sb., *O.Fr.* maniere; *manner, sort, custom,* LAȜ. 18983*; A. R. 6; P. L. S. xvii. 301; manere [maneere] C. L. 663; an manere of fissce HOM. I. 51; of manere CH. C. T. *F* 546; maner doctrine CH. C. T. *B* 1669; maner thing 3951; maner sergeant *E* 519; **maners** (*pl.*) manars ALEX. (Sk.) 2515, 4223.

    manēr-lī, adv., *courteously,* WILL. 5008; ALEX. (Sk.) 3953.

manēred, adj., *mannered,* LANGL. *B* ii. 27.

mang, sb. ?=Ȝe-**mang**; *mixture*; mong A. R. 384; ich nabbe no mong ... wiþ þe world HOM. I. 185; wiþ false godes Ȝe make mong MAN. (F.) 7384; mang (*for* a mang) wimmankin ORM. 239; on mang (*among*) wulfen MAT. x. 16; en mang SAX. CHR. 261; a mang ORM. 110; AYENB. 124; a mong LAȜ. 2229*; A. R. 2; he com a mong us HOM. I. 19; a mong manne O. & N. 553; þar a mong 497; Floriz siȝte & weop a mong FL. & BL. 431; a mong men H. H. 131; e mang PL. T. 793; emong TOWNL. 102; bi mong KATH. 200; SPEC. 35; bi mong wummen ['*inter mulieres*'] A. R. 100; **a manges** (*amongst*) PROCL. 7; SHOR. 21; PERC. 1148; a monges P. S. 324; MAND. 11.

    mong-corn, sb., *mixed corn,* PR. P. 342; PL. CR. 786.

mangen, v., *O.Fr.* mangier; *feast, eat*; **maunged** (*pret.*) LANGL. *A* xii. 71; **maunged** (*pple.*) vi. 250.

mangen, v., ? *mingle* : he mihte hire murþes monge wiþ blisse SPEC. 34.

mangen², v., *O.E.* (ge-)mangian, = *O. N.* manga; *trade*; mange PS. ci. 27; **manges** (*pres.*) A. R. 408*.

    mangere, sb., *O.E.* mangere, = *O.N.* mangari, *O.H.G.* mangeri, *mod.Eng.* (fish-)monger; *dealer, merchant,* ['*mercator*'] MAT. xiii. 45; mangare FRAG. 2; eir monger P. L. S. xii. 69; *comp.* **fisch-, hors-, wolle-mangere**.

    [**manging**, sb.; *comp.* **for-manging**.]

mang·ir, sb., *Fr.* mangeoire; *manger, stable,* WICL. E. W. 435; '*mansorium, presepium,*' PR. P. 325; **manjores** [mangers] (*pl.*) MAN. (F.) 11182.

mangerie, sb., *feast,* GAM. 345, 434; WICL. S. W. I. 4; mangeri YORK xxxi. 208; LANGL. *C* xiii 46; A. P. ii. 52, 1365; BARB. xx. 67*.

manglen, v.,=*M.L.G.* mangelen, *M.Du.* menghelen; *mingle*; **mongleð** (*pres.*) A. R. 338; **monglinde** (*pple.*) A. R. 116.

    monglunge, sb., = *M.L.G.* mangelinge.; **mingling**, A. R. 6.

mangonel, sb., *O.Fr.* mangonel; *machine for throwing stones,* P. S. 69; mangenel ROB. (W.) 11438; mongenel AYENB. 116; **mangunels** (*pl.*) AR. & MER. 2440.

manicle, sb., *O.Fr.* manicle; *manacle,* PR. P. 325; **maniclis** [*later ver.* manaclis] (*pl.*) WICL. PS. cxlix. 8.

manie, sb., ? *O.Fr.* manie; *madness,* CH. C. T. *A* 1374.

mánien, v., *O.E.* manian, manigean, monian, = *O.Fris.* monia, *O.L.G., O.H.G.* manōn; *remember, mention* : mid muþe monegen SAL. & SAT. 240 (REL. I. 182); **móneȝe** (*pres.*) SHOR. 89; maneð & meneȝeð HOM. I. 241.

manifest, adj., *O.Fr.* manifeste; *evident,* CH. BOET. iii. 10 (91).

maniȝ, adj., *O.E.* manig, maneg, mænig, monig, = *O.L.G.* manag, *O.H.G.* manag, manig, *Goth.* manags; *many*; maniȝ word ORM. 43; maniȝ what 7086; mani GEN. & EX. 696; mani man SHOR. 8; mani, moni O. & N. 1323, 1575; CH. C. T. *A* 3907; moni MARH. 1; moni (*many a one*) hit forlet HOM. I. 15; moni [mani] Ȝer LAȜ. 337; of moni ane londe 16189; on moni are wise 29131; moni enne deadne cniht 7993; moni dai A. R. 2; moni hwat 352; moni (*ms.* mony) on hungri hund MISC. 154; meni BEK. 654; meni a man TOR. 699; monie mus HOM. I. 53; **mǣnies** (*gen. m.*) P. L. S. viii. 18; monies HOM. I. 103; monies kunnes LAȜ. 1710; of mani (*dat.*) kine hewis ALEX. (Sk.) 3864; monies riches monnes sune 28932; **monine** (*acc. m.*) LAȜ. 3412; **manige** (*pl.*) GEN. & EX. 428; moniȝe HOM. I. 97; manie HOM. II. 27; AYENB. 26; monie A. R. 200; SPEC. 48; to monie tale O. & N. 257.

    mani-feald, adj., *O.E.* manigfeald,=*Goth.* managfalþs, *O.H.G.* managfalt; *manifold,* HOM. I. 233; manifald Iw. 2973; monivold A. R. 176; manifold GOW. I. 344.

    moni-volden, v., *O.E.* manegfealdian,=*O. H.G.* manigfaltōn; *multiply,* A. R. 402.

    meni-tüwe, adj., *O.E.* menigtywe; '*sollers*' FRAG. 2.

manige, manage, menige, sb., *O.E.*manegu, mænego, menigo, mengu,=*O.H.G.* managī; *Goth.* managei; *many, multitude,* MAT. iv. 25, v. 1; LK. viii. 40.

manjoure, manjure, *see* **manger**.

manke, sb., *O.E.* mancus; (*a coin*); goldes feale manke P. L. S. viii. 35.

mankin, v., *cf. M.Du.* manken; *maim,* '*mutilo,*' PR. P. 325; **mankit** (*pret.*) GOLAG. & GAW. 1013 (ANGL. II. 432).

manna, sb., *Lat.* manna, *O.Fr.* manne; *manna,*

GEN. & EX. 3330; manne HOM. II. 99; AYENB. 181.

mannish, see **mennisc.**

mannisse, mennisse, sb., *human nature,* HOM. II. 163, 165; mannesse [mennesse] KATH. 1118; see **mennisc.**

manouren, v., *O.Fr.* manoevrer, maneurer; *govern*; **manours** (*pres.*) ALEX. (Sk.) 837.

mänsien, v., *curse, excommunicate; for* amänsen; mansen (*ms.* mannsenn) ORM. 10522; **mänsed** *for* amänsed MISC. 154; LANGL. *B* iv. 160; A. P. ii. 774.

mänsing, sb., *for* amänsing; O. & N. 1312; ROB. (W.) 9686; upe mansinge 10368.

mansiōn, sb., *O.Fr.* mansion; *mansion (in astrology)*, CH. C. T. *A* 1974; *F* 50.

mantel, sb., *O.Fr.* mantel; *mantle*, LA3. 14755; BEK. 2100; CH. C. T. *A* 2163; mentel PR. P. 333; REL. I. 129; GEN. & EX. 2026; AYENB. 188; mantill ALEX. (Sk.) 3236; mantle (*dat.*) LA3. 15274; **mantles** (*pl.*) MARH. 7.

mantelet, sb., *little mantle*, CH. C. T. *A* 2163.

mapel, sb., *O.E.* mapel; *maple*; [mapil, -ul] CH. C. T. *A* 2923; mapul *'acer'* VOC. 181; PR. P. 325.

mappe-mounde, sb., *O.Fr.* mappemonde, *Lat.* mappa mundi; *map of the world*, GOW. III. 102.

marble, marbre, see **marmon.**

marbrin, adj., *of marble*, ALEX. (Sk.) 4353.

March, pr. n., *Lat.* Martius; *March*, ORM. 1891; Mearch JUL. 79; Marz HAV. 2559.

[**march,** sb., *O.Fr.* marche; *march.*]

march-gat, sb., *line of march*; march-gats (*pl.*) ALEX. (Sk.) 5076.

marchant, sb., *Fr.* marchand; *merchant*, CH. C. T. *A* 270; marchaunt FL. & BL. 42; LANGL. *B* iv. 132 (marchaund *A* iv. 115); marchands (marchaundez) (*pl.*) ALEX. (Sk.) 1557.

marchandise, sb., *Fr.* marchandise; *merchandise*, ROB. 99; H. H. 96 (98); LANGL. *B* PROL. 63.

marche, sb., *O.E.* mearc, = *O.L.G.* marca, Goth. marka, *O.H.G.* marcha; *march, boundary, 'confinium,'* PR. P. 325; LANGL. *B* xv. 438; þe marche of Wales TREV. VIII. 187; þe king of þe Marche P. L. S. xiii. 32; merke GEN. & EX. 440; **marches** (*gen.*) men *men of that region* ALEX. (Sk.) 2540; **marches** (*pl.*) WILL. 2214; MAND. 292; markes AYENB. 223; merkes TRIST. 2234.

marchen, v., = *O.H.G.* marchōn; *adjoin*: þe kingdom of Hungarie þat **marcheþ** (*pres.*) to þe lond of Polaine MAND. 6; monei and marchandise marchen to gideres LANGL. *B* PROL. 63.

Marchisc, adj., *cf. Ger.* Märkisch; *Mercian*: Marchisce monnen LA3. 30996.

máre, sb., *O.E.* mara, mera, = *O.N.* mara, *M.H.G.* mare; *nightmare*; *goblin*; *incubus*; *'ep(h)ialtes,'* PR. P. 326; mirie ge singeð ðis mere REL. I. 221; *comp.* niht-mare.

máre, see **mére.** **märe,** see **mēre** *under* **mā.**

mareis, sb., *O.Fr.* marois, mares; *morass*, LANGL. *B* xi. 344; CH. BOET. iii. 11 (97); mares AYENB. 250; mares CATH. 227; marras ALEX. (Sk.) 3893; marrass BARB. vi. 55.

mären, v., = *O.H.G.* mērōn; *increase, augment*: morin PR. P. 343; mori AYENB. 45; mōreþ (*pres.*) LIDG. M. P. 243; mōred (*pple.*) GOW. II. 181.

marescal, sb., *O.Fr.* marescal; *marshal*, PROCL. 9; mareschal ROB. 491; marschal BEK. 796; LANGL. *B* iii. 200.

margarite, sb., *Lat.* margarita; *pearl*, SPEC. 26; WICL. PROV. xxv. 12; margrite ALEX. (Sk.) 4901; **margarits** (*pl.*) ALEX. (Sk.) 3669.

margerie, sb., *pearl*, A. P. i. 1037, ii. 556; **margaris** (*pl.*) A. P. i. 199.

margine, sb., *Ital.* margine; *margin*, PR. P. 326; LANGL. *B* vii. 18; margins ['*margines'*] (*pl.*) TREV. I. 41.

margon, sb., *cf. Sc.* murgeon (*grumbling*); *muttering*: to sitt doune in margon and nolle ALEX. (Sk.) 628.

[**margrote,** sb., *O.E.* meregrot, *cf. O.L.G.* merigriota, *O.H.G.* marigrioz; *pearl.*]

margrote-stān, sb., *pearl*; **margrote-stänes** (*pl.*) ORM. 7407.

mar3, see **mear3.** **mar3en,** see **mor3en.**

mari, see **mear3.**

mariäge, sb., *O.Fr.* mariage; *marriage*, ROB. 31; AYENB. 220; CH. C. T. *A* 212; A. P. i. 414.

marien, v., *O.Fr.* marier; *marry*: marie ROB. 30; A. P. ii. 52; marissi AYENB. 220; mariede (*pret.*) TREV. IV. 9.

marinere, sb., *O.Fr.* marinaire; *mariner*, CH. C. T. *A* 1627; **mariners** (*pl.*) MAN. (F.) 14467.

marjoran, sb., *marjoram*, GOW. III. 133.

mark, sb., *O.E.* marc, = *M.Du.* mark, merk *O.Fris.* merk, *O.N.* mörk; *mark (coin)* *'marca,'* PR. P. 326; sixti mark of golde LA3. 22392; marc ROB. 393; mark LANGL. *A* v. 31; for marke ne for punde P. L. S. viii. 149.

marke, see **marche, mearke.**

market, sb., *Lat.* mercātus; *market*, SAX. CHR. 252; REL. I. 219; MISC. 16; ALIS. 1515; marcat AYENB. 36; ALEX. (Sk.) 421.

market-bētere, sb., *batteur de marché*, CH. C. T. *A* 3936.

market-daschare, sb., *'circumforaneus,'* PR. P. 326.

markis, marqvis, sb., *O.Fr.* marchis; *marquis*, CH. C. T. *E* 64, 453.

markisesse, sb., *marchioness,* CH. C. T. *E* 283, 294, 1014.

marl, sb.,=*M.Du.* margel, mergel, *O.H.G.* mergil; *marl,* PR. P. 327; TREV. II. 15.
　marle-pit, sb., *marl pit,* CH. C. T. *A* 3460.

marlin, v., *cf. Du.* marlen; *marl, manure with marl; whitewash; 'illaqueo,'* PR. P. 327; imarled (*pple.*) TREV. II. 15.

[marmon] marble, marbre, sb., *Lat.* marmor, *Fr.* marbre; *marble;* marble ALEX. (Sk.) 1330.

　marmon-, marme-stān, sb., *O.E.* marman-, mearmstān; *marble,* LA3. 1317, 7622; marbrestan KATH. 1489; marbreston LA3. 1317*; ROB. 476; marbelston HOM. II. 145; BEK. 2142; GREG. 946.

marmoset, sb., *? O.Fr.* marmoset; *some ape;* **marmozettes** (*pl.*) MAND. 210.

marou, sb., *prov.Eng.* marrow; *companion, mate;* (*v. r.* màru) *'socius'* PR. P. 327; maro TOWNL. 110.

marren, *see* merren. **marte,** *see* martre.

martinet, martnet, sb., *Fr.* martinet; *martlet,* PR. P. 327.

Martinmesse, sb., *Martinmas,* MAN. (H.) 230; martimes BARB. ix. 127.

martir, sb., *O.E.* martyr, *Lat.* martyr; *martyr,* LA3. 24283; KATH. 2217; **martiren** (*dat. pl.*) HOM. I. 239.

　martir-dōm, sb., *martyrdom, slaughter,* LA3. 10120.; martirdome BARB. vi. 289.

martre, sb., *Fr.* martre, marte; *marten;* martres (*gen.*) P. L. S. viii. 182; *comp.* fūl-mart.

martren, v., *martyr, kill;* martri ROB. (W.) 1601; martur D. TROY 12985; **martrede** (*pret.*) LA3. 10901; **imartred,** imartired (*pple.*) ROB. 81, 82; martrid ALEX. (Sk.) 1268, 3644.

maschin, v., *mash, 'misceo,'* PR. P. 328; *see* mǣschen.

masculin, adj., *Lat.* masculīnus; *male,* CH. BOET. ii. 3 (37).

máse, sb., *? O.E.* mase (*gurges*); *? maze, error, bewilderment,* ROB. 498; P. L. S. xxiii. 14; LANGL. *A* i. 6; CH. TRO. v. 468; A. P. ii. 395.
　máse-līche, adv., (*? for* másedlīche) A. R. 272.

māse, sb., *O.E.* māse,=*M.Du.* meese, *O.H.G.* meisa; *titmouse,* LIDG. M. P. 203; mose O. & N. 69; *comp.* col-, tit-māse.

másed, pple., *see* másen.
　másed-līche, adj., *stupidly,* A. R. 272*.
　másednesse, sb., *stupor,* CH. C. T. *E* 1061.

masel, sb., *M.Du.* masel; *measle;* masil *'serpedo'* PR. P. 328; **maseles** (*pl.*) VOC. 161.

maselin, sb., *O.Fr.* mazelin; *maple-bowl,* L. N. F. 219; CH. C. T. *B* 2042.

másen, v., *maze, make or become dizzy;* i

mase WILL. 438; ye mase CH. C. T. *E* 2387; **mǎsed** (*pple.*) GOW. III. 6; CH. C. T. *B* 526; mased of wine PS. lxxvii. 65; masid SACR. 653; masid and mad MIR. PL. 170; *comp.* a-, bi-mǎsen.

maser, sb., *O.N.* mösurr,=*M.Du.* maser, *O. H.G.* maser, masar, *O.Fr.* maser, mazer; *mazer, maple-wood, maple-bowl, 'mur*(h)*a,'* VOC. 198; PR. P. 328; GREG. 890; W. & I. 35; hire nap of mazere REL. I. 129 (HOM. II. 163); masers (*pl.*) MAN. (F.) 11418.

maske, sb., *O.E.* max,=*M.Du.* masche, *O.H.G.* masca, *O.N.* möskvi; *mesh, 'macula,'* PR. P. 329.

maskel, sb., *cf. M.Du.* maschel; *spot, flaw;* **maskle** (*dat.*) A. P. ii. 556; **mascles** (*pl.*) ALEX. (Sk.) 4989, 5138.
　maskel-lēs, adj., *spotless,* A. P. i. 743.

masken, *see* malskren.

maslin, *see* mæstling.

masoun, sb., *O.Fr.* maçon, machon, maschun; *mason,* P. L. S. xii. 58; mascun FL. & BL. 326; **machūnes** [machuns] (*pl.*) LA3. 15478; masons LANGL. PROL. *A* 101; masouns TRIST. 2829; masonis BARB. xvii. 937.

masonerie, sb., *Fr.* maçonnerie; *masonry,* PR. P. 329.

masporie, sb., *some kind of gem,* A. P. i. 1018.

masse, sb., *O.Fr.* masse; *mass, lump,* PR. P. 328.

masse, *see* mæsse. **mast,** *see* mæst.

mǎst, *see* mā. **master,** *see* maister.

mastif, *see* mestif.

mastik, sb., *Fr.* mastic; *mastic,* PR. P. 329.

mastilyon, mastling, *see* mæstling.

mat, adj., *O.Fr.* mat; *dejected, weary,* A. R. 382; C. L. 1205; WILL. 2441; mate Iw. 427; RICH. 6749; FLOR. 771; TOR. 679; to mourne nowe full mate and full madde YORK xlvi. 4; with mate chere BARB. xvii. 794; **máte** (*pl.*) KATH. 2015.

máte, sb.,=*M.L.G.* mate, *M.Du.* maet, *O.H.G.* gimazo (*conviva*); *mate, 'socius,'* PR. P. 329; ROB. 236; FER. 1372.
　mait-lēs, adj., *matchless,* D. TROY 7861.

[māte, adj., *?]*
　mōte-wōk, adj., *? moderately weak,* PR. P. 345.

máten, v., *O.Fr.* mater; *checkmate, overcome; become faint;* A. R. 98; mate A. P. i. 613; **máted** (*pple.*) C. L. 1324; no more mate ne dismaied GAW. 336.
　mæting, sb., *mate, checkmate:* at ilka matting (*v. r.* mating) þei seide chek MAN. (F.) 11399.

materas, sb., *O.Fr.* materas; *mattress;* matteras PR. P. 329; matras VOC. 260.

materie, sb., *O.Fr.* matere, *Lat.* materia; *matter,* A. R. 270; AYENB. 49; SHOR. 8; matere WILL. 711; CH. C. T. *B* 322; matir, mater BARB. iii. 301, iv. 216; matier *wood* PALL. iii. 282.

mathematique, adj., *O.Fr.* mathematique ; *mathematic*, GOW. III. 87.

matines, sb., pl., *Fr.* matines ; *matins*, R. S. vii ; MAN. (F.) 15136 ; matins [matines] LANGL. *A* v. 233 ; matins BRD. 10.

matrimoine, sb., *O.Fr.* matrimoine ; *matrimony*, LANGL. *C* iii. 149 ; CH. C. T. *A* 3095 ; matermoin HAMP. PS. l. 6*.

matte, sb., *O.E.* meatte, *Lat.* matta ; *mat*, VOC. 260 ; PR. P. 329 ; TREV. VII. 379.

mattok, sb., *O.E.* mattoc, mættoc, mattuc ; *mattock*, '*ligo*,' PR. P. 330 ; VOC. 234 ; ['*bidente*'] TREV. V. 129.

maturitē, sb., *slowness, deliberation*, BARB. xi. 583.

máþe, sb., *O.E.* maða,=*Goth.* maþa, *O.L.G.* matho, *O.H.G.* mado ; *insect, maggot*, '*cimex*,' FRAG. 3 ; maþe (*printed* maye), mathe '*tarmes, cimex*' PR. P. 230 ; máþen (*pl.*) AR. & MER. 484 ; meaċen HOM. I. 251.

maðek, sb., *cf. O.N.* maðkr, *Dan.* maddik, māk ; *maggot* ; mak '*tarmes, cimex*,' PR. P. 321 ; mauk '*cimex*' VOC. 190 ; maðekes (*pl.*) HOM. I. 326.

maðelen, v., *O.E.* maðelian, mæðlan, meðlan, *cf. Goth.* maþljan (λαλεῖν) ; *see* **mælen** ; *speak, discourse* ; medle LANGL. *A* xi. 93* ; melle A. P. i. 796 ; TOWNL. 163 ; maðeleð (*pres.*) A. R. 74 ; maðelinde (*pple.*) A. R. 86 ; **medelede** [melled] (*pret.*) LANGL. *B* iii. 36.

maðelere, sb., *O.E,* maċelere ; *one who discourses*, A. R. 88*.

maðelild, sb., *one fond of talking*, A. R. 88.

maðelunge, sb., *O.E.* maċelung ; *discoursing*, A. R. 76.

māðem, sb., *O.E.* māðum, māðm,=*O.L.G.* methom, *O.N.* *meiðm* (*pl.* meiðmar), *Goth.* maiþms ; *treasure* ; **māðmes** (*pl.*) LAȝ. 896 ; madmes REL. I. 174 ; (*ms.* maddmess) ORM. 6471.

maucht, *see* **meaht**.

maugrē, prep., *O.Fr.* mal-, maugrē ; *in spite (of)*, HAV. 1128 ; AYENB. 69 ; WICL. WISD. xix. 14 ; mau-, malgre LANGL. *C* xxi. 84 ; magre BARB. i. 453 ; magre his *in spite of him* ii. 124, iv. 194 ; maugre, maugref A. P. ii. 44, 54 ; maugref GAW. 1565 ; malegrefe, malegreve ALEX. (Sk.) 2782, 1747.

maugrē, sb., *ill-will, displeasure*, BARB. xvii. 60 ; LANGL. *B* vi. 242, ix. 153.

mauitē, sb., *O.Fr.* mauté ; *wickedness, evil intent*, BARB. iv. 730 ; v. 524 ; vi. 212 ; xix. 235 ; xix. 235*.

maumet, *see* **mahimet**.

maumetri, sb., *from* **mahimet** ; *idolatry* ; maumentri ALEX. (Sk.) 4486 ; HAMP. PS. xcvi. 7*, 8* ; maumetri MAN. (F.) 1337.

mavis, sb., *Fr.* mauvis ; *mavis, thrush*, R. R. 619 ; mavice PR. P. 330.

mawe, *see* **maȝe**.   **māwe**, *see* **mǣw, māȝe**.

---

māwen, v., *O.E.* māwan,=*M.Du.* maejen, *M.H.G.* mæjen, māwen, mǣn ; *mow (hay)* ; mawen, mowen HOM. I. 131, 137 ; mawe AYENB. 314 ; mowen P. L. S. viii. 11 ; MISC. 108 ; mowe O. & N. 1040 ; **mēowen** [mewen] (*pret.*) LAȝ. 1942.

māwer, sb.,=*M.Du.* maejer ; *mower*, VOC. 213 ; moware PR. P. 345.

mé, pron., *O.E.* me (*dat.*), mec, me (*acc.*),=*O.Fris., O.L.G.* mi (*dat. acc.*), *O.H.G.* mir (*dat.*), mih (*acc.*), *O.N.* mer (*dat.*), mik (*acc.*), *Goth.* mis (*dat.*), mik (*acc.*), *Lat.* mihi (*dat.*), me (*acc.*), *Gr.* μοί (*dat.*), μέ (*acc.*), *Ir.* me, *Welsh, Gael.* mi (*dat. acc.*) ; *me*, KATH. 470 ; A. R. 26 ; C. L. 330 ; arh ich was me self HOM. I. 277 ; lærest me to cwenken in me galnesses fule stinch ORM. 1191 ; i me self sah godes gast 12592 ; ic . . . have me hid þat he þe helpe . . . ase he wolde GEN. & EX. 358 ; me selve FL. & BL. 146 ; me, mi HAV. 2204, 2576 ; S. S. (Web.) 135, 383 ; help me LAȝ. 1199 ; þurh me [mi] seolfne 8333 ; wise mi 1200 ; mi self ich w(u)lle teo . . . to telde þas kinges 791 ; þat ich me draȝe to mine cunde O. & N. 273 ; bi mi seolve 835 ; mi (*ms.* my) selve so to bere SPEC. 93.

me, conj.,=? *O.Fris., M.Du.* men ; *but*, KATH. 327 ; me þu heċene hund þe hehe healent is min help MARH. 6 ; me wenestu . . . þet ich chuḷle leapen on him A. R. 54.

me, *see* **man**.

meaht, sb., *O.E.* meaht, meht, mæht, micht, myht, miht, *cf. O.Fris.* macht, mecht, *O.L.G., O.H.G.* maht, *Goth.* mahts *f.*, *O.N.* māttr *m.* ; *might, power* ; maht SPEC. 34 ; maȝt A. P. iii. 112 ; maght IW. 3621 ; mauȝt TRIST. 2791 ; maucht BARB. ii. 421, xii. 534 ; mahte HOM. I. 59 ; cuð þi mahte MARH. 7 ; mehte (mihte) HOM. I. 113 ; miht P. P. 250 ; miht [miȝt] C. L. 1174 ; miȝt LANGL. *B* PROL. 113 ; if it lai in his might [miht, miȝt] CH. C. T. *A* 538 ; o mihte A. R. 26 ; mihte & mein JUL. 54 ; none mihte [miȝte] O. & N. 1670 ; miȝte AYENB. 15 ; mihte, mighte PR. P. 337 ; miste REL. I. 184 (MISC. 133) ; **mæhte** [mihte] (*dat.*) LAȝ. 6810 ; mihte (*pl.*) HOM. I. 47 ; P. L. S. viii. 39 ; þine mihte R. S. iv (MISC. 162) ; mæhte [mihtes] LAȝ. 3242 ; mihtes A. R. 298 ; mahtes ORM. 4936 ; mihta (*gen. pl.*) HOM. I. 101 ; alre mahte rote ORM. 4976 ; mehten, mihten (*dat. pl.*) LAȝ. 23016, 28701 ; mihten A. R. 134 ; *comp.* ȝe-mæht, un-miht. miȝt-ful, adj., *mighty*, C. L. 588 ; migtful GEN. & EX. 100.

mightful-nes, sb., *strength*, PR. C. 754.

miht-lēs, adj., *O.E.* mihtelēas ; *powerless*, HOM. I. 111.

meaht², adj., *O.E.* meaht ; *from* **mǣi** ; *mighty, powerful* ; might SACR. 182 ; min migte name GEN. & EX. 3038 ; *comp.* al-miȝt, un-maht.

meahtiȝ, adj., *O.E.* meahtig, mæhtig, mehtig, mihtig,=*O.H.G.* mahtīg, *Goth.* mahteigs, *O.N.*

māttugr; *mighty*; mahtiȝ ORM. 806; mæhti [mihti] LAȝ. 2839; magti, migti GEN. & EX. 983, 3797; mihti A. R. 182; miȝti BRD. 16; AYENB. 103; mighti [mihti] CH. C. T. *A* 108; mehtiȝan (*dat. m.*) HOM. I. 129; comp. al-, un-mahtiȝ (-mihti).

miȝti-hēd, sb., *cf. O.H.G.* mahtīgheit; ['*potentatus*'] WICL. ECCLUS. X. 11.

maȝti-lī, adv., *O.E.* mihtiglīce; *mightily*, GAW. 2262.

mihtinesse, sb., *mightiness*, S. A. L. 83.

mēal, *see* mǣl.

mēane, mēanen, *see* mǣne, mǣnen.

mearȝ, sb., *O.E.* mearg, mearh,= *O.L.G.* marg, *O.Fris.* merg, *O.N.* mergr, *O.H.G.* marg, marc; *marrow*: þet meari JUL. 58; marȝ, mari, merȝ; merou, merouȝ, WICL. GEN. xlv. 18, JOB xxi. 24, PS. lxv. 15, EX. xvii. 3; mari, marou, maruh, marugh PR. P. 326; mari [meri] CH. C. T. *C* 542; margh PALL. iv. 477; merowe (*dat.*) APOL. 91.

mari-bōn, sb., *marrow-bone*, CH. C. T. *A* 380; LIDG. M. P. 165.

merewi, adj., *marrowy*, WICL. IS. xxxiv. 6.

meari, *see* mearȝ.

mearke, sb., *O.E.* mearc,= *M.H.G.* marc, *M.Du.* mark, merk, *O.N.* mark, merki; *mark, sign, boundary mark, mark for shooting*; marke LAȝ. 19099; CH. C. T. *D* 619; S. & C. I. xxxv; marke, merke MARH. 5, 13; merke PR. P. 334; A. R. 300; ORM. 17982; LANGL. *B* xv. 343; MAN. (H.) 129; merk GEN. & EX. 1003; PR. C. 4402; M. H. 57; as to a merke schote to him P. L. S. xviii. 44; marken, mærkes [marke, markes] (*pl.*) LAȝ. 18869, 29854; merkes PL. CR. 177; merken A. R. 364; comp. hēre-mǣrke, kine-merke.

merk-schot, sb., *distance between archery butts*, BARB. xii. 33.

mearkien, v., *O.E.* mearcian, *cf. O.N.* marka, merkja, *M.H.G.* merchen, merken; *mark*; markian HOM. I. 127; merkin '*signo*' PR. P. 334; merken CH. BOET. i. 4 (16); if i miȝt merke (*reach*) to Messedone ALEX. (Sk.) 5404; merke (*pres.*) LUD. COV. 308; markeþ SHOR. 7; markede (*pret.*) LAȝ. 5642; markedest MARH. 10; markid (*penetrated*) ALEX. (Sk.) 3674; mærcoden LAȝ. 26309; marked (*pple.*) LANGL. *B* xii. 186; markit S. S. (Wr.) 2595; merked A. P. ii. 1617; comp. ȝe-mearkien.

mærkung, sb., *O.E.* mearcung; *marking*, FRAG. I.

mearu, adj., *O.E.* mearu,= *O.H.G.* marawi, marwo; *tender, delicate*: mare MK. xiii. 28; mereuh MISC. 94; merou, merugh HALLIW. 550; meruwe (*pl.*) A. R. 378.

meaw, *see* mǣw. mecche, *see* mǣcche.

meche, sb., *O.Fr.* meche; *match*, '*lichinus*,' VOC. (W. W.) 754.

mēche, sb., *O.E.* mēce,= *O.L.G.* māki, *O.N.*

mǣkir, *Goth.* mēkeis (*acc.* mēki); *sword*; mǣche (*dat.*) LAȝ. 7495; mēchen (*dat. pl.*) LAȝ. 178.

mechel, *see* müchel. mēd, *see* mād.

méde, sb., *O.E.* meodu, medu,= *O.Fris.* mede, *O.H.G.* medo, meto, *O.N.* miöðr; *mead*' *mulsum*,' PR. P. 331; LAȝ. 6928; SPEC. 88; SHOR. 9; mede, meode L. H. R. 138, 139; meþe VOC. 178; CH. C. T. *A* 3261; meeth PALL. ii. 282.

mēde, sb., *O.E.* mēd,= *O.L.G.* mēda, mieda, *O.Fris.* mēde, meide, mīde, *O.H.G.* mēta, meata, miata, mieta; *meed, reward,* '*praemium, merces*,' PR. P. 331; LAȝ. 17646; A. R. 94; ORM. 4381; GEN. & EX. 1419; HAV. 102; BEK. 34; SPEC. 42; AYENB. 43; WILL. 2135; RICH. 5404; WICL. JOHN iv. 36; PR. C. 96; meede [mede] LANGL. *A* iii. 203; CH. C. T. *A* 3380; ower mede ['*merces*'] is michel MAT. v. 12; to mede for hire hokeres MARH. 18; mēden (*pl.*) KATH. 1646; A. R. 240; comp. kine-mēde.

mēde-ful, adj., *meritorious*, ROB. (W.) 8975.

mēd-ȝeorn, adj., *desirous of meed*, P. L. S. viii. 129; med (-iern) HOM. I. 175.

mēde², sb., *O.E.* mǣd,= *?O.Fris.* mēde; *mead, meadow*, O. & N. 438; FL. & BL. 434; ALIS. 8; CH. C. T. *A* 89; S. S. (Wr.) 1120; mede, medwe LEG. 200, 201; medewe '*pratum*' PR. P. 337; MAND. 312; mēdewe (*gen.*) MISC. 93; mēdewe (*dat.*) LAȝ. 30251; medewe [mede] P. S. 332 (p. 33); medue MAN. (H.) 195; mǣdwe (*pl.*) SAX. CHR. 254; medewen, medewes LAȝ. 1942, 24263; mede ROB. 187; medewes GOW. II. 327; medes MISC. 108.

mēdwe-grēne, adj.. *green as a meadow*, GOW. II. 266.

mēd-wurt, sb., *meadow-wort*, VOC. 139.

[mēde³, adj., *from* mōd; -*minded,* -*hearted*; comp. ēad-, hrēu-, wā-mēde.]

medeme, adj., *O.E.* medume, meodume, *cf. O.H.G.* metemo; *moderate*: medeme mel REL. I. 132 (HOM. II. 13).

[medemien, v., *O.E.* medemian, = *O.H.G.* metemen; *moderate*; comp. ȝe-medemien.]

mēden, v., *cf. O.L.G.* mēdean, miedōn, *O.H.G.* mietan; *reward, recompense*: he bihet to meden ham wið swiče hehe mede KATH. 415; medin PR. P. 331; medi AYENB. 146; mēde D. TROY 5124; mēdeþ (*pres.*) P. 17; LANGL. *B* iii. 215; mēdede (*pret.*) TREV. III. 421; meded WILL. 4646; imēaded (*pple.*) HOM. I. 243.

mēdewe, *see* mēde.

mediacioun, sb., *Fr.* mediation; *mediation*, CH. C. T. *B* 684.

medicine, sb., *O.Fr.* medecine; *medicine*, HOM. I. 187; SPEC. 88; TRIST. 1204; medisine MAN. (F.) 13896; medessing PALL. i. 797.

medil, *see* middel.

meditaciōn, sb., *O.Fr.* meditacion ; *meditation* ; **meditaciũns** (*pl.*) A. R. 44 ; MED. I.

[medle, sb., *O.Fr.* mesle (*mespilum*) ; *medlar*.] **medle-trē**, sb., *medlar tree*, BEV. 52 ; meletre VOC. 192.

medlē, sb., *O.Fr.* medlee, meslee ; *medley, mixture, fight,* ' *mixtura,*' PR. P. 331 ; melle IW. 505 ; BARB. vi. 361, x. 184 ; a medlee cote CH. C. T. *A* 328 ; **medlēs** (*pl.*) AYENB. 41 ; melleis BARB. xvii. 120.

medlen, v., *O.Fr.* medler, mesler ; *mix, meddle,* HOCCL. i. 176 ; medlin ' *misceo*' PR. P. 331 ; melle D. ARTH. 938 ; LUD. COV. 21 ; medliþ (*pres.*) SHOR. 28 ; and sir Pilate medill him YORK xxxiv. 327 ; mellis *cohabit* ALEX. (Sk.) 5430 ; **medled**, melled (*pret.*) WILL. 1709, 2323 ; **medled** (*pple.*) LANGL. *B* ix. 3 ; MAND. 76 ; melled PR. C. 9431 ; LIDG. M. P. 23 ; imedled [imelled] TREV. III. 469 ; imelled BRD. 13.

meddlinge, sb., *mixture, fighting,* WICL. JOHN xix. 39 ; melling TREV. I. 387 ; II. 159 ; melline BARB. v. 406.

medlen, *see* maðelen.

megge, sb. ; al þus wiþ megge [*v. r.* menging] & wiþ monge bitwixt hem wax þer werre stronge MAN. (F.) 5889.

mēȝhe, *see* mǣȝe.

mégre, adj., *O.Fr.* maigre ; *meagre,* LANGL. *B* v. 128 ; R. R. 218 ; A. P. ii. 1198 ; **megire** [meger] (*sb.*) *leanness* ALEX. (Sk.) 1164.

mēhe, *see* mǣȝe. **mehte**, *see* meaht.

mei, *see* mæi. **meiden**, *see* mæiden.

mein, *see* mæin. **meinē**, *see* mainē.

meineal, *see* maineal. **meister**, *see* maister.

meið, *see* mæið. **mēk**, *see* mēoc.

mel, sb., *O.N.* meˣal, miˣil ;=**middel :** o **mell(e)**, *prep.* [*O.N.* ā miˣli, milli] ; *between, among,* IW. 1436 ; o mell L. H. R. 90 ; a melle, e melle TOWNL. 55, 56 ; i melle [*O.N.* ī milli, miˣli], L. C. C. 24.

mēl, *see* mǣl. **mel**, *see* male.

melancoli, adj., *melancholy* ; malicoli ALEX. (Sk.) 2382, 2741.

melancolie, sb., *O.Fr.* melancolie ; *melan-choly, dudgeon,* CH. D. BL. 23 ; malencolie PR. P. 322 ; malancoli BARB. xvi. 128 ; melincoli PALL. iv. 883 ; malicoli ALEX. (Sk.) 1981.

melancolike, adj., *O.Fr.* melancolique ; *melancholy,* CH. C. T. *A* 1375.

melche, *see* milche.

melden, v., *O.E.* meldian,=*O.H.G.* meldōn, *O.L.G.* meldōn ; *announce, bear news of* : meld(e) REL. II. 210 ; yef þou þar of me melde M. H. 166 ; *comp.* bi-, for-melden.

mel-ðēu, sb., *O.E.* meledēaw,=*O.H.G.* militou ; *cf.* Goth. miliþs (*honey*) ; *mildew,* ' *uredo,*' PR. P. 337 ; HOM. I. 269 ; meldew(e) [' *ure-dine*'] WICL. GEN. xli. 6.

méle, sb., *O.E.* melo, meolo, mealo=*O.H.G.* melo, *M.Du.* mele, *O.N.* miöl ; *meal, flour,* ORM. 1552 ; HAV. 780 ; LANGL. *B* xiii. 261 ; MAND. 189 ; CH. C. T. *A* 3995 ; A. P. ii. 226 ; þet mele AYENB. 93 ; melu [meele] WICL. EXOD. xii. 34 ; **melewes** (*gen.*) MAT. xiii. 33. **méle-sek**, sb., *meal sack,* IW. 2032.

mēle, sb., *O.E.* mēle,=*M.L.G.* mēle, *M.H.G.* miol ; *cup, bowl* ; (*Midl. text* bolle) C. M. 3306 ; **mēles** (*pl.*) HALLIW. 548.

mēlen, *see* mǣlen.

melencolie, *see* melancolie.

melk, *see* milc. **melle**, *see* mail, mülne.

mellen, *see* medlen, maðelen.

mellere, *see* mülnere. **melline,** *see* medling.

melodie, sb., *O.Fr.* melodie ; *melody,* L. H. R. 28 ; AYENB. 151 ; CH. C. T. *A* 9.

melten, v., *O. E.* meltan, myltan (*pret.* mealt, melte) ; *melt, liquefy,* MARH. 6 ; ORM. 17417 ; REL. I. 185 ; GEN. & EX. 99 ; melte GOW. II. 37 ; melte [multe] TREV. VI. 335 ; þi mersi mai malte þi meke to spare A. P. ii. 776 ; **melteð** (*pres.*) A. R. 270 ; melteþ LANGL. *B* xvii. 226 ; melt HOM. I. 159 ; melte (*subj.*) AYENB. 171 ; **malt** (*pret.*) GEN. & EX. 1017 ; S. S. (Wr.) 2043 ; LAUNF. 740 ; GAW. 2080 ; milte P. L. S. xv. 204 ; þet . . . melten al of teares A. R. 110 ; ȝif ha ne mealte i teares H. M. 17 ; **mólten** (*pple.*) P. S. 253 ; LANGL. *B* xiii. 82 ; PR. C. 7126 ; M. H. 111 ; molte CH. TRO. v. 10 ; *comp.* for-, ȝe-, to-melten. **meltinge**, sb., *O.E.* meltung ; *melting,* PR. P. 332.

melu, adj., *mellow* ; melwe, melowe ' *maturus*' PR. P. 332.

membre, sb., *O.Fr.* membre ; *member,* PR. P. 332 ; P. L. S. xiv. 19 ; to na licherous lustes leeve ve (*r.* we) oure **membris** (*pl.*) ALEX. (Sk.) 4328 ; menbris 4544.

memorial, sb., *memorial,* GOW. II. 19.

memorie, sb., *O.Fr.* memorie ; *memory,* AYENB. 107 ; A. D. 263 ; AL. (L.[1]) 53.

menáce, sb., *O.Fr.* menace, manace ; *menace* ; manace CH. BOET. i. 4 (12) ; PR. C. 4350 ; manauce BARB. iii. 608 ; menas ALIS. 843.

menácen, v., *O.Fr.* menacer ; *menace* ; menace A. P. iii. 422 ; **manácen** (*pres.*) LANGL. (Wr.) 10898 ; **manausit** (*pret.*) BARB. ii. 68. **manásing**, sb., *menacing,* BARB. viii. 408 ; manasinge CH. C. T. *A* 2035.

menciōn, sb., *O.Fr.* mention ; *mention,* ARCH. LII. 37 ; LANGL. *C* ix. 247 ; mencioun [mensioune] CH. C. T. *A* 893.

mende, *see* münde.

menden, v.,=amenden ; *mend,* WILL. 647 ; **mendes** (*imper.*) 845 ; **mendid** (*pple.*) ALEX. (Sk.) 464.

mendinant, sb., *O.Fr.* mendinant ; *beggar* ; **mendinantz** (*pl.*) CH. C. T. *D* 1912.

méne, v., *? for* deméne, D. TROY 1750.

mēne, adj. & sb., *O.Fr.* moien ; *mean, middle* :

ine mene time AYENB. 36; in þe mene while WILL. 1148; WICL. JOHN iv. 31; meene 'medium' PR. P. 332; mene LANGL. B i. 158; CH. BOET. iv. 7 (146); wiþ treble, mene (tenor) & burdoun MAN. (F.) 11263.

mēne, mēnen, see mǣne, mǣnen.

meneȝen, see münegen.

menestral, sb., O.Fr. menestrel; minstrel, ROB. 272; menestrals (pl.) AYENB. 192; menestraus A. R. 84; ministrals. [minstrals, menstreles] CH. C. T. F 78.

menestralcie, sb., minstrelsy, MAN. (F.) 11267; menstralcie CH. C. T. A 2671.

menever, see menuver.

mengen, v., O.E. mengan, mǣngan, = O.Fris. nienga, O.L.G. mengian, M.H.G. mengen; from mang; mingle, disturb, trouble: wult þu balwe menge(n) LAȝ. 5016; his mod him gon mengen (be troubled) 3407; no durste heo nævere mængen imong Englisce monnen 31911; mèngin 'misceo' PR. P. 332; menge TREAT. 140; mengeþ (pres.) A. D. 252; wraþþe meinþ þe heorte blōd O. & N. 945; mengde (pret.) GEN. & EX. 3581; mengde [meinde] LAȝ. 15530; meingde [mengde] A. R. 326; menged (pple.) PS. cv. 35; PR. C. 6738: mengid APOL. 87; meind SPEC. 24; comp. ȝe-mengen.

　menginge, sb., mingling, 'mixtura,' PR. P. 332; TREAT. 136; menging PR. C. 4705.

mengen, see münegen.

meni, see maniȝ.　menien, see münien.

menisōn, sb., O.Fr. menison; dysentery, ROB. 568; menisoun LANGL. B xvi. 111.

mennisc, adj., O.E. mennisc, = O.L.G. mennisc, O.H.G. mennisc, O.N. menskr, Goth. mannisks; from man; human: mennesc flesc HOM. I. 91; in his menniske kinde ORM. DEDIC. 218; mennisc ?mankind HOM. I. 229; mennish, mannish II. 81.

　mennisc-leȝc, sb., humanity, ORM. 85.

　menske-līche, adv., honourably, A. R. 316; menskli GAW. 1312, 1983.

　menniscnesse, sb., humanity, HOM. I. 99; ORM. 1373.

mennisc, sb., O.E. mennisc, = O.H.G. mennisco, mannisco; man, human being; þese fower mannisshe HOM. II. 39; see mannisse.

menovin, menowe, sb., minnow; menawe VOC. (W.W.) 704; menouins (pl.) BARB. ii. 557.

menske, sb., O.E. mennisc (ALFR. P. C. 39); O.N. menska (humanity, virtue, honour), cf. O.L.G. menniski (humanity); dignity, honour, KATH. 135; A. R. 38; ROB. 33; WILL. 313; AM. & AMIL. 690; TRIST. 2118; A. P. i. 162; PERC. 1423; L. C. C. 12; mensce, menske LAȝ. 2681, 3360; with menske and with manhede DEGR. 83.

　menske-ful, adj., honourable, A. R. 358; menskful SPEC. 51; menscful M. H. 10.

　menskful-lī, adv., honourably, BARB. xix. 86.

mensken, v., dignify, honour, HOM. I. 281; menskin JUL. 7; menske WILL. 4834; LANGL. A. iii. 177; menskeð (pres.) H. M. 23; menskez A. P. ii. 141; menske (imper.) KATH. 2008; mensked (pple.) JOS. 146.

mentel, see mantel.

menuse, sb., O.Fr. menuise; minnow, B. B. 168; menuce PR. P. 333.

menusen, v., O.Fr. menuser; make less: ȝif he withdrawe or menuse (subj.) [v. r. amenuse, amenusith] CH. C. T. I 377.

menuver, sb., O.Fr. menu ver; miniver; (printed meniuier) FL. & BL. 110; menever [meniver] LANGL. B xx. 137; menever MAN. (F.) 11194.

mēoc, adj., O.N. miūkr, = M.Du. muik, Goth. mūks; meek: meoc & milde & softe ORM. 667; meok P. L. S. xviii. 6; meoke A. R. 280; SPEC. 73; mek & milde ROB. 287; meke HAV. 945; LANGL. A i. 147; MAND. 132; CH. C. T. A 69; PR. C. 395; meoke & milde men ORM. 3606; meik BARB. i. 390; mēokest (superl.) MARH. 4; comp. un-mēoc.

　mēok-hēde, sb., meekness, ROB. 389.

　mēoc-leȝc, sb., O.N. miūkleikr; meekness, ORM. 1546; mekeleic KATH. 1240.

　mēoke-līche, adv., O.N. miūkliga; meekly, MARH. 14; meoclike ORM. 11392; muekliche AYENB. 65; mukli FER. 1945; mek-, mekeliche WILL. 408, 808; mekeliche TRIST. 168; mekeli CH. C. T. E 548.

　mēocnesse, sb., meekness, ORM. 10720; mieknesse P. S. 335; mueknesse AYENB. 65.

mēoken, v., render meek, A. R. 276; meoken, meken ORM. 9385, 13950; meeken WICL. I PARAL. xviii. 1; mekin 'humilio' PR. P. 331; a man to meke him ALEX. (Sk.) 1746; meke LANGL. A v. 52; R. R. 3541; PR. C. 172; ȝif þei ... hem mēked (pret.) WILL. 1276.

mēos, sb., O.E. mēos, = O.H.G. mios, mies; moss; mēse (dat.) ORF. 246.

mēr, see mǣr.

mercenarie, sb., O.Fr. mercenaire; mer-cenary, CH. C. T. A 514.

mercer, sb., O.Fr. mercier; mercer, A. R. 66; LANGL. C vii. 250.

[merche, sb., O.E. merce (parsley); comp. stan-, wude-merche.]

Merche, sb., O.E. Merce, Myrce; Mercians; Merchene (gen. pl.) LAȝ. 30935; Merkine riche CHR. E. 373.

merci, sb., O.Fr. merci; mercy; thanks, A. R. 30; O. & N. 1092; GEN. & EX. 1241; have merci on me HOM. I. 209; MARH. 5; SPEC. 59; in þe merci be ido be fined ROB. (W.) 11155.

　merci-vol, adj., merciful, AYENB. 188.

merciable, adj., O.Fr. merciable; having mercy, HOM. I. 211; A. R. 30; JUL. 53; MISC. 226; PL. CR. 629; WICL. NUM. xiv. 19.

mercien, v., _thank_; **merciede** (_pret._) LANGL. _C_ iv. 21; mercied _B_ iii. 20.

mercement, sb., _amercement, fine_, LANGL. _C_ ii. 159; merciment _B_ i. 160; CATH. 235.

Mercuri, pr. n. & sb., _Mercury; quicksilver_; CH. C. T. _A_ 1385, _G_ 774; Marcure ALEX. (Sk.) 4498.

merdale, sb., _O.Fr._ merdaille; _camp followers_, BARB. ix. 249.

mére, _see_ máre.

mére, sb., _O.E._ mere _m._, _cf._ _O.L.G._ meri _f._, _O.H.G._ meri, mari _m._, _n._, _Goth._ marei _f._, _Lat._ mare _n.; lake, mere, sea;_ '_mare_,' PR. P. 333; C. L. 671; A. P. i. 158; D. TROY 10924; mere LA3. 21739; þisne muchelne mære [mere] 21959; **méres** (_pl._) PS. cxiii. 8*.

mére-maiden, sb., _mermaid_, R. R. 682; mermaidin '_siren_' PR. P. 334; mermaide [mermeidin] CH. C. T. _B_ 4460.

mére-man, sb., _merman_, REL. I. 221.

mére-minne, sb., = _O.H.G._ meriminni, -menni, _M.L.G._ merminne; _siren_; **mereminnes** (_gen._) KATH. (E.) 1490; **mereminnen** [mereminne] (_pl._) LA3. 1337.

mére-neddre, sb., _O.E._ merenæddre; '_muraena_,' FRAG. 3.

mére-swīn, sb., _cf._ _M.L.G._ merswīn; _dolphin_, D. ARTH. 1091; mersvin M. H. 25.

mére[2], sb., _O.E._ mere, myre, = _O.H.G._ merhe, marhe; _mare_, HAV. 2449; AYENB. 185; PERC. 713; mere, mare CH. C. T. _A_ 541, _H_ 78; meere PR. P. 333; mure BEK. 1173; mare REL. II. 211; TREV. V. 447; MAND. 250; _comp._ **cart-, stōd-mere.**

mére-wōd [marewod], O. & N. 496.

mēre, _see_ **mǽre.**

merel, sb., _O.Fr._ merel; _? counter or 'man' in the game of merils; ? ticket in a lottery_; under the clerkes lawe men seen the merel al misdrawe GOW. I. 18; merel III. 201.

mer3, _see_ **mear3.**

merghid, adj., _full of marrow_, HAMP. PS. lxv. 14.

meridian, adj., _meridian_, CH. ASTR. PROL. (3).

meridional, adj., _O.Fr._ meridional; _southern_, CH. C. T. _F_ 263.

merie, _see_ **mürie.**

merien, v., _O.E._ (a-)merian; _purify_; imered (_pple._) AYENB. 94; of merede golde MISC. 96; _comp._ **a-merien.**

merit, sb., _O.Fr._ merite; _merit_, A. R. 160; merite AYENB. 134; CH. C. T. _G_ 33; merite [merit] LANGL. _A_ i. 157; merote ALEX. (Sk.) 5226; merite GOW. III. 187.

merke, _see_ **mearke, marche, mirke.**

merling, sb., _O.Fr._ merlanc; _whiting_, '_gam-_ (_m_)_arus_,' PR. P. 334; riht als sturjoun etes merling M. H. 136.

merliōn, sb., _O.Fr._ esmerillon; _merlinhawk_, P. L. S. xxv. 9; merlioun WICL. LEV. xi. 13; merlion [merilioun, emerlion] CH. P. F. 339.

merren, v., _O.E._ merran, myrran, = _O.L.G._ merrian, _O.H.G._ merren, marren (_hinder_), _Goth._ marzjan (σκανδαλίζειν), _mod. Eng._ mar; _hinder, injure_: he walde merrin hire mei3-had MARH. 4; merre H. M. 17; marre PL. CR. 66; **marre** (_pres._) A. P. ii. 279; merre3 KATH. 1780; merres N. P. 39; a maide marreþ me SPEC. 29; marres ALEX. (Sk.) 2040; **marrande** (_pple._) YORK i. 93; **merren,** mearren (_subj._) JUL. 34, 35; **mǽrde** (_pret._) LA3. 1903; merden H. M. 9; and merden (morde) Irisc folc LA3. 22345; marrid ALEX. (Sk.) 3546; **marred** (_pple._) WILL. 664; _comp._ **a-, 3e-merren.**

merring, sb., _O.E._ myrring; _marring_, PR. C. 6114; marring WILL. 4362.

mersch, sb., _O.E._ mersc _m._, _cf._ _M.L.G._ mersch _f.; marsh_; þane merss AYENB. 251; mershe PS. cvi. 34; **mershe** [mershe] (_dat._) O. & N. 304; mershe WICL. GEN. xli. 18.

merschi, adj., _marshy_, CH. C. T. _D_ 1710; mershi WICL. GEN. xli. 2.

merþe, _see_ **mürhðe.   meru,** _see_ **mearu.**

merveile, sb., _O.Fr._ merveille; _marvel_, CH. BOET. iv. 5 (132); mervaill, mervale, mervaile ALEX. (Sk.) 549, 1061, 1814.

merveilen, v., _O.Fr._ merveiller; _marvel_, CH. BOET. ii. 5 (46); me merveilled (_pret._) LANGL. _B_ xi. 342; **mervalled** [mervailled] (_pple._) ALEX. (Sk.) 3218.

merveilous, adj., _O.Fr._ merveillus; _marvellous_; [marveillous, -velous] LANGL. _A_ PROL. 11; mervelous A. P. i. 1165; mervailous MAN. (H.) 174.

merveilos-like, adv., _marvellously_, MAN. (F.) 1691.

mes, sb., _O.Fr._ mes, _Lat._ missus; _mess_, MAP 337; TREV. V. 459; DEGR. 1202; vele mes AYENB. 55; mees LANGL. _B_ xiii. 52; messe PR. P. 334; thei were served of **messes** (_pl._) GAM. 467; messes A. P. ii. 637.

mēs, _see_ **mēos.**

mesaise, sb., _O.Fr._ mesaise; _want of ease_, HOM. I. 279; meseise ROB. 34; mes-, meoseise A. R. 108, 220; mes-, miseise LANGL. _A_ i. 24.

mesaise, adj., _diseased_; meseise ROB. (W.) 11985.

mesaistē, sb., _poverty_; miseiste, miseste WICL. JOB v. 21; MK. iv. 19; xii. 44; 2 COR. viii. 14.

mesaventūr(e), sb., _O.Fr._ mesaventūre; _misadventure_, HORN (L.) 326; mes-, misaventure CH. C. T. _B_ 616.

mescheance, sb., _O.Fr._ mescheance; _mischance_, ROB. 278; meschance CH. C. T. _B_ 602.

mēschen, _see_ **mǽschen.**

mescheven, v., _O.Fr._ meschever; _come to bad fortune_; **mischeeve** (_pr. subj._) P. R. L. P. 195; mischeve PALL. i. 614.

[**meschevous,** adj., _O.Fr._ meschevous; _unfortunate._]

meschevous-lī, adv., *unfortunately*, MAN. (F.) 14107.

meschief, sb., *O.Fr.* meschief, *cf. Span.* menoscabo; *ill-luck*; mischeef LANGL. PROL. *A* 64; meschef WILL. 1045; MAN. (F.) 2508.

mesel, adj. & sb., *O.Fr.* mesel, *Lat.* misellus; *leper*, ROB. 86; mesel, misel *'leprosus'* PR. P. 339; TRIST. 3175; (his hande) was mesele C. M. 5824; meseles (*pl.*) PL. CR. 623.

mesellerie, sb., *O.Fr.* mesellerie; *leprosy*: meselrie *['lepra']* TREV. V. 119.

mésen, v., *for* amésen; *moderate, soothe*; mese A. P. ii. 764; TOWNL. 175; meese YORK xxvi. 64, xliii. 238.

meson-deu, sb., *O.Fr.* maison Dieu; *hospital*, LANGL. *A* viii. 28; maison dewe CATH. 229; meson dieux (*pl.*) LANGL. *C* x. 30.

messe, sb.?: þe Jewes . . . sall noзt be merked with þat messe YORK xi. 162.

messāge, sb., *O.Fr.* message; *message*, MAN. (F.) 15027; (*messenger*) A. P. ii. 454; messāges (*pl.*) AYENB. 122.

messager, messanger, sb., *O.Fr.* messager; *messenger*; messager A. R. 190; AYENB. 211; JOS. 324; messagere [messinger] ALEX. (Sk.) 1690; messangers (*pl.*) 897.

messe, *see* mæsse. mēst, *see* mā.

mester, sb., *O.Fr.* mester, *from Lat.* ministerium; *office, trade; need, necessity*; A. R. 72; O. & N. 924; MAN. (F.) 586; GEN. & EX. 536; HAV. 823; C. L. 478; mestier P. L. S. xv. 44; þet mestier AYENB. 187; mester [mister] CH. C. T. *A* 613; mister, misteir BARB. xvii. 435, 938; mestire ALEX. (Sk.) 1774; mister CATH. 241; PR. C. 3476; C. M. 15661; mistere, mistire HAMP. PS. iii. 7*, *page* 514; mistir YORK viii. 52; BARB. xi. 452, xvii. 743, 753.

mestif, sb., *O.Fr.* mestif, mestiz; *mastiff*, PR. P. 329; mastif MAN. (H.) 189; mastives (*pl.*) TREV. VIII. 187.

mestling, *see* mæstling.

mestrin, v., *need, require*; us mistris (*pres.*) ALEX. (Sk.) 4281; we mister no sponis here at oure manging TOWNL. 90; what mistris þe YORK vii. 54.

mesuāge, sb., *O.Fr.* mesuage; *messuage*, CH. C. T. *A* 3979.

mesurable, adj., *Fr.* mesurable; *measurable*, LANGL. *A* i. 19.

mesūre, sb., *O.Fr.* mesure; *measure*, A. R. 372; H. M. 41; LANGL. *A* i. 33; PR. C. 7690; mesour BARB. xvi. 323; at all mesure x. 281.

mesūren, v., *O.Fr.* mesurer; *measure*, CH. BOET. iii. 2 (65); mesuri AYENB. 252.

met, sb., *O.E.* met, *cf. O.L.G.* (gi-)met, *O.H.G.* mez; *from* méten; *measure*, *'mensura,'* PR. 335; REL. I. 131; H. M. 19; GEN. & EX. 439; L. H. R. 79; LANGL. *B* xiii. 359; MAN. (F) 14978; met & mæþ ORM. 4584;

mett ALEX. (Sk.) 25; méte (*dat.*) CH. C. T. *I* 799; *comp.* зe-, ofer-met.

[met-fæst, adj., *O.E.* (ge-)metfæst; *modest*.]

metfastnesse, sb., *O.E.* gemetfæstniss; *modesty*, ORM. 2524.

met-зerde, sb., *measuring yard*, ['*ulna*'] TREV. VIII. 105.

[met-līc, adj., *O.E.* gemetlīc, *cf. O.Fris.* metlīk, *O.H.G.* mezlīch; *moderate*; *comp.* зe-metlīc.]

met-wand, sb., *measuring rod*, '*ulna*,' PR. P. 336.

metal, sb., *O.Fr.* metal; *metal*, ROB. 28; AYENB. 26; MAN. (F.) 4483; metel PR. P. 335.

méte, sb., *O.E.* mete, *cf. O.Fris.* mete, *O.L.G.* meti, mat, *O.N.* matr, *Goth.* mats (*gen.* matis) *m.*, *O.H.G.* maz *n.*; *meat, food, 'cibus,'* PR. P. 335; A. R. 412; ORM. 3213; O. & N. 107; GEN. & EX. 573; HAV. 459; CH. C. T. *A* 1615; mete, mæte LAЗ. 658, 4466; þane mete HOM. I. 237; AYENB. 29; meite BARB. iii. 393; no schal i never ete mete GREG. 82; mete ne drink DEGR. 1739; hui wenden to heore mete S. A. L. 153; at þe firste mete EGL. 1275; metes (*pl.*) LAЗ. 3558; *comp.* flēsc-, hwīt-, morзen-, nōn-mete.

méte-bord, sb., *table*, ['*mensa*'] TREV. III. 67; mete-bordes (*pl.*) LAЗ. 3638*.

mæte-cün, sb., *kind of food*, LAЗ. 941.

méte-cüsti, adj., *hospitable, liberal*, LAЗ. 348.

méte-зevare, sb., *host, entertainer*, '*dapsilis, dapaticus*,' PR. P. 335.

méte-lēs, adj., *O.E.* meteléas; *without food*, ROB. 170; LANGL. *B* x. 65.

méte-nīþing, sb., *O.N.* matnīðingr; *food-niggard*; metenīþinges (*pl.*) HOM. II. 227.

méti-süpe, sb., *O.E.* metescipe; *provision of food*, REL. I. 131; meteship WICL. TOB. ii. 1; C. M. 12565.

méte-sēl, sb., *dining hall*, MAN. (H.) 334.

mēte, *see* mǣte. mētels, mēten, *see* mǣ-.

méten, v., *O.E.* metan, = *O.L.G.* metan, *O.Fris., O.N.* meta, *Goth.* mitan, *O.H.G.* mezan; *measure*, HOM. II. 213; A. P. i. 1031; mete PS. cvii. 8; méteð (*pres.*) A. R. 232; mon hine met mid one зerde FRAG. 5; ge meteð MAT. vii. 2; HOM. I. 137; зe meten [mete] LANGL. *A* i. 151; met (*imper.*) FL. & BL. 328; mat (*pret.*) WICL. 2 KINGS viii. 2; met TOR. 701; EGL. 328; PR. C. 7695; heo meeten [mete] L. H. R. 30, 31; méten (*pple.*) GEN. & EX. 2701; REL. I. 46; *comp.* a-, зe-méten; *deriv.* met, māte, mæte.

métinge, sb., *O.E.* metung; *measuring*, '*mensuratio*,' PR. P. 336.

mēten, *see* зe-mǣten.

mēten, v., *O.E.* mētan, = *O.Fris.* mēta, *O.N.* mœta (mǣta), *O.L.G.* mōtian; *from* mōt; *meet, come together*, GEN. & EX. 2828; metin '*obvio*' PR. P. 335; mete WILL. 815; he

igon mete þreo cnihtes LAȝ. 18127; mete wid
him P. S. 324; mete [meete] LANGL. *B* xv. 246;
CH. C. T. *C* 693; þei meeten MAND. 164;
mĕtte (*pret.*) S. S. (Wr.) 1297; CH. C. T. *C*
713; metten GEN. & EX. 1790; ALIS. 5697; hio
metten þæt child MAT. ii. 11; hi metten wiþ
Ailmar king HORN (L.) 155; mette BRD. 14;
wher mette ȝe ou A. D. 259; *comp.* ȝe-mēten.

mētinge, sb., *O.E.* meting; *meeting,* PR. P.
336; BARB. iii. 15; viii. 242.

métre, sb., *Fr.* metre; *metre* : in prose...and
ech in meetre CH. C. T. *B* 3171; of metre,
of rime and of cadence GOW. II. 82; metre
ALEX. (Sk.) 3464; metir MAN. (F.) 196.

[**metsien,** v., *O.E.* metsian; *feed.*]

metsunge, sb., *O.E.* metsung; *repast,* LAȝ.
31769.

mette, sb., *O.E.* gemetta; *companion,* LANGL.
C xvi. 55.

metz, sb., *pity,* À. P. ii. 215.

mēðe, *see* **mǣðe. méven,** *see* **móven.**

mewe, *see* **mue. mex,** *see* **mix. mi,** *see* ·**mé.**

miche, sb., *O.Fr.* miche; *small loaf,* REL. II.
192; **micches** (*pl.*) R. R. 5585.

michel, *see* **müchel. mīchen** *see* **mūchen.**

mid, prep. & adv., *O.E.* mid, miȝ,=*O.L.G.*
mid, midi, *O.Fris.* mith, mithe, *Goth.* miþ,
O.N. meȝ, *O.H.G.* mit, miti; *with,* KATH.
1953; ORM. 11077; SPEC. 93; ALIS. 1513;
E. G. 359; mon hine met mid one ȝerde FRAG.
5; mid godere heorte HOM. I. 3; swingeð
him miȝ smele twige 149; þa claȝes ... þe
þe rapes weren mide biwunden 51; þeo wimon
wasmidchilde LAȝ. 266; mid sweorde tohewen
1557; he wes swike mid þan meste 2547;
weder mid þan bezsten 11966; he wolde ...
fleon a wæi mid his cnihten 7948; forfaren
mid hungre 23614; and lai mid (*sec. text* bi)
me seolven 25926; ich wulle wunien mid þe
29630; þe mide him weoren 2831; mid alle
mihte A. R. 4; þet heo mei weopen & menen
... mide þe salmwuruhte 274; valleð a dun
mid þeos gretunge 32; imeind mid spire and
grene segge O. & N. 18; ich alle blisse mid
me bringe 433; þe ule ... mid þisse worde
hire eȝen abrad 1044; mid sw(e)orde ...
fihte 1068; he mot mid [*v. r.* wiþ] me holde
1680; þar he and oþer mide gr(e)owe 136;
þat he mihte heom ilome beo mide 1768;
hom rǒd Ailmar þe king and mid him his
funding HORN (L.) 220; a tale mid the beste
474; mit te 628; schenk us mid þe furste
HORN 1154; mid him he hadde a stronge
axe ROB. 17; lat me speke mid mi broþer
289; i ne mai mid ȝou wende BRD. 23; mid
þe furste he (a)manseþ me BEK. 1942; mid
hem C. L. 439; mit te 399; him mide 1340;
þe povere alle mid idone (? *added thereto*)
S. A. L. 153; fifti mid idone L. N. F. 222;
Crist is mid ous SHOR. 5; mid gode riȝte
129; mid, mide AYENB. 5, 50; þat menskful
maide þat þere mid þe lies WILL. 3143; þat

maide him mide 2133; þou art mid childe A.
D. 259; mid [*v. r.* with] hem LANGL. *B* PROL.
147; mid, mit, mide MARH. 4, 13·; we willeð
gan miȝ þe JOHN xxi. 3; þe him mide ferden
MK. xv. 41; mide R. S. v (MISC. 178); GEN.
& EX. 2478.

mid-liggunge, sb., *lying with, together,*
HOM. II. 13.

mid-þólinge, sb., *compassion,* AYENB. 157.

mid-wīf, sb., *midwife,* '*obstetrix,*' PR. P.
337; PL. CR. 78; mid-, midewif MIRC 87,
98; midewif VOC. 143; mid-, medewiif WICL.
GEN. xxxviii. 27; medewif SHOR. 12; *cf.*
Span. comadre.

mid, adj., *O.E.* mid, *cf. O.L.G.* middi, *Goth.*
midjis, *O.N.* miȝr, *O.H.G.* mitt; *mid* : of **mid-
ḍen** (*dat.*) þine boseme ['*de medio sinu* (*tuo*)']
A. R. 146; a midden ane wælde LAȝ. 10001;
in þe midde scheld WILL. 3605; at mid(d)e
somer 1464; to **middere** (*dat. fem.*) nihte
['*media nocte*'] MAT. xxv. 6; on midre sæ ['*in
medio mari*'] MK. vi. 47; to midder niht LAȝ.
25714; se hælend stod heom on midden LK.
xxiv. 36; he stod a midde heom alle MISC. 54;
a midden alle his fon C. L. 333; his bodi to
barst o mid(de) heppes MARH. 10.

mid-dei, sb., *O.E.* middæg=*O.Fris.* middei,
O.H.G. mittitag; *midday,* A. R. 34; middai
BRD. 10; LANGL. *B* v. 500.

mid-festen, sb., *O.E.* midfesten (SAX. CHR.
175); *cf. M.Du.* midvasten; *middle of a fast,*
LAȝ. 22256.

mid-lenten, sb., *O.E.* midlengten; *mid-Lent,*
LANGL. *B* xvi. 172.

mid-morwen [**midmareȝen**], sb., *cf. O.H.G.*
mittimorgan; *middle of the morning,* A. R.
24; midmorn GAW. 1073; midmore H. V. 83.

midmorwe-tide, sb., *middle of the morning,* LANGL. *A* ii. 42.

mid-niht, sb., *cf. O.H.G.* mittinaht; *midnight,* LAȝ. 5766; A. R. 236; midniȝt TREAT.
132.

mid-rede, sb., *O.E.* midhriðe=*O.Fris.* mid-
rede; *midriff,* '*diaphragma,*' VOC. 208.

mid-rif, sb., *O.E.* midhrif, *cf. O.Fris.* mid-
ref; *midriff,* '*diaphragma,*' PR. P. 337; mid-
ref VOC. 183.

mid-sīde, sb., *middle of the side*; stonden in
water to midside REL. I. 222; to þe midside
MAP 338.

mid-somer, sb., *O.E.* midsumer, = *M.Du.*
midsomer; *midsummer,* VOC. 273; ROB. 302;
LANGL. *B* xiv. 160; MAN. (H.) 137; mis-
somer LEG. 85; midsumer dæi SAX. CHR. 259.

mid-þēih, sb., *middle of the thigh,* MISC. 150.

mid-wei, sb., *midway*; midwei betweonen
þet and ester A. R. 412.

mid-winter, sb., *O.E.* midwinter (SAX. CHR.
140), = *M.L.G.* midwinter; *midwinter,* HOM.
II. 55; midwinter dæi SAX. CHR. 266; midwinter ROB. 452; P. L. S. xxxiv. 12.

midde, sb., *middle*; **on midden** [*O.E.* on middan, *O.L.G.* on middian, middion], *amid*; on midden þere se HOM. I. 87; þe stæf tobræc a midden LAȝ. 8154; kærf þis lond a midde 4836; mi nest is ... rum a midde O. & N. 643; a midde þe hevene TREAT. 132; a midde þe strete AYENB. 143; a middes (? *for* midden) *amidst* JOS. 602; LANGL. *B* xiii. 82; DEP. R. PROL. 3; þæt wif stod þær on middes (*earlier text* middan) JOHN viii. 9; in middes þe se PR. C. 2938; in þe middis DEP. R. iv. 78.

midden-eard, sb., *O.E.* middaneard; *middle world* (*between heaven and hell*), MK. viii. 36; middenerd HICKES I. 168; P. L. S. viii. 98; LAȝ. 24778; MISC. 52.

midde-night, sb., = *M. Du.* middennacht; *midnight,* S. S. (Wr.) 2541.

midde-ward, adj., *O.E.* middanweard; *midst*: inne middewardere (*printed* midde warȝe) helle HOM. I. 43; noght þe biginninge na þe midward (*sb.*) HAMP. PS. *page* 503.

mide-winter, sb., *O.E.* middanwinter; *midwinter,* BRD. 11; **midewinteres** (*gen.*) LAȝ. 22905; midewinteres A. R. 412.

middel, adj., *O.E.* midla (*superlat.* midlesta), = *O.Fris.* middel, *O.H.G.* mittil; *middle,* AYENB. 122; þe middel broþer LAȝ. 12909*; þe middel place MAND. 2; medil S. & C. ii.; ISUM. 170; **midlest** (*superlat.*) GEN. & EX. 710; midleste A. R. 370; CHR. E. 112; AYENB. 122; þe midleste broȝer LAȝ. 2116.

middel-eard, sb., = **midden eard,** *earth,* HOM. I. 59; middelærd LAȝ. 11167; ORM. 3980; middelerd R. S. iv; GEN. & EX. 106; HAV. 2244; SPEC. 22; ALIS. 214.

middel-ȝard, sb., *O.L.G.* middilgard, *O.H.G.* mittelgart, *O.N.* meȝalgarȝr; *middle of the garden*; (*ms.* middel yard) CHEST. I. 67.

middel-niht, sb., *O.E.* middelneaht; *midnight,* LAȝ. 20607; a middelnihte O. & N. 325; abute middelniȝte HORN (L.) 1297.

Middel-sæx, pr. n., *O.E.* Middelseaxe; *Middlesex,* LAȝ. 15391; Middelsex ROB. 3.

middel-wei, sb., = *Ger.* mittelweg; *mid way, mean,* HOM. I. 255; A. R. 336.

middel-wereld, sb., *world,* ORM. 17538; middelwerld GEN. & EX. 98.

middel, sb., *O.E.* middel, = *M.Du.* middel, *M.H.G.* mittel; *middle,* A. R. 180; HOM. II. 85; HAV. 2092; þe middel of þis lond ROB. 229; hire middel smal SPEC. 35; medille TOWNL. 112; on heora middele ['*in medio eorum*'] MK. ix. 36; iveng me bi þan midle LAȝ. 28069; bi þe midle [middel] LANGL. *A* v. 202; in middel of his live S. S. (Web.) 1729.

middemiste, adj., *midmost,* LEECHD. III. 112.

midding, sb., *Dan.* mögdynge; *midding, dunghill,* '*sterquilinium,*' PR. C. 628; PALL. i. 750; miding TOWNL. 30.

mide, *see* mid.

mīȝen, v., *O.E.* mīgan, = *M.L.G.* mīgen, *O.N.* mīga; *make water*; **mǣh** [meh] (*pret.*) LAȝ. 17726.

migge, *see* **mügge.**

migge, sb., *O.E.* micge, *cf. M.H.G.* mīge, *M.Du.* mijghe; *urine,* A. R. 402; migga LEECHD. III. 132.

migrene, migreime, sb., *Fr.* migraine; *megrim, headache,* PR. P. 337.

Mihel, pr. n., *Michael,* KATH. 710.

Mihel-masse, sb., *Michaelmas,* ROB. 463; Mighel-messe HOCCL. ii. 14.

miht, *see* meaht.

mik, sb., *man,* C. M. 2807; **mikes** (*pl.*) *free labourers* A. P. i. 572.

mike, sb., *some part of a boat,* A. P. ii. 417.

mikel, *see* müchel. **mil,** *see* mel.

milc, sb., *O.E.* meoluc, meolc, milc, = *O.N.* miolk, *O.Fris.* melok, *Goth.* miluks, *O.H.G.* miluh; *milk,* LAȝ. 1182; ORM. 12662; O. & N. 1009; milk A. R. 320; HAV. 643; SPEC. 36; CH. C. T. *F* 614; milk [melk, melke] LANGL. *B* v. 444; melk AYENB. 137; melke ROB. 361; SHOR. 90; mulc BRD. 7; milche GEN. & EX. 2788.

milc-hwīt, adj., *O.E.* meolchwīt; *milk-white,* FRAG. 2; milcwhit LAȝ. 15938; melkwhit OCTOV. (W.) 1679; milkeqvite ALEX. (Sk.) 1579, 3776, 4533, 5468.

milc-rēm, sb., *cream,* HICKES I. 227.

milk-soppe, sb., *milksop,* CH. C. T. *B* 3100.

milce, milcien, *see* mildse, mildsien.

milche, adj., = *M.L.G.* melk, *O.H.G.* melch; *milch*: sche was melche LAI LE FR. 196; milche cou PR. P. 337; milche kie E. W. 57.

mílde, adj., *O.E.* milde, = *O.L.G.* mildi, *O.N.* mildr, *Goth.* milds, *O.H.G.* milti; *mild,* P. L. S. viii. 13; LAȝ. 30809; A. R. 120; ORM. 668; ROB. 287; SPEC. 62; milde of herte AYENB. 133; þat mild (*sb.*) he besechis ALEX. (Sk.) 5097; **mildere** (*dat. f.*) LAȝ. 13032; onswerede him ... mildere stevene HOM. I. 45; **mílden** (*dat. pl.*) LAȝ. 1192; **míldre** (*compar.*) A. R. 268; O. & N. 1775; **mildest** (*superl.*) LANGL. *B* xix. 250; *comp.* un-mílde.

mílde-hēde, sb., *clemency,* AYENB. 133.

míld-heorte, adj., *O.E.* mildheort, *O.H.G.* miltherzi; *benign,* LAȝ. 16813; HOM. II. 121.

mildhert-leȝc, sb., *clemency,* ORM. 1476.

mildheortnesse, sb., *clemency,* A. R. 120; ORM. 2896; mildhertnes PS. lxxxvii. 12.

míld-heorted adj., *mild-hearted,* HOM. II. 45.

míld-lích, adj., *mild*; mid **míldlíche** (*pl.*) worden LAȝ. 8832; **míldeliche** (*adv.*) A. R. 40; BRD. 3; WILL. 1898; mildeliche LANGL. *A* iii. 21; mildelike GEN. & EX. 1371.

míldenesse, sb., = *O.H.G.* miltnissa; *mildness,* SPEC. 73; AYENB. 65.

mílde-scipe, sb., *meekness,* LAȝ. 17146;

mildeschipe HOM. I. 269; mildschipe H. M. 41; mildshipe HOM. II. 49.

mílde², sb., *O.N.* mildi, = *O.H.G.* miltī; *clemency*, P. R. L. P. 167.

   milde-ful, adj., *mild, merciful*, PS. cxiv. 5; mildful JUL. 55.

milde, sb., *millet*, PALL. i. 556.

mildien, v., *cf. O.H.G.* milten; *grow mild*; mildi AYENB. 177.

mildse, sb., *O.E.* milds, milts; *clemency*; milze, milce, mildze, mildce, milzce LAȝ. 6616, 11055, 16831, 21889, 31391; milts(e) FRAG. 8; milce P. L. S. viii. 107; KATH. 297; A. R. 30; ORM. 1476; GEN. & EX. 3728; HAV. 1361; ROB. 57; C. L. 350; milse SHOR. 44; SPEC. 58; milse [milce] and ore O. & N. 1083, 1404; milse, milce L. H. R. 18, 19.

   mils-ful, adj., *merciful*, A. R. 264; C. L. 543; PS. cxiv. 5; milzful JUL. 56.

mildsien, v., *O.E.* mildsian, miltsian; *show clemency*; milcien HOM. I. 29; milcen ORM. 1041; milce MISC. 141; **milse**(*pres.*) ['*miseror*'] PS. cxiv. 5; **milce** (*imper.*) KATH. 2419; milce hore soulen A. R. 428; *comp.* ȝe-mildsien.

míle, sb., *O.E.* mil,= *O.N.* mīlą, *O.H.G.* mīla, milla, *Lat.* mille; *mile*: ane mile LAȝ. 5819; fif mile SAX. CHR. 256; fourti mile TREAT. 134; twenti mile weies ALIS. 4446; fiftene **milen** (*pl.*) MARH. 2.

milers, sb., pl., *thousands*, MAN. (F.) 13527.

milfoil, sb., *Fr.* mille-feuille; *milfoil*, REL. I. 36.

milioun, millioun, sb., *Fr.* million; *million*, CH. C. T. *D* 1685.

milkin, v., *? O.E.* meolcian, *cf. M.L.G.* melken, *O.N.* miolka, *O.H.G.* melchan; *milk*, '*mulgeo*,' PR. P. 338; melke TREV. I. 359; **milked** (*pret.*) MAND. 71; imelked (*pple.*) P. L. S. xiii. 234.

mille, millere, milne, milnere, see **mülne, mülnere**.

milse, milsien, see **mildse, mildsien**.

milte, sb., *O.E.* milte,= *O.Fris.* milte, *? O.N.* milti, *O.H.G.* milzi *n.; milt, spleen*, FRAG. 6; R. S. v (MISC. 178).

min, adv., *O.N.* minnr,= *O.Fris., O.H.G.* min, *Goth.* mins, *Lat.* minus; *less*, L. C. C. 22.

   minne, adj., *O.N.* ·minni,= *O.Fris.* minnera, minra, *O.H.G.* minniro, *Goth.* minniza, *Lat.* minor; *lesser*, PS. ix. 6; þe minne GAW. 1881; þe more and þe minne FLOR. 549; PERC. 1608.

mín, adj., *O.E.* mīn,= *O.N., O.L.G.* mīnn, *O.H.G.* mīn, *Goth.* meins; *mine, my*: min feder HOM. I. 125; min hirde MARH. 12; min seoruwe FRAG. 6; min heed LAȝ. 7289; CH. C. T. *A* 1144; min child ORM. 2801; min hus O. & N. 623; min heorte 37; min loverd GEN. & EX. 1625; þe viht is min A. R. 266; sone min REL. I. 186; fader min C. L.

325; mi hlaford MAT. xxiv. 48; mi lare JOHN vii. 16; mi fader KATH. 466; mi sone HAV. 387; mi deþ CH. C. T. *A* 1566; mi sawle MARH. 3; mi soulė A. R. 134; mi tunge O. & N. 37; mi bearn LAȝ. 2276; **mínes** (*gen. m. n.*) LAȝ. 3588; MISC. 54; mines federes A. R. 406; mine lustes brune 318; mine lives MARH. 13; mi laverdes 12; mine songes O. & N. 1358; mine wives HAV. 698; to mi lives ende AYENB. 1; **mínre**, mire (*gen. f.*) LK. xiv. 24, xv. 12; ANGL. VII. 220; mire LAȝ. 8407; þu ert mire soule light HOM. I. 191; **mínan** (*dat. m. n.*) ANGL. VII. 220; minan, minen, mine MAT. viii. 6, x. 22; JOHN xvi. 26; mine LAȝ. 698; A. R. 148; O. & N. 46; AYENB. 89; min ORM. 10998; of mi necke MARH. 13; of mi londe BEK. 440; to . . . mi broþer CH. C. T. *A* 1161; **mire** (*dat. f.*) HOM. I. 13; LAȝ. 7879; O. & N. 1741; bi mi miȝte BEK. 1597; of mi soule LANGL. *A* iv. 123; **mínne** (*acc. m.*) HOM. I. 113; A. R. 306; MISC. 53; minne, mine LAȝ. 789, 3384; mine O. & N. 36; min, mi ORM. 2957, 17725; mi broˇer MARH. 12; **míne** (*acc. f.*) A. R. 38; O. & N. 1188; min GEN. & EX. 277; min heorte KATH. 2163; mi hwile 765; mi mede HAV. 119; mi soule BEK. 356; **mín** (*acc. n.*) GEN. & EX. 2882; S. & C. I. iii; min iwit O. & N. 1188; min heed CH. C. T. *A* 782; mi lond LAȝ. 8349; BEK. 39; **míne** (*nom. acc. pl.*) LAȝ. 492; MARH. 8; A. R. 62; ORM. DEDIC. 52; O. & N. 605; HAV. 385; BEK. 914; ȝif ȝe beon mine KATH. 1774; mine [min] eiȝen LANGL. *A* v. 90; mi gees iv. 38; min GEN. & EX. 1760; SPEC. 47; min [mi] fet M. H. 18; **míre** (*gen. pl.*) LAȝ. 21985; **mínan** (*dat. pl.*) LK. i. 44; minen LAȝ. 2946; mine MARH. 20; O. & N. 364; BEK. 478; in alle mine neoden HOM. I. 205; mine, min A. R. 62, 234; min SPEC. 61.

mincen, v., *O.Fr.* mincer; *mince*; **mince** (*imper.*) C. B. 16, 29, 110; **mincid** (*pple.*) L. C. c. 18; mincid, mencid, minced C. B. 14, 15, 76.

minchen, see **münechene**.

minde, minden, mindi, see **münde, münden, mündi**.

míne, sb., *mineral*, PALL. i. 374; mine *mine* GOW. II. 87.

mínen, v., *Fr.* miner; *dig a mine*, MAND. 267; mini AYENB. 108; **míne** (*imper.*) PALL. iii. 334; **mínede** (*pret.*) TREV. III. 269; mineden ['*suffoderunt*'] WICL. GEN. xlix. 6; *comp.* under-mínen.

mingen, see **münegen. minien**, see **münien.**

ministraciōn, sb., *service*, ALEX. (Sk.) 3554.

ministral, see **menestral**.

ministre, sb., *O.Fr.* ministre; *minister, servant*, CH. C. T. *G* 1; **ministris** (*pl.*) ALEX. (Sk.) 1537.

ministren, v., *O.Fr.* ministrer; *minister*, LANGL. *B* xii. 54; ministre PR. C. 5958.

[**minne**, sb., *comp.* mére-minne.]

minne², sb., *O.N.* minni; *memory, remem-*

brance, C. M. 8835 ; on þi pouer þen have þou minne MIRC 1965.

minnen, v., *O.N.* minna (*remind*), minnaz (*remember*) ; *have in mind, remember* ; minne A. P. i. 582 ; C. M. 112 ; ANT. ARTH. xviii ; þat i mai minne on þe mon GAW. 1800 ; of one thinge i wolle iou minne M. ARTH. 169 ; me **minnis** [*O.N.* mik minnir] (*pres.*) TOWNL. 225 ; me minneþ (*ms.* mineþ) þat ic seȝde ȝuw ORM. 1817 ; me minez A. P. ii. 25 ; *see* **münien.**

   minning, mining, sb., *O.N.* minning ; *memory*, C. M. 13183.

minour, sb., *Fr.* mineur ; *miner*: as a minour sekith gold hid WICL. GL. PROV. ii. 4 ; minur MISC. 97 ; **minours** (*pl.*) LANGL. PROL. *A* 101 ; GOW. II. 198.

minster, *see* **münster. mint,** *see* **münet.**

minte, sb., *mite* (*insect*), *small piece, 'mica,'* VOC. (W. W.) 767 ; **mintis** (*pl.*) '*bibiones* (*vermes*)' 623.

minte², sb., *O.E.* minte, = *O.H.G.* minza, *Lat.* menta, mentha ; *mint, 'menta,'* VOC. 190 ; PR. P. 338 ; minte [mente] WICL. MAT. xxiii. 23 ; *comp.* **brōc-, hors-minte.**

minten, *see* **münten.**

minūte, sb., *Fr.* minute ; *minute*, PR. P. 338 ; in a minute [minte] while LANGL. *B* xi. 372, *C* xiv. 200 ; **minutes** (*pl.*) CH. ASTR. i. 7 (5).

miracle, sb., *O.Fr.* miracle ; *miracle, wonder*, A. R. 158 ; HAV. 500 ; þat miracle CHR. E. 764 ; **miracles** (*pl.*) SAX. CHR. 263 ; AYENB. 56.

mīre, sb., *O.E.* mȳre, mīre, = *M.Du.* miere, *Dan.* mȳre, *Swed.* mȳra ; *ant*, MISC. 8, 9 ; *comp.* **pisse-mīre.**

mīre, *see* **mūre. mirie,** *see* **mürie.**

mirke, adj., *O.E.* mirce, myrce, *O.L.G.* mirki, *O.N.* myrkr ; *dark, 'obscurus,'* PR. P. 339 ; REL. I. 210 ; ANT. ARTH. vi ; þe mirke nith HAV. 404 ; mirc GEN. & EX. 284 ; mirk MAN. (H.) 221 ; PR. C. 456 ; M. H. 98 ; merke R. R. 5339 ; þe merke dale LANGL. *B* i. 1 ; merke night N. P. 51 ; **merk** (*sb.*), *darkness*, A. P. ii. 894 ; in merke LANGL. *B* xvii. 240 ; in mirke H. S. 2164.

   mirknesse, sb., *darkness*, GEN. & EX. 3104 ; merk-, merkenesse LANGL. *B* xviii. 136 ; merk- nes APOL. 98.

mirken, v., *O.N.* myrkja ; *make dark*, HORN CH. 81 ; merke LUD. COV. 207 ; he sent mirknes and he **mirkid** (*pret.*) HAMP. PS. civ. 26 ; **mirkid** (*pple.*) lxxiii. 21.

mirour, sb., *O.Fr.* mireor ; *mirror*, AN. LIT. 91 ; AYENB. 158 ; LANGL. *B* xi. 8.

mirous, adj., *wonderful*, PALL. iv. 358.

mirre, sb., *O.E.* myrre, *O.Fr.* mirre, myrre ; *myrrh*, MAT. ii. 11 ; A. R. 372 ; MISC. 26.

mirt, sb., *Lat.* myrtus ; *myrtle*, PALL. i. 568.

mirþe, *see* **mürhðe. mirþren,** *see* **mürðren.**

mis, adv., *O.N.* mis ; *? from* **mīþen** ; *badly* : iȝeven mis & inumen mis HOM. I. 205 ; heo wrencheð hore muð mis A. R. 212 ; on elpi word þet tu mis iherest 296 ; seið & deð so much mis (*ill*) 86 ; ȝif him mis biveolle 200 ; þu ert mis biþouht MISC. 45 ; ich (h)abbe mis ido ROB. 339 ; mis bilevinde AYENB. 69 ; mis for to donne H. M. 17 ; to (a)mende boþe þe mis MAN. (H.) 303 ; **a mis** (*O.N.* ā mis) *amiss* O. & N. 1365 ; ȝe þencheþ a mis BRD. 8 ; he dude a mis LEG. 14 ; is faren o mis MIR. PL 152 ; misse . . . schaped WILL. 141.

[**mis-**, prefix, *O.E.* mis-,= *O.N.* mis-, *Goth.* missa-, *O.H.G.* missi- ; *wrongly, amiss.*]

mis-, *see* **mes-.**

mis-bēode, v., *O.E.* misbēodan, = *O.N.* mis- biōða, *M.H.G.* missebieten ; *abuse, offend*, O. & N. 1541 ; **misbēode** (*imper.*) LANGL. *A* vii. 45 [misbede *B* vi. 46] ; **misbēd** (*pret.*) MAN. (F.) 2088 : **misbóden** (*pple.*) GAW. 2339 ; who haþ ȝou misboden or offended CH. C. T. *A* 909.

mis-bére, v., *bear amiss*, TREV. III. 275 ; ȝif hit is **misborn** (*pple.*) H. M. 33 ; þeȝ þe him hadde þer misbore BEK. 1248.

mis-bilēave, sb., *misbelief, unbelief*, KATH. 348 ; misbileve AYENB. 13 ; S. S. (Wr.) 2302 ; misbiliefe PR. C. 5521.

mis-bilēved, adj., *infidel* ; misbilevede men ROB. (W.) 2442.

mis-beleven, v., *believe amiss*, GOW. III. 152.

mis-bóde, sb., *abuse, offence*, LAȝ. 11095 ; HOM. II. 79.

mis-cas, sb., *mishap*, ROB. (W.) 10047.

mis-chēosen, v., *choose ill* ; **mischēs** (*pret.*) GEN. & EX. 190.

mischeroun, sb., *gnat*, MAN. (F.) 14706.

mis-comforten, v., *discomfort* ; thai **mis- comfort** (*pres.*) HAMP. PS. cxix. 4* ; **mis- comfortand** (*pple.*) 4.

mis-cwēmen, v., *displease* ; **miscwēmeð** (*pres.*) A. R. 182 ; misqvēme (*subj.*) PL. T. 595 (P. P. 323).

mis-cweþen, v., *O.E.* miscweðan ; *speak ill* ; **miscweþen** (*pple.*) FRAG. 1.

mis-dēde, sb., *O.E.* misdǣd,= *Goth.* missa- dēds, *O.H.G.* missitāt ; *misdeed*, A. R. 124 ; ORM. 7830 ; O. & N. 231 ; **misdēde** (*pl.*) C. L. 394 ; **misdēden** (*dat. pl.*) LAȝ. 18389.

mis-dēme, v., *misjudge*, WICL. NUM. xiv. 11 ; **misdēmeþ** (*pres.*) CH. C. T. *E* 2410.

mis-departen, v., *divide amiss* ; **misdepar- teth** (*pres.*) CH. C. T. *B* 107.

mis-dōn, v., *O.E.* misdōn,= *O.H.G.* missituon ; *misdo, do amis*, O. & N. 1489 ; nalde na mon misdon wið oðre HOM. I. 15 ; misdo LANGL. *A* iii. 118 ; to misdone WILL. 2581 ; **misdēð** (*pres.*) HOM. I. 65 ; A. R. 124 ; **misdide** (*pret.*) KATH. 1207 ; misdede GEN. & EX. 1847 ; HAV. 992 ; MAP 336 ; misdiden ORM. 15154 ; **mis- dōn** (*pple.*) A. R. 98 ; SPEC. 72.

mis-dōere, sb., *misdoer*; **misdōeres** (*pl.*) AYENB. 8; MAND. 9.

mis-drawe, v., *misdraw*, SAINTS (Ld.) xlv. 168; **misdrawe** (*pple.*) GOW. I. 18.

misericorde, sb., *Fr.* misericorde ; *mercy*, A. R. 30; CH. A. B. C. 25; miserecorde A. P. i. 366.

miserie, sb., *O.Fr.* miserie ; *misery, servitude*, CH. C. T. *B* 3196; ALEX. (Sk.) 1774, 3550.

misentente, v., *cf. O.Fr.* mesentendre ; *misunderstand* ; **misetente** (*pple.*) A. P. i. 257.

mis-fal, sb., = *M.Du.* misval; *mischance* ; misval AYENB. 30.

mis-fallen, v., = *M.Du.* misvallen ; *turn out ill* ; misvalle (*pres. subj.*) AYENB. 193.

mis-fangen, v., *take amiss* : þah he beo god me hine mai misfonge O. & N. 1374.

mis-fáre, sb., = *O.N.* misför ; *ill faring*, L. H. R. 118; C. M. 315.

mis-fáren, v., *O.E.* misfaran, = *O.N.* misfara, *O.H.G.* missefaran ; *go astray, fare ill*, GEN. & EX. 1911 ; misfáre (*pple.*) WILL. 995 ; AMAD. (R.) xxi.

mis-fēren, v., *go astray, fare ill* ; misfērde (*pret.*) LAȜ. 26229 ; HAV. 1869 ; S. S. (Wr.) 2765 ; misferden KATH. 93 ; misferde WILL. 2999.

mis-gān, v., = *M.Du.* misgaen ; *go astray, err*: þou misgōs (*pres.*) HAV. 2707 ; (he) misgeþ AYENB. 94.

mis-gang, sb., *trespass*, C. M. 17235 ; misgong S. A. L. 163.

mis-ȝēmen, v., *O.E.* misgȳman ; *neglect* ; misseȝeme A. P. i. 322; misȝēmed (*pple.*) A. R. 344.

mis-gelt, sb., = gült, WILL. 1541.

mis-gīen, v., *misguide* ; misgīed (*pple.*) CH. C. T. *B* 3723.

mis-governaunce, sb., *misconduct*, CH. C. T. *B* 3202.

mis-gülten, v., = gülten : what have we þe misgelt (*pple.*) S. S. (Web.) 1697.

mis-hap, sb., *mishap*, PR. P. 339 ; mishep A. R. 180.

mis-happen, v., *happen unfortunately* ; mishappe (*pres. subj.*) CH. C. T. *A* 1646.

mis-hópien, v., = *M.L.G*, mishopen ; *despair*; mishópie (*pres.*) HOM. I. 213 ; mishópand (*pple.*) HAMP. PS. xliii. 20*.

mīsi, sb., *cf.* **mēos** ; *fen*, GAW. 749.

mis-kenninge, sb., ' *mespris par oi u de fet*,' REL. I. 33.

mis-lǽden, v., *O.E.* mislǽdan ; *mislead*; mislĕd(d)en (*pret.*) CH. TRO. iv. 48.

mis-lēren, v., *seduce, deceive* ; mislere P. L. S. xiv. 6 ; mislērede (*pret.*) LAȜ. 4311*.

mis-lōre, v., *O.E.* mislār ; *false doctrine*, HOM. II. 29.

mis-lēve, sb., = *M.L.G.* mislōve; *unbelief*, HOM. II. 73.

mis-lēven, v., *discredit* ; mislēveð (*pres.*) A. R. 416; misleveþ AYENB. 180 ; mislēfde (*pret.*) HOM. II. 137.

mis-līch, adj., *O.E.* mislīc, = *O.L.G.* mislīc, *O.N.* mislīkr, *O.H.G.* misselīch, *Goth.* missaleiks ; *various*, MARH. 1 ; A. R. 180; ȝiven ham misliche nomen KATH. 271 ; mislīche (*adv.*) HOM. I. 119 ; mistlice HOM. I. 230; heo misliche foren LAȜ. 6270; þat ich . . . of þat matere so misseliche þenke WILL. 711.

mis-līche, adv., *badly*, HOM. I. 281 ; messeli WILL. 207.

mis-līken, v., *O.E.* mislīcian, = *O.N.* mislīka, *O.H.G.* misselīchen ; *mislike, be displeasing*, A. R. 338 ; tis maȝ þe misliken ORM. 18287 ; mislikeð (*pres.*) P. L. S. viii. 7 ; mislikeþ O. & N. 344 ; mislīke (*subj.*) HORN (R.) 670 ; mislīkede (*pret.*) GEN. & EX. 1728 ; it mislikede me WILL. 2039.

mislīkunge, sb., *misliking*, A. R. 180; misliking SPEC. 72 ; IW. 537 ; PR. C. 9028.

mis-limpen, v., *O. E.* mislimpan ; *befall amiss* : him mai sone mislimpe HOM. I. 243.

mis-nimen, v., = *O.H.G.* misseneman ; *mistake* : misnimeð (*pres.*) A. R. 46 ; misnimþ AYENB. 83 ; þu misnōme (*pret.*) KATH. 455 ; misnumen (*pple.*) GEN. & EX. 3091 ; misnume O. & N. 1514.

misniming, sb., *malappropriation*, ROB. (W.) 10465.

mis-nótien, v., *abuse, misuse* ; misnóteð (*pres.*) A. R. 130.

mispaien, v., *O.Fr.* mespaier ; *displease* ; mispaie YORK v. 64; HAMP. PS. xcix. 4 ; MAN. (F.) 7811 ; mispaies (*pres.*) HAMP. PS. xxxvi. 37*; mispaie (*pres. subj.*) A. R. 218 ; mispaied (*pret.*) YORK iii. 399 ; mispaide (*pple.*) GOW. I. 178.

mis-proud, adj., *unwisely proud*, LANGL. C viii. 96; misproude (*pl.*) ALIS.² (Sk.) 312 ; misseproude WILL. 2944.

mis-rǽden, v., *O.E.* misrǽdan, = *M.H.G.* misserāten ; *give bad advice* ; misrede O. & N. 1063 ; HORN (R.) 298 ; misrĕt (*pres.*) AYENB. 184 ; misrǣden (*subj.*) LAȜ. 13130.

mis-réken, v., *? go astray* for sumeres tide is al to wlonc and doþ misreken monnes þonk O. & N. 490 ; and sone mai a word misreke þar muþ schal aȝen h(e)orte speke 675.

mis-rempen, v., *? go wide of the mark* : ȝef þe þincþ þat i misrempe (*pres.*) O. & N. 1787.

mis-reulen, v., *misgovern* ; misreuleth (*pres.*) LANGL. B ix. 59 ; misreule GOW. III. 170.

mis-sawe, sb., *evil-speaking*, A. R. 124.

mis-scheppen, v., = *M.Du.* misscheppen ; *deform* ; misschápen (*pple.*) *misshapen* A. P. ii. 1355 ; misshape LANGL. B vii. 95 [misshapen *A* viii. 79*].

misse, sb., = *M.Du.*, *M.H.G.* misse, *O.N.* missa; *loss, privation, defect, injury*, M. ARTH. 3677 ; of hete hi habbeð misse P. L. S.

viii. 118 ; be misse never so huge A. P. iii. 420 :
to (a)mende mi misse WILL. 532 ; i wol be
vengid of this gret(e) misse LUD. COV. 43 ;
misse (dat.) REL. I. 49 ; wiþute misse (fail)
ASS. 112 ; wiþoute misse SHOR. 135.

mis-seggen, v.,=M.Du. misseggen ; speak
evil : missegge SHOR. 98 ; miszigge AYENB.
189 ; misseist (pres.) KATH. 457 ; misseiᶛ
A. R. 34 ; misseie (subj.) E. G. 89 ; missaid
(pret.) MAN. (F.) 3485 ; misseid (pple.)
HAV. 1688.

mis-ziggere, sb., slanderer, AYENB. 256.

mi(s)-sēmand, pple., unseemly, HAMP. PS.
lxxii. 15*.

missen, v., O.E. missan,=O.H.G. missen,
O.Fris., O.N. missa ; miss : he sulden missen
hine GEN. & EX. 3336 ; missin 'careo' PR. P.
340 ; misse SPEC. 57 ; WILL. 1016 ; PR. C.
5266 ; A. P. i. 329 ; LUD. COV. 50 ; heo . . .
moten misse þ(e)rof KATH.651 ; huewenden . . .
of huere live to misse HORN (R.) 126 ; misseᶛ
(pres.) A. R. 364 ; (ye) misse (subj.) MISC.
40 ; miste (pret.) GEN. & EX. 3872 ; BEK.
50 ; an an swa ich þe miste LAȝ. 18817 ;
misten ORM. 8919 ; comp. ȝe-missen.

missing, sb., want, lack, ALEX. (Sk.) 4595 ;
YORK i. 48.

mis-spēche, sb.,=M.Du. misspraeke ; slander ; missespeche WILL. 1523.

mis-spēden, v., succeed ill ; misspēdde
(pret.) MAN. (F.) 6912.

mis-spenden, v., misspend ; misspéne (pres.)
LAȝ. 13483*.

mis-st(e)orte, -stürte, v., start amiss, O. &
N. 677.

mis-swáre, sb., ? perjury, HOM. I. 205.

mist, sb., O.E. mist, = M.Du. mist ; mist,
'nebula,' PR. P. 340 ; TREAT. 136 ; REL. I.
211 ; ALIŚ. 5761 ; LANGL. A PROL. 88 ; miste
(dat.) P. L. S. viii. 9.

mis-tǣchen, v., O. E. mistǣcan ; teach
falsely ; mistāgte (pres.) GEN. & EX. 475.

mis-táken, v., O. N. mistaka ; mistake ;
mistákeþ (pres.) R. R. 1540.

misti, adj., O.E. mistig,=M.Du. mistigh ;
misty, 'nebulosus,' PR. P. 340 ; REL. I. 265 ;
mistier (compar.) LANGL. B x. 181.

misti-hēde, sb., mystery, CH. COMP. M. 224.

misti-lī, adv., mistily, PR. C. 4364.

mistinesse, sb., mistiness, ['caligine'] WICL.
PROV. vii. 9.

mis-tīden, v., O.E. mistīdan ; turn out ill ;
mistīde (pres. subj.) O. & N. 1501.

mistil, sb., ? O.E. mistel, = O.H.G. mistil ;
? mistletoe ; peire bedis of mistill w. & I. 16.

mis-tīmen, v., O.E. mistīman ; turn out ill :
mistīmeᶛ (pres.) A. R. 200.

mis-trēouþe, sb., unbelief ; mistrauthe A. P. ii
996.

mistrin, see mestrin.

mis-trōst, sb.,=M.Du. mistroost ;=O.H.G.
missetrōst ; mistrust, LUD. COV. 126.

mis-trowe, sb.,=M.L.G. mistruwe ; diffidence, WILL. 3314.

mis-trowen, v., O.N. mistrūa,=M.Du. mis-
trouwen, M.H.G. missetriuwen ; mistrust ;
mistrowande (pres. pple.) WICL. BAR. i. 17 ;
missetrowede (pret.) WILL. 1480 ; mis-
trowet (pple.) A. R. 68*.

mis-trum, ? adj., ? infirm, meagre (diet) : of
mistrume (dat.) mel A. R. 262*.

mis-trūsten, v., mistrust ; mistriste CH. C.
T. C 369 ; PARTEN. 4107 ; mistrōstende
(pple.) WICL. BAR. i. 17.

mis-turnen, v., turn amiss, pervert ; mis-
turne WICL. LAM. iii. 36 ; misturnes (pres.)
PR. C. 1617.

mis-þenchen, v., have evil thoughts ; mis-
ᶛenche (pres. subj.) A. R. 62.

mis-þunchen, v., O.E. misþyncan ; seem
amiss : þat te misþuncheᶛ (pres.) KATH. 982.

mis-wenden, v., O.E. miswendan ; turn
amiss ; miswendeþ (pres.) AYENB. 40 ; mis-
went (pple.) FER. 1963.

mis-witen, v., be heedless, A. R. 202.

mis-wīven, v., wive amiss, GEN. & EX. 540.

mis-word, sb., = M.H.G. miswort ; harsh
word, slander, A. R. 190.

mis-wrīten, v., O.E. miswrītan ; miswrite ;
miswritene (pple. pl.) FRAG. 1.

mis-wune, sb., evil habit, REL. I. 132 (HOM.
II. 13).

mis-würchen, v., do ill ; miswerche WILL.
5148 ; miswroȝt (pple.) ASS. 187 ; mis-
wroght PR. C. 1993.

mit, see mid.

mīte, sb., O.E. mīte ['tamus'],=M.Du. mijte
(tick), O.H.G. mīza (gnat) ; mite (insect) ;
mītes (pl.) CH. C. T. D 560.

mīte², sb.,=M.L.G. mīte, M.Du. mijte ; mite,
minute object, 'minutum,' PR. P. 340 ; WILL.
2017 ; LANGL. B xx. 178 ; for . . . i nolde noȝt
ȝive a mite FER. 1579 ; weie a mite GOW. II.
275 ; nought worþ a mite CH. C. T. A 1558.

miteine, sb., Fr. mitaine ; mitten, glove, 'mitta,
mancus,' PR. P. 340 ; miteines (pl.) PL. CR.
428 ; mittens PALL. i. 1167.

mitigaciōn [mitigacioun], sb., Fr. mitigation ; mitigation, LANGL. B v. 477.

mīting, sb., mite, trifle, little one, TOWNL.
96 ; YORK xviii. 113 ; mighting xxxi. 110 ;
mi miting meede YORK xxxii. 305.

mitre, sb., Fr. mitre ; mitre, JOS. 293 ; ALEX.
(Sk.) 1541.

mitred, adj., mitred, DAV. DR. 79.

mīþen, v., O.E. mīᶜan, = O.L.G. mīthan,
O.H.G. mīdan ; avoid, conceal, SPEC. 24 ;
miþe A. P. i. 359 ; HORN CH. 825 ; his sorwe
he couþe ful wel miþe HAV. 948 ; mīᶛe (pr.
subj.) GEN. & EX. 3807.

F f

mix, sb., *O.E.* mix, meox,=*Fris.* miux, *cf. O.H.G.* mist, *Goth.* maihstus; *dunghill, dung,* KATH. 204; SHOR. 109; **mixe** (*dat.*) REL. I. 183; mexe HOM. I. 113; **mixes** (*pl.*) *vile men* MISC. 140.

mixen, sb., *O.E.* mixen, myxen, meoxen; *dunghill,* R. R. 6496; **mixene** (*dat.*) LK. xiv. 35; (*ms.* mixenne) A. R. 140; mixne JUL. 41.

mixen, v., *O.E.* miscan,=*O.H.G.* miskan; *cf. Lat.* miscēre; *mix*; **mixid** (*pple.*) S. & C. II. vi.

mō, *see* **mā.**

mobard, sb., *clown*; **mobardis** (*pl.*) YORK xxxviii. 137; mobbardis xliv. 74.

moche, mochel, *see* **müche, müchel.**

mōd, sb., *O.E.* mōd *n., cf. O.L.G.* mōd, *Goth.* mōds (ὀργή, θυμός), *O.N.* mōðr, *O.H.G.* muot *m.*; *mood, mind, courage,* ALIS. 102; mod [mood] CH. C. T. *A* 1760; þi wreððe & þi mod HOM. I. 67; his mod him gon mengen LAȝ. 3407; whar þu þat mod nime 24777; þat uvele mod O. & N. 8; ðin milde mod GEN. & EX. 3602; let þi mod overgo MAN. (F.) 3214; **mōdes** (*gen.*) ORM. 9386; **mōde** (*dat.*) O. & N. 661; mid clene mode HOM. I. 65; hit com him on mode LAȝ. 11; þa (*r.* wa) hire wes on mode 3105; þu woldest beon of oþer mode MISC. 84; he arn of one mode REL. I. 216; he þohte on his mode HORN (R.) 287; wiþ glade mode SPEC. 80; wiþ . . . dreri mode ISUM. 186; mude PR. C. 2391; *comp.* **over-mōd.**

　mōd-kare [**mōdcare**], sb., *O.E.* mōdcearu; *anxiety,* LAȝ. 3115.

　mōd-ful, adj., *ambitious,* LAȝ. 31464.

　mōd-lēste, sb., *O.E.* mōdlēast = *discouraged condition,* HOM. I. 111.

　mōd-sorȝe, sb., *O.E.* mōdsorg; *sorrow of mind,* LAȝ. 8692.

mōd², adj., *O.N.* mōðr; *courageous; -minded*; **mōde** (*pl.*) MISC. 91; *comp.* **aðel-, ān-, blīðe-, drēoriȝ-, ēad-, hrað-, lang-, over-, or-, sāri-, þōle-, wēa-mōd.**

mōd, *see* **mād.**

moder, sb., *prov.Eng.* (*Norf.*) mauther; *young girl,* PR. P. 341.

mōder, sb., *O.E.* mōdor, mōdur, = *O.L.G.* mōdar, *O.Fris.* mōder, *O.N.* mōðir, *O.H.G.* muoter; *mother,* MAT. xii. 46; LAȝ. 210; MARH. 2; BEK. 153; LANGL. *B* xiv. 264; moder [mooder] CH. C. T. *B* 276; mooder '*mater, genitrix*' PR. P. 341; JOS. 98; modir PERC. 445; moodir LIDG. M. P. 243; **mōder** (*gen.*) LAȝ. 11058; ORM. 168; WILL. 1177; GOW. I. 352; PL. CR. 270; PR. C. 447; on his moder kne C. M. 11681; **mōder** [*O.E.* mēder] (*dat.*) MAT. ii. 11; KATH. 931; ORM. 248; **mōder** (*pl.*) MK. x. 30; modres P. L. S. xviii. 25; modren AYENB. 67; *comp.* **eald-, fōster-, god-, stēop-mōder.**

　mōder-bern, sb., =*M.H.G.* muoterbarn; *mother's child,* MARH. 2.

mōder-burh, sb., *metropolis,* KATH. 46.

mōder-child, sb., *O.E.* mōdorcild; *mother's child,* HORN (H.) 664; P. L. S. xxi. 93.

mōder-chirche, sb., = *M.H.G.* müterkirche; *mother church,* CHR. E. 923.

mōder-hēde, sb., *motherhood,* SHOR. 117.

mōder-lēs, adj., = *M.H.G.* muoterlōs; *motherless,* KATH. 78.

mōder-lich, adj., *O.E.* mōdorlīc; *motherly*: i þi moderliche herte HOM. I. 285.

mōder-sune, sb., *mother's son,* HOM. I. 269; modersone GREG. 705.

mōdi, adj., *O.E.* mōdig, = *Goth.* mōdags, *O.N.* mōdugr, *M.H.G.* muotec, *mod.Eng.* moody; *full of passion, courageous, proud,* '*superbus,*' FRAG. 3; LAȝ. 8344; O. & N. 500; P. S. 220; OCTAV. (H.) 771; MIN. vi. 42; prud . . . & modi HOM. I. 43; modi & bold GEN. & EX. 2728; wel modi & wel murne HORN (L.) 704; modiȝ ORM. 8245; **mōdieste** (*superl.*) KATH. 1247; *comp.* **over-mōdi.**

　mōdiȝ-leȝc, sb., *pride,* ORM. 73.

　mōdiȝ-līke, adv., *courageously,* ORM. 2035; modiliche LANGL. *B* iv. 173.

　mōdiȝnesse, sb., *O.E.* mōdignes; *courage, passion, pride,* ORM. 12040; modinesse HOM. I. 103; O. & N. 1405; prude and modinesse MISC. 74.

mōdien, v., *O.E.* mōdegian, mōdgian; *be proud*; to modienne HOM. I. 219; he **mōdigað** (*pres.*) 103.

modifie, v., *Fr.* modifier; *modify,* CH. C. T. *A* 2542.

mōdrie, sb., *O.E.* mōdrie, mōdrige, = *? O.Fris.* mōdire, *? O.H.G.* muotera; *aunt* (*ms.* moddrie) LAȝ. 30644.

moeble, adj., *O.Fr.* moeble, moble; *moveable,* CH. ASTR. i. 21 (47); meble S. A. L. 48; **moebles** (*sb. pl.*), *moveable goods, furniture,* CH. C. T. *E* 1314; mobles D. ARTH. 666.

mōȝ, mōȝe, *see* **māȝ, māȝe.**

moȝen, *see* **muȝen.** **moȝte,** *see* **mohðe.**

mohðe, sb., *O.E.* mohðe, moððe, = *? M.Du.* motte; *moth*: mohðe fret te claðes H. M. 29; mogðe LK. xii. 33; mouȝþe [mouȝte] WICL. MAT. vi. 19; mouþe '*tinea*' REL. I. 6; mouþe [mothþe] LANGL. *C* xiii. 217; moþe VOC. 177; mouȝte '*tinea*' PR. P. 346; moughte VOC. 223; **mothes** [moþþes, mouhtes] (*pl.*) CH. C. T. *D* 560; moȝtes TREV. V. 119.

moiler, *see* **mulier.**

moisōn, sb., *O.Fr.* moison, muison; *measure,* REL. I. 192; alle þe **musouns** (*pl.*) in musike LANGL. *B* x. 172 [musons *A* xi. 128].

moiste, adj., *O.Fr.* moiste; *moist,* LANGL. *B* xvi. 68; CH. C. T. *A* 457.

moistin, v., *O. Fr.* moistir; *moisten,* '*humecto,*' PR. P. 341; **moisted** (*pret.*) PARTEN. 3574; **moisted** (*pple.*) MAND. 3.

moistūre, sb., *O. Fr.* moistour, moisteur ; *moisture, 'humor,'* PR. P. 341 ; TREV. III. 65.

móke, sb., *? net* : some can sette the moke awrie REL. I. 248.

móke², sb., ?=maðek, mauk ; *moth, 'tinea,'* VOC. 190, 223.

mokkin, v., *O.Fr.* moquer ; *mock, 'ludifico,'* PR. P. 341.

mokren, *see* mukren. **mōl,** *see* māl.

molaine, sb., *? some ornament of a shield* ; **molaines** (*pl.*) GAW. 169.

mólde, sb., *O.E.* molde, = *O.Fris.* molde, *O.N.* mold, *O.H.G.* molta, *Goth.* mulda ; *mould, ground, earth,* HOM. II. 258 ; SHOR. 137 ; AYENB. 95 ; FER. 361 ; of þære mólde (*dat.*) FRAG. 5 ; of þar eorþe molde MISC. 142 ; on molde SPEC. 33 ; WILL. 917 ; LANGL. *A* PROL. 64 ; A. P. ii. 1114 ; under molde MISC. 93 ; HORN (L.) 317 ; AN. LIT. 91 ; CHR. E. 743.

   móld(e)-ale, sb., *funeral banquet,* PR. P. 341.

   mōld-werp, sb.,=*M.H.G.* moltwerf ; *mole,* ['*talpa*'] WICL. LEV. xi. 30 ; moldewarp REL. II. 83 ; APOL. 57 ; PALL. i. 924.

mólde², sb., *cf. Span.* molde, *O.Fr.* modle ; *mould, model,* FER. 4939 ; LANGL. *B* xi. 341 ; mone of a mold(e) TRIST. 942 ; from þe molde (*? neck*) to þe nolle [' *a cervice usqve ad occipitium* '] TREV. V. 369 ; ?mólden (*pl.*) A. R. 84.

mōlen, *see* mālen.

molle, sb.,=*M.L.G.,M.Du.* mol; *mole,'talpa,'* PR. P. 342 ; VOC. 220.

molle², sb., *labour, trouble,* ALEX. (Sk.) 628, 4446.

molsh, adj., *soft,* PALL. ii. 141.

mōme, sb.,=*M.L.G.* mōme, mōne (*aunt, mother*), *M. Du.* moeme, *O. H. G.* muoma *aunt*), *O.N.* mōna (*mother*) ; *? aunt, 'matertera,'* PR. P. 342 ; mone GOW. I. 97.

momelen, v.,=*M.Du.* momelen, *M.L.G.* mummeln; *chatter, discourse*; mameli [*v. r.* momele, mamelen] LANGL. *B* v. 21 ; momeleþ (*pres.*) P. S. 238 ; mamelit REL. I. 179 ; momellis YORK xxvii. 106.

   mammlere, sb., *babbler* : Marcure (was) a mammlere of wordis ALEX (Sk.) 4498.

momurdotes, sb., pl., *sulks,* D. TROY 9088.

mon, *see* man. **mōn,** *see* mān.

monchen, v., *cf.* mouchen ; *munch* ; monche [muche, mucche, meche] hire mete CH. TRO. i. 907.

monde, sb., *O.Fr.* monde ; *world* : in þis mounde MAN. (F). 11974.

mone, *see* mune.

mōne, sb., *O.E.* mōna, = *O.L.G., O.H.G.* māno, *O.N.* māni, *Goth.* mēna, *Gr.* μήνη ; *moon, 'luna,'* PR. P. 342 ; ORM. 13843 ; HAV. 403 ; TREAT. 133 ; C. L. 103 ; S. S. (Wr.) 332 ; þe

mone, þene mone A. R. 166 ; mone [moone] CH.C.T.*A* 3352 ; mōnen (*gen.*) dæig *Monday* SAX. CHR. 258 ; monen dai ROB. 495 ; P. S. 340 ; mone dæi LAƷ. 13935 ; mone dai TREAT. 133 ; monen moruwe ARCH. LII. 38 ; mone niht SAX. CHR. 259 ; atte Janus mones (*month's*) idus PALL. ii. 29 ; mōnen, mone (*dat.*) LAƷ. 9128, 30498 ; mone KATH. 272 ; MAND. 51 ; under mone HAV. 2791 ; mōnen, mone (*pl.*) REL. I. 264, 268.

mōne, *see* māne, mōme.

moneie, sb., *O.Fr.* monoie, *Lat.* monēta ; *money,* LANGL. *B* i. 44 ; CH. C. T. *A* 703 ; mone MAN. (F.) 8996 ; PR. C. 5570 ; *see* münet.

moneƷen, *see* manien. **monek,** *see* munec.

monen, *see* munen.

mōnen, *see* mānen, mǣnen.

monesten, v., *O.Fr.* monester ; *admonish* ; moniss BARB. xii. 383 ; **monestis** (*pres.*) ALEX. (Sk.) 2592 ; **monishit** (*pret.*) ALEX. (Sk.) 1379* ; monest 3127, moneste [monishitt] 1173 ; þai monist BARB. xii. 379.

   monesting, sb., *admonition,* WICL. I COR. xiv. 3.

mōneð, sb., *O.E.* mōnað, mōnoð, mōnð *m., n.* = *O.Fris.* mōnath, *O.N.* mānaðr, *O.H.G.* mānod, *Goth.* mēnoþs *m., M.H.G.* mānot, mōnet *m., n. ; month,* LK. i. 36 ; GEN. & EX. 619 ; þane moneð LAƷ. 7220 ; moneþ ROB. 59 ; WILL. 5074 ; PL. CR. 248 ; monþ BEK. 1850 ; mōnþes (*gen.*) TREAT. 134 ; mōneþe (*dat.*) MAND. 49 ; monþe PROCL. 7 ; LANGL. (Wr.) 1648 ; mōnðes (*pl.*) JOHN iv. 35 ; LAƷ. 7771 ; monethes PR. C. 4988 ; monithis TOR. 1860 ; monþes, monþ P. L. S. xvi. 155, 157 ; moneð HOM. I. 33 ; tweolf moneð A. R. 218 ; fif moneþ ORM. 433 ; mōneþe (*dat. pl.*) C. M. 11220 ; *comp.* hervest-mōneþ.

mong, *see* mang. **mongere,** *see* mangere.

moni, *see* maniƷ. **monien,** *see* manien.

moniment, sb., *Lat.* monumentum ; *document* ; **monumentes** (*pl.*) BARB. xx. 44*.

monstre, sb., *O.Fr.* monstre ; *monster,* MAND. 47 ; CH. BOET. i. 4 (18).

mont, *see* munt.

mopisch, adj., *foolish,* BEK. 78 ; **moppische** (*pl.*) men BRD. 59.

moppe, sb., *cf. Du.* mop ; *mop, 'pupa,'* PR. P. 342 ; *? fool* S. S. (Web.) 1414 ; mop YORK xxxi. 196 ; **moppis** (*pl.*) DEP. R. iii. 276.

moppen, v., *bewilder, stupefy* ; **mapped** (? mopped) (*pple.*) and mased L. H. R. 216.

mōr, sb., *O.E.* mōr *m.*=*M.Du.* moer, *M.H.G.* muor *n. ; moor, 'palustre,'* PR. P. 342 ; PERC. 1127 ; mōre (*dat.*) GREG. 719 ; to þan more O. & N. 818 ; mōres (*pl.*) LAƷ. 4817 ; ALIS. 6074 ; P. S. 70.

   mōr-hen, sb., *mocr-hen,* P. S. 158.

   mōr-land, sb., *O.E.* mōrland ; *moor-land* ; (*ms.* moreland) GEN. & EX. 2968.

mōr-ven [mōrfen], sb., *moor-fen*, LAȜ. 20164.
mōr², sb., *O.E.* mōr ; *mulberry* ; mours (*pl.*(
HAMP. PS. lxxvii. 52.

mōr-bēam, sb., *O.E.* mōrbeam ; *mulberry-
tree*, FRAG. 3.

mūr-berie, sb., *mulberry*, REL. I. 37.

mōr-tree, sb., *mulberry-tree*, WICL. LK.
xvii. 6.

móre, sb., *O.E.* moru (*gen.* moran) (*see* ANGL.
VI. 176),=*M.Du.* more, *O.H.G.* morha, mo-
raha (*carrot*) ; *root, stump, ' radix,'* VOC.
181 ; HOM. I. 103 ; O. & N. 1328 ; ROB. 352 ;
LANGL. *B* xvi. 5 ; S. S. (Wr.) 1726 ; of þare
more HOM. II. 217 ; móren (*pl.*) HOM. II.
139 ; mores BRD. 13 ; bi moren and bi rote(n)
LAȜ. 31885 ; *comp.* wal-more.

mōre, *see under* mā.

moreine, sb., *O.Fr.* morine ; *murrain*, LANGL.
C iv. 97 ; morein PR. P. 343 ; morin HAMP.
PS. i. 1*.

móren, v., *uproot; root* ; mórede (*pret.*) took
root REL. I. 129 (HOM. II. 163) ; *uprooted* ROB.
(W.) 10263 ; móred (*pple.*) 10264 ; hure love
is mored (*rooted*) on þe ful vaste FER. 2834 ;
imored L. H. R. 28.

moren, *see* morȝen.　mōren, *see* māren.

morȝen, sb., *O.E.* morgen, mergen,=*O.L.G.*,
O.H.G. morgan, *M.Du.* morgen, margen, mergen, *O.Fris.* morn, *O.N.* morginn, *Goth.*
maurgins ; *morning, morrow* ; þane morȝen
AVENB. 46 ; morgen GEN. & EX. 247 ; morwin LUD. COV. 257 ; morwin, morwe PR. P.
344 ; morwe ALIS. 1241 ; CH. C. T. *A* 1492 ;
good morwe LANGL. *B* v. 306 ; moroun GAW.
1208 ; þe þridde morewe SPEC. 82 ; morn P. L. S.
xxix. 20 ; A. P. ii. 493 ; more RICH. 6977 ; bi
þe morghen, moreghen MISC. 33 ; a þe marȝen
HOM. I. 79 ; a morȝen, mærȝe, mærwe, marwen
[morwe] LAȜ. 1694, 18474, 20019, 26854 ; on
morgen, morwen GEN. & EX. 1161, 2305 ; til a
moregen HOM. II. 75 ; a morȝe O. & N. 432 ;
a morwen oþer a niht A. R. 22 ; to morwen 278 ;
on þe morwen HAV. 811 ; a morewen SPEC. 26 ;
a morwe BEK. 771 ; WILL. 1296 ; at morwhen
[morihen, mørn] PS. xxix. 6 ; at morwen
MAND. 164 ; at morne PR. C. 2668 ; to morȝen, mærȝen, marȝen [morwe] LAȜ. 16066,
23661, 26393 ; ær to marwen [morewe] eve
17732 ; to marhen KATH. 645 ; he sal to morwen þole deþ S. S. (Wr.) 647 ; to marȝan HOM.
I. 21 ; to morwe BEK. 769 ; LANGL. *A* ii. 26 ;
to morne DEGR. 1695 ; from morwe till eve
WILL. 793 ; *comp.* ǽr-, mid-morwen.

morgen-give, sb., *O.E.* morgengifu,=*O.H.*
G. morgangeba ; *a present made on the morrow
after marriage*, GEN. & EX. 1428 ; morhȝive
[marhenȝive] A. R. 94 ; marheȝive H. M. 39 ;
mor-, mærȝeve LAȜ. 14394, 31090 ; morive
PR. P. 343.

morgen-hwīle, sb., *morning* ; morgewile
HOM. II. 39 ; mornwhile D. ARTH. 2001 ;
morgenqvile GEN. & EX. 3275.

morȝen-līc, adj., *O.E.* morgenlīc ; *belonging
to the morrow* : se morgendliche daig MAT.
vi. 34.

morȝen-līht [morelīht], sb., *O.E.* morgen-
lēoht ; *morning light*, LAȜ. 17946 ; morewe
liȝt WICL. I KINGS xxv. 34.

morȝe-mete, sb., *O.E.* morgenmete, = *O.N.*
morginmatr ; *morning meal*, HOM. I. 237.

moren-milk, sb., = *Ger.* morgenmilch ;
morning milk, SPEC. 36.

morwe-song, sb., *morning song*, CH. C. T.
A 830.

morowe-slēp, sb., *morning sleep*, CR. K.
10 ; morgesclep MISC. 131.

morwe-spēche, sb., = *M.L.G.* morgen-
sprāke ; *morning meeting, ' crastinum collo-
quium,'* PR. P. 344 ; E. G. 54 ; mornspeche 45 ;
þis gilde shal have iiij. morwe-spēces (*pl.*)
be þe ȝere 60.

morwin-sterre, sb., *O.E.* morgensteorra ;
morning star, PR. P. 344.

morgen-tīd, sb., *O.E.* morgentīd ; *morning
time*, GEN. & EX. 59 ; moreȝentīd FL. & BL.
570 ; morwetiö FER. 2895 ; moretid ALIS.
4106.

morȝeninge, sb., *morning* : in þare morȝen-
(i)nge [moreweninge] O. & N. 1718 ; morweninge P. L. S. xvii. 584 ; morewening SPEC. 60 ;
morweninge, morninge ' *diluculum, matuti-
num'* PR. P. 344 ; LANGL. PROL. *A* 5 ; CH. C.
T. *A* 1062.

morken, *see* murken.

morknen, v., *O.N.* morkna ; *rot* ; mourkne
(*pres.*) A. P. ii. 407.

mormal, sb., *cf. O.Fr.* mal mort (*espèce de
lèpre*) ; *? mortification, ' malum mortuum,'*
PR. P. 343 ; CH. C. T. *A* 386.

morn, *see* morȝen.　mornen, *see* murnen.

morsel, sb., *O.Fr.* morsel ; *morsel*, AYENB.
248 ; GAW. 1690 ; CH. C. T. *A* 128 ; mossel
ROB. 342 ; mussel WICL. GEN. xviii. 5 ; mor-
sellis (*pl.*) BARB. ix. 398.

morsūre, sb., *O.Fr.* morsure ; *biting*, ALEX.
(Sk.) 4088.

mortal, mortel, adj., *O.Fr.* mortel ; *mortal*,
CH. C. T. *A* 1590.

morteis, sb., *Fr.* mortaise ; *mortise, ' gum-
phus, incastratura,'* PR. P. 344 ; MAND. 10.

morteisen, v., *mortise* ; morteised (*pple.*)
YORK xxvi. 163.

mortēr, sb., *Fr.* mortier, *Lat.* mortārium ;
mortar, ROB. 128 ; LANGL. *B* vi. 144 ; ALIS.
332 ; mortier AYENB. 116 ; a morter fast is
made aboute the tree PALL. iv. 113.

mortifie, v., *O.Fr.* mortifier ; *mortify*, CH.
C. T. *G* 1431.

mortrels [mortreus, mortreux], sb., pl., *O.
Fr.* mortreux ; *a sort of stew*, LANGL. *B* xiii.
91 ; mortreus [mortreux] CH. C. T. *A* 384
mortrewes C. B. 14, 19, 90.

morð, sb., *O.E.* morð,= *O.N.* morð, *O.L.G.*, *O.Fris.* morth, *O.H.G.* mord *n.*, *related to Lat.* mors (*gen.* mortis) *f.; death, murder*; þat morð LAӠ. 28715; morþ [murth] C. M. 1121; morth ALEX. 1279; mid morðe aqvellen LAӠ. 19739.

morþ-dēde, sb., *O.E.* morðdæd,= *M.H.G.* morttāt; *murder*, FRAG. 7.

morð-gomen, sb., *murderous play, murder*, LAӠ. 22908.

morð-slaӠe, sb., *O.E.* morðslaga; *murderer*; morðslaӠa (*pl.*) HOM. I. 29.

morð-spel, sb., *murder*, LAӠ. 19654.

morþer, morder, sb., *O.E.* morðor,= *Goth.* maurþr *n.; murder*, CH. C. T. *B* 1766 (Wr. 14987); PL. CR. 635; þane morþer (*ms.* morþre) LAӠ. 28715*; morþre (*dat.*) ROB. 144; murðre A. R. 310.

morþren, see murðren.

morwen, morweninge, see morӠen, morӠeninge.

mos, sb., *O.N.* mosi,= *M.Du.*, *O.H.G.* mos; *moss, 'muscus,'* PR. P. 344; IW. 2040; TRIAM. 392; mose PALL. iii. 365.

mos², sb., *O.N.* mosi,= *O.H.G.* mos, *M.Du.* mose; *moss* (*bog*), M. T. 15.

[mōs, sb., *O.E.* mōs,= *M.L.G.* mōs, *M.Du.* moes, *O.H.G.* muos; *pulp; comp.* appelmōs.]

mosardri, sb., *O.Fr.* musarderie; *indolence*, ALEX. (Sk.) 4486.

mōse, see māse.

mosel, sb., *O.Fr.* musel; *muzzle*, CH. C. T. *A* 2151.

most, see must. mōst, see mā.

mōste, v., *O.E.* mōste,= *O.Fris.* mōste, *O.L.G.* mōsta, *Goth.* (ga-)mōsta, *M.H.G.* muoste; *from* mōt² ; (*pret.*); *must*, KATH. 1397; O. & N. 665; P. L. S. xiii. 249; C. L. 220; CH. C. T. *A* 712; (*ns.* mosste) ORM. 7602; ne moste þer na mon in cumen LAӠ. 6712; of al þe brode eorðe ne moste he habben a grot for te deien uppon A. R. 260; moste, muste HOM. I. 9, 27; muste GEN. & EX. 2624; mostes þu FRAG. 7; þu mostest BEK. 1214; mostist MAP 336; alle mosten to helle te H. H. 8; moste WILL. 1052; and bede . . . þat hi moste þane wei mid hem go BRD. 5.

mot, sb., *? O.E.* mot,= *Du.* mot; *mote, atom*, WICL. LK. vi. 41; þet mot AYENB. 175; mote GAW. 2209; REL. I. 27; móte (*dat.*) LANGL. *B* x. 263; A. P. i. 725; mótes (*pl.*) CH. C. T. *D* 868; moten ELL. ROM. I. 317.

móte-lēs, adj., *spotless*, A. P. i. 899.

mot², sb., *Fr.* mot; *trumpet-note, bugle-note, 'classicum,'* PR. P. 344; blue bigli in buglez þre bare mote GAW. 1141; strankande ful stoutli moni stif mótez (*pl.*) 1364.

mōt, sb., *O.E.* mōt,= *O.N.* mot; *meeting, assembly*, O. & N. 468; REL. I. 219; gewitene mot SAX. CHR. 253; þat mot MISC. 45; þat meidene mot (*assembly*) KATH. 2458; þis meidenes mot (*dispute*) 1321; wormes holdeþ here mot A. D. 237; mōte (*dat.*) HOM. II. 83; LAӠ. 12885; GAW. 910; *comp.* Ӡe-mōt, wardmōt.

moot-halle, sb., *assembly-hall*, P. S. 336; LANGL. *B* iv. 135; GAM. 717.

mōt-hūs, sb., *meeting-house*, MISC. 46.

mōt², v., *O.E.* mōt,= *O.L.G.*, *O.Fris.* mōt, *Goth.* (ga-)mōt (χωρεῖ), *O.H.G.* muoz; (*pret.-pres.*) *must*, LAӠ. 3494; KATH. 1919; A. R. 98; O. & N. 672; GEN. & EX. 1304; SHOR. 11; WILL. 548; MIRC 1578; PL. CR. 557; A. P. i. 31; PR. C. 4207; mot [moot] WICL. LK. xix. 5; moot CH. C. T. *A* 735; HOCCL. i.75; þer fore ic mot Ӡu telle TREAT. 136; þa hwile þe he mot libbe P. L. S. viii. 17; þu most LAӠ. 9854; A. R. 102; BRD. 1; K. T. 446; we moten GREG. 36; Ӡe moten TRIST. 1754; motin CH. P. F. 546; Ӡe ne mote bileve her no leng BRD. 3; heo moten LAӠ. 478; ha moten MARH. 14; þei mote LANGL. *B* ix. 108; mōte (*subj.*) LAӠ. 4481; MARH. 5; O. & N. 52; GEN. & EX. 1621; C. L. 332; SPEC. 57; H. H. 151; as evere mote i [i moote] drinke win or ale CH. C. T. *A* 832; Ӡeornest tat tu mote sket up cumen intil heofne ORM. 1266; al so mote bitide þe S. S. (Wr.) 1185; welcome mot(e) he be MAN. (H.) 192; as mut(e) i þrive TOR. 1443; we moten HAV. 18; heo moten A. R. 298; see mōste.

móte, sb., *O. Fr.* mote; *moat*, MISC. 97; LANGL. *B* v. 595; A. P. i. 142; PR. C. 8896; (*castle*) ALEX. (Sk.) 3831, 4353; (*palace*) 3218, 3324, 5602.

móte, see mot. mōːewōk, see māte.

mōtien, v., *O.E.* mōtian; *moot, dispute, plead*, HOM. I. 43; moten KATH. 586; MARH. 17; mootin '*discepto*' PR. P. 345; mote LANGL. *A* i. 150; GOW. III. 296; IW. 3328; MIR. PL. 30; aӠens þe nile i not moote P. R. L. P. 202; mōtest (*pres.*) LAӠ. 1443.

mōtare, sb., *pleader, 'disceptator, placitàtor,'* PR. P. 345; mōteres (*pl.*) KATH. 725.

mōtild, sb., *female advocate*, HOM. I. 205; KATH. 397.

mōtinge, sb., *pleading, advocacy*, LAӠ. 6559; MISC. 39; LANGL. *B* vii. 58.

motif, sb., *motive, suspicion*, CH. C. T. *B* 628.

motlee, adj., *O.Fr.* mattelé; *motley*, CH. C. T. *A* 271; motle '*stromaticus*' PR. P. 345; mottelai CATH. 244.

motōn, sb., *O.Fr.* mouton, multon; *sheep, mutton; gold coin so called from having a figure of a sheep impressed upon it; 'ovilla, moto,'* PR. P. 345; motoun of golde LANGL. *B* iii. 24 [mutoun *A* iii. 25].

moþe, moþþe, see mohðe.

mouchen, v., *cf.* monchen; *? munch, cat*; mouchid (*pret.*) TOWNL. 320.

moue, sb., *O.Fr.* moe ; *grimace* ; (*ms.* mǫwe) PR. P. 346; mouwe P. S. 339; make mowes (*pl.*) on þe mone YORK xxxv. 286.

mouȝte, mouhte, *see* mohðe.

moule, sb., *Fr.* mule ; *chilblain*, '*pernio*,' PR. P. 346.

moulen, *see* muwlen. **mount**, *see* munt.

mountūre, sb., *saddle horse*, GAW. ,1691.

mournen, *see* murnen. **mous**, *see* mūs.

mousteren, v., *O.Fr.* mostrer, mustrer ; *muster* ; musterin PR. P. 349; **mustred** (*pple.*) PARTEN. 3003.

moustre, sb., *O.Fr.* moustre, mostre ; *show, appearance, muster*, LANGL. B xiii. 362; WICL. 3 KINGS v. 13 ; **mustours** (*pl.*) *dials, clocks* ALEX. (Sk.) 130.

moutin, v., = *O.L.G.* (ge-)mūtōn, *O. H. G.* mūzōn ; *moult*, '*plumeo, deplumeo*,' PR. P. 347 ; moute HALLIW. 564 ; **moutes** (*pres.*) PR. C. 781.

mouþ, *see* mūð. **mouwe**, *see* mūȝe.

móven, v., *O. Fr.* movoir, mevoir ; *move* ; move [moeve, meve] LANGL. B xvii. 194; move CH. P. F. 150; moevi BRD. 31 ; meove TREV. IV. 157 ; moife YORK xvii. 48 ; **méven** (*pres.*) MAND. 13 ; moffes YORK v. 2 ; **meffid** (*pret. pl.*) ALEX. (Sk.) 2403 ; **móved** (*pple.*) PL. CR. 826 ; meved WILL. 4285.

mowe, v., *from* moue; *make grimaces*, YORK xxxvi. 78.

mōwe, *see* mǣw, mǣȝe, moue.

mowen, *see* muȝen. **mōwen**, *see* mǣwen.

muance, sb., *O.Fr.* muance ; *change* ; mowence BARB. i. 134.

muche, *see* muchel.

müchel, muchel, adj. & adv., *O.E.* mycel, micel, = *O.L.G.* mikil, *O.N.* mikill, *Goth.* mikils, *O.H.G.* michil ; *much, great, large*, P. L. S. viii. 107 ; wes þet folc swa muchel [mochel] LAȝ. 1998 ; muchel seolver 7283 ; muchel dæl 9436 ; he lette makien ænne dic þe wæs . . . muche 10350 ; muche londe he him ȝef 136 ; to muchel *too much* HOM. I. 213 ; muchel luvede he us A. R. 292 ; a muche flod 74 ; þet so muche [muchel] drohen MARH. 2 ; þah he muche þolie KATH. 229 ; ich con muchel more O. & N. 1207 ; muche strengþe 764 ; þer couþe he muchel [mochil] helpe CH. C. T. A 258 ; mochel MAND. 181 ; HOCCL. i. 370; mochel AYENB. 181 ; moche 7 ; mechel SHOR. 147 ; meche 139 ; michel PL. CR. 55 ; þat michel can TRIST. 1528 ; michel, mikel GEN. & EX. 26, 1209 ; mikel PS. xxi. 26 ; M. H. 2 ; PR. C. 925 ; mekill, mekil ALEX. (Sk.) 927, 69 ; mikel blisse ORM. 788 ; þe ston was mikel HAV. 1025 ; he was mike 960 ; mikil '*multus*' PR. P. 337 ; mukel A. P. ii. 366 ; mekil S. & C. I. liii; mekel qvat *many various things* ALEX. (Sk.) 130, 5468; much qvat

GAW. 1280; muche wes þe sorewe P. S. 193 ; qvare on muse ȝe sa mekill · ALEX. (Sk.) 268 ; a muche [moche] man LANGL. *A* viii. 70 ; he was . . . so moche P. L. S. xv. 147 ; miche WICL. GEN. xxix. 7 ; PERC. 11 ; meche TOR. 270 ; E. G. 355 ; ALEX. (Sk.) 3306 ; miche 5602 ; þeo muchel blisse A. R. 38 ; þe mochele see BRD. 1 ; þat micle god ORM. 3900 ; of hire micle selþe 2634 ; **mücheles** (*gen. n.*) A. R. 102, 368 ; MISC. 86 ; mucheles þe mare HOM. I. 45 ; **müchelen**, muclen (*dat. m. n.*) LAȝ. 13276, 26626 ; michelen LK. xxi. 27 ; **müchelere**, muchlere (*dat. f.*) HOM. I. 87, 109 ; muchelere, muchelre, mochelere, muchere [mochere] LAȝ. 516, 3688, 8284, 28379 ; michelere stefne '*voce magna*' LK. iv. 33 ; mid michelere stefne xix. 37 ; **müchelne** (*acc. m.*) HOM. I. 15 ; muchelne, muchene, LAȝ. 8722, 24835 ; **müchele** (*acc. f.*) HOM. I. 123 ; **müchele** (*instr.*) A. R. 186 ; O. & N. 906 ; muchele mare P. L. S. viii. 194 ; KATH. 1246 ; muchele [moche] þe balder LAȝ. 16508; micle lahre ORM. 2664 ; **müchele** (*pl.*) KATH. 37 ; micle ORM. 8002 ; muche [moche, miche, meche] and lite CH. C. T. *A* 494 ; **müchelen** (*dat. pl.*) LAȝ. 5256.

müchel-hēde, sb., *greatness*, FL. & BL. 51 ; mochelhede AYENB. 93 ; mikelhede PS. viii. 2.

michelnesse, sb., *O.E.* mycelness, = *O.H.G.* michilnessi ; *greatness, magnitude*, HOM. II. 135 ; michelnes WICL. EX. ix. 24 ; mikelnes PS. xxviii. 4.

müchele, sb., *O.E.* mycelu, = *O.H.G.* michelī ; *magnitude* : of one mochel(e) & miȝte LANGL. B xvi. 182.

müchelin, v., *O.E.* myclian, miclian, = *O.N.* mikla, *Goth.* mikiljan, *O.H.G.* michilan ; *magnify, increase*, MARH. 15 ; mucli JUL. 19 ; **mücheleð** (*pres.*) A. R. 182 ; **mikled** (*pple.*) PS. xix. 6.

müchen, v. *cf. O.H.G.* mūhhan (*rob*) ; *steal, pilfer* ; michin '*manticulo*' PR. P. 337.

müchere, sb., *cf. O.H.G.* mūhhāri ; *pilferer* ; michare '*manticularius, furunculus*' PR. P. 336; ALEX. 3542 ; micher VOC. 213 ; R. R. 6541 ; ED. 827 ; TOWNL. 236 ; **müchares** [mucheres] (*pl.*) A. R. 150.

mud, sb., = *M.L.G.* mudde, mod, modde ; *mud*, '*limus*,' PR. P. 347 ; P. L. S. xxxi. 2 ; mod RICH. 4360 ; **mudde** (*dat.*) A. P. ii. 407.

mud-lī, adj., *muddy*, HAMP. PS. *page* 494.

müe, sb., *O.Fr.* mue, *cf. mod.Eng.* mews ; *coop to fatten poultry ; cage, prison* ; KN. L. 85 ; (*ms.* mwe) '*saginarium*' PR. P. 350; muwe, meue (*ms.* mewe) CH. C. T. *A* 349; meuwe WILL. 3336 ; mewe PALL. iv. 583 ; **mewes** (*pl.*) i. 526.

müen, v., *Fr.* muer ; *moult* ; **imūwed** (*pple.*) FER. 1738.

müge, sb., *O.Fr.* muge (*muguet*) ; *? musk*

plant, SPEC. 26 ; *? comp.* note-**migge** *under* **knute**.

mūȝe, sb., *O.E.* mūga, = *O.N.* mūgi ; *heap (of corn)* ; mughe, mowe ' *moie* ' VOC. 154 ; WICL. RUTH iii. 7 ; moghe ALEX. (Sk.) 4434 ; þourȝe felde or corn mowe or stak GEN. & EX. 6760 ; mow BARB. iv. 117 ; mūȝen (*pl.*) LAȝ. 29280.

mugen, v., *become cloudy* ; **muged** (*pret.*) GAW. 142.

muȝen, v., *O.E.* mugan, *cf. O.H.G.* mugan, magan, *M.Du.* moghen, *Goth.* magan, *O.N.* mega ; *from* maȝ ; *be able*, GEN. & EX. 1818 ; muȝhen ORM. 2959 ; moȝe AYENB. 21 ; mowen CH. TRO. ii. 1594 ; mowe WICL. GEN. xiii. 16 ; moun (*ms.* mown) PR. P. 346 ; **mahte** [mihte] [*O.E.*meahte,mihte,= *O.L.G.*, *O.H.G.* mahta, mohta, *Goth.* mahta, *O.N.* mätti], (*pret.*) *might*, KATH. 61 ; mahte MARH. 7 ; mahte, mehte, mihte, michte HOM. I. 33, 129, 161 ; mahte, mæhte, mihte, michte (*printed* mithte), micte (*printed* mitte), mohte LAȝ. 403, 1908, 3148, 5535, 9352, 24673 ; mihte ORM. 130 ; mihte, miȝte O. & N. 394 ; michte MISC. 27 ; miȝte BEK. 62 ; AYENB. 31 ; mihte, muhte A. R. 58, 94 ; micte HORN (H.) 10 ; micte, micthe, mithe,mouchte, moucte, mouchte, mouthe HAV. 42, 145, 147, 233, 356, 376, 1030 ; moht M. H. 7 ; moȝt GAW.84 ; moght Iw.226 ; MIN. vi. 58 ; mught PR. C. 282 ; moute AN. LIT. 3 ; mihtest ORM. 4682 ; H. H. 95 ; miȝtest P. L. S. xxiv. 206 ; AYENB. 104 ; mæhten LAȝ. 12294 ; mihten FRAG. 6 ; migten GEN. & EX. 573 ; muhten A. R. 58 ; mouȝten JOS. 23 ; **might** (*pple.*) MAND. 298 ; CH. TRO. iii. 654.

mūȝen, mūwen, v., *mow, stack (hay)* ; mughe ' *archoniare* ' CATH. 245 ; mowe other mowen *mow or stack* LANGL. *C* vi. 14.

mügge, sb., *O.E.* mycg, = *O.L.G.* muggia, *M. Du.* mugge, *O.H.G.* mucca ; *midge, gnat* ; mig(g)e VOC. 223.

muggel, sb., *tail* [vocatur . . cauda ab indigenis, patria lingva,· mughel FORDUN. SCOTICHRON. i. 138] ; **muggles** [moggles] (*pl.*) LAȝ. 29588.

mugling, sb., *tailed man* ; **muglinges** (*pl.*) LAȝ. 29590.

mug-wurt, sb., *O.E.* mucgwyrt ; *mug-wort*, ' *artemisia*,' FRAG. 3 ; mugw(u)rt REL. I. 36 ; mugwort PR. P. 347.

muk, sb., *O.N.* mykr *f.* (FRITZ.) ; *muck, dung, manure*, HAV. 2301 ; GOW. II. 290 ; LANGL. (Wr.) 4081 ; WICL. 3 KINGS xiv.10 ; PR. C. 9008 ; muc GEN. & EX. 2557 ; mok P. L. S. i. 20 ; mok (*v. r.* wose) LANGL. *C* xiii. 229.

muk-hille, sb., *muck-hill*, ' *sterquilinium*,' PR. P. 348 ; **mukhilles** [mokhulles] (*pl.*) TREV. VI. 213.

mukren, v., *? heap up* ; mukre CH. TRO. iii. 1375 ; **mokeren** (*pres.*) CH. BOET. ii. 5 (45).

mukrere, sb., *one who heaps up* ; **mokereres** (*pl.*) CH. BOET. ii. 5 (45) ; MISC. 214.

mul, sb., *cf. O.E.* myl, *M.L.G.*, *M.Du.* mul ; *dust, mould, rubbish ; ' pulvis*,' PR. P. 348 ; HAV. 6200 ; GOW. II. 204 ; A. P. i. 904.

mul-rein, sb., *mizzling shower, drizzle*, ' *plutina, pluviola*,' PR. P. 348.

[**mūl**, sb., ? = **mūr, mōr**.]

mūl-beri, sb., = *M.H.G.* mūlbere ; *mulberry*, ' *morum*,' PR. P. 348 ; molberi PS. lxxvii. 47.

mülc, see **milc**.

mūle, sb., *Fr.* mule, *Lat.* mūla ; *mule*, HOM. I. 5 ; AYENB. 223 ; muile LEG. 95 ; ALIS. 175 ; **mūles** (*pl.*) ROB. 189.

mulier, sb., *O.Fr.* moillier, *Lat.* mulier ; *woman, wife* : mede is moliere [*v. r.* muliere, mulirie] of amendes LANGL. *B* ii. 118 ; moillere, moiller *C* iii. 120.

mülle, see **mülne**.

mullin, v., *cf. O.H.G.* mullen, *O.N.* mylja ; *grind to powder*, ' *pulveriso*,' PR. P. 348.

mullok, sb., *refuse, manure*, CH. C. T. *A* 3873.

mülne, sb., *O.E.* myln, = *O.N.* mylna, *M.Du.* molen, *O.H.G.* mulin, muli, *? Lat.* molina ; *mill*, A. R. 72 ; P. S. 69 ; GAW. 2203 ; milne VOC. 235 ; WICL. MAT. xxiv. 41 ; mulle ROB. 547 ; mille PR. P. 337 ; MAND. 189 ; melle AYENB. 24 ; CH. C. T. *A* 3923 ; at mulne dure O. & N. 778 ; at mulne 86 ; **milnes** (*pl.*) MAN. (H.) 173 ; *comp.* **wind-milne**.

milne-hous, sb., *mill*, HAV. 1967 ; millehous PR. P. 337.

milne-stōn, sb., *O.E.* mylnstān ; *mill-stone*, REL. I. 81 ; milne-, milnstoon WICL. JUDG. ix. 53 ; milston VOC. 180 ; melston TREAT. 136 (mulleston SAINTS (Ld.) xlvi. 580) ; milleston E. G. 358.

mil-ward, sb., *O.E.* mylnweard ; *miller*, REL. I. 85 ; mel(l)eward P. L. S. xxiv. 6 ; **milwardes** [' *molendinarii*'] (*pl.*) TREV. IV. 319.

mülnere, sb., = *O.N.* mylnari, *O.H.G.* mulnāri, mulināri ; *miller*, LANGL. *A* ii. 80 ; milner, milner *C* iii. 113 ; *B* ii. 111* ; milner VOC. 212 ; millere CH. C. T. *A* 3150.

mülten, see **melten**.

mültestre, sb., *O.E.* myltestre, miltestre ; *harlot*, ' *meretrix*,' FRAG. 5 ; **miltistran** (*pl.*) MAT. xxi. 31.

multiplicatioun, sb., *Lat.* multiplicātiō ; *multiplying*, CH. C. T. *G* 849.

multiplie, v., *Fr.* multiplier ; *multiply*, CH. C. T. *C* 365 ; **multiplieþ** (*pres.*) AYENB. 190.

multitūde, sb., *Fr.* multitude ; *multitude*, PR. P. 348 ; ALEX. (Sk.) 69, 104, 927.

mummin, v., = *M.Du.* mommen ; *mutter*, ' *mutio*,' PR. P. 348 ; mum YORK xi. 175 ; **imummed** (*pple.*) DEP. R. iii. 371.

mun, see **man**.

múnd, sb., *O.E.* mund,=*O.L.G., O.N.* mund, *O.H.G.* munt; *palm of the hand, protection,* LA3. 25569; **múnde** (*dat.*) MISC. 197; he wende toward Bruges . . . wiþ swiþe gret(e) mounde P. S. 189; michel of mounde GREG. 645; a knight of mochel mounde LAUNF. 597; feouwer & sixti munden [mundes] LA3. 21994; *comp.* **schaft-monde.**

 múnd-breche, sb., *O.E.* mundbryce; ['*lae-sio majestatis*'] TREV. II. 95; mundebriche '*trespas vers seignur*' REL. I. 33.

münde, sb., *O.E.* (frēond-)mynd,=*O.N.* mynd (*form*), *Goth.* (ga-)munds, *O.H.G.* (gi-)-munt (*memory*), *Lat.* mens (*gen.* mentis), *Gr.* μῆτις; *mind, memory,* MARH. 21; TREAT. 134; WILL. 4123; habbe on ine munde A. R. 66; for his fader munde [minde] ['*pro patris memoria*'] TREV. VII. 315; muinde P. L. S. xxvi. 38; minde ORM. 17577; GOW. II. 82; LANGL. *B* v. 288; PS. cviii. 15; PR. C. 59; a man þat is out of his minde CH. C. T. *C* 494; as þe bok makiþ mind(e) FLOR. 2168; wox so mikel minde (*? number*) of flies MAN. (H.) 2625; tur(t)les michel mind GEN. & EX. 3676; mende '*mens, memoria*' PR. P. 332; LUD. COV. 19; hi ne conne mende have of þilke holi gode SHOR. 42; *comp.* **?3e-münde, wurð-münd.**

 minde-dai, sb., *commemoration day,* TREV. II. 413; ENG. ST. I. 311; E. W. 109.

 mind-lēs, adj., *mindless,* PR. C. 2288.

münde, sb., =3emünde; *memory*; minde REL. I. 186, 216; SPEC. 82; MAP 336; H. S. 727.

münden, v., =3emünden; *mind, bear in mind, remember*; minde (*pres.*) M. T. 134; mindende (*pple.*) WICL. PROL. ROM.

[**mündi,** adj.,=3emündi; *mindful.*]

 mindi3nesse, sb., *memory,* ORM. 11508.

múndian, v., *O.E.* mundian,=*O.L.G.* mun-dōn, *O.H.G.* muntōn; *protect,* HOM. I. 115.

[**mundlunge,** adv.; *comp.* un-mundlunge.]

mune, sb., *cf. O.E.* myne,=*M.H.G.* mun, *O.N.* munr, *Goth.* muns; *mind, memory*: do þu þis mid gode mune (*r. w.* sune) HOM. I. 57; he hadde . . . in mone (*r. w.* sone) P. L. S. x. 35.

 mune-dei, sb., *anniversary*; **munedawes** (*pl.*) A. R. 22.

müne, sb., *O.E.* myne, mene,=*O.N.* men, *O.L.G.* meni; *necklace,* '*monile,*' FRAG. 2.

munec [monek], sb., *O.E.* munec, munuc, *Lat.* monachus; *monk,* LA3. 12906; munuch HOM. I. 199; A. R. 318; monek ROB. 417; monk LANGL. *A* v. 233; **munekes** (*pl.*) O. & N. 729; **muneken** (*dat. pl.*) LA3. 29718.

 munec-clāð, sb., *monk's garment,* LA3. 12984.

 munec-līf, sb., *O.E.* munuclīf; *monastery,* LA3. 29717; ORM. 6292.

münechene, sb., *O.E.* mynecen; *nun,* LA3. 28476; monchen LANGL. *C* vii. 128; minchin TREV. VI. 53.

münegen, v., *O.E.* mynegian, myngian,= *O.H.G.* (bi-)munigōn; *bear in mind, remind,* A. R. 320; hare ahne deð ant drihtines mune-gin MARH. 15; mune gie [mune3i] LA3. 24027; mini3i 2033*; mungen C. L. 1193; munge, minge, menge WILL. 831, 1422, 1624; mengen LANGL. *B* vi. 97; munge *A* vii. 88; minge SPEC. 95; MIRC 1555; OCTAV. (H.) 7; EM. 926; A. P. i. 854; **münge** (*pres.*) P. L. S. xxv. 167; mune3eð HOM. I. 145; A. R. 144; munegeð REL. I. 130; mene3eð HOM. I. 241; **mene3i**-(*subj.*) SHOR. 104; **müne-gede** (*pret.*) HOM. I. 151; mune3ede hine to fusen LA3. 16648; *comp.* 3e-münezen.

 mune3ing, sb., *remembrance, commemora-tion,* HOM. I. 45; muniging GEN. & EX. 1623; munegunge A. R. 26.

munekien, v., *make a monk,* LA3. 12904.

müngen, *see* münegen.

munen, v., *O.E.* munan,=*Goth.* *munan (*pple.* munands), *O.N.* muna, *prov. Eng.* mun; *from* man; *think, remember, have in mind, intend; be about to, have to*; GEN. & EX. 687; ðe hertes costes we ogen to munen REL. I. 217; monen II. 8; monin '*com-memoro*' PR. P. 342; mone (*r. w.* sone) H. S. 1118; he wolde mone þat he was his broþer sone MAN. (F.) 4811; 3ho **munde** (*pret.*) . . . to dæþe ben istaned ORM. 1967; he ne monede hit nout long S. S. (Web.) 1782; kind him mond (*v. r.* wolde) forbede C. M. 1105; *comp.* 3e-munen.

münet, sb., *O.E.* mynet,=*O.N.* mynt, *O.H.G.* muniza, *Lat.* monēta; *mod. Eng.* mint; *money, coin,* '*numisma*' FRAG. 2; menet MAT. xxii. 19; AYENB. 241; mint MIRC 1775; PALL. iii. 1069; *see* moneie.

 minitere, sb., *O.E.* mynetere; *minter, moneyer,* SAX. CHR. 253; **müneteres** (*pl.*) LEB. JES. 853.

münien, münnen, v., *O.E.* mynian, mynnan, *have in mind, bear in mind, call to mind,* LA3. 2033; ma wundres ich habbe iwraht þene ich mahte munien JUL. 41; ma murhðen þen alle men mihten . . . munnen KATH. 1714; munne REL. II. 211; of hem we ohte munne SPEC. 112; i wolde nemne hire to dai ant i dorste hire munne 113; munne [min(n)e] C. L. 268; mene D. TROY 1799; minne MIRC 1397; **münie** (*pres.*) REL. I. 171; munne 125; munnest KATH. 972; þat þou mines [menes] of him ['*quod memor es ejus*'] PS. viii. 5; þe ofte munneð mi nome ant munegeð MARH. 21; 3e minnen [menen] LANGL. *B* xv. 454; mene (*imper.*) PR. C. 5740; he minede (*pret.*) (*admonished*) alle men to forleten here sinnes HOM. II. 139; þai be **mined** ['*remi-niscentur*'] (*pple.*) PS. xxi. 28; *see* **minnen**; *comp.* 3e-münnen.

mening, sb., *memento,* HAMP. PS. cxxxiv. 13, cxlviii. 1*.

münnen, *see* munien.

münster, sb., *O.E.* mynster, *Lat.* monastērium; *minster,* LAȝ. 13028; **münstre** (*dat.*) ROB. 419; minstre ORM. 1017; menstre E. G. 38.

munt, sb., *O.E.* munt, *cf. O.Fr.* munt, mont; *mount,* LAȝ. 18336; A. R. 106; KATH. 2498; ORM. 5374; GEN. & EX. 1744; munt, mont MAT. xiv. 23; JOHN vi. 3; mount ROB. 203; MAND. 12; mont HOM. I. 87; TREV. III. 227; **munte** (*dat.*) HOM. I. 5; **muntes** (*pl.*) SAX. CHR. 258.

münt, sb., *intention,* A. P. i. 1161; mint C. M. 463.

muntaine, sb., *O.Fr.* montaigne; *mountain*; montaine LAȝ. 1282; mountaine BRD. 24; **muntaines** (*pl.*) WILL. 2619.

muntance, sb., *O.Fr.* montance; *amount*; mountance WILL. 2391.

munten, v., *O.Fr.* munter, monter; *mount*; mounteþ (*pres.*) LANGL. *A* PROL. 64.

münten, v., myntan; *think, intend; point* : mede i mot munten P. S. 151; mintin PR. P. 338; minte TRIAM. 1513; **mente** (*pret.*) SHOR. 151; þou mintest GAW. 2274; al se he ment(e) to don SAX. CHR. 263; toward a mighti mountaine him mintid with his finger ALEX. (Sk.) 1089; **münt** (*pple.*) REL. I. 131 (HOM. II. 11); mint S. S. (Wr.) 1660; TOR. 1616; PERC. 1667; *comp.* ȝe-münten.

mūr, *see* mōr.

murcnen, v., *O.E.* murcnian; *murmur,* ORM. 7785; murcneden (*pret.*) LK. v. 30.

mure, *see* mére.

mūre, sb., *O.E.* mȳre (-næddre), *O.N.* mȳrr *f.; mire,* BEV. 2023; mire '*palus*' PR. P. 338; RICH. 6939; CH. C. T. *A* 508; MAN. (H.) 70; GAW. 749; miere REL. I. 7; **mūres** (*pl.*) P. S. 216; mures, mires WILL. 2619, 3507; *comp.* lei-mūre.

mūren, v., *Fr.* murer; *wall,* MAND. 278.

mürȝe, mürghe, sb., *? O.E.* myrg; *joy, mirth,* FER. 3350, 3440.

mürȝen, v., *O.E.* (a-)myrgan; *make merry*; **mürgeþ** (*pres.*) *ms.* miryhes REL. I. 124; SPEC. 45; miryhes (*printed* mirþhes), miryes PS. xxxi. 11, xlvi. 2.

mürhðe, sb., *O.E.* myrhð; *mirth, amusement,* P. L. S. viii. 184; MARH. 11; KATH. 2382; murhðe, muruhðe A. R. 92, 162; murhðe, murehðe, murðe [murhþe, murthe] LAȝ. 1794, 5111, 24268; murȝþe, mureȝþe, mureþe O. & N. 341, 355; merhþe, merȝþe ANGL. I. 16, 29; mireȝþe FL. & BL. 682; muriðe HOM. II. 185; murþe ALIS. 1575; murþe [merþe] C. L. 42; merþe WILL. 31; murþe LANGL. *A* iii. 11; mirthe [*v. r.* merþe] CH. C. T. *A* 773; **mürhðhe** (*ms.* murðhe) (*pl.*) HOM. I. 13; merthes GAM. 783.

murhþen, v., *give mirth* : mirthe LANGL. *B* xvii. 240 [murthen *C* xx. 206].

mūri, adj., *muddy, miry*; miri M. T. 128.

mürie, adj., *O.E.* myrge, merge; *merry,* LAȝ. 10147; KATH. 317; O. & N. 345; BRD. 14; K. T. 288; murie, merie, mirie WILL. 1148, 1905, 2853; murie [merie] CH. C. T. *A* 235; merie (*ms.* merye) SHOR. 148; FLOR. 1766; mirie GEN. & EX. 212; miri '*laetus, jucundus, hilaris*' PR. P. 338; S. S. (Wr.) 261; PR. C. 904; **mürie** (*adv.*) HORN (R.) 592; merie LANGL. *B* PROL. 10; mirie HOCCL. i. 237; **mürie** (*pl.*) FRAG. 7; murie wordes A. R. 390; **mürgre** (*compar.*) LAȝ. 24964; muriere song BRD. 10; **mürieste** [meriest(e)] (*superl.*) LANGL. *B* xv. 211.

müri-, meri-līche, adv., *merrily,* CH. C. T. *B* 1300.

mirinesse, sb., *mirth,* CH. BOET. iii. 2 (66).

murken, v., *O.E.* murcian; *murmur*; murkeden [morkeden] (*pret.*) PS. cv. 25.

murne, adj., *O.E.* (un-)murne; *sad, mournful,* LAȝ. 16159; HORN (L.) 704; ART. & MER. 8338.

murne[2], sb., = *O.H.G.* morne; *sorrow* : mourne i make A. D. 251.

murnen, v., *O.E.* murnan, *cf. O.L.G.* mornian, *O.H.G.* mornēn, *Goth.* maurnan; *mourn,* A. R. 310; GEN. & EX. 2053; murnin MARH. 14; murne MAN. (H.) 20; mournen SPEC. 69; mornin PR. P. 344; mourne [morne] WICL. JOHN xvi. 20; **morne** (*pres.*) CH. C. T. *A* 3704; mornes(t) S. S. (Wr.) 1862; we murne M. H. 36; **murne** (*subj.*) HORN (L.) 964; murnende (*pple.*) LAȝ. 18183; **murnede** (*pret.*) A. R. 366; murnede [mornede] LAȝ. 13090; mornede WILL. 825; mornede LANGL. *A* iii. 163 [morned *B* iii. 169]; *comp.* bi-murnen.

murnunge, sb., *O.E.* murnung; *mourning,* A. R. 342; murninge O. & N. 1598; BEK. 23; murning GEN. & EX. 3205; PR. C. 1846; mourning SPEC. 54; morning WILL. 746.

murnif, adj., *mournful*; mornif A. P. i. 262.

murten, v., *break, crush*; murte A. P. iii. 150.

mürðe, *see* mürhðe.

mürðren, murðren, v., *O.E.* myrðrian, = *Goth.* maurþrjan; *murder,* LAȝ. 21516; murðrin HOM. i. 247; murþri P. L. S. xxiii. 21; morþere LANGL. *A* iv. 42; mirþren ORM. 8124; morþeren PL. CR. 666; **mürðredest** (*pret.*) A. R. 310; **mürþered** (*pple.*) WILL. 1774; morþered CH. C. T. *E* 725; imurðred A. R. 244; imorþred ROB. 110; *comp.* a-mürðrin.

murþrere, sb., *murderer*; mordrere MAND. 292; murþereris ⌊morþereres⌋ (*pl.*) LANGL. *B* vi. 275.

mūs, sb., *O.E.* mūs, = *O.N., O.H.G.* mūs, *Lat.* mūs; *mouse,* HOM. I. 53; mous SPEC. 111;

LANGL. *B* PROL. 12 ; CH. C. T. *A* 144; muis P. S. 326 ; **mūs** [*O.E.* mȳs] (*pl.*) *mice* O. & N. 87; muis TREV. VII. 297; mis LANGL. *B* PROL. 147 ; mise MAND. 291 ; S. & C. II. lviii; **mūse** [*O.E.* mūsum] (*dat. pl.*) O. & N. 591 ; *comp.* **fēld-, flinder-, rēre-mous.**

  mūs-don, adj., *dun coloured,* PALL. iv. 812.

  mūs-ēre, sb., *mouse-ear* (*plant*), '*pilosella*,' VOC. 140.

  mous-falle, sb., *O.E.* mūsfealle, *O.H.G.* mūsfalla ; *mouse-trap,* '*muscipula*,' PR. P. 347.

  mūs-hēred, adj., *dun haired,* PALL. iv. 896.

  mūse-stock, sb., *mouse-trap,* '*muscipula*,' VOC. 132 ; (*ms.* musestoch) HOM. I. 53.

muscherōn, sb., *Fr.* mousseron ; *mushroom,* '*bletus*,' PR. P. 349.

muschil, sb., *O.E.* muscle, *cf. O.H.G.* muscula, *from Lat.* musculus ; *muscle,* PR. P. 348 ; muscle CH. C. T. *D* 2100.

muscle, sb., *mussel,* ALEX. (Sk.) 5469.

mūsen, v., *O.Fr.* muser ; *muse* ; muse PR. C. 6266 ; mose ALEX. (Sk.) 333 ; **mūse** (*pres.*) LANGL. *B* x. 181 ; mowsin '*muso, musso*' PR. P. 347 ; **mūsi** (*subj.*) AYENB. 104 ; **mūsed** (*pret.*) CH. C. T. *B* 1033.

mūsen², v., *cf. M.H.G.* mūsen ; *catch mice* ; mousin '*muricapio*' PR. P. 347 ; **mūseþ** (*pres.*) REL. I. 180.

musike, sb., *Fr.* musique ; *music,* GEN. & EX. 460 ; ALEX. (Sk.) 2238.

musk, sb., *Fr.* musc ; *musk,* PR. P. 349.

muskit, sb., *O.Fr.* moschet ; *musket, male sparrow-hawk* ; muskitte PR. P. 349 ; muskett '*capus*' CATH. 247 ; muskitt VOC. 220.

musōn, *see* **moisōn.**

must, sb., *O.E.* must, = *O.H.G.* most ; *from Lat.* mustum ; *must, new wine,* VOC. 257 ; WICL. JOB xxxii. 19; must [most] LA3. 8723 ; must, most LANGL. *B* xviii. 368, *C* xxi. 415. ·

mustarde, sb., *O.Fr.* moustarde ; *mustard,* PR. P. 349; mustard AN. LIT. 9 ; mostard AYENB. 143.

musti, adj., *moist, new*: use this ferment for musti brede PALL. xi. 525.

mustre, *see* **moustre.**

mūte, sb., *O.Fr.* mute, muete, meute ; *a pack of hounds,* WILL. 2192 ; GAW. 1451, 1720.

muteren, v., *mutter* ; moteringe ['*mussitantes*'] (*pple.*) WICL. 2 KINGS xii. 19.

mūð, sb., *O.E.* mūð, = *O.N.* mūðr, munnr, *Goth.* munþs, *O.Fris., O.L.G., O.H.G.* mund ; *mouth,* A. R. 60 ; REL. I. 211 ; muþ FRAG. 5 ; ORM. 13868 ; O. & N. 673 ; mudh MISC. 30 ; hold þi mouþ BEK. 2059 ; þane mouþ AYENB. 27 ; **mūðes** (*gen.*) REL. I. 131 ; **mūþe** (*dat.*) FL. & BL. 11 ; mouþe S. S. (Wr.) 2320 ; **mūðes** (*pl.*) LA3. 14823 ; GEN. & EX. 2316.

mūðe, sb., *O.E.* mūða, = *O.N.* munni ; *mouth, estuary,* MAT. viii. 18 ; mouþe ROB. 20.

mūþen, v., *cf. M.H.G.* munden ; *speak, utter*: mouþen LANGL. *B* iv. 115 ; **mouthed** (*pret.*) *C* xxi. 154.

[**muwel**] **moul,** sb., *cf. Dan.* mul, *Swed.* mögel *n.; mould, mildew,* '*mucor*,' PR. P. 346.

mūwen, *see* **mū3en.**

muwlen, sb., *cf. O.N.* mygla ; *mould, become mouldy,* A. R. 344 ; moulen CH. C. T. *B* 32 ; moulin PR. P. 346 ; **mouled** (*ms.* moweld) (*pple.*) PR. C. 5570 ; þi moulid mete P. R. L. P. 181 ; moulid bred REL. I. 85.

n.

nā, adv., *O.E.* nā, nō, = *O.H.G.* neo, nio, *O.N.* nei ; *for* ne ā ; *no, nay, not,* IW. 716 ; na swa swa heore bokeras ['*non sicut scribae eorum*'] MAT. vii. 29 ; þat he ne mihte na (*neither*) east na west SAX. CHR. 260 ; ne nom he na alle þa þe þer inne weren HOM. I. 123 ; na lengre ORM. 13163 ; na more HAV. 2363 ; H. H. 64 ; na þe mo AYENB. 41 ; na þe les WILL. 1751 ; CH. C. T. *B* 94 ; MAND. 14 ; ne beo þe dai na swa long LA3. 1327 ; nulle ich na so don 6239 ; na mare 15737 ; no durste þær bilæven na þæ vatte no þe læne 19444 ; na, no, neo þe les 141, 8033, 25611 ; nea swa strong 1552 ; he ne blakede no 7524 ; nouþer heort no hinde no mihte(n) heo … ifinde 30568 ; nam mare [*earlier text* na mare] MK. xv. 5 ; nam more (*ms.* namore) FRAG. 5 ; nam more [na mare] A. R. 246 ; betere is þo þene no [*ms. Oxon.* '*melius est tunc qvam nunqvam*'] 340 ; no he seið 334 ; noa he seiðe 222 ; na þe mo ROB. 257 ; nere þe voreward no so strong 379 ; no more 273 ; 3ef þu nult no JUL. 10 ; 3ef þu ne dest no MARH. 7 ; neo de les 13 ; no [neo] ðe les KATH. 1023 ; na mo O. & N. 564 ; na, no, neo þe les 374, 827, 1297 ; no hwat scholde ich a mong heom do 997 ; nis it no so derne idon P. L. S. viii. 39 ; no mo BEK. 24 ; no leng BRD. 3 ; no but it were 3oven to þe from above WICL. JOHN xix. 11 ; soster no broþer P. S. 205 ; no peis no griþ GREG. 580 ; no *no* C. L. 1099 ; WILL. 67, 2701 ; MAN. (H.) 41.

nā, *see* **nān.**

nab, sb., *projecting point of a hill*: nabb ALEX. (Sk.) 5494.

naciōn [**nacioun**], sb., *Fr.* nation ; *nation,* CH. C. T. *B* 268 ; nacioun TREV. V. 363 ; BARB. x. 331.

naddre, sb., *O.E.* nædre, næddre, *cf. O.L.G.* nadra, *O.H.G.* natra, natara *f., O.N.* naðra *f.,* naðr *m., Goth.* nadrs *m.; adder, viper,* LEG. 54 ; SHOR. 104 ; naddre [neddre] CH. C. T. *E* 1786 ; nedre HOM. I. 153 ; neddre A. R. 82 ; REL. I. 211 ; GEN. & EX. 323 ; GOW. III. 118 ; MAND. 205 ; nedire '*serpens*' VOC. 222 ;

neddire ALEX. (Sk.) 4757; **nedren** (*gen.*)
HOM. I. 223; neddre A. R. 222; on neddre
liche HOM. II. 59; **næddren** (*acc.*) MAT. vii.
10; **naddren** (*pl.*) P. L. S. viii. 138; neddren
A. R. 214; neddren, neddreHOM. I. 43, 53;
nadres SAX. CHR. 262; nedres ROB. 43;
naderes VOC. 255; **neddre** (*gen.*) streon *offspring of vipers* ORM. 9265; *see* **addre**;
comp. **mére-, water-neddre.**

   neder-copp, sb., *for* ǎttercop; *spider*;
VOC. (W. W.) 766.

næfre, adv., *O.E.* nǣfre; *for* ne ǣfre; *never*,
ORM. 807; nævre, nævere LAȝ. 3104, 26253;
nevre O. & N. 209; REL. I. 209; BRD. 3;
AYENB. 26; nevre more HAV. 672; neavere
KATH. 126; nevere, never LANGL. *A & B*
PROL. 12; never þe lesse A. R. 218; never
the latter ['*verumtamen*'] HAMP. PS. xxx.
7*, xlviii. 16; ne beo þe song never so murie
O. & N. 345; never mo BEK. 2039; never te
never yet MISC. 35; AYENB. 99; þou he be
nevir so wilde EGL. 708; ner SPEC. 36, 102;
ner, nere WICL. LEV. vi. 13, JOHN ix. 21.

nēh, *see* **nēh.**

næil, sb., *O.E.* nægl, nægel,= *O.H.G.* nagal,
nagel, *O.N.* nagl, *O.Fris.* neil; *nail,* '*clavus,*'
FRAG. 2; neil A. R. 404; nail '*clavus, unguis*'
PR. P. 350; CHR. E. 629; nal PALL. ii. 199;
neiles (*pl.*) MARH. 19; KATH. 2151; nailes
LAȝ. 21879; HAV. 2163; LANGL. *A* vii. 56;
CH. C. T. *D* 769; **neilen** (*dat. pl.*) HOM. I.
121; *comp.* **finger-neil (-nail).**

   neil-cnīf, sb., *nail knife*; **neilcnīves** (*pl.*)
JUL. 56.

næil-sex, sb., *O.E.* nægelseax; *nail knife,*
FRAG. 5; nailsax [nailsex] LAȝ. 30578.

næilen, v., *O.E.* næglian, *cf. Goth.* (ga-)nagljan,
O.H.G. nagalen, negilen, *O.N.* negla; *nail*;
nailin '*clavo*' PR. P. 351; nailen LEG. 186;
naȝlen ORM. 2100; **nailed** (*pple.*) HOM. II.
21; ineiled A. R. 114: inailed BRD. 5.

næiðer, pron. & conj., *for* ne ǣiðer; *neither*,
nor; here neiˣer GEN. & EX. 1276; neiþer
LANGL. *B* iv. 32, 130; neiþer, neither CH. C. T. *A*
946, 1135; neither (*ms.* neyther) knith ne
knave HAV. 458; me liste not to whispre
neither roune LIDG. TH. 695; neithir '*neuter*'
PR. P. 352.

næniȝ, adj., *for* ne ǣni; *none*; nenig (*ms.*
nengi) LEECHD. III. 98; ·naniȝ man ORM. 59;
neng (*not*) sullen ... ah ȝefen HOM. I. 135;
neng bi mine wrihtę at for his milde wille
HOM. II. 217.

náfe, *see* **náve.**

nagge, sb., *? M.Du.* negghe; *nag, horse,*
'*bestula, equillus,*' PR. P. 350; D. TROY 7727.

naht, sb., *O.E.* neaht, næht, neht, nyht, niht
(*gen.* nihte), nihtes,= *O.L.G., O.H.G.* naht
(*gen.* naht, nahtes), *Goth.* nahts (*gen.* nahts),
O.N. nătt, nŏtt (*gen.* năttar, nǣtr), *Lat.* nox
(*gen.* noctis), *Gr.* νύξ (*gen.* νυκτός); *night,* HOM.

I. 57; SPEC. 34; PS. ciii. 24*; naȝt A. P. ii.
807; naght Iw. 3897; nauht MISC. 90; naht,
niht ORM. 972, 1904; nagt, nigt GEN. & EX.
76, 1678; þa burh born alle niht LAȝ. 29307;
to niht þu scalt faren 709; niht & dai H. M.
15; a niht KATH. 1443; þu singest a niht
[niȝt] O. & N. 219; niht, nith HAV. 404, 2669;
nicht REL. I. 177; MISC. 27; nict HORN
(H.) 131; wiste his bodi niȝt and dai P. L. S.
x. 40; to niȝt AYENB. 51; **nahtes,** nihtes
(*gen.*) ORM. 6492, 16897; nihtes HOM. I. 7;
KATH. 1078; þu flihst nihtes [niȝtes] O. & N.
238; be nihtes (*for* nihte) SAX. CHR. 261; on
nighter tale [*? O.N.* ā nāttar þeli] *in the middle
of night* C. M. 6126; bi nither tale HAV. 2025;
bi nighter tale CH. C. T. *A* 97; HOCCL. i.
306; bi niȝter tale ALEX. 324; **nihte** (*dat.*)
P. L. S. viii. 39; in þere nihte LAȝ. 7576;
bi nihte, niht (*ms.* nith) 2082, 2373; bi nihte
A. R. 20; bi nihte [niȝte] O. & N. 365; niht
(*pl.*) A. R. 278; CH. E. 157; seoven niht LAȝ.
20192; feowertene niht 25675; nigt GEN. &
EX. ·583; niȝt JOS. 6; seve niȝt HORN (L.)
448; þre niȝt ROB. 170; sefen naht ORM.
545; Marches nahtes 1901; nihtes O. & N.
523; **nihte** (*gen. pl.*) MK. i. 13; LAȝ. 4506;
s(e)ove nihte blisse HOM. II. 169; **nihten,**
nihte (*dat. pl.*) LAȝ. 5457, 11929; *comp.* **Frī-,
mid-, middel-, Sater-niht.**

   niht-, niȝt-, night-cappe, sb., *night-cap,*
CH. C. T. *B* 1853.

   night-crake, sb., '*nicticorax*' (νυκτικόραξ);
night-bird, VOC. 188.

   niȝt-crōwe, sb., *night-crow,* WICL. LEV. xi.
16; nightcrawe CATH. 254.

   niht-fuel, sb.,= *M.H.G.* nahtvogel; *night-
bird,* ['*nictocorax*'] A. R. 142.

   nihte-gale, sb., *O.E.* nihtegale, = *O.H.G.*
nahta-, nahtigala; *nightingale,* SPEC. 92;
niȝtegale REL. I. 241; nihte-, niȝtingale O. &
N. 13, 1512; nightingale CH. C. T. *A* 98.

   niht-longe, adv.,= *M.H.G.* nahtlanc; *through
the night*: þe wal ... ne moste nihtlonges
[nihtlonge] ... istonden LAȝ. 15504.

   niht-līc, adv., *O.E.* nihtlīc; *nightly,* PS. xc.
5*; **nihtlīche** (*pl.*) deden REL. I. 131.

   night-mare, sb.,= *M.H.G.* nahtmare; *night-
mare,* '*ephialtes,*' PR. P. 356; niȝtmare SAINTS
(Ld.) xlv. 228.

   night-raven, sb., *O.E.* nihthræfn; *night-
raven,* CATH. 254; ['*nictocorax*'] HAMP.
PS. ci. 7.

   niht-þeoster, sb., *night-darkness,* HOM. II.
171.

   niht-wecche, sb., *O.E.* nihtwæcce; *night-
watch,* HOM. II. 39; nightwacche D. TROY
7352.

nāht, *see* **nāwiht.**

nāhwēr, adj., *for* ne āhwēr; *nowhere*;
nawer AYENB. 210; nawer, nowher, neouwer
[nohware] LAȝ. 753, 3301, 31800; nohwer

KATH. 1731; no-, nouhwar A. R. 134, 160; nowhar (*ms.* nowwhar) ORM. 3566; nowhar MARG. 113; nowar O. & N. 1168; nouȝwher LANGL. *A* ii. 193 [nowhere *B* ii. 217]; nohwere MISC. 85; nour ROB. 233; noure HAMP. PS. i. 1.

noure-whāre, adv.,=nāhwēr; HAMP. PS. iv. 5*; ix. 17*; nourwhare xxxi. 13*; nowre-whare cxlix. 1*; noure where cxviii. 45.

nāhwider,adv., *for* ne āhwider; *no-whither, to no place*; nohwider MARH. 9; nowider wardes SAX. CHR. 262; nouhwuder A. R. 424; noweder MAP 338.

nai, adv., *O.N.* nei,=nā; *no, nay, not,* A. R. 370; MARH. 18; O. & N. 464; WILL. 916; (*ms.* nay) SHOR. 159; GREG. 421; IW. 1152; MAND. 292; CH. C. T. *A* 1667; KATH. 777; naȝ ORM. 10285; nai, næi LAȝ. 13132, 23575; it is no nai CH. C. T. *B* 1956, *E* 817, 1139; GAM. 34; withoute nai GAM. 26; this is no nai 433; nei HOM. I. 27.

nail, *see* næil.

nais, adj., *O.N.* neiss; *ashamed*; nakid and nais M. H. 52.

nait, adj., *? O.N.* neytr; *useful, vigorous,* D. TROY 3878; *comp.* un-nait.

nait-lī, adv., *dexterously, quickly,* A. P. ii. 480; D. TROY 2427; naiteli ALEX. (Sk.) 2896.

naiten, v., *O.N.* neyta; *use*; naite A. P. ii. 531; D. TROY 776; PERC. 185; note D. TROY 402; naites (*pres.*) 2518, 2968.

naiten², v., *O.N.* neita; *say no*; naitin '*nego*' PR. P. 351; naiteth (*pres.*) CH. BOET. i. 1 (4)*; naitid (*pple.*) APOL. 77.

nāked, adj., *O.E.* nacod,=*O.H.G.* nacot, nachot, *O.N.* nöktr, nökkvi̇̆r, *Goth.* naqaþs; *naked, unprovided with defensive armour,* A. R. 316; ORM. 6164; BRD. 24; LANGL. *B* xii. 162; S. S. (Wr.) 152; agane armit men to ficht mai nakit men haff litill micht BARB. xiii. 98; his nakede sweord LAȝ. 686; þai sterte al nakede til hure FER. 2436; nakid *bare body* D. TROY 6403; *comp.* steort-nāked.

nāked-hēd, sb., *nakedness*, ALIS. 7056; nakidhed WICL. JER. ii. 25.

nāked-līche,adv.,*nakedly,* A. R. 316; AYENB. 174.

nākednesse, sb., *nakedness,* CH. C. T. *E* 866.

nāken, adj., *naked,* LANGL. *C* xxi. 51*.

nāken, nākenen, v., *? for* *nākeden, *O.E.* (ge-)nacodian,=*O.H.G.* (ki-)nachatōn; *strip naked*; nakin '*nudo*' PR. P. 351; nácnes (*pres.*) HOM. I. 283; nákens (*imper.*) ALEX. (Sk.) 4959; nákide (*pret.*) WICL. JOB xx. 19.

nakere, sb., *O.Fr.* nacaire; *a sort of drum,* DEGR. 1085; nakeres (*pl.*) MAND. 281; nakeris GAW. 1016; nakers CH. C. T. *A* 2511; nakrin (*gen. pl.*) GAW. 118.

nakenen, *see* nákin. nal, *see* næil.

[nāme, sb., *from* nimen; *comp.* arf-nāme.]

nāme, sb., *O.E.* nama, noma,=*O.Fris.* nama, *O.L.G., O.H.G.* namo *m.*, *Goth.* namo, *O.N.* nafn *n.*; *name*, '*nomen*,' PR. P. 351; LK. i. 5; ORM. 1831; GEN. & EX. 3497; ROB. 16; AYENB. 120; LIDG. M. P. 67; nome, noma [name] LAȝ. 1397, 1951; HOM. I. 59; O. & N. 1762; JOS. 156; þene nome A. R. 50; námen [name] (*dat.*) MAT. vii. 22; nomen LAȝ. 10136; bi nome MARH. 2; nómen (*acc.*) LAȝ. 7125; námen (*pl.*) MAT. x. 2; HOM. I. 221; namen, nomen, noma [names] LAȝ. 1802, 2067, 29417; nomen A. R. 204; KATH. 271; C. L. 615; names HOM. II. 91; *comp.* bī-, ēke-, tō-náme.

náme-cund, adj., ?=námecūð; *renowned*: Balaam was an ful namecund prophete ORM. 6863.

nóme-cūð, adj., *O.E.* namcūð; *renowned, famous,* KATH. 537; A. R. 334; namecouþ AL. (L.) 44; namekouþ P. S. 327; namekouth D. TROY 2630.

námecouþ-hēde, sb., *renown,* AYENB. 25.

náme-lēs, adj., *nameless,* CH. BOET. iv. 5 (131).

náme-līche, adv., *namely, especially,* BEK. 21; AYENB. 21; WILL. 1203; CH. TRO. ii. 212; nomelīche KATH. 21; A. R. 18.

námin, v., *O.E.* (ge-)namian; *name, '*nomino,*' PR. P. 351; námed (*pple.*) WICL. 1 MACC. x. 1; *comp.* ȝe-námen.

nāming, sb., *naming,* HORN (H.) 216.

nān, adj., *for* ne ān; *none, no,* ORM. 16164; nes ter nan þat mihte . . . wrenchen KATH. 124; na þing 227; na mon 925; nan sunne HOM. I. 35; nam (*for* nan) man LK. x. 22; nan [no] cniht LAȝ. 8593; næs þer nan [non] oðer 25468; na [no] king 25378; nan mon, no man O. & N. 274, 1539; non fis GEN. & EX. 1124; noan ne nime PROCL. 5; no þing A. R. 120; no man BEK. 272; nānes (*gen. m. n.*) LAȝ. 26754; nanes weis KATH. 975; nanes kinnes ORM. DEDIC. 274; none mannes SHOR. 93; nānre (*gen. f.*) HOM. I. 245; nānun (*dat. m. n.*) LK. iv. 26; nane [none] LAȝ. 24624; nonen AYENB. 68; none A. R. 68; nāre (*dat. f.*) LAȝ. 19468; nore MISC. 56; nǣnne (*acc. m.*) ORM. 165; nænne, nenne [nanne] LAȝ. 9862 & 19847; nenne A. R. 96; AYENB. 8; nanne P. L. S. viii. 60; O. & N. 812; nōne (*acc. f.*) P. L. S. viii. 119; O. & N. 536; MISC. 49; AYENB. 29; nāne (*pl.*) KATH. 106; i nane depe sinnes ORM. 12839; nane (none) LAȝ. 15287; none A. R. 168; BRD. 29.

nāniȝ, *see* nēniȝ.

ẖnap, sb., *O.E.* hnæp,=*O.L.G.* nap, *O.H.G.* hnapf, *O.N.* hnappr (*button*); *cup, bowl*: nap REL. I. 129; MAP 343; þene nap LAȝ. 14333; nep '*patera*' FRAG. 4; nep A. R.

344; **neppe** (*dat.*) A. R. 214; **nappes** (*pl.*)
R. S. V.

náp̃e, sb., *nape,* ' *cervix,*' PR. P. 351 ; ' *haterel*'
VOC. 144 ; ÁLIS. 1347.

náp̃e², v. : take a troute and **náp̃e** (*imper.*)
him C. B. 102; take a tenche and nape him 105.

napet, sb.,=**napkin**, PR. P. 351.

napkin, sb., *dimin. of* *nappe, *Fr.* nappe,
Lat. nappa ; *napkin* : napet or napekin
' *napella* ' PR. P. 351.

***h*nappen**, v., *O.E.* hnappian, hnæppian; *have
a nap, sleep* : nappin ' *dormito* ' PR. P. 351 ;
nappi LA3. 1219* ; nappe LANGL. *B* v. 393 ;
neppe (*pres.*) REL. II. 211 ; nappeð A. R.
324 ; HOM. II. 201 ; nappeþ CH. C. T. *H* 9;
nappide (*pret.*) WICL. PS. cxviii. 28 ;
nappid (*pple.*) TOWNL. 98.

nappinge, sb., *O.E.* hnappung ; *napping,
sleeping,* ' *dormitatio,*' PR. P. 351 ; napping
WICL. PROV. xxviii. 29.

naprōn, sb., *Fr.* napperon ; *apron,* BER. 33;
naprun ' *limas* ' PR. P. 351 ; VOC. (W. W.)
660.

narcotike, sb., *O.Fr.* narcotique ; *narcotic*;
narcotikes [nercotikes] (*pl.*) CH. C. T. *A*
1472 ; narcotiks CH. L. G. W. 2670.

nard, sb., *O.E.* nard,=*O.Fr.* nard, *Lat.*
nardus ; *spikenard, ointment* ; **narde** (*dat.*)
WICL. JOHN xii. 3.

nare, *see* **nēor** *under* **nēh.**

naru, adj., *O.E.* nearu,=*O.L.G.* naru; *narrow,*
ORM. 3687; TRIST. 1942; (*ms.* narw) PS. lxxxiii.
4 ; nare, naru SAINTS (Ld.) xlv. 157, 163;
nareu SAX. CHR. 262 ; nare P. L. S. viii.
174 ; TREAT. 139 ; nara MAT. vii. 14 ; narou,
narwe MAND. 45, 69 ; narwe (? naru), narowe
(? narou) ' *strictus* ' PR. P. 351 ; neruh A. R.
144 ; þæt narewe geat MAT. vii. 13 ; nærewe
LK. xiii. 24 ; þene narewe wei P. L. S. viii.
171 ; **narewe** (*dat. m.*) LA3. 5511 ; nearowe
JUL. 35 ; **narwe** (*pl.*) ORM. 9202 ; þauh 3e þe
neruwure (*comp.*) beon A. R. 430 ; forð and
narwere GEN. & EX. 3965 ; þe parluris [þur-
les] lest & **nerewest** [narewest] (*superlat.*)
A. R. 50.

narwe (*adv.*) LANGL. *B* xviii. 371 ; and
heeld hir narwe in cage CH. C. T. *A* 3224;
wel narewe þe biledet O. & N. 68; narewe
herted HOM. II. 29; neruwe A. R. 268.

narou-hēde, sb., *narrowness,* ' *strictura,*'
PR. P. 351.

neruh-līche, adv., *narrowly,* A. R. 334.

naruh̃e, sb., *narrowness* ; neruh̃e A. R. 378.

narwen, v., *O.E.* nearwian ; *make narrow* :
narweþ (*pres.*) TREV. I. 57; narweþ MAN.
(F.) 4577 ; narwand (*pple.*) [' *affligens*']
PS. xxxiv. 5.

náse, nóse, sb., *O.E.* nasu, nosu, *cf. O.H.G.*
nasa, *O.Fris.* nose, *O.N.* nös, *M.L.G., M.Du.*
nese *f., Lat.* nāsus *m.* ; *nose,* SHOR. 6;

AYENB. 154 ; GOW. I. 98 ; D. TROY 11 ; AMAD.
(R.) vii ; nose BEK. 2201 ; CH. C. T. *A* 152;
S. S. (Wr.) 2070; LUD. COV. 137 ; nease
HOM. I. 251 ; neose 127; nesa 23; neose
[nose] LA3. 8181; neose FRAG. 5 ; MARH.
9 ; A. R. 100; SPEC. 34; nese, nose PR. P.
354 ; MAN. (F.) 1821; nese HAV. 2450;
MAND. 204 ; IW. 260; PR. C. 626 ; D. ARTH.
2248; his nese smel REL. I. 208.

neose-þürl, sb., *O.E.* nasþyrl ; *nostril,* R. S.
v (MISC. 182); nosethirl, -þril CH. C. T. *A*
557; nose-, neseþirl, -þril WICL. 2 KINGS
xxii. 9; nesethirl PR. P. 354.

nóselingis, adv., *laying on the back,* ' *supinus,*' PR. P. 358.

nasel, sb., *O.Fr.* nasel ; *nose-piece (of a
helmet),* [*Wace* ' *nasaus* ' 9518] MAN. (F.)
10043.

natif, adj., *Lat.* nātīvus ; *native, slave* ;
TREV. VIII. 39.

nativitē, sb., *O.Fr.* nativité ; *nativity,* A. R.
412 ; nativitee CH. C. T. *B* 3206, *F* 45.

natte, *Fr.* natte,=**matte** ; *mat,* VOC. 197 ;
' *matta, storium* ' PR. P. 351.

natūre, sb., *O.Fr.* nature ; *nature,* MISC.
35 ; (*natural power*) ALEX. (Sk.) 3379* ;
natour ALEX. (Sk.) 4027.

naturel, adj., *O.Fr.* naturel ; *natural,* AYENB.
18; CH. C. T. *D* 1144 ; naturill brether,
brether naturell D. TROY 6770, 7786.

naturel-līche, adv., *naturally,* MISC. 30.

naule, *see* **navele. nāut**, *see* **nāwiht.**

nāuþer, pron. & conj., *for* ne āuþer ; *neither,
nor,* M. H. 18; C. M. 11679; (*ms.* nawþer) A. P. i.
1086; nowðer MARH. 5; nowþer (*ms.*
nowwþerr) ORM. 3124; nouðer A. R. 52, 350;
nouðer, neouðer LA3. 6972, 8723; nouþer·
C. L. 37; (*ms.* nowþer) MIRC 386 ; ISUM.
149; nouþer of ham Iw. 3230; nouther,
nother PR. C. 167, 465; nouþer of gold nor of
silver MAND. 239; noþer O. & N. 465;
noþer of flesch ne of blod TREAT. 134; noþer
... nor WILL. 1675; noiþer GREG. 712; LAI
LE FR. 153; nor CH. C. T. *A* 946.

náve, sb., *O.E.* nafu,=*O.N.* nöf, *M.Du.* nave,
O.H.G. naba ; *nave (of a wheel),* ' *modiolus,*'
PR. P. 351; WICL. 3 KINGS vii. 33; CH. C. T.
D 2266; nafe VOC. 278.

(**ná**)**ve-gār**, sb., *O.E.* nafogār,=*O.H.G.*
nabagēr ; *auger,* FRAG. 4; nauger VOC. 170;
nagere CHEST. I. 107.

navee, navie, sb., *O.Fr.* navee, navie ; *navy,
fleet,* ' *classis,*' PR. P. 352; WICL. 3 KINGS ix.
26 ; PARTEN. 5673; navie MAN. (F.) 2152;
CH. H. F. 216; navi ALEX. (Sk.) 1160; if ani
nave [navi] *ship* 3376.

navele, sb., *O.E.* nafola,=*O.N.* nafli, *O.Fris.*
naula, *O.H.G.* nabalo ; *navel,* LEG. 164;
SHOR. 44; navele [naule] LANGL. *B* xiv. 242;
WICL. S. SOL. vii. 2; navele [navel] CH. C. T.

A 1957; navil Iw. 1992; navil, novil PR. P.
360; novil LIDG. M. P. 199; naule A. P. i.
459; noule REL. I. 221.

navie, *see* **navee.**

nāwiht, nōht, nŏht, nŏt, pron. & adv., *O.E.*
nā-, nōwiht, nāuht, nāht, nōht, *mod. Eng.* naught
(nought), not; *for* ne āwiht; *nothing, not;*
nawiht heardes HOM. I. 257; nawiht, nawt,
naut MARH. 5, 7; nawiht, naht LA3. 473, 6129;
heo nefden noht ane moder 209; ne beo he
noht swa lohiboren 22041; nolde he nawiht
leȝhen ORM. 10351; noht (*ms.* nohht) tat
Crist maȝ cwemen 13117; god ne mihte noht
þa belles heren ringen 910; ne don we
nauht þus HOM. I. 111; nowiht, nowt, nout
[nawt, naut] A. R. 2, 86, 144; nout mucheles
102; nawiht, naht, nawt, nowiht, nowiȝt, nouht,
noht, noȝt, nout O. & N. 58, 884, 928, 1275, 1324,
1426, 1480, 1620; nacht, nocht MISC. 27, 28;
naȝt AYENB. 18; naught RICH. 5678; naught,
nouht, nought, noght, not CH. C. T. *A* 462,
744, 756, 789; nowicht (*ms.* nowicth), nouht
(*ms.* nouth) HAV. 97, 249; noht H. H. 60,
109; noȝt BRD. 3; for noȝt ROB. 449; nouȝt
[noȝt] LANGL. *B* iii. 268; be it nought so lite
GOW. I. 174; nouȝt, noght, nout S. S. (Wr.)
53, 209, 210; noght forþi *nevertheless* HAMP.
PS. xxxviii. 8; nocht for thi BARB. vii. 220;
nat forþi H. S. 5885; nought withstonding
notwithstanding GOW. II. 181; **nōhtes** (*gen.*)
LA3. 13947; ne beoð ha riht nohtes JUL. 22;
nōuhte [nouȝte] (*dat.*) C. L. 34; noȝte BEK.
864; bringe . . . to noȝte SHOR. 163; *deriv.*
nōuhti.

nāwþer, *see* **nāuþer.**

ne, adv., *O.E.* ne, = *O.N.* ne, *O.Fris., O.L.G., O.*
H.G., Goth. ne, ni, *Lat.* ne; *not,* GEN. & EX.
554; HAV. 49; he ne mihte speden LA3. 403;
leode ne beoð þar nane 1244; ne aras [naros]
he never mare 1555; þet we ne bicumen
prude A. R. 232; no wunne ne schal wonten
þe 398; þat hit ne forwurðe naut MARH. 5;
nefde (= ne hefde) ha buten iseid swa þet an
engel ne com JUL. 68; noht ne wære he
þanne god ORM. 11637; wenst þu þat ich ne
cunne singe O. & N. 47; hi ne seȝe no þing
BRD. 11; hi ne knewe hire speche noȝt BEK.
66; hou miȝte heo iseo qvelle hire child þat
hire hurte ne brac a tvo P. L. S. x. 15; þer
ne was LANGL. *B* PROL. 177; whi ne had
god mad us swa PR. C. 6228; i ne wot, i not
WILL. 903, 911; i ne hadde [nadde] TREV.
VIII. 69; i ne am [nam] CH. C. T. *A* 1122;
nich = ne ich JOHN xviii. 17; nah = ne ah
LA3. 28461; nam = ne am A. R. 38; ORM.
10281; nis = ne is JOS. 66; GREG. 177;
nabbe = ne habbe LA3. 7876; KATH. 1765;
nas = ne was ORM. 311; O. & N. 114; C. L.
6; nes A. R. 38; KATH. 307; nat = ne wat
I know not LA3. 6243; ORM. 2445; HOCCL. i.
329; not BEK. 163; not [noot] CH. C. T. *A*
284; þauh þu hit nute [ne wite] nout A. R.
58; nulle = ne wille LA3. 1447; MARH. 22;

nulle, nelle LANGL. *A* v. 238; nelle O. & N.
452; nille S. S. (Wr.) 1768; ne (*? for* na)
mare FRAG. 7; HOM. I. 167; A. R. 96*;
ne ðe las LA3. 1346; ne ðe les HOM. II. 79;
GEN. & EX. 3853; ne þe les HAV. 1108;
RICH. 3812; GOW. I. 4; WICL. LK. xii. 31;
ne beo þe þarm ne so smel MISC. 151; ne for
thi *never the less* C. M. 11619.

ne², conj., *O.E., O.N.* ne, = *O.L.G.* ne, ni, Goth.
nih, *O.H.G.* noh; *neither, nor,* MARH. 5;
HAV. 44; C. L. 37; JOS. 593; þei we it nusten
ne iseien P. L. S. viii. 51; ne dei ne niht HOM.
I. 43; ne wurðe nan cniht swa wod ne kempe
swa wilde LA3. 8593; ne ne mai na þing
wiðstonden his wille KATH. 227; þat ȝho ne
være shamed . . . ne shend ORM. 1992;
noþer griþ ne sibbe O. & N. 1005; ne for
sorewe ne for wo BEK. 63; ne mo ne les
AYENB. 118; ne (*v. r.* nor) i LANGL. *B* v.
640; þat þei ne be hurt ne harmed MAND.
245; he ne had nouther strenthe ne might
PR. C. 465; noþer gin ni monnes strengþe
ROB. 7; ni lene to piler ni to wal MIRC
271.

nēad, sb., *O.E.* nēad, nīed, nēd, nȳd, *cf. O.Fris.*
nēd, *O.L.G.* nōd, *O.H.G.* nōt, *O.N.* nauð,
nauðr, Goth. nauþs *f.,* *M.H.G.* nōt *f. m.;*
need, MAT. xi. 12; neod xviii. 7; wher hit
neod [neode] weore LA3. 5282; þe king isæh
þe neode [nede] 9526; neod (nod), neode
(node) HOM. I. 3, 11, 83; O. & N. 466, 638,
906; neod BEK. 874; ȝif hit neod is A. R.
80; hwar se ȝe habbeð neode 424; neode
C. L. 19; nede, neode LANGL. *B* vii. 68, *A*
viii. 70; ned, nede ORM. 906, 12965; WILL.
119, 3210; i hadde al þat me was ned MAP
335; nede '*necessitas*' PR. P. 352; MAND. 82;
(h)is nede (*call of nature*) vor to do ROB. 310;
nēdes [= *M.H.G.* nōtes], *of necessity* WILL.
1043; CH. C. T. *A* 1290; nedes costes *neces-*
sarily ANGL. III. 61; ARCH. LII. 33; at þare
neode (node) O. & N. 529; at nede HAV. 93;
nede [neede] CH. C. T. *E* 461; nied, niede
HOM. II. 121, 228; AYENB. 63, 95; nied
BARB. i. 254; niede ANGL. I. 23; HOM. I.
221; MISC. 32; **nēade** [*cf. O.E.* nȳde, nēde,
M.H.G. nōte], *necessarily* SHOR. 138; nede
ORM. 13558; O. & N. 636; P. L. S. xvii. 392;
JOS. 230; WILL. 3922; MIRC 154; he mot
nede beien LA3. 1051; **nēode** (*pl.*) FRAG. 8;
nēoden (*dat. pl.*) HOM. I. 205; A. R. 246;
neden REL. I. 181.

nēod-ful, adj., *needful,* A. R. 260; nedful
GEN. & EX. 2130; niedvol AYENB. 166; nud-
ful LANGL. *C* ii. 21; **nēdfulere** (*comp.*) HAMP.
PS. ix. 9.

nēod-lēs, adj., *needless,* BEK. 1662.

nēd-līche, adv., *O.E.* nȳdlīce; *necessarily,*
AL. (V.) 116; nedli Iw. 1184; PR. C. 2864;
nedeli CH. BOET. iii. 9 (84).

nēd-swōt, sb., *sweat of anguish,* A. R.
110.

neid-wai, adv., *of necessity*, BARB. xix. 156;
neidwais *necessarily* v. 242.

nēadunga, nēadlunga, adv., *O.E.* nēadunga;
necessarily; nedunga HOM. I. 123; nedunge,
nedlunge MARH. 22; nedlunge A. R. 190; ned-
linge HOM. II. 199; nedelingis M. ARTH. 753;
TOWNL. 290; nedelonges AMAD. (R.) xii.

[**nēan**, adv., *O.E.* nēan; *near*; comp. for-
nēan.]

nēat, nout, sb., *O.E.* nēat, *O.N.*naut,=*O.Fris.*
nāt, *O.H.G.* nōz; *from* nēoten; *neat, cow, ox*;
neet '*bos*' PR. P. 354; LANGL. *C* xxii. 266;
OCTOV. (W.) 927; net GEN. & EX. 940; nout
PS. cxliii. 14; **nētes** (*gen.*) HAV. 781; **nēte**
(*dat.*) IW. 252; **nāt** (*pl.*) SAX. CHR. 259;
neet GEN. & EX. 2097; PALL. i. 506; neet
[nete] CH. C. T. *A* 597; nete MAN. (H.) 117;
naute AMAD. (R.) xv; nowt ORM. 1298; **nēte**
[niete] (*gen. pl.*) LA3. 369; nēte (*dat. pl.*)
MIRC 353.

   neet-hirde, sb., *cowherd*, '*bubulcus*,' PR. P.
354; neetherde WICL. AMOS vii. 14; net-
herdes [neethurdes] (*gen.*) CH. C. T. *B* 2746.

neb, sb., *O.E.* nebb,=*M.L.G.*, *M.Du.* nebbe,
O.N. nef; *beak, face*, '*rostrum*,' PR. P. 352;
REL. I. 124; FL. & BL. 615; AL. (L.³) 514;
þet neb HOM. I. 121; his neb & his neose
LA3. 8181; scheau þi neb to me '*ostende
mihi faciem*' A. R. 90; neb [nebbe] ERC.
439; nebbes (*gen.*) HOM. II. 181; nebbe
(*dat.*) A. R. 276; nebbe (*pl.*) HOM. I. 43;
nebbes LA3. 4163.

   neb-schaft, sb., *face*, KATH. 448; neb-
scheaft JUL. 55; nebscheft A. R. 94.

nēce, sb., *O.Fr.* niece; *niece*, ROB. 203; A. P.
i. 233; CH. C. T. *B* 1290.

necessarie, adj., *O.Fr.* necessaire; *necessary*,
CH. C. T. *C* 681; necessari ALEX. (Sk.) 125.

necessitē, sb., *O.Fr.* necessité; *necessity*,
P. L. S. xxxi. 85; CH. C. T. *F* 593.

hnecke, sb., *O.E.* hnecca,=*O.Fris.* hnecka,
O.N. hnakki; *neck*; necke LA3. 687; MARH.
12; A. R. 322; O. & N. 122; BEK. 2202;
nekke '*collum*' PR. P. 352; LANGL. *B* v. 135;
MAND. 48; CH. C. T. *A* 238; þane nhicke
AYENB. 132; nicke HORN (R.) 1248.

   nekke-bōn, sb., *neckbone, collarbone*: (he)
brak his nekkebon in two LANGL. *C* i. 114*.

nēd, see **nēad.**   **neddre**, see **naddre.**

nēde, see **nēad, nēode.**

nēden, v., *O.E.* nēdan, nȳdan,=*O.N.* ney∂a,
Goth. nauþjan, *O.H.G.* nōten; *from* nēad;
need; compel; HOM. II. 75; ORM. 11820; þe
veond ne mei neden nenne mon to don sunne
A. R. 304; nedin '*egeo*' PR. P. 352; neid
BARB. xiii. 46; **nēde∂** (*pres.*) A. R. 72; REL.
I. 213 (MISC. 7); nedeþ, neodeþ, nudeþ LANGL.
A xi. 50, *B* x. 63, *C* xii. 48; al þat hem nedeþ
MAND. 241; þer of nedeþ nouht to speke CH.
C. T. *A* 462; net MAT. v. 41; A. R. 338;
nēde (*subj.*) HOM. I. 15; **nĕdde** (*pret.*) HOM.

I. 121; A. P. i. 1043; nedden LA3. 4048;
comp. bi-, 3e-nēden.

nēdi, adj., *O.E.* nēdig,=*O.N.* nau∂igr, *O.H.G.*
nōtag; *needy*, TREV. III. 383; P. L. S. v (b.)
58; LANGL. *B* vii. 71; CH. BOET. iv. 5 (131);
LUD. COV. 127; neodi HOM. I. 135.

nēdle, sb., *O.E.* nǣdl,=*O.Fris.* nēdle, *Goth.*
nēþla, *O.H.G.* nādla, nādela, *O.N.* nāl;
needle, '*acus*,' PR. P. 352; ORM. 6341; ROB.
99; TREV. IV. 211; LANGL. *B* xii. 162;
anne nedle eage LK. xviii. 25; nelde A. R.
324; PALL. i. 602.

   nēdle-, neelde-werk, sb., *needle-work*,
WICL. EX. xxvi. 1.

   nēdeler, sb., = *M.H.G.* nādelǣre; *needle-
seller*, LANGL. *B* v. 318; neldere *A* v. 161.

nēdlunge, see **nēdunga.**

nedre, see **naddre.**   **neet**, see **nēat.**

hnéfe, sb., *O.N.* hnefi; *fist*; nefe PERC. 2087;
neve HAV. 2405; D. TROY 13889; neeve
FLOR. 1636; neefe [nave, new] BARB. xv.
129; **neffes** (*pl.*) YORK xxix. 370; nevis
BARB. xx. 257.

nefe, see **neve.**

ne3, see **nēh.**   **ne3en**, see **ni3en.**

[**nē3en**, v., *? O.E.* nēgan (*satisfy*); *? comp.*
a-nē3en.]

nē3en, see **nēhen.**

hnē3en, v., *O.E.* hnǣgan; *neigh*; ne3en WICL.
IS. xxiv. 14; neyin PR. P. 352; neie (*ms.* neye)
FER. 3669; TREV. III. 179; **nēide** (*pret.*)
ROB. 459.

   nēiing(e), sb., *neighing*, WICL. JOB xxxix.
19; JER. xiii. 27; **neiingis**, ne3ingus (*pl.*)
JER. viii. 16, xiii. 27.

neghen, see **ni3en.**

negligence, sb., *O.Fr.* negligence; *negligence*; necligence CH. C. T. *A* 1881.

negligent [**necligent**], adj., *O.Fr.* negligent;
negligent, CH. C. T. *D* 1816.

nēh, adj. & adv., *O.E.* nēh, neāh, nīh,=*O.Fris.*
nēi, *O.L.G.*, *O.H.G.* nāh, *O.N.* nā, *Goth.* nēhw;
near, MARH. 6; ORM. 13537; SPEC. 34;
nech HOM. I. 141; al þat he neh com LA3.
1568; neh him 1989; þa Irisce men weoren
nakede neh þan 22340; nǣh cun 10260;
neih 1609 (*ms.* nehi 4996); neh, nehi, nei
(*ms.* ney) O. & N. 1252, 1267; þu ham neih
list FRAG. 6; ledde after him neih þan (*quite*)
al his ofspreng HOM. II. 33; hit is . . . ful
neih idon mid ham þet kume∂ so neih to
gederes A. R. 60; ancre þet nave∂ nout neih
hond hire vode 424; of neih 250; he is neih
(*ms.* neyh) honde (*near at hand*) MISC. 85; þat
stood me neih C. L. 370; hit eode hire herte
swiþe neih [nei3] 320; neg GEN. & EX. 1234;
hit was ne3 eve HORN (L.) 464; ne3 þan ende
BRD. 33; hi were ne3 what at te see BEK.
1958; nei3 LANGL. *B* iii. 144; marred nei3
honde *almost* WILL. 884; nei3[nigh]alle TREV.

VIII. 350; it was neigh ded LAI LE FR. 198;
nehi MAN. (H.) 203 ; nei (*ms.* ney) HAV. 464;
hit is nei (*ms.* **ney**) vif ʒer ROB. 195 ; (h)is
men nei wat (*nearly*) alle 80 ; neʒ [niʒ] WICL.
PROV. vii. 8 ; neigh [nigh] CH. C. T. *A* 1526;
neoh LK. xix. 11 ; nieʒ AYENB. 212 ; nigh
MAND. 44 ; S. S. (Wr.) 3185 ; ni PR. P. 355;
comp. **a-nēh.**

nēor, adv. (*compar.*), *near, nearer,* HOM. I.
137 ; LAʒ. 6484; come neor and neor
ALIS. 599 ; neor, ner O. & N. 386, 923,
1260; nier HORN (L.) 771 ; AYENB. 234;
ner ORM. 15235; TREAT. 133; GOW. III.
212 ; ner [neer] LANGL. *B* xvi. 69 ; CH. C. T.
A 839 ; ner [*O.N.* nǣr], *for* neh, ORM. 9638;
nare YORK xxii. 52.

     nĕrre, adj. (*compar.*), *nearer,* ORM. 15691;
nerre þe half TREAT. 133 ; narre, nerre
YORK ix. 62, xxxi. 321.

     nēst, nĕst, adj. & adv. (*superl.*), *next,* '*pro-
ximus,*' PR. P. 354; H. M. 23; ORM. 1054; (*ms.*
nesst) 4987 ; GEN. & EX. 3921 ; HOM. I. 247 ;
necst, next, nest O. & N. 688, 700; next þe
grounde TREAT. 132; his nexte cun LAʒ.
22837 ; þe nexte wike LANGL. *B* xiii. 154;
þe nexte wai M. ARTH. 278 ; love þine nixte
(*neighbour*) ase þi selve AYENB. 145.

     nēst-fald, adj., *nearest*: mi nestfalde cun
JUL. 33.

     nēih-ebūr, sb., *O.E.* neāhgebūr, = *O.H.G.*
nāhgipūr ; *neighbour,* A. R. 368 ; neghebur
PR. C. 5863; neʒʒebour, neʒebour, neʒibor
AYENB. 10, 36, 38 ; neihʒebor LANGL. *A* v.
74 ; neighbore *B* v. 94 ; neighebour [niʒhe-
bour] CH. C. T. *A* 535 ; **nēhgebūres** (*gen.*)
HOM. II. 95.

     nēighbour-hēde, sb., *neighbourhood,* PR. P.
352.

     nēʒebūr-rēde, sb., *vicinity*; (*ms.* neʒebur-
redde) HOM. I. 137 ; nehebo(r)reden II. 83.

     nēih-hende, adv. & prep., *near at hand*:
þe dom is us neihhende MISC. 142 ; nerhand,
nerand, nerehand HAMP. PS. i. 1, lxx. 12*,
xciii. 17, cxviii. 87.

     nēh-lēchen, v., *O.E.* neāhlǣcan ; *approach,*
HOM. II. 57 ; JUL. 8 ; **nēhlēcheð** (*pres.*)
HOM. II. 35 ; A. R. 60; neihlecheð REL. I.
131 ; **nēhlēhte** (*pret.*) LAʒ. 5267 ; neihleihte
MISC. 84.

     nēihlēchunge, sb., *approach,* A. R. 196.

     nēh-weste, sb., *O.E.* neāhwest, = *O.H.G.*
nāhwist, *O.N.* nāvist; *proximity, vicinity*:
heo comen hire a neweste (=*prov.Eng.*
anewst) LAʒ. 3507 ; a neoweste þan stane
17427 ; he com a neuste 25752 ; ga a neouste
26107.

nēhen, v., ? *O.E.* nēhwan, = *Goth.* nēhwjan,
O.H.G. nāhen ; *approach, come near to* ;
nehe M. H. 17 ; neihen A. R. 134 ; neʒhen,
nehʒhen ORM. 4491, 12571 ; neʒen [neighen]
LANGL. *B* xvii. 58 ; neggen REL. I. 212 ; neʒe

GAW. 1575 ; neghe PR. C. 1208 ; neiʒen MAP
345 ; neiʒe WICL. MAT. iii. 2 ; neiʒhe WILL.
3230; neighe MAND. 241 ; newhen HAV.
1866; nighe M. ARTH. 2832 ; **nēihede**
(*pret.*) A. R. 134; neiʒede WICL. MAT. viii.
5 ; neiʒhede [neiʒide] TREV. V. 249 ; neʒede
[neghed] LANGL. *B* xx. 231 ; neghid S. S.
(Wr.) 950; nighed TRIAM. 1512.

nei, *see* **nā.**   **nēi,** *see* **nēh.**

nēien, *see* **hnēʒen.**   **nēiʒ, nēih,** *see* **nēh.**

nēiing, *see under* **hnēʒen.**

neil, *see* **næil.**   **nēiðer,** *see* **næiðer.**

nekar(d), *see* **nigard.**

neked, sb., *little or nothing*: & of þat ilk
nuʒere bot neked nou wontez & i wolde loke
on þat lede GAW. 1062 ; bot to dele you for
drurie þat dawed bot neked 1805 ; þai noied
bot a nikid ALEX. (Sk.) 3935.

nekke, *see* **hnecke.**   **nēlde,** *see* **nēdle.**

[**nēme,** adj., = *M.H.G.* (an-)nǣme ; *comp.* **ūt-
nēme.**]

nemel, nemen, *see* **nimel, nimen.**

nemlen, v., = **nemnen** ; *name*; nemelin ·PR.
P. 353 ; **nemeld(e)** (*pret.*) GEN. & EX. 3533;
nemeled (*pple.*) P. R. L. P. 23.

nemnen, v., *O.E.* nemnan, = *O.H.G.* nemnan,
nemman, nennan, *Goth.* namnjan, *O.N.*
nefna ; *from* **nāme** ; *name, call by name,*
tell, relate, A. R. 204 ; ORM. 157 ; nemnen
[nemni] LAʒ. 2910 ; nemmen A. R. 316;
nemne SPEC. 113 ; PL. CR. 472 ; nemni
TREAT. 132 ; AYENB. 57 ; nempnen C. L.
299 ; nempnen [nempne] LANGL. *A* i. 21 ;
nemne, nevene MAN. (F.) 2291, 7357 ; LUD.
COV. 173 ; nevene (*pres.*) ISUM. 27 ; nemneð
A. R. 10; nevenes GAW. 10; ?nenis YORK
xxxii. 185 ; nem (*imper.*) A. R. 398 ; neven
ALEX. (Sk.) 833* ; **nemne** (*subj.*) A. R. 340;
nemnede (*pret.*) LAʒ. 7100 ; BRD. 3; H. S.
2287 ; nemnede, nemde REL. I. 131 (HOM. II.
11); nemned WILL. 368 ; nemde A. R. 138;
MISC. 53 ; AR. & MER. 9379; nevenid ALEX.
(Sk.) 76 ; **nemned** (*pple.*) ORM. 468 ; WICL.
HEBR. iii. 13 ; *comp.* **a-, ʒe-nemnen.**

     nemnunge, sb., = *O.H.G.* nemnunga ; *name,*
naming; nemmunge A. R. 290; þi neve-
ning (*ms.* þin evening) HORN (L.) 266.

neng, nēnig, *see* **næniʒ.**

nēod, nēode, *see* **nēad.**

nēode, sb., ? *O.E.* nēod, = *O.L.G.* niud,
O.H.G. niot ; ? *desire*: a þe is ure neode
LAʒ. 25686; *comp.* **un-nēode.**

nēod-ful, adj., *desirous,* A. R. 400.

nēod-līche, adv., *O.E.* nēodlīce ; *diligently*:
heo ferden alle niht neodliche swiðe LAʒ.
26923 ; needeli ALIS. (Sk.) 747; ant heom
i folhi **nēodelükest** (*superl.*) MARH. 13.

neoʒen, *see* **niʒen.**   **neomen,** *see* **nimen.**

neose, *see* **nāse.**

*h*nēosen, v.,=*O.N.* hniōsa, *O.H.G.* niusan ;
sneeze ; nesin '*sternuto*' PR. P. 354.

nēsinge, sb.,=*M.H.G.* niesunge ; *sneezing*,
'*sternutatio*,' PR. P. 354 ; nesing, neesing
WICL. JOB xli. 9 (*v. r.* fnesinge).

[nēoten, v., *O.E.* nēotan,=*O.L.G.* niotan,
O.N. niōta, *Goth.* niutan, *O.H.G.* niozan ;
use, enjoy ; deriv. nēat, nūten, nóte, nótien,
nüt, nütte, nütten.]

neoðer, neoðerien, *see* niðer, niðerien.

nēowe, *see* nēwe.   nēowel, *see* nīwel.

nēoweste, *see* nēhweste *under* nēh.

nep, *see* *h*nap.

nēp, sb., *O.E.* nēp(-flōd), *mod.Eng.* neap-
(-tide) ; *scanty ; low* : to serve the pouer
people of halfpeny worthes in the neep se-
sons E. G. 425.

nēpe, sb., *O.E.* nǣp,=*O.N.* nǣpa, *from Lat.*
nāpus ; *turnip*, VOC. 191 ; PR. P. 353 ; D.
TROY 3076.

nepte, sb., *O.E.* nepte, nǣpte, *Lat.* nepeta ;
catmint, PR. P. 353.

nēr, *see* nǣfre.   nercotik, *see* narcotike.

nēre, sb., *O.N.* nȳra,=*M.L.G.* nēre, *O.H.G.*
niero ; *kidney*, '*ren*,' VOC. 186 ; L. C. C. 52 ;
neere PR. P. 353 ; neeres (*pl.*) ['*renes*'] PS.
xxv. 2 ; neres P. P. 264 ; *comp.* kide-nēre.

nerfe, sb., *O. Fr.* nerf ; *nerve, physical
strength* ; nerf CH. TRO. ii. 642.

nĕrre, *see under* nēh.   neru, *see* naru.

*h*nesche, adj., *O.E.* hnesce,=*Goth.* hnasqws ;
prov.Eng. nesh ; *soft, tender* ; nesche A. R.
134 ; TREAT. 138 ; ALIS. 915 ; SHOR. 146 ; S. S.
(Web.) 732 ; TREV. I. 133 ; MAND. 303 ;
nesche and softe O. & N. 1546 ; in hard & in
nesche WILL. 495 ; neshe (*ms.* nesshe) ORM.
995 ; PR. C. 614 ; L. C. C. 33 ; nesch A. P. i.
605 ; nesh TOWNL. 113 ; nesh, neis HAV.
217, 2743 ; neschere (*compar.*) JUL.² 134.

ness-hēde, sb., *tenderness*, AYENB. 267.

nesse-liche, adv., *softly*, ROB. 435.

*h*neschen, v., *O.E.* hnescian ; *make tender,
mollify* ; neschin PR. P. 353 ; neshen ORM.
15909 ; nhesseþ (*pres.*) AYENB. 94 ; neissis
HAMP. PS. cxli. 6* ; nesched (*pple.*) PS. liv.
22 ; neissid ere his wordis aboven oile HAMP.
PS. liv. 24 ; nessid xc. 3*.

nése, *see* náse.   nēsin, *see* *h*nēosen.

nest, sb., *O.E.*,=*M.L.G.*, *O.H.G.* nest ; *nest*,
O. & N. 100 ; WILL. 83 ; CH. C. T. *B* 1749 ;
þet nest A. R. 134 ; neste (*dat.*) O. & N. 92 ;
nist (*ms.* nyst) (*pl.*) MAT. viii. 20 ; nestes
A. R. 132.

. nestling, sb., *nestling* ; nestlinges (*pl.*)
DEP. R. iii. 75.

nēst, *see under* nēh.

nestien, v., *O.E.* nistian, *cf. O.H.G.* nistēn ;
make a nest ; nesteð (*pres.*) A. R. 132 ;
nǣstieð LAȝ. 21753.

nestlen, v., *O.E.* nestlian, nistlian,=*M.L.G*

nestelen ;   *nestle, build nests*, HOCCL. i.
288 ; nestlin PR. P. 354 ; nestleþ (*pres.*)
LAȝ. 21753* ; nestild (*pret.*) ALEX. (Sk.)
506.

net, sb., *O.E.* nett,=*O.L.G.*, *O.N.* net, *Goth.*
nati, *O.H G.* nezi ; *net*, P. L. S. xvii. 157 ;
(*ms.* nett) ORM. 13474 ; nette (*dat.*) HOM. II.
209 ; AYENB. 170 ; net (*pl.*) MAT. iv. 22 ;
nettes LAȝ. 29251 ; A. R. 334 ; *comp.* fish-
net.

nēt, *see* nēat.   nete, *see* *h*nite.

nēten, *see* nūten.

netle, sb., *O.E.* netle, netele,=*M.L.G.* netele, *O.H.G.* nezila ; *nettle, 'urtica*,' FRAG.
3 ; VOC. 139 ; TREV. VI. 461 ; LIDG. M. P.
171 ; blind(e) netele LEECHD. III. 340 ;
netlen (*pl.*) AYENB. 230 ; a mong þe netle O.
& N. 593 ; netles H. S. 7515.

netlin, v.,=*Du.* netelen, *Ger.* neszeln ; *nettle*,
PR. P. 355 ; nettild (*pple.*) with ire ALEX.
(Sk.) 737.

neðen, neðer, *see* niðen, niðer.

nethring, *see under* niðerien.

nēuste, *see* nēhweste *under* nēh.

neuth, *see* niðen.

néve, sb., *O.E.* nefa, neofa,=*O.Fris.* neva,
O.H.G. nefo, *O.N.* nefi ; *nephew, grandson,
'nepos*,' PR. P. 355 ; GEN. & EX. 724 ; nefe,
neve SAX. CHR. 251, 255.

néve, *see* *h*néfe.

nevelinge, *see* nivelen.

nevenen, *see* nemnen.

neveu, sb., *O.Fr.* neveu ; *nephew, grandson*,
ROB. 19 ; DEGR. 26 ; CH. H. F. 617 ; CH.
L. G. W. 1440, 1442, 2659 : neveu [nevou]
MAN. (F.) 4920 ; nevu AYENB. 48.

nēvre, *see* nǣfre.

nēwe, adj., *O.E.* nēowe, nīowe, nīwe,=*O.L.G.*
niwi, nigi, *O.Fris.* nie, *O.H.G.* niuwi, *Goth.*
niujis, *O.N.* nȳr, *Lat.* novus, *Gr.* νέος ; *new*,
P. L. S. viii. 156 ; KATH. 2137 ; GEN. & EX.
1286 ; newe C. L. 780 ; AYENB. 97 ; LANGL.
A v. 171 ; CH. C. T. *E* 857 ; S. S. (Wr.)
1489 ; of newe E. G. 387 ; YORK xi. 141
(new TOWNL. viii. 141) ; o newe *anew* GAW.
65 ; newe, nowe O. & N. 1129 ; newe,
niwe ROB. 217, 517 ; newe, new, neow ORM.
3447, 7245, 11030 ; neowe HOM. I. 143 ; A. R.
8 ; neowe, niwe [neuwe] LAȝ. 2675, 23110 ;
niwe FL. & BL. 296 ; SHOR. 163 ; niewe MK.
i. 27 ; HICKES I. 223 (ANGL. I. 26) ; niwe,
nue L. H. R. 56 ; nue TREAT. 133 ; neu C. M.
1281 ; nēwes (*gen. n.*) GEN. & EX. 250 ;
nēwene (*acc. m.*) AYENB. 162.

nēow-cumen, adj., *newly come*, LAȝ. 8562.

nēwe-fangel, adj., *mod.Eng.* newfangled ;
eager for novelty, GOW. II. 273 ; CH. C. T. *F*
618 ; newfangel S. B. W. 35.

nēwefangelnes, sb., *fondness for novelty*,
CH. C. T. *F* 610.

nēw-ʒēr, sb., *new year*; **nēwʒēres** (*gen.*) daʒ ORM. 4230.

nēowe-līche, adv., *O.E.* nīwlīce; *newly,* A. R. 218; neweliche AR. & MER. 4658; neuliche CH. BOET. iv. 3 (122); neuli MAN. (H.) 67; REL. I. 146.

nēwlingis, adv., *newly, soon,* BARB. xiv. 86.

nēwen, adv., *O.E.* nēowan, nīwan; *recently; anew, afresh,* ORM. 173, 7069; neowen, neouwen LAʒ. 3591, 20683; n(e)owen, newe O. & N. 1129; newe P. S. 246; WILL. 1354; CH. C. T. *A* 3221; TRIST. 188.

nēwen[2], v., *O.E.* nēowian, nīwian, = *O.H.G.* niuwōn, *cf. Goth.* (ana-)niujan; *make new, renew;* newe CH. TRO. iii. 305; **nēweð** (*pres.*) REL. I. 210; neweþ ['*innovat*'] WICL. WISD. vii. 27; **nēwed** (*pret.*) PR. C. 7460; þo niwede here sorwes ANGL. I. 308; **nēwed** (*pple.*) JOS. 588; MIRC 642; *comp.* **a-, ʒe-nēwen.**

nēwlingis, *see under* **nēwe.**

ni, *see* ne[2]. **niatt,** *see* **nēat.**

nīce, adj., *O.Fr.* nice, *mod.Eng.* nice; *foolish,* '*iners*,' PR. P. 355; ROB. 106; SHOR. 46; AYENB. 59; S. S. (Web.) 412; HOCCL. i. 204; nice siʒtes LANGL. *B* xvi. 33; þat we ben wise and no þing nice CH. C. T. *D* 938; ? **nisechere** (*comp.*) TREV. VIII. 151.

nīcen, v., *become foolish*; **nīsen** (*pres. pl.*) GAW. 1206.

nicetē, sb., *O.Fr.* niceté; *folly, cowardice, licentiousness,* CH. C. T. *G* 495; nisste BARB. vii. 379; nisete DEP. R. iii. 144; nisete, niste TREV. I. 281, III. 317, IV. 51, 115, V. 227, VIII. 303.

nich, adv., *for* ne ich; *not I, no,* JOHN xviii. 17; O. & N. 266.

nicke, *see* **hnecke.**

nicken, v., ? *from* **nich**; *but cf. Sw.* neka; *deny, say no*; **nicke** (*pres.*) SPEC. 32; nickeð A. R. 308; nickes WILL. 4145; nikkes PERC. 1024; **nikked** (*pret.*) GAW. 2471; al nikked him with nai GAW. 706; **nekid** (*pple.*) ALEX. (Sk.) 1460.

nīed, *see* **nēad.** **nien,** *see* **niʒen.**

nīet, *see* **nēat.** **neiþe,** *see* **niʒeðe.**

nif, conj., ? *for* ne **if**; *cf. O.H.G.* nibu, *Goth.* niba; *unless,* GAW. 1769; A. P. ii. 424.

nifle, sb., ? *O.Fr.* nifle; *triviality* : he served hem wiþ **nifles** (*pl.*) & wiþ fables CH. C. T. *D* 1760.

nifte, sb., = *O.E.* nift, = *O.Fris., O.H.G.* nift, *O.N.* nipt, *Lat.* neptis; *niece, step-daughter,* PR. P. 355; GEN. & EX. 1386; **niftes** (*pl.*) MAN. (F.) 11011.

nig, adj., ? *niggardly* : nig and hard in al (h)is live ARCH. LII. 36.

nīʒ, *see* **nēh.**

nigard, adj., *niggard, miser,* LANGL. *B* xv. 136; WICL. ECCLUS. xiv. 3; CH. C. T. *D* 333;

nigart R. R. 1175; nekar, nekard ALEX. (Sk.) 1743.

nigardie, sb., *stinginess,* CH. C. T. *B* 1362.

niʒen, card. num., *O.E.* nigon, neogon, = *O. Fris., O.L.G.* nigun, *Goth., O.H.G.* niun, *O.N.* niu, *Ir., Gael.* naoi, *Lat.* novem; *nine,* HOM. I. 219; TRÍST. 364; niʒen, nihen [niʒe] LAʒ. 1188, 5149; niʒen [niʒene] and þritti 2804; þer æhte þer niʒene 26502; niʒhen ORM. 1051; nien IW. 402; niʒe SPEC. 37; nie þinges A. R. 326; neʒen AYENB. 45; neghen PR. C. 729; kniʒtes neʒene FER. 2720; neghe OCTOV. (W.) 650; neoʒe TREAT. 134; ʒif ʒe siggeð niene A. R. 30; nine LANGL. *B* i. 119; nine and twenti CH. C. T. *A* 24; nene D. TROY 2638.

niʒen-tēne [neʒentēne], card. num., *O.E.* nigontȳne; *nineteen,* LAʒ. 1850; neoʒentene P. L. S. xiii. 82; neʒentene FER. 2699.

neoʒen-tēoþe, ord. num., *O.E.* nigontēoða; *nineteenth,* P. L. S. xiii. 351.

nigen-ti, card. num., *ninety,* GEN. & EX. 1027.

niʒeðe, ord. num., *O.E.* nigoða, = *O.L.G.* nigunda, *Goth.* niunda, *O.N.* niundi; *ninth,* P.L.S. viii. 170; nigeðe HOM. II. 137; niʒeðe, nieðe A. R. 198, 236; neoʒeþe BEK. 617; niʒhende ORM. 4488; neʒende AYENB. 10; SHOR. 100; neghend PR. C. 3988; ninþe LANGL. (Wr.) 9545; H. S. 2906

niʒt, niht, *see* **naht.**

nigromancie, sb., *O.Fr.* nigromance; *necromancy,* LANGL. *A* v. 158; nigromance CATH 255; nigramansi BARB. iv. 747; nigromaunc MAN. (F.) 2241; nigremansi TREV. I. 231 negremauncie LUD. COV. 189.

nigromanciere, sb., *necromancer,* CATH. 255 nigramansour BARB. iv. 242; nigromansere nigromancier TREV. V. 345, VI. 19, VIII. 135; **nigramancers** (*pl.*) APOL. 95.

nigun, sb., *niggard, miser,* H. S. 5578; niggoun [nigon] GAM. 323.

niker, sb., *O.E.* nicor, = *M.Du.* nicker, necker, ? *O.N.* nykr, ? *O.H.G.* nichus; *a water-sprite*; nikir '*siren*' PR. P. 356; **nikeres** (*pl.*) LAʒ. 21746; MAN. (F.) 1447; nikeren AYENB. 61; *see* **iker.**

nikid, *see* **neked.**

nimel, adj., ? *O.E.* nemol; *nimble*; nimil '*capax*' PR. P. 356; nemel LIDG. M. P. 168; nemil L. H. R. 113; TREAT. OF FISH. 8.

nem(l)**-lī,** adv., *nimbly,* D. TROY 1226, 10940; nemeli YORK xxix. 219, xxxv. 120.

nimen, v., *O.E.* niman, nyman, neoman, = *O.L.G., Goth.* niman, *O.H.G.* neman, *O.Fris.* nima, *O.N.* nema; *take, seize,* SAX. CHR. 254; FRAG. 7; HOM. I. 9; KATH. 2148; ORM. 2910; C. L. 547; S. S. (Web.) 1430; LANGL. *B* xiii. 373; al heora god we sculen nimen [nime] LAʒ. 993; æt Dovre he þohte nimen lond (*disembark*) 9737; nimen wreche

GEN. & EX. 1042; nimin PR. P. 356; nime LEECHD. III. 96; O. & N. 1097; TREAT. 139; AYENB. 37; niman HOM I. 17; neoman 29; nemen 241; nemen JOHN vi. 21; neme RICH. 3876; neomen MARH. 3; hu durre ȝe neomen ow to cristes icorne JUL. 46; to nimene LAȝ. 30118; AYENB. 9; nimeð (*pres.*) A. R. 6; nimeþ FL. & BL. 9; he nimþ þane diaþ AYENB. 30; nim (*imper.*) LAȝ. 13156; A. R. 324; numeð HOM. I. 75; nime (*subj.*) O. & N. 359; AYENB. 55; we numen HOM. l. 23; nam (*pret.*) FER. 257; GOW. I. 212; S. S. (Wr.) 461; (*ms.* namm) ORM. 916; his licham of erðe he nam GEN. & EX. 200; hunger him nam 1490; ¬eðen he nam (*went*) to mirie dale 745; þe admiral hire nam to qvene FL. & BL. 791; þat men fro him his birþene nam HAV. 900; þe erl . . . nam until his lond 2930; þat nam he most to herte WILL. 1203; nam, nom KATH. 910, 1023; þe king nom [nam] þat writ on hond LAȝ. 484; þer he þa sæ nom (*il prit la mer*) 4966; nom HOM. I. 31; C. L. 917; BEK. 3; AYENB. 45; DEGR. 126; he nom leave A. R. 230; (þou) nome LAȝ. 5048; MISC. 139; HORN (L.) 1173; S. S. (Web.) 277; namen SAX. CHR. 261; MAN. (H.) PREF. 98; nomen HOM. I. 59; GEN. & EX. 1016; WILL. 1309; S. S. (Wr.) 317; MIN. ix. 53; nome P. L. S. xiii. 237; GREG. 269; nomen [nemen] LAȝ. 6723; nomen, neme HAV. 1207, 2790; neme HORN (L.) 60; nome (*subj.*) ROB. 66; nome [neme] LAȝ. 7093; numen (*pple.*) ORM. 6940; GEN. & EX. 409; nume HOM. II. 59; nomen JOS. 405; CH. TRO. v. 514; TRIAM. 1181; manrede of alle havede he nomen HAV. 2265; nome S. S. (Wr.) 2680; comp. **a-, bi-, for-, ȝe-, mis-, of-, ofer-, under-nimen**; *deriv.* **nāme, nume.**

nimere, sb., = *O.H.G.* nemāri; *one who seizes*, AYENB. 248.

nimunge, sb., = *O.H.G.* nemunga; *act of taking*, A. R. 38; niminge AYENB. 164.

nimphe, sb., *O.Fr.* nymphe, *Lat.* nympha; *nymph*; **nimphes** [nimphus] (*pl.*) CH. C. T. B 2928.

nip, sb., *? from* **nippen**; *? place of 'nipping'* cold: out of þe nippe [nipe] of þe norþ LANGL. B xviii. 162, C xxi. 168.

hnipen, v., *O.E.* hnipian, = *M.H.G.* nipfen; *slumber*: þa sunne gon to nipen LAȝ. 31734; þa nipeden (*pret.*) hio ealle & slepen MAT. xxv. 5.

nīpin, v., = *M.Du.* nijpen; *'premo, stringo,'* PR. P. 357.

nippen, v., *? = L.Ger.* knippen; *nip*: **nippe** (*imper.*) of the hed TREAT. OF FISH. 24; nipping (*pple.*) hus lippes LANGL. C vii. 104.

hnite, sb., *O.E.* hnitu, = *M.Du.* nete, *O.H.G.* niz; *nit*; **nite** *'lens'* PR. P. 357; VOC. 190; nete 177; niteȝ (*pl.*) PR. C. 651.

nīte, v., *O.N.* nīta; *deny*, L. H. R. 121; C. M. 883; **nitte** (*pret.*) M. H. 50; nit BARB. i. 52; nite (*pple.*) ALEX. (Sk.) 1460.

nitelen, v., *toil* : (þai) niteled (*pret.*) þer alle þe niȝt for noȝt at þe last A. P. ii. 888.

nitten, *see* **nütten.**

nīð, sb., *O.E.* nīð *m.* (*contention, envy*), *cf.* *O.L.G.* nīth *m.*, *O.H.G.* nīd *m.*, *O.N.* nīð *n.*, *Goth.* neiþ *n.* (φθόνος); *contention, envy, malice*, A. R. 404; nīð and win REL. I. 213; nīð and strif GEN. & EX. 373; nīð, niþ HOM. I. 153; niþ ORM. 123; niþ and wrake O. & N. 1194; niþ and onde 1401; nīþe (*dat.*) FRAG. 7; LAȝ. 3934*; SHOR. 156; S. S. (Web.) 1205; AM. & AMIL. 208; of niþe ant of onde P. S. 212.

nīð-craft, sb., *malice*, LAȝ. 7116.

nīð-ful, adj., *envious*, HOM. I. 57; GEN. & EX. 369; niþful ORM. 18254; þurh nīðfulne (*acc.*) craft LAȝ. 10219; þe nīðfulle (*pl.*) P. L. S. viii. 138*.

nīð², adj., *? O.E.* nīð, *cf.* *Goth.* (anda-)neiþs; *? dire*; niþ hellepine ORM. 13677.

[**nīðe**, adv., = *O.H.G.* nida, *M.H.G.* nide (*down*).]

nīðe-weard, adv., *O.E.* nyðe-, neoðeweard; *downwards*, ['*deorsum*'] MK. xv. 38; fro neþewarde HOM. II. 165.

niþemest, adj., *O.E.* niðemest; *lowest*, TREAT. 138; FER. 3257; neþemest C. M. 9926; þe neþemeste HOM. II. 219; þan neoþemeste pinan I. 117.

nīðen, adv., *O.E.* niðan, neoðan, neoðane, = *O.N.* neðan, *O.L.G.* nithana, *O.H.G.* nidana; *beneath*; neðen *from beneath* HOM. II. 105; ge sint nyðene ['*de deorsum*'] JOHN viii. 23; neuth BARB. xi. 538; comp. **bi-neoðen.**

nīðen, v., *O.N.* nīða, = *O.H.G.* nīden; *oppose, envy*; niþe ['*aemulari*'] PS. xxxvi. 8; nīðede (*pret.*) GEN. & EX. 1521.

nīðer, adv., *O.E.* niðor, neoðor, = *O.L.G.* nithar, *O.N.* niðr, *O.H.G.* nidar; *downwards*, LAȝ. 8182; falle nīðer REL. I. 223; ne warp þu me nawt neoðer into helle MARH. 17; neoðer, neðer HOM. II. 111; sal gliden on his brest(e) neðer GEN. & EX. 370.

neoðer-stīenge, sb., *descent*, HOM. II. 111.

nīðer-ward, adv., = *M.H.G.* niderwert; *downwards*, REL. I. 221; n(e)oþerward, neþerward O. & N. 144; netherward D. TROY 7717.

niþere, adj., *O.E.* niðera, neoðera, = *O.Fris.* nithere, *O.N.* neðri, *O.H.G.* nidaro; *nether, lower*, MARG. 160; þe niþer(e) ende BRD. 24; he beð neðer þanne he.er was HOM. II. 103; to þe niðer(e) wunienge 173; a niðer(e) helle grunde 229; his neþer(e) chavel TRIST. 1468; þe neþere lippe VOC. 146; neoðere A. R. 332; nimes of ȝour nethir glove (i.e. *boots*) & nakens ȝoure leggis ALEX. (Sk.) 4959;

neþire gloves 2767; in **neþerere** (*compar.*) of erth ['*in inferioribus terrae*'] HAMP. PS. cxxxviii. 14; **neþereste** (*superl.*) [=*O.H.G.* nidarŏsta]; *lowest*, CH. BOET. i. 1 (5).

niꝺerien, v., *O.E.* niꝺerian,=*O.N.* niꝺra, *O.H.G.* niderran, nidiran; *make low, debase, humilitate*; (*ms.* nyꝺerien) LK. vi. 37; niþren (*ms.* niþþrenn) ORM. 13966; neoꝺered (*pple.*) LAꝣ. 29992; *comp.* a-, ꝫe-**niꝺerien**.

　nething, sb., *abasement*, BARB. xix. 155*.

niꝺing, sb., *O.E.* niꞇing,=*O.N.* niꞇingr; *wretch, villain*, LAꝣ. 690; niþing HORN (R.) 204; AYENB. 139; ALIS. 2054; ISUM. 23; niþinge (*dat.*) MISC. 30; *comp.* **mete-niþing.**

nivelen, v., *Fr.* (re-)nifler; *snuffle*: nivelen & makien sure & grimme chere A. R. 240; **nifle** (*pres.*) REL. II. 211; nꝭvelinge [nevelinge] (*pple.*) wiþ þe nose LANGL. B v. 135.

niwe, *see* **newe.**

niwel, adj., *O.E.* nēowol, nȳwol, nēol; *deep, right down*: and nuel feol to grunde LAꝣ. 16777*.

　niuelinge, neuelinge, adv., *flat down*, '*resupinus*,' TREV. II. 203.

　niwelnisse, neowelnesse, sb., *O.E.* nēowel-, nȳwolness; *depth, abyss*, HOM. I. 225, 233.

nō, *see* **nā, nān.**

nōble, adj., *O.Fr.* noble; *noble, good*, A. R. 166; P. L. S. xiii. 119; nobill YORK xxv. 300; **nōbleste** (*superl.*) AYENB. 92; noblost ROB. (W.) 6346.

　nōble-līche, adv., *nobly, excellently*, AYENB. 55; nobliche P. L. S. xvii. 423; nobli ALEX. (Sk.) 260.

　nōble-man, sb., *person of high birth*; nōblemen (*pl.*) HOM. I. 273.

nōble², sb., *noble (coin worth 6s. 8d.*), LANGL. C iv. 47; nōbles (*pl.*) CH. C. T. C 907, G 1365.

nobleie, ·sb., *O.Fr.* noblei; *splendour, grandeur, nobility*, ROB. 46; MAN. (F.) 105; CH. C. T. F 77; noblei CH. C. T. G 449; noblai BARB. viii. 211; HAMP. PS. cviii. 2*; nobiꞁai BARB. ix. 95.

nōblen, v., *ennoble*; **nōbledest** (*pret.*) [*trans. Dante's* '*nobilitasti*'] CH. C. T. G 40.

noblesse, sb., *O.Fr.* noblesse; *nobility, magnificence*, AYENB. 20; CH. BOET. ii. 3 (37); noblesce A. R. 166.

nōbletē, sb., *O.Fr.* nobleté; *nobility*, TREV. I. 235, IV. 195, VII. 487, VIII. 15.

nōces, sb. pl., *O.Fr.* noces; *wedding*, A. R. 78.

nocke, sb.,=*M.Du.* nocke, *Dan., Swed.* nokke; *notch*; nokke PR. P. 357; nokkis (*pꞁ.*) P. R. L. P. 17.

nocturne, sb., *Fr.* nocturne; *nocturn* (*prayer at 3 ꜳ m.*), A. R. 270.

nodden, v., *nod*; nodde CH. C. T. H 47.

noddil, v., *strike with the fist*: dose noddil on him with neffes YORK xxix. 370.

nodile, sb., *noddle, top of the head*, '*occiput*,' VOC. 206; nodil, nodle PR. P. 357.

nōꝫt, nōht, *see* **nāwiht.**.

nōhtunge, sb., *from* **nāwiht**; *depreciation*, H. M. 9, 31; sum luꞇer word oꞇer sum nouhtunge A. R. 426.

noi, *see* **nui. noien**, *see* **nuien.**

noious, adj., *hurtful, difficult*, WICL. 2 THESS. iii. 2; BARB. xix. 742; PALL. i. 485.

noise, sb., *O.Fr.* noise; *noise, quarrel*, A. R. 66; TREAT. 135; AYENB. 66; GREG. 615.

noisen, v., *make a noise*; nois (*pres.*) ALEX. (Sk.) 4744.

nōk, sb., *nook, corner, angle*, HAV. 820; **nōke** (*dat.*) GAW. 660; of þe noke ['*anguli*'] PS. cxvii. 22; he held þe lettre bi þe noke DEGR. 165; **nōkes** (*pl.*) MIN. vii. 5; C. M. 17675.

[**nōked**, adj., *cornered; comp.* **fēower-nōked.**]

nokke, *see* **nocke.**

hnol, sb., *O.E.* hnoll,=*O.H.G.* hnol; *head*; nol '*vertex*' FRAG. 2; nol ['*cervicem*'] WICL. 4 KINGS xvii. 14; nolle (*dat.*) P. S. 157; TREV. V. 369.

nolpe, sb., *blow, stroke*, D. TROY 14037.

nolpen, v., *strike*; nolpit (*pret.*) D. TROY 6613, 13889; nolpit (*pple.*) 1257.

nombles, sb., pl., *O.Fr.* nomble; *? from Lat.* lumbulus; *entrails of a deer*, GAW. 1347; TRIST. 491; nombles of veneson C. B. 70; nomblis, nombles 10; numbles 114.

nombre, nombren, *see* **numbre, numbren.**

nóme, *see* **náme, nume.**

nōn, sb., *O.E.* nōn, *from Lat.* nōna (*scil.* hōra); *noon*, LAꝣ. 14039; MISC. 50; ær none LAꝣ. 17063; after none HORN (L.) 358; at none 801; **nōnes** (*pl.*) *meal at noon* LANGL. B vi. 147, C vii. 429, ix. 146.

　nōne-mete, sb., *O.E.* nōnmete; *mid-day repast*, '*merenda*,' REL. I. 6; nunmete PR. P. 360.

　nōne-(s)chenche, sb., *mod.Eng.* luncheon; *noon drinking*, LANGL. IV. *page* 165.

　nōn-tīd, sb., *O.E.* nōntīd; *noon-tide*, SAX. CHR. 264; nontide MK. xv. 33.

nōn, *see* **nān.**

nōnes, *in phrase* for þe nōnes,=for þen ōnes; *see under* **ān.**

nonne, *see* **nunne.**

hnoppe, sb., *? O.E.* hnoppa,=*M.L.G., M.Du.* noppe; *nap (on cloth)*; noppe '*villus*' PR. P. 358; CATH. 256.

hnoppe, v., *take the nap off*; noppe CATH. 256.

hnoppi, sb.,=*M.Du.* noppigh; *with a nap, rough*; noppi '*villosus*' PR. P. 358.

·ꞁor, *see* **nāuþer. norice**, *see* **nurice.**

norissen, *see* **nurissen. nors,** *see* **nurice.**

norð, sb. & adv., *O.E.* norð (*adv.*),=*O.Fris.*
north (*adv., sb.*), *O.H.G.* nord (*sb.*) ; *north* :
Albanac hefde al þat norð LAȜ. 2134 ; (þat)
wunede norð [norþ] 3443 ; Hengest is ifaren
norð 16442 ; min sete norð on hevene·maken
GEN. & EX. 278 ; norþ O. & N. 921 ; MISC.
96 ; she lokede norþ HAV. 1255 ; þet norþ
AYENB. 124 ; north '*septentrio*' PR. P. 358 ;
['*aquilonem*'] PS. lxxxviii. 13 ; in þe norþe CH.
C. T. *A* 4015.

    norð-dēle, sb., *O.E.* norðdǣl ; *north part,*
HOM. I. 219 ; norþdale ORM. 16412.

    norð-ēast, sb., *north-east,* SAX. CHR. 249.

    norð-ende, sb., *O.E.* norðende ; *north
quarter,* LAȜ. 12729 ; norþende HAV. 734.

    Norþ-folc, pr. n., *Norfolk,* MISC. 146.

    norþ-half, sb., *O.E.* norðhealf ; *north part,*
TREAT. 137 ; SHOR. 149.

    Norð-humberlond, pr. n., *Northumberland,*
LAȜ. 6390.

    Norþ-see, pr. n., *North-sea,* P. L. S. xiii. 17.

    norð-sīde, sb., LAȜ. 24515 ; norþside P. L.
S. xvii. 350.

    norþ-ward, adv., *O.E.* norðweard ; *north-
ward,* BEK. 1127 ; CH. C. T. *A* 1909.

    Norð-weȝe, pr. n., *Norway* ; **Norþweie**
[*O.E.* Norðwegum] (*dat.*) ROB. 182 ; Norweȝe
[Norweie] LAȜ. 4474 ; Norweie O. & N. 909 ;
MAN. (F.) 2878.

    Norþ-wīch, pr. n., *O.E.* Norðwīc ; *Norwich,*
MISC. 145 ; P. L. S. xiii. 64 ; Norwic SAX.
CHR. 263.

norðe, ? adj., *north* ; bi norðen LAȜ. 3742 ; bi
norðe 21043 ; bi norþe P. L. S. x. 1.

[**norþer,** adv., *O.N.* norðr ; *northernly.*]

    norþer-man, sb., *man of the north* : bot
northir men (*north*) wald no þing swa BARB.
xvii. 846.

norþer², adv., *O.N.* norðar ; *more northward* :
þo ferde he norþer LAȜ. 2674*.

norðerne, adj., *O.E.* norðerne ; *northern:*
þat norðerne volc LAȜ. 31333 ; a northerne
wind LEG. 165 ; norþerne wind SPEC. 51 ;
the norþern light CH. C. T. *A* 1987.

nóse, *see* **náse. nŏt,** *see* **nāwiht.**

nōt, *for* ne wāt, *see* ne.

hnot, adj., *O.E.* hnot ; *cleanly shaved* : a not
hed had he CH. C. T. *A* 109 ; not of his
nolle DEP. R. iii. 46.

notable, adj., *O.Fr.* notable ; *remarkable,*
CH. C. T. *B* 1865.

notarie, sb., *O.Fr.* notaries ; *notary, solicitor,
scribe,* LANGL. *C* xvii. 192 ; notaries (*pl.*)
AYENB. 40 ; notories LANGL. *C* iii. 185.

nóte, sb., *O.E.* notu ; *from* nēoten ; *use, occupation, enjoyment,* PR. P. 359 ; P. L. S. xix. 53 ; CHR.
E. 927 ; SHOR. 36 ; AYENB. 159 ; S. S. (Web.)
992 ; MIRC 1484 ; RICH. 2651 ; CH. C. T.

A 4068 ; H. S. 2073 ; A. P. ii. 381 ; for ure
note HOM. I. 195 ; is in þe eni oþer note bute
þu havest schille þrote O. & N. 557 ; note and
frame SPEC. 71 ; him selve to note HOM. II.
183 ; note, noote YORK xix. 268, xxxvi. 383.

    nóte-ful, adj., *useful,* WICL. E. W. 343 ;
CH. BOET. i. 1 (7) ; C. M. 8473.

    notsum, adj., *profitable* : heore nutene
neotsume (? *for* notsume) weren LAȜ. 342.

nóte², sb., *O.Fr.* note ; *note* : a note of þe
nihtegale SPEC. 26 ; þis was a meri note
YORK xv. 65 ; note *tenor of a letter* ALEX. (Sk.)
1719 ; nótes (*pl.*) AYENB. 105 ; A. P. i. 883.

note, *see* **hnute.**

nŏte, adj., *Lat.* nōtus ; *noted, excellent* : now
nar ȝe not fer fro þat note place GAW. 2091 ;
noied of þare note men ALEX. (Sk.) 1227 ;
our note men 4870.

nóten, v., *Fr.* noter ; *note, observe, indite,* A. R.
158 ; nótis (*pres.*) ALEX. (Sk.) 2795 ; noteþ
DEP. R. iv. 54 ; nóte (*imper.*) ALEX. (Sk.)
5655 ; so was nóted (*pret.*) þe note A. P. ii.
1651 ; notid ALEX. (Sk.) 4569.

nótien, v., *O.E.* notian ; *make use of, enjoy,*
HOM. II. 189 ; A. R. 106 ; noten GEN. & EX.
3144 ; notie SHOR. 21 ; note JOS. 588 ; H. S.
2602 ; ich nótie (*pres.*). . . mine þrote O. & N.
1033 ; note RICH. 5604 ; þu notest ORM.
12228 ; noteð JUL. 45 ; noteþ AYENB. 260 ;
þe king nuste whæt he inóted (*pple.*) hafde
LAȜ. 30603 ; *comp.* bi-nótien.

nōþe, adj., *O.E.* nōð,=*O.H.G.* nand ; *? boldness* : and lete al þat to noþe (*r. w.* soþe)
SHOR. 65 ; ȝaf he let to noþe 111.

nou, *see* **nu.**

nouche, sb., *O.Fr.* nouche, *from O.H.G.* nusca ;
ouch, necklace, collar, jewel, 'monile,' PR.
P. 359 ; ['*fibulam*'] WICL. 1 MACC. x. 89 ; a
nouche of golde CH. C. T. *D* 743 ; nouchis
(*pl.*) CH. H. F. 1350 ; nouches [ouchez] ALEX.
(Sk.) 3134.

noucin, sb., *O.N.* nauðsyn ; *necessity, distress* ;
(*printed* noncine) C. M. 5372 ; nocin 5802 ;
nowcin MARH. 1 ; KATH. 1698.

nōuht, *see* **nāwiht.**

nōuhti, adj., *having naught, penniless, worthless* : men . . . þat nedi ben and nouȝti [naȝti]
LANGL. *B* vi. 226 (Wr. 4245) ; noȝti A. P. ii.
1359 ; more nedier and nāuȝtier (*comp.*)
LANGL. *B* vii. 72*.

noule, *see* **navele.**

noumpere, sb., *O.Fr.* nonper ; *umpire,*
S. A. L. 210 ; noun-, noumpere LANGL. *A* v.
181 (nompeir *C* vii. 388) ; noumpere WICL.
PROL. ROM. ; oumpere PR. P. 360.

nounāge, sb., *A.Fr.* nonaage ; *nonage, mi-
nority* ; **nounāges** (*pl.*) D. RICH. iv. 6.

nouncertein, sb., *uncertainty,* CH. COMP. V.
46.

nout, *see* **nēat. nōut,** *see* **nāwiht.**

nouthe, *see* **nu.**

nōuðer, *see* **nāuþer. novele,** *see* **navele.**

novellis, sb., pl., *news*: whens evere this barne mai be that shewes þer novellis nowe YORK xx. 106.

novelrie, sb., *O.Fr.* novelerie; *novelty*; novellerie GOW. II. 259; **novelreis** (*pl.*) BARB. xix. 394; CH. C. T. *F* 619; H. F. 686.

noveltē, sb., *O.Fr.* noveleté : *novelty, news*: for wonder of novelte of his doinge TREV. III. 67; slike novelte mai noght disease YORK xv. 127; what novelte xxv. 118; **noveltees** (*pl.*) CH. C. T. (T.) 10933.

Novembre, sb., *Lat.* November; *November,* CH. ASTR. i. 10 (6).

novis, sb., *O.Fr.* novice; *novice,* CH. C. T. *B* 3129; VOC. (W. W.) 681.

nowel, sb., *O.Fr.* noël *from Lat.* nātālem; *Christmas,* CH. C. T. *F* 1255.

nowele, *for* **an owele** (*owl*); an nowele YORK xxix. 119.

nowt, *see* **nēat. nōwðer,** *see* **nāuþer.**

nu, ? nū, adv., *O.E.* nu,=*O.L.G., O.H.G., O.N., Goth.* nu, *Gr.* νύ; *now,* O. & N. 46; P. S. 257; til nu ORM. 14066; nu ʒu (*ms.* nuʒʒu) | *cf. O.H.G.* nu ju] 9281; and was redi to slon him nu ge GEN. & EX. 1328; nu ðen MARH. 13, 21; nu þen MISC. 85; nu ða HOM. I. 9; nu ðe LAʒ. 29863; KATH. 2120; nu þe FL. & BL. 12; þat me ofþinchet nu þe P. L. S. viii. 5; ich beo nu (a)n an ilaht JUL. 72; nu (a)n on A. R. 270; nu, nou HAV. 120, 2421; nou TREAT. 135; AYENB. 87; JOS. 29; nou þe CH. C. T. *A* 462; A. P. iii. 414.

nūd, nūden, *see* **nēad, nēden.**

nue, *see* **nēwe. nūel,** *see* **nīwel.**

[**nūht,** sb., *O.E.* nyht; *abundance.*]

    nūhtsom [neghtsom], adj.,=ʒenūhtsom ; *abundant,* PS. xcviii. 8.

    nūhtsomen, v., *O.E.* genyhtsumian; *abound*; nuhtsome PS. lxiv. 14.

    neghtsomnes, sb., *abundance,* PS. cxxix. 4.

nui, sb., *for* **anui**; *annoyance, hurt* : es ther any **noies** (*pl.*) of newe YORK xi. 386; ar ther any noies new TOWNL. viii. 386; noi LANGL. *C* iii. 554; *see* **anui.**

    noi-ful, adj., *hurtful,* WICL. PS. xxxvi. 2.

nuiaunce, sb., *trouble*; noiaunce MAN. (F.) 3444.

nuien, v., *for* **anuien**; *annoy*; noi YORK i. 71; ALEX. (Sk.) 676; **noiis** (*pres.*) ALEX. (Sk.) 4182; **noied** (*pret.*) ALEX. (Sk.) 1227; HAMP. PS. cvi. 18*; noiede WICL. JUD. xi. 1, xvi. 7; **noied** (*pple.*) LANGL. *C* iii. 19.

nuisance, sb., *O.Fr.* nuisance, *mod.Eng.* nuisance; *annoyance, damage*; noisance PARTEN. 401.

nuisen, v., *O.Fr.* nuisir; *injure*; nusi SHOR. 20.

numbre, sb., *O.Fr.* nombre; *number,* P. L. S. xvii. 223; nombre CH. C. T. *A* 716; nounbre ALEX. (Sk.) 89, 215, 449, 3641.

numbren, v., *O.Fr.* nombrer ; *number, enumerate* ; noumbre ROB. 61 ; ALIS. 1396 ; nombre [noumbre] LANGL. *B* i. 115.

nume, sb., *O.N.* (her-)numa; *from* nimen ; *act of taking; pledge, hostage* ; nome REL. I. 227 (MISC. 25) ; GEN. & EX. 2268.

nümen, *see* **nimen. nunan,** *see* **nu.**

nunne, sb., *O.E.* nunne, *O.Fr.* nonne, *Lat.* nonna; *nun,* A. R. 316; M. H. 164; nunne [nonne] LAʒ. 15651; nonne ROB. 222; GOW. III. 281; CH. C. T. *A* 118; **nunnen** (*pl.*) MISC. 77; nonne (*dat. pl.*) P. L. S. ix. 149.

nunnerie, sb., *O.Fr.* nonnerie; *nunnery,* P. L. S. xxxv. 148; nonnerie LAʒ. 15642*; P. L. S. xvi. 150; nunrie ROB. (W.) 2728.

nurceri, sb., *nursery, nursing*; norceri ROB. (W.) 8938*.

nurice, sb., *O.Fr.* nurrice, norrice; *nurse,* A. R. 82; norice P. L. S. xiii. 135; AYENB. 161; nors PALL. iv. 35; norise MIR. PL. 141; nursche WICL. I THESS. ii. 7.

nurissen, v., *O.Fr.* nurrir, norrir; *nourish*; *nurse*; norissi AYENB. 154; norischi BEK. 1874; **norisseþ** (*pres.*) SHOR. 99; **norisse** (*subj.*) CH. BOET. iii. 6 (79); **nurist** (*pple.*) PR. C. 4198.

    norisschinge, sb., *nursing,* ROB. (W.) 8938*.

nuritūre, sb., *O.Fr.* norreture, nurture; *nurture*; nurture MAN. (F.) 4295; nurtur D. TROY 7885; noriture PARTEN. 3837; norture PR. P. 358; AYENB. 113; GAM. 4.

nurnen, v., *bring forward, proffer; propose; utter, say*: he nolde not . . . nurne hir aʒainez GAW. 1661; to norne on þe same note 1669; i wil no giftez . . . i haf none you to norne 1823; nou (*? ms.* how) norne ʒe youre riʒt nome 2443; **nurned** (*pret.*) A. P. ii. 669; (þat prince) nurned him so neʒe þe þred GAW. 1771.

nurri, sb., *O.Fr.* nouri; *foster-child,* ['*alumnus*'] TREV. VII. 139; norie, norrei, norri, nori TREV. VI. 43, 79, VII. 379, 393, 519; nori WILL. 1511.

nurtren, v., *from* **nuritūre**; *nurture*; **nurtrid** (*pple.*) ALEX. (Sk.) 3177.

nurð, sb., *? noise,* A. R. 92*; leaveð ower nurð ant ower ladliche nere MARH. 21; nurð, nurhð HOM. I. 247; H. M. 31.

nürvil, nirvil, sb., *? for* **cnürvil**; *dwarf,* '(*homo*) *pusillus*,' PR. P. 357, 361.

nüt, adj., *O.E.* nytt,=*O.Fris.* nette, *O.N.* nytr, *Goth.* (un-)nuts, *O.H.G.* nuzzi; *from* **nēoten**; *useful,* LAʒ. 9470; *comp.* **un-nüt.**

***h*nute,** sb., *O.E.* hnutu (*pl.* hnyte), *cf. O.N.* hnot, *O.H.G.* hnuz; *nut*; nute HOM. I. 79; C. M. 18833; ane nhote AYENB. 143; note PR. P. 359; ALIS. 3293; he ne yaf a note of hise oþes HAV. 419; **nutes** (*pl.*) GEN. & EX.

3840 ; notes BEK. 1203 ; *comp.* **hasel-, pin-, wal-note.**

nute-brūn, adj., *nut-brown*, C. M. 18846.

nut-hak(k)e, sb., *nuthatch*, SQ. L. DEG. 55.

note-migge, sb., *nutmeg*, PR. P. 359 ; note-mugge ALIS. 6792 ; **notemigges** (*pl.*) C. B. 109.

note-schale, sb., = *M.H.G.* nuzschal ; *nutshell*, TREV. IV. 141 ; **nutescalen** (*pl.*) LA3. 29265.

nüten, sb., *O.E.* nȳten, nēten, nīeten ; *cattle*, FRAG. 3 ; nēten ['*jumentum*'] LK. x. 34 ; nŭtenu (*pl.*) HOM. I. 105 ; nutene LA3. 341.

nŭtte, sb., *O.E.* nytt, *cf. O.N.* nyt, *M.L.G.* nutte, *O.H.G* nuzzī *f.; use* : neoren noht to nuttes (*for* nutte) LA3. 13428.

nütten, v., *O.E.* nyttian, = *M.L.G., M.Du.* nutten, *O.H.G.* nuzzan ; *make use of, enjoy* : nutten hote spices A. R. 370 ; hu man birþ weoreldþing nitten ORM. 5543 ; nitte HAV. 941 ; **nütteð** (*pres.*) REL. I. 132 ; ȝef ha nutteð hare nome MARH. I ; **nütted** (*pple.*) HOM. II. 23.

O.

o, *see* on, of.

ō, interj., = *Goth., M.H.G.* ō ; *oh,* LA3. 17126 ; KATH. 1453 ; A. R. 54 ; AYENB. 93.

ō, *see* ā.

obedience, sb., *O.Fr.* obedience ; *obedience,* HOM. I. 213 ; AYENB. 140.

obediencer, sb., *a certain officer in a monastery,* LANGL. *C* vi. 91.

obedient, adj., *O.Fr.* obedient ;. *obedient,* A. R. 424.

obeiaunce, sb., *obedience,* ALEX. (Sk.) 5106.

obeiin, obeishe, v., *O.Fr.* obeir (obeiss-) ; *obey, reverence,* PR. P. 361 ; obeie [obeishe] WICL. GEN. xli. 40 ; obesche [obei] ALEX. (Sk.) 2416 ; **obes** (*pres.*) A.P. i. 886 ; obeschen ALEX. (Sk.) 1937 ; **obeising** (*pple.*) *obedient* CH. L. G. W. 1266 ; **obeide** (*pret.*) WICL. DEEDS vi. 7 ; **obéched** (*pret.*) A. P. ii. 745 ; obeschid [obeid] *made obeisance to* ALEX. (Sk.) 1620 ; obeisit BARB. ix. 304.

obeisant, adj., *O.Fr.* obeissant ; *obedient,* ROB. (W.) 10355 ; CH. C. T. *E* 66 ; obeisand BARB. iv. 603, viii. 10 ; obeissiant ALEX. (Sk.) *page* 279.

obeisaunce, sb., *obedience, obeisance* ; obeissance CH. C. T. *A* 2974 ; *E* 24, 230, 502 ; obeishaunce, obeisaunce WICL. I KINGS xv. 22, ECCLES. iv. 17 ; **obeisances** (*pl.*) CH. C. T. *F* 515 ; obeisaunces CH. L. G. W. 149.

obelei, *see* oblē.

obite, sb., *Lat.* obitus ; *death,* WICL. GEN. xxv. 11 ; R. W. 285 ; (*anniversary of death*) TREV. VI. 261.

obitte, pple., *from Lat.* obitus ; *dead* : i schulde be obitte YORK xxxvii. 269.

object, pple., *Lat.* objectus ; *placed in the way,* PALL. iv. 743, xii. 124.

oblē, ovelēte, sb., *O.Fr.* oublee, *O.E.* oflæte, *from med.Lat.* oblāta ; *offering ; sacramental wafer ; thin cake* ; HALLIW. 585 ; obelei VOC. (W. W.) 598 ; ovelete HOM. II. 97 ; **oblēs** (*pl.*) L. C. C. 22 ; obles and offrandis ['*oblaciones et holocausta*'] HAMP. PS. l. 20 ; obleies C. B. 73.

obligen, v., *O.Fr.* obligier ; *pledge* : ich wille obligi me to þe ROB. (W.) 280 ; to thi service i **oblissh** (*pres.*) me YORK xiv. 151 ; oblisheth WICL. PROV. xiii. 13 ; **obleged** (*pret.*) ROB. (W.) 6775 ; obleged MAN. (F.) 15266 ; oblishid, oblisht WICL. NUM. xxx. 4 ; WICL. PS. xix. 9.

obliviōn, sb., *O.Fr.* oblivion ; *oblivion,* GOW. II. 23.

obscūre, adj., *O.Fr.* obscur ; *obscure,* R. R. 5351.

obsequies, sb., pl., *O.Fr.* obseque, obseques, *A.Fr.* obsequies ; *obsequies,* CH. C. T. *A* 993 ; ['*civilitates*'] TREV. VII. 191.

observaunce, sb., *Fr.* observance ; *observance, duty,* CH. C. T. *A* 1045 ; **observaunces** (*pl.*) A.R. 24 ; CH. C. T. *F* 516 ; CH. L. G. W. 150.

observe, v., *O.Fr.* observer, *from Lat.* observāre ; *favour,* CH. C. T. *B* 1821.

obstacle, sb., *O.Fr.* obstacle ; *obstacle,* CH. C. T. *E* 1659.

obstinaci, sb., *obstinacy,* GOW. II. 117.

obstinacion, sb., *Lat.* obstinātiō ; *obstinacy,* TREV. VII. 371.

obstinate, adj., *Lat.* obstinātus ; *obstinate,* GOW. II. 117.

obstinaunce, sb., *O.Fr.* obstinance ; *obstinacy,* TREV. VII. 371.

oc, conj., *O.N.* ok ; *and* ; (*ms.* occ) ORM. 1216.

ōc, conj., *O.N.* auk ; = ēac ; *also,* GEN. & EX. 54 ; & oc it makeð his egen brigt REL. I. 210 ; ok HAV. 187.

ōc, *see* āk.

occasioun, sb., *O.Fr.* occasion ; *occasion,* CH. L. G. W. 994 ; **occasiōns** (*pl.*) of daliaunces CH. C. T. *C* 66.

occident, sb., *O.Fr.* occident, *from Lat* occidentem ; *west,* ALEX. (Sk.) 1045 ; CH. C. T. *B* 297.

occisoune, sb., *Lat.* occīsio ; *slaughter,* BARB. xiv. 220.

occupaciōn, sb., *O.Fr.* occupation ; *occupation,* GOW. II. 50.

occupie, v., *O.Fr.* occuper ; *occupy, dwell, make use of,* TREV. IV. 109 ; occupien him LANGL. *B* xvi. 196 ; okupien *C* xix. 207 ; **occupied** (*pple.*) *B* v. 409.

ocean, sb., *O.Fr.* occean ; *ocean* ; occean [occian] CH. C. T. *B* 505 ; occian BRD. 1 ; ALEX. (Sk.) 5503 ; occiane 2328.

ochen, v., *O.Fr.* ocher, hocher ; *notch, nick* ;

oches (*pres.*) D. ARTH. 2565 ; **ochede** (*pret.*) 4245.

Octōber, pr. n., *Lat.* Octōber ; *October*, CH. ASTR. i. 10 (6).

od, *see* ord.   **ōd,** *see* ād.

odde, adj., *O.N.* oddi ; *odd; unique; distinguished, special;* '*impar*,' PR. P. 362 ; A. P. ii. 426 ; odd(e) or even GOW. III. 138 ; odde (*irregular*) weddinge MIRC 198 ; ALEX. (Sk.) 2121, 4750 ; odde 27, 2631* ; od 94 ; **oddest** (*superl.*) 2008 ; oddist(e) 189, 1751.

    od, adv., *singly; specially,* MAN. (F.) 4614 ; D. TROY 7466, 9597, 10839.

    od-lī, adv., *curiously, excellently,* ALEX. (Sk.) 275 ; D. TROY 6859.

    od-men, sb., pl., *distinguished men, chieftains,* ALEX. (Sk.) 3783.

odious, adj., *Fr.* odieux ; *odious,* CH. C. T. D 2190.

odūr, sb., *O.Fr.* odor ; *odour,* P. L. S. xxxv. 76 ; odour TREV. III. 467 ; CH. C. T. *F* 913 ; odor [odour] C. M. 3701.

oerre, sb.,=irre ; *anger,* HOM. II. 228.

of, prep. & adv., *O.E.* of, æf,=*O.Fris.* of, af, ofe, *O.N.*, *Goth.* af, *O.L.G.* af, ava, *O.H.G.* ab, aba, abe, *Lat.* ab, *Gr.* ἀπό ; *of, off:* alused . . . of bende P. L. S. viii. 68 ; ared me . . . of eche deaðe HOM. II. 169 ; heo wolden kecchen of þe al þet þu hefdest A. R. 324 ; heo tok . . . of hire finger a riche ring FL. & BL. 2 ; Wawain hem tok þe knave of AR. & MER. 7826 ; þa veol þe king of horse LAȝ. 28790 ; þe king lihte of his stede HORN (R.) 51 ; þe leves fallen of þe tree ELL. ROM. I. 269 ; þe stude þet we of feollen MARH. 17 ; þe weoren icumen of (*sec. text* fram) Rome LAȝ. 5580 ; faren of his londe 1028 ; þa he sculde of live wende 6851 ; & braid hine of (*sec. text* ut of) þere scæðe 8177 ; isend . . . of (*v. r.* from) heovene KATH. 1586 ; þu most of londe fleo O. & N. 1304 ; heo beþ of londe idrive C. L. 444 ; the blod ran of his sides HAV. 1850 ; he aros of deaðe HOM. II. 23 ; he borwede of me LANGL. *A* iv. 40 ; þat . . . kun þe we beoð of [ove] icomene LAȝ. 451 ; to clensen hem of (*ms.* off) sinne ORM. 4111 ; þis gode prest þat we nu mælen ofe (*ms.* offe) 462 ; a man of god(e) inwit of alle þulke him mai skere TREAT. 133 ; þe dome þat ic eow er of sede P. L. S. viii. 79 ; þeo þreo sunnen þet ich spec of A. R. 56 ; þe child þat Isaie of tolde C. L. 646 ; þane gode mon þat of so feole þinge con O. & N. 1772 ; swulc he weore of witte LAȝ. 8226 ; of live i wolde be A. D. 252 ; if he were brouct of live HAV. 513 ; bringe of [o] live C. M. 11570 ; þat of his slepe awaked CH. C. T. *A* 2523 ; bicom of engel ate(l)lich deovel A. R. 52 ; he wrohte win of water ORM. 11081 ; þat ever wes mad of blod(e) ant bon(e) SPEC. 33 ; to warnie wummen of hore fol eien A. R. 54 ; he riht noht of hem ne rohte ORM. 9024 ; hi neren

aferede of nane licamliche þinunge HOM. I. 97 ; men beoþ of þe wel sore ofdrad O. & N. 1150 ; þus wes Marlin . . . iboren of his moder LAȝ. 15793 ; ich biȝet hit iwriten of þe writere MARH. 2 ; ich wolde þet heo weren of alle . . . iholden A. R. 48 ; Adam was biswike of Eve ALIS. 7709 ; itend of wraððe KATH. 156 ; i love mi lorde of (*for*) all YORK xviii. 216 ; of attere ifulled LAȝ. 14987 ; deien of hungre A. R. 342 ; þe king of þam londe HOM. I. 87 ; þat god of þisse londe LAȝ. 999 ; þe ord of þan sworde 18094 ; þe ende of þe tale A. R. 72 ; tvo of ous AYENB. 219 ; þat weorc is of stane LAȝ. 17180 ; eawles of irne KATH. 2209 ; o drope of blode P. L. S. xix. 294 ; þonkeð him of his ȝeove A. R. 44 ; heo is gulti of þe bestes deaðe 58 ; to bere witnisse of þis falshede BEK. 1117 ; þer of ne yaf he nouth a stra HAV. 315 ; have merci of me 491 ; for love of þe BEK. 39 ; for love of me SPEC. 69 ; a . . . mon of þriti ȝeren LAȝ. 377 ; a king of muche miht C. L. 275 ; ȝung of ȝeres A. R. 158 ; he was . . . clene of sinne ORM. 3171 ; riche of welðe GEN. & EX. 1355 ; of love he wes ful trewe SPEC. 91 ; ful of witte C. L. 400 ; liht of fote C. M. 3730 ; it was ten of [*v. r.* at] þe clokke CH. C. T. *B* 14 ; i ron of [*v. r.* on] blode H. H. 53 (55) ; smiten of HOM. I. 29 ; & fret of þe oðres earen 251 ; of (h)acken HOM. II. 139 ; his heved of swippen LAȝ. 878 ; sloh him of þat hæveð 3856 ; and þat þih him of smat 26071 ; scheken of A. R. 144 ; swipte hire of þat heaved KATH. 2485 ; here . . . . hevedes he of bot GEN. & EX. 2926 ; deþ him wile of drive MISC. 93 ; of weve BEK. 953 ; hi strupten his cloþes of 2225 ; his croune was al of ismite 2327 ; kerve of P. L. S. xiv. 19 ; (h)is heved of to smite *to smite off his head* A. D. 130 ; stripe of þe skin TREV. III. 173 ; of stripe RICH. 3399 ; ȝif þin hefet were ofe (*ms.* offe) HOM. I. 29 ; þe he specð ofe (*ms.* offe) HOM. II. 93 ; hwonne þeos rinde is ofe (*ms.* offe) A. R. 150 ; a tere af þin ye HALLIW. 24 ; on æsthalf o þis middelærd ORM. 3432 ; nu wot Adam sum del o wo GEN. & EX. 353 ; bringe o dawe LUD. COV. 291 ; some wolde have him a dawe RICH. 973.

of [2], conj., ?=*O.Fris.*, *M.Du.* of ; *? or* : this nis nought romaunce of skof ALIS. 667.

ōf, *see* ōð.

ofærne, *see* ofrennen.

of-āsken, v., *inquire;* ofaxen AL. (L.) 362 ; he ofeschte (*pret.*) what men hit were P. L. S. xiii. 343 ; ofācsed (*pple.*) AYENB. 152.

of-cálden, v., *make cold;* ofcóld (*pple.*) ROB. 322.

of-clépen, v., *call forth;* ofclépiþ (*pres.*) ALIS. 1810.

of-cwellen, v., *for* acwellen ; *kill,* ORM. 6897 ; ofcwalde (*pret.*) 8037.

of-dawen, v., *awake,* ALIS. 2265.

of-drǽden, v., *O.E.* ofdrǽdan, *for* ondrǽden ; *terrify* ; **ofdrēde** (*pres.*) HORN (L.) 291 ; **ofdrěd** (*pple.*) LAȝ. 8425 ; A. R. 178 ; ORM. 7925 ; GEN. & EX. 3955 ; ne beo þu nawiht ofdred KATH. 674 ; ofdrad O. & N. 1150 ; S. S. (Web.) 903.

of-draȝen, v., *attract* : to ofdrawen hire heorte A. R. 392.

of-druncnen, v., *drown* ; ofdruncneþ (*pres.*) ORM. 14611 ; þat was in þe sæ **ofdrunened** (*pple.*) 14852.

of-drunken, v., *drown* : al mankin was . . . ofdrunked (*pple.*) ORM. 6793.

ofe, *see* of.

of-earnen, v., *earn*, KATH. 2167 ; **ofearneð** (*pres.*) A. R. 194 ; oferned (*pple.*) HOM. II. 189.

ofen, *see* oven.

of-ēode, v. (*pret.*), *deserved* : þu . . . ofeodest FRAG. 7.

ofer, prep. & adv., *O.E.* ofer, = *O.Fris.*, *M.Du.* over, *Goth.* ufar, *O.N.* yfir, *O.H.G.* ubar ; *over* : ic wille senden flod ofer alne middenard HOM. I. 225 ; ðe halie gast com ofer þa apostlas 93 ; þa wes hit cuð over al þe burh 3 ; ofer erþe ORM. 14567 ; ich am duc ofer heom LAȝ. 461 ; þe heo weoren wældinde over 8386 ; over þene wal heo clumben 9420 ; over sæ varen 10512 ; over al *everywhere* 4868 ; þe herdes þe wakeden over here oref HOM. II. 31 ; he leapeð over ham A. R. 380 ; ofearneð . . . hure over hure 354 ; ich holde her hetel sweord over þin heaved 400 ; over (*above*) alle þing him luvien 92 ; wende over lond BEK. 1178 ; over þe nabbe i no mihte C. L. 1110 ; muche blisse þer is over al 732 ; þet no viend ous vondi over oure miȝte AYENB. 170 ; over þre mile AR. & MER. 6648 ; over al RICH. 2940 ; uvel over uvel A. R. 52 ; over night CH. TRO. ii. 1549 ; over *through* þe ȝer L. C. C. 33 ; he is lovird over lef MISC. 105 ; over alle þe wode þei hur soght TRIAM. 349 ; þe see toeode & þet i(s)raelisce folc wende over HOM. I. 141 ; nule he nout . . . wenden over A. R. 266 ; let lust over gan *go away* 118 ; it was over gon GEN. & EX. 3031 ; þo let Ubbe and al his care sowre over fare HAV. 2063 ; þe þroli þouȝt . . . he let over slide WILL. 3519 ; or PALL. i. 1032.

ofer-ǽt, sb., *O.E.* oferǽt, = *M.L.G.* overāt, *M.H.G.* überāz ; *over-eating* ; **overētes** (*pl.*) REL I. 131.

ofer-bēoden, v., *O.N.* yfirbiōða ; *overburden with commands* ; **oferbēden** (*pres. subj.*) ORM. 6233.

ofer-bīden, v., *O.E.* oferbīdan ; *remain over, last out, survive* : overbide CH. C. T. *D* 1260 ; C. M. 22687 ; þe fader mai boþe overbide ANGL. IV. 198 ; overbōd (*pret.*) S. S. (Web.) 1731 ; ur bost . . . is sone overbide (*pple.*) P. L. S. xxix. 92.

ofer-brǽdan, v., *O.E.* oferbrǽdan ; *spread*

over, cover : wes . . . mid palle **overbrǽd** (*pple.*) LAȝ. 19045.

ofer-cáld, sb., *excessive cold* ; overcolde PALL. xi. 54.

ofer-cark, v., *trouble* : shal noþer king ne kniȝt overcark þe comune LANGL. *C* iv. 472.

ofer-casten, v., *overcast* ; overkesten A. R. 274 ; **overcaste** (*pret.*) P. L. S. xvii. 354 ; overkeste D. ARTH. 3932 ; **overcast** (*pple.*) GOW. III. 310.

ofer-clōsen, v., *overshadow* ; **overclōseth** (*pres.*) LANGL. *C* xxi. 140.

ofer-clōþen, v., *cover* ; **overclōþeth** (*pres.*) LANGL. *C* xxi. 140*.

ofer-cumen, v., *O.E.* ofercuman, = *M.L.G.* overkomen ; *overcome*, ORM. 6275 ; overcumen LAȝ. 5535 ; þeo þet nimeð more an hond þen heo mei overcumen A. R. 198 ; overcume AYENB. 169 ; LANGL. *A* iv. 69 ; **overcōm** (*pret.*) CHR. E. 410 ; overcomen MARH. 1 ; **overcumen** (*pple.*) KATH. 133 ; GEN. & EX. 2108 ; overcume O. & N. 542 ; overcomen TRIST. 2795.

  ovir-cǫmere [overcomere], sb., *conqueror*, ALEX. (Sk.) 1610 ; ovircomer (*miswritten* ovircomers) [overcommer] 1903.

ofer-cwatie, v., *?satiate* ; overqvatie O. & N. 353.

ofer-dēde, sb., = *M.L.G.* overdād ; *exaggeration, intemperance* ; over dede HOM. II. 55 ; O. & N. 352 ; AYENB. 55.

ofer-dēre, adv., *over-dear* ; over dere ROB. 389.

ofer-dōn, v., *overdo* : ever ich þing me mei þauh overdon A. R. 286 ; **overdoon** (*pple.*) CH. C. T. *G* 645.

ofer-drinc, sb., *excessive drinking* ; **overdrinke** (*dat.*) HOM. I. 153.

ofer-drīven, v., *O.E.* oferdrīfan ; *pass over* : ȝe sall thame overdriff *survive* BARB. iv. 661 ; **ovirdrāfe** (*pret.*) ALEX. (Sk.) 1505 ; **overdrive** (*pple.*) *far gone in illness* MIRC 1813 ; ourdrivin *brought to an end* BARB. xix. 481.

ofer-ēake, sb., *O.E.* oferēaca ; *surplus* ; overeake HOM. I. 31.

ofer-ēode, v. (*pret.*), *O.E.* oferēode ; *went over ; passed away* : þe hunger þat Egipte overȝode C. M. 10524 ; overȝede ALEX. (Sk.) 350 ; þis ȝol overȝede GAW. 500.

ofer-etel, sb., *O.E.* oferetol ; *given to over-eating* ; '*edax*' FRAG. 5.

ofer-fallen, v., *O.E.* oferfeallan, = *M.H.G.* übervallen ; *fall over* : & tær fel dun þat hus . . . & oferfel (*pret.*) hem alle ORM. 4799.

ofer-fáren, v., *O.E.* oferfaran ; *pass over* : and **overváreþ** (*pres.*) veole þeode O. & N. 387.

ofer-fealden, v., *fold over* ; **ovirefólden** (*pple.*) ALEX. (Sk.) 5463.

ofer-flēoten, v., = *O.H.G.* ubarfliozan ; *flow*

over; **overflēot** (*pres.*) C. L. 849; **overflēt** (*pret.*) GEN. & EX. 586.

ofer-flōwen, v., *O.E.* oferflōwan; *overflow*; **oferflōweþ** (*pres.*) ORM. 10721.

overflōwendnesse, sb., *superfluity*, HOM. I. 115.

ofer-fülle, sb., *O.E.* oferfyllo,=*Goth.* ufarfullei; *redundance*; overfulle A. R. 160; O. & N. 354.

ofer-füllen, v.,=*Goth.* ufarfulljan, *M.H.G.* übervüllen; *overfill*; **overfüllet** (*pple.*) H. M. 19.

ofer-gangen, v., *O.E.* ofergangan, = *Goth.* ufargaggan, *O.H.G.* ubargān; *go over; surpass*; ORM. 10228; þanne sal þi child þi forbod overgangin MISC. 129; overgange HAV. 2587; overgan O. & N. 952; þat me ne mai mid swikedome overgan LA3. 15183; **oviregōs** (*pres.*) ALEX. (Sk.) 2534; þeo luve . . . overgeð ham alle voure A. R. 394; overgeþ O. & N. 947; AYENB. 34; **overgān** (*pple.*) LA3. 6258; MARH. 16; C. M. 4029; þa he hefde al þat lond overgan KATH. 520.

ofer-gard, adj. & adv., *? excessive*; overgard REL. II. 226; þe world was so overgart (*? arrogant*) P. S. 341; overgart proud C. L. 993; overgart gret WILL. 1069.

ofer-gart, sb., *? arrogance*, ORM. 8163; overgart MARH. 16.

ofer-grēat, adj., = *M.L.G.* overgrōt; *overgreat, too great*; overgreet CH. C. T. *G* 648.

ofer-gülden, v., *gild over*; **overgüldeð** (*pres.*) A. R. 182; **overgüld** (*pple.*) MARH. 9; ofergilded ORM. 2612; overgilt PR. C. 8902; overgilte LANGL. *B* xv. 121.

ofer-hardi, adj., *too daring*; overhardi LANGL. *C* iv. 300.

ofer-hāt, adj., *over-hot, too hot*; overhoot CH. C. T. *G* 955.

ofer-hebben, v., *O.E.* oferhebban; *? neglect*; **overhaf** (*pret.*) MAP 341; MAN. (F.) 14728.

ofer-hēh, adj., *O.E.* oferhēah,=*M.H.G.* überhōch; *over-high, too high*, ORM. 12061.

ofer-hélen, v., *O.E.* oferhelan; *cover*; overhelen WICL. GEN. ix. 14; **overhéleð** (*pres.*) HOM. II. 73; **overhéled** (*pple.*) PALL. ii. 16.

ofer-hélden, v., *overturn*; **overhéldis** (*pres.*) ALEX. (Sk.) 726.

ofer-hēren, *O.E.* oferhēran, -hȳran,=*M.H.G.* überhœren; *neglect*; **overhēr(e)d** (*pple.*) LEG. 17.

ofer-hi3ien, v., *overtake*; ourhi BARB. iii. 737; vi. 598.

ofer-hippen, v.,=*M.H.G.* überhüpfen; *leap over, omit*; overhippin PR. P. 372; **overhippis** (*pres.*) MAN. (F.) 64; o worde þei overhuppen LANGL. *B* xiii. 68.

ofer-ho3e, sb., *contempt*; overhohe H. M. 43; overhowe A. R. 276.

ofer-ho3ien, v., *O.E.* oferhogian; *contemn,*

despise; **overhoheð** (*pres.*) O. & N. 1406; oferhoweþ REL. I. 184.

ofer-hópe, sb., *over-confidence, presumption*; overhope HALLIW. 595; H. V. 68.

ofer-hóven, v., *hover over*; **overhóveþ** (*pres.*) LANGL. *B* xviii. 169; overehoveþ *C* iv. 300.

ofer-hu3ien, v., *contemn*; oferhugeð (*pres.*) LK. x. 16; oferhuged (*pple.*) MK. ix. 12.

ofer-hwǣr, adv., *everywhere*; overwhere MAN. (F.) 107.

ofer-hláden, v.,=*O.H.G.* ubarhladan; *overload*; **overláden** (*pple.*) A. R. 368; overlade LIDG. TH. 4557.

ofer-lǣden, v., *O.E.* oferlǣdan; *lead over; domineer*; ovirleden '*opprimo*' PR. P. 373; overlede LANGL. *B* iii. 314; **overlēde** (*pr. subj.*) PALL. x. 101.

ofer-lǣven, v., *O.E.* oferlǣfan; *leave over*: 3e shulen overleeven ['*transmittetis*'] WICL. LEV. xxv. 46.

ofer-lang(e), adj., *over-long*; overlange PR. C. 7274; overlonge O. & N. 450.

ofer-láte, adj., *over-late*; overlate PR. C. 1988; PERC. 1956.

ofer-hlēap, sb.,=*M.L.G.* overlōp; *omission, skip*; overlop [overlepe] M. H. 32.

ofer-hlēapen, v., *O.E.* oferhlēapan,=*M.L.G.* overlōpen; *outrun, catch*; overlepe H. S. 2916; **overlēapeð** (*pres.*) A. R. 380; **overlēep** (*pret.*) LANGL. *B* PROL. 150 (Wr. 299).

ofer-leggen,v.,=*M.H.G.* uberlegen; *overlay*: þei shulen overleie us wiþ stoonus ['*lapidibus nos obruent*'] WICL. EXOD. viii. 26; **overlaid** (*pple.*) D. ARTH. 3253.

over-leiing, sb., *pressure, trouble*, WICL. LK. xxi. 25.

ofer-lenden, v., *pass beyond*; ovirlende ALEX. (Sk.) 5069.

ofer-lifa, sb., *O.E.* oferlȳfa; *? excess*: þe oferlifa on hete (*r.* ete) & on wete HOM. I. 101.

ofer-liggen, v.,=*M.H.G.* uberligen; *overlie*: þeos ilke ehte þe þeos þus **overliggeð** (*pres.*) HOM. I. 53; in hire slep(e) þat o womman her owen child overlai (*pret.*) S. C. D. J. 35.

ofer-litil, adj., *too little*; ovirlitil PR. P. 373.

ofer-livien, v., = *O.H.G.* ubarleben, *O.Fris.* urlibba; *outlive, survive*; **overliveþ** (*pres.*) WICL. EXOD. xxi. 22; **overlevede** (*pret.*) TREV. V. 305.

ofer-lōken, v., *look over, peruse*; overlōked (*pple.*) CH. D. BL. 232.

ofer-macchen, v., *overmatch*; overmacche CH. C. T. *E* 1220; **overmacched** (*pple.*) WILL. 1216.

ofer-maistren,v.,*overmaster*; **overmaistrith** (*pres.*) LANGL. *B* iv. 176.

ofer-met, sb., *O.E.* ofermet; *excess*, ORM. 10720.

ofer-mēte, adj., *O.E.* ofermǣte ; *excessive,* FRAG. 7 ; A. R. 296.

ofer-mōd, sb., *O.E.* ofermōd,=*O.H.G.* ubermuot ; *pride*: for his overmode (*?ms.* overmod, *printed* overmoð) HOM. I. 9.

ofer-mōd², adj., *O.E.* ofermōd ; *proud*: ich was to overmod ANGL. III. 279 ; þe over-mōde (*pl.*) MISC. 81.

ofer-mōdi, adj., *O.E.* ofermōdig, = *O.H.G.* ubermuotīg ; *proud* ; overmodi HOM. I. 5.

   overmōdinesse, sb., *O.E.* ofermōdigness ; *pride,* HOM. I. 19.

ofer-mōre, adv., *besides, moreover* ; overmore LANGL. *C* ix. 35.

ofer-muche, adj., *overmuch* ; overmoche SHOR. 35 ; overmiche CH. BOET. iii. 7 (79).

ofer-müchel, adj., *overmuch* ; overmuchel A. R. 178 ; overmikel PR. C. 6662 ; ovirmikil '*nimius*' PR. P. 373.

ofer-nimen, v., *O.E.* oferniman ; *apprehend* ; overnome (*pple.*) P. 1.

ofer-nōn, sb., *afternoon* ; overnon ROB. (W.) 7302, 7487.

ofer-plentē, sb., *excess* ; overplente LANGL. *C* xiii. 234, *B* xiv. 73.

ofer-prüte, sb., *pride* ; overprute (*sec. text* orgul prude) REL. I. 180 (MISC. 120).

ofer-rǣchen, v., *overreach, cheat* ; overreche LANGL. *B* xiii. 374 ; overrought (*pret.*) MAN. (F.) 1562.

ofer-rēden, v., *read over* ; overrede S. A. L. 81.

ofer-rennen, v.,=*M.H.G.* uberrennen ; *over-run* ; orrennus (*pres.*) ANT. ARTH. xxi.

ofer-rīden, v., *O.E.* oferrīdan ; *ride over* ; override WILL. 4147 ; ourraid (*pret.*) BARB. ix. 513 ; overriden (*pple.*) AR. & MER. 9101 ; CH. C. T. *A* 2022.

ofer-rinnen, v.,=*M.H.G.* uberrinnen ; *run over* ; overeorninde (*pple.*) met H. M. 19.

ofer-scāwian, v., *survey* ; overscawian HOM. I. 117.

ofer-schēoten, v., *overshoot* ; **oversheet** (*pres. subj.*) ALEX. (Sk.) 1767*.

ofer-seilien, v., *sail across* ; oursaile BARB. iii. 686.

ofer-sēmen, v., *O.E.* ofersȳman ; *overload* ; oversēmd (*pple.*) HOM. II. 65.

ofer-sēn, v., *O.E.* ofersēon, = *O.H.G.* uber-sehan ; *oversee, superintend* ; oversen LANGL. *A* vi. 115 ; LUD. COV. 197 ; oversihð (*pres.*) P. L. S. viii. 38 ; oversah (*pret.*) LAȝ. 28023 ; and hi bih(e)old and overseȝ O. & N. 30 ; overseȝen LAȝ. 1946 ; oversein (*pple.*) GOW. III. 304.

ofer-setten, v., *O.E.* ofersettan ; *upset, con-quer* ; ovirsettin PR. P. 373 ; **oversette** (*pret.*) TREV. V. 199 ; þat londfolc hem oversette mid felefelde pine HOM. II. 51.

ofer-sitten, v., *O.E.* ofersittan ; *gain pos-session of*: we m(a)ȝen oversitten þis lond LAȝ. 8035.

ofer-skippen, v., *skip over, omit (in reading)* ; overskippin PR. P. 373 ; **overskipte** (*pret.*) CH. D. BL. 1208 ; overskipped (*pple.*) LANGL. *B* xi. 298.

   over-skipper, sb., *one who omits in reading* ; overskippers (*pl.*) LANGL. *B* xi. 302, *C* xiv. 123.

ofer-sleaht, sb., *top of a doorway* ; oversleaht HOM. I. 87.

ofer-sleie, sb., *O.E.* oferslege ; *top of a door-way,* '*(super)limen*,' FRAG. 4 ; ovirslai '*super-liminare*' PR. P. 374.

ofer-slope, sb., *O.E.* oferslop, = *O.N.* yfir-sloppr ; *over-slop, over-jacket* ; overslope, -sloppe CH. C. T. *G* 633.

ofer-sprǣden, v., *O.E.* ofersprǣdan ; *over-spread* ; oversprǣden LAȝ. 14188 ; **over-sprēdde** (*pret.*) A. R. 54 ; overspradde LANGL. *B* xix. 201 ; CH. C. T. *A* 2871.

ofer-steden, v. ; **overstad** (*pple.*) *over-whelmed,* MAN. (F.) 12770.

ofer-stīȝen, v., *O.E.* oferstīgan,=*Goth.* ufar-steigan, *O.H.G.* ubarstīgan ; *surmount* ; over-stihen JUL. 22 ; **oferstīeð** (*pres.*) R. S. iii. oferstāh (*pret.*) HOM. I. 225.

ofer-streccan, v., *overstretch* ; overstreit (*pple.*) MAN. (F.) 13270 ; was ner overstreit (*pple.*) MAN. (F.) 13270.

ofer-strīden, v.,=*M.L.G.* overstrīden ; *stride over*: overstrīt (*pres.*) þe cnolles HOM. II. 111 ; ovirestride (*subj.*) ALEX. (Sk.) 5477.

ofer-strong, adj., *excessively strong,* A. R. 294.

ofer-swīfen, v., *conquer,* ORM. 1848.

ofer-swīþe, adv., *O.E.* oferswīðe,=*M.L.G.* overswinde ; *far too much* ; overswiþe O. & N. 1518 ; overswuðe A. R. 178.

ofer-tāken, v., = *O.N.* yfirtaka ; *overtake* ; overtaken LANGL. *B* xvii. 82 ; overtake IW. 665 ; ourta BARB. iii. 97, viii. 190 ; **over-tōc** (*pret.*) GEN. & EX. 1756 ; overtok HAV. 1816 ; ROB. 64 ; A. P. ii. 1213 ; overtoken A. R. 244* ; ourtáne (*pple.*) *condemned* BARB. xix. 55.

ofer-tēon, v., *O.E.* ofertēon,=*M.H.G.* über-zichan ; *overspread*: þanne ic ofertēo (*pres.*) hefenes mid w(o)lcne HOM. I. 225.

ofer-tilden, v., *O.E.* oferteldan ; *furnish with an awning* ; overtild (*pple.*) JUL. 9.

ofer-treden, v.,=*M.L.G.* overtreden ; *tread upon* ; ofertrad (*pret.*) ORM. 12493.

ofer-trēowien, v., *O.E.* ofertrūwian ; *trust presumptuously* ; **overtrēoweþ** (*? for* or-trēoweþ) (*pres.*) godes milce HOM. I. 21 ; overtrowinge (*pple.*) WICL. 1 COR. iv. 4.

ofer-trūst, sb., *presumption* ; untrust and overtrust ['*presumptionem*'] A. R. 332.

ofer-trūsten, v., *be presumptuous* ; overtrusten A. R. 332.

ofer-trūsti, adj., *presumptious* ; þe overtrusti A. R. 334.

ofer-tülten, v., *upset, tilt over* ; **overtilte,** overtulte (*pres.*) LANGL. *C* xxiii. 54, 135.

ofer-tumblen, v., *upset* ; **ourtummillit** (*pret. pl.*) BARB. xvi. 643.

ofer-turnen, v., *overturn* ; overturne LANGL. (Wr.) 11065 ; **overturneð** (*pres.*) A. R. 356 ; **overturnde** (*pret.*) LEB. JES. 857 ; overterned MAN. (F.) 4627.

over-turning, sb., *overturning,* LANGL. *C* xix. 164.

ofer-þōȝt, adj., *anxious* : alle he weren overðogt GEN. & EX. 2219.

ofer-þrāwen, v., *overthrow* ; overþrowe LANGL. *B* viii. 36 ; hie overþrēwe (*pret.*) SHOR. 124 ; overþrōwen (*pple.*) CH. BOET. i. 4 (21) ; overþrawe AYENB. 15 ; ovērþrowe ALIS. 1113.

ofer-þünchen, v., = ofþünchen ; *grieve, repent* : it him oferþühte (*pret.*) ORM. 19596 ; overþoughte MAN. (F.) 2350.

ofer-þwart, adv., *across* ; overthwart CH. C. T. *A* 1991 ; overthwert *across* MAN. (F.) 860 ; A. P. ii. 316; overþwerte *wrongly* MAN. (F.) 2318 ; ourthwort BARB. viii. 172.

ofer-wacche, adj., *awake too late at night,* DEP. R. iii. 282.

ofer-walten, v., *overflow* ; **overwaltez** (*pres.*) A. P. ii. 370.

ofer-ward, adv., *across* ; overward PALL. iii. 139.

ofer-weȝen, v., = *M.H.G.* überwegen ; *overweigh* : luve overweið (*pres.*) hit A. R. 386.

ofer-wenden, v. ; oferwente (*pret.*) overcame, GEN. & EX. 2285.

ofer-wēne, sb., = *O.H.G.* ubarwāni ; *presumption* : pride & overwene REL. I. 216.

ofer-wēnen, v., *overween* ; **overwēninde** (*pple.*) AYENB. 169.

overwēninge, sb., *overweening,* AYENB. 17 ; overwening MAN. (F.) 6289.

ofer-weorpen, v., *O.E.* oferweorpan, = *O.H. G.* ubarwerfan ; *overthrow, subvert* : þet uðen ne stormes hit ne overw(e)orpen (*pres. subj.*) A. R. 142 ; oferwarp (*pret.*) ORM. 15807*.

ofer-werc, sb., *superstructure,* ORM. 1035.

ofer-whelmin, v., *overwhelm* ; overwhelmin, ovirqwelmin PR. P. 374 ; overwhelme R. R. 3775 ; D. ARTH. 3261 ; overqwelmis (*pres.*) ALEX. (Sk.) 560.

ofer-whelve, v., *turn over, overwhelm* ; overwhelve PALL. i. 161, 781 ; **overwhelveþ** (*pres.*) þe see CH. BOET. ii. 3 (39).

ofer-winnen, v., *O.E.* oferwinnan, = *O.H.G.* ubarwinnan ; *vanquish* ; **overwonnen** (*pret.*) PS. cviii. 3.

ofer-wrēon, v., *O.E.* oferwrēon (*pres.* ofer-

wrīht ZEIT. XXI. 25) ; *clothe* ; overwrie PALL. i. 143 ; overwrīȝeþ (*pres.*) C. L. 716.

of-fal, sb., *O.N.* affall, = *M.Du.* ʻafval ; *offal,* PR. P. 362 ; L. C. C. 29.

offe, *see* of.

of-fehten, v., *fight out* : þei were weri offouȝten (*pple.*) (*exhausted with fighting*) JOS. 552.

offenden, v., *O.Fr.* offendre ; *offend* ; **offended** (*pple.*) CH. C. T. *A* 909.

offense, sb., *offence,* QUAIR 38 ; offence, offens D. TROY 9700, 13911.

offensioun, sb., *offence, damage,* CH. C. T. *A* 2416 ; offencioun WICL. 2 COR. vi. 3.

of-fēren, v., *frighten off,* A. R. 230 ; offere SHOR. 129 ; of-, **affeare** (*pres. subj.*) JUL. 12, 13 ; offæred (*pple.*) SAX. CHR. 259 ; offæred, -fered LAȝ. 5270, 15491 ; offered FRAG. 7 ; offeared KATH. 669 ; H. M. 43 ; offerd GEN. & EX. 2844 ; FL. & BL. 475.

offertoire, sb., *O.Fr.* offertoire ; *offertory,* CH. C. T. *A* 710.

office, sb., *O.Fr.* office ; *office,* AYENB. 122 ; offiz GEN. & EX. 2071.

officer, sb., *O.Fr.* officier ; *officer,* P. L. S. xxvi. 73 ; **officeres** (*pl.*) CH. C. T. *E* 190.

official, sb., *O.Fr.* official ; *official* ; **officials** (*pl.*) P. S. 332 ; AYENB. 37.

offrande, offrende, sb., *O.Fr.* offrende ; *offering,* GEN. & EX. 1298, 1312 ; offrende HAV. 1386 ; offrandis (*pl.*) HAMP. PS. l. 20.

offrien, v., *O.E.* offrian, *from Lat.* offerre ; *offer,* HOM. I. 87 ; offren A. R. 152 ; ORM. 1011 ; offirs (*pres.*) ALEX. (Sk.) 3658 ; **offrede** (*pret.*) AYENB. 193 ; offreden LAȝ. 8093.

offringe, sb., *O.E.* offrung ; *offering,* AYENB. 194 ; **offirings** (*pl.*) ALEX. (Sk.) 164.

of-füllen, v., *fill up* ; **offülled** (*pple.*) LAȝ. 20438.

of-fürhten, v., *frighten* ; offrihte (*pret.*) LAȝ. 30267 * ; offrüht (*pple.*) KATH. 669 ; offruiht MISC. 54 ; offrigt REL. I. 226 ; GEN. & EX. 3692 ; ofright MAN. (F.) 14570 ; offirhte MAT. viii. 26.

of-gān, v., *obtain, deserve, require* ; of gon P. L. S. xviii. 21 ; vor to ofgon þine heorte A. R. 390 ; ofgon [agon] LANGL. *B* ix. 106 ; ic ofga et þe . . . his blod HOM. I. 117 ; ofgeð A. R. 258 ; ofgōn (*pple.*) A. R. 386 ; ofguo AYENB. 13.

of-gästen, v., *terrify* : so sore hi were offgäste (*pple.*) P. L. S. xiii. 212.

of-ȝiten, v., *perceive,* LAȝ. 25777.

of-grámien, v., *irritate* : hie ben offgrámede (*pple.*) wið hem selfen HOM. II. 69.

of-grīsen, v., *horrify* ; ofgrisen (*pple.*) HOM. II. 135, 173.

of-hæten, v., *heat* : he wes swiðe ofhæt (*pple.*) LAȝ. 9314.

of-healden, v., *retain* : ofhealde, -hiealde AYENB. 24, 177.

of-hēren, v., *hear*; ofhĕrde (*pret.*) HORN (L.) 41.

of-hüngren, v., *suffer hunger*; ofhüngred (*pple.*) AM. & AMIL. 1908; ʒif þu ert ofhungred efter þe swete A. R. 376; offingred 404; ofhungret KATH. 1030; offingred LAʒ. 31804; afingred BRD. 19; P. S. 342; afingred LANGL. *B* x. 59, *C* xii. 43.

of-kennen, v., *get to know*; *beget*; ofkende (*pret.*) LAʒ. 1659*; R. S. iv.

of-langen, v., *affect with longing*; oflonged [*O.E.* oflongod] (*pple.*) HOM. II. 183; O. & N. 1587; æfter þe ic wes oflonged LAʒ. 19034; alonged L. H. R. 23; RICH. 3049.

of-liggen, v., *lie exposed*; ofligge (*pres. subj.*) O. & N. 1505; þa weoren weri oflæien (*pple.*) (*exhausted with lying, cf.* of-walked) LAʒ. 19300.

of-lüsted, pple., *O.E.* oflysted; *excited with desire*: þa wes he . . . oflust [alust] after deores flæsce LAʒ. 30554.

of-nimen, v., *O.E.* ofniman; *seize*; ofnom (*pret.*) ROB. 171; ofnomen LAʒ. 26669.

of-rēchen, v., *reach, attain to*; ofreche HORN (L.) 1283; ROB. 285; WILL. 3874; ofrähte (*pret.*) JUL. 57; ofrauʒte BEV. 867.

of-rēden, v., *surpass in counselling*, REL. I. 187.

of-rennen, v., *overtake*; ofærne LAʒ. 13149.

of-rīden, v., *surpass in riding*, REL. I. 187.

of-sáken, v., *O.E.* ofsacan,=*O.N.* afsaka (*excuse*); *deny*: i ne mai hit noʒt ofsake P. L. S. xv. 60; (heo) ofsōke (*pret.*) LEG. 97.

of-scápe, v., *for* escápe; *escape*, ROB. 20; ofscapie, ofskapie ROB. (W.) 459, 582; ofscápede (*pret.*) ROB. (W.) 1196, 2676; ofscápel (*pple.*) ROB. (W.) 8200.

of-schámed, pple., *ashamed*, SHOR. 160; heo was . . . of(s)chamed [ofschomed] O. & N. 934; ofssamed ROB. 342; hie . . . bieð swiðe ofshamede HOM. II. 173.

of-sēchen, v., *O.N.* ofsœkja; *seek out*; ofseche P. L. S. xi. 13; ofsēcheð (*pres.*) A. R. 232; ofsōʒt (*pple.*) ROB. 145; ofsouʒt WILL. 1676.

of-senden, v., *send for*, ALIS. 500; S. S. (Web.) 704; ofsende SHOR. 41; WILL. 5293; (h)is kniʒtes he let ofsende ROB. 122; ofsende (*pres. subj.*) LAʒ. 15748; ofsent (*pple.*) LANGL. *A* ii. 37.

of-sēon, v., *O.E.* ofsēon; *catch sight of*: lest eni segges . . . hem ofse schuld WILL. 2223; ofseʒ (*pret.*) FER. 3739; ofsei REL. II. 272.

of-serven, v., *for* deserven; *merit*; ofservie MISC. 75; ofserved (*pple.*) A. R. 238.

of-setten, v., *O.E.* ofsettan; *set around*; ofsette (*pple.*) WILL. 2648.

of-sitten, v., *O.E.* ofsittan; *besiege, set round*; ofsit (*pres.*) FRAG. 1; ofsitte (*subj.*) HOM. I. 115.

of-slēan, v., *O.E.* ofslēan,=*Goth.* afslahan (ἀποκτείνειν); *slay, kill*, LAʒ. 685; ofslōh (*pret.*) LAʒ. 2559; JUL. 63; ofslaʒen, -sclawen (ofslawe) (*pple.*) LAʒ. 554, 27127; ofslagen GEN. & EX. 4077; ofslahe, -slawe O. & N. 1611.

of-spring, sb., *O.E.* ofspring; *offspring*, P. L. S. viii. 105; ORM. 11034; BRD. 3; C. L. 1484; ospringe TREV. I. 203, 285; oxspring C. M. 11415.

of-springen, v., *spring*: of wam we beoþ ofspronge (*pple.*) LAʒ. 26418.

of-sprüng, sb., *offspring*, A. R. 54; MISC. 38; ST. COD. 97.

of-stingen, v., *O.E.* ofstingan; *cut through*; ofstinge (*pres.*) LAʒ. 5034; ofstong (*pret.*) 10653; ofstungen (*pple.*) 11441.

of-strengðen, v., *fortify*; ofstrengþede (*pret.*) ROB. (W.) 2968.

of-swinken, v., *procure by labour*: þat we miʒte ofswinke oure mete ROB. 40; ofswonk (*pret.*) L. H. R. 26.

oft, adv., *O.E.* oft,=*O.L.G.* oft, *O.H.G.* ofto, *Goth.* ufta, *O.N.* opt; *oft, often*, P. L. S. viii. 11; oft, ofte HOM. I. 109, 147; O. & N. 539, 1545; LANGL. *A* ii. 16; ofte LAʒ. 622; MARH. 15; A. R. 68; ORM. 9016; REL. I. 214 (MISC. 8); HAV. 226; BEK. 15; LIDG. M. P. 6; ofte [often] hadde ben CH. C. T. *A* 310, *B* 278; ofture (*compar.*) A. R. 284; ofter HOM. I. 43; WILL. 610; LANGL. *B* xviii. 378; MAND. 19; oftest (*superl.*) KATH. 114.

of-táken, v., *overtake*: Cadwalan him after wende ah oftaken he hine ne mahte LAʒ. 31330; oftake BEK. 53; AYENB. 43; WILL. 2198; OCTOV. (W.) 1625; oftōk (*pret.*) HORN (R.) 1241; oftook RICH. 4367; oftoken A. R. 244; oftáke (*pple.*) E. G. 355.

often, *see* oft.

of-tēonen, v., *vex, irritate*; oftēoned (*pple.*) O. & N. 254; oftiened AYENB. 66.

ofte-sīðen, adv., *oftentimes, frequently*, A. R. 418; oftesiþe P. L. S. x. 17; HOCCL. vi. 51; oftsithe PR. C. 7460; oftesīþes CH. C. T. *A* 485; ofteziþes AYENB. 45.

ofte-tīme, adv., *oft-times*, CH. C. T. *A* 52; TRIAM. 28.

of-treden, v., *O.E.* oftredan; *tread down*, ORM. 11650.

of-þrüsmen, v., *O.E.* ofþrysman; *suffocate*; ofþresmeð (*pres.*) MK. iv. 19.

of-þunchen, v., *O.E.* ofþyncan,=*M.Du.* afdunken; *displease; repent*; ofþunche HOM. II. 179; ofþinche P. L. S. viii. 103; HORN (L.) 106; lest it shulde oþenkin him ['*ne poeniteret eum*'] WICL. EX. xiii. 17; ofþinke HORN (L.) 972; aþinke (*for* afþinke) TREV. IV. 461; ofþuncheð (*pres.*) JUL. 17; ofþuncheð, -þincheð [aþincheþ] LAʒ. 3364, 13577; ofþencheð ANGL. I. 7; ofþincheþ MISC. 58; sore ofþinkeþ it me ROB. 54;

aþinkeþ LANGL. *B* xviii. 89; aþinkeþ, -þenkeþ CH. C. T. *A* 3170; ofþünche (*subj.*) A. R. 118; ofþuhte (*pret.*) HOM. I. 157; O. & N. 397; ofþuhte [ofþohte] LA3. 20903; ofþo3te ROB. 171.

of-þünchung, SD., *displeasure*, A. R. 110; ofþunchinge KATH. 1703.

of-þürst, pple., *O.E.* ofþyrsted; *athirst*, A. R. 240; HOM. II. 199; MISC. 151; afurst (*for* afþurst) LANGL. *B* x. 59, *C* xii. 43; aferst HALLIW. 29; þe beggeres beoþ ofþurste HORN (L.) 1120 [afurste (R.) 1120].

of-wáken, v., *awake* (*intr.*); he ofwōk (*pret.*) AR. & MER. 3810.

of-walked, pple., *exhausted with walking*: whan þou art weri ofwalked [*v. r.* forwalked] LANGL. (Wr.) 8434 (*B* xiii. 204).

of-wundren, v., *O.E.* ofwundrian; *amaze*; ofw(u)ndred (*pple.*) SAX. CHR. 261.

ō3, *see* āh. ō3en, *see* ā3en. ō3t, *see* āwiht.

ōh, *see* āh. ōht, *see* āht, āwiht.

oignement, sb., *O.Fr.* oignement; *ointment*, P. L. S. xxiii. 124; oinement AYENB. 93; WILL. 136; HOCCL. ii. 42; CH. C. T. *A* 631; oinementis (*pl.*) WICL. MK. xvi. 1.

oile, *see* éle. oinoun, *see* oniōn.

ointen, v., *anoint*, ANGL. I. 318; WICL. MK. xvi. 1; *see* enointin.

ointing, sb., *anointing*, HAMP. PS. *page* 515.

oistre, *see* ostire. ōk, *see* ōc, āk.

ōken, *see* áken.

ōker, sb., *O.N.* ōkr,=*O.E.* wōcor, *O.Fris.* wōker, *Goth.* wōkrs, *O.H.G.* wuocher; *usury*, A. R. 202; P. R. L. P. 236; MIRC 379; APOL. 111; okir '*fenus*' CATH. 259; PS. xiv. 5; okir, okur H. S. 2395, 2457; feih þet he 3iveð ... to okere A. R. 326.

ōkeren, v.,=*Swed.* ockra; *increase by usury*; ōkered (*pres.*) A. R. 326; okerin H. S. 2456.

ōkering, sb., *usury*, H. S. 5944; HAMP. PS. lxxi. 14.

ōkerere, sb.,=*Swed.* ockrare; *usurer*, H. S. 2563; okerer M. H. 142; okirere HAMP. PS. cviii. 10; ōkerars (*pl.*) TOWNL. 313.

óld, óldlī, *see* eald.

olfend, sb., *O.E.* olfend, = *O.H.G.* olbenta, *Goth.* ulbandus; *camel*, FRAG. 3; olfentes (*gen.*) hær ORM. 3208; olvente (*dat.*) HOM. II. 127; olvende (*gen. pl.*) MAT. iii. 4.

ólie, *see* éle.

oliet, sb., *O.Fr.* oeillet, oillet; *eyelet*, PR. P. 363; oiletis (*pl.*) WICL. EXOD. xxvi. 4.

olifant, sb., *O.Fr.* olifant; *elephant*, IW. 257; olifont AYENB. 84; elifans 224; elifaunt, olifaunt PR. P. 138; of olifantes (*gen.*) bane (*elephant's bon* ivory) LA3. 23778; eliphauntes teeth BARTH. xiv. 33; olifauns (*pl.*) ALIS. 854; olifuntz CH. BOET. iii. 8 (80).

olive, sb., *O.Fr.* olive; *olive*, HOM. II. 89; PS. cxxvii. 3.

olm, adj., *O.N.* olmr; *cruel, fierce*: Giwes weren... olme and of luþere chere S. A. L. 152.

ōlühnen, v., *from O.E.* ōlyht (*flattery*, BL. HOM. 99); *flatter*, A. R. 248; ūleð (? *for* olehð) (*pres.*) KATH. 1496; H. M. 3; ōlhnede (*pret.*) JUL. 53.

ōlühnunge [ōlhtninge], sb., *flattery*, A. R. 192; olhnunge JUL. 12; olhnung [olhtnunge] KATH. 1502.

omell, *see* mell.

omnipotent, adj., *O.Fr.* omnipotent; *omnipotent*, CH. C. T. *D* 423.

on, *see* an. on-, *see* an-, un-, and-.

ōn, *see* ān. ond-, *see* and-.

onde, onden, *see* ande, anden.

onder, *see* under. ōnen, *see* ānen.

onerable, adj., *burdensome*, TREV. II. 143.

onerous, adj., *O.Fr.* onereux; *burdensome*, R. R. 5636.

ōni, *see* æni.

onicle, sb., *O.Fr.* onicle; *onyx*, SPEC. 25.

oniōn, sb., *O.Fr.* oignon; *onion*, PR. P. 365; oingnun, oinon VOC. (W. W.) 555, 572; oniōns [oiniouns] (*pl.*) CH. C. T. *A* 634.

oof, *see* ōwef. op, *see* up.

ópen, adj., *O.E.* open,=*O.L.G.* opan, *O.N.* opinn, *O.H.G.* ofan; *open*, A. R. 60; ORM. 732; ope O. & N. 168; ópene (*pl.*) KATH. 1132; AYENB. 257.

ópe(n)-hēde, adj., *bareheaded*, ROB. (W.) 6967; open heved [openhede] 10938.

ópen-līc, adj., *O.E.* openlīc; *conspicuous*, M. H. 29; openlike bisne ORM. 2909; openlíche (*adv.*) *openly* A. R. 86; C. L. 556; SHOR. 62; openlike, openli3 ORM. 281, 2316; opeliche A. D. 263; ópenlüker (*compar.*) A. R. 8.

openien, v., *O.E.* openian,=*O.L.G.* opanōn, *O.N.* opna, *O.H.G.* ofanōn; *open*; *explain*; (*ms.* oppenien) LA3. 19486; openen A. R. 70; opnen (*ms.* oppnenn) ORM. 4114; ase þe eie openeð A. R. 94; openede (*pret.*) P. L. S. x. 50; AYENB. 96; opnede GEN. & EX. 3773; opnede [opened] LANGL. *B* xviii. 247; openeden LA3. 5773; opened (*pple.*) KATH. 2455; *comp.* 3e-openen.

openunge, sb., *O.E.* openung; *opening*, A. R. 60; openinge PR. P. 368.

operaciōn, sb., *O.Fr.* operation; *operation*, CH. C. T. *D* 1148; GOW. III. 128.

opie, sb., *O.Fr.* opie, *from Lat.* opium; *opium*, CH. C. T. *A* 1472.

opinioun, sb., *O.Fr.* opinion; *opinion*, CH. C. T. *A* 183; GOW. I. 267; WICL. MAT. iv. 24; opiniōns (*pl.*) AYENB. 69.

oppe, *see* uppe.

oppen, oppere, *see* hoppen, hoppere.

op(p)ortūne, adj., *O.Fr.* opportun; *seasonable*, LIDG. TH. PROL. 149.

opportunitē, sb., *O.Fr.* oportunité ; *opportunity*, WICL. MAT. xxvi. 16.

oppōsen, v., *O.Fr.* oposer ; *oppose, contradict in argument*, CH. C. T. *D* 1597 ; oppose GOW. I. 49 ; oposin, aposen PR. P. 13 ; aposen LANGL. *A* iii. 5 (appose *B* iii. 5, apose *C* iv. 5).

opposite, adj., *O.Fr.* opposite ; *opposite*, CH. C. T. *A* 1894.

oppositiōn, sb., *O.Fr.* opposicion ; *opposition*, CH. C. T. *F* 1057.

oppresse, v., *O.Fr.* opresser ; *interfere with, suppress*, CH. FORTUNE 60.

   opressing, sb., *oppressing*, ALEX. (Sk.) 5336.

oppressioun, sb., *O.Fr.* oppression ; *oppression*, CH. L. G. W. 1868.

or, *see* āuþer, ēower, ofer.

[**or-**, prefix, *O.E.* or-, = *O.Fris.*,*M.Du.*or-, *O.N.* or-, ur-, *O.L.G.* ur-, *O.H.G.* ur-, *Goth.* us- ; *out of, from, destitute of.*]

ōr, sb., *O.E.* ōr ; *beginning* : buten ore & ende P. L. S. viii. 91 ; MISC. 64 ; are HOM. I. 171.

ōr², sb., *O.E.* ōr ; *ore, metal*, A. R. 284* ; oor TREV. II. 79 ; **ōre** [oore] (*dat.*) CH. C. T. *D* 1064 ; *comp.* silver-ōre.

ōr, *see* ǣr.

oracle, sb., *Fr.* óracle ; *oracle* ; **oracles** (*pl.*) CH. H. F. 11.

orage, sb., *A.F.* orache, *Fr.* arroches ; *orach* ; C. B. 5.

oratorie, sb., *O.Fr.* oratoire ; *oratory*, S. A. L. 227 ; oratori ALEX. (Sk.) 1651 ; (he) doon make an auter and an oratorie (*closet for devotion*) CH. C. T. *A* 1905.

oratour, sb., *O.Fr.* orateur ; *orator*, CH. BOET. iv. 4 (129).

Orcaneiȝe, pr. n., *Orkney*, LAȝ. 9556.

orchard, sb., *O.E.* orceard, = *ortceard, ortgeard, *Goth.* aurtigards ; *orchard*, ' (*hor*)*tus,*' FRAG. 4 ; A. R. 378 ; FL. & BL. 271 ; ROB. 104 ; TRIST. 2058 ; orchærd LAȝ. 12955.

   orchard-weard, sb., *gardener*, ' *ortolanus,*' FRAG. 4.

or-cost, adj., *O.N.* orkostr ; *wealth* : ȝif þer is orcost oðer eni ahte KATH. 1724.

ord, sb., *O.E.* ord *n.*, *cf.* *O.L.G.*, *O.Fris.* ord *m.*, *O.H.G.* ort *n.*, *O.N.* oddr *m.* ; *point,* ' *mucro,*' FRAG. 4 ; P. L. S. viii. 43 ; LAȝ. 8596 ; CHR. E. 174 ; AR. & MER. 1177 ; ord & ende ORM. 6775 ; FL. & BL. 47 ; C. M. 7770 ; heo is ord & ende of alle uvele HOM. I. 103 ; od MAN. (F.) 4614 ; **orde** (*dat.*) A. R. 212 ; mid speres orde O. & N. 1068 ; HORN (R.) 1389 ; orde (*pl.*) (*ms.* ordes, *r. w.* worde) LAȝ. 20658 ; mid stelene orden LAȝ. 8703.

   ord-frume, sb., *O.E.* ordfruma ; *beginning*, MARH. 8.

or-dāl, sb., *O.E.* ordāl, = *O.Fris.* ordēl, *O.H.G.* urteil ; *ordeal, judgment*, CH. TRO. iii. 1046.

ordenaunce, sb., *O.Fr.* ordenance ; *ordinance, provision*, C. M. 11292 ; A. P. ii. 698 ; CH. P. F. 390 ; ordinance AYENB. 124 ; PR. C. 8438.

ordenē, ordinē, adj., *O.Fr.* ordené ; *set in order, well-ordered*, CH. BOET. iii. 12 (102) ; ordine AYENB. 153.

   ordenē-līche, adv., *orderly*, AYENB. 125.

ordénen, v., *O.Fr.* ordener ; *ordain, appoint* ; ordeinin PR. P. 368 ; ordeine ROB. 139 ; ordeini AYENB. 94 ; ordane [ordaine] ALEX. (Sk.) 3176 ; **ordans** [ordains] (*pres.*) 3160, 3408 ; (þu) ordandis YORK xlvii. 87 ; ordeigned (*pret.*) LANGL. *B* PROL. 119 ; ordand ALEX. (Sk.) 52 ; ordaint A. P. ii. 237 ; ordand (*pple.*) ALEX. (Sk.) 3680, 3787.

   ordáning, sb., *intent*, BARB. xix. 26.

ordenour, sb., *O.Fr.* ordeneur ; *ordainer, arranger*, CH. TRO. iii ; ordeinour BEK. 211.

ordinat, adj., *Lat.* ordinātus ; *well ordered, regular* : a life blisful and ordinaat CH. C. T. *E* 1284.

ordre, sb., *O.Fr.* ordre ; *order ; religious orders* ; A. R. 10 ; MISC. 57 ; ROB. 440 ; AYENB. 48 ; ordere ALEX. (Sk.) 27 ; he had so long in ordere ibe ROB. (W.) 2315 ; an house of monekes to holde hor ordre bet 5720 ; bi ordre *in order* CH. L. G. W. 2514.

ordren, v., *order* ; **ordreþ** (*pres.*) SHOR. 47 ; iordret (*pple.*) HOM. I. 261.

ordūre, sb., *O.Fr.* ordure ; *ordure*, CH. BOET. i. 6 (29) ; CH. C. T. *I* 428 ; A. P. ii. 1092.

ōre, *see* āre.

oreisoun, sb., *O.Fr.* oreisun ; *orison, prayer*, MARG. 294 ; oreisun A. R. 16 ; P. L. S. xxxv. 165 ; oresun FL. & BL. 579 ; **orisōns** (*pl.*) CH. C. T. *B* 596.

orenge, sb., *Fr.* orenge ; *orange* : as orenge & oþer frit A. P. ii. 1044 ; oronge PR. P. 371.

Orewelle, pr. n., *Orwell*, CH. C. T. *A* 277.

orf, sb., *O.E.* orf ; (*inheritance*), *cattle*, LAȝ. 15316 ; GEN. & EX. 795 ; ROB. 6 ; P. S. 342 ; GOW. I. 17 ; þat orf, oref HOM. II. 35, 39 ; **oreve**, orve (*dat.*) O. & N. 1157.

   orf-cwalm, sb., *O.E.* orfcwealm ; *cattle plague*, SAX. CHR. 259 ; orfqvalm HOM. II. 61.

orfeverie, sb., *O.Fr.* orfevrie, orfaverie ; *goldsmith's work*, QUAIR 48.

orfreis, sb., *O.Fr.* orfrois ; *orfrey, gold fringe*, ALIS. 179.

organe, sb., *O.E.* organ, organa, orgel, *O.Fr.* *organe *pl.* orgenes, *from Lat.* organum ; *organ, instrument of music*, PS. cl. 4 ; orgone ' *organum* ' PR. P. 369 ; her vois was murier than the murie orgon . . . in the church CH. C. T. *B* 4041 ; aungelles with instruments of **organes** (*pl.*) and pipes A. P. ii. 1081 ; organes MAN. (F.) 11266 ; we hangiden up oure orguns *harps* WICL. PS. cxxxvi. 2 ; orgins HAMP. PS. cl. 4 ; whil the orgues maden

melodie CH. C. T. *G* 134 ; orgles, timbres, al
maner gleo ALIS. 191 ; on the orgons plaide
the porpos REL. I. 81.

organer, sb., *O.Fr.* organeor; *organist,* PR.
P. 369.

organister, sb., *O,Fr.* organiste ; *organist,*
'*orgonista*' PR. P. 369.

orgeilūs, adj., *O.Fr.* orguillus, orgoillus ;
proud, MISC. 30 ; orgulous ALIS. 2006.

orguil, orgel, sb., *O.Fr.,* orgoil, orguel ;
pride, HOM. II. 43, 63 ; orʒel (orhel) A. R.
224* ; orhel MARH. 11.

　　orʒhel-mōd, sb., *pride,* ORM. 6262.

　　orgul-prŭde, sb., *pride,* REL. I. 180 (MISC.
121) ; orgelpride GEN. & EX. 3767.

oriel, sb., *O.Fr.* oriol, *med.Lat.* oriolum ; *oriel,*
boudoir, '*cancellus,*' PR. P. 369 ; oriall SQ. L.
DEG. 93 ; E. T. 307.

orient, sb., *O.Fr.* orient, *from Lat.* orientem ;
east, CH. C. T. *B* 3504 ; ALEX. (Sk.) 94, 1111,
3079.

orient, adj., *eastern,* ALEX. (Sk.) 5269.

oriental, adj., *O.Fr.* oriental; *oriental,* CH.
L. G. W. 221 ; **orientales** (*sb. pl.*) *sapphires*
LANGL. *B* ii. 14.

origine, sb., *O.Fr.* origine ; *race,* ALEX. (Sk.)
91.

orloge, orologe, *see* **horologe.**

or-mēte, adj., *O.E.* ormǣte ; *immense,* SAX.
CHR. 263 ; ORM. 238.

or-mōd, adj., *O.E.* ormōd,=*O.H.G.* urmuot ;
despondent : þet we on unilimpan to ormode
ne beon HOM. I. 105.

orne, adj., *? O.E.* orne ; *? anxious,* SHOR. 80 ;
was **ornure** (*comp.*) of mete & of drunche
þen þe twei oc̃re A. R. 370 ; *comp.* **un-orne.**

ornement, sb., *O.Fr.* ornement, aornement ;
ornament ; ournement WICL. GEN. ii. 1 ; **or-
nementes** (*pl.*) A. P. ii. 1799 ; ornamentes
[aorne-, aournementes] CH. C. T. *E* 258.

ornen, v., *O.N.* orna (*heat, grow hot*), *cf. Swed.*
orna ; *warm, enrage ; droop, sink, shrink* :
if Elinus . . . ournes (*pres.*) for ferde D.
TROY 2540 ; ournand, ournond (*pple.*) D.
TROY 2203, 13399 ; ournit (*pret.*) 6404 ; we
have ournit (*pple.*) him with angur D. TROY
4857.

　　ourning, sb., *shrinking, terror,* D. TROY
4767, 12711.

ornen², v., *Fr.* orner ; *adorn* ; ourne WICL.
GEN. xxiv. 47 ; ourneden (*pret.*) WICL. I
PET. iii. 5 ; ourned (*pple.*) WICL. APOC. xxi. 19.

　　ourning, sb., *adorning,* WICL. I PET. iii. 3.

or-ped, adj., *O.E.* orped (*full grown*) ;
brave, ['*strenuus*'] TREV. V. 231 ; AYENB.
183 ; AR. & MER. 2166 ; GOW. I. ʁ29 ; orpud
'*audax*' PR. P. 371.

　　orped-līche,adv. ['*strenue*'], *bravely,* TREV.
V. 231 ; AR. & MER. 1729 ; orpedli GAW.
2232 ; A. P. ii. 623.

orped-schipe, sb., *bravery,* ALIS. 1413.

orphelin, sb., *Fr.* orphelin, orphenin ; *orphan,* CH. BOET. ii. 3 (37).

orpiment, sb., *O.Fr.* orpiment; *orpiment,* CH.
C. T. *G* 759.

orpin, sb., *Fr.* orpin ; *orpine,* PR. P. 371.

or-rāþ, adj., *without counsel, perplexed,* ORM.
3150.

　　orrāþnesse, sb., *perplexity,* ORM. 3145.

or-reste, sb., *? O.N.* orrosta, orrasta ; *combat,
conflict,* SAX. CHR. 233 ; orrest ORM. 12539.

orrible, *see* **horrible.**

or-sorʒ, adj., *O.E.* orsorg,=*O.H.G.* ursorg ;
free from care ; '*securus,*' orseoruh FRAG. 4.

ort, sb., *for* *or-ǣt; *cf. O.Fris.* ort, *M.
Du.* oorete, ooraete ; *leavings of a meal* ;
ortus (*pl.*) PR. P. 371.

or-trowe, adj., *O.E.* ortrȳwe,=*O.H.G.* ur-
tri(u)wi ; *diffident,* ORM. 11589.

or-trowian, v., *O.E.* ortrūwian ; *suspect,*
HOM. I. 113 ; ortrowec̃ (*pres.*) A. R. 382 ;
ortrowed (*pret.*) ALIS. (Sk.) 738 ; ortro-
weden ['*suspicabantur*'] WICL. JUDG. viii.
11.

or-trowþe, sb.,=*O.H.G.* urtri(u)wida (*sus-
picio*) ; *diffidence,* ORM. 3145.

or-trūwe, ortrowe, sb., *O.E.* ortrēow ;
diffidence, HOM. II. 43, 73 ; in ortrou [ortrewe]
(*suspicion*) þou art more ROB. (W.) 7021.

[**ōs**, sb., *O.E.* ōs, *O.N.* āss (*god*).]

Ōsbern, pr. n., *O.E.* Ōsbeorn ; BEK. 501.

ōsel, sb., *O.E.* ōsle,=*O.H.G.* amsala ; *ousel*
(*bird*), VOC. 164 ; L. C. C. 36 ; osul TREV. I.
237 ; osulle B. B. 144.

ōsen, v., *? cf. M.Du.* oosen (*haurire, effun-
dere*), *?* = **wōsen** ; *ooze* ; ose ['*scatere*'] PALL.
ix. 116.

osiere, sb., *O.Fr.* osier ; *osier,* PR. P. 371.

ospringe, *see* **ofspringe.**

ossen, v., *cf. early mod.Eng.* osse (*omen*) ;
show ; **osses** [ossus] (*pres.*) ALEX. (Sk.) 2263 ;
he **ossed** (*pret.*) him bi unninges A. P. iii. 213 ;
what & hase þu ossed to Alexander ALEX.
(Sk.) 2307.

　　ossinge, sb., *showing, prophecy,* ALEX.
(Sk.) 868 ; **ossingis** (*pl.*) 732.

ost, *see* **host.**

ostire, sb., *O.E.* ostre, *from Lat.* ostrea ;
oyster, VOC. 189 ; oistre CH. C. T. *A* 182 ;
oestres (*pl.*) TREV. II. 181.

ostrice, sb., *O.Fr.* ostrusce (*autruche*), *Span.*
avestruz ; *ostrich,* A. R. 132* ; ostriche VOC.
220 ; WICL. LEV. xi. 16 ; **ostrigis** (*pl.*) WICL.
JOB xxx. 29.

ōte, *see* **āte.**

oter, sb., *O.E.* oter, otor, otr,=*O.N.* otr, *O.
H.G.* ottar, oter, ottir ; *otter,* VOC. 251; MISC.
70 ; BRD. 30 ; otir LIDG. M. P. 158.

ōð, conj., *O.E.* ōð; *until*; oð þat ic þe segge MAT. ii. 13; of (=oð) se claþ drige beon LEECHD. III. 90; þer abide of all(e) his ʒeferen were ʒegadered HOM. I. 231; a þe (*? for* oþþe) HOM. I. 5.

ōþ, *see* āþ, wōd.   ōþam, *see* āþum.

ōðer, adj. & adv., *O.E.* ōðer,=*O.Fris.* ōther, *O.L.G.* othar, *Goth.* anþar, *O.H.G.* ander, *O.N.* annarr; *other*; þe oðer LAʒ. 22758; A. R. 14; þe oðer (*second*) heste HOM. I. 11; þa corn þer an oðer sorʒe LAʒ. 31807; oþer ALIS. 51; þat oþer ʒer O. & N. 101; an oþer þing 583; an oþer (*ms.* a noþer) C. L. 626; þis oþer dai BRD. 25; til þe oþer dai HAV. 1755; oðer hwat *something else* A. R. 98; ʒet was oþer what for whi þeʒ wæren drihten laþe ORM. 9729; oþer what we mote do P. L. S. xxi. 137; ah hit ilomp an oþer LAʒ. 14028; al hit iwarð on oðer [iwarþ oþer] 21005; ich þe wulle an oþer segge O. & N. 903; Floriz þencheþ al on oþer FL. & BL. 32; (H)avelok þouthe al an oþer HAV. 1395; þo þis holi man iseʒ þat hit non oþer nolde beo P. L. S. xvii. 511; he wolde al seggen oþer (*otherwise*) viii. 75; ah al heo þohten oðer LAʒ. 5429; al oðer hit itidde 27898; wan it nolde oþer gon ROB. 161; oþer i ne kan REL. I. 102; ōþres (*gen. m.*) O. & N. 1499; he ssel more lovie his oʒene zaule þanne an oþres AYENB. 197; oðres [oþer] mannes fr(e)ond P. L. S. viii. 15; þe oðres earen HOM. I. 251; heore nan ne icnew oðres speche 93; oðres monnes 13; oðer monnes 9; ever ich on halt oðres hond A. R. 252; oðre monnes sunnen 108; oþeres TREV. VIII 187*; ōðren (*dat. m. n.*) MAT. vi. 24; to an oþren AYENB. 175; oðre HOM. I. 29; on oðre stede HOM. II. 89; oðere LAʒ. 12604; anne after oþer *one after another* O. & N. 802; ōðerne (*acc. m.*) A. R. 404; þet eni mon scal wið oðerne misdon HOM. I. 55; oðerne, oðren LAʒ. 3881, 4841; oþerne FER. 995; oþerne, oþren AYENB. 8, 162; ōðre (*pl.*) KATH. 1051; oþre O. & N. 1593; ORM. 692; hwan þe oþre sawen þat HAV. 2416; oðere GEN. & EX. 2199; oþere BEK. 571; ōðre (*gen. pl.*) HOM. I. 9; oþre manne wit AYENB. 136; ōðren (*dat. pl.*) LK. xxiv. 9; LAʒ. 2716; oþeren SHOR. 45; oþre O. & N. 1376.

oðer-līche, adv., *O.E.* ōðerlīce; *otherwise*; ōðerlüker (*compar.*) HOM. II. 97; MISC. 63.

ōþer, *see* āuþer.

ōþer-*h*wīle, adv., *occasionally*, (*ms.* oþerhuil) AYENB. 21, 30, 40; LANGL. *C* vii. 160, *B* PROL. 164; otherwhile ... otherwhile (*conj.*) *sometimes ... sometimes* TREV. I. 71; oþerwhiles LANGL. *C* xvii. 364; other whiles among *at intervals* TREV. VII. 341.

ōþer-weies, adv., *otherwise*, LAʒ. 14029*; otherweies CH. C. T. *E* 1072.

ōther-wīse, adv., *on any other condition*, CH. C. T. *F* 534.

ōþom, *see* āþum.

[oðð̄er, *? for O.E.* oðð̄e, *cf. M.H.G.* oder; *or*, LEECHD. III. 100.]

[ou-, prefix, *O.N.* ū-, = un-.]

ou, *see* ēow.   ouch, *see* nouche.

ōuht, *see* āwiht.   oul, *see* awel.

oule, *see* ūle.

ou-list, adj., *cf. O.N.* ūlyst (*bad appetite*), ūlystugr (*unwilling*); *listless*, '*deses*,' PR. P. 374.

oulist-hēde, sb., *listlessness*, PR. P. 374.

ou-mautin, v., *O.N.* ūmǣtta; *swoon*; (*ms.* owmawtyn) '*sincopiso*' PR. P. 374.

oumpere, *see* noumpere.   ounce, *see* unce.

our, *see* ēower.   oure, *see* ūre.

ournement, *see* ornement.

ournen, *see* ornen.   ous, *see* ūs.

out, *see* ūt.   ōut, *see* āwiht.

outrāge, sb., *O.Fr.* outrage, ultrage; *outrage, excess, conceit*: he dede non outrage in drinking TREV. VII. 367; outragie [outrage] ['*lasciviam*'] VI. 293; ['*insolentiam*'] VII. 339; outtrage IV. 131; P. L. S. xiii. 95; HOCCL. I. 114; utrage HAV. 2837; oultrage GOW. III. 206; outrāges (*pl.*) AYENB. 19.

outrāge-lī, adv., *superfluously, outrageously*, HAMP. PS. xxiv. 3; outragelich ['*insolenter*'] TREV. IV. 205.

outrageous, adj., *O.Fr.* outrageus; *outrageous, excessive*, HOCCL. iv. 14; BARB. vi. 126, viii. 270, xi. 32; (*adv.*) *extremely* vi. 19.

outrageus-li, adv., *excessively*, HAMP. PS. xxv. 7, xxxiv. 8.

outragin, v., *Fr.* outrager; *outrage*, PR. P. 375.

ōuþer(e), *see* āuþer.   ove, *see* of.

ovelēte, *see* oblē.   ovemest, *see* uvemest.

oven, sb., *O.E.* ofen,=*O.Fris.* oven, *O.H.G.* ofan, *O.N.* ofn, ogn, *Goth.* auhns; *oven*, '*for(nax)*,' FRAG. 4; '*furnus*' VOC. 201; '*furnus, fornax, clibanus*' PR. P. 372; HOM. I. 41; MISC. 148; PS. xx. 10; ovin L. C. C. 38; ove TREV. I. 405; ofne (*dat.*) JUL. 38; ORM. 993; O. & N. 292; ovene CH. C. T. *I* 856; ovenes (*pl.*) TREV. I 417.

oven, *see* uven.   over, *see* ofer.

ōver, sb., *O.E.* ōfer,=*M.L.G.* ōver, *M.Du.* oever; *shore*; ōvre [ofre] (*dat.*) LAʒ. 8584; on þe seis ovre HAV. 321; overe MAN. (F.) 4336.

overe, overeste, *see* uvere.

over-hand, -herre, -leþer, -ling, -lippe, -man, *see under* uver.

overte, adj., *O.Fr.* overt; *overt, open*; CH. H. F. 718.

ovese, *see* evese.

ōveste, sb., *O.E.* ōfost, ōfest, ēfest,=*O.L.G.* ōfst(-līco); *haste*: an oveste he wende ... LAʒ. 21493; mid efste MK. vi. 25.

ovet, sb., *O.E.* ofet,=*M.L.G.* *M.Du.* ovet, *O.H.G.* obaz; *fruit*: þet ovet of þine wombe AYENB. 262.

ow, *see* ēow.

ōwef, sb., *O.E.* ōweb, ōwef, -wif; *woof*; oof '*subtegmen*' PR. P. 362; WICL. LEV. xiii. 47.

owel, *see* awel.

ōwen, *see* āȝen. ower, *see* ēower.

oxe, sb., *O.E.* oxa, = *O.Fris.* oxa, *O.N.* oxi, uxi, *O.H.G.* ohso, *Goth.* auhsa; *ox*, '*bos*,' VOC. 177; PR. P. 376; A. R. 32; AYENB. 111; SHOR. 122; oxe tunge *ox tongue (a plant)* VOC. 162; oxen *(acc.)* LK. xiii. 15; oxen *(pl.)* LAȝ. 31814; ROB. 275; ALIS. 760; oxin BARB. x. 381; Oxene vord *Oxford* LAȝ. 26241; Oxne ford CHR. E. 1CO3.

oxe-land, sb., *land workable by one ox*, TREV. II. 97.

oxe-stal, sb., *ox-stall*, CH. C. T. *E* 207; oxestalle (*dat.*) P. S. 257.

ōxien, *see* ăskien.

oxspring, *see* ofspring.

oyas, v., *O.Fr.* oyez, *mod.Eng.* oh yes! *(call for silence)* ; *hear!* YORK xxx. 368.

p.

pā, sb., *O.E.* pāwa, pēa, = *O.N.* pā(-fugl), *from Lat.* pāvō ; *peacock*, TOWNL. 99 ; po P. S. 159; pō (*gen.*) LANGL. *B* xii. 257; poos (*pl.*) WICL. 2 PARAL. ix. 21.

pā-cok, sb., *peacock*, VOC. 189; pa-, po-, pecok LANGL. *B* xii. 241; pokoc AYENB. 258; poocok MAND. 48; po-, pecok CH. C. T. *A* 104.

pē-henne, sb., *pea-hen*, PR. P. 390; pohen [pehen], pohenne LANGL. *B* xii. 240, *C* xv. 175.

pace, *see* pas, passen.

pacche, sb., *patch*, ['*assumentum*'] WICL. MK. ii. 21; pacch(e), patche, pahche '*pittacium*' PR. P. 377.

pacience, sb., *Fr.* patience; *patience*, A. R. 180; AYENB. 167; LANGL. *B* xiv. 192; A. P. iii. 1, 36; *(as name of a plant)* C. B. 69.

pacient, adj., *Fr.* patient; *patient*, TREV. III. 283; CH. C. T. *D* 1984; H. V. 106; *(sb.)* LANGL. *C* xiv. 31; pacientes (*pl.*) *C* x. 178.

pacient-liche, adv., *patiently*, LANGL. *C* xiii. 147.

padde, sb., *O.N.* padda, = *M.L.G.* padde, pedde, *M.Du.* padde, podde; *frog*, LUD. COV. 185; paddis (*pl.*) ['*ranae*'] WICL. EX. viii. 8; pades SAX. CHR. 262; podes GEN. & EX. 2977.

paddoke, sb., *frog or toad*, '*rana*,' REL. I. 8; paddok '*bufo*' PR. P. 376; LUD. COV. 164; paddokes (*pl.*) ALIS. 6126; TOWNL. 325. paddok-stōle, sb., *toad-stool*, CATH. 265.

páen, *see* paien.

pæti, adj., *O.E.* pætig; *cunning*, '*astutus*,' FRAG. 3.

páge, sb., *O.Fr.* pages; *page, boy*, HAV. 1730; P. S. 239; S. S. (Web.) 2445; CH. C. T. *C* 688; MAN. (F.) 6840.

pagine, sb., *O.Fr.* pagene; *page; pageant*; A. R. 286; C. M. 21295; pagent '*pagina*' PR. P. 377; PARTEN. 79; pagens [pagentis] (*pl.*) WICL. JER. xxxvi. 23.

paie, sb., *O.Fr.* paie; *satisfaction, pleasure*; to paie LANGL. *B* v. 556, *C* viii. 189, *C* xiv. 160, xx. 186; A. P. i. 1, 1164.

paiement, sb., *Fr.* paiement; *payment*, CH. C. T. *D* 131; A. P. i. 598.

paien, adj. & sb., *O.Fr.* paien; *pagan* ; *(ms.* payen) HORN (R.) 45; MAN. (F.) 7382; paen ' AYENB. 12; páens (*pl.*) ROB. (W.) 4914, 8283.

paien², v., *O.Fr.* paier; *mod.Eng.* pay; *satisfy, pay*, A. R. 108; paie BEK. 315; AYENB. 39; LANGL. *B* iii. 62; paieþ (*pres.*) LANGL. *C* viii. 277; paide (*pret.*) A. R. 290; paid A. P. i. 1165; paid (*pple.*) GEN. & EX. 2215; ipaid LAȝ. 2340*; ipaied A. R. 124; *comp.* mispaien.

paienie, sb., *O.Fr.* paienie; *pagan country*, FER. 761.

paiere, sb., *O.Fr.* paiere; *payer*, LANGL. *A* vi. 41, *C* viii. 194.

paile, sb., *O.E.* pægel; *pail*, '*multrale, multrum vel multre*,' PR. P. 377.

paiment, *see* pávement.

pain, sb., *Fr.* pain; *bread*, LANGL. *B* ix. 80; paine puffe [*Fr.* pain pouffé] C. B. 61, 68; pain purdeu, purdeuz [*Fr.* pain perdu] C. B. 42, 83; pain fondeu [*Fr.* pain fondu] F. C. 33; pein reguson [*Fr.* pain ragusain] C. B. 112.

paindemayn, sb., *O.Fr.* paindemain, *Church Lat.* pānis dominicus; *finest bread*, CH. C. T. *B* 1915; painmain, paiman VOC. (W.W.) 657, 788; painemain C. B. 8, 11, 12; take a painmain *(cake)* C. B. 90.

painime, sb., *O.Fr.* paenisme; *paganism, country of the pagans*, HORN (L.) 803; ROB. 401; MISC. 28; apparailled as a painim (*pagan*) LANGL. *B* v. 523; painim *B* xi. 157, *C* viii. 161; painimes (*pl.*) WICL. ROM. PROL.

painimeri, sb., *paganism*, CATH. 266.

painten, *see* peinten. pair(e), *see* peir.

pairen, *see* peiren.

pais, sb., *O.Fr.* pais, peis, pēs; *peace*, SAX. CHR. 265; LAȝ. 480*; GEN. & EX. 8; BEK. 404; SHOR. 122; DEGR. 1569; peis A. R. 166; pais [pes] O. & N. 1730; pes HOM. I. 141; MAN. (F.) 1936; MAND. 11; pees CH. C. T. *A* 532; peas PS. xxvii. 3.

paise, *see* peis, peisin.

paisen, v., ? = apaisen; and paisi (*make peace*) wiþ Cesare LAȝ. 8839*; pese D. TROY 3809; peesid (*pple.*) WICL. PS. lxxxii. 2; pesed MAN. (F.) 4570.

paisibilitē, sb.; *? O.Fr.* paisibleté ; *calm* ; pesibilite WICL. LK. viii. 24.

paisible, adj., *O.Fr.* paisible ; *peaceable*, AYENB. 261 ; peisible CH. BOET. i. 5 (23) ; pesible WICL. JOB v. 23, viii. 6 ; pesable MAN. (F.) 4040.

pésiblenesse, pésibilnesse, sb., *calm, calmness*, WICL. MAT. viii. 26 ; MK. iv. 39.

paitrūre, sb., *cf.* peitrel ; *defence for the neck of a horse*, GAW. 168, 601.

pakke, sb., *O.N.* pakki, = *M.Du.* pack ; *pack*, '*sarcina*,' PR. P. 378 ; pakke [pak] LANGL. *B* xiii. 201 ; pak LUD. COV. 137 ; packes (*pl.*) A. R. 166 ; REL. II. 175.

  pake-neelde, sb., *pack-needle* ; [*v. r.* pacneld, pakke nedle] LANGL. *A* v. 126.

pakken, v., *cf. M.Du.* packen ; *pack* : and pakken hem to gideres LANGL. *B* xv. 184 ; pakkin '*sarcino, fardello*' PR. P. 378.

pākoc, *see under* pā.

pal, peal, sb., *O.E.* pæll, pell, = *O.N.* pell, *from Lat.* pallium ; *pallium; pall; a costly sort of cloth* ; LAȝ. 897, 1296 ; REL. I. 119 ; pal [pel] KATH. 1461 ; pel MISC. 97 ; LAI LE FR. 185 ; pelle PR. P. 391 ; pall D. TROY 435 ; palle (*dat.*) R. S. v ; GREG. 193 ; þat wæde . . . al was it of þe betste pall(e) ORM. 8173 ; pælles (*pl.*) LAȝ. 2368.

pál, sb., *O.Fr.* pal, *from Lat.* pālus ; *cf.* pāl ; *pale, stake*, '*palus*,' FRAG. 4 ; '*palus, vallus*' PR. P. 378 ; a litil paal ['*paxillus*'] WICL. EZ. xv. 3 ; pale (*boundary*) D. TROY 13874 ; páls (*pl.*) (*ramparts*) D. TROY 5610.

pāl, sb., *O.E.* pāl, *O.N.* pāll, *from Lat.* pālus ; *cf.* pál ; *pole* ; pole '*contus, pertica*' PR. P. 407 ; TREV. I. 369 ; pōle (*dat.*) AYENB. 203 ; LANGL. *B* xviii. 52.

palais, śb., *O.Fr.* palais ; *palace*, HORN (L.) 1256 ; ROB. 352 ; paleis WILL. 2838 ; PR. P. 378 ; palaice, palais A. P. ii. 1389, 1531.

palat, sb., *O.Fr.* palat ; *palate* : the tonge . . . cleved to his palat [*ear. ver.* palet] WICL. LAM. iv. 4 ; palet '*palatum*' PR. P. 378.

pále, adj., *O.Fr.* pale ; *pale, foaming*, GREG. 732 ; as pale as a pelet LANGL. *A* v. 61 ; pale [paal] *C* xxi. 59 ; pale CH. C. T. *A* 205 ; pale D. TROY 13874.

palefrai, sb., *O.Fr.* palefreid, *from late Lat.* paraveredus ; *palfrey*, HOM. I. 5 ; palefrei HAV. 2060 ; ROB. 490 ; palfrai LANGL. *B* x. 308.

pálen, v., *from* pál ; *fence or enclose with stakes* ; pale MAN. (F.) 1055 ; pálid (*pple.*) ['*vallata*'] WICL. 4 KINGS xxv. 2.

palesie, *see* paralisie.

palet, sb., *? O.Fr.* palet ; *head-piece*, '*pelliris, galerus*' PR. P. 378 ; a prevy pallette . . . to hille here lewde heed DEP. R. iii. 325.

pali, sb., *O.Fr.* paille ; *straw*, '*cantabrum*,' PR. P. 379.

paliet, sb., *Fr.* paillet ; *pallet*, '*lectica*,' PR. P. 379.

palin, v., *deck with palls* ; palit (*pple.*) D. TROY 8385.

palle, *see* pal.

pallen, adj., *O.E.* pællen ; *made of pall* : ænne pallene curtel LAȝ. 23762 ; pallen webis ALEX. (Sk.) 1517.

pallen, v., *beat, strike, knock* : i palle (*pres.*) him doun LANGL. *B* xvi. 30 ; polle *C* xix. 50* ; pallit (*pret.*) thurgh the persans D. TROY 10022 ; and proude doun pallede (*pple.*) JOS. 499.

pallen², v., *O.Fr.* pallir ; *become vapid; lose spirit* ; pallin '*emorior*' PR. P. 379 ; palle GOW. III. 13 ; palle (*pres.*) REL. II. 211 ; palled CH. C. T. *B* 1292.

pallioun, sb., *O.Fr.* pallion ; *pallium*, BEK. 248 ; palliun SAINTS (Ld.) xxvii. 306.

palme, sb., *O.E.* palm (VOC. 32), *O.Fr.* palme, paume ; *palm (of the hand); palmtree* ; '*palma*,' PR. P. 380 ; ase palme oþer ase cipres AYENB. 131 ; palm REL. II. 244 ; paume P. L. S. xvii. 232 ; þe paume is pureli þe hand LANGL. *B* xvii. 141 ; palmen (*pl.*) REL. I. 267 ; paumez *antlers* GAW. 1155 ; palme sun(n)e dai *Palm Sunday* MISC. 39.

  palme-trē, sb., *O.E.* palmtrēo ; *palm-tree*, GEN. & EX. 3305.

  palm-twīg, sb., *O.E.* palmtwīg ; *twig of palm-tree*, FRAG. 3 ; HOM. II. 89.

palmere, sb., *palmer*, HORN (L.) 1027 ; palmeres [palmers] (*pl.*) LANGL. *B* v. 106.

palmi, adj., *made of palms* ; palmi basket PALL. xi. 458.

palpable, adj., *? O.Fr.* palpable ; *palpable, evident*, LANGL. *C* xix. 235.

paltok, sb., *O.Fr.* paltoc ; *a sort of coat*, '*baltheus*,' PR. P. 380 ; LANGL. *B* xviii. 25.

páment, *see* pávement.

pampren, v., *pamper* : to pappe and pampe (*? read* pampre) her fleische REL. I. 41 ; þou pamprest (*pres.*) ANGL. III. 287 ; *comp.* forpampren.

pan, *see* páne.

pancēre, sb., *O.Fr.* panciere ; *coat of mail*, ALEX. (Sk.) 4960.

panche, sb., *O.Fr.* pance ; *paunch, coat of mail*, PARTEN. 5773 ; paunche LANGL. *B* xiii. 87 ; CH. P. F. 610 ; paunces, paunz (*pl.*) MAN. (F.) 13553, 10028.

páne, sb., *O.Fr.* pane, panne, pan, *from Lat.* pannus ; *pane (of glass); patch; piece; garment* ; '*pagina*,' PR. P. 380 ; '*pen(n)ula*' 381 ; MAN. (F.) 12463 ; IW. 204 ; GAW. 154 ; EGL. 858 ; PARTEN. 5654 ; a pane of menuver FL. & BL. 110 ; pane of riche skinne TRIST. 569 ; robes wiþ riche pane WILL. 5356 ; uch pane of þat place A. P. i. 1033 ; a pan þat was broken ĘER. 5188.

panel, sb., *O.Fr.* panel; *panel,* '*pagella, panellus,*' PR. P. 381; IW. 473; C. M. 14982; ne putte men in panell (*jury list*) LANGL. *C* iv. 472.

paneter, panter, sb., *Fr.* panetier; *keeper of the pantry,* LANGL. *C* xvii. 151; paniter panter ROB. 187, 203.

paneterie, sb., *Fr.* paneterie; *pantry*; panetrie A. D. 259; pantrie '*panetorium vel panitria*' PR. P. 382.

pani, *see* pening.

panier, sb., *O.Fr.* panier, *from Lat.* pānārium, *mod.Eng.* panier; *basket,* HAV. 813; TREV. V. 195; ÇH. C. T. *E* 1568.

[**panis,** *mistake for* pavis.]

panne, sb., *O.E.* panne,=*O.Fris.* panne, ponne, *O.N.* panna, *O.H.G.* phanna, *? from Lat.* patina; *pan; skull;* '*patella,*' PR. P. 381; AYENB. 23; CH. C. T. *D* 1614; L. H. R. 150; panne [ponne] LANGL. *A* iv. 64; ponne (*printed* poune) ALIS. 2770; irene ponne '*sartago*' FRAG. 4; erthen panne C. B. 54; *comp.* brain-, hēaved-, herne-, knee-panne.

pan-kake, sb., *pancake,* '*laganum,*' PR. P. 380; pancake C. B. 46.

pans, *see* pening.

panter, pantrie, *see* paneter, paneterie, peinten.

pantere, sb., *O.Fr.* panthere; *panther,* '*pant(h)era,*' PR. P. 381; ALIS. 6820; panter REL. I. 225.

pantere², sb., *O.Fr.* pantiere; *snare for birds,* '*laqveus, pedica,*' PR. P. 381; foules ... that of the pantere and the nette ben escaped CH. L. G. W. 131.

pantin, v., *O.Fr.* pantoier; *pant, take short breaths,* '*anhelo,*' PR. P. 381; pante TOWNL. 217.

pāpe, sb., *O.E.* pāpa, *Lat.* pāpa; *pope,* PR. C. 1886; pape [pope] LA3. 14807; pope O. & N. 746; BEK. 249; LANGL. *B* xv. 485; pōpe (*gen.*) LANGL. *A* ii. 18.

   pāpe-dōm, sb., *popedom,* SAX. CHR. 253.

   pōpe-hōli, adj., *holy as a pope, hypocritical,* LANGL. *B* xiii. 284.

papejai, sb., *O.Fr.* papegai; *popinjay,* SPEC. 26; A. P. ii. 1465; papejai [papenjai, popinjai, popingai] CH. C. T. *E* 2322; popejai [papengai] TREV. IV. 307; popegai MAND. 274.

papelard, sb., *Fr.* papelard; *deceiver,* AYENB. 26; this paperlarde preste GEST. R. 401.

papelote, sb., *mess of porridge*; paplote CATH. 268; papelotes (*pl.*) LANGL. *C* x. 75.

papin, sb., *boiled pudding*; papins (*pl.*) C. B. 9.

papir, sb., *Lat.* papȳrus; *paper,* E. G. 5; paper GOW. II. 8.

   papir-whīt, adj., *white as paper,* CH. L. G. W. 1198.

pappe, sb., *? Lat.* pappa, papa; *pap, breast,*

teat, ORM. 6441; PR. C. 6767; OCTAV. (H.) 442; **pappes** (*pl.*) H. M. 35; A. R. 330*; HAV. 2132.

pappen, v., *? make soft,* KEL. I. 41.

par-, *see* per-.

paradīs, sb., *Fr.* paradis, *from Gk.* παράδεισος; *paradise,* LA3. 24122; SHOR. 157; LANGL. *B* x. 463; parais A. R. 66; in paradise MISC. 50; peradis D. TROY 5496; **paraises** (*gen.*) KATH. (E.) 893.

paragal, sb., *O.Fr.* paragel; *companion*; DEP. R. i. 71.

parāge, sb., *O.Fr.* parage, pairage, *cf. mod. Eng.* peerage; *high birth, lofty dignity,* FL. & BL. (H.) 666; S. S. (Web.) 243; A. P. i. 419; LIDG. M. P. 26.

parail, sb., = **appareil**; *apparel, dress,* LANGL. *C* xiii. 121; paraille, paraile *B* xi. 228, 235*, *C* xi. 116*; ALEX. (Sk.) 4676.

parailen, v.,=**appareilen**; *array, apparel*; **parrails** [*v. r.* apperels] (*pres.*) ALEX. (Sk.) 765; **parailede,** parailed (*pret.*) LANGL. *C* i. 25, iii. 224; parailed (*pple.*) ALEX. (Sk.) 1552; parreld 480, 4208; paraillid 5285.

paralisie, sb., *Fr.* paralysie; *palsy*; parlesi PR. C. 2996; L. H. R. 130; palesie WICL. MAT. iv. 24.

paramour, sb., *? O.Fr.* par amour; *paramour, lover, concubine,* CH. C. T. *D* 454; LANGL. *C* xvii. 107; ALEX. (Sk.) 5222; **paramouris** (*adv.*) *as a lover* BARB. xiii. 485; paramours CH. L. G. W. PROL. *A* 260; **paramours** (*pl.*) LANGL. *C* vii. 186; ALEX. (Sk.) 3769, 4337.

paraunter, adv., *O.Fr.* per aventure; *perchance,* LANGL. *C* viii. 297, ix. 43; perauntre xvii. 50; ROB. (W.) 2018; paraventure *B* xii. 184; peraventure WILL. 234.

parboilin, v., *O.Fr.* parbouillir; *parboil,* '*semibullio, parbullio,*' PR. P. 382; **parboile** (*imper.*) C. B. 6, 100; parboiled (*pple.*) 100.

parc, sb., *O.E.* pearruc, *O.Fr.* parc; *cf. M.H.G.* pferrich; *park,* LA3. 1432*; park P. L. S. i. 5; WILL. 2845; GOW. II. 45; park, parrok PR. P. 384; **parkes** (*pl.*) ROB. I.

parceiven, *see* perceiven.

parcelle, sb., *Fr.* parcelle; *parcel, part,* CH. ASTR. i. 21 (13); parcel LANGL. *B* x. 63; parcele ALEX. (Sk.) 4496; **parcells** (*pl.*) *shares* ALEX. (Sk.) 4318.

   parcel-mēle, adv., *by bits,* LANGL. *C* xx. 28; percel mel *A* iii. 72.

parcenēr, sb., *O.Fr.* parcener, parconier; *sharer, partner* ['*particeps*'] WICL. PROV. xxviii. 24; parciner ROB. (W.) 6309; **parcenēres,** parceneris (*pl.*) WICL. PROV. v. 17; WISD. vii. 4; I COR. ix. 12.

parchementer, *see* perchementer.

parchen, v., *parch*; parche '*frigo*' PR. P. 382.

parchmen, *see* perchemin.

parclōs, sb., *O.Fr.* parclos ; *enclosure, screen* ; parcloos PR. P. 382.

pardōn, sb., *O.Fr.* pardon ; *pardon,* CHR. ENG. 314 ; pardoun LANGL. *B* ii. 222 [pardun *A* ii. 198] ; pardoun CH. C. T. *C* 926.

pardonēre, sb., *O.Fr.* pardonaire ; *pardoner,* LANGL. *B* v. 648 ; pardoner CH. C. T. *A* 669.

párement, sb., *O.Fr.* parement ; *adornment* : chambre of **páramentz** (*pl.*) CH. C. T. *F* 269 ; dauncing chambres ful of parements CH. L. G. W. 1106.

parfit, see **perfit**.

parget, sb., *plaster (of a wall)*, PALL. i. 414.

pargetin, v., *O.Fr.* pargeter ; *parget, plaster,* '*linio,*' PR. P. 384 ; **pargete** (*pres.*) WICL. Ez. xiii. 11.

párin, v., *Fr.* parer ; *pare,* S. & C. I. xlvi ; parie P. L. S. xxiv. 234 ; pare LANGL. *B* v. 243 ; **páred** (*pret.*) LANGL. *C* vii. 242 ; **páred** (*pple.*) *prepared, fitted up* ALEX. (Sk.) 4208.

paritorie, sb., *O.Fr.* paritoire, parietaire ; *parietary, pellitory,* CH. C. T. *G* 581 ; paratori '*colitropium*' VOC. (W. W.) 787.

park, see **parc**.

parkere, sb., *park keeper,* '*indagator,*' PR. P. 384 ; E. W. 8.

parle, v., *O.Fr.* parler ; *speak,* DEP. R. iv. 48 ; **parled** (*pret. pl.*) iv. 88 ; **parled** (*pple.*) LANGL. *B* xviii. 268, *C* xxi. 281.

parlement, sb., *O.Fr.* parlement ; *parliament,* HAV. 1006 ; ROB. 463 ; MAN. (F.) 1726 ; perlament D. TROY 2095 ; parlament YORK xxxii. 33.

parlour, sb., *O.Fr.* parloir ; *parlour, conversation-room,* MAN. (F.) 7066 ; '*colloqvotorium*' CATH. 269 ; parloure LANGL. *B* x. 97 ; **parlūrs** (*gen.*) A. R. 68.

parosche, sb., *O.Fr.* paroche ; *parish,* BEK. 1879 ; paroshe P. S. 157 ; paresche MIRC 17 ; parishe CH. C. T. *A* 491.

parisch-preest, sb., *parish priest,* LANGL. *C* xvi. 211 ; **parissheprestes**, parshepreestes (*pl.*) LANGL. *B* x. 268, *C* xxiii. 280.

paroschian, sb.,*O.F.* parochien; *parishioner,* A. R. 198 ; parishen LANGL. *B* xi. 67 ; **parisschens** (*pl.*) *A* PROL. 79 ; parisshiens *B* xx. 280.

parrai, sb., *O.Fr.* parroie ; *nobility,* ALEX. (Sk.) 4028.

parren, v., *enclose* ; **parred** (*pple.*) HAV. 2439 ; IW. 3228 ; YORK xxxiii. 33.

parroken, v., *cf. Fr.* parquer, *Ger.* pferchen ; *impark, enclose, shut in* ; **parroked** (*pple.*) LANGL. *B* xv. 281.

parson, see **persōne**.

parsonāge, sb., *parsonage, benefice,* LANGL. *B* xiii. 245.

part, sb., *O.Fr.* part ; *part,* FL. & BL. 522 ;

MARG. 308 ; P. S. 338 ; AYENB. 110 ; a part *apart* CH. C. T. *F* 252 ; **pars** (*pl.*) ALIS. 664 ; GREG. 378.

parten, v., *O.Fr.* partir ; *part, participate, share ; separate* ; parte FL. & BL. 387 ; CH. BOET. i. 3 (10) ; **parteþ** (*pres.*) SHOR. 76 ; partis ALEX. (Sk.) 5418 ; (þai) part 1931 ; **parti** (*subj.*) A. R. 406 ; **parted** (*pple.*) MISC. 91 ; HAV. 2962.

partinge, sb., *imparting ; departure* ; DEP. R. i. 171 ; LANGL. *A* xi. 303 ; parting FL. & BL. 684 ; GREG. 480 ; TRIST. 165.

parténen, v., *O.Fr.* (a-)partenir ; *pertain* ; **perténis** (*pres.*) ALEX. (Sk.) 4309 ; (þai) pertines [pertenis] 1772 ; **parténed** (*pret.*) WILL. 1419 ; partenit BARB. xx. 313.

partener [**partiner**], sb.,=**parcener** ; *partner,* TREV. III. 477 ; partiner ROB. 309 ; AYENB. 256.

parti, sb., *Fr.* parti ; *party ; side* : for to fiȝt for our parti C. M. 7470 ; þere þat partie pursueth LANGL. *B* xvii. 302 ; leve þe trewe partie LANGL. *C* ii. 95 ; parti *A* i. 7 ; **parties** (*pl.*) LANGL. *B* xiv. 268.

parti², adj., *different* : þat þi personale proporcion sa parti is to mine ALEX. (Sk.) 668.

particulēr, adj., *Fr.* particulier ; *particular,* CH. C. T. *E* 35.

partie, sb., *O.Fr.* partie ; *part, portion ; passage in a book* ; L. H. R. 48 ; LEG. 3 ; SHOR. 166 ; LANGL. *B* i. 7, *C* xvi. 157 ; E. G. 382 ; more partie *most part* DEP. R. ii. 37 ; a partie *partly* LANGL. *B* xv. 17, *C* xvii. 168 ; **partis**, partise (*pl.*) ALEX. (Sk.) 3764, 3992.

partriche, see **pertriche**.

parvis, sb., *Fr.* parvis ; *parvis, church close,* CH. C. T. *A* 310.

pas, sb., *O.Fr.* pas, *from Lat.* passus ; *pace, passage ; pass, path* ; BEK. 69 ; SHOR. 133 ; CH. C. T. *F* 388 ; paas *A* 825 ; pace A. P. i. 677 ; passe YORK xxx. 116 ; ALEX. (Sk.) 2978 ; þe pas of Altoun LANGL. *C* xvii. 139 ; pace PR. P. 376 ; passe *canto* ALEX. (Sk.) 2845.

paschen, **pasken**, v., *cf. Swed.* paska ; *dash* : pasken in þe water P. L. S. xx. 8 ; and al to duste **pashed** (*pret.*) kinges LANGL. *B* xx. 99 ; passhed YORK xlvi. 38 ; **passchet** (*pple.*) LANGL. *A* v. 16.

paske, sb., *Lat.* pascha, *Hebr.* פֶּסַח ; *Passover,* ORM. 15850 ; LANGL. *B* xvi. 139 ; pasche GEN. & EX. 3157 ; pasc, pass YORK xxvii. 11, 29.

paske-daȝ, sb., *Passover-day,* ORM. 15552.

pasnepe, sb., *parsnip,* '*pastinaca,*' CATH. 270 ; parsnepe PALL. x. 158.

passāge, sb., *O.Fr.* passage ; *passage, journey,* HORN (L.) 1323 ; ARCH. LII. 35 ; MAN. (F.) 14012 ; A. P. iii. 97 ; pasage ALEX. (Sk.) 32.

passagēr, sb., *traveller* ; **passagērs** (*pl.*) MAN. (F.) 16593.

passen, v., *O.Fr.* passer; *pass, surpass, escape,* A. R. 330; passi LA3. 1341*; BEK. 1148; passe HAV. 1376; **passeth** (*pres.*) CH. C. T. *F* 404; **passe** (*imper.*) *B* 1633; **passinge** (*pple.*) LANGL. *C* xxii. 266; passunde A. P. ii. 1389; passande men MAN. (F.) 3297; passing *surpassing* ALEX. (Sk.) 1750; he is a passing (*excellent*) man CH. C. T. *G* 614; for passing of witt ALEX. (Sk.) 45; er i **pace** (*subj.*) (*ere I die*) CH. C. T. *F* 494; **passed** (*pple.*) CH. C. T. *E* 610.

passing-lī, adv., *passingly,* ALEX. (Sk.) 2904; passandli 1999, 3455, 3596.

passiūn, sb., *Fr.* passion; *passion,* HOM. I. 119; KATH. 1163; MISC. 154; passioun AYENB. 142; LANGL. *B* xiii. 90; passion *passion week* ROB. (W.) 11330.

páste, sb., *O.Fr.* paste; *paste, pastry,* LANGL. *B* xiii. 250; C. B. 39; make faire cofins of fine paast 75; past 45; paaste 98.

pastē, sb., *O.Fr.* pasté; *pastry, pie,* CH. C. T. *A* 4346; **pastees** (*pl.*) HAV. 644.

pasteláde, sb., *a dish in cookery,* C. B. 59; pistelade, petelade 62.

pastour, sb., *Lat.* pastor; *shepherd, herdsman;* **pastours** (*pl.*) LANGL. *A* xi. 300, *B* xii. 149, *C* xii. 293.

pastūre, sb., *O.Fr.* pasture; *pasture, food,* MAN. (F.) 7348; c. M. 18445; D. RICH. iii. 14; **pastours** (*pl.*) ALEX. (Sk.) 1198.

pasturen, v., *pasture, feed*; **pasturde** (*pple.*) ALEX. (Sk.) 5425.

páte, sb., *pate, head,* P. L. S. xxiii. 83: P. S. 237; LIDG. M. P. 54.

paten, sb., *O.Fr.* patin; *patten*; (*ms.* patien) '*calopodium*' PR. P. 385.

patene, sb., *O.Fr.* patene; *paten,* '*patena,*' PR. P. 385; pateine SHOR. 53.

patente, sb., *Fr.* patente; *letter patent, open letter; indulgence, pardon*; CH. C. T. *C* 337; patent LANGL. *B* xiv. 191; **patentes** (*pl.*) *B* v. 194.

pateren, v., *patter, chatter:* me thinke he patris (*pres.*) like a pi YORK xxxv. 266.

patriarche, sb., *O.Fr.* patriarche; *patriarch,* HOM. I. 131; patriarke A. R. 154; patriarc ROB. (W.) 9869; **patriarkes** (*pl.*) LANGL. *C* viii. 88.

patrimoigne, sb., *patrimony,* LANGL. *B* xx. 233.

patrōn, sb., *O.Fr.* patron; *patron, pattern,* '*patronus, exemplar,*' PR. P. 386; ROB. 471; patroun P. S. 326; **patrōns** (*pl.*) E. G. 321.

paþ, sb., *O.E.* pæð, = *O.Fris.* path, *O.H.G.* phad; *path,* ALIS. 3219; paþ [path] LANGL. *B* xiv. 300; peþ '*semita*' FRAG. 3; AYENB. 127; peth BARB. xviii. 366; **paþe** (*dat.*) HAV. 2390; **paðes** (*pl.*) LA3. 17330; paþes O. & N. 380; HAV. 268; peðes HOM. II. 129.

paðeren, v., ? = puðeren; *poke about;*

pother; **paðereð** [*v. r.* puðeres] (*pres.*) A. R. 214.

paue, sb., *O.Fr.* poue, poe, *?L.G.* paute, pōte; *paw,* PR. P. 386; poue (*ms.* powe) AR. & MER. 1491; **paues** (*ms.* pawes) (*pl.*) B. DISC. 1997; ISUM. 181; poues (*ms.* powes) RICH. 1082.

paume, see **palme**. **paunce,** see **panche**.

pautenēr, sb., *O.Fr.* pautonier, paltonier; *vagabond,* ALIS. 1737; FER. 859; sa fell, sa pautener (*adj.*) BARB. i. 462, ii. 194.

pautenēr², sb., *purse,* P. S. 327; '*cassidile*' PR. P. 387.

pávement, sb., *O.Fr.* pavement; *pavement,* BEK. 2146; pavement [paviment, pament] CH. C. T. *B* 1867; paviment ROB. (W.) 9791; pament GEST. R. 81; paiment D. TROY 352.

páven, v., *O.Fr.* paver; *pave*; pave CH. C. T. *G* 626; **páved** (*pple.*) P. S. 190; ALEX. (Sk.) 3220.

pavilōn, sb., *O.Fr.* pavillon; *pavilion,* ROB. 272; LANGL. *A* ii. 43; PARTEN. 911; paviloun GREG. 565; ne pelour in hus paveilon (*coif*) LANGL. *C* iv. 452; **pavilōns** (*pl.*) MAN. (F.) 4645.

pavis, sb., *O.Fr.* pavois; *a sort of shield,* REL. II. 22; (*printed* panys) D. TROY 5722; pavice PR. P. 386; **paves** (*pl.*) ALEX. (Sk.) 2223.

pavis², adj.: a pavis pillion hatt D. ARTH. 3460.

pax, sb., *Lat.* pax (*peace*); *crucifix to be kissed*: kisse the pax LIDG. *in* L. F. M. B. 296; þo prest (þo) pax wil kis L F. M. B. 48.

pax-brede, sb., '*osculatorium,*' VOC. (W.W.) 756.

paxwax, -wex, sb., *pax-wax, tendon of the neck,* PR. P. 388.

pē, see **pā**.

peautre, sb., *O.Fr.* peautre, peutre; *pewter,* E. W. 102; peutir PR. P. 395.

péce, sb., *O.Fr.* piece; *piece,* '*pars, crater,*' PR. P. 388; ROB. 555; LANGL. *B* xiv. 48; MAND. 10; IW. 760; pece man ROB. (W.) 7314; piece P. S. 334; **péces** (*pl.*) *cups* LANGL. *B* iii. 89.

péce-mēl, adv., *piece-meal.* ROB. 22.

pechē, sb., *O.Fr.* pechet, *from Lat.* peccātum; *sin, fault*; pechche A. P. i. 841.

[**pēchen,** v., *O.E.* pæcan (*deceive*); *comp.* bi-pēchen.]

pecken, see **picken**. **pēcok(e),** see *under* **pā**.

pecunie, sb., *Lat.* pecūnia; *money,* LANGL. *C* iv. 393.

pecunious, adj., *rich, moneyed,* LANGL. *C* xiii. 11.

pedaille, see **pitaill**.

peddare, sb., *prov.Eng.* (*East Anglian*) pedder; *pedler,* '*calatharius,*' PR. P. 389; peoddare A. R. 66; **pedderis** (*pl.*) WICL. E. W. 12.

pedde, sb., *prov.Eng.* (*East Anglian*) ped ; *basket*, '*idem quod panere*,' '*calathus*,' PR. P. 390.

pedlare, sb., *pedler*, PR. P. 390 ; pedlere LANGL. *B* v. 258.

peel, *see* **péle. peer,** *see* **peir.**

peere, *see* **pére. peering,** *see under* **péren.**

pegge, sb., *peg*, '*cavilla*,' PR. P. 390.

peggen, v., *stuff* (*with food*) : to pegge us as a peni hoge ALEX. (Sk.) 4278.

peine, sb., *O.Fr.* peine, poene, poine, *from Lat.* poena, *Gr.* ποινή ; *pain, penalty* ; BEK. 481 ; SHOR. 38 ; SPEC. 81 ; LANGL. *B* i. 167 ; MAND. 11 ; CH. C. T. *A* 1319 ; ROB. (W.) 7742.

peineble, adj.; *taking pains, careful*, H. S. 5802 ; penible CH. C. T. *E* 714.

peinin, v., *O.Fr.* pener ; *give* or *cause pain, punish ; urge ;* '*crucio*,'. PR. P. 390 ; peine WILL. 2898 ; i peine (*pres.*) me CH. C. T. *C* 330 ; þat peined (*pret.*) him to deþe LANGL. *B* i. 169 (Wr. 800) ; peined (*pple.*) WICL. DEEDS xxii. 5 ; painit D. TROY 10336.

peinten, v., *from O.Fr.* peint, *pple. of* peindre ; *paint*; peinte CH. C. T. *C* 17 ; paint ALEX. (Sk.) 4427 ; peint (*pres.*) BEK. 2151 ; painted (*pret.*) LANGL. *C* xxii. 11 ; painted A. P. i. 750 ; (þai) paintid ALEX. (Sk.) 1704 ; painted (*pple.*) ALEX. (Sk.) 4149 ; ipeint O. & N. 76.

paintour, sb., *painter*, ALEX. (Sk.) 5145 ; panter VOC. (W. W.) 687.

peintunge, sb., *painting*, A. R. 392.

peintūre, sb., *O.Fr.* peinture ; *painting*, A. R. 242 ; WICL. ESTH. i. 6.

peir, peer, sb., *O.Fr.* pair, (*couple*) ; *pair ; peer, equal ;* '*par*,' PR. P. 391, 394 ; per HAV. 2241 ; P. L. S. xvii. 380 ; paire gloves LANGL *C* vii. 251 ; peire FL. & BL. 566 ; paire CH. C. T. *C* 623 ; pere LANGL. *A* iii. 198, xi 194 ; péres (*pl.*) LANGL. *B* vii. 16.

pér-lēs, adj., *peerless*, WILL. 933.

peirement, sb., *damage, destruction*, WICL. MAT. xvi. 26 ; 2 COR. vii. 9.

peiren, v., *make worse, impair* ; peire (? *for* empeire) C. M. 8407 ; paire YORK xxvi. 14, xxxiv. 256 ; PALL. iii. 964 ; peireth (*pres.*) LANGL. *A* iii. 123 ; peired (*pple.*) *B* iii. 127 ; peired MAN. (F.) 8716.

peiring, sb., *destruction*, WICL. LK. ix. 25.

peiren², v., *be peer to, compare to* ; périth (*pres.*) LANGL. *A* xii. 4 ; pere ALEX. (Sk.) 1842 ; peren [peeren] LANGL. xv. 410 ; pere *are like* ALEX. (Sk.) 4703.

peis, sb., *O.Fr.* pois ; *weight*, LANGL. *B* v. 243 ; MAN. (F.) 11283 ; in paise *in the balance* ALEX. (Sk.) 3260.

peis, *see* **pais.**

peisin, v., *O.Fr.* poiser ; *weigh*, '*pondero, libro, trutino*,' PR. P. 390 ; paise ALEX. (Sk.)

4618 ; peisede (*pret.*) LANGL. *A* v. 131 [poised *B* v. 217].

peitrel, sb., *O.Fr.* peitral ; *breastplate of a horse in armour*, CH. C. T. *G* 564.

peitrel, v., *furnish with a* peitrel, LANGL. *C* v. 23.

pekke, sb., *peck* (*measure*), '*batus*,' PR. P. 391 ; CH. C. T. *A* 4010.

pēkok, *see under* **pā. pel,** *see* **pal.**

[**pel,** sb., ? *O.Fr.* pel ; ? *skin*.]

pel-lēk, sb., ? *wild thyme*, '*serpillum*,' VOC. 266.

pēl, sb., *peel, fort* : (he) did mak a pele . . . wrouht of tre fulle welle MAN. (H.) 157 ; peill BARB. x. 137, 152, 193, 207 ; pēlis (*pl.*) x. 147 ; *cf.* **pīle.**

péle, v., *for* **apéle** ; *appeal*, '*appello*,' PR. P. 391 ; N. P. 3 ; pele [peel] LANGL. *B* xvii. 302.

pēle, *see* **pēl.**

pēle, sb., *O.Fr.* pēle ; *baker's shovel ;* '*pala*,' CATH. 273 ; A. S. (H.) 79 ; C. B. 51.

pelegrim, pelgrim, sb., *O.Fr.* pelegrin ; *pilgrim*, P. L. S. xiv. 29, 32 ; pelegrim, pilegrim LAȝ. 30730, 30744 ; pilegrim A. R. 348 ; pilgrim AYENB. 86 ; M. H. 54 ; *see* **peregrin.**

pelet, *see* **pelot.**

pelfir, sb., *O.Fr.* pelfre, *cf. mod.Eng.* pelf, pilfer ; *plunder*, '*spolium*,' PR. P. 391.

pelle, *see* **pal.**

pellen, v., *expel, throw out* : to morwen shall ich forth pelle HAV. 810 ; þe powere . . . pellid (*pret.*) doune his kniȝtis ALEX. (Sk.) 117.

pellican, sb., *Gr.* πελεκάν ; *pelican* ; A. R. 118 ; pellicans (*pl.*) ALEX. (Sk.) 5129.

pellūre, *Fr.* pelure ; *fur*, WILL. 53 ; pelure LANGL. *A* ii. 9 ; pellure LANGL. *B* xv. 7 ; pelloure [pelour] ALEX. (Sk.) 2768.

pelot, pelet, sb., *O.Fr.* pelote, pilote ; *pellet, stone ball*, '*pileus, vel piliolus, rudus*,' PR. P. 391 ; as pale as a pelet LANGL. *B* v. 78 ; swifte as a pellet out of a gonne CH. H. F. 1643 ; pelote GOW. II. 306.

pelrināge, *see* **pilegrimāge.**

pelt, sb., ? *M.H.G.* pelz, pelliz, *from med.Lat.* pellicia ; *pelt*, P. R. L. P. 16.

pelten, *see* **pülten.**

pen, sb., *from* **pennen** ; *pen* (*for sheep, etc.*) ; penez (*pl.*) A. P. ii. 322.

penne-féd, adj., *fed in a pen*, A. P. ii. 57.

penance, sb., *O.Fr.* peneaunce, penance ; *penance, suffering*, P. L. S. x. 26 ; penaunce SHOR. 5 ; LANGL. *C* iv. 101, vi. 84, 196 ; *B* x. 120 ; penaunces (*pl.*) LANGL. *B* xiii. 66 ; *C* i. 27 ; *see* **penitence.**

penaunce-lēs, . adj., *without penance or suffering*, LANGL. *B* x. 462, *C* xii. 296.

penancēr, sb., *O.Fr.* peneancier ; *one who imposes a penance, confessor*, C. M. 26341 ; penauncer LANGL. *B* xx. 317* ; *see* **peni-tancer.**

penant, penaunt, sb., *O.Fr.* penant; *penitent,* GREG. 844, 870; penaunt LANGL. *C* v. 130; penauntes (*pl.*) *C* xvi. 101; *see* penitaunt.

pencel, sb., *O.Fr.* pincel; *pencil, paint-brush,* CH. C. T. *A* 2049.

pencel[2], sb., *O.Fr.* penoncel; *small banner,* ALIS. 2688; pensel LANGL. *C* xix. 189; pensiles (*pl.*) LAƷ. 27183*; penceles MAN. (F.) 12511.

penciōn, sb., *O.Fr.* pension, *mod.Eng.* pension; *payment, reward,* LANGL. *A* viii. 48.

pendaunt, sb., *O.Fr.* pendant; *hanging ornament*: pendande of a belt '*pendulum*' CATH. 274; **pendauntes** (*pl.*) LANGL. *B* xv. 7; pendauntez GAW. 2038, 2431.

pende, penden, *see* pünde, pünden.

penden v.; *for* appenden; *belong to*: þe apparement þat **pented** (*pple.*) to þe kirke A. P. ii. 1270.

peni, *see* pening.

pening, sb., *O.E.* pening, penig,= *O.N.* peningr, *O.Fris.* penning, pennig, panning, pannig, *O.H.G.* phenning, phenting; *penny,* ORM. 3287; peni LAƷ. 31961; R. S. v; BEK. 1174; AYENB. 23; panig MK. xii. 15; pani AR. & MER. 3639; **penie** (*dat.*) P. L. S. viii. 34; **peniƷes, paneƷes, panewes** (*pl.*) LAƷ. 3544, 14684, 29460; penies HAV. 776; MAN. (H.) 146; OCTAV. (H.) 726; penis M. H. 18; pens LANGL. *B* v. 243; S. & C. II. lvii; panes, panewes ROB. (W.) 1392, 1397*; pans BEK. 332; AYENB. 190.

peni-ale, sb., *poor ale (at a penny per gallon),* LANGL. *A* v. 134, *B* xv. 310; B. B. 208.

peni-lēs, adj., *penniless,* DEP. R. iii. 196.

penni-stāne, sb., *flat stone used as a quoit,* BARB. xiii. 581.

peni-worþ, sb., *pennyworth,* LANGL. *B* iii. 256 [peni, *v. r.* peniworth, *A* iii. 243]; **peniworthes** (*pl.*) *goods* LANGL. *C* vii. 384.

penitancēr, sb., *O.Fr.* penitencier; *one who imposes a penance, confessor*; penitancer LANGL. *C* xxiii. 319; penitancere *B* xx. 317; **penitauncērs** (*pl.*) LANGL. *C* vii. 256; *see* penancēr.

penitaunt, sb., *O.Fr.* penitent; *penitent,* LANGL. *C* v. 130*; *see* penant.

penitence, sb., *O.Fr.* penitence; *penitence, penance,* A. R. 8; HOM. II. 61; *see* penance.

?penitote, sb., *for* *peritote; *O.Fr.* peridon *from med.Lat.* peritot, pelidor, periodus (DU C.); *peridot, chrysolite*; **penitotes** (*pl.*) A. P. ii. 1472.

penne, sb., *O.Fr.* penne; *pen,* LEG. 85; P. S. 156; TREV. IV. 215; MAN. (F.) 11885; A, P. ii., 1724; þe **pennes** (*feathers*) (*pl.*) of þe pecok LANGL. *B* xii. 247; pennes of an aungell ALEX. (Sk.) 4529.

pennen, v., *O.E.* (on-)pennian; *pen (an animal), impound, shut in*: **penned** (*pple.*) HOM. II. 43; *comp.* bi-pennen.

pennēre, sb., *pencase,* CH. C. T. *E* 1879; pennare '*pennarium, calamarium*' PR. P. 392.

penōn, sb., *O.Fr.* penon; *pennon,* '*bandum, pennum,*' PR. P. 392; penoun CH. C. T. *A* 978; **penounes** (*pl.*) PL. CR. 562; pennons ALEX. (Sk.) 3028.

pensel, *see* pencel.

pensif, adj., *O.Fr.* pensif; *thoughtful,* A. P. i. 246; LANGL. *C* x. 299; GOW. II. 65; (*adv.*) *A* viii. 133.

penta(u)ngel, sb., *figure of five points,* GAW. 620, 636, 664.

Pentecoste, sb., *O.Fr.* pentecoste; *Pentecost,* SAX. CHR. 251; HOM. I. 89.

pentice, sb., *O.Fr.* apentis; *penthouse,* '*appendicium,*' PR. P. 392; pentis VOC. 236.

Peohtes [Peutes], pr. n., pl.,= *O.E.* Peohtas; *Picts,* LAƷ. 9922; Peihtes MAN. (F.) 5697.

peolien, *see* pilien.

people, sb., *O.Fr.* pueple, peuple; *people,* OCTAV. (H.) 854; people [poeple, peple] CH. C. T. *E* 995; BOET. i. 4 (15); puple SHOR. 53; WILL. 499; peple, poeple LANGL. *A* i. 5 (*B* i. 5); peuple *C* xii. 21; peple MAND. 3.

peose, *see* pése.

peper, sb., *O.E.* pipor, *from Lat.* piper; *pepper,* LANGL. *A* v. 155; MAND. 168; L. C. C. 19; piper PALL. vi. 168.

peper-corn, sb., *peppercorn*; pepirqwerne '*fractillum*' PR. P. 393; **pepercornes** (*pl.*) ALEX. (Sk.) 2025.

pepin, v., *O.Fr.* pepin, pupin, poupin; *pippin, kernel,* C. M. 8504; pepin, pipin PR. P. 401; pepin [popin] WICL. NUM. vi. 4; C. B. 32.

péple, *see* people.

[per-, par-, *prefix, O.Fr.* par-, per-; *through.*]

pér, *see* peir.

perantre, peraventure, *see* parauntre.

per-brēken, v., *break through*; perebrake [perbrake] MAN. (F.) 7950.

perceiven, v., *O.Fr.* percevoir; *perceive*; persawe BARB. i. 82; parceiveþ (*imper.*) WICL. JUDG. v. 3; perceived (*pret. pl.*) MAN. (F.) 12320; **per-, parceived** (*pple.*) LANGL. *B* v. 143.

persáving, sb., *perception,* BARB. iv. 358.

percen, perchen, perisshen, sb., *O.Fr.* percer, percher, ?pertuissier; *pierce*; percen CH. BOET. iii. 7 (81); persen LANGL. *C* xii. 295; perche CATH. 276; to perische wiþ dart WICL. E. W. 348; thirlith or **perissheth** (*pres.*) GEST. R. 47; persith DEP. R. iii. 11; (thei) persen WICL. 2 TIM. iii. 6; **pershaunt** (*pple.*) LANGL. *C* ii. 257; the king . . **perisshed** (*pret.*) the harnes GENER. 3367; persched is (*r.* his) scheld FER. 941; percede ROB. 17; D. ARTH. 2075; parsed [perside] TREV. VIII. 85; **ipersshed** (*pple.*)

LANGL. *B* xvii. 189; his sherte that was pershed in .v. places KN. L. 143.

perche, sb., *O.Fr.* perche; *perch (fish),* '*perca,*' VOC. 189; '*percha, parcha*' PR. P. 393; REL. I. 85; C. B. 102.

perche ², sb., *O.Fr.* perche; *perch, rod, pole* : in his hond (*? ms.* hong) he bar a long perche, his staf as the3 hit were P. L. S. xv. 80; in his hond a long perche he bar SAINTS (Ld.) xl. 78; chauntecleer sat on his perche CH. C. T. *B* 4074; '*pertica*' PR. P. 393.

perchemin, sb., *O.Fr.* parchemin; *parchment,* LANGL. *B* xiv. 191; parchemin LEG. 85; P. S. 156; parchmen A. P. ii. 1134; perchement VOC. 210.

 parchementer, sb., *preparer of parchment,* CATH. 269.

pére, sb., *O.E.* peru,= *O.N.* pera, *M.Du.* pere, *O.H.G.* pira, bira *f.; from Lat.* pirum *n.; pear,* VOC. 192; '*pirum*' PR. P. 394; AYENB. 208; FER. 5722; MAND. 245; peere DEP. R. PROL. 73; **péren** (*dat. pl.*) P. L. S. xxiii. 89.

 pere-jonette, sb., *early pear*; **perejoñettes** (*pl.*) LANGL. *C* xiii. 221.

 per-trē, sb., *pear-tree,* VOC. 227; peretree LANGL. *A* v. 16; '*pirus*' PR. P. 394.

pére ², sb., *O.Fr.* perre, pierre (*stone*); *stone, pier,* '*pila,*' PR. P. 394; oppon ech pere þar stent a tour FER. 1684; **piers** (*pl.*) ALEX. (Sk.) 4356.

pére, *see* peir.

perē, sb., *O.Fr.* perē; *perry*; perre ENG. ST. VII. 105; pereie SHOR. 8; VOC. (W. W.) 603; perre '*piretum*' PR. P. 394; pirrei, pirre CATH. 281.

peregrin, adj., *O.Fr.* peregrin; *foreign*; faucon peregrin CH. C. T. *F* 428; *see* pelegrim.

péren, v., *O.Fr.* pareir (*paroir*); *appear*; LANGL. *B* PROL. 173; *see* apéren.

 peering, sb., *appearing,* LANGL. *C* xxii. 92*.

perfeccion, sb., *O.Fr.* perfection; *perfection,* AYENB. 79; perfeccioun CH. BOET. ii. 4 (42); perfecciun (*? printed* perfectiun) A. R. 372.

perfit, adj., *O.Fr.* parfez; *perfect*; parfit AYENB. 185; perfit CH. BOET. i. 2 (8); perfight, parfight, parfit, perfit CH. C. T. *A* 72, 338, 422, 532; **parfiter** (*comp.*) LANGL. *B* xii. 25; **parfitest** (*superl.*) LANGL. *C* xiv. 99.

 parfit-līche, adv., *perfectly,* LANGL. *C* xvi. 180; perfitliche *B* xvi. 220; perfitli, parfitli WICL. AMOS v. 10, LK. i. 45; **perfitliest** (*superl.*) HAMP. PS. cv. 24.

 parfitnesse, sb., *perfection,* LANGL. *B* x. 200, xvi. 184; (*perfect life*) *C* vi. 90.

performie, v., *O.Fr.* parfournir; *perform,* FER. 2894; performe, -fourni CH. BOET. i. 4 (18), iii. 2 (67); **parfourned** (*pret.*) LANGL. *B* v. 607 [performede *A* vi. 88].

peril, sb., *O.Fr.* peril; *peril,* A. R. 194; AYENB. 16; MAN. (F.) 4840, 10285; perill [perle] ALEX. (Sk.) 1783.

perilous, adj., *O.F.* perilleus; *perilous,* AYENB. 22; CH. C. T. *B* 1209; perlouse D. TROY 564; peralus ALEX. (Sk.) 530.

 perilos-lī, adv., *dangerously,* LANGL. *C* i. 170.

perimancie, *see* piromancie.

perishe, v., *O.Fr.* periss-, perir; *perish,* C. M. 8789; **perissiþ** (*pres.*) CH. BOET. iii. 11 (96); **perissi** (*subj.*) MISC. 33.

perisshen, *see* percen.

perken, v., *perk* : þe popejaies **perken** (*pres.*) & pruinen fol proude ANGL. I. 95.

perle, sb., *O.Fr.* perle, *cf. M.Du., O.H.G.* pèrla, perala; *pearl,* '*margarita, glaucoma,*' PR. P. 394; AYENB. 158; A. P. i. 1; **perlis,** peerlis (*pl.*) LANGL. *B* x. 9, *A* xi. 12.

permutaciōn, sb., *exchange,* LANGL. *A* iii. 243, *C* iv. 316.

permūte, v., *exchange,* LANGL. *B* xiii. 110; **permūten** (*pres.*) *C* iii. 185.

pernen, v., ? = proinen; *preen* : papjaiez painted **perning** (*pple.*) GAW. 611.

perpetuel, adj., *O.Fr.* perpetuel; *perpetual,* LANGL. *B* xvii. 126; perpetuall ALEX. (Sk.) 3312.

perplexitē, sb., *O.Fr.* perplexité; *danger,* BARB. xi. 619; perplexitee GOW. III. 348.

perquer, adv., *O.Fr.* per quer, per cuer; *by heart,* BARB. i. 238.

perrē, perree, sb., *? O.Fr.* perrée; *precious stones,* LANGL. *B* x. 12; CH. C. T. *D* 344; pirre ALEX. (Sk.) 3954, 4036; **? perreies** (*pl.*) (*? ms.* perrieris) D. TROY 1670.

perreie, sb., *Fr.* porrée; *stew,* C. B. 32; perre F. C. 39; porre of peson L. C. C. 44.

pers, adj., *a blue colour* : in sanguin and in pers CH. C. T. *A* 439.

persawe, *see* perceiven.

persecucioun, sb., *O.Fr.* persecucion; *persecution,* TREV. V. 111; persecucioune BARB. iv. 5.

perseverance, sb., *O.Fr.* perseverance; *perseverance,* AYENB. 168.

perseveren, v., *O.Fr.* perseverer; *persevere*; persevere CH. C. T. *D* 148.

persil, sb., *Fr.* persil, *cf. O.E.* petersilie; *parsley*; percil [*v. r.* percile, percelle, persoli] LANGL. *B* vi. 288 [persil *A* vii. 273]; perseli L. C. C. 18; parcelli, perceli C. B. 5, 6, 7, 81, 100.

persōne, sb., *O.Fr.* persone; *person; parson;* '*persona,*' PR. P. 395; A. R. 316; AYENB. 12; persone [persoun] CH. C. T. *A* 478; persoun P. L. S. ix. 139; persun A. R. 126; **parsōnes** (*pl.*) LANGL. *B* x. 268.

personele, adj., *personal,* ALEX. (Sk.) 5142; personale 668.

pert, adj., *? for* apert; *pert, plain, clever, bold,* LAUNF. 294; wis and pert GEN. & EX. 3292; proud and pert CH. C. T. *A* 3950; stout he was and pert B. DISC. 123; pert in plai K. T. 18; a pert man D. TROY 1462; pert (*adv.*) ALEX. (Sk.) 2917; so pert *quite publicly* 2295; perte DEP. R. iv. 88.

    pert-lī, adv., *quickly, boldly, pertly,* TRIAM. 746; D. TROY 1033; perteliche WILL. 53; LANGL. *B* v. 15; perteli to hem he rad TOR. 1504; YORK xxix. 136.

    pertnes, sb., *prettiness,* D. TROY 9205.

pertēnen, *see* partēnen.

pertriche, sb., *O.Fr.* pertris, perdris, perdriz (GACHET, GLOSS.); *partridge, 'perdix,'* PR. P. 395; partriche DEP. R. iii. 38; **pertriches** (*pl.*) PL. CR. 764.

perturben, v., *O.Fr.* perturber; *perturb,* CH. C. T. *A* 906.

pervenke, sb., *O. E.* perfince, *O.Fr.* pervenche; *periwinkle,* FRAG. 3; P. S. 218.

perverten, v., *O.Fr.* pervertir; *pervert*; perverted (*pple.*) CH. BOET. ii. 1 (30).

pés, *see* pais.

pesche, sb., *O.Fr.* pesche, peske, = *Ital.* pesca, persica, *cf. O.E.* persoc; *peach*; peche, peshe, peske *'pomum Persicum'* PR. P. 388, 395.

pése, sb., *O.E.* piose, *Lat.* pīsum; *pea, 'pisa,'* PR. P. 395; ALIS. 5959; MAND. 158; PALL. x. 64; pees LANGL. *B* vi. 171; peose A. D. 243; pise FER. 5847; **pésen** (*pl.*) AYENB. 120; LANGL. *B* vi. 198; RICH. 6004.

    pés-codde, sb., *peascod, 'siliqua,'* PR. P. 395; **pésecoddes** (*pl.*) LANGL. *A* vii. 279 [pesecoddes *B* vi. 294].

    pése-lōf, **peese-lōf**, sb., *loaf made of pea-meal,* LANGL. *B* vi. 181, *C* ix. 176.

pestel, sb., *O.Fr.* pestel; *pestle, 'pila,'* PR. P. 395; GAM. 122; WICL. EXOD. xvi. 14.

pestilence, sb., *O.Fr.* pestilence; *pestilence, 'pestilencia,'* PR. P. 395; CH. C. T. *B* 4600; pestilence time LANGL. *C* xi. 272; **pestilences** (*pl.*) LANGL. *A* v. 13.

pet, *see* püt.

petegreu, sb., *pedigree,* APPEND. ROB. 585; pedegru, petigru, *'stemma'* PR. P. 390.

Peter, pr. n., *Peter*; LANGL. *B* v. 544; **Petres bourh** *Peterborough* SPEC. 88.

petit, adj., *Fr.* petit; *petty, small*: poverte nis but a petit þing LANGL. *B* xiv. 242; peti *C* xvii. 84; peti fet *small stalks* PALL. iii. 902; pedi feet iv. 375.

pettaile, *see* pitaile. **peuple**, *see* people.

peþ, *see* paþ. **peuter**, *see* peautre.

pevard, sb., *O.Fr.* poivrade; *a sauce,* C. B. 71.

pēvisch, adj., *peevish*; peivesshe [pevische] schrewe LANGL. *C* ix. 151.

philosophe, **phisike**, *see* filosofe, fisike.

Phippe, sb., = Philip; *name for a sparrow,* LANGL. xi. 41.

piane, *see* pione. **pibbel**, *see* pobble.

pic, *see* pich, picke.

pīc, sb., *O.E.* pīc, *cf.* Swed. pīk *m.*, Gael. pīc, *M.Du.* pijke *f.*; *pike, spike, peak*: þe pic LAȜ. 30752; þene pic 30857; pik MAN. (F.) 15835; *pike* (*fish*) CH. C. T. *E* 1419; pike LANGL. *B* v. 482; pike *'cuspis, lucius'* PR. P. 396; VOC. 189; S. S. (Web.) 1253; GREG. 845; ISUM. 497; **pīkes** (*pl.*) FRAG. 8; KATH. 1946; MISC. 149; ROB. 51; FER. 4647; as ful as an illes pil is of pikes P. L. S. xviii. 47.

    pīk-staf, sb., *O.N.* pīkstafr; *pike-staff,* LANGL. *B* vi. 105.

Picardes, [**Picars**], sb., pl., *Picts,* ROB. (W.) 46, 956, 1700.

[**picche-**, **pic-**, *stem of* picchen.]

    pic-forke, sb., *pitchfork,* LAȜ. 21597; pik-forke *'merga'* PR. P. 397.

    pitche-longes, adv., *headlong,* PALL. vi. 42.

picchen, v., *cf. M.Du.* picken, *O.N.* pikka; *fix* (*a stake*) *in the earth; pitch* (*a tent*); *pierce or divide with a sharp point; set* (*jewels*), set with *jewels*; picche TREV. I. 387; WICL. NUM. ii. 3; picche a two þe rotes LANGL. *B* vi. 105; and piccheþ (*pres.*) first his boþe P. 45; picchinde (*pple.*) stake SPEC. 110; þer he pihte (*pret.*) his stæf LAȜ. 29653; piȝte TREV. VII. 349; stakes of irn . . . he piȝte in Temese gronde ROB. 51; and wiþ hir bek hir selven so sche pighte CH. C. T. *F* 418; piȝte 1627; A. P. i. 228; a spere þat is pight into þe erþe MAND. 183; paveluns were piȝte ANT. ARTH. xxxvii; ipiȝt AYENB. 199; P. L. S. ii. 94.

pich, sb., *O.E.* pic, = *O.H.G.* pech, bech, *O.N.* bik *m., from Lat.* pix *f.*; *pitch,* P. L. S. viii. 110; MISC. 149; ROB. 410; ST. COD. 54; S. S. (Web.) 1280; þat pich HOM. I. 251; pick, pik, *'pix, pissa'* PR. P. 396; pic M. H. 111; pik HOM. I. 269; REL. I. 53.

pichen, v., *cover with pitch*; pike HAV. 707; ipiched (*pple.*) BRD. 5.

picher, sb., *O.Fr.* picher; *pitcher,* LEG. 22; E. G. 354; PERC. 454; pecher H. S. 10749; C. B. 39; *see* biker.

picke, sb., *cf. M.Du.* picke, *M.H.G.* bicke; *pickaxe*: pikke *'ligo'* CATH. 278; pik VOC. 234; his pic and his spade AYENB. 108; pike H. S. 941.

picken, v., *O.N.* pikka, = *M.Du.* picken, bicken, *M.H.G.* bicken; *pick; peck; steal; adorn*; A. R. 84*; he did hewe tres and pikke MAN. (F.) 9939; þus pore men her part ai pickez (*pres.*) *choose* A. P. i. 573; pecken [picke] out WICL. PROV. xxx. 17; pikke [pekke] (*imper.*) CH. C. T. *B* 4157; pilours and plodders piked (*pret.*) þere goodes D. TROY 12862; þe pie haþ pecked (*pple.*) you

N. P. 43; þe portalez picked (*adorned*) of rich platez A. P. i. 1036.

pikare, sb., *petty thief, thief, 'furculus,'* PR. P. 395; pikers (*pl.*) LANGL. *C* vi. 17.

pike-herneis, sb., *plunderers of armour,* LANGL. *C* xxiii. 263; pikehernois *B* xx. 261.

picke-purse, sb., *pickpocket,* CH. C. T. *A* 1998; pikeporses (*pl.*) LANGL. vii. 370.

picois, sb., *O.Fr.* picois; *pickaxe,* WICL. I KINGS xiii. 20; pikois LANGL. *B* iii. 307; pikeis '*ligo*' PR. P. 397.

pīe, sb., *O.Fr.* pīe; *magpie,* O. & N. 126; S. S. (Web.) 2217; LANGL. *B* xi. 338.

pīe², sb., *pie,* '*artocrea(s),*' PR. P. 395; L. C. C. 51; ? pīes (*gen.*) hele, ? *pie crust* LANGL. *B* vii. 194; hote pīes (*pl.*) *A* PROL. 104; pies of Paris C. B. 75.

pigeōn, sb., *O.Fr.* pigeon; *pigeon,* CATH. 277; pijon '*columbella*' PR. P. 396; pejōns, pejouns, pejonis (*pl.*) C. B. 58, 67, 109.

pigge, sb., *cf. M.Du.* bigghe; *pig, young swine;* '*porcellus,*' VOC. 177; '*porcellus*' PR. P. 395; pigges (*pl.*) A. R. 204; MAND. 72.

pik, *see* picke. pīk, pīke, *see* pīc.

pīked, adj., *piked,* CATH. 278; '*cuspidatus*' PR. P. 397; pīkede (*pl.*) MISC. 73; L. N. F. 506; piked(e) schon MIRC 43.

pikeis, *see* picois.

piken, v., *cf. M.Du.* picken, ? *Fr.* piquer; *pick, hoe*: to piken it and to weden it LANGL. *B* xvi. 17; pikin '*purgo*' PR. P. 397; pike GOW. II. 351, III. 162; pikeþ (*pres.*) P. S. 150; ni at þe mete þi toþ þou pike (*imper.*) MAS. 746; (þei) piked (*pret.*) LANGL. *B* vi. 113; piked (*pple.*) A. P. ii. 1466; ipiked SAINTS (Ld.) xlv. 68.

piken, *see* picken.

pīken, v., *look, peep*; pike CH. TRO. iii. 60, 2909; N. P. 57.

pikerel, sb., *little pike,* PR. P. 397; CH. C. T. *E* 1419.

pikil, sb., *cf. M.L.G., M.Du.* pekel; *pickle,* '*picula,*' PR. P. 397; M. T. 126.

pil, sb., *Lat.* pilus; *peel*: pil and piþ ANGL. II. 247.

pīl, sb., *O.E.* pīl, = *M.Du.* pijl, *O.H.G.* phīl *m., from Lat.* pīlum *n.; pile, stake,* LANGL. *B* xvi. 86; longe pīles (*pl.*) & grete dide þei make faste in Temese dide þei hem stake MAN. (F.) 4611; piles (*printed* wiles) (*darts*) on ile FRAG. 8.

pilāge [pelāge], sb., *Fr.* pillage; *plunder,* ALEX. (Sk.) 3179.

pilche, sb., *O.E.* pylce; *pilch, flannel or fur garment,* '(*tunica*) *pellicia,*' PR. P. 397; A. R. 362; PL. CR. 243; LIDG. M. P. 154; pilches (*pl.*) GEN. & EX. 377; MAND. 247.

pilche-clūt, sb., *cloth,* HOM. I.253; A. R.212.

pīle, sb., *O.Fr.* pīle; *pile, building,* '*pila,*' PR. P. 398; LANGL. *C* xxii. 366; if i dwelle in mi

pile of ston TOR. 375; pile A. P. i. 686; pīlis (*pl.*) H. V. 64.

pilegrim, *see* pelegrim.

pilegrimāge, sb., *O.Fr.* pelerinage; *pilgrimage,* LEB. JES. 647; pelrinage BEK. 5; pilgrimage LANGL. *B* vi. 86.

pilēr, sb., *O.Fr.* piler; *pillar,* HOM. I. 281; P. L. S. xv. 206; LANGL. *B* v. 602; S. S. (Wr.) 2022; pileer MAND. 91; pelare ALEX. (Sk.) 4707; pilars (*pl.*) ALEX. (Sk.) 5068.

pilewhei [piriwhit], sb., *some kind of drink*; LANGL. *A* v. 134.

pilien, peolien, v., *O.Fr.* peler, *from Lat.* pilāre; *peel; rob,* A. R. 86; pillin '*decortico*' PR. P. 399; rushes to pilie LANGL. *C* x. 81; pile GOW. I. 17; þus me pileþ (*pres.*) þe pore and pikeþ ful clene P. S. 150; þat pileþ holi kirke LANGL. *B* xix. 439; pilede (*pret.*) WICL. 3 ESDR. i. 36; piled (*pple.*) WILL. 5123; piled [pilled] CH. C. T. *A* 3935; peled PARTEN. 2169; ipiled P. S. 337; *comp.* bi-pilien.

pillinge, sb., *pillage,* DEP. R. i. 13.

pilion, sb., *sort of cap*: ne puten no pilion on his pil(e)d pate PL. CR. 839; a pavis pillione hatt D. ARTH. 3460.

pilken, v., *cf. Ital.* piluccare; ? *peel, pluck*: pilken & peolien A. R. 86; wolde he . . . pileken 84.

pilori, sb., *Fr.* pilori; *pillory,* P. S. 345; pillori LANGL. *A* ii. 181 [pilorie *B* ii. 203]; pulleri *C* iii. 2.

pilour, sb., *Fr.* pilleur; *robber,* LANGL. *A* iii. 188; pilours (*pl.*) D. TROY 12862; MAN. (F.) 6682; piloures LANGL. *B* xix. 413.

pilten, *see* pülten. pilwe, *see* püle.

piment, sb., *O.Fr.* piment; *a spiced drink,* A. R. 404; HAV. 1728; AR. & MER. 3133; CH. BOET. ii. 5 (50).

pinacle, sb., *Fr.* pinacle; *pinnacle,* '*pinaculum, pinna,*' PR. P. 399; pinacles (*pl.*) P. P. 251; pinnacles A. P. ii. 1463.

pinade, sb., *see* pīnnote *under* pīne²; *a dish made with 'pine nuts'* C. B. 34; pinnonade F. C. 31.

pincardine, sb., *some gem*: pinkardenes (*pl.*) A. P. ii. 1472.

pinchen [pinche], v., *O.Fr.* pincer; *pinch; find fault*; CH. C. T. *A* 326; to pinchen *H* 74; pinche S. S. (Web.) 1243; pinched (*pret.*) LANGL. *B* xiii. 371; hir wimpul pinched (*pple.*) was CH. C. T. *A* 151.

pinde, pinden, *see* pünde, pünden.

pīne, sb., *O.E.* pīn, = *O.L.G., O.H.G., O.N.* pīna; *torture, pain,* ['*supplicium*'] MAT. xxv. 46; P. L. S. viii. 68; LAʒ. 2515; KATH. 1031; A. R. 24; ORM. 2987; O. & N. 1566; GEN. & EX. 955; HAV. 540; BEK. 8; SHOR. 83; ALIS. 1944; LANGL. *B* x. 388; CH. C. T. *A* 1324; HOCCL. i. 448; IW. 489; FLOR. 2087; MIN. 29; ISUM. 73; M. H. 55; pīnen, pine (*pl.*) HOM. I. 43; A. R. 134; AYENB. 130;

pines ROB. (W.) 10525; **pīnan** (*dat. pl.*) HOM. I. 43; *see* **peine.**

pīn-ful, adj., *painful*, A. R. 356; þis pinful(l)e gin KATH. 1980.

pīne [2], sb., *Lat.* pīnus; *pine*, L. H. R. 70.

pīn-appel, sb., *pine-apple*, ANGL. I. 95.

pī(n)-note, sb., *O.E.* pīnhnutu; '*nut*' *of the pignon pine*, '*pinum,*' PR. P. 400.

     pīnnote-trē, sb., *pignon pine*, S. S. (Web.) 544.

pingen, *see* **püngen.**

pīnien, v., *O.E.* pīnian, = *O.H.G.* pīnōn, *O.N.* pīna; *mod.Eng.* pine; *torture, cause pain; suffer torture*; FRAG. 8; pinan HOM. I. 35; pinen A. R. 216; R. R. 3511; pinen þær þi bodiȝ a wiþ chele & þrist & hunger ORM. 1614; pini AYENB. 130; pine HAV. 1958; GREG. 708; **pīneð** (*pres.*) KATH. 1824; SPEC. 81; **pīnede** (*pret.*) LANGL. *A* i. 145; þe levedi pinede so sore CHR. E. 565; pineden SAX. CHR. 263; C. L. 314; **pīned** (*pple.*) M. H. 144; ipined A. R. 114; AYENB. 263; *comp.* **for-pīnen.**

pīnunge, sb., *O.E.* pīnung, pīning, = *O.N.* pīning; *torture*, A. R. 368; pining SAX. CHR. 263.

pīning-stōle, sb., *stool of punishment, cucking-stool*; **pīning(e)stōles** (*pl.*) LANGL. *B* iii. 78, *C* iv. 79.

piniōn, sb., *Fr.* pignon; *pinion*, '*pennula,*' PR. P. 400; C. M. 12958.

pinken, v., *make figures* : heo **pinkes** (*pres.*) wiþ heore penne on heore parchemin P. S. 156.

pinne, sb., *? O.E.* pinn, = *M.Du.* pinne *f.,* O. *N.* pinni, *Gael.* pinne *m., M.H.G.* phinne *f.; pin, nail, needle*, VOC. 203; '*cavilla, spintrum*' PR. P. 399; ALIS. 6146; CH. C. T. *A* 196; S. S. (Wr.) 1410; ED. 117; ȝe hitte þe pinne LUD. COV. 138; **pinnes** (*pl.*) P. L. S. xxxv. 59; *comp.* **dúre-pin.**

pinner, sb., *pin-maker*, D. TROY 1591.

pinnin, v., *pin, fasten with a pin*, '*concavillo,*' *enclose*, PR. P. 400; pinne L. H. R. 131; D. ARTH. 4047; þai **pinez** (*pres.*) me in a prisoun A. P. iii. 79; **pinnede** (*pret.*) LANGL. *A* v. 127 [pinned *B* v. 213].

pinok, sb., *? Welsh* pinc; *? hedge sparrow*, VOC. 177; pinnuc O. & N. 1130.

pinsōn, sb., *Fr.* pinçon; *stocking or gaiter,* '*pedipomita,*' PR. P. 400; '*pedulus*' VOC. (W.W.) 601.

pinte, sb., *O.Fr.* pinte; *pint* (*measure*), '*pinta,*' PR. P. 401.

pintil, sb., *O.E.* pintel (VOC. 65); '*veratrum,*' VOC. (W.W.) 632; *comp.* **ȝēk-pintel.**

pione, sb., *Lat.* pione, = *Lat.* paeonia; *peony*; pioni, piani (*pionia, poenia*) PR. P. 395, 401; pioine REL. I. 37; **piane** [*v. r.* peinie, pianie] (*pl.*) LANGL. *A* v. 155; piones *B* v. 312.

pīpe, sb., *? O.E.* pīpe, = *O.N.* pīpa, *M.L.G.* pīpe, *O.H.G.* fīfa; *pipe, fife*, O. & N. 22; REL. I.

111; ALIS. 7769; OCTAV. (H.) 197; o pipe of bras FL. & BL. 225; a barel or a pipe PALL. ii. 381; **pīpen** (*pl.*) LAȝ. 5110; the pipes of his longes CH. C. T. *A* 2752; pipes PALL. ii. 389; pipis DEP. R. iii. 275; *comp.* **bagge-, water-pīpe.**

pīpin, v., = *M.L.G.* pīpen, *M.H.G.* pfīfen; *pipe, play the pipe*, PR. P. 401; pipen in an ivileef CH. C. T. *A* 1838; pipe P. S. 216; TREV. III. 207; **pīped,** pipede (*pret.*) LANGL. *B* xviii. 406, *C* xxi. 453.

     pīpare, sb., *O.N.* pīpari, = *O.H.G.* phīfari; *piper*, PR. P. 401.

     pīping, sb., *piping*, LAȝ. 5110; MAN. (F.) 1775; much piping þer repaires GAW. 1017.

pīpin [2], v., *squeak*, '*pipio,*' PR. P. 401; ac **pīpest** (*pres.*) al so doþ a mose O. & N. 503.

? pipined, adj., *furnished with pips*, PALL. iii. 72.

pi-poudre, sb., *court of Pie Poudre*; **pipoudris** (*sb., pl.*) *cases in the court of Pie Powder* DEP. R. iii. 319.

pippe, sb., *cf. M.Du.* pippe; *pip* (*disease of birds*), '*pituita,*' PR. P. 401; SACR. 525.

pīren, v., *peer*; **pīreþ** (*pres.*) GOW. III. 29; piriþ DEP. R. iii. 49.

pirie, sb., *O.E.* pirige; *pear-tree*, CH. C. T. *E* 2342; piries (*pl.*) LANGL. *A* v. 16 : **pirie** (*dat. pl.*) S. S. (Web.) 555.

piriwhit, *see* **pilewhei.**

pirne, sb., '*needle*' *of a pine-tree; roll of a weaver's loom;* '*panus,*' PR. P. 402; þai fande a ferli faire tre . . . withouten bark ouþir bast, full of bare **pirnes** (*pl.*) ALEX. (Sk.) 4981.

piromancie, sb., *pyromancy*, ALEX. (Sk.) 4612; perimancie LANGL. *A* xi. 158.

pisan, *see* **püsane.**

pisse, sb., *Fr.* pisse; *urine*, CH. C. T. *D* 729.

     pisse-mīre, sb., *ant*, MAND. 301; CH. C. T. *D* 1825.

     pis-pot, sb., '*maniodella, madula,*' PR. P. 402.

pissen, v., *O.Fr.* pisser; MAND. 249; pisse P. L. S. xiv. 66; **pissed,** pissede (*pret.*) LANGL. *B* v. 348, *C* vii. 399.

piste, sb., *spikenard*; piste Indik ['*spicae Indicae*'] PALL. xi. 411.

pistle, *for* epistle, sb., *epistle,* A. R. 350; WICL. DEEDS xxiii. 33; LANGL. *B* xii. 30; pistil *A* xi. 229; CH. C. T. *E* 1154; pistill ALEX. (Sk.) 1791; **pistilis** (*pl.*) WICL. DEEDS xxii. 5; pistils ALEX. (Sk.) 3420.

pit, *see* **püt.**

pitaill, sb., *A.Fr.* pitaille, *O.Fr.* pietaille; *infantry*, BARB. xiii. 229*; pitall xi. 420*; pedaille MAN. (F.) 14885.

pitaunce, sb., *O.Fr.* pitance, *mod.Eng.* pittance : *provision, share*, A. R. 114; LANGL. *B* v. 270, *C* xvi. 61; a good pitaunce CH. C. T. *A* 224 : pietancia PR. P. 402.

pitē, sb., *O.Fr.* pité; *pity*, A. R. 368; FL. & BL. 529; ROB. 143; MAN. (F.) 5210; CH.

PITY 1, 5, 44 ; M. H. 7 ; piete SPEC. 89 ; petie D. TROY 8686.

pithe, sb., *O.E.* piŧa, = *M.Du.* pitte ; *pith*, 'medulla,' PR. P. 402 ; piþ ANGL. II. 247 ; CH. C. T. *D* 475 ; pith PERC. 1640 ; (*strength, power*) GAW. 1456 ; BARB. iii. 599.

pitous, adj., *O.Fr.* pitous ; *piteous*, AYENB. 150 ; ROB. 491 ; WILL. 643 ; CH. C. T. *A* 143.

    pitous-līch, adj., *pitiable, piteous*, LANGL. *C* xxi. 59 ; pitouslich, pitousli (*adv.*) LANGL. *C* v. 94, ii. 77 ; pitevousli, piteouseliche, pituoslich ROB. (W.) 5884* ; pitosliche 5607 ; pitousli WILL. 1168.

pittit, adj., *from* püt ; *full of pits* ; pittit [pottit] BARB. xi. 388.

pláce, sb., *O.Fr.* place ; *place, house*, A. R. 258 ; BEK. 117 ; GREG. 910 ; ALEX. (Sk.) 952 ; **pláces** (*pl.*) LANGL. *C* xiii. 246.

pláge, sb., *Lat.* plăga ; *region*, WICL. GEN. iv. 16 ; '*clima*' CATH. 282 ; **pláges** (*pl.*) CH. C. T. *B* 543 ; the 4 principals plages or quarters of the firmament CH. ASTR. i. 5 (5) ; 4 plages principalx ii. 31 (41).

pláge², sb., *Lat.* plāga ; *plague* ; [*v. r.* wounde] CH. C. T. *I* 593 ; LUD. COV. 125 ; plages PALL. iii. 396 ; **pláges** (*pl.*) WICL. GEN. xii. 17.

plaȝe, ? **plāȝe,** sb., = *M.H.G.* phlāge ; *play, sport ; fight* ; GEN. & EX. 537 ; plawe P. S. 153 ; TRIST. 3101 ; þi moder luþer plawe ROB. 291 ; LEG. 26 ; þat trewe wes in uch plawe HORN (R.) 1094 ; at þare plawe L. N. F. 217 ; ploȝe LAȝ. 20843 ; A. R. 184* ; plaȝen (*pl.*) LAȝ. 29219 ; plahen [plohen] KATH. 106 ; plawes SPEC. 45 ; M. H. 115.

    plōw-efēre, sb., *play-fellow*, SPEC. 49 ; **plaȝe-ivēren** [*sec. text* pleiveres] (*pl.*) LAȝ. 15631.

plaȝen, v., *O.N.* plaga ; *play* ; plaigen GEN. & EX. 2016 ; seþe and plai(e) GREG. 1031 ; plawe . . . at þe bars 381 ; plawin '*bullio*' PR. P. 403 ; plawe HAV. 950 ; FLOR. 258 ; SACR. 664 ; plaiand (*pple.*) L. C. C. 37 ; plaȝede mid worden LAȝ. 17335.

plaice, sb., *Fr.* plaise ; *plaice (fish)*, '*pecten*,' PR. P. 402 ; pleise REL. I. 81 ; plais C. B. 103 ; plaices (*pl.*) HAV. 896.

plaid, plait, sb., *O.Fr.* plaid, plait ; *plea, debate*, O. & N. 5 ; plait AYENB. 39 ; plai E. G. 350 ; plei, plae '*placitum*' CATH. 283 ; ple WICL. HEB. vi. 16.

plaiden, v., *O.Fr.* plaidier ; *plead* ; plaidi O. & N. 184 ; P. L. S. xix. 77 ; plaiti AYENB. 99 ; plede C. L. 1024 ; LANGL. *B* vii. 42 ; plete CH. TRO. ii. 1468 ; DEP. R. iii. 382 ; **plédite** (*pret.*) LANGL. *C* xxii. 295* ; plededen *B* PROL. 210 ; pledit DEP. R. iii. 319.

    plaitere, sb., *Fr.* plaideur ; *pleader*, AYENB. 98 ; **plaidūrs** (*pl.*) MISC. 76 ; pletours *fighting-men*, ALEX. (Sk.) 1731.

plaiding, sb., *pleading*, O. & N. 12 ; pleding LANGL. *C* iv. 452.

plain, adj. & adv., *O.Fr.* plain ; *flat, even, clear* : no covert miȝt þei kacche, þe cuntre was so plaine WILL. 2217 ; hir sones plein entente CH. C. T. *B* 324 ; ful plat and ful plein *B* 3947 ; speketh so plein that we mai understonde *E* 19 ; to tellen short and plein *G* 360 ; plein MAN. (F.) 1772.

    plein-lī, adv., *plainly, frankly* ; CH. C. T. *A* 1733 ; planli BARB. ix. 512, x. 520, xix. 54.

plaine, sb., *O.Fr.* plaigne ; *plain*, WILL. 3770 ; a lusti plaine CH. C. T. *E* 59 ; pleine PR. P. 404.

plainen, v., *Fr.* plaindre ; *complain* ; plaine ROB. 22 ; plaini AYENB. 132 ; pleine LANGL. *B* iii. 167 [plaine *A* iii. 161] ; **pleine** (*subj.*) LANGL. *C* ix. 166 ; plénand (*pple.*) HAMP. PS. xii. 1* ; pleined (*pret.*) WILL. 1845 ; pleinede LANGL. *C* i. 81 ; plainede ROB. (W.) 765 ; plained YORK xlviii. 296.

    plaininge, sb., *complaint*, ROB. (W.) 9711.

plainte, sb., *O.Fr.* plainte, pleinte ; *complaint*, LEG. 57 ; E. G. 349 ; plainte, pleinte ROB. 252, 328 ; pleinte HAV. 134 ; MAN. (F.) 3224 ; **pleintis** (*pl.*) DEP. R. iii. 306.

plaintif, sb., *O.Fr.* plaintif ; *plaintiff*, E. G. 360.

plait, sb., *O.Fr.* pleit, ploit ; *plait*, '*plica*,' PR. P. 402.

plait, *see* plaid.   **plaiten,** *see* plaiden.

plaitin, v., *plait*, '*plico*,' PR. P. 402 ; **plétede** (*pret.*) LANGL. *A* v. 126 [plaited *B* v. 212].

pláne, sb., *Fr.* plane ; *plane*, PR. P. 402.

pláne², sb., *Fr.* plane ; *plane (tree)*, '*platanus*,' PR. P. 402 ; WICL. GEN. xxx. 37 ; plone ANGL. I. 95.

planēte, sb., *O.Fr.* planēte ; *planet*, TREAT. 132 (SAINTS (Ld.) xlvi. 418).

plánin, v., *O.Fr.* planer ; *plane, make smooth*, '(*com*)*plano*,' PR. P. 403 ; **ipláned** (*pple.*) C. L. 696.

planir, *see* plenēr.

planke, sb., *Fr.* planche, planke ; *plank*, WILL. 2778 ; H. S. 5261 ; ALEX. (Sk.) 3740.

plantain, plantein, sb., *O.Fr.* plantain ; *plantain*, CH. C. T. *G* 581 ; plantein '*plantago*' PR. P. 403.

plante, sb., *O.E.* plant, = *O.Fr.* plante ; *plant, tree*, CH. C. T. *D* 763 ; plante LANGL. *B* v. 591 [plonte *A* vi. 72] ; plontte A. P. i. 104 ; **plants** (*pl.*) ALEX. (Sk.) 4995.

planten, v., *O.E.* plantian, = *O.N.* planta, *O.H.G.* phlanzōn, cf. *O.Fr.* planter ; *plant*, MAND. 50 ; plonteþ (*pres.*) AYENB. 123 ; plantede (*pret.*) SAX. CHR. 263 ; plant [planted] ALEX. (Sk.) 1654, 3146, 5656.

plasche, v.,= *M. Du.* plasch; *plash, pool,* '*lacuna,*' PR. P. 403; D. ARTH. 2798; *see* **flasche.**

plaster, sb., *O.E.* plaster, = *O.N.* plaster, *O.H.G.* plaster, phlaster, *O.Fr.* plastre; *plaster,* WICL. IS. xxxviii. 21; plastre SPEC. 89; LANGL. *B* xx. 308; **plastres** (*pl.*) AYENB. 148; *see* **emplastre.**

plasterin, v., *O.Fr.* plastrer; *plaster,* '*cata-plasmo, gipso,*' PR. P. 402; plastre LANGL. *C* xxiii. 310; **plastreþ** (*pres.*) xxiii. 314; **plastred** (*pple.*) *B* xvii. 95.

plat, adj., *O.Fr.* plat; *flat,* A. P. ii. 1379; legge hire plat to grunde JUL.[2] 41; i wol þe telle al plat CH. C. T. *C* 648; i slood doun plat to þe erþe WICL. DAN. viii. 18; plate feet ALIS. 2001; *see* **flat.**

plat-ful, adj., *brimful,* A. P. ii. 83.

plat-lī, adv., *flatly,* '*plane,*' PR. P. 403.

pláte, sb., *O.Fr.* plate; *plate, piece,* '*lamina, vel lama,*' PR. P. 403; a fine hauberk . . . ful strong . . . of plate CH. C. T. *B* 2055; all pa. ʒestis of plate pure as þe noble ALEX. (Sk.) 3673; **plates** (*pl.*) GEN. & EX. 2370; a paire plates large CH. C. T. *A* 2121; þritti **pláten** (*pl.*) of selver REL. I. 144; in Peeres plates *armour* LANGL. *C* xxi. 24; plates A. P. i. 1036; platis [platis] *armour plates* ALEX. (Sk.) 2214, 2450.

pláte-rōfes, sb. pl., *roofs made of plates of gold,* ALEX. (Sk.) 5260.

pláte-werkis, sb. pl., *works of plate,* ALEX. (Sk.) 3223.

plater, sb., *platter, dish,* '*paropsis,*' VOC. 178; PR. P. 403; L. H. R. 137; A. P. ii. 638; platere C. B. 17, 27, 114; **platers** (*pl.*) D. ARTH. 182.

platten, v., *O.E.* plættan,=*M.Du.* platten, pletten, *Swed.* plätta, *M.H.G.* platzen, blatzen; *strike, break; throw down flat;* **platte** (*pres.*) LANGL. *C* xix. 34*; **plattinde** (*pple.*) HAV. 2282; **platte** (*pret.*) LANGL. *A* v. 45 (*B* v. 63, *C* vii. 3); þat on hem plat(te) AR. & MEṘ. 9747; to armes al so swiþe plette HAV. 2613; his heved of he plette 2626.

plawe, *see* **plaʒe.**

pléchen, v., *O.Fr.* plecier, plessier (*plier, entrelacer); pleach;* **plécheth** (*pres.*) PALL. iii. 330.

plecke, sb., *O.E.* plæcca,=*M.L.G., M.Du.* plecke; *piece of ground,* '*porciuncula,*' PR. P. 405; plek HALLIW. 631; A. P. ii. 1379.

plecken, v.,=*M.Du.* plecken; *mark with spots;* **plekked(e)** (*pple.*) stones '*guttatos lapides*' TREV. I. 429.

pléden, *see* **plaiden.**

pleʒe, plege, sb., *O.E.* plega *m.* (*game, fight*), ? = *O.Fris.* plega, *M.H.G.* pflege *f.* (*custom, care); play, sport, battle,* HOM. II. 55, 211; (*ms.* pleie) HOM. I. 193; monine seorh-

fulne pleiʒe LAʒ. 2282; plæʒe 15554; pleowe 8187; (*ms.* pleie) A. R. 344; pleouwe 318; (*ms.* playe) ISUM. 188; M. H. 85; childes plaie GREG. 612; **pleʒes** (*pl.*) AYENB. 207; pleies P. L. S. xvi. 68; *comp.* **bal-pleowe.**

plei-(i)fēre, sb., *play-fellow,* P. L. S. xvii. 64; OCTOV. (W.) 1232; PR. P. 404.

plei-ful, adj., *playful,* HOM. I. 205.

plege, plegge, sb., *O.Fr.* plege; *pledge,* '*obses,*' CATH. 283; plegge '*vas*' PR. P. 404; TREV. III. 129; E. G. 382; ALEX. (Sk.) 1783.

plegen, v., *from* plege; *pledge:* of all i plege (*pres.*) and pleine me YORK xviii. 170.

pleʒen, v., *O.E.* plegian (*play*),= ? *O.Fris.* plegia, *M.Du.* plegen, *M.H.G.* phlegen (*use*); *play;* pleien LAʒ. 8131; A. R. 94; H. M. 41; (*ms.* pleyen) P. L. S. xiii. 150; MAND. 50; pleien ant rage A. D. 237; pleien, pleie O. & N. 213, 486; (*ms.* pleie) WILL. 678; (*ms.* pleye) HAV. 951; GREG. 738; pleien, pleie, plaien CH. C. T. *A* 236, 758; pleie wiþ a plow(e) LANGL. *B* iii. 307; plaje M. H. 80; ʒe þet pleieð (? *have inter-course*) mit te worlde A. R. 76; we pleien H. H. 68; **pleiende** (*pple.*) KATH. 1691; pleiinge LANGL. *C* xix. 274; **pleiede** (*pret.*) A. R. 318; pleoʒede, pleuwede LAʒ. 6978, 29219; pleide FL. & BL. 31; pl(e)oʒeden HOM. I. 7; pleoweden [pleoiden] LAʒ. 8145.

pleiere, sb., *player,* TREV. IV. 297; pleiare '*lusor, ludibundus*' PR. P. 404.

plein, adj., *O.Fr.* pleins; *full,* WILL. 3158; MAN. (F.) 10615; LANGL. *A* viii. 87.

pleine, *see* **plaine.**

plenēr, adj. & adv., *O.Fr.* plenier; *full; in full:* þe feire is þere iliche plenere FL. & BL. 216 (H.) 618; P. L. S. xvii. 435; plenere time LANGL. *B* xvi. 103; whan þe peple was plenere comen *B* xi. 108; planir ALEX. (Sk.) 4138.

plenēr-lī, adv., *fully,* H. S. 5811.

plentē, sb., *O.Fr.* plenté; *plenty,* A. R. 194; HAV. 1729; AYENB. 161; LANGL. *B* xi. 323; C. M. 4038; CH. C. T. *B* 443; plenteð GEN. & EX. 3709.

plenteūs, adv., *O.Fr.* plentius; *plenteous, abundant,* ROB. 23; plentevous LANGL. *B* x. 80; plentivous ROB. (W.) 531; MAN. (F.) 1390, 6438; plaintiose D. TROY 3153.

plenteous-līche, adv., *plenteously,* WILL. 180; **plenteouslier** (*comp.*) WICL. HEB. vi 19.

plentive, adj., *O. Fr.* pleinteive, plentif; *fertile,* MAN. (F.) 6444.

plentuustē, sb., *plentifulness,* HAMP. PS. xxxv. 9.

plesaunce, sb., *O.Fr.* plaisance; *pleasure,* LANGL. *C* ix. 14; GAW. 1247; plesauns D. TROY 2311.

plesaunte, adj., *O.Fr.* plaisant; *pleasing*, LANGL. *B* xiv. 101; plesaunt GAW. 808.

plésen, v., *O.Fr.* plaisir; *please*; plese WILL. 5435; LANGL. *B* xiv. 220; pléses (*pres.*) LANGL. *B* xv. 152*; plesen *A* x. 209; plésed (*pret.*) *C* iv. 492; pleseden *A* iii. 98; plésed (*pple.*) ALEX. (Sk.) 593.

 plésing, sb., *pleasure*, LANGL. *A* iii. 237.

pléte *see* plaiden. **pletten**, *see* platten.

pleume, *see* plūme.

plevine, sb., *O.Fr.* plevine (*garantie*); *pledge, contract*, IW. 1253.

plicchen, v., ?=plucken; *pluck*: & pliȝte (*pret.*) him of (h)is sadel FER. 3029; he pliȝte [plighte] his hors aboute CH. C. T. *B* 15; and pliȝten him in ALIS. 5831; pliȝt (*pple.*) MED. 626; þre leves have i pliȝt out of his boke CH. C. T. *D* 790.

plíen, v., *Fr.* plier; *bend*; plie CH. C. T. *E* 1169; plíe (*imper.*) PALL. xii. 61; plíande (*pple.*) grevez A. P. iii. 439; plíed (*pple.*) ovir with pure gold ALEX. (Sk.) 5260.

pliht, sb., *O.E.* pliht,=*O.Fris.*, *M.Du.* plicht, *M.H.G.* phliht; *from O.E.* plīon (*pret.* plēah) (*risk*), *M.H.G.* phlegen (*pret.* phlac); *plight; danger*: þre dæȝes hit rinde blod ... þat wæs swuþe mochel pliht LAȜ. 3897; ærst heo pleoweden and seoððe pliht makeden 8146; he bad him maken siker pligt (*engagement*) GEN. & EX. 1269; no pliȝt (*fight*) seche GAW. 266; i telle it ou a pliht (*on my faith*) P. S. 218; a pliht [pliȝt] C. L. 304; Joseph was wel bliþe a pliȝt SHOR. 120; i sigge a plight AR. & MER. 3909; pliȝt, plight (*guilt*) C. M. 8396; Crist deied for mannes pliht M. H. 99; have i no more plith HAV. 2002; pliȝt (*condition*) A. P. i. 1074; tiȝende ... of swiȝe muchele plihte LAȜ. 13309; beon icnowen of his pliht 18567; Moises fastede ... to pligt (? *duly*) XL daiges and XL nigt GEN. & EX. 3611; a spangel good of plight S. S. (Wr.) 1448; in mi plit CH. C. T. *E* 2335.

 pliht-ful, adj., *risky, dangerous*, M. H. 117.

 pliht-líc, adj., *perilous*; **plihtlíche** (*pl.*) spelles LAȜ. 23528.

plihten, v., *O.E.* plihtan,=*M.H.G.* phlihten; *plight, engage, pledge*, LAȜ. 13071; pliȝte HORN (L.) 305; FER. 1281; plightin, plitin PR. P. 405; plihte (*pres.*) MISC. 83; plíȝte SHOR. 145; i plihte [pliȝte] þe mi trouþe LANGL. *A* vi. 35; plihten [plicten] (*pret.*) LAȜ. 5543, 6571; plihten, pliȝten LANGL. *A* PROL. 46; pligt (*pple.*) GEN. & EX. 1275; pliȝt SPEC. 30; GREG. 32; ipliȝt ROB. 184; ipluht A. R. 208.

plihti, adj., = *M.Du.* plichtigh, *M.L.G.* plichtich, *M.H.G.* phlihtic; *in danger, guilty*: he sal be plighti [*v. r.* gilti] for þis an C. M. 26842.

plít, *see* pliht.

plod, sb., ? *puddle*: in a foul plodde [pludde] (*dat.*) in þe stret ROB. 536.

plōȝe, *see* plaȝe.

plōh, sb., *O.E.* plōh, *O.N.* plōgr,=*M.Du.* ploeg, *O.H.G.* phluog; *plough*, A. R. 384*; ORM. 15902; plouh REL. I. 173; plouh AL. (L.³) 11 (T. plouȝ); plow LANGL. *B* vii. 119 [plouh *A* viii. 104]; plogh [plough, plou] CH. C. T. *B* 1478; plou (*ms.* plow) HAV. 1017; MAND. 250; S. & C. I. xi.

 plōu-bat, sb., *plough paddle*, LANGL. *A* vii. 96*, *C* xi. 64*.

 plōu-beem, sb., *plough-beam*, VOC. 169; PR. P. 405.

 plōuh-fōt, sb., *plough foot*, LANGL. *C* ix. 64; plou fote *B* vi. 105.

 plōu-lōde, sb., '*carucata,*' VOC. 270.

 plōu-lond, sb., pl., *ploughlands*, ROB. (W.) 7676.

 plōuȝ-mon, sb., *ploughman*, LANGL. *B* vii. 3 [plouman *A* vi. 3].

 plōuh-pote, sb., ? =plouh-fōte; LANGL. *A* vii. 96; plowpote *B* vi. 105*.

 plōuh-schare, sb.,=*M.H.G.* phluocschar; *ploughshare*, TREV. II. 353.

 plōu-staf, sb., *plough staff*, ROB. (W.) 2198.

 plōu-stert,=*M.L.G.* plōchstert, *Swed.* plōgstiert; *plough-tail*, '*stiva,*' PR. P. 405.

plomaile, *see* plumaile.

plom, plomere, *see* plumb, plumber.

plomet, sb., *O.Fr.* plommet; *plummet*, CH. ASTR. ii. 23 (33).

plont, *see* plante.

plot, sb., *O.E.* plot, ?=*Goth.* plats (ἐπίβλημα); *plot* (*of ground*), '*porciuncula,*' PR. P. 405; LUD. COV. 47; mani foule plottes (*pl.*) LANGL. *B* xiii. 318; *see* blot.

plōu, plōuh, *see* plōh.

plover, sb., *O.Fr.* plovier; *plover*, VOC. 221; '*pluviarius*' PR. P. 405; plovers (*pl.*) PL. CR. 764; D. ARTH. 182.

pluck, sb., *stroke*; **pluckis** (*pl.*) TOR. (A.) 1611.

plukkin, v., *O.E.* pluccian,=*M.Du.* plucken, *M.H.G.* phlücken; *pluck*, '*vellico, excatheriso,*' PR. P. 405; plucke LUD. COV. 146; plockien MAT. xii. 1; plocke (*imper.*) LANGL. *C* viii. 229; plukked (*pret.*) LANGL. *B* xi. 109; (þai) plucked DEP. R. ii. 32; plukked (*pple.*) LANGL. *B* xii. 249.

plumāge, sb., *Fr.* plumāge; *plumage*, CH. C. T. *F* 426.

plumaile, sb., ? *O Fr.* plumail; *plumage*; plomaile DEP. R. ii. 32.

plumb, sb., *Fr.* plomb; *plumb*, PR. P. 405.

 plom-reule, sb., *plumb-rule*, CH. ASTR. ii. 38 (46).

plumber [plomere], sb., *Fr.* plombie̅; *plumber*, PR. P. 406.

plūme, sb., _O.E._ plūme,=_L.G._ plūme, prūme, _O.N._ plōma, _from Lat._ prūnum; _plum_; ploume '_prunum_' PR. P. 405; **plōmes** (_pl._) LANGL. _C_ xiii. 221.

plŭm-trē, sb., _O.E._ plūmtrēow; _plum-tree,_ VOC. 192; PR. P. 406; **plŏmtrēs** (_pl._) LANGL. _A_ v. 16.

plŭme, sb., _O.Fr._ plume; _plumes, feathers,_ DEP. R. iii. 49.

plūmen, v., _pick feathers off the neck_; pleume DEP. R. ii. 163.

plump, adj., _plump;_ **plumpe** (_dat._) D. ARTH. 2199.

plumpen, v.,=_M.Du._ plompen; _plump; drop, throw_; plump (_imper._) L. C. C. 51; plom hem in a boiling potte C. B. 76; hi **plum(p)ten** (_pret._) doune as a doppe ALIS. 5776.

plunge, sb., _pool,_ ALEX. (Sk.) 5546.

plungen, v., _O.Fr._ plungier, plongier; _plunge_; ploungen CH. BOET. iii. 2 (65).

pluralitē, sb., _plurality,_ LANGL. _C_ iv. 33; **pluralities** (_pl._) _A_ xi. 197.

plurel, adj., _O.Fr._ plurel; _plural,_ LANGL. _B_ x. 237.

pluttide, pluccid, adj., _? plump,_ D. TROY 3078, 3837.

pō, _see_ **pā.**

pobel, sb., _? O.E._ pabol, papol‒ popol(-stān); _pebble_; (_ms._ pobble) A. P. i. 117; **publes** (_pl._) S. A. L. 158.

pibbil-stōn, sb., _pebble,_ WICL. PROV. xx. 17; publeston TREV. I. 353.

pocalips, sb., _for_ **apocalips**; _Apocalypse,_ LANGL. _B_ xiii. 90, _C_ xvi. 99.

pocke, sb., _O.E._ pocc,=_M.Du._ pocke; _(small)-pox, pustula_; pokke '_porrigo_' PR. P. 407; **pockes** (_pl._) '_viroles_' VOC. 161; pokkes LANGL. _B_ xx. 97; pokkis FLOR. 2024.

pōcok, _see_ **pācok** _under_ **pā.**

podde, _see_ **padde.**

podel, sb., _puddle,_ '_lacuna,_' PR. P. 406; MAN. (H.) 54; N. P. 53.

poding, sb., _O.Fr._ boudin; _pudding_; pudding LANGL. _B_ xiii. 106; pudding of purpaisse C. B. 42; **podinges** (_pl._) P. L. S. xxxv. 59.

poding-ale, sb., _thick ale,_ LANGL. _C_ vii. 226; pudding ale _B_ v. 220*.

poeple, _see_ **people.** **poēr,** _see_ **pouēr.**

poete, sb., _O.Fr._ poete; _poet,_ LANGL. _B_ xii. 260; poite D. TROY 306; **poetes** (_pl._) _C_ xxi. 453.

poetrie, sb., _O.Fr._ poeterie; _poetry,_ PR. P. 406.

poffen, _see_ **puffen.**

pohe, sb., _O.E._ pohha; _? =_ **póke**; pouhe LANGL. _A_ viii. 178.

poinant, _see_ **poniant.**

poinen, v., _O.Fr._ poindre; _prick, goad_;

poined (_pret._) MAN. (F.) 16218; **? poined** (_pple._) _? ornamented, trimmed_ A. P. i. 217.

poinӡe, _see_ **punӡe.**

point, sb., _O.Fr._ point; _point,_ A. R. 178; ROB. 395; AYENB. 33; in god point (_condition_) CH. C. T. _A_ 200; TREAT. 140; point A. P. i 891·; at point devis _exactly_ CH. C. T. _F_ 560; puint LANGL. _C_ xi. 94*.

pointel, sb., _O.Fr._ pointel; _style for writing on tablets,_ '_stilus,_' PR. P. 406; A. P. ii. 1533; CH. C. T. _D_ 1742; pointel(e) [pontel] C. M. 11087.

pointin, v., _Fr._ pointer; _point,_ '_puncto,_' PR. P. 407; **pointest** (_pres._) LANGL. _C_ ix. 298; **pointe** (_imper._) with venegre C. B. 29; princes **pointid** (_pret._) it [_a diving box_] with pik ALEX. (Sk.) 5546.

poisie, sb., _O.Fr._ poesie; _poetry,_ LANGL. _B_ xviii. 406.

poisonen, v., _poison_; poisoun LANGL. _B_ vi. 300; **poisoned** (_pple._) A. P. ii. 1095.

poiseninge, sb., '_intoxicacio,_' PR. P. 407.

poisoun, sb., _O.Fr._ poison; _poison,_ P. L. S. xiii. 99; puisun KATH. 2344; H. M. 33.

póke, sb., _O.E._ poca,=_M.Du._ poke, _O.N._ poki, _O.Fr._ poche; _bag,_ '_sacculus,_' PR. P. 407; HAV. 555; LANGL. _B_ vii. 191; CH. C. T. _A_ 4278; _see_ **pouche.**

póke-ful, sb., _bagful,_ LANGL. _B_ vii. 191.

póken, pucken, v.,=_M.L.G., M.Du._ poken; _poke, thrust, throw_; **póke** (_pres._) REL. II. 211; pukketh [_v. r._ pukkes, pucketh, pouketh, poketh] LANGL. _B_ v. 620; **pókede** [poukede] (_pret._) CH. C. T. _A_ 4169; pokid RICH. 3459; pukked [poukede, poked] LANGL. _B_ v. 643.

poket, sb., _Fr._ pochette, poquette; _pocket,_ '_sacculus,_' PR. P. 407; **pokets** (_pl._) CH. C. T. _G_ 808.

pōkoc, _see_ **pācok** _under_ **pā.**

pol, sb., _cf. M.L.G._ poll, _M.Du._ pol, bol; _poll,_ '_caput,_' PR. P. 407; MARG. 177; LANGL. _B_ xi. 57; MAN. (H.) 279; **polle** (_dat._) P. S. 237; ANT. ARTH. ix; pulden prestes bi ӡe polle A. P. ii. 1265.

pol-ax, sb., _cf. M.L.G., M.Du._ polexe; _pole-axe,_ '_bipennis,_' PR. P. 407; polax, pollax CH. C. T. _A_ 2544; pollax RICH. 6870; pollex VOC. 263; _see_ **boleax.**

pol-bere, sb., _kind of barley,_ '_trimensis,_' PR. P. 407.

pol-cat, sb., _pole-cat,_ CH. C. T. _C_ 855; pul-kat '_pecoides_' VOC. (W.W.) 601; PR. P. 407.

pol-hēved, _? adj._; **polehēvedes** (_sb., pl._) (_tadpoles_) and froskes GEN. & EX. 2977.

pol-wigle, sb., _? for_ *polwigge; _tadpole,_ PR. P. 408.

pōl, sb., _O.E._ pōl,=_M.L.G._ pōl, _M.Du._ poel, _M.H.G._ pfuol; _pool, stream,_ REL. II. 81; PERC. 682: pool '_stagnum_' PR. P. 407;

MISC. 149; pōle (*dat.*) LAȝ. 21748; FLOR. 1738; pōles (*pl.*) HAV. 2101; pulles PALL. i. 1032; *comp.* water-pōl; *deriv.* pullen².

polaile, sb., *O.Fr.* poulaille, polaille; *poultry*, PR. P. 407; polile A. P. ii. 57; pulaile BARB. xi. 120*.

pōle, v., *? put a guard on (a horse's) head* (SKEAT) : let peitrel him and pole him LANGL. *C* v. 23.

pōle, *see* pāl, pōl,

polein, sb., *O.Fr.* polein; *kneecap; pulley; 'troc(h)lea,'* PR. P. 407; poleins (*pl.*) MAN. (F.) 10027; polaines GAW. 576.

polete, *see* pulete.

policie, sb., *O.Fr.* policie; *policy*, CH. C. T. *C* 600.

polischen, v., *O.Fr.* poliss- ; *polish* ; polische WICL. I PARAL. xxii. 2 ; polische LANGL. *A* v. 257 [polsche *B* v. 482]; police A. P. ii. 1131 ; policed, polised (*pple.*) A. P. ii. 1068, 1134; pullishet D. TROY 4589.

polīve, sb., *? O.Fr.* poulie; *pulley* [*r.w.* drive] CH. C. T. *F* 184.

polk, sb., *small pool*, PR. P. 408; HAV. 2685; TRIST. 2865; MAN. (H.) 277.

poll, *see* pallen, pullen.

pollen, v., *poll, shave the head*; pollid (*pple.*) '*capitonsus*' PR. P. 407; ALIS. 216 ; WICL. JOB i. 20.

polment, *see* pulment.

pomade, sb., *O.Fr.* pommade (*ointment made from apples*; *cider*), *mod.Eng.* pomade; *cider* : no piement ne pomade ne presiouse drinkes LANGL. *C* xxi. 412.

pomel, sb., *O.Fr.* pomel; *pommel*, WICL. JUDG. iii. 22 ; MAN. (F.) 10037; pomels (*pl.*) PL. CR. 562.

pomeli, adj., *Fr.* pommelé ; *spotted like an apple*, CH. C. T. *A* 616; his hakenei that was al pomeli gris *G* 559; pomli PALL. iv. 809.

pomgarnade, sb., *Lat.* pōmum grānātum; *pomegranate*, LANGL. *A* v. 155*; A. P. ii. 1466.

pomice, sb., *cf. Ital.* pomice, *Fr.* ponce; *pumice* ; pomeis, pomice '*pomex*' PR. P. 408.

pompe, sb., *Fr.* pompe; *pomp*, SHOR. 110; LANGL. *B* iii. 66 ; CH. C. T. *A* 525.

pompous, adj., *O.Fr.* pompeus ; *pompous*, CH. C. T. *B* 3661.

pond, *see* púnd.

ponde, sb., *cf.* pünde ; *pond, 'stagnum, vivarium,'* PR. P. 408; P. R. L. P. 26 ; ponde ['*lacum*'] TREV. I. 69; mi net his (*r.* is) nei honde in a wel fair ponde HORN (H.) 1173; *comp.* fish-ponde.

ponderūs, adj., *Lat.* ponderōsus; *heavy*, GEST. R. 66.

poniant, adj., *Fr.* poignant ; *poignant, 'acutus, acer,'* PR. P. 408; poinant, poinaunt C. B. 6, 33.

poniet, sb., *Fr.* poignet; *wrist-cuff*, '*premanica*,' PR. P. 408.

ponne, *see* panne. pōpe, *see* pāpe.

popejai, *see* papejai.

popelere, sb., *a sort of duck*, PR. P. 408.

pōpen, v., *blow a wind-instrument*; pōped [pouped] (*pret.*) CH. C. T. *B* 4589.

poperen, v., *trot* : (Religioun) poperiþ (*pres.*) on a palfrey LANGL. *A* xi. 210.

popet, sb., *puppet*, ALIS. 335; CH. C. T. *B* 1891.

popi, sb., *O.E.* popig ; *poppy, 'papaver,'* PR. P. 409; GOW. II. 102.

poplande, pple., *rushing* : þe pure poplande hourle plaies on mi heved A. P. iii. 319; þe pure populand hurle [perle] passis it umbi ALEX. (Sk.) 1154.

popler, v., *Fr.* poplier; *poplar, 'populus,'* PR. P. 408; PALL. iii. 194.

poppere, sb., *dagger*, CH. C. T. *A* 3931.

[popul, sb., *Fr.* peuple, *Lat.* pōpulus ; *poplar.*] popul-trē, sb., '*populus,*' PR. P. 408.

popul², sb., *a weed*, '*lolium,*' VOC. 201; popil '*gith, nigella*' CATH. 286. popil-mele, sb., *meal made of the seeds of this plant*, REL. I. 53.

porail, *see* pouraille.

porc, sb., *O.Fr.* porc; *pork, swine*, S. A. L. 156; pork RICH. 3049; porke D. TROY 3837; porkes (*pl.*) D. ARTH. 3121.

porche, v., *Fr.* porche; *porch*, ROB. 271; AYENB. 135; LANGL. *B* xvi. 225.

porciōn, sb., *O.Fr.* porcion; *portion*, CHR. E. 352; MAN. (F.) 14845; porcioun CH. C. T. *B* 1246; C. B. 18.

pōre, sb., *Fr.* pore; *pore (of the skin)*, '*porus,*' PR. P. 409.

pōre, *see* povre.

pōren, v., *cf. M.Du.* poren; *pore, look*; pure HORN (L.) 1092; pore [poure] and prie CH. C. T. *D* 1738; pouri AYENB. 177; pōred (*pret.*) ALIS. (Sk.) 1080.

pōren, v., *pour*; porin PR. P. 409; pore C. B. 16; pouren (*pres.*) CH. C. T. *G* 670; pourede (*pret.*) LANGL. *A* v. 134; the water on his hede she poured out GOW. I. 302.

poret, *see* porrette. porfil, *see* purfil.

pork, *see* porc.

porknell, sb., *little pig* ; Polidarius the porknell D. TROY 6368.

porpeis, sb., *O.Fr.* por(c)peis; *porpoise* ; purpeis '*foca, vitula marina, suillus*' PR. P. 417; C. B. 105; porpeis in broth F. C. 53.

porren, v., *thrust*; porris (*pres. pl.*) ALEX. (Sk.) 5560.

porrette, sb., *cf. Ital.* porretta ; *sort of leek* ; poret LANGL. *B* vi. 3co; PR. P. 409; porettes (*pl.*) LANGL. *B* vi. 288.

porett-plontes, sb., pl., *leeks*, LANGL. *C* ix. 310.

porse, *see* purse.

I i

port, sb., *O.E.*, *O.Fr.* port, *Lat.* portus ; *port, harbour,* FRAG. 4 ; AYENB. 86 ; **Portes** mouþe *Portsmouth* ROB. (W.) 5223 ; **porz** (*pl.*) LA3. 24415 ; pors ROB. (W.) 1169.

  Port-chæstre [**Portchestre**], pr. n., *Porchester,* LA3. 11239.

  port-dogge, port-hound, sbs., *town dog,* SAINTS (Ld.) xlv. 267, 274.

  port-rēve, sb., *port-reeve,* ROB. 540 ; GREG. 501 ; **portrēven** (*pl.*) ROB. (W.) 11205.

  port-toun, sb., *port-town, busy town,* SAINTS (Ld.) xlv. 267.

port [2], sb., *O.Fr.* port, *cf. Ital.* porto ; *deportment, bearing,* '*gestus,*' PR. P. 409 ; CH. C. T. *A* 69 ; porte LANGL. *B* xiii. 278, *C* vii. 30.

portal, sb., *O.Fr.* portal ; *portal :* þe **portalez** (*pl.*) piked of rich plates A. P. i. 1036.

portatif, adj., *O.Fr.* portatif ; *portable, light,* LANGL. *B* i. 155, *C* ii. 154.

portehors, sb., *O.Fr.* portehors ; *breviary* ; portous [porthors, porthous, porthos, portos] LANGL. *B* xv 122 ; poortos '*portiforium, breviarum*' PR. P. 410 ; on mi porthos i make an oath CH. C. T. *B* 1321.

porter, sb., *O.Fr.* portier ; *porter, doorkeeper,* FL. & BL. 241 ; ROB. 159 ; portour LANGL. *C* xxiii. 330 ; **porteres** (*pl.*) *B* v. 628 [porters *A* vi. 108 ; portours *C* vii. 370].

Porteshām, -hōm, pr. n., *Portsham,* O. & N. 1752, 1791.

portoure, sb., *O.Fr.* porteor ; *porter, carrier,* '*portitor, portator, gestor, calo,*' PR. P. 410.

portreie, porveance, *see* **pur-.**

póse, sb., *O.E.* geposu (*pl.*) (LEECHD. I. 148), *catarrh,* '*catarr(h)us,*' PR. P. 410 ; P. L. S. ix. 92 ; CH. C. T. *A* 4152 ; (he) fneseth faste, and eek he hath the pose *H* 62.

pósen, v., *O.Fr.* poser ; *suppose,* LANGL. *B* xviii. 293, *C* xx. 275.

poshote, sb., *posset,* C. B. 15, 36 ; poshotte 36 ; possot '*balducta*' PR. P. 410.

posnet, sb., *O.Fr.* poçonet, *prov.Eng.* posnet (*saucepan*) ; *small pot ;* '*urceus,*' PR. P. 410 ; E. W. 46 ; possenet C. B. 23, 72 ; **posnettis** (*pl.*) WICL. 2 PARAL. xxxv. 13.

posséde, v., *O.Fr.* posseder ; *possess,* ALEX. (Sk.) 2841.

possessiōn, possessioun, sb., *O.Fr.* possession ; *possession,* AYENB. 149, 150 ; possessioun CH. C. T. *A* 2242.

possessioner, sb., *beneficed clergyman* ; **possessioneres** (*pl.*) LANGL. *B* v. 144.

possibilitē, sb., *O.Fr.* possibilité ; *possibility,* CH. C. T. *A* 1291.

possible, adj., *O.Fr.* possible ; *possible,* CH. C. T. *E* 956 ; A. P. i. 452.

posson, v., *O.Fr.* poulser ; *push,* '*trudo,*' PR. P. 410 ; posse HORN (L.) 1011 ; posshen LANGL *C* vii. 96 ; **possed** (*pret.*) LANGL. *B* PROL. 151 ; puste HORN (R.) 1079 ; **possid** (*pple.*) LANGL. *B* PROL. 151.

post, sb., *O.E.* post, *from Lat.* postis ; *post,* TREAT. 135 ; **postes** (*pl.*) LA3. 28032 ; LANGL. *B* xvi. 54.

postel, sb., *O.Fr.* postel ; *post, stake,* C. M. 14980 ; **postles** (*pl.*) HOM. I. 127 ; LA3. 1316 ; JUL. 56.

posterne, sb., *O.Fr.* posterne, posterle, *Lat.* posterula ; *postern,* ROB. 19 ; WILL. 1752 ; **posternes** (*pl.*) LANGL. *B* v. 628.

postle, sb., *for* apostel ; *apostle, preacher* ; **postles,** posteles (*pl.*) LANGL. *B* vi. 151, xvi. 159.

pot, sb., *cf. L.G., Du., Dan., Fr.* pot ; *pot,* A. R. 368 ; AYENB. 58 ; IW. 759 ; potte LANGL. *B* xiii. 255 ; **puttes** (*pl.*) PALL. ii. 253 ; pottis *round deep holes* BARB. xi. 364, 385 ; *comp.* **pis-, water-pot.**

  pot-ful (? **pot ful**), sb., *potful,* LANGL. *B* vi. 189.

  pot-schoord, sb., *potsherd,* VOC. 171 ; pot scarth HAMP. PS. xxi. 15*.

  pot-spōn, sb., *ladle,* PR. P. 411.

potāge, sb., *O.Fr.* potāge ; *pottage,* A. R. 412 ; ROB. 404 ; CH. C. T. *B* 3623 ; L. C. C. 17 ; potage ware *potherbs* PALL. vii. 57 ; **potāges** (*pl.*) LANGL. *C* xvi. 47.

potagēr(e), sb., *Fr.* potager ; *pottage-maker,* LANGL. *B* v. 157, *C* vi. 132 ; '*leguminarius*' CATH. 288.

potel, sb., *O.Fr.* potel ; *pottle,* '*laguncula,*' PR. P. 411 ; WICL. IS. x. 33 ; E. G. 59 ; potel, potell LANGL. *B* v. 348, *A* v. 192, *C* vii. 399.

poten, *see* **puten.**

potente, sb., *cf. M.Du.* potente ; *tipped staff,* LANGL. *B* viii. 96 ; CH. C. T. *D* 1776.

pottare, sb., *potter,* '*ollarius,*' PR. P. 411 ; **potteres** (*gen.*) C. M. 16536

pouche, sb., ? = **póke** ; *pouch,* '*marsupium,*' PR. P. 411 ; CH. C. T. *A* 3931.

pouderon, v., *Fr.* poudrer ; *grind to powder,* '*condio,*' PR. P. 411 ; **pūdrid** (*pple.*) P. L. S. xxxv. 110 ; poudered A. P. i. 36 ; pouderd *salted* C. B. 14.

poudre, sb., *O.Fr.* poudre ; *powder,* P. L. S. xvii. 225 ; SHOR. 162 ; poudre caste *covered with dust* PALL. viii. 12 ; **poudres** (*pl.*) AYENB. 148 ; pouderes LANGL. *C* xxiii. 359.

pouēr, sb., *O.Fr.* pouoir, pooir ; *power ; army;* AYENB. 71 ; A. P. ii. 1654 ; (*ms.* power) SHOR. 21 ; PR. C. 5884 ; poer TREAT. 133 (paoer SAINTS (Ld.) xlvi. 427) ; pore, poeir, poer ROB. (W.) 2049, 7639, 4523 ; PR. P. 411 ; pouwere [power] ALEX. (Sk.) 2056.

pouke, *see* **pūke.**

poun, sb., *O.Fr.* peōn ; *pawn at chess,* CH. D. BL. 661 ; aufins and **pounis** (*pl.*) GEST. R. 70.

pound, *see* **púnd.**

pounen, v., *O.E.* punian ; *pound* ; poune WICL. MAT. xxi. 44 ; *comp.* **to-ponen.**

pouraille (**? povraille**), sb., *O.Fr.* pouraille; *poor people*, P. S. 223; pôraile WILL. 5123; poveraill, poveralȝe, poverale BARB. viii. 275, 368; poraille MAN. (F.) 6664; **purraileis** (*gen.*) DEP. R. ii. 165; **porails** (*pl.*) WICL. PROV. xxx. 14.

pouren, *see* **póren**.

pous, sb., *O.Fr.* pous; *pulse*, LANGL. *B* xvii. 66, *C* xx. 66; pouce, veine '*pulsus*' PR. P. 411; MAN. (F.) 9011.

poustē, sb., *O.Fr.* poesté; *power, ability*, LEG. 12; H. H. 7; CH. BOET. iv. 5 (131); MAN. (F.) 1244; PR. C. 3996; **poustees** (*pl.*) *violent attacks* LANGL. *B* xii. 11.

pouten, v., *pout*; **poute** (*pres.*) REL. II. 211.

povertē, sb., *O.Fr.* poverté; *poverty, meanness*, HOM. I. 143; A. R. 260; P. L. S. xxv. 65; C. L. 854; AYENB. 132; LANGL. *B* viii. 116, *C* xi. 116; **povert** *A* ix. 111, *B* xi. 264.

povre, adj., *O.Fr.* povre; *poor*; (*ms.* poure) KATH. 50; A. R. 260; O. & N. 482; R. S. v; HAV. 138; povere LAȝ. 22715; C. L. 890; povere [pore] LANGL. *B* v. 257; pore BEK. 1120; pover (*r. w.* recover) GOW. I. 357; REL. I. 115; **povre** (*sb.*) *the poor* LANGL. *C* xx. 237; **poverer** (*comp.*) ROB. (W.) 7617*.

povre-hēde, sb., *poverty*, AYENB. 130.

povre-līke, adv., *poorly*, HAV. 323; poure-liche CH. C. T. *E* 213, 1055; she praiet him pourli (*beseechingly*) with hir pure hert D. TROY 11553.

povernesse, sb., *poverty*, MISC. 75.

practif, adj., *experienced*, ALEX. (Sk.) 1582.

practisoure, sb., *practitioner, man of experience*, LANGL. *B* xvi. 107; **practisirs** ALEX. (Sk.) 1582.

prai, *see* **preie**.   **praiabill**, *see* **preiable**.

praiel, sb., *O.Fr.* praiel, prael; *meadow*, '*pratellum*,' PR. P. 411.

praiere, sb., *O.Fr.* praerie; *meadow*, GAW. 768.

prancen, v., *prance*, GOW. III. 41; praunce GAW. 2064; **prauncede** (*pret.*) FER. 5341.

prāne, sb., *prawn*, '*stingus*,' PR. P. 411.

pranglen, v., *cf. Goth.* praggan; *press, fetter*: i was þer with so harde **prangled** (*pple.*) HAV. 639.

prank, sb., *fold*; '*plica*,' PR. P. 411.

prankid, pple., *flounced*, '*plicatus*,' PR. P. 411; prankid gounes TOWNL. 312.

prat, adj., *O.E.* prætt, *cf. M.Du.* prat (*haughty, arrogant*); *cunning*; (*ms.* pratt) ORM. 6652; mid præt [pret] wrenchen (? prætwrenchen) LAȝ. 81, 5302.

práten, v., =*Du.* praten, *Dan.* prate, *Swed.* prata; *prate*; prate LIDG. M. P. 155; LUD. COV. 353.

prati, adj., *O.E.* prætig, prættig (*cunning*); *pretty*, '*formosus*,' PR. P. 411; D. TROY 13634; LUD. COV. 104; S. & C. I. lxi.

prati-lich, adv., *prettily*, REL. I. 63; pratili LUD. COV. 82.

precep [**precept**], sb., *O.Fr.* precep, precept; *instruction*, ALEX. (Sk.) 982.

prēchen, v., *O.Fr.* prêcher, *cf. O.E.* prēdician; *preach, speak, utter*, A. R. 70; preche MISC. 44; **prēcheþ** (*pres.*) LANGL. *C* xxiii. 277; prechen, precheþ *A* iii. 216, *C* iv. 279; **prēchinge** (*pple.*) *C* i. 57; prechande A. P. ii. 942; **prēchede** (*pret.*) C. L. 1416; LANGL. *C* vi. 115; **prēchede** (*pple.*) LANGL. *C* i. 57.

prēchinge, sb., *preaching*, MISC. 56; AYENB. 191; LANGL. *C* viii. 286; prechingue SAINTS (Ld.) xxvii. 1933.

prechūr, sb., *Fr.* precheur; *preacher*, A. R. 160; prechour SHOR. 48; **prechoures** (*pl.*) LANGL. *B* xii. 19.

precious, adj., *O.Fr.* precius; *precious*, CL. M. 82; AYENB. 112; SPEC. 103; precios A. P. i. 4; **preciousest** (*superl.*) LANGL. *A* ii. 12.

prēde, *see* **prūde**.

predecessour, sb., *one going before*; predecessour of princes *superior of all princes* ALEX. (Sk.) 1723.

predicatioun, sb., *O. Fr.* predicatiun; *preaching*, BEK. 1957.

preiable, adj., *to be entreated*; praiabill HAMP. PS. lxxxix. 15.

preie, sb., *O.Fr.* preie; *prey*, ROB. 270; praie MAN. (F.) 4513; prai ALEX. (Sk.) 1335; C. L. 1054; bi **praies** (*pl.*) ['*de praedis*'] TREV. I. 127.

preien, v., *O.Fr.* preier; *pray*; preie HORN (L.) 763; GREG. 861; **preie** (*pres.*) HAV. 1440; preie SPEC. 58; **preied** (*pret.*) LANGL. *B* i. 80 [priede *A* i. 78]; preieden WICL. JOHN iv. 31; **praid** (*pple.*) ALEX. (Sk.) 3183.

preiēre, sb., *O.Fr.* preiere; *prayer*, SAINTS (Ld.) xiv. 42; LANGL. *B* xiii. 135; CH. C. T. *A* 1204; MAN. (F.) 1373; preȝere AL. (V.) 304; **prairis** [praiers] (*pl.*) ALEX. (Sk.) 1483.

preis, sb., *O.Fr.* preis; *praise*, CH. TRO. ii. 1585; praise A. P. i. 301; pres A. P. 419; C. M. 6358.

preisen, v., *O.Fr.* preiser; *praise; appraise, value*; A. R. 64; preise HAV. 60; LANGL. *A* v. 174; CH. TRO. ii. 1583; praisi AYENB. 135; **preisede** (*pret.*) BEK. 1360; preiseden LANGL. *A* v. 177; preised (*pple.*) MAN. (F.) 3119; praised ALEX. (Sk.) 2666.

prejudice, sb., *O.Fr.* prejudice; *prejudice*, SHOR. 36.

prekien, *see* **prikien**.

prelāt, sb., *O.Fr.* prelat; *prelate*, LAȝ. 24502; CH. C. T. *A* 204; **prelāȝ** (*ʒl.*) A. R.

10; prelas AVENB. 39; prelates [prelatus]
LANGL. *A* i. 101.

prengen, v.,=*M.H.G.* phrengen, *M.Du.*
prangen; *press*; **preinte** (*pret.*) LANGL.
B xiii. 112; prengte *C* xvi. 121*; Richard
preinte (h)is eʒe (*winked*) oppon . . . FER.
4507.

prente, sb., *O.Fr.* (em-)preinte; *print, im-
print*, LEB. JES. 390; preinte [*v. r.* preente,
prente] LANGL. *C* xviii. 73; prente [*v. r.*
printe] CH. C. T. *D* 604; priente AYENB.
81; preente '*effigies, impressio*' PR. P. 412;
print *signature* ALEX. (Sk.) 3162*.

prentin, v., *print, 'imprimo,'* PR. P. 412;
with pine printed (*pple.*) in hert D. TROY 195.

prentis, sb., *apprentice, pupil*, LANGL. *C* iii.
224, *B* xiii. 393, *A* v. 116; **prentises** (*pl.*)
C xxii. 231; preentices *B* xix. 226; *see*
aprentis.

prentis-hōde, sb., *apprenticeship*, LANGL.
B v. 256, *C* vii. 251.

prēon, sb., *O.E.* prēon,=*L.G.* prēn, *M.H.G.*
phriem; *pin*, A. R. 84; **prēne** (*dat.*) ANT.
ARTH. xxix; **prēones** (*pl.*) KATH. 1947;
prenes GEN. & EX. 1872; *comp.* ēar-**prēon**.

prēonen, v., *stick with a pin*; prene ['*con-
figere*'] WICL. I KINGS xix. 10; SACR. 467;
prēoneð (*pres.*) R. S. v; þurgh his herte he
prēned (*pret.*) him MED. 859; **prēned** (*pple.*)
ALIS. (Sk.) 420; iprened in an clout ST. COD.
98; *comp.* bi-**prēonen**.

prēost, sb., *O.E.* prēost, *cf. O.Fr.* prestre;
priest, A. R. 16; BEK. 364; preost [prest]
LAʒ. 1; preost [prost], prest O. & N. 322,
902; pruest P. S. 159; prest GEN. & EX.
3922.

prēost-hood, sb., *O.E.* prēosthād; *priest-
hood*, TREV. IV. 105; LANGL. *C* xxii. 334.

presence, sb., *O.Fr.* presence; *presence*,
LANGL. *B* PROL. 173; CH. BOET. i. 3 (10);
i am here in your presente [*r. w.* tente] A. P. i.
389; & aprochen to his presens ii. 8.

present, sb., *O.Fr.* present; *present*, GEN. &
EX. 1831; AYENB. 189; þet present A. R.
152; present *signature* ALEX. (Sk.) 3162;
presands (*pl.*) 5466.

presenten, v., *O.Fr.* presenter; *present, represent*, PR. P. 412; presente MAN. (F.)
3219; thou shuld . . . present min estate
D. TROY 2190; i presend (*pres.*) ʒou . . . foure
hundreth fellis ALEX. (Sk.) 5138; (þai) pre-
sandis [present] *make presents* 1041; present
deliver 1791; **presentede** (*pret.*) CHR. E.
625; presentide LANGL. *C* xxii. 92; presented
A. P. 1217.

presōn, *see* prisōn.

press, sb., *from* pressen; *press; throng,
crowd*: vor me is loð presse A. R. 168; put
him in a presse LANGL. *B* v. 213; his presse
(*cupboard*) icovered with a falding reed CH.
C. T. *A* 3212; in grete presse P. L. S.

xxi. 30; as he eode in grete prece SAINTS
(Ld.) xxvi. 30; prees or throng '*pressura,*'
presse or pile of clothe '*pressorium,*' '*panni-
plicium,*' presse for grapis '*prelum*' PR. P.
412; pres, prees, *crowd* MAN. (F.) 369, 720.

pressin, v., *O.Fr.* presser; *press, 'premo,
comprimo, presso,*' PR. P. 412; prese ALEX.
(Sk.) 3483; i **presse** (*pres.*) win L. H. R. 136;
presis *hastens* ALEX. (Sk.) 954; **presed**
(*pret.*) MAN. (F.) 13811.

pressūre, sb., *O.Fr.* pressour, pressoir; *press*,
PR. P. 412; pressour WICL. APOC. xix. 15;
pressours (*pl.*) WICL. PS. viii. 1.

prest, adj., *O.Fr.* prest; *ready, prompt*, MARG.
98; WILL. 1598; LANGL. *B* xiv. 220; **prester**
[prestiore] (*comp.*) LANGL. *B* x. 289.

prest-lī, adv., *readily*, WILL. 1146; LANGL.
B vi. 95; prestliche *C* iv. 308.

prēst, *see* prēost. **pret**, *see* prat.

presumciōn, sb., *O.Fr.* presumpcion; *presumption, assumption*, AYENB. 17; presum-
ciun A. R. 208; presompcion LANGL. *C* xiv.
232; **presompcions** (*pl.*) *C* xii. 39.

presūmen, v., *O.Fr.* presumer; *presume*;
presūmen (*pres. pl.*) LANGL. *C* i. 135;
presūmit (*pret.*) BARB. i. 572, xi. 143.

pretenden, v., *O.Fr.* pretendre; *intend*, CH.
TROI. iv. 922; (þes Jewes) pretende (*pres.*)
me to take YORK xxviii. 52.

pretten, v., *cf. M.Du.* pratten; *take pride*: ief
þe pokoc him **prette** (*pret.*) vor his vaire taile
AYENB. 258.

préven, *see* próven.

pricasour, sb., *hard rider*, CH. C. T. *A* 189.

pricche, = **prike**, sb.; *? prick*; lecheries
pricches (*pl.*) A. R. 60.

pricchen,=**prikien**,v., *? prick*; priʒte (*pret.*)
HOM. II. 257.

prīde, prīden, *see* prūde, prūden.

prīen, v., *pry*; prie P. S. 222; WILL. 5019;
CH. C. T. *A* 3458; **prīed** (*pret.*) LANGL. *B*
xvi. 168.

prike, sb., *O.E.* prica, pricca,=*M.Du.* prick;
prick, spike, spur, mark on a target, MAT. v.
18; prike, prikke '*spintrum vel spinter,
cavilla, broccus, meta*' PR. P. 413; prikke
FRAG. 5; A. R. 228; GOW. I. 283; prikke
AYENB. 71; CH. H. F. 907; C. T. *B* 119; prick
ALEX. (Sk.) 45; (*point of time*) 4630.

priket, sb., *young buck,* '*capriolus,*' PR. P. 413.

priket[2], sb., *spike of a candlestick,* '*stiga,*'
PR. P. 413.

prikien, v., *? O.E.* prician,=*M.Du.* pricken;
prick, spur, FRAG. 8; priken HAV. 2639;
prikie P. L. S. xxi. 131; prike WILL. 2382;
prikie [prike prekie] LANGL. *B* xviii. 11;
prikin, prikkin '*pungo, stimulo*' PR. P. 413;
prikke S. & C. I. li.; prikeð (*pres.*) GEN. & EX.
3964; neddre . . . attreð hwat heo prikeð
HOM. II. 191; prikeþ CH. C. T. *A* 1043;

(heo) prikieỡ A. R. 244 ; prikieþ, prekieþ AYENB. 230, 257 ; **prikiende** (*pple.*) FRAG. 8 ; prikinde A. R. 134 ; prekande PERC. 605 ; **priked** (*pret.*) S. S. (Wr.) 2551 ; PR. C. 5338 ; he priked his stede K. T. 1080 ; **priked** (*pple.*) TREV. V. 371 ; proudli prikid all in prose ALEX. (Sk.) 5074.

prikere, sb., *horseman* : a proud prikere of Fraunce LANGL. *A* x. 8 ; a prikere on a palfrai *B* x. 308 ; prikiere LANGL. *C* xi. 134 ; prickker (*misprinted* pukken) PALL. iv. 845.

prikinge, sb., *pricking* ; prikiunge A. R. 234 ; pricking (*remorse*) in hert D. TROY 2183.

prikil, sb., *O.E.* pricle,=*M.Du.* prikel, prekel ; *prickle,* '*aculeus,*' PR. P. 413 ; H. S. 8486 ; prikel MAN. (F.) 16218.

prille, sb., *peg-top,* '*gyraculum,*' PR. P. 413.

primāte, sb., *O.Fr.* primat ; *primate, chief* : of þissen londe he wæs primate LAȜ. 29736 ; þe primate in blisse HAMP. PS. xxiii. 6 ; MAN (F.) 15302.

prīme, sb., *O.Fr.* prīme, *from Lat.* prīma (hōra) ; *prime (six o'clock in the morning),* A. R. 24 ; HORN (L.) 966 ; IW. 2304 ; CH. C. T. *B* 1396 ; bitvene þe prime and þe none GREG. 778 ; heiȝ prime LANGL. *A* vii. 105.

prīme-tīde, sb., *prime,* HORN (L.) 857.

prīme², adj., *prime, vigorous,* LANGL. *A* xii. 60 ; DEP. R. iii. 34.

primēr, sb., *primer,* '*primarius,*' PR. P. 413 ; CH. C. T. *B* 1707.

primerole, sb., *O.Fr.* primerole ; *primrose,* SPEC. 26 ; CH. C. T. *A* 3268.

primseinen, v., *O.N.* primsigna ; *sign (a person) with the cross, make a catechumen* ; primseȝnest (*pres.*) ORM. 1542 ; Martin yet nou i-primséned (*pple.*) AYENB. 188 ; primseȝned ORM. 16560.

prince, sb., *O.Fr.* prince ; *prince,* MARH. 2 ; BEK. 1342 ; SHOR. 15 ; prins LANGL. *C* xiii. 176.

prins-hōd(e), sb., *princely dignity,* WICL. MK. x. 42, JUDE 6.

princesse, sb., *Fr.* princesse ; *princess,* '*principisca,*' PR. P. 413 ; princes ALEX. (Sk.) 5099.

principal, adj., *O.Fr.* principal ; *principal,* AYENB. 106 ; þe palais principale A. P. ii. 1531.

    principal-līche, adv., *principally,* AYENB. 26 ; principaliche LANGL. *B* xiv. 194.

principalitē, sb., *O.Fr.* principalité ; *principality, chief place* ; '*principalitas,*' PR. P. 413 ; A. P. ii. 1672 ; principalete, principalte [principalite] ALEX. (Sk.) 648, 1737, 2311.

printe, *see* **prente.**

prior, sb., *O.Fr.* prior, priour, priur ; *prior* ; priour '*prior*' PR. P. 413 ; LANGL. *B* xi. 156 ; MAN. (F.) 7065 ; **prioures** (*pl.*) LANGL. *B* x. 267.

prioresse, sb., *O.Fr.* prioresse ; *prioress,* CH. C. T. *A* 118.

priorie, sb., *priory,* HAV. 2522.

prīs, sb., *O.Fr.* prīs, *cf. M.Du.* prijs, *O.N.* prīss *m. ; prize, price, high esteem,* A. R. 392 ; GEN. & EX. 326 ; HAV. 283 ; FL. & BL. 750 ; BEK. 150 ; AYENB. 234 ; CH. C. T. *A* 815 ; A. P. i. 272 ; AV. ARTH. xxxiv ; he berþ þat pris MISC. 98 ; þe pris for te winne SPEC. 103 ; *see* **preis.**

prīs-hēde, sb., *valour,* D. TROY 2907.

prīs-neet, sb., *prize ox,* LANGL. *B* xix. 261.

prīs-tale, sb., *excellent tale,* WILL. 161.

prīsin, v., *O.Fr.* prīsier, *cf. M.Du.* prijsen *O.N.* prīsa ; *appraise, value ; praise, extol* : '*taxo, metaxo,*' PR. P. 414 ; priss BARB. vi. 505, vii. 99 ; **prīsit** (*pret.*) BARB. xvi. 672 ; prīsit (*pple.*) x. 776, xvi. 525.

prīsare, sb., *appraiser,* '*metaxarius,*' PR. P. 413.

prisōn, sb., *O.Fr.* prison ; *prison,* LANGL. *A* ix. 94 ; prison. [prisoun] C. L. 368 ; prisoun BEK. 7 ; (*prisoner*) WILL. 1251 ; prisun A. R. 126 ; GEN. & EX. 2040 ; preson DEP. R. iii. 303 ; in ane prisune HOM. I. 33 ; to prisune (*prisoner*) 13 ; of prisune LAȜ. 1016* ; **prisōns** (*pl.*) captives LANGL. *C* viii. 277.

prisonēr, sb., *O.Fr.* prisonier ; *prisoner,* WILL. 1267 ; prisuner (*prison keeper*) GEN. & EX. 2042 ; presonere A. P. ii. 1217 ; **prison-ēres** (*pl.*) LANGL. *B* iii. 136.

privē, adj., *O.Fr.* privé ; *privy, secret, intimate,* LAȜ. 6877* ; A. R. 168 ; ROB. 25 ; privi CH. C. T. *G* 1452 ; previ DEP. R. iii. 111 ; **previest** (*superl.*) *C* xix. 98* ; **privēs** (*sb., pl.*) *intimates* MAN. (F.) 9299.

    privē-liche, adv., *secretly,* ROB. 104 ; AYENB. 225 ; previliche DEP. R. ii. 122 ; preveili LANGL. *C* xvi. 153.

prīven, v., *Fr.* priver ; *deprive* ; privin '*privo*' PR. P. 414 ; prife HAMP. PS. lxxvii. 65, lxxxiii. 13 ; **prīvid** (*pret.*) ci. 18.

privetē, sb., *O.Fr.* priveté ; *secrecy, silence,* HOM. I. 185 ; privete, privite PR. C. 3775, 4651 ; privite A. R. 152 ; CH. C. T. *G* 701 ; privettee *G* 1052, 1138 ; in prevate ALEX. (Sk.) 4997.

privilege, sb., *O.Fr.* privilege ; *privilege,* A. R. 160 ; AYENB. 15 ; **privilegies** (*pl.*) SAX. CHR. 263.

proces, sb., *Lat.* processus ; *narrative,* CH. C. T. *B* 3511 ; proces holde (*keep close to my story*) *F* 658 ; ALEX. (Sk.) 4259 ; **proses** (*pl.*) D. TROY 247, 13774.

processiōn, sb., *O.Fr.* procession ; *procession,* HOM. II. 89 ; processiun LAȜ. 18223 ; processioun P. L. S. xiii. 304 ; prosessioun A. P. i. 1096.

procuracie, sb., *procuracy, proxy* ; prokecie '*procuracia*' PR. P. 414 ; **procuracies** (*pl.*)

P. L. S. xvii. 320 ; **procurases** (*pl.*) *agencies* (*misprinted* prouianses) LANGL. *B page* vi.

procuratour, sb., *procurator, proctor,* CH. C. T. *D* 1596; prokeratour, -ketour '*procurator*' PR. P. 414.

procūren, sb., *O.Fr.* procurer; *procure*; procuri BEK. 1276; (to) proker hir pes D. TROY 11555; **procūre** (*pres.*) MAN. (F.) 7462.

   prokūring, sb., *procuring* : thurgh his prokuring (*intervention*) D. TROY 13766.

prodigalitē, sb., *O.Fr.* prodigalité; *prodigality,* AYENB. 21.

proēme, sb., *O.Fr.* proeme *from Lat.* proœmium; *proem, preface*; proheme CH. C. T. *E* 43.

profer, sb., *offer, proffer*; proffer D. TROY 262; þo **profers** (*pl.*) were made 250.

proferin, v., *O.Fr.* proferer; *proffer,* '*offero,*' PR. P. 414; profre LANGL. *B* xvii. 140; to proffer (*risk*) our persons D. TROY 3139; **profrest** (*pres.*) LANGL. *A* viii. 27 ; **profered** (*pret.*) WILL. 1267 ; profrede LANGL. *C* v. 91; proffrede *C* v. 67 ; **profert** (*pple.*) A. P. ii. 1463.

profess, sb., *profession* : þe abit of nonne heo tok ac me nolde hire profes noȝt make anone wise ROB. (W.) 8944.

professiōn, -fessioun, sb., *O.Fr.* profession, professiun; *profession,* LANGL. *B* i. 98; professiun A. R. 6.

profit, sb., *O.Fr.* profit; *profit,* AYENB. 99 ; LANGL. *B* PROL. 169; CH. BOET. i. 4 (15); prophete DEP. R. iv. 10, 48 ; **profectez** (*pl.*) ALEX. (Sk.) *page* 280.

profitable, adj., *O.Fr.* profitable ; *profitable,* AYENB. 185 ; LANGL. *A* i. 120.

profitin, v., *O.Fr.* profiter; *profit, grow,* PR. P. 414 ; profiti AYENB. 126 ; **profitide** (*pret.*) WICL. LK. ii. 52 ; profet ALEX. (Sk.) 2370.

profound, adj., *Fr.* profond ; *profound,* P. L. S. xvii. 221.

progenie, sb., *O.Fr.* progenie ; *progeny,* S. A. L. 145; WICL. GEN. xliii. 7 ; ALEX. (Sk.) 4021.

proinen, v., *? OFr.* progner, *mod.Eng.* prune, preen ; *preen* : **proineþ** [pruneþ] (*pres.*) him and pikeþ CH. C. T. *E* 2011 (Wr. 9885) ; he pruneþ him and pikeþ as doþ an hauke GOW. III. 75 ; pruinen ANGL. I. 95.

projecte, sb., *design,* ALEX. (Sk.) 3331.

proker, see **procūren.**

prókien, v., *cf. L.G.* proken; *stimulate*; prokie H. M. 47 ; **prókeð** (*pres.*) A. R. 204 ; **prókie** (*subj.*) A. R. 238 ; **prókede** (*pret.*) P. S. 343.

   prókinge, sb., *goading*; prokinnge A. R. 266.

prokkin, v., *beg, arrogate,* '*procor,*' PR. P. 414 ; **proches** [prokes] (*pres.*) ALEX. (Sk.) 1926.

prolouge, sb., *prologue, announcement,* ALEX. (Sk.) 2730, 5066.

prollin, v., *prowl,* '*scrutor,*' PR. P. 415 ; **prolle** (*pr. subj.*) CH. C. T. *G* 1412.

pro-, purlongin, v., *O.Fr.* prolonguer, *mod. Eng.* purloin ; *remove far off,* '*prolongo,* alieno,*' PR. P. 417 ; **proloined** (*pple.*) GEST. R. 135.

promissiōn, sb., *O.Fr.* promission ; *promise,* GEN. & EX. 4131 ; promissioun GEST. R. 134.

promōciōn, sb., *O.Fr.* promotion ; *promotion,* '*promocio,*' PR. P. 415.

promptin, v., *Lat.* promptāre ; *prompt,* '*promo, insenso,*' PR. P. 415.

pronge, sb., *cf. M.L.G.* prange, *M.Du.* pranghe ; *pang,* '*aerumna,*' PR. P. 415 ; **prongis** (*pl.*) LUD. COV. 287.

pronounce, sb., *O.Fr.* pronuncer ; *pronounce,* CH. C. T. *G* 1299; **pronounsand** (*pple.*) ALEX. (Sk.) 3395.

prophecie, sb., *O.Fr.* prophecie ; *prophecy,* A. R. 158; ROB. 132 ; TREV. I. 421 ; propheci C. M. 18988 ; **profecies** (*pl.*) ii. 1158.

prophecīen, v., *prophesy*; **propheciede** (*pret.*) TREV. VIII. 35 ; **prophesīd** (*pple.*) ALEX. (Sk.) 1896 ; *see* **prophetīsen.**

prophēte, sb., *O.Fr.* prophete ; *prophet,* ORM. 5195 ; profete 5801 ; A. P. i. 797.

prophetīsen, v., *O.Fr.* prophetizer, *from late Lat.* prophētīzāre ; *prophecy* ; **prophetised** (*pret.*) MAN. (F.) 16606 ; *see* **propheciēn.**

proporciōn, sb., *O.Fr.* proporcion ; *proportion,* ALEX. (Sk.) 668.

proppe, sb., *cf. M.Du.* proppe ; *prop,* '*contus,*' PR. P. 415.

propre, adj., *O.Fr.* propre ; *proper ; separate ; handsome*; PL. CR. 569 ; of propre kind CH. C. T. *F* 610 ; thou art a propre man *C* 309 ; here is propre service LANGL. *B* xiii. 51 ; **propurest** (*superl.*) ALEX. (Sk.) 3331 ; **propre** (*sb.*) MAN. (F.) 2380.

   propre-līche, adv., *properly,* A. R. 98 ; AYENB. 34 ; propreliche, propreli LANGL. *C* xvi. 153, *B* xiv. 274.

propreté, sb., *O.Fr.* propreté ; *property, peculiarity* ; propirte DEP. R. iii. 38; so passis þi propurti prete wemen all D. TROY 626 ; propurte 2530 ; **propertēs** (*pl.*) ALEX. (Sk.) 4257 ; propurtes DEP. R. iii. 65.

prōse, sb., *Fr.* prose ; *prose,* LEG. 60 ; CH. C. T. *B* 2124 ; sais þe prose (*prose story*) ALEX. (Sk.) 2062.

prospectives, sb., pl., *perspective glasses,* CH. C. T. *F* 234.

prosperitē, sb., *O.Fr.* prosperité ; *prosperity,* A. R. 194 ; ALEX. (Sk.) 1860.

prou, sb., *O.Fr.* prou ; *profit, advantage,* P. L. S. xvii. 311 ; SHOR. 147 ; AYENB. 78 ;

MIRC 548 ; LANGL. *C* i. 145* ; MAN. (F.)
1477 ; pru 8820.

proud, *see* prūd.

prouesse, sb., *O.Fr.* proesse; *prowess, valour,*
ROB. 112 ; AYENB. 90 ; pruesse HORN (L.)
556 ; MAN. (F.) 3119 ; proues GAW. 912, 1249.

prous, preus, adj., *O.Fr.* prous, preus;
valiant, AYENB. 83.

prout, *see* prūd.

próve, sb., *O.Fr.* prove, proeve ; *proof,* SHOR.
62 ; preove A. R. 52 ; preve CH. P. P. 497.

próven, v., *O.Fr.* prover; *prove,* AR. & MER.
3011 ; proven, prove, provi BEK. 788, 791,
1006 ; prove HORN (L.) 545 ; provie SHOR.
137 ; provi AYENB. 158 ; proven LANGL. *B*
viii. 120 [preven *A* ix. 115]; preoven A. R.
390 ; preve MAND. 50 ; próve ['*probet*']
(*imper.*) ech man him selven HOM. II. 93 ;
he próvede (*pret.*) his wepne JOS. 500.

provende, sb., *O.Fr.* provende; *provender;*
prebend; MAN. (F.) 11188 ; provende [pro-
vender] LANGL. *B* xiii. 243 ; **provendres**
(*pl.*) *C* iv. 32 ; provenders *A* PROL. 80* ; pro-
vendris WICL. E. W. 419.

provendren, v., *provide with provender* ;
provendreþ (*pres.*) LANGL. *C* iv. 187, *B* iii.
149.

provendrēres, sb., pl., *men holding pre-
bends,* LANGL. *A* iii. 5 ; provendereris WICL.
S. W. III. 211.

provēour, *see* purveiour.

proverbe, sb., *O.Fr.* proverbe ; *proverb,*
LANGL. *C* xviii. 51 ; WICL. JOHN xvi. 29 ;
proverbes (*pl.*) LANGL. *C* ix. 265.

providens, sb., *O.Fr.* providence ; *foresight,*
providence, ALEX. (Sk.) 3990, 4062.

province, sb., *O.Fr.* province ; *province,* A. P.
ii. 1300 ; ALEX. (Sk.) 187.

províne, v., *? O.Fr.* provigner ; *propagate by*
slips, PALL. xii. 38.

provisour, sb., *Fr.* proviseur ; *provisor,*
LANGL. *B* iv. 133, *C* iii. 182.

provókin, v., *O.Fr.* provoquer ; *provoke,*
'*provoco,*' PR. P. 415.

provost, sb., *O.Fr.* provost (*for* prevost);
provost, WILL. 2265 ; CH. BOET. i. 4 (14).

prūd, prūt, adj., *O.N.* prūðr, *O.E.* prūt; *proud;*
prud HOM. I. 43 ; R. S. v ; GEN. & EX. 1414 ;
FL. & BL. 241 ; HAV. 302 ; prud, prut A. R.
176, 276 ; prut [prout] LAȝ. 8828 ; prut MISC.
114 ; HORN (L.) 1389 ; proud SPEC. 104 ;
AYENB. 258 ; prout TREAT. 138 ; DEGR. 1559 ;
prūde (*dat.*) KATH. 578 ; proude LANGL. *B*
iii. 178 ; **prūttere** (*compar.*) MISC. 193 ; **prū-
dest** (*superl.*) A. R. 296 ; pruttest LAȝ. 20870.

 proude-herte, adj., *proud of heart,* LANGL.
B v. 63.

 prūd-līche, adv., *proudly,* KATH. 577 ; ANT.
ARTH. xxviii ; proudliche AYENB. 168 ; prudli
A. P. ii. 1379.

prūde, prūte, sb., *O.N.* prȳði, *O.E.* prȳt
(*gay attire*) ; *pride* ; prude A. R. 140 ; MISC.
38 ; pruide MAP 342 ; MIRC 1107 ; he heo
lette scruden mid unimete prude (*splendour*)
LAȝ. 14292 ; & nomen þer muchele prude
(*? treasure*) prude 11715 ; Bruttes hafden . . .
unimete prute [prude] 19409 ; prute MISC. 95 ;
ROB. 163 ; TREAT. 139 ; somdel of pruite
SAINTS (Ld.) xlvi. 682 ; pride MISC. 11 ;
LANGL. *B* x. 75 ; MAND. 3 ; S. S. (Wr.) 585 ;
OCTAV. (H.) 560 ; prede HOM. I. 221 ; MISC.
33 ; SHOR. 110 ; AYENB. 16.

 prīde-lēs, adj., *without pride,* CH. C. T. *E*
930.

prūden, prūden, v., *cf. O.N.* prȳða ; *be proud,*
pride one's self, REL. I. 188 ; pridin '*superbio*'
PR. P. 413 ; he ne ssel him not prede AYENB.
258 ; **proudeþ** (*pres.*) P. L. S. xxviii. 18 ;
WICL. JOB xv. 20 ; **prūden** (*subj.*) A. R. 232* ;
iprūd (*pple.*) *adorned* KATH. 1460.

prudence, sb., *O.Fr.* prudence ; *prudence,*
AYENB. 125.

prudent, adj., *O.Fr.* prudent ; *prudent,* CH.
C. T. *C* 110.

pruinen, prūnen, *see* proinen.

prūst, *see* prēost. **prūt,** *see* prūd.

psalm, sb., *Lat.* psalmus, *Gr.* ψαλμός ; *psalm,*
A. R. 52 ; **salmes** (*pl.*) LAȝ. 23754.

 salm-bōc, sb., *psalter,* HOM. II. 69.

 salm-song, sb., *psalm,* FRAG. 6.

 psalm-, salm-wrihte, sb., *psalmist,* H. M.
3 ; salmwurhte HOM. I. 135.

psalmen, v., *sing psalms* ; salme PS. ciii. 80.

psautēr, sautēr, sb., *O.Fr.* psaltier, sautier ;
psalter, TREV. IV. 39 ; psauter LANGL.
A iii. 277 ; (*psalmist*) *A* viii. 55, 107 ; sauter
H. M. 3.

psautrie, sb., *O.Fr.* psalterie ; *psaltery,*
WICL. AMOS vi. 5 ; sautrie CH. C. T. *A* 3213.

psautrien, v., *play on the psaltery* ; sautrien
LANGL. *C* xvi. 208.

publicān, sb., *Lat.* publicānus ; *publican* ;
pupplicānes (*gen.*) A. R. 328 ; **publicānes**
(*pl.*) ORM. 10147.

publissen, v., *Fr.* publier ; *publish* ; **pub-
lisshe** (*imper.*) LANGL. *C* xiii. 38 ; publice *B*
xi. 101 ; **publissed** [published] (*pple.*) CH. C.
T. *E* 415.

puding, *see* poding.

pūe, sb., *O.Fr.* puie, *Lat.* podium ; *pew, seat*
in church ; **pūwes** (*pl.*) LANGL. *C* vii. 144.

pūen, v., *O.Fr.* puier ; *lean, support one's*
self ; pue TREAT. 139 ; **pūed** (*pret.*) M. T.
260.

puf, sb., *cf. Dan.* puf, *Ger.* puf, buf ; *puff* : a
windes puf A. R. 122 ; **puffis** (*pl.*) LANGL. *C*
xx. 66*.

puffen, v., *cf. Dan.* puffe, *Ger.* puffen ; *puff,* A.
R. 272 ; poffe SAINTS (Ld.) xlv. 34 ; powȝe

[powe] ROB. (W.) 6394*; **puffed** (*pple.*) LANGL. *B* v. 16; *see* buffen.

pugge, sb.; *refuse*; **pugges** (*pl.*) ['*vilia excrementa*'] of . . . corne PALL. iii. 1079.

puint, *see* **point.**

pūke, sb., *O.N.* pūki, *cf. Ir.* pūca, *Welsh* pwca; *puck, demon, devil, hobgoblin,* MISC. 76; pouke LANGL. *B* xiv. 190; RICH. 566; **poukes** (*gen.*) LANGL. *C* xix. 282.

[**pukeres,** *? mistake for* puberes (*Lat.*), SHOR. 63.]

pukken, *see* **póken.**

pul, sb., *pull, trick*: pul or draȝte '*tractus*' PR. P. 416; CH. P. F. 164; MAN. (F.) 1809; mani il pul MAN. (F.) 3960.

pūle, sb., *O.E.* pyle,=*M.Du.* pulwe, *O.H.G.* phuluwi; *pillow,* '*pulvinar*,' FRAG. 4; pelewe [pile, pule] TREV. VII. 421; pilwe PR. P. 399; **pilwes** (*pl.*) GOW. I. 142; LIDG. M. P. 118.

pilwe-bere, sb., *pillow-case,* CH. C. T. *A* 694.

pulete, sb., *O.Fr.* polete; *pullet*; **poletes,** pulettis, pultis (*pl.*) LANGL. *A* vii. 267, *B* vi. 282.

pullen, v., *O.E.* pullian; *pull,* CH. C. T. *E* 2353; pullin '*traho*' PR. P. 416; pulle LANGL. *B* xvi. 73; A. P. ii. 68; let him þer inne pulle (*thrust*) L. H. R. 60; *comp.* **to-pullen.**

pullen[2], v., *from* pōl; *form pools*; **pulle** (*subj.*) ['*stagnet*'] PALL. i. 89.

pulment, sb., *O.Fr.* pulment, polment; *pottage,* GEN. & EX. 1490; C. M. 3532; polment A. P. ii. 628.

pulpit, sb., *Lat.* pulpitum; *pulpit,* PL. CR. 661.

puls, sb., *Lat.* puls; *pulse* (*plants*), *pease,* PALL. i. 723.

pült, sb., *knock, push,* LEG. 16.

pülten, v., *mod.Eng.* pelt; *push, thrust, knock*: hit wule pulten on him A. R. 366; pulte ROB. 71; to deþe . . . pulte REL. II. 244; pelte AR. & MER. 2926; **pülte** (*pres.*) O. & N. 873; pilteð HOM. II. 197; **pülte** (*pret.*) LAȜ. 7528*; FER. 774; hii pulte hem vorþ bivore þe oþere ROB. 479; pilte, pelte HORN (R.) 1433 (L. 1415); **pült** (*pple.*) C. L. 207; WILL. 4219; pilt GEN. & EX. 2214; S. S. (Wr.) 670; RICH. 4085; P. L. S. iii. 56; ipult LAȜ. 10839*; ARCH. LII. 36; heo weren ipult ut of paradise HOM. I. 129.

pültunge, sb., *beating,* A. R. 366.

pulter, sb., *O.Fr.* poulletier; *poulterer,* '*gallinarius,*' PR. P. 416.

pultèr[2], sb., *cf. Swed.* paltor (*pl.*); *rag,* DEP. R. ii. 165.

pultrie, sb., *poultry,* CH. C. T. *A* 598; E. G. 353.

pumpe, sb., *Fr.* pompe; *pump,* '*hauritorium,*' PR. P. 416.

punchin, v., ?=**bunchen**; *punch,* '*tundo,*' PR. P. 416; punche LUD. COV. 355; **punchiden** (*pret.*) WICL. EZ. xxxiv. 21.

punchoun, sb., *O.Fr.* poinchon; *punch, pricker,* '*stimulus, punctorium,*' PR. P. 416.

púnd, sb., *O.E.* pund,=*O.Fris., O.N., Goth.* pund, *O.H.G.* phunt *n.; pound* (*weight*); **púnde** (*dat.*) P. L. S. viii. 34; **pónd** (*pl.*) AYENB. 190; pundes FRAG. 6; **púnde** (*gen. pl.*) LAȜ. 4782; an hundred punde O. & N. 1101; an hundred pund HAV. 1633; pounde PL. CR. 410; GREG. 594; **púnden** (*dat. pl.*) LAȜ. 5118.

pound-mēle, adv., *pound by pound,* LANGL. *A* ii. 198.

pünde, sb., *from* pünden; (*?pound,enclosure*): *? pond*; min net liht her wel hende wiþinne a well feir pende HORN (R.) 1138; *see* **ponde.**

pin-fóld, sb., *pinfold, pound,* VOC. 239; '*inclusorium*' PR. P. 400; **pondfólde,** pun-, pon-, pinfolde (*dat.*) LANGL. *B* v. 633; þe poukes poundfalde LANGL. *C* xix. 282.

pundelan, sb., *warrior, hero,* BARB. iii. 159.

pünden, sb., *O.E.* pyndan; *impound, dam up* (*water*); orf to puinde L. N. F. 215; pinde '*includere*' CATH. 280; HALLIW. 625; in prison . . . pende PL. T. 598; þe water hwon me pünt (*pres.*) hit A. R. 72; (ȝe) pundeð IBID.; *comp.* **ȝe-pünden.**

pinder, sb., *pinder,* '*inclusor,*' VOC. 214; pindare PR. P. 400.

punge, sb., *O.E.* pung,=*O.N.* pungr, *Goth.* puggs, *O.H.G.* fung; *purse,* ALIS. 1728.

püngen, pungen, v., *O.E.* pyngan; *prick; thrust*; punge LANGL. *A* ix. 88; pinge FER. 1248, 2430; puinde [pungde] (*pret.*) LAȜ. 23933; he pu(n)gde him ofer þe brigge HORN (H.) 1117; **pungid** (*pple.*) WICL. PROV. xii. 18; *comp.* **to-püngen.**

punginge, sb., *pricking,* HAMP. PS. xxi. 5.

punȝē, sb., *? O.Fr.* poignée; *skirmish,* BARB. xii. 373; punȝhe [poinȝe] xvi. 307.

punischen, v., *O.Fr.* puniss-, punir; *punish,* LANGL. *B* iii. 78; CH. BOET. i. 4 (15); punishen, pun(s)chin PR. P. 416; punissi AYENB. 148.

pūple, *see* **people.**

[**pur-?**].

pur-blind, adj.,=*Swed.* purblind (RIETZ); *purblind,* '*luscus,*' PR. P. 416; ROB. 376; pureblinde ['*luscos*'] WICL. EX. xxi. 26.

pūr, adj., *O.Fr.* pūr; *pure, simple*; (*r.w.* fur) ROB. 8; CH. E. 628; for pure [puire] teone LANGL. *B* vii. 116; puir(e) LANGL. *A* v. 13, *B* xiii. 166; **pūrest** (*superl.*) ROB. 106.

pūr-līche, adv., *purely, wholly,* BEK. 984; LANGL. *C* xvi. 226; purli CH. D. BL. 5; LANGL. *B* xvii. 141; he pureli forȝetes A. P. ii. 1660; pureli ALEX. (Sk.) 187.

purchace, sb., *O.Fr.* pourchas; *purchase*;

earnings; endeavour; purchas CHR. E. 511;
CH. C. T. *A* 256; BARB. v. 534.

purchacen, v., *O.Fr.* pourchacier; *purchase;
earn; procure;* porchasen CH. C. T. *G* 1405;
por chaci AYENB. 9; porchase ROB. 16; pur-
chasede (*pret.*) LANGL. *A* vii. 3; **purchesand**
(*pple.*) BARB. ii. 188.

purchesing, sb., *purchasing, acquisition,*
CH. C. T. *A* 320; BARB. ii. 579.

purchaceour, -chasour, sb., *?earner of fees,*
CH. C. T. *A* 318.

pured, adj., *furred,* GAW. 154.

pūred, pple., *purified,* LANGL. *A* iv. 82*, *C* v.
91*; GAW. 633, 912, 1737; MAND. 286.

puren, *see* poren.

pūretē, sb., *O.Fr.* pureté; *purity,* A. R. 4;
purte AYENB. 202; purite A. P. ii. 1074.

purfil, sb., *purfle, border;* porfil A. P. i. 216;
purfile LANGL. *B* v. 26; purfil '*limbus*' PR. P.
416.

purfilen, v., *Fr.* pourfiler; *border;* **purfiled**
(*pple.*) with pelure LANGL. *B* ii. 9; purfiled
CH. C. T. *A* 193.

purfire, sb., *porphyry,* ALEX. (Sk.) 4682;
porfurie [porphurie] CH. C. T. *G* 775.

purgacioun, sb., *O.Fr.* purgacion; *purifi-
cation,* TREV. VIII. 7; WICL. NUM. vi. 9.

purgatorie, sb., *O.Fr.* purgatoire; *purgatory,*
A. R. 126; SHOR. 5; LANGL. *A* ii. 71, *B* xi.
128, *C* x. 11.

purgen, v., *O.Fr.* purgier; *purge, cleanse,*
LANGL. *B* xv. 529; purgi BEK. 371; AYENB.
132; purge WICL. LK. iii. 17; **purgis** (*pres.*)
ALEX. (Sk.) 4682; **purged** (*pple.*) PR. C.
6398.

pūrifīin, v., *O.Fr.* purifier; *purify,* PR. P. 417.

purpos, sb., *O.Fr.* propos; *purpose;* porpos
ROB. 121; AYENB. 220; purpos CH. C. T. *A*
3981.

purposin, v., *O.Fr.* proposer; *propose, purpose,* PR. P. 417; **purposede** (*pret.*) CH.
BOET. v. 6 (176).

purpre, sb., *O.Fr.* purpre; *purple,* KATH.
1461; REL. I. 119; purpre & pal he droh
A. D. 245; pourpre AYENB. 229; MAN. (F.)
4744; mid purpren MK. xv. 17; **purpres**
(*pl.*) LA3. 5928.

purpris, sb., *O.Fr.* porpris; *goods and
chattels,* A. D. 236.

purpuresse, sb., *seller of purple,* WICL.
DEEDS xvi. 14.

purse, sb., *O.E.* purs, *from Lat.* bursa; *purse,*
R. S. vii; purs P. S. 338; purs, burs '*bursa*'
PR. P. 417; purs, pors AYENB. 52; porse
E. G. 357; pors P. 64; **purses** (*pl.*) A. R.
168; porses LA3. 5927*.

purs-berer, sb., *treasurer,* P. L. S. xxiii. 114.

purs-kervere, sb., *cut-purse,* TREV. VIII.
181.

pursen, v., *put into a purse;* **porse** (*imper.*)
LANGL. *C* xiii. 164.

purseute, sb., *O.Fr.* poursuite; *succession*:
twelve in pourseut (*printed* poursent) i con
asspie A. P. i. 1035.

pursūin, v., *O.Fr.* por-, poursuir; *pursue,
prosecute,* '*prosequor, insequor*,' PR. P. 417;
pursūeþ (*pres.*) PL. CR. 664; porsueþ LANGL.
C xx. 284, xxii. 432; pursueth '' xvii. 302;
persewede (*pret.*) D. ARTH. 1476; porsuede
LANGL. *C* xviii. 167; porsewed (*pple.*) WILL.
2474.

purtenaunce, sb., *appurtenance,* LANGL. *B*
xv. 184*; portinaunce *C* iii. 108; **purten-
aunces** (*pl.*) *B* ii. 103.

purtreie, v., *O.Fr.* portraire; *portray;* portreie CH. C. T. *A* 96; **portreieþ** (*pres.*)
LANGL. *C* xvii. 320; purtraied (*pple.*) A. P. ii.
1465; **portreid** (*pple.*) WILL. 445.

purveiaunce, sb., *A.Fr.* purveaunce, *O.Fr.*
porvëance; *providence; provision; equipment*: al the purveiance that themperour . .
hath shapen for his doughter CH. C. T. *B* 247;
his prudent purveiance *B* 483; he schal make
with temptacioun also purveiaunce WICL. I
COR. x. 13; porveance ROB. 457; purveaunce
CHR. E. 498; WILL. 1598.

purveien, v., *O.Fr.* porveir, -veoir; *purvey,
provide;* purveie LANGL. *B* xiv. 28; porveie
ROB. 305; porvai A. P. iii. 36; **porveiþ** (*pres.*)
AYENB. 145; porvaies A. P. ii. 1502; pur-
veied (*pple.*) WILL. 1605; CH. BOET. i. 4 (21).

purveiour, sb., *Fr.* pourvoyeur; *purveyor,
provider;* purveour C. M. 11003; pur-, pro-
viour TREV. VIII. 147; proweour, prowor, -er
LANGL. *B* xix. 255, *C* xxii. 260.

pūsane, sb., *? O.Fr.* (gorgerette) pisainne;
gorget: thro3he ventaille and pusane ANT.
ARTH. xlv; pesan ALEX. (Sk.) 4960; pesane
D. ARTH. 3458; pisan GAW. 204.

pussen, *see* **possen.**

put, sb., *throw,* '*putting*' *of a stone,* HAV.
1055; H. V. 73; *see* but.

pūt, sb., *O.E.* pytt,=*O.N.* pyttr, *O.H.G.*
puzzi, phuzzi, *O.Fris.* pet; *from Lat.* puteus;
pit, LA3. 15961; A. R. 58; TREAT. 132; ALIS.
717; JOS. 4; þes put HOM. I. 49; pit MAND.
94; pet LK. vi. 39; **pütte** (*dat.*) A. R. 116;
R. S. v; LANGL. *B* x. 370; pette REL. I. 160;
AYENB. 207; *comp.* clēi-, col-püt.

püt-falle, sb., *pitfall,* P. S. 193; pitfalle
VOC. 264; PR. P. 402.

puten, v., *cf. O.Fr.* boter; *throw, push,
thrust,* A. R. 116; pute R. S. v; puten, putten HAV. 1033, 1051; puten [putte] LANGL.
B iii. 84; putten MAND. 227; putte BRD. 8;
MIRC 1660; puttin out S. S. (Wr.) 2068;
putte [poten] WICL. MK. v. 10; poti AYENB.
135; potte MAN. (F.) 8885; puitten LANGL. *A*
ix. 95; **pottist** (*pres.*) MAP 336; puteˈ REL.

I. 224 ; putteþ P. 16 ; heo puteð . . . . vorð
A. R. 328 ; þai pottin AUD. 35 ; put in . . .
þi sweord MISC. 43 ; puiteþ LANGL. *A* vi. 100 ;
(*pl.*) xi. 42 ; **putte** (*pret.*) LAȝ. 18092 ; MAND.
230 ; CH. C. T. *A* 988 ; puttest HOM. I. 15 ;
put (*pple.*) LUD. COV. 29 ; iput MARH. 22 ;
see **butten.**

putting(e), sb., *pushing, instigation,* HAMP.
PS. xii. 5 ; xxxv. 12 ; potting MAN. (F.) 8891.

puttok, sb., *a sort of hawk,* '*milvus*' PR. P.
418 ; potok VOC. 252.

puðeren, v., *pother* ; puðeres (*pres.*) A. R.
214*.

q.

qvá, *see* **hwá.** **qvād,** *see* **cwēd.**

qvadrant, sb., *O.Fr.* cadrant ; *quadrant,*
'*qvadrans,*' PR. P. 418 ; ? qvadrentes (*pl.*)
(? *ms.* in adrentes) ALEX. (Sk.) 129.

qvaier, sb., *O.Fr.* qvaier ; *quire (of paper),*
book, '*qvaternus*' PR. P. 418 ; qvaer A. R.
282 ; the qvair maid be King James, QUAIR
inscription.

qvaile, sb., *O.Fr.* qvaille, caille ; *quail,* VOC.
177 ; '*qvalia*' PR. P. 418 ; A. P. i. 1084.

qvailen, v., *O.Fr.* coaillier ; *curdle,* '*coagulo,*'
PR. P. 418 ; qvaile C. B. 27.

qvainen,=hwēnen ; *grieve* ; qvained (*pret.*)
C. M. 10495.

qvaintan, sb., *O.Fr.* qvintaine ; *quintain,* D.
TROY 1627.

qvaintese, *see* **cointese.**

qvákien, *see* **cwákien.** **qval,** *see* **hwal.**

qvále, *see* **cwále.**

qvalitē, sb., *O.Fr.* qvalité ; *quality,* TREAT.
133 ; ALEX. (Sk.) 4660 ; qvalitēs (*pl.*) AYENB.
153.

qvalm, *see* **cwalm.**

qvan, qwanne, *see* **hwanne.**

qvant, whant, sb., *pole,* '*contus,*' PR. P. 418.

qvantitē, sb., *O.Fr.* qvantité ; *quantity,* P. S.
334 ; P. L. S. xxx. 40 ; LANGL. *C* xxii. 376.

qvappen, v., *tremble* ; qvappe ['*palpitare*']
WICL. TOB. vi. 4 ; CH. TRO. iii. 57 ; qvappid
[cvappid] (*pret.*) ALEX. (Sk.) 2226.

qvarel, sb., *O.Fr.* qvarrel ; *quarrel, square
bolt,* ROB. 491 ; AYENB. 71 ; qwarel C. L. 826.

qvariere, sb., *O.Fr.* qvarrier, '*lapidicida,*' PR. P.
419.

qvarrē, adj., *O.Fr.* qvarré ; *square* ; *stout of
body,* LEG. 46 ; FER. 1072 ; MAN. (F.) 10310 ;
qvarri '*corpulentus*' PR. P. 419.

qvarrēre, sb., *O.Fr.* qvarriere ; *quarry* ;
qvarēre '*lapidicina*' PR. P. 419 ; WILL. 2232.

qvart, sb., *O.Fr.* qvart ; *quart,* '*quarta,*'
PR. P. 419 ; CH. C. T. *A* 649 ; C. B. 35.

qvartēr, sb., *O.Fr.* qvarter ; *quarter,* P. S.
341 ; CH. C. T. *D* 1963.

qvarterin, v., *cut into quarters* ; qvarter
(*imper.*) C. B. 18.

qvarterne, *see* **cwarterne.**

qvarteroun, sb., *O.Fr.* qvarteron ; *quarter,*
[*v.r.* qvartron, qvarterone, qvartroun] LANGL.
B v. 217.

qvaschin, v., *O.Fr.* qvasser ; *quash,* '*quas-
sare,*' PR. P. 419 ; cwesse, qveisse O. & N.
1388 ; qvassed (*pret.*) MAN. (H.) 209 ;
qvaschede [qvashte] LANGL. *C* xxi. 64.

qvási, adj., *queasy,* M. T. 333 ; it is a qvasi mete
TREAT. OF FISH. 24 ; ? qvaisi P. R. L. P. 215.

qvávien, *see* **cwávien.**

qvecchen, *see* **cwecchen.**

qvēd, *see* **cwēd.** **qveint,** *see* **coint.**

qveint, sb.,=cunte ; TRIST. 2254 ; CH. C. T.
D 444.

qveisen, v., *? squeeze* ; qveise (*imper.*) out
the jus REL. I. 302.

qvélen, *see* **cwélen.** **qvellen,** *see* **cwellen.**

qvēme, qwēme, *see* **cwēme.**

qvēn, qwēn, *see* **cwēn.**

qvence, sb., *? Fr.* cognasse ; *quince,* '*coc-
tonum,*' PR. P. 420 ; qvinces (*pl.*) C. B. 27, 97.

qvenchen, *see* **cwenchen.**

qver, sb., *O.Fr.* cuer ; *choir,* '*chorus,*' PR. P.
420 ; ROB. 224 ; MAND. 70 ; LIDG. M. P.
258 ; qveor P. L. S. xvii. 436 ; TREV. V. 183.

qverē, *see* **qvirē.**

qverēle, sb., *O.Fr.* qverēle ; *quarrel* : qverē-
les (*pl.*) AYENB. 83 ; CH. BOET. iii. 3 (70).

qvéren, v., *for* enqvéren ; *inquire* ; qvíris
[enqvirez] (*pres.*) ALEX. (Sk.) 1703.

qverister, sb., *chorister,* PR. P. 420 ; qver-
istres (*pl.*) P. L. S. xxvi. 9.

qverne, *see* **cwerne.**

qvert, whert, adj., *safe and sound* ; '*incolu-
mis, sospes,*' PR. P. 420 ; qvert S. S. (Web.)
771 ; LUD. COV. 202 ; hol and qvert LIDG.
M. P. 38 ; qvert MAN. (F.) 9990.

qvert, sb., *sound health,* MAN. (F.) 9990 ;
L. H. R. 108 ; in qverte FER. 325 ; his wif
was not in qverte E. T. 821.

qvest, sb., *? O.Fr.* qveste ; *inquiry, inquest,
jury, verdict,* GAM. 786 ; qveste '*duodena*'
PR. P. 420.

qvest-monger, sb., *inquest-holder* ; qvest-
mongeres (*pl.*) LANGL. *B* xix. 367.

qvestioun, sb., *O.Fr.* question ; *question,*
SHOR. 136.

qvéðen, *see* **cwéðen.** **qvī,** *see* **hwī.**

qvibibe, sb., *O.Fr.* cubebe ; *cubeb,* '*qvi-
parum,*' PR. P. 421 ; SPEC. 27 ; ALIS. 6796 ;
qvibibis (*pl.*) L. C. C. 16 ; C. B. 6, 37, 46.

qvic, *see* **cwic.** **qvicchen,** *see* **cwecchen.**

qvide, *see* **cude.**

qviete, sb., *O.Fr.* qviete ; *quiet,* CH. C. T.
E 1395.

qvik, *see* cwic. qvīle, *see* hwīle.

qville, sb., *? L.G.* qviele (WƠSTE, VOLKS-ÜBERL. 104); *quill,* '*calamus,*' PR. P. 421.

qvilte, sb., *O.Fr.* cuilte, coulte; *quilt,* '*culcitra,*' VOC. 178; '*culcitra*' PR. P. 421; qwiltes (*pl.*) E. G. 350; coultes MAP 334.

qvin, *see* hwin.

qvinade, sb., *pottage made of quinces,* C. B. 27.

qvir-boilli, sb., *? O.Fr.* cuir boilli; *boiled leather* (*for making armour*), CH. C. T. *B* 2065; qwirbolle BARB. xii. 22.

qvirē, sb., *O.Fr.* cuiree (*curée*), *quarry, captured game*; qvirre TRIST. 499; qverre GAW. 1324.

qvisseux, sb., pl., *O.Fr.* cuissel; *armour for the legs,* MAN. (F.) 10027.

qvistrōn, sb., *O.Fr.* qvistron, coistron; *vagabond, beggar*: ther nas knave no qwistron ALIS. 2511; qwistounes (*pl.*) ALEX. (Sk.) 4660.

qvīt, *see* hwīt.

qvitaunce, sb., *O.Fr.* qvitance; *quittance, payment,* '*acqvietancia,*' PR. P. 421; cwitaunce A. R. 126.

qvite (? qvīte), adj., *O.Fr.* qvite; *quit,* MISC. 46; FL. & BL. 724; M. ARTH. 490; cwite A. R. 6; qvit AYENB. 137; qvit and skere C. L. 1142; qvite (*adv.*) quite MAN. (H.) 50.

qvit-claim, v., *renounce,* D. TROY 1763, 13086; to qvit claim all qverels 1763; thai qvite claimit the qverell 13086.

qviten, v., *O.Fr.* qviter; *quit; requite, pay for;* R. S. vii (MISC. 190); PL. CR. 351; qvite WILL. 325; GAW. 2244; PR. C. 3920.

qviter, sb., *cf. L.G.* kwater, kwader, *H.G.* koder; *pus, suppuration,* REL. I. 192; qvetor I. 302; qvitere ['*saniem*'] WICL. JOB ii. 8; qviture [qwetour, qvetoure] ROB. 435 (W. 8956*); qvitoure P. L. S. xvii. 159; whitour PR. P. 525.

qviteren, v., *suppurate*; whitourin PR. P. 525; qviterende (*pple.*) WICL. WISD. V. 25.

qviver, sb., *O.Fr.* cuivre; *quiver,* '*pharetra,*' PR. P. 421; D. TROY 1730.

qvó, *see* hwá. qvoin, *see* coin.

qvoint, *see* coint. qvōn, *see* hwōn.

r.

rā, sb., *O.E.* rāh, = *O.N.* rā, *O.H.G.* rēh, rēch; *roe,* PR. C. 8938; roa [ra] MARH. 3; ro REL. I. 121; roo '*capreolus*' PR. P. 435; D. ARTH. 922; EGL. 261.

rā-buk, sb., = *O.H.G.* rēchboch; *roebuck,* VOC. 219.

roa-dēor, sb., *O.E.* rāhdēor; *roebuck,* '*capreolus,*' FRAG. 3.

rabbische, adj., *O.Fr.* rabi (*mad*); *unruly,* TREV. VIII. 85.

rabbisch-līch(e), adv., *hastily,* TREV. VIII. 121, 135.

rabbischnesse, sb., *rashness,* TREV. VIII. 147.

rabbe, *see* rápe.

rabel, sb., *pack* (*of hounds*), GAW. 1899.

rabet, sb., *O.Fr.* rabot; *rabbet, groove, grooving-plane,* '*runcina, runctura,*' PR. P. 421.

rabet², sb., *young rabbit,* '*cunicellus,*' PR. P. 421.

rabite, sb., *O.Fr.* arabi; *arab* (*horse*), GUY² (Z.) 10955; rabett 2222; rabetis (*pl.*) ALEX. (Sk.) 1320.

rablen, v., = *M.Du.* rabelen; *mutter;* rable (*imper.*) HALLIW. 661.

rabútin, *see* rebútin.

rac, sb., = *? M.Du.* rack; *? from* réken; *rack,* MAP 335.

racche, sb., *O.E.* rǣcc, *cf. O.N.* rakki; *sleuthhound;* racches (*pl.*) ORM. 13505; MISC. 92; MAP 339; GOW. II. 274; D. ARTH. 3999.

rácen, v., *? for* erácen; *eradicate*: to race out pride LIDG. M. P. 162; of race PR. C. 6704.

ráchen, v., *?* = rákien; *prepare;* ráchez (*pres.*) him radli to ride ALEX. (Sk.) 2031*.

racke, sb., *? M.Du.* racke; *rack;* rakke '*praesepe*' PR. P. 422; HALLIW. 665; at racke & at manger WICL. E. W. 435; a rake of iren for to rost on his eiren E. W. 102; rekke and manger PARTEN. 913; a peire rakkes (*pl.*) E. W. 56; rakkes and brandernes of erne 57.

racoillen, *see* recoillen.

hrǎd, adj., *O.N.* hrǣddr; *afraid*; rad ORM. 2170; Iw. 481; GAW. 251; A. P. ii. 1543; M. H. 73; TOWNL. 102; redd ALEX. (Sk.) 1040, 2510; rǎddest (*superl.*) 2510*.

rǎdnesse, sb., *terror,* D. ARTH. 120.

hrad, *see* hrað. rād, *see* rǣd.

rāde, sb., *O.E.* rād (*journey, road*), = *O.N.* reið (*riding*), *M.Du.* reede (*naval station*), *M. H.G.* reite (*riding*); *from* rīden; *road, journey,* TRIST. 955; on rade toward Rome JUL. 76; rade [rode] C. M. 11427; rode MAP 339; roode PR. P. 435; D. TROY 1045.

rood-hors, sb., = *M.H.G.* reitros; *saddle horse*; rood horsis (*pl.*) (*earlier ver.* ridinge hors) WICL. 3 KINGS iv. 26.

rǎdegound, *see* rědgound *under* rēad.

radevōre (? radenōre), sb., *piece of tapestry*: and weven in her stole the radevore (*? ms.* radeuore, ? radiuore, raduor) CH. L. G. W. 2352; Penelapie renewed her work in the raduore (*for* radeuor) *in* CH. L. G. W. (Sk.) *page* 188.

radish, sb., *Fr.* radis; *radish,* PALL. ii. 201.

rǣchen, v., *O.E.* rǣcan, rǣccan, = *O.Fris.* rēka, rētsa, rēsza, *O.H.G.* reichan; *reach,* LAꝫ. 21412; rechen LANGL. *B* xi. 353; reche

PR. C. 3814; M. H. 61; **ræche** (*pres.*) JOHN
xiii. 26; þe schadewe ... recheþ to Lempne
MAND. 17; **rēcheð** (*imper.*) forð mid boðe
honden A. R. 338; **rāhte** (*pret.*) LK. xxiv. 30;
raht SPEC. 42; M. H. 162; raȝt A. P. ii. 1691;
raghte [raughte] CH. C. T. *A* 136; raght
OCTAV. (H.) 977; me him to rehte [reahte]
anne sceld LAȝ. 23775; reiȝte P. L. S. xx. 95;
rāuȝt (*pple.*) WILL. 4823; PL. CR. 733; raught
LIDG. TH. 158; *comp.* a-, at-, ȝe-, of-, ofer-
ræchen.

hrææd, see hrað.

rēd, rāð, sb., *O.E.* ræd *m.*, *O.N.* rāð *n.*, *cf. O.
Fris.* rēd, *O.L.G.* rād, *O.H.G.* rāt *m.*, *counsel,
advice*; ræd, read [red] LAȝ. 610, 866 (*miswritten* rad 4411); hit þuhte him swiðe hærd
ræd (*fate*) 8164; red HOM. II. 61; O. & N. 682;
GEN. & EX. 309; ROB. 26; BEK. 100; C. L.
1001; SHOR. 127; P. P. 252; JOS. 491;
S. S. (Wr.) 1260; PR. C. 2014; ANT. ARTH.
viii; þesne red HOM. I. 63; he scholde
nimen his red (*resolution*) FL. & BL. 798;
reuthful is mi red A. D. 247; mi red is taken
TRIST. 139; red [reed] LANGL. *B* xiii. 374;
CH. C. T. *A* 665; red '*consilium*' PR. P. 426;
SPEC. 104; S. B. W. 118; HOCCL. i. 108; red,
read A. R. 6, 66; ræd, raþ ORM. 1415, 18719;
red, rath HAV. 75, 1194; read, reað, rað
KATH. 6, 579, 2000; he wes **rēdes** (*gen.*)ful LAȝ.
129; rædes men *counsellors* PROCL. 2; reades
mon A. R. 224; reaðes mon [*O.N.* rāðs maðr]
KATH. 573; **rēde** (*dat.*) LAȝ. 19238; heo
nomen heom to ræde (*ms.* ræden) [reade] þat
aȝæin heo wolden riden 20210; oure heuene
9; hwat scal us to rede *what shall help us*
HOM. I. 165; if we ... doþ al bi his rede
MISC. 37; what is me to rede WILL. 903; rade
AR. & MER. 2834; **rēdes** (*pl.*) LAȝ. 2090;
redes O. & N. 1222; *comp.* cün-, hī(w)-, un-,
wan-rēd.

rēd-ful, adj., *wise,* LAȝ. 5292; (*ms.* redeful)
C. L. 612.

rēd-ȝive, rēd-ȝeve, sb., *O.E.* rædgifa, = *O.
H.G.* rātgebo; *counsellor,* LAȝ. 11615, 24888.

rēd-lēas, adj., *O.E.* rædlēas, = *O.H.G.* rāte-
lōs; *without counsel,* HOM. I. 211; redles O.
& N. 691; WILL. 2915; TOWNL. 225.

rēd-man, sb., = *M.H.G.* rātman; *counsellor*;
readmen (*pl.*) KATH. 573*.

rēd-wīs, adj., *wise in counsel*; **rēadwīsest**
(*superl.*) MARH. 13.

[**rēd**[2], adj.; *comp.* ān-, fast-, lat-, sam-, twī-
rēd, (-rēd, -rēde).]

[**rēde,** adj., *O.E.* ræde, = *O.Fris.*, *M.L.G.* rēde,
M.H.G. reite; (*ready*).]

rede-lichē, adv., *O.E.* rædlīce; *readily,* O.
& N. 1281; WILL. 5467; PL. CR. 811; reade-
liche A. R. 344.

rædels, sb., *O.E.* rædels, = *M.L.G.* rēdelse,
M.H.G. rætsal; *riddle, enigma*; redels VOC.
160; WICL. JUDG. xiv. 16; redels [ridels]

TREV. III. 181; redels [redeles,rideles] LANGL.
B xiii. 184.

[**rēden,** sb., *O.E.* rēden (*condition*); *comp.*
brōðer-, cün-, cūð-, fēlaȝ-, fēr-, folc-,
frēond-, háte-, hēow-, luf-, man-,nēhebūr-,
sib-rēden (-rēden, -rāden).]

rēden[2], **rāðen,** v., *O.E.* rēdan, *O.N.* rāða,
= *Goth.* (ga-)rēdan, *O.Fris.* rēda, *O.L.G.*
rādan, *O.H.G.* rātan; *counsel ; explain ;
rule*; ræden LAȝ. 2330; reden CH. TRO. i.
668; reden, readan HOM. I. 37, 115; rede O.
& N. 1697; ROB. 65; BEK. 831; AYENB. 154;
GOW. I. 151; IW. 2153; P. L. S. iii. 60; PERC.
1248; ANT. ARTH. xliii; he kan rede þe a
riȝt FL. & BL. 142; rede redels LANGL. *B*
xiii. 184; þat dremes couþe rede A. P. ii. 1578;
raþen ORM. 2948; rathe, rothe, rede HAV.
104, 1335, 2817; **rēde** (*pres.*) ORM. 18336;
rede S. S. (Wr.) 1689; HOCCL. i. 382; ic rede
ðat ge flen GEN. & EX. 3118; reade A. R. 24;
redeþ C. L. 943; who se riht redeþ SPEC. 26;
read H. M. 9; yef me him wel ret AYENB. 22;
rēd (*imper.*) LAȝ. 18718; (ȝe) rædeð, reade
LAȝ. 875, 881; **rēdde** (*pret.*) ORM. 6496;
FL. & BL. 761; HAV. 1353; BEK. 467; redde
GEN. & EX. 3436; AYENB. 184; red L. H. R.
64; raddest O. & N. 159; ræden LAȝ. 11414;
comp. a-, bi-, for-, ȝe-rēden.

rēdere, sb., *O.E.* rēdere; *counsellor*; **rē-
deres** (*pl.*) AYENB. 184.

rēden[3], v., *O.E.* rēdan, (*originally the same
word as* **rēden**[2]); *read, say, speak*; ræden
[reade] LAȝ. 4497; A. R. 286; REL. I. 224,
263; HAV. 244; SHOR. 47; ne mahte hit
na mon rikenin ne redan (*v. r.* tellen) JUL.
51; clerkes þat conne reden C. L. 1241;
no tonge ne mihte reden ne þouȝt þenken
his mihtful deden 1359; rede L. H. R.
108; me hit mai in boke rede O. & N. 350;
hwa schal unker speche rede 1782; whi þei
wolde nought of him rede MAN. (F.) 10598;
þer herd(e) i rede in roune who Tristrem gat
and bar TRIST. 3; rede and sai(e) PR. C.
6288; **rēde** (*pres.*) GEN. & EX. 34; redeþ
ORM. 17286; redeð, ret A. R. 170, 244; ret
HOM. I. 125; AYENB. 231; þei reden MAND.
83; **rēdde** (*pret.*) REL. I. 59; redde [radde]
LANGL. *B* iii. 334; radde LAȝ. 10; redden
CH. C. T. *F* 713; **rēd** (*pple.*) ORM. 6870;
rad GOW. I. 194; MAND. 132; DEGR. 962;
ired MARH. I; irad FL. & BL. 578; P. L. S.
xvii. 431; *comp.* a-rēden.

rēdere, sb., *O.E.* rēdere; *reader,* MARH. 20;
BEK. 1068; SHOR. 47.

rēdunge, sb., *O.E.* rēding; *reading,* HOM.
I. 93; A. R. 286; on rædinge FRAG. I.

rēden[4], v., *O.E.* (ge-)rēdan, = *M.L.G.* rēden,
M.Du. reeden, *Goth.* raidjan, *M.H.G.* reiten;
get ready, prepare; **rēdde** (*pret.*) MAN. (F.)
14088; and large roum about hem redde AR.
& MER. 7906; þet setle swa h(e)o radden
HOM. I. 61; *comp.* a-, ȝe-rēden.

rēdi [readi], adj., *cf. M.L.G., Swed.* rēdig; *ready, prepared,* LAȝ. 8651; rædiȝ ORM. 2527; redi R. S. ii; MISC. 42; GEN. & EX. 1066; JOS. 42; L. H. R. 139; GREG. 344; LANGL. *A* iv. 155; WICL. JOHN vii. 6; MAND. 236; C. M. 7469; a redi wei TOR. 868; readi HOM. I. 277; rēdie (*pl.*) HOM. II. 191; þe rēdieste (*superl.*) answere TREV. III. 181; *comp.* ȝe-rædi.

rēdi-līche, adv., *readily,* WILL. 1226; redili TRIST. 1523; CH. C. T. *C* 667.

rēdinesse, sb., *readiness,* PR. P. 426.

rēdien, v.,=? *M.L.G.* rēdegen; *direct, prepare; discourse;* readien HOM. I. 249; redie MAND. 185; rēdies (*pres.*) IW. 1956; rēdi (*imper.*) the D. ARTH. 4137; and rēdied (*pret.*) hem forþ to wende C. M. 5040; ? *comp.* bi-rēdien.

rēf, rēfen, see rēaf, rēaven.

ræft, sb., *O.N.* raptr, *mod.Eng.* raft; *beam, plank;* raft AV. ARTH. xxv.

ræfter, sb., *O.E.* ræfter,=*M.Du., M.L.G.* rafter; *rafter;* refter 'tignum' FRAG. 4; 'trabs' VOC. 260; þane refter AYENB. 175; rafter CH. C. T. *A* 990; ræftres [refteres] (*pl.*) LAȝ. 7829; rafters [refters] TREV. VII. 349.

rēh, see *h*rēoh.

hræȝel, hrægel, sb., *O.E.* hrægl,=*O.H.G.* hregil, *O.Fris.* hreil; *robe, neckerchief;* reȝel O. & N. 562; reil HICKES I. 167; rail JOHN xiii. 4; ræglen [*'pannis'*] LK. ii. 12; *comp.* set-ræiȝol.

ræil-hūs, sb., *O.E.* hrægelhūs; *vestry,* ' *vestiarium*,' FRAG. 4.

reil-þein, sb., *O.E.* hrægelþēn; *valet,* SAX. CHR. 260.

hræilen, v., *cf. O.H.G.* (ki-)hregilōd (*adorned*); *clothe;* railed (*pple.*) GAW. 745; real(l)i railed (*ms.* railled) wiþ wel riche cloþes WILL. 1618; railid D. ARTH. 3263.

rēm, see *h*rēam.

rēmen [rēmen], v., *O.E.* (a-)rēman,=*M.H.G.* rēmen; *stretch out, extend; contend;* LAȝ. 4128, 16751; he gon ræmien (? *for* ræmen *or* ramien) [remi] *he began to stretch himself* 25991; holde everich his owene mester & nout ne reame (*imper.*) oðres A. R. 72; Brutus him rēmde [remde] (*pret.*) to LAȝ. 682; remed *yawned* LANGL. *C* viii. 7; heo . . . ræmden to gadere LAȝ. 623.

rǣren [reare], v., *O.E.* rǣran, *from* rīsan; *rear, raise,* LAȝ. 17458; rere DREAM 468; up rere P. L. S. xii. 41; þei lete rere a halle S. S. (Wr.) 138; rēre (*imper.*) PS. l. 11; rǣrde (*pret.*) LAȝ. 22111; and ivele laȝe rerde (*v. r.* arerde) P. L. S. viii. 86; ROB. 218; rēred (*pple.*) P. S. 249; rerid WICL. GEN. xxxiii. 20; *comp.* a-rǣren.

rǣs, sb., *O.E.* rǣs,=*O.N.* rās, *M.Du.* raes; *from* rīsen; *race, rush, course;* res O. & N. 512; MISC. 93; JOS. 491; he was wroþ and

maked a res S. S. (Web.) 2391; res [rees] GAM. 547; rees GOW. I. 335; TREV. VIII. 348; E. T. 1080; AV. ARTH. xxii; res, reas M. ARTH. 1957, 2961; moni grimne reas LAȝ. 2283; rase PR. C. 8938; rǣse, rese (*dat.*) LAȝ. 21367, 31236; on a rese PERC. 1366; A. P. ii. 1782; to þe bischope in a ras he ran M. H. 141; rǣsen (*dat. pl.*) LAȝ. 5200 (*miswritten* rasen 683); *comp.* fēond-rǣs, in-rās.

rǣsen, v., *O.E.* rǣsan, *cf. O.H.G.* reisōn, *O.N.* rāsa, *M.Du.* rāsen; *race, rush, run,* LAȝ. 1004; rese IPOM. 1831; rase GOW. II. 264; Saladin began to rase for ire RICH. 3633; rǣsde (*pret.*) MK. x. 50; ræsde [resde] LAȝ. 6496; ant te drake resde to hire MARH. 10; resede TREV. VI. 377; resden to þan castle LAȝ. 1679; *comp.* ȝe-rǣsen.

raf, sb., *O.Fr.* raffe, *cf. mod.Eng.* riffraff; *rubbish,* '*purgamenta*,' PALL. i. 827.

rafles, sb., pl., *O.Fr.* rafle; *a game at dice;* tables and rafles CH. C. T. *I* 793.

raft, rafter, see ræft, ræfter.

ráge, sb., *O.Fr.* rage; *rage, folly, madness,* ROB. 216; C. L. 197; AYENB. 142; ALIS. 980; a rage and suche a veze CH. C. T. *A* 1985.

rágerie, sb., *wantonness,* CH. C. T. *D* 455.

ragge, sb., *O.N.* rögg (*tuft*); *rag,* '*panniculus*,' PR. P. 421; SHOR. 110; ragges (*pl.*) P. S. 150; GOW. I. 100.

ragge-man, sb., *ragman*: ragman, or he that goith with jaggid clothis PR. P. 421; þat rageman [raggeman] (*sc. the devil*) þat first man deceivede LANGL. *C* xix. 122; rageman (*for* rageman-rolle) *a contemptuous name for a document with many seals*: (he) rauhte with his ragemon (*papal bull*) [ragman] LANGL. *A* PROL. 73.

ragged, adj., *ragged, shaggy,* MAP 338; a ragged colt ALIS. 684; raggede cloþes LANGL. *B* xi. 33.

rágin, v., *O.Fr.* ragier; *rage; be wanton,* '*rabio*,' PR. P. 421; rage ST. COD. 94; pleien ant rage A. D. 237; to rage & to pleie CH. C. T. *A* 3273; if any of his feris ráged (*pret.*) ALEX. (Sk.) 638; þou has rágid (*pple.*) 460.

rai, sb., *for* arrai; *array,* DEP. R. iii. 125.

rai[2], sb., *Fr.* raie; *striped cloth,* '*stragulum*,' VOC. 238; the raie is turned overthvert that sholde stonde adoun P. S. 336; raies (*pl.*) LANGL. *A* v. 125, *B* v. 211, *C* vii. 217.

rai-cloth, sb., *striped cloth,* WICL. PROV. xxxi. 22.

raie, sb., *Fr.* raie; *ray (a fish),* '*ragadia*,' VOC. 222; raiȝe ROB. (W.) *page* 821.

raien, v., *for* arraien; *set in array;* rai YORK xxvi. 246; rais (*pres.*) him ALEX. (Sk.) 2031; raied (*pret.*) ROB. (W.) 4386*.

raike, sb., *O.N.* reik *n., course, path,* D. ARTH. 2985; reike '*vagatio*' PR. P. 427.

raiken, v., *O.N.* reika; *wander;* raike PR. C. 4891; M. ARTH. 3373; M. H. 58; reike

MAP 342 ; **raikes** (*pres.*) A. P. iii. 89 ; **rai-kinde** (*pple.*) A. R. 140* ; **raiked** (*pret.*) GAW. 1727 ; reikid FLOR. 1648.

rail, sb., *cf. M.L.G., Swed.* regel, *O.H.G.* rigel ; *rail, 'paxillus,'* PR. P. 422 ; GOW. III. 75 ; **railes** (*pl.*) PALL. iv. 287.

railen, v., ? = *M.H.G.* regelen ; *arrange in a row* : raile vinis PR. P. 422 ; **railed** (*pple.*) opon a rowe MIN. iv. 83 ; lettres railed a right M. ARTH. 3531.

railen, *see hrǽilen,* reilen.

raimen, v., ? = **rāmien** ; *arrange; guide; direct* : raimen (*? govern*) þe realmes LANGL. *A* i. 93 ; and ham doþ raimi and kveadliche lede AYENB. 44 ; raime ALEX. 2488 ; falser men might no man raime E. T. 431 ; reime MAN. (H.) 185 ; al þat þe riche mai reime LANGL. *C* xiv. 96 ; **raimeþ** (*pres.*) P. S. 150 ; reimeþ P. R. L. P. 231 ; **reimed** (*pret.*) MAN. (H.) 29 ; raimed D. ARTH. 100 ; *comp.* a-reimen.

rain, *see hrān.*

raines, sb., *from* Rheims, pr. n. ; *fine linen of Rheims* ; serkis of raines ALEX. (Sk.) 4339 ; reines 1550.

raisen, v., *O.N.* reisa, = *Goth.* raisjan, *O.E.* (a-)rāsian ; *raise, rear* ; (*ms.* reȝȝsenn) ORM. 15599 ; reise CH. C. T. *D* 2102 ; WICL. JOHN xi. 11 ; nu raise (*pres.*) þai up þe rode HOM. I. 283 ; **reise** (*imper.*) SPEC. 100 ; **reisede** (*pret.*) REL. I. 224 ; *comp.* a-reisen.

raisin, *see* reisin.

raisōn, sb., *O.Fr.* raison, raisun ; *reason, talk* ; raisoun A. P. 191 ; reisun A. R. 78 ; resun KATH. 2248 ; P. L. S. III. 19 ; reson ROB. 196 ; reson [resun] C. L. 1080.

raisonable, adj., *O.Fr.* raisonable, resnable ; *reasonable, eloquent* ; resonable [*v. r.* res-nable, renable] LANGL. *C* i. 176 ; renable ROB. 414 ; AYENB. 95.

rak, sb., *? O.N.* rek ; *? cloud,* HALLIW. 662 ; A. P. iii. 176 ; D. TROY 1984 ; upon rak rises þe sunne GAW. 1695.

[**ráke**, sb., *from* ŕáken ; *comp.* ǽrend-réke.]

ráke², sb., *O.E.* race, = *M.L.G., M.Du.* rake *f.* ; *from* réken ; *rake, 'rastrum,'* PR. P. 422 ; VOC. 201 ; R. S. v ; CH. C. T. *A* 287.

ráke-stéle, sb., *rake-handle* : that tale is nat worth a rake stele CH. C. T. *D* 949.

ráke³, sb., *? region, path,* GAW. 2144 ; out of þe rake [rakke] of riȝtwisnes renne suld he nevire ALEX. (Sk.) 3383 ; þe rake on þe riȝt hand 5070.

ŕráke, sb., *O.E.* hraca, = *M.L.G.* rake, *O.H.G.* racho ; *throat* ; rake MARH. 11 ; R. S. v (MISC. 180) ; of þe feondes rake KATH. 919.

rāke, sb., ? = **rāike** ; (*r. w.* lake) TOWNL. 218.

rakel, sb., *hasty, rash,* CH. C. T. *H* 278 ; A. P. iii. 526 ; rakil HOCCL. i. 83 ; LUD. COV. 24 ; **rakele** (*pl.*) LEG. 37.

rakelnesse, sb., *rashness,* CH. C. T. *H* 283.

rakente, sb., *O.E.* racente, = *O.N.* rekendi, *O.H.G.* rahchinza ; *chain* ; **rekenthis** (*pl.*) ALEX. (Sk.) 5542.

raken-tēie, sb., *O.E.* racentēag (*for* racent-tēag) ; *chain,* BEV. 1636 ; raketeȝe LAȝ. 16752 ; raketiȝe ROB. (W.) 3001* ; raketeie P. L. S. viii. 141 ; JUL.² 112 ; raketehe HOM. I. 249 ; JUL. 46 ; **raketēgen** (*pl.*) MK. v. 4 ; rake-teien LEG. 162.

raket, sb., *O.Fr.* rachette, rasquette ; *racket* ; pleien raket to and fro, netle in, dokke out, now this, now that CH. TRO. iv. 460.

rákien, v., *O.E.* racian, = *M.L.G., M.Du.* raken, *O.N.* raka (*scrape*), *Swed.* raka (*scrape, run*), *Goth.* ufrakjan (ἐκτείνειν), *mod.Eng.* rake ; *wander, run, go* ; *scrape* ; rakin PR. P. 422 ; raken GEN. & EX. 2132, 3324 ; raken PL. CR. 72 ; rake AR. & MER. 8048 ; TOWNL. 167 ; Horn to halle **rákede** (*pret.*) HORN (R.) 1084 ; mid sweorden heom to rakeden LAȝ. 18058.

rákere, sb., *scavenger,* LANGL. *A* v. 165 ; racare *'fimarius'* PR. P. 421.

rakke, *see* racke.

ram, sb., *O.E.* ramm, = *M.L.G., O.H.G.* ram ; *ram, 'aries,'* VOC. 177 ; *'vervex'* PR. P. 422 ; [*'arietem'*] WICL. GEN. xv. 9 ; IW. 3019 ; (*ms.* ramm) ORM. 1136 ; rom *'aries'* FRAG. 4.

ramal, sb., *old wood* ; (*misprinted* rainal) PALL. iii. 292.

ramblen, v., *ramble* ; **romblinge** (*pple.*) LANGL. *C* vi. 11*.

Ramesēie, pr. n., *O.E.* Rammes ēge ; *Ramsey,* ROB. 283 ; Ramesǽie SAX. CHR. 266.

rāmien, v., = *M.L.G., M.Du.* rāmen, *O.H.G.* rāman, rāmen (*direct one's way*) ; *roam, direct one's course* ; romen CH. TRO. ii. 516 ; rome HAV. 64 ; into þe toun he moste rome AL. (L.³) 290 ; and rome fro home LANGL. *B* xi. 124, *C* xiii. 63 (Wr. 6836) ; roume [rombe] LANGL. *B* xi. 109, *C* xiii. 48* ; **rōmeþ** (*pres.*) ALIS. 7207 ; **rōminde** (*pple.*) S. S. (Web.) 1429 ; **rōmed** (*pret.*) WILL. 1608 ; þa Rom-leoden rameden ȝeond uþen LAȝ. 7854 ; *see* **rēmen, raimen** ; *der.* ramblen.

rōmere, sb., *wanderer, roamer,* LANGL. *B* x. 306* ; **rōmares** (*pl.*) LANGL. *C* iv. 120.

ramin, *see* ramne.

rammin, v., *ram, 'trudo, pilo'* PR. P. 422 ; **rammed** (*pret.*) AR. & MER. 533.

rammin, *see* ramne.

rammisch, adj., *like a ram,* CH. C. T. *G* 887.

ramne, sb., *Lat.* rhamnus ; *thorn, bramble,* [*'rhamnum'*] WICL. JUDG. ix. 14 ; rammin [ramin] HAMP. PS. lvii. 9.

rampe, sb., *? mod.Eng.* romp ; *common scold,* KN. L. 25.

rampen, v., *? O.Fr.* ramper ; *scratch with paws* : raumpe on him PR. C. 2225 ; she **rampeþ** (*pres.*) in mi face CH. C. T. *B* 3094 ; **rampend** (*pple.*) *rampant,* GOW. III. 74.

*h*ramsen, sb. pl., *O.E.* hramsa, = *M.L.G.* ramese, *Swed.* rams; *wild garlic*; ramsis PR. P. 422.

ran, sb., *song*, S. S. (Web.) 2723; ron SPEC. 33; CL. M. 2; a luve ron MISC. 93; ronnes (*pl.*) KATH. 108.

[**?** ran², sb., *O.N.* rann (*house*).]

   ran-saken, v., *O.N.* rannsaka (*search the house*), *from O.N.* rann, *see* ern; *ransack*, GEN. & EX. 2323; ransake CH. C. T. *A* 1005; REL. I. 8; ransakand (*pple.*) PS. vii. 10.

*h*rān, sb., *O.E.* hrān,= *O.N.* hreinn; *reindeer*: þe ronke racches þat ruskit þe ron MISC. 92.

   rain-dēr, sb., *reindeer*; raindere D. ARTH. 922.

ranc, adj., *O.E.* ranc, = *O.N.* rakkr; *rank, strong, brave, proud*, GEN. & EX. 2105; he folc & ranc (*ms.* rannc) ORM. 9622; rank MAN. (F.) 13805; H. S. 5095; ISUM. 200; ronk GAW. 513; A. P. i. 843; ranke (*pl.*) P. S. 341; ronke A. R. 268*; þe ronke racches MISC. 92.

rancle, sb., *festering sore*, REL. I. 52, 53.

ranclen, v., *rankle*: his flesch gan ranclen & rebelle BEV. (K.) 2832.

rancour, sb., *O.Fr.* rancœur, rancuer; *rancour*, CH. C. T. *A* 2732; ALEX. (Sk.) 2701.

rand, sb., *O.E.* rand, rond, *cf. O.H.G.* rand *m.*, *O.N.* rönd *f., margin, border*; rande (*dat.*) GAW. 1710; randes (*pl.*) PL. CR. 763; randez A. P. i. 105.

randen, v.? = *O.Fris.* randa; *lacerate*; ronden KATH. 1998; rondin ant rendin MARH. 6.

randōn, sb., *O.Fr.* randon (*course*); *mod.Eng.* (at) random; *onrush, force; torrent of words, harangue*; PARTEN. 1727; '*haringa*' PR. P. 423; furre fleeth in o randun P. L. S. xxxv. 132; randoun M. ARTH. 2888; randoum MAND. 238; in a randoune BARB. xvii. 694; intill a randoune BARB. xix. 596.

rangale, sb., *rabble, camp-followers*, BARB. xii. 474; rangall xiii. 341; rangald viii. 198.

rank, *see* ranc.

ranken, v., *? grow rank*: er hit ronke (*subj.*) in rote ANGL. IV. 193.

ransaken, *see under* ran².

ransūn, sb., *O.Fr.* raençon, raenchon, *from Lat.* redemptio; *ransom*; raunsun A. R. 124; raunson MAN. (F.) 3332; raunsoun C. L. 514; raimson ROB. (W.) 6046*; raunceoun LANGL. *B* viii. 35; ransōns [raunsons] (*pl.*) ALEX. (Sk.) 1665.

ransunen, v., *ransom*; (i) ransoun (*pres.*) ... mine lige LANGL. *C* xxi. 398 (raunceon *B* xviii. 347); raunsoned (*pple.*) *B* xvii. 301.

rap, sb., *cf. Dan.* rap, *Swed.* rapp; *rap*; rappe '*ictus*' PR. P. 423; rappes (*pl.*) EM. 660; ALIS. (Sk.) 348; rappis OCTOV. (W.) 334.

hrap, adj., *cf. Dan., Du.* rap; *swift*; rape GAM. 101; rápe (*adv.*) R. R. 6516.

   rap-lī, adv., *quickly*, WILL. 3179; FER. 384; LANGL. *A* v. 176; M. H. 32; MIN. vi. 67.

rāp, sb., *O.E.* rāp, *cf. M.L.G.* rēp *m., Goth.* raip, *O.N.* reip *n., O.H.G.* reif *m.; rope*, ORM. 15818; enne rap [rop] LA3. 20333; rop FRAG. 1; ROB. 448; FER. 2902; roop PR. P. 436; TREV. VII. 423; rāpe (*dat.*) REL. II. 282; rope SHOR. 110; rāpes (*pl.*) LA3. 1099; MIN. viii. 68; ropes GREG. 481; *comp.* sti(3)-rōp.

rápe, ráve, sb., *Lat.* rāpa, *O.Fr.* rabe, rave; *rape (the plant)*, '*rapa*,' VOC. 191; PR. P. 423; rabben (*pl.*) ALIS. 4983; wilde raves PALL. v. 170.

*h*rápe, sb., *haste*; rape PR. P. 423; HORN (L.) 1418; S. S. (Web.) 1631; REL. I. 115; H. S. 2143; rápe (*dat.*) LANGL. *B* v. 333; GOW. I. 296; over þeo table he leop a rape ALIS. 4239; in rape L. H. R. 135.

*h*rápen, v., *O.N.* hrapa; *rush, hasten*; rapin PR. P. 423; rapen GEN. & EX. 2376; rape GOW. I. 335; SACR. 659; rape and renne CH. C. T. *G* 1422; rape þe to shrifte LANGL. *B* v. 399; rápede (*pret.*) GEN. & EX. 1221; þe folk þat escaped to Scotland þam raped MAN. (H.) 90.

rāpere, sb.,= *M.L.G.* rēpere; *rope maker or seller*; ropere VOC. 212; TREV. V. 181; LANGL. *A* v. 166.

rappin, v.,= *Swed.* rappa; *rap, beat, '*pulso*,* PR. P. 423; rappe LUD. COV. 183; rappe a doune LANGL. *C* ii. 91.

rār, sb., *O.E.* (ge-)rār; *roar*; a rore (*printed* arore) MISC. 216; rōre (*dat.*) GOW. III. 74; rāris (*pl.*) GOL. & GAW. 85 (ANGL. II. 411).

rārin, v., *O.E.* rārian,= *M.L.G.* rāren, *M.Du., O.H.G.* rēren; *roar*, JUL. 49; HAMP. PS. lxxvi. 1; rare IW. 242; PR. C. 7341; rare, rore C. M. 12530; rair, rar BARB. iv. 423, x. 685; roren WICL. JOB vi. 5; roorin PR. P. 437; rore P. L. S. xix. 241; WILL. 86; EGL. 327; rārinde (*pple.*) MARH. 17; rārede (*pret.*) D. ARTH. 784; rorede OCTOV. (W.) 1739; he rorede als a bole HAV. 2438.

rārunge, sb., *O.E.* rārung; *roaring*, HOM. I. 253.

rās, *see* rǣs.

rascaile, rascalie, sb., *? O.Fr.* rascaille, *mod. Eng.* rascal; *refuse, worthless people*, PR. P. 423, 424; rascaille MAN. (F.) 6784; raskaille D. ARTH. 2881; rascaile, rasskaile *lean deer* DEP. R. ii. 119, 129.

rasch, adj., *O.N.* röskr,= *O.H.G.* rasc, *M.Du.* rasch; *rash, sharp, swift*; rasch & ronk A. P. i. 1166; rashe (*adv.*) L. C. C. 18.

raschen, v., *? O.N.* raska; *? rush*: riche stedes rependez and rasches (*pres.*) one armes D. ARTH. 2107.

rāsen, *see* rǣsen.

rásin, v., *O.Fr.* raser; *scrape*, PR. P. 424;
rásith (*pres.*) WICL. WISD. xiii. 11.

rasken, *see* raxen.

rásour, sb., *O.Fr.* rasour; *razor*, AYENB. 66;
CH. C. T. *B* 3246; rasure HAMP. PS. li. 2;
rásours (*pl.*) P. L. S. xix. 221.

raspen, v., *cf. Du.* raspen, *Dan.* raspe, *Swed.*
raspa; *rasp, scrape*: rospen & raken GEN.
& EX. 2132; rasped (*pple.*) A. P. ii. 1545.

rasse, sb., *raised mound, top*: a rasse of a rok
A. P. ii. 446.

raste, *see* reste.

*h*ratelen, v., = *M.Du.* ratelen, *M.H.G.* razzeln;
rattle, strike together; ratelen REL. I. 65;
ratled (*pret.*) AR. & MER. 7858.

ráten, v., *? Swed.* rata; *rate, scold*: he schal
be ráted (*pple.*) CH. C. T. *A* 3463; *comp.*
a-ráten.

raton, sb., *O.Fr.* raton; *rat*, LANGL. *B* PROL.
158; CATH. 300; ratones (*pl.*) LANGL. *C* i.
165, 198.

ratoner, sb., *rat-catcher*, LANGL. *A* v. 165;
ratonere *B* v. 322.

ratte, sb., = *M.Du.* ratte, *M.L.G.* rotte; *rat*;
rattes (*pl.*) LANGL. *B* PROL. 200; MAND.
250; rotte VOC. 177.

ratte², sb., *rag*: i rattes (*pl.*) & i clutes HOM.
I. 277.

ratten, v., *cf. M.H.G.* ratzen; *lacerate, tear*;
ratted (*pple.*) A. P. ii. 144; *comp.* to-ratten.

*h*rað, adj., *O.E.* hræð, hræd, *O.N.* hraðr, = *O.
H.G.* hrad; *swift*; rath PERC. 98; rad
GEN. & EX. 617; to cheste rad O. & N. 1043;
þar to he was ful rad S. A. L. 150; ræd
['*promptus*'] MAT. xxvi. 41; LAʒ. 12318;
rathe mon ANT. ARTH. xxxiv; rathe men D.
ARTH. 2550; grucching and luring him beoþ
rade O. & N. 423; heo beoþ to rad(e) SPEC.
45; ráðe (*adv.*) *quickly, soon*, LAʒ. 4338;
KATH. 554; ʒis sonde hem overtakeð raðe
GEN. & EX. 2313; raþe ORM. 13766; HAV.
358; AYENB. 152; MAN. (H.) 87; M. H. 26;
raþe oþer late O. & N. 1147; raþe [rathe]
LANGL. *A* iii. 56; whi rise ʒe so raþe [rathe]
CH. C. T. *A* 3768; reaðe, reðe [raðe] A. R.
54, 86; reðe HOM. I. 91; hraþur (*compar.*)
rather LEECHD. III. 120; raðer LAʒ. 3539;
A. R. 190, 384; oþer raðer oþer later ANGL.
I. 15; raþer BEK. 1652; SHOR. 160; CH. C.
T. *E* 2302; reðer HOM. I. 115; (h)is raþere
wif ROB. 285; þe raþere toun MAND. 46;
ráþest (*superl.*) LANGL. *B* xiv. 203.

raô-, read-līche, adj., *O.E.* hræð-, hræd-
līce; *quickly, promptly*, A. R. 422; reaðliche
HOM. I. 247; redliche 3; radliche LAʒ. 25603;
rathli L. H. R. 84; MIN. vii. 91; MIR. PL. 151;
radli JOS. 629; GAW. 367.

red-mōd, adj., *O.E.* hrædmōd; *hasty, irri-
table*, HOM. I. 105.

rāþ, rāþen, *see* rād, rāden.

hrāu, adj., *O.E.* hrēaw, = *O.L.G.* hrā, *O.N.*
hrār, *O.H.G.* rōw; *raw*; rau LEG. 36; P. S.
237; (*ms.* raw) '*crudus*,' PR. P. 424; ALIS.
4932; TREV. V. 431; MAND. 203; rauʒ
SAINTS (Ld.) xlv. 152; ra VOC. 200.

rāu-hēde, sb., *rawness*, '*cruditas*,' PR P.
424.

rāunesse, sb., *O.E.* hrēawness; *rawness*,
PR. P. 424.

rauncesūn, *see* ransūn.　ráve, *see* rápe.

*h*ráven, sb., *O.E.* hræfn, hræm, = *O.N.* hrafn,
M.L.G. raven, *O.H.G.* hraban; *raven*; raven
REL. I. 218; A. P. ii. 455; reafen [reven] A.
R. 84; reven TREAT. 133; revin VOC. 177;
reafnes (*gen.*) A. R. 84; remes LAʒ. 30392;
réfnes (*pl.*) LK. xii. 24; ravenes S. S. (Wr.)
3147.

réven-fōt, sb., *a plant*, REL. I. 36.

rávin, v., *O.Fr.* raver, *from Lat.* rabere; *rave*,
'*deliro*,' PR. P. 424; rave GOW. I. 282; MIRC
1326; LUD. COV. 44; ráve (*pres.*) HOCCL. i.
394; we rave CH. C. T. *G* 959; ráved (*pret.*)
LANGL. *B* xv. 10.

ravine, sb., *O.Fr.* ravine; *rapine*, CH. P. F.
323; PR. C. 9448.

ravinour, sb., *plunderer*, CH. BOET. iv. 3
(121); ravenour TREV. VII. 445; raviners
(*pl.*) CH. BOET. i. 4 (12).

ravissen, v., *O.Fr.* ravir; *ravish*: þou . . .
ravisest (*pres.*) ROB. 194; shrewes ravissen
CH. BOET. iv. 5 (131); ravischede (*pret.*)
LANGL. *A* iv. 36; ravisched MAN. (F.) 436.

ravissour, sb., *Fr.* ravisseur; *ravisher*;
ravischers (*pl.*) wiþ ravisschours MISC. 225.

rāw, *see* hrāu.

rāwe, sb., *O.E.* rāw, rǣw, = *? M.L.G.* rēge; *row*,
line, series, KATH. 1954; TRIST. 3095; A. P.
i. 544; AV. AR. vi; a rawe FER. 4605; rowe
'*series, linea*' PR. P. 438; HORN (R.) 1086;
AM. & AMIL. 1900; TOR. 818; on rowe
MIRC 123; for þre niʒtes a rowe he seiʒ
þat same siʒt ED. 68; on a rowe SPEC. 35;
rowe, rewe SHOR. 93, 95; rewe BEK. 2201;
GOW. I. 50, III. 308; CH. C. T. *A* 2866;
DEP. R. PROL. 54; a rewe A. R. 90; bi reawe
336; a rewe ROB. 252; bi rewe DAV. DR. 31;
rāwen (*pl.*) JUL. 21; *comp.* dai-rāwe (-rēwe).

rāwen, v., *? beam, ? dawn, run*: hertes
gonne rowe REL. I. 120; þe dai rōweþ (*pres.*)
LANGL. *C* ii. 114; er þe dai rēwe (*subj.*)
P. S. 239; rōwed (*pret.*) LANGL. *B* xviii. 123.

raxen, v., *stretch*; raske H. S. 4282; raskit
(*pres.*) VOC. 152; raxes him REL. II. 80;
raxed [roxed, roskid] (*pret.*) LANGL. *B* v.
398.

raxlen, v., *stretch*; roxle (*pres.*) REL. II.
211; raxlede (*pret.*) LAʒ. 25992; raxled A.
P. i. 1173; raxlede [rascled] & remed LANGL.
C viii. 7.

[re-, prefix, *O.Fr.* re-, *Lat.* re-; *again.*]

hrēac, sb., _O.E._ hrēac,=_O.N._ (torf-)hraukr ; _rick,_ '_acervus,_' FRAG. 2 ; rek REL. II. 80 ; reek PR. P. 428 ; **rēkes** (_pl._) WICL. EX. xxii. 6.

rēad, adj. & sb., _O.E._ rēad,=_O.Fris._ rād, _O.L. G._ rōd, _O.H.G._ rōt, _Goth._ rauds, _O.N._ rau∂r ; _red,_ A. R. 112 ; ræd [read] LA3. 15940 ; red HOM. I. 83 ; R. S. v ; red [reed] CH. C. T. _A_ 153 ; reed MAND. 57 ; WICL. GEN. xxxviii. 27 ; HOCCL. i. 159 ; ∂e reade se GEN. & EX. 2670 ; Mars yaf to her coroun reed _redness_ CH. L. G. W. 533 ; **rēades** (_gen. m._) A. R. 402 ; **rēade** (_dat. m. n._) REL. I. 173 ; mid his reade blode FRAG. 6 ; of reade [rede] golde LA3. 1181 ; **ræde** [rede] (_pl._) LA3. 1890 ; **rǎddore** (_compar._) C. L. 719 ; raddere P. S. 330 ; _comp._ **fūr-, in-rēad.**

rĕd-brēst, sb., _redbreast,_ PR. P. 426.

rĕd-gound, sb., _mod.Eng._ redgum ; '_scrophulus,_' PR. P. 426 ; **rǎdegounes** (_pl._) LANGL. _C_ xxiii. 83.

rĕdnesse, sb., _O.E._ rēadness ; _redness,_ PR. P. 426.

read, readen, see **rēd, rǣden.**

rēaden, v., _O.E._ rēadian, _cf. O.H.G._ rōtēn, rōten ; _make red_ ; **rēdes** (_pres._) SPEC. 34 ; **irēaded** (_pple._) A. R. 356.

rēaf, sb., _O.E._ rēaf,=_O.Fris._ rāf, _M.Du._ roof, _O. H.G._ roub ; _robe, garment; spoil; plunder;_ HOM. I. 225 ; warp he an his rugge a ræf [reaf] swi∂e deore LA3. 23760 ; **rēve** (_dat._) O. & N. 458 ; AR. & MER. 8396 ; læten þa ræf liggen LA3. 8612 ; **rēafe** (_gen. pl._) MAT. xxiii. 5 ; _comp._ **bed-, messe-rēaf** (-**rēf**).

rēaf-lāc, sb., _O.E._ rēaflāc ; _rapine,_ HOM. I. 39 ; ræflac LA3. 9939 ; reflac A. R. 202 ; GEN. & EX. 436 ; refloc HOM. II. 79.

rēal, adj., _O.Fr._ real ; _royal,_ WILL. 1597 ; real [ial, roial] CH. C. T. _A_ 1018.

rēal-līche, adv., _royally,_ TREV. III. 171.

rēalme, sb., _O.Fr._ realme, reaume, roiaume ; _realm, kingdom_ ; rialme GAW. 691 ; realme DREAM 250 ; roialme. MAN. (F.) 14763 ; reaume BEK. 948 ; WILL. 1964 ; reame LANGL. _C_ i. 192 ; reume _B_ viii. 105 ; reem TREV. I. 115, 121 ; reume WICL. MK. iii. 24 ; **rēalmes** (_pl._) LANGL. _A_ i. 93 ; reomes MAN. (F.) 5316 ; remes CH. C. T. _B_ 4326.

rēaltē, sb., _O.Fr._ realté ; _royalty_ ; rialte P. L. S. xxx. 88 ; realte, reaute WILL. 1926, 5006.

rēam, sb., _? O.E._ rēam, = _L.G._ raum, _Du._ room ; _cream_ ; rem AR. & MER. 1455 ; L. H. R. 146 ; _comp._ **milc-rēm.**

hrēam, sb., _O.E._ hrēam, ? = _O.N._ hreimr (? hreymr) ; _cry_ ; ream HOM. I. 209 ; A. R. 110* ; ræm [ream] LA3. 11280 ; ræm ORM. 8137 ; rem O. & N. 1215 ; REL. I. 223 ; GEN. & EX. 1962 ; **rēames** (_pl._) KATH. 164

rēame, see **rēalme. reamen,** see **rǣmen.**

rearde, sb., _O.E._ reord (_for_ *reard),=_O.H. G._ rarta, _Goth._ razda, _O.N._ rödd ; _voice,_

K k

sound; saying ; AYENB. 60 ; reord, rerd ORM. 9566, 16664 ; reorde [rorde] O. & N. 311 ; rorde ['_sonus_'] PS. xviii. 5 ; rerd S. S. (Web.) 910 ; IW. 2073 ; TOWNL. 307 ; he him kneu wel bi his rerde REL. II. 274 ; rerid ALEX. (Sk.) 387, 488 ; _comp._ **3e-reard,** _deriv._ **reordien.**

[? **reardi,** adj., _comp._ **el-reordi.**]

reas, see **rǣs. rea∂e,** see _hrá∂e._

rēaume, see **rēalme.**

rēaven, v., _O.E._ rēafian, _cf. O.Fris._ rāvia, _M. Du._ rōven, _O.N._ raufa, reyfa, _O.H.G._ roubōn, _Goth._ (bi-)raubōn ; _from_ **rēaf** ; _rob, plunder,_ KATH. 1229 ; A. R. 286 ; ræfen ORM. 2015 ; reven GEN. & EX. 2802 ; HAV. 480 ; reven, reve H. H. 120, 132 ; revin '_spolio_' PR. P. 431 ; reave SHOR. 46 ; reve CH. C. T. _A_ 4011 ; APOL. 48 ; **rēveþ** (_pres._) HOCCL. v. 21 ; reves PR. C. 251 ; revis S. S. (Wr.) 2348 ; if he rēve (_subj._) me mi ri3t LANGL. _B_ xviii. 274 ; **rǣvede** (_pret._) SAX. CHR. 261 ; LA3. 4038 ; reved MIN. iii. 122 ; refde MARH. 13 ; refte HAV. 2223 ; reveden MISC. 151 ; **rēved** (_pple._) WILL. 2755 ; WICL. JER. l. 37 ; reft TOR. 2176 ; raft S. S. (Wr.) 1015 ; _comp._ **bi-, 3e-rēavien.**

rēvere, sb., _O.E._ rēafere. = _O.N._ raufari, reyfari, _O.H.G._ roubāre ; _plunderer, robber,_ ['_praedo_'] WICL. JER. iv. 7 ; **rēaferes** (_pl._) HOM. I. 243 ; reaveres H. M. 29 ; reavares A. R. 150 ; ræveres SAX. CHR. 262 ; LA3. 14059 ; reveres HOM. I. 15 ; R. S. vii ; LANGL. _B_ xiv. 182.

rēverie, sb., = _Ger._ räuberei ; _spoliation,_ ROB. 194 ; MAN. (F.) 5827.

rēvunge, sb., _O.E._ rēafung ; _spoliation,_ HOM. I. 35 ; ræving [reving] LA3. 2647 ; reving P. L. S. viii. 128.

rebáte, v., _O.Fr._ rabatre ; _abate,_ LUD. COV. 76.

rebel, adj. & sb., _O.Fr._ rebelle ; _rebellious; rebel_ : þo hii were rebel ROB. 72 ; SHOR. 101 ; AYENB. 68 ; ['_rebellantem_'] TREV. III. 41.

rebellin, v., _Fr._ rebeller ; _rebel,_ PR. P. 425 ; **rebelland** (_pple._) BARB. x. 129*.

rebelliōn, sb., _O.Fr._ rebellion ; _rebellion,_ WICL. 3 KINGS xi. 27.

reboiting, see _under_ **rebútin.**

rebounden, v., _O.Fr._ rebondir ; _rebound,_ WICL. PROV. III. 10.

rebours, _in phr._ **at rebours,** _O.Fr._ a rebours (_backward_) ; _evasively_ : Bretons . . . ar bot avaunturs & manace mikel at rebours MAN. (F.) 12652 ; (he) answered him al at reburs 5165.

rebūke, v., _O.Fr._ rebouquer ; _rebuke,_ D. ARTH. 1333 ; **rebūked** (_pret._) LANGL. _B_ xi. 419.

rebútin, v., _O.Fr._ rebouter ; _repulse_ ; **rebútit** (_pple._) BARB. ii. 468 ; vii. 617 ; rabutit xii. 168 ; reboitit BARB. xii. 84.

reboiting, sb., _repulse,_ BARB. xii. 339.

recchen, v., *O.E.* reccan, reccean,=*O.H.G.*
recchen, *O.L.G.* rekkean, *O.N.* rekja ; *from*
réken ; *rule, direct; reach; expound, tell* :
ich þe wulle ræcchen deorne runen LAȜ.
14079 ; if he can rechen ðis dremes wold
GEN. & EX. 2122 ; wen hoe shulden þidere
recche REL. II. 278 ; recche out TREV. IV. 317;
þe sunne reccheð (*pres.*) hire rune MARH. 9 ;
reccheð ever abuten A. R. 164 ; ricchis, riches
GAW. 8, 1873 ; he recte (*printed* rette) (*pret.*)
heom þa get oðer bispell MAT. xiii. 31 ; and
rahte ut his tunge MARH. 10 ; þe king his hand
up rauȝte M. T. 7 ; out of his slepe he raught
M. ARTH. 3191 ; reiȝte MED. 642 ; rehten
['*narraverunt*'] MK. v. 16 ; ut of scipe heo
rehten (*went*) LAȜ. 25646 ; heo muche runen
ræhten heom bitweonen 25124 ; here armes
whan hi upward reiȝte P. L. S. xx. 95 ; **rāuht**
(*pple.*) MISC. 90 ; i was rauht on roode tre
P. L. S. xxv. 63 ; *comp.* **a-, for-, ȝe-recchen.**

[**recchend,** pple., *O.E.* reccend (*ruling*).]

rechen-dōm, sb., *O.E.* reccenddōm ; *re-gency, government,* FRAG. 1.

recching, sb., *interpretation* ; reching GEN.
& EX. 2058.

rĕcchen, *see* **rĕchen.**

receite, sb., *O.Fr.* recete (*recette*) ; *receipt, prescription,* CH. C. T. *G* 1353.

receive, v., *O.Fr.* recevir ; *receive,* H. S.
236 ; receiveth (*pres.*) LANGL. *C* iv. 501 ;
received (*pple.*) PR. C. 5436.

recet, sb., *O.Fr.* recet ; *receptacle, refuge, harbour,* ROB. 98 ; MAN. (F.) 4464 ; reset
BARB. v. 415 ; x. 139.

recetten, v., *from* recet ; *take harbour, take refuge* ; recetteþ (*pres.*) LANGL. *C* iv. 501 ;
were recetted (*pple.*) *had taken refuge,* ROB.
(W.) 4360, 4635.

recettor, sb., *harbourer,* LANGL. *C* iv. 501.

rēch, *see* **rēk.**

[**rēche,** sb.,=*M.L.G.* rōke, *M.H.G.* ruoche
(*care*) ; *care.*]

rēche-lēs, adj., *O.E.* rēcelēas,=*M.L.G.* rōke-lōs, *M.H.G.* ruochelōs ; *reckless,* HOM. II.
39 ; CH. C. T. *A* 179 ; LIDG. TH. 1039 ;
reckelæs ORM. 932 ; rechelese hinen HOM. I.
245.

    rēchelēs-līche, adv., *carelessly* ['*negligen-ter*'], TREV. VIII. 89.

    rēchelēsnesse, sb., *recklessness,* LANGL. *C*
ix. 259 ; rechelesnes *C* xii. 195 ; recchelesnes
B xi. 33 ; reklesnes PR. C. 3909.

    rēchelēste, sb., *O.E.* rēcelēst ; *heedlessness,*
HOM. II. 45.

rēchelen, v., *cf. O.E.* rēcelsian ; *offer incense* :
to **rēchelende** (*dat. inf.*) þe alter HOM. II. 133.

rēchels, rĕchels, sb., *O.E.* rēcels, rȳcels, rīcels,
cf. O.N. reykelsi ; *incense* ; recheles HOM. II. 45 ;
recheles, rechles A. R. 216, 376 ; rekels C. M.
11499 ; rekils TÔWNL. 125 ; recles M. H. 105 ;
(*ms.* recless, reccless) ORM. 1744, 6475 ; rek-

les PS. cxl. 2 ; riche(l)s LAȜ. 8091 ; richelis
PR. P. 433 ; rikels REL. I. 53.

rēchel-fat, sb., *O.E.* recelsfæt ; *censer,* HOM.
II. 133 ; reclefat ORM. 1736 ; recle-, reklefat
GEN. & EX. 3761, 3782.

rechen, *see* **recchen.**

rĕchen, v., *O.E.* rēcan, rēccan,=*O.N.* rœkja
(rækja), *O.L.G.* rōkian, *M.H.G.* ruochen ;
reck, care ; rehchen LAȜ. 18042 ; ye nolden
of me recchen (*? r. w.* wrecchen) MISC.
82 ; recche O. & N. 803 ; LANGL. *A* iv. 51
(65) ; rekken ORM. 16165 ; PR. P. 428 ;
recche (*pres.*) P. L. S. viii. 112 ; ne recche
ich noht his londes LAȜ. 17051 ; of oþer
þing ne recche i noht SPEC. 69 ; i recche
never who it here AUD. 51 ; recche, rekke
CH. C. T. *A* 1398, 2257 ; me ne reccheð A.
R. 104 ; reccheth [rekkeþ] LANGL. *B* xv. 172 ;
recþ O. & N. 491 ; reches A. P. ii. 465 ; ȝe
recheþ ALIS. 7319 ; þei recchen MAND. 64 ;
reche (*subj.*) REL. I. 225 ; whan so hit recche
(*r. w.* fecche) HORN (L.) 352 ; recke HAV.
2047 ; **rōhte, rŏhte** (*pret.*) LAȜ. 11482 ; (*ms.*
rohhte) ORM. 9024 ; þe nefre . . . nanes godes
ne rohte HOM. I. 9 ; roȝte AV. AR. xxv ; as him
no þing ne roȝte BEK. 1053 ; rouhte A. R. 60 ;
roghte [roughte] CH. C. T. *A* 3772 ; route
REL. II. 277 ; **rōuhte** [roȝte] (*subj.*) O. & N.
427.

rēchen, *see* **rǣchen.**

reclaime, sb., *reclaim* (*term in falconry*),
DEP. R. ii. 182.

reclaimen, v., *O.Fr.* reclaimer ; *reclaim* (*in
falconry*) : he wole . . . reclaime thee, and
bringe thee to lure CH. C. T. *H* 72.

    reclaiming, sb., *enticement,* CH. L. G. W.
1371.

reclūs, adj. & sb., *Fr.* reclus ; *recluse* : the
reclus frere E. W. 7 ; **reclūses** (*pl.*) A. R. 10.

reclūsen, v., *shut up* ; religious outriders
reclūsed (*pple.*) in here cloistres LANGL. *C*
v. 116.

recoilen, v., *O.Fr.* reculer ; *recoil,* A. R. 294.

recoillin, v., *O.Fr.* recueiller ; *bring together* ;
(þei) racoillede (*pret.*) MAN. (F.) 5287.

recomanden, v., *O.Fr.* recommander ; *com-mend* ; recomende CH. C. T. *G* 544 ; recom-andeth (*pres.*) CH. C. T. *B* 279.

recompense, v., *O.Fr.* recompense ; *reward,*
GOW. II. 278.

reconforten, v., *O.Fr.* reconforter ; *comfort* ;
reconforte CH. C. T. *A* 2852 ; **reconforted**
(*pple.*) LANGL. *B* v. 287.

    reconforting, sb., *comfort,* BARB. xi. 499.

reconissance, sb., *O.Fr.* recoignisance ;
recognisances (*law term*), CH. C. T. *B* 1520.

reconsilen, v., *O.Fr.* reconciler, *from Lat.*
reconciliāre ; *recover, reconcile* ; **recounse-linge** (*pple.*) WICL. 2 COR. v. 19 ; **recoun-selide** (*pret.*) WICL. 2 COR. v. 18 ; recoun-

seiled (*pple.*) WICL. JUDG. xix. 3 ; MAND. (PROL.).

recouncelere, sb., *reconciler,* WICL. DEUT. v. 5.

recounseling, sb., *reconciliation,* WICL. ECCLUS. xliv. 17.

record, sb., *record, witness,* LANGL. *C* iv. 474, *B* xviii. 85.

recorden, v., *O.Fr.* recorder ; *report, repeat,* A. R. 256 ; recordi AYENB. 21 ; **recorde** (*pres.*) CH. C. T. *A* 829.

recours, sb., *O.Fr.* recours ; *recourse,* CH. C. T. *F* 75.

recoverer, sb., *recovery,* A. P. ii. 394 ; re-covrere LANGL. *B* xvii. 67 ; recover *C* xx. 67.

recovri, v., *O.Fr.* recovrer, *from Lat.* recu-perāre ; *recover,* AYENB. 32 ; recovere KN. L. 179 ; recovere, -coevre LANGL. *B* xix. 239 ; rekevere WICL. ECCLUS. ii. 6 ; **recuvered** (*pret.*) WILL. 3874 ; rekowered ALIS. 5835.

recraien, v. ; **recraied,** recreiȝede (*pple.*) *recreant,* LANGL. *B* iii. 257 ; *A* iii. 244.

recreant, adj., *O.Fr.* recreant ; *recreant, defeated,* PARTEN. 4781 ; recreaunt GAW. 456; LANGL. *B* xviii. 100.

recreatiōn, sb., *Fr.* recreation ; *amusement,* GOW. III. 100.

rectour, sb., *Fr.* recteur ; *rector* ; **rectours** (*pl.*) LANGL. *C* iii. 184.

red, *see* hraðð.

[rēd, ? sb. ; *comp.* dai-rēd.]

rēd², sb., *? trench* ; **rēdes,** reedes (*pl.*) ['*fossae, fossulae*'] PALL. iv. 219, xii. 73.

rēd, *see* ræd, rēad, hrēod. **redd,** *see* hrad.

hredden, v., *O.E.* hreddan, = *O.Fris.* hredda, *O.H.G.* rettan ; *rid ; escape; deliver* ; redden ORM. 8126 ; ridde GAW. 2246 ; **riddes** (*pres.*) HOM. I. 273 ; **redde** (*pret.*) HOM. II. 19 ; ORM. 19316 ; þe children þer wiþ fram depe he redde M. T. 133 ; ruddest JUL. 75 ; red (*pple.*) AR. & MER. 8606 ; *comp.* a-**redden.**

reddour, *see* **redour.**

[rede ?, *comp.* mid-rede.]

rēde, sb., = *O.H.G.* rōtī ; *redness,* CH. C. T. *B* 356.

rēde, *see* rǣde. **rēdel,** *see* rǣdels.

redel, *see* ridel.

redempciōn, sb., *redemption,* WICL. LK. i. 68.

rēden, *see* rǣden. **rēdi,** *see* rǣdi.

redic, sb., *O.E.* rǣdic, = *O.H.G.* ratich, retich ; *radish,* FRAG. 3 ; redich REL. I. 36.

redingking, sb., (? *corruption of* *rīding-cniht* ; *cf. O.E.* rādcniht) ; *feudal retainer,* LANGL. *B* v. 323 ; **redingkinges** (*pl.*) LANGL. *C* iii. 112.

Rēdinges, pr. n., *O.E.* Rēadingas ; *Reading,* ROB. 424.

redolent, adj., *? O.Fr.* redolent ; *fragrant,* BER. 2765.

redour, sb., *O.Fr.* reidur (*roideur*) ; *stiffness, violence* ; reddour WILL. 2953 ; P. R. L. P. 213 ; D. ARTH. 1456 ; reddur PR. C. 6091 ; riddour [raddour] ALEX. (Sk.) 2329 ; **redūrs** (*pl.*) D. TROY 1805.

redoutable, adj., *O.Fr.* redoubtable ; *terrible, venerable,* CH. BOET. iv. 5 (131).

redouting, sb., *reverence,* CH. C. T. *A* 2050.

redressen, v., *O.Fr.* redrecier ; *redress,* ANGL. I. 325 ; redresse TREV. IV. 191.

rēf, *see* rēaf, hrēof.

rēfe, sb., = ȝerēfe ; *reeve,* MK. xv. 43 (*earlier text* gerefa) ; HOM. I. 163 ; reve '*praepositus*' PR. P. 431 ; LAȝ. 15597 ; MARH. 6 ; HAV. 1627 ; BEK. 49 ; reve [reeve] CH. C. T. *A* 542 ; **rēves** (*gen.*) LANGL. *B* v. 427 ; **rēven** (*pl.*) KATH. 1975 ; AYENB. 37 ; *comp.* **burh-, chirche-, port-, rēp-, schir-rēve.**

rēf-schipe, sb., *office of a reeve,* KATH. 11.

reeve-rolles, sb., pl., *reeue-rolls,* LANGL. *C* xxii. 465.

hrēfen, v., *O.E.* (ge-)hrēfan ; *from* hrōf ; *roof* ; refen SAX. CHR. 263 ; *comp.* ȝe-rēfen.

refēten, v., *feed, refresh* ; refēte (*pres. pl.*) ALEX. (Sk.) 4587.

reflāren, v., *O.Fr.* reflairer ; *blow back* ; **reflars** (*pres.*) YORK xli. 367.

reflexiōn, sb., *reflexion* ; **reflexiōns** (*pl.*) CH. C. T. *F* 230.

reformen, v., *Fr.* reformer ; *reform* ; reforme GOW. I. 273 ; reformeþ (*pres.*) AYENB. 81.

refrein(e), sb., *O.Fr.* ? refrain ; *refrain (of a song),* CH. TRO. ii. 1571.

refreinen, v., *O.Fr.* refrener ; *refrain, abstain,* '*refreno*,' PR. P. 427 ; refreine TREV. III. 187.

[**refreinen**², v., *sing a refrain.*]

refraining, sb., *singing of a refrain,* R. R. 749.

refreschen, v., *O.Fr.* refreschir ; *refresh* ; re-fresche CH. C. T. *A* 2622 ; **refreschede** (*pret.*) REL. I. 40.

reft, *see* rēaven. **refter,** *see* rǣfter.

refuge, sb., *O.Fr.* refuge ; *refuge,* CH. C. T. *A* 1720.

refūsin, v., *O.Fr.* refuser ; *refuse,* PR. P. 427.

refūte, sb., *O.Fr.* refute, *cf.* fuite ; *refuge,* PR. P. 427 ; refuit WICL. DEUT. xix. 12 ; refut(e) CH. C. T. *B* 546.

reg, *see* hrūg.

regard, sb., *O.Fr.* regvard ; *regard, respect* ; at regard of CH. C. T. *I* 788 ; *see* **reward.**

regale, sb., *O.Fr.* regale ; *royalty* : al þe regal of Rome WILL. 282 ; regall PL. T. xix (P. P. 309) ; **regals** (*pl.*) CH. L. G. W. 2128,

[**reȝel,** sb., *O.E.* regol ; *rule.*]

reȝhel-bōc, sb., *O.E.* regolbōc; *book of canons,* ORM. DEDIC. 8.

reȝol-sticke, sb., *rule,* '*regula,*' FRAG. 3.

hreȝel, *see* hræȝel.

regiben, v., *O.Fr.* regiber (*regimber*); *kick*; **regibbeð** (*pres.*) A. R. 138.

regioun, sb., *O.Fr.* region; *region,* A. P. i. 1177; CH. C. T. *C* 122.

registre, sb., *list, catalogue,* LANGL. *C* xxiii. 271; **registres** (*pl.*) *C* xii. 274.

registrēr, sb., *registrar, accountant,* LANGL. *C* xxii. 259; **registrēres** (*pl.*) LANGL. *B* ii. 173.

reȝn, *see* rein.

regne, sb., *O.Fr.* regne; *reign,* AYENB. 107; MAN. (F.) 7115; rengne HORN (L.) 901, 908.

regnen, v., *O.Fr.* regner; *reign*; regne LANGL. *A* iii. 271; reinin PR. P. 428; **regneþ** (*pres.*) P. S. 340; regne (*subj.*) HAV. 2586; regnede (*pret.*) MAN. (F.) 7287.

regraterie, sb., *O.Fr.* regraterie (*regratterie*); *act of regrating,* LANGL. *B* iii. 83.

regratour, sb., *regrator,* E. G. 353; regratour LANGL. *A* v. 140 [regratere *B* v. 226]; **regrateres** (*pl.*) *B* iii. 90.

regulēr, adj., *O.Fr.* regulier; *according to rule*; canouns **regulērs** (*pl.*) R. R. 6696.

regverdōn, sb., *from O.Fr.* regverdonner; *reward,* GOW. II. 206.

rēh, rēi, *see* hrēoh.

rehed, *see* hrēod.

reherce, v., *O.Fr.* rehercer; *rehearse,* PR. C. 2390; reherce, -herse CH. C. T. *B* 89; reherce (*imper.*) LANGL. *B* i. 22 [rehersen *A* i. 22.

rehercing, sb., *rehearsal,* CH. C. T. *A* 1650.

rehersaille, sb., *rehearsal,* CH. C. T. *G* 852.

rehéten, v., *O.Fr.* rehaitier; *refresh, cheer, encourage*; rehaite A. P. ii. 127; **rehétes** (*pres.*) ALEX. (Sk.) 3999, 5320; HAMP. PS. xxii. 2*.

rehétinge, sb., *comfort,* HAMP. PS. xxii. 2.

rehéten[2], v., *cf.* ráten; *rebuke*; rehete LANGL. *C* xiii. 35*.

reie, sb., *cf. M.Du., M.L.G., M.H.G.* reie; *round dance*; **reies** (*pl.*) CH. H. F. 1236.

reie, *see* rüȝe.

reihe, rouhe, sb., *O.E.* reohhe; *ray (fish),* PR. P. 427, 438; **rehȝen,** rihȝen (rohȝe) (*pl.*) LAȝ. 29557, 29583; righe MAN. (F.) 15196.

reil, *see* hræȝel.

reilen, v.; *run, roll*; raileth (*pres.*) LIDG. M. P. 220; railing (*pple.*) P. R. L. P. III; *cf.* **roilen.**

reimen, *see* raimen.

rein, sb., *O.E.* regn, rēn,=*O.N.* regn, *O.L.G.* regin, regan, *O.H.G.* regen, regan, *O.Fris.* rein, *Goth.* rign; *rain,* '*pluvia,*' PR. P. 428;

A. R. 246; GEN. & EX. 3265; TREAT. 136; LANGL. *B* xiv. 66; rein, rain LAȝ. 3898, 19745; reȝn ORM. 8622; ren HOM. I. 225; a reïnes drope P. L. S. xvii. 369; **reine** (*dat.*) R. S. i (MISC. 156); rene AYENB. 130.

rein-bowe, sb., *O.E.* rēnboga,=*O.H.G.* reginbogo; *rainbow,* VOC. 273; GEN. & EX. 637; P. L. S. ii. 170; renboge HOM. I. 225.

rein-drope,sb.,=*M.H.G.*regentrophe; *rain-drop,* S. B. W. 11.

rein-foul, sb., '*picus,*' PR. P. 428.

rein-water, sb.,=*O.H.G.* regenwazer; *rain-water,* HOM. II. 151.

reine, *see* rēne.   **reinen,** *see* regnen.

reines, sb., *O.Fr.* reins, *Lat.* rēnes; *reins, kidneys,* WICL. WISD. i. 6; renis LEECHD. III. 140.

reinin, v., *O. E.* regnian, rēnian (*pret.* rēnode, rān BL. HOM. 260), *cf. O.H.G.* regonōn, *O.N.* regna, rigna, *Goth.* rignjan; *rain,* PR. P. 428; reȝnen ORM. 8694; reine HOM. II. 99; rine LAȝ. 19745; AYENB. 49; **reineþ** (*pres.*) MAND. 45; reineþ LANGL. *B* xvii. 333; reineþ [*v. r.* rineþ, regneþ, raineþ] *C* xx. 315; rineþ ALIS. 4976; **reine** (*subj.*) A. R. 98*; rine TREAT. 136; **rīnde** [reinede] (*pret.*) LAȝ. 3895; rinde MAT. vii. 27; reinede, rainde, ron, roon LANGL. *B* xiv. 66, *C* xvi. 270; roon TREV. II. 239; *comp.* bi-reinen.

reisen, *see* raisen.

reisin, sb., *O.Fr.* reisin; *bunch of grapes; raisin*; ['*racemus*'] WICL. JUDG. viii. 2; **reisins** (*pl.*) ALIS. 5193; raisins of Corauns *currants* L. C. C. 16.

reisūn, *see* raisōn.

rejoissen, v., *O.Fr.* rejoiss-, rejoïr; *rejoice*; rejoisen [rejoisshen] LANGL. *C* xviii. 198; rejoische WILL. 4102; his broþer **rejoisede** (*pret.*) (*held*) þe regalte MAN. (F.) 5344.

rēk, sb., *O.E.* rēc,=*O.Fris.* rēk, *O.N.* reykr, *M.Du.* rook, *O.H.G.* rouch; *from* rēoken; *vapour, smoke,* ['*fumus*'] PS.. xvii. 9; reek PR. P. 428; rek [reech] C. M. 2744; **rēke** (*dat.*) PR. C. 9431; D. ARTH. 1041.

rēk, *see* hrēac.

réke, sb., ? *from* réken; ? *haste, hurry* [? *or* =rēk]; in a grete reke MED. 821; wiþ gret(e) reke AR. & MER. 7904.

rēkels, *see* rēchels.

réken, v., ? *O.E.* recan (*rule, go*), = *M.Du.* reken, *O.N.* reka (*conduct*), *Goth.* rikan (σωρεύειν), *M.H.G.* rechen; *extend; direct one's way, go*: þis kniȝtes come reken in BEK. 2091; reke FER. 1249; PALL. i. 194; þe no mon mai at reke HOM. II. 258; ich wule forþur reke O. & N. 1606; to her sone sche gan to reke OCTOV. (W.) 182; let me in reke BEV. 394; in erþe þi bodi reke AR. & MER. 1027; þat . . . al mai reke C. M. 11221; **rak** (*pret.*) FER. 2177; BEV. 3360; his spere to his heorte rac (*ms.* rack) LAȝ.

9320*; **réke** (*pple.*) BEV. 1686; in oure ashen olde is fir ireke CH. C. T. *A* 3882; *comp.* **mis-réken**; *deriv.* **ráke, rákien, recchen.**

reken, adj., *O.E.* recen,=*M.L.G.* reken; *prompt, apt,* GEN. & EX. 3485; SPEC. 27; A. P. i. 5; **rekeneste** (*superl.*) D. ARTH. 4081; *comp.* **un-reken.**

reken-lī, adv., *promptly,* GAW. 251; A. P. ii. 127.

rēken, v., *O.E.* rēcan, *cf. O.N.* reykja, *O.H.G.* rouchan; *from* **rēk**; *reek, steam*; reke PS. cxliii. 5; LIDG. M. P. 114; **rēkes** (*pres.*) AV. ARTH. xv.

rekenen, v., *O.E.* (ge-)recenian, = *O.Fris.* rekenia, reknia, *O.H.G.* rechenōn; *reckon, enumerate*; reknin '*computo*' PR. P. 428; reknen GOW. III. 121; rekene TREAT. 133; WILL. 1934; LANGL. *B* ii. 61; MAND. 213; B. DISC. 2109; WICL. MAT. xviii. 24; rekene [rekne] CH. C. T. *A* 1954; rekeni MISC. 193; AYENB. 19; rikenen A. R. 210; JOS. 76; rikenin JUL. 51; þine peines rikene hit were long SPEC. 68; **rekenede** (*pret.*) P. L. S. xxiii. 71; **recned** (*pple.*) ORM. 2055; *comp.* **ӡe-rekenen.**

rekenere, sb., *reckoner*; reknare '*computator*' PR. P. 428; **rikenares** (*pl.*) A. R. 214.

rekeninge, sb.,=*M.L.G.* rekeninge, *O.H.G.* rechenunga; *reckoning,* AYENB. 18; LANGL. *B* v. 434; CH. C. T. *A* 600; rikening JOS. 444; rekninge '*ratio, computatio*' PR. P. 428; rekning LIDG. M. P. 240; recning P. L. S. i. 23.

rekenthi, *see* **rakente.**

rekke, sb., *cf. M.Du.* recke (*pertica*); *? fetters, irons*: hevie **rekkes** (*pl.*) binde to hire fet P. L. S. xv. 192.

rēkken, *see* **rēchen.**

rēl, *see* **hrēol.**

relaciōn, sb., *O.Fr.* relation; *relation,* LANGL. *C* iv. 344, 346, 363.

relatif, sb., *O.Fr.* relatif; *relative* (*in grammar*), LANGL. *C* iv. 357.

reléf, sb., *O.Fr.* relief; *relief; remains of a meal*; A. R. 168; C. L. 1277; relefe GOW. III. 23; CATH. 303; releef '*reliqvie, fragmentum*' PR. P. 428; relif WICL. EX. xxix. 34; **relifs, relives** (*pl.*) WICL. EX. viii. 3, xxix. 34.

relente, v., *from O.Fr.* relent; *melt,* CH. C. T. *G* 1278.

relés, sb., *O.Fr.* reles; *release,* C. L. 509; TREV. III. 421; ROB. (W.) Q 2.

relessen, v., *O.Fr.* relesser, -laisser; *release*; **relesse** (*pres.*) CH. C. T. *F* 1613; relesi (*subj.*) ROB. (W.) 10297; **relessed** (*pple.*) SHOR. 65; LANGL. *B* iii. 58.

reléven, v., *O.Fr.* relever; *relieve*; releve LANGL. *B* xviii. 141; **reléved** (*pple.*) CH. C. T. *A* 4182.

relíen, v., *O.Fr.* relier, ralier; *rally*; reli BARB. iii. 34; **relíed(e)** (*pret.*) LANGL. *B* xx. 147, *C* xxiii. 148; releit BARB. ii. 401; releit (*pple.*) BARB. vii. 91.

religioun, sb., *O.Fr.* religion; *religion,* P. L. S. xvii. 253; religiun A. R. 8, 74; HOM. II. 49.

religiūs, adj. & sb., *O.Fr.* religious, religius; *religious; bound by monastic vows,* A. R. 74; religiouse wommen *nuns* E. W. 7; of þo ilke **religiouses** (*pl.*) (*monks*) fond seint Austin mani houses MAN. (F.) 15281.

relike, sb., *O.Fr.* relique; *relic, remainder,* ROB. 274; relikes (*pl.*) AYENB. 64; relikes WICL. NUM. xxiv. 19.

rēlin, *see* **kreolen. rem,** *see* **hráven.**

rēm, *see* **realme, hrēam, rēam.**

remanant, sb., *O.Fr.* remanant; *remnant,* FER. 3273; remenant CH. C. T. *A* 724; remenaunt MAN. (F.) 6988; ALIS. 5707; remenont AYENB. 100.

réme, *see* **rime.**

remedie, sb., *O.Fr.* remede; *remedy,* A. R. 124; AYENB. 207; CH. C. T. *A* 1216.

remembrance, sb., *O.Fr.* remembrance; *remembrance,* CH. C. T. *B* 3908; MAND. 252.

remembren, v., *O.Fr.* remembrer; *remember*; CH. C. T. *I* 1063; i remembred (*pret.*) me A. P. iii. 326.

rēmen, *see* **rǣmen, rūmen.**

hrēmen, v., *O.E.* hrēman, hrýman; *from* **hrēam**; *cry out*; remen KATH. 2371; R. S. iv; remen heo schule and grede MISC. 77; reme H. S. 7858; A. P. i. 1180; **rēmeð** (*pres.*) REL. I. 223; hrimð MAT. xii. 19; **rēmde** (*pret.*) HOM. I. 95; ӡif me remde lude fur A. R. 242; remden MAT. viii. 29; LAӡ. 5795; MARH. 18; **irēmd** (*pple.*) A. R. 2; *comp.* **bi-rēmen.**

re-mēne, v., *interpret,* WICL. 2 ESDR. xi. 13.

remēner, remēnour, sb., *interpreter,* WICL. GEN. xl. 22; I ESDR. PROL.

remēning(e), sb., *interpretation,* WICL. JUDG. vii. 15.

remena(u)nt, *see* **remanant.**

remeue, *see* **remüen.**

remissiōn, sb., *O.Fr.* remission; *remission,* A. R. 346; remissioun CHR. E. 634.

remorden, v., *O.Fr.* remordre; *vex, cause remorse*; **remordiþ** (*pres.*) CH. BOET. iv. 6 (140).

remounten, v., *? O.Fr.* remonter; *cause to rise again*: thow hast **remounted** (*pple.*) and norisshed me CH. BOET. iii. 1 (63).

rempen, v., *O.E.* rempian; *? run*: þei **rempede** (*pret.*) þem to reste MAN. (F.) 3492; *? comp.* **mis-rempen.**

remüen, v., *O.Fr.* remuer; *stir, move,* CH. BOET. ii. 6 (52); remue ROB. 533; AYENB. 104; remeue TRO. i. 691; **remeueþ** (*imp.*) CH. C. T. *G* 1008; **remüed** (*pret.*) WILL. 1326.

rēn, *see* **rein.**　**rēnable,** *see* **raisonable.**

renden, v., *O.E.* rendan, = *O.Fris.* renda; *rend, tear, lacerate,* KATH. 1999; rende WILL. 1851; LANGL. *B* PROL. 198; LIDG. M. P. 202; rende & rive C. M. 7507; rende (*pret.*) JUL. 70; HORN (R.) 727; ROB. 289; rente CH. C. T. *A* 990; rendden LA3. 7849; rended (*pple.*) M. H. 144; irend A. R. 148; *comp.* to-renden.

renden², v., *cf. O.N.* renna; *run into shape, melt*; rendid (*pret.*) HAMP. PS. cv. 9*; *? see* **rennen.**

rendren, v., *O.Fr.* rendre; *render, restore; translate*; LANGL. *C* xviii. 322; rendreþ (*pres.*) *C* xi. 88; rendred (*pret.*) *C* vii. 217; rendred, rendret *A* ix. 82, *B* viii. 90.

rēne (? rēn), sb., *O.N.* rein *f.*, *cf. M.Du.* reen, *M.H.G.* rein *m.*; *border,* PALL. i. 159.

rēne², sb., *Fr.* rēne, *cf. Ital.* redine; *rein, 'habena,'* PR. P. 429; reine CH. C. T. *A* 4083; reines (*pl.*) TREV. IV. 77.

renegat, sb., *renegade,* CH. C. T. *B* 932; CH. L. G. W. (*earlier prol.*) 401.

reneie, v., *O.Fr.* reneier (*renier*); *deny, renounce,* AYENB. 57; LANGL. *B* xi. 120; CH. C. T. *B* 3751.

re-nēwen, v. (*for* anēwen), *renew*; renēwede (*pret.*) TREV. IV. 119; renēwid (*pple.*) WICL. 2 COR. iv. 16.

[**renewlen,** v., *O.Fr.* renoveler, *from med.Lat.* renovellāre; *renew.*]

　　renüling, sb., *renewal,* WICL. MACC. xii. 17.

reng, sb., *O.Fr.* renc, rang, *cf. Welsh* rheng, rhenc. *Bret.* renk, *Ir.* rang, ranc; *rank, row*: on a reng AR. & MER. 1508; renk BRD. 12; out of þe renge he com ride BEV. 3631; renges [ringes] (*pl.*) CH. C. T. *A* 2594; *see* **hring.**

rengen, v., *O.Fr.* renger; *range, wander, roam*; renge MAN. (F.) 8257; rengeð (*pres.*) & reccheð ever abuten A. R. 164; his batail renged PARTEN. 2224; *see* **ringen.**

rēnien, v., *O.E.* rēnian, (ge-)regnian; *repair*; rēniende (*pple.*) ['*reficientes*'] heora nett MAT. iv. 21.

renisch, adj., *cf.* **runisch**; *furious, loud,* ALEX. (Sk.) 2943, 387.

　　renisch-lī, adv., *vehemently,* ALEX. (Sk.) 4931.

renk, *see* **reng, rink.**

renke, sb., *O.E.* renc (*pl.* renca); *from* ranc; *pride*; rencas (*pl.*) HOM. I. 7.

renne, sb., *cf. O.L.G., O.N.* renne, *O.H.G.* rinne; *from* rinnen; *run, course*; ren '*cursus*' PR. P. 429; rimes ren GEN. & EX. i; at a renne CH. C. T. *A* 4079.

rennels, sb., = *? O.H.G.* rennisal; *rennet*; renlis '*coagulum*' PR. P. 429.

rennen, v., *O.E.* rennan, ærnan, = *O.H.G.*

rennen, *O.Fris., O.N.* renna, *Goth.* (ur-)rannjan; *from* rinnen; *run, make to run*; ærnen LA3. 8129; erne O. & N. 1204; eorneð and eærne(ð) LA3. 6138; rende (*pret.*) HORN (H.) 1274; ernde HORN (R.) 1239; he ærnde to are hævene LA3. 9362; arnde ROB. 140; summe ærnden heore stede LA3. 8134; *comp.* for-, þurh-rennen.

rennare, sb., = *M.H.G.* rennære; *runner, 'cursor,'* PR. P. 429; rennere LANGL. *A* xi. 208; urnare CHR. E. 900.

renning, sb., *stream, course*: the renningis (*pl.*) of watris WICL. PS. i. 3.

rennen, *see* **rinnen.**

renomē, sb., *O.Fr.* renumee; *renown,* GOW. II. 43; renomme BARB. iv. 774; renownee BARB. viii. 290.

renoun, sb., *O.Fr.* renoń; *renown,* A. P. i. 985; CH. C. T. *A* 2240; MAN. (F.) 14753; men of grete renouns (*pl.*) MAN. (F.) 13774.

renouncen, v., *? O.Fr.* renoncer; *renounce*; renounced (*pple.*) GOW. I. 258.

rental, sb., *rental,* LANGL. *B* vi. 92 (*C* ix. 99).

rente, sb., *O.Fr.* rente; *rent, revenue,* O. & N. 1773; CH. C. T. *A* 256; rentes (*pl.*) SAX. CHR. 263; renten A. R. 168.

renten, rente, v., *O.Fr.* renter; *provide with rents, endow,* LANGL. *B* vii. 32, *A* viii. 35.

renüling, *see under* **renewlen.**

rēod, adj., *O.E.* rēod, = *O.N.* rioðr; *red* (? *for* rēad) LA3. 3528, 19890.

hrēod, rēod, sb., *O.E.* hrēod, = *M.Du.* ried, *O.H.G.* riot; *reed, reed-bed,* MAT. xi. 7, xii. 20; reod REL. I. 37; ALIS. 6433; red VOC. 226; reed PR. P. 426; M. H. 36; rēode (*dat.*) LA3. 20168; reedes (*pl.*) MAND. 189.

　　rēd-3erde, sb., *reed-sceptre,* HOM. I. 281.

　　rēden, reeden, adj., *made of reed,* TREV. VII. 487.

　　rēodi, adj., *reedy,* TREV. VII. 487*; reedi place ['*carectum*'] WICL. JOB viii. 11.

hrēof, adj., *O.E.* hrēof, = *O.N.* hriūfr, *O.H.G.* riob; *rough*; ref GEN. & EX. 3726.

hrēofel, adj., *O.E.* hrēofel; *leprous*; rēofle (*pl.*) LK. xii. 12; hrēoflin (*dat. pl.*) HOM. I. 229.

hrēoh, adj., *O.E.* hrēoh, = *O.L.G.* hrē; *fierce, bold*; reh LA3. 24784; rei REL. I. 188; reu weder MAT. xvi. 3; rǣhere (*dat. f.*) LA3. 7934; rǣhne (*acc. m.*) LA3. 3884; rǣhere [re3ere] (*compar.*) LA3. 4062; rēhest (*superl.*) MARH. 13; *comp.* **wel-rēow.**

　　rēh-līche, adv., *roughly,* LA3. 9324.

　　rǣh-scipe, sb., *roughness, fierceness,* LA3. 24943.

rēoken, v., *O.E.* rēocan, = *O.N.* riūka, *O.H.G.* riuchan; *give forth a vapour*: rēkeþ (*pres.*) CH. L. G. W. 2609; *deriv.* rēk, rēken, rōke.

hrēol, sb., *O.E.* hrēol; *reel* (*for thread*); reel,

rel, '*girgillus, alabrum*' VOC. 180, 269; PR. P. 428; REL. II. 81.

*h*rēolen, v., *wind on a reel; reel, stagger*; relin PR. P. 429; reole [rele, reli] LANGL. *C* x. 81; rēled (*pret.*) GAW. 304; A. P. iii. 147.

rēopen, *see* rīpen.   reorde, *see* rearde.

reordien, v., *O.E.* reordian; *speak, say*, LA3. 22174.

*h*rēosen, v., *O.E.* hrēosan; *fall, be ruined*; reosen LA3. 15487; þe wal rēoseð (*sec. text* falleþ) (*pres.*) 15887; rēas (*pret.*) LK. i. 12; rees LA3. 15518*; ruren þer to grunde 27986; *comp.* a-, to-rēosen; *deriv.* *h*rūre.

*h*reoðer, sb., *O.E.* hrīðer, = hryðer, *O.Fris.* hrīther, *O.H.G.* hrind (*pl.* hrindir); *ox, bovine animal*; reoðer A. R. 140*; reoþeres [roþeres] (*gen.*) ALIS. 4719; reþeres P. S. 220; reoðeren (*pl.*) KATH. 60; reðeren HOM. II. 37; reþren LEB. JES. 853; reþeren, riþeren, ruþeren TREV. III. 205, IV. 439; ruðeren [roþere] LA3. 8106; roþeren ROB. 52; PL. CR. 431.

    reoðer-heorde, sb., *O.E.* hryðerhyrde; *neatherd*, HOM. I. 97.

*h*reoþeren, adj., *O.E.* hrīþeren; *of cattle, bovine*: reþerne tounge. *bugloss* HALLIW. 680.

[*h*rēou, adj., *O.E.* hrēow, = *O.N.* hryggr; *sad, sorrowful.*]

    rēu-fol, adj., *rueful, pitiful*, ROB. (W.) 6709.

    rēou-līc, adj., *O.E.* hrēowlīc, = *M.Du.* rouwe-līc; *sad, compassionate*, FRAG. 8; reowlic HOM. I. 39; reoulich LA3. 15080; reulich ROB. 296; reouli (*ms.* reowli) ALIS. 6907; rew-, reuli GEN. & EX. 1162, 1968; reuli HORN (R.) 1057; MAP 338; (*ms.* rewli) S. S. (Web.) 910; **rēowlīche** (*adv.*) FRAG. 6; MARH. 4; reuliche AN. LIT. 10; WILL. 86; AR. & MER. 788; reuli M. H. 143 (*spelt* rewli 59); ruliche P. L. S. xii. 61.

    ? rū-mēde, sb., ? *sadness*, LA3. 12971.

    rēounesse, sb., *O.E.* hrēowness, = *M.Du.* rouwenisse; *grief, misery*, A. R. 144*; reunesse HAV. 2227 (*spelt* rewnesse 502).

[*h*rēous, sb., *grief.*]

    rēus-ful, adj., R. S. vii (MISC. 186).

*h*rēousien, v., *O.E.* hrēowsian, = *M.H.G.* riuwesen; *grieve, sorrow*; reusi(e)n HOM. I. 27; **rēowsiende** (*pple.*) MK. viii. 12; rēousede on heorte LA3. 22173; *comp.* bi-rēowsen.

    rēowsund, sb., *repentance*, ORM. 8799.

    rēowsunge, sb., *repentance*, LK. x. 13; ORM. 5563.

*h*rēouðe, sb., = *O.N.* hrygð; *ruth, repentance, sorrow, grief, misery*; reouðe A. R. 32, 150; r(e)ouþe HOM. I. 79; reuþe 149; reouðe, reowðe [reuþe] LA3. 10869, 12116; reowðe KATH. 2372; rewðe GEN. & EX. 2339; reuþe, rewþe ROB. 103, 175; reuþe O. & N. 1445*; BEK. 541; AYENB. 186; WILL. 3270; S. S. (Wr.)

1389; MAN. (H.) 71; reuþe LANGL. *A* iv. 95; CH. C. T. *A* 914; reuthe PR. C. 6729; ruþe HORN (L.) 673; ruthe PR. P. 439.

    rēouð-ful, adj., *ruthful*, A. R. 116; reouþful ALIS. 6501; reowðful HOM. I. 253; reuþful C. L. 378; REL. I. 121.

    rēowðful-līche, adv., *ruthfully*, MARH. 4.

    rēuþe-lēs, adj., *ruthless*, P. S. 255.

reoven, *see* rīven.

*h*rēowe, sb., *O.E.* hrēow, = *O.H.G.* riuwa, *M.L.G.* rūwe, *M.Du.* rūwe, rouwe; *sorrow*; reowe O. & N. 1445; rewe BEK. 1051; in rew(e) GEN. & EX. 3151.

    rēu-ful, adj., *rueful*, REL. I. 235; P. L. S. xvi. 25; GREG. 150; LANGL. *B* xiv. 148; M. H. 26; C. M. 14301; AUD. 7.

    rēowfulnesse, sb., *ruefulness*, H. M. 41; reoufulnesse A. R. 368.

[*h*rēowsum, adj., = *M.L.G.* rūwesam; *repentant.*]

    rēowsumnesse, sb., *repentance*, HOM. I. 21.

*h*rēowen, v., *O.E.* hrēowan, = *O.L.G.* hreuwan (*ms.* hreuuan), *O.H.G.* riuwan, *M.Du.* rouwen, *O.N.* hryggva, hryggja; *rue, grieve, repent; pity, be merciful*: ævere hit wule þe reouwen [reuwe] LA3. 16047; reowe REL. I. 184; SPEC. 37; ALIS. 3944; rewen (*printed* repen) REL. I. 221; riewe (*v. r.* reowe) ANGL. I. 8 (MISC. 59); rewe HORN (L.) 378; TRIAM. 1544; þe cǫk bigan of him to rewe HAV. 967; þat hii ssolde on him rewe ROB. 449; reue AUD. 42; ruwin '*poeniteo, compatior*' PR. P. 439; r(ē)oweþ (*pres.*) HOM. I. 149; it reoweþ him ORM. 3976; he reweþ me C. L. 541; me reweth sore for hende Nicholas CH. C. T. *A* 3462; he rewes ['*miseretur*'] PS. xxxvi. 26; (hie) rieweð HOM. II. 63; reewe (*imper.*) HOCCL. i. 412; rēw, reu (*pret.*) GEN. & EX. 1166, 1828; reu HOM. II. 147; C. M. 4972; god ræw of man ORM. 65; rewede HAV. 503; rewide WICL. EST. x. 12; rewed M. H. 15; reude BEK. 298; *comp.* a-, bi-rēowen.

rēp, *see* rīp.

repaire, v., *O.Fr.* repairer; *from Lat.* repatriāre; *restore to one's country*, CH. C. T. *B* 1516; TRIST. 2735.

reparailen, v., *O.Fr.* repareiller; *repair*; reparaild (*pple.*) HAMP. PS. ii. 9*, xxi. 24.

    reparailinge, sb., *restoration*, HAMP. PS. ci. 19*.

repárin, v., *O.Fr.* reparer; *repair, mend*, '*reparo*,' PR. P. 430.

repast, v., *O.Fr.* repast; *repast*, LANGL. *C* x. 148.

rēpe, sb., ? *from* rīpen; *cf.* rīp; *sheaf*, '*manipulus*,' VOC. 201, 233; WICL. DEUT. xxiv. 19; '*javele*' REL. II. 80; reepe TOWNL. 13.

repeir, sb., *resort, return*: umwhile to þe erþe þei make repeir MAN. (F.) 8078.

repel, sb., ?=**ripel**; ? *cudgel*; **repples** (*pl.*)
HOM. I. 231.

repéle, v., ? *repeal* : þer mai no man þis doom
repele ANGL. II. 243.

*h*repen, v., *O.E.* hrepian, hreppan,=*M.L.G.*,
M.Du. reppen, *O.N.* hreppa, *O.Fris.* reppa;
touch; repen & rinen A. R. 128; repie [reppe]
S. A. L. 150; **hrepe** (*imper.*) HOM. I. 221;
repie (*subj.*) FRAG. 8; **repede** (*pret.*) MAT.
viii. 3; *comp.* a-repen.

rēpen, *see* **rīpen.**

repentailles, sb., pl., *O. Fr.* repentailles;
penalty; penalties for breach of contract,
MAN. (F.) 11838.

repentance, sb., *O.Fr.* repentance; *repentance*, CH. C. T. *B* 680; repentaunce SHOR. 5.

repentant, adj., *O.Fr.* repentant; *penitent,*
ROB. (W.) 5917.

repente, v., *O.Fr.* repentir; *repent,* CH. C. T.
C 850; repenti ROB. 350; SHOR. 40; repenteþ
(*pres.*) AYENB. 238.

replenissen, v., *O.Fr.* repleniss-; *replenish*;
replenissed (*pple.*) CH. BOET. i. 4 (20).

replēte, adj., *O. Fr.* replet; *quite full*;
repleet CH. C. T. *B* 4147; WICL. PHIL. iv. 18.

replíe, v., *O.Fr.* replier; *reply,* CH. L. G. W.
343.

reportin, v., *O.Fr.* reporter; *report,* PR. P.
430; reported (*pple.*) CH. C. T. *B* 152.

reportour, sb., *reporter,* CH. C. T. *A* 814.

reprehende, v., *Lat.* reprehendere; *blame,*
CH. TRO. i. 510.

representen, v., *O.Fr.* representer; *represent*; **represented** (*pret.*) R. R. 7404.

represse, v., *repress,* GOW. III. 166.

reprévable, adj., *reprehensible,* WICL. PROV.
xxv. 10; GAL. ii. 11.

repréven, *see* **repróven.**

repromissioun, sb., *Lat.* reprōmissiōnem;
promise, WICL. HEB. xi. 39.

repróve, sb., *reproof,* WILL. 652; repreef
WICL. JUD. vii. 16; repreve CH. C. T. *C* 595.

repróven, repréven, v., *O.Fr.* reprover, *mod.
Eng.* reprove, reprieve; *reprove,* PR. C. 5314;
repreven WICL. PROV. xxv. 10; repreve
MAN. (F.) 11665.

repréving, sb., *reproof*; **reprévinges** (*pl.*)
MAND. 1.

repugne, v., *O.Fr.* repugner; *fight against,*
WICL. 1 KINGS xv. 23.

reputacioun, sb., *Fr.* reputation; *reputation,*
CH. C. T. *C* 602.

reqvére, v., *O.Fr.* reqverre; *require,* CH. C. T.
D 1052; **reqvíre** (*pres.*) GAW. 1056.

reqveste, sb., *O.Fr.* reqveste; *request,* A. P. i.
281; CH. C. T. *E* 104.

*h*rēr, adj., *O.E.* hrēr (LEECHD. II. 272), *mod.
Eng.* rare (*slightly cooked*); *tender; slightly
cooked*; rer (*ms.* rere) ' *dimollis* ' PR. P. 430.

rerāge, sb., *arrears*; **rerāges** (*pl.*) LANGL.
B v. 246; *see* **arerāge.**

rerbras, sb., *armour for the back of the arm,*
MAN. (F.) 10030.

rerde, *see* **rearde.** **rere**, *see* **hrüre.**

[**rēre**, adv., *O.Fr.* riere; *backward; comp.*
arēre.]

rēre-warde, sb., *rearguard,* D. ARTH.
1430; GOW. I. 220.

[*h*rēre-, ? *from* hrōre.]

rēre-mous, sb., *O.E.* hrēremūs (VOC. 77),
hrȳremūs (ZEIT. XXI. 43); *bat,* VOC. 164;
WICL. LEV. xi. 19; **rēremīs** (*pl.*) DEP. R. iii.
272.

*h*rēren, v., *O.E.* hrēran,=*O.N.* hrœra, *O.L.G.*
hrōrian, *O.H.G.* hruorian; *move*; **rēre** (*pres.*)
D. ARTH. 2810; *see* **hrōrien.**

rēren, *see* **rǣren.** **rerid**, *see* **rearde.**

rēs, *see* **rǣs.**

resalger, sb., ? *O.Fr.* resalgar (*réalgar*), *Sp.*
rejalgar; *realgar,* CH. C. T. *G* 814.

resche, *see* **rusche.**

reschen, v.? hire leofliche lich **reschte** (*pret.*)
of þe leie MARH. 18.

rescoue, v., *O.Fr.* rescourre; *rescue,* CH.
BOET. iv. 5 (133); MAN. (F.) 4966; D. ARTH.
4131.

rescous, sb., *O.Fr.* rescousse (*deliverance*);
rescue, CH. C. T. *A* 2643; rescu LUD. COV. 114.

resemble, v., *O.Fr.* resembler; *resemble,* CH.
C. T. *D* 90.

rēsen, *see* **hrisien.** **rēsen**, *see* **rǣsen.**

reset, *see* **recet.**

reserven, v., *O.Fr.* reserver; *keep private*;
reserved (*pple.*) CH. C. T. *A* 188.

residence, sb., *O.Fr.* residence; *residence,*
CH. C. T. *G* 660.

residue, sb., *O.Fr.* residu; *residue,* LANGL.
A vii. 93, *C* ix. 109.

resien, *see* **hrisien.**

resignen, v., *O.Fr.* resigner; *resign*; **resigne**
(*pres.*) CH. C. T. *B* 780.

resild, pple., ? *from O.Fr.* resel (*net*); ? *re-
ticulated*, ? *wrinkled* : resild as a resch ALEX.
(Sk.) 4126.

resine, sb., *Fr.* resine; *resin*; recine WICL.
JER. li. 8.

resistence, sb., *O.Fr.* resistence; *resistance,*
CH. C. T. *G* 909.

resolven, v., *O.Fr.* resoudre; *loosen; melt*;
resolved (*pple.*) CH. BOET. iv. 5 (133).

rēsōn, *see* **raisōn.**

resort, sb., *O.Fr.* resort (*ressort*); *resort,*
CH. TRO. iii. 135.

resounen, v., *O.Fr.* resoner; *resound*; re-
souneþ (*pres.*) CH. C. T. *A* 1278.

respit, sb., *O.Fr.* respit; *respite,* BEK. 631;
AYENB. 39.

respiten, v., *O.Fr.* respiter ; *respite*, CH. C. T. *F* 1582 ; respite S. S. (Web.) 1005.

respounen, v., *O.Fr.* respondre, responre ; *answer* ; **respoune** (*pres.*) MAN. (F.) 4238.

respounse, sb., *O.Fr.* respons ; *response*, BEK. 825.

reste, sb., *O.E.* rest, *cf. O.H.G.* resti, *O.L.G.* resta, rasta, *? O.N.* röst, *? Goth.* rasta ; *rest*, A. R. 166 ; O. & N. 281 ; GEN. & EX. 252 ; C. L. 90 ; CH. C. T. *E* 1855 ; none reste FRAG. 6 ; reste & ro ORM. 4972 ; to reste [raste] eode þa sunne LAȝ. 28328 ; to eche reste KATH. 2373 ; whan þei were a reste LANGL. (Wr.) 2939 ; *comp.* lich-raste, un-reste.

    reste-daȝ, sb., *O.E.* restedæg ; *day of rest*, ORM. 4175.

    reste-lees, adj., *O.E.* restlēas ; *restless*, CH. C. T. *C* 728.

resten, v., *O.E.* restan, = *O.H.G.* restan, *O.L. G.* restian ; *rest*, MARH. 6 ; A. R. 260 ; ORM. 9598 ; GEN. & EX. 1369 ; BRD. 6 ; LANGL. *B* iii. 235 ; þer þu scalt resten LAȝ. 17231 ; and ich me wulle ræsten 19038 ; reste CH. C. T. *D* 1736 ; resteþ (*pres.*) SHOR. 133 ; AYENB. 31 ; reste (*pret.*) LAȝ. 3602 ; he þat reste him on þe rode SPEC. 52 ; *comp.* ȝe-resten.

rēsti, adj., *prov.Eng.* reesty ; *rancid*, '*rancidus*,' PR. P. 431 ; REL. II. 29.

restif, adj., *O.Fr.* restif ; *restive*, PALL. x. 73.

rēstin, v., *become rancid* ; reestin '*ranceo*' PR. P. 431.

restitūtiōn, sb., *O.Fr.* restitution ; *restitution*, LANGL. *B* v. 235.

restōre, v., *O.Fr.* restorer ; *restore*, ROB. 194 ; WILL. 2953.

restreine, v., *O.Fr.* restreindre ; *restrain*, CH. C. T. *B* 3777 ; restreigne GOW. III. 206.

rēsŭn, rēsŭnable, *see* raisōn.

resurreccioun, sb., *O.Fr.* resurrection ; *resurrection*, BRD. 17 ; SHOR. 126.

retenaunce, sb., *O.Fr.* retenaunce, *from med.Lat.* retinentia ; *retinue, company* ; **retenauntz** (*pple.*) MAN. (F.) 15985.

retenue, sb., *O.Fr.* retenue ; *retinue*, CH. C. T. *A* 2502.

hrēðe, adj., *O.E.* rēðe, hrēðe ; *fierce, severe* : þe weder als in somer smeþe son began be ruȝ and reþe M. H. xviii ; ræþe FRAG. 1 ; for þa reða dome HOM. I. 15 ; þa wæren swiþe reþe ['*saevi*'] MAT. viii. 28. [rēðe ? *comp.* wand-rēðe.]

rēþer, *see* hreoðer.

rethor, sb., *Lat.* rhētōr ; *orator*, CH. C. T. *F* 38.

rethorike, sb., *rhetoric*, CH. C. T. *E* 32 ; retorik(e) LANGL. *B* xi. 98, *C* xiii. 35.

rētin, v., = *L G.* rauten, rōten, *? Swed.* röta ; *soak*, '*rigo, infundo*,' PR. P. 431.

retournen, v., *O.Fr.* retourner ; *return*, CH. C. T. *A* 2095 ; R. R. 382.

retrograd, adj., *Lat.* retrogradus ; *retrograde*, CH. ASTR. ii. 4 (19).

rettin, v., *O.Fr.* reter ; *account, impute*, '*reputo*,' PR. P. 431 ; rette WILL. 461 ; rette (*pres.*) CH. C. T. *A* 726 ; **rettid** ['*imputavit*'] HAMP. PS. xxxi. 2 ; *comp.* ? a-retten.

rēu, *see* hrēou.

reule, sb., *O.Fr.* reule, riule, *from Lat.* rēgula ; *rule*, P. L. S. ix. 46 ; AYENB. 150 ; CH. C. T. *A* 173 ; riule A. R. 82* ; riwle 2 ; **riulen** (*pl.*) 428 ; rule *line, row* PALL. iv. 526.

reulen, v., *O.Fr.* riuler, *from Lat.* rēgulāre ; *rule*, WICL. PROV. iii. 6 ; reule LANGL. *A* iv. 9 ; riwlen A. R. 4 ; **reuleþ** (*pres.*) AYENB. 124.

reume, *see* rēalme. **rēusien**, *see* hrēousien.

rēuþe, *see* hrēouðe.

[**rēve**, sb., *? = O.H.G.* reba (*vitis*) ; *comp.* hai-, hurd-reve.]

rēve, *see* rēfe.

revel, sb., *O.Fr.* revel ; *revel*, CH. C. T. *A* 2717 ; this revel . . . lasteth a fourtenight CH. L. G. W. 2251 ; **reveles** (*pl.*) WILL. 1953 ; LANGL. *B* xiii. 442.

revelen, v., *O.Fr.* reveler ; *revel* ; revele P. L. S. xxx. 15.

revelour, sb., *reveller*, CH. C. T. *A* 4371.

revelrie, sb., *revelry*, R. R. 720.

réven, *see* ráven. **rēven**, *see* rēaven.

reverence, sb., *O.Fr.* reverence ; *reverence*, LANGL. *B* xiv. 204 ; SAINTS (Ld.) lii. 44 ; ROB. 553 ; AYENB. 20 ; CH. C. T. *A* 141.

revers, adj., *O.Fr.* revers ; *opposite*, GOW. I. 167.

reversen, v., *from* revers ; *reverse* ; **reversed** (*pple.*) GOW. I. 3.

revesten, v., *O.Fr.* revestir ; *clothe* ; (hii) **revested** (*pret.*) ROB. 406 ; **revested** (*pple.*) WILL. 1959.

revilen, v., *see* avīli ; *revile* ; reville MAN. (F.) 11677 ; **reviled** (*pret.*) MAN. (F.) 6361 ; **reviled** (*pple.*) PR. C. 8544 ; GOW. III. 247.

reward, sb., *O.Fr.* regvard· ; *reward, regard*, WILL. 3339 ; LANGL. *B* xvii. 265 ; PR. C. 1880 ; *see* regard.

rewarde, v., *O.Fr.* regvarder ; *reward, regard ; agree* ; WILL. 3840 ; WICL. JOB xxxiii. 24 ; **rewardieþ** (*pres.*) FER. 3463 ; **rewarded** (*pret.*) LANGL. *B* xi. 361 ; **rewarded** (*pple.*) FER. 312.

    rewarding, sb., *reward*, LANGL. *C* iv. 340*.

rēwe, *see* rāwe, hrēowe.

rēwen, rēwðe, *see* hrēowen, hrēowðe.

rīal, *see* rēal.

rīalme, rīaltē, *see* rēalme, rēaltē.

ribald, sb., *O.Fr.* ribaud ; *ribald, worthless creature* ; ribaude [ribalde] LANGL. *B* xvi. 151 ; ribaud *C* xix. 170 ; ribaude (*the devil*) *B* xiv. 203 ; ribaud AYENB. 51 ; **ribalds** (*pl.*) LANGL.

C xvii. 46; ribaudes *C* vii. 435; rebaldiṣ BARB.
xvi. 137; ribauz HOM. I. 279.

ribaldaille, sb., *O.Fr.* ribaudaille; *rabble*;
rebaldaill BARB. i. 103.

riban, sb., *O.Fr.* riban, *cf. Bret.* ruban, *Ir.*
ribin, *Gael.* ribean; *ribbon,* REL. II. 19; riband
PR. P. 432; rebant VOC. 268; **ribanes** (*pl.*)
LANGL. *B* ii. 16; rebans ANT. ARTH. ii.

ribaud, *see* **ribald**.

ribaudie, sb., *O.Fr.* ribaudie; *ribaldry,* P. L.
S. xii. 59; AYENB. 128; LANGL. *A* PROL. 44;
CH. C. T. *C* 324; *see* **ribaudrie**.

ribaudour(e), sb., *profligate fellow,* LANGL.
A vii. 66, *B* vi. 75.

ribaudrie, sb., *O.Fr.* ribauderie; *ribaldry,*
LANGL. *C* i. 45; *see* **ribaudie**.

ribbe, sb., *O.E.* ribb *n., cf. M.Du.* ribbe, *O.H.
G.* ribbi, rippi *f., O.N.* rif *n.;* *rib, 'costa,'*
VOC. 148, 226; PR. P. 432: S. S. (Web.) 1572;
rib GEN. & EX. 227; of þan ribbe HOM. I.
223; **ribbes** (*pl.*) LAȝ. 1603; HAV. 1900;
CH. C. T. *D* 506; **ribben** (*dat. pl.*) LAȝ. 1599;
comp. **dōu-ribbe**.

   ribbe-wort, sb., *hounds-tongue (herb),* '*lan-
ciola,*' PR. P. 433.

ribben, v., *L.G.* ribben; *dress with a rib*;
ribbin flax '*metaxo*' PR. P. 433; to ribbe
(*v. r.* rubbe) LANGL. *C* x. 81*.

ribíbe, ribíbour, *see* **rübíbe, rübíbour.**

ricchen, *see* **recchen**.

rīche, adj., *O.E.* rīce, = *O.L.G.* rīki, *Goth.*
reiks, *O.H.G.* rīche; *rich, powerful,* '*opu-
lentus,*' PR. P. 433; LAȝ. 128; KATH. 2360;
A. R. 66; ORM. 6384; BEK. 341; SHOR. 57;
RICH. 5899; rike M. T. 15; C. M. 1796; **rīches**
(*gen. m.*) LAȝ. 28932; **richere** (*dat. f.*) LAȝ.
764; **rīchne,** richene (*acc. m.*) LAȝ. 6593,
23791; þine riche cloþes A. D. 238; **riche**
(*gen. pl.*) menne ROB. (W.) 8839; **rīcchere**
(*compar.*) HOM. I. 271; P. L. S. xiii. 92; P. S.
331; ricchere [richere] LAȝ. 9911; richere GEN.
& EX. 1280; richore CHR. E. 409; **rīcheste**
(*superl.*) HOM. I. 125; richcheste [richeste]
LAȝ. 5732; riccheste LANGL. *A* iii. 201;
rihchest LAȝ. 18929; ricchest MISC. 96; *comp.*
efen-rīke.

   riche-dōm, sb., = *O.L.G.* rīkidōm, *O.H.G.*
rīchetuom; *wealth,* LAȝ. 3328; H. M. 3.

   riche-līche, adv., *O.E.* rīclīce; *richly,* LAȝ.
2728; AM. & AMIL. 689; TRIST. 1553; riche-
like GEN. & EX. 2442; **rīchlīer** (*compar.*)
WILL. 1934.

rīche², sb., *O.E.* rīce, = *O.L.G., O.N.* rīki,
Goth. reiki, *O.H.G.* rīchi *n.;* *reign, realm,
dominion,* MAT. iii. 5; A. R. 208; KATH. 47;
O. & N. 357; HORN (R.) 20; TREAT. 132;
SPEC. 94; AYENB. 66; GOW. II. 268; A. P. i.
600; þat riche HOM. II. 67; ORM. 7011; þe
king sende swa wide swa ileste his riche LAȝ.
595; inne godes riche P. L. S. viii. 179; rike
HAV. 290; H. H. 176 (178); IW. 142; PS.

xxi. 29; FLOR. 1809; **rīche** (*pl.*) MAT. iv.
8; riche, richen LAȝ. 5396, 7899; *comp.*
**abbod-, bischop-, eorð-, ēst-, heoven-,
kine-, king-, weoreld-riche.**

rĭchels, *see* **rēchels.**

rīchen, v., = *M.Du.* rijken, *O.H.G.* rīchan;
enrich; become rich; riche WILL. 3014; C. M.
7481; **rīcheþ** (*pres.*) WICL. I KINGS ii. 7;
LANGL. *A* iii. 74.

richesse, sb., *O.Fr.* richesse; *riches,* '*opu-
lentia,*' PR. P. 433; ROB. 113; C. L. 1016;
WILL. 5057; **richesses** (*pl.*) A. R. 168;
AYENB. 24; ritchessis WICL. MK. x. 23.

richt, rict, *see* **riht.** **ridden,** *see* **hredden.**

ride, sb., = *Ger.* rit; *from* **rīden**; *ride,* GEN. &
EX. 3950.

ridel, sb., *O.Fr.* ridel (*rideau*); *curtain,* FER.
2537; redell ALEX. (Sk.) 4930.

hridel, sb., *riddle, sieve;* ridil '*cribrum*' PR.
P. 433.

[**rīdel**, sb., *one who rides;* *comp.* **for-rīdel.**]

riden, *see* **rüden.**

rīden, v., *O.E.* rīdan, = *O.Fris.* rīda, *O.N.*
rīða, *O.H.G.* rītan; *ride, go,* LAȝ. 432; A. R.
266; on horse riden HAV. 370; ridæn FRAG.
6; ridan HOM. I. 5; ride BEK. 1179; ALIS.
460; hi leten þat schup ride HORN (L.) 136;
rīdeþ (*pres.*) ORM. 6966; O. & N. 494; P. S.
222; rit H. M. 11; rīd A. D. 250; LANGL.
(Wr.) 339; **rīd** (*imper.*) LAȝ. 26551; **rīdende**
(*pple.*) LAȝ. 26795; WILL. 1954; ridinde A.
R. 216; **rād** (*pret.*) HOM. I. 5; LAȝ. 7240;
TRIST. 179; M. H. 135; PERC. 496; he . . .
rad ruglinge into helle MARH. 17; rod HORN
(L.) 630; TOR. 505; rode LANGL. *B* xvii. 99
(rood (Wr.) 11567); CH. C. T. *A* 390; þu ride
LAȝ. 26527; riden ALIS. 1038; FER. 1548;
CH. C. T. *A* 825; ridden RICH. 4025; **riden**
(*pple.*) CH. C. T. *A* 48; *comp.* **bi-, ȝe-rīden;**
deriv. **ride, rāde.**

   rīdere, sb., = *M.L.G.* rīder, *M.H.G.* rīter;
rider, LAȝ. 9288; LANGL. *A* x. 306; **rīdæres**
[ridares] (*pl.*) LAȝ. 19860.

   rīding, sb., *riding;* **rīðinges** (*pl.*) ORM.
9213; AYENB. 24.

hridlen, v., *riddle, sift;* ridlen A. R. 234;
ridelin '*cribro*' PR. P. 433; ridlide (*pret.*)
WICL. DAN. xiv. 13.

hridren, v., *O.E.* hridrian, = *M.L.G.* redern,
O.H.G. hritarōn; *riddle, sift;* **riddrede**
(*pret. subj.*) LK. xxii. 31.

rie, *see* **rüȝe.**

rif, sb., *O.N.* rif, = *M.Du.* rif; *reef (of a sail),*
GOW. III. 341.

[**hrif,** sb., *O.E.* hrif, = *O.H.G.* href, *O.Fris.* rif;
belly, womb; *comp.* **mid-rif.**]

rīf, sb., *O.E.* rīf, = *M.Du.* rijf, *O.N.* rīfr, *mod.
Eng.* rife; *abundant, frequent, openly known,*
GEN. & EX. 1252; SHOR. 112; HOCCL. i.
427; of riches þou art so rif EGL. 1041; rif

(rife) PS. xxvi. 1*; rife GOW. I. 213; rive H.
M. 29; P. L. S. xvii. 52; P. S. 195; WILL.
5414; balu þer wes rive LAȝ. 20079; his craftes are so rife PERC. 560; rīve (adv.) KATH.
2513; MAP 350; þat folc þat deide so rive
ROB. 252.

rīf-līche, adv., greatly, WILL. 1472; rifli
REL. II. 7; M. H. 42.

riflin, v., O.Fr. rifler; rifle, 'spolio,' PR. P.
433; riflede (pret.) LANGL. B v. 234; rí-
fild (pple.) MIN. ii. 16.

riflinge, sb., plunder, LANGL. B vii. 238.

rifloure, sb., plunderer, PR. P. 433; rifeler
LANGL. C vii. 316.

rift, sb., O.E. rift, ryft, reft,= O.N. ript, ? O.
H.G. reft; garment, veil: me hire hafd bi-
wefde mid ane hali rifte LAȝ. 28475; comp.
wāȝe-rift.

rifte, sb., from rīven; rift, 'rima,' PR. P.
433; reft R. R. 2661.

riften, v., rift; rifte 'ructo' CATH. 308;
riftes (pres.) PS. xviii. 3; rifted (pret.) PS.
xliv. 2.

riftere, sb., O.E. riftere; reaper, 'messor,'
FRAG. 2.

hrig, sb., ? O.N. hregg; ? storm; rig A. P. ii. 382.

rig(ge), see hrüg. righe, see reihe.

rigour, sb., O.Fr. rigour; rigour, CH. C. T.
F 775.

riht, adj., O.E. riht, ryht,= O.L.G. reht, O.
H.G. reht, O.N. rēttr, Goth. raihts, Lat.
rectus; right; straight; just; A. R. 8; (ms.
rihht) ORM. 15362; mi riht arm LAȝ. 28040;
richt REL. I. 23; MISC. 33; rith HAV. 1812;
MAP 336; rist REL. I. 49; riht (subst.) LAȝ.
2511; vor te don alle men riht A. R. 286;
riht, riȝt O. & N. 950, 969; rigt GEN. & EX.
3714; riȝt AYENB. 8; A. P. i. 621; a riht
aright, LAȝ. 17631; A. R. 76; ȝif þu hit const
a riht bilegge O. & N. 904; Jacob Ɣus him
bimeneð o rigt GEN. & EX. 2226; þe righte
wai MAND. 6; mid riȝte BRD. 3; SHOR. 60;
rihtere (dat. f.) LAȝ. 29092; rihtne (acc. m.)
LAȝ. 4768; O. & N. 1238; rihte (adv.) HOM.
I. 167; rihte, riht [riȝt] O. & N. 80, 1246; he
fl(e)ah dun rihte [riht] LAȝ. 25613; forð riht
heo wenden 7807; righte as þou were a god
MAND. 295; righte whare scho lai ISUM.
194; þat him riht luvieð KATH. 1741; riht
biȝeten ahte ORM. 1645; riht sone LIDG. M.
P. 29; an on riht, rihtes A. R. 18, 42; al
rihtes swa HOM. I. 133; & slowe doun riȝtes
WILL. 1165; comp. ȝe-, un-, up-riht.

riht-ful, adj., just, L. H. R. 144; riȝtful
LANGL. B PROL. 127; it is riȝtful WICL.
GEN. xxx. 30; riȝtvolore (comp.) ROB. (W.)
5391.

riȝtvol-līche, adv., rightfully, AYENB. 196.

rihtfulnesse, sb., righteousness, SPEC. 53;
riȝtfulnesse WICL. GEN. xviii. 19; riȝtfulnes
H. S. 600; riȝtvolnesse AYENB. 29.

riht-lēchen, v., O.E. rihtlǣcan; correct,
rectify, HOM. I. 17; REL. I. 130 (HOM. II.
9); rihtleche MISC. 86; riȝtleche WILL. 1310.

riht-līche, adv., O.E. rihtlīce; rightly, HOM.
II. 27; riȝtli WILL. 232.

rihtnesse, sb., O.E. rihtness, rihtnyss;
rectitude, LAȝ. 14; riȝtnesse AYENB. 154;
riȝtnisse BEK. 1627.

riht-wīs, adj., O.E. rihtwīs; righteous, HOM.
I. 35; rihtwis LAȝ. 6537; C. L. 63; P. L. S. xxv.
38; M. H. 1; rihtwis A. R. 268; ORM. 2880;
riȝtwis WICL. GEN. vi. 9; A. P. i. 674; rightwis S. S. (Wr.) 2754; PR. C. 9154; D. ARTH.
3989; ðe rigtwise GEN. & EX. 1043.

rigtwīs-hēd, sb., righteousness, justice,
GEN. & EX. 3740.

rihtwīs-leȝc, sb., righteousness, justice,
ORM. 2531.

rihtwis-līche, adv., righteously, KATH.
753; riȝtwisli A. P. i. 708.

rihtwīsnesse, sb., O.E. rihtwīsness; righteousness, A. R. 304; ORM. 10916; rihtwisnesse LAȝ. 6554; riȝtwisnesse LANGL. B xviii.
197; rightwisnesse MAND. 294.

rihten, v., O.E. rihtan,= O.L.G. rihtian, O.
H.G. rihtan, O.N. rētta, prov.Eng. right;
straighten, direct, correct, JUL. 9; to rihten
eo(u)re leoden LAȝ. 6254; his ban rihten
19502; to rihten here lif ORM. 18148; rigten
GEN. & EX. 3423; rihte MISC. 52; righte
GOW. III. 170; rithe HAV. 2611; þe riwle
þet rihteð (pres.) þe heorte A. R. 410;
rihtes JOS. 451; riht (imp.) JUL. 31; righte
['dirige'] PS. xxiv. 5; rihteþ ORM. 9201; rihteð
ou up arise A. R. 18; rihte (pret.) MARH. 20;
rihten LAȝ. 25732; here arewes riȝte P. L. S.
xviii. 43; rihted (pple.) ORM. 9208; comp.
a-, ȝe-rihten.

rihtnen, v., = rihten, make right, correct,
ORM. 10361.

rīke, see rīche. rīkels, see rēchels.

rikenen, see rekenen.

rīkien, v., O.N. rīkja; reign; rīked (pret.)
PS. xcii. 1.

rīm, sb., O.E. rīm, cf. O.Fris., O.N. rīm n., O.
H.G. rīm, hrīm m.; rhyme, number, song;
(ms. rym) HAV. 23; he þat haveþ þis rim
(ms. rym) iwriten MISC. 57; þurh tale & rime
of fowertiȝ ORM. 11248; rīmes (gen.) GEN.
& EX. 1; rīme (pl.) C. M. 14922; and seide
þes rime HORN (L.) 804; lat him rimes make
CH. C. T. B 96; rīme (dat. pl.) SHOR. 165;
HOCCL. i. 247; in prose ne in rime LEG. 60;
writen o rime REL. I. 224.

hrīm, sb., O.E. hrīm,= O.N. hrīm, O.H.G. rīm,
M.Du. rijm; rime, hoar frost; rim LAȝ.
28525; (ms. ryme) 'pruina' PR. P. 434.

rim-frost, sb., hoar-frost, 'pruina,' VOC.
239; rimforst SAINTS (Ld.) xlvi. 627.

rime, sb., O.E. rima, reoma; rim, border,
margin, VOC. 150; þe rime of his eȝen WICL.

TOB. xi. 14; rim of a whele PR. P. 434; rim, reme (*integument of the foetus*) PR. C. 520; **rimez** (*pl.*) GAW. 1343; *comp.* **dai-, sǣ-rime.**

rīmen, v., *O.E.* rīman,=*O.H.G.* (gi-)rīman; *rhyme, enumerate,* ORM. 11217; rime GOW. II. 173; CH. C. T. *B* 2122; MAN. (H.) 71; LIDG. M. P. 55; ich nelle eou noþer rede ne rime of king ne of eorl S. A. L. 148; **rīmeþ** (*pres.*) SHOR. 104; *comp.* **a-, ȝe-rīmen.**

rīmen, *see* **rūmen.**

hrimple, sb., *O.E.* hrympele,=*M.L.G.*, *M.Du.* rimpel; *ripple, wrinkle*; rimple, rimpil '*ruga*' PR. P. 434.

hrimplen, v., *cf. M.Du.* rimpelen; *wrinkle, ripple*; **rimplid** (*pple.*) PR. P. 434; R. R. 4495; LIDG. M. P. 200.

rīmþe, *see* **rūmðe.**

rīn, sb., *O.N.* rīn, = *M.Du.* rijn; *stream,* HALLIW. 685.

Rīn, pr. n., *O.E.* Rīn,=*O.N.*, *O.H.G.* Rīn; *Rhine*; Ruin LANGL. *A* PROL. 108; **Rīne** (*dat.*) TREV. V. 257.

rincen, v., *O.Fr.* rincer; *rinse*; rince REL. I. 7; **rinsede** (*pple.*) coupes D. ARTH. 3375.

rínde, sb., *O.E.* rind,=*M.Du.* rinde, *O.H.G.* rinda; *rind, shell,* O. & N. 602; L. H. R. 24; C. L. 1308; AYENB. 96; MAN. (H.) 333; þeos rinde A. R. 150.

    rínd-lēas, adj., *without rind,* A. R. 150.

rinele, *see* **rünele.**

hrīnen, v., *O.E.* hrīnen, = *O.L.G.*, *O.H.G.* hrīnan; *touch*; **rīneð** (*pres.*) A. R. 320; al þat ha rineð to JUL. 56; **rīn** (*imper.*) A. R. 408; **rīne** (*subj.*) CL. M. 65; **rān** (*pret.*) ORM. 15518; *comp.* **a-, æt-, ȝe-rinen.**

rīnunge, sb., *O.E.* hrīnung; *touching,* A. R. 408.

rīnen, *see* **reinen.**

hring, sb., *O.E.* hring,=*O.L.G.*, *O.H.G.* hring, *O.N.* hringr; *ring*; ring A. R. 420; PERC. 425; þis ring whil he is þin FL. & BL. 4; þisne ring P. L. S. xxii. 9; ænne ring LAȝ. 30805; in a ringe ALIS. 1112; i widewene ring H. M. 21; **ringes** (*pl.*) GEN. & EX. 1872; HAV. 2740; **ringe** (*dat. pl.*) LAȝ. 30803; *comp.* **ēre-, góld-, hōp-, kine-ring.**

    ring-worm, sb., *ring-worm,* '*impetigo,*' VOC. 255; ringwirm '*serpigo*' PR. P. 434.

hringen, v., *O.E.* hringan (*pret.* hringde),= *O.N.* hringja, *M.Du.* ringhen; *ring, sound*; ringen LAȝ. 16929; ORM. 901; HAV. 242; ringe CH. C. T. *A* 3896; **ringden** (*pret.*) SAX. CHR. 259; rang Iw. 1397; rong ROB. 509; LAI LE FR. 181; CH. C. T. *A* 3215; bellen þer ringeden [rongen] LAȝ. 24486; rungen M. T. 15; rongen LANGL. *B* xviii. 425; ronge E. T. 319; **rungen** (*pple.*) HAV. 1132; irungen LAȝ. 29441; irunge HORN (L.) 1016.

    ringinge, sb., *ringing,* P. L. S. xii. 45.

hringen², v., *O.E.* hringian, *cf. O.N.* hringja, *O.H.G.* (ge-)hringen; *form a ring, encircle, adorn with a ring*: let us alle aboute him ringe HALLIW. 686; **ringinde** (*pple.*) (? = renginde) abuten A. R. 140; fingres richeliche **iringed** (*pple.*) LANGL. *C* iii. 12.

Rīnisch, adj.,=*M.H.G.* rīnisch; *Rhenish,* D. ARTH. 203.

rink, sb., *O.E.* rinc,=*O.L.G.* rinc, *O.N.* rekkr; *man, warrior,* WILL. 1193; REL. I. 78; A. P. iii. 216; renk LANGL. *B* PROL. 192; GAW. 691; renke DEP. R. ii. 31; **rinkas** (*pl.*) LAȝ. 5188.

rinnen, v., *O.E.* rinnan, irnan, yrnan, eornan, = *O.L.G.*, *O.H.G.*, *Goth.* rinnan, *O.Fris.*, *O.N.* rinna, renna, *M.Du.* rinnen, rennen, runnen; *run*; rinne C. M. 10054; PERC. 1662; rinne [renne] CH. P. F. 247; rennen LANGL. *B* xv. 220; renne HAV. 1161; irnen, urnen, eornen LAȝ. 8130, 21229, 24696; eornen A. R. 86; KATH. 2300; ORM. 1336; eorne [urne] O. & N. 638; urne HORN (L.) 878; ROB. 514; (*ms.* yerne) AYENB. 173; OCTOV. (W.) 965; erne SPEC. 81; AR. & MER. 1228; **rinnes** (*pres.*) IW. 3245; renneð KATH. 2511; renneþ MAND. 32; renniþ TOR. 1298; irneð LAȝ. 29664; as weter þet eorneð JUL. 74; eorneþ [urneþ] O. & N. 375; eorneþ [erneþ] C. L. 730; eorneþ [ȝerneþ] TREV. I. 105; ȝernes GAW. 498; his eghen rinnes PR. C. 781; **runnande** (*pple.*) M. H. 114; eorninde LEG. 33; **ran** (*pret.*) GEN. & EX. 1009; HAV. 216; SPEC. 83; MAND. 41; CH. C. T. *B* 661; TOR. 1050; PR. C. 5297; (*ms.* rann) ORM. 1364; ron LANGL. *A* v. 43; ȝarn, ȝorn *B* xi. 59, *C* xiii. 13; ron HOM. I. 207; HICKES I. 227; AL. (V.) 404; euch weoved . . . ron of þat baleful blod KATH. 205; ran (*r. w.* barn) AR. & MER. 2587; earn LK. xv. 20; (*ms.* yarn) AYENB. 191; orn LAȝ. 18806; A. R. 188; ROB. 208; BEK. 88; SPEC. 58; þat blod orn a dun MISC. 42; runnen WICL. JOHN xx. 4; AV. ARTH. xxv; runne FER. 2438; ronnen CH. C. T. *A* 2925; urnen HOM. I. 3; II. 39; LAȝ. 4578; A. R. 112; urne BRD. 22; ȝornen A. P. ii. 881; **urne** (*subj.*) A. R. 164; JUL. 30; **runnen** (*pple.*) A. P. i. 26; ronnen, rone LANGL. *B* viii. 90, *A* ix. 82; ronnin '*coagulatus*' PR. P. 436; ironne CH. C. T. *A* 8; iorne HORN (L.) 1146; LEG. 23; *comp.* **æt-, bi-, of-, over-rinnen;** *deriv.* **renne, rennen, rūne.**

ríote, sb., *O.Fr.* riote; *riot, dissipation,* CH. C. T. *A* 4392; non wisure read ne mei bringen hire ut of hire riote A. R. 198; he ne heþ none hede of longe riote of tales AYENB. 99; men . . . seide in olde riote BEV. 1192.

ríoten, v., *O.Fr.* rioter; *make a riot*; riote CH. C. T. *A* 4414; he oft **ríot** (*pret.*) to the land *harried* BARB. v. 181.

ríotour, sb., *rioter,* CH. C. T. *C* 692; **ríotoures** (*pl.*) CH. C. T. *C* 661.

ríotous, adj., *O.Fr.* rioteus; *riotous*, CH. C.
T. *A* 4408; D. ARTH. 363.

ríotrie, sb., *rioting*, MAN. (F.) 2406.

*h*rip, sb., *O.N.* hrip; *peat-basket*; rippe (*dat.*)
HAV. 893.

rīp, sb., *O.E.* rīp, rȳp *n.*, *from* rīpen; *cf.* rēpe;
harvest ['*messis*'], MAT. xiii. 39; rip [ripe]
TREV. VIII. 185; rip [rep] WICL. 2 KINGS
xxi. 9; JOB xviii. 16; **ripis** (*pl.*) *sheaves*
['*manipulos*'] HAMP. PS. cxxv. 8; *comp.*
bed-rēp, ȝe-rīp.

rīp-man, sb., *reaper*, MAT. ix. 37; repman
S. B. W. 246; PALL. vii. 18.

rēpe-rēve, sb., *head-reaper*, LANGL. *C* vi.
15.

rēp-tīme, sb., *O.E.* rīptīma; *harvest*, WICL.
PROV. xxvi. 1.

rīpe, adj., *O.E.* rīpe, = *O.L.G.* rīpi, *O.H.G.*
rīphi, rīñ; *ripe*, '*maturus*,' PR. P. 434; FRAG.
6; O. & N. 211; BRD. 33; AYENB. 28; ALIS.
5757; LANGL. *B* xvi. 71.

[**ripel**, sb., = *M.L.G.* repele, *M.H.G.* rifel; *that
which rips*.]

ripil-stok, sb., *instrument for cleaning flax*,
VOC. 269.

ripelen, v., = *M.L.G.* repelen, *M.H.G.* rifeln;
scratch, tear: he **repulde** (*pret.*) his face
GUY² (Z.) 9617.

ripen, v., = ? *Swed.* ripa (*repa*); *rip, tear open;
examine*: ripe þaire ware C. M. 4893; ripe
up D. ARTH. 1877; **ripe** (*imper.*) TOWNL.
112; **ripande** (*pple.*) A. R. ii. 592; **riped**
(*pret.*) M. H. 143.

rīpen, v., *O.E.* rīpan, rȳpan, rēpan, = *Goth.*
raupjan; *reap*, P. L. S. viii. 11; ripe AYENB.
214; TREV. VIII. 185; reopen JUL. 75; repen
HOM. I. 161; MISC. 59; repe P. S. 152; PS.
cxxv. 5; reepe H. V. 70; **rēpiþ** (*pres.*) WICL.
JOHN iv. 36; **rēp** (*pret.*) LANGL. *B* xiii. 374;
repen HOM. I. 241; LAȜ. 10033; LEB. JES.
358; ropen LANGL. *B* xiii. 374; WICL. RUTH
i. 22; **rópen** (*pple.*) CH. L. G. W. 74; rope
PALL. x. 127; *deriv.* rīp, rēpe.

rīpere, sb., *O.E.* rīpere; *reaper*; repare PR.
P. 430; **rīperen** (*dat. pl.*) MAT. xiii. 30.

rīpinge, sb., *reaping*, ['*messis*'] WICL.
JUDG. xv. 1.

rīpen², v., *O.E.* rīpian, = *O.H.G.* rīfan; *from*
rīpe; *ripen*, HOM. II. 220; LANGL. *B* xvi.
39; ripin '*maturo*' PR. P. 434; **rīpest** (*pres.*)
HOCCL. II. 33; **rīpede** (*pret.*) HOM. I. 241;
rīpid (*pple.*) WICL. JOEL iii. 13.

rippen, *see* rüppen.

Rīpun, pr. n., *O.E.* Rīpum, Hrīpum; *Ripon*,
MISC. 146.

*h*rire, *see* *h*rüre.

rīs, sb., *cf. Fr.* riz, *Gr.* ὄρυζα; *rice*; (*ms.* ryȝs)
MAND. 310.; ris C. B. 22, 114.

*h*rīs, sb., *O.E.* hrīs, = *O.N.*; *O.H.G.* hrīs; *twig*;
ris SPEC. 26; LAUNF. 937; on blowe ris O. &

N. 1636; **rīse** (*dat.*) CH. C. T. *A* 3324; þer
he under rise liȝ LAȜ. 740; rīs (*pl.*) O. & N.
586; hulen of ris(e) & of leaves A. R. 100;
þe hare þat bredus in þe rise AV. ARTH. ii;
comp. wode-rīs.

rische, *see* rüsche.

[**rīse**, sb., *rise; comp.* up-rīse.]

*h*risel, *see* *h*rüsel.

rīsen, v., *O.E.* rīsan, = *O.L.G.* rīsan (*rise*),
O.H.G. rīsan (*flow, fail*), *M.Du.* rijsen (*rise*),
O.Fris., *O.N.* rīsa, *Goth.* (ur-)reisan; *rise*,
HOM. II. 103; ORM. 4197; GEN. & EX. 4039;
risen [rise] LANGL. *B* xix. 142; rise WICL.
NUM. xxviii. 24; S. S. (Wr.) 1603; **rīs** (*imper.*)
HAV. 584; **rās** (*pret.*) ORM. 2741; raas PS.
iii. 6; ros REL. I. 209; GEN. & EX. 261;
ALIS. 4378; WILL. 1193; roos LANGL. *B* v.
234; WICL. GEN. xix. 1; risen S. S. (Web.)
2089; GREG. 274; rise MAN. (H.) 222; resin
TOR. 2558; reson EGL. 284; **risen** (*pple.*)
ORM. 11552; CH. C. T. *A* 1499; risin S. S.
(Wr.) 211; *comp.* a-, bi-, ȝe-risen; *deriv.*
rist, **rāsen, raisen, ?rāren**.

*h*risien, v., *O.E.* hrisian, hrysian, = *O.L.G.*
hrisian, *Goth.* hrisjan; *quake, move*: þe eorðe
gon to rusien LAȜ. 15946; resie AYENB. 23;
rese CH. C. T. *A* 1986; **rüsed** (*pret.*) LANGL.
B xvi. 78*; riseden burnen [rusede wepne]
LAȜ. 26917; *comp.* a-resien.

rison, sb., *prov. Eng.* (*Chesh.*) rizzom; *head
of oats*; **risonis** (*pl.*) ALEX. (Sk.) 3060.

rist, sb., *O.E.* (a-)rist, = *Goth.* (ur-)rists; *from*
rīsen; *resurrection, rising*: þe sonne rist
ALIS. (Sk.) 791; *comp.* a-, up-rist (-riste).

ritten, v., ? = *O.H.G.* rizzan; *split*: for to ritte
and for to flo HAV. 2495; **ritte** (*pret.*) GAW.
1332; TRIST. 479; þat hure haberkes ritte
FER. 5030; *comp.* to-ritten.

rīþ, sb., *O.E.* rīð, = *O.L.G.* rīth, *M.L.G.* rīde f.;
small stream, '*rivus*,' FRAG. 3.

rīþer, *see* *h*reoðer.

rīþþe, *see* *h*rüþþe.

rīve, sb., = *M.L.G.* rīve, *M.Du.* rijve; *from*
rīven; *rake*, '*rastrum*,' PR. P. 435.

rīve², sb., *O.Fr.* rīve; *bank*, HORN (L.) 132.

rīve, *see* rīf.

rivel, sb., *wrinkle*, TREV. I. 257; **rivelis**
(*pl.*) ['*rugae*'] WICL. JOB xvi. 9.

rivelen, v., *wrinklen*; **rivele** (*pres.*) REL. II.
211; **riveling** (*pple.*) R. R. 7214; **riveleden**
(*pret. pl.*) LANGL. *B* v. 193*; **riveled** (*pret.*
pple.) LEG. 218; GOW. III. 370.

riveling, sb., *O.E.* rifling; *a sort of shoe*,
MAN. (H.) 282; (*nickname for Scotch*) MIN. ii.
19; his knichtis werid **revelinis** (*pl.*) off
hidis WINT. viii. 4421.

rīven, v., *O.N.* rīfa (*break*), = *M.Du.* rijven
(*rub*), *O.H.G.* rīban (*rub*); *rive, tear, break*;
rivin PR. P. 435; rive CH. C. T. *C* 828; PR.
C. 888; SACR. 713; a sondir rive LIDG. M.
P. 189; rife TOWNL. 11; **rīvez** (*pres.*) GAW.

1341 ; **rāf** (*pret.*) IW. 2615 ; PERC. 2157 ; þat spere þurh raf [rof] LAȝ. 23943 ; roof WICL. 2 KINGS ii. 23 ; ref C. M. 7809 ; REL. I. 1 ; SACR. 48 ; **riven** (*pple.*) MAN. (H.) 148 ; his robes riven were TRIST. 582 ; rifen IW. 3539 ; reven S. & C. I. xlix ; *comp.* to-riven ; *deriv.* rīve, rifte.

rivēre, sb., *O.Fr.* riviere ; *river*, HORN (L.) 230 ; ALIS. 5142 ; river FER. 3967 ; CH. C. T. *D* 2083.

rīxien, v., *O.E.* rīcsian, rīxian, = *O.H.G.* rīchisōn ; *reign* ; rixan SAX. CHR. 265 ; **rīxeð** (*pres.*) HICKES I. 224 ; rīxede (*pret.*) MAT. ii. 22.

rīxlien, v., *reign*, LAȝ. 19323 ; rixlen HOM. I. 271 ; ORM. 2237 ; let hit a rixlie (? *ms.* arixlye) MISC. 130 ; rixle HOM. II. 27 ; **rīxleð** (*pres.*) MARH. 19 ; A. R. 80 ; rixles D. TROY 2726 ; rīxlede (*pret.*) LAȝ. 6907.

   rīxlunge, sb., *reign*, HOM. I. 111 ; A. R. 248 ; rixlinge KATH. 44.

rō, sb., *O.E.* rōw, = *O.N.* rō, *M.Du.* roe, *O.H.G.* ruowa, rāwa ; *quiet, repose*, MARH. 20 ; ORM. 7042 ; C. L. 90 ; P. S. 149 ; MAP 336 ; M. H. 14 ; ro ant rest REL. I. 116 ; roo FLOR. 840 ; RICH. 7135 ; M. ARTH. 3614 ; TOWNL. 222 ; *comp.* un-rō.

   rō-lēs, adj., *restless* ; rooles ase the roo SPEC. 42.

rō, *see* rā.

robard, sb., *robber*, YORK vii. 47.

robben, v., *O.Fr.* rober ; *rob, spoil*, HAV. 1958 ; PL. CR. 459 ; **robbeð** (*pres.*) A. R. 86 ; robbeþ AYENB. 39 ; **robbede** (*pret.*) ROB. 16 ; **irobbed** (*pple.*) H. M. 15 ; rebbe ROB. (W.) 6041*.

robbeour, sb., *O.Fr.* robeor ; *robber*, ROB. 389 ; robbour MAND. 250 ; LANGL. *A* v. 242 ; robbere AYENB. 79 ; robbare A. R. 150.

rόbe, sb., *O.Fr.* robe ; *róbe*, MISC. 39 ; ROB. 313 ; AYENB. 119 ; **róbes** (*pl.*) *clothes* LANGL. *C* xvi. 202.

róbed, adj., *robed*, LANGL. *C* xi. 1.

roberie, sb., *O.Fr.* roberie ; *robbery*, HOM. II. 61 ; MISC. 30 ; AYENB. 9 ; robberie BEK. 396.

robous, sb., ? *O.Fr.* robeux ; *rubbish*, ' *petrosa*,' PR. P. 435.

roc, sb., *O.E.* rocc, = *O.N.* rokkr, *O.H.G.* roch (*tunica*) ; ' *toral*,' FRAG. 4.

hrōc, sb., *O.E.* hrōc, = *O.N.* hrōkr, *M.Du.* roec, *O.H.G.* hruoh ; *rook* ; roc ' *graculus* ' FRAG. 3 ; rok O. & N. 1130 ; rook PR. P. 436.

roche, sb., *O.Fr.* roche, roke, = *Ital.* rocca ; *rock*, GEN. & EX. 256 ; ROB. 22 ; AYENB. 142 ; WILL. 2367 ; rokke ' *rupes* ' PR. P. 436 ; CH. C. T. *F* 1061 ; **rockes** (*pl.*) GOW. I. 314.

roche [2], sb., *cf. Dan.* rokke, *Swed.* rocka ; *roach*, VOC. 189 ; PR. P. 435 ; REL. I. 85 ; *see* reihe.

rochet, sb., *Fr.* rochet ; *rochet*, ST. R. 501 ; rochette R. R. 4757.

rochi, adj., *rocky*, ROB. (W.) 2499.

rocke, sb., *M.Du.* rocke, *O.N.* rokkr, *O.H.G.* roccho ; *distaff*, VOC. 157 ; KN. L. 79 ; rokke PR. P. 436 ; rok REL. I. 4 ; MIR. PL. 122.

rocken, *see* rucken.

rŏdde, sb., *originally same word as* rōde ; SAINTS (Ld.) xl. 123 ; O. & N. 1123, 1646 ; *comp.* līm-rŏd(de).

rode, *see* rude.

rōde, sb., *O.E.* rōd (*pole, cross*), = *O.L.G.* rōda (*rod*), *O.Fris.* rōde (? *gallows*), *O.H.G.* ruota (*rod*) ; *rood, cross, crucifix*, LAȝ. 11165, 22101 ; KATH. 928 ; A. R. 26, 60 ; ORM. 5608 ; R. S. vii ; GEN. & EX. 386 ; HAV. 431 ; SPEC. 85 ; AYENB. 114 ; WILL. 1669 ; A. P. i. 704, iii. 270 ; O. & N. 1382 ; roid BARB. xii. 256* ; holi rōde (*gen.*) dai (*May* 3) ROB. (W.) 1932 ; rōde, roode (*dat.*) LANGL. *B* ii. 3 ; HOM. I. 121 ; *see* rŏdde.

   rūde-ēvin, sb., *eve of the rood*, BARB. xviii. 634.

   rōde-pīne, sb., *pain of the cross*, ORM. 2018.

   rōde-tāken, sb., *sign of the cross*, MARH. 10.

   rōde-trēo, sb., *cross, crucifix*, ORM. 5602.

rōde, *see* rāde, **rodi**, *see* rudi.

hrōf, sb., *O.E.* hrōf, = *O.Fris.* hrōf ; *roof, cover* ; rhōf ORM. 11351 ; rof FRAG. 6 ; HAV. 2082 ; ROB. 416 ; REL. II. 216 ; S. S. (Wr.) 2168 ; þe rof MISC. 179 ; þene rof LAȝ. 2894 ; rōfe (*dat.*) CH. C. T. *A* 3623 ; rove FRAG. 6 ; A. R. 152 ; hrōfen (*dat. pl.*) MAT. x. 27 ; *comp.* hūs-rōf.

rogge ?, sb. : and let(e) that losinger go on the roge (*r. w.* dogge) CHEST. II. 94 ; that ever i regnede on þir rog D. ARTH. 3272.

roggen, *see* ruggen. **rohȝe**, *see* reihe.

roi, sb., *Fr.* roi ; *king*, D. ARTH. 1670.

roial, *see* rēal.

roilin, v., ? *O.Fr.* roeler ; *wander*, ' *vagor*,' PR. P. 436 ; roile aboute CH. C. T. *D* 653 ; **roileþ** (*pres.*) LANGL. *A* xi. 206 ; CH. BOET. i. 6 (29) ; **roilend** (*pple.*) REL. II. 175 ; roillede, roiled (*pret.*) TREV. I. 145 ; rueled A. P. ii. 953.

roinouse [roinissche], adj., *O.Fr.* roigneus (*rogneux*) ; *dirty, scabby*, LANGL. *B* xx. 82, *C* xxiii. 83*.

rōk, sb., *Fr.* roc ; *rook* (*at chess*), MAN. (F.) 11397 ; rook PR. P. 436.

róke, sb., *cf. M.Du.* roke (*odor*) ; *from* rēoken ; *fog, vapour, cloud*, ' *nebula*,' PR. P. 436 ; GEN. & EX. 1163 ; BEV. 2471.

Rokes-buru, pr. n., *Roxburgh*, HAV. 139.

róki, adj., *foggy*, ' *nebulosus*,' PR. P. 436.

rokke, *see* rocke, roche.

rolle, sb., *O.Fr.* rolle, *from Lat.* rotula ; *roll*, A. R. 344 ; P. S. 157 ; CH. C. T. *C* 911.

rollen, v., *O.Fr.* roller ; *roll*, CH. TRO. ii. 659 ; rollede (*pret.*) LAȝ. 22287*.

rom, *see* **ram**.   **rōm**, *see* **rūm**.

romance, sb., *Fr.* romance ; *romance*, ROB. 487 ; romaunce CH. D. BL. 48.

Romanisc, adj., *Roman* : þis Romanisce folc LAӠ. 5628 ; þe Romanishe king ORM. 8327.

romant, sb., *O.Fr.* romant (*roman*) ; *romance* ; romanz (*pl.*) HAV. 2327 ; SPEC. 34.

romblen, *see* **ramblen**, **rammelin**.

Rōme, pr. n., *O.Fr.* Rōme ; *Rome*, LAӠ. 5238 ; O. & N. 1016.

   Rōme-burh, pr. n., *city of Rome*, LAӠ. 5347 ; ORM. 7010.

   Rōm-lēode, sb., *men of Rome*, LAӠ. 7187.

   Rōme-rīche, pr. n., *empire of Rome*, ORM. 8305.

   Rōme-scot, sb., *Peter's pence*, SAX. CHR. 250.

   Rōme-toun, pr. n., *town of Rome*, S. S. (Web.) 551.

   Rōme-þēod, pr. n., *Romans*, LAӠ. 9046.

   Rōm-ware, sb., *dwellers in Rome*, LAӠ. 7936.

rōmen, v., *? cf. Swed.* râma ; *bellow, rumble* ; romi AV. ARTH. xii ; HAMP. PS. xxxvii. 20 ; **rōmand** (*pple.*) ['*rugientes*'] PS. ciii. 21 ; rumiand HAMP. PS. xxi. 12 ; **rōmede** (*pret.*) D. ARTH. 784 ; romid ['*rugiebam*'] HAMP. PS. xxxvii. 8.

   rōming, sb., *commotion* : fisshes . . . sal . . . mak swilk roming (*v. r.* roring) PR. C. 4772.

rōmen, *see* **rāmen**.

Romes-ēie, pr. n., *O.E.* Rumes ēg (īg); *Romsey*, ROB. 377.

romour, *see* **rumour**.

ron, *see* **ran**.   **rōn**, *see* **hrān**.

rōnd, adj., *O.Fr.* roond, *Lat.* rotundus ; *round*, AYENB. 1 ; round ROB. 414 ; CH. C. T. *A* 3934 ; a round *around* BEK. 2152 ; MAN. (F.) 10536 ; a ronde BEV. 1373 ; a **rōnde** (*sb.*) [rounde] of bacon LANGL. *C* x. 148 ; te grene bowes beoð . . . forwurðen to druie hwite rondes A. R. 148.

rōndel, sb., *O.Fr.* rondel (*rondeau*) ; *roundel*, SAINTS (Ld.) xlvi. 452 ; rundel TREAT. 133 ; roundel CH. C. T. *A* 1529 ; **rōndels** (*pl.*) L. G. W. 423.

ronden, *see* **randen**.

rōnen, *for* *rōen, *cf. Swed.* rōa : *console* ; **rōned** (*pret.*) me þou es ['*consolatus est me*'] PS. lxx. 60.

ronge, *see* **hrunge**.   **ronk**, *see* **ranc**.

rop, sb., *O.E.* ropp (LEECHD. II. 230),=*M.Du.* rop ; *intestine*, A. P. iii. 270 ; **roppes** (*pl.*) AYENB. 62 ; *comp.* **ars-rop**.

rōp, *see* **rāp**.

hrōp, sb., *O.E.* hrōp,=*O.H.G.* hruof; *clamour* : þer wes wop, þer wes rop (*sec. text* cri) LAӠ. 12540 ; **rōpe** (*dat.*) S. S. (Web.) 1185.

hrōpen, v., *O.E.* hrōpan (*pret.* hrēop), *cf. O.L. G.* hrōpan, *O.H.G.* hruofan, *Goth.* hrōpjan ;

cry out : rope and rare IW. 242 ; **rōpeð** (*pres.*) A. R. 330*.

rōr, *see* **rār**.

hrōre, sb.,=*O.L.G.* hrōra, *O.H.G.* ruora ; *uproar, commotion* ; rore '*commotio*' PR. P. 436.

rōren, *see* **rāren**.

hrōrien, v., *cf.* hrēren ; *move* ; roorin PR. P. 437.

hrōs, sb., *O.N.* hrōs ; *praise* ; ros ORM. 4910 ; C. M. 11948 ; M. H. 43 ; rous H. S. 5160.

rōse, sb., *O.E.* rose,=*O.H.G.*, *Lat.* rosa ; *rose*, KATH. 1434 ; **rōsen** (*pl.*) A. R. 276.

   rōse-garlond, sb., *rose-garland*, CH. H. F. 135.

   rōse-lēf, sb., *rose-leaf* ; **rōselēves** (*pl.*) CH. L. G. W. 228.

   rōse-rēd, adj., *red as a rose*, HORN (L.) 16 ; rose reed CH. C. T. *G* 254.

hrōsen, v., *O.N.* hrōsa ; *praise, glorify* ; rosen ORM. 4906 ; rose A. P. ii. 1371 ; TOWNL. 10 ; **rōses** (*pres.*) M. H. 49 ; **rōsed** (*pret.*) C. M. 2417.

   rōsing, sb., *praising, glorifying*, ORM. 4902 ; PR. C. 7070.

rosēr, sb., *Fr.* rosier ; *rose bush*, HAV. 2919.

rospen, *see* **raspen**.

[**hrōst**, sb., *O.E.* hrōst,=*O.L.G.* hrōst, *M.Du.* roest, *mod.Eng.* roost.]

rōst, sb., *O.Fr.* rost (*rôt*), *O.H.G.* rōst ; *process of roasting; roast meat* ; LANGL. *A* PROL. 108 ; þenne mot ich habbe hennen a rost (*O.Fr.* en rost) P. S. 151.

   rōst-īrin, sb.,=*O.H.G.* rōstīsan ; '*craticula*,' PR. P. 437.

rōsten, v., *O.Fr.* rostir, *O.H.G.* rōsten ; *roast*, L. H. R. 59 ; roste P. L. S. xv. 203 ; P. S. 191 ; roste [rooste] CH. C. T. *A* 383 ; **rōsted** (*pple.*) IW. 757 ; irost LEG. 8.

rot, sb., *cf. M.Du.* rot ; *rot*, ['*putredo*'] WICL. PROV. xii. 4 ; A. P. ii. 1079 ; TOWNL. 84.

[**rōt**, sb., *O.E.* rōt ; *joy, mirth, hilarity ; comp.* un-rōt.]

rōte, sb., *O.Fr.* rote ; *psaltery*, C. L. 7408 ; SQ. L. DEG. 1071 ; TRIST. 1853 ; pleien on a rote CH. C. T. *A* 236 ; **rōtes** (*pl.*) A. P. ii. 1082.

rōte, *see* **route**.

rōte, sb., *O.N.* rōt ; *root*, A. R. 54 ; SPEC. 57 ; SHOR. 109 ; AYENB. 34 ; P. L. S. ii. 98 ; WILL. 638 ; H. S. 7619 ; rote [roote] LANGL. *B* xv. 99 ; CH. C. T. *A* 2 ; roote C. L. 830 ; **rōtes** (*pl.*) HOM. II. 161 ; ORM. 3213 ; MAND. 190 ; **rōten** (*dat. pl.*) KATH. 2153 ; rote LAӠ. 31885 ; *comp.* **wode-rōte**.

   rōt-fest, adj., *O.N.* rōtfastr ; *established, secure*, SAX. CHR. 256.

rotelen, v.,=*M.Du.* rotelen, *M.L.G.* roteln, ruteln ; *ruttle* : his þrote shal rotelen REL. I. 65 ; **rotled** (*pret.*) ALIS. 930.

rottilling, sb., *disturbance*, ALEX. (Sk.) 943.

roten, adj., *O.N.* rotinn ; *rotten, putrid,* LANGL. *B* xv. 99 ; CH. C. T. *A* 3875 ; H. S. 6765 ; rotin A. R. 84*.

roters, sb., pl., *from med.Lat.* rutārii, rotārii ; *mercenary soldiers*, ROB. (W.) 6032.

rotien, v., *O.E.* rotian, *cf. M.Du.* roten, *O.H.G.* rozzēn ; *rot, become putrid*, FRAG. 7 ; A. R. 116 ; MAP 347 ; roten ORM. 4773 ; MAND. 49 ; rotie MISC. 92 ; SPEC. 101 ; AYENB. 32 ; rote HAMP. PS. xv. 10 ; **rotie** [rote] (*pres. subj.*) CH. C. T. *A* 4407 ; **rotede** (*pret.*) GEN. & EX. 3342 ; BEK. 2304 ; **roted** (*pple.*) A. R. 84 ; WILL. 4124 ; *comp.* for-rotien.

rotunge, sb., *O.E.* rotung ; *rotting*, H. M. 13.

rōtin, v., *O.N.* rōta ; *root, 'radico,'* PR. P. 437 ; **rōted** (*pple.*) P. L. S. xxv. 157 ; iroted A. R. 386 ; AYENB. 26.

roþer, *see* hreoþer.

rōþer, sb., *O.E.* rōðer, = *O.H.G.* ruodar ; *rudder, oar, 'remus,'* FRAG. 2 ; AYENB. 160 ; GOW. I. 243 ; A. P. ii. 419 ; MAN. (F.) 6576 ; P. R. L. P. 22.

rouelle, sb., *O.Fr.* rouelle, roelle (*petite roue*) ; *rowel*, D. ARTH. 3262.

rouȝ, *see* rūh. **rouhe**, *see* reihe.

rouken, *see* rūken.

roum, roumen, *see* rūm, rūmen.

roun, sb., *O.N.* hrogn, = *O.H.G.* rogan ; *roe (of fish)* ; (*ms.* rowne) PR. P. 438.

rounci, *see* rūnci. **round**, *see* rōnd.

roune, rounen, *see* rūne, rūnen.

rouste, sb., *O.N.* raust ; *voice*, ALEX. 488 ; (*ms.* rowwst) ORM. 9197.

route, sb., *O.Fr.* route, rote ; *route, rote, way* : halt forð his rute A. R. 350 ; bi rote *by rote* REL. II. 245 ; CH. C. T. *B* 1712 ; be pure rote PL. CR. 377.

route[2], sb., *O.Fr.* route, = *Ital.* rotta (*rupture*) ; *rout, company*, P. L. S. xxi. 14 ; SHOR. 135 ; WILL. 1213 ; GREG. 600 ; LANGL. *B* PROL. 146 ; CH. C. T. *A* 622 ; uppen one route (*first text* weorede) of wolves LAȝ. 2598* ; rute A. R. 92* (*v. r.* verd).

routen, v., *? Fr.* router ; *assemble* ; route ROB. 39 ; WILL. 5478 ; in al þe lond durste no cristen route CH. C. T. *B* 540 ; route, *?* rute C. M. 14618.

routen[2], v., *O.N.* rauta ; *prov. Eng.* rawt ; *roar* ; route HAV. 1911 ; AV. ARTH. xii.

hrouten, *see* hrūten. **rouwen**, *see* rūhen.

rovaisoun, sb., *O.Fr.* rovaison ; *rogation* ; **roveisouns** (*pl.*) P. L. S. xvii. 348.

[**rōve** ? sb., *comp.* wude-rōve.]

Rovecestre, pr. n., *O.E.* Hrofeceaster ; *Rochester*, MISC. 145 ; Roucestre, Rouchestre ROB. 4, 6.

róvare, sb., ? = rēavere ; *rover, 'pirata,'* PR. P. 437.

rōwe, *see* rāwe.

rōwen, v., *O.E.* rōwan, = *O.N.* rōa ; *row*, JUL. 77 ; rowen HORN (R.) 122 ; rouwen [rowe] LAȝ. 7813 ; rogen REL. I. 174 ; **rōweþ** (*pres.*) P. S. 254 ; **rēowen** (*pret.*) LK. viii. 26 ; rewen BRD. 5 ; TRIST. 1655 ; *comp.* bi-, ümbe-rōwen.

rōwere, sb., *rower, 'remex,'* CATH. 312 ; roware PR. P. 437.

rōwen[2], v., *? cf. O.H.G.* ruowan (*rest*) ; *rest* ; rowe BER. 284.

rōwen, *see* rāwen.

rowst, *see* rouste. **roxen**, *see* raxen.

rū, *see* rūh, hrēou.

rubbin, v., *rub, 'frico,'* PR. P. 438 ; rubbe LANGL. *C* x. 81 ; **rubbeþ** (*pres.*) CH. C. T. *A* 3747 ; **rubbede** [rubbed] (*pret.*) LANGL. *B* xiii. 99.

rūbi, sb., *O.Fr.* rubi ; *ruby*, SPEC. 25 ; LANGL. *B* ii. 12 ; rubee CH. H. F. 1362 ; **rūbis** (*pl.*) AYENB. 76.

rūbibe [ribibe, rebibe], sb., *O.Fr.* rubebe ; *violin*, CH. C. T. *A* 3331 ; ribibe *old woman* CH. C. T. *D* 1377 ; ribibe PR. P. 433 ; HALLIW. 682.

rūbibour, sb., *player on the rubibe* ; ribibor, ribibour LANGL. *A* v. 165, *B* v. 322.

rubriche, sb., *O.Fr.* rubriche ; *rubric*, CH. C. T. *D* 346.

rucken, v., = *M.L.G.*, *M.Du.* rucken, *Swed.* rucke, rocke, *O.H.G.* rucchen ; *rock, agitate* ; rokken CH. C. T. *A* 4157 ; rokke ROB. 98 ; rocki AYENB. 116 ; **rockeð** (*pres.*) A. R. 82 ; **rokked** (*pret.*) LANGL. *B* xv. 11 ; rokked of þe roust of his riche bruni GAW. 2018 ; heo ruckeden (? *ms.* ruokeden) burnen LAȝ. 22287 ; *see* ruggen.

rŭcken, *see* rūken.

rucul, sb., *Lat.* eruca ; *cankerworm*, PALL. i. 855 ; (*rocket*) PALL. i. 854.

rŭdden, *see* hredden.

ruddok, sb., *O.E.* rudduc VOC. 29 ; *robin redbreast, 'fri(n)gilla,'* PR. P. 438 ; CH. P. F. 349 ; SQ. L. DEGR. 46 ; roddoc VOC. 164.

rude, sb., *O.E.* rudu *f.*, *cf. O.N.* roði *m.* ; *red, redness, ruddy*, A. R. 330 ; O. & N. 443 ; MISC. 193 ; rudde GOW. III. 27 ; E. T. 200 ; rudde [rode] CH. C. T. *A* 3317 ; rode SPEC. 26 ; ST. COD. 60 ; ALIS.[2] (Sk.) 178 ; M. ARTH. 179.

[**rŭde**, adj., *O.E.* gerȳde (*opportune*) ; *comp.* un-rŭde.]

rŭde[2], adj., *O.Fr.* rude ; *rude, rough ; undressed (of cloth)*, WILL. 1851 ; CH. C. T. *B* 3998 ; ruide D. ARTH. 1049.

rŭde-liche, adv., *rudely*, CH. C. T. *A* 734 ; ruidli, rudli BARB. ii. 349, ix. 750*.

rūdenesse, sb., *rudeness*, CH. C. T. *E* 397.

[**rudel**, sb., *ruddle*.]

rodel-wort, sb., LEECHD. III. 342.

ruden, v., *? O.N.* roða ; *become red, make red* ; ruddon '*frico*' PR. P. 438 ; necke & heved ... he **ruddede** (*pret.*) ... wiþ his here P. L. S. xvii. 172 ; as rodi as a rose ruddede [roddede] hus chekes LANGL. *C* xvi. 108 ; **iruded** [irudded] (*pple.*) & ireaded A. R. 50, 356.

rüden, v., *O.N.* ryðja, hryðja (*empty*) ; *rid, clear* : þe shal ruden þine wei ['*qui praeparabit viam tuam*'] HOM. II. 133 ; **rid** (*pret.*) D. TROY 1533 ; **irüd** (*pple.*) C. L. 1227.

rudi, adj., *ruddy*, A. R. 330 ; H. M. 35 ; **ruddi** PR. P. 438 ; **rodi** ALIS. 7833 ; LANGL. *B* xiii. 99 ; CH. C. T. *F* 385 ; WICL. MAT. xvi. 2 ; **rudie** (*pl.*) KATH. 1431.

rudnin, v., *O.N.* roðna ; *become red*, JUL. 27 ; þe reve **rudnede** (*pret.*) al o grome MARH. 19.

rue, sb., *O.Fr.* rue ; *rue*, '*ruta*,' REL. I. 36 ; WICL. LK. xi. 42.

ruet, sb., *trumpet*, '*lituus*,' CATH. 313 ; **ruet** [ruwet] LANGL. *B* v. 349 ; **ruwet** ALIS. 3699.

ruffelin, v., *ruffle*, PR. P. 439 ; **ruffeld** (*pple.*) C. M. 26391.

*h*rüg, sb., *O.E.* hrycg, = *O.N.* hryggr, *M.L.G.* rugge, *O.H.G.* hrucki, *prov. Eng.* rig ; *ridge, back* ; **rug** H. M. 17 ; TREAT. 139 ; þene rug A. R. 264 ; **rug** [rugge] LAȝ. 1912 ; **rig** HAV. 1775 ; IW. 1833 ; **rigge** ALIS. 5722 ; M. ARTH. 2178 ; þane reg AYENB. 116 ; **rügge** (*dat.*) LAȝ. 8157 ; O. & N. 775 ; LANGL. *B* xiv. 212 ; **hrigge** LEECHD. III. 120 ; **rigge** E. G. 354 ; **rügges** (*pl.*) LAȝ. 540 ; **rigges** PALL. i. 1151 ; **rigges** noþer vores FER. 1565 ; **rüggen** (*dat. pl.*) LAȝ. 27421.

   rig-bōn, sb., *O.E.* hrycgbān ; *backbone*, VOC. 245 ; LANGL. *C* vii. 400.

rüȝe, sb., *O.E.* ryge (*gen.* ryges), = *O.N.* rugr, *M.Du.* rogge, *O.H.G.* rocco ; *rye*, '*secale*,' P. S. 152 ; (*ms.* reye) CH. C. T. *D* 1746 ; **rie** PR. P. 433.

rüȝen, v., *cf. O.H.G.* (gi-)rūhen ; *from rüh* ; *roughen* ; **rouwe** (*pres.*) REL. II. 211.

rugged, adj., *cf. Swed.* rugget (RIETZ) ; *rugged*, AR. & MER. 1501 ; **ruggid** '*hispidus*' PR. P. 439.

ruggen, v., *O.N.* rugga ; = **rucken** ; *rock, agitate* ; roggin PR. P. 435 ; rogg(e) þam in sonder PR. C. 1230 ; **ruggede** (*pret.*) N. P. 41 ; roggede D. ARTH. 784 ; **rogged** LANGL. *B* xvi. 78 ; LIDG. M. P. 40.

ruggi, adj., *cf. Swed.* ruggig ; *hairy*, CH. C. T. *A* 2883 ; **roggi** PALL. xi. 86.

*h*rüglunge, adv., *cf. O. H. G.* ruckelingen ; *backwards* ; **ruglunge** FRAG. 6 ; JUL. 49 ; **ruglinge** MARH. 17.

rŭh, adj., *O.E.* rūh, = *M.Du.* rū, rūgh, rouw, *O.H.G.* rūh ; *rough, hairy* ; (*ms.* ruhh) ORM. 9211 ; **rugh** TOWNL. 100 ; **ru** GEN. & EX. 1539 ; rouȝ P. P. 216 ; WICL. GEN. xxv. 25 ; **rough** ALIS. 6261 ; MAND. 285 ; CH. C. T. *A* 3738 ; his ruhe necke MARH. 12 ; þet ruwe vel A. R. 120 ; **ro(u)we** (*pl.*) MAP 338 ; WILL.

4778 ; mid ruȝe felle O. & N. 1013 ; **rŭhure** (*compar.*) A. R. 284.

ruine, sb., *O.Fr.* ruine ; *ruin*, CH. C. T. *A* 2463.

ruit, sb., *O.Fr.* ruit ; *? noise* : riot and ruit L. H. R. 132.

rüke, sb., *cf. O.Swed.* rūka, *? O.N.* hrūga ; *heap* ; **rüken** (*pl.*) A. R. 214.

rukelen, v., *heap up*, A. R. 214 ; þu schalt rukelen on his heaved bearninde gleden 406.

rüken (*? h*rüken), v., *cf. Swed.* rūga (RIETZ), *Dan.* rūge (*brood*), *? O.N.* hrūga (*heap up*) ; *? mount on the back; crouch* ; *huddle together* : ne mei he nouþer on hire ne ruken ne riden A. R. 266 ; and in þi bedde to rouken (*v. r.* jouken) þus CH. TRO. v. 409 ; þe neddir þat on þam sal rouke PR. C. 6765 ; ruckin '*incurvor*' PR. P. 439 ; þe scheep þat roukeþ (*pres.*) in þe folde, CH. C. T. *A* 1308 ; þei rucken in hire neste GOW. II. 57 ; **rukking** (*pple.*) LIDG. M. P. 118 ; this shep rukking in his folde P. R. L. P. 19 ; *deriv.* rūke, *? h*rēac.

rúle, *see* **reule**.

rūm, adj., *O.E.* = *O.Fris.* rūm, *O.N.* rūmr, *O.H.G.* rūm, *Goth.* rūms ; *spacious, large*, O. & N. 643 ; **roum** ROB. 303 ; AR. & MER. 6936 ; CH. C. T. *A* 4126 ; heofnes rume riche ORM. 3689 ; in roume stede PS. xxx. 9 ; **roume** (*pl.*) A. D. 294 ; roume landes D. ARTH. 432 ; **roumer** (*compar.*) CH. C. T. *A* 4145.

   rūm-handed, adj., *liberal*, HOM. II. 29.

   rūm-hende, adj., *liberal*, LAȝ. 6538.

   rūm-līche, adv., *largely*, LAȝ. 2452.

rūm[2], sb., *O.E.* rūm, *cf. O.L.G.*, *O.H.G.* rūm *m.*, *O.N.* rūm *n.*, *Goth.* rūms *m.* ; *room, space*, LAȝ. 1003 ; þe laferd hafde litel rum in al þat micle riche ORM. 8489 ; gede on rum *aside, apart* GEN. & EX. 4000 ; roum S. S. (Web.) 599 ; CH. L. G. W. 1999 ; A. P. ii. 96 ; PR. C. 9168 ; ȝiveþ me roum AL. (L.[2]) 481 ; roum '*spacium*' PR. P. 438 ; rom S. S. (Wr.) 615 ; **roume** (*dat.*) AR. & MER. 8065 ; a roume (*at a distance*) he stod TRIST. 2355 ; a roume he hovid RICH. 464 ; stand on roume TOWNL. 235.

[**rūm-**, *stem of* **rūmen**.]

   rūm-hūs, sb., *latrina*, O. & N. 652.

rūmen, rümen, v., *O.E.* rȳman, rūmian, *cf. O. N.* rȳma, *O.L.G.* rūmian, *O.H.G.* rūmman ; *from rūm* ; *make room; extend; empty; yield* ; LAȝ. 10640 ; wanne he sal henne rimen REL. I. 175 ; rime MAN. (F.) 9868 ; reme AR. & MER. 4408 ; þi lond to reme HORN (L.) 1272 ; roume LANGL. *C* i. 181 ; **rēm** (*imper.*) LK. xiv. 9 ; remeþ AL. (T.) 505 ; **rūme** (*subj.*) C. M. 14922 ; and þene wæi **rūmde** (*pret.*) LAȝ. 28323 ; rumede ROB. 536 ; roumede JOS. 597 ; *comp.* ȝe-rūmen.

rūmien, *see* **rōmen**.

rummelin, rumlin, v., *cf. Dan.* rumle, *Du.* rommelen ; *rumble*, PR. P. 439 ; **romblen** CH. L. G. W. 1216.

rumour, sb., *O. Fr.* rumeur; *rumour*; romour YORK xxvi. 34; **rumours** (*pl.*) CH. BOET. ii. 7 (59).

rumpe, sb., *cf. Dan.* rumpe, *Swed.* rumpa; *rump*, '*cauda*,' PR. P. 439.

rūmðe, sb., *O.E.* rȳmð,=*M.Du.* ruimte; *from* (*rūm; room; space*; rimthe PR. P. 434; a rumðe *at large* LAȝ. 27492.

[**rün?** sb., *comp.* **cün-rün.**]

*h*rün, sb., *?*O.N. hryn; *noise*: ruten forð wið swuch rune KATH. 2031.

rúnci, sb., *O.Fr.* roncin (*cheval de service*); *stallion, hack,* HAV. 2569; rounci IW. 252; CH. C. T. *A* 390; rounce GAW. 303; rounse '*mannus*' VOC. 187; rouncin P. S. 190; **roun-sies** (*pl.*) MAN. (F.) 11422.

rūndel, see **rōndel.**

rüne, sb., *O.E.* ryne,=*O.Fris.* (blōd-)rene, *O.H.G.* run, *Goth.* runs *m.*; *from* **rinnen**; *running, course,* HOM. I. 207; MARH. 7; O. & N. 1156; hwon þe tunge is o rune A. R. 74; rine LK. viii. 44; *comp.* **blōd-rüne.**

rūne, sb., *O.E.* rūn,=*O.N.* rūn, *O.L.G., O.H.G., Goth.* rūna; *colloquy, conversation, counsel; language; letter*; O. & N. 1170; elche rune he ihurð P. L. S. viii. 45; ofte heo heolden rune LAȝ. 25332; rune, run ORM. 18719, 18786; on Sexisce runen (*r.* rune) LAȝ. 32000; roune ALIS. 806; TRIST. 3; ne mai no man wiþ þe holden roune AN. LIT. 4; herkne to mi roune SPEC. 29; wiþ briddes roune (*song*) 43; on Englische roun(e) MAN. (F.) 13757; **rünes** (*pl.*) KATH. 574; godes derne runes A. R. 96; þat he write runen LAȝ. 25340; *comp.* **lēod-, som-, sunder-rūne.**

rūn-stæf, sb., *letter*; **rūnstaven** (*pl.*) LAȝ. 9961.

rünel, sb., *O.E.* rynel; *stream,* '*cursor*,' FRAG. 3; rinel D. TROY 5709.

*h*runge, sb., *O.E.* hrung,= *M.H.G.* runge, *Goth.* hrugga (ῥάβδος); *rung (of a ladder)*: ronge of a ledder, of a carte CATH. 311; **ronges** (*pl.*) '*rideles*' REL. II. 83; LANGL. *B* xvi. 44; þe ronges and þe stalkes CH. C. T. *A* 3625.

rungen, v., *?rouse*; **rung** (*imper.*) up & sture þe A. R. 290.

rüni, adj., *O.E.* rynig, *?*=*M.H.G.* rünec; *?fleet*: þe runie wulf LAȝ. 20123 (*printed* rimie 1545).

rūnien, v., *O.E.* rūnian,=*O.H.G.* rūnen, *M. Du.* rūnen, ruinen; *whisper, talk,* HOM. II. 107; runen, runan (rouni) LAȝ. 2331, 32116; rounen P. S. 326; rounin '*susurro*' PR. P. 438; roune CH. C. T. *D* 1572; M. ARTH. 3423; rouneþ (*pres.*) in his ere LANGL. *B* iv. 13; heo runeþ to gaderes R. S. vii; rounede (*pret.*) BEK. 1200; H. S. 6931; rouned ALIS. 7614; A. P. iii. 64.

rūninge [rouni(n)ge], sb., *conversation,* LAȝ. 14070.

runisch, adj., *cf.* **renisch**; *?furious, terrible,* GAW. 457; A. P. ii. 1545.

runkel, see **wrunkel.**

rŭnken, v., *?from* **rūnien,** *as* **hĕrkien** *from* **hēren**; runk(e) or roune TOWNL. 68.

rūpere, sb., *O.E.* rȳpere; *robber*; **rūperes** (*pl.*) HOM. I. 15.

rüppen, v., *?O.E.* ryppan,=*M.H.G.* rupfen, rüpfen; *seize*: to rippen hem & ræfen ORM. 10212; rüpten (*pret.*) LAȝ. 10584.

*h*rüre, sb., *O.E.* hryre; *from* **hrēosen**; *ruin, fall*; rure O. & N. 1154; rire (*ms.* ryre) MAT. vii. 27; rere MK. v. 13.

rüsch, v., *?*=*M.H.G.* rüsch; *rush*: & take hem at one rusche (*ms.* russche) FER. 2888.

rusche, rüsche, sb., *O.E.* risce, resce, =*M.L.G.* riske, rische, *M.H.G.* rusche; *rush,* '*(s)cirpus*,' REL. I. 6; rische, rusche PR. P. 435; rusche LANGL. *A* iii. 137; rusche [reshe] ['*scirpus*'] WICL. JOB viii. 11; rishe R. R. 1701; resche '*juncus, sirpus*' VOC. 191; GOW. I. 160; S. S. (Wr.) 2884; LUD. COV. 170; resse AYENB. 253; **risches** (*pl.*) RICH. 6038; *comp.* **bul-rische.**

rüschen, v., *cf. M.L.G.* rüschen, *? M.H.G.* rüschen, riuschen; *rush*: & sau þe red(e) blod ruschen (*ms.* russchen) out FER. 497; rusche D. ARTH. 1339; **rŭshes** (*pres.*) AV. ARTH. iv; **rŭshinge** (*pple.*) CH. C. T. *A* 1641; **rŭsched** (*pret.*) GAW. 2204; A. P. ii. 368; windis blewen and rusheden into þat hous WICL. MAT. vii. 25.

*h*rüsel, sb., *O.E.* hrysel,=*O.L.G.* rusel; *grease,* '*adeps*,' FRAG. 2; rusel, risel LEECHD. III. 112, 124.

rüsien, see *h*risien.

rusken, v., *?*=*Swed.* ruska (*shake*): þe ronke racches þat **ruskit** (*pret.*) þe ron MISC. 92.

russet, sb., *O.Fr.* russet, rousset; *russet, red cloth,* MISC. 92; LANGL. *A* viii. 1.

rust, sb., *O.E.* rust,=*M.L.G.* rust, *O.H.G.* rost; *rust,* '*rubigo*,' PR. P. 439; A. R. 160; P. R. L. P. 183; PR. C. 5570; M. H. 105; roust GAW. 2018.

rusten, v., *O.E.* rustian, *cf. O.H.G.* rostēn; *rust,* A. R. 344; rustin '*rubigino*' PR. P. 439; ruste CH. C. T. *A* 502; **roustez** (*pres.*) SAINTS (Ld.) lv. 120; **irusted** (*pple.*) A. R. 160.

[*h*rüsten, v., *O.E.* hrystan, hyrstan,=*O.H.G.* hrusten; *comp.* **ȝe-rüsten.**]

rusti, adj.,=*M.H.G.* rostic; *rusty,* '*rubiginosus*,' PR. P. 439; CH. C. T. *A* 618.

rúte, see **route.**

rūten, v., *? rush, dart*: þai seȝe þe waȝes of þe se harde to gadre route FER. 1343; **rēat** (*pret.*) JUL. 58; ruten forð KATH. 2031; *comp.* **æt-rūten.**

*h*rūten, v., *O.E.* hrūtan, *cf. O.Fris.* hrūta, *M. Du.* rūten, *O.N.* hriōta, *O.H.G.* riozan; *snore*;

routin '*sterto*' PR. P. 438; route BEV. 1180;
þe wind so loude began to route (*roar*) CH.
TRO. iii. 743; RICH. 4304; routeþ (*pres.*)
['*stertit*'] WICL. PROV. x. 5; CH. C. T.*A* 3647;
rŭtte (*pret.*) *snored* LANGL. *B* v. 398.

rūþe, *see* *h*rēouðe. rūðer, *see* rēoðer.

rüþþe, adj., *O.E.* ryðða, = *M.H.G.* rüde ;
mastiff; ? rith A. P. ii. 1543.

s.

sǟ, sb., *O.N.* sār ; *tub, cask, '* *tina*,' VOC. 200 ;
so HAV. 933 ; soo PR. P. 462.

sǻ, *see* swǻ. saaf, *see* sauf.

sabat, sb., *O.Fr.* sabat ; *sabbath*, AYENB. 7 ;
sabbat C. M. 11997; sabote WICL. MK. ii. 27 ;
sabothis (*pl.*) 23 ; sabaz H. M. 17.

sabatūn, sb., *cf. Sp.* zabaton, *Ital.* ciabattone
(*large boot*); *steel shoe*; sabatouns (*pl.*) GAW.
574; sabatons MAN. (F.) 10026.

sáble, sb., *O.Fr.* sable ; *sable fur; sable
colour*; PR. P. 440; CH. COMPL. M. 284;
ALEX. (Sk.) 3946; sabille D. ARTH. 771;
enamelede of sable D. ARTH. 2027.

sabeline, sb., *O.Fr.* sabeline ; *sable, fur of
sable*, HOM. II. 231 ; sabline MISC. 70.

sac, sb., *O.E.* sæcc, = *O.H.G.* sach, *O.N.* sekkr,
Goth. sakkus, *Lat.* saccus ; *sack*, HOM. II.
139; P. L. S. xvii. 186; sak CH. C. T. *A*
4017; sek PR. P. 451 ; IW. 2032; PS. xxix.
12 ; a sek ful PR. C. 566; seck GEN. & EX.
2309; sech LK. x. 4 ; zech AYENB. 81 ; sackes
(*pl.*) LEG. 92 ; seckes HAV. 2019.

sachel, sb., ? *from O.Fr.* sache (*wise*) ; *philosopher*, ALEX. (Sk.) 716.

sachel², sb., *Lat.* saccellus ; *satchel*, WICL.
LK. x. 4 ; sachels (*pl.*) CH. BOET. i. 3 (12).

sacken, v., *cf. O.N.* sekka ; *put into a sack* ;
sakked (*pple.*) CH. C. T. *A* 4070.

sacrament, sb., *O.Fr.* sacrement ; *sacrament*,
A. R. 268; sacrement AYENB. 14 ; TREV. V.
231.

sácren, v., *O.Fr.* sacrer ; *consecrate ; take a
solemn oath* ; JOS. 302; sácreð (*pres.*) A. R.
208; sacreþ AYENB. 235 ; sacrieþ ROB. (W.)
7209; sácred (*pret.*) ROB. (W.) 9147; sácred
(*pple.*) *sacred* A. P. i. 1139; isacred TREV.
II. 115.

sácringe, sb., *consecration*, TREV. II. 115;
VI. 411; ['*sacrificium*'] TREV. V. 231 ; BEK.
1152.

sacrifíce, sb., *O.Fr.* sacrifice; *sacrifice*, AYENB.
187; sacrifis, sacrifese ALEX. (Sk.) 4461,
1486*; sakerfise A. P. i. 1064.

sacrificen, v., *sacrifice* ; sacrifísed (*pret.*)
LANGL. *B* xii. 118.

sacrifien, v., *O.Fr.* sacrifier; *sacrifice* ; sacrifie
WICL. EX. x. 11.

sacrilege, sb., *O.Fr.* sacrilege ; *sacrilege*,

AYENB. 34; sacrelege TREV. V. 265 ; sacrilag
ALEX. (Sk.) 4561.

sacristane, sb., *Fr.* sacristain, *mod.Eng.*
sacristan, sexton ; *sexton*, CATH. 315 ; sextein
CH. C. T. *B* 3126.

sad, adj., *O.E.* sæd, = *O.L.G.* sad, *O.N.* saðr,
Goth. saþs, *O.H.G.* sat ; *sad; sated ; resolute ;
quiet; solid; 'solidus*,' PR. P. 440; WILL.
1463; MAND. 159; Childric . . . is sad of
mine londe LA3. 50830 ; selden i am sad þat
semli for te se SPEC. 29 ; of worldes winne
sad MAP 341 ; hit was god and sad SHOR.
146 ; strong sad and sound GOW. III. 92 ;
sad and trewe CH. C. T. *B* 135 ; sad, sæd,
sed HOM. II. 75, 232 ; sed P. L. S. viii. 195 ;
ANGL. I. 31 ; zed and stable AYENB. 83 ;
þe sadde mon LANGL. *A* ix. 23 ; sad man *B*
viii. 28 ; mani sadd hundreth ALEX. (Sk.)
3883; sade (*pl.*) O. & N. 452 ; sadder
(*compar.*) LANGL. *B* v. 4 ; saddest (*superl.*)
LANGL. *C* xi. 49.

sad-lī, adv., *sadly, firmly, seriously*, WILL.
469; CH. C. T. *E* 1100; A. P. iii. 442 ; ALEX.
(Sk.) 347, 2568, 1139; sadloker (*comp.*)
LANGL. *A* v. 4.

sadnes (sadnesse), sb., *steadiness, steadfastness, discreetness*, ALEX. (Sk.) 1017; sadnesse CH. C. T. *E* 452 ; SHOR. 52.

sadel, sb., *O.E.* sadol, sadel, = *O.N.* söðull,
O.H.G. satul, satol, satal, satel ; *saddle*, GEN.
& EX. 3949 ; LANGL. *B* iv. 19 ; IW. 422 ;
CH. C. T. *A* 2162 ; GAW. 437 ; sadele (*dat.*)
LA3. 6473 ; sadeles (*pl.*) LA3. 3408 ; sadles
AR. & MER. 3881.

sadel-bowe, sb., *O.E.* sadelboga, = *O.H.G.*
satelpogo; *saddle-bow*, REL. I. 176 ; AR. &
MER. 8158.

sadelien, v., *O.E.* sadelian, = *M.L.G.* sadelen,
O.H.G. satalōn, *O.N.* söðla ; *saddle*, LA3. 13530;
BEV. 757 ; sadelede (*pret.*) HORN (L.) 715 ;
sadulde OCTAV. (H.) 1198 ; sadilt BARB. ii.
141 ; sadled (*pple.*) LANGL. *B* ii. 169 [sadelet
A ii. 147].

sadien, v., *O.E.* sadian, = *M.L.G.* saden, *O.H.*
G. satōn ; *make solid ; confirm ; be serious* ;
sade P. L. S. xxx. 4 ; saddin '*solido*' PR. P.
440; to sadde us in bileve LANGL. *B* x. 242 ;
comp. a-sadien.

sadiler, sb., = *M.L.G.* sadeler, *O.H.G.* satilāri,
O.N. söðlari ; *saddler*, VOC. 212.

sǣ, sb., *O.E.* sǣ *m. f., cf. O.N.* sǣr, siōr, siār,
O.L.G. sēu, sēo, *O.H.G.* sēo, sē (*gen.* sēwes),
Goth. saiws *m.* ; *sea, lake*, REL. I. 128 (HOM.
II. 161) ; ORM. 14796; sæ, sea [see] LA3.
123, 12005; þe sea [see] KATH. 1800 ; se
REL. I. 220 ; see TREAT. 137 ; þe reade see
A. R. 330 ; sǣs (*gen.*) MAT. xviii. 6 ; sees
REL. I. 220 ; seis HAV. 321 ; þære sæ strond
LA3. 7241 ; bi þere sæ side 25661 ; innan þan
sea HOM. I. 43 ; in þere sea 51 ; i ðer see A.
R. 230 ; bi þare see O. & N. 1754; *comp.*
Norþ-see.

see-calf, sb., *sea-calf*; **seecalves** (*pl.*) TREV. II. 13.

sǣ-clif, sb., *O.E.* sǣclif; *sea-cliff*, LAȝ. 12638.

sǣ-farinde, pple., *sea-faring*, REL. I. 128.

sǣ-fisc, sb., *O.E.* sǣfisc; *sea-fish*, LAȝ. 22550; seefisch TREV. I. 335.

sǣ-flōd, sb., *O.E.* sǣflōd; *sea*, LAȝ. 2630.

sǣ-flot(e), sb., *O.E.* sǣflota; *navy*, LAȝ. 4530.

see-froth, sb., *seaweed*, ['*algam*'] PALL. iv. 335.

sē-grúnd, sb., *O.E.* sǣgrund; *bottom of the sea*, REL. I. 220 (MISC. 16).

cee-hound, sb., *sea-dog*, ALIS. 5669.

sǣ-līðende, sb., pl., *travellers by sea*, LAȝ. 7821.

sǣ-man, sb., *O.E.* sǣmann; *seaman*; sǣmon LAȝ. 1165.

sē-māwe, sb., *sea-mew*; '*fulica*,' VOC. 189; semow(e) '*alcedo*' PR. P. 452; semewe LIDG. M. P. 202.

sǣ-rime, sb., *O.E.* sǣrima; *seashore*, LAȝ. 6216.

sǣ-sīde, sb., *seaside*, LAȝ. 9746.

sē-sond, sb., *sea-sand*, REL. I. 220 (MISC. 16).

sea-strēam, sb., *O.E.* sǣstrēam; *sea*, LAȝ. 326; MARH. 9.

sǣ-strond, sb., *O.E.* sǣstrand, -strond; *seashore*, '*litus*,' FRAG. 3; LAȝ. 9235; seestrond TREV. V. 11; bi þe seestronde P. S. 188.

sea-þistel, sb., *sea thistle*, '*tribulus marinus*' REL. I. 37.

see-water, sb., *O.E.* sǣwæter; *sea-water*, HOM. I. 159.

sǣ-wēri, adj *O.E.* sǣwērig; *sea weary*, LAȝ. 4619.

sē-wolf, sb., *sea-wolf*, ROB. 132.

sǣc, sǣclien, *see* **sēoc, sēclien.**

sǣd, sb., *O.E.* sǣd, *cf. O.Fris.* sēd, *O.L.G.* sād, *O.N.* sāð *n.*, *O.H.G.* sāt *f.*; *seed*, MAT. xiii. 3; sed ORM. 5070; O. & N. 1041; TREAT. 138; seed *children* LANGL. *C* xi. 251; þat sed JUL. 75; seed P. S. 152; LANGL. *A* iii. 261; seid *issue* BARB. i. 63; seð MISC. 9; þet zed AYENB. 143; sed, sad HOM. II. 161, 163; sǣde (*gen. pl.*) MAT. xiii. 32; *comp.* hemp-, līn-sǣd.

sǣd-āte, sb., *oat seed*; **sǣdāten** (*pl.*) SAX. CHR. 252.

sēd-foul, sb., *bird living on seeds*, CH. P. F. 512.

sǣd-hwǣte, sb., *wheat-seed*, SAX. CHR. 252.

seed-lēp, sb., *seed-basket*, PR. P. 451; sede lepe REL. I. 7; seed leep LANGL. *B* vi. 63*; sēdlǣpas (*pl.*) SAX. CHR. 252.

seed-tīme, sb., *O.E.* sǣdtīma; *seed-time*, TREV. I. 131.

sǣden, v., *sow seed, plant*, '*semento*;' sedin PR. P. 451; sede PALL. vi. 71; seeden *beget offspring* LANGL. *C* xi. 251.

sǣdere, sb., *O.E.* sǣdere, = *O.H.G.* sǣter; *sower*, MK. iv. 3.

sǣȝen, v., *O.E.* sǣgan, = *M.H.G.* seigen; *from* sīȝen; *sink, go*; sēȝes (*pres.*) ALEX. (Sk.) 4333; segeð GEN. & EX. 2232; (þai) seȝen 1481; *comp.* bi-sǣȝen.

sǣl, sēl, adj., *O.E.* sǣl, sēl, = *O.N.* sǣll, *Goth.* sēls (χρηστός, ἀγαθός); ·*happy, timely, good*, LAȝ. 1234, 4071; sel H. M. 47; sēlere (*dat. f.*) LAȝ. 21654; sēle (*pl.*) ALIS. 7430; sēlere (*gen. pl.*) LAȝ. 18011; sēlen (*dat. pl.*) LAȝ. 25162; sēlre, selere (*compar.*) LAȝ. 67, 21166; ge sind selren MAT. x. 31; sēlest (*superl.*) LAȝ. 918; þanne sælesten dæl ['*optimam partem*'] LK. x. 42; *comp.* un-sǣl.

sǣl-līche, adj., *fortunate*, LAȝ. 7863.

sǣl², sēl², sb., *O.E.* sǣl *m. f.*, *cf. O.N.* sǣla, *Goth.* sēlei *f.*; *happiness; fit season*; LAȝ. 12875, 24104; ORM. 14304; GEN. & EX. 417; M. T. 69; LUD. COV. 275; S. & C. I. lxxiv; mi sel mi saule hele HOM. I. 183; sele A. P. iii. 5; L. H. R. 72; AV. ARTH. lxiii; sǣle, sele (*dat.*) LAȝ. 734, 1310; sele O. & N. 953; selden sal he ben on sele REL. I. 180; a sele HOM. II. 183; at sume sele 185; it turned him to sele C. M. 4432; i am sette þus out of seill YORK vii. 136; sele YORK ii. 13; *comp.* barli-, méte-, un-sēl.

sǣlehðe, sb., (?=sǣlðe); *happiness, blessedness*; selehðe LAȝ. 25136; seluhðe A. R. 354; selhðe KATH. 895; H. M. 39; *comp.* ȝe-, un-sēlhðe.

sǣlen, v., *? O.E.* gesǣlan; *befit*: of kunde me ne sēlde (*pret.*) þe to spouse welde HORN (R.) 425.

sǣli, adj., *O.E.* (ge-)sǣlig, = *O.L.G.* sālig, *O.H. G.* sālig, *mod. Eng.* silly; *happy, good, innocent* (*often in contemptuous sense*); seli PR. P. 452; LAȝ. 1484*; KATH. 1421; A. R. 352; PS. ii. 13; WICL. ECCLUS. xiv. 2; A. P. i. 658; CH. C. T. *D* 1702; PR. C. 5810; seli child is sone ilered BEK. 158; REL. I. 110; seli wif what eileþ þe AN. LIT. 10; seli HAV. 477; ROB. 33; P. S. 194; CH. C. T. *A* 3601; PL. CR. 442; a seli litil clout MAND. 293; þe sili man S. S. (Wr.) 1361; *comp.* ȝe-, un-sēli.

sēli-līche, adv., *happily*, A. R. 184; seliliȝ ORM. 17318.

sēlinesse, sb., *happiness*, CH. TRO. iii. 813.

sǣlþe, sēlþe, sb., *O.E.* sǣlð, = *O.L.G.* sālða, *O.H.G.* sālida; *blessedness; felicity*; selþe REL. I. 181; HAV. 1338; SPEC. 33; (*ms.* sellþe) ORM. 3851; *comp.* ȝe-, un-sǣlðe.

sǣr, *see* **sār.** **sǣre**, *see* **schēre.**

sǣte, sb., *O.E.* sǣte, = *O.N.* sǣti, *M.Du.* sāte, *O.H.G.* gesǣze *n.; from* sitten; *seat*, ORM. 11961; sete HOM. II. 33; R. S. iv; MISC. 88; SHOR. 149; PS. xliv. 7: PR. C. 9318; DEGR. 1463; seete A. P. ii. 92; LIDG. M. P. 98; sēte (*pl.*) MISC. 73: *comp.* kine-sǣte.

[sǣte², sb., *O.E.* sǣta, = *O.H.G.* sāzo ; *dweller* ; *comp.* **Dor-, Sumer-sēte.**]

[sǣte ³, adj., *O.E.* sǣte ; *comp.* **and-sǣte.**]

Sæverne, pr. n., *O.E.* Sæfern ; *Severn,* LAȝ. 9579 ; Severne ROB. 2 ; MAN. (F.) 1960.

sǣw, *see* **sēau.**

sæx, *see* **sax. sáf,** *see* **sauf.**

saffing, *see under* **sauven.**

saffran, sb., *Fr.* safran ; *saffron,* REL. I. 129 ; saffroun CH. C. T. *B* 1920.

saffron, v., *colour with saffron,* CH. C. T. *C* 345 ; **saffrond** (*pple.*) ALEX. (Sk.) 4600.

saft, *see* **schaft.**

ságe, adj. & sb., *wise, wise person,* ALEX. (Sk.) 1649, 4704 ; LANGL. *B* x. 379 ; **ságes** (*pl.*) LANGL. *B* xiii. 423 ; DEP. R. iii. 257.

ságe-lī, adv., *wisely,* ALEX. (Sk.) 3359.

saȝe, sb., *O.E.* sage, = *M.L.G.* sage, *O.H.G.* saga, *O.N.* sög ; *saw* ; *sawe 'serra'* VOC. 181 ; PR. P. 441 ; PL. CR. 753.

saȝe ², sb., *O.E.* sagu, = *M.L.G.* sage, *O.N.,* *O.H.G.* saga, *mod.Eng.* saw ; *saying* ; þeo saȝe HOM. I. 133 ; ælc his saȝe sæide LAȝ. 26345 ; sage GEN. & EX. 4153 ; saȝe, sahe, sawe A. R. 56 ; sahe MARH. 8 ; sawe CH. C. T. *A* 1163 ; SACR. 393 ; such wonder nas never iherd in sawe C. L. 619 ; alle seide at o sawe WILL. 1112 ; **saghes** (*pl.*) PS. xviii. 4 ; sahen KATH. 358 ; sawen (sawes) LAȝ. 749 ; sawen REL. I. 171 ; Salomones sawes LANGL. *B* vii. 137 ; *comp.* **bī-, ȝein-, on-, wið-saȝe,** (**-sahe, -sawe**).

[**saȝel,** adj., *O.E.* sagol ; *comp.* **sōð-saȝel.**]

sāȝel, sb., *O.E.* sāgol, = *M.H.G.* seigel (*round of a ladder*) ; *? club, cudgel :* ælc bær an honde ænne saȝel stronge LAȝ. 12280 ; sowel *' fustis '* FRAG. 4 ; mid **sāhlen** (*dat. pl.*) [' *fustibus* '] MAT. xxvi. 47.

saȝen, v., = *M.L.G.* sagen, *O.H.G.* sagōn, *O.N.* saga ; *saw* ; *sawin ' serro '* PR. P. 441 ; **sageð** (*pres.*) REL. I. 223 ; **sahede** (*pret.*) MARH. 22 ; **isahet** (*pple.*) JUL. 38.

sawer, sb., *sawyer,* VOC. 212.

saȝen, *see* **seggen.**

saggard, sb., *? sluggard,* YORK xxxvi. 82.

saggin, v., *cf. L.G.* sacken, *Swed.* sacka ; *sag, sink down,* PR. P. 440.

sahe, *see* **saȝe.**

saht, adj., *O.E.* sæht (SAX. CHR. 215), = *O.N.* sāttr ; *reconciled, at peace,* P. S. 214 ; saȝt A. P. i. 52 ; saght Iw. 3898 ; saught TRIST. 273 ; MAN. (F.) 2490 ; saut AN. LIT. 8 ; sæht LAȝ. 5114 ; sahte ORM. 5731 ; SPEC. 47 ; saught (*? ms.* saughe) YORK iv. 34 ; **sauhte** (*pl.*) MISC. 97 ; bote mi sustren ben sauȝt(e) and some C. L. 520 ; *comp.* **ȝe-, un-, wrang-saht.**

sahtnesse, sb., *O.E.* sahtnyss (SAX. CHR. 203) ; *reconciliation, peace,* ORM. 3515 ; sauht-nesse [sauȝtnesse] C. L. 474 ; sahtnesse, sæht-

nesse, sehtnesse LAȝ. 2809, 8262, 30137 ; seihtnesse A. R. 120.

sahte, sb., *O.N.* sātt, sætt ; *reconciliation, peace,* A. R. 258* ; sæhte [sahte] LAȝ. 9844 ; sæhte SAX. CHR. 261 ; seihte A. R. 250 ; saughte D. ARTH. 1007.

sahten, v., *cf. O.E.* sehtian, *O.N.* sætta ; *reconcile, make peace* ; sauhten C. L. 546 ; **sauȝte** LANGL. *A* iv. 2* ; sauhte [sauȝte, saughte] GAM. 150 ; saute AN. LIT. 8 ; seite (*r. w.* eiȝte) ROB. 533.

sahtlen, v., *O.E.* sahtlian ; *reconcile, make peace,* ORM. 351 ; saghtlin, sauȝtle LANGL. *A* iv. 2* ; **saghtle** (*pres.*) PR. C. 1470 ; sauȝtle DEGR. 1757 ; **saȝtill** [saghtill] (*imper.*) ALEX. (Sk.) 865 ; **sahtleden,** sæhtleden (*pret.*) SAX. CHR. 264, 265 ; **sahtled** (*pple.*) ORM. 7976 ; saȝtled A. P. ii. 1139 ; saghteld Iw. 3952.

saughtling, sb., *reconciliation,* MAN. (F.) 3256.

ɛahtnien, v., *reconcile, make peace,* HOM. I. 39 ; sauȝtne LANGL. *A* iv. 2 ; sæhtnien [sehtne] LAȝ. 8776 ; seihtni [sachtni] A. R. 28 ; *comp.* **ȝe-sahtnien.**

saie, sb., *O.Fr.* saie ; *silk, serge,* WICL. EX. xxvi. 9 ; sai ALEX. (Sk.) 4600.

saien, v., *for* **assaien** ; *try* ; **saie** (*imper.*) YORK xxx. 99 ; **saied** (*pret.*) MAN. (F.) 823.

saien, *see* **seggen. saiff,** *see* **sauven.**

sail, *see* **seil.**

sailen, v., *for* **asailen** ; *assail* ; **sailede** (*pret.*) ROB. (W.) 382* ; (þai) sailid ALEX. (Sk.) 5559.

sailen, *see* **saliin.**

saim, sb., *cf. Ital.* saime, *O.Fr.* sain (*graisse*) ; *fat, lard,* HALLIW. 702 ; seim A. R. 412.

sain, sb., *? O.E.* segen, sægen (*? saying, speech*) : leve we þis for sain (*ms.* sayn) *? for truth* MISC. 91.

saine, *see* **seine.**

saint, seint, adj. & sb., *O.Fr.* saint ; *holy, saint,* AYENB. 13 ; seint LAȝ. 32 ; P. L. S. ix. 1 ; sant TOWNL. 101 ; sent Austin MAN. (F.) 15403 ; sanct BARB. i. 353 ; v. 336 ; seinte, seint, sein ROB. (W.) 5614, 9848, 10180 ; **sontes** (*pl.*) SPEC. 96 ; þe seintes legende of Cupide CH. C. T. *B* 61.

saint, *see* **samit.**

sainten, v., *Fr.* saintir ; *make a saint of* isonted (*pple.*) A. R. 350 ; sanctit BARB. xvii. 286, 875.

saintuarie, sb., *O.Fr.* saintuaire ; *sanctuary,* seintuarie WICL. EX. xv. 17 ; CH. C. T. *C* 953 ; seintewaire LANGL. *C* vi. 79 ; saintware ALEX. (Sk.) 1567.

saisen, v., *O.Fr.* saisir ; *seize, hold, take possession of* ; saise HAV. 251 ; seise LANGL. *B* xiii. 375 ; seise [sese] MAN. (F.) 7142 ; seisi BEK. 699 ; E. G. 362 ; sese WILL. 5391 ; sess BARB. x. 108 ; **sésand** (*pple.*) x. 774 ; **séside** (*pret.*) D. ARTH. 3067 ; sesed GAW. 822 ; sesit

BARB. vi. 447; xiv. 130; **seised** (*pple.*) HAV. 2513.

seising, sb., *seizing, conquest*, ALEX. (Sk.) 4396; sesing BARB. vi. 496.

saisine, sb., *O.Fr.* saisine; *seizin*, AYENB. 144; seisine ROB. 314; C. L. 237; sesine D. ARTH. 3588.

[**sáke**, sb., *O.E.* (ge-)saca, = *O.H.G.* sacho, *from* **sáken**; *comp.* **wiþer-sáke**.]

sáke², sb., *O.E.* sacu,=*O.L.G.* saca, *O.H.G.* sacha,*O.N.* sök; *sake; cause; litigation; injury*; FRAG. 7; GEN. & EX. 1392; TRIST. 2138; MAN. (H.) 135; A. P. i. 799; M. H. 130; þat he wid Romleode summe sake arerde LA3. 26290; sake & sinne ORM. 1335; cheste and sake O. & N. 1160; wiþute sake 1430; for hire sake A. R. 4; in hare senvolle sake SHOR. 66; withouten sake ['*sine causa*'] PS. iii. 8; saca & socne ANGL. VII. 220; sak YORK xii. 195; **sáken** (*pl.*) him weoren laˇe LA3. 31327.

sac-ful, adj., *O.E.* sacfull; *guilty*, HOM. I. 109.

sac-lǣs, adj., *O.E.* saclēas,=*O.N.* saklauss; *innocent*; (*ms.* sacclæs) ORM. 1900; sacles GEN. & EX. 916; sakles IW. 2526; MAN. (H.) 182; MIR. PL. 146; **saclēse** (*pl.*) HOM. II. 171.

[**sáken**, v., *O.E.* sacan, = *O.L.G.* sacan (*fight, chide*), *Goth.* sakan (μάχεσθαι), *O.H.G.* sahhan (*litigate, chide*); *comp.* **æt-, for-, of-, wið-sáken**.]

sal, sb., *O.E.* sæl (*gen.* sales)=*O.H.G.* sal *n.*, *hall, vestibule; tent*; in 3our sall YORK xxxiii. 87; in his sale ALEX. (Sk.) 502; **sále** (*dat.*) HORN (L.) 1107; SPEC. 26; AM. & AMIL. 444; GAW. 197; A. P. ii. 107; PERC. 1586; L. C. C. 10; **sáles** (*pl.*) ALEX. (Sk.) 4016.

sal, *see* **schal**.

sāl, sb., *O.E.* sāl,=*O.L.G.* sēl, *M.Du.* seel, *O.N., O.H.G.* seil *n.*; *rope*; sol MISC. 151; sool PR. P. 463.

salōde, sb., *? O.Fr.* salade; *helmet*, DREAM 1554.

salamandre, sb., *Fr.* salamandre; *salamander*, AYENB. 167.

salarie, sb., *O.Fr.* salaire; *salary*, LANGL. B v. 433; E. W. 31; salleri DEP. R. iv. 46.

sále, *see* **sáwle**.

sále, sb., *O.E.* sala, *cf. O.N., O.H.G.* sala; *sale, 'venditio,'* PR. P. 440; W. & I. 31; or set hem up to ani sale PL. T. iii. 63 (P. P. 340).

salere, sb., *O.Fr.* saliere; *saltcellar*, B. B. 7; salure GAW. 886.

salfe, sb., *O.E.* sealf,=*O.H.G.* salba; *salve, ointment*, ORM. 6477; þeos sealfe MK. xiv. 5; salvè A. R. 124; O. & N. 888; P. R. L. P. 173; CH. TRO. iv. 944; sealve SHOR. 2; *comp.* **ē3e-salfe**.

salfen, v., *O.E.* sealfian, = *Goth., O.H.G.* salbōn; *salve, anoint*, ORM. 9427; salve

(*imp.*) MARH. 5; **salvede** (*pret.*) LANGL. B xvi. 109; isalved (*pple.*) A. R. 274.

salving, sb., *salving*, YORK x. 334.

salhe, sb., *O.E.* sealh,=*O.H.G.* salha, salaha, *Lat.* salix, *Gr.* ἕλιξ; *sallow (tree), willow*; saluhe (*ms.* salwhe) '*salix*' PR. P. 441; salghe CATH. 317; sali PALL. xii. 139; **salyhes** (*pl.*) PS. cxxxvi. 2*; salwes CH. C. T. *D* 655; salewis WICL. LEV. xxiii. 40.

saliin, v., *O.Fr.* saillir; *dance*, PR. P. 441; saille LANGL. B xiii. 233; **sailede** (*pret.*) ROB. (W.) 5633.

salm, *see* **psalm**.

salmari, sb., *med.Lat.* salmāria; *baggage*, ALEX. (Sk.) 126.

salmen, *see* **psalmen**.

salmōn, sb., *O.Fr.* saumon; *salmon*, VOC. 189; salmond BARB. xix. 664; saumoun E. G. 354; **samoun** [samouns] (*pl.*) TREV. I. 369, 467.

salpetre, sb., *Fr.* salpetre; *saltpetre*, CHR. E. 183; CH. C. T. *G* 808.

sals, *see* **sause**.

salt, sb., *O.E.* sealt, salt,=*O.L.G., O.N., Goth.* salt, *O.H.G.* salz; *salt*, MAT. v. 13; A. R. 138; ORM. 1002; salt saveth catel LANGL. B xv. 421; Iw. 2047; þet zalt AYENB. 242; **saltes** (*gen.*) ORM. 1653; **sealte** (*dat.*) SHOR. 9.

salt-cote, sb., *saltcellar, 'salina,'* PR. P. 441.

salt², adj., *O.E.* sealt,=*O.Fris.* salt, *O.N.* saltr; *salt, salted, 'salsus,'* PR. P. 441; salt water P. L. S. viii. 126; þe salte se HAV. 1305; **sealte** (*dat.*) SHOR. 9; **saltne** (*acc. m.*) LA3. 6116; **salte** (*pl.*) ORM. 13849; MAND. 156; salte teres CH. C. T. *A* 1280.

salt, *see* **saut**.

salten, silten, v., *O.N.* salta, *O.E.* syltan; *season with salt*; saltin '*salio*' PR. P. 441; salte MAND. 149; **salte**, selte, silte (*pret.*) L. H. R. 58, 59.

salter, sb., *O.E.* sealtere; *salter, salt dealer*; '*salinator,*' VOC. 213; saltare PR. P. 441.

salŭen, v., *Fr.* saluer; *salute*; **salŭeþ** (*pres.*) CH. C. T. *B* 1284; salus YORK xxii. 184; **salŭede** (*pret.*) WILL. 4017; salusit BARB. iv. 509.

salŭing, sb., *salutation*, CH. C. T. *A* 1649.

saluh, salou, adj., *O.E.* salu,=*O.N.* sölr, *M. Du.* saluwe, *O.H.G.* salo; *sallow*, PR. P. 441.

salŭs, sb., *O.Fr.* salut; *salutation*, ALEX. (Sk.) 4647; **salŭtis** (*pl.*) 3088.

salvacioun, *see* **sauvaciūn**. **salve**, *see* **salfe**.

salven, *see* **salfen, sauven**.

salwe, *see* **salhe**.

sam, conj., *O.E.* sam; *whether; or*: sam ... sam HOM. II. 107.

[**sam-**, pref., *O.E.* sam-,=*O.N.* sam-, *? Gr.* ἅμα.]

[**sam-rǣd**, adj., *O.E.* samrǣd, *O.N.* sem-rāða; *harmonious.*]

somrēdnesse, sb., *O.E.* samrǣdness; *concord*, A. R. 254.

som-rūne, sb., *colloquy*, LAȝ. 5479.

sam-tal, adj., *agreed*: to ben **samtále** (*pl.*) & sahte ORM. 1535.

[**-sam**, *see* -sum.]

[**sām**, adj., *O.E.* sām-,= *O.L.G.* sām-, *O.H.G.* sāmi-, *Lat.* sēmi-, *Gr.* ἡμι- ; *half-*.]

sām-cweoc, adj., *half alive*, LK. x. 30.

sām-dēd, adj., *half dead*, ROB. 163.

sām-rēde, adj., *half red, half ripe*, LANGL. *C* ix. 311.

sām-rīpe, adj., *half ripe*, LANGL. *C* ix. 311*.

sambue, sb., *O.Fr.* sambue (*housse*), *O.H.G.* sambūch ; *housings*; sambu of silk ALIS. 176 (*see page* 373) ; **saumbues** (*pl.*) of the same threde M. ARTH. 2360 ; sambus LAUNF. 950 ; sambutes (*for* sambuces) ANT. ARTH. ii.

samburi, sb., *? from* sambue ; *litter*; saumburi LANGL. *C* iii. 178.

sáme, adj., *O.N.* samr, *cf.* Goth. sama, *O.H.G.* sama, *? Gr.* ὁμός ; *same*, P. L. S. xxv. 178 ; LANGL. *B* iii. 54 ; þe same seg WILL. 3435 ; þe same niȝt TREV. III. 131 ; þat same lond MAN. (H.) 136 ; of þe same tree MAND. 11 ; of þa same stanes ORM. 9914 ; some REL. II. 281 ; in the **sámin** (*dat.*) tim BARB. i. 252 ; on the sammin viss BARB. vii. 140 ; samine x. 563 *.

sáme, *see* **scháme**.

samed, adv., *O.E.* samod, somod, samed, somed,= *O.L.G.* samad, samod, samed, *Goth.* samaþ, *?O.H.G.* samant ; *together, once* ; þe ba somed læsinge speken LAȝ. 3068 ; gulcheð al ut somed A. R. 88 ; þet halt þe gode somed 254 ; somet MARH. 15.

samen, adv., *O.N.* saman,= *O.L.G.*, *O.H.G.* saman, *Goth.* samana ; *together, in company*, GEN. & EX. 40 ; HAV. 467 ; WILL. 1288 ; IW. 1417 ; MAN. (H.) 88 ; A. P. ii. 400 ; þeȝ baþe samen cwemden god ORM. 377 ; bunden samen C. M. 8778 ; sam, same, samin YORK viii. 126, xiii. 301, x. 235 ; same DEGR. 1396 ; somen H. M. 43 ; AV. ARTH. xxvii. ; ha somen (*v. r.* somet) seiden KATH. 532 ; sammin BARB. v. 72 ; vii. 513 ; x. 257 ; saṃin BARB. ii. 239.

samen-tale, adj., *harmonious*, C. M. 683 ; somentale A. R. 426 *.

samit, sb., *O.Fr.* samit ; *a rich silk-stuff*, S. A. L. 215 ; samet [samite] R. R. 873 ; B. DISC. 833 ; saint (*? ms.* saynt) GAW. 2431.

samne? in phrases; **at somne** [*O.E.* æt samne (somne), = *O.L.G.* at samne (samna)]; *at one*, *together*, LAȝ. 24146 ; **to somne**, tó sumne [*Ō.E.* to somne,= *O.L.G.* te samne, *O.H.G.* ze samana] *together* LAȝ. 61, 1393 ; to samen ORM. 649 ; to same HOM. II. 23 ; **in same** *in common, together* RICH. 4386 ; B. DISC.

2097 ; P. R. L. P. 141 ; i same FER. 1188 ; i some ROB. 40 ; C. L. 1418.

samnien, v., *O.E.* samnian, somnian,= *O. Fris.* samna, samena, *O.L.G.* samnōn, *O.H.G.* samanōn ; *gather up, congregate, call together, join together* ; samnen ORM. 3285 ; samne MISC. 105 ; A. P. ii. 53 ; somnien, sumnien LAȝ. 19183, 30628 ; somnen MISC. 104 ; sompnin MARH. 15 ; sumni (? = somoni) ROB. 181 ; **samme** (*imper.*) YORK xliv. 87 ; **same-nand** (*pple.*) PS. xxxii. 7 ; **somnede** (*pret.*) LAȝ. 6398 ; BEK. 1879 ; samned MAN. (H.) 3 ; somnedest HOM. I. 209 ; **samned** (*pple.*) AM. & AMIL. 415 ; sammed YORK xxxiv. 43 ; samened HAV. 2890 ; *comp.* ȝe-**samnien**.

samninge, sb., *O.E.* samnung, somnung,= *O.H.G.* samanunga ; *meeting, assembly, congregation*, HOM. II. 215 ; REL. I. 23 ; samning sampninge MAN. (F.) 3464, 6718 ; samening GEN. & EX. 458 ; somnunge H. M. 9.

sample, sb., *for* asample ; *sample, example* ; saumple LANGL. *C* xii. 288*.; sampill ALEX. (Sk.) 5306.

samplēre, sb., *sampler*, PARTEN. 2947 ; saumpler WICL. DEUT. xvii. 18 ; saumplarie *instructor* LANGL. *C* xv. 47.

sanct, *see* **saint**.

san, prep., *O.Fr.* san, sen ; *without* : saun ANGL. I. 304 ; GREG. 575 ; saunez PALL. iii. 1122 ; saun faile [san faille, samfaile] ROB. (W.) 8360.

sanctifīen, v., *O.Fr.* sanctifier ; *consecrate* ; **sanctified** (*pple.*) GOW. III. 234.

sand, sb., *O.E.* sand, sond, *n.*, *cf. O.L.G.* sand, *O.N.* sandr, *O.H.G.* sant, *m.; sand, land*, ORM. 14802 ; sond A. R. 402 ; GEN. & EX. 2718 ; HAV. 708 ; FER. 997 ; þat sond LAȝ. 123 ; SAINTS (Ld.) xlv. 162; be sand & be wattir ALEX. (Sk.) 4299 ; **sonde** (*dat.*) CH. C. T. *B* 509 ; *comp.* **sē-sond**.

sand-chisel, sb., *O.E.* sandceosel ; *sand*, ['*arenam*'] MAT. vii. 26.

Sand-wich, pr. n., *O.E.* Sandwīc ; *Sand-wich*, ROB. 300.

sandale, sb., *sandal*; **sandalies** (*pl.*) WICL. MK. vi. 9.

sande, sb., *O.E.* sand, sond,= *M.L.G.* sande *f.* (*mission*) ; *mission, message, messenger; dish at table* ; M. H. 8 ; ðis sonde hem over-takeð raðe GEN. & EX. 2313 ; he sent his sande fra toune to toune ISUM. 689 ; sand TRIST. 2351 ; saande, seande YORK x. 244 ; xiii. 235 ; sande, sonde LAȝ. 3125, 4971 ; sonde A. R. 190 ; H. H. 150 ; LANGL. *B* iii. 349 ; S. S. (Wr.) 340 ; A. P. i. 942 ; sone se hire sonde com aȝain KATH. 153 ; ai welcome be þi sonde CH. C. T. *B* 826 ; sonde '*daps*' FRAG. 4 ; sond '*missio*' PR. P. 464 ; gon on **sond** IPOM. 2283 ; of ever ilc sonde . . . most **and** best he gaf Benjamin GEN. & EX. 2295 ; **sondes** (*gen.*) mon *messenger* LAȝ. 13595 ; A.

R. 190; KATH. 518; sandis man, sandis men,
sendes men ALEX. (Sk.) 4234, 2399, 1170*;
sander man (*for* sandes man) ORM. 19383;
sander men SAX. CHR. 249; sandir men C. M.
21408; sonder man GEN. & EX. 1410; sander
bode HOM. II. 89; **sonden** (*pl.*) LA3. 4651;
A. R. 246; hwer beoð þine disches mid þine
swete sonde R. S. v; of alle his sonde PR. C.
3535; *comp.* on-sande.

sond-mon, sb., *messenger,* LA3. 12747.

sandi, adj., *O.E.* sandig, = *O.N.* söndugr,
sandy; sondi TREV. I. 333; MAND. 34.

sandiver, sb., *Fr.* suin de verre; *glass-gall*;
saundiver A. P. ii. 1036.

sang, sb., *O.E.* sang, song,= *O.H.G.* sang,
O.N. söngr, *Goth.* saggws; *from* **singen**;
song; ORM. 3923; REL. II. 193; IW. 396;
PR. C. 9254; MIR. PL. 151; sange ALEX.
(Sk.) 253; zang AYENB. 68; song LA3. 22701;
MARH. 2; **songes** (*gen.*) O. & N. 196; **songe**
(*dat.*) O. & N. 46; on songe no on spelle LA3.
12093; O. & N. 220; ðesne song HOM. I. 199;
zonge AYENB. 105; **songes,** songe (*pl.*) LA3.
5109, 9539; **songen** (*dat. pl.*) LA3. 19575;
songe O. & N. 82; *comp.* ēven-, lof-, morwe-,
uht-sang.

song-bōc, sb., *O.E.* sangbōc; *song-book*;
song bōkes (*pl.*) FRAG. 1.

songere, sb., *O.E.* sangere,= *O.H.G.* san-
gāri, *O.N.* söngari; *singer,* HOM. II. 117.

sangvin, adj., *O.Fr.* sangvine; *blood red*: of
his complexion he was sangvin CH. C. T. *A* 333.

sanke, v., *O.N.* sanka, samka; *assemble,* C.
M. 13840.

sannen, v., *O.N.* sanna (*from* sannr *sooth*);
demonstrate, ORM. 11289.

sant, *see* saint.

santrelle, sb., *stroller* (*term of reproach*): to
take Jesus, þat sauntrelle YORK xxviii. 190;
sauterell xxxi. 310, xxxii. 91, 274.

santren, v., *? mod.Eng.* saunter; *? hesitate*;
santred (*pret.*) and doubted PARTEN. 4653.

sauntering, sb.; *sauntering,* YORK xxxv.
70, 150.

sap, sb., *O.E* sæpp,= *M.L.G.*, *M.Du.* sap, *O.H.*
G. saph, saf; *sap,* P. P. 218; þet zep AYENB. 96.

sāpe, sb., *O.E.* sāpe,= *M.Du.* sēpe, *O.H.G.*
seipha; *soap,* HOM. I. 53; sqpe A. R. 66;
ROB. 6; LANGL. *B* xiv. 6.

sāper, sb., *soap dealer*; sōpers (*pl.*) LANGL.
C vi. 72.

saphir, sb., *Fr.* saphir; *sapphire,* MISC. 96;
SPEC. 25; P. L. S. xxxv. 89; safir AYENB. 82;
safferes [saphirs, saphires] (*pl.*) LANGL. *B* ii.
13.

sapience, sb., *O.Fr.* sapience, *Lat.* sapientia;
wisdom, CH. C. T. *G* 101; ALEX. (Sk.) 1022;
sapiences (*pl.*) CH. C. T. *G* 338.

sappi, adj.,= *M.Du.* sappigh; *sappy*; (*ms.*
sapy) PR. P. 441.

sār, sb., *O.E.* sār= *O.N.* sār, *Goth.* sair, *O.L.*
G., *O.H.G.* sēr; *sore, grief; wound; disease*;
HOM. I. 121; sar, sær LA3. 7998, 8477;
nowðer sar ne sorhe KATH. 1170; sor A. R.
376; O. & N. 1234; HAV. 234; P. S. 159;
ALIS. 4397; A. P. i. 130; þat sor P. L. S. x.
56; sor and blein GEN. & EX. 3027; **sāre**
(*dat.*) LA3. 12511; sore WILL. 891; D. ARTH.
932; þare sare ['*dolorum*'] anginne MAT.
xxiv. 8; *comp.* ēie-sōr.

sār[2], adj., *O.E.* sār,= *O.N.* sārr, *O.L.G.* sēr, *O.*
H.G. sēr; *sore, grievous, sad,* H. M. 35;
sar, sær LEECHD. III. 108; þe king was on
mode sar [sor] LA3. 638; þer fore is min
herte sær 7289; sor A. R. 208; sek and ser
MAP 338; **sāre** (*adv.*) MARH. 17; ORM.
3809; PR. C. 7402; PERC. 1114; sare [sore]
þu hit salt abuggen LA3. 8158; wepen alle
swiþe sare HAV. 401; him hungrede swiþe
sore 654; sore FRAG. 8; O. & N. 885; GEN.
& EX. 1166; C. L. 314; WILL. 593; CH. C.
T. *A* 148; sere ST. COD. 96; of hire sore
mode O. & N. 1595; **sārne** (*acc. m.*) LA3. 10423;
his leomes þat beoð sare LA3. 19501; woundes
sore SPEC. 68; **sārure** (*compar.*) LA3. 149;
sarre A. R. 112; H. M. 27; MIRC 1565;
WILL. 2025; zorer AYENB. 238; sararre
YORK xi. 160; **sōrest** (*superl.*) A. R. 382;
HOM. II. 173; sarrest JOS. 620.

sār-līc [sōrlīch], adj., *O.E.* sārlīc; *sore,*
LA3. 28457; **sōrlīche** (*adv.*) FRAG. 5.

Sarasene, sb., *O.Fr.* Sarazin; *Saracen, heathen,*
LANGL. *B* xi. 151; **Saracenes,** Sarasenes (*pl.*)
B iii. 325, xi. 115; Sarazins HORN (L.) 68, 607.

sardein, sb., *? O.Fr.* sardine; *sardine*; **sar-
deines** (*pl.*) C. B. 24.

sardoine, sb., *O.Fr.* sardoine; *sardonyx*;
sardoines (*pl.*) and ... calsidoines FL. & BL.
285 (H. 700).

Sares-büri, pr. n., *O.E.* Searobyrig; *Salis-
bury,* MISC. 145; Salesburi LA3. 15290.

sarge, sb., *O.Fr.* sarge; *serge,* WICL. EX.
xxvi. 9; CH. C. T. *A* 2568.

sāri, adj., *O.E.* sārig, *cf. O.H.G.* sērag; *sorry,*
miserable, ORM. 8945; sari HOM. I. 81;
MARH. 17; PR. C. 3468; PERC. 158; sari,
særi [sori] LA3. 166, 1476; sori '*tristis*' PR.
P. 465; A. R. 88; O. & N. 994; GEN. & EX.
974; HAV. 477; ROB. 52; P. S. 203; LANGL.
B x. 75; a sori þou3t WILL. 3696; ? seri
GEN. & EX. 408; **sārine** (*acc. m.*) HOM. I.
111; **sōrie** (*pl.*) A. R. 32; **sārier** (*compar.*)
IW. 2126; soriure A. R. 310; soriere P. L. S.
xxiii. 105; WICL. GEN. xl. 7; **sær3est** (*superl.*) LA3. 28459.

sāri-līche, adv., *sorrowfully,* MARH. 11;
sarili IW. 431; soriliche FRAG. 6; A. R. 224.

sāri-mōd, adj., *O.E.* sārigmōd; *sorry of*
mood, LA3. 29791; sorimod O. & N. 1218;
GEN. & EX. 3520.

sārinesse, sb., *O.E.* sārigness; *sorrow,*
LA3. 27560; sorinesse MISC. 76; HORN (L.)

922 ; sorinisse BEK. 1943 ; **sārinesse** (*pl.*) HOM. I. 103.

sārigen, v., *O.E.* sārgian ; *grieve,* MK. xiv. 33.

sarmōn, *see* **sermōn. sarp,** *see* **scharp.**

sarpe, sb., *O.Fr.* sarpe (*serpe*) ; *pruning hook* ; WICL. I KINGS xiii. 20.

sarpelēre, sb., *O.Fr.* serpeliere; *woolpack, bag,*LIDG. M. P. 204; **sarplērs** (*pl.*) FER. 4371.

sars, sb., ?*O.Fr.* saas (*sas*) ; *sieve* ; sarce PR. P. 441 ; CATH. 318 ; F. C. 67 ; W. & I. 82.

sarsin, v., *O.Fr.* sasser ; *strain* ; **sarce** (*imper.*) it smothe PALL. xi. 414; sarse it thrugh a sarce CATH. 318*.

sat, *see* **schat.**

Satanas, pr. n., *Lat.* Satanas ; *Satan,* D. ARTH. 3812.

Satern, pr. n., *Saturn* ; **Saternes** (*gen.*) daig [*O.E.* Sæternes dæg], *Saturday,* MK. xvi. I ; Sateres dai LAȝ. 13933*.

Sater-daȝ, pr. n., *O.E.* Sæterdæg, = *M.Du.* Saterdag ; *Saturday,* ORM. 4350 ; Saterdai P. L. S. ix. 199 ; Sater-, Seterdai LANGL. *A* v. 14; Sætterdæi LAȝ. 13933.

Sater-niȝt, pr. n., *night before Saturday,* ROB. 557.

satin, sb., *Fr.* satin ; *satin* ; **satins** (*pl.*) CH. C. T. *B* 137.

satisfac(c)iōn, sb., *Fr.* satisfaction ; *satisfaction,* AYENB. 32 ; satisfaccioun LANGL. *B* xiv. 94.

sattlen, *see* **setlen.**

sauf, adj., *Fr.* sauf ; *safe,* AYENB. 36 ; WILL. 868 ; sauf, sauve BEK. 434, 435 ; saf LANGL. *C* xv. 112 ; saaf WICL. MK. v. 23 ; save ROB. 54.

sauf-līche, adv., *safely,* WILL. 3051 ; saveli YORK xxxviii. 307 ; saufli BARB. x. 484 ; savelich TREV. VII. 533 ; **sáveloker** (*comp.*) TREV. IV. 163.

sauge, sb., *O.Fr.* sauge ; *sage* (*herb*), '*salvia,*' PR. P. 441 ; SPEC. 26 ; SACR. 585 ; *see* **sáve.**

saughe, *see* **saht, sēon.**

sauht, *see* **saht. sāule,** *see* **sāwle.**

saulee, *see* **suvel. saumburi,** *see* **samburi.**

saumoun, *see* **salmōn. saunez,** *see* **san.**

sause, sb., *O.Fr.* sause, sauce ; *sauce,* AYENB. 55 ; sause [sauce] CH. C. T. *B* 4024 ; sauce LANGL. *A* vii. 249 ; sals BARB. iii. 540.

saut(e), sb., *for* asaut ; *assault* ; ROB. (W.) 11869 ; HAMP. PS. xii. 4* ; ciii. 23* ; saute TREV. IV. 429 ; ALEX. (Sk.) 2221 ; **saltis** (*pl.*) BARB. xviii. 68.

sauten, v., *for* asauten ; *assault* (*a city*) ; **sauted** (*pret.*) TREV. VIII. 552.

sauter, *see* **psalter.**

sauterien, *see* **psalterien.**

sauturoure, sb., *O.Fr.* saultoir ; *saltire* (*in heraldry*), GAW. 4182.

sauvaciūn, sb., *O.Fr.* sauvatiōn ; *salvation,* A. R. 242 ; sauvacion SHOR. 29 ; salvacioun CH. COMPL. M. 213.

sauvāge, adj. & sb., *O.Fr.* sauvāge ; *savage* ; leounes sauvage ST. COD. 94 ; a colt savage ALIS. 767 ; liouns savage A. D. 237 ; **savāgius** (*pl.*) ALEX. (Sk.) 3914.

sauven, v., *O.Fr.* sauver, salver ; *save,* A. R. 98 ; sauven [salvin] KATH. 1025 ; salven HOM. I. 202 ; saven C. L. 554 ; sauvi ROB. (W.) 1260 ; sovi AYENB. 98 ; saiff YORK iv. 12 ; **sáveþ** (*pres.*) SPEC. 26.

sáfing, sb., *salvation* ; saffing YORK xiv. 103.

sauveour, sb., *O.Fr.* sauveour ; *saviour,* SHOR. 136 ; saveour MAN. (F.) 5358 ; saviour LANGL. *C* viii. 121.

sáve, sb., *Lat.* salvia ; *sage* (*herb*), CH. C. T. *A* 2713 ; *see* **sauge.**

sáve², prep. & conj., *O.Fr.* sauf *from Lat.* salvus ; *save, except,* CH. C. T. *G* 1355 ; LANGL. *A* PROL. 77, ii. 210, vii. 24.

saveine, sb., *O.E.* safine; *savine* (*shrub*), GOW. III. 130.

savel, sb., *O. Fr.* sable ; *sand* ; **savelles** (*pl.*) PALL. i. 353.

sávetē, sb., *O.Fr.* sauveté ; *safety,* C. L. 354 ; saufte, savite BARB. iii. 183, iv. 536.

savour, sb., *O.Fr.* savour, savor ; *savour, smell,* SPEC. 87 ; CH. C. T. *D* 2196 ; A. P. ii. 510 ; PR. C. 656 ; savor WILL. 638 ; savur A. R. 102 ; saver PALL. i. 751.

saver-lī, adv., *tastily, carefully,* YORK xxix. 80 ; (he) kisses him ... saverli GAW. 1937 ; and hade ben sojourned saverli 2048.

savouren, v., *O.Fr.* savourer ; *savour* ; savoure CH. C. T. *D* 171 ; **savoreþ** [savoureth] (*pres.*) LANGL. *B* viii. 108 ; **severed** (*pret.*) TREV. VIII. 17 ; savourd ALEX. (Sk.) 4821 ; flouris weill savourit (*pple.*) BARB. xvi. 70.

savouri, adj., *savoury* ; **saveriour** (*comp.*) LANGL. *C* xix. 65.

sawe, sawen, *see* **saȝe, saȝen.**

sāwen, sāwe, v., *O.E.* sāwan, sæwan (ALFR. P. C. 427) (*pret.* sēow), = *O.L.G.* sājan, sēhan (*pret.* sēu, saida), *O.H.G.* sāan, sāhan, sāwan (*pret.* sāta), *O.Fris.* sēa [*pple.* (e-)sēn], *O.N.* sā (*pret.* sēri, sādi) *Goth.* saian, saijan (*pret.* saisō) ; *sow* (*seed*), HOM. I. 133, 135 ; zawe AYENB. 214 ; sowen GEN. & EX. 2347 ; sowen ROB. 496 ; sowen [sowe] LANGL. *A* v. 548 ; sowe O. & N. 1039 ; **sāwen** (*pres.*) ORM. 5071 ; sawen PS. cxxv. 5 ; sēow (*pret.*) HOM. I. 133 ; siew, sew II. 151 ; seu ROB. 470 ; MAP 347 ; M. H. 145 ; sew LANGL. *C* vii. 271 ; sewe *B* xiii. 375 ; seowen P. L. S. viii. 11 ; JUL. 75 ; seowen [sewen] LAȝ. 1941 ; siewe ANGL. I. 8 ; sewe ROB. 21 ; **sāwen** (*pple.*) PR. C. 445 ; sowen CH. C. T. *C* 375 ; *comp.* bi-, ȝe-**sāwen** ; *deriv.* **sǣd.**

sāwere, sb., *sower*, HOM. I. 133'; sawer TREV. VIII. 469.

sōwing, sb., *sowing*, LANGL. *A* PROL. 21.

sāwle, sb., *O.E.* sāwel, sāwul, sāwl, sāul,=*O.N.* sāla, *O.L.G.* sēola, *O.H.G.* sēla, *Goth.* saiwala; *soul, mind*, MARH. 3; KATH. 215; ORM. 1555; sawle A. P. ii. 1599; sawle [saule] MK. xiv. 34; LK. i. 46; saule LA3. 27634; SHOR. 41; FER. 318; ISUM. 733; ANT. ARTH. xvii; sowle, soule FRAG. 5; soule A. R. 10; GEN. & EX. 486, 2525; TREAT. 139; CH. C. T. *E* 1134; þe soule loveþ þe bodi so þat nevere heo nule him wende fro C. L. 1169; sale ALEX. (Sk.) 1640; sōwle (*gen.*) HOM. II. 27; saule HOM. II. 77; mi saule hele HICKES I. 167; oure soule leche A. D. 262; **sāule** (*pl.*) ANGL. I. 24; saule, saulen HOM. I. 119, 129; saulen LA3. 18320; MISC. 56; zaulen AYENB. 1; soule P. L. S. ix. 181; soulen A. R. 30; alle soulen dai *All Souls' Day* BEK. 1149; þam wrecche saule HOM. I. 41.

sāwle-healethe, sb., *healing, salvation, safety*, TREV. I. 365, 371, II. 253.

sāwlen, v., *cf. M.H.G.* gesēlen; *endow with a soul*; **sōuled** (*pple.*) CH. C. T. *G* 329.

sax [sex], sb., *O.E.* seax,=*O.N.* sax, *O.H.G.* sahs; *knife, dagger*, LA3. 15214; sex '*cultellus*' FRAG. 4; MAN. (F.) 7874; **sæxe** (*dat.*) LA3. 5034; sexes [seaxes] (*pl.*) LA3. 15252; saxes, sexes ROB. 125, 144; **sæxen** (*dat. pl.*) LA3. 16148; *comp.* flint-, hond-, næil-sex.

Saxe, pr. n., *O.E.* Seaxa; *Saxon*; **Saxes** (*pl.*) LA3. 15219.

Sex-lond [Saxlond], pr. n., *land of the Saxons*, LA3. 15083.

Sex-lēode, pr. n., *Saxon people*, LA3. 15137.

Sæx-þēode, pr. n., *Saxon people*, LA3. 14494.

Saxisc, adj., *Saxon*: a Sexisc wimmon LA3. 14143; Rouuenne (spæc) Saxisc [Saxisse] 14979.

Saxon, pr. n., *Fr.* Saxon; *Saxon*; **Saxons** (*pl.*) LA3. 1976.

sayen, *see* seggen.

scabbe, sb., *O.E.* sceabb,=*Swed.* skabb; *scab, sore*, '*scabies*,' PR. P. 442; scabbe [shab] WICL. LEV. xxii. 22; scab MISC. 31; scabbe (*dat.*) CH. C. T. *C* 358; **scabbes** (*pl.*) LANGL. *B* xx. 82*; shabbes P. S. 239.

scabbed, pple., *scabbed*, HAV. 2449; schabbid LEG. 119; schabbede (scabbide) schep LANGL. *A* viii. 17.

scaberge, sb., *scabbard, sheath;* '*fourrel*,' PARTEN. 2790; scauberk, scaubert '*vagina*', PR. P. 443; scaubert ROB. 273; CHR. E. 628; **scaberke** [scabarge] (*dat.*) TREV. V. 373.

scafald, sb., *O.Fr.* eschafaut; *scaffold, warlike engine*, CATH. 320; skaffold [scafold] CH. C. T. *A* 2533; **scaffaldis** [scaffalis] (*pl.*) BARB. xvii. 343.

scaft, *see* schaft.

scailen, *see* schailin.

scaived, adj., *? wild*; skaived GAW. 2167.

scal, *see* schal.

scalc, sb., *scalp*; skalke [skalk] HAMP. PS. lxvii. 23; (he) shal hew downe the **skalkis** (*pl.*) of sinful HAMP. PS. cxxviii. 4.

scáld, sb., *O.N.* skáld; *poet*; **scáldes** (*pl.*) ORM. 2192.

scalden, v., *O.Fr.* *escalder, eschauder; cf. Ital.* scaldare, *Lat.* excaldāre; *scald*, HICKES I. 229; PR. C. 6576; **scoldeþ** (*pres.*) AYENB. 66; **schaldinde** (*pple.*) A. R. 246; scaldand PS. lxxxii. 10; **scalded** (*pret. pple.*) C. M. 15988; iscalded CH. C. T. *A* 2020.

scále, scāle, sb., *O.E.* scealu (*husk*), *O.N.* skál (*cup, scale-pan*), *cf. M.L.G., M.Du.* schale (? schāle), *O. H. G.* scala (? scāla) (*bowl, shell*),=*M. H. G.* schale, schāle, *mod. Eng.* shale; *scale; shell; bowl;* '*sqvama*,' PR. P. 442; REL. I. 7; ane scale of rede golde LA3. 5368; scale [shale] LANGL. *C* xiii. 145; schale of a not '*testula*' PR. P. 443; L. C. C. 30; scole '*lanx*' PR. P. 449; a bolle oþer a scole A. P. ii. 1145; a disch ine his one hond & a scoale [skale, schale] in his oþer A. R. 214; milc wes i þere scale [scole] LA3. 1182; in cupp(e) and schal(e) M. H. 120; **scálen** (*pl.*) LA3. 21327; scales GOW. I. 275; *comp.* hnute-, wæ3e-scale.

scále [schāle], sb., *O.N.* skáli; *shanty*, C. M. 8592.

scálen, *see* schailen.

scálin, schálin, v., *shell, peel*; '*exqvamo*,' PR. P. 442, 443; þe peces faste gunne schali FER. 3282.

scalle, sb., *cf. O.N.* skalli (*skull*); *scab*, '*glabra*,' VOC. 179; '*s(q)vama*' 222; '*glabra*' PR. P. 442; CH. ADAM 3; to kepe his heved . . . fro þe scalle HALLIW. 959; scalle [skalle] C. M. 11819; scalle PALL. vi. 138.

scallid, pple., *scalled;* '*glabrosus*,' PR. P. 442; scalled browes CH. C. T. *A* 627; þe scallede AYENB. 224.

scalōn, sb., *onion, shallot*; **scalōnes** (*pl.*) LANGL. *C* xi. 310; scalons (*printed* stalons) PALL. iv. 267.

scalop, sb., *O.Fr.* escalope; *scallop*, PR. P. 442; **skalopis** (*pl.*) D. ARTH. 3474.

scáme, *see* scháme.

scamoine [scamoiene], sb., *Lat.* scammonia; *scammony*, LA3. 17740.

scandle, sb., *O.Fr.* escandle, esclandre; *scandal, slander*, A. R. 12; schandle 380; shaundre TREV. III. 421; *see* sclaundre.

scant, adj., *scant*, '*parcus*,' PR. P. 442; skant (*adv.*) *scarcely* MIR. PL. 78; BARB. xx. 434*.

scantilōn, sb., *O.Fr.* eschantillon; *? mod. Eng.* scantling; *mason's rule*, C. M. 2231; scantlion '*mensura*' PR. P. 442; schauntillun FL. & BL. 325.

scápen, *see* **escápen.**

scaplori, sb., *Church Lat.* scapuláre; *scapular, sort of scarf*; CATH. 321 ; PR. P. 442 ; **scapeloris** (*pl.*) A. R. 424* ; chapolories PL. CR. 550.

scar, sb., *? O.Fr.* escar (*mépris, raillerie*); *mockery* : ure frenden to scare LAȝ. 5835.

scarlat, sb., *O.Fr.* escarlate ; *scarlet,* MISC. 22 ; LEG. 40 ; scarlet ROB. 313 ; LANGL. *A* ii. 13 ; scharlette D. ARTH. 3459.

scarle, *see* **skerrel.**

scarmichen, v., *O.Fr.* escarmoucier; *skirmish* ; scarmish PARTEN. 2079 ; skirmishe TREV. IV. 399.

   scarmeshing, sb., *skirmish,* CH. L. G. W. 1910.

scarmoche, sb., *Fr.* escarmouche ; *skirmish,* skarmysch [scarmich, scharmus] CH. TRO. ii. 611 ; skarmoch A. P. ii. 1186.

scarn, sb., *O.Fr.* escarn, escharn, eschern, *cf. O.H.G.* scern, *M.Du.* scherne ; *scorn, derision,* LAȝ. 17307 ; skarn ORM. 4402 ; scorn AYENB. 22 ; schorn [scarn] A. R. 108, 290 ; scorn '*derisio*' PR. P. 450 ; skorne MAND. 212 ; **scorne** (*dat.*) HOM. II. 169 ; schorne S. & C. II. xii.

   schorn-leihter, sb., *scornful laughter,* A. R. 344.

   scorn-líche, adv., *scornfully,* BEK. 710.

scarnen, v., *O.Fr.* escarnir, escharnir, eschernir, *cf. O.H.G.* scernōn, *M.Du.* schernen ; *scorn, deride* : scornin '*derideo*' PR. P. 450 ; scorni AYENB. 211 ; to scorne and to scolde LANGL. *B* ii. 81 ; **scarneð** (*ms.* scarned) (*pres.*) REL. I. 176 ; schorneð A. R. 248 ; **scornede** (*pret.*) BEK. 80 ; scorned MAN. (H.) 278 ; schorned WILL. 554 ; DEP. R. iii. 236 ; **skarned** (*pple.*) ORM. 7397 ; scorned MAND. 14.

   scorner, sb., *scorner*; LANGL. *B* xix. 279 ; scornere *C* xxii. 284 ; scornare '*derisor*' PR. P. 450.

   scærninge [scorninge], sb., *scorning, derision,* LAȝ. 2791 ; scornunge HOM. I. 207 ; schornunge A. R. 200 ; skorning [schorning] C. M. 18231.

scarp, *see* **scharp.**

scarre, sb., *O.Fr.* escarre ; *scar* ; '*rima,*' PR. P. 442 ; scar ['*cicatricem*'] WICL. LEV. xxii. 22 ; *see* **sker.**

scars, adj., *O.Fr.* escars ; *scarce,* BEK. 274 ; scars and chinche S. S. (Web.) 1244 ; scarse AYENB. 54.

   scars-líche, adv., *scarcely,* AYENB. 34 ; TREV. III. 171 ; scarseliche ROB. (W.) 10614.

   scarsnesse, sb., *scarceness,* AYENB. 159.

scarsitē, sb., *OFr.* escarsité ; *scarcity,* TREV. IV. 87 ; skarste TREV. III. 465 ; scarsete CH. C. T. *G* 1393.

[scarþ, sb., *O.N.* skarð ; = **scheard** ; *comp.* pot-scarþ.]

---

scáte, sb., *O.N.* skata ; *skate* (*fish*), '*sqvatus,*' PR. P. 443 ; '*ragadia, scatus*' CATH. 322 ; REL. I. 81 ; schate VOC. 254.

scaterin, v., = *M.Du.* scheteren; *scatter, shatter,* '*spargo,*' PR. P. 443 ; **schaterande** (*pple.*) GAW. 2083 ; **scatered** (*pret.*) SAX. CH. 261 ; skatered ['*dissipavit*'] PS. xvii. 15 ; **scatered** (*pple.*) CH. C. T. *G* 914 ; schaterid APOL. 81 ; ischatred AR. & MER. 553 ; *comp.* **to-scaterin.**

scáðe, sb., *O.E.* scaða, sceaða, *cf. O.L.G.* scatho, *O.N.* skaði, *O.H.G.* scado, *mod.Eng.* scaith ; *injury, wound, loss, odium ; enemy* ; REL. I. 221 ; GEN. & EX. 302 ; ne doð heo noht muchel scaðe [scaþe] LAȝ. 15784 ; þer wuneð þe scaðe (*sec. text* feond) inne þa (*r.* þe) scendeð þas leode 25691 ; scaþe HAV. 2006 ; S. A. L. 159 ; RICH. 5002 ; A. P. ii. 21 ; AVOW. A. xvi. ; hit is scaþe GAW. 674 ; scaþe, scathe '*damnum*' PR. P. 443 ; scaþe LANGL. *A* iv. 83 ; [skaþe *B* iv. 96] ; scathe [skaþe] CH. C. T. *A* 446 ; skaþe AN. LIT. 8 ; MAN. (F.) 5179; schaþe WILL. 3084 ; scaðɥ ['*latrones*'] MK. xv. 27 ; *comp.* **feond-, wáld-scaðe.**

   scáðe-dēde, sb., *harmful deed,* LAȝ. 29578.

   skáþe-lǣs, adj., *O.N.* skaðalauss ; *scathless,* ORM. 11356; scaþeles ANT. ARTH. xxxvii.

   skáþe-lī, adv., *with injury,* ALEX. (Sk.) 642.

   scáðe-werc, sb., *harm* ; (*sec. text* harm) LAȝ. 1547.

scapel, adj., = *Goth.* skapuls, *O.H.G.* scadel ; *harmful, noxious,* A. P. iii. 155 ; skathil D. TROY 4067 ; schathill [schatell] ALEX. (Sk.) 2992 ; skathill 4802.

[scáði, adj., *O.E.* scæðig ; *noxious* ; *comp.* **un-sháþig.**]

scáþin, v., *O.E.* sceaðian, = *O.N.* skaða, *O.H.G.* scadōn, *cf. Goth.* (ga-)skaþjan ; *harm, wound,* PR. P. 443 ; i schal scaþie hem FER. 759; skathe TOWNL. 213 ; **scáþed** (*pret.*) A. P. ii. 1776; skaþed DEP. R. ii. 105 ; **skáþed** (*pple.*) ORM. 4964.

scauten, v., *cf. O.N.* skota ; *push, thrust* ; **skautand** (*pple.*) ALEX. (Sk.) 4200.

sceaft, *see* **schaft. sceal,** *see* **schal.**

scēap, *see* **schēp. scéld.** *see* **schíld.**

scēnen, *see* **shǣnen.**

sceonien, *see* **schunien. sceort,** *see* **schort.**

sceptre [ceptre, septre], sb., *O.Fr.* sceptre, ceptre ; *sceptre,* CH. C. T. *B* 3334 ; septre PR. C. 4098 ; septer, septour ALEX. (Sk.) 502, 2324.

scēt, *see* **schēat.**

schabbe, *see* **scabbe. schád,** *see* **scheåd.**

scháde, schadewe, sb., *O.E.* sceadu *f.,* scead, scæd *n., cf. O.L.G.* scado *m.,* *M.L.G.* schede *m., n., O.H.G.* scato, *Goth.* skadus *m.*; *shade, shadow* : schade ROB. 107 ; P. L. S. i. 38 ; schade, schadue WILL. 22, 754 ; CH. C. T. *B*

7; schadwe HICKES I. 167; sceadewe HOM. I.
29; scheadewe A. R. 190; shadewe HOM. II.
29; schadewe MISC. 94; MAND. 16; scha-
dowe PR. P. 443; schadue ALIS. 2628; schadu
HORN CH. 526; ssed AYENB. 95.

schadowen, v., *O.E.* sceadwian, *cf. Goth.*
(ufar-)skadwjan, *O.L.G.* scadowan, *O.H.G.*
scatewan; *shadow, shade, 'umbro,'* PR. P.
443; ssedvi AYENB. 97; **schadeweþ** (*pres.*)
MAND. 157; **schadewede** (*pret.*) C. L. 875;
schadowed A. P. i. 42; *comp.* **bi-, üm-**
schadewen.

schǣd, *see* **scheád.**

schǣnen, v., *O.E.* scǣnan,=*O.H.G.* sceinan,
O.N. skeina; *from* **schīnen**; *break*: sceldes
gunnen scenen LAȝ. 31234; **scǣnden** (*pret.*)
LAȝ. 5186; *comp.* **to-schǣnen.**

schǣþe, sb., *O.E.* scǣð, sceáð,=*O.N.* skeið,
O.H.G. sceida; *sheath*; shǣþe ORM. 14675;
scǣðe, scaþe [seþe, seaþe] LAȝ. 8177, 23211;
scheþe ROB. 135; MAND. 86; CH. C. T. *B*
2066; schethe, schede PR. P. 444; schede
S. & C. I. lxi.; W. & I. 40.

schǣwen, *see* **scheáwen.**

schaft, sb.,=*O.E.* sceaft,=*O.L.G., O.H.G.* scaft,
O.N. skapt; *? from* **scheppen**; *shaft* (*arrow;
pole; spear-handle*), *'hastile,'* PR. P. 443;
B. DISC. 930; schaft [shaft] CH. C. T. *A*
1362; saft GEN. & EX. 3899; ssaft ROB.
419; scæft [scaft] LAȝ. 6494; schaft [scheft]
TREV. III. 449; **schafte** (*dat.*) PERC. 52;
schaftes (*pl.*) AMAD. (R.) lv; sceaftes, scaftes
[saftes] LAȝ. 4228; schafte FER. 1594.

　　schaft-monde, sb., *O.E.* sceaftmund; *palm,*
D. ARTH. 2546, 3843, 4232.

schafte, sb., *O.E.* sceaft,=*O.H.G.* (ga-)scaft,
Goth. (ga-)skafts; *from* **scheppen**; *creation,
creature,* KATH. 239; shafte LANGL. *B* xi.
387; safte GEN. & EX. 3628; schaft ['*figmen-
tum*'] PS. cii. 14; þulke schaft to underfonge
C. L. 661; **schafte** [shafte] (*pl.*) O. & N. 788;
god þat alle shafte (*ms.* shaffte) wrohte ORM.
PREF. 58; sceafte [scefte] P. L. S. viii. 42;
schefte SHOR. 35; **schefte** (*gen. pl.*) MARH.
11; *comp.* **frum-, ȝe-, neb-scheaft (-sceafte).**

schaȝe, sb., *O.E.* sceaga,=*Fris.* skage, *O.N.*
skagi　*m.* (*low cape*); *shaw, grove*;
schawe FLOR. 1504; **schaȝe** (*gen.*)
GAW. 2161; **schaȝe** (*dat.*) A. P. iii. 452;
schawe GOW. II. 45; DEGR. 1652; as goldfinch in þe schawe [shawe] CH. C. T. *A* 4367;
shawe REL. I. 244; **schawes** (*pl.*) ALIS. 6109;
WILL. 178; MIN. xi. 2; *comp.* **wude-scaȝe.**

schailin, scheilin, v., *cf. Swed.* skiäla (*go to
pieces*)·(RIETZ 588); *disperse, break up; 'dis-
gredi'* PR. P. 443; scale BARB. vi. 575; scaill
xvii. 99; **shailande** [skailand] (*pple.*) C. M.
18836; schailande *? straddling wide,* D. ARTH.
1098; **scaild,** skailed (*pret.*) C. M. 2524; scalit
BARB. vi. 428, ix. 429; **scálit** (*pple.*) vi. 28,
xvi. 211.

schak, sb., *shock; charge*; ALIS. 232; FER.
2663; schakke D. ARTH. 1759.

scháken, v., *O.E.* sceacan, scacan,=*O.L.G.*
*skacan (*pret.* skōc), *O.N.* skaka; *shake,
gallop, move*: he wole shaken his berd P. S.
324; schake ALIS. 4255; FER. 928; FLOR.
1978; þe maister him gan hom schake LEG.
41; shake SPEC. 110; schake, ssake ROB. 25,
218; ssake AYENB. 130; scake PR. C. 5410;
ich muhte ... scheken ham ... of me A. R.
344; Corineus com sceki (*ms.* scecky) LAȝ.
1536*; heo **schékeð** (*pres.*) hire spere A. R.
60; schek (*imper.*) A. R. 206; **sácande** (*pple.*)
MISC. 21; **schōk** [shook] (*pret.*) CH. C. T. *A*
2265; schok his heved S. S. (Web.) 1069; out
of þe sadil he schok PERC. 694; schok, ssok
ROB. 24, 208; nes þer nan biscop þat forð on
his wæi ne scoc LAȝ. 13246; schoke MAN.
(H.) 39; **scháke** [shake] (*pple.*) CH. C. T.
A 406; ischake LEG. 46; *comp.* **a-, at-, to-**
scháken (-scéken).

schakle, sb., *O.E.* sceacul, scacul (*pillory*),
M.Du. schakel (*ring*), *O.N.* skökull; *? from*
scháken; *shackle, 'numella,'* PR. P. 433;
scheakeles (*pl.*) A. R. 94.

schaklin, v., *shackle,* PR. P. 443.

schal, v., *O.E.* sceal, scel,=*O.L.G.* scal, *O.N.,
Goth.* skal, *O.H.G.* scal, scol, sal, sol; (*pret.-
pres.*); *shall, owe,* KATH. 403; C. L. 354;
MAND. 26; hit schal þunche þe swete A. R.
136; i schal [shal] slee ['*interficiam*'] WICL.
GEN. vi. 17; schal, shal, scal, sal O. & N. 346,
530, 960, 1199; shal (*ms.* shall) ORM. 155; it
shal be knowe GOW. I. 229; scheal, schel
SHOR. I, 108; scal, scæl, scel [sal] LAȝ. 1247,
5449, 5964; scal, sæl [sal] P. L. S. viii. 13,
45; (sal REL. I. 132; GEN. & EX. 12; MISC.
27; S. S. (Wr.) 268; MAN. (H.) 37; PR. C.
34; M. H. 11; sal, sel HICKES) I. 223; sel
ANGL. I. 8; voryef me þet ich þe ssel AYENB.
115; xal LUD. COV. 216; þu scalt HOM. I. 7;
schalt KATH. 516; þu schalt falle O. & N. 956;
þou schalt hit sone finde LANGL. *A* iv. 57;
shalt ORM. 157; xalt S. & C. I. i; schulen C. L.
57; WICL. GEN. vi. 20; ȝe schulen beon (*as
future*) A. R. 166; schulen ORM. 159; schullen PROCL. 2; sculen FRAG. 7; HOM. I. 7;
sculen, scullen [sollen] LAȝ. 702, 5417; sulen
GEN. & EX. 308; shule [schulle] O. & N. 1673;
schulle TREAT. 133; WILL. 3690; MAND. 6;
sulle ANGL. I. 8; sholen HAV. 621; schole
ALIS. 30; ssolle AYENB. 5; ssolle [schelle]
ROB. (W.) 7205; **shule** (*subj.*) ORM. 1832;
schulle [shulle] O. & N. 442; lest ihc schulle
hit forgo FL. & BL. 182; þat ha schulen
lasten a KATH. 279; *deriv.* **scholde, schuld.**

schále, *see* **scále.**

schalk, sb., *O.E.* scealc, =*O.L.G.* scalc, *O.
Fris.* skalk, schalk, *O.H.G.* scalc, scalh, *O.N.*
skalkr, *Goth.* skalks (δοῦλος); *servant, man,*
ALEX.[2] (Sk.) 20; GAW. 160; A. P. ii. 1029;
scalc (*sec. text* cniht) LAȝ. 19126; schalke
YORK xxx. 295; **scalkes** (*pl.*) LAȝ. 4219.

schalke, sb., *for* **chalk** ; *chalk,* D. ARTH.
1226, 1363.
schalowe, adj., *shallow, thin; 'bassus,'* PR.
P. 447 ; TREV. III. 131 ; shalou PARTEN. 739.
scháme, sb., *O.E.* sceamu, scamu, sceomu,
scomu, = *O.L.G., O.H.G.* scama, *O.N.* skömm;
shame, 'verecundia, ignominia,' PR. P. 443;
BEK. 11 ; SHOR. 35 ; MIRC 905 ; PR. C. 7145;
shame HOM. II. 173 ; ORM. 7284; HAV. 1939;
schame, schome, shome O. & N. 50, 363, 1483;
scame P. L. S. viii. 84 ; scame, scome HOM. I.
35, 59 ; scame, scome, sceome [same] LAȝ.
1434, 2294, 3493 ; same GEN. & EX. 234 ;
ssame AYENB. 8 ; schome KATH. 91 ; schome
ant schonde CHR. E. 542 ; shome H. H. 51 ;
scheome A. R. 60 ; ALIS. 7819; ha haveð us
alle scheome idon JUL. 72 ; þi stefne goþ . . .
to schome (*perit*) O. & N. 522 ; *comp.* léod-
scome.

sham-fast, adj., *O.E.* scamfæst; *modest,* (*ms.*
shammfasst) ORM. 2175; scham-, schamefast
CH. C. T. *A* 2055 ; schamefast '*verecundus*'
PR. P. 443 ; shamefast HOCCL. i. 431 ; LUD.
COV. 124 ; ssamvest AYENB. 222 ; sceomefest
FRAG. 5 ; **schómevaste** (*pl.*) MISC. 80.

shamfestnesse, sb., *modesty, 'verecundia,'*
HOM. II. 71 ; schamfestnesse CH. C. T. *A*
840.

scheóme-ful, adj., *shameful,* A. R. 302 ;
schamful CH. C. T. *C* 290.

ssamvol-líche, adv., *shamefully,* AYENB.
181.

schóme-léas, adj., *O.E.* scamléas, = *O H.G.*
scamalōs ; *shameless,* A. R. 170 ; schomeles
R. S. vi ; shameles LANGL. *C* iv. 47 [shamlees
B iii. 44].

scámelēst, sb., *O.E.* sceamléast; *shameless-
ness,* MK. vii. 22.

scheóme-lich, adj., *O.E.* sceamlíc, = *O.H.G.*
scamalīh ; *shameful,* A. R. 116 ; schamli WILL.
556 ; schómelíche (*adv.*) H. M. 31 ; R. S. vii;
schamliche TRIST. 1474 ; shameliche P. R. L.
P. 229 ; shamelike HAV. 2825 ; mid scome-
liche witen LAȝ. 20462.

schamel, sb., *O.E.* sceamul, scamul, sceomol,
= *O.L.G.* scamel, *O.H.G.* scamal ; *mod. Eng.*
shambles ; *stool, bench,* ['*scabellum*'] PS. cix.
1 ; schamil (*printed* sthamil) REL. II. 176;
schamel [scheomel] A. R. 166* ; shamill HAMP.
PS. xcvii. 5 ; *comp.* fót-scamel.

schámien, v., *O.E.* sceamian, scamian, *cf. O.
H.G.* scamen, *Goth.* skaman, *O.N.* skamma ;
shame ; scamien P. L. S. viii. 83 ; shamien
HOM. II. 69 ; schamin '*verecundor*' PR. P.
443 ; and schamen al mi kinrede CH. C. T. *F*
1565 ; it shal ne shamen þe ['*non te pudebit*']
WICL. IS. liv. 4 ; ssamie AYENB. 229 ; schame
PR. C. 7159 ; scomien [samie] LAȝ. 25209;
scheomien MARH. 7 ; scheomen A. R. 312 ;
eu schal sore schomie (*ms.* schomye) MISC.
41 ; shámeþ (*pres.*) ORM. 18284; **schámie**
[schomie](*imper.*) þe O. & N. 161; þai **scháme**

['*erubescant*'] (*subj.*) PS. vi. 11 ; þet eow
sceamie HOM. I. 35 ; **schómedest** (*pret.*)
LANGL. *A* iii. 183 [shamedest *B* iii. 189] ;
ham þas scamede HOM. I. 223 ; him swiče
scomede þat he swa iscend wes LAȝ. 4851 ;
shámed (*pple.*) ORM. 1986 ; HAV. 2745 ;
schamid and schent AUD. 1 ; schamede ['*con-
fusi*'] REL. I. 8 ; *comp.* a-, for-, of-**schámien.**

schámous-lī, adv., *shamefully,* YORK xxxii.
143.

schande, sb., *O.E.* sceand, scand, sceond,
scond, = *Goth.* skanda, *O.H.G.* scanda, scanta :
disgrace, ignominy ; shande ORM. 11956 ;
schonde O. & N. 1733 ; HORN (L.) 714; CHR.
E. 329 ; ROB. 65 ; P. S. 221 ; WILL. 555 ;
duden him muchele schonde MISC. 49 ; he
makede to sconde (*disgraced*) LAȝ. 7032 ;
sonde REL. I. 221.

scond-ful, adj., *disgraceful,* HOM. I. 31.

schand-lich, adj., *O.E.* scandlíc ; *shameful,* AR. & MER. 4286 ; s(c)andlic HOM. I.
239 ; wið scondliche deaðe LAȝ. 2274.

schandle, *see* **scandle.**

schanke, sb., *O.E.* sceanca, scanca, sceonca,
sconca ; *shank, 'crus,'* PR. P. 444 ; MAN. (H.)
189 ; sceonke FRAG. 2 ; **scanken** (*pl.*) JOHN
xix. 31 : sconken LAȝ. 1876 ; schonken JUL.
49 ; shankes (*pl.*) ORM. 4775 ; HAV. 1903 ;
schankes CH. C. T. *B* 1392 ; schonkes MIRC
892 ; schonken (*dat. pl.*) A. R. 258.

schantillōn, *see* **eschantillōn.**

schap, sb., *O.E.* (ge-)sceap, = *M.Du.* schap,
O.N. skap, *M.H.G.* (ge-)schaf ; *from* **scheppen** ; *shape, 'forma,'* PR. P. 444 ; WILL.
2885 ; TREV. III. 399 ; CH. C. T. *A* 1889 ;
shap & hiu HOM. II. 99 ; þe shap(pe) ne þe
shaft(e) LANGL. (Wr.) 7346 (*B* xi. 387) ; shap
membrum genitale ORM. 5937 ; REL. II. 28 ;
schape KATH. 449 ; JUL. 20; þi shape dide
þat hit ne sholde HOM. II. 67 ; **scheápe** (*dat.*)
A. R. 424 ; et þe schape þe d(e)ovel smuȝeð
in derneliche hwenne hit bið ȝaru to ga(l)-
liche deden HOM. I. 153.

schap-, scap-lēs, adj., *shapeless,* C. M.
350.

schap-lī, adj., *shapely, fit,* CH. C. T. *A*
372 ; WICL. GEN. xxiv. 16.

shaplinesse, sb., *beauty,* WICL. PS. xliv. 5.

schápien, v., *O.E.* sceapian, = *O.N.* skapa,
O.H.G. scafōn ; *shape ;* schapen [shapen]
CH. C. T. *A* 2541 ; shape LANGL. *B* ix. 131 ;
for to schape (*make*) þe chamberleines linnen
cloþes TREV. VII. 269 ; **shápeþ** (*pres.*)
ORM. 17583 ; schapeþ MAND. 160 ; **schépieð**
(*imper.*) A. R. 420 ; **schéped** (*pret.*) A. P. iii.
247 ; **sháped** (*pple.*) HAV. 424 ; H. H. 194 ;
comp. for-, ȝe-**schéapien.**

schápare, sb., *O.N.* skapari, = *O.H.G.* sca-
fari ; *shaper, 'formator, creator,'* PR. P. 444 ;
shapere WICL. IS. li. 13.

schápinge, sb., = *O.H.G.* skafunga; *shaping,* 'formatio,' PR. P. 444.

schar, sb., *O.E.* scear, = *O.H.G.* scar ; *from* **schéren** ; *ploughshare; shears; 'vomer,'* VOC. 169; WICL. DEUT. xxxi. 3 ; schar [schare] CH. C. T. *A* 3763; **scháre** (*dat.*) LANGL. *B* iii. 306 ; REL. I. 7 ; **ssáres** (*pl.*) ROB. 335 ; *comp.* **plōh-schar.**

scháre, sb., *O.E.* scearu ; *share-bone, groin,* 'ingven, pubes,' CATH. 333; TREAT. 139 (SAINTS (Ld.) xlvi. 727) ; schore VOC. 148 ; PR. P. 448; schere VOC. 246 ; A. R. 272.

scháren, v., *from* **scháre** ; *cut, shear*; **scharde** (*pret.*) DEGR. 1630 ; **schard** (*pple.*) *formed* ALEX. (Sk.) 4675 ; like schepe þat were scharid awai schall ʒe schake YORK xxviii. 141.

scharlette, *see* **scarlat.**

[**scharn,** sb., *O.E.* scearn, = *O.N.* skarn, *O. Fris.* skern ; *dung.*]

    sharn-bude, sb., *cf. mod.Eng.* shornbug ; *dung beetle,* GOW. I. 173; scar(n)bude VOC. 255 ; scearn-budda FRAG. 3; sharn-bodde PALL. ix. 60 ; **ssarnbodes** (*pl.*) AYENB. 61.

scharp, adj., *O.E.* scearp, = *O.L.G.* skarp, *O.Fris.* skerp, scherp, *O.N.* skarpr, *O.H.G.* scarf, sarf ; *sharp,* MARH. 9; ALIS. 6536 ; P. P. 216; MAND. 50; sharp ORM. 9211 ; scarp wes þe pic LAʒ. 30752; scarp hunger 20728; sarp GEN. & EX. 3577; ssarp AYENB. 165; scherp A. R. 212; P. L. S. xxi. 149; ane wiæxe . . . swiᵭe scærpe LAʒ. 4592 ; mid scearpe mire eaxe 2310; þe sharpe spere SPEC. 70; scharpe salve LANGL. *B* xx. 304; **scharpe** (*adv.*) O. & N. 141 ; schearpe wordes A. R. 82.

    scharp-līche, adv., *sharply,* WILL. 178; TREV. V. 171.

scharpnesse, sb., *sharpness,* PR. P. 444; TREV. III. 455 ; ssarpnesse AYENB. 165.

scharpin, v., *cf. O.E.* scerpan, scyrpan, = *M. Du.* scherpen ; *sharpen,* '*acuo,*' PR. P. 444; his nese shal sharpen REL. I. 65 ; sharpe WICL. WISD. v. 21 ; **scharpeþ** (*pres.*) CH. C. T. *A* 3763; him scerpeþ þe neose FRAG. 5; **scharped** (*pret.*) PS. lxiii. 4.

schat, sb., *O.E.* sceatt, = *O.L.G.* scat, *O.N.* skattr, *Goth.* skatts, *O.H.G.* scaz; *money, treasure*; sat GEN. & EX. 795.

schateren, *see* **scateren.**

scháþe, *see* **scáᵭe.** **schaþel,** *see* **scaþel.**

schaud, *see* **scheald.**

schavadri [**chevaldri**], sb., *baseness,* ALEX. (Sk.) 3371.

schavaldour, sb., *wanderer*; schaveldoure '*discursor, vaca'undus*' PR. P. 444.; **schavaldouris** (*pl.*) BARB. v. 205; schaveldours WICL. E. W. 249.

scháve, sb., *O.E.* sceafa, scafa, = *O.H.G.* scaba ; *scraper,* '*scalprum,*' PR. P. 444.

scháven, v., *O.E.* scafan, = *O.N.* skafa, *O.H. G.* scaban, scapan, *Goth.* scaban, *Lat.* scabere ; *shave, scrape*; schavin '*rado*' PR. P. 444 ; schafe PERC. 2095 ; **schóf** (*pret.*) IPOM. 1640; schoof WICL. I PARAL. xix. 4; scaft [safde] LAʒ. 22293; **scháven** (*pple.*) MAND. 33 ; R. R. 941 ; A. P. ii. 1134; shaven GEN. & EX. 2120; schavin REL. I. 51 ; ischaven A. R. 422*.

    scháving, sb., *shaving; slice cut off*; shaving CH. C. T. *G* 1239.

schawe, *see* **schaʒe.**

schāwen, *see* **scheāwen.**

schē, *see* **schēo.**

scheád, sb., *O.E.* (ge-)sceád, scád, = *O.H.G.* sceit, *mod. Eng.* shed ; *from* **scheáden,** *distinction, discrimation* : schead ba of god & of uvel KATH. 240; schad HOM. I. 255 ; nis bitwenen ʒunc & hem nan shæd ORM. 6229 ; shæd & skil 5534 ; *comp.* **sceád.**

scheáde, sb., *O.E.* scáde, *? cf. M.H.G.* scheide; *division; parting (of the hair), top of the head*; schade '*discrimen*' VOC. 183 ; shede D. TROY 3023 ; schode AR. & MER. 1480; schode [shode] CH. C. T. *A* 2007.

[**scheádel,** sb., *? = O.H.G.* sceitila; *division*; *comp.* **gate-schēdel (-schādil).**]

scheáden, v., *O.E.* scēaden (*pret.* scēod), = *O.L.G.* scēᵭan, skēdan, *O.H.G.* sceidan (*pret.* scied), *Goth.* skaidan (*pret.* skaiskaid); *shed, separate, divide,* A. R. 270; shæden ORM. 1209; schede O. & N. 197; MAN. (H.) 159; A. P. i. 411; schode GOW. I. 101 ; **shōdeᵭ** (*pres.*) HOM. II. 67 ; **shádde** (*pret.*) ORM. 3200; schadden MAN. (F.) 991 ; **shåd,** sad (*pple.*) GEN. & EX. 58, 148 ; *comp.* **for-ʒe-, to-scheáden.**

    shǣding, sb., *? = O.H.G.* sceidunga ; *division, separation,* ORM. 16863 ; scheding TOR. 518 ; schodinge PR. P. 447.

scheáf, sb., *O.E.* scēaf, = *M.L.G., M.Du.* schōf, schoof, *O.N.* skauf, *O.H.G.* scoub ; *from* **schūven** ; *sheaf*; shef LANGL. *B* iii. 324; schef [sheef] CH. C. T. *A* 104; scheef TREV. VIII. 335; scheef, schof '*merges*' PR. P. 444; **shǣfes** (*pl.*) ORM. 1481 ; sheves WICL. JUD. viii. 3; **schēve** [shēve] (*dat. pl.*) O. & N. 455.

    scēaf-mælen, adj., *O.E.* scēafmǣlum ; *by sheaves,* MAT. xiii. 30.

scheal, *see* **schal.**

scheáld, adj., *thin, shallow* : þe water was scheld HALLIW. 730; schold '*bassus*' PR. P. 447; schoolt [schoold] TREV. III. 131 ; inne senne so schealde SHOR. 93; **schaudest** [shaldest] (*superl.*) BARB. ix. 354.

scheap, *see* **schap.** **schēap,** *see* **schēp.**

scheard, sb., *O.E.* sceard, = *O.N.* skarᵭ; *from* **schéren** ; *shred, fragment*; sherd LIDG. M. P. 114; scherd, schord '*testa*' PR. P. 445; scherd [shord] WICL. IS. xxx. 14; **scherdes** (*pl.*) LEG. 34 ; TREV. IV. 151 ;

a dragon whos scherdes schinen as þe sonne GOW. III. 68 ; *comp.* **pot-schord.**

schēat, sb., *O.E.* scēat, *cf. O.Fris.* skāt, *M.Du.* schoot, *Goth.* skauts *m., O.N.* skaut *n., O.H. G.* scōz *m. f.; sheet*; scet '*mantele*' FRAG. 4 ; scat HOM. II. 231 ; sciet (*? printed* scier) P. L. S. viii. 183* ; opene þe shet ANGL. IV. 195 ; schene under schete REL. I. 180 ; *see* **schēte.**

schēawe, sb., *O.E.* scēaw,=*M.H.G.* schouwe ; *show* ; schewe '*monstratio*' PR. P. 446 ; schew DEP. R. iv. 56 ; schawe YORK xxx. 56.

shāw-erne, sb., *show-house*, [*interpretation of* '*Ephratah*' MIC. v. 2] ORM. 7025.

schēawen, scheauwen, v., *O.E.* scēawian,= *O.Fris.* skowia, *O.L.G.* scawōn, *O.H.G.* scau-wōn, scouwōn, *cf. Goth.* (us-)skawjan ; *show, point out,* A. R. 154, 210 ; scheawen, sceawen HOM. I. 49, 203 ; scheawen HOM. II. 73 ; schawen MARH. 15 ; shæwen ORM. 105 ; schewen MAND. 279 ; schewen GOW. I. 50 ; schewin PR. P. 446 ; scheawe SHOR. 152 ; sseawi, ssewi AYENB. 44, 56 ; schewi [sewi] O. & N. 151 ; schewe P. L. S. ix. 152 ; PS. iv. 6 ; schewe [shewe] LANGL. *A* i. 2 ; of his craft som what we wille hem schewe [shewe] CH. C. T. *D* 283 ; so fair it was to shewe H. F. 1305 ; shewe LIDG. M. P. 96 ; ssewe ROB. 210 ; schawe GAW. 27 ; A. P. ii. 1599 ; M. H. 96 ; for to se and for to shawe HAV. 2784 ; schowen C. L. 35 ; **sēaweþ** (*pres.*) MISC. 33 ; **schēau** (*imper.*) A. R. 292 ; schewe WICL. JOHN xiv. 8 ; **schēawede,** scheawude (*pret.*) A. R. 154, 160 ; scæwede, scewede, scawede [sewede] LAȝ. 1405, 2020, 7241 ; seawede MISC. 27 ; sseawede AYENB. 13 ; schewede MISC. 54 ; MAND. 86 ; sceawede sceaude HOM. I. 41 ; scheaude KATH. 915 ; sewede GEN. & EX. 2661 ; showeden A. D. 262 ; **schēwed** (*pple.*) MAN. (H.) 8 ; *comp.* ȝe-, ófer-schēawen.

shēawere, sb., *O. E.* scēawere, = *O.H.G.* scauwāri ; *pointer out ; mirror* ; HOM. II. 29 ; sseawere, ssewere AYENB. 84, 202 ; scheau-ware A. R. 90 ; schewere ALIS. 18 ; shewer P. 16 ; scawere ['*speculator*'] HOM. I. 117 ; *mirror* 53 ; **shēweres** (*pl.*) WICL. IS. iii. 23.

sēawinge, v., *O.E.* scēawung ; *vision,* MISC. 26 ; sseawinge AYENB. 14 ; **schēauwinges** (*pl.*) A. R. 268.

schēawles, *see* **schēoules.**

sched, *see* **scháde.**

scheden, v.,=*O.Fris.* schedda (*shake*), *M.L. G., M.Du.* schudden, *O.H.G.* scuttan, scutan (*pret.* scutta, scutida) ; *shed, pour,* MISC. 47 ; schedin PR. P. 444 ; schede P. L. S. xx. 69 ; LEG. 87 ; þou schalt schede ['*fundes*'] WICL. EX. xxix. 7 ; **schedde** (*pres.*) O. & N. 1616 ; shedeð, shat REL. I. 128 (HOM. II. 161) ; schet HOM. I. 159 ; (hio) schedeð A. R.

166 ; **schedde** (*pret.*) SHOR. 4 ; WICL. EX. xxiv. 6 ; god schedde his blod for alle men A. R. 312 ; schedde, scedde HOM. I. 157 ; scedde R. S. vii. (MISC. 186) ; sce(d)de [sadde] LAȝ. 5187 ; ssedde AYENB. 107 ; shedde [shadde] LANGL. *B* xvii. 288 ; mi brain schadde on þe ground BEK. 1578 ; hi . . . schedden his . . . blod MISC. 39 ; **shed** (*pple.*) H. H. 37 ; schad SHOR. 20 ; xad LUD. COV. 275 ; isched A. R. 402 ; ischad ALIS. 2772 ; *comp.* a-, bi-scheden.

schedunge, sb., *shedding,* A. R. 262.

schēden, *see* **scheāden.** **schēf,** *see* **schēaf.**

schefte, *see* **schafte.** **schēi,** *see* **schēoh.**

schéken, *see* **scháken.**

schel, *see* **schal, schil.**

schelchene, sb., = *M.H.G.* schelkin ; *from* **schalk** ; *female servant,* MISC. 45 ; schelchine A. R. 12.

schéld, *see* **scheald, schíld.**

schelfe, sb., *O.E.* scylf (BL. HOM. 27),=*M.Du.* schelf, schelve, *M.L.G.* schelf ; *shelf,* PR. P. 445 ; **schelves** (*pl.*) CH. C. T. *A* 3211.

schelle, sb., *O.E.* scell, sciell, scyll,=*O.N.* skel, *M.L.G., M.Du.* schelle, *Goth.* skalja (κέραμος) ; *shell ; drinking-vessel ; anything hollow ;* '*testa,*' VOC. 230 ; PR. P. 444 ; ALIS. 571 ; GOW. III. 76 ; shelle LANGL. *B* xi. 252 ; schelle [schille] TREV. III. 397 ; schille '*testa*' VOC. 254 ; skell YORK ii. 65 ; **schelles** (*pl.*) MAND. 193 ; schelles oþer coppes TREV. V. 449 ; schellis TREV. II. 215 ; schelles [skellis] MAN. (F.) 14683 ; schelles *hollows in the ground* [*O.E.* scel] ALEX. (Sk.) 4049 ; *comp.* āi-schelle.

schel-drake, sb., *sheldrake,* PR. P. 444 ; REL. I. 84.

shel-fish, sb., *O.E.* scelfisc,=*O.N.* skelfiskr ; *shellfish,* CH. BOET. ii. 5 (50).

schellen, v.,=*M.L.G.* schellen, *M.Du.* schol-len, schillen ; *shell* ; schillin '*excortico*' PR. P. 446.

schellen[2], v.,=*M.H.G.* schellen, *O.N.* skella ; *from* **schillen** ; *make a harsh noise:* to þe knight sche gan to skille OCTOV. (W.) 326 ; **schilled** (*pret.*) AR. & MER. 4750 ; þai schulde so doþ þe þonder FER. 631 ; schilde 727.

schemeren, *see* **schimeren.**

schench, sb., *O.E.* scenc,=*M.H.G.* schanc ; *draught (of wine) ; cup* ; scench S. S. (Web.) 562 ; **scenche** (*dat.*) P. L. S. viii. 167 ; schenches (*pl.*) LAȝ. 13461 ; *comp.* nōn-, water-, win-scenc.

schenchen, v., *O.E.* scencan,=*O.N.* skenkja, *O.H.G.* scenchan ; *pour out* ; scenchan LAȝ. 14962 ; shenchen REL. I. 113 ; schenche HORN (L.) 370 ; ALIS. 7581 ; þe drink for to schenche ROB. 118 ; and water on him gan schenche S. S. (Web.) 2247 ; schenkin PR. P. 445 ; **shenkest** (*pres.*) ORM. 15403 ; schen-

chiþ [schenke] CH. C. T. *E* 1722 (Wr. 9596);
schenche (*subj.*) MISC. 175; weoren þa bernes
iscængte (*pple.*) mid beore LA3. 8124.

[**schende,** sb.,=*M.H.G.* schende (*disgrace*);
? = **schande.**]

schend-ful, adj., *ignominious,* A. R. 356;
REL. I. 180; WICL. 1 KINGS xx. 30; CH. C.
T. *C* 290*; schindful P. L. S. xiii. 366.

schendful-līche, adv., *disgracefully,* A. R.
460; P. L. S. xi. 10; schend-, shenfulliche
LANGL. *A* iii. 261.

schendfulnesse, sb., *vileness,* A. R. 322.

schend-lāc, sb., *ignominy,* KATH. 1285; A.
R. 106.

schend-līc, adj.,=*M.H.G.* schentlīch; *ignominious*: silde fram sindliche (*first text*
scondliche) deaþe LA3. 2274*.

schendnesse, sb., *O.E.* scendnyss; *disgrace,*
MISC. 45; LEG. 36; ssendnesse ROB. 342;
schendnes TREV. V. 245; schindnisse BEK.
1302; *comp.* 3e-scendnesse.

schend-schüpe, sb., *ignominy, hurt,* JOS.
496; schendschepe PR. C. 7146; schenschepe
PR. P. 445; schenschip WILL. 556; WICL.
LEV. xviii. 17; schenchipe D. ARTH. 2435.

schenden, v., *O.E.* scendan, scindan, scyndan,
=*M.Du.* schenden, *O.H.G.* scendan, scentan;
from **schande**; *disgrace, revile, confound,* A.
R. 316; schenden ure wiþerwin(n)e MISC. 79;
shenden LANGL. *B* xi. 416; CH. H. F. 1016;
schendin '*confundo*' PR. P. 445; scenden
HOM. I. 21; scenden [sende, synde] LA3.
14167, 23682; schende O. & N. 274; MISC. 52;
HORN (L.) 680; BEK. 763; MIRC 1646;
WILL. 556; S. S. (Wr.) 501; WICL. ECCLUS.
xiii. 8; GAW. 2266; LUD. COV. 52; shende
HAV. 1422; H. H. 130; scende JUL.² 222; to
ssende and to destrue AYENB. 28; schinde
FER. 523; **schendest** (*pres.*) MARH. 7;
schent A. R. 298; H. M. 31; hou þis man me
schent BEK. 973; **schende** (*pret.*) P. L. S.
xix. 74; O. & N. 285; schente HORN (L.) 322;
shente LANGL. *B* xvii. 288; scenden LA3. 29787;
shended, shend (*pple.*) ORM. 1985, 4965;
shent GEN. & EX. 754; *comp.* **a-, 3e-schenden.**

schending, sb.,=*M.H.G.* schendunge; *dis-
gracing,* PL. CR. 94; sending REL. I. 218.

schene, *see* **schine.**

schēne, adj., *O.E.* scēne, scȳne, scēone,=*O.
Fris.* skēne, scōne, *O.L.G.* scōni, *O.H.G.*
scōni, *Goth.* skauns, *mod.Eng.* sheen; *beautiful, splendid,* KATH. 448; A. R. 100; FL. &
BL. 263; WILL. 2396; A. P. i. 1144; ISUM.
550; ANT. ARTH. vi; as schene as schininde
sunne MARH. 19; maiden schene C. L.´ 1191;
schene, scene HORN (H.) 97, 174; schene
[sheene] CH. C. T. *A* 115; shene LANGL. *B*
xviii. 409; M. ARTH. 51; sheene LIDG. TH.
1801; scene MISC. 196; feir & scene P. L. S.
viii. 171; scene, sceone, scone [scene] LA3.
2094, 2299, 3098; shene, scone ORM. 3431,

15665; **schēnre** (*compar.*) KATH. 1661; A.
R. 100.

schēnen, *see* **schænen.**

schenken, *see* **schenchen.**

schēo, pron., *she,* ALIS. 151; scheo,
scho MAN. (F.) 435; sheo E. W. 74; scheo,
sche FER. 1201, 2133; scæ SAX. CHR. 264;
sche WILL. 836; TRIST. 1669; MAND. 90;
sche, she, sge GEN. & EX. 335, 1444, 1925;
sche, she CH. C. T. *A* 121; sche [she], scho
LANGL. *A* i. 10 (*v. r.* heo); sche, she, scho,
sho HAV. 126, 192, 1232, 1721; sche, se MED.
843, 1034; scho MIRC 95; Iw. 198; S. S.
(Wr.) 295; GAW. 1259; PERC. 157; M. H. 15;
sho PR. C. 583; scho [sco] C. M. 5611.

schēoh, adj., *O.E.* scēoh,=*M.H.G.* schiech,
Dan. skȳ, *Swed.* skygg, *cf. O.N.* styggr; *shy,
fearful, timid*; scheouh A. R. 242; schei, skei
PR. P. 444; **schēowe** (*pl.*) A. R. 242; *see*
skig.

scheóme, *see* **scháme.**

scheort, *see* **schort.**

schēoten, v., *O.E.* scēotan,=*O.N.* skiōta, *O.
Fris.* skiata, schiata, *O.H.G.* sciozan; *shoot,
throw, send, fall, rush*; sceoten LA3. 313; he
wile . . . sheten in his heorte ORM. 3839; þoh
þat him sholde sheten to þolen for his soþe
word grimme dæþes pine 19952; scheten
GEN. & EX. 474; PL. CR. 773; scheten wiþ
bowe MAND. 154; schetin PR. P. 445; sheten
[schete] WICL. PS. lxiii. 6; schete HORN (L.)
939; WILL. 2399; schete [sheete] CH. C. T.
A 3928; schete, ssete ROB. 11, 377; schute
P. L. S. xvi. 139; FER. 3254; **schēte** (*pres.*)
S. S. (Web.) 1982; heo scheot þe earewen A.
R. 60; þe sunne gleam þe scheot from est
into west HOM. I. 265; schet GOW. I. 258;
hit schut forþ TREAT. 135; scheoteð forð sum
word KATH. 812; **scheat** (*pret.*) JUL. 71;
sceat, scæt [set] LA3. 254, 5081; schet GEN.
& EX. 475; CHR. E. 57; FER. 3962; MAN.
(F.) 1522; S. & C. x; schet, sset ROB. 11,
537; ssat AYENB. 45; schet (*ms.* schete),
schotte WICL. 4 KINGS xiii. 17; (?schot)
Iw. 1664; GAW. 317; PERC. 2114; scuten
HAV. 2431; scuten [soten] LA3. 12574; schoten GREG. 586; shoten LANGL. *B* xx. 224;
and shoten on him so don on bere dogges
HAV. 1838; schote ALIS. 2791; ssote ROB.
263; **schute** (*subj.*) HOM. I. 225; **schoten**
(*pple.*) JUL. 73; shote A. D. 291; *comp.* **æt-, 3e-
to-, þurh-schēoten**; *deriv.* **scheat, schēte,
schot, schote, schótien, schute, schütten.**

ssīetere, sb., *shooter, hurler,* AYENB. 174;
sheter WICL. GEN. xxi. 20; **ssētares** (*pl.*)
[schetors, scheoteres] ROB. (W.) 7482*.

schēotunge, sb., *shooting,* A. R. 60.

schēoules, sb., *cf. M.H.G.* sch(i)ūsel,=*M.Du.*
schouwsel; *scarecrow*; sheules, scheawles
O. & N. 1128, 1648.

[**schēowelen,** v., *frighten*; *comp.* **a-schēwelen.**]

schēowen, v., ? cf. M.L.G. schūwen, M.Du. scūwen, O.H.G. schūhhan; *from* schēoh; *shy, go astray, avoid* ; scheuen or eschuen ' *vito* ' REL. I. 7 ; skewe D. ARTH. 1562 ; *see* eschewen.

schēp, sb., O.E. scēp, scēap,=O.L.G. scāp, O.H.G. scāf; *sheep*, A. R. 122 ; shep ORM. 988 ; sep GEN. & EX. 940 ; þet ssep AYENB. 140 ; scep HOM. I. 121 ; sceap 245 ; shēpes (*gen.*) skin ORM. 3210 ; scēp (*pl.*) HICKES I. 225 ; scep[sceap] LA3.25681 ; shep HOM. II. 37 ; ORM. 3760 ; scheep WICL. JOHN ii. 14 ; sep REL. I. 209 ; seop MISC. 41 ; scēpe (*gen. pl.*) MAT. xviii. 12 ; scēpan (*dat. pl.*) MAT. xv. 24 ; scheapen LA3. 1546.

schēp-cot, sb., *sheep-cot*, VOC. 236 ; schep-cote PR. P. 445.

Ssēp-ēie, pr. n., O.E. Scēapēg, -īg ; *Sheppey*, ROB. 259.

schēp-fálde, sb., *sheep-fold*, CATH. 335.

shēp-hirde, sb., *shepherd*, ORM. 3587 ; schep-, sheepherde CH. C. T. A 514 ; schephurde BEK. 2134.

schepe, *see* schipe.

schēpe, sb., O.E. *scēapa (scēpa), *prov. Eng.* (*Linc.*) shep; *shepherd*; scheep [shepe] LANGL. A PROL. 2 [shepherde *C* i. 2] ; a chēpis (*gen.*) croke LIDG. (*in Skeat's Langl.*).

schēpish, adj., *meek as a sheep* ; shepisshe ORM. 6654.

scheppen, v., O.E. sceppan, scyppan,=O. Fris. skeppa, scheppa, O.N. skepja, O.H.G. sceffan, scaffan, Goth. (ga-)skapjan ; *form; create; make ready*; scheppeþ (*pres.*) SHOR. 5 ; ssepþ AYENB. 209 ; bote þou ... schippe (*subj.*) þe an on to fi3te FER. 542 ; schōp [O.E. sceōp, scōp,=O.L.G. (gi-)scōp, O.Fris. skōp, schōp, O.N. skōp, Goth. (ga-)skōp, O.H.G. scuof] (*pret.*) A. R. 138 ; SHOR. 147 ; CH. C. T. D 1780 ; AMAD. (R.) xxxix ; schop þe & al þe world KATH. 219 ; scheop R. S. v (MISC. 184) ; scop P. L. S. viii. 42 ; David þe þe salm scop HOM. I. 7 ; scop [sop] LA3. 14877 ; scupte [sipte] 1951 ; shop ORM. 1411 ; WICL. Is. xlv. 8 ; shoop LANGL. (Wr.) 5288 ; ssop AYENB. 87 ; shupte P. S. 238 ; þou shoope EM. 2 ; þou me shope [shuptest] of eorþe H. H. 156 ; schuptest MARH. 20 ; shuptest H. H. 158 ; schopen ALIS. 6970 ; þe frend shopen þe child(e) name HOM. II. 87 ; schope PS. cxviii. 73 ; shápen [O.E. sceapen,=Goth. (ga-)skapans, O.H.G. scafan] (*pple.*) shapen HOM. II. 105 ; ORM. 3551 ; schapen WILL. 126 ; CH. C. T. A 1392 ; *comp.* for-, 3e-, mis-scheppen ; *deriv.* schap, schafte, scōp.

sceppende, sb., O.E. sceppend, scyppend ; *creator*, HOM. I. 227 ; scheppande REL. I. 219 ; scuppend HOM. I. 129 ; schuppend KATH. 305 ; schuppinde A. R. 260 ; shippend ORM. 346.

scheppare, sb.,=M.L.G. schepper, M.H.G.

schepfære ; *creator*, REL. I. 282 ; sseppere AYENB. 7 ; scheppere [shepper] LANGL. B xvii. 167 ; shuppere REL. II. 229 ; schuppare A. R. 138 ; C. L. 1510 ; shuppare P. S. 238.

scheppinge, sb., = M. H. G. schepfunge ; *creation*; sseppinges (*pl.*) AYENB. 158.

schepstre, sb., cf. M.Du. schepper (*tailor*) ; *sempstress*; schipster TREV. VII. 269 ; a shepster (*gen.*) shere LANGL. B xiii. 331 (shappesters *C* vii. 75).

schepþe, sb.,=M.H.G. geschepfede ; *creature, form* ; sseppe AYENB. 81, 92.

scherd, *see* scheard. schére, *see* schāre.

schēre, sb., O.Fris. skēre, schēre,=M.L.G. schēre, O.H.G. scēra, M.H.G. schære ; *pair of shears;* ' *forfex* ' PR. P. 445 ; schere [shere] CH. C. T. A 2417 ; shere LANGL. B xiii. 331 ; s(c)ǣres (*pl.*) LA3. 14215 ; sh(e)res HAV. 857.

schēre, *for* chēre, sb., *mien*, GAW. 334.

schēre, *see* skēre.

schéren, v., O.E. sceran, = O.H.G. sceran, O.Fris. skera, schera, O.N. skera ; *shear*; sceren [seren] LA3. 20307 ; sheren GEN. & EX. 2347 ; sheren shep REL. I. 267 ; schere ROB. 150 ; RICH. 6038 ; PS. lxxxiii. 13 ; GAW. 213 ; schar (*pret.*) TRIST. 1493 ; AM. & AMIL. 2298 ; M. H. 55 ; of all(e) þe met(e) þat she schar DEGR. 801 ; shar HAV. 1413 ; he scar [sar] his crune LA3. 17663 ; scher MARH. 22 ; sceren LA3. 22291 ; schare MAN. (H.) 221 ; schóren (*pple.*) GEN. & EX. 1200 ; schorn RICH. 3001 ; schore TREV. V. 25 ; *comp.* 3e-, ümbe-schéren ; *deriv.* schére, schar, scháre, scóre, scheard.

schérer, sb.,=O.H.G. scerāri ; *shearer,* ' *messor*,' VOC. 213.

schéren[2], v., ? =M.L.G. scheren ; *? tremble*: for doute les te he valle he shoddreþ and shéreþ (*pres.*) SPEC. 110

scherp, *see* scharp. scherte, *see* schürte.

schēt, *see* schēat.

schēte, sb., O.E. scēte, scȳte ; *from* schēoten; *sheet*, ' *linteum*,' PR. P. 445 ; LEB. JES. 696 ; scheete [shete] CH. C. T. G 879 ; shete MED. 955 ; scete HICKES I. 224 ; ssete ROB. 435 ; scheetis (*pl.*) [' *linteamina* '] WICL. JOHN xx. 5 ; schetes LANGL. B xiv. 233 ; *see* schēat.

schetel, *see* schütel. schēten, *see* schēoten.

schetten, *see* schütten. schēþe, *see* schǣþe.

schēuen, schēules, *see* schēowen, schēoules.

scheve, sb.,=M.L.G. scheve, M.H.G. schebe; *? from* scháven ; *shaving* ; schevis (*? ms.* chewis) (*pl.*) VOC. 269 ; schifes of line CATH. 337.

schēven, *see* schüven.

schevere, scheveren, *see* schivere, -en.

schēw(e),schēwen, *see* schēawen,schēawe.

M m

schīde, sb., *O.E.* scīde,=*O.Fris.* skīd, *O.N.*
skīð, *M.H.G.* schīt ; *splinter, split wood, lath,
'assula'* PR. P. 446 ; ALIS. 6421 ; **schīdes**
(*pl.*) AR. & MER. 531 ; RICH. 1385 ; schides
[shides] and bordes LANGL. *B* ix. 131 ;
GOW. I. 314 ; cleven shides HAV. 917.

scīd-wal [sīdwal], sb., *O.E.* scīdeweall ;
wall of stakes, LAȝ. 10354.

[schīdren, v.,=*Ger.* scheitern ; *go to pieces;*
comp. **to-schīdren.]**

schift, sb., *O.E.* scift, *O.N.* skipt ; *shift, trick ;
at a schift *suddenly* M. H. 26 ; skiffte YORK
xxvi. 130.

schiften, v., *O.E.* sciftan, scyftan, *O.Fris.*
skifta, *M.Du.* schiften, *O.N.* skipta ; *mod.Eng.*
shift ; *change, move away; assign, divide ;*
schiftin PR. P. 446 ; leten riht scuften [sufte]
LAȝ. 4131 ; to schede and to schifte TREV. III.
477 ; schifte [shifte] CH. C. T. *D* 104 ; schifte,
skifte D. ARTH. 1213, 1643 ; skifte AMAD.
(Web.) 656 ; **scift** (*ms.* scyft) (*pres.*) HOM. I.
237 ; **shifte** (*pret.*) LANGL. *B* xx. 166 ;
(*printed* shiste) PARTEN. 2792 ; shiftede
GEN. & EX. 1732 ; shifteden ORM. 470 ;
schifted (*pple.*) AR. & MER. 1482 ; skifted
GAW. 19 ; comp. **to-schiften.**

schiftinge, sb., *change, turn,* PR. P. 446 ;
shifting ORM. 467.

schiggen, v., *shake :* schigge clothis '*excutio*'
PR. P. 446 ; he come **schig(g)inge** (*pple.*)
ayene DEGR. 345.

schil, adj., *O.E.* scyll,=*M.Du.*, *M.H.G.* schel ;
shrill, sonorous, WART. II. 41 ; ORF. 478 ;
shil PARTEN. 1997 ; schel LUD. COV. 180 ;
schille harpe O. & N. 142 ; **schille [shille]** (*pl.*)
CH. C. T. *B* 4585 ; sille REL. I. 221 ; **schille**
(*adv.*) WILL. 37 ; OCTOV. (W.) 563 ; MAN.
(H.) 30 ; TRIST. 3284 ; EGL. 300 ; schille and
loude GREG. 879 ; schille [schulle] O. & N.
1683 ; LANGL. *C* vii. 46 ; schulle HORN (L.)
207 ; cf. **schril.**

schild, sb., *O.E.* scild, scyld, sceld,=*O.L.G.*
scild, *O.Fris.* skeld, *Goth.* skildus, *O.N.*
skiöldr, *O.H.G.* scilt, skilt, schilt ; *shield,* EGL.
730 ; scheld A. R. 52 ; scheld [sheld] O. & N.
1022 ; sheld HAV. 624 ; sseld AYENB. I ;
sceld HOM. I. 69 ; þene sceld LAȝ. 7535 ;
sceold HOM. I. 243 ; **schilde** (*dat.*) OCTAV.
(H.) 825 ; schelde KATH. 810 ; PERC. 52 ;
scéldes, sculdes (*pl.*) LAȝ. 4187, 4193 ;
schélden (*dat. pl.*) LAȝ. 6699.

sceld-trume, sb., *O.E.* scildtruma ; *troop,
band,* LAȝ. 16371 ; scheldtrome OCTOV. (W.)
1505 ; BEV. 993 ; scheltrome RICH. 5629 ; scheltrone '*acies*' VOC. 240 ; scheltrun [shiltroun]
['*aciem*'] WICL. I KINGS iv. 2.

schilden, v., *O.E.* scildan, scyldan,=*O.N.*
skilda ; *shield, defend,* A. R. 82 ; scilden
[silde] wi ́ scondliche deaðe LAȝ. 2274 ;
sculden 5745 ; shilden ORM. 3794 ; silden
GEN. & EX. 214 ; schilde K. T. 537 ; shilde
hem from sunne P. S. 153 ; shilden [shelden]

LANGL. *B* x. 407 ; schulde BRD. 26 ; **schilt**
(*pres.*) A. R. 392 ; sculde ́ P. L. S. viii. 174 ;
schild (*imper.*) O. & N. 163 ; scild LAȝ. 1072 ;
scheld SHOR. 84 ; **schilde** (*subj.*) O. & N.
1253 ; schilde, shilde GEN. & EX. 2525, 4157 ;
shilde HAV. 16 ; Crist shilde us alle þer wi ́
HOM. II. 43 ; silde 11 ; ssilde AYENB. 271 ;
schilde (*pret.*) HOM. I. 259 ; scilde, scelde
LAȝ. 8431, 20155 ; comp. **ȝe-schilden.**

schélder, sb., *shielder,* PS. xvii. 7.

schildiȝ, *see* **schüldi.**

schile, sb., = *M.L.G.* schele, *M.Du.* schil,
O.N. skil ; *skill ; distinction, discrimination ;
reason, excuse ;* LIDG. M. P. 229 ; schil E. G.
30 ; schil, skil A. R. 204, 306 ; skil HOM. I.
61 ; ORM. 1652 ; GEN. & EX. 1425 ; H. S. 113 ;
skil, skile C. L. 489, 1074 ; skill YORK xliii.
113 ; scule HOM. II. 13 ; skele SHOR. 154 ;
skele, scele AYENB. 7, 11 ; **skiles** (*gen.*) A. R.
288 ; **skile** (*dat.*) O. & N. 186 ; **skiles**, skilles
(*pl.*) LANGL. *B* x. 301, xvii. 330.

skil-ful, adj., *skilful,* PR. P. 457 ; skilfull
YORK iii. 22 ; scelvol AYENB. 169.

skelvol-līche, adv., *skilfully,* AYENB. 6.

skilfulnesse, sb., *skilfulness,* PR. P. 457.

skil-lǣs, adj., *ignorant,* ORM. 3715.

scil-wīs, **skilwīse**, adj., *reasonable,* HAMP.
PS. xxxvii. 11* ; lvii. 11* ; comp. **un-scilwīs.**

scilwīs-lī, adv., *reasonably,* HAMP. PS.
xxxi. 6*.

schilien, v., *O.N.* skilja,=*M.Du.* schillen
(*differ*), *M.L.G.* schelen ; *divide, distinguish ;
be skilful ;* schillin '*segrego*' PR. P. 446 ;
schille M. H. 152 ; skilede (*pret.*) HOM. II.
119 ; **skiled** (*pple.*) ORM. 16860.

schille, schillen, *see* **schelle, schellen.**

schillen, v., = *O.H.G.* scellan (*pret.* scal),
O.N. skella (*pret.* skal) ; *clang, sound ;* schille
ORF. 270 ; wi ́ **schillinde** (*pple.*) stefne
MARH. 19 ; schillinge TREV. VII. 331 ; *deriv.*
schellen.

schilling, sb., *O.E.* scilling,=*O.H.G.* scilling,
O.N. skillingr, *Goth.* skilliggs ; *shilling,*
TRIST. 304 ; M. H. 141 ; shilling IW. 3058 ;
an hundred **schillinge** (*gen. pl.*) FL. & BL. 126.

schim, adj., cf. *O.E.* scima (*brightness*) ;
bright, A. P. i. 1077.

schimeren, v.,=*Ger.* schimmern, *M.Du.* schemeren, *Swed.* skimra ; *shine, dazzle ;* **schim-**
ere ́ (*pres.*) HOM. I. 326 ; **schimerande**
(*pple.*) H. M. 21 ; schemerande YORK i. 69,
schemered (*pret.*) GAW. 772.

schimeringe, sb., = *O.H.G.* scimeringe ;
shining, brightness, CH. C. T. *A* 4297 ; A. P. i.
80 ; skimering [skemering] YORK xvii. 123.

schimien, v., *O.E.* scimian, cf. *O.H.G.* sci-
man, *O.N.* skima ; *shimmer, glitter ;* **schim-**
inde (*pple.*) JUL. 55 ; **schimede** (*pret.*)
MARH. 44.

[schīn, sb., *O.E.* scīn ; *magic ;* comp. **dēofel-**
shīn.]

schinde, schinden, *see* schende, schende.

schine, shine, sb., *O.E.* scin, = *O.H.G.* scina, *M.L.G.*, *M.Du.* schene; *shin*, O. & N. 1060; shine P. R. L. P. 220; scine '*tibia*' FRAG. 2; schine [schene] CH. C. T. *A* 386; schene VOC. 184; shines (*pl.*) LANGL. *B* xi. 423.

schin-bande, sb., *?shin-plate*, (*?ms.*-bawde) D. ARTH. 3846; schinbandes (*pl.*) ANT. ARTH. (R.) xxxi.

schin-, schine-boon, sb., *O.E.* scinbān, = *O.H.G.* schinebein; *shinbone*, TREV. VIII. 65.

schīnen, v., *O.E.* scīnan, = *O.H.G.* scīnan, *O.L.G.* skīnan, *O.Fris.* skīna, schīna, *O.N.* skīna, *Goth.* skeinan; *shine*, A. R. 364; MISC. 142; schine BEK. 1409; scine LAȝ. 16644; ssine AYENB. 188; scineð, schineð (*pres.*) HOM. I. 83; shineþ ORM. 2138; schīne [shine] (*subj.*) O. & N. 963; schīninde (*pple.*) A. R. 224; schān (*pret.*) MARH. 2; C. M. 12574; shan ORM. 16169; PR. C. 6243; shane HAMP. PS. xcvi. 4; scean HOM. I. 43; scean, scan [son] LAȝ. 20608, 28773; schon [shoon] CH. C. T. *A* 198; scinen (*pl.*) LAȝ. 27361; shinen GOW. III. 68; sinen (*pple.*) REL. I. 209; *comp.* bi-schīnen; *deriv.* schǣnen.

schīning, sb., *lightning*, WICL. PS. cxliii. 6.

schīning-lī, adv., *splendidly*, WICL. LK. xvi. 19.

schingle, sb., *? for* *schindle, = *O.H.G.* scindela; *shingle*; schingil '*scindula*' PR. P. 446; ALIS. 2210; scingles (*pl.*) P. L. S. xxxv. 57.

schinglen, v., *? cf. Ger.* schindeln; *cover with shingles*: in þi schinglede [*v. r.* sengle] (*pple.*) schup LANGL. *A* x. 170 [shingled *B* ix. 141].

schip, sb., *O.E.* scip, scyp, = *O.Fris.* skip, schip, *O.L.G.*, *O.N.*, *Goth.* skip, *O.H.G.* scif, scef; *ship*, A. R. 142; þat schip AL. (T.) 190; MAN. (H.) 124; scip, schip [sip] LAȝ. 4536, 4753; a schip ful þer of ROB. 70; ssip 542; scip JUL². 215; ssip AYENB. 112; ship, schup LANGL. *A* ix. 131; schup HORN (L.) 597; schipes (*gen.*) A. R. 142; schipe (*dat.*) LAȝ. 3500; schipe, schepe WILL. 5088, 5212; to schipe he wende BEK. 1825; ssipe MISC. 32; scipe (*ms.* scype) (*pl.*) MK. iv. 36; scipe, schipe, scipen [sipes] LAȝ. 100, 6182, 6184; schipes O. & N. 1205; ALIS. 1511; scipen (*dat. pl.*) LAȝ. 14248; mide alle þa scipe SAX. CHR. 201.

ship-breche, sb., = *M.Du.* schipbreke; *shipwreck* ['*naufragium*'] WICL. 2 COR. xi. 25.

schip-brüche, sb., = *M.H.G.* schifbruch; *shipwreck*, TREV. II. 369.

shippe-craft, sb., *shipbuilding*, YORK viii. 67.

schip-fare, sb., *voyage*, TRIST. 926; schip-fair BARB. iii. 686.

scip-ferde, sb., *O.E.* scipfyrd; *navy*, LAȝ. 2156.

scip-gume, sb., *sailor*; scipgumen (*pl.*) LAȝ. 4560.

schip-hīre, sb., *passage money*, '*naulum*,' PR. P. 446; WICL. JONAH i. 3.

schip-man, sb., *O.E.* scipmann; *sailor; pirate*; CH. C. T. *A* 388; scipmen (*pl.*) LAȝ. 28308; ssipmen AYENB. 140.

schip-wriȝt, sb., *O.E.* scipwyrhta; *shipwright*, ALIS. 3665.

schipe, schepe, sb., *O.E.* scipe, VOC. 20; *wages, reward*, CH. C. T. *I.* 568; ssepe AYENB. 33.

[-schipe, suffix, *O.E.* -scipe, -scype, *cf. O.Fris.* -skipe, *O.L.G.* -scipi, -skepi, *O.N.* skapr, *O.H.G.* -scaf, -skeph; *from* scheppen; -*ship*; *comp.* āht-, bisi-, cnïht-, cwēd-, düsi-, earȝ-, fēlāȝ-, fēond-, frēond-, gāl-, gōd-, ȝēap-, hard-, hēh-, hende-, kēn-, lāferd-, mād-, man-, méte-, rēf-, schend-, sōð-, trēow-, tūn-, þral-, war-, wïld-, wit-, wurð-schipe (-schepe, -schüpe).]

schipen, v., *O.E.* scipian, = *M.L.G.* schepen, *M.H.G.* schiffen; *take ship, navigate*; shipen WICL. DEEDS xx. 3; schipeþ (*pres.*) ALIS. 1495; shipede (*pret.*) HORN (R.) 978; schipeden S. A. L. 152; schipped, i-schipped (*pret.*) TREV. VII. 95, 99.

schipping, sb., *shipping*: till his schipping [*v. r.* schippes] is he gane BARB. xvi. 16; schippine iii. 400.

schipien, v., *? from* schipe; *? remunerate, reward*: scipien heom mid londe LAȝ. 20012.

scipinge, sb., *wages*, LAȝ. 13656.

schipne, *see* schüpne.

schippen, *see* scheppen.

schīr, adj., *O.E.* scīr, = *O.L.G.* skīr, *O.N.* skïrr, *Goth.* skeirs, *M.H.G.* schīr; *clear, pure*, A. R. 144; LAUNF. 247; schir, scir, sir GEN. & EX. 518, 1835, 3848; a lith ful shir HAV. 588; shir water REL. I. 87; shir atter ORM. 15383; of galnesse skir & fre 8015; schir(e) schome KATH. 1286; schīre (*dat. n.*) PR. C. 6934; schīre (*pl.*) A. P. ii. 1278; schīre (*adv.*) FLOR. 98; þei brenne ... shire H. S. 12439; þat schinist so schire YORK xlvi. 202; schīrer (*compar.*) GAW. 956.

Schīre-burne, pr. n., *O.E.* Scīre burna; *Sherborne*, MISC. 145; P. L. S. xiii. 50.

schīr-līche, adv., *purely*, A. R. 154; schirli GAW. 1880.

schīrnesse, sb., *purity*, A. R. 386; shirnes HAMP. PS. vi. 2.

schīre, sb., *O.E.* scīr (*charge, province*), = *O.H.G.* scīra (*charge*); *shire, province*, PR. P. 447; scur '*provinciave l pagus*' FRAG. 3; anemuchele schire A. R. 334; schire [shire] CH. C. T: *A* 584; shire P. 71; schīren (*pl.*) ROB. 3; sciren LAȝ. 31995; schiren, schire MISC. 146.

schīr-rēve, sb., *O.E.* scīrgerēfa; *sheriff*, MARH. 2; MISC. 60; CH. C. T. *A* 359; scir-

reve P. L. S. viii. 25; shirreve P. S. 338;
serreve ANGL. I. 10; schirreff BARB. xvi.
583; ssereve ROB. (W.) 11601; · scīrerēvan
(*pl.*) ANGL. VII. 220.

schīren, v., *O.E.* scīran, = *M.L.G.* schīren, *O.N.*
skīra, *Goth.* (ga-)skeirjan; *make bright; make
clear, declare*; nes þer nan swa hæh mon þat
durste word sciren (*utter*) LAȝ. 16822; schīreð
(*pres.*) & brihteð þe heorte A. R. 384; god
almigtin ꝥe soꝥe shīre (*subj.*) GEN. & EX.
2036; *comp.* ȝe-schīren.

schirte, *see* schürte.

schirkin, v., *from* schīren; *ciear, brighten*;
schirkind (*pple.*) ALEX. (Sk.) 4816.

schismatike, adj. & sb., *Lat.* schismaticus;
schismatic: sarrasins and scismatikes (*pl.*)
LANGL. *C* xiii. 54.

schisme, sb., *O.Fr.* cisme; *schism*; scisme
GOW. I. 15.

schīte, sb., = *M.L.G.* schīte, *M.H.G.* schīze;
excrement, diarrhœa; schit TREV. VII. 51.

schīt-word, sb., = *M.L.G.* schītword; *foul
language,* O. & N. 286.

schitel, *see* schütel.

schīten, sb., *O.E.* scītan, = *M.L.G.* schīten,
O.N. skīta, *M.H.G.* schīzen; *cacare*; schitin
PR. P. 447; schite REL. II. 176; TREV. IV. 329;
shīteþ (*pres.*) ALIS. 5670; schoot (*pret.*)
TREV. V. 153; schite [ischete] (*pple.*) TREV.
IV. 329; *comp.* bi-schīten.

schitte, sb., *O.E.* scitta; *from* schīten;
excrement; skitte PR. P. 458.

schitten, *see* schütten.

schīve, sb.. = *M.L.G.* schīve, *M.Du.* schijve,
M.H.G. schībe; *scrap, slice,* '*lesca, scinda,*'
PR. P. 447; A. R. 416*; schife VOC. 241;
schīves (*pl.*) L. C. C. 17.

schīven, *see* schüven.

schivere, sb., = *M.H.G.* schivere, schevere,
M.Du. schevre; *shiver, splinter,* PR. P. 447;
CH. C. T. *D* 1840; shever IW. 3234; schiver
(*slice of bread*) B. B. 322; scifren, scivren
(*pl.*) LAȝ. 4537, 27785; schevires TOR. 177.

schiverin, v., = *M.Du.* schevren; *shiver,
splinter,* PR. P. 447; ther schiveren schaftes
CH. C. T. *A* 2605; scheverede (*pret.*) D.
ARTH. 1813; schivered, schevered (*pple.*)
WILL. 3411, 3616; scheverid RICH. 5309;
comp. to-schiveren.

schō, sb., *O.E.* sceō, sceōh, = *O.L.G.* scōh,
O.N. skōr, *Goth.* skōhs, *O.H.G.* scuoh, scuoch;
shoe, VOC. 197; R. S. i; schoo PR. P. 447;
scho [sho] CH. C. T. *A* 253; sho ORM. 10438;
sso AYENB. 220; schu REL. I. 83; sceōs (*pl.*)
HOM. I. 37; scheon A. R. 362; schon WILL.
14; schoon MAND. 59; shon HAV. 860; E.
G. 359; shoon LANGL. (Wr.) 9581; son GEN.
& EX. 2781; scōne (*gen. pl.*) MK. i. 7; schōn
(*dat. pl.*) MISC. 193; *comp.* hors-schō.

shō-þwang, sb., *O.E.* sceōþwang; *shoe-
thong,* ORM. 10387; shoþvong HOM. II. 137.

schōde, schōden, *see* scheāde, scheāden.

schoderen, *see* schuderen.

schōf, *see* scheāf.

schoggin, v., *cf. M.Du.* schocken, *M.H.G.*
schoggen, schucken; *shog, jog, shake up and
down,* '*agito,*' PR. P. 447; schogs (*pres.*)
ALEX. (Sk.) 5018; thei schokke D. ARTH.
4114; was schoggid ['*jactabatur*'] (*pple.*)
WICL. MAT. xiv. 24; ishogged PALL. xi.
322; *cf.* schiggen.

schōin, v., *O.E.* sceōian, = *O.H.G.* scuohōn;
shoe; '*calceo,*' PR. P. 447; sho HAV. 1138;
scheōinde (*pres. part.*) A. R. 16; scōiden
(*pret.*) LAȝ. 22291; ischood ·(*pple.*) LANGL.
A ii. 134; shodde [*v. r.* shoed, ischoud, schod]
B ii. 163; iscod LAȝ. 7831; ischode (*pl.*)
MISC. 91.

schōing, sb., *shoeing,* ['*calceamentum*']
WICL. EXOD. iii. 3, 5; ssoinge AYENB. 154.

schokke, sb., = *M.Du.* schocke, *M.H.G.* scho-
che; *shock (of corn),* PR. P. 447; VOC. 264.

schokken, v., = *M.Du..* schocken, *M.H.G.*
shochen; *heap up*; schokkin, PR. P. 447; þei
wille schokken hem to gidre MAND. 252.

schokken, *see* schoggen.

schōld, *see* scheāld.

scholde, v., *O.E.* sceolde, scolde, = *O.L.G.*
scolda, scolde, *Goth.* skulda, *O.N.* skuldi, skyldi,
O.H.G. scolta, solta; (*pret.*) *should; owed;
ought*; SHOR. 28; MAND. 6; i ne scholde do
no þing aȝen his wille þat he ne scholde sigge
þat ich wolde holi churche aspille BEK. ·1602;
ne scholde [sholde] he . . . so don O. & N.
381.; til þat godes sune . . . him sholde on
eorþe shæwen ORM. 268; scolde, scholde,
sculde [solde] LAȝ. 2079, 2084, 3746; scolde,
sculde HOM. I. 9; cride on him þat he ssolde
helpe hire ROB. 237; ȝif ure loverd demde
him al efter rihtwisnesse . . . wo scholde him
iwurðen A. R. 332; þe stronge . . . pinen þet
he schulde drien (*which he was to endure*)
112; sholde [shulde] LANGL. *B* iii. 202; þu
scholdest . . . bisechen A. R. 102; schuldest
KATH. 459; WILL. 5194; we scolden P. L. S.
viii. 164; þei sholden HAV. 1195; ssolden
AYENB. 6; sulden GEN. & EX. 957.

scholdre, *see* schuldre.

schóme, *see* scháme.

schonde, *see* schande.

schonien, *see* schunien.

schonke, *see* schanke.

schoppe, sb., *O.E.* sceoppa ['*gazophylacium*'
LK. xxi. 1]; *shop, tent, hut,* PR. P. 448; schoppe
[shoppe] CH. C. T. *A* 4376; shoppe E. G. 358;
ssoppe ROB. 541; **shope** (*pl.*) (*v. r.* shoppes)
LANGL. *B* ii. 213 [schoppes *A* ii. 189].

schor, sb., *cf. O.Sw.* skorra, *O.N.* skara;
menace, clamour; schoir BARB. vi. 621;
schour MAN. (F.) 6820.

schord, *see* scheard.

schóre, sb.,=*M.Du.*, *M.L.G.* schore ; *?from*
schéren ; *shore*, GAW. 2161 ; A. P. i. 230.

schóre², sb.,=*? M.Du.* schore ; *prop, support,*
PR. P. 448.

schóre³, adj., *steep, sheer*, BARB. x. 600 ; schoir
x. 22.

schóre, *see* scóre, scháre.

schóren, v.,=*M.Du.* schoren ; *shore up, prop* ;
issóred (*pple.*) AYENB. 207 ; *comp.* under-
schóren.

　schorier, sb., *support, prop* ; shoriers (*pl.*)
LANGL. *C* xix. 20.

schorn, *see* scarn.

schort, adj., *O.E.* sceort, = *O.H.G.* scurz ;
short, '*curtus, brevis*,' PR. P. 448 ; BEK. 776 ;
schort, short CH. C. T. *A* 306 ; short ORM.
16279 ; his minde es short ·PR. C. 774 ; short
[scort] O. & N. 73 ; ssort AYENB. 81 ; scheort
A. R. 146 ; an sceort [sort] bat LA3. 28624 ;
at the schorte soon D. ARTH. 1325 ; schortest
(*superl.*) CH. C. T. *A* 836.

　ssort-hēde, sb., *shortness*, AYENB. 99.

　scheort-līche, adv., *shortly, briefly*, A. R.
308 ; schortliche GREG. 962 ; shortlike ORM.
12788 ; shortli PR. C. 4848.

　schortnesse, sb., *O.E.* sceortness ; *shortness*, PR. P. 448.

schortin, v., *O.E.* sceortian ; *shorten, 'brevio,*'
PR. P. 448 ; schorten MAN. (H.) 175 ; schorte
[shorte] CH. C. T. *A* 791 ; scorteð (*pres.*)
HOM. I. 25 ; i-schorted (*pple.*) TREV. III.
259 ; *comp.* a-schortien.

schorved, *see* scurved.

schot, sb., *O.E.* (ge-)sceot,=*O.Fris.* scot, *O.
N.* skot, *M.L.G.* schot, *O.H.G.* scoz *n.; from*
schēoten ; *shot, missile*, TREV. IV. 431 ;
schot [shot] CH. C. T. *A* 2544.

　schotte-men, sb., *shooters*, D. ARTH. 2467.

　schot-windowe, sb., *sliding window* ; CH.
C. T. *A* 3358.

schot, *see* scot.

schote, sb., *cf. M.Du.*, *M.L.G.* schote *m.*;
?=schüte ; *shot, act of shooting*, S. & C. I.
xxxv ; schote '*tetanus*' PR. P. 448 ; at o schote
MAN. (F.) 1738 ; schotis (*pl.*) D. ARTH. 3627 ;
with stanis, schot and othir thing BARB. xvii.
35 ; *comp.* bowe-schote.

schótien, v., *O.E.* scotian, = *O.N.* skota ;
shoot : scotien mid heore flan LA3. 16555 ;
schote WILL. 178 ; PERC. 213 ; schote PR. C.
1906 ; schotte D. ARTH. 1992 ; schóteð
(*pres.*) H. M. 15 ; þei schoten MAND. 190.

schotten, *see* scotten.

schour, *see* schūr, schor.

schouren, v., *hasten* : hit is beter þat we to
heom schoure (*subj.*) ALIS. 3722.

schoute, *see* schūte.

schoutin, v., *shout*, '*vocifero*,' PR. P. 448 ;
shoute CH. TRO. ii. 614 ; TOR. 1854 ; MIR.
PL. 165 ; *see* schūte.

schouven, *see* schūven.

schovele, sb., *O.E.* sceofol, sceofl,=*M.L.G.*,
M.Du. schufel, schuffel ; *? from* schūvən ;
shovel, ROB. 99 ; schovele, schoule L. H. R.
42, 43 ; schoveles [shoveles] (*pl.*) LANGL. *B*
vi. 192 ; *comp.* stēor-scofle.

　schovelle-fōtede, adj., D. ARTH. 1098.

schovelin, v., *shovel, shuffle*, PR. P. 448 ;
shoveling (*pple.*) forþ [*v.r.* stumblende]
WICL. TOB. xi. 10*.

　schoveler, sb., *shoveler*, PR. P. 448.

schoven, v., *O.E.* scofian,=*M.H.G.* schoben ;
from schūven ; *shove, push* ; to shove hit up
LANGL. *C* xix. 20 ; ssofþ (*pres.*) at his dore
AYENB. 174.

schōwen, *see* schēawen.

schragge, sb., *scrag, jagged end* ; schragges
(*pl.*) D. ARTH. 3473.

schraggen, v., *trim, lop off* ; schragge PR. P. 448.

　schreggare, sb., *trimmer*, PR. P. 449.

schrápien, v.,=*M.L.G., M.Du.* schrapen, *M.
H.G.* schrapfen ; *scrape* : schreapien A. R.
116 ; shrapin LIDG. M. P. 184 ; schrapin,
scrapin PR. P. 450 ; skrape H. S. 7045 ;
scrápeþ (*pres.*) P. S. 239 ; LANGL. *B* xi. 423 ;
screapeð A. R. 344 ; scrapiþ N. P. 3 ; LUD.
COV. 41 ; scrépe (*subj.*) AYENB. 98 ; scrápe
(*imper.*) PALL. iv. 608 ; scrápede (*pret.*)
TREV. VIII. 213 ; scraped TRIAM. 392 ; ha
beoð iscrepte (*pple.*) ut of lives writ H. M. 23.

schrēade, sb., *O.E.* scrēade,=*M.L.G.* scrōde,
M.Du. schroode, *O.H.G.* (a-)scrōta ; *shred,
cutting* ; screade SHOR. 30 ; schrede PR. P.
448 ; AR. & MER. 1540 ; shrede HAV. 99 ;
schrēaden (*pl.*) A. R. 416.

schrēaden, v., *O.E.* screādian, = *M.L.G.*
scrōden, schrōden (*pple.* geschroden), *M.
Du.* schrooden, *O.H.G.* scrōtan (*pret.* screot) ;
tear up, cut ; shræden ORM. 8118 ; of shre-
den Gow. I. 138 ; schredin PR. P. 448 ; (heo)
schrēdeþ (*pres.*) MISC. 83 ; scrādieð (*imper.*)
LA3. 5866 ; schrēd (*pret.*) CH. C. T. (Wr.)
8103 ; þei schrede DEGR. 293 ; ANT. ARTH.
xliv ; schredde WICL. 4 KINGS iv. 39 ; schrede
(*pple.*) D. ARTH. 2688 ; *comp.* to-schrēden.

schreapien, *see* schrápien.

schrēawe, sb., *cf. O.E.* scrēawa (*barn mouse*) ;
shrew, evil person, VOC. 24 ; schrewe '*pravus*'
PR. P. 449 ; TREAT. 133 ; WILL. 4643 ; S. S.
(Wr.) 1705 ; CH. C. T. *D* 291 ; schrewe
levedi P. L. S. xiii. 202 ; shrewe, screwe P. S.
153, 154 ; shrewe LANGL. *B* x. 437 ; ssrewe
AYENB. 32 ; schrēawes (*pl.*) SHOR. 108 ;
schrewes MIRC 1481 ; schrewis AMAD. (R.)
xxxix ; schrewen BRD. 22 ; ne lust me wiþ þe
screwen chide O. & N. 287.

　schrēu-hēde, sb., *depravity*, P. L. S. xxiv.
31 ; SHOR. 158.

　schrēward, ssrēward, sb., *villain*, ROB.
(W.) 5441.

schrēawen, v., *deprave ; curse* ; schrewin '*de-*

pravo' PR. P. 449 ; i schrewe ʒou CH. C. T. *D* 446 ; **shrēwede** (*pret.*) LANGL. *C* vii. 75 ; **schrēwed** (*pple.*) [*mod.Eng.* shrewd] TREV. IV. 179 ; schrewed folk CH. BOET. i. 4 (18) ; schrewede havenes MAND. 46 ; schrewede aventures RICH. 2678 ; *comp.* **bi-schrēwen.**

schrēwed-hēde, sb., *wickedness,* ROB. (W.) 5676*.

schrēwednesse, sb., *corruption, fierceness,* TREV. I. 29, 153, V. 23.

schrēde, *see* **schrēade.**

schrēden, *see* **schrēaden, schrūden.**

schrenche, sb., ?=*M.Du.* schrenke (*cancellus*) ; *shed* : men iseoth ofte liʒtingue brenne hous and schrenche SAINTS (Ld.) xlvi. 588.

schrenchen, v., *O.E.* screncan,=*O.H.G.* screnchen ; *supplant, deceive,* JUL. 34 ; schrenche P. L. S. viii. 167 ; screnken ORM. 11861 ; **shrenche** (*pres. subj.*) HOM. II. 209 ; **schrencte** (*pret.*) KATH. 1189 ; *comp.* **a- schrenchen.**

schrēwed, pple., *see* **schrēawen.**

schrīchen, v.,=*O.L.G.* scrīcōn, *Swed.* skrīka ; *screech* ; schrichen FLOR. 454 ; M. T. 99 ; schrikin '*vagio*' PR. P. 449 ; shrike P. S. 158 ; TOWNL. 26 ; sriche GREG. 297 ; skriche REL. II. 192 ; S. S. (Web.) 1290 ; skrike PR. C. 7347 ; **schrīchest** (*ms.* schirchest) [schrichest] (*pres.*) O. & N. 223 ; shrikeð HOM. II. 181 ; sriken ANT. ARTH. x ; **scrīkede** (*pret.*) ED. 1671 ; schriched CH. C. T. *D* 4590 ; shrighte ALIS. 5738 ; GOW. III. 321 ; CH. C. T. *A* 2817 ; *comp.* **bi- schrīchen.**

schrīden, *see* **schrūden.**

schrift, sb., *O.E.* scrift *m.* (*confession*), *cf. O.H. G.* scrift (*writing*), *O.N.* skript *f.* (*script, picture, ecclesiastical discipline*) ; *from* **schrī- ven** ; *shrift, confession,* A. R. 8 ; his schrift ihalden HOM. I. 9 ; scrift LAʒ. 11391 ; soþne scrift FRAG. 8 ; schrifte '*confessio*' PR. P. 449 ; SHOR. 33 ; shrifte ORM. 9262 ; ssrifte AYENB. 14.

schrift-fader, sb., *confessor,* P. L. S. x. 69 ; schriftfader [schrefader] TREV. VI. 457 ; shriftfader P. 18.

schrifte, sb., *confessor* ; scrifte HOM. I. 19. **schriftes** (*gen.*) A. R. 418.

schrīken, *see* **schrīchen.**

schril, adj., *cf. L.G.* schrell ; *shrill* ; schrille (*adv.*) LANGL. *C* vii. 46* ; ORF. 102 ; (*ms.* schirlle) GREG. 415 ; **schrille** [shrille] (*pl.*) CH. C. T. *B* 4585 ; *see* **schil.**

schrillen, v., *sound shrill* ; **schrilliþ** (*pres.*) ALIS. 777 ; scrilles and scrikes ANT. ARTH. xlii ; *see* **schillen.**

schrimp, sb., *shrimp,* PR. P. 449 ; **schrimpes** (*pl.*) CH. C. T. *B* 3145 ; schrimpe (*applied to a dragon*) D. ARTH. 767.

schrīn, sb., *O.E.* scrīn,=*O.N.* skrīn, *O.H.G.* scrīne, *Lat.* scrīnium ; *shrine,* ALIS. 4670 ;

schrin, shrin FER. 2116, 2134 ; shrin A. D. 258 ; **schrīne** (*dat.*) BEK. 2434.

schrīnin, v., *lay in a shrine,* PR. P. 449 ; **ischrīned** (*pple.*) BEK. 2426.

schrinken, v., *O.E.* (for-)scrincan,=*M.Du.* schrinken, *O.Swed.* skrinka ; *shrink, contract* ; schrinke MISC. 80 ; S. A. L. 161 ; þanne his senewes gonne to schrinke TREV. III. 411 ; scrinkin LAʒ. 2278 ; **scrinkeþ** (*pres.*) P. S. 158 ; him scrinkeþ þa lippen FRAG. 5 ; **schrank** (*pret.*) GAW. 2313 ; schronk CH. BOET. i. 1 (5) : **shrunken** (*pple.*) GOW. I. 98 ; your shrunken lippis AN. LIT. 84 ; shrunke LIDG. M. P. 201 ; *comp.* **for-schrinken.**

schrippe, *see* **scrip.**

schrīven, v., *O.E.* scrīfan (*prescribe penance*), =*O.Fris.* skrīva (*write, prescribe penance*), *O.N.* skrīfa, *O.H.G.* scrīban (*write*) ; *from Lat.* scrībere ; *shrive, prescribe penance, confess* : ich chulle schriven me A. R. 340 ; he shal shrifen þe ORM. 6128 ; scriven LAʒ. 32074 ; schrive (*ms.* schryve) MISC. 79 ; SHOR. 33 ; shrife PR. C. 3508 ; **schrif** (*imper.*) A. R. 266 ; **scrāf** (*pret.*) LAʒ. 32192 ; schraf M. H. 10 ; schrof A. R. 68 ; schrof [shrof] LANGL. *A* iii. 44 ; **scrive** (*subj.*) HOM. I. 25 ; **schriven** (*pple.*) CH. C. T. *D* 1440 ; shriven HAV. 364 ; *comp.* **ʒe-schrīven** ; *deriv.* **schrift.**

ssrīvere, sb., *confessor,* AYENB. 174.

schrob, sb., *O.E.* scrobb ; *shrub* ; **schrobbes** (*pl.*) TREV. II. 61 ; shrobbis [schrubbes] LANGL. *C* i. 2.

Schrop-schīre, pr. n., *O.E.* Scrobscīre ; *Shropshire,* P. L. S. xiii. 30.

schrogge, *see* **scrogge.**

schrōp, sb., *? from root of* **schrāpien** ; *? scraping* ; schroup DEP. R. ii. 154.

schrūd, sb., *O.E.* scrūd,=*O.N.* skrūd ; *mod. Eng.* shroud ; *garment,* R. S. v (MISC. 172) ; shrud ORM. 17591 ; HAV. 303 ; scrud P. L. S. viii. 183 ; srud GEN. & EX. 795 ; þet ilke ssroud AYENB. 258 ; **schrūde** (*dat.*) A. R. 300 ; schroude A. P. ii. 47 ; **scrūd** [srud] (*pl.*) LAʒ. 10180 ; schrudes MARH. 19 ; shroudes LANGL. *B* PROL. 2 ; **scrūden** [scrude] (*dat. pl.*) LAʒ. 5362 ; *comp.* **here-, kine-scrūd.**

schrūden, v., *O.E.* scrȳdan,=*O.N.* skrȳða ; *enshroud, clothe,* A. R. 214 ; scruden LAʒ. 8945 ; shrude P. S. 153 ; shriden ORM. 3676 ; sriden GEN. & EX. 351 ; schride MAP 337 ; shride HAV. 963 ; schrede AM. & AMIL. 934 ; shrede HORN (R.) 718 ; ssrede AYENB. 90 ; **schrūdde** (*pret.*) KATH. 912 ; A. R. 302 ; schrudden MISC. 81 ; **shrid** (*pple.*) ORM. 137 ; HAV. 978 ; schred GREG. 535 ; *comp.* **ʒe-schrūden.**

schruggin, v., *cf. Dan.* skrugge, skrukke ; *mod.Eng.* shrug ; *shiver,* '*frigulo,*' PR. P. 449.

schucke, sb., *O.E.* scucca, sceucca ; *adversary; devil,* MARH. 7 ; A. R. 316, 326 ; H. M. 41 ; scucke LAʒ. 276 ; **schucken** (*pl.*) JUL. 56.

schudde̍, sb., *shed*, PR. P. 449.

schuderen, v., *cf. M.Du.* schuderen; *shudder*, KATH. 809; shoddreþ (*pres.*) SPEC. 110; schudrinde (*pple.*) MARH. 15; schoderide (*pret.*) D. ARTH. 2106; ischodred þen SHOR. 92.

　schudering, shodering, sb., *crashing*, ALEX. (Sk.) 2624.

schüften, *see* schiften.

schühten, v., *? cf. M.H.G.* schiu(h)zen; *terrify*; schüchteð (*pres.*) A. R. 312*.

schül, *see* schil.

schüld, sb., *O.E.* scyld, *cf. O.N.* skuld, skyld, *O.H.G.* scult; *from* schal; *debt, fault*; sculd FRAG. 5.

schulde, *see* scholde.

schülden, *see* schilden.

schüldig, adj., *O.E.* scyldig, = *O.H.G.* sculdīg; *from* schüld; *guilty*; sculdig HOM. I. 113; sculdi FRAG. 5; schuldi KATH. 2296; A. R. 206; H. M. 35; sceldig MAT. xxiii. 16; *comp.* dǣþ-shildiȝ.

schuldre, sb., *O.E.* sculdor, = *O.Fris.* sculder, *M.L.G.* schulder, *O.H.G.* scultera; *shoulder*; (*ms.* shulldre) ORM. 4776; schulder WILL. 3290; A. P. ii. 981; shulder PR. C. 5206; schuldre (*dat.*) MISC. 49; BEK. 2234; shuldre HAV. 604; sculdre [soldre] LAȜ. 19127; shuldre (*pl.*) SPEC. 52; schuldren JUL. 49; ssoldren ROB. 313.

　schuldir-bōn, sb., = *M.H.G.* schulterbein; *shoulder-bone*, PR. P. 449; scholderbon OCTOV. (W.) 1139.

　schulder-blad, sb., = *M.H.G.* schulterplat; *shoulder-blade*, IW. 2614.

schuldren, v., *shoulder*; shuldreden (*pret. pl.*) HAV. 1056.

schulen, *see* sculen.

schulle, sb., = *Ger.* scholle, *Swed.* skolla; *plaice*, HAV. 759.

schunchen, v., *terrify*, JUL. 34; we schuncheð (*pres.*) hine veor a wei A. R. 312; *comp.* a-schunchen.

schunien, v., *O.E.* scunian, sceonian; *shun, abhor, avoid*, KATH. 811; A. R. 82; H. M. 35; scunien, sceonien LAȜ. 14605, 14872; shunen ORM. 4502; sunen REL. I. 217; GEN. & EX. 1864; shonie LANGL. *B* PROL. 174; mannes mone shonie AL. (T.) 161; schone S. A. L. 159; MAP 336; JOS. 496; D. ARTH. 1719; schonie TREV. VI. 27; schunest (*pres.*) O. & N. 590; schoneþ TREAT. 133; schones ['*reprobat*'] PS. xxxiii. 10; we schoneþ '*devitamus*' TREV. III. 459; men þe schonieþ O. & N. 792; hi shoneþ ALIS. 4919; schunie (*subj.*) A. R. 92; scunede (*pret.*) HOM. I. 79; scunede [sonede] LAȜ. 3112; ich wolde þet oðre schuneden ase ȝe doð gederunge A. R. 286; *comp.* an-, bi-schunien.

schunten, v., *avoid*; schount ALEX. (Sk.) 180; þe skerre hors . . . þet schuntes (*pres.*)

A. R. 242*; schuntes, schountes D. ARTH. 1055, 3816; schunt(e) (*pret.*) GAW. 1902.

schüp, *see* schip.

schüpene, sb., *O.E.* scypen; *shed, stable*, LEG. 88; schipne [shepne] CH. C. T. *A* 2000; shipnes (*pl.*) *D* 871.

schüppen, schüppere, *see* scheppen, scheppere.

schūr, sb., *O.E.* scūr, *cf. O.H.G.* scūr *m.*, *O.N.* skūr *f.*; *shower, rain*; schour WILL. 4514; TRIST. 1936; B. DISC. 1156; schour [shour] CH. C. T. *A* 3520; shūres (*pl.*) HOM. II. 175; sures MISC. 9; schoures MAN. (H.) 333; shoures LANGL. *B* xviii. 409; mi sharpe shoures smerte CH. M. P. (Sk.) xxii. 66.

schūr², sb., *O.N.* skūrr; *shed*; schūrris (*pl.*) ALEX. (Sk.) 4049.

schürte, sb., *O.N.* skyrta, = *M.L.G.*, *M.Du.* schorte; *shirt*, BEK. 260; shurte HOM. II. 139; scurte [seorte] LAȜ. 23761; schurt(e) [schirte, scherte] ['*camisia*'] TREV. V. 445; schirt(e) '*interula*' PR. P. 447; schirte [sherte] ['*subucula*'] WICL. LEV. viii. 7; shirte HAV. 768; scherte [sherte] CH. C. T. *A* 1566; sherte LANGL. *B* xv. 330; sserte AYENB. 191.

　shürte-lappe, sb., *shirt-lap*, HORN (R.) 1209.

schürten, v., *? cf. M.H.G.* schürzen; *amuse*; schürteð (*pres.*) ou to gederes A. R. 422.

schüte, scüte, sb., *O.E.* scyte, = *O.Fris.* schet, *O.H.G.* scuz *m.*; *from* schēoten; *shoot, shot*, A. R. 60, 62; þene scute LAȜ. 1461; ech ȝeres scute (*sprout*) L. H. R. 28; such shute com in þe woman(n)es hed A. D. 262; ssute ROB. 537; *comp.* ūt-schüte.

schūte, sb., = *M.Du.* schuite, *O.N.* skūta; *flat bottomed boat*; shoute M. T. 120; schoute (*pl.*) ALIS². (Sk.) 484; schoutes RICH. 4785.

schütel, sb., *O.E.* scytel; *from* schēoten; *shuttle*; sc(h)itil, webstaris instrument PR. P. 447; schetil VOC. 235.

schütel², adj., *O.E.* scytel; *headstrong, unstable*; schitil '*praeceps*' PR. P. 447.

schütels, sb., *O.E.* scytels; *bolt*; scutles HOM. I. 127.

schütte, sb., *O.E.* scytta, = *M.L.G.*, *M.Du.* schutte, *O.N.* skyti, *O.H.G.* scuzzo; *shooter, archer*; scütten (*pl.*) LAȜ. 27046.

schüttel, sb., *O.E.* scyttel; *bolt*; schittil '*pessulus*' PR. P. 447; tvo ssetteles AYENB. 94.

schütten, v., *O.E.* scyttan, *cf. M.Du.*, *M.L.G.* schutten, *O.Fris.* sketta, *M.H.G.* schützen; *shut*; shutte [shette] LANGL. *B* PROL. 105; schitte WICL. IS. xxii. 22; schitte þe dore FER. 2027; shetten ALIS. 5821; ssette AYENB. 179; ssete (*imper.*) AYENB. 210; schutteð al þet þurl to A. R. 96; schette [shette] (*pret.*) CH. C. T. *A* 3634; shitte PARTEN. 3295; schutten ALIS. 2640; shetten GEN. & EX. 1078; schetten WILL. 3267; schet (*pple.*)

A. P. iii. 452; ischet S. S. (Web.) 2455; FER. 1952; *comp.* bi-, for-schütten.

schüven, schüven, v., *O.E.* scūfan, scēofan, *M.L.G.* schüven, *O.Fris.* skúva, *O.H.G.* sciuban, *mod.Eng.* shove; *push,* A. R. 314; scuven [seve] LA3. 17396, 21590; ure loverd . . . him . . . wile shufe fro him HOM. II. 53; shivin, shive LANGL. *C* xix. 20*; schive . . . a wei ['*abigere*'] WICL. JUDG. xvi. 19*; schouve [shouve] CH. C. T. *A* 3912; scheve BEV. 1407; schēf (*pret.*) MISC. 51; BRD. 19; FER. 1369; þæ wile þe he þa scipen ut scæf [sef] LA3. 9366; schof ALIS. 4250; S. S. (Wr.) 1411; shof HAV. 892; GOW. I. 165; CH. P. F. 154; scufen, scuven [soven] LA3. 7859, 20925; shoven ALIS. 5889; schove BEK. 2241; shoven (*pple.*) GOW. III. 202; schove [shove] CH. C. T. *F* 1281; iscoven HOM. I. 129; *deriv.* schēaf, schoven.

shouving, sb., *pushing,* CH. C. T. *H* 53.

science, sb., *Fr.* science; *knowledge,* CH. C. T. *G* 896; PR. C. 5946; LANGL. *A* xi. 145.

sciment, sb., *for* ciment; PALL. vi. 190.

scip, *see* schip. sclabbe, *see* slabbe.

sclæht, *see* slaht.

sclákien, *see* slákien. sclat, *see* slat.

sclaundre, sb., *O.Fr.* esclaundre; *slander; scandal;* ROB. 333; sclaunder WILL. 4045; sclondre AYENB. 6; sclaunder, slaunder PR. P. 458; slaundre FER. 132; *see* scandle.

sclaundren, v., *slander; cause to offend;* sclaundre CH. C. T. *G* 998; schal sclaundre ['*scandalizaverit*'] WICL. MK. ix. 41; sclaundrid (*pple.*) LUD. COV. 7.

sclaundrer, sb., *slanderer;* sclaunderers (*pl.*) PR. C. 7042.

sclavine, sb., *O.Fr.* esclavine; *pilgrim's cloak,* HORN (L.) 1054; sclavene VOC. (W.W.) 773; slavine ISUM. 497; slavein DEP. R. iii. 236; slaveine PR. P. 458; slavin CATH. 343.

scleire, *see* sleir. sclender, *see* slender.

sclēpen, *see* slēpen. sclíce, *see* slíce.

sclūse, sb., *O.Fr.* escluse; *sluice,* AYENB. 255.

scō, *see* schō.

scof, sb., = *O.Fris.* schof, *cf. O.H.G.* scopf; *scoff,* MIRC 902; bourd and scof MAN. (F.) 7586; kof ALIS. 6986; scoffes (*pl.*) AYENB. 128.

scoffing, sb., *scoffing,* LANGL. *B* xiii. 277.

scofle, *see* schovele.

scōgh, sb., *O.N.* skōgr; *wood,* C. M. 15826; scōghe (*dat.*) ALEX. (Sk.) 3915; scōghes (*pl.*) ANT. ARTH. v.

scólde, sb., *scold, blamer,* PR. P. 449; REL. I. 188; LANGL. *B* xix. 279; E. T. 653; skolde P. S. 335; ?skalde Iw. 69.

scólden, v., *scold, blame, reprimand;* scolde LANGL. *B* ii. 81.

scōle, sb., *O.E.* scōl (ÆLFR. 304), *cf. O.N.* skōli, *O.H.G.* scuola; *school,* LA3. 9897; A. R.

422; AYENB. 34; LANGL. *B* xx. 271; CH. C. T. *A* 125; drawing [oxen] bi the horne is noo goode scole (*? method, practice*) PALL. ii. 14.

scōle-fēre, sb., *school-fellow,* TREV. III. 449; VII. 397.

scōle-maistre, sb., *schoolmaster,* KATH. 522; scolmeistre A. R. 422.

scōle, *see* scale (scāle).

scoleie, v., *attend school, study,* CH. C. T. *A* 302; scolaid (*pret.*) ALEX. 645.

scōler, sb., *O.E.* scōlere, = *O.H.G.* scuolāre; *scholar,* MIRC 845; CH. C. T. *A* 3190; scōlers (*pl.*) AYENB. 39.

scolle, *see* sculle. scōm, *see* scūm.

scomfiten, v., *for* discomfiten; *discomfit;* sconfited [scomfitede] (*pret.*) TREV. III. 375; sckonfet ALEX. (Sk.) 4802.

sconde, *see* schande. scōne, *see* schēne.

sconse, sb., *O.Fr.* esconse; *sconce, candlestick fixed in a wall,* '*absconsus*,' VOC. (W. W.) 649; sconce '*absconsa*' PR. P. 450.

scōp, sb., *O.E.* scōp, sceōp, = *O.H.G.* scōf; *from* scheppen; *poet;* scōpes (*pl.*) LA3. 30615; *comp.* lēoð-scōp.

scōpe, sb., = *Swed.* skōpa, *M.H.G.* schuofe; *scoop,* '*alveolus*,' PR. P. 450; scōpes (*pl.*) MAN. (F.) 8168.

scōpen, v., *lade out (water),* A. P. iii. 156; do scope þis water MAN. (F.) 8164.

scorchen, v., *from* scóren; *score, cut;* scortche (*imper.*) B. B. 80.

scorchen[2], v., = scorcnen; *scorch*: þe peoples þat þe violent wind notus scorchiþ [*v. r.* scorklith] (*pres.*) CH. BOET. ii. 6 (55); scorched (*pple.*) PARTEN. 3678.

scorclen, v., = scorcnen, *as* brütlen = brütnen, drunklen = drucnen; *scorch*; scorklin '*ustulo*' PR. P. 450; scorklith [scorchiþ] (*pres.*) CH. BOET. ii. 6 (55).

scorcnen, v., *for* *scorpnen [*O.N.* skorpna] *as* droukin *for* droupnen; *crack, furrow*: þat te land was . . . scorcned (*pple.*) þurh þe druhþe ORM. 8626.

scóre, sb., *O.E.* scor, = *O.N.* skor; *from* schéren; *notch; number; score, twenty;* PR. P. 450; AR. & MER. 3109; LAUNF. 419; four score BRD. 14; ten score LANGL. *B* x. 180; skore oþer writ E. G. 362; schore DEP. R. ii. 42; ani shore or hol or reft R. R. 2660; foure schore WILL. 1102; score (*rank, manner*), MAN. (F.) 5028; score (*track*), MAN. (F.) 3377, 13694.

scorel, sqverel, sb., *O.Fr.* escurel, esqvirel; *squirrel,* PR. P. 450.

scóren, v., = *O.N.* skora (*cut*), = *M.L.G., M. Du.* schoren (*split*); *score*; scorin taliis '*tallio dico*' PR. P. 450; score (*printed* seore) PALL. vi. 119; on his 3erde skore shalle he alle messis in halle þat servet be B. B. 312; *? comp.* wiþ-scóren.

scóren², v.,=*M.H.G.* schoren ; *? thrust*: (þet) softe was iwend to þe sulven & efre þet scerpe scóred (*pple.*) me touwar(d) FRAG. 8.

scorf, *see* scurf.

scorge, sb., *O.Fr.* escorgie ; *scourge*, PR. P. 450 ; scourge WICL. JOHN ii. 15 ; schurge A. R. 418 ; scurges (*pl.*) MISC. 140 ; scourgen MARG. 121 ; sqvorges ['*flagella*'] PALL. iii. 113.

scorgen, v., *scourge* ; scourged(e) (*pret.*) ROB. 263.

scorn, scornen, *see* scarn, scarnen.

scorpioun, sb., *O.Fr.* scorpion, scorpiun ; *scorpion*, AYENB. 62.; LANGL. *B* xviii. 153 ; scorpiun A. R. 198.

scort, *see* schort.

scot, sb., *O.E.* scot, sceot, *O.Fris.* scot, schot, *M.H.G.* schoz ; *from* schēoten ; *scot, tribute, payment*, ROB. 296 ; P. S. 71 ; OCTOV. (W.) 282 ; ['*symbola*'] WICL. PROV. xxiii. 21 ; þet scot AYENB. 51 : schot S. & C. II. lxxiv.

Scot, sb.; *Scot*, '*Scotus*,' PR. P. 450 ; Scot-tes [*O.E.* Scottas, *O.N.* Skotar] (*pl.*) LA3. 10808.

   Scot-lond, pr. n., *O.E.* Scotland ; *Scotland*, LA3. 2130 ; O. & N. 908.

   Scot-lēode, sb., pl., *Scottish people* (*pl.*), LA3. 20047.

   Scot-þeode, pr. n., *Scottish nation*, LA3. 20417.

scotifer, sb., *Lat.* scūtifer ; *shieldbearer*; skotiferis, skottefers (*pl.*) D. ARTH. 3034, 2468.

scotele, *see* scutele. scotien, *see* schotien.

scotten, v.,=*M.H.G.*schozzen; *pay scot*, HOM. I. 201 ; 3e schulen scotten (*participate*) mid him of his blisse ine heovene A.R. 348; 3e schot-teð (*pres.*) IBID. ; ne gabbe þu ne schotte (*imper.*) ne chid þu wiþ none sotte REL. I. 183 (MISC. 126).

Scottisch, adj., *cf. O.E.* Scyttisc ; *Scottish* ; Scottishe (*pl.*) P. S. 222.

scoulen, *see* sculen.

scourin, v., *cf. M.L.G., M.H.G.* schuren ; *? from O.Fr.*escurer; *scour; '*erugino, verbero*'* PR. P. 450 : scoure L. C. C. 9.

scourge, *see* scorge.

scoute, sb., *O.Fr.* escoute ; *scout*, S. S. (Wr.) 2218 ; LUD. COV. 136.

   scoute-wach, sb., *guard*, A. P. ii. 838 ; skoutte waches (*pl.*) D. ARTH. 2468.

scoute ², sb., *O.N.* skūti ; *cave formed by projecting rocks* ; scoutes (*pl.*) GAW. 2167.

scouten, v., *O.Fr.* escouter, ascouter ; *scout, pry* ; skoute (*pres.*) A. P. 483.

scouter. ², v., *O.N.* skūta ; *project* ; scūtis (*pres. pl.*) ALEX. (Sk.) 4865.

scoverour, sb., *for* discoverour ; *explorer*,

scout ; scoverours (*pl.*) D. ARTH. 3118 ; scurreours BARB. xiv. 487*.

scragen, adj., *scraggy* : scragen & unefne A. R. 4* ; *see* scroggi.

scrápe, scrápien, *see* schrápien.

scrappe, sb., *scrap* ; scrappes (*pl.*) TREV. I. 15.

scratte, sb., *O.N.* skratti,=*O.H.G.* scraz ; *wizard ; monster ; '*armifraudita*'* VOC. (W. W.) 695 ; skratt VOC. 217.

scrattin, v., *cf. Dan.* kratte, *Swed.* kratta, *scratch*, '*scalpo, scabo*,' PR. P. 450; scratte PR. C. 7378 ; scratteð (*pres.*) A. R. 186*.

scrēade, *see* schrēade.

scrēamen, v., *scream*, H. M. 37 ; scrēmeþ (*pres.*) P. S. 158.

scrén, sb., *O.Fr.* escren, escran ; *screen*, VOC. 197 ; PR. P. 450.

screnchen, screnken, *see* schrenchen.

scrēwe, *see* schrēawe.

scrībe, sb., *Lat.* scrība ; *scribe*, WICL. MAT. viii. 19 ; scrībes (*pl.*) LANGL. *C* xxi. 26.

scrīchen, scrīken, *see* schrichen.

scrift, *see* schrift.

scrimming, sb., *from O.Fr.* escrimer ; *see* skirmen² ; *skirmishing*, BARB. xix. 521.

scrinken, *see* schrinken.

scrippe, sb., *O.N.* skreppa ; *scrip, bag*, '*pera*,' PR. P. 450; HORN (R.) 1069 ; TREV. III. 309 ; CH. C. T. *D* 1737 ; scrippe LANGL. *B* v. 542 ; schrippe *A* vi. 26.

scriptūre, sb., *O.Fr.* escripture ; *writing* ; i have put in it scripture CH. BOET. i. 4 (17) ; scripture is hir name LANGL. *B* x. 150.

scrit, sb., *O.Fr.* escrit ; *writing* : bi skore oþer bi scrit E. G. 357 ; scrites (*pl.*) MAN. (F.) 8071.

scrið, sb., *O.E.* scriðe, *O.N.* skriðr (*gliding*) ; *urging, entreaty*, GEN. & EX. 1419.

scrīðen, v., *O.E.* scrīdan,=*O.N.* skrīða, *O. L.G.* scrīdan, *O.H.G.* scrītan ; *glide ; escape*: þa com Scottene king scriðen to hirede LA3. 10799; skrith(e) MIN. v. 68 ; scrād (*pret.*) LA3. 4109 ; Tarbis him scrod (*urged*) GEN. & EX. 2695 ; scriðen (*pl.*) LA3. 8405 ; *comp.* þurh-scrīðen.

scrivein, sb., *O.Fr.* escrivain, *mod.Eng.* scrivener ; *scribe*, WICL. IS. xxxvi. 22 ; scrivener PR. P. 450 ; scriveins (*pl.*) AYENB. 44 ; scrivaines LANGL. *C* xii. 97.

scrīven, *see* schrīven.

scrobben, *see* scrubben.

scrof, adj., *cf.* scurf ; *rough* ; (*ms.* strof) A. P. ii. 1546.

scrof, *see* scurf.

scrog, sb., *cf. Dan.* skrog, *? O.N.* skrokkr, *Scotch* scrog ; *shrub, brushwood* ; skrogges (*pl.*) D. ARTH. 1641 ; shrog(g)es TOWNL. 110.

scroggi, adj., *covered with brushwood* ; þe

wei was stoni, thorni and scroggi GEST. R.
(H.) 19.

scroue, sb., *O.Fr.* escroue, escroe (*cédule*);
scroll, PR. P. 450; scrowe A. R. 282; **scrowis**
(*pl.*) WICL. MAT. xxiii. 5.

scrubben, v., = *L.G.* schrubben, *Dan.*
skrubbe; *scrub*; **scrobbeþ** (*pres.*) ALIS. 4310.

scrūd, scrūden, *see* **schrūd, schrūden.**

scruf, *see* **scurf.**

scrüple, sb., *? O.Fr.* scrupule; *scruple*; scriple
of herte WICL. 1 KINGS xxv. 31; (*the weight
so called*) PALL. ii. 418; **scripilles** (*pl.*)
[*'obolos'*] WICL. EX. xxx. 13.

scucke, *see* **schucke.**

scūe, sb., *O.E.* scūwa, scūa, *cf. O.N.* skuggi;
shadow; scu PR. P. 450; **skūez** (*pl.*) GAW.
2167.

scuften, *see* **schuften.**

[**scül**, adj., *cf. O.H.G.* scelah, *O.N.* skialgr
(*oblique*); *deriv.* **scülen.**]

    scül-ēiȝed, adj., *O.E.* sceolēged; *squinting*,
FRAG. 3.

scüld, *see* **schüld.**   **scülden**, *see* **schílden.**

scüle, *see* **schile.**

scülen, v., *O.E.* (be-)scylian, = *O.H.G.* scile-
hen; *squint at*; schulen A. R. 210.

scülen, v., = *Dan.* skūle, *Swed.* skūla, *M.L.G.*
schülen, *M.Du.* schuilen; *scowl*: skoul(e)
and stare PR. C. 2225; scoule '*oboculo*' PR.
P. 450.

scülken, v., *cf. Dan.* skulke, *L.G.* schulken;
from **scülen**, *as* **lŭrken** *from* **lūren**; *sculk,
lie hid; loiter*; sculke PS. xxxviii. 12; **skülkeþ**
(*pres.*) GOW. II. 93; sculke PR. C. 1788;
skŭlked (*pret.*) MAN. (F.) 6976.

    scouking, sb., *skulking, cowardice*, BARB.
viii. 140; into scouking *traitorously* vii. 130.

    scülkere, sb., *skulker*, REL. I. 133; skulker
D. ARTH. 3119; sculcare PR. P. 451; **scöl-
kers** (*pl.*) TREV. VII. 491.

sculkeri, sb., *ambush*; skoulkeri D. ARTH.
1644.

sculle, sb., *skull*, '*cranium*,' PR. P. 450; BEK.
2166; sculle [skulle] CH. C. T. *A* 3935;
schulle A. R. 296; skulle [scolle] TREV. V.
371; scolle VOC. 179; ROB. 16; FER. 353.

scūm, scōm (? **scŭm, scŏm**), sb., *O.N.*
skūm, = *O.H.G.* scūm; *scum*, '*spuma*,' PR. P.
449; scom REL. I. 52; **scōme** (*dat.*) AYENB.
44.

scūmen, v., = *O.H.G.* scūman; *foam*; (*ms.*
scummyn) PR. P. 450; (h)a **skūmede** (*? ms.*
skuntede) (*pret.*) als a bore FER. 3888.

scumer, sb., *O.Fr.* escumeur; *pirate, rover*;
scummar` BARB. xiv. 375; **skumers** [sku-
meres] (*pl.*) TREV. IV. 175.

scunien, *see* **schunien.**

scunneren, v., *Scotch* scunner; *loathe, avoid*;
skounrand (*pple.*) BARB. v. 201*; **skun-
nirrit** [scounrit] (*pret. pl.*) BARB. xvii. 651.

scurf, sb., *O.E.* scurf, scruf (LEECHD. I. 316);
M.L.G. schorf, *O.H.G.* scorf; *mod.Eng.*
scurf; *rubbish*, PR. P. 451; C. M. 11823;
scorf PALL. vi. 138; scrof VOC. 179; þe
schroff and schroup DEP. R. ii. 154.

scurge, *see* **scorge.**   **scürte**, *see* **schürte.**

scurved, adj., *O.E.* scurfed, = *M.L.G.* schor-
vet; *scabby* : þe ssorvede (*printed* ssornede)
AYENB. 224.

scut, adj., *short;* '*curtus, brevis*' PR. P. 451.

scut[2], sb., *short coat*, '*nepticula*,' PR. P. 451.

scüte, *see* **schüte.**

scutele, sb., *O.E.* scutel, = *M.L.G.* schutel,
schotel, *O.H.G.* scuzela; *dish*; scotile
'*scuttella*' VOC. 257.

scūven, *see* **schüven.**

sé, dem. prn. & def. art., *O.E.* se, *O.N.*, *Goth.*
sa, *Gr.* ὁ; *that; the*; se deofel MAT. iv. 5; se
man, se, sa gealle LEECHD. III. 84; se king,
se burch, se fir SAX. CHR. 249; se fader
HOM. I. 219; se man HOM. II. 223; se tiX-
inge 31; se king MISC. 26; se þe *he who*
221; se Xe HICKES I. 222; ze þet AYENB.
117; fram se biscop SAX. CHR. 250; **sie**
[*O.E.* sēo, sīo, *cf. O.N.* sia, sū, *Goth.* sō], (*fem.*)
MK. xv. 40; si modor HOM. I. 233; si mirre
MISC. 28; zi þet AYENB. 102; si [*?* = *Goth.* si,
O.L.G., O.H.G. siu, sia, sie] (*she*) ȝeleste sume
wile HOM. I. 235; **ses** (*? gen., for* þes) he-
lendes IBID.; **sa** (*? pl.*) handa LEECHD. III.
112; abutan sa earan 94.

se, *see* **swá.**

sē, sb., *O.Fr.* sed; *see* (*of a bishop*); *seat,
throne*; CHR. E. 363; hus se þar he sat LANGL.
C i. 114; ȝe were sette in ȝoure se DEP. R. i.
86; oure sire in his see above þe sevene sterris
iii. 352; þe king sat adoun in is (*r.* his) sce
SAINTS (Ld.) xxvii. 815; see BEK. 809.

sē, sea, *see* **sǣ.**   **sealfe**, *see* **salfe.**

sēam, sb., *O.E.* sēam, = *O.Fris.* sām, *M.L.G.*
sōm, *O.H.G.* saum, soum, (*border*), *O.N.*
saumr (*seam*); *seam;* seem '*sutura*'
PR. P. 452; þe coote was wiþout(e) seem
WICL. JOHN xix. 23; **sēames** (*pl.*) HOM. I.
225; MAND. 9.

sēam[2], sb., *O.E.* sēam, = *O.L.G.* sōm, *O.H.G.*
saum, soum, *from Lat.* sagma, *Gr.* σάγμα;
horse load, burden; seem PR. P. 452; a
seem of whete LANGL. *A* iii. 40; sæm
ORM. 3724; **sēames** (*pl.*) '*sarcinas*' LK.
xi. 46; semes O. & N. 775; GEN. & EX.
1368; WILL. 2554.

sēamære, sb., *O.E.* sēamere; *one who
sews*, '*sartor*,' FRAG. 2.

sēamestre, sb., *O.E.* sēamestre; *sempstress*;
sĕmsteris (*pl.*) D. TROY 1585.

sēar, sb., *O.E.* sēar, *cf. M.L.G.* sōr; *sere, dry,
arid*; seer '*aridus*' PR. P. 453; MAN. (H.)
18; seere braunches CH. R. R. 4749.

seernesse, sb., *aridity*, PR. P. 453.

seare, sb., *O.E.* searo, *cf. O.N.* sörvi, *Goth.* sarwa, *O.H.G.* (ge-)sarwe, serwe ; *armour, apparel* ; sere HOM. I. 33.

sēarien, v., *O.E.* sēarien, = *M.L.G.* sōren ; *become sere ; wither* ; seerin '*aresco*' PR. P. 453 ; sere ALIS. 796 ; **sērid** (*pple.*) LIDG. M. P. 241 ; *comp.* **a-sēarien.**

sēað, sb., *O.E.* sēað, = *O.Fris.* sāth, *M.L.G.* sōd, *M.H.G.* sōt ; *from* **sēoðen** ; *grave, pit* ; **sēaþe**, sæþe (*dat.*) FRAG. 6, 7 ; **sēaðen** (*dat. pl.*) LAȝ. 841.

sēau, sb., *O.E.* sēaw, = *O.H.G.* sou ; *juice, pottage, savoury dish* ; seu (*ms.* sew) VOC. 266 ; PR. P. 454 ; REL. I. 81 ; seeu (*ms.* seew) WICL. GEN. xxvii. 4 ; sewe GOW. II. 325 ; sǣw ORM. 994, 1470 ; **sēwe** (*dat.*) GAW. 892 ; L. C. C, 43 ; **sēwes** (*pl.*) CH. C. T. *F* 67 ; *comp.* **éle-sǣw.**

sēc, *see* **sēoc.**

sēchen, v., *O.E.* sēcan, sēcean, = *O.Fris.* sēka, *O.N.* sœkja, *O.L.G.* sōkean, *Goth.* sōkjan, *O.H.G.* suochan ; *seek*, A. R. 164 ; C. L. 26 ; sechan LK. xii. 29 ; ANGL. VII. 220 ; sechen, sæchen LAȝ. 12863, 17321 ; seche O. & N. 1759 ; ROB. 209 ; WILL. 223 ; MIRC 651 ; GOW. I. 267 ; MAND. 59 ; TOR. 1013 ; sechen to chirche JUL. 45 ; he scholde þider seche JOS. 528 ; zeche AYENB. 94 ; siche MARG. 28 ; FER. 1935 ; seche [seke] LANGL. *A* PROL. 47 ; CH. C. T. *A* 17, 784 ; seken ORM. 2718 ; GEN. & EX. 3598 ; sekin PR. P. 451 ; seke ISUM. 130 ; **sēche** (*pres.*) MARH. 14 ; secst HOM. II. 29 ; secheð HOM. I. 173 ; secheþ SPEC. 26 ; secð HOM. II. 226 ; sekþ MISC. 65 ; (ȝe) secheþ MISC. 42 ; (hio) secheð MAT. vi. 32 ; **sēch** (*imper.*) A. R. 336 ; HOM. II. 9 ; **sēchinde** (*pple.*) HOM. I. 127 ; sechinde [sechinge] LAȝ. 1383 ; **sōhte** (*pret.*) KATH. 976 ; SPEC. 41 ; (*ms.* sohhte) ORM. 2942 ; souhte A. R. 130 ; soȝte HORN (L.) 39 ; sohten LAȝ. 11168 ; **sōht** (*pple.*) SPEC. 53 ; (*ms.* sohht) ORM. 6454; sogt, sowt GEN. & EX. 848, 2870 ; *comp.* **a-, æt-, an-, bi-, ȝe-, of-, þurh-, under-sēchen.**

sēkere, sb., *seeker*, WICL. GEN. xxxi. 35.

sēching, sb., *seeking*, WILL. 2190.

sēclien, v., *O.E.* sȳclian ; *to be ill* ; **sēcli** (*pres. subj.*) A. R. 50 ; **sǣclede** (*pret.*) SAX. CHR. 266 ; **isǣcled** (*pple.*) LAȝ. 30549 ; iseclid BER. 2583.

sēcnien, v., *sicken, become ill* ; sekenin PR. P. 451 ; **sēcneð** (*pres.*) A. R. 368 ; **siiknede** (*pret.*) WICL. 4 KINGS xx. 1 ; **sēcned** (*pple.*) ORM. 4771.

secrē, adj. & sb., *O.Fr.* secroi ; *secret, private*, CH. C. T. *G* 178, 643 ; secree *B* 3211 ; secre *a prayer in the Mass* ROB. (W.) 12044 ; L. F. M. B. 143 ; secre of secretes *name of a book* CH. C. T. *G* 1447.

secrē-lī, adv., *secretly*, CH. C. T. *E* 763.

secrēnesse, sb., *secrecy*, CH. C. T. *B* 773.

secrēte, sb., *O.Fr.* secrete ; *secret, private*

prayer, '*oracio*,' CATH. 327 ; **secrētes** (*pl.*) CH. C. T. *G* 1447.

secte, sb., *Fr.* secte ; *sect, set, company, class of mankind, likeness*, LANGL. *B* xi. 237 ; TREV. IV. 175 ; CH. C. T. *E* 1171, *F* 17 ; in that secte our saveoure saved al mankind LANGL. *B* xiv. 258 ; secte *form C* viii. 130.

sectour, sb., *for* **executor** ; *executor*, CATH. 327 ; sektour D. ARTH. 665 ; seketoure PR. P. 451 ; **secturs** (*pl.*) TOWNL. 326 ; secatours REL. I. 314 ; secutors LANGL. *C* xvii. 277.

seculēr, adj. & sb., *O.Fr.* seculer ; *secular*, BEK. 2229 ; CH. C. T. *E* 1322 ; (*layman*) BARB. iv. 12 ; (*secular priest*) LANGL. *B* ix. 177.

secunde, adj. & sb., *O.Fr.* second ; *second*, ROB. 282 ; SHOR. 45 ; TREV. IV. 275.

sed, *see* **sad. sēd**, *see* **sǣd.**

sedeful, *see* **sideful. sedewal**, *see* **zedewal.**

sedill, sb., *? from* *side *cf. O.N.* siðugr ; *? modest* ; sedill douvis ALEX. (Sk.) 3937.

sēdin, *see* **sǣdin.**

see, *see* **sē. sefen**, *see* **seofen.**

seg, sb., *O.E.* secg, = *O.L.G.* segg, *O.N.* seggr ; *messenger, man*, WILL. 772 ; seg, sæg LAȝ. 4443, 8015 ; segge LANGL. *B* iii. 63 ; A. P. ii. 398 ; ANT. ARTH. xxiii ; **segges** (*pl.*) LAȝ. 20854 ; of segge werke ALEX. (Sk.) 4473.

seg[2], sb., *O.E.* secg, = *M.Du.* segghe ; *related to* **saȝe** ; *sedge*, '*carex*,' FRAG. 3 ; segge PR. P. 451 ; VOC. 191 ; **segge** (*dat.*) O. & N. 18 ; **segges** (*pl.*) PALL. i. 525 ; seggis WICL. GEN. xli. 18.

sēge, sb., *O.Fr.* sēge, siege ; *seat, throne ; siege* ; A. R. 238 ; ROB. 132 ; JOS. 292 ; MAN. (F.) 5082 ; sege *seat* ALEX. (Sk.) 236, 1872, 5182 ; (*siege*) 1029, 3020 ; (*camp*) 2442 ; seige siege 1282 ; **sēgis** (*pl.*) ['*sedes*'] HAMP. PS. cxxi. 5.

sēgen, v., *lay siege to* ; **sēgande** (*pple.*) BARB. xvii. 511 ; **sēgid** (*pple.*) ALEX. (Sk.) 4296 ; **sēgit** (*pret.*) BARB. xi. 114.

seggen, v., *O.E.* secgan, sæcgan, secgean, sagian, = *O.L.G.* seggean, *O.Fris.* sedza, sidza, *O.N.* segja, *O.H.G.* sagen, sagan ; *say, speak, tell*, MAT. xi. 7 ; KATH. 635 ; ORM. 55 ; FL. & BL. 281 ; segge O. & N. 393 ; ROB. 17 ; SHOR. 23 ; seggen, segen HOM. II. 27 ; seggen, siggen FRAG. 7, 8 ; seggen, siggen, suggen HOM. I. 19, 127, 133 ; LAȝ. 512, 983, 1164 ; siggen A. R. 24 ; sigge MISC. 28 ; TREAT. 136 ; ALIS. 4198 ; S. S. (Wr.) 1708 ; zigge AYENB. 5 ; suggen R. S. iv ; REL. I. 245 ; sugge C. L. 420 ; seggen, seie LANGL. *A* iii. 166, *B* iii. 172 ; sægen SAX. CHR. 263 ; (*ms.* seien) H. M. 3 ; REL. I. 128 ; SPEC. 84 ; (*ms.* seyen) HAV. 2886 ; seien, seigen GEN. & EX. 1139, 2494 ; sein PR. P. 451 ; S. & C. I. xlix ; sein, seie CH. C. T. *A* 779, 1151 ; seie WILL. 1279 ; MAND. 2 ; WICL. GEN. xxxvii. 20 ; H. S. 37 ; seie, seiȝe JOS. 142, 161 ; sagen REL. I. 186 ;

(*ms.* saien) AN. LIT. 3; (*ms.* sayen) SPEC. 36;
MAP 347; sain RICH. 1339; to seggene MK.
ii. 9; **segge** (*pres.*) HOM. I. 21; sugge [segge]
LA3. 6321; seist KATH. 392; O. & N. 265; sei𐑂
HOM. I. 21; A. R. 48; hit sei𐑂 in þere tale LA3.
22889; seiþ, sæiþ FRAG. 5, 7; (hio) segge𐑂
MAT. v. 11; HOM. I. 29; seggeþ FRAG. 8;
sigge𐑂 A. R. 52; **seie** (*imper.*) A. R. 238; LEG.
32; seie, sæi3e, saie [sei] LA3. 2269, 2270,
30283; seie, sei O. & N. 217, 556; sugge𐑂
[seggeþ] LA3. 865; **segge** (*subj.*) O. & N. 844;
sægde (*pret.*) LK. xiv. 16; saigde, saide MK.
vi. 31, xvi. 15; se3de ORM. 149; seide SAX.
CHR. 249; A. R. 72; KATH. 154; C. L. 860;
LANGL. *A* iii. 48; CH. C. T. *A* 776; WICL.
GEN. ii. 18; seide, sæide, saide [seide, saide]
LA3. 672, 1256, 3379; seide, sede O. & N. 9,
217; HORN (L.) 271, 309; sede FL. & BL. 3;
BEK. 130; segden MK. ii. 18; seiden, seden
HOM. I. 91, 97; heo was hohful and erede
hwat heo þar after hire **séde** (*should say*) O.
& N. 1296; **seid** (*pple.*) GEN. & EX. 2425;
SPEC. 105; H. H. 15; *comp.* bi-, 3e-**seggen.**

[**seggere**, sb., *sayer; comp.* sôþ-siggere.]

seignior, sb., *O.Fr.* seigniour; *prince, lord*;
seniour YORK xxx. 73; seniour(e) 2634, 3073;
seneiours, seneours (*pl.*) 1614, 2487.

seignorie, sb., *O.Fr.* seignorie; *seigniory,*
lordship, ROB. 284; seignurie MAN. (F.) 2648;
seniourie [senori] ALEX. (Sk.) 1913, 1936;
sen3ori, sen3hori, sen3eroi, sen3houri BARB. v.
232, i. 97, xv. 324, i. 151.

[**sehen,** *? pret. pple. of* sêon.]

sehen-lĭch, adj., *O.E.* (ge-)sewenlĭc; *visible;*
sehelĭche (*pl.*) MARH. 11; KATH. 250;
comp. 3e-, un-**sehenlĭch** (-sewenlĭch).

seht, sehtnien, *see* saht, sahtnien.

seien, *see* **seggen**.

seil, sb., *O.E.* segl *m., n., cf. O.N.* segl *n.,*
O.L.G. segel, *O.H.G.* segal *m.; sail, 'velum,'*
FRAG. 2; PR. P. 451; LA3. 4595; HAV. 711;
BEK. 1837; **seile** (*dat.*) CH. C. T. *A* 696;
saile WILL. 568; **seiles** (*pl.*) LA3. 1339.

seil-clā𐑂, sb., *sail-cloth,* (*ms.* -clæ𐑂, -cloþ)
LA3. 4549.

seil-3erd, sb., *O.E.* segelgyrd; *sail-yard,*
'*antenna,*' FRAG. 2; PR. P. 451.

seil [2], sb., *O.E.* sigle, (in-)segele, = *O.N.* sigli,
Goth. sigljô; *seal,* H. M. 11; seel PROCL. 7;
MAND. 231; sel AYENB. 62; sehel LANGL. *B*
vii. 23*; *comp.* in-**seil**.

seilien, v., *O.E.* seglian, = *O.H.G.* segelen,
O.N. sigla; *sail,* LA3. 25525; seiglien LK.
viii. 22; seilen [seile] WICL. WISD. xiv. 1;
seile MAND. 305; saile WILL. 2673; **seilinge**
(*pple.*) LANGL. *C* xxi. 344; seillinge *B* xviii.
304; sailand ALEX. (Sk.) 161; sailande
flowing GAW. 865; **seileden** (*pret.*) L. N. F.
217; ALIS. 5640; seilide TREV. V. 265.

sailer, sb., *sailor*; **sailers** (*pl.*) ALEX. (Sk.)
4359.

seilien [2], v., *O.E.* (in-)seglian, *cf. Goth.* sigljan;
seal; seelin '*sigillo*' PR. P. 452; sele LUD.
COV. 340; **iseilet** (*pple.*) MARH. 5; isealed
A. R. 388; selid ALEX. (Sk.) 1170.

seeling, sb., *sealing,* WICL. JOB xxxviii. 14.

seim, *see* saim.

seine, sb., *O.E.* segn, = *O.Fr.* seigne, signe;
sign, LA3. 9282*; JOS. 197; MAN. (F.) 10162;
signe BEK. 411; WILL. 3213; singe, senge
YORK ix. 290, xi. 100; seni, sine PR. P. 453,
456; **signes** (*pl.*) A. R. 70; singes YORK xi.
156.

seine [2], sb., *O.E.* segne, sagene, = *O.L.G.* segina,
O.Fr. seine, *from Lat.* sagēna; *fishing net;*
snare; VOC. 159; HORN (H.) 700; **saines**
(*pl.*) ALEX. (Sk.) 4270.

seinian, v., *O.E.* segnian, = *O.H.G.* segenōn,
O.N. signa, *O.Fr.* seignier; *sign, mark with*
a sign; bless; HOM. I. 127; PERC. 287; i sal
saine ['*benedicam*'] PS. lxii. 5; signe þer wiþ þi
foreheved P. L. S. xvii. 66; **saine** (*subj.*)
FLOR. 297; and **seinede** [signed] (*pret.*)
him LANGL. *B* v. 456, *C* viii. 63; he sanit
him (*crossed himself*) BARB. vii. 98; **iseined**
(*pple.*) wiþ ure seel PROCL. 7; iseinet (*printed*
isemet) MARH. 23.

seint, *see* saint, ceint.

seis, *see* sês. **seisen**, *see* saisen.

sek, sb., = *O.H.G.* sech; *ploughshare, 'vomer,'*
VOC. 234.

sêk, *see* sêoc.

sêken, sîken, v., *cf. M.L.G.* sûken, *O.H.G.*
siuchan; *be ill*; seeken ['*aegrotare*'] WICL.
I KINGS xxx. 13; seke PALL. xi. 117; **siikide**
(*pret.*) WICL. 4 KINGS xiii. 14.

sêken, *see* sêchen. **sekir**, *see* siker.

sêl, *see* sâêl, seil. **selc**, *see* silk.

selcū𐑂, *see under* sêld [2].

séld, sb., *O.E.* seld; ? = **setl**; *seat,* HOM. I. 93;
sélde (*dat.*) LA3. 25988.

séld [2], adj., *O.E.* seld; *seldom, few*; **sélde**
(*pl.*) time CH. C. T. *E* 146; **sélde** (*adv.*) *E*
427; **seldere** (*compar.*) HOM. II. 207.

sél-cū𐑂, adj. & sb., *O.E.* sel-, seldcū𐑂; *rare,*
strange, wonderful; wonder, GEN. & EX.
3972; sel-, sulcu𐑂 LA3. 280, 3894; selku𐑂 A.
R. 8; selcuþ ORM. 19217; selcouþ ALIS. 154;
selkouþ LANGL. (Wr.) 7309; þe selcouþe si3t
WILL. 2329; ænne sélcū𐑂ne (*acc. masc.*) mon
LA3. 19059; sélcouthe (*pl.*) maners PR. C. 1518.

sélkouthen, v., *make wonderful*; **sélkoath**
(*imper.*) HAMP. PS. xvi. 8; **sélkouthid** (*pret.*)
HAMP. PS. xxx. 27; **sélkouthid** (*pple.*) HAMP.
PS. iv. 4.

sélcū𐑂-lîche, adv., *seldom,* LA3. 10301; sel-
cuþlike ORM. 2586; selcouþli WILL. 2650;
selcutheli ALEX. (Sk.) 3076.

séld-hwonne, adv., *O.E.* seldhwonne; *sel-*
dom, A. R. 428; sildwhanne (*ms.* silde-) TREV.
I. 333.

sel-līc, adj & adv., *O.E.* sel-, syl-, seldlīc,= *O.L.G.* seldlīc, *Goth.* sildaleiks ; *strange, marvellous, illustrious,* GEN. & EX. 466 ; sellich AN. LIT. 96 ; sel-, sillich [sul-, seollich] LA3. 6438, 7328 ; a selli man TRIST. 83 ; S. S. (Web.) 248 ; Iw. 107 ; MAN. (H.) 174 ; sellic (*a wonder*) FRAG. 8 ; P. L. S. viii. 92 ; LA3. 23050 ; sullic HOM. I. 171 ; sullich JUL. 55 ; of selliche [seollliche] wisdome O. & N. 1299 ; **sellichne** (*acc. m.*) LA3. 18847.

séld-sēne, adj., *O. N.* sialdsēnn ; *rarely seen, rare* ; seltsene H. M. 27 ; our speche schal beon seldcene A. R. 80.

séld-spēche, sb., *taciturnity,* A. R. 76.

ɛélden, adj., *O.E.* seldan, seldon, seldum,= *O.N.* sialdan, *O.H.G.* seltan ; *rare* ; (*ms.* seldenn) ORM. 8468 ; REL. I. 180 ; GOW. I. 30 ; OCTAV. (H.) 72 ; HOCCL. i. 165 ; PR. C. 260 ; selden [seldom] LANGL. *B* vii. 137 ; seldum GEN. & EX. 2181 ; REL. I. 214 ; seldom PR. P. 452 ; seldoum ALEX. (Sk.) 4220 ; selde HOM. II. 73 ; LA3. 8018 ; A. R. 72 ; O. & N. 944 ; FL. & BL. 462 ; CH. C. T. *A* 1539 ; sielde S. A. L. 157 ; P. S. 330.

[**séle,** sb., *O.E.* sele,= *O.L.G.* seli, *O.N.* salr ; *house, hall* ; comp. lēf-, wun-sele.]

séle², sb., *O.E.* seolh,= *O.N.* selr, *O.H.G.* selach ; *seal (animal),* PR. P. 452 ; HAV. 755.

seel-skin, sb., *seal-skin,* REL. I. 51.

séle³, sb.,= *M.L.G.* sele, *O.N.* seli, *O.H.G.* silo *m.* ; *seal, signet,* PR. P. 452 ; sile GEN. & EX. 2978.

sēlen, *see* **sǣlen.** **seler,** *see* **celler.**

self, adj., *O.E.* self, seolf, sielf, silf, sylf,= *O.L.G.*, *O.Fris.* self, *O.N.* sialfr, *O.H.G.* selbēr, *Goth.* silba ; *self, himself, same* : he is god self KATH. 1093 ; heo was hire self þer imong 1579 ; þe biscop self ORM. 1022 ; self his kinde GEN. & EX. 1806 ; self (? *for* selfe) ῗe fon 2610 ; god self ... seiþ SHOR. 23 ; he self [*v. r.* him self] was on þe feld C. M. 4059 ; what man self es PR. C. 128 ; 3e self H. S. 6301 ; 3e [3ou] self C. M. 14691 ; ne beo þe levre þan þi self þi mei ne þi mo3e P. L. S. viii. 15 ; zelf AYENB. 59 ; þu seolf LA3. 3192 ; he [him] seolf wende 14201 ; al swa þe king sulf 4261 ; heo seolf (*for* seolfe) 5990 ; þeih heo seolf nabbe non MISC. 128 ; sulf A. R. 56 ; þu sulf O. & N. 497 ; þo com our loverd silf P. L. S. xix. 290 ; þe seolfe [seolve] coc O. & N. 1679 ; þe selve duk WILL. 1368 ; þe sulve moder A. R. 328 ; **selfem** (*dat. m.*) JOHN i. 22 ; seolfan LA3. 828 ; selven KATH. 1130 ; seidest þe selven (*for* selve) 631 ; of mi selven ISUM. 120 ; me seolven HOM. I. 253 ; of þe seolven MARH. 4 ; sulven A. R. 296 ; selve B. DISC. 75 ; at þe selve huse FL. & BL. 21 ; seolve ALIS. 2323 ; þe seolve hit turneþ to grome O. & N. 1284 ; þe kat ful wel him sulve liveþ 810 ; **sülfne** (*acc. m.*) P. L. S. viii. 7 ; hine seolfne HOM. I. 95 ; seolfne, sulfne, seolfan, seolven, sulven [seolve] LA3. 493, 1865,

2227, 8333, 8457 ; seolfen, seolven JUL. 26, 27 ; þe selfen ORM. 4469 ; þurh god selfen 4131 ; and make him selven wood CH. C. T. *A* 184 ; sulfen FRAG. 6 ; he huld him silve forlore BEK. 663 ; **seolve** (*pl.*) LA3. 25381 ; þise selve word(e)s WILL. 889 þe sulve stottes O. & N. 495 ; **selven** (*dat. pl.*) KATH. 1298 ; beon eow selven riche LA3. 5802 ; selven [selve] LANGL. *A* v. 35 ; to us sulven A. R. 302 ; **selven** (*acc. pl.*) KATH. 362 ; us selfen ORM. 12970 ; seolven LA3. 5199 ; þuruh us sulven A. R. 248 ; eow selfe HOM. I. 15 ; þo heo hadde ... heo selve forsineged HOM. II. 143 ; ou selve SPEC. 101 ; 3ou self LANGL. *B* ii. 38 ; for ous silve BEK. 1668.

seolf-cwale, sb., *O.E.* sylfcwala ; *suicide,* HOM. I. 103.

silf-grēne, sb., *house leek,* VOC. 265.

self-wil, sb., *O.E.* selfwill ; *self-will* ; **self-willes** ['*ultro*'] MK. iv. 28 ; wilt þou silfwilles lete þe slen FER. 221.

self-wille, sb., *O.N.* sialfvili,= *O.H.G.* selbwillo ; *self-will,* P. L. S. xxxi. 195.

sel(f)-willi, adv.,= *M.Du.* selfwilligh ; *following his own will only,* PR. P. 452.

sēli, *see* **sǣli.**

sēlin, *see* **seilien²**.

selk, *see* **silk.**

selle, sb., *O.Fr.* selle ; *seat* ; **sellis** (*pl.*) WICL. 2 MACC. xiv. 21.

selle, *see* **celle.**

sellen, v., *O.E.* sellan, syllan,= *O.H.G.* sellan, *O.L.G.* sellian, *O.Fris.* sella, *O.N.* selja, *Goth.* saljan (θύειν) ; *from* **sále** ; *sell, give,* ORM. 6345 ; sellen [sille] WICL. GEN. xlvii. 22 ; sellin PR. P. 452 ; selle APOL. 9 ; zelle AYENB. 36 ; selle LANGL. *B* iii. 195 [sulle *A* iii. 189] ; selle [sulle, sille] TREV. VIII. 105 ; seollen, sullen LA3. 29057, 31053 ; sullen HOM I. 135 ; A. R. 190 ; H. M. 27 ; sulle ROB. 223 ; P. S. 151 ; sillen LK. xi. 7 ; sille P. L. S. xxiii. 130 ; to sellenne MAT. xx. 23 ; **sülest** (*pres.*) H. M. 27 ; selleþ, sulleþ E. G. 355 ; sulleῗ A. R. 398 ; silῗ MAT. xiii. 44 ; **sele** (*imper.*) me drinken ['*da mihi bibere*'] JOHN iv. 10 ; sule A. R. 290 ; sule, sile LEECHD. III. 106, 108 ; **sealde** (*pret.*) SHOR. 47 ; salde, sælde [solde] LA3. 10020, 13437 ; salde Iw. 1703 ; PR. C. 4849 ; solde A. R. 398 ; GEN. & EX. 1843 ; BRD. 25 ; sealden [salden] HOM. I. 91 ; salden ORM. 15557 ; **sóld** (*pple.*) LANGL. (Wr.) 11089 ; comp. 3e-sellen.

sellere, süllere, sb.,= *O.Fris.* seller ; *dealer, giver,* E. G. 355 ; sullere HOM. II. 213 ; **selleres** (*pl.*) CH. C. T. *A* 248 ; MAND. 86 ; sellers [silleris] WICL. EZ. xxvii. 22 ; sullares LEB. JES. 633.

sellinge, sb., *selling,* '*venditio*,' PR. P. 452.

seller, *see* **celler.**

sellīc, *see under* **séld²**.

selten, *see* **salten.** **sélþe,** *see* **sǣlþe.**

selūre, *see* **celūre.** **selver,** *see* **silver.**
sēm, *see* **sēam.**

semblable, adj., *O.Fr.* semblable; *like,*
LANGL. *B* x. 367, xviii. 10, *C* iv. 337, xix. 213.

semblance, sb., *appearance,* ALEX. (Sk.) 5192;
sembalaunce 4098.

sembland, adj., *like,* LANGL. *B* x. 367*.

semblant, sb., *Fr.* semblant; *semblance,*
show, aspect, HOM. I. 247; ROB. 129; WILL.
228; semblaunt A. R. 90; FL. & BL. 50;
sembland YORK xvii. 93.

semblē, sb., *for* **assemblē**; *meeting, army,*
ALEX. (Sk.) 797, 3978; semle 1573; LANGL.
A PROL. 97; DEP. R. iv. 85.

semblen, v., *Fr.* sembler; *seem;* **sembeles**
(*pres.*) M. H. 136.

semelen, v., *Fr.* sembler (*? for* **assemblen**);
assemble, GEN. & EX. 3865; semelin E. G. 47;
semble D. ARTH. 63; semple ALEX. (Sk.)
2796; **semblis** (*pres.*) *attacks* ALEX. (Sk.)
1333; hi sembleþ AYENB. 176; **sembled**
(*pret.*) ALEX. (Sk.) 617; (þai) sembled DEP.
R. iii. 357; **sembled** (*pple.*) WILL. 2147;
semblid iv. 32.

sembling, sb., *assembling,* ALEX. (Sk.) 769.

sēme, adj., *O.N.* sœmr; *befitting, decent,*
GAW. 1085; A. P. ii. 549; *comp.* **ȝe-sēme.**

sēme-lïch, adj., *O.N.* sœmiligr; *seemly,* B.
DISC. 849; semlich A. R. 94*; GOW. III. 299;
semeli *'decens'* PR. P. 452; PR. C. 73; sem-,
semeli CH. C. T. *A* 751; semli ISUM. 268;
simli TOR. 126; þi semliche schape KATH.
449; þat semliche child WILL. 49; **sēme-**
liche (*adv.*) TREV. VIII. 87; semelike GEN.
& EX. 1504; semeli ALEX. (Sk.) 281, 424;
sēmloker (*compar.*) GAW. 83; **sēmlokest**
(*superl.*) SPEC. 27; semelokest ANT. ARTH.
vi.

sēmelīnesse, sb., *seemliness,* '*decentia,*' PR.
P. 452.

sēmelitē, sb., *seemliness,* YORK xxv. 116.

sēmen, v., *O.E.* sēman, = *O.N.* sœma;
befit, be becoming, suit; seem; SPEC. 32;
preostes heo þer setten as þer to mihte(n)
semen LAȝ. 10207; fendes made hem semen
to ben so hole MAND. 284; seme O. & N.
187·; **sēmeþ** (*pres.*) ORM. DEDIC. 66; LANGL.
A PROL. 32; as hit best semeð HOM. I. 271;
þat semiþ þe wel N. P. 56; semiþ non to hur
but he EGL. 112; semes HAV. 2916; me
semes WILL. 620; þine eien semen dede CH.
TRO. iv. 1092; **sēmde** (*pret.*) A. R. 112*;
WILL. 2880; it was nevere man . . . þat so
wel semde (*ms.* semede) king or caiser for
to be HAV. 976; semed [semid] ALEX.
(Sk.) 2108; semden MARH. 9; *comp.* **bi-, ȝe-**
sēmen.

sēming, sb. : to mi seminge *as it appears to*
'me CH. C. T. *B* 1838.

sēmen², v., *O.E.* sēman, sȳman; *from* **sēam²**;

burden; press upon; ge sēmeð ['*oneratis*']
(*pres.*) LK. xi. 46; semeþ HOM. II. 93; **sēmde**
(*pret.*) SHOR. 85; semdest me mid sunne
FRAG. 6; *comp.* **over-sēmen.**

[**semi-,** prefix, *Lat.* sēmi- ; *half.*]
semi-cōpe, sb., *short cope,* CH. C. T. *A* 262.
sen, *see* **siþþen, seofen.** **sēn,** *see* **sēon.**

senāt, sb., *Fr.* senat; *senate,* CH. BOET. i. 4
(18); (*ms.* senaht) LAȝ. 25388.

senatour, sb., *Fr.* senateur; *senator,* ROB.
89; D. ARTH. 227; CH. C. T. *B* 1044; **sena-**
tūrs (*pl.*) LAȝ. 25337.

senchen, v., *O.E.* sencan, = *O.L.G.* (bi-)-
senkian, *O.H.G.* senkan, senchan; *from* **sin-**
ken; *make to sink,* JUL. 79; in sunne ant
sorewe i am **seint** (*pple.*) SPEC. 24; *comp.*
a-, bi-, for-senchen.

sencinge, sb., *for* **censinge**; PR. P. 452.
sendal, *see* **cendal.**

senden, v., *O.E.* sendan, = *O.L.G.* sendean,
O.N. senda, *O.H.G.* sentan, *Goth.* sandjan;
from **sande**; *send,* LAȝ. 27879; A. R. 422;
GEN. & EX. 1683; CH. C. T. *A* 2976; þenne
sende (*shall send*) ic eou rihte widerunge
HOM. I. 13; sendeþ BRD. 13; he sent [sendeþ]
þe his sonde LAȝ. 26367; sent HOM. I. 7; A.
R. 246; TREAT. 136; (heo) sendeð LAȝ. 476;
send (*imper.*) MARH. 3; sende (*subj.*) O. &
N. 1570; sende (*pret.*) KATH. 48; ORM.
1861; HAV. 136; P. S. 248; sende, sente MK.
vi. 27, ix. 37; þou sent ALEX. (Sk.) 2022;
sendest H. H. 209 (211); heo senden LAȝ.
662; **send** (*pple.*) ORM. 97; sent LANGL. *B*
xvii. 146; *comp.* **a-, bi-, ȝe-, of-senden.**

sendere, sb., *sender,* HOM. II. 111.

sendinge, sb., *sending,* '*missio,*' PR. P. 452.
senden, *see* **schenden.**

sendren, *see* **sindren.**

sēne, sb., *O.E.* sien, sȳn, = *O.L.G.* siun, *O.N.*
siōn, sȳn, *Goth.* siuns; *from* **sēon**; *sight*
(*act or power of seeing*), HOM. II. 25; ORM.
9394; ich habbe god sene O. & N. 368; *comp.*
an-, ēah-sēne.

sēne², adj., *O.E.* (ge-)sēne, sȳne, = *M.L.G.*
sūne, *O.N.* sēnn, sȳnn; *visible,* '*mani-*
festus,' PR. P. 452; ORM. 2173; GEN. & EX.
1173; HAV. 656; SPEC. 58; LANGL. *B* i. 147;
CH. C. T. *A* 134; *? truthful* GAW. 148, 341;
comp. **bī-, ēð-, ȝe-, sēld-sēne.**

seneȝen, *see* **sŭneȝen.**

senevel, sb., *O.Fr.* seneveil (*sénevé*); *mus-*
tard seed, WICL. MAT. xiii. 31; senvei PALL.
viii. 149.

senewe, *see* **sinewe.** **senge,** *see* **seine.**

sengin, v., *O.E.* (be-)sengan, = *M.H.G.* sengen; *from* **singen**; *singe,* '*ustulo,*' PR. P.
453; senge CH. C. T. *D* 349; zengþ (*pres.*)
AYENB. 229; seind (*pple.*) CH. C. T. *B* 4035;
comp. **bi-sengen.**

seniour(e), *see* **seignior.**

senne, *see* sünne.

sens, sb., = cense *for* encens ; *incense*, ALEX. (Sk.) 4184 ; sence CATH. 330 ; sense LANGL. *B* xix. 82, *C* xxi. 86.

sensitife, adj., *Fr.* sensitif ; *sensitive*, ALEX. (Sk.) 4381.

sensour, sb., *for* encenser ; sensours [censours] (*pl.*) ALEX. (Sk.) 1565.

sent, sb., *for* assent ; WILL. 1983 ; sente YORK xxxii. 144.

sent², sb., *? from O.Fr.* sentir ; *scent*, D. ARTH. 1040 ; BARB. vi. 500.

sentement, sb., *O.Fr.* sentement ; *feeling*, CH. L. G. W. 69.

sentence, sb., *O.Fr.* sentence ; *sentence*, *opinion, verdict*, A. R. 348 ; P. L. S. xvii. 518 ; CH. C. T. *B* 4167 ; sentens *I* 58.

senuwe, *see* sinewe. senvei, *see* senevel.

sēo, *see* am.

sēoc, sēc, adj., *O.E.* sēoc, sīoc, = *O.L.G.* seoc, sioc, *O.Fris.* siek, siak, *O.N.* siūkr, *Goth.* siuks, *O.H.G.* sioch ; *sick, ill*, ORM. 6165, 8073 ; seoc, sec, sæc [seac] LA3. 2794, 6667, 6781 ; siec ANGL. I. 19 ; sec HOM. I. 23 ; sek GEN. & EX. 1175 ; AN. LIT. 7 ; PR. C. 772 ; M. ARTH. 158 ; seek PR. P. 451 ; MAND. 11 ; PALL. iii. 939 ; SACR. 556 ; sec, sic A. R. 176 ; sik [sek] CH. C. T. *D* 1592 ; siik [seek] WICL. MAT. xxv. 39 ; sik MISC. 31 ; HORN (L.) 272 ; BEK. 892 ; seik BARB. ix. 112 ; zik AYENB. 148 ; þe sike man SHOR. 29 ; sücne (*acc. m.*) LA3. 17682 ; sēoke (*pl.*) LA3. 29541 ; heo beoð boðe seke A. R. 364 ; alle sike 32 ; seeke HOCCL. i. 409 ; sike LANGL. *B* xv. 568 ; sēoken (*dat. pl.*) MAT. ix. 12 ; sēccure (*compar.*) A. R. 46.

sīk-lēwe, adj., *unhealthy*, TREV. I. 257, III. 303.

sēk-lī, adj., *sickly*, WILL. 1505 ; seekli HOCCL. i. 15.

sēocnesse, sb., *sickness*, LA3. 19303 ; seocnysse LEECHD. III. 126 ; seknesse MAND. 89 ; secnesse, sicnesse A. R. 178, 182 ; sicnesse HOM. II. 167 ; sekenes TREV. III. 265 ; seiknes BARB. iv. 191 ; siknisse BRD. 14.

seofen, seofe, card. num., *O.E.* seofon, seofone, = *M.L.G.* seven, soven, *Goth.* sibun, *O.H.G.* sibun, siben, *O.Fris.* siugun, sigun, *O.N.* siau ; *seven*, HOM. I. 41, 43 ; he haveð sefene 27 ; seofen, sefen ORM. 545, 8573 : seofan, seofen, sefen, seofe MAT. xii. 45, xv. 36 ; MK. viii. 20 ; seoven, sovene LA3. 463, 716 ; seoven, sovene, seovene A. R. 22, 62, 278 ; seovene FRAG. 8 ; MISC. 59 ; seven KATH. 1680 ; seven LANGL. *A* iv. 73 [sevene *B* iv. 86] ; seaven MIR. PL. 7 ; sevene HICKES I. 222 ; HAV. 2125 ; P. S. 157 ; SHOR. 45 ; FER. 1493 ; sevene and fifti TREV. I. 45 ; soven HOM. I. 13 ; sove TREAT. 132 ; þare seofene MAT. xxii. 28.

seoven-fáld, adj., *O.E.* seofonfeald, = *O.N.* siaufaldr, *O.H.G.* sibunfalt ; *sevenfold*, HOM.

I. 26i ; sefenfald ORM. DEDIC. 267 ; seovevald MARH. 23 ; seovevold A. R. 38 ; sevefeald HOM. II. 171.

seoventēne, card. num., *O.E.* seofontȳne, = *O.N.* siautian, *M.H.G.* sibenzehen ; *seventeen*, LA3. 2140 ; seventene AR. & MER. 8905 ; seventenþe HAV. 2559.

seventēþe, ord. num., *O.E.* seofontēoða ; *seventeenth*, AL. (L.²) 325.

seofenti3, card. num., *O.E.* seofontig, = *O. H.G.* sibunzug ; *seventy*, ORM. 4319 ; seoventi A. R. 62 ; seventi GEN. & EX. 706.

seventiþe, ord. num., *O.E.* seofontigoða ; *seventieth*, ROB. 282.

seofeðe, ord. num., *O.E.* seofoða, = *O.H.G.* sibendo, *O.Fris.* siugunda, *O.N.* siaundi ; *seventh*, HOM. I. 39 ; seoveðe MARH. 8 ; A. R. 14 ; seveþe HORN (R.) 927 ; JOS. 577 ; A. D. 240 ; soveþe TREAT. 138 ; sefþe P. L. S. ii. 97 ; sefte LUD. COV. 83 ; sevenþe HAV. 1825 ; LANGL. (Wr.) 9531 ; CH. C. T. *A* 1462 ; seveneth LANGL. *B* xiv. 306 ; seofende MAT. xxii. 26 ; sefende ORM. 4464 ; sevende GEN. & EX. 445 ; SHOR. 52 ; sevend PR. C. 362.

sēoke, sb., = *M.L.G.* sūke, *O.H.G.* siuchī ; *illness* ; for sike [seke] þei mi3te uneþe stonde CH. C. T. *D* 394 ; in sik(e) and sor(e) GEN. & EX. 1239.

seolf, *see* self. seolk, *see* silk.

seolver, *see* silver.

sēon, sēn, v., *O.E.* sēon, sīon, = *O.Fris.*, *O.N.* sia, *O.L.G.*, *O.H.G.* sehan, *Goth.* saihwan ; *see*, ORM. 47, 318 ; seon LANGL. *A* i. 146, see *B* i. 170 ; sien ANGL. I. 16 ; suen SPEC. 100 ; sen GEN. & EX. 279 ; RICH. 129 ; sen, seen CH. C. T. *A* 1709 ; seo, se H. H. 219 ; to seonne JUL. 47 ; zienne AYENB. 108 ; te sene SPEC. 96 ; sende HOM. II. 139 ; sixt (*pres.*) TREAT. 134 ; C. L. 8 ; SHOR. 140 ; (*ms.* sixst) R. S. vi (MISC. 184) ; sixt [suxt] LANGL. *C* xi. 158 ; TREV. VIII. 219 ; sext S. S. (Web.) 362 ; sichst, sist (*ms.* syst) O. & N. 242 ; sist JUL. 31 ; A. D. 289 ; sest HOM. II. 137 ; KATH. 1074 ; HAV. 534 ; sihþ (*ms.* syhþ), siþ O. & N. 950 ; si3þ TREAT. 133 ; zi3þ AYENB. 11 ; siht HOM. I. 29 ; sið LA3. 4380 ; seoð, seð H. M. 15 ; seoþ, seþ ORM. 670, 3829 ; we seeþ [seth] LANGL. *B* iii. 216 ; 3e seoþ BEK. 897 ; (hi) seoþ, soþ O. & N. 884 ; seh (*imper.*) A. D. 258 ; seah (*pret.*) MAT. iv. 21 ; sah (*ms.* sahh) ORM. 148 ; sag GEN. & EX. 26 ; sa3 FER. 4563 ; sa3, sau3 HORN (L.) 125, 167 ; sauh MAN. (H.) 31 ; saugh MAND. 24 ; TOR. 1718 ; sauh LANGL. *A* v. 9, 10 ; saw [*v. r.* sei3, sey, sau3], say LANGL. *B* v. 9, 10 ; saugh, seigh CH. C. T. *A* 193, 850 ; sau MAP 336 ; REL. I. 61 ; sau, saw HAV. 1251, 2410 ; saw PERC. 873 ; seh MARH. 10 ; KATH. 173 ; SPEC. 41 ; se3 BEK. 28 ; A. P. i. 158 ; se3, sei L. H. R. 36 ; seih C. L. 369 ; (*ms.* seyh) MISC. 53 ; sei3 JOS. 58 ; (*? printed* seigh) ALIS. 5718 ; sei3, si3 TREV. III. 127 ; sei DEGR. 1567 ;

sai EGL. 380; se AUD. 78; see S. S. (Wr.)
480; S. & C. II. lxxiv; sigh Gow. I. 46; (þu)
sehe HOM. I. 259; seȝe P. L. S. x. 56; SHOR.
83; sei (*ms.* sey) [seie, seiȝ, sawe] LANGL.
B viii. 75; seȝe *A* ix. 66; þeȝ A. D. 263; FER.
4017; seghen MISC. 27; sæȝhen ORM. 3342;
sehen KATH. 280; seȝen P. L. S. viii. 49;
seghe ISUM. 17; sawen MAND. 91; sogen,
sowen GEN. & EX. 3108, 3522; sowen HAV.
957; sēhe (*subj.*) H. M. 17; sæȝe LAȝ. 6275;
seȝe TREAT. 135; seiȝe LANGL. *B* v. 86; soge
REL. I. 220; sowe HAV. 1323; zeȝen AYENB.
204; seȝhen (*pple.*) ORM. 3335; seȝen MISC.
229; seiȝen [sein] LANGL. *B* x. 68; seien,
seie WILL. 264, 5003; sein MAND. 70; CH.
C. T. *B* 172; EGL. 450; seiȝe P. P. 216;
sewen GEN. & EX. 1195; *comp.* bi-, for-,
ȝe-, of-, ofer-, þurh-sēon; *deriv.* sēne, siht,
sihðe.

sēere, sb., = *Ger.* seher; *seer, prophet,* WICL.
2 KINGS xv. 27.

sēon[2], v., ?= *O.H.G.* sīn, *M.Du.* sijn; ? be; hu
mai ðis sen GEN. & EX. 1923.

sēon, *see* sihen. seorȝe, *see* sorȝe.

sēoðen, v., *O.E.* sēoðan, = *O.N.* sioða, *O.Fris.*
siatha, *O.H.G.* siodan; *seethe, boil, cook*;
sethin '*coqvo*' PR. P. 454; seþe ROB. 404;
RICH. 1493; CH. C. T. *A* 383; P. 43; sēþen
(*pres.*) MAND. 129; sēþe (*subj.*) A. P. ii. 631;
sēoþinge (*pple.*) MARG. 251; sēð (*pret.*)
GEN. & EX. 1487; seþ L. H. R. 60; IW. 1699;
seeþ CH. C. T. *E* 227; suden LAȝ. 20978;
soden BRD. 8; soden [soþen] LANGL. *B* xv.
288; soden (*pple.*) RICH. 3069; MAND. 251;
soþen IW. 1701; sode WILL. 1849; isoden
[isothen] LANGL. *B* xv. 425; isode BRD. 8;
LEG. 8; *deriv.* sēað.

sēoþing, sb., *cooking,* TREV. IV. 439.

seoðð̆en, *see* siþþen.

seoven, seoveð̆e, *see* seofen, seofeð̆e.

sēowen, v., *O.E.* sēowian, sīwian, *O.H.G.*
siuwan, sīwan, *cf. Goth.* siujan, *Lat.* suere;
sew; sewe EM. 59; DEP. R. iii. 138; sowe
LANGL. *B* vi. 9; souwe [sewe] *A* vii. 9;
souwe JOS. 427; sowe (*pres.*) '*suo*' PR. P.
466; seweð̆ MK. ii. 21; sēouweð̆ (*imper.*)
A. R. 420; sēwede (*pret.*) LEG. 74; sewide,
souwide WICL. JOB xvi. 16: sēwed (*pple.*)
WILL. 3060; sewed [sowed] CH. C. T. *A* 685;
comp. bi-, ȝe-sēowen.

sēwer, sb., *sewer,* CATH. 331; sēweris (*pl.*)
LANGL. *A* xi. 301; sewars APOL. 106; sowers
DEP. R. iii. 165.

sēowestre, sb., *stitcher, sempstress*; sewestre
[siwestere] LANGL. *C* vii. 362; sewstare, sow-
stare, soware '*sutrix*' PR. P. 454.

sep, *see* sap. sēp, *see* schēp.

septre, *see* sceptre.

septemtrioun, sb., *O.Fr.* septentrion; *north,*
CH. C. T. *B* 3657.

sepulcre, sb., *O.Fr.* sepulcre; *sepulchre,*

HOM. II. 21; A. R. 170; ROB. 411; SHOR.
124; sepulcre MISC. 52.

sepultūre, sb., *Fr.* sepulture; *sepulture,*
sepulchre, ROB. 166; SHOR. 128; CH. C. T.
C 558.

sēr, adj., *O.N.* sēr; *several, particular,* ORM.
18653; AV. ARTH. x; sēre (*pl.*) FLOR. 331;
GAW. 822; P. R. L. P. 139; PR. C. 48; seere
MAN. (H.) 19.

sēre-lĕpi, adj., *separate, various*: sekenes
of serelepi kendis ALEX. (Sk.) 4440; a serelepi gifte 4521; sērlĕpes (*adv.*) ORM.
513; GAW. 501; LANGL. *B* xvii. 164.

sēr-līche, adv., = *Dan.* særlig; ?*particularly,* WILL. 2149; sereli YORK xliv. 24.

sērnes, sb., *variety,* ['*varietate*'] HAMP.
PS. xliv. 11; sērnesis [sernessis] (*pl.*) xliv.
15.

sērtē, sb., *variety*; sērties (*pl.*) ALEX. (Sk.)
4654.

sēr, *see* sēar, sār.

sercle, *for* cercle; *circle,* ALEX. (Sk.) 3736;
WICL. PROV. xi. 22.

sēre, *see* seare. serewe, *see* sorȝe.

sergant, sb., *O.Fr.* sergant, serjant; = servant;
*sergeant, servant, man-at-arms; serjeant-at-
law; overseer of labourers*; MISC. 33; sergant,
serjaunt (*v. r.* servaunt) C. M. 15915; serjant
BEK. 687; serjaunt P. S. 327; sergeaunt PR. C.
6084; D. ARTH. 1173; sergont, serjont AYENB.
32, 188; serganz (*pl.*) HAV. 2088; sergeants
CH. C. T. *G* 361.

serge, sb., = cerge; *wax candle,* CATH. 330;
serges [cirges] (*pl.*) C. M. 20701; serges A.
P. ii. 1489.

serȝe, *see* sorȝe.

sergeantie, sb., *men at arms*; serjauntie
MAN. (F.) 11979.

serie, sb., *Lat.* series; *series,* CH. C. T. *A* 3067.

serious, adj., *O.Fr.* serieux; *serious*: serious,
sad and feithefulle PR. P. 453.

cerious-lī, adv., *seriously,* CH. C. T. *B* 185.

serke, sb., *O.E.* serce, syrce, = *O.N.* serkr;
shirt, LANGL. *B* v. 66; P. R. L. P. 128; EM.
501; serk HAV. 603; serke (*dat.*) S. S. (Wr.)
1391; wiþoute serke GAM. 259.

sermōn, sb., *O.Fr.* sermon, sermun; *sermon,
talk,* AYENB. 20; sermun MISC. 187; sarmoun
MAN. (F.) 9240; sarmōns (*pl.*) LANGL. *C* xv.
201.

sermonen, v., *O.Fr.* sermoner (*sermonner*);
talk, preach, HOM. I. 81; sermone CH. C. T. *C*
879; sermonie MISC. 77; sermōnes (*pres.*)
YORK xxx. 302.

serpent, sb., *O.Fr.* serpent; *serpent,* P. L. S.
iii. 26; CH. C. T. *D* 1994; serpentis (*pl.*)
ALEX. (Sk.) 3707.

serrai, adj., ? *O.Fr.* seré; *close*; sarrai (*adv.*)
BARB. viii. 296.

serri-lī, adv., *closely*; sarrali BARB. viii. 222;
ix. 140; xvii. 96; xviii. 195; sarreli MAN. (F.)
13536.

sertane, adj., *for* certein; *certain* : a sertane
folk ALEX. (Sk.) 3956; sertein MAN. (F.)
8151; a sertan (*sb.*) of giftes ALEX. (Sk.)
5121.

sertes, adv., *for* certes; *certainly*, MAN. (F.)
1223; sertis ALEX. (Sk.) 4371.

sērtie, *see uuder* sēr.

servāge, sb., *Fr.* servage; *servitude, service*,
ROB. (W.) 257, 263, 1059; CH. C. T. *E* 147;
D. BL. 769; PR. C. 1157.

servant, sb., *O.Fr.* servant; *servant*, A. R.
428; servaunt A. P. i. 698; servand, sirvand
ALEX. (Sk.) 2779, 1962; servandis (*pl.*)
2753; her servant (i. e. *lover*) for to be CH. L.
G. W. 1957.

serven, v., *for* deserven; *deserve*; serve
(*pres.*) HAMP. PS. iv. 5; served (*pple.*)
earned ALEX. (Sk.) 3426.

servien, v., *O.Fr.* servir; *serve*, A. R. 12;
serven GEN. & EX. 5; HAV. 1230; SPEC.
74; serfen ORM. 471; servie BEK. 156;
serveð (*pres.*) KATH. 2104; serveþ O. & N.
1579.

   servinge, sb., *serving*, LAȝ. 8097; serving
SPEC. 69.

servisable, adj., *useful, obliging*, CH. C. T.
G 1014; MAN. (F.) 3139.

servise, sb., *O.E.* serfise,=*O.Fr.* servise, service; *service*, LAȝ. 8071; A. R. 312; GEN. &
EX. 1672; BEK. 136; service H. H. 244; here
is propre service (*serving*) LANGL. *B* xiii. 51;
godes service to hiure *C* x. 227; servise (*duty*)
C iv. 451; serves ALEX. (Sk.) 918.

servitour, sb., *O.Fr.* serviteur; *attendant*;
servitūrs (*pl.*) MAN. (F.) 11300.

servitūte, sb., *O.Fr.* servitut; *servitude*,
CH. C. T. *E* 798.

serwe, *see* sorȝe.

sēs, sb., *for* *cēs, *O.Fr.* ces; *cessation, conclusion*, MAN. (F.) 182; AR. & MER. 3188;
seis YORK viii. 19.

sēsen, *see* saisen, cessen.

sēsonen, v., *season*; sēsonde (*pple.*) ALEX.
(Sk.) 2923.

sēsoun, sb., *O.Fr.* saison; *season*, WILL. 29;
seson, sesun LANGL. *A*, *B* PROL. 1; seson,
[sesoun] CH. C. T. *G* 1343; cesoun WICL.
ECCLES. viii. 6.

sessen, *see* cessen.

sester, sb., *Lat.* sextārius; *a measure of capacity*, PALL. viii. 148.

set, sb., *O.E.* set, = *O.N.* set, *O.H.G.* sez;
from sitten; *seat*; sēte (*dat.*) P. P. 218;
sette A. R. 358; sēten (*dat. pl.*) LAȝ. 30841.

   set-ræiȝel, sb., *O.E.* sethrægel; *chair-cover*,
'tapete,' FRAG. 4.

set², sb., *from* setten; *trap for game*; setis
(*pl.*) BARB. iii. 479.

sēte, adj., *proper, ?suitable*, L. C. C. 8; PALL.
ii. 420; sete, seete SPEC. 89, 114; i dide
al þat þe was sete MAP 337; sete for mannis
bodi E. G. 397; sete qvile *fitting time* ALEX.
(Sk.) 3081; *comp.* un-sēte.

sēte, *see* sǣte.

setel, sb., *O.E.* setol (ZEIT. XXI. 25), setl, *n.*, *cf.*
O.H.G. sezal; *mod.Eng.* settle; *seat, throne*;
['*sedem*'] MAT. xxv. 31; ['*cathedra*'] PS. i. 1;
setil PR. C. 6122; þet setle (*printed* secle)
HOM. I. 61; setle (*dat.*) HOM. II. 59; O. & N.
594; setle (*pl.*) MAT. xxi. 12; *comp.* dōm-,
hēah-, kine-, mēl-, þrüm-setel (-seotel).

   setel-gang, sb., *O.E.* setlgang; *down-going* :
fram sonne-springe to setel-gang PS. xlix. 1.

Seterdai, *see* Saterdaȝ *under* Satern.

setlen, v., *O.E.* setlan; *settle, place together;*
sink down; satelin '*basso*' PR. P. 440; setliþ
(*pres.*) ALIS. 484; sattles (*pres.*) YORK 248;
setlede (*pret.*) FER. 3281; setled (*pple.*)
ORM. 14049; WILL. 2452.

setnesse, sb., *O.E.* setness; *constitution,
statute*, HOM. I. 261; ORM. 16837; zetnesse
AYENB. 104; sætnesse LAȝ. 4258; *comp.* ȝe-
setnesse.

setten, v., *O.E.* settan,=*O.L.G.* settean, *O.N.*
setja, *Goth.* satjan, *O.H.G.* sezzan; *from*
sitten; *set, place*, FRAG. 1; LAȝ. 5309;
MAND. 72; setten a chirche JUL. 77; gif he
wile setten us over (*carry over*) þat michele
water HOM. II. 43; þe swel schal setten (*subside*) A. R. 274; til þat to sette bigan þe sunne
HAV. 2671; on kneus þei schule hem sette
MIRC 272; (we) setteð (*pres.*) HOM. I. 19; ac
men setten nat bi (*value*) songewarie LANGL.
C x. 302; sete (*imper.*) LAȝ. 3699; MISC. 85;
LEG. 58; sete (*pret.*) MARH. 19;
BEK. 1593; CH. C. T. *B* 329; & sette þas laȝe
HOM. I. 17; þe king sette to fleonne LAȝ. 1570;
and sette treen GEN. & EX. 1278; Havelok
sette him dun HAV. 927; schi(r)-reves he
sette (*constituted*) 266; (þu) settest JUL. 61;
setten P. L. S. xxxi. 296; P. S. 215; setten a
sertein dai WILL. 1462; set (*pple.*) SPEC. 31;
þa setten heo biscopes LAȝ. 10200; (*ms.* sett)
ORM. DEDIC. 41; *comp.* a-, bi-, for-, ȝe-,
of-, ofer-, ümb-setten.

seþen, *see* seoðen. seþþen, *see* siþþen.

seūr, adj., *O.Fr.* seür; *sure*, S. S. (Web.)
2033; HOCCL. i. 320; sur WILL. 973; suir
P. P. 216.

   seūr-lī, adv., *surely*; seurli, sureli ALEX.
(Sk.) 1986*, 1833.

seūrement, sb., *assurance*; surement ALEX.
(Sk.) 2748.

seūrtē, sb., *O.Fr.* seurté; *surety*, WILL. 1463;
seurtee CH. C. T. *B* 243.

seute, sb., *O.Fr.* sieute, suite; *mod.Eng.*
suit, suite; following, company, retinue,

N n

WILL. 1080; siute ROB. 36; seute [suite, sute] LANGL. *B.* v. 504; suite CH. C. T. *A* 2873; soite, soute D. ARTH. 81, 3931, 3941.

seven, *see* seofen.

severen, v., *O.Fr.* severer, sevrer; *sever*; **severes** (*pres.*) GAW. 1797; **severed** (*pple.*) TREV. IV. 325.

sēw, *see* sēau.

sewe, sb., *from* sewen; *pursuit*, PALL. v. 184.

sewen, v., *O.Fr.* sewir, sivir, suir; *follow*; sewe (?seue) MAND. 226; sewe, seuwe WILL. 581, 2821; sewe [suwe] LANGL. *B* xi. 21 (Wr.) 6615; siwi LA3. 1387*; suin PR. P. 483; sue WICL. GEN. xxiv. 8; **seweð** (*pres.*) HOM. II. 85; siweþ SPEC. 24; siweþ [siveþ] O. & N. 1526; suweð A. R. 208; **suwede** (*pret.*) C. L. 1274.

siwinge, sb., *carrying out orders*, ROB. (W.) 10320.

sēwen, *see* schēawen, sēowen.

sewen², v. : to sewe at þe mete *set upon the table 'deponere'* CATH. 331; '*ferculo*' PR. P. 454.

sewer, sb., *waiter (at table)*; '*depositor*,' CATH. 331; ['*dapifer*'] TREV. VI. 435; A. P. ii. 639.

sex, *see* sax, six.

sextein, *see* sacristain. **shaft,** *see* schaft.

shalmie, sb., *O.Fr.* chalemie; *shawm, trumpet*; **shalmies** (*pl.*) CH. H. F. 1218.

shaundre, *see* scandle.

si, *see* se, si3e. **sī,** *see* am.

sib, adj. & sb., *O.E.* sibb, sib, =*O.Fris.* sib, *O.H.G.* sipp, *Goth.* (un-)sibjis, *O.N.* sifi; *related, relative,* ORM. 307; GEN. & EX. 228; TOWNL. 162; 3ef ho were sib to the MIRC 1353; sib or fremde 1558; sibbe LANGL. *B* v. 634 [sib *A* vi. 113]; sibbe '*consangvineus*' PR. P. 455; HORN (R.) 68; AR. & MER. 8914; TRIST. 722; **sibbe** (*pl.*) P. L. S. viii. 17; A. R. 204; HAV. 2277; SHOR. 69; sibbe men LA3. 1360; sibbe & fremde MAN. (F.) 6484; *comp.* god-, 3e-sib.

sib-man, sb., *relative,* '*affinis*;' **sibmen** (*pl.*) LA3. 1360*.

sib-rēden, sb., *O. E.* sibrǣden; *kinship, affinity,* SAX. CHR. 255; sibredin D. ARTH. 691; sibrede PR. P. 455; ROB. 492; GOW. III. 284.

sib-sum, adj., *O.E.* sibsum, =*O.H.G.* sibbisam; *amicable, peaceable* : þa **sibsume** (*pl.*) ['*pacifici*'] MAT. v. 9; *comp.* 3e-sibsum.

sibsumnesse, sb., *O.E.* sibsumness; *peace, amity,* HOM. I. 91; MISC. 54.

sib, sibbe, sb., *O.E.* sibb, =*O.Fris.* sibbe, *Goth.* sibja, *O.H.G.* sippa, sibba, ? *O.N.* sif (*pl.* sifjar); *peace; kindred*; MAT. x. 13; HOM. I. 193, 243; sibbe O. & N. 1005; P. L. S. v (b.) 56; ALIS. 5982; S. S. (Web.) 288; sæhte and sibbe he luvede LA3. 6096; of Daviþes kin & sibbe ORM. 3315.

sibnesse, sb., *O.E.* (ge-)sibness; *peace; kindred*; HOM. I. 275; MISC. 92; LEG. 53; H. H. 204 (206).

sīc, sb., *sigh,* JUL. 21; sik SPEC. 92; **sīke** (*dat.*) CH. C. T. *A* 1117; **sīkes** (*pl.*) A. R. 284; siches HOM. II. 83.

sīc, *see* sēoc. **sīcer,** *see* sīder.

sīche, sīchen, *see* sīc, sīke, sīken.

sicomore, sb., *O.Fr.* sicomore; *sycamore*; **sicomoris** (*pl.*) WICL. 3 KINGS x. 27; sichomures ALEX. (Sk.) 4973.

sīd, adj., *O.E.* sīd, =*M.L.G.* sīd, sīt, *O.N.* sīðr; *wide, ample, far,* PR. C. 1534; siid '*talaris*' PR. P. 455; þe side coote WICL. GEN. xxxvii. 23; **sīde** (*pl.*) H. S. 3227; side robes SPEC. 37; **side** (*adv.*) LA3. 29902; wide & side ORM. 5900; **sidder** (*compar.*) LANGL. *B* v. 193.

[**side,** sb., *O.E.* sido, =*O.L.G.* sidu, *Goth.* sidus, *O.H.G.* situ, *O.N.* siðr ; *custom*.]

side-ful, adj., *well-behaved, moral ; 'pudicus,'* FRAG. 5 ; sedeful ORM. 2175.

sīde, sb., *O.E.* sīde, =*O.L.G.* sīda, *O.N.* sīða, *O.H.G.* sīta ; *side,* A. R. 294 ; ORM. 4777 ; C. L. 1154 ; SPEC. 88 ; side bi side LA3. 19824 ; C. M. 1786 ; **siden** (*dat.*) JOHN xx. 25 ; on ælchere siden [side] LA3. 621 ; al so þu dost on þire side O. & N. 429 ; Horn stant bi þi side HORN (H.) 1007 ; a **sīde** *aside* GOW. II. 372 ; CH. C. T. *A* 896 ; stoude a side FER. 1826 ; bi **sīden,** bi side, bi sides *beside, besides* LA3. 5498, 12426, 24411 ; bi side þe buregh HOM. II. 31 ; bi sides hem IBID. ; þar bi side O. & N. 25 ; bi side þe urþe TREAT. 134 ; **sīden** (*pl.*) A. R. 392 ; side HOM. I. 147 ; *comp.* ēst-, norþ-, sūþ-, west-, wiðer-side.

sīde-benche, sb., *side-bench,* WILL. 4565.

sīd-bord, sb., *side-table,* GAW. 115 ; sideborde LANGL. *B* xiii. 36.

sideling, adv., = *M.H.G.* sītelingen ; *sideways,* D. TROY 7320 ; sidling, sidlinges MAN. (F.) 10348, 10869 ; sidelinges GENER. 206 ; sidlinges D. ARTH. 1039.

sīd-wā3, sb., *side-wall ;* **sidwages** (*pl.*) GR. 18 ; sidwowes FRAG. 6.

sīder, sb., *O.Fr.* sidre, *low Lat.* sicera, *Gr.* σίκερα, *Heb.* שֵׁכָר ; *cider, strong drink,* LEG. 81 ; sidir '*sicera*' PR. P. 455 ; cider [siþer, sicer, ciser] CH. C. T. *B* 3245 ; sider [cidre, siþer, ciser, cisar, sidur] WICL. JUDG. xiii. 14 ; sicher, sycher VOC. 178 ; cithir WICL. JUDG. xiii. 14 ; C. M. 10982, 12679.

sie, *see* sē. **siec,** *see* sēoc.

sīen, *see* si3en, sīhen. **sife,** *see* sive.

sifte, sb., = *M.Du.* sifte ; *sieve ;* cifte [cive] '*cribrum*' PR. P. 78.

siften, v., *O.E.* siftan, = *M.Du.* siften ; *sift,* C. M. 15523 ; siftin '*cribro*' PR. P. 455 ; **sifted** (*pple.*) CH. C. T. *G* 941.

siʒ (? sīʒ=sīc), sb., *sigh*; sighe (*dat.*) CH. C.
T. *A* 1117; sighes [sikes] (*pl.*) P. F. 248.

sigaldren, *see under* siʒe².

siʒe, sb., *O.E.* sige, *cf. O.H.G.* sigi, *m., Goth.*
sigis *n.*; *victory*, HOM. I. 13; LAʒ. 23896;
ORM. 5461; si (*ms.* sy) JUL. 10.

   siʒe-fast, adj., *victorious*, ORM. 16958;
siʒefest (*printed* siʒefeit) FRAG. 4.

[siʒe², ?.]

   siʒe-craft, sb., *? magic art*, LAʒ. 15501.

   si-galdren, v., *enchant*, A. R. 208; sigaldride
(*pret.*) H. S. 503.

   sigaldrie, sb., *enchantment*, A. R. 208*;
ALIS. 7015; CHEST. II. 69.

siʒen, v., *from* siʒ; *cf.* sīken; *sigh*; sighen
GOW. II. 319; S.S. (Wr.) 1860; sihghin, sihin,
sigh, '*suspiro*,' PR. P. 455; sihin LUD. COV.
391; siʒh(e) (*pres.*) WILL. 909; sihð O. & N.
1587; sighes PERC. 1064; sigande (*pple.*)GEN.
& EX. 1436; siʒede [siʒhede] (*pret.*) LANGL.
B xviii. 263; siʒede PL. CR. 442; siʒide WICL.
JOSH. xv. 18; siʒed TOR. 2388; sighed EGL.
832; sīked [sighed] MAN. (F.) 9073.

   siʒing, sb., *sighing*, FER. 1040; sighinges
(*pl.*) ALEX. (Sk.) 441.

sīʒen, v., *O.E.* sīgan, = *O:L.G., O.H.G.* sīgan,
O.Fris., O.N. sīga; *glide, fall, sink*: siʒen to
helle LAʒ. 14589; forð heo gunnen siʒen
29071; sihen KATH. 2353; sieʒe ALEX. (Sk.)
716; sie CH. TRO. v. 182; PALL. xi. 326;
sīheð (*pres.*) H. M. 47; hie arist anes a dai
and eft sigeð HOM. II. 109; sieð R. S. iii;
sīʒe (*subj.*) SHOR. 3; sah, seh (*pret.*) LAʒ.
2918, 10255; doun bi (h)is chin (h)it seʒ
FER. 589; siʒen LAʒ. 8682; sihen JUL. 76;
comp. a-, an-, ʒe-, to-sīʒen.

siggen, *see* seggen.   signe, *see* seine.

signet, sb., *O.Fr.* signet; *signet*; singettez
(*pl.*) A. P. i. 838.

signifiaunce, sb., *Fr.* signifiance; *signifi-
cance*, WILL. 2958; signefiance MISC. 28.

signifien, v., *Fr.* signifier; *signify*; signifie
ROB. 154; signifieth (*pres.*) ALIS.² (Sk.)
853; signified (*pret.*) ALEX. (Sk.) 515.

sīʒt, *see* siht.

sīhe, sb., = *O.H.G.* sīha, *M.Du.* sijghe, *O.N.*
sīa; *strainer, colander*, '*colum*,' PR. P. 79*.

sīhen, sēon, v., *O.E.* sēon, = *O.H.G.* sīhan,
M.Du. sijghen, *O.N.* sīa; *strain, run*; sī
(*imper.*) L. C. C. 7; þet rede blod sēh (*pret.*)
ut HOM. I. 121.

sīhen, *see* sīʒen.

siht, sb., = *O.H.G.* siht *f.; ? =*sihðe; *from*
sēon; *sight, vision*, LAʒ. 20929; R. S. v;
SPEC. 25; sigt GEN. & EX. 1626; siʒt SHOR.
33; WILL. 762; GREG. 1048; sight MAN.
(H.) 174; S. S. (Wr.) 2414; sith LUD. COV.
117; sihte P. L. S. viii. 184; HOM. II. 61;
CHR. E. 769; siʒte MARG. 155; *comp.* an-,
bi-, fore-, ēaʒe-, ēiʒe-(ie), in-siht (-sihte,
-siʒt).

sighte-lēs, adj., *sightless*, GEN. & EX. 1528.

siht², sb., *sigh*; siʒt GREG. 13; siʒtes (*pl.*)
WILL. 924.

sihten, v., *O.E.* siccettan; *sigh*; sighte REL.
I. 71; s. & C. II. lxii; sihte (*pret.*) HOM.
II. 169; (*ms.* sihtte) L. N. F. 217; sihte [siʒte]
O. & N. 1291; siʒte FL. & BL. 59; BRD. 28;
TREV. VIII. 227; sighte CH. C. T. *B* 1035;
DEGR. 209.

sihti, adj.,= *M.H.G.* sihtic; *visible*; sighti
'*visibilis*' PR. P. 455.

sihðe, sb., *O.E.* (ge-)sihð *f.; from* sēon; *vision*,
MK. xvi. 8; A. R. 48; KATH. 497, 1620;
sigðe (*ms.* sigðhe) GEN. & EX. 1630; sihþe
ORM. 12670; ziʒþe AYENB. 47; *comp.* an-,
ēaʒe (ēh-), ʒe-sihðe.

sīk, *see* sīc.

sīke, sb., *O.E.* sīc,= *O.N.* sīk, sīki *n.*; *trench,
rill*, '*rivus*,' VOC. 195; sīche (*dat.*) SHOR.
123; sikes (*pl.*) MAN. (F.) 8165; sikes, sikis
BARB. xix. 742, xi. 300.

sīke, *see* sēoke.

sikel, sb., *O.E.* sicol,= *O.H.G.* sichila, *from
Lat.* secula; *sickle*, KATH. 827; TREAT. 133;
sikel [sikul] LANGL. *B* iii. 306 (Wr. 1983); sikil
'*falx*' PR. P. 455; REL. I. 7.

sikel, adj., *? sickly*: Lazarus was swiþe sikel
a man S. A. L. 151.

sīken, v., *O.E.* sīcan (*pret.* sāc); *sigh*, A. R.
110; H. M. 27; siken . . . & suhʒhen ORM.
6924; sike HORN (L.) 426; HAV. 291; WILL.
691; CH. C. T. *A* 1540; sike & grone P. L. S.
xix. 266; siken, siche(n) LAʒ. 12772, 13626;
sike, siche BEK. 536, 1446; siche MISC. 50;
REL. I. 274; TREV. VII. 535; LAUNF. 249;
sīkeð (*pres.*) HOM. I. 43; sikeþ O. & N.
1352; sīkande (*pple.*) A. P. ii. 715; sīkede
(*pret.*) LANGL. *B* xviii. 263; siked MAN.
(F.) 9073; sīked (*pple.*) SPEC. 92; *comp.*
bi-sīchen.

   sīking, sb., *sighing*, HAV. 234; SPEC. 53;
WILL. 5451.

sīken, *see* sēken.

siker, adj.,= *O.Fris.* siker, sikur, *O.L.G.* sikur̄,
O.H.G. sichur, *from Lat.* secūrus; *secure*, P.
L. S. viii. 20; LAʒ. 15092; KATH. 25; ORM.
4844; GEN. & EX. 869; ROB. 116; WILL. 2361;
CH. C. T. *F* 1139; PR. C. 8558; sekir TOR.
1001; sikerure (*compar.*) A. R. 164; sikerer
LANGL. *B* xii. 162.

   siker-hēde, sb., = *O.H.G.* sichurheit; *security*, O. & N. 1265.

   siker-lēc, sb., *security*, MARH. 14.

   siker-līche, adv.,= *O.H.G.* sichurlīcho; *securely*, A. R. 62; O. & N. 1139; sikerlike GEN.
& EX. 1500; HAV. 422; sekirli PERC. 2002.

   sikernesse, sb., *security*, A. R. 342; HAV.
2856; JOS. 623; sikernisse BEK. 800; sekir-
nes PERC. 1204.

sikerien, v., *O.E.* sicerian; *trickle*: þat

sikeriez (? *printed* sikeniez) (*pres.*) out of þe se SAINTS (Ld.) xlvi. 641.

sikerin, v.,=*O.Fris.* sikura, sikeria, *O.L.G.* sicorōn, *O.H.G.* sicherōn ; *make sure,* '*securo,*' PR. P. 455 ; sikeri ROB. 545 ; FER. 2041 ; sikere S. S. (Wr.) 47 ; **sikerede** (*pret.*) MAN. (F.) 14716 ; **sikered** (*pple.*) WILL. 1463.

sikil, *see* swikel. **sǐknen,** *see* sēcnen.

sílden, *see* schílden, **sile,** *see* séle.

sīlen, v., *glide ; go* ; sile DEGR. 343 ; YORK xviii. 196 ; **sīles** (*pres.*) he a doune AV. ARTH. xvi ; silis, silez ALEX. (Sk.) 111, 2922* ; & **sīled** (*pret.*) firre A. P. ii. 131.

sīlen [2], v., *Swed.* sīla ; *strain* ; sīle '*colare*' CATH. 339 ; **sīled** (*pple.*) HALLIW. 743 ; silud L. C. C. 21.

silence, sb., *O.Fr.* silence ; *silence,* A. R. 22.

silf, *see* self.

silk, sb., *O.E.* seolc,=*O.N.* silki ; *silk,* SPEC. 36 ; P. L. S. i. 11 ; MAND. 212 ; CH. C. T. *F* 613 ; silke ALEX. (Sk.) 4016 ; selc FL. & BL. 536 ; selk B. DISC. 223 ; selke SHOR. 34 ; **seolke** (*dat.*) LAȝ. 22764 ; A. R. 420.

 silk-þrēd, sb.,=*O.N.* silkiþrāčr ; *silk thread,* WILL. 4430.

 selk-werk, sb., *embroidery of silk,* JOS. 427.
 silk-wirm, sb., *O.E.* seolcwyrm ; *silk-worm,* PR. P. 456.

silken, adj., *O.E.* seolcen ; *silken,* LIDG. M. P. 104 ; silkin EM. 377 ; selkin ALIS. 278 ; mid **seolkene** (*dat. pl.*) þrede MISC. 77.

sillable, sb., *O.Fr.* sillabe ; *syllable,* PR. P. 455 ; CH. C. T. *E* 101 ; silipp YORK x. 26.

sille, *see* sülle. **sillen,** *see* sellen.

silour, *see* celūre.

silte, sb., *silt* ; cilte '*glarea*' PR. P. 77.

silten, *see* salten.

silver, sb., *O.E.* silofor, seolfor, sylfor,=*O.N.* silfr, *O.H.G.* silbar, silabar, *Goth.* silubr ; *silver,* BEK. 1472 ; silfer ORM. 15796 ; selver [silver] LANGL. PROL. *A* 78, *B* 81 ; CH. C. T. *A* 115 ; selfer MAT. x. 9 ; selver KATH. 270 ; R. S. vii ; WILL. 2554 ; P. 5 ; zelver AYENB. 35 ; seolver HOM. I. 9 ; A. R. 152 ; O. & N. 1366 ; seolver, sulver LAȝ. 884, 3570 ; suelfer P. L. S. viii. 133 ; **selfre** (*dat.*) HOM. I. 227 ; seolvre MISC. 89 ; selvre 28 ; seolvere, selvere LAȝ. 2451, 4388 ; *comp.* qvik-silver.

 silver-ōre, sb., *silver-ore,* P. S. 338.

silveren, adj., *O.E.* sylofren, seolfren, silfren, = *O.H.G.* silberīn, *Goth.* silubreins ; *made of silver,* WICL. GEN. xliv. 2 ; mid selvrene stikke ANGL. IV. 194 ; bi þe selverne biȝe L. H. R. 29 ; **seolverne** (*pl.*) LAȝ. 22783 ; silverene D. ARTH. 1949.

silverin, v., *cover with silver,* '*deargento,*' PR. P. 456.

siment, *see* ciment.

simfan, sb., *O.Fr.* symphans, sinfanie ; *musical instrument* ; ['*symphonies*'] MAN. (F.) 11387.

similacre [semilacre], sb., *O.Fr.* simulacre, *Lat.* simulācrum ; *image,* ALEX. (Sk.) 2997 ; **semilacris** (*pl.*) 5637 ; simolacries 4460.

simile, sb., *simile;* LANGL. *C* xix. 299.

similitūde, sb., *Lat.* similitūdō ; *comparison, proposition,* CH. C. T. *G* 431 ; LANGL. *C* xx. 160*.

simle, adv., *O.E.* simle, simble,=*O.L.G.* simbla, simla. *O.H.G.* simble ; *always,* HOM. I. 239.

simnel, sb., *O.Fr.* simenel ; *fine sort of bread,* VOC. 198 ; PR. P. 456 ; **simenels** (*pl.*) HAV. 779.

simonde, sb., *for* ciment ; *cement,* YORK viii. 102.

simonie, sb., *O.Fr.* simonie ; *simony,* A. R. 202 ; P. L. S. xvii. 145 ; simonie LANGL. *A* PROL. 83.

simple, adj., *O.Fr.* simple ; *simple. ignorant,* A. R. 128 ; REL. I. 226 ; BEK. 1058 ; AYENB. 134 ; WILL. 714 ; sempill ALEX. (Sk.) 4404.

 simple-liche, adv., *simply,* AYENB. 134.

 simpilnesse, sb., *simplicity,* PR. P. 456 ; ALEX. (Sk.) 4040, 4051.

simplesse, sb., *O.Fr.* simplesse ; *simplicity,* AYENB. 140 ; LANGL. *A* xi. 121*.

simpletē, sb., *O.Fr.* sempleté ; *simplicity,* LANGL. *B* x. 165 ; simplite *A* xi. 121*.

simulaciōn, sb., *O.Fr.* simulacion ; *simulation,* AYENB. 23.

[**sin-**, *O.E.*=*O.L.G.* *O.H.G.* sin-, *O.N.* sī- ; *always, perpetually.*]

 sin-fulle, adj., *O.E.* sinfulle ; *endless,* REL. I. 38.

 sin-grēne, sb., *O.E.* singrēne, = *M.H.G.* singrüene ; *house leek,* '*Jovis barba,*' REL. I. 37 ; '*semperviva*' PALL. i. 853 ; sengrene PR. P. 251.

[**sin-hīwen,** sb., pl., *O.E.* sinhīwan ; *husband and wife* ; *comp.* ȝe-sinhēowen.]

sinagins, sb., pl., ? *monsters,* ALEX. (Sk). 5452.

sinagoge, sb., *synagogue,* ALEX. (Sk.) 1058.

sinder, sb., *O.E.* sinder,=*M.L.G.* sinder, *O.N.* sindr, *O.H.G.* sintar ; *cinder,* '*scoria,*' CATH. 340 ; sindir PR. P. 456.

sinder, *see* sünder.

sindren, v., *L.G.* sin(d)ern ; *purify* : þe ilke welle is zvo clier and zvo izendred (*pple.*) AYENB. 251.

sinegen, *see* süneȝen.

sinewe, sb., *O.E.* sinu, seonu,=*O.Fris.* sine, *M.L.G.* sene, *O.N.* sin, *O.H.G.* senewa ; *sinew,* '*nervus,*' PR. P. 456 ; S. S. (Web.) 1048 ; sinue (*ms.* sinwe), senewe WICL. GEN. xxxii. 25 ; senwe GEN. & EX. 1805 ; **sinuen** (*pl.*) P. L. S. xv. 194 ; senuwen LAȝ. 6498 ; seonewen MARH. 7 ; sina, sinan LEECHD. III. 88, 120 ; senewes SPEC. 101 ; senous, sinnous YORK xxxv. 108, 133 ; *comp.* hōuȝ-senu.

sinowi, adj., *sinewy* ; senowi PALL. iv. 684.

singable, adj., *singable* ; singabil [*'cantabiles'*] HAMP. PS. cxviii. 54.

singe, *see* **seine.**

singen, v., *O.E.* sigan,=*O.L.G.*, *O.H.G.* singan, *Goth.* siggwan, *O.N.* syngva, syngja ; *sing,* A. R. 44 ; LANGL. *B* xv. 219 ; zinge AYENB. 22 ; **singeþ** (*pres.*) ORM. 1725 ; SHOR. 23 ; þu **singe** (*subj.*) O. & N. 226 ; **singende,** **singinge** (*pple.*) LA3. 26946, 29714 ; **sang** [song] (*pret.*) CH. C. T. *A* 2212 ; song O. & N. 20 ; songe LANGL. *B* xix. 206 ; þu sunge O. & N. 1049 ; sungun, sungen HOM. I. 5, 7 ; sungen SAX. CHR. 249 ; MARH. 22 ; ORM. 3373 ; GEN. & EX. 3288 ; O. & N. 1663 ; sunge GOW. II. 176 ; sunge, songen BRD. 17, 20 ; songen MAND. 278 ; **sunge** (*subj.*) O. & N. 1026 ; sunge [songe] LA3. 17435 ; **sunge,** songe (*pple.*) CH. C. T. *A* 266 ; isungen [isonge] LA3. 18520 ; *deriv.* **sang, sengen.**

    singare, v., *singer,* PR. P. 456.

    singstere, sb., *female singer,* WICL. I ESD. ii. 65.

sin3en, *see* **süne3en.**

single, adj.,*O.Fr.* single, sengle; *single* ; sengle SHOR. 65 ; AYENB. 48.

singnette, *see* **signet.**

singuler, adj., *O.Fr.* singuler ; *singular,* TREV. IV. 17.

sinke, sb.,=*M.Du.* sinke ; *sink ; well of a lamp ; 'exceptorium, mergulus'* PR. P. 456.

sinken, v., *O.E.* sincan,=*O.L.G.* sincan, *O. H.G.* sinkan, *Goth.* sigqan, *O.N.* sökkva ; *sink, glide, fall,* LA3. 20156 ; MARH. 7 ; CH. C. T. *A* 951 ; **sinkeþ** (*pres.*) ORM. 13381 ; **sank** (*pret.*) GOW. I. 229 ; LANGL. *B* xviii. 67 ; sunken LA3. 4582 ; JUL. 79 ; GEN. & EX. 3775 ; sonken MAP 339 ; i **sonke** (*ms.* songe, *r. w.* lonke) (*subj.*) P. S. 156 ; **sunken** (*pple.*) ORM. 14569 ; sonken WILL. 4111 ; MAND. 101 ; isunken LA3. 28485 ; *comp.* **a-, bi-, for-sinken** ; *deriv.* **senchen.**

sinne, *see* **sünne.**

sinoper, sb.,*O.Fr.*sinople ; *red earth,* 'sinopis,' CATH. 341 ; sinopir PR. P. 456.

sinoþ, sb., *O.E.* synoð, *Lat.* synodus, *Gr.* σύνοδος ; *synod,* FRAG. 2 ; sinað LA3. 25338 ; sinod TREV. V. 231.

sioun, *see* **ciūn. sip,** *see* **schip.**

siphre, sb., *Arab.* çifr ; *cipher,* DEP. R. iv. 53.

sippin, v.,=*M.Du.* sippen, *M.H.G.* supfen ; *sip,* PR. P. 456 ; sippe CH. C. T. *D* 176.

sire, sb., *O.Fr.* sīre ; *sire, sir,* LA3. 22485 ; A. R. 86 ; R. S. vii ; HAV. 310 ; sire bischop BEK. 997 ; ser YORK xxii. 151 ; sire soile *fatherland* ALEX. (Sk.) 5021 ; **sires** (*pl.*) WILL. 2248 ; CH. C. T. *A* 3909.

sirurgien, *see* **cirurgian.**

sirupe, sirope, sb., *? O.Fr.* syrop ; *syrup,* CATH. 341 ; sorip PR. P. 465.

sis, sb., *O.Fr.* sis ; *six of dice,* CH. C. T. *B* 3851

sīse, sb., *for* asīse ; *mod.Eng.* size ; *session, meeting,* MAP 337 ; *due measure* MIRC 1282 ; *'dome of lond'* PR. P. 456.

sīse², sb., *size* : sise for bokis limininge PR. P. 456.

sīsen, v.,*for* asīsen ; *constitute* ; sīsed (*pret.*) ALEX. (Sk.) 4654.

sīsour, sb., *for* asīsour, *? juryman, ? witness* : a sisoure and a sompnoure LANGL. *B* iv. 167 ; **sīsoures** [sisours] GAM. 881 ; fals men þat bein sisours H. S. 1335.

sissen, v., *cf. Du.* sissen ; *hiss* ; **cisses** (*pres.*) VOC. 152.

sīser, *see* **sīder.**

siste, *see* **sixte. sister,** *see* **süster.**

sīte, sb., *Fr.* sīte ; *position,* CH. ASTR. ii. 17 (28).

sīte², sb., *O.N.* sȳti ; *sorrow, pain,* L. H. R. 63 ; ALEX. 546 ; TRIST. 1940 ; MIN. vii. 65 ; M. H. 149 ; ANT. ARTH. xvii : sorowe and site MAN. (H.) 5 ; FLOR. 1631 ; ser3he & sit ORM. 4852 ; in site and care C. M. 1410.

sīten, v., *O.N.* sȳta ; *be sorrowful, be anxious* ; **sīte** (*pres.*) C. M. 11675.

sitole, sb., *for* **citole** ; *guitar,* A. P. i. 91.

sitten, v., *O.E.* sittan,=*O.L.G.* sittean, *O.N.* sitja, *Goth.* sitan, *O.H.G.* sizzan, *Lat.* sīdere ; *sit,* LA3. 23035 ; KATH. 1574 ; A. R. 22 ; ORM. 14086 ; LANGL. *B* ii. 96 ; sitte BRD. 24 ; to sittenne MAT. xx. 23 ; þu **sittest** (*pres.*) O. & N. 89 ; (he) sit A. R. 332 ; MARH. 17 ; K. T. 699 ; (we) sitteþ O. & N. 1682 ; heo sitteð LA3. 14118 ; **site** (*imper.*) HOM. I. 91 ; REL. I. 186 : AN. LIT. 1 ; sitteð a dun LA3. 24865 ; **sitte** (*subj.*) A. R. 10 ; þet mon ... er timan to his borde ne sitte HOM. I. 105 ; ne sitte hit hire se uvele H. M. 7 ; **sittinde** (*pple.*) A. R. 16 ; **sittende** [sittinge] LANGL. *B* xviii. 48 ; inne wel sittende schon MISC. 193 ; **sat** [*O.E.* sæt,=*O.L.G.*, *O.N.*, *Goth.* sat, *O.H.G.* **saz**] (*pret.*) ORM. 5807 ; O. & N. 939 ; RICH. 3454 ; CH. C. T. *A* 615 ; PERC. 968 ; LIDG. M. P. 12 ; he sat at te bordes ende BEK. 1186 ; he sat a dun 1172 ; þer he sæt [sat] ... an his kinebenche LA3. 9692 ; þe swike set [sat] a dun 12958 ; set A. R. 156 ; KATH. 139 ; seet L. N. F. 217 ; FER. 1200 ; (þu) sete FRAG. 6 ; A. R. 238 ; FER. 1403 ; LANGL. (Wr.) 12653 ; **sæten** (*pl.*) ORM. 15560 ; heo sæten [seten] stille LA3. 25121 ; heo seten to borde 14950 ; seten A. R. 258 ; JOS. 432 ; RICH. 3471 ; CH. C. T. *F* 92 ; seeten MISC. 229 ; seten [saten] WICL. JOSH. viii. 9 ; sete BRD. 6 ; TRIST. 549 ; **sæte** (*subj.*) BEK. 1213 ; WILL. 1622 ; **seten** [*O.E.* seten, = *O.N.* setinn, *Goth.* sitans, *O.H. G.* (ge-)sezen] (*pple.*) HOM. II. 103 ; HAV. 738 ; S. S. (Web.) 2649 ; MIRC 1179 ; WICL. LK. vii. 37 ; seten [siten] CH. C. T. *A* 1452 ; **siten** PS. cxxvi. 2 ; *comp.* **æt-, bi-, for-, 3e-, wiþ-sitten** ; *deriv.* **set, setel, setten, sæte.**

sittunge, sb.,=*M.H.G.* sitzunge ; *sitting,* A. R. 156.

siþ,adv., *O.E.* siþ,*cf.O.N.* sið,*O.H.G.* sīd, *Goth.* (þana-)seiþs ; *?after,* Gow. I. 104 ; siþ whanne LANGL. *B* xx. 186 ; *see* **siþþen.**

sið, sb., *O.E.* sið, *cf. O.L.G.* sīth, sīð, *O.H.G.* sind, *Goth.* sinþs *m., O.N.* sinn *n.* ; *journey* ; *time ; chance* ; H. M. 9 ; hwet unseli sið JUL. 47 ; Ebri(u)s (h)adden seli sið GEN. & EX. 2546 ; siþ AN. LIT. 9 ; þane seoruhful(l)e siþ FRAG. 5 ; dreiȝen ... wrecche siþ 6 ; oðer siðe HOM. I. 37 ; þridde siðe *thirdly* HOM. II. 95 ; sum siþe ORM. 5372 ; ðis one siðe GEN. & EX. 3093 ; þe þridde siþe O. & N. 325 ; hu þe beon (*r.* beo) on siðe *in what condition art thou* LAȝ. 30284 ; aris ... to þine fæie siðe (*death*) 26040 ; at sume siþe O. & N. 293 ; in unker siþe 993 ; of here liwes (*r.* lives) siþe L. N. F. 217 ; wanede hire sīðes (*pl.*) LAȝ. 25847 ; his wrecche siðes HOM. II. 169 ; þreo siðes REL. I. 130 ; ten siðes GEN. & EX. 1731 ; siþes LANGL. *B* v. 431 ; MAND. 291 ; seofen siðan, siðen (*ms.* syðan, syðen) ['*septies*'] LK. xvii. 4 ; fif siðen A. R. 18 ; siþen P. S. 151 ; siðe LAȝ. 1103 ; fif siðe KATH. 794 ; siþe HAV. 778 ; FL. & BL. 212 ; SPEC. 134 ; CH. C. T. *B* 1155 ; feole siþe K. T. 470 ; mani þousand siþe [siþes] WILL. 103, 1696 ; feill siss *many times* BARB. v. 178 ; *comp.* **bále-, cwále-, déaþ-, earfeþ-, forð-, heðen-, líf-, sorh-, wā-, wēn-sīð.**

sīþe, sb., *O.E.* siðe,=*M.L.G.* sigde, segede *f.,* *O.N.* sigðr *m.; scythe,* HAV. 2553 ; LANGL. *B* iii. 306 ; RICH. 6788 ; GAW. 2202 ; sithe '*falx*' VOC. 277 ; PR. P. 457.

sīþid, adj., *armed with a scythe,* ALEX. (Sk.) 3598, 3821.

siþen, siþenes, *see* **siþþen.**

sīþien, sīþian, v., = *O.E.* sīðian,=*O.L.G.* sīthōn, *O.H.G.* sindōn, *O.N.* sinna ; *from* **sīð** ; *make a journey* ; FRAG. 8 ; þet ure saule moten eft sīðian to him HOM. I. 119 ; siðen REL. I. 224 ; forð siðen LAȝ. 21279 ; (heo) **sīþieþ** (*pres.*) FRAG. 8.

siþre? in phr. **of siþre,** *?*=*O.N.* of sidir ; *of late,* ORM. 322.

siþþen, adv., *O.E.* siððan, siþþon, seoððan, syððan ;=**siþ** þan ; *since, ago,* ORM. 231 ; siþþen [sithen, siþþe, sithe] þat þe world bigan CH. C. T. *A* 2102 ; siþ [sin] þat i hadde a wif *E* 1545 ; siðþe MISC. 163 ; siþþe TREAT. 132 ; siþþe [seþþe] he burede him L. H. R. 26 ; siþþe [sith] LANGL. *B* v. 441 ; siþen [sithhen, siþenes, sitthenes] vi. 65 ; siþe [sith, sin] xx. 320 ; siþþe a longe time PL. CR. 158 ; siþe Criste deide 353 ; siððan, seoþþan HOM. I. 43, 109 ; seoððen A. R. 146 ; R. S. v ; seoþþen his deaȝes beoþ igon FRAG. 5 ; seoððen, suððen LAȝ. 1725, 3915 ; seþþen C. L. 46 ; TRIST. 109 ; seþþen, seþþe WILL. 104, 433 ; seoþþe [soþþe], seþþe O. & N. 324, 1402 ; seððe LK. vii. 45 ; þat furst is feir ant seþþe unsete SPEC.

23 ; zeþþe AYENB. 14 ; siððen (*?* siðen) [*O.N.* siðan], siðen, seðen HOM. II. 3, 21, 183 ; siðen GEN. & EX. 84 ; REL. I. 210 (MISC. 3) ; siþen HAV. 399 ; JOS. 4 ; MAND. 75 ; MAN. (H.) 2 ; A. P. i. 13 ; ISUM. 146 ; sithe M. ARTH. 126 ; seþen RICH. 5562 ; seþin ANT. ARTH. xx ; sithen, sen PR. C. 731, 2212 ; sin HOCCL. i. 71 ; sen DEGR. 1648 ; sen, sin YORK xxv. 66, i. 139 ; sethens E. G. 414.

sive, sb., *O.E.* sife,=*M.Du., M.L.G.* seve, *O.H.G.* sib ; *sieve,* '*cribrum,*' PR. P. 457 ; GOW. I. 294 ; CH. C. T. *G* 940 ; sife FRAG. 4 ; VOC. 201 ; *comp.* **hēr-seve.**

siveþe, sb., *O.E.* sifoða ; *bran,* '*acus,*' sifethe FRAG. 4 ; **sivedis** (*pl.*) '*furfur*' PR. P. 457.

sivin,v.,=*M.Du.* seven, *M.H.G.* siben ; *strain, sift,* '*colo,*' PR. P. 457.

siwen, *see* **sewen.**

six, card. num., *O.E.* six, seox, siex, sex,=*O.N.* sex, *O.L.G.* sehs, *O.H.G.* sehs, sehse, *Goth.* saihs, *Lat.* sex, *Gr.* ἕξ ; *six,* BRD. 28 ; six, sixe A. R. 30, 298 ; CH. C. T. *F* 391 (Wr. 10705) ; six hundred LAȝ. 613 ; þa sixe 25387 ; sixe HAV. 2788 ; LANGL. *B* v. 431 ; siexe P. S. 157 ; sex GEN. & EX. 577 ; MAN. (H.) 66 ; C. M. 18877 ; LIDG. M. P. 34 ; E. G. 355 ; þe sexe daȝhes ORM. 4166.

six-tēne, card. num., *O.E.* sixtēne, -tȳne ; *sixteen,* LAȝ. 1103 ; A. R. 298 ; sextene ORM. 572 ; AR. & MER. 3948.

six-tēoþe, ord. num., *O.E.* sixtēoða ; *sixteenth,* P. L. S. xii. 81 ; sixteþe ROB. 261 ; sixtenðe JUL. 78.

six-ti, card. num., *O. E.* sixtig, = *O. L. G.* sehstig ; *sixty,* LAȝ. 27446 ; BEK. 1851 ; sextiȝ ORM. 7675 ; sexti PR. C. 4525.

sixti-fáld, adj., *sixty-fold,* H. M. 23.

zixtiaȝte, ord. num., *O.E.* sixtigeða ; *sixtieth,* AYENB. 234.

sixte, ord. num., *O.E.* sixta,=*O.L.G.* sehsta, *O.H.G.* sehsto, *Goth.* saihsta, *Lat.* sextus ; *sixth,* A. R. 14 ; BEK. 601 ; sexte ORM. 4322 ; GOW. III. 121 ; S. S. (Wr.) 83 ; PR. C. 3982 ; siste, seste HOM. I. 39, 43 ; seste A. D. 240.

skaffaut, *see* **scafald.**

skalk, *see* **scalc.**

skar, *see* **sker. skarn,** *see* **scarn.**

skec, sb., *plundering,* AR. & MER. 4736 ; skek [skec, scek, shekke, checke, cheke] ROB. (W.) 5131.

skēi, *see* **schēoh.**

skeine, sb., *O.Fr.* escaigne ; *skein,* PR. P. 457.

skek, *see* **skec. skele,** *see* **schile.**

skell, *see* **schelle. skellen,** *see* **schellen.**

skelp, sb., *blow, stroke* ; schath of **skelpis** (*pl.*) YORK xxxiii. 35.

skelpen, v., *prov.Eng.* skelp ; *beat, flog* ; skelp ALEX. (Sk.) 1924 ; he **skelpte** (*pret.*) oute of score YORK xxvi. 81.

skelten, v. : scoleres skelten (*pret.*) þeratte þe

skil for to find A. P. ii. 1554; skete skarmoch
skelt (*pple.*) much skathe lached 1186.

  skelting, sb. : for skelting of harme D. TROY
1089, 6042.

skemering, *see* **schimeringe** *under* **schimeren.**

skemten, v., *O.N.* skemta ; *amuse, delight* :
and skente hi mid mine songe O. & N. 449 ;
skente (*pret.*) 1085.

  skemting, sb., *amusement, pleasure*, ORM.
2165 ; skempting REL. I. 218 ; skenting LAȝ.
30625 ; skentinge O. & N. 446.

skenten, *see* **skemten.**

skeppe, sb., *O.N.* skeppa (*measure*); *skep,
carrying-basket,* '*sporta,*' PR. P. 457 ; PALL.
iii. 209.

sker, sb., *O.N.* sker *n.; projecting rock* ;
scarre ['*scopulus*'] WICL. 1 KINGS xiv. 5 ;
undir a skerre HALLIW. 709.

sker [2], adj., *O.N.* skiarr ; *scared, timid* ; skar
TOWNL. 198 ; þe skerre hors A. R. 242*.

skēre, adj., *O.E.* scǣre,=*O.N.* skǣr ; *sheer;
clear, pure* ; C. L. 1073 ; LEG. 85 ; qvit and
skere [schere] LAUNF. 429, 881 ; **schēre**
Þūrs dai *Holy Thursday* [*cf. Dan., Swed.*
skær Tōrsdag] BRD. 16 ; MIRC 640 ; schere
Þors dai MAND. 19 ; scere Þors dai REL. I.
144 ; shere Þurs dai SACR. 398 ; sheer Þurs
dai LIDG. M. P. 253 ; **skēre** (*adv.*) B. DISC.
1914 ; habbeð iqveðen us scere LAȝ. 12752 ;
of blisse ȝe beoþ skere R. S. iv ; makie we us
clene and skere MISC. 73 ; **skĕrre** (*compar.*)
A. R. 314.

skēren, v., *cf. O.Dan.* skǣre, *O.Swed.* skǣra ;
purify, A. R. 308 ; skere ROB. 334 ; REL. I.
241 ; ALIS. 3295 ; S. S. (Wr.) 3398 ; þar of
þu . . . most þe skere O. & N. 1302 ; of scaþe
i wol me skere P. S. 156 ; skere, schere MISC.
190, 191.

sker(r)el, sb., *? scarecrow,* '*larva,*' PR. P. 457 ;
scarle CATH. 321.

skerren, v., *cf. O.N.* skirra, *Swed.* skiarra ; *from*
sker ; *scare,* ORM. 676 ; skeren a wei '*abigo*'
PR. P. 457 ; **scarrez** (*pres.*) A. P. ii. 598 ;
skerrid (*pret.*) ALEX. (Sk.) 4802 ; skerrit
D. TROY 13404 ; *? comp.* **a-scürren.**

skēt, adj., *O.E.* scēot, *O.N.* skiōtr ; *quick* ;
skēte (*adv.*) AR. & MER. 294 ; GAM. 187 ;
TRIST. 896 : GAW. 19 ; A. P. iii. 195 ; skete,
skeet ALIS. 3047, 5637 ; sket ORM. 1266 ;
HAV. 1926 ; B. DISC. 484.

  skēt-lī, adv., *suddenly,* ALEX. (Sk.) 5040.

skēwed, adj., *? from* **skīe** ; *piebald* : the
skewed ['*varii*'] goos PALL. i. 703 ; skewed
horses CHEST. II. 142.

skēwen, *see* **schēowen.**

skīe, sb., *O.N.* skȳ *n., cf. O.L.G.* skio, sceo *m.* ;
sky, cloud, '*nubes,*' PR. P. 457 ; ALIS. 318 ; GOW.
II. 50 ; LIDG. M. P. 161 ; ðat brigte skie biforen
hem fleg GEN. & EX. 3643 ; þat it ne
lefte not a skie in al þe welken CH. H. *F* 1600 ;

skēwes (*pl.*) C. L. 1494 ; A. P. ii. 1206 ; skewis
ALEX. 561 ; skiwes REL. I. 262 ; skies 210.

skiffte, skiften, *see* **schift, schiften.**

skig, adj.,=*M.L.G.* schügge, *Swed.* skygg,
? O.N. styggr ; *?*=**schēoh** ; *timid, careful,*
A. P. ii. 21.

skil, skile, *see* **schile.**

skilli, *?* adj. : þen watz a skilli skivalde, qven
scaped alle the wilde A. P. ii. 529.

skin, sb., *O.E.* scinn,=*O.N.* skinn ; *skin,* CH.
C. T. *A* 3811 ; LIDG. M. P. 133 ; þet skin
AYENB. 81 ; scin 230 ; skinne '*pellis, cutis*'
PR. P. 457 ; skinne (*dat.*) EM. 954 ; **skinnes**
(*pl.*) MAND. 216 ; skinnes DEP. R. ii. 32 ;
comp. **barm-, bēre-skin.**

skinden, v., *O.E.* scyndan,=*O.N.* skynda,
O.H.G. scuntan ; *hasten* : ðe chapmen skinden
(*printed* skiuden) (*pret.*) here fare GEN.
& EX. 1989.

skine, *see* **schine.**

skinnere, sb., *skinner,* VOC. 181 ; skinnare
PR. P. 457.

skinneri, sb., *skins, furs,* B. B. 180.

skip, sb., *hop, jump,* '*saltus,*' PR. P. 458.

skippin, v., *skip, jump,* '*salto,*' PR. P. 458 ;
skippe HORN (R.) 1361 ; ALIS. 768 ; WICL.
1 KINGS x. 6 ; CH. C. T. *A* 3259 ; HOCCL. I.
120 ; skipte (*pret.*) LANGL. *B* xi. 103 ; MAN.
(F.) 11365 ; *comp.* **over-skippen.**

  skippare, sb., *skipper,* '*saltator,*' PR. P. 458 ;
skipperes (*pl.*) GEN. & EX. 3087.

skīr, *see* **schir.**

skirmen, v., *scream* ; **skirmand** (*pple.*) ALEX.
(Sk.) 5157.

skirmen [2], v.,=*O.H.G.* scirman, *M.L.G.* schermen
; *fence,* A. R. 212 ; bigunnen mid sceldes
to scurmen LAȝ. 8144 ; skirme ALIS. 662 ;
schirme O. & N. 306 ; **skirmden** (*pret.*) LAȝ.
8406.

  skirming, sb., *fencing,* HAV. 2323.

skirmishen, *see* **scarmishen.**

skirpen, v., *O.N.* skirpa (*spit*) ; *spit out ; reject* :
þat unfæle folc þat skirpeþ (*pres.*) þar
ongænes ORM. 7389.

skirt, sb., *skirt,* PR. P. 458 ; MAN. (F.) 7884 ;
skirtes (*pl.*) MAND. 221.

skirwit, sb., *parsnip,* '*pastinaca,*' PR. P. 449.

skitte, *see* **schitte.**

skivalde? sb. : þen watz a skilli skivalde,
qven scaped alle þe wilde A. P. ii. 529.

sklaire, *see* **sleir.**

sklither, sklitheringe, *see* **slith-.**

skof, *see* **scof.** **skrat,** *see* **scrat.**

skrīchen, skrīken, *see* **schrīchen.**

skŭlken, *see* **scŭlken.**

skunniren, *see* **scunneren.**

slā, sb., *O.E.* slāhe,=*O.H.G.* slēha, *M.Du.*
slee ; *sloe* ; slo '*prunum*' PR. P. 459 ; HAV.

2051 ; R. R. 928 ; þis lives blisse nis wurð a slo R. S. ii (MISC. 160) ; **slōn** (*pl.*) VOC. 163 ; ALIS. 4983.

slōe-thorn, sb., *O.E.* slāhþorn, = *Ger.* schlēhdorn ; *sloethorn*, VOC. 163.

slabbe, sb., *slab; plate, sheet* : a slab [ane sclabbe] of ire TREAT. 135 ; SAINTS (Ld.) xlvi. 531 ; **slabbes** (*pl.*) FER. 3313.

slabben, v., *cf. Swed.* slabba (RIETZ) ; *? wallow* : hou ist thet hi ine helle **slabbeþ** (*pres.*) SHOR. 151.

slaberen, *see* **slaveren**.

slac, sb., *ravine*, ANT. ARTH. xxiii ; slak DEG. 333 ; **slake** (*dat.*) ISUM. 622.

slac², adj., *O.E.* slæc, sleac, = *O.L.G.* slac, *O.N.* slakr, *O.H.G.* slach ; *slack, loose, lax, 'piger,'* FRAG. 2 ; slac, slak AYENB. 32, 141 ; slak *'laxus'* PR. P. 458 ; ALIS. 1252 ; þe slake skin CH. C. T. *E* 1849.

slacnesse, sb., *O.E.* sleacness ; *slackness*, AYENB. 33 ; slaknesse PR. P. 458 ; CH. C. T. *I* 680.

slácien, *see* **slákien**.

slæd, sb., *O.E.* slæd ; *valley* ; **slæde**, slade (*dat.*) LAȝ. 8585, 26887 ; slade GAW. 2147 ; A. P. i. 141 ; PALL. ix. 176 ; **sládes** (*pl.*) LAȝ. 28365 ; GOW. II. 93.

slæn, *see* **slēan**.

slēp, **slāp**, sb., *O.E.* slǣp, = *O.Fris.* slēp, *Goth.* slēps, *O.L.G.* slāp, *O.H.G.* slāf ; *sleep*, ORM. 1903, 3148 ; slep A. R. 144 ; AYENB. 31 ; sleep LANGL. *A* PROL. 45 ; slepe *B* PROL. 45 ; CH. C. T. (Wr.) 1392 (sleepe *A* 1390) ; **slǣpes** (*gen.*) ORM. 2971 ; **slǣpe** (*dat.*) ORM. 8352 ; heo weren on slæpe LAȝ. 1159 ; slepe TOR. 927 ; slape LK. ix. 32 ; HOM. II. 77 ; HORN (R.) 1315 ; a slape WILL. 1995 ; on slope FLOR. 1632.

slēp-ern, sb., *O.E.* slǣpern ; *sleeping room, bed room,* ' *dormitorium,*' FRAG. 4.

slēp-lēs, adj., *O.E.* slǣplēas ; *sleepless*, CATH. 344.

slēplēaste, sb., *O.E.* slǣplēast ; (*? printed* slop-), *sleeplessness*, FRAG. 3.

slēp-wurt, sb., *lettuce*, ' *lactuca,*' REL. I. 37.

[**slæpel**] **slāpel**, sb., *O.E.* slāpol ; *one fond of sleep*, FRAG. 3.

slǣpen [**slēpe**], v., *O.E.* slǣpan, slēpan (*pret.* slēp, slǣpte) (ÄLFR. P. C. 101), = *Goth.* slēpan (*pret.* saislēp), *O.Fris.* slēpa, *O.L.G.* slāpan, *O.H.G.* slāfan (*pret.* slief) ; *sleep*, LAȝ. 733 ; slepen A. R. 4 ; slepin PR. P. 459 ; sclepen REL. I. 177 ; slepe AYENB. 29 ; FLOR. 1657 ; sclepe LUD. COV. 41 ; slapen HOM. II. 7 ; slape S. S. (Web.) 929 ; AR. & MER. 814 ; **slēpeð** (*pres.*) A. R. 212 ; slepð MISC. 192 ; **slēpinde** (*pple.*) A. R. 224 ; **slēp** (*pret.*) GEN. & EX. 967 ; MISC. 24 ; ROB. 151 ; **slǣp** [sleap], **slēpte** LAȝ. 4005, 26009 ; slep [sleep, sleepte] LANGL. *B* v. 382 ; sleep [slepte] CH. C. T. *A* 98 ; slepte A. R. 236 ; MAND. 222 ; (*ms.* sleppte)

ORM. 2484 ; þu sleptest FRAG. 7 ; slepen HAV. 2128 ; **slēpe** (*subj.*) ROB. 213 ; **slēped** [slept] (*pple.*) LANGL. *B* v. 4 ; *comp.* a-, ȝe-**slǣpen**.

slēpere, sb., *sleeper*, TREAT. 138 ; slepare A. R. 258.

slēping, sb., *sleeping*, H. S. 5739.

slēpi, adj., *O.E.* slǣpig, = *M.L.G.* slāpich, *O.H.G.* slāfag ; *sleepy* ; slepi GEN. & EX. 871 ; GOW. II. 113 ; **slēpie** (*dat.*) A. R. 272.

slǣten, v., *incite* (*dogs*) *; hunt* ; ORM. 13485 ; him to slete mid grete houndes REL. II. 278 ; heo . . . **slǣtten** [sleatten] (*pret.*) him wið hundes JUL. 52 ; a slat (*pple.*) swin P. S. 154.

slǣting, sb., *inciting* (*dogs*), *hunting*, [*later text* honting] LAȝ. 12304 ; bole slating ALIS. 200.

[**slaȝe**, sb., *O.E.* slaga, = *O.H.G.* slago ; *from* **slēan** ; *slayer* ; *comp.* āȝen-, man-slaȝe.]

slaht, sb., *O.E.* (wæl-)sleaht, slyht *m.*, *cf. O.H.G.* slaht *f.* ; *from* **slēan** ; *slaughter, death*, KATH. 200 ; slaȝt ROB. 56 ; FER. 5519 ; slaȝt, sclaȝt A. P. i. 800, ii. 56 ; slauht JOS. 266 ; slaught ALIS. 1910 ; GOW. I. 348 ; slaht, slæht, sleȝht, sclæht LAȝ. 2544, 11252, 17951, 28730 ; sleight AR. & MER. 4936 ; ELL. ROM. I. 279 ; ofte the mannes sleȝte (*printed* sieȝte) arift were man hiȝt weneth wel litel SHOR. 98 ; *comp.* fēond-, man-, wæl-slaht.

slahter, sb., *O.N.* slātr *n.*, *slaughter*, M. H. 38 ; slaghter PR. C. 3367 ; slauhter MAN. (H.) 91 ; slaughter AR. & MER. 3918 ; slautir PR. P. 458.

slaȝþe, sb., *?* = **slaht** ; *slaughter*, AYENB. 90.

slaie, sb., *O.E.* slahæ, slae ; *sley* (*of a loom*), '*pecten,*' VOC. 217, 234 ; PR. P. 458.

slak, *see* **slac**.

slákien, v., *O.E.* (a-)sleacian (*from* slac), = *M. Du.* slaken, *O.L.G.* slekian ; *loose, set free,. slacken, slake*, A. R. 134 ; JUL. 26 ; þet . . . his sunbendes nule slakien HOM. I. 51 ; sclakien P. L. S. viii. 19 ; slakie . . . his bendes LAȝ. 23345* ; þe pride of sir David bigon fast to slaken MIN. ix. 49 ; slakin '*laxo*' PR. P. 458 ; slake CH. C. T. *E* 802 ; S. S. (Wr.) 1210 ; PR. C. 6224 ; LIDG. M. P. 53 ; TOWNL. 117 ; his wraþþe for to slake WILL. 728 ; ne mi þurst slake LANGL. *B* xviii. 366 ; to slake his lust M. H. 80 ; in slep(e) i **slákie** (*pres.*) SPEC. 54 ; fasteð til his fel him slakeð REL. I. 211 ; slakeþ SHOR. 30 ; swa þat i slakie to ofearnen hevenriche KATH. 2166 ; and **sláked** (*pret.*) his spere PERC. 1696 ; *comp.* a-slákien.

slān, *see* **slēan**.

slante? sb. ; *in phrase* on slante [slonte] ; *aslant*, ANT. ARTH. xlviii ; o slante D. ARTH. 2254 ; a slonte '*oblique*' PR. P. 6.

slāp, **slāpen**, *see* **slǣp**, **slǣpen**.

slappe, sb., = *Ger.* slappe ; *slap, blow*, PALL. iv. 763.

slaschen, v.,=*Dan.* slaske, *? Swed.* slaska ; *lash* ; **slascht** (*pple.*) WICL. 3 KINGS v. 18.

slat, sclat, sb., *? O.Fr.* esclat ; *slate, tile,* PR. P. 458 ; **slattes** (*pl.*) MAP 350 ; sclattes TREV. I. 399 ; WICL. IS. ix. 10.

slatten, v., *? O.N.* slatta, sletta (*dab*) ; *hang down ; fall :* heo **sleateð** [sclattes, sletteð] (*pres.*) a dun boa two hore earen A. R. 212 ; þat heo to grunde ... **sletten** (*pret.*) ALIS. 2262.

slāu, adj., *O.E.* slāw,=*M.Du.* sleuw, *O.L.G.* slēu, *O.N.* slǣr, sliōr, *O.H.G.* slēw ; *slow :* slau & let HOM. II. 183 ; (*ms.* slaw) '*hebes, tardus*' PR. P. 458 ; slaw H. M. 37 ; ORM. 9885 ; slou R. S. vi ; TREAT. 138 ; slouh A. R. 258 ; slouȝ end dul TREV. V. 255 ; **slāwe** (*pl.*) PR. C. 5546 ; LUD. COV. 251 ; slowe CH. C. T. *D* 1816 ; slouwe P. S. 325 ; **slōweste** (*superl.*) SPEC. 176.

    slāw-līche, adv., *slowly* (*printed* slapliche), REL. I. 131 (slawliche HOM. II. 11) ; slauli PR. C. 3193.

    slāunesse, sb., *slowness,* PR. P. 458.

    slō(w)-wurm, sb., *O.E.* slāwwyrm ; *slow-worm,* '*stellio*,' FRAG. 3.

slauht, *see* **slaht.** **slaunder,** *see* **sclandre.**

slāuþe, *see* **slēwðe.**

slaveren, v., *cf. M.Du.* slabberen, *Swed.* slabbra ; *slaver* ; **slavers** (*pres.*) PR. C. 784 ; **slaverit** (*pret.*) '*bave*' VOC. 143 ; *comp.* **bi-slaberen.**

slavine, *see* **sclavine.** **slāw,** *see* **slāu.**

slāwe, sb., *sloth, slowness* ; witoute slow(e) L. H. R. 214.

slāwen, v., *O.E.* slāwian ; *be slow,* HOM. I. 161.

slē, *see* **slēh.**

slēan, v., *O.E.* slēan (*pret.* slōh),=*O.N.* slā, *O. L.G., O.H.G., Goth.* slahan ; *strike, beat ; slay, kill* ; A. R. 138 ; slæn [slean] LAȝ. 3952 ; slen HORN (L.) 85 ; P. L. S. xiii. 98 ; ALIS. 1735 ; slen, sleen '*occido*' PR. P. 459 ; slen LANGL. *A* iii. 267 ; sle TOR. 240 ; slee FER. 55 ; slen [sleen, sle, slee] CH. C. T. *A* 661, 1645 ; slæn, slan ORM. 4450, 8040 ; slea, slaȝe AYENB. 8, 223 ; slaa ISUM. 302 ; slon MISC. 27 ; GEN. & EX. 1328 ; SPEC. 91 ; sclon S. & C. II. xliii ; slo HAV. 1364 ; PERC. 925 ; to slenne A. R. 130 ; **slēað** (*pres.*) A. R. 118 ; slað H. M. 29 ; sloð REL. I. 218 ; heo slaȝeð HOM. I. 53 ; **slēa** (*imper.*) A. R. 206 ; **slōh** (*pret.*) LAȝ. 767 ; ORM. 3590 ; HORN (R.) 611 ; sloȝ, slouȝ, slou ROB. 19, 20, 23, 175 ; slogh EGL. 448 ; fir ... he slogh IW. 2039 ; slouh A. R. 118 ; slouȝ BEK. 2356 ; slough [slou] CH. C. T. *A* 987 ; slou HAV. 501 (*spelt* slow 2633) ; LANGL. *B* xx. 149 ; S. S. (Wr.) 1028 ; A. P. i. 1221 ; TOR. 698 ; slog, slug GEN. & EX. 2668, 3913 ; slooȝ, sleu (*ms.* slew) WICL. LK. xv. 27 ; sleu (*for* slu) TRIAM. 1436 ; þu sloȝe LAȝ. 10999 ; slowe HAV. 2069 ; P. L. S. xxiv. 189 ; heo sloȝen (*pitched*) heore teldes wide ȝeond þa feldes LAȝ. 7865 ; sloȝ-

hen ORM. 13782 ; sloghe PR. C. 5526 ; slowen A. R. 270 ; slowen MAND. 280 ; **slōȝe** (*subj.*) HOM. I. 39 ; slowe FER. 467 ; **slaȝen** (*pple.*) ORM. 4458 ; slagen GEN. & EX. 509 ; slawen HAV. 2000 ; SPEC. 96 ; sclawen HICKES I. 228 ; slawe K. T. 1055 ; S. S. (Wr.) 1646 ; slain MAND. 280 ; CH. C. T. *A* 992 ; slan PERC. 555 ; slon TOR. 2128 ; sleie WILL. 379 ; *comp.* a-, ȝe-, of-**slēan** ; *deriv.* **slaȝe, sleȝe, slegge, slaht, slahter.**

    slēere, sb., *slayer,* TREV. V. 373.

slēc, adj., *sleek, smooth,* KATH. 1677 ; sleek PALL. ii. 152.

slecken, v., *O.E.* (ge-)sleccan,=*Swed.* släcka ; *from* **slac** ; *slack-, slake, extinguish :* to slecken fir & cwenken ORM. 10126 ; slekken wel þin þirst 14484 ; slekkin '*extinguo*' PR. P. 459 ; slecche M. T. 200 ; **slecked** (*pple.*) ORM. 5689.

slede, sb.,=*M.Du., M.L.G.* slede, slide, *O.N.* sleði, *O.H.G.* slito ; *from* **slīden** ; *sledge ;* '*traha,*' PR. P. 458 ; VOC. 232 ; **sledis** (*pl.*) WICL. 1 PARAL. xx. 3.

sleȝ, *see* **slēh.**

sleȝe, sb., *O.E.* slege, slæge, *cf. O.L.G.* slegi *m. ; from* **slēan** ; *slaughter :* ure drihten wes iled to sleȝe al swa me dede a scep HOM. I. 121 ; *comp.* ofer-sleie.

slegge, sb., *O.E.* slecg, *O.N.* sleggja ; *from* **slēan** ; *sledge (hammer),* FER. 1308 ; TREV. VI. 199 ; **slegges** (*pl.*) PARTEN. 3000.

slēh, adj., *O.N.* slǣgr (slǣgr) ; *sly, clever,* LAȝ. 14366* ; ORM. 13498 ; sleȝ ASS. 144 ; BEK. 2050 ; FER. 1446 ; L. C. C. 19 ; sleeȝ WICL. 2 PARAL. ii. 12 ; sle BARB. iv. 512 ; slegh MIRC 856 ; scleȝ B. DISC. 351 ; sleih (*ms.* sleyh) MISC. 88 ; slei (*ms.* sley) HAV. 1084 ; MAN. (F.) 9849 ; sleigh [sli] CH. C. T. *A* 3201 ; sliȝ ALIS. 9 ; sligh LIDG. M. P. 29 ; sli '*astutus, callidus*' PR. P. 459 ; **slēȝe** (*pl.*) AYENB. 265 ; sleghe PR. C. 7570 ; (*ms.* sleye) SPEC. 86 ; **slēiere** [slihere] (*compar.*) CH. C. T. *D* 1322 ; slear BARB. xvii. 244 ; **slēhest** (*superl.*) MARH. 12 ; sleast BARB. xvii. 435.

    slēh-līche, adv., *O.N.* slǣgliga ; *slily,* LANGL. *C* vii. 107 ; sleiliche WILL. 637 ; sleiȝ-, slili CH. C. T. *A* 1444 ; sleli YORK xxx. 8 ; BARB. xix. 538.

    slīnesse, sb., *slyness,* '*astucia*' PR. P. 459.

    slēih-schüpe, -schipe, sb., *? skilfulness,* C. L. 801.

    slēhþe, sb., *O.N.* slǣgð ; *sleight, contrivance,* LAȝ. 17212* (*first text* liste) ; sleȝþe AYENB. 18 ; FER. 2405 ; sleghþe MIRC 364 ; sleghte YORK xxii. 88 ; sleiȝþe WILL. 2151 ; sleiȝþe [sleiþe] TREV. IV. 93 ; sleithe '*calliditas, astutia*' PR. P. 458 ; sleihte MAN. (H.) 129 ; sleithe, sleiȝte LANGL. *B* xix. 94, *C* xxii. 98 ; sleithe [sleighte] CH. C. T. *A* 604 ; sleighte MAND. 301 ; slicht BARB. v. 105, 488.

slēi, slēih, *see* **slēh.**

sleir, sb.,=*M.H.G.* sleier; *veil*: a ladi in a sklaire LANGL. *B* vi. 7 (skleir *A* vi. 7).

sleithe, *see* slēhþe.   **slēk**, *see* slēc.

slēkin, v., *?polish*,'*lucibricennulo*,' PR. P. 459.

sleknen, v., *slacken*; sleken PR. C. 6558; **slekenid** (*pple.*) APOL. 19.

slēn, *see* slēan.

slender, adj., *cf. M.Du.* slinder; *slender*,'*gracilis*,' PR. P. 459; RICH. 3530; slendir LIDG. M. P. 50; sclender MAND. 291; CH. C. T. *A* 587.

slengen, v., *cf. O.N.* slöngva; *from* slingen; *sling*; slenge HAV.2435; þre hondred hevedes of (h)a **slende** (*pret.*) BEV. 248; Crist on crois was sleint (*pple.*) P. R. L. P.248; grete slabbes of stel(e) & ire to þe walles þo wer(e)n islente FER. 3313.

slente, sb.,?=**slante**; *slope*: bi slente oþer slade A. P. i. 141.

slenten, v., *cf. ? Swed.* slinta; *glide, fall*: into þe erþe hit **sclente** (*pret.*) HALLIW. 711; he sleint o wai AM. & AMIL. 2279; *comp.* to-slenten.

slenting, sb., *shooting*: slenting of arwes GAW. 1160.

slǣp, slǣpen, *see* slǣp, slǣpen.

slēt, sb., *cf. M.H.G.* slōz (*hail*); slete; *sleet*, HICKES I. 231 (P. L. S. xxxv. 39); sleet '*nicula*' PR. P. 459; CH. C. T. *F* 1250; L. G. W. 1220; **slēte** (*dat.*) IW. 375; GAW. 729; TOWNL. 99.

slēten, *see* slǣten.   **sletten**, *see* slatten.

sleðrende, adj., *falling like sleet*, HOM. II.99.

sleuth, *see* slōþ.   **slēuþe**, *see* slēwðe.

slēve, sb., *O.E.* slēfe, slȳfe, slȳf; *sleeve*, '*manica*,' PR. P. 459; GREG. 790; OCTAV. (H.) 1027; sleve [sleeve] CH. C. T. *G* 1224; **slēfes** (*pl.*) M. H. 111; **slēven** (*dat. pl.*) A. R. 56; P. S. 156.

[**slēwe**, sb., = *M.H.G.* slewe; *from* slāu; *sloth, languor*.]

slēu-vol, adj., *slothful*, AYENB. 32.

slēuvel-liche, adv., *slothfully*, AYENB. 32.

slēwðe, sb., *O.E.* slǣwð; *sloth*, '*desidia*,' HOM. I. 103; sleuþe ROB. 195; AYENB. 31; SHOR. 34; SPEC. 49; TREV. VI. 235; sleuȝthe LANGL. *A* PROL. 45; sleuthe *B* PROL. 45, *C* i. 45; slauðe HOM. I. 107; slauþe LAȝ. 27039; PL. CR. 91; A. P. ii. 178; LUD. COV. 404; slouþe CH. C. T. *B* 530; slouhðe A. R. 144.

slēuhþen, v., *be slow*, ANGL. I. 9; *comp.* for-slēuþen.

slī, *see* slēh.

slīce, sb., *O.Fr.* esclice; *slice, splinter; spatula*; '*spat(h)a*,' PR. P. 459; þei braken speris to **sclīces** (*pl.*) ALIS. 3833.

slich, sb.,=*M.L.G.* slik, *M.H.G.* slich; *slime, wet mud*, MIR. PL. 4; slik BARB. xiii. 352.

slīden, v., *O.E.* slīdan, = *M.H.G.* slīten; *slide*, *glide*, A. R. 252; uppe þi breste þou schalt sliden (*ms.* slyden) S. A. L. 222; slide O. & N. 1390; CH. TRO. v. 351; LIDG. M. P. 65; **slīdeð**, slit (*pres.*) A. R. 74; slit AYENB. 149; SPEC. 110; **slōd** (*pret.*) WILL. 792; L. H. R. 136; LAUNF. 214; A. P. i. 59; slod [slood] TREV. VII. 237; slood WICL. LAM. iii. 53; slaid BARB. iii. 701, x. 700; slide (*subj.*) LEG. 169; **sliden** (*pple.*) WICL. PROV. xxiv. 10; islide O. & N. 686; hit is forþ islide P. L. S. xxix. 88; *comp.* a-slīden; *deriv.* slede, slider.

slider, adj., *O.E.* slidor; *from* slīden; *slippery*, LEG. 168; TREV. I. 63; GOW. III. 14; CH. C. T. *A* 1264; AM. & AMIL. 1843; þe wei is slider O. & N. 956; slidir '*lubricus*' PR. P. 459; sledir H. S. 5262; sclider '*labilis*' CATH. 322; sklither HAMP. PS. xxxiv. 7, xxxviii. 1*.

slidirnesse, sb., *slipperiness*, WICL. PS. xxxiv. 6; PR. P. 459.

slideri, adj., *cf. Swed.* sliddrig; *slippery*, VOC. 160; WICL. PS. xxxiv. 6; slidri LEG. 221; sliddri A. R. 74.

sliderin, v., *O.E.* sliderian (ALFR. P. C. 276), = *M.Du.* slideren; *slip*, '*labor*,' PR. P. 459; slidre GENER. 4152.

sliþering, sb., *slipperiness*; sklithiringe [*v.r.* sclitering] HAMP. PS. cxiv. 8*.

sliȝ, *see* slēh.

sliȝt, adj.,=*M.Du.* slicht, slecht, *O.H.G.* sleht, *O.N.* slēttr, *Goth.* slaihts; *slight, smooth*, A. P. i. 190; sleght TOWNL. 145.

slih, *see* slēh.   **slīk**, *see* swülc.

slike, adj., *?from* slīken; *smooth, sleek*, PR. P. 459; HAV. 1157; R. R. 542; PALL. i. 689.

slīken, v.,=*M.L.G.* slīken, *O.H.G.* slīchan; *glide*; **slikes** (*pres.*) ANT. ARTH. xlviii; *comp.* a-slīken.

slikien, v., *from* slike; *polish, make smooth*: he can so wel his wordes slike GOW. II. 365; **slikeð** (*pres.*) R. S. i; til sleuþe and sleep sliken his sides LANGL. *B* ii. 98 (Wr. 1080); isliked (*pple.*) O. & N. 841; AYENB. 99; isliket KATH. 1675.

slim, sb., *O.E.* slīm,=*O.N.*, *M.L.G.*, *M.H.G.* slīm; *slime*, A. R. 276; SHOR. 105; **slīme** (*dat.*) GOW. III. 96; PS. lxviii. 3; WICL. GEN. ii. 7.

slīmi, adj., *slimy*, LANGL. *B* v. 392.

slinge, sb.,=*O.Fris.*, *M.Du.* slinge, *O.H.G.* slinga; *sling*, '*funda*,' PR. P. 459; SHOR. 132; LANGL. *B* xx. 162; sclinge REL. I. 6; **slinges** (*pl.*) ALIS. 1191; *comp.* staf-slinge.

slingen, v., *O.E.* slingan, = *O.H.G.* slingan, *M.Du.* slinghen, *O.N.* slyngva; *sling*; slingin PR. P. 459; **slang** (*pret.*) SHOR. 132; slong JUL. 63; S. S. (Web.) 1316; and out at þe dore him slong [sclong] CH. C. T. *H* 306; slungen RICH. 5227; slonge ROB. 362; **slongen**

(*pple.*) PERC. 672; sloungen MAP 338; *deriv.* slinge, slengen.

slingare, sb., *slinger*, PR. P. 459; slinger OCTOV. (W.) 1599; **slengers** (*pl.*) ALEX. (Sk.) 2219.

slinken, v., *slink*; sclink BER. 3334.

slip, sb.,=*M.L.G.* slip, *mod. Eng.* slip; *from* slīpen; *wet clay*, '*limus*,' PR. P. 459.

slip², sb., ?=*M.Du.*, *M.L.G.* slippe (*lacinia, sinus vestis*)*; mod.Eng.* slip; *lappet of a garment*, '*lacinia*,' PR. P. 459.

slip³, sb., *blow*, A. P. ii. 1264.

slīpen, v., *O.E.* (to-)slīpan,=*M.L.G.* slīpen (*glide, sharpen*), *M.Du.* slijpen (*polish, sharpen*), *O.H.G.* slīfan (*glide*); *glide*; slipe F. C. 15; **slīpeþ** (*pres.*) GOW. II. 347; *deriv.* slip, sliper, slippen.

sliper, adj., *O.E.* slipur, = *M.L.G.* slipper, *M.H.G.* slipfer; *from* slīpen; *slippery*; (*v. r.* cliper) P. P. 215; PS. xxxiv. 6; WICL. PROV. xxvi. 28; D. TROY 11295.

slippe, sb.,=*M.H.G.* slipf; *slip, descent*, E. G. 374.

slippen, v.,=*M.Du.*, *M.L.G.* slippen, *M.H.G.* slipfen; *from* slīpen; *slip, glide, escape*; slippe (?=sluppe) HUNT. 114; er þai slippe (*escape*)·miȝt A. P. ii. 1785; **slippus** (*pres.*) ANT. ARTH. xlviii; **slipte** (*pret.*) GOW. II. 72; **slipped** (*pple.*) GAW. 244.

slite, sb., *O.E.* slite,=*O.N.* slit, *O.H.G.* sliz; *from* slīten; *slit, crack, cleft; opening in a garment*; ['*scissura*'[MK. ii. 21; stones hi doþ in heore slitte O. & N. 1118; **slitte** (*dat.*) GOW. I. 15; þu most habbe … twenti marc ine þi slitte FL. & BL. 348; and tok þe messanger bi þe slit(e) AR. & MER. 1406.

slīten, v., *O.E.* slītan,=*O.L.G.* slītan, *O.Fris.*, *O.N.* slīta, *O.H.G.* slīzan; *slit, split, break*; slitin PR. P. 459; **slītende** (*pple.*) MK. ix. 26; **slāt** (*pret.*) MAT. xxvi. 65; **slitin** (*pple.*) PR. P. 459; *comp.* to-slīten; *deriv.* slite, slitten, ? slæten.

slitten, v.,=*M.H.G.* slitzen; *from* slīten; *slit, pierce, split*; slitte PS. lxxxviii. 24*; CH. C. T. *F* 1260; þe longage of þe Norþhumbres … is scharp slitting (*pple.*) & froting & unschape TREV.II. 163; slitte(*pret.*) OCTOV.(W.)1664; TREV. V. 39; islit (*pple.*) LAȜ. 14221; *comp.* to-slitten.

. **slittinga**, sb., *slitting*, HOM. I. 33.

sliþer, *see* slider.

slīven, v., *O.E.* (to-)slīfan; *cut, cleave*; slivin '*findo*' PR. P. 459.

sliver, sb., *from* slīven; *part, portion*: al hool or of hem sliver CH. TRO. iii. 1013.

slō, *see* slā, slōh.

slober, sb., '*faeces immundae*,' PR. P. 459; D. TROY 2529.

slobren, v.,=*M.Du.* slobberen, ? *Swed.* slub-

bra (RIETZ 628), *mod.Eng.* slobber; ? *wallow*: þei slober in þe mere S. & C. II. lvi.

? **sloder**, sb., = ? *M.H.G.* sloter(-gruobe); *puddle*: hie secheð to þe fule sloddre (? *ms.* floddri) & þar on waleweð HOM. II. 37.

slogardie, *see* slugardie.

slōh, sb., *O.E.* slōh, slōg; *slough, bog*, KATH. 1677; slogh TRIAM. 366; slough CH. C. T. *B* 3988; slo O. & N. 1394; **slōghe** (*dat.*) ISUM. 403; slowe ALIS. 6075; WICL. 2 PET. ii. 22; **slōwes** (*pl.*) DEGR. 1656.

sloknin, v., *O.N.* slokna; *quench, extinguish, stop*, '*extingvo*,' PR. P. 459; slokin TOWNL. 117; HAMP. PS. i. I; **slokenes** (*pres.*) M. H. 37; **slekind**, slokind (*pple.*) HAMP. PS. xii. 4, xvii. 10.

slomen, slomer, *see* slumen, slumer.

slōn, *see* slēan.

slop, sloppe, sb., *O.E.* slop, *O.N.* sloppr *m.*; *from* slūpen; *slop, robe*, PR. P. 460; slope H. S. 522; **sloppes** (*pl.*) CH. C. T. *I* 422; *comp.* over-slope.

slop², sb., *gap*, BARB. viii. 274; **sloppis** (*pl.*) BARB. viii. 179.

slop³, sb., *pool*; **sloppes** (*pl.*) D. ARTH. 3923.

slópe, sb., *slope*; a slope *aslope* R. R. 4464; H. V. 54.

slor, sb., ? *mod.Eng.* slur; *mud*, '*coenum, limus*,' PR. P. 460.

slot, sb., *O.Fris.* slot, *O.H.G.* sloz; *mod. Eng.* slot; *bar, bolt*, '*vectis*,' VOC. 237; '*pessulus*' PR. P. 460; **slottes** (*pl.*) ['*vectes*'] PS. cvi. 16; slot *pit of the stomach* GAW. 1330, 1593; slote D. ARTH. 2254.

sloterin, v., ? =*M.Du.* slodderen; *bespatter*, '*maculo*,' PR. P. 460; **slotered** (*pple.*) GENER. 7066; PR. C. 2367.

slōþ, sb., *O.N.* slōð; *mod.Eng.* sleuth(-hound); *track*, ORM. 1194; C. M. 1254; sleuth BARB. vii. 21, 44.

    sleuth-húnd, sb., *sleuth-hound*, BARB. vi. 484, 669; sluthe hunde CATH. 345.

slōu, *see* slāu.

slouh, sb., ? =*M.H.G.* slūch; *slough, skin*, PR. C. 520*; slughe ALEX. (Sk.) 4456; **slūghe** [slouȝe] (*dat.*) C. M. 745; **slōȝis** (*pl.*) ALEX. (Sk.) 5085.

slouþe, *see* slēwðe.

slovein, sb., ? *from M.Du.* slof; *sloven*, LUD. COV. 218.

sluchched, adj., *cf.* slich; *muddy*, A. P. iii. 341.

slude, sb., *sludge*; sloude D. ARTH. 3179.

sluggardi, sb., *sloth*; sloggardie, slogardie [slogardrie] CH. C. T. *A* 1042, *G* 17.

slugge, sb., *sluggard*, '*deses, segnis*,' PR. P. 460.

sluggen, v., *be sluggish*; sluggin '*desidio, torpeo*,' PR. P. 460; *comp.* for-sluggen.

sluggi, adj., *lazy*, '*desidiosus, torpidus*,' PR. P. 460; A. R. 258; REL. I. 13; LIDG. M. P. 52;

sluggi [sloggi] CH. C. T. *1706*; sloggi P. R. L. P. 26.

slume, sb., *O.E.* sluma; *slumber, sleep*; sloumbe A. P. iii. 186.

slumen, v., *cf. M.H.G.* slummen, *M.L.G.* slomen; *slumber, sleep*; slumen, slume, slumme LA3. 17995, 18408, 32058; **slomande** (*pple.*) ALEX. 176.

slumer, sb., = *M.H.G.* slummer; *slumber*; **slomere** [slumber(e), slomber(e)] (*dat.*) CH.C. T. *A* 3816; slomoure D. ARTH. 3221.

slumeren, v., *O.E.* slumerian, *cf. M.H.G.* slumern, slummern; *slumber, sleep*; slomire D. ARTH. 4044; **slumeren** (*pres.*) REL. I. 221; **slumberde** (*pret.*) LANGL. *A* PROL. 10; slombered *B* PROL. 10.

slumi, adj., *sleepy, drowsy*: slummi & sluggi A. R. 258.

[**slūpen,** v., *O.E.* slūpan, *cf. M.L.G.* slūpen, *Goth.* sliupan, *O.H.G.* sliufan; *slip, glide; comp.* **æt-slūpen**; *deriv.* **slüppen, slop.**]

[**slüppen,** v., = *M.H.G.* slüpfen; *from* slūpen; *slip; comp.* **a-slüppen.**]

slutte, sb., *? cf. M.Du.* slodde; *slut, sloven*, PR. P. 460; LUD. COV. 218; crabbe is a slutte to kerve B. B. 158.

slutti, adj., *dirty*, '*coenulentus*,' PR. P. 460; a slutti coppe HALLIW. 761.

sluttish, slottisch, adj., *slovenly*, CH. C. T. *G* 636.

smac, sb., *O.E.* smæcc (smæc), = *M.L.G.* smak, *O.Fris.* smek, *O.H.G.* smach, smac; *smack, taste, flavour*; (*ms.* smacc) ORM. 1653; smac, smak AYENB. 33, 177; smak '*gustus*' PR. P. 460; O. & N. 823; smach M. T. 199; smech P. L. S. viii. 140; A. R. 94; **smakke** (*dat.*) P. L. S. xxxv. 77; **smackes** (*pl.*) AYENB. 112; **smecche** (*gen. pl.*) MARH. 9.

smech-lēas, adj., *tasteless*, A. R. 138.

smæl, sb., *O.E.* smæll, smell, = *Swed.* smäll; *blow, shock of battle*; **smællen** (*dat. pl.*) LA3. 27052.

smæte, adj., *? O.E.* smæte; *from* smīten; *? polished, pure*; smeate KATH. (E.) 1655; **smeatest** (*superl.*) MARH. 11.

smakin, v., *? O.E.* smacian, = *O.Fris.* smakia, *M.Du.; M.L.G.* smaken; *smack, taste*, PR. P. 460; (J)osep dede his lich ... swete smaken GEN. & EX. 2443; a word þat of god smakeþ (*pres.*) SHOR. 48; al þet hi zieþ and smackeþ of þe guodes ... of þise worlde AYENB. 92; smake (*subj.*) REL. I. 208; *comp.* 3e-smakien.

smal, adj., *O.E.* smæl, = *O.Fris.* smel, *O.L.G.* smal, *O.H.G.* smal, *Goth.* smals; *small, slender, thin*, HOM. I. 85; O. & N. 73; TREAT. 138; REL. I. 120; CH. C. T. *A* 153; smal folc LA3. 15367; ne smal ne grat AYENB. 137; mi middel smal & long A. D. 246; smel A. R. 278; **smále** (*pl.*) GEN. & EX. 656, 2107; MAND. 49; **smálere** [smelre] (*compar.*) A. R. 314*.

smal-līche, adv., *small*, AYENB. 111.

smalnesse, sb., *smallness*, TREV. II. 181.

smálin, v., *cf. M.L.G., M.Du.* smalen; *become less*, '*minoro*,' PR. P. 460.

smaragde, sb., *O.Fr.* smaragde; *emerald*, MISC. 98; *see* **emeraude.**

smaragdone, sb., *emerald*, ALEX. (Sk.) 3356; **smaragdans,** smaragdens (*pl.*) 3678, 5424.

smart, *see* **smerte.**

smateren, v., *? cf. L.G.* smaddern; *defile*; **smaterid** (*pple.*) wiþ smoke REL. I. 240.

smateren[2], v., *cf. Swed.* smattra, *M.H.G.* smeteren; *chatter, prate*; smatter S. & C. II. lxxii.

smēa3, adj., *O.E.* smēa; *from* smū3en; *subtle*; smegh HOM. II. 195.

smēih-līche, adv., *subtlety*, HOM. II. 71.

smēhnesse, sb., *cunningness*, HOM. II. 205.

smēa3an, v., *O.E.* smēagan; *ask, think, consider*, HOM. I. 219; smeagen MK. viii. 11; **smēade** (*pret.*) HOM. I. 219.

smeate, *see* **smæte.**

smēc, sb., *O.E.* smēc, smȳc, smīc, = *M.Du.* smook; *fume, vapour*, ORM. 1088; smek '*fumus*' PR. P. 460; M. H. 104; S. & C. II. xlvii; smech P. L. S. viii. 140; MISC. 28; AYENB. 66; **smēke** (*dat.*) P. L. S. viii. 9 (smike HOM. I. 161; smeche ANGL. I. 8; ZUP. ÜBUNGSB. 52); HOM. II. 220; SHOR. 96; smiche HOM. II. 258; MISC. 75; **smēkis** (*pl.*) AV. ARTH. xv.

smecchen, v., *O.E.* smeccan, = *O.Fris.* smekka, *O.H.G.* smecchan; *from* smac; *smack, taste*: swa swoteliche he **smeccheð** (*pres.*) me KATH. 1537; **smeihte** [smachte] (*pret.*) A. R. 106; simauhte, smau3te LANGL. *B* v. 363, *C* vii. 414; **smaught** (*pple.*) BER. 3122; *comp.* 3e-smecchen.

smecchunge, sb., *tasting*, A. R. 64; H. M. 13; smechunge HOM. I. 245.

smech, *see* **smac.**

smēch, smēk, *see* **smēc.**

smēkin, v., *O.E.* smēcan, = *M.Du.* smōken; *give off vapour*, '*fumo*,' PR. P. 460; **smē-kende** (*pple.*) MAT. xii. 20; **smēkide** (*pret.*) WICL. ECCLUS. xxiv. 21; **smēkid** (*pple.*) REL. I. 240.

smel, *see* **smal.**

smel, sb., *smell,* '*odor*,' PR. P. 460; HOM. I. 53; A. R. 104; O. & N. 822; REL. I. 208; ROB. 43; CH. C. T. *A* 2427; þane smel AYENB. 178; smel, smul L. H. R. 26, 27; smeal KATH. 1600*; smeol ALIS. 2573; smul HOM. II. 99; smil BRD. 4; TREV. IV. 137; **smelles** (*pl.*) A. R. 104; **smelle** (*gen. pl.*) KATH. 614.

smellen, v., *smell*, A. R. 88; CH. C. T. *A* 3691; smelle HOM. I. 53; smellin PR. P. 460; smeallen MARH. 4; smullen HOM. II. 35; smille P. L. S. xii. 148; **smelleð** (*pres.*) KATH. 1537; REL. I. 225; smelleþ LANGL. *B* xi. 426; smul-

leþ SPEC. 88; smulliþ ALIS. 6793; smilleþ
FER. 2546; **smellinde** (*pple.*) A. R. 340;
smellinge MAND. 11; **smelde** (*pret.*) CH. H.
F. 1685; C. M. 16562; smelde, smilde, smulde
L. H. R. 26, 27; S. A. L. 150; smilde P. L. S.
xii. 146; **ismelled** (*pple.*) HOM. I. 189.

smellunge, sb., *smelling*, A. R. 48; smellinge C. L. 1176; smeallunge HOM. I. 245; H.
M. 13.

smelt, sb., *O.E.* smelt, smylt; *smelt (fish)*;
smelte PR. P. 460.

smeorte, smeorten, *see* **smerte, smerten.**

smeoþen, *see* **smiðien.**

[**smer,** sb., ? = *M.H.G.* smier (*laughter*);
mockery; *comp.* **bī-smer.**]

smére, adv., ? *scornfully*: (heo) louȝ smere a
non LEG. 34; gan to lawe smere FER. 386;
he smere loh LAȝ. 14981*; & wel smere louȝ
P. L. S. xxi. 152; smere he lou REL. II. 272;
smare TRIST. 2870.

smére[2], sb., *O.E.* smeru, smeoru, = *O.H.G.*
smero, *O.N.* smiör *n.*; *anointing, ointment,*
ORM. 13244; GEN. & EX. 1573; E. G. 356.

smér-gavel, sb., *tax on unguents,* E. G. 359.

smérien, v., *O.E.* smerian, smirian, smyrian,
= *M.Du.* smeren, *M.H.G.* smeren, smeiren,
smirwen; *smear, anoint*; smeren GEN. & EX.
2442; smerie [smirie] L. H. R. 18; smurien A.
R. 372; smirie P. L. S. xix. 183; **sméreþ**
(*pres.*) SHOR. 14; hi smerieþ AYENB. 60;
smurieð HOM. I. 53; **smíre,** smure (*imper.*)
LEECHD. III. 86, 114; **smérede** (*pret.*) AYENB.
187; S. S. (Web.) 1151; smered PS. xliv. 8;
smereden KATH. 1613; **sméred** (*pple.*) ORM.
994; *comp.* **bi-, ȝe-smérien** (**-smerwen**).

méringe, sb., *smearing, anointing*; **smér-
inges** (*pl.*) AYENB. 148.

smerles, sb., *O.E.* smerels, smyrels, = *O.N.*
smyrsl, *M.Du.* smersel; *ointment,* GEN. &
EX. 2454; smerlis M. H. 97; þe smerieles ne
is naȝt worþ AYENB. 217; smirles KATH.
1612; H. M. 13; smuriles A. R. 372; smerl
C. M. 11504.

smerte, sb., = *M.Du.*, *M.L.G.* smerte, *O.H.G.*
smerza; *smart, grief,* P. L. S. viii. 57; A. R.
294; C. L. 1153; REL. I. 111; AL. (L.) 281;
CH. C. T. *A* 3813; smert PR. P. 460; S. S.
(Web.) 777; HOCCL. i. 40; smierte HOM. II.
223.

smerte[2], adj., *smart, sharp, rough,* A. R. 294;
HAV. 2055; smerte & kene MARG. 121; oure
sorewe þat is so smerte MAN. (F.) 6922;
smearte HOM. I. 243; one smerte ȝerde A. D.
298; wiþ a ȝerde smerte CH. C. T. *A* 149;
esi fir and smart CH. C. T. *G* 768; **smerte**
(*adv.*) B. DISC. 601; smit . . . so smerte
LANGL. *B* xi. 426; to gaa smert *promptly*
ALEX. (Sk.) 5575; **smerte** (*pl.*) gier HOM. II.
61; þornes smerte H. V. 93; **smærte** [smorte]
(*dat. pl.*) LAȝ. 20318.

smert-liche, adv., *smartly, quickly*: smit

smertliche KATH. 2016; biddiþ him smeortli
ALIS. 1911.

smerten, v., = *M.L.G.*, *M.Du.* smerten, *O.H.G.*
smerzan (*pret.* smarz); *smart, be sorrowful,*
HOM. I. 83; HOM. II. 207; smertin PR. P.
460; smeorten A. R. 238; smerte SPEC. 70;
S. S. (Web.) 2642; MIRC 1260; þe dint bigan
ful sore to smerte HAV. 2647; smurte TREAT.
140; **smerteð** (*pres.*) H. M. 31; smertes PR.
C. 1317; **smerte** (*subj.*) A. D. 295; CH. C. T.
A 230; smeorte REL. I. 177; **smeart** (*pret.*)
HOM. II. 21; þo him smart [smert] so sore
LEG. 186, 187; smurte P. L. S. xvii. 116;
smerten ALIS. 5848.

smertinge, sb., *smarting,* HOM. I. 83; HOM.
II. 165; smeortung A. R. 294.

smeþ, *see* **smið.**

smēðe, smōðe, adj. & adv., *O.E.* smēðe,
cf. M.L.G. smōde; *smooth, flat*; smeðe
A. R. 2; smeþe ROB. 424; PS. liv. 22*;
WICL. GEN. xxvii. 11; smeþe & softe
waȝe ORM. 9666; smethe A. D. 240; PR.
C. 6349; L. C. C. 47; smoþe A. P. i. 6;
smoþe [smothe] CH. C. T. *A* 690; smothe
PR. P. 461; LIDG. M. P. 245; **smēðere**
(*compar.*) HOM. II. 189.

smēðe-liche, adv., *smoothly,* KATH. 356;
smeþeliche TREV. VII. 259.

smeþe[2], sb., *level surface, 'planities,'* PR. P.
460.

smeþen, *see* **smiðien.**

smēþien, smōðien, v., *O.E.* smēðian, *cf. L.G.*
smœden; *render smooth*; smeþien HOM. I. 31;
smeðen A. R. 4; **ismēðet** (*pple.*) KATH. 1675;
ismoþed AYENB. 57.

smēthnes, sb., *O.E.* smēðnyss; *flatness,
level surface, 'planities,'* PR. P. 460.

smīc, smīch, *see* **smēc.**

smiker, adj., *O.E.* smicer, = *O.H.G.* smechar;
elegant, ORM. 13679.

smil, *see* **smel.**

smīl, sb., *cf. Dan.* smiil, *M.H.G.* smiel; *smile,*
BRD. 4.

smīlin, v., = *Dan.* smile, *Swed.* smila, *M.H.G.*
smielen; *smile, 'subrideo,'* PR. P. 461;
smile FER. 452; CH. C. T. *D* 1446; MAN. (H.)
185; DEGR. 804; **smiland** (*pple.*) WILL.
991.

smilere, sb., *smiler,* CH. C. T. *A* 1999.

smirien, *see* **smerien.**

smirken, v., *O.E.* smercian; *smirk, smile
constrainedly*; **smirkende** (*pple.*) KATH.
356.

smirles, *see* **smerles.**

smite, sb., = *M.L.G.* smite, smete, *M.Du.* smete,
O.H.G. smiz; *from* **smīten**; *blow, stroke,
throw,* GEN. & EX. 2990; S. S. (Wr.) 1959;
þane smite AYENB. 140; smete HALLIW. 762;
smiten (*dat. pl.*) HOM. II. 207; smite LAȝ.
535.

smīten, v., O.E. smītan,=O.Fris. smīta, M.L.G. smītan, Goth. (bi-, ga-)smeitan, O.H.G. smīzan ; smite, throw, LAȝ. 9204 ; A. R. 324 ; ORM. 14677 ; O. & N. 78 ; for to smite þis holi man his swerd he drouȝ BEK. 2122 ; children . . . he wolde smite and bete P. L. S. xxiii. 49 ; smite fur BRD. 30 ; smït (pres.) A. R. 60 ; TREAT. 136 ; CH. C. T. E 122 ; ut of his ᷄rote it smit an onde REL. I. 220 ; (heo) smiteᷥ LAȝ. 20172 ; smīt (imp.) GEN. & EX. 3360 ; smāt (pret.) MAT. xxvi. 68 (earlier text slōh) ; MARH. 9 ; IW. 377 ; Brutus heom smat on LAȝ. 534 ; uppen þene helm he hine smat [smot] 7512 ; smot GEN. & EX. 2925 ; ROB. 185 ; HAV. 1676 ; AYENB. 48 ; B. DISC. 1185 ; þu me smite [smete] bi þon rugge LAȝ. 8157 ; smiten REL. I. 101 ; GOW. II. 72 ; heo smiten [smete] to gædere LAȝ. 5183 ; smite ROB. 217 ; smeten RICH. 3988 ; smite (subj.) C. L. 1255 ; AN. LIT. 10 ; smiten (pple.) MAN. (H.) 178 ; wiᷥ lepre smiten GEN. & EX. 3690 ; smeten FLOR. 1620 ; comp. bi-, ȝe-smīten ; deriv. smite, smitten.

smītare, sb., smiter, A. R. 156.

smitten, v.,=M.L.G. smitten, M.Du. smetten, M.H.G. smitzen ; from smīten ; contaminate, pollute ; smitted (pple.) PS. cv. 39 ; TREV. I. 359* ; smittid WICL. E. W. 436 ; ismittet H. M. 13 ; comp. bi-smitten.

smiᷥ, sb., O.E. smiᷥ,=O.N. smiᷥr, O.Fris. smeth, O.H.G. smid ; smith, workman, A. R. 52 ; GEN. & EX. 466 ; smiᷥ [smiþ] LAȝ. 21131 ; smith CH. C. T. A 2025 ; smiþes (pl.) O. & N. 1206 ; SPEC. 86 ; smeþes REL. I. 240 ; comp. gōld-, mæstling-smiᷥ.

smiᷥien, v., O.E. smiᷥian, = O.N. smiᷥa, Goth. (ga-)smiþōn, O.H.G. smidōn ; forge ; smiᷥie, smiᷥeȝe LAȝ. 30743, 30749 ; smiþie LANGL. B iii. 305 ; smeoᷥien A. R. 284 ; smethe 'fabricare' CATH. 346 ; ALEX. (Sk.) 5515 ; smeoᷥiᷥ [smiᷥes] (pres.) A. R. 52 ; smeoᷥede [smiþede] (pret.) LAȝ. 1563 ; smithed CH. C. T. A 3762.

smiᷥᷥe, sb., O.E. smiᷥᷥe,=O.Fris. smitha, O.N. smiᷥja, O.H.G. smidda ; smithy, forge, A. R. 284 ; smiþþe P. L. S. ix. 60 ; smithi PR. P. 461 ; smeþi [smeþei] C. M. 23238.

smoc, sb., O.E. smocc, smoc, O.N. smokkr, cf. O.H.G. smoccho ; smock, chemise, REL. I. 129 (HOM. II. 163) ; smok 'interula' PR. P. 461 ; A. D. 106 ; GOW. I. 115 ; PL. CR. 79 ; CH. C. T. A 3238 ; schetis ether smockis (pl.) WICL. IS. iii. 22*.

smoc-lēs, adj., without a smock, CH. C. T. E 875.

smod, sb., cf. Sc. smot, mod.Eng. smut ; filth, A. P. ii. 711.

smok, see smoc.

smóke, sb., O.E. smoca ; smoke, 'fumus,' PR. P. 461 ; MARH. 9 ; BRD. 23 ; ALIS. 4352 ; LANGL. B xvii. 321 ; MAND. 247 ; CH. C. T. D 278.

smóki, adj., smoky, PR. P. 461.

smókien, v., O.E. smocian ; smoke, LAȝ. 25734 ; and smókeþ (pres.) as þei a fere were TREV. I. 119 ; smókede (pret.) BRD. 23 ; & smoked heom mid ful smoke SAX. CHR. 262.

smólder, sb., smoulder, PALL. i. 929 ; smoke and smolder MAP 345 ; LANGL. B xvii. 321.

smóldren, v., smoulder, suffocate : in smolderande (pple.) smoke A. P. ii. 955.

smólt, adj., O.E. smolt,=M.Du. smolt ; clear, serene, quiet, ['serenus'] MAT. xvi. 2 ; GAW. 1763 ; A. P. ii. 732.

smóre, sb., dense smoke, LANGL. C xx. 303*.

smóren, v., O.E. smorian,=M.L.G., M.Du. smoren ; suffocate ; smore C. M. 5573 ; smore 'fumigo' PR. P. 461 ; smóred (pple.) PR. C. 7601.

smorten, v., cf. L.G. smorten, smurten ; ? smart ; smourte (pret.) ROB. 322.

smorᷥer, sb., smother, thick smoke, HOM. I. 43 ; smorþer (ms. smorþre) LANGL. C xx. 303 ; smorther ROB. 407* ; D. TROY 911 ; smoþer MAP 339 ; to helle smurᷥre A. R. 272.

smorᷥren, v., smother, suffocate ; smorᷥrinde (pple.) HOM. I. 251 ; smeorᷥrinde [smoᷥrinde] smoke MARH. 9.

[smoter ?]

smoter-lích, adj., dirty, smutty : she was somdel smoterlich CH. C. T. A 3963.

[smoteren, v., cf. Du. smoderen ; comp. bismoteren.]

smotten, v., ? cf. M.H.G. smutzen ; ? smut ; ismotted (v. r. smitted) (pple.) TREV. I. 359.

smoþe, see smeᷥe. smoþer, see smorᷥer.

[smudden, v., cf. L.G. smudden ; besmut, soil ; comp. bi-smudden.]

[smudelen, v., cf. L.G. smuddren ; pollute ; comp. bi-smudelen.]

smūȝen, v., O.E. smūgan, cf. O.N. smiūga, M.H.G. smiegen ; creep, glide ; smūȝeᷥ (pres.) HOM. I. 153 ; smuȝᷥ II. 191 ; deriv. smēaȝ.

smül, smüllen, see smel, smellen.

smülting, sb., O.E. smylting ; mixed metal, 'electrum,' FRAG. 4.

smürien, see smerien.

snacchen, v., cf. M.Du. snacken ; snatch, seize ; snecchen A. R. 324 ; snacches (pres.) MAN. (F.) 13889 ; snache (subj.) ALIS. 6559.

snāde, sb., cf. O.E. snǣd,=O.N. sneiᷥ ; from snīᷥen ; piece, bit, C. M. 15387 ; snode 'offa' FRAG. 4 ; AYENB. 77 ; mid þer ilke snode MISC. 40 ; snede HOM. II. 181 ; snōdes (pl.) ['frusta'] PS. cxlvii. 17.

[snǣden, v., O.E. snǣdan,=O.N. sneiᷥa, M.H.G. sneiten ; cut ; comp. to-snǣden.]

snǣsen, v., O.E. snǣsan, = O.N. sneisa ; strike ; snesen A. R. 212* ; comp. a-snǣsen.

snaile, sb., *O.E.* snegl, snægl *f.*, *cf. M.L.G.* snegel, sneil, *O.N.* snigill *m.; snail,* '*limax,*' PR. P. 461; LUD. COV. 209; snawile P. L. S. xxxv. 40; snele VOC. 190; snile 223; HALLIW. 766; **snailes** (*pl.*) O. & N. 87; MAND. 169.

[snaipe, sb., *O.N.* sneypa; *disgrace.*]

snaipe-lī, adv., *? O.N.* sneypiliga; *? disgracefully,* ANT. ARTH. vii.

snaipen, v., *O.N.* sneypa (*outrage*); *nip; check*; snaipe C. M. 13027; **snaiped** (*pret.*) GAW. 2003; ALEX. (Sk.) 3633, 3995.

snáke, sb., *O.E.* snaca, = *M.L.G.*, *M.Du.* snake, *O.N.* snakr; *snake,* '*anguis,*' PR. P. 461; GEN. & EX. 2805; Gow. III. 118; **snáken** (*pl.*) LK. x. 19; P. L. S. viii. 138; MISC. 148; ALIS. 5972; snakes SAX. CHR. 262; PS. xiii. 3.

snakeren, v., *? sneak*: .hwon he **snakereð** (*pres.*) toward ou A. R. 380; **snakerinde** (*pple.*) 290.

snápe, sb., *cf. O.N.* snöp; *? winter pasture,* ALEX. (Sk.) 1560.

snapren, v., *trip, stumble*; snapre ERC. 381; þi foot schal not snapere ['*pes tuus non impinget*'] WICL. PROV. iii. 23; **snapirs** [snappers] (*pres.*) ALEX. (Sk.) 847; **snaperid** (*pret.*) TREV. VII. 187.

snáre, sb., *O.E.* snear, = *M.Du.*, *M.L.G.* snare, *O.N.* snara; *snare, noose, halter,* '*laqveus, pedica,*' PR. P. 461; '*pedica*' REL. I. 7; P. S. 192; CH. C. T. *A* 1490; PS. ix. 16; LIDG. M. P. 168.

snárin, v., *cf. O.Swed.* snærja; *set a snare, ensnare,* '*illaqueo,*' PR. P. 461.

snarl, sb., = *Swed.* snarel; *noose, halter*: (he) heng him self wiþ a snarl (*v. r.* snare) TREV. V. 209; **snarles** (*pl.*) TREV. II. 385.

snarlin, v., *entangle,* '*illaqueo,*' PR. P. 461; **snarlede** (*pret.*) TREV. VII. 431.

snart, adj., *severe,* ALEX. (Sk.) 3633; (*adv.*) GAW. 2003.

snatiren, v., *stumble*; **snatirs** (*pres.*) ALEX. (Sk.) 3995.

snattid, adj., *snub-nosed,* '*simus,*' PR. P. 461; snatted nose TREV. III. 285; (*printed* fuatted) ALIS. 6447.

snāu, sb., *O.E.* snāw, = *Goth.* snaiws, *M.Du.* sneeuw, *O.L.G.*, *O.H.G.* snēo (*gen.* snēwes), *O.N.* snær, sniār, sniōr; *from* **sniwen**; *snow,* LA3. 27459; C. M. 18498; (*ms.* snaw) AYENB. 267; IW. 375; A. P. ii. 222; PR. C. 1440; snow HOM. II. 99; snou VOC. 160; O. & N. 430; (*ms.* snow) PR. P. 462; TREAT. 136; MAND. 130; CH. C. T. *B* 3942; snou3 LEG. 165; snouh (*ms.* snowh), snou3 (*ms.* snow3) C. L. 722; **snāwe** [snowe] (*dat.*) LA3. 20125; snowe O. & N. 413.

snāw-water, sb., *snow-water,* HOM. I. 159.
snāw-hwīt, adj., *O.E.* snāwhwīt; *snow-* white, MARH. 18; snauwhit LA3. 24521; snouhwit A. R. 314.

snāwen, v., *snow*; snowin PR. P. 462; **snāwes** (*pres.*) VOC. 201.

snāwi, adj., = *M.H.G.* snēwic, *O.N.* snæugr, sniougr; *snowy,* HOM. I. 251.

snecchen, *see* **snacchen.** **snēde,** *see* **snáde.**

snegge, sb., = *O.H.G.* sneggo, snecco, *M.L.G.* snigge; *snail*; þane snegge AYENB. 32.

snekke, sb., *prov.Eng.* sneck; *latch, lock,* '*pessulus,*' PR. P. 461; snek VOC. 237; TOWNL. 106; snekk 346.

snel, adj., *O.E.* snell, snel, = *O.L.G.* snel, *O. H.G.* snell, *O.N.* sniallr; *quick, active,* LA3. 28860; REL. I. 132 (HOM. II. 13); O. & N. 829; MISC. 97; MIRC 121; **snelle** (*pl.*) S. S. (Wr.) 316; þreo snelle sunen LA3. 7064; kni3tes swiþe snelle HORN (L.) 1463; þine cokes snelle MAP 334; **snelle** (*adv.*) ALIS. 2670; EM. 309; PERC. 2170.

snēle, *see* **snaile.**

snellen, v., *cf. M.H.G.* snellen; *incite*; **snelles** snellus (*pres.*) ANT. ARTH. vii.

snēp, adj., *cf. O.N.* snāpr; *foolish*: hit þincheþ boþe wise and snepe noht þat þu singe ac þat þu wepe O. & N. 225.

snerchen, v., *O.N.* snerkja; *sputter*: þe hude . . . as hit **snarchte** (*pret.*) [snercte] ant barst MARH. 18.

snéren, v., *sneer*; sal snere ['*subsannabit*'] PS. ii. 4; **snéred** (*pret.*) ['*deriserunt*'] xxxiv. 16.

snēsen, v., *sneeze*; **snēseth** (*pres.*) TREV. V. 389; CH. C. T. *H* 62 (*v. r.* fneseþ).

snēsinge, sb., *sneezing,* TREV. V. 389.

snēsen, *see* **snǣsen.**

snevel, sb., *snivel*: snevel of þe nose '*pus nasi*' VOC. (W. W.) 631.

snevelen, v., *? =* **snüvelen**; *snivel*: snevelinge (*v. r.* nevelinge) wiþ þe nose LANGL. *B* v. 135*.

snevien, v., *? cf. Dan.* snive; *sniff*: & **snevieð** (*pres.*) avre fule HOM. II. 37; & alse mid nose **sneved** (*pret.*) 207.

snēwen, *see* **snīwen.**

snibbin, v., *cf. Dan.* snibbe, *Swed.* snebba; *snub, reprove,* '*reprehendo,*' PR. P. 461; snibbe [snebbe] CH. C. T. *A* 523; **snibbed** (*pret.*) PS. ix. 6; M. H. 38; C. M. 18228; snibid HAMP. PS. xv. 7; **snibbid** (*pple.*) APOL. 6; LIDG. M. P. 256; *see* **snubben.**

snibinge, sb., *reproof,* HAMP. PS. xv. 7*.

sni3en, v., *creep*: þan **sni3es** (*pres.*) þar out of þat snith hill ALEX. (Sk.) 4095.

snīken, v., *O.E.* snīcan, = *Dan.* snīge, *Swed.* snīga; *creep*;. **snīkeð** (*pres.*) HOM. I. 251.

snīle, *see* **snaile.**

snīpe, sb., *cf. O.N.* (mȳri-)snīpa; *snipe,* PR. P. 461; REL. I. 82.

snīte, sb., *O.E.* snīte (VOC. 29, 62) ; *? snipe ; 'ibis,'* VOC. 177 ; HALLIW. 767 ; WICL. IS. xxxiv. 11 ; (*spelt* snighte) LIDG. M. P. 192.

snīten, *see* snūten.

snīteren, v., *? drive* ; **sniterand** (*pple.*) ANT. ARTH. vii ; þe snau **snitered** (*pret.*) GAW. 2003.

snīþ, adj., *from* **snīþen** ; *? smooth* : þat snīþ hill ALEX. (Sk.) 4095.

snīðen, v., *O.E.* snīðan,=*O.L.G.* snīthan, *O.Fris.* snītha, *O.N.* snīða, *Goth.* sneiþan, *O.H.G.* snīdan ; *cut* ; **snīþ** (*imper.*) ORM. 14666 ; **snāþ** (*pret.*) ORM. 1338 ; *deriv.* **snāde.**

snivelen, *see* snüvelen.

snīwen, v., *O.E.* snīwan,=*O.H.G.* snīwan, *O.N.* *snīva, *snȳja (*pres.* snȳr, *pple.* snivinn), *M.Du.* sneeuwen ; *snow* ; snewe ORF. 245 ; hit **snīwiþ** (*pres.*) ALIS. 6450 ; sniuþ [sniwe(þ)] O. & N. 620 ; sneweth LAUNF. 293 ; **snēwe** (*imper.*) *sprinkle* PALL. xi. 332 ; **snēuwinge** (*pple.*) C. L. 722 ; **snēu** (*ms.* snew) (*pret.*) MAN. (F.) 13551 ; snewede CH. C. T. *A* 345 ; *comp.* **bi-snīwen** ; *deriv.* snāu.

snobben, v., *hiccup* ; **snobbe** (*pres.*) REL. II. 211 ; weping and **snobbinge** (*pple.*) ED. 1978.

snochinge, sb., *speaking through the nose* : mi barein speche, hosnes and snochinge TREV. I. 11.

snōd, sb., *O.E.* snōd ; *Sc.* snood ; *chaplet, 'vitta,'* FRAG. 2.

snōde, *see* snāde.

snoff, *see* snuff.

snōk, sb., *promontory* ; snuk BARB. i. 188.

snōken, v., = *L.G.* snōken, *Swed.* snōka ; *sniff, smell, 'nicto,'* PR. P. 462.

snóre, sb., *O.E.* snoru, *cf. O.Fris.* snore, *O.N.* snör, snor, *O.H.G.* snur, snor ; *daughter-in-law, 'nurus,'* FRAG. 2 ; LK. xii. 53.

snórin, v.,=*L.G.* snoren ; *snore, 'sterteo,'* PR. P. 462 ; **snóreth** (*pres.*) CH. C. T. *B* 790*.

snorkil, sb., *wrinkle, blemish,* HAMP. PS. cxlvii. 5*.

snorten, *see* snurten.

snot, sb.,=*M.L.G.* snot, snotte, *O.Fris.* snotte, *M.Du.* snot, snut, *? M.H.G.* snuz ; *nasal mucus, 'mucus,'* VOC. 245 ; snotte VOC. 186 ; PR. P. 462 ; ED. 1281.

snóter, adj., *O.E.* snotor, *O.N.* snotr,=*Goth.* snutrs ; *wise, prudent, 'prudens,'* FRAG. 3 ; (*ms.* snoterr) ORM. 7087 ; **snóterne** (*acc. m.*) HOM. I. 117.

snoternesse, sneternesse, sb., *O.E.* snotor-ness ; *prudence, wisdom,* HOM. I. 95, 97.

Snoting-hām, pr. n., *O.E.* Snotinga hām ; *Nottingham,* MISC. 146.

snotte, *see* snot.   **snōu,** *see* snāu.

snoute, *see* snūte.

snubben, v., *O.N.* snubba ; *snub, reprove* ;

snübe [snib] HAMP. PS. lviii. 17* ; **snuband** (*pple.*) ii. 1 ; *see* snibben.

snuff, sb., *candle snuff* ; **snoffes** (*pl.*) WICL. EX. xxv. 38*.

snoff-dish, sb., *snuff-tray,* WICL. EX. xxv. 38.

snuffen, v., *snuff* ; **snuffid** (*pple.*) WICL. EXOD. xxv. 38.

snuffers, sb., *snuffers,* WICL. xxxvii. 23.

snūk, see snōk.

snurtin, v., *cf. L.G.* snurten ; *snort, snore,* PR. P. 462 ; he **snorteþ** (*pres.*) in his slepe CH. C. T. *A* 4163.

snūte, sb.,=*M.L.G.* snūte, *M.Du.* snuite ; *snout, nose,* HORN (L.) 1082 ; REL. I. 224 ; snoute '*rostrum*' PR. P. 462 ; HORN (R.) 1088 ; ALIS. 6534 ; GOW. I. 283 ; CH. C. T. *B* 4095.

snūtel, sb., *snuffers* ; **snītels** (*pl.*) WICL. NUM. iv. 9 (*later ver.*).

snūten, v., *O.N.* snȳta,=*M.Du.* snuiten, *M.L.G.* snūten, *O.H.G.* snūzan ; *blow (the nose)* ; snitin '*emungo*' PR. P. 461 ; þi nese snite MAS. 745 ; **snēting,** snitinge (*pple.*) B. B. 13, 134 ; **snĭtte** (*pret.*) P. L. S. ix. 85 ; **snĭtid** (*pple.*) WICL. EXOD. xxxvii. 23 ; isnit P. L. S. ix. 91.

snūter, sb., *snuffers* ; **snīters** (*pl.*) WICL. NUM. iv. 9 (*earlier ver.*).

snūvelen, v., = *L.G.* snüfeln, *Swed.* snöfla, *Dan.* snœvle ; *snivel* ; snivele VOC. 173 ; snivelle CATH. 347 ; **snüvelinde** (*pple.*) nose SAINTS (Ld.) xlvii. 678 (TREAT. 138) ; *see* snevelen.

snūven, v.,=*M.L.G.* snūven, *Swed.* snūfva, *M.Du.* snuiven, *M.H.G.* snūben ; *pant* ; to **snüvende** (*pple.*) HOM. II. 191.

só, *see* swā.   **sō,** *see* sā, schō.

sobbin, v., *sob, 'singulto,'* PR. P. 462 ; sobbe LIDG. TH. 3380 ; **sobbeþ** (*pres.*) GOW. I. 289 ; þei sobbe LUD. COV. 105 ; **sobband** (*pple.*) ALEX. (Sk.) 3 · **sobbed** (*pret.*) LANGL. *B* xiv. 326.

sobbing, sb., *sobbing,* HAV. 234 ; zobbinge AYENB. 211.

sóbre, adj., *O.Fr.* sobre ; *sober,* AYENB. 254 ; LANGL. *B* xiv. 53 ; CH. C. T. *B* 97.

sóbre-līche, adv., *soberly,* AYENB. 248 ; sobirli ALEX. (Sk.) 2356, 4643, 5340 ; sobire ALEX. (Sk.) 4266.

sóbretē, sb., *O.Fr.* sobrieté ; *sobriety,* LANGL. *C* xvi. 187, *B* xiii. 217 ; sobirte *A* xi 121*.

sócien, v., *form an alliance with* ; **sócied** (*pret.*) TREV. VIII. 147 ; he socied to him þe emperour VIII. 333 ; **isócied** (*pple.*) III. 133 ; socied VII. 533.

socke, *see* sōk.

sōcne, sb., *O.E.* sōcn, *cf. O.N.* sōkn, *Goth.* sōkns, *O.H.G.* sōhni ; *from* sáken ; *inquiry ; refuge ; territory ; custom* ; ANGL. VII. 220 ;

inne swīx̆e feire (r. ferre) stude from socne þes folkes LAȜ. 2365; sokne CH. C. T. A 3987; sookne PR. P. 463; sokene LANGL. B ii. 110; comp. bi-, chirch-, frið-, hāmsōcne.

sodain, sb., O.Fr. soudain, from Lat. subitāneus; sudden, PR. C. 1951; sodein CH. BOET. i. 3 (10); C. T. B 421.

    sodain-līche, adv., suddenly, AYENB. 64; sodeinliche CH. C. T. A 1575; sodanli ALEX. (Sk.) 1052.

sode, sb., cf. M.L.G. sode, O.Fris. satha; sod, turf : a grene sod L. C. C. 6.

soffren, see suffren.

sŏfte, adj. & adv.,=O.E. sēfte adj., sōfte adv., O.H.G. samft, semft, ? M.L.G. sachte; soft, mild, warm, KATH. 1539; A. R. 304; O. & N. 6; GEN.& EX. 335; C. L. 957; LANGL. A PROL. 1; (r. w. ofte) HAV. 991; SPEC. 62; (ms. sŏffte) ORM. 1307; heo þuhten (r. þuhte) him wel softe LAȜ. 14897; sŏfte [O.E. sōfte,= O.L.G. sāfto] (adv.) GOW. II. 45; wer he læi softe LAȜ. 4004; sitte softe P. S. 154; sŏftre (compar.) LAȜ. 16109; sŏftest (superl.) MARH. 11.

    sŏft-gern, adj. & sb., luxurious : þe sŏftgern (pl.) HOM. II. 75.

    sŏft-gerne, sb., luxury, HOM. II. 75.

    sŏfte-līche, adv., softly, A. R. 368; LEG. 6; softeli LANGL. A iii. 37; softli ALEX. (Sk.) 2952.

    sŏftnesse, sb., O.E. sōftness; softness, LAȜ. 25549; A. R. 196.

sŏftin, v.,?=M.L.G. sachten; make soft, 'mollio,' PR. P. 462; softi LAȜ. 12042*; sŏfteð (pres.) A. R. 244; sŏfte (imper.) MARH. 5; sŏftid (pple.).WICL. PS. liv. 22.

sŏftnen, v., soften; softne HOCCL. iv. 11; softene D. ARTH. 2691.

soȝe, see suȝe.   **sōȝen,** see swōȝen.

soht, see suht.

soile, sb., O.Fr. soile; soil, country. ' solum,' PR. P. 463; A. P. ii. 1039; ALEX. (Sk.) 3728.

soilen, v., for assoilen; absolve; i soile (pres.) YORK xxxii. 361.

soilen², v., O.Fr. soillier (souiller); soil; soileð (pres.) A. R. 84; soiled (pple.) L. H. R. 143; soiled [suiled] LANGL. B xiv. 2; isoilled P. L. S. xvii. 10.

soilūre, sb., O.Fr. soillure; soil, pollution, ROB. (W.) 8501.

sojournen, v., O.Fr. sojourner; sojourn; sojourne MAN. (F.) 3560; sojorne CH. L. G. W. 2476; suggeourns (pres.) D. ARTH. 54; sojournede (pret.) P. L. S. ix. 104; sudjornet (pret.) BARB. xvi. 47.

    sudjorning, sb., staying, BARB. vi. 26.

sok, sb., O.E. soc; from sūken; suck, WICL. Is. xi. 8; seser childes of her sok A. P. iii. 391.

sok², sb., O.E. socc,=O.H.G. soc, soch, O.N. sokkr, M.Du. socke, Lat. soccus; sock, VOC. 259; socke PR. P. 462; sockes (pl.) P. S. 330; sokkes P. 26.

sōke, sb., O.E. sōc; ?=sōcne; soke, jurisdiction, territory : in þe sok(e) E. G. 350.

soket, sb., O.Fr. soket; socket, PR. P. 463; ALIS. 4415.

sókin, v., soak, PR. P. 463.

sol, adj., soiled, dirty, REL. I. 129 (HOM. II. 163); a sol cloð A. R. 324.

sōl, see sāl.

solace, sb., O.Fr. solaz; solace, A. P. i. 130; PR. C. 3729; solas ROB. 16; LEG. 12; AYENB. 72; WILL. 1550; solauce D. ARTH. 239, 659.

solacen, v., O.Fr. solacier; solace, LEG. 91; solaci AYENB. 63; ROB. (W.) 11511; solace ALIS. 433; solaseth (pres.) LANGL. B xiii. 453; solaced (pple.) B xix. 22.

solaciūs, adj., agreeable, BARB. x. 290.

solain, adj. & sb., O.Fr. solain, mod.Eng. sullen; alone; solitary person; solitary meal; PARTEN. 5431; solein ' solitarius' PR. P. 463; GOW. III. 6; CH. P. F. 607; WICL. JOB iii. 14; LIDG. TH. 249; sulaine D. ARTH. 2592.

sólde, sb., Fr. solde; wages, hire, O. & N. 764; soud MAN. (F.) 14234; soude ' stipendium' WICL. 2 COR. xi. 8.

sóldiour, sb., O.Fr. soldoier, soudoier; soldier; soudiour WILL. 3954; saudeor, saudiour TREV. VI. 437, VII. 339; soudeurs (pl.) AYENB. 146; saugeours, sougeours ALEX. (Sk.) 2172, 2828.

sóle, adj., O.Fr. sol; sole, solitary, GOW. I. 320.

sóle², sb., O.E. sole,=O.H.G. sola, Goth. sulja, Lat. solea; sole (of the foot), 'planta, solea' PR. P. 463; þe sóles (ms. sooles) of his feet TREV. IV. 351.

sole³, sb., Fr. sole, Lat. solea; sole (fish), PR. P. 463.

solemne, adj., O.Fr. solemne; solemn, famous, A. P. ii. 1171; solempne CH. C. T. A 209; TREV. I. 95, IV. 379, V. 299, VI. 37.

    solempne-lī, adv., with pomp, CH. C. T. B 317, 691, G 272; TREV. VII. 271.

solempnetē, sb., O.Fr. solemnité; solemnity, P. S. 249; solempnitee D. ARTH. 514.

soler, sb., O.E. solere, O.Fr. solier; upper chamber, summer room, ' solarium,' PR. P. 464; P. L. S. xiii. 340; LEG. 31; CHR. E. 847; WICL. JOSH. ii. 6; soler, solere TREV. III. 285, VI. 277.

solȝ, see suluh.

soli, adj., ?=solwi; VOC. 171.

soli², sb., ? Lat. solium; throne; soli(e) A. P. ii. 1171, 1678.

solide, adj., O.Fr. solide; solid, CH. ASTR. i. 17 (9).

solien, v., *O.E.* solian,=*M.L.G.* solen, *Goth.* (bi-)sauljan; *sully, soil, defile*: nis... noht so hwit þat hit ne soleþ (*pres.*) O. & N. 1276.

solitarie, adj., *O.Fr.* solitaire; *alone,* LANGL. *C* xviii. 7.

solitūde, sb., *O.Fr.* solitude; *solitude,* CH. COMPL. M. 65.

solp, *see* sulpen.

solsecle, sb., *Lat.* soiseqvium; *turnsol,* SPEC. 6.

solstice, sb., *Fr.* solstice; *solstice*; **solstices** (*pl.*) GEN. & EX. 150.

soluȝ, *see* suluh.

soluciōn, sb., *O.Fr.* solutıon; *discharge,* GOW. II. 86.

solwi, adj., *dirty,* VOC. 171; TRIST. 1777.

solwin, solowin, v., *cf. M.Du.* soluwen; *soil, 'maculo,'* PR. P. 464; **solowede** (*pret.*) H. S. 9153; **solwid** (*pple.*) C. M. 22491.

som, *see* sum. **som-,** *see* sam-.

sōm, adj., *O.E.* (ge-)sōm, *cf. O.L.G.* sōmi, *Goth.* samjan sis (εὐπροσωπῆσαι); *convenient, accordant*; **sōme** (*pl.*) LAȝ. 9883; C. L. 520; some & sauhte MISC. 97; *see* sēme; *comp.* ȝesōm.

  sōm-līch? adj., *fit*: hit is riht & somlich (*v. r.* semlich) A. R. 94.

sōm², sb., *cf. O.N.* saumr; *trace of a cart*; soim BARB. x. 180; *see* sēam.

sóme, *see* sáme.

sōme, sb., *O.E.* sōm; *concord,* LAȝ. 2552; A. R. 426; mid sib and mid some MISC. 89.

somed, somen, *see* samed, samen.

somer, sb., *O.Fr.* somier, sumer (*sommier*); *sumpter horse,* PR. P. 464; summer BARB. xix. 746; **somers** (*pl.*) MAP 334; someris ALIS. 850.

somer, *see* sumer. **somnien,** *see* samnien.

somonen, v., *O.Fr.* somoner, sumuner, semoner; *summon*; somoni MISC. 26; AYENB. 87; somone MAN. (F.) 3265; somoni, someni(?) ROB. 377, 568; somni, sumni(?) BEK. 397, 737; somne CH. C. T. *D* 1361; sompne LANGL. *B* iii. 314; **somondis** (*pres.*) HAMP. PS. vii. 12*; **somnede** (*pret.*) TREV. III. 201; **sommed,** somned (*pple.*) TREV. VII. 529, VIII. 151.

somonour, somnour, sb., *O.Fr.* semoneor; *apparitor,* CH. C. T. *A* 543; somonour [sompnour] LANGL. *B* iv. 167.

somounce, sb., *O.Fr.* sumunse, semonce; *summons,* L. H. R. 38; somons MAN. (F.) 977; somouns D. ARTH. 91.

son, sb., *O.Fr.* son; *sound,* WILL. 39; PR. C. 4971; soun TREAT. 135; ALIS. 772; CH. C. T. *D* 2273.

sond, *see* sand.

sonde, *see* sande, schande.

sonder, *see* sunder. **sone,** *see* sune.

sōne, adv., *O.E.* sōna,=*O.Fris.* sōn, *O.L.G.* sāna, sāno, sāne, sān; *soon, forthwith,* LAȝ. 200; O. & N. 1058; TREAT. 132; AN. LIT. 11; S. S. (Wr.) 362; PR. C. 68; so sone so he haveð overkumen A. R. 374; sone se (*as soon as*) ich seh þe leome KATH. 477; þa seȝde Zacarias þus til godes engel sone ORM. 198; & sone swa þat steorne stod 6450; sone so Loth ut of Sodome cam GEN. & EX. 1109; sone [soone] CH. C. T. *E* 1617; þen(n)e he asoilede hire sone [soone] LANGL. *A* iii. 47; so sone so *B* xvii. 63; sune MIN. v. 5; eft sones god seide to Abraham WICL. GEN. xvii. 9; **sōnre** (*compar.*) A. R. 58; **sōnest** (*superl.*) A. R. 392; sonnest LANGL. *B* i. 70.

sonet, sb., *O.Fr.* sonete; *a musical instrument,* A. P. ii. 1516; symbales & **sonetez** (*pl.*) 1415.

song, *see* sang.

songewarie, sb., *from O.Fr.* songe (*dream*); *interpretation of dreams,* LANGL. *C* x. 302, *B* vii. 148, 150.

sonne, *see* sunne. **sont,** *see* saint.

sope, sb., *O.E.* sopa (LEECHD. II. 134),=*M.Du.* sope, *O.N.* sopi; *from* sūpen; *sup, small quantity (of water,* &c.), MISC. 152; A. P. ii. 108; *comp.* grund-sope.

sōpe, *see* sāpe. **sopen,** *see* soupen.

sophime, sb., *sophism,* CH. C. T. *E* 5; **sophimes** (*pl.*) *F* 554.

sophister, sb., *teacher, professor,* TREV. V. 175; sophistre LANGL. *C* xviii. 311.

sophistrie, sb., *O.Fr.* sophisterie; *sophistry,* AYENB. 65; CH. L. G. W. 137; LANGL. *B* xix. 343; sofistri CATH. 348.

soppe, sb., = *M.L.G., M.Du.* soppe, *O.N.* soppa; *from* sūpen; *sop, steeped morsel,* PR. P. 465; soppe [sop] LANGL. *B* xv. 175; CH. C. T. *A* 334; sop GAW. 1135; MAN. (F.) 7547; zop AYENB. 107; *comp.* milk-soppe.

soppe², sb., *crowd, troop,* D. ARTH. 1493, 2818, 3729; sop BARB. iii. 47, vii. 567; **soppis** (*pl.*) *heap* BARB. viii. 326.

sōr, *see* sār.

sorcer, sb., *O.Fr.* sorcier; *sorcerer*; **sorsers** (*pl.*) A. P. ii. 1579.

sorceress, sb., *sorceress*; **sorceresses** (*pl.*) CH. H. F. 1261.

sorcerie, sb., *O.Fr.* sorcerie; *sorcery, magic,* LANGL. *A* x. 210; sorceri MIRC 973; sorsoir A. P. 1576.

Sóre, pr. n., *Soar,* MAN. (F.) 2270.

sóre, sb., *O.Fr.* sore (*blonde*); *sore (buck of the fourth year)*: founs, **soures** (*pl.*), bukkes CH. D. BL. 429.

sorel, sb., ? *Fr.* surelle; *sorrel, 'acetosa,'* PR. P. 465; L. C. C. 54.

sóren, v., *Fr.* essorer; *soar, mount aloft*; soor CH. C. T. *F* 123.

[**sorȝ?** adj.; *comp.* or-sorȝ.]

sorȝe, sb., *O.E.* sorg, sorh,=*O.N.* sorg, *O.L.G.* sorga, *O.H.G.* sorga, suorga, *Goth.* saurga (λύπη, μέριμνα); *sorrow,* P. L. S. viii. 98; SHOR. 32; zorȝe AYENB. 27; sorge, sorwe GEN. & EX. 68, 268; sorȝe, seorwe [sorwe] LAȝ. 2919, 31807; sorȝe, sorwe, sorewe, seorhe, seorwe, serewe O. & N. 431, 884, 1599; soreȝe FL. & BL. 528; sorhe MARH. 6; þat wurᵗeð al to sorhe & to care H. M. 27; to sorhe BEK. 1516; MAND. 25; CH. C. T. *D* 1079; in sorwe and pine ligge HAV. 1374; ham to sorwe and teene REL. I. 243; sorowe '*dolor, moeror*' PR. P. 465; serwe LANGL. *A* ii. 89; sorwe *B* ii. 120; seorȝe R. S. iii; seorwe FRAG. 5; seoruwe A. R. 88; serȝhe ORM. 4852; serwe JOS. 705; K. T. 250; serwe and deol C. L. 110; serewe REL. I. 262; soru [sorou] C. M. 1281; **sorȝe** (*pl.*) ANGL. I. 17; sorȝe, sorȝen [sorewe] LAȝ. 12332, 27129; sorouse YORK xii. 7; **sorȝen** (*dat. pl.*) HOM. I. 71; sorghen MISC. 32; sorewen H. H. 44; *comp.* **mōd-sorȝe.**

sorh-, **soruful** [sorhfol], adj., *O.E.* sorg-, sorhfull,=*O.H.G.* sorgfoll; *sorrowful,* LAȝ. 167, 325; seorh-, seoruhful FRAG. 5, 6; seoruhful A. R. 88; soruful TREV. V. 433; soru (*ms.* sorw)-, sorful' HAV. 1248, 2541; sorful A. D. 235.

seorhful-līche, adv., *sorrowfully,* H. M. 17; seoruhfulliche A. R. 400.

sorh-līche, adv., *O.E.* sorglīce; *sorrowfully,* LAȝ. 21883; sorh-, seoruhliche FRAG. 5.

sorh-sīð, sb., *misfortune*; **sorhsīðes** (*pl.*) LAȝ. 11109.

sorȝien, v., *O.E.* sorgian, *cf. O.L.G.* sorgōn, *O.H.G.* sorgen, *Goth.* saurgan; *sorrow, grieve*; zorȝi AYENB. 171; sorhen H. M. 27; sorwin '*doleo, lugeo*' PR. P. 465; LUD. COV. 361; sorwi SHOR. 32; sorwe WILL. 691; sorowe APOL. 7; seoruwen A. R. 308; serȝhen ORM. 8950; **sorewe** (*pres.*) REL. I. 122; serewe SPEC. 93; sorweþ CH. C. T. *A* 2652; **sorȝede** (*ms.* sorȝeden) [sòrewede] (*pret.*) LAȝ. 5078.

sorȝinge, sb., *O.E.* sorgung; *sorrowing*; sorewing SPEC. 53; serwinge C. L. 1390.

sorhe, *see* **sorȝe.** **sōri,** *see* **sāriȝ.**

sorname, *see* **surnoun.** **sort,** *see* **schort.**

sort, sb., *O.Fr.* sort; *lot, destiny, chance,* CH. C. T. *A* 844; sort(e) WICL. LK. i. 5, EPH. i. 11.

sortelegie, sb., *Lat.* sortilegium; *divination by lots,* TREV. I. 411.

sorwe, *see* **sorȝe.** **soster,** *see* **süster.**

sot, adj. & sb., *O.E.* sot, *O.Fr.* sot; *foolish; fool,* LAȝ. 1442; A. R. 66; sotte songes KATH. 107; lat **sottes** (*pl.*) chide O. & N. 297; sottes LANGL. x. 256.

sot-hēde, sb., *folly,* O. & N. 1375; sothed C. M. 18235.

sot-līch, adj., *foolish*: heora sotliche cure LAȝ. 1970; **sotlīche** (*adv.*) KATH. 359; sotlice SAX. CHR. 261.

sot-scipe, sb., *folly,* SAX. CHR. 260; LAȝ. 23178; sotschipe A. R. 362.

ɛōt, sb., *O.E.* sōt,=*O.N.* sōt, *M.Du.* soet; *soot,* ALIS. 6636; sot [soot] TREV. VII. 379; soot '*fuligo*' PR. P. 465; sote P. S. 195.

sōte, *see* **swēte.**

sotel, *see* **sutel.** **sotelen,** *see* **sutilen.**

soten, v., *O.Fr.* (a-)soter; *besot; toxicate*; **soted** (*pple.*) CH. C. T. *G* 1341.

sōti, adj., *O.E.* sōtig,=*O.N.* sōtugr; *sooty,* '*fuliginosus,*' PR. P. 465; O. & N. 578; soti [sooti] CH. C. T. *B* 4022.

sōð, adj. & sb., *O.E.* sōð,=*O.L.G.* sōd, *O.N.* saᵗr, sannr; *sooth, true,* KATH. 873; A. R. 52; GEN. & EX. 1032; soþ ORM. 313; WILL. 2799; PL. CR. 841; S. S. (Wr.) 2470; soth PR. C. 7687; A. P. i. 481; **sōð,** soþ (*subst.*) *truth* O. & N. 217, 950; soþ CH. C. T. *A* 845; soᵗe KATH. 155; GEN. & EX. 74; þat soᵗe P. L. S. xx. 39; for soᵗe A. R. 88; for soþe JOS. 3; vor zoþe AYENB. 16; to soᵗe LAȝ. 2177; þenne segge ic eou to soᵗe HOM. I. 9; þat wit tu wel to soþe ORM. 234; as me mai to soþe iseo TREAT. 132; **sōðes** (*gen.*) A. R. 102; H. M. 17; **sōðere** (*dat. f.*) HOM. II. 191; soþere LAȝ. 673; **sōðne** (*acc. m.*) HOM. I. 227; soþne FRAG. 8; þa(n) soᵗen ileafan HOM. I. 97; **ɛōðere** (*gen. pl.*) LAȝ. 20734; **sōðen** (*dat. pl.*) LAȝ. 25204; on soᵗe sagen GEN. & EX. 14; **sōþer** (*compar.*) SHOR. 31; **sōþest** (*superl.*) LANGL. *B* x. 441.

sōð-cnāwes, adv., *? aware of the truth*: beo nu soð cnawes KATH. 1079; beo soð cnawes JUL. 54.

sōð-cwide, sb., *true saying*; **sōᵗqvides** (*pl.*) LAȝ. 9524.

sōð-fest, adj., *O.E.* soᵗfæst; *truthful,* LAȝ. 6535; A. R. 26; soþfest FRAG. 3; soþfast C. L. 639; H. H. 18; sothfast WICL. JOHN iii. 33.

sōþfastlīke, adv., *truthfully,* ORM. 2995.

sōᵗfestnesse, sb., *O.E.* sōᵗfæstnyss; *truthfulness,* HOM. I. 119; soþfastnesse ORM. 17850.

zōþ-hēde, sb., *truth,* AYENB. 105.

sōð-līche, adv., *O.E.* sōᵗlīce; *truly, in truth,* HOM. I. 15; A. R. 12; H. M. 9; soþlike ORM. 6445; sothlike PS. ii. 6; soþli CH. C. T. *A* 1199; sothli WICL. JOHN iii. 2; soothli HOCCL. iii. 42.

sōþnesse, sb., *truthfulness,* MISC. 47; soþenesse LANGL. *B* iii. 24; soþnisse BEK. 1629.

sōþ-saȝe, sb., *true saying,* O. & N. 1038.

sōð-sagel, adj., *O.E.* sōᵗsagol; *truth-speaking,* HOM. II. 131; soþsawel '*veridicus*' FRAG. 3.

sōþ-schüpe, sb., *truth,* C. L. 1020.

sōþ-siggere, sb., *speaker of truth*; **zōþ-ziggeres** (*pl.*) AYENB. 256.

sōðien, v., *O.E.* (ge-)sōᵗian, *cf. O.N.* sanna; *prove true,* LAȝ. 8491; *comp.* **ȝe-sōðien.**

souchen, v., *O.Fr.* souchier, suscher; *suspect*; souche WILL. 1983; **souches** (*pres.*) PR. C. 788; **souched** (*pret.*) TOWNL. 319.

soudan, sb., *O.Fr.* soldan; *sultan*, CH. C. T. *B* 177.

soudanesse, sb., *sultaness*, CH. C. T. *B* 358.

soude, *see* **sólde.**

souden, v., *O.Fr.* souder, *from Lat.* solidāre; *strengthen*; **soudid** (*pple.*) WICL. DEEDS iii. 7 (*later ver.*).

souen, v., *gall, grieve* : soure [sure] suld him soue ALEX. (Sk.) 2313, 5348.

soufre, sb., *O.Fr.* soufre; *sulphur*, A. P. ii. 954.

sough, sb., ?=*M.L.G.* sō, *mod.Eng.* sough; *sewer*, PALL. i. 515.

souȝt, *see* **suht.** **souken,** *see* **súken.**

sōule, *see* **sáwle.** **soun,** *see* **son.**

sound, *see* **súnd.**

sounen, v., *O.Fr.* soner, suner; *sound*; soune CH. BOET. ii. 3 (37); **sūne** (*imper.*) HORN (L.) 209; **souned** (*pret.*) PARTEN. 4718.

soupen, *see* **súpen.**

soupen, v., *O.Fr.* souper, soper; *sup, eat supper*; soupe HAV. 1766; soupi AYENB. 52; **souped** (*pple.*) WILL. 2997; soped GREG. 1037.

souper, sb., *O.Fr.* super, soper; *supper*; super HAV. 1762; soper FL. & BL. 23; BRD. 6.

souple, adj. *O.Fr.* souple; *supple*, ROB. 223; CH. C. T. *A* 203.

sour, souren, *see* **súr, súren.**

sour, sb., *O.N.* saurr; *dirt, 'coenum,'* PR. P. 466.

souri, adj., *O.N.* saurugr; *dirty, 'coenosus,'* PR. P. 466.

sournoun, *see* **surnoun.**

ȝours, sb., *O.Fr.* surse; *rising; source*; CH. C. T. *E* 49; D. ARTH. 1978, 2511.

soute, *see* **seute.**

soutere, *see* **sútere.**

souþ, sb., *O.N.* sauþr; *sheep*; **sowþes** (*pl.*) ORM. 15565.

souþ, *see* **súð.** **sovel,** *see* **suvel.**

soven, *see* **seoven.**

soverain, adj. & sb., *O.Fr.* soverain; *sovereign, superior*, SHOR. 110; AYENB. 189; WILL. 3954; sov17ein ROB. 15; sufferaine YORK xiv. 46; our suffraind sire YORK x. 163; þe soveraine sire ALEX. (Sk.) 1724; **sovereignes,** -eines (*pl.*) *persons of consequence* LANGL. *C* xii. 269, *B* xii. 200; **soverainest** (*superl.*) ALEX. (Sk.) 1913; soverinest 3097.

soverainetē, sb., *O.Fr.* souveraineté; *sovereignty*, ALEX. (Sk 1859; sovereinte CH. C. T. *D* 1038.

sowe, *see* **suȝe.** **sowel,** *see* **suvel.**

sowen, *see* **séowen.** **sōwen,** *see* **sáwen.**

---

sōwenen, *see* **swōȝnen.** **sōwle,** *see* **sáwle.**

sowþ, *see* **souþ.**

spā, sb., *O.N.* spā; *prophecy*, C. M. 14526.

spā², v., *O.N.* spā; *prophesy* : wit propheci to spa C. M. 18988.

spac, adj., *O.N.* spakr (*quiet, wise*); *active, ready; wise; quiet, gentle*; spac (*? for* sprac) to uvel & slaw to god HOM. I. 305; spac & hwat II. 183; spak H. S. 319; K. T. 774; **spakest** (*superl.*) A. P. iii. 169.

   spac-līche, adv., *O.N.* spakliga; *readily, speedily*, SPEC. 37; spakliche (*v. r.* spracliche) LANGL. *B* xvii. 81, xviii. 12; spakli WILL. 3357; A. P. iii. 338.

spáce, sb., *O.Fr.* espace; *space*, CH. C. T. *A* 1896; PR. C. 3933; spess BARB. xv. 285.

spáde, sb., *O.E.* spadu (*pl.* spadan),=*O.L.G.* spado, *O.Fris.* spada, *? Gr.* σπάθη; *spade,* '*vanga,*' FRAG. 4; PR. P. 467; A. R. 384; R. S. v; AYENB. 108; IW. 3225; CH. C. T. *A* 553.

spǽche, sb., *O.E.* spǽc, sprǽc, = *O.Fris.* sprēke, sprītze, *O.L.G.* sprāca, *O.H.G.* sprācha, sprāhha; *from* **spéken**; *speech, oration, decision* : him was his spæche . . . bīræfed ORM. 2831; þare spæche FRAG. 1; spæche, speche, speke LAȝ. 1971, 4018, 13076; spræche JOHN xviii. 32; speche PR. P. 467; KATH. 808; O. & N. 13; BEK. 66; C. L. 22; AYENB. 21; CH. C. T. *A* 307; S. S. (Wr.) 445; on ure speche HOM. I. 89; gief he him set a speche HOM. II. 179; uvel speche A. R. 80; þat wiþ þe holde speche MISC. 86; al was on speche ðor biforen GEN. & EX. 665; speke HAV. 946; **sprǽche** (*pl.*) MAT. xix. 1; spechen C. L. 25; heo cuþen alle spechen HOM. I. 93; his spechen weoren gode LAȝ. 30161; spæches ORM. 16057; *comp.* **fóre-, morwe-spéche.**

   spǽc-hūs, sb.,=*M.H.G.* sprāchhūs; *audience-room*; spǽchūse (*dat.*) LAȝ. 13036.

spær, *see* **spar.**

spǽten, v., *O.E.* spǽtan; *from* spāt; *spit*, MK. xiv. 65; speten O. & N. 39; J. T. 45; WICL. DEUT. xxv. 9; spete TREV. IV. 327; **spēteð,** spet (*pres.*) A. R. 80, 82; **spétte** (*pret.*) A. R. 106; P. L. S. iii. 59; WICL. JOHN ix. 6; spatte LEB. JES. 1; TREV. III. 317; & spetten him on MK. xv. 19; spatten MISC. 45; *comp.* **bi-spéten.**

Spaine, pr.n., *Fr.* Espagne; *Spain*, LAȝ. 1351.

spainel, sb., *O.Fr.* espagneul; *spaniel*, CH. C. T. *D* 267; H. V. 91; spangel S. S. (Wr.) 1448.

Spainisc, adj., *Spanish* : of Spainisce (e)ard(e) LAȝ. 30703.

spáke, sb., *O.E.* spāca,=*M.L.G.* spēke, *O.H.G.* speicha; *spoke (of a wheel),* '*radius,*' VOC. 202; spoke PR. P. 469; **spáken** (*pl.*) JUL. 56; spoken MISC. 149; spokes P. R. L. P. 238; CH. C. T. *D* 2257.

spákien, v., *from* **spac**; *? hasten* : he mot

spakie to donde sunne a wei MISC. 192; **spákid** (*pret.*) *soothed, quieted* ALEX. (Sk.) 237.

spalde, sb.,=*M.H.G.* spalte; *chip, splinter*; spalle [spolle] '*assula*' PR. P. 467.

spalden, v.,=*M.L.G.* spalden, *O.H.G.* spalten (*pret.* spielt); *chip, split*; **spaldid** (*pple.*) D. ARTH. 3699.

spále, sb.,=*M.H.G.* spale (*rung of a ladder*), *? O.N.* spölr (*plank*); *splinter*, O. & N. 258; **spális** (*pl.*) GOLAG. 629 (ANGL. II. 424).

spān, see **spōn**.

spánin, v., *O.E.* spanan,=*O.H.G.* spanan (*entice*), *M.Du.* spanen (*wean*); *wean*, '*ablacto*,' PR. P. 467; and his ibedde from him spanne O. & N. 1490; **spáned** (*v. r.* wenid) (*pple.*) & taken fro milke HAMP. PS. cxxx. 4; spenned A. P. i. 53; *comp.* **for-spánen**.

spangel, sb., = *M.H.G.* spengel (*brooch*); *spangle*, '*lorale*,' PR. P. 467.

spanne, sb., *O.E.* spann, sponn,=*O.H.G.* spanna, *O.N.* spönn; *span*, PR. P. 467; LEG. 67; FER. 1607; CH. C. T. *A* 155; D. ARTH. 2060; sponne lengore SPEC. 35.

spannen, v., *O.E.* spannan (*pret.* spēonn, spenn),=*O.H.G.* spannan (*pret.* spien); *span, stretch, fold*: þe king spans (*pres.*) his spere AV. ARTH. xiii; mi honde i spenned (*pret.*) A. P. 49.

[**spar**, sb., *? O.E.* spær,=*M.H.G.* spar; *spar*.] **spær-stōn**, sb., '*gypsum*,' FRAG. 4.

spar[2], adj., *O.E.* spær,=*O.N.* sparr, *O.H.G.* spar; *spare, parsimonious*: upon spare wise GAW. 901; the schortte ribbis aboune the spare (*waist*) D. ARTH. 2060.

spar-hende, adj., *O.E.* spærhende; *niggardly*, '*parcus*,' FRAG. 3.

spar-lēs, adj., *plentiful*, ALEX. (Sk.) 5467.

[**spar-līche**, adv., *O.E.* spærlīce, *O.N.* sparliga; *sparingly; comp.* **un-sparlīche**.]

spar-lire, sb., *O.E.* spær-, spearlira; *calf of the leg*, OCTOV.(W.)330; BEV. 2311; sparlir GAW. 158; sperlire FRAG. 2; VOC. 148; sparliuer (*printed* sparlyver) TREV. V. 355; **sparliuers** [*v. r.* -luris] (*pl.*) WICL. DEUT. xxviii. 35.

sparcle, see **spearcle**.

spáre, sb.,=*O.H.G.* sparī; *parsimony*, C. M. 2909.

sparewe, see **sparwe**.

spárien, v., *O.E.* sparian,=*O.N.* spara, *O.H.G.* sparen; *spare*, LAȝ. 27487; ȝif heo mei sparien eni povre schreaden A. R. 416; sparen MAN. (H.) 114; sparie ROB. 428; ALIS. 2624; spari AYENB. 157; þe to telle null(e) i spare for whi hit is þat i care A. D. 258; **spáreþ** (*pres.*) C. L. 419; sparieð KATH. 808; **spáre** (*subj.*) PR. C. 3928; **spárede** (*pret.*) HOM. I. 121; LAȝ. 21066; BEK. 62; **spáred** (*pple.*) GEN. & EX. 3587; ispared A. R. 364; SPEC. 70; *comp.* **a-spárien**.

sparke, see **spearke**.   **sparowe**, see **sparwe**.

sparplin, v., *O.Fr.* esparpeiller (*éparpiller*); *sprinkle, scatter*, '*dispergo*,' PR. P. 467; **sparpled** (*pret.*) MAN. (F.) 8488; sperpolid ALEX. (Sk.) 4162; **sparplid** (*pple.*) WICL. DEEDS v. 36.

sparre, sb., *O.N.* sparri *m.*,=*O.H.G.* sparro, *M.Du.* sparre, sperre; *spar, beam, rafter*, '*tignum*,' PR. P. 467; VOC. 236; sparre and rafter CH. C. T. *A* 990; **sparres** (*pl.*) WICL. ECCLUS. xxix. 29.

spar(r)el, sb., *bolt*: sperel or closel '*firmaculum*' PR. P. 469.

sparren, v., *O.E.* sparrian, *cf. O.H.G.* sparran, sperren, *O.N.* sperra; *close, bar*; sperin '*claudo*' PR. P. 469; speren GEN. & EX. 2194; sperre MAN. (H.) 240; **sperriþ** (*pres.*) APOL. 34; spers PR. C. 3835; **spareð** (*imper.*) A. R. 70*; **sparred** (*pret.*) FLOR. 1774; sperrid TOR. 364; sperid M. ARTH. 2997; **sperred** (*pple.*) HOM. I. 285; IW. 2979; sperd ORM. 4122; isperred [ispered] LANGL. *B* xix. 162*; he ferde to þe tour þer he woren sperde HAV. 448; *comp.* **bi-, un-sparren, (-sperren)**.

sparþe, sb., *O.N.* sparða; *halberd*, GAW. 209; CH. C. T. *A* 2520; sparthe '*bipennis*' PR. P. 467; **sparrethis** [sparthis] (*pl.*) ALEX. (Sk.) 2458.

sparwe, sb., *O.E.* spearwa, spearuwa,=*Goth.* sparwa, *O.H.G.* sparo, *O.N.* spörr; *sparrow*, PS. x. 1; sparwe [sparowe] CH. C. T. *A* 626; sparuwe, sparewe A. R. 142, 152; sparewe BEK. 1098; sparowe '*passer*' PR. P. 467; **sparewen** (*pl.*) LK. xii. 6; sparewen LAȝ. 29258.

spar-, sper-hauk, sb., *O.N.* sparrhaukr; *sparrow-, sparhawk*, '*nisus*,' PR. P. 468; C. M. 1789; sparhauk LANGL. *B* vi. 199 (Wr.) 4191); sperhauk CH. P. F. 338.

spāt, sb., *cf. O.E.* spēd; *spittle*; spot C. L. 1147; **spōte** (*dat.*) SHOR. 88.

spāte-, spēte-, speat-wil, adj., *inclined to spit*, MARH. 9, 12.

spátel, sb., *O.E.* spātl, spādl,=*O.Fris.* spēdel; *spittle*, HOM. I. 279; spotel TREV. I. 195; spotle '*sputum*' PR. P. 469; A. R. 288; WICL. JOHN ix. 6; **spōtle** (*dat.*) LEB. JES. 2.

spátlen, v., *O.E.* spātlian, spǣtlian; *spit upon*; **spátle** (*pres.*) REL. II. 211; **spátled** (*pple.*) MAN. (F.) 8196.

spōtlunge, *O.E.* spātlung; *spitting*, A. R. 188; spateling HOM. I. 279.

spaunin [**spánin**], v., *spawn* (*of fish*), PR. P. 467.

spauninge, sb., *spawning*, PR. P. 467.

spaveine, sb., *O.Fr.* esparvain; *spavin* (*in farriery*), PR. P. 467.

spearcle, sb., *spark*; sparcle TREV. VII. 379; CH. C. T. *B* 2095; WICL. IS. i. 31; sparkle K. T. 194; **sparkles** (*pl.*) LEG. 194; MAN. (F.) 8544.

spearclen, v.,=*M.Du.* sparkelen ; *sparkle* ; sparklin '*scintillo*' PR. P. 467 ; **sperclinde** (*pple.*) A. R. 34 ; sparclinge [sparklinge] CH. C. T. *A* 2164 ; of his spetewile muᵹ(e) **sperklede** (*pret.*) fur ut MARH. 9.

spearke, sb., *O.E.* spearca,=*M.Du.*, *M.L.G.* sparke ; *spark* ; sparke, sperke A. R. 96, 296 ; sparke HAV. 91 ; GOW. I. 258 ; CH. C. T. *B* 2095 ; sparc, spærc LAȝ. 21482, 23508 ; **spearken** (*pl.*) AYENB. 137.

sparkin, v., *O.E.* spearcian,=*M.L.G.* sparken ; *give out sparks ;* '*scintillo*,' PR. P. 467 ; **sparkede** (*pret.*) HAV. 2144.

spéce, *see* **spíce. spēche,** *see* **spǣche.**

special, adj., *Fr.* special ; *special*, ROB. 422 ; AYENB. 15 ; ALIS. 7609 ; A. P. i. 235 ; speciall BARB. v. 501.

special-līche, adv., *specially*, AYENB. 7 ; specialli CH. C. T. *A* 15.

specialtē, sb., *partiality*, BARB. vii. 246.

specke, sb., *O.E.* specca ; *speck, spot* ; spec A. P. ii. 551 ; spekke '*pittacium*' PR. P. 468 ; **speckes,** speches (*pl.*) A. R. 288*.

speckid, adj., *specked*, WICL. GEN. xxx. 32 ; spekked TREV. I. 189.

spectacle, sb., *O.Fr.* spectacle ; *spectacle*, CH. C. T. *D* 1203 ; HOCCL. vi. 57.

spēde, sb., *O.E.* spēd,=*O.L.G.* spōd, *O.H.G.* spuot ; *speed, hurry ; issue, success :* he that was ur moste spede P. P. 215 ; thurgh helpe and spede of praier PR. C. 2882 ; sped AN. LIT. 6 ; ahtes sped (*abundance*) ORM. 12252 ; ᵹat hem sal bringen ivel sped GEN. & EX. 310 ; erneþ toward te sæ wiþ mikel sped ORM. 18094 ; **spēde** (*dat.*) LAȝ. 23417* ; *comp.* ? ȝe-, un-, **wan-spēde.**

spēd-ful, adj., *speedy*, WICL. TOB. iii. 6 ; CH. BOET. iv. 4 (125) ; spedeful TREV. III. 11 ; speidfull BARB. v. 486.

spēd-lī, adv., *O.E.* spēdlīce ; *speedily*, ALIS. 3451 ; WILL. 5468 ; A. P. ii. 1729.

spēden, v., *O.E.* spēdan,=*M.Du.* spoeden, *O.H.G.* (ge-)spuotōn ; *speed, hasten, follow after ; prosper ;* LAȝ. 403 ; ORM. 12317 ; GEN. & EX. 2303 ; þei mighte not speden in hire viage MAND. 305 ; spede HORN (L.) 461 ; HAV. 1634 ; LANGL. *B* xvii. 81 ; S. S. (Wr.) 1965 ; PR. C. 3585 ; **spēt** (*pres.*) JUL. 42 ; O. & N. 763 ; spēdeþ (*imper.*) P. S. 240 ; spedeþ ȝou MAN. (F.) 8907 ; god spede ȝou CH. C. T. *A* 2558 ; **spēdde** (*pret.*) LAȝ. 13214 ; homward he him spedde CH. C. T. *A* 1217 ; heo spedde O. & N. 1792 ; hi ne spedde noȝt þere BRD. 22 ; *comp.* ȝe-**spēden.**

spēdar, sb., *helper*, YORK i. 110.

spēdi, adj., *O.E.* spēdig,=*O.H.G.* spuotīg ; *speedy* ; spēdiast (*superl.*) BARB. vi. 591 ; *comp.* **wan-spēdi.**

spēdi-lī, adv., *speedily*, CH. C. T. *B* 1442.

speir, sb., *the opening in a garment,* '*cluniculum*,' PR. P. 468 ; **spaire** (*dat.*) D. ARTH. 2060.

speir[2], sb., *for* **espeir** ; *hope*, MAN. (F.) 2360.

[**spéke,** sb., *O.E.* sprec, ? =*M.H.G.* sprech (*speaking*).]

spéke-ful, (? **spēke-**), adj., *talkative*, A. R. 100.

spéke-man, sb., *spokesman*, AYENB. 99 ; spékemen (*pl.*) 60.

spēke, *see* **spǣche.**

spéken, v., *O.E.* specan, spreacan,=*O.L.G.* sprecan, *O.Fris.* spreka, *O.H.G.* sprechan ; *speak*, HOM. I. 35 ; LAȝ. 2977 ; KATH. 315 ; A. R. 48 ; ORM. 2733 ; GEN. & EX. 2027 ; speke O. & N. 261 ; ROB. 199 ; CH. C. T. *F* 652 ; speoken HOM. I. 89 ; MARH. 16 ; spreken MAT. vi. 7 ; to spekene LAȝ. 24722 ; AYENB. 94 ; to spekende HOM. II. 35 ; **spékest** (*pres.*) O. & N. 1282 ; spext BEK. 768 ; REL. I. 265 ; spekeð HOM. I. 73 ; A. R. 82 ; specþ O. & N. 1072 ; (heo) specað HOM. I. 89 ; **spec** (*imper.*) LAȝ. 12945 ; spek A. D. 293 ; speke we *let us say* A. R. 104 ; **spékende** (*pple.*) MAT. xv. 31 ; spekinde FRAG. 8 ; **spǣc,** spac (*pret.*) LAȝ. 121, 2841 ; spac SAX. CHR. 264 ; HOM. II. 11 ; A. D. 263 ; ORM. 224 ; O. & N. 396 ; GEN. & EX. 925 ; BEK. 28 ; spak HAV. 2389 ; ALIS. 231 ; WICL. JOHN ix. 29 ; CH. C. T. *A* 124 ; spec A. R. 56 ; P. S. 248 ; spek KATH. 311 ; AYENB. 251 ; speek C. L. 458 ; þu **spēke** O. & N. 554 ; MISC. 139 ; þou speke M. T. 83 ; spæken, spræken, LK. iv. 36, ix. 30 ; spæken, speken LAȝ. 3248, 8249 ; spæken ORM. 1027 ; speken GREG. 294 ; CH. C. T. *B* 214 ; speeken LANGL. *A* ii. 201 (spoke *B* ii. 225) ; spoken GEN. & EX. 2913 ; **spǣke** (*subj.*) ORM. 16260 ; speke BRD. 21 ; **spéken** [*O.E.* sprecen, *cf. O.L.G.* (gi-)sprocan, *O.H.G.* (ge-)sprochen] (*pple.*) HOM. II. 51 ; HAV. 2369 ; AN. LIT. 8 ; speke WILL. 4605 ; spoken MAND. 40 ; MAN. (H.) 77 ; spoken, spoke CH. C. T. *A* 31, *B* 58 ; spoke GOW. I. 60 ; *comp.* **bī-, for-, ȝe-spéken ;** *deriv.* **spǣche.**

spékere, sb.,=*O.H.G.* sprechăre ; *speaker*, TREV. IV. 141 ; speker H. S. 8292.

spékinge, sb., *speaking*, AYENB. 50.

speklen, v.,=*M.Du.* speckelen, spickelen ; *speckle* ; þei ben ... alle **spekelede** (*pple.*) MAND. 290.

spel, sb., *O.E.* spell,=*O.L.G.*, *O.H.G.* spel, *O.N.* spiall, *Goth.* spill (μῦθος) ; *mod.Eng.* spell ; *speech, narrative*, LAȝ. 16397 ; MIRC 170 ; (*ms.* spell) ORM. 10025 ; spel & leoð REL. I. 129 (HOM. II. 163) ; spel ne song SPEC. 68 ; **spelles** (*gen.*) ORM. 13474 ; **spelle** (*dat.*) O. & N. 1794 ; HAV. 338 ; CH. C. T. *B* 2083 ; M. H. 63 ; of godes spelle AYENB. 11 ; listeþ to þis spelle FER. 4002 ; with speche and spelle LUD. COV. 123 ; **spelles** [*r.w.* helle, tellen], (*pl.*) LAȝ. 8100, 26538 ; spellen 22052 ;

spelles A. R. 120; *comp.* bī-, gŏd-, lār-, lāð-, lēod-, morð-, wil-spel.

spélde, sb., *O.E.* speld (*torch*), = *M.H.G.* spelte, spilte *f.* (*splinter*); *spill*, LEG. 162; ofte gret fuir . . . wext of a luitel spielde S. A. L. 157; **spélden** (*pl.*) LEG. 166; speldes WILL. 3392.

spelder, sb., = *M.L.G.* spelder, *M.H.G.* spelter, spilter; *splinter*; **spildurs** (*pl.*) AV. ARTH. xiii.

speldren, v., *spell*; ʒif þat tu canst speldren hem, Adam þou findest **speldred** (*pple.*) ORM. 16440.

spélien, v., *spare*, HOM. II. 213; al þat tu miht spelen ORM. 10133; spelie AL. (T.) 208; FER. 458; spele MAP 339; MAN. (F.) 12428; spele and spare LANGL. *C* xiv. 77; gef þu ani þing **spélest** (*pres.*) & levest HOM. II. 31; **ispéled** (*pple.*) S. S. (Web.) 542.

spelke, sb., *O.N.* spialk; *splinter, chip*, PR. P. 468.

spellien, v., *O.E.* spellian, = *M.Du.*, *M.H.G.* spellen, *O.N.* spialla, *Goth.* spillōn; *spell, speak, say, narrate* : þat folc gan to spel-(l)ien Irlondes speche LAʒ. 10068; spellen P. S. 240; LUD. COV. 157; þe (a)postles foren sone an an til hæþen folc to spellen ORM. 8528; spellin '*syllabico*' PR. P. 468; spelle LEG. 3; GREG. 611; GOW. II. 20; no tonge ne mai hit telle . . . ne mouþ spelle C. L. 692; speke and spelle LANGL. *B* xv. 600; **spelle** (*pres.*) HAV. 15; speleð A. R. 170; **spellede** (*pret.*) SHOR. 133; hio spelleden LK. xxiv. 15; speleden I.Aʒ. 4051; **spelled** (*pple.*) ORM. 5747.

spellunge, sb., *story, narration*, A. R. 64; spelling H. S. 10976; spellinge '*syllabicatio*' PR. P. 468.

spelonk, sb., *Lat.* spelunca; *cave*, ALEX. (Sk.) 5392; **spelonkes** (*pl.*) LANGL. *B* xv. 270.

spenden, spenen, v., *O.E.* (a-, for-)spendan, *O.N.* spenna, ? = *O.H.G.* spentōn, *from Lat.* expendere; *spend*; spendin '*expendo*' PR. P. 468; spende CH. C. T. *A* 4135; spene P. S. 151; P. L. S. i. 16; speneþ (*pres.*) O. & N. 1525; þat have no **spending** (*pple.*) silver LANGL. *B* xi. 278; **spende** (*pret.*) LAʒ. 13657; **spended** (*pple.*) WILL. 4324; LANGL. *B* v. 380; spende PALL. iv. 541; ispended AYENB. 171; ispend P. L. S. viii. 6; ispened A. R. 322; *comp.* a-, for-spenden.

spendere, sb., = *M.H.G.* spendære; *spender, spendthrift*, AYENB. 190; spendour LANGL. *C* vi. 28.

spendinge, sb., *spending*, AYENB. 21; LANGL. *B* xiv. 197.

spenen, *see* spenden.

spenne, sb., ? *O.N.* spenna; *space, interval; quickset hedge*; GAW. 1074, 2316; spene ALEX. (Sk.) 4162.

spennen, v., = *M.H.G.* spennen, *O.N.* spenna; *from* spannen; *stretch out, embrace; fasten*; spain BARB. iii. 582; up spende (*pret.*) his feder tunge A. R. 158; spainit BARB. iii. 583; spenned (*pple.*) A. P. i. 53; spend GAW. 587; *comp.* un-spennen.

spense, sb., *O.Fr.* dispense; *expense; provision-room*; A. R. 350; P. L. S. xvii. 31; boteri or celere PR. P. 468; spence, spens YORK xxxvi. 241, xxxii. 134; **spencis** (*pl.*) D. ARTH. 3163; spences LANGL. *C* xvii. 40.

spenser, sb., *O.Fr.* despensier; *steward*, TREV. IV. 331.

speoren, *see* spürien.

spēowen, *see* spīwen. sper, *see* spar.

sperclen, *see* spearclen.

spére, sb., *O.E.*, spere = *M.Du.* spere, *O.H.G.* sper, *O.N.* spiör *n.*, *Lat.* sparum, sparus; *spear*, '*hasta*,' PR. P. 468; O. & N. 1022; HAV. 347; CHR. E. 631; CH. C. T. *A* 2894; OCTAV. (H.) 853; PR. C. 5292; þat spere LAʒ. 18082; **spéres** (*gen.*) LAʒ. 8596; spére, speren [speres, speares] (*pl.*) LAʒ. 5864, 27462; speres A. R. 110; *comp.* bōr-, wal-spere.

spér-man, sb., *spearman*, ROB. 378.

spére-scæft, sb., *spear-shaft*, LAʒ. 14752.

spére-wurt, sb., *spear-wort*, REL. I. 37; sperewort PR. P. 469.

spēre, sb., *sphere*, PR. P. 468; PR. C. 4887; PARTEN. 6509; TREV. I. 227.

sperel, *see* spar(r)el.

spéren, v., *spear*; **spérid** (*pret.*) ALEX. (Sk.) 3649.

speren, *see* spüren.

sperhauke, *see* sparhauk *under* sparwe.

sperke, *see* spearke.

sperling, spirling, sb., = *M.L.G.* spirling; *sparling* (*small fish*), '*cammarus*,' VOC. 189, 222; sperling M. H. 136; **sperlinges** (*pl.*) L. C. C. 54.

sperme, sb., *O.Fr.* sparme; *sperm*, CH. C. T. *B* 3199.

sperren, *see* sparren. spete, *see* spite.

spēten, *see* spǣten. spēwen, *see* spīwen.

spic, sb., *O.E.* spic (VOC. 82), = *O.N.* spik, *O.H.G.* spec, spech; *lard; bacon fat;* '*lardum*,' FRAG. 4; LAʒ. 24437; spik PR. P. 469; MAN. (F.) 12345; **spiche** (*dat.*) MISC. 151.

spice, spéce, sb., *O.Fr.* espice, espece; *spice; species*; A. R. 78, 208; spice SHOR. 90; JOS. 193; **spíces** (*pl.*) A. R. 370; WICL. GEN. xxxvii. 25; speces (*printed* spetes) MAP 334.

spícen, v., *embalm*, TREV. V. 287.

spícer, sb., *O.Fr.* espicier; *grocer*, VOC. 212; PR. P. 469; **spícers**, spiceres (*pl.*) LANGL. *C* iii. 235; *B* ii. 225; spiserez A. P. ii. 1038.

spicerie, sb., *O.Fr.* espicerie ; *grocery, spices,*
P. S. 333 ; PL. CR. 301 ; D. ARTH. 162.

spie, espie, sb., *cf. M.Du.* spie, *O.Fr.* ẹspie ;
spy; 'explorator,' PR. P. 469 ; FL. & BL. 332 ;
ALIS. 3556 ; ISUM. 249 ; espie GOW. I. 81 ;
spíes (*pl.*) GEN. & EX. 2169 ; aspies ROB.
557 ; espies D. TROY 13424.

spien, espien, v., *cf. M.Du.* spien, *O.N.* speja,
O.H.G. spehōn, *O.Fr.* espier, *Lat.* spẹcere ;
spy, espy, GEN. & EX. 2172 ; spiin '*exploro*'
PR. P. 469 ; spie A. P. ii. 780 ; spien [espien]
CH. C. T. *D* 316 ; espie LIDG. M. P. 64 ; **aspíen**
MAND. 66 ; aspie WILL. 774 ; **aspíeþ** (*pres.*)
AYENB. 253 ; **aspíeden** (*pret.*) A. R. 196.

　spíere, sb., *spy,* WICL. ESTH. viii. 5 ;
spíares (*pl.*) LAȝ. 1492*.

spigot, sb., *spigot,* PR. P. 469 ; WICL. JOB
xxxii. 19.

spīk(e), sb., *Lat.* spīca ; *ear of corn,* LANGL. *C*
xiii. 180.

spīkenard, sb., *spikenard,* LEB. JES. 770 ;
['*nardi spicati*'] WICL. MK. xiv. 3.

spilder, see **spelder.**

spile, sb., ?=*O.N., O.L.G., O.H.G.* spil, *O.
Fris.* spil, spel ; *sport, game*; ˘under and
levene made spile (*r. w.* qvīle) GEN. & EX.
3462 ; froskes & podes spile GEN. & EX.
2977.

　spil-qverne, sb., *spinning-top, 'gyraculum,'*
REL. I. 9.

spilen, v., *O.E.* spilian, =*O.N.* spila, *O.L.G.,
O.H.G.* spilōn ; *play, sport,* GEN. & EX. 2532 ;
spilede (*pret.*) LAȝ. 13816 ; SHOR. 120 ;
spiled (*pple.*) GEN. & EX. 3183.

spille, sb., *cf. M.Du.* spelle (*acicula*) ; *pin*: hit
nis noȝt worþ a spille BEK. 850 (*v. r.* fille
SAINTS (Ld.) xxvii. 856).

spillen, v., *O.E.* spillan, spildan, =*O.N.* spilla,
M.L.G., M.Du. spillen, spilden, *O.H.G.* spildan ; *spill, lose; kill; be destroyed*; LAȝ.
880 ; JUL. 24 : spillin '*effundo*' PR. P. 469 ;
spille BEK. 204 ; C. L. 903 ; WILL. 966 ;
LANGL. *B* xix. 298 ; GOW. II. 114 ; MAN.
(H.) 279 ; PR. C. 1320 ; ISUM. 193 ; al his
hwile he scholde spille O. & N. 1020 ; late ye
nouth mi bodi spille HAV. 2422 ; **spilleð**
(*pres.*) HOM. II. 213 ; spilþ AYENB. 182 ;
spille (*subj.*) HOM. I. 17 ; PALL. viii. 164 ;
spilde (*pret.*) REL. I. 48 ; spilden LAȝ. 28863 ;
þai spilden mani a man TRIST. 40 ; **spilled**
(*pple.*) ALIS. 1062 ; spilt GOW. I. 270 ; *comp.*
a-, for-, ȝe-spillen.

　spillinge, sb., *spilling,* PR. P. 469.

spindle, sb., *O.E.* spinl, =*O.H.G.* spinnela ;
spindle, 'fusus,' FRAG. 4 ; LEG. 105 ; spindele
VOC. 157.

spink, sb., *cf. Swed.* (gul)-spink ; *prov.Eng.*
spink ; *? chaffinch,* VOC. 189.

spinnen, v., *O.E.* spinnan, =*Goth., O.H.G.*
spinnan, *O.N.* spinna ; *spin*; spinnin '*neo*'
PR. P. 469 ; spinne LANGL. *A* v. 130 ; spinnen

B v. 216 ; he **spinnes** (*pres.*) him out a
grete space ALEX. (Sk.) 3033 ; þe **spun-
niande** (*pple.*) aspaltoun A. P. ii. 1038 ; **span**
(*pret.*) LEG. 74 ; LIDG. M. P. 90 ; sponne CH.
TRO. iii. 734 ; **sponnen** (*pple.*) WICL. JUDG.
xvi. 9 ; *comp.* **ȝe-spinnen.**

spinnere, sb., =*M.L.G.* spinnere ; *spinner,
spider,* REL. I. 219 ; spinnare '*filatrix,
aranea,*' PR. P. 469 ; spiþre (*for* spinþre)
AYENB. 164.

spinnestere, sb., =*Du.* spinster ; *mod.Eng.*
spinster ; *woman who spins*; **spinnesteres**
(*pl.*) LANGL. *B* v. 216 ; spinsters *A* v. 130.

spīr, sb., *O.E.* spīr, *cf. M.L.G.* spīr *n., O.N.*
spīra *f.*; *sprout,* PALL. iii. 1034 ; shal
nevere spir springen up LANGL. *C* xiii. 180 ;
spire '*hastula*' PR. P. 469 ; as an ok comeþ
of a litle spire CH. TRO. ii. 1335 ; imeind
mid spire and grene segge O. & N. 18.

spirakle, sb., *Lat.* spirāculum ; *breath, spirit,*
A. P. ii. 408.

spiren, see **spürien.**

spīrin, v., *sprout, 'spico,'* PR. P. 469 ; **spīred**
(*pple.*) '*germée*' REL. II. 81.

spirit, sb., *O.Fr.* espirit, esprit ; *spirit,* GEN.
& EX. 203 ; CH. C. T. *A* 2765 ; sprit RICH.
394.

spiritualtē, sb., *O.Fr.* esperitualité ; *spirituality,* LANGL. *C* vii. 125.

spirituel, adj., *O.Fr.* spirituel ; *spiritual,*
LANGL. *C* vii. 125*.

spirling, see **speiling.**

spísen, v., *for* despísen ; *despise*; spise
(*pres.*) ALEX. (Sk.) 2931 ; spiseth LANGL. *C*
vii. 122*.

spit, see **despít.**

spite, sb., *cf. O.E.* spitu *f., M.Du.* spit, spet
O.H.G. spiz *m.; spit, 'veru,'* FRAG. 4 ; ROB.
207 ; OCTOV. (W.) 122 ; B. DISC. 582 ; spite,
spete PR. P. 469 ; spete VOC. 256 ; **spetes**
(*pl.*) MAP 334 ; spiten *? fins* LAȝ. 21329.

[**spitel,** sb., *small spike.*]

　spitil-forke, sb., *spiked fork*; spitill
forkes (*pl.*) TUND. 739.

　spitel-staf, sb., *spiked-staff,* A. R. 384*.

spitel², sb., =hospital ; *hospital.*

　spitel-üvel, sb., *leprosy,* A. R. 148.

spitien, sb., =*M.Du.* spiten, speten, *M.H.G.*
spizzen ; *from* **spite** ; *put on a spit* ; **spite**
(*pres.*) PR. P. 469 ; **spiteden** (*pret.*) KATH.
2210* ; **ispited** (*pple.*) LAȝ. 26522 ; ROB.
207 ; *comp.* þurh-spitien.

spitous, spetous, adj., *spiteful, angry,
terrible,* ALEX. (Sk.) 2458, 4567 ; spetos
GAW. 209 ; *see* **despitous.**

spitten, v., *O.E.* spittan ; *spit,* JUL. 49 ; **spit**
(*pres.*) H. M. 17 ; spitte CH. C. T. *C* 421 ;
spitten and spewen LANGL. *B* x. 40 ; spit
(*imper.*) A. R. 290 ; **spitte** (*subj.*) MIRC 890 ;

spitte (*pret.*) ALIS. 890; spitten C. M.
16635; spitte L. H. R. 189; *comp.* bi-spitten.

spiþre, *see* spinnere.

spīwen, v., *O. E.* spīwan (*pret.* spāw,
spēowde),=*O.L.G.* spīwan, *O.H.G.* spīwan,
spīan, *M.H.G.* spīwen spīen (*pret.* spei,
spīwete), *O.Fris.* spīa, *Goth.* speiwan, *M.Du.*
spēeuwen, *O.N.* spȳja, *Lat.* spuere ; *spew,*
vomit ; speowen A. R. 240; JUL. 49; spuwe
TREV. VII. 503; spewen out WICL. JOB xx.
15; spewe P. S. 336; CH. C. T. *B* 2606;
spēoweð (*pres.*) HOM. I. 251; spuwþ
LEECHD. III. 140; speweð HOM. II. 37;
REL. I. 211; spewen LANGL. *B* x. 40; spī
(*imper.*) A. R. 310; ispēwed [ispūwed] (*pple.*)
TREV. IV. 439; *comp.* a-spēowen.

splatten, v., *cut open* ; to splatt(e) þe bor
EGL. 490; splatte (*imper.*) (*? for* platte) *press*
flat, PALL. ii. 123.

splecken, ? =plecken ; splekked [plekked]
(*pple.*) stones ['*guttatos lapides*'] TREV. I.
429.

spleien, v., *display, spread* ; splai PALL. i.
625; COMP. BL. KN. 33; splaied (*pple.*) LUD.
COV. 242; *cf.* displeien

splēn, sb., *Lat.* splēn ; *spleen*, PR. P. 469.

splent, sb.,=*M.L.G.* splente, splinte; *splinter;*
small plate of armour; REL. II. 84; D. ARTH.
2061; splentes (*pl.*) of steel RICH. 4979.

splentide, adj., *adorned with small plates,*
D. ARTH. 3264.

[splot, sb., *O.E.* splot ; *spot.*]

splotti, adj., *spotty*, WICL. GEN. xxx. 35.

spoile, sb., *booty, spoil* ; spoile ALIS. 986;
spuilis (*pl.*) WICL. GEN. xlix. 27.

spoilin, v., *Fr.* spolier; *unclothe; rob; 'spolio,'*
PR. P. 470; to spoile helle MIRC 509; 3e shulen
spoile [*spoliabitis*'] WICL. EX. iii. 22; spoilis
(*pres.*) ALEX. (Sk.) 4962; spulȝeit (*pple.*)
BARB. xiii.

spōke, *see* spāke.

spōle, sb.,=*M.Du.* spoele, *O.H.G.* spuole ;
spool, 'panus,' VOC. 180; PR. P. 470.

spōn, sb., *O.E.* spōn,=*O.Fris.* spōn, *O.N.*
spönn, spānn, *O.H.G.* spān, *M.Du.* spaen ;
chip, splinter; spoon; VOC. 232; PR. P. 470;
TRIST. 2050; spoon S. & C. I. xlix; spōne
(*dat.*) PERC. 2250; L. C. C. 51; spōnes (*pl.*)
['*astulis*'] TREV. V. 455; spones [spoones]
CH. C. T. *C* 908.

spōn-nēowe, adj., *quite new*, ALIS. 4055;
spannewe HAV. 968; CH. TRO. iii. 1665.

sponge, sb., *O.Fr.* esponge ; *sponge*, A. R.
262; spunge MK. xv. 36.

[spor, sb., *O.E.* spor,=*O.N.*, *O.H.G.* spor;
track, foot-print; comp. hwēl-spōr.]

spore, *see* spure. spornen, *see* spurnen.

sport, *see* disport.

spot, sb., *cf. M.Du.* spotte; *spot, 'macula,'* VOC.

206; PR. P. 470; þe spot AYENB. 228; þane
spot 237; spottes (*pl.*) REL. I. 225; LANGL.
B xiii. 315.

spōt, spōtel, *see* spāt, spātel.

spotti, adj., *spotty*, AYENB. 192; L. H. R. 213;
WICL. GEN. xxx. 35.

spotton, v.,=*M.Du.* spotten ; *spot, 'maculo,'*
PR. P. 470; spotted (*pple.*) GEN. & EX. 1721;
spottid WICL. GEN. xxx. 35; *comp.* bi-
spotten.

spous, *see* spūs.

spoute, spouten, *see* spūte, spūten.

[sprac, adj., *cf. O.N.* sprǣkr (*lively*), ? *prov.*
Eng. (*Berks.*) sprack; *energetic, lively; see*
spac.]

sprac-, sprak-līche, adj., *sprightly, lively*,
LANGL. *C* xxi. 10; *B* xviii. 12*.

sprǣche, *see* spǣche.

sprǣden, v., ? *O.E.* sprǣdan,=*M.Du.* sprēden,
O.H.G. spreitan ; *spread* ; sprǣde [sprede]
LAȝ. 14203; spreden A. R. 400; spredin
'*expando*' PR. P. 470; srede O. & N. 437;
LANGL. *B* iii. 308; S. S. (Wr.) 622; MAN.
(H.) 38; þat writ an an he gan sprede AL.
(T.) 450; sprēt (*pres.*) HOM. I. 77; A. R.
98; AYENB. 17; sprēdde (*pret.*) P. L. S. xiii.
118; spradde LAȝ. 1215; GOW. I. 182; A. P.
iii. 365; spredden GEN. & EX. 2567; sprēd
(*pple.*) ORM. 1015; sprad GREG. 566; LUD.
COV. 373; *comp.* bi-, ȝe-, over-, to-sprǣden.

sprai, sb., *cf. Dan.* sprag;=sprec; *spray,*
sprout, sedge ; (*ms.* spray) FL. & BL. 275; ROB.
552; SPEC. 27; CH. C. T. *B* 1960; spraies
(*pl.*) TREV. IV. 157.

spraien, v.,=*Dan.* spraga, ? *Swed.* spraka ;
sprout ; spraie (*ms.* spraye) ERC. 335; TOWNL.
145; sprai S. & C. II. xlii.

spranke, sb., = *M.Du.* spranke ; *spark* ;
spronke LEG. 159; *sprout* PALL. xii. 116.

spraulin, v., *O.E.* sprēawlian ; *sprawl, 'pal-*
pito,' PR. P. 470; spraule WICL. TOB. vi.
4; spraulend (*pple.*) GOW. II. 5; sprau-
lid (*pret.*) LUD. COV. 186; sprauleden HAV.
475.

sprec, sb., *O.N.* sprek; ? *sprout*; sprékes
(*pl.*) MARH. 15.

sprecchen, v., *O.E.* (on-)spreccan ; *sprout*:
ant sprechi in ham sprekes of lustes MARH.
15.

sprēden, *see* sprǣden. spréken, *see* spéken.

sprenge, sb.,=? *O.H.G.* springa (*trap*);
? *springe*: þu schalt hwippen on a sprenge
O. & N. 1066.

sprengen, v., *O.E.* sprengan,=*O.H.G.* sprengan, *O.N.* sprengja; *from* springen ; *make*
to spring; sprinkle; WICL. IS. xxxviii. 25;
sprengen [springen] CH. C. T. *B* 1183; spren-
geþ (*pres.*) GAM. 503; sprengeð ou mid hali
water A. R. 16; sprengde (*pret.*) LEG. 180;

spreinde 181; spreind(e) ALIS. 341; sprende
LEG. 156; spreinden JOS. 314; **sprenged**
(*pple.*) TREV. V. 7; misbileve . . . a mong
men was isprenged ROB. 119; spreind
[ispreind] CH. C. T. *A* 2169, *B* 422; sprent
ALEX. (Sk.) 743; **isprengde**, isprinde (*pl.*)
A. R. 92; *comp.* **bi-sprengen.**

sprenkelin, v.,=*M.Du.* sprenkelen; *sprinkle,*
PR. P. 470; **sprancleth** [sprankeleþ] (*pres.*)
TREV. I. 319.

sprenten, v.,=*M.H.G.* sprenzen; *leap, run*:
sparkes of fire þat obout sal sprent PR. C. 6814;
sprenten ALEX. (Sk.) 786*; **sprente** (*pret.*)
OCTAV. (H.) 473; to the chambirdore he
sprente M. ARTH. 1846; sprent GAW. 1896;
PERC. 1709.

sprenþinge, sb., *leaping,* TREV. I. 369.

sprēot, sb., *O.E.* sprēot,=*M.L.G.* sprēt, *M.Du.*
spriet; *from* **sprūten**; *sprit, pole,* ALIS. 858;
spret '*contus*' PR. P. 470; OCTOV. (W.) 601;
WILL. 2754; **sprēte** (*dat.*) A. P. iii. 104;
comp. **bō3-sprēt.**

sprēþ(e), adj., ?=*Ger.* spröde; ? *fragile :* þa3
ich be spreþ(e) (*r. w.* dēþe) SHOR. 103.

sprigge, sb., ? *cf.* *M.L.G.* sprick *n.* ; *sprig,*
LANGL. *C* vi. 139.

sprind, adj., *O.E.* sprind; *lively :* be a man
never so sprind SHOR. 2.

springal, sb., = *M.L.G.,* *M.H.G.* springal,
O.Fr. espringalle; *an engine for casting stones* ;
springols (*pl.*) FER. 3310; spri[n]galdis
(springaltez) ALEX. 1419; springaldis BARB.
xvii. 247.

springe, sb., *O.E.* springe, spring,=*O.L.G.*
(aha-)spring, *O.H.G.* (ur-)spring *m.* ; *spring*
(*fountain, vernal season*), *upspringing ;*
dawn ; twig, rod ; '*scaturigo, planta,*'
PR. P. 470; GEN. & EX. 581; C. M. 11699;
who so spareþ þe springe [spring] ['*qui parcit*
virgae'] LANGL. *B* v. 41; on þe spring(e) of
þe dai FER. 3513; *comp.* **dai-, of-, wel-**
spring.

spring-flood, sb., *spring-tide,* CH. C. T. *F*
1070.

springen, v., *O.E.* springan,=*O.L.G., O.H.G.*
springan, *O.N.* springa; *spring, jump, rise,*
A. R. 282; springe C. L. 593; þe blostme
ginneþ springe and sprede O. & N. 437; þis
word bigon to springe AL. (V.) 223; þe dai
bigan to springe CH. C. T. *A* 822; TOR.
1958; a weile . . . ðat **springeð** (*pres.*) ai
REL. I. 210; **sprang** (*pret.*) KATH. 2488;
ORM. 10258; ALIS. 5539; sprong GEN. & EX.
60; LANGL. (Wr.) 9041; K. T. 194; of hire
wisdome sprong [sprang] þat word wide LA3.
6302; sprungen HOM. I. 141; senwe sprungen
fro ðe lið GEN. & EX. 1804; **sprungen** (*pple.*)
ORM. 511; HAV. 1131; sprunge HORN (L.)
1015; sprongen GOW. III. 249; *comp.* **a-,**
at-, 3e-, to-springen ; *deriv.* **springe,**
sprengen, sprung.

springing, sb., *springing :* in springing
time ['*verno tempore*'] TREV. I. 65.

sprintel, sb., *twig :* ofte druie **sprintles**
(*pl.*) bereþ winberien A. R. 276.

sprit, *see* **spirit.** **spronke,** *see* **spranke.**

sprōte, sb., *O.E.* sprote,=*M.Du.* sprote, *O.N.*
sproti, *O.H.G.* sprozo ; *from* **sprūten** ; *sprig,*
sprout, TOWNL. 14 ; i ne have stikke i ne have
sprote HAV. 1142; **sprōtes** (*pl.*) ALEX. 790.

sprotte, sb.,=*L.G.* sprotte ; *sprat,* VOC. 222 ;
M. T. 119.

sprouten, *see* **sprūten.**

sprung, sb.,=*O.H.G.* sprung ; *from* **springen** ;
jump, rising, KATH. 322; **sprunges** (*pl.*)
JUL. 50; CHR. E. 195; *comp.* **of-sprung.**

sprūte, sb.,=*M.L.G.* sprūte, *M.Du.* spruite,
sprout ; **sproutes** (*pl.*) PS. lxxvii. 51.

sprūten, v., *cf.* *O.E.* sprēotan, *O.Fris.* sprūta,
M.Du. spruiten, *M.H.G.* spriuzen; *sprout,*
germinate, HOM. II. 217; sprute C. M. 11216;
sproutin '*pullulo*' PR. P. 471; **sprūteð**
(*pres.*) H. M. 11; **sproutand** (*pple.*). PS.
lxiv. 11; *deriv.* **sprūte, sprēot, sprōte,**
sprütten.

sprütten, v., *O.E.* spryttan,=*M.H.G.* sprüt-
zen; *sprout :* þe wiði þet **sprütteð** (*pres.*)
ut A. R. 86.

spudde, sb., *spud, digging-knife,* '*cultellus,*'
PR. P. 471.

spure, sb., *O.E.* spura, spora,=*O.H.G.* sporo,
O.N. spori ; *spur,* '*calcar,*' VOC. 221; O. & N.
777; spore PR. P. 470; HAV. 2569; ROB.
396; **spuren** [spores] (*pl.*) LA3. 23772;
sporen HOM. I. 243; ALIS. 817; spores
LANGL. *B* xviii. 12; CH. C. T. *A* 473.

spore-lēs, adj., *without a spur,* P. S. 71.

spurgen, v., *O.Fr.* espurger ; *ferment* ; spor-
gen '*spumo*' PR. P. 470; **spourgide** (*pret.*)
GEST. R. 403.

spurien, v., *cf.* *O.H.G.* sporōn ; *spur, urge*
on, incite ; spurie [sporie] LA3. 21354;
spureð (*pres.*) H. M. 13; GEN. & EX. 3970;
spurede (*pret.*) JUL. 59; spored PARTEN.
4214.

spürien, v., *O.E.* spyrian,=*O.N.* spyrja, *O.*
H.G. spurian; *track ; inqvire* ; spire WILL.
4594; GOW. I. 198; MAN. (H.) 112; M. H.
95; sperin '*percunctor, inqviro*' PR. P. 469;
spere '*investigo*' REL. I. 6; FLOR. 293;
TRIAM. 592; spire [spere] (*pres.*) LANGL.
B xvii. 1; speore *C* xx. 1*; **spüred** (*pple.*)
GAW. 901; spired HAV. 2620.

spirring, sb., *questioning,* B. B. 2; **spir-**
ringes (*pl.*) YORK xxxiii. 64.

spurn, sb.,=*O.H.G.* (ana-)spurn ; *stumbling-*
block : his fot sal finde a spurn C. M. 4324.

spurnen, v., *O.E.* spurnan, spornan, *cf.*
O.N. spyrna, *O.L.G., O.H.G.* spurnan ; *spurn,*
kick, A. R. 188; spurnin '*calcitro*' PR. P.
471; þei schulen not spurne þer inne ['*non*

impingent in ea'] WICL. JER. xxxi. 9; þe wiket op spurne HORN (H.) 1115; in speche i sporne REL. II. 211; þe ground he sporneþ GOW. II. 72; þat þou ne spurn(e) þi fot til stan [*' ne offendas ad lapidem pedem tuum'*] PS. XC. 12; **spurnde** (*pret.*) ROB. 341; *comp.* **æt-spurnen.**

spūs, sb., *O.Fr.* espus, espous, espos; *spouse, husband,* A. R. 10; **spūse** (*O.Fr.* espūse) *wife* A. R. 98; O. & N. 1527; spouse AYENB. 118.

    spūs-hād, sb., *marriage; marriage vow,* HOM. I. 143; spushod II. 45; spoushed TREV. VII. 245.

    spūsāge, sb., *wedlock;* spousage TREV. IV. 173.

spousaile, sb., *O.Fr.* espousaille; *spousal,* MIRC 532; spousaile [spousaille] CH. C. T. *E* 115; **spōsailes** (*pl.*) AYENB. 189; spousails TREV. III. 365.

spūse, sb., *marriage vow;* breke spuse O. & N. 1334.

    spous-breche, sb., *adulterer,* SHOR. 64.

    spous-brekere, sb., *adulterer,* TREV. VIH. 25.

    spūs-brüche, sb., *adultery,* A. R. 56; spousbruche ROB. 220; spousbreche AYENB. 37.

spūsen, v., *O.Fr.* espouser; *espouse, betroth,* HAV. 1123; spouse P. L. S. xvi. 11; spousi AYENB. 225; **spoused** (*pple.*) SPEC. 72; ispoused JUL.² 10.

    spūsing, sb., *espousal,* O. & N. 1336; HAV. 1164.

spūte, sb., = *M.Du.* spuite (*siphon*); *spout;* spoute PR. P. 470.

spúten, *see* **dispúten.**

spūten, v., *cf.* M.Du. spuiten, O.N. spȳta, M. H.G. spiutzen; *spout, vomit;* spoute ANGL. I. 81; PALL. i. 1097; **spouted** (*pple.*) CH. C. T. *B* 487; MAN. (F.) 8196.

sputten, v., ?=putten; *urge:* ich ... ham ... sputte (*pres.*) to more MARH. 14; þet flesch sput (*v. r.* put) ... touward swetnesse A. R. 196*; as þe feond sputte (*v. r.* spurede) (*pret.*) ham te don JUL. 58.

sqvachen, v., *O. Fr.* esqvacher; *squash;* sqvacche [swacche] PS. cvi. 6*; **sqvached** (*pple.*) LEG. 224.

sqvaimus, *see* **sqveimous.**

sqváre, adj., *O.Fr.* esqvarre; *square,* CH. C. T. *A* 1076; PALL. ii. 107; sqware '*qvadrus*' PR. P. 471; sware A. P. i. 837.

sqvárin, v., *O.Fr.* esqvarir, *cf. Ital.* squadrare; *square;* '*qvadro,*' PR. P. 471.

sqvames, sb., pl., *from Lat.* sqvama; *scales,* CH. C. T. *G* 759.

sqvatten, v., ?*O.Fr.* esqvatir; *squat;* sqwat-(te) PS. cix. 6; **sqvat** (*pple.*) WICL. 2 KINGS xxii. 8; L. H. R. 142; *comp.* **to-sqvatten.**

sqveimous [sqvaimous], adj., *cf. A.Fr.* escoimous; *squeamish, disdainful,* CH. C. T.

A 3337; sqveimous PR. P. 419; sqvaimus (*v. r.* scaimes, scoimes) TREV. VII. 461; scoimous A. P. ii. 21.

sqvēlen, v., *cf. Swed.* sqväla; *squeal;* **sqvē-lande** (*pple.*) C. M. 1344.

    sqvēling, sb., *squealing,* M. H. 167; sweling ALEX. (Sk.) 4112.

sqviari, sb., *O.Fr.* escuyerie; *company of esquires,* BARB. xx. 320.

sqvier, sb., *O.Fr.* esqvier, escvier; *squire, esquire,* CH. C. T. *A* 79; WICL. 1 KINGS xiv. 6; sqwier D. ARTH. 1179; scwier P. S. 219; swiere ALEX. (Sk.) 4507; **swíars** (*pl.*) ALEX. (Sk.) 1184.

sqville, sb., *Fr.* squille; *squill* (*herb*), PR. P. 471.

sqviler, sb., *O.Fr.* sculier; *scullion;* H. S. 5913; sqvillare, dische wescheare '*lixa*' PR. P. 471.

sqvinacie, sqvinanci, sb., *Fr.* esquinancie: *quinsy,* TREV. III. 335; sqvinacie PR. P. 471.

sqvint, ?=a sqvint, adv., *asquint,* A. R. 212.

sqvire, sb., *O.Fr.* esqvire, esqvierre (*équerre*), *cf. Ital.* squadra; *square,* FL. & BL. 325; C. M. 2231; **sqvires** (*pl.*) CH. AST. i. 12.

sqvīre, *see* **swire.**

sqvirel, sb., *O.Fr.* escurel; *squirrel,* S. S. (Web.) 2777; scurel VOC. 251; **sqvirelles,** sqverels (*pl.*) CH. D. BL. 431; P. F. 196.

sriden, *see* **schrūden. srūd,** *see* **schrūd.**

ssaft, *see* **schaft.**

stabbe, sb., *stab,* '*stigma,*' PR. P. 471.

stáble, sb., *O. Fr.* estable; *stable,* O. & N. 629; ROB. 280.

stáble², adj., *O.Fr.* estable; *stable, firm,* ROB. 54; AYENB. 83; JOS. 245; stabill YORK i. 62.

    stábel-(l)ī, adv. *firmly,* YORK xvii. 6, 140; stabilli BARB. xiii. 635.

stáblen, v., *O.Fr.* establer; *put in stable;* **stáblede** (*pret.*) CHR. E. 310; **istábled** (*pple.*) TREV. IV. 179.

stáblen², stablissen, v., *O.Fr.* establir, (-iss-); *establish,* LEG. 71; stablen [stablische] WICL. PS. xx. 12; stablisse LANGL. *B* i. 120*.

stac, sb., *O.N.* stakkr; *stack,* HAV. 814; stack '*acervus*' PR. P. 471; **stakkes** (*pl.*) MAN. (F.) 14690; *comp.* **corn-, hēi-stak.**

stacion, v., *O.Fr.* stacion; *station,* PR. P. 471.

stacionere, sb., *mod.Eng.* stationer; *bookseller,* '*bibliopola,*' PR. P. 471.

stackin, v., = *Dan.* stakke, *Swed.* stacka; *stack;* stakkin PR. P. 471; **stacke** (*imper.*) VOC. 154.

stad, sb., ?*position, status;* mannes stad SHOR. 158; in what stad his fader were S. S. (Wr.) 3325.

stad, *see* **steden.**

stæf, staf, sb., *O.E.* stæf,= *O.Fris.* stef, *M. Du.* staf, *O.N.* stafr, *O.H.G.* stab; *staff; letter, verse*; LAȝ. 8154, 22105; stef FRAG. 1; staf GEN. & EX. 3149; ROB. 126; SPEC. 48; CH. C. T. *A* 495; PARTEN. 6555; (*ms.* staff) ORM. 4312; **stáve** (*dat.*) LAȝ. 8152; O. & N. 1167; H. V. 85; steave A. R. 292; **stáfes** (*pl.*) ORM. 16391; staves ROB. 126; MAND. 247; steves AYENB. 156; **stáven** (*dat. pl.*) LAȝ. 21154; *comp.* **bōc-, candel-, cart-, di(s)-staf.**

stef-creft, sb., *O.E.* stæfcræft; *grammar*, HOM. I. 235.

[**Staf-ford,** pr. n., *O.E.* Stæfford; *Stafford*.]

Stafford-schire, pr. n., *O.E.* Stæffordscire; *Staffordshire*, MISC. 146; P. L. S. xiii. 26.

staf-ful, adj., *quite full*, GAW. 494.

stef-ifēȝ, sb., *O.E.* stæfgefēg; *alphabet*; **stefifēiȝe** (*dat.*) FRAG. 1.

staf-līc, adj., *O.E.* stæflīc; *literal*; staflike drinch ORM. 14478.

stef-liþere, sb., *O.E.* stæfliðer; *sling*, '(*f*)*undibulum*,' FRAG. 4.

staf-slinge, sb., *staff-sling*, C. M. 7528; CH. C. T. *B* 2019; **staffslingis** (*pl.*) BARB. xvii. 344.

stæffing, sb., *thrusting*; staffing BARB. xvii. 785.

stæir, adj., *cf. L.G.* stēger, *M.H.G.* steigel; *from* **stiȝen**: *steep*: so staire & so stepe ALEX. (Sk.) 4828; staire A. P. i. 1022.

stæire, sb., *O.E.* stæger,= *M.Du.* stēger *m.*; *from* **stiȝen**; *stair, step, ladder*; steire HOM. II. 165; A. R. 248; CH. COMPL. M. 129; steir L. H. R. 134; staire (*dat.*) LEG. 31.

stæiren, v., *ascend*; stairis (*pres.*) ALEX. (Sk.) 4834; stairand (*pple.*) ALEX. (Sk.) 3923.

stæle-wurðe, stǎlwurðe, adj., *O.E.* stǎlwyrðe, *?for* staðolwyrðe, *see* **staðelwurðe** *under* **staðel**; *stalwart, strong*; stelewurðe HOM. I. 25; stealewurðe JUL. 45; stalewurðe MARH. 15; staleworþe LAȝ. 3812*; ROB. 39; stalwurþe ANT. ARTH. lv; stalworþe HAV. 904; stalworþ WILL. 1950; TRIST. 90; APOL. 108; stalworth PR. C. 689; stalword ALEX. (Sk.) 3937.

stǎlward-hēde, sb., *courage*, ROB. (W.) 4337.

stǎlwurþ-līȝ, adj., *stalwart, strong*; (*ms.* stall-) ORM. 11947; stalworthli, stalwartli ALEX. (Sk.) 1149, 2625.

stēne, sb., *O.E.* stæna; *waterpot*; stene MISC. 85; ['*hydria*'] WICL. 3 KINGS xviii. 12; PALL. iv. 666; **stēnen** (*pl.*) LEG. 58; stenes ['*urnas*'] TREV. V. 165.

stēned, adj., *built with stones*; staned stren-this ALEX. (Sk.) 4352.

stēnen, adj., *O.E.* stænen,= *O.H.G.* steinīn, *Goth.* staineins; *from* **stān**; *of stone*; stonen

PR. P. 477; WICL. GEN. xxxv. 14; stænene [stonene] wal LAȝ. 9241; ine stonene þruh A. R. 378; stanene fetles ORM. 14029.

stēnen [2], v., *O.E.* stænan, *cf. O.H.G.* steinōn, *Goth.* stainjan; *stone*, JOHN x. 32; stene AYENB. 213; stonin PR. P. 477; stonie LANGL. *B* xii. 77; **stēnde** (*pret.*) REL. I. 144; stenede A. R. 122; **istēnet** (*pple.*) JUL. 40; staned, istaned ORM. 1968, 1978.

stēp, *see* **stēap. stæpe,** *see* **stápe.**

stærc, *see* **starc.**

[**stæðe ?**]

stæðe-lī, adj.,= *O.Fris.* stēdelīk, *? M.H.G.* stætelīch; *strong, mighty*: nes þer nan swa stæðeli þat lengore mihte stonden LAȝ. 1600.

staf, *see* **stæf.**

stáge, sb., *O.Fr.* estage; *stage*, FL. & BL. 255; ALIS. 7685; PARTEN. 4925; A. P. i. 410; **stáges** (*pl.*) AYENB. 122.

stágen, v., *erect, build*; stage MAN. (F.) 3090.

staien, v., *O.Fr.* estaier; *stay, remain*, PR. P. 473; staid (*pret.*) LIDG. M. P. 103.

stair, *see* **stēir.**

stáke, sb., *O.E.* staca,= *O.Fris.* stake; *from* stéken; *stake, 'sudes*,' PR. P. 471; HAV. 2830; WILL. 1723; CH. C. T. *A* 2552; FLOR. 1665; E. G. 362; to ane stake binde LAȝ. 16684; **stáke** (*pl.*) SPEC. 110; *comp.* **ále-stáke.**

stáken, v.,= *M.Du., M.L.G.* staken; *stake*; stake MAN. (F.) 1852.

stakerin, v., *O.N.* stakra, *cf. M.Du.* staggeren; *stagger, 'vacillo, titubo*,' PR. P. 471; stakir YORK xxx. 84; stakirs [staker] ALEX. (Sk.) 845; **stakereþ** (*pres.*) CH. L. G. W. 2687; **stakered** (*pret.*) MAN. (F.) 12377; stakerd TOWNL. 308.

stal, sb., *O.E.* steall,= *O.Fris., M.Du., O.H.G.* stal, *O.N.* stallr; *stall, station*, H. M. 5; H. S. 11874; E. G. 353; heo stal makeden LAȝ. 1671; þer tok he stal MAN. (F.) 14144; þu nevre mon . . . stal ne stode O. & N. 1632; steal HOM. I. 263; stel MISC. 99; þe bone . . . no stel ne schel him stonde SHOR. 28; **stalle** (*dat.*) REL. I. 219; ALIS. 1885; P. S. 188; MAN. (H.) 327; IW. 695; ANT. ARTH. xxxv; i stude & i stalle KATH. 682; oxe a stalle O. & N. 629; i stall þat heȝhest is in heofne ORM. 2145; **stalles** (*pl.*) LANGL. *B* xvi. 128; stallis *choir seats* ALEX. (Sk.) 4543.

stal-feht, sb., *stubborn fight*, LAȝ. 1841.

stále, stála, sb., *O.E.* stalu,= *O.H.G.* stala; *from* stélen; *theft*, HOM. I. 13, 39; stale P. L. S. viii. 128; AYENB. 9; bi stale *stealth* HOM. I. 249; **stále** (*pl.*) MAT. xv. 19; stales *conspiracies* YORK xxxi. 75.

stále [2], sb.,= *M.L.G.* stale, *M.Du.* stael; *stalk of a plant; rung of a ladder; handle*; A. P. i. 1001; PALL. xi. 194; a ladel . . . with a long stale

LANGL. *C* xxii. 279*; þeos two **stálen** (*pl.*) of þisse leddre A. R. 354; stales SHOR. 3.

stále [3], adj., *cf.M.Du.* stel; *stale,* '*defaecatus,*' VOC. 198; PR. P. 472; CH. C. T. *B* 1954; wiþ stale ale REL. I. 52.

stálin, v., '*defaeco,*' PR. P. 472.

stalke, sb., *diminut. of* **stále** [2]; *stalk, reed,* '*calamus,*' VOC. 210; PR. P. 472; LANGL. *C* xix. 39; R. R. 1701; P. R. L. P. 2; to climben bi þe ronges and þe **stalkes** (*pl.*) CH. C. T. *A* 3625.

stalkin, v., *cf. Dan.* stalke, *mod.Eng.* stalk; *go softly,* PR. P. 472; FER. 2389; stalke and crepe GOW. II. 351; **stalkeþ** (*pres.*) CH.C.T. *A* 1479; and stalke ye þeder ful stille LIDG. M. P. 112; **stalked** (*pret.*) TRIST. 2578; A. P. i. 152; stalkeden WILL. 2728.

stalking, sb., *stalking,* YORK xxx. 157.

stallen, v., *O.E.* steallian,=*M.H.G.* stallen; *place in a stall; locate;* stalle A. P. i. 188; þar holi soulen stalleþ SHOR. 91; stall (*imper.*) ALEX. (Sk.) 589; **stallid** (*pret.*) ALEX. (Sk.) 195; **stalled** (*pple.*) A. P. ii. 1334; stallid LIDG. M. P. 168; istalled CH. H. F. 1364; *comp.* **for-stallen.**

stalōn, sb., *O.Fr.* estalōn; *stallion,* VOC. 187; PALL. iv. 782; stalun P. L. S. xxxv. 167.

stalðe, sb., *? Swed.* stöld; *stealth,* GEN. & EX. 1767; stalþe ROB. 197.

stǎlworthe, *see* **stǽlewurðe.**

stam, *see* **stem.**

stameren, v.,=*M.L.G., M.Du.* stameren, *O. H.G.* stammalōn; *stammer,* REL. I. 65; **stamered** (*pret.*) AR. & MER. 2864.

stamin, *see* **stemin.**

stamin, sb., *O.Fr.* estamine; *linsey-woolsey,* A. R. 418; BEK. 2228; '*stamina*' PR. P. 472; CH. C. T. *I* 1052; TREV. VII. 307; (*v. r.* stames) CH. L. G. W. 2360.

stampin, v., *cf.M.L.G., M.Du.* stampen, *O.N.* stappa, *O.H.G.* stamfōn; *stamp, pound,* PR. P. 472; stampe WICL. Is. xiv. 23; **stampe** (*pres.*) CH. C. T. *C* 538; **stamped** (*pple.*) ALIS. 332.

stān, sb., *O.E.* stān,=*O.L.G., O.Fris.* stēn, *O.H.G.* stein, *O.N.* steinn, *Goth.* stains; *stone,* KATH. 1260; ORM. 9879; IW. 361; þene stan HOM. I. 141; stan [ston] LA3. 2315; ston GEN. & EX. 1604; þes deorewurðe ston A. R. 134; al so stille als a ston HAV. 928; þane ston SHOR. 124; stoon MAND. 90; WICL. JOHN xi. 31; **stānes** (*pl.*) *testes* SAX. CHR. 253; stanes [stones] LA3. 5694; stones O. & N. 1118; AYENB. 140; **stānen** (*dat. pl.*) MAT. iii. 9; LA3. 626; stane HOM. I. 9; þe rode stond in stone SPEC. 86; *comp.* **fīr-, 3im-, marmon-, milne-stān** (-stōn).

stōn-bowe, sb., '*arcuballista,*' PR. P. 477.

stōn-croppe, sb., *O.E.* stāncrop; *stonecrop* (*a herb*), '*crassula minor,*' PR. P. 477.

stān-cün, sb., *class of stone;* (*ms.* stæncun) LA3. 2847.

stān-deed, adj., *stone-dead,* TREV. I. 205; standed HAV. 1815; PARTEN. 115.

stān-grāf, sb., *stone-pit;* **stāngræfen** (*? ms.* graffen, *sec. text* greaves) (*dat. pl.*) LA3. 31881.

stōn-hard, adj.,=*Ger.* steinhart; *hard as stone,* L. H. R. 139; stonharde *fast* A. P. ii. 884.

Stān-henge, pr. n., *Stonehenge,* LA3. 15191; Stonhenge ROB. 149.

stōn-hēp, sb.,=*Ger.* steinhaufe; *stone heap,* WICL. 4 KINGS x. 8.

Stān-lē3, pr. n., *Stanley;* **Stānlēghe** (*dat.*) P. L. S. xvii. 387.

stān-marche, sb., *O.E.* stānmerce; *a herb,* '*macedonia, alexandria,*' PR. P. 472.

stōn-stille, adj., *stone-still,* A. R. 414; stane still MIN. ii. 32.

stān-wall, sb., *O.E.* stānweall; *stone-wall,* LA3. 15846.

stanc, sb., *O.Fr.* estanc; *pool, tank, reservoir,* A. P. ii. 1018; staunke ALEX. (Sk.) 3854, 3923; **stankes** (*pl.*) MAND. 209; stangis HAMP. PS. cvi. 35.

stanchen, v., *O. Fr.* estanchier; *stanch,* TREV. VII. 289; staunchen CH. BOET. iii. 3 (7); staunche GOW. I. 138; astaunche LIDG. M. P. 30; stonchi AYENB. 73.

stand, sb., *cf. M.Du.* stand, *M.H.G.* stant; *stand, position,* C. M. 1694.

standard, sb., *O.Fr.* estendard; *standard,* ROB. 303.

stande, sb., *O.E.* stand,=*M.Du.* stande, *O.H.G.* stanta; *from* **standen**; *cask;* stonde '*futis*' PR. P. 477; R. S. v; ['*cupa*'] PALL. i. 1051; **stondis** (*pl.*) WICL. JER. lii. 19.

standen, v., *O.E.* standan, stondan,=*O.L.G., Goth.* standan, *O.N.* standa, *O.Fris.* stonda, *O.H.G.* stantan; *stand,* ORM. 649; standen in 2149; stonden FRAG. 6; LA3. 25938; A. R. 22; GEN. & EX. 1607; HAV. 689; CH. C. T. *A* 88; stondin PR. P. 477; stande, stonde LANGL. *B* xvii. 54 (Wr. 11478); **stondest** (*pres.*) HOM. I. 35; stonst A. R. 236; mare eie stondeð men of monne HOM. I. 161; stondeþ C. L. 693; for þe me stondeþ þe more rape HORN (L.) 554; stondeð, stant HOM. II. 103, 220; stondeð, stont, stond, stant, stænt [stend] LA3. 107, 1397, 8214, 18850, 21321; stant GOW. II. 136; stont O. & N. 623; MAND. 73; hu stont ham A. R. 80; stont, stond SPEC. 35, 86; stont, stent P. L. S. viii. 10; stent MAT. xii. 25; HOM. I. 221; P. L. S. x. 10; FER. 1147; 3e stondeð KATH. 632; (heo) stondeð LA3. 9475; A. R. 366; (þu) **stonde** (*subj.*) ORM. 5008; þat he to dai stonde us bi MAN. (F.) 9223; **stondende** (*pple.*) GEN. & EX. 3149; stondinde A. R. 16; **stōd** (*pret.*) LA3. 1175; A. R. 352; ORM. 141; TOR. 241; te king(e) stod eie of him HOM. II. 139; al.Engelond of him stod awe HAV. 277; heo stód up MARH. 21; he stod up & seide ROB. 141; (þu) stode HOM. I. 195; O. & N. 1632; stoden GEN. & EX. 3543; GOW. III. 146; **standen** (*pple.*) L. H. R. 116; MAN. (H.) 191; A. P. i. 518; standen, stonden CH. C. T. *E* 1494 (Wr. 9368); stonden E. T. 322; *comp.* **a-, æt-, bi-, ȝe-, ümbe-, wið-standen**; *deriv.* **stand, stande.**

stānen, *see* **stǣnen. stang,** *see* **stanc.**

stange, sb., *cf. M.Du.* stange, *O.H.G.* stanga, *? O.N.* stöng; *from* **stingen**; *pole,* GAW. 1614.

stange[2], sb., *? from* **stingen**; *sting*: þe scorpion forbare is stange C. M. 693.

stangen, v., *O.N.* stanga; *sting, prick*; **stanged** (*pret.*) C. M. 12528; PR. C. 5293; L. H. R. 117.

   stanging, sb., *stinging*; stangine HAMP. PS. xlviii. 13.

stāni, adj., = *O.H.G.* steinīg, *Goth.* stainahs; *stony,* H. M. 17; M. H. 52; stoni '*lapidosus*' PR. P. 477.

stápe, sb., *O.E.* stæpe, stepe, = *M.Du.* stap, *O.H.G.* staph; *from* **stápen**; *step,* '*gradus*;' stæpe FRAG. 3; stape [steape] O. & N. 1592; stape ROB. 338; AYENB. 47; FER. 3989; **steapes** (*pl.*) HOM. I. 187.

   stápe-fole, adj., *high,* A. P. iii. 122.

stapel, sb., *O.E.* stapul, = *O.Fris.* stapul, stapel, *M.L.G., M.Du.* stapel, *O.H.G.* stafol, stafel; *staple; step;* FER. 2181; the staple to be reduced from Mirbonrach to Caleis TREV. VIII. 488; the marchauntes of the staple 571; thei borowed .. of the staple 582; ech stapel of his bed S. S. (Web.) 201; stapil, stapul PR. P. 472; C. M. 8288; schet wiþ a stapil and a clasp RICH. 4084; þe steire of fiftene **stoples** (*pl.*) REL. I. 130; (HOM. II. 165).

stápen, v., *? O.E.* stapan, = *O.L.G.* stapan, *? O. Fris.* stapa; *proceed, go*; stape FER. 5793; stepen HOM. I. 207; MARH. 15; **stápeth** (*pres.*) TREV. VII. 527; **stépe** (*subj.*) FL. & BL. 303; **stōp** [*O.E.* stōp, = *O.L.G., O.Fris.* stōp] (*pret.*) LAȝ. 23861; L. N. F. 213; stap ROB. 338; TRIST. 2865; step KATH. 714; P. L. S. xv. 82; ALIS. 4448; (þu) stepe JUL. 62; stopen MK. xv. 29; stopen, stepen LAȝ. 9235, 21035, 28408; **stápen** (*pple.*) CH. C. T. *E* 1514; *deriv.* **stápe, stapel, steppe, steppen.**

staplen, v., *O.E.* (under-)staplian, = *M.Du.* stapelen; *furnish with a staple*; **stapled** (*pple.*) GAW. 606.

star, sb., *O.N.* störr; *sedge,* '*carex*,' PR. P. 472; HAV. 939.

starc, adj., *O. E.* stearc, = *O.Fris.* sterk, *O.L.G.* starc, *O.N.* starkr, styrkr, *O.H.G.* starch; *stark, strong, severe,* MARH. 9; starc & hard ORM. 1472; starc and strong O. & N. 5; starc and stor 1473; stark '*rigidus*' PR. P. 472; HAV. 341; SPEC. 87; ALIS. 5527; CH. C. T. *E* 1458; MAN. (H.) 174; stærc, sterc [starc] LAȝ. 9197, 10905; sterc HOM. I. 5; sterk ne sterch ne kene MISC. 156, 157; S. S. (Web.) 2123; FER. 3241; þe sterke dom A. R. 144; **starkere** (*dat. f.*) LAȝ. 5568; **stærcne** (*acc. m.*) LAȝ. 21227; **sterke** (*adv.*) LAȝ. 16683; heoræ sceaftes weoren starke (*pl.*) LAȝ. 4228.

stark-blínd, adj., *stark blind,* TREV. III. 97.

stark-dēd, adj., JOS. 567; IW. 1880.

starc-líche, adv., *harshly,* KATH. 718; stærcliche LAȝ. 26973; starkli LUD. COV. 124.

starche, sb., *cf. M.H.G.* sterke; *starch,* PR. P. 472.

stáre, sb., *O.E.* stær, = *O.N.* stari, *O.H.G.* stara; *starling,* '*sturnus*,' HALLIW. 798; TREV. iv. 307; CH. P. F. 348; LIDG. M. P. 150; ster '*turdus*' FRAG. 3.

[**stáre**[2], sb.]

   stáre-blind, adj., *O.E.* stareblind, = *O.Fris.* stareblind, *O.H.G.* starablind; *? quite blind,* O. & N. 241.

stárin, v., *O.E.* starian, = *O.H.G.* staren, *O.N.* stara; *stare; glitter, gleam, shine;* '*patentibus oculis respicere, niteo,*' PR. P. 472; stare GOW. II. 63; CH. C. T. *B* 1887; A. P. i. 149; PR. C. 7426; **stárest** (*pres.*) O. & N. 77; stareþ R. S. v; **stárinde** (*pple.*) HAV. 508; staringe, starinde TREAT. 140 (SAINTS (Ld.) xlvi. 798); **stáred** (*pret.*) LANGL. *B* xvi. 168; *comp.* **bi-stáren.**

stark, *see* **starc.**

starken, v., *O.E.* stearcian, *cf. O.H.G.* starchēn, sterchen; *strengthen; become rigid,* P. R. L. P. 224; þe stormes **starked** (*pret.*) C. M. 1845.

starling, sb., *from* **stáre**; *starling,* '*sturnus*,' PR. P. 472; starling [sterling] C. M. 1789; sterling VOC. 188; SQ. L. DEG. 56.

start, *see under* **steort.**

stát, sb., *O.Fr.* estat; *state, estate,* ROB. 43; C. L. 1206; AYENB. 48; stat, æstat A. R. 178, 204; estat H. M. 13; E. G. 6; estat [astat] CH. C. T. *G* 1388; state YORK xxvi. 23; **astáz** (*pl.*) A. R. 160.

   státe-lī, adv., *in proper position,* YORK xxvi. 82.

statue, sb., *O.Fr.* statue; *statue,* A. P. ii. 995; CH. C. T. *B* 3349.

statūre, sb., *O.Fr.* stature; *stature,* PR. P. 472; PR. C. 4980.

statūt, sb., *O.Fr.* statut; *statute,* BEK. 697; GAW. 1060; **statūz** (*pl.*) P. S. 188.

stáþe, sb., *O.E.* stæð *n. m.* (?), *cf. O.Fris.* sted, *O.L.G.* stath, *O.H.G.* stad, stat, *Goth.* staþs *m.*; *landing-place, bank,* '*statio*.'

PR. P. 473; steþ *'ripa'* FRAG. 3; uppen
Sevarne staþe LA3. 7.

staðel, sb., *O.E.* staˠol,=*O.N.* stöðull, *N.L.
G.* stadel, *O.H.G.* stadal; *foundation, base,*
LA3. 15911; stathel VOC. 264; ·buldeð ower
boldes uppon treowe **staðele** (*pl.*) JUL. 72.

staðel-fest, adj., *stable,* KATH. 71; **staðel-
væste** (*acc. f.*) LA3. 9819; *comp.* **un-staþelfest.**

[**staðelfæsten,** v., *make stable; comp.* 3e-
staðelfæsten.]

staþelnesse, sb., *stability,* PS. lxviii. 3.

staðel-wurþe, adj., *firm,* A. R. 272*.

staðel², adj., *stable;* þurh **staðele** (*dat.*) . . .
3efe LA3. 401.

staðelien, v., *O.E.* staˠolian; *make stable,
establish;* **staðelede** (*pret.*) HOM. II. 127;
staþeled PS. viii. 4*; *comp.* 3e-**staðelien.**

steal, *see* stal.

stēam, sb., *O.E.* stēam,=*M.Du.* stoom; *steam;
flame;* steem *'vapor, flamma'* PR. P. 473;
stem H. S. 2526; a stem als it were a sunnebem HAV. 591.

stēap, adj., *O.E.* stēap,=*O.Fris.* stāp; *steep;
difficult; bright:* moni steap mon LA3. 1532;
stæp & heh ORM. 11379; þe paþ . . . was
narwe and step ALIS. 7041; þe strem step
AR. & MER. 7910; wiþ vois ful stepe RICH.
5985; **stēpne** (*acc. m.*) LA3. 19815; in uche
steppe i3e A. P. ii. 583; stepe staired stones
1396; **stēpe** [steepe] (*pl.*) CH. C. T. *A* 201;
ei3en stepe K. T. 15; ehnen **stĕappre** (*comp.*)
þene steorren MARH. 9.

stēpnesse, sb., *steepness,* '*elevacio,*' PR. P.
474.

steape, *see* stápe. **stecche,** *see* stücche.

stede, sb., *O.E.* stede, styde, *cf. O.N.* staðr,
Goth. staþs (*gen.* stadis) *m., O.H.G.* stat *m.,*
state *f., M.H.G.* stat, state, stete, *O.L.G.*
stad, stedi, *M.Du.* stad, stede, *O.Fris.*
sted *f.; stead, place;* (*ms.* stēde) ORM.
10101; (*ms.* stede) 13786; TREAT. 136;
ISUM. 511; PR. C. 5002; þane stede HOM. I.
221; he toc him on sunes stede GEN. &
EX. 2637; in stede MAND. 227; stande in
stede *stand in stead* P. P. 252; stonde in
stede MIRC 328; strengþe stont us in no
stide P. L. S. xxix. 90; stede, stude O. & N.
966, 1654; stude MARH. 9; we stude hatte
Camelford LA3. 28534; i þan stude (*there*)
he hine wolde slæn 6370; þene stude A. R.
6; in hore stude *instead of them* 426; i
stude H. M. 23; toward þan ilke stude MISC.
49; ure loverd heom onswerede an on ine þe
stude 39; he heveden a stude (*place, estate*)
þer bi side A. D. 259; in þe stude S. S. (Wr.)
1738; an oon in the stude LEG. 15; stede,
stude (*pl.*) O. & N. 590; studen A. R. 342;
stedes HAV. 1846; AYENB. 140; **stüden**
(*dat. pl.*) LA3. 10214; in menie stede P. L. S.
xii. 42; *comp.* **bed-stede,** *h*lēow-stüde.

stede-fæst, adj., *steadfast,* PROCL. 3; stede-
fast ORM. 1597; BEK. 38; SHOR. 159;
GOW. III. 115; stedevest AYENB. 84; stude-
vest A. R. 302; JUL. 75; stidefast LIDG. M.
P. 65; *comp.* **un-stedefast.**

stedfast-hēde, sb., *stability,* S. B. W. 286.

stedefæst-līche, adv., *steadfastly,* PROCL.
4; stedefastliche MISC. 28.

stedefastnesse, sb., *steadfastness,* '*stabili-
tas,*' PR. P. 473; MISC. 6; stidefastnesse CH.
BOET. iii. 11 (97).

stüdefast-schipe, sb., *stability,* C. L. 282.

stēde, sb., *O.E.* stēda; *steed, horse,* HAV.
1675; ALIS. 692; OCTAV. (H.) 718; PERC.
605; þene stede LA3. 6496; **stēden** (*dat.*)
LA3. 26796; **stēden** (*pl.*) ROB. 185; CHR.
E. 621; SPEC. 48; A. D. 236; AR. & MER.
3372.

steden, v., *cf. M.L.G., M.Du.* steden, *O.N.*
steðja (*pple.* staddr); *stand, place:* i schall
nott stedde in no stede YORK xlvi. 94; ne
stüdeð (*pres.*) (*avails*) h(e)om nawiht HOM.
I. 77; **stedd** (*imper.*) ALEX. (Sk.) 3977;
stedid (*pple.*) P. 13; sted TOWNL. 25; o
mang þi fas her sted ertou IW. 2987; stad
JOS. 397; GAW. 644; in sorowe was he stad
FLOR. 1313; *comp.* **bi-, wið-steden.**

stedi, adj., *O.E.* stedig,=*M.Du.* stedigh, *M.
L.G.* stedich, *M.H.G.* stetic; *steady, stable:*
stunt & stidi3 ORM. 9885.

stee, *see* stī3e. **steem,** *see* stefne.

stef, *see* stæf.

stēf, adj., *cf. O.Fris.* stēf, *Swed.* stȳf; ?=**stīf;**
rigid, strong, WILL. 2984; strong and stef
GREG. 574; a stef strem WICL. PROV. xviii.
4; stef on stede TRIST. 3079; **stēvere** (*compar.*) WICL. DEUT. ix. 14.

stēf-hēde, sb., *rigour,* AYENB. 263.

stēf-līche, adj., *rigidly,* AYENB. 258.

stefne, sb., *O.E.* stefn, stemn,=*O.N.* stefna,
Goth. stibna, *O.Fris., O.L.G.* stemna, *O.H.G.*
stimna; *voice; constitution, appointment;*
MAT. ii. 18; FRAG. 1; HOM. II. 43; MARH.
21; A. R. 82; ORM. 10680; stefne, stevene
KATH. 717; stefne, stevne, stevene O. & N.
317, 522, 898; stefne, stefene, stevene [stemne]
LA3. 11994, 19525, 20790; stevene GEN. &
EX. 622; HAV. 1275; SHOR. 128; GOW. II.
30, 253; H. S. 10062; ISUM. 36; PR. C. 4559;
wiþ merie stevene CH. C. T. *A* 2562; þei
setten stevene 4383; bi her bothe assent was
set a steven CH. COMP. M. 52; stevine
TRIAM. 12; over gestes it has þe steem
(*preeminence*) MAN. (F.) 98; **stefne** (*gen.
pl.*) MARH. 20.

stefnen, v., *O.E.* ·(ge-)stefnian, (a-)stemnian,
= *O.N.* stefna, stemna; *give voice for,
appoint;* stevene P. R. L. P. 113; TRIST. 2937;
stémes (*pres.*) ALEX. (Sk.) 2960; þan stemes
he with *then he considers with himself* ALEX.

(Sk.) 5301 ; **stemmed** (*pret.*) *resolved* GAW.
230 ; *comp.* ʒe-stefnen.

stevening, sb., *shouting, appointment,* SPEC.
46 ; stevening YORK xxxii. 6.

stefneten, v., *? stop* ; hwi . . . **steven(e)tiǒ**
(*pres.*) se stille KATH. (E.) 1265.

steiin, v.,= *M.Du.* steghen, *M.H.G.* stegen ;
from **stiʒen** ; *climb,* '*scando,*' PR. P. 473 ;
steied (*pret.*) HALLIW. 804.

steike, sb., *O.N.* steik ; *steak,* '*carbonella,*
frixa, assa,' PR. P. 473 ; **stēkis** (*pl.*) of venson
or bef C. B. 40.

steil, sb., *cf. O.H.G.* stiagil ; *from* **stiʒen** ; *step,*
'*gradus,*' PR. P. 473.

steinen, *see* **disteinen.** **stēire,** *see* **stǣire.**

stéken, v.,= *O.L.G.* stecan (*pret.* stac), *O.H.G.*
stechan (*pret.* stach) ; *prick, fix, fasten ; close* ;
A. R. 62 ; let hem steken in an hus ORM.
8087 ; steke H. S. 11225 ; þe dore to steke
faste BEK. 683 ; **stékis** (*pres.*) ALEX. (Sk.)
5485 ; steke 2139 ; **stak** (*pret.*) R. R. 458 ;
Iw. 699 ; steken A. P. ii. 884 ; **stéke** (*pple.*)
S. S. (Wr.) 1360 ; RICH. 4282 ; isteke TRIST.
2999 ; stiken wiþ þe spere JOS. 273 ; stoken
ALIS. 1132 ; IW. 695 ; GAW. 782 ; stoke GOW.
I. 60 ; **istékene** (*pl.*) A. R. 50 ; *comp.* **bi-,**
to-stéken ; *deriv.* **stáke.**

stekien, *see* **stikien.** **stel,** *see* **stal.**

stēl, sb., *O.E.* stȳle, *cf. O.N.* stāl, *M.Du.* stael,
O.H.G. stāl, stahal ; *steel,* LAʒ. 25879 ;
A. R. 160 ; C. L. 1254 ; SHOR. 132 ; steel
MAND. 190 ; CH. C. T. *E* 2426 ; stiel TRIST.
3324 ; mid **stēles** (*gen.*) egge LAʒ. 9799 ; **stile**
(*dat.*) FER. 4433 ; PARTEN. 2259.

stēl-boʒe, sb. : ? Arthur stop a stel boʒe and
leop an his blancke LAʒ. 23899.

stēl-gere, sb., *armour,* GAW. 260.

stéle, sb., *O.E.* stel,= *M.L.G.* stel, stil, *M.Du.*
stele, *O.H.G.* stil *m. ; stake, stalk, handle ;*
'*ansa,*' PR. P. 473 ; stele ['*ligno*'], stile PALL.
iii. 770, xii. 77 ; **stéle** (*dat.*) CH. C. T. *A* 3785 ;
a ladel . . . with a longe stele LANGL. *B* xix.
274 ; *comp.* .**ráke-stele.**

stélen, v., *O.E.* stelan,= *O.L.G., O.H.G.* stelan, *O.Fris., O.N.* stela, *Goth.* stilan (κλέπτειν) ;
steal, hide, GEN. & EX. 1035 ; stelen swa we
wolden LAʒ. 736 ; steolen (*go stealthily*) ut
of hirede 2353 ; stelen [stele] CH. C. T. *A*
562 ; stele AYENB. 79 ; **stéle** (*pres. subj.*)
ORM. 4467 ; GEN. & EX. 3511 ; **stal** (*pret.*)
LANGL. *B* xiii. 367 ; CH. C. T. *A* 3995 ; LIDG.
M. P. 255 ; S. & C. I. lxxii ; Marie . . . stal a
wei fram hire kunne S. A. L. 160 ; stel ROB.
564 ; þu stele O. & N. 105 ; (heo) stelen P. L.
S. viii. 81 ; stolen LANGL. *B* xix. 151 ; **stólen**
(*pple.*) GEN. & EX. 1748 ; istolen HOM. I. 31 ;
istole BEK. 810 ; *comp.* **bi-, for-stélen** ;
deriv. **stále, stalǒe.**

stēlen, adj., *O.E.* stȳlen, = *M.L.G.* stēlen ; *made*
of steel, ALIS. 1980 ; WILL. 3535 ; stilen
PARTEN. 256 ; þe stelene brond LAʒ. 7634.

stēlen [2], v., *O.E.* stȳlan,= *O.N.* stǣla ; *steel,*
sharpen ; provide with steel ; istēled (*pple.*)
C. L. 1248 ; þet istelede irn JUL. 59.

stellen, v., *O.E.* stellan,= *M.L.G., M.Du.* stellen, *O.H.G.* stellan ; *from* **stal** ; *set, establish* ;
hire nome . . . þe me ærst hire on **stálde**
(*pret.*) LAʒ. 7132 ; stolde MISC. 92 ; riwle þet
mon stolde A. R. 8 ; **istáld** (*pple.*) A. R. 6 ; H. M.
19 ; *comp.* **a-stellen.**

stem, sb., *O.E.* stemn, stefn, stæln,= *O.N.*
stafn, *O.L.G.* stamn, *O.H.G.* stam ; *stem,*
MAN. (H.) 296 ; stam *foreship* D. ARTH.
3664 ; one the **stamine** (*dat.*) D. ARTH. 3658 ;
on stamin ho stod A. P. ii. 486.

stēm, *see* **stēam.**

stēmin, v., *O.E.* stēman, stȳman ; *from* **stēam** ;
steam ; blaze ; '*flammo,*' PR. P. 473 ; eien . . .
þat **stēmed** (*pret.*) as a forneis of a leed CH. C.
T. *A* 202.

stemmen, v., = *M.H.G.* stemmen, *? O.N.*
stemma ; *stop* ; **stemme** (*pres. subj.*) A. P. ii.
905.

stemne, *see* **stefne.**

stench, sb., *O.E.* stenc,= *O.L.G.* stanc, *O.H.*
G. stanc, stanch ; *from* **stinken** ; *stench,*
smell, odour, MARH. 11 ; A. R. 216 ; MISC.
63 ; BRD. 23 ; AYENB. 248 ; **stenche** (*dat.*)
HOM. II. 167 ; R. S. v (MISC. 174) ; TREV. V.
99 ; **stenches** (*pl.*) HOM. I. 193.

stenchen, v., *O.E.* stencan ; *annoy with stench* :
þe smoke þere of hem ssolde boþe stenche &
blende ROB. 407 ; *comp.* **a-stenchen.**

stēnen, *see* **stǣnen.**

steng, sb., *O.E.* steng *m.,*= *M.Du.* stenge *f. ;*
from **stingen** ; *? crowbar,* '*vectis,*' FRAG. 4.

stente, sb., *O.Fr.* estente ; *valuation,* '*taxacio,*'
PR. P. 474 ; **stentes** (*pl.*) ROB. (W.) 7678*.

stenten, *see* **stünten.**

steolen, *see* **stélen.** **stēon,** *see* **stiʒen.**

[stēop ?]

stēop-bern, sb., *O.E.* stēopbearn,= *O.N.*
stiūpbarn ; *step-child ; orphan* ; HOM. I. 115.

stēop-brōthir, sb.,= *O.H.G.* stiefbruoder ;
step-brother ; stepbrothir PR. P. 474.

stēop-child, sb., *O.E.* stēopcild ; *step-child* ;
stēpchild (*pl.*) JOHN xiv. 18 ; stepchildre
PS. xciii. 6.

stēop-doghter, sb., *O.E.* stēopdohtor,=
O.N. stiūpdōttir, *O.H.G.* stieftohter ; *step-*
daughter ; stepdoghter VOC. 205 ; stepdouʒter
['*privignam*'] TREV. V. 103.

stēop-fader, sb., *O.E.* stēopfæder,= *O.Fris.*
stiap-, stiepfeder, *O.N.* stiūpfaᵕir, *O.H.G.*
stiuf-, stiof-, stieffater ; *step-father* ; stepfader
TREV. I. 93 ; M. H. 123.

stēop-mōder, sb., *O.E.* stēopmōder,= *O.N.*
stiupmōᵕir, *O.H.G.* stiufmuoter ; *step-mother,*
LAʒ. 14421 ; stepmoder WILL. 2640 ; step-
moodir LIDG. M. P. 219 ; stipmoder P. L. S.
xii. 88.

stēop-sune, sb., *O.E.* stēopsunu, = *O.N.* stiūpsonr, *O.H.G.* stiufsun ; *step-son* ; stepsune LAȝ. 32138 ; stepsone '*privignus*' PR. P. 474 ; ROB. 61.

stēop-sistir, sb., = *O.H.G.* stiefswester ; *step-sister* ; stepsister PR. P. 474.

stēor, sb., *O.E.* -stēor, = *O.H.G.* stior, *Goth.* stiur ; *steer, young ox* ; ster [steer] CH. C. T. *A* 2149.

stēor², sb., *O.E.* stēor, *cf. M.Du.* stier, stuir, stuer, *M.H.G.* stiur, *O.N.* stȳri *n.; helm, rudder* ; ster ORM. 15258 ; mi seil and ek mi stere [steere] CH. C. T. *B* 833 ; steere WICL. PROV. xxiii. 34 ; **stēores** (*gen.*) man ORM. 2135 ; steores mon [steres man] LAȝ. 11985 ; **stēre** (*dat.*) P. L. S. xxxv. 154 ; stere [steere, stiere] LANGL. *B* viii. 35 ; þu most to stere (*boat, by synecdoche*) HORN (L.) 101.

stēr-bord, sb., *O.E.* stēorbord, = *M.Du.* stier-, stuirbord ; *starboard* ; **stērburde** (*ms.* stere-) (*dat.*) D. ARTH. 3665.

stēr-lēs, adj., *rudderless*, CH. C. T. *B* 439.

stēor-(mo)n, sb., *O.E.* stēormann, = *M.H.G.* stiurman ; *steersman, 'gubernator,'* FRAG. 2 ; sterman VOC. 274 ; TOWNL. 31 ; **stēormen** (*pl.*) LAȝ. 28436.

stēor-scofle, sb., *O.E.* stēorsceofol ; *rudder*, FRAG. 2.

steer-staf, sb., *tiller*, WICL. PROV. xxiii. 34.

stēre-trē, sb., *tiller*, TOWNL. 31.

stēore, sb., *O.E.* stēor, stȳr, = *O.H.G.* stiura *f.; government, discipline*, HOM. I. 117 ; stere GEN. & EX. 3418.

stēor-lēas, adj., *O.E.* stēorlēas ; *without discipline*, HOM. I. 117.

stēore², sb., *O.E.* stēora, = *O.H.G.* stiuro, *O.N.* stiōri *m.; governor* ; stere GEN. & EX. 3420 ; he þat is lord of fortune be þi stere [steere] CH. C. T. *B* 448.

stēoren, v., *O.E.* stēoran, stīeran, stȳran, *cf. O.H.G.* stiuran, *O.Fris.* stiura, stiora, *O.N.* stȳra, *Goth.* stiurjan (ἱστάναι) ; *steer, govern ; guide* ; ORM. 1559 ; stere FLOR. 825 ; PS. ii. 9 ; TRIST. 2571 ; A. P. iii. 27 ; þu **stēorest** (*pres.*) te sea stream þet hit fleden ne mot fir þan þu markedest MARH. 10 ; stereth LANGL. *B* viii. 47 ; stureth *A* ix. 42 ; **stēor** (*imper.*) me ant streng me JUL. 33 ; þin herte nu þu stere (*govern, restrain*) HORN (L.) 434 ; Maxence **stēorede** (*pret.*) þe refschipe KATH. 11 ; steered P. P. 216 ; *comp.* ȝe-**stēoren**.

stēringe, sb., *steering*, MISC. 18.

stēoren, *see* **stēren**.

steorne, sb., *O.E.* steorn, *O.N.* stiorn, = *O.Fris.* stiorne, stiarne, *mod.Eng.* stern ; *rudder (of a ship)* ; LANGL. *A* ix. 30 ; sterne CH. H. F. 437 ; M. T. 129 ; þe helme & þe sterne A. P. iii. 149 ; steerne WICL. PROV. xxiii. 34.

steorne, *see* **steorre**.

steorre, **steorne**, sb., *O.E.* steorra *m.*, *O.N.* stiarna *f.*, *cf. O.L.G.* sterro, *O.H.G.* sterro,

ster 10 *m.*, *Goth.* stairnō *f.; star* : steorre KATH. 1663 ; þanne steorre MAT. ii. 10 ; þe steorre LAȝ. 17870 ; ænne ... sterre [st(e)orre] 17865 ; sterre PR. P. 474 ; GEN. & EX. 132 ; SHOR. 123 ; LANGL. *B* xviii. 231 ; CH. C. T. *A* 2061 ; MAND. 70 ; si sterre MISC. 26 ; steorne ORM. 3464 ; sterne C. M. 11391 ; MAN. (F.) 9031 ; M. H. 95 ; þat **steorne** (*gen.*) leom ORM. 6536 ; **steorran** (*pl.*) MAT. xxiv. 29 ; steorren HOM. I. 143 ; MARH. 9 ; ALIS. 134 ; steorrene SAINTS (Ld.) xlvi. 402 (sterren TREAT. 132) ; sterren AYENB. 267 ; steorre O. & N. 1329 ; **sterren** [storre] (*dat. pl.*) LAȝ. 9127 ; *comp.* dai-, ēven-, lōde-, morwen-sterre.

st(e)orre-wīs, adj., *learned in the stars*, O. & N. 1318.

ster-wort, sb., *starwort*, LEECHD. III. 345.

steorred, **steorned**, pple., *cf. O.H.G.* gestirnōt ; *starred* ; sterned PR. C. 993 ; þe stirrede bur MARH. 22.

steorri, adj., *starry* ; sterri CH. BOET. ii. 2 (36).

steort, sb., *O.E.* steort, = *O.Fris.* stert, stirt, *O.H.G.* sterz ; *tail ; plough handle ; stalk* ; stert PR. P. 474 ; REL. I. 208 ; HAV. 2823 ; **stortes** (*pl.*) ['*acutis surculis*'] PALL. iv. 387 ; *comp.* plōu-stert.

start-blīnd, adj., *quite blind*, TREV. VI. 235 ; start- [*v. r.* streiȝt-, stark-]blind III. 97.

steort-nāket, adj., *quite naked*, MARH. 5 ; JUL. 16, 17 ; steort-, stertnaked (*printed* steorc-, stercnaked) A. R. 148, 260 ; start-naked C. L. 431 ; (*printed* starcnaked) P. S. 336.

steorve, sb., *O.E.* steorfa, = *O.H.G.* sterbo ; *pestilence*, MARH. 12 ; JUL. 49 ; steorfa HOM. I. 13.

steorven, v., *O.E.* steorfan, = *O.Fris.* sterva, *M.L.G.* sterven, *O.H.G.* sterban, *mod.Eng.* starve ; *die, perish* ; steorve S. A. L. 153 ; sterven GOW. II. 28 ; CH. C. T. *E* 1991 ; stervin PR. P. 474 ; sterve AYENB. 70 ; LIDG. M. P. 32 ; PALL. i. 931 ; **steorveð** (*pres.*) A. R. 222 ; sterveþ S. S. (Web.) 348 ; sterft MISC. 31 ; sterve (*subj.*) LANGL. *B* xi. 422 ; sterven HOM. I. 71 ; **starf** (*pret.*) GEN. & EX. 481 ; ALIS. 579 ; CH. C. T. *A* 933 ; LIDG. M. P. 149 ; sterf A. R. 360 ; (þou) storve REL. II. 275 ; sturven SAX. CHR. 262 ; sturfe HOM. I. 233 ; storven GEN. & EX. 2975 ; ALIS. 5082 ; storven, storve AYENB. 12, 67 ; CH. C. T. *C* 888 ; **storve** (*subj.*) GEN. & EX. 1958 ; storven (*pple.*) GEN. & EX. 3162 ; S. S. (Web.) 1126 ; storve WILL. 1515 ; istorven A. R. 308 ; istorve CH. C. T. *A* 2014 ; þe **storvene** (*sb.*) KATH. 1043 ; *comp.* a-steorven ; *deriv.* steorve, sterven.

stervinge, sb., = *M.H.G.* sterbunge ; *dying* AYENB. 95.

stēowien (*? for* stēawien), v., *cf. M.L.G., M. Du.* stouwen, *? M.H.G.* stouwen, stöuwen ;

coerce, restrain: he sette stronge lawen to steo-
wien [stewe] his folk LAȝ. 6266; steowi MISC.
193; stew þine unwittie wordes MARH. 6; stou
LUD. COV. 217; **stēwede** (*pret.*) D. ARTH.
1489; **stēwed**, stowed,stouwet (*pple.*) LANGL.
A v. 39, *C* vi. 146 (*B* v. 48 ruled); istewet
KATH. 657; *comp.* **wiŏ-stēwen.**

stĕp, *see* stēap, stēop.

stĕpel, sb., *O.E.* stĕpel, stȳpel; *steeple, high
tower*, ROB. 528; LAI LE FR. 152; anne stepel
HOM. I. 93; stepul VOC. 193; stipul FLOR.
1887; **stēples** (*pl.*) AYENB. 23.

stĕpen, *see* stápen.

stĕpin, v., *O.N.* steypa (*pour*); *steep, soak,
'infundo,'* PR. P. 474; stepe PALL. ii. 281;
stēped (*pple.*) L. C. C. 6.

steppe, sb.,=*M.Du.* steppe; *from* stápen;
step, PR. P. 474; **steppes** (*pl.*) C. L. 740;
TREV. IV. 153; MAND. 81; IW. 2889; *comp.*
fōt-steppe.

steppen, v., *O.E.* steppan, stæppan,=*O.Fris.*
steppa, *O.H.G.* steffan; *step, go*, LAȝ. 18420;
steppin PR. P. 474; steppe ROB. 336; LANGL.
B v. 352; **steppeŏ** (*pres.*) REL. I. 208; **stapte**
(*pret.*) OCTOV. (W.) 1435.

ster, *see* stáre.   **stēr**, *see* stēor.

sterc, *see* starc.

stĕre, adj., ? =*O.H.G.* stiuri; *? strong, steadfast*, HORN (L.) 1344; GUY² (Z.) 662.

   stĕr-līche, adv., *? strongly; harshly*; (*v. r.*
sterneliche) CH. TRO. iii. 677; swa hwet swa
þe laverd speke to his men sterliche HOM.
I. 111.

stĕre, *see* stēore.

stĕren, v., *O.E.* stēran; *from* stōr; *offer
incense*: me schal ham steoren mid guldene
chelle HOM. I. 193.

stĕren², v., *O.E.* stēran,=*O.Fris.* (to-)stēra,
M.H.G. stœren; *move*: ne stēreŏ (*imper.*)
ge nogt of ŏe stede REL. I. 217 (MISC. 13);
comp. **a-stēren.**

stĕren, *see* stēoren.   **sterien**, *see* stūrien.

sterk, *see* starc.

sterling, sb., *cf. M.H.G.* sterlinc; *sterling
penny*, LANGL. *B* xv. 342; **sterlinges** (*pl.*)
ROB. 294; CH. C. T. *C* 907; an hondred
schillinges sterlinges TREV. VIII. 167.

sterling, *see* starling.

sterne, *see* steorre, stūrne.

sterne, *see* steorne.   **sterre**, *see* steorre.

stert, sb., *start; moment*, HAV. 1873; CH. C.
T. *A* 1705; C. M. 14298; stirt '*saltus, momentum*' PR. P. 476.

   stert-hwūle, sb., *moment*, A. R. 336.

stert, *see* steort.

sterten, v., *O.N.* sterta,=*M.H.G.* sterzen;
start, leap; sterte CH. C. T. *A* 1044; stirtin
PR. P. 476; stirte MAN. (F.) 5027; bulluc ster-
teþ (*pres.*) A.S. iv; **sterte** (*pret.*) WILL. 2277;
S. S. (Web.) 1472; DEGR. 1205; M. ARTH.

2989; heil maister he seid(e) and to him
sterte MED. 421; sturte ROB. 212; sturte
[storte] LAȝ. 23951; stirte HAV. 566; AR.
& MER. 3338; Pharaon stirte up GEN. & EX.
2931; stirte forth HAV. 873; *comp.* **a-, æt-,
mis-sterten.**

stertil, adj., *hasty*; stirtil or hasti, '*praeceps*'
PR. P. 476.

stertlen, v., *rush, stumble along*; **stertlinde**
(*pple.*) MAP 335; a courser stertling as the fir
CH. L. G. W. 1204.

[**sterven**, v., *O.E.* (a-)sterfan, styrfan,=*M.L.
G.* sterben, *M.H.G.* sterben; *from* **steorven**;
put to death; comp. **a-sterven.**]

sterven, *see* steorven.

stĕten, v., *cf. O.Fris.* stēta, *O.N.* steyta; *beat*;
stĕtten (*pret.*) AR. & MER. 3322, 3817.

stevene, *see* stefne.   **stewe**, *see* stūve.

stĕwe, sb., ?=*M.L.G.* stouwe; *fish-pond*, PALL.
i. 769; stewe, stue '*vivarium*' PR. P. 481;
stēwe (*dat.*) CH. C. T. *A* 350; stiewe M. T. 119.

stĕwien, *see* stēowien.

stī, *see* stiȝ, stiȝe.

stiburn, stoburn, adj.. *stubborn, 'austerus,'*
PR. P. 475; stiburn LIDG. M. P. 168; stiborn
CH. C. T. *D* 456.

   stiburnesse, sb., *stubbornness*, PR. P. 475.

sticche, *see* stücche.

sticchen, v.,=*M.L.G., M.Du.* sticken, *O.H.G.*
sticchan, stecchan;=**stikien**; *stitch, stick*;
stichin, stikkin up '*succingo*' PR. P. 475;
sticheŏ (*pres.*) H. M. 9; **stiȝte** (*pret.*) HOM.
II. 257; **stiȝt** (*pple.*) WILL. 4425; istiht (*ms.*
istihd) A. R. 424; *comp.* **þurh-sticchen.**

stiche, sb., *O.E.* stice,=*O.Fris.* steke, *Goth.*
stiks, *O.H.G.* stich *m.; stitch*, PR. P. 475; A.
R. 110; H. M. 35; stik(e) IW. 3053; **stiches**
(*pl.*) PL. CR. 553.

   stich-wurt, sb., *stitch-wort*, REL. I. 37.

sticke, sb., *O.E.* sticca,=*M.L.G.* sticke, *O.H.
G.* steccho, stecko *m.; stick, fragment*; stike
[steke] ALEX. (Sk.) 1311; FRAG. 3; O. & N.
1625; enne sticke A. R. 370; stikke PR. P.
475; LANGL. *B* xii. 14; CH. C. T. *G* 1265;
stikkes (*pl.*) ORM. 8651; *comp.* **candel-,
fīr-, reȝol-sticke.**

stide, stidiȝ, *see* stede, stedi.

stīe, stīen, *see* stiȝe, stiȝen.

stif, adj., *O.E.*,=*M.L.G.* stif, *M.Du.* stijf;
stiff, strong, valiant; 'rigidus, fortis, robustus,'
PR. P. 475; FRAG. 5; BEK. 637; WILL. 3535;
OCTOV. (W.) 1041; IW. 31; GAW. 107; stif [stiif]
C. M. 18140; stif he wes on þonke LAȝ. 2110;
stif and starc O. & N. 5; þe wind was strong
and stif inouȝ BRD. 21; stif hit stod up riht
A. D. 254; stif and toght CH. C. T. *D* 2267;
a stif spere ALIS. 2745; stiffe ALEX. (Sk.)
1149, 4758; stive here HOM. II. 139; **stive**
(*pl.*) LEG. 92; **stīvest** (*superl.*) ST. COD.
58; LANGL. *B* xiii. 294; þe stiffeste P. S. 337.

stīf-līche, adv., *valiantly, strongly*, BEK. 171 ; stifli Iw. 3186.

stīfnesse, sb., *stiffness*, PR. P. 475.

stīȝ, sb., *O.E.* stīg, = *O.H.G.* stīg, *O.N.* stīgr *m.; from* stīȝen ; *path, way* ; stih ORM. 12916 ; sti '*semita*' PR. P. 475 ; REL. I. 213 ; GEN. & EX. 3958 ; HAV. 2618 ; S. S. (Web.) 712 ; WILL. 212 ; WICL. JOB xvi. 23 ; IW. 599 ; HORN CH. 172 ; ISUM. 40 ; **stīȝe** (*dat.*) A. P. iii. 402 ; **stīȝes** (*pl.*) HOM. I. 7 ; rihteþ . . . drihtines narwe stiȝhes ORM. 9202 ; stihes ['*semitas*'] PS. viii. 9 ; sties MAN. (F.) 8432 ; **stīȝen** (*dat. pl.*) LAȝ. 16366 ; *comp.* **up-stīȝ**. [**stīȝe** ? *from* **stīȝen**.]

stī-rōp, sb., *O.E.* stigrāp, = *O.N.* stigreip, *M.L.G.* stegerēp, *O.H.G.* stegereif ; *stirrup*, FRAG. 6 ; PR. P. 476 ; HORN (L.) 758 ; BEK. 190 ; AR. & MER. 3260 ; CH. C. T. D 1665 ; LAUNF. 977 ; GAW. 2060 ; **steropes** (*pl.*) ALEX. (Sk.) 840* ; steropes 792*.

stīȝe, sb., *O.E.* stīg, = *O.H.G.* stīga *M.Du.* stijghe *f., prov.Eng.* stee ; *ladder; rising, ascent; sty* ; sti S. S. (Web.) 3295 ; stie '*hara*' VOC. 178 ; L. H. R. 215 ; sti PR. P. 475 ; A. R. 128 ; CH. C. T. D 1829 ; sti an ye (*stye in the eye*) PR. P. 475 ; stee [stegh] ALEX. (Sk.) 2481 ; **stīes** (*pl.*) ALEX. 1437 ; *comp.* **swīn-stī**.

stī-ward, sb., *O.E.* stīweard, = *O.N.* stīvarðr ; *steward*, PR. P. 476 ; A. R. 386 ; GEN. & EX. 1991 ; HAV. 666 ; MAND. 236 ; S. S. (Wr.) 1508 ; DEGR. 1577 ; stiward, -wærd LAȝ. 1475, 13571.

stīȝele, sb., *O.E.* stigel, = *M.L.G.* stegel, *M.H.G.* stigele *f.; stile* ; stile '*scansile*' PR. P. 475 ; CH. C. T. C 1772 ; MIN. i. 88.

stīȝen, v., *O.E.* stīgan, = *O.L.G.*, *O.H.G.* stīgan, *O.Fris.*, *O.N.* stīga, *Goth.* steigan, *Gr.* στείχειν ; *rise, ascend*, HOM. I. 149 ; LEB. JES. 206 ; stiȝhen ORM. 2753 ; stihen KATH. 1012 ; stien A. R. 40 ; CH. BOET. iii. 9 (88) ; stige HOM. II. 111 ; stiȝe P. R. L. P. 200 ; stiȝe [stie] TREV. IV. 411 ; stie MAND. 134 ; steiȝen C. L. 164 ; steȝe, sti (*r. w.* be) SHOR. 126 ; þer þu wenest heȝest to steo R. S. i (MISC. 158) ; sten REL. I. 49 ; stihð (*pres.*) A. R. 216 ; stiȝþ O. & N. 1405 ; þe stench þet of þi muð stiheð MARH. 13 ; steþ C. L. 1490 ; stighen PS. xxi. 30 ; stig (*imper.*) GEN. & EX. 4100 ; stāh [*O.E.* stāg, stāh] (*pret.*) ORM. 16700 ; steaȝ AYENB. 13 ; steah KATH. 338 ; steah, steh HOM. II. 3, 23 ; steg HOM. II. 111 ; GEN. & EX. 319 ; steȝ P. L. S. xx. 94 ; stegh MIRC 518 ; steegh ALIS. 5827 ; steh MARH. 1 ; steih A. R. 250 ; MISC. 88 ; steiȝ PL. CR. 810 ; steiȝ [stiȝ] TREV. III. 125 ; steigh MAND. 96 ; stei ROB. 322 ; PR. C. 4603 ; (þu) stihe, stuhe JUL. 62, 63 ; stehe SPEC. 69 ; (heo) stiȝen LAȝ. 26005 ; stihen MARH. 10 ; stiȝe A. P. ii. 389 ; stiȝhen (*pple.*) ORM. 2783 ; stigen GEN. & EX. 4130 ; stoȝen FER. 5027 ; *comp.* **a-, ȝe-, over-stīȝen** ; *deriv.* **stīȝ**, **stīȝe**, **stīȝe**, **stīȝele**, **steil**, **stǣire**.

stēghere, sb., *climber*, HAMP. PS. *pages* 506, 509.

stēhinge, sb., *mounting*, HAMP. PS. cxix. 1.

stiggin, v., *cf. O.N.* styggja ; *start* ; **stiggis** (*pres.*) ALEX. (Sk.) 5301.

stīh, stīhen, see **stīȝ, stīȝen**.

stihten, v., *O.E.* stihtan, = *M.Du.* stichten, *O.H.G.* stiftan ; *dispose, destine* : æfter þam þe him for stihted (*ms.* stihteð) (*pple.*) wæs LK. xxii. 22 ; stiȝt ALEX. (Sk.) 4897 ; stight *written* ALEX. (Sk.) 2693*.

stihtlen, v., *dispose, order, rule* ; stiȝtle PL. CR. 315 ; ? stigh(t)ill YORK xxxi. 75 ; **stiȝtlez** (*pres.*) GAW. 2213 ; stighillis (? *for* stightillis) ALEX. (Sk.) 755* ; sti(h)tlede [stiȝtlide] (*pret.*) LANGL. C xvi. 40 ; stiȝtled WILL. 1199 ; ALEX. 195 ; **stiȝtled** (*pple.*) A. P. ii. 90.

stike, see **stiche**.

stikel, adj., *O.E.* sticol, = *M.L.G.* stekel, *O.H.G.* stechal ; *difficult*, PARTEN. 5848.

stikil-līche, adv., *with difficulty*, ALIS. 219.

stikelinde, stikelunge, adv., *? intently*, H. M. 17.

stikeling, sb., = *M.Du.* stekelingh, *M.H.G.* stichelinc ; *stickling (fish)*, '*silurus*,' PR. P. 475 ; stikling VOC. 222.

stikien, v., *O.E.* stician, *cf. Gr.* στίζω ; *stick, stab, prick* ; stikie LEG. 189 ; stekie LANGL. B i. 121 ; **stikeþ** (*pres.*) TREAT. 140 (stikez SAINTS (Ld.) xlvi. 782) ; earewe þet . . . stikeð i ðe heorte A. R. 60 ; **stikinde** (*pple.*) H. M. 35 ; **stikede** (*pret.*) BEV. 828 ; þat sweord stike(de) [stekede] feste LAȝ. 7533 ; ful he stikede of arewen P. L. S. xviii. 48 ; he stiked fast S. S. (Wr.) 1246 ; heo stikeden [stekede] mid cnifes LAȝ. 20962 ; **istiked** (*pple.*) CH. C. T. A 1565 ; istiked ase swin P. S. 190.

sticking, sb., *stabbing*, ALEX. (Sk.) 2623.

stikil, adj., *? piercing* : þe stikill stormes ALEX. (Sk.) 4186.

stikke, see **sticke**.

stil, sb., *Lat.* stilus ; *stylus, pen* ; **stīle** (*dat.*) TREV. V. 297.

stīle, see **stīȝele**. **stīl**, see **stēl**.

stillatorie, sb., *still (for distilling)* ; '*stillatorium*' PR. P. 475 ; **stillatories** (*pl.*) CH. C. T. G 580.

stille, adj., *O.E.* stille, = *M.Du.* stille, *O.H.G.* stilli ; *still*, '*qvietus, tranqvillus*,' PR. P. 475 ; KATH. 373 ; A. R. 116 ; ORM. 1177 ; C. L. 1030 ; APOL. 5 ; CH. C. T. F 1472 ; Joram wes stille (*silent*) LAȝ. 15890 ; beo nu stille O. & N. 261 ; he lai þer stille ROB. 11 ; **stille** (*adv.*) LAȝ. 6747 ; GEN. & EX. 2428 ; HAV. 2997 ; PR. C. 3782 ; ȝif þis child dvelle stille (*yet*) here GREG. 77.

stille-hēde, sb., = *M.H.G.* stilheit ; *quietude, rest*, AYENB. 142.

stil-lich, adj., *O.E.* stillīc ; *quiet* : mid stilliche ginne LAӠ. 2374 ; **stillīche** (*adv.*) A. R. 82 ; R. S. i ; ALIS. 1566 ; H. S. 2432 ; stilleliche ROB. 19 ; stilli he mournes ALEX. (Sk.) 1136.

stilnesse, sb., *stillness,* HOM. I. 115 ; A. R. 414 ; WICL. JOHN xi. 28.

stille [2], sb., = *O.H.G.* stillī ; *calm, quiet* : þet makest stille efter storme A. R. 376.

stillen, v., *O.E.* stillan, = *O.H.G.* stillan, *M. Du.* stillen, *O.N.* stilla ; *still,* GEN. & EX. 3924 ; stillin '*pacifico*' PR. P. 475 ; stille HORN (L.) 676 ; FLOR. 831 ; his wraþþe for to stille BEK. 467 ; **stilleð** (*pres.*) his teares A. R. 186 ; *comp.* Ӡe-stillen.

stillin, v., *drop, trickle,* '*stillo,*' PR. P. 475.

stilte, sb., = *M.L.G., Dan.* stylte, *Swed.* stylta, *M.Du.* stelte, *O.H.G.* stelza ; *stilt,* '*calepodium, lignipodium*' PR. P. 475 ; REL. I. 86.

stilðe, sb., = *O.H.G.* stillida ; *quiet, tranquillity,* A. R. 156 ; H. M. 41.

stīme, sb., *? particle* : ne he iwis might se a stime C. M. 19652.

stinc, sb., *? O.E.* stinc ; *smell, stench,* GEN. & EX. 2975 ; stink '*foetor*' PR. P. 475 ; MISC. 228 ; CH. C. T. *D* 2274, *B* 3810 ; LUD. COV. 275 ; stinch ORM. 1192 ; **stinche** (*dat.*) BEK. 2406.

stingen, v., *O.E.* stingan, = *O.N.* stinga, *Goth.* (us-)stiggan, *Lat.* (in-)stingvere ; *sting, puncture ; vex* ; PL. CR. 648 ; **stingeð** (*pres.*) A. R. 82 ; stingeþ MISC. 149 ; **stinginde** (*pple.*) A. R. 82 ; **stang** (*pret.*) S. S. (Web.) 759 ; stong LAӠ. 11363 ; GEN. & EX. 3896 ; stungen ORM. 17441 ; stongen MAND. 286 ; **stunge** (*subj.*) HOM. II. 209 ; **stungen** (*pple.*) GEN. & EX. 3901 ; PS. iv. 5* ; stongen CH. C. T. *A* 1079 ; E. T. 645 ; istungen HOM. I. 121 ; istunge LAӠ. 27597 ; O. & N. 515 ; istongen PL. CR. 553 ; istonge SPEC. 84 ; *comp.* of-, to-, þurh-stingen ; *deriv.* stange, steng.

stinken, v., *O.E.* stincan (*smell*), = *M.L.G.* stinken, *O.H.G.* stinchan (*smell*), *Goth.* stigqan (κόπτειν) ; *cf. O.N.* stökkva (*run, leap*) ; *smell,* ORM. 4781 ; heo hit ne muwen stinken (*smell*) A. R. 86 ; stinkin '*foeteo, oleo*' PR. P. 475 ; þeo þet rotieð and **stinkeþ** (*pres.*) A. R. 84 ; & stinkeð fule ORM. 1201 ; **stinkinde** (*pple.*) AYENB. 32 ; þat stinkende lic ORM. 8195 ; stinkinge ROB. 214 ; **stanc** (*pret.*) ORM. 8077 ; stank LANGL. (Wr.) 10753 ; MAND. 97 ; CH. C. T. *B* 3807 ; H. S. 8313 ; stonc A. R. 326 ; stonk BEK. 2404 ; C. L. 1284 ; stunken A. R. 230 ; þat stunken swiþe swete ORM. 8194 ; **stunken** (*subj.*) A. R. 86 ; **stonken** (*pple.*) MAND. 10 ; *comp.* Ӡe-stinken ; *deriv.* stinc, stench, stünch.

stinten, *see* stünten. **stīpel,** *see* stēpel.

[**stīpen,** v., = *L.G.* stīpen ; *prop up ; comp.* under-stīpen.]

stiper, sb., *cf. L.G.* stiper ; *prop, support* : þe stipre þat is under þe vine set L. H. R. 135.

[**stipren,** v., *prop up ; comp.* under-stipren.]

stirien, *see* stürien.

stirk, sb., *O.E.* stirc, stўric ; *bullock,* '*juvencus,*' VOC. 204.

stirke, sb., = *M.L.G.* sterke, *M.H.G.* stirke, sterke ; *heifer,* '*juvenca,*' PR. P. 476.

stirne, *see* stürne. **stirred,** *see* steorred.

stirt, *see* stert. **stirten,** *see* sterten.

stirtil, *see* stertil.

stith, sb., *O.N.* steði ; *anvil, stithy,* HAV. 1877 ; CH. C. T. *A* 2026 ; stithe '*incus*' PR. P. 476 ; steþi C. M. 23237.

stīð, adj., *O.E.* stīð, = *O.Fris.* stīth ; *hardy, strong, brave,* LAӠ. 10083 ; stīð & strong GEN. & EX. 1591 ; stiþ ALIS.[2] (Sk.) 91 ; stith MAN. (H.) 194 ; strong and steith PALL. iv. 892 ; **stīðne** (*acc.*) dom HOM. I. 95 ; **stīþe** (*adv.*) M. H. 4 ; aӠein þe stiþ(e) i stod SPEC. 99 ; **stīþe** (*pl.*) PS. x. 7 ; stīðe men LAӠ. 24886 ; stedes stiþe K. T. 338 ; kniӠtes stiþe on stede TRIST. 66 ; stiþe stormes AMAD. (R.) xlviii ; **stīther** (*compar.*) PR. C. 3173.

stīð-imained, adj., *strong in might,* LAӠ. 25820.

stīð-imōded, adj., *strong in mind,* LAӠ. 26022.

stīþ-lī, adj., *bravely,* GAW. 431.

stīven, v., = *M.L.G.* stīven, *M.Du.* stijven, *O. Fris.* stīva ; *from* stīf ; *stiffen, make robust ;* stive TRIST. 1169 ; **stīved** (*pple.*) WILL. 3033.

stob, sb., = *M.Du.* stobbe, *? O.N.* stobbi ; *stub, stump* ; stòb ant stokke REL. I. 168 ; **stobbus** (*pl.*) ED. 4326.

stoble, *see* stuble. **stoburn,** *see* stiburn.

stoc, sb., *O.E.* stocc, = *O.Fris.* stok, *O.N.* stokkr, *O.H.G.* stoc, stoch ; *stock, trunk,* FRAG. 3 ; stoc, stok O. & N. 25, 1113 ; WICL. JOB xiii. 27 ; stok VOC. 236 ; SAINTS (Ld.) xlv. 233 ; AR. & MER. 3865 ; PR. C. 676 ; over stok and over ston ORF. 332 ; stok, stock SHOR. 1, 109 ; stock MAND. 10 ; **stocke** (*dat.*) AYENB. 19 ; **stockes** (*pl.*) '*çeps*' VOC. 163 ; LAӠ. 5694 ; MARH. 1 ; stokkes LANGL. *A* iv. 95, v. 585 ; FLOR. 1150 ; **stocken** (*dat. pl.*) LAӠ. 626 ; *comp.* cāl-, herth-, mūse-stoc.

stok-douve, sb., = *M.Du.* stockduive ; *stockdove ;* (*ms.* -dowe) PR. P. 476.

stok-fish, sb., = *M.H.G.* stocvisch ; *stockfish,* PR. P. 476.

stoc [2], sb., *O.E.* stoc(-weard) ; *place ;* **stókes** (*pl.*) (*ms.* stokess) ORM. 15694.

stōd, sb., *O.E.* stōd, *cf. O.N.* stöð *n.,* *O.H.G.* stuot *f.;* *stud, stable,* HICKES I. 231 ; stode PERC. 326 ; stood GOW. III. 204 ; **stōde** (*dat.*) O. & N. 495 ; þou come of liþer stode P. S. 201.

stōd-mere, sb., *O.E.* stōdmyre ; *stud-mare,* A. R. 316 ; PERC. 367.

stóde, sb., _O.E._ stod,= _O.N._ stoð ; _post_ ; stoþe
PR. P. 478 ; **stoodes** (_pl._) WICL. ECCLUS. xxix.
29 ; _comp._ **dor-stóde**.

stode, stodul, _see_ **stude, studul.**

stoffe, sb., _stuff (cloth)_ ; stuffe ALEX. (Sk.)
2980*.

stoffen, v., _O.Fr._ estoffer; _stuff; suffocate_ ;
stuffin PR. P. 481 ; **stoffes** (_pres._) JOS. 601;
stuffede (_pret._) D. ARTH. 3616; **stoffed**
(_pple._) GAW. 606 ; stuffed ['_suffocatus_'] TREV.
VI. 449.

stofnen, v., _O.N._ stofna ; _institute_ ; **stofned**
(_pple._) ORM. 14561.

stok, _see_ **stoc.**

stóken, v.,= _M.L.G., M.Du._ stoken ; _stab_ ;
stoke CH. C. T. _A_ 2546; & **stókeþ** (_pres._)
him bitwene (h)is browes FER. 4615; stokes
D. ARTH. 2554.

stoken, v.,= _M.Du.,M.L.G.,M.H.G._ stocken;
put in the stocks ; stokken '_coigner_,' VOC. 163;
stockid (_pret._) him an a stol WART. II. 106;
stocked in prison CH. TRO. iii. 380.

stoki, adj.,= _Ger._ stockig ; _stock-like_ ; stoki L.
H. R. 148.

stól, sb., _O.E._ stol,= _O.L.G., O.Fris._ stōl, _O.N._
stōll, _Goth._ stōls, _O.H.G._ stuol ; _stool, seat_, A.
R. 166 ; MISC. 145 ; S. S. (Web.) 1889 ; stool
'_scabellum_' PR. P. 476 ; **stóle** (_dat._) AYENB.
239 ; LANGL. _B_ v. 394 ; **stóles, stooles** (_pl._)
CH. C. T. _D_ 288 ; stoles TREV. IV. 99 ; _comp._
bischop-, dóm-, kine-stól.

stóle, sb., _Lat._ stola ; _stole_, TREV. IV. 365.

stomak, sb., _Fr._ estomac ; _stomach_, PR. P.
476 ; stomake ALEX. (Sk.) 4436.

stomeren, _see_ **stumren.**

stomlen, _see_ **stumlen.**

stompe, _see_ **stumpe.   stón,** _see_ **stán.**

stónd, _see_ **stúnde.**

stonde, stonden, _see_ **stande, standen.**

stónen, _see_ **stǽnen.   stonien,** _see_ **stunien.**

stoppe, sb., _O.E._ stoppa, _cf. O.E._ stēap,=
_O.N._ staup, _mod.Eng._ stoup ; _bucket, wooden
mug,_ '_situla_,' PR. P. 477.

stoppel, sb., _stopper (of a bottle)_, '_ducillus_,'
PR. P. 477 ; CHEST. I. 142.

stoppin, v., _O.N._ stoppa,= _M.L.G., M.Du._
stoppen, _M.H.G._ stopfen ; _stop,_ '_obturo_,' PR.
P. 477 ; stoppen CH. TRO. ii. 804 ; stoppen
here mouþ P. S. 326 ; P. 10 ; stoppi AYENB.
257 ; **stoppeð** (_pres._) A. R. 72 ; **stoppede**
(_pret._) ISUM. 448 ; & stoppede (h)is wounde
FER. 215 ; he-stopped heore wai ALIS. 1224 ;
stopped (_pple._) PR. C. 7368 ; istopped SHOR.
156 ; _comp._ **for-stoppen.**

stór, sb., _O.E._ stōr, ? = _O.H.G._ stūr ; _incense,_
VOC. 140 ; LEG. 96 ; MISC. 26 ; AYENB. 211 ;
(_printed_ scor) SHOR. 123.

stór², sb., _O.Fr._ estōr, _low Lat._ staurum ;
store ; farm stock, ROB. 395 ; JOS. 456 ; A. P.
i. 846 ; stoor PR. P. 477 ; CH. C.T._A_ 598 ; **stóre**
(_dat._) DEGR. 72.

stōr³, sb., _? O.Fr._ estor, estour, _?from O.H.G._
(ki-)stōr ; _tumult, battle; army_ ; FLOR. 1659 ;
stour IW. 2633 ; PR. C. 1838 ; M. ARTH. 3064;
PARTEN. 2231 ; stoure ALEX. (Sk.) 785, 3501;
stour BARB. ii. 355 ; bale stour _death-pang_
A. P. iii. 427 ; **stōre** (_dat._) ALIS. 2110 ; S. S.
(Web.) 956 ; stoure WILL. 4214 ; EGL. 9;
ANT. ARTH. xliii ; **stoures** (_pl._) CH. C. T.
_B_ 3560 ; stures M. H. 23.

stōr⁴, adj., _O.E._ stōr,= _O.Fris._ stōr, _O.N._ stōrr;
great, strong, LA3. 85 ; O. & N. 1473 ; GEN.
& EX. 842 ; HAV. 2383 ; stor ant stark CHR.
E. 464 ; stoor '_austerus, rigidus_' PR. P. 477 ; þe
strimes urneþ store FL. & BL. 228 ; **stōre** (_pl._)
GAW. 1923 ; ANT. ARTH. (R.) lv ; his mæhte
weren store LA3. 3242 ; store windes IW. 373.

 stōr-līc, adj., _great, strong_ : þat feht wes
swiðe storlic [storlich] LA3. 10647 ; **stōrlīche**
(_adv._) KATH. 1274.

storben, v., _for_ disturben ; _trouble_ ; **storbis**
(_pres._) ALEX. (Sk.) 667 ; **storbet** (_pple._)
3605 ; stourbed 934.

stordi, _see_ **sturdi.**

stōren, v., _O.Fr._ estōrer, _Lat._ (in-)staurāre ;
store ; store P. S. 70 ; astore (?=enstore) ROB.
107 ; ALIS. 5817 ; astori AYENB. 136 ; he
stōrede (_pret._) þem wiþ corn MAN. (F.) 2916 ;
istōred (_pple._) LA3. 13412*.

storial, adj., _historical_ ;   storial mirour
[='_Speculum historiale_'] CH. L. G. W. PROL.
_A_ 307 ; storial sooth 702 ; _see_ **historial.**

storie, sb., _O.Fr._ istoire, estoire ; _story, history,
legend_, A. R. 154 ; CH. C. T. _G_ 35 ; store ALEX.
(Sk.) 3854, 4828 ; stori 2050 ; **stories** (_pl._)
LANGL. _B_ vii. 73 ; _see_ **histoire.**

stōrien, v.,= _M.Du., O.H.G._ stōren ; _move_ :
þat folk gan to storie LA3. 23756* ; sturie(n)
. . . ne storen KATH. 362 ; store E. T. 755 ;
w. & I. 78 ; hwen ha . . . ne stōrið (_pres._)
ham seolf MARH. 14 ; loke ye store not of þat
stede GUY² (Z.) 3869 ; **stōrede** (_pret._) LA3.
9334*.

stork, sb., _O.E._ storc,= _M.L.G._ stork, _O.N._
storkr, _O.H.G._ storch, storach; _stork,_ '_ciconia_,'
PR. P. 477 ; MAN. (F.) 14574 ; CH. P. F. 361.

storm, sb., _O.E._ storm,= _M.L.G.,M.Du._ storm,
_O.N._ stormr, _O.H.G._ sturm ; _storm, tumult,_
['_tempestas_'] HOM. II. 43 ; SHOR. 161 ; CH. C.
T. _A_ 1980 ; M. H. 136 ; **storme** (_dat._) HOM. I.
143 ; LA3. 25587 ; A. R. 376 ; **stormes** (_pl._)
JUL. 79 ; GOW. I. 143 ; **storme** (_dat. pl._)
HOM. II. 43.

stormi, adj., _O.E._ stormig,= _M.H.G._ stürmec ;
stormy, LIDG. M. P. 2.

stort, _see_ **steort.**

stot, sb., _O.E._ stotte,= _M.Du._ stutte (_mare_) ;
horse, '_caballus_,' PR. P. 477 ; CH. C. T. _A_
615 ; **stottes** (_pl._) O. & N. 495.

stot², sb., _cf. Swed._ stut (_bullock_) ; _bullock_ ;
(_ms._ stott) '_buculus_' VOC. 218 ; stotte ISUM.
92 ; **stottes** (_pl._) LANGL. _B_ xix. 262.

stot[3], sb., *stoat*, *(ms.* stott) LUD. COV. 218.

stotin, v., *stutter; hesitate, pause; 'titubo,'*
PR. P. 477; **stote** (*pres.*) A. P. i. 149; **stote**
(*pres. subj.*) VOC. 173; **stotiand** (*pple.*)
REL. I. 291; þere þei **stotede** (*pret.*) (*stayed*)
a stound DEGR. 226.

stoþe, *see* **stode**.   **stouke**, *see* **stūke**.

stounde, *see* **stúnde**.   **stoupen**, *see* **stūpen**.

stour, *see* **stōr**, **stūr**.

stout, adj., *cf. M.Du.* stout, *O.N.* stoltr, *O.H.G.*
stolz, *? O.Fr.* estout (*hardi, téméraire*);
stout, bold, proud, SPEC. 52; SHOR. 150;
GREG. 883; CH. C. T. *A* 545; LAI LE FR.
249; stout ... and fers BEK. 512; so proud
& so stout S. A. L. 224; stout and kene K. T.
936; **stoute** (*pl.*) P. S. 255; **stoutar** (*comp.*)
BARB. xv. 524; **stoutest** (*superl.*) BARB. xi. 470.

     stout-līche, adv., *stoutly, proudly,* WILL.
1950; stoutli ALEX. (Sk.) 279.

     stoutnes, sb., *stubbornness,* BARB. vii. 356.

stout, *see* **stūt**.

stoven, sb., *O.N.* stofn; *stalk, stem,* C. M.
8037.

stovēr, sb., *O.Fr.* estouvoir (*provision*); *food,*
S. S. (Web.) 2606; TRIST. 1149.

stōwe, sb., *O.E.* stōw,=*O.Fris., O.N.* stō,
mod.Eng. -stow *in place-names; place,* MAT.
xiv. 15; an oðre stowe HOM. I. 219; stou
SPEC. 98; to þere stowe [stouwe] LAȝ. 1174;
stōwen (*dat. pl.*) MK. i. 45; HOM. II. 207;
comp. **cwalm-, earding-, leir-stōwe**.

stōwien, *see* **stēowien**.

stōwin, v., *from* **stōwe**; *stow away; place,*
'*loco,*' PR. P. 478; **stauez** (*pres.*) A. P. ii. 480;
stōwide (*pret.*) ['*collocavit*'] TREV. III.
277; wel wern þai **stōwed** (*pple.*) A. P. ii.
113; staued A. P. ii. 352; *comp.* **bi-stōwen**.

strā, *see* **strāu**.

strāc, sb.,=*M.H.G.* streich; *from* **strīken** ;
stroke; stroc ROB. 17; strok MAND. 260;
TRIAM. 1229; GAW. 287; þane strok AYENB.
34; strok [strook] CH. C. T. *A* 1709; **strākes**
(*pl.*) IW. 640; strakis ALEX. (Sk.) 1015;
strokis EGL. 310.

strǣte, sb., *O.E.* strǣt,=*O.L.G.* strāta, *O.H.*
G. strāza; *road, lane, street,* LAȝ. 4823; ORM.
7358; strete KATH. 1671; O. & N. 962; strate
HOM. II. 227; MIN. vi. 56; **strǣte** (*pl.*) LK.
x. 10; (*ms.* strætte) LAȝ. 4843; **strǣten** (*dat.*
pl.) LAȝ. 4839.

Strĕt-ford, pr. n., *Stratford,* DAV. DR. 113.

strai, sb., *cf. mod.Eng.* stray (*large open com-*
mon): *wandering,* '*vagacio,*' PR. P. 478; on
strai *astray* BARB. xiii. 195; o strai C. M.
6827; a straie A. P. 1161.

straiin, v., *O.Fr.* estraier; *stray, wander,*
'*vago,*' PR. P. 478; straie LUD. COV. 74;
straie (*pres.*) M. H. 52; **astraied** (*pple.*)
GOW. II. 132.

strail, sb., *O.E.* strǣgl, *from Lat.* strāgulum ;

bedclothing ['*stratum*'] PS. vi. 7*; straile
'*stragulum*' PR. P. 478.

strait, *see* **streit**.

straives, sb., pl., *strays*: waives and straives
LANGL. *C* i. 92.

strākien, v., *O.E.* strācian, = *M.H.G.* strei-
chen; *from* **strīken**; *stroke*; strōke CH. C.
T. *F* 162; þei over lond **strākeþ** (*pres.*) (*run*)
PL. CR. 82; **strōked** (*pret.*) TRIST. 3108;
sterin stevin upon strake straked (*? sounded*)
þar trumpis ALEX. (Sk.) 1386; þire traitours
trouthis has **strākid** (*pret.*) (*have made a*
covenant) 3192.

strāl, sb., *O.E.* strǣl *m. f.* (*arrow*), *cf. M.Du.*
straele (*arrow*), *O.H.G.* strāla *f.* (*arrow,*
shot); *? arrow*; **strāles** (*pl.*) LAȝ. 5695.

strand, sb., *O.E.* strand,=*M.H.G.* strant *m.,*
M.Du. strande *n.* (*shore*), *O.N.* strönd *f.*
(*shore, river*); *strand, shore, river,* C. M. 8191 ;
strond GEN. & EX. 2717; ALIS. 1981; GREG.
281; þat strond LAȝ. 19916; stronde '*litus*'
PR. P. 480; ['*flumen*'] TREV. I. 429; ['*tor-*
rentem'] WICL. DEUT. ix. 21; **strande** (*dat.*)
ORM. 19450; stronde HORN (R.) 115; P. L.
S. xxiii. 31; A. P. i. 152; CH. C. T. *B* 825;
framward þan stronde [þare stronde] LAȝ.
9408; up to þe nekke in þe stronde ANGL. I.
306; **strandis** ['*torrentes*'] (*pl.*) HAMP. PS.
xvii. 5; *comp.* **sǣ-strond**.

strang, adj., *O.E.* strang, strong,=*O.L.G.*
strang, *O.N.* strangr, *O.H.G.* strang, streng ;
strong, severe, powerful, brave, SHOR. 14;
AYENB. 32; IW. 2386; S. S. (Wr.) 2655; PR.
C. 881; strang & mihti HOM. I. 231; strang
& hard ORM. 6326; strong LAȝ. 173; MARH.
12; A. R. 6; O. & N. 5; strong and stark
HAV. 608; ALIS. 5527; strong and stif BRD.
21; strong & stor P. R. L. P. 101; **strongen**,
stronge (*dat. m. n.*) LAȝ. 11733, 27551; me
wolde to stronge deþe him bringe BEK. 1076;
strangere (*dat. f.*) LAȝ. 27773*; **strangne**
(*acc. m.*) AYENB. 227; strongne LAȝ. 6392;
stronge[strang(e)] (*acc. f.*)LAȝ.1567; **strange**
(*adv.*) S. S. (Wr.) 197; stronge O. & N. 254;
BRD. 24; **strange** (*pl.*) ORM. 19963; stronge
GEN. & EX. 3713; AYENB. 83; gares swiðe
stronge LAȝ. 26282; stronge (*distressing*)
tidinge 5271; **strongen** (*dat. pl.*) LAȝ. 17395;
strengre (*compar.*) MARH. 12; A. R. 326;
strengor K. T. 657; strenger CH. C. T. *C* 825;
strengeste (*superl.*) LAȝ. 2109; KATH. 734;
þe strengeste HORN (L.) 823 (H. 852); ROB.
111; strengest A. R. 280; GOW. III. 147;
comp. **un-strang**.

     strang-līche,adv.,*O.E.*stranglīce; *strongly,*
P. L. S. xvii. 295; strongliche MARH. 14;
strongliche [stra(n)gliche] LAȝ. 7843; **strong-**
lüker (*compar.*) H. M. 15; **stronglükest**
(*superl.*) A. R. 218.

strange, adj., *O.Fr.* estrange; *strange,* ROB.
16; strange [straunge, stronge] TREV. IV.
13; straunge '*extraneus*' PR. P. 479.

strangen, v., _become strange_; **strangeþ**
(_pret._) GOW. II. 264.

stranger, sb., _O.Fr._ estrangier; _stranger_;
straunger PR. P. 479.

strangien, v., _O.E._ strangian,=_O.H.G._ strangen; _become strong, make strong_: þet . . .
eower feond stronȝian HOM. I. 13; **strangede**
[strongede] (_pret._) LAȝ. 4461; _comp._ ȝe-, un-
strangien.

stranglen, v., _O.Fr._ estrangler; _strangle_;
astrangli AYENB. 50; **strangled** (_pple._) HAV.
640; astrangled ROB. 342.

strapil, sb., _?O.E._ strapul; _leggings, ' tibiale,'_
VOC. 259; **strapeles** (_pl._) A. R. 420; straples
BEK. 1477; strapeles [straples] TREV. V.
355.

strāu, sb., _O.E._ strēaw, strāw,=_O.N._ strā,
O.Fris. strē, _O.L.G._ strō, _O.H.G._ strau, strou,
strō; _straw, (ms._ straw) LANGL. _B_ xiv. 251,
_C_ xvii. 93; stra Iw. 2655; þer of ne yaf he
nouth a stra HAV. 315; stro BARB. iii. 320*;
streuw LEECHD. III. 114; strea A. R. 296;
stre GOW. I. 282; stre [stree] CH. C. T. _A_
2918; stree MAND. 253; _comp._ **bed-strāu.**

strāu-beri, sb., _O.E._ strēawberie; _straw-_
berry, ' fragum,' PR. P. 478; streberi VOC.
141.

strawen, v., _O.E._ strēowian,=_O.Fris._ strewa,
_O.H.G._ strewen, _M.H.G._ streuwen, strauwen,
strouwen, _O.N._ strā, _Goth._ straujan (_pret._
strawida), _cf. Lat._ sternere (_perf._ strāvi);
strew, FL. & BL. 436; strawen (_ms._ strawwenn)
ORM. 8193; strowin PR. P. 480; **strewes**
(_pres._) PS. cxlvii. 16*; strewe [strawe] (_subj._)
CH. C. T. _F_ 613; **strewed** (_pret._) TRIST.
2195; strowede MAT. xxi. 8; **strawed**
(_pple._) ALIS. 1026; WILL. 1617; EGL. 376;
strawid S. S. (Wr.) 2692; strowed LIDG. M. P.
186; istrawed FER. 421; _comp._ **bi-strēwen.**

strē, strēa, _see_ **strāu.**

strēam, sb., _O.E._ strēam,=_O.Fris._ strām, _O._
_L.G._ strōm, _O.H.G._ straum, stroum, _O.N._
straumr; _stream, river,_ MARH. 5; AYENB.
72; stream, stræm, strem LAȝ. 2849, 6116,
19756; striem P. L. S. viii. 126; strem GEN.
& EX. 2096; S. S. (Wr.) 3191; streṁ [streem]
CH. C. T. _A_ 3895; **strēame** (_dat._) ANGL.
VII. 220; swimmen . . . mid þe streme
HOM. I. 51; **strēmen** (_dat. pl._) LAȝ. 31229;
strēames (_pl._) A. R. 188; strimes FL. & BL.
228; _comp._ **sēa-, water-strēam.**

strecche, sb., _?compass, grasp_: bið swa
mihtles on his modes streche HOM. I. 111.

strecchen, v., _O.E._ streccan,=_M.Du._ streck-
en, _O.H.G._ strecchan; _stretch, extend,_ JUL.
27; strecche WILL. 219; MAND. 2; CH. C.
T. _G_ 469; strechche AYENB. 103; **streccheð**
(_pres._) A. R. 280; streccheþ MISC. 101; þe
over Germania streccheþ [' _se extendit_'] bi
sides Alpes to . . . TREV. I. 255; strekis
[straighit] ALEX. (Sk.) 1953; strekez D.

ARTH. 1229; **streche** (_imper._) KATH. 2265;
streahte, stræhte, strehte [strahte] (_pret._)
LAȝ. 1910, 17886, 21227; streihte A. R. 280;
streiȝte P. L. S. xvii. 345; MED. 641; TREV.
II. 107; he strahte forþ his riht earm HOM.
I. 189; he strahte him MARH. 9; straughte
CH. C. T. _A_ 2916; strehten MK. xi. 8; and
a wei streiȝten JOS. 456; **streiht** (_pple._) JOS.
519; streiȝt P. R. L. P. 252; William **streiȝt**
(_adv., straightway, forthwith_) went hem to
WILL. 3328; streȝt ALEX. (Sk.) 1758; strauȝt
CH. BOET. v. 5 (170); straught OCTOV. (W.)
959; unto þe helle straught he went GOW.
III. 36; streȝt on _near_ ALEX. (Sk.) 1574;
comp. **a-, ȝe-strecchen.**

strecour, sb., _dog for the chase,_ BARB. vi.
487.

streek, _see_ **strike.**

strēȝen, v., _O.E._ strēgan; _strew_; strie PALL.
iii. 10.

streȝten, v., _march_; **streiȝtiş** (_pres._) ALEX.
(Sk.) 2032.

streiȝt, _see under_ **strecchen.**

streikin, _see_ **streken.**

streinin, v., _O.Fr._ estreindre; _strain, stretch;_
constrain; restrain; ' stringo,' PR. P. 479;
streine CH. C. T. _E_ 1753; **stránes** (_pres._)
out ALEX. (Sk.) 840*; straini (_subj._) AYENB.
263; **streine** (_imper._) L. C. C. 43; **streined**
(_pple._) ALEX. (Sk.) 3121; strenȝeit BARB. xii.
248; istreind BEK. 1477.

streinour, sb., _strainer, ' colatorium, con-_
strictorium,' PR. P. 479; L. C. C. 16; C. B. 90.

streinðe, _see_ **strengðe.**

streit, adj., _O.Fr._ estreit; _strait, ' strictus,'_
' angustus, artus,' PR. P. 479; LAȝ. 22270;
BEK. 260; CH. C. T. _A_ 174; streit þat is to
seie narou MAND. 45; strait AYENB. 54;
streiȝt blind TREV. VI. 235; till a **stráte**
(_sb._) thai held thair wai BARB. iv. 458;
strátest (_superl._) BARB. vi. 463.

streit-, strait-līche, adv., _straitly,_ AYENB.
7, 34.

straitnes, sb., _narrowness,_ BARB. xii. 430.

streitin, v., _straiten,_ PR. P. 479; **streitid**
(_pple._) WICL. JOB xviii. 7.

streke, adv. (_from O.E._ strec _adj._),=_M.Du._
strek, strak, _M.H.G._ strac; _straight,_ D. ARTH.
1792; strek, strik PR. C. 2623, 3378.

strik-lī, adv., _straight,_ PR. C. 3288.

streke, _see_ **strike, strecchen.**

stréken, v., _cf. Swed._ streka; _stretch_; REL.
I. 65; streke PS. lxxxiv. 6; streikin ' _ex-_
tendo' PR. P. 479; **strékez** (_pres._) GEN. & EX. 481;
stréked (_pple._) MAN. (F.) 12703; ?striked
WILL. 1617.

strékinge, sb., _stretching,_ HAMP. PS. xxi. 17.

strēken, _see_ **strīken. strēṁ,** _see_ **strēam.**

strēmen, v., _O.N._ streyma; _from_ **strēam**;

stream, flow; welthes if þai **strēmen** (*pres.*) ['*affluant*'] PS. lxi. 11 ; **strēminge** (*pple.*) JOS. 560; **strēmden** [streamden] (*pret.*) A. R. 188 *; **strēmed** (*pple.*) WICL. PROV. v. 16.

strēmere, sb., *streamer*: stremere of fane, '*cherucus*' PR. P. 479.

strēn, *see* **strēon**.

strench P, sb. : alle he (*sc.* deaþ) riveþ in one strench R. S. i (MISC. 156).

strencðe, *see* **strengðe**.

strēnen, *see* **strēonen**.

streng, sb., *O.E.* streng, = *M.L.G.* streng, *O.N.* strengr, *O.H.G.* stranc ; *string*, GEN. & EX. 479; ROB. 456; TREV. III. 211 ; þene streng LAȝ. 1454; streng [string] CH. C. T. *D* 2067; string '*funiculus*' PR. P. 480; **strenge** (*dat.*) O. & N. 1230; **strenges** (*pl.*) P. L. S. xvii. 156; stringes MAP 339; ALIS. 208; PS. xxxii. 2 ; **strengen** (*dat. pl.*) JOHN ii. 15 ; LAȝ. 17983.

strenge, sb., *O.E.* strengu, = *O.L.G.* strengīn, strengī, *O.H.G.* strengī; *from* **strang**; *strength, force; strong place*; REL. I. 130 ; GEN. & EX. 714; ROB. 302; GREG. 238; (? *ms.* strenȝe) LAȝ. 26690; L. H. R. 44*; strenghe REL. I. 185; **strenghes** (*pl.*) D. ARTH. 2242.

strengen, v., *O.E.* strengan, = *O.H.G.* (er-)strengan, *O.N.* strengja ; *make strong, fortify*, KATH. 942*; ORM. 2614; **strengeð** (*pres.*) MARH. 14; strengeþ SHOR. 26 ; strengith TOR. 6; **streng** (*imper.*) JUL. 33 ; **strenged** (*pple.*) ORM. 2748; strenghed PS. xxvi. 14*; *comp.* ȝe-, un-strengen.

strengðe, sb., *O.E.* strengðu, = *O.H.G.* strengida ; *strength, fortitude*, KATH. 647; H. M. 11 ; GEN. & EX. 581; strengðe, strengða, strenče [strengþe] LAȝ. 1540, 3727, 17213; strengče, strencðe, strenče A. R. 140, 280; strengþe C. L. 801; SHOR. 17; strenghe, strencþe, strenþe O. & N. 173, 781, 1713; strenčče MARH. 19; strencþe ORM. 5519; streinče HOM. I. 69; streinþe SPEC. 57; strenþe A. P. ii. 1155.

strencþe-lēs, adj., *strengthless*, ORM. 12530.

strengðen, v., *strengthen*; strenȝče KATH. 942; strengþi AVENB. 86; **strengþeþ** (*pres.*) LANGL. *B* viii. 47; strenčeð A. R. 140; **strengþede** (*pret.*) ROB. 140; & streinþede (*exerted*) him . . . to serve god A. D. 257; gestrænčþed (*pple.*) LK. i. 80; istrengþed MISC. 30.

strenken, v., *sprinkle*, ORM. 1099; **strenked** (*pple.*) ORM. 1771.

strenkelin, v., *sprinkle, scatter; disperse, destroy; '*aspergo*,'* PR. P. 479; strenkle A. P. ii. ȝ07; strenkil HAMP. PS. lx. 9.

strenkilinge, sb., *sprinkling*, HAMP. PS. l. 8*.

strenkil, sb., *aspergillum*; haliwater stik, '*aspersorium, isopus*' PR. P. 479; PS. l. 9; **strencles** (*pl.*) ORM. 1095.

strenðe, *see* **strengðe**.

strēon, sb., *O.E.* (ge-)strēon (*treasure*), *cf.* *O.L.G.* (ge-)striuni, *O.H.G.* (ka-)striuni (*gain*); *progeny*, H. M. 3 ; þat holi streon MISC. 153 ; þe streon (*embryo*) a midde þe eiȝe SAINTS (Ld.) xlvi. 396; and biȝete on him a steorne streone ALIS. 511; streon, stren ORM. 27, 16396; C. L. 155; stren SAL. & SAT. 230; SHOR. 64; stren [streen] CH. C. T. *E* 157; **strēne** (*dat.*) HOM. II. 19; **strēones** (*pl.*) A. R. 208; *comp.* ȝe-strēon.

strēonen, **strēonien**, v., *O.E.* strēonan, strȳnan, *cf.* *O.H.G.* striunan, *O.L.G.* (ge-)striunian; *beget*, LAȝ. 18844, 18846; strenen REL. I. 222; **strēoneð** (*pres.*) A. R. 234; streoneþ ALIS. 7057; strenes HAV. 2983; strinen PS. lxxii. 27; **strēonede** (*pret.*) H. M. 9; C. L. 1388; enne sune on hire he streonede [strenede] LAȝ. 2571; strende HOM. II. 19; **strēoned** (*pple.*) ORM. 33; *comp.* ȝe-strēonen; *deriv.* **strūnd**.

strēonunge, sb., *procreation*, H. M. 33; streoninge C. L. 1389.

strēpen, *see* **strūpen**.

stresse, sb., *for* **destresse**; *force*: take noȝt be stresse his wiffe YORK xx. 188.

strēte, sb., *O.Fr.* estrēte ; *estreat; rate-book*; strete, catchepol bok to gader by mercimentis PR. P. 480.

strēte, *see* **strǣte**.

strewen, *see* **strawen**.

stride, sb., *cf.* *M.L.G.* strede, *M.H.G.* strit; *from* **strīden** ; *stride, step*, PR. P. 480; ALIS. 4447; FER. 3221; **strídes** (*pl.*) HICKES I. 168 (HOM. II. 111); ten stride TRIST. 1488; an hundred stride FER. 4643.

strīden, v., ? *O.E.* (be-)strīdan, = *M.L.G., M. Du.* strīden ; *stride, step*, LAȝ. 17982; striden on stede AN. LIT. 96; stride C. M. 10235; IW. 1552; **strīt** (*pres.*) SPEC. 110; **strīd** (*imper.*) SPEC. 111; **stridende** (*pple.*) HICKES I. 168; **strād** (*pret.*) IW. 3193; *comp.* **bi-, over-strīden** ; *deriv.* **stride**.

strie, sb., *O.Fr.* estrie ; *hag, hideous woman*, HAV. 98; stri TOWNL. 148.

strien, *see* **struien**.

strīen, *see* **strēȝen**.

strīf, sb., *O.Fr.* estrīf ; *strife*, KATH. 680; A. R. 200; LAȝ. 24966*; GEN. & EX. 373; BEK. 1642; C. L. 273; CH. C. T. *A* 1187; **strīvis** (*pl.*) ALEX. (Sk.) 4251.

strike, sb., *O.E.* strica ; *from* **strīken** ; *streak, stroke, line*: this chapitre . . . that . . . with obels, that is, with a stric, we han befor notid WICL. ESTH. x. 3*; a gret strike MAN. (F.) 1420; streke PR. P. 479; **strikes** (*pl.*) TREV. III. 249; (*lines*) CH. ASTR. i. 7 (5).

strike [2], sb., = *L.G.* strike, streke; *from* **strīken** ; *hank; strickle*; '(h)ostorium,' VOC. 201; strek(e) PR. P. 479; stric of line '*lini*

manipulus' VOC. 217; streek of flax PR. P. 479;
a strike of flex CH. C. T. *A* 676.

strikile, sb., *cf. M.Du.* strekel; *strickle*,
'(*h*)*ostorium*,' VOC. (W. W.) 726.

striken, v., *O.E.* strīcan, = *M.L.G.* strīken, *M.
Du.*strijken,*O.H.G.*strīchen,?*cf.O.N.*striūka;
strike,stroke,rub, LA3. 20303; strikeS. S.(Web.)
1254; PR. C. 7018; A. P. i. 1124; strike þer on
blak sope REL. I. 108; robes... strike (*iron*)
MAN. (F.) 11192; strekin PR. P. 479; **strīkeð**
(*pres.*) KATH. 2514; strem þat strikeþ (*runs*)
stille SPEC. 44; **strīk** (*imper.*) A. R. 408;
strāc (*pret.*) LA3. 9318; MARH. 5; strac in
after godes folc ORM. 14810; strak APOL. 3;
a mous... strok [strook] forþ LANGL. *B*
PROL. 183, *C* i. 197; he strok his berd S. S.
(Web.) 142; & loveli strek (h)is mane FER.
244; strek (*fell*) into a studie WILL. 4038;
þe strunden þe striken a dun of þine... fet
HOM. I. 187; striken men þiderward MARH.
17; hue striken seil HORN (R.) 1023; **striken**
(*pple.*) JOS. 519; MAP 341; strekin PERC.
1371; *deriv.* strike, strikile, strāc, strāken.

strīnd, *see* strūnd. **strīnen**, *see* strēonen.

string, *see* streng.

strīpe, sb., *cf. M.L.G., M.H.G.* strīfe *m.*,
M.Du. strijpe; *stripe*, '*vibex*,' PR. P. 480;
VOC. 267; S. & C. II. lxxiv.

strīpen, *see* strūpen.

strīþþe, strīþe, sb., *position of legs when
placed firmly*, GAW. 846, 2305.

strīvin, v., *O.Fr.* estriver; *strive*, '*contendo*,'
PR. P. 480; strive HORN (R.) 729; CH. C. T.
A 3040; WICL. JOHN xviii. 36; strife M. H.
48; **strīveð** (*pres.*) A. R. 84; R. S. v (MISC.
172); **strīvede** (*pret.*) BEK. 1576; strived
WILL. 4099; striveden LEG. 31; ALIS. 2870;
strof A. R. 398; CH. C. T. *A* 1038.

strōc, *see* strāc.

strogelin, v., *struggle*, '*colluctor*,' PR. P. 480;
strugle [strogle] CH. C. T. *E* 2374.

stroien, v., *see* struien.

strōken, *see* strāken. **strond**, *see* strand.

strong, *see* strang. **strōt**, *see* strūt.

stroþe, sb., ? *O.N.* storð; *small wood*; bi a
strothe rande GAW. 1710.

stroþe[2],adj.?; qven stroþe men slepe A. P. i. 115.

stroupe, *see* strūpe.

strout, -en, *see* strūt, -en.

strowen, *see* strawen.

strūcion, sb., *Lat.* strūthiō; *ostrich*, A. R.
132*; WICL. LEV. xi. 16.

struien, v., *for* destruien; *destroy*; struien
JOS. 507; strien A. P. ii. 307, 1768; stroien
LANGL. *B* xv. 587; **struieð** (*pres.*) HOM. II.
161.

strumpet, sb., ? *from O.Fr.* strupe, stupre;
strumpet, '*meretrix*,' PR. P. 481; S. P. S 153;
strompet TREV. III. 333; LANGL. *C* xv. 42;
strumpetis (*pl.*) CH. BOET. i. 1 (6).

strūnd, sb., *O.E.* strȳnd; *from* **strēonen**;
generation; strend PS. xxi. 32; ne boჳ **ne**
strind O. & N. 242; **strūnde** (*dat.*) JUL. 55; of
heore strund(e) [*v. r.* streone] LA3. 2736;
strinde FLOR. 2174; **strinds** (*pl.*) ALEX. (Sk.)
5104; **strūnden** (*dat. pl.*) A. R. 28*.

strūnde, sb., *stream*; **strūnden** (*pl.*) HOM. I.
187; A. R. 188*; H. M. 35; þi strivande
stremeჳ of strindeჳ so moni A. P. iii. 311.

strūpe, sb., *O.N.* strūpi; *throat*; stroupe
PR. P. 480; ANGL. I. 83; MAN. (H.) 190.

strūpen, v., *O.E.* (be-)strȳpan, = *M.Du.* stroo-
pen, *O.H.G.* stroufen; *strip*, KATH. 1548;
stripe MARG. 238; hem for to stripe [strepe]
of harneis and of wede CH. C. T. *A* 1006;
strēpeþ (*pres.*) AYENB. 105; **strīpe** (*imper.*)
L. C. C. 48; **strūpte** (*pret.*) P. L. S. xx. 81;
strepte HOM. II. 195; struptest JUL. 63;
heo strepten of heore cloþes LEB. JES. 812;
i**strūped** (*pple.*) A. R. 148*; istrupt C. L. 431;
comp. be-strūpen.

strūt, sb., = *M.H.G.* strūz; *swelling, contention*;
H. S. 3350; C. M. 3461; P. S. 334; he maden
mikel strout HAV. 1039; stintst of thi strot
A. P. i. 353; strot ne strif 848; a strut [strout]
'*turgide*' PR. P. 480; a strout P. S. 336 (P.
55); his eghne stode on strout ISUM. 620.

strūten, v., *cf. M.H.G.* striuzen; *strut; swell
out*; stroutin '*turgeo*' PR. P. 480; strout
contend HAV. 1779; **stroutende** (*pple.*) REL.
II. 15; **strūted** (*pret.*) AR. & MER. 233; his
heer (was) strouted (*pple.*) as a fanne CH. C.
T. *A* 3315.

stūbbe, stubbe, sb., *O.E.* stybb, = *M.L.G.*
stubbe, *O.N.* stubbi; *stock of a tree*, GAW.
2293; IPOM. 1270; **stubbes** (*pl.*) CH. C. T.
A 1978; a mong þe stubbe O. & N. 506.

stuble, sb., *O.Fr.* estuble, *cf. M.Du.* stoppel
M.H.G. stupfel; *stubble* ['*stipula*'] WICL.
JOSH. ii. 6; stubble PR. C. 3185; stubbil,
stobul PR. P. 481; stouple ROB. 223.

stubbel-, stobel-goos, sb., *stubble-goose*,
CH. C. T. *A* 4351.

stūcche, sb., *O.E.* stycce, = *M.L.G.* stucke,
O.H.G. stucchi *n.*; *piece, fragment*, '*frustum*,'
FRAG. 4; 'MISC. 64; a sticche of ure brede
P. L. S. viii. 96; stecche ANGL. I. 18; **stüc-
chen** (*pl.*) KATH. 2032; stucchen [sticches]
LA3. 16703; stechches AYENB. 111.

stūcchen, sb., *small piece*, A. R. 428; **stüc-
chenes** (*pl.*) A. R. 14; KATH. 2032*.

stude, sb., *O.E.* studu (*destina*), = *M.H.G.*
stud (*post*); *stud, button*; stode '*bulla*'
VOC. 175; **stodis** (*pl.*) E. W. 46.

stüde, *see* stede.

studie, sb., *O.Fr.* estudie; *study, thought*,
P. L. S. xvii. 217; AYENB. 78; WILL. 4056;
stodi ALEX. (Sk.) 263.

studien, v., *cf. O.H.G.* (ga-)studian, *O.N.*
styðja; *prop, support; stop*; heo ne **studeð**
[stut] (*pres.*) never ancre wununge A. R. 142;

hwi studgi [studiȝe] ȝe nu KATH. (E.) 1264 ;
ne studgeȭ ah sturieȭ MARH. 9 ; studis
ALEX. (Sk.) 2960 ; ine þise vour virtues ham
studede (*pret.*) þe iealde filozofes AYENB. 126.

studien², v., *O.Fr.* estudier ; *study* ; studie
P. L. S. xvii. 279 ; AYENB. 24 ; stodi ALEX.
(Sk.) 2480 ; **astudieȭ** (*imper.*) A. R. 200 ;
studieden (*pret.*) LANGL. *B* xv. 587.

studul, stodul, sb., = *O.H.G.* studil, *O.N.*
stuȭill ; *support, prop,* ' *telarium*,' PR. P. 481.
studel-fast ?, adj., *steadfast,* JUL. 74.

stue, *see* **stēwe, stüve.**

stüf-bæþ, *see under* **stüve.**

stuff, stuffen, *see* **stoffe, stoffen.**

stūke, sb., *L.G.* stūke, *prov. Eng., Sc.* stook ;
pile of sheaves ; stouke ' *arconius* ' CATH.
366 ; his hede is like a stouke TOWNL. 313.

stull, sb., *great piece* : stuffis so ȝour stomake
with stullis (*pl.*) ALEX. (Sk.) 4436.

stulpe, sb., *cf. ON.* stolpi ; *peg, post,* ' *paxillus,*' PR. P. 481 ; stulpes (*pl.*) PALL. i. 1054.

stumlen, stumblen, v., = *M.Du.* stomelen ;
stumble, PL. CR. 591 ; stummelin, stomelin
' *caespito* ' PR. P. 476, 481 ; stumble VOC. 143 ;
REL. I.6 ; stomble LANGL. *A* ix. 27 ; **stomble**
(*pres.*) REL. II. 211 ; stomblen CH. C. T. *A*
2613 ; **stomling** (*pple.*) TOR. 661 ; **stumbleȭ,**
stombled (*pret.*) MAN. (F.) 12435, 13050 ;
stomelid M. ARTH. 115.

stombling, stomling, sb., *stumbling,* ALEX.
(Sk.) 2623.

stumpe, sb., = *M.Du.* stompe, *M.L.G.* stump,
M.H.G. stumpf ; *stump, trunk,* PR. P. 481 ;
TOWNL. 308 ; **stompe** (*dat.*) JOS. 681 ; EGL.
739 ; **stumpes** (*pl.*) LIDG. M. P. 30 ; stompus
TRIAM. 1561.

stumpen, v.; *stumble* : þat stumpeþ (*pres.*)
at þe flesches more O. & N. 1392.

stumren, v., *O.N.* stumra ; *stumble* ; **stomere**
(*pres.*) REL. II. 211.

stunch, sb., = *O.H.G.* stunc ; *from stinken* ;
odour, smell, HOM. I. 43 ; A. R. 104 ; ARCH.
LII. 36 ; **stunche** (*dat.*) MISC. 77.

stúnde, sb., *O.E., O.N.* stund, = *O.Fris.* stunde,
stonde, *O.L.G.* stunda, *O.H.G.* stunt, stunta ;
moment, hour, time, KATH. 1269 ; HORN (L.)
739 ; ane lutele stunde LAȝ. 3440 ; sume
stunde O. & N. 1353 ; stund GEN. & EX. 2041 ;
stunt HAMP. PS. li. 5* ; stonde ALEX. (Sk.)
2831 ; oþer stund [= *O.H.G.* andera stunt]
ORM. 996 ; stounde SAINTS (Ld.) xlvi. 605 ;
LANGL. *B* viii. 65 ; M. ARTH. 114 ; he þolede
mani a biter stounde S. S. (Web.) 849 ; a longe
stounde A. D. 262 ; **stúnde** (*dat.*) A. R. 190 ; O.
& N. 802 ; HAV. 2614 ; stounde A. P. i. 658 ;
help nòu in þis stounde K. T. 1076 ; **stúndum**
(*dat. pl.*) *at times* HAMP. PS. xliii. 11*.

stúnd-mēle, adv., *O.E.* stundmælum ; *moment by moment,* HICKES I. 168 (HOM. II. 113) ;
stoundemele WILL. 736 ; stoundmeel WICL.
NUM. x. 7 ; stoundemele CH. TRO. v. 674.

stúnden, v., = *M.L.G.* stunden, *O.Swed.*
stunda ; *? remain* : no þing he 'ne stinte ne
stounded (*pret.*) MAN. (F.) 10902 ; ȭor he
stunden (*? for* stundeden) til helpe cam GEN.
& EX. 1987.

stountinge, sb., *delay,* D. ARTH. 491.

stunien, v., *stun, astonish* ; stoniin ' *percello,
stupefacio* ' PR. P. 476 ; **stonede** (*pret.*) AUD.
78 ; stouned GAW. 301 ; stonaid ALEX. (Sk.)
2589 ; **stoned** (*pple.*) PARTEN. 2940 ; stunaid
men HAMP. PS. *page* 510 ; *comp.* **a-stunien.**

stoniing, sb., *astonishment,* WICL. MK. v. 42 ;
stoneinge WICL. 4 KINGS xxii. 19 ; stoininge
PR. P. 477.

stunt, adj., *O.E.* stunt, = *O.N.* stuttr, *M.H.G.*
stunz ; *obtuse, foolish,* ' *stultus,*' FRAG. 1 ;
stunt & dil ORM. 3714.

stunt-līc, adj., *stupid,* HOM. I. 109.

stuntnesse, sb., *O.E.* stuntnyss ; *folly,* HOM.
I. 117.

stünten, v., *O.E.* (a-)styntan, = *O.N.* stytta ;
from stunt ; *stint, cease, stop* : þe qvale gon
to stunte LAȝ. 31891 ; stinten ORM. 12844 ;
to stinten al oure strif LIDG. M. P. 95 ; stintin
PR. P. 475 ; stinte WILL. 1042 ; LANGL. *B* i.
120 ; AV. ARTH. xxviii ; stinten, stenten CH.
C. T. *A* 903, 2732 ; **stünteþ** (*pres.*) C. L. 894 ;
stinteþ MIRC 897 ; stunt A. R. 202 ; (hi) stinteþ
P. L. S. xxxv. 99 ; **stünte** (*pret.*) A. D. 247 ;
stinte E. T. 241 ; er he stinte in eni stede
BEK. 1126 ; stente M. ARTH. 1844 ; **stünt,**
istunt (*pple.*) REL. I. 123 ; stunt is al mi
plawe A. D. 251 ; *comp.* **a-, æt-stünten.**

stinting, sb., *stoppage,* BARB. vii. 40, xii. 14.

stūpen, v., *O.E.* stūpian, = *O.N.* stūpa, *M.Du.*
stuipen ; *stoop,* JUL. 72 ; þat mon ne mæi
mid strengȭe stupen [stoupe] hine to grunde
LAȝ. 25950 ; stoupin ' *inclino* ' PR. P. 478 ;
stoupe ' *nutare* ' REL. I. 6 ; LANGL. *B* v. 394 ;
B. DISC. 322 ; over þe table he gon stoupe
ALIS. 1103 ; stope FER. 4065 ; **stoupeþ** (*pres.*)
TREV. II. 185 ; CH. C. T. *E* 2348 ; **stoupi**
(*subj.*) AYENB. 151 ; **stūpede** (*pret.*) MISC.
53 ; FL. & BL. 697 ; he stoupede doun H. S.
5615 ; IW. 3255.

stūr, adj., ? = *M.L.G.* stūr ; *rigid, strong*; stour
SPEC. 87 ; H. S. 11473 ; sture BARB. x. 158,
xii. 92 ; *? cf.* **stōr.**

sturdi, adj., *O.Fr.* estourdi, estordi ; *sturdy,*
' *contumax,*' PR. P. 481 ; REL. I. 6 ; sturdi
[stordi] CH. C. T. *E* 698 ; stordi ALIS. 1332 ;
GOW. III. 289 ; stordi, stourdi ROB. 157, 186.

sturdi-lī, adv., *sturdily,* BARB. ii. 363.

sturdinesse, sb., *sturdiness,* ' *contumacia,*'
PR. P. 481 ; REL. I. 7 ; CH. C. T. *E* 700.

Stūre, pr. n., *Stour,* P. L. S. viii. 126 ; Ailmar
rod bi Sture [Stoure] HORN (L.) 685 (R. 688) ;
Stoure LAȝ. 2472.

stūren, *see* **stēoren.**

sturgiūn, sb., *O.Fr.* esturgeon ; *sturgeon,*
HAV. 753.

stürien, v., *O.E.* styrian ; *stir, move, excite,* LAȝ. 17403 ; KATH. 1273 ; A. R. 130 ; sturie S. A. L. 159 ; FER. 876 ; sture HORN (R.) 1445 ; stiren ORM. 2810 ; AR. & MER. 2842 ; stirin PR. P. 476 ; stire, stere LANGL. *B* xvii. 54, 220 ; CH. C. T. *C* 346 ; sterie AYENB. 173 ; steren MAND. 22 ; stere GOW. II. 13 ; stere strif AUD. 35 ; stireð (*pres.*) REL. I. 209 ; ne he sake ne sturað HOM. I. 113 ; he . . . sturieð aa mare MARH. 9 ; sturieð leihtres A. R. 198 ; stüre (*imper.*) A. R. 290 ; ohtliche eou sturieð LAȝ. 15254 ; stüriinde (*pple.*) A. R. 152 ; stürede (*pret.*) KATH. 2146 ; ROB. 17 ; JOS. 567 ; sturede his tunge LAȝ. 17434 ; he stired þe coles CH. C. T. *G* 1278 ; stureden þa leoden LAȝ. 10717 ; stired (*pple.*) GEN. & EX. 3961 ; stired ['*motus*'] PS. xii. 15 ; istured LAȝ. 8118 ; istured hidere KATH. 797 ; *comp.* **a-, an-, bi-, to-**stürien.

    stürunge, sb., *O.E.* styrung ; *stirring,* HOM. I. 189 ; stiring [stering] WICL. MAT. viii. 24 ; steringe P. R. L. P. 251.

stürmen, v., *O.E.* styrman,= *O.N.* styrma, *O.H.G.* sturmen ; *from* **storm** ; *storm ; be stormy ; agitate* ; **storminge** (*pple.*) CH. BOET. i. 6 (29) ; stürmden (*pret.*) LAȝ. 18327 ; þa Freinsce weoren istürmede (*pple.*) LAȝ. 1670.

stürne, adj., *O.E.* styrne ; *stern, severe, austere,* A. R. 268 ; FL. & BL. 701 ; MAP 336 ; ROB. 27 ; P. S. 153 ; SPEC. 31 ; þat feiht was swiðe sturne LAȝ. 2473 ; he wes . . . sturne [sterne] wið þa dusie 6586 ; þe sturne hit bi-gripte GAW. 214 ; stirne ORM. 15514 ; sterne PR. P. 474 ; AYENB. 130 ; B. DISC. 402 ; sterne and stout DEGR. 105 ; steorne ALIS. 2390 ; sterine, sterinne D. ARTH. 157, 735, 3622 ; sterinneste (*superl.*) D. ARTH. 3872 ; **stürne** [sterne] (*adv.*) O. & N. 112.

    sterene-fulle, sterinfulle, adj., *fierce,* D. ARTH. 2692, 3824.

    stürn-hēde, sb., *severity,* ROB. 132.

    stürn-liche, adv., *O.E.* styrnlīce ; *sternly,* LAȝ. 27461 ; sturneliche ROB. 321 ; sterneliche LANGL. (Wr.) 5603 ; sterinli D. ARTH. 745.

sturte, sb., *impetuosity* : be noȝt to sturten with þi sturte ALEX. (Sk.) 3758.

sturten, adj., *impetuous, quarrelsome,* ALEX. (Sk.) 3758 ; to sterin or to sturtin 4257.

sturten, *see* sterten.

stüt, sb., *O.E.* stüt ; *midge,* '*culex,*' FRAG. 3 ; gnattes and stoutes (*pl.*) TREV. V. 159.

stuten, v., *O.Fr.* estoutir ; *become foolish* ; stotais (*pres.*) D. ARTH. 1435, 3467, 4271.

stutten, v.,= *M.L.G.* stutten, *M.H.G.* stutzen ; *cease, stay* : þah monie estterten us summe schulen stutten JUL. 51 ; þa ne cuðen ha neaver stutten [*v. r.* stunten] hare cleppen A. R. 72* ; stitte P. L. S. xiii. 239 ; stutteð (*pres.*) HOM. I. 267 ; stute (*imper.*) KATH. 1540 ; stuttinge (*pple.*) WICL. IS. xxxii. 4 ; stutte

(*pret.*) JUL. 70 ; þe stefne stutte MARH. 21 ; *comp.* **æt-stutten.**

stüve, sb., *O.Fr.* estuve ; *hot bath ; brothel* ; stue, steue (*ms.* stewe) '*stupha, terme*' PR. P. 481 ; **stuives,** stives, stiues (*ms.* stiwes), steues (*ms.* stewes), stuwes, stues (*pl.*) LANGL. *A* vii. 65, *B* vi. 72, *C* ix. 71 ; stives CH. C. T. *D* 1332 ; steues (*ms.* stewes) GOW. III. 291 ; MAND. 131 ; stues PL. CR. 631.

stüf-bæþ, sb., *hot bath,* LEECHD. III. 92.

stuwin, stúin, v., *O.Fr.* estuver ; *stew, give a hot bath,* '*stupho, balneo*' PR. P. 481.

suágin, v., *for* aswágen ; *assuage,* '*mitigo,*' PR. P. 481 ; suage PALL. iv. 883 ; **suágede** (*pret.*) TREV. V. 285 ; suáged (*pple.*) MAN. (F.) 4570.

subact, adj., *Lat.* subactus ; *well worked* : in lande subact PALL. xii. 216 ; *lying low* TREV. I. 145 ; *subject* II. 103.

subjeccioun, sb., *subjection, service,* CH. C. T. *B* 270 ; COMP. M. 32.

sublimatories, sb., pl., *vessels for sublimation,* CH. C. T. *G* 793.

sublīmen, v., *Lat.* sublīmāre ; *sublimate* ; sublīmed (*pple.*) CH. C. T. *G* 774.

    sublīming, sb., *sublimation,* CH. C. T. *G* 770.

submitten, v., *Lat.* submittere ; *submit* ; summitte TREV. III. 125 ; **submittede** (*pret.*) CH. BOET. i. 4 (19).

subsidie, sb., *? A.Fr.* *subsidie (subside) ; *subsidy,* CR. K. 36.

substance, sb., *O.Fr.* substance ; *substance,* AYENB. 113 ; CH. C. T. *C* 539.

substanciel, adj., *O.Fr.* substanciel ; *substantial,* AYENB. 113.

substantif, sb., *substantive* : adjectif and substantif LANGL. *C* iv. 345, 363.

suburb, sb., *Lat.* suburbium ; *suburb,* PR. P. 482 ; **suburbes** (*pl.*) CH. C. T. *G* 657.

subverten, v., *Lat.* subvertere ; *subvert,* WICL. TIT. iii. 11.

sūc, *see* sēoc.

succedent, sb., *Lat.* succedens ; *follower,* CH. ASTR. ii. 4 (19) ; (he) maketh to crafte nature a succedent PALL. iii. 1124.

successōr, sb., *O.Fr.* successur ; *successor,* TREV. IV. 14 ; CH. C. T. *E* 138 ; successoure ALEX. (Sk.) 4286.

such, *see* swülc.

suclen, v., = *L.G.* sukeln ; *suckle* ; **suclid** (*pple.*) WICL. JOB iii. 12.

sucre, sb., *O.Fr.* sucre ; *sugar,* AYENB. 83 ; SPEC. 26 ; sucre [sugre] CH. C. T. *F* 614 ; sucre, sugre, suger LANGL. *A* v. 100, *B* v. 122.

sucuren, v., *O.Fr.* succurre ; *succour* ; socoure WILL. 1186 ; MAN. (F.) 7469 ; soucouri AYENB. 186 ; socori ROB. 399 ; **sucurede** (*pret.*) MISC. 32.

sucurs, sb., *O.Fr.* sucurs; *succour*, A. R. 244; socurs MAN. (F.) 5166; socours ST. COD. 95; sucur FL. & BL. 674; socour CH. C. T. *B* 664; places of socour [soker] ['*xenodochia*'] TREV. IV. 137.

sudarie, sb., *Lat.* sūdārium; *napkin*, YORK xxxvi. 387; TREV. V. 189.

sudeakne, sb., *O.Fr.* subdiacne; *subdeacon*, SHOR. 50.

sudēn, sb., *subdean*; **sudēnes** (*pl.*) LANGL. *A* ii. 150.

sudūen, v., *O.Fr.* souduire; *subdue*; **sodewed** [sudewid] (*pret.*) TREV. III. 19; **sodūed** [sudued] (*pple.*) TREV. III. 123.

suen, see **sēon.**

suet, sb., *cf. O.Fr.* seu; *suet*; suet(e) '*liquamen*' PR. P. 483.

suffisant, adj., *O.Fr.* suffisant, *Lat.* sufficiens; *able, sufficient*, CH. C. T. *B* 243, *C* 932; sufficiant ALEX. (Sk.) 4396; sufficiand BARB. i. 368.

suffīse, v., *O.Fr.* soufire; *suffice*; souffise HOCCL. i. 356; **suffīce** (*subj.*) ALEX. (Sk.) 3196.

suffrable, adj., *tolerable, merciful*, MAN. (F.) 16513; suffrabil HAMP. PS. cvi. 29*.

suffraunce, sb., *O.Fr.* sufrance; *sufferance, permission, patience*, TREV. V. 135; LANGL. *C* i. 124, *A* iii. 93, *B* xi. 370; suffrance *B* vi. 146; unsittinge suffraunce *C* v. 189.

suffraunt, adj., *patient*, CH. D. BL. 1010.

suffren, v., *O.Fr.* sufrir, soffrir; *suffer*, LANGL. *C* xxii. 68; soffri LAȝ. 24854*; i **suffre** (*pres.*) LANGL. *A* iv. 1; soffreþ AYENB. 139; **suffride** (*pret.*) LANGL. *C* xiv. 17*; **suffred** (*pple.*) WILL. 1014.

suȝe, sb., *O.E.* sugu, = *M.L.G.* suge, soge; *sow*; suwe A. R. 204; zoȝe AYENB. 61; soghe TOWNL. 91; sowe LANGL. *B* xi. 333; CH. C. T. *A* 2019; sowe [souwe] WICL. LEV. xi. 7, PROV. xi. 22; sowe [*cf. med.Lat.* sus, scrofa] *a warlike engine* ROB. (W.) 8480; sow BARB. xvii. 597, 621.

suge-þistel, sb., *sow-thistle*, LEECHD. III. 346; southistile PR. P. 466.

suget, adj. & sb., *O.Fr.* subgiet; *subject*: i am simpill sugett of thine YORK xiv. 64; soget WILL. 473; PL. CR. 650; þe citee was lower and sogett to þe temple TREV. I. 111; soget 287; þerh is suget Devenschire . . . þe Est-Saxons . . . were **sogettis** (*pl.*) to þe bisshop of Londoun TREV. II. 123; subgetz CH. C. T. *E* 482.

sugetten, v., *subject*; **sugettide** (*pret.*) WICL. GEN. xlvii. 20.

sugge, sb., *O.E.* sucge; *name of a bird*, '*curruca*,' PR. P. 483; *comp.* **hei-sugge**.

süggen, see **seggen.**

suggestioun, sb., *O.Fr.* suggestioun; *criminal charge*, LANGL. *B* vii. 67; CH. C. T. *B* 3607.

suhȝhen, v., = *Dan.* sukke, *?Swed.* sucka; *sigh, pant*, ORM. 7924; **suggeþ** (*pres.*) ST. COD. 54; **suggeden** (*pret.*) REL. I. 224.

suht, sb., *O.E.* suht, = *O.L.G.*, *O.H.G.* suht, *O.N.* sótt; *sickness*; soght C. M. 14157; *comp.* ȝalou-, gol-souȝt.

suite, see **seute.**

[süke? sb.; *from* süken; *comp.* huniȝsouke.**]**

süken, v., *O.E.* sūcan, sūgan, = *O.H.G.* sūgan, *O.N.* sūga, *cf. Lat.* sūgere; *suck*, LAȝ. 13194; souken C. L. 78; ȝeeven . . . souken ['*lactaverunt*'] WICL. LAM. iv. 3; þou shalt souke ['*suges*'] þe milc Is. lx. 16; souke GREG. 109; LANGL. *B* xi. 116; ȝeveþ zouke AYENB. 60; **sükeð** (*pres.*) H. M. 39; REL. I. 220; souketh PALL. xi. 16; soukes PR. C. 6767; neddren heore breosten sukeþ MISC. 151; **sükende** (*pple.*) LAȝ. 20973; sukinde HOM. I. 5; soukend GOW. I. 268; **sēc** (*pret.*) A. R. 330; sek AR. & MER. 2861; sæc [soc] LAȝ. 12981; sek S. A. L. 159; sok OCTOV. (W.) 555; P. L. S. iv. 39; S. & C. II. xlvi; þu suke LK. xi. 27; þa titles þæt þu suke [soke] LAȝ. 5026; soke AR. & MER. 2954; AL. (L.[1]) 1045; soken ALIS. 6119; **soken** (*pple.*) LUD. COV. 28; soke H. S. 6024; isoke TREV. III. 267; *deriv.* **sok.**

[sukle, *?* sb., *from* süken; *mod.Eng.* (honey-) suckle.**]**

sokil-blome, sb., *? honeysuckle*; '*locusta*' VOC. (W. W.) 787.

sukling, sokelinge, sb., *a herb*, '*locusta*,' PR. P. 463.

sükling, sokling, sb., = *M.H.G.* süglinc; *suckling*; sokelinge '*sububer*' PR. P. 463.

sulaine, see **solain.**

sulc, see **swülc.** **sülf**, see **self.**

sulȝe, see **suluh.**

sülien, v., *O.E.* sylian, = *O.L.G.* sulian, *M.H.G.* suln; *soil, defile, pollute*; **süleð** (*pres.*) H. M. 35; sulieþ O. & N. 1240; **isüled** (*pple.*) PL. CR. 752; isuled, isuiled A. R. 158, 396; isulet MARH. 3; *comp.* **bi-sülien.**

sülle, sb., *O.E.* syll, = *O.N.* sylla, *M.Du.* sulle, *Goth.* sulja; *sill, base, platform*, '*basis*,' FRAG. 5; sille PR. P. 456; sille (*dat.*) GAW. 55; selle CH. C. T. *A* 3822.

süllen, see **sellen.** **süllīc**, see **sellic.**

sulpen, v., *? pollute*; solp ALEX. (Sk.) 4292; sulpande (*pple.*) A. P. i. 725; sulped (*pret.* *pple.*) A. P. ii. 15; *comp.* **bi-sulpen.**

suluh, sb., *O.E.* sulh *f.*; *plough*, A. R. 384; sulȝe LAȝ. 31811; soluȝ LEG. 46; solouȝ [solowe] TREV. VII. 535; solou VOC. 180; zuolȝ AYENB. 242; þritti solh (*pl. carucates*) of londe LAȝ. 18779; twenti sulhene [solȝene] lond LAȝ. 13176; sulȝene [solwene] 18789.

sülver, see **silver.**

sum, adj. & pron., *O.E.* sum, = *O.L.G.* sum, *O.N.*

sumr, *O.H.G.* sum, *Goth.* sums ; *some, certain,* '*aliqvis,*' PR. P. 484 ; MAND. 265 ; sum blind man O. & N. 1237 ; sum was king GEN. & EX. 834 ; swilc oðer sum 686 ; er he sum þing þer of wiste P. L. S. xiii. 134 ; sum man [*'homo qvidam'*] WICL. LK. xv. 11 ; HOCCL. i. 261 ; sum time WILL. 780 ; oþer sum *others* C. M. 18967 ; te sea sencte him on his þriture sum (*him with thirty others*) JUL. 79 ; sum [som] LAȝ. 12055 ; som CH. C. T. *C* 409 ; som time ROB. 5 ; his fiftend som TRIST. 817 ; sumes (*gen. m. n.*) HOM. II. 217 ; ORM. 18702 ; sumes weis A. R. 354 ; summes cunnes LAȝ. 21765 ; sume kunnes FL. & BL. 415 ; sum kin BARB. x. 519 ; sumen (*dat. m.*) MAT. xxv. 15 ; sume O. & N. 293 ; et sume time A. R. 48 ; summere (*dat. f.*) LAȝ. 16842 ; sumne (*acc. m.*) LAȝ. 10139 ; A. R. 28 ; FL. & BL. 318 ; somne (*printed* soume) reed REL. II. 276 ; sume (*acc. f.*) A. R. 390 ; sume (*pl.*) ORM. 6574 ; sume men HOM. II. 37 ; O. & N. 879 ; A. R. 8 ; summe [some] LANGL. *A* PROL. 20 ; some SHOR. 24 ; hii clupede me samdede king some ROB. (W.) 3415 ; þe angvisses some 7204 ; bigonne to fle some 8171 ; summen (*dat. pl.*) LAȝ. 24111.

sum-dēl, adv., *somewhat* ; sumdeill grai BARB. i. 383 ; SAINTS (Ld.) xlv. 168 ; she was som del deef CH. C. T. *A* 446.

sum-tīme, adv., *sometimes*, LANGL. *C* xviii. 99* ; '*aliqvando*' CATH. 371.

sum-whær, adv., *somewhere*, ORM. 6929 ; sumwher SPEC. 110 ; sumwar P. L. S. xii. 78.

sum-what, pron. & adv., *somewhat*, KATH. 507 ; A. R. 44 ; sumwhat ORM. 958.

sum², conj., *cf. Dan., Swed.* som, *O.N.* sem ; *O.E.* some, same, = *O.H.G.* sama ; *so, as, soever,* C. M. 1149 ; al swa hit sum iwarð LAȝ. 1497 ; swa sum (*for so*) þe godspel kiþeþ ORM. 302 ; whær sum (*wheresoever*) we finden 1827 ; whare sum þai ȝede Iw. 30 ; sum (*for* swa sum) i þe telle AMAD. (R.) lxix ; what som evere SHOR. 103 ; what som ever EGL. 56 ; hou som it be IW. 1507.

[-sum, suffix, *O.E.* -sum, = *O.L.G.*, *O.H.G.* -sam, *O.N.* -samr, *Goth.* -sams ; *-some ; comp.* ang-, bēi-, būh, folȝ-, glad-, hāl-, hand-, hēr-, lāð-, luf-, lust-, not-, nüht-, sib-, wan-, wlat-, wün-sum (-som, -sam).

sumer, sb., *O.E.* sumor, *cf. O.Fris.* sumur, *O.L.G., O.H.G.* sumar, sumer *m.*, *O.N.* sumarr *m.*, sumar *n.; summer,* MAT. xxiv. 32 ; A. R. 20 ; ORM. 11254 ; REL. I. 220 ; sumer [somer] LAȝ. 22250 ; somer CH. C. T. *A* 394 ; M. H. 22 ; sumeres (*gen.*) LAȝ. 2861 ; sumeres tide O. & N. 489 ; þe someris dai P. L. S. xxxv. 151 ; sumere (*dat.*) FRAG. 2 ; O. & N. 416 ; zomere AYENB. 131 ; *comp.* mid-sumer.

somer-dai, sb., = *M.H.G.* sumertac ; *summer-day,* GOW. I. 184.

somer-game, sb., *summer game,* LANGL. *B* v. 413.

sumer-līch, adj., *O.E.* sumorlīc ; *summer-like,* KATH. 1678.

Sumer-sēte, pr. n., *O.E.* Sumorsǣte ; *Somerset,* LAȝ. 21013 ; Somersete ROB. 3 ; Sumersete scire SAX. CHR. 249.

sumer-tīd, sb., = *O.H.G.* sumerzīt ; *summer time,* GEN. & EX. 1224 ; somertide CH. C. T. *F* 142.

somer-tīme, sb., *O.N.* sumartīmi ; *summer time,* LANGL. *B* xv. 94.

sumer²? adj., *? certain* : in one sumere dale O. & N. 1.

summe, sb., *O.Fr.* somme ; *sum, amount, number,* BEK. 332 (SAINTS (Ld.) xxvii. 386) ; somme AYENB. 260 ; D. ARTH. 448 ; soumme A. P. iii. 509.

summer, *see* somer.

summitee, sb., *O.Fr.* sommette ; *summit* ; PALL. iv. 240 ; summite BARB. iii. 706.

sumne, sumnien, *see* samne, samnien.

sumpter, *see* somer.

súnd, sb., *O.E.* = *O.L.G.* (ge-)sund, *O.H.G.* (ge-)sunt ; *sound, healthy,* LAȝ. 15762 ; ORM. 14818 ; FL. & BL. 364 ; REL. I. 210 ; sound HORN (R.) 580 ; CHR. E. 749 ; P. L. S. xii. 109 ; SPEC. 89 ; Iw. 2740 ; sauf and sound LANGL. *B* viii. 34 ; sond REL. I. 161 ; sounde (*adv.*) OCTAV. (H.) 72 ; *comp.* ȝe-súnd.

súnd-fullen, v., *O.E.* gesund-fullian ; *be healthy, prosper* ; soundfulle PS. i. 3.

súnd², sb., *O.E.* sund, = *O.N.* sund ; *? for* swund ; *sound (of the sea), strait, frith,* LAȝ. 21326 ; C. M. 621 ; sound HORN (R.) 628 ; PR. P. 466.

súnde, sb., *cf. O.Fris.* sunde, *O.H.G.* (gi-)suntī ; *health* ; mid sunde LAȝ. 4967 ; wiþ sounde LAI LE FR. 86 ; JOS. 675 ; in sounde GAW. 2489 ; *comp.* ȝe-súnde.

sunder, adj. & adv., *O.E.* sundor, = *O.L.G.* sundor, sundar, *O.N.* sundr, *O.H.G.* suntar ; *separately* ; **in sonder** [= *O.N.* ī sundr, *O.H.G.* in suntar] *in sunder* PR. C. 1787 ; EGL. 389 ; in sondir TRIAM. 200 ; in sinder P. P. 216 ; on sunder *asunder* [*'seorsum'*] MAT. xvii. 1 ; on sunder, o sunder GEN. & EX. 116, 3909 ; a sundir H. S. 1671 ; a sonder CH. C. T. *A* 491 ; on sundren [*O.E.* on sundran, sundrum, = *O.L.G.* an sundron] MAT. xvi. 22 ; a sundren [*'seorsum'*] MK. iv. 34.

sunder-blēo, sb., *divers colour* ; **sunder-blēs** (*gen.*) GEN. & EX. 1729.

sunder-hälȝe, sb., *O.E.* sundorhālga ; *pharisee,* HOM. I. 245.

sünder-līch, adj., *O.E.* synderlīc, = *O.H.G.* sunterlīch ; *separate, divers, particular* ; sunderliche þinges A. R. 14 ; sünderliche (*adv.*) H. M. 3 ; sinderliche C. L. 1508.

sunderlinge, adv., = *O.Fris.* sunderlinge ;

separately: to uch one sunderling he ȝaf a dole C. L. 290.

sunder-lipe, -lipes, adv., *? O.E.* sundorlēpe, -lēpes,=*O.Fris.* sunderlēpis ; *separately,particularly*, HOM. I. 11, 137 ; sunderlepes 261 ; A. R. 90* ; sunderlupes HOM. II. 5 ; A. P. iii. 12 ; sonderlipes MAN. (F.) 3879.

sunder-rēd, sb., *divers counsel*, GEN. & EX. 3808.

sunder-rūne, sb., *private conversation, particular counsel*, LAȝ. 31414 ; HOM. II. 29 ; sunderrun ORM. 16978 ; GEN. & EX. 991.

sunder², sb., *O.E.* suner (LIND. MT. viii. 30) ; *herd of swine*, GAW. 1440.

sundren, v., *O.E.* sundrian, syndrian,=*O.N.* sundra, *O.H.G.* suntarōn ; *sunder, separate*, GEN. & EX. 468 ; sundren god from uvele A. R. 270 ; nan ne mei sundren fram oðer KATH. 1794 ; sondre PR. C. 4789 ; sundreð (*pres.*) H. M. 23 ; sundren REL. I. 224 ; **sundrede** (*pret.*) A. R. 414 ; sundride WICL. DEUT. xxxii. 8* ; **sundred** (*pple.*) GAW. 659 ; *comp.* ȝe-**sundren.**

sundring, sb.,=*M.H.G.* sunderunge ; *sundering*, GEN. & EX. 458.

sundri, sündri, adj., *O.E.* syndrig,=*O.N.* sundrugr, *O.H.G.* suntrīg ; *sundry, separate, diverse*, GEN. & EX. 1985 ; sundri, sindri LAȝ. 2688, 11832 ; sondri Gow. II. 261 ; CH. C. T. *B* 181.

sune, sb., *O.E.* sunu,=*O.L.G.* sunu, *O.Fris.* sunu, sune, sun, *O.H.G.* sunu, sun, *O.N.* sunr, sonr, *Goth.* sunus ; *son*, A. R. 26 ; ORM. 494 ; GEN. & EX. 46 ; MISC. 29 ; sune, suna MAT. i. 20, ii. 15 ; suna LEECHD. III. 82 ; sune [sone] LAȝ. 246 ; sone HAV. 660 ; SHOR. 132 ; WICL. JOHN xvii. 1 ; CH. C. T. *A* 79 ; **sune** [*O.E.* suna] (*gen.*) MAT. xxiv. 27 ; HOM. I. 123 ; JUL. 48 ; MISC. 191 ; sune [sones] LAȝ. 9630 ; sunes MAT. xxii. 2 ; MARH. 2 ; GEN. & EX. 1984 ; sone ROB. 479 ; **sune** (*dat.*) HOM. I. 41 ; sune [sone] LAȝ. 309 ; **sune** [*O.E.* suna] (*pl.*) HOM. I. 55 ; sunan SAX. CHR. 248 ; sunen A. R. 270 ; sunen [sones] LAȝ. 2541 ; sunes, sunas MAT. xx. 21, xxi. 28 ; sunes ORM. 488 ; HOM. I. 225 ; sunes, sunen GEN. & EX. 540, 2175 ; sone, sones L. H. R. 18, 20 ; **sune** (*gen. pl.*) MAT. xxvii. 56 ; **sunan** (*dat. pl.*) LAȝ. 2538 ; *comp.* **god-, mōder-, stēop-sune.**

sünegen, v., *O.E.* syngian,=*M.L.G.* sundigen ; *sin*, A. R. 56 ; suneȝan HOM. I. 103 ; singen GEN. & EX. 172 ; MIRC 1073 ; sunegie MISC. 78 ; sunegi LEG. 154 ; sunge 178 ; seneȝi SHOR. 32 ; zeneȝi AYENB. 20 ; sinewi P. L. S. xvii. 450 ; **süneȝeð** (*pres.*) HOM. I. 153 ; sinȝheþ ORM. 3970 ; singeþ, singeþ, sinneþ LANGL. *A* ix. 17, *B* viii. 22, *C* xi. 23 ; we sunegieð HOM. I. 17 ; **sünge** (*imper.*) H. V. 110 ; **sünegie** (*subj.*) A. R. 58 ; sunegi O. & N. 928 ; **süneȝede** (*pret.*) P. L. S. viii. 142* ; sunegede JUL. 60 ; sinied P. L. S. iii.

24 ; þu sungedest HOM. I. 9 ; sunegeden MISC. 68 ; sungeden C. L. 1381 ; singeden P. R. L. P. 243 ; *comp.* **for-, ȝe-süneȝen.**

zeneȝere, sb., *sinner*, AYENB. 113.

sünegild, sb., *female sinner*, H. M. 43.

sünegunge, sb., *sinning*, A. R. 52 ; suneginge MISC. 141.

sunien, v., *see* **schunien.**

sunne, sb., *O.E.* sunne, = *O.L.G.* sunna, sunne, *O.N.*, *O.H.G.* sunna, *Goth.* sunnō ; *sun*, MARH. 9 ; A. R. 140 ; ORM. 7273 ; GEN. & EX. 132 ; WILL. 3073 ; ALIS. 639 ; AV. ARTH. lxv ; LUD. COV. 21 ; sunne, sonne [sonne] LAȝ. 7239, 8122 ; sunna SAINTS (Ld.) xlvi. 436 ; C. L. 101 ; CH. C. T. *A* 7 ; **sunnen** (*gen.*) **dai** (*Sunday*) M. H. 5 ; sun-(n)en dei A. R. 412 ; sun(n)en daȝ ORM. 4360 ; sunnen, sunne dei HOM. I. 139 ; sonnen dai ROB. 495 ; sun(n)en niht A. R. 22 ; after sunna upgange LEECHD. III. 98 ; þe sunne hete HOM. II. 151 ; ase is þe sunne gleam A. R. 94 ; it malt at ðe sunne sine (*sunshine*) GEN. & EX. 3337 ; on one sun(n)e niȝte MISC. 162 ; þe sunne upriste HORN. (L.) 1436 ; **sunnen** (*dat.*) LK. xxi. 25 ; under sunnan LAȝ. 108 ; under þere sunnen 24982 ; under sunne A. R. 250 ; MARH. 15.

sunne-bǣm, sb., *O.E.* sunnebeam ; *sunbeam*, ORM. 7278 ; sunnebem HAV. 592 ; sonnebem SPEC. 33 ; zonnebiam AYENB. 108.

sünne, sb., *O.E.* synn, sinn, senn, *O.N.* synd, *O.L.G.* sundia, sundea, *O.H.G.* sunta ; *sin*, KATH. 1177 ; A. R. 10 ; H. M. 35 ; sunne, sinne C. L. 232 ; sinne ORM. 816 ; GEN. & EX. 182 ; HAV. 536 ; MIRC 71 ; LANGL. *B* i. 147 ; MAND. 249 ; senne MAT. xii. 31 ; MISC. 28 ; SHOR. 32 ; zenne AYENB. 16 ; **sünne** (*pl.*) HOM. I. 33 ; O. & N. 1395 ; sinne C. L. 828 ; sinne, senne, sinnen MAT. ix. 6 ; MK. ii. 5 ; LK. v. 20 ; sinne, senne, sinnen, sennen, sinnes HOM. II. 5, 11, 69, 73 ; FRAG. 8 ; sunne, sinnes L. H. R. 30 ; sunnen (*r. w.* wunne) LAȝ. 24116 ; sunnen MARH. 26 ; A. R. 144 ; REL. II. 276 ; **sünna** (*gen. pl.*) HOM. I. 37 ; HOM. I. 29 ; **sünnen** (*dat. pl.*) O. & N 858 ; sunnan, sunne HOM. I. 35, 127 ; sinnen MAT. i. 21 ; HOM. II. 87 ; *comp.* **hēaved-sünne.**

sün-bōte, sb., *expiation of sin*, HOM. I. 51 ; sinbote HOM. II. 83.

sün-ful, adj., *O.E.* syn-, synnfull ; *sinful*, A. R. 56 ; R. S. ii ; ALIS. 4627 ; sinful ORM. 12048.

sinful-līke, adv., *sinfully*, ORM. 16155.

sinful-hēd, sb., *sinfulness*, GEN. & EX. 180 ; sunfolhēde S. A. L. 50.

sinfulnesse, sb., *sinfulness*, PR. P. 456.

sinne-lēas, adj., *O.E.* synlēas ; *sinless*, JOHN viii. 7 ; sinnelæs ORM. 5742 ; senneles SHOR. 3.

sünni, adj., *O. E.* synnig ; *sinful* ; sinni
PARTEN. 5218.

sūpen, v., *O.E.* sūpan,=*M.L.G.* sūpen, *O.N.*
sūpa, *O.H.G.* sūfan ; *sup, drink up, suck up,
sop* ; soupen LANGL. *B* ii. 96 ; me wille soupen
win ST. COD. 100 ; soupe P. S. 334 ; soupe
'sorbeo' PR. P. 466 ; **sōp** (*pret.*) ANGL. I.
314 ; soop up [*'absorbuit'*] WICL. APOC. xii.
16 ; **sopen** (*pple.*) WICL. PS. cxxiii. 4 ;
deriv. **sope, soppe.**

superfluitē, sb., *O.Fr.* superfluité ; *superfluity,* ALEX. (Sk.) 4277.

supowell, sb., *extraneous aid,* ALEX. (Sk.)
4300 ; suppowale BARB. xvi. 139.

supowellen, v., *support* ; suppowelle D.
ARTH. 2818.

supplement, sb., *Lat.* supplēmentum ; *patch,*
WICL. MK. ii. 21.

supplíen, v., *supplicate* ; **supplíed** (*pret.*)
ALEX. (Sk.) 163.

suppōsen, v., *O.Fr.* supposer ; *suppose* ; suppose CH. C. T. *D* 786 ; PR. C. 3776.

suppōsinge, sb., *supposition,* CH. C. T. *E*
1041.

suppriour, sb., *sub-prior,* LANGL. *C* vii.
153.

suprísen, v., *from O.Fr.* surpris-, surprendre ;
take by surprise ; suprise ALEX. (Sk.) 2390 ;
suppriss BARB. vi. 37 ; **supprísede** (*pple.*) D.
ARTH. 2616 ; supprisside 1420 ; supprised
R. R. 3235 ; supprisit BARB. xviii. 426.

sūr, adj., *O.E.* sūr,=*M.L.G.* sūr, *O.N.* sūrr, *O.
H.G.* sūr ; *sour,* HOM. I. 129 ; ORM. 15208 ; O. &
N. 866 ; MISC. 77 ; sour SPEC. 114 ; a soure
lof LANGL. *B* xiii. 48 ; **sūrre** (*compar.*) A. R.
114 ; **sūre** (*adv.*) O. & N. 1082 ; *deriv.* **sūren.**

sūr-dāgh, sb.,=*M.H.G.* sūrteic ; *'fermentum,' sour dough, leaven,* VOC. 201 ; sourdouȝ
WICL. MAT. xiii. 33.

sūr, *see* seūr.

sūrance, sb., *for* assūrance ; *assurance* ;
surrauns D. ARTH. 1381.

surcharge, sb., *Fr.* surcharge ; *additional
load,* BARB. xvi. 458.

surcote, sb., *O.Fr.* surcote ; *surcoat,* CH. C.
T. *A* 617 ; surcott(e) D. ARTH. 2434, 3252.

[**sūre,** sb., *O.E.* sūre,=*O.H.G.* sūre, *O.N.*
sūra (*acid*) ; *comp.* **wode-sūre.**]

sūren, v., *O.E.* sūrian,=*O.H.G.* sūren ; *turn
sour* ; sourin *'aceso'* PR. P. 466 ; zoureþ
(*pres.*) AYENB. 205 ; sourid (*pple.*) WICL.
MAT. xiii. 33.

sūren, v., *for* assūren ; *promise* ; **sūred** [en-
surid] (*pple.*) ALEX. (Sk.) 2633.

sūretē, sb., *O.Fr.* seureté ; *careless confidence,*
CH. AN. 215.

surfēt, sb., *O.Fr.* surfait ; *surfeit, 'excessus'*
PR. P. 484 ; LANGL. *A* v. 210 ; surfeet TREV.
IV. 329.

surgerie, sb., *O.Fr.* cirurgie (?*cirurgerie) ;
surgery, CH. C. T. *A* 413 ; surgerie [surgenrie]
LANGL. *B* xvi. 106.

surgien, *see* cirurgian.

surmísen, v., *blame* ; **surmísing** (*pple.*)
TREV. VIII. 526.

surmounten, v., *O.Fr.* sur-, sour-, sormon-
ter ; *surmount* ; sourmounten CH. BOET. iii.
7 (80) ; **surmontes** (*pres.*) ALEX. (Sk.) 4449 ;
surmountid (*pret.*) 2361.

surnoun, sb., *O.Fr.* sur-, sornom ; *surname* ;
sournoun CHR. E. 982 ; surname TREV. III.
265 ; sorname LANGL. *C* iv. 369.

surplis, sb., *O.Fr.* sur-, sorpliz, -peliz ; *sur-
plice, 'superpellicium,'* VOC. 231 ; CH. C. T.
A 3323 ; surplice PR. P. 485 ; **surples** (*pl.*)
ALEX. (Sk.) 1550.

surqviderie, sb., *O.Fr.* surqviderie (*pré-
somption*) ; *presumption,* A. R. 56 ; surqvidrie
CH. C. T. *I* 403 ; surqvidri, surqvitri ALEX.
(Sk.) 4293, 4254 ; surqvidre GAW. 2457 ;
sorqvidriȝe A. P. i. 309 ; succudri BARB. xi.
11, xvi. 327 ; sukudri BARB. xi. 11*.

surqvidour, sb., *proud man* ; **sorqvidours**
(*pl.*) LANGL. xx. 341.

surtraien, v., *strain* ; surtrai PALL. iii. 1097.

surtraiten?, v., *withdraw* ; surtreet (*imper.*)
PALL. iv. 460.

surveiance, sb., *inspection,* CH. C. T. *C* 95.

[**surveien,** v., *A.Fr.* surveer ; *survey, oversee.*]

surveior, sb., *surveyor,* E. W. 54.

sus, adv., ?=*O.L.G., O.H.G.* sus ; *thus :* þo
þe sus (*? for* þus) biggeð and sulleð HOM. II.
215.

suspecīon, sb., *O.Fr.* souspection ; *suspicion,
expectation,* MAN. (F.) 2905 ; LANGL. *C* xviii.
315 ; suspecioun CH. C. T. *D* 306 ; suspec-
tioun TREV. III. 117.

suspect, adj., *O.Fr.* suspect ; *open to suspicion,* CH. C. T. *E* 905.

suspenden, v., *O.Fr.* suspendre ; *suspend* ;
suspendede (*pret.*) ROB. 563 ; **sospendiez**
(*pres.*) SAINTS (Ld.) xxvii. 856.

sustenance, sb., *O.Fr.* soustenance ; *sustenance,* P. L. S. xvi. 70 ; sustenaunce P. S.
340 ; SHOR. 90 ; sustinance ALEX. (Sk.)
4269.

susténe, v., *O.Fr.* sus-, sous-, sostenir ; *sus-
tain,* ROB. 108 ; sustene-, -teine LANGL. *B* ix.
108 ; susteene CH. C. T. *B* 160 ; sostieni
AYENB. 56 ; **sousteined** (*pret.*) ROB. (W.)
2386.

susteininge, sb., *support, sustenance,* TREV.
II. 219.

süster, sb., *O.E.* sweostor, swustor,=*O.Fris.*
swester, suster, *O.N.* systir, *O.L.G., O.H.G.*
svester, *Goth.* swistar, *Russ.* сестра, *Lat.*
soror ; *sister,* H. M. 17 ; WILL. 2643 ; MIRC
831 ; suster LANGL. *A* iii. 54 ; sustre *B* iii.
63 ; CH. C. T. *A* 871 ; E. G. 9 ; suster [soster

LAȝ. 25534; C. L. 358; soster P. L. S. xiii.
144; P. S. 205; zoster AYENB. 89; sister
GEN. & EX. 766; PR. P. 457; **suster** [soster]
(*gen.*) LAȝ. 3813; þere qvene suster sunen
3757; **swüstre** [*O.E.* sweostor] (*pl.*) MAT.
xiii. 56; sustre HOM. I. 7; sustren A. R. 2;
C. L. 462; MAND. 102; sustren [sostres] LAȝ.
3032; sustres ORM. 6382; *comp.* **half-, stēop-
süster.**

 süster-sone, sb., = *M.H.G.* swestersun;
(? suster sone) ROB. 37; we are **siȝtersones**
(*pl.*) PERC. 1441.

sūtare, sb., *O.N.* sūtari, = *O.H.G.* sūtāri; *bootmaker,* A. R. 324; soutere PL. CR. 752; E. G.
358; soutare '*sutor*' PR. P. 466; **zouteres**
(*gen.*) AYENB. 66; a souter son TREV. VII.
135.

 souteresse, sb., *female shoe-dealer,* LANGL.
B v. 315.

sutel, adj., *O.E.* sweotol, swutol, sutol; *manifest, evident,* KATH. 381; A. R. 154; ORM.
18862; sutel [sotel] LAȝ. 1519; sotel SPEC.
23; swutel MK. vi. 14.

 sutel-līche, adv., *O.E.* sweotollīce; *plainly,*
HOM. I. 41; KATH. 1340.

sutelin, v., *O.E.* sweotulian; *manifest, make
evident,* JUL. 18; **soteleþ** (*pres.*) SPEC. 23;
sutelie (*subj.*) A. R. 154; **sutelede** (*pret.*)
KATH. 1036; *comp.* **ȝe-sutelien.**

suteltē, sb., *O.Fr.* sotilleté; *subtilty, skill,
ingenious contrivance,* PR. C. 5903; subtilte
CH. C. T. *G* 844; sotelte L. C. C. 5; C. B.
68, 69; sotilte TREV. I. 231; sutelte BARB.
iii. 611; sutell BARB. xix. 32.

suti (? **sūti**), adj., *sooty, foul,* A. R. 228; H.
M. 35; mi saule þet is suti HOM. I. 185; hu
swart þing ant hu suti is sunne MARH. 15;
sutti OCTAV. (H.) 885.

sutil, adj., *O.Fr.* soutil; *subtle;* sotil AYENB.
24; LANGL. *B* xv. 12; sotil [soutil, subtil]
CH. C. T. *A* 2049; sotel [sotil] TREV. III.
369.

 sotel-(l)iche, adv., *cleverly, skilfully,* TREV.
IV. 141, 391; ȝuttili YORK viii. 77.

sutilen, v., *make subtle; reason subtly;*
sutils (*pres.*) HAMP. PS. ii. 2*; suteleþ LANGL.
C xxii. 459; **sotelide** (*pret.*) xxi. 336.

sūð, sb. & adv., *O.E.* sūð, = *O.H.G.* sund, *O.N.*
sunnr; *south,* GEN. & EX. 829; REL. I. 211;
Merlin ferde riht suð [suþ] LAȝ. 18898; suþ
ORM. 12125; MISC. 96; suth, south HAV.
434, 1255; wheþer i be souþ oþer west SPEC.
59; south '*auster*' PR. P. 466; þet zouþ
AYENB. 124.

sūþ-dāle, sb., *O.E.* sūðdæl; *south part,*
ORM. 16418.

sūð-ende, sb., *south quarter,* LAȝ. 3372.

Sūþ-fólk, pr. n., *O.E.* Sūðfolc; *Suffolk,*
MISC. 146; Souþfolk P. L. S. xiii. 64.

sūð-ȝæt [sūþ-ȝeat], sb., *south gate,* LAȝ.
27932.

suð-half, sb., *O.E.* sūðhealf; *south quarter,*
LAȝ. 15937; souþhalf LEG. 192.

Sūð-hǽmtūn, pr. n., *Southampton,* LAȝ.
19917.

sūð-lond, sb., *O.E.* sūðland; *southern land,*
LAȝ. 2111.

Sūð-sǽxe, pr. n., *O.E.* Sūðseaxe; *Sussex,*
LAȝ. 15368; Suþsexe MISC. 146; Souþsex P.
L. S. xiii. 51.

souþ-side, sb., *south side,* TREV. V. 297;
MAND. 73.

sūð-ward, adv., *O.E.* sūðweard; *southward,*
LAȝ. 20193; souþward ROB. 113.

South-werk, pr. n., *Southwark,* CH. C. T.
A 20.

souþ-western, adj., *south-western:* þe souþ-
westerne wind LANGL. *A* v. 14.

sūþ, see **swīð.**

sūðe, adj., *O.E.* sūða (dat. sūðan); *south:*
bi suðen (*ms.* sūðen) [suþe] LAȝ. 30214;
bi souþe AR. & MER. 3117; bi souþe þe boru
HAV. 2828.

sūðen, adv., *O.E.* sūðan, = *O.N.* sunnan (*ab
austro*), *O.H.G.* sundan (*meridies*); *from the
south,* GEN. & EX. 1167.

 sūðen-wind, sb., *south wind,* GEN. & EX.
3084; southenwind PS. lxxvii. 26; **souþin-
windes** (*pl.*) M. T. 128.

[**sūðer,** sb., *O.N.* suðr, = *O.H.G.* sundar (*south,
south wind*).]

 Souþer-ēie, pr. n., *O.E.* Sūðrīge; *Surrey,*
P. L. S. xiii. 53.

[**sūþer²,** adj., = *M.L.G.* süder (*southern*).]

 suþer-wude sb., (? *for* suþerne wude), '*abrotonum,*' FRAG. 3.

sūðerne, adj., *O.E.* sūðerne, = *M.H.G.* sundern;
southern, sūðerne LAȝ. 32038; a souþerne wind BRD.
22; southerne wode [*O.E.* sūðerne wudu]
southernwood, '*abrotonum*' PR. P. 467.

sūððen, see **siþþen.**

suvel, sb., *O.E.* sufol, sufl, = *O.N.* sufl, *prov.
Eng.* (*Pem.*) sowli; *savoury food,* P. L. S. viii.
23; A. R. 192; sovel LANGL. *C* ix. 286; saulee
['*edulium*'] *B* xvi. 11; (*ms.* sowel, *r. w.* covel)
HAV. 767; souel (? *ms.* sowel) ['*pulmentum*']
WICL. GEN. xxvii. 4; souel (*printed* sonel)
WICL. S. W. I. 63; sowlle TOWNL. 87; sowel
'*edulium*' VOC. (W. W.) 579.

suwe, see **suȝe.** **suwen,** see **sewen.**

sūwien, see **swīȝien. svá,** see **swá.**

swá, adv., *O.E.* swa (? swā), swæ, swe
(? swē), se (? sē), = *O.N.* svā, svo, so, su,
Goth. swa (οὕτως), swē (ὡς), *O.L.G., O.H.G.*
sō; *so, as,* PERC. 1463; PR. C. 28; swa [so]
strong LAȝ. 600; a swa hende gome 3812;
swa [so] sone swa [so] heo mihten 25645;
wha swa [se] *whoever* 4841; wer swa [so]
heo wolleð 479; hu swa his riche men ræden
him wolden 30227; swo 2348; swe he dude

29805; sa [so] me scal lacnien his leomes 19500; swa hwet swa *whatsoever* HOM. I. 111; swa se hi mihten 235; hwen se *whensoever* 85; swa, se H. M. 5, 7; ORM. 63, 6450; swa wraˇe workes KATH. 173; swa)as *so as* 1055; so kene 181; se wide se þe lond was 49; se lengre se mare 1720; swa þat heo ches him MARH. 2; so lengre so leovere IBID.; sone so heo icumen wes 4; riht swo [so] hi weren ipeint mid wode O. & N. 76; þah þu iseo þe steorre al swa (*r. w.* mo) 1329; so sone so þu sittest a brode 518; sva, sa M. H. 5, 38; swo wel swo ANGL. I. 11; for to friˇen hise geste swo GEN. & EX. 1070; so fagen so fueles arn 15; swo P. S. 150; swo we mowe sigge MISC. 28; wo so hath beleave IBID.; zvo (*ms.* zuo) AYENB. 14; so C. L. 104; SHOR. 3; LANGL. A PROL. 10; so hit al iwearþ FRAG. 8; so so *so that* FRAG. 1; hwo so 7; so sone so A. R. 374; hwam so he luveˇ 184; hu woc so hit ever beo 138; o hwuche wise se heo ever wule 8; hwon se ȝe ever willeˇ 412; se lengre se betere 8; so whit so eni lilie flour HORN (R.) 15; so þat . . . inome hi were at te laste BEK. 6; so god a mon SPEC. 74; to þe simple so as to þe riche WILL. 338; whit so feþer of swon K. T. 12; what so him list CH. C. T. D 1291; sa ALEX. (Sk.) 146, 269; sa gates *in such a manner* YORK x. 30; sa gat BARB. vii. 368*.

swac, adj., = M.L.G. swak, M.H.G. swach; *weak*, GEN. & EX. 1528; *comp.* un-swac.

swǽlen, v., O.E. swǽlan, ? = O.N. svǽla (*smoke*); *sweal, burn*; swele BER. 2349; swēlde (*pret.*) LAȝ. 25594; iswēled (*part.*) ['*ustulatum*'] TREV. VIII. 143; *comp.* for-swǽlen.

swǽt, *see* swāt.

swǽten, v., O.E. swǽtan, = M.L.G. swēten, O.N. sveita, M.H.G. sweizen; *from* swāt; *sweat, perspire*; sweten LAȝ. 19797; sweten A. R. 362; sveten SPEC. 70; swete O. & N. 1716; swete HORN (L.) 1407; SAINTS (Ld.) xlvi. 596; swete [sweete] CH. C. T. G 522; swēt (*pres.*) A. R. 360; swētte (*pret.*) A. R. 110; swette MAND. 96; CH. C. T. G 560; swatte FER. 5753; WICL. ECCLES. ii. 19; *comp.* bi-swǽten.

swǽting, sb., *sweating*, LAȝ. 17763.

swaif, sb., = O.H.G. sveib; *from* swīven; *blow*: þe swaif of þe sworde A. P. ii. 1268; swaife ALEX. (Sk.) 806.

swain, *see* swān.

swaiven, v., O.E. swǽfan, *cf.* O.H.G. sveibōn; *go swiftly*; swaives (*pres.*) to þe see boþem A. P. iii. 253.

swále, sb., ? O.N. svöl (*pl.* svalar) (*covered walk*); *shed, awning*, '*umbra, umbraculum*,' PR. P. 481; M. T. 43.

swálen, v.: *burn up, dry up*: heo heom letten swalen inne swærte fure LAȝ. 10188; swáleˇ

(*pres.*) heore bures 6147; þei swáliden ['*aestuaverunt*'] (*pret.*) WICL. MAT. xiii. 6.

swalewe, sb., O.E. swealewe, swealwe, = M.Du. swaluwe, O.H.G. swalawa, swaluwa, swalewa, O.N. svala; *swallow*, ALIS. 3787; swaluwe S. A. L. 158; swalowe PR. P. 481; swalwe CH. C. T. A 3258; swalu TRIST. 1366; swaio VOC. 188; swolwe FER. 4232.

swalȝ, sb., = M.L.G. swalg, swalch *m.; from* swelȝen; *whirlpool*; swalgh D. TROY 13299.

swalm, sb., *swelling*; (*v. r.* swel) A. R. 274*; svalm ? *suffering*, ? *swoon*, C. M. 20758.

swalmen, v., *swell*: soche a sweme his harte can swalme FLOR. 770.

swalterin, v., *faint*, '*ex(h)alo, sincopizo*,' PR. P. 481.

swalu, swalwe, *see* swalewe.

swalwen, *see* swelȝen.

swan, sb., O.E. swan, = M.H.G. swan, O.N. svanr; *swan*, CH. C. T. A 206; B. DISC. 1367; swon SPEC. 28; swanes (*pl.*) ALEX. (Sk.) 4276.

swān, swein, sb., O.E. swān, O.N. sveinn, = O.H.G. svein; *swain, countryman, young man*; swon PALL. iii. 1086; swein SAX. CHR. 257; swein HAV. 273; P. S. 189; GREG. 684; swein, swain LAȝ. 3505, 3530; CH. C. T. A 4027; swain S. S. (Wr.) 2131; OCTAV. (H.) 1241; sweine (*dat.*) LAȝ. 3576; sweines (*pl.*) ROB. 53; þreo cnihtes & heore sweines LAȝ. 18128; sweinen (*dat. pl.*) LAȝ. 22776.

swange, sb., O.N. svangi; *loins*, GAW. 138; D. ARTH. 1129.

swǎp, sb., O.E. swǎp, = O.N. sveipr, ? M.H.G. sweif; *swoop, stroke*, TOWNL. 206; LUD. COV. 8; swap, sweip PR. P. 482; swappe (*dat.*) CH. H. F. 543; sqwappe ANT. ARTH. xlii.

swāpen, v., O.E. swāpan (*pret.* swēop), = O.N. sveipa (*pret.* sveip), M.H.G. sweifen (*pret.* swief); *swoop; sweep*; swopen R. S. v (MISC. 176); swope [swoope] þe floor CH. C. T. G 936 (Wr. 12864); zvope AYENB. 109; swōpeˇ (*pres.*) A. R. 314; swōpeþ (*imper.*) P. L. S. xv. 184; swēp (*pret.*) RICH. 6929; swaines ful swiþe swepen þer til A. P. ii. 1509; swōpen (*pple.*) HOM. II. 87; swopen TRIST. 2193; isvope ROB. 338; *comp.* bi-swāpen.

swappen, v., *from* swǎp; *cf.* O.N. sveipa (*pret.* sveipta), ? M.H.G. sweifen (*pret.* sweifte); *strike; move quickly*: his heed for to swappe CH. C. T. E 586; swappes (*pres.*) D. ARTH. 4244; swapte (*pret.*) L. H. R. 142; Beofs to him swapte LAȝ. 26775*; he swapte his hed undir the watir GEST. R. 3; sweppede D. ARTH. 1795; swapped (*pple.*) WILL. 3609.

swār, adj., O.E. swār, swǣr, = O.L.G. svār, O.N. svárr, O.H.G. svār, O.Fris. swēr, Goth. swērs (ἔντιμος); *heavy, sore*, FLOR. 90; GAW. 138; for sware ungriþ ORM. 16280; swēre (*adv.*) JUL. 46.

swērnes, sb., *O. E.* swǣrnyss ; *sadness,* APOL. 107.

swarde, sb., *O.E.* sweard, *cf. O.Fris., M.Du.* swarde, *M.H.G.* swarte *f., O.N.* svörᵍr *m. ; sward, skin,* ' *cespis, coriana* ' PR. P. 482 ; **swarthe** (*dat.*) D. ARTH. 1466.

[**swáre**, sb., *one who takes an oath ; comp.* **mān-sware.**]

swáre ², sb., *O.E.* (and-, āᵍ-)swaru, = *O.N.* svara ; *oath,* A. R. 344 ; HOM. II. 163 ; false sware HOM. II. 259 ; sware LAȝ. 10893* ; P. S. 247 ; LEG. 35 ; sware [? *O.N.* svar] *response, answer,* ORM. 2422 ; *comp.* **and-, mis-swáre.**

swáren, v., *O.E.* (and-)swarian, = *O.N.* svara ; *answer, respond,* ORM. 8938 ; **swáred** (*pret.*) GAW. 2011.

swarm, sb., *O.E.* swearm, = *M.H.G.* swarm, *O.N.* svarmr (*swarm, tumult*) ; *swarm,* ' *examen,*' VOC. 223 ; PR. P. 482 ; swarm of bees CH. C. T. *B* 4582 ; his breste is bored wiþ deþes **swarmes** (*pl.*) L. H. R. 135.

swarmin, v., = *M. Du.* swermen ; *swarm,* ' *examino,*' PR. P. 482 ; swarmen RICH. 5751 ; swarmen [swermen] CH. C. T. *D* 1693.

swart, adj., *O.E.* sweart, = *O.Fris., O.L.G.* svart, *O.N.* svartr, *Goth.* swarts, *O.H.G.* svarz, swarz ; *swart, black,* MARH. 15 ; GEN. & EX. 286 ; swart P. L. S. xvii. 354 ; RICH. 465 ; L. H. R. 223 ; DREAM 1864 ; svart ROB. 490 ; þe swarte pich MISC. 149 ; **swarte** (*dat. m.*) LAȝ. 28053 ; **swarture** (*comp.*) A. R. 284.

swartnesse, sb., *swartness,* PR. P. 482.

swarten, v., *O.E.* sweartian, *cf. O.N.* svarta, *O.H.G.* svarzēn ; *become swarthy* ; **swartede** (? *ms.* swartete) (*pret.*) MARH. 18 ; *comp.* **for-swarten.**

swartish, adj., *darkish* ; swartish (*adv.*) red CH. H. F. 1647.

swāt, sb., *O.E.* swāt, *cf. O.Fris., O.L.G.* svēt, *O. H.G.* sveiz, *O.N.* sveiti ; *sweat, perspiration,* HOM. I. 281 ; ORM. 1616 ; swat (? *ms.* swæt) [swot] LAȝ. 2281 ; swot A. R. 110 ; swot HAV. 2662 ; C. L. 200 ; swot [swoot] TREV. III. 227 ; WICL. LK. xxii. 44 ; CH. C. T. *G* 578 ; swoot LUD. COV. 30 ; zvot AYENB. 96 ; sot [soot] FER. 719 ; swete PR. P. 483 ; **swāte** (*dat.*) LAȝ. 17803 ; swote SHOR. 162 ; *comp.* **nēd-swōt.**

swāti, **swōti**, adj., *O. E.* swātig, = *O. N.* sveitugr, *M.H.G.* sweizic ; *covered with perspiration,* A. R. 104 ; swoti D. TROY 2366.

[**swáþe**, sb., *swathe, bandage.*]

swaþe-bend, sb., *bandage, swaddling-band* ; **swáthebendes** (*pl.*) SHOR. 121.

swaþ-clūt, sb., *swaddling cloth,* VOC. 143.

swáþe ², sb., *O.E.* swaᵒu (*vestigium*), = *M.Du.* swade (*striga*) ; *foot-steps, track ; row of mown grass,* PR. P. 482 ; ' *andeine* ' VOC. 154 ; cam him no fieres swaᵒe ner GEN. & EX. 3786 ; **swáthes** (*pl.*) D. ARTH. 2508.

swáþin, v., *cf. O.E.* (be-)sweᵒian ; *swathe, bind,* ' *fascio,*' PR. P. 482 ; **swáþe** (*pres.*) MED. 976 ; **swáþed** (*pret.*) C. M. 11236 ; **swéthed** (*pple.*) PALL. vi. 19 ; iswaþid VOC. 143.

[**swaþel**, sb., *O.E.* swæᵒil, sweᵒel, = *M.Du.* swadel ; *swaddle, bandage.*]

sveþel-band, sb., *swaddling cloth,* C. M. 1343.

sweþel-clout, sb., *swaddling cloth,* M. H. 91.

swaþel ², adj., *tightly bound* : swaþel & toȝt AL. (T.) 116.

swaþlen, v., *swaddle, bind* ; **swetheled** (*pret.*) C. M. 11236 ; **swaþild** (*pple.*) VOC. 203 ; sweþled GAW. 2034.

swaᵒrien, v., *O.E.* swaᵒrian ; *sleep, dream, swoon* : he bigan to swoudri as a slep him nome SAINTS (Ld.) lxiii. 268 ; (*printed* swondrie) P. L. S. xvii. 257.

swoddringe, sb., *sleep, dream* : a svoddringe him nom ROB. 264 ; þo þoȝte him in his swoudringe SAINTS (Ld.) lxiii. 269 (*printed* swondringe P. L. S. xvii. 258).

swebben, v., *O.E.* swebban, = *O.N.* svefja, *O.H.G.* (in-)sweppan, *Lat.* sōpire ; *sleep, swoon* ; he **sweveᵒ** (*pres.*) HOM. I. 233 ; **swevede** (*pret.*) LAȝ. 25548 ; he swefede þe mid þen sweiȝe FRAG. 7 ; *comp.* **a-, ȝe-swebben.**

swēȝ, sb., *O.E.* swēg ; *from* **swōȝen** ; *sound* ; sweig LK. xv. 25 ; sweih FRAG. 1 ; swei HOM. I. 87 ; **swēiȝe** (*dat.*) FRAG. 7.

swēȝen, v., *O.E.* swēgan ; = *L.G.* swögen (*groan*); *sound* ; sweiȝen FRAG. 1 ; (*ms.* sweieᵒ) & singeᵒ HOM. I. 193 ; **swēȝand** (*pple.*) ALEX. (Sk.) 5019 ; it **swēied** (*pret.*) so murie LANGL. *B* PROL. 10.

swēying, sb., *noise,* YORK xxx. 371.

sweȝer, sb., *O.E.* sweger, = *O.H.G.* swiger, *cf. Goth.* swaihrō, *Lat.* socrus, *Gr.* ἑκυρά ; ' *so*(*crus*),' FRAG. 2 ; swegre, swigre (*ms.* swygre) [' *socrum* '] MAT. viii. 14, x. 35.

sweiȝ, sb., *O.N.* sveigr ; *sway, movement* CH. BOET. ii. 1 (32) ; sweght of our swappes YORK xxxiii. 362.

swēiȝ, **swēiȝen**, *see* **swēȝ**, **swēȝen.**

sweiȝen, v., *O.N.* sveigja ; *sway, move* ; **swēied** (*ms.* sweyed) (*pret.*) GAW. 1429 ; þe sail sweied on þe see A. P. iii. 151.

swiere, *see* **sqvíer.**

swein, *see* **swan.** **sweip**, *see* **swap.**

swel, sb., *? O.E.* swell, = *M.L.G., M.Du.* swel ; *swelling, tumour,* S. S. (Web.) 1566 ; þene swel A. R. 274.

swēlen, *see* **swǣlen.**

swelȝ, sb., = *M.Du.* swelgh, *M.L.G.* swelch, *O.N.* svelgr ; *whirlpool, pit* ; zvelȝ AYENB. 55, 82 ; sweluȝ WICL. 3 KINGS xi. 27 ; swoluȝ WICL. PROV. xiii. 15 ; sweluh ' *vorago* ' PR. P. 482 ; swelu, sveluh (*ms.* swelw, svelhu) MAN. (F.)

1453; swelogh MAND. 33; swolȝ A. P. iii.
250; swolouȝ ['*vorago*'] TREV. V. 139;
swolowe (*dat.*) CH. L. G. W. 1104; **swelowes**
[swolwes] (*pl.*) TREV. I. 65.

swelȝen, v., *O.E.* swelgan, sweolgan, = *O.L.G.*
(far-)svelgan, *O.N.* svelga (*pret.* svalg), *O.H.*
G. swelgan, swelhan; *swallow; overwhelm*;
swelghe PR. C. 6232; swelwen LIDG. TH.
4073; swelwin '*glutio*' PR. P. 482; swelewe
P. R. L. P. 237; swelowe [swolwe, swolewe]
TREV. VIII. 241; swolhen (*for* sweolhen)
JUL. 74; swolȝhen ORM. 10224; swolwe CH.
C. T. *H* 36; swolewe BEK. 2192; zvelȝþ
(*pres.*) AYENB. 123; swelewith TREV. I. 119;
swolegeð HOM. II. 181; swoluweð A. R. 8;
swoleweþ C. L. 1129; swelighis ['*devorant*']
PS. xiii. 4; **swelge** (*subj.*) HOM. II. 43;
swalh (*pret.*) ORM. 14592; **swolgen** (*pple.*)
GEN. & EX. 1976; swolȝed A. P. iii. 363;
swolihed PS. xliii. 25*; *comp.* a-, for-, ȝe-
swelȝen; *deriv.* swelȝ, swalȝ.

swelle, adj., *tumid, proud* ; bolde and swelle
OCTAV. (H.) 1557.

swellen, v., *O.E.* swellan, = *M.L.G.* swellen,
O.H.G. swellan, *O.N.* svella ; *swell,* LAȝ.
19800; swelle MISC. 80; LANGL. *B* xix. 278;
CH. C. T. *A* 2752; swelleð (*pres.*) H. M. 31;
swelle (*subj.*) A. R. 274; **swal** [sval] (*pret.*)
O. & N. 7; swal CH. C. T. *D* 967; for hunger
ich swal MISC. 82; þer hit up swal A. D. 293;
swollen (*pple.*) CH. C. T. *E* 950; swolle
P. L. S. xxv. 162; *comp.* **to-swellen**; *deriv.*
swel.

 swellinge, sb., = *M.H.G.* swellunge; *swelling,* LANGL. *B* v. 122.

swelm, sb., = ? *O.H.G.* swilm (*sleep*) ; *heat,
glow*; (*ms.* swelme) ALEX. 750; þe swelme
leþe A. P. iii. 3.

swelten, v., *O.E.* sweltan, = *O.L.G.* sweltan,
O.N. svelta, *Goth.* swiltan, *O.H.G.* swelzan;
become faint; die; ['*mori*'] MAT. xxvi. 35;
MARH. 7; ORM. 915; swelten ich schal &
beornen HOM. I. 197; swelten LAȝ. 19801;
JOS. 377; swelte LANGL. *B* v. 154; LIDG.
M. P. 38; D. ARTH. 813; i **swelte** (*pres.*)
and swete CH. C. T. *A* 3703; **swealt** (*pret.*)
HOM. I. 225; swælt LAȝ. 26566; swalt ORM.
31; swalt A. P. i. 815; swelt WILL. 1494;
swulten LAȝ. 6071; swulten ORM. 5321;
comp. a-, for-swelten.

swelten[2], v., *O.N.* svelta (*pret.* svelta) ; *kill*;
swelt (*pple.*) ALIS. 7559*; *comp.* **for-swelten.**

swelwen, *see* **swelȝen.**

swēm, sb., *O.N.* sveimr (*tumult*), = ? *M.H.G.*
sweim; *sorrow, care,* GEN. & EX. 391; swem(e)
FLOR. 770; LUD. COV. 72, 109; sweem
'*tristitia, molestia*' PR. P. 482; (*trance*) CR. K.
29; for verai **swēme** (*dat.*) LIDG. M. P. 38.

 swēm-ful, adj., *sorrowful,* SACR. 798; LUD.
COV. 72; swemeful LIDG. M. P. 38.

swēm-lī, adj., *swooning*: a swemli swouh
L. H. R. 135.

swēmen, v., *O.E.* (a-)swæman, = *M.L.G.*
swēmen, *M.H.G.* sweimen (*wander*), ? *O.N.*
sveima (*swim, soar, wander about*); *swoon,
fall into a trance, grieve*: he scal
alle þa swiken swemen (? *printed* swe-
nien) mid eiȝe LAȝ. 16099; ne sweamen hire
heorte mid wernunge A. R. 330; **swēmith**
(*pres.*) LUD. COV. 148; **swēmande** (*pple.*)
A. P. ii. 563; **isweamed** (*pret. pple.*) H. M.
17.

swemilen, *see* **swimilen.**

swenchen, v., *O.E.* swencan, swencean,
from **swinken**; *fatigue; torment, afflict,*
HOM. I. 13; A. R. 134; swenche P. L. S.
viii. 125; swenchen, swenken ORM. 8942,
12216; **swenchest** (*pres.*) H. M. 35; þu
swenchest te to swiðe MARH. 5; monine
mon on swevene ofte heo swencheð LAȝ.
15787; swencten (*pret.*) HOM. I. 101; *comp.*
ȝe-swenchen.

sweng, sb., *O.E.* sweng, = *O.Fris.* sweng;
from **swingen**; *beat, stroke,* O. & N. 799;
sweng SAINTS (Ld.) xlv. 173; **swenges** (*pl.*)
MARH. 14; A. R. 80; **swenge** (*dat. pl.*) O. &
N. 803.

swengen, v., *O.E.* swengan, = *O.Fris.*
swenga; *swing; beat, lash*; swengin PR. P.
482; **sweng** (*imper.*) A. R. 290; swengeð
LAȝ. 22839; **sweinde** (*pret.*) A. R. 280;
sweinde LAȝ. 8183, 21138; swende MARH. 10.

swengil, sb., = *M.Du.* swenghel, ? *M.H.G.*
swengel; *rod, scourge,* PR. P. 482.

sweor, sb., *O.E.* sweor, *cf. O.H.G.* swehur,
sweher, *Goth.* swaihra, *Lat.* socer, *Gr.* ἑκυρός;
father-in-law, '*socer,*' FRAG. 2.

sweor[2], sb., *O.E.* sweor; '*columna*;' (*printed*
speer) FRAG. 3.

sweord, sb., *O.E.* sweord, sword, swyrd,
swurd, = *O.Fris.* swerd, swird, *O.L.G.* swerd,
O.N. sverð, *O.H.G.* swert; *sword,* MARH. 5;
A. R. 212; sweord ALIS. 147; þat sweord
[swerd] LAȝ. 1558; swerd ORM. 16284; GEN.
& EX. 1307; swerd CH. C. T. *A* 112; WICL.
JOHN xviii. 10; E. 265; W. & I. 41; þat
swerd FER. 743; sverd ROB. 207; sword
MAND. 291; þet zvord AYENB. 148; swird
FLOR. 585; **sweordes** (*gen.*) A. R. 60;
sw(e)orde [swerde] (*dat.*) O. & N. 1068;
sweord [sweordes] (*pl.*) LAȝ. 22813, 24472;
swerd [swerde, swerdes] LANGL. *A* i. 97;
sweordan [sweorden] (*dat. pl.*) MAT. xxvi. 47,
55; sweorden LAȝ. 6699.

 swerd-berare, sb., *sword-bearer,* PR. P.
483.

 sweord-brōðer, sb., = *M.H.G.* swertbruo-
der; *comrade in arms*; **sweord brotheren**
(*pl.*) LAȝ. 30523.

 swerd-man, sb., *swordsman, gladiator,*
TREV. V. 23.

sweore, *see* **swíre. sweoven,** *see* **sweven.**

swēp, sb., *O.E.* swǣp; *scope, meaning*: ꝥes dremes swep ne wot he nogt GEN. & EX. 2112; swepe ALEX. (Sk.) 248.

swépe, sb., *O.E.* swipe, sweope,= *O.N.* svipa, *M.Du.* swepe; *whip, scourge*, ORM. 15562; sweipe PR. P. 482; **swépis** (*pl.*) TOWNL. 227; **swépen** (*dat. pl.*) HOM. I. 231.

swépen, v., *seize*; swepe A. P. iii. 341.

swēpen, v., *cf. O.N.* sōpa (*for* svōpa); *sweep, glide*; swepin PR. P. 482; swepe CH. C. T. *E* 978; FLOR. 138; þe water con (*r.* gon) swepe A. P. i. 111; **swēped** (*pple.*) PR. C. 4947.

swēpare, sb., *sweeper*, PR. P. 482.

sweper, *see* **swiper. swerd,** *see* **sweord.**

swēre, *see* **swáre, swíre.**

swérien, v., *O.E.* swerian,= *O.L.G.* swerian, *O.H.G.* swerian, swerran, *O.N.* swerja, *cf.Goth.* swaran; *swear*, LAȝ. 5403; swerien A. R. 70; swerie ROB. 346; zverie AYENB. 6; swerie [swere] TREV. VIII. 139; swere LANGL. *A* i. 97; CH. C. T. *A* 1821; **swérie** (*pres.*) MARH. 21; swereð A. R. 98; (hi) swerieþ BEK. 2022; **swōr** [*O.E.* swōr,= *Goth., O.L.G., O.H.G.* swōr, *O.N.* svōr, sōr] (*pret.*) *swore*, LAȝ. 3446; HAV. 398; BEK. 719; swor GEN. & EX. 1338; swor ROB. 347; swor P. S. 192; swor LANGL. *A* ii. 146; zvor AYENB. 45; swar IW. 521; M. H. 39; PERC. 381; swer EGL. 430; TOR. 889; þu swore LAȝ. 5041; BEK. 1015; sware þou PS. lxxxviii. 50; (heo) sworen LAȝ. 6170; LANGL. *B* xv. 586; swore CH. C. T. *A* 1826; swore MAN. (H.) 28; **swóren** [*O.E.* (for-)sworen,= *O.H.G.* (gi-)-sworan, *O.N.* svarinn] (*pple.*) *sworn*, GEN. & EX. 1525; sworen HAV. 439; LANGL. *B* v. 376; sworn CH. C. T. *B* 1331; swore GREG. 788; *comp.* **for-, ȝe-swérien**; *deriv.* **swáre, swōr.**

    swérare, sb., *swearer*, PR. P. 482.

swerken, v., *O.E.* sweorcan,= *O.L.G.* swerkan; *be obscure*; **swærkeð** (*pres.*) LAȝ. 22030; **swurken** (*pret.*) LAȝ. 11973.

swermen, *see* **swarmen.**

swerven, v., *O.E.* sweorfan,= *O.N.* sverfa (*scour*), *O.Fris.* swerva, *M.Du.* swerven (*wander*), *O.H.G.* swerbạn, *Goth.* (af-)svairban; *swerve*: þat it mai swerve to no side GOW. III. 92; **swarf** (*pret.*) FER. 743; þe dint swarf AR. & MER. 9369; heo swerf to Criste KATH. 2212.

swēte, adj., *O.E.* swēte,= *O.Fris.* swēte, *O. L.G.* swōti, *O.N.* scētr, *Goth.* sūtis, *O.H.G.* suozi, *M.H.G.* süeze, *Lat.* svāvis (*for* svād-vis), *Gr.* ἡδύς; *sweet, pleasant*, HAV. 2927; ROB. 244; MAND. 57; PR. C. 4915; swete, swete, svete O. & N. 358, 866; sweote HOM. I. 53; swet ORM. 1258; GEN. & EX. 3302; sqvete ANT. ARTH. xxv; swite S. S. (Wr.) 2080; swete, swote KATH. 613, 1600; A. R. 80, 98;

swete, swote SPEC. 23, 57; swete, swote, sote CH. C. T. *A* 3205, 3206; swete [sote] WICL. NUM. xxviii. 13; swote MARH. 4; þet swete FRAG. 7; HOM. I. 215; mi swete (*dear one*) ALEX. (Sk.) 2826; lose þe swete (*scil.* life) A. P. iii. 364; ALEX. (Sk.) 3068; has the swete levede D. ARTH. 3360, 3703; **swōte** (*adv.*) FRAG. 7; HOM. I. 53; A. R. 238; swote L. H. R. 24; **swěttere** (*compar.*) HOM. II. 33; H. M. 29; svettere SPEC. 68; swettere WICL. JUDG. xiv. 18; swetture MISC. 97; swetter PR. C. 3699; swottre MARH. 11; **swětteste** (*superl.*) REL. I. 220; swotest KATH. 614.

    swēt-līc, adv., *O.E.* swētlīc,= *sweetly*; a svetli svire SPEC. 52; **swēt-līke** (*adv.*) ORM. 1647; sweteliche KATH. 673; A. R. 264; R. S. iv; sweteliche WILL. 1329; swoteluche H. M. 41.

    swētnesse, sb., *O.E.* swētness,= *O.H.G.* suoznissi; *sweetness*, H. M. 7; swetnesse C. L. 582; swetnesse, swotnesse A. R. 80, 102; swotnesse, svotnesse L. H. R. 28, 29.

swēte, sb., ? = *O.N.* scēti, *O.H.G.* suozī; *? sweetness*: a bende ... in swete to were GAW. 2518.

swēte, *see* **swát.**

swēten, v., *O.E.* swētan,= *O.H.G.* suozan; *sweeten*; swetin '*dulcoro*' PR. P. 483; salt þat ure mete **swēteþ** (*pres.*) ORM. 1649.

swēten, *see* **swǣten.**

swēting, sb., *sweet one, darling*, HOM. I. 271; sweting WILL. 916; TOWNL. 96; sveting SPEC. 51; swetting (*loved one*) YORK xl. 40.

sweþel, sweþen, *see* **swaþel, swaþen.**

sweven, sb., *O.E.* swefen, swefn *n.*, *cf. O.N.* svefn, *O.L.G.* sweban (*dat.* swefne), *Lat.* somnus, *Gr.* ὕπνος *m.*; *sleep, dream, vision*, KATH. (E.) 1560; JUL. 75; sweven HORN (L.) 679 (R.) 681); GOW. I. 24; S. S. (Wr.) 1936; a sweven LAȝ. 25552; þat sworen 25553; a selco(u)þe sweven ... sche mette WILL. 2869; swevin OCTAV. (H.) 158; TOWNL. 108; sweven [swevene, swefne] CH. C. T. *B* 3930; swevene LANGL. *A* PROL. 11; þe PR. P. 483; swevene P. L. S. xiii. 147; **swefne** (*dat.*) A. R. 224; swefne [swevene] LAȝ. 1221; swevene GEN. & EX. 225; **swefnes** (*pl.*) A. R. 268; **swefnen** (*dat. pl.*) MAT. ii. 22; swefnen LAȝ. 1158.

swevenen, v., *O.E.* swefnian; *dream*; **swevenеþ** (*pres.*) WICL. IS. xxix. 8; **swevenid** (*pret.*) LANGL. *A* PROL. 10*.

    swevenere, sb., *dreamer*; **sweveneres** (*pl.*) (*later ver.* dremeris) WICL. JER. xxvii. 9.

swévet, sb., *O.E.* sweofot; *sleep, dream*; **swévete,** sweovete (*dat.*) LAȝ. 17773, 17802; sweovete KATH. 1438.

swīc, sb., *O.E.* swīc,= *M.H.G.* swīch; *from* **swīken**; *deception*; **swīke** (*dat.*) REL. I. 220; swike C. M. 11556; D. TROY 11837; swiche HOM. II. 258.

swīc-ful, adj., *crafty*, LA3. 8022 ; swikfull HAMP. PS. lxxiii. 14.

swich, *see* swūlc.

swīe, adj., *O.E.* swīge ; *silent* : þe devel com to þis maide swie MARG. 157.

swī-messe, sb., *low mass*, HOM. II. 97.

swī-wike, sb., *Holyweek*, A. R. 70*.

swīen, *see* swī3en.

swift, adj., *O.E.* swift ; *? from* swīven ; *swift*, ORM. 6972 ; swift MISC. 94 ; REL. I. 179 ; AR. & MER. 535 ; swifte (*pl.*) LA3. 5902 ; CH. C. T. *A* 190 ; A. P. i. 570 ; swifte wateres A. R. 252 ; swiftre [swiftere] (*compar.*) LA3. 26068 ; swifture A. R. 94.

zvift-hēde, sb., *swiftness*, AYENB. 78.

swift-līche, adv., *O.E.* swiftlīce ; *swiftly*, KATH. 689 ; swiftliche LEG. 41 ; sviftlīker (*compar.*) HALLIW. 826.

swiftnesse, sb., *O.E.* swiftness ; *swiftness*, A. R. 94 ; swiftnes PR. C. 9029.

swift-schipe, sb., *swiftness*, A. R. 398.

swī3en, sūwien, v., *O.E.* swīgian, sūgian, sūwian, *cf. O.Fris.* swīgia, *O.L.G.* svīgōn, *O. H.G.* svīgēn ; *be silent* ; swīgeð (*pres.*) HOM. II. 103 ; sūwinde (*pple.*) A. R. 256 ; swīede (*pret.*) HOM. II. 101 ; swi3eden LA3. 16820.

swike, sb., *O.E.* swica, = *O.N.* sviki ; *from* swīken ; *deceiver, traitor*, A. R. 98 ; swike HAV. 423 ; SPEC. 46 ; D. TROY 11833 ; swike ROB. 221 ; sweoke H. M. 45 ; swiken (*gen.*) LA3. 22858 ; swiken (*pl.*) LA3. 3816 ; MISC. 61 ; þe swichen (*printed* þes wichen) HOM. II. 223 ; *comp.* lāverd-, lēod-swike.

swike, adj., *O.E.* swice ; *deceptive, treacherous*, LA3. 14865 ; swike GEN. & EX. 2845.

swike-dōm, sb., *O.E.* swicdōm, = *O.N.* svik-dōmr ; *treachery*, HOM. I. 55 ; ORM. 3997 ; GEN. & EX. 2883 ; swikedom LA3. 8310 ; P. S. 220 ; swike-, svikedom O. & N. 167 ; svike-dom ROB. 36.

swike-hēde, sb., *treachery*, O. & N. 162*.

swike, sb., *O.E.* swice ; *trap*, '*decipula*,' VOC. 221 ; RICH. 4081 ; IW. 677 ; he bindeð uppon þa swike chese HOM. I. 53.

swikel, adj., *O.E.* swicol, = *O.N.* svikall ; *deceptive, treacherous*, HOM. I. 43 ; swikel HAV. 1108 ; FER. 4589 ; PS. v. 7 ; MAN. (F.) 3828 ; svikel CHR. E. 791 ; þe swikele king LA3. 15026 ; þe swikele wimon REL. I. 144 ; sikil HAMP. PS. xlii. 1* ; swikelne (*acc. m.*) REL. I. 183 ; swikele (*pl.*) P. L. S. viii. 127 ; swi-kelure (*compar.*) A. R. 180 ; swikelest (*superl.*) LA3. 15221.

svikel-dōm, sb., *deception*, O. & N. 163 ; swikeldom ROB. 110*.

svikel-hēde, sb., *deception*, O. & N. 162.

swikel-(l)īche, adv., *deceptively*, H. M. 39 ; swikelli '*dolose*' PS. xiii. 3 ; swikilli HAMP. PS. v. 11.

swikelnesse, sb., *deception*, MISC. 74.

swīken, v., *O.E.* swīcan, = *O.L.G.* swīcan, *O.N.* svīkja (*pret.* sveik), *O.H.G.* swīchan ; *cease, fail, deceive*, REL. I. 223 (MISC. 20) ; 3if he nule nefre swiken (*cease*) HOM. I. 23 ; swiken [swike] LA3. 4101 ; swīke (*pres.*) O. & N. 1459 ; swike SPEC. 48 ; zvikeþ AYENB. 157 ; swīc (*imper.*) REL. I. 213 (MISC. 7) ; swīke (*subj.*) MARH. 5 ; swiken (*pret.*) HOM. I. 43 ; his men him sviken SAX. CHR. 264 ; *comp.* a-, bi-, 3e-swīken ; *deriv.* swīc, swīke, swike, swikel.

swīking, sb., *O.E.* swīcung ; *deception*, REL. I. 222.

[swīkende? *pres. pple. of* swīken.]

swicande-līche, adv. (*printed* swican̄-liche) ; *deceptively*, HOM. I. 25.

swilc, *see* swūlc.

swilen, v., *O.E.* swilian ; *swill, wash* : dishes swilen HAV. 919 ; swile H. S. 5828 ; þe flume shal hit swile A. D. 239.

swīme, sb., *O.E.* swīma, = *O.Fris.* swīma, *O.N.* svīmi ; *dizziness*, MAN. (F.) 8400 ; H. S. 11287 ; DEGR. 1211 ; D. ARTH. 4246 ; TOWNL. 8 ; swime [svime] C. M. 14201.

swimel, sb., *giddy motion* ; swimbel (*v. r.* rumbel) CH. C. T. *A* 1979.

swimilen, v., *feel dizzy* ; swemile (*pres.*) ALEX. (Sk.) 156.

swimmen, v., *O.E.* swimman, = *O.H.G.* swim-man, *O.N.* svimma, *M.Du.* swimmen, swem-men ; *swim*, HOM. I. 51 ; swimmen LANGL. *B* xii. 163 ; swimme HORN (R.) 1432 ; CH. C. T. *A* 3575 ; swemme LA3. 28078* ; swim (*imper.*) HOM. I. 51 ; swam (*pret.*) HOM. I. 51 ; swam WILL. 2760 ; GOW. II. 272 ; swam, svam BRD. 8, 31 ; swam, swom FER. 3958, 3965 ; swummen HOM. I. 129 ; swom-men WICL. DEEDS xxvii. 42 ; svommen [swomme] LA3. 1342 ; swomme LEG. 166.

swimmere, sb., = *M.H.G.* swimmer ; *swim-mer*, LANGL. *B* xii. 167.

swīn, sb., *O.E.* swīn, = *O.Fris.* swīn, *O.N.*, *O.L.G.* swīn, *O.H.G.* swīn, *Goth.* swein ; *swine*, A. R. 128 ; swin CH. C. T. *C* 556 ; þat wilde swin LA3. 468 ; þet zvin AYENB. 255 (*spelt* zuyn 179) ; swīnes (*gen.*) HOM. I. 169 ; swunes P. L. S. viii. 73 ; swīn (*pl.*) A. R. 230 ; ORM. 7410 ; swin LA3. 25682 ; MAND. 248 ; svin REL. I. 62 ; swīne (*gen. pl.*) LK. viii. 32 ; swīnen (*dat. pl.*) MK. v. 16 ; HOM. I. 135 ; *comp.* mast-, mer-swīn.

swiin-kote, sb., *swine-cote*, PR. P. 483.

swīn-flēsch, sb., *pork*, VOC. 200.

swiin-herd(e), sb., *swine-herd*, PR. P. 483 ; swinhirde CATH. 374.

swīn-stī, sb., *O.N.* svīnstī, = *M.Du.* swijn-stije, *O.H.G.* swīnstīge ; *pig-sty*, PR. C. 9002.

swinc, sb., *O.E.* (ge-)swinc ; *from* swinken ; *labour*, A. R. 94 ; ORM. 6102 ; swinc [swinch] LA3. 2297 ; swink HAV. 770 ; C. L. 200 ;

TRIST. 1116; CH. C. T. *A* 188; PR. C. 755; LUD. COV. 30; zvinch AYENB. 83; **swinkes** (*gen.*) HOM. I. 163; KATH. 806; **swinke** (*dat.*) A. R. 220; swinche P. L. S. viii. 186; swinche BEK. 9; **swinkes** (*pl.*) A. R. 240; *comp.* ȝe-**swinch**.

    swinc-ful, adj., *laborious,* HOM. I. 7; ORM. 2621.

    swincfulnesse, sb., *laboriousness,* ORM. 2526.

swínden, v., *O.E.* swindan, = *M.Du.* swinden. *O.H.G.* swintan; *vanish, decrease, consume,* P. L. S. viii. 29; swinde P. S. 150; *comp.* a-**swínden**.

swing, sb., *O.E.* swing; *swing, blow with a sword*; **swinge** (*dat.*) D. TROY 1271.

swingen, v., *O.E.* swingan, = *O.L.G.*, *O.H.G.* swingan; *swing, vibrate, beat,* MAT. xxvii. 26; ORM. 6362; swingen hit swiftliche a-buten JUL. 58; swinge HAV. 214; **swingeð** (*pres.*) HOM. I. 149; (þai) **swengen** (*rush out*) A. P. ii. 109; **swenged** (*pret.*) A. P. ii. 667; **swing** (*imper.*) L. C. C. 11; **swang** (*pret.*) A. P. i. 1058; swong MAN. (F.) 13054; he swong hire al abuten his swire MARH. 9; with swerdes swonge þei to gider WILL. 3856; **sw(u)ngen** (*pple.*) HAV. 226; swungen [swongen] PS. lxxii. 5; swongen with swepis TOWNL. 227; swonge L. H. R. 142; *comp.* **bi-, ȝe-, to-swingen**; *deriv.* **swing, sweng, swengen, swingle**.

    swinginge, sb., *beating,* HOM. II. 57; swing-ing ORM. 5527.

swingle, sb., *O.E.* swingele, = *M.Du.* swing-hel; *rod, whip,* VOC. 156.

swinglen, v., = *M.Du.* swinghelen; *whip, beat*; *swingle* (*flax*); swingle VOC. 156; i bete and **swingile** (*pres.*) flex REL. II. 197.

swīnisch, adj., = *M.H.G.* swīnisch; *swinish*: swinisse men HOM. II. 37.

swink, see **swinc**.

swinken, v., *O.E.* swincan; *labour; travel*; HOM. I. 19; A. R. 358; GEN. & EX. 3778; swinken after mete ORM. 6100; swinken for til halȝhen 15760; swinke ROB. 99; MIRC 1346; CH. C. T. *A* 186; ISUM. 396; **zvinke** (*pres.*) AYENB. 171; swinkeð H. M. 39; **swanc** (*pret.*) ORM. 17699; GEN. & EX. 2014; swank HAV. 788; SHOR. 165; swonc A. R. 110; swonc LAȜ. 7488; þat ich . . . hider swonk |swonc] O. & N. 462; swunken GEN. & EX. 1656; swunken [swonke] LAȜ. 17408; swunke P. L. S. viii. 128; swonken LANGL. *A* PROL. 21; swonke RICH. 3762; swunke (*subj.*) HOM. II. 229; swunche we P. L. S. viii. 160; swunke [swonke] LAȜ. 17909; **swunken** (*pple.*) ORM. 6103; iswunken A. R. 404; i-swonke FER. 152; *comp.* **a-, bi-, of-swinken**; *deriv.* **swinc, swenchen**.

    swinkere, sb., *worker, toiler,* CH. C. T. *A* 531; **zvinkeres** (*pl.*) AYENB. 90.

    swincunge, sb., *labour,* HOM. I. 69.

swipe, sb., *O.N.* svipr, = *M.H.G.* (nider-, umme-)swif; *stroke,* PR. P. 482; þat of þen ilke sweorde enne swipe [swip] hefde LAȜ. 7648.

swiper, adj., *O.E.* swipor; *quick,* ANGL. I. 154; TREV. III. 361; swipir '*agilis*' PR. P. 484.

    sweper-lī, adv., *quickly,* D. ARTH. 1128, 1465.

swippen, v., *O.E.* swipian, = *O.N.* svipa; *vibrate, beat*: & lette hit a dun swippen LAȜ. 16510; ich wulle mid swerde his heved of swippen 878; **swippes** (*pres.*) PR. C. 2196; **sqvippand** (*pple.*) ANT. ARTH. v; **swipte** (*pret.*) MAN. (F.) 12117; and his sweord . . . swipte mid maine LAȜ. 23978; swipte hire of þat heaved KATH. 2485; heo bið sone iswipt forð A. R. 228.

swird, see **sweord**.

swire, sb., *O.E.* swira, swyra, swiora, sweora, = *O.N.* sviri; *neck,* MARH. 9; swire HAV. 311; SPEC. 28; FER. 997; B. DISC. 230; A. P. ii. 1744; PERC. 790; MIN. viii. 68; sqvire ANT. ARTH. xl; swire, sweore A. R. 58, 394; swiere ANGL. I. 15; HOM. II. 224; þene swure, sweore [swere] LAȜ. 4012, 26565; sweore, swore HOM. I. 49, 169; sweore, sweore, swere O. & N. 73, 1125; swere P. L. S. viii. 73; swere ALIS. 2000; S. S. (Web.) 461; GOW. II. 30; svere ROB. 389; zvere AYENB. 155; **sweoren** (*dat.*) MK. ix. 42; **sweoren** (*dat. pl.*) LAȜ. 22786.

    swire-bān, sb., *O.E.* swir-, sweorbān; *neck-bone,* D. ARTH. 2959.

    sweor-bēah, sb., *O.E.* sweorbēah; *neck-lace,* '*monile*,' FRAG. 2.

    sweor-cops, sb., *O.E.* sweorcops; '*boja*,' FRAG. 5.

swīð, adj., *O.E.* swīð, swȳð, = *O.L.G.* swīth, *Goth.* svinþs, *M.H.G.* swind, *O.N.* svinnr; *strong*; **swīðe** (*adv.*) *strongly, greatly, quickly,* MARH. 5; GEN. & EX. 1009; swiþe wel ORM. 2157; swiþe HAV. 111; C. L. 320; WILL. 41; TRIST. 165; LANGL. *B* v. 456; CH. C. T. *B* 730; S. S. (Wr.) 2; TOR. 1956; swiþe fele SHOR. 104; high þe swiþe PS. xxx. 3; þis mai ran til hir moder swiþe M. H. 39; sviþe BRD. 2; CHR. E. 341; MAN. (H.) 134; swithe PR. C. 5713; sqviþe ANT. ARTH. xiv; swiðe, swuðe A. R. 58, 236; swiðe, swuðe LAȜ. 1349, 4170; swiþe, sviþe, swuþe, suþe O. & N. 2, 376, 1245, 1591; swuþe FRAG. 5; suþe FL. & BL. 355; **swiþer** (*compar.*) LAȜ. 1580; swiþþer FER. 816; þe swiðre (*printed* spiðre) H. M. 39; on þar swiðeran (*right*) halfe HOM. I. 229; mid his swiðren honde LAȜ. 21424; **swiðest** (*superl.*) KATH. 734; swa we hit swiðest maȝen don LAȜ. 25794.

    swīðe-līche, adv., *O.E.* swīðlīce; *quickly,* LAȜ. 4421.

swĩðen, v., *O.N.* svĩ̆̆a ; *burn, light up* ; swiþe PS. CXX. 6* ; swĩðeð (*ms.* swideð) (*pres.*) REL. I. 210 (MISC. 3) ; swiþez A. P. iii. 478 ; swãth (*pret.*) ['*combussit*'] PS. CV. 18 ; *comp.* for-swĩðen.

swĩven, v., *O.E.* swĩfan, = *O.N.* svĩfa, *O.Fris.* swĩva ; (*? move quickly*) ; *have sexual intercourse with* : yon wenche wol i swive CH. C. T. *A* 4178 ; *comp.* ofer-swĩfen ; *deriv.* swaif, swaiven.

swĩving, sb., P. S. 69 (A. D. 98).

swó, *see* swá. swodrien, *see* swaðrien.

swŏft, swŏfte, sb., *sweepings*, MISC. 176, 177.

swŏȝ, sb., ?=swēȝ ; *soughing (of the wind)*, *sound; swoon*; GEN. & EX. 484 ; þe swogh of þe see D. ARTH. 759 ; fel . . . doun in swogh IW. 824 ; swogh, swough CH. C. T. *A* 1979, 3619 ; swouȝ BEV. 1563 ; swough AR. & MER. 7142 ; M. ARTH. 903 ; swouh L. H. R. 135 ; swou (*ms.* swow) RICH. 796 ; sou EGL. 374 ; swŏȝe (*dat.*) ALEX. 5020 ; swoghe ISUM. 89 ; Clement lai in swoghe OCTAV. (H.) 900 ; he com with a swowe MAN. (H.) 170 ; fel doun on swowe WILL. 87 ; þe swoughes and þe cries LIDG. TH. 4348.

swŏȝen, v., *O.E.* swŏgan, = *O.L.G.* swŏgan, *Goth.* (ga-)swŏgjan (στενάζειν) ; *sough (as the wind); sound; swoon*; swowe ALIS. 7874 ; swowe or swelte LANGL. *B* v. 154 ; swŏwinde (*pple.*) A. R. 288 ; swēȝ (*pret.*) GAW. 1796 ; þe soun . . . swei in his ere A. P. iii. 429 ; swowed [swouned] LANGL. *B* xiv. 326 ; soghe, soghed, soughid PARTEN. 1944, 2890 ; souȝed A. P. iii. 140 ; iswoȝen, iswowen (*pple.*) LAȝ. 3074, 4516 ; a dun he feol iswoȝe *dead* HORN (L.) 428 ; isvoȝe BRD. 1 ; iswowen JOS. 203 ; iswowe ROB. 290 ; ALIS. 2438 ; iswowe, isowe TREV. VI. 477 ; *deriv.* swŏg, swēȝ, swēȝen.

swŏhinge, sb., *O.E.* (ge-)swŏwung ; *groaning, fainting*, HORN (H.) 464 ; swoȝing ALEX. (Sk.) 4385.

swŏȝne, sb., *swoon*; swoune (*ms.* swowne) GOW. I. 268 ; PR. C. 7289 ; sche fel in swoune TRIAM. 375 ; she fil (on) swoune [a swoune] CH. C. T. *C* 245.

swŏȝnen, v., *swoon*; swounin ' *bilbio, syncopo*' PR. P. 484 ; swowene MAN. (F.) 1841 ; swoune & swelte CH. COMPL. M. 216 ; swŏghened (*pret.*) ALIS. 5857 ; swounede WICL. ESTH. xv. 18 ; swouned WILL. 2098 ; LANGL. *B* xx. 104 ; DEGR. 360 ; swounid OCTAV. 983 ; sowenede, sounede FER. 1080, 4221 ; sowened EM. 780 ; sounid TOR. 1886 ; swŏuned (*pple.*) CH. C. T. *A* 913.

swŏȝning, sb., *swooning*, HORN (L.) 444 ; svoweninge S. A. L. 156 ; swouninge JOS. 543 ; soȝening, sowening FER. 1134, 2585.

swolȝ, swolȝen, *see* swelȝ, swelȝen.

swon, *see* swan. swŏn, *see* swãn.

swŏpen, *see* swãpen.

swŏr, sb., = *M.H.G.* swuor ; *from* swérien ; *oath* ; swŏre (*dat.*) MIRC 1067.

sword, *see* sweord. swŏt, *see* swãt.

swŏte, *see* swēte.

swŏu, swŏuȝ, swŏuh, *see* swŏȝ.

swŏwen, *see* swŏȝen.

swülc, adj., (adv., conj.), *O.E.* swylc, swilc, swelc, = *O.L.G.* sulic, *O.H.G.* su-, solich, *Goth.* swaleiks, *O.N.* slĩkr ; *such; like as, as if* : a swulc mon LAȝ. 31585) swilc 1375 ; he ferde . . . sulc (*sec. text* al se; he walde awede 6486 ; sulch 4085 ; such [soch] werc 491 ; swilc ORM. 1632 ; nis no loverd swilc se is crist P. L. S. viii. 40 ; swilc tiding GEN. & EX. 407 ; metal swilc he wolde 3620 ; swilk HAV. 1118 ; RICH. 6019 ; PR. C. 155 ; ISUM. 11 ; swilch REL. I. 131 ; swilch, swich HICKES I. 222 ; swuch MARH. 16 ; KATH. 690 ; A. R. 12 ; H. M. 7 ; swuch man O. & N. 1496 ; amansed swuch [such] þu art 1307 ; svich worð bold non swuch þer nas iseiȝe LEG. 48 ; swic E. G. 360 ; swich [such] CH. C. T. *A* 3 ; a such kniȝt ROB. 217 ; such is evel ant elde SPEC. 48 ; soch EGL. 30 ; sich PERC. 159 ; TOWNL. 27 ; sech TOR. 2241 ; selk AN. LIT. 5 ; slik IW. 141 ; MIN. viii. 35 ; swiche drede WILL. 781 ; swülches (*gen. m. n.*) LAȝ. 20337 ; swülc(h)ere [solchere] (*gen. f.*) LAȝ. 487 ; swülchen (*dat. m.*) LAȝ. 2087 ; swuche A. R. 318 ; swülchere (*dat. f.*) LAȝ. 4449 ; mid sucher sorȝe SHOR. 33 ; swulcne (*acc. m.*) LAȝ. 18934 ; swilcne HOM. I. 37 ; nenne swuchne mon A. R. 96 ; swichne ANGL. I. 32 ; swülche (*pl.*) LAȝ. 6564 ; swilke [swiche, suche] LANGL. *A* PROL. 32 ; suche BEK. 174 ; DEGR. 112 ; oþere suche TREV. VIII. 327 ; soche AUD. 81 ; zvichen (*dat. pl.*) AYENB. 37.

swüre, *see* swíre. swüð, *see* swĩð.

t.

tá, *see* táken.

tã, sb., *O.E.* tá, = *O.N.* tã, *O.H.G.* zēha ; *toe*, C. M. 5932 ; taa PR. C. 1910 ; to HAV. 1743 ; BEK. 1478 ; P. L. S. vi. 13 ; too PR. P. 495 ; S. S. (Wr.) 1131 ; tŏn (*pl.*) SAINTS (Ld.) xlv. 68 ; PL. CR. 426 ; toon CH. C. T. *B* 4052 ; tase YORK xxxv. 180 ; tãn (*dat. pl.*) JUL. 59.

tabard, sb., *O.Fr.* tabard, tabart ; *tabard, short coat*, CH. C. T. *A* 541 ; REL. I. 62 ; tabard [tabart] LANGL. *A* v. 111 ; tabarde LANGL. *C* vii. 203 ; tabbard ' *colobium*' PR. P. 485 ; taberd VOC. (W. W.) 734.

taber, *see* tabour.

tabernacle, sb., *O.Fr.* tabernacle ; *tabernacle*, GEN. & EX. 3174 ; ROB. 20.

táble, sb., *O.Fr.* table ; *table*, HOM. I. 11 ; þe kniȝtes ȝeden to table HORN (L.) 587 ; table *food* DEP. R. i. 58 ; (*picture*) TREV. V. 399 ;

table mele ['*tabulatim*'] PALL. iii. 148;
tábles (*pl.*) GEN. & EX. 3578; tables of ston
AYENB. 5; gemenes of des and of tables
backgammon 45; ROB. (W.) 3965; tables
garden beds ['*areæ*'] PALL. i. 810; *see* **tavel.**

tábille-man, sb., *piece used in playing at
draughts, dice, chess, etc.,* '*seaccus, calculus,*'
CATH. 376.

tabler, sb., *O.Fr.* tablier (*board for games
with moveable pieces*), MAN. (F.) 11395.

tablette, sb., *O.Fr.* tablette; *tablet*; tabu-
lette PALL. vi. 195; **tablettes** (*pl.*) SHOR.
92.

tabour, sb., *O.Fr.* tabour; *tabour, small
drum,* HAV. 2329; CH. C. T. *D* 2268; taboure '*timpanum, terrificium*' PR. P. 485;
tabor '*tympanum*' REL. I. 6; ROB. (W.)
8166; **tabours** (*pl.*) WILL. 3813; taberes
drummers LANGL. *A* ii. 79.

tabre, v., *play on the tabour,* LANGL. *C* xvi.
205, *B* xiii. 230.

taburn, sb., *tambourine, small drum,* CATH.
376; **taburnes** (*pl.*) [*v. r.* taburs] ALEX. (Sk.)
1385; timbres & tabornes A. P. ii. 1414.

taburner, sb., *player on the tambourine,*
CATH. 376; taberner VOC. (W. W.) 688.

taburnistre, sb., *female tambourine player*;
taburnistirs (*pl.*) HAMP. PS. lxvii. 27.

tache, sb., *touchwood,* LANGL. *C* xx. 211;
tacche *B* xvii. 245*.

tache [2], sb., *O.Fr.* tache, teche; *manners,
quality*; teche, tecche '*mos, conditio*' PR. P.
487; **tacches** (*pl.*) LANGL. *B* ix. 146; tecches
AYENB. 32; wikked tecches MAN. (F.) 3899.

tache [3], adj., *touchy; redoubtable*; es nane so
teche of þi time ALEX. (Sk.) 663.

tache, *see* **takke.**

tachen, v., *? for* **atachen**; *attach*: to tache
& teie MAN. (F.) 12056; **tacchis** (*pres.*)
ALEX. (Sk.) 5065; tasselez þerto **tacched**
(*pple.*) GAW. 219; þer hit onez is tachched,
twinne wil hit never 2512; in me weore
tacched sorwes two L. H. R. 142; tached oþer
tiʒed A. P. i. 464.

tācnen, v., *O.E.* tācnian,= *O.N.* tākna, teikna,
cf. Goth. taiknjan, *O.H.G.* zeichenan, zeich-
nan; *from* **tāken**; *betoken, signify,* ORM.
1639; toknin PR. P. 495; **tōkeneð** (*pres.*)
GEN. & EX. 638; þet tokeneþ holi þinges
SHOR. 8; WILL. 2937; **tācnede** (*pret.*) LAʒ.
2832; **tākned** (*pple.*) PS. iv. 7; *comp.* **bi-,
ʒe-tācnen.**

tācninge, sb., *O.E.* tācnung, *cf. O.H.G.* zei-
chenunga; *betokening, signification,* LAʒ.
15974; O. & N. 1213; tokning GEN. & EX.
1624; tokeninge C. L. 557.

tāde, tadde, sb., *O.E.* tādie; *toad*; tade '*bufo*'
VOC. 190; tadde '(*rana ru*)*beta*' FRAG. 3; HOM.
I. 53; P. s. 238; tode MAP 339; toode '*bufo*'
PR. P. 495; **tadden** (*pl.*) A. R. 214; tadden,

tadde HOM. I. 51, 53; toden LEG. 188; tades
PR. C. 6900; todes P. L. S. xxv. 56; MAND. 61.

tæchen, v., *O. E.* tæcan, tæcean; *teach,
show,* ORM. 3468; (*ms.* tachen, *r. w.*
sæchen) LAʒ. 17320; teachen 2419; ic þe
wulle tache Bruttisce spæche 26833; techen
A. R. 210; GEN. & EX. 2792; LANGL.
B xvii. 40; teche O. & N. 1021; CH. C. T.
A 308; PR. C. 5548; ihc schal þe teche a
trewe ifere ASS. 46; teache P. L. S. viii. 152;
t(e)achen HOM. II. 17; **tæcheþ** (*pres.*) FRAG.
5; techeð, tecð HOM. I. 95, 109; techeð A. R.
220; tekþ AYENB. 6; hi techeþ 8; **tēch** (*imper.*)
HOM. I. 51; DEGR. 914; **tæhte,** tahte,
taute [tehte] (*pret.*) LAʒ. 804, 3705, 10240;
tehte, tahte HOM. I. 89, 107; tahte KATH.
1821; CHR. E. 105; (*ms.* tahhte) ORM. 1071;
tachte MISC. 35; me taʒte hire þe wei BEK.
75; tauʒte LANGL. *A* i. 74; tauʒte þai sche
was free FER. 1391; he taughte him sone
to þe kiste PERC. 2109; taute AN. LIT. 8;
taite REL. I. 187; teihte A. R. 144; teiʒte P.
L. S. xvii. 47; toʒte AYENB. 96; **taht** (*ms.*
tahht) (*pple.*) ORM. 18741; tauʒt LANGL. *B*
x. 223; *comp.* **bi-, ʒe-tæchen.**

tēchere, sb., *teacher,* ALIS. 17; TREV. III.
219; **tēchours,** techers (*pl.*) LANGL. *C* xvii.
246, xxiii. 120.

tēchinge, sb., *O.E.* tǣcung; *teaching,* MISC.
44; AYENB. 17; teching LANGL. *B* x. 151.

tǣlen, v., *O.E.* tǣlan,= *O.N.* tǣla; *blame, reprove,* LK. xiv. 29; LAʒ. 3334; ORM. 2033;
telen, tele O. & N. 1377, 1415; tele AN. LIT.
96; **tēleþ** (*pres.*) REL. I. 176; MISC. 116;
tǣlden (*pret.*) LAʒ. 3801; telden ['*vitupera-
verunt*'] MK. vii. 2; **tǣled** (*pple.*) ORM.
16743.

[**tǣse,** adj., *O.E.* tǣse (*mild*); *comp.* **ʒe-tǣse.**]

tǣsel, sb., *O.E.* tǣsel,= *O.H.G.* zeisala; *tea-
sel*; tesel VOC. 141; tasil '*carduus*' VOC. 191;
PR. P. 487; WICL. IS. xxxiv. 13; **tāseles** (*pl.*)
LANGL. *B* xv. 446.

tǣsen, tāsen, v., *O.E.* tǣsan,= *M.Du.* tēsen,
Swed. tēsa, *M.H.G.* zeisen; *pull about; card
(wool)*; *tease*; tese wolle CATH. 380; tosin
'*carpo*' PR. P. 497; toose REL. II. 197; þei
tōse (*pres.*) and pulle GOW. I. 17; **taised**
(*pple.*) GAW. 1169; *comp.* **to-tāsen.**

taffata, sb., *Fr.* taffetas; *taffeta,* CH. C. T. *A*
440; taffata (*v. r.* tafeta) ALEX. (Sk.) 1515.

taggen, v., *see* **toggen**; *? pull*: he shal be
taggud [togged] (*pple.*) wundersare ERC. 437.

tail, sb., *O.E.* tægel, tægl,= *O.H.G.* zagel *m.*
(*tail*), *O.N.,* *Goth.* tagl *n.* (θρίξ); *tail; retinue*; HOM. II. 197; BRD. 8; AYENB. 61;
MAND. 291; CH. C. T. *D* 466; þe teil, þene
teil A. R. 208; **teile** (*dat.*) A. R. 254; **tailes**
(*pl.*) LAʒ. 29557; tailles LANGL. *B* v. 19.

taille-end, sb., *tail-end,* LANGL. *B* v. 395.

tail-lēs, adj., *tailless,* LEG. 151.

tail-rōp, sb., '*avaluer,*' VOC. 168.

taile, talie, sb., *O.Fr.* taille; *tally,* '*talea*,' PR. P. 486; taile [taille] CH. C. T. *A* 570; be taile and be score LAUNF. 419; taile oþer scrit E. G. 362; tale (*v. r.* tailȝe, tailyie) *agreement* BARB. xx. 134; tailyie *tax* xii. 320; tail *man* MAN. (F.) 15363; taile LANGL. *B* iii. 130; **tailes** (*taxes*) AYENB. 38.

tailed, pple., *having a tail,* MAP 338; þe tailede sterre ROB. 416; *comp.* ȝe-teiled.

tailed², pple., *entailed*; londes tailede and not tailede TREV. VIII. 502, 503.

tailende, sb., *for* **taillinge**; *reckoning by tally,* LANGL. *B* viii. 82.

tailin, taliin, v., *O.Fr.* tailler; *mark on a tally; set down; agree upon; tax*; PR. P. 485, 486; taile MAN. (F.) 16550; itailed (*pple.*) SHOR. 165; ȝif i bigge . . . auȝt but ȝif it be itailed [itailled] i forȝete it ȝerne LANGL. *B* v. 429; talit [tailȝeit] BARB. xix. 188.

taillāge, sb., *O.Fr.* taillage; *tallage,* BEK. 343; LANGL. *B* xix. 37; taliage WILL. 5124; **talāges** (*pl.*) LANGL. *C* xx. 37*.

tailōn, sb., *O.Fr.* taillon; *slip of a tree*; **taliōns** (*pl.*) ['*taleas*'] PALL. iii. 991.

tailour, sb., *O.Fr.* tailleor; *tailor*; **tailōrs** (*pl.*) ROB. 313; taillours LANGL. *A* PROL. 100; *B* PROL. 220; tailours C i. 223; tailloures (*gen. pl.*) LANGL. *B* xv. 447.

taisen, *see* **tæsen.**

tait, adj. & sb., *O. N.* teitr (*cheerful*), = *O. H.G.* zeiz (*tender*); *joyous, lively; cheerfulness*: tait . . . & qvoint A. P. iii. 871; **tait(e)** (*pl.*) bestes GAW. 1377; þe laddes were kaske and teite HAV. 1841; with taite at þaire hertes ALEX. (Sk.) 1208; taite *sport, conflict* ALEX. (Sk.) 3979.

takel, sb., *cf. M.L.G., Du.* takel; *tackle, gear, implement, arrow,* GEN. & EX. 883; CH. C. T. *A* 106; A. P. iii. 233; ur takel ur tol WART. II. 109; takil MAN. (F.) 12081; TOR. 1404.

takelen, v., *from* **takel**; *ensnare, catch*; **takild** (*pple.*) with sum luv HAMP. PS. *page* 512.

takelinge, sb., *tackling*: takellinge for theire shippes TREV. IV. 63.

taken, táken, v., *O.N.* taka (*touch, seize*), *M. L.G.* taken (*touch*), *cf. Goth.* tēkan; *take; touch, seize, give,* ORM. 85; GEN. & EX. 1318; to þan fehte taken LAȝ. 23688; hu heo mihten taken on þat þe scucke weore fordon 25965; take MISC. 95; GREG. 85; MAND. 167; S. S. (Wr.) 240; take, tan TRIST. 881, 1418; on þe sakles he suld ta wrake C. M. 11554; **tákeþ** (*pres.*) SHOR. 99; he takeþ nu to fulhtnen ORM. 18269; tais BARB. ii. 146; tac (*imper.*) SPEC. 106; þet ne mai naȝt þolie þet me him take AYENB. 22; tōc (*pret.*) HOM. II. 167; LAȝ. 7976; JUL. 70; þat Adam god forlet & toc him to þe deofel ORM. 356; ȝho toc wel wiþ godes word

2457; tok ALIS. 3813; ure lord . . . tok (*touched*) his lepre MISC. 31; þe gailer him tok (*gave*) an appel P. L. S. xxiv. 231; took LANGL. (Wr.) 10735; tuk ISUM. 189; (þou) toke HAV. 1216; JOS. 438; CH. D. BL. 483; þei token S. S. (Wr.) 1463; token to ȝeien KATH. 2091; he token leve GEN. & EX. 2200; ant token him a kine ȝerde P. S. 215; hii toke (*first text* bitahten) him . . . al þe borh LAȝ. 24039*; he tōke (*subj.*) GEN. & EX. 1531; **taken** (*pple.*) ORM. 1150; C. L. 202; CH. C. T. *A* 1439; tan GAW. 490; LUD. COV. 15; HOM. I. 27; itaken LAȝ. 680; itake BEK. 619; *comp.* **a-, bi-, of-, over-, under-táken.**

táking, sb., *catch,* ['*captio*'] WICL. PS. xxxiv. 8.

tāken, sb., *O.E.* tācen, tācn, *cf. O.Fris.* tēken, *O.L.G.* tēcan, *O.N.* tākn, teikn, *O.H.G.* zeichan *n., Goth.* taikns *f.*; *token, sign,* MAT. xii. 38; KATH. 195; ORM. 3335; PS. lxxxv. 17; þat taken LAȝ. 17956; token GEN. & EX. 646; tocne A. R. 316; AYENB. 226; tokne PR. P. 495; tācne (*dat.*) LAȝ. 17900; in tokne þat pais scholde be bitvext god and manne SHOR. 131; **tāken, tacne** [*O.E.* tācen, tācnu] (*pl.*) MAT. xxiv. 24; LK. xxi. 11; tacna, tacne HOM. I. 91; tacne LAȝ. 32151; **tācnen** (*dat. pl.*) LAȝ. 1158; *comp.* fore-tāken.

takke, sb., *clasp,* '*fibula*,' PR. P. 485; tak L. H. R. 145; tache CATH. 376*.

takkin, v., *tack, sew,* PR. P. 485; **tacked** (*pple.*) PARTEN. 4802; *see* **tachen.**

tal, adj., *O.E.* (ge-)tæl (*pl.* getale), (un-)tala (*North.* MAT. xxvii. 23), = *Goth.* (un-)tals (πειθής, παιδευτός), *? O.H.G.* (gi-)zal (*agilis*); *tall, seemly; docile;* '*elegans*,' PR. P. 486; she made him . . . so humble and talle (*? obsequious*) CH. COMPL. M. 38; þer is no bagpipe half so tal LIDG. M. P. 200; a talle man LUD. COV. 215.

tal-līche, adv., *becomingly*: talliche hire atired WILL. 1706; talli '*eleganter*' PR. P. 486; D. TROY 8813.

tále, sb., *O.E.* talu, = *O.Fris.* tale, *O.N.* tala, *O.H.G.* zala; *tale, speech, narration; number*; LAȝ. 7397, 12765; KATH. 637, 1293; A. R. 68, 320; ORM. 4310; HAV. 2026; AYENB. 234; LIDG. M. P. 44; iherde ich holde grete tale O. & N. 3; and tolde Eve a tale GEN. & EX. 321; sevene ger bi tale 1673; on Engel(e) tale 2526; þe palmere seide on his tale HORN (H.) 1072; folc wiþoute tale ROB. 393; he ȝaf answere and tale SHOR. 123; þe tale (*v. r.* noumbre) of tvelve C. M. 18910; taill BARB. xvii. 835; **tále** (*pl.*) O. & N. 257; talen LAȝ. 15869.

tále-teller, sb., *tale-bearer,* TREV. I. 337; LANGL. *B* xx. 297; **táletellours** (*pl.*) C xxiii. 299.

tále-wīs, adj., *skilled in speech; loquacious, slanderous*; LANGL. *A* iii. 126; talwis *B* iii,

130; talewiis B. B. 12; talewise men HOM. II. 193.

tále, tæl, sb., *O.E.* tāl, tæl (*reproach, calumny*) =*O.N.* tāl (*injury*), *O.H.G.* zāla (*danger*); *calumny*, LK. iii. 14; tel HAV. 191; teil H. S. 2042; tole SHOR. 36.

tāl-līch, adj., *O.E.* tāl-, tællīc; *blasphemous*: talliche word MAT. xv. 19.

talent, sb., *talent* (*sum of money*); **talentes**, talentis (*pl.*) ALEX. (Sk.) 1666, 3154.

talent[2], sb., *O.Fr.* talent; *desire, appetite, purpose*, LEG. 12; '*appetitus*' PR. P. 486; BARB. iii. 694; YORK xxi. 69.

talentif, adj., *desirous*, GAW. 350.

talevas, sb., *O.Fr.* talevas (*sorte de bouclier*); *a sort of shield*; **talevaces** (*pl.*) HAV. 2323.

talʒ, sb.,=*M.L.G.* talg, talch, *M.Du.* talgh *Swed.* talg; *tallow*; talgh REL. I. 53; PALL. i. 444; taluʒ (*ms.* talwʒ) WICL. LEV. vi. 12; E. G. 359; talugh RICH. 1552; taluh PR. P. 486; *deriv.* talwi.

talʒen, v., *tallow*; talwin '*sepo*' PR. P. 487; **talwid** (*pple.*) IBID.; taloghid ALEX. (Sk.) 4208.

talie, see taile.

tálien, v., *O.E.* talian, = *O.Fris.* talia, *O.N.* tala, *O.H.G.* zalōn; *tell a tale*, KATH. 795; talen CH. C. T. *A* 772; tale WILL. 160; tale ni telle ED. 3677; **tálie** (*pres.*) A. R. 356; hunten þar talieð LAʒ. 20857; þeo mariners crieþ and taleþ ALIS. 1415; **táliinde** (*pple.*) AYENB. 207; **tálede** (*pret.*) KATH. 1827; taleden LAʒ. 3800; **táled** (*pple.*) GOW. III. 329.

talkin, v., *talk*, '*fabulo, sermocino*,' PR. P. 486; MARH. 13; talke LANGL. *B* xvii. 82; **talkeð** (*pres.*) A. R. 422; H. M. 17; **talkie** (*subj.*) LAʒ. 788; **talkede** (*pret.*) TREV. IV. 359; A. P. ii. 132; PERC. 1526; talkeden WILL. 3077.

talkinge, sb., *talking*, LEG. 3.

talmen, v., *O.N.* talma (*hinder*), = *M.L.G.*, *Du.* talmen (*delay*); *?fail*; talme FLOR. 769; til mi tonge talmes (*pres.*) REL. I. 292; mi harte talmes D. ARTH. 2581.

talōn, sb., *O.Fr.* talon; *talon, claw*: **talōns** (*pl.*) ALEX. (Sk.) 5454; taloundes TREV. I. 83.

taluʒ, see talʒ.

talwi, adj., *from* talʒ; *tallowy*, PR. P. 486.

táme, adj., *O.E.* tam, tom,=*M.L.G.* tam, *O.N.* tamr, *O.H.G.* zam; *from root of Goth.* (ga-)-timan, *O.H.G.* zeman (*be fitting*); *tame*, GEN. & EX. 174; SPEC. 71; GOW. I. 144; tom KATH. 1318; **támere** (*dat. f.*) MAT. xxi. 5; **tóme** (*pl.*) O. & N. 1444; P. S. 194; **tommure** (*compar.*) A. R. 144.

támę-hēd, sb., *tameness*, GEN. & EX. 1485.

támen, v., *for* attámin; *broach* (*a cask*); *pierce*; **támed** (*pple.*) ALEX. (Sk.) 2622.

támien, témien, v., *O.E.* temian, *cf. M.L.G.* temen, *O.Fris.* tema, *Goth.* (ga-)tamjan, *O.N.*

temja, *O.H.G.* zeman; *from* **táme**; *tame*; tamin PR. P. 486; tame MIRC 1728; temien HOM. II. 63; A. R. 138; temie LAʒ. 25231*; teme MAP 335; **témeþ** (*pres.*) REL. I. 245; **témede** (*pret.*) A. R. 176; TREV. I. 187; **témed** (*pple.*) P. S. 214; *comp.* a-, ʒe-támien (-témien).

táminge, sb., *taming*, PR. P. 486.

tán, see **táken**.

tange, sb., *O.E.* tang (ÆLFR. 67),=*O.Fris.*, *M.L.G.* tange, *M.Du.* tanghe, *O.N.* töng, *O.H.G.* zange; *pair of tongs*, '*forceps*,' VOC. 232; FER. 1308; tonge PR. P. 496; O. & N. 156; P. L. S. ix. 79; **tangen** (*dat. pl.*) BRD. 22.

tange[2], sb., *O.N.* tangi; *sting; dagger; '*pugio*,' VOC. 221; tonge '*aculeus*' PR. P. 496.

tangid, pple., *stung*, ALEX. (Sk.) 3637, 3886, 4798.

tangil, adj.: tanggil or froward and angri PR. P. 486.

tankard, sb., *O.Fr.* tanqvart; *tankard*, '*amphora*,' VOC. 178; PR. P. 486.

tanni, tauni, adj., *cf. Du.* tanig; *tawny*, PR. P. 486; tauni LANGL. *B* v. 196; taunde (*for* taune) ALEX. (Sk.) 4335.

tannin, v., *cf. M.Du.* tannen, tanen, *O.Fr.* tanner, taner; *tan*, PR. P. 486; itanned (*pple.*) E. G. 358.

tannere, sb.,=*M.Du.* taner; *tanner*, E. G. 359; **tanneris** (*pl.*) LANGL. *A* PROL. 100.

tape, see tappe[2], tappen.

taper, sb., *O.E.* tapur; *taper*, HOM. II. 47; P. L. S. ix. 12; SHOR. 49; taper, tapur, tapir TREV. VII. 425; **tapre** (*dat.*) LANGL. *B* xvii. 203; **taperes** (*pl.*) MARH. 18; BEK. 1883.

tapet, sb., *O.Fr.* tapit; *cloth, tapestry*, '*tapetum*,' PR. P. 486; tapit '*tapete*' VOC. 242; GAW. 568; **tapetis** (*pl.*) WICL. 2 KINGS xvii. 28; tapites GAW. 77.

tapicēr, tapecēr, sb., *O.Fr.* tapicier; *tapestry-worker*, CH. C. T. *A* 362; tapecer, tapesere PR. P. 486.

tapicerie, sb., *tapestry;* tapecerie HALLIW. 850.

tapissinge, sb., *hangings*, HAMP. PS. xvii. 13.

tappe, sb., *O.E.* tæppa,=*M.L.G.* tappe, *O.H.G.* zapho; *tap*, PR. P. 486; CH. C. T. *A* 3890; teppe AYENB. 27.

tappe[2], sb., *O.E.* tæppe; *tape*, '*taenia*,' VOC. 196; **tapes** (*pl.*) CH. C. T. *A* 3241.

tappe[3], sb., *tap, blow*, GAW. 2357.

tappen, v.,=*M.L.G.* tappen; *tap, beat*: your foot ye tappin (*pres.*) AN. LIT. 86; **tep** (*imper.*) A. R. 296; tape GAW. 406.

[**tappen**[2], v., *draw liquor*.]

tæppare, sb., *O.E.* tæppere,=*O.Fris.* tapper; '*caupo*,' FRAG. 2.

tappestere, sb., *O.E.* tæppestre ; *tapster*, CH. C. T. *A* 241 ; **tapesters** (*pl.*) *barmaids* LANGL. *A* ii. 79.

táre, sb., *tare (weed)*, CH. C. T. *A* 4000 ; **táren** (*pl.*) AR. & MER. 7364 ; taris [*'zizania'*] WICL. MAT. xiii. 25.

tarettes, sb., pl., *from O.Fr.* teride ; *transport vessel*, MIN. iii. 80.

targe, sb., *charter*, *'carta,'* PR. P. 486.

targe², sb., *O.E.* targe,=*O.N.* targa, *O.Fr.* targe ; *small shield*, P. P. 217 ; CH. C. T. *A* 471 ; PARTEN. 4212 ; **targes** (*pl.*) ROB. 361.

targen, v., *O.Fr.* targier ; *retard* ; targi MISC. 36 ; targe ALIS. (Sk.) 211 ; PALL. iii. 1075 ; **targede** (*pret.*) P. L. S. xiii. 179.

tarȝen, *see* terȝen.

target, sb., *O.Fr.* targette ; *target, shield*, PR. P. 487 ; **targetes** [targettes] (*pl.*) ALEX. (Sk.) 2622.

tarien, *see* terȝen.   tarne, *see* terne.

tarse, sb., *silk of Tartary*, LANGL. *B* xv. 163 ; tars ALEX. (Sk) 1515, 4673.

tartarine, sb., ?*O.Fr.* tartarin ; *silk of Tartary*, LANGL. *B* xv. 224.

tarte, sb., *O.Fr.* tarte ; *tart, small pie*, PR. P. 487 ; **tartes** (*pl.*) R. R. 7041.

tartre, sb., *Fr.* tartre ; *tartar*, CH. C. T. *G* 813.

tas, sb., *O.Fr.* tas ; *heap* ; (*ms.* taas) CH. C. T. *A* 1005 ; tasse PR. P. 487.

tāsel, *see* tæsel.

tásen, *see* teisen.   tāsen, *see* tæsen.

taske, sb., *O.Fr.* tasque, tasche ; *assessment, task*, *'taxa,'* PR. P. 487 ; C. M. 5872.

tasker, sb., *thresher*, VOC. (W.W.) 697 ; taskar BARB. v. 318.

tassel, sb., *O.Fr.* tassel ; *tassel*, PR. P. 487 ; C. M. 4389.

tást, sb., *O.Fr.* tast ; *taste*, TREV. III. 467 ; CH. P. F. 160 ; *'gustus, sapor'* PR. P. 487 ; **tástes** (*pl.*) *investigations* LANGL. *B* xii. 131.

tásten, v., *O.Fr.* taster ; *feel, touch, taste ; kiss* ; taastin PR. P. 487 ; taste FLOR. 109 ; he gan taste aboute þe mouþ LANGL. *B* xiii. 346 ; **tásteþ** (*pres.*) AYENB. 245 ; **táste** (*imper.*) CH. C. T. *G* 503 ; **tástede** (*pret.*) LANGL. *C* xx. 122 ; tasted S. S. (Web.) 1048 ; architriclin tastide þe watir WICL. JOHN ii. 9 ; tastit BARB. ix. 388.

tásting, sb., *groping*, LANGL. *C* xx. 122*.

[tater, sb., *O.N.* töturr,=*L.G.* tater ; *rag.*]

tatered, adj., *tattered, lacerated*, PL. CR. 753 ; tatird PR. C. 1537 ; TOWNL. 4.

taterin, v.,=*M.L.G., M.Du.* tateren ; *babble*, *'blatero,'* PR. P. 487.

tateringe, sb., =*M.Du.* tateringhe ; *'garritus,'* PR. P. 487 ; WICL. E. W. 192.

[taþ, sb., *O.N.* tað ; *dung, manure.*]

taþin, v., *'stercoro,'* *manure*, PR. P. 487.

taunen, v.,=*M.L.G.* tōnen, *M.Du.* toonen, *M.H.G.* zounen ; *point out*, GEN. & EX. 1022 ; **taunede** (*pret.*) REL. I. 226.

tauni, *see* tanni.

tavel, sb., *O.E.* tæfel,=*O.N.* tafl, *O.H.G.* zabel ; *game played with moveable pieces* : somme pleoide mid tavel LAȝ. 8133*.

tævel-bred, adj.,=*O.H.G.* zabelbret ; *game board*, LAȝ. 8133.

tavelin, v., *O.E.* tæflan,=*O.N.* tefla, *M.H.G.* zabelen ; *play at tables (backgammon, draughts, etc.)*, MARH. 13 ; **taveleþ** (*pres.*) O. & N. 1666 ; teveli (*subj.*) KATH. 822.

teveling, sb., *sport* ; (*printed* teneling) GAW. 1514.

taverne, sb., *O.Fr.* taverne ; *tavern*, ROB. 195 ; AYENB. 56 ; CH. C. T. *A* 3334 ; **tavernes** (*pl.*) LANGL. *C* iii. 98.

tavernēr, sb., *Fr.* tavernier ; *innkeeper*, CH. C. T. *C* 685 ; **taverniers** (*pl.*) AYENB. 44.

tawe, sb., *O.E.* (ge-)taw, = *M.L.G.* tawe, tauwe, touwe, *M.Du.* touwe, *M.H.G.* (ge-)zouwe *n.* ; *tackle* ; towe PALL. vii. 36 ; teu (*ms.* tew) *'piscalia'* PR. P. 490 ; (*tawing of leather*) *'frunicio'* 489.

tawen, sb., *O.E.* tēawian, tāwian,=*M.L.G.* tawen, tauwen, touwen, *M.Du.* touwen, *Goth.* taujan (*pret.* tawida) (ποιεῖν), *O.H.G.* zawan, zowan, *mod. Eng.* taw ; *prepare (leather)* ; *dress (hemp)* ; *scourge* ; (*ms.* tawwenn) ORM. 15908 ; tewin PR. P. 490 ; tewe MAN. (F.) 12453 ; **tawiþ** (*pres.*) REL. II. 175 ; *comp.* ȝe-tawen.

tawer, sb., *O. E.* tawere ; *tawer*, [*'coriarius'*] WICL. DEEDS ix. 43 ; teware PR. P. 490.

tax, sb., *Fr.* taxe ; *tax*, P. S. 151.

taxāciōn, sb., *Fr.* taxation ; *taxation*, P. S. 337.

taxen, v., *tax* ; taxeþ (*pret.*) LANGL. *C* ii. 159.

taxour, sb., *assessor of a fine*, LANGL. *C* ix. 37 ; taxoure B vi. 40.

te, *see* tō.   te-, *see* to-.

tēafor, sb., *O.E.* tēafor ; *red paint*, *'minium ;'* (*ms.* teapor) FRAG. 2.

tealt, sb., ? *O.E.* tealt (*vacillans*) ; *unstable condition* : cristninge stant te tealte SHOR. 9.

tēam, sb., *O.E.* tēam (*progeny*), = *O.Fris.* tām (*bridle, progeny*), *M.L.G.* tōm, *M.Du.* toom, *O.N.* taumr, *O.H.G.* zaum, zoum (*reins*) ; *from* tēon ; *team ; progeny* FRAG. 8 ; weox swa his team JUL. 60 ; þene team A. R. 336 ; tæm ORM. 2415 ; tem ROB. 261 ; ALIS. 2350 ; PS. cxxxvi. 7 ; teem CHR. E. 147 ; a tem of foure grete oxen LANGL. *B* xix. 257 ; **tēames** (*gen.*) HOM. II. 133 ; **tēme** (*dat.*) TOR. 2022 ; **tēmes** (*pl.*) O. & N. 776 ; P. L. S. xxi. 129 ; **tēame** (*gen. pl.*) H. M. 41 ; *comp.* barn-tēam.

tēar, sb., *O.E.* tēar, tēagor, *cf. O.Fris.* tār *m.*,

O.N. tār, *Goth.* tagr *n.*, *O.H.G.* zaher *m.*, *Ir.*
dear, deor, deur, *Gael.* deur, *Welsh* dagr, *Gr.*
δάκρυ *n.; tear*; ter HAV. 285; MAP 338;
teer LUD. COV. 213; þe ter, tere HOM. I.
159; tere [teere] CH. C. T. *B* 3251; tēares
(*pl.*) KATH. 2361; A. R. 110; teares [teres]
LAȝ. 5075; tæres ORM. 13849; teres O. & N.
426; GEN. & EX. 2356; GOW. I. 143; tiares
AYENB. 173; tieres, tires HORN (L.) 654,
960; tēaren (*dat. pl.*) LK. vii. 38; HORN
(R.) 970; tere S. A. L. 163; *comp.* huni-tiar.

teche, *see* tache. tēchen, *see* tǣchen.

tedir, *see* teþer. teen, *see* tēon.

tēȝe, sb., *O.E.* tēge, tȳge *m.*, teag,=*O.N.*
taug *f.; tie, chain*; teiȝe SAINTS (Ld.) xiv.
301; tēis (*pl.*) JOS. 504; tēȝen (*dat. pl.*)
LAȝ. 20998; *comp.* rake-tēȝe.

    tēȝ-doggue, sb., *chained dog* (*ms.* doggue)
SAINTS(Ld.)xlv.301; teidoggeWICL.E.W.252.

teȝele, sb., *O.E.* tigele, *from Lat.* tegula ; *tile*,
AYENB. 167; tigel GEN. & EX. 2552; tile
CH. C. T. *D* 2105; *der.* tīlette, tīlin, tīlare.

    teghel-stān, sb., *tile, brick*, REL. I. 54; tiel-
stoon WICL. IS. xvi. 11.

tēȝen, v., ? *O.E.* tēgan, tȳgan ; *tie, bind*,
HOM. II. 257; teien [tiȝe] LAȝ. 20997; teiȝen
LANGL. *A* i. 94; tien *B* i. 96; (*ms.* teye)
P. L. S. xxi. 130; PALL. iv. 752; tēide (*pret.*)
A. R. 140; tiȝed [tied] C. L. 407; tēiȝed, teied
(*pple.*) WILL. 3226, 3232; tiȝed A. P. i. 464;
tighed ALIS. 779; teid CH. C. T. *E* 2432;
comp. ȝe-, un-tēȝen.

teie, sb., *O.Fr.* teie, toie, *from Lat.* thēca ;
coffer, (*ms.* teye) '*theca*' PR. P. 487; tie GOW.
II. 246; S. S. (Wr.) 2951; DEGR. 552; biloken
in hire teye R. S. vii (MISC. 190).

tēie, *see* tēȝe. teil, *see* tail.

tein, sb., *O.E.* tān,=*M.Du.* teen, *O.N.* teinn,
Goth. tains, *O.H.G.* zein ; *thin plate of metal* :
a tein of silver CH. C. T. *G* 1225; teines
(*pl.*) 1337.

teint, adj., *O.Fr.* taint ; *tinged, red*, MAN.
(F.) 10903.

teinten, v.; *for* ateinten ; *attaint* ; tainte
YORK xxvi. 6.

teisen, v., *O.Fr.* teser, toiser (*string a bow*);
poise a blow ; teisande (*pple.*) MAN. (F.)
10906; he tásit (*pret.*) þe vire *fixed the bolt
on the cross-bow* BARB. v. 623; teised MAN.
(F.) 12368.

teit, *see* tait.

tek, sb., *cf. M.Du.* tick, *M.H.G.* zic; *tick* ; tek,
or litille touche '*tactulus*' PR. P. 487.

teke, *see* tike.

tel, sb., *O.E.* (ge-)tæl,=*O.N.* tal, *O.L.G.* (gi-)-
tal ; *number*, A. R. 372; *comp.* ȝe-tel.

tēl, *see* tāle.

tēld, sb., *O.E.* teld,=*M.Du.* telde, *O.N.* tiald,
O.H.G. zelt ; *from* tēlden ; *covering, tent*,
LAȝ. 31384; M. ARTH. 2624; teld, tgel

(=tield) GEN. & EX. 2025, 3769; telt PR. P.
488; telde MAN. (F.) 649; tild ALEX. (Sk.)
1343, 3860; tēldes [tealdes] (*gen.*)· LAȝ.
26335; télde (*dat.*) IW. 2053; A. P. ii. 866;
téld [teldes] (*pl.*) LAȝ. 16462; telde (*ms.*
teldes, *r. w.* velde) (*gen. pl.*) LAȝ. 17491; (*ms.*
telden) 24436; télden (*dat. pl.*) LAȝ. 17367.

tílden, v., *O.E.* (be-, ofer-)teldan (*pple.* tolden),=*O.N.* tialda (*pret.* tialdaða) ; *pitch* (*a
tent*) ; *erect; dwell; set* (*a trap*) ; telde ALIS.
5067; tilden his musestock HOM. I. 53;
tillen þe nettes A. R. 334; he téldeð (*pl.*) þe
grune HOM. II. 211; tildeð A. R. 334*; þei
tílddeden ['*tetenderunt*'] (*pret.*) Absalon a
tabernacle WICL. 2 KINGS xvi. 22; télded
(*pple.*) GAW. 884; telded on lofte A. P. ii.
1342; teldit [iteldid, itilled] LANGL. *A* ii.
44; itælded LAȝ. 17489; a paviloun iteld he
sigh LĄUNF. 263; itild PALL. iv. 164; *comp.*
bi-, over-tílden ; *deriv.* téld.

    tildunge, sb., *setting of traps*, A. R. 278.

téle, sb., *cf. M.Du.* teelingh (*qverqvedula*) ;
teal (*bird*), '*cercele*,' VOC. 165 ; '*turcella, tur-
bella*' PR. P. 487; SQ. L. DEG. 320.

téle², sb., ?=*tile, sorcery* : with cha(r)mes &
with tele he is ibroȝt aȝein to hele MIRC 368.

tēlen, *see* tǣlen. telien, *see* tilien.

[telle, adj., *comp.* earfeð-telle.]

tellen, v., *O.E.* tellan,=*O.Fris.* tella, *O.L.G.*
tellian, *O.N.* telja, *O.H.G.* zellan ; *from* tále ;
tell,narrate; number; account; LAȝ.14; ORM.
4550; GEN. & EX. 497; CH. C. T. *B* 247; telle
O. & N. 293; HAV. 2615; AYENB. 17; OCTAV.
(H.) 575; here nomes i shal telle SPEC. 104;
him i telle (*pres.*) a loverd þat þus con bete
bales A. D. 240; (þu) telest FER. 1578; telest,
tellest JUL. 54, 55; (he) tele ð LK. xiv. 28; tele
(*imper.*) HOM. I. 249; hit telleð A. R. 74;
teileþ [telþ] O. & N. 340; tel 133; tel [telle]
LAȝ. 26089; tealde (*pret.*) HOM. II. 31;
SHOR. 52; AYENB. 239; talde KATH. 1318;
ORM. 5372; IW. 359; talde [tolde] LAȝ. 1350;
tolde A. R. 66; LANGL. *B* v. 252; mo þen ten
siþen told(e) i mi tax P. S. 151; twentı pund he
þem tolde OCTAV. (H.) 587; he ne tolde þer of
noȝt (*thought nothing of it*) ROB. (W.) 8636;
tolde [telde] WICL. GEN. xl. 9; telde S. S.
(Web.) 798; tolden GEN. & EX. 2221; þene
mahum þe heo tolden (*accounted*) for god LAȝ.
231; táld (*pple.*) PR. C. 213; told A. R. 356; he
was for a kempe told HAV. 1036; teld RICH.
5345; *comp.* bi-, ȝe-tellen.

    tellere, sb., *teller*, WILL. 334; *comp.* tále-
teller.

    tellinge, sb., *numbering, speaking*, AYENB.
1; LANGL. *C* xxiii. 8.

telwin, v., *O.N.* telgja ; *cut, chop*, '*reseco*,' PR.
P. 488.

tēm, *see* tēam.

téme, sb., *O.Fr.* tesme, *Gr.* θέμα ; *theme, instruction*, PR. P. 488; LANGL. *A* iii. 86;
tēmes (*pl.*) ALEX. (Sk.) 2519.

tēmen, v., *O.E.* tēman, tȳman, *O.N.* teyma ;
from **tēam** ; *lead, bring forth* : to witnesse
. . . temen (*take*) P. L. S. viii. 54 ; teman FRAG.
8 ; þer to þu scalt teman (*draw*) LAȝ. 1265 ;
he hehte Tennancius to Cornwale temen 7174 ;
teamen H. M. 33 ; ȝho ne mihte tæmen (*conceive*) ORM. 130 ; teme O. & N. 499 ; H. S.
9546 ; A. P. iii. 316 ; innes for to teme & take
MAN. (F.) 11177 ; i nul nout teme (*? contend*)
SPEC. 32 ; timen GEN. & EX. 982 ; PL. CR.
742 ; tēmeð (*pres.*) A. R. 220 ; tŭmde (*pret.*)
LAȝ. 27919 ; he temde (*ms.* temed) him to þe
king *he appealed to the king* TRIST. 431 ;
hine to hærre temden 1956.

    tēminge, sb., *childbirth*, H. V. 4.

tēmin, v., *O.N.* tœma ; *from* tōm ; *prov.Eng.*
teem ; *make empty ; pour out ; 'vacuo,'* PR. P.
488 ; **tēmez** (*pres.*) D. ARTH. 1801 ; þei teme
sadils HALLIW. 857.

Temese, pr. n., *O.E.* Temese ; *Thames,*
ROB. 2 ; MAN. (F.) 4331 ; þa Temese LAȝ.
7404 ; Temse TREV. IV. 185.

témien, *see* támien.

temperance, sb., *Fr.* temperance ; *temperance,* AYENB. 124.

tempeste, sb., *O.Fr.* tempeste ; *tempest,
plague,* MISC. 32 ; SAINTS (Ld.) xlv. 195 ;
AYENB. 73 ; REL. I. 265 ; MAN. (F.) 2984.

temple, sb., *O.Fr.* temple, *O.E.* tempel ;
temple, PR. P. 488 ; LAȝ. 1137 ; KATH. 1489 ;
þæt temple MAT. xii. 6.

templērs, sb., pl., *knights templar,* LANGL.
C xviii. 209.

temporal, sb., *O.Fr.* temporel ; *temporary,
short lived,* WICL. MK. iv. 17.

temporalitē, sb., *O.Fr.* temporalité; *revenues
of the church* ; LANGL. *C* xxiii. 128 ; **tem-
peraltēs** (*pl.*) LANGL. *B* xx. 127.

temprē, adj., *modified, temperate,* HAMP. PS.
l. 1* ; cxxxvii. 5*.

tempren, v., *O.E.* temprian, *O.Fr.* temprer ;
temper, moderate, PL. CR. 743 ; tempre WICL.
EZ. xxvi. 9 ; tempire ALEX. (Sk.) 3466 ; tem-
preþ (*pres.*) AYENB. 254 ; tempred (*pple.*)
ORM. 2893 ; REL. I. 111.

temprūre, sb., *due proportion ; moderation* :
þe elementis . . . so travailed out of temperoure
ALEX. (Sk.) 543.

tempten, *see* tenten.

temptour, sb., *O.Fr.* tempteor ; *tempter,* CH.
C. T. *D* 1655.

temse, sb., cf. *M.L.G.* temes, *M.Du.* tems ;
temse, sieve, VOC. 200 ; PR. P. 488.

temsin, v., *O.E.* temsian,=*M.Du.* temsen ;
strain; pass through a sieve ; PR. P. 488.

tēn, card. num., *O.E.* tēn, tīen, tȳn, tēne, tȳne,
=*O.Fris.* tian, tien, *O.N.* tiu, tio, *O.L.G.* tehan,
Goth. taihun, *O.H.G.* zehan, zehani, *Lat.*
decem, *Gr.* δέκα, *Ir.,* *Gael.* deich, *Welsh*
deg ; *ten,* AYENB. 5 ; GREG. 115 ; ten, tene,

A. R. 200, 244 ; SHOR. 92 ; ten siþen P. S.
151 ; bi tene 188 ; ten, tgen (=tien) GEN. &
EX. 1955, 3413 ; ten ȝer LAȝ. 2514 ; tene [ten]
beoð inohȝe 3388 ; tene KATH. 794 ; BEK.
1851 ; teon, tien (*ms.* tyen) MAT. xviii. 24,
xxv. 28 ; tien HOM. I. 219 ; *comp.* **eahte-,
fēower-, fīf-, niȝen-, seven-, six-, þrēo-
tēne.**

    tēn-fóld, adj., *tenfold,* HOM. II. 135.

tēn, *see* tēon.

tenaunt, sb., *Fr.* tenant ; *tenant,* LANGL. *B*
vi. 39 ; **tenauntes** (*pl.*) *C* xviii. 45*.

tenche, sb., *O.Fr.* tenche ; *tench,* VOC. 189 ;
PR. P. 488 ; REL. I. 51.

tēnde, *see* tēoþe.

tenden, v., *O.E.* (on-)tendan, = *Goth.* tand-
jan ; *set on fire, burn,* HOM. I. 81 ; tende
[tiende] TREAT. 135 (SAINTS (Ld.) xlvi. 523) ;
he shal teenden ['*incendet'*] WICL. EXOD.
xxx. 7 ; tende PARTEN. 2136 ; tende (*pres.
subj.*) A. R. 296 ; tende (*pret.*) S. S. (Wr.)
2183 ; hi tende here liȝt P. L. S. ix. 16 ; ten-
deden LANGL. *B* xviii. 238 ; itend (*pple.*)
KATH. 197 ; TREV. III. 395 ; tend lowe
ALEX. (Sk.) 4179 ; *comp.* a-, on-tenden.

tenden [2], v., *Fr.* tendre ; *attend* ; tende (*? for*
atende) L. H. R. 120 ; ȝe tendiþ (*pres.*) to
idelnes ['*vacatis otio'*] WICL. EX. V. 17.

tender, sb., *O.E.* tender ; *tinder,* LANGL. *C*
xx. 211* ; *see* tinder, tunder.

tendoir, v., *Fr.* tendoir ; *stretching frame* ;
tentoure PR. P. 489.

tendre, adj., *Fr.* tendre ; *tender, delicate,* A.
R. 372 ; HAV. 217 ; AYENB. 31.

    tender-līche, adv., *tenderly,* TREV. II. 307.

tendren, v., *O.N.* tendra ; *set on fire* : & fenġ
his neb to rudnin ant tendrin JUL. 29 ; **tendreð** (*pres.*) H. M. 31.

tendrōn, sb., *Fr.* tendron ; *tender shoot, bud,*
PR. P. 488 ; PALL. iii. 774.

tēne, *see* tēone.

teneis, sb., *tennis ; 'teniludus,'* PR. P. 488.

tēnel, sb., *O.E.* tǣnel ; *basket, 'tenella, car-
tallus,'* PR. P. 489.

tenement, sb., *Fr.* tenement ; *tenement,* PR.
P. 488 ; E. G. 362.

tēnen, *see* tēonen.

Tenet, pr. n., *O.E.* Tenet ; *Thanet,* ROB. 122 ;
TREV. II. 43.

tenge, adj., *O.E.* (ge-)tenge,=*O.H.G.* (gi-)-
zengi ; *near to* : cludes h(e)ovene tenge (*ms.*
tinge, *r. w.* genge) O. & N. 1001.

tenoun, sb., *Fr.* tenon ; *tenon, 'tenaculum,'*
PR. P. 489.

tenour, sb., *meaning, contents,* ALEX. (Sk.)
3566, 4239 ; '*tenor'* PR. P. 489.

tenserie, sb., *? for* * censerie ; *extraordinary
impost,* SAX. CHR. 262.

tentāciūn, sb., *O.Fr.* tentation ; *temptation, trial,* A. R. 232.

tente, sb., *O.Fr.* tente, tende, *cf. Ital.* tenda ; *tent,* ' *tentorium,*' PR. P. 489 ; hor **tentes** (*pl.*) and hor pavilons ROB. 203 ; tentes LANGL. *A* ii. 44.

tente², sb., *O.Fr.* tente, *cf. Ital.* tenta ; *probe* : tente of a wounde or a soore ' *tenta* ' PR. P. 489.

tente³, sb., *for* en**tente** ; *intention* : tac tente [' *attende*'] WICL. PROV. vii. 24 ; DEP. R. ii. 92 ; tent P. L. S. xxv. 8 ; LIDG. M. P. 34 ; HAMP. PS. liv. 7.

tenten, v., *O.Fr.* tenter, tempter ; *stretch ; try, tempt,* A. R. 236 ; tente (' *extendo*') cloth PR. P. 489 ; **temptide** (*pret.*) WICL. DAN. i. 14 ; **itented,** itempted (*pple.*) A. R. 226, 228 ; itouked and itented LANGL. *B* xv. 447.

tentoure, *see* tendoir.

tēnþe, *see* tēoþe.

teolie, teolien, *see* tilie, tilien.

tēon, v., *O.E.* tēon, = *O.L.G.* tiohan, *Goth.* tiuhan, *O.Fris.* tia, *O.N.* *tiuga (*pple.* toginn), *O.H.G.* ziohan, ziuhan, *Lat.* dūcere ; *lead, draw ; go, mount* ; LAȝ. 1831 ; C. L. 821 ; ALIS. 6954 ; teon [teo] KATH. 2129 ; teo O. & N. 1232 ; ALIS. 729 ; a þousend men ne mowe hire enes of þe stede teo P. L. S. xxi. 112 ; ten AR. & MER. 1435 ; towarde Egipte he gunne ten GEN. & EX. 1953 ; he wulde ꝥat he sulde hem ten ꝥat he wel ꝥewed sulde ben 1913 ; ⸱tee ROB. 202 ; RICH. 5137 ; LUD. COV. 33 ; te ORF. 280 ; M. ARTH. 1015 ; tuen H. H. 234 ; **tēo** (*pres.*) LEG. 48 ; tihþ O. & N. 1435 ; hit hine tið to þan bittre deꝥe HOM. I. 27 ; teoð, teð HOM. II. 35, 37 ; **tīh** (*imper.*) LAȝ. 17416 ; **tīe** (*subj.*) JOHN vi. 44 ; ten REL. I. 216 ; **tēah,** teh (*pret.*) HOM. II. 139, 185 ; teh I. 129 ; teah (*ms.* tah), tæh, teih LAȝ. 640, 805, 21616 ; teȝ FL. & BL. 617 ; teg GEN. & EX. 320 ; teiȝ JOS. 57 ; AL. (T.) 449 ; teiȝe ROB. (W.) 6579* ; tei REL. II. 278 ; tigh GOW. II. 318 ; hio tugen LK. v. 11 ; tuȝen, tuwen LAȝ. 1834, 2619 ; tuhen MARH. 22 ; **togen** (*pple.*) GEN. & EX. 3647 ; towen GAW. 1093 ; A. P. i. 251 ; *comp.* a-, bi-, ȝe-, ofer-, to-, wið-tēon ; *deriv.* tēȝe, tēam, toȝe, toȝen, toggen, tüȝe, tüȝel, tuggen, tuken, tucken, tüht, tühten.

tēon², v., *O.E.* tēon, = *Goth.* teihan, *O.H.G.* zīhan, *Lat.* dīcere ; *accuse* : holi churche ... þat me tīȝþ (*pres.*) on BEK. 1180.

tēon, *see* tēn.

tēone, sb., *O.E.* tēona, = *O.L.G.* tiono ; *? from* **tēon** ; *vexation, injury,* A. R. 114 ; BEK. 829 ; ALIS. 2980 ; A. D. 291 ; þene teone FRAG. 6 ; ne do he þe nevre swa muchelne teone HOM. I. 15 ; wiðute teone & treie 193 ; for to don him teone MISC. 49 ; hi hedden teone and seorewe 89 ; teone [tone] and schame O. & N. 50 ; teone, tuone LAȝ. 176, 6013 ; teone [tene] KATH. (E) 402 ; tene ORM. 19866 ; GEN. & EX. 2992 ; HAV. 729 ; SHOR.

130 ; LANGL. *B* vi. 119 ; S. S. (Wr.) 1797 ; MAN. (F.) 8212 ; TRIAM. 345 ; PERC. 1345 ; M. ARTH. 1449 ; i lede mi lif wiþ tene and kare AN. LIT. 7 ; he livede in tene & wo L. H. R. 18 ; ȝif i told(e) him treuli mi tene and min anger WILL. 552 ; tene [teene] CH. C. T. *A* 3106 ; tiene AYENB. 31 ; teine BARB. xviii. 233 ; tuene REL. I. 263 ; **tēonen** (*dat.*) LAȝ. 4362 ; **tēonen** (*acc.*) MAT. xx. 13 ; **tēonen** (*pl.*) LAȝ. 11689 ; H. M. 7.

tēon-ful, adj., *O.E.* tēonfull ; *painful,* ' *injuriosus,*' FRAG. 2 ; LAȝ. 4585 ; tenful WILL. 2666.

tēnful-lī, adv., *sorrowfully,* A. P. ii. 160.

tēonen, v., *O.E.* tēonian, tȳnan, = *O.L.G.* (ge-)tiunean, *O.Fris.* tiona, tiuna ; *harm, irritate* ; tenin PR. P. 489 ; tene GAW. 2002 ; S. S. (Wr.) 2434 ; H. S. 7443 ; EGL. 444 ; **tēne** (*pres.*) A. P. ii. 759 ; teoneð [teoneð] KATH. (E.) 550 ; heo teoneð A. R. 118 ; alle wordes him tieneþ and greveþ AYENB. 142 ; teneþ LANGL. *B* xv. 412 ; þu **tēonedest** (*pret.*) MISC. 139 ; **tēned** (*pple.*) WILL. 1992 ; tenid REL. II. 196 ; DEP. R. iii. 81 ; *comp.* a-, ȝe-, of-tēonen.

tēonðe, *see* tēoþe.

teorien, v., *O.E.* teorian, = *O.L.G.* (far-)terian, *? M.H.G.* zern ; *tire* ; tire TOWNL. 126 ; him teoreþ (*pres.*) his miht FRAG. 5 ; *comp.* a-teorien.

tēoþe, tēnde, sb., *O.E.* tēoða, tēogoða, *O.N.* tiundi, tiondi, = *O.L.G.* tehando, *Goth.* taihunda, *O.H.G.* zehanto ; *tithe, tenth* ; teoþ BEK. 619 ; teþe SHOR. 101 ; teithe MIRC 347 ; tigeðe, tieðe HOM. II. 83, 137 ; tiþe LANGL. *B* xv. 480 ; A. P. ii. 216 ; tigðe, tende GEN. & EX. 597, 895 ; tend TOWNL. 9 ; c. M. 1062 ; tiþe, tenþe WILL. 4715, 5346 ; tende ORM. 2715 ; AYENB. 2 ; tend PR. C. 3990 ; teonðe HOM. I. 219 ; tenþe H. S. 2926 ; **tēndis** (*pl.*) and offrandis HAMP. PS. lxxviii. 1* ; tendes W. & I. 78.

tēoþien, v., *O.E.* tēoðian, tīogoðian ; *tithe, give a tenth* ; teoþe (*ms.* theoþe) MISC. 77 ; teðien HOM. II. 215 ; teþen [tiþen] LANGL. *C* xiv. 73 ; teithe MIRC 349 ; tithin PR. P. 495 ; tende CATH. 379 ; tēþeȝede (*pret.*) ROB. 261 ; itēoþeged (*pple.*) A. R. 28.

tēoþere, sb., *tither* ; **tīþeres** (*pl.*) CH. C. T. *D* 1312.

tēoþing, sb., *O.E.* tīoðung ; *tithing, title,* P. L. S. xii. 40 ; teþinge ROB. 267 ; E. G. 361 ; tiðinge HOM. II. 129 ; tiþing [' *decimationem*'] WICL. TOB. i. 7 ; teondunge LK. xviii. 12 ; **tēndingis** (*pl.*) GEST. R. 17.

teppe, *see* tappe.   **tēr,** *see* tēar.

tercel, adj. & sb., *O.Fr.* tercel, terce ; *? male hawk, ? male eagle* ; tercel egle CH. P. F. 392, 449 ; the tercel gan she calle CH. P. F. 405 ; tercel hawk ' *tercellus* ' PR. P. 489 ; tercel. tercelle VOC. (W. W.) 616, 701 ; terselle CATH. 380 ; these egles **tercels** (*pl.*) CH. P. F. 540.

tercelet, sb., *Fr.* tiercelet ; *small hawk,* CH. C. T. *F* 504 ; **tercelets** (*pl.*) CH. P. F. 659.

tercian, adj., *Lat.* tertiāna ; *tertian* (*fever*), LANGL. *A* xii. 80.

tere, sb., *O.E.* teru, teoru, = *M.Du.* tere, *? O.N.* tiara ; *tar,* VOC. 279 ; terre PR. P. 489 ; ter GEN. & EX. 662 ; tarre LANGL. *C* x. 262.

tēre, adj., *difficult, tiresome,* ALEX. 150 ; DEGR. 1409 ; ANT. ARTH. (R.) x ; to tere ALEX. (Sk.) 4767 ; *see* **tōr.**

tēre [2], adj., *? Fr.* (*Pic.*) tere (*tendre*); *fine*; tere (teer) flour ['*similae*'] TREV. III. 9 ; a sqvier tere HALLIW 859.

teren, v., = *M.L.G.* teren ; *tar, cover with tar* ; tere HAV. 707 ; terrin PR. P. 489 ; **terred** (*pple.*) GEN. & EX. 2596.

tēren, v., *O.E.* teran, = *Goth.* (ga-)tairan, *O.H.G.* zeran ; *tear, tire, lacerate,* P. L. S. xxv. 150 ; tere ALIS. 5985 ; CH. C. T. *B* 1326 ; DEGR. 1688 ; tere FL. & BL. 736 ; tereþ (*pres.*) MISC. 67 ; he tireð on his ket REL. I. 218 ; tar (*pret.*) K. T. 100 ; AL. (T.) 146 ; and tar hire bi þan ere LAȝ. 25850* ; sche tar hire her S. S. (Web.) 472 ; teren MAND. 81 ; ED. 3036 ; tiere LAȝ. 24843* ; some . . . þe holi mannes clothes tere AL. (T.) 326 ; tóren (*pple.*) S. S. (Web.) 782 ; tore L. H. R. 143 ; *comp.* to-téren.

terflen, v., *O.E.* tearflian ; *wallow,* ['*volutabatur*'] ; **terflede** (*pret.*) MK. ix. 20.

terȝen, v., *O.E.* tergan, tyrgan, = *M.L.G.* tergen, *M.Du.* terghen (*vex, irritate*); *delay, hinder ; provoke ;* teriin '*irrito, moror*' terwin '*fatigo*' PR. P. 489 ; gef him . . . bilimpeð for to tirgen REL. I. 216 ; targe ROB. (W.) 2363 ; i wol not tarien you CH. C. T. *F* 73 ; tarie ALIS. 2010 ; tarie (*pres.*) MAND. 160 ; þat ȝe terren him to wraþþe ['*ut eum ad iracundiam provocetis*'] WICL. DEUT. iv. 25 ; he tarȝede a lute while P. L. S. xiii. 179 ; **taried** (*pret.*) ['*irritaverunt*'] PS. cv. 16 ; itaried ['*fatigatus*'] REL. I. 9.

    terring, sb., *provocation, delaying* ; **terringes** (*pl.*) WICL. 4 KINGS xxiii. 26 ; targinge ROB. (W.) 4216.

terme, sb., *O.Fr.* terme ; *term, period,* A. R. 338 ; FL. & BL. 432 ; AYENB. 33 ; **termes** (*pl.*) *expressions* LANGL. *B* xii. 237 ; termis, teermes *ends* WICL. DAN. iv. 19, MAT. xxiv. 31.

terminen, sb., *O.Fr.* terminer ; *determine, limit* ; **termineth** (*pres.*) WICL. HEB. iv. 7 ; **termined** (*pple.*) WICL. I KINGS xx. 33 ; iterminet LANGL. *A* i. 95 ; termenid *C* ii. 93.

terminour, sb., *termination* ; **terminours** (*pl.*) LANGL. *C* iv. 409*.

terne, sb., *O.N.* tiörn ; *tarn, lake,* A. P. ii. 1041 ; D. TROY 11187 ; tarne AV. ARTH. x.

ternen, *see* **turnen.**   **terre,** *see* **tere.**

terren, *see* **terȝen.**

Tervagant, pr. n., *O.Fr.* Tervagant ; *an idol,* LAȝ. 13911.

terven, v., *O.E.* (ge-)tyrfian, = *O.H.G.* zerben ; *roll :* truit & treget to helle schal terve L. H. R. 207.

testament, sb., *Fr.* testament ; *testament,* AYENB. 191.

testēre, sb., *O.Fr.* testiere ; *tester ; head-piece ; helmet ;* teester of a bed PR. P. 489 ; testre ALEX. (Sk.) 4914 ; beddis **testēris** (*pl.*) WICL. E. W. 434 ; testeres CH. C. T. *A* 2499.

testif, adj., *testy, headstrong,* CH. C. T. *A* 4004.

testifien, v., *testify, preach,* LANGL. *B* xiii. 93.

teter, sb., *O.E.* teter, *cf.* *O.H.G.* zitaroch (*impetigo*) ; *tetter,* VOC. 267 ; **tetres** [teteres] (*pl.*) TREV. II. 61.

    teter-wert, sb., *celandine,* LEECHD. III. 346.

tette, *see* **titte.**

tēþe, tēðien, *see* **tēoþe, tēoþien.**

teþer, sb., *O.N.* tioðr, = *O.Fris.* tiader, tieder ; *tether ;* tedir VOC. 234.

teþren, v., *O.N.* tioðra, *cf. L.G.* tudern ; *tether ;* **teþrid** (*printed* teyryd) (*pple.*) ERC. 437.

tew, tewen, *see* **tawe, tawen.**

texte, sb., *O.Fr.* texte ; *text, saying* ; text CH. C. T. *B* 45 ; CATH. 380 ; text LANGL. *C* ii. 202 ; *B* ii. 121 ; tixte LANGL. *B* x. 270.

textual, adj., *? O.Fr.* textuel ; *literal,* CH. C. T. *I* 57.

tīar, *see* **tēar.**

ticchen, sb., *O.E.* ticcen ; *cf. ? O.H.G.* ziḳki ; *kid,* ' *hoedus,*' FRAG. 3 ; A.R.100 ; (*ms.* tycchen) LK. xv. 29 ; **ticchenan** (*dat. pl.*) MAT. xxv. 32.

tīd, tīt, adv., *cf. O.N.* tiðr (*frequent*) *neut.* titt (*frequently, quickly*) ; *quickly* ; tid, tit WILL. 1013, 4167 ; tit GOW. II. 320 ; H. S. 1764 ; TOR. 2063 ; PR. C. 1914 ; M. ARTH. 3713 ; *also* tīd, als tit, as tit *immediately* A. P. ii. 64 ; LANGL. *B* xiii. 319 ; AMAD. (R.) lvi ; **tittire** (*comp.*) ALEX. (Sk.) 2519.

tīd-līche, adv., *O.N.* tiðliga ; *quickly,* JUL. 58 ; KATH. 1956* ; tidlike GEN. & EX. 1231 ; titli LANGL. *C* ii. 92*, xxi. 469*.

tīde, sb., *O.E.* tīd, = *O.L.G., O.Fris.* tīd, *O.N.* tīð *f., O.H.G.* zīt *f. n. ; tide, time, season, hour,* FRAG. 1 ; MARH. 18 ; O. & N. 489 ; WILL. 859 ; PR. C. 379 ; þeos tide HOM. I. 89 ; þ(e)o tid 87 ; þe tide is wel neih icume MISC. 42 ; þe tide is ebbid H. V. 69 ; eche tide . . . of þe dai BRD. 10 ; al tide PS. lxxviii. 4 ; sum tide MIN. i. 17 ; it chaungeþ as þe tide CH. C. T. *B* 1134 ; tid ORM. 13402 ; ðe holie tid þat me clepeð advent HOM. II. 3 ; it sal ben ðe laste tid GEN. & EX. 263 ; **tīde** (*dat.*) GOW. I. 133 ; AM. & AMIL. 281 ; an are tide LAȝ. 14924 ; in þe same tide LIDG. TH. 3409 ; **tīde** (*pl.*)

HOM. I. 89; O. & N. 26; hadde he ibeon þer anne dai oþer twa bare tide P. L. S. viii. 70; þare tide ['*temporum*'] MAT. xvi. 3; **tīden** (*dat. pl.*) A. R. 22; *comp.* bürð-, ēven-, hē3-, herfest-, lenten-, mēl-, mor3en-, nōn-, sumer-; ühten-, undern-, winter-tīde.

tīde-ful, adj., *seasonable* : in tideful time HAMP. PS. xxxi. 7.

tīdfulnes, sb., *times of necessity,* HAMP. PS. ix. 9*.

tīden, v., *O.E.* tīdan, *O.N.* tīða; *betide, happen,* JOS. 392; CH. C. T. *B* 337; tidin PR. P. 493; tide WILL. 3017; HOCCL. vi. 39; tīdeþ (*pres.*) BEK. 1928; tid LANGL. *C* xiv. 213*; tit MARG. 308; S. S. (Web.) 292; P. S. 334; Aþulf tit no wounde HORN (R.) 1352; him ne tit non oþer mede P. L. S. xii. 48; **tīde** (*subj.*) A. D. 294; tidde (*pret.*) CHR. E. 542; WILL. 198; *comp.* bi-, 3e-, mis-tīden.

tider, see þider.

tīdi, adj., = *M.Du.* tijdigh, *M.L.G.* tīdich, *O.H. G.* zītīg, *mod.Eng.* tidy; *seasonable; honest*; GEN. & EX. 2105; LANGL. *B* ix. 104; WICL. JAS. v. 7; PARTEN. 5722; þe tidi child WILL. 160; *comp.* un-tīdi.

tīdinde, tīdende, tīðinde, sb., *O.N.* tīðindi; *message,* LA3. 2052, 3734, 5153; tiþende ORM. DEDIC. 158.

tīdinge, sb., = *Du.* tijding, *Ger.* zeitung : *event, tidings, message,* LA3. 24907; H. M. 45; S. S. (Web.) 423; tiding GEN. & EX. 407; bringe heom l(e)ove tidinge [tiþinge] O. & N. 1035; tiding, tiþing ROB. 79, 172; tiðinge A. R. 70; tiþinge BEK. 1527; tīðinge (*pl.*) HOM. II. 33.

tidive, sb., *? some small bird*; tidives (*pl.*) CH. C. T. *F* 648.

tidren, see tüdren. **tīe,** see tē3e.

tiel, see tü3el. **tiele,** see te3ele.

tien, see tēn. **tīen,** see tē3en.

tīene, see tēone. **tīer,** see tēar.

tiffen, sb., *O.Fr.* tiffer (*orner*); *adorn, carve,* ALIS. 4109; GAW. L129; tiffed (*pple.*) WILL. 1725; ALEX. (Sk.) 4465.

tiffung, sb., *finery,* A. R. 420*.

[**-ti3,** *suffix,* *O.E.* -tig, = *O.N.* -tigr, *Goth.* *tigus (*pl.* tigjus), *O.H.G.* -zig; -*ty*; *comp.* ahte-, fēower-, fīf-, ni3en-, seven-, six-, þrī-ti3 (-ti).]

ti3e, see tü3e. **ti3el,** see tü3el.

ti3ele, see te3ele. **tī3en,** see tē3en.

tīgre, sb., *Fr.* tigre; *tiger,* CH. C. T. *A* 1657; tīgirs, tigris (*pl.*) ALEX. (Sk.) 3850, 3573.

ti3ten, see tihten.

tiht, sb., *O.E.* tiht; *from* tēon; *accusation*; ? ti3t M. T. 163.

tiht, see tüht.

tīht, adj., = þīht; *thick, dense* : a wod(e) þat wase full tight TOR. (A.) 589.

ti3t-lī, adv., *tightly,* WILL. 66.

tihten, v., *? Ger.* tichten, dichten (*think*); *devise, ?intend*; ti3t (*pres.*) GAW. 2483; þe sones of Israel . . . ti3ten shiltron WICL. JUDG. xx. 33; þe foli þat his breþeren ti3t(e) C. M. 4124; tiht (*ms.* thit) (*pple.*) HAV. 2990; ti3t P. R. L. P. 177; tight OCTOV. (W.) 1476; AM. & AMIL. 1697; IW. 111; iti3t PL. CR. 168; iti3t he hadde his paviloun GREG. 565; *see* dihten.

tihten, *see* tühten.

tike, sb., *? O.E.* ticia, *cf.* *M.L.G.*, *M.Du.* teke, *O.H.G.* zeche; *tick* (*insect*), PR. P. 493; P. S. 238; teke VOC. 255.

tīke, sb., *O.N.* tīk, *prov. Eng.* tike; *dog; rustic*; tīkes (*pl.*) LANGL. *B* xix. 37; D. ARTH. 3642.

tikel, adj., *ticklish*; *frail, wanton*; LANGL. *A* iii. 126; A. D. 237; CH. C. T. *A* 3428; MAN. (F.) 13413; tikil '*titillosus*' PR. P. 493; tekil LUD. COV. 134.

tikelin, v., *tickle,* '*titillo,*' PR. P. 493; tikled [tikeled] (*pret.*) CH. C. T. *D* 395; tikelid HOCCL. i. 204.

til, prep., *O.N.* til, = *O.Fris.*, til; *till, to,* GEN. & EX. 85; HAV. 141; SPEC. 30; LANGL. *A* i. 95; WICL. JOHN xvi. 24; M. H. 3; til hi iafen up here castles SAX. CHR. 261; he se3de þus til him ORM. 803; goþ til him WILL. 266; til Athenes CH. C. T. *A* 2964; if he be til god bousom PR. C. 85; til þat men com KATH. 719 [(E.) aðet me come]; til þat he scholde to hevene wende C. L. 44; til þat it be night GOW. II. 109; til ORM. 10229; *comp.* in-, on-, until.

tilden, *see* telden.

tile, sb., = *O.H.G.* zila (*studium*); *cultivation; produce; goods* : so mikel wex his tile GEN. & EX. 1519; *see* tēle.

til-man, sb., = *M.Du.* teelman; *cultivator,* '*cultor,*' VOC. 218; ANGL. I. 314; tilmon D. TROY 2462; tilmen (*pl.*) C. M. 4696.

Tile, pr. n., *O.E.* Tile; *Thule,* TREV. I. 325; CH. BOET. iii. 5 (77).

tīle, *see* te3ele.

tīlette, sb., *small tile,* PALL. vi. 195.

til-hēwen, v., *for* to-hēwen; *hew, cut up*; till hēwit (*pret.*) BARB. ii. 381; tillhēwin (*pple.*) BARB. xx. 367.

tilie, sb., *O.E.* tilia; *cultivator,* HOM. II. 181; teolie HOM. I. 133; tilien (*pl.*) MK. xii. 7; A. R. 416; REL. I. 129; (HOM. II. 163); *comp.* eorðe-, lond-tilie.

tiliēn, v., *O.E.* tilian, teolian (*study, strive, cultivate*), *cf.* *O.L.G.* tilian, *O.Fris.* tilia, *M.L.G.*, *M.Du.* telen (*cultivate*), *O.H.G.* zilēn (*study, strive*); *till,* LA3. 2618; A. R. 384; PL. CR. 743; tilien [tulien] þe eorþe LANGL. *A* viii. 2, *B* vii. 2; tilie ROB. 21; tilen REL. I. 210; GEN. & EX. 363; teolien S. A. L. 149; teolie LEB. JES. 245; tulien HOM. II. 155; to teoliende efter istreone I. 133; tileð (*pres.*)

A. R. 78 ; þei tilen not þe lond MAND. 64 ;
ilede (*pret.*) SAX. CHR. 262 ; ure loverd
more þen two and þritti ȝer tiled(e) efter hore
luve A. R. 404 ; tileden LAȜ. 1940 ; no mete
ne tiliden [teleden] LANGL. *B* xiv. 67 ; tilede
here liflode ROB. 41 ; **tiled** (*pple.*) SAX. CHR.
262 ; MAN. (F.) 1854 ; *comp.* ȝe-tilien.

   tiliere, sb., = *M.L.G.* teler ; *tiller, farmer,*
GEN. & EX. 1482 ; DEP. R. i. 54 ; teoliare
LEB. JES. 589 ; *comp.* erþe-tiliere.

   tilunge, sb., *O.E.* tilung, teolung, = *M.L.G.*
telinge, *M.Du.* teelinghe ; *husbandry ; culture ; sorcery* ; A. R. 296 ; wichecraft and
telinge MIRC 360 ; teliinge SHOR. 95 ; tuliinge
LANGL. *B* xiv. 63 ; **telinges** (*pl.*) TREV. III.
265 ; teolunges A. R. 208.

   tilin, v., *from* teȝele ; *tile*, PR. P. 494.

   tilare, sb., *tiler*, PR. P. 494.

tillen, v., *O.E.* (ge-)tillan, = *O.H.G.* *zillan
(*pret.* zilta) ; *touch, reach, extend* ; tille TREV.
III. 131 ; alle þat he miȝt(e) tille FER. 59 ; to
fraude wild(e) he never tille MAN. (H.) 128 ;
tilde (*pret.*) TREV. V. 193 ; þe leome þat tilde
westward ROB. 152 ; þe niþer(e) ende tilde to
his chinne BRD. 24.

tillen, *see* tüllen, té den.

tilten, v., *mod.Eng.* tilt ; *be overthrown*: þis
ilk(e) toun schal tilte to grounde A. P. iii.
361 ; tiltis (*pres.*) ALEX. (Sk.) 1303 ; feole
temples . . . tülten (*pret.*) to þe eorþe JOS.
100.

tilðe, sb., *O.E.* tilð ; *tilth, culture, toil ; produce* ; A. R. 78 ; tilþe P. L. S. viii. 29 ; CHR. E.
20 ; GOW. II. 190 ; LANGL. *B* xix. 430 ; þe tilþe
of riȝtfulnesse schal be stilnesse WICL. IS.
xxxii. 17 ; *comp.* ȝe-tilðe.

timber, sb., *O.E.* timber, = *O. Fris.* timber,
O.N. timbr, *O.H.G.* zimbar ; *timber, material,*
LAȜ. 22929 ; CH. C. T. *A* 3666 ; timbre LANGL.
B xix. 316.

timbre, sb., *O.Fr.* timbre ; *tambourine*, A. P.
ii. 1414 ; ['*tympanum*'] WICL. IS. v. 12 ;
timbir, litil taboure '*tempanillum*' PR. P. 494 ;
timbre [tembre] *helmet-crest* ALEX. (Sk.)
1230 ; **timbres** (*pl.*) ALIS. 191.

timbrien, v., *O.E.* timbrian, = *O.N.* timbra,
O.H.G. zimberēn ; *build*, LAȜ. 5940 ; timbren
A. R. 124 ; ORM. 13368 ; timbre LANGL. *B* xi.
352 ; **timbred** (*pple.*) WILL. 2015 ; *comp.*
ȝe-timbrien.

   timbrunge, sb., *O.E.* timbrunge ; *building*,
HOM. I. 93 ; A. R. 124 ; timbringe HOM. I.
227.

tīme, sb., *Lat.* thȳmus ; *thyme*, PR. P. 494.

tīme [2], sb., *O.E.* tīma, = *O.N.* tīmi ; *time*, ORM.
63 ; HAV. 1714 ; SHOR. 83 ; SAINTS (Ld.) xlvi.
606 ; CH. C. T. *A* 35 ; þa þe time com LAȜ.
291 ; þene time REL. I. 175 (MISC. 112) ; sum
time A. R. 92 ; er tīman (*dat.*) HOM. I. 103 ; an
þan timen LAȜ. 9668 ; an ane timen 30068 ; on

ane time 30064 ; on a time KATH. 2 ; on rihte
time HOM. I. 133 ; **bī tīme** (*betimes*) he aros
P. L. S. xii. 118 ; in olde time AL. (L.[1]) 2 ; be
time GENER. 522 ; tīmen (*pl.*) C. L. 1403 ;
time HOM. II. 3 ; fele times LANGL. *B* xiii.
330 ; *comp.* æven-, hervest-, lif-, sumer-,
un-, winter-tīme.

   tīm-līch, adj., *timely, seasonable*, AYENB.
44 ; **tīmlīche** (*adv.*) HOM. I. 25 ; LAȜ.
31369 ; JUL. 9 ; A. D. 261 ; timeli JOS. 415 ;
tīmlüker (*compar.*) KATH. 2117.

tīmen, v., *O.E.* (ge-)tīmian, = *O.N.* tīma ;
happen ; befall : us sal timen ðe betre sped
GEN. & EX. 3820 ; hit timeð H. M. 35 ; so me
wel time WILL. 3570 ; **timede** (*pret.*) D.
ARTH. 3150 ; *comp.* bi-, ȝe-tīmen.

   tīminge, sb., *accident, event*, GEN. & EX.
31 ; timing HORN (H.) 166.

tīmen, *see* tēmen.

timpan, sb., *Lat.* tympanum ; *drum*, WICL.
EX. xv. 20 ; timpan C. M. 21309.

tin, sb., *O.E.* tin, = *O.N.* tin, *M.Du.* tin, ten,
O.H.G. zin ; *tin*, REL. I. 129 ; ROB. 6 ; PL.
CR. 195 ; E. G. 358.

tinclen, v., *O.E.* tinclan (*tinkle*) ; *tinkle, tingle* ;
boþe his eeris shulen tinclen WICL. I KINGS
iii. 11 ; **tenclis** (*printed* tenelis) [tinkill]
(*pres. pl.*) ALEX. (Sk.) 1385.

tind, sb., *O.E.* tind, = *O.N.* tindr, *M.L.G.* tinde ;
M.H.G. zint ; *tine, tooth*, TRIST. 507 ; **tínde**
(*dat.*) A. P. i. 78 ; LIDG. M. P. 203 ; **tíndes**
(*pl.*) A. R. 354 ; tindes of harowis ALEX. 3908 ;
tindis TRIAM. 1085.

tinder, sb., *O.E.* tynder ; *tinder*, LAȜ. 29267 ;
see tunder.

tine, adj., *tiny* : a litil tine egg ALEX. (Sk.)
507 ; littell tine child LUD. COV. 414.

tīnen, v., *O.N.* tȳna ; *lose* ; tine WILL. 299 ;
LANGL. *B* i. 112 ; APOL. 43 ; MIN. x. 18 ;
AMAD. (R.) lxiv ; PERC. 911 ; M. H. 73 ;
þu tīnes (*pres.*) GEN. & EX. 3518 ; he tines
PR. C. 2027 ; þei tine MAN. (F.) 4514 ; tīne
(*subj.*) PS. xxv. 9 ; **tīnte** (*pret.*) HAV.
2023.

tīnen, *see* tūnen.

tinglen, v., *cf. M.Du.* tinghelen ; *tingle, tinkle* :
hise eeris tingle ['*tinniant*'] (*subj.*) WICL.
JER. xix. 3 ; **tinglinge** (*pple.*) I COR. xiii. 1*.

tinken, v. ; *make a tinkling noise* ; **tinkinge**
(*pple.*) WICL. I COR. xiii. 1.

   tinkere, sb., *tinker*, LANGL. *A* v. 160 ; tin-
kare '*tintinarius*' PR. P. 494.

tinnen, adj., *O.E.* tinen, tinnen, = *O.H.G.*
zinīn ; *made of tin*, PALL. vi. 99.

tinnin, v., *tin, cover with tin*, PR. P. 494 ;
tinned (*pple.*) CH. H. F. 1482.

tīnsel, sb., *loss, ruin*, C. M. 916 ; MAN. (F.)
9836* ; tinsale BARB. v. 455 ; tinsill HAMP.
PS. lxxxvii. 12.

tinte, sb., *O.N.* tinta (*pint, small bottle*); *a vessel holding half a bushel,* PR. P. 494.

tintreȝe, sb., *O.E.* tintrege; *torment*; tin- trehe, tintreohe KATH. 404, 620; **tintreohen** (*pl.*) HOM. I. 261.

tintreȝen, v., *O.E.* tintregian; *torment*; tin- traȝeð (*ms.* tintraȝed) (*pres.*) HOM. I. 13.

tip, sb., *cf. Du., Dan.* tip, *M.H.G.* zipf; *tip; extremity*: tip of the nese '*pirula*' PR. P. 494; vort þe nede tippe A. R. 338.

tippen, v., *tip*: a staf tipped (*pple.*) wiþ horn CH. C. T. *D* 1740; þose traitoures arn tipped (*consummate*) schrewes A. P. iii. 78.

tippen[2], v., *cf. Sw.* tippa; *overthrow*: tipe doun ȝonder toun A. P. iii. 506.

tippet, sb., *? O.E.* tæppet; *tippet,* MAN. (H.) 280; tipett '*liripipium*' PR. P. 494; tipet [tepet] CH. C. T. *A* 233; tipit VOC. 238.

tir, sb., *for* atír; *attire, equipment,* WILL. 1725; tire ROB. (W.) 1188*.

tīr, sb., *O.E.* tīr, tȳr, = *O.L.G.* tīr, *O.N.* tīrr; *glory, authority,* LAȝ. 2051, 4237.

　　tīr-ful, adj., *mighty*: þe tirfulle feond LAȝ. 2893.

tīr, *see* tēar.

tirannie, sb., *O.Fr.* tirannie; *tyranny,* CH. C. T. *A* 941.

tirant, sb., *O.Fr.* tirant; *tyrant,* ROB. 374; tiraunt LANGL. *C* xxiii. 60; **tirauntes** (*pl.*) DEP. R. i. 54; tirauns LANGL. *C* iii. 211.

tirantrie, sb., *tyranny*; terauntrie PR. P. 488; (*ms.* tirauntire: *? read* tirauntrie *or* tirauntise) A. P. ii. 187.

tirdil, sb., *O.E.* tyrdel (LEECHD. II. 408); *diminut. of* tord; *dung, ? dunghill,* '*rudus,*' PR. P. 494; **trideis** (*pl.*) REL. I. 53.

tiren, *see* teren.

tirf (*? tirfe*), sb., *turning up of a hat or sleeve,* PR. P. 494.

tirȝen, *see* terȝen.

tirnen, *see* turnen.　**tirpeil,** *see* trepeil.

Tīsdæi, *see under* Tīu.

tísen, v., *O.Fr.* tiser; *entice*; tise TRIAM. 107; tísen (*pres.*) LANGL. *C* viii. 91*; **tísed** (*pret.*) MIRC 1424.

tisik, sb., *O.Fr.* tisique, *from Lat.* phthisica; *phthisis, consumption,* PR. P. 494; tisike 389.

[**tit,** sb., = *O.N.* tittr; *small bird.*]

　　tit-māse, sb., *titmouse,* VOC. 188; titemose VOC. 165; PR. P. 494.

tīt, *see* tīd.

[**titelen,** v., *tattle.*]

　　titelere, sb., *tattler*; **titeleres** (*pl.*) LANGL. *B* xx. 297; tituleris DEP. R. iv. 57.

[**titeren,** v., *cf. O.N.* titra, *M.H.G.* zitern (*tremble*), *mod.Eng.* titter.]

　　titerere, sb., *tattler*; **titereres** (*pl.*) LANGL. *B* xx. 297, *C* xxiii. 299.

title, sb., *O.Fr.* title; *title, claim; tittle*; LANGL. *C* xiv. 106; titel [titil] ['*apex*'] WICL. MAT. v. 18.

titler, sb., *? hound*; **titleres** (*pl.*) GAW. 1726.

titte, sb., *O.E.* titt, = *M.L.G., M.Du.* titte, *M.H.G.* zitze; *teat*; tette GEN. & EX. 2621; tete '*uber*' PR. P. 489; GOW. I. 268; CH. C. T. *A* 3704; **tittes** (*pl.*) LAȝ. 5025; KATH. 2129; A. R. 330; SPEC. 35; **titten** (*dat. pl.*) LAȝ. 11936.

titte[2], sb., *pull, tug,* PR. C. 1915; tit WINT. viii. 2037.

titten, v., *pull tightly*: in strang pains be streined and titted (*pple.*) PR. C. 7216.

tīðe, adj., *O.E.* tīða; *sharing in*: ben (þere) bene tīðe HOM. II. 27.

tīþe, *see* tēoþe.

tīðen, v., *O.E.* tīðian, tȳðian, = *O.L.G.* tvīthōn, *M.H.G.* (ge-)zwīden; *concede, grant*; tiþe ROB. 114; tīþe (*pres.*) LUD. COV. 35; tiþeþ ORM. 5365; tuðe me mine bone HOM. I. 207; **tiðed** (*pple.*) HOM. II. 135; tid C. M. 10966; *comp.* ȝe-tūðen.

tīþien, *see* tēoþien.　**tīðinde,** *see* tīdinde.

tīðinge, *see* tīdinge.

Tīu, sb., *O.E.* Tīw, = *O.N.* Tȳr; *Mars*; **Tīwes dai** [*O.E.* Tīwes dæg, = *O.Fris.* Ties dei, *O.N.* Tȳs dagr, *O.H.G.* Zies tag] *Tuesday* BEK. 2462; AR. & MER. 8787; Tiwes niȝth (*r.* niȝt) DAV. DR. 43; Tisdæi, Tisdei LAȝ. 13924*, 13932.

[**to-, te-,** prefix, *O.E.* to-, *O.Fris.* to-, te-, ti-, *O.L.G.* te-, ti-, *O.H.G.* zar-, zer-, zir-, za-, ze-, zi-, *? Goth.* twis-; *Lat.* dis-; *to-, asunder, apart.*]

tō, tǒ, prep., & adv., *O.E.* tō, = *O.L.G., O.Fris.* tō, te, *O.H.G.* zuo, zo, za, ze; *to*; ȝif eni man seið eawiht to eou HOM. I. 3; þeo þat ham to (*v. r.* to heom) luteð KATH. 1781; gað to bedde LAȝ. 711; þat gærsume þe scolde þe to cume 10509; Sæxes him sette to 15276; cum to me A. R. 98; valleð to þer eorðe 18; godes engel com him to ORM. 143; his soule mote cume to hevene MISC. 155; he ... ȝede to londe HORN (L.) 1022; þan he was to þe erþe brouth HAV. 248; he ... steigh to hevene MAND. 133; þeo men þe ic þene herm to dude (*to whom I did the harm*) HOM. I. 31; þær it to bilimpeþ ORM. 1657; make hine to kinge *make him king* LAȝ. 11468; þa hine to monne iber 65; preoste ihoded O. & N. 1311; me ches him to kinge ROB. 302; cheosen him to lemmon SPEC. 62; þeo þet is iwend to wulvene A. R. 120; demde hire te deaðe MARH. 19; þa he to (*for*) bisne nom LAȝ. 30; habben to wife 145; ȝeve ... to wive CH. C. T. *A* 1860; that i might borow(e) it to this dede M. ARTH. 172; he hem gaf to andswere HOM. II. 81; tac þe rode to þi staf SPEC. 106; werp nu to token dun ðat wand GEN. & EX. 2803; him to luve, to wurþscipe HOM. I. 5; mare hit him deð to herme þenne to gode 27 ; and warni men to h(e)ore note O. & N. 330 ; ever ich brewestere . . . þat breweþ to (*for*) sale E. G. 354 ; þat þai be sold . . . to (*at*) as hie pris as hit mai E. W. 70 ; efre forð to domes dei HOM. I. 45 ; heo him wolden beon liðe a to heore live LAȝ. 10017 ; ævere to his live 10396 ; hi is zik to þe diaþe AYENB. 197 ; to (*for* to þam *until*) we be gon TOWNL. 64 ; tō dei *to-day* HOM. I. 3 ; to dai A. R. 278 ; te dai MISC. 33 ; to dæi, to daie LAȝ. 5442, 25908 ; tō morgen *to-morrow* GEN. & EX. 3017 ; tō niȝt *to-night* BEK. 1574 ; to middelnihte *at midnight* O. & N. 731 ; teven (=te even) *at even* AYENB. 51 ; to ȝere *this year* LAȝ. 8039 ; P. S. 214 ; þe king gon to spekene LAȝ. 24722 ; hwat heo haveð to donne A. R. 52 ; to luvien JUL. 5 ; ta like 37 ; þat is te cumen HOM. II. 5 ; þa . . . iwende godes engel to HOM. I. 87 ; alle heo hiȝeden to LAȝ. 2317 ; stercliche (heo) to stopen 9798 ; þa Bruttes þo ræsden 26645 ; it was . . . toiled to and fro MAP 338 ; hie tuneð to hire fif gaten HOM. II. 181 ; schutteð al þet þurl to A. R. 96 ; and stek to þe dore S. S. (Wr.) 1398 ; to (*too*) softe ORM. 2899 ; to longe P. L. S. viii. 178 ; to strang AYENB. 51 ; to *too, also* MAT. v. 40 ; WILL. 11 ; too badde is mi witte 5024 ; *comp.* in-, on-, un-tō.

tō, *see* tā, twā.

tō-ayēns, prep., *against*, AYENB. 10.

to-bēaten, v., *O.E.* tobēatan ; *beat severely* ; tobēteþ (*pres.*) LAȝ. 3308 ; O. & N. 1610 ; tobēot (*pret.*) P. L. S. xxiii. 53 ; ase me tobeot his cheoken A. R. 106.

to-belle, v., *swell extremely*, BEV. 2656 ; tobollen (*pple.*) A. R. 122 ; him wæren fet & þeos tobollen & toblawen ORM. 8080 ; tobollen, -bolle LANGL. B v. 84.

to-bēren, v., *O.E.* toberan ; *bear asunder* ; tebeoreð (*pres.*) H. M. 31 ; tobar (*pret.*) GEN. & EX. 2146 ; hou his sustren hem tobeeren C. L. 522 ; tobóren (*pple.*) C. L. 49.

tō-bilimpen, v., *belong to* ; tōbilimpeþ (*pres.*) ORM. 1657.

to-bīten, v., =*M.L.G.* tobīten, *M.H.G.* zerbīzen ; *bite in pieces* ; tobōt (*pret.*) ANGL. I. 318.

to-blāsten, v., *blast asunder* ; toblăst(e) (*pret.*) MAN. (F.) 9293.

to-blāwen, v., *O.E.* toblāwan ; *inflate* ; toblāwen (*pple.*) ORM. 8080 ; toblowen & tobollen A. R. 122.

to-brǣden, v., *O.E.* tobrǣdan ; *spread out* ; hio tobrēdeð (*pres.*) MAT. xxiii. 5 ; tobrĕddest (*pret.*) PS. xvii. 37.

to-breiden, v., *O.E.* tobregdan ; *draw asunder* ; tobraid (*pret.*) GOW. II. 53 ; WICL. LK. ix. 42 ; tobreid AL. (L.¹) 396 ; tobróde (*pple.*) O. & N. 1008.

to-brḗken, v., *O.E.* tobrecan, =*O.Fris.* to-, tebreka, *O.H.G.* zebrechan ; *break in pieces* : he hine up bræid swulc he hine tobreken

wolde LAȝ. 16520 ; tobreocan HOM. I. 127 ; tobreke O. & N. 1554 ; S. S. (Wr.) 301 ; WICL. LK. xx. 18 ; H. V. 29 ; tobrékeð (*pres.*) A. R. 164 ; tobrekeþ AYENB. 64 ; tobrac (*pret.*) LEG. 12 ; tobrak WILL. 3237 ; ALIS. 570 ; GOW. III. 296 ; tobróke (*pple.*) LANGL. B vii. 28.

to-bresten, v., *O.E.* toberstan, =*O.L.G.* tebrestan, *O.H.G.* za-, ze-, ziprestan ; *burst asunder*, ORM. 16147 ; tobreste DEGR. 1509 ; tobersteð (*pres.*) A. R. 214 ; toberste (*subj.*) O. & N. 122 ; tobrast (*pret.*) SACR. 48 ; tobrast, -barst WICL. DEEDS i. 18 ; tobarst LAȝ. 1921 ; MARH. 10 ; WILL. 374 ; ALIS. 2325 ; toborsten JOS. 509 ; toborste LAȝ. 5926* ; (*printed* toberste) MARG. 243 ; tobrosten (*pple.*) CH. C. T. *A* 2691.

to-brūsen, v., *O.E.* tobrȳsan ; *bruise in pieces* ; tobrisen ORM. 12032 ; tobrūsede (*pret.*) WICL. 4 KINGS xviii. 4 ; tobrīsed (*pple.*) MAT. xxi. 44 ; HAV. 1950.

to-brūtnen, v., *break up* ; tobrittenith (*pres.*) PARTON. 596 ; tobritned (*ms.* tobrittnedd) (*pple.*) ORM. 9468.

to-brütten, v., *O.E.* tobryttan ; *break up* ; tobrütte (*pret.*) LAȝ. 1602.

to-būnen, v., *beat severely* : and þe totorveþ and tobūneþ (*pres.*) O. & N. 1166 ; so tobete and so tobōned (*pple.*) SHOR. 85.

to-būsten, v., *?beat, bruise greatly* : and me tobüsteþ [toburste(þ)] (*pres.*) and tobeteþ O. & N. 1610.

to-chēowen, v., *O.E.* tocēowan ; *chew in pieces* : tochēoweð (*pres.*) & tovret A. R. 202 ; (heo) tocheoweð HOM. I. 251.

to-chīnen, v., *O.E.* tocīnan ; *split up* ; tochine O. & N. 1565 ; tochineð (*pres.*) HOM. II. 199 ; tochān (*pret.*) HOM. I. 141 ; þa heorte tochan [tochon] LAȝ. 21235 ; tochon FER. 3001 ; þe roche tochon MISC. 92 ; toc(h)oon ALIS. 573 ; tochine (*pple.*) ORF. 260.

to-clateren, v., *shatter in pieces* ; toclatirs (*pres.*) ALEX. (Sk.) 799 ; toclatered (*pple.*) WILL. 2858 ; þai had hit al toclatrid FER. 897.

to-clēoven, v., *O.E.* toclēofan, =*O.H.G.* zechliuban ; *cleave asunder* ; tocleve AYENB. 56 ; LANGL. B xii. 141 ; CH. TRO. v. 613 ; toclēef (*pret.*) ORM. 14798 ; toclef ROB. 186 ; GOW. III. 296 ; þe naddre toclef LEG. 54 ; tocleef TREV. VII. 463 ; tocluven LAȝ. 1920 ; toclove L. H. R. 137 ; toclóven (*pple.*) JOS. 516 ; L. H. R. 142.

to-combren, v., *discomfit* ; tocombirs (*pres.*) ALEX. (Sk.) 1302.

tō-comen, v., *come together* ; tōcomen (*pret. pl.*) LANGL. *C* xxii. 343.

tō-coming, adj., *future*, ['*ventura*'] WICL. PS. xxi. 34.

to-cruschen, v., *crush in pieces* ; tocrusshe HAV. 1992 ; þe walles tobreke & al tccrusch-ede (*pple.*) FER. 5153.

tō-cüme, sb., *O.E.* tōcyme; *coming*, HOM. I. 89; tocom(e) M. H. 8.

tō-cwēme, adv., *agreeably*, ORM. 1087.

to-dǣl, sb., *O.E.* todāl; *division*, LK. xii. 51; þet forme todol FRAG. I.

to-dǣlen, v., *O.E.* todǣlan,=*O.Fris.* todēla, *O.L.G.* tedēlian, *O.H.G.* za-, ze-, ziteilen; *divide, separate*, LA3. 9519; ORM. 10495; todelen FRAG. 6; todealen A. R. 186; todele AYENB. 80; TREV. VIII. 151; todēleþ (*pres.*) C. L. 845; TREV. I. 133; todǣlde [todelde] (*pret.*) LA3. 772; todǣled (*pple.*) FRAG. 7.

   to-dēlinge, sb., *division, separation*, AYENB. 72.

to-daschen, v., *dash in pieces*; todaschte (*pret.*) P. L. S. xxiii. 84; C. L. 1342; todashed (*pple.*) TREV. III. 63; ba weoren todascte LA3. 1469.

tōde, *see* tāde.

to-dōn, v., *O.E.* todōn; *separate*, LA3. 2945; todede (*pret.*) LA3. 6507.

to-drǣven, v., *O.E.* todrǣfan; *disperse*; todreven A. R. 254; todrēfde (*pret.*) HOM. I. 93; todrēfed, todreved (*pple.*) LA3. 330, 16141; todreved MISC. 37; C. L. 446; groundes of hilles todreved are PS. xvii. 8; todreaved KATH. 92.

to-dra3en, v., *draw asunder; draw out fully*; LA3. 1506; todragen GEN. & EX. 191; todra3e HORN (L.) 1492; todrawe WICL. GEN. xl. 19; AMAD. (R.) xvi; todrauhð (*pres.*) A. R. 122; todrōh (*pret.*) LA3. 25618; todrogh IW. 823; todrough ALIS. 4659; todrowe P. L. S. xix. 249; LANGL. *B* x. 35; todra3e (*pple.*) O. & N. 1462; todrawen HAV. 2001; todrawe WILL. 1564; S. S. (Wr.) 877.

to-drēosen, v., *fall to pieces, decay*: he schal todreosen so lef on bouh MISC. 94; todrese LA3. 9245*; todrōren (*pple.*) ST. COD. 98; todrore MISC. 152.

to-drīven, v., *O.E.* todrīfan,=*O.Fris.* todrīva, *O.H.G.* zetrīban; *drive asunder*, LA3. 26048; al todrive SAINTS (Ld.) xlvi. 580; C. L. 862; P. L. S. ii. 122; todrāf, -drōf (*pret.*) LA3. 549, 1604; L. H. R. 141; todriven (*pple.*) KATH. 2079; todrifen ORM. 16397.

to-dünnen, v., *strike with a sounding blow*; todünnet (*pple.*) HOM. I. 281.

to-ēode, v., pret., *O.E.* toēode; *went apart, separated*, HOM. I. 141; to3eode LA3. 23980*.

to-fallen, v., *O.H.G.* tofeallan, = *O.L.G.* tefallan, *O.H.G.* zifallan; *fall in pieces*, LA3. 18867; tofalleþ (*pres.*) TREV. I. 133; tovalþ AYENB. 184; tofēol (*pret.*) LA3. 16496; tofēlle (*subj.*) ORM. 16185; tofallen (*pple.*) LA3. 22092; tofalle FER. 5011.

to-fáren, v., *O.E.* tofaran,=*O.L.G.* tefaran, *O.H.G.* zefaran; *go asunder*; the folk . . . shall tofare ANGL. III. 546.

to-fēren, v., *O.E.* tofēran; *go apart, separate*; tofērden (*pret.*) HOM. I. 93.

to-filchen, v., *seize*; tofilched (*pret.*) GAW. 1172.

to-flēon, v.,=*O.H.G.* zefliugan; *fly asunder*; toflo3en (*pple.*) LA3. 28668.

to-flēoten, v., = *O.H.G.* zafliozan; *be dissipated*; toflēoteð (*pres.*) A. R. 76; tovlēoten (*subj.*) A. R. 72.

tō-flight, sb.,=*M.L.G.* tovlucht, *O.H.G.* zuofluht; *refuge*, PS. ix. 10, xvii. 3.

tō-foren, prep. & adv., *O.E.* tōforan,=*O.L.G.* teforan; *before*, MAT. xxv. 32; HOM. I. 121; LA3. 14071; KATH. 51; of þan toforen iseide redes men PROCL. 3; toforn S. S. (Web.) 304; GREG. 191; tofore TREAT. 133; WILL. 142; LANGL. *B* v. 457; tofore þe kinge O. & N. 1728; tovore AYENB. 7.

to-freten, v., *corrode*; tovret (*pres.*) A. R. 202.

to-fruschen, v., *break in pieces*; tofrusshe HAV. 1993; tofrusched (*pret.*) RICH. 5032.

toft, sb., *O.E.* toft, *O.N.* topt; *toft, piece of ground, 'campus,'* PR. P. 495; LANGL. *A* PROL. 14.

to-gān, v., *O.E.* togangan,=*O.L.G.* tagangan, *O.H.G.* zegān; *go asunder*; togāð (*pres.*) HOM. I. 239; togō (*subj.*) SHOR. 29.

[**to3e**, sb., *from* tēon; *leader*; *comp.* here-to3e.]

tō-3ein, prep., *O.E.* tōgegn, tōgegnes,=*O.L.G.* tegegnes; *against, opposite*, A. R. 130; þeo þet stalewurðe beoð ant starke to3ein me MARH. 15; to3æn, -3eines, -3enes, -3ænes LA3. 1439, 3586, 3626, 9792; to3eines HOM. I. 121; JUL. 35; HORN (R.) 1328; C. L. 386; þat scholde me socouri to3en min enimis FER. 172; (*ms.* toyeynes) MISC. 42; to-, tegenes HOM. II. 9; to3ænes ORM. 8632; to3eanes P. L. S. viii. 175; toyens AYENB. 112.

to3en, v., *O.N.* toga,=*M.L.G.* togen, *M.Du.* toghen, *O.Fris.* toga,=*M.H.G.* zogen; *tow*; towe A. P. iii. 100; towen (*pres.*) JOS. 374; to3ede (*pret.*) LA3. 7536*.

to-3esceōden, v., *separate*, HOM. I. 237.

toggen, v.,=*M.L.G.*, *M.Du.* tocken, ? *O.H.G.* zocchōn; *tug, pull*: ne toggen (*dally*) mid him ne pleien A. R. 424; toggin PR. P. 495; toggið (*pres.*) MARH. 14; toggiþ P. L. S. xxxii. 5; toggid (*pret.*) ALIS. 2305; REL. I. 59; togged (*pple.*) MAP 339; *comp.* to-toggen.

toglen, v.,=*M.H.G.* zoglen;=**toggen**; toggle A. R. 424*.

to-glīden, v., *O.E.* toglīdan, = *O.L.G.* teglīdan; *glide asunder*; toglide MISC. 94; toglād [toglōd] (*pret.*) LA3. 18083.

to-gna3en, v., *gnaw in pieces*; tognawe ALIS. 4629; PR. C. 863; tognōwe (*pret.*) P. L. S. xix. 248; ALIS. 6119; tognawe (*pple.*) REL. I. 167.

to-gnīde, v., *crush, grind*, PS. xvii. 43*; tognōd (*pret.*) PS. civ. 33*.

to-grínden, v., *grind in pieces*; **togrint** (*pres.*) LANGL. *C* xii. 62; **togrounde** (*pple.*) PALL. i. 1135.

toȝt, *see* **toht.**

tōh, adj., *O.E.* tōh, *cf. M.L.G.* tā, *O.H.G.* zāhi; *tough, tenacious,* M. H. xv; touh, tou MAN. (F.) 13038; touȝ WICL. GEN. xi. 3; tough R. R. 1726; tough, tou PR. P. 498; tou ROB. 175; REL. II. 29; tōȝen (*dat. n.*) LAȝ. 9319; towe ROB. (W.) 5890; tōȝe (*pl.*) LAȝ. 5865*; toghe OCTOV. (W.) 1084; **tōwer** (*comp.*) LANGL. *C* xiii. 187.

   tōughnesse, tōunesse, sb., *toughness,* PR. P. 498.

to-hacken, v.,=*O.Fris.* tohakia, *M.H.G.* zer-hacken; *hack in pieces*; tohakke ROB. 141; **tohakked** [tohacked] (*pple.*) TREV. VI. 445.

to-harwen, v., *harrow completely*; **toharewide** (*pret.*) LANGL. *C* xxii. 268.

to-hēawen, v., *O.E.* tohēawan,=*O.Fris.* ta-hāwa, *M.H.G.* zerhouwen; *hew in pieces*: þaȝ me ssolde hine al toheawe AYENB. 178; tohewe HORN (R.) 1324; GAW. 1853; to-hēwen (*pres.*) CH. C. T. *A* 2609; tohēu (*pret.*) P. L. S. xix. 228; (þai) tohewe FER. 897; **tohēawen** (*pple.*) LAȝ. 178; tohewen HAV. 2001; tohewe WILL. 3412.

to-hélden, v., *glide away*; **tohélden** (*pret.*) LAȝ. 27464.

tō-hópe, sb., *O.E.* tōhopa,=*O.L.G.* tōhopa; *hope,* HOM. I. 155, 191; II. 149.

toht, adj., *mod.Eng.* taut; *tough, tight, firm, binding*; toȝt ROB. 22; A. P. i. 521; þe kniȝt so toȝt GAW. 1869; he makeþ his mawe touht P. S. 331; toght [touht, touȝt, tought] CH. C. T. *D* 2267; tout P. 28; ROB. (W.) xx. 150; mi tohte rude MISC. 193; þe tohte ilete O. & N. 1446; ifillid toȝte (*adv.*) FER. 4390.

to-hurlen, v., *dash in pieces*; **tohurles** (*pres.*) JOS. 533.

to-hurren, v., *? hurry apart,* A. R. 426*.

to-hwiðeren, v., *whirl in pieces*: þat alle þise fowr hweoles **tohwideren** (*pres. subj.*) to stucches KATH. 2018; **tohwiðered** (*pple.*) o hweoles A. R. 362.

toil, sb., *toil, labour*; **toile** (*dat.*) D. ARTH 1802.

toilen, v., *? O.Fr.* toiller, toiller; *mod.Eng.* toil; *pull about, harass; labour*: to toilen wiþ þe erþe PL. CR. 742; it was ... reuliche **toiled** (*pple.*) to and fro MAP 338; tore and toiled L. H. R. 143.

tōken, *see* **tāken.**

to-kerve, v., *O.E.* toceorfan,=*O.Fris.* to-kerva; *carve in pieces,* A. P. ii. 1700.

tokker, *see* **touker** *under* **tuken.**

tōknen, *see* **tācnen.**

tol, sb., *O.E.* toll,=*O.L.G.* tol, *O.N.* tollr, *O. H.G.* zol; *toll, 'vectigal,'* PR. P. 495; LAȝ.

13316; AYENB. 192; S. S. (Web.) 2050; LANGL. *C* xiv. 73; **tolle** (*dat.*) P. S. 237; **tolles** (*pl.*) ANGL. VII. 220.

   tol-bōþe, sb., *custom-house, 'telonium,'* VOC. 274; WICL. MAT. ix. 9.

   tol-gaderer, sb., *toll-gatherer,* WICL. MAT. PROL.

   tol-hous, sb., *custom-house, 'telonium,'* PR. P. 496.

tōl, sb., *O.E.* tōl,=*O.N.* tōl; *tool, instrument,* WILL. 2243; tol and takel GEN. & EX. 469; tool PR. P. 496; tole ALIS. 815; GAW. 413; toile ALEX. (Sk.) 4708; **tōlen** (*pl.*) LAȝ. 29253; tooles LANGL. *A* xi. 133; toles *B* x. 177; **tōle** (*dat. pl.*) AR. & MER. 507; *comp.* **egge-tōl.**

tōle, *see* **tāle.**

toli, sb., *scarlet dye,* ALEX. (Sk.) 4335; tuli *'puniceus'* PR. P. 505.

tō-liggen, v., *pertain to*; **tōlið** (*pres.*) ORM. 1408.

to-limien, v., *dismember*: **tolimeð** (*pres.*) A. R. 84; **tolimede** (*pret.*) JUL. 58; **tolimet** (*pple.*) H. M. 21.

to-liðien, v., *O.E.* toliðian,=*M.L.G.* toleden, *M.H.G.* zerliden; *dismember*; **tolideden** (*pret.*) LAȝ. 4226; **toleðed** (*pple.*) LAȝ. 25929.

tollen, v., *O.N.* tolla, = *Du.* tollen, *M.H.G.* zollen; *pay or take toll,* PR. P. 496; LANGL. *C* xiv. 51; CH. C. T. *A* 562.

   tollere, sb., *O.E.* tollere,=*M.H.G.* zoller; *toll collector,* H. S. 5572; tollare *'telonarius'* PR. P. 496; **tolleres** (*pl.*) LANGL. *B* PROL. 220.

tollen [2], v., *cf.* **tüllen**; *draw, allure, entice*: drawen or tollen CH. BOET. ii. 7 (56); tollin *'incito'* PR. P. 496; tolli O. & N. 1627; SAINTS (Ld.) xlv. 307; þis **tolleð** (*pres.*) him touward þe A. R. 290; tolleð (*misprinted* colleð) men to him REL I. 221; ha tollið to gederes ant toggið MARH. 14; tolls ALEX. (Sk.) 3640; **tollede** (*pret.*) TREV. VI. 149; tollid his oune wif a wai S. S. (Wr.) 3052; tolled LANGL. *B* v. 214; *comp.* **to-tollen.**

to-luggen, v., *drag about*; **tologged** (*pple.*) LANGL. *A* ii. 192; tolugged *B* ii. 216.

to-lūken, v., *O.E.* tolūcan; *tear asunder,* KATH. 2123; **tolēac** (*pret.*) MARH. 7; to-luken LAȝ. 2602; JUL. 78; **toloken** (*pple.*) MARH. 6.

tom, *see* **táme.**

tōm, adj., *O.E.* tōm, = *O.N.* tōmr, *cf. O.L.G.* tōmi; *empty, unoccupied*; toom *'vacuus'* PR. P. 496; FLOR. 144; MAN. (H.) 192; tome saule ['*animam inanem*'] PS. cvi. 9*; toim, tume BARB. v. 642, xvii. 735; **tōme** (*pl.*) C. M. 17798; TOWNL. 113.

tōm [2], sb., *O.N.* tōm; *ease, leisure,* WILL. 3778; A. P. i. 134; toom PR. P. 496; i have no toom [tome] to telle LANGL. *A* ii. 160; tome ROB.

557; S. S. (Web.) 3; GREG. 275; MAN. (F.) 3409; PR. C. 6248; TOWNL. 176.

to-marred, pple., *defiled*, A. P. ii. 1114.

tombe, sb., *O.Fr.* tombe, tumbe; *tomb*, TREV. I. 35; tombe [toumbe] CH. C. T. *F* 518; tumbe LAȝ. 6080*; P. L. S. xii. 141.

tomben, tomblen, see **tumben, tumblen.**

tomberel, sb., *O.Fr.* tomberel; *tumbrel, dung-cart*; tomerel, tumrel PR. P. 496, 506.

to-melte, v., *dissolve*, HOM. I. 269.

tō-náme, sb.,=*M.L.G.* tōname, *M.H.G.* zuo-name; *surname*, HOM. II. 143; LANGL. *C* xiii. 211; WICL. ECCLUS. xlvii. 19; MAN. (F.) 7000; tonome [toname] LAȝ. 9383.

tonge, see **tange, tunge.**

tonne, see **tunne.**

top, sb., *O.E.* topp, top,=*O.Fris.*, *M.Du.* top, *O.N.* toppr, *M.H.G.* zopf; *top, tuft of hair, head*, PR. P. 496; ALIS. 1417; CH. C. T. *A* 590; top ne more O. & N. 1328; top and tail AR. & MER. 8136; top over tail WILL. 2776; a top of heer VOC. 144; a top of flax REL. II. 78; **toppe** (*dat.*) JUL. 28; O. & N. 1422; BRD. 16; LANGL. *A* iii. 135; P. 11; TRIAM. 764; D. ARTH. 1144; bi þone toppe he hine nom LAȝ. 684; teon seiles to toppe 1339; bi toppes and bi here O. & N. 428; *comp.* **fore-top.**

top², sb., *cf. M.Du.* dop, top, *O.H.G.* toph; *top, teetotum*, PR. P. 496; ALIS. 1727.

to-parti, v., *for* disparti; *depart*, MISC. 57; **toparted** (*pple.*) P. S. 332.

to-ponen, v., *pound completely*; **toponede** (*pret.*) ['contrivit'] WICL. PS. civ. 16.

toppe, see **tōuppen.**

toppin, v., = *M.Du.* toppen; *wrestle, 'colluctor,'* PR. P. 496; hi **toppede** (*pret.*) ofte P. L. S. xxiv. 15.

topping, sb., *? mane, ? head*: þe tail and his topping GAW. 191.

topte, ?adj.: a topte sail (*topsail*) DEP. R. iv. 72.

to-pulle, v., *pull in pieces*, MAN. (F.) 10210; **topullid** (*pple.*) WICL. IS. xviii. 7.

to-pungen, v., *prick*; **topungid** (*pple.*) WICL. PS. xxix. 13.

to-qvaschin, v., *shake asunder*; **toqvashte** (*pple.*) LANGL. *C* xxi. 259.

tor, sb., *tor, hill-top*; tor, toure ALEX. (Sk.) 2109.

tōr, adj., *O.E.* tor, = *O.N.* tor-; *O.H.G.* zur-; *? Goth.* tuz-; *difficult, strong*, GAW. 165; tor for to paien A. R. 108*; tor & hefiȝ lif ORM. 6350; tor, toor WILL. 1428, 5066; *see* **tēre.**

tōr-fēre, sb., *O.N.* torfœra; *hardship, difficulty*, C. M. 8662; D. ARTH. 3451; ALEX. (Sk.) 3729; torfere [torfer] 1261; torfar 1193; torfoir YORK xl. 160, 174.

to-rácen, v., *tear to pieces*, CH. C. T. *E* 572; **toráced** (*pple.*) GAW. 1168.

to-rǣsen, v., *? fly to pieces*: & reat to þet hweol swa þet hit al toresde (*? printed* to refde) JUL. 58.

to-ratten, v., *rend asunder*; torattis (*pres.*) D. ARTH. 2235.

torche, sb., *O.Fr.* torche; *torch*, FL. & BL. 238; LANGL. *B* xvii. 203; torches (*pl.*) JOS. 191; GAW. 1119.

tord, sb., *O.E.* tord, = *M.Du.* tord; *excrement, 'stercus,'* VOC. 208; O. & N. 1686; CH. C. T. *B* 2120; toord PR. P. 497; þou art not worþ a tord N. P. 16; *deriv.* **tirdel.**

to-rēaven, v., *take completely away*; **torēveþ** (*pres.*) LANGL. *C* iv. 203.

to-renden, v., *O.Fris.* to-, terenda; *rend in pieces*, FRAG. 6; torende MARG. 132; LANGL. *B* x. 112; torendeð (*pres.*) A. R. 362; torente (*pret.*) S. S. (Wr.) 484; torent (*pple.*) REL. I. 46.

to-rēosen, v., *O.E.* tohrēosan; *fall to pieces*, LAȝ. 9245; torǣs (*pret.*) LAȝ. 9426.

torf, see **turf.**

to-ritten, v., *tear in pieces*; toritte (*pret.*) ORF. 79*.

to-rīven, v., *tear, break up, rend asunder*, MAN. (H.) 170; torōf (*pret.*) LAȝ. 7844*; HAV. 1792; GOW. III. 296; AMAD. (R.) xlviii; LANGL. *C* xxi. 63; toriven (*pple.*) HAV. 1953; A. P. i. 1197; torive L. H. R. 138.

torment, sb., *O.Fr.* torment, turment; *tempest; torment, suffering*; AYENB. 79; torment, turment PR. P. 497; CH. C. T. *A* 1298; tourment P. L. S. xx. 58; turment LEG. 3.

tormenten, v., *Fr.* tourmenter; *torment*; turmentin PR. P. 506; **tormenteþ** (*pres.*) AYENB. 53; itourmented (*pple.*) P. L. S. xvii. 170.

tormentinge, sb., *torture*, CH. C. T. *E* 1038.

tormentise, sb., *torture*, CH. C. T. *B* 3707.

tormentour, sb., *tormentor*, CH. C. T. *B* 818; turmentour S. S. (Web.) 498.

torne, tornen, see **turne, turnen.**

torpelnesse, sb., *distraction, whirl*, A. R. 322.

torplen, v., *fall headlong*: torplen into helle A. R. 322; torpleð (*pres.*) A. R. 324.

tortle, see **turtle.**

tortu, sb., *O.Fr.* tortue; *tortoise*, KN. L. 15; *see* **tortuce.**

tortuce, sb., *? med.Lat.* tortūca; *tortoise*, PR. P. 497.

tortuous, adj., *oblique* (*in astrology*) CH. C. T. *B* 302.

to-rŭschen, v., *shatter*; torŭsshed (*pple.*) and tobroke ['confracta'] TREV. IV. 399.

torvien, v., *O.E.* torfian; *throw*; torfede (*pret.*) MK. xii. 41; torvede LAȝ. 16703*; *comp.* **to-torvien.**

to-scailen, v., *scatter*; **to-scailed** (*pret.*) ALEX. (Sk.) 4150.

to-scateren, v., *scatter asunder* : toscatere
we ['*dissipemus*'] WICL. JER. v. 5 ; it al
toschatird ALEX. (Sk.) 4150 ; **toscatered**
(*pple.*) CH. C. T. *D* 1969.

tosch, see tusch.

to-schǣden, v., *O.E.* tosceādan, *? O.H.G.*
za-, zisceidan ; *separate* ; toshǣdan ORM.
19862 ; **toschēdeþ** (*pres.*) TREV. I. 133 ;
toscǣdde (*pret.*) LAȝ. 30262 ; **toschǎd**
(*pple.*) TREV. III. 241.

to-schǣnen, v., *O.E.* toscǣnan ; *break in
pieces* ; toscǣne [to sene] LAȝ. 2309 ; to-
schēneþ, -sheneþ (*pres.*) O. & N. 1120 ;
toscǣnde, -sceande (*pret.*) LAȝ. 2315, 16492 ;
? þat floc þat was toskeȝned (*pple.*) *scattered*
(? *ms.* toskeȝȝredd) ORM. 1498.

to-scháken, v., *O.E.* tosceacan ; *shake to
pieces* ; toshake ANGL. III. 546 ; PALL. ii. 240 ;
toschákeð (*pres.*) O. & N. 1647 ; **toschōk**
(*pret.*) BEV. 742 ; **tosháken** (*pple.*) WICL.
IS. xxiv. 20.

to-schellen, v., *shell, peel* ; **to-schullen**
(*pple.*) LANGL. *B* xvii. 191.

to-schēoten, v., *O. E.* toscēotan ; *break
asunder* ; toshĕtt (*pret.*) ALIS. (Sk.) 1008.

to-schideren, v., *shiver in pieces* : þer speres
can toschider GUY² (Z.) 1468.

to-schifte, v., *shift asunder, divide*, SHOR.
27 ; TREV. I. 97 ; **toschift** (*pple.*) SHOR. 27.

to-schivere, v., *shiver in pieces*, ALIS. 2728 ;
toshivere HAV. 1993 ; **toshivrede** (*pret.*)
HICKES I. 168 ; **toschivered** (*pple.*) WILL.
3603 ; toshivered CH. P. F. 493.

to-schrēaden, v., *shred in pieces* ; **toschrēde**
(*pres.*) CH. C. T. *A* 2609.

tōsen, see tǣsen.

to-sīȝen, v., *fall into decay* : þe bodi schal tosie
ARCH. LII. 33 (*printed* tofye SPEC. 101 ; A. D.
225).

to-skilen, v., *divide* ; toskiled (*pple.*) ORM.
18652.

to-slenten, v., *? fly to pieces* : þat al þat
heved toslente (? *printed* toflente) (*pret.*)
FER. 4940.

to-slīten, v., *O.E.* toslītan,=*O.H.G.* zislīzan ;
break asunder ; **tosliten** (*pres. subj.*) ['*di-
rumpant*'] MAT. vii. 6.

to-slitten, v., = *M.H.G.* zerslitzen ; *slit in
pieces* ; **toslitte** (*pret.*) JUL.² 146 ; toslit
(*pple.*) O. & N. 694.

to-snǣden, v., *cut in two* ; tosnǣdde (*ms.*
tosnæde) (*pret.*) LAȝ. 4015 ; tosnadde (*ms.*
tosnaðde) 28050.

to-soillen, v., *soil all over* ; tosoilled (*pple.*)
MAN. (F.) 12345.

to-sprǣden, v.,=*O.H.G.* zispreitan ; *spread
about, scatter* : þe schep wolleþ **tosprēde**
(*printed* to sprede) BEK. 2134 ; **tosprĕt**
(*pres.*) A. R. 402 ; **tosprĕd** (*pple.*) HOM. I.
147 ; tospred, -sprad HOM. II. 21, 205 ; to-
sprad ROB. 134.

to-springen, v.,=*M.L.G.* tospringen, *O.H.G.*
zispringen ; *burst asunder* ; tospringe AL. (L.¹)
1020 ; er him ouȝte þe herte tospringe
(*printed* to springe) C. L. 593 ; **tosprunge**
(*pret.*) H. S. 10673.

to-sqvatten, v., *bruise, crush* : shal tosqvatte
['*conteret*'] þin heed WICL. E. W. 461 ; **tosqvat**
(*pple.*) S. A. L. 16 ; treuþ(e) is al tosqwat P. 72.

tóst, adj., *toast* ; toost '*tostus*' PR. P. 497.

to-stéken, v., = *M.H.G.* zerstechen ; *prick* :
tostéke (*pret.*) hi his swete hefed with one
thornene coroune SHOR. 85.

tósten, v., *toast*, PR. P. 497.

to-stere, v., *move completely*, WICL. PS. xxviii. 8.

to-stingen, v., *sting severely* ; **tostingeþ**
(*pres.*) MISC. 152.

to-stonai, v., *astound completely*, BARB. xviii.
547.

to-swellen, v., *O.E.* toswellan,=*O.H.G.* zi-
svellan ; *swell to excess* : toswelle shul his
flodes ['*intumescent fluctus ejus*'] WICL. JER.
v. 22 ; toswal (*pret.*) MAN. (F.) 10876 ;
toswollen (*pple.*) KATH. 842 ; A. R. 282 ;
toswollen LAȝ. 17815 ; toswolle LEG. 54 ; S. S.
(Web.) 1588.

to-swingen, v.,=*O.L.G.* tesvingan ; *swing in
pieces* ; **toswungen** (*pple.*) LAȝ. 8026.

to-swinken, v., *labour greatly* ; **toswinke**
(*pres. pl.*) CH. C. T. *C* 519.

tō-tagge, sb., *appendage*, A. R. 318.

to-tāsen, v.,=*M.H.G.* zerzeisen ; *pull to
pieces* ; totose O. & N. 70.

tote, sb.,=*M.Du.* tote (*apex*), *O.H.G.* zota
(*tuft*) ; *heap* : of a tote ['*cumuli*'] TREV.
V. 163*.

tote², sb., *? toe* ; totez (*pple.*) A. P. ii. 41.

[tōte- ? *stem of* tōten.]

tōte-hil, sb., *mount of observation*, '*specula*,'
PR. P. 497 ; toothil ['*speculam*'] WICL. IS. xxi.
5 ; MAND. 312.

totelen, see tutelen.

tōten, v., *O.E.* tōtian ; *peep, look, view*, HOM.
II. 211 ; A. R. 52 ; and bad me toten on þe
tree LANGL. *B* xvi. 22 ; tote GOW. II. 143 ;
tōtes (*pres.*) GAW. 1476 ; totz A. P. i. 513 ;
tōte (*imper.*) P. L. S. xxix. 27 ; **tōtede** (*pret.*)
HAV. 2106 ; his ton toteden out PL. CR. 425.

tootere, sb., *one who looks*, WICL. IS. xxi. 6.

tōtild, adj., *fond of peeping* : totilde ancre
A. R. 102.

to-tēon, v., *O.E.* totēon,=*O.H.G.* ziziuhan ;
draw away, HOM. I. 9.

to-téren, v., *O.E.* toteran ; *tear in pieces*,
GEN. & EX. 2089 ; teteren A. R. 84 ;
toteoren JUL. 12 ; totere HAV. 1839 ;
totire TOWNL. 143 ; **totéreþ** (*pres.*) MISC.
153 ; tetireð REL. I. 218 (MISC. 13) ; **totar**
(*pret*). CH. C. T. *B* 3801 ; AL. (L.³) 530 ;
totere MARG. 126 ; **totóren** (*pple.*) LAȝ. 4994 ;
SPEC. 70 ; totorn LANGL. *B* v. 197.

toteron, v., *cf. M.Du.* touteren; *totter,* '*vacillo*,' PR. P. 498; **totrede** (*pret.*) *swung*, TREV. II. 387.

toti, adj., *dizzy*, N. P. I; min heed is toti CH. C. T. *A* 4253.

totir, sb., *cf. M.Du.* touter; *see-saw, swing*, '*oscillum*,' PR. P. 498; **totres** (*pl.*) TREV. II. 387.

to-toggen, v., *pull in pieces*; tetoggeð (*pres.*) & tetireð hem mid hire teð REL. I. 218 (MISC. 13).

to-tollen, v., *drag to pieces*; **totolled** (*pple.*) P. 59.

to-torvien, v., *O.E.* totorfian; *throw away*; (*ms.* totorvion) HOM. I. 9; **totorveð,** -torveþ (*pres.*) O. & N. 1119.

to-tréden, v., = *O.H.G.* zitretan; *tread in pieces*, A. R. 166; **totrad** (*pret.*) ['*conculcavit*'] WICL. PS. lv. 2; totreden LAȝ. 26771; totraden ALIS. 3946; **totréden** (*pple.*) HOM. I. 133; LAȝ. 380; totrede SHOR. 29.

to-tühten, v., *stretch asunder*; tospred & **totüht** [totiht] (*pple.*) HOM. II. 205.

tō-turn, sb., *refuge*, MARH. 11.

to-tüsen, v., = *Ger.* zerzausen; *rumple thoroughly*; **totüsed** (*pple.*) HAV. 1948.

to-twǽmen, v., *O.E.* totwǽman; *divide, separate*; **totwĕmde** (*pret.*) MARH. 17; totwemde (*printed* totweinde), -tweamede A. R. 396; ær heo totwemden LAȝ. 26593.

to-twicchen, v., *twitch in pieces*; totwiccheð (*pres.*) HOM. I. 53; totwiccheþ O. & N. 1647; **totwiȝt(e)** (*pret.*) WILL. 2097; **totwiht** (*pple.*) P. S. 337; totwiȝt P. 59.

to-twinnen, v., *divide asunder*; totwinneþ (*pres.*) ORM. 19060; **totwinnede** (*pret.*) A. R. 396*.

tōð, sb., *O.E.* tōð, *cf. O.Fris.* tōth, *O.L.G.* tand, *O.N.* tönn. *Goth.* tunþus, *O.H.G.* zand, zan, *Welsh* dant, *Lat.* dens, *Gr.* ὀδούς; *tooth*, A. R. 218; GEN. & EX. 4148; toþ Gow. II. 252; teð (*pl.*) KATH. 194; A. R. 288; teþ FRAG. 8; O. & N. 1538; S. S. (Wr.) 2592; unðonc his teð *in spite of his teeth* A. R. 236*; betvene his teþ AYENB. 67; tōþe (*gen. pl.*) MAT. xxv. 30; tōþan, toþa, toþe (*dat. pl.*) LEECHD. III. 102; toðen HOM. II. 183; toðen, toþen LAȝ. 21384, 27631; *comp.* wang-tōþ.

tooth-draware, sb., *tooth-drawer*, '*edentator*,' PR. P. 498.

tooth-lēs, adj., *toothless*, PR. P. 498.

tōþid, adj., *toothed*; '*dentatus*' PR. P. 498.

to-þrüsten, v., *thrust in pieces*; toþrüste (*pret.*) HOM. I. 131.

toú, sb., *? O.E.* tow, =*M.Du.* touw, *O.N.* tō; *tow*; (*ms.* tow) GOW. II. 315; CH. C. T. *A* 3774; towe LANGL. *B* xvii. 245.

tōu, see **tōh.**

towaille, sb., *O.Fr.* touaille; *towel*, FL. & BL. 563; MAND. 250; CH. C. T. *B* 3935;

towaille, touwaile SHOR. 50, 51'; (*ms.* twaille) ED. 64; **tuailes** (*ms.* twailes) (*pl.*) JOS. 285; towailes MIRC 1871.

touch, sb., *touch, contact*, A. P. iii. 252; *request*, GAW. 1301; **touches** (*pl.*) *sounds* 120; (*covenants*) 1677.

touchen, v., *O.Fr.* toucher; *touch*, C. L. 1309; touche CH. C. T. *D* 87; **touchend** (*pple.*) WILL. 1383; **touchede** (*pret.*) LANGL. *B* xvii. 147, *C* xx. 122.

tōuȝ, see **tōh.** **touken,** see **tuken.**

toun, see **tūn.**

tō-uppen, adv. & prep., *on the top of, above*: toup alle oþren SHOR. 127; toppe alle þing AYENB. 106.

tour, see **tūr.** **tourment,** see **torment.**

toute, see **tūte.** **tow,** see **tou.**

to-wǣven, v., =*O.H.G.* zeweiban; *waft away*; **toweaved** (*printed* to weaved) (*pple.*) A. R. 148.

to-walten, v., *overflow*; **towalten** (*pret.*) A. P. ii. 428.

tō-ward, adj. & prep., *O.E.* tōweard, =*O.L.G.* tōward; *toward*, HOM. I. 3; KATH. 412; ORM. 5038; O. & N. 1254; toward SAINTS (Ld.) 470; TREV. III. 99; sum ben devout holi and toward (*well disposed*) P. L. S. xxxi. 315; toward, -wardes LAȝ. 515, 566; towarde *at hand* LANGL. *C* i. 214.

to-warplen, v., *? scatter*: hwon alle þe leaves schulen beon **towarpled** (*pple.*) A. R. 322.

to-wāwen, v., = *O.H.G.* ziwāhan; *scatter by blowing*: þe wind hem wolde towowen ANGL. III. 279.

towe, see **tawe.** **towen,** see **toȝen.**

to-wenden, v., *O.E.* towendan; *turn aside, divert; go to pieces*: mid þusendfeld wrenches he þe harte **towendeð** (*pres.*) HOM. II. 191; towent A. R. 324; hit al **towende** (*? ms.* to wode) (*pret.*) to scifren LAȝ. 4537; and towende . . . a muchelere wraððe 29429.

to-werpen, v., *O.E.* toweorpan, =*O.L.G.* te-werpan; *throw away*; **towerpeð** (*pres.*) HOM. I. 109; **toworpen** (*pple.*) ORM. 16277.

to-winden, v., *go to pieces*: al to pieces he **towond** (*pret.*) FER. 2568.

to-wrǣsten, v., *wrest asunder*; **towrǎsten** (*pret.*) MAP 338.

to-wringen, v., *distort*; **towrong** (*pret.*) HORN (L.) 1074.

to-wríþen, v., *O.E.* towríðan; *distort*; towriþen TRIST. 3179.

to-wurðen, v., *perish*: for hungere **towurðen** (*pple.*) LAȝ. 20744.

tráce, sb., *Fr.* trace; *trace, track*, PR. C. 4349; trace of a wei over a felde '*trames*' PR. P. 498; **traiss** BARB. vi. 553; **traise** YORK xxx. 118; **tras** A. P. i. 1113.

trácin, v., *Fr.* tracer; *trace out*, PR. P. 499; what wei we **tráce** (*pres.*) CH. P. F. 54.

[**traht**, sb., *O.E.* traht ; *exposition.*]

trahtnen, v., *O.E.* trahtnian ; *treat of, expound*; **trahtned** (*pple.*) ORM. 11680.

traice, sb., *trace, horse harness,* '*restis,*' PR. P. 499 ; trais CH. C. T. *A* 2139 ; traise CATH. 391.

traie, *see* treȝe.

traien, traissen, v., *O.Fr.* trair, traiss- ; *betray* ; traie S. S. (Web.) 1526 ; WICL. MAT. xxvi. 46 ; traise JOS. 624 ; **traiest** (*pres.*) MISC. 42 ; **traiid** (*pret.*) P. L. S. iii. 42 ; **traied** (*pple.*) JOS. 102 ; traised CHR. E. 830 ; traissed (*printed* traisted) WILL. 2075.

traile, sb., *Fr.* traille ; *trail*, PR. P. 499.

trailed, adj., *from O.Fr.* treille (*trellis*) ; *intertwined; fenced around* ; A. P. ii. 1473 ; DEP. R. i. 47.

trailen, v., *Fr.* trailler ; *trail* ; traile PALL. iii. 289 ; þe geaunt to lond for to traille TOR. 1316 ; **trailende** (*pple.*) REL. II. 15 ; trailinge WICL. ESTH. xv. 7.

    trailliṅg, sb., *dragging*, LANGL. *B* xii. 242.

train, sb., *O.Fr.* trahin ; *enticement, stratagem*, D. ARTH. 4192 ; BARB. xix. 360 ; trane viii. 440 ; TOR. 803, 1455 ; traine YORK x. 102 ; '*proditio*' PR. P. 499.

traine, sb., *O.Fr.* trahine ; *race* : God haþ taken in þe his flexhli trene L. H. R. 147.

trainin, v., *O.Fr.* trahiner ; *draw on; delay; entice; lay a stratagem* ; PR. P. 499 ; traine D. ARTH. 1683.

traisōn, sb., *O.Fr.* traison ; *treason*, HAV. 444 ; traisoun ROB. 107 ; AYENB. 37 ; A. P. i. 187 ; treison LANGL. *C* xxi. 326 ; treisun A. R. 56 ; tresone MAN. (F.) 14034 ; D. ARTH. 878 ; tresun HOM. I. 279 ; **traisōns** (*pl.*) MAN. (F.) 7128.

traisonable, adj., *treasonable* ; tresonabill BARB. v. 550.

traissen, *see* traien.

traiþe-lī, adv., *? surely*, A. P. ii. 907 ; & entises him to tene more traiþli þen ever 1137.

traiterous-lī, adv., *treacherously* ; treterousli MAN. (F.) 14360.

traitorie, sb., *treachery*, CH. C. T. *B* 781 ; traiterie MAN. (F.) 9698.

traitre, sb., *O.Fr.* traitre (*acc.* traitor) ; *traitor*, traitour P. L. S. xiii. 106 ; C. L. 1333 ; treitur HOM. I. 279 ; **treitres** (*pl.*) H. M. 9 ; treitours [tretours] LANGL. *C* xxi. 425.

tramaile, sb., *Fr.* tramail ; *trammel,* '*tragum,*' PR. P. 499.

tramme, sb., *machine, contrivance*, A. P. iii. 101 ; **traimmes** (*pl.*) ALEX. (Sk.) 127.

tramountaine, sb., *O.Fr.* tramontaigne, *late Lat.* transmontāna (stella) ; *pole star* : i (*sc.* Lucifer) schal telde vp mi trone in þe tramountaine A. P. ii. 211.

trampelin, v.,=*Du.* trampelen, *L.G.* trampeln ; *trample*, PR. P. 499 ; **trampliþ** (*pres.*) WICL. PROV. vi. 13.

trampin, v.,=*M.L.G.* trampen, *Dan.* trampe ; *tramp*, PR. P. 499 ; **trampiþ** (*pres.*) WICL. PROV. vi. 13.

transcrit, sb., *O.Fr.* transcrit ; *transcript*, BEK. 547.

transe, sb., *Fr.* transe ; *trance* ; traunce CH. C. T. *A* 1572.

transgressour, sb., *O. Fr.* transgresseur ; *transgressor* ; **transgressōrs** (*pl.*) LANGL. *B* ii. 92*.

translacioun, sb., *O.Fr.* translation ; *translation*, TREV. IV. 37.

transmigraciōn, sb., *O.Fr.* transmigration ; *deportation*, ROB. 9.

transporten, sb., *Fr.* transporter ; *transport*, CH. BOET. i. 4 (19).

transversen, v., *transgress* ; **transverseþ** (*pres.*) LANGL. *C* iv. 449 ; **transversede** (*pret.*) LANGL. *C* xv. 209 ; *see* traversen.

trant, sb., *? cf. M.Du., Swed.* trant (*step*) ; *trick*, TOWNL. 145 ; trante YORK xxix. 234, xxxii. 252 ; **trantes** (*pl.*) P. P. 265 ; trantis YORK xlii. 168.

tranten, v., *cf. M.Du.* tranten, *Swed.* trant (*walk*) (RIETZ) ; *deceive ;*. **trantes** (*pres.*) GAW. 1707.

trappe, sb., *O.E.* treppe, = *M.Du., O.Fr.* trappe ; *trap,* '*decipula,*' PR. P. 499 ; VOC. 264 ; ORM. 12301 ; CH. C. T. *A* 145.

trappe[2], sb., *trappings*, ALIS. 3421 ; **trappis** (*pl.*) RICH. 1516.

trappen, v., *O.E.* (be-)træppan, = *M.Du.* trappen ; *trap* ; **trappid** (*pple.*) '*illaqueatus, deceptus*' PR. P. 499 ; YORK xxvi. 267 ; trapped LIDG. TH. 1830 ; *comp.* bi-trappen.

trappen[2], v., *furnish with trappings* ; trappin PR. P. 499 ; **trappid** (*pple.*) '*phaleratus,*' PR. P. 499 ; trapped CH. C. T. *A* 2890 ; trappede stedes D. ARTH. 1757.

trappure, sb., *trappings of a horse*, PR. P. 499 ; **trappures** CH. C. T. *A* 2499.

trasch, sb., *? trousers* : with rent cockrez at þe kne & his clutte **trashes** (*pl.*) A. P. ii. 40.

trāste, *see* trūst.

tratte, sb., *trot* ; trate FER. 1370 ; **trattes** (*pl.*) WILL. 4769 ; trattis TOWNL. 150.

travail, v., *Fr.* travail ; *toil, labour, trouble, travel*, MISC. 33 ; CHR. E. 487 ; AYENB. 130 ; FER. 2123 ; travaille CH. C. T. *E* 1210 ; travel WICL. 1 COR. xv. 58 ; travailȝe BARB. i. 23 ; **travailes** (*pl.*) LANGL. *C* x. 234.

travailin, v., *O.Fr.* travailer ; *travail, labour; travel; '*laboro,*' PR.P. 499 ; travaili AYENB. 33 ; travaile PR. C. 5942 ; i wolde **travaille** (*pres.*) ... twenti hundred mile LANGL. *B* xvi. 10 ; þu travaillest þer aboute noȝt P. L. S. xiii. 163 ; traveilist WICL. MK. v. 15 ; travaland BARB. xi. 369 ; **travalit** (*pret.*) BARB. vi. 27 ; travalit (*pple.*) vii. 376.

    travailinge, sb., *labour*, LANGL. *A* v. 235.

travaillour, sb., *O.Fr.* travailleur; *labourer*; **travaillours** (*pl.*) LANGL. *B* xiii. 239.

travailous, adj., *laborious, troublesome*; travailous, travelous, traveilous WICL. EX. vi. 6, JOB vii. 3, 1 TIM. iii. 1.

　travailous-li, adv., *laboriously,* WICL. WISD. xv. 7.

tráve, sb., *frame in which unruly horses are confined,* CATH. 391; CH. C. T. *A* 3282; (*ms.* trawe) PR. P. 500.

travers, sb., *something placed across; drawback*; (love) hath ever some travers GOW. III. 384; travas (*? screen*) '*transversum*' PR. P. 499; trifled a traverce wer alle A. P. ii. 1473.

traversen, v., *O.Fr.* traverser; =transversen; *transgress*; **traversed** (*pret.*) LANGL. *B* xii. 284; traversit BARB. xvii. 532.

trē, *see* **trēo.**

treble, adj., *O.Fr.* treble; *treble,* PR. P. 501; tribill BARB. xviii. 30*; treble song '*precentus*' PR. P. 501.

trecherie, *see* **tricherie.**

tréde, sb., =*M.L.G., M.Du.* trede, *M.H.G.* trit *m.; tread, step,* FLOR. 1882; **tréden** [treoden] (*pl.*) A. R. 380; treoden H. M. 15.

tréden, v., *O.E.* tredan, =*O.Fris.* treda, *M.L.G.* tredan, *O.H.G.* tretan, *Goth.* trudan, *O.N.* troča (*pret.* trač, träčum, *pple.* tročinn); *tread,* ORM. 2571; trede ROB. 132; tréde (*pres.*) CH. C. T. *A* 3022; trad (*pret.*) ORM. 2561; TREV. III. 347; þo þou into hevene trede REL. I. 49; treden [traden, troden] WICL. LK. xii. 1; trede ED. 2940; troden LANGL. *B* xi. 347; trod *engendered* LANGL. *C* xiv. 166; tréden (*pple.*) ORM. 4416; FER. 2297; treden [troden] CH. C. T. *C* 712; troden WILL. 3402; itrede O. & N. 501; *comp.* for-, of-, to-**tréden.**

tredil, sb., *O.E.* tredel; *treadle, step,* '*gradus*,' PR. P. 501.

tree, *see* **trēo.**

trēen, adj., *O.E.* trēowen; *wooden,* WICL. EX. vii. 19; C. M. 12392; PALL. iv. 916; trein BARB. x. 361.

treȝe, sb., *O.E.* tryge, = *? L.G.* trügge; *tray*; **treies** (*pl.*) RICH. 1490.

treȝe [2], sb., *O.E.* trega, *cf. O.L.G.* trego, *O.N.* tregi *m., Goth.* trigo *f.* (λύπη); *affliction, grief,* ANGL. IV. 195; þer is blisse a buten treȝe P. L. S. viii. 187; treie HOM. I. 193; AN. LIT. 6; WILL. 2073; REL. I. 113; (*ms.* treye) GREG. 736; PR. C. 7327; al þat whilom was murþe is turned to treie and tene P. S. 340; beo ibroȝt out of treie P. L. S. xix. 172; traie S. S. (Web.) 1681.

treȝen, v., *O.E.* tregian, =*O.N.* trega; *afflict, grieve*: ðu **tregest** (*pres.*) me GEN. & EX. 3975; **traied** (*pret.*) PS. v. 11; traid S. S. (Web.) 523; war for þi herte treide HORN (H.) 1313; **traied** (*pple.*) ALIS. 3046; *comp.* a-treȝen.

treget, sb., *O.Fr.* tresgiet (*magie*); *deceit*: þat never did treget ne truite L. H. R. 198.

tregettin, v., *O.Fr.* tresgetter; *juggle,* '*praestigio*,' PR. P. 501.

tregettour, sb., *O.Fr.* tresgetteres (*magicien*); *juggler,* C. M. 12247; CH. H. F. 1277; **tregetours** (*pl.*) CH. C. T. *F* 1141, 1143.

treget(t)rie, sb., *piece of trickery,* R. R. 6376, 6384.

treie, *see* **treȝe.**

treie, sb., *O.Fr.* trei, treis; *from Lat.* trēs; *three* (*throw at dice*), CH. C. T. *C* 653.

trein, *see* **traien.**

treisten, *see* **trūsten.**

trelest, adj., *trelliced,* ALEX. (Sk.) 3343.

trelis, sb., *Fr.* treillis; *trellis,* PR. P. 501.

tremblen, v., *O.Fr.* trembler; *tremble,* CH. C. T. *I* 598; tremble H. V. 80; tremblede, **trembled** (*pret.*) LANGL. *A* ii. 211; tremled LANGL. *B* v. 357*.

tremelinge, sb., *trembling,* PR. P. 501.

trenchaunt, adj., *O.Fr.* trenchant (*tranchant*); *cutting, sharp,* FER. 537; trenchaunt or pliant '*versatilis*' PR. P. 501.

trenche, sb., *O.Fr.* trenche; *trench, hollow walk,* CH. C. T. *F* 392.

trenchūr, sb., *O.Fr.* tranchoire; *trencher; carving knife*; P. S. 204; trenchoure PR. P. 501; trenchere VOC. 178.

trenden, v., =*M.L.G.* trenden; *roll*: rollen & trenden CH. BOET. iii. 11 (100); and in his ármes **trende** (*pret.*) HORN (H.) 452 [trente (R.) 434]; and hire bokes went(e) and trent(e) S. S. (Web.) 2370; **itrent** (*pple.*) FER. 5881; aboute itrent SHOR. 137; *comp.* bi-, un-**trenden.**

trendil, sb., *O.E.* trendel, =*M.H.G.* trendel, *M.L.G.* trendel, trindel; *roll, roller, wheel,* ['*sphaera*'], WICL. IS. xxix. 3; '*troc(h)lea*' PR. P. 502; '*insubulum*' HALLIW. 887; trendul D. TROY 453.

trendled, adj., *? O.E.* tryndeled; *rounded*: trendled als a wel REL. I. 225 (MISC. 23).

trendlen, v., *O.E.* (a-)trendlian, = *M.H.G.* trendeln, *M.L.G.* trendelen, trindelen; *roll*: letten teares trendlen KATH. 2361; trendlin, trendelin PR. P. 502; þeȝ appel **trendli** (*pres. subj.*) from þon tr(e)owe O. & N. 135; **trendeled** (*pret.*) A. P. i. 41; trendelid RICH. 4506; be trendlid ['*volvi*'] WICL. JUDG. vii. 13; þe hedde trendild on þe borde GUY [2] (Z.) 3712.

tréne, *see* **traine.**

trenket, sb., *O.Fr.* trenchet (*tranchet*); *shoemaker's knife,* PR. P. 502.

trennen, v., =*? M.Du., M.H.G.* trennen (*separate*): uch toþ fram oþer is **trent** (*pple.*) REL. II. 212.

trental, sb., *thirty masses for the dead*; **trentalis** (*pl.*) LANGL. *B* xi. 146*.

trēo, sb., *O.E.* trēo, trēow, trēu,=*O.L.G.* trio, *Goth.* triu (*gen.* triwis), *O.Fris.*, *O.N.* trē, *cf. Gr.* δόρυ; *tree,* KATH. 1192; FL. & BL. 291; þat treo LA3. 26061; þet treou, treo A. R. 150, 392; trew GEN. & EX. 3301; tree LANGL. *B* xvi. 4; tre '*arbor, lignum*' PR. P. 500; HAV. 1022; S. S. (Wr.) 599; trou SHOR. 157; þet trau AYENB. 26; trēowes, trowes (*gen.*) HOM. I. 221; treouwes A. R. 150; trēowe, trowe (*dat.*) O. & N. 135; trowe SHOR. 159; trewe AR. & MER. 1599; treuwe HOM. II. 107; trēowa (*pl.*) LK. xxi. 29; treowe HOM. I. 5; treowen, treon LA3. 1835, 25978; treon SAINTS (Ld.) xlvi. 587; treon, tron O. & N. 1133, 1201; two treon '*duo ligna*' A. R. 402; treos ORM. 14; trowes ALIS. 6762; trēowe (*gen. pl.*) MAT. iii. 10; trēowen (*dat. pl.*) LA3. 511; trewen MK. xi. 8; *comp.* appel-, bōc-, box-, cheri-, dóre-, galwe-, per-, plŭm-, rōde-, ulm-, war3-, wīntrēo.

trēo-werk, sb., *woodwork,* LA3. 22899.

trēosien, trousien, v., *O.E.* trēowsian, trȳwsian; *have faith, trust,* LA3. 8315, 8489; þe king him trēousede (*pret.*) on 9308.

trēou, *see* **trēo.**

trēouðe, trēowðe, trēuweþe [trēuþe], sb., *O.E.* trēowð,=*O. H. G.* (ga-)triuwida, *O.N.* .trygð; *truth, faith,* LA3. 4340, 9819, 25471; treowþe ALIS. 4115; on ure treowþe PROCL. 3; treouðe HOM. I. 205; treouþe S. A. L. 153; trewðe REL. I. 130; GEN. & EX. 1524; trewþe HORN (L.) 305; AYENB. 221; gaf treuth believed BARB. iv. 223; treuþe ROB. 172; SPEC. 30; SHOR. 60; WILL. 2006; FER. 1281; LANGL. *A* i. 12; WICL. JOHN iv. 23; ure treuþe [trouþe] and ure hope C. L. 1294; treuthe SAX. CHR. 261; truþe FL. & BL. 396; trouðe A. R. 310; treuþe, trouthe CH. C. T. *A* 46; trouthe PR. P. 503; PR. C. 4388; trowþe (*ms.* trowwþe) ORM. 1350; trawþe A. P. i. 495; *comp.* or-, un-, wan-trēuþe (-trouþe).

trow**þe-lǽs,** adj., *truthless,* ORM. 188.

trēowe, sb., *O.E.* trēow,=*O.L.G.* treuwa, *O. Fris.* treuwa, triuwa, trouwa, *O.H.G.* triuwa, trūwa, *O.N.* trū, *Goth.* triggwa; *fidelity; agreement; truce, respite*; trewe CH. TRO. iv. 1312; MAN. (H.) 195; TRIAM. 1251; D. ARTH. 879; trewe, truwe LANGL. *B* vi. 332, *C* ix. 355; *comp.* mis-, or-trūwe.

trow(e)-**fest,** adj., *O.E.* trēowfæst; *faithful,* HOM. I. 89.

trēw-, **trēofestnesse,** sb., *fidelity,* HOM. I. 99, 109.

trēu-fol, adj., *O.E.* (ge-)trēowfull; *faithful,* A. R. 203.

trēounesse, sb., *faithfulness,* A. R. 294; trewnesse FL. &. BL. 500.

trēowe, trig, adj., *O.E.* trēowe, trȳwe, *O.N.* trūr, tryggr, =*O.L.G.* triuui (=triuwi), *O.Fris.* triuwe, trouwe, *O.H.G.* (ge-)triuwi, trūwi, *Goth.* triggws; *true, faithful*; treowe HOM. I. 143;

KATH. 231; A. R. 100; treowe, treouwe [trewe] LA3. 8851, 31068; trewe GEN. & EX. 720; trewe HAV. 1756; SHOR. 124; TRIST. 110; LANGL. *A* i. 86; EGL. 262; trewe [truwe] TREV. V. 447; treue ST. COD. 95; triewe LIDG. M. P. 55; triwe ROB. 488; FER. 3502; truwe PR. P. 503; true P. L. S. xvi. 125; trig (*ms.* trigg) & trowe (*ms.* trowwe) ORM. 6177; trēowe (*pl.*) PROCL. 6; trēowere, treoure [treuwere, treuere] (*comp.*) LA3. 8932, 18840; trewere luve HOM. I. 275; trēoweste (*superl.*) LA3. 25487; *comp.* or-, un-, 3e-trēowe (-trēwe.)

trēu-lāc, sb.,=*O.N.* trūleikr; *fidelity,* HOM. I. 215.

trēow-līche, adv., *O.E.* trēowlīce; *truly,* MARH. 13; treouliche A. R. 218; treou-, treuliche LA3. 5545, 26403; treuliche ROB. 348; treu-, treweliche LANGL. I. 153; treuli MAND. 48.

trēow-scipe, sb., *fidelity,* HOM. I. 109; treow-, treouscipe [treou-, treusipe] LA3. 6541, 9818; treoweschipe A. R. 8.

trēowen, v., *O.E.* trēowian, trūwian, *cf. O.H.G.* triuwen, trūwen, trūen, *O.N.* trūa, *Goth.* trauan; *trust, believe*; trowen HOM. I. 67; (*ms.* trowwenn) ORM. 214; treuwen (*printed* treuþen) feondes lore MISC. 196; trowe HAV. 1656; LANGL. *B* xvii. 162; PR. C. 3776; 3e mowe us wel trowe WILL. 4840; þou shalt not trowe ['*non credes*'] WICL. DEUT. xxviii. 66; trowe (*pres.*) CH. C. T. *A* 691; H. S. 483; trouwe JOS. 216; troue ANT. ARTH. xxii; true REL. I. 22; þu treowest hire mid muchel wouh MISC. 94; treoweð KATH. 1334; trewið GEN. & EX. 2037; (heo) treoweð [troueþ] LA3. 3413; troweþ (*imper.*) MAND. 6; trowande (*pple.*) A. P. i. 662; trowede (*pret.*) LA3. 2351; ROB. 110; trawed A. P. i. 282; troud (*printed* croud) (*pple.*) HAV. 2338; *comp.* mis-, or-trowian (-trowen).

trouabile, adj., *credible,* HAMP. PS. xcii. 7.

trēowes, sb., *? plur.* of **trēowe**; *truce, faith,* ALIS. 2808; trewes RICH. 3207; TRIST. 111; B. DISC. 537; MAN. (F.) 7743; triws A. R. 286; triwes ROB. 488; truwes K. T. 215; truwes [trues] ['*foedus*'] TREV. V. 433; trewis, trowis BARB. xiv. 96, xv. 102; truwis, truis PR. P. 503; trues P. L. S. xvi. 23.

trēowðe, *see* **trēouðe.**

trepeil, sb., *O.Fr.* trepel; *agitation, disturbance*; tirpeil MAN. (H.) 98; turpel, turpeil, turpail MAN. (F.) 1665, 6866, 15422.

trepget, sb., *O.Fr.* trebuchet; *trap; engine for casting stones*; LANGL. *A* xii. 86; trebgot PR. P. 501; tribochetes (*pl.*) ALEX. (Sk.) 1296.

tresaunce, sb., *corridor,* '*transcencia*' PR. P. 502.

tresche, sb., *O.Fr.* tresche (*danse, bal*), *Ital.* tresca (*spezie di ballo*); *a sort of dance*; in a tresche ROB. (W.) 7067.

trēsōne, *see* **traisōn.**

tresōr, sb., *O.Fr.* tresor; *treasure, money,*
SAX. CHR. 261; A. R. 126; SPEC. 89; tresore
tresour LANGL. *A* i. 81, *B* i. 83, *C* ii. 79;
tresur MISC. 97.

tresorēr, sb., *O.Fr.* tresorer; *treasurer,* TREV.
VIII. 209; tresourer P. L. S. xvii. 382; tre-
soriere AYENB. 231; tresorere, treserour
LANGL. *B* xx. 259, *C* xxiii. 260.

tresorie, sb., *O.Fr.* tresorie; *treasury,* ROB.
274; LEB. JES. 448; A. P. ii. 1317; ALEX.
(Sk.) 1666.

trespas, sb., *O.Fr.* trespas; *trespass, sin,*
P. L. S. ix. 196; MAP 348 (A. D. 241); MAN.
(F.) 209; BARB. xii. 485.

trespassen, v., *O.Fr.* trespasser; *trespass;*
trespasseþ (*pres.*) LANGL. *A* iii. 274; **tres-**
passide (*pret.*) TREV. III. 263; trespassit
BARB. xi. 553; **trespassed** (*pple.*) LANGL.
B v. 375; trespast LANGL. *C* vii. 426*.

trespas(s)our, sb., *trespasser,* PR. P. 502;
trespassours (*pl.*) LANGL. *C* ii. 92*.

tresse, sb., *O.Fr.* tresce; *tress, plait of hair,*
CH. C. T. *A* 1049; **tresses** (*pl.*) ARCH. LII.
35; S. S. (Web.) 478; trisses ALEX. (Sk.) 3450.

tressour, sb., *head-dress,* GAW. 1739.

trēst, *see* **trūst.**

trestel, sb., *O.Fr.* trestel; *trestle;* **trestlis**
(*pl.*) WICL. EX. xxvi. 20; trestes A. P. ii. 832;
GAW. 884, 1648.

trēsten, *see* **trūsten. trēsūn,** *see* **traisōn.**

trētable, adj., *O.Fr.* traitable; *tractable,* PR.
P. 502; AYENB. 94.

tretē, sb., *Fr.* traité; *treaty,* PR. P. 502; CH.
C. T. *A* 1288; H. V. 78.

trēten, v., *Fr.* traiter; *treat; handle; manage:*
to treten of folie CH. C. T. *C* 64; trete WICL.
COLOSS. ii. 21; **trēteþ** (*pres.*) AYENB. 142;
trētit (*pret.*) BARB. iii. 741; (*made a treaty*)
BARB. iv. 172.

trēting, sb., *negociation,* BARB. xiv. 8.

tretis, adj., *O.Fr.* traitis (*allongé, bien fait*);
Lat. tractīcius; *well made, pretty,* ENG. ST.
VII. 103; fair & tretis FER. 5883; it was
gentil and tretis R. R. 1216; tretis BARB.
xi. 35, x. 125; tretiss xix. 145.

tretis[2], sb., *treatise; short poem; treaty,*
TREV. III. 369; MAND. 314; CH. ASTR.
PROL.; CH. C. T. *E* 331; DEP. R. PROL. 51.

treu, sb., *O.Fr.* treü, *Lat.* tribūtum; *tribute;*
(*ms.* trew) FER. 4393.

treuāge, sb., *O.Fr.* treuage, truage; (*ms.* trew-
age) HORN (L.) 1498; truage LAꝫ. 7189*;
ROB. 46; truwage FER. 1731.

trēuþe, *see* **trēouðe. trēw,** *see* **trēo.**

trēwe, trēwes, trēwðe, *see* **trēowe, trēowes,**
trēouðe.

triacle, sb., *O.Fr.* triacle, *mod.Eng.* treacle;
antidote to poison, sovereign remedy, AYENB.
17; ALIS. (Sk.) 198; CH. C. T. *B* 479; love
is triacle of hevene LANGL. *B* ii. 146.

tribochet, *see* **trepget.**

tribulaciōn, sb., *Fr.* tribulation; *tribulation,*
PR. C. 4133; **tribulaciūns** (*pl.*) A. R. 402.

tribut, sb., *Fr.* tribut; *tribute,* LANGL. *B* xix.
37; tribute ALEX. (Sk.) 888, 1044.

tributarie, v., *Fr.* tributaire; *tributary,* CH.
C. T. *B* 3866.

trichen, v., *O.Fr.* trichier, trechier; *trick,* P.
S. 69; **itricchet** (*pple.*) H. M. 9.

tricherie, sb., *O.Fr.* tricherie, trecherie, tri-
querie; *treachery, trickery,* A. R. 202; LEB.
JES. 826; SPEC. 45; tricherie, trecherie SHOR.
162, 163; tricherie, treccherie LANGL. *A* i.
172, *B* i. 196; trecherie HAV. 2988; P. L.
S. xiii. 107.

tricherous, adj., *treacherous,* MAN. (F.)
16519; trecherous PR. C. 4232.

trichōr, sb., *O.Fr.* tricheor; *traitor,* ROB. 455;
trichour SPEC. 45; **trichūrs** (*pl.*) MISC. 153.

tridel, *see* **tirdel.**

triē, adj., *Fr.* trié; *tried, choice,* SHOR. 165;
WILL. 761; LANGL. *B* xvi. 4.

trien, v., *O.Fr.* trier; *try,* DEP. R. ii. 83;
triin '*eligo, discerno*' PR. P. 502; **trīede**
(*pret.*) D. ARTH. 1947; **trīed** (*pple.*) LANGL.
B i. 205.

trīed, adj., *choice,* LANGL. *B* xvi. 4; **trīedest**
(*superl.*) *A* i. 126.

trīed-līche, adv., *excellently;* (*v. r.* trielich)
LANGL. *B* PROL. 14.

triennels, sb., pl., *masses said for three years,*
LANGL. *B* x. 330.

trifle, *see* **trūfle.**

trifled, pple., *? ornamented with trefoils;* so
trailed and trifled A. P. ii. 1473.

trig, *see* **trēowe.**

triklen, v., *trickle;* **trikilen** (*pres.*) P. R. L.
P. 207; **trikland** (*pple.*) IW. 1558; **trikled**
(*pret.*) CH. C. T. *B* 1864; trikild ALEX. (Sk.)
4974.

trillin, v., *cf. Dan.* trille, *Swed.* trilla; *twirl;*
'*volvo*,' PR. P. 502; trille A. P. i. 78; ꝫe mote
trille a pin CH. C. T. *F* 316; **trillid** (*pret.*)
OCTAV. (H.) 269.

trimen, *see* **trūmen.**

trīnd, adj., *cf. O.Fris., Dan., Swed.* trind, *M.*
L.G. trint, trent (*round*): umbe trind (*? ms.*
trin) [= *M.L.G.* umme trint, trent] *around,*
ORM. 17563.

trindil, sb., *? O.E.* tryndel; *roller, pulley,*
'*troc(h)lea*,' VOC. 217; *see* **trendil.**

trindlen, v., *?* = **trendlen** : *roll* : his hevid
trindeld (*pret.*) on þe sand IW. 3259.

trīne, adj., *O.Fr.* trīne; *threefold;* trine compas *earth, sea, and heaven,* CH. C. T. *G*
45.

trīnen, v., = *Swed.* trīna (*pret.* trān) (RIETZ
750), *Dan.* trīne (*pret.* treen); *go, march;* trine
YORK i. 5; **trīne** (*pres.*) D. ARTH. 1757; trines

ALEX. (Sk.) 5171, 5195, 5231; **trīnande** (*pple.*) A. P. ii. 976; **trōn** (*pret.*) A. P. ii. 132.

trīnen [2], *? for* **atrīnen**; *touch*, LANGL. *C* xxi. 87; he ne dorste hem trine AL. (T.) 429.

trinitē, sb., *O.Fr.* trinité; *trinity*, P. L. S. xvii. 236; LANGL. *C* xix. 211, 264.

trípe, sb., *O.Fr.* tripe; *tripe*, PR. P. 502; VOC. 208; (*? morsel*) CH. C. T. *D* 1747.

tripet, sb., *? O.Fr.* tripout (*mauvaise manœuvre, complot*); *evil scheme*; tripet ne truit L. H. R. 132; truit and tripet 147.

trippe, sb., *flock, troop*: a trippe of gaite PERC. 186; trip MAN. (H.) 203.

trippin, v., = *M.Du.* trippen, *Dan.* trippe, *Swed.* trippa; *trip*, 'caespito,' PR. P. 503; trippe and daunce CH. C. T. *A* 3328; thei **trippe** (*pres.*) on trappede stedis D. ARTH. 3713.

trīse, sb., *cf. L.G.* trisse, *Swed.* trissa, *Dan.* tridse; *windlass, roller*; triiste, triis 'troc(h)-lea' PR. P. 503.

trīsen, v., *cf. M.L.G.* tritzen, *Dan.* tridse; *trice, hoist up*: out of his sete i wol him trise CH. C. T. *B* 3715; thei **trīsen** (*pres.*) up þaire sailez D. ARTH. 832.

trīst, trīsten, *see* **trūst, trūsten**.

tristre, sb., *tryst* (*station in hunting*), *rendezvous*, A. R. 332; trister GAW. 1712; i stand at mi tristur TOWNL. 310; triste MAN. (F.) 858; trist BARB. vii. 23; set trist *appointed* BARB. vii. 235; **tristeres** (*pl.*) GAW. 1170; tristors 1146.

trīwe, *see* **trēowe**.

troched, adj., *? ornamented*; troched toures A. P. ii. 1383.

trod, sb., *O.E.* trod; *from* **tréden**; *step*; **trodes** (*pl.*) A. R. 380*; **trodus** (*pl.*) ED. 513.

trodden, v., *tread, trace out*; **troddeð** (*pres.*) A. R. 232; **trodde** (*subj.*) A. R. 380.

troȝ, sb., *O.E.* trog, = *O.H.G.* trog *m.*, *O.N.* trog *n.; trough*; trogh VOC. 200; trogh, trough CH. C. T. *A* 3627; trouȝ WICL. JOSH. iii. 15; trough, trou PR. P. 503; trou 'auge' VOC. 155; **trowes** (*pl.*) PALL. iv. 916; *comp.* dōu-trough.

Troie, pr. n., *Gr.* Τροία; *Troy*, LAȝ. 205.

Troiisc, Troinisc, adj., *cf. M.H.G.* troisch; *Trojan*: of þan Troiscen monnen LAȝ. 410; þa Troinisce men LAȝ. 1955.

troilen, v., *O.Fr.* troiller (*enchanter*); *beguile*; þou troiledest (*pret.*) LANGL. *C* xxi. 321; **troiled** (*pple.*) 334.

trokien, *see* **trukien**.

trollin, v., *L.G.* trullen; *troll, roll; walk, wander; '* volvo,' PR. P. 503; þus haþ he trolled (*pple.*) forþ þise two and þritti winter LANGL. *B* xviii 296, *C* xxi. 334.

trome, *see* **trume**.   **trompe**, *see* **trumpe**.

tronchōn, sb., *O.Fr.* tronchon; *truncheon, staff,* M. ARTH. 3071; tronchoun CH. C. T. *A* 2615; trunchone PR. P. 504; trunsioune BARB. xvi. 129.

tronchōn [2], sb., *intestinal worm*; trunchon PR. P. 504.

tróne, sb., *O.Fr.* trone; *throne*, A. R. 40; HAV. 1316; LIDG. M. P. 12.

trónen, v., *enthrone*, LANGL. *A* i. 122*; **tróneþ** (*pres.*) LANGL. *B* i. 131.

tropel, sb., *O.Fr.* tropel; *troop*; **tropellis** (*pl.*) BARB. xiii. 275.

trōst, trōsten, *see* **trūst, trūsten**.

trot, sb., *O.Fr.* trot; *trot*, A. P. ii. 976; CH. C. T. *G* 575; C. M. 15872.

trotevale, sb., *idle talk*, H. S. 47, 5972, 8081, 9245; (*printed* trotenale) MAP 337.

trotton, v., *O.Fr.* troter; *trot*, PR. P. 503; **trotted** (*pret.*) LANGL. *A* ii. 135, *B* ii. 164.

trou, trouȝ, *see* **troȝ**.

troute, sb., *O.Fr.* truite; *trout,* 'tructa,' VOC. 222; PR. P. 503; troute [trute] C. M. 11884; **troutis** (*pl.*) BARB. ii. 577.

trouðe, *see* **trēouðe**.

trowe, trowen, *see* **trēowe, trēowen**.

truāge, *see* **treuāge**.

truandise, sb., *O.Fr.* truandise (*imposture*); *begging*; mid iseli **truwandise** (*dat.*) ['*trutannisatione*'] A. R. 330.

truant, sb., *O.Fr.* truant, truand; *truant*, GOW. II. 13; truont AYENB. 174; trouaunt '*trutannus*' PR. P. 503.

truanten, v., *O.Fr.* truander; *play the truant*; trouantin '*trutannizo*' PR. P. 503.

truble, sb., *O.Fr.* tourble, trouble; *trouble*, trubuil H. M. 29; turble, torble PR. P. 497.

trublen, turblen, v., *O Fr.* trubler, turbler; *trouble*, A. R. 268; **troubleþ** (*pres.*) AYENB. 150; **turblide** (*pret.*) WICL. MK. ix. 19; trobled *stumbled* LANGL. *C* vii. 408; **tribled** (*pple.*) HAMP. PS. ix. 22*; troblid DEP. R. PROL. 15.

truele, trulle, sb., *O.Fr.* truele, *Lat.* truella, trulla; *trowel*, WICL. AMOS vii. 7.

trūfle, sb., *O.Fr.* trufle; *trifle, nonsense*, ROB. 417; FER. 3459; LANGL. *B* xii. 140, xviii. 147; trifle PR. P. 502; trefele LANGL. *C* xv. 83*; **trūfles** (*pl.*) A. R. 106; AYENB. 58; trifles GAW. 108.

trūflen, v., *O.Fr.* trufler; *trifle, beguile*; trufli AYENB. 214; trifelin PR. P. 502; **trofle** (*pres. subj.*) D. ARTH. 2932.

    trifflour, sb., *trifler*, DEP. R. iii. 118.

truit, sb., *O.Fr.* trut (*ruse*); *evil design*; L. H. R. 132, 147.

truke, sb., *? lack*: ðat he ne be dead for truke of ðin helpe GEN. & EX. 3508.

truken, v., *Fr.* troquer, *cf. Span.* trocar; *truck,*

exchange; trukkon *'cambio'* PR. P. 503;
trukie (*pres. subj.*) H. M. 5; þet for ani worldliche luve his luve trukie A. R. 408.

trukien, v., *O.E.* trucian; *fail, be lacking,*
KATH. 1814; eaver se þu mare haves se þe
schal mare trukien (trukie) H. M. 7; trukien
[trokie] LAȝ. 17171; truke MISC. 96; troken
GEN. & EX. 105; **trukeþ** (*pres.*) FRAG. 5; þat
heo trukieð ... treoðe to halden LAȝ. 16861;
trukie (*subj.*) HOM. I. 53; trukede [trokede]
(*pret.*) LAȝ. 16416; ȝif bileave him trukede
A. R. 230; *comp.* ȝe-, **wan-trukien.**

    trukunge, sb., *failing, defection,* A. R. 12;
trukinge HOM. I. 79.

[**trum,** adj., *O.E.* trum; *firm, robust; comp.*
mis-, un-trum.]

trume, sb., *O.E.* truma; *troop, army,* H. M.
21; MISC. 149; GEN. & EX. 1829; trume
[trome] LAȝ. 28352; trome HAV. 8; SHOR.
108; AR. & MER. 5108; FER. 2372; D. ARTH.
3592; *comp.* **schild-trume.**

trümen, v., *O.E.* trymian, trymman; *from*
trum; *? become strong*; ge sal of a sune
trimen (*become pregnant*) GEN. & EX. 1024;
trimede (*pret.*) 1198; *comp.* **bi-, ȝe-trümen.**

trumpe, sb., *O.Fr.* trompe, *cf. O.N., O.H.G.*
trumba; *trumpet, 'tuba,'* PR. P. 503; VOC.
202; trumpe [trompe] CH. C. T. *A* 674;
trompe MAN. (F.) 9916; **trumpes** (*pl.*)
ALIS. 185; trumpen A. P. ii. 1402.

trumpen, v., *trumpet*; trumpon PR. P. 504;
gert trump BARB. viii. 293; **trumpand** (*pple.*)
ix. 137; **trumpit** (*pret.*) xvii. 356.

    trumper, sb., *trumpeter, 'tubicen, bucci-
nator,'* VOC. 218; trumpoure PR. P. 504.

trumpet, sb., *Fr.* trompette; *small trumpet,*
PR. P. 504.

trunk, sb., *Fr.* tronc, *mod.Eng.* trunk : trunke
for kepinge of fische *'gurgustium, nassa'* PR.
P. 504.

trunken, v., *truncate*; trunke PALL. iv. 86.

trunsioune, see **tronchōn.**

trusse, sb., *O.Fr.* trousse; *truss, pack, bundle,*
PR. P. 504; PARTEN. 720; D. ARTH. 3592;
trous SPEC. 110; **trusses** (*pl.*) A. R. 168*.

trussel, sb., *O.Fr.* troussel (*trousseau*); *bundle;*
trusselle D. ARTH. 3655.

trussen, v., *O.Fr.* trusser, trousser; *pack up;*
be off; go away, A. R. 322; trussin PR. P.
504; trusse HAV. 2017; LANGL. *A* ii. 194;
trossi ROB. 487.

trūst, trŭst, sb., *cf. O.N.* traust (*reliance*), Goth.
trausti *n.* (διαθήκη), *O.Fris.* trāst, *O.H.G.* trōst
m. (*consolation*); *trust,* HOM. I. 187; A. R. 202;
MAND. 283; have trust on his help H. M. 11;
trist P. S. 304; E. T. 550; P. L. S. v. 5; EGL.
546; trist to longe lif HOM. II. 75; trist
[trost] WICL. IS. xxxi. 6; trest FL. & BL. 408;
DEP. R. i. 47; trost PR. P. 503; *comp.* **mis-,
over-, un-, wan-trŭst.**

trŭst, trŭst[2], adj., *cf. O.N.* traustr (*trusty*),
O.H.G. (gi-)trōst (*confident*); *trusty, trustful* :
þat ȝe arn trust on A. R. 66*; trest AR. &
MER. 271; trost MAN. (H.) 60; E. 6. 46.

trŭst-līche, adv., *trustfully,* HOM. II. 9;
trustli MED. 1107; trostl(i)ke REL. I. 223;
trostli *'confidenter'* PR. P. 503; WICL. PROV.
x. 9; traistli HAMP. PS. lxvii. 1*.

    traistnes, sb., *confidence,* HAMP. PS. xxix. 7.

trŭsten, trŭsten, v., *O.N.* treysta (*confide*), =
O.H.G. trōsten (*console*); *trust, confide,* HOM.
I.213; MAND. 226; traste LANGL. *A* viii. 166*;
truste ROB. 33; tristen HAV. 253; triste
SHOR. 118; GOW. I. 176; E. T. 230; tresten
REL. I. 181; trosten, triste WICL. PS. cxvii.
8; trosten, tresten PL. CR. 237; trosti AYENB.
242; **trīste** (*pres.*) O. & N. 760; tresteþ LAȝ.
17941*; tristeþ MISC. 94; treistes MAN. (H.)
119; traistes PR. C. 1091; HAMP. PS. ii. 13;
we truste CH. C. T. *A* 501; **trŭste** [triste]
(*subj.*) O. & N. 1273; þet alle meidenes ...
trustin on þe MARH. 7; ȝif þei ... trusten in
god almiȝti LANGL. *B* xi. 279; **trōstende**
(*pple.*) E. G. 53; **trŭste** (*pret.*) KATH. 2222;
ROB. 204; triste BRD. 5; TREV. III. 343;
comp. **mis-, over-trŭsten.**

    traistinge, sb., *confidence,* HAMP. PS. lxx. 4*.

trŭsti, trŭsti, adj., = *Dan.* trōstig (*confident*);
trusty, faithful, MARH. 13; A. R. 334; SPEC. 47;
LANGL. *B* viii. 82; trusti [tristi] TREV. V. 443;
tristi DEP. R. ii. 103; A. P. ii. 763; tristi and
trew(e) WILL. 596; trosti PR. P. 503.

    trīsti-līche, adv., *trustfully,* WICL. GEN.
xxxiv. 25; trustili *confidently* WILL. 3904.

trŭþ, sb., *O.E.* trŭð, = *O.N.* trŭðr (*histrio*);
trumpeter, 'liticen,' FRAG. 2.

trŭþe, see **trēouðe. trŭwe,** see **trēowe.**

tubbe, sb., *L.G.* tubbe, *Du.* tobbe; *tub,* PR.
P. 504; CH. C. T. *A* 3621.

tucken, v., = *M.L.G.* tucken, *O.H.G.* zucchen;
tuck, draw : tucken [tukken] up PR. P. 504;
tucked [tukked] (*pple.*) CH. C. T. *D* 1737.

tuder, sb., *O.E.* tudor, tuddor; *product,* HOM.
II. 177.

tüdren, v., *O.E.* tydran, tyddran; *produce,
beget* : to tidren (*ms.* tiddrenn) & to tæmen
ORM. 18307; **tüderið** (*pres.*) HOM. II. 177;
tüdered (*pple.*) GEN. & EX. 630.

    tüderende, tüderinde, sb., *? for* tüderinge,
production, HOM. II. 55, 177.

tŭel, sb., *O.Fr.* tuel (*tuyau*); *pipe, tube,* L.
C. C. 38; tuwel CH. C. T. *D* 2148.

tuen, see **tēon.**

tŭft, sb., *? O.Fr.* tuffe; *tuft* : a tuft [toft] of
heres CH. C. T. *A* 555; tufte (*dat.*) TREV. V.
163; **tuftes** (*pl.*) TREV. I. 83.

tüȝe, sb., *? O.E.* tyge, = *O.H.G.* zug; *from*
tēon; *draught, pull*; tige *draught of liquor*
HOM. II. 67.

tü-brügge, sb.,=*Ger.* zugbrücke; *draw-bridge*; ROB. 543; P. S. 222.

tüȝel, sb., *O.E.* tygel,=*O.N.* tygill, *O.H.G.* zugil; *from* tēon; *rein, traces*; tiȝel '*tractorium*' FRAG. 4; þe reines oþer þe tiels (*pl.*) TREV. IV. 77.

tuggen, v., *cf.* toggen, tucken; *tug, draw*; tugge H. S. 9287; **tugge** (*pres.*) MED. 441; ituggid (*pple.*) HOCCL. i. 197.

tüht, sb., *O.E.* tyht,=*M.Du.* tucht, *O.H.G.* zuht; *from* tēon; *discipline, conduct, usage*, HOM. I. 247; abide .. in that tiȝt HALLIW. 878; what for laughinge & oþer tihtes (*pl.*) MAN. (F.) 9307; *comp.* un-tüht.

tühten, v. *O.E.* tyhtan, = *M.L.G.* tuchten, *M.H.G.* zühten; *draw; conduct, persuade, instruct; discipline*: tuhten & teachen HOM. I. 267; tuhten ne chasten A. R. 268; tiȝt(e) A. P. i. 717; teȝt (*pres.*) SHOR. 97; to Mahune heo tuhteð (*draw to*) LAȝ. 27321; tihten (*ms.* tihhtenn) & turnen hæþen folc . . . to lefen uppo Criste ORM. 7048; **tühte** (*subj.*) HOM. II. 247; **tühte** (*pret.*) FRAG. 7; þe deofel heom tuhte to þan werke HOM. I. 121; tihte II. 228; tuhten LAȝ. 810; to Charlis host aȝen þai tiȝt(e) FER. 1015; tiȝt (*pple.*) BEV. 3047; Alisaunder þo to reste is tight (*gone*) ALIS. 4485; *comp.* bi-, ȝe-, to-tühten.

tuihting, sb., *O.E.* tyhtung; *leading*, HOM. II. 29; tihtinge I. 229.

tühtle, sb., *discipline, usage*: for þere ilke tuhtle (*later text* þinge) LAȝ. 24675; *comp.* un-tühtle.

tuken, v., *O.E.* tucian; *pluck, tease; full (cloth)*; **tukest** (*pres.*) O. & N. 63; tukeð KATH. 550; & tukeð hire al to wundre H. M. 17; **tuked** (*pple.*) HOM. II. 21; he was . . . so scheomeliche ituked and so seoruhfu(l)-liche ipined A. R. 366; itouked [itukked] (*tucked up*) CH. C. T. (Wr.) 7319 (*D* 1737); itouked LANGL. *B* xv. 447; *cf.* tucken.

    touker, sb., *fuller*, '*fullo*,' VOC. 181; WICL. MK. ix. 2; **tokkeris** [toukers, toucheris] (*pl.*) LANGL. *A* PROL. 100.

tülien, *see* tilien.

tulk, tolk, sb., *O.N.* tulkr (*interpreter*); *man*, GAW. 3, 1966; tolk A. P. ii. 687.

tüllen, v., *O.E.* (for-)tyllan; *draw, entice*; tille S. S. (Wr.) 1563; as muche place as mid a þvong ich mai aboute tille ROB. (W.) 2492; to þe scole him to tille C. M. 12175; to tille þis yong man to foli M. H. 113; **tülleþ** (*pres.*) R. S. vii (MISC. 188); tilles PR. C. 1183; **tülde** (*pret.*) A. R. 320; he havede . . . al þe folk tilled (*pple.*) intil his hond HAV. 438.

tumbe, *see* tombe.

tumben, v., *O.E.* tumbian,=*O.N.* tumba; *jump, dance*; **tumbede** (*pret.*) MAT. xiv. 6; **tombede** (*pple.*) [tombede] TREV. IV. 365.

    tumbestere, sb., *female dancer*; Herodias douȝter .. was a tumbestre HALLIW. 894;

tombester TREV. IV. 15; **tumbesteris** [tombesteres] (*pl.*) CH. C. T. *C* 477.

tumblen, v.,=*M.L.G.* tumelen, *Dan.* tumle, *Swed.* tumla (*stagger, wallow*); *tumble, dance*; tumlin '*voluto*' PR. P. 506; tomblen ALIS. 2465; tummill BARB. ix. 452; **tumbleþ** (*pres.*) SHOR. 34; **tumblide** (*pret.*) ['*saltavit*'] WICL. MAT. xiv. 6; tomblede TREV. IV. 365; his heved . . . tomblede on þe sond FER. 997; tumlit BARB. iv. 182; tombled MAN. (H.) 70; sche . . . tombled over þe hacches WILL. 1776; tumlit (*pple.*) BARB. iv. 229.

    tumbler, sb., = *M.L.G.* tumeler; *tumbler*, '*saltator*,' VOC. 218; WICL. ECCLUS ix. 4; tumlare PR. P. 506.

tümen, *see* tēmen.   **tumrel**, *see* tomberel.

tün, sb., *O.E.* tūn,=*O.L.G.* tūn, *M.Du.* tuin, *O.H.G.* zūn m., *O.N.* tūn n.; *enclosure, farm, town*, ORM. 7016, 8629; REL. I. 217; GEN. & EX. 713; HAV. 764; toun GREG. 569; þen toun ROB. 184; i have bouȝt a toun ['*villam*'] WICL. LK. xiv. 18; toun and tour CH. C. T. *E* 2172; **tūnes** [tounes] (*gen.*) LAȝ. 1639; **tūne** (*dat.*) A. R. 418; O. & N. 1169; Averil eode of tune LAȝ. 24196; somer com to toune (*first text* to londe) 24242*; lenten is come . . . to toune SPEC. 43; **tūnes** [tounes] (*pl.*) LAȝ. 1639; tounes AYENB. 30; **tūnen** (*dat. pl.*) LAȝ. 31894; *comp.* æppel-, burȝ-, leih-, wert-, wike-tūn.

tün-folk, sb., *townfolk*, ROB. 70.

tün-mon, sb., *townsman*, '*villanus*,' FRAG. 4; tounemen (*pl.*) DEP. R. ii. 41.

tün-scipe, sb., *O.E.* tūnscipe; *township*, SAX. CHR. 262.

tunder, sb., *O.N.* tundr,=*M.L.G.* tunder, *M.H.G.* zunder; *tinder*, REL. I. 220; MAN. (F.) 14683; tundir PR. P. 506; tunder [tonder] LANGL. *B* xvii. 245; *see* tinder, tender.

tŭne, sb., *tune*, LANGL. *C* xxi. 470*; '*tonus, modulus*' CATH. 396.

tünen, v., *O.E.* tȳnan,=*M.L.G.* tūnen, *M.Du.* tuìnen, *O.Fris.* (be-)tēna, *O.H.G.* zūnen; *from* tün; *enclose*, A. R. 62; tuinen MIRC 490; tinin PR. P. 494; tine CL. M. 69; P. R. L. P. 167; tün (*imper.*) A, R. 104; **tünden** (*pret.*) LAȝ. 15320; *comp.* bi-, un-tünen.

tunge, sb., *O.E.* tunge,=*O.L.G.* tunge, *O.N.* tunga, *Goth.* tuggō, *O.H.G.* zunga, *Lat.* lingva *for* *dingva; *tongue*, '*lingva*,' tunge of a bocle '*lingula*,' tunge of a balance '*examen*' PR. P. 506; KATH. 194; A. R. 70; ORM. 4869; GEN. & EX. 372; GOW. I. 64; tunga LEECHD. III. 102; tunge [tonge] O. & N. 37; L. H. R. 26; tonge [tunge] CH. C. T. *A* 265; tonge ROB. 12; AYENB. 24; LANGL. *A* viii. 80; PERC. 628; his tunge (*gen.*) bend MK. vii. 35; **tunga** [tunge] (*dat.*) LEECHD. III. 102; **tungen** (*pl.*) A. R. 410; tongen SHOR. 100; **tungen** (*dat. pl.*) HOM. I. 93; tunge MISC. 56.

tunicle, sb., *O.Fr.* tunicle ; *tunic,* LANGL. *B* xv. 163 ; CATH. 396.

tunne, sb., *O.E.* tunne,=*O.N.,* *O.H.G., Ir., Gael.* tunna ; *tun* : tun cask A. R. 214 ; R. S. v ; tonne AYENB. 27 ; GREG. 86 ; CH. C. T. *A* 3894 ; to are tunne LAȝ. 14957 ; **tunnen** (*pl.*) 30672 ; tonnes of bras MAN. (F.) 2246 ; *comp.* **wīn-tunne.**

   tun-hōve, sb., *ground ivy,* ' *hedera terrestris,*' PR. P. 506.

tuppe, sb., *tup, ram,* '*vervex,*' VOC. 219 ; PALL. viii. 78 ; **tupis** (*pl.*) ALEX. (Sk.) 5566.

tūr, sb., *O.Fr.* tur, tor, *Lat.* turris ; *tower,* SAX. CHR. 234 ; A. R. 226 ; GEN. & EX. 661 ; FL. & BL. 222 ; ane tur [anne tour] LAȝ. 7763 ; tour HAV. 2073 ; H. H. 31 ; LANGL. *A* PROL. 14 ; tor A. P. i. 965 ; **tūres** (*gen.*) A. R. 228 ; **tūres,** turres [toures] (*pl.*) LAȝ. 7781, 23888.

turble, turblen, *see* **truble, trublen.**

turbot, sb., *O.Fr.* tourbot ; *turbot,* VOC. 189 ; turbut PR. P. 506 ; HAV. 754.

turet, sb., *O.Fr.* tourette ; *turret,* FER. 2441 ; toret CH. C. T. *A* 1909 ; toret '*turricula*' PR. P. 497 ; SAINTS (Ld.) xlv. 15.

turet², sb., *O.Fr.* touret ; *ring on a dog's collar* ; **tourettes** (*pl.*) CH. C. T. *A* 2152.

turf, sb., *O.E.* turf (*dat.* tyrf), *cf. O.H.G.* zurba *f., O.N.* torfa *f.,* torf *n., O.Fris.* turf *m. ; turf, sod,* LAȝ. 15395 ; tórf TREV. I. 339 ; mid ... turf and clute O. & N. 1167 ; **turves** (*pl.*) HAV. 939 ; turves [torves] CH. C. T. *E* 2235.

turken, turknen, v., *cf. O.Fr.* torquer ; *turn, twist* ; he **torkans** [torkis] with *turns towards* ALEX. (Sk.) 2967.

turment, *see* **torment.**

turn, sb., *O.Fr.* tourn ; *turn, trick,* A. R. 280 ; H. M. 47 ; GEN. & EX. 63 ; a freendes turn [torn] CH. C. T. *C* 815 ; tourn SAINTS (Ld.) xlv. 176 ; torn FER. 2019 ; **turnes** (*pl.*) KATH. 853 ; *comp.* **ȝein-, tō-turn.**

turnai, v., *Fr.* tornei, tornoi (*tournoi*) ; *tourney,* ALEX. (Sk.) 5429 ; tournei BEK. 213.

turnaien, v., *turn, wheel, engage in a tournament* ; **tornaieez** (*pres.*) GAW. 1707 ; **tournaied** *pret.*) GAW. 41.

turneiment, tournement, sb., *O.Fr.* tornoiement ; *tournament,* CH. C. T. *B* 1906 ; turnement A. R. 390 ; tornement AYENB. 101 ; **turnemens** (*pl.*) ROB. (W.) 11041*.

turnen, türnen, v., *O.E.* tyrnan, türnian,= *O.N.* turna, *O.H.G.* turnen, *from O.Fr.* torner, *Lat.* tornāre ; *turn,* ORM. 169 ; ich ule turnen me a wei A. R. 76 ; turne riht to wronge A. D. 237 ; turȝe [teorne] LAȝ. 12734 ; tuirne LEG. 51 ; tornen C. L. 1211 ; torne [tourne] LANGL. *B* xx. 46 ; terne AR. & MERL. 7940 ; FER. 869 ; MIRC 2034 ; B. DISC. 581 ; tirne of þe hide A. P. ii. 630 ; **turneð** (*pres.*) R. S. i ; turneð

giu to me HOM. II. 59 ; turninde (*pple.*) A. R. 356 ; **turnde** (*pret.*) LAȝ. 4092 ; O. & N. 1090 ; BRD. 24 ; turnden MARH 19 ; tirnden (*ms.* tirneden) HAV. 603 ; **iturnd** (*pple.*) HOM. I. 53 ; R. S. v ; of Latin iturnd into Englisch JUL. 2 ; iterned HORN (H.) 460 ; *comp.* **a-, bi-turnen.**

   turnare, sb., *turner,* PR. P. 507 ; **turners** (*pl.*) D. TRÓY 1586.

   turninge, sb., *turning,* PR. P. 507 ; **turninges** (*pl.*) MAN. (F.) 12860.

turren, v., *butt* (*as a ram*) ; **turred** (*pret.*) ALEX. (Sk.) 5567.

turtle, turtre, sb., *O.E.* turtle, *O.Fr.* turtre ; *turtle-dove,* ORM. 989 ; AL. (L².) 129 ; tortle SPEC. 38 ; **turtres** (*gen.*) REL. I. 224 ; **turtles** (*pl.*) MAND. 87 ; tortors GAW. 612 ; **turtlen** (*dat. pl.*) HOM. II. 49.

   turtle-brid, sb., *young turtle,* HOM. II. 47.

   turtel-douf(e), sb., = *O. H. G.* turteltūba ; *turtle-dove,* M. H. 159.

tusch, sb., *O.E.* tusc, tux, = *O.Fris.* tusk, tusch ; *tusk,* EGL. 383 ; tosch PR. P. 497 ; **tuskes** (*pl.*) JUL. 68 ; REL. I. 8 ; tusches S. S. (Web.) 914 ; toskes PS. lvii. 7 ; tuȝes A. R. 280 ; ALIS. 6547.

   tusked, adj., *tusked,* CH. C. T. *F* 1254 ; tuskid, toschid PR. P. 497.

tuschel, sb., *O.E.* tuxel ; *tusk* ; **tuxlis** (*pl.*) OCTOV. (W.) 929.

[**tūsen,** v., *Ger.* zausen ; *comp.* **to-tūsen.**]

tūte, sb., *backside, buttocks* ; toute P. L. S. xxxv. 136 ; CH. C. T. *A* 3812 ; TOWNL. 9 ; **touten** (*pl.*) ENG. ST. I. 102.

tutel, sb., *beak* ; A. R. 74 ; **tuteles** (*pl.*) A. R. 80.

tutelen, v., *whisper* ; totelin '*susurro*' PR. P. 498 ; **tuteleð** (*pres.*) A. R. 212.

   tutelere, sb., *tittler, tattler* ; **tutelers** (*pl.*) LANGL. *B* xx. 297.

   tutling, sb., *tooting* : a tutling of his horne herd thai BARB. xix. 664.

tutour, sb., *O.Fr.* tutor ; *guardian,* LANGL. *A* i. 54* ; *B* i. 56 ; ii. 52.

tūðen, *see* **tīðen.**

[**tūwe ?** *comp.* **maniȝ-tūwe.**]

tüwel, *see* **tüel.** **tux,** *see* **tusch.**

twā, *see* **twēȝen.**

twæmen, v., *O.E.* twæman ; *divide, separate* ; twemen LAȝ. 2948 ; twemen (*printed* tweinen) H. M. 27 ; tweamen A. R. 252* ; tweamin MARH. 5 ; *comp.* **to-twæmen.**

[**twecche,** sb., *O.E.* twecca ; *twitch* ; *comp.* **angel-twæcche.**]

twēȝen, card. num., *O.E.* twegen,= *O.Fris.* twēne, *O.L.G.* tvēne, *O.H.G.* zvēne, zwēne, *Goth.* tweihnai, twai, *O.N.* tveir, *Lat.* duo, *Gr.* δύο ; *twain, two* (*masc.*), HOM. I. 229 ; ORM. 12722 ; tweigen MAT. xviii. 20 ; heo tweien

FRAG. I; HOM. I. 41; twein 81; tweien [tweie] cnihtes LAȝ. 9462; tweiȝe [twei] 1113; twein 2530; þei tweine WILL. 1528; hem tweie 2147; tweien (ms. tweyen), tweie, tweine LANGL. B v. 32; tweie deles TREV. III. 7; tweiȝe, twei E. G. 350, 359; tveie [twei] men O. & N. 795; twei P. L. S. x. 41; tvaie boȝes AYENB. 67; twei A. R. 370; twei kinges ROB. 38; a schaft or twei S. S. (Wr.) 747; tweine CH. C. T. A 1134; PR. P. 504; hise tweȝen (for twa) dohtres ORM. 6386; of tweȝen (for twam) prestes 483; twā (? twa) [O.E. twā f., twa, tu n.,= O. Fris. twā f., twa (? twā) n., O.L.G. tvā f., tve n., O.N. tvǣr f., tvö n., Goth. twōs f., twa n., O.H.G. zvā, zvö f., zvei n., Lat. duae f., duo n.]; two (fem., neut.), H.M.9; twa sunne HOM. I. 33; twa gildene crunes ORM. 8179; tweamin a twa (in two) MARH. 5; hire twa sustren LAȝ. 3067; tva [two] niht LAȝ. 1113; twa children 7145; two dolen A. R. 10; two & þrittuȝe HOM. II. 47; tvo ȝer ROB. 225; two & fourti 60; tvo & tventiþe 440; tvo ziden AYENB. 153; a two JOS. 103; DEGR. 1620; breke a two SPEC. 40; to smite þe corde a two CH. C. T. A 3569; two, to GEN. & EX. 377, 423; on two, on to HAV. 471, 1823; tvo, to WILL. 1698, 2877; to P. S. 329; E. G. 30; PR. P. 495; twa (for tweȝen) men SAX. CHR. 262; twa prestes ORM. 487; twa daies S. S. (Wr.) 3180; tweigre, tweire (gen. m. f. n.) MAT. xviii. 16, xxi. 31; tweire A. R. 406; tweire cunne O. & N. 888; tweire LAȝ. 17560; FER. 311; twam (dat.) HOM. I. 31; P. L. S. viii. 155; of twam þinge O. & N. 1477; twam L. N. F. 217; ich habbe . . . iȝeven hit mine twam [two] dohtren LAȝ. 3167; bitweone twom (ms. twō) monnen 22968; twan HOM. I. 241; bituhen unc twa MARH. 8; of þa twa prestes ORM. 490.

twā-fáld, adj., twofold, HOM. I. 151; ORM. 14034; twovold A. R. 50.

tweien, twein, see twēȝen.

tweinen, see twinen.

twelf, card. num., O.E. twelf, twelfe,= O.Fris. twelf, twelef, twilif, O.L.G. tvelif, twelifi, Goth. twalif, O.N. tolf, O.H.G. zvelif, zvelife; twelve, KATH. 1844; twelfmoneth LANGL. B xiii. 337; tvelf TREAT. 139; AYENB. 11; twelf [twelve] winter ·LANGL. B v. 196; twelf, ·twælf, twealf LAȝ. 1168, 8106, 11065; twelfe iferan 1621; þas twælfe 25275; tweolve of þine witiȝ(e)n 4368; (ms. twellfe, twellfte) ORM. 537, 11069; tweolf, tweolve A. R. 200, 218; twelve O. & N. 836; tvelve AR. & MER. 934; an of þam twelfen MAT. xxvi. 14.

twelfte, ord. num., O.E. twelfta, = O.H.G. zwelifto; twelfth, ORM. 11047; twelfte PR. C. 4802; LUD. COV. 84; tvelfte AYENB. 14; tweolfte A. R. 412.

twēmen, see twæmen.

twengen, v., = O.H.G. zvengan, dwengan; from

twingen; twinge, press tightly; þat me ne twenge þine hude O. & N. 1114; þu twengest (pres.) þar mid so doþ a tonge 156; he tvengde (pret.) and schok hire bi þe nose P. L. S. ix. 81.

twěntiȝ, card. num., O.E. twēntig, = O.L.G. tvēntig, O.H.G. zveinzig, Goth. twai tigjus; twenty; (ms. twenntiȝ) ORM. 1894; twenti SAX. 252; KATH. 2502; tventi HAV. 259; TREAT. 139; SAINTS (Ld.) xlvi. 716.

twěntüðe, ord. num., twentieth, MARH. 23; tventeoþe [twentiþe] TREAT. 139 (SAINTS (Ld.) xlvi. 714); twentiþe WILL. 5354.

tweolf, see twelf.

[twēon, O.E. twēon; see bi-twēonen.]

twēonian, v., O.E. twēonian, twȳnian; doubt, debate, be at variance; HOM. I. 109; twēoneð (pres.) MK. xi. 23; twēoneden, twineden (pret.) LAȝ. 907, 3791.

twēoninge, sb., O.E. twēonung, twȳnung; doubt: and weren a tweninge (ms. at we-nynge) MISC. 54.

[twī-, prefix, O.E. twī-, = O.Fris. twī-, O.N. tvī-, O.H.G. zvī-; twi-; two.]

twī-bil, adj., O.E. twībill; two-edged axe, 'bipennis,' VOC. 196; PR. P. 505; SPEC. 110; N. P. 13; in brade axe and twibile HAMP. PS. lxxiii. 7.

twī-feald, adj., O.E. twīfeald,= O.Fris. twī-fald, O.N. tvīfaldr, O.H.G. zvīfalt; twofold, double, HOM. II. 187; twifald ORM. 4997; twi-, tvifald M. H. 19, 37; twifold REL. I. 131; twifold PS. cviii. 29.

twī-light, sb., = Ger. zwīlicht; twilight, 'diluculum, crepusculum,' PR. P. 505.

twī-rǣd, adj., O.E. twīrǣd; of different opinion: weoren alle twirǣde [twireade] LAȝ: 19416.

twī-spěche, sb., O.E. twigsprǣc; equivoca-tion, double meaning, REL. I. 129 (HOM. II. 163).

twicchen, v., ? O.E. twiccan,= M.H.G. zwic-ken; twitch: twichin, twikkin PR. P. 505; tvicche P. L. S. xxi. 131; twik (imper.) PALL. vi. 26; twiȝt(e) (pret.) WART. II. 258; hure swerdes out þai twiȝte FER. 1596; twight(e) M. ARTH. 1038; twight [twiȝt] (pple.) CH. C. T. D 1563; þe bord he fond of tviȝt TRIST. 1952; comp. to-twicchen.

twie, adj., O.E. twiwa,= O.Fris. twīa, M. L.G. twīe, twige; twice, GEN. & EX. 808; twie HORN (L.) 1452; tvie CHR. E. 247; twige, twewe MK. xiv. 30, 72; twie, twien A. R. 23, 36; twien, tweien [twie] LAȝ. 7908, 10504; twiȝen HOM. I. 37.

twies, adv.,= M.L.G. twīes, twiges, M.H.G. zwīes; twice, HOM. I. 227; A. R. 70; twies LANGL. C viii. 29; MAND. 261; tvies AYENB. 35; twiges SAX. CHR. 248; twiȝes (ms. twiȝ-ȝess) ORM. 566; twiȝes WILL. 3721; E. G. 357·

twīg, twĭg, sb., *O.E.* twīg,=*M.Du.* twijgh, *M.L.G.* twīch, *O.H.G.* zvīg *n.; twig, branch,* CH. C. T. *I* 391 ; tvig AYENB. 22 ; twigge PR. P. 505 ; twi MAT. xxiv. 32 ; twigga (*pl.*) HOM. I. 5 ; twigges PS. lxxix. 11 ; DEP. R. iii. 87 ; twīge (*dat. pl.*) HOM. I. 149 ; *comp.* **līm-, palm-twīg.**

twiȝe, *see* **twíe.**

[**twih,** adv., *O.E.* twih ; *comp.* **bi-twih.**]

twin, sb., *O.N.* tvinnr, tvennr ; *twin, two* : þis twinne seolþe ORM. 8769 ; twinne srud GEN. & EX. 2367 ; **in twinne** *in two* S. S. (Wr.) 1499 ; **a twinne** *asunder* JOS. 49 ; o twinne REL. I. 214.

   twinling, sb., ? = *O.H.G.* zvinelinc ; *twin,* '*gemellus,*' VOC. 213 ; PR. P. 505 ; **twinlingis** (*pl.*) WICL. GEN. xxv. 24.

twīn, sb., *O.E.* twīn,=*M.Du.* twijn ; *twine, doubled thread,* PR. P. 505 ; a **twines** (*gen.*) þræd LAȝ. 14220 ; twines þreed CH. C. T. *A* 2030.

twinclen, v., *O.E.* twinclian ; *twinkle* ; **twincleþ** (*pres.*) WICL. PROV. vi. 13 ; **twinkled** (*pret.*) CH. C. T. *A* 267.

   twinclere, sb.,=*M.H.G.* zwinkeler ; *twinkler,* WICL. ECCLUS. xxvii. 25.

   twinkling, sb., *twinkling,* CH. COMPL. M. 222 ; tvinkling REL. II. 14 ; twinkelinge PR. P. 505 ; N. P. 50.

twingen,v.,=*O.Fris.*twinga,dwinga,thwinga, *O.L.G.* thvingan, ? *O.N.* þvinga, *O.H.G.* dwingan ; *press, oppress* ; **twinges** (*pres.*) ['*affligit*'] PS. xli. 10 ; **twungen** (*pple.*) PS. xxxvii. 9* ; *deriv.* **twengen.**

twīnien, *see* **twēonien.**

twīnin, v.,=*M.Du.* twijnen ; *twine, twist,* '*torqueo,*' PR. P. 505 ; **twīneþ** (*pres.*) REL. I. 240 ; **twīnande** (*pple.*) A. P. ii. 1691 ; **twined** (*pple.*) LAȝ. 14220* ; LANGL. *B* xvii. 204.

twinkin, v.,=*M.H.G.* zwinken ; *wink,* '*nicto,*' PR. P. 505 ; **twinke** (*pres.*) GAM. 453.

twinne, sb., *O.E.* getwinn *m.; twin,* '*gemellus, geminus,*' PR. P. 505 ; **twinnes** (*pl.*) TREV. I. 211 ; *comp.* ȝe-**twinne.**

twinnen, v., *twin, divide, separate,* A. R. 332 ; H. M. 27 ; twinnen PL. CR. 496 ; twinne WILL. 1572 ; LIDG. M. P. 247 ; P. R. L. P. 109 ; twinne [tvinne] C. M. 9634 ; **twinne** (*pres. subj.*) HOCCL. i. 17 ; er ȝe twinne CH. C. T. *G* 182 ; þa twin(n)eden here þonkes (= *quarrelled*) LAȝ. 3791 ; **tvinned** (*pple.*) TRIST. 2694 ; *comp.* **a-, to-twinnen.**

   twinnunge, sb., *O.N.* tvenning ; *separation,* A. R. 396 ; twinninge H. M. 27.

twinsien, v., ? *retreat* ; **twinseden** (*pret. pl.*) cnihtes LAȝ. 4236.

twint, sb., ?*M.Du.* twint ; *jot* : had(de) nat a twint BER. 433 ; twinte DEP. R. iii. 81.

twisel, adj.,=*O.H.G.* zvis(e)l ; *double* : twisil tunge WICL. ECCLES. v. 16.

twiselen, v., *divide, fork* ; tunge **twiselende** (*pple.*) *cloven tongues* (*of fire*) HOM. II. 117.

twist, sb.,=*M.Du.* twist ; *twist, branch,* PR. P. 505 ; **twiste** (*dat.*) CH. C. T. *E* 2349.

twisten, v.,=*M.L.G., M.Du.* twisten ; *strip boughs; twist* ; twiste CH. C. T. *F* 566 ; '*defrondare*' CATH. 399 ; **twisteð** (*pres.*) *is nauseated* HOM. II. 213 ; hi tvisteþ ine tvo AYENB. 159.

twister, sb., *stripper of boughs,* CATH. 399.

twiteren, v.,=*O.H.G.* zvizarōn ; *twitter* : **twitereþ** (*pres.*) TREV. I. 237 ; þilke brid . . . twitriþ [twiterith] CH. BOET. iii. 2 (68).

twix, adv., *O.E.* (be-)twix,=*O.L.G.* tvisc, *O.H. G.* zvisk ; *among* : tvix (? *for* atvix) þe thorns C. M. 3179 ; *comp.* **a-, bi-twix.**

twō, *see* **twēȝen.**

<hr>

þ.

þā, *see under* **þe.**

þā, adv., *O.E.* þā,=*O.N.* þā, *O.Fris.* thā, *O. L.G.* thō, *O.H.G.* dō ; *then, when,* KATH. 24 ; ORM. 453 ; þa ȝet seiþ þeo soule FRAG. 7 ; þa ȝet HOM. I. 31 ; þa þe heo comen 87 ; þa ða he wolde 219 ; þa ða 223 ; þa ðe 227 ; þaa PERC. 497 ; þa wes Turnus sari LAȝ. 166 ; þa [þo] al þis wes idon 849 ; þæ (? *for* þa) 1253 ; þeo 3037 ; þo HOM. II. 5 ; FL. & BL. 53 ; HAV. 291 ; BRD. 10 ; AYENB. 141 ; WILL. 1865 ; HOCCL. i. 12 ; L. C. C. 11 ; þo stod on old stoc þar bi side O. & N. 25 ; þo heo bet do ne mihte 1070 ; þoo, þo LANGL. *A* ii. 119, *B* ii. 148 ; þo [tho] CH. C. T. *A* 993 ; ðo REL. I. 209 ; GEN. & EX. 717 ; tho M. ARTH. 249 ; þo, þeo, þoa, þeoa A. R. 52, 78, 236, 314.

þæ, *see* **þe.**   **þær,** *see* **þār.**

þēw, *see* **þēau.**   **þēwen,** *see* **þēwen.**

þáflen, *see* **þávien.**

þāh, adv. & conj., *O.E.* þēah, þēh,=*O.Fris.* thāch, *O.L.G.* thōh, *O.H.G.* dōh, *Goth.* þauh, *O.N.* þō ; *though,* MK. xiv. 31 ; SPEC. 29 ; no wonder þah (*ut*) me be wo 81 ; þah he muche þolie KATH. 229 ; as þah (*tamquam*) þe almihti ne nihte . . . 987 ; þah MARH. 8 ; þe(h)hweðere 11(47) ; as hi þah ledað to deðe HOM. I. 119 ; þah hweðre his saule wes in helle 131 ; ase þah he saide 189 ; þach 147 ; þaȝh 15 ; þauh 203 ; þech 171 ; þah, þaih, þeah, þeih [þoh, þeh] LAȝ. 244, 2345, 2513, 4039 ; al se þæh (? þæh) he ne mihte libben 6702 ; nute we . . . þeh he heo nabbe to wife 25678 ; þah, þaih, þeh, þeih, þeȝ O. & N. 811, 1235, 1425, 1724 ; þaȝ MIRC 214 ; GAW. 350 ; AUD. 52 ; þaȝ þer bi ȝome bronches þet ne bieþ naȝt diadlich(e) zenne AYENB. 9 ; þaȝ es þe wone is kveadvol 6 ; þagh L. C. C. 5 ; heo hit mai don þauh A. R. 6 ; þauh hit þunche attri hit is þauh healuwinde 190 ; ase þauh hit were 338 ; þauh [þauȝ] C. L. 1296 ; þauh

[þei3] LANGL. *A* i. 10; þau, þei MAN. (F.)
4878, 4879; þe3 FL. & BL. 62; TREAT. 134;
þe3h ANGL. I. 7; þe3h, þei3h, þei WILL. 451,
919, 1563; þei ROB. 30; E. G. 350; SAINTS
(Ld.) xlvi. 493; hit nis no wonder þei me be
wo ANGL. II. 254; nis no sellic þei heom
beo wo P. L. S. viii. 92; þoh wethere SAX.
CHR. 261; it was þoh (*ms.* þohh) ful mikel
riht ORM. 23; þoh swa þeh (*ms.* þehh) *notwithstanding* 249; þoh wheþre 310; ᵹog
GEN. & EX. 4; þogh EGL. 727; thogh [þough,
þouh] CH. C. T. *A* 253; þou3 E. G. 75; þou
HAV. 299; TOR. 1020; SACR. 702.

þak, sb., *O.E.* þæc,=*O.N.* þak, *O.H.G.* dach;
thatch, roof, VOC. 237; PR. P. 490; þakkes
(*pl.*) MAN. (F.) 14689.

þāke, adj., ? *O. E.* þāce; *loose-textured,
flabby*; thoke '*insolidus*' PR. P. 491.

þakken, v., *O.E.* þaccian; *pat, stroke*; þak-
keþ (*pres.*) P. L. S. xxxv. 142; þakked
(*pple.*) CH. C. T. *A* 3304.

þakkin, v.,=*M.H.G.* dachen; *thatch*; thakkin
PR. P. 490.

þan, conj., *O.E.* þonne, þanne (SAX. CHR. 178),
ᵹon (LK. xii. 23)=*O.L.G.* than, *O.H.G.*
danne, denne; *than,* ORM. 1761; þe leste
steorre is ... more þane al þe eorþe SAINTS
(Ld.) xlvi. 398; ᵹan HOM. I. 223; þan, þanne
SAX. CHR. 264, 265; þan, þon, þane, þene O. &
N. 24, 39, 505, 564; ᵹan, ᵹanne GEN. & EX.
142,144; þane, þanne, þonne, þene, þenne [þan,
þane] LA3. 3014, 3453, 6515, 6903, 8916; þanne
MAT. vi. 25; AYENB. 16; þanne, þænne P. L.
S. viii. 1; þenne HOM. I. 27; biᵹ ... fulre
þene he wes 25; er þonne, þon 37, 93; þen
FRAG. 7; KATH. 170; SPEC. 38; JOS. 592;
er þen C. L. 492; þen, þene, þenne A. R. 8,
84, 108.

þan², conj. & adv., *O.E.* þon, þonne, þanne,
þænne,=*O.L.G.* than, *Goth.* þan, *O.H.G.*
danne, denne; *then, when*; Iw. 1804;
PR. C. 4712; hie lieᵹ þan (*when*) hie Crist
loverd clepieᵹ HOM. II. 21; þan, þane, þanne
ORM. 469, 557, 8401; þan, þanne HAV. 156,
232; ᵹan sal him (god) almightin luven GEN.
& EX. 9; ᵹan (*when*) man hem telleᵹ soᵹe
tale 17; ᵹanne 999; þane, þon, þonne, þenne
[þanne] LA3. 711, 1546, 3413, 3616, 25799;
þane (*when*) he cumeᵹ HOM. II. 5; þanne
SAX. CHR. 265; P. L. S. viii. 3, 20; SAINTS (Ld.)
xlvi. 397; AYENB. 128; WICL. JOHN xii. 16;
þanne, þonne, þenne O. & N. 508, 822; þonne
FRAG. 5; þanne [þenne] LANGL. *A* i. 56;
CH. C. T. *G* 1193, 1211; þenne HOM. I.
133; SPEC. 37; H. H. 80; þenne, þen KATH.
373, 562; þene, þeone, þenne, þeonne A. R.
42, 46, 108, 218.

þanc, sb., *O.E.* þanc, þonc,=*O.L.G.* thanc,
Goth. þank, *O.H.G.* danch; danc; *thanks,
thought, favour,* P. L. S. viii. 36, 45; he cuᵹe
him ᵹer of wel gret ᵹanc (*ms.* ᵹhanc) GEN. &
EX. 1659; þanc, þonc LA3. 4360, 5068; þank

AYENB. 18; GOW. I. 66; MAND. 285; þank
[thank] CH. C. T. *A* 1808; god þank HAV.
2005; þank have 3e TRIST. 2081; þonc A. R.
222; ROB. 485; A. P. i. 900; nenne þonc
HOM. I. 137; hire þonc wenden JUL. 14;
gode ᵹonc *thanks to God* HOM. II. 11; þonc,
þonk O. & N. 461, 490; þonk REL. I. 109; GAW.
1380; his þankes [=*O.H.G.* sines danches]
(*gen.*) *willingly* CH. C. T. *A* 1626; his
þonkes HOM. I. 31; BEK. 291; hire þonkes
O. & N. 70; here þankes SAX. CHR. 265;
MAT. xvi. 8; on gode þanke HOM. II. 5;
habben in his þonke HOM. I. 29; wende on
his þonke þat hit were for unᵹeawe LA3. 3063;
þankes (*pl.*) MAT. xv. 19; þonkes LA3. 3791;
AYENB. 18; gode þonkes A. D. 287; *comp.* 3e-,
ümbe-, un-þanc.

[**þanc-ful**, adj., *O.E.* þancfull; *thankful.*]
þankful-līche, adv., *thankfully,* P. S. 156.

þanc, see þan.

þanen, adv., *O.E.* þanon, þanone,=*O.Fris.*
thana, *O.L.G.* thanan, *O.H.G.* danan, dannan;
thence, MAT. iv. 21; þanen hit was ibroht up
into heofene HOM. I. 241; þanene, þonene,
þeonene, þonne, þeonne, þenne [þanene, þanne]
LA3. 235, 654, 1297, 5971, 16257, 31362;
þanene LEG. 49; þanene, þanne ROB. 288;
377; þat comen þanne P. L. S. viii. 71; þonne,
þenne, O. & N. 132, 1726; þenen, þeonen SAX.
CHR. 249, 251; þenene REL. I. 282; hwa
hefde ... þat licome ilad þeonne KATH. 2233;
þenne HAV. 1185; H. H. 5; JOS. 25; RICH.
2947; þeonne LANGL. *A* i. 71; þennes LANGL.
B i. 73; þannes ANGL. I. 15; BEK. 1141;
AYENB. 12; þennes WILL. 2191; CH. C. T.
B 510.

þene-ward, adv., *thence,* H. M. 43; þeone-
ward A. R. 296.

þank, see þanc.

þankien, v., *O.E.* þancian, þoncian, *cf. Goth.*
þagkjan, *O.L.G.* thancōn, *O.H.G.* danchōn,
O.N. þakka; *thank*; þanken CH. TRO. ii.
848; (*ms.* þannkenn) ORM. 3900; þanke S. S.
(Wr.) 2421; þonkien HOM. I. 5; we sculan
þonkian him þere (*thank him for the*) muchele
mildheortnesse HOM. I. 121; þonkie MISC. 81;
þonke WILL. 3522; þankie (*pres.*) LA3. 3534;
ich hit þonkie þe HOM. I. 191; þonki KATH.
2415; BRD. 3; AYENB. 6; þe ich þonki ...
alle þine deden JUL. 61; ᵹanc (*imper.*) GEN.
& EX. 1320; þonkeᵹ A. R. 430; þonke (*subj.*)
A. R. 256; þankede (*pret.*) ISUM. 501; he
þonkede hire LA3. 1261; iþonked (*pple.*)
LA3. 20827.

þanking, sb., *O.E.* þancung; *thanking,* PR.
C. 7842; þankingis (*pl.*) ['*gratias*'] WICL.
2 KINGS viii. 10.

þanne, see þan, þanen.

þar, see þarf.

þār, adv., *O.E.* þǣr, þār,=*O.Fris.* thēr, *O.N.,
Goth.* þār, *O.L.G.* thār, *O.H.G.* dār, dāra;

there, MISC. 154; ALIS. 2130; PR. C. 361;
EGL. 341; þar, þare, þær, þære SAX. CHR.
251, 255, 261, 263; þar, þare, ᵹere HOM. II.
5, 7, 171; þar, þear, þær, þer, þare, þære, þere,
LAȝ. 8, 123, 259, 607, 716, 1830, 3378, 27332;
þar, þare O. & N. 26, 295; þar, þor, þer FER.
102, 150, 544; þær FRAG. 5; ORM. 55; þær,
þære JOHN xi. 31, 32; þer HORN (L.) 523;
C. L. 491; ROB. 69; AYENB. 1; þer þe heo
beoᵹ HOM. I. 9; & wunede sum hwile þer
[þear] KATH. 8; þer, þere P. L. S. viii. 22, 50;
A. R. 68, 80; WILL. 216, 1627; CH. C. T. A
43; *C* 689; ᵹor GEN. & EX. 211; þor, þore
HAV. 922, 1044; ᵹore REL. I. 211; þore H.
H. 63; RICH. 316; MAN. (F.) 2930; **þēr**
aboute A. D. 263; WILL. 972; **þēr æfter**
thereafter LAȝ. 1220; þar after O. & N. 393;
þer after LANGL. *A* v. 137; ᵹor after GEN. &
EX. 146; þrafter KATH. 191; **þēr an** *thereon*
HORN (R.) 573; þer an [þar on] LAȝ. 7275;
þar Rome nou on stondeᵹ 107; þer on A. R.
178; þar on, one O. & N. 104, 1240; **þær**
one ORM. 957; þer one HOM. II. 89; **þær**
onȝæn ORM. 5304; þer aȝen BEK. 294; H. H.
86; þer a mong R. S. ii; þer ate FL. & BL.
138; **þēr bī** *thereby* A. R. 160; H. M. 23;
þe abbot þat þer bi stod CHR. E. 763; þer
bifore P. S. 221; þer bivore MISC. 43; þer
buve FL. & BL. 294; **þær fore** *therefore*
ORM. 2431; þer fore LAȝ. 316; KATH. 301;
þere for HOM. I. 135; þar fore O. & N. 758;
þer vore AYENB. 6; ᵹor fore GEN. & EX.
1215; **þer fram** BRD. 24; þar from O. & N.
137; ᵹor gen GEN. & EX. 2797; þerien HAV.
2271; **þær inne** *therein* ORM. 1651; þer
inne A. R. 352; TREAT. 138; þat folc þer inne
LAȝ. 642; þer inne, þrinne SHOR. 9; þrinne
JUL. 30; þrin LAȝ. AYENB. 34; **þar mid** O. & N.
81; þer mid P. L. S. xix. 169; þer mide A. R.
150; C. L. 374; WILL. 5358; ᵹor mide GEN.
& EX. 2656; **þār nēh** HOM. I. 43; **þēr of**
thereof FRAG. 6; þar of O. & N. 120; þer
of H. M. 25; LANGL. *B* xiii. 420; þer offe
HOM. II. 167; ᵹor of GEN. & EX. 234; þrof
KATH. 818; SHOR. 6; **þār over** O. & N.
1136; þer over BRD. 27; **þōr til** HAV. 1443;
þēr tō *thereto* A. R. 6; BEK. 37; AYENB. 8;
þar to þu stele in o dai O. & N. 103; ne schal
þar nevre cume to 611; **þēr tōȝeines** A. R. 80;
þer teyens AYENB. 11; **þær þurh** ORM. 2325;
þar þurh O. & N. 1558; þer ufenen LAȝ. 17696;
þēr under *thereunder* S: S. (Wr.) 2055; ᵹor
under GEN. & EX. 3184; þer **uppe** BRD. 6;
þer uppe [þruppe] A. R. 42; þer oppe SHOR.
2; þer uppon HOM. I. 53; þer upon MARH.
21; **þēr ūti** MK. iii. 31; þer ute LAȝ. 1179;
Peter stod þer ute MISC. 43; þar oute Iw.
2156; **þēr wiᵹ** *therewith* MARH. 15; þer
wiþ BEK. 272; LANGL. *A* i. 16; **þēr wiᵹ-**
inne H. M. 11.

þāre, *see* **þār.**

þarf, v., *O.E.* þearf, = *O.N.*, *Goth.* þarf, *O.L.G.*
tharf, *O.H.G.* darf, (*pret. pres.*) *it is required;*

it behoves; (he) needs; HOM. II. 69; H. M. 5;
REL. I. 174; O. & N. 803; TRIST. 3053; (*ms.*
þarrf) ORM. 12886; þarf, þerf P. L. S. viii. 22,
23; þerf A. R. 192; þarf [þar] MAN. (F.) 7237;
darf ANGL. I. 9; FL. & BL. 315; SHOR. 64;
derf HOM. I. 187; þar CH. C. T. *A* 4320;
thar GAW. 2355; PR. C. 2167; TOWNL. 14;
ne þar [dar] he seche non oþer leche C. L.
733; þer P. L. S. xvii. 516; dar OCTOV. (W.)
1337; þearft þu FRAG. 6; þerft HOM. I. 37;
þērf(t) tu, þer(t) tu A. R. 136; þart LEG. 40;
dert [þert] LAȝ. 22923; þat tu wenen ne þarf
[þerf] KATH. 1160; þarst GOW. II. 61; þerst
BRD. 29; tharst þou [darstou] nevere care
LANGL. *B* xiv. 55; þurfen ȝe LAȝ. 18042;
þurven A. R. 6; þurve HOM. I. 253; (we)
þorven LEG. 211; (ȝe) þore P. L. S. xxi. 158;
þurfe (*subj.*) ORM. 7766; þurve A. R. 172;
MISC. 75; þat tes unseli ne þurve nawt seggen
JUL. 69; *comp.* **be-þarf;** *deriv.* **þarfe, þurfen.**

þarfe, sb., *O.E.* þearf, = *O.L.G.* therva, *O.N.*
þörf; *need:* alle þa þat hafden ned & þarfe
to þin heipe ORM. 12247; nir hit nan þerf
HOM. I. 9.

þarfe², sb., *O.E.* þearfa; *poor person;*
þærfa (*printed* wærfa) FRAG. 2; ic em
þarva & wrecche HOM. I. 115; **þearfen** (*dat.*
pl.) MAT. xix. 21.

þarm, sb., *O.E.* þearm, = *O.N.* þarmr, *O.Fris.*
therm, *O.H.G.* darm; *bowel,* MISC. 151;
tharm VOC. 247; PR. P. 490; **þærmes, þer-**
mes [þarmes] (*pl.*) LAȝ. 318, 18451; þermes
FRAG. 6; þarmes H. M. 35; þarmis H. S. 702.

þarnen, v., *O.N.* þarna, þarfna; *lack, lose:*
þat ilke þing þat tu ful wel ne miht te self
noht þarnen (*ms.* þarrnenn) ORM. 10142;
þarne HAV. 2835; tharne PR. C. 8509; AV.
ARTH. lxvi; TOWNL. 126; **þarnes** (*pres.*)
ALEX. (Sk.) 2709; **tharned** (*pret.*) ALEX.
(Sk.) 3071.

þat, *see under* **þe.**

þávien, v., *O.E.* þafian; *permit;* þeavien
(*ms.* þeauien) JUL. 18; ᵹaven GEN. & EX.
3139; þave HAV. 2696; **þávieᵹ** (*pres.*)
ant þolieᵹ MARH. 15; ne þave þu þat storm
me duve HOM. II. 43; þat god ne þole noht
ne þafe ORM. 5457; *comp.* **ȝe-þávien.**

þáfunge, sb., *O.E.* þafung; *permission,* A.
R. 344*.

þau, þauȝ, þauh, *see* **þāh.**

þāwe, sb., *O.N.* þá; *thaw, melting;* thowe
PR. P. 492.

þāwen, v., *O.E.* þāwian, *cf. O.N.* þeyja, *M.L.*
G. dāwen, douwen, *O.H.G.* do(u)wan, de(u)-
wan (*digest*); *thaw, melt;* thowin PR. P.
492; **thōwes** (*pres.*) VOC. 201.

þe, dem. adj., dem. & rel. pron., *O.E.* þe, = *O.L.*
G. the, *O.Fris.* the, thi, *O.H.G.* der; *the, that,*
who, which, HAV. 9; AYENB. 5; þeman LEECHD.
III. 98; þe creft FRAG. 1; þe man þe wule
siker ben P. L. S. viii. 20; þe ᵹe he *who* HOM.

I. 109; þe king LAȝ. 126; þæ 6153; þa [þe] dai 1327; þe is al so federleas A. R. 10; þe þe 86; Moises . . . þe havede þe lawe to ȝeme H. H. 24; ꝺe GEN. & EX. 47; þerl (=erl) GREG. 1; his broꝺer þa wæs eorl of Norꝺhamtune(s)scire SAX. CHR. 253; þa (*for* þeo) wæte LEECHD. III. 82; þe (*for* þat) land SAX. CHR. 262; þe bred HOM. II. 97; þe child LAȝ. 295; þe lond KATH. 49; þe water A. R. 72; þe liht O. & N. 734; þe folc ORM. 141; þe child ROB. 11; after þan þe he haved idon P. L. S. viii. 87; þe [þat] we swa take him on LAȝ. 3333; wiꝺ þon þa [þe] he mote libben 886; and wiste wel . . . þe wraþþe binimeþ monnes red O. & N. 941; þe (*for* þes) kinges SAX. CHR. 264; þe mannes HOM. II. 73; þe deofles HOM. I. 223; A. R. 268; þe gostes O. & N. 1398; þe (*for* þere) assa fet HOM. I. 5; bi þe tunge 41; to þe (*for* þam) kinge KATH. 1307; ine þe londe HOM. I. 151; of þe blod ORM. 1070; þæt bed þe se lame on laig MK. ii. 4; þe (*for* þene) SAX. CHR. 262; ·KATH. 117; bihealden þe forbodene appel HOM. II. 35; to þan ilke weie þe he ful ȝeare wuste LAȝ. 525; þe dom ORM. DEDIC. 75; me taȝte hire þe wei BEK. 75; þene fule onkume . . . þa þe d(e)ovel haveꝺ in (e)ow ibroht HOM. I. 149; þe (*for* þa) teþ LEECHD. III. 104; þe wilde deor HOM. II. 35; þa þe þæt word gehereꝺ MK. iv. 18; þe þeo sunne wrohten FRAG. 7; we þe . . . misdoꝺ HOM. I. 173; alle þa þe wolden . . . LAȝ. 20479; bi þe (*for* þam) eȝen HOM. I. 41; þe hi hit bitechan willaꝺ ANGL. VII. 220; þe þe was to l(e)of wre(c)che men to swenchen HOM. I. 175.

þēo (*fem.*) [*O.E.* þēo (=*O.L.G.* thiu, *O.H.G.* diu)] JOHN xix. 20; þeo, þe FRAG. 5; þæs deofles lore þeo (*qvae*) þe likede wel 7; þeo, þo, þe HOM. I. 83, 87, 89; þeo, þe, þæ LAȝ. 4010, 9815, 12005; þeo deꝺ al so A. R. 52; þeo heorte 282; þe (? þē) heorte 50; þ(e)o, þe ule O. & N. 26, 187; þe qvene ROB. 26; þe þrote AYENB. 14; a . . . chaumber þe clerli was peinted WILL. 4422; to þe (þe ? *for* þere) eorꝺe MARH. 9; to þe soule biheve A. R. 388; þe (þe ? *for* þa) speche O. & N. 13.

þæt (*nom. acc. n.; also as conj.*) [*O.E.* þæt, = *O.N.* þat, *O.L.G.* that, *Goth.* þata, *O.H.G.* daz, *mod.Eng.* that] LK. i. 21; PROCL. 2; þet SAX. CHR. 249; þæt, þat P. L. S. viii. 4, 7; þat LAȝ. 297; þet 1239; þæt [þat] weder 7843; þæt [þat] he cume swiꝺe 11500; hit is feole ȝere þat heore þrættes comen here 26294; þat KATH. 70; H. M. 3; O. & N. 1259; þat fur R. S. iv; þat heved ROB. 186; þat folc MARG. 250; þat child HAV. 575; SHOR. 12; þat gode hors FER. 244; þou hast duere aboht þat þou levedest me noht H. H. 60; þet FRAG. 6; JUL. 39; AYENB. I; þet beoꝺ ure eȝan HOM. I. 127; al þet me eaver deꝺ A. R. 4; þet

water 314; þat (*for* þe) gode man ORM. 2105; þat oþer knight CH. C. T. *A* 1014; mon . . . þet [þat] þeos boc rede LAȝ. 58; þe þet is idel A. R. 212; þat (*for* þeo) burde WILL. 683; þat (*for* þes) folkes ORM. 1689; after þat (*for* þam) 6792; alle þet (*for* þeo) luvieꝺ þe A. R. 282; þe þat seggeþ SHOR. 99; þut lond ROB. (W.) 5077, 12014; þut water 5354; þut scoble 4237; þut on, þe oþer 10671; he was glad of þut cas 6773.

þæs (*gen. m. n.*) [*O.E.* þæs, = *O.L.G.* thes, *O.N.* þess, *Goth.* þis, *O.H.G.* des, *Gr.* τοῦ], FRAG. 5; þæs þe ma MK. x. 48; an man þæs name wæs Jairus LK. viii. 41; þas folkes i. 10; þæs þes SAX. CHR. 249; þæs kinges LAȝ. 806; þes [þe] kinges sune 332; wa wes Lumbardisce folc þes [þas] 2745; þas (? *for* þæs) kinges ferde 713; þeos [þes] sweordes 7560; þes HOM. I. 5, 9; A. R. 62; MARH. 2; þes te mare ORM. 444; þes wateres MISC. 146; þes þe bet SAL. & SAT. 234; þes, þas O. & N. 338, 822, 882, 1442; wel was him þas P. L. S. xii. 122; Oliver wax hol sone þas FER. 1387.

þēre (*gen. f.*) [*O.E.* þǣre] HOM. I. 5; LAȝ. 331; þare LK. i. 10; þare ule O. & N. 28; þer saule SHOR. 4; þer hvile AYENB. 217.

þam (*dat. m. n.*) [*O.E.* þǣm, þam, þan] MAT. ii. 8; after þan LK. x. 1; þam HOM. I. 37; þan 17; þon 41; to þa deꝺe 121; þan PROCL. 5; on þan londe LAȝ. 127; **under þan** (*meanwhile*) 9660; nas he noht to þan iboren 30779; hit was . . . iseid þon kinge (*to the king*) 8001; **wiꝺ þon** þe þu him ȝeve griꝺ 8253; he redde þæn [þan] kæisere 9266; þen kinge 1626; **for þan** *because, therefore*, REL. I. 128; for þan þe HOM. I. 225; for þan com ic of hefne dun ORM. 17621; mid þan [þon] O. & N. 801; at þan [þen] ende 1288, 1514; with þan þu wilt his child take HAV. 532; fram þan time SHOR. 123; **er þan** AM. & AMIL. 2398; **bī þan** he com bi þat barn WILL. 220; ief to god be þan þet he heþ þe AYENB. 195; inne þo time 12; at þo daie 14; to ꝺan GEN. & EX. 2792; bi ꝺan sal Sarra selꝺe timen 1023; for ꝺo 1046; on þen flore FRAG. 5; of þen epple A. R. 66; wiꝺ þen þet 284; þat reafde þen [þe] riche Job his ahte JUL. 40, 41; to þen inne AN. LIT. 3; at þen ende TRIST. 3287; at ten (*for* at þen) ende H. M. 7; R. S. vii; atte (*for* at þe) LANGL. *A*, *B* PROL. 42; at ten ale CH. C. T. *D* 1349; to þen unwiht MARH. 12; wiꝺ þon þat 5; after þon *afterwards* CHR. E. 303; bi þon C. L. 556; under þon M. T. 107; in þa tune SAX. CHR. 256; in þa time MISC. 29; þo tides of þo daie 34; for þo þe HOM. II. 25; or þo CH. D. BL. 234.

þēre (*dat. f.*) FRAG. 6; þære, þare LAȝ. 93, 1233; on þere helle grunde P. L. S. viii. 149; on þere alde laȝe HOM. I. 87; bi þer heorte 41; in þar sæ 143; þare JOHN xvii. 11; to þare wunda LEECHD. III. 86; of þare ule O. & N. 31 ; in þare tide SHOR. 86 ; þere A. R. 54 ; to þer eorðe 18.

þē (*instr. neut.*), [*O.E.* þē, þȳ, *Goth.* þe, *O.L.G.* thiu, *O.H.G.* diu, *O.N.* þvi] ; þe, þæ bet LAȝ. 701, 30597 ; for þi 26134 ; þe bet A. R. 58 ; for þi, þui 6, 152 ; þe wurs O. & N. 34 ; nartu þe wisere 1330 ; for þe, þi 65, 69 ; ich rede þi þat men beo ware 860 ; þe bet BRD. 7 ; þe leng þe more BEK. 46 ; þe bet CH. C. T. *D* 1951 ; for þi A. 1841 ; for þui (*printed* forþ *in*) FRAG. 2 ; þi HOM. I. 93 ; þi les 117 ; to ði 223 ; þi læs MAT. xv. 32 ; þi sathanas . . . þe saule wule derie MISC. 76 ; for þi KATH. 85 ; ORM. 361 ; HAV. 1194 ; C. L. 85 ; PR. C. 189 ; DEGR. 29 ; ANT. ARTH. xxxiv ; for ði GEN. & EX. 1581 ; wi(þ) þi MAN. (F.) 8997.

þene (*acc. m.*) [*O.E.* þæne, þane, þone] FRAG. 1 ; HOM. I. 31 ; P. L. S. viii. 171 ; S. A. L. 157 ; þene, þen A. R. 10, 52 ; MARH. 11 ; KATH. 1188, 1189 ; þane, þene, þen LAȝ. 133, 418, 3253 ; þane, þene O. & N. 249, 1093 ; þane PROCL. 7 ; SHOR. 5 ; AYENB. 7 ; þane wei BEK. 701 ; þen toun ROB. 184 ; þen appel H. H. 10 ; þan FER. 2419 ; inne þane (*for* þam) fehte LAȝ. 214 ; to þone kinge SAX. CHR. 252.

þā - (*acc. f.*) JOHN xix. 17 ; HOM. I. 37 ; þa mol(de) FRAG. 5 ; he nom þa Englisca boc þa makede seint Beda LAȝ. 31 ; þo MISC. 29 ; þa (*for* þeo) drane SAX. CHR. 256 ; þa laȝe HOM. I. 9 ; þa qvene LAȝ. 198 ; on þa (*for* þere) ealde laȝe HOM. I. 9.

þā, þēo (*nom. acc. pl.*) [*O.E.* þā, þæge, = *O.Fris.* thā, *O.L.G.* thia, thea, *Goth.* þai, *O.N.* þeir, *O.H.G.* dia, die, *mod.Eng.* they] FRAG. 1 ; þa oðre men HOM. I. 5 ; þa songes þa we nu singeð 125 ; þeo þe ihereð 47 ; þa, þæ, þea, þo, þeo, þe, þaie, þai LAȝ. 2009, 2020, 3068, 3638, 6420, 7095, 7121, 7789, 19542, 21868 ; þa kinges ORM. 6451 ; þa þat wæren gode men 53 ; þeȝ (*ms.* þeȝȝ) ledden heore lif 125 ; þa [þai] wandes C. M. 1423 ; þa, þai PR. C. 152, 1253 ; þo ANGL. I. 13 ; HOM. II. 3 ; MISC. 27 ; AYENB. 6 ; þo þe luve-den unriht P. L. S. viii. 47 ; þ(e)o O. & N. 843 ; þeo þe haveþ . . . 1675 ; þoa þinges KATH. 360 ; þeo þat 501 ; þeo, þie HOM. II. 107 ; þo, þei WILL. 318, 1757 ; MAND. 10, 20 ; þo wordes CH. C. T. *A* 1123 ; þei [þei] were 407 ; þo, þe HAV. 1841, 2044 ; þo hote baþes ROB. 28 ; þe twei breþeren 38 ; þei were agaste 50 ; þeo BEK. 421 ; CHR. E. 492 ; þeo, þe A. R. 2, 166 ; þe fet MARH. 18 ; men þe ani god cunnen REL. I. 130 ; þei LANGL. *A* i. 8 ; þei þat nolden on me leven H. H. 235 ; alle þei þat god drede LUD. COV. 84 ; þai doȝti(e) men FER. 458 ; þa (*for* þam) ORM. 2796 ; of þa oðre SAX. CHR. 252 ; bi þa hon-den HOM. I. 41 ; of þa men KATH. 145 ; al þa Grickes þea heo neih comen LAȝ. 581 ; non of þo SPEC. 29 ; oon of þoo CH. C. T. *A* 2351 ; to þeo þet tin nome munnið MARH. 21.

þāra (*gen. pl.*) [*O.E.* þāra, *O.N.* þeirra] HOM. I. 221 ; þare LK. i. 4 ; ANGL. VII. 220 ; O. & N. 1584 ; þare, þere LAȝ. 1776, 3346 ; ðere monne HOM. I. 135 ; þair PR. C. 52 ; þeȝre ORM. 127 ; eȝþer þeȝres 2508.

þām (*dat. pl.*) [*O.E.* þām, þǣm] MK. v. 2 ; P. L. S. viii. 134 ; PR. C. 202 ; ISUM. 148 ; þam, þan HOM. I. 93 ; þa ilke wepne þa þe apostel spekð of 155 ; þeȝm ORM. 1142 ; þan MAT. v. 44 ; O. & N. 1762 ; PROCL. 3 ; bi þan dagen HOM. II. 47 ; þan, þon, þen LAȝ. 714, 747, 7806 ; þan, þon, þe AYENB. 11, 30, 139 ; þon HOM. I. 175 ; to þen eien A. R. 50 ; ich am on of þe A. D. 253.

þe [2], conj., *O.E.* þe, = *O.L.G.* the, *O.Fris.* tha, *Goth.* þau ; *either, or,* O. & N. 824, 1064 ; weþer est þe west ROB. 220 ; þe, þa LAȝ. 1418, 16812.

þe [3], conj., *? O.E.* þe, = *O.Fris.* tha ; *for þan, than,* HOM. II. 119 ; if ȝe beoð strengre þe heo I. 151 ; na mo þe [*v. r.* þene] O. & N. 564.

þe [4], conj., *? O.E.* þe ; *for þā ; then, when* : þe þe (*earlier text* ða þā) he was '*cum esset*' FRAG. 1 ; þe ȝet þa ȝet 7 ; þe ȝet LAȝ. 263 ; MARH. 1 ; O. & N. 1624 ; þe [*sec. text* þō] Dunwale havede isæd LAȝ. 4150 ; þa þe HOM. I. 87 ; þo þe he him shop HOM. II. 35 ; þa þæ he hine beseag MK. viii. 24.

þe [5], adv., *? O.E.* þe ; *for þār* ; þer ðe (*where*) nevre deað ne com HOM. I. 193.

[þē-, *stem of* þēon.]

þē-dōm, þēodōm, sb., *prosperity, thrift,* LANGL. *A* x. 105, *C* viii. 53 ; thedom S. S. (Web.) 587 ; evil thedom LUD. COV. 139 ; þedam '*vigencia*' PR. P. 490.

þē, *see* þe, þeoh.   þēah, *see* þāh.

þearf, *see* þarf.

þēau, sb., *O.E.* þēaw, = *O.L.G.* thau, *O.H.G.* dau ; *manner, virtue* : þes þeau A. R. 278 ; þeau, þeu HOM. II. 47 ; theu (*ms.* thew) '*mos*' PR. P. 490 ; PS. lxiv. 7 ; þæw JOHN xix. 40 ; þurh haliȝ þæw ORM. 6754 ; of maine and of þēauwe (*dat.*) LAȝ. 6361 ; þēawes (*pl.*) A. R. 240 ; H. M. 3 ; ivele þeawes HOM. II. 71 ; þeawes AYENB. 17 ; þeauwes, þewes, þæwes [þeuwes] LAȝ. 2147, 6899, 7161 ; gode þewes O. & N. 1017 ; þewes C. L. 763 ; WILL. 189 ; MIRC 1482 ; A. P. ii. 1436 ; CH. C. T. *E* 409 ; fele þewes REL. I. 109 ; thewes PR. C. 1883 ; þewis LIDG. M. P. 256 ; M. ARTH. 1081 ; AMAD. (R.) xxxix ; *comp.* fēond-, lēod-, man-, un-þēau.

þēau-ful, adj., *moral, virtuous* : þ(e)awfulle mihtes H. M. 45 ; mid þeauful(l)e talen A. R. 422 ; mid þeufulle worden LAȝ. 1797 ; *comp.* un-þeufol.

þēu-lēs, adj., *immoral,* (*ms.* þeweles) P. S. 255.

   thowlěsnes, sb., *heedlessness,* BARB. i. 333.

þēawed, adj., *mannered ; virtuous ; so* boner

& þewed A. P. ii. 733; wel ðewed GEN. & EX. 1914; *comp.* un-þēwed.

þecchen, v., *O.E.* þeccan,=*O.N.* þekja, *O.H.G.* decchan; *thatch, cover*; þecche LANGL. *B* xix. 232; *comp.* bi-þecchen.

þeker, sb., *O.E.* þecere,=*O.H.G.* dechari; *thatcher, coverer*, VOC. 212.

þechene, sb., *O.E.* þecen; *covering*, ['*tectum*'] LK. vii. 6.

þechene [2], *for* þechele, sb., *O.E.* þecele, þæcele =fæcele, *O.H.G.* fachala, *O.L.G.* fakla, *from Lat.* facula; *torch*, LA3. 8084.

þēde, sb., *prov.Eng.* (*Norf.*) thead, fead; *brewer's wicker strainer*: hatz þou bro3t beverage in þede A. P. ii. 1717; thede, breuares instrument '*qvalus*' PR. P. 490.

þēde, *see* þēode. **þeder**, *see* þider.

þēdom, *see under* þē-.

ðēf, sb., *O.N.* þefr; *taste*, GEN. & EX. 3340.

þēf, **þĕfte**, *see* þēof, þēofðe. **þēh**, *see* þāh.

þēh, **þēi3**, *see* þēoh. **þei, þei3, þeih**, *see* þāh.

þēfe, sb., *O.E.* þýfe; *leafy branch, twig*; theve, brusch (*brushwood*), PR. P. 490.

þēfe-þorne, sb., *O.E.* þēfeþorn; *buckthorn*, ['*rhamnus*'] HAMP. PS. lvii. 9*; þeveþorn WICL. JUDG. ix. 14; þefþorn REL. I. 37.

þein, sb., *O.E.* þegen, þegn, þēn,=*O.N.* þegn, *O.L.G.* thegan, *O.H.G.* degan; *thegn, thane; man, soldier; 'minister, satrapa'*, FRAG. 2, 4; LA3. 1584; thein HAV. 2466; þeign, þen MK. ix. 35, x. 43; þeinas (*pl.*) ANGL. VII. 220; þeignes MAT. viii. 8, 9; þeines MARH. 17; R. S. v; P. S. 217; CHR. E. 583; ðeines, þeignes SAX. CHR. 250, 255; þeigne (*gen. pl.*) HOM. I. 229; þeignen (*dat. pl.*) MAT. xxviii. 12; HOM. I. 231; LA3. 3349; *comp.* bur-, disc-, lār-, reil-þein.

þeinen, v., *O.E.* þegnian; *be a thegn; minister, serve*; LA3. 30786; þeine (*pres. subj.*) REL. I. 181; þeignede (*pret.*) MAT. viii. 15; þeineden LA3. 24595; *comp.* 3e-þeinen.

þēninge, sb., *O.E.* þegnung; *ministration*, HOM. I. 233.

þelc, *see* þulc. **þellich**, *see* þullic.

þen, *see* þan. **þēn**, *see* þēon, þein.

þenchen, v., *O.E.* þencan, þencean,=*O.L.G.* thenkean, *O.N.* þenkja, *Goth.* þagkjan, *O.H.G.* denchan; *from* þanc; *think*, HOM. I. 15; KATH. 1736; A. R. 204; þenchen on his saule LA3. 18139; þenchen, þenche C. L. 17; þenche AYENB. 47; RICH. 4574; CH. C. T. *A* 3253; þenken ORM. 1761; þenke WILL. 4908; MAND. 278; thenkin PR. P. 490; þenche (*pres.*) O. & N. 485; whan i þenke [þinke] þer on LANGL. *B* v. 609; thenkeste (=thenkest þu) HAV. 578; þencheþ FL. & BL. 32; þenkes HAV. 306; 3e þencheþ BRD. 8; ðenkeð (*ms.* ðhenkeð) GEN. & EX. 2028; þenc, þench (*imper.*) LA3. 8555, 8782; A. R. 120, 122; þench on me MISC. 50; SPEC. 92; þenk on me AL. (T.) 173; þenke

(*subj.*) WILL. 711; þōhte (*pret.*) KATH. 137; SPEC. 41; (*ms.* þohhte) ORM. 2377; þohte, þoute LA3. 1255, 14410; ðogte GEN. & EX. 333; þouthe, þoucte HAV. 504, 1869; þōht (*pple.*) HOM. II. 71; (*ms.* þohht) ORM. 2364; *comp.* bi-, 3e-, ümbe-þenchen.

þenchinge, sb., *thinking*, AYENB. 6.

þenken, *see* þenchen.

þenne, *see* þan, þanen, þünne.

þennen, v., *O.E.* þennan, þenian,=*O.L.G.* thennian, *O.N.* þenja, *O.H.G.* dennan, *Goth.* (us-)þanjan, *Gr.* τείνειν; *stretch*; þinne ['*extendam*'] PS. cvii. 10.

þēo, *see* þe, þēoh.

þēode, sb., *O.E.* þēod,=*O.L.G.* thioda, *O. Fris.* thiada, *Goth.* þiuda (ἔθνος), *O.N.* þióð, *O.H.G.* diota, diot; *people, nation; country*; ['*gens*'] LK. xxi. 10; LA3. 5218; C. L. 20; K. T. 995; þeod ORM. 3438; of þissere þeode LA3. 5417; ine unkuðe þeode A. R. 250; wo þere þeode R. S. vi (MISC. 184); into þare þeode O. & N. 1583; in þeode SPEC. 23; þiode HOM. I. 237; þede HAV. 105; MAN. (H.) 19; PERC. 1255; A. P. i. 482; M. ARTH. 61; Babiloine þeo riche þede ALIS. 7959; in þe heþen þede OCTAV. (H.) 615; þēode (*pl.*) MAT. xxviii. 19; HOM. I. 15; þeode [þode] O. & N. 387; ase fele þede ase fele þewes REL. I. 109; ðeden GEN. & EX. 2302; alle þēde (*gen.*) spæches ORM. 16057; *comp.* kine-þēode.

[þēodi, adj., *from* þēode; *comp.* el-þēodi.]

þēodísc, adj.,=*Goth.* *þiudisks, *O.H.G.* diutisc; *belonging to a nation or country*: þa þeodisce men LA3. 5838; *comp.*el-þeodisc.

þēof, sb., *O.E.* þēof,=*O.L.G.* thiof, *O.Fris.* thiaf, *O.N.* þiófr, *Goth.* þiubs, *O.H.G.* diob, diub; *thief*, A. R. 174; H. M. 17; BEK. 384; þu fule þeof HORN (L.) 323; þief AYENB. 37; þuef SPEC. 106; ðef GEN. & EX. 1773; þef [thef] LANGL. *B* xii. 206; þif S. S. (Wr.) 512; þēofes (*pl.*) MAT. vi. 19; þeoves (*ms.* þeoues), þeves HOM. I. 15, 29; þeoves A. R. 174; O. & N. 1372; þiefes HOM. II. 61; thiafes ANGL. VII. 220; þefes MAND. 250; thefes PR. C. 5210; þēovene (*for* þeove) (*gen. pl.*) MISC. 39; þeove [þeve] (*dat. pl.*) P. L. S. viii. 22; þieve ANGL. I. 9.

þēof-līche, adv., *stealthily*, BRD. 13; ALIS. 4002; þeeflīche TREV. VIII. 155; thevelich, þeflīche LANGL. *B* xviii. 336 ((Wr.) 12755).

þēofðe, sb., *O.E.* þēofð,=*O.N.* þýfð, *O.Fris.* thiufthe; *theft*, HOM. I. 13; LA3. 4263; A. R. 202; þeofþe BEK. 396; þiefþe AYENB. 9; þufþe ROB. 503; þefte MISC. 31; CH. C. T. *A* 4395; H. S. 2073; ðefte GEN. & EX. 3512.

þēoh, sb., *O.E.* þēoh,=*O.Fris.* thiach, *O.N.* þio, *O.L.G.* thio, *O.H.G.* dioh, ? *from* þēon; *thigh*: þeo O. & N. 1496; þe3 B. DISC. 476; þe HAV. 1950; thee D. ARTH. 1046; þeh LA3. 30581; þat þih 26071; þei3 [þi3] TREV.

IV. 185; þih FRAG. 2; þi ROB. 244; thi PR.
P. 490; þēȝ (pl.) HOM. II. 258; þih 211;
þeos ORM. 8079; þies MISC. 51; þeȝes SPEC.
52.

þeon, v., *O.E.* þēon, þīon, = *O.L.G.* thīhan,
thion, *Goth.* þeihan, *O.H.G.* dīhan; *increase,
prosper, flourish* : ꝺen GEN. & EX. 803: þen
LUD. COV. 340; so mot i þen AR. & MER.
1048; theen '*vigeo*' PR. P. 490; þe CH. C.
T. *D* 2207; þee EGL. 430; þīeꝺ (*pres.*) HOM.
II. 177; þīch (*imper.*) REL. I. 22; þēoinge
(*pple.*) BEK. 149; þēah (*pret.*) LK. ii. 52;
þeagh REL. I. 129; his welꝺe ꝺeg GEN. & EX.
2012; he . . . ꝺogen wel 2542; þogen (*pple.*)
HOM. II. 127; þowen SPEC. 23; *comp.* ȝe-
þeon; *deriv.* þicke, þīht.

þeorf, adj., *O.E.* þeorf, = *O.N.* þiarfr, *M.Du.*
derf, *O.H.G.* derb; *prov.Eng.* tharf (cake);
unfermented, ORM. 997; þerf bred MAND.
19; þerf breed WICL. GEN. xix. 3; þerf brede
TREV. V. 9; therf ['*azymus*'] PR. P. 490;
þerve (*pl.*) A. P. ii. 635.

þeorfling, sb., *O.E.* þeorfling; *anything unfermented*, ORM. 1590.

þēos, see **þes**.

þēoster, þēster, sb., *O.E.* þēoster, þȳster
(*pl.* þēostru); *darkness*; þuster O. & N. 230;
þeostra, þeostre ['*tenebrae*'] MAT. vi. 23;
LK. xi. 35; þustre SHOR. 121; þeostron
(*dat. pl.*) LK. xii. 3; þeostran HOM. I. 131;
comp. niht-þeoster.

þēoster-ful, adj., *O.E.* þēoster-, þȳsterfull;
dark, MAT. vi. 23.

þēosternesse, sb., *O.E.* þēosterness; *ob-
scurity, darkness*, A. R. 142; ORM. 3787;
þiesternesse REL. I. 131; þusternesse [þester-
nesse] O. & N. 369; ꝺisternesse GEN. & EX.
58; þisternesse HAV. 2191; þesternesse
LANGL. *B* xvi. 160.

þēostre, þēstre, adj., *O.E.* þēostre, þȳstre,
= *O.L.G.* thiustri, *O.Fris.* thiuster, ?*M.H.G.*
diuster; *obscure, dark*; þeostre LK. xi. 34;
þiestre AYENB. 159; þester AR. & MER. 1705;
JOS. 160; (*ms.* þessterr) ORM. 13426; þuster
REL. I. 89; þister SHOR. 140; þe þestere niht
LAȜ. 7563; þustere 9802; bi þeostre nihte
O. & N. 1432; þūstre (*pl.*) P. L. S. viii. 38;
ꝺiestre HICKES I. 222; þūstrore (*compar.*)
MISC. 150.

þiester-līche, adv., *darkly*, AYENB. 244.

þēostren, þēstren, v., *O.E.* þēostrian; *become
dark*; þeostren HOM. I. 143; þester JOS.
235; þēostreꝺ (*pres.*) A. R. 94; þēostrede
[þustrede] (*pret.*) LAȜ. 4575; þa þestrede þe
dæi SAX. CHR. 260; *comp.* a-þēostren.

[**þēote**, sb., *O.E.* þēote; *comp.* weter-þēote.]

þēoten, v., *O.E.* þēotan, = *O.N.* þiota, *O.H.G.*
diozan; *howl*, MARH. 22; þuten ORM. 2034;
þēoteꝺ (*pres.*) A. R. 120; þēotinde (*pple.*)
KATH. 163.

þēow, adj., *O.E.* þēow; *servile* : þeow wummon MARH. 4; þeu HAV. 2205; þeu, þu
BEK. 648, 652; þēwe (*dat.*) WILL. 5514;
þēowe [þeue] (*pl.*) LAȜ. 334; þewe & freo
AL. (L.³) 2.

þēow-dōm, sb., *O.E.* þēowdōm; *servitude*,
H. M. 5; (*ms.* þeowwdom) ORM. 3617; þeow-
dom [þeudom] LAȜ. 454; þeoudom A. R. 218;
þedom C. L. 247.

þēou-līc, adj., = *M.H.G.* dielīch; *servile* :
þewlike dede ORM. 4177.

þū-man, sb., *O.E.* þēow-, þēomann; *serf*,
ROB. (W.) 9655; þēwemen (*pl.*) ROB. (W.)
10317.

þēow, sb., *O.E.* þēow, = *Goth.* þius (*pl.* þiwōs);
slave, servant, FRAG. 6; þeow, þew (*ms.*
þeoww, þeww) ORM. 31, 7454; þeou LAȜ.
29390; þeu HOM. II. 181; *comp.* lād-tēow,
lār-þēu.

þēowe, sb., *O.E.* þēowa; *slave*, MAT. xviii.
28; MARH. 1; A. R. 372; þewe C. L. 245;
þeowen (*acc.*) LK. xiv. 17.

þēowe[2], sb., *O.E.* þēowe, = *Goth.* þiwi, *O.L.G.*
thiu, *O.N.* þȳ, *O.H.G.* diu; *female slave*,
MARH. 4; *comp.* ēꝺel-þēowe.

þēowien, v., *O.E.* þēowian; *serve*, LAȜ. 10015;
þēowest (*pres.*) HOM. I. 25.

þēowten, v., *from O.E.* þēowet (*service*); *do
service*, ORM. 44.

þer, see **þarf**.   **þēr**, see **þār**.

þerf, see **þarf**, **þeorf**.

þerfling, see *under* **þeorf**.

þerl, þerlen, see **þurl**, **þurlen**.

þerm, see **þarm**.

þērne, sb., = *O.N.* þerna, *O.L.G.* thiorna, *O.H.*
G. diorna, dierna; *girl*, HAV. 298; SHOR.
63; H. S. 7354; þierne AYENB. 129.

þes, prn., *O.E.* þes, þys, = *O.L.G.* these, *O.
Fris.* this, thisse, *O.N.* þessi, *O.H.G.* des,
dis; *comp. of* þe *and* se; *this*, HOM. I. 33;
KATH. 230; FER. 1146; þes nome A. R. 170;
þes dai O. & N. 259; þes boȝ AYENB. 41; ꝺes
GEN. & EX. 3967; þes [þis] cniht LAȜ. 398;
þus [þes] dom 16937; þis gode man ORM.
461; þis king C. L. 307; þēos [*O.E.* þēos,
= *O.L.G.* thius], (*fem.*) A. R. 84; þeos weorld
HOM. I. 33; þeos (þos) ule O. & N. 41, 1667;
þeos, þes LAȜ. 7083, 10110; þies JOHN vii.
36; þis [*O.E.* þis, þys] (*neut.*) LAȜ. 6287; O.
& N. 113; þis beoꝺ godes wordes A. R. 144;
þis water 246; þis mot O. & N. 468; þis, þus
KATH. 1047; þis (*for* þeos) lare H. M. 3;
þis (*for* þisses) folkes LAȜ. 824*; þis lifes
ORM. 2708; to þis (*for* þisen) lande SAX.
CHR. 254; of þis (*for* þisse) dede MARH. 22;
þis (*for* þisne) ard LAȜ. 13473; þisses [*O.E.*
þisses, þises] (*gen. m. n.*) HOM. I. 97; LAȜ.
823; þisses hweolpes A. R. 198; þisse [*O.E.*
þisse, þusse] (*gen. f.*) MAT. xiii. 22; þisse
HOM. I. 21; þissere 103; þisan, þise [*O.E.*
þisum, þysum, þissum] (*dat. m. n.*) HOM. I.
225, 227; þisen, þissen MAT. viii. 9, xvii. 20;

þissen, þisse FRAG. 7, 8 ; in þissen [þisse] londe LAȝ. 10391 ; et tissen one cherre A. R. 266 ; to ꝥise lande SAX. CHR. 253 ; þisse HOM. I. 5 ; in þisse live BRD. 30 ; þisse [*O.E.* þisse, þysse] (*dat. f.*) A. R. 320 ; on þisse wise MARH. 13 ; þisse, þissere MAT. xii. 32, xiii. 22 ; HOM. I. 9, 89 ; þissere SHOR. 53 ; i þissere [þisse] burh LAȝ. 5320 ; þisser, þesser, þeser HOM. I. 235 ; þisne [*O.E.* þisne] (*acc. m.*) P. L. S. xiii. 141, xxii. 14 ; þisne, þesne HOM. I. 5, 27 ; þisne, þusne LAȝ. 827, 4081, 10453 ; þesne A. R. 58 ; þerne AYENB. 94 ; SHOR. 161 ; þās [*O.E.* þās] (*acc. f.*) HOM. I. 235 ; LAȝ. 2044 ; þas (*for* þeos) weorld HOM. I. 35 ; þas burh LAȝ. 2061.

þās [*O.E.* þās, þǣs,=*O.L.G.* these, *O.H.G.* dise, dese, *mod.Eng.* those] (*nom. acc. pl.*) ; *those, these,* PS. xxii. 5 ; þas, þos, þeos, þes HOM. I. 11, 49, 53, 163 ; þas, þæs, þes, þeos, þus LAȝ. 476, 1038, 1250, 2219, 3816 ; þas, þir PR. C. 257, 491 ; þos P. L. S. viii. 21 ; ALIS. 6477 ; þes HORN (L.) 804 ; H. M. 9 ; ꝥes GEN. & EX. 1643 ; þeos A. R. 94 ; MARH. 5 ; þeos, þos O. & N. 139 ; þos, þise AYENB. 7, 10 ; þese REL. I. 132 (HOM. II. 13) ; þese, þise LANGL. *B* PROL. 184 ; þise ORM. 4573 ; WILL. 889 ; þus, þuse FER. 660, 1174 ; þuse TREAT. 135 ; þeos SAINTS (Ld.) xlvi. 513) ; þir M. H. 9 ; in þeos (*for* þissen) wordes A. R. 158 ; þisse, þissere [*O.E.* þissa] (*gen. pl.*) LAȝ. 2463, 7180 : þissere MAT. xiii. 22 ; þisen [*O.E.* þissum] (*dat. pl.*) MAT. iii. 9 ; þisan xxvii. 21 ; þissen FRAG. 1 ; LAȝ. 10461 ; þisse HOM. I. 11 ; bi þisse twam worde P. L. S. viii. 155.

þēstre, *see* þeostre.

þeþen, adv., *O.N.* þaꝧan ; *thence,* ORM. 1098 ; HAV. 2498 ; PR. C. 5831 ; ꝧeꝧen REL. I. 225 ; summe for pride fellen ꝧeꝧen GEN. & EX. 65 ; ꝧeꝧen (*ms.* ꝧeden) ut comen vii neet 2097.

þēu, þēw, *see* þēau, þēow.

þēwe, *see* þēowe.

þēwen, v., *O.E.* þēowan, þēwan, þȳwan, þȳn, = *M.H.G.* diuhen, dūhen, diuwen, *M.L.G.* dū-wen, *M.Du.* douwen ; *oppress* : þat he miȝte þat liþere folc so þewe P. L. S. xxiv. 57 ; to þæwen ȝunker childre ORM. 6217 ; ne þeawe þine servanz A. R. 268* ; *comp.* under-þēwen.

þī, *see* þē *under* þe, þeoh, þīn.

þicke, adj., *O.E.* þicce,=*O.L.G.* thikki, *O.N.* þiokkr, þykkr, *O.H.G.* dicchi ; *from* þihen ; *thick,* LAȝ. 12578 ; A. R. 382 ; O. & N. 580 ; þikke LANGL. *B* xix. 398 ; CH. C. T. *A* 1056 ; thikke '*spissus, densus*' PR. P. 490 ; thurgh thikke & thurgh thenne CH. C. T. *A* 4066 ; þicke (*adv.*) FRAG. 8 ; BEK. 77 ; þat folk cam wel þicke aboute o(u)re loverd LEB. JES. 411 ; he þankit him þicke (*often*) D. TROY 9972 ; ꝧicke GEN. & EX. 2988 ; isawe al to þikke SHOR. 27 ; þiccure (*compar.*) A. R.

50 ; þickure MISC. 148 ; þenne Crist assoiled þicker (*oftener*) men WICL. E. W. 344.

þicke-liste, adj., *? hard of hearing,* HOM. II. 129.

þiknesse, sb., *O.E.* þicness,=*O.H.G.* dicnissi ; *thickness,* TREV. I. 45 ; thikkenesse PR. P. 491.

þicke [2], sb., = *? O.H.G.* dicchī ; *thickness* ; thikke D. ARTH. 3755 ; ine þe þicke O. & N. 1626.

þicken, v., *O.E.* þiccian ; *thicken* ; thikkin PR. P. 491 ; þicke GOW. II. 327 ; þickeþ (*pres.*) TREAT. 139.

þider, adv., *O.E.* þider, þyder,=*O.N.* þaꝧra ; *thither,* A. R. 128 ; ORM. 1700 ; O. & N. 719 ; HAV. 850 ; BEK. 73 ; CH. C. T. *A* 1263 ; MAN. (H.) 6 ; þider in LAȝ. 544 ; þider ut 31559 ; nulle we þuder [þider] wende 12625 ; þider, þeder WILL. 33, 2235 ; þuder ROB. 539 ; AL. (T.) 202 ; þeder P. 32 ; þedur EGL. 432 ; tider MAN. (F.) 12423.

þider-ward, adv., *thitherward,* O. & N. 143 ; þiderward LAȝ. 1662 ; PR. C. 7539 ; þiderward, -wardes KATH. 160, 2059.

þief, *see* þeof. þierne, *see* þerne.

þiestre, *see* þeostre. þifel, *see* þuvel.

þiggen, v., *O.E.* þicgan,=*O.N.* þiggja, *O.L.G.* thiggean, *O.H.G.* diggen ; *take* (*food, drink*) ; *receive* ; *beg* ; thiggin PR. P. 28 ; þigge D. TROY 13549 ; mi mete to þigge HAV. 1373 ; þiggieꝧ (*pres.*) MK. vii. 5 ; þiȝe (*imper.*) LEECHD. III. 92 ; þet mon ... to muchel ne þigge on ete & on wete þigge HOM. I. 105 ; þiggand (*pple.*) PS. xxxix. 18.

þigging, sb., *begging,* '*mendicacio*' PR. P. 490.

þih, *see* þeoh.

þiht, adj., = *M.H.G.* dihte, *O.N.* þéttr ; *from* þeon ; *firm, solid,* '*solidus,*' PR. P. 491 ; *see* tīht.

þihtin, v., *make solid,* '*soiido,*' PR. P. 491.

þikke, *see* þicke. þilc, *see* þulc.

þild, *see* þuld.

þille, sb., *shaft of a cart,* '*temo,*' PR. P. 491 ; þilles '*timons*' REL. II. 83 ; PALL. vii. 38.

thil-hors, sb., *shaft-horse,* VOC. 202.

þimbel, *see* þumel.

þīn, adj. & pron., *O.E.* þīn,=*O.L.G.* thīn, *O.N.* þīnn *m.*, þīn *f.*, þītt *n.*, *Goth.* þeins *m.*, þeina *f.*, þein *n.*, *O.H.G.* dīnēr *m.*, dīniu *f.*, dīnaz *n.* ; *thine, thy* : þin feder HOM. I. 125 ; þin name AYENB. 107 ; þin moder JOHN xix. 27 ; þin blisse FRAG. 6 ; þin wif ORM. 156 ; þin herte 1460 ; þin heved O. & N. 74 ; þi bodi 73 ; he is þin FL. & BL. 4 ; al þat þin is SPEC. 42 : þi fader LAȝ. 2292 ; þi man HAV. 2173 ; þi wille BEK. 341 ; þi broþer CH. C. T. *A* 1131 ; þi moder A. R. 54 ; þi tunge R. S. v ; þines (*gen. m. n.*) FRAG. 6 ; HOM. I. 11 ; þines [þine] cunnes LAȝ. 16546 ; þine songes O. & N. 896 ; þine wifes SHOR. 69 ; þīre (*gen. f.*) LAȝ.

28104 ; MISC. 88 ; of þi luve MARH. 18 ; þīnan
(dat. m. n.) JOHN xvii. 12 ; þinen, þine MAT.
iv. 10, vi. 6 ; þine LAӠ. 1247 ; A. R. 26 ; ORM.
1251 ; O. & N. 221 ; SPEC. 62 ; SHOR. 69 ; of
þine gode FRAG. 6 ; of þin owe bodi BEK.
846 ; þīnre (dat. f.) MK. vii. 29 ; þire FRAG.
6 ; HOM. I. 33 ; LAӠ. 1576 ; O. & N. 915 ;
þīnne [þine] (acc. m.) LAӠ. 5074 ; þinne, þine
A. R. 38, 106 ; þine HOM. I. 17 ; O. & N. 339 ;
AYENB. 8 ; þine flor FRAG. 7 ; þin ORM. 1319 ;
þi KATH. 444 ; BEK. 141 ; þīne (acc. f.) A. R.
48 ; O. & N. 258 ; þine sunne R. S. i ; þine
saule SHOR. 87 ; þine strengþe AYENB. 54 ;
for þine sake SPEC. 28 ; þine [þin] lare LAӠ.
697 ; þurh þi muchele mihte KATH. 654 ;
neih þine (for þinre) heorte FRAG. 7 ; bi þine
tale A. R. 334 ; þīn (acc. n.) R. S. v ; þin fule
hold FRAG. 6 ; ðin GEN. & EX. 397 ; þi swerd
LAӠ. 5072 ; þi (þin) sweven 25577 ; þīne (nom.
acc. pl.) LAӠ. 3093 ; A. R. 100 ; ORM. 6727 ;
HAV. 620 ; for þe and þine freende SHOR. 28 ;
þine fon BEK. 1648 ; þin LAӠ. 25878* ; þin
fet SPEC. 80 ; þin armes (ms. þi narmes) WILL.
666 ; for þi children BEK. 424 ; þīnre (gen.
pl.) MAT. v. 29 ; þinra HOM. I. 111 ; þire LAӠ.
22448 ; þīne (dat. pl.) FRAG. 6 ; BEK. 979 ;
SHOR. 90 ; of þine fan KATH. 688 ; don þine
[þin] uniwinen wa LAӠ. 14466 ; for þine
wounden SPEC. 82.

þinchen, see þunchen.

þing, sb., O.E. þing, = O.N. þing, O.L.G., O.
Fris. thing, O.H.G. ding ; thing, ORM. 1839 ;
heo ne seide na þing soð LAӠ. 3013 ; and na
ðing ne rohten 6274 ; evrich þing O. & N.
229 ; he þat þer of no þing (v. r. nowiht) not
1247 ; þe þouӠt no þing i not C. L. 1050 ; wis-
likes þinges (gen.) certainly ORM. 3186 ; for
godes þing(e) (dat.) for God's sake HOM. I. 67 ;
for þine þinge LAӠ. 5033 ; for hire þinge O. & N.
1597 ; þing (pl.) HOM. I. 75 ; C. L. 8 ; MAN.
(H.) 158 ; M. H. 1 ; þurh alle þing LAӠ. 2722 ;
O. & N. 771 ; over alle þing JUL. 75 ; ORM.
3640, 8644, 13664 ; þinge H. M. 7 ; þing,
þinges A. R. 52, 286 ; MARH. 8, 9 ; þing,
þinge, þinges, þinge (gen. pl.) LK. i. 1 ; KATH.
254 ; A. R. 398 ; þingen (dat. pl.) HOM. I.
133 ; þinge LAӠ. 16042 ; O. & N. 1540 ; SHOR.
100 ; in alle þinge ORM. 1655 ; PROCL. 3 ;
comp. brūd-, hŭs-, wif-, wonder-þing.

[þingen, v., O.E. þingan ; grow, increase;
comp. ӠeӠ-þingen.]
þingen², v., O.E. þingian, = O.N. þinga, O.L.G.
thingōn, O.H.G. dingōn ; conciliate, HOM. II.
43 ; to þingen (ms. þingenn) us wiþ ure god
ORM. 8997.

þinken, see þunchen.

þinne, see þunne.

þinnesse, see under þunne.

þīode, see þēode. þirde, see þridde.

þirl, þirlen, see þurl, þurlen.

þirs, see þurs.

þirst, þirsten, see þurst, þursten.

þis, see þes.

þistel, sb., O.E. þistel, = O.N. þistill, O.H.G.
distill ; thistle ; thistil (ms. thystylle) 'carduus' PR. P. 491 ; þistles (pl.) REL. I. 264 ;
comp. hors-, sea-, suӠe-þistel.

þistre, see þēostre. þīwien, see þēowien.

þixel, sb., = O.H.G. dehsala, dehsela, M.Du.
diechsel ; adze, ['ascia'] PS. lxxvii. 6 ; thixil
PR. P. 491 ; VOC. 234.

þō, sb., O.E. þōhe, = Goth. þāhō, O.H.G. dāha ;
clay ; (printed yo, r. w. tough) PALL. i. 402.

þō, see þā.

þode [þodde], sb., O.E. þoden ; wind, LAӠ.
27645.

þōӠt, see þōht. þōh see þāh.

þōht, þóht, sb., O.E. (ge-)þōht, = O.N. þōttr,
O.H.G. (ge-)dāht ; thought, LAӠ. 18549 ;
KATH. 512 ; P. S. 220 ; (ms. þohht) ORM.
2577 ; þoht [þoӠt] O. & N. 492 ; þoӠt FL. &
BL. 34 ; BEK. 861 ; ðogt GEN. & EX. 1558 ;
þouht A. R. 62 ; þouht [þouӠt] C. L. 6 ; þouӠt
CH. D. BL. 4 ; þouӠt, þout WILL. 4054, 4116 ;
þouth HAV. 1190 ; þout S. S. (Web.) 688 ;
þōhte (dat.) O. & N. 940 ; ich am in grete
þohte SPEC. 113 ; þoӠte AYENB. 6 ; þōhtes
(pl.) H. M. 5.

þōht-ful, adj., thoughtful, ORM. 3423 ;
þoӠtful FL. & BL. 168.

þōke, see þāke.

þol, sb., O.E. þoll, = O.N. þollr, M.Du. dol,
dolle, M.L.G. dolle ; peg, 'cavilla ;' (ms.
tholle) PR. P. 492.

þóle, sb., O.N. þol n. ; patience : min ðole is
long GEN. & EX. 3496.

þóle-bürde, adj., patient, HOM. II. 79.

ðólebürdnesse, sb., patience, HOM. II. 53.

þóle-mōd, sb., patience, HOM. I. 69.

þóle-mōd, adj., O.E. þolemōd, cf. O.N. þolin-
mōðr ; patient, LAӠ. 3141 ; KATH. 177 ; A. R.
158 ; C. L. 854 ; SPEC. 72 ; H. S. 10925 ; hwi
nule we . . . beo þolemode SPEC. 91 ; thole-
mode HAMP. PS. lxxii. 14*.

þólemōd-līche, adv., patiently, A. R. 46 ; P.
R. L. P. 240 ; tholemodeli HAMP. PS. xxxvi. 11.

þólemōdnesse, sb., patience, A. R. 8 ; H. M.
41 ; C. L. 985 ; AYENB. 68 ; tholemodnes
HAMP. PS. ix. 19*.

þólien, v., O.E. þolian, cf. O.N. þola, O.L.G.
tholean, tholōn, O.H.G. dolan, dolen, dolōn,
Goth. þulan, Lat. *tulere (perf. tuli) ; bear,
suffer, KATH. 1006 ; A. R. 6 ; þolien deþ C. L.
410 ; þolien sore SPEC. 28 ; þene deað þolien
[þolie] LAӠ. 284 ; þolen ORM. 897 ; ALIS.
5361 ; þolen hunger HOM. II. 35 ; ðolen SAX.
CHR. 256 ; GEN. & EX. 3664 ; þolie MISC.
148 ; BEK. 36 ; AYENB. 22 ; þolie, þole LANGL.
A iv. 71 ; þole WILL. 918 ; Gow. II. 355 ;
S. S. (Wr.) 647 ; H. S. 1449 ; APOL. 56 ; A. P.
ii. 190 ; (þu) þólest (pres.) LAӠ. 4322 ; (he)
þoleð 8285 ; Ӡe þolieð A. R. 188 ; þóle (imper.

AYENB. 117 ; ich þólie (*subj.*) A. R. 352 ; þólede (*pret.*) TREAT. 136 ; SHOR. 25 ; PL. CR. 90 ; þóled (*pple.*) WILL. 4514 ; CH. C. T. *D* 1546 ; *comp.* 3e-þólien.

þóling, sb., *O.E.* þolung, = *O.H.G.* dolunga ; *sufferance,* APOL. 5.

þŏmbe, þŏme *see* þūme. þon, *see* þan.

þonc, *see* þanc. þonder, þoner, *see* þuner.

þong, *see* þwang. þonkien, *see* þankien.

þōr, *see* þār.

Pōresdai, *see under* þuner.

þorh, *see* þurh.

þorn, sb., *O.E.* þorn, thorn, = *O.N.* þorn, *O.L.G.* thorn, *O.H.G.* dorn, *Goth.* þaurnus ; *thorn,* CH. C. T. *A* 2923 ; hou Moises ise3 a þorn berne BRD. 16 ; þorne LANGL. *B* xii. 228 ; þorne (*dat.*) ISUM. 103 ; þornes (*pl.*) A. R. 134 ; ORM. 9212 ; O. & N. 586 ; H. S. 7514 ; ðornes GEN. & EX. 1334 ; þornen (*dat. pl.*) MAT. vii. 16 ; LA3. 649 ; *comp.* ha3e-, hwīt-, slō-, þēfe-þorn.

   thorn-bak, sb., *thornback,* VOC. 189 ; PR. P. 492 ; þornbake HAV. 832.

   þorn-hog, sb., *hedgehog,* AYENB. 66.

þornen, v., *O.E.* þyrnen, = *O.Fris.* thornen, *O.H.G.* durnīn ; *made of thorns :* þe þornene krune A. R. 258 ; þernene helm MK. xv. 17.

þorni, adj., = *M.H.G.* dornec ; *thorny,* A. R. 134 ; M. H. 52.

þorp, sb., *O.E.* þorp, = *O.N.* þorp, *O.L.G.*, *O.Fris.* thorp, *Goth.* þaurp (ἀγρός), *O.H.G.* dorf ; *mod.Eng.* thorp(e) (*in local names*) ; *village, town* ; thorp, throp '*oppidum*' PR. P. 492 ; þrop HOM. II. 89 ; CH. C. T. *E* 199 (Wr. 8057) ; þorpe (*dat.*) LIDG. M. P. 140 ; þorpes (*pl.*) A. P. ii. 1178 ; þropes WILL. 2141.

þorst, *see* þurst.

þoru, þoru3, þorw, *see* þurh. þos, *see* þus.

þost, sb., *O.E.* þost, = *O.H.G.* dost ; *dung,* ['*stercus*'] PS. lxxxii. 11* ; TREV. IV. 423 ; WICL. ECCLUS. ix. 10 ; thost PR. P. 492 ; PALL. iv. 348 ; þoste (*dat.*) P. S. 237.

þou, *see* þū. þou, þou3, þouh, *see* þāh.

þou3t, þouht, *see* þōht. þoume, *see* þūme.

þout, *see* þōht. þouten, *see* þūten.

þōwe, þōwen, *see* þāwe, þāwen.

þrā, adj., *O.N.* þrār ; *bold ; severe, strong ;* IW. 3570 ; C. M. 14392 ; M: H. 124 ; þro MAP 336 ; TRIST. 37 ; K. T. 1078 ; OCTAV. (H.) 547 ; M. ARTH. 1525. ; þe dupe river þat wilde was & thro FER. 3968 ; his þro þo3t GAW. 645 ; men þat þro were to fi3t(e) WILL. 3264 ; hur peinis were so þroo TRIAM. 405.

   þrā-lī, adj., *O.N.* þrāligr, = *O.E.* þrēalīc, *O.H.G.* drōlīh ; *bold ; eager :* þe þroli þou3t WILL. 3518 ; þrālī (*adv.*) L. H. R. 110 ; ANT. ARTH. xv ; þroli into þe develez þrote man þringez A. P. ii. 180 ; þroli LANGL. *A* ix. 107.

þrā ², sb., *O.N.* þrā, ? *O.E.* þrēa *n. ; cf.* þrāwe ;

struggle, victory, mastery ; þro H. S. 10570 ; A. P. ii. 754 ; ani werre or ani þro MAN. (F.) 10204 ; þe knight bihaldez him in þroo PERC. 673.

þrǣd [þrēd], sb., *O.E.* þrǣd, = *O.Fris.* thrēd, *O.N.* þrāðr, *O.H.G.* drāt ; *from* þrāwen ; *thread,* LA3. 14220 ; þrēde (*dat.*) MISC. 77 ; WILL. 4430 ; þrēdes (*pl.*) CH. BOET. i. 1 (5) ; *comp.* gōld-, silk-þrēd.

   þrēd-bare, adj., *threadbare,* LANGL. *A* v. 113 ; CH. C. T. *A* 290.

þrǣpen, *see* þrēapen.

þrǣst, sb., *crowd, host* ; þrĕste (*dat.*) AYENB. 121 ; þrast(e) GAW. 1443 ; thonere thrăstis (*pl.*) ALEX. (Sk.) 554.

þrǣsten, v., *O.E.* þrǣstan ; *press, force :* mine cnihtes balde scullen þræsten [þreaste] biforen me LA3. 23373 ; þreste ALIS. 3326 ; CH. C. T. *A* 2612 ; TRIAM. 774 ; þurles þer þet water þrest in A. R. 314 ; thrĕste (*imper.*) PALL. xi. 86 ; þrăste (*pret.*) LA3. 27644 ; CH. C. T. *C* 260 ; þreaste ... smoke ut MARH. 9 ; heo þresten (*printed* wresten) in uppon me A. R. 220 ; þrasten [þreste] ut of telden LA3. 26318 ; þraste MAP 338 ; þrast (*pple.*) P. L. S. xxx. 69 ; þo he was at þe 3ate out þrast A. D. 260 ; *comp.* for-, 3e-þrǣsten.

þrǣt, *see* þrēat.

þrā3e, sb., *O.E.* þrāg (HOLTZM. I. 176) ; *course or space of time,* HOM. I. 35 ; þra3he ORM. 3475 ; þrawe ALIS. 3836 ; RICH. 5062 ; MAN. (H.) 180 ; M. H. 142 ; in a lite þrawe HAV. 276 ; þro3e HORN (L.) 336 ; lat me nu habbe mine þro3e [þrowe] O. & N. 260 ; and blisse mid heom sume þrowe 478 ; þrowe LA3. 640 ; WILL. 679 ; CH. C. T. *D* 1815 ; P. S. 343 ; one þrowe MISC. 79 ; none þrowe P. L. S. xvii. 292 ; umbe þrowe SPEC. 25 ; thre dais out a thrawe ALEX. (Sk.) 2046.

þrā3en, *see* þrēan.

þral, adj., *O.E.* þearl, = *O.Fris.* thral, *M.L.G.*, *M.Du.* dral ; *? severe,* REL. II. 118 ; was in hert(e) thral GENER. 3947.

þral ², sb., *O.E.* þræl, *cf. O.N.* þrǣll ; *thrall, slave,* MARH. 12 ; R. S. vi (MISC. 184) ; HAV. 527 ; ROB. 198 ; C. L. 242 ; SPEC. 89 ; SHOR. 102 ; þral [thral] and bonde CH. C. T. *D* 1660 ; ðral GEN. & EX. 2881 ; þral [þrel] H. M. 7 ; þrel LA3. 14852 ; A. R. 370 ; AYENB. 19 ; thrill BARB. i. 243 ; þralles (*gen.*) C. L. 547 ; þralle (*dat.*) HORN (L.) 419 ; þrales (*pl.*) HOM. II. 121 ; þreles P. L. S. viii. 95 ; þralles LA3. 492 ; H. M. 5 ; þrelles A. R. 130 ; þrallen (*dat. pl.*) LA3. 10014 ; þralle AL. (T.) 252.

   þral-dōm, sb., *O.N.* þrældōmr ; *servitude,* LA3. 29156 ; ROB. 12 ; LANGL. (Wr.) 12281 ; LIDG. M. P. 63 ; PR. C. 8005 ; threldome, thirldome BARB. i. 265 ; i. 236.

   þral-hōd, sb., *servitude,* HORN (L.) 439 ; þralhede HORN (R.) 443 ; ROB. 47.

   þral-shipe, sb., *servitude,* HOM. II. 37.

þrel-weorc, sb., *O.N.* þrælverk; HOM. I. 47; þrelwerc LAȝ. 455.

þrallāge, sb., *slavery*; thrillage BARB. i. 101, 275.

þralle, sb., *slavery* : that helpeth soules out of þralle P. R. L. P. 91.

þrallen ?, v.,=*O.N.* þrǣla; *make thrall, enslave*, R. R. 882; þrallede (*pret.*) LAȝ. 11205; þralled MAND. 2.

þrang, sb., *O.E.* (ge-)þrang,=*M.Du.* drangh, *M.H.G.* dranc; *from* þringen; *throng, crowd*, ALIS. 3409; lw. 839; TOR. 1057; thrang PR. C. 4704; þrong A. P. ii. 135.

þrangien, v., *throng, press*; thronge (*pres.*) D. ARTH. 3755.

þráve, sb., *O.N.* þrefi *m.; bundle, number*, HALLIW. 867; thrave [þreve] LANGL. *B* xvi. 55; thrafe TOWNL. 12.

þrāwe, sb., *O.E.* þrēa,=*O.N.* þrā, *O.H.G.* dra(u)wa, dro(u)wa, droa *f.; from O.E.* þrēowan; *throe*, ALIS. 616; throwe '*ærumna*,' PR. P. 493; þrōwe, þrowes (*pl.*) HOM. II. 181; þrowe (*ms.* þrowes, *r. w.* blowe) S. A. L. 156; þrawes M. H. 181; þrowes ROB. 12; TRIAM. 411; on his last(e) þrowe ASS. *B* 533; in al his harde þrowe L. H. R. 150; *comp.* dēþ-þrāwe.

þrāwe, *see* þrāȝe.

þrāwen, v., *O.E.* þrāwan, *cf. O.H.G.* drājan (*pret.* drāta); *throw; turn round*; þrawe PS. cxxxix. 11; and ssel þrawe þet chef into þe vere AYENB. 139; þrowe R. S. i; þrowe TREV. III. 171; LANGL. *B* xvi. 131; þrēou (*pret.*) LAȝ. 807; (*ms.* þreow) ALIS. 2427; þreu HORN (L.) 1076; AYENB. 133; (*ms.* þrew) LANGL. *B* xx. 163; þu þrewe HORN (L.) 1172; þa cheorles up þreowen [þreuwen] LAȝ. 12321; þrāwen (*pple.*) GAW. 194; þrauwen mid winde LAȝ. 27359; was þrowen ['*jactabatur*'] WICL. MAT. xiv. 24; þrowe CH. C. T. *F* 326; iþrowen C. L. 739; mine lo(c)kes were iþrowe ANGL. III. 279; *comp.* for-, over-þrāwen; *deriv.* þrǣd.

þrē ? sb.,=þrā: þureh his þre creftes (? þrecreftes) HOM. II. 167.

þrē, *see* þrī.

þrēan, v., *O.E.* þrēan, þrēawian, þrēagean, *cf. O.L.G.* githrōōn, *O.H.G.* dra(u)wan, dro(u)wan; *castigate, threaten*; þrāghand ['*castigans*'] PS. cxviii. 18*; þrēde (*ms.* þredde) (*pret.*) ['*increpavit*'] MAT. xvii. 18; þrād (*pple.*) A. P. ii. 751; *comp.* ȝe-þrēan.

þrēap, sb., *contradiction*, ROB. (W.) 2391; þrep (*ms.* þrepe) D. TROY 9845; a þrep MISC. 149; wituten threpe C. M. 13310; witouten þrep A. P. ii. 350.

þrēapen, v., *O.E.* þrēapian (*rebuke*); *speak against, contradict* : him birþ þræpen . . . onȝænes alle sinnes ORM. 5744; þrepe H. S. 6067; D. TROY 12134; threpe PS. xciii. 10; TOWNL. 102; þrēapeð (*pres.*) aȝain þe KATH. 1939;

þreap . . . & þreate 1509; thrēpide (*pret.*) D. ARTH. 930.

þrēping, sb., *strife*, A. P. ii. 183.

þrēapnen, v., *threaten*; þreapni AYENB. 84.

þrēapninge, sb., *threatening*, AYENB. 65.

þrēat, sb., *O.E.* þrēat,=*M.L.G.* drōt; *from O.E.* þrēotan; *crowd; oppression; threat*; HOM. II. 61; þe þræt wes þa mare LAȝ. 9791; þræt þas kaiseres 22582; ðret GEN. & EX. 2021; þrat A. P. iii. 55; þrēte (*dat.*) O. & N. 58; LEG. 24; þrēates (*pl.*) KATH. 40; þrættes [þretes] LAȝ. 26294; nule heo forgo Robin for al heore þrete R. S. vii (MISC. 190).

þrēaten, v., *O.E.* þrēatian,=*O.N.* þreyta, *M.L.G.* drōten; *oppress, threaten*, KATH. 623; A. R. 248; þretien LAȝ. 17300; þrete GREG. 396; CH. L. G. W. 754; H. S. 6398; A. P. i. 560; þrēteþ (*pres.*) O. & N. 1609; BEK. 1553; SPEC. 43; þreat H. M. 17; (heo) þretia ð LAȝ. 493; þrete ['*arguas*'] PS. vi. 2; þrēatede (*pret.*) ['*increpavit*'] LK. ix. 55; treatede (*printed* þreateð), LAȝ. 641; þrætede [þretede] 7644; ðreated GEN. & EX. 4125; þrette A. R. 366; HAV. 1163; ðrette GEN. & EX. 2023; þreteden WICL. MK. x. 13; summe þrætteden [þrettede] heore veond LAȝ. 27131; þrēted (*pple.*) GAW. 1725; þrat P. S. 158; *comp.* ȝe-þrēaten.

þrēatunge, sb., *O.E.* þrēatung; *menace*, A. R. 156; þretinge LAȝ. 19369*; LANGL. *B* xviii. 279; þreting MISC. 156.

þrēatnen, v.,=þrēaten, *threaten*; þrētneþ (*pres.*) ROB. 457; þrētnede (*pret.*) BEK. 1804; LEG. 31.

þrētninge, sb., *threatening*, P. L. S. xiii. 248; (*ms.* þretningue) S. A. L. 152.

þrēd, *see* þrǣd. þrel, *see* þral.

þreng, sb.,=*M.H.G.* gedrenge; *from* þringen; *crowd*; AR. & MER. 6109; a mong þe þrenge LAȝ. 2229*.

þrengen, v., ? *O.E.* þrengan, = *O.N.* þröngva, *M.H.G.* drengen; *press*; þrengden (*pret.*) ORM. 16182; (hi) þrengde þe man þær inne SAX. CHR. 262.

þrēo, *see* þrī. þreoie, þreowe, *see* þríe.

þrēowien, v., *O.E.* þrēowian; ? =þrowien : he þrēowede (*pret.*) longe LAȝ. 11389.

þrēp, þrēpen, *see* þrēap, þrēapen.

[þresch-, *stem of* þreschen.]

þresch-wold, sb., *O.E.* þresc-, þrex-, þerscwald, -wold, þerscold, ? *O.N.* þreskiöldr; *threshold, ' limen,'* PR. P. 492; þreshwold CH. C. T. *A* 3482; thressewolde LANGL. *B* v. 357 [threswolde *A* v. 201]; þreoxwold FRAG. 4; þriswald IW. 3222; þerswald VOC. 170.

þreschen, v., *O.E.* þrescan, þerscan,=*O.N.* þreskja, *Goth.* þriskan, *O.H.G.* dresken; *thrash, beat*; threschin '*trituro*' PR. P. 492; þressin AN. LIT. 84; þresche REL. I. 6; þresche [thresche] CH. C. T. *A* 536; þreshe [thresche] (*pres.*) LANGL. *B* v. 553 (Wr. 3596); þu þreshest (*ms.* þresshesst) tine shæfes ORM. 1481 ; þreosche (*subj.*) A. R. 306 ; þroshen (*pple.*) ORM. 1530 ; iþroschen A. R. 186 ; iþor(s)chen HOM. I. 85 ; iþorsse AYENB. 139.

þreschare, sb., = *M. H. G.* drescher ; *thrasher*, PR. P. 492.

þreschwold, *see under* þresch-.

þrĕst, *see* þrĕst.

þrĕsten, *see* þrĕsten, þrūsten.

þrĕt, þrētien, *see* þrēat, þrēaten.

þréve, *see* þráve.

þrī, card. num., *O.E.* þrī, þrȳ *m.*, þrēo *f. n.*, = *O. Fris.* thrē *m.*, thria *f.*, thriu *n.*, *O.L.G.* thria, thrie *m. f.*, thriu *n.*, *O.H.G.* drī *m.*, drio *f.*, driu *n.*, *O.N.* þrīr *m.*, þriar *f.*, þriū *n.*, *Goth.* þreis *m.*, þrija *n.*, *Lat.* trēs, trīs *m. f.*, tria *n.*, *Gr.* τρεῖς *m. f.*, τρία *n.*, *Welsh, Ir., Gael., Bret.* trī ; *three* : þri dages MK. viii. 2 ; þri reaferes HOM. I. 243 ; godes þri SHOR. 142 ; þri hestes AYENB. 7 ; þri þing 170 ; þru dæges LEECHD. III. 134 ; þreo hlafes LK. xi. 5 ; þreo men LA3. 22144 ; þreo kinges ORM. 6526 ; þreo sunnen A. R. 56 ; þreo 3er BEK. 147 ; þreo ni3t JOS. 6 ; 3our þreo HORN (L.) 815 ; þreo & twentuˇe KATH. 2213 ; þrie kinges MISC. 27 ; þrie þing HOM. II. 27 ; þre niht LEECHD. III. 134 ; þre dagas SAX. CHR. 256 ; þre daies HAV. 655 ; þre sones CHR. E. 107 ; þre siþes MAND. 291 ; we þre K. T. 810 ; þre and twenti ROB. 62 ; þre & þrittiþe 441 ; þrē (*gen. pl.*) sum *three at a time* BARB. iii. 420 ; þrēom (*dat.*) HOM. I. 99 ; LA3. 11517 ; æfter þreom dagen MK. viii. 31 ; þrem MAT. xxvi. 61.

þrēo-fáld, adj., *O.E.* þrie-, þrȳfeald, = *O. Fris.* thrīfald, *O.H.G.* drīfalt ; *threefold*, MARH. 11 ; þréovold A. R. 82 ; þrefald ORM. 14034.

þrēo-hād, sb., *trinity*, HOM. I. 267 ; JUL. 78.

þrēo-tēne, þrītēne, card. num., *O.E.* þrēotȳne ; *thirteen* ; þreottene LA3. 7771 ; A. R. 234 ; þrittene ORM. 11071 ; þrettene BEK. 2435 ; þrettene LANGL. *A* v. 128 [threttine *C* vii. 220].

þrettēþe, ord. num., *thirteenth*, ROB. 232 ; þreotteoþe MISC. 146 ; threttende PR. C. 7173.

þrī-ti, þritti, card. num., *O.E.* þrītig, þrittig, = *O.L.G.* thrītig, *O.H.G.* drīzug ; *thirty*, LA3. 377, 2688 ; þritti SAX. CHR. 256 ; FRAG. 1 ; A. R. 46 ; BEK. 814 ; H. H. 72 ; MAN. (H.) 6 ; þritti3 ORM. 3207 ; thretti, thirti PR. P. 492 ; after þan þrittie LA3. 26631.

þritti-fáld, adj., *thirtyfold*, H. M. 23.

þrittŭðe, ord. num., *O.E.* þrītigoˇča ; *thirtieth*, H. M. 23 ; KATH. 43 ; HOM. II. 47 ; þrittiþe ROB. 441 ; þritta3te AYENB. 234.

þricche, *see* þrücche.

þridde, ord. num., *O.E.* þridda, = *O.L.G.* thriddio, *Goth.* þridja, *O.N.* þriči, *O.H.G.* dritto,

Lat. tertius ; *third*, A. R. 14 ; BEK. 583 ; LANGL. *B* xvi. 188 ; CH. C. T. *C* 836 ; MAND. 48 ; L. C. C. 55 ; ˇridde GEN. & EX. 761 ; thredde PR. C. 4210 ; þirde S. S. (Wr.) 49.

þrīe, *see* þrī.

þrie, adv., *O.E.* þriga, þriwa, *O.Fris.* thrīa, thrija, *O.L.G.* thrijo, thriwo ; *thrice*, H. M. 23 ; ROB. 191 ; ALIS. 1263 ; S. S. (Wr.) 2179 ; CHEST. II. 25 ; þreowe MK. xiv. 30 ; þrie, þrien, þreie, þreoien LA3. 14338, 14352, 17432, 26066 ; þrien REL. I. 144.

þries, adv., = *M.H.G.* drīes ; *thrice*, LA3. 26066 ; A. R. 106 ; AYENB. 35 ; SHOR. 11 ; þries, thries CH. C. T. *A* 63 ; ˇries REL. I. 209 ; þri3es ORM. 1149.

þrif, sb., *O.N.* þrif ; *increase, fortune* ; þref FER. 2017.

þrīfen, v., *O.N.* þrifa ; *thrive* ; þrifen . . . & waxen ORM. 10868 ; þrive (*ms.* þriue) HORN (L.) 620 ; HAV. 280 ; LANGL. *B* v. 284 ; TOR. 1155 ; þrive [thrive] CH. C. T. *A* 3675 ; thrife TOWNL. 8 ; if þretti þrivande be þrad A. P. ii. 751 ; fele þrivande þonkkez GAW. 1980 ; þrāf (*pret.*) ORM. 3182 ; PERC. 212 ; M. H. 109 ; þrof ROB. 11 ; **þriven** (*pple.*) SPEC. 23 ; iþriven REL. I. 121 ; mani threvin dukis ALEX. (Sk.) 1326.

þrift, sb., *O.N.* þrift ; *from* þrīfen ; *thrift, prosperity*, P. L. S. xiv. 70 ; TRIST. 3060 ; þrift, thrift CH. C. T. *G* 1425 ; **þriftes** (*pl.*) SPEC. 47.

þrifti, adj., *prosperous*, CH. C. T. *B* 138 ; thrifti '*vigens*' PR. P. 492 ; D. ARTH. 317.

þrīle, adj., *cf. M.H.G.* drilich ; *threefold*, A. R. 26 ; þrile i þreo hades MARH. 11.

þrille-hōd, sb., *trinity*, C. L. 1239.

þrillen, *see* þürlen.　**þrim**, *see* þrüm.

þrimnesse, sb., *cf. O.E.* þrīness, þrȳness, *O.H. G.* drīnissa ; *trinity* ; (*ms.* þrimmnesse) ORM. 11177 ; þrumnesse A. R. 160 ; MARH. 11 ; (*ms.* þrunesse) HOM. I. 259 ; þremnesse HOM. II. 137 ; þreomnesse HOM. I. 99.

þrin, adj., *O.N.* þrinnr ; *threefold, triple* : þrinne lac ORM. 1144 ; on þrinne wis(e) M. H. 163 ; his sones þrinne (*for* þri) HAV. 716.

þring, sb., *O.E.* geþring, = *O.L.G.* gethring ; *throng, crowd, press*, BEV. 1365 ; þringe (*dat.*) MISC. 86 ; ALIS. 2533 ; atforen al þan dringe [þringe] LA3. 14966.

þringen, v., *O.E.* þringan, = *O.L.G.* thringan, *O.H.G.* dringan ; *throng, press* ; þringe MISC. 42 ; ALIS. 2388 ; **þringeþ** (*pres.*) HOM. I. 237 ; þringez A. P. ii. 180 ; þringen WICL. LK. viii. 45 ; þringe (*subj.*) O. & N. 796 ; **þringing** (*pple.*) R. R. 656 ; **thrang** (*pret.*) PS. lxxvii. 59 ; þrong LA3. 10652 ; þe blod þurch brini þrong TRIST. 3264 ; þrough her herte it þrong GOW. II. 262 ; þrungen MK. v. 24 ; JUL. 67 ; LANGL. *B* v. 517 ; binnen heo þrungen [þronge] LA3. 9421 ; þrongen A. P. ii. 1775 ; þronge EM. 659 ; **thrungen**

(*pple.*) PS. lxxii. 22 ; WICL. E. W. 319 ; *comp.* bi-, for-, ȝe-þringen ; *deriv.* þring, þrang, þreng, þrengen, þrung.

þripel, adj., *for* *triple ; *triple*, ORM. 6770.

þrist, *see* þürst.

þrīste, adj., *O.E.* þrīste,=*O.L.G.* thrīsti ; *bold*, HOM. I. 117 ; LAȝ. 356 ; O. & N. 171 ; arȝe we beoþ to done god to uvele al to þriste P. L. S. viii. 10.

þrīste-lēchen, v., *O.E.* þrīstlǣcan ; *make bold*, HOM. I. 25.

þristen, *see* þürsten. þrīsten, *see* þrūsten.

þrīven, *see* þrīfen. þrō, *see* þrā.

þrobben, v., *throb* ; þrobbart (*pple.*) LANGL. A xii. 48.

þrōȝe, þrōȝen, *see* þrāȝe, þrāȝen.

þroh, *see* þruh. þrōlī, *see* þrālī *under* þrā.

þrom, *see* þrum.

þrōn, v., *O.N.* þrōa-sk, = Ger. drūhen ; *grow, increase* ; þro M. H. 112.

þrong, *see* þrang. þrop, *see* þorp.

þrostle, sb., *O.E.* þrostle,=*M.H.G.* trostel ; ?=þruschel ; *throstle*, O. & N. 1659 ; þrostel GOW. I. 54 ; þrostil H. S. 7481 ; þrustle SPEC. 26 ; þrustele WILL. 820.

þrostel-coc, sb., *throstle-cock* ; (?*ms.* þrestel-) SPEC. 43 ; þrostel-, þrustelcok CH. C. T. B 1959.

þrōte, sb., *O.E.* þrotu (*gen.* þrotan),=*O.H.G.* droza, ? *cf. O.Fris.*, *M.Du.*, *M.L.G.* strote ; *throat*, A. R. 216 ; O. & N. 558 ; AYENB. 14 ; ALIS. 5952 ; MAND. 290 ; S. S. (Wr.) 2070 ; PS. v. 11 ; GAW. 1740 ; throte 'guttur' PR. P. 492 ; SAX. CHR. 262 ; þrota LEECHD. III. 106.

þrōte-bolle, sb., *O.E.* þrotbolla ; *larynx*, R. S. v (MISC. 178) ; CH. C. T. A 4273 ; REL. II. 78.

þrotlen, v.,=*Ger.* droszeln ; *throttle, strangle* ; throtlet (*pret.*) D. TROY 12752.

þrouȝ, *see* þruh.

þrōwe, *see* þrāȝe, þrāwe.

þrōwen, *see* þrāwen.

þrowin, v., *O.E.* þrowian, = *O.H.G.* drōēn, trōēn, druoēn, trūēn ; *bear, suffer, endure*, KATH. 1140 ; þrowede (*pret.*) HOM. II. 101 ; þrouwede I. 17 ; he ᵹrowede and ᵹolede GEN. & EX. 1180.

þrowunge, sb., *O.E.* þrowung, þrowing,= *O.H.G.* druunga ; *suffering*, HOM. I. 119 ; A. R. 372 ; þrowinge ORM. 15205 ; þrowinge MISC. 38 ; ᵹrowing GEN. & EX. 1317.

þrublen, v., *press* ; þrublande (*pple.*) A. P. ii. 504 ; þrobled (*pret.*) A. P. ii. 879 ; threpild [threpelitt] (*congregated*) ALEX. (Sk.) 1476.

þrücche, sb., *O.E.* (of-)þrycce,=*M.H.G.* druc ; *push, rush* ; thricche D. TROY 12752 ; þrich GAW. 1713.

þrücchen, v., *O.E.* þryccan,=*O.H.G.* drucchan ; *push, rush* : a dun þrucche LAȝ. 19483 ; þriȝt (*pret.*) GAW. 1443 ; þriȝt (*pple.*) A. P. ii. 135.

þruh, sb., *O.E.* þruh,=*O.H.G.* druha, truha, *M.H.G.* truhe, *O.N.* þrō ; *coffin*, HOM. I. 51 ; KATH. 2515 ; þeos þruh A. R. 378 ; one þruh of stone MISC. 51 ; thrugh TOWNL. 290 ; þroh CHR. E. 747 ; þrouȝ P. L. S. xvi. 168 ; SHOR. 4 ; TREV. VII. 535 ; H. V. 13 ; þrowe (*dat.*) REL. I. 161 ; throghes (*pl.*) PS. lxxvii. 6 ; thurghis ALEX. (Sk.) 4452.

þruh, *see* þurh.

þrum, sb.,=*O.H.G.* drum (*finis, stirps*), M. *Du.* drom (*licium*) ; *thrum* ; thrum 'licium,' PR. P. 493 ; throm VOC. 235.

þrum, þrüm, sb., *O.E.* þrymm ; *troop, power, glory* : on a thrum [þrom] MAN. (F.) 13459 ; þrom C. M. 7423 ; on a throm *in a heap* ALEX. (Sk.) 3642.

þrüm-ferde, sb., *troop*, LAȝ. 1356.

þrim-setel, sb., *O.E.* þrymsetl ; *throne*, MAT. v. 34 ; þrimsetles (*pl.*) HOM. I. 219.

þrumlen, v.,=*Ger.* drumeln ; *stumble* : he thromlide [thrumbled] (*pret.*) at þe þreshe-fold LANGL. C vii. 408 ; thrompelde A v. 201.

þrummen, v.,=*M.Du.* drommen ; *compress* : þa þre boc þrumde (*pret.*) to are LAȝ. 54.

þrümnesse, *see* þrimnesse.

þrung, sb., *from* þringen ; *throng, crowd*, A. R. 162 ; þrunge (*dat.*) LAȝ. 27524.

þrungen, v., *throng, press close* : heo þrungeð (*pres.*) alle to gederes A. R. 252.

þruppe, sb., ? *passage, defile* : Brutus hefede gode cnihtes ... ilead to þære þruppe (*later text* to gadere) LAȝ. 531 ['*trespas*' WACE 271.]

þrüsche, sb., *O.E.* þrysce, *cf. O.H.G.* drosca ; *thrush*, O. & N. 1659 ; WILL. 820 ; þruisse P. L. S. xxxv. 96.

þrushil, sb., = *M.H.G.* droschel ; *thrush* ; (*ms.* thrushil) '*merula*' PR. P. 493 ; *see* þrostle.

[þrüsmen, v., *O.E.* (a-)þrysman, = *O.Fris.* t(h)resma ; *suffocate*; *comp.* a-, for-, of-þrüsmen.]

þrüst, *see* þürst.

þrüsten, v., *O.N.* þrȳsta,=? *O.H.G.* drūsten ; *thrust* ; þruste LEG. 22 ; thristin '*premo*' PR. P. 491 ; i shal þristen uth þin heie (*v.* eie) HAV. 1152 ; þat he mouthe in seckes þriste HAV. 2019 ; þrīstiþ [þrestiþ] ['*premit*'] (*pres.*) WICL. PROV. xxx. 33 ; þrüste (*pret.*) CHR. E. 671 ; þruste [þriste, þreste] CH. C. T. E 2003 (Wr. 9877) ; ᵹrist(e) GEN. & EX. 2110 ; þurch his bodi he þreste TRIST. 2391 ; hi þresten out hare eȝen AYENB. 204 ; þriste MAN. (F.) 8886 ; *comp.* to-þrüsten.

þrustle, *see* þrostle.

þu, þū, pron., O.E. þu, þū, = Goth. þū, O.N. þū, O.L.G., O.Fris. thu (? thū), O.H.G. du, Ir., Gael., Lat. tū, Gr. τύ (σύ); thou, FRAG. 6; KATH. 217; ORM. DEDIC. 11; O. & N. 33; HORN (L.) 91; BEK. 40; ðu GEN. & EX. 360; þu, þou, þeou LAȝ. 475, 482, 2978; HAV. 527; þou C. L. 326; AYENB. 5; GREG. 37; LANGL. A i. 25; þou [thou] CH. C. T. A 1094; wil(t) te (for tu) HAV. 528; ne þarf þe (for þu) H. M. 11; þin [O.E., O.N. þīn, O.L.G. thīn, Goth. þeina, O.H.G. dīn] (gen.) HAV. 1128; þe [O.E. þe, = O.L.G., O.Fris. thi, O.N. þer, O.H.G. dir, Goth. þus], (dat.) thee KATH. 210; A. R. 12; BEK. 39; CH. C. T. A 1129; hider am ic send to þe þis blisse þe to kiþen ORM. 210; hwi leavestu ham þe ane JUL. 22; ȝif þu þe self wel nimest gom ORM. 4162; of þi sulf A. R. 276; þou wost þi selve CH. C. T. A 1174; þē [O.E. þec, þe, = O.L.G. thi, thic O.Fris. thi, O.N. þik, Goth. þuk, O.H.G. dih, dich] (acc.) LAȝ. 474; KATH. 219; ORM. 670; O. & N. 34; þou sselt þe resti AYENB. 7; for to preise þe selve FER. 158; þi selven LANGL. A i. 131; þi zelve AYENB. 145.

þū, see þēow. þū, see þē under þe.

þücke, sb., = O.N. þykkr (? indignation, offence); trick, injury: vor te don þe eft swuche þucke A. R. 326.

þüden, v., O.E. þȳdan; press, push; þŭdde (pret.) LAȝ. 1898; MARH. 12; comp. ȝe-, under-þüden.

þüder, see þider. þuef, see þēof.

þuften, sb., ? female servant, A. R. 4; þuftenes (gen.) H. M. 45.

þülc, adj., O.E. þylc, = O.N. þvilīkr; such, that; þulke time ROB. 182; þulke dom C. L. 180; þilke stevene FL. & BL. 54; þilke text CH. C. T. A 182; þelke sone SHOR. 140; þilke (pl.) H. H. 135; WILL. 3530; þilke daies LAȝ. 1284*.

þüld, sb., O.E. (ge-)þyld, = O.L.G. (gi-)thuld, O.H.G. dult; patience; þild ORM. 2613; PS. ix. 19.

þülde-līche, adv., O.E. þyldelīce, = O.H.G. (ka-)dultlīcho; patiently, A. R. 106; JUL. 29; þildeliȝ ORM. 1186; comp. un-þüldelīche.

þüldi, adj., O.E. (ge-)þyldig, = O.H.G. dultīg; patient, KATH. 177.

þüllīch, adj., O.E. þyllīc, þys-, þuslīc; such, of this kind, KATH. 849; A. R. 8*; þullich þulli H. M. 9, 25; þulh [þulli] mot MARH. 7; þellic MK. ii. 12; þe oþer heste is þellich AYENB. 6; þellīcne (acc. m.) MAT. xviii. 5; þüllīche (pl.) A. R. 84*; þelliche þinges AYENB. 27; of þulliche nesche wepnen HOM. I. 255.

þüme, sb., O.E. þūma, = O.Fris. thūma, O. H.G. dūmo; thumb, A. R. 18; þoume REL. I. 190; SAINTS (Ld.) xlv. 356; AYENB. 43; thoumbe PR. P. 492; þome VOC. 184; þombe, thombe LANGL. B v. 439; CH. C. T. A 563.

þūmel, sb., O.E. þȳmel (LEECHD. II. 150), = O.N. þūmall; thimble; thimbil 'theca' (pollicis) PR. P. 491; N. P. 13.

[þun ?.]

þun-wange, sb., O.E. þunwange, -wenge, -wonge, = O.N. þunnvangi, O.H.G. dunwengi; temple, 'tempus,' REL. I. 54; thunwonge PR. P. 493; thonwange HALLIW. 866.

þünchen, v., O.E. þyncan, = O.L.G. thunkean, O.N. þykkja, O.H.G. dunchan, Goth. þugkjan; seem, KATH. 691; A. R. 122; hit walde me þunchen þet softeste beð HOM. I. 35; þunche P. S. 194; þinchen P. L. S. viii. 31; HOM. I. 33; þinche O. & N. 346; ðinken GEN. & EX. 234; þenche MISC. 35; þinchest (pres.) MISC. 84; þuncheð KATH. 346; hit þuncheð fisc LAȝ. 1325; swa heom bezst þincheð 22578; þuncheþ C. L. 720; heom þuncheþ FRAG. 5; me þuncheþ H. H. 138 (140); þincheð HOM. II. 220; þinkeþ ORM. 15667; FL. & BL. 42; þincheþ, þuncþ, þincþ, þinkþ O. & N. 255, 541, 1672; þincþ P. L. S. viii. 3; þinkþ MISC. 58; SHOR. 30; me þinkeþ, þinkes WILL. 430, 839; þingð HOM. I. 145; ase me þingþ AYENB. 18; (weies) . . . þe monnen þuncheð rihte HOM. I. 119; þünche (subj.) LAȝ. 13584; KATH. 278; þinke LANGL. B xviii. 249; þühte, þühte (pret.) KATH. 85; (ms. þuhhte) ORM. 8936; hit þuhte read blod A. R. 112; þuhte HOM. I. 129; þuhte, þucte (printed þutte), þuste [þohte] LAȝ. 770, 4435, 5268; þuhte, þuȝte O. & N. 21, 1661; þuȝte FL. & BL. 54; BEK. 56; ðugte (ms. ðhugte) GEN. & EX. 407; þuste ASS. 226; þoȝte, þouȝte ROB. 14, 181; þouȝte LANGL. B xi. 47; him þouhte [þouȝte, thoughte] CH. C. T. A 954; comp. bi-, for-, ȝe-, mis-, of-þünchen.

þuner, sb., O.E. þunor, = O.H.G. donar; thunder; þoner Iw. 370; PS. lxxvi. 19; thoner TOWNL. 63; ðunder GEN. & EX. 1108; þonder MAND. 292; CH. C. T. D 732; þunres (gen.) KATH. 2024; (printed wunres) HOM. I. 43; þonres ST. COD. 54; þunres dæi [O.E. þunres dæg, = O.Fris. Thunres dei, M.H.G. Dunres, Donres tag, O.N. þōrs dagr] Thursday LAȝ. 13929; þurs dei A. R. 40; þurs, þors dai LANGL. B xvi. 140; þores dai ROB. 297; C. L. 1417; SHOR. 126; þunre (dat.) HOM. I. 43; þondre ROB. 308; AYENB. 130.

þoner-blăst, sb., peal of thunder, Iw. 339; þonderblast FLOR. 1643.

þonder-clap, sb., thunder-clap, CH. C. T. I 174.

þonder-dent, sb., thunder-clap, CH. C. T. A 3807.

þuneren, v., O.E. þunerian, þunrian, = O.H.G. donarōn; thunder; thunderin 'tonat' PR. P. 493; þoneres (pres.) VOC. 201; þunrede (pret.) JOHN xii. 29; þonord HAMP. xvii. 15; thunret D. TROY 3691; thonered PS. xvii. 14; þonderde JOS. 235; þonrid S. S. (Wr.) 2213.

þunne, adj., *O.E.* þynne,= *O.N.* þunnr, *O.H.G.* dunni, *cf. Lat.* tenuis, *Gr.* τανύς; *thin,* A. R. 144; REL. I. 120; MAP 336; SPEC. 47; þinne CH. C. T. *E* 1682; þenne SHOR. 99; þünne (*adv.*) O. & N. 1529.

þin-hēde, sb., *thinness,* '*tenuitas,*' PR. P. 491.

þinnesse, sb., *O.E.* þynnyss; *thinness,* '*tenuitas,*' PR. P. 491.

þünnen, v., *O.E.* (ge-)þynnian; *make thin*; thinnin '*tenuo,*' PR. P. 491.

þurfe, adj., *O.N.* þurfi; *needful*: al þat hem was þurfe (*ms.* þurrfe) ORM. 9628.

þurfen, v., ? *O.E.* þurfan (*pret.* þorfte), ?= *O.N.* þurfa (*pret.* þurfti), *Goth.* þaurban (*pret.* þaurfta), *O.H.G.* (be-)durfan (*pret.* dorfti); *from* þarf; *need*; þurfte (*pret.*) A. R. 336; P. S. 338; (*ms.* þurrfte) ORM. 16164; þurte HOM. II. 35; M. H. 40; armed ... as semli ... as ani segges þurte WILL. 3788; ꝥurte GEN. & EX. 234; thurte TOWNL. 317; þorte FL. & BL. 253; ne þorte us have friȝt MAP 338; *comp.* bi-þurfen.

þurh, prep., *O.E.* þurh, þyrh, þerh,= *O.L.G.* thurh, thuru, *O.H.G.* durh, duruh, *O.Fris.* thruch, *Goth.* þairh (διά); *mod.Eng.* through, thorough; *through; by means of; by*: þurh muchele ȝeoven KATH. 37; þah he were deadlich þurh þat he mon was 1117; as ha wes iwisset þurh þen engel JUL. 39; isahet þurh & þurh IBID.; þurh Cristes helpe ORM. DEDIC. 26; he heold þe stronge castles þurh [þorh] ... his fader ȝefe LAȝ. 401; þurh elleoven ȝere þe king wunede þere 31871; þorh 283; þurch (*ms.* þurhc), þur SAX. CHR. 263, 264; þurh, þurch HOM. I. 223; þurh þuruh A. R. 92; and ich so do þurȝ niht and dai O. & N. 447; þurȝ [þureh] ginne castel and burȝ me mai iwinne 765; þurh, þurch 1401; þurw, þur 1405; þurch MISC. 30; þurch godes grace GREG. 55; ꝥurg GEN. & EX. 195; þurȝ A. P. i. 669; ichosen þurȝ us PROCL. 2; þurȝh, þurch (? *printed* burth) WILL. 635, 655; þurgh [thurgh] CH. C. *A* 920; thurgh PR. C. 1428; thurght HAMP. PS. iv. 6*; ix. 14*; lxx. 10*; þureȝ FL. & BL. 312; þuregh HOM. II. 33; þorȝ SHOR. 14; MIRC 502; þet iziȝþ þorȝ þane wal AYENB. 81; þorgh MAND. 3; þoruh, þoruȝ, þoru, þurf L. H. R. 18, 19; þoru HAV. 631; (*spelt* þorw 367); þoru, þorw, þorw ROB. 1, 3, 84; þoru (*ms.* þorw) DEGR. 1263; þoru (*ms.* þorw) deþ he scholde þe lif forlete C. L. 178; þourh SPEC. 31; boþe were maked þourh me H. H. 94 (96); þorouȝ LANGL. *A* ii. 123; thorw *B* ii. 152; þorou APOL. 30; þurf, þoruȝ TREAT. 132 (SAINTS (Ld.) xlvi. 424); þruh REL. I. 102; þur LEECHD. III. 94.

þurh-bóren, v.,= *O.H.G.* durhþorōn; *bore through*; þoroubore MAN. (F.) 16184.

þurh-costnen, v., *completely provide*; þurh-costned (*pple.*) LAȝ. 25440.

þurh-drīven, v., *O.E.* þurhdrīfan; *drive*

through, KATH. 1943; þurhdriven (*pple.*) KATH. 1204.

þurh-ēode, v. (*pret.*), *passed through*: heo þurheoden [þorhȝeode] Francene þeode LAȝ. 5217.

þurgh-fare, sb., *thoroughfare,* CH. C. T. *A* 2847; thurghfare PR. P. 493.

þurh-fēren, v., *go through*; þurhfērde (*pret.*) KATH. 1147.

þurh-gān, v., *O.E.* þurhgangan; *go through,* ORM. 12860; þurhgon LAȝ. 19645.

þurh-gengen, v., *go through, spread over,* LAȝ. 1207.

þurh-leasten, v., *endure,* HOM. I. 251.

þurh-lōken, v., *look through,* ORM. DEDIC. 68.

þurh-rennen, v.,= *M.H.G.* durchrennen; *run through*; þurhærnen LAȝ. 16657; þurh-arnden (*pret.*) LAȝ. 12129.

þorou-rīde, v.,= *M.H.G.* durchrīten; *ride through,* MAN. (F.) 14516.

þurh-schēoten, v., *O.E.* þurhscēotan; *shoot through*; þorghschoten (*pret.*) MAN. (F.) 4373.

þurh-scrīden, v., *go through,* LAȝ. 10887.

þurh-sēchen, v., *seek through*; þurhseken ORM. 242; þurhsēcheð (*pres.*) HOM. II. 191; þurhsicheþ HOM. I. 165; þorwsōuȝte (*pret.*) ROB. 151; þurhsōht (*pple.*) KATH. 520; þourhsoht SPEC. 99; þurchsouȝt GREG. 74.

þurh-sēon, v., *O.E.* þurhsēon; *see through, perceive,* A. R. 50; þurhsihð (*pres.*) HOM. II. 222; ꝥurhsiȝð HICKES I. 223.

þurh-spitien, v., *cover with spikes*: and þurh spitien hit [þe hweol] al wið irnene gadien JUL. 57.

þurh-sticchen, v., *stitch through*; þuruh-stihten (*pret.*) A. R. 272.

þurh-stingen, v., *sting through*; þourh-stong (*pret.*) SPEC. 70; þurhstungen (*pple.*) HOM. I. 147.

þurh-ūt, adv. & prep., *throughout,* MARH. 22; ȝif he wile heo þurhut forleten HOM. I. 23; þurhut al his kinelond LAȝ. 4826; þurh-, þurȝut O. & N. 879; ꝥurgut GEN. & EX. 3704; þuruhut A. R. 212; þourhout al France P. S. 193.

þurh-waxen, v.,= *M.H.G.* durchwahsen; *grow over*: wes þe munt þurhwexen (*pple.*) mid ane wude feiren (*r.* feire) LAȝ. 18338.

þurh-wúnden, v., *wound through*: sinnes þat ... þurhwúnden (*pres.*) all þat bodiȝ ORM. 17443.

þurh-wunian, v., *O.E.* þurhwunian; *continue,* LAȝ. 1384; þurhwunest (*pres.*) KATH. 662.

þürl, sb., *O.E.* þyrel; *perforation, window*: þet þurl A. R. 96; ꝥirl REL. I. 211; þürle (*dat.*) A. R. 344; þürles (*pl.*) A. R. 50; þurles, thirles C. M. 18687; þerles AYENB. 204; *comp.* ēh-, nóse-þürl.

þŭrlen, v., *O.E.* þyrlian, *from O.E.* þyrel, = *O. H.G.* durchil (*pierced*); *mod.Eng.* thrill; *drill, pierce*; A. R. 392; C. L. 1152; thirlin, thrillin '*perforo*' PR. P. 491; þirle [þrille] WICL. NUM. xxiv. 8; þrulle TREV. I. 339; þirles (*pres.*) WILL. 612; þirle (*imper.*) H. V. 26; þŭrlede (*pret.*) JUL. 40; TREV. VII. 349; þerlede L. H. R. 223; þurleden A. R. 292; JOS. 509; þurleden, þirled LANGL. *A* i. 148, *B* i. 172; þŭrled (*pple.*) SPEC. 62; thirled CH. C. T. *A* 2710; thirlid MIR. PL. 151; iþurled MISC. 140; iþurled [iþorled] LA3. 4541.

þŭrlunge, sb., *O.E.* þyrelung; *perforation*, HOM. I. 207.

þŭrrok, sb., ? *O.E.* þurruc, = *M.Du.* durk; *sink*, CH. C. T. *I* 363; thurrok '*sentina*' PR. P. 493.

þŭrs, sb., *O.E.* þyrs, = *O.N.* þurs, *O.H.G.* durs; *gigantic spectre*, KATH. 1880; MARH. 11; thurs D. ARTH. 1100; þirs WICL. IS. xxxiv. 15.

þŭrst, sb., *O.E.* þurst, þyrst, = *O.L.G.* thurst, *O.H.G.* durst; *thirst*, A. R. 114; ALIS. 5058; MISC. 37; CH. C. T. *B* 100; þurst [þorst] LA3. 6224; þurst [þrust] KATH. 1702; þurst, furst, þrist, þrest LANGL. *B* xviii. 366, *C* xxi. 413; þorst AYENB. 73; þurst, þirst HOM. II. 35, 75; þirst, þrist ORM. 5682, 14484; thirst, thrist PR. P. 491; þerst FER. 2810; þrist LIDG. M. P. 53; thrist PR. C. 6204; LUD. COV. 105; thrist, threst PR. C. 3254, 6204; þrest IW. 2974; MAND. 230; þŭrste (*dat.*) LA3. 18951.

þŭrst-lēu, adj., *thirsty*, LIDG. M. P. 75.

þŭrsten, v., *O.E.* þyrstan, = *O.N.* þyrsta, *O.H. G.* dursten; *thirst*: ne schal him þurste nevere MISC. 85; þristen [þirste] WICL. ECCLUS. xxiv. 29; thristin, thirstin PR. P. 491; þŭrsteþ (*pres.*) GOW. II. 135; me þursteþ [þristeþ] LANGL. *B* xviii. 365; þirsteþ ORM. 14485; (hio) þirsteð MAT. v. 6; þŭrste (*pret.*) A. R. 188; *comp.* a-, for-, of-þŭrsten.

þŭrsti, adj., *O.E.* þurstig, ðrystig, = *O.H.G.* durstag; *thirsty*; þristi3 ORM. 6163; thristi PR. P. 491; thresti PR. C. 6165.

þŭs, adv., *O.E.* þus, = *O.L.G.*, *O.Fris.* thus; *thus*, FRAG. 7; A. R. 10; ORM. 237; C. L. 610; LANGL. *B* ix. 92; MAN. (H.) 164; þus, thus PR. P. 493, 536; þus he hine bipecheð HOM. II. 217; þus [þos] hafeð Modred idon LA3. 28136; þus sone 29625; þus þriste O. & N. 758; þus old A. D. 253; þos SHOR. 3; þos, þous, þus AYENB. 52, 71, 189.

þŭs, see þes.

þŭsend, sb., *O.E.* þūsend, = *O.N.* þūsund, *Goth.* þūsundi, *O.H.G.* dūsunt; *thousand*, A. R. 236; þusend, þusund [þousend] LA3. 465, 24613; a þusend mile HORN (L.) 319; an þusende ORM. 1316; þousend(e) TREAT. 134 (SAINTS (Ld.) xlvi. 492); þousend [þousand] CH. C. T. *A* 1954 (Wr. 1956); þousond AYENB. 71; þŭsende, þusunde (*pl.*) LA3. 545, 1669; twa þusende MK. v. 13; feower þusend manne

MAT. xv. 38; fowr þusend KATH. 2037; mid teon þusenden LK. xiv. 31.

þŭsend-fald, adj., *thousandfold*, KATH. 2323; þusendfeld HOM. II. 191.

þŭstre, see þēostre.

þūten, v., *cf. M.H.G.* duzen; *say '*thou*' to a person*; thoutin PR. P. 492.

þūten, see þēoten.

þŭvel, sb., *O.E.* þȳfel; *shrub*: þifel '*frutex*' FRAG. 3; þŭvele (*dat. pl.*) O. & N. 278.

þwang, sb., *O.E.* þwang, þwong, = *O.N.* þvengr; *thong*, ORM. 10439; þwang IW. 3160; þwong [þwang] LA3. 14219; þwong WICL. JOHN i. 27; GAW. 194; þong ROB. 116; REL. I. 115; þvanges (*pl.*) M. H. 10; þwonges [þwanges] LA3. 22295; þwonges [þonges] TREV. 369; þonges WILL. 1720; þwange MK. i. 7; *comp.* schō-þwang.

þwanged, adj., *thonged*: þwongede [þongede] scheon A. R. 362.

þwarle, adj., ? *intricate*: a þwarle knot GAW. 194.

þweorh, adj., *O.E.* þweorh, = *Goth.* þwairhs (ὀργίλος), *O.N.* þverr, *O.H.G.* dverh (*transverse, perverse*); *perverse*: þweore cneores MAT. xvii. 17.

þwert, adv., *O.N.* þvert *acc. neut. of* þverr; *athwart, across*: over þwert HAV. 2822; over þvert MAN. (H.) 241; a thirt *athwart* PARTEN. 169; þwert over MARH. 10; A. R. 82; þwert over þe ilond TREV. V. 225; his herte ðo wurð ðwert (*for* þwer *perverse*) GEN. & EX. 3099.

þwert-ūt, adj., *completely*, ORM. 194; þvertut forlore HOM. II. 123.

þwerten, v., *thwart, hinder, prevent*; ðwerted (*pret.*) GEN. & EX. 1324.

þwitel, sb., *whittle, jack-knife*, TREV. IV. 329; CH. C. T. *A* 3933.

þwīten, v., *O.E.* þwītan; *cut up*; thwitin PR. P. 488; þwiten (*pple.*) R. R. 933.

u.

ŭch, see 3ehwilc.

ŭder, sb., *O.E.* ūder, ūdr, = *M.Du.* uider, *M. H.G.* ūter, *Gr.* οὖθαρ; *udder*; uddir, iddir '*uber*' PR. P. 258.

ūfel, see ŭvel. ufere, see uvere.

ug, sb., *O.N.* uggr; *fear*; ugge (*dat.*) HOM. I. 209.

ug-lī, adj., *O.N.* uggligr; *mod.Eng.* ugly; *horrid, frightful*, '*horribilis*,' PR. P. 509; S. S. (Web.) 2782; CH. C. T. *E* 673; IW. 1730; PR. C. 6683; MIR. PL. 164; an uglike snake GEN. & EX. 2805.

ŭglines, sb., *ugliness*, PR. C. 2364.

ugsome, adj., *frightful*, D. TROY 877.

uggin, v., *O.N.* ugga; *shudder, feel horror,* '*horreo*,' PR. P. 509; ugge PR. C. 6419; þat ow **uggi** (*pres. subj.*) wið ham A. R. 92*; **uggid** (*pple.*) APOL. 109.

ugging, sb., *horror,* GEN. & EX. 950; HAMP. PS. *page* 514.

ūhte, sb., *O.E.* ūhte,=*O.L.G.* ūhta, *O.H.G.* ūhta, uohta, *Goth.* ūhtwō, *O.N.* ōtta; *early morning*: on uhten ORM. 2484; hi sloȝen and fuȝten þe niȝt and þe uȝten HORN (L.) 1376 [(R.) 1424; ouȝten (H.) 1415]; upon uȝten A. P. ii. 893.

ūht-song, sb., *O.E.* ūhtsang,=*O.H.G.* ūhtisang, *O.N.* ōttusöngr; *matins,* A. R. 20; uhtensang ORM. 6360.

ūhten-tīd, sb., = *O.N.* ōttutīd; *morning,* ORM. 5832.

ülde, see **elde.**

ūle, sb., *O.E.* ūle,=*O.H.G.* ūla, uwila, *O.N.* ugla; *owl,* '*bubo, strix*,' FRAG. 3; O. & N. 26; oule AYENB. 27; GOW. I. 100; CH. P. F. 343.

ülke, see **ilke.**

[**ulm,** sb., *Lat.* ulmus; *elm.*]

ulm-tree, sb., *O.E.* ulmtrēow; *elm-tree,* WICL. IS. xli. 19.

ümbe, umbe, prep. & adv., *O.E.* ymbe, ymb, embe,=*O.L.G.* umbi, *O.H.G.* umpi, umbi, umbe, *O.N.* umb, um, *Lat.* ambi, amb-, *Gr.* ἀμφί; *round, about, after,* SPEC. 49; GAW. 589; A. P. ii. 879; PALL. vii. 106; beo umbe me MARH. 6; is umbe (*ms.* vmbe) . . . þet heo him luvie A. R. 218; umbe stunde 344; umbe stount BARB. vii. 398; umbe hwile *sometimes* KATH. 12; umbe (*ms.* ummbe) trin(d) [=*M.L.G.* umme trint] *around* ORM. 17563; þat we nu mælen umbe 304; umbe, embe HOM. I. 51, 95; umbe fif winter LAȝ. 6617; he þohte embe uvel 6563; umbe þrowe SPEC. 25; umbe, umbi ALEX. (Sk.) 1154, 2209, 3250; ævre um wile SAX. CHR. 262; um while LANGL. *B* v. 345; MAN. (H.) 241; um qvile ALEX. (Sk.) 23, 3079, 4745; embe MAT. xvii. 12; REL. I. 131; to speke so embe noȝt P. L. S. xix. 164.

ümbe-breid, sb., *reproach*; umbraid ALEX. (Sk.) 1800; **umbreides** (*pl.*) MAN. (F.) 3485.

ümbe-breiden, v., *address; reproach*; **umbreide** (*subj.*) MAN. (F.) 8004; **umbraide** (*pret.*) A. P. ii. 1622.

ümbe-casten, v., *throw round*; umbecast BARB. v. 552; **umbekesten** (*pret.*) GAW. 1434; **umbecast** (*pple.*) WILL. 4693.

ümbe-clappen, v., *surround*; umbiclappis (*pres.*) ALEX. (Sk.) 4171; (þai) umclappis 2473; **umbiclappid** (*pple.*) 3451.

ümbe-clōsen, v., *enclose*; **umbclōsed** (*pret.*) MAN. (F.) 4080.

ümbe-clüppen, v., *O.E.* ymbclyppan; *embrace*; **umbeclipped** (*pret.*) GAW. 616; umclipped PS. cxviii. 61.

ümbe-delven, v., *dig round*; **umbidelve** (*imper.*) PALL. iv. 324.

ümbe-fálden, v., *fold around*; **umbefóldes** (*pres.*) GAW. 181; **umfáldin** (*pple.*) *enclosed* ALEX. (Sk.) 4717.

ümb-ēode, v., *O.E.* ymbēode; *go round*; umyhode ['*circumivi*'] PS. xxvi. 6; umbi-yeden (*ms.* un-) HAV. 1842.

ümbe-gang, sb., *O.E.* ymbegang,=*O.H.G.* umbi-, umbegang, *O.N.* umgangr; *circuit, circumference*; umgang ['*circuitus*'] PS. xi. 9; C. M. 9192.

ümbe-gangen, v., *O.E.* ymbegangan, ymbgān, =*O.H.G.* umbigangan, umbegān; *go about, go round*; umga HAMP. PS. lviii. 7, 16; umbigonge (*pple.*) PALL. iv. 197, 435; umbigoon HALLIW. 899.

umganginge, sb., *circuit,* HAMP. PS. cxii. 3*.

ümbe-ȝiven, v.,=*O.H.G.* umbigeban; *place round*; **umgivand** (*pple.*) PS. iii. 7.

ümbe-grōwen, v., *grow around*; **umbegrōwen** (*pple.*) A. P. ii. 488.

ümbe-hēden, v., *guard about*; **umhēde** (*imper.*) ALEX. (Sk.) 731*.

ümbe-hilen, v., *cover up*; **umhild** (*pret.*) HAMP. PS. xliii. 21.

ümbe-hoȝe, sb., *O.E.* ymbehoga; *care, anxiety*; **embhugan** (*dat.*) MAT. vi. 34 (*earlier text* ymbehogan).

ümbe-hwīle, see under **ümbe.**

ümbe-keorvunge, sb., *circumcision,* HOM. I. 207.

ümbe-lappe, v., *wrap around, embrace, envelop,* CATH. 402; **umlapped** (*pret.*) PS. xxxix. 13; **umbilapped** (*pple.*) WICL. HEB. v. 2; umlappe HAMP. PS. xxvi. 5; **umlappis** (*pres.*) HAMP.PS. i. 1*; umlappid HAMP. PS. xxxix. 16.

ümbe-leggen, v., *lay around*; **umbeleid** (*pple.*) MAN. (H.) 187.

ümbe-liggen, v.,=*M.H.G.* umbeligen; *lie around*; umbeliȝe A. P. i. 836.

ümbe-lōken, v., *look around*; umbiloke C. M. 8468.

ümbe-lūken, v., *shut round, enclose*; umlouke HAMP. PS. cxliv. 3.

ümbe-rōwen, v., *row around*; **umbirōwen** (*pple.*) LAȝ. 114.

ümbe-scháden, v., *shade around*; umshade TOWNL. 75.

ümbe-schéren, *cf. Dan.* omskære; *circumcise* (*ms.* ummbesherenn) ORM. 4138; **umbeshǽren** (*pret.*) ORM. 4084; **umbeshoren** (*pple.*) ORM. 4080.

ümbe-schīnen, v., *O.E.* ymbscīnan; *shine around*; **umbeschōn** (*pret.*) A. P. iii. 455.

ümbe-sēgen, v., *besiege*; **umsēged** (*pret.*) HAMP. PS. xxi. 11.

ümbe-sēon, v., *look about*; **umsē** (*imper.*) ALEX. (Sk.) 3728.

ümbe-setten, v., *O.E.* ymbsettan ; *surround*; umsette (*pret.*) C. M. 195 ; umset (*pple.*) with sere enmis PR. C. 1250.

ümbe-snīþen, v., *O.E.* ymbsnīðan ; *circumcise* ; embsniþen HOM. I. 81 ; embscniȼen LK. i. 59.

ümbe-standen, v., *O.E.* ymbe-, ymbstandan ; *stand around*; umbistōde (*pret.*) HAV. 1875.

ümbe-strīden, v., *bestride, mount (a horse)* : he umstrād (*pret.*) a nobil stede IW. 1302.

ümbe-wealden, v., *surround*; ümbe-walt (*pres.*) A. P. ii. 1181.

ümbe-sweien, v., *surround*; umbesweied (*pple.*) wiþ seven grete wateres A. P. ii. 1380.

ümbe-þanc, sb., *O.E.* ymbeþanc ; *thought* ; embeþonke (*dat.*) HOM. II. 87.

umbe-þenken, v., *meditate*, ORM. 1240 ; umbi-, umthinke the TOWNL. 4, 251 ; umbiþōghte (*pret.*) ISUM. 426 ; umþoght M. H. 79.

ümbe-ūten, prep., *O.E.* ymbūtan ; *round*; embeuten MK. xiv. 47.

ümbe-wæven, v., *enclose* ; umbewēved (*pret.*) GAW. 581.

[ümbri, sb., *O.E.* ymbren ; *Ember fast.*]
ümbri-dei, sb., *O.E.* ymbrendæg, *cf. O.N.* imbrudagr ; *Ember-day*, A. R. 70 ; embirdai PR. P. 139.

imbri-wike, sb., *Ember-week*, (*ms.* ymbri-) A. R. 70*.

[un-, on-, prefix, *O.E.* un-,=*O.L.G.*, *O.H.G.*, *Goth.* un-, *O.N.* ū-, ō-, *Lat.* in-, *Gr.* ἀν- ; *expressing negation.*]

[un²-, prefix, *O.E.* un-,=*O.Fris.* un-, und-,= *and ; expressing reversal of an action.*]

[un³-, prefix,=*O.Fris.*, *Goth.* und-, *? O.L.G.* und-, unt- ; *until.*]

un-abásit-lī, adv., *boldly*, BARB. vi. 20.

un-ablī, adv., *without ability*, ALEX. (Sk.) 2308.

un-achteled, adj., *innumerable*, GEN. & EX. 796.

un-æled, adj., *unburnt* ; uneled PALL. ix 103.

un-afüllendlīc, adj., *O.E.* unafyllendlīc ; *insatiable* : unafillendliche gredinesse HOM. I. 103.

un-aȝeten [onaȝete], adj., *unnoticed*, LAȝ. 25797.

un-aneomned, adj., *unnamed*, HOM. I. 43.

un-armed, adj., *unarmed* ; onarmed AYENB. 170 ; unarmit BARB. vii. 552.

un-aserved, adj., *undeserved*, ROB. (W.) 1256*.

un-äsked, adj., *unasked*, A. R. 338.

un-avanced, adj., *unprompted*, GOW. II. 205.

un-bain, adj., *cf. O.N.* ūbeinn ; *not ready*, C. M. 17735 ; unbein ANGL. IV. 186.

un-báld, adj., *O.E.* unbeald ; *timid*, C. M. 15914 ; unbolde REL. I. 120 ; unbálde [onbolde] (*pl.*) LAȝ. 995 ; unbolde SPEC. 100.

un²-bálden, v., *discourage, enfeeble* ; unbálded [onbalded] (*pple.*) LAȝ. 11547.

un-beden, adj., *O.E.* unbeden ; *unbidden* ORM. 17081.

un-bélden, v., *discourage* ; ounbelde ANGL. V. 280.

un²-benden, v., *unbend*, ALEX. (Sk.) 1974* ; unbende (*pret.*) GOW. I. 108 ; unbente GEN. & EX. 483.

un-bēne, adj., *comfortless* : in moni a bonk unbene GAW. 710.

un-berind, adj., *O.E.* unberende,=*Goth.* unbairands ; *unbearing*, HOM. II. 125.

un-bermed, adj., *unfermented*, ORM. 1591.

un-bibüried, adj., *unburied*, KATH. 2275.

un-bicomelīch, adj., *uncomely* ; he makede him unbicomelich HORN (L.) 1065.

un-biden, adj., *unbidden*, C. M. 14912.

un-bigged, adj., = *Swed.* obyggd ; *uncultivated*, ORM. 3199.

un-bigunnen, adj., *not begun*, ORM. 18574.

un-bihēve, adj., *unprofitable*, HOM. II. 7 ; unbihēfre (*compar.*) HOM. I. 265.

un-bihēve², sb., *disadvantage*, HOM. II. 121.

un-bihōf, sb., *disadvantage* ; unbihōve (*dat.*) LAȝ. 8576.

un-bilēave, sb., *for* *unȝelēave, *O.E.* ungelēafa ; *unbelief*, KATH. 261 ; unbileve HOM. II. 81 ; unbelēfen (*acc.*) MK. xvi. 14.

unbilēve-ful, adj., *unbelieving*, WICL. JUDG. xx. 5 ; WISD. x. 7.

unbilēvefulness, sb., *unbelief*, WICL. MK. ix. 23.

un²-binden, v., *O.E.* unbindan, = *O.Fris.* undbinda ; *unbind*, HOM. I. 7 ; ORM. 3682 ; unbinden [unbinde] LANGL. *B* PROL. 101 ; onbinde AYENB. 172 ; unbond (*pret.*) P. L. S. viii. 95 ; GEN. & EX. 2223 ; heo unbunde LAȝ. 5926 ; unbounden HAV. 601 ; unbúnden (*pple.*) HOM. I. 7 ; MISC. 55 ; unbounde ALIS. 85 ; M. T. 103.

un-bischoped, pple., *unconfirmed* : and longe beon unbishoped A. R. 208.

un-bisehenesse, sb., *inattention*, A. R. 344*.

un-bisorȝelīch, adj., *negligent* ; unbisorȝelīche (*adv.*) HOM. I. 43.

un-blisse, sb., *unhappiness*, FRAG. 5.

un-blessid, adj., *unblessed*, WICL. ECCLUS. xxvii. 24 ; onblissede (*pl.*) AYENB. 41.

un-blīþe, sb., *O.E.* unblīðe,=*O.H.G.* unblīdi ; *sad, joyless*, O. & N. 1585 ; HAV. 141 ; SPEC. 30 ; A. P. ii. 1017.

un²-bocle, v., *unbuckle*, CH. C. T. *F* 555.

un-boȝsam, adj., *disobedient* ; onboȝsam AYENB. 21 ; unbousom PR. C. 8596.

unboȝsamnesse, sb., *disobedience*; onboȝ-samnesse AYENB. 33.

un-boht, adj., *unbought, unatoned for*, ANGL. I. 10; A. D. 296; unbocht HOM. I. 163; un-bouht MISC. 60.

un-boren, adj., *unborn*, ORM. 17327; unbore REL. I. 184.

un-bōtlīc, adj., *irremediable*; unbotliche lure H. M. 17.

un-bowed, adj., *unbowed*, CH. BOET. iv. 7 (148).

un-brasten, adj., *unburst*, A. P. ii. 365.

un-brīche, adj., *O.E.* unbrȳce; *profitless*, HALLIW. 900.

un-būhsum, adj., *disobedient*, REL. I. 184 (MISC. 185); onbuxum PR. P. 364.

    unbūxum-hēd, sb., *disobedience*, GEN. & EX. 345.

    unbūxsumnes, sb., *disobedience*, HAMP. PS. lxxvii. 64*.

un-būried, adj., *unburied*, A. R. 352.

unc, pron., *O.E.* unc, = *O.L.G.* unc, *Goth.* ugk, ugkis, *O.N.* okkr (*dat. acc. dual*); *us two*, FRAG. 7; LAȝ. 25901; MARH. 5; REL. I. 186; GEN. & EX. 1776; (*ms.* unnc) ORM. DEDIC. 27; bitweonen unc (*v. r.* us twa) KATH. 1526: **unker** [*O.E.* uncer, *O.H.G.* unker, *O.N.* okkar (*gen. dual*), LAȝ. 23665; (*v. r.* us tvæ) hweþer unker O. & N. 151.

un²-cacchen, v., *let loose*; uncacchid (*pret. pl.*) hertes ['*timuerunt*'] *lost courage* ALEX. (Sk.) 2588.

un-callid, adj., *uncalled for*, ALEX. (Sk.) 832.

únce, ounce, sb., *O.Fr.* once, *Lat.* uncia; *ounce*, CH. C. T. G 1121.

un-chaunce, sb., *mischance*, ALEX. (Sk.) 822*.

ünche, sb., *O.E.* ynce, *from Lat.* uncia; *inch, ounce*, LEG. 66; inche PR. P. 261; enchen (*pl.*) FER. 4660; seven inche MAN. (F.) 10039; enches BEV. 746; feouwer unchene long LAȝ. 23970.

un²-clāþen, v., *unclothe*; uncloþe WICL. EZ. xlix. 19; uncleþe ALEX. (Sk.) 5505; **un-clōþede** (*pret.*) HAV. 659.

uncle, sb., *O.Fr.* uncle; *uncle*, ROB. 58; WICL. PARAL. xxvii. 32; uncles (*pl.*) 3443.

un²-clenchen, v., *unclench*; unclainte (*pret.*) þe barres ALIS.² (Sk.) 1172.

un-clēne, adj., *O.E.* unclǣne; *unclean*, HOM. I. 27; ORM. 1712; O. & N. 91; GEN. & EX. 1867; onclene PR. P. 364; a draught of un-clene (*poison*) ALEX. (Sk.) 1106.

    onclēn-līch, adj., *uncleanly*, AYENB. 42; vor unclēnliche (*dat.*) cause IBID.

    unclǣnnesse, sb., *uncleanness*, ORM. 2168; onclennesse AYENB. 203; unclannesse A. P. ii. 30; unclennes ALEX. (Sk.) 4218.

un-clēnsed, adj., *uncleansed*, ORM. 10617.

un-cnāwen, adj., *unknown*; unknawen PR. C. 337; ALEX. (Sk.) 3715; unknowen CH. C. T. D 1397.

un-cnāwing, sb., *ignorance*; unknawinge HAMP. PS. cxlvii. 5*.

un²-cnütten, v., *O.E.* uncnyttan; *unknit, loosen*; unknitten CH. BOET. v. 3 (154); un-knette TREV. II. 43; **uncnütte** (*subj.*) HOM. II. 137; uncnette MK. i. 7; **unknitte** (*pret.*) WICL. ECCLUS. xxviii. 18; uncnüt (*pple.*) KATH. 1156.

un-coint, adj., *unwise*; unqvaint HAMP. PS. ciii. 26*.

un-comberen, v., *cease from encumbering*, PALL. vi. 51.

un-conninde, adj., = *Goth.* unkunnands; *ig-norant*, AYENB. 59; unconand HAMP. PS. page 4.

    onconninde-hēde, sb., *ignorance*, AYENB. 33.

    unconand-lī, adv., *ignorantly*, HAMP. PS. lxxiv. 2*.

un-conninge, sb., *ignorance*, BEK. 1024; on-conninge AYENB. 131; unkunning PR. C. 169.

un-corrigible, adj., *incorrigible*, WICL. PROV. xxiv. 19.

un-corrumpid, adj.; *uncorrupted*; uncor-umpid ALEX. (Sk.) 4334.

un-corsaid, adj., *unridden*, ALEX. (Sk.) 3775.

un-cost, sb., *vice*, REL. I. 213.

un-covenable, adj., *unsuitable*; uncunable HAMP. PS. lxxi. 9*.

    uncunabil-lī, adv., *unsuitably*, HAMP. PS. xxxviii. 1*.

    unconabilnes, sb., *unreasonableness*, HAMP. PS. cv. 31*.

un²-coveren, v., *uncover*; unkevered (*pret.*) JOS. 559.

un-coverlīch, adj., *? irrecoverable*, H. M. 27.

un-cristen, adj., *unchristian*; **uncristene** (*pl.*) LANGL. A i. 91 [uncristne B i. 93].

un-cumelīch, adj., *uncomely*, H. M. 25; on-comeli LANGL. B ix. 160 [uncomeli A x. 180].

un-cunabiltē, sb., *inconvenience*; **uncon-abiltēs** (*pl.*) HAMP. PS. lxxii. 14*.

un-cünde, adj., *O.E.* uncynde; *unnatural, unkind*, TREAT. 135; unkuinde JOS. 242; un-kinde GEN. & EX. 449; TRIST. 2758; onkinde PR. P. 365; onkende AYENB. 188.

    unkünde-līch, adj., *unkindly, unnatural*, A. R. 116; **unkíndelīke** (*adv.*) HAV. 1250; unkindeli, -lich, GOW. I. 348.

un-cūrabil, adj., *incurable*, HAMP. PS. cxxxix. 3*.

un-curteis, adj., *uncourteous*; uncortoise A. P. i. 303.

un-cüsti, adj., *O.E.* uncystig; *illiberal*, '*par-cus*,' FRAG. 3.

un-cūð [oncūþ], adj., *O.E.* uncūð, = *Goth.*

unkunþs, *O.N.* ūkunnr; *uncouth, strange, foreign*, LAȜ. 6624; uncuþ ORM. 228; uncouþ C. L. 586; into uncuþe londe HORN (L.) 729; unkūðe (*pl.*) A. R. 348; uncouþe PERC. 1047; unkouþe LANGL. *B* vii. 155.

uncūþ-līȝ, adj., *strangely*, ORM. 14341.

un-cūðð̄e, sb., *O.E.* uncyðð̄u; *foreign country*, A. R. 140.

un-cwēme, adj., *disagreeable*, ORM. 1527; unqveme HOM. II. 9.

un-cwenked, adj., *unquenched*, ORM. 10491.

un-dampned, adj., *uncondemned*, WICL. DEEDS xvi. 37.

un-dēadlīch, adj., *O. E.* undēadlīc; *immortal*, KATH. 965; H. M. 39; undedlich MARH. 10; undeedli WICL. 1 TIM. i. 17.

undēadlīchnesse, sb., *immortality*, KATH. 1122; undeðlicnesse HOM. II. 33; undeadlinesse WICL. WISD. iii. 4.

un-dēad, adj., *not dead*: undede ALEX. (Sk.) 158.

un-defouled, adj., *not trodden down, undefeated*, ALEX. (Sk.) 2630.

un-dēmed, adj., *unjudged*, ORM. 16725.

un-dēore, adj., *O.E.* undēore; *not dear*, A. R. 408.

un-dēp, adj., = *O.H.G.* untiuf; *not deep*, SAX. CHR. 262.

under, prep. & adv., *O.E.* under, = *O.L.G.*, *Goth.* undar (ὑπό), *O.N.* undir, *O.Fris.* under, onder, *M.Du.* onder, *O.H.G.* untar, *Lat.* inter; *under; between, among; during*; under (*ms.* vnder) þere sunnen LAȜ. 24982; under [onder] þan wude 4734; liggen under laðest mon H. M. 31; under (*ms.* unnderr) water dippest ORM. 1551; þat sit at mulne under cogge O. & N. 86; under ure wede C. L. 657; under an holu ok WILL. 295; under a brode banke LANGL. *A* PROL. 8; under mire onwalde HOM. I. 13; under godes warde H. M. 7; þat holi churche under fote were so BEK. 2019; te veorðe valleð under þe uttre A. R. 222; þe wes under wedlac iboren LAȜ. 395; under eou alle nis þar nan 915; under are stunde 16501; under þan comen tiðende 9660; under þis com þe þurs KATH. 1880; he þolede diaþ onder Pouns Pilate AYENB. 12; þat þou ne go nauȝt onder (*intereas*) SHOR. 163; *comp.* an-under.

under-beren, v., *O.E.* underberan; *support*; underbern WICL. ECCLUS. xii. 14; onderbere AYENB. 84.

under-delvin, v., *dig under*, 'suffodio,' PR. P. 511; underdelve WICL. ECCLUS. xii. 18.

under-dōn, v., = *O.H.G.* untartuōn; *put under, subject*, GEN. & EX. 4041.

under-ēode, v., (*pret.*) *went under*; onderȝede SHOR. 122.

un-derf, adj., = *O.N.* ūdiarfr; *infirm*: þat underve flesch KATH. 1174.

under-fangen, v., *seize, receive*, ORM. 11728; undervongen A. R. 190; underfongin PR. P. 511; underfange LAȜ. 10141*; underfonge HORN (R.) 335; C. L. 661; A. D. 257; WILL. 5259; ondervonge AYENB. 14; undervongest (*pres.*) HOM. I. 51; underfangð HOM. I. 239; undervongeð A. R. 190; undirfangid (*pret.*) ALEX. (Sk.) 910.

under-fōn, v., *O.E.* underfōn, = *O.H.G.* untarfāhan; *seize, receive*, HOM. I. 65; LAȜ. 3222; HOM. II. 96; KATH. 702; ORM. 3956; GEN. & EX. 1679; CHR. E. 994; undervon A. R. 14; underfo MAP 343; to underfonne MARH. 20; underfēs, -vest (*pret.*) KATH. 983; underfeð H. M. 41; A. R. 190*; (ha) undervoð H. M. 19*; undervōh (*imper.*) LAȜ. 16880; (he) underfō (*subj.*) HOM. I. 119; underfēng (*pret.*) LAȜ. 3271; KATH. 1102; GEN. & EX. 480; underveng A. R. 40; underfonge [underfeng] LANGL. *B* i. 76; onderving AYENB. 5; underfongen, -fonge (*pple.*) LANGL. *A* v. 635; underfonge BEK. 1391; undervonge LAȜ. 26257; underfon [onderfon] 3431; underfon KATH. 702.

under-gangen, v., *O.E.* undergangan; *undergo*, ORM. 10661; undergon GEN. & EX. 1147; undergo P. L. S. xiii. 280; hi undergōþ (*pres.*) SHOR. 11.

under-ȝiten, v., *O.E.* undergitan; *perceive, understand*, A. R. 150; underȝite P. L. S. xxiii. 54; underȝete S. C. Ð. J. 41; underȝat (*pret.*) O. & N. 1055; FL. & BL. 35; underȝat, -ȝæt [onderȝeat] LAȜ. 15028, 26988; underȝet ROB. 62; undirgat S. S. (Wr.) 3151; þu underȝete P. L. S. xxiv. 200; (heo) underȝeten LAȜ. 1811; underȝete P. L. S. xiii. 291; underȝeten (*pple.*) LAȜ. 26869; underȝete O. & N. 168.

under-ȝōke, v., = *Ger.* unterjochen; *subjugate*, WICL. JUDITH ii. 3.

under-king, sb., *viceroy*, LAȜ. 31340.

under-lein, v., *O.E.* underlecgan, = *O.H.G.* untarleccan; *underlay*, WICL. JER. xxvii. 11; underleiden (*pret.*) GEN. & EX. 3388; underleid (*pple.*) FRAG. 6.

under-liggen, v., = *O.H.G.* untarliggan; *lie under*; underligge WICL. EX. xxi. 31.

underling, sb., *inferior, subject*, LAȜ. 3657; A. R. 198; underlinge LANGL. *B* vi. 47; ALEX. (Sk.) 1861; onderlinges (*pl.*) AYENB. 39.

under-lūten, v., *O.E.* underlūtan; *submit*; underlūte, -loute (*pple.*) C. M. 7163; undirloute WICL. GEN. xxxvii. 8.

under-mīne, v., *undermine*, WICL. MAT. vi. 20.

undern, sb., *O.E.* undern, = *O.L.G.* undorn, undern, *O.N.* undorn, *Goth.* undaurns, *O.H.G.* untarn *m.; the time from nine to twelve o'clock in the morning*, A. R. 24; CH. C. T. *E* 260; at undren and at middai MISC. 33; undurn WICL. JOHN iv. 6; undorne

ALEX. (Sk.) 3853; **underne** (*dat.*) BEK. 2465; at þon heie undarne MISC. 56; undurne MAND. 163; DEGR. 1209; aboute onderne S. A. L. 160; at ondre ЬHOR. 84.

under-mēl, sb., *O.É.* undernmǣl; *morning meal*, CH. C. T. *D* 875.

under-tīd, sb., *O.É.* underntīd; *morning time*, HOM. I. 91; A. R. 400; undertide ORF. 179.

undern-tīme, sb., *morning time*, ORM. 19458; undirtime M. ARTH. 2807.

un-derne, adj., *O.E.* underne; *not secret*, JUL. 75.

under-néþen, adv., *underneath*, PL. CR. 695; undirneþe TREV. V. 123; CH. BOET. iii. 5 (75).

under-nimen, v., *O.E.* undernyman, = *O.H.G.* untarneman; *seize, catch; undertake; perceive; take to, agree to; blame*; KATH. 123; A. R. 202; (*ms.* undernymen) MAT. xix. 12; ondernime AYENB. 83; underneomen H. M. 19; undernemen MAND. 139; underneme '*reprehendo*' PR. P. 511; **undernimeþ** (*pres.*) LANGL. *B* v. 115; **undernam** (*pret.*) GEN. & EX. 1553; undirnam S. S. (Wr.) 3236; undernomen LA₃. 8067; C. L. 598; **undernumen** [ondernome] (*pple.*) LA₃. 26734; undernomen MAP 336; undirnomen WICL. JOHN iii. 20; undernome P. S. 323; undirnome S. S. (Wr.) 2858.

under-putten, v., *place under, subjugate*; underputt (*pple.*) ALEX. (Sk.) 5402.

under-schóren, v., = *M.Du.* onderschoren; *shore or prop up*; undershóred (*pple.*) LANGL. *C* xix. 47.

under-sēchen, v., = *Ger.* untersūchen; *examine*; onderzēkþ (*pres.*) AYENB. 184.

under-settin, v., = *M.L.G.* undersetten, *M.Du.* ondersetten; *underset, prop, 'suffulcio,'* PR. P. 511; undersette S. S. (Web.) 2101; underset (*pres.*) A. R. 254; REL. I. 223; undersette (*pret.*) MAN. (F.) 284.

under-standen, v., *O.E.* understandan, = *O. N.* undirstanda, *O.Fris.* understonda; *understand, perceive; receive*; ORM. 1763; understonden HOM. I. 35; A. R. 50; understonde P. L. S. viii. 97; O. & N. 1262; HAV. 2814; C. L. 953; onderstonde SHOR. 24; AYENB. 14; understond (*r. w.* weñd) (*pres.*) O. & N. 1463; understōd (*pret.*) LA₃. 4484; KATH. 2145; WILL. 877; Josep al it understod GEN. & EX. 2210; Josep wel faire him understod (*received*) 2293; understoden (*received*) him mid procession HOM. II. 89; understanden (*pple.*) PR. C. 1681; HOCCL. iii. 4.

understondinge, sb., *understanding*; onderstandinge AYENB. 24; þe xij **understandings** (*pl.*) ('*intelligences,*' in astrology) ALEX. (Sk.) 279.

under-stipen, v., = understipren; A. R. 142*.

under-stipren, v., *cf. M.H.G.* understivelen; *shore up, support*, A. R. 142.

under-táken, v., *undertake*, ORM. 10314; WICL. PS. xlvii. 4; **undertōc** (*pret.*) SPEC. 41; undiretuke ALEX. (Sk.) 2967; **undertōke** (*subj.*) HAV. 377.

under-þēode, sb., *subjects* : (he) scal spenen among al his underþede HOM. I. 85.

undur-þēwe, v., = *Du.* onderdouwen; *subject*; underþewe ALIS. 1406.

under-þūden, v., *O.E.* underþȳdan, -þeodan; *subject*; underþeden SAX. CHR. 260; alle þ(e)o þet him b(e)oð **underþēde** (*pple.*) HOM. I. 85.

under-wīten, v., = *O.L.G.* undarwītan; *perceive*; underwāt (*pret.*) O. & N. 1091.

under-wrōte, v., *O.E.* (under-)wrōtan; (ÆLFR.); *root up, undermine*, MISC. 97.

un-digne, adj., *unworthy*, CH. C. T. *E* 359.

un-discreet, adj., *undiscerning*, CH. C. T. *E* 996.

un-distreined, adj., *uncompelled*, ALEX. (Sk.) 2779.

un-distrobbed [-tourblett], adj., *undisturbed*, ALEX. (Sk.) 3418.

un²-dōn, v., *O.E.* undōn, = *O.Fris.* un-, unndūa; *undo*, GR. 39; GEN. & EX. 2114; didden hem dereli undo *cut them up* (*hunting term*) GAW. 1327; ondo SHOR. 78; AYENB. 106; undüde (*pret.*) HOM. I. 121; undede WILL. 4846; LAI LE FR. 183; undōn (*pple.*) FRAG. 7; LA₃. 19206; ondon SHOR. 61.

un-drē₃, adj., *impatient*; undreh SPEC. 41; undrēghe (*pl.*) HALLIW. 901.

un-dühti, adj., = *M.H.G.* untühtic; *unprofitable, unworthy*, JUL. 4.

un²-dütten, v., *unfasten*, S. B. W. 254; undütte (*pret.*) KATH. 1821; undüt (*pple.*) LEG. 36.

un-efen, adj., *O.E.* unefen; *uneven*: unefne chaunge A. R. 312; unevene (*pl.*) MISC. 75; unevene (*adv.*) MISC. 86.

unefen-lích, adv., *unevenly*, A. R. 410; unevenli ['*iniqve*'] WICL. GEN. xvi. 5.

un-ended, pple., *O.E.* ungeended; *infinite*, GEN. & EX. 3518.

un-endlíche, adv., *infinitely*, A. R. 398.

un-erned, adj., *unearned*, HOM. II. 33.

un-ēsen, v., *make uneasy*; unēseth (*pres.*) PALL. iii. 562; unēsid (*pple.*) ALEX. (Sk.) 5054.

un-ēðe, adj., *O.E.* uneāðe; *difficult*, LA₃. 2259; uneāðe (*adv.*) *with difficulty, scarcely*, HOM. II. 225; uneaþe (*ms.* unneaþe) O. & N. 1605; unæþe ORM. 16289; uneþe TREAT. 136; MAND. 282; GAW. 134; oneþe PL. CR. 217; unēðes GEN. & EX. 2341; uneþes GOW. III. 6; unneths HAMP. PS. iv. 6*.

un-fǣle,-vēle, adj., *O.E.* unfǣle; *improper, ungracious*, LAȝ. 22018, 23868; his unfǣle þeowes ORM. 8034; unfele gast MK. vi. 49; þe unfele (*wicked*) man HOM. II. 79; unfele, -vele O. & N. 1003, 1381; unvele MISC. 73; unfeale A. R. 198*; onvele S. A. L. 153.

un-fain, adj., *O.E.* unfǣgen; *not desirous, displeased*, M. ARTH. 2691; M. H. 90; unfein REL. I. 113; unfawe LAUNF. 732.

un-fair, adj., *O.E.* unfæger, = *Goth.* unfagrs; *unfair, frightful*; unfaire (*pl.*) A. P. ii. 1801; ALEX. (Sk.) 4864; unfair (*adv.*) *extremely; cruelly; horribly*, ALEX. (Sk.) 1189, 2041, 1224, 837, 555.

un-fāken, adj., *O.E.* unfǣcne; *not deceptive*, ORM. 4149.

un²-fálden, v., *O.E.* unfealdan; *unfold*; unvolden A. R. 100; unfolde LANGL. *B* xvii. 176; unfáldis (*pres.*) 3027; unfēold [onfeold] (*pret.*) LAȝ. 10544; unfeld LK. iv. 17; unfeelde LANGL. *C* iii. 73; unfólde (*pple.*) M. ARTH. 1044.

un-fanded, adj., *untried*; unvonded A. R. 232.

un²-fasten, v., *unfasten*; unvesten A. R. 218; onfestin PR. P. 365.

un²-fastnen, v., *unfasten*, WICL. IS. xiv. 27; unvestnen A. R. 252.

un-fēawe, adj., *not few*; unfæwe ORM. 792.

un-fēre, adj., *O.E.* unfēre, = *O.N.* ūfœrr; *indisposed*, LAȝ. 6780; unfer GEN. & EX. 2810; unfēre (*dat. pl.*) L. H. R. 114.

un-fēre², sb., *infirmity*, C. M. 3556.

un-festlīch, adj., *unfestive*, CH. C. T. *F* 366.

un²-feteren, v., *unfetter*; unfetere LANGL. *A* iii. 134 [unfettre *B* iii. 138].

un-fīled, adj., *unpolluted*, HOM. II. 133.

un-flichand, adj., *from O.Fr.* flechir; *unflinching*, HAMP. PS. ii. 9*.

un-forȝólden, adj., *O.E.* unforgolden; *unrequited*, HOM. I. 41, 163; unforȝolde AN. LIT. 91.

un-forgúlt, adj., *without guilt*, H. M. 43.

un-fráme, sb., *disadvantage*, GEN. & EX. 1566.

un-frē, adj., *unfortunate*, A. P. ii. 1129.

un-freined, adj., *unasked*, A. R. 338*.

un-fréme, sb., *O.E.* unfremu; *disadvantage, mischief*, P. L. S. viii. 114*; HOM. II. 195.

un-frēond, sb., = *M.H.G.* unvriunt; *hostile person*; onfrēondes (*pl.*) LAȝ. 17612*.

     unfrīnd-schip, sb., *enmity*, ALEX. (Sk.) 2722.

un-freten, adj., *not devoured*, FRAG. 7.

u(n)-frigt, adj., *O.E.* unforht; *intrepid*, GEN. & EX. 3713.

un-frið, sb., *O.E.* unfrið, = *O.N.* ūfriðr; *discord, dissension*, LAȝ. 2557.

un-fruitous, un-fructuous, adj., *unfruit-*

ful, WICL. EX. xxiii. 26; JOB xxiv. 20; EPH. v. 11; TIT. iii. 14.

un-fulhtned, adj., *not baptized*, ORM. 16895.

un-fūllable, adj., *insatiable*; unfilabil HAMP. PS. c. 7*.

un²-ȝarken, v., *unfasten*; unȝarkid, unȝarked (*pret.*) ALEX. (Sk.) 2147, 3209.

un-garnist, adj., *unadorned*, A. P. ii. 137.

un-gāstlī, adv., *unspiritually*, ALEX. (Sk.) 4430.

un-ȝeare, adv., *O.E.* ungeare; *unexpectedly*; unghere GEN. & EX. 3047.

un-ȝearu, adj., *O.E.* ungearu; *unprepared*, HOM. I. 103.

un-gein, adj., *perilous* : þes gates are ungaine FLOR. 1421.

     ungein-liche, adv., *ungainly*, MARH. 9.

un-ȝerīm, adj., *O.E.* ungerīne; *innumerable*, ORM. 18993.

un-glad, adj., *O.E.* unglæd, = *O.N.* ūglaðr; *not happy*, SPEC. 29; WILL. 747; A. P. iii. 63.

     ungled-liche, adv., *sorrowfully*, A. R. 338.

un-gōd, adj., *O.E.* ungōd, = *O.H.G.* unguot; *not good*, ORM. 16739; P. L. S. xxiii. 22; A. D. 235; sum ungōd (*sb.*) O. & N. 1364.

un²-ȝóken, v., *unyoke*; unȝóked (*pple.*) TREV. V. 367.

un-greiþ, adj., = *O.N.* ūgreiðr; *unprepared*, SPEC. 99.

un-grēte, sb., *smallness*, O. & N. 752.

un-griþ, sb., *O.E.* ungrið; *dissension*, ORM. 16280.

un-gülti, adj., *O.E.* ungyltig; *not guilty*; ungilti WICL. GEN. xxxvii. 22; ongulti PR. P. 365.

un-gürde, adj., *ungirt*, TREV. VIII. 213; ungert GAM. 215.

un²-hādien [on-hōdi], v., *deprive of clerical functions, 'unfrock,'* LAȝ. 13169.

un-hǣle, sb., *O.E.* unhǣlu; *infirmity, mischief*, LAȝ. 11546; ORM. 4779; unhele HOM. I. 7; CH. C. T. *C* 116; FLOR. 1340.

un-hǣled, adj., *unhealed*; unhealed A. R. 328.

un-hǣlðe, sb., *O.E.* unhǣlð; *sickness*; unhelðe HOM. II. 35; unhelþe P. L. S. viii. 8; MISC. 108.

un²-hǣren, v., *deprive of hair*; unheerid (*pple.*) ['*depilatus*'] WICL. EZ. xxix. 18.

[un-haȝer, adj., *not apt.*]

     unhaȝher-liȝ, adv., *cf. O.N.* ūhagligr; *unskilfully*, ORM. 425.

un-hāl, adj., = *O.N.* ūheill, *O.H.G.* unhail, *Goth.* unhails; *not whole; sick, ill*; ORM. 4778; M. H. 129; unhol A. R. 370; unhāle [onhole] (*pl.*) LAȝ. 19647; unhole MISC. 82.

un-hālsum, adj., = *O.N.* uheilsamr; *unwholesome*, ORM. 7177; onholsum PR. P. 365; unhalesome ALEX. (Sk.) 4387.

un-hap, sb.,=*O.N.* ūhapp ; *unhap, mischance, misfortune,* H. M. 29 ; O. & N. 1267* ; GAW. 438 ; A. P. ii. 143 ; unhappe ALEX. (Sk.) 4554 ; onhap PR. P. 365 ; unhep A. R. 180.

un-happen, adj., *bad* ; unhappen glette A. P. ii. 573.

un-happi, adj., *unhappy,* CH. C.ᵢ T. (Wr.) 4726 ; LIDG. TH. 821 ; onhappi PR. P. 365 ; **unhappiest** (*superl.*) ALEX. (Sk.) 713.

un²-hardelen, v., *from O.Fr.* hardelle (*troupe*) ; *disperse* ; **unhardeled** (*pret.*) GAW. 1697.

un-hardi, adj., *not bold,* LANGL. *B* PROL. 180.

un²-haspen, v., *disclose* ; unhaspe A. P. ii. 688.

un-heled, adj., *unhidden, open,* TREV. I. 367 ; unhelid ALEX. (Sk.) 3450.

un²-helien, v., *O.E.* unhelian ; *uncover, reveal, make known* : unhilen ORM. 12944 ; unhile WICL. RUTH iii. 4 ; **unhelen** (*pr. subj.*) HOM. II. 77 ; **unheled** (*pple.*) A. R. 150 ; TREV. VIII. 161 ; ounhelid FER. 586 ; unhiled LANGL. *B* xvii. 319.

un-hemed [**un-hemmid**], adj., *unrestrained,* ALEX. (Sk.) 2835.

un-hende, sb., *discourteous, rude,* A. R. 204 ; H. M. 9 ; P. R. L. P. 191 ; IW. 2943 ; unhende [onhende] LA3. 28826.

    unhende-līche, adv., *rudely,* KATH. 2148 ; ROB. (W.) 8540.

un-hent, adj., *uncaught,* WILL. 1671.

un-hērlī, adj., *O.E.* unhīerlīc (BL. HOM. 203), unhȳrlīc (B. D. D. 11),=*O.N.* ūhȳrligr ; *monstrous,* M. H. 129.

un-hērsamnesse, sb., *O.E.* unhērsumness ; *disobedience,* HOM. I. 235 ; (*ms.* unnherrsummnesse) ORM. 4277.

un-hēwen, adj., *unhewn,* ALEX. (Sk.) 1945.

un²-hūden, v., *uncover ; make public* ; **unhĭd** (*pple.*) ALEX. (Sk.) 3437.

un-hóld, adj., *O.E.* unhold,=*O.L.G.* unhold, *O.H.G.* unhold ; *not friendly, hostile* ['*inimicus*'] MAT. xiii. 28 ; **unhólde** (*pl.*) SPEC. 24 ; unholde HOM. I. 161 ; A. R. 222.

un-honourable, adj., *dishonourable,* ALEX. (Sk.) 2950.

un-hópe, sb., *despair,* HOM. I. 251 ; A. R. 8.

un-hühtlīc, adj., *without joy,* LA3. 5101.

un²-hülen, v., *uncover* ; unhillen GEN. & EX. 1912 ; onhillin PR. P. 365 ; **unhülede** (*pret.*) A. R. 58* ; **unhüled** (*pple.*) JOS. 515.

un-hurt, adj., *unhurt,* JUL. 31 ; ALEX. (Sk.) 5530.

un²-hūsen, v., *unhouse* ; **unhoused** (*pple.*) JOS. 455.

un-hwarfed, adj., *unchanged* ; unwharfed ORM. 18794.

un-hwáte, sb., *misfortune,* O. & N. 1267.

ūnicorne, sb., *O.Fr.* unicorne : *unicorn,* A. R. 120 ; **ūnicornes** (*pl.*) ALEX. (Sk.) 3593.

un-icunde [**onicunde**], adj., *O.E.* ungecunde ; *foreign,* LA3. 18429.

un-ifēie, adj., *enormous,* LA3. 5573.

un-ifēle, adj., *O.E.* ungefēle ; *? insensible* : mid unifele þingen LA3. 21744.

un-ifōh, [**onifōh**], adj., *O.E.* ungefōg,=*O. Fris.* unefōg ; *enormous,* LA3. 8674 ; unifouh FRAG. 7.

un-ihĕrsamnesse, sb., *disobedience,* HOM. I. 221.

un-ihōded, adj., *not ordained,* O. & N. 1178.

un-ilīc, adj., *O.E.* ungelīc,=*O.H.G.* ungelīch ; *unlike,* LA3. 9919 ; unilich JUL. 60 ; **unilīche** (*pl.*) P. L. S. viii. 179.

un-ilīke, sb., *unlike* : þin unilike O. & N. 806.

un-ilimp, sb., *O.E.* ungelimp (SAX. CHR. 219) ; *misfortune,* HOM. II. 117 ; unilimpe REL. I. 174.

un-iloȝe, adj., *?=Ger.* ungelogen ; *without a lie* ; ouniloȝe FER. 511*.

un-imáke, sb., *what is unlike* : elches wurmes unimake [onimake] LA3. 17961.

un-imeað, sb., *want of moderation,* JUL. 5.

un-imet, sb., *O.E.* ungemet ; *excess* ; **uniméte** (*dat.*) A. R. 74.

un-imet², adj., *O.E.* ungemet,=*O.H.G.* ungimeȝ ; *immense* : in his unimete (? unimēte) blisse A. R. 40 ; **uniméte** (*adv.*) A. R. 102 ; H. M. 19.

un-imēte, adj. & adv., *O.E.* ungemæte, *O. H.G.* ungemāȝ ; *immoderate, immense ; immensely* ; HOM. I. 101 ; MISC. 73 ; unimete [onimete] LA3. 4964.

    unimēte-līche, adv., HOM. I. 281 ; A. R. 398.

un-iqvēme, adj., *inconvenient,* REL. I. 184.

un-irīmed, adj., *innumerable,* LA3. 433.

un-irūde, adj., *O.E.* ungerȳde, *enormous, cruel* : unirude duntes HOM. I. 253 ; *see* **un-rūde.**

    unirēd-līce, adv., *roughly* ; uniredlice underfangeth HOM. I. 239.

un-isaht, adj., *unreconciled, at enmity* ; **unisahte** (*pl.*) HOM. I. 39.

[**un-iseȝen,** adj., *unseen.*]

    uniseȝen-, **un-isewenlīch,** adj., *O.E.* ungesewenlīc ; *invisible,* HOM. I. 97.

un-isēle, adj., *unhappy, unfortunate,* P. L. S. viii. 101 ; LA3. 26446 ; O. & N. 1004.

    unisēli, adj., *O.E.* ungesælig ; *unhappy, unfortunate,* LA3. 4010 ; A. R. 68.

    unisē(l)-līche, adv., *unhappily,* LA3. 7022.

un-isĕlðe, sb.,*O.E.* ungesælð; *infelicity,* HOM. I. 171 ; LA3. 2545.

un-isibbe, adj., *unfriendliness,* LA3. 9845.

un-isōm, adj., *at variance* ; **unisōme** (*pl.*) O. & N. 1522.

un-isúnde, sb., *unhealthy*, LAȝ. 18452.

un-itáld, adj., *untold*, HOM. I. 233.

ūnitē, sb., *O.Fr.* unité; *unity*, CH. C. T. *E* 1334.

ūniversitē, sb., *Fr.* université; *university*, P. L. S. xvii. 245.

un-iwar, adv., *unawares*, ROB. 88.

un-iwasse, adj., = *M.H.G.* ungewaschen; *unwashen*, HOM. I. 237.

un-iweald, sb., = *O.Fris.* unewald, *M.H.G.* ungewalt; *impotence*, HOM. II. 63.

un-iwealde, adj., *impotent*; uniwælde, LAȝ. 5901.

un-iwider, sb., *O.E.* ungewider (BL. HOM. 125),= *O.L.G.* ungiwideri, *O.H.G.* ungiwitiri; *tempest*; uniwidere (*dat.*) HOM. I. 115.

un-iwil, sb., *unwillingness*, HOM. I. 69.

un-iwrench, sb., *evil trick*: for þin(e) uniwrenche MISC. 174, 175.

unker, adj., *O.E.* uncer,= *O.N.* okkarr; *see* unc; *of us two*, LK. xii. 13; FRAG. 7.; ORM. DEDIC. 80; O. & N. 1689; unkere (*dat. pl.*) LAȝ. 8891.

un²-kevelen, v., *take off the gag*; unkeveleden (*pret.*) HAV. 601.

un-lǣred, adj., *O.E.* unlǣred; *unlearned, untaught*, ORM. 17117; unlered PR. C. 5947; unlerid WICL. 1 COR. ẋiv. 16.

un-laȝe, sb., *illegality*, GEN. & EX. 1762; unlawe BEK. 602.

[**unlaȝe-ful**, adj., *unlawful*.]

u(n)lāȝe-līche, adv., *illegally*, HOM. I. 115; unlaȝhelike ORM. 15867; unlahfulliche SPEC. 53.

un-lēaf, adj., *O.E.* ungelēaf; *unbelieving*: þu art unlef (*thou disbelievest*) mine worde HOM. II. 125.

un-lēafful, adj., *unlawful*, '*illicitus*'; onleefful PR. P. 366; H. V. 111; unlefful D. TROY 13686; unleveful CH. C. T. *I* 593.

un-lēaflīch, adj., *O.E.* ungelēaflīc,= *O.H.G.* ungloublīch; *incredible*: unleflich HOM. II. 125; un-lēflīche (*pl.*) KATH. 345.

un-lēde, adj., *O.E.* unlǣde,= *Goth.* unlēds; *poor, miserable*; (*ms.* vnlede) O. & N. 1644; MISC. 122 (*printed* vulede REL. I. 182); AL. (T.) 333.

un-lengþe, sb., *brevity*, O. & N. 752.

un-lēod, sb., *foreign people*; unlēoden [onleode] (*pl.*) LAȝ. 6939; onlede SHOR. 109.

un-lēof, adj., *O.E.* unlēof,= *O.H.G.* unliub, *Goth.* unliubs; *not dear*; unlef and unqveme HOM. II. 189; unl(e)ofne breð HOM. I. 153.

un-līc, adj., *cf.O.Fris.* unlīk, *O.N.* ūlīkr,= un-ilīc; *unlike*, ORM. 16859; unlich JUL. 14; unliche (*adv.*) P. L. S. v. (b.) 55.

unlik-lī, adv., *unlikely*, ALEX. (Sk.) 5552.

un²-līden, v.; *uncover*: unlīden (*pres.*) A. R. 58*; unlūded (*pret.*) 58*.

un-ligel, adj., *for* unligen, ? *O.E.* ungelygen: *truthful*, HOM. II. 131.

un²-līkien? v.; *?feel dislike*; **onlīkede** (*pret.*) LAȝ. 3266*.

un²-līmen, v., *unfasten, separate*, A. R. 256.

un-limp, sb.,=**unilimp**; *misfortune*, HOM. II. 61; A. R. 274.

un-litel, adj., *O.E.* unlytel; *not little*, ORM. 726.

un-loked, adj., *not guarded*, GREG. 964.

un²-lósen, v., *unloose*, LANGL. *B* xvii. 139.

un-lōðnesse, sb., *innocence*, A. R. 340.

un²-louken, v., *O.E.* unlūcan; *unlock, unfasten*, C. L. 77; LANGL. *B* xii. 112; onlouken SHOR. 55; onlēak (*pret.*) AYENB. 67; unlek S. S. (Web.) 2251; B. DISC. 1816; **unloke** (*pple.*) MISC. 147; TREV. VI. 203.

un-lovelīch, adj.,*unlovely*, LANGL. *B* xv. 114.

un-lust, sb., *O.E.* unlust,= *O.H.G.* unlust, *Goth.* unlustus; *displeasure*, H. M. 35; ORM. 2623; MAP 336; idelnesse and unlust CH. C. T. *I* 680; unlust HOCCL. i. 189.

un-lusti, adj., = *M.H.G.* unlustig; *idle, slothful*; ȝemeleas & unlusti HOM. I. 205; þe onlosti þat bieþ slacke to godes service AYENB. 170; onlisti PR. P. 366; unlustie (*pl.*) H. M. 43.

unlusti-lī, adv., *slothfully*, P. L. S. xxv. 143.

un-lǔved(e), adj., *unlawful*, HOM. II. 71.

un-mád, adj., *unfinished*, GEN. & EX. 671.

un-maȝe, adj., *O.E.* unmaga,= *O.H.G.* unmag; *impotent*; on-, ounmawe FER. i. 2658, 2766.

un-maht, adj., *impotent*; **unmahte** (*dat. m.*) A. D. 297.

un²-mensken, v., *disgrace*; **unmenskeð** (*pres.*) MARH. 14.

un-mēoc, adj., *not meek*, ORM. 9880; unmek C. M. 11815.

un-merred, adj., *unmarred*, MARH. 10.

un-mēte, adj., *O.E.* unmǣte,= *O.H.G.* unmāzi; *not meet*; *immoderate, immense*; HOM. I. 103; SPEC. 23; B. DISC. 1629; GAW. 208; A. P. i. 758; D. ARTH. 4070; unmete [unmeete] C. L. 634; unmaite HAMP. PS. lxxii. 14*.

unmēte-lī, adv.,*unfittingly*,ALEX. (Sk.) 321.

un-mēþ, sb., *immoderateness*, FL. & BL. 675; mid unmeþe O. & N. 352.

un-mēþ², adj., *immoderate*, A. R. 50; unmeð muchel hird JUL. 4; unmeth C. M. 11815.

unmeaðe-liche, adv., MARH. 15; unmeð-lüker (*compar.*) A. R. 238.

unmēðschipe, sb., *immoderateness*, A. R. 122.

un-mévable, **unmóvable**, adj., *immoveable*, WICL. EX. xv. 16; 1 PARAL. xvi. 30; HEB. vi. 18; unmoebles (*sb., pl.*) LANGL. *B* iii. 267.

un-miht, sb., *O.E.* unmiht,=*O.H.G.* un-maht, *Goth.* unmahts; *impotence*; (*ms.* vnmyþt, *r. w.* liht) A. D. 290; unmight MAN. (F.) 15564; unmihte KATH. 1022; **unmiȝte** (*dat.*) BEK. 1441.

un-mihti, adj., *O.E.* unmehtig,=*O.H.G.* un-mahtīg·; *powerless*, HOM. II. 35; SPEC. 22; unmiȝti CH. BOET. i. 4 (13); WICL. WISD. xi. 18.

un-milde, adj., *O.E.* unmilde,=*O.H.G.* un-milt; *ungentle*, ORM. 9880; O. & N. 61; BEK. 1494.

un-mind(e), adj., *unmindful*, C. M. 1572.

un-mündlunge, adv., *O.E.* unmyndlinga; *unexpectedly*, HOM. I. 249; (*ms.* unmunlunge) A. R. 280.

un-mürie, adj., *not merry*, O. & N. 346.

un-nait, adj., *useless*, ['*inania*'] PS. ii. 1; i sall not make unnait ['*non faciam irrita*'] HAMP. PS. lxxxviii. 34; **unneite** (*pl.*) A. R. 130*; **unnaite** (*adv.*) C. M. 5976.

un-nēd, adj., *not compelled*, ORM. 11457; un-net A. R. 340.

unnen, v., *O.E.* unnan,=*O.H.G.* unnan, *O.N.* unna; *from* an; *favour, grant*, A. R. 282; HOM. II. 79; ich **unne** (*pres.*) hire wel SPEC. 40; ȝif þu hit wel unnest A. R. 282; (heo) unneð A. R. 22; **ūðe,** ūþe (*pret.*) LAȝ. 193, 13035; uþe ORM. 3451; **ūðe** (*subj.*) A. R. 90; ouþe MAP 336; **unned** (*pple.*) H. M. 13; *comp.* ȝe-**unnen.**

　unnunge, sb., = *M. H. G.* (g)unnunge; *favour, sign*, A. R. 282; he ossed him (? *read* hem) bi **unninges** (*pl.*) A. P. iii. 213.

un-nēode, sb., *? displeasure*: to his aȝre un-neode [onneode] LAȝ. 308.

un-neomelīch, adj., *incapable*, KATH. 1185.

un-noiandnes, sb., *innocence, harmlessness*, HAMP. PS. xxv. 11; xl. 13.

un-nombirable, adj., *inmumerable*, ALEX· (Sk.) 2365.

un-nüt, adj., *O.E.* unnytt,=*Goth.* unnuts, *O.N.* ūnytr, *O.H.G.* unnuz; *unprofitable*, P. L. S. viii. 3*; MISC. 192; unnet A. R. 82; REL. I. 129; unnit HOM. II. 83; ORM. 4921; **unnüt** (*sb.*) A. R. 352; unnit ORM. 8059; on unnet HOM. I. 107; unnutte speche HOM. II. 129; unnitte 65; **unnette** (*pl.*) JUL. 45.

un-ofserved, adj., *undeserved*, ROB. (W.) 1256.

un-orne, adj., *O.E.* unorne; *feeble, frail, plain, rude*, A. R. 108; O. & N. 317, 1492; HORN (R.) 338; ure unorne fleis HOM. I. 85; his fode was unorne ORM. 828; unorne læfe 16809.

　unorne-like, adj., *basely, meanly*, HAV. 1941; unorneliȝ ORM. 3750.

un-páred, adj., *unpared*, P. L. S. xxiv. 232.

un-perfeccioun, sb., *imperfection*, WICL. ECCLUS. xxxviii. 31.

un-perfit, unparfit, adj., *imperfect*, WICL. PS. cxxxviii. 16.

un-pīned, adj., *not pained*, ORM. 1367; P. L. S. xvii. 173.

un -pinnen, v., *unpin*: þe ȝate unpinne A. P. i. 727; **unpinnede** (*pret.*) LANGL. *B* xi. 108, *C* xiii. 47.

un-pliȝt, sb., *? peril*; that mi soule have no unpliȝt ASS. *B* 194.

un-possible, adj., *impossible*, ALEX. (Sk.) 4249.

un-profitable, adj., *unprofitable*, (*spelt* un-prophetable) ALEX. (Sk.) 3560.

un-próvednes, sb., *inexperience*, ALEX. (Sk.) 1019.

un-punissinge, sb., *impunity*, HAMP. PS. xciii. 11*.

un-rǣd, sb., *O.E.* unrǣd,=*O.N.* ūrāð; *bad counsel, imprudence, ill-fate*, LAȝ. 6517·; unrað [onread] 3038; unred O. & N. 1464; GEN. & EX. 1906.

un-rēde, sb., *? O.E.* unrǣden; *imprudence; folly*, O. & N. 1355.

un-rēdi, adj., *unready*, WICL. 2 COR. ix. 4; PR. C. 1990.

un-reken, adj., *unready, not apt*, GEN. & EX. 2817; SPEC. 100.

un-reste, sb., =*M.L.G.* unreste, -raste; *unrest*, CH. TRO. iv. 879; WICL. JUDITH xiv. 9.

un-resti, adj., *restless*; unristi HAMP. PS. cxl. 10*.

un-rīde, *see* unrūde.

un-riht, adj., *O.E.* unriht,=*O.L.G.* unreht, *O.H.G.* unreht; *wrong*; unriȝt BEK. 330; WICL. ECCLUS. v. 10; **unriht** (*sb.*) P. L. S. viii. 47; LAȝ. 6553; unrigt GEN. & EX. 1276; unright RICH. 2125.

　unriht-ful, adj., *wrongful*; **onriȝtvolle** (*pl.*) AYENB. 39.

　unrihtfulnesse, sb., O. & N. 1742.

　unriht-wīs, adj., *O.E.* unrihtwīs; *un-righteous*, HOM. I. 115; ORM. 390; unrigtwis GEN. & EX. 2014.

un-rīpe, adj., *O.E.* unrīpe,=*O.H.G.* unrīfi; *unripe*, O. & N. 320.

un-rō, sb., = *O.N.* ūrō, *O.H.G.* unrāwa; *disquiet*, C. M. 7438; unroo PERC. 362.

un-rōt, adj., *O.E.* unrōt; *sad, sorrowful*, ['*tristis*'] MAT. xxvi. 38.

un-rūde, adj., *enormous, cruel*, HOM. I. 249; unride ORM. 4784; REL. I. 220; HAV. 964; AR. & MER. 886; K. T. 142; MAN. (F.) 3435; PERC. 1131; TOWNL. 84; ALEX. (Sk.) 871; a geaunt unride TRIST. 2712; ounride FER. 747; þe **unrīdeste** (*superl.*) HAV. 1985.

　unrūde-līche, adv., *cruelly*, JUL. 54; unrideli HOM. I. 281; GAW. 1432; ALEX. (Sk.) 638; unruidli 566.

un-sad, adj., *O.E.* unsæd; *light, unsteady,* CH. C. T. *E* 995.

un-sæhte, sb.,=*O.N.* ūsātt; *discord, dissension,* LAȝ. 11459; unseihte FRAG. 7.

unsahtnesse, sb., *discord,* ORM. 7187.

un-sæl, adj., *O.E.* unsæl, *O.N.* ūsæl; *unhappy*; unsel LAȝ. 30541; unsele REL. I. 113; usel (*ms.* usell) ORM. 3668.

onsēl-liche, adv., *unhappily,* LAȝ. 7022*.

un-sæl², sb.,*infelicity, misfortune*; unsel MISC. 149; O. & N. 1263*; unsele ALEX. (Sk.) 1106; on unsele (*inopportunely*) REL. I. 131.

un-sælhðe, sb., *infelicity*; unselhðe A. R. 86; JUL. 47.

un-sæliȝ, adj., *O.E.* unsælig,=*O.H.G.* unsālig; *unhappy*; unseliȝ ORM. 4812; unseli KATH. 1811; A. R. 174; unceli GAW. 1562; GEN. & EX. 1073; JOS. 704; GOW. II. 142; CH. C. T. *G* 468; IW. 2939.

un-sælðe, sb., *O.E.* unsælð; *infelicity,* LAȝ. 4748; unselþe ORM. 1561; O. & N. 1263; unselðe GEN. & EX. 3026.

un-samen, adv., *not together,* ALEX. (Sk.) 605.

un-sauht, pple.,=*O.N.* ūsāttr; *irreconciled, at enmity,* JOS. 64; unsaught GOW. III. 153; unsahte (*pl.*) LAȝ. 3930.

un-scáþeful, adj., *harmless*; unskaþeful ORM. 1176.

un-schámefast, adj., *immodest* ['*impudens*'] WICL. DAN. viii. 23; onschamefast PR. P. 367.

un-schent, adj., *unharmed,* ALEX. (Sk.) 2143.

un-schōd, adj., *O.E.* unscōd; *without shoes, unshod,* WICL. IS. xx. 3.

un-schaplīch, adj., *unshapely*: unshaplich REL. I. 129.

un-scháþiȝ, adj., *innocent,* (*ms.* unnshaþiȝ) ORM. 2889.

un-shápiȝnesse, sb., *innocence,* ORM. 58.

un-schrivel, adj., *?neglectful of confession*; onssrivel AYENB. 32.

un-schriven, adj., *unshriven,* A. R. 314.

un²-schütten, v., *open*; unschette (*pret.*) CH. C. T. *E* 2047; unschüt (*pple.*) MISC. 228; unshet IW. 63.

un-seȝȝend-līc, adj., *inexpressible,* ORM. 2823.

un-sehen, adj., *unseen, invisible,* MARH. 10; unseie SAINTS (Ld.) xlv. 66; unseiene [unsehene] gostes A. R. 312; unsene *unique* ALEX. (Sk.) 1026.

unseȝhen-līc adj., *invisible,* ORM. 17296; unsehelich KATH. 255.

un²-seilen, v., *unseal*; unsélid (*pret.*) MISC. 216.

un-sēmlī, adj.,=*O.N.* ūsǣmiligr; *unseemly,* MAP 335; SPEC. 31; D. ARTH. 1044.

un-sēne, adj., *unseen,* HOM. II. 47; GEN. & EX. 2878.

un-sēouwed, adj., *not sewed; without seam*; A. R. 344; unsued P. L. S. xxiv. 169.

un²-sēowen, v., *unsew, slit open*; unsowen LANGL. *B* v. 66.

un-sēsōn, sb.; in unseson *out of season* ALEX. (Sk.) 4439.

un-sēte, adj.,=*M.Du.* onsāte; *?improper,* SPEC. 23, 31, 49; his sorwe was unsete TRIST. 1238; unsete mete TREV. IV. 11.

un-sib, adj., *not related*; unsibbe (*dat. pl.*) ORM. 2474.

un-sibbe, sb., *O.E.* unsibb,=*O.H.G.* unsippe; *enmity*; onsibbe LAȝ. 9845*.

un-siker, adj., *insecure,* A. R. 144; TREV. VIII. 327; PR. C. 1089.

unsikernes, sb., *insecurity,* PR. C. 9049.

un-sīþ, sb., *misfortune*; unsīþe (*dat.*) O. & N. 1164.

un-skil, sb.,=*O.N.* ūskil; *indiscretion,* GEN. & EX. 3506; MISC. 148; (*ms.* unnskill) ORM. 427; unskile HOM. I. 65; P. S. 333.

un-skīlwis, adj., *unreasonable*; unscilwis HAMP. PS. lxxxv. 10*.

unscilwīs-lī, adv., *unwisely,* HAMP. PS. xxvi. 4*; xxxv. 7*.

un-slagen, adj., *unslain,* GEN. & EX. 1332; unslaine ALEX. (Sk.) 2475.

un-slēgh, adj.,=*O.N.* ūslœgr; *not sly,* PR. C. 1938; unsleiȝ WICL. PROV. xxiii. 28; þe unsleie MISC. 82.

un-smēþe, adj., *O.E.* unsmēðe; *rough,* ORM. 9209.

un-sode, adj., *O.E.* unsoden; *unsodden,* O. & N. 1007.

un-sŏfte, adj., *O.E.* unsōfte; *hard, severe,* MISC. 91; CH. C. T. *E* 1824.

un-sōm, adj., *discordant*; unsōme (*pl.*) LAȝ. 3931.

un-sōuht, adj., *unsought,* A. R. 324.

un-soumid, adj., *unnumbered,* ALEX. (Sk.) 1991.

un-sparlīche, adv.,=*O.N.* ūsparliga; *unsparingly,* JUL. 59; unspareli GAW. 979; D. ARTH. 3160.

un-spēde, sb., *O.E.* unspēd; *misfortune,* C. M. 15420.

un-spēdful, adj., *unsuccessful,* HAMP. PS. cxxviii. 4.

un²-spennen, v., *unbind, loose*; unspennede (*pret.*) A. R. 158*.

un²-sperren, v., *unbolt, open,* ORM. 12158; unsperre, -spere LANGL. *B* xviii. 259, *C* xxi. 272; unspere LIDG. M. P. 54.

un²-spurne, v., *push open,* HORN (L.) 1074.

un-stáble, adj., *unstable,* A. R. 122; P. L. S. xxiv. 183.

un-staþelfest, adj., *O.E.* unstaðolfæst; *un-*

stable, [*inconstans'*] HOM. I. 151; unsta-ꝧelvest A. R. 208.

un-stedefast, adj., *unsteadfast*, P. L. S. viii. 122; HOM. II. 61; TREV. I. 357.

   unstedefastnesse, sb., *unsteadfastness*; onstedefastnesse PR. P. 367.

un²-stéke, v., *unfasten*, HOM. II. 258; LEG. 26.

un-stirabil, adj., *immovable*, HAMP. PS. *page* 506.

un-strang, adj., *O. E.* unstrang; *infirm, weak*, ORM. 7911; unstrong LAȝ. 10474; MARH. 12; A. R. 6; O. & N. 561; **unstrengre** (*compar.*) A. R. 222.

un²-strangien, v., *enfeeble*; unstronge REL. I. 119.

un²-strengen, v., *enfeeble*; **unstrengeꝺ** (*pres.*) JUL. 44.

un-strengꝺe, sb., *infirmity*, KATH. 1027; unstrengþe O. & N. 751; unstrencꝺe A. R. 232; unstrencþe ORM. 16915.

un²-strengꝺen, v., *enfeeble*; unstrencꝺen A. R. 138; **unstrengꝺet** (*pple.*) KATH. 1275.

un-strēoned, adj., *not begot*, LAȝ. 18882.

un-suffrabil, adj., *insufferable*, ['*intolerabilem'*] HAMP. PS. cxxiii. 4.

un-suget, adj., *not subject*, WICL. HEB. ii. 8.

un-súnd, adj., *unsound*; unsond M. ARTH. 2859.

un-súnde, sb., *mortality, death*, LAȝ. 39315; unsounde A. D. 263.

un-súre, adj., *insecure*, ALEX. (Sk.) 2136.

un-swac, sb., *? = Ger.* unschwach; *displeasing*, GEN. & EX. 1212.

un-tāght, adj., *untaught*, PR. C. 5873.

un-táld, adj., *untold*, PR. C. 7447; ALEX. (Sk.) 2677; untold OCTOV. (W.) 821.

un-talelīch, adj., *= M.Du.* ontallik, *M.H.G.* unzallich; *innumerable*, HOM. I. 251; A. R. 410.

un²-tēȝen, v., *O.E.* untȳgan; *untie*; untiȝe BEV. 3384; **untēigeꝺ** (*imper.*) MK. xi. 2.

un-témid, adj., *untamed*, WICL. ECCLUS. xxx. 8; untamed HAMP. PS. xxiv. 11*.

un-tīdi, adj., *= M.L.G.* untīdich, *O.H.G.* unzītig; *mod.Eng.* untidy; *unseasonable, improper, mean, dishonest*, KATH. 2433; WILL. 1455; LANGL. *B* xx. 118; *C* iv. 87.

un-tiht, -tiȝth, ꜳ ᷤᴊ ᷣ., *= M.H.G.* unzuht; *vice*, MAP 336, 342.

un³-til, adv., *until*, LANGL. *B* PROL. 227; PR. C. 555; ISUM. 243; ontil HAV. 761.

un-tiled, adj., *untilled*, LANGL. *B* xv. 451; untuled ROB. 372.

un-tīme, sb., *O.E.* untīma, *= O.N.* ūtīmi; *unseasonableness*, REL. I. 132; in untime A. R. 344; LANGL. *B* ix. 186; CH. C. T. *I* 1051; ontime LEG. 14.

un-tīmelīche, adj., *untimely*, REL. I. 131.

un³-tō, prep., *= O.L.G.* unto; *unto, until*, HAV. 2399; CH. C. T. *A* 488; FLOR. 1335; PERC. 1273; M. H. 37; a while unto i cum ogain IW. 930; onto LUD. COV. 126.

un-todǣled, adj., *undivided*, ORM. 11518.

un-todēlinde, adj., *indivisible*; ontodelinde AYENB. 266.

   untodēlend-līch, adj., *indivisible*, HOM. I. 99.

un-towen, adj., *= M.L.G.* untogen, *M.H.G.* ungezogen; *untrained, rude*, A. R. 372; untohe H. M. 31; untoun SPEC. 32; **untohene** (*pl.*) HOM. I. 245.

   untohe-līche, adv., *= M.L.G.* untogelīke; *rudely*, HOM. I. 247; H. M. 17.

   untoweschipe, sb., *rudeness*, A. R. 170.

un²-trenden, v., *unroll*; untrende (*pres. subj.*) MISC. 99.

un-trēwe, adj., *= M.L.G.* untrūwe; *untrue*, HORN (R.) 645; SPEC. 46; ontrewe AYENB. 18; PR. P. 368.

   untrēwnesse, sb., *untruthfulness*, HOM. II. 228; untreunesse P. L. S. viii. 134.

un-trīst, *see* untrūst.

un-trouþe, sb., *O.E.* untrēowꝺ; *untruth*, CH. C. T. *B* 687; ontreuþe AYENB. 17.

un-trum, adj., *O.E.* untrum; *infirm*, MK. xiv. 38; **untrume** (*pl.*) MARH. 22; þa untrummen men HOM. I. 91.

   untrumnesse, sb., *O.E.* untrumness; *infirmity*, HOM. I. 103; ORM. 72; untrumnisse JOHN xi. 14.

un²-trumen, v., *O.E.* untrumian; *enfeeble*; untrumed [ontromed] (*pple.*) LAȝ. 15037.

un-trūst, sb., *distrust, diffidence*, A. R. 332; CH. C. T. *E* 2206; untrust, -trist WICL. ROM. iv. 20.

un-trūsti, -trīsti, adj., *untrusty*, TREV. III. 265; ontrusti, ontristi PR. P. 368.

un-tuderi, adj., *barren*, GEN. & EX. 964.

un-tühtle, sb., *want of discipline*, LAȝ. 24655.

un²-tūnen [on-tūne], v., *O.E.* untȳnan; *unenclose*, LAȝ. 18949; **untīneꝺ** (*pres.*) HOM. II. 115; **untūnden** (*pret.*) LAȝ. 9781.

un-twēmet, -tweamet, adj., *undivided*, JUL. 54, 55.

unꝺe, *see* ūꝺe.

un-þank, sb., *O.E.* unþanc, *= O.H.G.* undanc, -danch; *ingratitude, displeasure*, P. S. 327; unthank CH. C. T. *A* 4082; unꝺonc LAȝ. 22370; A. R. 202; unþonk A. P. ii. 183; onþank þan AYENB. 69; his **unþankes** (*gen.*) *unwillingly* ORM. 7194; here unþankes SAX. CHR. 265; hire unꝺonkes [onþonkes] LAȝ. 4502; min unþonkes P. L. S. xxiii. 102; hit is þe an **unꝺonke** (*dat.*) *it displeases thee* LAȝ. 11769.

un-þēaw, sb., *O.E.* unþeaw; *bad manner, vice*, H. M. 9; unꝺeau A. R. 152; unþeu O. &

N. 194; unðēawe [onþeue] (*dat.*) LAȝ. 3064; unþǣwes (*pl.*) ORM. 17782; unþewes REL. I. 110.

unþēu-fol, adj.; *vicious*; (*printed* un-thenfol) P. S. 159.

un-þēwed, adj., *ill-mannered*, MISC. 185; unðewed GEN. & EX. 2555; unthewed PR. C. 5873; unþǣwed ORM. 2186.

un-ðēode, sb., *strangers*, A. R. 312.

un-þólelīch, adj., *intolerable*, HOM. I. 251.

un-þóliinde, adj., *intolerable*; onþoliinde vol of brene AYENB. 264.

un-þrift ?, sb., *folly*; unþrifte (*dat.*) A. P. ii. 516.

un-þrifti, adj., *unthrifty*; onthrifti PR. P. 367.

un-þrīvande, adj., *uncourteous*, GAW. 1499.

un-þüldelīche, adv., *impatiently*, KATH. 163.

un-waker, adj., *unwatchful*, A. R. 272.

un-wáldes, adv., O.E. unwealdes; *from weakness*, HOM. I. 23.

un-war, adj., O.E. unwær; *unwary*, WICL. PROV. xxiii. 28; HOCCL. i. 41; unwar. [on-war] LAȝ. 7810; þe unware soule A. R. 274.

unwar-līche, adv., O.E. unwǣrlīce; *un-warily*, HOM. II. 191; unwarli ALEX. (Sk.) 5329.

un-warned, adj., *unwarned*, ROB. 51; WICL. 2 MACC. viii. 6.

un-waschen, -weaschen, adj., *unwashen*, HOM. I. 187, 202; unwaschen A. P. ii. 34.

un-wedded, pple., *unwedded*, LEG. 30; un-weddede (*pl.*) H. M. 13.

un-weder, sb.,=M.L.G. unweder, M.Du. on-weder, M.H.G. unweter; *bad weather, tempest*, GEN. & EX. 3058.

un-wélde, adj., *impotent*, GEN. & EX. 347; his limes arn unwelde MISC. 3; unwelde GOW. I. 312; R. R. 359; TOWNL. 76.

un-wéldi, adj., = M.L.G. unweldich; *un-wieldy*, CH. C. T. H 55.

un-wéli, adj., *poor*, PS. lxxviii. 8*.

un-wemmed, adj., *spotless*, A. R. 10; ORM. 14735; CH. C. T. B 924.

un-weote, sb., O.E. unweòta; *ignorant person*, A. R. 8; unwiten, unweoten (*pl.*) KATH. 1054; unweoten MARH. 6.

unwitenesse, sb., *ignorance*, A. R. 278; un-weotenesse HOM. I. 255.

unwit-schipe, sb., *ignorance*, HOM. I. 275.

un-wĕpned [onwĕpned], adj., *unarmed*, LAȝ. 5654; unwĕpnede (*pl.*) HOM. II. 191.

un-wĕrȝed, adj., *unwearied*; unweried ALEX. (Sk.) 3622; unwĕrȝede (? *printed* unwerȝeð) (*pl.*) HOM. I. 261.

un-wīd, adj., *not wide*; unwīde (*pl.*) C. M. 8667.

un-wiht, sb., *monster*, LAȝ. 15734; unwiht A. R. 238; H. M. 41; O. & N. 33.

un-wiht², adj., *frightful*, O. & N. 339.

un-wil, sb., O.E. unwill (*gen.* unwilles);

? *displeasure*: min unwil [unwil] *against my will* MARH. 13; hire unwilles (*gen.*) JUL. 6.

un-wille, sb., O.E. unwilla,=O.H.G. unwille; *displeasure*; his unwille HOM. II. 123.

un-wille², adj., ? *displeasing*: ever ich blisse him is unwille O. & N. 422; mid unwille heorte A. R. 238.

un-willi, adj.,=O.H.G. unwillīg; *unwilling*, '*invitus*,' CATH. 418.

un²-wíndin, v., O.E. unwindan; *unwind*; onwindin PR. P. 368; unwond (*pret.*) LAI LE FR. 189.

un-wine, sb., O.E. unwine; *enemy*, KATH. 1228; A. R. 178; ORM. 19838; unwines (*pl.*) LAȝ. 1628.

un-winne, adj., *invincible*, TRIST. 1235.

un-wīs, adj., O.E. unwīs,=O.L.G. unwīs, O. H.G. unwīs, Goth. unweis; *unwise*, ORM. 16954; unwis SPEC. 42; unwis [onwis] LAȝ. 16022.

unwīs-dōm, sb., *folly*, A. R. 278; unwisdom [onwisdom] LAȝ. 3383.

unwīs-līche, adv., O.E. unwīslīce; *unwisely*, A. R. 338.

un-wit, sb., O.E. ungewit; *folly*, JUL. 22; (*ms.* unwitt) ORM. 6003; unwit CH. COMPL. M. 271; onwit AYENB. 82.

un-witer, adj., O.N. ūvitr; *ignorant, un-knowing*, LAȝ. 16023.

unwitter-lī, adv., *unwittingly*, A. R. 294.

un-witinde, adj.,=O.L.G. unwitandi, O.N. ūvitandi, M.H.G. unwizzende; *unwitting*: of healde oþre manne þinges . . . onwitinde and wiþoute wille of þe lhorde AYENB. 37; un-witinge, -wetinge CH. C. T. G 1320.

unwetand-lī, adv., *secretly*, ALEX. (Sk.) 134.

un-witti, adj.,=O.N. ūvitugr, O.H.G. unwiz-zig; *foolish*, LAȝ. 786; WICL. WISD. iii. 12.

unwitti-lī, adv., *foolishly*, LANGL. A iii. 101; B iii. 105.

un-wrǣst [on-wrěst], adj., O.E. unwrǣst; *infirm, invalid*, LAȝ. 16307; unwreast HOM. I. 237; unwrest A. R. 274; unwrest CHR. E. 921; ALIS. 620; B. DISC. 2118; unwrǣste ORM. 4889; unwreste C. L. 1149; unwraste HAV. 2821; þe unwreste herd ['*iners pastor*'] HOM. II. 39; unwraste dede S. S. (Web.) 1917; unwrĕste (*pl.*) O. & N. 178; unwreaste men KATH. 1266.

unwrĕst-līche, adv., *wickedly*, C. L. 1468; unwreastliche A. R. 294.

unwrĕstschipe, sb., *wickedness*, A. R. 304; unwrestschupe C. L. 1143.

un²-wrappen, v., *unwrap*, CH. BOET. iv. 6 (133).

un-wrench, sb., *evil turn, trick, sin*, A. R. 268; unwrenches (*pl.*) HOM. II. 79; unwrenche (*dat. pl.*) O. & N. 169; *cf.* uniwrenche.

un²-wrīen, v., ? O.E. *unwrīhan, -wrēon; *dis-

cover, reveal, A. R. 328; unwre CH. TRO. i. 860; ounwrie FER. 1849; unwreon' ALIS. 336; unwreo MISC. 91; BEK. 2300; **unwrīhð** (*pres.*) A. R. 84; heo unwreoð 88; **unwrīh** (*imper.*) A. R. 316; **unwrēah** (*pret.*) KATH. 1769; unwreag MAT. xvi. 17; unwreih FRAG. 5; A. R. 56; unwreiʒ AL. (T.) 434; unwrien A. R. 58; **unwríen** (*pple.*) P. L. S. viii. 81; unwrien, -wroʒen, -wroʒe, -wrowe O. & N. 162, 848; onwriʒe AYENB. 88; unwrogen LK. xvii. 30.

un-wúnded, adj., *unwounded,* ORM. 14735; unwondid ALEX. (Sk.) 1235.

un-wune, adj., *unaccustomed;* unwone C. M. 10139.

un²-wunien, v., *disuse; discontinue;* **onwoneþ** (*pres.*) AYENB. 32.

un-wünne, sb., = *O.H.G.* unwunna; *sadness,* P. L. S. viii. 105; (*printed* unwune) HOM. I. 71; unwunne SPEC. 47; unwinne P. L. S. vii. 6; unwin ALEX. (Sk.) 531.

unwin-lī, adj., *sadly,* D. ARTH. 955.

un-wurð, adj., *O.E.* unweorð; *not worth,* A. R. 94; unwurþ FRAG. 6; ORM. 16163; O. & N. 770; REL. I. 181; unworð [onworþ], unwurðe LAʒ. 3464, 24656; unworþ C. L. 1112; onworþ AYENB. 132.

  onworþ-hēde, sb., *unworthiness,* AYENB. 17.

  unwurð-līch, adj., *unworthy,* H. M. 33; onworþlich AYENB. 132; unworthli [unworth-eli] ALEX. (Sk.) 869.

  onworþnesse, sb., *unworthness,* AYENB. 19.

un-wurð², sb., *?unworthliness:* ne schalt (t)u ... wite þe wið unworð H. M. 33.

un-wurði, adj., = *O. H. G.* unwirdig; *un-worthy,* HOM. I. 281; onwurþi PR. P. 368; unworþi BEK. 1397; SPEC. 73.

un²-wurðien, v., *O.E.* unweorðian; *disdain, dishonour;* onworþi AYENB. 22; **unwurðeð** (*pres.*) HOM. II. 181; unwurþeþ ORM. 18285; **geunwurðed** (*pple.*) HOM. I. 227.

up, prep. & adv., *O.E.* up, = *O.L.G.* up, upp, *O. Fris.* up, op, *O.N.* upp, *cf. Goth.* iup (ἄνω), *O. H.G.* ūf; *up; upon:* þe beoð up on (*earlier text* uppan) munt aset MAT. v. 14; þe prophete stod ... in þe venne up to his muʒe HOM. I. 47; up arisen FRAG. 7; walkeð ... up & dun P. L. S. viii. 121; heo wunden up seiles LAʒ. 1101; he þene streng up braid 1454; þa hit alles up brac 3077; he aras up 28008; he stod up 11422; þat maide dronc up (*sec. text* ut) þat win 14349; hebben up [up hebbe] 24781; þa postes ... þa heolden up þa halle 28033; nu raise þai up þe rode HOM. I. 283; up an heih A. R. 130; þet we holden hit up 138; ʒiven oðrʒ strencðe & up holden ham 140; ʒelden up 266; hebben up 264; stien up to þer heovene 356; ant rihte hire up MARH. 20; kesten kang eien up on ʒunge wummen A. R. 56; up o þe rode 24; biheold

after help up' toward hevene KATH. 745; ʒeven an an up hare ʒeomere bileave 1830; up (*ms.* upp) cumen ORM. 1267;' ʒho ras up 2741; up springen 10543; up hofen 12148; & bereþ up his hæfed 1297; up in heofnes ærd 3886; up to þe chinne O. & N. 96; hong up þin ax 658; ariseþ up MISC. 42; corn ðat was up sprungen GEN. & EX. 3050; slep ðor non ðe ða ne up waked 3466; mi dore he broken up HAV. 1960; bere up ROB. 7; up walle 28; bulde up 67; up in þe lifte TREAT. 136; if þe deu is up idrawe 137; oure riʒtes up to holde BEK. 1283; up risen C. L. 578; up holden (*uphold*) 609; up rered 1394; he us up nom 1488; heve up WILL. 348; up drawe ALIS. 2633; B. DISC. 637; up breke OCTOV. (W.) 190; WICL. GEN. xix. 9; up rapen S. S. (Web.) 1620; up lifte RICH. 4474; up reise MAN. (H.) 78; ful boldeli shal trouthe hire heed up bere HOCCL. i. 286; up swal CH. C. T. *B* 1750; he ʒaf up þe gost *B* 1862; turned up swa doune PR. C. 7230; up so doun(e) LANGL. *B* xx. 53; APOL. 19; ROB. (W.) 6831*; up so doun CH. C. T. *A* 1377; up stonde OCTAV. (H.) 362; up stande ISUM. 324; & stiʒen up (*sec. text* uppe) þan hulle LAʒ. 26005; and sat up one vaire boʒe O. & N. 15; me mai up (*v. r.* uppe) one smale sticke me sette 1625; to honte up þe kinges lond(e) ROB. 16; up qvam ðu it findes GEN. & EX. 2320; up erthe LANGL. *B* ix. 99; up peril of mi lif CH. C. T. *D* 2271; op winde REL. II. 273; op ʒelde FER. 4015; op nime *aspire* AYENB. 83; þer op wexeþ alle guodes 75.

[up-ahefed, pple., *elated.]*

  upahefednesse, sb., *elation,* HOM. I. 115.

up-an, prep., = *O.N.* uppā; *upon,* M. T. 97; up-on KATH. 131; R. S. vii; HAV. 2061; BEK. 1502; S. S. (Web.) 190; þei leveden upon him LANGL. *B* i. 117; þu bohtest us upon þe rode SPEC. 58; upo loft 114; evrich upon oþer rideþ O. & N. 494; and hupte uppon on blowe ris 1636; habbeð attir uppon (? = uppen) heore heorte HOM. I. 51; trust uppon godes strencðe A. R. 280; þat muche wo us brouʒte uppon C. L. 1482; uppon, uppo ORM. DEDIC. 69, 105; þeo hwile ðet ich truste uppo mon HOM. I. 213; sorhe upo sorhe H. M. 37; opan, oppon SHOR. 56, 86; opon MAN. (H.) 37; oppon AN. LIT. 10.

up-arīsinge, sb., *resurrection;* oparisinge AYENB. 213.

up-ariste, sb., *resurrection,* HOM. I. 207.

up-astīhunge, sb., *ascension,* MARH. 1.

up-bræid [upbreid], sb., *reproach,* LAʒ. 26036; upbraide ALEX. (Sk.) 1800; upbreid 'AL. (L.³) 155; oupbreid ST. COD. 961.

up-breiden, v., *upbraid:* and flesches lustes hire upbreide O. & N. 1414.

  up-breidinge, sb., *upbraiding;* upbreid-inges (*pl.*) LAʒ. 19117.

up-brixle, sb., *reproach,* ORM. 4871.

up-brüd, sb., *reproach*, H. M. 33 ; þet upbrud A. R. 108.

upen, *see* uppen.

up-fieringe, sb., *? upper chamber* : on witte-sunnedeie . . . com ferliche muchel swei of heofne and fulde al þa upfieringe (*? printed* upfleunge) mid fure HOM. I. 89.

up-flōr, sb., *upper story, 'solarium,'* FRAG. 4 ; **upflōre** (*dat.*) HOM. I. 89.

up-gang, sb., *O.E.* upgang,=*O.N.* uppgangr, *O. H. G.* ūfgang ; *rise* ; **upgange** (*dat.*) LEECHD. III. 98.

up-háld, sb., =*M.L.G.* upholt, *O.N.* upphald ; *uphold, support* ; (*ms.* upphald) ORM. 9217.

up-hēping, sb., *accumulation*, CH. BOET. ii. 3 (37).

up-hóldere, sb., *smallware dealer, 'velaber,'* PR. P. 512 ; **uphólderes** (*pl.*) LANGL. *A* v. 168.

up-land, sb., *upland*, MAN. (F.) 10453.

up-londisch, adj., *uplandish*, PR. P. 512.

up-līc, adj., *O.E.* uplīc ; *upper* ; **upplican** (*dat.*) HOM. I. 41.

uppe, adv., *O.E.* uppe,=*O.L.G.* uppe, *O.N.* uppi, *cf. Goth.* iupa ; *above, up*, ORM. 1169 ; ROB. 185 ; ӡe schulen habben þer uppe þe brihte sihte of godes nebscheft A. R. 94 ; he was uppe JOS. 234 ; heo stiӡen uppe on . . . treowe HOM. I. 5 ; uppe on his heved LAӡ. 17495 ; uppe on þe rode MISC. 167 ; uppe in þe lufte SAINTS (Ld.) xlvi. 599 ; his dore is uppe (*open*) CH. C. T. *F* 615 ; drink it al oppe HORN (H.) 1161 ; ief hi vindeþ þe gate oppe AYENB. 255.

uppen, prep., *O.E.* uppan,=*O.L.G.* uppan, *O.Fris.* uppa, oppa, *O.H.G.* ūfen, üffen ; *over*, HOM. I. 5 ; þe holie gast wile cumen uppen þe HOM. II. 21 ; uppen Sevarne staþe LAӡ. 7 ; uppen æstre 22309 ; ich þe wulle swerien uppen mine sweorde þat . . . 1078 ; uppe [uppen] þere Tambre 28544 ; uppen [uppe] lif & uppen leome 500 ; treo uppen treo 20719 ; uppen eorðe REL. I. 178 ; uppen, uppe A. R. 242, 286 ; uppen, upe O. & N. 733, 1683 ; upen HOM. I. 243 ; rideþ uppe stede R. S. iv ; uppe a lasse hul ROB. 204 ; upe þe doune 174 ; upe BRD. 11 ; leide upe (*v. r.* up) þis dede man L. H. R. 44 ; ope þe heved Jesum he smot LEG. 49 ; ope þe steple AYENB. 180 ; a glose ope þe sautere·187 ; ssete þe dore ope þe 210 ; *comp.* **an-, tō-uppen.**

üppen, v., *O.E.* yppan,=*O.N.* yppa ; *disclose*, A. R. 146 ; **ippen** (*pres.*) REL. I. 129 (HOM. II. 165) ; **üppede** (*pret.*) A. R. 146.

üppinge, sb., *O.E.* ypping ; *disclosure*, A. R. 148.

upper, adj.,=*M.L.G.* uppere ; *upper*, ALIS. 5691.

up-riht, adj., *O.E.* upriht,=*O.H.G.* ufreht ; *upright*, A. R. 266 ; C. L. 695 ; i ne mai stonde upriht A. D. 246 ; uprigt GEN. & EX. 3248 ; opriӡt AYENB. 56.

up-rīse, v., *O.N.* upprīsa ; *uprise*, C. M. 18571.

uprīsinge, sb., = *M.L.G.* oprīsinge ; *uprising*, HORN (L.) 844 ; ROB. 379 ; uprising REL. I. 57 ; oprisinge AYENB. 227.

up-rist, sb., *resurrection, rising*, S. S. (Web.) 1649 ; REL. I. 161 ; upriste (*dat.*) CH. C. T. *A* 1051 ; RICH. 3194 ; tofore þe sunne upriste HORN (L.) 1436 ; opriste LEB. JES. 704 ; SHOR. 124.

up-stīӡ, sb., *O.E.* upstīg,=*O.H.G.* ufstīc *m.*; *ascension* ; **upstīӡe** (*dat.*) HOM. I. 89.

up-ward, adv., *O.E.* upweard ; *upward, up-wards*, LAӡ. 15244 ; upward KATH. 1989 ; ORM. 2056 ; upward, -wardes A. R. 72, 122 ; uppard JUL. 74.

ürchōn, *see* irchoun.

ūre, pron., *O.E.* ūre, ūser,=*O.L.G.*, *O.Fris.* ūser, *O.H.G.* unser, *Goth.* unsara (*gen. pl.*) ; *our, of us*, MISC. 142 ; HORN (L.) 815 ; ure an HOM. I. 21 ; ure fifti LAӡ. 16311 ; hwuch ure KATH. 803 ; ure alre moder A. R. 52 ; ure nan ORM. 7766 ; ure elre land 7491 ; ure eiþer O. & N. 185 ; ur GEN. & EX. 2262 ; oure CH. C. T. *A* 823 ; our FER. 2629 ; our TRIST. 1022 ; our on WILL. 3388.

ūre[2], adj. pron., *O.E.* ūre, ūser, *O.L.G.*, *O.Fris.* ūse, *Goth.* unsar, *O.H.G.* unser ; *our*, A. R. 90 ; HAV. 13 ; JOS. 32 ; þe bur is ure O. & N. 958 ; in ure londe LAӡ. 735 ; for ure lufe ORM. 3883 ; ure (*earlier text* urne) hlaf MAT. vi. 11 ; ure (*for* ures) forme fader gult HOM. I. 171 ; ures hlafordes HOM. I. 235 ; **ūren** (*dat. m.*) LK. i. 73 ; **ūrne** (*acc. m.*) LK. xvii. 5 ; þine aldren and ure LAӡ. 7356 ; ure godes KATH. 458 ; ure seckes GEN. & EX. 2261 ; **ūren** (*dat. pl.*) LK. i. 71 ; mid ure honden LAӡ. 975 ; oure BEK. 132 ; AYENB. 6 ; WILL. 876 ; LANGL. *B* PROL. 102 (Wr. 204) ; she shal ben oure CH. TRO. iv. 539 ; þat is oures CH. C. T. *B* 1463 ; **ūres** (*gen. m.*) HOM. I. 235 ; P. L. S. viii. 99.

ūre[3], sb., *O.Fr.* ūre, hūre, ōre, hōre ; *hour*, MISC. 34 ; ure, oure, hour(e), hore C. M. 15664 ; oure AYENB. 19 ; houre TREV. V. 191 ; **ūres** (*pl.*) A. R. 6 ; oures [houres] CH. C. T. *C* 671.

ūre, sb., *O.Fr.* eur, eur, *from med. Lat.* *agurium, Lat.* augurium ; *destiny, luck, fate*, BARB. i. 312 ; vi. 17, 377 ; ix. 68 ; xv. 376 ; CT. LOVE 634.

ūri, adj., *? O.E.* ūrig, *? O.N.* ūrigr ; *dirty* : his ouri wed M. H. 88.

urinal, sb., *O.Fr.* urinal ; *urinal*, LAӡ. 17725* ; S. S. (Web.) 1049.

urīne, sb., *O.Fr.* urīne ; *urine* ; P. S. 333 ; CH. C. T. *D* 121 ; urin ALEX. (Sk.) 3826.

urisoun, sb., *O.Fr.* hourson ; *an appendage to a helmet*, GAW. 608.

urnement, sb., *O.Fr.* ournement ; *ornament* ; **urnements** (*pl.*) A. P. ii. 1284.

urnen, *see* **rinnen. ürre,** *see* **irre.**
ürþe, *see* **eorðe.**

ürþe, sb., *O.E.* yrþ; *act of ploughing* : erþe BARTH. XVII. 18.

  ürþling, sb., *O.E.* yrþling; *husbandman,* '*arator,*' FRAG. 2.

ūs, ŭs, pron., *O.E.* ūs, ūsic (*acc.*),=*O.L.G.* *O.Fris.* ūs, *O.N.* oss, *Goth.* uns, unsis, *O.H. G.* uns (*dat.*), unsih (*acc.*); (*dat. acc.*) *us,* KATH. 402; GEN. & EX. 1620; LANGL. B PROL. 154; CH. C. T. A 748; nan of us HOM. I. 51; swa us w(u)rse bið LA3. 972; us selve we habbet cokes 3315; ous 1509; þat blisse als us (*ms.* uss) se3þ seþ boc ORM. 1064; ous SPEC. 73; SHOR. 5; AYENB. 6.

ŭs, sb., *O.Fr.* ūs; *use,* A. R. 16; ROB. 458; AYENB. 55; ALEX. (Sk.) 2950.

ūsāge, sb., *O.Fr.* usāge; *usage,* ALIS. 1286; CH. C. T. E 1706; PR. C. 3790.

ūsaunce, sb., *O.Fr.* usance; *custom, habit,* FER. 2217; CH. P. F. 674.

ūsaunt, adj., *accustomed,* FER. 3296.

uscher, sb., *O.Fr.* uissier; *usher,* '*ostiarius,*' VOC. 210; PR. P. 512; ARCH. LII. 38.

ūsen, v., *O.Fr.* user; *use,* PL. CR. 63; usi AYENB. 48; use MIRC 1144; **ūseð** (*pres.*) HOM. I. 207; **ūsede** (*pret.*) LA3. 24293*.

ūsle, sb., *O.E.* ysle,=*M.H.G.* usele, üsele, *O. N.* usli; *embers,* A. P. ii. 747; isil PR. P. 266; **ūseles** (*pl.*) TREV. IV. 431; **iselen** (*dat. pl.*) HOM. II. 65.

ūsūre, sb., *O.Fr.* usūre; *usury,* MIRC 372; WICL. LEV. xxv. 37.

ūsurēr, sb., *O.Fr.* usurier; *usurer,* PR. P. 513.

ūt, prep. & adv., *O.E.* ūt,=*O.L.G.,O.N.,Goth.* ūt, *O.H.G.* ūz; *out (of)* : me sculde leten ut þe king of prisun SAX. CHR. 264; to bringen þe ut of huse ... ut æt þire dure FRAG. 6; & þe ... drinkes ut speweð HOM. II. 37; ne mihte he þat sweord ut dra3en LA3. 7537; driven ut 4982; of Spaine ich wes ut idriven 6213; ibrout ut of þon qvarterne 726; out 286; sched ut ['*effunde*'] A. R. 320; ut of þe heorte 404; þet tet blod barst ut MARH. 5; ut of teone 29; ti neb ... tendreð ut of teone H. M. 31; ut bresten of þe deofles band ORM. 61; ut lesen 3619; mankin ... was lesed ut of helle DEDIC. 166; w(e)orp hit ut O. & N. 121; ne mai me hem ut driven R. S. v; get ne migte 'ðis folc ut gon GEN. & EX. 3021; qvan he weren ut tune went GEN. & EX. 2311; ut lede HAV. 89; he drou ut ... his gode swerd 2733; he wenten ut of halle HORN (L.) 71; out te ROB. 214; seche out 127; icome out Irlond 103; al out *altogether* 54, 121; hi ... fonde hit out P. L. S. xiii. 288; þe angel ... þat drof me out at þe 3ate L. H. R. 22; out of londe heo mot fle C. L. 440; of on fulnesse heo weren out riht 283; out þisse londe CHR. E. 246; þare cam an naddre out þe gras LEG. 54; þou shalt never out wende H. H. 129; out inome (*except*) þe dede of spous-

hod AYENB. 221; king Alisaundre is out iriden ALIS. 4110; out scrape WILL. .448; out breide RICH. 4523; out breke FER. 2996; out louke PS. li. 7*; out taken GOW. I. 146; MAND.250; i shal out drie ['*exsiccabo*'] WICL. Is. xlii. 15; he goþ ... out at þe dore CH. C. T. E 367; out wele *choose* MAN. (F.) 7340; comp. þurh-ūt.

ūt-bīwiste, ? *for* **ūt-wiste**; *exile*; min utbi-wiste is ... longe ituþed HOM. I. 157.

ūt-cume, sb., *outcome, issue,* A. R. 80.

ūte, adv., *O.E.* ūte,=*O.L.G.* ūte, *O.N.* ūti, *Goth.* ūta, *O.H.G.* ūze; *out,* A. R. 150; REL. I. 184; hwen he beoð ute H. M. 31; her ute sitteð six men LA3. 19704; þe folc þær ute stod ORM. 141; oute P. L. S. i. 42; MAN. (F.) 12419; þat hii ne solde oute wende ROB. 170; stonde þer oute P. S. 324; whan þe candel is oute 329.

ūte-wið, adv., *beyond,* A. R. 38; H. M. 31.

ūtemest, adv., *O.E.* ūte-, ȳtemest; *utmost, last*; utmest ORPH. 347; outemest RICH. 2931; þe utemæste LA3. 11023; þe utmeste ende TREV. VI. 359.

uten, v., *O.E.* utan, uton, wutan, wutun, = *O.L. G.* wita, *M.Du.* weten: uten don elmessen *let us do alms* HOM. I. 107; uten we heom to liðe LA3. 20635; ute we (*misprinted* þe) us biwerien P. L. S. viii. 168; ute we þah to him fare O. & N. 1779.

ūten, adv. & prep., *O.E.* ūtan, ūton,=*O.L.G., O.N.* ūtan, *Goth.* ūtana, *O.H.G.* ūzan; *out, beyond, without,* LA3. 5699; uten childre GEN. & EX. 653; uten erd 2406; uten herdes (*r.* erdes) 2410; comp. bi-, ümbe-, wið-ūten.

  ūten-lădde, sb., *foreigner,* HAV. 2153.

  outen-land, sb., *foreign land,* PS. cxxxvi. 4.

  ūten-stede, sb., *foreign place,* GEN. & EX. 1741.

  ūten-wiþ, adv., *beyond,* ORM. 4778.

ūten[2], v., *O.E.* ūtian,=*M.Du.* ūten, *O.H.G.* ūzōn; *put out*; outen ENG. ST. I. 308; CH. C. T. G 834.

ūt-gang, sb., *O.E.* ūtgang,=*O.H.G.* ūzgang; *outgoing, exit,* C. L. 878; outgang PS. cvi. 35; out3ong C. L. 878.

ūt-gangel, adj., *fond of going out* : if þi loverd is neufangel ne be þou nout for þi outgangel ANGL. IV. 197.

ūt-hēese, sb., *hue and cry*; outhees CH. C. T. A 2012; uthest, uheste O. & N. 1683, 1698.

ūt-halve, adv., ? *on the outside* : he isez ... his nest ifuled uthalve O. & N. 110.

ūt-lēete, sb., *outlet* : bi þare see in ore utlete O. & N. 1754.

ūt-la3e, sb., *O.E.* ūtlaga,= *O.N.* ūtlagi; *outlaw,* LA3. 10568; utlage GEN. & EX. 431; utlawe FRAG. 2; outlawe CH. C. T. H 231; **ūtla3en** (*pl.*) LA3. 1121; utlagen HOM. II.

ūt-la3ien, v., *O.E.* ūtlagian; *outlaw*; out-

lawin PR. P. 375; **outlawed** (*pple.*) REL. I.
267; TREV. V. 175.

ūt-land, sb., *O.E., O.N.* ūtland; *foreign land*;
outlandes (*pl.*) MAN. (F.) 3212; outlondis
D. ARTH. 3697.

ūtlandisch, adj., *O.E.* ūtlendisc; *foreign*:
outlandische kinges MAN. (F.) 11127; of
outlondisse manne HORN (H.) 613.

ut-līche, adv., *utterly*; utleche ROB. 66.

ūt-nēme, adj., *exceptional, extreme*, C. M.
4827.

ūt-nimen, v., *take out*; **ūtnumen** (*pple.*)
exceptional, extraordinary: he shal ben ut
numen man ORM. 163; hire ut num(e)ne
(*exquisite*) feire JUL. 7.

ūtren, v.,=*L.G.* ütern, *H.G.* äuszern; *utter*;
outre PARTEN. 1563.

ūt-rīder, adj., *rider out*; an outridere that
loved venerie CH. C. T. *A* 166.

ūt-schūte, sb., ? *O.E.* ūtscyte; *excess*: loþ
me beoþ wives utschute O. & N. 1468.

ŭtter, adv., *O.E.* ūtor, ūttor,=*O.N.* ūtar;
outwards, out, MAT. xx. 28.

ŭttre, adj., *O.E.* ūtera, ūttera, ȳtra,=*O.Fris.*
ūtera, *O.N.* ȳtri, *O.H.G.* ūzero, ūzzero; *utter,
outer*, A. R. 4; utter CH. TRO. iii. 664; PR. C.
4815.

ŭtter-līch, adj.,=*M.L.G.* ūterlīk, *M.H.G.*
ūzerlich; *outward*; **ŭtterlīche** (*dat. pl.*) [*v. r.*
openliche] sunnen A. R. 344*.

ŭtterlīche, adv., *outwardly; utterly, to the
uttermost*; A. R. 206; utterlike ORM. 16510;
utter-, outerli CH. C. T. *D* 664.

ŭttereste, adj.,=*O.Fris.* ūtersta, ūtrosta, *O.H.
G.* ūzarōsto; *last, extreme*, 'extremus,' PR. P.
513; L. H. R. 69; uttereste, outereste CH.
BOET. i. 1 (7), ii. 6 (55); uttermest (*for* ute-
mest) *uttermost*, TREV. I. 75.

ūt-ward, adv., *O.E.* ūtweard; *outward*, A. R.
100; utwardes 92; outward BEK. 2189;
TREV. III. 469.

ūt-wiþ, adv. & prep., *on the outside; out of;
beyond*; ORM. 13116; outwiþ JOS. 186; out-
with PR. C. 6669.

ūðe, sb., *O.E.* ȳð, =*O.N.* ūðr, unnr, *O.L.G.*
ūthia, *O.H.G.* unda; *wave*, HOM. II. 143;
ūþe (*pl.*) HOM. I. 43; uðen LAȝ. 4578; þet
uðen [unðes] ne stormes hit ne overw(e)orpen
A. R. 142; iþes A. P. iii. 147; itheȝ D. TROY
1992; ithis ALEX. 63.

ūðien, v., *O.E.* ȳðian; *flow*; ūþiende (*pple.*)
FRAG. 2.

ūþ-wite, sb., *O.E.* ūðwita; *philosopher*; ūþ-
wites (*pl.*) (*ms.* uþwītess) ORM. 7083.

[**uve,** adv.,=*O.Fris.* ova, *O.H.G.* oba, obe
(*above*).]

uve-ward, adj., *O.E.* ufeweard; *above*:
fram ufewearden MAT. xxvii. 51; to uvewarde
REL. I. 130 (MISC. 165).

üvel, adj. & sb., *O.E.* ȳfel,=*O.Fris., M.Du.*

evel, *O.H.G.* ubil, *Goth.* ubils; *evil*; (*ms.* vuel)
P. L. S. viii. 47; A. R. 52; C. L. 5; (*ms.* uuel)
FL. & BL. 441; an uvel ræd LAȝ. 3777; his
uvel wille BEK. 2291; ufel FRAG. 5; HOM. I.
35; ifel SAX. CHR. 252; ORM. 1742; ivel
GEN. & EX. 3718; S. S. (Web.) 1878; TRIST.
1682; efel MAT. vii. 17; evel ANGL. I. 14;
eovel SPEC. 75; ALIS. 753; þat ufel *malady*
LAȝ. 17598; þeo þet ou eni uvel [eil] doð A.
R. 186; þat uvel P. L. S. xii. 108; ivel HAV.
114; el (*r. w.* del = devel) MIRC 365;
üvelne (*acc. m.*) REL. I. 182 (MISC. 122);
uvelne [uvele] ræd LAȝ. 2541; his uvele
deden A. R. 86; evele willes AYENB. 66; to
oþre uvele dede O. & N. 1376; *comp.* **lond-
üvel.**

üfele [üvele], adv., LAȝ. 1903; þat heo him
uvel walden 7686; uvele [el] iheowed A. R. 368;
uvele bicom him BEK. 1179; ivele LANGL.
(Wr.) 2808; ivle FER. 2557; evile MAND.
10; el [*v. r.* ille] heowet HOM. I. 249.

üvel-dēde, sb., *O.E.* yfeldǣd, = *O.H.G.*
ubeltāt; *evil deed*, HOM. I. 247; iveldede GEN.
& EX. 502; eveldede TREV. III. 143.

evel-dōer, sb., *evil doer*, TREV. IV. 221.

üfelnesse, sb., *O.E.* yfelnesse,=*M.H.G.*
übelnisse; *evil*, HOM. I. 95.

üvelien, v., *O.E.* yfelian, yflian; *do evil;
hurt; become evil; fall sick* : ne scal us na
mon uvelien (*ms.* uuelien) þer vore HOM. I.
15; men . . . þat eveleþ nouȝt TREV. I. 81;
üvelede (*pret.*) ROB. (W.) 7162; eyilde H.
S. 8032; uvelede, evelled ROB. (W.) xx.
429; iüveled (*pple.*) LAȝ. 31774.

uvemest, adj., ? *O.E.* ufemest, yfemest; *top-
most*: þe uvemeste bou HOM. II. 219; an alre
uvemeste A. R. 328; þe ovemeste TREAT. 132;
ovemast C. L. 789; an ovemast 715.

uven, adv., *O.E.* ufan, ufane, = *O.N.* ofan,
O.H.G. obana; *above, from above*: oð ðe ge
seon ufene gescredde LK. xxiv. 49; *comp.* **an-,
bi-uven.**

uvenan, adv., *O.E.* ufenan (*from above*
(JOHN iii. 31), ufenan, ufenon (*above*) (B. D. D.
144, 212); *O.H.G.* obenan (*above*); *above*: he
cnelede þar ufenain (? ufen an) LAȝ. 1217;
and Walwain heom uvenon [ovenon] 27706;
oveno (*ms.* oueno, *printed* oneno) þe sherte
HORN (R.) 1485.

uver, adv., *O.E.* ufor, *O.N.* ofar; *higher up*,
ORM. 1715.

uvere, adj., *O.E.* ufera,=*M.L.G.* overe, *O.N.*
öfri, efri, *O.H.G.* obero; *upper*, A. R. 332; þe
ufere [overe] hond habben LAȝ. 1520; heore
uvere (*ms.* vuere) breih MISC. 150; overe
WICL. LEV. xi. 29; þe overe lippe VOC. 146;
þe overe side SAINTS (Ld.) xlvi. 463; his
overe cheoke MARG. 159; in þe over side
TREV. I. 127; on þe ovir ende FLOR. 600;
an uvere daȝe(n) LAȝ. 27794.

overeste, adj., = *M.Du.* overste, *O.H.G.*

oberōsto; *highest, uppermost,* CH. C. T. *A* 290; overest WICL. EX. xxxix. 21.

over-hand, sb.,=*M.Du.* overhand, *M.H.G.* oberhant; *upper hand, superiority,* ORM. 5458; til Josue wan þe overhand C. M. 6956; overhond ROB. 83.

over-herre, sb.,=*M.H.G.* oberherre; *overlord,* H. M. 29; MIRC 1226.

over-leþer, sb., = *Ger.* oberleder; *upper-leather,* VOC. 181; ovirlethir PR. P. 373.

overling, sb., *superior,* AYENB. 141; D. ARTH. 289; overlingis (*pl.*) HAMP. PS. lxv. 10.

over-lippe, sb.,=*Ger.* oberlippe; *upper-lip,* GOW. II. 140; CH. C. T. *A* 133.

over-man, sb.,=*M.Du.* overman, *M.H.G.* oberman; *overman, superior officer,* GEN. & EX. 3424; C. M. 6968.

v.

v-; *for many words beginning with this letter, see* f-, *also* w-.

vacaunt, adj., *Fr.* vacant; *vacant,* PR. P. 507; vacant ALEX. (Sk.) 4774; vacauns ROB. (W.) 9697.

vache, sb., *O.Fr.* vache; *cow, beast,* CH. TRUTH 22.

vacherie, sb., *O.Fr.* vacherie; *dairy,* '*vaccaria,*' PR. P. 507.

vaile, v., *O.Fr.* valoir; *be well, aid, avail,* LEG. 49; MAN. (F.) 12580; vailes (*pres.*) ALEX. (Sk.) 103.

vain, adj., *O.Fr.* vain; *vain,* MAN. (F.) 10706; vaine ALEX. (Sk.) 389.

vair, sb., *O.Fr.* vair, *Lat.* varius; *a sort of fur,* TRIST. 1381; veir MAN. (F.) 615; veir ant gris REL. I. 121.

vaires, sb.: in vaires (*O.Fr.* en voire), *in truth,* GAW. 1015.

val, sb., *O.Fr.* val; *valley,* L. H. R. 26*; vaal 27; vaile, vale ALEX. (Sk.) 1205, 3980, 4164.

val, *see* fal.

valē, sb., *O.Fr.* valée; *valley,* ASS. *B* 590; BARB. vii. 4; valeie L. H. R. 18, 26; valaiis (*pl.*) BARB. xi. 185.

vallauntise, sb., *O.Fr.* vaillantise; *bravery;* vaillauntise MAN. (F.) 12193.

valliant, adj., *Fr.* vaillant; *valiant,* MAN. (F.) 6952; vailʒeand BARB. xvii. 218.

valour, sb., *O.Fr.* valour, valur; *valour:* wordes . . . of greet valour LAUNF. 984; valour ALEX. (Sk.) 2493.

value, sb., *O.Fr.* value, *cf. Ital.* valuta; *valúe, rank,* LANGL. *C* xiv. 202; CH. C. T. *B* 1361; valou MAN. (F.) 4911.

vampe, sb., *Fr.* avant-pied; *vamp, front of a boot or hose,* PR. P. 508; vaumpez (*pl.*) A. R. 420.

vanissen, v., *Lat.* vānescere; *vanish,* CH. BOET. iii. 4 (74); vanischiþ (*pres.*) TREV. VIII. 157; vanesched (*pret.*) LANGL. *B* xii. 293.

vanitē, sb., *O.Fr.* vanité; *empty space, vanity,* H. M. 27; PR. C. 7228; it was bot vacant & wide as vanite it were ALEX. (Sk.) 4774; vanitēs (*pl.*) AYENB. 77.

[vant, *for* avant.]

vant-warde, sb., *Fr.* avant-garde; *vanguard,* ROB. 362; vauntwarde LANGL. *B* xx. 94; vaward ALEX. (Sk.) 3617; vangard BARB. xi. 164.

vant², sb., *cf. Ital.* vanto; *vaunt:* he dorste make a vant [vaunt] (*? printed* avant, avaunt) CH. C. T. *A* 227.

vanten, v., *O.Fr.* vanter; *vaunt;* vaunton PR. P. 508; vant (*imper.*) ALEX. (Sk.) 2713.

vapour, sb., *O.Fr.* vapour; *vapour,* CH. C. T. *F* 393.

variacioun, sb., *O.Fr.* variation; *variation,* CH. C. T. *A* 2588.

variance, sb., *variance,* PR. C. 1446; variaunce PR. P. 508.

váriin, v., *O.Fr.* varier; *vary,* PR. P. 508; váriande (*pple.*) PR. C. 1447.

[vassal, sb., *O.Fr.* vassal; *vassal.*]

vassalāge, sb., *O.Fr.* vassalage; *prowess, valour,* FER. 1671; CH. C. T. *A* 3054; BARB. xvi. 4; MAN. (F.) 12331.

vath, interj., *fie!* WICL. MAT. xxvii. 40; vah WICL. Is. xliv. 16.

vaumbras, sb., *Fr.* avant-bras; *armour for the front of the arm,* MAN. (F.) 10030.

vavasour, sb., *O.Fr.* vavassour; *vavasor, sub-vassal,* FER. 430; vavasours (*pl.*) MAN. (F.) 10996.

vax, *see* fax.

vēāge, sb., *O.Fr.* veiāge, viage; *voyage, journey,* (*v. r.* veiage) ROB. (W.) 4112; viage MAN. (F.) 8840; CH. C. T. *A* 77; BARB. v. 207; LIDG. TH. 1311.

vecchen, *see* fecchen.

vecke, sb., *old woman,* R. R. 4286, 4495.

veile, sb., *O.Fr.* veile; *veil,* PR. P. 508; CH. C. T. *A* 695; veil GEN. & EX. 3616; veiles (*pl.*) A. R. 420.

veilen, v., *O.Fr.* veler; *veil;* veilede (*pret.*) TREV. V. 305.

vein, *see* fein. veir, *see* vair.

vel, *see* fel.

vēl, veel, sb., *O.Fr.* veel; *veal,* CH. C. T. *E* 1420; veel PR. P. 508.

velim, sb., *Fr.* velin; *vellum,* PR. P. 508.

velvet, velwet, sb., *Ital.* velluto; *velvet,* PR. P. 508; felvet LAUNF. 950.

ven, *see* fen.

vencuse, v., *O.Fr.* vaincre; *vanquish,* MAN. (F.) 7396; WILL. GL. 1 COR. i. 25; venqvise

D. ARTH. 1984; **venkisched** [venkised] (*pple.*)
LANGL. *C* xxi. 106; venqvissed, venqvished
CH. C. T. *B* 291; fenked ALIS.[2] (Sk.) 111;
vencust, vencuste ALEX. (Sk.) 3122, 3875.

vendāge, sb., *O.Fr.* vendenge; *vintage,*
LANGL. *B* xviii. 367 (*C* xxi. 414).

vēne, sb., *O.Fr.* vēne; *vein,* TRIST. 2214;
veine PR. P. 508; **vánis** (*pl.*) BARB. vii. 173.

veneisūn, sb., *O.Fr.* venoison; *venison,* HAV.
1726; veneson ROB. 243; venison MAN. (F.)
1500.

venemostē, sb., *O.Fr.* venimosité; *poison,*
LANGL. *C* xxi. 161.

venerie, *see* **venorie**.

vengeance, sb., *O.Fr.* venjance; *vengeance,*
ROB. 334; PR. C. 4852; vengance ALEX.
(Sk.) 1484; vengeans BARB. vi. 506; ven-
jaunce LANGL. *A* iii. 245.

vengeaunt, adj., *avenging,* HAMP. PS. xcviii.
9.

vengin, v., *O.Fr.* venger; *avenge, vindicate,*
PR. P. 508; venge WILL. 5197; LANGL. *A* v.
106; MAN. (F.) 6501; PERC. 568; **vengid**
(*pple.*) ALEX. (Sk.) 2090; vengit BARB. xix.
151.

venial, adj., *O.Fr.* venial; *venial,* AYENB. 16.

venie, sb., *Lat.* venia (*pardon*); *request for
pardon,* A. R. 46, 258, 426.

venim, sb., *O.Fr.* venin; *venom, poison,*
AYENB. 22; MAN. (F.) 9689; HOCCL. i. 211;
fenim SHOR. 105; *charmers and venim
makers* HAMP. PS. lvii. 5*.

venimin, v., *O.Fr.* envenimer; *envenom,* PR.
P. 508; **venimed** (*pple.*) WICL. WISD. xvi. 10.

venimous, adj., *O.Fr.* venimeus; *venomous,
malarious,* BEK. 440; AYENB. 203; veni-
mouse MAN. (F.) 16594.

venorie, sb., *Fr.* venerie; *game, hunting,*
WILL. 1685; venerie CH. C. T. *A* 166; MAN.
(F.) 856.

ventousen, v., *O.Fr.* ventouser; *cup* : **ven-
toused** (*pple.*) HALLIW. 908.

   ventūsinge, sb., *cupping,* CH. C. T. *A* 2747.

veor, *see* **feor**.

ver-, = **for-**, *see under this prefix.*

verai, adj. & adv., *O.Fr.* verai (*vrai*); *true,
very,* A. P. i. 1184; AUD. 11; verrai HOCCL.
i. 71; verrei JOS. 341; LANGL. *B* xvii. 289;
verrei [verri] WICL. LK. xvi. 11.

   verai-li, adv., *verily,* GAW. 866; verreiliche
JOS. 351; verrali ALEX. (Sk.) 2928.

verdegrese, sb., *O.Fr.* vert de gris; *verdigris;*
verte grece '*viride graecum*' PR. P. 509;
verdegrece VOC. (W. W.) 619; verdegrees CH.
C. T. *G* 791.

verdit, sb., *O.Fr.* verdit; *verdict,* ROB. 141;
CH. C. T. *A* 787.

verdūre, sb., *verdure,* ALEX. (Sk.) 4979.

verge, sb., *O.Fr.* verge; *verge* : verge in a
writis werke, '*virgata*' PR. P. 508.

vergeous, sb., *O.Fr.* verjus; *verjuice,* L. H. R.
137.

vergier, sb., *O.Fr.* verger; *orchard,* ALIS.
1920.

vermiliōn, sb., *Fr.* vermillon; *vermilion,*
PR. P. 508; ALEX. (Sk.) 4336; vermeon 3945.

vermine, sb., *O.Fr.* vermine; *vermin,* CH.
C. T. *E* 1095; **vermins** (*gen. pl.*) ALEX.
(Sk.) 4797.

vernāge, sb., *O.Fr.* vernage; *Italian white
wine,* PR. P. 509.

vernicle, sb., *copy of the handkerchief of St.
Veronica,* LANGL. *B* v. 530.

vernisch, sb., *O.Fr.* vernis; *varnish,* PR. P.
509.

vernischin, v., *varnish,* PR. P. 509.

verre, sb., *O.Fr.* voirre, verre; *glass,* PR. P.
508; ver WICL. PROV. xxiii. 31; ALEX. (Sk.)
4351; **verris** (*pl.*) ANT. ARTH. xxxvi.

vers, sb., *O.E.* fers, *O.Fr.* vers; *verse,* BRD.
10; vers, fers FRAG. 1; fers ORM. 11943.

verset, sb., *O.Fr.* verset (*verset*); *short verse,*
A. R. 16.

vers(e)len, *O.Fr.* verseiller; *say versicles,* A.
R. 44; **versalie** (*pres. subj.*) 120; **versled**
(*pret.*) MAN. (F.) 16472.

versifiin, v., *O.Fr.* versifier; *versify,* PR. P.
508.

verti, *for* averti, adj., *prudent,* BARB. xviii.
439.

vertū, sb., *O.Fr.* vertu, vertut; *virtue, valour,*
H. M. 13; MARG. 316; ALEX. (Sk.) 5324;
vertūz (*pl.*) A. R. 340; vertous ALEX. (Sk.)
4410; vertuis BARB. x. 295.

vertuous, adj., *O.Fr.* vertuous; *virtuous,
precious,* CH. C. T. *A* 251; A. P. ii. 1280.

verveine, sb., *O.Fr.* verveine; *vervain,* GOW.
II. 262; vervein VOC. (W. W.) 711.

vessel, sb., *O.Fr.* vessel; *vessel,* JOS. 298;
MIRC 114; CH. C. T. *D* 100; fessel SHOR.
56; *make we na vessal of virre* ALEX. (Sk.)
4351; vessail A. P. ii. 1713.

vesselment, sb., *vessels,* A. P. ii. 1280, 1288.

vest, *see* **fest**.

vestiment, sb., *O.Fr.* vestiment; *vestment,*
PR. P. 509; vestement C. M. 3701; **vestimenz**
(*pl.*) A. R. 418.

vestrie, sb., *O.Fr.* vestiaire; *vestry,* PR. P.
509.

vestūre, sb., *O.Fr.* vesture; *vesture, clothing,*
LANGL. *B* i. 23; vestoure ALEX. (Sk.) 1539.

vet, *see* **fet**.

veuter, sb., *one who tracks deer by the fuite;*
veuters (*pl.*) GAW. 1146.

vexin, v., *O.Fr.* vexer; *vex,* PR. P. 509.

viage, *see* **vēāge**.

viaunce, sb., *food,* ALEX. (Sk.) 4121.

viaundour, sb., *hospitable person,* MAN. (F.)
4076.

vicari, sb., *Fr.* vicaire, *Lat.* vicārius ; *vicar,* SHOR. 54 ; vicair [viker] CH. P. F. 379 ; viker LANGL. *B* xix. 477 ; vicar PR. C. 3837.

vice, sb., *O.Fr.* vice ; *vice, defect,* ROB. 195 ; AYENB. 27 ; viss BARB. vi. 355.

vicious, viciūs, adj., *O.Fr.* vicious, vicieus ; *vicious,* CH. C. T. *B* 3653.

victorie, sb., *O.Fr.* victorie ; *victory,* SHOR. 148 ; ALIS. 7663 ; LANGL. *B* iii. 348 ; CH. C. T. *G* 34.

vie, sb., *O.Fr.* vie ; *life (biography),* JUL. 2 ; ASS. *B* 879.

víen, v., *for* envíen ; *vie :* to vie [*v. r.* envie] who miȝt slepe best CH. D. BL. 173.

vīen, *see* fēȝen.

vigile, sb., *Fr.* vigile ; *vigil* ; **vigiles** (*pl.*) A. R. 412.

vigour, sb., *O.Fr.* vigour ; *vigour,* FER. 961.

vīl, adj., *O.Fr.* vīl ; *vile,* TREAT. 138 ; C. L. 1112 ; AYENB. 82 ; vil, fil P. L. S: i. 3, 4 ; fil ROB. 47.

  vīl-hēde, sb., *vileness,* AYENB. 130.

  vīl-līche, adv., *vilely,* P. L. S. xi. 9 ; AYENB. 133.

vilans, adj., *? O.Fr.* vilains ; *vile, base, ugly* ; (*printed* vilaus) D. TROY 527 ; a velans vale ALEX. (Sk.) 4164.

vīle, *see* fīle.

vilein, sb., *O.Fr.* vilain ; *villain,* AYENB. 18.

vileinie, sb., *O.Fr.* vilainie ; *villainy,* A. R. 216 ; CH. C. T. *B* 4477 ; vilainie S. S. (Web.) 1794 ; velani ALEX. (Sk.) 4550.

villāge, sb., *Fr.* villāge ; *village,* CH. C. T. *C* 687.

viltē, sb., *O.Fr.* vilté ; *meanness,* A. R. 354 ; ROB. 519 ; BEK. 763 ; S. S. (Web.) 1220.

vínden, *see* fínden.

vīne, sb., *O.Fr.* vine, vigne ; *vine,* LEB. JES. 244 ; AYENB. 96 ; vigne SHOR. 30 ; **vinyhes** (*pl.*) PS. civ. 33 ; **vīnes** ALEX. (Sk.) 4899.

vinegre, sb., *Fr.* vinaigre ; *vinegar,* PR. P. 510 ; WICL. MK. xv. 36.

vinēre, sb., *vineyard,* HAMP. PS. lxxix. 9 ; **vinērs** (*pl.*) lxxvii. 52.

viniter, sb., *vintner,* ROB. 542.

viniterie, sb., *vintry,* ROB. 542.

viole, *see* fiole.

violence, sb., *O.Fr.* violence ; *violence,* PR. P. 510 ; CH. C. T. *G* 908.

violent, adj., *O.Fr.* violent ; *violent,* PR. P. 510 ; CH. C. T. *C* 867.

violet, vialet, sb., *O.Fr.* violette ; *violet,* PR. P. 509.

vīre, sb., *O.Fr.* vire ; *cross bolt,* BARB. v. 595.

virelai, sb., *sort of rondeau* ; **virelaies** (*pl.*) CH. C. T. (T.) 11260 ; balades, roundels, virelaies CH. L. G. W. 423.

virgine, sb., *O.Fr.* virgine ; *virgin,* SPEC. 88 ;

vergine A. P. i. 1099 ; **virgines** (*pl.*) KATH. 2342.

virginitē, sb., *O.Fr.* virginité ; *virginity,* CH. C. T. *D* 62 ; verginte, verginite A. P. i. 767, ii. 1071.

virnin, v., *Fr.* vironner ; *environ ;* ' *vallo, circumdo* ' PR. P. 510.

virole, sb., *O.Fr.* virole ; *ferrule,* PR. P. 510.

virre, *see* verre.

vīs, sb., *O.Fr.* vis ; *look, face,* ALIS. 267 ; B. DISC. 60 ; MAN. (F.) 3747 ; vice *fine appearance* ALEX. (Sk.) 1539.

vīs², sb., *O.Fr.* vis ; *screw, spiral staircase,* [' *cochleam* '] WICL. 3 KINGS vi. 8.

[**vīs³,** sb., *O.Fr.* vise ; =**devīs** ; *device.*]

visāge, sb., *O.Fr.* visāge ; *visage,* AYENB. 45 ; ALIS. 5652 ; CH. C. T. *E* 949 ; fisage FER. 1079.

visēre, sb., *O.Fr.* visiere ; *visor,* PR. P. 511 ; viser MAN. (F.) 8552.

visioun, sb., *O.Fr.* vision, visiun ; *vision,* P. L. S. xvii. 226 ; visioun [visiun] C. M. 4454 ; vision ALEX. (Sk.) 1508.

visiten, v., *O.Fr.* visiter ; *visit,* MAN. (F.) 3866 ; visite CH. C. T. *A* 493 ; visiti MISC. 28 ; **visitis** [viseten] (*pres.*) ALEX. (Sk.) 1964 ; **visitede** (*pret.*) A. R. 154.

vitaille, sb., *O.Fr.* vitaille ; *from Lat.* victuālia ; *victuals,* MAN. (F.) 10786 ; vitaille [vitaille] CH. C. T. *A* 248 ; **vitailes** (*pl.*) WILL. 1121.

vītal, adj., *Fr.* vital ; *vital,* CH. C. T. *A* 2802.

vitremite, sb., *woman's cap,* CH. C. T. *B* 3562.

vitriol, sb., *Fr.* vitriol ; *vitriol,* CH. C. T. *G* 808.

vittalit, pple., *victualed, stored,* BARB. iv. 63 ; vetelid TOR. 2188.

vittelleris, sb. pl., *foragers,* BARB. xiv. 407 ; vittelouris xiv. 429 ; vitaillers LANGL. *B* ii. 60.

vivēre, sb., *O.Fr.* vivier, *Lat.* vīvārium ; *pool ;* (*printed* vinere) C. M. 13764 ; *see* **wiwere.**

vlēon, *see* flēon.

vocat, sb., *for* **advocat** ; *advocate, pleader ;* vokite VOC. (W.W.) 680 ; **vocates** (*pl.*) LANGL. *B* ii. 60.

voide, adj., *O.Fr.* void, vuid ; *void,* TREV. III. 329 ; CH. BOET. ii. 5 (50) ; PR. C. 390.

voidin, v., *O.Fr.* vuidier ; *make void,* PR. P. 512 ; voide CH. C. T. *A* 2751 ; MAN. (F.) 5388 ; HOCCL. v. 25 ; **voidis** (*pres.*) doun þe levis ALEX. (Sk.) 4145.

voiz, sb., *O.Fr.* voiz, vois ; *voice,* HAV. 1264 ; BEK. 1097 ; vois JUL.² 73 ; voice ALEX. (Sk.) 1000 ; voce BARB. xi. 407 ; xii. 200.

volāge, adj., *O.Fr.* volage ; *giddy ;* thought volage R. R. 1284.

volageous, adj., *giddy,* BARB. viii. 455.

volatile, sb., *fowl, birds,* ALEX. (Sk.) 4637;
volatilis (*pl.*) [*' altilia '*] WICL. MAT. xxii. 4.

volc, *see* **folc.**

volume, sb., *Fr.* volume; *volume,* CH. C. T.
D 681.

volupēr, sb., *woman's cap,* CH. C. T. *A* 3241;
volupeer 4303; voleper CATH. 404.

vomen, v., *Fr.* vomir; *vomit;* **vome** (*pres.
subj.*) WICL. LEV. xviii. 25.

voming, sb., *vomiting,* WICL. JER. xlviii. 26.

vomit, sb., *vomit, ' vomitus,'* PR. P. 512; CH.
C. T. *A* 2756.

vor, *see* **for.**

vou, sb., *O.Fr.* vou; *vow,* SHOR. 75; C. M.
15256.

vouchen, v., *O.Fr.* voucher; *vouch :* i vouche
hur safe AMAD. (R.) liii; i vouche it save
LUD. COV. 336; vouche sauf þat his sone hire
wedde WILL. 1449; **vouchid** (*pret.*) safe
ALEX. (Sk.) 303

vouin, v., *O.Fr.* vouer; *vow,* PR. P. 512.

vouren, v., *O.Fr.* vorer; *devour;* **voure**
(*pres.*) MAN. (F.) 10318.

voute, sb., *O.Fr.* voute, volte; *vault,* PR. P.
512.

voutrière, sb., *for* **avoutrière**; *adulteress,*
ALEX. (Sk.) 4532.

voutūr, sb., *O.Fr.* voutour, voltour; *vulture,*
WICL. JOB xxviii. 7; **voutres** (*pl.*) ALEX.
(Sk.) 3945.

vrēo, *see* **frēo.** **vūst**, *see* **fūst.**

w.

wá, *see* **hwá.**

wā, wǣ, wei, interj., sb., adv., & adj., *O.E.* wā
(*adv., interj.*), wēa (*sb.*), *O.N.* vei, *cf. O.L.G.,
O.H.G.* wē, *Goth.* wai, *Lat.* vae; *woe;
calamity, illness; sorrowful;* wa la wa *alas*
FRAG. 8; wei la 6; wa [wo] la wa LAȝ. 7971;
wa [we] le 28092; wæi [wei] la wæi 8031; we
[wo] la 3456; wo la wo A. R. 88; wei la wei
50; wo la wo O. & N. 412; wai 120, 220; wo
la wo SPEC. 74; cried he neiþer wo ne wai
MAN. (F.) 15879; wai HICKES I. 223; R. S.
v; wai la wai ISUM. 140; wai [wei] la wai
[wei] CH. C. T. *B* 1308; wei HOM. I. 33;
MARH. 16; wei la wei GEN. & EX. 2088; wei
HAV. 462; WILL. 935; a wei C. L. 188; wa
is me HOM. I. 35; him was ful wa TRIST.
2769; then wes he wa BARB. i. 348; wa
worth(e) þe PR. C. 7396; wa [wo] wes him
LAȝ. 317; wa [wo] worþe þan monne 3359;
wa (*printed* þa) 3105, 4281; wao 316; wæ
28329; wo me þet ic libbe FRAG. 5; wo þere
þeode R. S. vi; Jacob was wo GEN. & EX.
1833; him was waȝ (*ms.* waȝȝ) ORM. 11904;
inoh wa þu havest idon me MARH. 11; þat us
wa deð JUL. 51; þolen wa ORM. 897; in wele
and waa ISUM. 305; wa ALEX. (Sk.) 528,

2721, 3075, 5606; þeone is al þet wo iwurðen
to wunne A. R. 218; wo & weane 80; wirchen ... wo HAV. 510; woa HOM. I. 213;
wee P. S. 152; **wæs** (*pl.*) ALEX. (Sk.) 4592;
wāer (*comp.*) BARB. xvi. 245*; he þe **wāest**
(*superl.*) of þe werd ALEX. (Sk.) 2004.

wā-dæi, sb., = *M.H.G.* wētac; *day of woe,*
LAȝ. 8750.

wō-fare, sb., *sorrow,* H. S. 6482.

wō-ful, adj., *woeful,* CH. C. T. *A* 2056.

wā-mēd(e), adj., *angry,* LAȝ. 6368; wemed(e)
HICKES I. 167.

wei-mēre, sb., ? = *Goth.* wajamērei (*blasphemy*); *lamentation*; ah schulen **weimēres**
(*gen.*) leod ai mare (singen) in helle H. M. 21.

wēa-mōd, adj., *O.E.* wēamōd; *angry,* A. R.
118; wemod HOM. I. 5; wemod TREAT. 138.

wēmōdnesse, sb., *O.E.* wēamōdness;
anger, [*' ira '*] HOM. I. 103.

wā-sīð, sb., *time of woe,* H. M. 37; wosið
HOM. II. 209; wosith MAN. (F.) 15712; wei-
sið LAȝ. 25846; **wēasīþes** (*pl.*) FRAG. 8.

wā-word, sb., *sentence of woe,* A. R. 306*.

waast, *see* **wást.**

wac, adj., *cf. M.Du.* wack; *feeble,* GEN. & EX.
1197.

wāc, waik, adj., *O.E.* wāc, *O.N.* veikr, = *O.L.G.*
wēc, *O.H.G.* weich; *from* **wīken**; *weak,* HOM.
I.185; ORM. 6185; wac LAȝ. 10775; woc A. R.
12; min wlite is wan & min herte woc REL. I.
186 (MISC. 135); wooc GEN. & EX. 1874; wok
REL. I. 123; MISC. 93; waik LANGL. *C* vi. 23;
weik '*debilis, imbecilis, lentus,*' PR. P. 520;
ant teos wake world MARH. 1; þis waike
[weike] womman CH. C. T. *B* 932; **wāke**
(*pl.*) ORM. 13671; woke A. R. 268; waike
HAV. 1012; **wācre** (*compar.*) KATH. 1267;
wacre, wackere LAȝ. 4539, 23593; *comp.* (liþ-),
leoþ-wōk, mōte-wok.

wōc-līc, adj., *O.E.* wāclīc; *weakly,* ' *vilis,*'
FRAG. 4; **wāclīche** (*adv.*) H. M. 9; wocliche
A. R. 294.

wācnesse, sb., *weakness,* HOM. I. 273; woc-
nesse A. R. 66; waiknes PR. C. 9026.

wacche, sb., *O.E.* wæcce, = *O.H.G.* wacha;
from **wāken**; *watch, vigil,* MISC. 139; FER.
1696; GOW. II.96; LANGL. *B* ix.96; M. ARTH.
2605; wecche A. R. 114; ORM. 1451; **wecche**
[*'vigilias'*] (*pl.*) HOM. II. 95; wecches ORM.
1617; **wacchis** (*pl.*) *watchmen,* ALEX. (Sk.)
5215; BARB. iii. 187; (*scouts*) xix. 442;
wecchen (*dat. pl.*) A. R. 138; KATH. 1766;
comp. **niht-wache**.

wacche-man, sb., *watchman*; **wacchemen**
(*pl.*) ALEX. (Sk.) 5164.

wacchen, v., *O.E.* wæccan (BL. HOM. 137);
watch, keep a vigil; wacche GOW. I. 162;
HOCCL. i. 305; PARTEN. 5375; wecche LUD.
COV. 280; **wecchinde** (*pple.*) MARH. 15.

wachet, waget, sb., *sort of blue cloth,* CH. C. T. *A* 3321.

wacsen, *see* **waxen.**

wād, wōd, sb., *O.E.* wād,=*O.Fris., M.Du.* weed, *O.H.G.* weit ; *woad (dye)* ; PR. P. 513; wod FRAG. 3 ; wod REL. I. 36 ; wood CH. F. AGE 17 ; **wōde** (*dat.*) O. & N. 76.

wāden, v., *O.E.* wadan,=*O.Fris.* wada, *O.N.* vaða, *O.H.G.* watan ; *wade,* FRAG. 6 ; GEN. & EX. 1799; waden CH. C. T. 1684; wadin '*vado*' PR. P. 513 ; wade HAV. 2645 ; ROB. 99; AM. & AMIL. 1364; AL. (L¹.) 548; heo **wādeð** (*pres.*) ine wete A. R. 74 ; **wōd** (*pret.*) LAȝ. 18095 ; P. L. S. xv. 92 ; FER. 591 ; GAW. 787.; ISUM. 178 ; he wod into þe water P. S. 287; þe knotten wode in his flesch BEK. 1479 ; woude BARB. ix. 388.

wæde, sb., *O.E.* wæde, *cf. O.Fris.* wēde, *O.L. G.* wādi *n., O.H.G.* wāt *f., mod.Eng.* (widow's) weeds ; *garment ; dress, armour ;* þat wæde ORM. 8171 ; wede '*vestimentum*' PR. P. 519 ; LAȝ. 26754* ; HAV. 94; TRIST. 1512; WILL. 3535 ; OCTAV. (H.) 600 ; A. P. ii. 793 ; wede GEN. & EX. 1972 ; **wæde** (*pl.*) FRAG. 6 ; wede PERC. 1252 ; weden C. L. 547 ; all his garisons in glissinand weais ALEX. (Sk.) 3015 ; in ani here wedis 1010 ; weaden A. R. 424; **wēden** [wede] (*dat. pl.*) LAȝ. 23773 ; *comp.* ȝe-**wēde.**

wēdle, sb., *O.E.* wædl ; *poverty,* LAȝ. 1002.

wēdle ², sb., *O.E.* wædla ; *poor person,* MK. x. 46 ; ORM. 5638 ; þe **wēdlen** (*pl.*) LAȝ. 5872 ; (*spelt* weaðlen 427 ; weðlen 497).

wēdlien, v., *O.E.* wædlian ; *be poor ; beg ;* þe king wædlien (*? ms.* wæilien) agon LAȝ. 28880; me scamed (*r.* scameð) þæt ic **wēdlie** ['*mendicare erubesco*'] (*pres. subj.*) LK. xvi. 3.

wēfeles, sb., *O.E.* wæfels ; *coat* ['*pallium*'] MAT. v. 40.

wǣȝe, sb., *O.E.* wǣg, = *O.N.* vāg, *O.H.G.* wāga ; *weighing machine, balance ; wey* ; weie A. R. 60 ; TREV. VI. 267 ; (*ms.* weye) LANGL. *B* v. 93 ; waie AYENB. 256 ; weihe WILL. 947 ; WICL. IS. xl. 12 ; **wēien** (*pl.*) A. R. 372 ; twelf weien (makieþ) on foþir REL. I. 70 ; breken brǣde weiȝes [weies] (*some offensive weapon ; ? a different word*) LAȝ. 30982 ; mid spæren and mid grǣte waȝen 21505.

    wēie-scale, sb., *O.E.* wǣgscalu,=*M.H.G.* wāgeschale ; *scale of a balance* ; (*ms.* weye-) ARCH. LVII. 313.

wǣȝen, v., *O.E.* wǣgan (*afflict, deceive*),= *O. H.G.* weigen (*afflict, harass*) ; *? waye* SHOR. 159.

wǣi, *see* **wei.**

wēken, v., *O.E.* wēcan ; *from* **wāc** ; *weaken, enfeeble* ; weken CH. TRO. iv. 1144.

wæl, sb., *O.E.* wæl,=*O.N.* val, *O.H.G.* wal ; *death, slaughter ;* þat wæl, wal, wel [wal] LAȝ. 4111, 6427, 10830 ; **wæles** (*gen.*) 812 ; on wǣle liggen 9497.

wæl-kempe, sb., *mighty warrior,* LAȝ. 23810.

wal-kirie, sb., *O.E.* wælcyrige, = *O.N.* valkyrja ;. *necromancer* ; wichez & **walkiries** (*pl.*) A. P. ii. 1577.

wel-rēow, sb., *O.E.* wæl-hrēow ; *cruel, murderous,* HOM. I. 229.

wæl-slaht, sb., *O.E.* wæl-sleaht ; *slaughter,* LAȝ. 1369.

wal-spere, sb., *O.E.* wæl-spere ; *spear,* LAȝ. 28577.

wæl, *see* **wel.**

wæl, sb., *O.E.* wæl, = *M.L.G.* wael, *M.Du.* wæl ; *whirlpool,* '*gurges,*' FRAG. 3 ; wel MISC. 149; helle wel MAP 339; weel ['*torrente*'] PS. xxxv. 9 ; wele C. M. 2903.

wæld, *see* **wāld.**

wǣling, *?* adj., *?* *wanton* : unþæwful word & wæling word ORM. 2192.

wællen, *see* **wellen.**

wēne, wāne, sb., *misery,* LAȝ. 2198, 5655 ; weane JUL. 46; H. M. 9; weane & teone A. R. 114; wene HOM. I. 169; in wo & wane II. 224; wan(e) and wrake K. T. 66 ; ant wenden of þeos weanen to weolen MARH. I.

wān-līch, adj. *? miserable* ; wanliche iberen LAȝ. 30288; wanliche weoren þa sonden 25990.

wān-sīð, sb., *misery, woe* ; **wan-, wensīðe** (*dat.*) LAȝ. 539, 3088.

wēn-slaht, sb., *slaughter,* LAȝ. 9520.

wānsum, adj., *miserable,* GEN. & EX. 1099 ; wansom PERC. 1065.

wēne, *see* **wēne.**

wǣpen, sb., *O.E.* wæpen,=*O.Fris.* wēpen, *Goth.* *wēpn (*pl.* wēpna) (ὅπλα), *O.N.* vāpn, *O.L.G.* wāpan, *O.H.G.* wāfan ; *weapon* ; þat it muȝe ben til us god wapen ȝæn þe deofel ORM. 2616 ; wepen CH. C. T. *A* 1591; DEGR. 1606 ; wepne *membrum virile* LANGL. *B* ix. 180; wapen MAN. (H.) 187; PR. C. 1707 ; wopen GEN. & EX. 469 ; **wǣpen** (*pl.*) ORM. 4556 ; wæpne LK. vi. 22 ; wæpne, wapen [wepne] LAȝ. 499, 1702 ; wepne HOM. I. 23; O. & N. 1369; wepne HAV. 89; FER. 171 ; LANGL. *B* xii. 107 ; wapne HOM. II. 11 ; **wēpne** (*gen. pl.*) MARH. 22 ; **wēpnen** (*dat. pl.*) P. L. S. viii. 169 ; A. R. 60 ; wepnen, wapnen [wepne] LAȝ. 367, 947 ; wapnen, wapne HOM. II. 9, 230.

wēp-man, sb., *O. E.* wæpmann, wæpenmonn ; *man, male,* MAT. xix. 4 ; (*ms.* wepmann) ORM. 7998 ; wepmon O. & N. 1379 ; wepmon MISC. 85 ; P. S. 153 ; wep-, wapmon LAȝ. 1868, 19055 ; wapman HOM. II. 21 ; GEN. & EX. 1001 ; **wēpmen** (*pl.*) KATH. 2355 ; wep-, weopmen A. R. 10, 54 ; wapmen MISC. 151.

wǣpmon-cün, sb., *male sex,* LAȝ. 498 ; wepmankin ORM. 4092.

wēpen-take, sb., *O.E.* wæpengetace, -ge-

tæce, *O.N.* vāpnatak; *wapentake*, TREV. II.
87.

wǣpnien, v., *O.E.* wǣpnian, = *O.Fris.* wēpna,
O.N. vāpna, *O.H.G.* wāfenen; *provide with
weapons, arm*; wæpnien LAȝ. 20347; **wēpne**
(*imper.*) LAȝ. 17945; **wēpnede** (*pret.*) KATH.
191; **wæpned** (*pple.*) ORM. 677; wopened
GEN. & EX. 3373; iwæpned LAȝ. 17375; wa-
penned bernes ALEX. (Sk.) 1250.

wær, *see* **war.** **wærf,** *see* **hwarf.**

wǣsand, sb., *O.E.* wǣsand, = *O.Fris.* wā-
sende, *O.H.G.* weisunt; *windpipe*; wesand
VOC. 207; '*isophagus*' VOC. (W.W.) 676;
BARB. vii. 584*; wesaunt GAW. 1336.

wæstme, *see* **wastum.**

wǣt, adj., *O.E.* wǣt, = *O.Fris.* wēt, *O.N.* vātr;
wet; wes ich al wet LAȝ. 28080; MAN. (F.)
15576; mid wet eien A. R. 278; wet GEN. &
EX. 2356; weet C. L. 1433; MAND. 100; þis
halwende wēt (*sb.*) HOM. I. 187; in wate
and drie PR. C. 7611; the pure fettres . . .
weren . . . **wēte** (*pl.*) CH. C. T. *A* 1280; þai
wate war BARB. iv. 380; wete DEGR. 824.

wēt-, wāt-shōd, adj., *wetshod*; LANGL. *B*
xiv. 161, *C* xvii. 14.

wǣte, sb., *O.E.* wǣta *m.*, wǣte *f.*, *O.N.* vǣta
f., *moisture, drink*, ORM. 7852; wete A. R.
164; HOM. II. 123; REL. I. 210; wete TREAT.
138; **wēte** (*dat.*) HOM. I. 103; **wǣten** (*acc.*)
LK. viii. 6; þane wæten LEECHD. III. 120;
wǣten, wætun (*pl.*) LEECHD. III. 82, 86;
wēten (*gen. pl.*) LAȝ. 19769.

wǣten, v., *O.E.* wǣtan, = *O.N.* vǣta; *wet,
moisten*; wete ROB. 321; MAN. (F.) 10340;
wētes (*pres.*) SPEC. 36; weeten WICL. JOB
xxiv. 9; **wēte** (*subj.*) MAND. 158; **wǎtte**
[wette] (*pret.*) ROB. 322; **wĕt** (*pple.*) SPEC.
30; *comp.* **bi-wēten.**

wæter, *see* **water.**

wǣven, v., *O.E.* (be-)wǣfan, = *O.N.* veifa,
Goth. (bi-)waibjan, *O.H.G.* (ze-)weiban; *wave,
twist, shake; move; go*; weve ROB. 64; A. P. i.
318; into þe cloistre . . . hi gonne weve BEK.
2077; **wēveð** (*pres.*) HOM. II. 85; wonderliche
it þe weves WILL. 922; and smot of Modred
his hefd þat hit **wēfde** (*pret.*) a (þene) feld LAȝ.
28049*; his brond aboute he wevede B. DISC.
505; & weved up a window(e) WILL. 2978;
and wefden [wefde] up þa castles ȝæte LAȝ.
19003; wafte A. P. ii. 422, 453; **wēved**
(*pple.*) ENG. ST. I. 102; he wolde his heved
fro þe bodi have iweved ALIS. 3807; hii were
a doun iweved BEV. 954; *comp.* **bi-, to-,
ümbe-wǣven** (-**wēven**); *deriv.* **wǣfeles.**

wafre, sb., *O.Fr.* waufre, gaufre; *wafer,
thin cake*; **wafres** (*pl.*) LANGL. *B* xiii. 271;
HOCCL. i. 146.

wafrere, sb., *maker* or *seller of wafer-cakes,
confectioner*, LANGL. *B* xiii. 266; waferer *A*
vi. 120; **wafereres,** waifrers (*pl.*) CH. C. T.
C 479.

wafrestre, sb., *female maker* or *seller of
wafer cakes,* LANGL. *C* viii. 285.

waȝ, *see* **wai.**

waȝ, sb., *O.E.* wāg, wāh, wǣg, = *O.Fris.* wach,
M.Du. weegh, *O.N.* veggr, *Goth.* waddjus;
wall; wah LAȝ. 25887; wah oðer wal A. R.
104*; wagh PS. lxi. 4; waghe [wagh] *bank
of snow* ALEX. (Sk.). 1757; as till waghe
heldid and till wall down put ['*tanquam
parieti inclinato et macerie depulse*'] HAMP.
PS. lxi. 3; woȝ AYENB. 72; wouh JOS. 204;
wough PALL. i. 785; **waghe** (*dat.*) PR. C.
6619; woȝe GAW. 858; woughe WICL. PS.
lxi. 4; wawe, wowe HAV. 474, 2078; wowe
P. L. S. xvii. 293; **waȝes** [wowes] (*pl.*) LAȝ.
10182; waȝhes ORM. 6825; wahes HOM. I.
247; H. M. 31; wowes O. & N. 1528; wowes,
woawes A. R. 172, 346; wowes LANGL. *B* iii.
61; MAND. 247; **wōȝe** (*dat. pl.*) HORN (L.)
970; *comp.* **hēle-, sīd-waȝ.**

waȝhe-rift, sb., *O.E.* wāhrift; *wall covering,
veil*, ORM. 1014; wahreft '*velum*' FRAG. 2.

wáȝe, sb., ? = *M.L.G.* wage *f.*, *M.Du.* waeghe,
M.H.G. wage, *cf.* *wēi; flood, wave*, AYENB.
207; wawe GOW. I. 184; M. T. 130; **waȝes** (*pl.*)
LAȝ. 11977*; waghes D. TROY 270; waȝes,
wawes FER. 1343, 1345; wawes BRD. 24;
TRIST. 371; LANGL. viii. 40; CH. C. T. *B*
468; LIDG. M. P. 63; waughes TOWNL 31.

wáge, sb., *O.Fr.* wage, gvage, gage; *wage;
pledge; 'stipendium,'* PR. P. 513; ALIS. 904;
HOCCL. i. 119; D. ARTH. 302; **wáges** (*pl.*)
LANGL. *B* xi. 283.

wágen, v., *O.Fr.* wager, gvager, gager; *wage,
engage*; wage TREV. VII. 321; LANGL. *A*
iv. 84; D. ARTH. 547; i wagene mine hevede
2445.

wágeour, sb., *hired soldier*; **wágeouris**
(*pl.*) BARB. ix. 48*.

wageoure, sb., *O.Fr.* gageure; *wager*, P. S.
218; wajoure H. S. 5598, 5601.

waggin, v., *cf. Swed.* vagga, ? *M.H.G.* wacken;
wag, shake, move, '*moveo, vacillo,*' PR. P. 513;
wagge '*vacillare, titubare*' REL. I. 6; he
wole wagge his urine in a vessel P. S. 333;
wagge aboute þe cloistre 332; wepne wagge
HAV. 89; **waggeð** (*pres.*) A. R. 374; waggeþ
LANGL. *B* xvi. 41; P. 44; he waggeþ þe lippen AYENB. 211; **wagginge** (*pple.*) as a reed
TREV. VII. 321; **wagged** (*pret.*) PL. CR. 226;
waged [wagged] ALEX. (Sk.) 968.

waghe, *see* **waȝ.**

waȝien, v., *O.E.* wagian, *cf. Goth.* wagjan, *O.H.
G.* wagan; *move*; waȝeȝen LAȝ. 26941; he
ne mai . . . wawie fot ne hond SAINTS (Ld.)
xiv. 234; wawe ROB. 207; ALIS. 2634;
WILL. 19; **wawes** (*pres.*) JOS. 52; wagiende
(*pple.*) HOM. II. 175; **wawede** (*pret.*) TREV.
VI. 427; **wawid** (*pple.*) WICL. LK. vii. 24.

wagren [**wageren**], v., ? *O.N.* vagra; *reel!,*

stumble, stagger, ['nutare'], WICL. ECCLES.
xii. 3; þe wagerinde (pple.) wind H. V. 86.

waȝn, see wain. wāh, see wāȝ.

wai, see wā, wei.

waif, sb., O.Fr. gaif; waif; waives, weives
(pl.) LANGL. B PROL. 94, C i. 92.

waik, see wāc.

wailen, v., cf. Swed. veila, vaila, O.N. vǣla;
wail, lament; waile AR. & MER. 2573; CH. C.
T. A 1295; weilen WICL. MAT. ix. 15; weilin
PR. P. 520; wailend (pple.) GOW. I. 144;
wailede (weiled) (pret.) LANGL. B xiv. 332;
weiled WILL. 1515; comp. bi-wailen.

waimentin, v., O.Fr. gvaimenter; mourn,
wail, 'lamento,' PR. P. 513; waimente
(imper.) ANGL. I. 306.

waimentinge, sb., lamentation, CH. C. T. A
995; HAMP. PS. xxxiii. 21.

wain, sb., O.E. wægn, wæn,=O.Fris. wain,
wein, O.N. vagn, O.L.G. wagon, O.H.G.
wagan; wain, waggon, 'plaustrum,' PR. P.
513; MAP 338; ROB. 416; C. M. 11653;
wain [wein] WICL. 1 KINGS vi. 7; wein
FRAG. 4; waȝn ORM. PREF. 21; wane BARB.
xi. 25; waines (pl.) GEN. & EX. 2362;
waines AR. & MER. 4722; TREV. VII. 355;
MAND. 250; comp. croude-wain.

wain², sb., O.Fr. gaaing; gain, booty, AYENB.
43; waine SHOR. 80; MAN. (F.) 5953.

wainoun, sb., O.Fr. waignon (chien); a term
of abuse, SPEC. 47.

waiour, sb., O.Fr. gayoir; horse-pond; (ms.
wayoure) PR. P. 513; waiers (pl.) MAN. (F.)
11186.

waite, sb., O.Fr. waite, gvaite, gaite; mod.
Eng. wait; watchman, spy, AYENB. 121; D.
TROY 352; waites (pl.) S. S. (Web.) 1490;
C. M. 11541.

wairingle, see wariangel.

waiten, v., O.Fr. waiter, gaitier; watch,
wait, heed, HAV. 512; WICL. 2 ESDR. xii.
24; CH. C. T. B 246; waitin 'observo' PR. P.
513; his profit to waite LANGL. B v. 202;
waiteþ (pres.) AYENB. 179; lokeþ and
waiteþ wane he come to londe LAȝ. 23077*;
waites PR. C. 1243; waiteand (pple.) ALEX.
(Sk.) 3835; waitand BARB. xiii. 598; waited,
waitid (pret.) ALEX. (Sk.) 1764, 5621;
vatit BARB. v. 36, 640; waiteden HOM. II.
87; weiteden A. R. 196*; comp. a-waiten.

waitere, sb., watchman, ['speculator']
WICL. 4 KINGS ix. 17.

waiþe, see wāþe.

waiven, v., ? A.F. weiver, from O.N. veifa,
=Goth. (bi-)waibjan, O.H.G. (zi-)weiban; see
wǣven; waive, move, send: to waiven [waive]
up þe wiket LANGL. B v. 611; and of hire herte
alle zenne to waivie (ms. wayuye, printed
waynye) AYENB. 18; waife ALEX. (Sk.) 297;
waife (wafe) (pres.) 723; waives 4656; waived
(pres.) (printed wayned) from þe beres WILL.

2386; and waived a wei wanhope LANGL. B
xx. 867; waived his berd GAW. 396; fro þennes
art þou weived (pple.) CH. C. T. B 308;
waifid ALEX. (Sk.) 822.

wáke, sb., O.E. (niht-)wacu, = O.N. vaka, M.
L.G. wake; wake, watch, vigil, A. R. 314;
O. & N. 1590; comp. līch-wake.

wáke-man, sb., watchman; wakemon A.R.
14; wak(e)man S. S. (Wr.) 1443; wakemen
(pl.) (ms. wǎkemenn) ORM. 3812.

wáken, v., O.E. *wacan (pret. wōc); wake, be
awake; keep a vigil; REL. I. 221; (ms.
wakenn) ORM. 7484; þat æver alc god mon
to niht wákie (subj.) LAȝ. 23732; leste ...
sake wo & wraþþe bitwene hem wake MIRC
1654; wọc (pret.) GEN. & EX. 2111; þo he woc
(first text awoc) of sleape LAȝ. 25566*; wok
TRIST. 2616; aboute þe middel of þe nith
wok Ubbe HAV. 2093; wok [wook] CH. C.
T. A 1393; wook E. T. 778; hirdes woke
ORM. 3752; wọken LANGL. B xiv. 69; woke
BEK. 681; wáken (pple.) MIN. ix. 34; wakin
TOR. 280; comp. a-wáken.

waker, adj., O.E. wacor, = O.N. vakr, O.H.
G. wachar; watchful, A. R. 142; REL. I. 132;
waker LANGL. C x. 259; P. R. L. P. 187;
waker and snel MISC. 97; wakir 'pervigil'
PR. P. 514; wakere (sb.) LANGL. A v. 223*;
wakere (pl.) MARH. 17; comp. un-waker.

wákien, v., O.E. wacian, wacigan, cf. Goth.
wakan, O.N. vaka, O.L.G. wacōn, O.H.G.
wachēn; awake, watch, MAT. xxvi. 40; A. R.
6; H. M. 37; ne mihtestu one tide wakien
mid me MISC. 42; wakegen HOM. II. 41; i
shal yemen þe and waken (for biwaken)
HAV. 630; waki AYENB. 52; wake WILL.
2007; wauch him BARB. vi. 95; wák
(printed walk) (pres.) BARB. xvii. 930;
wákieþ (imper.) MISC. 42; wákiinde (pple.)
A. R. 144; wákeden (pret.) LAȝ. 9859; þe
herdes þe wakeden over here oref HOM. II.
31; al to late þeih wakeden P. S. 343; wakit
(printed walkit) BARB. xvii. 324; wáked
(pple.) GEN. & EX. 2516; þat haveth ... fele
nihtes waked HAV. 2999; wakid OCTAV.
(H.) 30; wakit (printed walkit) BARB. xiv.
455; comp. a-, an-, bi-, for-, ȝe-wákien.

wákinge, sb., waking, watch, PR. P. 514;
waking WICL. ECCLUS. xxxviii. 27; wacunge
HOM. I. 69; wákingis (pl.) WICL. WISD.
xiii. 23.

wǎkien, v., O.E. wācian; from wāc; weaken,
soften, moisten: his heorte gon to wakien
[wokie] LAȝ. 2798; wakie [wokie] ... wikkede
hertes LANGL. C xv. 25; water to woke with
Themes LANGL. B xv. 332; wōkeþ (pres.)
MISC. 101; and with warme water ... wokeþ hit
LANGL. C xvii. 332; wǎkede (pret.) LAȝ. 2938.

waknen, v., O.E. wacnian, wæcnan, O.N.
vakna, = Goth. (ga-)waknan; be awakened:
wakenen H. M. 31; wakne HAV. 2164; wac-
neþ (ms. waccneþþ) (pres.) ORM. 5845;

wakeneð MARH. 11 ; wakeneþ SPEC. 60 ; þer wakeneþ in þe world wondred ant wee P. S. 152 ; wakens ALEX. (Sk.) 2222 ; **wakned** (*pret.*) M. H. 134 ; wakened P. L. S. xxv. 148 ; *comp.* **a-wakenen.**

wal, sb., *O.E.* weall,=*O.Fris.*, *O.L.G.* wal, *Lat.* vallum ; *wall*, A. R. 104 ; þane wal LAȝ. 7085 ; FL. & BL. 273 ; ROB. 98 ; CH. C. T. E 1047 ; **walle** (*dat.*) REL. I. 272 ; **walles** (*pl.*) LAȝ. 1135 ; O. & N. 767 ; walles ORM. 14801 ; walles, walle P. L. S. viii. 21 ; HOM. I. 163 ; **wealle** (*pl.*) SAX. CHR. 123 ; *comp.* **burh-, fore-, grúnd-, stán-wal.**

wal-wurt, sb., *O.E.* wealwyrt ; *wall-wort,* '*ebulum,*' REL. I. 36.

[**wal-,** *? stem of* **wallen.**]

wal-hāt, adj., *boiling hot,* JUL. 31 ; ORM. 14196.

wal, *see* **wæl, wel, walh.**

walc, sb., *O.E.* (ge-)wealc,=*M.H.G.* walc, *O.N.* valk ; *from* **walken** ; *walk, march* ; was nouthire waldis in þar walke ALEX. (Sk.) 3799 ; walc and win LAȝ. 2542 ; **walke** (*dat.*) REL. I. 223 ; walke CH. C. T. *A* 1969.

walcnien, v., *? =* **walkien ; walcni^ð** (*pres.*) MARH. 9.

wáld, wæld, sb., *O.E.* weald, = *O.L.G.*, *O. Fris.* wald, *O.H.G.* wald, walt, *O.N.* völlr (*plain*) ; *weald, wold, woodland,* LAȝ. 21339, 31216 ; wold GEN. & EX. 938 ; wold LUD. Cov. 16 ; **wálde** [wolde] (*dat.*) LAȝ. 20842 ; wolde O. & N. 1724 ; REL. I. 222 ; wold MISC. 38 ; **wáldes** (*pl.*) MARH. 10 ; woldes LAȝ. 20138 ; walde 16265 ; **wálden** (*dat. pl.*) 12832.

wáld-scaðe, sb., *destructive monster,* LAȝ. 25859 ; wældscæðe 6446.

wáld², sb., *O.E.* (ge-)weald *n. m.* (*?*) *cf. O.Fris.* wald *m.*, *O.N.* vald *n.*, *O.L.G.* (gi-)wald, *O. H.G.* (ge-)walt *n. f.; power, dominion, realm* : i . . . ah it wald ORM. 11815 ; wald C. M. 18044 ; he . . . haveð his soule weald HOM. II. 79 ; wold GEN. & EX. 2000 ; wold FER. 3324 ; **wáldes** (*gen.*) H. M. 27 ; willes & woldes A. R. 6 ; **wálde** (*dat.*) ORM. 38 ; walde L. H. R. 93 ; þet alle þing haveð on¹ wealde ANGL. I. 31 ; þer wes swa mochel folc . . . þat a nevere nane walde [wolde] ne mihte hit al halde LAȝ. 5253 ; wolde PERC. 2006 ; M. ARTH. 745 ; his herte she hadde in wolde EM. 399 ; gif þu havest welþe a wold(e) REL. I. 174 ; *comp.* **an-, ȝe-wáld.**

wáld³, *?* adv., *O. E.* weald (*perchance*) ; *perhaps* : and wald he hit forȝete SHOR. 34.

wáld⁴, sb., *O.N.* valdr ; *governor* ; wold GEN. & EX. 3412.

walde, sb., *cf. Sp.* gualda, *Port.* gualde, *Fr.* gaude ; *weld* (*the plant reseda luteola*) ; wolde LEECHD. III. 349 ; wolde, welde '*sandix*' PR. P. 520, 532 ; welde CH. F. AGE 17.

wald-ēȝed, adj., *O.N.* valdeygðr ; *wall eyed ; with glaring eyes,* ALEX. (Sk.) 608 ; wawil eȝed [waugheeghed] 1706.

wálden, wélden, v., *O.E.* wealdan, waldan, *O.L.G.* waldan, *Goth.* waldan, *O.Fris.* walda, *O.H.G.* waltan ; *have in one's power, govern, rule* : to walden kineriche LAȝ. 2966 ; wealden HOM. II. 79 ; H. M. 39 ; welden (*r. w.* ihealden) P. L. S. viii. 28 ; HOM. I. 163 ; welde HORN (R.) 313 (H. 318) ; wolde HORN (L.) 308 ; P. L. S. xv. 10 ; CH. C. T. *D* 271 ; MAN. (H.) 264 ; **wealdeð** (*pres.*) P. L. S. viii. 195 ; wealdeð, weldeð JUL. 4, 5 ; wialdeð ANGL. I. 11 ; drihte þe alle domes waldeð LAȝ. 16378 ; walt 32049 ; walt HOM. II. 195 ; wealdeð ['*dominantur*'] MAT. xx. 25 ; ȝoure tongen ȝe wealde SHOR. 100 ; **wealdinde** (*pple.*) KATH. 934 ; wældinde LAȝ. 8386 ; **wield** (*pret.*) HOM. II. 169 ; heo welden LAȝ. 183 ; *comp.* **a-, ȝe-, on-wálden** ; *deriv.* **wáld, wélde, wélden.**

wéldere, sb., *possessor,* WICL. ECCLES. vii. 13.

wáldend, wáldende [wéldende], sb., *O.E.* wealdend ; *governour, ruler,* LAȝ. 25568, 28205 ; weldende HOM. II. 17 ; weldende HOM. I. 75.

wálding [wélding], sb., *O.E.* wealdung ; *power, possession,* LAȝ. 19011 ; weldinge HAMP. PS. cxxxiv. 4.

[**wále,** sb. ; *comp.* **wude-wale.**]

wále², sb., *O.N.* val *n.*, *cf. O.H.G.* wala *f.; choice, option,* D. TROY 11952 ; ic þin wale iwearþ FRAG. 8 ; to wale C. M. 5375 ; ANT. ARTH. xxvii.

wále³, adj., *cf. Goth.* walis (γνήσιος, ἠγαπημένος) ; *noble, choice, famous* : þe wale burde GAW. 1010 ; þi wale regne A. P. ii. 1734 ; feres wale GEN. & EX. 888 ; wale ALEX. (Sk.) 75, 2261, 3561, 4772 ; waile qvile *good time* 4597.

wále⁴, sb., *O.E.* walu,=*? O.Fris.* walu, *O.N.* völr (*round stick*), *Goth.* walus (ῥάβδος) ; *wale, weal, wheal ; gunwale ;* MAN. (F.) 12062 ; D. ARTH. 740 ; wale or stripe, wale of a schippe PR. P. 514 ; *comp.* **wort-wale.**

wále, *see* **wéle.**

wálen, v., *choose* ; wale SPEC. 33 ; ALEX. (Sk.) 1667, 4655 ; D. TROY 127 ; **wále** (*imper.*) ALEX. (Sk.) 1014 ; **wáled** (*pple.*) GAW. 1276 ; *see* **wélen.**

walet, sb., *?=***watel** ; *wallet,* LANGL. *C* xi. 269 ; CH. C. T. *A* 686.

walewen, *see* **walwen.**

walh, adj., *O.E.* walh, *cf. M.L.G.* walg-(-haftich) ; *that has a sickly taste,* H. M. 35 ; walhwe swete PR. P. 515.

Walh, sb., *O.E.* Wealh,=*O.H.G.* Walh, *O.N.* *Valr (*pl.*) Valir) ; *Welshman, foreigner, slave* : ælc wælh wurðe ivreoid LAȝ. 14852.

Wáles, pr. n., *Wales,* LAȝ. 2121.

wal-more, sb., *O.E.* wealmore ; *parsnip, carrot,* '*pastinaca,*' FRAG. 3.

wal-note, sb., *O.N.* valhnot; *foreign nut,* *walnut,* PR. P. 514; LANGL. (Wr.) 7082.

walnote-shale, sb., *shell of walnut,* CH. C. H. F. 1281.

walnot-trē, sb., *walnut-tree,* VOC. 192.

Waling-ford, pr. n., *O.E.* Wealinga ford; *Wallingford,* ROB. 299.

Walisc, Welisc, adj., *O.E.* Weallisc, Wælisc, = *O.N.* Valskr, *O.H.G.* Walhisc; *Welsh, foreign,* LA3. 13021, 31632; þe Walsche LANGL. *A* v. 167 [Walsche *B* v. 324]; **Wælsce,** Welsce [Walse] (*pl.*) LA3. 2120, 2124.

[**walk-,** *stem of* **walken.**]

walk-milne, sb., = *M.L.G.* walkemole, *M. H.G.* walkmüle; *fulling-mill,* TOWNL. 313.

walk, *see* **walc, wakien.**

walken, v., *O.E.* wealcan, = *M.H.G.* walken, walchen; *walk, roll, toss,* WILL. 2129; CH. C. T. *A* 2309; wide walken LANGL. *B* viii. 14; **walkeð** (*pres.*) HOM. I. 173; he walkeþ & wendeþ FRAG. 5; walkeþ GREG. 846; **walkende** (*pple.*) HOM. II. 51; **wēlc** (*pret.*) KATH. 916; welk PR. C. 4390; PERC. 209; AMAD. (R.) xliv; þa ... scipen þa 3eond þa sæ weolken LA3. 12040; welken GEN. & EX. 568; **walke** (*pple.*) HORN (L.) 953; *comp.* 3e-**walken**; *deriv.* **walc.**

walkere, sb., *? O.E.* wealcere, = *O.H.G.* walchāre, *mod.Eng.* walker; *fuller*; ['*fullo*'] WICL. MK. ix. 2; walker '*fullo*' VOC. 212; at a **walkeres** (*gen.*) hous BEK. 1135.

walkien, v., *O.N.* valka; *walk,* A. R. 4; **walkede** (*pret.*) C. L. 1376; walkide WICL. JOHN vii. 1; walkeden GEN. & EX. 3882; walkede LA3. 12040* (*first text* weolken); **walked** (*pple.*) LANGL. *B* v. 537; cloth ... **iwalked** [*v. r. for* itouked] (*fulled*) *B* xv. 447*.

walkne sb., ? = **wolcne**; *welkin,* GEN. & EX. 103, 136; tagte fuel on walkene his fligt 161.

walle, sb., *? O.E.* weall, = *O.Fris.* walla *m.; well,* MARH. ſ1; FL. & BL. 291; L. H. R. 32; at ore walle MISC. 84; **wallen** (*pl.*) HOM. I. 189; wallen REL. I. 267.

wallen, v., *O.E.* weallan, = *O.L.G.*, *O.H.G.* wallan, *O.Fris.* walla; *well up, boil,* ORM. 10507; REL. I. 101; **walleð** (*pres.*) A. R. 118; HOM. II. 227; þar· melke and honi walleþ SHOR. 90; **weallinde** (*pple.*) A. R. 216; wallinde hat *boiling hot* JUL. 70; wallinde MISC. 149; wallande A. P. i. 365; walling ALIS. 1622; **wēol** (*pret.*) KATH. 1925; JUL. 58; þe blod out of his wounde wel K. T. 1087; wul HOM. II. 167; **weolle** (*subj.*) JUL. 70; *comp.* for-**wallen**; *deriv.* **walle, wel, welle, wellen, walm, welm.**

wallen², v., *surround with a wall; build a wall;* walle ROB. 51; **wallede** (*pret.*) GEN. & EX. 435; **wallit** (*pple.*) BARB. i. 107, ii. 220; **walled** (*pple.*) PL. CR. 164; *comp.* bi-**wallen.**

wallare, sb., *cf. prov.Eng.* (dry-)waller; *stonemason,* PR. P. 514.

walm, sb., *? O.E.* wælm, = *O.H.G.* walm: *from* **wallen**; *well, spring; bubbling up*; S. S. (Wr.) 2390; **walme** (*dat.*) LA3. 22124; **walmes** (*pl.*) HOM. I. 141; *see* **welm.**

walm-hāt, adj., *boiling hot,* JUL. 68.

walop, sb., *O.Fr.* galop; *gallop,* WILL. 1770.

Walne, pr. n., *O.E.* Wyln, = *M.H.G.* Welhin; *Welshwoman, foreigner, slave*; wealne MAT. xxvi. 71.

walopen, v., *O.Fr.* galoper; *gallop*: **walopande** (*pple.*) D. ARTH. 2827; waloping PARTEN. 4827; **walopit** (*pret.*) BARB. ii. 440.

walten, v., *O.E.* wealtan, = *M.H.G.* walzen (*pret.* wielz), *O.N.* velta (*pret.* valt, *pple.* oltinn); *roll, overturn*; walt(e) D. TROY 909; **waltez** (*pres.*) A. P. ii. 1037.

walterin, weltrin, v., *cf.M.L.G.* waltern, woltern, weltern, *M.Du.* welteren; *welter, roll about, overturn, 'voluto,'* PR. P. 521; waltre TREV. VII. 203; weltiren BARB. xi. 25; **waltrand** BARB. iii. 719; **waltered** (*pret.*) A. P. ii. 415; waltrid DEP. R. ii. 187; **walterid** (*pple.*) LUD. COV. 342; he was waltrid ['*volvebatur*'] WICL. JUDG. v. 27; weltrit BARB. iii. 719.

walwen, v., *O.E.* wealwian, *cf.* Goth. (af-, at-)-walwjan, *Lat.* volvere; *turn, wallow, roll*; walwe LEG. 221; welwin PR. P. 521; i **walwe** (*pres.*) and winde CH. C. T. *D* 1102; i walowe YORK xxxix. 10; walew
ð HOM. II. 37; þeos walewið in wurðinge H. M. 13; waleweþ AYENB. 126; walweth aboute LANGL *B* viii. 41 [waleweth *A* ix. 46]; **walloing** (*pple.*) TOR. (A.) 189; walowand ALEX. (Sk.) 4064; **walewede** (*pret.*) LEB. JES. 227; D. ARTH. 3838; walewide WICL. JUD. xiii. 10; weole-wede JUL. 41; þe wawes walwede agein þe wal FER. 1345; **walwed** [iwalwed] (*pple.*) TREV. VI. 301; *comp.* bi-**walwen.**

walewing, sb., *wallowing, rolling,* A. R. 294; walewing [walwing] WICL. 2 PET. ii. 22.

wámbe, sb., *O.E.* wamb, womb, = *O.N.* vömb, *Goth.* wamba; *belly, womb,* ORM. 2471; wambe PS. xvi. 14; PR. C. 4161; wombe FRAG. 7; A. R. 78; wombe LA3. 19800; AYENB. 50; **wómbe** (*gen.*) HOM. II. 11; **wómbe** (*dat.*) MAND. 157; to þere wombe HOM. II. 181.

wamlin, v., *cf. Dan.* vamle, *Swed.* vâmla (*vàlta*); *belch, vomit, feel sick; ? roll about; 'nauseo,'* PR. P. 515; to wamel at his hert A. P. iii. 300; he **womblede** (*pret.*) & tomblede ED. 3213.

[**wan,** adj., *O.E.* wan, won, = *O.Fris.* wan, won, *O.L.G.* wan, *O.H.G.* wan, *O.N.* vanr, *Goth.* wans; *wanting.*]

wan-belēve, sb., *want of faith, 'diffidentia,'* PR. P. 515.

wan-belēvenesse, sb., *'perfidia,'* PR. P. 515.

wan-bode, sb., *one who bids a low price, 'invalidus licitator,'* PR. P. 515.

wan-hope, sb., *cf. M.Du.* wanhope; *despair,* ROB. 333; AYENB. 29; LANGL. *B* ii. 99; MAND. 285; CH. C. T. *A* 1249; TOWNL. 184; SACR. 67; wonhope C. L. 951.

wan-hopien, v., *cf. M.Du.* wanhopen; *despair*; wanhope (*pres. subj.*) ANGL. I. 68.

wan-mōl, adj., *not eloquent,* GEN. & EX. 2817.

wan-rēd, sb., (*? for* wandrēd); *? misfortune*: wanne hie segen men wanred þolien HOM. II. 147.

wan-shápe, sb., *deformity,* ANGL. I. 81.

wan-spēde, sb., *O.E.* wanspēd; *ill fortune,* D. TROY 9327.

wan-spēdi, adj., *O.E.* wanspēdig; *unfortunate*; wanspedie men HOM. II. 177.

wan-toȝen, adj., *wanton; irregular*; wantowen LANGL. *C* iv. 143; wantowe, -toun *'insolens'* PR. P. 515; wantoun CH. C. T. *A* 208; wanton werkis ALEX. (Sk.) 12.

wantoun-hēde, sb., *wantonness,* PR. P. 515; wantonhede D. TROY 2911.

wantounesse, sb., *wantonness,* LANGL. *A* iii. 120.

wan-trowþe, sb., *incredulity,* ORM. 3148.

wan-trukien, v., *fail, be wanting*; bote þe wantrokie of live SHOR. 34.

wan-trūst, sb., *cf. M.Du.* wantroost; *one without trust,* CH. C. T. *H* 281.

wan-wit, sb., *O.E.* wanwit; *folly,* L. H. R. 180.

wan², adj., *O.E.* wann, wonn; *wan, pale, faint,* P. S. 191; P. L. S. ii. 43; TREV. III. 371; AM. & AMIL. 2445; K. T. 285; CH. C. T. *A* 2456; WICL. EX. xxi. 25; min wlite is wan REL. I. 186 (MISC. 135); wan, won SPEC. 54, 83, 93; won H. M. 43.

wannesse, sb., *wanness,* [*'livor'*] WICL. PROV. xx. 30.

wan, *see* **hwan.**

wān, sb., =**wēne**; *hope; store, quantity*: al mi worldli wan REL. II. 96; wane IW. 1429; PERC. 422; MIN. iii. 93; BARB. xvii. 249; won FER. 3571; AN. LIT. 6; P. S. 341; CH. TRO. iv. 1134; riche won FL. & BL. 386; god won HAV. 1791; S. S. (Web.) 2420; of god(e) corn gret won (*plenty*) ROB. 2; þo he sai non oþer won (*chance*) 12; nis þer non oþer won P. S. 149; þis worldes won SPEC. 24; hit is wicked won REL. I. 263; woon LANGL. *B* xx. 170; wone, woone *'copia'* PR. P. 532; wone HOCCL. i. 294; ne con ich me no wone L. N. F. 215; bi mine wone (*opinion*) HAV. 1711; *comp.* ȝe-wān.

wān-lēs, adj., *hopeless,* MARH. II.

wand, adj. & sb., *O.N.* vandr; *evil*: þu

schalt in þe putte wunie mid þe **wonde** (*sb.*) *devil,* MISC. 175.

wand-lich, adj., *? O.N.* vandligr; *wicked*; enne swiȝe wandliche sune LAȜ. 6358.

wand-rāþ, sb., = ? **wandrēðe**; *suffering,* ORM. 4846.

wond-rēd, sb., =**wandrēðe**; P. S. 152.

wand-rēðe, sb., *O.N.* vandræði; *misery, distress, peril,* LAȜ. 12511; þat wandreȝe JUL. 22; wandreþe C. L. 1116; wandrethe TOWNL. 137; wandreth M. H. 21; ALEX. (Sk.) 528; wondreȝe MARH. 4; A. R. 156.

wandsom-lī, adv., *sorrowfully*; weri and wandsomdli D. ARTH. 4012.

wand, sb., *O.N.* vöndr, =*Goth.* wandus; *wand, rod,* ORM. 16178; wand TRIST. 909; PR. C. 5880; LUD. COV. 95; wond GEN. & EX. 2715; wond MAP 336; **wandes** (*pl.*) C. M. 1418; *comp.* elle-, met-wand.

wande, sb., *? O.N.* vandi (*difficulty*); *hesitation, doubt, refusal*; wiþouten wande C. M. 8465, 11518.

wandelard, sb., *A.Fr.* wandelard; *? vagabond, ? criminal*: þise men lift þer standard agein David wandelard MAN. (H.) 115.

wandien, v., *O.E.* wandian; *from* **winden**; *cf.* **wenden**; *flee, hesitate, fear*; wande TRIAM. 1526; wonde P. S. 335; MIRC 384; JOS. 399; WILL. 4071; GOW. I. 332; RICH. 228; AM. & AMIL. 550; CH. L. G. W. 1185; to love nul i noht wonde SPEC. 29; wondi SHOR. 103; wonde (*pres.*) HORN (L.) 337 (R. 343); þu ne wandest for nane men MAT. xxii. 16; wand [waned] (*pret.*) *flinched* ALEX. (Sk.) 1411; vaindist [wandist] BARB. xii. 109; wonded A. P. ii. 855.

wandlessour, *see* **wanlasour**.

wandrien, wondrien, v., *O.E.* wandrian, = *M.L.G., M.Du.* wanderen; *wander,* LAȜ. 12044, 14796; wandren LANGL. *B* vi. 304; wandrin, wanderen *'vagor'* PR. P. 515; schal wandre [*'ambulaverit'*] WICL. JOHN xi. 9; wandreð (*pres.*) HOM. II. 85; wondrinde (*pple.*) MARH. 11; wondrede (*pret.*) LAȜ. 21969; P. S. 240.

wanderare, sb., *wanderer,* PR. P. 515.

wandringe, sb., =*M.L.G.* wanderinge; *wandering,* LANGL. *A* PROL. 7; wanderinge PR. P. 515.

wáne, sb., *O.E.* wana, =*O.N.* vani; *wane, decline, decrease, lack*; GEN. & EX. 1028, 3353; no god nis him wane P. L. S. viii. 185; wane PR. P. 515; SHOR. 121; P. S. 341; AR. & MER. 3132; GOW. II. 307; MAN. (F.) 8329; E. G. 30; wone H. M. 29; to muche wone HOM. I. 213; vor wone of witnesse A. R. 68; two wone of twenti KATH. 67; wone REL. I. 171; C. L. 229; ful lutel þer wæs wone LAȜ. 1905; wone of wit SPEC. 30; *comp.* and-wane.

wane, *see* **wune.** **wāne,** *see* **wǣne.**

wanelasour, *see* **wanlasour.**

wang, sb., *O.E.* wang, wong;=*O.L.G.* wang, *O.N.* vangr, *Goth.* waggs; *plain, field*; wong *'territorium'* PR. P. 532; **wonges** (*pl.*) HAV. 397.

wange, sb., *O.E.* wange, wonge, wenge, *cf. O. L.G., O.H.G.* wanga *n., O.N.* vangi *m.; cheek,* LK. vi. 29; wænge MAT. v. 39; **wongen** (*pl.*) LAȝ. 30268; wonges SPEC. 28; TRIST. 732; JOS. 647; *comp.* þun-wange.

 wang-tōth, sb., *cheek-tooth, grinder,* CH. C. T. *B* 3234; PR. P. 515; wang-, wongtooþ WICL. JUDG. xv. 19; wongtoth VOC. 207.

wangere, wongere, sb., *O.E.* wangere,= *O.H.G.* wangāri, *Goth.* waggareis; *pillow,* CH. C. T. *B* 2102.

wánien, v., *O.E.* wanian, wonian,=*O.Fris.* wania, wonia, *O.H.G.* wanōn, *O.N.* vana; *wane, lessen; make pale* : Bruttes gunnen wonien [wanien] LAȝ. 26991; heo w(u)lleð wonien us 982; wanie LANGL. *B* vii. 5 [wonien *A* viii. 59]; wanie FER. 1645; CH. C. T. *A* 2078; MAN. (F.) 1255; wane GOW. III. 109; MAND. 44; **wóneþ** (*pres.*) ROB. 42; þe mone waxeð & woneð A. R. 166; wanieð hire rihtes HOM. II. 177; wannes ALEX. (Sk.) 4142, 4627; (hi) wanieþ BRD. 15; wonieð KATH. 2218; **wánand** (*pple.*) YORK ix. 204; **wániand** (*used as sb.*) *the waning moon, unlucky time* : in þe waniand xvi. 37; waneand vii. 45; **wánede** (*pret.*) ROB. 410; wonede C. L. 232; *comp.* ȝe-**wánien.**

 wánunge, sb., *O.E.* wanung, wonung; *waning*; wiðute wanunge [wonunge] KATH. 922; wonunge HOM. II. 35; woning C. L. 228.

wānien, v., *O.E.* wānian,=*M.Du.* weenen, *O.H.G.* weinōn, *O.N.* veina; *wail*; wanen ORM. 5653; wepen and weinen [woni] LAȝ. 25827; wonie and grede O. & N. 975; heo woneþ & groneþ MISC. 152; **wōniende** (*pple.*) FRAG. 5; **wānede** (*pret.*) LAȝ. 25847.

 wānunge, sb., *O.E.* wānung; *wailing,* H. M. 37; waning LAȝ. 17796; woninge [wanunge] & wop P. L. S. viii. 117; woaning FRAG. 5; woninge MISC. 74; woning O. & N. 311; REL. I. 227.

wankel, adj., *O.E.* wancol,=*O.L.G.* wancol, *O.H.G.* wanchal; *unstable,* REL. I. 221; wankill M. P. 138.

wanlace, sb., *winding in the chase,* HALLIW. 915; **wanlaces** (*pl.*) MAN. (F.) 12860.

wanlasour, sb., *one who drives game* ; wanelasour *'alator'* VOC. (W. W.) 562; **wandlessours** (*pl.*) IPOM. 387.

wanne, *see* **hwanne.**

wansin, v., *O.E.* wansian; *become less,* '*decresco*' PR. P. 515; Marches nahtes wansen (*ms.* wannsenn) ORM. 1901; **wansit** (*pret.*) P. R. L. P. 234.

want, adj. & sb., *O.N.* vant *n.; lacking, deficient; want, deficiency* : hem was want (*ms.* wannt) gastlic insiht ORM. 14398; wante P. S. 341; þet tu hevedest wonte þer of A. R. 284.

 wantsum, adj., *poor,* ORM. 14824; M. H. xviii.

wanten, v., *O.N.* vanta; *want, be lacking* : o word ne schal þer wonten A. R. 344; **wanteþ** (*pres.*) LANGL. *B* xiv. 173; wonteþ C. L. 649; nawiht ne wonteð ham KATH. 1685; all þat wanteþ Cristes hald ORM. 13380; **wantede** (*pret.*) GEN. & EX. 1233; wantede HAV. 1243; wanted PR. C. 6194.

wánunge, *see under* **wánien.**

wāpen, *see* **wǣpen.**

wāpman, *see under* **wǣpen.**

wappe, sb., *blow,* D. TROY 6405; whapp YORK xxiii. 199; wap ALEX. (Sk.) 5318; at a wapp *in a moment* 3040.

wappen, v., *wrap up; beat* : wappin in clothis '*involvo*' PR. P. 515; **wappond** (*pple.*) *lashing about* D. TROY 9513; þe ȝonge men . . . upon þe wiket **wapped** (*pret.*) *struck, beat* A. P. ii. 882; wappit [*v.r.* swappit] out the stane BARB. xvii. 691; in wrathenesse ar wapped YORK xxxi. 12; his bodi is **wappid** (*pple.*) al in wo S. & C. II. xxxiii. *comp.* a-, at-, bi-**wappen.**

wappin, v., *bark,* '*nicto,*' PR. P. 515.

war, adj., *O.E.* wær,=*O.L.G.* war, *O.N.* var, *O.H.G.* (gi-)war, *Goth.* wars; *ware, wary,* A. R. 48; GEN. & EX. 721; ich ulle makien þe war of alle mine wiheles MARH. 16; ȝif man mihte wurþen war (*perceive*) þat ȝho wiþ childe wære ORM. 1963; war '*cautus*' PR. P. 516; HAV. 788; CHR. E. 477; WILL. 1201; FER. 4635; IW. 12; PR. C. 2022; þat no man ne scholde . . . war of him beo BEK. 1151; war & wis SPEC. 30; CH. C. T. *A* 309; war, wær LAȝ. 1486, 2967; war, wer, wear O. & N. 170, 1638; wer S. S. (Web.) 410; **ware** (*pl.*) O. & N. 860; **warre** (*compar.*) HOM. I. 253; warre P. S. 339; **wareste** (*ms.* warreste) (*superl.*) LAȝ. 2108; *comp.* ȝe-, un-**war** (-wær).

 war-līche, adv., *O.E.* wærlīce; *warily,* LAȝ. 12300; warliche A. R. 138; KATH. 82; wearliche HOM. I. 245; warli ANT. ARTH. xxxviii.; LUD. COV. 334.

 warnesse, sb., *O.E.* wærness; *wariness,* WICL. DEUT. xxxii. 28.

 war-scipe, sb., *O.E.* wærscipe; *caution,* LAȝ. 5603; warschipe A. R. 252; H. M. 41; warsipe REL. I. 218 (MISC. 14).

war [2], interj. & sb., *hunters' cry* ; with hai & war GAW. 1158.

war [3], sb., *cf. Swed.* var; *pus, humour* ; war & wirsen ORM. 4782; ware [wore] C. M. 11835.

[wăr, sb., *O.E.* wǣr, *O.N.* vār (*pl.* vāra), *O.H.G.* wāra ; *faith.*]

wăr-loghe, sb., *O.E.* wǣrloga,= *O.L.G.* wār-logo ; *liar, traitor, sorcerer,* D. TROY 4439 ; warlow ALEX. (Sk.) 1706 ; wărlowes (*gen.*) A. P. iii. 258 ; warlowe YORK xxx. 258 ; wăr-laȝes (*pl.*) ALEX. (Sk.) 3795, 4425 ; warlaus *devils* C. M. 23250.

wār, *see* hwār.

warant, sb., *O.Fr.* wărant, gvarant, garant ; *guarantee, safeguard,* MARH. 8 ; warant HAV. 2067 ; waraunt MISC. 94 ; warand BARB. 502 ; held his wai his warand (*place of safety*) till BARB. xix. 679 ; warrand x. 247 ; warrande xiii. 434 ; to warrand *in safety* xiii. 710.

warantie, v., *O.Fr.* warandir, gvarantir, garantir ; *defend, save,* MISC. 77 ; warante MAN. (F.) 5919 ; CH. C. T. *C* 338 ; warand BARB. ii. 504 ; warantie (*pres. subj.*) MISC. 89.

warantise, sb., *O.Fr.* warentise, garantise ; *guarantee ;* warantizez (*pl.*) D. ARTH. 1614.

warbrace, sb., *Fr.* gardebras ; *bracer,* PR. P. 516.

warch, warchen, *see* werk, werken.

ward, sb., *O.E.* weard,= *O.L.G.* ward, *Goth.* wards, *O.H.G.* wart, *O.N.* vörðr ; *warden, keeper, guard :* set ilk man sine till his ward BARB.xvii. 627 ; wardes(*gen.*) men (*for O.E.* weardmen) wardsmen LAȝ. 19304 ; warde (*dat.*) LANGL. *B* xviii. 320 ; wardes (*pl.*) ALIS. 1977 ; till thar wardis thai went BARB. xvii. 349 ; wearden (*dat. pl.*) MAT. xxvii. 66 ; *comp.* bére-, dúre-, erf-, ȝáte-, hei (hai-), mül-, stī-, wude-ward.

ward², adv., *O.E.* weard (TREAT. 15),= *O.L.G., O.Fris.*-ward, *Goth.*-wairþs, *O.N.*-verðr, *O.H.G.* -wert, -wart ; *towards :* a bac ward LAȝ. 20086 ; onȝein wærd 1673 ; a dune ward A. R. 130 ; a dune ward, dune ward [wardes] KATH. 1992, 2022 ; a wei ward O. & N. 376 ; til hevene ward GEN. & EX. 3025 ; to gode ward MISC. 28 ; to me ward HORN (L.) 1118 ; to hevene ward WILL. 102 ; a wei wardes 2188 ; *comp.* after-, bac-, ēast-, fore-, forð-, fram-, ȝeond-, hām-, heone-, hider-, hinde-, hwider-, in-, niðe-, niðer-, norð-, sūð-, tō-, þene-, þider-, up-, ūt-, uve-, west-, wiðer-ward.

warde, sb., *O.E.* weard,= *M.L.G.* warde, *O.H.G.* warta ; *guardianship ; district ;* A. R. 6 ; warde '*custodia*' PR. P. 516 ; LAȝ. 19402 ; MARG. 35 ; WILL. 376 ; MAN. (H.) 132 ; under godes warde H. M. 7 ; his air, but ward, releif or taill BARB. xii. 320 ; ward *guard* ALEX. (Sk.) 5614 ; (*troop*) 3040 ; warde *charge* ALEX. (Sk.) 77, 5172 ; wardes *watch towers* FER. 1344 ; PR. C. 9083 ; *comp.* bac-, efter-, for-warde.

ward-mōt, sb., *meeting of the ward,* LANGL. *B* PROL. 94.

wardecorce, sb., *Fr.* gardecorps ; *cloak,* PR. P. 516.

wardein, sb., *O.Fr.* gvardains ; *guardian,* A. R. 272 ; wardein P. L. S. xiii. 105 ; WILL. 1104 ; wardan ALEX. (Sk.) 75 ; wardane BARB. xiv. 512 ; wardanis (*pl.*) *regents* xiv. 33 ; wardeines *umpires* GAM. 279.

wardanri, sb., *wardenship,* BARB. viii. 362*.

wardien, v., *O.E.* weardian,= *O.Fris.* wardia, *O.L.G.* wardōn, *cf. O.H.G.* wartēn ; *guard ;* wardie P. L. S. xii. 24 ; warde LEG. 19 ; GREG. 980 ; ALEX. (Sk.) 5374 ; wardeð (*pres.*) A. R. 182 ; wardiþ ALIS. 906 ; wardede (*pret.*) ROB. 27 ; wardide WICL. 1 MACC. iv. 61 ; warded (*pple.*) WILL. 101.

wardone, sb., *sort of pear,* '*volemum*' PR. P. 516.

wardrere, sb., *club,* ALEX. (Sk.) 838 ; warder '*bacillus*' PR. P. 516.

wardrop, sb., *O.Fr.* garderobe ; *wardrobe,* D. ARTH. 4203 ; wardrope D. ARTH. 901, 2622 ; warderope '*vestiaria*' PR. P. 516 ; wardrobe *latrina* CH. C. T. *B* 1762.

warderopere, sb., '*vestiarius,*' PR. P. 516.

wăre, sb., *O.E.* waru,= *O.N.* vara, *M.Du.* ware ; *wares, goods, merchandise,* P. L. S. viii. 346 ; A. R. 66 ; GEN. & EX. 1990 ; ware LAȝ. 11356 ; ALIS. 7077 ; A. D. 279 ; LANGL. *A* ii. 189 ; CH. C. T. *D* 522 ; M. T. 204 ; IW. 2992.

wăre², sb., *O.E.* waru,= *O.H.G.* wara ; *caution ;* be on war(e) SPEC. 46.

war-lok, sb., *prison,* '*sera compeditalis,*' PR. P. 517 ; A. P. iii. 80.

wăre³, sb., pl., *O.E.* waru ; *the inhabitants of a place ; host, collection ;* waters ware, windes ware PS. xvii. 11, 16 ; al englene were HOM. I. 195 ; *comp.* burh-, Cant-, eorðe-, helle-, heoven-, Rōm-ware.

wăre, *see* wére.

Wárehām, pr. n., *Wareham,* ROB. 288.

wareine, sb., *O.Fr.* warene ; *warren,* PR. P. 516 ; LANGL. *B* PROL. 163.

wáren, v., *lay out, spend ;* ware H. S. 5798 ; wair '*commutare*' CATH. 408 ; *comp.* bi-wáren.

Warewik, pr. n., *Warwick,* PROCL. 10.

warf, *see* hwarf.　warh, *see* wari.

wari, sb., *O.E.* wearg, werg, werig,= *O.H.G.* warch, *O.N.* vargr, *Goth.* (launa-)wargs ; *villain, felon,* MARH. 4 ; weri [wari] A. R. 352 ; warien (*gen. pl.*) LAȝ. 28215.

warȝ-trēo, sb., *O.N.* vargtrē ; *gallows,* BEK. 2216 ; warhtreo HOM. I. 283 ; waritreo A. R. 122 ; MISC. 51 ; waritreo [weritreo] LAȝ. 5714.

wari², adj., *O.E.* warig (*dirty*),= *O.H.G.* warag ; *? Swed.* varig (*purulent*) ; *dirty ;* wori P. L. S. viii. 72 ; A. R. 386 ; wori BRD. 12 ; in worie watere HOM. I. 29.

wariangel, sb., *? O.E.* weargincel,= *O.H.G.*

warcheṅgil, *M.L.G.* wargingel; *dimin.· of* wari; *butcher-bird*; wairingle ALEX. (Sk.) 1706; **wariangles** (*pl.*) CH. C. T. *D* 1408.

warien, v., *O.E.* wergian, wyrgean,=*O.H.G.* (for-)wergen, *Goth.* (ga-)wargjan (κατακρίνειν); *curse, condemn,* A. R. 70; wariin PR. P. 516; warie H. S. 1288; HOCCL. i. 63; curse or warie WICL. ROM. xii. 14; weri PR. C. 7422; **warie** (*pres.*) CH. C. T. *B* 372; werieð HOM. I. 109; wergeð, weregeð MAT. xv. 4; MK. vii. 10; **warie** (*subj.*) HAV. 433; **weried** (*pret.*) PS. lxi. 5; PR. C. 4202; **waried** (*pple.*) KATH. 203; GEN. & EX. 544; waried Gow. II. 144; TOR. 1540; weregede REL. I. 131; þe weregede gastes HOM. I. 239; *comp.* a-, for-, 3e-warien (-war3en, -wer3en).

wariunge, sb., *O.E.* wyrigung; *malediction,* A. R. 200; warienge HOM. II. 179; wariinge PR. P. 516.

wárien, v., *O.E.* warian,=*O.Fris.* waria, *O.N.* vara, *O.L.G.* warōn, *O.H.G.* (be-)warōn; *beware,* A. R. 418; waren GEN. & EX. 2154; warie ALIS. 4083: ware ROB. 115; FLOR. 405; **wáre** (*imper.*) Gow. II. 388; ware þe for wanhope LANGL. *A* 225; war þe S. & C. I. xxxv.

warissen, warisōn, *see* garissen, garisōn.

warke, *see* werk.

wărla3e, -loghe, *see under* wǎr.

warm, adj., *O.E.* wearm,=*O.L.G.* warm, *O.N.* varmr, *O.H.G.* warm; *warm,* ORM. 10146; O. & N. 622; warm CH. C. T. *D* 1827; **warme** (*pl.*) A. R. 418; **warmer** (*compar.*) LANGL. *B* xviii. 410.

warmen, wermen, v., *O.E.* wearmian, wyrman, *cf. Goth.* warmjan, *O.H.G.* warmen; *warm*; warmen ORM. 2711; warmen [wormie] LA3. 16209; warme S. S. (Wr.) 2526; **warmen** (*pres.*) REL. I. 220; **wurme** (*imper.*) LEECHD. III. 114; **wermende** (*pple.*) MK. xiv. 67; **wermde** (*pret.*) JOHN xviii. 18; wermede MISC. 43.

warmþe, sb.,=*M.L.G.* wermede; *warmth*; wermþe HOM. I. 37.

warne, sb., *O.E.* wearn; *? denial*; wiþouten more warne C. M. 11133.

warne[2], conj., *for* wǽre ne, *cf. O.H.G.* neware, *O.L.G.* ni wári; *were it not that,* HAMP. PS. xli. 8.

warnere, sb., *from* wareine; *warrener,* PR. P. 517.

warnestūre, sb., *O.Fr.* garnesture, warnesture; *garniture, garrison,* ROB. 94.

warnien, v., *O.E.* wearnian, warnian,=*O.N.* varna, *O.L.G.* warnōn; *warn, admonish*: ic wulle ... warnie ow wið herme P. L. S. viii. 115; Vor to warnie wummen of hore fol eien A. R. 54; to warnie þe beforn FER. 1808; wearnen H. M. 37; warnen GEN. & EX. 1581; warne ROB. 45; **warni** (*pres.*) O. & N. 330; þat warneþ men men from wo SPEC. 35; heo hi wernað wið drunkenesse HOM. I. 111;

warnie (*subj.*) A. R. 270; **warne** (*imper.*) MK. i. 44; **warnede** (*pret.*) LANGL. *A* iv. 55; he warnede alle his cnihtes LA3. 7984; he warnede hine seolven 28499; tvei serjantz... warnede him þat me wolde to stronge deþe him bringe BEK. 1076; **warned** (*pple.*) C. L. 390; CH. C. T. *A* 3535; *comp.* 3e-warnen.

warninge, sb., *O. E.* wearnung,=*O.H.G.* warnunge; *warning, admonition,* A. R. 256*; warninge LEG. 95; MIRC 1247.

warnischen, *see* garnischen.

warp, sb., *O.E.* wearp,=*M.L.G.* warp, *M.Du.* werp, *O.H.G.* warf; *from* weorpen; *warp,* 'stamen,' VOC. 218; PR. P. 517; **werpe** (*dat.*) WICL. LEV. xiii. 48.

warpin, v., *O.N.* varpa; *throw, bend,* 'curvo,' PR. P. 517; warpe ['ordiri'] ... a webbe TREV. V. 365; **warpes** (*pres.*) JOS. 257; all þe wordis at (=þat) he þaim **werpid** (*pret.*) uttered ALEX.) 202; unethis werped [warpid] he þat worde 709; wid open werped [warpid] he þe 3atis 1526; thou art **warpid** (*pple.*) alle in wo TOWNL. 226; warpid in sonder ALEX. (Sk.) 798*; a ... worde has **werpid** (*pple.*) & spoken 243; *cf.* **weorpen.**

warre, sb., *O.E.* wearr (*wart*), *cf. O.H.G.* werra (*varix*); warre, knotte of a tre PR. P. 516.

warrok, sb.,*girth,* 'sirentorium,' VOC. (W.W.) 612.

warroke, v., *girth, fasten with a girth,* LANGL. *C* v. 21, *A* iv. 19; warrok *B* iv. 20; ye **warrok** (*imper.*) YORK xxx. 525.

warte, sb., *O.E.* wearte,=*O.N.* varta, *M.Du.* warte, wrate, *O.H.G.* warza; *wart,* 'verruca,' VOC. 207; werte 179, 267; CH. C. T. *A* 555; werte, wrete PR. P. 533.

warþ, sb., *O. E.* wearoð, wearð,=*O. H. G.* warid; *shore,* A. P. iii. 339; **waruðe** (*dat.*) MAT. xiii. 2; warþe GAW. 715.

was, *see* wesen.

wascheles, sb., *? washing vessel,* JOS. 288.

waschen, weaschen, weschen, v., *O.E.* wascan,=*O.L.G., O.H.G.* wascan, wescan (*pret.* wōsc, wasc); *wash,* HOM. I. 159, 189, 202; washen (*ms.* wasshen) HOM. II. 57; (*ms.* wasshenn) ORM. 2711; washen HAV. 1233; waschen Gow. II. 137; wascen [wassen] LA3. 10182; wassen GEN. & EX. 2291; wasche BRD. 30; wasche, wasse FL. & BL. 304, 564; weschen SHOR. 4; wesche CATH. 415; wesse AYENB. 371; woshen A. D. 262; **wascheð** (*pres.*) A. R. 300; wosheþ REL. I. 264; **wōsch** (*pret.*) P. L. S. xxiii. 125; ARCH. LVII. 241; MAND. 91; wosch, wesch HOM. I. 79, 157; weosch A. R. 300; wesh (*ms.* wessh) ORM. 1103; wesch SHOR. 86; woschen LANGL. *A* ii. 196; wosche BRD. 12; CH. C. T. *A* 2283; weshen LANGL. *B* ii. 220; wessen HOM. II. 65; wesche GREG. 887; wesse MISC. 29; **waschen** (*pple.*) WILL. 5070; GREG. 1004; MAND. 92; waschen, iweschen HOM. I.

157, 159; iwaschen A. R. 288; iwasche BEK.
2223; *comp.* a-weschen.

waschestre, sb., *female washer*; washes-
tren (*dat. pl.*) HOM. II. 57.

waschunge, sb., *washing*, A. R. 332.

wáse, sb., *cf.* M.L.G., M.Du. wase (*bundle,
torch*), Dan., Swed. vase (*fascis*) m.; *torch*;
wase '*stupa*' VOC. 180 (W. W. 627); þe lie of
þe fur stod on heȝ as hit a was(e) were BRD.
23; a brenning wase BER. 2351.

wāse, waise, sb., *? O.E.* wās, = *O.N.* veisa
(*pool, cesspool*), O.Fris., M.L.G. wāse (*slime*);
slime; wase, waise CATH. 409; waiȝe MAN.
(H.) 70; wose PR. P. 532; AYENB. 87; BER.
1742; as weodes wexen in wose and in donge
LANGL. C xiii. 229.

waspe, sb., *O.E.* wæps, = *O.H.G.* wafsa, wefsa,
Lat. vespa; *wasp,* VOC. 190, 222; PR. P. 517;
PL. CR. 648; ALEX. (Sk.) 738; waspis (*pl.*)
3011.

wassail, *see under* wesen.

wást, sb., = *O.H.G.* wahst, *Goth.* wahstus;
from waxen; *stature; waist*; PR. P.. 517;
GAW. 144; ANT. ARTH. xlv; waast CH. C. T.
B 1890; wáste (*dat.*) GOW. II. 373; wacste
HOM. I. 77.

wást[2], adj., *O.Fr.* wast, gast, *Lat.* vastus;
waste, void, solitary, ROB. 372; WILL. 690;
he haþ maad mi covenaunt wast ['*pactum
meum irritum fecit*'] (*later ver.* voide) WICL.
GEN. xvii. 14; in ore waste [vaste] þicke hegge
O. & N. 17; waste (*wasteful expenditure*) CH.
C. T. *C* 593; waste. wahes H. M. 31; waist
BARB. vii. 151*.

wáste, sb., = M.H.G. waste; *waste, desert*,
HOM. II. 163; waste E. T. 451; GAW. 2098.

wastel, sb., *O.Fr.* wastel, gastel (*gâteau*);
cake made of finest flour, '*libellus,*' PR. P. 517;
HAV. 878; LANGL. *B* v. 293.

wastel-breed, sb., *bread*, CH. C. T. *A* 147.

wásten, v., *O.Fr.* waster, gvaster, gaster, *Lat.*
vastāre; *lay waste, waste away*, LAȝ. 22575;
waste MAND. 239; gaste SPEC. 90; wásteð
(*pres.*) A. R. 138; H. M. 29; wasteþ AYENB.
19; wástede (*pret.*) ROB. 136; wásted
(*pple.*) WILL. 2620.

wastine, sb., *O.Fr.* wastine, gvastine; *waste,
desert*; in þe wastine HOM. I. 141; in the
wastin(e) ALIS. 7121.

wastme, *see* wastum.

wástour, sb., *waster*, LANGL. *B* vi. 176;
wastoure ALEX. (Sk.) 5310.

wastum, sb., *O.E.* wæstum, westem, wæstm, =
O.L.G. wastom; *from* waxen; *fruit; growth;
stature*; MARH. 2; wæstme ['*fructus*'] MAT.
xxi. 19; wastme ORM. 1939; westum KATH.
(E.) 63, 310; westme (*ms.* vestme) (*dat.*)
LAȝ. 15698; wastme HOM. II. 127; wæstmes
(*pl.*) MAT. vii. 17; westmes HOM. I. 13;
wastmes LAȝ. 32108; wæstme (*gen. pl.*)

MAT. xxi. 34; westman (*dat. pl.*) MAT. vii.
16; westme SAX. CHR. 252.

wastme-lǣs, adj., *O.E.* wastemlēas; *without fruit*, ORM. 13858.

wat, *see* hwat. wāt, *see* wǣt.

wáte, *see* hwáte.

watel, sb., *O.E.* watel; *hurdle woven with
twigs; basket*: a watel (*? v. r.* walet) ful of
nobles LANGL. *C* xi. 269.

watel-ful, sb., *basketful*, LANGL. *C* xi. 269.

watelen, *see* watlen.

water, sb., *O.E.* wæter, = *O.Fris.* weter, *O.L.
G.* water, *O.H.G.* wazar, *Gr.* ὕδωρ; *water,
river, sea*, A. R. 314; ORM. 3212; GEN. & EX.
749; þat water AL. (T.) 190; water, wæter
LAȝ. 113, 5263; weter HOM. I. 129; ANGL.
I. 15; MARH. 19; þet weter AYENB. 106;
weater JUL. 31; wattir BARB. ix. 683; watres
(*gen.*) GEN. & EX. 1246; watre (*dat.*) LANGL.
B xvii. 228; watere HOM. I. 225; wætere
(*pl.*) JOHN iii. 23; watere MAT. xiv. 28;
watres SAX. CHR. 122; wateres LAȝ. 2007;
wateres A. R. 402; wateren [watere] (*gen.
pl.*) LAȝ. 24262; *comp.* rein-, see-, welle-
water.

water-brēþ, sb., *vapour*, TREAT. 137 (SAINTS
(Ld.) xlvi. 629).

water-bulge, sb., *water-bag*, H. M. 35.

water-cresse, sb., *water-cress*; water kresse
ALIS. 5767; waterkirs(e) VOC. 190; water-
crasses (*pl.*) LANGL. *C* vii. 292.

water-drinch, sb., *water-drink*, ORM. 14482

water-fet, sb., *O.E.* wæterfæt; *water-pot,*
['*hydriam*'] JOHN iv. 28; waterfate (*pl.*)
ii. 6.

water-fetles, sb., *water-pot*; (*ms.*) waterr
fetless (*pl.*) ORM. 14411.

water-flōd, sb., *O.E.* wæterflōd; *water-
flood*, ORM. 18088.

water-foul, sb., = M. H. G. wazzervogel;
water-fowl, CH. P. F. 504.

water-gate, sb., *water-gate*, FER. 4651.

watir-lili, sb., *water-lily*, PR. P. 518.

watir-nedire, sb., *water-snake,* '(h)ydrus,'
VOC. 223.

water-pípe, sb., *water-pipe*, VOC. 202.

water-pōl, sb., *water-pool*, ROB. 131.

water-pot, sb., *water-pot*, CH. C. T. *E* 290;
watirpot WICL. JOHN iv. 28.

water-püt, sb., *O.E.* wæterpytt; *water-pit,*
TREV. III. 401.

water-scenc, sb., *draught of water*, LAȝ.
19695.

water-strǣm, sb., *water-stream*, ORM.
18092.

weter-þeote, sb., *O.E.* wæterþeote; *torrent*;
weterþeotan (*pl.*) of þer micele niwelnisse
HOM. I. 225.

wateren, v., *O.E.* wæterian, *cf.* M.L.G. wateren,

weteren, *M.H.G.* wezzern; *water*; (*ms.* wattrenn) ORM. 13848; for to wattren here sep GEN. & EX. 2745; watere TREV. III. 415; wettrien (*printed* wectrien) HOM. I. 9; weteri AYENB. 98; **watereden** (*pret.*) LEB. JES. 305; his eȝen watreden LANGL. *A* vii. 162 [wattered *B* vi. 177].

    watringe, sb.,=*M.L.G.* wateringe, weteringe, *M.H.G.* wezzerunge; *watering*, PR. P. 518.

wateri,adj., *O.E.* wæterig,=*M.L.G.* waterich; *watery*, *like water* : þe heorte þet was wateri A. R. 376; wateri [watri] cloudes LANGL. *B* xviii. 410.

watlen, v., *cover with hurdles*; **watlede** (*pret.*) LANGL. *B* xix. 323; watelide *C* xxii. 328.

Watlinge-strēte, pr. n.,˙ *O.E.* Wætlinga strǣte; *Watling-street*, ROB. 8.

wattri, adj., *venomous*, M. H. 138; *see* attrie.

wāþe, sb., *O.E.* wāð, wǣð, *cf. O.H.G.* weida, *O.N.* veiðr; *hunting*; waithe D. TROY 2350; waith GAW.1381; ANT. ARTH. xxxiv; TOWNL. 33.

wāþe², sb., *O.N.* vāði; *peril, hurt*, MAN. (F.) 10352; wathe PR. C. 4558; ALEX. (Sk.) 119, 1411, 3523, 5586; waþe, woþe GAW. 222, 2355; ALEX. (Sk.) 1103; waith BARB. vii. 305; **wāþes** (*pl.*) AV. ARTH. xiv; wathes [woþes] ['*pericula*'] PS. cxiv. 3.

    wōthe-lī, adv., *O.N.* vāðaliga; *perilously*, D. TROY 8827.

wáven, v., *O.E.* wafian,=*O.N.* vafa, *M.H.G.* waben; *wave, fluctuate, vacillate*; wave HOCCL. i. 399; waff YORK xii. 95; **wáfe** (*pres. subj.*) TOWNL. 312; **wáwand** (*pple.*) BARB. ix. 245*; **wávid** (*pret.*) wiþ eche wind LIDG. M. P. 256; wavid YORK xxxii. 317.

wāvernesse, wēvernesse, sb., *O.E.* wæferness; *?splendour*: vorsien þisne midelard mid his wovernesse ANGL. I. 31.

waveron, v., *O.N.* vafra,=*M.H.G.* waberēn; *waver*, '*vacillo*,' PR. P. 518; **weverinde** (*pple.*) SHOR. 16; wawerand BARB. xii. 185; wawering vi. 584; **wawerit** (*pret.*) vii. 41.

waveschen, v.,=**weiven**; *put away*; **waveschid** (*pple.*) ALEX. (Sk.) 822.

wawe, *see* waȝe.

wāwe,sb.,*O.E.* wāwa, ? wāwe,*cf.O.H.G.* wēwa *f.*, wēwo *m.; woe, misery*, HOM. I. 73; wowe REL. I. 130; O. & N. 414; wowe LAȝ. 6268*; SHOR. 166; þis worldis wel(e) nis bot wowe P. L. S. v. (a) 3; þeo weowe FRAG. 5; of þan wowe HOM. II. 197; **wāwen** (*pl.*) ORM. 13349; wowe MISC. 139; wowes A. R. 198; **wāwen** (*dat. pl.*) HOM. I. 87.

[**wāwen**, v., *O.E.* wāwan, = *O.H.G.* wāhan, wājan, *Goth.* waian; *blow; comp.*to-wōwen.]

wawien, *see* waȝien.

wax, sb., *O.E.* weax,=*O.N.* vax, *O.Fris.* wax,

O.H.G. wahs; *wax*, '*cera*,' PR. P. 518; RICH. 782; M. H. 153; LUD. COV. 341; wax [wex] CH. C. T. *A* 675; wex HOM. II. 47; wex LAȝ. 2370; P. L. S. xiii. 121; LANGL. *B* xvii. 204.

    wax-bred, sb., *O.E.* weaxbred; *wax tablet*: (god) wrate his himself in stenene **waxbredene** (*dat. pl.*) HOM. I. 235.

wax², sb.,=*M.H.G.* wahs; *increase*, C. M. 1430.
    wax-dōm, sb.,=*M.Du.* wasdōm; *increase*; waxdam MAN. (F.) 6546*.

[**waxe?** *comp.* wode-wexe.]

waxen,adj., *cf. M.H.G.* wehsin; *made of wax*, GAW. 1650.

waxen²,v., *O.E.* weaxan (*pret.* wēox),=*O.Fris.* waxa (*pret.* wōx), *O.N.* vaxa (*pret.* vōx, ōx), *O.L.G.* wahsan (*pret.* wōhs), *O.H.G.* wahsan (*pret.* wuohs); *wax, grow, become*, FRAG. 6; ORM. 3935; GEN. & EX. 1128; hu muche god mihte of inker streon waxen H. M. 3; heore volc gon waxen LAȝ. 26990; waxin PR. P. 518; wexen HOM. II. 69; wexe HORN (L.) 441; AYENB. 95; WILL. 124; GOW. II. 109; MAND. 44; HOCCL. ii. 5; **waxe** (*pres.*) SPEC. 54; þou ... wext REL. I. 265; waxeð MARH. 11; waxeð, wacseð A. R. 54, 74; wext AYENB. 119; west O. & N. 689; (heo) waxað HOM. I. 35; wexeð LK. xii. 27; **wexinde** (*pple.*) REL. I. 184 (MISC. 128); **wēox** (*pret.*) A. R. 258; weox, wex KATH. 12, 19; weox, wex LAȝ. 1995, 25516; Lucifer ... wox so proud C. L. 97; wex ORM. 7694; GEN. & EX. 584; wex HAV. 281; GREG. 159; DEGR. 619; ANT. ARTH. xxvi; wex [weex] LANGL. *B* iii. 328; wex, wax ROB. 11, 322; wex [weex], wax CH. C. T. *B* 3868, 3936; weax, wæx SAX. CHR. 251, 264; wax WILL. 630; MAND. 303; TOR. 73; (þu) weoxe JUL. 63; (heo) weoxen LAȝ. 7165; woxen JOS. 433; GOW. II. 152; woxen [wexen] LANGL. *B* ix. 32; **waxen** (*pple.*) ORM. 3190; waxen HAV. 791; IW. 1738; *comp.* ȝe-, þurh-**waxen**.

waxen,v.,= *O.H.G.* wahsan; *fasten with wax*; waxin '*cero*,' PR. P. 518; **wexede** (*pret.*) CH. ASTR. ii. 40 (49).

waxt, *see* wast.

we (?**wē**), pron.; *O.E.* we,=*O.L.G.* wi, *O.N.* ver, *O.H.G.* wir, *Goth.* weis; *we*, MARH. 2; we LAȝ. 364; we (*printed* þe) 4128; we, we O. & N. 1690; we, we HAV. 621, 1058; weo HOM. I. 55.

wē, wēa, *see* **wā**.   **wealden**, *see* **wálden**.

wealle, weallen, *see* **walle, wallen**.

weane, *see* **wǣne**.   **wearnien**, *see* **warnien**.

weaschen, *see* **waschen**.

web, sb., *O.E.* webb, *cf. O.L.G.* (godu-)web, *M.L.G.* webbe (HOLTZM. I. 170), *O.H.G.* weppi *n.*, *O.N.* (guð-)vefr *m.; from* **wéven**; *web*, REL. I. 115; TREAT. 139; AYENB. 164; WICL. JOB vii. 6; weob A. R. 322; web LANGL. *A* v. 92 [webbe *B* v. 111]; webbe '*tela*' PR. P. 519; **webbe** (*dat.*) LAȝ. 19947;

webbis (*pl.*) *fabrics, woven cloths* ALEX. (Sk.) 4911, 5295 ; webis [webbes] 1577 ; *comp.* **cop-, gold-webbe.**

web-bēm, sb., *O.E.* webbēam, = *O.H.G.* weppeboum ; *weaver's beam, 'liciatorium'* VOC. 218.

webbe, sb., *O.E.* webba ; *weaver,* CH. C. T. *A* 362 ; **webbes** (*pl.*) P. S. 188.

webbe [2], sb., *O.E.* webbe ; *female weaver,* LANGL. *B* v. 215.

webbon, v., *O.E.* webban ; *weave,* PR. P. 519.

webbare, sb., *weaver,* PR. P. 519.

webstere, sb., *O.E.* webbestre ; (*female*) *weaver,* WICL. JOB vii. 6 ; webstar PR. P. 519 ; **websters** (*pl.*) D. TROY 1587 ; *comp.* **wolle-webstre.**

wecche, *see* **wacche.**

wecchen, v., *O.E.* weccan, = *O.H.G.* wecchan, *O.L.G.* wekkian, *O.N.* vekja, *Goth.* (us-)wakjan ; *wake up, rouse, excite,* HOM. II. 137 ; wechche S. S. (Web.) 1628 ; **weccheð** (*pres.*) a mong hem flite REL. I. 128 ; **weccheð** (*imper.*) LAȝ. 798 ; **wæht(e)** [wehte] (*pret.*) LAȝ. 16216 ; weiȝte BRD. 21 ; *comp.* **a-wecchen.**

wed, sb., *O.E.* wedd, = *O.Fris.* wed, *O.N.* veð, *Goth.* wadi (ἀρραβών), *O.H.G.* wetti ; *pledge, compact,* A. R. 394 ; wed SPEC. 110 ; ALIS. 884 ; MIRC 1290 ; GOW. I. 94 ; E. G. 8 ; þet beste wed AYENB. 113 ; wed [wedde] LANGL. *B* v. 244 ; wedde PR. P. 519 ; TRIST. 320 ; **weddes** (*gen.*) HOM. I. 225 ; **wedde** (*dat.*) GEN. & EX. 2198 ; wedde MAND. 13 ; AMAD. (R.) xxxiii ; mi lond ich wulle sette to wedde LAȝ. 25172 ; his nekke liþ to wedde CH. C. T. *A* 1218.

wed-brek(e), sb., *adulterer,* PS. xlix. 18.

wed-brōðer, sb., *O.N.* veðbrōðir ; *pledged brother,* SAX. CHR. 30 ; wedbrōðer LAȝ. 14469 ; wedbroþer HORN (H.) 295.

wed-lāc, sb., *O.E.* wedlāc ; *wedlock, marriage ;* LAȝ. 395 ; MISC. 150 ; wedlac ORM. 2499 ; wedlok '*matrimonium*' PR. P. 520 ; wedlaik MAN. (H.) 254 ; PR. C. 8261 ; **wedlāke** (*dat.*) A. R. 206 ; **wedlōkes** (*pl.*) LANGL. *B* ix. 152.

wed-lowe, sb., *O.E.* wedloga ; *pledge breaker,* FRAG. 7.

wed-setten, v., *mortgage, pledge ;* wedsette MAN. (F.) 11796 ; YORK xxxii. 346.

wēd, *see* **wēod.**

wedden, v., *O.E.* weddian, = *M.Du.* wedden, *O.N.* veðja, *Goth.* (ga-)wadjōn, *M.H.G.* wetten ; *wed, pledge,* ORM. 10407 ; GEN. & EX. 1090 ; wedden CH. C. T. *B* 223 ; weddi SHOR. 67 ; weddi wif ROB. 331 ; wedde REL. I. 196 ; ich **wedde** (*pres.*) boþe mine eres LANGL. *A* iv. 129 ; **weddede** (*pret.*) LAȝ. 4432 ; LEG. 81 ; (hio) weddeden LK. xxii. 5 ; **wedded** (*pple.*) ORM. 1942 ; wedded HAV. 2770 ; weddid breþrin M. T. 2 ; iwedded A. R. 394 ; LAȝ. 9568 ; ALIS. 4400 ; *comp.* **bi-wedden.**

wedding, sb., *O.E.* weddung, = *M.Du.* weddinghe ; *wedding,* GEN. & EX. 1428 ; wedding P. L. S. xvii. 92.

wēde, *see* **wǣde.**

wēden, v., *O.E.* wēden, = *O.H.G.* wuoten, wuaten ; *from* **wōd** ; *be mad, rage,* A. R. 264 ; KATH. 1263 ; ORM. 14140 ; wede ROB. 53 ; TRIST. 1049 ; IW. 2632 ; OCTAV. (H.) 339 ; M. ARTH. 787 ; **wēdined** (*pple.*) KATH. 379 ; *comp.* **a-wēden.**

weder, sb., *O.E.* weder, = *O.L.G.* wedar, *O.N.* veðr, *O.H.G.* wetar ; *weather ; bad weather, storm ;* GEN. & EX. 3055 ; FL. & BL. 70 ; TREAT. 136 ; AYENB. 129 ; LANGL. *B* xviii. 410 ; A. P. ii. 444 ; M. H. 135 ; þet weder wes swa wilde LAȝ. 4679 ; a derk weder þer aros ROB. 560 ; **wederes** (*gen.*) LAȝ. 9734 ; **wedere** (*dat.*) LAȝ. 4603 ; of þe hote weder WILL. 2440 ; **wedirs** (*pl.*) PR. C. 1424 ; **wedere** (*gen. pl.*) LAȝ. 25638 ; *comp.* **un-weder.**

weder-coc, sb., *weather-cock,* AYENB. 180.

weder-wīs, adj., *weather-wise,* LANGL. *B* xv. 350.

wederin, v., *O.E.* wedrian, = *O.N.* viðra, *M.H.G.* witeren ; *expose to weather ; wither ; 'auro,'* PR. P. 519 ; i widder (*pres.*) a wai TOWNL. 21.

wederinge, sb., *O.E.* wederung, *cf. Ger.* witterung ; *weathering, seasoning, 'temperies,'* PR. P. 519 ; p. 76 ; wedering E. G. 23 ; widerunge HOM. I. 13.

Wĕdnes dai, *see* **Wōden.**

wedir, *see* **weþer.** **wee,** *see* **wīȝe.**

weed, *see* **wēod.**

weet, weeten, *see* **wǣt, wǣten.**

wēf, sb., *? from* **wǣven** ; *mod.Eng.* whiff ; *breath, gust,* (ms. weffe) '*vapor*' PR. P. 520 ; fro þe comeþ awikke wef MAP 335 ; wef (? ms. weffe) RICH. 5291.

weft, sb., *O.E.* weft, *O.N.* veftr ; *from* **wéven** ; *weft,* '*trama*,' VOC. 218 ; WICL. EXOD. xxxix. 3 ; TOWNL. 18.

Wēȝe, pr. n., *Wye,* LAȝ. 29943 ; Weie (ms. Weye) ROB. 275 ; SPEC. 26 ; MAN. (F.) 8304.

wēȝe, *see* **wǣȝe.**

weȝen, v., *O.E.* wegan, = *O.H.G.* wegan, *O. Fris., O.N.* vega, *Goth.* (ga-)wigan, *Lat.* vehere ; *weigh, bear ; set free ;* P. L. S. viii. 32 ; weiȝen HOM. II. 222 ; weien HOM. II. 213 ; A. R. 336 ; weiin PR. P. 520 ; to weȝe boþe scheld and spere O. & N. 1022 ; weȝe AYENB. 44 ; weie TREV. IV. 207 ; (ms. weye) TREAT. 132 ; GOW. II. 275 ; **weie** (*pres.*) HOCCL. vi. 49 ; weeihð, weieð A. R. 232, 332 ; þai weȝen her ankres A. P. iii. 103 ; **waiȝ** [weiȝ, wei] (*pret.*) TREV. IV. 7 ; wai (ms. way) REL. II. 277 ; weȝ BEV. 1424 ; heo weȝe on heore honde feouwer sweord LAȝ. 24471 ; weie TREV. VI. 267 ; þe child swa hevi wogh [= *M.H.G.* wuoc *weighed*] HALLIW. 937 ; weiede, weide LAȝ. 24478, 26279 ; **weien**

(*pple.*) LANGL. *A* i. 152 ; iweȝe AYENB. 152 :
iweie TREV. III. 129 ; E. G. 356 ; wowin
'*libratus,*' PR. P. 533 ; *comp.* a-, over-weȝen ;
deriv. **wei, waȝe, wǣȝe, waȝn, weggen,
wiht.**

weiere, sb., = *M.L.G., M.H.G.* weger ;
weigher : this same cercle is cleped also the
weiere, *eqvator,* of the dai CH. ASTR. i. 17 (9.)

wegge, sb., *O.E.* wecg, = *O.N.* veggr, *M.L.G.*
wegge, wigge, *M.Du.* wegghe, wigghe, *O.H.
G.* weggi, wekki ; *wedge,* '*cuneus,*' VOC. 203 ;
CH. ASTR. i. 14 (8) ; PALL. ii. 246 ; wegge
[wedge], wigge PR. P. 520, 526.

weggen, v., *O.E.* wecgan, = *M.L.G.* weggen,
M.Du. wegghen, *O.H.G.* wegan ; *move, agitate* ; weieð (*ms.* weieð) (*pres.*) LAȝ. 20137 ;
iweid (*pple.*) SHOR. 14.

weht, *see* wiht[2].

wei, sb., *O.E.* weg, = *O.L.G., O.H.G.* weg,
O.Fris. wei, *O.N.* vegr, *Goth.* wigs, *related
to Lat.* via, (*for* *veha) ; *from* weȝen ;
way, JOHN xiv. 5 ; weig MAT. iii. 3 ; wei
FRAG. 3 ; HOM. I. 119 ; GEN. & EX. 1429 ;
wuderward hie sullen wei holden REL. I. 128 ;
wei HAV. 772 ; JOS. 32 ; (*ms.* wey) WILL.
205 ; LANGL. *B* v. 540 ; EGL. 274 ; and to
him þane wei nom BEK. 707 ; wei, wæi, wai
LAȝ. 1348, 15548, 28430 ; wei, wai O. & N.
249, 956 ; half wai midmorewe S. S. (Web.)
1626 ; **a weȝ** [*O.E.* a, on weg] *away,* ORM.
3196 ; a wei A. R. 238 ; a wei GOW. III. 226 ;
wende a wei WILL. 2207 ; faren a wæi [wei]
LAȝ. 10050 ; **a wei ward** he halde 8878 ;
he gengþ . . . a wei [wai] ward O. & N. 376 ;
a wif þat is a wai ward (*wayward*) WILL.
3985 ; o wai TRIST. 316 ; **alne wai** *always*
MISC. 148 ; AYENB. 6 ; al wei CH. C. T. *A*
275 ; **weiȝes,** weies (*gen.*) LEG. 156, 181 ;
þe oðer twa turnden anes weis KATH. 1986 ;
nānes weies, *no ways* LAȝ. 11216 ; nones
weis A. R. 86 ; hwuches weis JUL. 42 ; summes weis H. M. 9 ; oþer weies (*ms.* weyes)
MIRC 602 ; oðer weies, weis LAȝ. 10199, 16329 ;
oðer weis HOM. I. 31 ; oþer weis C. L. 626 ;
alles weis *always* A. R. 4 ; H. M. 27 ; alle
weies MAND. 21 ; **weȝe** (*dat.*) ORM. 6569 ; weie
O. & N. 820 ; weie LAȝ. 524 ; HORN (L.) 759 ;
weiȝes [weies] (*pl.*) LAȝ. 26915 ; weiges GEN.
& EX. 3244 ; weiȝes, weies WILL. 1224, 2131,
2207 ; weoȝes, weies HOM. I. 5, 7 ; weies
A. R. 78 ; ALIS. 2903 ; *comp.* **bi-wei.**

wei-brēde, sb., *O.E.* wegbrǣde, = *M.L.G.*
wegebrēde, *O.H.G.* wegebreita ; *waybread*
(*herb*), '*plantago,*' VOC. 265 ; PR. P. 520.

wai-fare, sb., *course,* D. ARTH. 1797.

wei-farende, pple., *wayfaring,* MAN. (F.)
3659.

wei-fērende, sb., *O.E.* wegfērende ; *way-
farer,* MAT. xxvii. 39 ; weiverinde A. R. 350 ;
waiverinde AYENB. 39.

wei-lokker, sb., *? from M.Du.** weghlokker ;
enticer away ; weilokere B. B. 19.

wæi-witer, sb., *guide* ; wæiwitere (*pl.*)
LAȝ. 12860.

wei-witti, sb., *guide* ; weiwittie (*pl.*) LAȝ.
12860*.

wei, *see* **wā.**

[**wēi,** sb., *O.E.* wǣg, = *Goth.* wēgs, *O.Fris.* wēi,
O.L.G. wēg, wāg, *O.H.G.* wāg, *O.N.* vāgr ;
water, comp. **hāle-wēi.**]

wēie, *see* **wǣȝe.** **weien,** *see* **weȝen.**

weiht, *see* **wiht**[2]. **weik,** *see* **wāc.**

weilen, *see* **wailen.** **wein,** *see* **wain.**

weiven, *see* **waiven.**

weivērinde, *see* **weifērinde** *under* **weī.**

wēke, sb., *O.E.* weoca, = *M.L.G.* weke, *M.Du.*
wieke, *O.H.G.* wieche ; *wick,* . VOC. 231 ;
LANGL. *C* xx. 164, 171 ; ED. 1277 ; weke (*ms.*
wueke) HOM. II. 47 ; weike 206 ; wicke LANGL.
C xx. PR. P. 520 ; LANGL. *B* xvii. 205.

wēke, *see* **wike.** **wēken,** *see* **wǣken.**

wel, adv., *O.E.* wel, = *O.L.G.* wel, *O.Fris.*
wel, wal, wol, *O.N.* vel, val, *O.H.G.* wela,
wola, *Goth.* waila ; *well, very,* MARH. 3 ;
ORM. 1033 ; þu ert wel don man HOM. II. 29 ;
þe mon þe wel deð he wel ifehð HOM. I. 131 ;
wel is us A. R. 190 ; wel oft O. & N. 36 ; wel
[vel] 95 ; wel PROCL. 2 ; BEK. 118 ; AYENB.
22 ; MAND. 41 ; hit wes wel neih middai
MISC. 50 ; wel sone HORN (L.) 42 ; a wel old
cherl WILL. 4 ; wel mai him be SPEC. 59 ;
þou seidist wel WICL. JOHN iv. 17 ; wel
wurðe þe LAȝ. 13079 ; and hine wæl lere
23121 ; wæl la *well done !* 12805 ; welle [wele]
29622 ; wel [weel] CH. C. T. *A* 1826 ; wel,
wol GEN. & EX. 229, 1995 ; he shulde yéme
hire wel HAV. 209 ; wel sixti 1747 ; wele
ALEX. (Sk.) 30, 44 ; wele [wella] 1970 ;
wol fair 185 ; wol RICH. 1280 ; S. & C. I. liii.
wil S. S. (Wr.) 332.

wel-dǣde, -dēd, sb., *O.E.* weldǣd, = *Goth.*
wailadēds, *O.H.G.* wolatāt ; *benefit,* LAȝ.
3306, 8052 ; weldede HOM. I. 133 ; **weldēdes**
(*pl.*) LANGL. *A* iii. 62.

wel-fare, sb., = *M.L.G.* wolvare ; *welfare,*
C. L. 189 ; WILL. 2076 ; GOW. II. 116 ; CH.
C. T. *B* 1529 ; weillfair BARB. xii. 156 ; weilfar
viii. 377*.

wel-wil(l)inde, adj., *O.E.* wel willende ;
benevolent, AYENB. 112 ; welwillinge PR. P.
521.

wel[2], sb., = *? O.H.G.* giwel (*globe*) ; *ball* ;
trendled als a wel REL. I. 225.

wel[3], sb., *O.E.* well, wyll ; *well, fountain,* A. R.
72 ; O. & N. 917 ; æt þam welle JOHN iv.
6 ; þane wel LAȝ. 19812* ; welles (*gen.*)
LAȝ. 19756 ; wille (*dat.*) LAȝ. 19810* ; welles
(*pl.*) TREV. I. 399 ; *see* **welle.**

wel-cresse, sb., *watercress,* VOC. 226 ;
'*nasturcium*' VOC. (W.W.) 712 ; *cf.* **welle-
carse.**

wil-spring, sb., *O.E.* well-, wyllspring ; *well-*

spring, P. L. S. xiii. 293 ; welsprung HOM. I. 195 ; **wellspringes** (*pl.*) HOM. I. 225 ; *cf.* **welle-spring.**

wel⁴, *see* **wæl, wil.**

wēl, sb., *O.E.* wēl (GREIN), *? ornament* : al þat wæl & al þat gold LAȝ. 8111.

wēl, *see* **wāl.**

welch(e), adj. & sb. ? = **walisc** ; *? Welsh cloth, ? flannel* ; she sholde nouȝte have walked on þat welche [welsche] so was it thredbare LANGL. *B* v. 199 (*C* v. 205).

welcome, *see under* **wil.**

wélde, adj., *O.E.* wylde ; *from* **wálden** ; *dominant, powerful* ; freo of heorte of wisdom wilde MISC. 96 ; **weldre** (*compar.*) HOM. I. 105 ; *comp.* **ȝe-wélde.**

wélde², sb., *O.E.* (ge-)wylde, = *M.L.G.* welde, *O.N.* veldi ; *dominion, power*, P. L. S. xiii. 52 ; C. M. 462 ; R. R. 395 ; M. H. 109 ; þat i mote þe seo on cristen mannes welde FER. 3716.

wélde, *see* **wálde.**

wélden, v., *O.E.* (ge-)weldan, wyldan, = *O.N.* valda (*pret.* voldi, olli, *pple.* valdit, voldit) ; *wield, dominate, rule*, A. R. 358 ; MARH. 5 ; GEN. & EX. 2143 ; to welden (*ms.* weldenn) al his kinedom ORM. 8159 ; welden REL. I. 174 ; LANGL. *B* xi. 72 ; to welden al þis worldes winne C. L. 183 ; welden, wælden LAȝ. 1250, 3335 ; welde HAV. 129 ; ROB. 536 ; H. H. 106 (108) ; MIRC 237 ; FLOR. 765 ; PR. C. 5777 ; wilde FER. 5179 ; CR. K. 96 ; **wéldeð** (? = wealdeð) (*pres.*) HOM. I. 153 ; welt H. M. 7 ; wildis *possess* ALEX. (Sk.) 4481 ; **wélde** (*pret.*) HOM. II. 119 ; wealde SAX. CHR. 250 ; wildid [weldid] *governed* ALEX. (Sk.) 2303 ; walde, wolde [welde, wolde] LAȝ. 2440, 5986, 24134 ; welde [walde] C. L. 978 ; welde JOS. 600 ; welten GEN. & EX. 840 ; **wólde** (*subj.*) REL. I. 183 ; **wealt** (*pple.*) KATH. 190 ; welt WILL. 856 ; til he was on the rode wold GEN. & EX. 255 ; þat wold (*suppressed*) HAV. 1932 ; *comp.* **a-, bi-, ge-wélden(-wílden).**

wíldar [**wéldar**], sb., *governor*, ALEX. (Sk.) 3166 ; wildire [welder] 1608.

wélden, *see* **wálden.**

[**wéldi**, adj., = *M.L.G.* weldich ; *powerful* ; *comp.* **un-wéldi.**]

wéle, sb., *O.E.* welạ, weola, = *O.L.G.* welo, *O.H.G.* wolo ; *weal, wealth, happiness*, KATH. 1511 ; LANGL. *B* xx. 39 ; *C* xiii. 236 ; PR. C. 1002 ; wele or wo GOW. I. 46 ; wele, weole HOM. I. 145 ; wele, weole HOM. I. 145 ; wele, weole LAȝ. 7732, 10394 ; wele [weole] O. & N. 1273 ; weole A. R. 192 ; weole (*v. r.* wele) P. P. 250 ; þene eorþliche weole FRAG. 6 ; weole and wunne R. S. ii ; weole C. L. 504 ; SPEC. 32 ; K. T. 967 ; weole after wowe REL. I. 174 ; in welȝe and wale GEN. & EX. 809 ; weolan (*gen.*) FRAG. 6 ; weolæn (*pl.*) FRAG. 6 ; weolen MARH. 1 ; **wélene,**

welena (*gen. pl.*) HOM. I. 33, 111 ; **wélen** (*dat. pl.*) LK. viii. 14 ; weolen FRAG. 6 ; *comp.* **weoreld-wéle.**

weole-ful, adj., *wealthy*, HOM. I. 259.

wélefulnesse, sb., *weal*, CH. BOET. i. 4 (21).

wélsom, adj., *prosperous*, WICL. GEN. xxiv. 21.

wele, *see* **wel.**

[**weleȝen**, v., *O.E.* welegian, welgian ; *make rich* ; *comp.* **a-, ȝe-weleȝen.**]

wélen, v., *O.N.* velja, = *O.H.G.* wellen ; *Goth.* waljan ; *cf.* **wálen** ; *choose* ; wele IW. 2507 ; out wele MAN. (F.) 7340.

welewen, *see* **welhen.**

welhen, v., *? from* **walh** ; *wither, fade, dry up* ; shal welewen ['*marcescet*'] WICL. IS. xix. 6 ; **weolewe** (*pres.*) SPEC. 50 ; welweþ TREV. VII. 477 ; welewith P. R. L. P. 173 ; welyhes PS. lxxxix. 6* ; **welwed** (*pret.*) A. P. iii. 475 ; *comp.* **for-welhen.**

welewunge, sb., *fading away*, H. M. 35.

wéli, adj., *O.E.* welig, weleg, = *O.H.G.* welagi ; *wealthy, 'dives,'* FRAG. 2 ; MAT. xxvii. 57 ; weoli LAȝ. 13904 ; þa weoleȝen 427 ; **wéliest** (*superl.*) C. M. 7879 ; *comp.* **un-wéli.**

welk, sb., *O.E.* weoloc, wiluc ; *whelk* ; wilk (*ms.* wilke) '*concha*' PR. P. 528 ; VOC. 189, 254 ; **welkes** (*pl.*) L. C. C. 17.

welkin, v., *cf. M.Du.* welken, *O.H.G.* welchen ; *wither, fade, dry up, 'marceo,'* PR. P. 521 ; welke REL. I. 6 ; **welkeþ** (*pres.*) TREV. I. 77 ; GOW. I. 35 ; welkes PS. lxxxix. 6 ; PR. C. 707 ; **welkede** (*pret.*) GEN. & EX. 2107 ; welked CH. C. T. *C* 738 ; *comp.* **for-welken.**

welkne, *see* **wolcne.**

welle, sb., *O.E.* wella, wylla *m.*, wylle *f.*, *O.N.* vella (*boiling heat, torrent*), *cf. O.H.G.* wella *f.* (*wave*) ; *from* **wallen** ; *well, fountain, flood*, VOC. 239 ; PR. P. 520 ; SPEC. 94 ; WILL. 5535 ; MAND. 169 ; S. S. (Wr.) 1382 ; welle A. R. 156 ; ane welle HOM. I. 41 ; þa welle LAȝ. 19812 ; þan welle [wille] 19771 ; wælle [welle] 17025 ; þeos welle LEB. JES. 322 ; welle, wulle, wille SHOR. 119 ; wille P. L. S. xiii. 295 ; **welle** (*dat.*) ORM. 19314 ; þere welle [wille] LAȝ. 19750 ; swulche fisces in wælle (*ms.* walle, *r. w.* kulle) 21334 ; **wellen** (*pl.*) A. R. 282 ; wellen BRD. 12 ; AYENB. 80 ; weallen (*sec. text* welles) LAȝ. 1240 ; at **Welle** (*Wells*) MISC. 145.

welle-carse, sb., *cf. O.E.* wylle cærse ; *water cress* ; **welle carses** (*pl.*) LANGL. *C* vii. 292.

welle-spring, sb., *well spring*, GEN. & EX. 1243.

wælle-strēam, sb., *O.E.* wylle-strēam ; *well-stream* ; (*ms.* walle-) LAȝ. 2849.

welle-water, sb., *well water*, HOM. I. 159 ; wellewater LAȝ. 19792 ; wulle wæter LEECHD. III. 90.

wellen, v., *O.E.* wellan, wyllan, *O.N.* vella, *cf.*
M.L.G. wallen, *O.H.G.* wellōn; *mod.Eng.* well,
weld ; *well up, bubble, flow ; pour forth ; boil,
melt ; weld* ; REL. I. 53 ; welle M. H. 29 ; eli letten
ho welle MARG. 60 ; þei schulen welle to gidere [' *conflabunt*'] WICL. IS. ii. 4 ; **welle**
(*pres.*) SPEC. 40 ; welle '*fundo, coagulo*' PR.
P. 520 ; he welleð of þe h(e)orte HOM.
I. 159 ; **welle** (*subj.*) L. C. C. 51 ; **welland**
(*pple.*) PR. C. 7126 ; **welden** (*pret.*) LEG. 191 ;
welled (*pple.*) L. H. R. 59 ; iwelled A. R. 284 ;
iwelled JUL². 54.

wellen², v., *O.E.* wyllan, = *M.H.G.* wellen ;
? roll : to þan scipen **wælden** (*? carried*) [*sec.*
text ladden] LAȝ. 1131 ; heo weopen . . . and
heore væx fæire wælden to volde 21874.

welm, sb., *O.E.* welm, wylm ; *boil, tumour ;
fountain ;* **welmes** (*pl.*) MAP 355 ; in þe
welmes [' *in scatebris* '] TREV. I. 429 ; *? comp.*
fōt-welm.

welmen, v., *boil* ; he **welmeþ** (*pres.*) up so he
were wod FL. & BL. (H.) 719.

welp, *see* **hwelp.**

welten, v., *O.E.* wyltan, *O.N.* velta (*pret.* velta),
= *M.H.G.* welzen ; *from* **walten** ; *overturn ;*
welte (*pret.*) D. ARTH. 3152 ; welt A. P. iii.
115 ; **iwelt** (*pple.*) SHOR. 162 ; *comp.* **a-**
welten.

weltren, *see* **walteren.**

welðe, sb., = *M.L.G.*, *M.Du.* welde ; *wealth,
happiness, pleasure,* GEN. & EX. 796 ; welþe
REL. I. 174 ; WILL. 2076 ; LIDG. M. P. 69 ;
ISUM. 699 ; ANT. ARTH. xvii ; welthe PR. C.
1307 ; welþe, weolþe LANGL. *A* i. 53 ; **welþes**
(*pl.*) [' *divitias* '] PS. xxxvi. 16* ; welthis (*as
sing.*) ALEX. (Sk.) 3290, 4467.

welþi, adj., = *M.Du.* weldich ; *wealthy,* H. V.
115.

welwen, *see* **welhen.**

wem, sb., *from* **wemmen**, *cf.* *O.E.* wamm,
womm *n. m.*, = *O.N.*, *O.H.G.* wamm, vamm
n., *O.L.G.* wam *m. ; spot, stain, crime,* A. R.
10 ; wem P. L. S. xviii. 96 ; LANGL. *B* xviii.
131 ; WICL. PROV. ix. 7 ; CH. C. T. *F* 121 ;
M. H. 126 ; wemm RICH. 1090 ; wem, wembe
HAMP. PS. xviii. 14 ; **wemme** (*dat.*) MISC.
98 ; A. P. i. 1002 ; PARTEN. 466.

wem-lēs, adj., *spotless, faultless,* C. M. 5748 ;
wemlees P. R. L. P. 211 ; wemmeles MISC.
139 ; CH. C. T. *G* 47.

wemmen, v., *O.E.* wemman, = *O.H.G.* (gi-)-
wemman, *Goth.* (ana-)wammjan ; *stain, defile ;*
wemmi ROB. 206 ; **wemde** (*pret.*) HOM. I.
83 ; **wemmed** (*pple.*) ORM. 2326 ; wemmed
JOS. 678 ; WICL. DEUT. xii. 15 ; *comp.* **a-,**
ȝe-wemmen.

wemmunge, sb., *pollution,* H. M. 13.

wen, *see* **wenne.**

wenchen, v., *O.Fr.* gvenchir, gvencir, gvinchir,
O.H.G. wenkan, wenchan ; *kick out (as a*

horse) ; *go back* ; wincen LIDG. M. P. 162 ;
to winse LANGL. *C* v. 22 ; wincin ' *calcitro* '
PR. P. 528 ; **winche** (*pres.*) D. ARTH. 2104 ;
wendeð ou ant **wencheð** (*imper.*) frommard
him A. R. 98.

wenchel, sb., *? O.E.* wencel (*pl.* winclo), *mod.
Eng.* wench ; *child, boy or girl ; girl, young
woman* ; were & wif & wenchel A. R. 334* ;
ȝuw is boren nu . . . an wenchel (*ms.* wenn-
chell) ORM. 3356 ; wenche [' *puella* '] WICL.
MAT. ix. 24 ; ' *ancilla* ' PR. P. 521 ; MISC.
100 ; P. L. S. xvii. 116 ; WILL. 1901 ; AR. &
MER. 894 ; GOW. I. 263 ; LANGL. *A* v. 208 ;
CH. C. T. *A* 3254 ; þa **wænclen** (*ms.* wan-
clen) (*pl.*) LAȝ. 31834 ; **wenclen** (? wenchen)
(*dat. pl.*) P. L. S. xvii. 98.

wend, sb., *O.E.* wend, = *O.Fris.* wend ; *path ;*
went CH. TRO. ii. 815 ; PALL. ii. 96.

[**wende ?** adj., *comp.* **hāl-, hwīl-, luve-**
wende.]

wenden, v., *O.E.* wendan, = *O.L.G.* wendean,
O.Fris. wenda, *O.N.* venda, *O.H.G.* wentan,
wendan, *Goth.* wandjan, *mod. Eng.* wend ;
from **winden** ; *turn, direct one's course, go ;
change* ; A. R. 110 ; ha willeð alle wenden to
Criste KATH. 693 ; to wenden ðus here ðoght
GEN. & EX. 4061 ; (*ms.* wendenn) ORM. 3441 ;
wenden [wende] LAȝ. 712 ; leves wenden
[wende] O. & N. 1326 ; wenden, wende LANGL.
A v. 144 ; CH. C. T. *A* 21 ; wende DEGR.
1567 ; to hevene wende H. H. 248 ; wende
into helle AYENB. 13 ; to wendende MISC. 192 ;
wende (*pres.*) O. & N. 288 ; wendeð LAȝ.
20864 ; went A. R. 104 ; þe wedercoc . . . þet
him went mid eche winde AYENB. 180 ; wend
O. & N. 1464 ; **wend** (*imper.*) A. R. 100 ; wend
þe hider LAȝ. 24177 ; **wendinde** (*pple.*) LEG.
36 ; **wende** (*pret.*) FL. & BL. 17 ; BEK. 69 ; þa
leaf wende LAȝ. 46 ; wende hire þiderward
KATH. 160 ; he wende heowes JUL. 38 ;
wente GEN. & EX. 321 ; wente MAND. 224 ;
(þu) wendest LAȝ. 5057 ; (hie) wenden REL.
l. 129 ; wenden, wenten MK. iii. 22 ; LK.
xxiv. 33 ; wenten JOS. 191 ; wend (*ms.* wennd)
(*pple.*) ORM. DEDIC. 113 ; wend HAV. 2138 ;
went S. S. (Wr.) 1485 ; his douȝter . . . was
went a wai WILL. 1984 ; *comp.* **a-, æt-, bi-,**
ȝe-, to-wenden.

wendinge, sb., *O.E.* wendung, wending ;
wending, turning, AYENB. 70 ; wending [' *ver-
sura* '] PALL. ii. 12.

wendi, adj., = *M.L.G.* wendich, *M.H.G.* wen-
dec ; *averse* : þu art windi [wundi] of me
JUL. 10, 11 ; windi of wisdom KATH. 376.

wēne, sb., *O.E.* wēn, *cf.* *O.Fris.* wēn, *Goth.*
wēns, *O.N.* vān *f.*, *O.H.G.* wāni *f.*, wān *m.*, *O.L.*
G. wān *m. ; thought, doubt, supposition,* FL. &
BL. 651 ; þa ȝet hit weore a wene whar þu heo
mihtes aȝe LAȝ. 18752 ; of þine kume nis na
wene 28141 ; hit bið a muchele wæne whær
ȝe iseon me avere mare 13503 ; an oþer þing
me is a wene O. & N. 239 ; wen is þat he was

fórdred ORM. 7152 ; vain [wane] BARB. vii. 2 ;
wēne (*dat.*) A. R. 390* ; wene MIRC 381 ;
þenne is þat folc buten wene þat reouðe heom
is to cumene LAȝ. 21763 ; withouten wene
PERC. 1987 ; M. ARTH. 548 ; wiþuten wen
ORM. 4326 ; wēne (*pl.*) REL. I. 173 (MISC.
108) ; *comp.* over-wēne.

wēne², adj., *O.E.* (or-)wāna, = *Goth.* (us-)wēna
O.N. vænn, *O.H.G.* (ur-)wāni ; *hopeful ;
beautiful ;* wēner (*compar.*) þen Wenore (=
Guenever) GAW. 945 ; *comp.* ǽr-wēne.

   wēn-līch, adj., *O.E.* wēnlīc, = *O.N.* vænligr ;
hopeful ; fine : of atel(l)iche to wenliche
HOM. II. 83 ; he mai beon . . . wenliche
(*good*) lorþeu REL. I. 173 (MISC. 108) ; **wēn-**
lūkest (*superl.*) HOM. II. 29.

wēne, *see* wǣne.

wénen, v., *O.E.* wenian, = *O.N.* venja, *M L.G.*
wenen, wennen, *M.Du.* wennen, *O.H.G.* wen-
nan ; *wean (an infant) ;* wene '*ablacto*' PR. P.
522 , **wénide** (*pret.*) WICL. PS. cxxx. 2 ;
comp. for-, ȝe-wénen.

   wéning, sb., *weaning,* WICL. GEN. xvi. 8.

wēnen, v., *O.E.* wēnan, = *O.Fris.* wēna, *Goth.*
wēnjan, *O.N.* vǣna, *O.L.G.* wānian, *O.H.G.*
wānan ; *ween, hope, suppose,* MAT. v. 17 ; A.
R. 106 ; ORM. 9826 ; wenen LAȝ. 4196 ; MAND.
252 ; þanne mawe we wenen þat he wule us
w(u)rþie REL. I. 183 ; mon mai longe lives
wene MISC. 156 (*printed* thene R. S. i) ;
wene WILL. 554 ; PR. C. 2154 ; wēne (*pres.*)
O. & N. 237 ; GEN. & EX. 315 ; wene HORN
(L.) 663 ; HAV. 655 ; SPEC. 92 ; þu. wenest
HOM. I. 7 ; wanst O. & N. 1644 ; weneþ SHOR.
67 ; AYENB. 21 ; moni mon weneþ . . . longes
lives REL. I. 174 (MISC. 112) ; wenis BARB. ii.
288 ; wēn (*imper.*) JUL. 12 ; wěnde (*ms.*
wennde) (*pret.*) ORM. 1993 ; wende BEK.
1758 ; TRIST. 33 ; DEGR. 1591 ; AMAD. (R.)
xxxvii. ; M. ARTH. 1160 ; þet heo þer wende
vinde MISC. 53 ; his knape wende it were a
der GEN. & EX. 477 ; þu wendest FRAG. 7 ;
wenit BARB. iv. 771 ; wenden MARH. 4 ; heo
wenden [wende] þat his sawen soðe weren
LAȝ. 749 ; wěnde (*subj.*) LANGL. *B* xiii. 280 ;
comp. ȝe-, over-wēnen.

   wēninge, sb., = *M.H.G.* wǣnunge ; *weening,*
AYENB. 113 ; CH. BOET. v. 6 (172) ; wening
TRIST. 1730.

wēnen, *see* hwēnen.

weng, sb., *O.N.* vengr, vængr ; *wing,* '*ala,*'
VOC. 233 ; PR. P. 522 ; whenge (*dat.*) LANGL.
B xii. 263* ; wenges (*pl.*) H. M. 47 ; ORM.
8024 ; PS. xvi. 8 ; MAND. 48 ; CH. C. T.
A 1964* ; *see* winge.

wenne, sb., *O.E.* wenn (*pl.* wennas), = *M.L.G.*
wene ; *wen,* '*verruca,*' PR. P. 522 ; VOC. 267 ;
wen AYENB. 262.

wenne, *see* wūnne, hwenne.

Wěnsdaie, *see* Wōden. **went,** *see* wend.

wēod, sb., *O.E.* wēod, wīod, = *O.L.G.* wiod *n. ;*

weed, ALIS. 796 ; weed '*herba nociva*' PR. P.
519 ; wed REL. I. 214 ; þat wed SHOR. 32 ;
wede ALEX. (Sk.) 413 ; wēode [wode] (*dat.*)
O. & N. 320 ; wiedes (*pl.*) HOM. II. 129 ;
weoden ROB. 404 ; *comp.* doke-, knop-
weed.

   wēod-hook, sb., *hoe,* ['*sarculum*'] WICL.
IS. vii. 25 ; wiedhoc AYENB. 121 ; wedhoc
VOC. 232.

wēoden, v., *O.E.* wēodian, = *M.Du.* wieden,
M.L.G. wēden ; *weed, free from weeds ;* weede
H. V. 77 ; wede '*runco*' PR. P. 519 ; **weede**
(*imper.*) PALL. ii. 289.

[weoh, adj., *O.E.* weoh, wīh (*sacred*), = *O.L.G.*
wīh (*temple*).]

   weofed, sb., *O.E.* weofed, weo-, wī-, wīgbed,
(-bed = bēod) ; *altar :* þat weofed [wefd] LAȝ.
1189 ; weofd 28750 ; weoved A. R. 346 ;
KATH. 203 ; weved ROB. 224 ; PS. xxv. 6 ;
wieved AYENB. 236 ; **wǽfde** [wefde] (*dat.*)
LAȝ. 8089 ; wēvedes (*printed* wenedes) (*pl.*)
TREV. I. 161.

weolcne, *see* wolcne.

weole, weoli, *see* wéle, wéli.

weolewen, *see* welwen.

weopen, *see* wēpen.

weorc, sb., *O.E.* weorc, worc, = *O.L.G.* werc,
O.Fris. werk, wirk, *O.N.* verk, *O.H.G.* werch,
Gr. ἔργον ; *work,* HOM. I. 93 ; weorc, werc
P. L. S. viii. 54, 64 ; ORM. 1833, 6522 ; weorc,
werc, worc LAȝ. 491, 2574, 8709 ; werc A. R.
118 ; weork S. A. L. 150 ; werk '*opus*' PR. P.
522 ; HAV. 866 ; C. L. 3 ; SPEC. 42 ; MAND.
93 ; CH. C. T. *A* 479 ; ðat werk GEN. & EX.
3902 ; work TREAT. 133 ; werc HOM. I. 223 ;
werc (*pl.*) MAT. xxiii. 5 ; REL. I. 131 (HOM.
II. 11) ; weorc, werkes HOM. I. 23, 99 ;
werkes LAȝ. 5979 ; werkes A. R. 118 ; wurkes
MARH. 6 ; **werke** (*gen. pl.*) HOM. I. 9 ;
weorcan, wercan (*dat. pl.*) HOM. I. 107,
109 ; wercan MAT. xxiii. 3 ; weorken LAȝ.
7106 ; wurken MISC. 193 ; werke FRAG. 7 ;
HOM. II. 220 ; werke MISC. 58 ; weorche
[werche] P. L. S. viii. 6 ; *comp.* and-, ȝe-
wunder-weorc.

   werk-beeste, sb., *beast of burden,* WICL. PS.
lxx. 23.

   werk-dai, sb., *O.N.* verkdagr, = *M.H.G.*
werctac ; *work-day,* PR. P. 522 ; MAS. 270 ;
werkedei A. R. 20 ; warke daȝ ORM. 11315 ;
werkedai LEG. 74 ; MIRC 1005.

   work-ful, adj., *O.E.* weorcfull ; *industrious ;*
work-vol AYENB. 199.

   werk-hous, sb., *O.E.* weorchūs ; *workshop,*
'*artificina, opificium,*' PR. P. 522.

   weorc-man, sb., *O.E.* weorcmann ; *workman,* LAȝ. 22892 ; wercmon A. R. 404 ; werk-
man LANGL. *B* xiv. 137.

   werkman-shipe, sb., *workmanship, work,*
LANGL. *B* ii. 91 ; werkmanshup *C* iii. 96, xx.
141 ; werkemanship *B* x. 288.

weored, sb., *O.E.* weorud, werud, werod, =*O.L.G.* werod; *host, troop*, LAȝ. 19922; wered ['*cohortem*'] MK. xv. 16; werd ['*turba*'] LK. vi. 17; werd REL. I. 188; wird GEN. & EX. 1790; **werod** (*pl.*) HOM. I. 219; weoredes A. R. 30*; weordes [wordes] MARH. 22; *comp.* **cniht-, mon-weored.**

wored-strencðe, sb., *army*, LAȝ. 509.

[**weored**[2], adj., *O.E.* weorod, wered; *sweet.*]

weorednesse, sb., *O.E.* weredness; *sweetness* : salt ȝiveð mete wordnesse A. R. 138.

weoreld, sb., *O.E.* weorold, worold, weoruld, woruld, world, =*O.L.G.* werold, *O.N.* veröld, *O.H.G.* weralt, werolt, worolt, werelt, werlte ; *?a compound of* wer *and* alde; *world; age, eternity*; ORM. 10515; weoreld, weoruld, woreld LAȝ. 5028, 8116, 9969 ; þeos woreld, weorld HOM. I. 19, 33 ; woreld, wereld HOM. II. 99, 161 ; world A. R. 92; KATH. 189 ; world C. L. 570; WICL. JOHN iii. 16 ; world, worlde LANGL. *A* PROL. 19; *B* PROL. 19; wordle TREAT. 132; SHOR. 95 ; AYENB. 7 ; wordle wiþouten ende P. L. S. xii. 109 ; werld GEN. & EX. 42 ; werld FER. 123 ; in werld ['*in seculum*'] PS. xliii. 9 ; werlde PERC. 284 ; werde '*mundus, seculum*' PR. P. 522; werd HAV. 1290; MAN. (F.) 222; ALEX. (Sk.) 18, 4142, 4136; þe litill werde *the miȝrocosm* 4494 ; **weorlde** (*gen.*) HOM. I. 21 ; weorlde, worlde MAT. xiii. 22, 39 ; worlde O. & N. 476 ; **weorlde** [worle] (*dat.*) LAȝ. 23081 ; a þere ilke worlde þa þis wes iwurðen 23425 ; on ðere eche weorlde HOM. I. 135 ; he beo ever iheied from worlde to worlde A. R. 430 ; from worlde into worlde MARH. 22 ; werelde MISC. 105 ; **worulde** (*gen. pl.*) LAȝ. 9072 ; in alre worlde world KATH. 663.

weoreld-ǽhte, sb., *O. E.* woruldǽht ; *worldly property*, ORM. 12079 ; worldaihte REL. I. 183.

weorld-ȝelp, sb., *worldly boast, praise*, HOM. I. 105.

weoreld-king, sb., *O.E.* woruldcyning ; *king of the world*, LAȝ. 6328.

world-lích, adj., *O.E.* woruldlīc ; *worldly*, A. R. 92; H. M. 29 ; worldlich C. L. 983 ; werdli [wordli] ALEX. (Sk.) 3262.

werdlinesse, sb., *worldliness*, PR. P. 522.

weoreld-lif, sb., *O.E.* weoruld-, woruldlíf ; *life of the world*, ORM. 2978.

weorld-mon, sb., *O.E.* weoruld mann ; *man of the world*, LAȝ. 28131 ; worldmon KATH. 486.

weoreld-ríche, sb., *O. E.* woruldríce ; *worldly kingdom*, ORM. 11800 ; weorldríche LAȝ. 15179.

weorld-scóme, sb., *shame of the world*, LAȝ. 8323.

weorld-sēli, adj., *happy; fortunate*; weorldseli men LAȝ. 11043.

weoreld-shipe, sb., *O. E.* woruldscipe ; *worldliness*, ORM. 6322.

weoreld-þing, sb., *thing of the world*, ORM. 2966; worldþing HOM. I. 105 ; wereldþing II. 127.

woruld-wele, sb., *O. E.* weoruldwela ; *worldly wealth*, P. L. S. viii. 183 ; woreldwele HOM. II. 29.

world-wis, adj., *worldly wise*, LAȝ. 15496.

world-witti, adj., *worldly wise*, KATH. 488.

world-wünne, sb., *worldly joy* : worldwunne wiðuten poverte HOM. I. 143.

weorien, *see* **wérien.**

weorpen, werpen, worpen, v., *O.E.* weorpan, worpan, wurpan, =*O.L.G.* werpan, *O.N.* verpa, *Goth.* wairpan, *O.H.G.* werfan, werphan ; *throw, twist*, LAȝ. 2488, 6428, 17429; weorpen, worpen A. R. 166, 404 ; werpen GEN. & EX. 3358; to werpen dun þe deofel ORM. 3575 ; worpe [werpe] of horse O. & N. 768 ; wurpen LK. xx. 19 ; **werpe** (*pres.*) PS. ii. 3 ; werpeð ðus hire web & weveð on hire wise REL. I. 219; werp up ALEX. (Sk.) 557 ; werpis 798 ; **worp** (*imper.*) A. R. 478 ; **werpinde** (*pple.*) HOM. II. 175 ; **wearp**, warp (*pret.*) KATH. 831, 894 ; warp ORM. 1095; O. & N. 125 ; GEN. & EX. 2640 ; warp LAȝ. 4518; HAV. 1061 ; PS. lxxvii. 67 ; GAW. 2025 ; on þat us warp from wo SPEC. 32 ; warp, werp HOM. I. 41, 129 ; weorp JUL. 38; werp A. R. 52 ; þu wurpe FRAG. 6 ; JUL. 60; (hie) wurpen REL. I. 129; (MISC. 161) ; KATH. 1832 ; wurpen LAȝ. 4033; MISC. 151; **wurpe** (*subj.*) A. R. 124 ; **worpen** (*pple.*) ORM. 8136; GEN. & EX. 1943 ; *comp.* **a-, bi-, for-, ȝe-, over-, to-weorpen (-werpen)** ; *deriv.* **warp, wurp.**

[**weorpere**, sb., *thrower; comp.* **knif-w(e)orpare.**]

weorre, *see* **werre.**

weorþien, *see* **würðien.** **weote**, *see* **wite.**

weoðelen, v., =*M.H.G.* wedelen; *waver, hesitate*; **weoðeleden** [wiþeleode] (*pret. pl.*) his fluhtes LAȝ. 2885.

weoved, *see* **weofed.**

wēp, sb., =*wōp* ; *weeping*, GEN. & EX. 2328 ; woep AN. LIT. 90.

wēp-lī, adj., *lamentable*, CH. BOET. i. 1 (5).

wēpen, v., *O.E.* wēpan, (*pret.* wēop), =*O. Fris.* wēpa, *O.N.* œpa (*pret.* œpta), *O.L.G.* wōpian (*pret.* weop), *Goth.* wōpjan (*pret.* wōpida) (φωνεῖν), *O.H.G.* wuofen (*pret.* wiof) ; *weep*, ORM. 5653 ; wepen LAȝ. 11980 ; wepe BEK. 1425 ; WILL. 310 ; weopen A. R. 274 ; **wēpe** (*pres.*) O. & N. 876 ; wepþ AYENB. 93 ; **wēpinde** (*pple.*) ROB. 338 ; weopinde FRAG. 5 ; **wēop** (*pret.*) A. R. 106 ; weop HORN (L.) 69 ; SPEC. 70 ; weop [wep] LAȝ. 6650 ; weop, wiop JOHN xx. 11 ; wiep HOM. II. 149 ; wep R. S. vii ; GEN. & EX. 4149 ; wep BRD. 9; IW. 2773 ; LIDG. TH. 1002 ; weep CH. C. T. *D* 588 ; wip S. S. (Wr.) 2508 ; wep, wepte WILL. 33, 50 ; wept LANGL. *B* xviii. 91 ; wepte

CH. C. T. *A* 148; wepen HAV. 152; wepe ROB. 338; wepten ORM. 8140; wōpen (*pple.*) AM. & AMIL. 2281; CH. C. T. *F* 523; wept *E* 1544; *see* ēpen; *comp.* bi-, 3e-wēpen.

wēpinge, sb., *weeping*, PR. P. 522; LA3. 5970*; SPEC. 30.

wēpen, *see* wǣpen.

wĕpman, *see under* wǣpen.

wer, sb., *O.E.* wer,=*O.L.G.*, *O.H.G.* wer, *O.N.* verr, *Goth.* wair, *Lat.* vir; *man* MAT. i. 19; were P. L. S. viii. 16; JUL. 15; H. M. 31; ORM. 4604; O. & N. 1522; GEN. & EX. 3977; were MISC. 85; ver GAW. 866; **weres** (*pl.*) LA3. 17333; *comp.* burh-wer.

    wer-wolf, sb., *O.E.* werewulf, = *M.H.G.* werwolf; *werewolf*, λυκάνθρωπος; WILL. 15; **werwolves** (*pl.*) PL. CR. 459.

wer², sb., *O.E.* wer *m.* (ALFR. P. C. 278); *weir, dam, pond*: þis strem 3ou ledeth to þe sorouful were CH. P. F. 138; **wéres** (*pl.*) ['*stagna*'] PS. cvi. 35; M. T. 121; wæres and feonnes SAX. CHR. 122; *comp.* fisch-wer.

[**wer** (? wēr), sb., ? *M.Du.* wier (*seaweed*).]
    wer-cok, sb., ? *pheasant*, L. C. C. 36.

wer, *see* war, werre.

wēr, *see* hwār, hweþer.

werble, sb., *O.Fr.* werble; *warble*; **werbles** (*pl.*) GAW. 119; werbeles TREV. I. 355.

werblen, v., *O.Fr.* werbler; *mod.Eng.* warble; *blow, sound*; þe **werbelande** (*pple.*) winde GAW. 2004; **werbild** (*pple.*) trompis ALEX. (Sk.) 2222.

werc, *see* weorc.   **werchen**, *see* wurchen.

werd, *see* weored, weoreld.

werde, *see* würde.

werden, **weorden**, v., *O.E.* werdan, wyrdan, = *O.L.G.* (a-)wardean, *O.H.G.* wartan, wertan; *harm, injure*, ORM. 5185, 6249.

wére, sb., *doubt*: boute were P. L. S. xxxv. 21; mi witte is in a were YORK v. 1; were LANGL. *B* xi. 111; *B* xvi. 3; weer *C* xiii. 50; wehere MAN. (H.) 306; weir BARB. xvii. 496; wer ii. 43.

wére², v., *cf. O.Fris., M.Du., M.L.G.* were, ? *O.H.G.* weri; ? *defence, protection*: he wurð ðane Egiptes were GEN. & EX. 2680; al englene were (*host*) HOM. I. 195; were GAW. 1628; þe Inglismen put þam to were ful baldli MIN. iii. 95; that falles unto the were TOWNL. 218.

were, *see* wer, werre.

wēre, sb., *cf. O.N.* vár, *Sw.* vår, *Latin* ver; *spring*, BARB. v. 1; ware HAMP. PS. lxxiii. 18; wair CATH. 408; veer PALL. iv. 251.

wéred, *see* weored.   **wereld**, *see* weoreld.

wéren, v.,=*M.Du., O.H.G.* weren; ? *remain*; **wérende** (*pple.*) on worlde REL. I. 184.

wer3en, *see* warien, wur3en.

wērgen, *see* wērien.   **werhte**, *see* wurhte.

---

weri, *see* wari.

wēri, adj., *O.E.* wērig,=*O.L.G.* (sīth-)wōrig, *O.H.G.* wōrag; *weary*; ['*fatigatus*'] JOHN iv. 6; KATH. 439; GEN. & EX. 975; weri '*fessus*' PR. P. 522; LA3. 1328; SHOR. 85; AYENB. 84; weri so water in wore A. D. 148; GOW. I. 318; CH. C. T. *A* 4234; EGL. 595; so weri he was of þe wei BEK. 1171; **wérie** (*dat. m.*) HOM. II. 7; **wérie** (*pl.*) LA3. 18406; **wēri** (*adv.*) HOM. II. 227; *comp.* for-, sǣ-wēri.

    wēri-hēde, sb., *weariness*, AYENB. 33.

    wērinesse, sb., *O.E.* wērignesse; *weariness*, ROB. (W.) 7384; REL. I. 130; werinisse P. L. S. ix. 61.

wérien, v., *O.E.* werian, wergan, = *O.L.G.* werean, *O.H.G.* werian, *O.N.* verja, *Goth.* warjan; *defend*, P. L. S. viii. 162; A. R. 52; werien [werie] KATH. 788; ic eou wulle werien wið elcne herm HOM. I. 13; weren 69; werien PROCL. 4; SHOR. 5; þe while he mai, . . . i compe hine werien [werie] LA3. 8288; weorien heom mid wepnen 21289; werie REL. I. 172; MISC. 44; HORN (L.) 785; ROB. 390; AYENB. 182; ALIS. 3656; weren GEN. & EX. 1272; þu miht weren þe fra þe3m ORM. 1406; weren PL. CR. 435; were HAV. 2152; CH. C. T. *A* 2550; MAN. (F.) 5957; IW. 3136; TOR. 527; AV. ARTH. xxxix; **wéreð** (*pres.*) A. R. 312; wereþ O. & N. 834; (heo) werieð A. R. 246; **wére** (*imper.*) þe . . . a3ean me A. R. 400; **wéreden** (*pret.*) LA3. 5696; *comp.* bi-wérien.

wérien², v., *O.E.* werian,=*O.H.G.* werian, *O.N.* verja, *Goth.* wasjan; *wearing* (*cloth*), A. R. 418; to werien moni a feir schrud R. S. v (MISC. 172); werien [werie, were] LANGL. *B* xiv. 329; were CHR. E. 642; were, weore SPEC. 37; **wérieþ** [wereþ] (*pres.*) O. & N. 1174; **wérede** (*pret.*) BEK. 1475; REL. I. 121; wered GOW. I. 33; CH. C. T. *A* 75; LIDG. M. P. 49; wereden R. S. iv (MISC. 164); wereden WICL. 3 ESDR. iii. 2; beien heo wereden anes kunnes iweden LA3. 30081; *comp.* for-wérien.

wériung, **werunge**, sb., *wearing*, A. R. 8.

werien, *see* warien.

wērien, v., *O.E.* wērigean; *make or become weary*; werie AYENB. 99; **wērgeð** [weries] (*pres.*) A. R. 252; ne wergeð he neaver to wurchen ow al þat wandreðe JUL. 22; it werieþ me CH. C. T. *G* 1304; (3e) wergið ow seolven MARH. 18; **wērgede** (*pret.*) JUL. 54; *comp.* forwērien.

    wērgunge, sb., *wearying*, A. R. 252.

werk, sb., *O.E.* wærc,=*O.N.* verkr; *pain, grief*, SPEC. 31; REL. I. 51; warch A. R. 282*; *comp.* heed-werk.

werk, **werken**, *see* weorc, wurchen.

werkin, v., *O.E.* wærcian (LEECHD. II. 318), *cf. O.N.* verkja, virkja; *pain, ache*; '*doleo*'

PR. P. 522; **warchende** (*pple.*) A. R. 360*; moni warchond wound D. TROY 10035.

werkinge, sb., *aching*; warkinge HAMP. PS. lxviii. 31; '*cephalia*' PR. P. 523; **werkingis** xxxvii. 2.

werld, *see* **weoreld**. **werm**, *see* **wurm**.

wermen, *see* **warmen**.

wermōd, sb., *O.E.* wermōd,=*O.H.G.* weri-muote; *wormwood*, '*absinthium*,' VOC. 139; WICL. LAM. iii. 19; PALL. xi. 344.

wernard, sb., *O.Fr.* gvernart (*trompeur*); *deceiver, liar*, LANGL. *B* iii. 179; **wernardes** (*pl.*) *A* ii. 98, *B* ii. 128, *C* iii. 142.

wernen, v., *O.E.* wyrnan,=*O.L.G.* wernian, *O.H.G.* warnen; *refuse, deny*, A. R. 330; wer-nen [werne] LA3. 30310; wernin, wearne KATH. 770; ne mai ich mine songes werne O. & N. 1358; werne HAV. 1345; OCTOV. (W.) 1937; MIRC 841; GOW. I. 162; CH. C T. *D* 333; HOCCL. i. 442; dorst he nou3t werne þe wille of his lord WILL. 305; werne, wurne HORN (L.) 916, 1086; TREV. VII. 329; wurne ROB. 367; **werneþ** (*pres.*) LANGL. *B* xx. 12; **werne** (*subj.*) GEN. & EX. 2797; **wernde** (*pret.*) A. R. 248; wernde AYENB. 189; wornde P. L. S. xvi. 127; and warnden [wornde] him in3eong LA3. 28370; **werned** (*pple.*) GEN. & EX. 3171; iwerned TREV. III. 101; *comp.* for-**wernen**.

wernunge, sb., *refusal*, A. R. 330.

wernien, *see* **warnien**.

werp, *see* **warp**. **werpen**, *see* **weorpen**.

werre, sb., *O.Fr.* werre, gverre *f.*, *cf. M.Du.*, *M.L.G.* werre *n. m.*, *O.H.G.* werra *f.* (*strife*), *M.H.G.* werre *m. f.*; *war*: þa þe ledden here lif in werre & in winne HOM. I. 175; werre C. L. 442; AYENB. 30; OCTOV. (W.) 1621; P. P. 217; WILL. 1083; LANGL. *B* xi. 323; CH. TRO. v. 1393; þer efter wæx svythe micel werre (*ms.* uuerre) betvyx þe king & Randolf SAX. CHR. 264; weorre A. R. 72; KATH. 13; weorre.MISC. 91; halde weorre [werre] LA3. 18660; werre, were GOW. I. 107, 214; H. S. 462, 10571; S. S. (Wr.) 2858, 5860; PARTEN. 3172, 3659; were GEN. & EX. 1788; were MAN. (H.) 22; DEGR. 393; PR. C. 2296, 4088; weir, wer BARB. xvi. 178, viii. 495; justit of wer *in a warlike manner* xix. 787.

werre², adj., *O.N.*verri,=*O.Fris.* werra, wirra; *worse*, ORM. 4898; GEN. & EX. 3951; werre H. S. 474; APOL. 55; M. H. 90; worre MIRC 1242; GAW. 1588; *see* **wurs**.

werreien, v., *O.Fr.* gverroier; *make war upon, wage war*, CH. C. T. *A* 1544; werrai HAMP. PS. xvi. 14; werrai ALEX. (Sk.) 2689; warrai BARB. v. 220, xiv. 39, xx. 522; warra ix. 646; varrai viii. 24; **werrais** (*pres.*) ALEX. (Sk.) 2495; varraiis BARB. xii. 363; **warraiand** (*pple.*) BARB. i. 140; **war-rait** (*pret.*) BARB. ix. 650, ix. 744; werraid ALEX. (Sk.) 3730.

werreour, sb., *O.Fr.* gverreur; *warrior*, MAN. (F.) 6106; ALEX. (Sk.) 5310; werraiour ALEX. (Sk.) 3211; BARB. v. 85*; **werreours** (*pl.*) 2643; werriours BARB. xx. 416; weorreur A. R. 62.

werrien, v., *cf. M.L.G.*, *M.Du.* werren; *make war, fight against*: David . . . toc to verrien him SAX. CHR. 261; werren, weorren KATH. 32, 1357; werren BEK. 2380; weorren H. M. 17; werre SHOR. 133; WILL. 1070; werri þe grace of þe holi gost AYENB. 29; **weorreð** (*pres.*) A. R. 60; weorreð a3ein me MARH. 8; weorreþ MISC. 89; þat he weorri mi wit MARH. 3; **weorrede** [werrede] (*pret.*) LA3. 20191; werrede BEK. 1588.

werring, sb., *cf. M.Du.* werringhe; *warring, dissension*, SPEC. 23.

wers, *see* **wurs**. **werstillare**, *see* **wrǎstlare**.

wert, *see* **wurt**. **werte**, *see* **warte**.

werþ, *see* **wurð**.

wérunge, *see under* **wérien**.

werve, sb., *O.E.* weorf, weoruf; *plough*, HOM. I. 85.

werven, *see* **hwerven**.

werwolf, *see under* **wer**.

wēsand, *see* **wǣsand**.

weschen, *see* **waschen**.

wésele, sb., *O.E.* wesle,=*M.L.G.* wesele, *O.H.G.* wisela; *weasel*, '*belette*,' VOC. 166; CH. C. T. *A* 3234; wesile '*mustela*' PR. P. 523.

[**wesen**, v., *O.E.* wesan,=*O.L.G.*, *O.H.G.* wesan, *O.Fris.* wesa, *Goth.* wisan, *O.N.* vesa, vera; *be*:] **wæs** (*imper.*) hæil *hail!* LA3. 14309; hale wese ge MAT. xxviii. 9; **wæs** (*pret.*) was FRAG. 7; wæs [wes] P. L. S. viii. 1; wæs, wes, weas SAX. CHR. 249; wæs [was] LA3. 513; heore wes þat wurse 26997; wes MARH. 2; seoðþan þe cristindom wes HOM. I. 5; efter þet he wes arisen 141; was A. R. 8; ORM. 540; GEN. & EX. 38; was, wes O. & N. 1; hire was boþe stronge and sure 1082; ou nas (=ne was) never icunde þar to 114; was HORN (L.) 13; LANGL. *A* PROL. 1; of his oth ne was him nouth HAV. 313; such an anker he was bicome BEK. 1155; wes H. H. 4; AYENB. 7; TRIST. 24; þu were O. & N. 1059; were BEK. 774; WICL. JOHN i. 48; CH. C. T. *B* 3850; wæron, wæran *were* SAX. CHR. 250; wæran JOHN xvii. 6; wæren MK. i. 32; wæren ORM. 53; weren HOM. I. 15; A. R. 72; heo weren [were] þeowe LA3. 334; hu heo ivaren weoren *how they had fared* 26610; weren [weoren] C. L. 448; were LANGL. *B* PROL. 55; CH. C. T. *A* 18; weren, ware, wore HAV. 156, 237. 400; waren HOM. II. 31; waren MISC. 33; ware MAN. (H.) 90; PERC. 1041; woren GEN. & EX. 347; wore EM. 410; þu **wǣre** (*subj.*) ORM. 5620; were O. & N. 53; vere H. H. 131; axede hire what hire were FL. & BL. 467; in muche murþe he were SPEC. 39; þat hire nevere sæl **nere** (*should be*) LA3. 12875; weore LANGL. *A* i.

10 [were *B* i. 10] ; ware TRIST. 376 ; war n(e) (*were it not that*) som hope ware PR. C. 7264 ; wore HAV. 504 ; *deriv.* **wiste.**

wēsen, v., *from* **wōs** ; *ooze, suppurate* ; wese N. P. 65.

wesseil, sb., *for* **wes heil** ; *see* **wesen** ; *wassail,* HAV. 1246.

wesseilen, v., *wassail* ; **wesseilen** (*pres.*) HAV. 2098 ; fele siþes haveden **wosseiled** (*pple.*) 1737.

west, sb. & adv., *O.E.* west,= *O.Fris.* west *n.* ; *west ; in the west* ; (*ms.* wesst) ORM. 12125 ; west MISC. 96; west *towards the west* LAȝ. 1278; i shal walk(e) west MIR. PL. 172 ; west '*occidens*' PR. P. 523 ; AYENB. 124 ; þet west LAȝ. 1231; þe west A. R. 94 ; toward þane west BRD. 3.

west-dāle, sb., *O.E.* westdǣl ; *west part,* ORM. 16406.

west-ende, sb., *O.E.* westende ; *west quarter,* A. R. 244.

west-half, sb., *O.E.* westhealf ; *west side* : an westhalve LAȝ. 29287.

West-mara-lond, pr. n., *Westmoreland,* MISC. 146.

West-münster, -minster, pr. n., *Westminster,* LANGL. *A* ii. 131.

West-sæx, pr. n., *O.E.* Westseaxe ; *Wessex,* LAȝ. 15390 ; Westsex P. L. S. xiii. 45.

west-sīde, sb., *west part,* BEK. 1410 ; CH. C. T. *E* 57 ; PR. C. 5127.

west-ward, adv., *O.E.* westweard ; *westward,* CH. C. T. *A* 2581 ; westwarde LANGL. *B* xvi. 169.

west-wind, sb., *O.E.* westwind ; *west wind,* PR. P. 523.

weste, ? sb., *west* ; bi westen A. R. 232 ; KATH. 591 ; bi westen [weste] Sævarne LAȝ. 2136 ; bi westen HORN (H.) 5 (bi weste L.) ; bi weste P. L. S. xii. 9 ; SHOR. 137 ; fro westen GEN. & EX. 3096.

wĕste, adj., *O.E.* wēste,= *O.Fris.* wōste, *O.H.G.* wuosti ; *waste, desert,* REL. I. 128 ; O. & N. 1528 ; (*ms.* wesste) ORM. 1391 ; weste paþes LAȝ. 17330.

wĕste[2], sb.,= *O.H.G.* wuostī ; *desert,* HOM. II. 127 ; O. & N. 1000 ; (*ms.* wesste) ORM. 17409.

westen, adv., *O.E.* westan,= *O.N.* vestan, *O.L.G.* westan, westane, *O.H.G.* westana ; *from the west,* GEN. & EX. 3915 ; þa com þer westene winden LAȝ. 25591.

wĕsten, sb., *O.E.* wēsten *m. n.,* cf. *O.Fris.* wōstene, wēstene, wēstenie, *O.L.G.* wōstinnea, *O.H.G.* wōstinna *f.* ; *desert,* '*desertum,*' FRAG. 3 ; **wĕstene** (*dat.*) HOM. I. 245.

wĕsten[2], v., *O.E.* wēstan,= *M.Du.* woesten, *M.H.G.* wüesten ; *lay waste,* LAȝ. 20941 ; **wĕsten** (*pret.*) LAȝ. 1754 ; **wĕst** (*pple.*) LAȝ. 1123; REL. I. 124 ; *comp.* a-, ȝe-**wĕsten.**

western, adj., *western* ; westren CH. BOET. i. 2 (8) ; *comp.* souþ-**western.**

wĕsterne, sb., *O.E.* wēstern ; *desert,* HOM. II. 129.

wĕsti, adj., *O.E.* wēstig ; *desert,* HOM. I. 277 ; **wĕstiȝe** (*pl.*) paþes LAȝ. 1120.

westme, *see* **wastum.** **wēt,** *see* **wǣt.**

wete, *see* **wite.** **wēte,** *see* **hwǣte.**

weten, *see* **witen.** **wēten,** *see* **wǣten.**

weter, *see* **water.**

wēðe, adj., *O.E.* wēðe,= *O.L.G.* wōthi, *Goth.* wōþis ; *sweet, pleasant* : a wind so wiþe (*ms.* wyþe) & so col A. P. iii. 454.

weðer, sb., *O.E.* weðer,= *O.N.* veðr, *O.L.G.* wither, *Goth.* wiþrus (*lamb*), *O.H.G.* widar ; *wether,* GEN. & EX. 3998 ; weþer '*vervex*' VOC. 177 ; CH. C. T. *A* 3249 ; MAN. (F.) 11490 ; weþir ['*arietem*'] WICL. GEN. xv. 9 ; wedir '*aries, vervex*' PR. P. 519 ; **weþeres** (*pl.*) ROB. 52 ; weþres PS. lxiv. 14 ; wethris ALEX. (Sk.) 4476.

wéved, *see* **weofed.** **wével,** *see* **wivel.**

wéven, v., *O.E.* wefan (*pret.* wæf),= *O.N.* vefa (*pret.* vaf, vōf, ōf), *M.Du.* weven, *O.H.G.* weban (*pret.* wab), *mod.H.G.* weben (*pret.* wob) ; *weave* ; wevin '*texo*' PR. P. 523 ; **wéve** (*pres.*) LANGL. *B* v. 555 ; weveð REL. I. 219 ; waf (*pret.*) LEG. 74 ; GOW. II. 320 ; weven A. P. i. 71 ; wof ASS. *B* 836 ; woven WICL. IS. lix. 5 ; *comp.* ȝe-**wéven** ; *deriv.* **web, weft, wivel.**

wévere, sb.,= *M.Du.* wever, *O.H.G.* weberi ; *weaver,* WICL. JOB vii. 6 ; **wéveris** (*pl.*) LANGL. *A* PROL. 99 ; weveres *B* PROL. 219.

wével, *see* **wivel.** **wēven,** *see* **wǣven.**

wex, *see* **wax.** **wexen,** *see* **waxen.**

wh-, *see* **hw-.**

wī, interj., *cf. M.H.G.* wī ; *woe,* HOM. I. 165 ; HOM. II. 183.

wī, *see* **hwī, wīȝ.**

wīc, sb., *O.E.* wīc *n.f., cf. O.L.G.* wīc *n., M.L.G.* wīk *n. f., O.Fris.* wīk *f., O.H.G.* wīch ; *from Lat.* vīcus, *mod.Eng.* wick *in local names; dwelling,* '*castellum,*' FRAG. 4 ; ORM. 8512 ; wike ALIS. 4608 ; REL. II. 93 ; **wīke** (*dat.*) H. H. 175 (177) ; C. M. 7917 ; on þere ilke wike LAȝ. 31960 ; **wīke** (*pl.*) O. & N. 604.

wīke-tūn, sb., *O.E.* wīctūn ; *court,* O. & N. 730.

wicche, sb., *O.E.* wicca (VOC. 60) ; *wizard,* JUL. 40 ; wicche TREV. III. 401 ; LANGL. *B* xviii. 69 ; wichche AYENB. 41 ; witche, wiche '*magus*' PR. P. 526 ; **wicches** (*pl.*) GEN. & EX. 3028 ; wiches PR. C. 4214.

wicche[2], sb., *O.E.* wicce (VOC. 74) ; *witch,* FRAG. 2 ; wicche P. L. S. xix. 282 ; SPEC. 38 ; MAND. 159 ; wiche PERC. 826 ; **wichen** (*pl.*) AYENB. 19.

[**wicche-,** *? stem of* **wicchen.**]

wicche-creft, sb., *O.E.* wicce-, wiccræft ; *witchcraft,* HOM. I. 115 ; KATH. 1052 ; A. R.

268; wicchecraft ORM. 7077; wicchecraft WILL. 118; CH. C. T. *D* 1305.

wicchen, v., *O.E.* wiccian, = *M.L.G.* wicken; *bewitch*; wicchin, witchin, wichin '*hariolor, incantor*' PR. P. 527; wicche WILL. 2539; wiche AN. LIT. 11; **wicchand** (*pple.*) PS. lvii. 6; *comp.* **bi-wicchen.**

wicching, sb., *bewitching,* HOM. II. 213; wicchinge P. L. S. xxi. 128.

wiche, sb., *O.E.* wice; *wych elm,* '*ulmus,*' PR. P. 526.

wicke, adj., *from* **wiken**; *wicked; feeble, contemptible*; A. R. 358*; ȝif þin macche is wis & god & tu witlæs & wicke ORM. 6197; wicke and feble was here ȝogt GEN. & EX. 1072; wicke ROB. 208; WILL. 4599; GOW. I. 194; a wicke dede HAV. 688; wicke [wikke] CH. C. T. *A* 1087; wikke REL. I. 222; SHOR. 152; LANGL. *B* v. 229; MAN. (H.) 288; wikke cloþes HAV. 2458; wik lose HAMP. PS. lxvii. 33; wicci SAX. CHR. 264.

wik-hals, sb., *rogue,* MAN. (H.) 267.

wik-hēde, sb., *wickedness,* P. L. S. xvi. 34.

wickenesse, sb., *wickedness,* '*iniqvitas,*' WICL. PROV. iv. 17; wiknesse PS. v. 7.

wicked, adj., *wicked,* AYENB. I; WILL. 3507; wicked, wikked SPEC. 30, 31; wikked MAND. 249; MAN. (H.) 66; S. S. (Wr.) 294; wikkid '*malus*' PR. P. 527; þes wickede wifman LAȝ. 14983*; **wikkeder** (*compar.*) FER. 2187.

wicked-hēde, sb., *wickedness,* AYENB. 43; C. M. 18251.

wikked-līche, adv., *wickedly,* LANGL. *A* v. 143; wickedli WILL. 3860.

wickednesse, sb., *wickedness,* P. S. 220; AYENB. 31; wikkednesse LANGL. *B* v. 290.

wid, *see* **wiþ.**

wīd, adj., *O.E.* wīd, = *O.L.G., O.Fris.* wīd, *O. N.* vīðr, *O.H.G.* wīt; *wide,* GEN. & EX. 565; wiþ & siþ ORM. 9174; wid LEG. 166; PL. CR. 164; CH. C. T. *A* 491; wiid '*latus, amplus*' PR. P. 526; þat wide water LAȝ. 113; þe wide world GOW. III. 242; **widne** (*acc. m.*) A. R. 56; **wīde** (*pl.*) GREG. 769.

wīde, adv., *widely,* GEN. & EX. 672; ich wende from heom wide O. & N. 288; wide hwear *widely spread* SAX. CHR. 249; wide whar ORM. 8943; wide where LANGL. *B* viii. 62; wide GREG. 846; þa scipen foren wide LAȝ. 100; wide and side 29902; þat me miȝte ihure wide BRD. 22; wide and side S. S. (Web.) 1687; AR. & MER. 200; wide whare P. L. S. i. 43; BEV. 2141; PERC. 1481.

widnesse, sb., *width,* MARH. 17; wiidnesse PR. P. 526.

wīden, v., *cf. M.L.G.* wīden, *M.Du.* wijden, *M.H.G.* wīten; *widen*; widin '*dilato*' PR. P. 526; **wīde** (*imper.*) PALL. iii. 923.

widene, adv., = *M.H.G.* wītene, wīten; *afar, widely*: widene cuþ LAȝ. 161; widen iwalken 112; wende i widene LANGL. *A* PROL. 4.

wider, *see* **hwider.** **wideren,** *see* **wederen.**

widewe, sb., *O.E.* widuwe, widwe, wuduwe, wudwe, = *O.L.G.* widowa, *Goth.* widuwō, *O.H. G.* wituwa, witwa, *cf. Lat.* vidua; *widow,* A. R. 10; widewe [widwe] CH. C. T. *A* 253; widwe ORM. 7651; widue PR. P. 526; P. L. S. xiii. 221; wodewe AYENB. 48; **widewen** (*pl.*) LAȝ. 9689; **widewene** (*gen. pl.*) REL. II. 276.

widewe-hād, sb., *widowhood,* H. M. 23; widewehod HOM. II. 45.

widewe-schrūd, sb., *widow's dress,* A. R. 300.

widuer, sb., = *M.Du.* weduwer, *M.H.G.* witewære; *widower,* S. S. (Wr.) 3438; widuare '*virbius*' PR. P. 526; **wideweres** (*pl.*) LANGL. *A* x. 194; widwers *B* xi. 174.

wīe, *see* **wīȝe.** **wīed,** *see* **wēod.**

wīel, *see* **wīȝel.** **wieved,** *see* **weofed.**

wīf, sb., *O.E.* wīf, = *O.N.* vīf, *O.L.G.* wīf, *O.H.G.* wīb; *wife, woman,* ORM. 301; wif and were O. & N. 1522; ðat wif GEN. & EX. 231; wif H. H. 178 (180); wif and man HAV. 1713; þet kveade wif AYENB. 181; wiif WILL. 375; **wifes** (*gen.*) HOM. I. 129; wives C. L. 195; **wīfe** (*dat.*) REL. I. 178; wife, wive O. & N. 1334; wive CH. C. T. *A* 1860; þat wold(e) hir have to wive inome GREG. 524; **wif** (*pl.*) HOM. I. 225; wif, wives LAȝ. 25, 1507; wive SPEC. 89; wifes SAX. CHR. 258; HOM. I. 49; wives HAV. 2; **wīfe,** wiven (*gen. pl.*) LAȝ. 25496, 28477; wifen MAT. xi. 11; wivene A. R. 158; **wīfan** (*dat. pl.*) HOM. I. 111; wifan, wifen MAT. xv. 38; wifen LAȝ. 31923; *comp.* **hūs-, mid-wīf.**

wif-hōd, sb., *O.E.* wīfhād; *wifehood,* CH. C. T. *E* 699; wiifhood PR. P. 526.

wif-kin, sb., *O.E.* wīfcynn; *womankind,* GEN. & EX. 656.

wiif-lēs, adj., *wifeless,* PR. P. 526.

wif-lī, adj., *O.E.* wīflīc; *wifely,* CH. C. T. *E* 429.

wif-man, sb., *O. E.* wīfman, wimman; *woman,* AYENB. 11; wifman, wimman ORM. 291, 2031; wifmon, wimmon LAȝ. 152, 1869; wimman GEN. & EX. 375; wimman HAV. 1168; SHOR. 80; PERC. 1683; wimmon, wimman, wummon O. & N. 1357, 1359, 1413; wummon A. R. 12; MARH. 3; wumman RICH. 3863; wman HAV. 281; **wifmen** (*pl.*) AYENB. 10.

womman-hēde, sb., *womanhood,* CH. C. T. *A* 1748; HOCCL. vi. 30.

wifman-kin, sb., *womankind,* ORM. 3058.

wummon-līch, adj., *womanly,* A. R. 274; wommanlich WICL. 3 KINGS xiv. 24.

wif-ðing, sb., *O.E.* wīfþing; *nuptials,* LAȝ. 31128.

wifle (? **wifle**), sb., *O.E.* wīfel (LEO); ? = ***wiȝ- bil**; *battle-axe,* '*bipennis,*' PR. P. 526; **wifles** (*pl.*) MAN. (F.) 4383.

wig, sb., *O.E.* wicg,=*O.N.* vigg ; *? beast of burden,* HOM. II. 89.

wiʒ, sb., *O.E.* wīg,=*O.N.* vīg, *O.L.G.*, *O.H.G.* wīg ; *battle* ; wi GEN. & EX. 3220 ; **wiʒe** (*dat.*) LAʒ. 4728, 25365.

    wi-eax, -æx, -ax, sb., *battle-axe,* LAʒ. 1567, 2264, 28029.

wiʒe, sb., *O.E.* wīga ; *soldier, man,* GAW. 131 ; A. P. ii. 545 ; wiʒh WILL. 565 ; wie LANGL. *B* xii. 291 ; wie DEGR. 563 ; wee ALEX. (Sk.) 134, 477, 5317 ; **wiʒes** (*pl.*) WILL. 2036 ; wies ALEX. (Sk.) 1030*.

wiʒel, sb., *O.E.* (stēor-)wīgel (*pl.* wīglu) ; *sorcery, deceit,* LAʒ. 19250 ; **wigeles** (*pl.*) REL. I. 131 ; wiʒeles, wieles A. R. 92, 300 ; wiheles MARH. 13 ; wieles FRAG. 8.

    wiʒel-ful, adj., *deceitful,* LAʒ. 31659.

wigelen, v., *cf. M.H.G.* wigelen, *M.Du.* wighelen ; *from* weʒen ; *totter, reel* : wigeleð (*pres.*) ase vordrunken mon A. R. 214.

[wiʒelien, v., *O.E.* wīglian,=*M.Du.* wijchelen ; *practise sorcery* ; *comp.* bi-wiʒelien.]

    wielare, sb., *O.E.* wīgelere, wīglere,=*M.Du.* wijcheler ; *sorcerer, 'augur,'* FRAG. 2 ; A. R. 106.

    wiʒelunge, sb., *O.E.* wīgelung ; *sorcery,* HOM. I. 115 ; wigeling LAʒ. 15791.

[wigge? sb. ; *comp.* ēar-wigge.]

wigge, sb., *cf. M.Du.* wegghe ; *prov.Eng.* wig ; *a sort of cake,* PR. P. 526.

wigge, *see* **wegge.**

wiht, sb., *O.E.* wiht, wuht *f. n., cf. O.L.G.*, *O.H.G.* wiht *f. n. m.* (?), *Goth.* waihts *f., O.N.* vætt, vættr *f., mod.Eng.* whit ; *'wight, creature, thing,* ORM. 1761 ; þet awariede wiht MARH. 7 ; a lute wiht A. R. 72 ; fæire wiht LAʒ. 25869; þat svete wiht SPEC. 45 ; wiht, wiʒt O. & N. 434 ; wiht LANGL. *A* i. 61 ; wiʒt LANGL. *B* i. 63 ; wiʒt WILL. 407 ; A. P. i. 338 ; a litel wiʒt HORN (L.) 503 ; GREG. 601 ; wight MAND. 130 ; M. ARTH. 608 ; everi wight CH. C. T. *A* 842 ; a litel wight MAN. (F.) 8041 ; ʒif þou me lovest ani wight (*at all*) S. S. (Web.) 293 ; of one mihtie wihte MISC. 86 ; **wihte,** wiʒte, wihtes, wiʒtes (*pl.*) O. & N. 87, 204, 431, 598 ; alle qvike wihte P. L. S. viii. 40 ; feole cunne wihte LAʒ. 15775 ; wiʒtes *genii* ROB. 130 ; elves and . . . wightes [wiʒtes] CH. C. T. *A* 3479 ; **wihte** (*gen. pl.*) MARH. 22 ; **wihte** (*dat. pl.*) C. L. 1449 ; *comp.* ā-, un-wiht.

wiht², sb., *O.E.* (ge-)wiht,=*M.Du.* wicht, *M.H.G.* (ge-)wiht, wihte, *O.N.* vētt, vætt ; *from* weʒen ; *weight,* LAʒ. 30835 ; CHR. E. 503 ; C. L. 638 ; wiʒt FL. & BL. 650 ; wiʒt, weiʒte TREV. V. 397 ; wigte GEN. & EX. 439 ; wiʒte E. G. 356; wecht BARB. xvii. 693 ; **wihte** (*dat.*) P. L. S. viii. 191 ; wiʒte AYENB. 44 ; wighte CH. TRO. ii. 1385 ; weihte, weiʒte BOET. i. 2 (8) ; weght(e) PR. C. 7690 ; **weʒtes** [weghtes] (*pl.*) LANGL. *B* xiv. **292 ; wehte**

(*dat. pl.*) ORM. 7812 ; *comp.* ʒe-wiht, aʒēn-wiʒte.

wiht³, adj., ?=*M.L.G.* wicht (*heavy*) ; *? from* weʒen ; *brave, valiant,* LAʒ. 20588 ; (*ms.* wyht) A. D. 252 ; strong and wiht K. T. 996 ; wiʒt WILL. 2877 ; ALIS. 1390 ; TRIST. 90 ; PR. C. 689 ; M. ARTH. 460 ; hardi and wight EGL. 8 ; wicht BARB. x. 531, xvii. 740 ; wicth, with HAV. 344, 1064 ; a wiʒte man LANGL. *B* ix. 21 [wiht mon *A* x. 20] ; þe wiʒte FER. 456 ; wihte (*pl.*) LAʒ. 777 ; **wiʒtere** (*compar.*) WILL. 3441 ; **wihteste** (*superl.*) LAʒ. 31498 ; *comp.* ʒe-, un-wiht.

    wiht-līche, adv., *vigorously, nimbly,* LANGL. *A* vii. 22 ; wihtli *A* vi. 29 ; wiʒtliche FER. 5384 ; wiʒtliche, -li WILL. 65, 92 ; wihtli HAMP. PS. vi. 10*.

Wiht, pr. n., *O.E.* Wiht ; *Isle of Wight,* MISC. 146 ; Wiʒt ROB. 2.

wike, sb., *O.E.* wiocu, wucu (*gen.* wiecan, wucan)=*O.Fris.* wike, *O.N.* vika, *Goth.* wikō, *O.H.G.* wecha ; *week,* A. R. 70 ; wike TREAT. 133 ; MAN. (F.) 9353 ; Iw. 3058 ; E. G. 9 ; i þere wike LAʒ. 13927 ; wike [weke] CH. C. T. *A* 1539 ; weke GOW. III. 116 ; wuke ORM. 4188 ; wouk BARB. xiv. 132 ; woke *'septimana'* PR. P. 532 ; AYENB. 110 ; MAND. 261 ; H. S. 293 ; E. G. 18 ; **wiken,** wike (*pl.*) LAʒ. 3919, 3921 ; wike BRD. 10 ; wuke HOM. II. 3 ; weke AMAD. (R.) xvii ; woken SHOR. 121 ; woke OCTOV. (W.) 612 ; **wukes** GEN. & EX. 2473 ; wokes ISUM. 153 ; wikene [wekene] (*gen. pl.*) LAʒ. 22931 ; **wiken** [wyke] (*dat. pl.*) LAʒ. 24815 ; wuken HOM. II. 3.

    wuke-mālum, adv., *by weeks,* ORM. 536.

wike, sb., *O.E.* wīc,=*O.N.* vīk, *M.L.G.* wīk (wyk), *M.Du.* wijk *f. ; from* wīken ; *corner, angle* : wike of þe eghe CATH. 417 ; **wīkez** (*pl.*) GAW. 1572.

wike², sb., *O.E.* wīce ; *office,* HOM. II. 91 ; JUL. 24 ; wike LAʒ. 29752 ; ALIS. 4608 ; **wiken** (*pl.*) HOM. I. 137 ; HOM. II. 183 ; mine wike beoþ wel gode O. & N. 605 ; wike SHOR. 25.

wīke, *see* **wīc.**

wiken, sb., *office, charge* ; (*ms.* wikenn) ORM. 7208.

wīken², v., *O.E.* wīcan,=*O.L.G.* wīcan, *O.N.* vīkja (*pret.* veik), *O.H.G.* wīchan, *Gr.* εἴκειν ; *go, fall* ; wīke (*ms.* wyke) (*pres.*) SPEC. 87 ; *deriv.* wīke, wāc.

wikir, sb.,=*Swed.* vikker, vekker ; *wicker, 'vimen,'* PR. P. 527.

wiket, sb., *O.Fr.* gvichet ; *wicket,* HORN (L.) 1074 ; LANGL. *B* v. 611 ; A. P. ii. 857 ; PERC. 490.

wīkien, v., *O.E.* wīcian, *dwell,* LAʒ. 18102.

    wīkinge, sb., *domicile* ; (*ms.* wickinge) LAʒ. 30453.

wikke, *see* **wicke.**

wīknere, sb., *O.E.* wīcnere; *officer*; **wīc-neres** (*pl.*) ANGL. VII. 220; wikenares LAȝ. 18175; **wikeneren** (*dat. pl.*) LAȝ. 6704.

wīkeninge, sb., *domicile*, LAȝ. 30453*.

wil, adj., *O.E.* (druncen-)will (ALFR. P. C. 120); *voluntary, desirable, ?agreeable*: ȝif ou is wilre KATH. 571; *comp.* **ān-, drunc-, frete-, hērc-, ȝe-, spāt-, wrēch-wil.**

wil-cume, adj., *O.E.* wilcuma,=*O.H.G.* willicomo; *welcome*: wilkume schaltu beon me A. R. 394; wil-, wulcume [wil-, welcome] LAȝ. 4901, 8528; welcume HOM. I. 259; O. & N. 1600; GEN. & EX. 1830; welcume H. H. 149; LANGL. *A* ii. 208; wolcome BEK. 1265; PR. P. 532.

wil-cumen, v., *O.E.* wil-, wylcumian; *welcome*: to wulcumen Mærlin LAȝ. 17098; **wolcumeþ** [welcomeþ] (*pres.*) O. & N. 440; welcumieð MAT. v. 47; **wilcumede** [wilcomede] (*pret.*) LAȝ. 10957; welcumede GEN. & EX. 1396; welcomede BRD. 3.

wil-dæi, sb., *O.E.* wildæg; *day wished for*; **wildaȝes** (*pl.*) LAȝ. 1798.

wil-gomen, sb., *pleasant sport*, LAȝ. 20944.

wil-ȝeove, adj., *?voluntarily given*, MARH. 16; A. R. 368.

Wil-helm, pr. n., *? O.E.* Wilhelm, Willelm (SAX. CHR. 211), *O.H.G.* Wilihelm; *William*; Willam A. R. 340; Williem BEK. 511; William WILL. 70.

wil-līche, adv., *O.E.* willīce; *voluntarily*, LEG. 8.

wil-schrift, sb., *voluntary confession*, A. R. 340.

wil-spel, sb.,=*O.L.G.* wilspel; *good news*, LAȝ. 1350.

wil-sum, adj., *O.E.* wilsum; *agreeable*, WILL. 5394.

wil-tīdende, sb., *joyful news*, LAȝ. 17090.

wil², sb., *O.E.* (ge-)will,=*O.N.* vil *n.; will, pleasure;* HOM. I. 61; A. R. 12; GEN. & EX. 194; lives wil HOM. I. 193; ȝef þi wil is MARH. 7; mi wil & mi weol(e) KATH. 2139; idel wil REL. I. 132; wil LAȝ. 2793; (*ms.* wyl) SPEC. 37; þat wil SHOR. 16; **willes** *willingly* A. R. 6; willes & waldes H. M. 27; *comp.* **ȝe-, self-, un-wil.**

wil-cwēme, *?adj., well pleased, content*: ich am wilcweme JUL. 32.

wil-ful, adj., *wilful, willing, desirous*, R. S. vi (MISC. 184); wilful BEK. 1309; wilfull BARB. ii. 354, xi. 266.

wilful-līche, adv., *wilfully; with good will*: wilfulliche þolien HOM. I. 279; wilfulliche LANGL. *C* xxii. 373; wilvolliche AYENB. 140; wilfulli '*voluntarie*' PR. P. 528; BARB. ii. 172, 386, iii. 404.

wilfulnesse, sb., *wilfulness*, '*pertinacia*,' HOM. II. 73; CH. C. T. *A* 3057.

wilful-shipe, sb., *wilfulness*, HOM. II. 205.

wil, *see* **wille, wilde.**

wilde, adj., *O.E.* wilde, *O.N.* villr,=*O.Fris.* wilde, *O.H.G.* wildi, *Goth.* wilþeis; *wild, savage, uncultivated ; self-willed ; bewildered, having lost one's way*; ORM. 6191; wilde flod O. & N. 946; wilde LAȝ. 785; SHOR. 121; SPEC. 48; wilde fur ROB. 410; (h)is wilde lore A. D. 246; wilde fir RICH. 5229; wilde of red(e) FLOR. 35; wild HORN (L.) 252; wilde and wod Iw. 1650; will of rede 379; wil and weri GEN. & EX. 975; BARB. i. 348, xiii. 477; will ALEX. (Sk.) 1272; wil HAV. 863; wille REL. I. 209; wilde der GEN. & EX. 169; in a wode þai were gon wille ISUM. 159; in wilde studes ROB. 130; mid wilde deoren MK. i. 13; **wildore** (*compar.*) A. D. 253; **wildest** (*superl.*) SPEC. 48.

wil-drēam, sb., *delusive dream*; **wil-drēmes** (*pl.*) A. P. iii. 473.

wil-gate, sb., *going astray*, PR. P. 527.

wildnesse, sb., *wildness*, PR. P. 528.

wild-scipe, sb., *wildness*, LAȝ. 20845.

wild-som, adj., *O.N.* villusamr; *wild, unpleasant; wandering, bewildered*: þe wildsome wai TOR. 508; moni wilsum wai he rod GAW. 689; wilsom '*dubius*' PR. P. 528; TOWNL. 268; ALEX. (Sk.) 4076, 5565.

wild-hēdid, adj., *wild-headed*, ALEX. (Sk.) 12.

wilde², sb., *?=M.H.G.* wilt *n.; wild deer*, GAW. 1150; D. TROY 2347; **wilde** (*dat.*) LAȝ. 1129.

wilde, wilden, *see* **wélde, wélden.**

wilderne, sb., *wild place, wilderness*, REL. I. 130 (HOM. II. 165); A. R. 160* (*v. r.* wildernesse); þar is wilderne muchel LAȝ. 1238; to þan wilderne 523.

wildernesse, sb., *cf. M.Du.* wildernisse; *wilderness*, A. R. 158; wildernesse LAȝ. 30335; wildernisse [wildernesse] O. & N. 1000; wildernisse P. L. S. xv. 131.

wīle, sb., *O.E.* wīl, *cf.* **wīȝel**; *wile, trick, artifice*, '*astutia*,' PR. P. 528; CH. C. T. *A* 3403; M. H. 2; **wiles** (*pl.*) SAX. CHR. 257; KATH. 893; ORM. 10317; wiles PR. C. 1360; wilis S. S. (Wr.) 2731.

wīle², sb., *? O.E.* *wīgela*; *see* **wīȝelien;** *sorcerer*: nes i never wicche ne wile (*ms.* wyle) SPEC. 38.

wīle, *see* **hwīle.**

wilȝe, sb., *O.E.* wilig, welig,=*M.L.G.* wilge, *M.Du.* wilghe, wilighe; *willow*; wilghe HALLIW. 931; wilwe CH. C. T. *A* 2922; wilwe, wilowe '*salix*' PR. P. 528; weloghe VOC. 228.

wīli, adj., *wily, artful,* '*astutus*,' PR. P. 528; WILL. 2764; C. M. 11807; CH. C. T. *B* 3130; MAN. (F.) 9849.

wilie, sb., *O.E.* wilige ; *basket* ; **wilien** (*pl.*) MK. vi. 43.

wilk, *see* welk.

wille, sb., *O.E.* willa, = *O.H.G.* willo, *O.L.G.* willeo, *Goth.* wilja, *O.N.* vili *m.; will, longing, pleasure,* MAT. xviii. 4 ; MARH. I ; A. R. 60 ; ORM. 258 ; wille HAV. 528 ; LANGL. *B* PROL. 200 ; MAN. (H.) 92 ; PR. C. 2394 ; ʒif hit þi wille weore LAʒ. 29058 ; so gret wille him com to BEK. 121 ; þis wiide wille went a wai SPEC. 23 ; his laste wille TREV. IV. 11 ; gret wille (*desire*) heo hadde LEG. 6 ; mid englene wille HOM. I. 193 ; þe wrenne . . . fale manne song a wille O. & N. 1722 ; þou shalt have . . . weder after wille A. D. 298 ; **willen** (*dat.*) HOM. I. 89 ; after heore willen LAʒ. 31650 ; wind (stod) an willen [at wille] 1102 ; weder heom stod on wille 7845 ; **willen** (*acc.*) MAT. vii. 21 ; willen LAʒ. 1270 ; REL. I. 183 ; *comp.* ʒe-, self-, un-wille.

wille [2], adj., *O.E.* (ān-)wille, = *O.H.G.* (ein-)-willi ; *desirable, agreeable* : wille oðer qveme HOM. II. 213 ; ʒif hit þe weore wille an heorte LAʒ. 20816 ; *comp.* ān-, un-wille.

wille-līche, adv., *voluntarily,* HOM. I. 41 ; A. R. 396.

wille, *see* welle.

willed, adj., *willed* ; averouse and evel willed LANGL. *C* ii. 189.

willen, v., *O.E.* *willan (*pple.* willende), = *O.H.G.* wellan, *O.L.G.* willian, wellian, *Goth.* wiljan, *O.N.* vilja, *Lat.* velle, (*see* ZEIT. XIX. 158) ; *will* ; (*ms.* wīlenn) ORM. 5297 ; wile [*O.E.* wile, wyle, wille, wylle (*volo, vult, velim, velit*), = *O.L.G.* willeo (*volo*), wili (*vult*), willie (*velim, velit*), *O.H.G.* willu, wille, wile (*volo*), wili, wile (*vult*), wilja (*velim*), *Goth.* wiljau (*volo*), wili (*vult*)] (1, 3, *pres. ind. & subj.*) ; *will* HOM. II. 11 ; ORM. 105 ; SEN. & EX. 277 ; wile HAV. 3 ; AYENB. 57, 101 ; wille REL. I. 208 ; SHOR. 59 ; MAN. (H.) 105 ; he wile KATH. 491 ; ich ulle 485 ; wille, wulle HOM. I. 13 ; wile [wile], wille, wulle O. & N. 214, 262, 553, 903 ; wille, wulle [wolle, wole] LAʒ. 697, 1387 ; wule P. L. S. viii. 79 ; MARH. 3 ; A. R. 60 ; ich ulle C. L. 38 ; wole BEK. 39 ; wolle, wole H. H. 78, 138 ; wol WILL. 281 ; GOW. I. 107 ; LANGL. *A* PROL. 38 ; CH. C. T. *A* 42 ; **wilt** *wilt* (2 *pres. ind.*) KATH. 1359 ; wilt, wult A. R. 90, 98 ; O. & N. 165, 1409 ; wult MARH. 4 ; wult, wlt [wolt] LAʒ. 694, 3189 ; **wille** *will* (2 *pres. subj.*) O. & N. 1289 ; wille SHOR. 103 ; wuŀle HOM. I. 57 ; beute ʒif þu wulle [wolle] icnawen beo LAʒ. 26433 ; **wille,** wile [wile], wule *will* O. & N. 185, 188, 1360 ; wulle H. M. 31 ; **willeð** [*O.E.* willað, wyllað *will*] (*pres. ind. pl.*) KATH. 692 ; willeþ AYENB. 16 ; willeþ, wulleþ MISC. 59, 61 ; wulleð A. R. 34, 48 ; O. & N. 1257 ; wulleð, wlleð [wolleþ] LAʒ. 481, 3753 ; wolleþ BEK. 403 ; **wilen** [*O.E.* willen, wyllen] *will* (*pres. subj. pl.*) GEN. & EX. 2304 ; on hwilche

halve h(e)o wilen falle (*cadent*) HOM. I. 153 ; willen (*for* willeþ) PROCL. 6 ; **wolde** [*O.E.* wolde, = *O.H.G.* wolte, *O.L.G.* welda, *O.N.*, *Goth.* wilda] (*pret.*) *would* A. R. 8 ; ORM. 150 ; wolde (*ms.* uuolde) SAX. CHR. 260 ; wolde HAV. 367 ; BEK. 17 ; AYENB. 16 ; CH. C. T. *A* 255 ; wolde, walde LAʒ. 358, 18911 ; wolde, wulde GEN. & EX. 214, 912 ; wulde REL. I. 219 ; walde HOM. I. 5 ; MARH. 2 ; walde PR. C. 4395 ; (þu) woldest LAʒ. 6230 ; (heo) wolden [wolde] 362 ; þat heo him uvel walden [wolde] 7686 ; wolde [walde] cumen O. & N. 1678 ; **wold** [= *Ger.* gewollt] (*pple.*) LANGL. *B* xv. 258.

willi, adj., = *O.H.G.* willīg ; *willing, ready ; voluntary, 'voluntarius,'* CATH. 418 ; C. M. 26351 ; þat þou be willi in þi witt ALEX. (Sk.) 2689.

willien, v., *O.E.* willian (*pret.* willode) ; *will, long for* : willen hire to wif FL. & BL. 587 ; wille MIRC 1074 ; willest LANGL. *B* xii. 221 ; willieþ AYENB. 142 ; **willede** (*pret.*) ROB. 12.

wilnien, wilniæn, v., *O.E.* wilnian, = *O.N.* vilna ; *desire,* LAʒ. 5955, 29877 ; wilnen A. R. 60 ; to wilnen & to ʒeornen ORM. 12152 ; wilni AYENB. WILL. 3563 ; CH. C. T. *A* 2114 ; **wilni** (*pres.*) MARH. 8 ; wilnest MAND. 295 ; wilneþ C. L. 385 ; LANGL. *B* iii. 110 ; we wilniað LAʒ. 1073 ; wilnieþ BRD. 23 ; **wilnede** (*pret.*) BEK. 201 ; he . . . wilnede þeos mæidenes LAʒ. 3202 ; heo wilneden after w(e)orre 2626 ; *comp.* ʒe-wilnien.

wilnunge, sb., *O.E.* wilnung ; *desire,* A. R. 148 ; wilninge LAʒ. 3160* ; AYENB. 22 ; willing CH. C. T. *E* 319.

Wiltūn-scīre, pr. n., *O.E.* Wiltūnscire ; *Wiltshire,* LAʒ. 21017 ; Wiltoneschire MISC. 146 ; Wilteschire P. L. S. xiii. 47.

wilwe, *see* wilʒe.

wimbil, sb., *cf.* Dan. vimmel, *M.L.G.* wiemel, wemel ; *wimble, boring tool, 'terebrum,'* VOC. 234 ; PR. P. 528 ; N. P. 13 ; wimble PALL. xi. 85.

wimman, *see* wīfman *under* wīf.

wimpel, sb., *cf. M.Du.* wimpel, *O.N.* vimpill ; *wimple,* A. R. 420 ; wimpel ROB. 338 ; wimpel [wimpil, wimpul] CH. C. T. *A* 151 ; wimpil '*peplum*' PR. P. 528.

wimpel-lēas, adj., *without wimple,* A. R. 420.

wimplin, v., *M.Du.* wimpelen ; *cover with a wimple,* A. R. 420* ; **wimpleþ** (*pres.*) CH. BOET. ii. 1 (3⅟) ; *comp.* bi-wimplen.

wimplinge, sb., *covering with a wimple* ; (*ms.* wimlunge) A. R. 420*.

win, sb., *O.E.* winn, = *M.L.G.* win *n.; from* **winnen** ; *labour ; contention ; acquisition* ; REL. I. 128 ; GEN. & EX. 598 ; (*ms.* winn) ORM. 6118 ; win LAʒ. 404 ; **winne** (*dat.*) HOM. I. 175 ; O. & N. 670 ; and ʒelpen of þan winne LAʒ. 12072* ; *comp.* ʒe-win.

wīn, sb., *O.E.* wīn, = *O.H.G.* wīn, *O.N.* vīn,

Goth. wein, _Lat._ vīnum, _Gr._ οἶνος ; _wine_, A. R.
114 ; win LAȝ. 8723 ; HAV. 1729 ; CH. C. T.
_A_ 637 ; þet guode win AYENB. 167 ; **wīnes**
(_gen._) ORM. 11118 ; **wīne** (_dat._) LAȝ. 14299 ;
wīnes (_pl._) A. R. 376.

wine-ballis, sb., pl., _balls of tartar_, PR. P.
529.

wīn-berie, sb., _O.E._ wīnberige ; _grape_, A. R.
276 ; **wīnberian** (_pl._) LEECHD. III. 114.

wīn-draf, sb., _lees of wine_, PALL. iii. 162.

wīn-drunken, adj., _O.E._ wīndruncen ; _drunk
with wine_, LAȝ. 8126 ; windrunke REL. I.
178 ; windronken TREV. III. 207.

wīn-ȝeard, sb., _O.E._ wīngeard ; _vineyard_,
A. R. 294 ; winyard MISC. 34 ; winiærd SAX.
CHR. 263 ; winȝord SPEC. 41.

wīn-scench, sb., _cf. M.H.G._ wīnschanc ;
draught of wine, LAȝ. 6932.

wīn-trēow, sb., _O.E._ wīntrēow ; _vine_, '_vi-
tis_,' FRAG. 3 ; wintre GEN. & EX. 2059 ; wintre
C. M. 4465.

wīn-tunne, sb., _wine-tun_, LAȝ. 30677.

wincen, see **wenchen**.

winche, sb., _O.E._ wince ; _winch, windlass_,
PALL. i. 426.

Winchecumbe, pr. n., _Winchcomb_, P. L. S.
xiii. 7.

winchen, see **wenchen**.

wínd, sb., _O.E._ wind, =_O.L.G._ wind, _O.H.G._
wint, _O.N._ vindr, _Goth._ winds, _Lat._ ventus ;
wind, LAȝ. 19347 ; TREAT. 138 ; þene wind
A. R. 282 ; **wíndes** (_gen._) LAȝ. 28239 ; **wínde**
(_dat._) A. R. 282 ; mid þan winde LAȝ. 236 ;
wíndes (_pl._) MARH. 10 ; **wíndon** (_dat. pl._)
LK. viii. 25 ; winden MK. xiii. 27 ; _comp._
ēst-, **west-wínd**.

wínd-milne, sb., _windmill_, REL. I. 7 ; wind-
mulle ROB. 547.

wind-ōge, sb., _O.N._ vindauga ; _window_,
GEN. & EX. 602 ; windohe A. R. 50* ; windowe
'_fenestra_' PR. P. 529 ; LANGL. _B_ iii. 49 [win-
dow _A_ iii. 48] ; MAND. 216 ; wondowe FER.
1361.

[**wínde**, sb., _cf. O.N._ vinda (_hank_), _M.Du._,
_M.H.G._ winde (_winding instrument, windlass; twining plant, convolvulus_) ; _comp._
wiþ-wínde.]

wind-ǎs, sb., _O.N._ vindāss, = _M.Du._ wind-
asse ; _windlass_, '_trochlea_,' PR. P. 529 ; RICH.
71 ; CH. C. T. _F_ 184 ; MAN. (F.) 12087 ; P. P.
iii. 103.

wínd-bēm, sb., '_lacunar_,' PR. P. 529.

wínde-clūt, sb., _winding cloth_, ORM. 3320.

[**windel**, sb., = _M.Du._ windel ; _winder ; comp._
ȝarn-windel.]

windel[2], sb., _fan, winnowing basket_ ; (_ms._
winndell) ORM. 10550 ; windil VOC. 201 ; PR.
P. 529.

winden, v., _O.E._ windan, = _O.L.G._ windan,

Goth. (bi-, us-)windan, _O.N._ vinda, _O.H.G._
wintan ; _wind, turn, twist_, GEN. & EX. 2448 ;
stanes heo letten . . . winden LAȝ. 27461 ;
over sæ winden (_go_) 20818 ; winde HAV.
221 ; ne shalt þou nevere henne winde H. H.
146 ; **wínde** (_pres._) CH. C. T. _D_ 1102 ; win-
deð, wint A. R. 296, 314 ; **wínd** (_imper._)
SHOR. 34 ; **wand** (_pret._) ORM. 3320 ; to þe
grounde he wand TRIST. 2349 ; wond MISC.
44, 51 ; P. L. S. xxiv. 126 ; GOW. II. 359 ; and
smat an Arðures sceld þat he wond a þene
feld LAȝ. 23964 ; into lef reste his sowle wond
GEN. & EX. 4136 ; Oliver . . . til him wond
FER. 892 ; wunden into widen (_r._ wide) sæ
LAȝ. 25541 ; and wonden [wounden _A_ ii. 196]
him in cloutes LANGL. _B_ ii. 220 ; wounden
EGL. 859 ; aboute þe bodi a rop þei wonde H.
S. 8056 ; in þe sadel wounde AR. & MER.
9162 ; **wúnden** (_pple._) ORM. 3326 ; wnden
HAV. 546 ; wonden TOWNL. 232 ; _comp._ **at-**,
bi-, **ȝe-**, **un-wínden** ; _deriv._ **wínde**, **wen-
den**.

wíndinge, sb., _winding_, '_obvolucio, tortura_,'
PR. P. 529.

wínden, v., = _M.Du._ winden, _O.H.G._ wintōn ;
winnow, A. R. 270 ; windin '_ventilo_' PR. F.
529 ; _cf._ **windwen**.

wíndi, adj., _O.E._ windig ; _windy_, LIDG. M. P.
2.

windi, see **wendi**.

Windles-ōver, pr. n., _O.E._ Windles ōfer ;
Windsor ; **Windlesōvre** (_dat._) SAX. CHR.
255 ; Windelsore ROB. 341.

windwen, v., _O.E._ windwian ; _winnow_, ORM.
10483 ; windwin HOM. I. 85 ; windwe MAND.
107 ; windewe, winewe WICL. JER. xlix. 36 ;
winwin '_ventilo_' PR. P. 530 ; windweð (_pres._)
JUL. 79 ; windwede (_pret._) A. R. 270.

windwere, sb., _winnower_ ; **windeweris**
[wineweres] (_pl._) WICL. JER. li. 2.

windwunge, sb., _winnowing_, A. R. 270.

wine, sb., _O.E._ wine, = _O.L.G._, _O.H.G._ wini,
_O.N._ vinr, vin ; _friend_, LAȝ. 2289 ; wine REL.
I. 217 ; **wines** (_pl._) P. L. S. viii. 111 ; _comp._
un-wine.

wine-mǣi, sb., _O.E._ wine(-mæg) ; _friend_,
LAȝ. 5831.

winge, sb., _wing_, LANGL. _B_ xii. 263 ; S. S.
(Wr.) 2196 ; **wingen** (_pl._) BRD. 9 ; AYENB.
217 ; ALIS. 485 ; hwingen A. R. 130 ; hwingen
LAȝ. 29263 ; winges CH. C. T. _A_ 1964 ; _see_
weng.

wink, sb., _sleep, nap_, LANGL. _A_ v. 3, 212.

winken, v., = _M.H.G._ winken (_pret._ wanc) ;
wink : hwon þe heorte . . . foð on ase to
winken A. R. 288 ; **wank** (_pret._) M. T. 79.

winkin, v., _O.E._ wincian, = _M.H.G._ winken
(_pret._ wincte) ; _wink_, '_nicto_,' PR. P. 530 ;
winke PR. C. 4970 ; a **winking** (_pple._) wrath
LANGL. _B_ xi. 4 ; wisdome winked (_pret._)
uppon mede LANGL. _B_ iv. 154.

winkung, sb., *winking* : lokinge wiðuten winkunge HOM. I. 145 ; winkinge *slumber* LANGL. *B* v. 3, *C* xii. 167.

[**winne**, sb., *O.E.* winna, = *O.H.G.* winno ; *fighter ; comp.* ʒe-, wiðer-winne.]

[**winne** ? adj. ; *comp.* arveð-, ēð-winne.]

winne, see wünne.

winnen, v., *O.E.* winnan, = *O.L.G.*, *O.H.G.* winnan, *O.N.* vinna, *Goth.* winnan (πάσχειν) ; *mod.Eng.* win ; *strive, contend ; acquire, win* ; MISC. 195 ; to winnen heofnes kinedom ORM. 801 ; wunnen aʒean A. R. 238 ; winne TREAT. 138 ; AYENB. 17 ; A. P. i. 578 ; winneð (*pres.*) HOM. II. 51 ; ðanne sumer and winter winnen REL. I. 220 ; **wan** (*pret.*) HOM. II. 187 ; wan GOW. III. 137 ; MAND. 228 ; he wan ... to William WILL. 2498 ; wan LANGL. *B* vi. 98 [won *A* vii. 89] ; wan to wax *attained his full growth* ALEX. (Sk.) 3986 ; wan awai *got away* BARB. xvi. 555 ; wan our *got over* ix. 405 ; wunnen ORM. 10175 ; wonnen LANGL. *B* PROL. 22 ; **wonne** (*subj.*) GREG. 238 ; **wunnen** (*pple.*) ORM. 6112 ; wunne RICH. 1348 ; wonnen MAND. 46 ; CH. C. T. *A* 877 ; *comp.* a-, bi-, ʒe-, over-winnen.

winner, sb., *winner* ; winners (*pl.*) LANGL. *C* i. 222.

winstere, sb., *tradeswoman*, LANGL. *A* v. 129*.

winninge, sb., *winning, 'lucrum,'* PR. P. 530 ; (*ms.* wynnygge) AYENB. 23 ; winning WICL. I TIM. iii. 8.

winter, sb., *O.E.* winter *m. n., cf. O.L.G., O. H.G.* winter, *Goth.* wintrus, *O.N.* vitr, vetr *m.; winter*, A. R. 20 ; þat winter LAʒ. 9887 ; þene winter 22246 ; **wintres** (*gen.*) LAʒ. 2861 ; winteres O. & N. 458 ; **wintre** (*dat.*) MK. xiii. 18 ; wintre LANGL. *B* xvii. 226 ; wintre, wintere O. & N. 533 ; **winter** (*pl.*) ORM. 15594 ; GEN. & EX. 567 ; winter LAʒ. 194 ; HAV. 417 ; H. H. 45 ; CH. C. T. *D* 1651 ; **wintre** (*gen. pl.*) P. L. S. viii. 105 ; REL. I. 130 ; twenti wintre LAʒ. 9028 ; vele wintre MISC. 41 ; **wintren** (*dat. pl.*) LAʒ. 9695 ; wintre HOM. I. 159 ; HOM. II. 220 ; ANGL. I. 6 ; *comp.* mid-winter.

winter-stal, sb., *winter resort*, MISC. 148.

winter-tīd, sb., = *M.H.G.* winterzīt ; *winter time*, MAN. (H.) 240.

winter-tīme, sb., *winter time*, LANGL. *C* xiii. 189.

winter-wele, sb., *winter pleasure*, SPEC. 43.

winter-wō, sb., *winter woe*, SPEC. 43.

wintred, pple., *O.E.* gewintrad ; *advanced in years, aged*, ORM. 453.

winwen, see windwen.

wīpe, sb., *Swed.* vipa ; *? lapwing, plover, 'upupa,'* VOC. 188 ; PR. P. 530 ; *'van(n)el(l)e'* VOC. 165.

wīpin, v., *O.E.* wīpian ; *wipe, 'tergo,'* PR. P. 530 ; wipe MAN. (F.) 11187 ; PR. C. 7977 ; Horn gan his swerd gripe & on his arme wipe HORN (L.) 606 ; wipi AYENB. 161 ; **wipeð** (*pres.*) A. R. 230 ; wipeþ (*ms.* wypeþ) P. S. 214 (A. D. 127) ; **wīpede** (*pret.*) HOM. II. 145 ; wipede P. L. S. xxiii. 125 ; wiped LANGL. *B* ii. 220 ; MAND. 97 ; heo wipeden hors leove mid dinnene claðe LAʒ. 22289.

wippen, v., *cf. L.G., Du.* wippen, *Dan.* vippe, *Swed.* vippa ; *tremble ; flap* : þu schalt wippen [hwippen] on a sprenge O. & N. 1066 ; **wipping** (*pple.*) MAN. (F.) 8197.

wips, see wisp.

wīr, sb., *O.E.* wīr, = *O.N.* vīr, *M.L.G.* wīre ; *wire, 'filum aereum vel ferreum,'* PR. P. 530 ; Iw. 2967 ; **wīre** (*dat.*) TREV. I. 355 ; LANGL. *B* ii. 11 ; *comp.* gōld-wīr.

wirchen, see wurchen. **wird**, see weored.

wirde, see wurde.

Wire-chestre, pr. n., *O.E.* Wigera ceaster ; *Worcester*, PROCL. 8 ; Wirecestre MISC. 146 ; Wircestre ROB. 2.

Wirhale, pr. n., *Wirral*, SPEC. 26 ; B. DISC. 1014.

wirien, see wurʒen. **wirken**, see wurchen.

wirling [wirling], sb., *Scotch* wirl, wurl ; *dwarf*, ALEX. (Sk.) 1706 ; wirlinges (*pl.*) 1733.

wirm, see wurm. **wirste**, see wriste.

wirstill, see wræstlen. **wirwen**, see wurʒen.

wis, adj. & adv., *O.E.* wiss(-līce), = *O.L.G.*, *O.Fris.* wis, *Goth.* (un-)wis, *O N.* viss ; *from* witen ; *certain, certainly*, HOM. I. 187 ; A. R. 38 ; (*ms.* wiss) ORM. 19 ; al so wis so he god is R. S. vii ; wis MISC. 88 ; AM. & AMIL. 1292 ; to wisse HORN (L.) 121 ; SHOR. I ; WILL. 3397 ; þat wite þu to wisse KATH. 1543 ; mid wisse HOM. II. 25 ; *comp.* ʒe-wis.

wis-līche, adv., *O.E.* wisslīce ; *certainly*, REL. I. 130 ; wislike, -liʒ ORM. 928, 10330 ; wislike HAV. 274 ; wisli CH. C. T. *E* 2175 ; i wot wisli WILL. 2947.

wīs, adj., *O.E.* wīs, = *O.L.G.* wīs, *O.N.* vīss, *O. H.G.* wīs, *Goth.* (un-)weis ; *wise*, A. R. 64 ; O. & N. 192 ; GEN. & EX. 100 ; ben wis *be wise* ORM. 2279 ; wurþen wis 11609 ; wis BEK. 166 ; SPEC. 103 ; CH. C. T. *A* 68 ; wis and war LAʒ. 26000 ; ALIS. 2129 ; wis mon REL. I. 183 ; of alle þewes was she wis HAV. 282 ; þe wise O. & N. 176 ; Merlin þe wise LAʒ. 32178 ; þe wise mon REL. I. 183 ; þam **wisen** (*dat.*) were MAT. vii. 24 ; **wīsne** (*acc. m.*) LAʒ. 8758 ; **wīsre**, wisure (*compar.*) A. R. 198, 338 ; wisure, wisere O. & N. 1250, 1330 ; **wīseste** (*superl.*) LAʒ. 2107 ; *comp.* ʒe-, sceād-, laʒe-, rǣd-, riht-, steorre-, tále-, un-, weder-, wrang-wīs.

wīs-dōm, sb., *O.E.* wīsdōm, = *O.L.G.* wīsdōm, *O.H.G.* wīstuom ; *wisdom*, KATH. 487 ; A. R. 26 ; O. & N. 772 ; þene wisdom HOM. I.

123; wisdom LAȝ. 26240; (ms. wissdom)
ORM. 15986.

wīs-hēde, sb.,=*O.H.G.* wīsheit; *wisdom*,
AYENB. 68.

wīs-līche, adv., *O.E.* wīslīce; *wisely*, A. R.
104; KATH. 82; wisliche AYENB. 94; wislike
ORM. 6113; GEN. & EX. 1091; **wīslūker**
(*compar.*) A. R. 234; wisloker P. S. 194.

wīsnesse, sb., *O.E.* wīsness; *wisdom*,
C. L. 292.

wisard, sb., *O.Fr.* gvisart; *wizard*, PR. P.
530.

wisch, wischen, *see* **wüsch, wüschen.**

wīse, sb., *O.E.* wīse; *stalk*, '*fragus*,' PR. P.
531; HALLIW. 934; stræberie wise FRAG. 3.

wīse², sb., *O.E.* wīse, wīs,=*O.L.G.*, *O.H.G.*
wīsa; *wise, manner*: on one wise A. R. 6;
heore wode wise O. & N. 1029; on alle wise
ORM. 2626; on fendes wise GEN. & EX. 2961;
wise REL. I. 176; PROCL. 6; on none wise
C. L. 573; on heore londes wise LAȝ. 25426;
on þis wise PR. C. 3622; in none wise BEK.
1212; in alle wise (*ms.* wyse) A. D. 257; in
þis wise WILL. 485; CH. C. T. *A* 1446; oþere
wise *in other manner* SHOR. 42; oþer wise
WILL. 396; on oþre wisen FRAG. I; **wīsen**
(*dat. pl.*) HOM. I. 109; A. R. 318; in vele
wisen AYENB. 62; on monie wise FRAG. I;
on two wise HOM. II. 9; þu forleost al þine
wise (? *melody*) O. & N. 519.

wisenen, v., *O.E.* wisnian, *cf. O.N.* visna, *O.H.
G.* wesnēn, wesenēn; *wizen, dry up*; **wise-
ned** (*pret.*) HALLIW. 934.

wīsien, wissien, v., *O.E.* wīsian, wissian, *cf.
O.L.G.* wīsean, *O.N.* vīsa, *O.H.G.* wīsan, wīssan;
show, guide; point out; wissien HOM. I. 13;
wissien [wissi]LAȝ. 5280; wissen ORM. 10823;
wisen C. L. 297; wissen TREAT. 133; i wol
wissen [wisse] ȝou þe wei LANGL. *A* v. 562;
wissin '*dirigo*' PR. P. 530; wisi O. & N. 915;
wise ROB. 524; CHR. E. 499; AUD. 49; wise
[wisse] CH. C. T. *D* 1008; wisse HAV. 104;
WILL. 2110; PR. C. 9304; M. H. 51; wisse and
rede GREG. 975; þu **wīsest** (*pres.*) [visest] O.
& N. 973; wiseþ FRAG. 8; wise me LAȝ. 1200;
wisse me to þi deore sone SPEC. 93; Crist þe
wisse (*direct*) HORN (L.) 1457; **wisede** (*pret.*)
LEG. 98; wissede LAȝ. 1365*; *comp.* ȝe-
wīsien.

wīsegend, sb., *O.E.* wīsiend; *director*, HOM.
I. 115.

wissunge, sb., *cf. O.E.* wīsung; *instruction,
direction*, HOM. I. 73; KATH. 190; wissing
ORM. 11830; wissinge SHOR. 96; *comp.* ȝe-
wissunge.

wisk, sb., *cf. Sw.* viska; *whisk, swift stroke*,
BARB. v. 641.

wisp, sb., *wisp*, '*torqves*,' PR. P. 530; TOWNL.
18; a wisp [wips] of hei LEG. 91; **wispe**
[wips(e)] (*dat.*) LANGL. *A* v. 195; *comp.*
ars-wisp.

wissien, *see* **wīsien.**

[**wiste**, sb., *O.E.* wist, = *O.N.* vist, *O.H.G.* wist,
Goth. wists (φύσις); *from* **wesen**; '*existence;*
comp. **bī-, ȝe-, nēh-wiste(-weste).**]

wistinge, sb., *learning*, HAMP. PS. liv. 7.

wit, sb., *O.E.* witt, wit,=*O.N.* vit, *O. Fris.*
wit, *O.L.G.* (gi-)wit, *M.Du.* wite, wete,
Goth. (un-)witi (*gen.* witjis), *O.H.G.* wizzi *n.*,
wit, intelligence, mind, O. & N. 681; þet wit
HOM. I. 123; wit & skil ORM. 1652; wit C.
L. 1080; AYENB. 11; CH. C. T. *A* 746; his
wit he forlæs LAȝ. 1661; wit and wisdom A.
D. 288; SHOR. 138; **witte** (*dat.*) P. L. S.
xiii. 220; ælc bi his witte wisdom sæiden
(*r.* sæide) LAȝ. 25627; ase heo were of witte
HORN (L.) 1084; out of witte MARG. 245;
OCTOV. (W.) 1667; **wittes** (*pl.*) TREAT.
139; PR. C. 5518; þe vif wittes A. R. 14;
comp. **for-, ȝe-, in-, un-, wan-wit.**

wit-ful, [witfol], adj., *witty*, LAȝ. 911;
witvol AYENB. 150.

wit-lēas, adj., *O.E.* gewitlēas; *witless*, A.
R. 256; witles O. & N. 692; witles BEK.
1940; AN. LIT. 91; **witlēase** (*pl.*) AYENB.
86.

wit, pron., *O.E.* wit,=*O.L.G.* wit, *O.N.* vit,
Goth. wit (νώ); *we two*, JOHN xvii. 11; FRAG.
7; HOM. I. 33; KATH. 1523*; GEN. & EX.
1775; (*ms.* witt) ORM. 201; þe bet wit (*sec.
text* we) mawen libben LAȝ. 9515; wit tweie
23653.

wit, *see* **wiðð.**

[**wīt?** sb.; *comp.* **ed-wīt.**]

wite, sb., *O.E.* wita, wiota, weota, wute (ALFR.
P. C. 2),=*O.Fris.* wita, *Goth.* (un-)wita, *O.H.
G.* wizo; *one who knows; witness*; ORM.
8672; beon weote & witnesse þer of A. R.
204*; wet his (*r.* is) mistike ne mei non wete
… afonde SHOR. 24; **witene** (*gen. pl.*)
LAȝ. 11545; *comp.* **ȝe-, un-wite(-weote).**

wit-scipe, sb., *O.E.* (ge-)witscipe,=*O.Fris.*
witskipe, *O.L.G.* (ge-)witscepi, *O.H.G.* (gi-)-
wizscaf; *knowledge, testimony*, HOM. I. 25;
witshipe ORM. 5709.

wite-word, sb., *O.E.* witword,=*O.N.* vitorð;
testimony; testament; ['*testamentum*'] PS.
xxiv. 14; MAN. (H.) 152; HAMP. PS. xxiv. 11.

wīte, sb., *O.E.* wīte,=*O.L.G.* wīti, *O.N.* vīti,
O.H.G. wīzi *n.; punishment, penalty, torment*,
HOM. I. 103; A. R. 4; ORM. 3295; GEN. &
EX. 2035; C. M. 884; CH. C. T. *G* 953; FLOR.
1637; LIDG. M. P. 32; uppe wite of feowerti
wunden (*r.* wunde) LAȝ. 5118; **wīten** (*dat.
pl.*) LAȝ. 1046; *comp.* **blōd-, ferd-, helle-
wīte.**

wīteȝan, v., *O.E.* wītigian, wītgian,=*O.Fris.*
wītgia, *M.Du.* wittighen, *O.H.G.* wīzagōn,
wīzzagōn, wizegōn, wizigōn; *prophesy*, HOM.
I. 91; **witeȝede** (*pret.*) HOM. I. 7; witegeden
MAT. vii. 22; *comp.* **ȝe-wīteȝen.**

wītegunge, sb., *O.E.* wītegung,=*O.H.G.*

wīzegunga; *prophecy*, HOM. I. 97; wite-
hunge ORM. 15149.

wīteʒe, sb., *O.E.* wītega, wītiga, wītga,=*O.N.*
vitki (*for* vitgi, *see* HOLTZM. 105), *O.H.G.*
wīzago, wizzago, wīzago, wīzigo; *wise man,
sage, prophet*, HOM. I. 43; H. M. 5; witeʒe,
witie LAʒ. 9094, 17415; witege HOM. II. 83;
witiʒe I. 233; **wītegen** (*gen.*) MK. i. 2;
wītegen (*dat. pl.*) MAT. vii. 15; witeʒen
KATH. 484; tweolve of þine witiʒ(e)n [wittie]
LAʒ. 4368.

witeʒen, *see* **witien.**

[**witel**, adj., *? O.E.* witol; *knowing; comp.*
ʒeare-witel.]

witen, v., *O.E.* witan, wytan (*pret.* wiste,
wisse), = *O.L.G.* witan, *Goth.* witan (*pret.*
wissa) (εἰδέναι), *O.N.* vita, *M.Du.* witen, weten, *O.H.G.* wizan (*pret.* wista, westa, wissa);
know; take care of, guard; ORM. 199; GEN.
& EX. 328; witen S. S. (Web.) 1459; CH. C. T.
E 1740; þu schalt . . . witen hure wille HORN
(L.) 288; witen, wuten A. R. 4, 96; witin, wetin
'*scio*' PR. P. 531; witen [wite] C. L. 1256;
H. H. 71 (69); wite O. & N. 1139; wite FL.
& BL. 555; ROB. 8, 101; ALIS. 3951; wat
is þi wille let me wite AN. LIT. 3; to wite þe
fram þe fende SHOR. 90; witte PR. C. 4734;
wete APOL. 37; PERC. 320; AMAD. (R.) iii;
K. L. 67; to witene LAʒ. 3163; þet is to
witene AYENB. I; dooþ me to witene LANGL.
B viii. 13; to witinge TREV. I. 347; to
wetinge E. G. 349; witen he wolde . . . wat
þing hit were LAʒ. 271; Brutus hine lette
witen (? = witien) 1854; **witand** [wittand]
(*pple.*) PS. lxxxvi. 4; **wiste** (*pret.*) ORM. 521;
GEN. & EX. 779; (ha) wiste . . . hire foster-
moderes ahte MARH. 2; wiste AYENB. 98;
WILL. 145; A. P. i. 376; PERC. 1146; so
foul he him wiste P. S. 220; wiste, wuste O. &
N. 10, 147; wiste [wuste] LANGL. *A* PROL. 12;
to þan ilke weie þe he ful ʒeare wuste [wiste]
LAʒ. 525; Brutus wes i þon castle & hine
wel wuste 1693; wuste A. R. 110, 270; wuste
ROB. 461; heo wisten HOM. I. 19; wusten
LAʒ. 18679; **wiste** (*subj.*) SPEC. 39; **wist**
(*pple.*) H. H. 49; WILL. 2705; LANGL. *B*
xviii. 203; WICL. MAT. x. 26; CH. C. T. *F*
260; *comp.* **bi-, ʒe-witen.**

[**witende**, pple., *witting, knowing; comp.*
un-witinde.]

witinde-līche, adv.,=*M.H.G.* wizzentlīche;
wittingly, AYENB. 8; witingli TREV. III.
463.

witende, sb., *O.N.* vitand,=*M.H.G.* wiz-
zende; *knowledge*: be his witinde AYENB. 6;
wiþoute hare witende 37; wiþouten witinge
CH. C. T. *A* 1611.

witen, v., *O.E.* wītan,=*O.L.G.* wītan, *Goth.*
(fra-, in-)weitan, *O.H.G.* wīzan, *cf. Lat.* vidēre,
Gr. ið-; *see; keep; impute*; ALIS. 1725; S. S.
(Wr.) 349; to witen ant to welden MARH.
2; schal he his mishap witen [wite] me O. &
N. 1249; to witen us wiþ þan unwihte MISC.

72; two lawen Adam scholde . . . witen and
holden C. L. 168; witin '*imputo*' PR. P.
531; wite LIDG. TH. 1042; M. ARTH. 1153;
TOWNL. 15; whom schal i it wite WILL. 458;
wīte (*pres.*) SPEC. 39 (A. D. 161); CH. C.
T. *B* 3860; þu witest A. R. 3ʹ4; wel þu witest
(? witest) ham JUL. 51; witeþ ANT. ARTH.
xvii; wites M. H. 124; witez on his lire
GAW. 2050; witeð, wite JUL. 74, 75; he
witeð [wit] & wealdeð alle þing P. L. S.
viii. 42; ANGL. I. 11; wit HOM. II. 123;
þe deofel . . . wit (*perceives*) heo HOM. I. 21;
þet wit (*guards*) & wereð us A. R. 312; þe
vif wittes þet witeð þe heorte al se wakemen
14; witeþ BEK. 2111; AYENB. 69; if we us
witeþ (*ms.* wyteþ) from heved sunne MISC.
37; and witeþ þis castel so wel C. L. 825;
wit (*imper.*) þat þe selve FER. 5127; witeþ
(*keep*) me fro schaþe WILL. 3008; wite (*look*)
ʒe þet ʒe ʒemen þenne (*r.* þene) halie sunne
dei HOM. I. 11; þet tu wite me wið ham A.
R. 28; god . . . wite þet he us lende P. L. S.
viii. 61; Crist . . . wite his soule fro helle
pine HAV. 405; **wāt** [*O.E.* wāt,=*O.L.G.*
weit, *O.N.* veit, *Goth.* (in-)wait, *O.H.G.* weiz,
Gr. οἶδα, *mod.Eng.* wot] (*pret.-pres.* 1, 3 *sing.*)
wot, know, knows, ORM. 12107; þen ha wat
(*knew*) hire woh KATH. 562; wat [wot] LAʒ.
15622; wat DEGR. 1727; i wat noght hu he on
þam hitte C. M. 7152; wat, wot A. R. 52, 312;
ich wat, wot O. & N. 61, 1179; he . . . þat
hine ful wot (*who knows that he is foul*)
236; wot GEN. & EX. 487; wot HAV. 1345;
AYENB. 9; MAND. 98; D. TROY 536; wot,
LANGL. *A* PROL. 43; CH. C. T. *A* 389; woot
WICL. GEN. iv. 9; ichot (=ich wot) ROB.
431; SPEC. 25; þu wast MARH. 17; ORM.
11259; wost A. R. 96; wost P. S. 151; SHOR.
165; S. S. (Wr.) 1103; ANT. ARTH. xx; þu
hit wost LAʒ. 15836; witen MK. xii. 14;
MARH. 17; ORM. 7932; GEN. & EX. 74;
witen CH. C. T. *A* 1794, *D* 1890; WICL. LK.
xx. 21; ʒit witen LAʒ. 5627; (heo) witen
[witeþ] 15060; we wuten LEB. JES. 286; ʒe
wuten A. R. 236; wuteð (*for* wuten) 252;
hie wuten HOM. II. 161; ʒe witeþ (*for* witen)
P. L. S. xii. 149; þei witeþ [weteþ] TREV. III.
283; **wite** (*subj.*) ORM. 5710; wite Crist
[=*M.H.G.* wizze Krist] HOM. I. 29; wite þu
know A. R. 90; god hit wute 250; wite P. S.
327; LANGL. *B* v. 297; S. S. (Wr.) 1032; wite
LAʒ. 15090; wite (*keep*) mine Bruttes 28604;
wite þu MISC. 47; god it wite HAV. 517;
witen (*pple.*) ORM. 8222; witen C. M. 10793;
she wende no man . . . hadde witen of hore
dede DEGR. 1592; *comp.* **æt-, bi-, ʒe-
under-witen.**

wīten², v., *O.E.* wītan; *go, depart*: þe wolf
to wīteþ (*pres.*) LAʒ. 21311* (*first text*
iwiteð); ne **wīte** (*imper.*) þou noght fra me
['*ne discesseris a me*'] PS. xxi. 12; witeð ['*ite*']
HOM. II. 5; *comp.* ʒe-wīten.

witer, adj., *O.N.* vitr; *knowing, wise; evident;*

LAȝ.9600; C.L.75; SPEC. 28; witer taken ORM. 4013; witter and war GEN. & EX. 1308; witter ALEX. (Sk.) 629; **witereste** (*superl.*) LAȝ. 15204; *comp.* un-, wei-witer.

witter-hēd, sb., *wisdom,* GEN. & EX. 3667.

witer-līche, adv., *O.N.* vitrliga; *surely,* LAȝ. 17563; H. S. 5; witerliche KATH. 283; witterliche A. R. 70; witerlike ORM. 785; witerlike HAV. 671; witterlike GEN. & EX. 769; witerli AN. LIT. 8; WILL. 305; LANGL. *A* i. 72; witterli LANGL *B* i. 74.

witeren, v., *O.N.* vitra; *make wise, make sure;* witere (*imper.*) LAȝ. 1200; wite me & were & witere JUL. 33; **witered** (*pple.*) JOS. 466; A. P. ii. 1587.

witering, sb., *information;* vittering BARB. iv. 562, v. 342.

witi, adj., *O.E.* witig, wittig, = *O.L.G.* witig, witag, *O.N.* vitugr, *O.H.G.* wizīg, wizzīg; *witty, skilful, wise,* O. & N. 1189; witti KATH. 317; witti P. L. S. xiii. 219; WICL. DEUT. i. 13; þis childes witige ['*prophetalis*'] gost HOM. II. 127; þe witeȝe [wittie] wurhte LAȝ. 21134; mine witie men 15829; **wittiest** (*superl.*) KATH. 533; *comp.* un-witti.

witti-hēde, sb., *wisdom,* SHOR. 138.

witti-lī, adv., *wittily,* WILL. 2602.

wītie, wīti̯ȝe, *see* witeȝe.

witien, v., *O.E.* witian, weotian, witigan, = *O.N.* vitja; *keep, guard, preserve,* LEB. JES. 529; witeȝen wel þa leoden LAȝ. 23122; witie 1854*; witien heom frame sunne S. A. L. 154; witie TREAT. 133; AYENB. 166; witie, wetie E. G. 350, 357; he wéteð (*pres.*) Peteres hus LAȝ. 32155; þat he ... witeȝe (*subj.*) me wið [witie me fram] sconde 23738; **witede** (*pret.*) LAȝ. 4608*; wited WILL. 176; *comp.* bi-witien.

wītien, v., *O.E.* (ed-)wītian, = *O.N.* vīta, *Goth.* (fair-, id-)weitjan, *O.H.G.* (it-)wīzōn; *impute, blame:* of al þis gilt ihc am to wite FL. & BL. 723; þe whiche two of al þis wo i wīte (*pres.*) CH. C. T. *B* 3860; hare overherren witið ham & wraððeð H. M. 29; al he **wītede** (*pret.*) [witte] hit win̄ LANGL. *A* i. 31; **wīted** (*pple.*) WILL. 519; witid WICL. JUDG. iv. 9; *comp.* ed-wīȝien.

witinge, *see* witende *under* witen.

witnen, v., *O.N.* vitna; *attest, testify;* **witneð** (*pres.*) A. R. 30; **wittened** (*pret.*) WILL. 3462.

witnesse, sb., *O.E.* (ge-)witness, = *M.Du.* wetenisse, *O.H.G.* (gi-)wiznessi; *witness, testimony,* P. L. S. viii. 57; A. R. 68; GEN. & EX. 3843; witnesse LAȝ. 13231; PROCL. 7; AYENB. 10; witnisse BEK. 830; *comp.* ȝe-witnesse, un-witenesse.

witnessen, v., *witness;* **witnisseþ** (*pres.*) LANGL. *B* PROL. 191.

[**witted,** adj., *witted;* comp. dul-witted.]

witter, witti, *see* witer, witi.

---

wið, prep., *O.E.* wið, = *O.N.* við, *O.L.G.* wið, *O.Fris.* with; *with, against:* þa spæken twegen weres wið hine LK. ix. 30; neb wið (*to*) neb HOM. I. 61; of þan icompe þe ure drihten hefde wið þene feond 129; he spec wið ðene halie mon 133; muð wið muðe HOM. II. 105; wið, wid, wit P. L. S. viii. 114, 115, 151; breoste wið [wiþ] breoste LAȝ. 1874; ne scal þe nan man scilden wið (*sec. text* fram) scondliche deaðe 2274; to feahten wið þon keisere 5532; þa spæc ælc wið oðer 22941; wið þan þe [wiþ þat] (*provided that*) þu me helpe 8485; we sendeð wið and wið (*at once*) 20747; ȝif þe king me stont wið 23127; wit [wid] 3002; þe him wit feohten (*r.* fohten) 3711; hwo mei wið þeos witen him A. R. 278; wið þen þat 284; beateð hire bare bodi wið bittre besmen MARH. 5; þet ich ... mahe stonden wið him 10; wið þon þet ich mote meidene-mede habben IBID.; wið (*v. r.* mid) rihte KATH. 770; to wurþen god wiþ bedes and wiþ lakes ORM. 905; ben wiþ childe 2446; swa bilæf ȝho þær wiþ him 3160; & her ic wile wiþ & wiþ þa seofne seolþes shæwen 5628; al þat folc toc niþ wiþ him 10267; wið [wiþ, wiþ] me wroð O. & N. 1087, 1608; wroþ wiþ [wit] his bridde 111; wiþ þat (= wiþ þon þat) Caim fro him fleg wið wif and hagte (*r.* agte) GEN. & EX. 431; wit rigt 56; ðis fis wuneð wið ðe segrund REL. I. 220; iseined wiþ ure seel PROCL. 7; þu schalt wiþ me to bure gon HORN (L.) 286; to speke wiþ him ROB. 104; iwrouȝht wel with þe beste LEG. 48; biset ic am wiþ liþere men MARG. 56; wiþ rihte C. L. 298; never schal fo him stonde wiþ 701; w(i)t 28; cloþes ... such(e) as i mai weore wiþ winne SPEC. 37; war be wiþ þe swike 46; idemd wiþ wrong AYENB. 12; face wiþ (*to*) face 244; biloved wiþ (*by*) riche & wiþ pore WILL. 1060; wiþ þat 3161; with traisoun wit feloni HAV. 1090; wit (*printed* þit) 997; lat him ... bringe wid him silver P. S. 325; wit, wid C. M. 338; heremites ... with hoked(e) staves LANGL. *B* PROL. 53; þough he were wounded with his enemi LANGL. *B* xvi. 105; wiz MISC. 194; *comp.* in-, ūt-wið.

wið-æften, adv., *O.E.* wiðæftan; *back,* MAT. ix. 20; wiðefte MK. v. 27.

wið-breiden, v., *? snatch away; turn away;* bute þu wiðbreide (*subj.*) þe H. M. 9.

wið-büggen, v., *redeem;* wiþbegge AYENB. 186.

wið-būwen, v., *decline, avoid,* A. R. 116; wiðbuhe (*printed* wiðhuhe) H. M. 37.

with-clépin, v., *call back, recall, 'revoco,'* PR. P. 530; wiþclepie AYENB. 215; **wiþclépe** (*pres.*) ALIS. 1301; wiþclépid (*pple.*) TREV. V. 209.

wiþ-cwéðen, v., *O.E.* wiðcweðan; *contradict,* FRAG. 1; **wiðqvað** (*pret.*) HOM. II. 137; wiðcwæðen SAX. CHR. 250.

wið-draȝen, v., *withdraw*; wiþdraȝe SHOR.
153; AYENB. 9; wiþdrawe H. S. 12492; wið-
drawe∂ (*pres.*) A. R. 230; wiðdrōg (*pret.*)
GEN. & EX. 599; witdrou HAV. 502; þe tveie
. . . wiþdrowe hem BEK. 1914.

wiþelen, *see* weo∂elen.

wiðer, adv.; **wiðere,** adj., *O.E.* wi∂er, wi∂re,
cf. O.L.G. wither, wi∂er, wi∂ar, withere, *O.
Fris.* wither, *O.H.G.* widar, widari, *O.N.* vi∂r,
Goth. wiþra (πρός); *against; hostile*: þe
wi∂er [wiþer] wes an compe LAȝ. 9287; weȝe
and wiþer weȝe AYENB. 137; wethire halfe
wrong side ALEX. (Sk.) 3355; *comp.* an-
wiðer(-wiðre).

wiþer-blench, sb., *attack*; þat mai ago
deaþes wiþerblench R. S. i (MISC. 156).

wiðer-craft, sb., = *M. H. G.* widerkraft;
craft; **wiðercraften** (*pl.*) LAȝ. 19629.

wiðer-dēde, sb., *hostile deed*; **wiðerdēden**
(*pl.*) LAȝ. 21086.

wiðer-feht, sb., *battle*; **wiðervehte** (*dat.*)
LAȝ. 28669.

wiþer-gāst, sb., *hostile spirit*, ORM. 11389.

wiðer-gome [wiþergame], sb., *contest*,
LAȝ. 24700.

wiðer-hap, sb., *adverse fortune*; **wiðer-
heppes, -happes** (*pl.*) LAȝ. 405, 9269.

wiðer-iwinne, sb., *adversary*, LAȝ. 4535.

wiðer-laȝe, sb., = *O.Fris.* witherlaga; *ruf-
fian*, LAȝ. 10968; wi∂erlahe MARH. 5.

wiþer-rǣs, sb., *hostile attack*; **wiþerreases**
(*pl.*) LAȝ. 25096*.

wiðer-sake, sb., *O.E.* wi∂ersaca, = *O.H.G.*
widersacho; *adversary*, LAȝ. 12620.

wiðer-sīde, sb., *opposite side*, LAȝ. 11972.

wiþer-strencþe, sb., *hostile strength*: awiht
of wiþerstrencþe ORM. 6905.

wiþer-þēod, sb., *hostile people*, ORM. 10227.

wiðer-ward, adj. & adv., *O.E.* wi∂erweard,
= *O.H.G.* widarwart, -wert,*Goth.* wiþrawairþs;
adverse, contrary, LAȝ. 6875; wi∂erward GEN.
& EX. 2935; wiþerward ORM. 9667; DAV.
DR. 20; witerwarde ALEX. (Sk.) 4247.

wiðer-winne, sb., *O.E.* wi∂erwinna, = *O.H.
G.* widarwinno; *adversary, enemy*, MAT. v. 25;
wiþerwinne MISC. 74; wi∂erwinne, -wine REL.
I. 226, 235; wi∂erwine A. R. 196; KATH.
1197; wiþerwine ROB. 325; RICH. 6012.

wiðer², sb., *O.E.* wi∂re; *adversity, resistance*:
wi∂er [wiþer] com toȝeines me LAȝ. 4678;
ȝif þe king wolde wi∂ heom wi∂er heolden
[wiþer holde] 9175.

wiðer-ful, adj., *hostile, valiant*, HOM. II.
51; wi∂erfulle [wiþerfolle] cheorles LAȝ.
21520.

wiðerfulnesse, sb., *hostility*, ['*impietatis*']
HOM. II. 63.

wiðerin, v., *O.E.* wi∂erian, = *O.H.G.* widarōn,
M.Du. wederen; *oppose, resist*, MARH. 14;

wiðereð (*pres.*) A. R. 238*; wiþreþ ORM.
1181; wi∂erie∂ HOM. II. 123; wi∂eren REL.
I. 219.

wiþering, sb., *adversary*, HORN (L.) 148;
wetheruns (*pl.*) D. TROY 5048.

wiþerling, sb., *O.E.* wi∂erling; *adversary*,
HORN (R.) 154, (H.) 156.

wið-geonde, prep., *O.E.* wi∂geondan; *beyond*;
[*mistranslation of* ' *circum* '] MAT. iii. 5.

wiþ-hálden, v., *withhold*, CH. BOET. iv. 6
(142); withhalde A. P. ii. 740; **wiþháldest**
(*pres.*) ALIS. 1310; þet ȝe wi∂holden ou from
vlesliche lustes A. R. 348; **wiðhēld** (*pret.*)
GEN. & EX. 2033; withheld HAV. 2356.

wiði, wiþþe, sb., *O.E.* wi∂ig, wi∂∂e, *cf. O.
Fris.* withthe, *O.N.* vi∂ja, vi∂, *O.H.G.* wīda,
Gr. ἰτέα; *withy, willow; halter, fetter*; wi∂i
A. R. 86; wiþi '*salix*' VOC. 181; FRAG.
3; wiþþe '*circulus*' 2; wiþi, wiþe PERC.
423, 444; do∂ wi∂∂e an his sweore LAȝ.
22833; wiþþe AYENB. 135; ALIS. 4714; H. S.
11553; withe, witthe, withth(e) '*boia*' PR. P.
531; **wiþþes** (*pl.*) ORM. 15563; wiþies WICL.
LEV. xxiii. 40; wiþþis WICL. JUD. vi. 9; iteied
to somne mid wi∂en LAȝ. 25973.

wiþi-bond, sb., *woodbine*; **weþebondes**
(*gen.*) LANGL. *A* vi. 9.

wiþ-winde, sb., *O.E.* wi∂owinde, = *M.L.G.,
M.Du.* wedewinde; *cf.* wudebinde; *woodbine*; **wiðwindes** (*gen.*) LANGL. *B* v. 525.

wið-innen, prep. & adv., *O.E.* wi∂innan (SAX.
CHR. 244); *within*, A. R. 4; REL. I. 215; wereden þene tun wi∂innen [wiþine] LAȝ. 18300;
wi∂innen þan fif nihte 30600; wiþinnen ORM.
1020; C. L. 717; wiþinnen, -inne FRAG. 7;
wiþinne TREAT. I 39; AYENB. 10; biþinne FL.
& BL. 244; E. G. 355 (*see* ENG. ST. V. 371).

wiþ-nimen, v., *blame, reprove*; wiþnime
AYENB. 137; **wiþnome** (*pple.*) SHOR. 146.

wið-sahe, sb., *contradiction*, A. R. 288*;
wiþsawe C. M. 5877.

wið-sáken, v., *O.E.* wi∂sacan; *deny, renounce*, A. R. 88; wi∂saken LAȝ. 10898;
wiðsōc (*pret.*) LAȝ. 13000; þu wiþsoke FRAG.
8; wiþsáken (*pple.*) JOS. 178.

wiþ-scóren, v., *retrench, cut short*; **wiþ-
scóre** (*imper.*) and wiþdraȝ þine willes AYENB.
254.

wiþ-seggen, v., = *O.Fris.* withsedza; *contradict*, ORM. 7646; wiþsegge ROB. 106; wi∂-
siggen A. R. 86; wiþzigge AYENB. 175; wi∂-
suggen [wiþsegge] LAȝ. 13237; wiþseie CH.
C. T. *A* 805; HOCCL. vi. 47; wiþsegge (*pres.*)
HORN (L.) 1276; **wiþseide** (*pret.*) BEK. 526.

wið-sigginge, sb., *contradiction*, A. R. 288.

with-settin, v., *resist*, ' *obsisto*,' PR. P. 530;
þei him wiþsette (*pret.*) MAN. (F.) 2931.

with-sitten, v., *resist*, HAV. 1683; withsitte
EGL. 1022; **wiþsat** (*pret.*) AR. & MER.
9065.

wiþ-standen, v., *O.E.* wiðstandan,=*O.Fris* withstonda ; *withstand, resist,* ORM. 7646 : wiðstonden KATH. 228 ; A. R. 218 ; wiðstonden LAӠ. 1419 ; wiþstonde LANGL. *B* PROL. 156 ; **wiþstondeþ** (*pres.*) AYENB. 265 : **wiðstōd** (*pret.*) GEN. & EX. 2649 ; **wiþstonde** (*pple.*) CH. BOET. i. 4 (14).

wiþ-steden, v., *withstand, stop up*; **wiþsted** (*pple.*) MAN. (F.) 13503.

wið-stēwen, v., *? check* : heore uvel . . . þu aӡest to hetiene & wiðstewen HOM. I. 15.

with-táke, v., *reprove,* HAMP. PS. xlix. 9 ; **withtōke** (*pret.*) c. 6.

    with-tákere, sb., *reprover,* HAMP. PS. ix. 24*.

wið-tēon, -tīen, v., *O.E.* wiðtēon ; *draw back,* HOM. II. 79, 139 ; **wiðtēoð** (*pres.*) HOM. I. 143 ; **wiðtēo** (*subj.*) REL. I. 132.

wið-ūten, prep. & adv., *O.E.* wiðūtan (SAX. CHR. 244) ; *without, beyond,* GEN. & EX. 503 ; hu me schal beren him wiðuten A. R. 4 ; wiðuten leave 230 ; wiþuten galle ORM. 1253 ; wiðuten dore LAӠ. 2382 ; wiðute þan volke 26215 ; wiþuten, wiþ-, witute O. & N. 183, 863 ; wiðute (*v. r.* buten) live KATH. 252 ; wiþute FL. & BL. 511 ; with-, wituten HAV. 179, 425 ; wiþouten C. L. 717 ; JOS. 316 ; wiþouten ende A. D. 263 ; WICL. JOHN viii. 35 ; wiþoute BEK. 267 ; AYENB. 6 ; biþute FL. & BL. 218 ; biþoute E. G. 349.

wið-ward, prep., *contrary to,* KATH. (E.) 1958.

wivil, wévil, sb., *O.E.* wifel,=*O.L.G.* (gold-)wivil, *M.L.G., M.Du.* wevel, *O.N.* (tord-)yfill, *O.H.G.* wibil ; *? from* **wéven** ; *weevil,* '*curculio,*' PR. P. 523, 531.

wīvin, v., *O.E.* wīfian,=*M.L.G.* wīven, *M.Du.* wijven ; *take a wife,* PR. P. 531 ; wivi (*ms.* wyui) AYENB. 225 ; FER. 2096 ; wive ROB. 35 ; TRIST. 2896 ; CH. C. T. *E* 140 ; **wiveþ** (*pres.*) REL. I. 115 ; **wīvede** (*pret.*) A. R. 308 ; *comp.* ӡe-**wīven.**

wivre, gvivre, sb., *O.Fr.* guivre, *mod.Eng.* wyvern ; *viper* : jalousie þat wikked wivere CH. TRO. iii. 1010 ; addres, **gvivres** (*pl.*) (printed quinres) ALIS. 5609.

wīwere, sb., *cf. O.H.G.* wīwāri ; = **vīvere** ; '*vivarium*' VOC. 195.

wlach, adj., *O.E.* wlæc,=*M.L.G.* wlack ; *tepid* ; (*printed* wlath) GEN. & EX. 3300 ; wlach, wlak SAINTS (Ld.) xlvi. 695 (TREAT. 138) ; wlech A. R. 202 ; JUL. 31 ; **wlache** (*dat.*) HOM. II. 151.

wlaffen, *see* **blaffen.   wlak,** *see* **wlach.**

wlanc, wlonc, adj., *O.E.* wlanc, wlonc,=*O.L. G.* wlanc ; *proud, splendid,* REL. I. 174, 180 ; wlonc O. & N. 489 ; prud & wlonc HOM. I. 35 ; wlonk A. P. iii. 486 ; PALL. iii. 449 ; in wlanke wede M. H. 42 ; of al mi weole wlonke P. S. 156 ; of his wlonke murþe WILL. 1634 ; wlanke deor P. L. S. xxv. 3 ; wlonke wordes KATH. 844 ; þe **wlonkest** (*superl.*) wedes GAW. 2025.

wlonk-hēde, sb., *pride,* O. & N. 1400.

wlaunknesse, sb., *pride,* P. S. 330 ; wloncnesse (*ms.* wlongnesse) HOM. I. 9.

wlanke, sb., *cf. O.E.* wlenco ; *? pride* ; for wlaunke P. S. 341.

wlappen, *see* **lappen.**

wlate, sb., *O.E.* wlætta ; *disgust,* O. & N. 1506.

    wlate-ful, adj., *disgusting,* H. M. 25 ; wlatful TREV. II. 167 ; PS. xiii. 1 ; wlatvol AYENB. 241.

    wlatsum, adj., *disgusting,* '*abominabilis,*' WICL. LEV. xi. 11 ; P. R. L. P. 173 ; A. P. ii. 541 ; wlatsom CH. C. T. *B* 3814 ; PR. C. 459 ; wlathsum HAMP. PS. xiii. 2, lii. 2.

wlátien, v., *O.E.* wlatian,=*M.L.G.* wlaten ; *feel disgust* : ham wolde wlatien þer aӡean A. R. 86 ; overfulle makeþ wlatie O. & N. 354 ; wlate ['*abominari*'] PS. v. 7 ; wlath HAMP. PS. v. 7 ; **wlátiþ** (*pres.*) APOL. 92 ; him wlatis H. S. 3541 ; wlatez A. P. ii. 305 ; wlattis ALEX. (Sk.) 4277 ; **wlátede** (*pret.*) WICL. JER. xiv. 19 ; **wláted** (*pple.*) ALEX. (Sk.) 5634 ; wlethid HAMP. PS. cv. 37 ; *comp.* a-**wláten.**

    wlátunge, sb., *O.E.* wlatung ; *disgust, abomination,* A. R. 80 ; wlatinge SHOR. 26 ; MIRC 894 ; **wláthingis** (*pl.*) HAMP. PS. *page* 516.

wlecchen, v., *O.E.* wleccan ; *from* **wlach** ; *make tepid* ; **iwlaht** (*pple.*) JUL. 70.

    wlecchunge, sb., *? tepidity* ; H. M. 45.

wlech, *see* **wlach.**

wlenchen, v., *O.E.* (ge-)wlencan, = *O.L.G.* (gi-)wlenkian ; *from* **wlanc** ; *be proud, make proud* : þi lease wit þat tu wlenches (*pres.*) (*pridest*) te in KATH. 1010 ; wlencð his soule HOM. II. 189 ; *comp.* a-, for-**wlenchen.**

wlispen, v.,=*O.L.G.* wlispen, *M.Du., M.H.G.* lispen ; *lisp* ; *lisþ* ; lispin PR. P. 306 ; **wlispit** (*pret.*) BARB. i. 393 ; lisped [lipsed] CH. C. T. *A* 264.

wlite, sb., *O.E.* wlite,=*O.L.G.* wliti, *Goth.* wlits, *O.N.* litr ; *from* **wlīten** ; *face, form* ; *beauty* ; KATH. 1463 ; ORM. 666 ; GEN. & EX. 2288 ; LAӠ. 19205 ; REL. I. 185 ; PS. xliv. 5 ; þe lilie mid hire faire wlite O & N. 439.

wlīten, v., *O.E.* wlītan,=*O.N.* līta ; *look, see* ; **wlīteþ** (*pres.*) SPEC. 43 ; *deriv.* **wlite, wlīti.**

wlíti, adj., *O.E.* wlitig,=*O.L.G.* wlitig ; *beautiful,* KATH. 313.

wlō, sb., *O.E.* wlōh ; *? nap, ? hem, fringe* : cloþes ful feble . . . þe **wlōn** (*pl.*) offe PL. CR. 736.

wlonc, *see* **wlanc.   wó,** *see* **hwá.**

wō, *see* **wā.   wōc,** *see* **wāc.**

wōd, adj., *O.E.* wōd,=*Goth.* wōds, *O.H.G.* wuot. *O.N.* ōðr ; *mad, furious,* A. R. 96 ; KATH. 31 ; ORM. 4676 ; O. & N. 566 ; wod LAӠ. 1714 ; HAV. 508 ; BEK. 1939 ; A. P. ii. 204 ; wood CH. C. T. *A* 184 ; TOR. 643 ; wood, ooth PR. P. 372, 531 ; woud ALEX.

(Sk.) 5428; **wōde** (*pl.*) PR. C. 99; wode houndes AYENB. 70; **wōden** (*dat. pl.*) LA3. 19092; **wŏddre** (*compar.*) A. R. 264.

wōd-hēde, sb., *fury, madness*, AYENB. 18; wodhed GEN. & EX. 3539.

wōd-līche, adv., *furiously*, HALLIW. 938; wodli WILL. 550; **wōdlokere** (*compar.*) LA3. 3201.

wōdnesse, sb., *O.E.* wōdnyss, = *M.Du.* woedenisse, *O.H.G.* wōtnissa; '*furor*,' PR. P. 531; AL. (L.²) 474; woodnesse CH. C. T. *A* 3452; SACR. 502.

wōd-schipe, sb., *fury*, A. R. 120.

wōd, *see* **wād**. **wode**, *see* **wude**.

Wōden, pr. n., *O.E.* Wōden, = *O.L.G.* Wōden, *O.H.G.* Wuotan, *O.N.* Ŏðinn; *name of a god*, LA3. 13903; Mercuri is on oure langage Woden MAN. (F.) 7372; **Wŏdnes** (*gen.*) dei *Wednesday* A. R. 70; Wodnes, Wednes dai LANGL. *B* xiii. 154; Wednes dai KATH. 2215; Wednes-, Wodenes-, Wedones-, Wodnes-, Wensdai(e) ROB. (W.) 2432.

woder, *see* **hwider**.

wōdien, v., *? O.E.* wōdian; *cf.* **wēden**; *rage*; woodeþ (*pres.*) CH. BOET. iv. 4 (123); woodedist (*pret.*) WICL. IS. xxxvii. 29.

[wō3-, *stem of* wō3en.]

wōh-lāc, sb., *wooing*, A. R. 388*.

wōuh-lēche, sb.; *wooing*, A. R. 96.

wō3, *see* wā3, wōh.

wō3en, v., *O.E.* wōgian; *woo*; wo3e HOM. I. 187; HORN (L.) 546; wowen A. R. 388; wowen AR. & MER. 772; wowin '*proco(r)*' PR. P. 533; wowe P. 11; S. S. (Wr.) 230; EGL. 1064; **wōweþ** (*pres.*) SPEC. 46 (A. D. 167); CH. C. T. *A* 3372; **wōwende** (*wo-winge*] (*pple.*) WICL. PROV. vii. 13; **wōwude** (*pret.*) A. R. 390; wowede LANGL. *B* iv. 74; wowid TRIAM. 67; woghit D. TROY 482.

wōwere, sb., *O.E.* wōgere; *wooer*, VOC. 176; woware A. R. 90; woware PR. P. 532; **wōweres** (*pl.*) LANGL. *B* xi. 71.

wōwunge, **wōuhinge**, sb., *wooing*, A. R. 116, 204; wowinge PR. P. 533; wowing SPEC. 28.

wōh, adj. & sb., *O.E.* wōh, = *O.L.G.* wāh, *Goth.* (un-)wāhs; *crooked; bad, evil*; A. R. 2; **wōh** (*subst.*) ORM. 5555; 3if þu me dest woh HOM. I. 33; þat woh R. S. v (MISC. 184); woh LA3. 4333; H. H. 52; wo3h WILL. 544; wogh OCTAV. (H.) 561; ISUM. 95; A. P. i. 621; woche D. TROY 5050, 6722, 7756; wouh A. R. 54; wouh [wou3] C. L. 385; wou3 BRD. 13; wou REL. I. 243; **wōhe** (*dat.*) KATH. 1243; wo3e LA3. 24811; mid wowe ne mid rihte MISC. 49; mid wo3e dome HOM. II. 179; **wō3e**, wowe (*pl.*) O. & N. 815; wo3he dedes ORM. 1375; wowe domes FRAG. 7.

wōhnesse, sb., *O.E.* wōhnyss; *depravity*, HOM. I. 103.

wōhlāc, *see under* wō3-.

woke, *see* **wike**. **wōkien**, *see* **wākien**.

wol, *see* **wel**, **wil**.

wolc, sb., *some bird*: þe wilde laveroc ant wolc & þe wodwale SPEC. 26 (A. D. 145).

wolcne, sb., *O.E.* wolcen, wolcn *n.*, *m.*, *cf.* *O.L.G.* wolcan, *O.H.G.* wolchan *n.*; *welkin*, *cloud, sky*; se wolcne SAX. CHR. 249; þa wolcne gon to dunien LA3. 27452; weolcne '*nubes*' FRAG. 3; þere weolcne [þare wolkne] he was swiðe nih LA3. 2883; **wolcne**, weolcne (*pl.*) LA3. 4575, 23947; wolkne SHOR. 137; to beholde þe wolcne [welkene] and þe sterres of hevene TREV. III. 459; wolkne [welkne] LANGL. *B* xvii. 160; weolcne MARH. 7; A. R. 246; wolcnen (*dat. pl.*) MAT. xxvi. 64; wolcnen, weolcnen LA3. 102, 25592; under weolcne [welkne] *in the open air* O. & N. 1682; in welkne SPEC. 114; on the welkne CH. C. T. *E* 1124.

wōld, *see* **wāld**. **wolf**, *see* **wulf**.

wolle, *see* **wulle**. **wollen**, *see* **willen**.

wómbe, *see* **wámbe**. **won**, *see* **wan**.

wōn, *see* **wān**. **wond**, *see* **wand**.

wónde, **wóndien**, *see* **wúnde**, **wúndien**.

wonder, *see* **wunder**.

wondrien, *see* **wandrien**, **wundrien**.

wóne, *see* **wáne**, **wune**. **wōne**, *see* **wēne**.

wonene, *see* **hwanene**. **wong**, *see* **wang**.

wonge, *see* **wange**.

wonien, *see* **wanien**, **wunien**.

wōnien, **wōning**, *see* **wānien**, **wānunge**.

wonne, *see* **wünne**, **hwanne**.

wont, sb., *mole*, VOC. 177; **wontes** (*pl.*) TREV. I. 339.

wonten, *see* **wanten**.

wōp, sb., *O.E.* wōp, = *O.L.G.* wōp, *O.N.* ōp, *O.H.G.* wuof; *weeping, lamentation*, P. L. S. viii. 117; A. R. 110; KATH. 2364; ORM. 7931; O. & N. 878; wop ROB. 34; SHOR. 147; AYENB. 265; wop, weop LA3. 5970, 11991.

wōpen, *see* **wǣpen**.

wōpi, adj., *O.E.* wōpig; *tearful*; **wōpie** (*pl.*) A. R. 376.

wōr, sb., *? O.E.* wōr, *? = M.H.G.* wuor; *? dam, pool*; **wōre** (*dat.*) REL. I. 122; weri so water in wore SPEC. 28 (A. D. 148); i wake as water in wore A. D. 249.

worc, *see* **weorc**. **wŏrchen**, *see* **wurchen**.

word, sb., *O.E.* word, = *O.L.G.* word, *Goth.* waurd, *O.H.G.* wort, *O.N.* orð; *word; saying, report*; A. R. 88; ORM. 282; O. & N. 300; þat word BEK. 537; þat word com to Dunwale LA3. 4118; þat weord [word] þat ich þe sende 8311; þer com word to þe king(e) ROB. 203; ich . . . sende him word P. L. S. xii. 102; word, wurd GEN. & EX. 46, 736; wurd H. S. 1240; **word** (*pl.*) P. L. S. viii. 5: word, wordes A. R. 58, 76; word, weord, worde, wordes HOM. I. 47; 57, 65, 93; word, wordes LAȝ. 7333, 3702 ; word, worde, wordes O. & N. 139, 178, 1715 ; worde HORN (H.) 857 ; **worde** (*gen. pl.*) LK. i. 4 ; **wordon** (*dat. pl.*) HOM. I. 227 ; worden LK. i. 20 ; worden PROCL. 11 ; worden, wurden [worde] LAȝ. 6675, 23632 ; worde P. L. S. viii. 6 ; O. & N. 158 ; mid worde & mid werke FRAG. 7 ; hie ne mihten mid worde here blisse tellen HOM. II. 115 ; *comp.* **bī-, bod-, hēre-, hux-, lof-word.**

   word-fest, adj., *O.E.* wordfæst ; *truthful, faithful* ; (*ms.* weordfeste) HOM. I. 111.

   word-lēs, adj., *wordless, ineffable* : wordles song ['*jubilus*'] HOM. II. 113.

word, see **weored.**

wordien, v., *utter words; speak* : þe king **wordede** (*pret.*) þus LAȝ. 18052 ; wordeden LANGL. *A* iv. 33.

wordle, woreld, see **weoreld.**

wōren, v., *O.E.* wōrian (*wander*) ; *wander, go, weary* : þis worldes won wiþ muchel unwinne us woren wolde SPEC. 24 ; this jurni to wore LUD. COV. 96 ; þis mong **wōreð** (*pres.*) so þe eien A. R. 386.

wori, see **wari. world,** see **weoreld.**

worm, see **wurm. worpen,** see **weorpen.**

worre, see **werre. wors,** see **würs.**

Wor-stede, pr. n., *Worstead,* CH. C. T. *A* 262.

wort, see **wurt.**

worþ, worþi, see **wurð, wurþiȝ.**

worþiȝ, sb., *O.E.* weorðig ; '*praedium,*' FRAG. 4.

wōs, sb., *O.E.* wōs, = *M.L.G.* wōs ; *juice,* HALLIW. 939 ; wose LEECHD. III. 345 ; ALEX. 413 ; wus ALIS. (Sk.) 712.

[**wōse,** sb., *O.E.* wāsa ; *satyr; comp.* **wode-wōse.**]

wōse, see **wāse.**

wōsen, v., *cf.* **wēsen** ; *ooze, exude* : **wōseth** (*pres.*) out ['*exundant*'] humours TREV. I. 63.

wōþe, see **wāþe²**.

wōu, wōuȝ, wōuh, see **wōh.**

wounde, see **wúnde. wōwe,** see **wāwe.**

wōwen, see **wōȝen.**

wrā, sb., *O.N.* (v)rā ; *angle, corner,* ALEX. 1585 ; wro P. L. S. xxviii. 26 ; M. T. 306 ; GAW. 2222 ; **wrās** (*pl.*) C. M. 18155 ; crepen in wros HAV. 68.

wrabbe, v., *? denounce* ; wrabbe [wrobbe] and wrie ERC. 38.

   wrabbere, sb., *? denouncer* : wreieres and **wrobberes** (*pl.*) HAV. 39.

wrac, adj., = *O.Fris., M.Du.* wrak ; *from* **wré-ken** ; *base* ; of heorte wrac LEG. 12.

wrac², sb., *cf. O.L.G.* wrak, *O.N.* vrek *n.; wreck* ; wrak CH. *C.* T. *B* 513 ; wrek PR. P. 533 ; wrec REL. I. 33.

wrǣcche, see **wrecche.**

wrǣche, sb., *O.E.* wrǣc, = *O.Fris.* wrētze,

O.L.G. wrēka, *Goth.* wrākja, *O.H.G.* rācha, *from* **wréken** ; *vengeance, punishment,* ORM. 19 ; wreche ['*vindicta*'] A. R. 186 ; ȝif þu hevedest wreche inumen HOM. I. 197 ; wreche and wrake GEN. & EX. 552 ; wreche LAȝ. 29581 ; MISC. 143 ; MAP 338 ; BEK. 1496 ; SHOR. 133 ; AYENB. 45 ; CH. C. T. *B* 679 ; S. S. (Wr.) 1775 ; he scholde tak(e) of heom wreche ALIS. 2858 ; do wreche FER. 4181 ; þe fende she miȝth do wreche AL. (L.¹) 216 ; þe fischers were adrad of wreche GREG. 299 ; wrǣche, wrache HOM. II. 29, 51 ; wrache CHR. E. 724 ; A. P. ii. 229 ; wreke '*vindicta*' PR. P. 533 ; wreke PS. lvii. 11 ; PR. C. 5538 ; TOWNL. 248.

   wrēch-ful, adj., *revengeful,* SAINTS (Ld.) xlv. 333.

   wrēch-wil, adj., *desirous of vengeance,* H. M. 47.

wrǣst, sb., *twist* ; **wrǣstes** (*pl.*) C. M. 3462.

wrǣst², adj., *O.E.* wrǣst ; *? from* **wriðen** ; *powerful* ; *wrest* (*? hard; or mistake for* unwrest) CL. M. 31 ; wrast GAW. 1423 ; *comp.* **un-wrǣst.**

   ? wrǣst-līch, adj. : an wræstliche þan folke (*? r.* þan unwrǣstliche folke) LAȝ. 29582.

wrǣste, sb. : wreste of an harpe, '*plectrum,*' PR. P. 533 ; wrast VOC. 240.

wrǣsten, v., *O.E.* wrǣstan ; *wrest, twist* ; wresten A. R. 374 ; wrestin '*plecto*' PR. P. 533 ; five kniȝtes miȝte not wreste þe rop out of his hond TREV. V. 181 ; **wrǣste** [wreste] (*pret.*) LAȝ. 7532 ; wraste PARTEN. 1377 ; hur fingres sche wrast(e) MAN. (F.) 3194 ; wrast(e) out min(e) iȝen A. P. iii. 80 ; **wrǣst** (*pple.*) M. T. 289 ; mani kniht of sadel was wrast MAN. (F.) 13571 ; *comp.* **to-wrǣsten.**

wrǣstlen, v., *O.E.* wrǣstlian, = *M.Du.* wras-telen ; *wrestle* ; wrestlen KATH. 2064 ; wrest-lin MARH. 14 ; wrastlen A. R. 80 ; wrastele ROB. 22 ; wrastle CH. C. T. *A* 3928 ; wirstill [wrastill] ALEX. (Sk.) 2276 ; **wrǣstele** (*pres.*) LUD. COV. 185 ; **wrǣstel** (*subj.*) LANGL. *B* xiv. 224 ; **wrǣstlede** (*pret.*) GEN. & EX. 1803 ; wrastlide WICL. GEN. xxxii. 24 ; wræstleden LAȝ. 24699.

   wrǣstlare, sb., *O.E.* wrǣstlere ; *wrestler,* A. R. 222 ; wrastlere TREV. II. 383 ; werstil-lare [wristiller] ALEX. (Sk.) 2287.

   wrǣstlunge, wrǣstlinge, sb., *wrestling,* A. R. 318, 374 ; wrastlinge LAȝ. 1871 ; O. & N. 795 ; wristilling ALEX. (Sk.) 2260.

wrǣðe, sb., *O.E.* wrǣð, = *M.H.G.* reide, *mod.Eng.* wreath ; *from* **wriðen** ; *ring* ; wreþe CH. C. T. *A* 2145 ; WICL. GEN. xli. 42 ; *fold* D. ARTH. 1093.

wrǣþen, v., = *O.L.G.* wrēthian, *O.N.* reiða, see **wrāðien** ; *make or become wroth* ; wreth JUL. 11 ; he wile wreþe wið þe HOM. I. 33 ; wreþi AYENB. 60 ; AM. & AMIL. 606 ; wrethin PR. P. 534 ; **wrēþeþ** (*pres.*) SHOR. 127 ;

wrēðede (*pret.*) HOM. I. 197; wreaðede JUL. 13; wrēthid (*pple.*) PS. ii. 12.

wrǽððe, sb., *O.E.* wrǽðo, wrǽððo (MK. iii. 21*), = *O.N.* reiði; *wrath* : mid þǽre wrǽððe [wreþþe] LAȝ. 3073; þurh his wraððe [wraþþe] 6379; wreððe HOM. I. 105; A. R. 120; GEN. & EX. 3863; wreððe, wraððe MARH. 9, 18; wraþþe ORM. 124; O. & N. 941; wraþþe ROB. 35; P. L. S. xxiv. 93; C. L. 407; WILL. 728; wraðe REL. I. 130; wrathe HAV. 2977; wraþe [wratthe] LANGL. *B* iv. 34; wreþe AYENB. 8; AM. & AMIL. 830; wrethe PR. P. 534.

wrēð-ful, adj., *wrathful*, HOM. I. 43; A. R. 118; wrethful PR. C. 5107; þe wreþvolle AYENB. 30.

wrǽþþe-lees, adj., *without wrath*, SPEC. 42 (A. D. 186).

wrāȝen, v., *? make angry*; wrāged (*pple.*) REL. I. 178 (MISC. 119); wrawid ALEX. (Sk.) 3167.

wrāh, adj., = *Swed.* vrå; *perverse, headstrong*: ne wrah ne mispaiet A. R. 416*; wrau (*ms.* wraw) '*perversus, protervus*' PR. P. 533; CH. C. T. *H* 46; wrau and gelous S. S. (Web.) 1742; so angri and so wrau N. P. 9; wroȝ ROB. 16, 24; wrōwe (*pl.*) REL. I. 241.

wrāhnesse, sb., *indignation*: he doþ alle þing wiþ anoi and wraunesse (*ms.* wrawnesse) CH. C. T. *I* 680.

wrak, see wrac.

wráke, sb., *O.E.* wracu, = *M.Du.*, *M.L.G.* wrake, *Goth.* wraka (διωγμός) : *from* wréken; *vengeance, persecution, injury*, HOM. II. 61; JUL. 50; GEN. & EX. 552; þa ilke wrake þe ic dude þe HOM. I. 9; wrake & sake 13; niþ and wrake O. & N. 1194; he þe wile wrake don MISC. 136; wrake LAȝ. 4040; SHOR. 141; MAP 352; FER. 1815; REL. I. 262; FLOR. 582; DEP. R. i. 130; LIDG. M. P. 125; M. ARTH. 935; LUD. COV. 200; he schal do ȝou no wrake ALIS. 7605; vengeaunce & wrake P. P. 252; werre and wrake P. 1; wo & wrake AL. (L.¹) 45; sorowe and wrake H. S. 3392; wele after wrake SAL. & SAT. 230; wiþouten wrake GREG. 338.

wráke-dōm, sb., *vengeance*, LAȝ. 76.

wráke-ful, adj., *revengeful*, SPEC. 22.

wrákeful-līche, adv., *vengefully*, KATH. 2076*.

wrang, adj., *O.N.* rangr, = *M.Du.* wrongh; *from* wringen; *crooked, twisted; wrong, evil, unjust*; AYENB. 159; Iw. 3134; wrang & crumb ORM. 9207; wrong REL. I. 210; SPEC. 31; WILL. 706; wrong '*curvus, oblongus*;' wrang (*sb.*) S. S. (Web.) 2900; wrong '*injuria*' PR. P. 534; BEK. 172; SPEC. 68; LANGL. *A* iii. 169; wrange (*dat.*) SAX. CHR. 252; wronge LAȝ. 27300; mid wronge P. L. S. viii. 106; hwo singeþ wronge (*adv.*) O. & N. 196; þeo ... habbeð þe nebbes wrong wende A. R. 254; wronger (*compar.*) WILL. 1176.

wrong-ful, adj., *wrongful*, P. S. 256; WICL. LK. xii. 58.

wrong-līche, adv., *wrongly*, AYENB. 8.

wrongnesse, sb., *crookedness*, '*curvitas*' PR. P. 534.

wrang-seht, adj., = *O. N.* rangsāttr; *at enmity*; wran(g)sehte (*pl.*) HOM. I. 241.

wrang-wīs, adj., *cf. Swed.* vrångvīs; *unjust*; wrongwise reven HOM. I. 175.

wrangwīs-lī, adv., *unjustly*, PR. C. 3865; LANGL. *A* iii. 76*.

wrangwīsnes, sb., *iniquity*, M. H. 136.

[wrangen, v., = *M.L.G.* wrangen; *strive, contend.*]

wranger, sb., = *M.L.G.* wranger; *one who strives*; (*ms.* wragger) TOWNL. 308.

wrangunge, sb., *strife*; (*ms.* wraggunge) A. R. 374*.

wranglen, v., *wrangle*; wrangle LANGL. *C* xvii. 80*.

wranglinge, sb., *wrangling*, LANGL. *B* iv. 34; wrangelunge (*? ms.* wragelunge) A. R. 374.

wrappen, v., *wrap*; wrappe CH. C. T. *E* 583; MED. 975; P. R. L. P. 226; wrapped (*pret.*) WILL. 745; wrappid RICH. 3094; wrapped (*pple.*) PL. CR. 435; MAND. 60; *comp.* bi-, for-, in-wrappen.

wrase, sb., *? = Swed.* vrase; *bundle* : one wrase of þornes MISC. 48; warse '*fasciculus*' CATH. 425.

wrāst, see wrǽst.

wrāð, adj., *O.E.* wrāð, = *O.L.G.* wrēth, *O.N.* reiðr, *O.H.G.* reid; *from* wrīðen; *wroth, angry; bad*; HOM. I. 15; þi les ðe god iwurðe wrað wið eou 117; wraþ ORM. 4814; wraþ MAP 360; Iw. 136; MIN. vii. 14; wrad, wroð [wroþ] LAȝ. 652, 3773 (*miswritten* wræð 8268); wroð A. R. 120; GEN. & EX. 1215; wroð, wroþ O. & N. 1043, 1608; wroþ ROB. 31; AYENB. 22; MAN. (H.) 86; S. S. (Wr.) 1798; wroþ and grim ALIS. 754; wroth or blithe CH. P. F. 504; wrooth HOCCL. i. 431; o wraðe time JUL. 57; wrāðer(e) (*dat. f.*) MARH. 10; wraðere [wroþere] LAȝ. 29556; wroðere A. R. 102; to wroþere hele MISC. 148; P. L. S. xxiv. 187; LANGL. *B* xiv. 120; wrothir haile ALEX. (Sk.) 1759; wrāðe (*pl.*) KATH. 173; wroþe HORN (L.) 348; wroþere (*compar.*) BEK. 361; wrāðest (*superl.*) LAȝ. 18583.

wrāþe, adv., *wrathfully, ill*, MISC. 75; wroðe HOM. II. 193; wroþe AYENB. 20; ived wroþe O. & N. 1529; þou ȝeldest nou mi love wroþe ROB. 31.

wrāð-līche, adv., *O. E.* wrāðlīce; *wrathfully*, LAȝ. 7379; wroþlice, -li WILL. 2074, 4598.

wrǽðe, see wrǽððe.

wrāðien, wrǽþþen, v., *O.E.* (ge-)wrāðian, *make wroth, become wroth*; wraþen WICL.

PROV. xviii. 14; wrathe [wraþþe] CH. C. T. *H*
80; wraththe, wratthe, wrathe LANGL. *A* ii.
85, *B* ii. 116, *C* iii. 118; þe king bigon to
wraðͤen KATH. 746; þat tu ne darst noht
drihtin wraþþen ORM. 5615; wraþþen AN. LIT.
3; wraþþe C. L. 594; wreͨðen A. R. 312;
wrōþe (*pres. subj.*) S. S. (Wr.) 1780; wrāðede
[wreþþede] (*pret.*) LAȝ. 4577; wraþed WILL.
981; MAN. (H.) 129; A. P. ii. 230; *comp.* a-,
ȝe-wrāðien.

wrᾰððe, see wræͨðͦe.

wrāu, wrāw, *see* wrāh.

wraxlien, v., *O.E.* wraxlian, = *O.Fris.* wraxlia;
strive, wrestle; wraxli LAȝ. 1858*; P. L. S.
xii. 70; wraxle [wraskle] LANGL. *C* xvii. 80;
wraxlede (*pret.*) LAȝ. 24699*.

    wraxlinge, sb., *O.E.* wraxlung; *strife, con-
tention,* LAȝ. 1871*.

wrec, *see* wrac.

wrecche, adj., *O.E.* wræcc; *from* wréken;
wretched, A. R. 56; ORM. 3878; R. S. ii
(MISC. 162); þu wrecche wiht O. & N. 556;
mi wrecche lif BRD. 23; a wrecche mon LAȝ.
3474; þat wrᾰcche volc 23571; þet wrᾰcche
lif FRAG. 5.

    wrecche-dōm, sb., *misery,* A. R. 232; JUL.
49; wrecchedom MISC. 152.

    wrecc-hēde, sb., *misery,* O. & N. 1219;
BEK. 1282; wrecehed SAX. CHR. 262.

    wrᾰcche-līche, adv., *miserably,* LAȝ. 30554;
wrecchelike, -liȝ ORM. 3326, 3708.

    wrechenesse, sb., *misery,* REL. II. 227.

wrecche², sb., *O.E.* wrᾰcca, wrecca, = *O.L.G.*
wrekkio, *O.H.G.* reccho; *wretch,* FRAG. 5;
HOM. I. 129; KATH. 2078; ORM. 10140; O.
& N. 534; wrecche SPEC. 38; CH. C. T. *A*
931; a slouȝ wrecche TREAT. 138; wrechche
AYENB. 31; wrᾰcche, wrehche [wrecche] LAȝ.
3860, 5932; wrecchen (*pl.*) P. L. S. viii. 142;
þa wrᾰcchen LAȝ. 20898; þes wrecche MISC.
75; wrecches GEN. & EX. 1074.

wrecched, adj., *wretched,* CH. C. T. *A* 921;
wreched PR. C. 557; wrecchede bali ST. COD.
99.

    wrecched-hēde, sb., *misery,* ROB. 102; PS.
xi. 6.

    wrecched-līche, adv., *wretchedly,* MAND.
251.

    wrecchednesse, sb., *wretchedness,* LANGL.
B xi. 44.

wrecchen, v., *O.E.* wreccan; *rouse* : and of
his eire briddes wraȝte [wrauhte] (*pret.*) O. &
N. 106; *comp.* a-wrecchen.

wrēche, *see* wrᾰche.

[wrēȝe, sb., = *M.L.G.* wrōge, *M.H.G.* rüege,
Goth. wrōhs; *? accusation.*]

    wrēi-ful, adj., *accusatory,* A. R. 302.

wrēȝen, wrēȝhen, v., *O.E.* wrēgan, = *O.L.G.*
wrōgian, *O.Fris.* wrōgia, wrēja, *Goth.* wrōhjan,
O.N. rœgja, *O.H.G.* ruogen; *accuse, betray,*

ORM. 416, 2889; wreigen LK. xxiii. 2; wreien
A. R. 304; wreghe PR. C. 5462; wreie FL. &
BL. 533; (*ms.* wreye) MISC. 39; CH. C. T. *A*
3503; wreie BEV. 1211; wraie AYENB. 175;
ȝif þu wrēiest (*pres.*) þe selfen HOM. I. 27;
wrēinde [wreȝinde] (*pple.*) A. R. 2; wrēide
(*pret.*) SAX. CHR. 260; BEK. 1226; wreiden
MK. xv. 3; wrēghed (*pple.*) IW. 2859;
comp. bi-, for-, ȝe-wrēȝen.

    wrēiere, sb., *O.E.* wrēgere, = *M.L.G.* wrō-
ger; *accuser,* HOM. I. 57; wrēieres (*pl.*)
HAV. 39.

    wrēiunge, sb., *O.E.* wrēging (ÆLFR. 317),
= *O.Fris.* wrōginge; *accusation,* A. R. 200;
wreinge REL. I. 267.

wréken, v., *O.E.* wrecan, = *O.L.G.* wrecan,
Goth. wrikan (διώκειν), *O.N.* reka, *O.H.G.*
rechan; *avenge; wreak*; ORM. 914; þet þu
wult for te wreken þe reaven god his strencͨe
A. R. 286; wreken HAV. 327; þat heo hine
wreken wolden LAȝ. 1627; wreke WILL.
1111; CH. C. T. *A* 961; PERC. 1065; TOWNL.
286; hvanne he his ssel wreke out of his velaȝrede AYENB. 189; þe devel fram hir for
to wreke GREG. 216; wrek þe on þat þuef SPEC.
106; wreoke (*subj.*) HOM. I. 209; wrak
(*pret.*) AR. & MER. 5972; TRIST. 3253;
GAM. 303; wrak AYENB. 215; wreken LAȝ.
13749; wréken [*O.E.* wrecen, = *O.N.* rekinn,
O.H.G. (gi-)rochan] (*pple.*) GEN. & EX.
2028; wreken HAV. 2368; TRIST. 3295; (h)a
wolde be wreken FER. 1949; wroke 5431;
wreke S. S. (Web.) 350; SACR. 212; wreken
LANGL. *A* ii. 169 [wroke *B* ii. 194]; wroken
GEN. & EX. 3191; wroken MAN. (F.) 8354;
wroken fro uch a woþe A. P. i. 375; wrokin
TOWNL. 115; *comp.* a-, ȝe-, on-wréken;
deriv. wrak, wráke, wrecche, wrᾰche.

    wrékere, sb., = *M.L.G., M.Du.* wreker;
avenger, CH. P. F. 361; wreker WICL. JOSH.
xx. 4.

wrench, sb., *O.E.* wrenc, = *M.H.G.* ranc;
wrench, guile, deceit, A. R. 338; R. S. i (MISC.
156); þeȝ he ne cunne wrench bute anne O.
& N. 811; wrench MISC. 36; REL. I. 175;
AYENB. 129; wrenche RICH. 4050; wrenk
M. H. 2; wrenke MAN. (H.) 58; wrenche
(*dat.*) LAȝ. 81; ROB. 55; wrenches (*pl.*) A.
R. 270; O. & N. 798; S. S. (Web.) 1756;
CH. C. T. *G* 1081; HOCCL. i. 378; wrenkes
and wiles PR. C. 1360; wrenche (*gen. pl.*)
O. & N. 813; wrenchen (*dat. pl.*) LAȝ. 1894;
wrenche P. L. S. viii. 127; þat is aferd of
plaites wrenche O. & N. 472; *comp.* un-
wrench.

    wrench-ful, adj., *deceitful, artful,* A. R. 268;
KATH. 892.

wrenchen, v., *O.E.* wrencan, = *M.H.G.*
renken; *wrench, twist*; A. R. 222; wrenchen
ut of þe weie MARH. 4; wrenche HOM. I.
281; wrinchand (*pple.*) PR. C. 1538; wreint
(*pple.*) P. S. 157; *comp.* ᾰt-, bi-wrenchen.

wrengðe, sb., *from* wrang ; *distortion; depravity,* REL. I. 210.

wrenne, sb., *O.E.* wrenna, wrænna ; *wren,* '*regulus,*' VOC. 177, 188, 221 ; PR. P. 533 ; GOW. III. 349 ; wrenne, wranne O. & N. 564.

wrēon, *see* wrīhen.

wreoðien, v., *O.E.* wreðian, wreoðian,=*O.L. G.* wrethian ; *lean upon, trust in* ; wreoðeð (*pres.*) A. R. 252 ; þat ʒe wreoðieð ow on KATH. 859.

wrēst, wrēsten, wrēstlen, *see* wræst, *etc.*

wrēþe, *see* wræðe, wræððe.

wrēðen, *see* wræðen. wrēððe, *see* wræððe.

wricken, v.,=*Du.* wrikken, *Dan.* vrikke, *Swed.* vricka ; *move* ; wrikkend (*pple.*) REL. II. 216 ; þe devel wrickede (*pret.*) her and þer P. L. S. ix. 82.

wriden, v., *O.E.* wridan ; *spread out, flourish* ; wride PALL. ii. 207.

wrīe (? wrīe), adj. = *M.L.G.* wrige ; *wry* ; a wrīe (*adv.*) *awry* REL. I. 248 ; R. R. 291.

wrien, v., *O.E.* wrigian ; *turn, twist* ; wrie CH. TRO. ii. 906 ; wrieð (*pres.*) R. S. iii ; hue wrieþ a wei SPEC. 48 ; wrí (*imper.*) MIRC 888 ; wried (*pret.*) CH. C. T. *A* 3283 ; wríed (*pple.*) GENER. 5957.

wrīen, *see* wrīhen.

wrigtful, wrigtelēs, *see under* wurht.

wrīheles, wrīeles, sb., *O.E.* wrīgels ; *veil,* A. R. 322, 420* ; wriels [wrielis] WICL. JOB xxiv. 8.

wrīhen, v., *O.E.* wrīon, wrēon,=*O.H.G.* (int-)-rīhan ; *cover, hide, veil :* wummon . . . schal wrihen hire heavet A. R. 420* ; to wrien & te helien þet gongþurl 84 ; wrihe HOM. I. 279 ; wri (*ms.* wry) AYENB. 258 ; wreon A. R. 27859 ; ALIS. 1606 ; wrīe (*pres.*) CH. C. T. *D* 1827 ; wrihð A. R. 84 ; wriʒþ, wrieþ AYENB. 61, 167 ; (heo) wreoð A. R. 88 ; wrīe (*imper.*) H. S. 1147 ; wreoð LAʒ. 17762 ; wrēih (*pret.*) A. R. 390 ; wreih [wrei] C. L. 918 ; wreigh AM. & AMIL. 2333 ; wriʒen [wreʒe] LAʒ. 17349 ; wrien MISC. 64 ; wreghen AR. & MER. 1774 ; wrugen P. L. S. viii. 81 ; wrīen (*pple.*) R. R. 912 ; iwriʒen [iwroʒe] LAʒ. 5192 ; iwrien [iwriʒen] A. R. 58 ; iwriʒe, iwreʒe AYENB. 66, 96 ; iwrie ROB. 55 ; LANGL. *B* xiv. 232 ; CH. C. T. *A* 2904 ; *comp.* bi-, over-, un-wrīen (-wrēon).

wriht, wrihte, *see* wurht, wurhte.

wringe, sb., *O.E.* (wīn-)wringe (MAT. xxi. 33) ; *press,* PALL. xi. 107.

wringen, v., *O.E.* wringan,=*M.L.G.* wringen, *M.Du.* wringhen, *O.H.G.* ringan ; *wring, twist, press,* LANGL. *B* xiv. 18 ; washen and wringen HAV. 1233 ; wringe REL. I. 115 ; OCTOV. (W.) 212 ; and gan wringe his lippe HORN (H.) 1105 ; (h)is honden gon he wringe P. S. 193 ; win to wringe L. H. R. 138 ; wringeþ (*pres.*) P. S. 188 ; CH. C. T. *E* 1553 ; wringeð ut A. R. 322 ; wringinde (*pple.*)

HORN (L.) 112 ; wrang (*pret.*) IW. 2773 ; wrong GEN. & EX. 2064 ; wrong HICKES I. 228 ; S. S. (Wr.) 2508 ; to grounde he him wrong TRIST. 3262 ; he wrungen hondes HAV. 152 ; wrongin ALIS. 7891 ; þe wrecchis wringet þe mok P. L. S. i. 20 ; wrungun [wrongun] (*pple.*) WICL. NUM. xxiv. 10.

wringer, sb., *wringer,* P. L. S. i. 21 ; wringeris (*pl.*) *winepress treaders* WICL. IS. xvi. 10.

wrinkil, sb., ? *O.E.* wrincle,=*M.Du.* wrinkel ; *wrinkle,* '*ruga,*' PR. P. 534 ; wrinclis (*pl.*) WICL. GEN. xxxviii. 14.

wriste, sb., *O.E.* wrist, *cf. M.L.G.* wrist, *O.Fris.* (hand-)wirst, *O.N.* rist *f.,* *M.H.G.* rist *m.,* riste *f.;* ? *from* wrīðen ; *wrist* ; wrist(e), wirste PR. P. 534 ; wirste VOC. 208 ; *comp.* hand-wriste.

writ, sb., *O.E.* writ,=*O.N.* rit ; *writ, writing, scripture,* A. R. 98 ; ORM. DEDIC. 331 ; GEN. & EX. 1974 ; writ PROCL. 7 ; HAV. 2486 ; C. L. 76 ; AUD. 10 ; þat writ LAʒ. 484 ; þet holi writ MISC. 36 ; writte of þe kingis coort, '*breve*' PR. P. 534 ; write (*dat.*) LAʒ. 25005 ; write (*pl.*) SAX. CHR. 250 ; writen LAʒ. 9131 ; writen (*dat. pl.*) LAʒ. 22981.

writ-rūne, sb., *letter, document* ; writrūnen (*pl.*) LAʒ. 5750.

wrīten, v., *O.E.* wrītan,=*O.L.G.* wrītan, *O.N.* rīta, *O.H.G.* rīzan (*tear*) ; *write* ; LK. i. 3 ; A. R. 42 ; ORM. 3554 ; writen LAʒ. 41 ; ALIS. 4793 ; wrīteð (*pres.*) LAʒ. 28869 ; writ FRAG. I ; HOM. I. 21 ; O. & N. 1756 ; wrīte (*subj.*) ORM. DEDIC. 97 ; wrāt (*pret.*) ORM. 5816 ; wrat LAʒ. 6317 ; FLOR. 2175 ; wrot A. R. 12 ; O. & N. 235 ; GEN. & EX. 462 ; wrot BEK. 164 ; wrote LANGL. *B* x. 169 [wrot *A* xi. 125] ; wroot ALIS. 4778 ; writen ORM. 5810 ; writen LANGL. *B* x. 428 ; write GOW. III. 85 ; write (*subj.*) MARH. 23 ; writen (*pple.*) ORM. 3085 ; writen CH. C. T. *G* 210 ; PR. C. 2042 ; write MISC. 41 ; *comp.* ʒe-writen.

writere, sb., *O.E.* writere ; *writer, scribe,* MARH. 2 ; wrīteres (*pl.*) MAT. ii. 4.

wrītunge, sb., *writing,* A. R. 80 ; writing AYENB. 8.

wrīþe, sb., *O.E.* wrīða ; *twist* ; wrīþen (*pl.*) ALIS. 5723 ; wrestes aiþer writh(e) C. M. 3462.

wrīðen, v., *O.E.* wrīðan,=*O.N.* rīða, *O.H.G.* (ki-)rīdan ; *writhe, twist, turn, bind,* LAʒ. 17394 ; wriþen a wei CH. BOET. v. 3 (154) ; of þe wai . . . ne wil he wriþe A. P. i. 350 ; wrīþinde (*pple.*) as a wond MAP 336 ; wrāð (*pret.*) LAʒ. 6729 ; wroþ S. S. (Web.) 1792 ; MAN. (F.) 13054 ; wroth GAW. 1200 ; wrooþ WICL. 2 KINGS xxiii. 21 ; writhen (*ms.* uurythen) SAX. CHR. 262 ; he wriþen him one crune MISC. 48 ; wriþen (*pple.*) LANGL. *B* xvii. 174 ; writhin, wreþin '*tortus*' PR. P. 534 ; a wrethin haft S. & C. I. lxi ; *comp.* ʒe-, to-wrīðen ; *deriv.* wrīðe, wrāð, wrææðe.

wrixlen, v., *O.E.* wrixlian ; *envelop; wrap* ;

exchange; wrixle D. TROY 3120; **wrixlis**, wrixles (*pres.*) D. TROY 445, 9327; **wrixlit** (*pret.*) 2061; wel bisemez þe wiʒe **wrūxled** (*pple.*) in grene GAW. 2191; *comp.* bi-**wrixlen.**

wrixlunge, sb., *O.E.* wrixlunge; *change*, HOM. I. 207.

wrō, *see* **wrā.** **wrōʒ**, *see* **wrāh.**

wrohte, *see* **wurhte.** **wrong**, *see* **wrang.**

wrōt, sb., *O.E.* wrōt; *snout*, LEG. 43; REL. II. 210.

wrōten, v., *O.E.* wrōtian, *cf. M.L.G.* wrōten, *M.Du.* wroeten, *O.H.G.* ruozen, *O.N.* rōta; *from* **wrōt**; *root, turn up with the snout*, REL. II. 216; wrotin PR. P. 534; wrote MIN. 23; ORF. 239; AV. ARTH. xii; **wrōteð** [wroteþ] (*pres.*) LAʒ. 469; wroteþ CH. C. T. *I* 157; (hie) wroteð HOM. II. 37; *comp.* under-**wrōten.**

wrōð, *see* **wrāð.** **wruhte**, *see* **wurhte.**

wrunkel, sb., = *M.Du.* wronkel; *cf.* **wrinkel**: *wrinkle*; runkel C. M. 18840.

wrunklen, v., = *M.Du.* wronkelen; *wrinkle*; rouncles (*pres.*) PR. C. 773; ronkled (*pple.*) GAW. 953.

wrusum, *see* **wursum.** **wu**, *see* **hwu.**

wū, interj., ? = *M.H.G.* wū; *woe*; wu me HOM. I. 157; (*ms.* wumme) ['*heu mihi*'] HOM. II. 149; MARH. 13.

wude, sb., *O.E.* wudu, widu (EP. GL.), = *M.L. G., M.Du.* wede, *O.H.G.* witu (GRAFF), *O.N.* viðr *m.; wood*, A. R. 402; ORM. 14568; GEN. & EX. 1306; wude (*printed* þude) LAʒ. 731; þene wude [wode] 8700; wode 1237; wode ['*nemus*'] WICL. GEN. xxi. 33; MAND. 50; CH. C. T. *A* 1422; þane wode AYENB. 95; **wude** [*O.E.* wuda] (*gen.*) HORN (L.) 1227; wode ROB. 88; JOS. 475; under a wilde wode side P. L. S. xxv. 1; under þe wode gore SPEC. 91; wudes [wodes] LAʒ. 8687; **wude** [wode] [*O.E.* wuda] (*dat.*) O. & N. 615; wudan ANGL. VII. 220; **wude** [wodes] [*O.E.* wudu] (*pl.*) LAʒ. 11772; wudes SAX. CHR. 256; MARH. 10; **wuden** (*dat. pl.*) LAʒ. 16126.

wude-bínde, sb., *O.E.* wudubind, -binde; *woodbind* (-*bine*), VOC. 140; wodebinde PR. P. 531; CH. C. T. *A* 1508.

wude-brūne, sb., '*buglossa*,' REL. I. 36.

wode-craft, sb., *woodcraft*, CH. C. T. *A* 110.

wode-douve, sb., *wood-dove*, CH. C. T. *B* 1960; wodedove '*palumba*' PR. P. 531.

wode-hake, sb., *? woodpecker*, '*picus*' PR. P. 531.

wude-hunig, sb., *O.E.* wuduhunig; *forest honey*; ['*mel silvestre*'] MAT. iii. 4.

wode-koc, sb., *O.E.* wuducoc; *woodcock*, VOC. 164; wodekok PR. P. 531.

wode-land, sb., *O.E.* wuduland; *woodland*, MAN. (F.) 1972; wodelond LAʒ. 1699.

wude-lēie [wodelēʒe], sb., *grove*; **wudelēi** (*dat.*) A. R. 96; wodeleie ST. COD. 58; under wodeleie HORN (R.) 1160; wudeliʒe HORN (L.) 1158.

wude-merch(e), sb., *O.E.* wudumerce; *wood-parsley*, '*saniculum*,' REL. I. 36.

w(u)de-minte, sb., '*origanum*,' REL. I. 37.

wode-rime, sb., *edge of the wood*, LAʒ. 739.

woode-línde, sb., *linden tree of a forest*, GAM. 676, 702.

wode-rōte, sb., *root of the wood*, LAʒ. 467.

wode-rīs, sb., *bushwood*, GAM. 771, 803; under þis wode ris DEGR. 172.

wude-rōve, sb., *O.E.* wudurōfe; *woodroof*, '*hastula regia*,' VOC. 140; woderove PR. P. 531; SPEC. 43 (A. D. 164).

wude-scaʒe [wodesaie], sb., *thicket of the wood*, LAʒ. 21561, 27367; wodeschawe B. DISC. 1112; GAM. 638; ISUM. 74; wodeshawe REL. I. 245.

wode-scaþe, sb., *monster of the wood*, LAʒ. 25859*.

wode-schīde, sb., *split wood*, '*teda*,' PR. P. 531.

Wude-stoc, pr. n., *O.E.* Wudestoc; *Wood-stock*; **Wudestoke** (*dat.*) SAX. CHR. 249; Wodestoke ROB. 439.

wode-sūre, sb., *O.E.* wudusūre; *wood-sorrel*, LEECHD. III. 349.

wude-wale, sb., = *M.L.G., M.Du.* wedewal (*galgulus, oriolus*), *M.H.G.* witewal (*oriolus*): *witwall* (*bird*), O. & N. 1659; wodewale '*picus*' PR. P. 531; R. R. 658.

wode-ward, sb., *forest-keeper*, VOC. 164; PR. P. 531; P. S. 149.

wode-wexe, sb., *O.E.* wuduweaxe; *some plant*; wodewexen (*dat. pl.*) E. G. 358.

wode-wōse, sb., *O.E.* wuduwāsa; *satyr, faun*, '*satyrus*,' PR. P. 531; ['*pilosus*'] WICL. IS. xxxiv. 14; GAW. 721; full of wod-wōse (*pl.*) ALEX. (Sk.) 1540.

wudi, adj., *woody*; wodi WICL. NUM. xiii. 21.

wuke, *see* **wike.** **wül**, *see* **wil.**

wulder, sb., *O.E.* wuldor *n., cf.* Goth. wulþus (*gen.* wulþaus) *m.* (δόξα); *glory*; (*ms.* wull-derr) ORM. 3379; alre wuldre FRAG. 7.

wulf, sb., *O.E.* wulf, = *Goth.* wulfs, *O.N.* ulfr, *O.H.G.* wolf; *wolf*, KATH. 31; A. R. 120; wulf H. S. 5523; LUD. COV. 68; wulf, wlf [wolf] LAʒ. 1545, 21305; wlf HAV. 573; wolf ROB. 280; CH. C. T. *A* 313; wulves fleit '*fungus*' REL. I. 37; **wulve** (*dat.*) A. R. 252; **wulfes** (*pl.*) LK. x. 3; wulves O. & N. 1008; wulfes MIR. PL. 150; wolves CHR. E. 732; **wlfan** (*dat. pl.*) LAʒ. 2599; *comp.* sē-, wer-**wolf.**

wulle, sb., *O.E.* wull, = *Goth.* wullu, *O.N.* ull,

O.H.G. wolla; *wool*, ORM. 12652; wulle
'*lana*' PR. P. 534; GOW. I. 17; wolle ROB.
351; P. L. S. xix. 249; AYENB. 137; LANGL.
A vi. 13; MAND. 208.

wul-lok, sb., *lock of wool*, PR. P. 534.

wolle-mongere, sb., *woolstapler*, E. G.
353; **wolmongers** (*pl.*) ROB. (W.) 11173.

wol-pack, sb., *woolpack*; **wolpackes** (*pl.*)
ROB. (W.) 11171.

wolle-ward, adj. & adv., *clothed in wool*:
wolleward and wete shoed LANGL. *B* xviii. 1;
wolward PL. CR. 788; wol warde PR. C. 3512.

wolle-webstere, sb., *wool weaver*, LANGL.
B PROL. 219.

wülle, *see* **welle**.

wullen, adj., *cf. O.E.* wyllen, = *O.H.G.* wullīn;
woollen; wollen ALIS. 4459; LANGL. *B* v.
215.

wüllen, *see* **willen**.

wülvene, sb., *O.E.* wylfen (LEECHD. I. 362), =
M.H.G. wulfinne; *she-wolf*, A. R. 120; wlvine
HAV. 573.

wumman, *see* **wifman**.

wúnd, adj., *O.E.* wund, = *O.L.G.* wnd, *Goth.*
wunds, *M.H.G.* wunt; *wounded*: þe wunde
studen A. R. 136.

wúnde, sb., *O.E.* wund, = *O.L.G.* wnda, *O.H.G.*
wunta, wnta, *O.N.* und; *wound*, A. R. 326;
ORM. 2218; wunde, wnde [wonde] LA3. 2285;
8604; wunde HAV. 1980; wnde (*r. w.* grunde)
MISC. 42; wonde AYENB. 174; FER. 502;
wounde C. L. 1198; CH. C. T. *A* 1010; **wúnde**
(*pl.*) MISC. 140; wounde SPEC. 97; BEV.
724; wunden LK. x. 34; A. R. 274; wunten,
wundes HOM. II. 19, 33; wunden [wundes]
LA3. 16589; wounden ROB. 223; wondes PR.
C. 5337; **wúnden** (*dat. pl.*) HOM. I. 75;
wounden SPEC. 82; wunde LEG. 210.

wunder, sb., *O.E.* wundor, = *O.L.G.* wndar,
M.Du. wonder, *O.H.G.* wuntar, wntar, *O.N.*
undr; *wonder, miracle, prodigy*, KATH. 152;
ORM. 218; GEN. & EX. 69; 3if him þunche∂
wunder A. R. 8; wunder hit is þat heo nawe-
deþ O. & N. 1384; nis wunder nan (*it is no
wonder*) þah he (n)abide 1389; wunder 361;
wunder, wnder [wonder] LA3. 1126, 4659;
wundur ANT. ARTH. xvi; wonder C. L. 619;
CH. C. T. *E* 337; þat wonder BEK. 86; no
wonder þah me be wo SPEC. 81; have∂ i-
schriven hire al to wundre (*marvellously*) A. R.
68; to tuken swa to wundre H. M. 27;
wunder (*pl.*) SAX. CHR. 262; wundre LK.
v. 26; HOM. I. 139; mihte þe luþer mon don
al þe wonder þat he con A. D. 298; wundres
A. R. 390; **wundre** [wnder] (*gen. pl.*) O. & N.
852.

wonder-blíþe, adj., *wonderfully blithe*,
WILL. 1895; wunderbli∂e LA3. 25542.

w(u)nder-craft [wondercraft], sb., *O.E.*
wundorcræft; *strange art, magic*; **wonder-
crafts** (*pl.*) LA3. 1147.

w(u)nder - crefti [wondercrafti], adj.,
strangely artful, LA3. 1154.

wonder-dēde, sb., *O.E.* wundordǣd, = *M.Du.*
wonderdaed, *M.H.G.* wundertät; *wonderful
deed*, ROB. 337.

wunder-feole, sb., *cf. M.H.G.* wundervil;
very many, LA3. 22900.

wunder-ful, adj., *cf. Ger.* wundervoll; *wonderful*, C. L. 62; wonderfol LA3. 280*;
wondervol AYENB. 15.

wunder-gōd, adj., = *O.H.G.* w(u)nterguot
marvellously good, LA3. 25475.

wunder-grēt, adj., = *Ger.* wundergrosz;
strangely great, '*immanis*,' PR. P. 534.

wunder-hēh, adj., *very high*, ORM. 12055.

wunder-kēne, adj., *very bold*, LA3. 7277.

wunder-līc, adj., *O.E.* wundorlīc, = *O.L.G.*
w(u)ndarlīc, *O.H.G.* wuntarlīch; *marvellous*,
LA3. 12308; wunderlich MAT. xxi. 42; A. R.
390; an wunderlike taken ORM. 15461;
wonderlīche (*adv.*) SHOR. 162; WILL. 345;
woundirliche S. S. (Wr.) 2678; wondirli
WICL. JOB x. 16; PERC. 2004; **wunder-
lüker** (*compar.*) HOM. I. 93; **wunder-
līcheste** (*superl.*) MISC. 60; wunderlikeste,
-lukeste P. L. S. viii. 34.

wonder-longe, adj., *very long*, LANGL. *B*
xv. 1; CH. C. T. *A* 1654.

wunder-mēre, adj., *? very gloriously*, LA3.
15641.

w(u)nder-müchel, adj., *very great*, LA3.
25078; wundermikel ORM. 7284.

wundernesse, sb., *marvel*, MISC. 85.

w(u)nder-rīche, adj., *very powerful*, LA3.
25295.

w(u)nder-strong, adj., *very strong*, LA3.
1744.

wonder-þing, sb., = *Ger.* wunderding; *wonderful thing*, BRD. 31; C. L. 644; GREG. 393.

wunder-wel, adj., *wondrous well*, LA3.
11361; wonderwel iwroht A. D. 158.

wonder - worc, sb., *O.E.* wundorweorc;
marvellous work, LA3. 17376*.

wúndien, v., *O.E.* wundian, = *O.H.G.* wuntōn; *wound*; wundie HOM. I. 15; wunden
A. R. 124; ORM. 12484; wondi, woundi
TREV. III. 459; **wúnde∂** (*pres.*) H. M. 15;
wóndede (*pret.*) BEK. 2125; wundeden
HOM. II. 33; wundeden [wondede] LA3.
1724; **wúnded** (*pple.*) GEN. & EX. 853;
wounded CH. TRO. ii. 533; *comp.* **for-, 3e-
wúndien.**

wundrel, sb., *miracle*, '*prodigium, ostentum*'
PR. P. 534.

wundren, v., *O.E.* wundrian, = *O.H.G.* wun-
tarōn; *wonder*, ORM. 15475; **w(u)ndri**
(*pres.*) O. & N. 228; ne w(u)ndre þou nawiht
þer fore LA3. 473; ne wundrie heo hire nowiht
A. R. 376; **wondrinde** (*pple.*) AYENB. 267;

wundrede (*pret.*) KATH. 377; hem wondrede alle BEK. 101; þe fend wondrede swiþe C. L. 1039; i wondred me R. R. 738; wundreden LA3. 21979; *comp.* a-, for-, of-wundren.

wundrunge,sb.,*O.E.* wundrung; *wondering*, HOM. I. 89.

wune, sb., *O.E.* (ge-)wuna,=*M.L.G.* wone, *O. L.G.* (gi-)wono, *O.H.G.* (gi-)wona, *O.N.* vani *m.; custom, habit; habitation, dwelling*; A. R. 266; GEN. & EX. 513, 1681; wune MISC. 87; REL. I. 217; FL. & BL. 557; he dude al se hit is wune [wone] LA3. 11184; ne mæi i noht . . . habben . . . þesne wone 13492; wone ROB. 287; C. L. 278; AYENB. 6; LANGL. *B* xv. 177; CH. C. T. *A* 335; H. F. 1166; MAN. (H.) 75; a3ens wone SHOR. 164; wile þou art in þis wreche wone P. L. S. ii. 164; wane PERC. 1347; woon CH. H. F. 1166; **wune** (*pl.*) HOM. II. 105; wunen GEN. & EX. 688; wones LANGL. *B* iii. 234; woanes [wanes] A. R. 416; wanes MARH. 21; *comp.* 3e-wune.

wun-sele, sb., *dwelling-hall, mansion*, LA3. 15703.

wune[2], adj., *O.E.* (ge-)wuna, = *O.H.G.* (gi-)-won, *O.N.* vanr; *accustomed*, GEN. & EX. 1530; wone HAV. 2297; OCTOV. (W.) 1062; MAN. (F.) 14514; ne were þou wone be god and mild S. S. (Web.) 672; *comp.* 3e-wune.

wunien, v., *O.E.* wunian, *cf. O.L.G.* w(u)nōn, wonōn, wanōn (*inhabit*), *O. H. G.* wonēn (*inhabit*), giwonēn (*be accustomed*); *dwell, inhabit; be accustomed*; A. R. 134; KATH. 921; wunian, wnian, wnien HOM. I. 13, 25, 63; wunien (*ms.* wunyen) MISC. 83; wunien, wonien [wonie] LA3. 478, 1205; þær he wunen sholde ORM. 8708; to wunen swa þe folc þer to 19541; wunen GEN. & EX. 306; wune HORN (L.) 731; wonien REL. I. 103; LEB. JES. 283; wonie HORN (R.) 735; (*ms.* wonye) TREAT. 138; AYENB. 220; wone LANGL. *A* ii. 200 [wonie *B* ii. 224]; wone HAV. 247; PS. lxiv. 5; GOW. I. 340; WICL. PROV. xiv. 9; PR. C. 16; M. ARTH. 2445; to wunienne FRAG. 7; HOM. I. 117; **wunest** (*pres.*) O. & N. 338; wuneð HOM. I. 21; wuniað 27; heo wunieþ on eorþe FRAG. 7; wonieþ AYENB. 87; **wune** (*imper.*) A. R. 162; wunieð ou to lutel drunch A. R. 412; **wuniende**, wune3ende (*pple.*) LA3. 19217, 29278; **wunede** (*pret.*) HOM. II. 85; (*printed* þunede) LA3. 5274; wonede ALIS. 5269; woneden CH. C. T. *A* 2927; **woned** (*pple.*) SPEC. 47; REL. II. 274; H. H. 46; CH. C. T. *E* 339; wunt or usid '*assuetus*' PR. P. 534; *comp.* 3e-, þurh-, un-wunien.

wununge, sb., *O.E.* wunung, *cf. O.H.G.* wononga; *habitation, mansion*, HOM. I. 105; A. R. 74; KATH. 2456; wuninge MISC. 144; wuning [woning] O. & N. 1760; woninge P. L. S. xv. 75; woning LA3. 7094*; SPEC. 102; CH. C. T. *A* 606.

woning-stede, sb., *dwelling-place*, REL. I. 167.

wünne, **wunne**, sb., *O.E.* wynn, wenn, wunn, *cf.O.L.G.* wnnea, *O.H.G.* wunna; *joy, pleasure, delight*, A. R. 192; KATH. 1511; H. M. 27; ? winne 39; hit is min hihte hit is mi wunne O. & N. 272; wunne SPEC. 47; wunne, winne [winne, wonne] LA3. 3099, 14160, 21474, 22732; winne HAV. 2965; ALIS. 3887; MAN. (F.) 14908; E. T. 1024; TRIAM. 533; M. ARTH. 3788; paradises winne C. L. 1312; wenne S. S. (Web.) 2581; **wünne** (*pl.*) MISC. 126; alle mine wunnen LA3. 22668; **wünnen**, wnnen (*gen. pl.*) LA3. 1124, 3421; **wünnen** (*dat. pl.*) MARH. 1; JUL. 79; *comp.* un-weoreld-wünne.

w(ü)n-folc, sb., *joyous folk*, LA3. 1385.

winne-halle, sb., *joy-hall*, GAW. 2456.

wün-līc, adj., *O.E.* wynlīc,=*O.H.G.* wunnilīh; *joyous, delicious*, LA3. 8090; winli WILL. 749; his wonliche cun LA3. 8772.

wünsum, sb., *O.E.* wynsum, = *O.L.G.* w(u)nsum, *O. H. G.* wunnisam; *winsome, delightful*, MARH. 19; wunsum [wonsom] LA3. 1187; winsum ['*suave*'] MAT. xi. 30; winsom ['*jucundus*'] PS. lxxx. 3; **wnsumre** (*compar.*) LA3. 905.

wünsumen, v., *O.E.* wynsumian; *be winsome*; **winsomes** (*pres.*) PS. cii. 3.

wunton, v., *from* **wunt** *pple. of* **wunien**; *accustom; frequent, practise, '*assuefacio, usito*'* PR. P. 534.

wurc, *see* weorc.

wurchen, sb., *O.E.* wyrcean, wyrcan, wircan, wercan, weorcean,=*O.L.G.* wirkean, *O.Fris.* werka, wirtsa, *O.H.G.* wurchan, wirchan, wirkan, *O.N.* yrkja, *Goth.* waurkjan; *work, do*, FRAG. 7; A. R. 6; wraðe workes wurchen KATH.174; wurchen, wrchen [werche, wirche] LA3. 1547, 5066; wrchen, wurche, wirche O. & N. 408, 722; wurche ROB. 232; he micte me wirchen michel wo HAV. 510; wirche [werche] CH. C. T. *A* 2759; worchen C. L. 17; worche MAND. 1; WICL. JOHN ix. 4; worche, wirche WILL. 307, 471; bigin to werchen god AN. LIT. 91; werchen [worche] LANGL. *A* i. 80; werche SHOR. 29; AYENB. 174; GOW. I. 63; AUD. 9; werche wo FER. 1545; chirches . . . werche sche dede GREG. 218; drihtines wille wirken ORM. 9342; wirke PR. C. 6905; ISUM. 398; wircan, wercan HOM. I. 221; werken GEN. & EX. 2799; to wurchene HOM. I. 117; **wrcheþ** (*pres.*) MISC. 148; (heo) wurcheð HOM. I. 139; **wurch** (*imper.*) JUL. 35; **worhte** [wrohte] (*pret.*) LA3. 12024; warhte HOM. I. 91; wrohte (*ms.* wrohhte) ORM. 2256; wrouhte A. R. 258; wrouhte [wrou3te] LANGL. *A* i. 80; wroute AN. LIT. 91; wrohten FRAG. 7; wrogten GEN. & EX. 529; wrou3ten WILL. 3873; **wroht** (*pple.*) SPEC. 31; wro3t MIRC 210; wrouht HAV. 2810; wrouht [wrou3t] C.

L. 3 ; LANGL. *A* iii. 101 ; WICL. GEN. xxxiv.
7 ; wrouht [wrouȝt, wroght] CH. C. T. *G*
1332 ; wrought LUD. COV. 28 ; *comp.* bi-,
for-, ȝe-wurchen.

wircher, sb.,=*M.H.G.* würkære ; *worker,*
CH. COMP. M. 261 ; worchere TREV. II. 177.

wirking, sb.,=*M.H.G.* wirkunge ; *working,*
PR. C. 4907.

wurd, *see* word.

wurde, sb., *O.E.* wyrd,=*O.L.G.* wrd, wrth,
O.H.G. wrt ; *from* wurčen ; *weird, fate* ;
wirde GAW. 2134 ; A. P. i. 249 ; ANT. ARTH.
xvi ; werd, werid, wird, wirid ALEX. (Sk.)
3247, 443, 689, 2260 ; wurdes (*pl.*) H. M.
33 ; wirdes LANGL. *C* xiii. 209 ; werdes
[wierdes] CH. BOET. i. 1 (4) ; HORN CH. 456 ;
ISUM. 202 ; werdes, wirdis ALEX. (Sk.) 270,
4950 ; werdis BARB. ii. 239 ; wordis, werdis
HAMP. PS. lxxiv. 5, cxl. 4.

wurȝen, v., *O.E.* (a-)wyrgan,=*O.Fris.* werga,
wirga, *M.L.G.* worgen, *M.Du.* worghen, *O.
H.G.* wrgan ; *mod.Eng.* worry ; *choke, throttle;
tear, worry* ; wirwin, worowen '*strangulo,
suffoco*' PR. P. 530 ; wiri(e) R. R. 6264 ; worowe
PR. C. 1229 ; wirhieþ [wirieþ] (*pres.*) LANGL.
C x. 226 ; wiriȝede (*pret.*) TREV. VII. 465 ;
wiriede P. R. L. P. 83 ; wiried (*pple.*) TREV.
VII. 534 ; wirwed, werewed HAV. 1915, 1921 ;
iworewid DEP. R. iii. 72 ; *comp.* a-wurien.

wurht, sb., *O.E.* (ge-)wyrht, *O.L.G.* (far-,
ge-)wrht ; *from* wurchen ; *thing done, merit* ;
wrihte (*pl.*) HOM. I. 69 ; bi mine wrihte
HOM. II. 217 ; after hise wrihte ORM. 8240 ;
comp. ȝe-wurht.

wrigt-ful, adv., *deservedly* ; wrigtful we in
sorwe ben GEN. & EX. 2204.

[wurht-lēas, adj.]

wrigtelēs-līke, adv., *wrongfully,* GEN. &
EX. 2076.

wurhte [wrohte], sb., *O.E.* wyrhta,=*O.H.
G.* wrhto, *O.L.G.* wrhteo ; *wright, workman,
carpenter,* LAȝ. 21134 ; wruhte MARH. 20 ;
JUL. 60 ; wrihte ORM. 18780 ; wriȝte LANGL.
B x. 401 ; wriȝte [wrihte, wrighte] CH. C. T.
A 614 ; wrihte, write PR. P. 534 ; wuruhte
(*gen.*) A. R. 284 ; wurhten (*pl.*) HOM. I. 93 ;
wurhten [wrohtes] LAȝ. 16969 ; werhten MK.
xii. 10 ; wirhten LK. x. 2 ; *comp.* gödspel-,
hwēol-, psalm-, schip-wurhte.

wurm, sb., *O.E.* wyrm, weorm,=*O.L.G.* wrm,
O.H.G. wurm, wrm, *Goth.* waurms, *M.Du.*
worm, *O.N.* ormr, *Lat.* vermis ; *worm, serpent,
dragon,* A. R. 138 ; ORM. 4870 ; wrm MISC.
149 ; worm TREAT. 139 ; C. L. 1129 ; GOW.
II. 299 ; wirm GEN. & EX. 2925 ; REL. I. 211 ;
wirm PR. P. 530 ; werm MISC. 28 ; AYENB.
215 ; LUD. COV. 29 ; wurmes (*gen.*) LAȝ.
17961 ; wurmes (*pl.*) FRAG. 6 ; MARH. 10 ;
wurmes [wormes] O. & N. 601 ; wormes BEK.
2237 ; LANGL. *B* xi. 320 ; wirmes, wirme
GEN. & EX. 178, 2982 ; wurmen (*gen. pl.*)
LAȝ. 15942 ; wurmene HOM. II. 35 ; H. M. 43 ;

wermene AYENB. 216 ; wurmen (*dat. pl.*)
FRAG. 6 ; LAȝ. 30500 ; *comp.* hond-, ring-,
silk-, wort-wirm(-werm).

werm-ēte, adj., = ? *M.H.G.* wurmæze ;
worm-eaten ; (*ms.* wermethe) AYENB. 229.

wurmen, *see* warmen.

wurnen, *see* wernen.

wurp, sb., *O.E.* wyrp,=*M.Du.* worp, *M.H.G.*
wurf ; *from* weorpen ; *cast, throw,* A. R. 56 ;
wrp MISC. 41 ; anes stanes werp LK. xxii. 41.

wurs, adj.,=*M.Du.* wers, wors ; *bad, wicked* ;
þe wurse, wrse [worse] *the devil* LAȝ. 1140,
2843 ; þe wurse, werse HOM. II. 187, 191 ;
wurs [*O.E.* wyrs, wirs,=*O.L.G.*, *O.H.G.* wirs,
Goth. wairs] (*compar.*) *worse* (*adv.*) P. L. S.
viii. 111 ; wurs, wors LAȝ. 3104, 3453 ; wurs,
wurs, wrs O. & N. 34, 793 ; wurs HORN (L.)
116 ; wors SHOR. 97 ; AYENB. 20 ; REL. II.
276 ; no þe wors hire nas MARG. 252 ; wers
M. T. 33 ; wurse [*O.E.* wyrsa,=*O.L.G.* wirsa,
Goth. wairsiza] *worse* (*adj.*) A. R. 82 ; a
wurse stede P. L. S. viii. 149 ; þe fal is wurse
H. M. 15 ; Bruttes hafden þat wurse LAȝ.
26712 ; worse LANGL. *B* xiv. 224 ; worse
[werse] CH. C. T. *A* 1224 ; werse ORM. 7395 ;
HAV. 1100 ; werse EGL. 293 ; ævre it was werse
& werse SAX. CHR. 262 ; wirse MAN. (F.)
8696 ; wurst [*O.E.* wyrst] (*superl.*) *worst*
(*adv.*) FRAG. 7 ; P. L. S. viii. 119 ; MARH. 14 ;
wurst HORN (L.) 68 ; werst ORM. 4250 ; werst
FER. 2809 ; wurste *worst* (*adj.*) A. R. 82 ;
wurste LAȝ. 29545 ; worste 28773* ; wrst(e)
SPEC. 99 ; werste GOW. I. 25 ; werst PR. P. 523.

wursien, v., *O.E.* wyrsian ; *render worse,*
HOM. I. 47 ; wersen ORM. 11845 ; worsi
AYENB. 33 ; wurseð (*pres.*) A. R. 326 ; wursi
(*subj.*) KATH. 2165 ; wursede [wersede]
(*pret.*) LAȝ. 18931 ; iwursed (*pple.*) HOM. I.
202 ; iworsed C. L. 811 ; iwersed PROCL. 6.

wursum, wrusum, sb., *O.E.* wyrms, worms,
worsm ; *purulent matter* ; wrusum [wursum]
['*sanies*'] A. R. 274 ; wrusum [wirsum] 322 ;
wirsen ORM. 4782 ; worsum and ware C. M.
11835.

wurt, sb., *O.E.* wyrt,=*O.L.G.* wrt, *O.N.* urt,
Goth. waurts, *O.H.G.* wurz, wrz *f.* ; *root* ; *herb* ;
GEN. & EX. 119 ; wurt, wrt REL. I. 175 ;
wert MAT. xiii. 26 ; wirte LK. xi. 42 ; wort
PR. P. 532 ; WICL. LK. xi. 42 ; wurtes (*pl.*)
VOC. 264 ; wortes GOW. III. 162 ; LANGL. *B*
vi. 310 ; CH. C. T. *E* 226 ; L. C. C. 54 ; wirte
(*gen. pl.*) MK. iv. 32 ; wurten (*dat. pl.*) LAȝ.
31884 ; *comp.* bān-, biscop-, blöd-, hul-, liþ-,
liver-, lung-, mēd-, mug-, ribbe-, slæp-,
spére-, stich-, wal-wurt.

wort-ȝerd, sb., *O.E.* wyrtgeard ; *kitchen
garden,* WICL. 3 KINGS xxi. 2.

wer(t)-tūn, sb., *O.E.* wyrttūn ; *garden,*
['*hortum*'] LK. xiii. 19.

worte-wale, sb., *O.E.* wyrtwalu,=*O.H.G.*
wurzala ; *root,* VOC. 172 ; S. & C. I. xxxvi.

wort-wirm, sb., *root worm*, '*eruca*,' PR. P. 532; worte-worm WICL. JOEL i. 4.

wurte, sb., *O.E.* (man-)wyrt (LEECHD. II. 96), = *M.Du.* worte, *M.L.G.* werte, *Swed.* vört, *M.H.G.* würze; *wort* (*in brewing*), '*idromellum*' VOC. 257; worte CH. C. T. *G* 813.

wurð, wurðe, adj. & adv., *O.E.* weorð, wyrð, wurð, wyrðe, wierðe, = *O.L.G.* werth, *O.N.* verðr, *Goth.* wairþs, *O.H.G.* werd ; *worth*, *worthy*, A. R. 38, 130; MARH. I, 5 ; þat is mare wurð KATH. 70; wurþ ORM. 1156; wurþ, wrþ, wrþ O. & N. 572, 1550; nis noht wurð [worþ] LAȝ. 26555 ; þat Vortiger weore wurðe [worþe] to walden þas þeode 13445; wurþ, wurþe ROB. 322, 458; wuirth LEG. 28; wurðe MAT. iii. 11; he bið deðes wurðe HOM. I. 109; he is wurðe þer to HOM. II. 93; wurþe CHR. E. 741; worþ, worþe BEK. 357, 2217; worþ SHOR. 95; worþ, worþe AYENB. 7, 74; his worþ(e) burde WILL. 2522; werth PS. xciii. 11 ; werþe S. S. (Web.) 2580; þæt hi beon . . . þare lande wurþe ANGL. VII. 220 ; **wurðer** (*compar.*) LAȝ. 30993 ; **wurðest** (*superl.*) FRAG. 6 ; *comp.* **ār-, dēore-, kine-, līc-, luve-, stal-, un-wurðe**.

wurð-līch, adj., *O.E.* weorð-, wurðlīc; = *O.H.G.* werdlīh; *worthy, valuable, dignified, stately*, HOM. II. 29 ; **wurðliche** (*adv.*) A. R. 40; REL. I. 130; wurþlike ORM. 1033; worli SPEC. 39 ; **wurðliche** (*pl.*) KATH. 1577; wurðliche wepnen LAȝ. 28923; **worþloker** (*compar.*) A. P. iii. 464; **wurðlükest** [worþlokest] (*superl.*) LAȝ. 25496.

wurð-münd, sb., *O.E.* weorð-, wurðmynd ; *dignity, honour*, KATH. 218 ; wurþmin ORM. 3379; wurðment HOM. I. 107; worðmunt LAȝ. 18851.

worþnesse, sb., *O.E.* weorðness ; *dignity, honour*, PROCL. 3; ALEX. (Sk.) 3163; worthenes 1825 ; *comp.* **un-worþnesse**.

wurðschipe, wurschipe, sb., *O.E.* weorð-, wyrðscipe ; *dignity, honour, worship*, A. R. 196, 278; wurþ-, wurschipe O. & N. 1288, 1344; þene wurðscipe LAȝ. 19534; worðschepe [worsipe] 3159 ; wurþschipe ORM. 726; worðssipe AYENB. 8 ; wurschipe HOM. I. 191; worshipe [worschip] LANGL. (Wr.) 2041 (*B* iii. 348) ; worshepe PR. C. 6217.

worþssip-vol, adj., *worshipful*, AYENB. 80.

worþssip-lich, adj., *honourable*, AYENB. 80; **worschiplich** (*adv.*) AL. (L.¹) 58.

wurðschipen, v., *worship, honour, revere*, KATH. 55 ; wurðsupen HOM. II. 5 ; worþ-, worssipie AYENB. 5, 212; worschipe MAND. 3.

worschipere, sb., *worshipper*, LEG. 216.

wurð², sb., *O.E.* weorð, worð, wurð, = *O.N.* verð, virð, *Goth.* wairþ, *O.H.G.* werd ; *worth, value, dignity, merit*, A. R. 290; H. M. 27 ; þet wurð HOM. I. 91; wurþ ROB. 373; worþ AYENB. 99 ; P. 47; worthe LANGL. *B* iv. 170; for monne weordes (*esteem*) ðinge HOM. I. 25 ; knightis of gret(e) worþe (*r. w.* eorþe) ALIS.

1652 ; wiðute mine **wurðes** (*pl.*) JUL. 65 ; *comp.* **un-worð**.

wurð-ful, adj., *O.E.* weorðfull ; *valuable, precious; worthy*, KATH. 1017 ; wurþful ORM. 5195 ; O. & N. 1481; þe worþvolle AYENB. 16.

wurðful-hēd, sb., *preciousness*, GEN. & EX. 3499.

wurðe, sb., *? O.E.* weorðu, (or-)wyrðu, = *O.H.G.* wirdī ; *dignity, honour* : never likede me mi werþe A. D. 291 ; drihtin to lofe & wurþe ORM. 1141.

wurðen, v., *O.E.* weorðan, worðan, wyrðan, wurðan, = *O.L.G.* werthan, *O.N.* verða, *Goth.* wairþan, *O.H.G.* werdan, *Lat.* vertere ; *become*, HOM. I. 63 ; KATH. 157 ; GEN. & EX. 41; wurðen, wrþan LAȝ. 1234, 12376; bliþe wurrþenn ORM. 160; hit schal wrþe wel isene O. & N. 846; worþe WILL. 327 ; lat hem worþe *let them alone* LANGL. *B* ii. 47; what shal worthe of us M. ARTH. 1817 ; **wurðest** (*pres.*) LAȝ. 16091 ; wurst HOM. I. 221 ; wurst HORN (L.) 324 ; worst P. L. S. xii. 109; S. S. (Wr.) 505 ; wurðeð A. R. 96 ; hit worþeþ al to win SPEC. 36 ; wurð MAT. xiii. 32 ; wurð . . . bi *will be* GEN. & EX. 2058 ; wurþ MISC. 120 ; HORN (L.) 460 ; wurþ, worþ O. & N. 405, 770; worþ AYENB. 90; to morwe worþ þe mariage imad LANGL. *A* ii. 22; worþ to noght Iw. 1642 ; we worþeþ BEK. 996; **wurð** (*imper.*) LAȝ. 3192; **wurþe** (*subj.*) FRAG. 8; wurþe [wrþe] O. & N. 400; wel wurðe [worþe] þe LAȝ. 13079; wurþe, worþe HAV. 1102, 2221; **warð** (*pret.*) LAȝ. 2927; warð KATH. 27; warþ ORM. 147; þo warþ þis monke swiþe wo A. D. 257; wearð A. R. 244; werð HOM. I. 133; þis maide werþ a slep P. L. S. xxi. 38; he werþ upe BEK. 1176; wurðe, wurð GEN. & EX. 272, 283; worþ LANGL. *B* v. 160; wurðen MAT. iii. 16; þeȝ wurden swiþe ofdredde ORM. 3343; wurðen LAȝ. 1112; **worðen** [*O.E.* worden] (*pple.*) HOM. II. 197; worþen A. P. i. 394; wurþen ORM. 1888; *comp.* **a-, for-, ȝe-, to-wurðen**.

wurðien, wurþian, v., *O.E.* weorðian, wyrðian, wurðian, = *O.N.* virða, *Goth.* wairþōn, *O.H.G.* werdōn ; *esteem, honour, revere, adore*, HOM. I. 11, 45; wurðien [weorþi] LAȝ. 9510; to lofen him & wurþen ORM. 208; wrþie REL. I. 172; worþ P. L. S. iv. 18; worth ALEX. (Sk.) 2124; **wurðeð** (*pres.*) REL. I. 129; **wurðede** [worþede] (*pret.*) LAȝ. 2614; **wurðed** (*pple.*) GEN. & EX. 1012; *comp.* **a-, ȝe-, un-wurðien**.

wurþiȝ, wurþi, adj., = *O.L.G.* wirthig, *O.N.* verðugr, *O.H.G.* wirdīg ; *worthy*, ORM. 2705, 4200; wurdi GEN. & EX. 1012; REL. I. 218; wurþi RICH. 6134; worþi BEK. 412; C. L. 1524; WILL. 2792; CH. C. T. *A* 43; þat ich þare to worþi be SPEC. 73; **wurthiest** (*superl.*) LUD. COV. 20; *comp.* **luve-, un-wurði**.

worþi-līche, adv., *worthily*, LANGL. (Wr.) 895.

worþinesse, sb., *worthiness*; CH. C. T. *A* 50.

wurðigen, v., = *M.L.G.* wordigen, *M.H.G.* wirdigen; *honour*, SAX. CHR. 31; þat tu wurge [wurȝgi] mi mawmez MARH. 18; þat ha ... wurȝehen (? *ms.* wurȝchen) [wurgin] KATH. 659; *comp.* ȝe-wurðigen.

wurðinge, wurðinge, sb., *O.E.* weorðung; *reverence, veneration*, HOM. I. 7, 109; wurð-ing GEN. & EX. 1774; worþinge SHOR. 96.

wurðing-dei, sb., *day of worship*, HOM. I. 9.

worþing-niȝth, sb., *? feast of the adoration*, DAV. DR. 87.

wurþinge, sb., *prov. Eng.* worthing; *mire, filth*, HOM. I. 85; wurȝinge MARH. 3; H. M. 13.

wurven, *see* hwerfen.   wūs, *see* wōs.

wŭsch, sb., *O.E.* wūsc, = *O.H.G.* wunsc, *M.Du.* wunsch, wensch, *O.N.* ōsk; *wish*, PR. P. 535; wisch MAND. 145.

wŭschin, wĭschin, v., *O.E.* wȳscan, = *O.H. G.* wunscan, *M.Du.* wunschen, wenschen, *O.N.* œskja; *wish, 'opto,'* PR. P. 530, 535; wische AR. & MER. 8732; MAND. 145; wŭsche (*pres.*) LANGL. *A* v. 92 [wisshe *B* v. 111]; wussheð REL. I. 130 (HOM. I.

165); wisheth CR. K. 58; wesseþ AYENB. 56; weschte (*pret.*) LANGL. *A* v. 195; weste AL. (T.) 335; wisten HOM. II. 3; GEN. & EX. 801.

wŭssinge, sb., *wishing*, HOM. II. 179; wis-sing REL. I. 216.

wuten, *see* witen.

y.

y- (*consonant*), *see* ȝ.

y-, ȳ- (*vowels*), *see* i-, ȝe-, ȝi-, ī-.

yā, *see* ȝā.   yald, *see* eald.

yem, *see* ēam.   yēme, *see* ȝēme.

yernen, *see* eornen.   yif, *see* ȝif.

ȳle, *see* īle.   yoie, *see* joie.

yok, *see* ȝoc.   yond, *see* ȝeond.

you, yow, *see* ēow.   yven, *see* ȝifen.

z.

z-, *see* s-.   zed, *see* sad.

zedewal, zeduale, sb., *O.Fr.* citoual; *ze-doary*, A. R. 370; sedewale SPEC. 27; ALIS. 6792; cetewale [cetuale] CH. C. T. *A* 3207.

ADDITIONS AND CORRECTIONS.

[New articles are denoted by the sign (*) prefixed. Articles to be substituted for those in the text are denoted by the sign (†).]

ǣr.

*ǣr-þēod, sb., *ancient people, antiquity*: hou Jesus him welk in areþede A. P. i. 711; arethede ISUM. 6; arthede DEGR. 7.

al.

al swa. Page 18, line 2, for 'elle' read 'telle.'

*alas, interj., *O.Fr.* a las; *alas*, A. R. 148; ROB. 34; allas CH. C. T. *B* 121; aḷace D. TROY 9716.

a-merien. Cancel the article; the example belongs to **amerran**.

a-merran. Add to examples of pple.: 'amerd AYENB. 125; (? *ms.* amered, *r. w.* iherd) S. S. (Web.) 2266.'

amirail. To senses add '*commander.*'

an.

†an ān, *mod. Eng.* anon; *continuously, straight forward, immediately*: (*ms.* anan) KATH. 31; þeȝ wisten sone an an (*ms.* anan) ORM. 225; fowertiȝ daȝhes aȝ on an (*ms.* onnan) 11331; *see further examples* under **ān.**

and, prep. The evidence for the existence of this word in Middle-English is questionable; the reading of most of the examples given being uncertain.

*[and-lang], andelang, endelong, endlang, anlong, along, adv. & prep., *O.E.* andlang, ollunc, *cf. O.L.G.* andlang, *O.Fris.* ondling, ondleng, *O.N.* endlangr, endilangr (*adj.*), *mod. Eng.* endlong, along; *lengthways; along; along the length of.* [*This word is by Stratmann regarded as a phrase; for examples of it* (*written as two words,* ande lang, *etc.*) *see* **and, lang, lange.**]

*apōsen, *see* **oppōsen.**

*aqvílen, enqvílen, v., *O.Fr.* aqvillir, anqvillir; *receive, obtain*; aqvile A. P. i. 699; enqvílen (*pres. pl.*) A. P. iii. 39; aqvild (*pple.*) (*ms.* aquylde) A. P. i. 966.

a-reimen. Insert: '*from O.Fr.* reimbre, *Lat.* redimere.'

bi-luvien.

†āre-thēde, *see* **ǣr-þeod** (*in Supplement*).

aretten. Insert '*O.Fr.* aretter.'

asaumple. Add at end '*see* **ensample, sample.**'

assīse. Add '*see* **sīse.**'

*astaunchen, *see* **stanchen.**

atel.

atel-līche. For '*cruel*' read '*dreadful, hideous, cruel.*'

a-weleȝen. Dele '*luxurious . . . 282.*'

bacin. For '*basin,*' read '*basin; helmet.*'

bēden. For '*O.L.G.*' read '= *O.L.G.*' Line 4, for 'bede' read 'béde (*pres.*).'

bélden. For 'forfren' read 'frofren.'

†beork, sb., *from* **beorken**; *bark (of a dog)* ANGL. IV. 197.

beren. Read **béren**, and afterwards **bēreð, bére, bérinde, bóren, bérere, béring.**

best. Read **bést, béstes, bést-līch.**

†bétil, *see* **bitel²** (*in Supplement*).

bēten. Line 2 should read '*O.L.G.* bōtian, *O.Fris.* bēta, *O.N.* bœ̄ta, *Goth.* bōtjan, *O.H.G.* puozan.'

betren. Read '*better, improve; prevail.*'

bi-berȝen. For '*protect*' read '*ward off.*'

bid. Cancel the entry.

bidden. Line 16, for '**bidde** (*subj.*)' read '**bidde.**'

bi-fóren.

bifore-gōere. For 'nise' read 'rise.'

bifore-ocupíen. For 'couegteth' read 'coveiteth.'

bi-gálen. For **bigolen** read **bigōlen.**

*bī-halve(s), *see* **half².**

bihōflen. See other forms under **bus,** which should have been included here.

bi-hrīnen. Cancel the article.

bi-lēoȝen. For 'biliveth' read 'bilieth (*pres.*).'

bi-luvien. For 'impers., *delight, please*'

read '*delight, please (impers.)* ; *approve, love.*'

binden. For **i, u** read **í, ú** in all the forms printed in clarendon type.

bi-strūpen. Insert rendering : '*strip away.*'

†**bitel** [2], **bétil,** sb., *O.E.* bitula, ? betel ; *beetle (insect)* ; bitil PR. P. 37 ; (*ms.* betylle) VOC. 225.

†**bi-tranten,** v., *deceive* : betrant (*printed* betraut) D. TROY 731 ; **betrantid** (*pret.*) 11767.

bi-þenchen. In line 4 for '*deþenche*' read '*beþenche.*'

bi-þünchen. For '*biðunched*' read '*biðüncheð.*'

blac [2]. Read '*sb.*'

blēchen. Add at end '*blaȝt white* A. P. 212 ; ALEX. (Sk.) 4925, 5482 ; blaught 1559*.'

blētsien. Dele quotation from RICH. 546.

blind, blindin. Read '**blínd, blíndin,**' and place the accent on the **í** in all the forms printed in clarendon type.

†**blismen,** *see* **blesmin.**

bōc [2].

    †**bŭk-mast,** sb., = *M.L.G.* bōkmast ; *beech mast,* HALLIW. 37.

borȝ. To the renderings add '*surety.*'

brā. For '*drench*' read '*dreuch.*'

brainid. Read '*adj.*'

brusch. For '*Ital.* brusco' read '*O.Fr.* brosse, *cf. Ital.* brusco.'

brusche. For '*Ital.* brusca' read '*O.Fr.* brosse, *cf. Ital.* brusca.'

bulken. For '*bulk* (*pres.*)' read '*bulk.*'

bulge. Dele '*bubble.*'

caft, keft. Add '*perhaps mistake for* cast, kest.'

cáld. At end of p. 99 read 'GREG. 798 ; **cáld** (*sb.*) *cold* HAV. 856 ; P. S. 330 ; WILL. 908.'

can [2]. For 'BRUCE' read 'BARB.'

cape [? cāpe]. Read '**cápe, cāpe,**' and dele '*cf.*'

cape [? cāpen]. Read '**cápen, cāpen.**'

chauceūre, sb., *O.Fr.* chauceure ; *hose of mail* : chauceore FER. 5631.

*****chēten** [2], v., *for* eschēten ; *escheat,* PR. P. 73.

chēse.

    †**chēs-lēpe,** sb., = chēs-lippe ; VOC. 222.

†**chevisance,** sb., *O.Fr.* chevissance ; *accomplishment, acquisition, booty ; success ; profit ; help ; loan on usury* ; CH. C. T. *B* 1519 ; MAN. (H.) 105, 181 ; GAW. 1390, 1406, 1678, 1939 ; LANGL. *C* xxiii. 16 ; GOW. II. 332 ; **chevesances** (*pl.*) LANGL. *B* v. 249.

child, childin. The i in all the forms printed in clarendon type, except **childre,** should have the acute accent. So also in the

derivatives and compounds of the sb., except children-lēas.

chīnen. Dele ' (*intrans.*).'

†**chippe,** sb., *O.E.* cipp ; *sharebeam of plough,* '*dentale,*' VOC. (W.W.) 628.

chippe. Read '**chippe** [2],' and dele '*? O.E.* cipp (*dentale*).'

chiteren. Insert '*twitter*' before '*chatter.*'

chois. Add at end '**chois** (*adj.*) *choice* ALEX.[2] 727 ; WILL. 400 ; choise D. TROY 490, 1356.'

clǣne. Read '*adj. & adv.,*' and add to the renderings '*clean, quite, entirely.*'

clēowen. Read '*sb.*'

colwie. Cancel the entry.

†**complī, complīn,** *O.Fr.* complie ; *compline, last church service of the day* ; compli BEK. 2078 ; cumpli A. R. 24, 428 ; cumpelie 22 ; complin CH. C. T. *I* 386 ; compline '*completorium*' PR. P. 109.

†**consence,** sb., *O.Fr.* consense ; *consent* : kunsence A. R. 288.

covert. Insert '*adj. & sb.*'

†**cucking,** sb., *cacatio,* PR. P. 106.

culrāge. Read '*O.Fr.* culrage.'

*****cúri,** *see* keweri.

cussin. Add '*qvischen*' ISUM. 579.

dalc. Dele '*dolc* . . . 3027.'

†**dalien,** v., *? O.Fr.* dallier (*railler*) ; *mod. Eng.* dally ; *play, sport ; converse, talk* : daliin '*fabulor, colloqvor*' PR. P. 112 ; dali GAW. 1253 ; daile MAN. (H.) 116 ; (*ms.* dayly, *r. w.* bayly, fayle, counsayle) A. P. i. 313 ; disours **dalie** (*pres.*) ALIS. 6991 ; **dailieden** (*pret.*) GAW. 1114.

dēawen. For '*touwen (touuen)*' read '*touwēn, touwōn.*'

defende. Read '*defend, prohibit.*'

†**defense,** sb., *O.Fr.* defense ; *defence ; prohibition* ; ROB. 253 ; defence GOW. I. 29 ; diffense R. R. 1142.

defoulen. Dele '*defoul.*'

†**degráden,** v., *O.Fr.* degrader, *Lat.* dēgradāre ; *degrade* ; **degráde** (*pret.*) TOWNL. 20 ; degrated D. TROY 12574 ; **degráded** (*pple.*) TREV. V. 35.

†**deinen,** v., *O.Fr.* deigner, *Lat.* dignārī ; *deign* ; (*impers.*) *appear worthy* : **daineþ** (*pres.*) AYENB. 18 ; deineþ CH. AN. 181 ; **deined** (*pret.*) LANGL. *B* vi. 310 ; ham ne dainede naȝt to do zenne AYENB. 76.

†**demeine,** *see* demaine.

descrīen. Read '*announce, describe ; descry* : descriin '*describo.*''

*****discomfiten,** v., *from O.Fr.* desconfit (*pple. of* desconfire) ; *discomfit, rout* : discoumfitin '*confuto, supero, vinco*' PR. P. 122 ; deʂconfite MAN. (F.) 13944 ; **deskumfit** (*pple.*) A. R. 250 ; *see* scomfiten.

distincte. Cancel the entry.

distincten. Add ' distincte R. R. 6199.'

douvre. Read ' *rabbit-burrow*.'

dwīnen. Read '*faint, wither*, [' *tabescere*'].'

earm [2]. Read ' adj.'

effraien. Read ' =affraien.'

eȝe.

    ei-fold. Cancel the entry.

eischste. Cancel the entry.

encombrous. For ' COMP. M. 339 ' read ' CH. COMP. V. 42.'

encenser. For '*censor*' read '*censer*.'

enclinaunt. For ' xxxiii ' read ' xxxii. (*misprinted* xxxiii *in Oxford edition*).'

ende. For ' anātis ' read ' anatis.'

*__endelong, endlang__ : *see* **andlang** (*in Supplement*).

endenten. Dele ' þen . . . anon.'

ender. Insert ' *O.N.* endr (*adv.*),= *O.E.* end (*formerly*).'

†**entēr,** adj., *entire, whole ; earnest, sincere ;* L. H. R. 196 ; entier LIDG. M. P. 100 ; intire 68.

    *__entēr-līche__, adv., *entirely*, MAS. 241 ; entereli *earnestly* ISUM. 434.

fāh. Read ' TRIST. 1220.'

far.

    far-cost. For '*see* **fercost**' read ' **fercostez** (*pl.*) D. ARTH. 743. '

fēle [2]. Read ' adj.'

fēon. For ' feid ' read ' feid (*pple.*).'

feorren. Near the end insert ' LAȜ.' before ' 5328.'

†**fercost,** *see* **farcost** *under* **far.**

feste, festin, festour. Read ' *féste*,' *etc.*

finden. Read **fínden,** and accent the í and ú in the forms in clarendon type.

fliteren. Cancel the entry. (A misprint in Southey : *see* **floteren.**)

floteren. Add at end ' fluttered (*misprinted* flitterid *in ed.* 1817) MAL. V. 4.'

fóldin. Read '*put in a fold*' (italics).

foltin. Read '*act like a fool ; make a fool of*.'

for-garen. For '*oppose*' read '*forfeit*.' For ' 14582 ' read ' 14584.'

for-ȝearwien. Dele this entry.

for-hōnen. Cancel the entry ; ' forhone ' is probably a mistake for ' forhoue,' see **for-hoȝien.**

for-lange. Read ' adv.'

for-liggen. Read '*lie illicitly with*.'

for-tühting. Read ' sb.'

forð-nimen. Read '*seize ; go away*.' For ' forðnam ' read ' forðnam (*pret.*).'

fōt. Line 8, ' fot ' should come after ' euch.'

frounter. Line 3, insert ' one ' after ' frikis.'

garison. Cancel the entry.

ȝe-dwilð. Cancel, and transfer examples to **ȝedwéld.**

ȝe-fēren. Dele ' (*pret.*).'

ȝe-gāst.

    igāst-līche. Read ' adv., *terribly*.'

ȝe-līc. Line 10, dele ' þe . . . beo.'

ȝemǣre. Read ' sb.'

genillere. For ' chaunceore' read 'chauceore.'

geomancie. For '*mensuration*' read '*geomancy*.'

†**geometrien,** sb., *geometer* ; **gemetriens** (*pl.*) LIDG. BOCH. [*ed.* 1554] I. ii. 200 [*ed.* 1558 *has* geometries].

ȝeonien. Line 5, read ' ȝeonie [ȝonie].'

ȝe-rennen. Cancel the entry.

†[**ȝe-scōle**] **i-scōle,** sb., *troop*, HOM. I. 243 ; *comp.* her(e)-iscōle.

geste, gesten [2]. Read ' géste, gésten.'

get [2]. For ' ges . . . 254 ' read '*see* **ges.**'

ȝirnen. For ' garennan ' read ' garinnan.' The rendering should be simply ' *run*.'

glēde. For ' berninde (*pple.*)' read ' berninde.'

gnēde. Last line, dele ' gnede . . . 154.' (Example from LANGL. belongs to **gnēdi.**)

gnēden. Add at end ' knede (? *be too scanty*) C. M. 2448.'

***gnēdi,** adj., =**gnēde** : LANGL. *C* xvi. 85.

gobben. Read ' gobbenen.'

god.

    gode-frigti-hēd. Last line, read ' godesfrigtihed 540.'

ȝoman. For ' AMAD. (L.)' read ' AMAD. (R.) l.' For ' AM.' read ' AMAD. (W.)'

góre. Dele ' gorred . . . 4645.'

***gōren,** v., *? from O.E.* gār (*spear*) ; *gore, stab* : gōred (*pple.*) (*ms.* gorred) ALEX. (Sk.) 4645.

grān. Read ' sb.'

grúnden. Dele ' grune . . . 338.'

haȝe. Dele ' hawch . . . 35,' and transfer ' *comp.* chirche-hawe ' from **haȝe** [2].

haste, hasten, hasti, hastif, hastives : read ' háste, hásten,' *etc.*

hèlle.

    helle-ware. For ' sb., pl.'·read ' sb. (collective).'

hellen [2]. Page 336, line 1, insert ' HAMP. PS.' before ' lxviii.'

hevenen. Add at end ' ? D. TROY 2083.'

heviin. Dele quotation from D. TROY.

hind, hinde, hinden. Read ' hínd,' *etc.*

hinder [2]. Read ' adv.'

hoferede. Read ' adj. (sb.) '

hori. Dele quotation from TREV.

hors.

    horse-charche. Read ' *horse-load*.'

†[hurfte, *misreading for* hürst TREAT. 132 ; cf. SAINTS (Ld.) xlvi. 394.]

hurlen. Read '*hurl ; rush violently.*'

hurst. Read 'hürst,' and insert ' (*r. w.* furst) SAINTS (Ld.) xlvi. 394.'

hürten. For '*offend*' read '*strike.*'

in-ȝetten. Read 'in-ȝeoten.'

in-ward.
    inward-lích. Read '*inward ; earnest.*'

irk. For '*distasteful*' read '*tired, dissatisfied.*'

irkin. Read '*be tired or dissatisfied.*'

keweri. Read ' *O.Fr.* qveuerie.'

læsten. Line 4, for '*þine*' read '*pine.*'

laȝe.
    *laȝe-líc, adj., *O.E.* lahlíc, *O.N.* lag-, lögligr; *lawful*: mid laweliche deden MISC. 106.

laȝe-líche. Cancel the last quotation.

láh.
    lóȝ-lí. Read 'adj. & adv., *O.N.* lágligr, -liga.'

léosen. Line 4, for '*lessen*' read '*leosen.*'

linde. Read '*linde.*'

*liniolf, sb., *O.Fr.* lignoel ; *shoemaker's waxed thread,* '*indula,*' PR. P. 306.

macche. Add '*see* meche.'

†maineal, adj. & sb., *mod.Eng.* menial ; *belonging to the house ; household servant ; intimate friend* ; meineal TREV. VII. 413, 419 ; WICL. ROM. xvi. 5.

maten. For '*mæting*' read '*máting.*'

meche. Read '*see* macche ; *match, wick,* '*lichinus.*''

merien. Dele '*comp.* a-merien.'

mel. Dele ' mel, *see* male.'

melden. Read '*mélden.*'

metz. Cancel the entry.

mid. Line 33, for '(*? added thereto*)' read '(*? at the same time, immediately*).' After line 34 insert 'mid idone S. S. 1368, 1442.'

*mid-idóne, *see* mid.

mis-lích. Add at end ' misseli (*mistakenly*) 207.' Cancel following entry.

moinel, sb., *O.Fr.* maynel ; *mullion of a window* ; moinelus (*pl.*) DEGR. 1459.

†móke, sb. : some can sette the moke awrie (*? be miserly*) REL. I. 248 ; þe wrecchis wringit þe moke so fast, up ham silf hi nul not spened P. L. S. i. 20.

mop. For '*mop*' read '*mod.Eng.* mop ; *rag doll.*'

naiten. Dele quotation from D. TROY.

nifte. Dele '*step-daughter.*'

†ofer-lang, adj., *over-long* : overlang PR. C 7274 ; overlonge (*adv.*) O. & N. 450.

ofer-láte. Read 'adv.'

of-stingen. Read '*pierce through.*'

*oliprance, sb., cf. *A.Fr.* orprance, *prov. Eng.* olliprance ; *vanity, ostentation,* A. P. ii. 1305 ; H. S. 4581, 4695.

pallen. Add '*see* pellen.'

pensif. Insert 'LANGL.' before ' *A.*'

pil. For ' *Lat.* pilus ' read '*from* pilien.'

píl. For '*darts*' read '*prickles.*'

pithe. Read 'piþe,' and transfer to alphabetical place.

pittit. Read '*pitted.*'

plat. Read 'adj. & adv.'

pouderon. Read '*sprinkle with powder, grind to powder.*'

qvilte. Add at end 'qviltus poined [*O.Fr.* coulte pointe, *Eng.* counterpane] DEGR. 1491.'

qviter. For '*cf* . . . koder' read ' *O.F.* cuiture (*burning, matter from a boil*).'

rac. For '*rack*' read '*? torment.*'

raimen. Dele '*comp.* a-reimen.'

rakel. Read 'adj.'

*recháte, sb., *recheat, signal of recall in hunting,* MAL. i. 52 ; rechase (*r.w.* trase) ANT. ARTH. (R.) v.

*recháten, v., *blow the recheat* ; rechátand (*pple.*) GAW. 1911 ; rechâted (*pret.*) 1466.

remüen. For 'TRO.' read 'CH. TRO.'

retenaunce. For 'retenauntz (*pple.*)' read ' retenauntz.'

hrēðe, rēðe, rēþer. These entries should come after rettin.

*revaien, v., *A.Fr.* riveier ; *hunt waterfowl, go hawking*: revei DEGR. 506 ; revaies (*pres.*) 807 ; irevaied (*pple.*) 659.

*rēverie, *see under* rēaven.

rivēre. For '*river*' read '*river ; hawking-ground for waterfowl (cf. Ger.* revier)' ; and add to examples 'PARTON. 178, 631 ; CH. TRO. iv. 385 ; revērez (*pl.*) A. P. i. 104.

rŏdde. Before the examples insert '*rod, staff.*'

rude. For '*ruddy*' read '*ruddiness.*'

ruel-bán, sb., *? walrus-bone* ; ruel bon DEGR. 1429 ; rewel boon CH. C. T. 2068 ; IPOM. (K.) A 6456.

*rūmēde, *see under* hrēou.

rünel. Insert '*runner*' before '*stream.*'

sēēl.
    †sēēl-líche, adv., *fortunately,* LAȝ. 7863.

samplēre. Read '*mod.Eng.* sampler ; *original, model.*'

†saulee, sb., *A.Fr.* saulee (*soûlée*) ; *satisfaction of appetite* [*v. r.* saule, soule, saulees] ; glossed '*edulium*' LANGL. xvi. 11.

*savenappe, sb., *napkin,* PR. P. 441* ; sanap, sanop 441 ; sanappus (*pl.*) DEGR. 1387.

*scalc, *see* schalk.

*schalmie, sb., *O.Fr.* chalemie ; *shawm, trumpet* ; shalmies (*pl.*) CH. H. F. 1218 ; (*ms.* scalmuse) DEGR. 1086.

schat. Insert as first example ' scat HOM. II. 231.'

scheād.	Read at end '*comp*. ᴣe-sceād.'
scheaf.
    scēaf-mǣlen.	Read 'adv.'
*scheltrone, *see* sceldtrume *under* schíld.
schote.	Line 3, for 'schote' read 'schote or crikke (*spasm*).'
schoutin.	Dele '*see* schūte.'
schrēawen.	The pple. schrēwed should be explained '*cursed, bad, wicked*.'
schunten.	Read '*avoid, swerve, flinch*.'
scof.	Line 3, for 'kof' read 'skof.'
séld.
    sel-cūð, and its derivatives should be written with e, not é.
*sengle, *see* single.
sēon.	Page 544, line 5, 'þeᴣ' should be removed to next line, before 'sæᴣhen,' and the quotation from P. L. S. (line 7) should be placed in line 5, after '*A* ix. 66.'
*sh-, *see* sch-.	*sīli, *see* sǣli.
singen.
    singare.	Read 'sb.'
snūten.	To rendering add '*snuff (a candle)*.'
sodain.	Read 'adj.'
spēde.	To rendering add '*aid*.'
spinnen.	Dele quotation from A. P.
sqvachen.	For 'cvi.' read 'cix.'
sqvatten.	Read '*mod.Eng.* squat; *break, throw down*.'
†sqvint, ? sb.; *cf. Du*. schuinte (*slope*): a sqvint (*adv*.) asquint, A. R. 212.
*ss-, *see* sch-.
stǣned.	Read 'stāned,' and remove to p. 574.
stále.	For '*rung*' read '*side*.'
stede.	Line 13, for 'we' read 'þe.'
stikel.	For '*difficult*' read '*steep, piercing*'; and add example given below under stikil.

stikel-līche.	For '*with difficulty*' read '*intently*.'
†stikil, *see* stikel.
†streᴣt, adj., *straight; here treated as pple. of* strecchen, *q.v.*
summitee.	For 'sommette' read 'sommité.'
súnd.	Read 'adj.'
suteltē.	Dele 'sutell BARB. xix. 32.'
suvel.	Dele 'saulee . . . 11.' (*See* saulee in *Supplement*.)
swengen.	Dele quotations from A. P.
swippen.	For '*vibrate, beat*' read '*brandish; move or strike swiftly*.'
targen.	Add 'targe ROB. (W.) 2363.'
    *targinge, sb., *delay*, ROB. (W.) 4216.
tēoþe.	Read 'ord. num. & sb.'
terᴣen.	Dele 'targe ROB. (W.) 2363.'
    terring.	Dele 'targinge ROB. (W.) 4216.'
*umbrēre, sb., *O.Fr.* ombriere; *visor*, PERC. 1522.
ut-līche.	Read 'ūt-līche.'
vīne.	For '*vine*' read '*vine, vineyard*.' Add to references 'A. P. i. 502, 507, 628.
wainen, v., ? *O.Fr.* waagnier (*gagner*); *gain*: ? waine A. P. ii. 1616; wainez (*pres.*) (*r. w.* frainez) ? *grants* i. 1504; wained (*pret.*) i. 1701; wained (*pple.*) (*ms.* vayned) ? *brought* i. 249.
warnestūre.	Add '(printed warme store) FLOR. 1168.'
warnestūren, v., *store, fortify*; CH. C. T. B 2487; warnestōred (*pple.*) CH. BOET. i. 4 (12)*.
*werle, sb., ? *garland*, A. P. i. 209.
*westernais, adv., ? *mistake for O.Fr.* bestorneis; *wrong side up, inside out*, A. P. i. 307.